Mechanical properties of selected alloys. See Chapter 3 for additional data.

Identification	Condition[a]	Yield strength MPa (ksi)	Ultimate tensile strength MPa (ksi)	Elongation in 50 mm %	Reduction in area %	Hardness
Alloy steels						
AISI 4130	Annealed	361 (52)	560 (81)	28	56	156 HB
	Q&T 205°C	1460 (212)	1630 (236)	10	41	467 HB
AISI 4340	Annealed	470 (68)	745 (108)	22	50	220 HB
	Q&T 315°C	1590 (230)	1720 (250)	10	40	486 HB
Carbon steels						
AISI 1020	Q&T 870°C	295 (43)	395 (57)	37	60	100 HB
AISI 1040	Annealed	350 (51)	520 (75)	30	57	150 HB
AISI 1060	Annealed	372 (54)	626 (91)	22	38	179 HB
AISI 1080	Q&T 800°C	380 (55)	615 (89)	25	30	255 HB
Cast irons						
ASTM 20	As cast	—	152 (22)	—	—	156 HB
ASTM 50	As cast	—	362 (52.5)	—	—	262HB
ASTM 60-40-18	Annealed	324 (47)	448 (65)	—	—	160 HB
ASTM 120-90-02	Q&T	827 (120)	965 (140)	—	—	325 HB
Stainless steels						
AISI 301	Annealed	205 (29.7)	515 (74.7)	40	—	RB 85
	1/8-hard	380 (55)	690 (100)	40	—	RC 20
	1/4-hard	515 (74.7)	860 (125)	25	—	RC 25
	1/2-hard	758 (110)	1034 (150)	18	—	RC 32
	Full-hard	965 (140)	1276 (185)	9	—	RC 41
AISI 304	Annealed	215 (31.2)	505 (73.2)	40	—	201 HB
Aluminum alloys						
1100	O	35 (5.07)	90 (13.0)	40	—	—
	H14	120 (17.4)	125 (18.1)	25	—	—
3004	O	69 (10)	179 (26)	20	—	—
	H38	234 (34)	276 (40)	6	—	—
5052	O	90 (13.0)	190 (27.6)	25	—	—
	H32	186 (27)	234 (34)	62	—	—
	H36	234 (34)	269 (39)	10		
7050	T7651	490 (71)	552 (80)	—	—	—
Copper alloys						
Red brass, 15% Zn	Annealed	91 (13)	280 (41)	47	—	64 HRF
	Hard	406 (59)	434 (63)	5	—	73 HRB
Phosphor bronze, 5% Sn	Annealed	175 (25)	350 (51)	55	—	40 HRB
High-silicon bronze	Annealed	210 (31)	441 (64)	55	—	66 HRB
Titanium alloys						
99.5% Ti	Annealed	240 (35)	330 (48)	30	—	—
Ti-6Al-4V	Annealed	925 (134)	1000 (145)	10	—	—
	Solution + age	1100 (159)	1175 (170)	10	—	—

Note: [a]Q&T, quenched and tempered.

Definitions and conversions.

Definitions		Conversion factors
Acceleration of gravity	1 g = 9.8066 m/s^2 (32.174 ft/s^2)	1 in. = 25.4 mm
Energy	1 kcal = 4187 J	1 lbm = 0.4536 kg
Length	1 mile = 5280 ft	1 ft = 0.3048 m
Power	1 horsepower = 550 ft-lb/s	1 lb = 4.448 N
Temperature	Fahrenheit: $t_F = \frac{9}{5}t_C + 32$	1 lb = 386.1 lbm-in./s^2
	Rankine: $t_R = t_F + 459.67$	1 ton = 2000 lb (short ton) or 2240 lb (long ton)
	Kelvin: $t_K = t_C + 273.15$	
Volume	1 ft^3 = 7.48 gal	1 ksi = 6.895 MPa
		1 ft-lb = 1.356 J
		1 kW =3413 BTU/hr
		1 quart = 0.946 liter
		1 kcal = 3.968 BTU

Online Resources.

Your new textbook provides 18-month access to online study resources that may include videos, code, web chapters, and more. Refer to the preface in the textbook for a detailed list of resources.

Follow the instructions below to register for the Companion Website for Serope Kalpakjian and Steven R. Schmid's *Manufacturing Processes for Engineering Materials,* Sixth Edition.

1. Go to www.pearsonhighered.com/engineering-resources
2. Either browse by author name or use the search feature to locate the resource page that corresponds with your textbook.
3. Click Companion Website.
4. Click Register and follow the on-screen instructions to create a login name and password.

Scratch here to reveal your access code.
This access code may only be used by the original purchaser.

Use a coin to scratch off the coating and reveal your access code.
Do not use a sharp knife or other sharp object as it may damage the code.

Use the login name and password you created during registration to start using the online resources that accompany your textbook.

IMPORTANT:

This access code can only be used once. This subscription is valid for 18 months upon activation and is not transferrable. If the access code has already been revealed it may no longer be valid. If this is the case you can purchase a subscription on the login page for the Companion Website.

For technical support go to http://247pearsoned.custhelp.com

Manufacturing Processes for Engineering Materials

SIXTH EDITION

Serope Kalpakjian
Steven R. Schmid

PEARSON

Boston Columbus Indianapolis New York San Francisco Hoboken
Amsterdam Cape Town Dubai London Madrid Milan Munich Paris Montreal Toronto
Delhi Mexico City Sao Paulo Sydney Hong Kong Seoul Singapore Taipei Tokyo

Vice President and Editorial Director, Engineering and Computer Science: Marcia J. Horton
Editor in Chief: Julian Partridge
Executive Editor: Holly Stark
Editorial Assistant: Amanda Brands
Field Marketing Manager: Demetrius Hall
Marketing Assistant: Jon Bryant
Managing Producer: Scott Disanno
Content Producer: Erin Ault
Project Manager: Louise Capulli, Lakeside Editorial Services L.L.C.
Operations Specialist: Maura Zaldivar-Garcia
Manager, Rights and Permissions: Ben Ferrini
Cover Designer: Black Horse Designs
Back Cover Images: Smartphone: rami_hakala/Fotolia and Video Screenshot: Courtesy of the Metal Powder Industries Federation.
Printer/Binder: LSC Communications
Cover Printer: LSC Communications
Composition: SPi Global
Typeface: 11/13 Sabon

Pearson Education Ltd., London
Pearson Education Singapore, Pte. Ltd
Pearson Education Canada, Inc.
Pearson Education –Japan
Pearson Education Australia PTY, Limited
Pearson Education North Asia, Ltd., Hong Kong
Pearson Educacion de Mexico, S.A. de C.V.
Pearson Education Malaysia, Pte.Ltd.
Pearson Education, Inc., Hoboken, New Jersey

Library of Congress Cataloging-in-Publication Data

Names: Kalpakjian, Serope, author. | Schmid, Steven R., author.
Title: Manufacturing processes for engineering materials / Serope Kalpakjian, Steven R. Schmid.
Description: Sixth edition. | Pearson Education. | Includes bibliographical references and index.
Identifiers: LCCN 2016022092 | ISBN 9780134290553 | ISBN 0134290550
Subjects: LCSH: Manufacturing processes.
Classification: LCC TS183 .K34 2016 | DDC 670–dc23 LC record available at https://lccn.loc.gov/2016022092

85 2023

ISBN-10: 0-13-429055-0
ISBN-13: 978-0-13-429055-3

Dedication

What can be said about Serope Kalpakjian?

He is a giant of manufacturing research, publishing, and education. ("But I am so small," he would say, "far too small to be a giant.")

He has written the world's most popular textbook for over three generations and has had more influence on training engineers than anyone. He has won the ASM and SME Gold Medals, indicating the high esteem that professional societies hold for him, yet he remains humble.

Universally beloved, Serope can work a room, whether it's filled with engineers, scientists, philosophers, or poets. (No one is perfect, he reminds us.) Everyone is his friend, just as he is everyone's friend.

He is a phenomenal speaker. I know of no other person that has been asked to give the commencement address at his own university. He was asked to give a Founder's Lecture at a NAMRC Conference, and told he could speak of whatever he wished: his life as a professor, his 40 years of writing, his decades of innovative teaching, his cutting-edge research; but he only had 30 minutes. ("30 minutes? I'll just speak slowly," he says.)

A devoted husband and father, interested and involved, his children and grandchildren have excelled in life—this is perhaps the greatest measure of man. ("I hated that he asked math questions at dinner," complains his daughter—a Professor at the University of Michigan.) Always patient and caring, he excelled at advising students and mentoring faculty. Especially coauthors. Especially me. ("Book authors make much less than those that throw a ball for a living," he lamented.)

There are some people, unfortunately a very few, that you meet in your life that you treasure their company and realize that they are genuinely great; people that have a lasting influence on your life; people that are your life-long friends. There are some people that you can hold as a role-model and not be disappointed.

That's Serope.

Sincerely,
Steven Schmid

Contents

3 Structure and Manufacturing Properties of Metals 86

4 Tribology, Metrology, and Product Quality 139

5 Casting Processes and Heat Treatment 196

Preface

Manufacturing has undergone a rebirth in its development and research activities since the mid-2000s. With the recognition that manufacturing adds value to national economies, governments around the world have been investing in their infrastructure, and are now partnering with industry to bring new manufacturing capabilities to the marketplace.

In view of the advances made in all aspects of manufacturing, the authors have continued to present a comprehensive, balanced and, most importantly, an up-to-date coverage of the science, engineering, and technology of manufacturing processes and operations. As in previous editions, this text maintains the same number of chapters while continuing to emphasize the interdisciplinary nature of all manufacturing activities, including the complex interactions among materials, design, and manufacturing processes.

Every attempt has been made to motivate and challenge students to understand and develop an appreciation of the vital importance of manufacturing in the modern global economy. The extensive questions and problems, at the end of each chapter, are designed to encourage students to explore viable solutions to a wide variety of challenges, giving them an opportunity to assess the capabilities as well as limitations of all manufacturing processes and operations. These challenges include economic considerations and the competitive aspects in a global marketplace. The numerous examples and case studies throughout the book also help give students a perspective on the real-world applications of the topics described in the book.

What's New in This Edition

- The text has been thoroughly *updated*, with numerous new materials and illustrations relevant to all aspects of manufacturing.
- A major addition to this revised text is the introduction of *QR codes*. Recognizing the proliferation of smart phones and tablets, and the inherent Internet browsing capability in these devices, the QR codes provide a readily available video insight into real manufacturing operations. (Please note that users must download a QR code reader to their smart device; data and roaming charges may also apply.)
- Each chapter now begins with a *list of variables* for the topics covered in that chapter.
- Wherever appropriate, *illustrations and examples* have been replaced to indicate *recent advances* in manufacturing.
- The text contains more cross-references to other relevant sections, tables, and illustrations in the book.
- The Questions, Problems, and Design problems at the end of each chapter have been significantly *expanded*.

- The *Bibliographies* at the end of each chapter have been thoroughly *updated*.
- A *Solutions Manual*, available for use by instructors, has been expanded; it now provides *MATLAB code* for numerous problems, allowing instructors to easily change the relevant parameters.

The following are the new or expanded topics in this edition:

Chapter	Topics
1	Economic multiplier; technology readiness level; a case study on three-dimensional printing of guitars; expansion of general trends in manufacturing.
2	Leeb hardness test; expansion of flow stress description.
3	Advanced high-strength steels; third generation steels; expansion of discussion on magnesium, chromium, and rare-earth metals.
4	Environmentally-friendly lubricants; validation of products and processes.
5	Strip casting; mold ablation; design of ribs in castings; computer modeling of casting; a case study.
6	Servo presses; electrically assisted forging; the Hall process; a case study on the production of aluminum foil.
7	Single-point incremental forming; age forming; hot stamping.
8	Expansion of tool-condition monitoring; through the cutting-tool cooling.
9	Expansion of laser machining; blue-arc machining; laser micro-jet; hybrid machining systems.
10	Electrically conductive polymers; big-area additive manufacturing; laser-engineered net shaping; friction stir modeling; The Maker Movement; design for additive manufacturing.
11	Expansion of metal injection molding; dynamic compaction of powders; combustion synthesis; pseudo-isostatic pressing; roll densification; graphene.
12	Expansion of friction welding; a case study on Blisks.
13	Wafer-scale integration; three-dimensional circuits; clean rooms; immersion lithography; pitch splitting; chip on board; system-in-package; roll-to-roll printing, including silver nanoparticles, inks, inkjet printing, gravure, flexographic, and screen printing; the MolTun process; photonic integrated circuits.
14	Intelligent robots and cobots; smart sensors; sensor validation.
15	ERP, ERP-II, and MES; manufacturing cell design; expansion of lean manufacturing; developments in communications; Internet of Things; Big Data; cloud storage; cloud computing.
16	Energy consumption in manufacturing; process energy demand and effects of workpiece materials.

Acknowledgments

We gratefully acknowledge below individuals for their contributions to various sections in the book, as well as for their reviews, comments, and constructive suggestions in this revision.

We are happy to present below as list of those individuals, in academic institutions as well as in industrial and research organizations who, in one way or another, have made various contributions to this and the recent editions of this book. Kent M. Kalpakjian, Micron Technology, Inc., was the original author of the sections on the fabrication of microelectronic devices.

D. Adams
G. Boothroyd
D. Bourell
J. Cesarone
A. Cinar
D.A. Dornfeld
M. Dugger
D.R. Durham
M. Giordano
M. Hawkins
K. Jones
R. Kassing
K. Kozlovsky
K.M. Kulkarni
M. Madou
H. Malkani
M. Molnar
S. Mostovoy
C. Nair
P.G. Nash
J. Neidig
C. Petronis
M. Prygoski
R. Shivpuri
K.S. Smith
B.S. Thakkar
J.E. Wang
K.R. Williams
P.K. Wright

We would also like to acknowledge the dedication and continued help and cooperation of our editor, Holly Stark, and Erin Ault, Content Producer.

We are grateful to many organizations that supplied us with numerous illustrations, videos, and various materials for the text. These contributions have specifically been acknowledged throughout the text.

SEROPE KALPAKJIAN
STEVEN R. SCHMID

About the Authors

Serope Kalpakjian is professor emeritus of mechanical and materials engineering at the Illinois Institute of Technology. He is the author of *Mechanical Processing of Materials* and co-author of *Lubricants and Lubrication in Metalworking Operations* (with E.S. Nachtman); both of the first editions of his textbooks *Manufacturing Processes for Engineering Materials* and *Manufacturing Engineering and Technology* have received the M. Eugene Merchant Manufacturing Textbook Award. He has conducted research in various areas of manufacturing, is the author of numerous technical papers and articles in handbooks and encyclopedias, and has edited several conference proceedings. He also has been editor and co-editor of various technical journals and has served on the editorial board of Encyclopedia Americana.

Among other awards, Professor Kalpakjian has received the Forging Industry Educational and Research Foundation Best Paper Award, the Excellence in Teaching Award from IIT, the ASME Centennial Medallion, the International Education Award from SME, A Person of the Millennium Award from IIT; the Albert Easton White Outstanding Teacher Award from ASM International, and the 2016 SME Gold Medal. The Outstanding Young Manufacturing Engineer Award, by SME, in 2001, was named after him. Professor Kalpakjian is a Life Fellow ASME, Fellow SME, Fellow and Life Member ASM International, Fellow Emeritus the International Academy for Production Engineering (CIRP), and is a founding member and past president of NAMRI. He is a graduate of Robert College (High Honor, Istanbul), Harvard University, and the Massachusetts Institute of Technology.

Steven R. Schmid is professor of Aerospace and Mechanical Engineering at the University of Notre Dame, where he teaches and conducts research in the general areas of manufacturing, machine design, and tribology. He received his B.S. degree from Illinois Institute of Technology (with Honors) and Master's and Ph.D. degrees from Northwestern University, all in mechanical engineering. He has received numerous awards, including the John T. Parsons Award from SME, the Newkirk Award from ASME, the Kaneb Center Teaching Award (three times), and the Ruth and Joel Spira Award for Excellence in Teaching.

Professor Schmid served as the President of the North American Manufacturing Research Institution (NAMRI, 2015–2016) and was appointed the first Academic Fellow at the Advanced Manufacturing National Program Office, US Department of Commerce, where he helped design the National Network for Manufacturing Innovation. Starting in 2016, he will serve as the Program Director for the Manufacturing Machines and Equipment program at the National Science Foundation. Dr. Schmid is the author of over 140 technical papers, and has co-authored the texts *Fundamentals of Machine Elements*, *Fundamentals of Fluid Film Lubrication*, and *Manufacturing Engineering and Technology*.

Introduction

CHAPTER 1

The objectives of this chapter are:

- Define manufacturing and describe the technical and economic considerations involved in manufacturing successful products.
- Explain the relationships among product design and engineering and factors such as materials and processes selection and the various costs involved.
- Describe the important trends in modern manufacturing and how they can be utilized in a highly competitive global marketplace to minimize production costs.

1.1 What Is Manufacturing?

As you read this Introduction, take a few moments to inspect various objects around you: pencil, paper clip, laptop computer, bicycle, and smartphone. You will note that these objects have been transformed from various raw materials into individual parts and then assembled into specific products. Some objects, such as nails, bolts, and paper clips, are made of one material; the vast majority of products are, however, made of numerous parts from a wide variety of materials (Fig. 1.1). A ball-point pen, for example, consists of about a dozen parts, a lawn mower about 300 parts, a grand piano about 12,000 parts, a typical automobile about 15,000 parts, and a Boeing 787 about 2.3 million parts. All are produced by a combination of processes, called manufacturing.

Manufacturing is the process of converting raw materials into products; it encompasses the design and manufacturing of goods using various production methods and techniques. Manufacturing began during 5000 to 4000 B.C. with the production of various articles, such as pottery, knives and tools, from wood, clay, stone, and metal (Table 1.1). The word *manufacturing* is derived from the Latin *manu* and *factus*, meaning made by hand; the word *manufacture* first appeared in 1567, and the word *manufacturing*, in 1683. The word **production** is also used interchangeably with the word manufacturing.

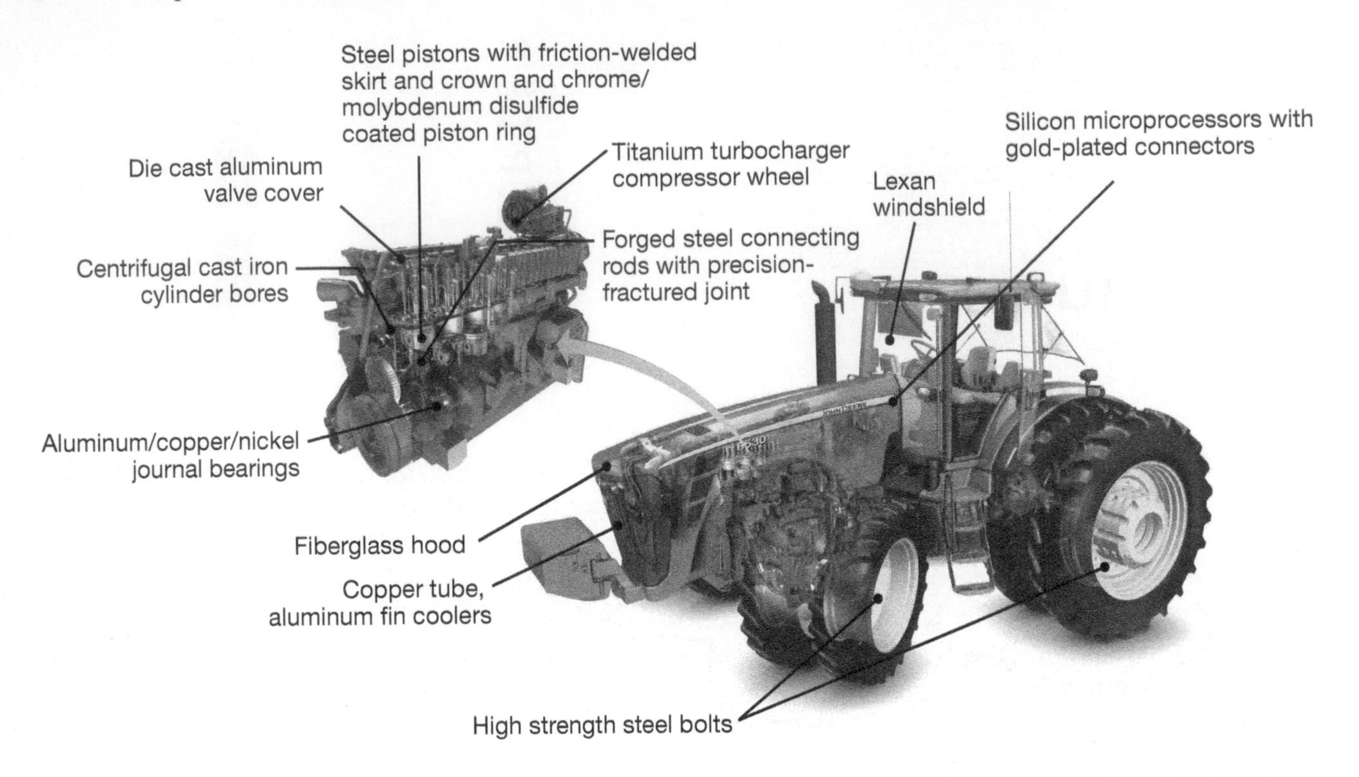

FIGURE 1.1 Model 8430 tractor, with detailed illustration of its diesel engine, showing the variety of materials and processes incorporated. *Source:* Courtesy of John Deere Company.

Manufacturing may produce *discrete products*, meaning individual parts, such as nails, rivets, bolts, and steel balls. On the other hand, wire, sheet metal, tubing, and pipe are *continuous products*, that may then be cut into individual pieces and thus become discrete products.

Because a manufactured item has undergone a number of changes whereby raw material has become a specific and useful product, it has **added value**, defined by its monetary worth. Clay, for example, has a certain value as mined; when the clay is used to make a ceramic dinner plate, a cutting tool, or an electrical insulator, value is added to the clay. Similarly, a wire coat hanger has added value over and above the cost of a piece of wire from which it is made.

Manufacturing is extremely important for national and global economies. Consider Fig. 1.2, which shows the economic multiplier of different sectors in the US economy. The economic multiplier indicates the amount of general activity in the economy generated from one dollar of activity in a given sector. Note that manufacturing has a multiplier over 1.5, and that it is higher than any other sector in the economy. This high economic multiplier has a number of implications, including:

1. The wealth of a country is closely tied to the level of its manufacturing activity, especially in advanced manufacturing or high value-added processes.

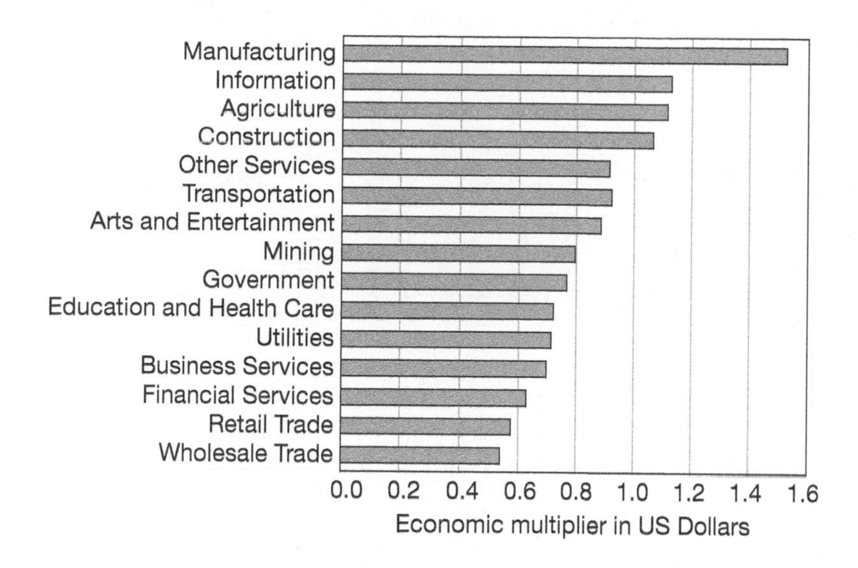

FIGURE 1.2 Economic multiplier for different sectors of the US Economy. *Source:* US Dept. of Commerce.

2. Although the emergence of global economies is often perceived as detrimental in the media, global trading leads to an increase in the wealth of each participating nation, due to the economic multiplier effect. However, to take advantage of this multiplier, a vibrant manufacturing sector is essential.
3. Nations with large Gross Domestic Product (GDP) growth have their economic activity concentrated on high value-added products, such as automobiles, airplanes, medical devices, computers, electronics, and various machinery. Other products, such as clothing, toys, and hand-held tools, are labor-intensive and thus are generally concentrated in countries where labor rates are lower.
4. Nations where labor rates are high can still compete by the application of automation and by continuous improvements in worker productivity.

Manufacturing is a complex activity, involving people who have a broad range of disciplines and skills, together with a wide variety of machinery, equipment, and tools with various levels of automation and controls, such as computers, robots, and material-handling equipment. Manufacturing activities must be responsive to several demands and trends:

1. A product must fully meet **design requirements, specifications,** and **standards.**
2. It must be manufactured by the most **economical** and **environmentally friendly** methods.
3. **Quality** must be **built into the product** at each stage, from design to assembly, rather than relying on quality testing after the product is made.
4. In a highly competitive and global environment, production methods must be sufficiently **flexible** to respond to constantly changing market demands, types of products, production rates and quantities, and **on-time delivery** to the customer.

TABLE 1.1 Historical development of materials and manufacturing processes.

Period	Dates	Metals and casting	Various materials and composites	Forming and shaping	Joining	Tools, machining, and manufacturing systems
Egypt: ~ 3100 B.C. to ~ 300 B.C. Greece: ~ 1100 B.C. to ~ 146 B.C. Roman Empire: ~ 500 B.C. to 476 A.D.	Before 4000 B.C.	Gold, copper, meteoric iron	Earthenware, glazing, natural fibers	Hammering		Tools of stone, flint, wood, bone, ivory, composite tools
	4000–3000 B.C.	Copper casting, stone and metal molds, lost-wax process, silver, lead, tin, bronze		Stamping, jewelry	Soldering (Cu-Au, Cu-Pb, Pb-Sn)	Corundum (alumina, emery)
	3000–2000 B.C.	Bronze casting and drawing, gold leaf	Glass beads, potter's wheel, glass vessels	Wire by slitting sheet metal	Riveting, brazing	Hoe making, hammered axes, tools for ironmaking and carpentry
	2000–1000 B.C.	Wrought iron, brass				
	1000–1 B.C.	Cast iron, cast steel	Glass pressing and blowing	Stamping of coins	Forge welding of iron and steel, gluing	Improved chisels, saws, files, woodworking lathes
Middle Ages: ~ 476 to 1492 Renaissance: 14th to 16th centuries	1–1000 A.D.	Zinc, steel	Venetian glass	Armor, coining, forging, steel swords		Etching of armor
	1000–1500	Blast furnace, type metals, casting of bells, pewter	Crystal glass	Wire drawing, gold- and silversmith work		Sandpaper, windmill-driven saw
	1500–1600	Cast-iron cannon, tinplate	Cast plate glass, flint glass	Water power for metalworking, rolling mill for coinage strips		Hand lathe for wood
	1600–1700	Permanent-mold casting, brass from copper and metallic zinc	Porcelain	Rolling (lead, gold, silver), shape rolling (lead)		Boring, turning, screw-cutting lathe, drill press

Industrial Revolution: ~ 1750 to 1850	1700–1800	Malleable cast iron, crucible steel (iron bars and rods)		Extrusion (lead pipe), deep drawing, rolling		
WWI	1800–1900	Centrifugal casting, Bessemer process, electrolytic aluminum, nickel steel, babbitt, galvanized steel, powder metallurgy, open-hearth steel	Window glass from slit cylinder, light bulb, vulcanization, rubber processing, polyester, styrene, celluloid, rubber extrusion, molding	Steam hammer, steel rolling, seamless tube, steel-rail rolling, continuous rolling, electroplating		Shaping, milling, copying lathe for gunstocks, turret lathe, universal milling machine, vitrified grinding wheel
	1900–1920		Automatic bottle making, bakelite, borosilicate glass	Tube rolling, hot extrusion	Oxyacetylene; arc, electrical-resistance, and thermit welding	Geared lathe, automatic screw machine, hobbing, high-speed steel tools, aluminum oxide, and silicon carbide (synthetic)
	1920–1940	Die casting	Development of plastics, casting, molding, polyvinyl chloride, cellulose acetate, polyethylene, glass fibers	Tungsten wire from metal powder	Coated electrodes	Tungsten carbide, mass production, transfer machines
WWII	1940–1950	Lost-wax process for engineering parts	Acrylics, synthetic rubber, epoxies, photosensitive glass	Extrusion (steel), swaging, powder metals for engineering parts	Submerged arc welding	Phosphate conversion coatings, total quality control
	1950–1960	Ceramic mold, nodular iron, semiconductors, continuous casting	Acrylonitrile-butadiene-styrene, silicones, fluorocarbons, polyurethane, float glass, tempered glass, glass ceramics	Cold extrusion (steel), explosive forming, thermochemical processing	Gas metal arc, gas tungsten arc, and electroslag welding; explosion welding	Electrical and chemical machining, automatic control.

(*continued*)

TABLE 1.1 Historical development of materials and manufacturing processes (*continued*).

Period	Dates	Metals and casting	Various materials and composites	Forming and shaping	Joining	Tools, machining, and manufacturing systems
Space Age	1960–1970	Squeeze casting, single-crystal turbine blades	Acetals, polycarbonate, cold forming of plastics, reinforced plastics, filament winding	Hydroforming, hydrostatic extrusion, electroforming	Plasma-arc and electron-beam welding, adhesive bonding	Titanium carbide, synthetic diamond, numerical control, integrated circuit chip
	1970–1990	Compacted graphite, vacuum casting, organically-bonded sand, automation of molding and pouring, rapid solidification, metal-matrix composites, semi-solid metalworking, amorphous metals, shape-memory alloys	Adhesives, composite materials, semiconductors, optical fibers, structural ceramics, ceramic-matrix composites, biodegradable plastics, electrically-conducting polymers	Precision forging, isothermal forging, superplastic forming, dies made by computer-aided design and manufacturing, net-shape forging and forming, computer simulation	Laser beam, diffusion bonding (also combined with superplastic forming), surface-mount soldering	Cubic boron nitride, coated tools, diamond turning, ultraprecision machining, computer-integrated manufacturing, industrial robots, machining and turning centers, flexible manufacturing systems, sensor technology, automated inspection, computer simulation and optimization
Information Age	1990–2000	Rheocasting, computer-aided design of molds and dies, rapid tooling	Nanophase materials, metal foams, high-temperature superconductors, machinable ceramics, diamond-like carbon	Additive manufacturing, rapid tooling, environmentally-friendly metalworking fluids	Friction stir welding, lead-free solders, laser butt-welded (tailored) sheet-metal blanks	Micro- and nanofabrication, LIGA, dry etching, linear motor drives, artificial neural networks, Six Sigma
	2000–2010s	TRIP and TWIP steels	Carbon nanotubes, graphene	Single point incremental forming, hot stamping, electrically assisted forming	Linear friction welding	Digital manufacturing, three-dimensional computer chips, blue-arc machining, soft lithography, flexible electronics

5. New developments in **materials, production methods,** and **computer integration** of both technological and managerial activities must constantly be evaluated with a view to their timely and economic implementation.
6. Manufacturing activities must be viewed as a large **system,** each part of which is interrelated. Such systems can be modeled in order to study the effect of various factors, such as changes in market demand, product design, materials, costs, and production methods, on product quality and cost.
7. A manufacturer must work with the customer for timely feedback for **continuous product improvement.**
8. **Global sourcing** of components and products requires adherence to quality systems, and the management of global supply chains.
9. A manufacturing organization must constantly strive for **higher productivity,** defined as the optimum use of all its resources: materials, technology, machines, energy, capital, and labor.

1.2 Product Design and Concurrent Engineering

Product design is a critical activity because it has been estimated that, generally, 70 to 80% of the cost of product development and manufacture, as well as environmental impact and energy consumption, are determined at the *initial* stages of product design. The design process for a particular product first requires a clear understanding of the functions and the performance expected of that product. The product may be new or it may be an improved model of an existing product. The market for the product and its anticipated uses must be defined clearly, with the assistance of sales personnel, market analysts, and others in the organization.

Product development generally follows the flow outlined in Table 1.2. *Technology Readiness Level* (TRL) and *Manufacturing Readiness Level* (MRL) are measures of a product's ability to be produced, marketed, and sold. In practice, all technologies must progress from some starting point up to a TRL and MRL of 9. A new scientific discovery or product idea begins at a TRL of 1 and it may or may not ever be suitable for commercial application. New versions of existing products may start at some higher TRL or MRL level, but the flow of development is always the same.

Note that each stage of a product development generally requires different skills and resources. Demonstrating a new concept in a laboratory environment (TRL 3) and demonstrating the concept in a new system in a real environment (TRL 7) are very different tasks. Similarly, producing a laboratory prototype (MRL 4) is very different from demonstrating manufacturing strategies for producing a product at scale (MRL 7), which is also very different from having a production facility in place.

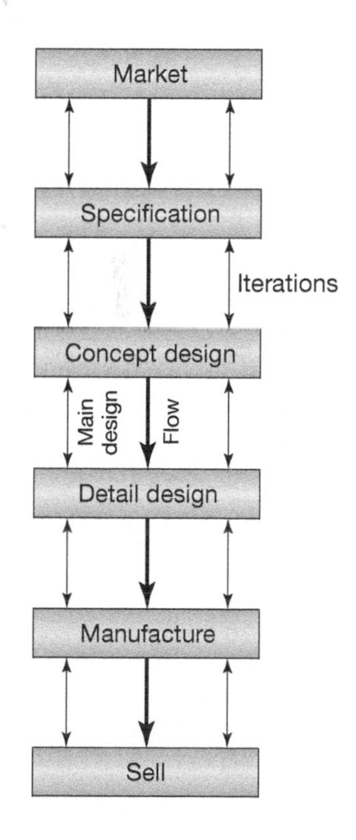

FIGURE 1.3 General product flow, from market analysis to selling the product, and depicting concurrent engineering. *Source:* After S. Pugh.

Another proposed product development process is shown in Fig. 1.3; it still has a general product flow, from market analysis to design and to manufacturing, but it contains deliberate iterations. Also, while not shown explicitly, it is recognized that all disciplines are involved in the earliest stages of product design. They progress concurrently, so that the iterations (which, by nature, occur from design changes or decisions

TABLE 1.2 Definitions of Technology Readiness Level (TRL) and Manufacturing Readiness Level (MRL).

TRL	Description	MRL	Description
1	Basic principles observed and reported	1	Manufacturing feasibility assessed
2	Technology concept and/or application formulated	2	Manufacturing concepts defined
3	Analytical and experimental critical function and/or characteristic proof of concept	3	Manufacturing concepts developed
4	Component and/or breadboard validation in a laboratory environment	4	Capability to produce the technology in a laboratory environment
5	Component or breadboard validation in a relevant environment	5	Capability to produce prototype components in a production relevant environment
6	System/subsystem model or prototype demonstration in a relevant environment	6	Capability to produce a prototype system or subsystem in a production relevant environment
7	System prototype demonstration in an operational environment	7	Capability to produce systems, subsystems or components in a production representative environment
8	Actual system completed and qualified through test and demonstration	8	Pilot line capability demonstrated; ready to begin low rate initial production
9	Actual system proven through successful mission operations	9	Low rate production demonstrated; capability in place to begin full rate production

between alternatives) result in less wasted effort and lost time. A key to this approach is the well-recognized importance of *communication* among and within different disciplines. While there must be communication among engineering, marketing, and service functions, there must also be avenues of interaction between engineering sub-disciplines, such as design for manufacture, design for recyclability, and design for safety.

Concurrent engineering, also called **simultaneous engineering**, is a systematic approach integrating the design and manufacture of products with the view toward optimizing all elements involved in the *life cycle* of the product (see Section 1.4). The basic goals of concurrent engineering are to minimize product design and engineering changes, as well as the time and costs involved in taking the product from design concept to production and introduction of the product into the marketplace.

Although the concept of concurrent engineering appears to be logical and efficient, its implementation can require considerable time and effort, especially when those using it either are not able to work as a team or fail to appreciate its real benefits. For concurrent engineering to succeed it must: (a) have the full support of an organization's top management; (b) have multifunctional and interacting work teams, including support groups; and (c) utilize all available state-of-the-art technologies.

A powerful and effective tool, particularly for complex production systems, is **computer simulation** in evaluating the performance of the product and the design of the manufacturing system to produce it. Computer simulation also helps in the early detection of design flaws, identifying possible problems in a particular production system, and optimizing manufacturing lines for minimum product cost. Several computer simulation software packages, using animated graphics and with various capabilities, are widely available.

Some steps in the production process will require a **prototype**, a physical model of the product. An important technique is **additive manufacturing** (Section 10.12), that relies on CAD/CAM and various manufacturing techniques (typically using polymers or metal powders) to rapidly produce prototypes of a part. These techniques are now advanced to such an extent that they can be used for low-volume economical production of actual parts.

During the prototype stage, modifications of the original design, the materials selected, or production methods may be necessary. After this phase has been completed, appropriate process plans, manufacturing methods (Table 1.3), equipment, and tooling are selected, with the cooperation of manufacturing engineers, process planners, and all those involved in production.

TABLE 1.3 Shapes and some common methods of production.

Shape or feature	Production method[a]
Flat surfaces	Rolling, planing, broaching, milling, shaping, grinding
Parts with cavities	End milling, electrical-discharge machining, electrochemical machining, ultrasonic machining, blanking, casting, forging, extrusion, injection molding, metal injection molding
Parts with sharp features	Permanent-mold casting, machining, grinding, fabricating[b], powder metallurgy, coining
Thin hollow shapes	Slush casting, electroforming, fabricating, filament winding, blow molding, sheet forming, spinning
Tubular shapes	Extrusion, drawing, filament winding, roll forming, spinning, centrifugal casting
Tubular parts	Rubber forming, tube hydroforming, explosive forming, spinning, blow molding, sand casting, filament winding
Curvature on thin sheets	Stretch forming, peen forming, fabricating, thermoforming
Openings in thin sheets	Blanking, chemical blanking, photochemical blanking, laser machining
Cross sections	Drawing, extrusion, shaving, turning, centerless grinding, swaging, roll forming
Square edges	Fine blanking, machining, shaving, belt grinding
Small holes	Laser or electron-beam machining, electrical-discharge machining, electrochemical machining, chemical blanking
Surface textures	Knurling, wire brushing, grinding, belt grinding, shot blasting, etching, laser texturing, injection molding, compression molding
Detailed surface features	Coining, investment casting, permanent-mold casting, machining, injection molding, compression molding
Threaded parts	Thread cutting, thread rolling, thread grinding, injection molding
Very large parts	Casting, forging, fabricating, assembly
Very small parts	Investment casting, etching, powder metallurgy, nanofabrication, LIGA, micromachining

Note:
[a] Rapid prototyping operations can produce all of these features to some degree.
[b] 'Fabricating' refers to assembly from separately manufactured components.

CASE STUDY 1.1 Three-Dimensional Printing of Guitars

The design flexibility of additive manufacturing is illustrated by the custom guitars produced by ODD, Inc. These guitars are designed in CAD programs, with full artistic freedom to pursue innovative designs; those in Figure 1.4 are only a selection of the many available. The CAD file is then sent to a three-dimensional printer, using the selective laser sintering process and produced from nylon (Duraform PA). As printed, the guitars are white. They are first dyed to a new base color, then subsequently hand-painted and sprayed with a clear satin lacquer. The customer-specified hardware (pickups, bridges, necks, tuning heads, etc.) are then mounted to produce the electric guitar.

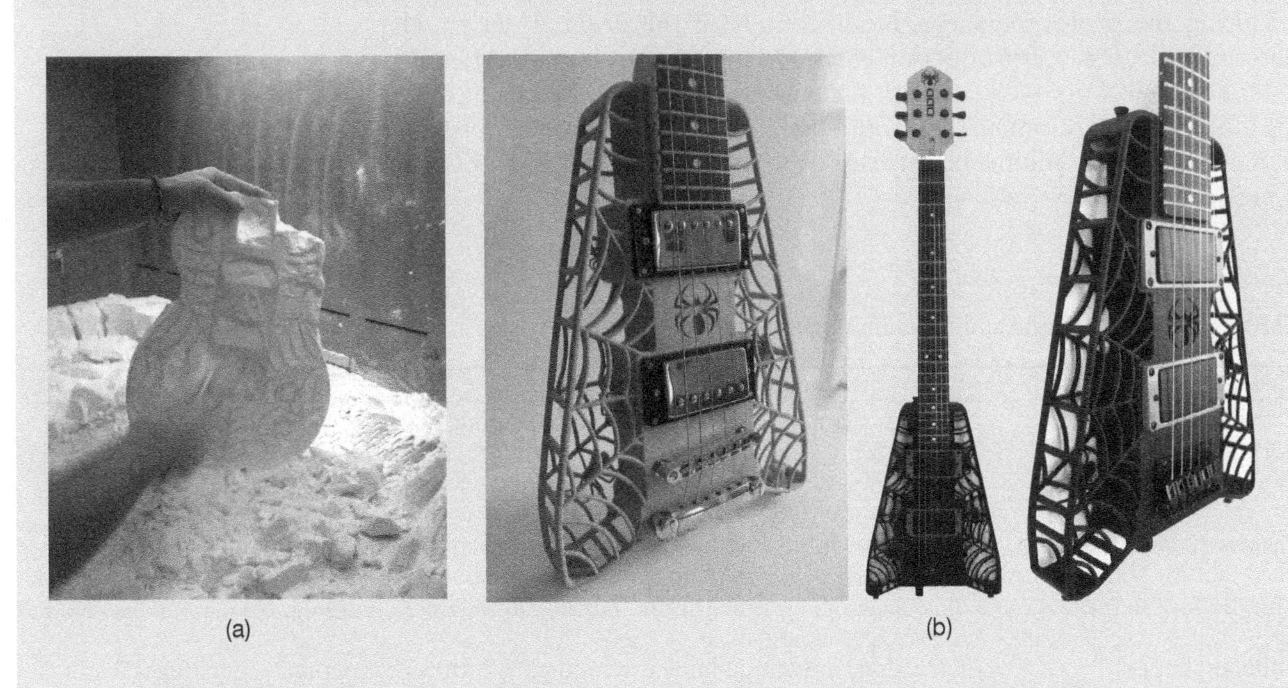

FIGURE 1.4 Guitars produced through additive manufacturing. (a) Spider design being removed from a powder bed. Note that the support material, or *cake*, has some strength and needs to be carefully removed; (b) finished Spider guitars. *Source:* Courtesy of O. Diegel, Massey University, New Zealand.

1.3 Design for Manufacture, Assembly, Disassembly, and Service

Design and manufacturing should never be viewed as separate disciplines and activities. Each part or component of a product must be designed so that it not only meets design requirements and specifications but also so it can be manufactured economically and with relative ease. This broad concept, known as **design for manufacture** (DFM), is a comprehensive approach to the production of goods. It integrates the product design process with materials, manufacturing methods, process planning, assembly, testing, and quality assurance.

Effective implementation of design for manufacture requires that designers acquire a fundamental understanding of the characteristics, capabilities, and limitations of materials, production methods, machinery, and equipment. Also included are such characteristics as variability in machine performance, dimensional accuracy and surface finish of the parts produced, processing time, and the effect of processing methods on part quality.

Designers and product engineers assess the impact of any design modifications on manufacturing process selection, tools and dies, assembly methods, inspection and, especially, product cost. Establishing quantitative relationships is essential in order to optimize the design for ease of manufacture and assembly at *minimum cost* (also called *producibility*). Computer-aided design, engineering, manufacturing, and process planning techniques, using powerful computer software, are indispensable to such analysis. They include **expert systems**, which are computer programs with optimization capabilities, thus expediting the traditional iterative process in design optimization.

After individual parts have been manufactured, they are assembled into a product. **Assembly** is an important phase of the overall manufacturing operation, requiring the consideration of the ease, speed, and cost of putting parts together (Fig. 1.5). Products should be designed for quick and easy **disassembly** so that they can be taken apart for maintenance, servicing, or recycling of their components.

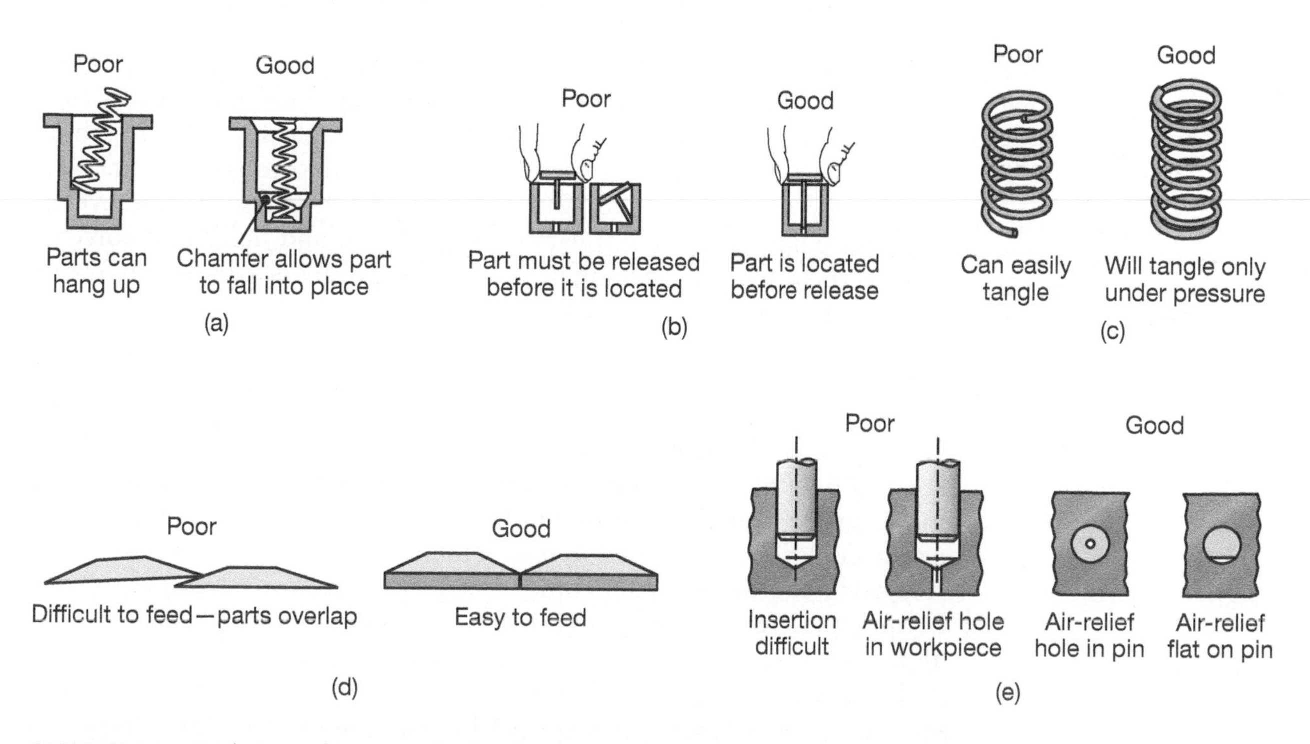

FIGURE 1.5 Redesign of parts to facilitate automated assembly.
Source: Boothroyd, G., Dewhurst, P., and Knight, W.A., *Product Design for Manufacture and Assembly*, 3rd ed., 2010, CRC Press.

Because assembly operations can contribute significantly to product cost, **design for assembly** (DFA), and **design for disassembly** are important aspects of manufacturing. **Design for service** is another important aspect, ensuring that individual parts in a product are easy to reach and service; if they are difficult to access for service, they are generally designed to be more robust. These activities have now been combined into **design for manufacture and assembly** (DFMA), which recognizes the inherent and important interrelationships among design, manufacturing, and assembly of products.

Design principles for economical production may be summarized as follows:

- Designs should be as simple as possible to manufacture, assemble, disassemble, service, and recycle.
- Materials should be chosen for their appropriate design and manufacturing characteristics as well as for their service life.
- Dimensional accuracy and surface finish requirements should be as broad as permissible.
- Secondary and finishing operations should be avoided or minimized, because they can add significantly to product cost.

1.4 Environmentally Conscious Design, Sustainable Manufacturing, and Product Life Cycle

In the United States alone, more than 30 billion kg of plastic products are discarded each year, and 62 billion kg of paper products. Every three months, US industries and consumers discard enough aluminum to rebuild the country's commercial air fleet. Globally, countless tons of automobiles, televisions, appliances, and computers are discarded each year. Metalworking fluids, such as lubricants and coolants, and fluids and solvents, such as those used in cleaning manufactured products, can pollute the air and water, unless recycled or disposed of properly.

Likewise, there are numerous byproducts from manufacturing plants: (a) sand with additives, used in metal-casting processes; (b) water, oil, and other fluids from heat-treating and facilities; (c) slag from foundries and welding operations; and (d) a wide variety of metallic and nonmetallic scrap produced in such operations as sheet forming, casting, and molding. Consider also the various effects of water and air pollution, acid rain, ozone depletion, hazardous wastes, landfill seepage, and global warming. Recycling efforts have gained increasing momentum over the years; aluminum cans, for example, are now recycled at a rate of 67% and plastics at around 9%.

The present and potential adverse effects of these activities, their damage to the environment and to the earth's ecosystem, and, ultimately, their effects on the quality of human life are now well recognized. In response, a wide range of laws and regulations have been and continue to be promulgated by local, state, and federal governments and international organizations. The regulations are generally stringent, and their implementation can have a major impact on the economic operation of manufacturing facilities. These efforts have been most successful when there is value added, such as in reducing energy requirements (and associated costs) or substituting materials that have both environmental design and cost benefits.

Major progress continues to be made regarding **design for recycling** (DFR) and **design for the environment** (DFE), or **green design**, indicating universal awareness of the challenges outlined above; waste has become unacceptable. This comprehensive approach anticipates the possible negative environmental impact of materials, products, and processes, so that they can be considered at the earliest stages of design and production.

Among various developments is **sustainable manufacturing**, which refers to the realization that our natural resources are vital to global economic activity, and that energy and materials management are essential to ensuring that resources will continue to be available for future generations. The basic guidelines to be followed are:

- Reducing waste of materials *at their source*, by refinements in product design and the amount of materials used.
- Reducing the use of hazardous materials in products and processes.
- Ensuring proper handling and disposal of all waste.
- Making improvements in waste treatment and in recycling and reuse of materials.

The **cradle-to-cradle** philosophy encourages the use of environmentally friendly materials and designs. By considering the entire life cycle of a product, materials can be selected and employed with minimal real waste. Environmentally friendly materials can be:

- Part of a *biological cycle*, where (usually organic) materials are used in design, function properly for their intended life, and can then be safely disposed of. Such materials degrade naturally and, in the simplest version, lead to new soil that can sustain life.
- Part of an *industrial cycle*, such as aluminum in beverage containers, that serve an intended purpose and are then recycled and the same material is re-used continuously.

Product Life Cycle (PLC). *Product life cycle* consists of all the stages that a product goes through, from design, development, production, distribution, use, and to its ultimate disposal and recycling. Typically, a product undergoes the following five stages:

1. Product *development* stage, which can involve much time and high costs,
2. *Market introduction* stage, in which the acceptance of the product in the marketplace is closely watched,
3. *Growth* stage, with increasing sales volume, lower manufacturing cost per unit, and hence higher profitability to the manufacturer,

4. *Maturation* stage, where sales volume begins to peak and competitive products begin to appear in the marketplace, and
5. *Decline* stage, with decreasing sales volume and profitability.

Product Life Cycle Management (PLCM). While life cycle consists of stages from development to the ultimate disposal or recycling of the product, *product life cycle management* is generally defined as the strategies employed by the manufacturer as the product goes through its life cycle. Various strategies can be employed in PLCM, depending on the type of product, customer response, and market conditions.

Guidelines for Green Design and Manufacturing. In reviewing the various activities described thus far, it can be noted that there are overarching relationships among the basic concepts of DFMA, DFD, DFE, and DFR. These relationships can be summarized as guidelines, now rapidly being accepted worldwide:

1. Reduce waste of materials, by refining product design, reducing the amount of materials used in products, and selecting manufacturing processes that minimize scrap, such as forming instead of machining.
2. Reduce the use of hazardous materials in products and processes.
3. Investigate manufacturing technologies that produce environmentally friendly and safe products and by-products.
4. Make improvements in methods of recycling, waste treatment, and reuse of materials.
5. Minimize energy use and, whenever possible, encourage the use of renewable sources of energy. Selection of materials can have a major impact on the latent energy in products, as described in Section 16.5.
6. Encourage recycling by using materials that are a part of either industrial or biological cycling, but not both in the same product. Ensure proper handling and disposal of all waste of materials used in products, but are not appropriate for industrial or biological cycling.

1.5 Selecting Materials

An increasingly wide variety of materials is now available, each having its own characteristics, composition, applications, advantages, limitations, and costs. The major types of materials used in manufacturing today are:

1. **Ferrous metals**: carbon steels, alloy steels, stainless steels, and tool and die steels (Chapter 3).
2. **Nonferrous metals and alloys**: aluminum, magnesium, copper, nickel, superalloys, titanium, refractory metals (molybdenum, niobium, tungsten, and tantalum), beryllium, zirconium, low-melting alloys (lead, zinc, and tin), and precious metals (Chapter 3).
3. **Plastics**: thermoplastics, thermosets, and elastomers (Chapter 10).
4. **Ceramics**: glass ceramics, glasses, graphite, and diamond (Chapter 11).

5. **Composite materials**: reinforced plastics, metal-matrix and ceramic-matrix composites, and honeycomb structures; these are also known as *engineered materials* (Chapters 10 and 11).
6. **Nanomaterials, shape-memory alloys, metal foams, amorphous alloys, superconductors,** and **semiconductors** (Chapters 3 and 13).

Material substitution. As new or improved materials continue to be developed, there are important trends in their selection and application. Aerospace structures, sporting goods, automobiles and numerous high-tech products have especially been at the forefront of new material usage. Because of the vested interests of the producers of different types of natural and engineered materials, there are constantly shifting trends in the use of these materials, driven principally by economics. For example, by demonstrating steel's technical and economic advantages, steel producers are countering the increased use of plastics in automobiles and aluminum in beverage cans. Likewise, aluminum producers are countering the use of various materials in automobiles (Fig. 1.6).

Some examples of the use or substitution of materials in common products are: (a) metal vs. plastic paper clips; (b) plastic vs. sheet metal light-switch plates; (c) wood vs. metal or reinforced-plastic handles for hammers; (d) glass vs. metal or plastic water pitchers; (e) plastic vs. copper pipes; (f) sheet metal vs. plastic chairs; (g) galvanized steel vs. copper or aluminum nails, and (h) aluminum vs. cast iron pans.

Material properties. In selecting materials for products, the first consideration generally involves **mechanical properties** (Chapter 2); typically, strength, toughness, ductility, hardness, elasticity, and fatigue and creep resistance. These properties can be modified by various heat treatment methods (Chapter 5). The *strength-to-weight* and *stiffness-to-weight ratios* of materials are also important considerations, particularly for aerospace

Robotically-applied, advanced arc-welding processes provide consistent, high-quality assembly of castings, extrusions and sheet components
Die-cast components are thin-walled to maximize weight reduction yet provide high performance
Advanced extrusion bending processes support complex shapes and tight radii
Strong, thin-walled hydroformed extrusions exhibit high ductility, energy absorption and toughness

(a) (b)

FIGURE 1.6 (a) The C7 Corvette sports car, using an all-aluminum chassis which is 45 kg lighter and 57% stiffer than the C6 Corvette; (b) The aluminum body structure, showing various components made by extrusion, sheet forming, and casting processes. *Source:* Courtesy of General Motors Corp.

and automotive applications. Aluminum, titanium, and reinforced plastics, for example, have higher strength-to-weight ratios than steels and cast irons. The mechanical properties specified for a product and its components should be appropriate for the conditions under which the product is expected to function.

Physical properties (Chapter 3), such as density, specific heat, thermal expansion and conductivity, melting point, and electrical and magnetic properties also have to be considered. **Chemical properties** play a significant role in hostile as well as normal environments. Oxidation, corrosion, general degradation of properties, and flammability of materials are among the important factors to be considered, as is toxicity; see, for example, the development of lead-free solders (Chapter 12). Physical and chemical properties are both important in advanced machining processes (Chapter 9). The **manufacturing properties** of materials determine whether they can be processed (cast, formed, shaped, machined, welded, or heat treated for property enhancement). Moreover, the specific methods used to process materials to the desired shapes should not adversely affect the product's final properties, service life, and its cost.

Cost and availability. The economic aspects of material selection are as important as the technological considerations. Cost and availability of raw and processed materials are a major concern in manufacturing. If these materials are not commercially available in the desired shapes, dimensions, tolerances, and quantities, substitutes or additional processing may be required; these steps can contribute significantly to product cost, as described in Chapter 16. For example, if a round bar of a specific diameter is required but is not commercially available, then a larger rod may be purchased and reduced in its diameter, as by machining, drawing though a die, or grinding.

Reliability of supply, as well as of demand, can significantly affect material costs. Most countries import numerous raw materials that are essential for their production of goods. The United States, for example, imports natural rubber, diamond, cobalt, titanium, chromium, aluminum, copper, and nickel from various countries. The geopolitical implications of such reliance on other countries are self-evident.

Various costs are involved in processing materials by different methods. Some methods require expensive machinery, others require extensive labor (*labor intensive*), and still others require personnel with special skills, high levels of formal education, or specialized training.

Service life and recycling. Time- and service-dependent phenomena, such as wear, fatigue, creep, and dimensional stability, are important considerations as they can significantly affect a product's performance and, if not controlled, can lead to failure of the product. The compatibility of the different materials used in a product is also important, such as the galvanic action between mating parts made of dissimilar metals, causing corrosion of the parts. Recycling or proper disposal of the individual components in a product at the end of its useful life is important, as is the proper treatment and disposal of toxic wastes.

CASE STUDY 1.2 Baseball Bats

Baseball bats for major and minor leagues are generally made of wood from the northern white ash tree, a wood that has high dimensional stability and a high elastic modulus and strength-to-weight ratio, and high shock resistance. Wooden bats can, however, break and may cause serious injury; this is especially true of the recent trend of using maple wood for bats. Wooden bats are made on semiautomatic lathes (Section 8.9.2), followed by finishing operations and labeling. The straight uniform grain, required for such bats, has become increasingly difficult to find, particularly when the best wood comes from ash trees at least 45 years old.

For the amateur market and school and college players, aluminum bats (top portion of Fig. 1.7) have been made since the 1970s as a cost-saving alternative to wood. The bats are made by various processes, described in Chapter 7. Metal bats are now mostly made from high-strength aluminum tubing, but can also incorporate titanium. The bats are designed to have the same center of percussion (the *sweet spot*) as wooden bats, and are usually filled with polyurethane or cork for improved sound damping and controlling the balance of the bat.

Metal bats possess desirable performance characteristics, such as lower weight than wooden bats, optimum weight distribution along the bat's length, and superior impact dynamics. Also, as documented by scientific studies, there is a general consensus that metal bats outperform wooden bats. Developments in bat materials include composite materials (Section 10.9), consisting of high-strength graphite fibers embedded in an epoxy resin matrix. The inner woven sleeve (lower portion of Fig. 1.7) is made of Kevlar fibers (an aramid), which add strength to the bat and dampen its vibrations.

FIGURE 1.7 Cross sections of baseball bats made of aluminum (top two) and composite material (bottom two).

Source: Mizuno Sports, Inc.

1.6 Selecting Manufacturing Processes

As can be seen in Table 1.3, a wide range of manufacturing processes are available to produce a wide variety of parts, shapes, and sizes. Note also that there is often more than one method of manufacturing a particular part from a given material (Fig. 1.8). The broad categories of processing methods can be listed as follows:

- **Casting:** Expendable mold and permanent mold (Chapter 5).
- **Forming and shaping:** Rolling, forging, extrusion, drawing, sheet forming, powder metallurgy, and molding (Chapters 6, 7, 10, and 11).
- **Machining:** Turning, boring, drilling, milling, planing, shaping, broaching, grinding, ultrasonic machining; chemical, electrical, and electrochemical machining; and high-energy beam machining (Chapters 8 and 9).
- **Joining:** Welding, brazing, soldering, diffusion bonding, adhesive bonding, and mechanical joining (Chapter 12).
- **Micromanufacturing and nanomanufacturing:** Surface micromachining, dry and wet etching, and LIGA (Chapter 13).
- **Finishing:** Honing, lapping, polishing, burnishing, deburring, surface treating, coating, and plating (Chapter 9).

Selection of a particular manufacturing process or a sequence of processes depends not only on the shape of the part to be produced but also on other factors. Brittle and hard materials, for example, cannot easily be shaped whereas they can be cast or machined. A manufacturing process usually alters the properties of materials. For example, metals that are formed at room temperature typically become stronger, harder, and less ductile. Characteristics such as *castability, formability, machinability*, and *weldability* of materials have to be considered.

Examples of manufacturing processes used, or substituted, for common products include: (a) forging vs. casting of crankshafts; (b) sheet metal vs. cast hubcaps (c) casting vs. stamping sheet metal for frying pans; (d) machined vs. powder metallurgy gears; (e) thread rolling vs. machining of bolts; and (f) casting vs. welding of machine tool structures.

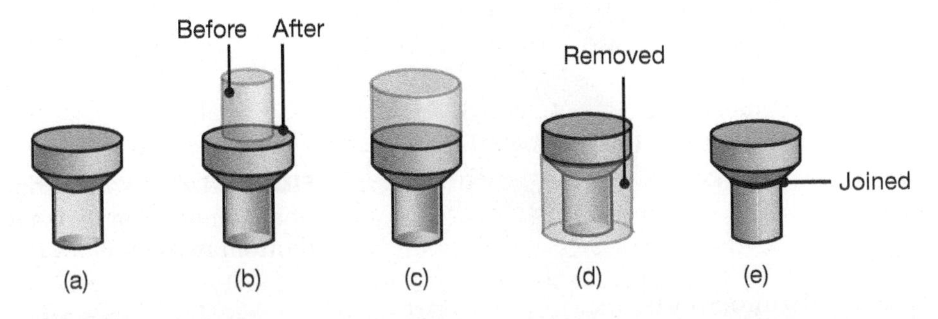

FIGURE 1.8 Various methods of making a simple part: (a) casting or powder metallurgy; (b) forging or upsetting; (c) extrusion; (d) machining; and (e) joining two pieces.

There are constant challenges to develop new solutions to production problems and finding new methods for significant cost reduction. For example, sheet-metal blanks have long been cut from large rolled sheets using traditional punches and dies. Although they are still in wide use, these operations are increasingly replaced by laser-cutting techniques. With computer controls, the laser path can automatically be controlled, producing a wide variety of shapes accurately, repeatedly, and economically.

Part size and dimensional accuracy. The size, thickness, and shape complexity of a part have a major bearing on the process selected. Complex parts, for example, may not be formed easily and economically, whereas they may be produced by casting, injection molding, and powder metallurgy methods, or may be fabricated and assembled from individual pieces. Likewise, flat parts with thin cross sections may not be cast properly. Dimensional tolerances and surface finish (Chapter 4) obtained in hot-working operations cannot be as fine as those in cold-working operations, because dimensional changes, warping, and surface oxidation occur during processing at elevated temperatures. Also, some casting processes produce a better surface finish than others, because of the different types of mold materials used. Moreover, the appearance of materials after they have been manufactured into products greatly influences their appeal to the consumer; color, surface texture, and feel are characteristics typically considered when making a purchasing decision.

The size and shape of manufactured products vary widely (Fig. 1.9). The main landing gear for the twin-engine, 400-passenger Boeing 777 jetliner, for example, is 4.3 m (14 ft) tall, and has three axles and six wheels; the main structure of the landing gear is made by forging, followed by several machining operations (Chapters 6, 8, and 9). At the other extreme is the manufacturing of microscopic parts and mechanisms. These components are produced through surface micromachining operations, typically using electron beam, laser beam, and wet and dry etching techniques on materials such as silicon.

Ultraprecision manufacturing techniques and related machinery have come into common use. For machining mirror-like surfaces, for example, the cutting tool is a very sharp diamond tip; the equipment has very high stiffness and is operated in a room where the temperature is controlled within one degree Celsius. Techniques such as molecular-beam epitaxy are being implemented to achieve dimensional accuracies on the order of the atomic lattice (nanometer).

Microelectromechanical systems (MEMS - see Chapter 13) are micromechanisms combined with an integrated circuit. They are widely used in sensors, cameras, and magnetic storage devices; they can potentially power microrobots to repair human cells, and produce microknives for surgery. Among the more recent developments are **nanoelectromechanical systems** (NEMS) that operate at the same scale as biological molecules (refer to Fig. 1.9). Much effort is being directed at using nanoscale materials, and to develop entire classes of new materials. Carbon nanotubes (Section 13.18) can now reinforce high-performance composite materials, assist in developing nanometer sized electronic devices, provide

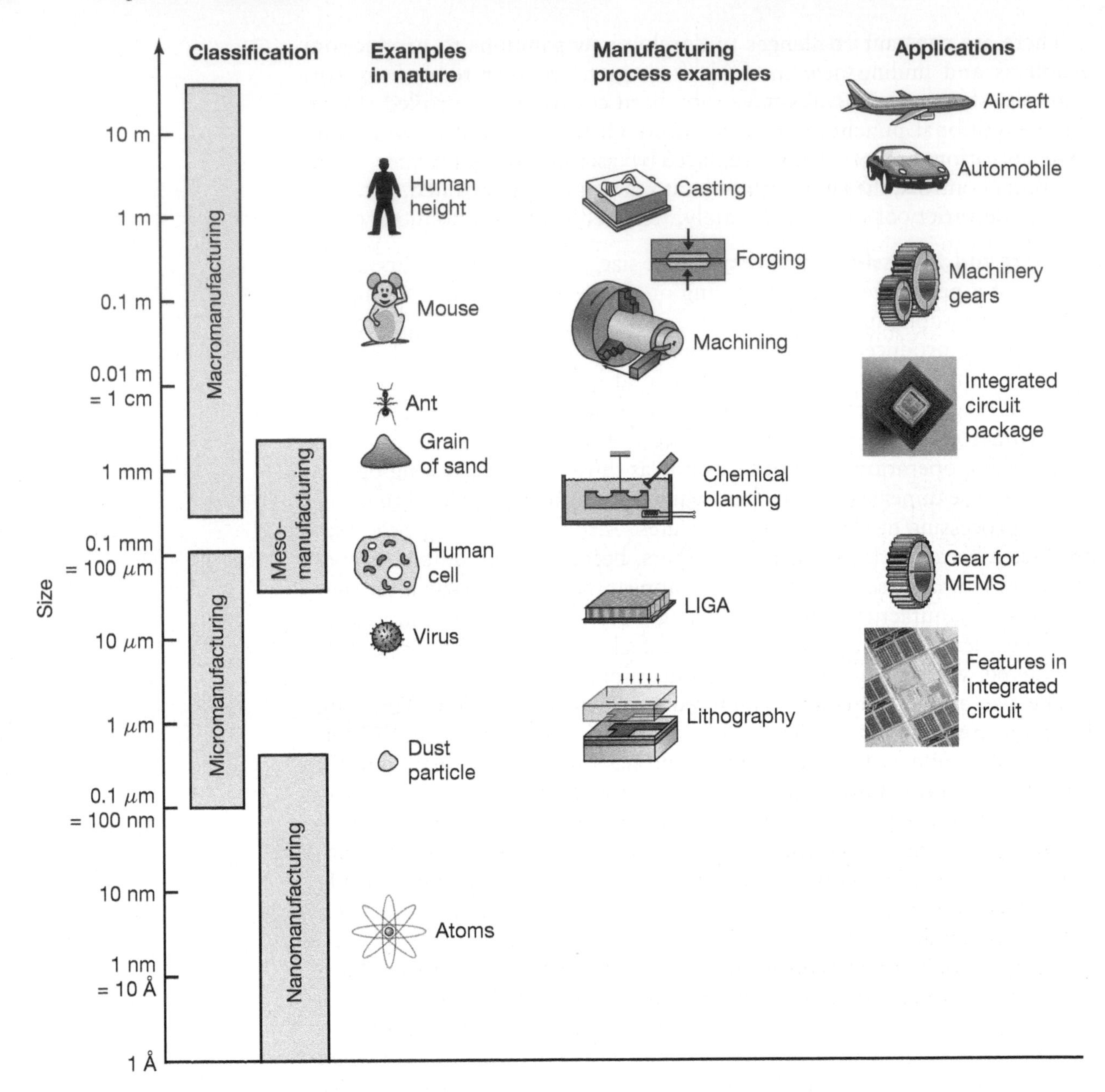

FIGURE 1.9 Illustration of the range of common sizes of parts and the capabilities of manufacturing processes in producing these parts.

shielding of electric circuits from magnetic fields, and store hydrogen in next-generation fuel cells.

Manufacturing and operational costs. Considerations of the design and cost of tooling, the lead time required to start production, and the effect of workpiece materials on tool and die life are of major importance (Chapter 16). For example, (a) tooling costs can be very significant, depending on tool size, design, and expected tool life; (b) a set of steel dies for stamping sheet metal fenders for automobiles, for example, may cost about $2 million or more; (c) for components made from expensive materials, such as titanium landing gear for aircraft or tantalum-based capacitors, the lower the scrap rate, the lower the production cost; and (d) because machining operations take longer time and waste material, they may not be as economical as forming operations, all other factors being equal.

The quantity of parts required and the desired production rate (pieces per hour) also are important factors in determining the processes to be used and the economics of production. Beverage cans or transistors, for example, are consumed in numbers and at rates much higher than such products as propellers for ships or large gears for heavy machinery. Availability of machines and equipment, operating experience, and economic considerations within the manufacturing facility are additional cost factors. If a certain part cannot be produced within a particular manufacturing facility, it has to be made by **outsourcing**.

The operation of some production machinery has significant environmental and safety implications. For example, chemical vapor deposition and electroplating processes involve the use of hazardous chemicals, such as chlorine gases and cyanide solutions, respectively. Metalworking operations often require the use of lubricants, the disposal of which has potential environmental hazards (Section 4.4.4). Unless controlled, these processes can cause significant air, water, and noise pollution.

Net-shape manufacturing. Because not all manufacturing operations produce finished parts to desired specifications, finishing operations may be necessary. A forged part, for example, may not have the desired dimensional accuracy or surface finish. Likewise, it may be difficult, impossible, or uneconomical to produce a part that has a number of holes in it by using only one manufacturing process. Also, the holes produced by a particular process may not have sufficient roundness, dimensional accuracy, or surface finish.

Because additional operations can significantly contribute to the product cost, *net-shape* or *near-net-shape manufacturing*, where the part is made in one operation and as close to the final desired dimensions, tolerances, and specifications as possible, has become an important goal. Typical examples are near-net-shape forging and casting of parts, powder-metallurgy techniques, metal-injection molding of metal powders, and injection molding of plastics and ceramics (Chapters 5, 6, 10, and 11).

1.7 Computer-Integrated Manufacturing

Computer-integrated manufacturing (CIM) has a very broad range of applications, including control and optimization of manufacturing operations, material handling, assembly, automated inspection and testing of products, inventory control, and various management activities. CIM has the capability for (a) improved responsiveness to rapid changes in market demand and product modifications; (b) more efficient use of materials, machinery, and personnel, and reduction in inventory; (c) better control of production and management of the total manufacturing operation; and (d) manufacturing high-quality products at low cost.

An outline of the major applications of computers in manufacturing (Chapters 14 and 15) is given below.

a. **Computer numerical control** (CNC) is a method of controlling the movements of machine components by direct insertion of coded instructions in the form of numerical data (Fig. 1.10). It was first implemented in the early 1950s and was a major advance in the automation of all types of machines.

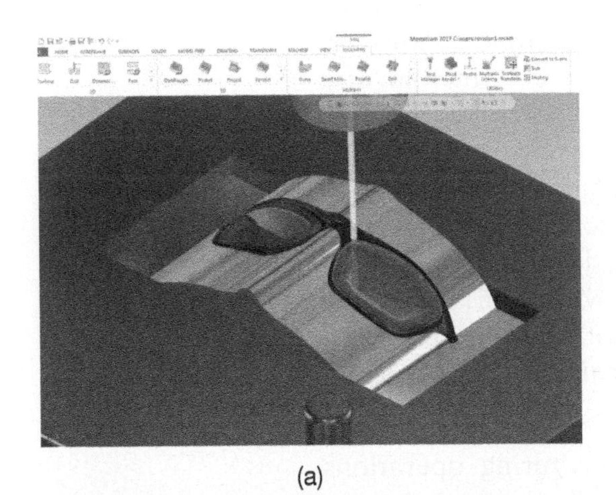

(a)

(b)

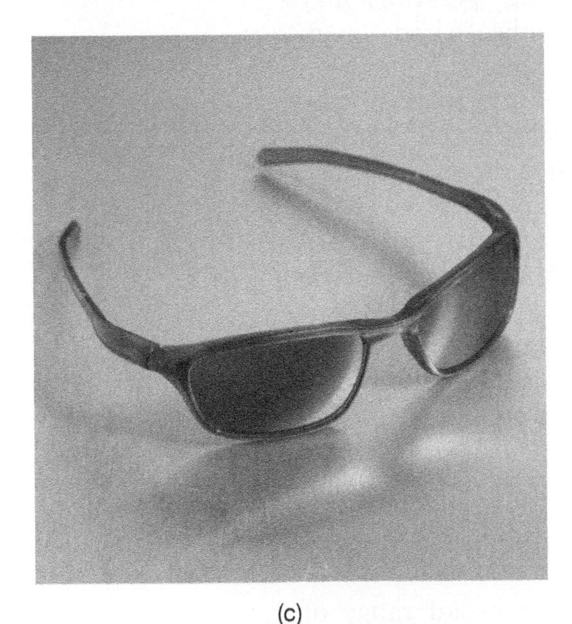

(c)

FIGURE 1.10 Machining a mold cavity for making sunglasses. (a) Computer model of the sunglasses as designed and viewed on the monitor. (b) Machining the die cavity using a computer numerical control milling machine. (c) Final product produced from the mold. *Source:* Courtesy of Mastercam / CNC Software, Inc.

b. In **adaptive control** (AC) the parameters in a manufacturing operation are adjusted automatically to optimize production rate and product quality and to minimize product cost. For example, forces, temperatures, surface finish, and dimensions of a part being machined are constantly monitored. If these parameters shift outside the acceptable range, the AC system automatically adjusts the process variables until these parameters fall within the acceptable range.

c. **Industrial robots.** Introduced in the early 1960s, industrial robots have rapidly been replacing humans in operations that are repetitive, boring, and dangerous, reducing the possibility of human error, decreasing variability in product quality, and improving productivity and safety. Robots with sensory-perception capabilities (*intelligent robots*) have been developed, with movements that simulate those of humans (Chapter 14).

d. In **automated handling**, computers make possible highly efficient handling of materials and products in various stages of completion (*work in progress*), such as when being moved from storage to machines, from machine to machine, and when products are at various points of inspection, inventory, and shipment (Section 14.6).

QR Code 1.1 Robot assembly. *Source:* Courtesy of the FANUC America Corporation.

e. **Automated and robotic assembly systems** are replacing costly assembly by human operators. Products must be designed or redesigned so that they can be assembled more easily by machines.

f. **Computer-aided process planning** (CAPP) is capable of improving plant productivity by optimizing process plans, reducing planning costs, and improving the consistency of product quality and reliability. Functions such as cost estimating and monitoring of work standards (time required to perform a certain operation) can now be incorporated into the system (Section 15.6).

g. **Group technology** (GT) is an approach by which all parts of a product are organized by classifying them into families and according to similarities in (a) design and (b) manufacturing processes employed to produce the part. Part designs and process plans can be standardized, and families of similar parts can be produced more efficiently and economically (Section 15.8).

h. In **just-in-time production** (JIT) supplies are delivered exactly when needed (just in time) to be ready for production, parts are produced just in time to be made into subassemblies and assemblies, and products are finished just in time to be delivered to the customer. Inventory-carrying costs are lower, defects are detected right away, productivity is increased, and high-quality products are made at lower cost (Section 15.12).

i. **Cellular manufacturing** consists of workstations, called *manufacturing cells*, that typically contain several machines, often controlled by a central robot, with each machine performing a different operation on the part (Fig. 1.11).

j. **Flexible manufacturing systems** (FMS) integrate manufacturing cells into a large unit, all computer-controlled. FMS have the highest level of efficiency, sophistication, and productivity among manufacturing

FIGURE 1.11 General view of a cellular manufacturing system, showing machines organized in functional groupings. *Source:* Courtesy of Advance Turning and Manufacturing.

systems. Although very costly, they are capable of efficiently producing parts in small quantities and rapidly changing production sequences on different parts. This flexibility enables FMS to respond rapidly to changes in market demand for a wide variety of products.

k. **Expert systems** are basically complex computer programs, with the capability of performing tasks and solving difficult real-life problems, much as human experts would.
l. **Artificial intelligence** (AI) is an important field involving the use of machines and computers to replace human intelligence. Computer-controlled systems are now capable of learning from experience and can make decisions that optimize operations and minimize costs.
m. **Artificial neural networks** (ANN) are designed to simulate the thought processes of the human brain. They have the capability of modeling and simulating production facilities, monitoring and controlling manufacturing operations, diagnosing problems in machine performance, conducting financial planning, and managing a company's manufacturing strategy.

The implementation of advanced technologies in manufacturing facilities requires significant expertise, time, and capital investment. They may be applied improperly or implemented on a too large or ambitious scale, involving major expenditures with questionable *return on investment* (ROI).

1.8 Lean Production and Agile Manufacturing

Lean production, also called *lean manufacturing* (Section 15.13), involves (a) a major assessment of each activity of a company regarding the effectiveness of its operations; (b) the efficiency of the machinery and equipment, while maintaining and improving product quality; (c) the number of personnel involved in a particular operation; and (d) a thorough analysis with the goal of reducing the cost of each activity in the production of goods, including both productive and nonproductive labor. Lean manufacturing may require a fundamental change in corporate culture, as well as require cooperation and teamwork between management and the work force. Lean production does not necessarily mean cutting back resources, but aims at *continuously improving efficiency and profitability*, by removing *all* types of waste in its operations (**zero-base waste**) and dealing with problems as soon as they arise.

Agile manufacturing aims to ensure *flexibility* (agility) in the manufacturing enterprise, so that it can rapidly respond to changes in product variety and to customers' needs. Agility is to be achieved through machines and equipment with built-in flexibility (**reconfigurable machines**), using *modular* components that can be arranged and rearranged in different ways in a manufacturing plant, thus reducing changeover time. The system uses advanced computer hardware and software and advanced communications systems.

1.9 Quality Assurance and Total Quality Management

Product quality is one of the most important considerations in manufacturing, as it directly influences the marketability and cost of a product and its customer satisfaction. Traditionally, quality assurance has involved inspecting parts after they have been manufactured, to ensure that they conform to a set of specifications and standards. Quality, however, cannot be inspected into a product after it is made; it must be *built into a product*, from the early stages of design through all subsequent stages of manufacturing and assembly. Because products typically are manufactured using a variety of processes, each of which can have significant variations in their performance, the **control of processes**, not the control of the products, is a critical factor in product quality.

Producing defective products can be very costly to the manufacturer, causing difficulties in assembly operations, necessitating repairs in the field and resulting in customer dissatisfaction. **Product integrity** is a term defined as the degree to which a product: (a) is suitable for its intended purpose; (b) fills a real market need; (c) functions reliably during its life expectancy; and (d) can be maintained with relative ease.

Total quality management (TQM) and *quality assurance* are the responsibility of everyone involved in the design and manufacture of a product. The technological and economic importance of product quality has been highlighted by pioneers such as Deming, Taguchi, and Juran (see Section 16.3). They emphasized: (a) the importance of management's commitment to product quality; (b) pride of workmanship at all levels of production; and (c) the use of powerful techniques such as **statistical process control** (SPC) and *control charts*, involving on-line monitoring of part production and rapid identification of the sources of quality problems (Chapter 4). The major goal is to *prevent defects* from occurring rather than detecting them after the product is made. As a consequence, computer chips, for example, are now produced such that only a few chips out of a million may be defective.

Quality assurance includes the implementation of **design of experiments**, a technique in which the factors involved in a production process and their interactions are studied *simultaneously*. Thus, for example, variables affecting dimensional accuracy or surface finish in a machining operation can now be readily identified, allowing appropriate actions to be taken during production.

Global competitiveness has created the need for international consensus regarding the establishment of quality-control methods. These efforts have resulted in the establishment of the International Organization for Standardization ISO 9000 series on Quality Management and Quality Assurance Standards and of QS 9000 (see Section 16.3.6). These standards are a *quality process certification* and not a product certification. A company's registration means that the company conforms to consistent practices, as specified by *its own quality system*. The standards, which are now the world standard for quality, have permanently influenced the manner in which companies conduct business in world trade.

1.10 Manufacturing Costs and Global Competitiveness

The cost of a product is often the overriding consideration in its marketability and general customer satisfaction. Manufacturing costs typically represent about 40% of a product's selling price, but it can vary considerably depending on the specific product and manufacturing methods used. The total cost of manufacturing a product consists of costs of materials, tooling, and labor, and includes fixed and capital costs; several factors are involved in each cost category (Chapter 16). Manufacturing costs can be minimized by analyzing the product design to determine, for example, whether part size and shape are optimal and the materials selected are the least costly while still possessing the desired properties and characteristics. The possibility of substituting materials is always an important consideration in minimizing product costs.

The economics of manufacturing have always been a major consideration, and it has become even more so as **global competitiveness** for high-quality products (**world-class manufacturing**) has become a necessity in worldwide markets. Beginning with the 1960s, the following trends have had a major impact on manufacturing:

- Global competition increased and markets became multinational and dynamic;
- Market conditions fluctuated widely;
- Customers demanded high-quality, low-cost products and on-time delivery; and
- Product variety increased substantially, products became more complex, and product life cycles became shorter.

TABLE 1.4 Approximate relative hourly compensation for workers in manufacturing in 2013 (United States = 100).

Country	Relative labor rate (US = 100)	GDP per hour worked
Norway	181	75.13
United States	100	67.32
Belgium	126	60.98
Netherlands	116	60.06
France	118	59.24
Germany	135	57.36
Ireland	116	56.05
Australia	130	55.87
Denmark	141	55.75
Sweden	141	55.28
United Kingdom	85	51.38
Spain	77	49.59
Italy	102	45.04
Japan	80	43.77
Singapore	66	41.46
Rep. Korea	60	32.31
India	5	3.40
China	8.6	—

Note: Compensation can vary significantly with benefits. Data for China and India are estimates, use different statistical measures of compensation, and are provided for comparison purposes only.
Source: US Department of Labor.

An important factor continues to be the wide disparity in labor costs, by an order of magnitude, among various countries. Table 1.4 shows the relative hourly compensation for production workers in manufacturing, based on a scale of 100 for the United States. These estimates are approximate because of such factors as various benefits and housing allowances that are not calculated consistently.

Outsourcing is the practice of taking internal company activities and paying an outside firm to perform them. With multinational companies, outsourcing can involve shifting activities to other divisions in different countries. Outsourcing of manufacturing tasks became viable only when the communication and shipping infrastructure developed sufficiently, a trend that started in the early 1990s. For example, an Indian software firm could not effectively collaborate with European or American counterparts until fiber optic communication lines and high-speed internet access became possible, even if the labor costs were lower.

The cost of a product is often the overriding consideration in its marketability and in general customer satisfaction. Keeping costs at a minimum is a constant challenge to manufacturing companies and crucial to their very survival. Labor-intensive products today are either made or assembled in countries where labor costs are among the lowest; they are, however, bound to rise as the living standards in those countries rise. Likewise, software development and information technology can be far more economical to implement in Asian countries than those in the West. However, countries with relatively high labor rates have begun to adapt by concentrating on advanced manufacturing technologies and by raising their worker productivity. There are several measures of *worker productivity*; Table 1.4 uses gross domestic product (GDP) per hour worked.

The approaches require that manufacturers **benchmark** their operations. Benchmarking means understanding the competitive position of a company with respect to that of others, and setting realistic goals for its future. It is a *reference* from which various measurements can be made and compared.

1.11 General Trends in Manufacturing

Several trends regarding various aspects of modern manufacturing are:

1. Product variety and complexity continue to increase.
2. Product life cycles are becoming shorter.
3. Markets continue to become multinational and global competition is increasing rapidly.
4. Customers are consistently demanding high-quality, reliable, and low-cost products.
5. Developments continue in the quality of materials and their selection for improved recyclability.
6. Machining is faster and able to achieve better tolerances because of innovative control strategies and suppression of chatter.
7. The most economical and environmentally friendly (green) manufacturing methods are being increasingly pursued. Energy management is increasingly important.
8. The desire to reduce energy consumption has driven designs that optimize structures and minimize weight. Weight savings research continues with the exploitation of materials with higher strength-to-weight and stiffness-to-weight ratios, particularly in the automotive, aerospace, and sporting industries.
9. Titanium, magnesium, aluminum, and fiber-reinforced polymers are increasingly seen as necessary technologies to meet energy efficiency goals.
10. Improvements are being made in predictive models of the effects of material-processing parameters on product integrity, applied during a product's design stage.
11. Developments continue in ultraprecision manufacturing, micromanufacturing, and nanomanufacturing, approaching the level of atomic dimensions.

12. Computer simulation, modeling, and control strategies are being applied to all areas of manufacturing.
13. Additive manufacturing has become pervasive, with a wide range of equipment availability and cost, and a broader range of materials and properties available. Advances in software sophistication have made three-dimensional modeling capability widely available. Additive manufacturing will be increasingly used for part production, and not just restricted to prototyping.
14. Advances in optimization of manufacturing processes and production systems are making them more agile.
15. Manufacturing activities are viewed not as individual, separate tasks, but as making up a large system, with all its parts interrelated.
16. It has become common to build quality into the product at each stage of its production.
17. Continuing efforts are aimed at achieving higher levels of productivity and eliminating or minimizing waste with optimum use of an organization's resources.
18. Software continues to expand into all aspects of manufacturing, from machinery controls to supply chain management.
19. Sensors of all types are being incorporated into machines, providing data for process validation and providing historical information that can be stored for a product. A term used to incorporate computer data into all parts of a product's lifecycle is the *digital thread*.
20. Lean production and information technology are being implemented as powerful tools to help meet global challenges. Machine tools are increasingly capable of communicating, giving plant managers real-time information about factory floor operations.
21. Advances in communication and sensors will lead to unprecedented access to data (as with cloud-based storage), improving control of the manufacturing enterprise, quality and efficiency, and management of complex global supply chains.

QR Code 1.2 Smart manufacturing. *Source:* Courtesy of the Smart Manufacturing Leadership Coalition.

CHAPTER

2 Fundamentals of the Mechanical Behavior of Materials

This chapter describes:

- The types of tests commonly used to determine the mechanical properties of materials.
- Stress–strain curves, their features, significance, and dependence on parameters such as temperature and deformation rate.
- Characteristics and the roles of hardness, fatigue, creep, impact, and residual stresses in materials processing.
- Yield criteria and their applications in determining forces and energies required in processing metals.

Symbols

A area, m^2
c one-half specimen depth, m
c_p specific heat, Nm/kgK
C strength coefficient, N/m^2
d diameter, m
D ball diameter, m
e engineering strain, m/m
$\dot{e}$ engineering strain rate, s^{-1}
E elastic modulus, N/m^2
E_p plastic modulus, N/m^2
G modulus of rigidity, N/m^2
HB Brinell hardness
HK Knoop hardness
HV Vickers hardness
I moment of inertia, m^4
k material shear strength, N/m^2
K strength coefficient, N/m^2
l length, m
l_o original or gage length, m
L diagonal length, m
m strain-rate sensitivity exponent
M bending moment, Nm
n strain-hardening exponent
N number of cycles
P force, N
r radius, m
s engineering stress, N/m^2
S stress amplitude, N/m^2
S_y yield strength, N/m^2
S'_y yield strength in plane strain, N/m^2
S_{ut} ultimate tensile strength, N/m^2
t thickness, m; time, s
T torque, Nm
u specific energy, Nm/m^3
v velocity; rate of deformation, m/s

W work done, Nm
Δ dilatation or volume strain
ΔT temperature rise, °C
ϵ true strain, m/m
$\overline{\epsilon}$ effective strain, m/m
$\dot{\epsilon}$ true strain rate, s^{-1}
η efficiency
ϕ angle of twist, radians
γ shear strain, m/m
ν Poisson's ratio
ρ density, N/m^3
σ true stress, N/m^2
$\overline{\sigma}$ effective stress, N/m^2
τ shear stress, N/m^2

Subscripts

1, 2, 3 principal directions
c compressive
f final, fracture, flow
m mean
o initial
t tensile

2.1 Introduction

The manufacturing methods and techniques by which materials can be shaped into useful products were outlined in Chapter 1. One of the oldest and most important groups of manufacturing processes involves **plastic deformation**, namely, shaping materials by applying forces on the workpiece by various means. Also known as **deformation processing**, it includes *bulk deformation processes* (such as forging, rolling, extrusion, and rod and wire drawing) and *sheet forming processes* (such as bending, deep drawing, spinning, and many other processes) that are generally referred to as *pressworking*.

In making a product such as a sheet metal automobile fender, the material is subjected to *tension*. In making a turbine disk, a solid cylindrical piece of metal is subjected to *compression*. A piece of plastic tubing is expanded by *internal pressure* to make a beverage bottle, subjecting the material to tension in more than one direction. In punching a hole in sheet metal, the material is subjected to *shearing* through its thickness. This chapter deals with fundamental aspects of the mechanical behavior of materials during such plastic deformation.

In all these processes, the material is subjected to one or more of the three basic modes of deformation, as shown in Fig. 2.1, namely, tension, compression, and shear. The amount of deformation to which the material is subjected is defined as **strain**. For tension or compression, the **engineering strain**, or **nominal strain**, is defined as

$$e = \frac{l - l_o}{l_o}. \tag{2.1}$$

Note that in tension the strain is positive, and in compression it is negative. In the deformation shown in Fig. 2.1c, the **shear strain** is defined as

$$\gamma = \frac{a}{b}. \tag{2.2}$$

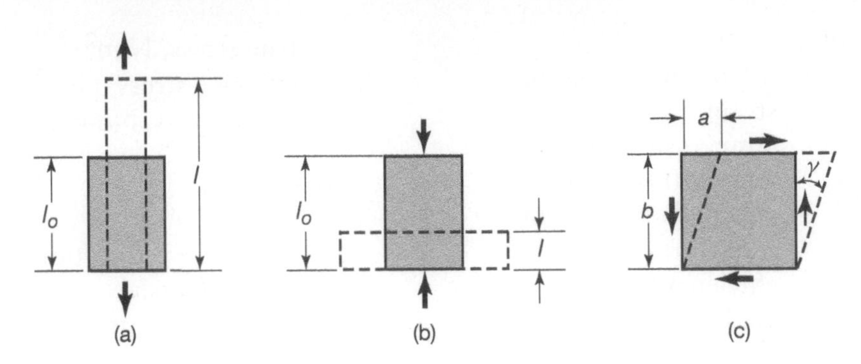

FIGURE 2.1 Types of strain. (a) Tensile. (b) Compressive. (c) Shear. All deformation processes in manufacturing involve strains of these types.

In order to change the shape of the elements shown in Fig. 2.1, *forces* must be applied to them, as shown by arrows. The determination of these forces as a function of strain is an important aspect in the study of manufacturing processes. A knowledge of these forces is essential in the design of proper equipment to be used, selection of tool and die materials for proper strength, and in determining whether a specific metalworking operation can be carried out on equipment of certain capacity.

2.2 Tension

QR Code 2.1 A tension test of steel. *Source:* Courtesy of Instron.

Because of its relative simplicity, the **tension test** is the most common for determining the *strength-deformation characteristics of materials*. It involves preparation of a test specimen (according to standard specifications) and testing it under tension on suitable testing equipment.

The test specimen has an original length l_o and an original cross-sectional area A_o (Fig. 2.2). Most specimens are solid and round; flat sheet or tubular specimens are also tested under tension. The original length is the distance between **gage marks** on the specimen or clips on an *extensometer* and typically is 25 or 50 mm (1 or 2 in.). Longer lengths may be used for larger specimens, such as for structural members, as well as shorter ones for specific small part applications.

Typical results from a tension test are shown in Fig. 2.3. The **engineering stress,** or **nominal stress,** is defined as the ratio of the applied load to the original area of the specimen,

$$s = \frac{P}{A_o}. \tag{2.3}$$

When the load is first applied, the specimen elongates proportionately to the load up to the **proportional limit;** this is the range of **linear elastic behavior** (Fig. 2.3). The material will continue to deform elastically, although not strictly linearly, up to the **yield point.** If the load is removed before the yield point is reached, the specimen will return to its original length. The **modulus of elasticity,** or **Young's modulus,** E, is defined as

$$E = \frac{s}{e}. \tag{2.4}$$

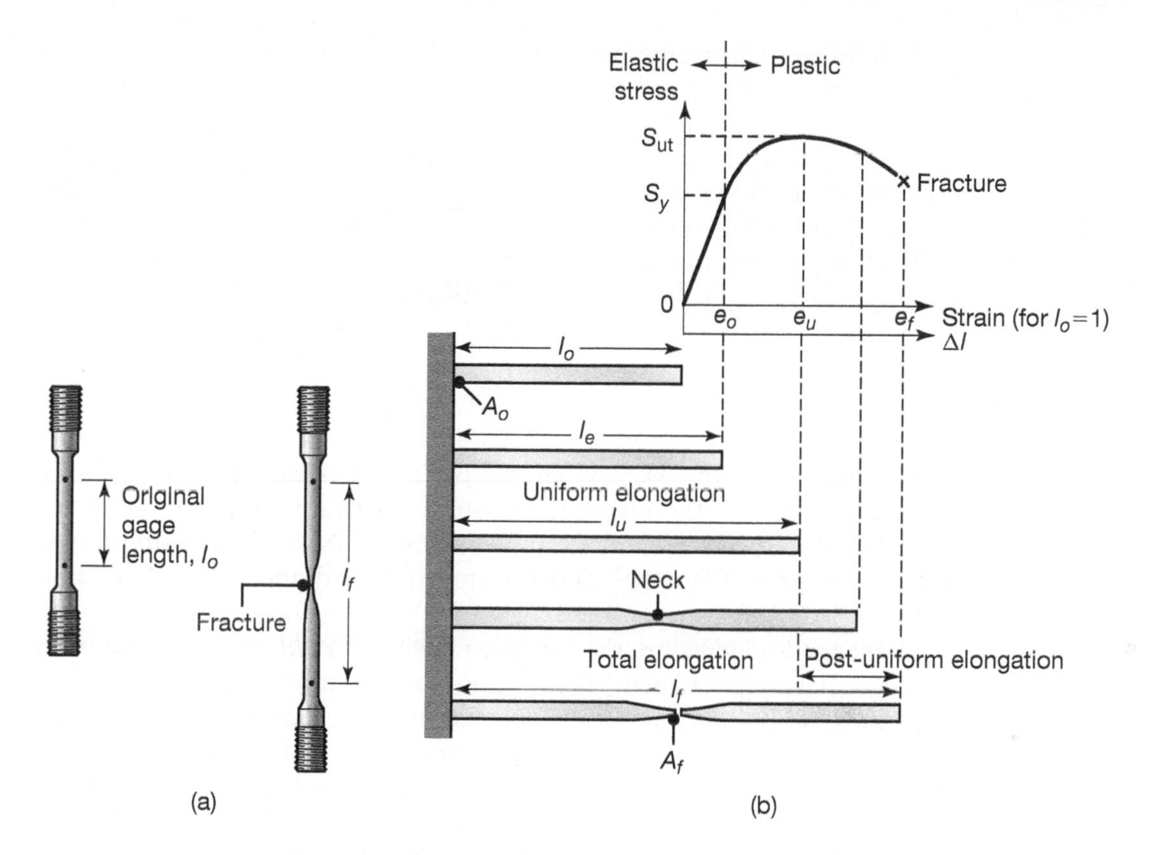

FIGURE 2.2 (a) Original and final shape of a standard tensile-test specimen. (b) Outline of a tensile-test sequence showing different stages in the elongation of the specimen.

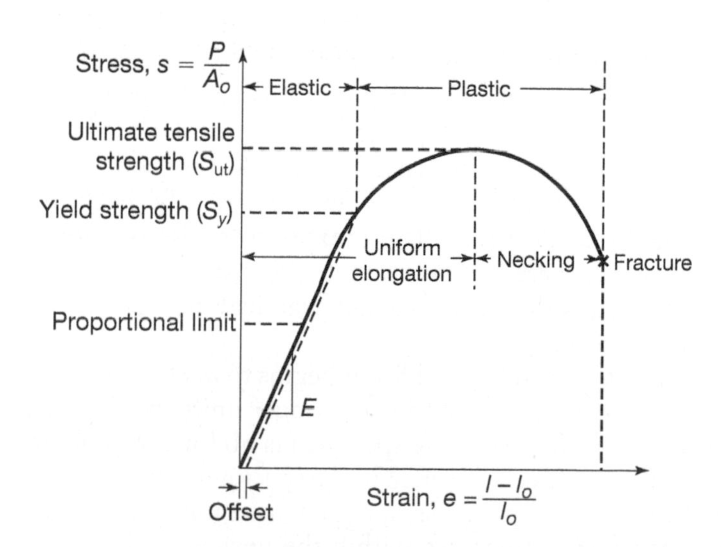

FIGURE 2.3 A typical stress–strain curve obtained from a tension test, showing various features.

This *linear relationship* between stress and strain is known as **Hooke's law**, the more generalized forms of which are given in Section 2.11. The elongation of the specimen is accompanied by a contraction of its lateral dimensions. The *absolute value* of the ratio of the lateral strain to longitudinal strain is known as **Poisson's ratio,** ν. Typical values for E and ν for various materials are given in the inside front cover.

The area under the stress–strain curve up to the yield point of the material is known as the **modulus of resilience:**

$$\text{Modulus of resilience} = \frac{S_y e_o}{2} = \frac{S_y^2}{2E}. \tag{2.5}$$

This property has the units of **energy per unit volume** and indicates the **specific energy** that the material can store elastically. Typical values for modulus of resilience are, for example, 2.1×10^4 Nm/m^3 (3 in.-lb/in^3) for annealed copper, 1.9×10^5 (28) for annealed medium-carbon steel, and 2.7×10^6 (385) for spring steel.

As the load is increased, the specimen begins to *yield*, that is, it begins to undergo **plastic (permanent) deformation;** the relationship between stress and strain is nonlinear. For most engineering materials the *rate of change* in the slope of the stress–strain curve beyond the yield point is very small, thus the determination of S_y can be difficult. The usual practice is to define the yield strength as the point on the curve that is *offset* by a strain of (usually) 0.2%, or 0.002 (Fig. 2.3). If other offset strains are used, they should be specified in reporting the yield strength of the material.

Recall that yielding is necessary in metalworking processes such as forging, rolling, and sheet metal forming operations, where materials have to be subjected to permanent deformation to develop the desired part shape. In the design of structures and load-bearing members, however, yielding is not acceptable at all since it causes their permanent deformation.

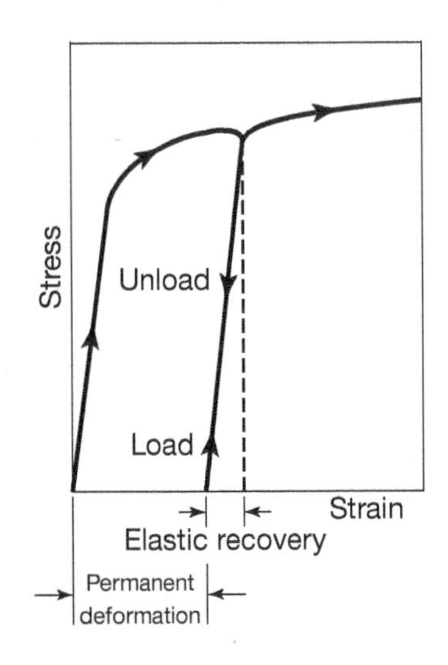

FIGURE 2.4 Schematic illustration of loading and unloading of a tensile-test specimen. Note that during unloading the curve follows a path parallel to the original elastic slope.

As the specimen continues to elongate under increasing load beyond S_y, its cross-sectional area decreases *permanently and uniformly* throughout its gage length. If the specimen is unloaded from a stress level higher than S_y, the stress–strain curve closely follows a straight line downward and parallel to the original elastic slope, as shown in Fig. 2.4. As the load, hence the engineering stress, is further increased, the curve eventually reaches a maximum and then begins to decrease. The maximum stress is known as the **tensile strength** or **ultimate tensile strength** (S_{ut}) of the material. Ultimate tensile strength, S_{ut}, is thus a simple and practical measure of the maximum strength of a material.

When the test specimen is loaded beyond S_{ut}, it begins to *neck* (Fig. 2.2a) and the elongation between the gage marks is no longer uniform. That is, the change in the cross-sectional area of the specimen is no longer uniform but is concentrated *locally* in a "neck" formed in the specimen (called **necking,** or *necking down*). As the test progresses, the engineering stress drops further and the specimen finally fractures within the necked region. The final stress level (marked by an $\times$ in Fig. 2.2b) at fracture is known as the **breaking** or **fracture stress.**

2.2.1 Ductility

The strain in the specimen at fracture is a measure of *ductility*, that is, how large a strain the material withstands before fracture. Note from Fig. 2.2b that until S_{ut} is reached, elongation is *uniform* and the strain up to the S_{ut} is thus called **uniform strain.** The elongation at fracture is known as the **total elongation,** and is measured between the original gage marks after the two pieces of the broken specimen are placed together.

Two quantities that are commonly used to define ductility in a tension test are *elongation* and *reduction of area.* **Elongation** is defined as

$$\text{Elongation} = \frac{l_f - l_o}{l_o} \times 100\%, \tag{2.6}$$

and is based on the total elongation (see the inside front cover for values). **Reduction of area** is defined as

$$\text{Reduction of area} = \frac{A_o - A_f}{A_o} \times 100\%. \tag{2.7}$$

Note that a material that necks down to a point at fracture, such as a glass rod at elevated temperature, has a reduction of area of 100%.

Elongation and reduction of area generally are related to each other for most engineering metals and alloys. Elongation ranges approximately between 10 and 60%, and values between 20 and 90% are typical for reduction of area for most materials. *Thermoplastics* (Chapter 10) and *superplastic* materials (Section 2.2.7) exhibit much higher ductility. Brittle materials, by definition, have little or no ductility; typical examples are common glass at room temperature and gray cast ion.

Necking is a *local* phenomenon. Put a series of gage marks at different points on the specimen, pull and break it, and then calculate the percent elongation for each pair of gage marks, it will be found that with decreasing gage length, the percent elongation increases. The closest pair of gage marks will have undergone the largest elongation, because they are closest to the necked and fractured region. Note, however, that the curves will not approach zero elongation because the specimen already has undergone some finite permanent elongation before it fractures. Consequently, it is important to include the gage length in reporting elongation data. Other tensile properties are generally found to be independent of gage length.

2.2.2 True Stress and True Strain

Because stress is defined as the ratio of force to area, *true stress* is likewise defined as

$$\sigma = \frac{P}{A}, \tag{2.8}$$

where A is the actual (hence true) or instantaneous area supporting the load.

The complete tension test may be regarded as a series of incremental tension tests where, for each succeeding increment, the specimen is a little longer than at the preceding stage. Thus, *true strain* (or *natural* or *logarithmic strain*), ϵ, can be defined as

$$\epsilon = \int_{l_o}^{l} \frac{dl}{l} = \ln\left(\frac{l}{l_o}\right). \tag{2.9}$$

Note that, for small values of engineering strain, $e = \epsilon$ since $\ln(1+e) = \epsilon$. For larger strains, however, the values rapidly diverge, as can be seen in Table 2.1.

TABLE 2.1 Comparison of engineering and true strains in tension.

e	ϵ
0.01	0.01
0.05	0.049
0.1	0.095
0.2	0.18
0.5	0.4
1	0.69
2	1.1
5	1.8
10	2.4

The *volume* of a metal specimen remains constant in the plastic region of the test (see *volume constancy*, Section 2.11.5). Thus, the true strain within the uniform elongation range can be expressed as

$$\epsilon = \ln\left(\frac{l}{l_o}\right) = \ln\left(\frac{A_o}{A}\right) = \ln\left(\frac{D_o}{D}\right)^2 = 2\ln\left(\frac{D_o}{D}\right). \tag{2.10}$$

Once necking begins, the true strain at any point between the gage marks of the specimen can be calculated from the reduction in the cross-sectional area at that point; thus, by definition, the largest strain is at the narrowest region of the neck.

It can be seen that at small strains the engineering and true strains are very close and, therefore, either one can be used in calculations. However, for the large strains encountered in metalworking operations, the true strain should be used because it is a better measure of the actual strain, as can be illustrated by the following two examples.

a. Assume that a tension-test specimen is elongated to twice its original length; this deformation is equivalent to compressing a specimen to one-half its original height. Using the subscripts t and c for tension and compression, respectively, it can be seen that $\epsilon_t = 0.69$ and $\epsilon_c = -0.69$, whereas $e_t = 1$ and $e_c = -0.5$. Thus, true strain is a better measure of strain.
b. Assume that a specimen 10 mm in height is compressed to a final thickness of zero; thus, $\epsilon_c = -\infty$, whereas $e_c = -1$. Note that the specimen will have deformed infinitely (because its final thickness is zero, and hence its diameter has become infinite); this is exactly what the value of the true strain indicates.

From these two simple examples, it can be seen that true strains are consistent with the actual physical phenomenon, whereas engineering strains are not.

2.2.3 True Stress–True Strain Curves

The relationship between engineering and true stress and strain can now be used to construct *true stress–true strain curves*, from a curve such as that shown in Fig. 2.3. A typical true stress–true strain curve is shown in Fig. 2.5a. For convenience, such a curve is often approximated by the equation

$$\sigma = K\epsilon^n. \tag{2.11}$$

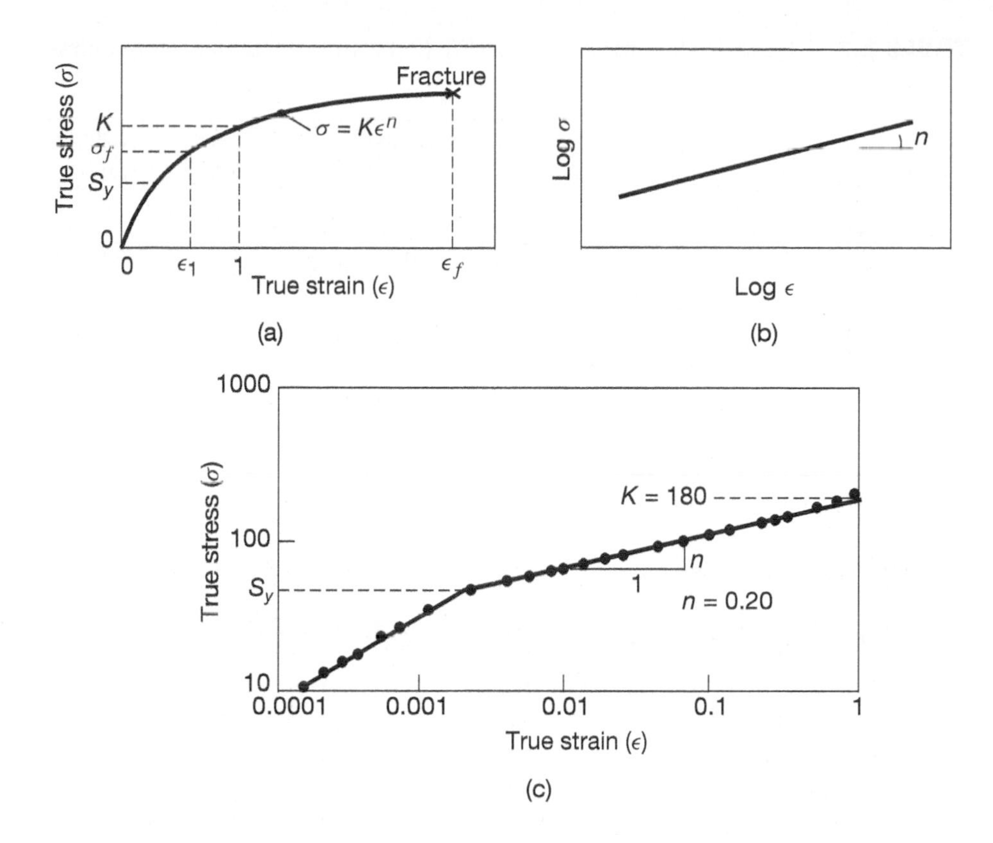

FIGURE 2.5 (a) True stress–true strain curve in tension. Note that, unlike in an engineering stress–strain curve, the slope is always positive and the slope decreases with increasing strain. Although in the elastic range stress and strain are proportional, the total curve can be approximated by the power expression shown. On this curve, S_y is the yield strength and σ_f is the flow stress. (b) True stress–true strain curve plotted on a log-log scale. (c) True stress–true strain curve in tension for 1100-O aluminum plotted on a log-log scale. Note the large difference in the slopes in the elastic and plastic ranges.

Note that Eq. (2.11) indicates neither the elastic region nor the yield point of the material, but these quantities are readily available from the engineering stress–strain curve. Since the strains at the yield point are very small, the difference between true yield strength and engineering yield strength is negligible for metals.

Equation (2.11) can then be rewritten as

$$\log \sigma = \log K + n \log \epsilon.$$

Plot the true stress–true strain curve on a log-log scale, obtain Fig. 2.5b. The slope n is the **strain-hardening exponent**, and K is the **strength coefficient.** Note that K is the true stress at a true strain of unity. Values of K and n for a variety of engineering materials are given in Table 2.2. The true stress–true strain curves for several materials are given in Fig. 2.6.

TABLE 2.2 Typical values for K and n in Eq. (2.11) at room temperature unless noted.

Material	Strength coefficient, K MPa	Strength coefficient, K psi	Strain-hardening exponent, n
Aluminum, 1100-O	180	2610	0.20
2024-T4	690	10,000	0.16
5052-O	210	3045	0.13
6061-O	205	3000	0.20
6061-T6	410	6000	0.05
7075-O	400	5800	0.17
Brass, 70-30, annealed	895	13,000	0.49
85-15, cold rolled	580	8400	0.34
Bronze (phosphor), annealed	720	10,400	0.46
Cobalt-base alloy, heat treated	2070	30,000	0.50
Copper, annealed	315	4570	0.54
Molybdenum, annealed	725	10,500	0.13
Steel, low carbon	530	7680	0.26
1045, hot rolled	965	14,000	0.14
1112, annealed	760	11,000	0.19
1112, cold rolled	760	11,000	0.08
4135, annealed	1015	14,700	0.17
4135, cold rolled	1100	16,000	0.14
4340, annealed	640	9280	0.15
17-4 P-H, annealed	1200	17,400	0.05
52100, annealed	1450	21,000	0.07
304 stainless, annealed	1275	18,500	0.45
410 stainless, annealed	960	13,900	0.10
Titanium			
Ti-6Al-4V, annealed, 20°C	1400	200	0.015
Ti-6Al-4V, annealed, 200°C	1040	150	0.026
Ti-6Al-4V, annealed, 600°C	650	94	0.064
Ti-6Al-4V, annealed, 800°C	350	51	0.146

Some differences between Table 2.2 and these curves exist because of different sources of data and test conditions employed.

Flow Stress. In Fig. 2.5a, σ_f is known as the **flow stress,** and is defined as the true stress required to continue plastic deformation at a particular true strain, ϵ_1. For strain-hardening materials, the flow stress increases with increasing strain. Note also from Fig. 2.5c that the elastic strains are much smaller than plastic strains. Consequently, and although both effects exist, elastic strains are generally ignored in calculations for forming processes throughout the rest of this text; the plastic strain will thus be the total strain that the material has undergone.

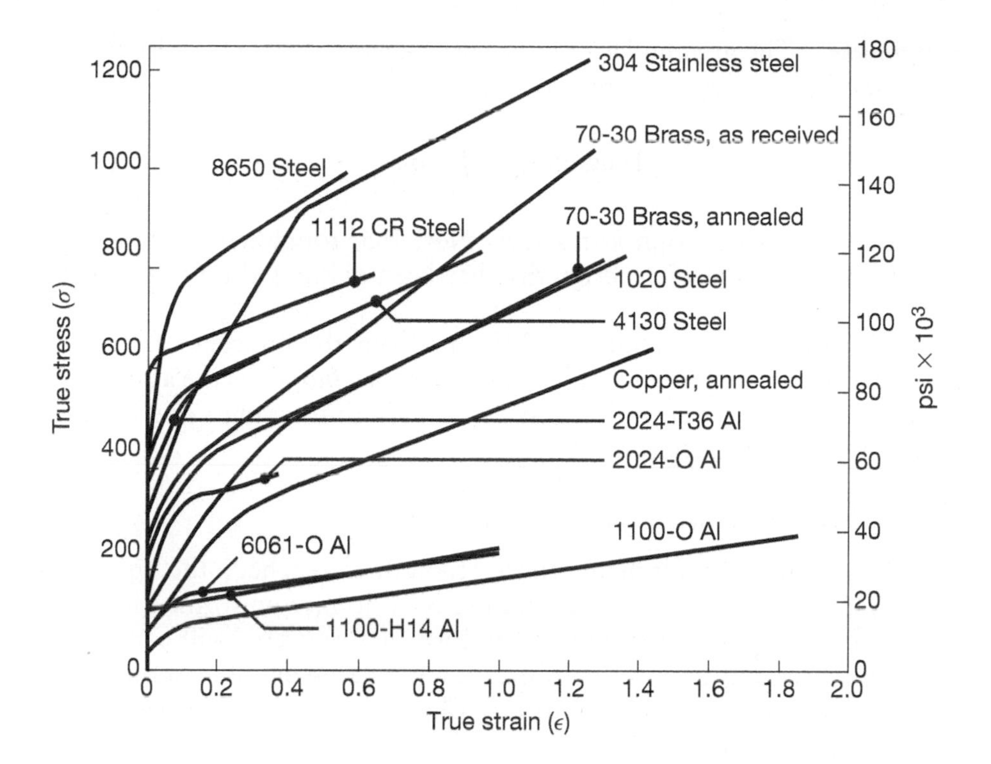

FIGURE 2.6 True stress–true strain curves in tension at room temperature for various metals. The point of intersection of each curve at the ordinate is the yield strength, S_y, thus the elastic portions of the curves are not indicated. When the K and n values are determined from these curves, they may not agree with those given in Table 2.2 because of the different sources from which they were collected. *Source:* S. Kalpakjian.

Sometimes it is desired to determine the average flow stress, or $\overline{\sigma}_f$, in a manufacturing operation. The average flow stress can be obtained in a number of ways, depending on the material's stress–strain curve:

- In general, a material's average flow stress up to a strain of ϵ_1 is given by

$$\overline{\sigma}_f = \frac{1}{\epsilon_1}\int_0^{\epsilon_1} \sigma\, d\epsilon. \tag{2.12}$$

- For a strain-hardening material with a power law behavior as given in Eq. (2.11), the average flow stress is then given by:

$$\overline{\sigma}_f = \frac{1}{\epsilon_1}\int_0^{\epsilon_1} K\epsilon^n d\epsilon = \frac{K\epsilon^n}{n+1}. \tag{2.13}$$

  Similar equations can be developed for other forms of stress–strain curves (see Section 2.2.5).

- If a material doesn't strain harden, then the average flow stress is merely the yield strength.

Toughness. The area under the true stress–true strain curve is known as **toughness** and can be expressed as

$$\text{Toughness} = \int_0^{\epsilon_f} \sigma \, d\epsilon, \tag{2.14}$$

where ϵ_f is the true strain at fracture. Note that toughness is the energy per unit volume (*specific energy*) that has been dissipated up to the point of fracture.

As defined here, toughness is different from the concept of *fracture toughness*, as treated in textbooks on *fracture mechanics* as the ability of a material containing a crack to resist fracture. Both toughness and fracture toughness are very important in design.

2.2.4 Instability in Tension

Recall that once the ultimate tensile strength is reached, the specimen will begin to neck and thus deformation is no longer uniform. This phenomenon is important because nonuniform deformation will cause part thickness variations and strain localization in processing of materials, particularly in sheet forming operations (Chapter 7) where the material is subjected to tension.

In this section it will now be shown that the true strain at the onset of necking is numerically equal to the strain-hardening exponent, n, given in Eq. (2.11). Note in Fig. 2.2 that the slope of the load-elongation curve at S_{ut} is zero (or $dP/d\epsilon = 0$). It is here that *instability* begins; that is, the specimen begins to neck and can no longer support the load because the cross-sectional area of the necked region is becoming smaller as the test progresses. Using the relationships

$$\epsilon = \ln\left(\frac{A_o}{A}\right), \quad A = A_o e^{-\epsilon}, \text{ and } P = \sigma A = \sigma A_o e^{-\epsilon}$$

determine $dP/d\epsilon$ by noting that

$$\frac{dP}{d\epsilon} = \frac{d}{d\epsilon}\left(\sigma A_o e^{-\epsilon}\right) = A_o\left(\frac{d\sigma}{d\epsilon}e^{-\epsilon} - \sigma e^{-\epsilon}\right).$$

Because $dP/d\epsilon = 0$ where necking begins, set this expression equal to zero. Thus,

$$\frac{d\sigma}{d\epsilon} = \sigma.$$

However, since

$$\sigma = K\epsilon^n,$$

thus,

$$nK\epsilon^{n-1} = K\epsilon^n,$$

and therefore

$$\epsilon = n. \tag{2.15}$$

Instability in a tension test can be viewed as a phenomenon in which two competing processes are taking place simultaneously:

a. As the load on the specimen is increased, its cross-sectional area decreases, which becomes more pronounced in the region where necking begins.
b. With increasing strain, however, the material becomes stronger due to strain hardening. Since the load on the specimen is the product of area and strength, instability sets in when the *rate of decrease* in cross-sectional area is greater than the *rate of increase* in strength; this condition is also known as **geometric softening.**

EXAMPLE 2.1 Calculation of Ultimate Tensile Strength

Given: A material has a true stress–true strain curve given by $\sigma = 700\epsilon^{0.5}$ MPa.

Find: Calculate the true ultimate tensile strength and the engineering ultimate tensile strength of this material.

Solution: The necking strain corresponds to the maximum load. The necking strain for this material is given as

$$\epsilon = n = 0.5.$$

Therefore, the true ultimate tensile strength is given by

$$S_{ut,true} = 700\,(0.5)^{0.5} = 495 \text{ MPa}.$$

The cross-sectional area at the onset of necking is obtained from Eq. (2.10) as

$$\ln\left(\frac{A_o}{A_{neck}}\right) = n = 0.5.$$

Consequently,

$$A_{neck} = A_o e^{-0.5},$$

and the maximum load P is

$$P = \sigma A = \sigma A_o e^{-0.5},$$

where σ is the true ultimate tensile strength. Hence,

$$P = (495)(0.606)(A_o) = (300 \text{ MN})A_o.$$

Since $S_{ut} = P/A_o$, it can be seen that $S_{ut} = 300$ MPa.

2.2.5 Types of Stress–Strain Curves

Each material has a differently shaped stress–strain curve, depending on its composition and various other factors that are described in detail later in this chapter. In addition to the power law given in Eq. (2.11), some

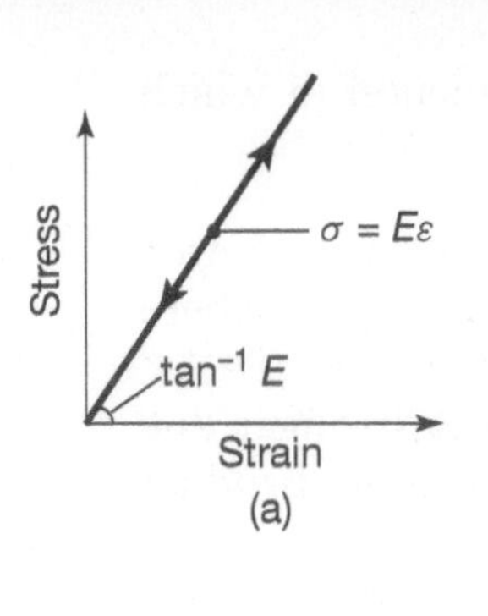

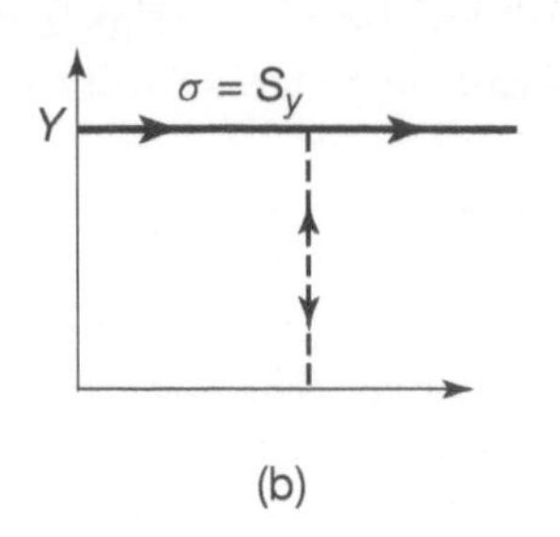

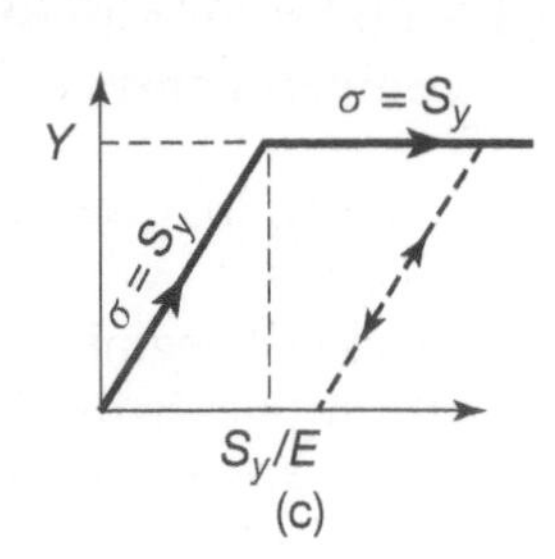

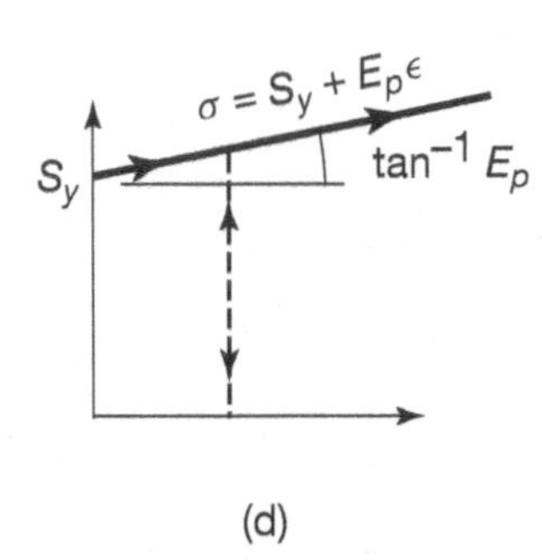

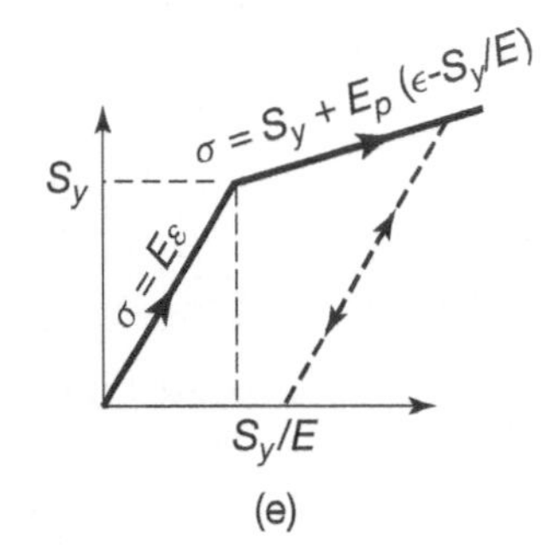

FIGURE 2.7 Schematic illustration of various types of idealized stress–strain curves. (a) Perfectly elastic; (b) rigid, perfectly plastic; (c) elastic, perfectly plastic; (d) rigid, linearly strain hardening; and (e) elastic, linearly strain hardening. The broken lines and arrows indicate unloading and reloading during the test.

of the major types of curves are shown in Fig. 2.7, with their associated stress–strain equations, and have the following characteristics:

1. A **perfectly elastic material** displays linear behavior with slope E. The behavior of brittle materials, such as common glass, most ceramics, and some cast irons, may be represented by the curve shown in Fig. 2.7a. There is a limit to the stress the material can sustain, after which it fractures. Permanent deformation, if any, is negligible.
2. A **rigid, perfectly plastic** material has, by definition, an infinite value of elastic modulus, E. Once the stress reaches the yield strength, S_y, it continues to undergo deformation at the same stress level; that is, there is no strain hardening. When the load is released, the material has undergone permanent deformation; there is no elastic recovery. (Fig. 2.7b).
3. The behavior of an **elastic, perfectly plastic** material is a combination of the first two: It has a finite elastic modulus, no strain hardening, and it undergoes elastic recovery when the load is released (Fig. 2.7c).
4. A **rigid, linearly strain-hardening** curve model assumes that the elastic strains are negligible, and that the strain-hardening behavior is defined by a straight line with a slope equal to the **plastic modulus,** E_p. Such materials have no elastic recovery upon unloading (Fig. 2.7d).
5. An **elastic, linearly strain-hardening** curve (Fig. 2.7e) uses an elastic portion with slope E, and a linear plastic portion with a plastic modulus, E_p.

Note that some of these curves can be expressed by Eq. (2.11) by changing the value of n (Fig. 2.8) or by other equations of a similar nature.

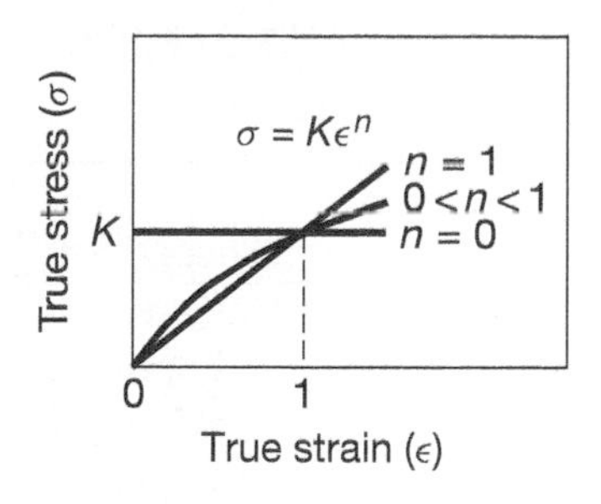

FIGURE 2.8 The effect of strain-hardening exponent n on the shape of true stress–true strain curves. When $n = 1$, the material is linear elastic, and when $n = 0$, it is rigid and perfectly plastic.

2.2.6 Effects of Temperature

Various factors that have an influence on the shape of stress–strain curves are described in this and subsequent sections. The first factor is temperature; although difficult to generalize, increasing temperature usually lowers the modulus of elasticity, yield strength, and ultimate tensile strength, and increases ductility and toughness (Fig. 2.9). Temperature also affects the strain-hardening exponent, n, of most metals, in that n decreases with increasing temperature. Depending on the type of material, its composition and level of impurities present, elevated temperatures can have other significant effects, as described in Chapter 3. The influence of temperature is best discussed in conjunction with strain rate for the reasons explained in the next section.

2.2.7 Effects of Strain Rate

Depending on the particular manufacturing operation and the characteristics of the equipment used, a workpiece may be formed at speeds that could range from very low to very high. Whereas the *deformation speed* or *rate* is typically defined as the speed at which, for example, a tension test is being carried out (such as m/s), the *strain rate* (such as 10^2 s^{-1}, 10^4 s^{-1}, and so on) is also a function of the specimen shape, as described below. Typical deformation speeds and strain rates in various metalworking processes are shown in Table 2.3.

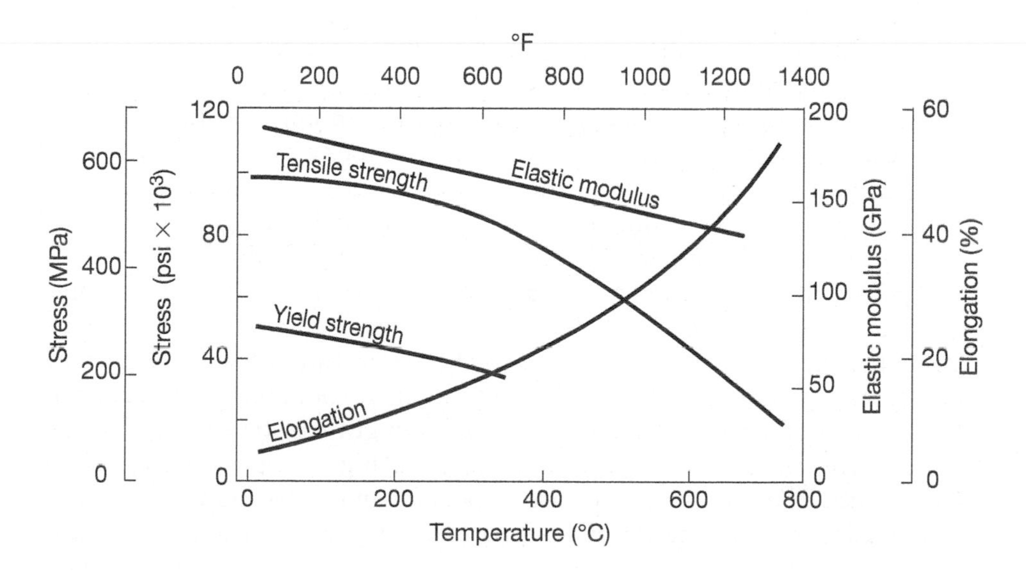

FIGURE 2.9 Effect of temperature on mechanical properties of carbon steel. Most materials display similar temperature sensitivity for elastic modulus, yield strength, ultimate strength, and ductility.

TABLE 2.3 Typical ranges of strain, deformation speed, and strain rates in metalworking processes.

Process	True strain ϵ	Deformation speed m/s	Deformation speed ft/min	Strain rate s^{-1}
Cold Working				
Forging, rolling	0.1–0.5	0.1–100	20–20,000	$1–10^3$
Wire and tube drawing	0.05–0.5	0.1–100	20–20,000	$1–10^4$
Explosive forming	0.05–0.2	10–100	2000–20,000	$10–10^5$
Hot working and warm working				
Forging, rolling	0.1–0.5	0.1–30	20–6000	$1–10^3$
Extrusion	2–5	0.1–1	20–200	$10^{-1}–10^2$
Machining	1–10	0.1–100	20–20,000	$10^3–10^6$
Sheet metal forming	0.1–0.5	0.05–2	10–400	$1–10^2$
Superplastic forming	0.2–3	$10^{-4}–10^{-2}$	0.02–2	$10^{-4}–10^{-2}$

The engineering strain rate, $\dot{e}$, is defined as

$$\dot{e} = \frac{de}{dt} = \frac{d\left(\frac{l - l_o}{l_o}\right)}{dt} = \frac{1}{l_o}\frac{dl}{dt} = \frac{v}{l_o}, \tag{2.16}$$

and the true strain rate, $\dot{\epsilon}$, as

$$\dot{\epsilon} = \frac{d\epsilon}{dt} = \frac{d\left[\ln\left(\frac{l}{l_o}\right)\right]}{dt} = \frac{1}{l}\frac{dl}{dt} = \frac{v}{l}, \tag{2.17}$$

where v is the rate of deformation, for example, the speed of the jaws of the testing machine.

It can be seen from the equations above that the engineering strain rate is constant for a tension test with a constant velocity, while the true strain rate, $\dot{\epsilon}$, is not. Therefore, in order to maintain a constant $\dot{\epsilon}$, the speed must be increased accordingly. (Note that for small changes in length of the specimen during a test, this difference is not significant.)

Typical effects of temperature and strain rate on the strength of metals are shown in Fig. 2.10. Note that increasing strain rate increases strength; the sensitivity of strength to the strain rate increases with temperature; and that this effect is relatively small at room temperature. From Fig. 2.10 it can be noted that the same strength can be obtained either at low temperature and low strain rate or at high temperature and high strain rate. These relationships are important in estimating the resistance of materials to deformation when processing them at various strain rates and temperatures.

The effect of strain rate on the strength of materials is generally expressed by

$$\sigma = C\dot{\epsilon}^m, \tag{2.18}$$

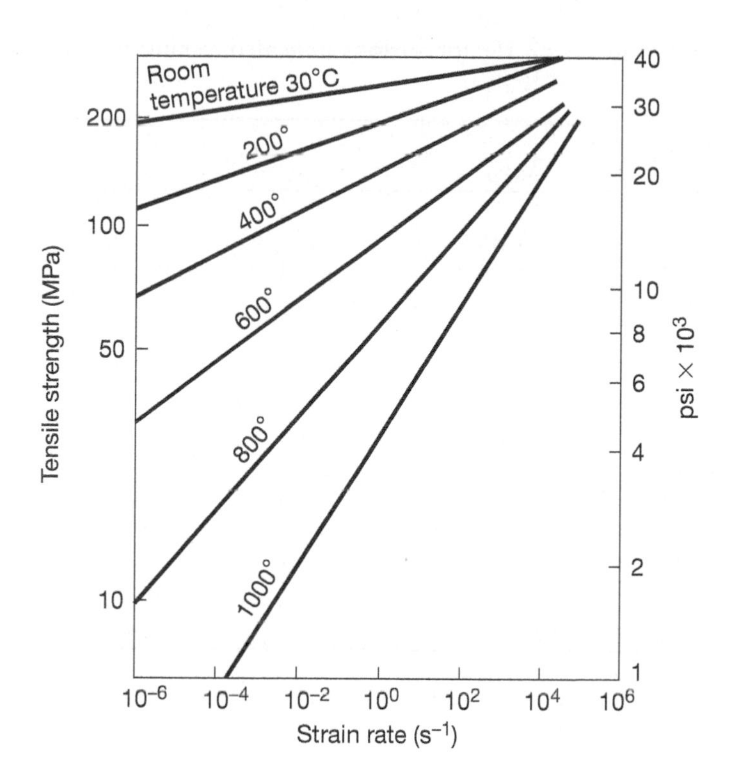

FIGURE 2.10 The effect of strain rate on the ultimate tensile strength of aluminum. Note that as temperature increases, the slope increases. Thus, tensile strength becomes more and more sensitive to strain rate as temperature increases. *Source:* After J.H. Hollomon.

where C is the **strength coefficient,** similar to K in Eq. (2.11), and m is the **strain-rate sensitivity exponent** of the material. A typical range of values for m is up to 0.05 for cold working and 0.05 to 0.4 for hot working of metals, and 0.3 to 0.85 for superplastic materials (see below). Some specific values for C and m are given in Table 2.4. It has been observed that the value of m decreases with metals of increasing strength.

The magnitude of m has a significant effect on necking in a tension test. Experimental observations have shown that with higher m values, a material stretches to a greater length before it fails, an indication that necking and/or fracture is delayed with increasing m. When necking is about to begin in a region of the specimen, that region's strength with respect to the rest of the specimen increases because of strain hardening. However, the strain rate in the necked region is also higher than in the rest of the specimen. Since the material in the necked region is becoming stronger as it is strained at a higher rate, this region exhibits a higher resistance to further necking.

Consider the effect of high strain rate sensitivity exponent. As a tension test progresses, necking becomes more *diffuse* and the specimen becomes longer before it fractures; hence, total elongation increases with increasing m. In addition, elongation after necking (*post-uniform elongation*) also increases with increasing m.

The effect of strain rate on strength also depends on the level of strain, because strain-rate effects increase with strain. Moreover, the strain rate also affects the strain-hardening exponent, n, because it decreases with increasing strain rate.

TABLE 2.4 Approximate range of values for C and m in Eq. (2.18) for various annealed metals at true strains ranging from 0.2 to 1.0, and at a strain rate of $2 \times 10^{-4}\ s^{-1}$.

	Temperature		C		
Material	°C	°F	MPa	ksi	m
Aluminum	200–500	390–932	82–14	12–2	0.07–0.23
Aluminum alloys	200–500	390–932	310–35	45–5	0–0.20
Copper	300–900	572–1650	240–20	35–3	0.06–0.17
Copper alloys (brasses)	200–800	390–1470	415–14	60–2	0.02–0.3
Lead	100–300	212–572	11–2	1.6–0.3	0.1–0.2
Magnesium	200–400	390–750	140–14	20–2	0.07–0.43
Steel					
Low carbon	900–1200	1650–2200	165–48	24–7	0.08–0.22
Medium carbon	900–1200	1650–2200	160–48	23–7	0.07–0.24
Stainless	600–1200	1100–2200	415–35	60–5	0.02–0.4
Titanium	200–1000	390–1800	930–14	135–2	0.04–0.3
Titanium alloys	200–1000	390–1800	900–35	130–5	0.02–0.3
Ti-6Al-4V	800–950	1470–1700	65–11	9.5–1.6	0.50–0.80
Zirconium	200–1000	3 90–1800	830–27	120–4	0.04–0.4

Note: As temperature increases, C decreases and m increases. As strain increases, C increases and m may increase or decrease, or it may become negative within certain ranges of temperature and strain. *Source:* After T. Altan and F.W. Boulger.

Because the formability of materials depends largely on their ductility (see Section 2.2.1), it is important to recognize the effect of temperature and strain rate on ductility; generally, higher strain rates have an adverse effect on the ductility of materials. The increase in ductility due to the strain-rate sensitivity of materials has been exploited in **superplastic forming of metals**, as described in Section 7.5.6. The term *superplastic* refers to the capability of some materials to undergo large uniform elongation prior to failure; elongation may be on the order of a few hundred percent to over 2000%. Examples of such behavior are hot glass, thermoplastic polymers at elevated temperatures, very fine-grain alloys of zinc-aluminum, and titanium alloys. Nickel alloys also can display superplastic behavior when in a nanocrystalline form (see Section 3.11.9).

2.2.8 Effects of Hydrostatic Pressure

Although most testing of materials is carried out at ambient pressure, experiments also can be performed under hydrostatic conditions, with pressures ranging up to 10^3 MPa (10^5 psi). Four important observations have been made concerning the effects of high hydrostatic pressure on the behavior of materials:

1. it substantially increases the strain at fracture (Fig. 2.11);
2. it has little or no effect on the general shape of the true stress–true strain curve, but extends it;

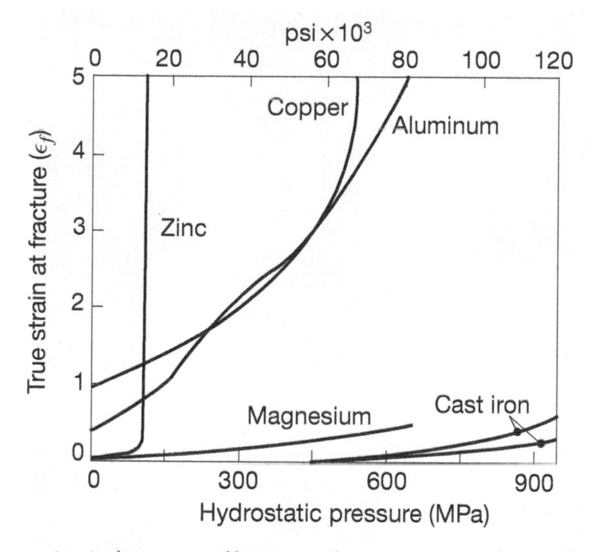

FIGURE 2.11 The effect of hydrostatic pressure on true strain at fracture in tension for various metals. Even cast iron becomes ductile under high pressure. *Source:* After H.L.D. Pugh and D. Green.

3. it has no effect on the strain or the maximum load at which necking begins; and
4. the mechanical properties of metals generally are not altered after being subjected to hydrostatic pressure.

The increase in ductility due to hydrostatic pressure has also been observed in compression and torsion tests. The increase has been observed not only with ductile metals, but also with brittle metals and nonmetallic materials. Various brittle materials such as grey cast iron can deform plastically when subjected to hydrostatic pressure. The level of the pressure required to impart or enhance ductility depends on the particular material.

2.2.9 Effects of Radiation

In view of the nuclear applications of various metals and alloys, studies have been conducted on the effects of radiation on material properties. Typical changes in the mechanical properties of steels and other metals exposed to high-energy radiation are: (a) higher yield strength; (b) higher tensile strength; (c) higher hardness; and (d) lower ductility and toughness. The magnitudes of these changes depend on the particular material and its processing history, as well as the temperature and level of radiation to which it is subjected.

2.3 Compression

In many operations in metalworking, such as forging, rolling, and extrusion, the workpieces are subjected to externally applied compressive forces. The **compression test**, in which the specimen is subjected to a compressive load as shown in Fig. 2.1b, can give useful information on the stresses required and the behavior of materials under compression. This test is typically carried out by compressing (*upsetting*) a solid cylindrical specimen between two flat *platens*. The deformation shown in Fig. 2.1b is

FIGURE 2.12 Barreling in compressing a round solid cylindrical specimen (7075-O aluminum) between flat dies. Barreling is caused by friction at the die–specimen interfaces, which retards the free flow of the material (See also Figs. 6.1 and 6.2). *Source:* After K.M. Kulkarni and S. Kalpakjian.

ideal; in practice, friction between the specimen and the two dies causes **barreling** (Fig. 2.12), because friction prevents the top and bottom surfaces from radially expanding freely.

This phenomenon makes it difficult to obtain relevant data and to properly construct a compressive stress–strain curve because (a) the cross-sectional area of the specimen changes along its height and (b) friction dissipates energy; this energy is supplied through an increased compressive force. However, with effective lubrication and other means (see Section 4.4.3), friction (and therefore barreling) can be minimized to obtain a reasonably constant cross-sectional area during the test.

The engineering strain rate, $\dot{e}$, in compression is given by

$$\dot{e} = -\frac{v}{h_o}, \tag{2.19}$$

where v is the speed of the die and h_o is the original height of the specimen. The true strain rate, $\dot{\epsilon}$, is given by

$$\dot{\epsilon} = -\frac{v}{h}, \tag{2.20}$$

where h is the instantaneous height of the specimen. Note that if v is constant, the true strain rate increases as the test progresses. In order to conduct tests at a constant true strain rate, a *cam plastometer* has been designed that, through a cam action, reduces the magnitude of v proportionately as the specimen height h decreases during the test.

The compression test can also be utilized to determine the ductility of a material, by observing any cracks that form on the barreled surfaces of the specimen (See Fig. 3.21d). Hydrostatic pressure has a beneficial effect in delaying the formation of these cracks. With a sufficiently ductile material and effective lubrication, these tests can be carried out to large uniform strains.

2.3.1 Plane-Strain Compression Test

The *plane-strain compression test* (Fig. 2.13) has been designed to simulate such bulk-deformation processes as forging and rolling (as described in Section 6.2). In this test, the die and workpiece geometries are such that the width of the specimen does not undergo any significant change during compression; thus the volume of material under the dies is in the condition of *plane strain* (see Section 2.11.3). The yield strength of a material in plane strain, S'_y, is given by

$$S'_y = \frac{2}{\sqrt{3}} S_y = 1.15 S_y, \tag{2.21}$$

according to the distortion-energy yield criterion (see Section 2.11.2).

As the geometric relationships in Fig. 2.13 indicate, the parameters for this test must be chosen properly to make the results meaningful. Moreover, caution should be exercised in the test procedures, such as preparing the die surfaces, aligning the dies, lubricating the contacting surfaces, and accurately measuring the load required.

When the results of tension and compression tests on the same material are compared, it is found that for most *ductile* metals, the true stress–true strain curves for both tests coincide (Fig. 2.14). However, this is not true for brittle materials, particularly with respect to ductility (see also Section 3.8).

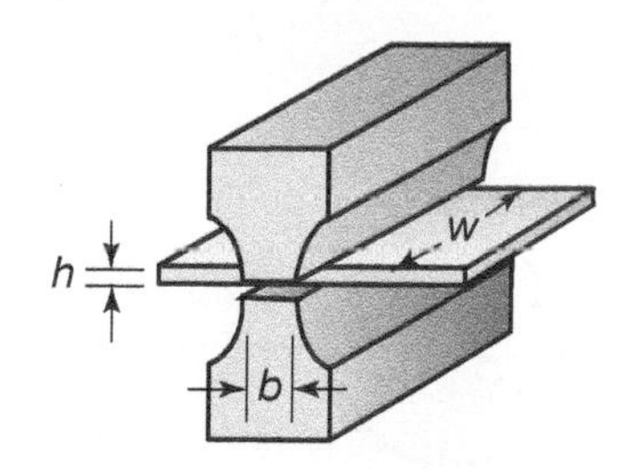

FIGURE 2.13 Schematic illustration of the plane-strain compression test. For this test to be useful and reproducible, it is recommended that $w > 5h$, $w > 5b$, and $2h < b < 4h$. This test gives the yield strength of the material in plane strain, S'_y.

2.3.2 Bauschinger Effect

In deformation processing of materials, a workpiece is sometimes first subjected to tension and then to compression, or vice versa. Three examples are (a) bending and unbending of sheet metal; (b) roller leveling in flattening sheet metal (see Section 6.3.4); and (c) reverse drawing in making deep-drawn cup-shaped parts (see Section 7.6.2).

It has been noted that when a piece of metal, with a tensile yield strength of S_y, is subjected to tension into the plastic range, and then the load is released and applied in compression, its yield strength in compression is found to be lower than that in tension (Fig. 2.15). This phenomenon, known as the *Bauschinger effect*, is exhibited to varying degrees by all metals and alloys and is also observed when the loading path is reversed, that is, compression followed by tension. Because of the lowered yield strength in the reverse direction of load application, this phenomenon is also called

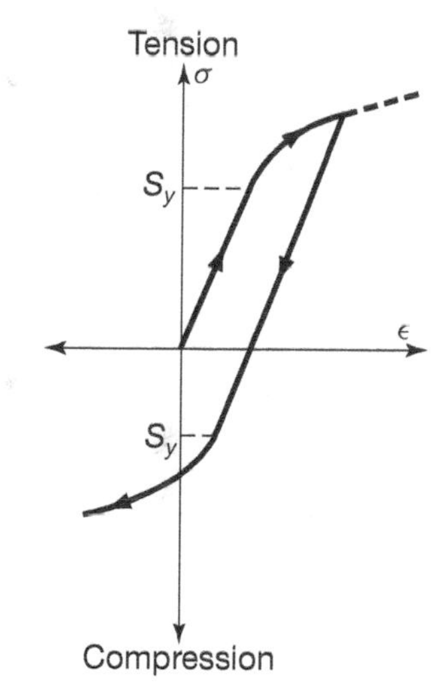

FIGURE 2.15 Schematic illustration of the Bauschinger effect. Arrows show loading and unloading paths. Note the decrease in the yield strength in compression after the specimen has been subjected to tension. The same result is obtained if compression is applied first, followed by tension, whereby the yield strength in tension decreases.

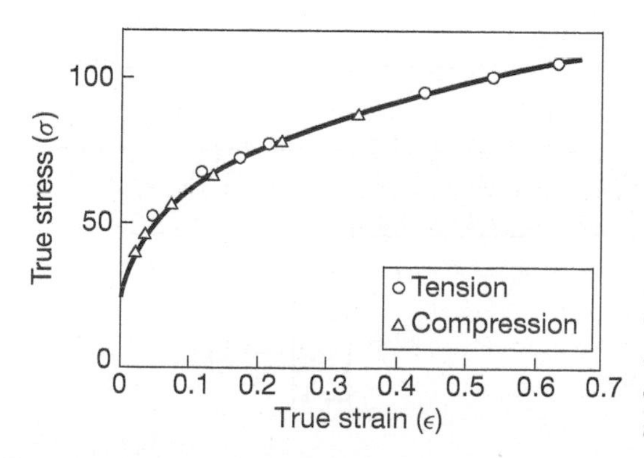

FIGURE 2.14 True stress–true strain curve in tension and compression for aluminum. For ductile metals, the curves for tension and compression are identical.

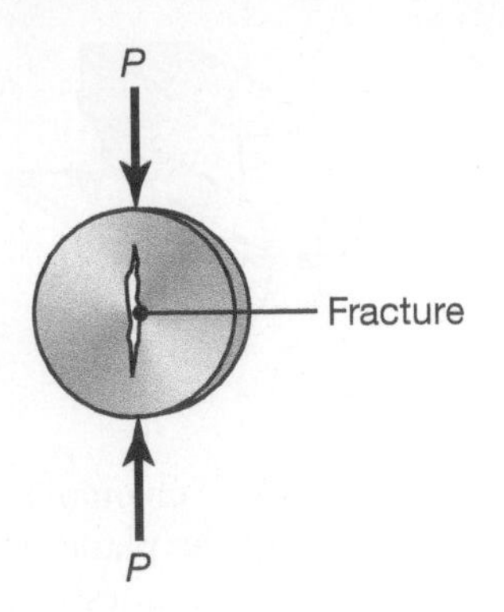

FIGURE 2.16 Disk test on a brittle material, showing the direction of loading and the fracture path. This test is useful for brittle materials, such as ceramics and carbides.

strain softening or **work softening,** a behavior that is also observed in torsion (Section 2.4).

2.3.3 The Disk Test

For brittle materials, such as ceramics and glasses, a **disk test** has been developed where a disk is subjected to compression between two hardened flat platens (Fig. 2.16). When loaded as shown, tensile stresses develop perpendicular to the vertical centerline along the disk, causing the disk to fracture and split in half vertically (see also *rotary tube piercing* in Section 6.3.5).

The tensile stress, σ, in the disk is uniform along the centerline and can be calculated from the formula

$$\sigma = \frac{2P}{\pi dt}, \tag{2.22}$$

where P is the load at fracture, d is the diameter of the disk, and t is its thickness. In order to avoid premature failure of the disk at the top and bottom contact points, thin strips of soft metal are placed between the disk and the platens. These strips also protect the platens from being damaged during the test.

2.4 Torsion

Another method of determining material properties is the **torsion test.** In order to obtain an approximately uniform stress and strain distribution along the cross section, this test is generally carried out on a tubular specimen with a reduced mid-section (Fig. 2.17). The *shear stress,* τ, can be determined from the equation

$$\tau = \frac{T}{2\pi r^2 t}, \tag{2.23}$$

where T is the torque applied, r is the mean radius, and t is the thickness of the reduced section in the middle of the tube. The shear strain, γ, is

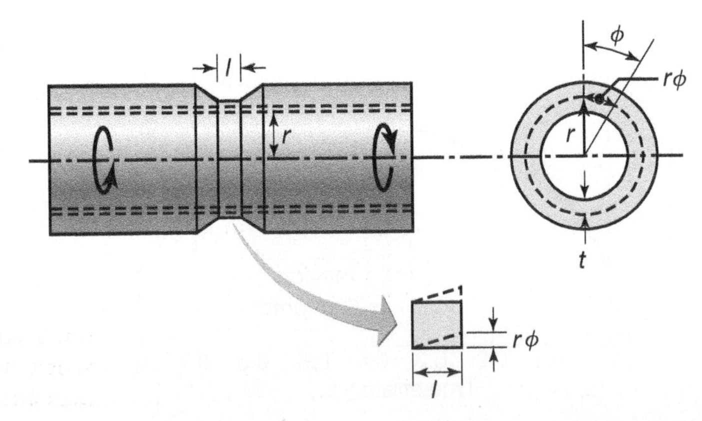

FIGURE 2.17 A typical torsion-test specimen. The specimen is mounted in a machine and is twisted. Note the shear deformation of an element in the reduced section.

determined from the equation

$$\gamma = \frac{r\phi}{l}, \tag{2.24}$$

where l is the length of the reduced section and ϕ is the angle of twist, in radians. With the shear stress and shear strain obtained from this test, the shear stress–shear strain curve of the material can be constructed.

In the elastic range, the ratio of the shear stress to shear strain is known as the **shear modulus** or the **modulus of rigidity**, G:

$$G = \frac{\tau}{\gamma}. \tag{2.25}$$

The shear modulus and the modulus of elasticity are related by the formula

$$G = \frac{E}{2(1+\nu)}, \tag{2.26}$$

which is based on a comparison of **simple shear** and **pure shear** strains (Fig. 2.18). Note from this figure that the difference between the two strains is that simple shear is equivalent to pure shear plus a rotation of $\gamma/2$ degrees.

Example 2.2 shows that a thin-walled tube does not neck in torsion, unlike the tension-test specimen described in Section 2.2.4. Consequently, changes in the cross-sectional area of the specimen are not a concern in torsion testing. The shear stress–shear strain curves obtained from torsion tests increase monotonically, just as they do in true stress–true strain curves.

Torsion tests can also be performed on solid round bars at elevated temperatures in order to estimate the *forgeability* of metals (see Section 6.2.5); the greater the number of twists prior to failure, the better the forgeability. Torsion tests can also be conducted on round solid bars that are compressed axially to produce a beneficial effect similar to that of hydrostatic pressure (see Section 2.2.8). The effect of compressive stresses in increasing the maximum shear strain at fracture has also been observed in metal cutting (see Section 8.2). The normal compressive stress has no effect on the magnitude of shear stresses required to cause yielding or to continue the deformation, just as hydrostatic pressure has no effect on the general shape of the stress–strain curve.

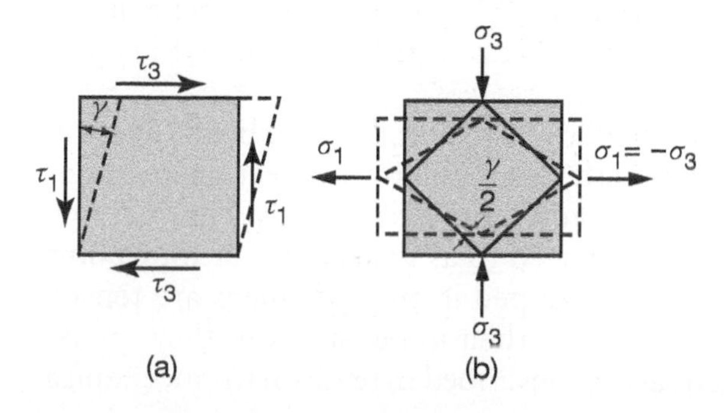

FIGURE 2.18 Comparison of (a) simple shear and (b) pure shear. Note that simple shear is equivalent to pure shear plus a rotation.

EXAMPLE 2.2 Instability in Torsion of a Thin-Walled Tube

Given: A thin-walled tube is made of a material whose true stress–true strain curve is given by $\sigma = K\epsilon^n$.

Find: Show that necking cannot take place in the torsion of this tube.

Solution: According to Eq. (2.23), the expression for torque, T, is

$$T = 2\pi r^2 t\tau,$$

where for this case the *shear stress*, τ, can be related to the normal stress, σ, using Eq. (2.39) for the distortion-energy criterion as $\tau = \sigma/\sqrt{3}$. The criterion for instability is

$$\frac{dT}{d\epsilon} = 0.$$

Because r and t are constant,

$$\frac{dT}{d\epsilon} = \left(\frac{2}{\sqrt{3}}\right)\pi r^2 \tau \frac{d\sigma}{d\epsilon}.$$

For a material represented by $\sigma = K\epsilon^n$,

$$\frac{d\sigma}{d\epsilon} = nK\epsilon^{n-1}.$$

Therefore,

$$\frac{dT}{d\epsilon} = \left(\frac{2}{\sqrt{3}}\right)\pi r^2 \tau nK\epsilon^{n-1}.$$

Since none of the quantities in this expression are zero, $dT/d\epsilon$ cannot be zero. Consequently, a tube in torsion does not undergo instability.

2.5 Bending

Preparing tension test specimens from such brittle materials as ceramics and carbides can be difficult because: (a) shaping and machining them to proper dimensions can be challenging; (b) brittle materials are sensitive to surface defects, scratches, and imperfections; (c) clamping brittle specimens for testing can be difficult; and (d) improper alignment of the test specimen may result in nonuniform stress distribution along the cross section of the specimen.

A commonly used test method for brittle materials is the **bend (flexure) test**, usually involving a specimen with rectangular cross section and supported at both ends (Fig. 2.19). The load is applied vertically, either at one or two points, hence these tests are referred to as **three-point** or **four-point** bending, respectively. The stresses developed in the specimens are tensile at their lower surfaces and compressive at their upper surfaces; they can be calculated using simple beam equations described in texts on the mechanics

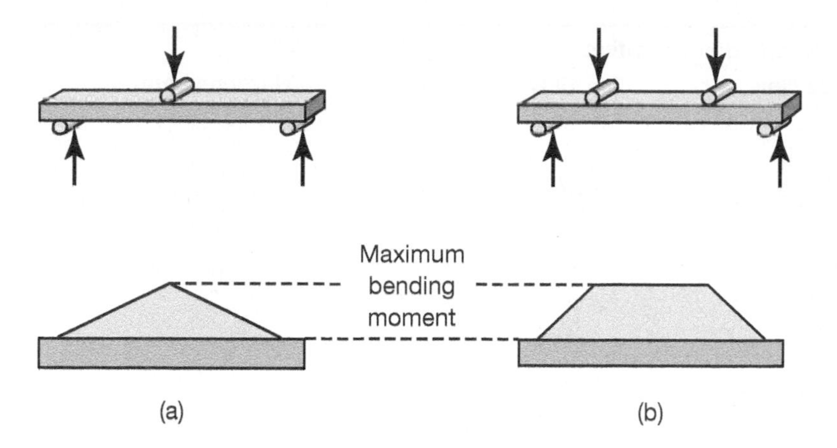

FIGURE 2.19 Two bend-test methods for brittle materials: (a) three-point bending and (b) four-point bending. The shaded areas on the beams represent the bending-moment diagrams, described in texts on the mechanics of solids. Note the region of constant maximum bending moment in (b), whereas the maximum bending moment occurs only at the center of the specimen in (a).

of solids. The stress at fracture in bending is known as the **modulus of rupture,** or **transverse rupture strength,** and is obtained from the formula

$$\sigma = \frac{Mc}{I}, \tag{2.27}$$

where M is the bending moment, c is one-half of the specimen depth, and I is the moment of inertia of the cross section.

Note that there is a basic difference between the two loading conditions in Fig. 2.19. In three-point bending, the maximum stress is at the center of the beam; in the four-point test, the maximum stress is constant between the two loading points. The stress magnitude is the same in both situations when all other parameters are maintained. There is, however, a higher probability for defects and imperfections to be present in the larger volume of material between the loading points in the four-point test than in the much smaller volume under the single load in the three-point test. This means that, as also verified by experiments, the four-point test is likely to result in a lower modulus of rupture than the three-point test. Also, as expected, the results of the four-point test also show less scatter than those in the three-point test.

QR Code 2.2 Instron 5544 3-point bend test. *Source:* Courtesy of Instron.

2.6 Hardness

One of the most common tests for assessing the mechanical properties of materials is the **hardness test.** *Hardness* of a material is generally defined as its *resistance to permanent indentation*; it can also be defined as its resistance to scratching or to wear (see Chapter 4). Several techniques have been developed to measure the hardness of materials, using various indenter geometries and materials. It should be noted that hardness is not a fundamental property, because resistance to indentation depends on the shape of the indenter and the load applied.

The most common standardized hardness tests are summarized in Fig. 2.20. Hardness tests are commonly performed on dedicated equipment in a laboratory, but portable hardness testers are also available. Portable hardness testers can generally perform any hardness test, and can be

Test	Indenter	Shape of indentation: Side view	Shape of indentation: Top view	Load, P	Hardness number
Brinell	10-mm steel or tungsten carbide ball	D; d	d	500 kg 1500 kg 3000 kg	$HB = \frac{2P}{(\pi D)(D - \sqrt{D^2 - d^2})}$
Vickers	Diamond pyramid	136°	L	1–120 kg	$HV = \frac{1.854P}{L^2}$
Knoop	Diamond pyramid	$L/b = 7.11$ $b/t = 4.00$ t	b; L	25 g–5 kg	$HK = \frac{14.2P}{L^2}$
Rockwell A, C, D	Diamond cone	120° $t = $ mm		60 kg 150 kg 100 kg	HRA, HRC, HRD $= 100 - 500t$
B, F, G	$\frac{1}{16}$ - in. diameter steel ball	$t = $ mm		100 kg 60 kg 150 kg	HRB, HRF, HRG $= 130 - 500t$
E	$\frac{1}{8}$ - in. diameter steel ball			100 kg	HRE $= 130 - 500t$

FIGURE 2.20 General characteristics of hardness testing methods. The Knoop test is known as a microhardness test because of the light load and small impressions.

specially configured for particular geometric features, such as hole interiors, gear teeth, and so on, or can be tailored for specific classes of material.

2.6.1 Brinell Test

In the **Brinell test**, a steel or tungsten carbide ball 10 mm in diameter is pressed against a surface with a load of 500, 1500, or 3000 kg. The *Brinell hardness number* (HB) is defined as the ratio of the load, P, to the curved area of indentation, or

$$HB = \frac{2P}{(\pi D)\left(D - \sqrt{D^2 - d^2}\right)} \text{ kg/mm}^2, \qquad (2.28)$$

where D is the diameter of the ball and d is the diameter of the impression, in millimeters.

Depending on the condition of the material tested, different types of indentations are obtained on the surface after a hardness test has been performed. As shown in Fig. 2.21, annealed materials generally have a rounded profile, whereas cold-worked (strain-hardened) materials have a sharp profile. The correct method of measuring the indentation diameter, d, for both cases is shown in the figure.

Because the indenter material has a finite elastic modulus, it also undergoes elastic deformation under the applied load, P; thus, hardness measurements may not be as accurate as expected. A common method of minimizing this effect is to use tungsten carbide balls, which, because of their high modulus of elasticity, deform less than steel balls. Tungsten carbide is generally recommended for Brinell hardness numbers higher than 500. In reporting the test results for these higher hardnesses, the type of ball used should be cited. Since harder workpiece materials produce very small impressions, a 1500-kg or 3000-kg load is recommended in order to obtain impressions that are sufficiently large for accurate measurements.

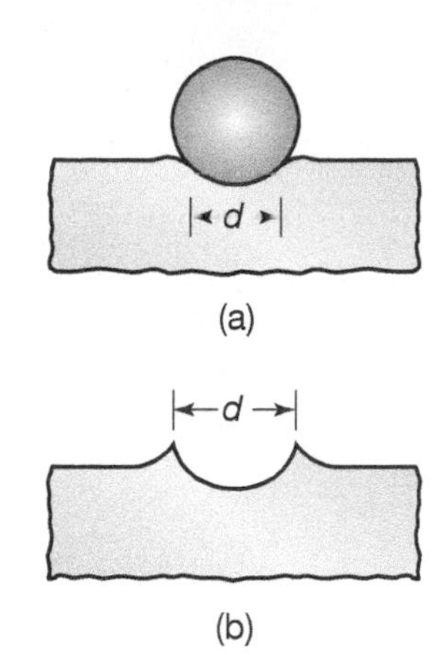

FIGURE 2.21 Indentation geometry for Brinell hardness testing: (a) annealed metal and (b) work-hardened metal. Note the difference in metal flow at the periphery of the impressions.

The Brinell test is generally suitable for materials of low to medium hardness. Because the impressions made by the same indenter at different loads are not geometrically similar, the Brinell hardness number depends on the load used. Consequently, the load employed should also be cited with the test results.

Brinelling is a term used to describe permanent indentations on a surface between contacting bodies such as a ball bearing indenting a flat surface under fluctuating loads or due to vibrations. Such loads commonly occur during transportation of equipment or machinery, or from dynamic loads associated with vibration during use.

2.6.2 Rockwell Test

In the **Rockwell test**, the *depth* of penetration is measured. The indenter is pressed on the surface, first with a *minor* load and then with a *major* load. The difference in the depth of penetration is a measure of the hardness. There are several Rockwell hardness scales that employ different loads, indenter materials, and indenter geometries. Some of the more common hardness scales and the indenters used are listed in Fig. 2.20. The Rockwell hardness number, which is obtained directly from the testing machine, is expressed as a value followed by an abbreviation for the test used; for example, 55 HRC refers to a Rockwell C scale with a value of 55. *Rockwell superficial hardness* tests have also been developed using lighter loads and the same type of indenters.

QR Code 2.3 Knoop and Vickers test equipment. *Source:* Courtesy of Instron.

2.6.3 Vickers Test

The **Vickers test**, formerly known as the *diamond pyramid hardness test*, uses a pyramid-shaped diamond indenter (See Fig. 2.20) with loads ranging from 1 to 120 kg. The *Vickers hardness number* (HV) is given by the formula

$$\text{HV} = \frac{1.854P}{L^2}. \tag{2.29}$$

The impressions are typically less than 0.5 mm on the diagonal. The Vickers test gives essentially the same hardness number regardless of the load, and is thus suitable for testing materials with a wide range of hardness, including very hard steels.

2.6.4 Knoop Test

The **Knoop test** uses a diamond indenter in the shape of an elongated pyramid (See Fig. 2.20), using loads generally ranging from 25 g to 5 kg. The *Knoop hardness number* (HK) is given by the formula

$$\mathrm{HK} = \frac{14.2P}{L^2}. \tag{2.30}$$

Because the size of the indentation is generally in the range of 0.01 to 0.10 mm, surface preparation is very important. The hardness number obtained depends on the applied load, thus test results should always cite the load applied. The Knoop test is a *microhardness* test because of the light loads utilized; hence it is suitable for very small or thin specimens, and for brittle materials such as gemstones, carbides, and glass. Because the impressions are very small, this test is also used to measure the hardness of individual grains in a metal.

2.6.5 Scleroscope

The *scleroscope* (from the Greek *skleros*, meaning "hard") is an instrument in which a diamond-tipped indenter (called *hammer*), enclosed in a glass tube, is dropped onto the specimen from a certain height. The hardness is related to the *rebound* of the indenter: the higher the rebound, the harder the material tested. An electronic version of a scleroscope, called a *Leeb*, or Equotip, test, has been developed. In this test, a carbide hammer impacts the surface, and incident and rebound velocities are electronically measured. A *Leeb number* is then calculated and usually converted to Rockwell or Vickers hardness. Because the instrument is portable, it is useful for measuring the hardness of large objects.

2.6.6 Mohs Test

The **Mohs test** is based on the capability of one material to scratch another. The Mohs hardness is expressed on a scale of 10, with 1 for talc and 10 for diamond (the hardest substance known); thus, a material with a higher Mohs hardness can scratch materials with a lower hardness. Soft metals have a Mohs hardness of 2–3, hardened steels about 6, and aluminum oxide 9. The Mohs scale generally is used by mineralogists and geologists, but not manufacturing; its crudeness and qualitative nature is not conducive to modern quality-control methods (see Section 4.9).

2.6.7 Shore Test and Durometer

The hardness of materials such as rubbers, plastics, and similar soft and elastic nonmetallic materials is generally measured by a Shore test, with an instrument called a *durometer* (from the Latin *durus*, meaning "hard"). An indenter is first pressed against the surface and then a constant load is rapidly applied. The *depth* of penetration is measured after one second; the hardness is inversely related to the penetration. There are two different scales for this test. Type A has a blunt indenter and an applied load of 1 kg; the test is typically used for softer materials. Type D has a sharper indenter and a load of 5 kg, and is used for harder materials. The hardness numbers in these tests range from 0 to 100.

2.6.8 Relationship Between Hardness and Strength

Since hardness is the resistance to permanent indentation, hardness testing is equivalent to performing a compression test on a small volume of a material's surface. Some correlation is therefore expected between hardness and yield strength, S_y, in the form of

$$\text{Hardness} = cS_y, \tag{2.31}$$

where c is a proportionality constant. The magnitude of c is not a constant for a material or even a class of materials because the hardness of a given material depends upon its processing history, such as cold or hot working and surface treatments. Theoretical studies, based on plane-strain slip-line analysis of a smooth flat punch indenting the surface of a semi-infinite body, have shown that for a perfectly plastic material of yield strength, S_y, the magnitude of c is about 3. This is in reasonable agreement with experimental data, as can be seen in Fig. 2.22. Note that cold-worked metals (which are close to being perfectly plastic in their behavior) show better agreement than annealed metals. The higher value of c for annealed materials is explained by the fact that, due to strain hardening, the average yield strength they exhibit during indentation is higher than their initial yield strength.

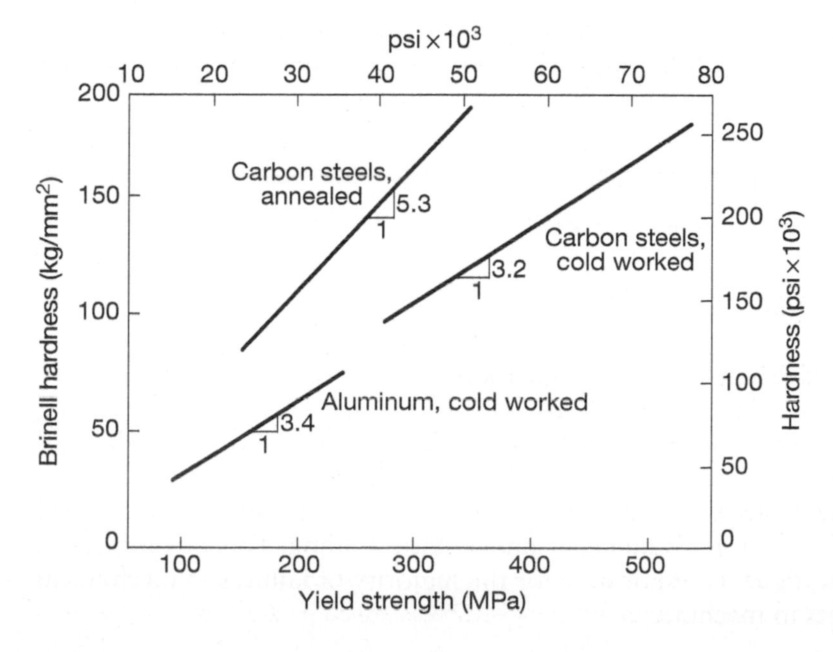

FIGURE 2.22 Relation between Brinell hardness and yield strength for aluminum and steels. For comparison, the Brinell hardness (which is always measured in kg/mm^2) is converted to psi units on the right scale.

FIGURE 2.23 Bulk deformation in mild steel under a spherical indenter. *Source:* After M.C. Shaw and C.T. Yang.

More practically, a relationship also has been observed between the ultimate tensile strength (expressed in MPa) and Brinell hardness number for steels:

$$S_{ut} = 3.5(\text{HB}). \tag{2.32}$$

Note that the depth of the deformed zone in a hardness test is about one order of magnitude larger than the depth of indentation (Fig. 2.23). For a hardness test to be valid, the material should be allowed to fully develop this zone. This is why thinner specimens require smaller indentations.

The **hot hardness** of materials is important in applications where the materials are subjected to elevated temperatures, such as in cutting tools in machining and dies for hot metalworking operations (see Section 6.8). The tests can be carried out using conventional testers with some modifications, such as enclosing the specimen and the indenter in a small electric furnace.

EXAMPLE 2.3 Calculation of Modulus of Resilience from Hardness

Given: A piece of steel is highly deformed at room temperature. Its hardness is found to be 300 HB.

Find: Estimate the modulus of resilience for this material.

Solution: Since the steel has been subjected to large strains at room temperature, it may be assumed that its stress–strain curve has flattened considerably, thus approaching the shape of a perfectly plastic curve. According to Eq. (2.31) and using a value of $c = 3$,

$$S_y = \frac{300}{3} = 100 \text{ kg/mm}^2 = 981 \text{ MPa}$$

The modulus of resilience is defined by Eq. (2.5),

$$\text{Modulus of resilience} = \frac{S_y^2}{2E}.$$

From the inside front cover, $E = 200$ GPa for steel. Hence,

$$\text{Modulus of resilience} = \frac{(981 \times 10^6)^2}{2 \times 200 \times 10^9} = 2.41 \times 10^6 \text{ Nm/m}^3.$$

2.7 Fatigue

Gears, cams, shafts, springs, and tools and dies are often subjected to rapidly fluctuating (cyclic) loads; these may be caused by fluctuating mechanical loads (such as in gear teeth, dies, and cutters), or by thermal stresses (such as in a room-temperature die coming into repeated contact with hot workpieces). Under these conditions, the part fails at a stress level below which failure would occur under static loading. This phenomenon, known as **fatigue**, is responsible for the majority of failures in mechanical components in machinery.

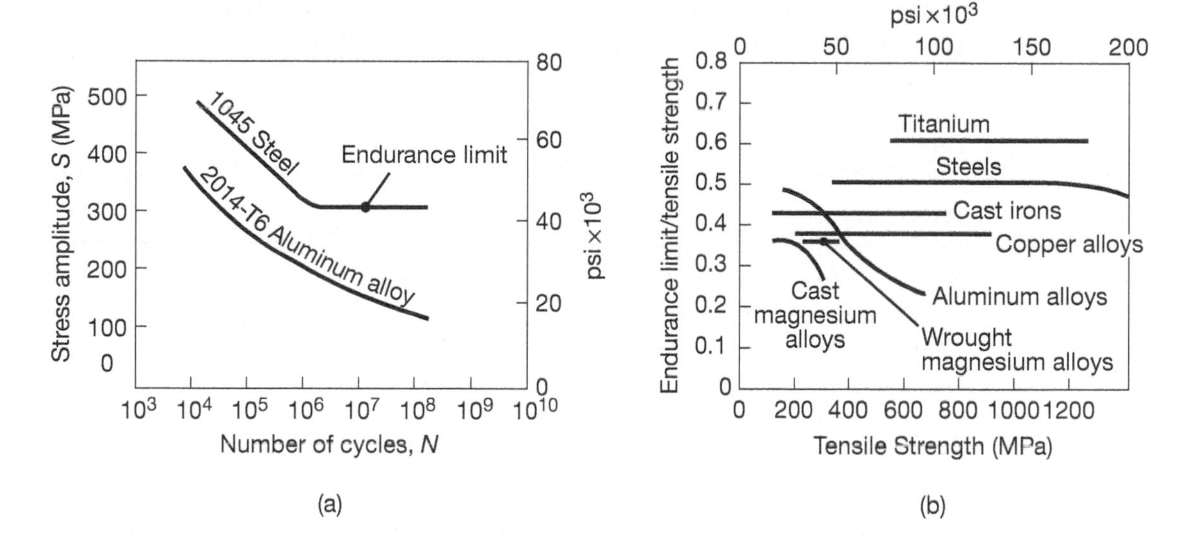

FIGURE 2.24 Fatigue in metals. (a) Typical *S-N* curves for two metals. Note that, unlike steel, aluminum does not have an endurance limit. (b) Ratio of fatigue strength to tensile strength for various metals, as a function of tensile strength.

In **fatigue tests**, specimens are repeatedly subjected to various states of stress, usually in a combination of tension, compression, or torsion. The test is carried out at various stress amplitudes (S), and the number of cycles (N) to cause total failure of the specimen is recorded.

A typical plot of the data obtained, known as an *S-N curve* or *Wöhler diagram*, is shown in Fig. 2.24a. These curves are based on complete reversal of the stress, that is, maximum tension, maximum compression, maximum tension, and so on, such as that obtained by bending a piece of wire alternately in one direction, and then in the other. The fatigue test can also be performed on a rotating shaft with a constant applied load, often using four-point bending (See Fig. 2.19b). The maximum stress to which the material can be subjected without fatigue failure regardless of the number of cycles is known as the **endurance limit** or **fatigue limit**.

The fatigue strength for metals has been found to be related to their ultimate tensile strength, S_{ut}, as shown in Fig. 2.24b. Note that for steels the endurance limit is about one-half their tensile strength. Although most metals, especially steels, have a definite endurance limit, aluminum alloys do not have one and the *S-N* curve continues its downward trend. For metals exhibiting such behavior (most face-centered cubic metals; see Fig. 3.3), fatigue strength is specified at a specific number of cycles, such as 10^7 as the useful service life of the component.

2.8 Creep

Creep is the permanent elongation of a material under a static load maintained for a period of time. It is a phenomenon of metals and some nonmetallic materials, such as thermoplastics and rubbers, and it can occur at any temperature. Lead, for example, creeps at room temperature under a constant tensile load. Also, the window glass in old houses has been found to be thicker at the bottom than at its top, the glass having undergone creep by its own weight over many years. For metals and their alloys, creep occurs at elevated temperatures, beginning at about 200°C (400°F) for aluminum alloys, and up to about 1500°C (2800°F) for refractory alloys.

The mechanism of creep at elevated temperature is generally attributed to *grain-boundary sliding*. Creep is especially important in high-temperature applications, such as gas-turbine blades and similar components in jet engines and rocket motors. High-pressure steam lines and nuclear-fuel elements also are subject to creep. Creep deformation also can occur in tools and dies that are subjected to high stresses at elevated temperatures during metalworking operations such as hot forging and extrusion.

The **creep test** typically consists of subjecting a specimen to a constant tensile load (hence constant engineering stress) at a certain temperature, and measuring the change in length that occurs over a period of time. A typical creep curve consists of primary, secondary, and tertiary stages (Fig. 2.25). The specimen eventually fails by necking and fracture, as in the tension test, and is called **rupture** or **creep rupture**. As expected, the creep rate increases with temperature and the applied load.

Design against creep usually requires knowledge of the secondary (linear) range and its slope, because the creep rate can be determined reliably when the curve has a constant slope. Generally, resistance to creep increases with the melting temperature of a material. Stainless steels, superalloys, and refractory metals and their alloys are thus commonly used in applications where creep resistance is required.

Stress relaxation is closely related to creep. In this phenomenon, the stresses resulting from external loading of a structural component decrease in magnitude over a period of time, even though the dimensions of the component remain constant. Common examples are rivets, bolts, guy wires, and parts under tension, compression, or bending; it is particularly common and important in thermoplastics (see Section 10.3).

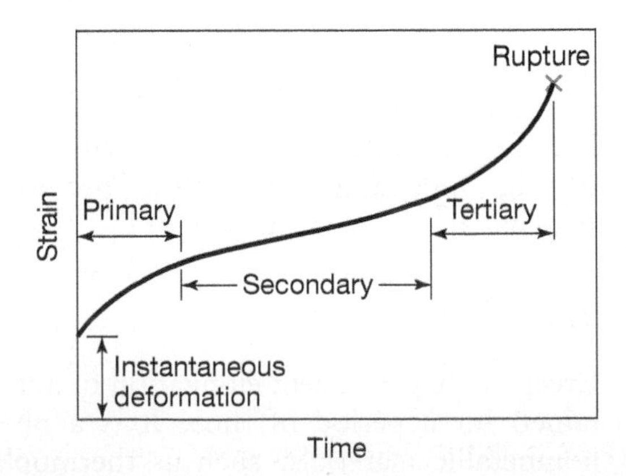

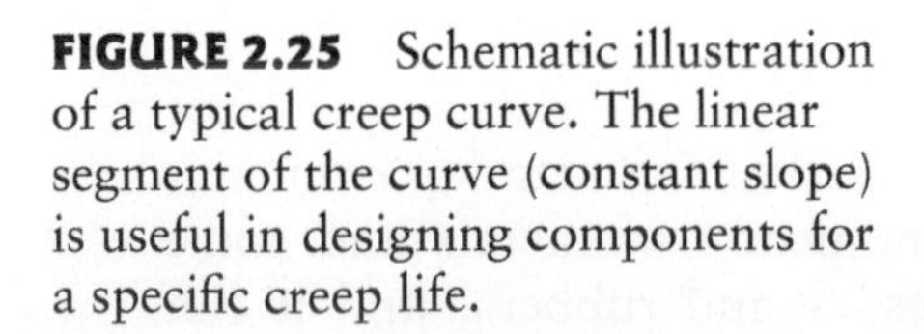
FIGURE 2.25 Schematic illustration of a typical creep curve. The linear segment of the curve (constant slope) is useful in designing components for a specific creep life.

2.9 Impact

In many manufacturing operations, as well as during their service life, various components are subjected to *impact* (or *dynamic*) *loading*. A typical **impact test** consists of placing a *notched specimen* in an impact tester and breaking it with a swinging pendulum (Fig. 2.26). In the **Charpy test** the specimen is supported at both ends, whereas in the **Izod test** it is supported at one end, like a cantilever beam. From the amount of swing of the pendulum, the *energy dissipated* in breaking the specimen is obtained; this energy is the **impact toughness** of the material.

Impact tests are particularly useful in determining the ductile-brittle *transition temperature* of materials (See Fig. 3.26). Generally, materials that have high impact resistance are also those that have high strength and high ductility. Sensitivity of materials to surface defects (**notch sensitivity**) is important, as it lowers their impact toughness.

2.10 Residual Stresses

Inhomogeneous deformation during materials processing leads to *residual stresses*; these are stresses that remain within a part after it has been shaped and all external forces have been removed. A typical example of inhomogeneous deformation is the bending of a beam (Fig. 2.27). Recall that the bending moment first produces a linear elastic stress distribution (Fig. 2.27a); as the moment is increased, the outer fibers begin to yield. For a typical strain-hardening material, the stress distribution is as shown in Fig. 2.27b. When the part is unloaded, it is bent permanently because it has undergone plastic deformation.

As previously shown in Fig. 2.4, all recovery is elastic; therefore, the moments of the areas *oab* and *oac* about the neutral axis in Fig. 2.27c

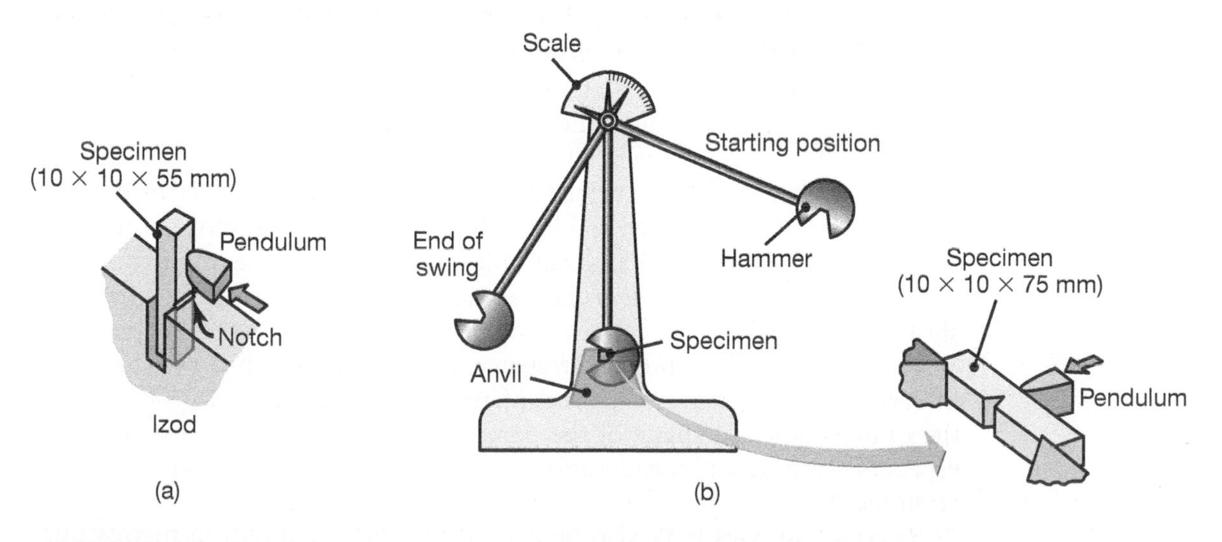

FIGURE 2.26 Impact test specimens: (a) Izod and (b) charpy.

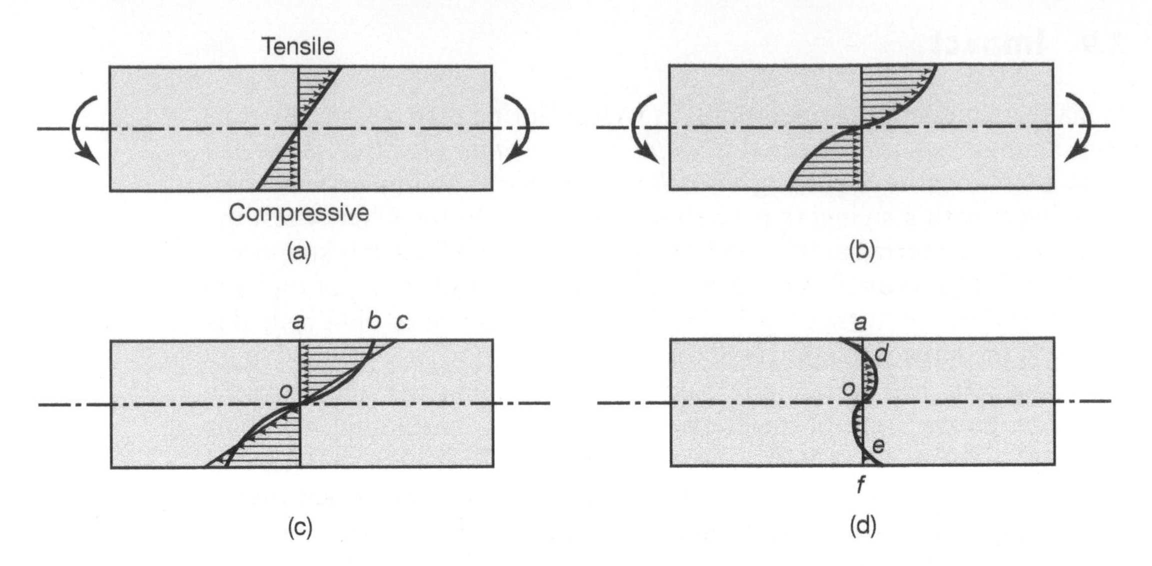

FIGURE 2.27 Residual stresses developed in bending a beam made of an elastic, strain-hardening material. Note that unloading is equivalent to applying an equal and opposite moment to the part, as shown in (b). Because of nonuniform deformation, most parts made by plastic deformation processes contain residual stresses. Note also that the forces and moments due to residual stresses must be internally balanced.

must be equal. (For purposes of this discussion, assume that the neutral axis does not shift.) The difference between the two stress distributions produces the residual stress pattern shown in the figure. Note that there are compressive residual stresses in layers *ad* and *oe*, and tensile residual stresses in layers *do* and *ef*. With no external forces present, the residual stresses in the beam must be in static equilibrium.

Although this example involves stresses in one direction only, in most situations relevant to manufacturing the residual stresses are three dimensional. When the shape of the beam is altered, such as in removing a layer of material by machining or grinding, equilibrium will be disturbed. The beam will then acquire a new radius of curvature in order to balance the internal forces. Another example of the effect of residual stresses is the drilling of round holes on surfaces of parts that have residual stresses. As a result of removing this material by drilling, the equilibrium of the residual stresses will be disturbed and the hole may now become elliptical. Such disturbances will lead to *warping*, some common examples of which are shown in Fig. 2.28.

The equilibrium of internal stresses may also be disturbed by *relaxation* of residual stresses over a period of time, which results in instability of the dimensions and shape of the part. These dimensional changes are an especially important consideration for precision machinery and measuring equipment.

Residual stresses may also be caused by *phase changes* in metals during or after processing, due to density variations among different phases,

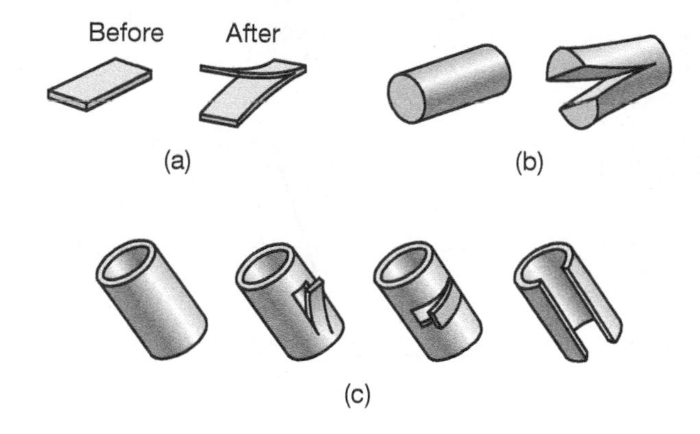

FIGURE 2.28 Distortion of parts with residual stresses after cutting or slitting: (a) rolled sheet or plate; (b) drawn rod; and (c) thin-walled tubing. Because of the presence of residual stresses on the surfaces of parts, a round drill may produce an oval-shaped hole because of relaxation of stresses when a portion is removed.

such as between ferrite and martensite in steels (see Section 5.6.1). Phase changes cause microscopic volumetric changes and thus result in residual stresses, a phenomenon that is important in warm and hot working of metals and in their heat treatment to improve their properties. Residual stresses can also be caused by *temperature gradients* within a part, as during the cooling cycle of a casting (Chapter 5) or in a grinding operation (Section 9.6).

2.10.1 Effects of Residual Stresses

Because they lower the fatigue life and fracture strength of a part, tensile residual stresses on the surface are generally undesirable. Note that a surface with tensile residual stresses will support lower additional tensile stresses (due to external loading) than a surface that is free from residual stresses. This phenomenon is particularly important for relatively brittle materials such as ceramics or cast irons, where fracture can occur with little or no plastic deformation. Tensile residual stresses can also lead to **stress cracking** or **stress-corrosion cracking** (Section 3.8.2) over a period of time. Conversely, compressive residual stresses on a surface are generally desirable. In fact, in order to increase the fatigue life of components, compressive residual stresses are intentionally imparted on surfaces using common techniques such as **shot peening** and **surface rolling** (see Section 4.5.1).

2.10.2 Reduction of Residual Stresses

Residual stresses may be reduced or eliminated either by **stress-relief annealing** (see Section 5.11.4) or by further *plastic deformation*. Given sufficient time, residual stresses may also be diminished at room temperature by *relaxation*; the time required for relaxation can be greatly reduced by increasing the temperature of the component. Because relaxation of residual stresses by stress-relief annealing is generally accompanied by warpage of the part, a *machining allowance* is commonly provided to compensate for dimensional changes during stress relieving. Machining allowance refers to a slightly oversized part, so that the final desired dimensions can be obtained through various material-removal operations (Chapter 8).

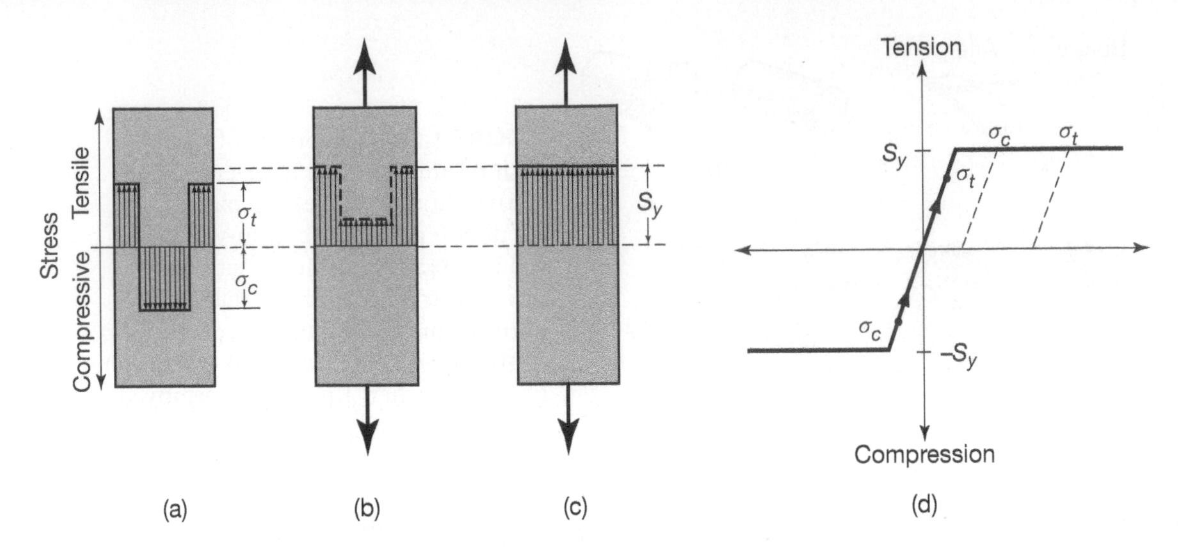

FIGURE 2.29 Elimination of residual stresses by stretching. Residual stresses can be also reduced or eliminated by thermal treatments, such as stress relieving or annealing.

The mechanism of residual stress reduction or elimination by plastic deformation can be described as follows: Assume that a piece of metal has the residual stresses shown in Fig. 2.29a, namely, tensile on the outside and compressive on the inside; these stresses are in equilibrium because there are no external forces acting on the part. Also assume that the material is elastic-perfectly plastic, as shown in Fig. 2.29d. For the levels of the residual stresses shown on the stress–strain diagram, note that the tensile and compressive stresses are both at a level below the yield strength, S_y; also note that all residual stresses are in the elastic range.

If a uniformly distributed tension is now applied to this part, points σ_c and σ_t in the diagram move up on the stress–strain curve, as shown by the arrows. The maximum level that these stresses can reach is the tensile yield strength, S_y. With sufficiently high loading, the stress distribution throughout the part eventually becomes uniform, as shown in Fig. 2.29c. If the load is now removed, the stresses recover elastically and the part is now free of residual stresses. Note also that very little stretching is required to relieve these residual stresses, because the elastic portions of the stress–strain curves for metals are very steep; hence the elastic stresses can be raised to the yield strength with very little strain (See also Fig. 2.6).

The technique for reducing or relieving residual stresses by plastic deformation, such as by stretching as just described, requires sufficient straining to establish a uniformly distributed stress within the part. Consequently, a material such as the elastic, linearly strain-hardening type shown in Fig. 2.7e can never reach this condition because the compressive stress, σ_c', will always lag behind σ_t'. If the slope of the stress–strain curve in the plastic region is small, the difference between σ_c' and σ_t' will be rather small and little residual stress will be left in the part after unloading.

EXAMPLE 2.4 Elimination of Residual Stresses by Tension

Given: Refer to Fig. 2.29a and assume that $\sigma_t = 140$ MPa and $\sigma_c = -140$ MPa. The material is aluminum and the length of the specimen is 0.25 m. Assume that the yield strength of the material is 150 MPa.

Find: Calculate the length to which this specimen should be stretched so that, when unloaded, it will be free from residual stresses.

Solution: Stretching should be to the extent that σ_c reaches the yield strength in tension, S_y. Therefore, the total strain should be equal to the sum of the strain required to bring the compressive residual stress to zero and the strain required to bring it to the tensile yield strength. Hence,

$$\epsilon_{total} = \frac{\sigma_c}{E} + \frac{S_y}{E}. \tag{2.33}$$

For aluminum, let $E = 70$ GPa, as obtained from the inside front cover. Thus,

$$\epsilon_{total} = \frac{140}{70 \times 10^3} + \frac{150}{70 \times 10^3} = 0.00414.$$

Hence, the stretched length should be

$$\ln\left(\frac{l_f}{0.25}\right) = 0.00414 \text{ or } l_f = 0.2510 \text{ m}.$$

As the strains are very small, engineering strains can be used in these calculations. Hence,

$$\frac{l_f - 0.25}{0.25} = 0.00414 \text{ or } l_f = 0.2510 \text{ m}.$$

2.11 Yield Criteria

In manufacturing operations where plastic deformation is involved, the material is generally subjected to biaxial or triaxial stresses. For example:

1. in the expansion of a thin-walled, spherical shell under internal pressure, an element in the shell is subjected to equal biaxial tensile stresses (Fig. 2.30a);
2. in deep drawing of sheet metal (Section 7.6) an element in the flange is subjected to a tensile radial stress and to compressive stresses on its surface and in the circumferential direction (Fig. 2.30c); and
3. in drawing of a rod or wire through a conical die (Fig. 6.59), an element in the deformation zone is subjected to tension in its longitudinal direction and to compression on its conical surface (Fig. 2.30b).

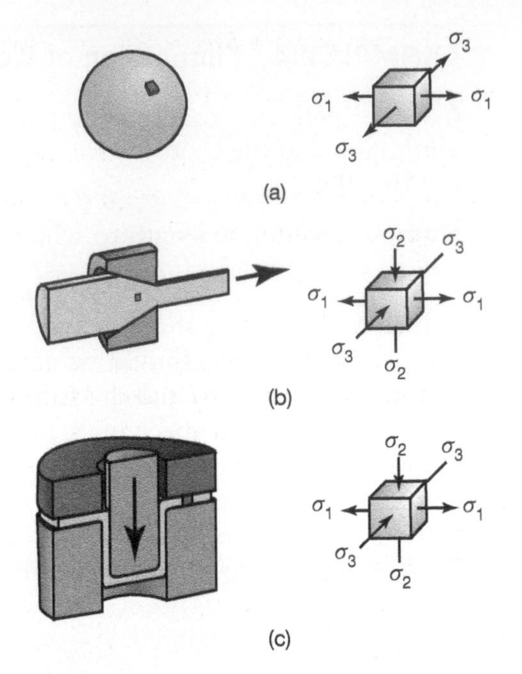

FIGURE 2.30 The state of stress in various metalworking operations. (a) Expansion of a thin-walled spherical shell under internal pressure. (b) Drawing of round rod or wire through a conical die to reduce its diameter (see Section 6.5). (c) Deep drawing of sheet metal with a punch and die to make a cup (see Section 7.6).

As shown in subsequent chapters, several other examples can be given in which the material is subjected to various normal and shear stresses during processing.

In the *elastic range*, the strains in the elements are represented by the *generalized Hooke's law* equations:

$$\epsilon_1 = \frac{1}{E}\left[\sigma_1 - \nu\left(\sigma_2 + \sigma_3\right)\right], \tag{2.34a}$$

$$\epsilon_2 = \frac{1}{E}\left[\sigma_2 - \nu\left(\sigma_1 + \sigma_3\right)\right], \tag{2.34b}$$

$$\epsilon_3 = \frac{1}{E}\left[\sigma_3 - \nu\left(\sigma_1 + \sigma_2\right)\right]. \tag{2.34c}$$

Thus, for simple tension where $\sigma_2 = \sigma_3 = 0$,

$$\epsilon_1 = \frac{\sigma_1}{E},$$

and

$$\epsilon_2 = \epsilon_3 = -\nu\frac{\sigma_1}{E}. \tag{2.35}$$

The negative sign indicates a contraction of the element in the lateral directions.

Recall that in a simple tension or compression test when the applied stress reaches the uniaxial yield strength, S_y, the material will deform *plastically*. For more complex state of stress, however, relationships have been developed among these stresses that will predict yielding. These relationships are known as *yield criteria*; the most common are the maximum-shear-stress criterion and the distortion-energy criterion.

2.11.1 Maximum-Shear-Stress Criterion

The **maximum-shear-stress criterion,** also known as the **Tresca criterion,** states that yielding occurs when the maximum shear stress within an element is equal to or exceeds a *critical value.* As shown in Section 3.3, this critical value is a *material property* and is called the **shear yield strength,** k. Thus, for yielding to occur,

$$\tau_{\text{max}} \geq k. \tag{2.36}$$

A convenient method of determining the magnitude of the stresses acting on an element is by using **Mohr's circles for stresses,** described in detail in textbooks on mechanics of solids. The **principal stresses** and their *directions* can easily be determined from Mohr's circle or from stress-transformation equations.

If the maximum shear stress is equal to or exceeds k, yielding will occur. It can be seen that there are many combinations of stresses (known as *states of stress*) that can produce the same maximum shear stress. From the simple tension test, note that

$$k = \frac{S_y}{2}, \tag{2.37}$$

where S_y is the uniaxial yield strength of the material.

From Mohr's circle for simple tension, the maximum-shear-stress criterion can be rewritten as

$$\sigma_{\text{max}} - \sigma_{\text{min}} = S_y. \tag{2.38}$$

For three-dimensional stresses, the maximum and minimum normal stresses produce the largest Mohr's circle of stress, hence the *largest* shear stress, and thus *the intermediate stress has no effect on yielding.* Note that the left-hand side of Eq. (2.38) represents the *applied stresses* and that the right-hand side is a *material property.* Also, assuming that (a) the material is *continuous, homogeneous,* and *isotropic* (it has the same properties in all directions), and (b) the yield strength in tension and in compression are equal (see *Bauschinger effect* in Section 2.3.2).

2.11.2 Distortion-Energy Criterion

The **distortion-energy criterion,** also called the **von Mises criterion,** states that yielding occurs when the relationship between the principal stresses and uniaxial yield strength, S_y, of the material is

$$(\sigma_1 - \sigma_2)^2 + (\sigma_2 - \sigma_3)^2 + (\sigma_3 - \sigma_1)^2 = 2S_y^2. \tag{2.39}$$

Note that, unlike the maximum-shear-stress criterion, the intermediate principal stress is now included in this expression. Here again, the left-hand side of the equation represents the applied stresses and the right-hand side a material property.

EXAMPLE 2.5 Yielding of a Thin-Walled Shell

Given: A thin-walled spherical shell is under internal pressure, p. The shell is 0.5 m in diameter and 2.5 mm thick. It is made of a perfectly plastic material with a yield strength of 150 MPa.

Find: Calculate the pressure required to cause yielding of the shell according to both yield criteria.

Solution: For this shell under internal pressure, the membrane stresses are given by

$$\sigma_1 = \sigma_2 = \frac{pr}{2t}, \qquad (2.40)$$

where $r = 0.25$ m and $t = 0.0025$ m. The stress in the thickness direction, σ_3, is negligible because of the high r/t ratio of the shell. Thus, according to the maximum-shear-stress criterion,

$$\sigma_{max} - \sigma_{min} = S_y,$$

or

$$\sigma_1 - 0 = S_y,$$

and

$$\sigma_2 - 0 = S_y.$$

Hence $\sigma_1 = \sigma_2 = 150$ MPa. The pressure required is then

$$p = \frac{2tS_y}{r} = \frac{(2)(0.0025)\left(150 \times 10^6\right)}{0.25} = 3.0 \text{ MPa}.$$

According to the distortion-energy criterion,

$$(\sigma_1 - \sigma_2)^2 + (\sigma_2 - \sigma_3)^2 + (\sigma_3 - \sigma_1)^2 = 2S_y^2,$$

or

$$0 + \sigma_2^2 + \sigma_1^2 = 2S_y^2.$$

Hence $\sigma_1 = \sigma_2 = S_y$. Therefore the answer is the same, or $p = 3.0$ MPa.

2.11.3 Plane Stress and Plane Strain

Plane stress is the state of stress in which one or two of the pairs of faces on an elemental cube are free from stress; in such a circumstance, all of the stresses act in a plane. An example is a thin-walled tube in torsion: There are no stresses acting normal to the inside or outside surfaces of the tube, thus the tube is in plane stress. Other examples are given in Fig. 2.31.

Plane strain is the condition where one of the pairs of faces on an elemental cube undergoes zero strain (Figs. 2.31c and d); in this case, all of the strains act in a plane. The **plane-strain compression test,** described earlier and shown in Fig. 2.13, is an example of plane strain. The width of the specimen is kept essentially constant during deformation, by properly

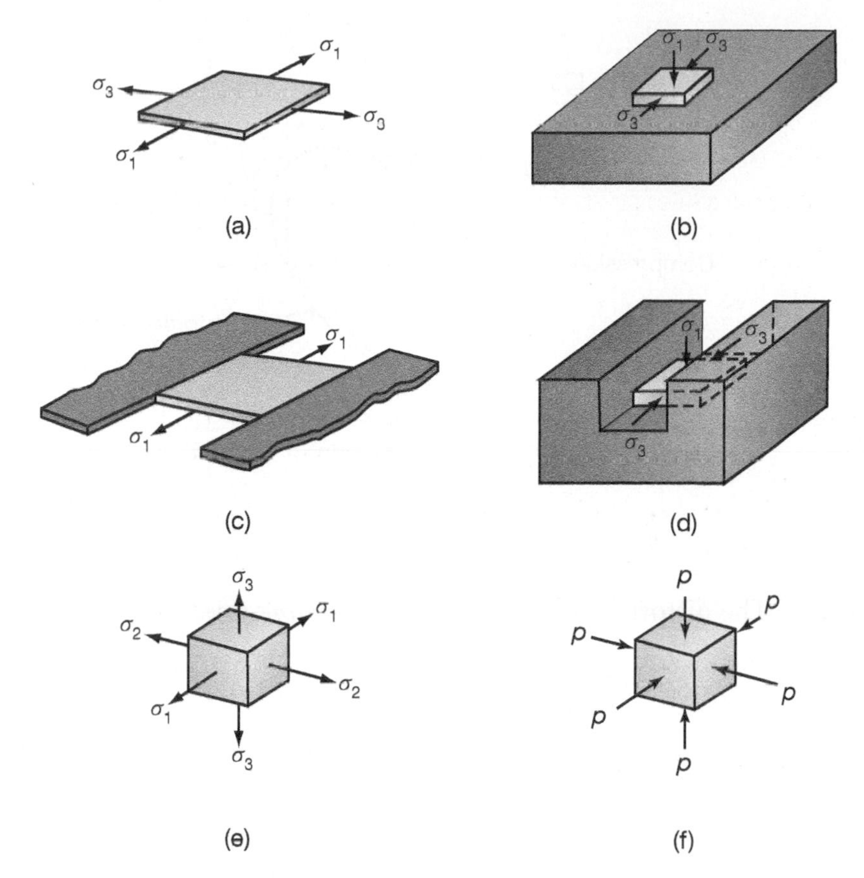

FIGURE 2.31 Examples of states of stress: (a) plane stress in sheet stretching; there are no stresses acting on the surfaces of the sheet. (b) plane stress in compression; there are no stresses acting on the sides of the specimen being compressed. (c) plane strain in tension; the width of the sheet remains constant while being stretched. (d) plane strain in compression (See also Fig. 2.13); the width of the specimen remains constant due to the restraint by the groove. (e) Triaxial tensile stresses acting on an element. (f) Hydrostatic compression of an element. Note also that an element on the cylindrical portion of a thin-walled tube in torsion is in the condition of both plane stress and plane strain.

selecting specimen dimensions. Note that an element does not have to be physically constrained on the pair of faces for plane-strain conditions to exist. An example is the torsion of a thin-walled tube in which it can be shown that the wall thickness remains constant.

A review of the two yield criteria just described will indicate that the plane-stress condition (in which $\sigma_2 = 0$) can be represented by the diagram in Fig. 2.32. The maximum-shear-stress criterion gives an envelope of straight lines. In the first quadrant, where $\sigma_1 > 0$ and $\sigma_3 > 0$, and σ_2 is always zero (for plane stress), Eq. (2.38) reduces to $\sigma_{max} = S_y$. The maximum value that either σ_1 or σ_3 can acquire is S_y; hence the straight lines, as shown in Fig. 2.32.

The same situation exists in the third quadrant because σ_1 and σ_3 are both compressive. In the second and fourth quadrants, σ_2 (which is zero for the plane-stress condition) is the intermediate stress. Thus, for the second quadrant, Eq. (2.38) reduces to

$$\sigma_3 - \sigma_1 = S_y, \tag{2.41}$$

and for the fourth quadrant, it reduces to

$$\sigma_1 - \sigma_3 = S_y. \tag{2.42}$$

Note that Eqs. (2.41) and (2.42) represent the 45° lines in Fig. 2.32.

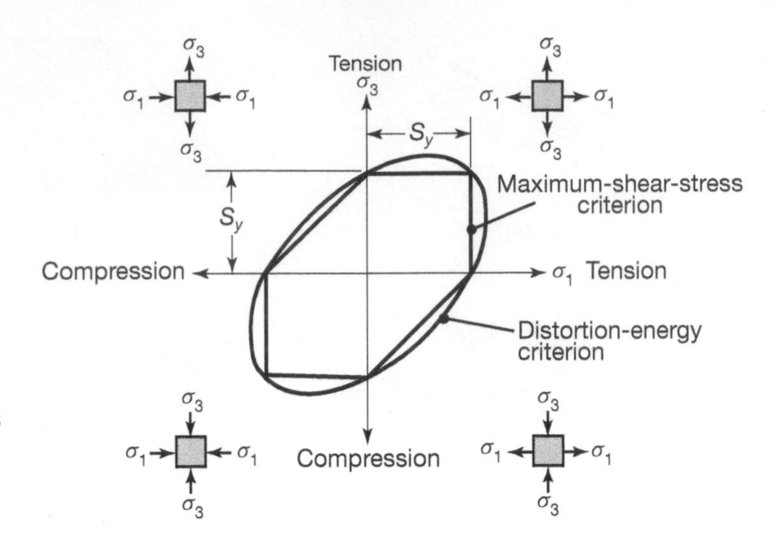

FIGURE 2.32 Plane-stress diagrams for maximum-shear-stress and distortion-energy criteria. Note that $\sigma_2 = 0$.

Plane stress. The distortion-energy criterion for *plane stress* reduces to

$$\sigma_1^2 + \sigma_3^2 - \sigma_1\sigma_3 = S_y^2 \tag{2.43}$$

and is shown graphically in Fig. 2.32. Whenever a point (with its coordinates representing the two principal stresses) falls on these boundaries, the element will yield.

The three-dimensional elastic stress–strain relationships were given by Eqs. (2.34a to c). When the stresses applied are sufficiently high to cause *plastic deformation*, the stress–strain relationships are obtained from **flow rules** (*Lévy-Mises equations*), described in detail in texts on plasticity. These relationships relate the stress and strain *increment* as follows:

$$d\epsilon_1 = \frac{d\bar{\epsilon}}{\bar{\sigma}}\left[\sigma_1 - \frac{1}{2}(\sigma_2 + \sigma_3)\right], \tag{2.44a}$$

$$d\epsilon_2 = \frac{d\bar{\epsilon}}{\bar{\sigma}}\left[\sigma_2 - \frac{1}{2}(\sigma_1 + \sigma_3)\right], \tag{2.44b}$$

$$d\epsilon_3 = \frac{d\bar{\epsilon}}{\bar{\sigma}}\left[\sigma_3 - \frac{1}{2}(\sigma_1 + \sigma_2)\right]. \tag{2.44c}$$

Note that these expressions are similar to those given by the elastic stress–strain relationships of the generalized Hooke's law, Eqs. (2.34).

Plane strain. For the *plane-strain* condition shown in Figs. 2.31c and d, $\epsilon_2 = 0$. Therefore, $d\epsilon_2 = 0$ and

$$\sigma_2 = \frac{\sigma_1 + \sigma_3}{2}. \tag{2.45}$$

Note that σ_2 is now an intermediate stress.

For plane-strain compression shown in Figs. 2.13 and 2.31d, the distortion-energy criterion (which includes the intermediate stress) reduces to

$$\sigma_1 - \sigma_3 = \frac{2}{\sqrt{3}} S_y = 1.15 S_y = S_y'. \tag{2.46}$$

Note that whereas for the maximum-shear-stress criterion $k = S_y/2$, for the distortion-energy criterion for the plane-strain condition, $k = S_y/\sqrt{3}$.

2.11.4 Experimental Verification of Yield Criteria

The yield criteria just described have been tested experimentally, typically using a specimen in the shape of a thin-walled tube under internal pressure and/or torsion as well as tension. Under such loading, it is then possible to generate different states of plane stress. Experiments using a variety of *ductile* materials have shown that the distortion-energy criterion agrees slightly better with the experimental data than does the maximum-shear-stress criterion. This is why the distortion-energy criterion is often used for the analysis of metalworking processes (see Chapters 6 and 7). The simpler maximum-shear-stress criterion, however, also can be used because the difference between the two criteria is negligible for most engineering applications.

2.11.5 Volume Strain

Summing the three equations of the generalized Hooke's law [Eqs. (2.34a to c)], the following is obtained:

$$\epsilon_1 + \epsilon_2 + \epsilon_3 = \frac{1 - 2\nu}{E} (\sigma_1 + \sigma_2 + \sigma_3), \tag{2.47}$$

where the left-hand side of the equation can be shown to be the **volume strain** or **dilatation**, Δ. Thus,

$$\Delta = \frac{\text{Volume change}}{\text{Original volume}} = \frac{1 - 2\nu}{E} (\sigma_1 + \sigma_2 + \sigma_3). \tag{2.48}$$

Note that in the plastic range, where $\nu = 0.5$, the volume change is zero. Thus, in plastic working of metals,

$$\epsilon_1 + \epsilon_2 + \epsilon_3 = 0, \tag{2.49}$$

which is a convenient means of determining a third strain if two strains are known. In the *elastic* range, where $0 < \nu < 0.5$, it can be seen from Eq. (2.48) that the volume of a tension-test specimen increases and that of a compression-test specimen decreases during the test.

The **bulk modulus** is defined as

$$\text{Bulk modulus} = \frac{\sigma_m}{\Delta} = \frac{E}{3(1 - 2\nu)}, \tag{2.50}$$

where σ_m is the *mean stress*, defined as

$$\sigma_m = \frac{1}{3}(\sigma_1 + \sigma_2 + \sigma_3). \tag{2.51}$$

2.11.6 Effective Stress and Effective Strain

A convenient means of expressing the state of stress on an element is the **effective stress** (or *equivalent stress*), $\overline{\sigma}$, and **effective strain,** $\overline{\epsilon}$. For the maximum-shear-stress criterion, the effective stress is

$$\overline{\sigma} = \sigma_1 - \sigma_3, \tag{2.52}$$

and for the distortion-energy criterion, it is

$$\overline{\sigma} = \frac{1}{\sqrt{2}}\left[(\sigma_1 - \sigma_2)^2 + (\sigma_2 - \sigma_3)^2 + (\sigma_3 - \sigma_1)^2\right]^{1/2}, \tag{2.53}$$

and is often referred to as the **von Mises stress.** The factor $1/\sqrt{2}$ is chosen so that, for yielding in simple tension, the effective stress is equal to the uniaxial yield strength, S_y.

The strains are likewise related to the effective strain. Thus, for the maximum-shear-stress criterion, the effective strain is

$$\overline{\epsilon} = \frac{2}{3}(\epsilon_1 - \epsilon_3), \tag{2.54}$$

and for the distortion-energy criterion, it is

$$\overline{\epsilon} = \frac{\sqrt{2}}{3}\left[(\epsilon_1 - \epsilon_2)^2 + (\epsilon_2 - \epsilon_3)^2 + (\epsilon_3 - \epsilon_1)^2\right]^{1/2}. \tag{2.55}$$

Again, the factors 2/3 and $\sqrt{2}/3$ are chosen to maintain consistency for the condition of simple tension. Note also that stress–strain curves may also be called effective stress–effective strain curves.

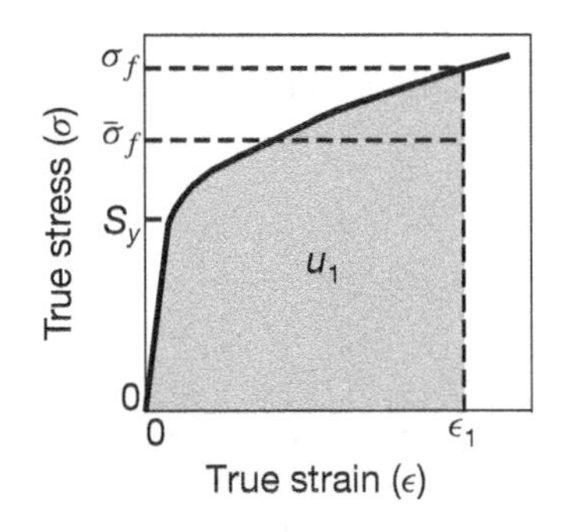

FIGURE 2.33 Schematic illustration of true stress–true strain curve showing yield strength S_y, average flow stress $\overline{\sigma}_f$, specific energy u_1, and flow stress σ_f.

2.12 Work of Deformation

Work is defined as the product of collinear force and distance; a quantity equivalent to work per unit volume is then the product of stress and strain. Because the relationship between stress and strain in the plastic range depends on the particular stress–strain curve of a material, the work is best calculated by referring to Fig. 2.33.

Note that the area under the true stress–true strain curve for any strain ϵ is the **energy per unit volume,** u, (**specific energy**) of the material deformed, and is expressed as

$$u = \int_0^{\epsilon_1} \sigma \, d\epsilon. \tag{2.56}$$

As described in Section 2.2.3, true stress–true strain curves generally can be represented by the simple expression

$$\sigma = K\epsilon^n.$$

Thus Eq. (2.56) can be written as

$$u = K\int_0^{\epsilon_1} \epsilon^n \, d\epsilon,$$

or

$$u = \frac{K\epsilon_1^{n+1}}{n+1} = \overline{\sigma}_f \epsilon_1, \tag{2.57}$$

where $\overline{\sigma}_f$ is the **average flow stress** of the material (see Section 2.2.3).

The energy represents the work expended in uniaxial deformation. For triaxial states of stress, a general expression is given by

$$du = \sigma_1 \, d\epsilon_1 + \sigma_2 \, d\epsilon_2 + \sigma_3 \, d\epsilon_3.$$

For a more general condition, the effective stress and effective strain can be used. The energy per unit volume is then expressed by

$$u = \int_0^{\epsilon} \overline{\sigma} \, d\overline{\epsilon}. \tag{2.58}$$

To obtain the work expended, u is multiplied by the volume of the material plastically deformed:

$$\text{Work} = (u)(\text{volume}). \tag{2.59}$$

The energy represented by Eq. (2.59) is the *minimum energy* or the *ideal energy* required for uniform (homogeneous) deformation. However, the energy required for actual deformation involves two additional factors: (a) the energy required to overcome *friction* at the die–workpiece interfaces, and (b) the **redundant work** of deformation, which is described as follows.

In Fig. 2.34a, a block of material is being deformed into shape by forging, extrusion, or drawing through a die (as described in Chapter 6). As shown in Fig. 2.34b, the deformation is uniform, or homogeneous. In reality, however, the material often deforms as in Fig. 2.34c because of the effects of friction and die geometry. The difference between (b) and (c) in Fig. 2.34 can be described by the fact that (c) has undergone additional shearing along horizontal planes.

Shearing requires expenditure of energy because additional plastic work has to be done in subjecting the various layers to undergo shear strains; this is known as *redundant work*. The word redundant indicates the fact that this additional energy does not contribute to the overall shape change of the part. Note also that grid patterns (b) and (c) in Fig. 2.34 have the same overall shape and dimensions.

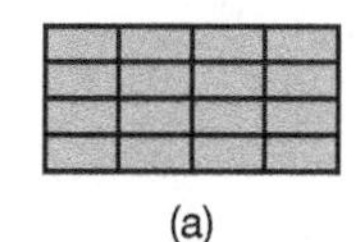

(a)

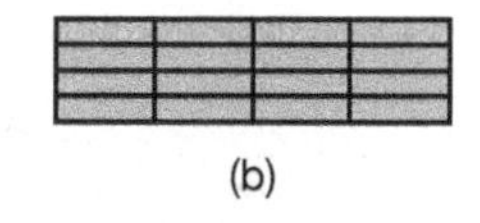

(b)

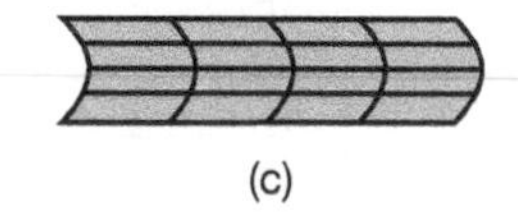

(c)

FIGURE 2.34 Deformation of grid patterns in a workpiece: (a) original pattern; (b) after ideal deformation; and (c) after inhomogeneous deformation, requiring redundant work of deformation. Note that (c) is basically (b) with additional shearing, especially at the outer layers. Thus, (c) requires greater work of deformation (See also Figs. 6.3 and 6.49).

The *total specific energy* required can now be written as

$$u_{\text{total}} = u_{\text{ideal}} + u_{\text{friction}} + u_{\text{redundant}}. \tag{2.60}$$

The *efficiency*, η, of a process is then defined as

$$\eta = \frac{u_{\text{ideal}}}{u_{\text{total}}}. \tag{2.61}$$

Depending on the particular process, frictional conditions, die geometry, and other process parameters, the magnitude of η varies widely; typical estimated values are 30 to 60% for extrusion and 75 to 95% for rolling.

EXAMPLE 2.6 Expansion of a Thin-Walled Spherical Shell

Given: A thin-walled spherical shell made of a perfectly plastic material of yield strength, S_y, original radius, r_o, and thickness, t_o, is being expanded by internal pressure.

Find: (a) Calculate the work done in expanding this shell to a radius of r_f. (b) If the diameter expands at a constant rate, what changes take place in the power consumed as the radius increases?

Solution: The membrane stresses are given by

$$\sigma_1 = \sigma_2 = S_y$$

(from Example 2.5), where r and t are instantaneous dimensions. The true strains in the membrane are given by

$$\epsilon_1 = \epsilon_2 = \ln\left(\frac{2\pi r_f}{2\pi r_o}\right) = \ln\left(\frac{r_f}{r_o}\right).$$

Because an element in this shell is subjected to equal biaxial stretching, the specific energy is

$$u = \int_0^{\epsilon_1} \sigma_1 \, d\epsilon_1 + \int_0^{\epsilon_2} \sigma_2 \, d\epsilon_2 = 2\sigma_1\epsilon_1 = 2S_y \ln\left(\frac{r_f}{f_o}\right).$$

Since the volume of the shell material is $4\pi r_o^2 t_o$, the work done is

$$W = (u)(\text{Volume}) = 8\pi S_y r_o^2 t_o \ln\left(\frac{r_f}{r_o}\right).$$

The specific energy can also be calculated from the effective stresses and strains. Thus, according to the distortion-energy criterion,

$$\bar{\sigma} = \frac{1}{\sqrt{2}}\left[(0)^2 + (\sigma_2)^2 + (-\sigma_1)^2\right]^{-1/2} = \sigma_1 = \sigma_2$$

and

$$\bar{\epsilon} = \frac{\sqrt{2}}{3}\left[(0)^2 + (\epsilon_2 + 2\epsilon_2)^2 + (-2\epsilon_2 - \epsilon_2)^2\right]^{1/2} = 2\epsilon_2 = 2\epsilon_1.$$

Note that the thickness strain $\epsilon_3 = -2\epsilon_2 = -2\epsilon_1$ because of volume constancy in plastic deformation, where $\epsilon_1 + \epsilon_2 + \epsilon_3 = 0$. Hence

$$u = \int_0^{\epsilon} \overline{\sigma}\,\overline{\epsilon} = \int_0^{2\epsilon_1} \sigma_1\, d\epsilon_1 = 2\sigma_1\epsilon_1.$$

Thus, the answer is the same. Power is defined as the rate of work; thus

$$\text{Power} = \frac{dW}{dt}.$$

Since all other factors in the expression are constant, the expression for work can be written as being proportional to strain

$$W \propto \ln\left(\frac{r}{r_o}\right) \propto (\ln r - \ln r_o).$$

Hence

$$\text{Power} \propto \frac{1}{r}\frac{dr}{dt}.$$

Because the shell is expanding at a constant rate, dr/dt = Constant. Hence the power is related to the instantaneous radius r by

$$\text{Power} \propto \frac{1}{r}.$$

2.12.1 Work, Heat, and Temperature Rise

The mechanical work in plastic deformation is converted into heat; however, the conversion is not 100% because a small portion of this energy is stored within the deformed material as elastic energy, known as *stored energy* (see Section 3.6). Typically, this energy represents 5 to 10% of the total energy input, although it may be as high as 30% in some alloys. In a simple plastic deformation process and assuming that all work of deformation is converted into heat, the *temperature rise*, ΔT, is given by

$$\Delta T = \frac{u_{\text{total}}}{\rho c_p}, \tag{2.62}$$

where u_{total} is the specific energy from Eq. (2.60), ρ is the density, and c_p is the *specific heat* of the material. It will be noted that higher temperatures are associated with large areas under the stress–strain curve of the material and smaller values of specific heat. Note that the rise in temperature should be calculated using the stress–strain curve at the appropriate strain rate (see Section 2.2.7). Because physical properties, such as specific heat and thermal conductivity also depend on temperature (see Sections 3.9.3 and 3.9.4), they should be taken into account in all calculations.

As an example, the theoretical temperature rise compressing a 27-mm-tall specimen down to 10 mm (hence a true strain of 1) can be estimated to be: for aluminum, 75°C (165°F); copper, 140°C (285°F);

low-carbon steel, 280°C (535°F); and titanium, 570°C (1060°F). Note that the temperature rise given by Eq. (2.62) is for an ideal situation where there is no heat loss during deformation; in actual operations, there is heat loss to the environment, to tools and dies, and to lubricants or coolants used. If deformation is carried out slowly, the actual temperature rise will be a small portion of the value calculated from this equation. Conversely, if the operation is performed very rapidly, these losses are relatively small. Under extreme conditions, an *adiabatic* state is approached, whereby the temperature rise is very high, leading to **incipient melting**.

EXAMPLE 2.7 Temperature Rise in Simple Deformation

Given: A cylindrical specimen 25 mm in diameter and 25 mm high is being compressed by dropping a weight of 500 N on it from a certain height. The material has the following properties: $K = 895$ MPa, $n = 0.5$, density = 8000 kg/m^3, and specific heat = 350 J/kg-K.

Find: Assuming no heat loss and no friction, calculate the final height of the specimen if the temperature rise is 150°C.

Solution: The expression for heat is given by

$$\text{Heat} = (c_p)(\rho)(\text{Volume})(\Delta T),$$

or,

$$\text{Heat} = (350)(8000)\left[\frac{(\pi)(0.025)^2}{4}\right](0.025)(150) = 5154 \text{ Nm}.$$

Also, ideally

$$\text{Heat} = \text{Work} = (\text{volume})(u) = \frac{\pi}{4}(0.025)^3\frac{(895 \times 10^6)\epsilon^{1.5}}{1.5},$$

so that

$$\epsilon^{1.5} = \frac{(5154)(1.5)(4)}{(\pi)(0.025)^3(895 \times 10^6)} = 0.704;$$

hence,

$$\epsilon = 0.79.$$

Using absolute values,

$$\ln\left(\frac{h_o}{h_f}\right) = \ln\left(\frac{0.025}{h_f}\right) = 0.79,$$

and therefore, $h_f = 0.011$ m, or 11 mm.

SUMMARY

- Many manufacturing processes involve shaping materials by plastic deformation; consequently, mechanical properties such as strength, elasticity, ductility, hardness, toughness, and the energy required for plastic deformation are important factors. The behavior of materials, in turn, depends on the particular material and its condition, as well as other variables, particularly temperature, strain rate, and the state of stress. (Section 2.1)

- Mechanical properties measured by tension tests include modulus of elasticity (E), yield strength (S_y), ultimate tensile strength (S_{ut}), and Poisson's ratio (ν). Ductility, as measured by elongation and reduction of area, can also be determined by tension tests. True stress–true strain curves are important for such mechanical properties as strength coefficient (K), strain-hardening exponent (n), strain-rate sensitivity exponent (m), and toughness. (Section 2.2)
- Compression tests closely simulate manufacturing processes such as forging, rolling, and extrusion. Properties measured by compression tests are subject to inaccuracy due to the presence of friction and barreling. (Section 2.3)
- Torsion tests are typically conducted on tubular specimens that are subjected to twisting. These tests model such manufacturing processes as shearing, cutting, and various machining processes. (Section 2.4)
- Bend or flexure tests are commonly used for brittle materials; the outer fiber stress at fracture in bending is known as the modulus of rupture or transverse rupture strength. The forces applied in bend tests model those incurred in such manufacturing processes as forming of sheet and plate, as well as testing of tool and die materials. (Section 2.5)
- A variety of hardness tests are available to test the resistance of a material to permanent indentation. Hardness is related to strength and wear resistance, but is itself not a fundamental property of a material. (Section 2.6)
- Fatigue tests model manufacturing processes whereby components are subjected to rapidly fluctuating loads. The quantity measured is the endurance limit or fatigue limit; that is, the maximum stress to which a material can be subjected without fatigue failure, regardless of the number of cycles. (Section 2.7)
- Creep is the permanent elongation of a component under a sustained static load. A creep test typically consists of subjecting a specimen to a constant tensile load at a certain temperature and measuring its change in length over a period of time. The specimen eventually fails by necking and rupture. (Section 2.8)
- Impact tests model some high-rate manufacturing operations as well as the service conditions in which materials are subjected to impact (or dynamic) loading, such as drop forging or the behavior of tool and die materials in interrupted cutting or high-rate deformation. Impact tests determine the energy required to fracture the specimen, known as impact toughness. Impact tests are also useful in determining the ductile-brittle transition temperature of materials. (Section 2.9)
- Residual stresses are those which remain in a part after it has been deformed and all external forces have been removed. The nature and level of residual stresses depend on the manner in which the part has

been plastically deformed. Residual stresses may be reduced or eliminated by stress-relief annealing, by further plastic deformation, or by relaxation. (Section 2.10)

- In metalworking operations, the workpiece material is generally subjected to three-dimensional stresses through various tools and dies. Yield criteria establish relationships between the uniaxial yield strength of the material and the stresses applied; the two most widely used are the maximum-shear-stress (Tresca) and the distortion-energy (von Mises) criteria. (Section 2.11)
- Because it requires energy to deform materials, the work of deformation per unit volume of material (u) is an important parameter and is comprised of ideal, frictional, and redundant work components. In addition to supplying information on force and energy requirements, the work of deformation also indicates the amount of heat developed and, hence, the temperature rise in the workpiece during plastic deformation. (Section 2.12)

SUMMARY OF EQUATIONS

Engineering strain, $e = \dfrac{l - l_o}{l_o}$

Engineering strain rate, $\dot{e} = \dfrac{v}{l_o}$

Engineering stress, $\sigma = \dfrac{P}{A_o}$

True strain, $\epsilon = \ln\left(\dfrac{l}{l_o}\right)$

True strain rate, $\dot{\epsilon} = \dfrac{v}{l}$

True stress, $\sigma = \dfrac{P}{A}$

Modulus of elasticity, $E = \dfrac{\sigma}{e}$

Shear modulus, $G = \dfrac{E}{2(1 + \nu)}$

Modulus of resilience $= \dfrac{S_y^2}{2E}$

Elongation $= \dfrac{l_f - l_o}{l_o} \times 100\%$

Reduction of area $= \dfrac{A_o - A_f}{A_o} \times 100\%$

Shear strain in torsion, $\gamma = \frac{r\phi}{l}$

Hooke's law, $\epsilon_1 = \frac{1}{E}[\sigma_1 - \nu(\sigma_2 + \sigma_3)]$, etc.

Effective strain (Tresca), $\bar{\epsilon} = \frac{2}{3}(\epsilon_1 - \epsilon_3)$

Effective strain (von Mises), $\bar{\epsilon} = \frac{\sqrt{2}}{3}\left[(\epsilon_1 - \epsilon_2)^2 + (\epsilon_2 - \epsilon_3)^2 + (\epsilon_3 - \epsilon_1)^2\right]^{1/2}$

Effective stress (Tresca), $\bar{\sigma} = \sigma_1 - \sigma_3$

Effective stress (von Mises), $\bar{\sigma} = \frac{1}{\sqrt{2}}\left[(\sigma_1 - \sigma_2)^2 + (\sigma_2 - \sigma_3)^2 + (\sigma_3 - \sigma_1)^2\right]^{1/2}$

True stress–true strain relationship (power law), $\sigma = K\epsilon^n$

True stress–true strain rate relationship, $\sigma = C\dot{\epsilon}^m$

Flow rules, $d\epsilon_1 = \frac{d\bar{\epsilon}}{\bar{\sigma}}\left[\sigma_1 - \frac{1}{2}(\sigma_2 + \sigma_3)\right]$, etc.

Maximum-shear-stress criterion (Tresca), $\sigma_{max} - \sigma_{min} = S_y$

Distortion-energy criterion (von Mises), $(\sigma_1 - \sigma_2)^2 + (\sigma_2 - \sigma_3)^2 + (\sigma_3 - \sigma_1)^2 = 2S_y^2$

Shear yield strength, $k = S_y/2$ for Tresca and $k = S_y/\sqrt{3}$ for von Mises (plane strain)

Volume strain (dilatation), $\Delta = \frac{1 - 2\nu}{E}(\sigma_1 + \sigma_2 + \sigma_3)$

Bulk modulus $= \frac{E}{3(1 - 2\nu)}$

BIBLIOGRAPHY

Ashby, M.F., *Materials Selection in Mechanical Design*, 4th ed., Pergamon, 2010.

ASM Handbook, Vol. 8: Mechanical Testing, ASM International, 2000.

ASM Handbook, Vol. 20: Materials Selection and Design, ASM International, 1997.

Beer, F.P., Johnston, E.R., DeWolf, J.T., and Mazurek, D., *Mechanics of Materials*, 7th ed., McGraw-Hill, 2014.

Brandt, D.A., and Warner, J.C., *Metallurgy Fundamentals*, 5th ed., Goodheart-Willcox, 2009.

Budinski, K.G., *Engineering Materials: Properties and Selection*, 9th ed. Prentice Hall, 2009.

Davis, J.R. (ed.), *Tensile Testing*, 2nd ed., ASM International, 2004.

Dowling, N.E., *Mechanical Behavior of Materials: Engineering Methods for Deformation, Fracture, and Fatigue,* 4th ed. Prentice Hall, 2012.

Herzberg, R.W., *Deformation and Fracture Mechanics of Engineering Materials*, 5th ed., Wiley, 2012.

Hosford, W.F., *Mechanical Behavior of Materials*, 2nd ed., Cambridge, 2009.

Popov, E.P., *Engineering Mechanics of Solids*, 2nd ed., Pearson, 1998.

QUESTIONS

2.1 Can you calculate the percent elongation of materials based only on the information given in Fig. 2.6? Explain.

2.2 Explain if it is possible for stress–strain curves in tension tests to reach 0% elongation as the gage length is increased further.

2.3 Explain why the difference between engineering strain and true strain becomes larger as strain increases. Is this phenomenon true for both tensile and compressive strains? Explain.

2.4 Using the same scale for stress, the tensile true stress–true strain curve is higher than the engineering stress–strain curve. Explain whether this condition also holds for a compression test.

2.5 Which of the two tests, tension or compression, requires a higher capacity testing machine than the other? Explain.

2.6 Explain how the modulus of resilience of a material changes, if at all, as it is strained: (a) for an elastic, perfectly plastic material; and (b) for an elastic, linearly strain-hardening material.

2.7 If you pull and break a tensile-test specimen rapidly, where would the temperature be the highest? Explain why.

2.8 Comment on the temperature distribution if the specimen in Question 2.7 is pulled very slowly.

2.9 In a tension test, the area under the true stress–true strain curve is the work done per unit volume (the specific work). Also, the area under the load-elongation curve represents the work done on the specimen. If you divide this latter work by the volume of the specimen between the gage marks, you will determine the work done per unit volume (assuming that all deformation is confined between the gage marks). Will this specific work be the same as the area under the true stress–true strain curve? Explain. Will your answer be the same for any value of strain? Explain.

2.10 The note at the bottom of Table 2.4 states that as temperature increases, C decreases and m increases. Explain why.

2.11 You are given the K and n values of two different materials. Is this information sufficient to determine which material is tougher? If not, what additional information do you need, and why?

2.12 Modify the curves in Fig. 2.7 to indicate the effects of temperature. Explain your changes.

2.13 Using a specific example, show why the deformation rate, say in m/s, and the true strain rate are not the same.

2.14 It has been stated that the higher the value of m, the more diffuse the neck is, and likewise, the lower the value of m, the more localized the neck is. Explain the reason for this behavior.

2.15 Explain why materials with high m values, such as hot glass and silly putty, when stretched slowly, undergo large elongations before failure. Consider events taking place in the necked region of the specimen.

2.16 Assume that you are running four-point bending tests on a number of identical specimens of the same length and cross section, but with increasing distance between the upper points of loading (see Fig. 2.19b). What changes, if any, would you expect in the test results? Explain.

2.17 Would Eq. (2.10) hold true in the elastic range? Explain.

2.18 Why have different types of hardness tests been developed? How would you measure the hardness of a very large object?

2.19 Which hardness tests and scales would you use for very thin strips of material, such as aluminum foil? Why?

2.20 List and explain the factors that you would consider in selecting an appropriate hardness test and scale for a particular application.

2.21 In a Brinell hardness test, the resulting impression is found to be an ellipse. Give possible explanations for this result.

2.22 Referring to Fig. 2.20, the material for testers are either steel, tungsten carbide, or diamond. Why isn't diamond used for all of the tests?

2.23 What role does friction play in a hardness test? Can high friction between a material and indenter affect a hardness test? Explain.

2.24 Describe the difference between creep and stress relaxation, giving two examples for each as they relate to engineering applications.

2.25 Referring to the two impact tests shown in Fig. 2.26, explain how different the results would be if the specimens were impacted from the opposite directions.

2.26 If you remove the layer *ad* from the part shown in Fig. 2.27d, such as by machining or grinding, which way will the specimen curve? (*Hint:* Assume that the part in diagram (d) is composed of four horizontal springs held at the ends. Thus, from the top down, there is compression, tension, compression, and tension springs.)

2.27 Is it possible to completely remove residual stresses in a piece of material by the technique described in Fig. 2.29 if the material is elastic, linearly strain hardening? Explain.

2.28 Referring to Fig. 2.29, would it be possible to eliminate residual stresses by compression? Assume that the piece of material will not buckle under the uniaxial compressive force.

2.29 List and explain the desirable mechanical properties for (a) an elevator cable; (b) a bandage; (c) a shoe sole; (d) a fish hook; (e) an automotive piston; (f) a boat propeller; (g) a gas-turbine blade; and (h) a staple.

2.30 Make a sketch showing the nature and distribution of the residual stresses in Figs. 2.28a and b before the parts were cut. Assume that the split parts are free from any stresses. (*Hint:* Force these parts back to the shape they were in before they were cut.)

2.31 It is possible to calculate the work of plastic deformation by measuring the temperature rise in a workpiece, assuming that there is no heat loss and that the temperature distribution is uniform throughout. If the specific heat of the material decreases with increasing temperature, will the work of deformation calculated using the specific heat at room temperature be higher or lower than the actual work done? Explain.

2.32 Explain whether or not the volume of a metal specimen changes when the specimen is subjected to a state of (a) uniaxial compressive stress and (b) uniaxial tensile stress, all in the elastic range.

2.33 It is relatively easy to subject a specimen to hydrostatic compression, such as by using a chamber filled with a liquid. Devise a means whereby the specimen (say, in the shape of a cube or a round disk) can be subjected to hydrostatic tension, or one approaching this state of stress. (Note that a thin-walled, internally pressurized spherical shell is not a correct answer, because it is subjected only to a state of plane stress.)

2.34 Referring to Fig. 2.17, make sketches of the state of stress for an element in the reduced section of the tube when it is subjected to (a) torsion only; (b) torsion while the tube is internally pressurized; and (c) torsion while the tube is externally pressurized. Assume that the tube is a closed-end tube.

2.35 A penny-shaped piece of soft metal is brazed to the ends of two flat, round steel rods of the same diameter as the piece. The assembly is then subjected to uniaxial tension. What is the state of stress to which the soft metal is subjected? Explain.

2.36 A circular disk of soft metal is being compressed between two flat, hardened circular steel punches of having the same diameter as the disk. Assume that the disk material is perfectly plastic and that there is no friction or any temperature effects. Explain the change, if any, in the magnitude of the punch force as the disk is being compressed plastically to, say, a fraction of its original thickness.

2.37 A perfectly plastic metal is yielding under the stress state σ_1, σ_2, σ_3, where $\sigma_1 > \sigma_2 > \sigma_3$. Explain what happens if σ_1 is increased.

2.38 What is the dilatation of a material with a Poisson's ratio of 0.5? Is it possible for a material to have a Poisson's ratio of 0.7? Give a rationale for your answer.

2.39 Can a material have a negative Poisson's ratio? Give a rationale for your answer.

2.40 As clearly as possible, define plane stress and plane strain.

2.41 What test would you use to evaluate the hardness of a coating on a metal surface? Would it matter if the coating was harder or softer than the substrate? Explain.

2.42 List the advantages and limitations of the stress–strain relationships given in Fig. 2.7.

2.43 Plot the data in the inside front cover on a bar chart, showing the range of values, and comment on the results.

2.44 A hardness test is conducted on as-received metal as a quality check. The results show that the hardness is too high, indicating that the material may not have sufficient ductility for the intended application. The supplier is reluctant to accept the return of the material, instead claiming that the diamond cone used in the Rockwell testing was worn and blunt, and hence the test needed to be recalibrated. Is this explanation plausible? Explain.

2.45 Explain why a 0.2% offset is used to obtain the yield strength in a tension test.

2.46 Referring to Question 2.45, would the offset method be necessary for a highly strained hardened material? Explain.

2.47 Explain why the hardness of a material is related to a multiple of the uniaxial compressive stress, since both involve compression of workpiece material.

2.48 Without using the words "stress" or "strain", define *elastic modulus*.

PROBLEMS

2.49 A strip of metal is originally 1.0 m long. It is stretched in three steps: first to a length of 1.5 m, then to 2.5 m, and finally to 3.0 m. Show that the total true strain is the sum of the true strains in each step, that is, that the strains are additive. Show that, using engineering strains, the strain for each step cannot be added to obtain the total strain.

2.50 A paper clip is made of wire 1.00 mm in diameter. If the original material from which the wire is made is a rod 15 mm in diameter, calculate the longitudinal and diametrical engineering and true strains that the wire has undergone during processing.

2.51 A material has the following properties: $S_{ut} = 350$ MPa and $n = 0.20$. Calculate its strength coefficient, K.

2.52 Based on the information given in Fig. 2.6, calculate the ultimate tensile strength of 304 stainless steel.

2.53 Calculate the ultimate tensile strength (engineering) of a material whose strength coefficient is 300 MPa and that necks at a true strain of 0.25.

2.54 A material has a strength coefficient $K = 700$ MPa. Assuming that a tensile-test specimen made from this material begins to neck at a true strain of 0.20, show that the ultimate tensile strength of this material is 415 MPa.

2.55 A cable is made of four parallel strands of different materials, all behaving according to the equation $\sigma = K\epsilon^n$, where $n = 0.20$. The materials, strength coefficients, and cross sections are as follows:

Material A: $K = 450$ MPa, $A_o = 7\ \text{mm}^2$
Material B: $K = 600$ MPa, $A_o = 2.5\ \text{mm}^2$
Material C: $K = 300$ MPa, $A_o = 3\ \text{mm}^2$
Material D: $K = 750$ MPa, $A_o = 2\ \text{mm}^2$

(a) Calculate the maximum tensile force that this cable can withstand prior to necking.
(b) Explain how you would arrive at an answer if the n values of the three strands were different from each other.

2.56 Using only Fig. 2.6, calculate the maximum load in tension testing of a 304 stainless-steel specimen with an original diameter of 6.0 mm.

2.57 Using the data given in the inside front cover, calculate the values of the shear modulus G for the metals listed in the table.

2.58 Derive an expression for the toughness of a material represented by the equation $\sigma = K(\epsilon + 0.2)^n$ and whose fracture strain is denoted as ϵ_f.

2.59 A cylindrical specimen made of a brittle material 50 mm high and with a diameter of 25 mm is subjected to a compressive force along its axis. It is found that fracture takes place at an angle of 45° under a load of 130 kN. Calculate the shear stress and the normal stress, respectively, acting on the fracture surface.

2.60 What is the modulus of resilience of a highly cold-worked piece of steel with a hardness of 280 HB? Of a piece of highly cold-worked copper with a hardness of 175 HB?

2.61 Calculate the work done in frictionless compression of a solid cylinder 40 mm high and 15 mm in diameter to a reduction in height of 50% for the following materials: (a) 1100-O aluminum; (b) annealed copper; (c) annealed 304 stainless steel; and (d) 70-30 brass, annealed.

2.62 A tensile-test specimen is made of a material represented by the equation $\sigma = K(\epsilon + n)^n$. (a) Determine the true strain at which necking will begin. (b) Show that it is possible for an engineering material to exhibit this behavior.

2.63 Take two solid cylindrical specimens of equal diameter, but different heights. Assume that both specimens are compressed (frictionless) by the same percent reduction, say 50%. Prove that the final diameters will be the same.

2.64 In a disk test performed on a specimen 50 mm in diameter and 2.5 mm thick, the specimen fractures at a stress of 500 MPa. What was the load on it at fracture?

2.65 In Fig. 2.29a, let the tensile and compressive residual stresses both be 70 MPa, and the modulus of elasticity of the material be 200 GPa with a modulus of resilience of 225 kNm/m^3. If the original length in diagram (a) is 500 mm, what should be the stretched length in diagram (b) so that, when unloaded, the strip will be free of residual stresses?

2.66 A horizontal rigid bar *c-c* is subjecting specimen *a* to tension and specimen *b* to frictionless compression such that the bar remains horizontal (see the accompanying figure). The force F is located at a distance ratio of 2:1. Both specimens have an original cross-sectional area of $1 \times 10^{-4}\ \text{m}^2$ and the original lengths are $a = 200$ mm and $b = 115$ mm. The material for specimen a has a true stress–true strain curve of $\sigma = (700\ \text{MPa})\epsilon^{0.5}$. Plot the true stress–true strain curve that the material for specimen b should have for the bar to remain horizontal.

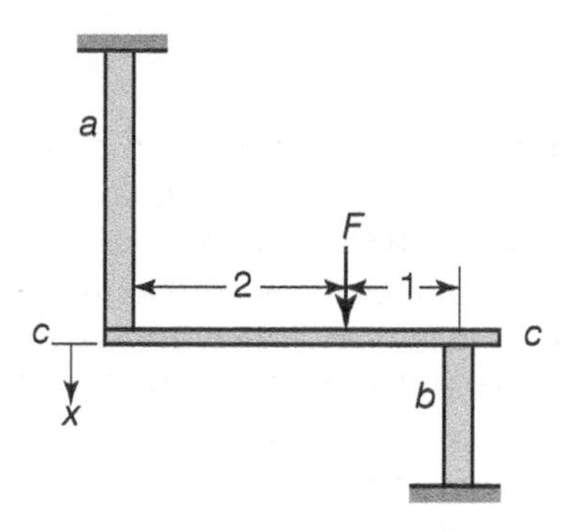

2.67 Inspect the curve that you obtained in Problem 2.66. Does a typical strain-hardening material behave in that manner? Explain.

2.68 Show that you can take a bent bar made of an elastic, perfectly plastic material and straighten it by stretching it into the plastic range. (*Hint:* Observe the events shown in Fig. 2.29.)

2.69 A bar 1 m long is bent and then stress relieved. The radius of curvature to the neutral axis is 0.50 m. The bar is 25 mm thick and is made of an elastic, perfectly plastic material with $S_y = 500$ MPa and $E = 207$ GPa. Calculate the length to which this bar should be stretched so that, after unloading, it will become and remain straight.

2.70 Assume that a material with a uniaxial yield strength S_y yields under a stress state of principal stresses $\sigma_1, \sigma_2, \sigma_3$, where $\sigma_1 > \sigma_2 > \sigma_3$. Show that the superposition of a hydrostatic stress p on this system (such as placing the specimen in a chamber pressurized with a liquid) does not affect yielding. In other words, the material will still yield according to yield criteria.

2.71 Give two different and specific examples in which the maximum-shear-stress and the distortion-energy criteria give the same answer.

2.72 A thin-walled spherical shell with a yield strength S_y is subjected to an internal pressure p. With appropriate equations, show whether or not the pressure required to yield this shell depends on the particular yield criterion used.

2.73 Show that, according to the distortion-energy criterion, the yield strength in plane strain is $1.15S_y$, where S_y is the uniaxial yield strength of the material.

2.74 What would be the answer to Problem 2.73 if the maximum-shear-stress criterion were used?

2.75 A closed-end, thin-walled cylinder of original length l, thickness t, and internal radius r is subjected to an internal pressure p. Using the generalized Hooke's law equations, show the change, if any, that occurs in the length of this cylinder when it is pressurized. Let $\nu = 0.25$.

2.76 A round, thin-walled tube is subjected to tension in the elastic range. Show that both the thickness and the diameter decrease as tension increases.

2.77 Take a long cylindrical balloon and, with a thin felt-tip pen, mark a small square on it. What will be the shape of this square after you blow up the balloon, (a) a larger square; (b) a rectangle with its long axis in the circumferential direction; (c) a rectangle with its long axis in the longitudinal direction; or (d) an ellipse? Perform this experiment, and, based on your observations, explain the results, using appropriate equations. Assume that the material the balloon is made up of is perfectly elastic and isotropic and that this situation represents a thin-walled closed-end cylinder under internal pressure.

2.78 Take a cubic piece of metal with a side length l_o and deform it plastically to the shape of a rectangular parallelepiped of dimensions l_1, l_2, and l_3. Assuming that the material is rigid and perfectly plastic, show that volume constancy requires that the following expression be satisfied: $\epsilon_1 + \epsilon_2 + \epsilon_3 = 0$.

2.79 What is the diameter of an originally 40-mm-diameter solid steel ball when the ball is subjected to a hydrostatic pressure of 2 GPa?

2.80 Determine the effective stress and effective strain in plane-strain compression according to the distortion-energy criterion.

2.81 (a) Calculate the work done in expanding a 2-mm-thick spherical shell from a diameter of 100 mm to 150 mm, where the shell is made of a material for which $\sigma = 200 + 50\epsilon^{0.5}$ MPa. (b) Does your answer depend on the particular yield criterion used? Explain.

2.82 A cylindrical slug that has a diameter of 25 mm and is 25 mm high is placed at the center of a 50-mm-diameter cavity in a rigid die (see the accompanying figure). The slug is surrounded by a compressible matrix, the pressure of which is given by the relation

$$p_m = 150\frac{\Delta V}{V_{om}} \text{ MPa}$$

where m denotes matrix and V_{om} is the original volume of the compressible matrix. Both the slug and the matrix are being compressed by a piston and without any friction. The initial pressure of the matrix is zero, and the slug material has the true stress–true strain curve of $\sigma = 600\epsilon^{0.4}$.

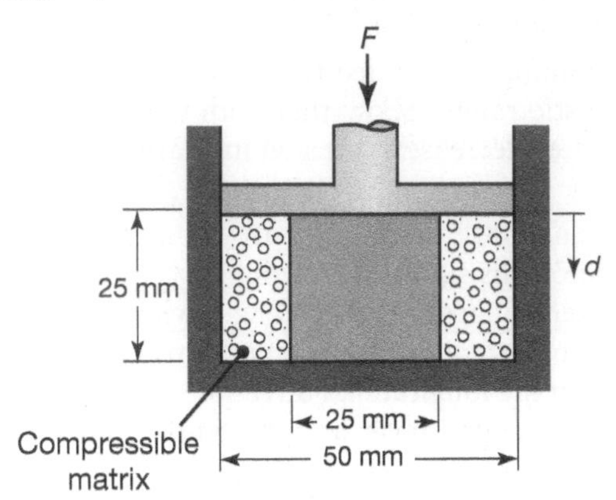

Obtain an expression for the force F versus piston travel d up to $d = 10$ mm.

2.83 A specimen in the shape of a cube 25 mm on each side is being compressed without friction in a die cavity, as shown in Fig. 2.31d, where the width of the groove is 30 mm. Assume that the linearly strain-hardening material has the true stress–true strain curve given by $\sigma = 70 + 30\epsilon$ MPa. Calculate the compressive force required when the height of the specimen is 3 mm, according to both yield criteria.

2.84 Obtain expressions for the specific energy for a material for each of the stress–strain curves shown in Fig. 2.7, similar to those shown in Section 2.12.

2.85 A material with a yield strength of 75 MPa is subjected to principal (normal) stresses of σ_1, $\sigma_2 = 0$, and $\sigma_3 = -\sigma_1/2$. What is the value of σ_1 when the metal yields according to the von Mises criterion? What if $\sigma_2 = \sigma_1/3$?

2.86 A steel plate is 100 mm × 100 mm × 10 mm thick. It is subjected to biaxial tension $\sigma_1 = \sigma_2 = 350$ MPa, with the stress in the thickness direction of $\sigma_3 = 0$. What is the change in volume using the von Mises criterion? What would this change in volume be if the plate were made of copper?

2.87 A 50-mm-wide, 1-mm-thick strip is rolled to a final thickness of 0.5 mm. It is noted that the strip has increased in width to 51 mm. What is the strain in the rolling direction?

2.88 An aluminum alloy yields at a stress of 50 MPa in uniaxial tension. If this material is subjected to the stresses $\sigma_1 = 25$ MPa, $\sigma_2 = 15$ MPa, and $\sigma_3 = -26$ MPa, will it yield? Explain.

2.89 A pure aluminum cylindrical specimen 25 mm in diameter and 25 mm high is being compressed by dropping a weight of 1000 N on it from a certain height. After deformation, it is found that the temperature rise in the specimen is 50°C. Assuming no heat loss and no friction, calculate the final height of the specimen. Use the following information for the material: $K = 205$ MPa, $n = 0.4$, $\rho = 7800$ kg/m^3, and $c_p = 450$ J/kg-K.

2.90 A ductile metal cylinder 100 mm high is compressed to a final height of 30 mm in two steps between frictionless platens. After the first step the cylinder is 70 mm high. Calculate both the engineering strain and the true strain for both steps, compare them, and comment on your observations.

2.91 Suppose the cylinder in Problem 2.90 has an initial diameter of 50 mm and is made of 1100-O aluminum. Determine the load required for each step.

2.92 Determine the specific energy and actual energy expended for the entire process described in the previous two problems.

2.93 A metal has a strain-hardening exponent of 0.22. At a true strain of 0.2, the true stress is 80 MPa. (a) Determine the stress–strain relationship for this material. (b) Determine the ultimate tensile strength for this material.

2.94 The area of each face of a metal cube is 5 cm^2, and the metal has a shear yield strength k of 140 MPa. Compressive loads of 40 kN and 80 kN are applied to different faces (say in the x- and y- directions). What must be the compressive load applied to the z-direction to cause yielding according to the Tresca yield criterion? Assume a frictionless condition.

2.95 A tensile force of 9 kN is applied to the ends of a solid bar of 7.0 mm diameter. Under load, the diameter reduces to 5.00 mm. Assuming uniform deformation and volume constancy, determine (a) the engineering stress and strain and (b) the true stress and strain. (c) If the original bar had been subjected to a true stress of 345 MPa and the resulting diameter was 5.60 mm, what are the engineering stress and strain for this condition?

2.96 Two identical specimens 20 mm in diameter and with test sections 25 mm long are made of 1112 steel. One is in the as-received condition and the other is annealed. (a) What will be the true strain when necking occurs, and what will be the elongation of the samples at that instant? (b) Find the ultimate strength for these materials.

2.97 During the production of a part, a metal with a yield strength of 200 MPa is subjected to a stress state σ_1, $\sigma_2 = \sigma_1/3$, and $\sigma_3 = 0$. Sketch the Mohr's circle diagram for this stress state. Determine the stress σ_1 necessary to cause yielding by the maximum-shear-stress and the von Mises criteria.

2.98 The following data are taken from a stainless steel tension-test specimen:

Load P (kN)	Extension Δl (mm)
7.10	0
11.1	0.50
13.3	2.0
16.0	5.0
18.7	10
20.0	15.2
20.5 (max)	21.5
14.7 (fracture)	25

Also, $A_o = 3.5 \times 10^{-5}$ m^2, $A_f = 1.0 \times 10^{-5}$ in.^{2}, $l_o = 50$ mm. Plot the true stress–true strain curve for the material.

2.99 A metal is yielding plastically under the stress state shown.

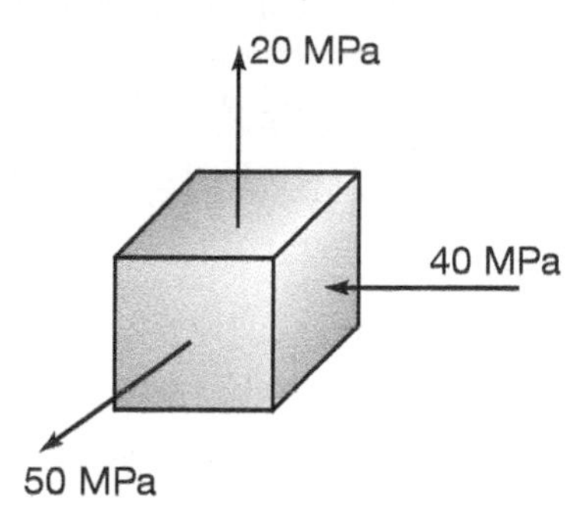

(a) Label the principal axes according to their proper numerical convention (1, 2, 3).

(b) What is the yield strength using the Tresca criterion?

(c) What if the von Mises criterion is used?

(d) The stress state causes measured strains of $\epsilon_1 = 0.4$ and $\epsilon_2 = 0.1$, with ϵ_3 not being measured. What is the value of ϵ_3?

2.100 Estimate the depth of penetration in a Brinell hardness test using 500 kg as the load when the sample is a cold-worked aluminum with a yield strength of 150 MPa.

2.101 It has been proposed to modify the von Mises yield criterion as:

$$(\sigma_1 - \sigma_2)^a + (\sigma_2 - \sigma_3)^a + (\sigma_3 - \sigma_1)^a = C,$$

where C is a constant and a is an even integer larger than 2. Plot this yield criterion for $a = 4$ and $a = 12$, along with the Tresca and von Mises criterion, in plane stress. (*Hint:* See Fig. 2.32.)

2.102 Assume that you are asked to give a quiz to students on the contents of this chapter. Prepare three quantitative problems and three qualitative questions, and supply the answers.

CHAPTER

Structure and Manufacturing Properties of Metals

This chapter describes the structure and properties of metals, and how they affect their manufacturing characteristics:

- Crystal structures, grains and grain boundaries, plastic deformation, and thermal effects.
- The characteristics and applications of cold, warm, and hot working of metals.
- The ductile and brittle behavior of materials, modes of failure, and the effects of various factors on fracture behavior.
- Physical properties of materials and their relevance to manufacturing processes.
- General properties and engineering applications of ferrous and nonferrous metals and alloys.

Symbols

a lattice spacing, m
b lattice spacing, m
d diameter, m
E elastic modulus, N/m^2
G shear modulus, N/m^2
k constant in Hall–Petch equation, N/m$^2\sqrt{\text{m}}$
n ASTM grain-size number
N number of grains per 0.0645 mm^2
S_f fracture strength, N/m^2
S_y yield strength, N/m^2
x Cartesian coordinate, m
γ shear strain, m/m
λ length, m
σ true stress, N/m^2
τ shear stress, N/m^2

3.1 Introduction

The structure of metals, that is, the arrangement of atoms, greatly influences their *properties* and *behavior*. Understanding these structures is important in order to evaluate, predict, and control their properties, as well as to make their appropriate selection for specific applications. For example, there are a number of reasons for the development of single-crystal turbine blades (Fig. 3.1) for use in jet engines, with properties that are better than those of polycrystalline blades made conventionally.

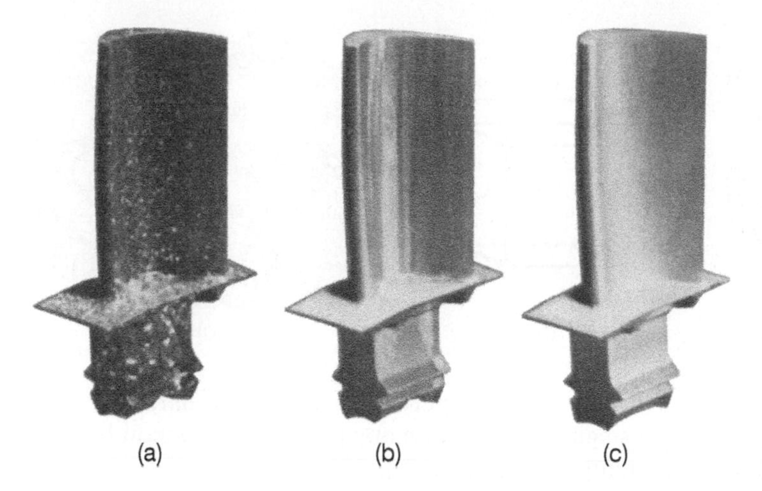

FIGURE 3.1 Turbine blades for jet engines, manufactured by three different methods: (a) conventionally cast; (b) directionally solidified, with columnar grains, as can be seen from the vertical streaks; and (c) single crystal. Although more expensive, single-crystal blades have properties at high temperatures that are superior to those of other blades. *Source:* United Technologies Pratt and Whitney.

This chapter begins with a general review of the crystal structure of metals and the role of grain size, grain boundaries, inclusions, and imperfections as they affect plastic deformation in metalworking operations. The failure and fracture of metals, both ductile and brittle, will then be reviewed, together with factors that influence fracture, such as state of stress, temperature, strain rate, and external and internal defects present in metals. The rest of the chapter provides a brief discussion of the general properties and applications of ferrous and nonferrous metals and alloys, including some data to guide product design and material selection.

3.2 The Crystal Structure of Metals

When metals solidify from a molten state, the atoms arrange themselves into various orderly configurations, called **crystals**; the arrangement of the atoms in the crystal is called **crystalline structure**. The smallest group of atoms showing the characteristic **lattice structure** of a particular metal is known as a **unit cell**; it is the building block of a crystal. The three basic patterns of atomic arrangement found in most metals are (Figs. 3.2–3.4):

1. **Body-centered cubic (bcc)**
2. **Face-centered cubic (fcc)**
3. **Hexagonal close packed (hcp)**

The reason why different crystal structures develop during solidification is due to the differences in the energy required to form these particular structures. Tungsten, for example, forms a bcc structure because it requires lower energy than would be required to form other structures. Likewise, aluminum forms an fcc structure. At certain temperatures, the same metal may form different structures. For example, iron forms a bcc structure below 912°C (1674°F) and above 1394°C (2541°F), but it forms an fcc structure between 912°C and 1394°C. The appearance of more than one type of crystal structure in a metal is known as **allotropism**, or **polymorphism**. These structural changes are an important aspect of the heat treatment of metals and alloys, as described in Section 5.11.

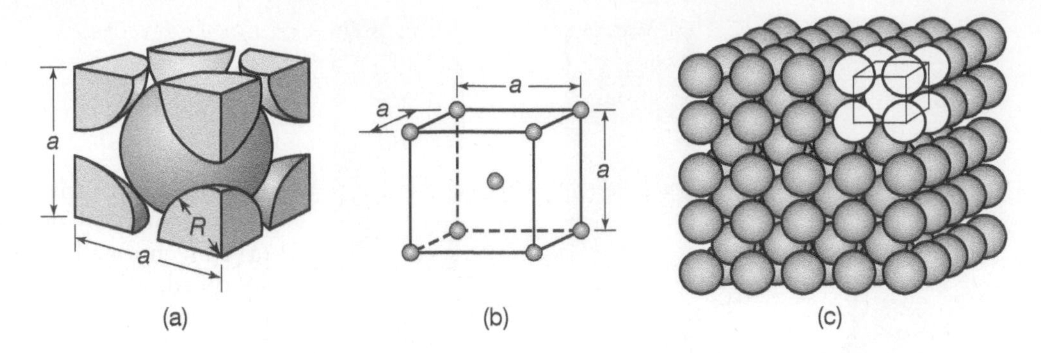

FIGURE 3.2 The body-centered cubic (bcc) crystal structure: (a) hard-ball model; (b) unit cell; and (c) single crystal with many unit cells. Common bcc metals include chromium, titanium, and tungsten.

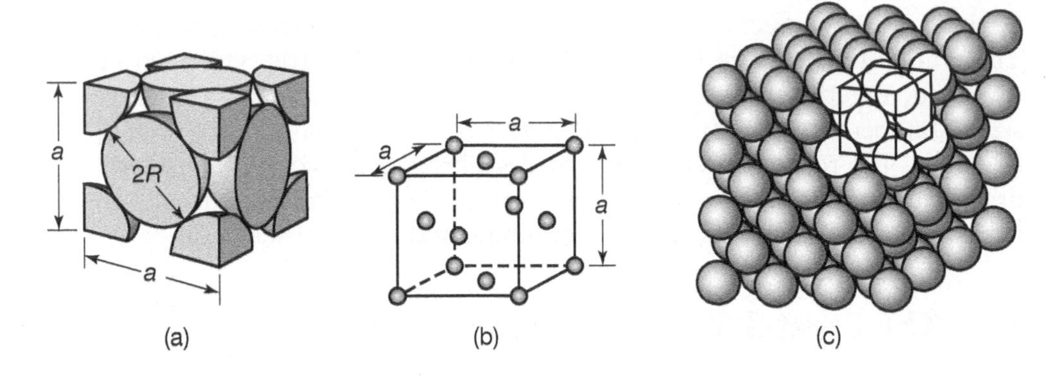

FIGURE 3.3 The face-centered cubic (fcc) crystal structure: (a) hard-ball model; (b) unit cell; and (c) single crystal with many unit cells. Common fcc metals include aluminum, copper, gold, and silver.

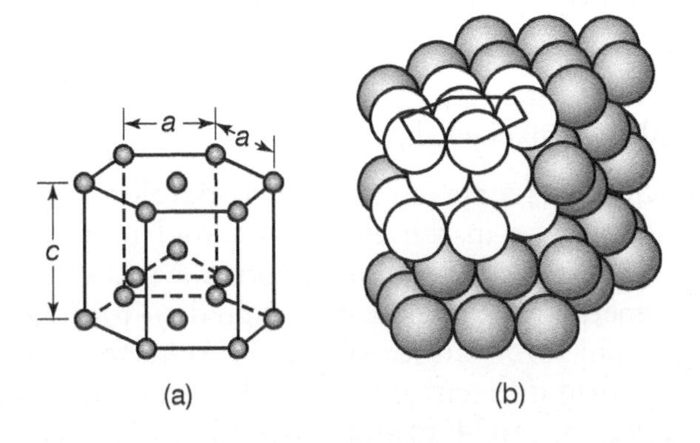

FIGURE 3.4 The hexagonal close-packed (hcp) crystal structure: (a) unit cell; and (b) single crystal with many unit cells. The top and bottom planes are called *basal planes*. Common hcp metals include zinc, magnesium, and cobalt.

Crystal structures can be modified by adding atoms of another metal or metals. Known as **alloying**, it typically improves the properties of the metal, as described in Section 5.2.

3.3 Deformation and Strength of Single Crystals

When a metal crystal is subjected to an external force, it first undergoes **elastic deformation**; that is, it returns to its original shape when the force is removed (see Section 2.2). An analogy to this type of behavior is a spring that stretches when loaded and returns to its original shape when unloaded. However, if the force on the crystal structure is increased sufficiently, the crystal undergoes **plastic (permanent) deformation**; that is, it does not return to its original shape when the force is removed.

There are two basic mechanisms by which plastic deformation may take place in crystal structures:

1. **Slip.** The slip mechanism involves the slipping of one plane of atoms over an adjacent plane (*slip plane*) under a shear stress, as shown schematically in Fig. 3.5a. This mechanism is much like the sliding of playing cards against each other. Just as it takes a certain amount of force to slide the cards against each other, so a crystal requires a

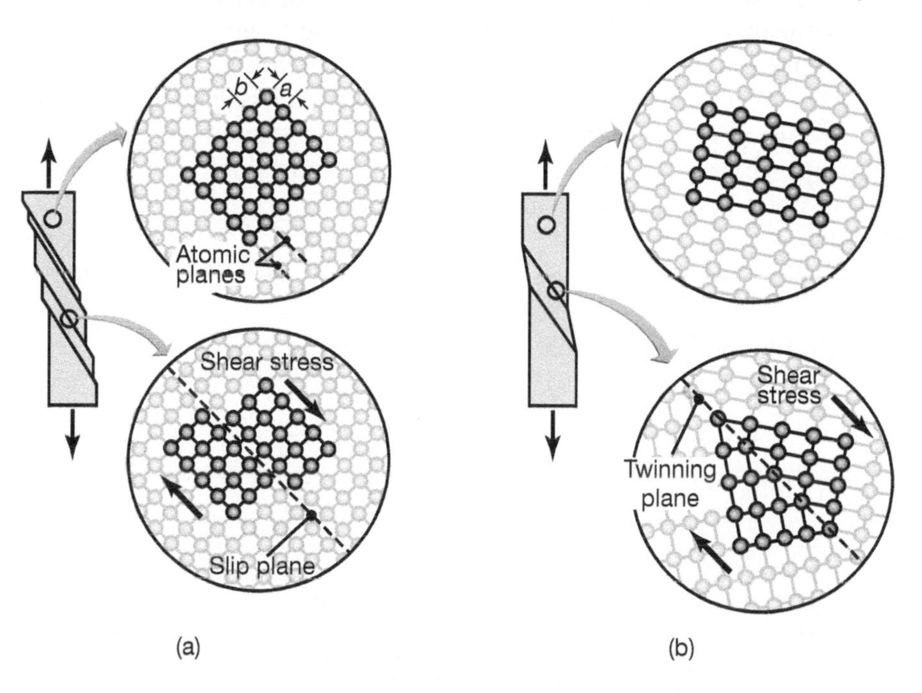

FIGURE 3.5 Permanent deformation of a single crystal under a tensile load. The highlighted grid of atoms emphasizes the motion that occurs within the lattice. (a) Deformation by slip. The b/a ratio influences the magnitude of the shear stress required to cause slip. Note that the slip planes tend to align themselves in the direction of pulling. (b) Deformation by twinning, involving generation of a "twin" around a line of symmetry subjected to shear. Note that the tensile load results in a shear stress in the plane illustrated.

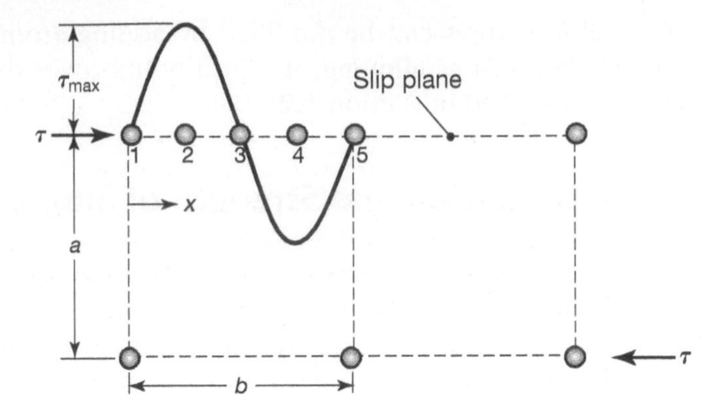

FIGURE 3.6 Variation of shear stress in moving a plane of atoms over another plane.

certain amount of shear stress (called **critical shear stress**) to undergo permanent deformation. Thus, there must be a shear stress of sufficient magnitude within a crystal for plastic deformation to occur.

The **maximum theoretical shear stress,** τ_{max}, to cause permanent deformation in a perfect crystal is obtained as follows. When there is no stress present, the atoms in the crystal are in *equilibrium* (Fig. 3.6). Under a shear stress, the upper row of atoms move to the right, where the position of an atom is denoted as x. Thus, when $x = 0$ or $x = b$ the shear stress is zero. Each atom of the upper row is attracted to the nearest atom of the lower row, resulting in *nonequilibrium* at positions 2 and 4; the stresses are now at a maximum, but opposite in sign. Note that at position 3, the shear stress is again zero, since this position is now symmetric. In Fig. 3.6, it is assumed that, as a first approximation, the shear stress varies sinusoidally. Hence, the shear stress at a displacement x is

$$\tau = \tau_{max} \sin \frac{2\pi x}{b}, \tag{3.1}$$

which, for small values of x/b can be written as

$$\tau = \tau_{max} \frac{2\pi x}{b}.$$

From Hooke's law in torsion [Eq. (2.25)],

$$\tau = G\gamma = G\left(\frac{x}{a}\right).$$

Hence,

$$\tau_{max} = \frac{Gb}{2\pi a}. \tag{3.2}$$

If it is assumed that b is approximately equal to a, then

$$\tau_{max} = \frac{G}{2\pi}. \tag{3.3}$$

It has been shown that with more refined calculations the value of τ_{max} is between $G/10$ and $G/30$.

From Eq. (3.2), it can be seen that the shear stress required to cause slip in single crystals is directly proportional to b/a. It can therefore be stated that slip in a crystal takes place along planes of maximum atomic density, or that slip takes place in closely packed planes and in closely packed directions. Because the b/a ratio is different for different directions within the crystal, a single crystal therefore has different properties when tested in different directions; thus, a crystal is **anisotropic.** A common example of anisotropy is woven cloth, which stretches differently when pulled in different directions, or plywood, which is much stronger in the planar direction than along its thickness direction (where it splits easily).

2. **Twinning.** The second mechanism of plastic deformation is *twinning*, in which a portion of the crystal forms a mirror image of itself across the *plane of twinning* (Fig. 3.5b). Twins form abruptly and are the cause of the creaking sound (called *tin cry*) heard when a tin or zinc rod is bent at room temperature. Twinning usually occurs in hcp and bcc metals by plastic deformation and in fcc metals by annealing (see Section 5.11.4).

3.3.1 Slip Systems

The combination of a slip plane and its direction of slip is known as a **slip system.** In general, metals with five or more slip systems are ductile, while those with less than five are not ductile (see Section 2.2.1). Note that each pattern of atomic arrangement will have a different number of potential slip systems:

1. In body-centered cubic crystals, there are 48 possible slip systems; thus, the probability is high that an externally applied shear stress will operate on one of the systems and cause slip. However, because of the relatively high b/a ratio, the required shear stress is high. Metals with bcc structures (such as titanium, molybdenum, and tungsten) have good strength and moderate ductility.
2. In face-centered cubic crystals, there are 12 slip systems. The probability of shear stress aligning with a slip system is moderate, and the required shear stress is low. Metals with fcc structures (such as aluminum, copper, gold, and silver) have moderate strength and good ductility.
3. The hexagonal close-packed crystal has three slip systems, and thus it has a low probability of slip; however, more systems become active at elevated temperatures. Metals with hcp structures (such as beryllium, magnesium, and zinc) are generally brittle.

Note in Fig. 3.5a that (a) the portions of the single crystal that have slipped have rotated from their original angular position toward the direction of the tensile force; and (b) slip has taken place along certain planes only. With the use of electron microscopy, it has been shown that what appears to be a single slip plane is actually a **slip band**, consisting of a number of slip planes (Fig. 3.7).

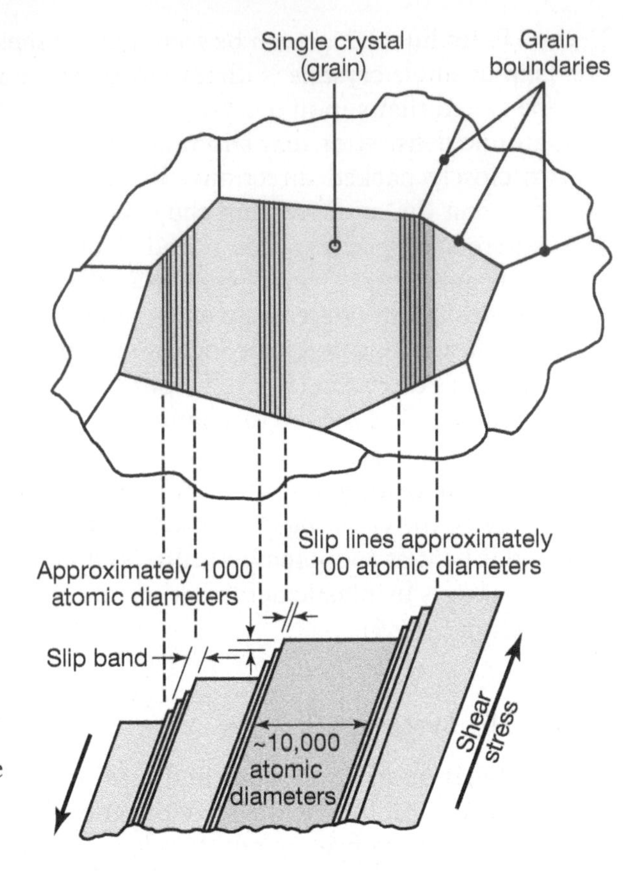

FIGURE 3.7 Schematic illustration of slip lines and slip bands in a single crystal subjected to a shear stress. A slip band consists of a number of slip planes. The crystal at the center of the upper drawing is an individual grain surrounded by other grains.

3.3.2 Ideal Tensile Strength of Metals

The ideal or theoretical tensile strength of metals can be obtained as follows: In the bar shown in Fig. 3.8, the interatomic distance is a when no external stress is applied. In order to increase this distance, a tensile force has to be applied to overcome the cohesive force between the atoms. The cohesive force is zero in the unstrained equilibrium condition.

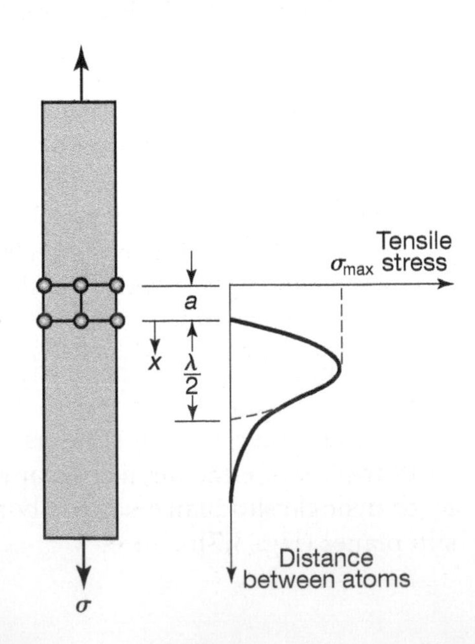

FIGURE 3.8 Variation of cohesive stress as a function of distance between a row of atoms.

When the tensile stress reaches σ_{max} the atomic bonds between two neighboring atomic planes break; σ_{max} is known as the *ideal tensile strength*.

The ideal tensile strength can be estimated by the following approach: Referring to Fig. 3.8, it can be shown that

$$\sigma_{max} = \frac{E\lambda}{2\pi a} \tag{3.4}$$

and that the work done per unit area in breaking the bond between the two atomic planes (i.e., the area under the cohesive force curve) is given by

$$\text{Work} = \frac{\sigma_{max}\lambda}{\pi}. \tag{3.5}$$

This work is expended in creating two new fracture surfaces, thus involving the surface energy of the material, γ. The total surface energy then is 2γ. Combining these equations, we find that

$$\sigma_{max} = \sqrt{\frac{E\gamma}{a}}. \tag{3.6}$$

When appropriate values are substituted into this equation, we have

$$\sigma_{max} \simeq \frac{E}{10}. \tag{3.7}$$

Thus, for example, the theoretical strength of iron would be on the order of 20 GPa (3×10^6 psi).

3.3.3 Imperfections

The actual strength of metals is approximately one to two orders of magnitude lower than the strength levels obtained from Eq. (3.7). This discrepancy has been explained in terms of **imperfections** in the crystal structure. Unlike the idealized models described previously, actual metal crystals contain a large number of imperfections and defects, which are categorized as follows:

1. **Point defects,** such as a **vacancy** (a missing atom), an **interstitial atom** (an extra atom in the lattice), or an **impurity atom** (a foreign atom that has replaced an atom of the pure metal) (Fig. 3.9);
2. **Linear,** or *one-dimensional, defects,* called **dislocations** (Fig. 3.10);
3. **Planar,** or *two-dimensional, imperfections,* such as **grain boundaries** and **phase boundaries;** and
4. **Volume,** or *bulk, imperfections,* such as **voids, inclusions** (nonmetallic elements such as oxides, sulfides, and silicates), other **phases,** or **cracks.**

A slip plane containing a dislocation requires lower shear stress to cause slip than does a plane in a perfect lattice (Fig. 3.11). This is understandable because, when a dislocation is present, slip requires far fewer atoms to

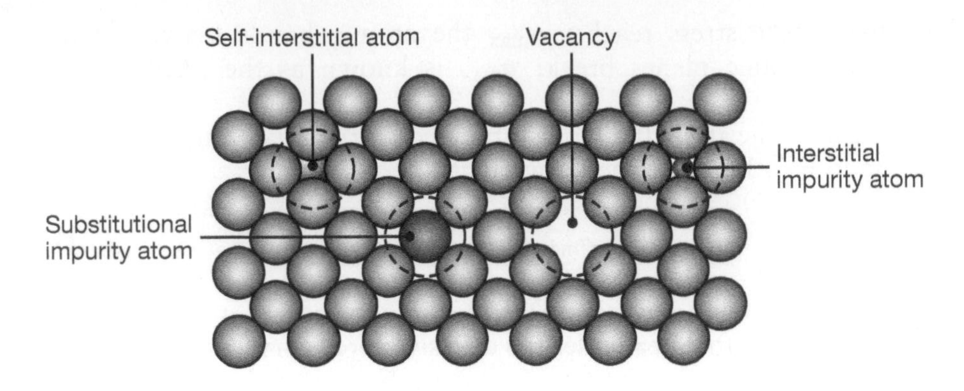

FIGURE 3.9 Various defects in a single-crystal lattice.

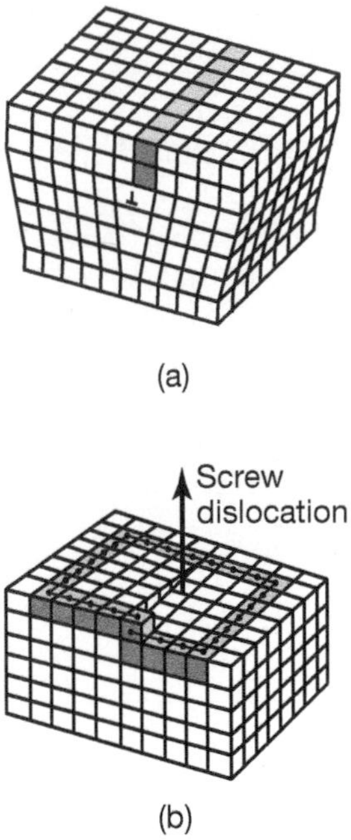

move to new lattice locations. One analogy that can be used to describe the movement of an edge dislocation is that of moving a large carpet by forming a hump at one end and moving the hump forward toward the other end. In this way, the force required to move a heavy carpet is much less than that required to slide the whole carpet along the floor.

The **density of dislocations** is the total length of dislocation lines per unit volume, ($mm/mm^3 = mm^{-2}$); it increases with increasing plastic deformation, by as much as 10^6 mm^{-2} at room temperature. The dislocation densities for some conditions are as follows:

1. Very pure single crystals: 0 to 10^3 mm^{-2};
2. Annealed single crystals: 10^5 to 10^6;
3. Annealed polycrystals: 10^7 to 10^8; and
4. Highly cold-worked metals: 10^{11} to 10^{12}.

The mechanical properties of metals, such as yield and fracture strength, as well as electrical conductivity, are affected by lattice defects and are known as **structure-sensitive properties.** On the other hand, physical properties, such as melting point, specific heat, coefficient of thermal expansion, and elastic constants are not sensitive to these defects and are thus known as **structure-insensitive properties.**

(a)

Screw dislocation

(b)

FIGURE 3.10 (a) Edge dislocation, a linear defect at the edge of an extra plane of atoms. (b) Screw dislocation, a helical defect in a three-dimensional lattice of atoms. Screw dislocations are so named because the atomic planes form a helical ramp.

3.3.4 Strain Hardening (Work Hardening)

Although the presence of a dislocation lowers the shear stress required to cause slip, dislocations can (a) become entangled and thus interfere with each other; and (b) be impeded by barriers, such as grain boundaries and impurities and inclusions in the material. Entanglement and impediments increase the shear stress required for slip.

The increase in the shear stress, and hence the increase in the overall strength of a metal, is known as **strain hardening,** or **work hardening** (see Section 2.2.3). The greater the deformation, the more the entanglements, thus increasing the metal's strength. Work hardening is used extensively to strengthen metals in metalworking processes at ambient temperature. Typical examples are (a) strengthening wire by drawing it through a die to reduce its cross section (Section 6.5); (b) producing the head on a bolt

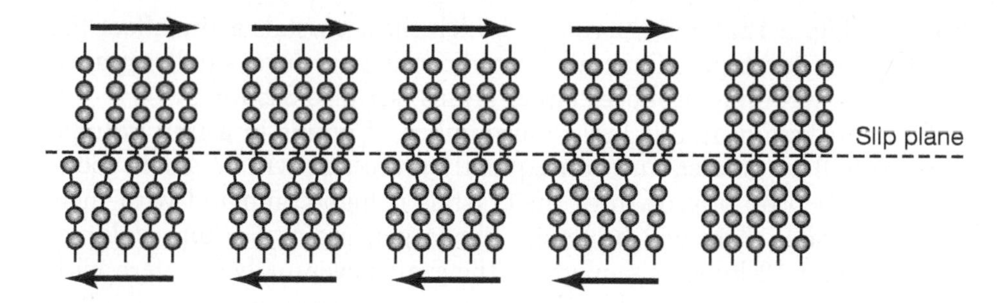

FIGURE 3.11 Movement of an edge dislocation across the crystal lattice under a shear stress. Dislocations help explain why the actual strength of metals is much lower than that predicted by atomic theory.

by forging it (Section 6.2.4); and (c) producing sheet metal for automobile bodies and aircraft fuselages by rolling (Section 6.3). As shown by Eq. (2.11), the degree of strain hardening is indicated by the magnitude of the strain-hardening exponent, n (See also Table 2.2). Among the three crystal structures, hcp has the lowest n value, followed by bcc, and then fcc, which has the highest.

3.4 Grains and Grain Boundaries

Metals commonly used for manufacturing various products are composed of many individual and randomly oriented crystals (*grains*), called **polycrystals.** When a mass of molten metal begins to solidify, crystals begin to form independently of each other, with random orientations and at various locations within the liquid mass (Fig. 3.12). Each of the crystals eventually grows into a crystalline structure, or a grain. The number and the size of the grains developed in a unit volume depend on the rate at which **nucleation** (the initial stage of formation of crystals) takes place. The *number* of different sites in which individual crystals begin to form (seven are

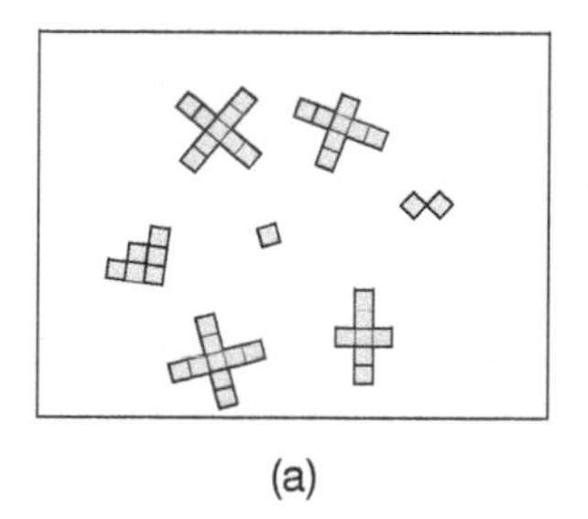

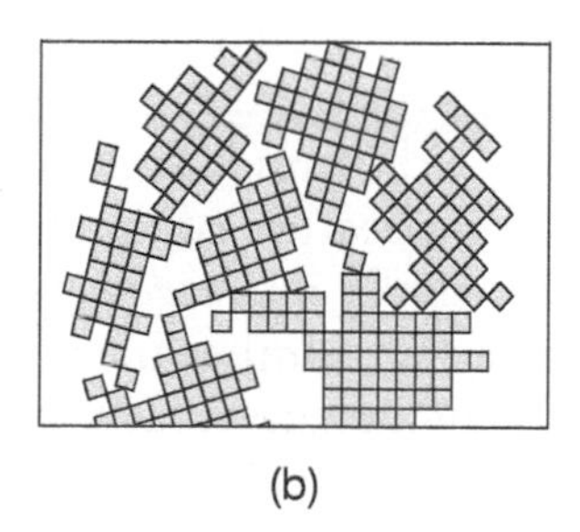

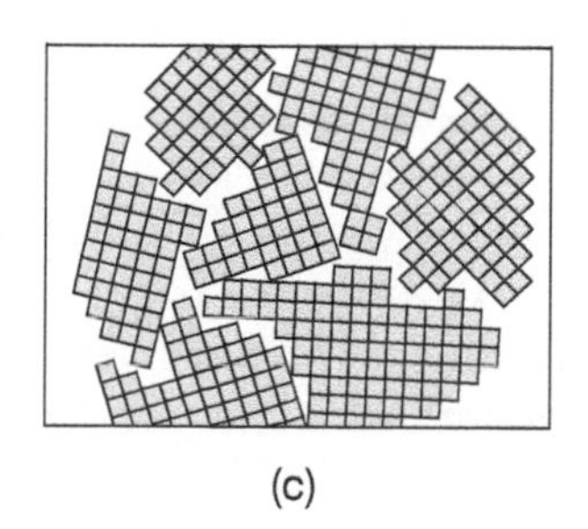

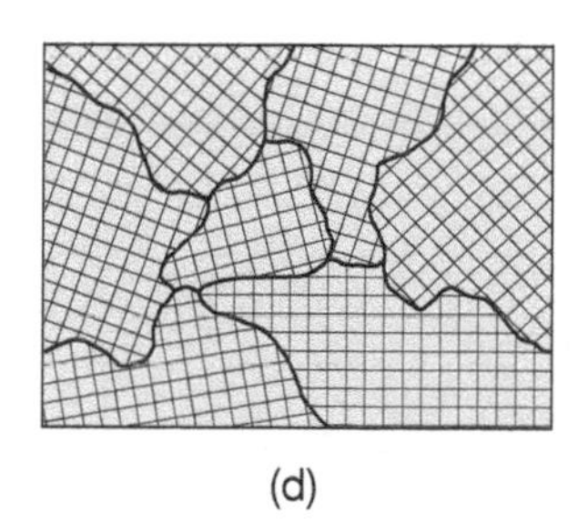

FIGURE 3.12 Schematic illustration of the various stages during solidification of molten metal. Each small square represents a unit cell. (a) Nucleation of crystals at random sites in the molten metal. Note that the crystallographic orientation of each site is different. (b) and (c) Growth of crystals as solidification continues. (d) Solidified metal, showing individual grains and grain boundaries. Note the different angles at which neighboring grains meet each other.

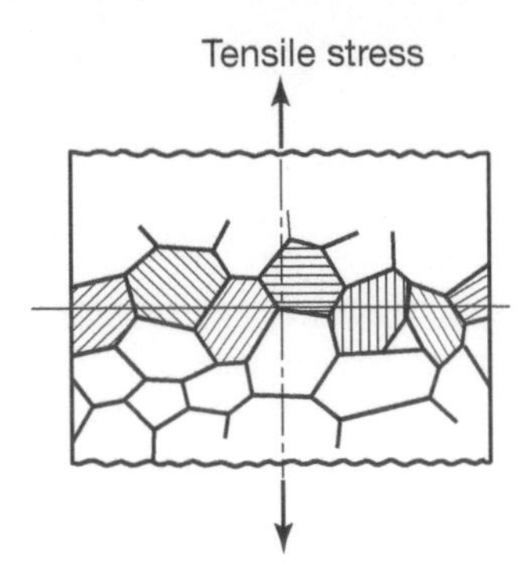

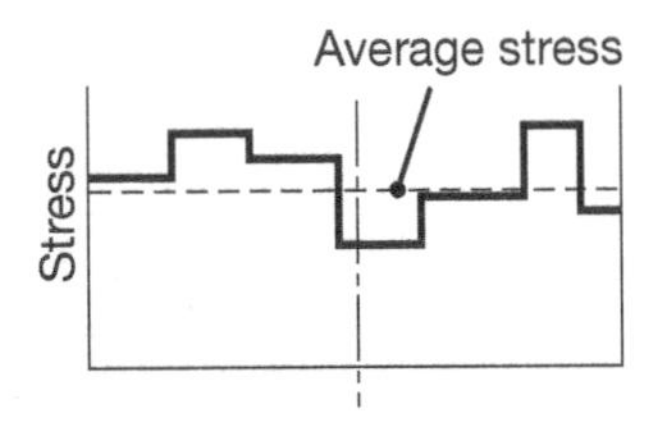

FIGURE 3.13 Variation of tensile stress across a plane of polycrystalline metal specimen subjected to tension. Note that the strength exhibited by each grain depends on its orientation.

shown in Fig. 3.12a) and the *rate* at which these crystals grow affect the size of grains developed. Generally, rapid cooling produces smaller grains, whereas slow cooling produces larger grains (see also Section 5.3).

If the nucleation rate is high, the number of grains in a unit volume of metal will be greater, and consequently grain size will be small. Conversely, if the rate of growth of the crystals is high compared with their nucleation rate, there will be fewer, but larger, grains per unit volume. Note in Fig. 3.12 how, as they grow, the grains eventually interfere with and impinge upon one another. The *surfaces* that separate the individual grains are called **grain boundaries**. Each grain consists of either a single crystal (for pure metals) or a polycrystalline aggregate (for alloys).

Note that the crystallographic orientation changes abruptly from one grain to the next across the grain boundaries. Recall from Section 3.3 that the behavior of a single crystal or a single grain is *anisotropic*. The behavior of a piece of polycrystalline metal (Fig. 3.13) is essentially *isotropic* because the grains have random crystallographic orientations, thus properties do not vary with the direction of testing. In practice, however, perfectly isotropic metals are rare, because the crystal structure usually is not equiaxed and cold working results in preferred orientation of the crystal (Section 3.5).

3.4.1 Grain Size

Grain size significantly influences the mechanical properties of metals. Large grain size generally is associated with low strength, low hardness, and high ductility. Also, as discussed in Section 3.6, large grains produce a rough surface appearance after being stretched (as in sheet metals) or compressed (as in forging and other metalworking processes). The yield strength, S_y, is the most sensitive property, and is related to grain size by the empirical formula (known as the *Hall–Petch equation*):

$$S_y = S_{yi} + kd^{-1/2}, \tag{3.8}$$

where S_{yi} is a basic yield strength (which can be regarded as the stress opposing the motion of dislocations in a very large grain), k is a constant indicating the extent to which dislocations are piled up at barriers (such as grain boundaries), and d is the grain diameter. Equation (3.8) is valid below the recrystallization temperature of the material.

Grain size (Table 3.1) is usually measured by counting the number of grains within a given area or the number of grains that intersect a given length of a line, randomly drawn on an enlarged photograph (taken under

TABLE 3.1 Grain sizes.

ASTM no.	−3	0	3	5	7	9	12
Grains/mm^2	1	8	64	256	1024	4096	32,800
Grains/mm^3	0.7	16	360	2900	23,000	185,000	4,200,000

a microscope) of the grains on a polished and etched specimen. Software is available to automate these tasks. Grain size may also be determined by referring to a standard chart. The American Society for Testing and Materials (ASTM) grain-size number, n, is related to the number of grains, N, per square inch at a magnification of 100× (equal to 0.0645 mm^2 of actual area) by the expression

$$N = 2^{n-1}. \tag{3.9}$$

Grains of sizes between 5 and 8 are generally considered fine grains. A grain size of 7 generally is acceptable for sheet metals used to make car bodies, appliances, and kitchen utensils. Grains can also be large enough to be visible to the naked eye, such as grains of zinc on the surface of galvanized sheet steel.

3.4.2 Influence of Grain Boundaries

Grain boundaries have a major influence on the strength and ductility of metals. Moreover, because they interfere with the movement of dislocations, they also influence strain hardening. The magnitude of these effects depends on temperature, rate of deformation, and the type and amount of impurities present along the grain boundaries. Grain boundaries are more reactive (i.e., they will form chemical bonds more readily) than the grains themselves, because the atoms along the grain boundaries are packed less efficiently and are more disordered than the atoms within the grains. For this reason, corrosion is most prevalent at grain boundaries.

When brought into close atomic contact with certain low-melting-point metals, a normally ductile and strong metal can crack under very low stresses, a phenomenon referred to as **grain-boundary embrittlement** (Fig. 3.14). Examples include (a) aluminum wetted with a mercury-zinc amalgam; and (b) liquid gallium and copper at elevated temperature wetted with lead or bismuth; these elements weaken the grain boundaries of the metal. The term **liquid-metal embrittlement** is used to describe such phenomena, because the embrittling element is in a liquid state. Embrittlement can also occur at temperatures well below the melting point of the embrittling element, a phenomenon known as **solid-metal embrittlement.**

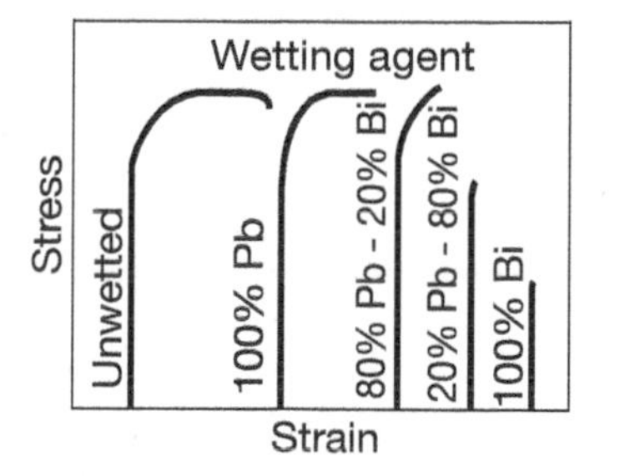

FIGURE 3.14 Embrittlement of copper by lead and bismuth at 350°C (660°F). Embrittlement has important effects on the strength, ductility, and toughness of materials.

Hot shortness is caused by local melting of a constituent or an impurity along the grain boundary at a temperature below the melting point of the metal itself. When such a metal is subjected to plastic deformation at elevated temperatures (*hot working*; see Section 3.7), the metal crumbles along its grain boundaries; examples of hot shortness include antimony in copper, and leaded steels and brass. To avoid hot shortness, the metal is usually worked at a lower temperature, which prevents softening and melting along the grain boundaries. **Temper embrittlement** is another form of embrittlement that occurs in alloy steels, and is caused by segregation (movement) of impurities to the grain boundaries.

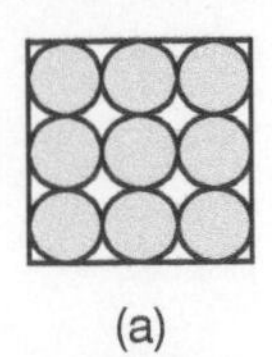

(a)

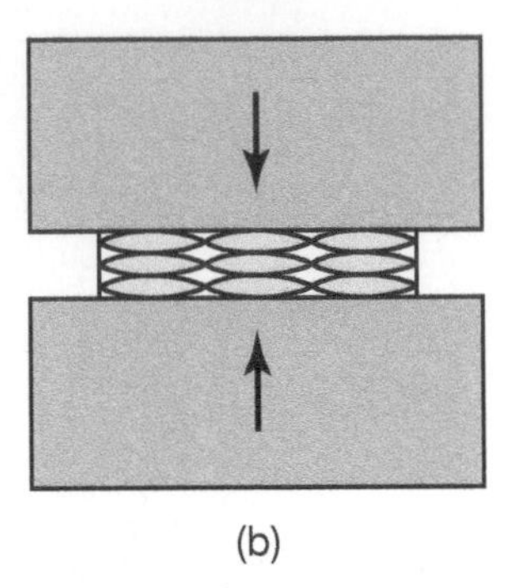

(b)

FIGURE 3.15 Plastic deformation of idealized (equiaxed) grains in a specimen subjected to compression, such as is done in rolling or forging of metals: (a) before deformation; and (b) after deformation. Note the alignment of grain boundaries along a horizontal direction.

3.5 Plastic Deformation of Polycrystalline Metals

When a polycrystalline metal, with uniform equiaxed grains (i.e., having equal dimensions in all directions, as shown in the model in Fig. 3.15a), is subjected to plastic deformation at room temperature (*cold working*), the grains become permanently deformed. The deformation process may be carried out either by compressing the metal (as in forging; Section 6.2) or by subjecting it to tension (as in stretching sheet metal; Chapter 7). The deformation within each grain takes place by the mechanisms described in Section 3.3.

After plastic deformation, the deformed metal exhibits higher strength because of the entanglement of dislocations with grain boundaries during deformation. The amount of increase in strength depends on the amount of deformation (strain) to which the metal is subjected; the greater the deformation, the stronger the metal becomes. Because they have a larger grain-boundary surface area per unit volume of metal, the increase in strength is greater for metals with smaller grains.

Anisotropy (texture). Due to plastic deformation, the grains become elongated in one direction and contracted in the other (Fig. 3.15b). As a result, the behavior of the metal becomes **anisotropic**, whereby its properties in the vertical direction in the figure are now different from those in the horizontal direction. The degree of anisotropy depends on how uniformly the metal is deformed throughout its volume. Note from the direction of the crack in Fig. 3.16, for example, that the ductility of the cold-rolled sheet in its vertical (transverse) direction is lower than in its longitudinal direction.

Anisotropy influences both the mechanical as well as the physical properties of metals. Sheet steel for electrical transformers, for example, is rolled in such a way that the resulting deformation imparts anisotropic magnetic properties to the sheet, thus reducing magnetic-hysteresis losses and improving the efficiency of transformers (see also the discussion of amorphous alloys in Section 3.11.9). There are two general types of anisotropy in metals:

1. **Preferred orientation.** Also called **crystallographic anisotropy**, *preferred orientation* can best be described by referring to Fig. 3.5. Note

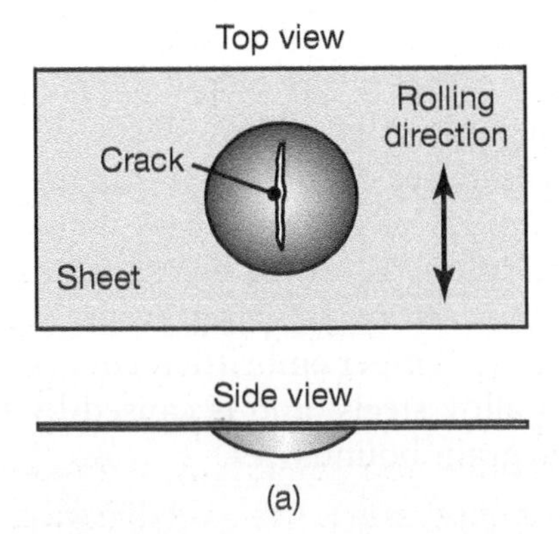

(a)

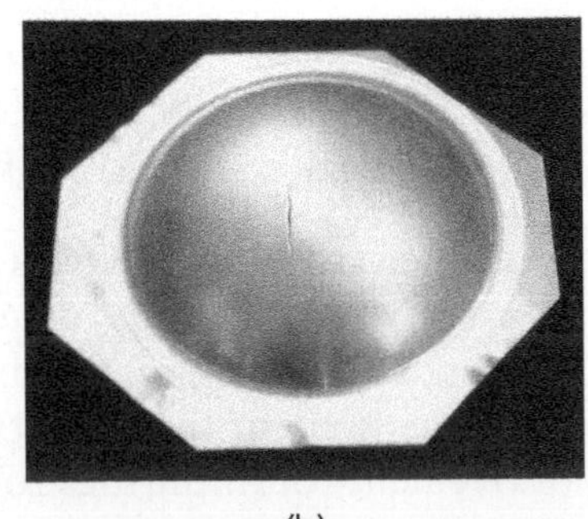

(b)

FIGURE 3.16 (a) Illustration of a crack in sheet metal subjected to bulging, such as by pushing a steel ball against the sheet. Note the orientation of the crack with respect to the rolling direction of the sheet. This material is anisotropic. (b) Aluminum sheet with a crack (vertical dark line at the center) developed in a bulge test. *Source:* J.S. Kallend, Illinois Institute of Technology.

that when a metal crystal is subjected to tension, the sliding blocks rotate toward the direction of pulling; thus, slip planes and slip bands tend to align themselves with the direction of deformation. For a polycrystalline aggregate, with grains in various orientations (Fig. 3.15), all slip directions tend to align themselves with the direction of pulling. Conversely, under compression, the slip planes tend to align themselves in a direction perpendicular to the direction of compression.

2. **Mechanical fibering.** Mechanical fibering results from the alignment of impurities, inclusions (stringers), and voids in the metal during deformation. Note, for example, that if the spherical grains (shown in Fig. 3.15a) were coated with impurities, these impurities would, after deformation, align themselves generally in a horizontal direction. Since impurities weaken grain boundaries, this metal would be weak and less ductile when tested in the vertical direction than in the horizontal direction. An analogy to this case would be plywood, which is strong in tension along its planar directions but peels off easily when tested in tension along its thickness direction.

3.6 Recovery, Recrystallization, and Grain Growth

Recall that plastic deformation at room temperature results in:

- the deformation of grains and grain boundaries;
- a general increase in strength;
- a decrease in ductility; and
- anisotropic behavior due to nonuniform deformation.

These effects can be reversed and the properties of the metal brought back to their original levels by heating the metal within a specific temperature range and period of time. The temperature range and the period of time required depend on the material and various other factors. Three events take place consecutively during the heating process:

1. **Recovery.** During recovery, which occurs at a specific temperature range below the **recrystallization temperature** of the metal (see below), the stresses in the highly deformed regions are relieved and the number of mobile dislocations is reduced. Subgrain boundaries (called **polygonization**) begin to form, with no significant change in such mechanical properties as hardness and strength, although with some increase in ductility (Fig. 3.17).
2. **Recrystallization.** The process of replacing grains with new, equiaxed and stress-free grains at elevated temperature is called *recrystallization*. The temperature for recrystallization ranges approximately between 0.3 and $0.5T_m$, where T_m is the melting point of the metal on the absolute scale. The recrystallization temperature is generally defined as the temperature at which complete recrystallization occurs within approximately one hour. Lead, tin, cadmium, and zinc recrystallize around room temperature. Recrystallization decreases

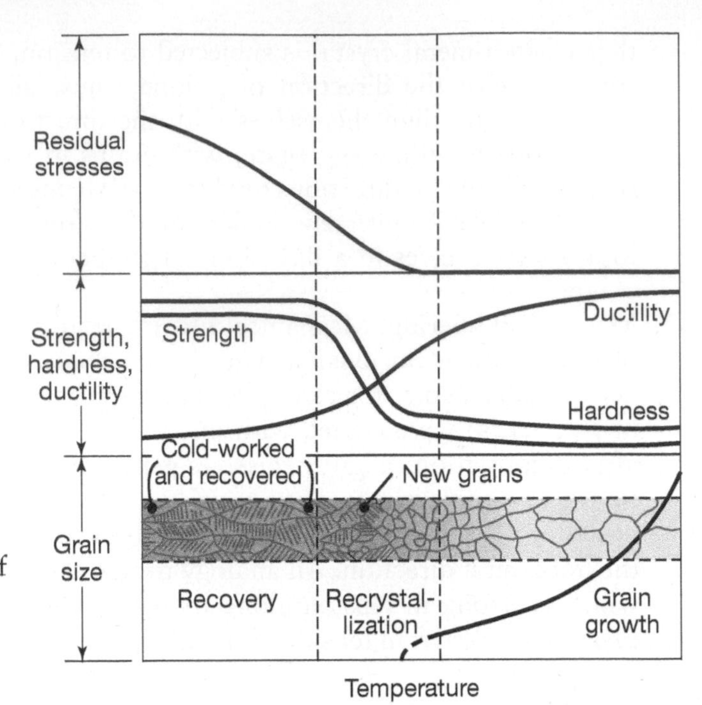

FIGURE 3.17 Schematic illustration of the effects of recovery, recrystallization, and grain growth on mechanical properties and shape and size of grains. Note the formation of small new grains during recrystallization. *Source:* After G. Sachs.

the density of dislocations and lowers the strength of the metal, but raises its ductility (Fig. 3.17).

Recrystallization depends on the degree of prior cold work (work hardening); the higher the amount of cold work, the lower the temperature required for recrystallization to occur. The reason for this inverse relationship is that as the amount of cold work increases, the number of dislocations and the amount of energy stored in the dislocations (*stored energy*) also increase. The stored energy supplies some of the energy required for recrystallization. Also, recrystallization is a function of time, because it involves *diffusion*, that is, movement and exchange of atoms across grain boundaries.

The effects on recrystallization of temperature, time, and reduction in the thickness or height of the workpiece by cold working can be summarized as follows (Fig. 3.18):

- For a constant amount of deformation by cold working, the time required for recrystallization decreases with increasing temperature.

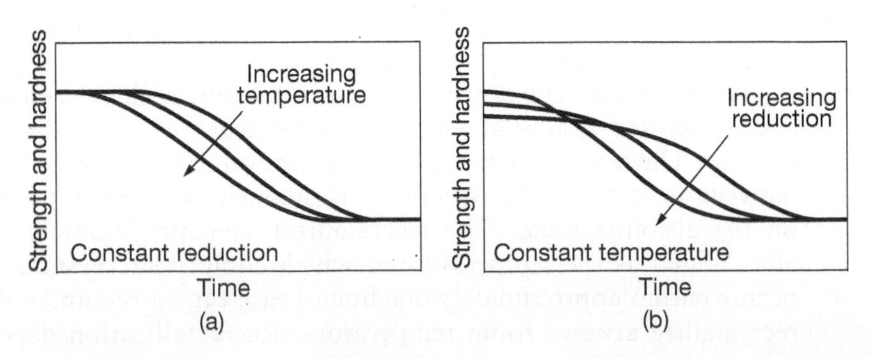

FIGURE 3.18 Variation of strength and hardness with recrystallization temperature, time, and prior cold work. Note that the more a metal is cold worked, the less time it takes to recrystallize, because of the higher stored energy from cold working due to increased dislocation density.

- The more the prior cold work, the lower the temperature required for recrystallization.
- The higher the amount of deformation, the smaller the resulting grain size after recrystallization (Fig. 3.19). Recrystallizing a deformed metal is a common method of converting a coarse-grained structure to one of fine grain, with improved properties.
- Anisotropy due to preferred orientation usually persists after recrystallization. To restore isotropy, a temperature higher than that required for recrystallization may be necessary.

3. **Grain growth.** With continued increase in the temperature of the metal, the grains begin to grow; their size may eventually exceed the original grain size. This phenomenon is known as *grain growth*, and it has an adverse effect on hardness and strength (Fig. 3.17). More importantly, however, large grains produce the **orange-peel effect,** resulting in a rough surface appearance, such as when sheet metal is stretched to form a part or when bulk metal is subjected to compression (Fig. 3.20), such as in forging operations.

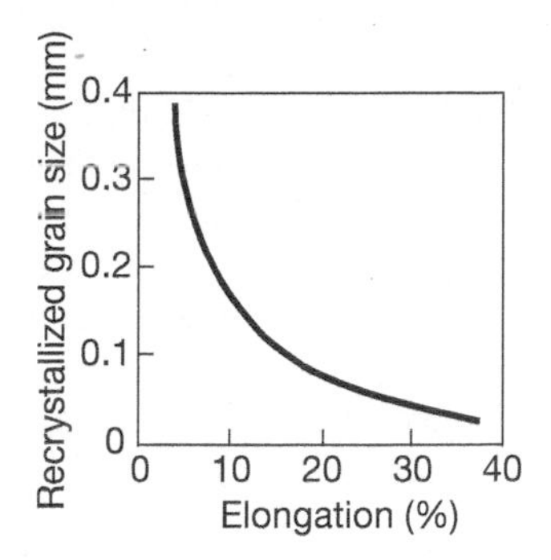

FIGURE 3.19 The effect of prior cold work on the recrystallized grain size of alpha brass. Below a critical elongation (strain), typically 5%, no recrystallization occurs.

3.7 Cold, Warm, and Hot Working

When plastic deformation is carried out above the recrystallization temperature of the metal, it is called **hot working,** and if it is done below its recrystallization temperature, it is called **cold working.** As the name implies, **warm working** is carried out at an intermediate temperature, thus warm working is a compromise between cold and hot working. The temperature ranges for these three categories of plastic deformation are given in Table 3.2 in terms of the **homologous temperature,** defined as the ratio of the working temperature to the metal's melting temperature, both on the absolute scale.

There are important technological differences in metal products that are processed by cold, warm, or hot working. For example, compared with cold-worked parts, hot-worked parts generally have (a) lower dimensional accuracy, because of uneven thermal expansion and contraction during processing of the part; and (b) a rougher surface appearance and finish, because of the oxide layer that usually develops. Other important manufacturing characteristics, such as formability, machinability, and weldability, are also affected by cold, warm, and hot working to different extents.

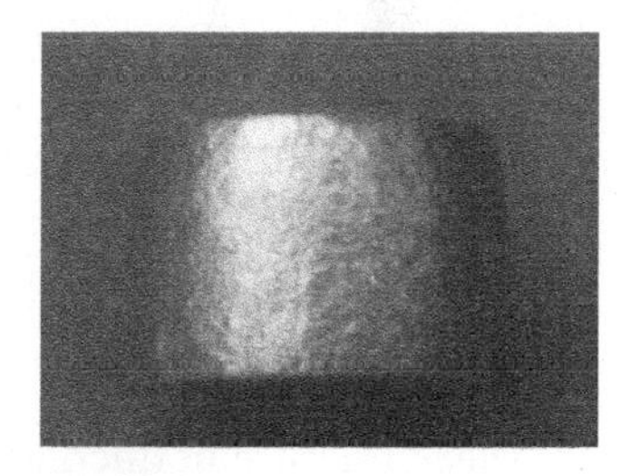

FIGURE 3.20 Surface roughness on the cylindrical surface of an aluminum specimen subjected to compression. *Source:* A. Mulc and S. Kalpakjian.

TABLE 3.2 Homologous temperature ranges for various processes.

Process	T/T_m
Cold working	< 0.3
Warm working	0.3 to 0.5
Hot working	> 0.6

3.8 Failure and Fracture

Failure is one of the most important aspects of a material's behavior because it directly influences its selection of a particular application, the methods of manufacturing, and the service life of the component. Because of the numerous factors involved, failure and fracture of materials is a complex area of study. In this section, we consider only those aspects of

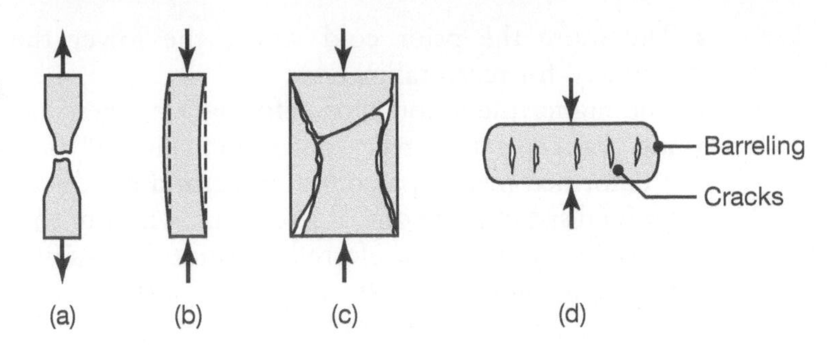

FIGURE 3.21 Schematic illustration of types of failure in materials: (a) necking and fracture of ductile materials; (b) buckling of ductile materials under a compressive load; (c) fracture of brittle materials in compression; (d) cracking on the barreled surface of ductile materials in compression (See also Fig. 6.1b).

failure that are of particular significance in the selection and processing of materials.

There are two general types of failure: (1) **fracture** and separation of the material, through either internal or external cracking; and (2) excessive deformation, including **buckling** (Fig. 3.21b). Fracture is further divided into two general categories: ductile and brittle (Fig. 3.22). It should also be noted that although failure of materials is generally regarded as undesirable, certain products are indeed designed to fail. Examples include (a) beverage cans with tabs or entire tops of food cans that are removed by tearing the sheet metal along a prescribed path; and (b) metal or plastic screw caps for bottles.

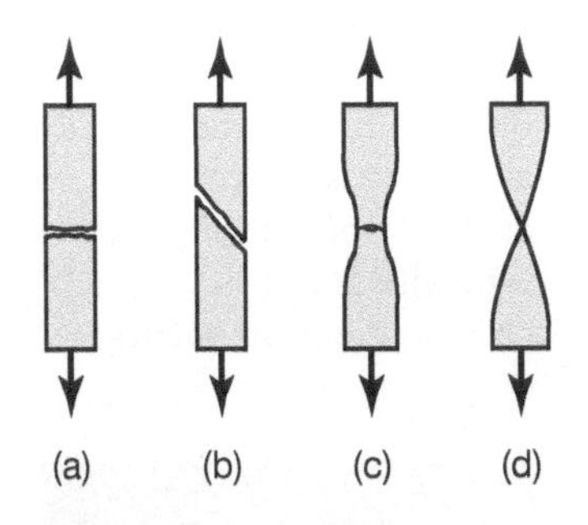

FIGURE 3.22 Schematic illustration of the types of fracture in tension: (a) brittle fracture in polycrystalline metals; (b) shear fracture in ductile single crystals (See also Fig. 3.5a); (c) ductile cup-and-cone fracture in polycrystalline metals (See also Fig. 2.2); and (d) complete ductile fracture in polycrystalline metals, with 100% reduction of area.

3.8.1 Ductile Fracture

Ductile fracture is characterized by plastic deformation that precedes failure of the part. In a tension test, for example, highly ductile materials, such as gold and lead, may neck down to a point and then fail (Fig. 3.22d). Most metals and alloys, however, neck down to a finite area and then fail. Ductile fracture generally takes place along planes on which the *shear stress is a maximum*. In torsion, for example, a ductile metal fractures along a plane perpendicular to the axis of twist, that is, the plane on which the shear stress is a maximum. Fracture in shear is a result of extensive slip along slip planes within the grains.

Upon close examination, the surface in ductile fracture (Fig. 3.23) indicates a *fibrous* pattern with *dimples*, as if a number of very small

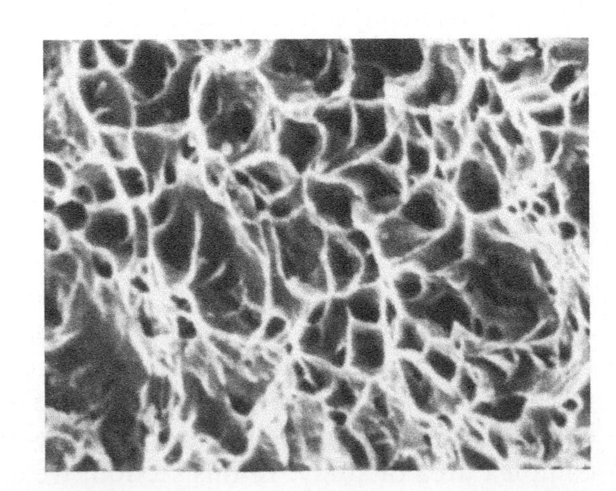

FIGURE 3.23 Surface of ductile fracture in low-carbon steel, showing dimples. Fracture is usually initiated at impurities, inclusions, or pre-existing voids in the metal. *Source:* After K.-H. Habig and D. Klaffke.

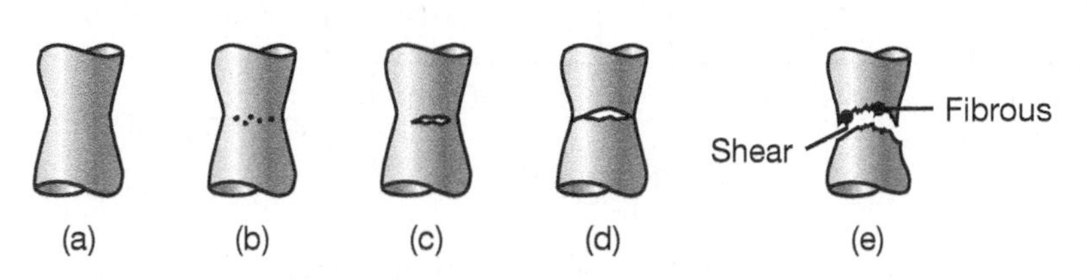

FIGURE 3.24 Sequence of events in necking and fracture of a tensile-test specimen: (a) early stage of necking; (b) small voids begin to form within the necked region; (c) voids coalesce, producing an internal crack; (d) rest of cross section begins to fail at the periphery by shearing; and (e) final fracture surfaces, known as cup-(top fracture surface) and-cone (bottom surface) fracture.

tension tests have been carried out over the fracture surface. Failure is initiated with the formation of tiny voids (usually originating around small inclusions or pre-existing defects), which then grow and coalesce, resulting in cracks that grow in size and lead to fracture. In a tension-test specimen, fracture typically begins at the center of the necked region from the growth and coalescence of cavities (Fig. 3.24). The central region becomes one large crack, which then propagates to the periphery of the necked region. Because of its appearance, this type of fracture is called a **cup-and-cone fracture.**

Effects of inclusions. Because they are nucleation sites for voids, *inclusions* have a major influence on ductile fracture and hence on the formability of materials. Inclusions may consist of impurities of various kinds and second-phase particles, such as oxides, carbides, and sulfides. The extent of their influence depends on such factors as their shape, hardness, distribution, and volume fraction. The higher their volume fraction, the lower will be the ductility of the material. Voids and porosity developed during processing, such as from casting (see Section 5.12) and some metalworking operations, also reduce the ductility of a material.

Two factors have a significant effect on void formation:

1. The strength of the bond at the interface of an inclusion and the matrix (the material around the inclusion). If the bond is strong, there will then be lower tendency for void formation during plastic deformation.
2. The hardness of the inclusion. If the inclusion is soft, such as manganese sulfide, it will conform to the overall change in shape during plastic deformation. If it is hard, such as a carbide or an oxide, it could lead to void formation. Also, because of their brittle nature, hard inclusions may also break up into smaller particles during deformation, as shown in Fig. 3.25.

The alignment of inclusions during plastic deformation leads to **mechanical fibering.** Subsequent processing of such a material must therefore involve considerations of the proper direction of working for maximum ductility and strength.

Transition temperature. Metals undergo a sharp change in ductility and toughness across a narrow temperature range. Called the *transition*

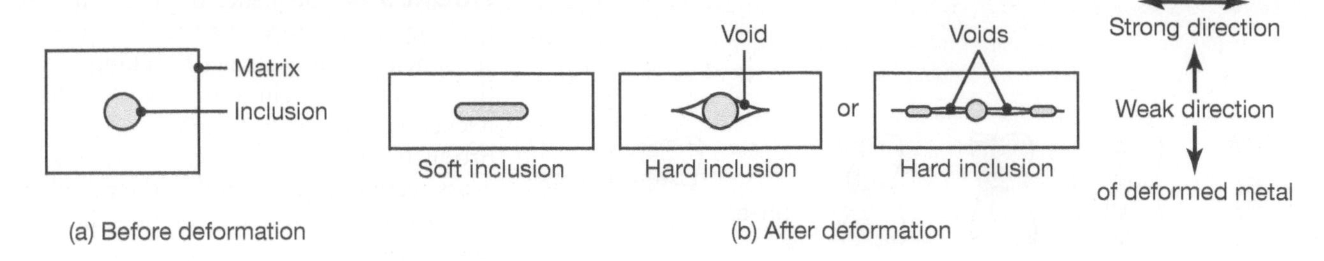

FIGURE 3.25 Schematic illustration of the deformation of inclusions and their effect on void formation in plastic deformation. Note that hard inclusions can cause voids because they do not conform to the overall matrix deformation.

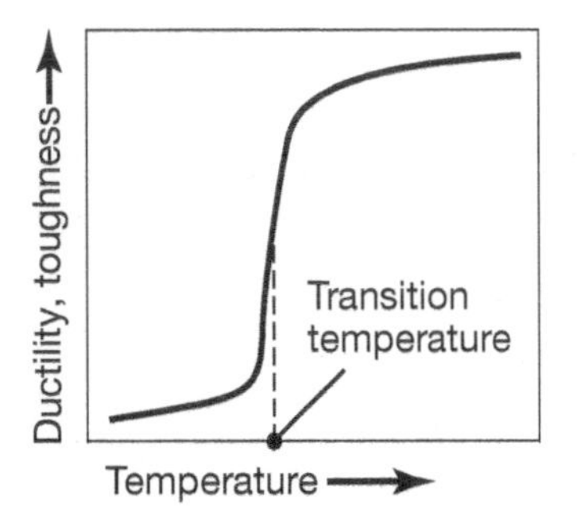

FIGURE 3.26 Schematic illustration of transition temperature. Note the narrow temperature range across which the behavior of the metal undergoes a major transition.

temperature (Fig. 3.26), this phenomenon typically occurs in body-centered cubic and some hexagonal close-packed metals; it is rarely exhibited by face-centered cubic metals. Transition temperature depends on factors such as composition, microstructure, grain size, surface finish and shape of the specimen, and rate of deformation. High rates, abrupt changes in shape, and surface notches raise the transition temperature.

Strain aging. *Strain aging* is a phenomenon in which carbon atoms in steels segregate to dislocations, thereby pinning them and thus increasing the steel's resistance to dislocation movement; the result is increased strength and reduced ductility. While usually taking place over several days at room temperature, strain aging can occur in just a few hours at a higher temperature; it is then called **accelerated strain aging**. For steels, the phenomenon is also called **blue brittleness** because of the color of the steel.

3.8.2 Brittle Fracture

Brittle fracture occurs with little or no plastic deformation preceding the separation of the material into two or more pieces. Figure 3.27 shows a typical example of the surface of brittle fracture. In tension, brittle fracture takes place along a crystallographic plane (called a **cleavage plane**) on which the normal tensile stress is a maximum. Body-centered cubic and some hexagonal close-packed metals fracture by cleavage, whereas

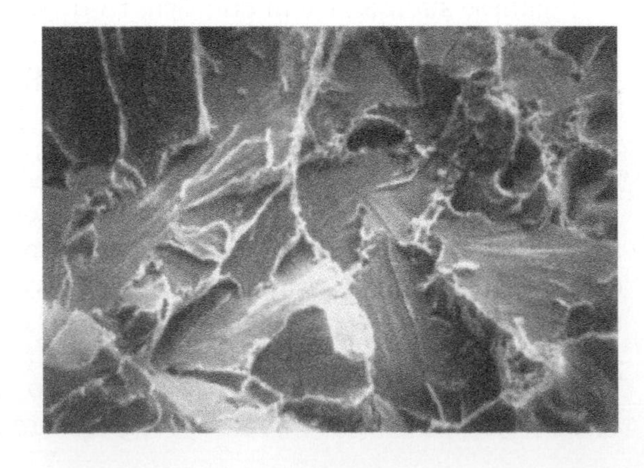

FIGURE 3.27 Typical fracture surface of steel that has failed in a brittle manner. The fracture path is transgranular (through the grains). Compare this surface with the ductile fracture surface shown in Fig. 3.23. Magnification: 200×. *Source:* Packer Engineering Associates, Inc.

face-centered cubic metals usually do not fail in brittle fracture. In general, low temperature and high rates of deformation promote brittle fracture.

In a polycrystalline metal under tension, the fracture surface has a bright granular appearance; this is because of the changes in the direction of the cleavage planes as the crack propagates from one grain to another. Brittle fracture in compression is more complex and, in theory, follows a path that is 45° to the direction of the applied force.

Materials that display fracture along a cleavage plane include chalk, gray cast iron, and concrete, which under tension typically fail in the manner shown in Fig. 3.21a. In torsion, they fail along a plane at 45° to the axis of twist, that is, along a plane on which the tensile stress is a maximum.

Defects. An important factor in fracture is the presence of such **defects** as scratches, external or internal cracks, and various other flaws. Under tension, the tip of a crack is subjected to high tensile stresses that propagate the crack rapidly, because the brittle material has little capacity to dissipate energy. It can be shown that the **fracture strength,** S_f, of a specimen with a crack perpendicular to the direction of pulling is related to the length of the crack, as follows:

$$S_f \propto \frac{1}{\sqrt{\text{Crack length}}}. \tag{3.10}$$

The presence of defects is essential in explaining why brittle materials are significantly weaker in tension as compared with their strength in compression. Under tensile stresses, cracks propagate rapidly, causing what is known as *catastrophic failure.* With polycrystalline metals, the fracture paths most commonly observed are **transgranular** (*transcrystalline,* or *intragranular*), meaning that the crack propagates *through* the grain. **Intergranular** fracture, where the crack propagates along the grain boundaries (Fig. 3.28), generally occurs when the grain boundaries are (a) soft; (b) contain a brittle phase; or (c) have been weakened by liquid-or solid-metal embrittlement (Section 3.4.2).

It can be shown that the maximum crack velocity in a brittle material is about 62% of the elastic wave-propagation (or acoustic) velocity of the material. This velocity is given by the formula $\sqrt{E/\rho}$, where E is the elastic

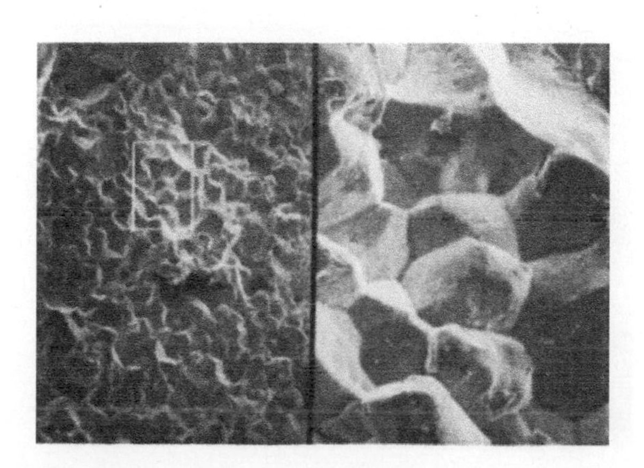

FIGURE 3.28 Intergranular fracture, at two different magnifications. Grains and grain boundaries are clearly visible in this micrograph. The fracture path is along the grain boundaries. Magnification: left, 100×; right, 500×. *Source:* Packer Engineering Associates, Inc.

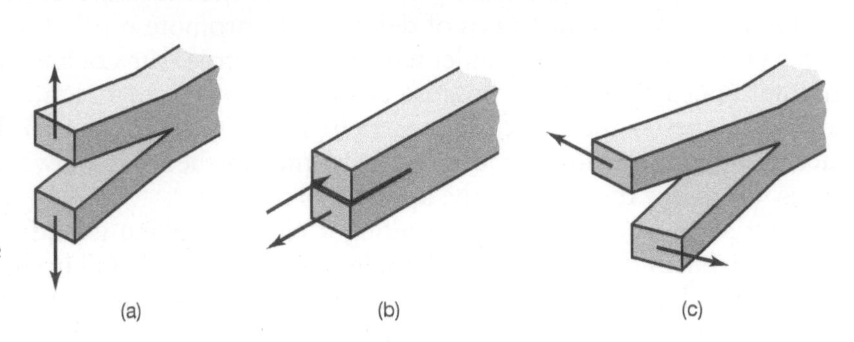

FIGURE 3.29 Three modes of fracture. (a) Mode I has been studied extensively, because it is the most commonly observed in engineering structures and components. (b) Mode II is rare, but is seen in some wear modes and packaging. (c) Mode III is the tearing process; examples include pull tabs on food cans, tearing a piece of paper and cutting materials with a pair of scissors.

modulus and ρ is the mass density; thus, for steel, the maximum crack velocity is 2000 m/s (6600 ft/s).

As shown in Fig. 3.29, cracks may be subjected to stresses in different directions. Mode I is tensile stress applied perpendicular to the crack. Modes II and III are shear stresses applied in two different directions.

Fatigue fracture. In this type of fracture, minute external or internal cracks develop at pre-existing flaws or defects in the material. The cracks then propagate throughout the body of the part and eventually lead to its total failure. The fracture surface in fatigue generally displays **beach marks**, as can be seen in Fig. 3.30. Under large magnification (higher than 1000×), a series of **striations** can be observed on fracture surfaces, where each beach mark consists of several striations.

Because of its sensitivity to surface defects, the fatigue life of a specimen or a part is greatly influenced by the method of preparation of its surfaces (Fig. 3.31).

The fatigue strength of manufactured parts can generally be improved by the following methods:

1. Inducing compressive residual stresses on surfaces, such as by shot peening or roller burnishing (see Section 4.5.1);
2. Surface (case) hardening by various means of heat treatment (see Section 5.11.3);

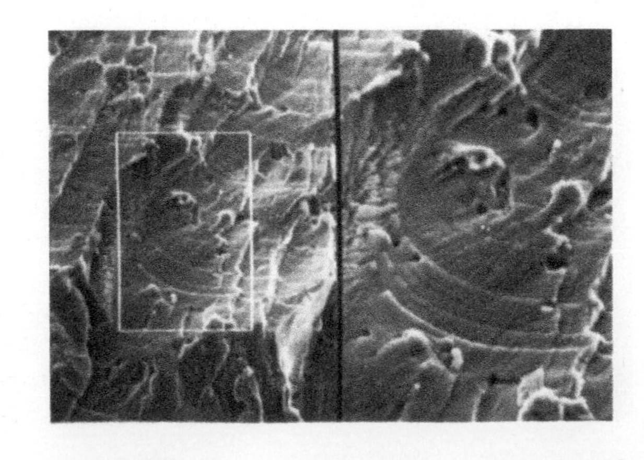

FIGURE 3.30 Typical fatigue fracture surface on metals, showing beach marks. Most components in machines and engines fail by fatigue and not by excessive static loading. Magnification: left, 500×; right, 1000×. *Source:* Packer Engineering Associates, Inc.

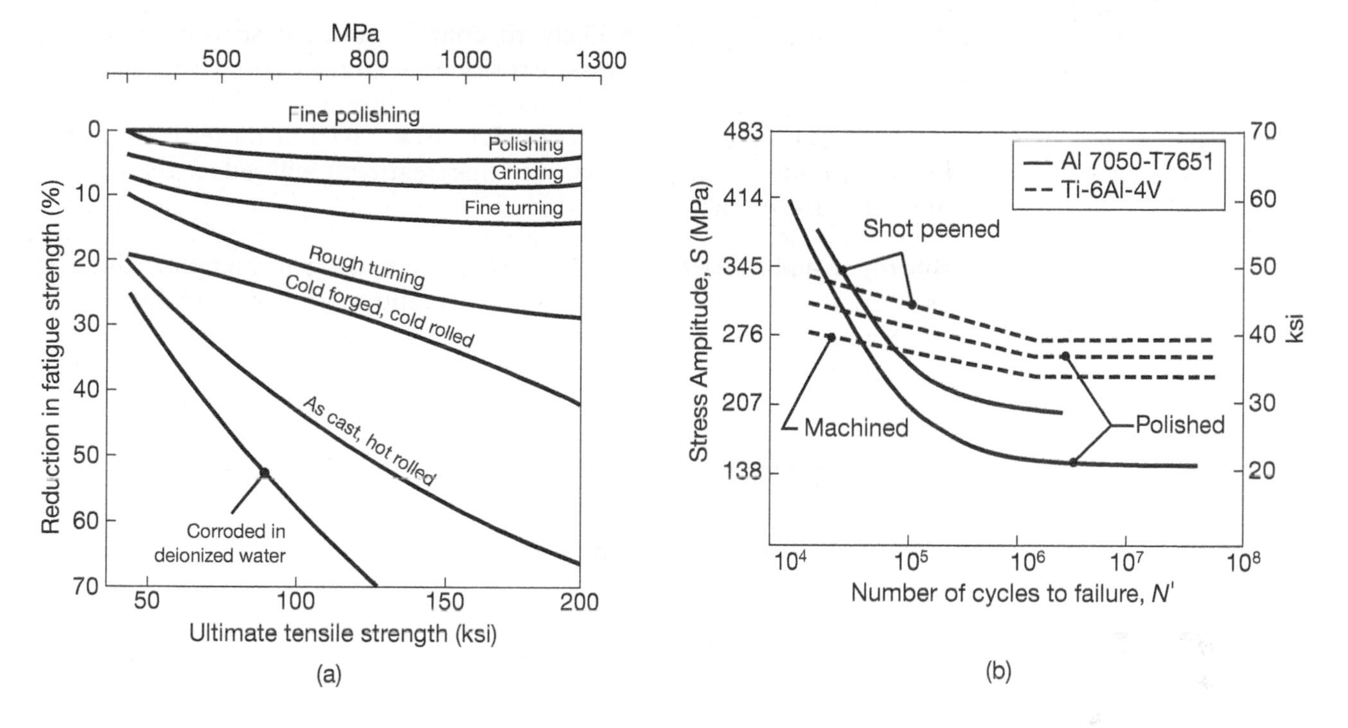

FIGURE 3.31 Reduction in fatigue strength of cast steels subjected to various surface-finishing operations. Note that the reduction is greater as the surface roughness and strength of the steel increase.

3. Providing a fine surface finish on the part, thereby reducing the effects of notches and other surface imperfections; and
4. Selecting appropriate materials and ensuring that they are effectively free of inclusions, voids, and impurities.

Conversely, the following factors and processes can reduce fatigue strength: decarburization, surface pits such as due to corrosion that act as stress raisers, hydrogen embrittlement (see below), galvanizing, and electroplating (Section 4.5.1).

Stress-corrosion cracking. An otherwise ductile metal can fail in a brittle manner due to stress-corrosion cracking (also called *stress cracking* or *season cracking*). Stress-corrosion cracking generally involves small concentrations of corrosive chemicals, in amounts lower than what would be considered corrosive. After forming, the parts may, either over a period of time or soon after it is made, develop numerous microscopic cracks due to chemical interactions with the environment. The resulting crack propagation may be intergranular or transgranular.

The susceptibility of metals to stress-corrosion cracking mainly depends on (a) the material, (b) the presence and magnitude of tensile residual stresses, and (c) the environment. Brass and austenitic stainless steels are among the metals that are highly susceptible to stress cracking. Environmental factors, such as salt water or other chemicals that could be corrosive to some metals, can increase stress-corrosion cracking. Metals that are cold worked are likely to contain residual stresses, and thus are more susceptible to stress-corrosion cracking as compared with hot-worked or annealed metals. The usual procedure to avoid stress-corrosion cracking is to *stress relieve* (Section 5.11.4) the part just after it is formed. Full annealing may also be done, but this treatment will reduce the strength of cold-worked parts.

Hydrogen embrittlement. The presence of hydrogen can reduce ductility and cause severe embrittlement in metals, alloys, and nonmetallic materials, leading to premature failure. Known as *hydrogen embrittlement*, this phenomenon is especially severe in high-strength steels (see Sections 3.10.1 and 3.10.2). Possible sources of hydrogen in metals are:

a. during melting of the metal;
b. *pickling*, that is, removal of surface oxides by chemical or electrochemical reaction;
c. through electrolysis in electroplating; and
d. from water vapor in the atmosphere or from moist electrodes and fluxes used during welding.

Oxygen also can cause embrittlement in metals, especially in copper alloys.

3.8.3 Size Effect

The dependence of the properties of a material on its size is known as *size effect*. Note from the foregoing discussions and from Eq. (3.10) that defects, cracks, imperfections, and the like are less likely to be present in a part as its size decreases. Thus, the strength and ductility of a part increase with decreasing size.

Although strength is related to cross-sectional area of a part, its length is also important, because the greater the length, the greater the probability for defects to exist. As a common analogy, a long chain is more likely to be weaker than a shorter chain, because the probability of one of the links being weak increases with the number of links, hence the length of chain.

3.9 Physical Properties

In addition to their mechanical properties, the **physical properties** of materials must also be considered in their selection and processing. Properties of particular interest in manufacturing are density, melting point, specific heat, thermal conductivity and expansion, electrical and magnetic properties, and resistance to oxidation and corrosion, as described next.

3.9.1 Density

The density of a metal depends on its atomic weight, atomic radius, and the packing of the atoms; alloying elements generally have a minor effect on density. The range of densities for a variety of materials at room temperature are given in the inside front cover.

Weight reduction is an important consideration in product design, particularly for aircraft and aerospace structures, automotive bodies and components, and various other products for which energy consumption and power limitations are major concerns. **Specific strength** (strength-to-weight ratio) and **specific stiffness** (stiffness-to-weight ratio) of materials and structures are important considerations (see also Section 10.9.1). *Substitution* of materials for weight and cost reduction and ease of manufacturing are major factors in competitive manufacturing.

Density also is an important factor in the selection of materials for high-speed equipment, such as the use of magnesium in printing and textile machinery, many components of which usually operate at very high speeds. The resulting light weight of the components in these applications reduces inertial forces that otherwise could cause lead to vibrations, dimensional inaccuracies, and even part failure over time. On the other hand, there are also applications for which higher densities, such as in using tungsten, are desirable. Examples include counterweights for various mechanisms (using lead and steel), flywheels, and components for self-winding watches (using high-density materials).

3.9.2 Melting Point

The melting point defines the temperature where a material's phase changes from solid to liquid. Unlike pure metals that have a definite melting point, melting of an alloy occurs over a wide range of temperatures, depending on the alloying elements (see Section 5.2). Since the recrystallization temperature of a metal is related to its melting point (Section 3.6), operations such as annealing, heat treating, hot working, and the selection of tool and die materials require consideration of the melting points of the metals involved. These considerations, in turn, influence the selection of tool and die materials in manufacturing operations.

Another major influence of the melting point is in the selection of the equipment and melting practice in casting operations; the higher the melting point of the material, the more difficult the operation becomes to perform (Section 5.5). In electrical-discharge machining (Section 9.13), laser machining, laser welding (Section 9.14), and electron beam melting (Section 10.12), as examples, the melting points of metals are important processing parameters.

3.9.3 Specific Heat

Specific heat is the energy required to raise the temperature of a unit mass of material by one degree. Alloying elements have a relatively minor effect on the specific heat of metals. The temperature rise in a workpiece, such as resulting from forming or machining operations, is a function of the work done and the specific heat of the workpiece material (see Section 2.12.1); thus, the lower the specific heat, the higher the temperature rise in the material.

3.9.4 Thermal Conductivity

Thermal conductivity indicates the rate at which heat flows within and through the material. Metals generally have high thermal conductivity, whereas ceramics and polymers have poor conductivity. Because of the large difference in their thermal conductivities, alloying elements can have a significant effect on the thermal conductivity of alloys.

When heat is generated by such means as plastic deformation or friction, the heat should be conducted away at a sufficiently high rate to prevent excessive temperature rise, which can result in high thermal gradients and thus cause inhomogeneous deformation in metalworking processes. One of the main difficulties in machining titanium, for example, is caused by its very low thermal conductivity (Section 8.5.2).

3.9.5 Thermal Expansion

When materials are heated, their dimensions increase, a phenomenon known as **thermal expansion**. Alloying elements have a relatively minor effect on the thermal expansion of metals. Examples where expansion or contraction is important include electronic and computer components, precision machined components, glass-to-metal seals, metal-to-ceramic subassemblies (see also Section 11.8.2), electronic packaging, struts on jet engines, and moving parts in machinery that require certain clearances for proper functioning.

Thermal stresses result from relative expansion and contraction of components or within the material itself, leading to *cracking, warping,* or *loosening* of components in an assembled product during its service life. Ceramic parts and tools and dies made of brittle materials are particularly sensitive to thermal stresses. Thermal conductivity, in conjunction with thermal expansion, plays the most significant role in causing thermal stresses, both in manufactured components and in tools and dies. To reduce thermal stresses, a combination of high thermal conductivity and low thermal expansion is desirable. Thermal stresses may also be caused by **anisotropy of thermal expansion** of the material, which is generally observed in hexagonal close-packed metals, ceramics, and composite materials.

Thermal fatigue results from thermal cycling. This phenomenon is particularly important, for example, (a) in a forging operation, when hot workpieces are placed over relatively cool dies, thus subjecting the die surfaces to thermal cycling; and (b) in interrupted cutting operations such as in milling (see Section 8.10). **Thermal shock** is the term generally used to describe development of cracks after a single thermal cycle.

The influence of temperature on dimensions and the modulus of elasticity are significant factors in precision instruments and equipment. A spring, for example, will have a lower stiffness as its temperature increases, because of reduced elastic modulus. Similarly, a tuning fork or a pendulum will have different frequencies at different temperatures. To address some of the problems of thermal expansion, a family of iron-nickel alloys that have very low thermal-expansion coefficients are available, and are known as **low-expansion alloys**; these alloys also have good thermal-fatigue resistance. Typical compositions are 64% Fe-36% Ni (Invar) and 54% Fe-28% Ni-18% Co (Kovar).

3.9.6 Electrical and Magnetic Properties

Electrical conductivity and dielectric properties of materials are important not only in electrical equipment and machinery, but also in such manufacturing processes as magnetic-pulse forming of sheet metals (Section 7.5.5) and in electrical-discharge machining and electrochemical grinding of hard and brittle materials (Chapter 9).

The **electrical conductivity** of a material can be defined as the degree to which a material conducts electricity. The units of electrical conductivity are mho/m or mho/ft, where mho is the inverse of ohm (the unit for electrical resistance). Materials with high conductivity, such as metals, are generally referred to as **conductors**. The influence of the type of atomic bonding on the electrical conductivity of materials is the same as that for thermal conductivity. Alloying elements have a major effect on the electrical conductivity of metals: the higher the conductivity of the alloying element, the higher is the conductivity of the alloy.

Electrical resistivity is the inverse of conductivity, and materials with high resistivity are referred to as **dielectrics**, or **insulators. Dielectric strength** is the maximum electrical field strength that can be withstood without breaking down (losing its electrical insulating properties) or it is defined as the voltage required per unit distance, in V/m (V/ft).

Superconductivity is the phenomenon of theoretically zero electrical resistivity that occurs in some metals and alloys below a critical temperature. The highest temperature at which superconductivity has to date been exhibited, at about −123°C (−190°F), is with an alloy of thallium, barium, calcium, copper, and oxygen; other material compositions are continuously being investigated. Developments in superconductivity are important in that the efficiency of such electrical components as large high-power magnets, high-voltage power lines, and various other electronic and computer components can be markedly improved.

Ferromagnetism is the large and permanent magnetization resulting from an alignment of magnetic moments between neighboring atoms (such as iron, nickel, and cobalt). It has important applications in electric motors, generators, transformers, and microwave devices. **Ferromagnetism** is the permanent and large magnetization exhibited by some ceramic materials, such as cubic ferrites.

The **piezoelectric** effect (*piezo* from the Greek, meaning *to press*), in which there is a reversible interaction between an elastic strain and an electric field, is exhibited by some materials, such as certain ceramics and quartz crystals. This property is utilized in making *transducers*, which are devices that convert the strain from an external force to electrical energy. Typical applications of the piezoelectric effect include force or pressure transducers, strain gages, sonar detectors, and microphones.

Magnetostriction is the phenomenon of expansion and contraction of a material when subjected to a magnetic field. Pure nickel and some iron-nickel alloys exhibit this behavior. Magnetostriction is the principle behind ultrasonic machining equipment (see Section 9.9).

3.9.7 Resistance to Corrosion

Corrosion is the deterioration of metals and ceramics; **degradation** is a similar phenomenon in plastics. *Corrosion resistance* is an important aspect of material selection, especially for applications in the chemical, food, and petroleum industries. Environmental oxidation and corrosion of components and structures also are a major concern in automobiles, aircraft, and various other transportation equipment.

Corrosion resistance depends on the composition of the materials and the particular environment. Chemicals (acids, alkali, and salts), the environment (oxygen, salts, and pollution), and water (fresh or salt) may all affect corrosion resistance. Nonferrous metals (excluding magnesium), stainless steels, and nonmetallic materials generally have high corrosion resistance. Steels and cast irons generally have poor resistance and must therefore be protected by various means, such as coatings and surface treatments (Section 4.5).

Corrosion can occur over an entire surface, or it can be localized, such as in **pitting**. It can occur along grain boundaries of metals as **intergranular corrosion** and at the interface of bolted or riveted joints as **crevice corrosion.** Two dissimilar metals may form a *galvanic cell* (two electrodes in an electrolyte in a corrosive environment, including moisture) and cause **galvanic corrosion**. Two-phase alloys (Section 5.2.3) are more susceptible to galvanic corrosion, because of the two different metals involved, than are single-phase alloys or pure metals. Heat treatment can have a significant influence on corrosion resistance.

Tool and die materials can be susceptible to chemical attack by lubricants and coolants (Section 4.4.4). The chemical reaction alters their surface finish and adversely influences the metalworking operation. An example is tools and dies made of carbides that have cobalt as a binder (Section 8.6.4); the cobalt can be attacked by elements in the metalworking fluid (see Sections 4.4.4 and 8.7), called **selective leaching**. Compatibility of the tool, die, and workpiece materials and the metalworking fluid is thus an important consideration in material selection.

Chemical reactions should not always be regarded as having only adverse effects. Some advanced machining processes, such as chemical machining and electrochemical machining, are indeed based on controlled chemical reactions (Chapter 9).

The usefulness of some level of oxidation is exhibited by the corrosion resistance of some metals. Examples include:

a. Aluminum develops a thin (a few atomic layers), strong, and adherent hard oxide film that protects the surface from further environmental corrosion.

b. Titanium develops a film of titanium oxide that protects against corrosion.
c. Stainless steels, because of the chromium present in the alloy, develop a protective film on their surfaces (see Section 4.2), known as **passivation**. If the protective film is scratched, thus exposing the metal underneath to the environment, a new oxide film forms in due time.

3.10 General Properties and Applications of Ferrous Alloys

By virtue of their very wide range of mechanical, physical, and chemical properties, **ferrous alloys** are among the most useful of all metals. Ferrous metals and alloys contain iron as their base metal and are variously categorized as carbon and alloy steels, stainless steels, tool and die steels, cast irons, and cast steels. Ferrous alloys are produced as sheet steel for automobiles, appliances, and containers; as plates for ships, boilers, and bridges; as structural members (such as I-beams); as bar products for leaf springs, gears, axles, crankshafts, and railroad rails; as stock for tools and dies; as music wire; and as fasteners, such as bolts, rivets, and nuts.

A typical US passenger car contains about 1000 kg (2200 lb) of steel, accounting for about 60% of its weight. As an example of their widespread use, ferrous materials comprise 70 to 85% by weight of virtually all structural members and mechanical components. Carbon steels are the least expensive of all metals, but stainless steels can be costly.

3.10.1 Carbon and Alloy Steels

Carbon and alloy steels are among the most widely used metals. The composition and processing of these steels are controlled in such a manner that makes them suitable for a wide variety of applications. They are available in various basic product shapes: plate, sheet, strip, bar, wire, tube, castings, and forgings.

Several elements are added to steels to impart various specific properties, such as hardenability, strength, hardness, toughness, wear resistance, workability, weldability, and machinability. Generally, the higher the percentages of these elements, the higher the particular properties that they impart to steels. Thus, for example, the higher the carbon content, the higher the hardenability of the steel and the higher its strength, hardness, and wear resistance. Conversely, however, ductility, weldability, and toughness are reduced with increasing carbon content.

Carbon steels. Carbon steels are generally classified as low, medium, and high (See Fig. 3.32):

1. **Low-carbon steel,** also called **mild steel,** has less than 0.30% C. It is generally used for common products such as bolts, nuts, sheet plates, tubes, and machine components that do not require high strength.

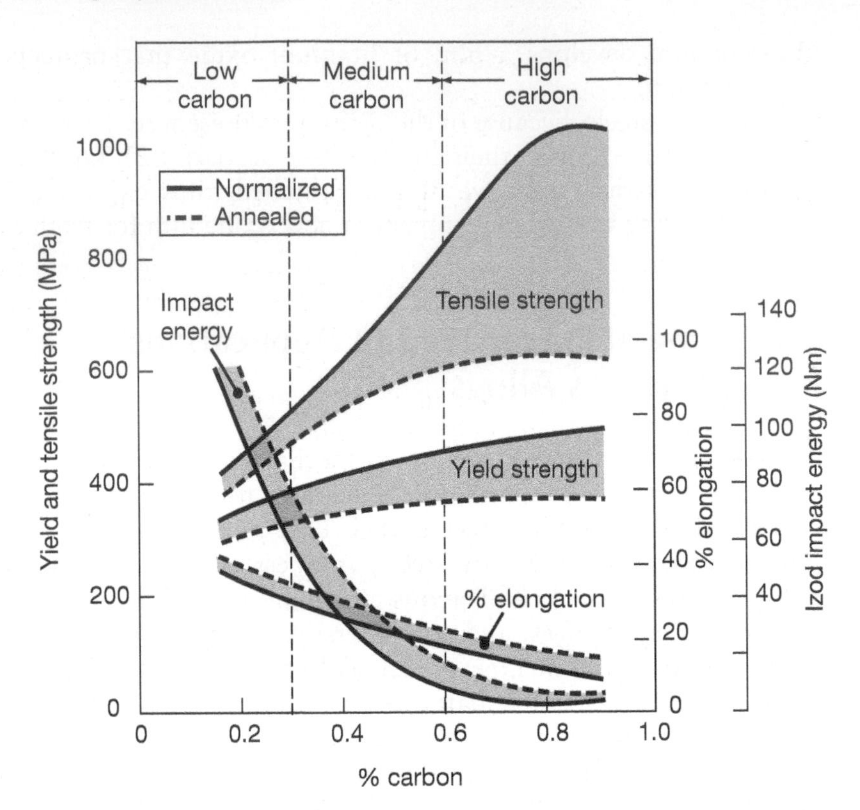

FIGURE 3.32 Effect of carbon content on the mechanical properties of carbon steel.

2. **Medium-carbon steel** has 0.30 to 0.60% C. It is generally used in applications requiring higher strength than those using low-carbon steels, such as machinery, automotive and agricultural equipment (gears, axles, connecting rods, crankshafts), railroad equipment, and metalworking machinery.
3. **High-carbon steel** has more than 0.60% C. It is generally used for parts requiring strength, hardness, and wear resistance; examples are springs, cutlery, cable, music wire, and rails. After being manufactured into appropriate shapes, the parts are usually heat treated and tempered. The higher the carbon content of the steel, the higher its hardness, strength, and wear resistance after heat treatment.
4. Carbon steels containing sulfur or phosphorus are known as **resulfurized** and **rephosphorized carbon steels**, with the major characteristic of improved machinability of the steels, as described in Section 8.5.1.

Alloy steels. Steels containing significant amounts of alloying elements are called *alloy steels*. **Structural-grade alloy steels** are used mainly in the construction and transportation industries, because of their high strength. Other types of alloy steels also are available for applications where strength, hardness, resistance to creep, fatigue, and toughness are required. These steels may also be heat treated to obtain the desired properties.

High-strength low-alloy (HSLA) steels. In order to improve the strength-to-weight ratio of steels, a number of *high-strength low-alloy* steels have

TABLE 3.3 Mechanical properties of selected advanced high-strength steels.

Steel	Minimum yield strength MPa (ksi)	Minimum ultimate strength MPa (ksi)	Elongation in 50 mm (%)	Strain hardening exponent, *n*
BH 260/370	260 (38)	370 (54)	32	0.13
HSLA 350/450	350 (50)	450 (65)	25	0.14
DP 350/600	350 (50)	600 (87)	27	0.14
DP 500/800	500 (72)	800 (116)	17	0.14
DP 700/1000	700 (101)	1000 (145)	15	0.13
TRIP 450/800	450 (65)	800 (116)	29	0.24
TRIP 400/600	400 (58)	600 (87)	30	0.23
CP 700/800	700 (101)	800 (116)	12	0.13
MART 950/1200	950 (138)	1200 (174)	6	0.07
MART 1250/1520	1250 (181)	1520 (220)	5	0.065
27MnCrB5, as rolled	478 (69)	967 (140)	12	0.06
hot stamped	1097 (159)	1350 (196)	5	0.06
37MnB4, as rolled	580 (84)	810 (117)	12	0.06
hot stamped	1378 (200)	2040 (297)	4	0.06

been developed. They have a low carbon content (usually less than 0.30%), and are characterized by a microstructure consisting of fine-grain ferrite and a hard second phase of carbides, carbonitrides, or nitrides. Mechanical properties for selected HSLA steels are given in Table 3.3; these steels have high strength and energy-absorption capabilities as compared to conventional steels.

Sheet products of HSLA steels typically are used in automobile bodies to reduce weight (and thus reduce fuel consumption), and in transportation, mining, and agricultural equipment. Plates made of HSLA steel are used in ships, bridges, and building construction, and such shapes as I-beams, channels, and angles are used in buildings and numerous other structures.

Dual-phase steels. Designated by the letter D, these steels are specially processed and have a mixed ferrite and martensite structure. Developed in the late 1960s, dual-phase steels have high work-hardening characteristics (i.e., a high *n* value; see Section 2.2.3), and thus possess good ductility and formability.

Microalloyed steels. Microalloyed steels possess superior properties and thus can often eliminate the need for heat treatment. These steels have a ferrite-pearlite microstructure with fine dispersed particles of carbonitride. A number of microalloyed steels have been developed, typically containing 0.5% C, 0.8% Mn, and 0.1% V. When subjected to carefully controlled cooling (usually in air), these materials develop improved and uniform strength. Compared to medium-carbon steels, microalloyed steels also can provide cost savings of as much as 10%, since the additional steps of quenching, tempering, and stress relieving are not required.

Nano-alloyed steels are also under continued development. These steels have extremely small grain sizes (10–100 nm), and are produced using metallic glasses (see Section 5.10.8) as a precursor. The metallic glass is subjected to a carefully controlled vitrification (crystallization) process, with a high nucleation rate, thus resulting in very fine nanoscale phases.

3.10.2 Ultra-High-Strength Steels

Ultra-high-strength steels are defined as those with an ultimate tensile strength higher than 700 MPa (100 ksi). There are five important types of ultra-high-strength steel: dual-phase, TRIP, TWIP, complex phase, and martensitic. The main application of these steels is for crashworthy design of automobiles. The use of stronger steels allows for smaller cross sections in structural components, thus resulting in weight savings and fuel economy without compromising safety. The significant drawbacks of all these steels are higher cost, tool and die wear, forming loads, and springback.

TRIP steels consist of a ferrite–bainite matrix and 5–20% retained austenite. During forming, the austenite progressively transforms into martensite. Thus, TRIP steels have both excellent ductility, because of the austenite and high strength after forming; as a result, these steels can be used to produce more complex parts than can be done using other high-strength steels.

TWIP steels (from *TW*inning-*I*nduced *P*lasticity) are austenitic and have high manganese content (17–20%). These steels derive their properties from the generation of twins during deformation (see Section 3.3), without a phase change, resulting in very high strain hardening and avoiding necking during processing. As can be seen in Fig. 3.33, TWIP steels combine high strength and high formability.

Complex-phase grades (CP grades) are very fine-grained microstructures of ferrite and a high volume fraction of hard phases (martensite and bainite). These steels can have ultimate tensile strengths as high as 800 MPa (115 ksi), and are therefore of interest for automotive crash applications, such as in bumpers and roof supports. *Martensitic grades* also are available, consisting of high fractions of martensite to attain tensile strengths as high as 1500 MPa (217 ksi).

Terminology has developed to refer to the generations of advanced high strength steels (AHSS), differentiated by their shading color, as shown in Fig. 3.33:

- Conventional high-strength steels include the traditional mild grades, bake hardenable, and HSLA grades.
- **First Generation AHSS** refer to dual phase, complex phase, TRIP, and martensitic steels.
- TWIP steels are part of a class of materials referred to as **Second Generation AHSS**.
- **Third Generation AHSS** are under development, and have started to become available. These materials combine the high strength of Second Generation AHSS with improved formability of First Generation AHSS, through careful control of microstructures and phases.

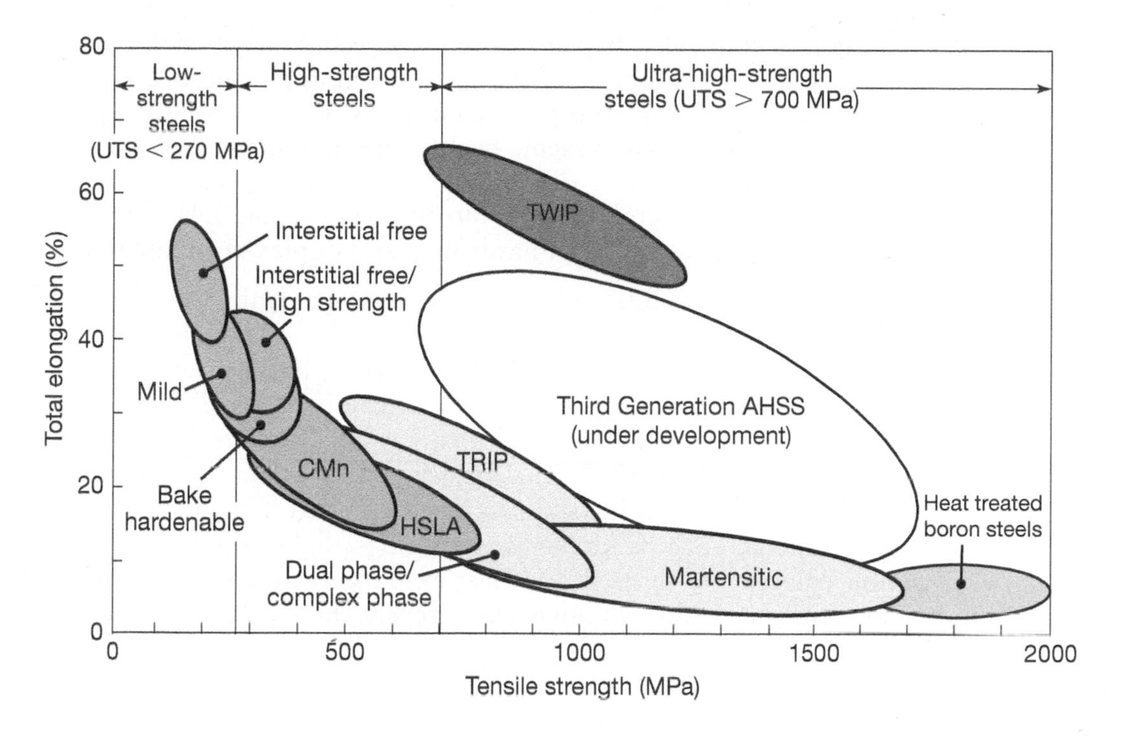

FIGURE 3.33 Comparison of advanced high-strength steels.

For example, a high strength phase such as martensite or ultra fine-grained ferrite may be mixed with a constituent that is highly formable and ductile, such as austenite.

3.10.3 Stainless Steels

Stainless steels are characterized primarily by their corrosion resistance, high strength and ductility, and high chromium content. They are called *stainless* because, in the presence of oxygen (air), they develop a thin, hard adherent film of *chromium oxide* that protects the metal from corrosion (*passivation*; see Section 4.2). This protective film builds up again if the surface is scratched. For passivation to occur, the minimum chromium content of the steel should be in the range of 10–12% by weight.

In addition to chromium, other alloying elements in stainless steels typically include nickel, molybdenum, copper, titanium, silicon, manganese, columbium, aluminum, nitrogen, and sulfur. The higher the carbon content, the lower is the corrosion resistance of stainless steels. The reason is that the carbon combines with the chromium in the steel and forms chromium carbide, which lowers the passivity of the steel. The chromium carbide introduces a second phase in the steel, which promotes galvanic corrosion. The letter *L*, indicating low, is used to identify low-carbon stainless steels.

Developed in the early 1900s, stainless steels are made by techniques basically similar to those used in other types of steelmaking (see Section 5.5), using electric furnaces or the basic-oxygen process. The level of impurities is controlled by various refining techniques. Stainless steels are available in a wide variety of shapes. Typical applications are in the chemical, food-processing, and petroleum industries, and for such products as cutlery, kitchen equipment, health care and surgical equipment, and automotive trim.

Stainless steels are generally divided into five types (Table 3.4): austenitic, ferritic, martensitic, precipitation-hardening, and duplex-structure steels.

1. **Austenitic steels** (200 and 300 series) are generally composed of chromium, nickel, and manganese in iron. They are nonmagnetic and have excellent corrosion resistance, but are susceptible to stress-corrosion cracking. These steels are the most ductile of all stainless steels and hence can be formed easily. They are hardened by cold working; however, with increasing cold work, their formability is reduced. Austenitic stainless steels are used in a wide variety of applications, such as kitchenware, fittings, lightweight transportation equipment, furnace and heat-exchanger components, welded construction, and parts subjected to severe chemical environments.
2. **Ferritic steels** (400 series) have a high chromium content: up to 27%. They are magnetic and have good corrosion resistance, but have lower ductility (hence have lower formability) than austenitic stainless steels. Ferritic stainless steels are hardened by cold working and are not heat treatable. They generally are used for nonstructural applications, such as kitchen equipment and automotive trim.
3. **Martensitic steels** (400 and 500 series) have a chromium content as high as 18% and they generally do not contain nickel. These steels

TABLE 3.4 Room-temperature mechanical properties and typical applications of annealed stainless steels.

AISI (UNS)	Yield strength MPa (ksi)	Ultimate tensile strength MPa (ksi)	Elongation (%)	Characteristics and typical applications
303 (S30300)	240–260 (35–38)	550–620 (80–90)	50–53	Screw-machine products, shafts, valves, bolts, bushings, and nuts; aircraft fittings; rivets; screws
304 (S30400)	240–290 (35–42)	565–620 (82–90)	55–60	Chemical and food-processing equipment, brewing equipment, cryogenic vessels, gutters, downspouts, and flashings
316 (S31600)	210–290 (30–42)	550–590 (80–85)	55–60	High corrosion resistance and high creep strength, Chemical and pulp-handling equipment, photo-graphic equipment, and brandy vats
410 (S41000)	240–310 (35–45)	480–520 (70–75)	25–35	Machine parts, pump shafts, bolts, bushings, coal chutes, cutlery, fishing tackle, hardware, jet engine parts, mining machinery, rifle barrels, and screws
416 (S41600)	550–585 (80–85)	620–720 (90–105)	18–15	Aircraft fittings, bolts, nuts, fire extinguisher inserts, rivets, and screws

are magnetic and have high strength, hardness, and fatigue resistance, and good ductility, but moderate corrosion resistance; they are hardenable by heat treatment. Martensitic stainless steels typically are used for cutlery, surgical tools, instruments, valves, and springs.

4. **Precipitation-hardening** (PH) **steels** contain chromium and nickel, along with copper, aluminum, titanium, or molybdenum. They have good corrosion resistance, good ductility, and high strength at elevated temperatures. Their main application is in aircraft and aerospace structural components.
5. **Duplex-structure steels** contain a mixture of austenite and ferrite. They have good strength and higher resistance to corrosion (in most environments) and to stress-corrosion cracking than the 300 series austenitic steels. Typical applications of duplex-structure steels are in water-treatment plants and heat-exchanger components.

3.10.4 Tool and Die Steels

Tool and die steels are specially alloyed steels and have high strength, impact toughness, and wear resistance at elevated temperatures. Various types of tool and die materials are widely used for a variety of important manufacturing processes.

1. **High-speed steels** (HSS) are the most highly alloyed tool and die steels, and maintain their hardness and strength at elevated operating temperatures. There are two basic types of high-speed steels: the **molybdenum type** (M series) and the **tungsten type** (T series). The M-series contain up to about 10% molybdenum, with chromium, vanadium, tungsten, and cobalt as other alloying elements. The T-series contain 12–18% tungsten, with chromium, vanadium, and cobalt as additional alloying elements. As compared with the T-series steels, the M-series steels generally have higher abrasion resistance, have less distortion during heat treatment, and are less expensive. The M-series steels constitute about 95% of all HSS produced in the United States. High-speed steel tools can be coated with titanium nitride and titanium carbide for better resistance to wear (see Section 4.4.2).
2. **Hot-work steels** (H series) are designed for use at elevated temperatures and have high toughness and high resistance to wear and cracking. The alloying elements are generally tungsten, molybdenum, chromium, and vanadium.
3. **Cold-work steels** (A, D, and O series) are used for cold-working operations. They generally have high resistance to wear and cracking. These steels are available as oil-hardening or air-hardening types.
4. **Shock-resisting steels** (S series) are designed for impact toughness; other properties depend on the particular composition. Applications for these steels include dies, punches, and chisels.

3.11 General Properties and Applications of Nonferrous Metals and Alloys

Nonferrous metals and alloys cover a very wide range of materials, from the more common metals such as aluminum, copper, and magnesium, to high-strength, high-temperature alloys such as tungsten, tantalum, and molybdenum. Although more expensive than ferrous metals, nonferrous metals and alloys have important applications, because of their wide range of mechanical, physical, and chemical properties and characteristics.

A turbofan jet engine for the Boeing 757 aircraft typically contains the following nonferrous metals and alloys: 38% titanium, 37% nickel, 12% chromium, 6% cobalt, 5% aluminum, 1% niobium (columbium), and 0.02% tantalum. Without these materials, a jet engine (Fig. 3.34) could not be designed, manufactured, and operated at the required energy and efficiency levels.

Typical examples of the applications of nonferrous metals and alloys include (a) aluminum for cooking utensils and aircraft bodies; (b) copper wire for electricity and copper tubing for water in residences; (c) titanium for jet-engine turbine blades and artificial joints; and (d) tantalum for rocket engines.

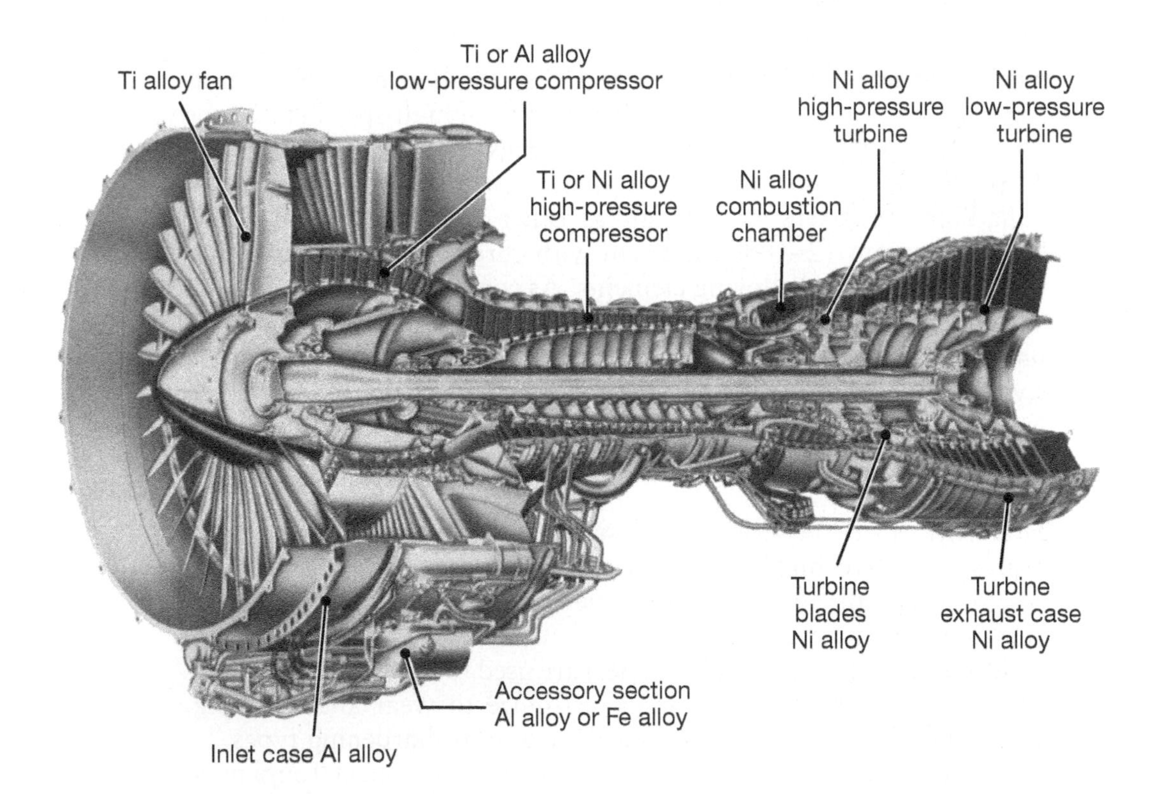

FIGURE 3.34 Cross section of a jet engine (PW2037) showing various components and the alloys used in making them. *Source:* After United Aircraft Pratt & Whitney.

3.11.1 Aluminum and Aluminum Alloys

Important factors in selecting *aluminum* (Al) and its alloys are their high strength-to-weight ratio, resistance to corrosion by many chemicals, high thermal and electrical conductivity, nontoxicity, reflectivity, appearance, and ease of formability and machinability. Also, they are nonmagnetic.

Principal uses of aluminum and its alloys include, in decreasing order of consumption, containers and packaging (aluminum beverage cans and foil), buildings and various other types of construction, transportation [aircraft and aerospace applications, buses, automobiles (See Fig. 1.6), railroad cars, and marine craft], electrical products (nonmagnetic and economical electrical conductors), consumer durables (appliances, cooking utensils, and outdoor furniture), and portable tools (Tables 3.5 and 3.6). Nearly all high-voltage transmission line wiring is made of aluminum cable (such as AISI 1350-H19), with steel reinforcement (known as ACSR cable).

The frame and the body panels of the Rolls Royce Phantom coupe are made of aluminum, improving the car's strength-to-weight and torsional rigidity-to-weight ratios. In its structural (load-bearing) components, 82% of a Boeing 747 aircraft and 70% of a Boeing 777 aircraft is aluminum. The Boeing 787 Dreamliner (first placed into service in late 2011) is well recognized for its carbon-fiber reinforced composite fuselage, although it still uses 20% aluminum by weight and 15% titanium.

As in all cases of material selection (see Section 1.5), each aluminum alloy has particular properties and its selection can be critical with respect to manufacturing characteristics and cost considerations. A typical aluminum beverage can, for example, consists of the following alloys: 3004 or 3104 for the can body, 5182 for the lid, and 5042

TABLE 3.5 Properties of various aluminum alloys at room temperature.

Alloy (UNS)	Temper	Yield strength MPa (ksi)	Ultimate tensile strength MPa (ksi)	Elongation in 50 mm (%)
1100 (A91100)	O	35 (5)	90 (13)	35–45
	H14	120 (17)	125 (18)	9–20
1350 (A91350)	O	30 (4.5)	85 (12)	23
	H19	165 (24)	185 (27)	1.5
2024 (A92024)	O	75 (11)	190 (27)	20–22
	T4	325 (47)	470 (68)	19–20
3003 (A93003)	O	40 (6)	110 (16)	30–40
	H14	145 (21)	150 (22)	8–16
5052 (A95052)	O	90 (13)	190 (27)	25–30
	H34	215 (31)	260 (38)	10–14
6061 (A96061)	O	55 (8)	125 (18)	25–30
	T6	275 (40)	310 (45)	12–17
7075 (A97075)	O	105 (15)	230 (33)	16–17
	T6	500 (72)	570 (83)	11
8090	T8X	400 (58)	480 (70)	4–5

TABLE 3.6 Manufacturing properties and typical applications of wrought aluminum alloys.

	Characteristics*			
Alloy	Corrosion resistance	Machinability	Weldability	Typical applications
1100	A	D–C	A	Sheet-metal work, spun hollow parts, tin-stock
2014	C	C–B	C–B	Heavy-duty forgings, plate and extrusions for aircraft structural components, wheels
2024	C	B–C	B–C	Truck wheels, screw machine products, aircraft structures
3003	A	D–C	A	Cooking utensils, chemical equipment, pressure vessels, sheet metal work, builders' hardware, storage tanks
5052	A	D–C	A	Sheet metal work, hydraulic tubes, and appliances; bus, truck, and marine uses
6061	B	D–C	A	Heavy-duty structural applications where corrosion resistance is needed; automotive structures; furniture, pipelines, hydraulic tubing
7005	D	B–D	B	Extruded structural members, large heat exchangers, tennis racquets, softball bats
7075	C	B–D	D	Aircraft and other structures; keys, hydraulic fittings
8090	A–B	B–D	B	Aircraft frames, helicopter structural components

*From A (excellent) to D (poor).

for the tab. (All sheet is in the H19 condition, the highest cold-worked state.) Similarly, aluminum-lithium alloys, such as 8090, are advantageous for aerospace applications because of their strength-to-weight and stiffness-to-weight ratios, but their high cost restricts their use elsewhere.

Aluminum alloys are available as mill products, that is, wrought product made into various shapes by such processes as rolling, extrusion, drawing, and forging. Aluminum ingots are available for casting, as are powder metals for powder metallurgy applications (Chapter 11). There are two types of wrought alloys of aluminum: (a) alloys that can be hardened by *cold working* (designated by the letter H) and are not heat treatable; and (b) alloys that are hardenable by *heat treatment* (designated by the letter T). The letter O indicates the annealed condition. Most aluminum alloys can be machined, formed, and welded with relative ease.

3.11.2 Magnesium and Magnesium Alloys

Magnesium (Mg) is the lightest engineering metal available; its alloys are used in structural and nonstructural applications where weight is of primary importance. Magnesium is also an alloying element in various nonferrous metals.

Typical uses of magnesium alloys include aircraft and missile components, material-handling equipment, portable power tools, luggage, bicycles, sporting goods, and general lightweight components. These alloys are available as either castings or wrought products, such as extruded bars

TABLE 3.7 Properties and typical forms of selected wrought magnesium alloys.

Alloy	Nominal composition	Condition	Yield strength MPa (ksi)	Ultimate tensile strength MPa (ksi)	Elongation in 50 mm (%)	Typical forms
AM50	4.5 AL, 0.27 Mn, 0.2 Zn	F	125 (18)	200 (29)	10	Die and sand castings
AM60	5.5 Al, 0.5 Si, 0.35 Cu	F	130 (22)	220 (32)	6	Die castings
AZ31B	3.0 Al, 1.0 Zn, 0.2 Mn	F	200 (29)	260 (38)	15	Extrusions
		H24	290 (42)	220 (32)	15	Sheet and plate
AZ80A	8.5 Al, 0.5 Zn, 0.2 Mn	T5	275 (40)	380 (55)	7	Extrusions and forgings
AZ91A	8.3 Al, 0.13 Mn, 0.35 Zn	F	150 (22)	230 (33)	3	Die castings
HK31A	0.7 Zr, 3 Th	H24	200 (29)	255 (37)	8	Sheet and plates
ZE10	1.0 Zn, 1.0 Ce	F	163 (24)	263 (38)	16	Sheet and plates
ZEK199	1.0 Zn, 0.3 Zr, 1.0 Ce	F	308 (45)	311 (45)	19	Extrusions and sheet
ZK60A	5.7 Zn, 0.55 Zr	T5	300 (44)	365 (53)	11	Extrusions and forgings

and shapes, forgings, and rolled plate and sheet. Magnesium alloys are also used in printing and textile machinery, to minimize inertial forces in high-speed components. Magnesium also has good vibration-damping characteristics.

Because it is not sufficiently strong in its pure form, magnesium is alloyed with various elements (Table 3.7) to impart certain specific properties, particularly high strength-to-weight ratios. A variety of magnesium alloys are available with good casting, forming, and machining characteristics. Magnesium alloys oxidize rapidly (they are *pyrophoric*), thus they are a potential fire hazard; precautions must be taken, especially when machining, grinding, or sand casting magnesium alloys. Products made of magnesium and its alloys are, however, not a fire hazard.

Magnesium is easy to cast but difficult to form. Efforts have been made to promote the increased use of magnesium in automobiles through improved welding and sheet formability. Alloys AM50, AM60, AZ31, and ZE10 are of high current interest.

3.11.3 Copper and Copper Alloys

First produced in about 4000 B.C., *copper* (Cu) and its alloys have properties similar to those of aluminum alloys. They are among the best conductors of electricity and heat, and also have good resistance to corrosion. They can be processed easily by various forming, machining, casting, and joining techniques.

Copper alloys often are attractive for applications where combined properties, such as electrical and mechanical properties, corrosion resistance, thermal conductivity, and wear resistance are required. Applications include electrical and electronic components, springs, cartridges for small

TABLE 3.8 Properties and typical applications of various wrought copper and brasses.

Type and UNS number	Nominal composition (%)	Yield strength MPa (ksi)	Ultimate tensile strength MPa (ksi)	Elongation in 50 mm (%)	Typical applications
Oxygen-free electronic (C10100)	99.99 Cu	70–365 (10–52)	220–450 (32–65)	55–4	Bus bars, waveguides, hollow conductors, lead in wires, coaxial cables and tubes, microwave tubes, rectifiers
Red brass (C23000)	85.0 Cu, 15.0 Zn	70–435 (10–63)	270–472 (39–68)	55–3	Weather stripping, conduit, sockets, fasteners, fire extinguishers, condenser and heat-exchanger tubing
Low Brass (C24000)	80.0 Cu, 20.0 Zn	80–450 (12–65)	300–850 (44–120)	55–3	Battery caps, bellows, musical instruments, clock dials, flexible hose
Free-cutting brass (C36000)	61.5 Cu, 3.0 Pb, 35.5 Zn	125–310 (18–45)	340–470 (50–68)	53–18	Gears, pinions, automatic high-speed screw-machine parts
Naval brass (C46400 to C46700)	60.0 Cu, 39.25 Zn, 0.75 Sn	170–455 (25–66)	380–610 (55–88)	50–17	Aircraft turnbuckle barrels, balls, bolts, marine hardware, valve stems, condenser plates

arms, plumbing, heat exchangers, and marine hardware, as well as some consumer goods, such as cooking utensils, jewelry, and other decorative objects (Table 3.8).

Copper alloys can have a wide variety of properties, and their manufacturing properties can be improved by the addition of alloying elements and by heat treatment. The most common copper alloys are brasses and bronzes; others are copper nickels and nickel silvers.

Brass is an alloy of copper and zinc; it was one of the earliest alloys developed and has numerous applications, including decorative objects. **Bronze** is an alloy of copper and tin (Table 3.9); other bronzes include (a) *aluminum bronze*, an alloy of copper and aluminum; (b) *tin bronze*; (c) *beryllium bronze* (a beryllium copper); and (d) *phosphor bronze*; the latter two have good strength and high hardness for such applications as springs and bearings.

3.11.4 Nickel and Nickel Alloys

Nickel (Ni), a silver-white metal discovered in 1751, is a major alloying element that imparts strength, toughness, and corrosion resistance to metals. It is used extensively in stainless steels and nickel-base alloys. These alloys are used for high-temperature applications, such as jet-engine components, rockets, and nuclear power plants, as well as in food-handling and chemical-processing equipment, coins, and marine applications. Because nickel is magnetic, its alloys are also used in electromagnetic applications, such as solenoids. As a metal, the principal use of nickel is in electroplating for resistance to corrosion and wear and for appearance.

Nickel alloys containing chromium, cobalt, and molybdenum have high strength and corrosion resistance at elevated temperatures. The behavior

TABLE 3.9 Properties and typical applications of various wrought bronzes.

Type and UNS number	Nominal composition (%)	Yield strength MPa (ksi)	Ultimate tensile strength MPa (ksi)	Elongation in 50 mm (%)	Typical applications
Architectural bronze (C38500)	57.0 Cu, 3.0 Pb, 40.0 Zn	140 (20)	415 (60)	30	Architectural extrusions, storefronts, thresholds, trim, butts, hinges
Phosphor bronze, 5% A (C51000)	95.0 Cu, 5.0 Sn, trace P	130–550 (19–80)	325–960 (47–140)	64–2	Bellows, clutch disks, cotter pins, diaphragms, fasteners, wire brushes, chemical hardware, textile machinery
Free-cutting phosphor bronze (C54400)	88.0 Cu, 4.0 Pb, 4.0 Zn, 4.0 Sn	130–435 (19–63)	300–520 (44–75)	50–15	Bearings, bushings, gears, pinions, shafts, thrust washers, valve parts
Low-silicon bronze, B (C65100)	98.5 Cu, 1.5 Si	100–475 (15–69)	275–655 (40–95)	55–11	Hydraulic pressure lines, bolts, marine hardware, electrical conduits, heat-exchanger tubing
Nickel-silver 65–18 (C74500)	65.0 Cu, 17.0 Zn, 18.0 Ni	390–710 (56–103)	170–620 (25–90)	45–3	Rivets, screws, zippers, camera parts, base for silver plate, nameplates, etching stock

of these alloys in machining, forming, casting, and welding can be modified by various other alloying elements. A variety of nickel alloys that have a range of strengths at different temperatures are shown in Table 3.10. *Monel* is a nickel-copper alloy, and Inconel is a nickel-chromium alloy. *Hastelloy*, a nickel-molybdenum-chromium alloy, has good corrosion resistance and high strength at elevated temperatures. *Nichrome*, an alloy of nickel, chromium, and iron, has high oxidation and electrical resistance and is commonly used for electrical-heating elements. *Invar*, an alloy of iron and nickel, has a low coefficient of thermal expansion, and has been used in precision scientific instruments and camera/optics applications (see Section 3.9.5).

3.11.5 Superalloys

Superalloys are important in high-temperature applications. They generally have good resistance to corrosion, mechanical and thermal fatigue, mechanical and thermal shock, creep, and erosion at elevated temperatures. Most superalloys have a maximum service temperature of about 1000°C (1800°F) for structural applications, but can be as high as 1200°C (2200°F) for non-load-bearing components.

Also known as **heat-resistant** or **high-temperature** alloys, major applications are in jet engines, gas turbines, reciprocating engines, rocket engines; tools and dies for hot working of metals; and in the nuclear, chemical, and petrochemical industries. They are generally identified by trade names or by special numbering systems, and are available in a variety of shapes. Three basic categories of superalloys are **iron base**, **cobalt base**, or **nickel**

TABLE 3.10 Properties and typical applications of various nickel alloys. All alloy names are trade names.

Alloy (condition)	Principal alloying elements (%)	Yield strength MPa (ksi)	Ultimate tensile strength MPa (ksi)	Elongation in 50 mm (%)	Typical applications
Nickel 200 (annealed)	None	100–275 (14–40)	380–550 (55–80)	60–40	Chemical- and food-processing industry, aerospace equipment, electronic parts
Duranickel 301 (age hardened)	4.4 Al, 0.6 Ti	900 (130)	1300 (190)	28	Springs, plastics-extrusion equipment, molds for glass
Monel R-405 (hot rolled)	30 Cu	230 (33)	525 (75)	35	Screw-machine products, water-meter parts
Monel K-500 (age hardened)	29 Cu, 3Al	750 (108)	1050 (150)	20	Pump shafts, valve stems, springs
Inconel 600 (annealed)	15 Cr, 8 Fe	210 (30)	640 (93)	48	Gas-turbine parts, heat-treating equipment, electronic parts, nuclear reactors
Hastelloy C-4 (solution treated and quenched)	16 Cr, 15 Mo	400 (58)	785 (114)	54	High-temperature stability, resistance to stress-corrosion cracking

base; they contain nickel, chromium, cobalt, and molybdenum as major alloying elements, with aluminum, tungsten, and titanium as additional elements.

1. Iron-base superalloys generally contain 32–67% iron, 15–22% chromium, and 9–38% nickel. Common alloys in this group are the *Incoloy* series.
2. Nickel-base superalloys are the most common of the superalloys and are available in a wide variety of compositions (Table 3.11). The range of nickel is from 38 to 76%; they also contain up to 27% chromium and 20% cobalt. Common alloys in this group include the *Hastelloy, Inconel, Nimonic, René, Udimet, Astroloy,* and *Waspaloy* series.
3. Cobalt-base superalloys generally contain 35–65% cobalt, 19–30% chromium, and up to 35% nickel. They are not as strong as nickel-base superalloys, but they retain their strength at higher temperatures than do nickel-base superalloys.

3.11.6 Titanium and Titanium Alloys

Titanium (Ti) was discovered in 1791, but was not commercially produced until the 1950s. Although it is relatively expensive, its high strength-to-weight ratio and its corrosion resistance at room and elevated temperatures make titanium very attractive for such applications as components for aircraft, jet engine, racing car, and marine craft; submarine hulls; chemical and petrochemical industries; and biomaterials, such as orthopedic implants (Table 3.12). Unalloyed titanium, known as

TABLE 3.11 Properties and typical applications of various nickel-base superalloys at 870°C (1600°F). All alloy names are trade names.

Alloy	Condition	Yield strength MPa (ksi)	Ultimate tensile strength MPa (ksi)	Elongation in 50 mm (%)	Typical applications
Astroloy	Wrought	690 (100)	770 (112)	25	Forgings for high-temperature applications
Hastelloy X	Wrought	180 (26)	255 (37)	50	Jet-engine sheet parts
IN-100	Cast	695 (100)	885 (128)	6	Jet-engine blades and wheels
IN-102	Wrought	200 (29)	215 (31)	110	Superheater and jet-engine parts
Inconel 625	Wrought	275 (40)	285 (41)	125	Aircraft engines and structures, chemical-processing equipment
Inconel 718	Wrought	330 (48)	340 (49)	88	Jet-engine and rocket parts
MAR-M 200	Cast	760 (110)	840 (122)	4	Jet-engine blades
MAR-M 432	Cast	605 (88)	730 (106)	8	Integrally cast turbine wheels
René 41	Wrought	550 (80)	620 (90)	19	Jet-engine parts
Udimet 700	Wrought	635 (92)	690 (100)	27	Jet-engine parts
Waspaloy	Wrought	515 (75)	525 (76)	35	Jet-engine parts

TABLE 3.12 Properties and typical applications of wrought titanium alloys.

Nominal composition (%)	UNS	Condition	Temp (°C)	Yield strength MPa (ksi)	Ultimate tensile strength MPa (ksi)	Elongation (%)	Typical applications
99.5 Ti	R50250	Annealed	25	240 (35)	330 (48)	30	Airframes; chemical, desalination,
			300	95 (14)	150 (22)	32	and marine parts; plate-type heat exchangers
5 Al, 2.5 Sn	R54520	Annealed	25	810 (117)	860 (125)	16	Aircraft-engine compressor blades
			300	450 (65)	565 (82)	18	and ducting; steam-turbine blades
6 Al, 4 V	R56400	Annealed	25	925 (134)	1000 (145)	14	Rocket motor cases; blades and
			300	650 (94)	725 (105)	14	disks for aircraft turbines and
			425	570 (83)	670 (97)	18	compressors; orthopedic implants;
			550	430 (62)	530 (77)	35	structural forgings
		Solution	25	1100 (160)	1175 (170)	10	
		+ age	300	900 (130)	980 (142)	10	
13 V, 11 Cr, 3Al	R58010	Solution	25	1210 (175)	1275 (185)	8	High-strength fasteners; aerospace
		+ age	425	830 (120)	1100 (160)	12	components; honeycomb panels

commercially pure titanium, has excellent corrosion resistance for applications where strength considerations are secondary. Aluminum, vanadium, molybdenum, manganese, and other alloying elements are added to titanium alloys to impart properties such as improved workability, strength, and hardenability. Titanium alloys are available for service at 550°C (1000°F) for long periods of time, and at up to 750°C (1400°F) for shorter periods.

The properties and manufacturing characteristics of titanium alloys are extremely sensitive to small variations in both alloying as well as residual elements. Proper control of composition and processing is thus essential, including prevention of surface contamination by hydrogen, oxygen, or nitrogen during processing, as they cause embrittlement of titanium, resulting in reduced toughness and ductility.

The body-centered cubic structure of titanium [beta-titanium, above 880°C (1600°F)] is ductile, whereas its hexagonal close-packed structure (alpha-titanium) is somewhat brittle and is susceptible to stress corrosion. A variety of other titanium structures (alpha, near alpha, alpha-beta, and beta) can be obtained by alloying and heat treating, such that the properties can be optimized for specific applications. **Titanium aluminide intermetallics** (TiAl and Ti_3Al) have higher stiffness and lower density than conventional titanium alloys, and can withstand higher temperatures.

3.11.7 Refractory Metals

Molybdenum, columbium, tungsten, and tantalum are referred to as *refractory metals*, because of their high melting point. These elements were discovered about two centuries ago and have been used as important alloying elements in steels and superalloys; however, their use as engineering metals and alloys did not begin until about the 1940s. More than most other metals and alloys, refractory metals and their alloys retain their strength at elevated temperatures. Consequently, they are of great importance and are used in rocket engines, gas turbines, and various other aerospace applications; in the electronics, nuclear power, and chemical industries; and as tool and die materials. The associated temperature range can be on the order of 1100°C to 2200°C (2000°F to 4000°F), where strength and oxidation of other materials are of major concern.

1. **Molybdenum.** *Molybdenum* (Mo), a silvery white metal, has high melting point, high modulus of elasticity, good resistance to thermal shock, and good electrical and thermal conductivity. Typical applications are in solid-propellant rockets, jet engines, honeycomb structures, electronic components, heating elements, and molds for die casting. Principal alloying elements in molybdenum are titanium and zirconium. Molybdenum, which is used in greater amounts than any other refractory metal, is also an important alloying element in cast and wrought alloys, such as steels and heat-resistant alloys, imparting strength, toughness, and corrosion resistance. A major disadvantage of molybdenum alloys is their low resistance to oxidation at temperatures above about 500°C (950°F) thus necessitating the use of protective coatings.
2. **Niobium.** *Niobium* (Nb), or *columbium* (after the mineral *columbite*), possesses good ductility and formability and has greater resistance to oxidation than do other refractory metals. With various alloying elements, niobium alloys can be produced with moderate strength and good fabrication characteristics. These alloys are

used in rockets; missiles; and nuclear, chemical, and superconductor applications. Niobium is also an alloying element in various alloys and superalloys.

3. **Tungsten.** *Tungsten* (W, from *wolframite*) was first identified in 1781, and is the most abundant of all refractory metals. Tungsten melts at 3410°C (6170°F), the highest melting temperature of any metal, and is characterized by high strength at elevated temperatures. However, it has high density, brittleness at low temperatures, and poor resistance to oxidation. Tungsten and its alloys are used for applications involving temperatures above 1650°C (3000°F), such as nozzle throat liners in missiles and in the hottest parts of jet and rocket engines, circuit breakers, welding electrodes, and spark-plug electrodes. The filament wire in incandescent light bulbs is made of pure tungsten, using powder metallurgy and wire-drawing techniques. Because of its high density, tungsten is also used in balancing weights and counterbalances in mechanical systems, including self-winding watches. Tungsten is an important element in tool and die steels, imparting strength and hardness at elevated temperatures. For example, tungsten carbide (with cobalt as a binder for the carbide particles) is one of the most important tool and die materials.
4. **Tantalum.** *Tantalum* (Ta) is characterized by its high melting point of 3000°C (5425°F), good ductility, and good resistance to corrosion; however, it has high density and poor resistance to chemicals at temperatures above 150°C (300°F). It is used extensively in electrolytic capacitors and various components in the electrical, electronic, and chemical industries, as well as for thermal applications, such as in furnaces and acid-resistant heat exchangers. A variety of tantalum-base alloys is available in various shapes for use in missiles and aircraft.

3.11.8 Other Nonferrous Metals

1. **Chromium.** Chromium is a hard and brittle metal that can be polished to a smooth, lustrous surface. The main applications of chromium are as an alloy in other metals, notably stainless steels (Section 3.10.3), high-speed steels (Section 3.10.4), and superalloys, such as Inconel (Section 3.11.4). Because of its aesthetic properties, as well as its inherent passivation abilities, chromium is commonly used as a coating, generally applied through electroplating.
2. **Beryllium.** Steel gray in color, *beryllium* (Be) has a high strength-to-weight ratio. Unalloyed beryllium is used in nuclear and X-ray applications (because of its low neutron absorption characteristics), and in rocket nozzles, space and missile structures, aircraft disc brakes, and precision instruments and mirrors. Its alloys of copper and nickel are used in such applications as springs (*beryllium-copper*), electrical contacts, and nonsparking tools for use in explosive environments, such as mines and in metal-powder production. Beryllium and its oxide are toxic and precautions must be taken in its production and processing.

3. **Zirconium.** *Zirconium* (Zr), silvery in appearance, has good strength and ductility at elevated temperatures, and has good corrosion resistance, because of an adherent oxide film on its surfaces. The element is used in electronic components and nuclear power reactor applications.
4. **Low-melting-point metals.** The major metals in this category are lead, zinc, and tin.
   a. **Lead.** *Lead* (Pb, after *plumbum*) has high density, good resistance to corrosion (by virtue of the stable lead-oxide layer that forms and protects its surfaces), very low hardness, low strength, high ductility, and good workability. Alloying with various elements, such as antimony and tin, enhances lead's properties, making it suitable for such applications as bearing alloys, cable sheathing, and lead-acid storage batteries. Lead is also used for damping sound and vibrations, radiation shielding against X-rays, weights and counterbalances, and in the chemical industries. The oldest lead artifacts were made around 3000 B.C. Lead is also an alloying element in solders, steels, and copper alloys and promotes corrosion resistance and machinability. Because of its toxicity, however, environmental contamination by lead continues to be a major concern, and it is being phased out in several countries (see, for example, *lead-free solders*, Section 12.14.3).
   b. **Zinc.** *Zinc* (Zn), which has a bluish-white color, is the fourth most utilized metal in industry, after iron, aluminum, and copper. Zinc has two major uses: (a) for galvanizing iron, steel sheet, and wire; and (b) as an alloy base for casting. In **galvanizing**, zinc serves as the anode and protects the steel (cathode) from corrosive attack, should the coating be scratched or punctured. Zinc is also used as an alloying element; brass, for example, is an alloy of copper and zinc.

      Major alloying elements in zinc are aluminum, copper, and magnesium; they impart strength and improve castability. Zinc-base alloys are used extensively in die casting for making such products as fuel pumps and grills for automobiles, components for household appliances (vacuum cleaners, washing machines, and kitchen equipment), machine parts, and photoengraving plates. Another use for zinc is in superplastic alloys (see Section 2.2.7), which have good formability characteristics by virtue of their capacity to undergo large deformation without failure. Very fine-grained 78% Zn-22% Al sheet is a common example of a superplastic zinc alloy that can be formed with methods commonly used for forming plastics or metals.
   c. **Tin.** Although used in small amounts, *tin* (Sn, after *stannum*), a silvery-white lustrous metal, is an important metal. Its most extensive use is as a protective coating on steel sheet (**tin plate**), used for making containers (tin cans) for food and various other products. Inside a sealed can, the steel is cathodic and

is protected by the tin (anode), so that the steel does not corrode. The low shear strength of the tin coatings on steel sheet also improves its performance in deep drawing and general pressworking operations.

Unalloyed tin is used in such applications as lining material for water-distillation plants and as a molten metal over which plate glass is made (see Section 11.11). Also called **white metals,** tin-base alloys generally contain copper, antimony, and lead, imparting hardness, strength, and corrosion resistance. Organ pipes, for example, are made of tin alloys. Because of their low friction coefficients, tin alloys are used as journal-bearing materials. These alloys, known as **babbitts**, consist of tin, copper, and antimony. **Pewter** is another alloy of this class. Developed in the 15th century, it is used for tableware, hollowware, and decorative artifacts. Tin is also an alloying element for type metals; dental alloys; and bronze (copper-tin alloy), titanium, and zirconium alloys. Tin-lead alloys are common soldering materials, with a wide range of compositions and melting points (see Section 12.14.3).

5. **Precious metals.** Also called *noble metals*, gold, silver, and platinum are the most important precious metals.
   a. **Gold** (Au, after *aurum*) is soft and ductile and has good corrosion resistance at any temperature. Typical applications include electric contact and terminals, jewelry, coinage, reflectors, gold leaf for decorative purposes, and dental work.
   b. **Silver** (Ag, after *argentum*) is a ductile metal and has the highest electrical and thermal conductivity of any metal; however, it develops an oxide film that adversely affects its surface appearance and properties. Typical applications include electrical contacts, solders, bearings, food and chemical equipment, tableware, jewelry, and coinage. *Sterling silver* is an alloy of silver and 7.5% copper, used for tableware and jewelry.
   c. **Platinum** (Pt) is a grayish-white, soft and ductile metal that has good corrosion resistance, even at elevated temperatures. Platinum alloys are used as electrical contacts, spark-plug electrodes, catalysts for automobile pollution-control devices, filaments, nozzles, dies for extruding glass fibers, thermocouples, the electrochemical industry, jewelry, and dental work.

3.11.9 Special Metals and Alloys

1. **Shape-memory alloys.** *Shape-memory alloys*, after being plastically deformed at room temperature into various shapes, return to their original shapes upon heating. For example, a piece of straight wire made of these alloys can be wound into a helical spring; when heated with a match, the spring uncoils and returns to its original straight shape. A typical shape-memory alloy consists of 55% Ni and 45% Ti; other alloys include copper-aluminum-nickel,

copper-zinc-aluminum, iron-manganese-silicon, and nickel-titanium. These alloys generally have good ductility, corrosion resistance, and high electrical conductivity.

The behavior of shape-memory alloys can be reversible, that is, the shape can switch back and forth repeatedly upon periodic application and removal of heat. Typical applications include temperature sensors, clamps, connectors, fasteners, and seals that are easy to install.

2. **Amorphous alloys.** *Amorphous alloys* are a class of metal alloys that, unlike ordinary metals, do not have a long-range crystalline structure (see Section 5.10.8). They have no grain boundaries, and the atoms are randomly and tightly packed. Because their structure resembles that of glasses (Section 11.10), these alloys are also called **metallic glasses**. Amorphous alloys are now available in bulk quantities, as well as wire, ribbon, strip, and powder, and continue to be investigated as an important and emerging material system.

   Amorphous alloys typically consist of iron, nickel, and chromium, alloyed with carbon, phosphorus, boron, aluminum, and silicon. These alloys exhibit excellent corrosion resistance, good ductility, and high strength. They also undergo very low loss from magnetic hysteresis, making them suitable for magnetic steel cores for transformers, generators, motors, lamp ballasts, magnetic amplifiers, and linear accelerators.

   The amorphous structure is obtained by extremely rapid cooling of the molten alloy. One such method is called *splat cooling* or *melt spinning* (See Fig. 5.29), in which the alloy is propelled at a very high speed against a rotating metal disk. Because the rate of cooling is on the order of 10^6 to 10^8 K/s, the molten alloy does not have sufficient time to crystallize. If, however, an amorphous alloy's temperature is raised and then cooled, the alloy develops a crystalline structure.

3. **Nanomaterials.** First investigated in the early 1980s and generally called nanomaterials, these materials have certain properties that are often superior to those of traditional materials. These characteristics include strength, hardness, ductility, wear resistance and corrosion resistance suitable for structural (load-bearing) and nonstructural applications, in combination with unique electrical, magnetic, and optical properties. Applications for nanomaterials include cutting tools, metal powders, computer chips, flat-panel displays for laptops, sensors, and various electrical and magnetic components (see also Sections 8.6.10, 11.8.1, and 13.18).

   Nanomaterials are available in granular form, fibers, films, and composites (containing particles that are on the order of 1 to 100 nm in size). The composition of nanomaterials may consist of any combination of chemical elements; among the more important compositions are carbides, oxides, nitrides, metals and alloys, organic polymers, and various composites. Methods for producing nanomaterials include inert-gas condensation, plasma synthesis, electrodeposition, sol-gel synthesis, and mechanical alloying or ball milling.

4. **Metal foams.** In metal foams, usually of aluminum alloys but also of titanium or tantalum, the metal itself consists of only 5–20% of the structure's volume. The foams can be produced by blowing air into molten metal and tapping the froth that forms at the surface, which solidifies into a foam. Alternative methods include: (a) chemical vapor deposition onto a polymer or carbon foam lattice; (b) slip casting metal powders onto a polymer foam; and (c) doping molten or powder metals with titanium hydride, which then releases hydrogen gas at the elevated casting or sintering temperatures. Metal foams have unique combinations of strength-to-density and stiffness-to-density ratios. Although these ratios are not as high as that of the base metal itself, the foams are very lightweight, thus making them an attractive material for aerospace applications; other applications include filters, lightweight beams, and orthopedic implants.
5. **Rare-earth metals.** Rare-earth metals are so named because they are generally difficult to mine and hence are available in small quantities. However, they are often more common in the Earth's crust than precious metals such as gold and platinum. Because of their unique magnetic, luminescent, and electrical properties, rare-earth metals are essential in many modern technologies. Among the more important rare-earth metals are the following:

   a. **Yttrium** is used in compact fluorescent lamps, light-emitting diodes, flat-panel monitors, laser technology, and superconductor applications.
   b. **Lanthanum** is used in catalytic converters and anode materials in high-performance batteries. Most hybrid automobiles depend on lanthanum anodes in their batteries; for example, it is estimated that each Toyota Prius uses 10–15 kg of lanthanum.
   c. **Dysprosium** is used in the production of lasers and commercial lighting, as well as in dosimeters to measure radiation exposure. When exposed to radiation, dysprosium emits light, which can be measured; it is also highly *magnetostrictive*, deforming under a magnetic field, thus making it useful for transducers and resonators.
   d. **Neodymium** is more common than cobalt, nickel or copper, and is the second-most common rare-earth metal (after cerium). Its alloy $Nd_2Fe_{14}B$ produces the strongest permanent magnets known and, as such, is used where small but powerful magnets are required, such as in-ear headphones, microphones, and hard disks.
   e. **Cerium** is mainly used in catalytic converters in automobiles for the oxidation of carbon monoxide and nitrous oxide, although it is also used in glass manufacture, permanent magnets, fuel cells, and in polishing optical components.
   f. **Samarium** is used in compounds of cobalt (usually $SmCo_5$ or $SmCo_{17}$) and is the second-strongest permanent magnet known,

next to neodymium magnets. However, samarium-cobalt magnets have better stability and can be used to temperatures as high as 700°C, whereas neodymium magnets are limited to 300°C or so.

g. **Terbium** is used as a dopant in solid-state electronic devices and in various sensors. It has the highest magnetostriction of any alloy.

Various other rare-earth metals have been used in industrial applications, usually for magnets (praseodymium, holmium), computer and portable electronic device displays (scandium, europium), and in radiation shielding (gadolinium, erbium) or generation (thulium).

SUMMARY

- Manufacturing properties of metals and alloys depend largely on their mechanical and physical properties. These properties, in turn, are governed mainly by their crystal structure, grain boundaries, grain size, texture, and various imperfections. (Sections 3.1, 3.2)
- Dislocations are responsible for the lower shear stress required to cause slip for plastic deformation to occur. However, when dislocations become entangled with one another or are impeded by barriers such as grain boundaries, impurities, and inclusions, the shear stress required to cause slip increases; this phenomenon is called work hardening, or strain hardening. (Section 3.3)
- Grain size and the nature of grain boundaries significantly affect properties, including strength, ductility, and hardness, as well as increase the tendency for embrittlement with reduced toughness. Grain boundaries enhance strain hardening by interfering with the movement of dislocations. (Section 3.4)
- Plastic deformation at room temperature (cold working) of polycrystalline materials results in higher strength of the material as dislocations become entangled. The deformation also generally causes anisotropy, whereby the mechanical properties are different in different directions. (Section 3.5)
- The effects of cold working can be reversed by heating the metal within a specific temperature range and for a specific period of time, through the sequential processes of recovery, recrystallization, and grain growth. (Section 3.6)
- Metals and alloys can be worked at room, warm, or high temperatures. Their overall behavior, force and energy requirements during processing, and their workability depend largely on whether the working temperature is below or above their recrystallization temperature. (Section 3.7)
- Failure and fracture of a workpiece when subjected to deformation during metalworking operations is an important consideration. Two types

of fracture are ductile fracture and brittle fracture. Ductile fracture is characterized by plastic deformation preceding fracture and requires a considerable amount of energy. Because it is not preceded by plastic deformation, brittle fracture can be catastrophic and requires much less energy than does ductile fracture. Impurities and inclusions, as well as factors such as the environment, strain rate, and state of stress, can play a major role in the fracture behavior of metals and alloys. (Section 3.8)

- Physical and chemical properties of metals and alloys can significantly affect design considerations, service requirements, compatibility with other materials (including tools and dies), and the behavior of the metals and alloys during processing. (Section 3.9)
- An extensive variety of metals and alloys is available, with a wide range of properties, such as strength, toughness, hardness, ductility, creep, and resistance to high temperatures and oxidation. Generally, these materials can be classified as (a) ferrous metals and alloys; (b) nonferrous metals and alloys; (c) superalloys; (d) refractory metals and their alloys; and (e) various other types, including amorphous alloys, shape-memory alloys, nanomaterials, and metal foams. (Sections 3.10, 3.11)
- Rare-earth metals are not actually rare, but are difficult to mine in large amounts. These useful metals have widespread applications, including permanent magnets and devices that exploit them, computer and portable electronic displays, batteries and fuel cells, and radiation shielding and generation. (Section 3.11)

SUMMARY OF EQUATIONS

Theoretical shear strength of metals, $\tau_{max} = \frac{G}{2}\pi$

Theoretical tensile strength of metals, $\sigma_{max} = \sqrt{\frac{E\gamma}{a}} \simeq \frac{E}{10}$

Hall–Petch equation, $Y = Y_i + kd^{-1/2}$

ASTM grain-size number, $N = 2^{n-1}$

Fracture strength vs. crack length, $S_f \propto \frac{1}{\sqrt{\text{Crack length}}}$

BIBLIOGRAPHY

Ashby, M.F., *Materials Selection in Mechanical Design*, 4th ed., Pergamon, 2010.

Ashby, M.F., and Jones, D.R.H., *Engineering Materials*, Vol. 1, *An Introduction to Their Properties and Applications*, 3rd ed., Pergamon, 2005; Vol. 2, *An Introduction to Microstructures, Processing and Design*, Pergamon, 2005.

ASM Handbook, various volumes ASM International.

ASM Specialty Handbooks, various volumes. ASM International.

Bhadeshia, H., and Honeycombe, R., *Steels: Microstructure and Properties*, 3rd ed., Butterworth-Heinemann, 2006.

Brandt, D.A., and Warner, J.C., *Metallurgy Fundamentals*, 5th ed., Goodheart-Wilcox, 2009.

Bryson, W.E., *Heat Treatment, Selection and Application of Tool Steels*, 2nd ed., Hanser Gardner, 2005.

Budinski, K.G., *Engineering Materials: Properties and Selection*, 9th ed., Prentice Hall, 2009.

Callister, W.D., Jr., and Rethwisch, D.G., *Materials Science and Engineering*, 9th ed., Wiley, 2013.

Dieter, G. E., and Schmidt, L. *Engineering Design: A Materials and Processing Approach*, 5th ed., McGraw-Hill, 2012.

Donachie, M.J. (ed.), *Titanium: A Technical Guide*, 2nd ed., ASM International, 2000.

Donachie, M.J., and Donachie, S.J., *Superalloys: A Technical Guide*, 2nd ed., ASM International, 2002.

Geddes, B., Leon, H., and Huang, X., *Superalloys: Alloying and Performance*, ASM International, 2010.

Hertzberg, R.W., and Vinci, R.P., *Deformation and Fracture Mechanics of Engineering Materials*, 5th ed., Wiley, 2012.

Hosford, W.F., *Physical Metallurgy*, 2nd ed., Taylor & Francis, 2010.

Kaufman, J.G., *Introduction to Aluminum Alloys and Tempers*, ASM International, 2000.

Krauss, G., *Steels: Processing, Structure, and Performance*, ASM International, 2nd ed., 2015.

Lagoudas, D.C. (ed.), *Shape Memory Alloys: Modeling and Engineering Applications*, Springer, 2008.

Leo, D.J., *Engineering Analysis of Smart Material Systems*, Wiley, 2007.

Liu, A.F., *Mechanics and Mechanisms of Fracture: An Introduction*, ASM International, 2005.

Lutjering, G., and Williams, J.C., *Titanium*, 2nd ed., Springer, 2007.

McGuire, M.F., *Stainless Steels for Design Engineers*, ASM International, 2008.

Reed, C., *The Superalloys: Fundamentals and Applications*, Cambridge University Press, 2008.

Revie, R.W. (ed.), *Uhlig's Corrosion Handbook*, 3rd ed., Wiley-Interscience, 2011.

Russel, A., and Lee, K.L., *Structure-Property Relations in Nonferrous Metals*, Wiley-Interscience, 2005.

Schaffer, J., Saxena, A., Antalovich, S., Sanders, T., and Warner, S., *The Science & Design of Engineering Materials*, 2nd ed., McGraw-Hill, 1999.

Schwartz, M., *Smart Materials*, CRC Press, 2008.

Shackelford, J. F., *Introduction to Materials Science for Engineers*, 8th ed., Macmillan, 2014.

Thermal Properties of Metals, ASM International, 2002.

Woldman's Engineering Alloys, 9th ed., ASM International, 2000.

QUESTIONS

3.1 What is the difference between a unit cell and a single crystal?

3.2 Explain why it is useful to study the crystal structure of metals.

3.3 What effects does recrystallization have on the properties of metals?

3.4 What is the significance of a slip system?

3.5 Explain what is meant by structure-sensitive and structure-insensitive properties of metals.

3.6 What is the relationship between nucleation rate and the number of grains per unit volume of a metal?

3.7 Explain the difference between recovery and recrystallization.

3.8 Is it possible for two pieces of the same metal to have different recrystallization temperatures? Is it possible for recrystallization to take place in some regions of a workpiece before other regions in the same workpiece? Explain your answers.

3.9 Describe why different crystal structures exhibit different strengths and ductilities.

3.10 Explain the difference between preferred orientation and mechanical fibering.

3.11 Name some analogies to mechanical fibering (e.g., layers of thin dough sprinkled with flour).

3.12 A cold-worked piece of metal has been recrystallized. When tested, it is found to be anisotropic. Explain the probable reason for its anisotropy.

3.13 Does recrystallization eliminate mechanical fibering in a workpiece? Explain.

3.14 Explain why the orange-peel effect may be a concern for metal surfaces.

3.15 How can one tell the difference between two parts made of the same metal, one shaped by cold working and the other by hot working? Explain the differences that might be observed. Note that there are several methods that can be used to determine the differences between the two parts.

3.16 Explain why the strength of a polycrystalline metal at room temperature decreases as its grain size increases.

3.17 Explain why steel is so commonly used in engineering applications.

3.18 What is the significance of metals such as lead and tin having recrystallization temperatures at about room temperature?

3.19 You are given a deck of playing cards held together by a rubber band. Which of the material-behavior phenomena described in this chapter could you demonstrate? What are the effects of increasing the number of rubber bands holding the cards together? Explain. (*Hint:* Inspect Figs. 3.5 and 3.7.)

3.20 Using the information given in Chapters 2 and 3, describe the conditions that induce brittle fracture in an otherwise ductile piece of metal.

3.21 Make a list of metals that would be suitable for a (a) paper clip; (b) bicycle frame; (c) razor blade; (d) food container; and (e) oil well drill bit.

3.22 Explain the advantages and limitations of cold, warm, and hot working of metals, respectively.

3.23 Explain why parts may crack when suddenly subjected to extremes of temperature.

3.24 From your own experience and observations, list three applications for each of the following metals and their alloys: (a) steel; (b) aluminum; (c) copper; (d) magnesium; and (e) titanium.

3.25 From your own experience and observations, list three applications that are not suitable for each of the following metals and their alloys: (a) steel; (b) aluminum; (c) copper; (d) magnesium; and (e) titanium.

3.26 Name products that would not have been developed to their advanced stages as we find them today if alloys with high strength and corrosion and creep resistance at elevated temperatures had not been developed.

3.27 Inspect several metal products and components and make an educated guess as to what materials they are made from. Give reasons for your guess. If you list two or more possibilities, explain your reasoning.

3.28 List three engineering applications for which the following physical properties would be desirable: (a) high density; (b) low melting point; and (c) high thermal conductivity.

3.29 Two physical properties that have a major influence on the cracking of workpieces or dies during thermal cycling are thermal conductivity and thermal expansion. Explain why.

3.30 Describe the advantages of nanomaterials over traditional materials.

3.31 Aluminum has been cited as a possible substitute material for steel in automobiles. What concerns, if any, would you have before purchasing an aluminum automobile?

3.32 Lead shot is popular among sportsmen for hunting, but birds commonly ingest the pellets (along with gravel) to help digest food. What substitute materials would you recommend for lead, and why?

3.33 What are metallic glasses? Why is the word "glass" used for these materials?

3.34 Which of the materials described in this chapter has the highest (a) density; (b) electrical conductivity; (c) thermal conductivity; (d) strength; and (e) cost?

3.35 What is twinning? How does it differ from slip?

3.36 What properties of titanium make it attractive for use in race-car and jet-engine components? Why is titanium not used widely for engine components in passenger cars?

3.37 How do stainless steels become stainless?

3.38 What are the main applications of rare-earth metals?

3.39 What property of neodymium makes it most suitable for in-ear headphones?

3.40 Identify several different products that are made of stainless steel, and explain why they are made of that material.

3.41 Explain why TRIP and TWIP steels have been developed.

3.42 Aluminum has a low density and high strength-to-weight ratio, and these are the main benefits of making cars from aluminum alloys. However, the average amount of steel in cars has increased in the past decade. List reasons to explain these two observations.

PROBLEMS

3.43 Calculate the theoretical (a) shear strength and (b) tensile strength for aluminum, plain-carbon steel, and tungsten. Estimate the ratios of their theoretical strength to actual strength.

3.44 A technician determines that the grain size of a certain etched specimen is 4. Upon further checking, it is found that the magnification used was 150, instead of 100, the latter of which is required by ASTM standards. What is the correct grain size?

3.45 Estimate the number of grains in a regular paper clip if its ASTM grain size is 7.

3.46 The natural frequency f of a cantilever beam is given by the expression

$$f = 0.56\sqrt{\frac{EIg}{wL^4}},$$

where E is the modulus of elasticity, I is the moment of inertia, g is the gravitational constant, w is the weight of the beam per unit length, and L is the length of the beam. How does the natural frequency of the beam change, if at all, as its temperature is increased?

3.47 A strip of metal is reduced in thickness by cold working from 30 mm to 15 mm. A similar strip is reduced from 25 mm to 10 mm. Which one of these strips will recrystallize at a lower temperature? Why?

3.48 A 2-m long, simply supported beam with a round cross section is subjected to a load of 25 kg at its center. (a) If the shaft is made from AISI 303 steel and has a diameter of 20 mm, what is the deflection under the load? (b) For shafts made from 2024-T4 aluminum, architectural bronze, and 99.5% titanium, respectively, what must the diameter of the shaft be for the shaft to have the same deflection as in part (a)?

3.49 If the diameter of the aluminum atom is 0.21 nm, estimate the number of atoms in a grain with an ASTM grain size of 5.

3.50 Plot the following for the materials described in this chapter: (a) yield strength versus density; (b) modulus of elasticity versus strength; and (c) modulus of elasticity versus relative cost.

3.51 The following data is obtained in tension tests of brass:

Grain size (μm)	Yield strength (MPa)
15	150
20	140
50	105
75	90
100	75

Does this material follow the Hall–Petch effect? If so, what is the value of k?

3.52 It can be shown that thermal distortion in precision devices is low for high values of thermal conductivity divided by the thermal expansion coefficient. Rank the materials in the inside front cover according to their suitability to resist thermal distortion.

3.53 An ISO equivalent to Eq. (3.9) is given by

$$N' = 8\left(2^{G_m}\right)$$

where G_m is the metric grain-size number and N' is the number of grains per square millimeter at 1× magnification. Can the ASTM grain-size number and the metric grain-size number be used interchangeably? Explain your answer.

3.54 Using strength and density data, determine the minimum weight of a 1-m. long tension member that must support a load of 4 kN, manufactured from (a) annealed 303 stainless steel; (b) normalized 8620 steel; (c) as-rolled 1080 steel; (d) 5052-O aluminum alloy; (e) AZ31B-F magnesium; and (f) pure copper.

3.55 Assume that you are asked to give a quiz to students on the contents of this chapter. Prepare three quantitative problems and three qualitative questions, and supply the answers.

Tribology, Metrology, and Product Quality

CHAPTER

EXAMPLES:

Several important considerations in materials processing are described in this chapter:

- Surface structures, texture, and surface properties as they affect processing of materials.
- The role of friction, wear, and lubrication (tribology) in manufacturing processes, and the characteristics of various metalworking fluids.
- Surface treatments to enhance the appearance and performance of manufactured products.
- Engineering metrology, instrumentation, and dimensional tolerances, and how they affect product quality and performance.
- Destructive and nondestructive inspection methods for manufactured parts.
- Statistical techniques for quality assurance of products.

Symbols

A_2 control chart constant
c electroplating constant, m^3/A-s
d_2 control chart constant
D_3 control chart constant
D_4 control chart constant
F friction force, N
I current, A
L sliding distance, m
l length of profile, m
k wear coefficient
m friction factor
N normal force between surfaces, N
n number of surface measures
p indentation hardness, N/m^2
R range
$\overline{R}$ average of range values
R_a arithmetic mean roughness, m
R_q root-mean-square roughness, m
S_{sy} shear strength, N/m^2
t time, s
$\overline{x}$ average
$\overline{\overline{x}}$ average of averages
V volume, m^3
y roughness height measure, m
μ coefficient of friction
σ normal stress, N/m^2 also, standard deviation
τ shear stress, N/m^2
τ_i interfacial shear stress, N/m^2

4.1 Introduction

The various mechanical, physical, thermal, and chemical effects induced during processing of materials have an important influence on the **surface** of a manufactured part. A surface generally has properties and behavior that can be considerably different from those of the part's interior. Although the **bulk material** generally determines a component's overall mechanical properties, the surfaces directly influence several important properties and characteristics of the manufactured part, including:

- Friction and wear properties of the part during subsequent processing, when it comes into direct contact with tools, dies, and molds, or during its service;
- Effectiveness of lubricants during manufacturing operations and during service;
- Appearance of the part;
- Impact of a surface on subsequent operations, such as painting, coating, welding, soldering, and adhesive bonding, as well as corrosion resistance;
- Initiation of cracks due to surface defects, such as roughness, scratches, seams, and heat-affected zones, which could lead to weakening and premature failure of the part by fatigue or other fracture mechanisms; and
- Thermal and electrical conductivity of contacting bodies; for example, a rough surface will have higher thermal and electrical resistance than a smooth surface, because of the fewer direct contact points between the two surfaces.

Tribology is the study of rubbing surfaces, and includes friction, wear, and lubrication and other *surface phenomena*. Consequently, (a) friction influences force and energy requirements and surface quality of the parts produced; (b) wear alters the surface geometry of tools and dies, which, in turn, adversely affects the quality of manufactured products and the economics of production; and (c) lubrication is an integral aspect of all manufacturing operations, as well as in the proper functioning of machinery and equipment.

The properties and characteristics of surfaces can be modified and improved by various **surface treatments**. Several mechanical, thermal, electrical, and chemical methods can be employed to improve tribological performance, including surface finish and appearance.

4.2 Surface Structure and Properties

Surfaces have considerable complexity. Referring to Fig. 4.1, the portion of the metal in the interior is called the **metal substrate** (the **bulk metal**); its structure depends on the composition and processing history of the metal.

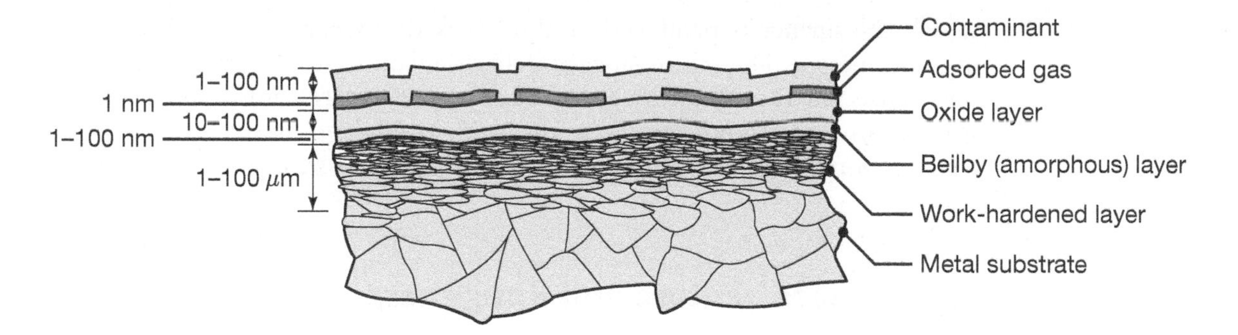

FIGURE 4.1 Schematic illustration of the cross section of the surface structure of metals. The thickness of the individual layers depends on processing conditions and the environment.

Above this bulk metal is a layer that usually has been plastically deformed and work hardened during processing. The depth and properties of the work-hardened layer (called the **surface structure**) depend on such factors as the processing history and the extent of frictional sliding to which the surface was subjected. If the surface was produced by machining with a dull tool or under poor processing conditions, or was ground with a dull grinding wheel, this layer will be relatively thick. Also, during processing, **residual stresses** in this work-hardened layer may develop, due to (a) nonuniform surface deformation; or (b) severe temperature gradients.

Additionally, a **Beilby**, or **amorphous, layer**, may develop on top of the work-hardened layer. Consisting of a structure that is microcrystalline or amorphous, this layer develops in some machining and surface finishing operations, where melting and surface flow, followed by rapid quenching, may have been encountered. Unless the metal is processed and kept in an inert (oxygen-free) environment, or it is a noble metal, an **oxide layer** usually develops on top of the work-hardened or amorphous layer.

Under normal environmental conditions, surface oxide layers are generally covered with *adsorbed* layers of gas and moisture. The outermost layer of the metal may be covered with *contaminants*, such as dirt, dust, grease, lubricants, cleaning-compound residues, and pollutants from the environment. It is thus apparent that surfaces generally have properties that are significantly different from those of the substrate.

Surface integrity. *Surface integrity* describes not only the geometric (topological) features of surfaces but also their mechanical and metallurgical properties and various characteristics. Surface integrity is an important consideration, because it can influence a product's fatigue strength, resistance to corrosion, and service life.

Several **defects**, produced during processing, can be responsible for lack of surface integrity. They are usually caused by a combination of such factors as (a) defects in the original material; (b) the method by which the surface is produced; and (c) lack of proper control of processing parameters. The major surface defects encountered in practice usually consist of one or more of the following: *cracks, craters, folds, laps, seams, splatter, inclusions, intergranular attack, heat-affected zones, metallurgical transformations, plastic deformation,* and *residual stresses.*

4.3 Surface Texture and Roughness

Regardless of the method of production, all surfaces have their own characteristics, generally referred to as **surface texture**. Although the description of surface geometry can be complex, certain guidelines have been established for identifying surface texture in terms of well-defined and measurable quantities (Fig. 4.2):

1. **Flaws**, or **defects**, are random irregularities, such as scratches, cracks, cavities, depressions, seams, tears, and inclusions.
2. **Lay**, or **directionality**, is the direction of the predominant surface pattern, usually visible to the naked eye.
3. **Waviness** is a recurrent deviation from a flat surface, much like waves on the surface of water. It is described and measured in terms of the space between adjacent crests of the waves (*waviness width*) and the height between the crests and valleys of the waves (*waviness height*). Waviness may be caused by (a) deflections of tools, dies, and of the workpiece; (b) warping from forces or temperature; (c) uneven lubrication; and (d) vibration or any periodic mechanical or thermal variations in the system during the manufacturing operation.
4. **Roughness** consists of closely-spaced irregular deviations, on a scale smaller than that for waviness. Roughness, which may be superimposed on waviness, is expressed in terms of its height, its width, and the distance on the surface along which it is measured.

Surface roughness. *Surface roughness* is generally described by two methods:

1. The **arithmetic mean value,** R_a, is based on the schematic illustration of a rough surface, as shown in Fig. 4.3; it is defined as

$$R_a = \frac{y_a + y_b + y_c + \cdots + y_n}{n} = \frac{1}{n}\sum_{i=1}^{n} y_i = \frac{1}{l}\int_0^l |y|\, dx, \qquad (4.1)$$

where all ordinates $y_a, y_b, y_c, \cdots$ are absolute values. The last term in Eq. (4.1) refers to the R_a of a continuous surface or wave, as is commonly encountered in analog signal processing, and l is the total profile length measured. Typically, 8000 height measurements are taken over one millimeter in length, depending on the characteristics of the surface being measured.

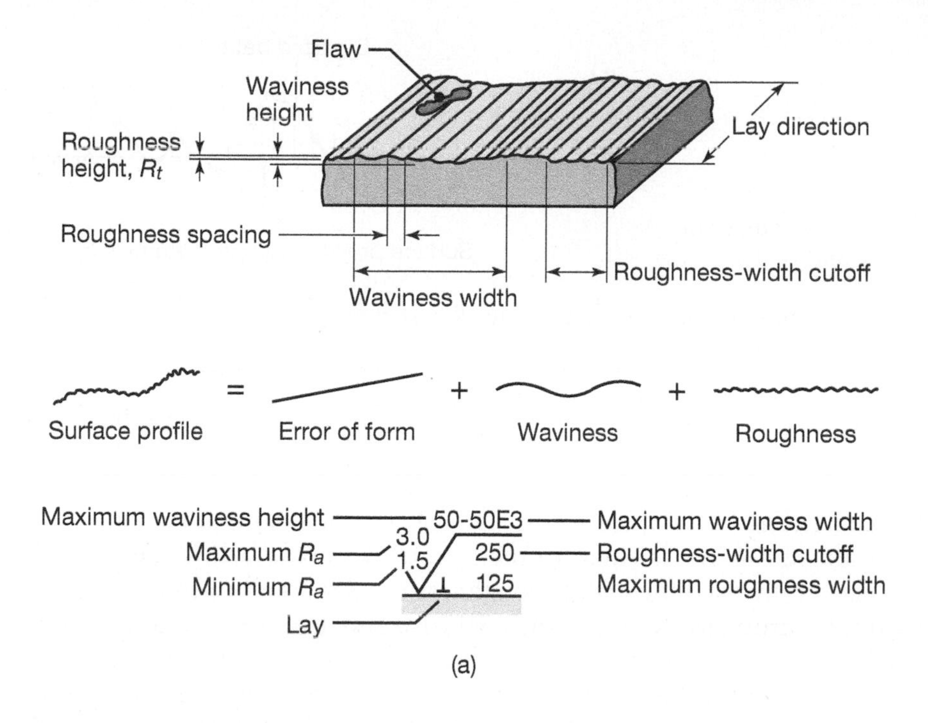

Lay symbol	Interpretation	Examples
=	Lay parallel to the line representing the surface to which the symbol is applied	√=
⊥	Lay perpendicular to the line representing the surface to which the symbol is applied	√⊥
X	Lay angular in both directions to line representing the surface to which symbol is applied	√X
P	Pitted, protuberant, porous, or particulate nondirectional lay	√P

(b)

FIGURE 4.2 (a) Standard terminology and symbols used to describe surface finish. The quantities are given in μm. (b) Common surface-lay symbols.

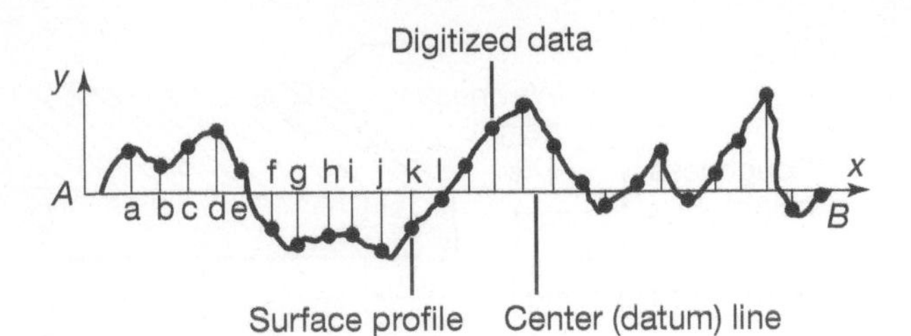

FIGURE 4.3 Coordinates used for measurement of surface roughness, used in Eqs. (4.1) and (4.2).

2. The **root-mean-square average,** R_q, is defined as

$$R_q = \sqrt{\frac{y_a^2 + y_b^2 + y_c^2 + \cdots + y_n^2}{n}} = \sqrt{\frac{1}{n}\sum_{i=1}^{n} y_i^2} = \sqrt{\frac{1}{l}\int_0^l y^2\,dx}. \quad (4.2)$$

The center line (*datum*) shown in Fig. 4.3 is located so that the sum of the areas above the line is equal to the sum of the areas below the line. The units generally used for surface roughness are μm (micrometer, or micron) or μin. (microinch), where 1 μm = 40 μin. and 1 μin. = 0.025 μm.

Comparing Eqs. (4.1) and (4.2), it can be seen that there is a relationship between R_a and R_q. Also, it can be shown that for a surface roughness in the shape of a sine curve, R_q is larger than R_a by a factor of 1.11. R_q is more sensitive to the largest asperity peaks and deepest valleys; these peaks and valleys are important for friction and lubrication. In practice, R_q is often used instead of R_a.

The **maximum roughness height,** R_t, may also be used as a measure of surface roughness. Defined as the height from the deepest trough to the highest peak, it is widely used in engineering practice because of its simplicity.

Symbols for surface roughness. Acceptable limits for surface roughness are specified on technical drawings by the symbols shown in the lower portion of Fig. 4.2; their values are placed to the left of the check mark. Symbols used to describe a surface specify only the roughness, waviness, and lay; they do not include flaws. Whenever flaws are important, a special note is included in technical drawings to describe the method to be used to inspect for surface flaws.

Measuring surface roughness. Several commercially available instruments, called **surface profilometers,** are used to measure and record surface roughness. The most common use a diamond stylus traveling along a straight line over the surface (Figs. 4.4a and b); the distance that the stylus travels can be specified and controlled (See Fig. 4.2). Profilometer traces are presented at an exaggerated vertical scale (a few orders of magnitude greater than the horizontal scale; Figs. 4.4c-f), called **gain** on the recording instrument. The recorded profile is thus significantly distorted, making the surface appear much rougher than it actually is. A profilometer compensates for any surface waviness and indicates only roughness, usually in terms of R_a or R_q. If desired, the profile data can be transferred to a computer for further analysis.

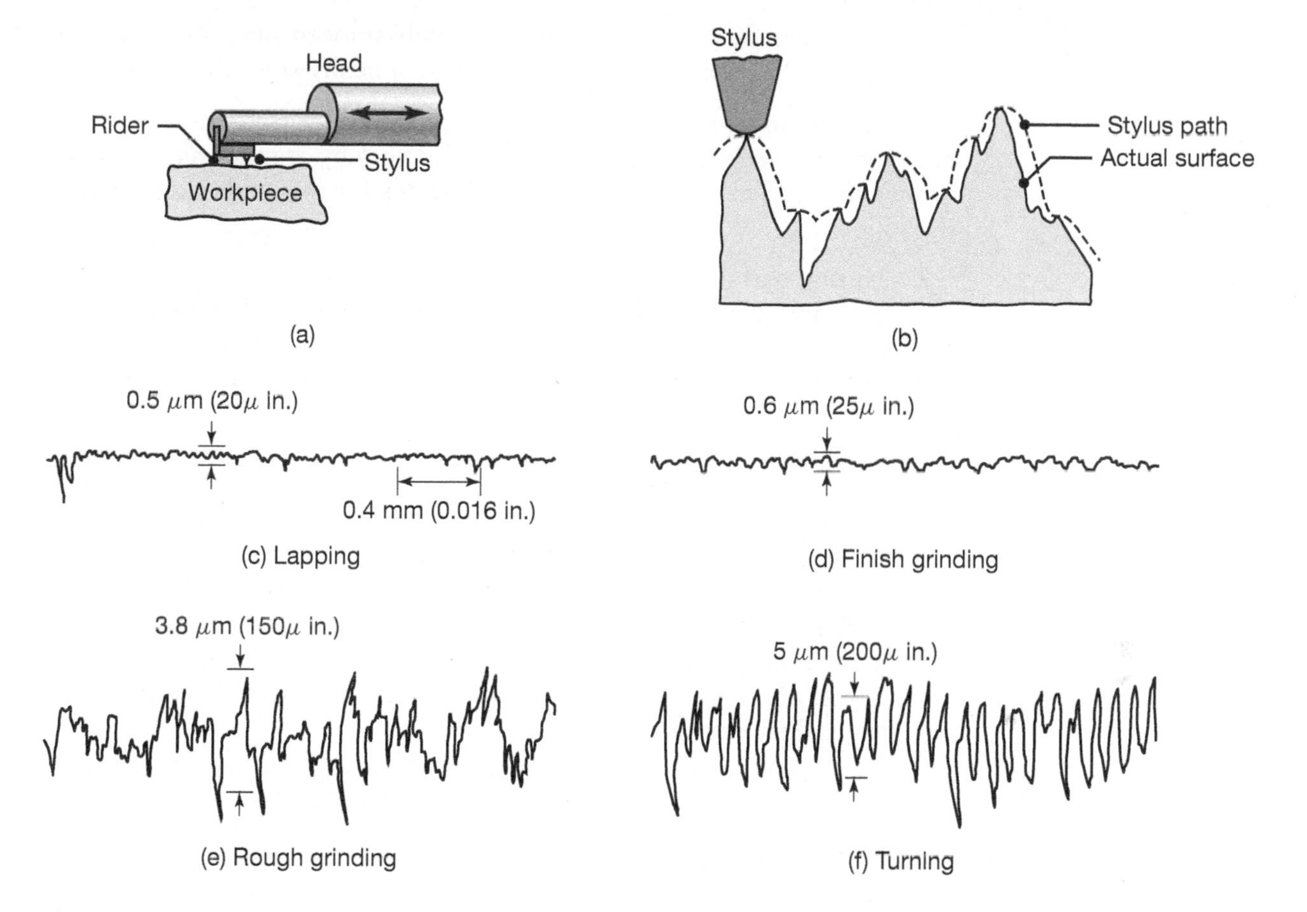

FIGURE 4.4 (a) Measuring surface roughness with a stylus. The rider supports the stylus and guards against damage. (b) Path of the stylus in measurements of surface roughness (broken line) compared with the actual roughness profile. Note that the profile of the stylus' path is smoother than the actual surface profile. Typical surface profiles produced by (c) lapping; (d) finish grinding; (e) rough grinding; and (f) turning processes. Note the difference between the vertical and horizontal scales.

Surface roughness can be observed directly through such means as (a) *interferometry*; and (b) optical, scanning-electron, laser or atomic-force *microscopy*. The microscopy techniques are especially useful for imaging very smooth surfaces for which features cannot be captured by less sensitive instruments. Stereoscopic photographs are particularly useful for three-dimensional views of surfaces; they can also be used to measure surface roughness.

There are three methods for taking **three-dimensional surface measurements.** An *optical-interference* microscope shines a light beam against a reflective surface and records the interference fringes that result from the incident and its reflective waves. *Laser profilometers* are used to measure surfaces either through interferometric techniques or by moving an objective lens to maintain a constant focal length over a surface; the movement of the lens is then a measure of the surface. An *atomic-force* *microscope* can be used to measure extremely smooth surfaces; it even has the capability of distinguishing atoms on atomically smooth surfaces.

Surface roughness in engineering practice. Surface roughness design requirements for engineering applications can vary by as much as two orders of magnitude. The reasons and considerations for this wide range include the following:

1. Mating surfaces require different *precision*. For example, ball bearings and gages require very smooth surfaces and high precision, whereas surfaces for gaskets and brake drums are much rougher.
2. Tribological considerations, including the effects of roughness on friction, wear, and lubrication.
3. Fatigue and notch sensitivity, because rougher surfaces usually have shorter fatigue life (See Fig. 2.24).
4. Electrical and thermal contact resistance, because the rougher the surface, the higher its contact resistance.
5. Corrosion resistance, because the rougher the surface, the greater will be the possibility of entrapping corrosive media.
6. Subsequent processing, such as painting and coating, where a certain degree of roughness generally results in better bonding (a phenomenon also called *mechanical locking*).
7. Depending on the application, a rougher surface may be preferred over a smoother one mainly for aesthetic reasons.
8. Cost considerations, because the finer the surface finish, the higher is the manufacturing cost (see Section 16.7).

4.4 Tribology: Friction, Wear, and Lubrication

4.4.1 Friction

Friction is defined as the resistance to relative sliding between two bodies in contact under a load. Metalworking processes are significantly affected by friction, because of the relative motion and the forces present between tools, dies, and workpieces. Friction dissipates energy, causing heat generation. The subsequent rise in temperature can have a major detrimental effect on the overall operation, such as by causing excessive heating in grinding, leading to *heat checks* (see Section 9.4.3). Furthermore, because it impedes free movement at the tool, die, and workpiece interfaces, friction significantly affects the flow and deformation of materials in metalworking processes.

It should be noted that friction is sometimes desirable. For example, friction is necessary in the rolling of metals to make sheet and plate (Section 6.3), for *friction welding* (Section 12.10) and for machine components such as brakes and clutches.

Adhesion theory of friction. The *adhesion theory of friction* is a common and simple explanation of friction, and is based on the observation that two clean, unlubricated metal surfaces contact each other at only a

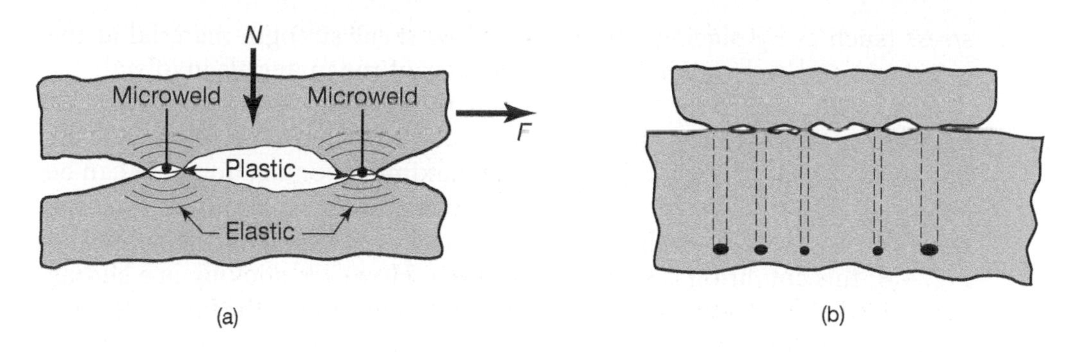

FIGURE 4.5 (a) Schematic illustration of the interface of two contacting surfaces, showing highly loaded asperities; and (b) illustration of the proportion of the apparent area to the real area of contact. The ratio of the areas can be as high as four to five orders of magnitude.

fraction of their apparent area of contact (Fig. 4.5). This observation holds regardless of their surface roughness. The load at the interface is thus supported by the contacting asperities. The sum of the contacting areas is known as the **real area of contact,** A_r.

With metals, very light loads are sufficient to cause plastic deformation at contacting asperities. Because the real area of contact is very small, the asperities are highly loaded and, as a result, they form **microwelds.** The intimate contact of asperities thus develop an **adhesive bond,** the nature of which involves *atomic interactions, mutual solubility* of the two bodies, and *diffusion.* The strength of the bond depends on the physical and mechanical properties of the metals in contact, the temperature, and the cleanliness of the surfaces, such as the nature and thickness of any oxide film or contaminants present.

Coefficient of friction. Sliding between two bodies under a normal load, N, requires the application of a tangential force, F. According to the adhesion theory, F is the force required to *shear the junctions,* and thus it is called the *friction force.* The *coefficient of friction,* μ, at the interface is defined as

$$\mu = \frac{F}{N} = \frac{\tau A_r}{\sigma A_r} = \frac{\tau}{\sigma}, \tag{4.3}$$

where τ is the shear strength of the junction and σ is the normal stress, which, for a plastically deformed asperity, is equivalent to the hardness of the material (see Section 2.6). Equation (4.3) is referred to as the **Coulomb Friction Law.** The coefficient of friction can now be defined as

$$\mu = \frac{\tau}{\text{Hardness}}. \tag{4.4}$$

It can be seen that the nature and strength of the interface is the most significant factor in the magnitude of the friction force. A strong interface requires a high friction force for relative sliding. Equation (4.4) indicates that the coefficient of friction can be reduced either by *decreasing the shear stress* (such as by placing a thin film of low shear strength material at the interface) and/or by *increasing the hardness* of the materials involved.

As the contact force increases and if there are no contaminant fluids trapped at the interface, the real area of contact will eventually reach the apparent area of contact; this area is the maximum contact area that can be achieved. The friction force will now become the force required to shear the entire interface, reaching a maximum and leveling off, as shown in Fig. 4.6; this condition is known as *sticking*. However, sticking in a sliding interface does not necessarily mean complete adhesion at the interface, as it would in welding or brazing operations; rather, it means that the frictional stress at the surface has reached a limiting value (which is related to the shear yield stress, k, of the material).

Once the real area of contact is equal to the apparent area of contact, further increases in the normal force, N, do not affect the friction force, F (See Fig. 4.6); hence, by definition, the coefficient of friction decreases. Because this situation is counterintuitive, a more useful approach is to define a **friction factor**, or **shear factor**, m, as

$$m = \frac{\tau_i}{S_{sy}}, \tag{4.5}$$

where τ_i is the shear strength of the interface and S_{sy} is the *shear yield strength* of the softer material in a sliding pair. The quantity S_{sy} is equal to $S_y/2$ according to the maximum-shear-stress criterion [Eq. (2.37)] and to $S_y/\sqrt{3}$ according to the distortion-energy criterion (Section 2.11.2). Equation (4.5) is often referred to as the *Tresca friction model*. Note that when $m = 0$ in this definition, there is no friction, and when $m = 1$, complete sticking or welding takes place at the interface.

The effects of load, temperature, speed, and environment on the coefficient of friction are difficult to generalize. The coefficient of friction in sliding contact has been measured as low as 0.02 and as high as 100. This wide range is not surprising in view of the numerous variables involved in friction. In metalworking operations that use lubricants, the range for μ is much narrower, generally around 0.05–0.1 for cold forming, and 0.2–0.7 for hot forming.

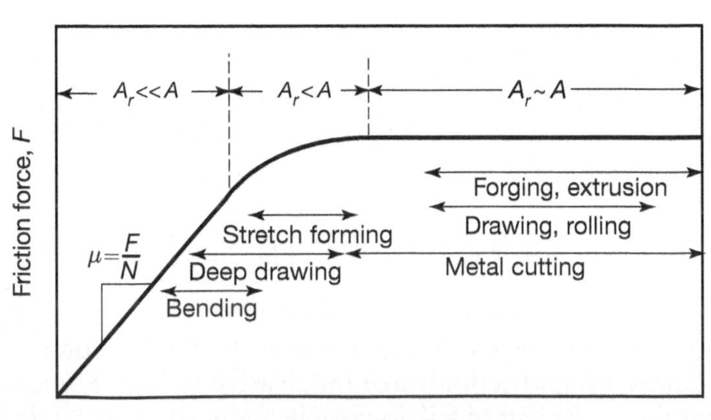

FIGURE 4.6 Schematic illustration of the relation between friction force, F, and normal force, N. Note that as the real area of contact approaches the apparent area, the friction force reaches a maximum and stabilizes at that level. At low normal forces, the friction force is proportional to normal force; most machine components operate in this region. The friction force is not linearly related to normal force in metalworking operations because of the high forces involved.

Abrasion theory of friction. If the upper body in the model shown in Fig. 4.5a is harder than the lower body, and/or its surface has hard protruding particles, then it will scratch and produce grooves in the softer body. This phenomenon, known as **plowing,** is an important aspect in abrasive friction; in fact, it can be a dominant mechanism for situations where adhesion between the two bodies is not particularly strong or contacting surfaces are aggressive. The plowing force can contribute significantly to friction and thus to the measured coefficient of friction at the interface.

Measuring friction. The coefficient of friction or friction factor is generally determined experimentally, either by simulated tests using various small-scale specimens, or at relevant scales in a manufacturing process. The techniques used generally involve measurements of either forces or dimensional changes in the test specimen. One test that has gained wide acceptance, particularly for bulk deformation processes such as forging, is the **ring compression test.** In this test, a flat ring is compressed plastically between two flat platens (Fig. 4.7). As its height is reduced, the ring expands radially outward, because of volume constancy. If friction at the platen–specimen interfaces is zero, both the inner and outer diameters of the ring will expand as if the ring were a solid disk. With increasing friction, however, the inner diameter decreases, because it takes less energy to reduce the inner diameter than to expand the outer diameter of the specimen.

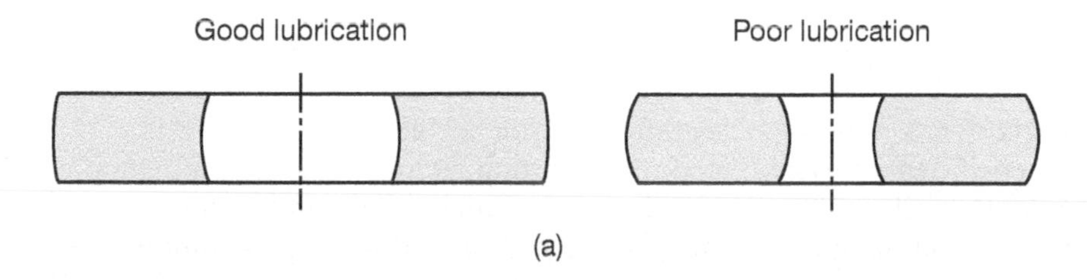

(a)

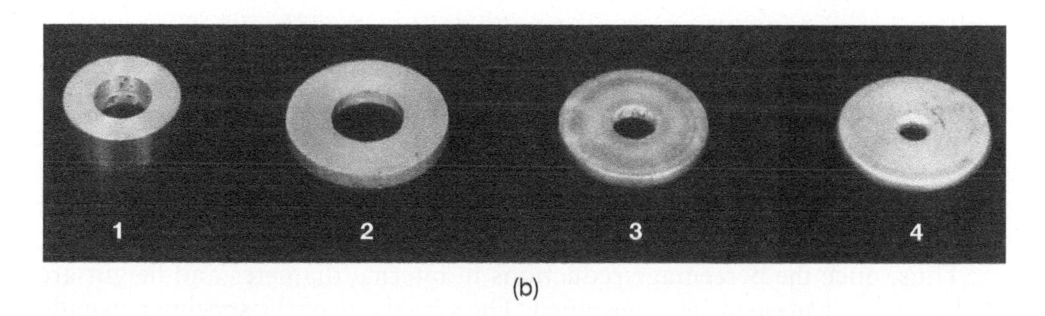

(b)

FIGURE 4.7 (a) The effects of lubrication on barreling in the ring compression test. (a) With good lubrication, both the inner and outer diameters increase as the specimen is compressed; and with poor or no lubrication, friction is high, and the inner diameter decreases. The direction of barreling depends on the relative motion of the cylindrical surfaces with respect to the flat dies. (b) Test results: (1) original specimen, and (2–4) the specimen under increasing friction. *Source:* After A.T. Male and M.G. Cockcroft.

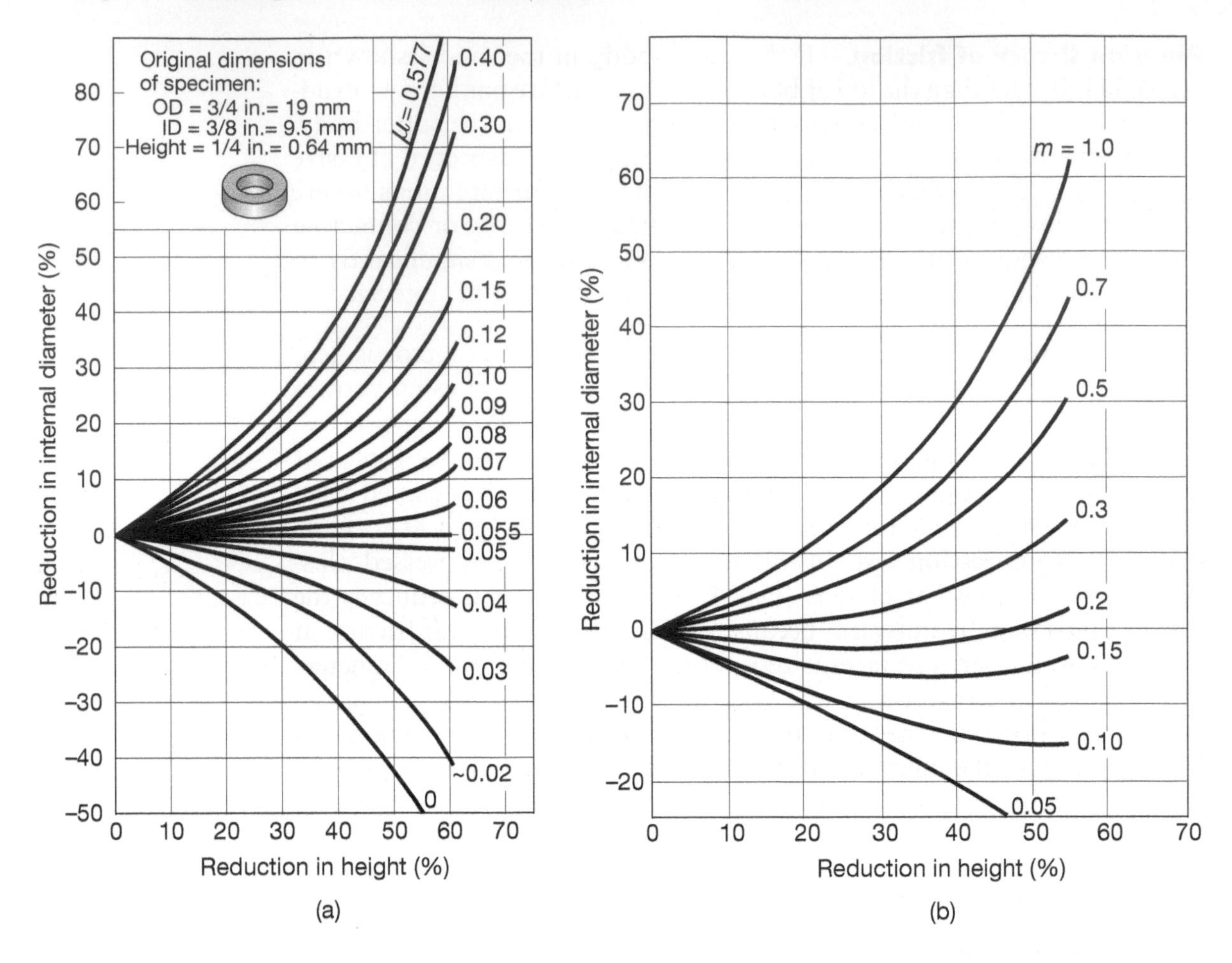

FIGURE 4.8 Charts to determine friction in ring compression tests: (a) coefficient of friction, μ; and (b) friction factor, m. Friction is determined from these charts from the percent reduction in height and by measuring the percent change in the internal diameter of the specimen after compression.

For a certain reduction in height, the internal diameter increases from the original if μ is low and decreases if μ is high. By measuring the change in the specimen's internal diameter and using the curves shown in Fig. 4.8 (which are developed through theoretical analyses), the average coefficient of friction or friction factor can be determined. The geometry of each ring has its own specific set of curves; the most commonly used is a specimen with proportions of outer diameter to inner diameter to height of 6:3:2. Thus, once the percentage reductions in internal diameter and height are known, μ can easily be determined. The actual size of the specimen usually is not relevant in these tests, except when conducted at elevated temperatures, since smaller specimens will cool faster than large ones as heat is transferred to the platens.

Temperature rise caused by friction. The energy dissipated in overcoming friction is converted into heat, except for a small portion that remains in the material as stored energy (see Section 2.12.1). If the frictional energy is high enough, the interface temperature can be increased sufficiently to soften and even melt the material on the surface. Several analytical expressions have been developed in metalworking operations to calculate the temperature rise; Eq. (9.9) is one such expression applicable to grinding operations.

Reducing friction. Friction can be reduced by:

1. Selecting materials that exhibit low adhesion, such as relatively hard materials like carbides and ceramics;
2. Using various lubricants; and
3. Applying various coatings on surfaces in contact, including soft solid films.

Friction in plastics and ceramics. Plastics (Chapter 10) generally possess low frictional characteristics, making them attractive for applications such as bearings, gears, seals, etc. The factors involved in metal friction described earlier are also generally applicable to polymers. However, since adhesion is low with polymers, the plowing component of friction is a significant factor, along with friction arising from *hysteresis losses*.

An important factor in plastics applications is the effect of temperature rise at interfaces caused by friction. Polymers quickly lose their strength and become soft as temperature increases; thus, their low thermal conductivity and low melting points are significant factors. If the temperature rise is not monitored and controlled, sliding surfaces can melt and/or undergo thermal degradation.

In the frictional behavior of ceramics (Section 11.8) the causes of friction are similar to that in metals, thus adhesion and plowing contribute to the friction force. However, because the asperities of ceramic surfaces usually remain elastic, the Coulomb law [Eq. (4.3)] generally works well.

EXAMPLE 4.1 Determining the Coefficient of Friction

Given: In a ring compression test, a specimen is 10 mm in height with outside diameter of 30 mm and inside diameter of 15 mm. It is reduced in thickness by 50%, and the outer diameter (OD) is measured as 39 mm after deformation.

Find: Determine the coefficient of friction, μ, and the friction factor, m.

Solution: First, the new ID needs to be determined; this diameter is obtained from volume constancy as follows:

$$\text{Volume} = \frac{\pi}{4}\left(30^2 - 15^2\right)10 = \frac{\pi}{4}\left(39^2 - \text{ID}^2\right)5.$$

From this equation, the new ID is found to be 13 mm. Thus,

$$\text{Change in ID} = \frac{15-13}{15} \times 100\% = 13\% \text{ (decrease)}.$$

For a 50% reduction in height and a 13% reduction in ID, Fig. 4.8 yields $\mu = 0.09$ and $m = 0.4$.

4.4.2 Wear

Wear is defined as the progressive loss or undesired removal of material from a surface. It has important technological and economic significance, especially because wear alters the shape of the workpiece, tool, and die interfaces, adversely affecting the manufacturing operations and the dimensions and the quality of the parts produced. At the same time, however, it can have an important beneficial effect of reducing the surface roughness by removing the peaks from asperities. Thus, under controlled conditions, wear may be regarded as a type of polishing operation. The *running-in* period for various components of machines and engines involves this type of wear.

Because they are expected to wear during their normal use, especially under high loads, some components of machinery are equipped with **wear plates** (or *wear parts*). Examples of wear parts include cylinder liners in extrusion presses and tool and die inserts in forging, drawing, and casting operations. These parts can easily be replaced when worn, without causing any damage to the rest of the machinery.

Adhesive wear. If a tangential force is applied to the model shown in Fig. 4.5, shearing of the junctions can take place either at the original interface of the two bodies or along a path below or above the interface (Fig. 4.9); this phenomenon is known as *adhesive wear*. The fracture path depends on whether or not the strength of the adhesive bond of the asperities is higher than the cohesive strength of either of the two bodies.

Based on the probability that a junction between two sliding surfaces will lead to the formation of a wear particle, the **Archard wear law** provides an expression for adhesive wear,

$$V = k\frac{LN}{3p}, \tag{4.6}$$

where V is the volume of material removed by wear from the surface, k is the *wear coefficient* (dimensionless), L is the length of travel, N is the normal load, and p is the indentation hardness of the softer body. It should be pointed out that, in some references, the factor 3 is deleted from the denominator and incorporated into the wear coefficient.

Table 4.1 lists typical values of k. Note that the wear coefficient for the same pair of materials can vary by a factor of 3, depending on particular test conditions and sample preparation. Mutual solubility of the mating

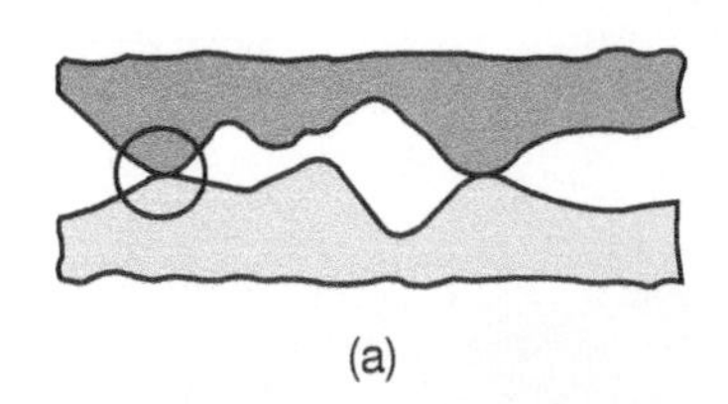

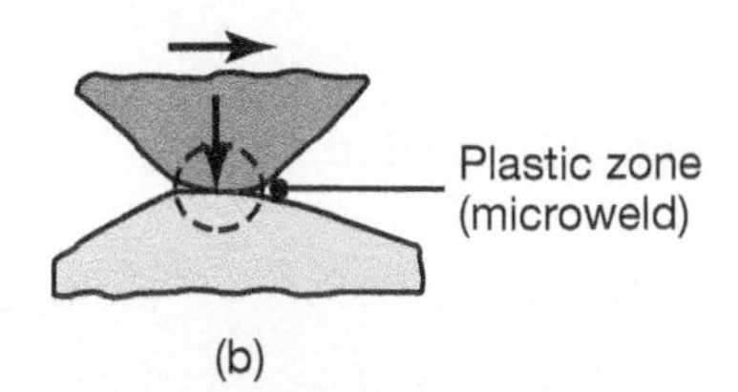

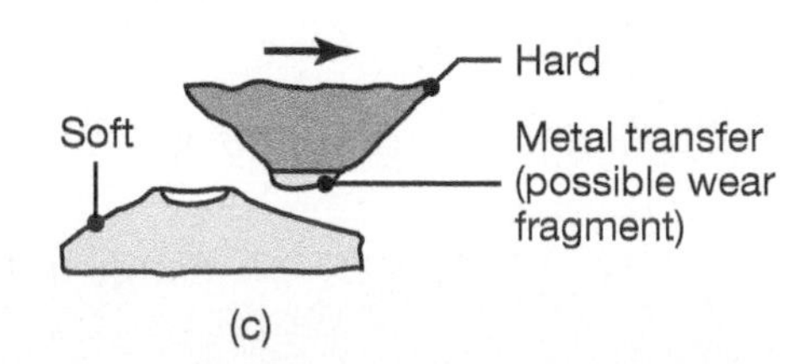

FIGURE 4.9 Schematic illustration of (a) asperities contacting, (b) adhesion between two asperities, and (c) the formation of a wear particle.

TABLE 4.1 Approximate order of magnitude for the wear coefficient, k, in air.

Unlubricated	k	Lubricated	k
Mild steel on mild steel	10^{-2} to 10^{-3}	52100 steel on 52100 steel	10^{-7} to 10^{-10}
60-40 brass on hardened tool steel	10^{-3}	Aluminum bronze on hardened steel	10^{-8} 10^{-9}
Hardened tool steel on hardened tool steel	10^{-4}	Hardened steel on hardened steel	10^{-9} 10^{-9}
Polytetrafluoroethylene (PTFE) on tool steel	10^{-5}		
Tungsten carbide on mild steel	10^{-6}		

bodies is a significant parameter in adhesion; thus similar metal pairs have higher wear coefficients than dissimilar pairs.

Thus far, the discussion on adhesive wear has been based on the assumption that the surface layers of the two contacting bodies are clean and free from contaminants; in such a case, adhesion is very strong and the adhesive wear rate can be very high, leading to **severe wear**. As shown in Fig. 4.1, however, metal surfaces are almost always covered with contaminants and oxide layers, which can have a profound effect on wear behavior.

Adhesive wear coefficients can be reduced by one or more of the following:

1. Selecting pairs of materials that do not form a strong adhesive bond, such as one of the pair being a hard material;
2. Using materials that form a thin oxide layer;
3. Applying a hard coating; and
4. Using a lubricant, or improving the lubrication condition.

EXAMPLE 4.2 Adhesive Wear in Sliding

Given: The end of a rod made of 60-40 brass is sliding over the unlubricated surface of hardened tool steel with a load of 500 N. The hardness of brass is 400 MN/m^2.

Find: What is the distance traveled to produce a wear volume of 1 mm^3 by adhesive wear of the brass rod?

Solution: The parameters in Eq. (4.6) for adhesive wear are as follows:

$$V = 1.0 \text{ mm}^3 = 1 \times 10^{-9} \text{ m}^3;$$

$$k = 10^{-3} \text{ (from Table 4.1)};$$

$$N = 500 \text{ N};$$

$$p = 400 \times 10^6 \text{ N/m}^2.$$

Therefore, the distance traveled is obtained from Eq. (4.6) as

$$L = \frac{3Vp}{kN} = \frac{(3)(1 \times 10^{-9})(400 \times 10^6)}{(10^{-3})(500)} = 2.4 \text{ m}.$$

Abrasive wear. *Abrasive wear* is caused by a hard and rough surface, or a surface with hard protruding features or embedded particles, sliding against another surface. This type of wear produces microchips or *slivers*, resulting in grooves or scratches on the softer of the mating surfaces (Fig. 4.10). Abrasive machining processes, described in Chapter 9, such as grinding, ultrasonic machining, and abrasive-jet machining, act in this manner, except that in these operations, the process parameters are controlled so as to produce desired surfaces, whereas abrasive wear is unintended and undesired.

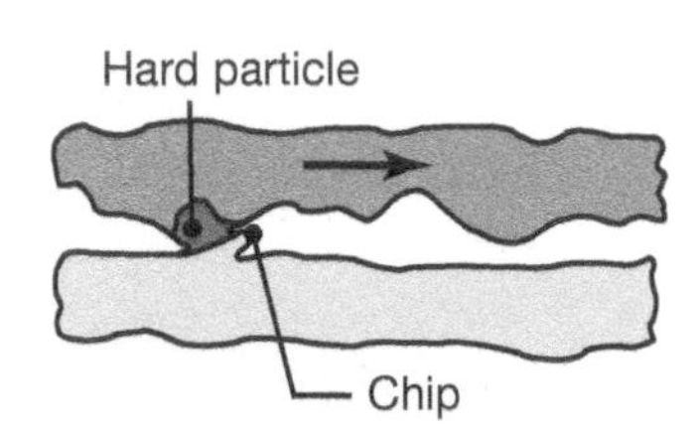

FIGURE 4.10 Schematic illustration of abrasive wear in sliding. Longitudinal scratches on a surface usually indicate abrasive wear.

The *abrasive wear resistance* of pure metals and ceramics is found to be directly proportional to their hardness, and the Archard wear law given in Eq. (4.6) applies. Thus, abrasive wear can be reduced by increasing the hardness of the softer of the materials or by modifying the processing parameters of load and sliding speed. However, very compliant materials, such as elastomers and rubbers, have high abrasive wear resistance; this is because they deform elastically and then recover after the abrasive particles have crossed over the surfaces. The best example is automobile tires, which have long lives (such as 70,000 km or 40,000 miles) even though they are repeatedly operated on abrasive road surfaces; note that even hardened steels would not last long under such abrasive conditions.

Fatigue wear. Also called **surface fatigue** or **surface fracture wear,** *fatigue wear* is caused by surfaces being subjected to cyclic loading, such as in forging and die-casting dies. Cracks are generated on the surface by cyclic mechanical or thermal stresses, such as cool die surfaces repeatedly contacting hot workpieces (*heat checking*). These cracks then join, and the surface begins to *spall.* Fatigue wear can be reduced by (a) lowering contact stresses; (b) reducing thermal cycling; and (c) improving the quality of materials, such as with fewer impurities, inclusions, and various other flaws that may act as local points for crack initiation (see also Section 3.8).

Other types of wear. Two other types of wear are important. **Fretting corrosion** occurs at interfaces that are subjected to very small movements, such as in vibrating machinery or collars on rotating shafts. **Impact wear** is the intentional removal of small amounts of material from a surface by impacting particles. Deburring by vibratory finishing and tumbling (see Section 9.8), and ultrasonic machining (Section 9.9) are two examples.

4.4.3 Lubrication

The interface between tools, dies, molds, and workpieces in manufacturing operations is subjected to a wide range of variables, including:

1. **Contact pressure,** ranging from elastic stresses to multiples of the yield strength of the workpiece material;

2. **Speed**, ranging from very low (such as in superplastic forming operations) to very high (such as in explosive forming, drawing of thin wire, and abrasive and high-speed machining operations); and
3. **Temperature**, ranging from ambient to near the tooling melting point (such as in hot extrusion and squeeze casting).

If two surfaces slide against each other under high pressure, speed, and/or temperature, friction and wear will be high; under these conditions, some form of lubrication may be necessary.

Lubrication regimes. Four regimes of lubrication are relevant to metalworking processes, as described below (Fig. 4.11):

1. **Thick film**: The two surfaces are completely separated by a fluid film, which has a thickness at least one order of magnitude greater than that of the surface roughness. Consequently there is no metal-to-metal contact between the surfaces. In this regime, the normal load is light and is supported by a *hydrodynamic fluid film* between the surfaces. The coefficient of friction is very low, typically ranging from 0.001 to 0.02, and wear is practically nonexistent.
2. **Thin film**: As the normal load increases or as the speed and/or viscosity of the fluid decrease (such as caused by temperature rise), the film thickness is reduced to between 3 and 10 times that of the surface roughness. There may be some metal-to-metal contact at the higher asperities. As with thick film lubrication, friction is very low and wear is practically nonexistent. Often no distinction between thin film and thick film is made, and either regime is referred to as *full film lubrication*.
3. **Mixed**: In this regime, a significant portion of the load is carried by the metal-to-metal contact of the asperities and the rest of the load is carried by a pressurized fluid film between the two surfaces.

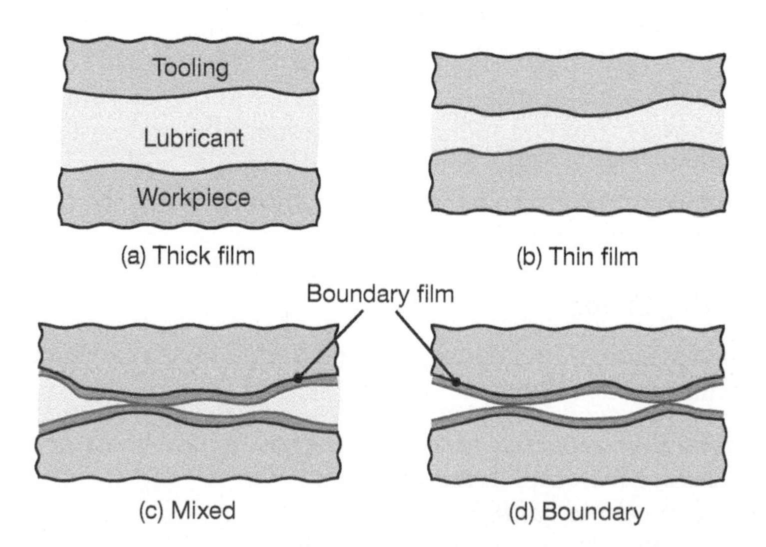

FIGURE 4.11 Regimes of lubrication generally occurring in metalworking operations.

The average film thickness is less than three times that of the surface roughness. With proper selection of lubricants, a strongly adhering *boundary film*, a few molecules thick, can develop over the asperities. This film prevents direct metal-to-metal contact, thus reducing wear. Depending on the strength of the boundary film and various other parameters, the friction coefficient in mixed lubrication may range up to about 0.4, but a value between 0.05 and 0.2 is more common.

4. **Boundary lubrication**: The load in this regime is supported by the contacting surfaces covered with a *boundary layer*. A lubricant may or may not be present in the surface valleys, but even if so, it will not be pressurized enough to support a significant portion of the load. Depending on the boundary-film thickness and its strength, the friction coefficient typically ranges from about 0.1 to 0.4. Common *boundary lubricants* are natural oils, fats, fatty acids, and soaps. Boundary films form rapidly on metal surfaces; however, the film can *break down*, it can be removed by being disturbed, rubbed off during sliding, or it *desorbs* due to high temperatures at the interface. Deprived of this protective layer, the clean metal surfaces contact each other and, as a consequence, severe wear and scoring can occur.

Surface roughness and geometric effects. Surface roughness, particularly in mixed lubrication, is important in that roughness can serve to develop local reservoirs or pockets for lubricants. The lubricant can then be trapped within the surface, and, because these fluids are incompressible, they can support a substantial portion of the normal load. In metalworking operations, it is generally desirable for the workpiece, and not the die, to have the rougher surface, as otherwise the workpiece surface will be damaged by the rougher and harder die surface.

Reduction of friction and wear by *entrainment* or *entrapment of lubricants* may not always be desirable. The reason is that with a thick lubricant film, the workpiece surface cannot come into full contact with the die surfaces; as a result, it does not acquire the shiny appearance that is generally desirable on the part. A thick film of lubricant results in a grainy, dull surface on the workpiece, called **orange peel**. In such operations as coining and forging (Section 6.2), trapped lubricants are especially undesirable, because they prevent precise shape generation. Furthermore, such operations as rolling require some controlled level of friction, so that the film thickness must not be excessive.

4.4.4 Metalworking Fluids

On the basis of the foregoing discussions, the functions of metalworking fluids may be summarized as follows:

- *Reduce friction*, thus reducing force and energy requirements and preventing excessive increases in temperature;
- *Reduce wear, seizure*, and *galling*;
- *Improve material flow* in dies, tools, and molds;

- *Act as a thermal barrier* between the workpiece and tool and die surfaces, thus preventing workpiece cooling in hot-working processes; and
- *Act as a release* or *parting agent* to help in the removal or ejection of parts from dies and molds.

Several types of **metalworking fluids** with a wide variety of chemistries, properties, and characteristics are available to fulfill these requirements:

1. **Oils** have high film strength, as evidenced by how difficult it can be to clean an oily surface. The sources of oils are **mineral** (petroleum or hydrocarbon), **animal**, or **vegetable**. For environmental reasons, there is significant continuing interest in replacing mineral oils with naturally degrading vegetable oil-base stocks. Oils may be **compounded** with a variety of additives, as discussed below.
2. An **emulsion** is a mixture of two immiscible liquids, usually mixtures of oil and water in various proportions, along with additives. Emulsions are of two types. In a *direct emulsion*, oil is dispersed in water as very small droplets, with average diameters typically ranging between 0.1 μm (4 μin.) and 30 μm (1200 μin.). In an *indirect* or *invert emulsion*, water droplets are dispersed in oil. Direct emulsions are important fluids, because the presence of water (typically 95% of the emulsion by volume) gives them high cooling capacity, yet they still provide good lubrication.
3. **Synthetic solutions** are fluids that contain inorganic and other chemicals, dissolved in water. Various chemical agents are added to impart different properties. **Semisynthetic solutions** are basically synthetic solutions to which small amounts of emulsifiable oils have been added.
4. Metalworking fluids are usually blended with various **additives**, including oxidation inhibitors, rust preventatives, detergents, odor control agents, antiseptics, and foam inhibitors.
5. **Soaps** are generally reaction products of sodium or potassium salts with fatty acids. Alkali soaps are soluble in water, but other metal soaps are generally insoluble. Soaps are effective boundary lubricants and can also form thick-film layers at die–workpiece interfaces, particularly when applied over conversion coatings (Section 4.5.1) used for cold metalworking operations.
6. **Waxes** may be of animal or plant (*paraffin*) origin and have complex structures. They have limited use in metalworking operations, except for copper and, as chlorinated paraffin, for stainless steels and high-temperature alloys.

Solid lubricants. Because of their unique properties and characteristics, several solid materials are used as lubricants in manufacturing operations:

1. **Graphite.** *Graphite*, described in Section 11.13, is weak in shear along its layers and hence has a low coefficient of friction in that

direction; thus, it can be a good solid lubricant, particularly at elevated temperatures. However, in a vacuum or an inert-gas atmosphere, graphite can be abrasive and result in high friction. Graphite may be applied either by rubbing it on surfaces or as a colloidal (dispersion of small particles) suspension in liquid carriers, such as water, oil, or alcohols.

2. **Molybdenum disulfide.** *Molybdenum disulfide* (MoS_2), another widely used lamellar solid lubricant, is similar in appearance to graphite and is used as a lubricant at room temperature. Unlike graphite, its friction coefficient is high in an ambient environment. Oils are common carriers for molybdenum disulfide, but they can also be rubbed onto the surfaces of a workpiece.
3. **Soft metals and polymer coatings.** Because of their low strength, thin layers of *soft metals* and *polymer coatings* can be used as solid lubricants. Suitable metals are lead, indium, cadmium, tin, and silver, and polymers include PTFE, polyethylene, and methacrylates. Soft metals are used to coat such high-strength metals as steels, stainless steels, and high-temperature alloys; for example, copper or tin is chemically deposited on the surface before the metal is processed.
4. **Glass.** Although a solid material at room temperature, *glass* becomes viscous at high temperatures and thus can serve as a liquid lubricant. Moreover, its poor thermal conductivity makes glass an attractive lubricant, because it acts as a thermal barrier between hot workpieces and relatively cool dies. Typical applications of glass include hot extrusion (Section 6.4) and forging (Section 6.2).
5. **Conversion Coatings.** Lubricants may not always adhere properly to workpiece surfaces, particularly under high normal and shear stresses. This characteristic is important in forging, extrusion, and wire drawing of steels, stainless steels, and high-temperature alloys. To improve lubricant retention, the workpiece surfaces are first transformed through chemical reaction with acids (hence the term *conversion*). The reaction leaves a somewhat rough and spongy surface, which acts as a carrier for the lubricant. Zinc phosphate conversion coatings are often used on carbon and low-alloy steels; oxalate coatings are used for stainless steels and high-temperature alloys (see Section 4.5.1).

4.5 Surface Treatments, Coatings, and Cleaning

After a part or a component is manufactured, all or specific areas of its surfaces may have to be further processed in order to impart certain specific properties and characteristics. *Surface treatments* may be necessary to:

- Improve surface roughness, appearance, dimensional accuracy, and frictional characteristics;
- Improve resistance to wear, erosion, and indentation (such as the slideways in machine tools; surfaces susceptible to wear; and shafts, rolls, cams, gears, and bearings);

- Control friction in contacting surfaces of tools, dies, bearings, and machine ways;
- Improve resistance to corrosion and oxidation, which is of particular concern with sheet metals for automotive or other outdoor uses, gas-turbine components, and medical devices;
- Improve fatigue resistance, as with bearings and shafts; and
- Impart decorative features, color, or special surface texture.

4.5.1 Surface Treatment Processes

Several processes are used for surface treatments, based on mechanical, chemical, thermal, and physical methods.

1. **Shot peening, water-jet peening,** and **laser shot peening.** In *shot peening*, the surface of the workpiece is impacted repeatedly with cast-steel, glass, or ceramic shot (small balls). Using shot sizes ranging from 0.125 to 5 mm (0.005 to 0.2 in.) in diameter, they make overlapping indentations on the surface, inducing plastic deformation of the surface, to depths up to 1.25 mm (0.05 in.). Because the deformation is not uniform throughout a part's thickness, the process imparts compressive residual stresses on the surface, thus improving the fatigue life of the component. Shot peening is used extensively on shafts, gears, springs, oil-well drilling equipment, and jet-engine parts, such as turbines and compressor blades.

   In *water-jet peening*, a water jet at pressures as high as 400 MPa (60 ksi) impinges on the surface of the workpiece, inducing compressive residual stresses, similar to those induced in shot peening. This method can be used effectively on steels and aluminum alloys.

   In *laser shot peening*, the surface is subjected to laser shocks from a high-powered laser (up to 1 kW). This method can be used successfully on jet-engine fan blades and on materials such as titanium and nickel alloys, inducing compressive surface residual stresses at depths more than 1 mm (0.04 in.).

   Ultrasonic peening uses a hand tool equipped with a piezoelectric transducer. Operating typically at a frequency of 22 kHz, the tool can have a variety of heads for different peening applications.

QR Code 4.1 Shot peening. *Source:* Courtesy of Electronics, Inc.

2. **Roller burnishing (surface rolling).** In *roller burnishing*, the component surface is cold worked using a hard and highly polished roller or a series of rollers (Fig. 4.12). Roller burnishing can be used on flat, cylindrical, or conical surfaces, and it improves the surface finish by removing such imperfections as scratches, tool marks, and pits. Corrosion resistance is also improved, because corrosive chemicals and residues cannot be entrapped in the surface. *Internal* cylindrical surfaces are burnished by a similar process, called **ballizing,** or **ball burnishing,** in which a smooth ball, slightly larger than the bore diameter, is pushed through the length of the hole.
3. In **explosive hardening,** the surface is subjected to high pressure by placing a layer of explosive sheet directly on the workpiece surface and detonating it; the contact pressures can be as high as 35 GPa

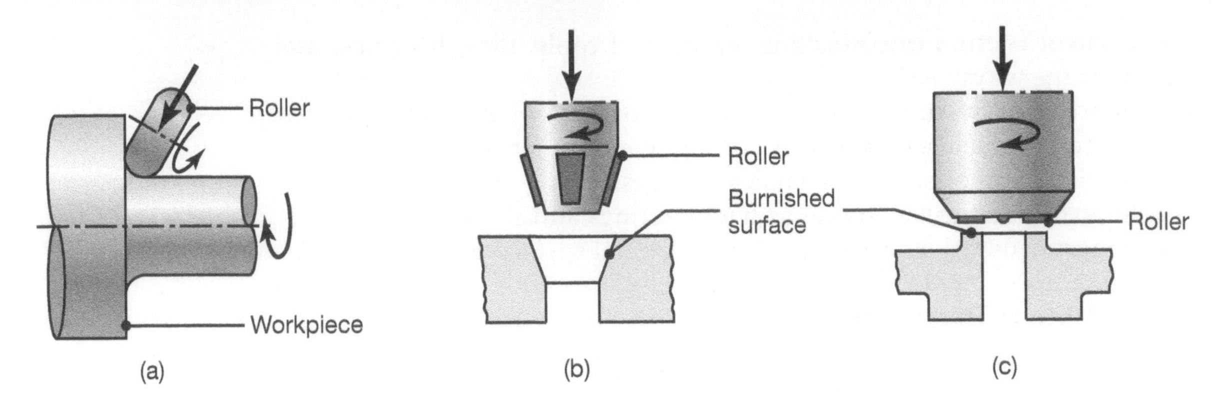

FIGURE 4.12 Examples of roller burnishing of (a) the fillet of a stepped shaft, (b) an internal conical surface, and (c) a flat surface.

(5×10^6 psi) and typically last about 2–3 μs. Large increases in surface hardness can be obtained by this method, with very little change (less than 5%) in the shape of the component. Railroad rail surfaces can, for example, be hardened by this method.

4. In **Cladding (clad bonding)**, metals are bonded with a thin layer of corrosion-resistant metal by applying pressure with rolls; multiple-layer cladding may be utilized in special applications. Cladding of aluminum (*Alclad*) is a common application, in which corrosion-resistant layers of pure aluminum are clad over an aluminum alloy core. Other applications include steels clad with stainless steel or nickel alloys. Cladding may also be performed using dies, as in cladding steel wire with copper. **Laser cladding** involves the fusion of a different material over a substrate. It can be applied to metals and ceramics for enhanced friction and wear behavior.
5. **Mechanical plating (mechanical coating, impact plating, peen plating).** In this process, fine metal particles on a surface are impacted with glass, ceramic, or porcelain spherical beads; the resulting high stress causes particles to bond onto to a workpiece surface. It is typically used for hardened-steel parts for automobiles, with plating thickness usually less than 0.025 μm (0.001 in.).
6. **Case hardening (carburizing, carbonitriding, cyaniding, nitriding, flame hardening, induction hardening).** These processes are described in Section 5.11.3 and summarized in Table 5.7. In addition to the common heat sources of gas and electricity, laser beams also are used as a heat source in surface hardening of both metals and ceramics. Case hardening, as well as some of the other surface-treatment processes, induce residual stresses on surfaces.
7. **Hard facing.** In this process, a relatively thick layer, edge, or point, consisting of wear-resistant hard metal, is deposited on the surface by any of the welding techniques described in Chapter 12. Several layers are usually deposited, called *weld overlay*. Hard coatings of tungsten carbide, chromium, and molybdenum carbide also can be deposited using an electric arc, variously called **spark hardening**, electric spark

hardening, or electrospark deposition. Hard-facing alloys are available as electrodes, rods, wire, and powder. Typical applications for hard facing include valve seats, oil-well drilling tools, and dies for hot metalworking operations. Worn parts are also hard faced to extend their use.

8. **Thermal spraying.** In *thermal spraying* (Fig. 4.13), also called **metallizing,** metal in the form of rod, wire, or powder is melted in a stream of oxyacetylene flame, electric arc, or plasma arc, and the droplets are sprayed onto a preheated surface at speeds up to 100 m/s (20,000 ft/min), using a compressed-air spray gun. The surfaces to be sprayed may be roughened to improve bond strength. Typical applications for this process include automotive components, steel structures, storage tanks, rocket-motor nozzles, and tank cars that are sprayed with zinc or aluminum, up to 0.25 mm (0.010 in.) in thickness.

   The energy sources for thermal-spraying processes are combustion (chemical) and electrical. Thus, thermal-spraying processes can be categorized as follows:

   a. **Combustion spraying:**
      - **Thermal wire spraying,** where an oxyfuel flame melts the wire (Fig. 4.13), and the gas stream entrains molten metal as particles that are deposited onto the target surface;
      - **Thermal metal-powder spraying:** metal powder is deposited on the surface using an oxyfuel flame;
      - **Detonation gun:** a controlled explosion takes place using an oxyfuel gas mixture; and
      - **High-velocity oxyfuel gas spraying,** which has a similar high performance as detonation gun spraying, but can be less expensive.

   b. **Electrical spraying:**
      - **Twin-wire arc.** In this process, an arc is formed between two consumable wire electrodes; melted metal is entrained in the gas spray.

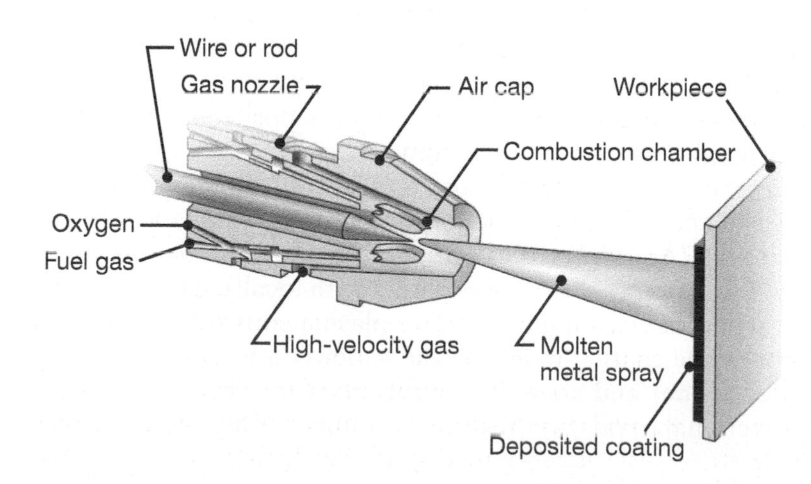

FIGURE 4.13 Schematic illustration of thermal wire spraying.

- **Plasma.** Plasma spraying produces temperatures on the order of 8300°C (15,000°F), so this approach is suitable for high-temperature metals. Bond strength is very good, with very low oxide content. **Low-pressure plasma spray** and **vacuum plasma spray** also produce coatings with high bond strength.

Cold spraying is a process in which the particles to be sprayed are not melted, thus oxidation is minimal. The spray jet is highly focused and has very high impact velocities.

9. **Surface texturing.** For technical, functional, optical, or aesthetic reasons, manufactured surfaces can be further modified by various secondary operations. Texturing methods generally consist of (a) etching, using chemicals or sputtering techniques; (b) electric arcs; (c) laser pulses; and (d) atomic oxygen, which reacts with surfaces and produces fine, cone-like surface textures. The possible adverse effects of any of these processes on the properties and performance of materials should be considered in their selection.
10. **Vapor deposition.** In this process, the workpiece is surrounded by chemically reactive gases that contain chemical compounds of the materials to be deposited. The deposited material, typically a few micrometers thick, may consist of metals, alloys, carbides, nitrides, borides, ceramics, or various oxides. The substrate may be metal, plastic, glass, or paper. Typical applications include coatings for cutting tools, drills, reamers, milling cutters, punches, dies, and wear surfaces (see also Section 13.5 on *semiconductor manufacturing*).

    There are two major vapor deposition process categories: *physical vapor deposition* and *chemical vapor deposition*.

    a. **Physical vapor deposition** (PVD). These processes are carried out in a high vacuum and at temperatures in the range of 200°C to 500°C (400°F to 900°F). Physical deposition processes involve some method of vapor generation without the use of chemical reactions; condensing the vapor onto a workpiece leads to the deposition of thin films. Common material coatings produced through PVD are titanium nitride, chromium nitride, and titanium aluminum nitride. Process variations are characterized according to their energy source:

    In **vacuum deposition**, the metal to be deposited is evaporated at high temperatures in a vacuum and is deposited on the substrate, which is usually at room or slightly higher temperature. Uniform coatings can be deposited on complex shapes. In **arc deposition** (Arc-PVD), the coating material (cathode) is evaporated by a number of arc evaporators with localized electric arcs. The arcs produce a highly reactive plasma consisting of ionized vapor of the coating material; the vapor condenses on the substrate (anode) and coats it. Applications for this process may be functional (oxidation-resistant coatings for high-temperature applications, electronics, and optics) or decorative (hardware, appliances, and jewelry).

In **sputtering**, an electric field ionizes an inert gas (usually argon). The positive ions bombard the coating material (cathode) and cause sputtering (ejecting) of its atoms; these atoms then condense on the workpiece, which is heated to improve bonding. In **reactive sputtering**, the inert gas is replaced by a reactive gas; if oxygen, the atoms are oxidized and the oxides are deposited. **Radio frequency (RF) sputtering** is used for nonconductive materials, such as electrical insulators and semiconductor devices. Sputtering is widely used in the semiconductor industry to deposit thin films in the manufacture of integrated circuits (Section 13.5).

Ion plating is a generic term describing the combined processes of sputter cleaning and deposition through vacuum evaporation. An electric field causes a glow discharge, generating a plasma; the plasma bombards the substrate, cleaning it and making it suitable for coating by the target material. **Ion-beam-enhanced (assisted) deposition** is capable of producing thin films as coatings for semiconductors, tribological, and optical applications. **Dual ion-beam-assisted deposition** or *dual ion-beam sputtering* (DIBS) is a coating technique that uses two ion beams: one cleans (or *bombards*) the substrate, and can help with the deposition process; the other sputters a target. The combined action of two ion beams result in good adhesion on metals, ceramics, and polymers. Ceramic bearings and dental instruments are two examples of its application.

b. **Chemical vapor deposition** (CVD) is a thermochemical process. In a typical application, such as for coating cutting tools with titanium nitride (TiN) (Fig. 4.14), the tools are placed on a graphite tray and are heated to 950°C to 1050°C (1740°F to 1920°F) in an atmosphere of inert gas. Titanium tetrachloride (a vapor), hydrogen, and nitrogen are then introduced into the chamber. The resulting chemical reactions form a thin coating of titanium nitride on the tool's surfaces. For coating with titanium carbide, methane is substituted for hydrogen and nitrogen gases. CVD coatings are usually thicker than

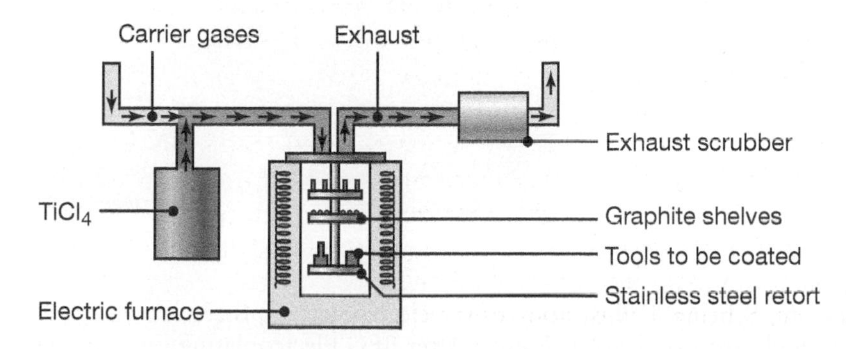

FIGURE 4.14 Schematic illustration of the chemical vapor deposition process.

those of PVD. **Medium-temperature CVD** (MTCVD) produces coatings that have higher resistance to crack propagation than CVD coatings.

11. In **ion implantation**, ions are accelerated toward a surface (in a vacuum) and penetrate the substrate to a depth of a few micrometers. This process (not to be confused with ion plating mentioned above) modifies surface properties, increasing the surface hardness and improving resistance to friction, wear, and corrosion. This process can be controlled accurately, and the surface can be masked to prevent ion implantation in unwanted places. When used in specific applications, such as semiconductors, the process is called **doping** (alloying with small amounts of various elements).
12. **Diffusion coating.** In this process, an alloying element is diffused into the surface, thus altering its properties; the elements can be in solid, liquid, or gaseous states. Diffusion coatings are given different names, depending on the diffused element (see also carburizing, nitriding, and boronizing, in Table 5.7).
13. **Electroplating.** The workpiece (cathode) is plated with a different metal (anode) while both are suspended in a bath containing a water-base electrolyte solution (Fig. 4.15). Although this process involves a number of reactions, the basic procedure is that the metal ions from the anode are discharged (under the potential from the external source of electricity), they combine with the ions in the solution, and are deposited on the cathode.

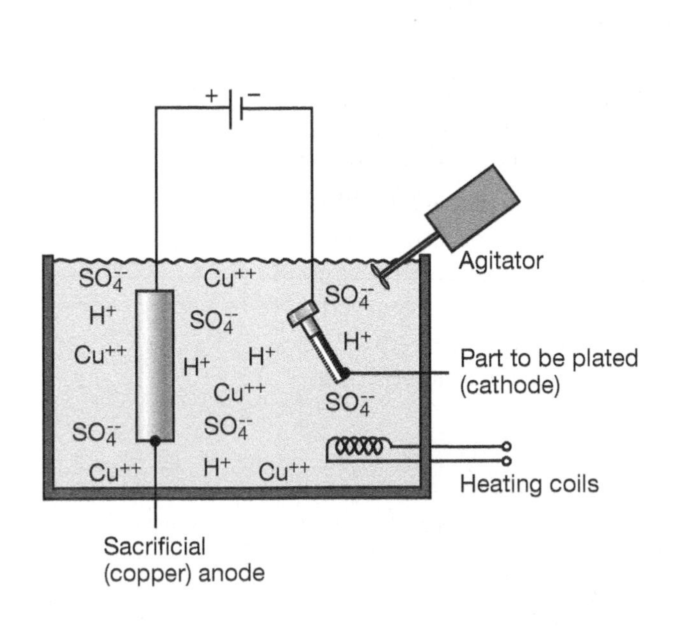

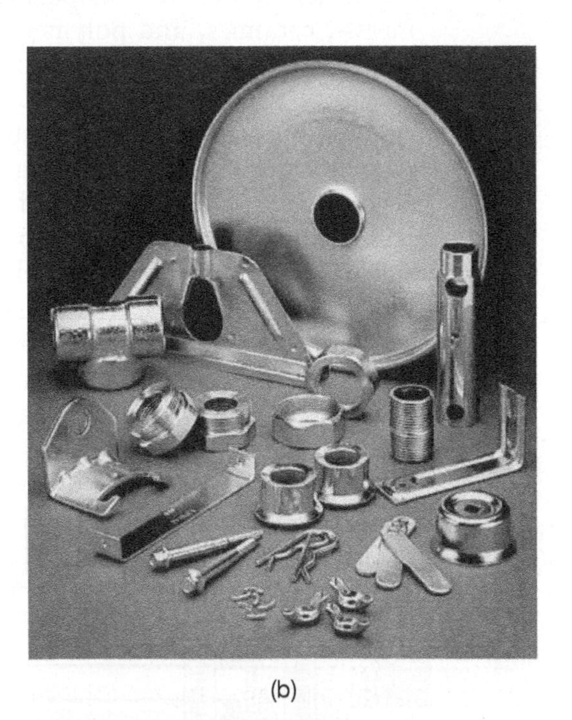

FIGURE 4.15 (a) Schematic illustration of the electroplating process. (b) Examples of electroplated parts. *Source:* After BFG Electroplating.

The volume of the plated metal can be calculated from the expression

$$\text{Volume of metal plated} = cIt, \tag{4.7}$$

where I is the current in amperes, t is time, and c is a constant that depends on the plated metal, the electrolyte, and the efficiency of the system, being typically in the range of 0.03 to 0.1 mm^3/amp-s. For the same volume of material deposited, the deposited thickness is inversely proportional to the surface area. The time required for electroplating is typically long, because the deposition rate is generally on the order of 75 μm/hr. Thin electroplated layers are typically on the order of 1 μm (40 μin.), and thick layers can be as much as 500 μm (0.02 in.).

The plating solutions are either strong acids or cyanide solutions. As the metal is plated from the solution, it has to be periodically replenished. This is accomplished through two principal methods: (a) salts of metals are occasionally added to the solution; or (b) a *sacrificial anode* of the metal to be plated is located in the electroplating tank, which dissolves at the same rate that metal is deposited. There are three main methods of electroplating:

a. In **rack plating**, the parts to be plated are placed on a rack, which is then moved through a series of processing tanks.
b. **Barrel plating** is used to plate small parts, such as bolts, nuts, gears, etc., that are placed in a container, or barrel, often constructed of polypropylene. Electrolytic fluid flows through the barrel and provides the plating metal. Electric current is provided by contacts in the barrel; current is provided to every part through contact with the barrel and other parts. During barrel plating, the barrel rotates to ensure plating uniformity.
c. In **brush processing**, the electrolytic fluid is pumped through a hand-held brush with metal bristles. It is suitable for plating of very large parts, and can be used to apply coatings on large equipment without the need for disassembly.

Common plating materials include chromium, nickel, cadmium, copper, zinc, silver, gold, and tin. **Chromium plating** is carried out by plating the metal first with copper, then with nickel, and finally with chromium. **Hard chromium plating** is deposited directly on the base metal, and has a hardness of up to 70 HRC.

Typical electroplating applications include copper plating of aluminum wire and phenolic printed circuit boards, decorative chrome plating of automotive parts, tin plating of copper electrical terminals (for ease of soldering), and plating of various components for enhanced appearance and resistance to wear and corrosion. Because they do not develop oxide films, noble metals (gold, silver, and platinum) are important electroplating materials for the electronics and jewelry industries. Plastics, such as ABS, polypropylene, polysulfone, polycarbonate, polyester, and nylon, can have metallic coatings applied through electroplating. Because they basically are not electrically conductive (although they can be made so; see Section 10.7.2), plastics must first be preplated by such processes as electroless nickel plating (see below). Parts to be electroplated may be simple or complex, and size is not a limitation; complex shapes may, however, develop varying plating thicknesses.

14. **Electroforming.** A process related to electroplating and especially useful for the manufacture of very small parts is *electroforming*. Metal is electrodeposited on a *mandrel* (also called a *mold* or a *matrix*), which is then removed; the coating itself thus becomes the product (Fig. 4.16). Both simple and complex shapes can be produced by electroforming, with wall thicknesses as small as 0.025 mm (0.001 in.). Parts may weigh from a few grams to as much as 270 kg (600 lb).

    Mandrels are made from a variety of materials, including: (a) metallic, such as zinc or aluminum; (b) nonmetallic, which can be made electrically conductive with the appropriate coatings; and (c) low-melting alloys, wax, or plastics, all of which can be melted away or dissolved with suitable chemicals.

    The electroforming process is particularly suitable for low production quantities or intricate parts (such as molds, dies, waveguides, nozzles, and bellows) made of nickel, copper, gold, and silver. The process is also suitable for aerospace, electronics, and electro-optics applications.

15. **Electroless plating.** This process is carried out by chemical reactions and without the use of an external source of electricity. In electroless

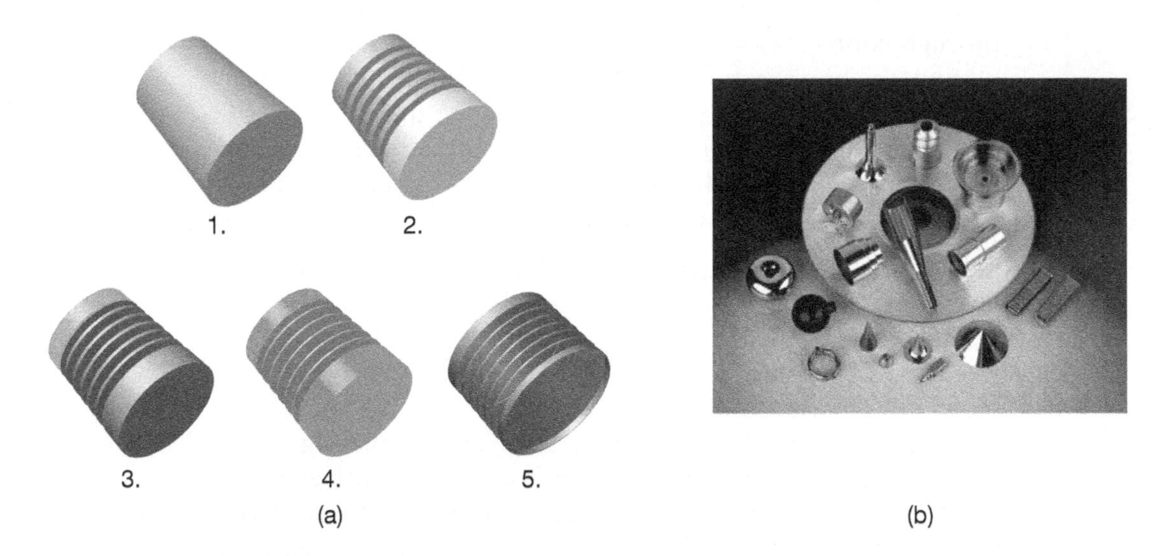

FIGURE 4.16 (a) Typical sequence in electroforming. (1) A mandrel is selected with the correct nominal size. (2) The desired geometry (in this case, that of a bellows) is machined into the mandrel. (3) The desired metal is electroplated onto the mandrel. (4) The plated material is trimmed if necessary. (5) The mandrel is dissolved through chemical machining (see Section 9.10). (b) A collection of electroformed parts. *Source:* After Servometer, Cedar Grove, NJ.

nickel plating, nickel chloride (a metallic salt) is reduced (with sodium hypophosphite as the reducing agent) to nickel metal, which is then deposited on the workpiece. The hardness of nickel plating ranges between 425 and 575 HV, and the plating can be heat treated to a hardness of 1000 HV. The coating has excellent wear and corrosion resistance.

16. **Anodizing.** Anodizing is an oxidation process (*anodic oxidation*) whereby the workpiece surfaces are converted to a hard and porous oxide layer that provides corrosion resistance and a decorative finish. The workpiece is the anode in an electrolytic cell immersed in an acid bath, resulting in chemical adsorption of oxygen from the bath. Organic dyes of various colors (typically black, red, bronze, gold, or gray) can be used to produce stable and durable surface films. Typical applications for anodizing include aluminum furniture and utensils, architectural shapes, automobile trim, picture frames, keys, and sporting goods. Anodized surfaces also serve as a good base for painting, especially for aluminum, which otherwise would be difficult to paint.
17. **Conversion coating.** In this process, also called **chemical-reaction priming**, a coating forms on metal surfaces as a result of chemical or electrochemical reactions. Various metals, particularly steel, aluminum, and zinc, can be conversion coated. Phosphates, chromates, and oxalates are used to produce conversion coatings, for such purposes as prepainting, decorative finishes, and protection against corrosion. An important application is in conversion coating of workpieces as a lubricant carrier in cold-forming operations (see Section 4.4.4).
18. **Coloring.** As the name implies, coloring involves processes that alter the color of metals and ceramics. The change is caused by the conversion of surfaces by chemical, electrochemical, or thermal processes into chemical compounds, such as oxides, chromates, and phosphates. In **blackening**, iron and steels develop a lustrous black-oxide film, using solutions of hot caustic soda.
19. **Hot dipping.** In this process, the workpiece (usually steel or iron) is dipped into a bath of molten metal, such as zinc (for galvanized-steel sheet and plumbing supplies), tin (for tin plate and tin cans for food containers), aluminum (*aluminizing*), and *terne* (lead alloyed with 10% to 20% tin). Hot-dipped coatings on discrete parts or sheet metal provide galvanized pipe, plumbing supplies, and numerous other products with long-term resistance to corrosion. The coating thickness is usually given in terms of coating weight per unit surface area of the sheet, typically 150–900 g/m^2 (0.5–3 oz/ft^2). The service life of hot-dipped parts depends on the thickness of the zinc coating and the environment to which it is exposed. Various **precoated sheet steels** are used extensively, such as in automobile bodies and food containers.
20. **Porcelain enameling.** Metals are coated with a variety of glassy (vitreous) substances to provide corrosion and electrical resistance

and for service at elevated temperatures. The coatings are generally classified as porcelain enamels, typically including enamels and ceramics (Section 11.8). The word **enamel** is also used for glossy paints, indicating a smooth and relatively hard coating. Porcelain enamels are glassy inorganic coatings consisting of various metal oxides. **Enameling** involves fusing the coating material on the substrate material by heating them both to 425°C to 1000°C (800°F to 1800°F), so as to liquefy the oxides. Depending on their composition, enamels have varying resistances to alkali, acids, detergents, cleansers, and water, and are available in a variety of colors.

Typical applications for porcelain enameling include household appliances, plumbing fixtures, chemical-processing equipment, signs, cookware, and jewelry, and as protective coatings on jet-engine components. The coating may be applied by dipping, spraying, or electrodeposition; the thickness is usually in the range of 0.05–0.6 mm (0.002–0.025 in.). Metals that are coated with porcelain enamel are typically steels, cast iron, and aluminum. **Glazing** is the application of glassy coatings on ceramic and earthenware to give them decorative finishes and to make them impervious to moisture. Glass coatings are used as lining for chemical resistance, with a thickness much greater than in enameling.

21. **Organic coatings.** Metal surfaces may be coated or *precoated* with a variety of organic coatings, films, and laminates to improve appearance and corrosion resistance. Coatings are applied to the coil stock on continuous lines, with thicknesses generally in the range of 0.0025–0.2 mm (0.0001–0.008 in.). These coatings have a wide range of flexibility, durability, hardness, texture, color, gloss, and resistance to abrasion and chemicals. Coated sheet metal is subsequently formed into such products as cabinets, appliance housings, paneling, shelving, siding for residential buildings, gutters, and metal furniture.

    Other applications of organic coatings include naval aircraft that are subjected to high humidity, seawater, rain, pollutants (such as from ship exhaust stacks), aviation fuel, deicing fluids, and battery acid, and parts that are impacted by abrasive particles, such as dust, gravel, stones, and deicing salts. For aluminum structures, organic coatings typically consist of an epoxy primer and a polyurethane topcoat, with a lifetime of four to six years.

22. **Ceramic coatings.** Ceramics, such as aluminum oxide and zirconium oxide, are applied to a surface at room temperature, usually by thermal-spraying techniques. These coatings serve as a thermal barrier, especially in such applications as hot-extrusion dies, diesel-engine components, and turbine blades.

23. **Painting.** Paints are classified as enamels, lacquers, and water-base paints. They have a wide range of characteristics and applications, and are applied by brushing, dipping, or spraying. In **electrocoating** (**electrostatic spraying**), paint particles are charged electrostatically

to make them attracted to surfaces, producing a uniformly adherent coating.

24. **Diamond coating.** Diamond coating of metals, glass, ceramics, and plastics involve the use of various chemical and plasma-assisted vapor deposition processes and ion-beam enhanced deposition techniques. *Free-standing diamond films* on the order of 1 mm (0.040 in.) thick can be made, including smooth and optically clear diamond films. Combined with such properties of diamond as hardness, wear resistance, high thermal conductivity, and transparency to ultraviolet light and microwave frequencies, diamond coatings have been applied most often to exploit diamond's high hardness. Growing diamond films on crystalline copper substrates is done by implantation of carbon ions is used in making computer chips (Chapter 13). Diamond can be doped to form *p*- and *n*-type ends on semiconductors in making transistors; its high thermal conductivity allows closer packing of chips than with silicon or gallium-arsenide chips, thus significantly increasing the speed of computers.
25. **Diamond-like carbon** (DLC). Using a low-temperature, ion-beam-assisted deposition process, this material is applied as a coating of a few nanometers in thickness, with a hardness of about 5000 HV. Less expensive than diamond films, DLC has important applications in tools and dies, gears, bearings, microelectromechanical systems, and small (micro-scale) probes, cutting tools (such as drills and end mills), measuring instruments, surgical knives, turbine blades, and fuel-injection nozzles.
26. **Surface texturing.** As described throughout this text, each manufacturing process produces a certain surface texture, finish, and appearance, which may be acceptable for its intended function. However, manufactured surfaces can be further modified by secondary operations for various reasons.

    Called *surface texturing*, these additional processes generally consist of the following techniques:

    a. Etching, using chemicals or sputtering techniques;
    b. Electric arcs;
    c. Lasers, using excimer lasers with pulsed beams; applications include molds for permanent-mold casting, rolls for temper mills, golf-club heads, and computer hard disks; and
    d. Atomic oxygen, which reacts with surfaces and produces a fine, conelike surface texture.

4.5.2 Cleaning of Surfaces

A clean surface can have both beneficial as well as detrimental effects. Although a contaminated surface reduces the tendency for adhesion and galling between mating parts, cleanliness generally is essential for more effective application of metalworking fluids, coating and painting, adhesive bonding, welding, brazing, soldering, reliability of machinery, and in assembly operations. *Contaminants* (also called soils) may consist of rust, scale, chips and other metallic and nonmetallic debris, metalworking fluids, solid lubricants, pigments, polishing and lapping compounds, and general environmental elements.

Cleaning involves the removal of solid, semisolid, or liquid contaminants from a surface. The type of cleaning process required depends on the type of contaminants to be removed. **Mechanical cleaning methods** consist of physically disturbing the contaminants, such as by wire or fiber brushing, dry or wet abrasive blasting, tumbling, steam jets, and ultrasonic cleaning.

In **electrolytic cleaning**, a charge is applied to the part to be cleaned in an aqueous solution, which results in the formation of bubbles of hydrogen or oxygen. The bubbles are abrasive and thus aid in the removal of contaminants from the surface.

Chemical cleaning methods are effective in removing oil, grease, and metalworking fluids from surfaces. They consist of one or more of the following operations:

1. *Solution*: the soil dissolves in the cleaning solution.
2. *Saponification*: a chemical reaction that converts animal or vegetable oils into a soap that is soluble in water.
3. *Emulsification*: the cleaning solution reacts with a liquid (such as oils) and forms an emulsion, so that the oil is suspended in the solution in the form of small droplets.
4. *Dispersion*: the concentration of solid contaminants on the surface is decreased by surface-active materials in the cleaning solution; these solid contaminants are then suspended into the cleaning solution in the form of solid particles.
5. *Aggregation*: lubricants are removed from the surface by various agents in the cleaning fluid, and collect as large dirt particles.

Cleaning fluids, including *alkaline solutions, emulsions, solvents, hot vapors, acids, salts*, and mixtures of *organic compounds*, can be used in conjunction with electrochemical processes for more effective cleaning. In **vapor degreasing**, a heated and evaporated solvent condenses on parts that are at room temperature. Dirt and lubricants are then removed and dispersed, emulsified, or dissolved in the solvent, which drips off the parts into the solvent tank.

Mechanical agitation of the surface can aid in removal of contaminants; the methods include wire brushing, abrasive blasting, abrasive jets, and ultrasonic vibration of a solvent bath (**ultrasonic cleaning**).

Cleaning parts with complex shapes can be difficult because interior corners are difficult to penetrate by mechanical or chemical means. Alternative designs may thus have to be considered, such as (a) avoiding deep blind holes; (b) providing appropriate drain holes in the part; or (c) producing the part from several smaller components instead of one large, difficult-to-clean component.

4.6 Engineering Metrology and Instrumentation

Engineering metrology is the measurement of dimensions, such as length, thickness, diameter, taper, angle, flatness, and profiles. In this section the characteristics of the instruments and the techniques used in engineering metrology are described. Numerous measuring instruments and devices are used in metrology. It is important to first describe briefly the quality of an instrument, as follows:

1. **Accuracy**: the degree of agreement between the measured dimension and its true magnitude.
2. **Precision**: the degree to which the instrument gives repeated measurements.
3. **Resolution**: the smallest dimension that can be read on an instrument.
4. **Sensitivity**: the smallest difference in dimensions that the instrument can detect or distinguish.

4.6.1 Measuring Instruments

1. **Line-graduated instruments.** Line-graduated instruments are used for measuring length (linear measurements) or angles (angular measurements), *graduated* meaning that it is marked to indicate a certain quantity. The simplest and most commonly used instrument for linear measurements is a *steel rule* (*machinist's rule*), bar, or tape with fractional or decimal graduations. Lengths are measured directly to an accuracy that is limited to the nearest division, usually 1 mm or $\frac{1}{64}$ in.

   Vernier calipers have a graduated beam and a sliding jaw with a *vernier*. Also called *caliper gages*, the two jaws of the caliper contact the part being measured and the dimension is read at the matching graduated lines. The vernier improves the sensitivity of a simple rule by indicating fractions of the smallest division on the graduated beam, usually to 25 μm (0.001 in.). Vernier calipers, which can be used to measure inside or outside lengths, are also equipped with *digital readouts*, which are easier to read and less subject to human error. Vernier *height gages* are vernier calipers with setups similar to that of a depth gage and have similar sensitivity.

   Micrometers have a graduated, threaded spindle and are commonly used for measuring the thickness and inside or outside diameters of parts. Circumferential vernier readings to a sensitivity of 2.5 μm (0.0001 in.) can be obtained. Micrometers are also available for measuring depths (*micrometer depth gage*) and internal diameters (*inside micrometer*) with the same sensitivity. The anvils on micrometers can be equipped with conical or ball contacts, used in measuring inside recesses, threaded rod diameters, and wall thicknesses of tubes and curved sheets.

   Diffraction gratings consist of two flat optical glasses with closely spaced parallel lines scribed on their surfaces. The grating on the shorter glass is inclined slightly; as a result, interference fringes develop when it is viewed over the longer glass. The position of these fringes depends on the relative position of the two sets of glasses.

2. **Indirect-reading instruments**. These instruments typically consist of calipers and dividers, without any graduated scales. They are used to transfer the size measured to a direct-reading instrument, such as a graduated rule. **Telescoping gages** are available for indirect measurement of holes or cavities.

   Angles are measured in degrees, radians, or minutes and seconds of arc. A **bevel protractor** is a direct-reading instrument similar to a common protractor, with the exception that it has a movable member. The two blades of the protractor are placed in contact with the part being measured, and the angle is read directly on the vernier scale. Another type of bevel protractor is the **combination square,** which is a steel rule equipped with devices for measuring 45° and 90° angles.

   Measuring with a **sine bar** involves placing the part on an inclined bar or plate and adjusting the angle by placing gage blocks on a surface plate. After the part is placed on the sine bar, a dial indicator (see below) is used to scan the top surface of the part. **Gage blocks** are added or removed as necessary until the top surface is parallel to the surface plate. The angle on the part is then calculated from geometric relationships. Angles can also be measured by using **angle gage blocks.** These blocks have different tapers that can be assembled in various combinations and used in a manner similar to that for sine bars.

3. **Comparative length-measuring instruments**. These instruments are used for measuring *comparative lengths*. Also called *deviation-type instruments*, they amplify and measure variations or deviations in distance between two or more surfaces.

   Dial indicators are simple mechanical devices that convert linear displacements of a pointer to rotation of an indicator on a circular dial. The indicator is set to zero at a certain reference surface, and the instrument or the surface to be measured (either external or internal) is brought into contact with the pointer. The movement of the indicator is read directly on the circular dial (as either plus or minus) to accuracies as high as 1 μm (40 μin.).

   Unlike mechanical systems, **electronic gages** sense the movement of the contacting pointer through changes in the electrical resistance of a strain gage or through inductance or capacitance. The electrical signals are then converted and displayed as linear dimensions. A common electronic gage is the **linear variable differential transformer** (LVDT), used extensively for measuring small displacements. Although they are more expensive than other types of gages, electronic gages have several advantages, such as ease of operation, rapid response, digital readout, less possibility of human error, versatility, flexibility, and the capability to be integrated into automated systems through microprocessors and computers.

4. **Measuring straightness, flatness, roundness, and profile**. These geometric features are important aspects of engineering design and manufacturing; consequently, their accurate measurement is critical.

Straightness can be checked with straight edges or dial indicators. **Autocollimators**, resembling a telescope with a light beam that bounces back from the object, are used for accurately measuring small angular deviations on a flat surface. Optical means, such as **transits** and **laser beams**, are used for aligning individual machine elements in the assembly of machine components.

Flatness can be measured by mechanical means, using a surface plate and a dial indicator; this method can also be used for measuring perpendicularity, which can be measured with the use of precision steel squares as well. Flatness can also be measured by **interferometry**, using an **optical flat**. The flat, a glass or fused quartz disk, with parallel flat surfaces, is placed on the surface of the workpiece. When a monochromatic (one wavelength) light beam is aimed at the surface at an angle, the optical flat splits it into two beams, appearing as light and dark bands to the naked eye. The number of fringes that appear is related to the distance between the surface of the part and the bottom surface of the optical flat. Consequently, a truly flat workpiece surface (that is, when the angle between the two surfaces is zero) will not split the light beam, and no fringes will appear. When the surfaces are not flat, fringes are curved. The interferometry method is also used for observing surface textures and scratches through microscopes.

Roundness is generally described as deviations from true roundness (mathematically, a circle), although the term **out of roundness** is actually more descriptive of the shape of the part. The various methods of measuring roundness fall into two basic categories. In the first method, the round part is placed on a V-block or between centers and is rotated, with the pointer of a dial indicator in contact with the surface. After a full rotation of the workpiece, the difference between the maximum and minimum readings on the dial is noted; this difference is called the **total indicator reading** (TIR), or **full indicator movement**. In the second method, called **circular tracing**, the part is placed on a platform, and its roundness is measured by rotating the platform.

Threads and **gear teeth** must be produced accurately for the smooth operation of gears, reduction of wear and noise level, and interchangeability. They are measured by means of thread gages of various designs that compare the thread produced with a standard thread. The gages used include threaded plug gages, screw-pitch gages (similar to radius gages), micrometers with cone-shaped points, and snap gages with anvils in the shape of threads. Gear teeth are measured with instruments that are similar to dial indicators, with calipers and with micrometers using pins or balls of various diameters.

QR Code 4.2 Coordinate measurement machine scanning technology. *Source:* Courtesy of Carl Zeiss.

Optical projectors, also called **optical comparators**, are now used for checking all profiles. The part is mounted on a table, or between centers, and a light source projects the image on a screen, at magnifications up to 100× or higher. Linear and angular measurements are made directly on the screen, which is equipped with reference lines and circles. The screen can be rotated to allow angular measurements as small as one minute, using verniers.

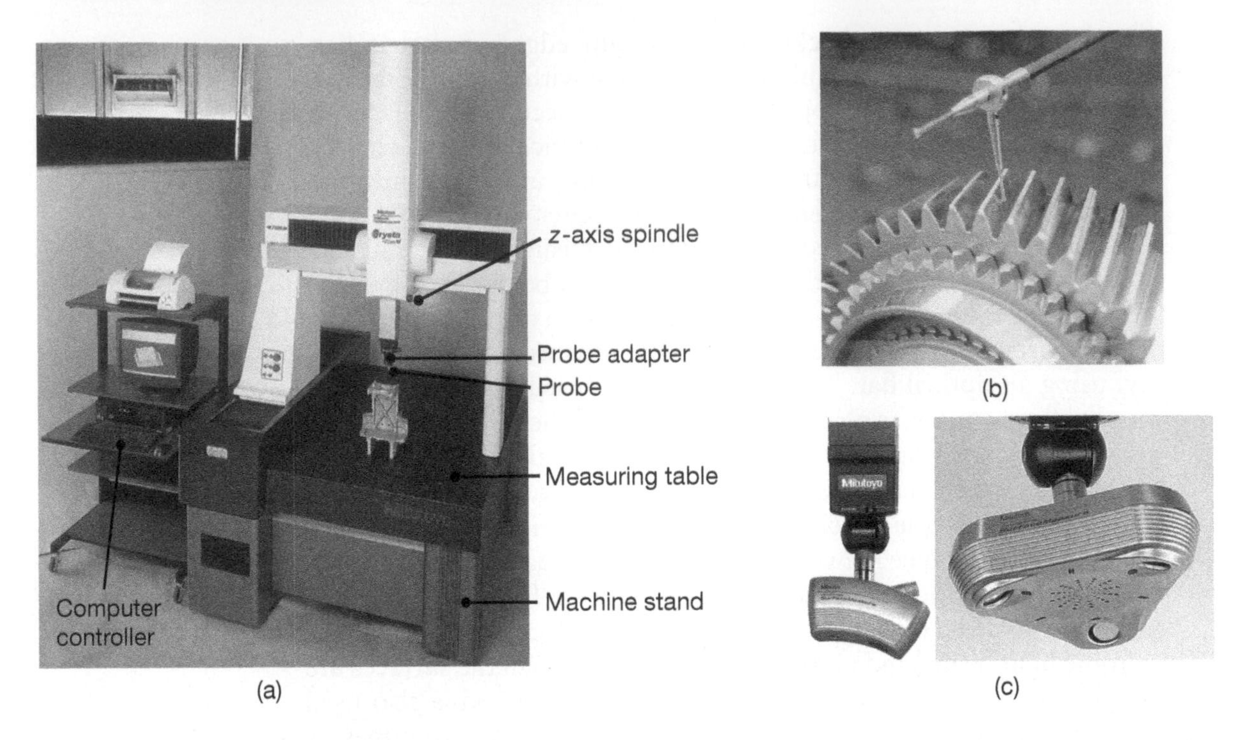

FIGURE 4.17 (a) A coordinate measuring machine with part being measured; (b) a touch signal probe measuring the geometry of a gear; and (c) examples of laser probes. *Source:* Mitutoyo America Corp.

QR Code 4.3 Large coordinate measurement machines.
Source: Courtesy of Carl Zeiss.

5. **Coordinate measuring machines** (CMMs, Fig. 4.17) are versatile in their capability to record measurements of complex profiles rapidly and with high sensitivity [0.25 μm (10 μin.)] They are generally computer controlled for on-line inspection of parts. They can be placed close to machine tools, or incorporated into manufacturing cells (see Section 15.9) for efficient inspection and rapid feedback for adjusting of processing parameters before the next part is made. The machines are made more rugged to resist environmental effects in manufacturing plants, such as temperature variations, vibration, and dirt.
6. **Gages**. Thus far, the word gage (or gauge) has been used to describe some types of measuring instruments, such as a caliper gage, depth gage, telescoping gage, electronic gage, strain gage, and radius gage. However, *gage* also has a variety of other meanings, such as in pressure gage; gage length of a tension-test specimen; and gages for sheet metal, wire, railroad rail, and the bore of shotguns. It should also be noted that, traditionally, the words *gage* and *instrument* have been used interchangeably.

   Gage blocks (or Johansson or *Jo blocks*) are individual square, rectangular, or round metal or ceramic blocks of various sizes. Their surfaces are lapped and are flat, and parallel within a range of 0.02–0.12 μm (1–5 μin.). The blocks can be assembled in many combinations to obtain desired lengths. Dimensional accuracy can be as high as 0.05 μm (2 μin.), but environmental-temperature control is important in using gages for high-precision measurements. Gage-block assemblies are commonly utilized in industry as an accurate reference length; angle blocks are made similarly and are available for angular gaging.

Fixed gages are replicas of the shapes of the parts to be measured. **Plug gages** are commonly used for holes. The *GO gage* is smaller than the *NOT GO* (or *NO GO*) *gage*, and slides into any hole whose smallest dimension is less than the diameter of the gage. The NOT GO gage must not go into the hole. Two gages, one GO gage and one NOT GO gage, are required for such measurements, although both may be on the same device, either at opposite ends or in two steps at one end (*step-type gage*). Plug gages are also available for measuring internal tapers (in which deviations between the gage and the part are indicated by the looseness of the gage), splines, and threads (in which the GO gage must screw into the threaded hole). It should be noted that such gages are not compatible with advanced quality-control techniques and lean manufacturing, and as such, are becoming less popular.

Ring gages are used to measure shafts and similar round parts, and **ring thread gages** are used to measure external threads. The GO and NOT GO features on these gages are identified by the type of knurling on the outside diameter of the rings. **Snap gages** are commonly used to measure external dimensions; they are made with adjustable gaging surfaces for use with parts that have different dimensions. Although fixed gages are inexpensive and easy to use, they only indicate whether a part is too small or too large, as compared with an established standard; they do not measure actual dimensions.

There are several types of **pneumatic gages**, also called **air gages**. The gage head has holes through which pressurized air, supplied by a constant-pressure line, escapes. The smaller the gap between the gage and the hole, the more difficult it is for the air to escape, and hence the back pressure is higher. The back pressure, sensed and indicated by a pressure gage, is calibrated to read dimensional variations of holes. The air gage can be rotated during its use to observe and measure any out-of-roundness of the hole. Outside diameters of such parts as pins and shafts also can be measured, where the air plug is in the shape of a ring slipped over the part. In cases where a ring is not suitable, a fork-shaped gage head (with the air holes at the tips) can be used.

Air gages are easy to use and the resolution can be as fine as 0.125 μm (5 μin.). If the surface roughness of parts is too high, the readings may be unreliable. The non-contacting nature and the low pressure of an air gage has the benefit of not distorting or damaging the part being measured, as could be the case with mechanical gages.

7. **Microscopes.** These are optical instruments used to view and measure very fine details, shapes, and dimensions on small and medium-sized tools, dies, and workpieces. The most common and versatile microscope used in tool rooms is the **toolmaker's microscope**, which is equipped with a stage that is movable in two principal directions and can be read to 2.5 μm (0.0001 in.). The **light-section microscope** is used to measure small surface details and the thickness of deposited films and coatings. A thin, light band is applied obliquely to the surface of the part, and the reflection is viewed at 90°, showing surface roughness, contours, and other features. The **scanning electron microscope** (SEM) has excellent depth of field, and as a result, all regions of a complex part are in focus and can be viewed and photographed to show extremely fine detail. SEM is particularly useful for studying surface textures and fracture patterns. Although expensive, such microscopes are capable of magnifications higher than $100,000\times$.

4.6.2 Automated Measurement

With advanced automation in all aspects of manufacturing processes and operations, the need for *automated measurement* (also called *automated inspection*, see Section 4.8.3) has become apparent. Flexible manufacturing systems and manufacturing cells, described in Chapter 15, have led to the adoption of advanced measurement techniques and systems. Automated inspection involves various on-line sensor systems that monitor the dimensions of parts *while being made*, and use these measurements as feedback in a closed-loop control system (see Section 14.3.3) to correct the process whenever necessary.

To appreciate the importance of on-line monitoring of dimensions, consider the following: If a machine has been producing a certain part with acceptable dimensions, what factors contribute to subsequent deviation in the dimension of the same part produced by the same machine? The major factors could be due to:

1. Static and dynamic deflections of the machine, because of vibrations and fluctuating forces, caused by variations such as in the properties and dimensions of the incoming workpiece material;
2. Deformation of the machine, because of thermal effects, including changes in the temperature of the metalworking fluids, machine bearings and components, or the environment; and
3. Wear of the tooling, dies, and molds.

Consequently, the dimensions of the parts produced will vary, thus necessitating monitoring of dimensions during production. **In-process workpiece control** is accomplished by special gaging and is used in a variety of applications, such as high-production machining and grinding.

Traditionally, part measurements have been made after it has been produced, a procedure known as postprocess inspection. In modern manufacturing practice, measurements are made while the part is being produced on the machine called *in-process, on-line,* or *real-time inspection*. Here, the term *inspection* means to check the dimensions of what is produced or being produced, and to observe whether it complies with the dimensional accuracy specified for that part.

4.7 Dimensional Tolerances

Dimensional tolerance is defined as the permissible or acceptable variation in the dimensions (height, width, depth, diameter, angles) of a part. Tolerances are unavoidable, because it is virtually impossible and not always necessary to manufacture two parts that have precisely the same dimensions. Furthermore, because close dimensional tolerances substantially increase the product cost, specifying a narrow tolerance range is undesirable economically. Tolerances become important only when a part is to be assembled, mated or interact with another part; surfaces that are free or not functional typically do not need close dimensional tolerance control.

Certain terminology has been established to clearly define geometric tolerances, such as the ISO system shown in Fig. 4.18a. Note that both the shaft and the hole have minimum and maximum diameters, respectively, the difference being the dimensional tolerance for each member. A proper

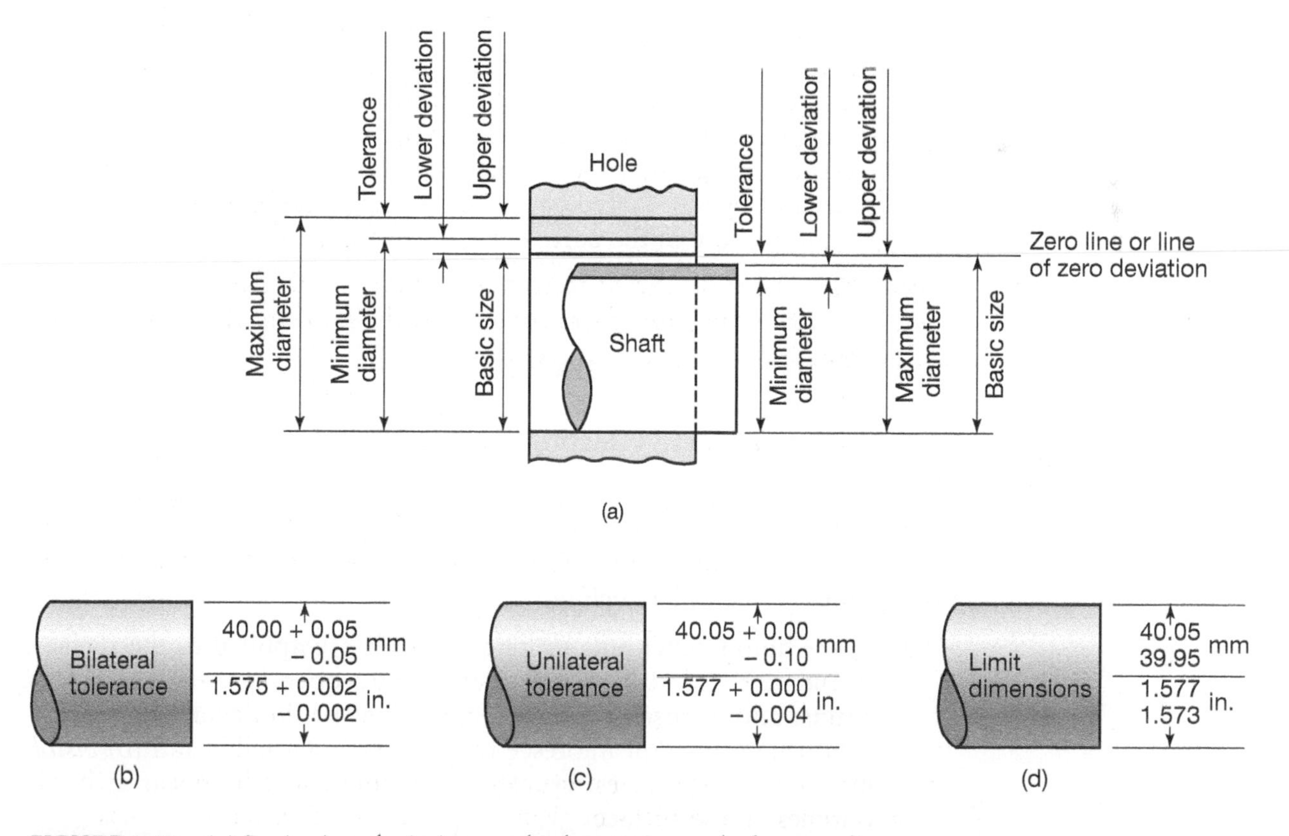

FIGURE 4.18 (a) Basic size, deviation, and tolerance on a shaft, according to the ISO system. (b)–(d) Various methods of assigning tolerances on a shaft.

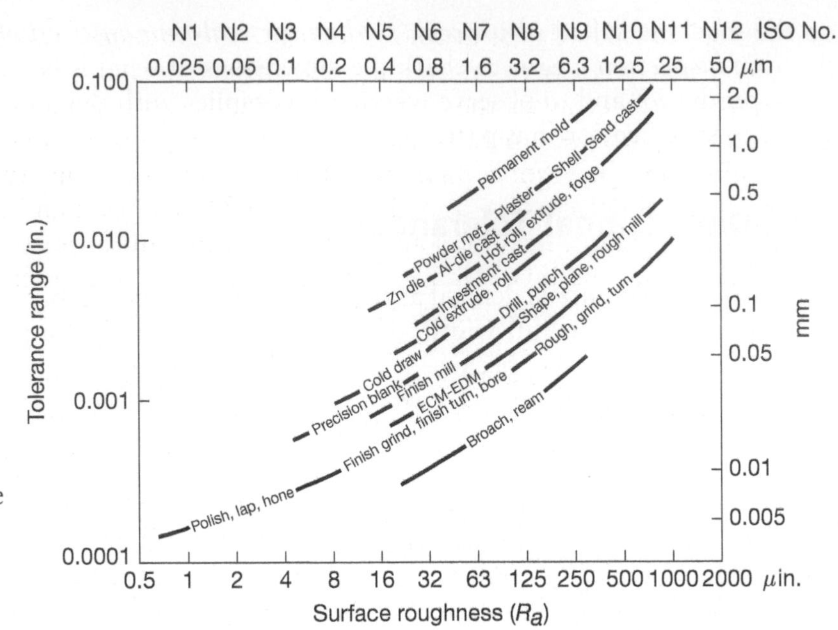

FIGURE 4.19 Tolerances and surface roughness obtained in various manufacturing processes. These tolerances apply to a 25-mm (1-in.) workpiece dimension.

engineering drawing must specify these parameters with numerical values, as shown in Fig. 4.18b.

The range of dimensional tolerances obtained in various manufacturing processes is given in Fig. 4.19. Note that there is a general relationship between tolerances and surface finish of parts manufactured by different processes. Also note the wide range of tolerances and surface finishes that can be obtained. Furthermore, the larger the part, the greater the obtainable tolerance range becomes. Experience has shown that dimensional inaccuracies of manufactured parts are approximately proportional to the cube root of the size of the part. Thus, for example, doubling the size of a part increases the inaccuracies by $2^{1/3} = 1.26$ times, or 26%.

4.8 Testing and Inspection

4.8.1 Nondestructive Testing Techniques

Nondestructive testing (NDT), as the name implies, is carried out in such a manner that the integrity and surface texture of the part are preserved. The basic principles of the more commonly used nondestructive testing techniques are described below.

1. In the **liquid-penetrants technique,** fluids are applied to the surfaces of the part and allowed to penetrate into openings, such as cracks, seams, and porosity. Two common types of liquids are (a) *fluorescent* penetrants that fluoresce under ultraviolet light; and (b) *visible* penetrants, using dyes, usually red in color, which appear as bright outlines on the surface.
2. The **magnetic-particle inspection technique** consists of placing fine ferromagnetic particles on the surface of the part. The particles can

be applied either dry or in a liquid carrier such as water or oil. When the part is magnetized with a magnetic field, a discontinuity (defect) on the surface causes the particles to gather visibly around it. The collected particles generally take the shape and size of the defect.

3. In **ultrasonic inspection**, an ultrasonic beam travels through the part. An internal defect, such as a crack, interrupts the beam and reflects back a portion of the ultrasonic energy. The amplitude of the energy reflected and the time required for return indicate the presence and location of the flaws in the part. The ultrasonic waves are generated by transducers of various types and shapes, called *probes* or *search units*. They operate on the principle of *piezoelectricity* (Section 3.9.6), using materials such as quartz, lithium sulfate, and various ceramics. *Couplants*, such as water, oil, glycerin, and grease, are used to transmit the ultrasonic waves from the transducer to the test piece. The ultrasonic inspection method has high penetrating power and sensitivity and its accuracy is higher than that of other nondestructive inspection methods.
4. The **acoustic-emission technique** detects signals (high-frequency stress waves) generated by the workpiece itself during such phenomena as plastic deformation, crack initiation and propagation, phase transformation, and rapid reorientation of grain boundaries. Bubble formation during boiling and friction and wear of sliding interfaces are other sources of acoustic signals. Acoustic emissions are detected by sensors consisting of piezoelectric ceramic elements. The acoustic-emission technique is typically performed by stressing elastically the part or structure, such as bending a beam, applying torque to a shaft, and pressurizing a vessel. It is particularly effective for continuous surveillance of load-bearing structures.
5. The **acoustic-impact technique** consists of tapping the surface of an object and listening to and analyzing the transmitted sound waves in order to detect discontinuities and flaws. The principle is basically similar to tapping walls, desktops, or countertops in various locations with fingers or a light hammer and listening to the sound emitted. The acoustic-impact technique can be instrumented and automated and is easy to perform; however, the results depend on the geometry and mass of the part, thus requiring a reference standard to identify flaws.
6. **Radiography** involves X-ray inspection to detect internal flaws or variations in the density and thickness of the part. In **digital radiography**, a linear array of detectors is used instead of film, and the data are stored in computer memory. Computer tomography is a similar system except that the monitor produces X-ray images of thin cross sections of the workpiece. **Computer-assisted tomography (catscan)** is based on the same principle and is used widely in medical practice.
7. The **eddy-current inspection method** is based on the principle of electromagnetic induction. The part is placed in, or adjacent to, an electric coil through which alternating current (exciting current) flows at frequencies ranging from 6 to 60 MHz. This current induces eddy currents in the part. Defects existing in the part impede and change

the direction of the eddy currents, causing changes in the electromagnetic field. These changes affect the exciting coil (inspection coil), whose voltage is monitored to detect the presence of flaws.

8. **Thermal inspection** involves observing temperature changes by contact- or noncontact-type heat-sensing devices, such as contact temperature probes (See Fig. 5.28c) or infrared scanners. Defects in the workpiece, such as cracks, poorly made joints, and debonded regions in laminated structures, cause a change in the temperature distribution. In **thermographic inspection**, heat-sensitive paints and papers, liquid crystals, and other coatings are applied to the surface of the part. Changes in their color or appearance indicate the presence of defects.
9. The **holography technique** produces a three-dimensional image of the part, using an optical system. This technique is generally used on simple shapes and highly polished surfaces, and the image is recorded on a photographic film. **Holographic interferometry** is used in inspecting various shapes and surface characteristics. Defects are revealed by using multiple-exposure techniques while the part is subjected to external forces or changes in temperature.

   In **acoustic holography**, information on internal defects is obtained directly from the image of the interior of the part. In **liquid-surface acoustical holography**, the part and two ultrasonic transducers (one for the object beam and the other for the reference beam) are immersed in a tank filled with water. The holographic image is then obtained from the ripples in the tank. In **scanning acoustical holography**, only one transducer is used and the hologram is produced by electronic-phase detection.

4.8.2 Destructive Testing Techniques

As the name indicates, in *destructive testing methods* the part no longer maintains its integrity, original shape, or surface texture. Recall that *mechanical testing methods*, described in Chapter 2, are all destructive, in that a sample has to be removed from the part in order to test it, and it cannot be used after the test. Other destructive tests include speed testing of grinding wheels to determine their bursting speed, high-pressure testing of pressure vessels to determine their bursting pressure, and formability tests for sheet metal (see Section 7.7.1). Hardness tests that leave large impressions (such as Brinell) also may be regarded as destructive testing, although microhardness tests are typically nondestructive, because only a very small permanent indentation is made. This distinction is based on the assumption that the material is not *notch sensitive* (Section 2.9).

4.8.3 Automated Inspection

Traditionally, individual parts and assemblies of parts have been manufactured in batches, sent to inspection in quality-control rooms, and, if approved, put in inventory. If products do not pass the quality inspection, they are either reworked, recycled, scrapped, or kept in inventory on the basis of a certain acceptable deviation from the standard. Such a system (**postprocess inspection**) is obviously inefficient as it tracks the defects after they have occurred and in no way attempts to prevent defects.

Automated inspection uses a variety of sensors that monitor the relevant parameters during the manufacturing process (**on-line inspection**). Using these measurements, the process then automatically corrects itself to produce acceptable parts; further inspection of the part at another location in the plant is thus unnecessary and inefficient. Parts may also be inspected immediately after they are produced (**in-process inspection**).

The use of appropriate sensors (see Section 14.8) and computer-control systems (Chapter 15) has enabled the integration of automated inspection into manufacturing operations. Such a system ensures that no part is moved from one manufacturing operation to another (such as a turning operation on a lathe followed by cylindrical grinding) unless the part is made correctly and meets the standards set for the first operation. Automated inspection is flexible and responsive to product design changes, less operator skill is required, productivity is increased, and parts have higher quality, reliability, and dimensional accuracy.

Sensors for automated inspection. Rapid advances in **sensor technology** (Section 14.8) have made feasible the real-time monitoring of manufacturing operations. Using various probes and sensors, it is possible to rapidly detect such parameters as dimensional accuracy, surface roughness, temperature, force, power, vibration, tool wear, and the presence of external or internal defects.

4.9 Quality Assurance

Quality assurance is the total effort by a manufacturer to ensure that its products conform to a detailed set of specifications and standards regarding, for example, surface finish, dimensional tolerances, composition, color, and mechanical, physical, and chemical properties of materials. The standards ensure proper assembly, using interchangeable, defect-free components, and the fabrication of a product that performs as intended by its designers.

The often-repeated statement that quality must be *built into a product* reflects the important concept that quality cannot be inspected into a finished product. Every aspect of design and manufacturing operations, such as material selection, production, and assembly, must be analyzed in detail to ensure that quality is truly built into the final product.

An essential approach is to control materials and processes in such a manner that the products are made correctly in the first place. Because 100% inspection may be impractical, several methods of inspecting smaller and statistically relevant sample lots are implemented. These methods all use statistics to determine the probability of defects occurring in the total production batch.

Inspection involves a series of steps:

1. Inspecting incoming materials to ensure that they meet specific property, dimension, and surface finish and integrity requirements;
2. Inspecting individual product components to make sure that they meet all specifications;
3. Inspecting the product to make sure that individual parts have been assembled properly; and
4. Testing the product to make sure that it consistently functions as designed and as intended for its use in the marketplace.

Inspections must continue throughout production, because there are always possibilities of variations in (a) the dimensions and properties of incoming materials; (b) the performance of tools, dies, and machines used; (c) possibilities of human error; and (d) errors made during assembly of the product. Consequently, no two products are made exactly alike. An important aspect of quality control is the capability to promptly analyze defects and eliminate them or to reduce them to acceptable levels. The totality of all these activities is called **total quality management** (TQM).

In order to control quality, one must be able to (1) *measure quantitatively* the level of quality and (2) *identify* all the material and process variables that can be controlled. The level of quality obtained during production can then be established by inspecting the product to determine whether it meets all the specifications for dimensional tolerances, surface finish, defects, and various other characteristics.

4.9.1 Statistical Methods of Quality Control

The use of *statistical methods* is essential in modern manufacturing operations, because of the large number of material and process variables involved. Events that occur randomly (without any particular trend or pattern) are called **chance variations**; those that can be traced to specific causes are called **assignable variations**. For example, results from four-point bending tests typically have a natural range of strength predictions, because of chance variations in material strength throughout its volume. These variations arise from the random distribution of flaws of various types and sizes within the material (see Section 2.5). Also, if the specimens are poorly prepared, so that some have one or more notches and others are notch-free, this can affect the test results. Although the existence of *variability* in production operations has been recognized for centuries, it was Eli Whitney (1765–1825) who first grasped its full significance when he found that *interchangeable* parts were indispensable to the mass production of firearms.

The terms commonly used in **statistical quality control** (SQC) are:

1. **Sample size**: The number of parts to be inspected, whose properties are studied to gain information about the whole population.
2. **Population** (also called the **universe**): The totality of individual parts of the same design from which samples are taken.

3. **Random sampling:** Taking a sample from a population or lot in which each item has an equal chance of being included in the sample.
4. **Lot size:** A subset of the population that may be treated as representative of the population.

Samples are inspected for certain specific characteristics and features, such as dimensional tolerances, surface finish, and various defects, using the techniques and instruments described earlier in this chapter. These characteristics fall into two categories: those that can be measured *quantitatively* (known as method of variables) and those that can be measured *qualitatively* (method of attributes).

The **method of variables** is the quantitative *measurement* of such characteristics as dimensions, tolerances, surface finish, and physical and mechanical properties. Such measurements are made for each of the members in the lot under consideration. The results are then compared with the specifications for that particular part. The **method of attributes** involves observing the *presence or absence* of qualitative characteristics, such as external or internal defects in machined, formed, or welded parts, or dents in sheet metal products, for each of the units in the lot under consideration. Sample size for attributes-type data is generally larger than that for variables-type data, because accurate qualitative measures are more difficult to obtain and variance is therefore higher.

Measurement results typically will vary during inspection. For example, when measuring the diameter of turned shafts as they are produced on a lathe (using a micrometer), it will be found that their diameters vary, even though it is ideally desirable for all the shafts to be exactly the same size. If the measured diameters of the turned shafts in a given population are listed and compared, it will be noted that one or more shafts have the smallest diameter, and one or more have the largest diameter. The majority of the turned shafts have diameters that lie between these extremes. The diameters can then be grouped and plotted on a *bar graph*, where each bar represents the number of parts in each diameter group (Fig. 4.20a). The bars show a **distribution** (also called a **spread** or **dispersion**) of the diameter measurements. The bell-shaped curve in Fig. 4.20a is called **frequency distribution**, and represents the frequency with which parts within each diameter group are being produced.

Data from manufacturing operations often fit curves that are represented by a mathematically derived **normal distribution curve** (Fig. 4.20b). These curves, also called *Gaussian curves*, are developed on the basis of probability. The bell-shaped normal distribution curve fitted to the data in Fig. 4.20b has two important features. First, it shows that the diameters of most shafts tend to cluster around an *average* value (**arithmetic mean**). This average is generally designated as $\bar{x}$ and is calculated from the expression

$$\bar{x} = \frac{x_1 + x_2 + x_3 + \cdots + x_n}{n}, \tag{4.8}$$

where the numerator is the sum of all measured values (diameters) and n is the number of measurements (the number of shafts).

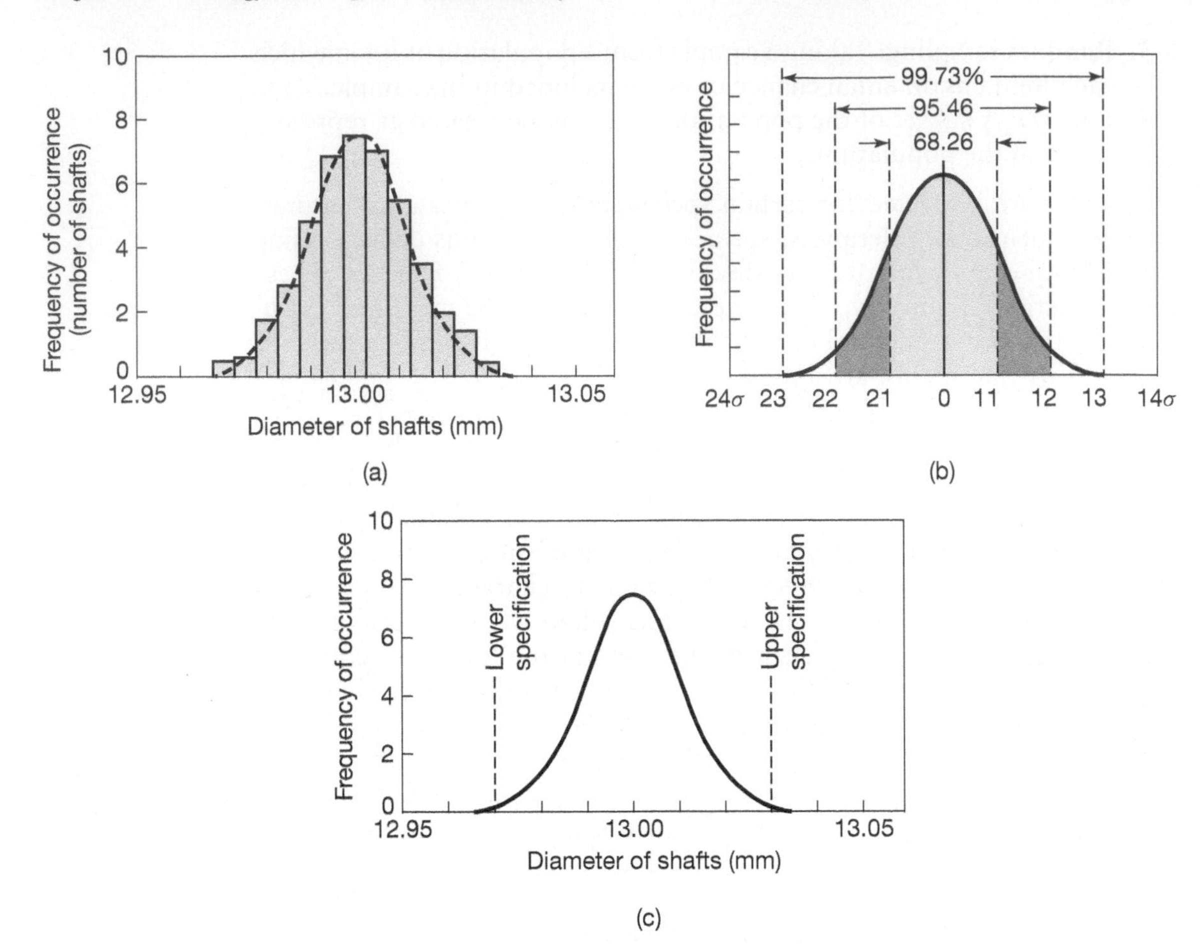

FIGURE 4.20 (a) A plot of the number of shafts measured and their respective diameters. This type of curve is called a *frequency distribution*. (b) A normal distribution curve indicating areas within each range of standard deviation. *Note:* The greater the range, the higher the percentage of parts that fall within it. (c) Frequency distribution curve, showing lower and upper specification limits.

The second feature of this curve is its width, indicating the *dispersion* of the diameters measured; the wider the curve, the greater is the dispersion. The difference between the largest value and smallest value is called the *range*, R:

$$R = x_{\text{max}} - x_{\text{min}}. \tag{4.9}$$

Dispersion is estimated by the **standard deviation**, which is generally denoted as σ and obtained from the expression

$$\sigma = \sqrt{\frac{(x_1 - \overline{x})^2 + (x_2 - \overline{x})^2 + (x_3 - \overline{x})^2 + \cdots + (x_n - \overline{x})^2}{n - 1}}, \tag{4.10}$$

where x is the measured value for each part. Note from the numerator in Eq. (4.10) that (a) as the curve widens, the standard deviation becomes greater; and (b) σ has units of linear dimension.

Since the number of parts that fall within each group is known, the percentage of the total population represented by each group can be calculated. Thus, Fig. 4.20b shows that the diameters of 99.73% of the turned shafts fall within the range of $\pm 3\sigma$, 95.46% within $\pm 2\sigma$, and 68.26% within $\pm 1\sigma$; only 0.2% fall outside the $\pm 3\sigma$ range.

Six sigma. An important practice in manufacturing operations, as well as in business and service industries, is the concept of *six sigma.* Note from the preceding discussion that three sigma in manufacturing would result in 0.27%, or 2700 parts per million defective parts. This is an unacceptable rate in modern manufacturing; in fact, it has been estimated that at the three sigma level, virtually no modern computer would function properly and reliably, and in the service industries, 270 million incorrect credit-card transactions would be recorded each year in the United States alone. It has further been estimated that companies operating at three to four sigma levels lose about 10–15% of their total revenue due to defects. Consequently, extensive efforts continue to be made to virtually eliminate all defects in products and processes.

Six sigma is a set of statistical tools, based on well-known total quality management principles to continually measure the quality of products and services. It includes such considerations as customer satisfaction, delivering defect-free products, and understanding process capabilities. The approach consists of a clear focus on defining the problem, measuring relevant quantities, and analyzing, improving, and controlling processes and activities. Six sigma has been combined with the approaches in lean manufacturing (see Section 15.13) to produce a hybrid methodology called **lean six sigma.** This approach sees the approach of six sigma (elimination of variation and design) and lean manufacturing (elimination of waste and development of value streams) as complimentary.

4.9.2 Statistical Process Control

If the number of parts that do not meet set standards increases during a production run, the cause must be determined (for example, variability in incoming materials, machine controls, degradation of metalworking fluids, operator boredom, etc.) and appropriate actions taken. It was only in the early 1950s that a systematic statistical approach was developed to guide machine operators. Known as *statistical process control* (SPC), this approach involves several elements: (a) control charts and setting control limits, (b) capabilities of the particular manufacturing process, and (c) characteristics of the machinery involved.

Control charts. The frequency distribution curve, given in Fig. 4.20a, shows a range of shaft diameters that may fall beyond the specified design tolerance range. The same bell-shaped curve is shown in Fig. 4.20c, which now includes the specified tolerances for the diameters of the turned shafts. *Control charts* graphically represent the variations of a process over a period of time; they consist of data plotted during production, and typically there are two plots. The quantity $\bar{x}$ in Fig. 4.21a is the average

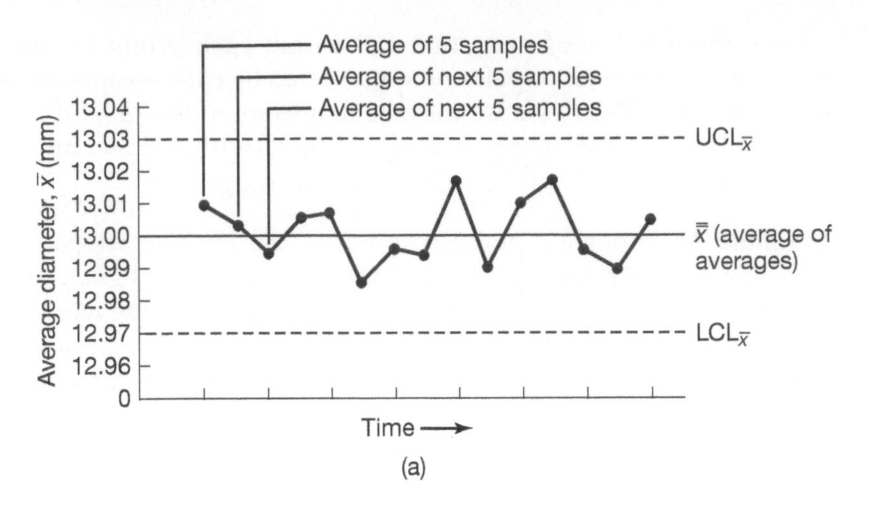

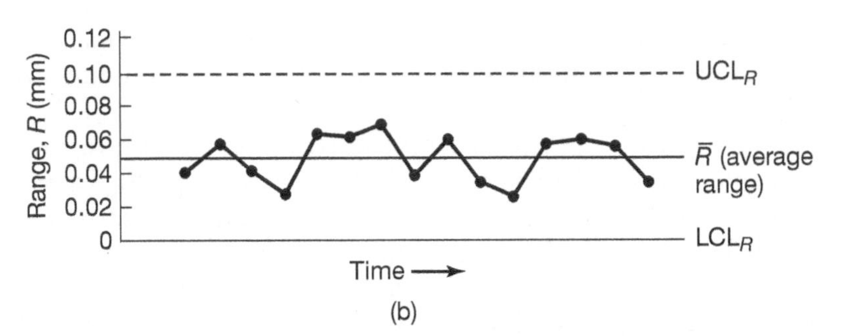

FIGURE 4.21 Control charts used in statistical quality control. The process shown is in good statistical control, because all points fall within the lower and upper control limits. In this illustration, the sample size is five, and the number of samples is 15.

for each subset of samples taken and inspected, say, a subset consisting of 5 parts. It has been observed that in manufacturing, a sample size of between 2 and 10 parts can be sufficiently accurate, depending on the standard deviation and provided that sample size is held constant throughout the inspection.

The frequency of sampling depends on the nature of the manufacturing process. Some processes may require continuous sampling, whereas others may require only one sample per day. Quality-control analysts are best qualified to determine the appropriate frequency for a particular situation. Because the measurements shown in Fig. 4.21a are made consecutively, the abscissa of these control charts also represents time. The solid horizontal line in this figure is the **average of averages (grand average)**, denoted as $\bar{\bar{x}}$ and represents the *population mean*. The upper and lower horizontal broken lines indicate the **control limits** for *the process*.

The control limits are set on these charts according to statistical-control formulas, designed to keep actual production within the acceptable $\pm 3\sigma$ range. Thus, for $\bar{x}$,

$$\text{Upper control limit } (\text{UCL}_{\bar{x}}) = \bar{x} + 3\sigma = \bar{\bar{x}} + A_2\bar{R} \tag{4.11}$$

and

$$\text{Lower control limit } (\text{LCL}_{\bar{x}}) = \bar{x} - 3\sigma = \bar{\bar{x}} - A_2\bar{R}, \tag{4.12}$$

where A_2 is obtained from Table 4.2, and $\bar{R}$ is the average of the R values.

TABLE 4.2 Constants for control charts.

Sample size	A_2	D_4	D_3	d_2
2	1.880	3.267	0	1.128
3	1.023	2.575	0	1.693
4	0.729	2.282	0	2.059
5	0.577	2.115	0	2.326
6	0.483	2.004	0	2.534
7	0.419	1.924	0.078	2.704
8	0.373	1.864	0.136	2.847
9	0.337	1.816	0.184	2.970
10	0.308	1.777	0.223	3.078
12	0.266	1.716	0.284	3.258
15	0.223	1.652	0.348	3.472
20	0.180	1.586	0.414	3.735

The major goal of statistical process control is to improve the manufacturing process via the aid of control charts to eliminate assignable causes. Control limits are calculated based on the historical production capability of the equipment itself. Generally not associated with design tolerance specifications or dimensions, they indicate the limits within which a certain percentage of measured values is normally expected to fall due to the inherent variations of the process itself.

The second control chart, given in Fig. 4.21b, shows the range R in each subset of samples. The solid horizontal line represents the average of R values, denoted as $\overline{R}$ in the lot; it is a measure of the variability in the samples. The upper and lower control limits for R are obtained from the equations

$$\text{UCL}_R = D_4\overline{R} \tag{4.13}$$

and

$$\text{LCL}_R = D_3\overline{R}, \tag{4.14}$$

where the constants D_4 and D_3 are read from Table 4.2. This table also includes values for the constant d_2 which is used in estimating the standard deviation from the equation

$$\sigma = \frac{\overline{R}}{d_2}. \tag{4.15}$$

When the curve of a control chart is similar to that shown in Fig. 4.21b, the process is said to be **in good statistical control**. In other words, (a) there is no clear and discernible trend in the curve; (b) the points (measured values) are random with time; and (c) they do not exceed the control limits. Note that curves such as those shown in Figs. 4.22a through c indicate certain trends; for example, about halfway in Fig. 4.22a, the diameter of the shafts increases with time, perhaps due to a change in one of the process variables, such as wear of the cutting tool. If, as in Fig. 4.22b, the trend is toward consistently larger diameters (hovering around the upper control limit), it could mean that the tool settings on the lathe may not be correct and, as a result, the shafts being turned are consistently too large.

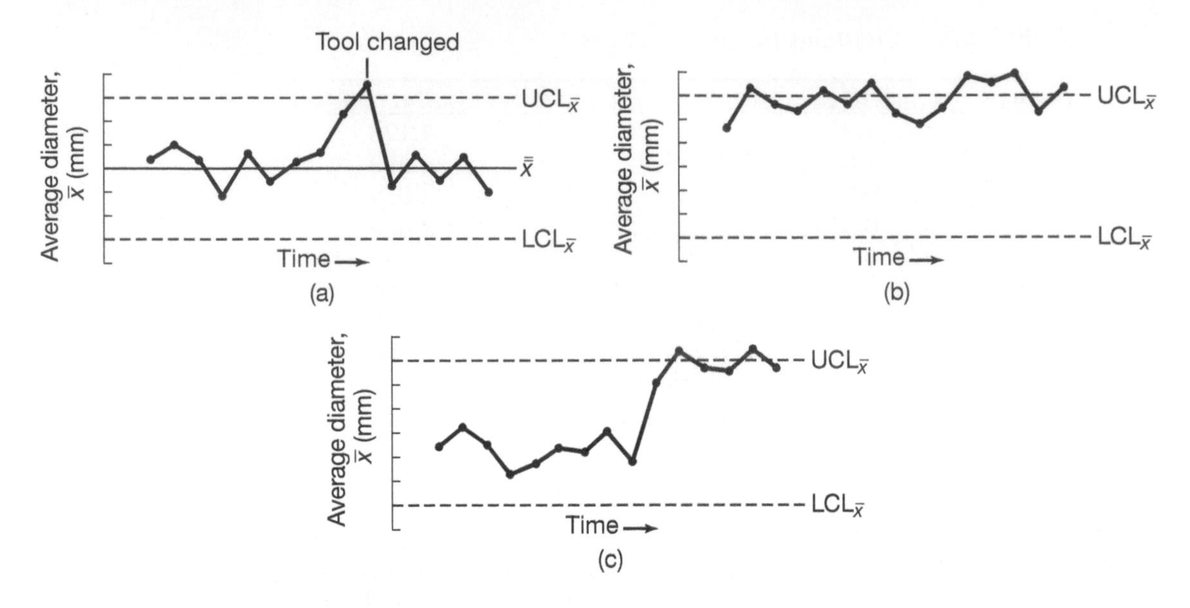

FIGURE 4.22 Control charts. (a) Process begins to become out of control, because of factors such as tool wear. The tool is changed, and the process is then in good statistical control. (b) Process parameters are not set properly; thus, all parts are around the upper control limit. (c) Process becomes out of control, because of factors such as a sudden change in the properties of the incoming material.

Figure 4.22c shows two distinct trends that may be caused by factors such as a change in the properties of the incoming material or a change in the performance of the cutting fluid (for example, by degradation). These situations place the process **out of control**.

Analyzing control-chart patterns and trends requires considerable experience in order to be able to identify the specific cause(s) of an out-of-control situation. A further reason for out-of-control situations is **overcontrol** of the manufacturing operation, that is, setting upper and lower control limits too close to each other, and hence setting a smaller standard-deviation range. In order to avoid overcontrol, control limits are set on process capability, rather than unrelated (and sometimes arbitrary) ranges.

It is evident that operator training is critical for successful implementation of SPC on the shop floor. Once process-control guidelines are established, operators should also be given some authority to make adjustments in processes that are becoming out of control, a task now made easier by the availability of a variety of software. For example, electronic measuring devices are integrated directly into a computer system or transmit their data to the *cloud* (remote servers for data storage where they can be accessed by other computers or devices) for real-time SPC. Figure 4.23 shows such a multifunctional computer system in which the output from a digital caliper or micrometer is analyzed in real time and displayed in several ways, such as by frequency distribution curves and control charts.

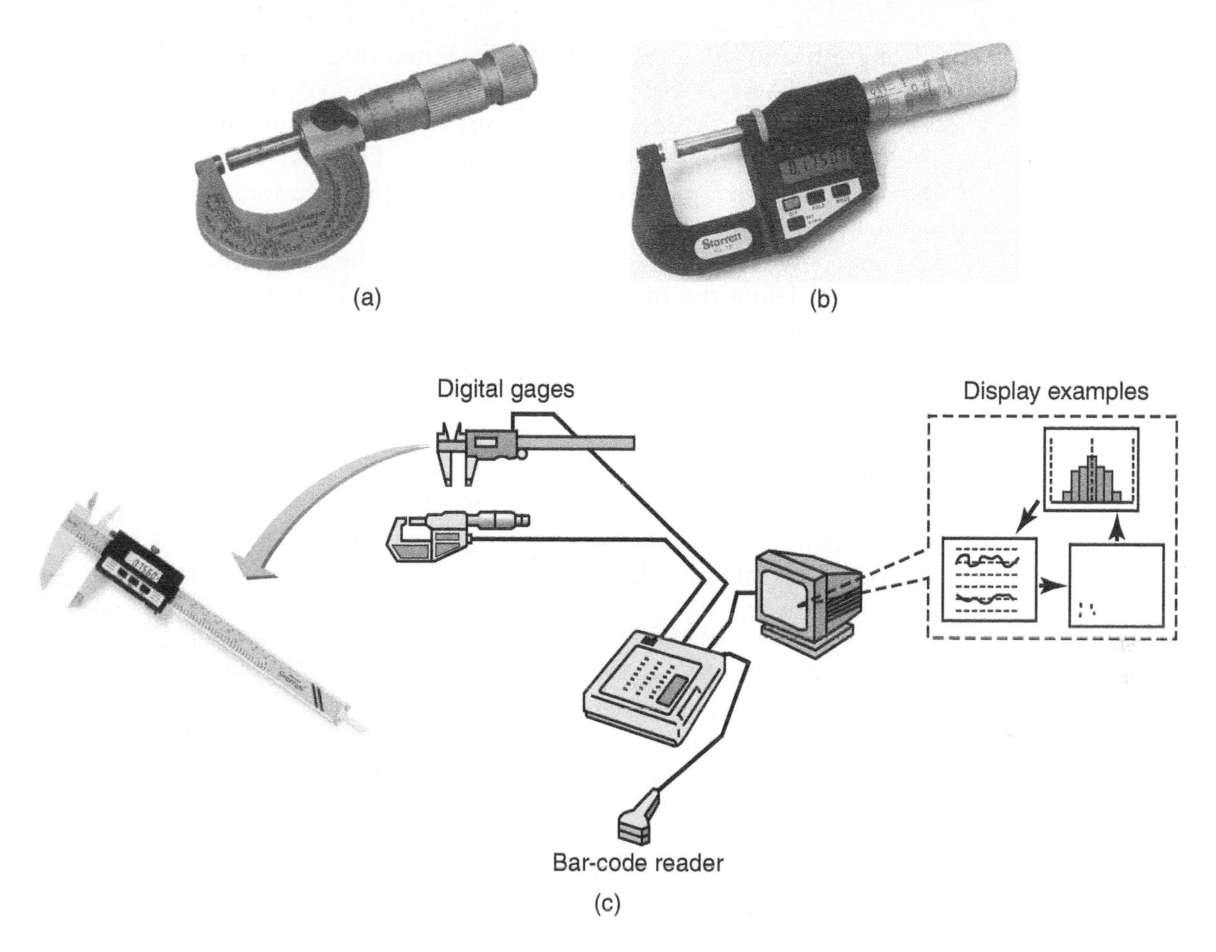

FIGURE 4.23 Schematic illustration showing integration of digital gages with a miniprocessor for real-time data acquisition and SPC/SQC capabilities. Note the examples on the CRT displays, such as frequency distribution and control charts. *Source:* L.S. Starrett Company.

Process capability. *Process capability* is defined as the limits within which individual measurement in a particular manufacturing operation would normally be expected to fall when only random variation is present. It tells us that the process is capable of producing parts within certain limits of precision. Since a manufacturing process involves materials, machinery, and operators, each of these aspects can be analyzed individually to be able to identify a problem when process capabilities do not meet part specifications.

4.9.3 Validation

Quality-control approaches described above are focused upon the repeatability and controllability of a manufacturing process. While it is important that design objectives be met by the manufacturing process, this is not the focus of many of the quality-control procedures described above. **Process validation** involves the analysis of data gathered throughout a product's development, including design and manufacturing. It has the goal of confirming that (a) quality is designed into a product; (b) a manufacturing process is properly designed to achieve the desired quality levels; and (c) that the designed process can reliably and robustly produce products that meet the design specifications. Process validation requires three essential steps: process design, process qualification, and continued process verification.

Process design involves collecting data from the design phase and using this data to define the manufacturing strategy. This stage involves such approaches as *design of experiments (DOE)* to identify the sensitivity of outcomes to process variations, optimization of processes, and identification of critical variables; determination of *critical quality attributes* (CQA) that are essential for determining product quality; determination of a strategy for process control; and *design space verification* to confirm that quality can be obtained within a range of input and process variables. The end result of successful process design will be a strategy for producing the product as intended by the designer, with inherent incorporation of quality standards.

Process qualification. In this step, the manufacturing strategy developed in process design is assessed to ensure that manufacturing targets can be attained reproducibly, and that manufacturing equipment can indeed confirm quality and output goals. Operating ranges should be shown to be capable of being held as long as would be necessary during routine production. It is important to verify that equipment operates in accordance with the process requirements in all anticipated ranges of operation. This verification should include challenging the equipment functions while under a load comparable to that expected during routine production. Process performance must be verified and qualified, based on a written protocol. Once process qualification has been completed, production can start.

Continued process verification (CPV) is the ongoing monitoring of production quality, including the procedures described above. Some federal agencies have codified particular CPV procedures and reporting requirements, notably in the pharmaceutical industry.

SUMMARY

- The surface of a workpiece can significantly affect its properties and characteristics, including friction and wear, lubricant effectiveness, appearance and geometric features, and thermal and electrical conductivity of contacting bodies. (Section 4.1)
- The surface structure of a metal has typically been plastically deformed and work hardened during prior processing. Surface roughness can be quantified by various techniques. Surface integrity can be affected by several defects. Flaws, directionality, roughness, and

waviness are measurable quantities that are used to describe surface texture. (Sections 4.2, 4.3)

- Tribology is the science and technology of interacting surfaces, and encompasses friction, wear, and lubrication. Friction may be desirable or undesirable, depending on the specific manufacturing circumstances. (Section 4.4)
- Wear, defined as the progressive loss or removal of material from a surface, alters the geometry of the workpiece and tool and die interfaces, and thus affects the manufacturing operation, dimensional accuracy, and the quality of parts produced. Friction and wear can be reduced by using various liquid or solid lubricants, as well as by applying ultrasonic vibrations. Four regimes of lubrication are relevant to metalworking processes. (Section 4.4)
- Several surface treatments are used to impart specific physical and mechanical properties to the workpiece. The techniques employed typically include mechanical working, physical and chemical means, heat treatments, and coatings. Cleaning of a manufactured workpiece involves removal of solid, semisolid, and liquid contaminants by various means. (Section 4.5)
- Parts made are measured by a variety of instruments with specific features and characteristics. In automated measurement, measuring equipment are linked to microprocessors and computers, for accurate in-process control of manufacturing operations. (Section 4.6)
- Dimensional tolerances and their specification are important factors in manufacturing, as they not only affect subsequent assembly of the parts made and the accuracy and operation of all types of machinery and equipment, but also can significantly influence product cost. (Section 4.7)
- Several nondestructive and destructive testing techniques are available for inspection of parts made. Inspection of each part being made is now possible using automated and reliable inspection techniques. (Section 4.8)
- Quality assurance is the total effort by a manufacturer to ensure that its products conform to a specific set of standards and specifications. Statistical quality-control and process-control techniques are now widely employed in defect detection and prevention. The total quality-control philosophy focuses on prevention, rather than detection, of defects. (Section 4.9)
- Process validation is a three-part effort (1) to ensure that a particular manufacturing system design is capable of achieving desired quality levels and design attributes; (2) that the manufacturing system developed is certified to meet production based on a design standard; and (3) that quality assurance procedures are used to confirm that desired quality is maintained during production. (Section 4.9)

SUMMARY OF EQUATIONS

Arithmetic mean value, $R_a = \dfrac{y_a + y_b + y_c + \cdots + y_n}{n} = \dfrac{1}{n}\sum_{i=1}^{n} y_i = \dfrac{1}{l}\int_0^l |y|\, dx$

Root-mean-square average, $R_q = \sqrt{\dfrac{y_a^2 + y_b^2 + y_c^2 + \cdots + y_n^2}{n}} = \sqrt{\dfrac{1}{n}\sum_{i=1}^{n} y_i^2} = \left[\dfrac{1}{l}\int_0^l y^2\, dx\right]^{1/2}$

Coefficient of friction, $\mu = \dfrac{F}{N} = \dfrac{\tau}{\text{Hardness}}$

Friction (shear) factor, $m = \dfrac{\tau_i}{k}$

Adhesive wear, $V = k\dfrac{lW}{3p}$

Arithmetic mean, $\bar{x} = \dfrac{x_1 + x_2 + x_3 + \cdots + x_n}{n}$

Range, $R = x_{\text{max}} - x_{\text{min}}$

Standard deviation, $\sigma = \sqrt{\dfrac{(x_1 - \bar{x})^2 + (x_2 - \bar{x})^2 + (x_3 - \bar{x})^2 + \cdots + (x_n - \bar{x})^2}{n-1}}$

Upper control limit, $\text{UCL}_{\bar{x}} = \bar{x} + 3\sigma = \bar{\bar{x}} + A_2\bar{R}$

Lower control limit, $\text{LCL}_{\bar{x}} = \bar{x} - 3\sigma = \bar{\bar{x}} - A_2\bar{R}$

BIBLIOGRAPHY

ASM Handbook, Vol. 5A: Thermal Spray Technology, ASM International, 2013.

ASM Handbook, Vol. 5B: Protective Organic Coatings, ASM International, 2013.

Bayer, R.G., *Mechanical Wear Fundamentals and Testing*, 2nd ed., Dekker, 2005.

Besterfield, D.H., *Quality Improvement*, 9th ed., Prentice Hall, 2012.

Bhushan, B., *Introduction to Tribology*, 2nd ed., Wiley, 2013.

Breyfogle, F., *Implementing Six Sigma: Smarter Solutions Using Statistical Methods*, 2nd ed., Wiley, 2003.

Campbell, R., *Integrated Product Design and Manufacturing Using Geometric Dimensioning and Tolerancing*, CRC Press, 2002.

Cogorno, G., *Geometric Dimensioning and Tolerancing for Mechanical Design*, 2nd ed., McGraw-Hill, 2012.

Curtis, M.A., *Handbook of Dimensional Measurement*, 5th ed., Industrial Press, 2014.

Hamrock, B.J., Schmid, S.R., and Jacobson, B.O., *Fundamentals of Fluid Film Lubrication*, 2nd ed, CRC Press, 2005.mon

Heiller, C., *Handbook of Nondestructive Evaluation*, 2nd ed., McGraw-Hill, 2013.

Hocken, R.J., and Pereira, P.H., (eds.), *Coordinate Measuring Machines and Systems*, 2nd ed., Dekker, 2012.

Hull, B., *Manufacturing Best Practices: Optimizing Productivity and Product Quality,* Wiley, 2011.

Krulikowski, A., *Fundamentals of Geometric Dimensioning and Tolerancing*, 3rd ed., Cengage, 2012.

Madsen, D.A., and Madsen, D.P., *Geometric Dimensioning and Tolerancing*, 9th ed., Goodheart-Wilcox, 2013.

Meadows, J.D., *Geometric Dimensioning and Tolerancing*, Dekker, 1995.

Montgomery, D.C., *Introduction to Statistical Quality Control*, 7th ed., Wiley, 2013.

Process Validation: General Principles and Practices. US Food and Drug Administration, 2011, at www.fda.gov.

Pyzdek, T., and Keller, P., *The Six Sigma Handbook*, 4th ed., McGraw-Hill, 2014.

Rathore, A., and Sofer, G. (eds.), *Process Validation in Manufacturing of Biopharmaceuticals*, 3rd ed., CRC Press, 2012.

Roy, M. (ed.), *Surface Engineering for Enhanced Performance Against Wear*, Springer, 2014.

Stachowiak, G.W., and Batchelor, A.W., *Engineering Tribology*, 4th ed., Butterworth-Heinemann, 2014.

Whitehouse, D.J., *Handbook of Surface and Nanometrology*, 2nd ed., CRC Press, 2011.

Williams, J.A., *Introduction to Tribology*, Cambridge University Press, 2006.

QUESTIONS

4.1 Explain what is meant by surface integrity. Why should we be interested in it?

4.2 Why are surface-roughness design requirements in engineering so broad? Give appropriate examples.

4.3 We have seen that a surface has various layers. Describe the factors that influence the thickness of these layers.

4.4 What is the consequence of oxides of metals being generally much harder than the base metal? Explain.

4.5 What factors would you consider in specifying the lay of a surface?

4.6 Describe the effects of various surface defects (see Section 4.3) on the performance of engineering components in service. How would you go about determining whether or not each of these defects is important for a particular application?

4.7 Explain why the same surface roughness values do not necessarily represent the same type of surface.

4.8 In using a surface roughness measuring instrument, how would you go about determining the cutoff value? Give appropriate examples.

4.9 What is the significance of the fact that the stylus path and the actual surface profile are generally not the same?

4.10 Give two examples each in which waviness of a surface would be (a) desirable and (b) undesirable.

4.11 Explain why surface temperature increases when two bodies are rubbed against each other. What is the significance of a temperature rise due to friction?

4.12 To what factors would you attribute the fact that the coefficient of friction in hot working is higher than in cold working, as shown in Table 4.1?

4.13 In Section 4.4.1, it was noted that the values of the coefficient of friction can be much higher than unity. Explain why.

4.14 Describe the tribological differences between ordinary machine elements (such as meshing gears, cams in contact with followers, and ball bearings with inner and outer races) and elements of metalworking processes (such as forging, rolling, and extrusion, which involve workpieces in contact with tools and dies).

4.15 Give the reasons that an originally round specimen in a ring-compression test may become oval after deformation.

4.16 Can the temperature rise at a sliding interface exceed the melting point of the metals? Explain.

4.17 List and briefly describe the types of wear encountered in engineering practice.

4.18 Explain why each of the terms in the Archard formula for adhesive wear, Eq. (4.6), should affect the wear volume.

4.19 How can adhesive wear be reduced? How can fatigue wear be reduced?

4.20 It has been stated that as the normal load decreases, abrasive wear is reduced. Explain why this is so.

4.21 Does the presence of a lubricant affect abrasive wear? Explain.

4.22 Explain how you would estimate the magnitude of the wear coefficient for a pencil writing on paper.

4.23 Describe a test method for determining the wear coefficient k in Eq. (4.6). What would be the difficulties in applying the results from this test to a manufacturing application, such as predicting the life of tools and dies?

4.24 Why is the abrasive wear resistance of a material a function of its hardness?

4.25 We have seen that wear can have detrimental effects on engineering components, tools, dies, etc. Can you visualize situations in which wear could be beneficial? Give some examples. (*Hint:* Note that writing with a pencil is a wear process.)

4.26 On the basis of the topics discussed in this chapter, do you think there is a direct correlation between friction and wear of materials? Explain.

4.27 You have undoubtedly replaced parts in various appliances and automobiles because they were worn.

Describe the methodology you would follow in determining the type(s) of wear these components have undergone.

4.28 Why is the study of lubrication regimes important?

4.29 Explain why so many different types of metalworking fluids have been developed.

4.30 Differentiate between (a) coolants and lubricants; (b) liquid and solid lubricants; (c) direct and indirect emulsions; and (d) plain and compounded oils.

4.31 Explain the role of conversion coatings. Based on Fig. 4.11, what lubrication regime is most suitable for application of conversion coatings?

4.32 Explain why surface treatment of manufactured products may be necessary. Give several examples.

4.33 Which surface treatments are functional, and which are decorative? Give examples.

4.34 Give examples of several typical applications of mechanical surface treatment.

4.35 Explain the difference between case hardening and hard facing.

4.36 List several applications for coated sheet metal, including galvanized steel.

4.37 Explain how roller-burnishing processes induce residual stresses on the surface of workpieces.

4.38 List several products or components that could not be made properly, or function effectively in service, without implementation of the knowledge involved in Sections 4.2 through 4.5.

4.39 Explain the difference between direct- and indirect-reading linear measurements.

4.40 Why have coordinate measuring machines become important instruments in modern manufacturing? Give some examples of applications.

4.41 Give reasons that the control of dimensional tolerances is important.

4.42 Give examples where it may be preferable to specify unilateral tolerances as opposed to bilateral tolerances in design.

4.43 Explain why a measuring instrument may not have sufficient precision.

4.44 Comment on the differences, if any, between (a) roundness and circularity; (b) roundness and eccentricity; and (c) roundness and cylindricity.

4.45 It has been stated that dimensional tolerances for nonmetallic stock, such as plastics, are usually wider than for metals. Explain why. Consider physical and mechanical properties.

4.46 Describe the basic features of nondestructive testing techniques that use electrical energy.

4.47 Identify the nondestructive techniques that are capable of detecting internal flaws and those that detect external flaws only.

4.48 Which of the nondestructive inspection techniques are suitable for nonmetallic materials? Why?

4.49 Why is automated inspection becoming an important part of manufacturing engineering?

4.50 Describe situations in which the use of destructive testing techniques is unavoidable.

4.51 Should products be designed and built for a certain expected life? Explain.

4.52 What are the consequences of setting lower and upper specifications closer to the peak of the curve in Fig. 4.22?

4.53 Identify factors that can cause a process to become out of control. Give several examples of such factors.

4.54 In reading this chapter, you will have noted that the specific term *dimensional tolerance* is often used, rather than just the word *tolerance*. Do you think this distinction is important? Explain.

4.55 Give an example of an assignable variation and a chance variation.

4.56 List the variables that can affect the magnitude of the temperature rise and its distribution throughout a workpiece in sliding contact.

PROBLEMS

4.57 Referring to the surface profile in Fig. 4.3, give some numerical values for the vertical distances from the center line. Calculate the R_a and R_q values. Then give another set of values for the same general profile, and calculate the same quantities. Comment on your results.

4.58 Calculate the ratio of R_a/R_q for (a) a sine wave; (b) a saw-tooth profile; and (c) a square wave.

4.59 Refer to Fig. 4.7b and make measurements of the external and internal diameters (in the horizontal direction in the photograph) of the four specimens shown. Remembering that in plastic deformation the volume of the rings remains constant, calculate (a) the reduction in height and (b) the coefficient of friction for each of the three compressed specimens.

4.60 Using Fig. 4.8a, make a plot of the coefficient of friction versus the change in internal diameter for a reduction in height of (a) 20%; (b) 40%; and (c) 60%.

4.61 In Example 4.1, assume that the coefficient of friction is 0.30. If all other initial parameters remain the same, what is the new internal diameter of the ring specimen?

4.62 How would you go about estimating forces required for roller burnishing? (*Hint:* Consider hardness testing.)

4.63 Estimate the plating thickness in electroplating a 50-mm solid metal ball using a current of 1 A and a plating time of 1 hour. Assume that $c = 0.08$ mm^3/amp-s.

4.64 Assume that a steel rule expands by 2% because of an increase in environmental temperature. What will be the indicated diameter of a shaft whose actual diameter is 30.00 mm?

4.65 Examine Eqs. (4.2) and (4.10). What is the relationship between R_q and σ? What would be the equation for the standard deviation of a continuous curve?

4.66 Calculate the control limits for averages and ranges for the following: number of samples = 9; $\overline{\overline{x}} = 50$; and $\overline{R} = 5$.

4.67 Calculate the control limits for the following: number of samples = 5; $\overline{\overline{x}} = 40.5$; and $\text{UCL}_R = 3.500$.

4.68 In an inspection with a sample size of 8, it was found that the average range was 10 and the average of averages was 60. Calculate the control limits for averages and ranges.

4.69 Determine the control limits for the data shown in the following table:

x_1	x_2	x_3	x_4
0.65	0.75	0.67	0.65
0.69	0.73	0.70	0.68
0.65	0.68	0.65	0.61
0.64	0.65	0.60	0.60
0.68	0.72	0.70	0.66
0.70	0.74	0.65	0.71

4.70 Calculate the mean, median, and standard deviation for all of the data in Problem 4.69.

4.71 The average of averages of a number of samples of size 7 was determined to be 125. The average range was 17.82, and the standard deviation was 5.85. The following measurements were taken in a sample: 120, 143, 124, 130, 105, 132, 121, and 127. Is the process in control?

4.72 Assume that you are asked to give a quiz to students on the contents of this chapter. Prepare three quantitative problems and three qualitative questions, and supply the answers.

CHAPTER

5 Casting Processes and Heat Treatment

This chapter describes the fundamentals of metal casting and the characteristics of casting processes, including:

- Mechanisms of the solidification of metals, characteristics of fluid flow, and the role of entrapped gases and shrinkage.
- Properties of casting alloys and their applications.
- Characteristics of expendable-mold and permanent-mold casting processes, their applications and economic considerations.
- Design considerations and simulation techniques for casting.
- Economic considerations.

Symbols

A	area, m^2
C	coefficient in Chvorinov's rule, s/m^2
C_L	weight composition of liquid phase
C_o	nominal weight composition
C_S	weight composition of solid phase
D	channel diameter, m
f	frictional loss
g	gravitational acceleration, 9.81 m/s^2
h	elevation, m
L	weight fraction of liquid
n	exponent
p	pressure, N/m^2
Q	volumetric flow rate, m^3/s
Re	Reynolds number
S	weight fraction of solid
T_L	liquidus temperature, °C
T_S	solidus temperature, °C
t	time, s
v	velocity, m/s
η	viscosity, Ns/m^2
ρ	mass density, kg/m^3

5.1 Introduction

As described throughout this text, several methods can be used to shape materials into useful products. **Casting** is one of the oldest methods and was first used about 4000 B.C. to make ornaments, arrowheads, and various other objects. Casting processes basically involve the introduction of molten metal into a mold cavity where, upon solidification, the metal takes the shape of the cavity. This process is capable of producing intricate shapes in a single piece, ranging in size from very large to very small, including those with internal cavities. Typical cast products are engine blocks, cylinder heads, transmission housings, pistons, turbine disks, railroad and automotive wheels, and ornamental artifacts.

All metals can be cast in, or nearly in, the final shape desired, often with only minor finishing operations required. With appropriate control of material and process parameters, parts can be cast with almost uniform properties throughout. As with all other manufacturing processes, a knowledge of certain fundamental aspects is essential to high quality production, with good surface finish, dimensional accuracy, strength, and lack of defects.

The important factors in casting operations are:

1. **Solidification** of the metal from its molten state, and accompanying shrinkage;
2. **Flow** of the molten metal into the mold cavity;
3. **Heat transfer** during solidification and cooling of the metal in the mold; and
4. **Mold material** and its influence on the casting operation.

5.2 Solidification of Metals

An overview of the solidification of metals and alloys is presented in this section. The topics covered are essential to an understanding of the structures developed in casting and the structure-property relationships obtained in the casting processes described throughout this chapter.

Pure metals have clearly defined melting or freezing points, and solidification takes place at a constant temperature (Fig. 5.1a). When the temperature of the molten metal is reduced to the freezing point, the latent heat of fusion is given off while the temperature remains constant. At the end of this isothermal phase change, solidification is complete and the solid metal cools to room temperature. The casting contracts as it cools, due to (a) contraction from a superheated state to the metal's solidification temperature; and (b) cooling, as a solid, from the solidification temperature to room temperature (Fig. 5.1b). A significant density change also can occur as a result of phase change from liquid to solid.

Unlike pure metals, **alloys** solidify over a *range* of temperatures. Solidification begins when the temperature of the molten metal drops below the **liquidus**, and is completed when the temperature reaches the **solidus**. Within this temperature range the alloy is in a *mushy* or *pasty* state, whose composition and state are described by the particular alloy's phase diagram.

5.2.1 Solid Solutions

Two terms are essential in describing alloys: solute and solvent. **Solute** is the *minor* element (such as sugar or salt) that is added to the **solvent**, which is the *major* element (such as water). In terms of the elements involved

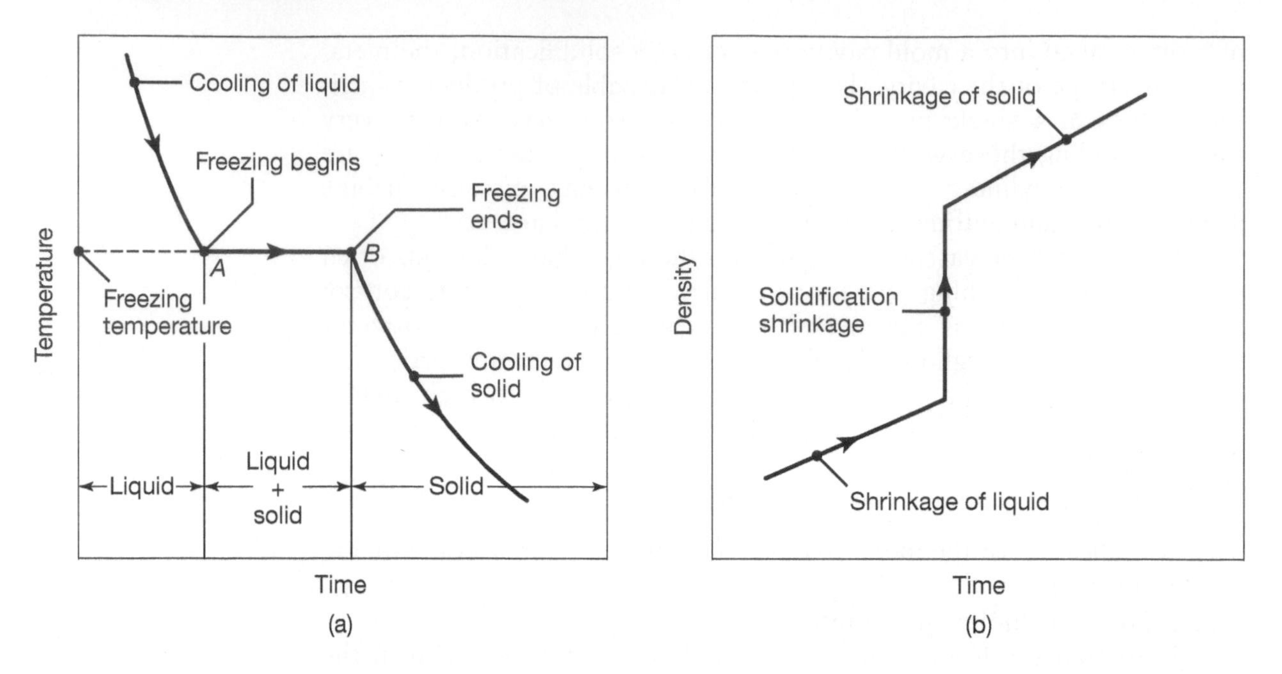

FIGURE 5.1 (a) Temperature as a function of time for the solidification of pure metals. Note that freezing takes place at a constant temperature. (b) Density as a function of time.

in a metal's crystal structure (see Section 3.2), the solute is the element (*solute atoms*) added to the solvent (*host atoms*). When the particular crystal structure of the solvent is maintained during alloying, the alloy is called a **solid solution**.

Substitutional solid solutions. If the size of the solute atom is similar to that of the solvent atom, the solute atoms can replace solvent atoms and form a *substitutional solid solution* (See Fig. 3.9). An example is brass, an alloy of zinc and copper, in which zinc (solute atom) is introduced into the lattice of copper (solvent atoms). The properties of brasses can thus be altered over a certain range by controlling the amount of zinc in copper.

Interstitial solid solutions. If the size of the solute atom is much smaller than that of the solvent atom, the solute atom occupies an interstitial position (as shown in Fig. 3.9) and forms an *interstitial solid solution*. A major example of interstitial solution is steel, an alloy of iron and carbon, in which carbon atoms are present in an interstitial position between iron atoms. As will be shown in Section 5.11, the properties of steel can thus be varied over a wide range by controlling the amount of carbon in iron. This is one reason that steel, in addition to being relatively inexpensive, is such a versatile and important material with a wide range of properties and applications.

5.2.2 Intermetallic Compounds

Intermetallic compounds are complex structures in which solute atoms are present among solvent atoms in specific proportions; some intermetallic compounds have solid solubility. The type of atomic bonds may range from metallic to ionic. Intermetallic compounds are strong, hard, and brittle. An example is copper in aluminum, where an intermetallic compound of $CuAl_2$ can be made to precipitate from an aluminum-copper alloy; this is an example of precipitation hardening (see Section 5.11.2).

5.2.3 Two-Phase Alloys

A solid solution is one in which two or more elements are soluble in a solid state, forming a single homogeneous material in which the alloying elements are uniformly distributed throughout the solid. There is, however, a limit to the concentration of solute atoms in a solvent-atom lattice, just as there is a limit to the solubility of sugar in water. Most alloys consist of two or more solid phases, and thus may be regarded as mechanical mixtures. Such a system with two solid phases is called a **two-phase system,** in which each phase is a homogeneous part of the total mass and has its own characteristics and properties.

A typical example of a two-phase system in metals is lead added to copper in the molten state. After the mixture solidifies, the structure consists of two phases: (a) one phase has a small amount of lead in solid solution in copper; and (b) another phase in which lead particles, approximately spherical in shape, are *dispersed* throughout the matrix of the primary phase (Fig. 5.2a). This copper-lead alloy has properties that are different from those of either copper or lead alone.

Alloying with finely dispersed particles (called *second-phase particles*) is an important method of strengthening alloys and controlling their properties. Generally, in two-phase alloys, the second-phase particles become obstacles to dislocation movement, which increases the alloy's strength (Section 3.3). Another example of a two-phase alloy is the aggregate structure shown in Fig. 5.2b; it contains two sets of grains, each with its own composition and properties. The darker grains may, for example, have a different structure than that of the lighter grains and be brittle, whereas the lighter grains may be ductile.

5.2.4 Phase Diagrams

A *phase diagram,* also called an **equilibrium diagram** or **constitutional diagram,** graphically illustrates the relationships among temperature, composition, and the phases present in a particular alloy system. *Equilibrium* means that the state of a system remains constant over an indefinite

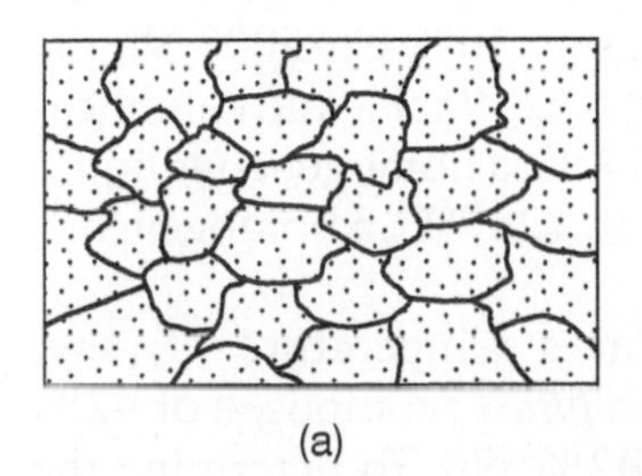
(a)

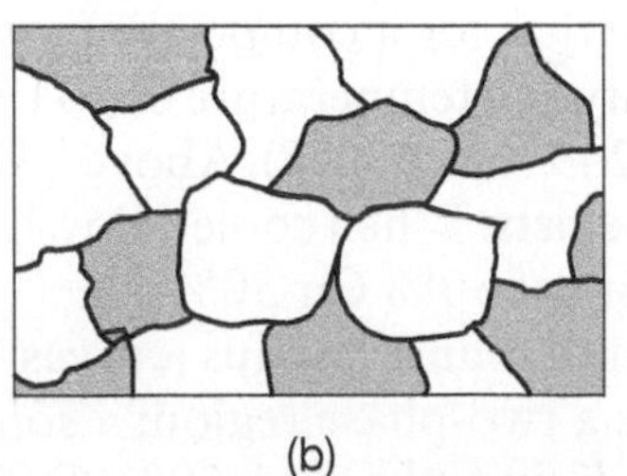
(b)

FIGURE 5.2 (a) Schematic illustration of grains, grain boundaries, and particles dispersed throughout the structure of a two-phase system, such as lead-copper alloy. The grains represent lead in solid solution of copper, and the particles are lead as a second phase. (b) Schematic illustration of a two-phase system, consisting of two sets of grains: dark and light. Dark and light grains have their own compositions and properties.

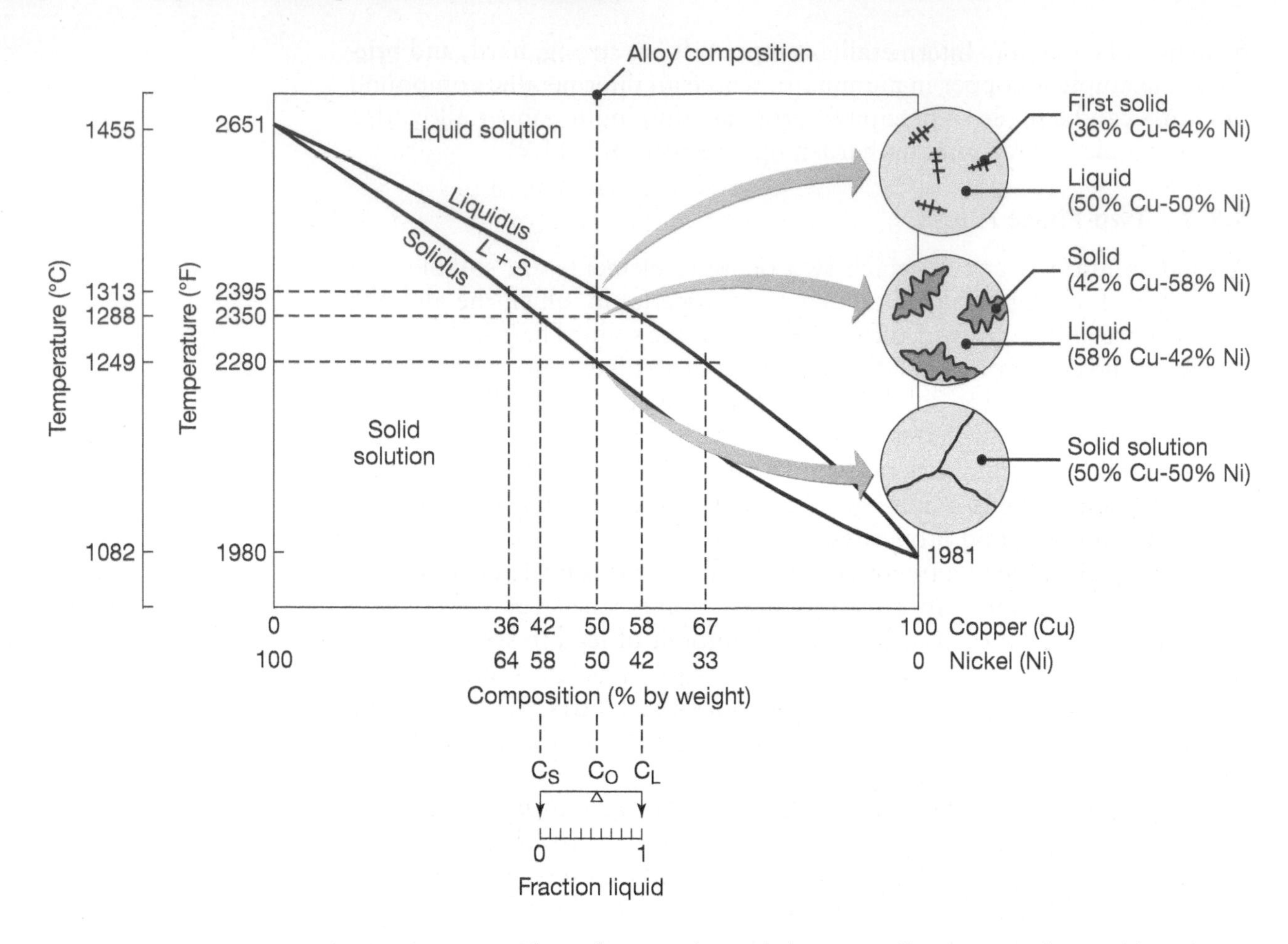

FIGURE 5.3 Phase diagram for nickel-copper alloy system obtained by a low rate of solidification. Note that pure nickel and pure copper each have one freezing or melting temperature. The top circle on the right depicts the nucleation of crystals; the second circle shows the formation of dendrites; and the bottom circle shows the solidified alloy with grain boundaries.

period of time. *Constitutional* indicates the relationships among structure, composition, and physical makeup of the alloy.

An example of a phase diagram is shown in Fig. 5.3 for the nickel-copper alloy. It is called a **binary phase diagram** because there are two elements (nickel and copper) in the system. The left boundary (100% Ni) of this phase diagram indicates the melting point of nickel, and the right boundary (100% Cu) indicates the melting point of copper. (All percentages are by weight.) Note that for a composition of, say, 50% Cu-50% Ni, the alloy begins to solidify at a temperature of 1313°C (2395°F), and solidification is complete at 1249°C (2280°F). Above 1313°C, a homogeneous liquid of 50% Cu-50% Ni exists. When cooled slowly to 1249°C, a homogeneous solid solution consists of 50% Cu-50% Ni.

Between the liquidus and solidus curves at a temperature of, say, 1288°C (2350°F), is a two-phase region: a *solid phase*, composed of 42% Cu-58% Ni, and a *liquid phase* of 58% Cu-42% Ni. To determine the solid composition, one must go left horizontally to the solidus curve and read down, to obtain 42% Cu-58% Ni. The liquid composition can be determined similarly by going to the right to the liquidus curve (58% Cu-42% Ni).

The completely solidified alloy in the phase diagram shown in Fig. 5.3 is a **solid solution**, because the alloying element (Cu, the solute atom) is completely dissolved in the host metal (Ni, the solvent atom), and each grain has the same composition. The mechanical properties of solid solutions of Cu-Ni depend on their composition; for example, by increasing the nickel content, the properties of copper are improved. The improvement is due to *pinning* (blocking) of dislocations at solute atoms of nickel, which may also be regarded as impurity atoms (See Fig. 3.9). As a result, dislocations cannot move as freely and, consequently, the strength of the alloy increases.

Lever rule. The composition of various phases in a phase diagram can be determined by a procedure called the *lever rule*. As shown in the lower portion of Fig. 5.3, a lever between the solidus and liquidus lines (called *tie line*) is first constructed, which is balanced (on the triangular support) at the nominal weight composition C_o of the alloy. The left end of the lever represents the composition C_S of the solid phase, and the right end of the composition C_L of the liquid phase. Note from the graduated scale in the figure that the liquid fraction is also indicated along this tie line, which ranges from 0 at the left (fully solid) to 1 at the right (fully liquid).

The lever rule states that the **weight fraction of solid** is proportional to the distance between C_o and C_L:

$$\frac{S}{S+L} = \frac{C_o - C_L}{C_S - C_L}. \tag{5.1}$$

Likewise, the **weight fraction of liquid** is proportional to the distance between C_S and C_o. Thus,

$$\frac{L}{S+L} = \frac{C_S - C_o}{C_S - C_L}. \tag{5.2}$$

Note that these quantities are fractions, and hence they must be multiplied by 100 to obtain percentages.

From inspection of the dashed line in Fig. 5.3 and for a nominal alloy composition of $C_o = 50\%$ Cu-50% Ni, it can be noted that because C_o is closer to C_L than it is to C_S, the solid phase contains less copper than the liquid phase. By measuring on the phase diagram and using the lever-rule equations, it can be seen that the composition of the solid phase is 42% Cu and of the liquid phase is 58% Cu (as stated in the middle circle at the right in Fig. 5.3).

Note that these calculations refer to copper. If the phase diagram in the figure is now reversed, so that the left boundary is 0% nickel (whereby nickel now becomes the alloying element in copper), these calculations give us the compositions of the solid and liquid phases in terms of nickel.

The lever rule is also known as the *inverse lever rule* because, as indicated by Eqs. (5.1) and (5.2), the amount of each phase is proportional to the length of the opposite end of the lever.

5.2.5 The Iron-Carbon System

The *iron-carbon binary system* is represented by the **iron-iron carbide phase diagram**, shown in Fig. 5.4a. Note that pure iron melts at a temperature of 1538°C (2800°F), as shown at the left in Fig. 5.4a. As it cools, it first forms δ-iron, then γ-iron, and finally α-iron. Commercially pure iron contains up to 0.008% C, steels up to 2.11% C, and cast irons up to 6.67% C, although most cast irons contain less than 4.5% C.

1. **Ferrite.** *Alpha ferrite*, or simply *ferrite*, is a solid solution of body-centered cubic iron and has a maximum solid solubility of 0.022% C at a temperature of 727°C (1341°F). *Delta ferrite* is stable only at very high temperatures and has no significant or practical engineering applications. Ferrite is relatively soft and ductile, and is magnetic from room temperature up to 768°C (1414°F). Although very little carbon can dissolve interstitially in bcc iron, the amount of carbon significantly affects the mechanical properties of ferrite. Also, significant amounts of chromium, manganese, nickel, molybdenum, tungsten, and silicon can be contained in iron in solid solution, imparting certain desirable properties.
2. **Austenite.** Between 1394°C (2541°F) and 912°C (1674°F) iron undergoes an *allotropic transformation* (see Section 3.2) from the bcc

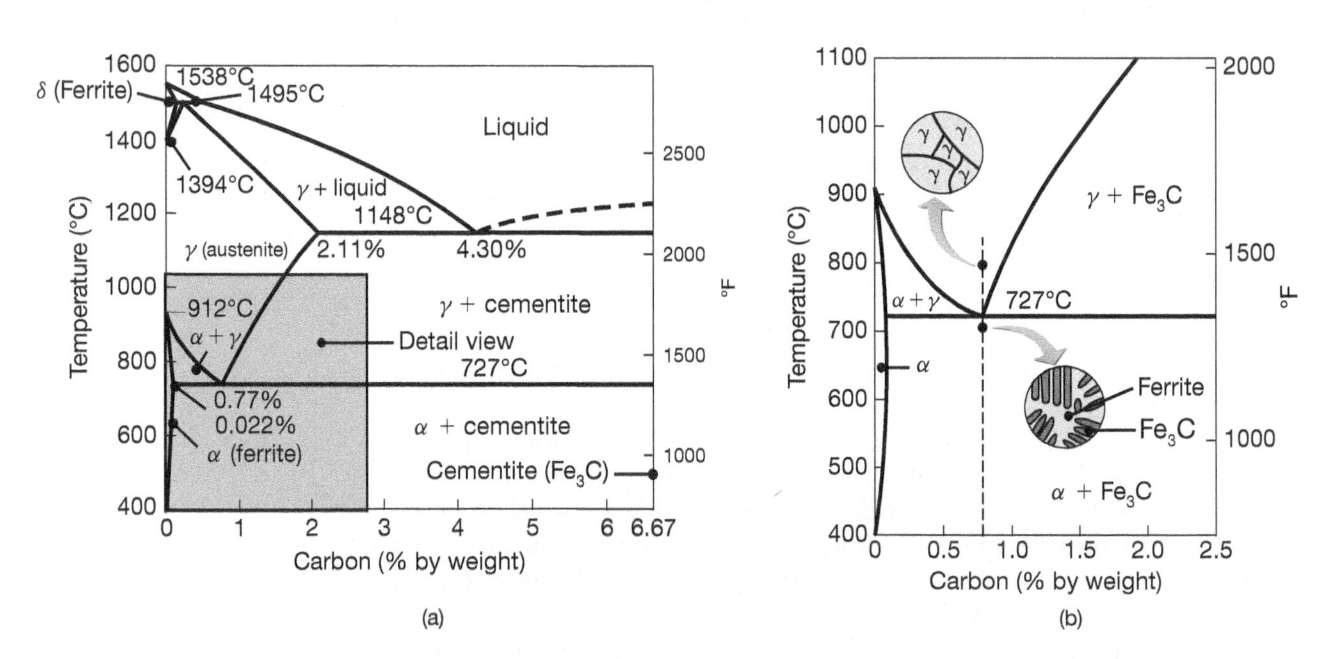

FIGURE 5.4 (a) The iron-iron carbide phase diagram. (b) Detail view of the microstructures above and below the eutectoid temperature of 727°C (1341°F). Because of the importance of steel as an engineering material, this diagram is one of the most important phase diagrams.

to fcc structure, becoming what is known as *gamma iron*, or more commonly, *austenite*. This structure has a solid solubility of up to 2.11% C at 1148°C (2098°F), which is about two orders of magnitude higher than that of ferrite, with the carbon occupying interstitial positions. (Note that the atomic radius of Fe is 0.124 nm and for C it is 0.071 nm.) Austenite is an important phase in the heat treatment of steels, described in Section 5.11. It is denser than ferrite, and its single-phase fcc structure is ductile at elevated temperatures, thus it possesses good formability. Large amounts of nickel and manganese can also be dissolved in fcc iron to impart various properties. Austenitic steel is nonmagnetic at high temperatures, and austenitic stainless steels are nonmagnetic at room temperature.

3. **Cementite.** The right boundary of Fig. 5.4a represents *cementite*, also called **carbide**, which is 100% iron carbide (Fe_3C) with a carbon content of 6.67%. (This carbide should not be confused with various other carbides used for tool and die materials, described in Section 8.6.4.) Cementite is a very hard and brittle intermetallic compound (Section 5.2.2), and significantly influences the properties of steels. It can be alloyed with elements such as chromium, molybdenum, and manganese for enhanced properties.

5.2.6 The Iron-Iron Carbide Phase Diagram

Various microstructures can be developed in steels, depending on its carbon content and the method of heat treatment (Section 5.11). For example, consider iron with a 0.77% C content that is being cooled *very slowly* from a temperature of 1100°C (2012°F) in the austenite phase. The reason for the slow cooling rate is to maintain equilibrium; higher rates of cooling are used in heat treating (see Section 5.11). At 727°C (1341°F) a reaction takes place in which austenite is transformed into alpha ferrite (bcc) and cementite. Because the solid solubility of carbon in ferrite is only 0.022%, the extra carbon forms cementite.

This reaction is called a **eutectoid** (meaning *eutecticlike*) **reaction**, indicating that at a certain temperature a single solid phase (austenite) is transformed into two solid phases, namely ferrite and cementite. The structure of the eutectoid steel is called **pearlite**, because, at low magnifications, it resembles mother of pearl. The microstructure of pearlite consists of alternate layers (*lamellae*) of ferrite and cementite (Fig. 5.4b); consequently, the mechanical properties of pearlite are intermediate between ferrite (soft and ductile) and cementite (hard and brittle).

In iron with less than 0.77% C, the microstructure formed consists of a pearlite phase (ferrite and cementite) and a ferrite phase. The ferrite in the pearlite is called *eutectoid ferrite*; the ferrite phase is called *proeutectoid ferrite* (*pro* meaning before), because it forms at a temperature higher than the eutectoid temperature of 727°C (1341°F). If the carbon content is higher than 0.77%, the austenite transforms into pearlite and cementite. The cementite in the pearlite is called *eutectoid cementite*, and the cementite phase is called *proeutectoid cementite*, because it forms at a temperature higher than the eutectoid temperature.

Effects of alloying elements in iron. Although carbon is the basic element that transforms iron into steel, other elements are also added to impart various desirable properties. The effect of these alloying elements on the iron-iron carbide phase diagram is to shift the eutectoid temperature and eutectoid composition (the percentage of carbon in steel at the eutectoid point). The eutectoid temperature may be raised or lowered from 727°C (1341°F), depending on the particular alloying element. Alloying elements always lower the eutectoid composition; that is, the carbon content becomes less than 0.77%. Lowering the eutectoid temperature means increasing the austenite range; thus, an alloying element, such as nickel, is known as an **austenite former**, because it has an fcc structure and thus tends to favor the fcc structure of austenite. Conversely, chromium and molybdenum have the bcc structure, causing these two elements to favor the bcc structure of ferrite; these elements are known as **ferrite formers.**

EXAMPLE 5.1 Determining the Amount of Phases in Carbon Steel

Given: A 10-kg, 1040 steel casting as it is being cooled slowly to the following temperatures: (a) 900°C; (b) 728°C; and (c) 726°C.

Find: Determine the amount of gamma and alpha phases in the steel casting.

Solution: (a) Referring to Fig. 5.4b, a vertical line is drawn at 0.40% C at 900°C. This is in the single-phase austenite region, so the percent gamma is 100 (10 kg) and percent alpha is zero. (b) At 728°C, the alloy is in the two-phase gamma-alpha field. When the phase diagram is drawn in greater detail, the weight percentages of each phase by the lever rule:

$$\text{Percent alpha} = \left(\frac{C_\gamma - C_o}{C_\gamma - C_\alpha}\right) \times 100\% = \left(\frac{0.77 - 0.40}{0.77 - 0.022}\right) \times 100\% = 49.5\%, \text{ or } 4.95 \text{ kg}$$

$$\text{Percent gamma} = \left(\frac{C_o - C_\alpha}{C_\gamma - C_\alpha}\right) \times 100\% = \left(\frac{0.40 - 0.022}{0.77 - 0.022}\right) \times 100\% = 50.5\%, \text{ or } 5.05 \text{ kg}$$

(c) At 726°C, the alloy will be in the two-phase alpha and Fe_3C region. No gamma phase will be present. Again, the lever rule is used to find the amount of alpha present:

$$\text{Percent alpha} = \left(\frac{6.67 - 0.40}{6.67 - 0.022}\right) \times 100\% = 94.3\%, \text{ or } 9.43 \text{ kg}$$

5.3 Cast Structures

The type of *cast structure* developed during solidification of metals and alloys depends on the composition of the particular alloy, the rate of heat transfer, and the flow of the liquid metal during the casting process. As described throughout this chapter, the structures developed, in turn, affect the properties of the castings.

5.3.1 Pure Metals

The typical grain structure of a pure metal that has solidified in a square mold is shown in Fig. 5.5a. At the mold walls the metal cools rapidly (*chill zone*) because the walls are at ambient or slightly elevated temperature, and as a result, the casting develops a solidified **skin** (*shell*) of fine *equiaxed grains*. The grains grow in the direction opposite to the heat transfer from the mold. Grains that have a favorable orientation, called **columnar grains,** grow preferentially (See middle of Fig. 5.5). Note that grains that have substantially different orientations are blocked from further growth.

5.3.2 Alloys

Because pure metals have limited mechanical properties, they are often enhanced and modified by **alloying.** The vast majority of metals used in engineering applications are some form of an **alloy,** defined as two or more chemical elements, at least one of which is a metal.

Solidification in alloys begins when the temperature drops below the liquidus, T_L, and is complete when it reaches the solidus, T_S (Fig. 5.6). Within this temperature range, the alloy is in a mushy or pasty state, with **columnar dendrites** (from the Greek *dendron* meaning akin to, and *drys* meaning tree). Note in the lower right of the figure the presence of liquid metal between the dendrite arms. Dendrites have three-dimensional arms and branches (*secondary arms*), which eventually interlock, as shown in Fig. 5.7. The width of the mushy zone (where both liquid and solid phases

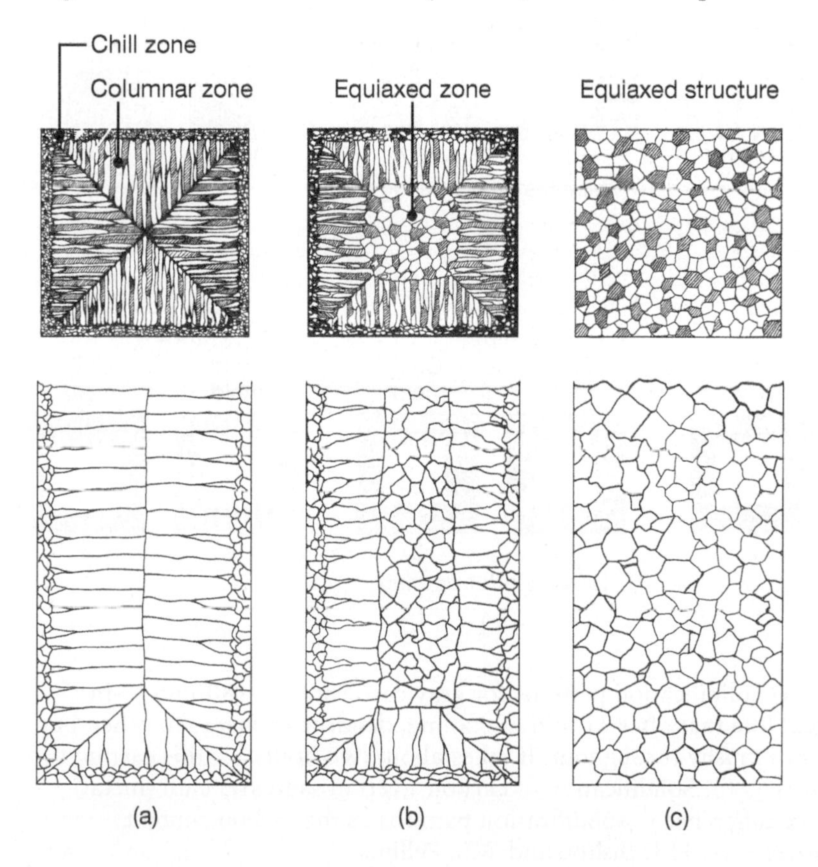

FIGURE 5.5 Schematic illustration of three cast structures of metals solidified in a square mold: (a) pure metals, with preferred texture at the cool mold wall. Note in the middle of the figure that only favorable oriented grains grow away from the mold surface; (b) solid-solution alloys; and (c) structure obtained by heterogeneous nucleation of grains.

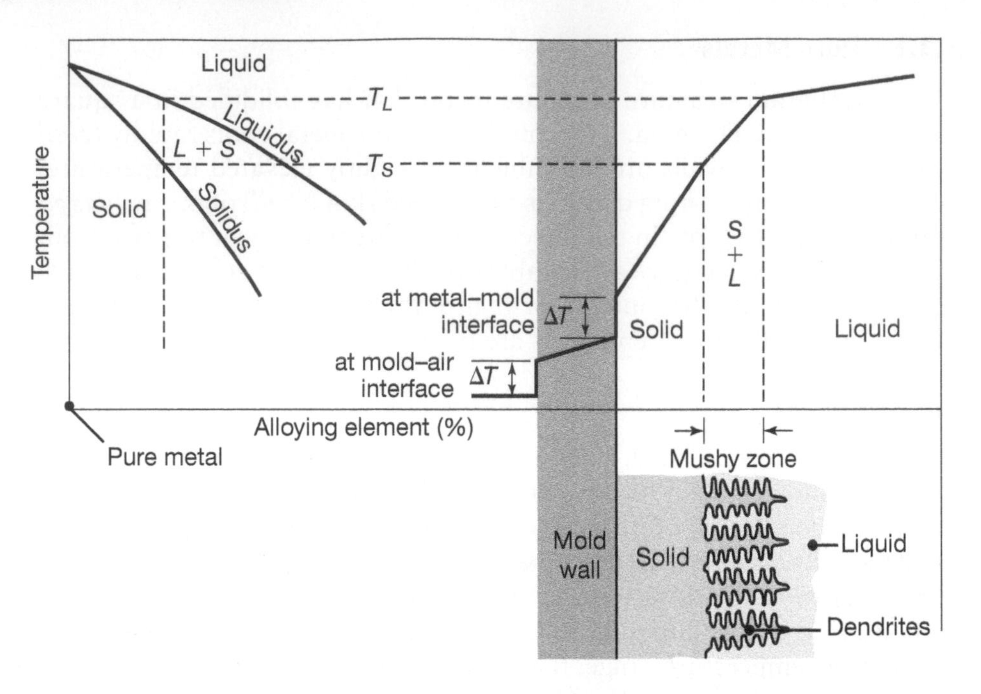

FIGURE 5.6 Schematic illustration of alloy solidification and temperature distribution in the solidifying metal. Note the formation of dendrites in the semi-solid (mushy) zone.

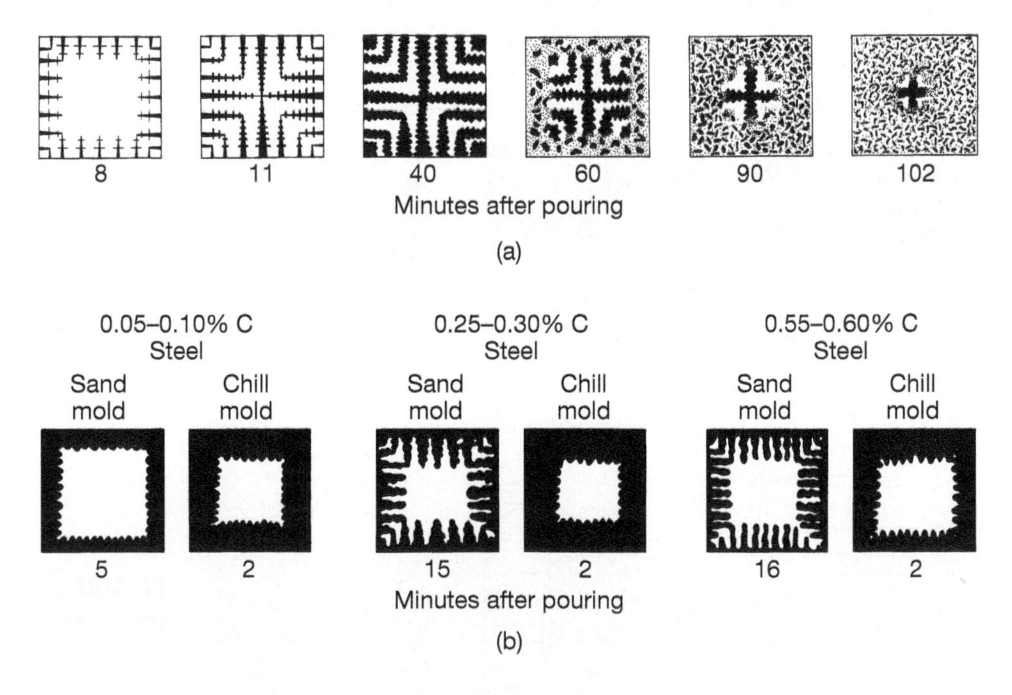

FIGURE 5.7 (a) Solidification patterns for gray cast iron in a 180-mm (7-in.) square casting. Note that after 11 min of cooling, dendrites reach each other, but the casting is still mushy throughout. It takes about two hours for this casting to solidify completely. (b) Solidification of carbon steels in sand and chill (metal) molds. Note the difference in solidification patterns as the carbon content increases. *Source:* After H.F. Bishop and W.S. Pellini.

are present) is important during solidification. This zone is described in terms of a temperature difference, known as the **freezing range**, as

$$\text{Freezing range} = T_L - T_S. \tag{5.3}$$

Note in Fig. 5.6 that pure metals have a freezing range that approaches zero, and that the *solidification front* moves as a plane front, without forming a mushy zone. Eutectics solidify in a similar manner, with an approximately plane front. The type of solidification structure developed depends on the composition of the eutectic. For example, for alloys with a nearly symmetrical phase diagram, the structure is generally lamellar, with two or more solid phases present, depending on the alloy system. When the volume fraction of the minor phase of the alloy is less than about 25%, the structure generally becomes *fibrous*.

A short freezing range for alloys generally involves a temperature difference of less than 50°C (90°F), and a long freezing range, greater than 110°C (200°F). Ferrous castings typically have narrow semi-solid (mushy) zones, whereas aluminum and magnesium alloys have wide mushy zones. Consequently, these alloys are in a semi-solid state throughout most of the solidification process, which is the main reason that thixocasting is feasible with these alloys (see Section 5.10.6).

Effects of cooling rate. Slow cooling rates (on the order of 10^2 K/s) or long local solidification times result in coarse dendritic structures, with large spacing between the dendrite arms. For higher cooling rates (on the order of 10^4 K/s) or short local solidification times, the structure becomes finer, with smaller dendrite arm spacing. For still higher cooling rates (on the order of 10^6 to 10^8 K/s), the structures developed are *amorphous* (meaning without any ordered crystalline structure, as described in Sections 3.11.9 and 5.10.8).

The structures developed and the resulting grain size, in turn, influence the properties of the casting. For example, as grain size decreases, (a) the strength and ductility of the cast alloy increase (see *Hall–Petch equation*, Section 3.4.1); (b) microporosity (interdendritic shrinkage voids) in the casting decreases; and (c) the tendency for the casting to crack (*hot tearing*) during solidification decreases. Moreover, lack of uniformity in grain size and distribution within castings results in anisotropic properties.

5.3.3 Structure-Property Relationships

Because all castings must possess certain specific properties to meet design and service requirements, the relationships between the properties and the structures developed during solidification are important considerations. This section describes these relationships in terms of dendrite morphology and the concentration of alloying elements in various regions of the casting.

The compositions of dendrites and of the liquid metal in casting are given by the phase diagram of the particular alloy. When the alloy is cooled very slowly, each dendrite develops a uniform composition. Under normal cooling rates typically encountered in practice, however, **cored**

dendrites are formed, which have a surface composition that is different from that at their centers (known as *concentration gradient*). The surface has a higher concentration of alloying elements than at the core of the dendrite, due to solute rejection from the core toward the surface during solidification of the dendrite (called **microsegregation**). The darker shading in the interdendritic liquid near the dendrite roots shown in Fig. 5.8 indicates that these regions have a higher solute concentration; consequently, microsegregation in these regions is much more pronounced than in others.

In contrast to microsegregation, **macrosegregation** involves differences in composition throughout the casting itself. In situations where the solidifying front moves away from the surface of a casting as a plane front (Fig. 5.9), lower-melting-point constituents in the solidifying alloy are driven toward the center (**normal segregation**); such a casting has a higher concentration of alloying elements at its center than at its surfaces. The opposite occurs in such dendritic structures as those for solid-solution alloys (Fig. 5.5b); that is, the center of the casting has a lower concentration of alloying elements (**inverse segregation**). The reason for this behavior is that the liquid metal (which has a higher concentration of alloying elements) enters the cavities developed from solidification shrinkage in the dendrite arms (that have solidified sooner). Another form

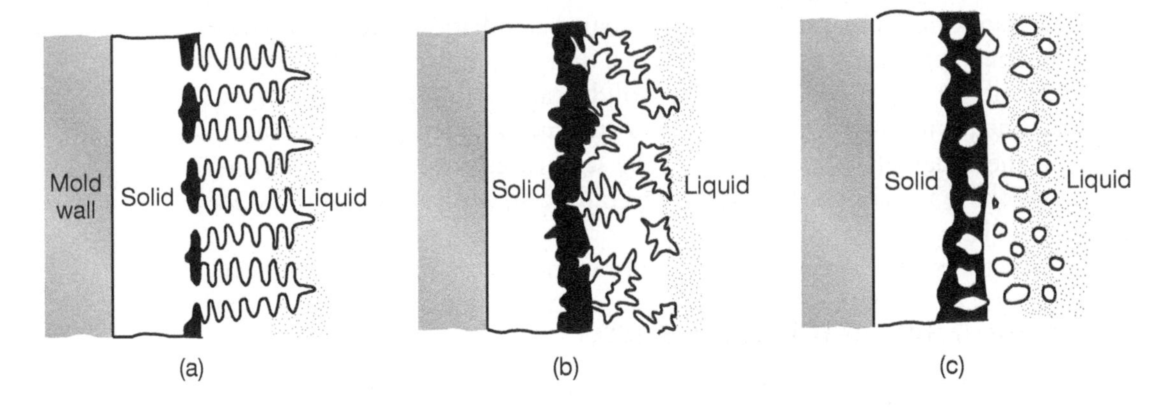

FIGURE 5.8 Schematic illustration of three basic types of cast structures: (a) columnar dendritic; (b) equiaxed dendritic; and (c) equiaxed nondendritic. *Source:* After D. Apelian.

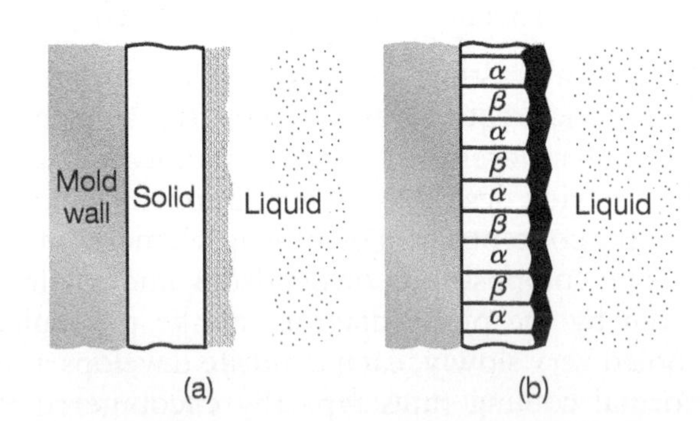

FIGURE 5.9 Schematic illustration of cast structures in (a) plane front, single phase; and (b) plane front, two phase. *Source:* After D. Apelian.

of segregation is due to gravity (**gravity segregation**), whereby higher density inclusions or compounds sink and lighter elements (such as antimony in an antimony-lead alloy) float to the surface.

A typical cast structure of a solid-solution alloy, with an inner zone of equiaxed grains, is shown in Fig. 5.5b. The inner zone can be extended throughout the casting, as shown in Fig. 5.5c, by adding a nucleating agent called **inoculant** to the alloy. The inoculant induces nucleation of grains throughout the liquid metal (**heterogeneous nucleation**). An example is the use of TiB_2 in aluminum alloys to refine grains and improve mechanical properties.

Because of the presence of thermal gradients in a solidifying mass of liquid metal and because of the presence of gravity (hence density differences), *convection* influences the cast structures developed. Convection promotes the formation of a chill zone (See Fig. 5.5), refines the grain size, and accelerates the transition from columnar to equiaxed grains. The structure shown in Fig. 5.8b can also be obtained by increasing convection within the liquid metal, whereby dendrite arms separate (**dendrite multiplication**). Conversely, reducing or eliminating convection results in coarser and longer columnar dendritic grains.

Convection can be enhanced by using mechanical or electromagnetic methods. Because the dendrite arms are not particularly strong, they can be broken up by agitation or mechanical vibration in the early stages of solidification (see **rheocasting,** Section 5.10.6). This action results in finer grain size, with equiaxed nondendritic grains that are distributed more uniformly throughout the casting (Fig. 5.8c). A side benefit is the *thixotropic* behavior of alloys (i.e., the viscosity decreases when the liquid metal is agitated), leading to improved castability of the metal. Another form of semisolid metal forming is **thixotropic casting,** where a solid billet is first heated to a semisolid state and then injected into a die-casting die (Section 5.10.3).

5.4 Fluid Flow and Heat Transfer

5.4.1 Fluid Flow

A basic gravity casting system is shown in Fig. 5.10. The molten metal is poured through a **pouring basin** (*cup*); it then flows through the **sprue** to the **well,** and into **runners** and to the mold cavity. **Risers,** also called **feeders,** serve as reservoirs of molten metal to supply the metal necessary to prevent shrinkage, which could lead to porosity. Although such a **gating system** appears to be relatively simple, successful casting requires proper design and control of the solidification process to ensure adequate fluid flow during casting. One of the most important functions of the gating system is to *trap contaminants* (such as oxides and other inclusions) in the molten metal, by having the contaminants adhere to the walls of the gating system, thereby preventing their reaching the mold cavity. Moreover, a properly designed gating system avoids or minimizes such problems as premature cooling, turbulence, and gas entrapment. Even before it reaches

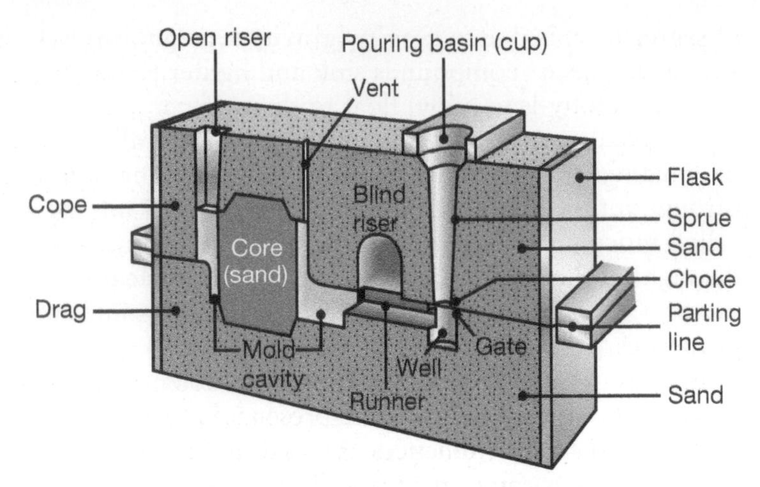

FIGURE 5.10 Cross section of a typical sand mold showing various features.

the mold cavity, the molten metal must be handled so as to avoid forming oxides on the molten metal surfaces (from exposure to the environment) or the introduction of impurities into the molten metal.

Two basic principles of fluid flow are relevant to gating design: Bernoulli's theorem and the law of mass continuity.

Bernoulli's theorem. *Bernoulli's theorem* is based on the principle of conservation of energy, and relates pressure, velocity, elevation of the fluid at any location in the system, and frictional losses. Thus,

$$h + \frac{p}{\rho g} + \frac{v^2}{2g} = \text{Constant}, \tag{5.4}$$

where h is the elevation above a certain reference plane, p is the pressure at that elevation, v is the velocity of the liquid at that elevation, ρ is the density of the fluid (assuming that it is incompressible), and g is the gravitational constant. Conservation of energy requires that at any particular location in the system, the following relationship be satisfied:

$$h_1 + \frac{p_1}{\rho g} + \frac{v_1^2}{2g} = h_2 + \frac{p_2}{\rho g} + \frac{v_2^2}{2g} + f. \tag{5.5}$$

where the subscripts 1 and 2 represent two different elevations, respectively, and f represents the frictional loss in the liquid as it travels downward through the gating system. The frictional loss includes such factors as energy loss at the liquid-mold wall interfaces and turbulence in the liquid.

Mass continuity. The mass continuity law states that for an incompressible liquid and in a system with impermeable walls, the rate of flow is constant; thus,

$$Q = A_1 v_1 = A_2 v_2, \tag{5.6}$$

where Q is the volumetric rate of flow (such as m^3/s), A is the cross-sectional area of the liquid stream, and v is the velocity of the liquid in that particular location. The subscripts 1 and 2 in Eq. (5.6) pertain to two different locations in the system. The permeability of the walls of the system is important because, otherwise, some liquid will permeate through (for example, in sand molds; Section 5.8.1) and the flow rate will decrease as the liquid travels through the system. Coatings are often used to inhibit such behavior in sand molds.

Sprue profile. An application of the two principles stated above is the traditional tapered design of sprues (Fig. 5.10), in which the shape of the sprue can be determined by using Eqs. (5.5) and (5.6). Assuming that the pressure at the top of the sprue is equal to the pressure at the bottom, and that there are no frictional losses in the system, the relationship between height and cross-sectional area at any point in the sprue is given by the parabolic relationship:

$$\frac{A_1}{A_2} = \sqrt{\frac{h_2}{h_1}}, \tag{5.7}$$

where, for example, subscript 1 denotes the top of the sprue and subscript 2 the bottom. Note that in a free-falling liquid, such as water from a faucet, the cross-sectional area of the stream decreases as it gains velocity downward. Thus, moving downward from the top, the cross-sectional area of the sprue must decrease. If a sprue is designed with a constant cross-sectional area, regions may develop where the molten metal loses contact with the sprue walls. As a result, **aspiration** may occur, whereby air will be sucked in or be entrapped in the liquid. Straight-sided sprues require a *choking* mechanism at their bottom, consisting of either a *choke core* or a *runner choke*, as shown in Fig. 5.10. A choking mechanism slows down the fluid flow so that aspiration will not occur.

Modeling of mold filling requires the application of Eqs. (5.5) and (5.6); see also Section 5.12.5. Consider the situation shown in Fig. 5.10 where molten metal is poured into a pouring basin. The metal then flows through a sprue to a gate and runner, and fills the mold cavity. If the pouring basin has a cross-sectional area that is much larger than the sprue bottom, the velocity of the molten metal at the top of the pouring basin will be very low. If frictional losses are due to viscous dissipation of energy, then f in Eq. (5.5) can be taken as a function of vertical distance, and is often approximated as a linear function. The velocity of the molten metal leaving the gate is then obtained from Eq. (5.5) as

$$v = c\sqrt{2gh}, \tag{5.8}$$

where h is the distance from the sprue base to the height of the liquid metal, and c is a friction factor. This factor ranges between 0 and 1, and for frictionless flow, it is unity. The magnitude of c varies with mold material, runner layout, and channel size, and can include energy losses due to turbulence and viscous effects.

If the liquid level has reached a height x, then the gate velocity is

$$v = c\sqrt{2g}\sqrt{h - x}. \tag{5.9}$$

The flow rate through the gate will be the product of this velocity and the gate area, according to Eq. (5.6). The shape of the casting will determine the height as a function of time. Equation (5.9) allows calculation of the flow rate, and dividing the casting volume by the mean flow rate then gives the mold fill time.

Simulation of mold filling helps designers specify the runner diameter, and the size and number of sprues and pouring basins. To ensure that the runners do not choke prematurely, the fill time must be a small fraction of the solidification time (see Section 5.4.4). However, the velocity must not be so high as to erode the mold material (known as *mold wash*) or to result in too high of a Reynolds number (see below), because it may result in turbulence and associated air entrainment. Several computational tools (see Section 5.12.5) are now available to evaluate gating designs and help determine the size of the mold components.

Flow characteristics. An important consideration in fluid flow in gating systems is the presence of **turbulence**, as opposed to **laminar flow** of fluids. The **Reynolds number**, Re, is used to characterize this aspect of fluid flow; Re represents the ratio of the inertia to the viscous forces in fluid flow, and is expressed as

$$\text{Re} = \frac{vD\rho}{\eta}, \tag{5.10}$$

where v is the velocity of the liquid, D is the diameter of the channel, and ρ and η are the density and viscosity, respectively, of the liquid.

The higher the Reynolds number, the greater is the tendency for turbulent flow. In ordinary gating systems Re ranges from 2000 to 20,000; Re values of up to 2000 represent laminar flow. Between 2000 and 20,000 the flow is a mixture of laminar and turbulent, and is generally regarded as harmless in gating systems for casting. However, Re values over 20,000 represent severe turbulence, resulting in air entrainment and *dross* formation. Dross is the scum that forms on the surface of the molten metal as a result of the reaction of the liquid metal with air and other gases. Techniques for minimizing turbulence generally involve avoidance of sudden changes in the flow direction and in the geometry of channel cross sections of the gating system.

Dross or *slag* (nonmetallic products from mutual dissolution of flux and nonmetallic impurities) can be eliminated only by *vacuum casting* (Section 5.8.5). The reduction and control of dross or slag is another important consideration in fluid flow. This can be achieved by skimming (using *dross traps*), properly designing pouring basins and gating systems, or using filters. Filters are usually made of ceramic, mica, or fiberglass. Their proper location and placement is important for effective filtering of dross and slag.

EXAMPLE 5.2 Design and Analysis of a Sprue for Casting

Given: The desired volume flow rate of the molten metal into a mold is 0.01 m^3/min. The top of the sprue has a diameter of 20 mm and its length is 200 mm.

Find: What diameter should be specified at the bottom of the sprue in order to prevent aspiration? What is the resultant velocity and Reynolds number at the bottom of the sprue if the metal being cast is aluminum and has a viscosity of 0.004 N-s/m^2?

Solution: A pouring basin is typically provided on top of a sprue so that molten metal may be present above the sprue opening; however, this complication will be ignored in this case. Note that the metal volume flow rate is $Q = 0.01$ m^3/min = 1.667×10^{-4} m^3/s. Let subscripts 1 and 2 refer to the top and bottom of the sprue, respectively. Since $d_1 = 20$ mm $= 0.02$ m,

$$A_1 = \frac{\pi}{4}d^2 = \frac{\pi}{4}(0.002)^2 = 3.14 \times 10^{-4}\ \text{m}^2.$$

Therefore,

$$v_1 = \frac{Q}{A_1} = \frac{1.667 \times 10^{-4}}{3.14 \times 10^{-4}} = 0.531\ \text{m/s}.$$

Assuming no frictional losses and recognizing that the pressure at the top and bottom of the sprue is atmospheric, Eq. (5.5) gives

$$0.2 + \frac{(0.531)^2}{2(9.81)} + \frac{p_{\text{atm}}}{\rho g} = 0 + \frac{v_2^2}{2(9.81)} + \frac{p_{\text{atm}}}{\rho g},$$

or $v_2 = 1.45$ m/s. To prevent aspiration, the sprue opening should be the same as that required by flow continuity, or

$$Q = A_2 v^2 = 1.667 \times 10^{-4}\ \text{m}^3/\text{s} = A_2(1.45\ \text{m/s}),$$

or $A_2 = 1.150 \times 10^{-4}$ m^2, and therefore $d = 12$ mm. The profile of the sprue will be parabolic, as suggested by Eq. (5.7). In calculating the Reynolds number, note from the inside front cover that the density of aluminum is 2700 kg/m^3. The density for molten aluminum will of course be lower, but not significantly, so this value is sufficient for this problem. From Eq. (5.10),

$$\text{Re} = \frac{vD\rho}{\eta} = \frac{(1.45)(0.012)(2700)}{0.004} = 11,745.$$

As stated above, this magnitude is typical for casting molds, representing a mixture of laminar and turbulent flow.

5.4.2 Fluidity of Molten Metal

A term commonly used to describe the ability of the molten metal to fill mold cavities is *fluidity*. This term consists of two basic factors: (1) characteristics of the molten metal; and (2) casting parameters. The following characteristics of molten metal influence fluidity:

1. **Viscosity**. Fluidity decreases as viscosity and the viscosity index (its sensitivity to temperature) increase.
2. **Surface tension**. A high surface tension of the liquid metal reduces fluidity. Oxide films developed on the surface of the molten metal have a significant adverse effect on fluidity; an oxide film on the surface of pure molten aluminum, for example, triples the surface tension.
3. **Inclusions**. As insoluble particles, inclusions can have a significant adverse effect on fluidity. This effect can be verified by observing the viscosity of a liquid, such as oil, with and without fine sand particles in it; it will be noted that the former will have lower viscosity.
4. **Solidification pattern of the alloy**. The manner in which solidification occurs, as described in Section 5.3, can influence fluidity. Moreover, fluidity is inversely proportional to the freezing range [see Eq. (5.3)]; thus, the shorter the range (as in pure metals and eutectics), the higher the fluidity becomes. Conversely, alloys with long freezing ranges (such as solid-solution alloys) have lower fluidity.

The following casting parameters influence fluidity and the fluid flow and thermal characteristics of the system:

1. **Mold design**. The design and dimensions of such components as the sprue, runners, and risers all influence fluidity to varying degrees.
2. **Mold material and its surface characteristics**. The higher the thermal conductivity of the mold and the rougher its surfaces, the lower is fluidity. Heating the mold improves fluidity, although it also increases the solidification time, resulting in coarser grains and hence lower strength.
3. **Degree of superheat**. Defined as the increment of temperature above an alloy's melting point, superheat improves fluidity by delaying solidification.
4. **Rate of pouring**. The lower the rate of pouring into the mold, the lower the fluidity, because the metal cools faster.
5. **Heat transfer**. Heat transfer directly affects the viscosity of the liquid metal, and hence its fluidity.

The term **castability** is generally used to describe the ease with which a metal can be cast to produce a part with good quality. Because this term also includes casting practices, the factors listed above have a direct effect on castability.

Tests for fluidity. Several tests have been developed to quantify fluidity, although none are accepted universally. In one such common test, the molten metal is made to flow along a channel that is at room temperature; the distance the metal flows before it solidifies is a measure of its fluidity. Obviously, this length is a function of the thermal properties of the metal and the mold, as well as of the design of the channel. Still, such fluidity tests are useful and simulate casting situations to a reasonable degree.

5.4.3 Heat Transfer

A major consideration in casting is the heat transfer during the complete cycle from pouring to solidification and cooling of the casting to room temperature. *Heat flow* at different locations in the system depends on many factors, relating to the casting material and the mold and process parameters. For instance, in casting thin sections, the metal flow rates must be high enough to avoid premature chilling and solidification. On the other hand, the flow rate must not be so high as to cause excessive turbulence, with its detrimental effects on the properties of the casting.

A typical temperature distribution in the mold-liquid metal interface is shown in Fig. 5.6. As expected, the shape of the curve will depend on the thermal properties of the molten metal and the mold material, such as sand, metal, or a ceramic. Heat from the liquid metal being poured is given off through the mold wall and the surrounding air. The temperature drop at the air-mold and mold-metal interfaces is caused by the presence of boundary layers and imperfect contact at these interfaces.

5.4.4 Solidification Time

During the early stages of solidification, a thin solidified skin begins to form at the cool mold walls; as time passes, the skin thickens. With flat mold walls, the thickness is proportional to the square root of time. Thus, doubling the time will make the skin $\sqrt{2} = 1.41$ times, or 41%, thicker.

The *solidification time* is a function of the volume of a casting and its surface area (**Chvorinov's rule**), and is given by

$$\text{Solidification time} = C\left(\frac{\text{Volume}}{\text{Surface area}}\right)^n, \tag{5.11}$$

where C is a constant that reflects the mold material and the metal properties, including latent heat and temperature. The parameter n typically has a value between 1.5 and 2, and is usually taken as 2. Thus, a large sphere solidifies and cools to ambient temperature at a much lower rate than does a smaller sphere, because the volume of a sphere is proportional to the cube of its diameter whereas the surface area is proportional to the square of its diameter. Similarly, it can be shown that molten metal in a cube-shaped mold will solidify faster than in a spherical mold of the same volume.

The effects of mold geometry and elapsed time on skin thickness and its shape are shown in Fig. 5.11. The unsolidified molten metal has been poured from the mold at different time intervals, ranging from 5 s to 6 min.

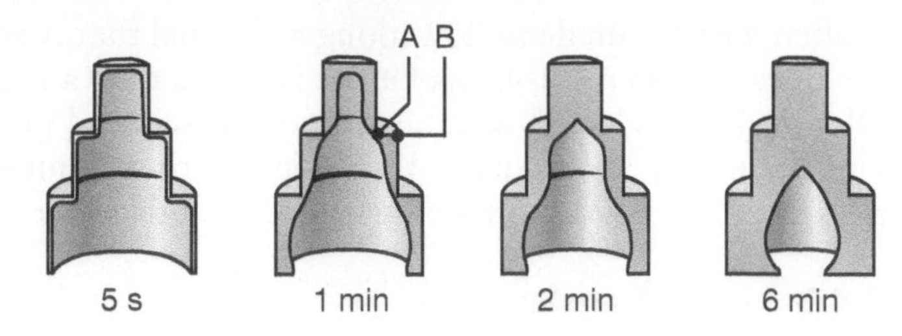

FIGURE 5.11 Solidified skin on a steel casting; the remaining molten metal is poured out at the times indicated in the figure. Hollow ornamental and decorative objects are made by a process called slush casting, which is based on this principle. *Source:* After H.F. Taylor, J. Wulff, and M.C. Flemings

Note that the skin thickness increases with elapsed time, and that the skin is thinner at internal angles (location A in the figure) than at external angles (location B). As expected, the metal cools slower at internal angles than at external angles. Note that this operation is very similar to making hollow chocolate candies in various shapes.

EXAMPLE 5.3 Solidification Times for Various Solid Shapes

Given: Three pieces being cast have the same volume but different shapes. One is a sphere, one a cube, and the other a cylinder with a height equal to its diameter.

Find: Which piece will solidify the fastest and which one the slowest? Use $n = 2$.

Solution: The volume is unity, so from Eq. (5.11):

$$\text{Solidification time} \propto \frac{1}{(\text{Surface area})^2}.$$

The respective surface areas are

$$\text{Sphere: } V = \left(\frac{4}{3}\right)\pi r^3, \quad r = \left(\frac{3}{4\pi}\right)^{1/3}, \quad \text{and} \quad A = 4\pi r^2 = 4\pi\left(\frac{3}{4\pi}\right)^{2/3} = 4.84$$

$$\text{Cube: } V = a^3, \quad a = 1, \quad \text{and } A = 6a^2 = 6.$$

$$\text{Cylinder: } V = \pi r^2 h = 2\pi r^3, \quad r = \left(\frac{1}{2\pi}\right)^{1/3}, \quad \text{and}$$

$$A = 2\pi r^2 + 2\pi rh = 6\pi r^2 = 6\pi\left(\frac{1}{2\pi}\right)^{2/3} = 5.54.$$

Therefore, the respective solidification times t are

$$t_{\text{sphere}} = 0.043C, \quad t_{\text{cube}} = 0.028C, \quad \text{and} \quad t_{\text{cylinder}} = 0.033C.$$

Therefore, the cube-shaped casting will solidify the fastest, and the sphere-shaped casting will solidify the slowest.

TABLE 5.1 Volumetric solidification contraction or expansion for various cast metals.

Contraction (%)		Expansion (%)	
Aluminum	7.1	Bismuth	3.3
Zinc	6.5	Silicon	2.9
Al – 4.5% Cu	6.3	Gray iron	2.5
Gold	5.5		
White iron	4–5.5		
Copper	4.9		
Brass (70–30)	4.5		
Magnesium	4.2		
90% Cu – 10% Al	4		
Carbon steels	2.5–4		
Al – 12% Si	3.8		
Lead	3.2		

5.4.5 Shrinkage

Metals generally shrink during solidification and cooling to room temperature, as shown in Fig. 5.1 and Table 5.1. *Shrinkage* in a casting causes dimensional changes and sometimes cracking and is a result of the following phenomena:

1. Contraction of the molten metal as it cools before it begins to solidify,
2. Contraction of the metal during phase change from liquid to solid (latent heat of fusion), and
3. Contraction of the solidified metal (the casting) as its temperature drops to ambient temperature.

The largest potential amount of shrinkage occurs during the phase change of the material from liquid to solid; this can be reduced or eliminated through the use of risers or pressure-feeding of molten metal. The amount of contraction during the solidification of various metals is shown in Table 5.1. Note that some metals, such as gray cast iron, expand. The reason is that graphite has a relatively high specific volume, and when it precipitates as graphite flakes during solidification of the gray cast iron, it causes a net expansion of the metal. Shrinkage, especially that due to thermal contraction, is further discussed in Section 5.12.2 in connection with design considerations in casting.

5.5 Melting Practice and Furnaces

Melting practice is an important aspect of casting operations, because it has a direct bearing on quality. Furnaces are charged with melting stock, consisting of liquid and/or solid metal, alloying elements, flux, and slag-forming constituents.

Fluxes are inorganic compounds that refine the molten metal by removing dissolved gases and various impurities. They have several functions, depending on the metal. For aluminum alloys, for example, there are cover fluxes (to form a barrier to oxidation), cleaning fluxes, drossing fluxes to protect the molten metal from oxidation, refining fluxes, and wall-cleaning fluxes (to reduce the detrimental effect that some fluxes have on furnace linings, particularly induction furnaces). Fluxes may be added manually or they can be injected automatically into the molten metal. To protect the surface of the molten metal against atmospheric reaction and contamination, as well as to refine the melt, the metal must be insulated against heat loss. Insulation typically is provided by covering the surface of the melt or mixing it with compounds that form a slag. In casting steels, for example, the composition of the slag includes CaO, SiO_2, MnO, and FeO.

The metal *charge* may be composed of commercially pure **primary metals**, which can include remelted or recycled scrap. Clean scrapped castings, gates, and risers may also be included in the charge. If the melting points of the alloying elements are sufficiently low, pure alloying elements are added for the desired composition in the melt. If the melting points are too high, the alloying elements do not mix readily with the low-melting-point metals; in this case, **master alloys (hardeners)** are often used. These usually consist of lower-melting point alloys with higher concentrations of one or two of the required alloying elements. Differences in specific gravity of master alloys should not be too high to cause segregation in the melt.

Melting furnaces. The melting furnaces commonly used in foundries are: electric-arc, induction, crucible, and cupolas.

Electric-arc furnaces are used extensively in foundries. They have such advantages as high rate of melting (hence high production rate), much less pollution than the other types of furnaces, and the ability to hold the molten metal for any length of time for alloying purposes.

Induction furnaces are especially useful in smaller foundries and produce composition-controlled smaller melts. The *coreless induction furnace* consists of a crucible, completely surrounded with a water-cooled copper coil through which high frequency current passes. Because there is a strong electromagnetic stirring action during induction heating, this type of furnace has excellent mixing characteristics for the purposes of alloying and adding new charge of metal. The *core* or *channel furnace* uses a low frequency (as low as 60 Hz) and has a coil that surrounds only a small portion of the unit.

Crucible furnaces have been used extensively throughout history; they are heated with various fuels, such as commercial gases, fuel oil, fossil fuel, as well as electricity. These furnaces may be stationary, tilting, or movable. Most ferrous and nonferrous metals are melted in these furnaces.

Cupolas are basically refractory lined, vertical steel vessels that are charged with alternating layers of metal, coke, and flux. Although they require major investments, cupolas operate continuously, have high melting rates, and produce large amounts of molten metal.

Levitation melting involves magnetic suspension of the molten metal. An induction coil simultaneously heats a solid billet and stirs and confines the metal, thus eliminating the need for a crucible, which could be a source of contamination of the melt with oxide inclusions. The molten metal then flows downward into an investment-casting mold (see Section 5.9.2). Investment castings made by this method are free of refractory inclusions and gas porosity, and have uniform fine-grained structure.

Foundries and foundry automation. The casting operations described in rest of this chapter are typically carried out in *foundries*. Modern foundries have automated and computer-integrated facilities for all aspects of their operations, and produce castings at high production rates, at low cost, and with excellent quality control.

Foundry operations basically involve three separate activities. (a) Pattern and mold making, utilizing computer-aided design and **additive manufacturing** techniques (see Section 10.12); (b) melting the metals while controlling their composition and impurities; and (c) various operations, such as pouring the molten metal into molds (carried along conveyors), shake-out, cleaning, heat treatment, and inspection. All operations are automated, including the use of industrial robots (Section 14.7).

5.6 Casting Alloys

Chapter 3 summarized the properties of wrought structures. The general properties of various casting alloys and processes are summarized in Fig. 5.12 and Tables 5.2 through 5.5.

5.6.1 Ferrous Casting Alloys

The term **cast iron** refers to a family of ferrous alloys composed of iron, carbon (ranging from 2.11% to about 4.5%), and silicon (up to about 3.5%). Cast irons are usually classified according to their solidification morphology, and by their structure (ferrite, pearlite, quenched and tempered, or austempered). Cast irons have lower melting temperatures than steels, which is why the casting process is so suitable for iron with high carbon content.

Cast irons represent the largest amount (by weight) of all metals cast and they can easily be cast into complex shapes. They generally possess several desirable properties, such as high strength, wear resistance, hardness, and good machinability (Chapter 8).

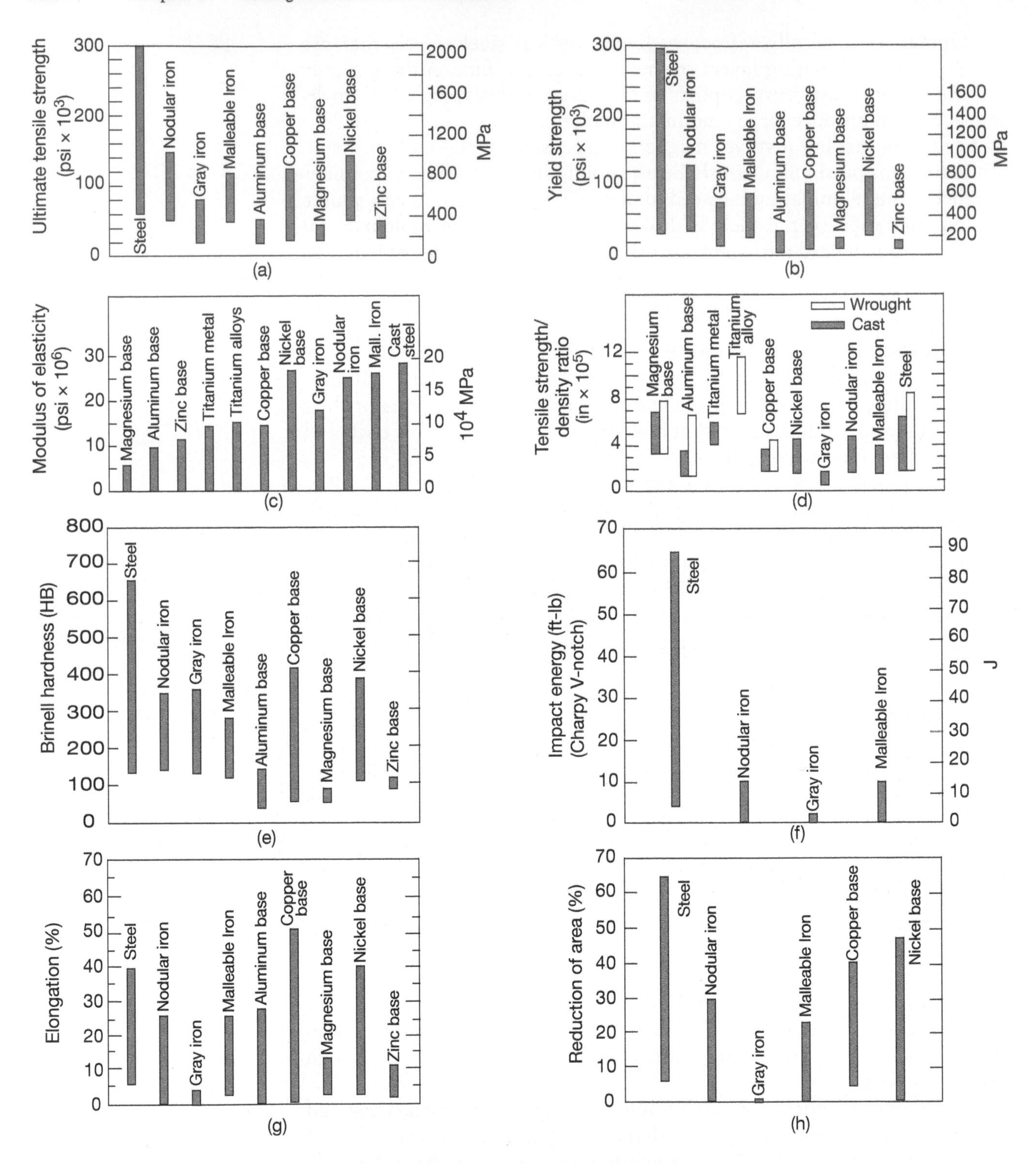

FIGURE 5.12 Mechanical properties for various groups of cast alloys. Compare with various tables of properties in Chapter 3. *Source:* After Steel Society of America.

TABLE 5.2 General characteristics of casting processes.

	Sand	Shell	Evaporative pattern	Plaster	Investment	Permanent mold	Die	Centrifugal
Typical materials cast	All	All	All	Nonferrous (Al, Mg, Zn, Cu)	All	All	Nonferrous (Al, Mg, Zn, Cu)	All
Mass (kg):								
minimum	0.01	0.01	0.01	0.01	0.001	0.1	<0.01	0.01
maximum	No limit	100+	100+	50+	100+	300	50	5000+
Typical surface finish (μm R_a)	5–25	1–3	5–25	1–2	0.3–2	2–6	1–2	2–10
Porosity[1]	3–5	4–5	3–5	4–5	5	2–3	1–3	1–2
Shape complexity[1]	1–2	2–3	1–2	1–2	1	2–3	3–4	3–4
Dimensional accuracy[1]	3	2	3	2	1	1	1	3
Section thickness (mm):								
minimum:	3	2	2	1	1	2	0.5	2
maximum:	No limit	—	—	—	75	50	12	100
Typ. dimensional tolerance	1.6–4 (0.25 for small)	±0.003		±0.005–0.010	±0.005	±0.015	±0.001–0.005	±0.015
Cost[1]								
Equipment	3–5	3	2–3	3–5	3–5	2	1	1
Pattern/die	3–5	2–3	2–3	3–5	2–3	2	1	1
Labor	1–3	3	3	1–2	1–2	3	5	5
Typical lead time[2]	Days	Weeks	weeks	Days	Weeks	Weeks	Weeks-months	Months
Typical production rate[2]	1–20	5–50	1–20	1–10	1–1000	5–50	2–200	1–1000
Minimum quantity[2]	1	100	500	10	10	1000	10,000	10–10,000

Notes:
1. Relative rating, 1 best, 5 worst. For example, die casting has relatively low porosity, mid- to low shape complexity, high dimensional accuracy, high equipment and die costs and low labor costs. These ratings are only general; significant variations can occur depending on the manufacturing methods used.
2. Approximate values without the use of rapid prototyping technologies.

TABLE 5.3 Typical applications for castings and casting characteristics.

Type of alloy	Castability*	Weldability*	Machinability*	Applications
Aluminum	G-E	F	G-E	Pistons, clutch housings, intake manifolds, engine blocks, heads, cross members, valve bodies, oil pans, suspension components
Copper	F-G	F	G-E	Pumps, valves, gear blanks, marine propellers
Gray Iron	E	D	G	Engine blocks, gears, brake disks and drums, machine bases
Magnesium	G-E	G	E	Crankcase, transmission housings, automotive panel beams, steering components
Malleable iron	G	D	G	Farm and construction machinery, heavy-duty bearings, railroad rolling stock
Nickel	F	F	F	Gas turbine blades, pump and valve components for chemical plants
Nodular iron	G	D	G	Crankshafts, heavy-duty gears
Steel (carbon and low alloy)	F	E	F-G	Die blocks, heavy-duty gear blanks, aircraft undercarriage members, railroad wheels
Steel (high alloy)	F	E	F	Gas turbine housings, pump and valve components, rock crusher jaws
White iron (Fe_3C)	G	VP	VP	Mill liners, shot blasting nozzles, railroad brake shoes, crushers and pulverizers
Zinc	E	D	E	Door handles, radiator grills, toys

*E, excellent; G, good; F, fair; VP, very poor; D, difficult.

TABLE 5.4 Properties and typical applications of cast irons.

Cast iron	Type	Yield strength MPa (ksi)	Ultimate tensile strength MPa (ksi)	Elongation in 50 mm (%)	Typical applications
Gray	Ferritic	140 (20)	170 (25)	0.4	Pipe, sanitary ware
	Pearlitic	240 (35)	275 (40)	0.4	Engine blocks, machine tools
	Martensitic	550 (80)	550 (80)	0	Wear surfaces
Ductile (Nodular)	Ferritic	275 (40)	415 (60)	18	Pipe, general service
	Pearlitic	380 (55)	550 (80)	6	Crankshafts, highly stressed parts
	Tempered Martensite	620 (90)	825 (120)	2	High-strength machine parts, wear resistance
Malleable	Ferritic	240 (35)	365 (50)	18	Hardware, pipe fittings, general engineering service
	Pearlitic	310 (45)	450 (65)	10	Couplings
	Tempered	550 (80)	700 (100)	2	Gears, connecting rods
White	Pearlitic	275 (40)	275 (40)	0	Wear resistance, mill rolls

TABLE 5.5 Typical properties of nonferrous casting alloys.

Alloy	Condition	Casting method*	Yield strength MPa (ksi)	Ultimate tensile strength MPa (ksi)	Elongation in 50 mm (%)	Hardness (HB)
Aluminum						
195	T6	S	220 (30)	280 (40)	2	—
319	T6	S	180 (25)	250 (36)	1.5	—
357	T6	S	296 (43)	345 (50)	2.0	90
380	F	D	165 (24)	331 (48)	3.0	80
390	F	D	241 (35)	279 (40)	1.0	120
Magnesium						
AZ63A	T4	S, P	95 (13)	275 (40)	12	—
AZ91A	F	D	150 (21)	230 (33)	3	—
EZ33A	T5	S, D	110 (16)	160 (23)	3	—
HK31A	T6	S, D	105 (15)	210 (30)	8	—
QE22A	T6	S	205 (30)	275 (40)	4	—
Copper						
Brass C83600	—	S	177 (26)	255 (40)	30	60
Bronze C86500	—	S	193 (28)	490 (71)	30	98
Bronze C93700	—	P	124 (18)	240 (35)	20	60
Zinc						
No. 3	—	D	—	283 (41)	10	82
No. 5	—	D	—	331 (48)	7	91
ZA27	—	P	365 (53)	425 (62)	1	115

*S, sand; D, die; P, permanent mold.

1. **Gray cast iron.** In this structure, graphite exists largely in the form of flakes (Fig. 5.13a). It is called *gray cast iron*, or *gray iron*, because when broken, the fracture path is along the graphite flakes and the surface has a gray, sooty appearance. The flakes act as stress raisers, thus greatly reducing ductility. Gray iron is weak in tension, although strong in compression, as are other brittle materials. On the other hand, the graphite flakes give this material the capacity to dampen vibrations by the internal friction (hence energy dissipation) caused by these flakes. Gray iron is thus a suitable and commonly used material for constructing structures and machine tool bases in which vibration damping is important (Section 8.12). Gray cast irons are specified by a two-digit ASTM designation. Class 20, for example, indicates that the material must have a minimum tensile strength of 20 ksi (140 MPa).

   The basic types of gray cast iron are ferritic, pearlitic, and martensitic (see Section 5.11). In **ferritic gray iron**, also known as fully gray iron, the structure consists of graphite flakes in an alpha ferrite matrix. **Pearlitic gray iron** has a structure consisting of graphite in a matrix of pearlite; although still brittle, it is stronger than gray iron. **Martensitic gray iron** is obtained by austenitizing a pearlitic gray iron, followed by rapid quenching, to produce a structure of graphite in a martensite matrix; as a result, it is very hard. Gray iron castings have relatively few shrinkage cavities and little porosity. Typical uses include engine blocks, machine bases, electric-motor housings, pipes, and wear surfaces of machinery.

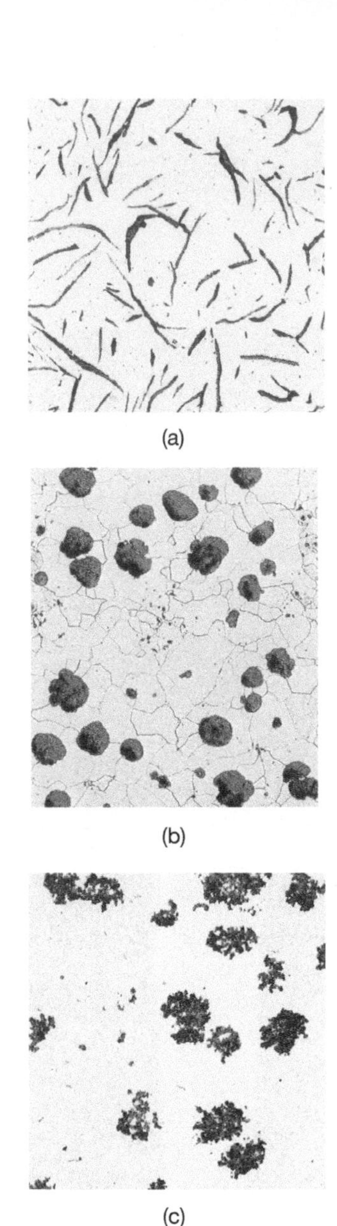

FIGURE 5.13 Cast iron microstructure. Magnification: 100×. (a) ferritic gray iron with graphite flakes; (b) ferritic nodular iron, (ductile iron) with graphite in nodular form; and (c) ferritic malleable iron, solidified as white cast iron, with the carbon present as cementite (Fe_3C), then heat treated to graphitize the carbon.

2. **Ductile iron (nodular iron).** In this structure, graphite is in *nodular* (*spheroid*) form (Fig. 5.13b), making ductile iron more shock resistant. The shape of graphite flakes can be modified into nodules (spheres) by small additions of magnesium and/or cerium to the molten metal prior to pouring. Ductile iron can be heat treated to make it ferritic, pearlitic, or to have a structure of tempered martensite. Typically used for machine parts, pipe, and crankshafts, ductile cast irons are specified by a set of two-digit numbers. Class or grade 80-55-06, for example, indicates that the material has a minimum tensile strength of 80 ksi (550 MPa), a minimum yield strength of 55 ksi (380 MPa) and a 6% elongation.
3. **White cast iron.** White cast iron is very hard, wear resistant, and brittle, because its structure contains large amounts of iron carbide instead of graphite. This structure is obtained either by rapid cooling of gray iron or by adjusting its composition by keeping the carbon and silicon content low. This cast iron is also called *white iron*, because the absence of graphite gives the fracture surface a white crystalline appearance. Because of its extreme hardness and wear resistance, white cast iron typically is used for liners for machinery that processes abrasive materials, rolls for rolling mills, and railroad-car brake shoes.
4. **Malleable iron.** Malleable iron is obtained by annealing white cast iron in an atmosphere of carbon monoxide and carbon dioxide, between 800°C and 900°C (1470°F and 1650°F) for up to several hours. During this process the cementite decomposes (*dissociates*) into iron and graphite. The graphite exists as *clusters* (Fig. 5.13c) in a ferrite or pearlite matrix, and thus has a structure similar to nodular iron, that imparts ductility, strength, and shock resistance, hence the term *malleable*. Typical uses include hardware and railroad equipment. Malleable irons are specified by a five-digit designation; for example, 35018 indicates that the yield strength of the material is 35 ksi (240 MPa), and its elongation is 18%.
5. **Compacted-graphite iron.** The compacted-graphite structure has short, thick, and interconnected flakes, with undulating surfaces and rounded extremities. The mechanical and physical properties of this cast iron are between those of flake graphite and nodular graphite cast irons. Compacted-graphite iron is easy to cast and has consistent properties throughout the casting. Typical applications include automotive engine blocks, crankcases, ingot molds, cylinder heads, and brake disks; it is particularly suitable for components at elevated temperatures and resists thermal fatigue. Its machinability is better than nodular iron.

6. **Cast steels.** The high temperatures required to melt steels (See Table 3.3) requires appropriate mold materials, particularly in view of the high reactivity of steels with oxygen. Steel castings possess properties that are more uniform (isotropic) than those made by mechanical working processes (see Chapter 6). Cast steels can be welded; however, welding alters the cast microstructure in the heat-affected zone (see Section 12.7), influencing the strength, ductility, and toughness of the base metal.
7. **Cast stainless steels.** Casting of stainless steels involves considerations similar to those for steels. Stainless steels generally have a long freezing range (see Section 5.3.2) and high melting temperatures, and develop various structures. Cast stainless steels are available in various compositions, and they can be heat treated and welded. The cast products have high heat and corrosion resistance. Nickel-based casting alloys, for example, are used for severely corrosive environments and service at very high temperatures.

5.6.2 Nonferrous Casting Alloys

Aluminum-based alloys. Cast alloys with an aluminum base have a wide range of mechanical properties, largely because of various hardening mechanisms and heat treatments to which they can be subjected. These castings are lightweight, hence also called *light-metal castings*, and have good machinability. However, except for alloys containing silicon, they generally have low resistance to wear and abrasion. Aluminum-based cast alloys have numerous applications, including automotive engine blocks, architectural, decorative, aerospace, and electrical components.

Magnesium-based alloys. Cast alloys with a magnesium base have good corrosion resistance and moderate strength (depending on the proper use of a protective coating system), and the particular heat treatment. They have very good strength-to-weight ratios and have been widely used for aerospace and automotive structural applications.

Copper-based alloys. Cast alloys with a copper base have the advantages of good electrical and thermal conductivity, corrosion resistance, nontoxicity (unless they contain lead), good machinability, and wear properties (thus suitable as bearing materials).

Zinc-based alloys. Cast alloys with a zinc base have good fluidity and good strength for structural applications. These alloys are widely used in die casting of structural shapes, electrical conduit, and corrosion-resistant parts.

High-temperature alloys. With a wide range of properties and applications, these alloys typically require temperatures of up to 1650°C (3000°F) for casting titanium and superalloys, and higher for refractory alloys. Typical applications include jet and rocket engine components, which otherwise would be difficult or uneconomical to produce by other means.

5.7 Ingot Casting and Continuous Casting

Traditionally, the first step in metal processing is the casting of the molten metal into a solid form (**ingot**) for further processing. In ingot casting, the molten metal is poured (*teemed*) from the ladle into ingot molds, in which the metal solidifies; this is basically equivalent to sand casting or to permanent mold casting. Molten metal is poured into the permanent molds, yielding the microstructure shown in Fig. 5.5. Gas entrapment can be reduced by bottom pouring, using an insulated collar on top of the mold, or by using an exothermic compound that produces heat when it contacts the molten metal (hot top). These techniques reduce the rate of cooling and result in a higher yield of high quality metal. The cooled ingots are then removed (*stripped*) from the molds and lowered into *soaking pits*, where they are reheated to a uniform temperature of about 1200°C (2200°F) for subsequent processing. Ingots may be square, rectangular, or round in cross section, and their weight ranges from a few hundred to tens of thousands of kilograms.

5.7.1 Ferrous Alloy Ingots

Several reactions take place during solidification of an ingot that have an important influence on the quality of the steel produced. For example, significant amounts of oxygen and other gases can dissolve in the molten metal during steelmaking, leading to porosity defects (see Section 5.12.1). Because the solubility limit of gases in the metal decreases sharply as its temperature decreases, much of these gases cavitate during solidification of the metal. The rejected oxygen combines with carbon and forms carbon monoxide, which causes porosity in the solidified ingot. Depending on the amount of gas evolved during solidification, three types of steel ingots can be produced.

1. **Killed steel** is fully deoxidized steel; that is, oxygen is removed and thus porosity is eliminated. In the deoxidation process, the dissolved oxygen in the molten metal is made to react with elements (typically aluminum) that are added to the melt; vanadium, titanium, and zirconium also are used. These elements have an affinity for oxygen and form metallic oxides; with aluminum, the product is called aluminum-killed steel. The term *killed* comes from the fact that the steel lies quietly in the mold.
2. **Semi-killed steel** is partially deoxidized steel. It contains some porosity, generally in the upper central section of the ingot, but otherwise has little or no pipe (see Section 5.12.1), and thus scrap is reduced. Piping is less because it is compensated for by the presence of porosity in that region. Semi-killed steels are economical to produce.
3. In a **rimmed steel**, which generally has less than 0.15% carbon, the evolved gases are only partially killed or controlled by the addition of elements, such as aluminum. Rimmed steels have little or no piping, with a ductile skin, and with good surface finish.

5.7.2 Continuous Casting

Conceived in the 1860s, *continuous casting*, or **strand casting**, was first developed for casting non-ferrous metal strip. It is now widely used for steel production and at low cost. A continuous casting system is shown schematically in Fig. 5.14a. The molten metal in the ladle (not shown) is cleaned and equalized in temperature, by blowing nitrogen gas through it for 5 to 10 min. The metal is then poured into a refractory-lined intermediate pouring vessel (**tundish**), where impurities are skimmed off.

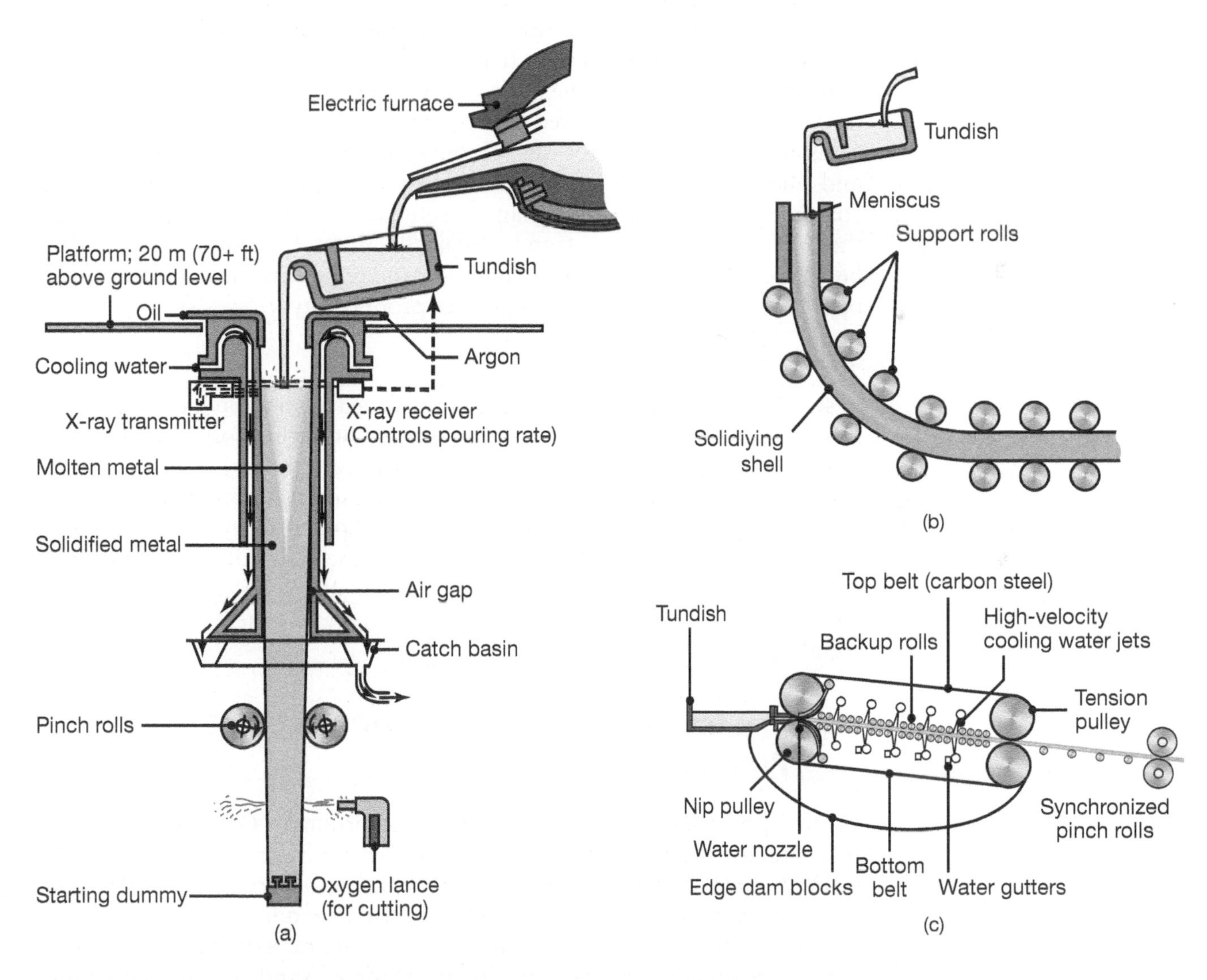

FIGURE 5.14 (a) The continuous-casting process for steel. Typically, the solidified metal descends at a speed of 25 mm/s (1 in./s). Note that the platform is about 20 m (65 ft) above ground level. *Source: Metalcaster's Reference and Guide*, American Foundrymen's Society. (b) Continuous casting using support or guide rollers to allow transition from a vertical pour zone to horizontal conveyors. (c) Continuous strip casting of nonferrous metal strip.

Before starting the casting process, a solid *starter*, or *dummy bar*, is inserted into the bottom of the mold. The molten metal is then poured and solidifies on the starter bar (See bottom of Fig. 5.14a). The bar is withdrawn at the same rate the metal is poured, along a path supported by rollers (called *pinch rolls*). The cooling rate is such that the metal develops a solidified skin (shell) to support itself during its travel downward, at speeds typically around 25 mm/s (1 in./s). The shell thickness at the exit end of the mold is about 12–18 mm (0.5–0.75 in.). Additional cooling is provided by water sprays along the travel path of the solidifying metal. The molds are generally coated with graphite, or a similar solid lubricant, to reduce friction and adhesion at the mold-metal interfaces. The molds also vibrated to further reduce friction and sticking.

The continuously cast metal is cut into desired lengths by shearing or torch cutting, or it may be fed directly into a rolling mill for further reductions in thickness and for shape rolling of products, such as channels and I-beams (Section 6.3.5). Continuously-cast metals have more uniform composition and properties than those made by ingot casting. Although the thickness of the steel strand is typically about 250 mm (10 in.), it can be 12 mm (0.5 in.) or less. The thinner strand reduces the number of rolling operations required and thus improves productivity.

5.7.3 Strip Casting

In strip casting, thin slabs or strips are produced from molten metal, which solidifies in manner similar to strand casting, but the hot solid is then rolled to form the final shape (Fig. 5.14b). The compressive stresses developed in rolling serve to reduce any porosity in the material, improving the properties. Strip casting eliminates a hot rolling operation in the production of metal strips or slabs. Final thicknesses on the order of 2–6 mm (0.08–0.25 in.) can be obtained for carbon, stainless, and electrical steels (used in making the iron cores of motors, transformers, and generators) and other metals.

5.8 Expendable-Mold, Permanent-Pattern Casting Processes

Casting processes are generally classified according to (a) mold materials; (b) molding processes; and (c) methods of feeding the mold with the molten metal (See also Table 5.2). The two major categories are *expendable-mold* and *permanent-mold* casting (Section 5.10). Expendable-mold processes are further categorized as *permanent pattern* and *expendable pattern* processes. Expendable molds typically are made of sand, plaster, ceramics, and similar materials, which generally are mixed with various **binders** or bonding agents.

5.8.1 Sand Casting

The *sand casting* process consists of (a) placing a pattern, having the shape of the desired casting, in sand to make an imprint; (b) incorporating a gating system; (c) filling the resulting cavity with molten metal; (d) allowing the metal to solidify; (e) breaking away the sand mold; and (f) removing the casting and finishing it. Examples of parts made by sand casting are engine blocks, cylinder heads, machine-tool bases, and housings for pumps and motors. While the origins of sand casting date to ancient times (See Table 1.1), it is still the most prevalent form of casting. In the United States alone, about 15 million tons of metal are cast by this method each year.

Sands. Sand, the product of the disintegration of rocks over extremely long periods of time, is inexpensive and suitable as mold material because of its resistance to high temperatures. Most sand-casting operations use silica sands (SiO_2). There are two general types of sand: *naturally bonded* (*bank sand*) and *synthetic* (*lake sand*). Because its composition can be controlled more accurately, synthetic sand is preferred by most foundries.

Several factors are considered in the selection of sand for molds. Sand having fine, round grains can be closely packed and forms a smooth mold surface. Good **permeability** of molds and cores allows gases and steam evolved during casting to escape easily. The mold should have good **collapsibility**, because the casting shrinks while cooling. Fine sand enhances mold strength but lowers mold permeability. Sand is typically conditioned before use. *Mulling machines* are used to uniformly mull (mixing thoroughly) sand with additives. Clay is typically used as a cohesive agent to bond the sand particles, giving the sand higher strength.

Types of sand molds. The major components of a typical sand mold are shown in Fig. 5.10. There are three basic types of sand molds: green-sand, cold-box, and no-bake molds. The most common mold material is *green molding sand*, the term green referring to the fact that the sand in the mold is moist or damp while the metal is being poured into it. Green molding sand is a mixture of sand, clay, and water, and is the least expensive in making molds.

In the **skin-dried** method, the mold surfaces are dried, either by storing the mold in air or drying it with torches; the molds may also be baked. These molds are generally used for large castings because of their high strength; also, they are stronger than green-sand molds and impart better dimensional accuracy and surface finish to the casting. However, distortion of the mold is greater, the castings are more susceptible to hot tearing (because of the lower collapsibility of the mold), and the production rate is lower because of the length of drying time required.

In the **no-bake** mold process, a synthetic liquid resin is mixed with sand, and the mixture hardens at room temperature. Because bonding of the mold takes place without heat, they are called *cold-setting processes*. The **cold-box** mold process uses organic and inorganic binders that are blended into the sand to chemically bond the grains for greater mold strength; no heat is used. These molds are dimensionally more accurate than green-sand molds but are more expensive to produce.

Patterns. *Patterns*, which are used to mold the sand mixture into the shape of the casting, may be made of wood, plastic, or metal; they can also use **additive manufacturing** techniques (Section 10.12). Patterns may also be made from a combination of materials to reduce wear in critical regions. The selection of a pattern material depends on the size and shape of the casting, the dimensional accuracy, the quantity of castings required, and the molding process to be used. Because patterns are repeatedly used to make molds, the strength and durability of the pattern material selected must be sufficient for the number of castings the mold is expected to produce. Patterns are usually coated with a **parting agent** to facilitate their removal from the molds.

Patterns can be designed with a variety of features for specific applications as well as for economic considerations: (a) **One-piece patterns** are generally used for simpler shapes and low-quantity production; they are typically made of wood, and are inexpensive; (b) **split patterns** are two-piece patterns, made so that each part forms a portion of the cavity for the casting, thus allowing the casting of complicated shapes; and (c) **match-plate patterns** are a common type of mounted pattern, in which two-piece patterns are constructed, by securing each half of one or more split patterns to the opposite sides of a single plate. In such patterns, the gating system can be mounted on the drag side of the pattern.

Cores. Cores are utilized for castings with internal cavities or passageways, such as in automotive engine blocks or valve bodies. They are placed in the mold cavity prior to pouring, and are removed from the casting during shakeout. Cores must possess certain strength, permeability, ability to withstand heat, and collapsibility. They are typically made of sand aggregates and are anchored by **core prints**, which are recesses in the mold to support the core and to provide vents for the escape of gases. To keep the core from shifting during casting, metal supports (called **chaplets**) may be used to anchor the core in place. Cores are shaped in core boxes, which are used much as patterns are used to form sand molds. The sand can be packed into the boxes with sweeps or blown into the box by compressed air from *core blowers*.

Sand-molding machines. The sand mixture is compacted around the pattern by molding machines. These machines eliminate arduous labor and produce higher-quality castings by manipulating the mold in a controlled manner. Mechanization of the molding process can further be assisted by **jolting** the assembly, in which the flask, the molding sand, and the pattern are all placed on a pattern plate, mounted on an anvil and jolted upward by air pressure at rapid intervals. The inertial forces compact the sand around the pattern.

In **vertical flaskless molding** the halves of the pattern form a vertical chamber wall against which the sand is blown and compacted. The mold halves are then packed horizontally, with the parting line oriented vertically, and moved along a pouring conveyor.

Sandslingers fill the flask uniformly with sand, under a stream of high pressure. *Sandthrowers* are used to fill large flasks, and are typically operated by machine. An impeller in the machine throws sand from its blades or cups at such high speeds that the sand is placed and compressed, so that additional compaction is unnecessary.

In **impact molding**, the sand is compacted by controlled explosion or instantaneous release of compressed gases, producing molds with uniform strength and good permeability. In **vacuum molding**, also known as the *"V" process*, the pattern is covered tightly with a thin plastic sheet. A flask is then placed over the pattern and is filled with sand. A second sheet of plastic is placed on top of the sand, and a vacuum action compresses and hardens the sand to allow the pattern to be withdrawn. Both halves of the mold are made this way and then assembled. During pouring of molten metal, the mold remains under vacuum but the casting cavity does not. When the metal has solidified, the vacuum is turned off and the sand falls away, releasing the casting.

The sand casting operation. The sequence of operations in sand casting is shown in Fig. 5.15. After the mold has been shaped and the cores have been placed in position, the two halves (*cope* and *drag*) are closed (Fig. 5.10), clamped, and weighted down (to prevent the separation of the mold sections under the pressure exerted when the molten metal is poured in). The design of the *gating system* is important for proper delivery of the molten metal into the mold cavity. Turbulence must be minimized, air and gases must be allowed to escape by vents or other means, and proper temperature gradients must be established and maintained to eliminate shrinkage and porosity. The design of *risers* also is important for supplying the necessary amount of molten metal during solidification of the casting. Note that the pouring basin may also serve as a riser for this purpose. After solidification, the casting is shaken out of its mold and the sand and oxide layers, adhering to the casting, are removed by vibration (using a shaker) or by sand blasting. The risers and gates are then removed by sawing, trimming in dies, abrasive disks, or by oxyfuel-gas cutting,

Almost all commercially used alloys can be sand cast. The surface finish largely depends on the mold materials. Although dimensional accuracy is not as good as that of other casting processes (See Table 5.2), intricate shapes can be cast, such as cast-iron engine blocks, transmission housings, machinery frames and very large propellers for ocean liners. Sand casting can be economical for relatively small as well as large production runs; equipment costs are generally low.

Mold ablation. Ablation has been used to improve the mechanical properties and production rates in sand casting. In this process, a sand mold is filled with molten metal, and the mold is then immediately sprayed with a liquid and/or gas solvent to progressively erode the sand. As the mold is removed, the liquid stream causes rapid and directional solidification of the metal. With properly designed risers, mold ablation results in significantly lower porosity than in conventional sand casting, leading to higher strength and ductility of the casting, and has therefore been applied to normally difficult-to-cast materials or metal-matrix composites. Since ablation speeds solidification and also removes cores, significant productivity improvements can also be achieved.

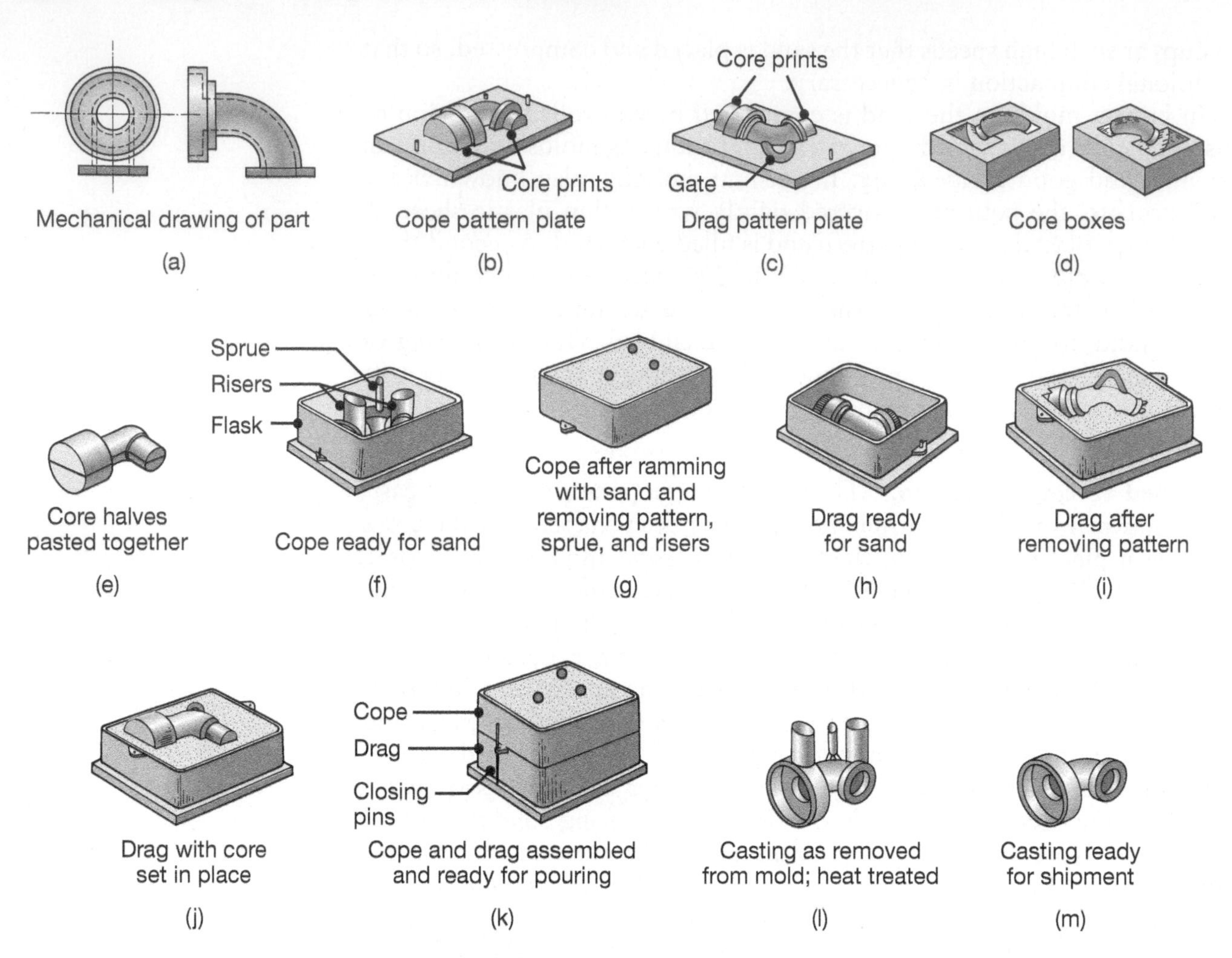

FIGURE 5.15 Schematic illustration of the sequence of operations in sand casting. (a) A mechanical drawing of the part, used to create patterns. (b-c) Patterns mounted on plates equipped with pins for alignment. Note the presence of core prints designed to hold the core in place. (d-e) Core boxes produce core halves, which are pasted together. The cores will be used to produce the hollow area of the part shown in (a). (f) The cope half of the mold is assembled by securing the cope pattern plate to the flask with aligning pins, and attaching inserts to form the sprue and risers. (g) The flask is rammed with sand and the plate and inserts are removed. (h) The drag half is produced in a similar manner. (j) The core is set in place within the drag cavity. (k) The mold is closed by placing the cope on top of the drag and securing the assembly with pins. (l) After the metal solidifies, the casting is removed from the mold. (m) The sprue and risers are cut off and recycled, and the casting is cleaned, inspected, and heat treated (when necessary). *Source:* After Steel Founders' Society of America.

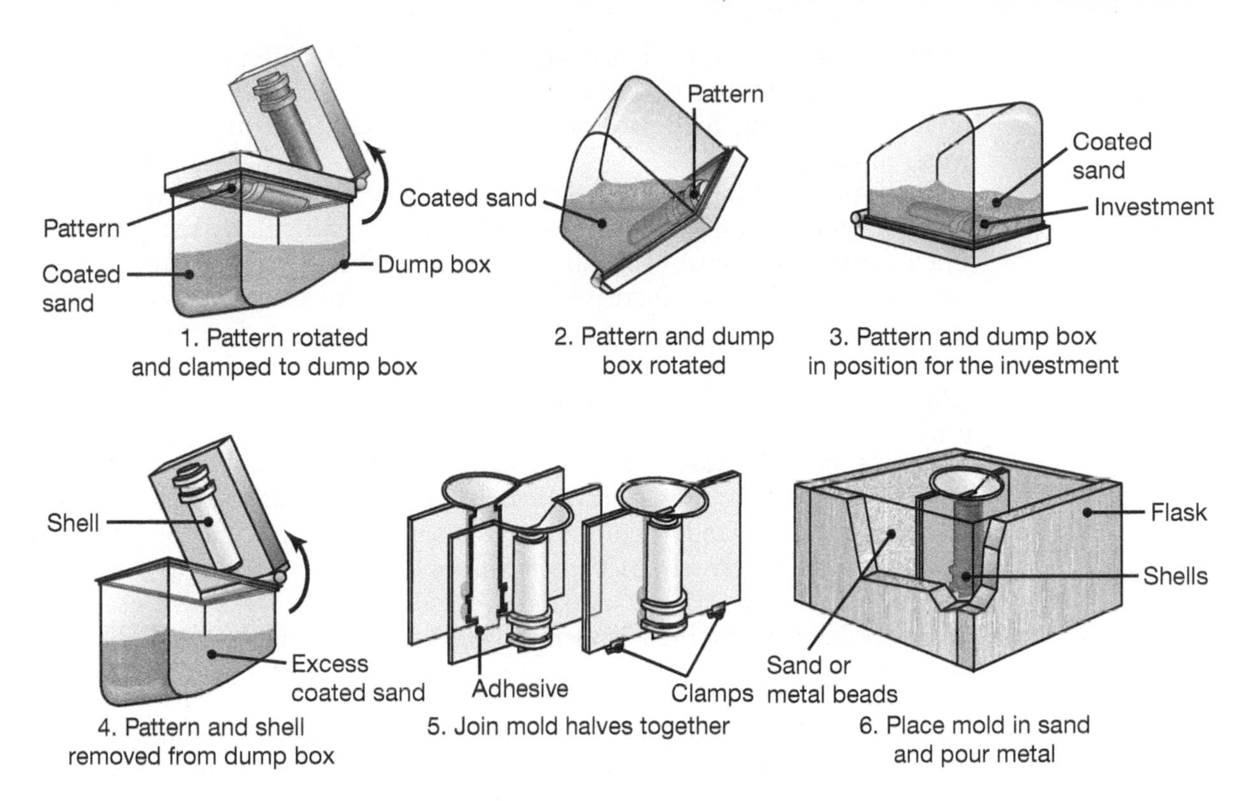

FIGURE 5.16 Schematic illustration of the shell-molding process, also called the *dump-box* technique.

5.8.2 Shell-Mold Casting

Shell-mold casting (Fig. 5.16) has grown significantly because it can produce many types of castings with close dimensional tolerances, good surface finish, and at a low cost. In this process, a mounted pattern, made of a ferrous metal or aluminum, is heated to 175°C to 370°C (350°F to 700°F), coated with a parting agent, such as silicone, and is clamped to a box or chamber containing a fine sand containing a 2.5–4% thermosetting resin binder (such as phenol-formaldehyde), which coats the sand particles. The sand mixture is then turned over the heated pattern, coating it evenly. The assembly is often placed in an oven for a short period of time to complete the curing of the resin. The shell hardens around the pattern and is then removed, using built-in ejector pins. Two half-shells are made in this manner and are bonded or clamped together in preparation for pouring.

The shells are light and thin, usually 5–10 mm, and thus their thermal characteristics are different from those for thicker molds. The thin shells allow gases to escape during solidification of the metal. The mold is generally used vertically and is supported by surrounding it with steel shot. The mold walls are relatively smooth, resulting in low resistance to molten metal flow and producing castings with sharper corners, thinner sections, and smaller projections than are possible in green-sand molds. Several castings can be made in a single mold, using a multiple gating system.

Applications include small mechanical parts requiring high precision, gear housings, cylinder heads, and connecting rods; the process is also widely used in producing high-precision molding cores, such as engine-block water jackets.

Shell-mold casting may be more economical than other casting processes, depending on various production factors, particularly energy cost. The relatively high cost of metal patterns becomes a smaller factor as the size of production run increases. The high quality of the finished casting can significantly reduce the costs for subsequent cleaning, machining, and other finishing operations. Complex shapes can be produced with less labor, and the process can be automated fairly easily.

Sodium silicate process. The mold material in this process is a mixture of sand and 1 to 6% sodium silicate (*waterglass*), or various other chemicals, as the binder for sand. The mixture is then packed around the pattern and hardened by blowing CO_2 gas through it. Also known as *silicate-bonded sand* or the carbon-dioxide process, this process is also used to make cores, reducing their tendency to tear, because of their flexibility at elevated temperatures.

Rammed graphite molding. Rammed graphite is used to make molds for casting reactive metals, such as titanium and zirconium, because these metals react vigorously with silica. The molds are packed, much like sand molds, then air dried, baked at 175°C (350°F), and fired at 870°C (1600°F). They are then stored under controlled humidity and temperature. The casting procedures are similar to those for sand molds.

5.8.3 Plaster-Mold Casting

The *plaster-mold casting* process and the ceramic-mold and investment-casting processes are known as **precision casting**, because of the high dimensional accuracy and good surface finish obtained. Typical parts made are lock components, gears, valves, fittings, tooling, and ornaments, weighing as little as 1 g. The mold is made of *plaster of paris* (gypsum, or calcium sulfate), with the addition of talc and silica flour to improve strength and control the time required for the plaster to set. These components are mixed with water and the resulting slurry is poured over the pattern.

After the plaster sets, usually within 15 min, the pattern is removed and the mold is dried. The mold halves are then assembled to form the mold cavity and are preheated to about 120°C (250°F) for 16 hrs; the molten metal is then poured into the mold. Because plaster molds have very low permeability, the gases evolved during solidification of the metal cannot escape; consequently, the molten metal is poured either in a vacuum or under pressure. The permeability of plaster molds can be substantially increased by the *Antioch* process. The molds are dehydrated in an autoclave (pressurized oven) for 6–12 hrs, then rehydrated in air for 14 hrs. Another method of increasing permeability is to use foamed plaster, containing trapped air bubbles.

Patterns for plaster molding are generally made of aluminum alloys, thermosetting plastics, brass, or zinc alloys; wood patterns are not suitable, due to the presence of water-based slurry. Because there is a limit to the maximum temperature that the plaster mold can withstand, generally about 1200°C (2200°F), plaster-mold casting is used only for casting aluminum, magnesium, zinc, and some copper-based alloys. The castings have fine detail and good surface finish. Since plaster molds have lower thermal conductivity than other types of molds, the castings cool slowly, resulting in more uniform grain structure, with less warpage and better mechanical properties.

5.8.4 Ceramic-Mold Casting

The *ceramic-mold casting* process, also called **cope-and-drag investment casting**, is another precision casting process; it is similar to the plaster-mold process, with the exception that it uses refractory mold materials suitable for high-temperature applications. The slurry is a mixture of fine-grained zircon ($ZrSiO_4$), aluminum oxide, and fused silica; they are mixed with bonding agents, and poured over the pattern (Fig. 5.17) which has been placed in a flask. The pattern may be made of wood or metal. After setting, the molds (ceramic facings) are removed, dried, burned off to remove volatile matter, and baked. The molds are then clamped firmly and are used as for all-ceramic molds. In the related *Shaw* process, the ceramic facings are backed by fireclay to make the molds stronger.

The high-temperature resistance of the refractory molding materials allows these molds to be used in casting ferrous and high-temperature alloys, stainless steels, and tool steels. Although the process is somewhat expensive, the castings have good dimensional accuracy and surface finish over a wide range of sizes and intricate shapes, with some parts weighing as much as 700 kg. Typical parts made are impellers, cutters for machining, dies for metalworking, and molds for making plastic or rubber components.

5.8.5 Vacuum Casting

The *vacuum-casting process*, also called *counter-gravity low-pressure* (CL) *process*, but not to be confused with the vacuum-molding process

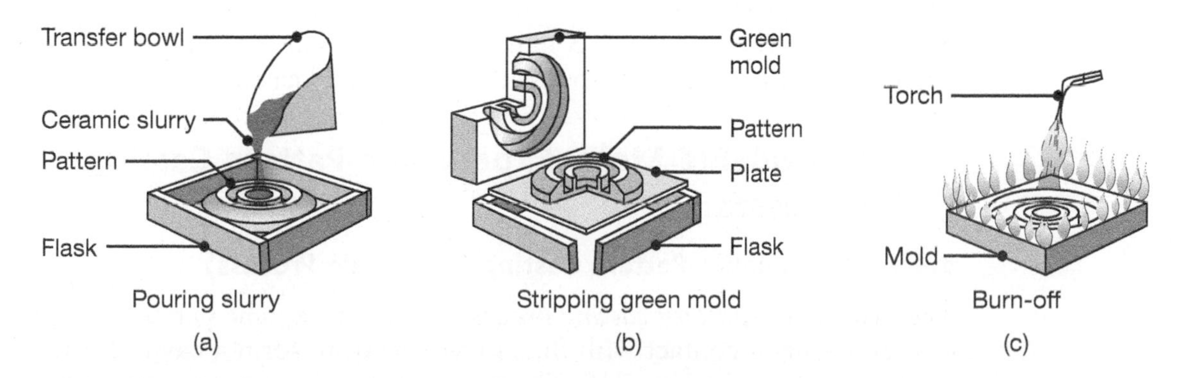

FIGURE 5.17 Sequence of operations in making a ceramic mold.

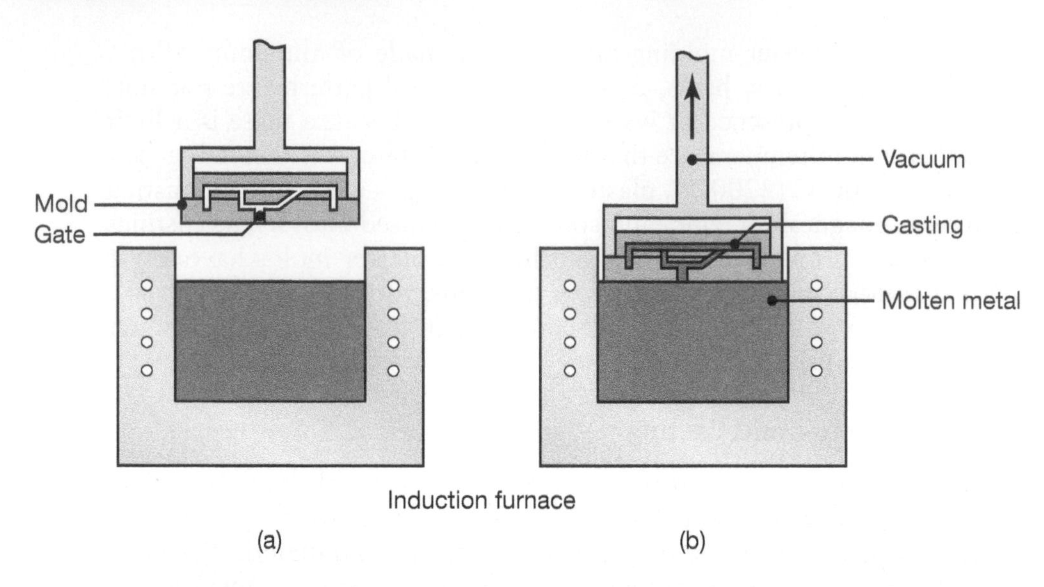

FIGURE 5.18 Schematic illustration of the vacuum-casting process. Note that the mold has a bottom gate. (a) before and (b) after immersion of the mold into the molten metal. *Source:* After R. Blackburn.

described in Section 5.8.1, is shown in Fig. 5.18. A mixture of fine sand and urethane is molded over metal dies and cured with amine vapor. The mold is then held with a robot arm and partially immersed into molten metal in an induction furnace. The metal may be melted in air (*CLA process*) or in a vacuum (*CLV process*). The vacuum reduces the air pressure inside the mold to about two-thirds of atmospheric pressure, drawing the molten metal into the mold cavities through a gate at the bottom of the mold. After the mold is filled, it is withdrawn from the molten metal. The furnace is at a temperature usually 55°C (100°F) above the liquidus temperature; consequently, the metal begins to solidify within a fraction of a second.

This process is an alternative to investment, shell-mold, and green-sand casting, and is particularly suitable for thin-walled (0.75 mm) complex shapes and with uniform properties. Carbon and low- and high-alloy steel and stainless steel parts, weighing as much as 70 kg, can be vacuum cast. These parts, which are often in the form of superalloys for gas turbines, may have walls as thin as 0.5 mm. The process can be automated and production costs are similar to those for green-sand casting.

5.9 Expendable-Mold, Expendable-Pattern Casting Processes

5.9.1 Expendable-Pattern Casting (Lost Foam Process)

The *expendable-pattern casting process* uses a polystyrene pattern, which evaporates upon contact with the molten metal, to form a cavity for the casting, as shown in Fig. 5.19. The process is also known as *evaporative-pattern* or *lost-foam casting*, and under the trade name *Full-Mold* process.

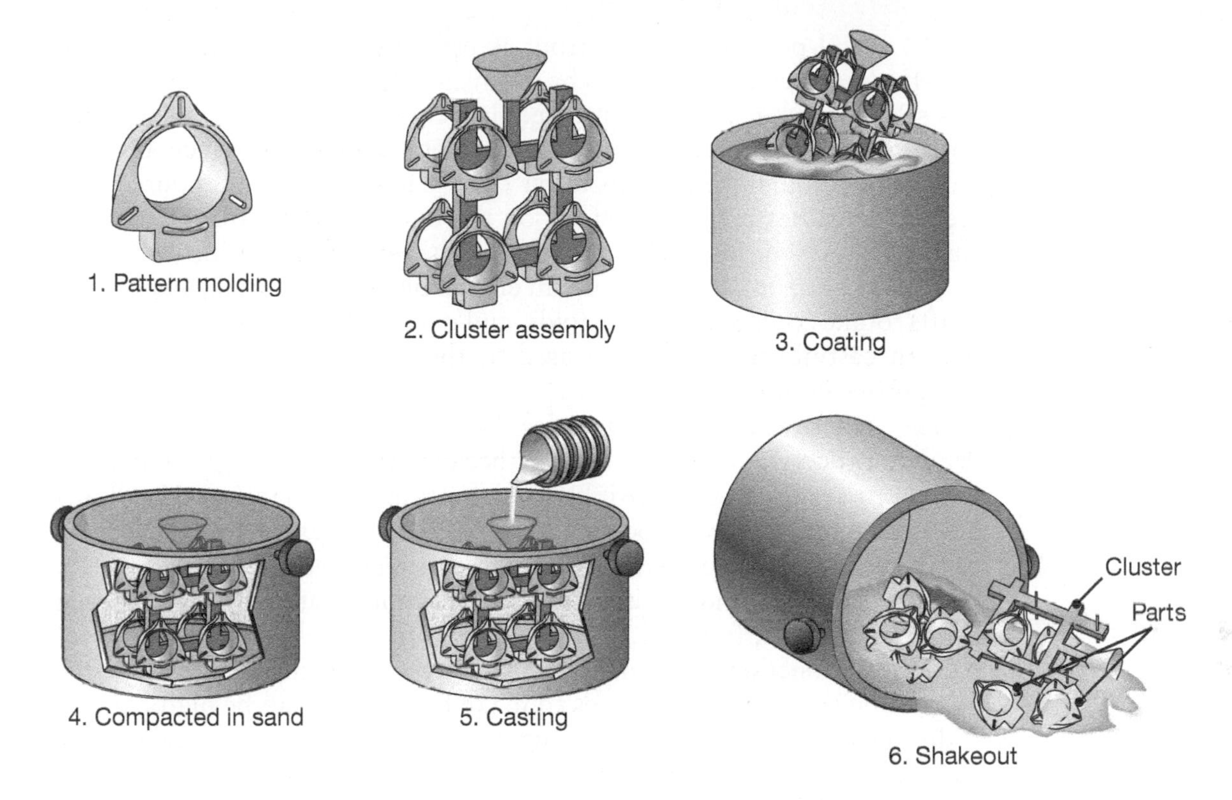

FIGURE 5.19 Schematic illustration of the expendable-pattern casting process, also known as lost-foam or evaporative-pattern casting.

It is one of the more important casting processes for ferrous and nonferrous metals, particularly for the automotive industry. First, raw expandable polystyrene (EPS) beads, containing 5–8% pentane (a volatile hydrocarbon), are placed in a preheated die, typically made of aluminum. The polystyrene expands and takes the shape of the die cavity; additional heat is applied to fuse and bond the beads together. The die is then cooled and opened, and the pattern is removed. Complex patterns may also be made by bonding various individual pattern sections, using hot-melt adhesive (Section 12.14.1).

The pattern is then coated with a water-based refractory slurry, dried, and placed in a flask. The flask is filled with loose fine sand, which surrounds and supports the pattern; it is periodically compacted by various means. Then, without removing the pattern, the molten metal is poured into the mold. This action immediately vaporizes the pattern and fills the mold cavity, completely replacing the space previously occupied by the polystyrene pattern. The heat degrades (depolymerizes) the polystyrene, and the degradation products are vented into the surrounding sand. In a modification of this process, the polystyrene pattern is surrounded by a ceramic sell (*Replicast® C-S process*; see below).

The molten metal flow velocity in the mold depends on the rate of degradation of the polymer. The flow is basically laminar, with Reynolds numbers in the range of 400–3000 (see Section 5.4.1). The velocity of the metal-polymer pattern front is estimated to be in the range of 0.1–1 m/s, and can be controlled by producing patterns with cavities or hollow sections. Because the polymer requires considerable energy to degrade, large thermal gradients are present at the metal-polymer interface. In other words, the molten metal cools faster than it would if it were poured into a cavity; this has important effects on the microstructure throughout the casting and also leads to directional solidification of the metal.

Typical parts made are aluminum engine blocks, cylinder heads, crankshafts, brake components, manifolds, and machine bases. The evaporative-pattern casting process is also used in the production of metal-matrix composites (Section 11.14).

The evaporative-pattern process has the following characteristics: (a) The process is relatively simple, because there are no parting lines, cores, or riser systems; hence it has design flexibility; (b) inexpensive flasks are satisfactory for this process; (c) polystyrene is inexpensive and can be easily processed into patterns with complex shapes, various sizes, and fine surface detail; (d) the casting itself requires minimum cleaning and finishing operations; (e) the operation can be automated and is economical for long production runs; and (f) the cost of producing the die used for expanding the polystyrene beads to make the pattern can be high.

In a modification of the evaporative-pattern process, called the *Replicast® C-S process*, a polystyrene pattern is surrounded by a ceramic shell; then the pattern is burned out prior to pouring the molten metal into the mold. The principal advantage of this process over investment casting (which uses wax patterns, Section 5.9.2) is that carbon pickup into the metal is avoided entirely. Further developments in evaporative-pattern casting include the production of metal-matrix composites (Section 11.14). During molding of the polymer pattern, fibers or particles are embedded throughout, which then become an integral part of the casting. Other techniques include the modification and grain refinement of the casting by using grain refiners and modifier master alloys.

5.9.2 Investment Casting (Lost Wax Process)

The *investment-casting* process, also called the *lost wax* process, was first used during the period 4000–3000 B.C. Typical parts made are mechanical components such as gears, cams, valves, and ratchets; parts up to 1.5 m (5 ft) in diameter and weighing as much as 1140 kg (2500 lb) have been successfully cast by this process. The sequence involved in investment casting are shown in Fig. 5.20.

The pattern is made by injecting *semisolid* or *liquid* wax or plastic into a metal die in the shape of the pattern, or is made through additive manufacturing methods (Section 10.12). The pattern is then removed and dipped into a slurry of refractory material, such as very fine silica and binders, ethyl silicate, and acids. After this initial coating has dried, the pattern is coated repeatedly to increase its thickness. (The term *investment* comes from investing the pattern with the refractory material.) Wax patterns require careful handling, because they are not strong enough to withstand the forces involved during mold making. The one-piece mold is dried in air

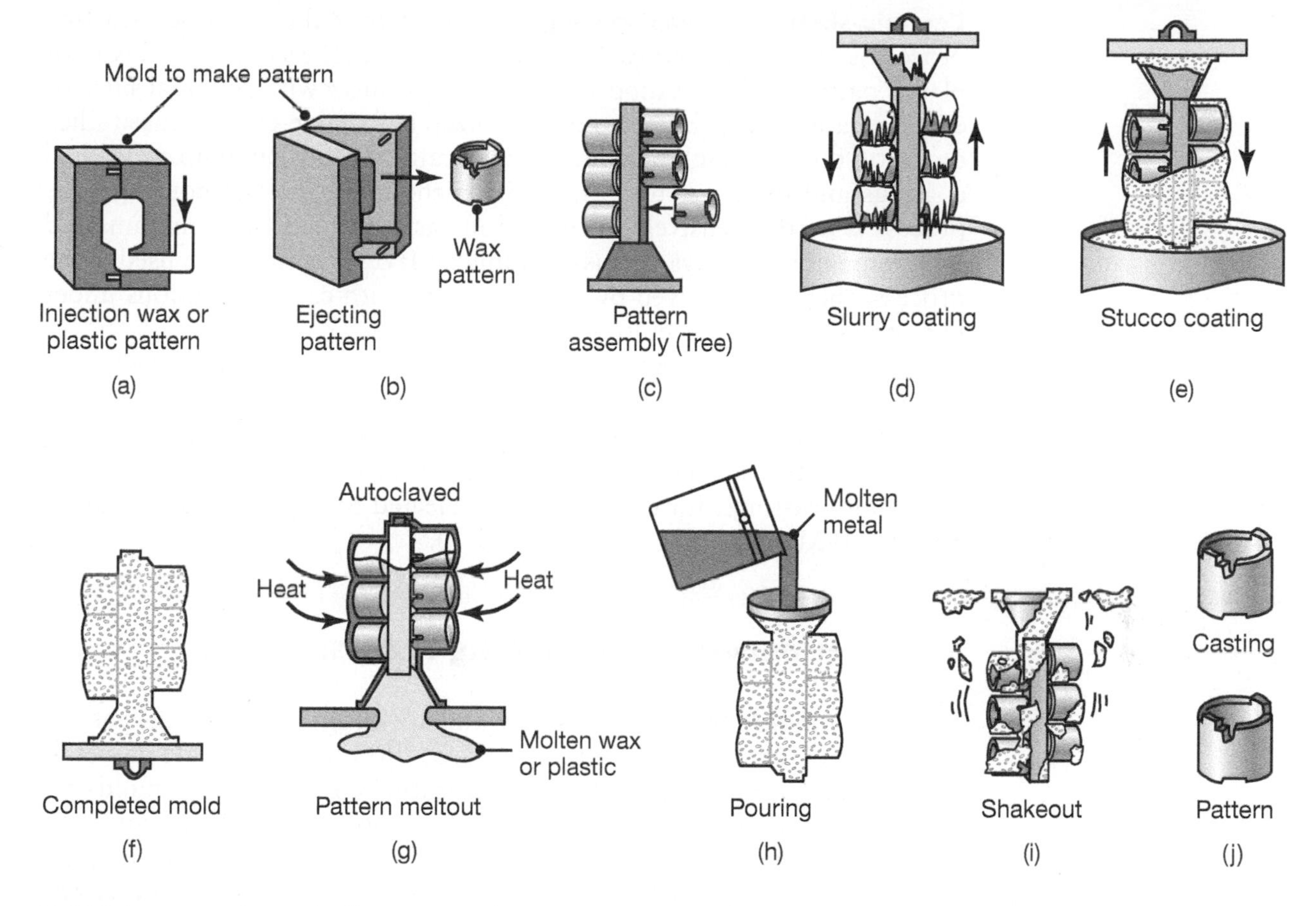

FIGURE 5.20 Schematic illustration of investment casting (lost wax process). Castings by this method can be made with very fine detail and from a variety of metals. *Source:* Steel Founders' Society of America.

and heated to a temperature of 90°C to 175°C (200°F to 350°F) for about 4 hrs (depending on the metal to be cast), to drive off the water of crystallization (chemically combined water). After the metal has been poured and has solidified, the mold is broken up and the casting is removed.

Several patterns can be joined to make one mold, called a **tree** (Fig. 5.20), significantly increasing the production rate. For small parts, the tree can be inserted into a permeable flask and filled with a liquid slurry investment. The investment is then placed into a chamber and evacuated, to remove air bubbles, until the mold solidifies. The flask is usually placed in a vacuum-casting machine, so that the molten metal is drawn into the permeable mold and onto the part, producing fine detail.

The labor and materials involved can make the lost-wax process costly, but little or no finishing is required. The process is capable of producing intricate shapes from a wide variety of ferrous and nonferrous metals and alloys, with parts generally weighing from 1 g to 100 kg and as much as 1140 kg. It is suitable for casting high-melting-point alloys, with a good surface finish and close dimensional tolerances.

Ceramic-shell investment casting. A variation of the investment-casting process is *ceramic-shell investment casting*. It uses the same type of wax or plastic pattern, which is dipped (a) first into a slurry with colloidal silica or ethyl silicate binder; (b) then into a fluidized bed of fine-grained fused silica or zircon flour; and (c) then into coarse-grain silica, to build up additional coatings and thickness to withstand the thermal shock of pouring. The process is used extensively for precision casting of steels, aluminum, and high-temperature alloys, and is economical. If ceramic cores are used in the process, they are removed by leaching them with caustic solutions under high pressure and temperature.

The molten metal may be poured in a vacuum, to extract evolved gases and reduce oxidation, further improving the quality of the casting. The castings made by this and as well as other processes may be subjected to hot isostatic pressing (see Section 11.3.3) to further reduce microporosity. Aluminum castings, for example, are subjected to a gas pressure of up to about 100 MPa (15 ksi) at 500°C (900°F).

5.10 Permanent-Mold Casting Processes

Permanent molds are used repeatedly and are designed so that the casting can be easily removed and the mold reused. The molds are made of metals that maintain their strength at high temperatures. Because metal molds are better heat conductors than the expendable molds, described above, the solidifying casting is subjected to a higher rate of cooling. This, in turn, affects the microstructure and grain size within the casting, as described in Section 5.3.

In permanent-mold casting, two halves of a mold are made from steel, bronze, refractory metal alloys, or graphite (*semipermanent mold*). The mold cavity and the gating system are machined into the mold itself, and thus become an integral part of the mold. If required, cores are made of metal or sand aggregate, and are placed in the mold prior to casting. Typical core materials are shell or no-bake cores, gray iron, low-carbon steel, and hot-work die steel. Inserts may also be used for various parts of the mold.

In order to increase the life of permanent molds, the surfaces of the mold cavity may be coated with a refractory slurry, or sprayed with graphite every few castings. The coatings also serve as parting agents and thermal barriers to control the rate of cooling of the casting. Mechanical ejectors, such as pins located in various parts of the mold, may be necessary for removal of complex castings. Ejectors usually leave small round impressions on castings and gating systems, as they do in injection-molding of plastics (see Section 10.10.2).

The molds are clamped together by mechanical means and are heated to facilitate metal flow and reduce damage (*thermal fatigue*, see Section 3.9.5) to the dies. The molten metal then flows through the gating system, fills the mold cavity, and, after solidification, the molds are opened and the casting is removed. Special methods of cooling the mold include water or the use of air-cooled fins. The casting operation is automated, especially for large production runs.

Used mostly for aluminum, magnesium, and copper alloys (because of their generally lower melting points), as well as steels cast in graphite or heat-resistant metal molds, the castings have good surface finish, close dimensional tolerances, and uniform and good mechanical properties. Typical parts made include automobile pistons, cylinder heads, connecting rods, gear blanks for appliances, and kitchenware. Because of the high cost of dies, permanent-mold casting is not economical for small production runs. Moreover, because of the difficulty in removing the casting from the mold, intricate shapes cannot be cast by this process, although intricate internal cavities can be cast using sand cores.

5.10.1 Slush Casting

It was noted in Fig. 5.11 that a solidified skin first develops in a casting and that this skin becomes thicker with time. Hollow castings with thin walls can be made by permanent-mold casting using this principle in a process called *slush casting*. The molten metal is poured into the metal mold, and after the desired thickness of solidified skin is obtained, the mold is inverted (slung) and the remaining liquid metal is poured out. The mold halves are then opened and the casting is removed. This process is suitable for small production runs, and is typically used for making ornamental and decorative objects and toys, from low-melting-point metals, such as zinc, tin, and lead alloys.

5.10.2 Pressure Casting

In the *pressure-casting* process, also called *pressure pouring* or *low-pressure casting* (Fig. 5.21), the molten metal is forced upward into a graphite or metal mold, by gas pressure which is maintained until the metal has completely solidified in the mold. In a similar system, the molten metal may also be forced upward by a vacuum, which also removes dissolved gases and produces castings with lower porosity.

5.10.3 Die Casting

The *die-casting* process, developed in the early 1900s, is an example of permanent-mold casting, in which the molten metal is forced into the die cavity at pressures ranging from 0.7 to 700 MPa. The European term *pressure die casting*, or simply die casting, described in this section, is not to be confused with the term *pressure casting*, described in Section 5.10.2. Typical parts made by die casting include transmission housings, valve bodies, motors, business machine and appliance components, hand tools, and toys. The weight of most castings ranges from less than 90 g to about 25 kg.

1. **Hot-chamber process.** The *hot-chamber* process (Fig. 5.22) involves the use of a piston, which traps a specific volume of molten metal and forces it into the die cavity through a gooseneck and nozzle. The pressures range up to 35 MPa, with an average of about 15 MPa.

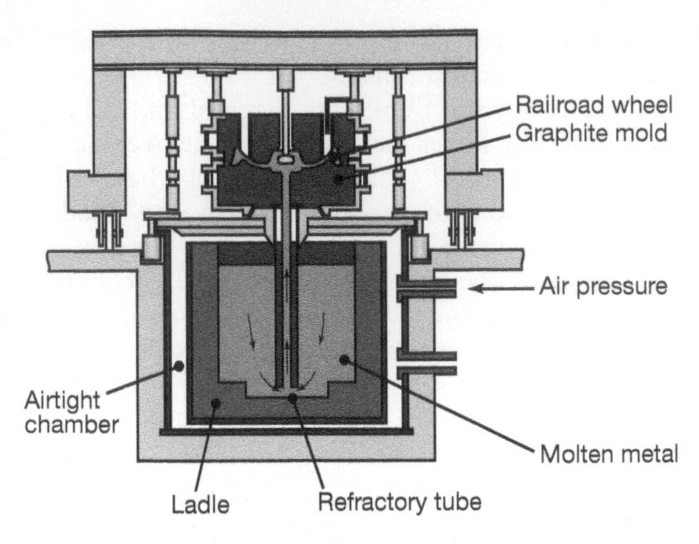

FIGURE 5.21 The pressure casting process, utilizing graphite molds for the production of steel railroad wheels. *Source:* Griffin Wheel Division of Amsted Industries Incorporated.

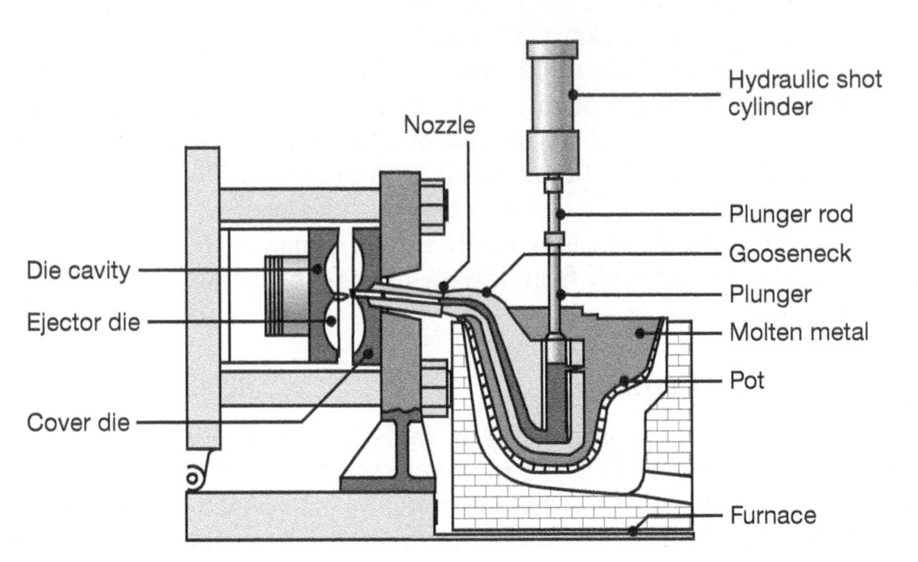

FIGURE 5.22 Schematic illustration of the hot-chamber die-casting process.

The metal is held under pressure until it solidifies in the die. To improve die life and to aid in rapid metal cooling (thus improving productivity), dies are usually cooled by circulating water or oil through various passageways in the die block. Low-melting-point alloys, such as zinc, tin, and lead, are commonly cast by this process. Cycle times typically range up to 900 shots (individual injections) per hour for zinc, although very small components, such as zipper teeth, can be cast at more than 300 shots per minute.

2. **Cold-chamber process.** In the *cold-chamber* process (Fig. 5.23) molten metal is introduced into the injection cylinder (*shot chamber*). The shot chamber is not heated, hence the term cold chamber. The metal is forced into the die cavity at pressures typically ranging from 20 to 70 MPa, although they may be as high as 150 MPa. The machines may be horizontal or vertical. Aluminum, magnesium, and

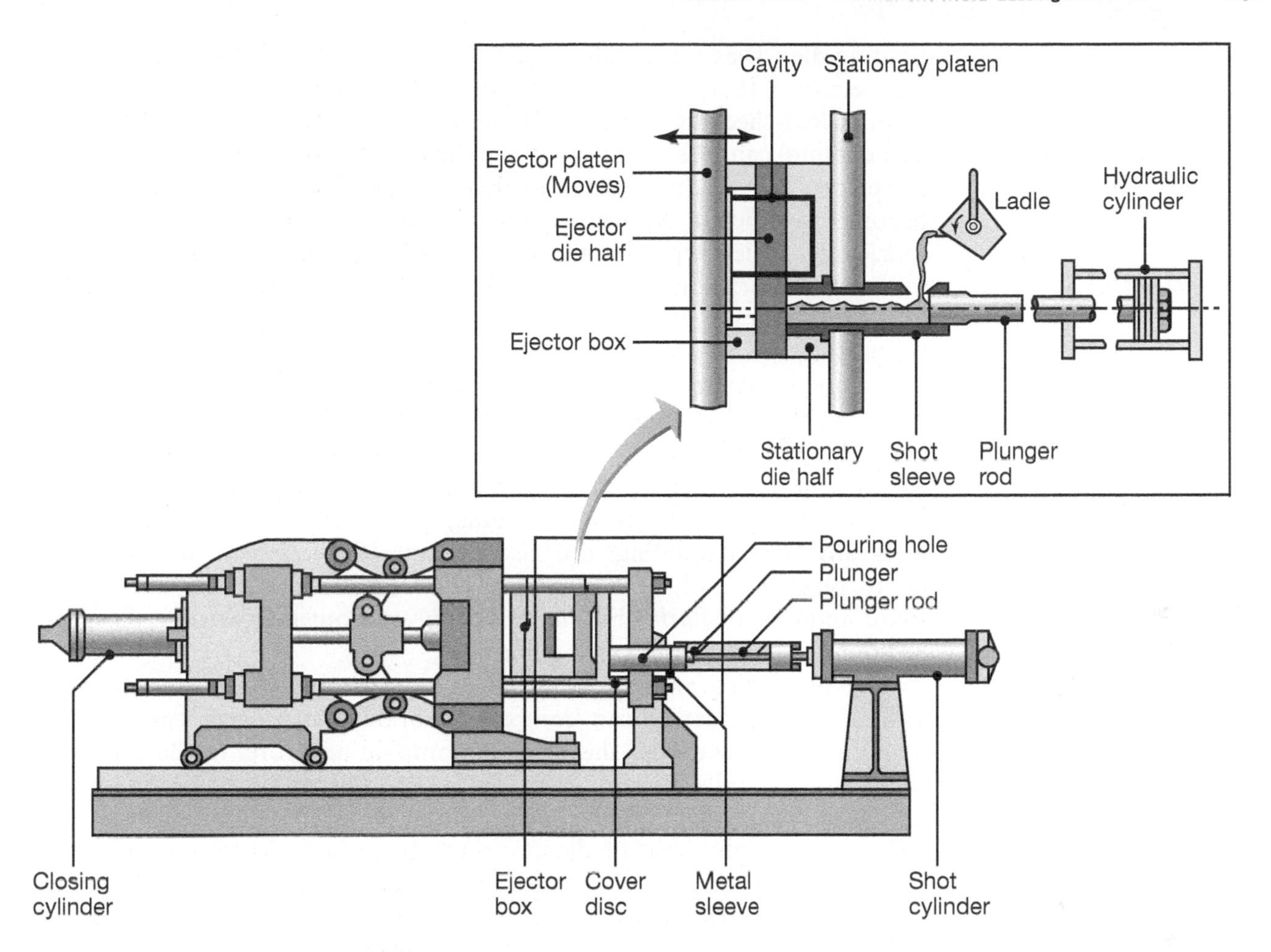

FIGURE 5.23 Schematic illustration of the cold-chamber die-casting process. These machines are large compared to the size of the casting, because high forces are required to keep the two halves of the die closed under pressure.

copper alloys are often cast by this method, although other metals, including ferrous metals, are also cast. Molten metal temperatures start at about 600°C (1150°F) for aluminum and magnesium alloys, and increase considerably for copper-based and iron-based alloys.

Process capabilities and machine selection. Because of the high pressures involved in die casting, the dies have a tendency to part unless clamped together tightly. The machines are rated according to the clamping force that can be exerted. The capacities of commercially available machines range from about 25 to 3000 tons. Other factors involved in the selection of die-casting machines include die size, piston stroke, shot pressure, and cost.

Dies may be made single cavity, multiple cavity (several identical cavities), combination cavity (several different cavities), or unit dies (simple small dies that can be combined in two or more units in a master holding die). Dies are usually made of hot-work die steels or mold steels (see Section 3.10.4). Die wear increases with the temperature of the molten metal. **Heat-checking** of the dies (surface cracking from repeated heating and cooling) can be a problem. When die materials are selected and maintained properly, dies may last more than half a million shots before die wear becomes significant. Alloys, except magnesium alloys, generally require lubricants, which typically are water base, with graphite or other compounds in suspension. Because of the high cooling capacity of water, these lubricants also are effective in keeping die temperatures low. Lubricants (as parting agents) are usually applied as thin coatings on die surfaces.

Highly automated, die casting has the capability for high production rates, with good strength, high-quality parts with complex shapes, and good dimensional accuracy and surface detail, thus requiring little or no subsequent machining or finishing operations (*net-shape forming*; see also Section 1.6). Components such as pins, shafts, and fasteners can be cast integrally (**insert molding**; see also Section 10.10.2) in a process similar to placing wooden sticks in popsicles. Ejector marks remain, as do small amounts of *flash* (thin material squeezed out between the dies; see Section 6.2.3) at the die parting line.

The properties and applications of die-cast materials are given in Table 5.6. Die casting can compete favorably with other casting and manufacturing methods, sheet metal stamping and forging. Because the molten metal chills rapidly at the die walls, the casting has a fine-grain and hard skin, with strength that is than in the center. Consequently, the strength-to-weight ratio of die-cast parts increases with decreasing wall thickness. With good surface finish and dimensional accuracy, die casting can produce bearing surfaces that would normally be machined. Equipment costs, particularly the cost of dies, are somewhat high, but labor costs are generally low because the process usually is semi- or fully automated.

TABLE 5.6 Properties and typical applications of common die-casting alloys.

Alloy	Yield strength MPa (ksi)	Ultimate tensile strength MPa (ksi)	Elongation in 50 mm (%)	Applications
Aluminum 380 (3.5 Cu–8.5 Si)	320 (45)	160 (23)	2.5	Appliances, automotive components, electrical motor frames and housings, engine blocks
Aluminum 13 (12 Si)	300 (44)	150 (22)	2.5	Complex shapes with thin walls, parts requiring strength at elevated temperatures
Brass 858 (60 Cu)	380 (55)	200 (29)	15	Plumbing fixtures, lock hardware, bushings, ornamental castings
Magnesium AZ91B (9 Al–0.7 Zn)	230 (33)	160 (23)	3	Power tools, automotive parts, sporting goods
Zinc No. 3 (4 Al)	280 (40)	—	10	Automotive parts, office equipment, household utensils, building hardware, toys
Zinc No. 5 (4 Al–1 Cu)	320 (45)	—	7	Appliances, automotive parts, building hardware, business equipment

Overcasting. A technology seen as critical for reducing vehicle weight is to *overcast* aluminum or magnesium, using a steel or copper insert to form a *hybrid casting*. This method allows the high strength of steel or corrosion resistance and heat transfer capabilities of copper to be locally applied, while exploiting the low density of aluminum and magnesium. Typically, an insert is placed in a die-casting mold (see also *insert molding*, Section 10.10.2), although permanent-mold casting is also often used for overcasting. Overcasting requires careful design of parts, as the adhesion of aluminum and magnesium on steel is poor, and mechanical features, such as grooves, bosses, and ribs are essential to ensure proper joint integrity.

5.10.4 Centrifugal Casting

The *centrifugal casting* process utilizes the inertial forces caused by rotation to force the molten metal into the mold cavities. There are three methods of centrifugal casting:

1. **True centrifugal casting.** Hollow cylindrical parts, such as pipes, gun barrels, and lampposts, can be produced by this technique, as shown in Fig. 5.24, in which molten metal is poured into a rotating mold. The axis of rotation is usually horizontal but can also be vertical for short parts. The molds are made of steel, iron, or graphite, and may be coated with a refractory lining to increase mold life. Mold surfaces can be shaped so that pipes with various outer patterns and shapes can be cast; the inner surface of the casting remains cylindrical because the molten metal is uniformly distributed by centrifugal forces. However, because of density differences in the radial direction, the lighter elements, such as dross, impurities, and pieces of the refractory lining, tend to collect on the inner surface of the casting.

   Cylindrical parts, ranging from 13 mm to 3 m (0.5 in. to 10 ft) in diameter and 16 m (53 ft) long, can be cast centrifugally; wall thicknesses typically range from 6 to 125 mm (0.25 to 5 in.). The pressure generated by the centrifugal force is high (as much as

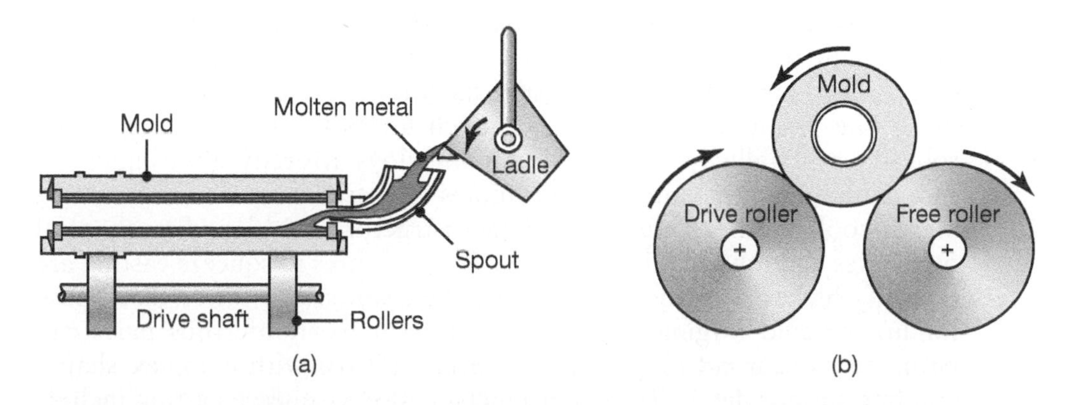

FIGURE 5.24 Schematic illustration of the centrifugal casting process. Pipes, cylinder liners, and similarly shaped hollow parts can be cast by this process.

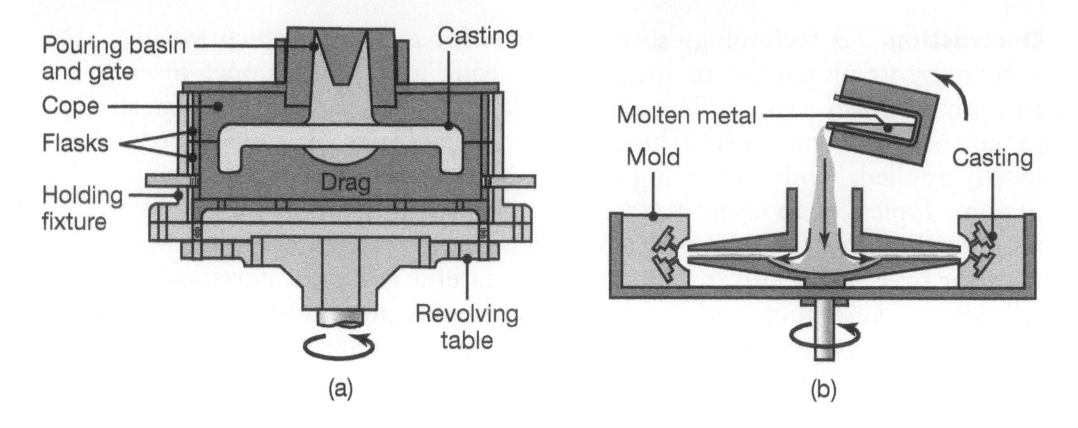

FIGURE 5.25 (a) Schematic illustration of the semicentrifugal casting process. Wheels with spokes can be cast by this process. (b) Schematic illustration of casting by centrifuging. The molds are placed at the periphery of the machine, and the molten metal is forced into the molds by centrifugal forces.

150 times gravitational acceleration), which is necessary for casting thick-walled parts. Castings of good quality, dimensional accuracy, and external surface detail are obtained by this process. In addition to pipes, typical parts made are bushings, engine cylinder liners, street lamps, and bearing rings, with or without flanges.

2. **Semicentrifugal casting.** An example of *semicentrifugal* casting is shown in Fig. 5.25a. This method is used to cast parts with rotational symmetry, such as a wheel with spokes.
3. **Centrifuging.** Also called *centrifuge casting*, mold cavities of any shape are placed at a distance from the axis of rotation. The molten metal is poured at the center and is forced into the mold by centrifugal forces (Fig. 5.25b). The properties within the castings vary by the distance from the axis of rotation, mainly because air bubbles in the molten metal are pushed toward the axis of rotation.

5.10.5 Squeeze Casting

The *squeeze-casting* process involves solidification of the molten metal under high pressure, thus it is a combination of casting and forging (Fig. 5.26). The machinery typically includes a die, punch, and ejector pins. The pressure applied by the punch keeps the entrapped gases in solution, especially hydrogen in aluminum alloys. Moreover, the high pressure at the die-metal interfaces promotes heat transfer, and the resulting higher cooling rate produces a fine microstructure, with good mechanical properties and limited microporosity. The pressures in squeeze casting are typically higher than those in pressure die casting, but lower than those for hot or cold forging (Section 6.2). Ferrous or nonferrous parts can be made to *near-net shape* (see also Section 1.6), with complex shapes and fine surface detail. Typical products made by squeeze casting include automotive wheels, mortar bodies (a short-barreled cannon), brake drums, and valve bodies.

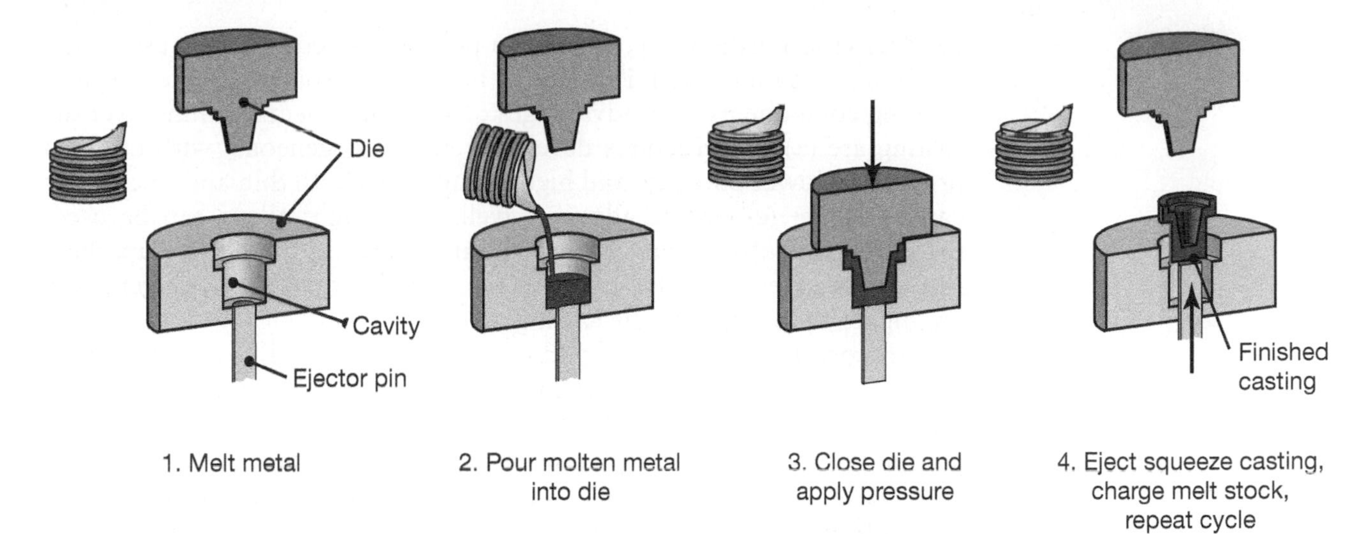

FIGURE 5.26 Sequence of operations in the squeeze-casting process. This process combines the advantages of casting and forging.

5.10.6 Semisolid Metal Forming (Thixocasting) and Rheocasting

In the *semisolid metal forming process* (developed in the 1970s and also called *semisolid metalworking*), the metal or alloy has a nondendritic and fine-grained structure when it enters the die or the mold. The alloy exhibits *thixotropic* behavior (its viscosity decreases when agitated), hence this process is also known as **thixoforming** or **thixocasting**. For example, at rest and above its solidus temperature, the alloy has the consistency of butter, but when agitated vigorously its consistency is more like machine oil.

Thixotropic behavior has been utilized in developing machines and technologies combining casting and forging of parts, with cast billets that are forged when 30–40% liquid, or in a mushy state. Processing metals in their mushy state also has led to developments in *mushy-state extrusion*, similar to injection molding (described in Section 10.2.2), *forging*, and *rolling* (hence the term *semisolid metalworking*). These processes are also used in making parts from specially designed casting alloys, wrought alloys, and metal-matrix composites (Section 11.14). They also have the capability for blending granules of different alloys, called *thixoblending*, for specific applications.

Thixotropic behavior has also been utilized in developing technologies that combine casting and forging of parts, using cast billets that are forged when the metal is 30–40% liquid. Parts made include automotive control arms, brackets, and steering components. Processing steels by thixoforming has not yet reached the same stage as with aluminum and magnesium, largely because of the high temperatures involved (that adversely affect die life) and the difficulty in making complex part shapes.

The main advantages of thixocasting are that the semisolid metal is just above the solidification temperature when injected into a die, resulting in reductions in solidification and cycle times, and thereby increased productivity. Furthermore, shrinkage porosity is reduced because of the lower superheat involved. Parts made include control arms, brackets, and steering components. The advantages of semisolid metal forming over die casting are (a) the structures developed are homogeneous, with uniform properties, lower porosity, and high strength; (b) both thin and thick parts can be made; (c) casting alloys as well as wrought alloys can be used; (d) parts can subsequently be heat treated; and (e) the lower superheat results in shorter cycle times; however, material and overall costs generally are higher than those for die casting.

In **rheocasting**, a slurry (a solid suspended in a liquid) is obtained from a melt furnace and is cooled, and then magnetically stirred prior to injecting it into a mold or die. Rheocasting has been successfully applied to aluminum and magnesium, and has been used to make engine blocks, crankcases (as for motorcycles or lawn mowers), and various marine applications.

5.10.7 Casting Techniques for Single-Crystal Components

Casting techniques can best be illustrated by describing the casting of gas turbine blades, which are generally made of nickel-based superalloys. The procedures involved can also be used for other alloys and components.

Conventional casting of turbine blades. The conventional casting process involves investment casting, using ceramic molds (See Fig. 5.17). The molten metal is poured into the mold and begins solidifying at the ceramic walls. The grain structure developed is polycrystalline, and the presence of grain boundaries makes this structure susceptible to creep and cracking along those boundaries under the centrifugal forces at elevated temperatures.

Directionally solidified blades. In the *directional solidification* process (Fig. 5.27a), the ceramic mold is prepared basically by investment casting techniques and is preheated by radiant heating. The mold is supported by a water-cooled chill plate. After the metal is poured into the mold, the assembly is lowered slowly, whereby crystals begin to grow at the chill-plate surface. The blade is directionally solidified, with longitudinal (but no transverse) grain boundaries. Consequently, the blade is stronger in the direction of centrifugal forces developed in the gas turbine, that is, in the longitudinal direction.

Single-crystal blades. In the process for single-crystal blades, first made in 1967, the mold is prepared by investment casting techniques. It has a constriction in the shape of a corkscrew (Figs. 5.27b and c), allowing only one crystal through it. As the assembly is lowered slowly, a single crystal grows upward through the constriction and begins to grow in the mold. Strict control of the rate of movement is necessary in this process. The solidified mass in the mold is a single-crystal blade. Although more expensive than other blades (the largest blades are over 15 kg and can

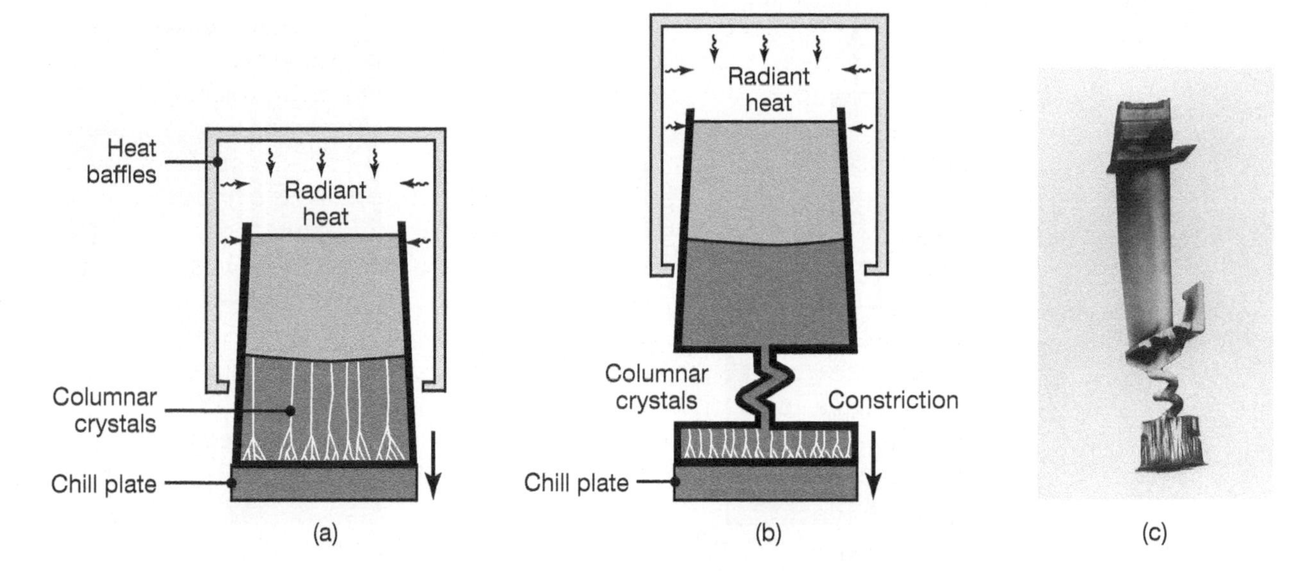

FIGURE 5.27 Methods of casting turbine blades: (a) directional solidification; (b) method to produce a single-crystal blade; and (c) a single-crystal blade with the constriction portion still attached. *Source:* (a) and (b) After B.H. Kear, (c) Courtesy of ASM International.

cost over \$6000 each), the absence of grain boundaries makes these blades resistant to creep and thermal shock, imparting a longer and more reliable service life.

Single-crystal growing. Single-crystal growing is a major activity in the manufacture of microelectronic devices (see Chapter 13). There are two basic methods of crystal growing. In the **crystal pulling** method, known as the *Czochralski process* (Fig. 5.28a), a seed crystal is dipped into the molten metal, and then pulled at a rate of about 10 μm/s while rotating at about 1 rev/s. The liquid metal begins to solidify on the seed, and the crystal structure of the seed is continued throughout the part. *Dopants* (alloying elements) may be added to the liquid metal to impart specific electrical properties. Single crystals of silicon, germanium, and various metals such as palladium, silver, and gold are grown by this process. Single-crystal ingots up to 400 mm in diameter and over 2 m in length have been produced by this technique; 200 and 300 mm cylinders are more common in the production of silicon wafers for integrated circuit manufacture (Chapter 13).

The second technique for crystal growing is the **floating zone** method (Fig. 5.28b). Starting with a rod of polycrystalline silicon, resting on a single crystal, an induction coil heats these two pieces while moving slowly upward. The single crystal grows upward while maintaining its orientation. Thin wafers are then cut from the rod, cleaned, and polished for use in microelectronic device fabrication (see Chapter 13). Because of the limited diameters that can be produced in this process, it is has largely been

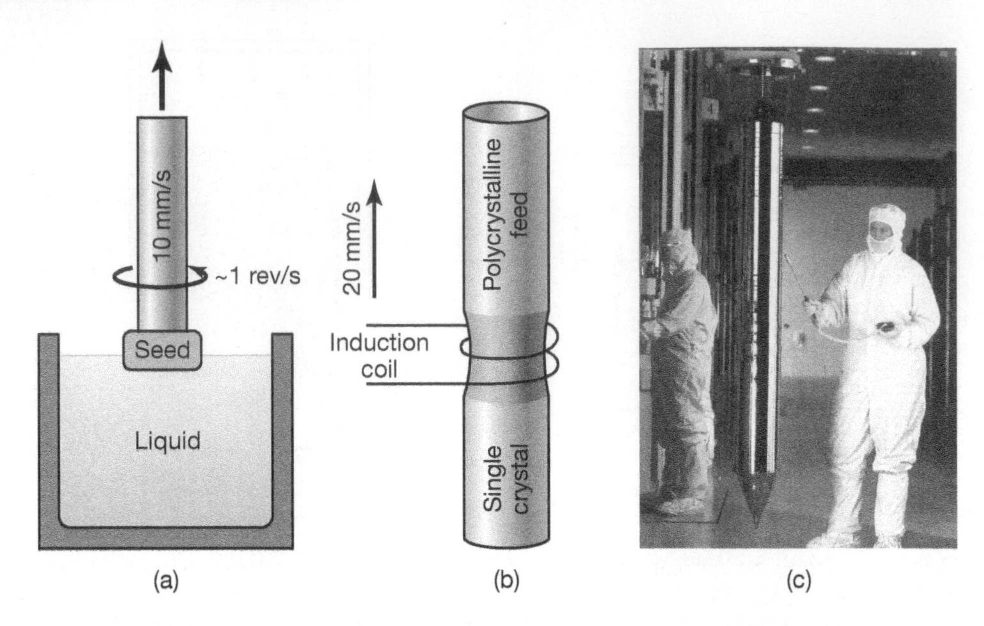

FIGURE 5.28 Two methods of crystal growing: (a) crystal pulling (Czochralski process) and (b) floating-zone method. Crystal growing is especially important in the semiconductor industry. (c) A single-crystal silicon ingot produced by the Czochralski process. *Source:* Courtesy of Intel Corp.

replaced by the Czochralski process (Fig. 5.28) for silicon, but is still used for cylinder diameters under 150 mm because it is a cost-effective method for producing single crystal cylinders in small quantities.

5.10.8 Rapid Solidification

First developed in the 1960s, *rapid solidification* involves cooling the molten metal at rates as high as 10^6 K/s, whereby it does not have sufficient time to crystallize. These alloys are called **amorphous alloys** or **metallic glasses**, because they do not have a long-range crystalline structure (see Section 3.2). They typically contain iron, nickel, and chromium, that are alloyed with carbon, phosphorus, boron, aluminum, and silicon. Among other effects, rapid solidification results in a significant extension of solid solubility, grain refinement, and reduced microsegregation.

Amorphous alloys exhibit excellent corrosion resistance, good ductility, high strength, very little magnetic hysteresis loss, high resistance to eddy currents, and high permeability. The last three properties are utilized in making magnetic steel cores used in transformers, generators, motors, lamp ballasts, magnetic amplifiers, and linear accelerators, with greatly improved efficiency. Another major application is rapidly-solidified superalloy powders, which are consolidated into near-net shapes for use in aerospace engines. Amorphous alloys are produced in the form of wire, ribbon, strip, and powder. In one process called **melt spinning** (Fig. 5.29), the alloy is melted (by induction in a ceramic crucible) and propelled, under high gas pressure, at very high speed against a rotating copper disk (chill block) where it chills rapidly (splat cooling).

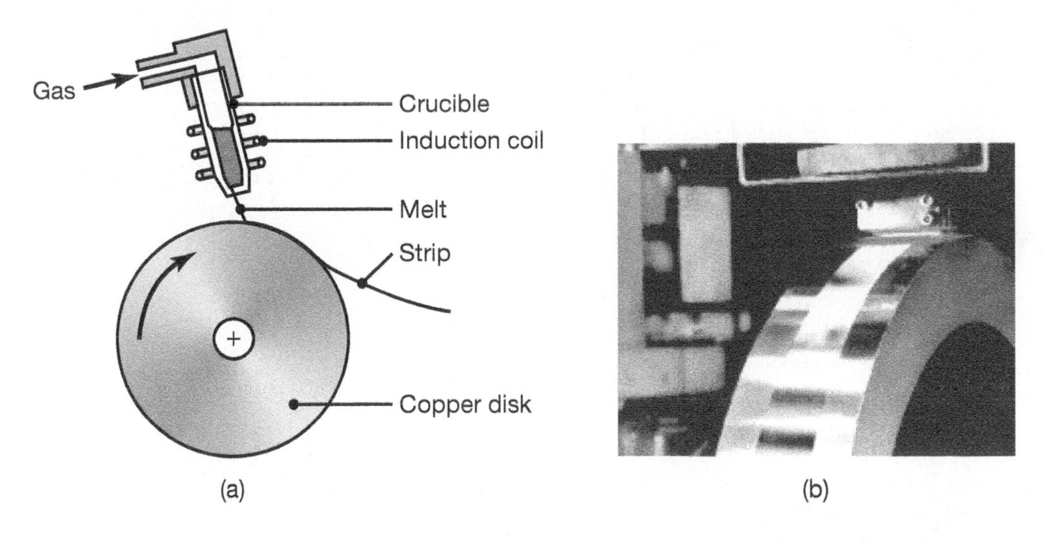

FIGURE 5.29 (a) Schematic illustration of the melt-spinning process to produce thin strips of amorphous metal. (b) Photograph of nickel-alloy production through melt-spinning. *Source:* Courtesy of Siemens AG.

5.11 Heat Treatment

The various microstructures developed during metal processing can be modified by *heat treatment* techniques, involving controlled heating and cooling of the alloys at various rates (also known as **thermal treatment**). These treatments induce phase transformations that greatly influence mechanical properties, such as strength, hardness, ductility, toughness, and wear resistance of the alloys. The effects of thermal treatment depend primarily on the alloy, its composition and microstructure, the degree of prior cold work, and the rates of heating and cooling during heat treatment.

5.11.1 Heat Treating Ferrous Alloys

The microstructural changes that occur in the iron-carbon system (Section 5.2.5) are described below.

Pearlite. If the ferrite and cementite lamellae in the pearlite structure (see Section 5.2.6) of the eutectoid steel are thin and closely packed, the microstructure is called **fine pearlite.** If the lamellae are thick and widely spaced, the structure is called **coarse pearlite.** The difference between the two depends on the rate of cooling through the eutectoid temperature, a reaction in which austenite is transformed into pearlite. If the rate of cooling is relatively high (as in air), the structure is fine pearlite is produced; if slow (as in a furnace), it is coarse pearlite.

The transformation from austenite to pearlite (and for other structures) is best illustrated by Figs. 5.30b and c. These diagrams are called **isothermal transformation (IT) diagrams** or **time-temperature-transformation (TTT) diagrams.** They are constructed from the data in Fig. 5.30a, which

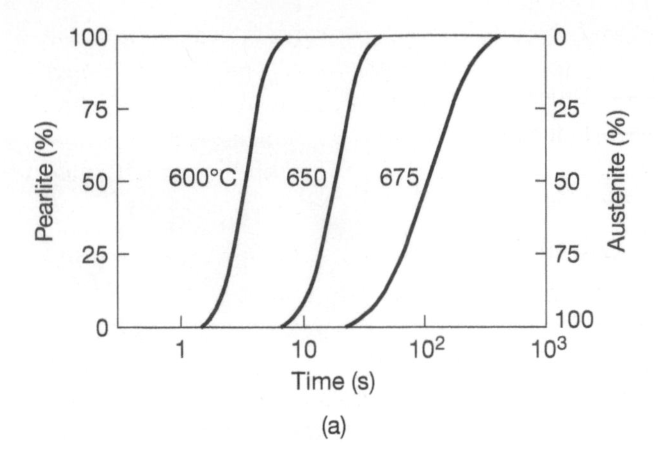

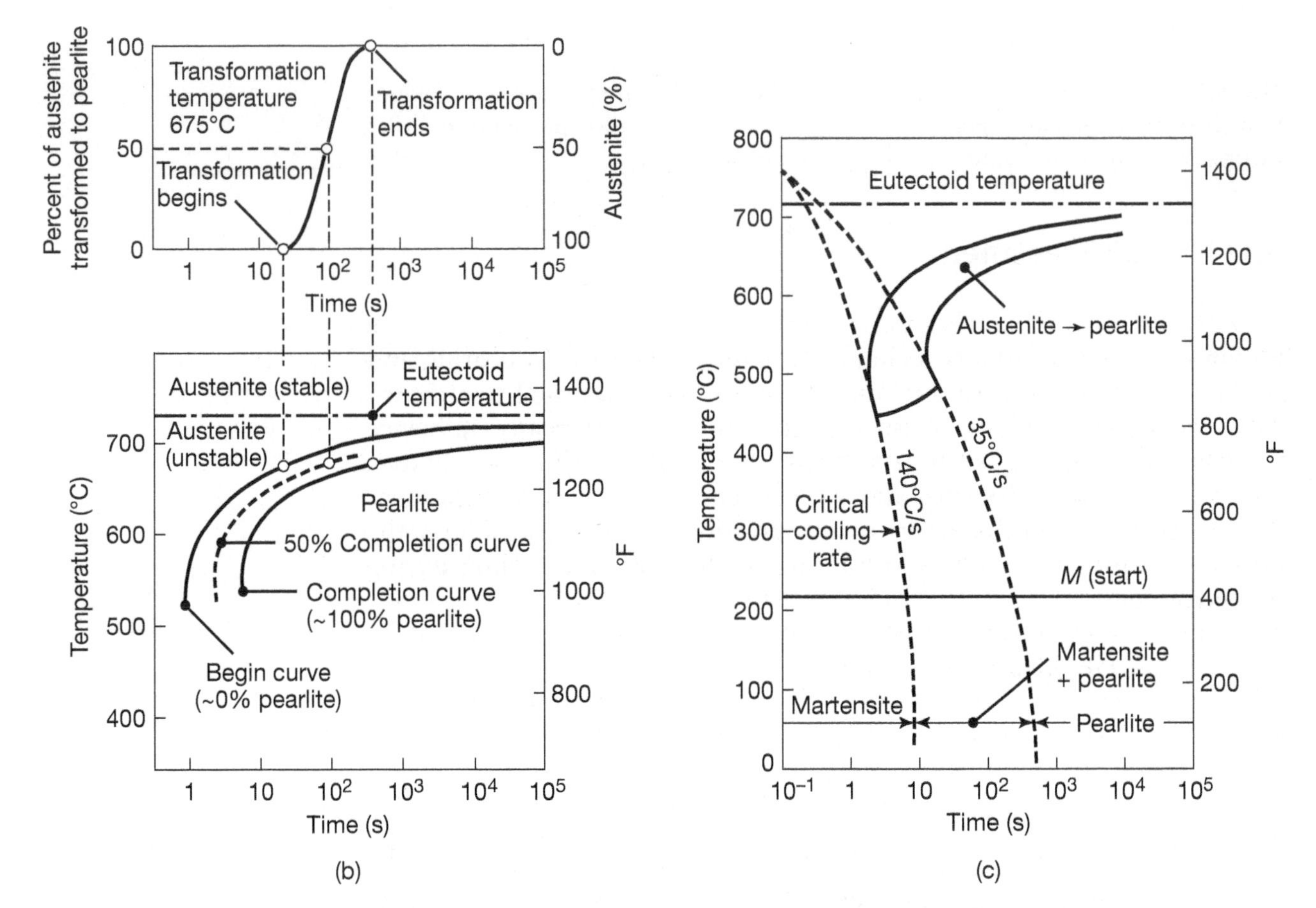

FIGURE 5.30 (a) Austenite to pearlite transformation of iron-carbon alloys as a function of time and temperature. (b) Isothermal transformation diagram obtained from (a) for a transformation temperature of 675°C (1247°F). (c) Microstructures obtained for a eutectoid iron-carbon alloy as a function of cooling rate.

shows the percentage of austenite transformed into pearlite as a function of temperature and time. The higher the temperature and/or the longer the time, the higher is the percentage of austenite transformed to pearlite. Note that for each temperature, a minimum time is required for the transformation to start and that sometime later, all the austenite is transformed to pearlite.

Spheroidite. When pearlite is heated to just below the eutectoid temperature and held at that temperature for a period of time, say for a day at 700°C (1300°F), the cementite lamellae (Fig. 5.4b) transform to *spherical* shapes (*spheroidites*). Unlike the lamellar shape of cementite, which acts as stress raisers, spheroidites are less conducive to stress concentration, because of their rounded shapes. Consequently, this structure has higher toughness and lower hardness than the pearlite structure. In this form it can be cold worked and the spheroidal particles prevent the propagation of any cracks within the material during working.

Bainite. Visible only under electron microscopy, *bainite* has a very fine microstructure, consisting of ferrite and cementite. It can be produced in steels by alloying and at cooling rates that are higher than those required for transformation to pearlite. This structure, called *bainitic steel*, is generally stronger and more ductile than pearlitic steel at the same hardness level.

Martensite. When austenite is cooled rapidly (such as by quenching in water), its fcc structure is transformed to a *body-centered tetragonal* (bct) structure. This structure can be described as a body-centered rectangular prism that is slightly elongated along one of its principal axes, called *martensite*. Because it does not have as many slip systems as a bcc structure and the carbon is in interstitial positions, martensite is extremely hard and brittle, lacks toughness, and therefore has limited use. Martensite transformation takes place almost instantaneously (Fig. 5.30c), because it does not involve the diffusion process (a time-dependent phenomenon that is the mechanism in other transformations).

Because of the different densities of the various phases in the structure, transformations involve volume changes. For example, when austenite transforms to martensite, its volume increases (hence density decreases) by as much as 4%. A similar but smaller volume expansion also occurs when austenite transforms to pearlite. These expansions and the resulting thermal gradients in a quenched part can cause the development of internal stresses within the body, which may cause parts to crack during heat treatment, as in **quench cracking** of steels, caused by rapid cooling during quenching.

Retained austenite. If the temperature at which the alloy is quenched is not sufficiently low, only a portion of the structure is transformed to martensite. The rest is *retained austenite*, which is visible as white areas in the structure along with dark needlelike martensite. Retained austenite can cause dimensional instability and cracking of the part and lowers its hardness and strength.

Tempered martensite. *Tempering* is a heating process that reduces martensite's hardness and improves its toughness. The body-centered tetragonal martensite is heated to an intermediate temperature, where it transforms to a two-phase microstructure, consisting of body-centered cubic alpha ferrite and small particles of cementite. Longer tempering time and higher temperature decrease martensite's hardness. The reason is that the cementite particles coalesce and grow, and the distance between the particles in the soft ferrite matrix increases as the less stable, smaller carbide particles dissolve.

Hardenability of ferrous alloys. The capability of an alloy to be hardened by heat treatment is called its *hardenability*; it is a measure of the depth of hardness that can be obtained by heating and subsequent quenching. (The term hardenability should not be confused with hardness, which is the resistance of a material to indentation or scratching.) Hardenability of ferrous alloys depends on their carbon content, the grain size of the austenite, and the alloying elements. The **Jominy test** has been developed in order to determine alloy hardenability.

Quenching media. Quenching may be carried out in water, brine (saltwater), oils, molten salts, or air, as well as caustic solutions, polymer solutions, and various gases. Because of the differences in the thermal conductivity, specific heat, and heat of vaporization of these media, the rate of cooling (**severity of quench**) also will be different. In relative terms and in decreasing order, the cooling capacity of several quenching media is: (a) agitated brine 5; (b) still water 1; (c) still oil 0.3; (d) cold gas 0.1; and (e) still air 0.02. Agitation is also a significant factor in the rate of cooling. In tool steels the quenching medium is specified by a letter (See Table 3.5), such as W for water hardening, O for oil hardening, and A for air hardening. The cooling rate also depends on the surface area-to-volume ratio of the part [see Eq. (5.11)]. The higher this ratio, the higher is the cooling rate; thus, for example, a thick plate cools more slowly than a thin plate with the same surface area.

Water is a common medium for rapid cooling; however, the heated metal may form a **vapor blanket** along its surfaces from water-vapor bubbles that form when water boils at the metal-water interfaces. This blanket creates a barrier to heat conduction because of the lower thermal conductivity of the vapor. Agitating the fluid or the part helps to reduce or eliminate the blanket. Also, water may be sprayed on the part under high pressure. Brine is an effective quenching medium because salt helps to nucleate bubbles at the interfaces, thus improving agitation. However, brine can corrode the part. **Die quenching** is a term used to describe the process of clamping the part to be heat treated to a die, which chills selected regions of the part. In this way cooling rates and warpage can be controlled.

5.11.2 Heat Treating Nonferrous Alloys and Stainless Steels

Nonferrous alloys and some stainless steels generally cannot be heat treated by the techniques for ferrous alloys, because nonferrous alloys

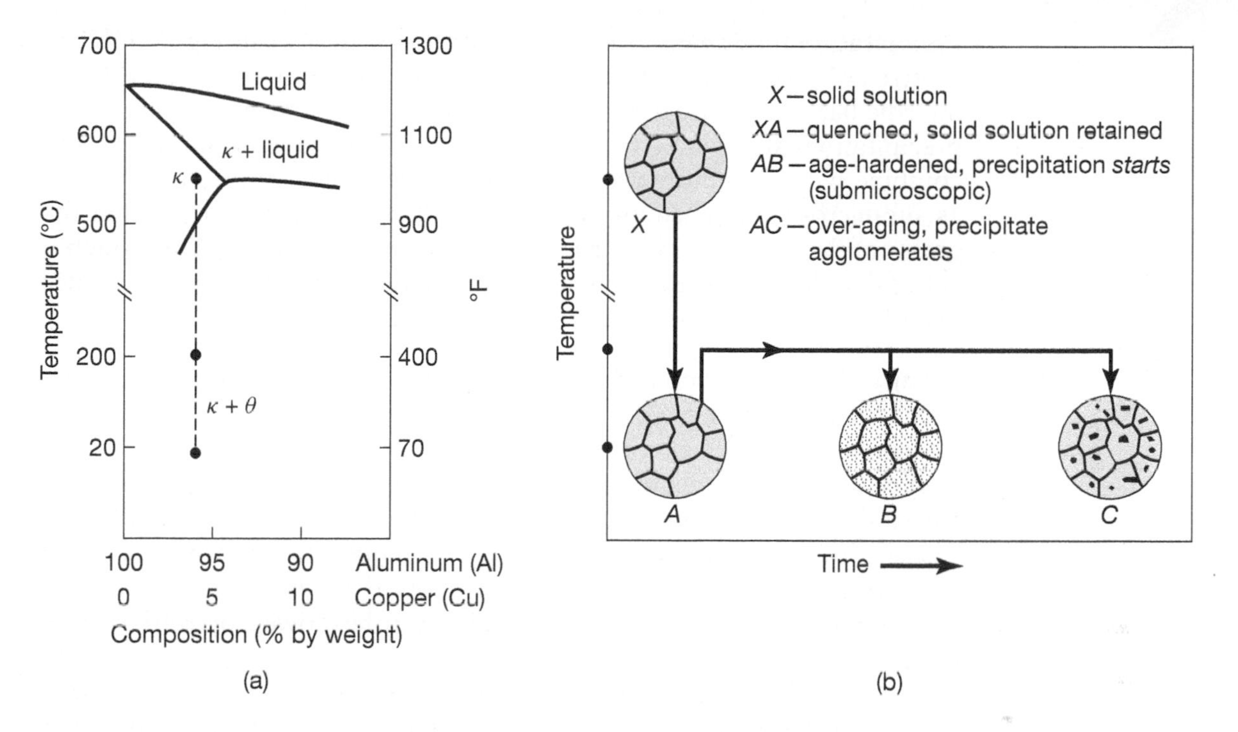

FIGURE 5.31 (a) Phase diagram for the aluminum-copper alloy system. (b) Various microstructures obtained during the age-hardening process.

do not undergo phase transformations as steels do. Heat-treatable aluminum alloys (see Section 3.11.1), copper alloys, and martensitic and precipitation-hardening stainless steels are hardened and strengthened by **precipitation hardening**. This is a technique in which small particles of a different phase (called *precipitates*) are uniformly dispersed in the matrix of the original phase (See Fig. 5.2a). Precipitates form because the solid solubility of one element (one component of the alloy) in the other is exceeded.

Three stages are involved in the precipitation-hardening process, which can best be described by referring to the phase diagram for the aluminum-copper system, given in Fig. 5.31. For an alloy with a composition of 95.5% Al-4.5% Cu, a single-phase (κ) substitutional solid-solution of copper (solute) in aluminum (solvent) exists between 500° and 570°C (930° and 1060°F). The κ phase is aluminum rich, has an fcc structure, and is ductile. Below the lower temperature, that is, below the lower solubility curve, two phases are present: κ and θ (a hard intermetallic compound of $CuAl_2$). This alloy can be heat treated and its properties modified by solution treatment or precipitation.

Solution treatment. In *solution treatment* the alloy is heated to within the solid-solution κ phase, say 540°C (1000°F), and cooled rapidly, such as by quenching in water. The structure obtained soon after quenching (*A* in Fig. 5.31b) consists only of the single phase κ; this alloy has moderate strength and very good ductility.

Precipitation hardening. The structure obtained in *A* in Fig. 5.31b can be strengthened by *precipitation hardening*. The alloy is reheated to an intermediate temperature and held there for a period of time, during which precipitation takes place. The copper atoms diffuse to nucleation sites and combine with aluminum atoms, producing the theta phase, which form as submicroscopic precipitates (shown in B by the small dots within the grains of the κ phase). This structure is stronger than that in *A*, although it is less ductile; the increase in strength is attributed to increased resistance to dislocation movement in the region of the precipitates.

Aging. Because the precipitation process is one of time and temperature, it is also called aging, and the property improvement is known as *age hardening*. If carried out above room temperature, the process is called **artificial aging**. Several aluminum alloys harden and become stronger over a period of time at room temperature, by a process known as **natural aging**. These alloys are first quenched and then, if required, are formed at room temperature into various shapes, and allowed to develop strength and hardness by natural aging. Natural aging can be slowed by refrigerating the quenched alloy.

In the precipitation process, as the reheated alloy is held at that temperature for an extended period of time, the precipitates begin to coalesce and grow. They become larger but fewer, as shown by the larger dots in C in Fig. 5.31b. This process is called **overaging**, which makes the alloy softer and less strong, although the part treated has better dimensional stability over time. There is an optimal time-temperature relationship in the aging process to develop desired properties. An aged alloy can be used only up to a certain maximum temperature in service, otherwise it will overage and lose some of its strength and hardness.

Maraging. This is a precipitation-hardening treatment process for a special group of high-strength iron-based alloys; the word *maraging* is derived from the words *mar*tensite and *aging*. In this process, one or more intermetallic compounds (Section 5.2.2) are precipitated in a matrix of low-carbon martensite. A typical maraging steel may contain 18% nickel and other elements, and aging is done at 480°C (900°F). Hardening by maraging does not depend on the cooling rate, thus full uniform hardness can be obtained throughout large parts and with minimal distortion. Typical uses of maraging steels are for dies and tooling for casting, molding, forging, and extrusion.

5.11.3 Case Hardening

The heat treatment processes described thus far involve microstructural alterations and property changes in the *bulk* of a part by *through hardening*. In many cases, however, alteration of only the *surface* properties of a part (hence the term *case hardening*) is desirable, particularly for improving resistance to surface indentation, fatigue, and wear. Typical applications include gear teeth, cams, shafts, bearings, fasteners, pins, automotive clutch plates, and tools and dies. Through hardening of these parts would not be desirable, because a hard part generally lacks the necessary toughness for these applications; a small surface crack, for example, can propagate rapidly through the whole part and cause total failure.

Several surface-hardening processes include (Table 5.7): **carburizing** (*gas, liquid,* and *pack carburizing*), **carbonitriding, cyaniding, nitriding, boronizing,** and **flame** and **induction hardening.** Basically, the component is heated in an atmosphere containing such elements as carbon, nitrogen, or boron, that alter the composition, microstructure, and properties of surfaces to various degrees.

For steels with sufficiently high carbon content, surface hardening takes place without the need for any of these additional elements; only the processes described in Section 5.11.1 are needed to alter the microstructures, usually by flame hardening or induction hardening. Laser beams and electron beams are also used effectively to harden both small and large surfaces and for through hardening of relatively small parts.

Because case hardening is a localized heat treatment, case-hardened parts have a hardness gradient. Typically, the hardness is greatest at the surface and decreases below the surface, and the rate of decrease depends on the composition of the metal and the process variables. Surface-hardening techniques can also be used for *tempering* (see Section 5.11.1), thereby modifying the properties of surfaces that have been subjected to heat treatment. Several other processes and techniques for surface hardening, such as shot peening and surface rolling, improve wear resistance and various other characteristics, as described in Section 4.5.1.

Decarburization. This is a phenomenon in which carbon-containing alloys lose carbon from their surfaces as a result of heat treatment or by hot working in a medium (usually oxygen) that reacts with the carbon. Decarburization is undesirable, because it adversely affects the hardenability of the surfaces of the part by lowering the carbon content; it also affects the hardness, strength, and fatigue life of steels by significantly lowering their endurance limit. Decarburization is best avoided by processing the alloy in an inert atmosphere or a vacuum, or by using neutral salt baths during heat treatment.

5.11.4 Annealing

Annealing is a general term to describe the restoration of a cold-worked or heat-treated part to its original properties, so as to increase its ductility (hence formability), reduce hardness and strength, or modify its microstructure. Annealing is also used to relieve residual stresses (Section 2.10) in a manufactured part for improved subsequent machinability and dimensional stability. The term annealing also applies to thermal treatment of glasses (Section 11.11.2) and weldments (Chapter 12).

The annealing process typically involves the following sequence: (1) heating the part to a specific range of temperature; (2) holding it at that temperature for a period of time (called *soaking*); and (3) cooling it slowly. The process may be carried out in an inert or controlled atmosphere, or is performed at low temperatures to prevent or minimize surface oxidation

TABLE 5.7 Outline of heat treatment processes for surface hardening.

Process	Metals hardened	Element added to surface	Procedure	General characteristics	Typical applications
Carburizing	Low-carbon steel (0.2% C), alloy steels (0.08–0.2% C)	C	Heat steel at 870°C to 950°C (1600°F to 1750°F) in an atmosphere of carbonaceous gases (gas carburizing) or carbon-containing solids (pack carburizing). Then quench.	A hard, high-carbon surface is produced. Hardness 55–65 HRC. Case depth <0.5–1.5 mm. Some distortion of part during heat treatment.	Gears, cams, shafts, bearings, piston pins, sprockets, clutch plates
Carbonitriding	Low-carbon steel	C and N	Heat steel at 700°C to 800°C (1300°F to 1600°F) in an atmosphere of carbonaceous gas and ammonia. Then quench in oil.	Surface hardness 55–62 HRC. Case depth 0.07–0.5 mm. Less distortion than in carburizing.	Bolts, nuts, gears
Cyaniding	Low-carbon steel (0.2% C), alloy steels (0.08–0.2% C)	C and N	Heat steel at 760°C to 845°C (1400°F to 1550°F) in a molten bath of solutions of cyanide (e.g., 30% sodium cyanide) and other salts.	Surface hardness up to 65 HRC. Case depth 0.025–0.25 mm. Some distortion.	Bolts, nuts, screws, small gears
Nitriding	Steels (1% Al, 1.5% Cr, 0.3% Mo), alloy steels (Cr, Mo), stainless steels, high-speed steels	N	Heat steel at 500°C to 600°C (925°F to 1100°F) in an atmosphere of ammonia gas or mixtures of molten cyanide salts. No further treatment.	Surface hardness up to 1100 HV. Case depth 0.1–0.6 mm (0.005–0.030 in.) and 0.02–0.07 mm for high speed steel.	Gears, shafts, sprockets, valves, cutters, boring bars
Boronizing	Steels	B	Part is heated using boron-containing gas or solid in contact with part.	Extremely hard and wear-resistance surface. Case depth 0.025–0.075 mm.	Tool and die steels
Flame hardening	Medium-carbon steels, cast irons	None	Surface is heated with an oxyacetylene torch, then quenched with water spray or other quenching methods.	Surface hardness 50–60 HRC. Case depth 0.7–6 mm. Little distortion.	Axles, crankshafts, piston rods, lathe beds, and centers
Induction hardening	Same as above.	None	Metal part is placed in copper induction coils and is heated by high frequency current, then quenched.	Same as above.	Same as above.

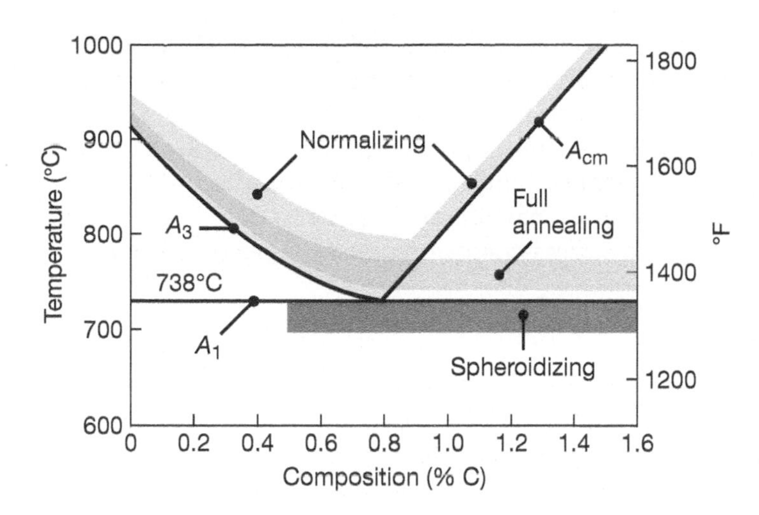

FIGURE 5.32 Temperature ranges for heat treating plain-carbon steels, as indicated on the iron-iron carbide phase diagram.

of the part. Annealing temperatures may be higher than the recrystallization temperature, depending on the degree of cold work (hence *stored energy*, see Section 2.12.1). The recrystallization temperature for copper, for example, ranges between 200°C and 300°C (400°F and 600°F), whereas the annealing temperature required to fully recover the original properties ranges from 260°C to 650°C (500°F to 1200°F), depending on the degree of prior cold work.

Full annealing is a term used for annealing ferrous alloys, generally low- and medium-carbon steels. The steel is heated to above A_1 or A_3 (Fig. 5.32), and cooling takes place slowly, say 10°C (20°F) per hour, in a furnace after it is turned off. The structure obtained in full annealing is coarse pearlite, which is soft and ductile, and has small uniform grains. Excessive softness in the annealing of steels can be avoided if the entire cooling cycle is carried out in still air. **Normalizing** is a process in which the part is heated to a temperature of A_3 or A_{cm} to transform the structure to austenite. The structure obtained is fine pearlite, with small uniform grains, and the process results in higher strength and hardness, and lower ductility than in full annealing. Normalizing generally refines the grain structure, produces uniform structure (*homogenization*), decreases residual stresses, and improves machinability.

Process annealing. In *process annealing*, also called *intermediate annealing*, subcritical annealing or *in-process annealing*), the part is annealed to restore its ductility, a portion or all of which may have been exhausted by work hardening during prior cold working. In this way, the part can be worked further into the final desired shape. If the temperature is too high and/or the time of annealing is too long, grain growth may result, with adverse effects on the formability of annealed parts.

Stress-relief annealing. Residual stresses may have been induced during forming, machining, or other shaping processes, or are caused by volume changes during phase transformations. To reduce or eliminate these stresses, the part is subjected to *stress-relief annealing*, or simply **stress**

relieving. The temperature and time required for this process depend on the material and the magnitude of residual stresses present. For steels, for example, the part is heated to below A_1 as shown in Fig. 5.32, thus avoiding phase transformations; cooling is slow, such as in still air. Stress relieving promotes dimensional stability in cases where subsequent relaxing of residual stresses may cause part distortion over a period of time; it also reduces the tendency for stress-corrosion cracking (Section 3.8.2).

5.11.5 Tempering

If steels are hardened by heat treatment, *tempering* reduces residual stresses and increases ductility and toughness. The steel is heated to a specific temperature, depending on composition, and then cooled at a prescribed rate. Alloy steels may undergo **temper embrittlement**, caused by the segregation of impurities along the grain boundaries, at temperatures between 480°C and 590°C (900°F and 1100°F). The term tempering is also used for glasses (see Section 11.11.2).

In **austempering**, the heated steel is quenched from the austenitizing temperature rapidly enough to avoid formation of ferrite or pearlite. It is held at a certain temperature until isothermal transformation from austenite to bainite is complete; it is then cooled to room temperature (usually in still air) at a moderate rate, to avoid thermal gradients within the part. The quenching medium most commonly used is molten salt, and at temperatures ranging from 160° to 750°C (320° to 1380°F).

Austempering is often substituted for conventional quenching and tempering, either to (a) reduce the tendency for cracking and distortion during quenching; or (b) improve ductility and toughness while maintaining hardness. Because of the relatively short cycle time, austempering is economical for numerous applications. In *modified austempering*, a mixed structure of pearlite and bainite is obtained. The best example of this practice is **patenting**, which provides high ductility and moderately high strength, such as patented wire (Section 6.5.3).

In **martempering (marquenching)**, steel or cast iron is quenched from the austenitizing temperature into a hot fluid medium (such as hot oil or molten salt). It is held at that temperature until the temperature is uniform throughout the part, and then cooled at a moderate rate (such as in air) to avoid temperature gradients within the part. The part is then tempered, because the structure thus obtained is primarily untempered martensite and thus it is not suitable for most applications. Martempered steels have a lower tendency to crack, distort, or develop residual stresses during heat treatment. For steels with lower hardenability, process that is suitable is *modified martempering*, in which the quenching temperature is lower and hence the cooling rate is higher.

In **ausforming**, also called **thermomechanical processing**, the steel is formed into desired shapes within controlled ranges of temperature and time, to avoid formation of nonmartensitic transformation products.

The part is then cooled at various rates to obtain the desired microstructures. Ausformed parts have superior mechanical properties.

5.11.6 Cryogenic Treatment

In *cryogenic tempering*, the temperature of steel is lowered from room temperature to −180°C (−300°F), at a rate of as low as 2°C per minute in order to avoid thermal shock. The part is then maintained at this temperature for 24 to 36 hrs; the conversion of austenite to martensite occurs slowly but almost completely, as compared to typically only 50 to 90% in conventional quenching. As a result, additional precipitates of carbon, with chromium, tungsten and other elements, form, the grain structure is refined, and the residual stresses are relieved. After the 24 to 36 hr soak time, the parts are tempered to stabilize the martensite.

Cryogenically treated steels have higher hardness and wear resistance than untreated steels. For example, the wear resistance of D-2 tool steels (see Section 3.10.3) can increase by over 800% after cryogenic treatment, although most tool steels show a 100–200% increase in tool life (see Section 8.3). Applications of cryogenic treatment include tools and dies, aerospace materials, golf club heads, gun barrels, and dental instruments.

5.11.7 Design for Heat Treating

In addition to the metallurgical factors described above, successful heat treating involves design considerations so as to avoid certain problems, such as cracking, warping, and development of nonuniform properties throughout the part. The cooling rate during quenching must be uniform, particularly with complex shapes of varying cross sections and thicknesses; this is to avoid severe temperature gradients within the part, which can lead to thermal stresses, cause cracking, residual stresses, and stress-corrosion cracking.

As a general guideline, (a) parts should have as nearly uniform thicknesses as possible or the transition between regions of different thicknesses should be smooth; (b) internal or external sharp corners should be avoided; (c) parts with holes, grooves, keyways, splines, and unsymmetrical shapes may be difficult to heat treat, because they may crack during quenching; (d) large surfaces with thin cross sections may warp; and (e) hot forgings and hot-rolled products may have a *decarburized skin*, and thus may not properly respond to heat treatment.

5.11.8 Cleaning, Finishing, and Inspection of Castings

After solidification and removal from the mold or die, castings are generally subjected to several additional processes. In sand casting, the casting is shaken out of its mold and the sand, and oxide layers adhering to the castings are removed by vibration or by sand blasting. Castings may also be cleaned electrochemically or by pickling them with chemicals to remove surface oxides, which could adversely affect their machinability (Section 8.5).

TABLE 5.8 Casting processes and their advantages and limitations.

Process	Advantages	Limitations
Sand	Almost any metal can be cast; no limit to size, shape or weight; low tooling cost	Some finishing required; somewhat coarse finish; wide tolerances
Shell mold	Good dimensional accuracy and surface finish; high production rate	Part size limited; expensive patterns and equipment required
Expendable pattern	Most metals cast with no limit to size; complex shapes	Patterns have low strength and can be costly for low quantities
Plaster mold	Intricate shapes; good dimensional accuracy and finish; low porosity	Limited to nonferrous metals; limited size and volume of production; mold making time relatively long
Ceramic mold	Intricate shapes; close tolerance parts; good surface finish	Limited size
Investment	Intricate shapes; excellent surface finish and accuracy; almost any metal cast	Part size limited; expensive patterns, molds, and labor
Permanent mold	Good surface finish and dimensional accuracy; low porosity; high production rate	High mold cost; limited shape and intricacy; not suitable for high-melting-point metals
Die	Excellent dimensional accuracy and surface finish; high production rate	Die cost is high; part size limited; usually limited to nonferrous metals; long lead time
Centrifugal	Large cylindrical parts with good quality; high production rate	Equipment is expensive; part shape limited

Finishing operations for castings may include straightening or forging them with dies, and machining or grinding, to obtain final dimensions.

Several methods are available for inspection of castings to determine their quality and the presence of any defects. Castings may be inspected visually or optically for surface defects. Subsurface and internal defects are investigated using nondestructive techniques, described in Section 4.8.1. Test specimens are removed from various sections of a casting and tested for strength, ductility and other mechanical properties, and to determine the presence and location of any internal defects.

Pressure tightness of such cast components as valves, pumps, and pipes is usually determined by sealing the openings in the casting and pressurizing it with water, oil, or air, and inspected for leaks.

5.12 Design Considerations

Certain guidelines and design principles pertaining to casting have been developed over many years. Although these principles were established primarily through practical experience, analytical methods and computer-aided design and manufacturing techniques (Chapter 15) are now in wide use, improving productivity and the quality of castings. Some of the advantages and limitations of casting processes that impact design are given in Table 5.8.

5.12.1 Defects in Castings

Depending on casting design and the practices employed, several defects can develop in castings. The International Committee of Foundry Technical

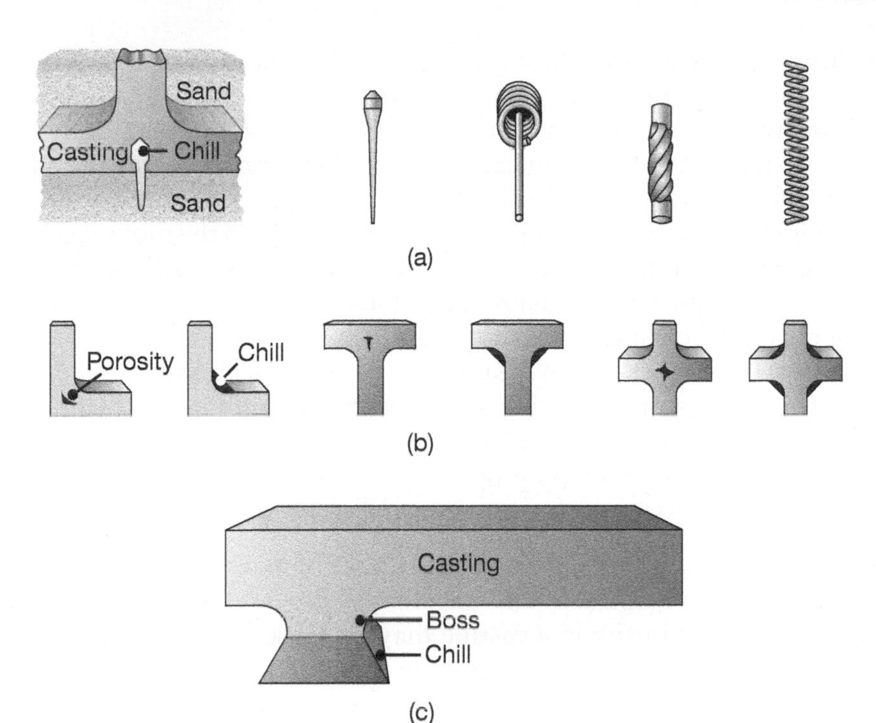

FIGURE 5.33 Various types of (a) internal and (b) external chills (dark areas at corners), used in castings to eliminate porosity caused by shrinkage. Chills are placed in regions where there is a larger volume of metal, as shown in (c).

Associations has developed standardized nomenclature, consisting of seven basic categories of casting defects:

1. **Metallic projections**, consisting of fins, flash, rough surfaces, or massive projections, such as swells.
2. **Cavities**, consisting of rounded or rough internal or exposed cavities, including blowholes, pinholes, and shrinkage cavities (See *porosity*, and Fig. 5.33).
3. **Discontinuities**, such as cracks, cold or hot tears, and cold shuts. If the solidifying metal is constrained from shrinking freely, cracking and tearing can occur. Coarse grains and the presence of low-melting segregates along the grain boundaries increase the tendency for hot tearing. *Cold shut* is an interface in a casting that lacks complete fusion, because of the meeting of two streams of partially solidified metal.
4. **Defective surface**, such as surface folds, laps, scars, adhering sand layers, and oxide scale.
5. **Incomplete casting**, such as misruns (due to premature solidification in a certain region in the casting), insufficient volume of metal poured, molten metal being at too low a temperature or pouring the metal too slowly, and runout (due to loss of metal from the mold after pouring).
6. **Incorrect dimensions or shape**, owing to such factors as improper shrinkage allowance, pattern mounting error, uneven contraction, deformed pattern, or warped casting. As discussed in Section 5.7, a pipe defect may result from metal shrinkage. A pipe refers to a surface that shrinks away from a mold, generally leaving a concave depression.

7. **Inclusions**, which form during melting, solidification, or molding. Inclusions may form (a) during melting because of reaction of the molten metal with the environment (usually oxygen) or the crucible material; (b) chemical reactions between components in the molten metal; (c) slags and other foreign material entrapped in the molten metal; (d) reactions between the metal and the mold material; and (e) spalling of the mold and core surfaces. All of these indicate the importance of maintaining melt quality and continuously monitoring the conditions of the molds.

   Generally nonmetallic, inclusions are regarded as harmful, because they act as stress raisers and thus reduce the strength of the casting. Hard inclusions (spots) in a casting may chip or damage cutting tools during subsequent machining operations. Inclusions can be filtered out during processing of the molten metal.

Porosity. *Porosity* is detrimental to the ductility of a casting and its surface finish, making it permeable and affecting the pressure tightness of a cast pressure vessel. Porosity in a casting may be caused either by *shrinkage* or *trapped gases*, or both. Porosity due to shrinkage is explained by the fact that thin sections in a casting solidify sooner than thick sections; as a result, the molten metal cannot enter into the thicker regions where the surfaces have already solidified, This condition leads to porosity in the thicker section, because the metal has to contract but is prevented from doing so by the solidified skin. **Microporosity** can also develop when the liquid metal solidifies and shrinks between dendrites and between dendrite branches (See Fig. 5.8).

Porosity due to shrinkage can be reduced or eliminated by various means, including the following:

1. Adequate liquid metal should be provided to prevent cavities caused by shrinkage.
2. Internal or external **chills** typically are used in sand casting (Fig. 5.33), to increase the solidification rate in thicker regions. Internal chills are usually made of the same material as the castings; external chills may be made of the same material or may be made of iron, copper, or graphite.
3. Making the temperature gradient steep, using, for example, mold materials that have high thermal conductivity.
4. Subjecting the casting to **hot isostatic pressing** (see Section 11.3.3); this is a costly method and is used mainly for critical components, as in aircraft parts.

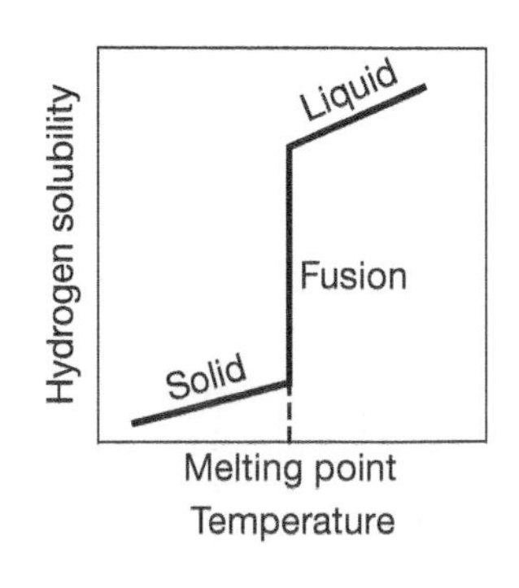

FIGURE 5.34 Solubility of hydrogen in aluminum. Note the sharp decrease in solubility as the molten metal begins to solidify.

Porosity due to gases is due to the fact that liquid metals have much greater **solubility** for gases than do solids (Fig. 5.34). When a metal begins to solidify, the dissolved gases are expelled from the solution, causing porosity. Gases may also be result of reactions of the molten metal with the mold materials; they either accumulate in regions of existing porosity, such as in interdendritic areas of the casting, or they cause microporosity, particularly in cast iron, aluminum, and copper.

Dissolved gases may be removed from the molten metal by flushing or purging it with an inert gas or by melting and pouring the metal in a vacuum. If the dissolved gas is oxygen, the molten metal can be *deoxidized*; steel is usually deoxidized with aluminum or silicon, and copper-based alloys, with copper alloy containing 15% phosphorus.

Whether microporosity is a result of shrinkage or is caused by gases may be difficult to determine. If the porosity is spherical and has smooth walls (much like the shiny surfaces of holes in Swiss cheese), it generally is from trapped gases. If the walls are rough and angular, porosity is likely from shrinkage between dendrites. Gross porosity (*macroporosity*) is from shrinkage, and is generally called **shrinkage cavities**.

5.12.2 General Design Considerations

There are two types of design issues in casting: (a) geometric features, tolerances, etc., that should be incorporated into the part; and (b) mold features that are required to produce the desired casting. Robust design (see also Section 16.2) of castings usually involves the following steps:

1. Design the part so that the shape is easily cast; several design considerations are described throughout this chapter.
2. Select a casting process and a material that is suitable for the part, its size, required production quantity, and mechanical properties. Often, steps 1 and 2 have to be specified simultaneously, which can be a demanding design challenge.
3. Locate the parting line of the mold or die.
4. Locate and design the gating system, to allow uniform feeding of the mold cavity with molten metal, including risers, sprue, and screens.
5. Select an appropriate runner geometry for the system.
6. Locate mold features, such as sprues, screens, and risers, as appropriate.
7. Ensure that proper controls and good practices are all in place.

Design of cast parts. The following considerations are important in designing castings:

1. **Corners, angles, and section thickness**. Sharp corners, angles, and fillets should be avoided as much as possible, because they act as stress raisers and may cause cracking and tearing of the metal, as well as of the dies, during solidification. Fillet radii should be selected so as to reduce stress concentrations and to ensure proper liquid-metal flow during pouring. The radii usually range from 3 to 25 mm, although smaller radii may be permissible in small castings and in specific applications. On the other hand, if the fillet radii are too large, the volume of the material in those regions is also large and the rate of cooling is lower.

   Section changes in castings should be smoothly blended into each other. The location of the largest circle that can be inscribed in a particular region is critical (Figs. 5.35a to c). Because the cooling rate in regions with larger circles is lower, they result in **hot spots**; these regions can then lead to **shrinkage cavities** and **porosity** (Fig. 5.35d). Cavities at hot spots can be eliminated using small cores. It is important to try to maintain uniform cross sections and wall thicknesses

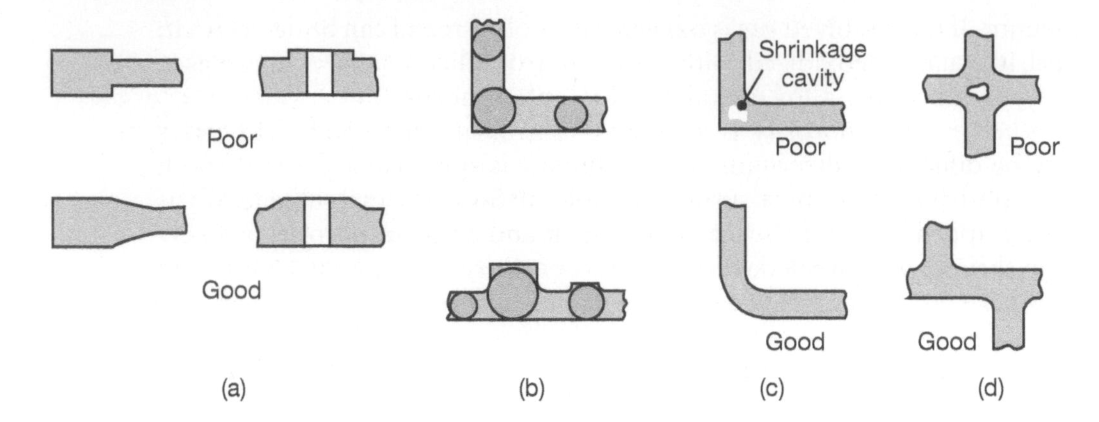

FIGURE 5.35 (a) Suggested design modifications to avoid defects in castings. Note that sharp corners are avoided to reduce stress concentrations; (b, c, d) examples of designs showing the importance of maintaining uniform cross sections in castings to avoid hot spots and shrinkage cavities.

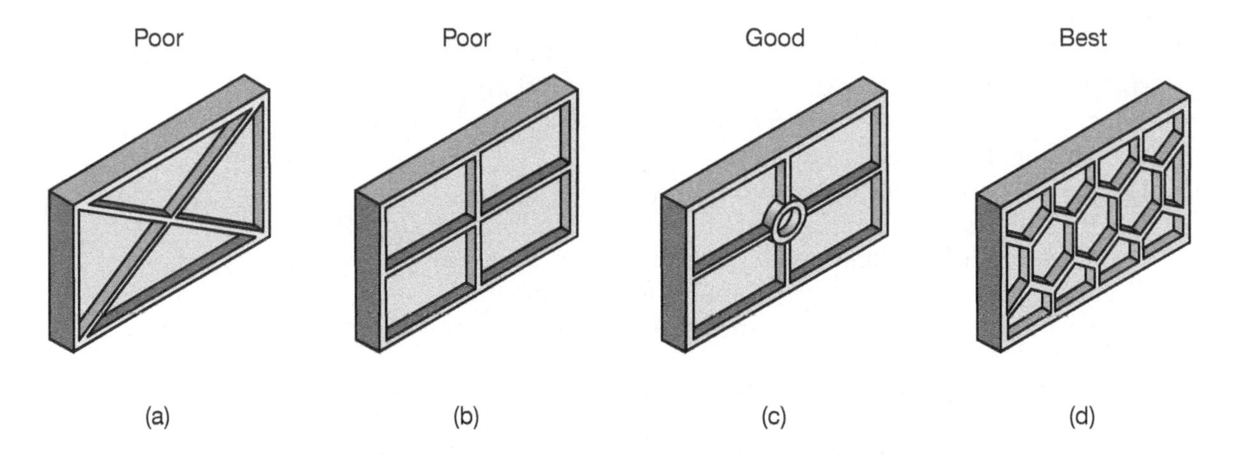

FIGURE 5.36 Rib designs for use on thin sections or flat surfaces to control or eliminate warping. Note the progression of designs: from left to right, the rib designs have improved castability and reliability.

throughout the casting so as to avoid or minimize shrinkage cavities. Although they increase the production cost, *metal chills* or *paddings* can eliminate or minimize hot spots (See Fig. 5.33).

2. **Flat areas.** Large flat areas (plain surfaces) should be avoided, because they may warp during cooling or develop poor surface finish, due to uneven flow of metal during pouring. A common technique is to break up flat surfaces with staggered ribs and serrations (see below).
3. **Ribs.** One method of producing parts with uniform thickness is to eliminate large, bulky volumes in the casting, as shown in Fig. 5.35. However, this can result in a loss in stiffness, and, especially with flat regions, it can lead to warping. One solution is to use ribs or support structure on the casting, as shown in Fig. 5.36; these are usually placed on the side that is less visible. Ribs should, in general,

have a thickness around 80% of the adjoining member thickness, and they should be deeper than their strut thickness. It is usually beneficial to have the ribs solidify before the members they adjoin. Ribs should not be used on both sides of a casting, and they should not meet at acute angles, because of complications in molding.

4. **Shrinkage.** To avoid cracking of the casting during cooling, allowances should be made for shrinkage. In castings with intersecting ribs, the tensile stresses developed can be reduced by staggering the ribs or by modifying the intersection geometry. Pattern dimensions should also provide for shrinkage of the metal during solidification and cooling. Allowances for shrinkage, known as **patternmaker's shrinkage allowances,** typically range from about 10 to 20 mm/m.
5. **Draft.** A small draft (taper) is typically provided in sand-mold patterns, to enable removal of the pattern without damaging the mold. Drafts generally range from 5 to 15 mm/m. Depending on the quality of the pattern, draft angles usually range from 0.5° to 2°. The draft angles on inside surfaces are typically twice this range; they have to be so because the casting shrinks inward toward the core.
6. **Dimensional tolerances.** Tolerances depend on the particular casting process, casting size, and type of pattern used. They should be as wide as possible, within the limits of good part performance, as otherwise the cost of the casting increases. In commercial practice, dimensional tolerances are usually in the range of ±0.8 mm for small castings, and may be as much as ±6 mm for large castings.
7. **Lettering and markings.** It is common practice to include some method of part identification in castings, such as letters, numbers, or corporate logos. These features can be depressions on the surface of the casting or they can protrude from the surface. In sand casting, for example, a pattern plate is produced by machining on a CNC mill (Section 8.10); it is simpler to machine letters into the pattern surface, resulting to sunken letters. In die casting, on the other hand, it is simpler to machine letters into the die, resulting to protruding letters.
8. **Finishing operations.** It is important to consider any subsequent machining and finishing operations that may be performed on castings. For example, if a hole is to be drilled, it is better to locate it on a flat surface than on a curved surface of the casting, to prevent the drill from wandering. An even better design would incorporate a small dimple, as a starting point for the drilling operations. Castings should also include features that allow them to be easily clamped into machine tools for subsequent finishing operations.

Parting line location. The location of the parting line is an important consideration, because it influences mold design, ease of molding, number and shape of cores required, method of support, and the gating system. A casting should be oriented in the mold such that its larger portion is relatively lower and that the height of the casting is minimized. Orientation of the casting also determines the distribution of porosity; consider casting of aluminum, for example, where porosity is due in part to hydrogen gas bubbles gases, and these will float upward (due to buoyancy), resulting in a higher porosity on the top regions. Recall also from Fig. 5.34 that hydrogen is soluble in liquid metal but not in solid; thus, critical surfaces should be oriented so that they face downwards.

A properly oriented casting can then have its parting line specified (See Fig. 5.10). In general, the parting line should be (a) along a flat plane, rather than be contoured; (b) at the corners or edges of castings, rather than on flat surfaces in the middle of the casting, so that the *flash* at the parting line (material squeezing out between the two halves of the mold) will not be as visible; (c) placed as low as possible relative to the casting for less dense metals, such as aluminum and magnesium alloys, and located at about the mid-height for denser metals, such as steels; and (d) in sand casting, it is typical that the runners, gates, and the sprue well are all placed in the drag on the parting line.

Gate design and location. Gates are the connections between the runners and the part to be cast. Some of the considerations in designing gating systems are as follows:

1. Multiple gates are often preferable, and are especially necessary for large castings. Multiple gates have the benefits of allowing lower pouring temperature, and also reduce the temperature gradients in the casting.
2. Gates should feed into thick sections of castings.
3. A fillet should be used where a gate meets a casting, so as to produce less turbulence than for abrupt junctions.
4. A gate closest to the sprue should be placed sufficiently away from it so that it can be easily removed; this distance may be as small as a few mm for small castings, and up to 500 mm for large parts.
5. The minimum gate length should be 3 to 5 times the gate diameter, on the metal being cast. Its cross section should be sufficiently large to allow filling of the mold cavity, and it should be smaller than the runner cross section.
6. Curved gates should be avoided; when necessary, a straight section in the gate should be located immediately adjacent to the casting.

Runner design. The runner is a horizontal distribution channel that receives molten metal from the sprue and delivers it to the gates. A single runner is used for simple parts, but two-runner systems can be specified for more complicated castings. Runners are also used to trap dross (a mixture of oxide and metal that forms on the surface of molten metals) and keep it from entering the gates and the mold cavity. Commonly, dross traps are placed at the ends of runners, and the runner projects above the gates, to ensure that the metal in the gates is tapped from below the surface.

Designing other mold features. The main goal in designing a *sprue* (described in Section 5.4.1) is to achieve the required metal flow rates while preventing aspiration or excessive dross formation. Flow rates are determined such that turbulence is avoided, but the mold is filled quickly as compared to the solidification time required. A *pouring basin* can be used to ensure that the metal flow into the sprue is uninterrupted. Also, if the level of molten metal is retained in the pouring basin during pouring, then the dross will float and will be prevented from entering the mold cavity. *Filters* are used to trap large contaminants; they also serve to slow the metal and make the flow more laminar. *Chills* can be used to speed solidification of metal in a particular region of a casting.

5.12.3 Design Principles for Expendable-Mold Casting

Expendable-mold casting processes have certain specific design considerations, that are mainly concerned with mold material, part size, and the casting method. Important design considerations are described below.

1. *Mold layout.* The features in the mold must be placed logically and compactly, including gates when necessary. Solidification should begin at one end of the mold and progresses in a uniform front across the casting, with risers solidifying last. Although traditionally mold layout has been based on experience, commercial computer programs are now available for this task (see Section 5.12.5). Based on finite-difference algorithms, these techniques allow the simulation of mold filling and the rapid evaluation of mold layouts.
2. *Riser design.* A major concern in the design of castings is the size and placement of risers (See Fig. 5.10). Risers are essential in controlling the progression of the solidification front across a casting, and are an important feature in mold layout. Blind risers are good design features, and retain heat longer than open risers. Risers are designed according to five basic rules:
   a. The metal in the riser must not solidify before the casting does; this is usually ensured by avoiding the use of small risers and by using cylindrical risers with small aspect ratios (height to cross section). Spherical risers are the most efficient shape, but are difficult to cast.
   b. The riser volume must be large enough to provide sufficient liquid metal necessary to compensate for shrinkage in the casting.
   c. Junctions between the casting and the riser should not develop a *hot spot*, where shrinkage porosity can occur.
   d. Risers must be placed such that liquid metal can be delivered to locations where it is most needed.
   e. There must be sufficient pressure to drive the liquid metal into locations in the mold where it is needed; risers are therefore not as useful for metals with low density, such as aluminum alloys, than for those with a higher density, such as steel and cast irons.
3. *Machining allowance.* Because most expendable-mold castings require some additional finishing operations, such as machining and grinding, allowances must be made in casting design. Machining

allowances, which are included in pattern dimensions, depend on the type of casting, and they increase with the size and section thickness of castings. Allowances usually range from about 2 to 5 mm (0.08–0.2 in.) for small castings, and to more than 25 mm (1 in.) for large castings.

5.12.4 Design Principles for Permanent-Mold Casting

General design guidelines and examples for permanent-mold casting are shown schematically for die casting in Fig. 5.37. Note that the cross sections have been reduced in order to decrease the solidification time and also save material. Special considerations are also involved in designing tooling for die casting. Although designs may be modified to eliminate the draft for better dimensional accuracy, a draft angle of 0.5° or even 0.25° is usually required, as otherwise galling (localized seizure or sticking of material) may occur between the casting and the dies, and cause part distortion.

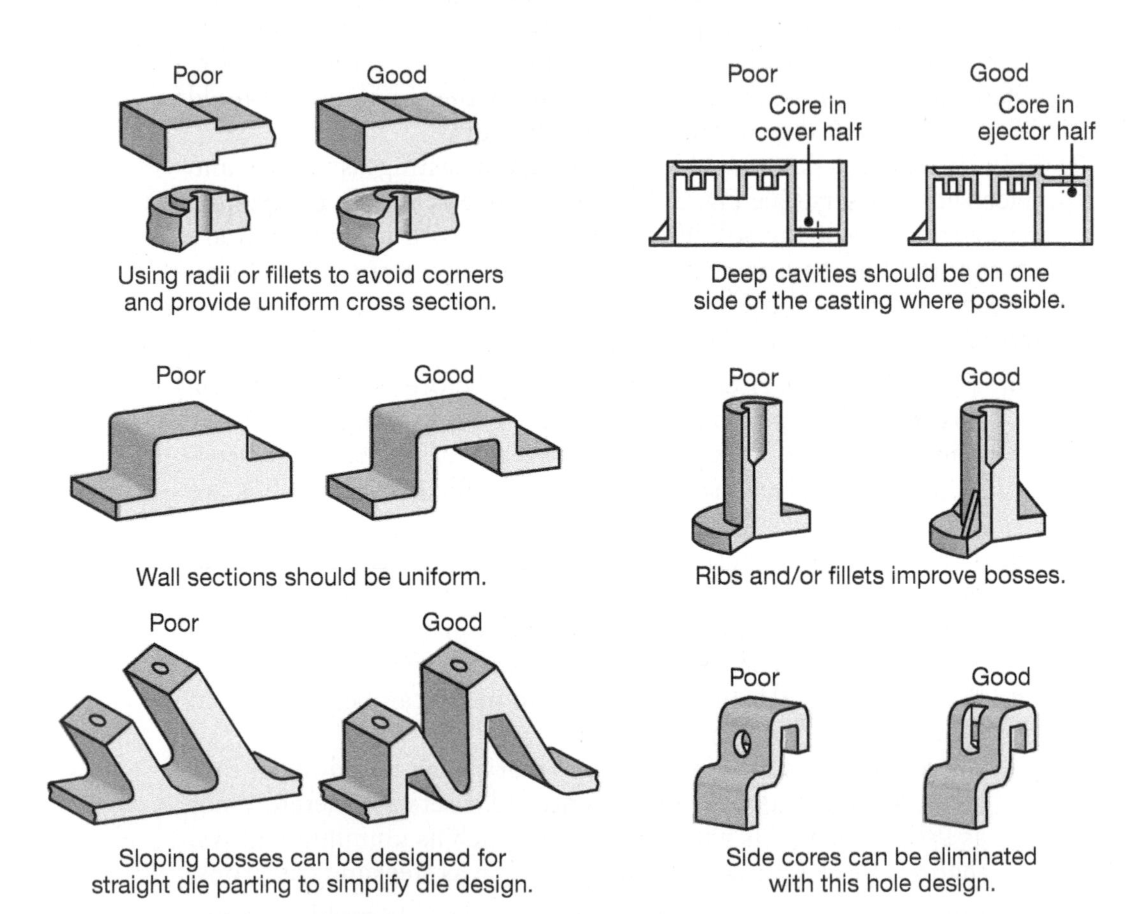

FIGURE 5.37 Suggested design modifications to avoid defects in castings. *Source:* After the North American Die Casting Association.

Die-cast parts are nearly-net shaped, typically requiring only the removal of gates and minor trimming, to remove flashing and other minor defects. The surface finish and dimensional accuracy of die-cast parts are very good (See Table 5.2), and, in general, they do not require a machining allowance.

5.12.5 Computer Modeling of Casting Processes

Casting processes involve complex interactions among material and process variables, a quantitative study of these interactions is thus essential to the proper design and the production of high quality castings. Advances in computer processing speeds and development of more powerful software have enabled modeling of various aspects of casting design and operations. These advances include studies in fluid flow, heat transfer, and the microstructures developed during solidification. The benefits of modeling include faster response to design changes, increased productivity, improved quality, and easier planning and cost estimating.

Modeling of *fluid flow* is based on Bernoulli's and the continuity equations, described in Section 5.4.1. The models predict the behavior of the hot metal during pouring into the gating system and its travel into the mold cavity, as well as velocity and pressure distributions in the whole system. Modeling of *heat transfer* in casting includes coupling of fluid flow and heat transfer, and the effects of surface conditions, thermal properties of the materials involved, and natural and forced convection on cooling of the casting. Recall that surface conditions vary during solidification, as a layer of air develops between the casting and the mold wall due to shrinkage. Similar studies involve modeling of the development of *microstructures* in casting. These studies encompass heat flow, temperature gradients, nucleation and growth of crystals, formation of dendritic and equiaxed structures, impingement of grains on each other, and movement of the liquid-solid interface during solidification.

Modeling is now capable of predicting, for example, the width of the mushy zone (See Fig. 5.6) during solidification, the development of grain size in castings, and the capability to calculate isotherms (giving insight into possible hot spots and subsequent development of shrinkage cavities). With advances in computer-aided design and manufacturing (see Chapter 14), modeling has become easier to implement. Several commercial software programs are now available on modeling and casting processes, including SOLIDCast, NovaCast, ProCAST, and Magmasoft.

5.13 Economics of Casting

It has been noted throughout this chapter that some casting processes require more labor than others, some require expensive dies and machinery, and some take much more time than others to complete the operation. Each of these factors, outlined in Table 5.2, affects, to various degrees, the overall cost of a casting operation. As described in greater detail in Chapter 15, the total cost of a product includes the costs of (a) materials, (b) dies and tooling, (c) equipment, and (d) labor. Preparations for casting a product include making molds and dies that require materials, machinery, time, and effort, which all contribute to costs. Although relatively little cost is involved in molds for sand casting, die-casting dies require expensive materials and a great deal of machining and preparation. Facilities also are required for melting and pouring the molten metal into the molds or dies, including furnaces and related equipment, their costs depending on the level of automation desired. Finally, costs are also involved in cleaning and inspecting castings.

The labor required in casting operations can vary considerably, depending on the particular process and level of automation. Investment casting, for example, requires a great deal of labor because of the large number of steps involved in this operation. Conversely, operations such as highly automated die casting can maintain high production rates with little labor required.

The cost of equipment per cast part (*unit cost*) decreases as the number of parts increases (Fig. 5.38); sustained high production rates can then justify the high cost of dies and machinery. If demand is relatively small, however, the cost per part increases rapidly. It then becomes more economical to manufacture the part by sand casting or by other manufacturing methods. Note that Fig. 5.38 can also include other casting processes suitable for making the same part. The two processes compared (for example, sand and die casting) produce castings with significantly different dimensional and surface-finish characteristics, thus not all manufacturing decisions should be based purely on economic considerations. In fact, parts can usually be made by more than one or two processes (See, for example, Fig. 1.8); thus, the final decision depends on both economic as well as technical considerations. The competitive aspects of manufacturing processes are described in greater detail in Chapter 15.

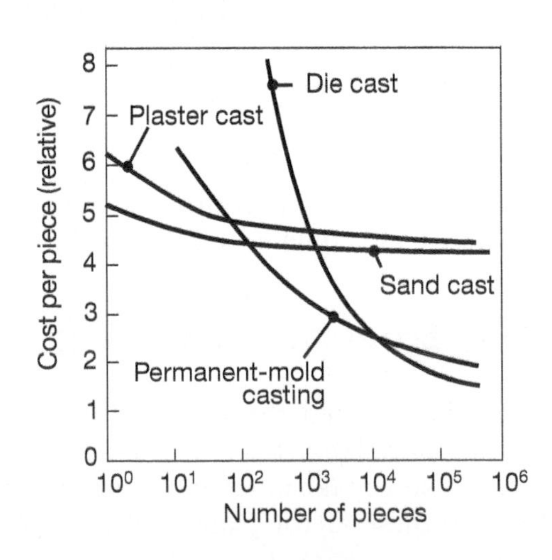

FIGURE 5.38 Economic comparison of making a part by two different casting processes. Note that because of the high cost of equipment, die casting is economical mainly for large production runs. *Source:* After the North American Die Casting Association.

CASE STUDY 5.1 Sand Casting of a Tractor Hitch

Farm implements such as mowers, aerators, and post hole diggers, are attached to tractors through a hitch; the Easy Hitch is a design by Greenwell Manufacturing that eases attachment of these implements to a standard tractor, easing the requirement that the tractor be precisely aligned with the implement for attachment. The Easy Hitch, shown in Fig. 5.39a, was originally constructed from a welded assembly, with five plate pieces that needed to be plasma cut (Section 9.14.2), formed (Section 7.5), machined (Chapter 8), and arc welded (Section 12.3).

Further, to allow proper assembly, all of the holes shown had to be machined and tapped (Section 8.9.4) to make sure that they would align properly for attachment to the tractor. Not surprisingly, this is a laborious process that also required considerable skill.

Dotson Iron Castings was approached by Greenwell Manufacturing to investigate the possibility of converting the welded assembly to a more simple design using sand casting. The final result is shown in Fig. 5.39b. To obtain this design:

1. Dotson combined the welded components into one cast part, so that the housing could be produced in one pour in a sand casting operation (Section 5.8.1).
2. The material selected was D80-55-06 ductile iron (pearlitic ductile cast iron from Table 5.4) to withstand the dynamic stresses encountered by the coupling.
3. Before finalizing the design, computer generated mold filling simulations were conducted to ensure the molten metal filled the mold with minimal turbulence and produced a casting without hot spots or excessive porosity.
4. Lettering and part identification that had been previously painted on (and were subject to damage) were incorporated into the casting.

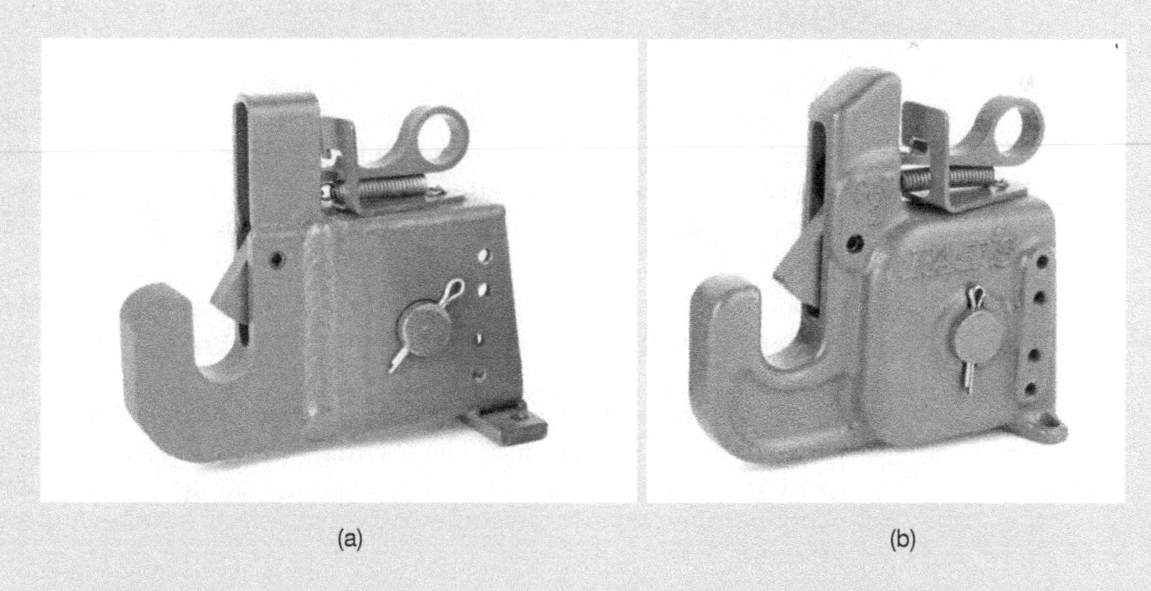

FIGURE 5.39 Casting of the Easy Hitch. (a) Original design, with a housing made from a welded assembly of five components and finish machining. (b) Redesigned Easy Hitch, constructed of a sand cast housing.

5. The mounting holes were reinforced with a boss and the holes in the as-cast condition were properly aligned, eliminating the need for drilling and tapping to ensure the hitch could be assembled properly.

The two videos show the design and construction steps for the pattern plate and a video of an actual pour. The resulting design confirmed the mold fill simulations, and high-quality parts were produced without any further design iterations.

In testing, the cast design demonstrated a 150kN capacity, which was an improvement of over 20 kN compared to the welded assembly. A larger version demonstrated a 200 kN capacity, representing an improvement of over 40 kN from the welded design.

Around 10,000 of the smaller hitch and 5,000 of the larger hitch are produced annually. Typical order quantities are 1,000, which are ideal for automated foundries. Dotson casts and finish grinds at a rate of 160 castings per hour.

Not only does the cast design outperform the welded design, but the customer realized a 50% reduction in cost compared to the labor-intensive fabrication.

Source: Courtesy of Dotson Iron Castings and Greenwell Manufacturing.

SUMMARY

- Metal casting is among the oldest and most common manufacturing processes. Solidification of pure metals takes place at a clearly defined constant temperature, whereas solidification of alloys occurs over a range of temperatures. Phase diagrams are important for identifying the solidification point or points for metals and alloys. (Sections 5.1 and 5.2)
- The composition and cooling rate of the melt affect the size and shape of grains and dendrites, and thus influence the properties of the casting. (Section 5.3)
- Most metals shrink during solidification, although gray cast iron and a few others actually expand. The resulting loss of dimensional accuracy and s cracking are difficulties that can arise during solidification and cooling. Several basic categories of casting defects have been identified. (Section 5.3)
- In casting, the molten metal or alloy typically flows through a variety of passages, including pouring basins, sprues, runners, risers, and gating systems, before reaching the mold cavity. Bernoulli's theorem, the

continuity law, and the Reynolds number are the analytical tools used in designing an appropriate system and in eliminating defects associated with fluid flow. Heat transfer affects fluid flow and solidification time in casting. Solidification time is a function of the volume of a casting and its surface area (Chvorinov's rule). (Section 5.4)

- Melting practices have a direct effect on the quality of castings; factors that affect melting include: (a) inorganic compounds or fluxes that are added to the molten metal to remove dissolved gases and various impurities; (b) the type of furnace used; and (c) foundry operations, including pattern and mold making, pouring of the melt, removal of castings from molds, cleaning, heat treatment, and inspection. (Section 5.5)
- Several ferrous and nonferrous casting alloys are available, with a wide range of properties, casting characteristics, and applications. Because castings are often designed and produced to be assembled with other mechanical components and into structures, various other considerations are also important, such as weldability, machinability, and surface conditions. (Section 5.6)
- The traditional ingot-casting process has been largely replaced for many continuous-casting methods for both ferrous and nonferrous metals. (Section 5.7)
- Casting processes are generally classified as either expendable-mold or permanent-mold casting. The most common expendable-mold methods are sand, shell-mold, plaster, ceramic-mold, and investment casting. Permanent-mold methods include slush casting, pressure casting, and die casting. Compared to permanent-mold casting, expendable-mold casting usually involves lower mold and equipment costs, but with lower dimensional accuracy. (Sections 5.8–5.10)
- Castings, as well as wrought parts made by other manufacturing processes, may be subjected to subsequent heat-treatment operations, to enhance various properties and service life. Several transformations take place in microstructures that have widely varying characteristics and properties. Important mechanisms of hardening and strengthening involve thermal treatments, including quenching and precipitation hardening. (Section 5.11)
- General principles have been established to aid designers to produce castings that are free from defects and meet dimensional tolerance and service requirements. Because of the large number of variables involved, close control of all parameters is essential, particularly those related to the nature of liquid metal flow into the molds and dies and the rate of cooling in different regions of castings. (Section 5.12)
- Within limits of good performance, the economic aspects of casting are as important as the technical considerations. Factors affecting the overall cost include the cost of materials, molds, dies, equipment, and labor, each of which varies with the particular casting operation. An important parameter is the cost per casting, which, for large production runs, can justify major expenditures typical of automated machinery and operations. (Section 5.13)

SUMMARY OF EQUATIONS

Bernoulli's theorem, $h + \frac{p}{\rho g} + \frac{v^2}{2g} = \text{Constant}$

Sprue contour, $\frac{A_1}{A_2} = \sqrt{\frac{h_2}{h_1}}$

Reynolds number, $\text{Re} = \frac{vD\rho}{\eta}$

Chvorinov's rule, Solidification time $= C\left(\frac{\text{Volume}}{\text{Surface area}}\right)^n$

Continuity equation, $Q = A_1 v_1 = A_2 v_2$

BIBLIOGRAPHY

ASM Handbook, Vol. 3: Alloy Phase Diagrams, ASM International, 2016.

ASM Handbook, Vol. 4: Heat Treating, ASM International, 1991.

ASM Handbook, Vol. 4C: Induction Heating and Heat Treatment, ASM International, 2014.

ASM Handbook, Vol. 15: Casting, ASM International, 1988.

ASM Specialty Handbook: Cast Irons, ASM International, 1996.

Bradley, E.F., *High-Performance Castings: A Technical Guide*, Edison Welding Institute, 1989.

Campbell, J., *Castings*, 2nd ed., Butterworth-Heinemann, 2003.

—, *Complete Casting Handbook: Metal Casting Processes, Techniques and Design*, Butterworth-Heinemann, 2011.

Dossett, J.L., and Totten, G.E., *ASM Handbook, Vol 4A: Steel Heat Treating - Fundamentals and Processes*, ASM International, 2013.

Glicksman, M.E., *Principles of Solidification: An Introduction to Modern Casting and Crystal Growth Concepts*, Springer, 2010.

Goodrich, G. (ed.) *Casting Defects Handbook: Iron & Steel*, 2nd ed., American Foundry Society, 2008.

—, *Investment Casting Handbook*, Investment Casting Institute, 1997.

Kaye, A., and Street, A.C., *Die Casting Metallurgy*, Butterworth, 1982.

Kirkwood, D.H., Suery, M., Kapranos, P., and Atkinson, H.V., *Semi-solid Processing of Alloys*, Springer, 2012.

Krauss, G., *Steels: Heat Treatment and Processing Principles*, ASM International, 1990.

Kurz, W., and Fisher, D.J., *Fundamentals of Solidification*, 4th ed., Trans Tech Pub., 1998.

Sahoo, M., and Sahu, S., *Principles of Metal Casting*, McGraw-Hill, 2014.

Steel Castings Handbook, 6th ed., Steel Founders' Society of America, 1995.

Totten, G.E., *Steel Heat Treatment Handbook*, 2nd ed., CRC Press, 2006.

Wieser, P.P. (ed.), *Steel Castings Handbook*, 6th ed., ASM International, 1995.

QUESTIONS

5.1 Describe the characteristics of (a) an alloy; (b) pearlite; (c) austenite; (d) martensite; and (e) cementite.

5.2 What are the effects of mold materials on fluid flow and heat transfer?

5.3 How can you tell whether cavities in a casting are due to porosity or to shrinkage?

5.4 How does the shape of graphite in cast iron affect its properties?

5.5 Explain the difference between short and long freezing ranges. How are they determined? Why are they important?

5.6 It is known that pouring molten metal at a high rate into a mold has certain disadvantages. Are there any disadvantages to pouring it very slowly? Explain.

5.7 Why does porosity have detrimental effects on the mechanical properties of castings? Which physical properties are also affected adversely by porosity?

5.8 A spoked hand wheel is to be cast in gray iron. In order to prevent hot tearing of the spokes, would you insulate the spokes or chill them? Explain.

5.9 Which of the following considerations are important for a riser to function properly? (a) Have a surface area larger than the part being cast; (b) be kept open to atmospheric pressure; (c) solidify first. Why?

5.10 Explain why the constant c in Eq. (5.9) depends on mold material, metal properties, and temperature.

5.11 Explain why gray iron undergoes expansion, rather than contraction, during solidification.

5.12 Is there porosity in a chocolate bar? In an ice cube? Explain.

5.13 Explain the reasons for hot tearing in castings.

5.14 Review Fig. 5.10 and make a summary, explaining the purpose of each feature shown and the consequences of omitting the feature from the mold design.

5.15 Would you be concerned about the fact that parts of internal chills are left within the casting? What materials do you think chills should be made of, and why?

5.16 Are external chills as effective as internal chills? Explain.

5.17 Do you think early formation of dendrites in a mold can impede the free flow of molten metal into the mold? Explain.

5.18 Is there any difference in the tendency for shrinkage void formation for metals with short freezing and long freezing ranges, respectively? Explain.

5.19 It has long been observed by foundry-men that low pouring temperatures, i.e., low superheat, promote equiaxed grains over columnar grains. Also, equiaxed grains become finer as the pouring temperature decreases. Explain these phenomena.

5.20 What are the reasons for the large variety of casting processes that have been developed over the years?

5.21 Why can blind risers be smaller than open-top risers?

5.22 Why are risers not as useful in die casting as they are in sand casting?

5.23 Would you recommend preheating the molds in permanent-mold casting? Also, would you remove the casting soon after it has solidified? Explain.

5.24 In a sand-casting operation, what factors determine the time at which you would remove the casting from the mold?

5.25 Recently, cores in sand casting have been produced from salt. What advantages and disadvantages would you expect from using salt cores?

5.26 Explain why the strength-to-weight ratio of die-cast parts increases with decreasing wall thickness.

5.27 We note that the ductility of some cast alloys is essentially zero (See Fig. 5.12). Do you think this should be a significant concern in engineering applications of castings? Explain.

5.28 The modulus of elasticity of gray iron varies significantly with its type, such as the ASTM class. Explain why.

5.29 List and explain the considerations for selecting pattern materials.

5.30 Why is the investment-casting process capable of producing fine surface detail on castings?

5.31 Explain why a casting may have a slightly different shape than the pattern used to make the mold.

5.32 Explain why squeeze casting produces parts with better mechanical properties, dimensional accuracy, and surface finish than expendable-mold processes.

5.33 Why are steels more difficult to cast than cast irons?

5.34 What would you do to improve the surface finish in expendable-mold casting processes?

5.35 You have seen that even though die casting produces thin parts, there is a limit to the minimum thickness. Why can't even thinner parts be made by this process?

5.36 What differences, if any, would you expect in the properties of castings made by permanent-mold versus sand-casting methods?

5.37 Which of the casting processes would be suitable for making small toys in large numbers? Why?

5.38 Why are allowances provided for in making patterns? What do they depend on?

5.39 Explain the difference in the importance of drafts in green-sand casting versus permanent-mold casting.

5.40 Make a list of the mold and die materials used in the casting processes described in this chapter. Under each type of material, list the casting processes that are used, and explain why these processes are suitable for that particular mold or die material.

5.41 Explain why carbon is so effective in imparting strength to iron in the form of steel.

5.42 Describe the engineering significance of the existence of a eutectic point in phase diagrams.

5.43 Explain the difference between hardness and hardenability.

5.44 Explain why it may be desirable for castings to be subjected to various heat treatments.

5.45 Describe the differences between case hardening and through hardening insofar as engineering applications are concerned.

5.46 *Type metal* is a bismuth alloy used to cast type for printing. Explain why bismuth is ideal for this process.

5.47 Do you expect to see larger solidification shrinkage for a material with a bcc crystal structure or fcc? Explain.

5.48 Describe the drawbacks to having a riser that is (a) too large; or (b) too small.

5.49 If you were to incorporate lettering on a sand casting, would you make the letters protrude from the surface or recess into the surface? What if the part were to be made by investment casting?

5.50 List and briefly explain the three mechanisms by which metals shrink during casting.

5.51 Explain the significance of the "tree" in investment casting.

5.52 Sketch the microstructure you would expect for a slab cast through (a) continuous casting; (b) strip casting; and (c) melt spinning.

5.53 The general design recommendations for a well in sand casting are that (a) its diameter should be twice the sprue exit diameter; and (b) the depth should be approximately twice the depth of the runner. Explain the consequences of deviating from these rules.

5.54 Describe thixocasting and rheocasting.

5.55 Sketch the temperature profile you would expect for (a) continuous casting of a billet; (b) sand casting of a cube; and (c) centrifugal casting of a pipe.

5.56 What are the benefits and drawbacks to having a pouring temperature that is much higher than the metal's melting temperature? What are the advantages and disadvantages in having the pouring temperature remain close to the melting temperature?

5.57 What are the benefits and drawbacks to heating the mold in investment casting before pouring in the molten metal? Are there any drawbacks? Explain.

5.58 Can a chaplet also be a chill? Explain.

5.59 Rank the casting processes described in this chapter in terms of their solidification rate. That is, which processes extract heat the fastest from a given volume of metal?

5.60 The heavy regions of parts typically are placed in the drag in sand casting and not in the cope. Explain why.

PROBLEMS

5.61 Using Fig. 5.3, estimate the following quantities for a 60% Cu-40% Ni alloy: (a) liquidus temperature; (b) solidus temperature; (c) percentage of nickel in the liquid at 1250°C (2550°F); (d) the major phase at 1250°C; and (e) the ratio of solid to liquid at 1250°C.

5.62 Determine the amount of gamma and alpha phases (Fig. 5.4b) in a 10-kg, 1060 steel casting as it is being cooled to the following temperatures: (a) 750°C; (b) 728°C; and (c) 726°C.

5.63 Derive Eq. (5.7).

5.64 A round casting is 0.3 m in diameter and 0.5 m in length. Another casting of the same metal is elliptical in cross section, with a major-to-minor axis ratio of 3, and has the same length and cross-sectional area as the round casting. Both pieces are cast under the same conditions. What is the ratio of the solidification times of the two castings?

5.65 Two halves of a mold (cope and drag) are weighted down to keep them from separating under the pressure exerted by the molten metal (buoyancy). Consider a solid, spherical steel casting, 225 mm in diameter, that is being produced by sand casting. Each flask (See Fig. 5.10) is 0.5 m by 0.5 m and 350 mm deep. The parting line is at the middle of the part. Estimate the clamping force required if the molten metal has a density of 8.96 g/cm^3 and the sand has a density of 1.2 g/cm^3.

5.66 Would the position of the parting line in Problem 5.65 influence your answer? Explain.

5.67 Plot the clamping force in Problem 5.65 as a function of increasing diameter of the casting, from 0.025 m to 0.50 m.

5.68 A cylindrical casting has a diameter of 300 mm and a length of 1 m. Another casting of the same metal is rectangular in cross section, with a width-to-thickness ratio of 3, and has the same length and cross-sectional area as the round casting. Both pieces are cast under the same conditions. The cylindrical casting solidifies in three minutes. How long does the rectangular cross section casting take to solidify?

5.69 Sketch a graph of specific volume versus temperature for a metal that shrinks as it cools from the liquid state to room temperature. On the graph, mark the area where shrinkage is compensated for by risers.

5.70 A 75-mm thick square plate and a right circular cylinder with a radius of 100 mm and height of 50 mm each have the same volume. If each is to be cast using a cylindrical riser, will each part require the same size riser to ensure proper feeding? Explain.

5.71 Assume that the top of a round sprue has a diameter of 100 mm and has a height of 300 mm. At the bottom, the sprue has a diameter of 25 mm. Based on Eq. (5.7), plot the profile of the sprue diameter as a function of its height.

5.72 The blank for the spool shown in the figure below is to be sand cast out of A-319, an aluminum casting alloy. Make a sketch of the wooden pattern for this part. Include all necessary allowances for shrinkage and machining.

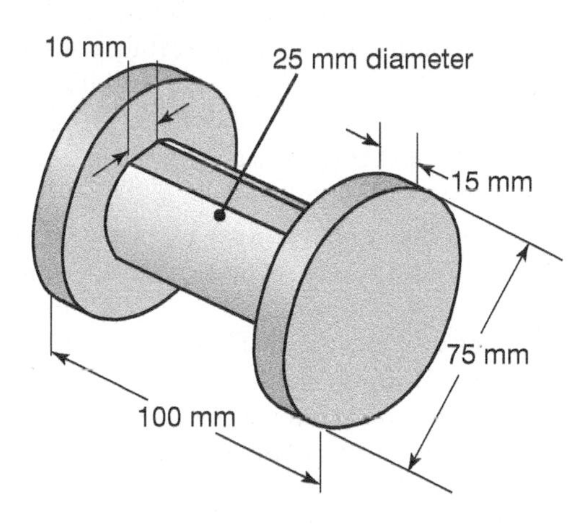

5.73 Repeat Problem 5.72, but assume that the aluminum spool is to be cast using expendable-pattern casting. Explain the important differences between the two patterns.

5.74 The optimum shape of a riser is spherical to ensure that it cools more slowly than the casting it feeds. Spherically shaped risers, however, are difficult to cast. (a) Sketch the shape of a blind riser that is easy to mold, but also has the smallest possible surface area-to-volume ratio. (b) Compare the solidification time of the riser in part (a) to that of a riser shaped like a right circular cylinder. Assume that the volume of each riser is the same, and that for each the height is equal to the diameter (see Example 5.2).

5.75 A cylinder with a diameter-to-height ratio of 1 solidifies in four minutes in a sand casting operation. What is the solidification time if the cylinder height is tripled? What is the time if the diameter is tripled?

5.76 Estimate the clamping force for a diecasting machine in which the casting is rectangular, with projected dimensions of 75 mm × 150 mm. Would your answer depend on whether or not it is a hot-chamber or cold-chamber process? Explain.

5.77 When designing patterns for casting, patternmakers use special rulers that automatically incorporate solid shrinkage allowances into their designs. Therefore, a 250 mm patternmaker's ruler is longer than 250 mm. How long is a patternmaker's ruler designed for the making of patterns for (a) aluminum castings; (b) pure silicon; and (c) high-manganese steel?

5.78 In sand casting, it is important that the cope mold half be held down with enough force to keep it from floating when the molten metal is poured in. For the casting shown in the figure below, calculate the minimum amount of weight necessary to keep the cope from floating up as the molten metal is poured in. (*Hint:* The buoyancy force exerted by the molten metal on the cope is related to the effective height of the metal head above the cope.)

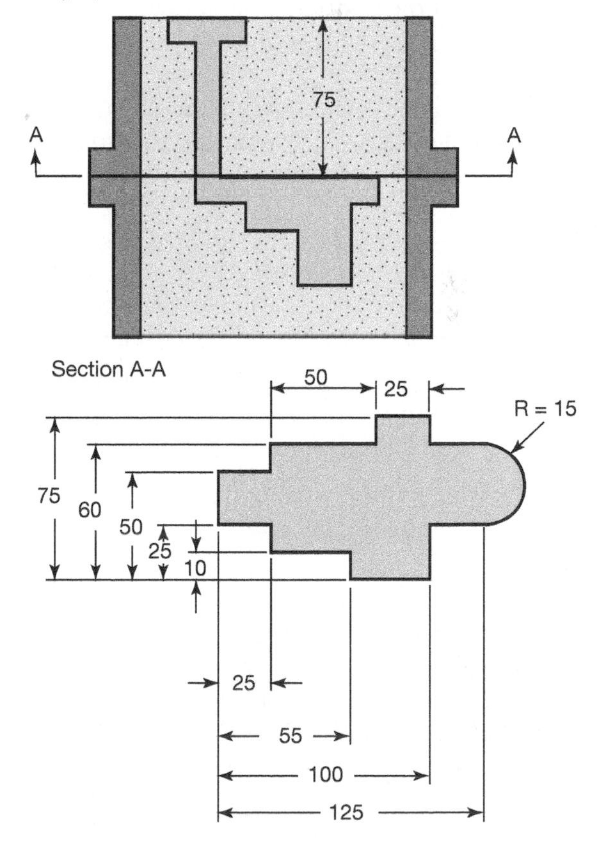

5.79 The part shown below is a hemispherical shell used as an acetabular (mushroom shaped) cup in a total hip replacement. Select a casting process for this part and provide a sketch of all patterns or tooling needed if it is to be produced from cobalt-chrome alloy.

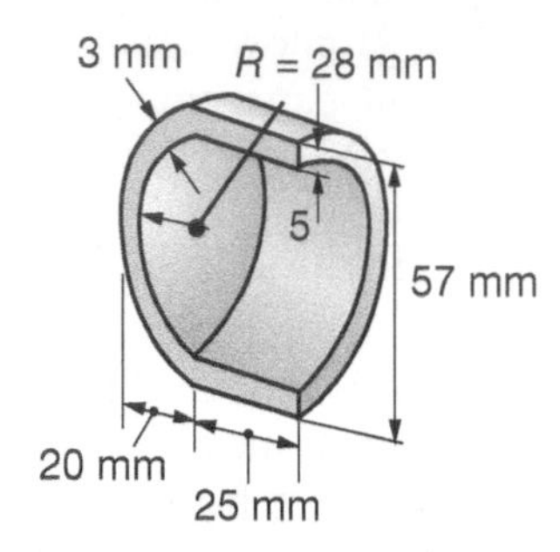

5.80 A sprue is 200 mm long and has a diameter of 125 mm at the top, where the metal is poured. If a flow rate of 60,000 mm^3/s is to be achieved, what should be the diameter of the bottom of the sprue?

5.81 Steel piping is to be produced by centrifugal casting. The length is 4 m, the diameter is 1 m, and the thickness is 15 mm. Using basic equations from dynamics and statics, determine the rotational speed needed to have the centripetal force be 70 times its weight.

5.82 A sphere with a diameter of 75 mm solidifies in three minutes in a sand-casting operation. Using the same equipment and metal, estimate the time required to produce the part considered in Problem 5.72.

5.83 Small amounts of slag often persist after skimming and are introduced into the molten metal flow in casting. Recognizing that the slag is much less dense than the metal, design mold features that will remove small amounts of slag before the metal reaches the mold cavity.

5.84 Pure aluminum is poured into a sand mold. The metal level in the pouring basin is 250 mm above the metal level in the mold, and the runner is circular with a 10 mm diameter. What is the velocity and rate of the flow of the metal into the mold? Is the flow turbulent or laminar? Use a viscosity of η = 0.0015 Ns/m^2.

5.85 For the sprue described in Problem 5.84, what runner diameter is needed to ensure a Reynolds number of 2000? How long will a 300,000 mm^3 casting take to fill with such a runner?

5.86 How long would it take for the sprue in Problem 5.84 to feed a casting with a square cross section of 50 mm per side and a height of 100 mm? Assume the sprue is frictionless

5.87 A rectangular mold with dimensions 100 mm × 200 mm × 400 mm is filled with aluminum with no superheat. Determine the final dimensions of the part as it cools to room temperature. Repeat the analysis for gray cast iron.

5.88 The constant C in Chvorinov's rule is given as 3 s/mm^2 and is used to produce a cylindrical casting with a diameter of 75 mm and a height of 125 mm. Estimate the time for the casting to fully solidify. The mold can be broken safely when the solidified shell is at least 20 mm. Assuming the cylinder cools evenly, how much time must transpire after pouring the molten metal before the mold can be broken?

5.89 A jeweler wishes to produce twenty gold rings in one investment-casting operation. The wax parts are attached to a wax central sprue with a 20 mm diameter. The rings are located in four rows, each 15 mm from the other on the sprue. The rings require a 3 mm diameter, 12 mm long runner to the sprue. Estimate the weight of gold needed to completely fill the rings, runners, and sprues. Assume a typical ring has a 25 mm outer diameter, 19 mm inner diameter and 5 mm width. The specific gravity of gold is 19.3.

5.90 Plot the solidification time of a cylinder with a volume of 15,000 mm^3 as a function of its diameter-to-length ratio. Assume that the solidification time for a 25-mm diameter sphere is three minutes.

5.91 If an acceleration of 100 g is necessary to produce a part in true centrifugal casting and the part has an inner diameter of 250 mm, a mean outer diameter of 350 mm, and a length of 6 m, what rotational speed is needed?

5.92 Assume that you are asked to give a quiz to students on the contents of this chapter. Prepare three quantitative problems and three qualitative questions, and supply the answers.

DESIGN

5.93 The figures below indicate various defects and discontinuities in cast products. Review each one and offer design solutions to avoid them.

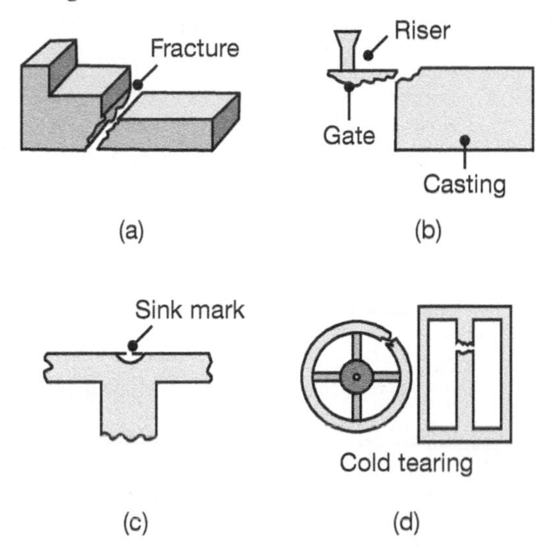

5.94 Porosity developed in the boss of a casting is illustrated in the figure below. Show that by simply repositioning the parting line of this casting, this problem can be eliminated.

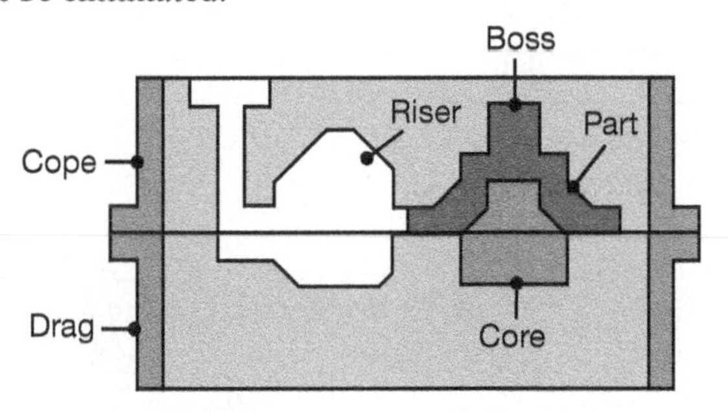

5.95 Design test methods to determine the fluidity of metals in casting (see Section 5.4.2). Make appropriate sketches and explain the important features of each design.

5.96 Design an experiment to measure the constants C and n from Chvorinov's Rule [Eq. (5.11)]. Describe the features of your design, and comment on any difficulties that might be encountered in running such an experiment.

5.97 For the wheel illustrated in the figure below, show how (a) riser placement; (b) core placement; (c) padding; and (d) chills may be used to help feed molten metal and eliminate porosity in the isolated hob boss.

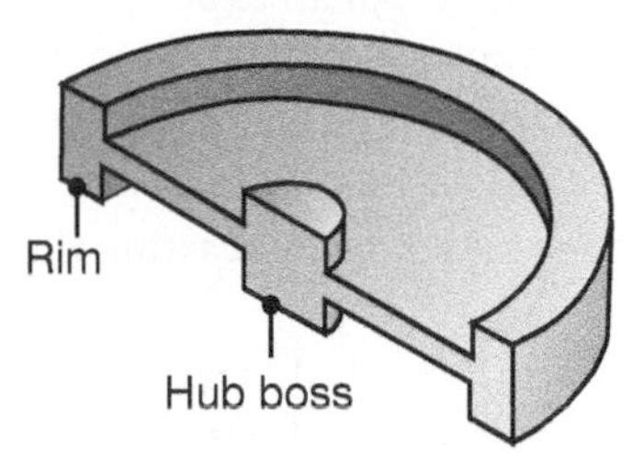

5.98 In the figure below, the original casting design shown in (a) was changed to the design shown in (b). The casting is round, with a vertical axis of symmetry. As a functional part, what advantages do you think the new design has over the old one?

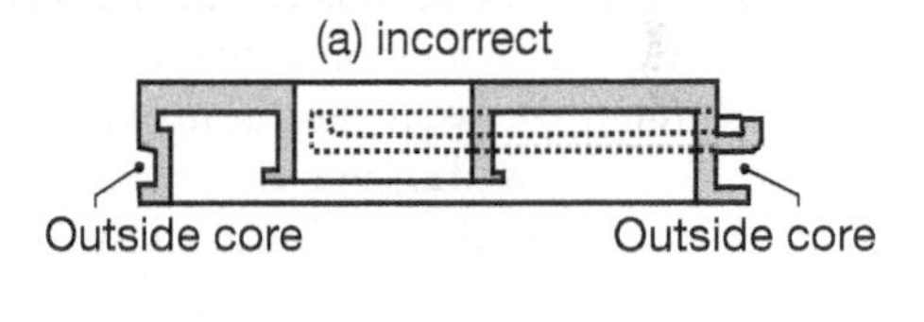

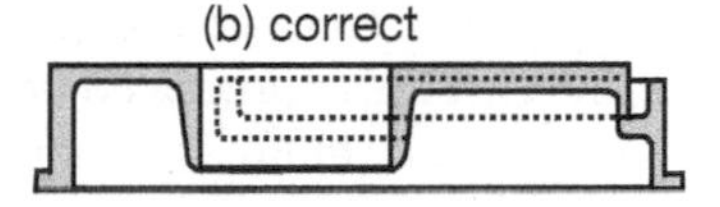

5.99 An incorrect and a correct design for casting are shown, respectively, in the figure below. Review the changes made and comment on their advantages.

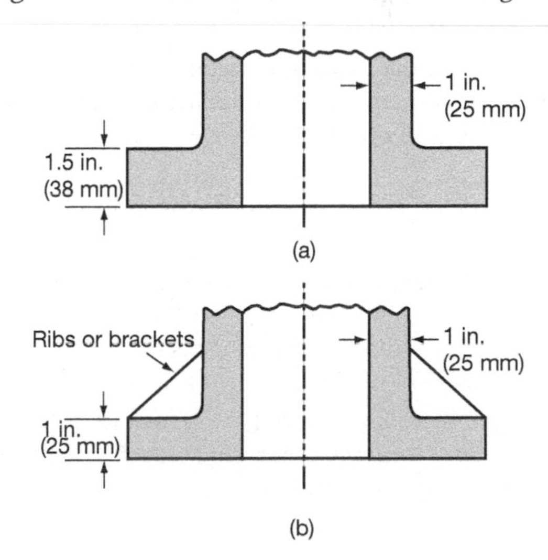

5.100 Write a brief report on the permeability of molds and the techniques that are used to determine permeability.

5.101 Utilizing the equipment and materials available in a typical kitchen, design an experiment to reproduce results similar to those shown in Fig. 5.11.

5.102 Design a test method to measure the permeability of sand for sand casting.

5.103 Reproduce Fig. 5.5 for a casting that is spherical in shape.

5.104 Explain how ribs and serrations are helpful in casting flat surfaces that otherwise may warp. How would you go about designing the geometry (thickness, depth, etc.) of a reinforcing rib?

5.105 A growing trend is the production of patterns and molds through rapid-prototyping approaches, described in Chapter 10. Consider the case of an injection molding operation, where the patterns are produced by rapid prototyping, and then hand assembled onto trees and processed in traditional fashion. What design rules discussed in this chapter would still be valid, and which would not be as important in this case?

5.106 Repeat Problem 5.105 for the case where (a) a pattern for sand casting is produced by rapid prototyping and (b) a sand mold for sand casting is produced.

5.107 It is sometimes desirable to cool metals more slowly than they would be if the molds were maintained at room temperature. List and explain the methods you would use to slow down the cooling process.

5.108 Describe the procedures that would be involved in making a bronze statue. Which casting process or processes would be suitable? Why?

5.109 Outline the casting processes that would be most suitable for making small toys. Explain your choices.

5.110 In casting metal alloys, what would you expect to occur if the mold were agitated (vibrated) aggressively after the molten metal had been in the mold for a sufficient period of time to form a skin?

5.111 Light metals commonly are cast in vulcanized rubber molds. Conduct a literature search and describe the mechanics of this process.

5.112 Three sets of designs for die casting are shown below. Note the changes made to original die design (1) and comment on the reasons.

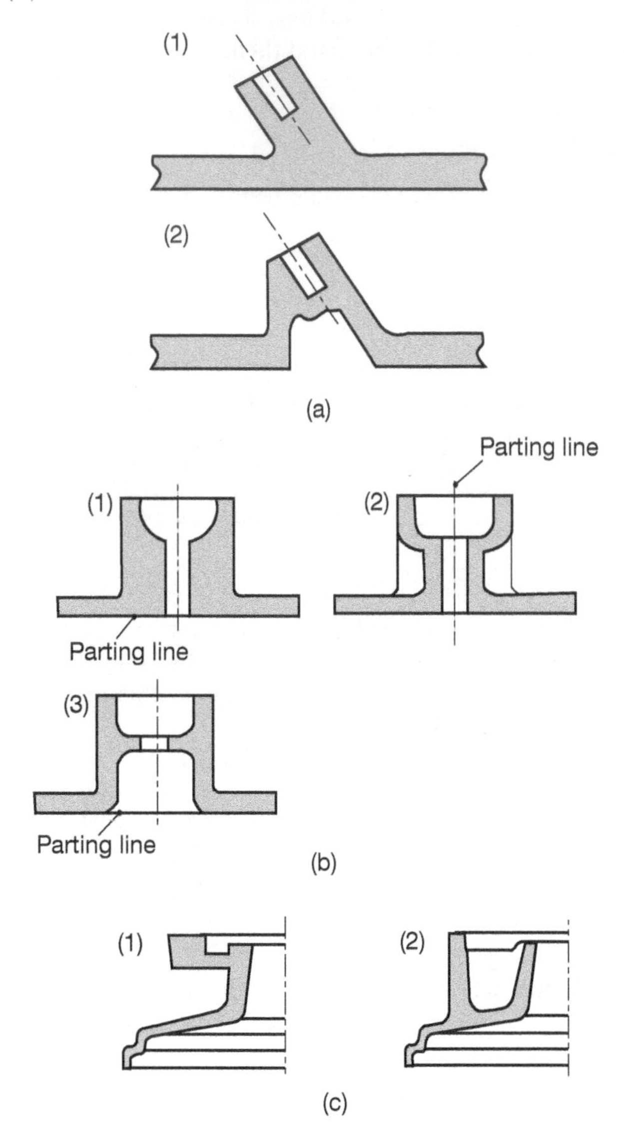

5.113 Note that in cast jewelry, gemstones are usually cast in place; that is, they are not attached after the ring is cast, but are incorporated into the ring. Design a ring with a means of securing a gemstone in the wax pattern, such that it will remain in the mold as the wax is being melted. Could such an approach be used in lost foam casting?

5.114 The following part is to be cast of 10% Sn bronze at the rate of 100 parts per month. To find an appropriate casting process, consider all the processes in this chapter, then reject those that are (a) technically inadmissible; (b) technically feasible but too expensive for the purpose; and (c) identify the most economical one. Write a rationale using common-sense assumptions about cost.

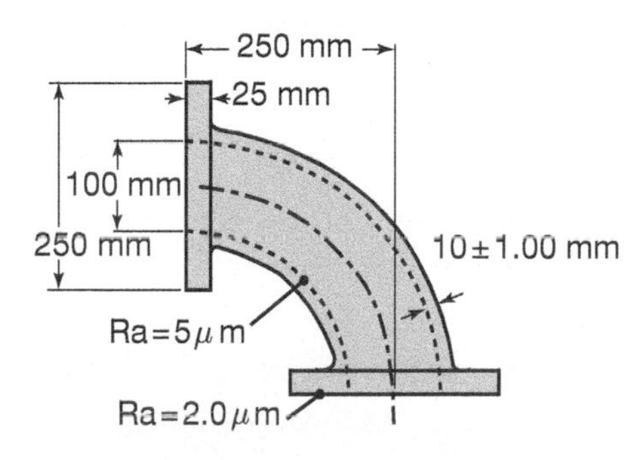

5.115 The two illustrations shown are proposed designs of a gating system for an aluminum low-power water turbine blade. The first uses a conventional sprue-runner-gate system, while the second uses a ceramic filter underneath a pouring cup, but without gates (direct pour method). Evaluate the two designs, and list their advantages and disadvantages. Based on your analysis, select a preferred approach.

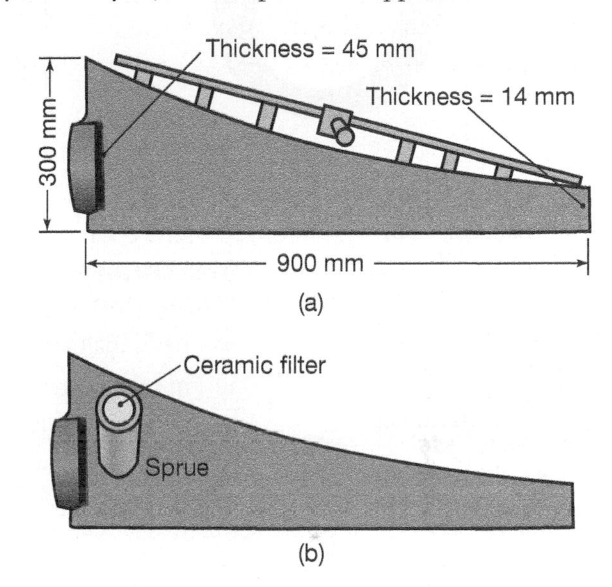

CHAPTER

6 Bulk Deformation Processes

This chapter describes the bulk-deformation processes of forging, rolling, extrusion, drawing, and swaging. It covers:

- The fundamental principles of these processes.
- Important processing parameters such as force and power requirements, temperature, formability, tool and die materials, and metalworking fluids.
- The characteristics of machinery and equipment used.
- Die manufacturing methods and die failures.

Symbols

a	width in plane strain, m
A	area, m^2
C	integration constant
D	diameter, m
e	engineering strain
$\dot{e}$	engineering strain rate, s^{-1}
F	force, N
F'	force per width, N/m
h	thickness, m
H	dimensionless angular position
k	yield strength in shear, N/m^2
K	strength coefficient, N/m^2
K_e	extrusion constant, N/m^2
K_p	force factor in forging
L	length of contact, m
N	angular speed, rpm
n	strain-hardening exponent
p	pressure, N/m^2
p_{av}	average pressure, N/m^2
r	radius, m
R	roll radius, m
R_e	extrusion ratio
R'	flattened roll radius, m
S_y	yield strength, N/m^2
S_y'	yield strength in plane strain, N/m^2
S_{ut}	ultimate tensile strength, N/m^2
T	torque, Nm
u	energy dissipated per volume, N/m^2
v	velocity, m/s
w	width, m
x	Cartesian coordinate, m
y	Cartesian coordinate, m
z	Cartesian coordinate, m
α	bite angle or die angle, deg
ϵ	true strain
$\dot{\epsilon}$	true strain rate, s^{-1}
$\bar{\dot{\epsilon}}$	average true strain rate, s^{-1}

ϕ	wedge angle, deg
Φ	inhomogeneity factor
μ	coefficient of friction
σ	normal stress, N/m^2
σ_d	draw stress, N/m^2
$\overline{\sigma}_f$	average flow stress, N/m^2
τ	shear stress, N/m^2
ω	angular velocity, deg/s

Subscripts

av	average
b	back
f	final, flow, front
max	maximum
n	neutral point
o	original
r	radial

6.1 Introduction

This chapter describes those metalworking processes where the workpiece is subjected to *plastic deformation*, under forces applied through dies and tooling. Deformation processes generally are classified by type of operation as either primary working or secondary working; they are further divided into the three categories of cold (meaning room temperature), warm, and hot working.

Primary-working operations involve taking a solid piece of metal (generally from a cast state) and breaking it down successively into *wrought* materials of various shapes by the basic processes of forging, rolling, extrusion, and drawing. **Secondary-working** operations typically involve further processing of the products from primary working into final or semifinal products, such as bolts, gears, and sheet metal parts.

In **bulk deformation** (Table 6.1), the parts being made have a relatively small surface-area-to-volume (or to thickness) ratio, and the entire part volume is deformed (hence the term *bulk*). Typical examples are hand tools, shafts, and turbine disks. In **sheet forming** (Chapter 7), the surface-area-to-thickness ratio of the products made is much higher, and the deformation can be localized. In sheet forming, thickness variations usually are not desirable, as they can lead to part failure during processing; examples are sheet metal automobile body panels and beverage cans.

The processes covered in this chapter concern only metals; the forming and shaping processes for plastics are described in Chapter 10, and for metal powders, ceramics, glasses, composite materials, and superconductors, in Chapter 11.

QR Code 6.1 What is forging? *Source:* Courtesy of the Forging Industry Association, www.forging.org.

6.2 Forging

Forging denotes a family of processes used to make discrete parts, in which plastic deformation is caused by compressive stresses applied through various dies and tooling. Forging is one of the oldest metalworking operations, dating back to 5000 B.C. (See Table 1.1). Historically, forgings could be produced with a heavy hammer and an anvil, using techniques practiced

TABLE 6.1 General characteristics of bulk deformation processes.

Process	General characteristics
Forging	Production of discrete parts with a set of dies; some finishing operations usually necessary; similar parts can be made by casting or powder-metallurgy techniques; usually performed at elevated temperatures; dies and equipment costs are high; moderate to high labor costs; moderate to high operator skill
Rolling	
Flat	Production of flat plate, sheet, and foil at high speeds, and with good surface finish, especially in cold rolling; requires very high capital investment; low to moderate labor cost
Shape	Production of various structural shapes, such as I-beams and rails, at high speeds; includes thread and ring rolling; requires shaped rolls and expensive equipment; low to moderate labor cost; moderate operator skill
Extrusion	Production of long lengths of solid or hollow products with constant cross sections, usually performed at elevated temperatures; product is then cut to desired lengths; can be competitive with roll forming; cold extrusion has similarities to forging and is used to make discrete products; moderate to high die and equipment cost; low to moderate labor cost; low to moderate operator skill
Drawing	Production of long rod, wire, and tubing, with round or various cross sections; smaller cross sections than extrusions; good surface finish; low to moderate die, equipment and labor costs; low to moderate operator skill
Swaging	Radial forging of discrete or long parts with various internal and external shapes; generally carried out at room temperature; low to moderate operator skill

by blacksmiths for centuries. Modern forgings use automated machinery and high production rates, and include bolts, gears, engine components, turbine disks, and structural components for a wide variety of applications.

Forging can be carried out at room or at elevated temperatures; depending on the temperature, a process can be called *cold*, *warm*, or *hot forging*. The temperature range for these categories is given in Table 3.2 in terms of the *homologous temperature*, T/T_m where T_m is the melting point (absolute scale) of the workpiece material. Note that the homologous recrystallization temperature for metals is about 0.5 (See also Fig. 3.17).

QR Code 6.2 Open-die forging operations. *Source:* Courtesy of the Forging Industry Association, www.forging.org.

6.2.1 Open-Die Forging

Open-die forging typically involves hot working and large deformations using simple tools such as flat dies, rounded sections, punches, or saddles. A common process, known as **upsetting**, involves placing a solid cylindrical workpiece (the blank) between two flat dies (platens) and reducing its height (Fig. 6.1a). Under ideal conditions, a solid cylinder deforms *uniformly*, as shown in Fig. 6.1a, known as **homogeneous deformation**. Because the volume of the cylinder remains constant in plastic deformation (see Section 2.11.5), any reduction in its height causes an increase in diameter. The reduction in height is defined as

$$\text{Reduction in height} = \frac{h_o - h_1}{h_o} \times 100\%. \tag{6.1}$$

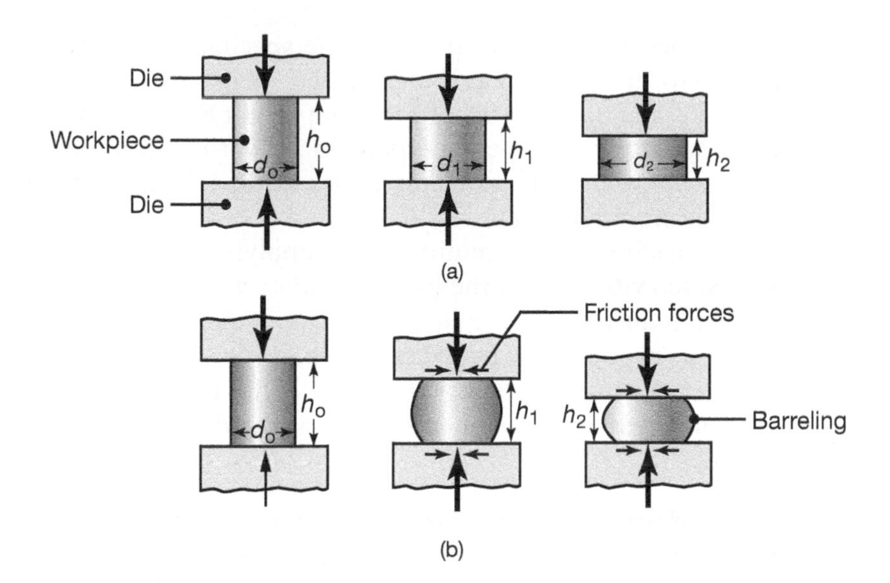

FIGURE 6.1 (a) Ideal deformation of a solid cylindrical specimen compressed between flat frictionless dies (platens), an operation known as upsetting. (b) Deformation in upsetting with friction at the die-workpiece interfaces. Note barrelling of the billet caused by friction.

From Eqs. (2.1) and (2.9) and using absolute values (as is often done for convenience with bulk-deformation processes), the engineering strain is

$$e_1 = \frac{h_o - h_1}{h_o}, \tag{6.2}$$

and the true strain is

$$\epsilon_1 = \ln\left(\frac{h_o}{h_1}\right). \tag{6.3}$$

QR Code 6.3 Animation of open-die forging. *Source:* Courtesy of the Forging Industry Association, www.forging.org.

With a relative velocity v between the platens, Eqs. (2.16) and (2.17) give the strain rate encountered by the specimen as

$$\dot{e}_1 = -\frac{v}{h_o} \tag{6.4}$$

and

$$\dot{\epsilon}_1 = -\frac{v}{h_1}. \tag{6.5}$$

If the specimen is reduced in height from h_o to h_2, the subscript 1 in the foregoing equations is replaced by subscript 2. Positions 1 and 2 in Fig. 6.1a may be regarded as instantaneous positions during an upsetting operation. Note also that the true strain rate, $\dot{\epsilon}$, increases rapidly as the height of the specimen approaches zero.

Barreling. Unlike the specimen shown in Fig. 6.1a, in actual practice the specimen develops a **barrel** shape during upsetting, as shown in Fig. 6.1b and in Fig. 2.12. Barreling is caused primarily by *frictional forces* that oppose the radially outward flow of the material at the die-workpiece interfaces. Barreling can also occur in frictionless upsetting of hot workpieces between cool dies; the reason is that the material at the die-specimen interfaces cools rapidly, whereas the rest of the specimen remains relatively hot. Since the strength of the material decreases with increasing temperature

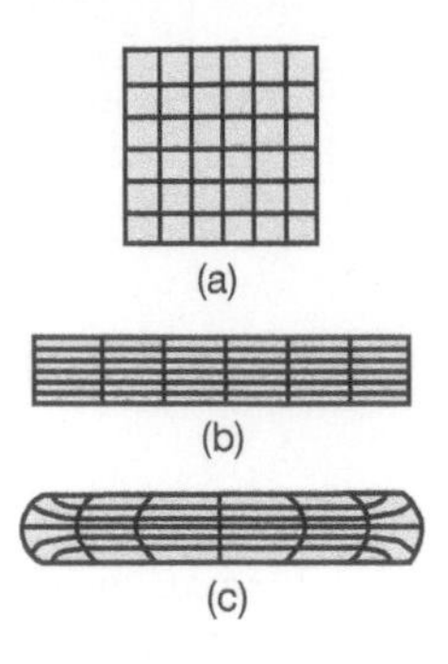

FIGURE 6.2 Schematic illustration of grid deformation in upsetting: (a) original grid pattern; (b) after deformation, without friction; and (c) after deformation, with friction. Such deformation patterns can be used to calculate the strains within a deforming body.

(Section 2.2.6), the upper and lower portions of the specimen exhibit a greater resistance to deformation than at the center.

As shown in Figs. 6.2 and 6.3, a result of barreling is that the deformation throughout the specimen becomes *nonuniform* or *inhomogeneous*. Note the approximately triangular stagnant zones (called *dead-metal zones*) at the top and bottom regions in Fig. 6.3.

Barreling caused by friction can be minimized by applying an effective lubricant or by **ultrasonic vibration** of the platens. In hot-working operations, barreling can be reduced by using heated dies or a thermal barrier at the interfaces, such as produced by a lubricant film.

Double barreling can occur in (a) slender specimens, with high ratios of height to cross-sectional area; and (b) when friction at the die-workpiece interfaces is very high.

Forces and work of deformation under ideal conditions. If friction at the workpiece-die interfaces is zero and the material is perfectly plastic, with a yield strength of S_y, the normal compressive stress on the cylindrical specimen is uniform and at the level of S_y. The force at any height h_1 is

$$F = S_y A_1, \tag{6.6}$$

where A_1 is the cross-sectional area and is obtained from volume constancy:

$$A_1 = \frac{A_o h_o}{h_1}.$$

The ideal work of deformation is the product of the specimen volume and the specific energy [see Eq. (2.59)] and is expressed as

$$\text{Work} = \text{Volume} \int_0^{\epsilon_1} \sigma \, d\epsilon, \tag{6.7}$$

FIGURE 6.3 Grain flow lines in upsetting a solid, cylindrical steel specimen at elevated temperatures between two flat cool dies. Note the highly inhomogeneous deformation and barreling, and the difference in shape of the bottom and top sections of the specimen. The latter results from the hot specimen resting on the lower die before deformation proceeds. The lower portion of the specimen began to cool, thus exhibiting higher strength and hence deforming less than the top surface.

where ϵ_1 is obtained from Eq. (6.3). If the material is strain hardening, with a true stress–true strain curve given by Eq. (2.11), or $\sigma = K\epsilon^n$, then the force at any stage during deformation becomes

$$F = \sigma_f A_1, \tag{6.8}$$

where σ_f is the flow stress of the material (see Section 2.2.3). The expression for the work done is

$$\text{Work} = (\text{Volume})\left(\overline{\sigma}_f\right)(\epsilon_1), \tag{6.9}$$

where $\overline{\sigma}_f$ is the average flow stress. Recall that for a work hardening material, the average flow stress is given by

$$\overline{\sigma}_f = \frac{K}{\epsilon_1}\int_0^{\epsilon_1} \epsilon^n \, d\epsilon = \frac{K\epsilon_1^n}{n+1}. \tag{6.10}$$

6.2.2 Methods of Analysis

There are several methods of analysis to theoretically determine stresses, strains, strain rates, forces, and local temperature rise in deformation processing. The most common approaches are:

Slab method. The *slab method* is one of the earlier and simpler methods of analyzing the stresses and loads in bulk-deformation processes. As described below, this method requires the selection of an element in the workpiece and identification of all the normal and frictional stresses acting on that element.

1. **Slab analysis of plane-strain forging.** Consider the case of simple compression with friction (Fig. 6.4). As the flat dies reduce the part thickness, it expands laterally, causing frictional forces to act in the opposite direction to the motion. The frictional forces are indicated

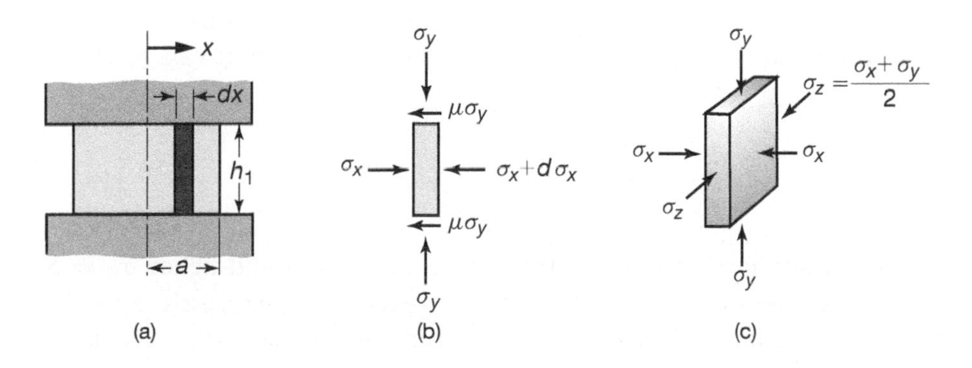

FIGURE 6.4 Stresses on an element in plane-strain compression (forging) between flat dies with friction. The horizontal stress σ_x is assumed to be uniformly distributed along the height, h. Identifying the stresses on an element (slab) is the first step in the slab method of analysis of metalworking processes.

by the horizontal arrows shown in Fig. 6.4. For simplicity, assume that the deformation is in *plane strain*; that is, the workpiece is not free to flow in the z-direction (see also Section 2.11.3). Also, consider a perfectly plastic material, so that the flow stress is the yield stress of the material (see Section 2.2.3).

An element showing all the applied stresses is shown in Fig. 6.4b. Note the difference between the horizontal stresses acting on the side faces, which is due to the frictional stresses on the element. Assume also that the lateral stress distribution, σ_x is uniform along the height, h, of the element.

The next step in this analysis is to balance the horizontal forces, because the element must be in static equilibrium. Assuming unit width,

$$(\sigma_x + d\sigma_x)\, h + 2\mu\sigma_y dx - \sigma_x h = 0,$$

or

$$d\sigma_x + \frac{2\mu\sigma_y}{h} dx = 0.$$

Note that there is only one equation but two unknowns, σ_x and σ_y. The required second equation is obtained from a yield criterion (Section 2.11), as follows: As shown in Fig. 6.4c, this element is subjected to triaxial compression. According to the distortion-energy criterion for plane strain,

$$\sigma_y - \sigma_x = \frac{2}{\sqrt{3}} S_y = S'_y. \tag{6.11}$$

Hence,

$$d\sigma_y = d\sigma_x.$$

Note that it has been assumed that σ_y and σ_x are *principal stresses*, although in the strictest sense, they cannot be principal stresses because a shear stress is also acting on the same plane. However, this assumption is acceptable for low values of the coefficient of friction, μ. Note also that σ_z in Fig. 6.4c is derived in a manner similar to that in Eq. (2.44). There are now two equations that can be solved by noting that

$$\frac{d\sigma_y}{\sigma_y} = -\frac{2\mu}{h} dx, \qquad \text{or} \qquad \sigma_y = Ce^{-2\mu x/h}. \tag{6.12}$$

The boundary conditions are at $x = a$, $\sigma_x = 0$ and hence $\sigma_y = S'_y$ at the edges of the specimen. (All stresses are compressive, so that negative signs for stresses are ignored for convenience.) Thus, the value of C becomes

$$C = S'_y e^{2\mu a/h},$$

and therefore,

$$p = \sigma_y = S'_y e^{2\mu(a-x)/h}, \tag{6.13}$$

and from Eq. (6.11),

$$\sigma_x = \sigma_y - S_y' = S_y'\left[e^{2\mu(a-x)/h} - 1\right]. \tag{6.14}$$

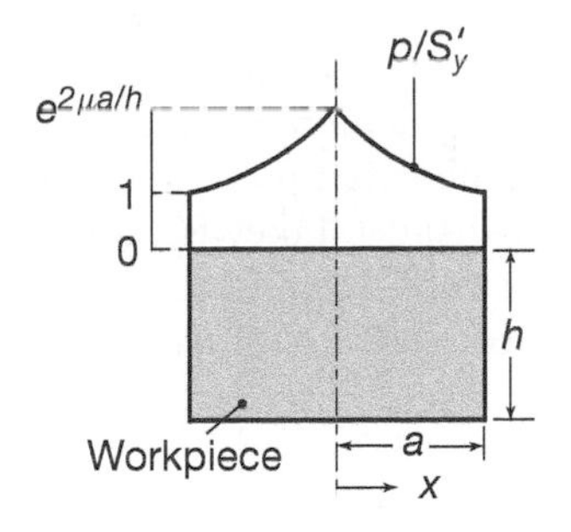

FIGURE 6.5 Distribution of die pressure, in dimensionless form of p/S_y', in plane-strain compression with sliding friction. Note that the pressure at the left and right boundaries is equal to the yield strength of the material in plane strain, S_y'. Sliding friction means that the frictional stress is directly proportional to the normal stress.

Equation (6.13) is plotted qualitatively and in dimensionless form in Fig. 6.5. Note that the pressure increases exponentially toward the center of the part, and that it increases with the a/h ratio and increasing friction. Because of its shape, the pressure-distribution curve in Fig. 6.5 is referred to as the *friction hill*. The pressure with friction is higher than it is without friction, and, with thin workpieces, significantly so. Since the work required to overcome friction must be supplied by the upsetting force, friction is increasingly important with high aspect ratios.

It can be seen that the area under the pressure curve in Fig. 6.5 is the upsetting **force per unit width** of the specimen. An approximate expression for the *average pressure*, p_{av}, is given by

$$p_{av} \simeq S_y'\left(1 + \frac{\mu a}{h}\right). \tag{6.15}$$

Again note the significant influence of the magnitude of a/h and friction on the pressure required, especially at high a/h ratios. Since the *forging force*, F, is the product of the average pressure and the contact area,

$$F = (p_{av})(2aw). \tag{6.16}$$

Note that the expressions for pressure are in terms of an *instantaneous height*, h, hence the force at any h during an upsetting operation must be calculated individually.

A rectangular specimen can be upset without being constrained on its sides (thus it is in a state of *plane stress*; see Fig. 2.31). According to the distortion-energy criterion, the normal stress distribution can then be given qualitatively by the plot in Fig. 6.6. Because the elements at the four corners are in uniaxial compression, the pressure there is S_y, and the friction hill along the edges of the specimen is as shown.

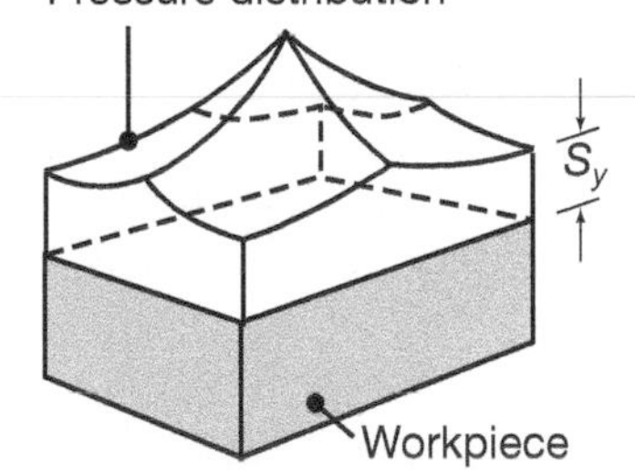

FIGURE 6.6 Die pressure distribution in compressing a rectangular workpiece with sliding friction and under conditions of plane stress, using the *distortion-energy criterion*. Note that the stress at the corners is equal to the uniaxial yield strength, S_y, of the material.

Figure 6.7 shows the lateral expansion (top view) of the edges of a rectangular specimen in actual plane-stress upsetting. Some typical results indicate that for an increase in specimen length of 40%, the increase in width (narrow dimension) is 230%. The reason for the significantly larger increase in width is because the material flows the most in the direction of least resistance (thus minimizing the energy dissipated). Because of its smaller magnitude, the width has less cumulative frictional resistance than does the length. Likewise, after upsetting, a specimen in the shape of a cube eventually acquires a pancake shape (with a barreled periphery), because the diagonal direction expands at a slower rate than in the other directions.

2. **Forging of a solid cylindrical workpiece.** The pressure distribution in forging of a solid cylindrical specimen (Fig. 6.8) can be determined

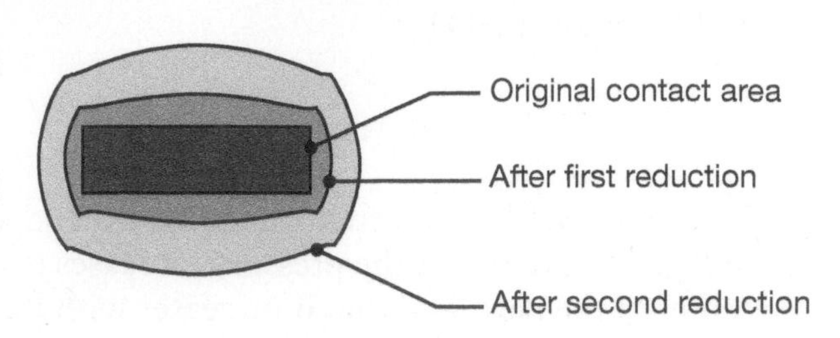

FIGURE 6.7 Increase in die-workpiece contact area of an originally rectangular specimen (viewed from the top) compressed between flat dies and with friction. Note that the length of the specimen (horizontal dimension) has increased proportionately less than its width (vertical dimension). Likewise, a specimen originally in the shape of a cube acquires the shape of a pancake after deformation with friction.

using the slab method of analysis. The approach is: (a) isolate a segment of angle $d\theta$ in a solid cylinder of radius r and height h; (b) take a small element of radial length dx; and (c) place on this element all the normal and frictional stresses acting on it. Following an approach similar to the analysis of plane-strain forging, an expression for the pressure, p, at any radius x can be obtained as

$$p = S_y e^{2\mu(r-x)/h}. \tag{6.17}$$

The average pressure, p_{av}, is given approximately as

$$p_{av} \simeq S_y\left(1+\frac{2\mu r}{3h}\right), \tag{6.18}$$

and thus the forging force is

$$F = (p_{av})\left(\pi r^2\right). \tag{6.19}$$

For strain-hardening materials, S_y in Eqs. (6.17) and (6.18) must be replaced by the flow stress, σ_f. The coefficient of friction, μ, in Eqs. (6.17) and (6.18) can be estimated as ranging from 0.05 to 0.1 for cold forging, and 0.1 to 0.2 for hot forging, depending on the effectiveness of the lubricant; it can be higher than these values, especially in regions where the workpiece surface is devoid of lubricant (see also Section 4.4.3).

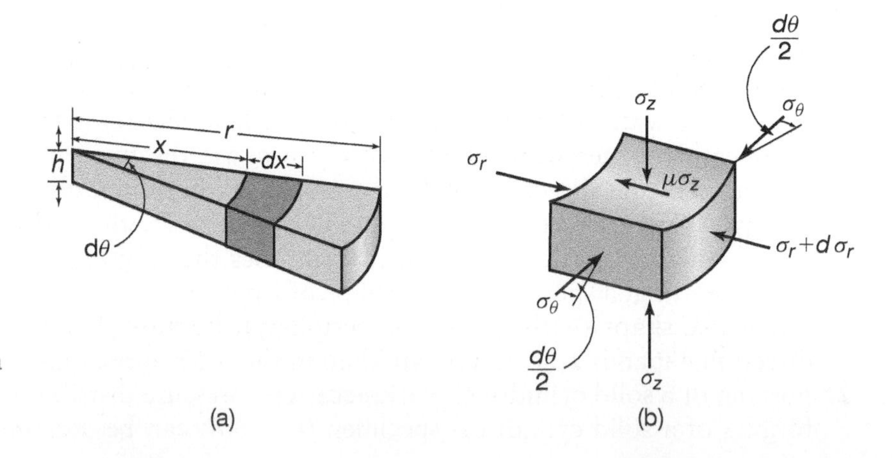

FIGURE 6.8 Stresses on an element in forging of a solid cylindrical workpiece between flat dies and with friction. Compare this figure and the stresses involved with Fig. 6.4.

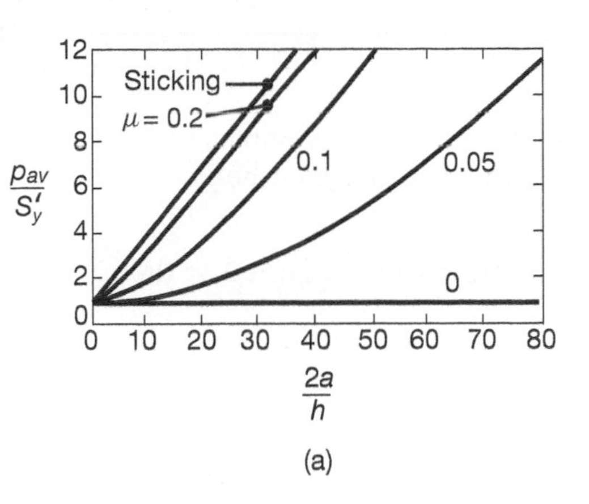

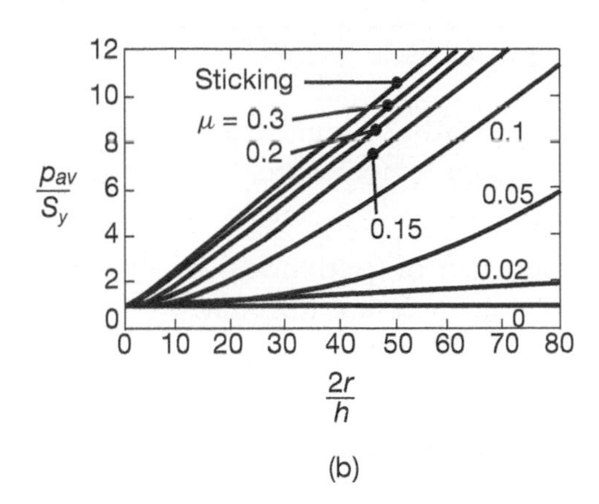

FIGURE 6.9 Ratio of average die pressure to yield strength as a function of friction and aspect ratio of the specimen: (a) plane-strain compression; and (b) compression of a solid cylindrical specimen. Note that the yield strength in (b) is S_y, and not S'_y as it is in the plane-strain compression shown in (a).

The effects of friction and the aspect ratio of the specimen (that is, a/h or r/h) on the average pressure, p_{av}, in upsetting are illustrated in Fig. 6.9. The pressure is given in a dimensionless form and hence can be regarded as a pressure-multiplying factor.

3. **Forging under sticking condition.** Note that the product of μ and p is the frictional stress (surface shear stress) acting at the workpiece-die interface at any location x from the center. As p increases toward the center, μp also increases; however, the value of μp cannot be higher than the shear yield strength, k, of the material. The condition when $\mu p = k$ is known as *sticking*. Note that in plane strain, the value of k is $S'_y/2$ (see Section 2.11.3). Sticking does not necessarily mean adhesion at the interface, but it reflects the fact that the material does not move relative to the platen surfaces.

For the sticking condition, the normal stress distribution in plane strain can be shown to be

$$p = S'_y \left(1 + \frac{a - x}{h}\right). \tag{6.20}$$

Note that the pressure varies linearly with x, as also shown in Fig. 6.10. The normal stress distribution for a cylindrical specimen under sticking condition can be shown to be

$$p = S_y \left(1 + \frac{r_o - r}{h}\right). \tag{6.21}$$

Again, the stress distribution is linear, as shown in Fig. 6.10 for plane strain.

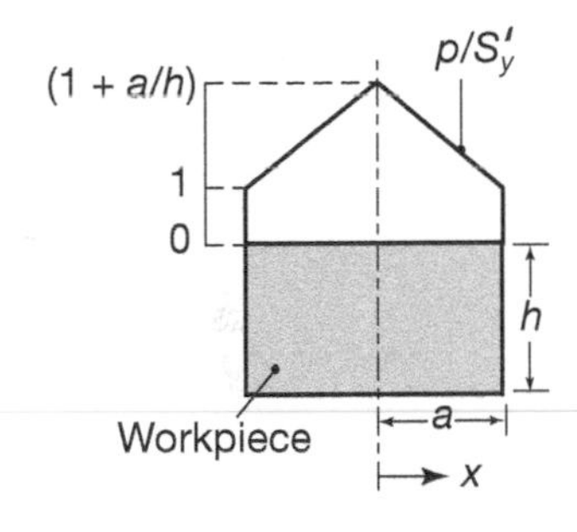

FIGURE 6.10 Distribution of dimensionless die pressure, p/S'_y, in compressing a rectangular specimen in plane strain and under sticking conditions. Note that the pressure at the edges is the uniaxial yield strength of the material in plane strain, S'_y.

EXAMPLE 6.1 Upsetting Force

Given: A cylindrical specimen made of annealed 4135 steel has a diameter of 150 mm and is 100 mm high. It is upset, at room temperature, by open-die forging with flat dies to a height of 50 mm.

Find: Assuming that the coefficient of friction is 0.2, calculate the upsetting force required at the end of the stroke. Use the average-pressure formula.

Solution: The average-pressure formula, from Eq. (6.18), is given by

$$p_{av} \simeq \sigma_f \left(1 + \frac{2\mu r}{3h}\right),$$

where S_y is replaced by σ_f because the workpiece material is strain hardening. From Table 2.2, $K = 1015$ MPa and $n = 0.17$. The absolute value of the true strain is

$$\epsilon_1 = \ln\left(\frac{100}{50}\right) = 0.693,$$

and therefore,

$$\sigma_f = K\epsilon_1^n = (1015)(0.693)^{0.17} = 953.6 \text{ MPa}.$$

The final height of the specimen, h_1, is 50 mm. The radius r at the end of the stroke is found from volume constancy:

$$\left(\frac{\pi 150^2}{4}\right) 100 = \left(\pi r_1^2\right)(50) \quad \text{and} \quad r_1 = 106.1 \text{ mm}.$$

Thus,

$$p_{av} \simeq 953.6\left[1 + \frac{(2)(0.2)(106.1)}{(3)(50)}\right] = 1223 \text{ MPa}.$$

The upsetting force is

$$F = \pi(1223)(0.1061)^2 = 43.25 \text{ MN}.$$

Note that Fig. 6.9b can also be used to solve this problem.

EXAMPLE 6.2 Transition from Sliding to Sticking Friction

Given: In plane-strain upsetting, the frictional stress cannot be higher than the shear yield strength, k, of the workpiece material. Thus, there may be a distance x in Fig. 6.4 where a transition occurs from sliding to sticking friction.

Find: Derive an expression for x in terms of a, h, and μ only.

Solution: The shear stress at the interface due to friction can be expressed as

$$\tau = \mu p.$$

However, the shear stress cannot exceed the yield shear stress, k, of the material, which, for plane strain, is $S_y'/2$. The pressure curve in Fig. 6.5 is given by Eq. (6.13); thus, in the limit,

$$\mu S_y' e^{2\mu(a-x)/h} = S_y'/2,$$

or

$$2\mu \frac{(a-x)}{h} = \ln\left(\frac{1}{2\mu}\right).$$

Hence,

$$x = a - \left(\frac{h}{2\mu}\right) \ln\left(\frac{1}{2\mu}\right).$$

Note that, as expected, the magnitude of x decreases as μ decreases. However, the pressures must be sufficiently high to cause sticking, that is, the a/h ratio must be high. For example, let $a = 10$ mm and $h = 1$ mm. Then, for $\mu = 0.2$, $x = 7.71$ mm, and for $\mu = 0.4$, $x = 9.72$ mm.

Finite-element method. In the *finite-element method*, the deformation zone in an elastic-plastic body is divided into a number of elements, interconnected at a finite number of nodal points. Next, a set of simultaneous equations is developed that represents unknown stresses and deformation increments subject to the boundary conditions. From the solution of these equations, actual velocity distributions and the stresses at the workpiece-die interfaces are calculated. Application of this technique requires such inputs as the stress–strain characteristics of the material as a function of strain rate and temperature, and frictional and heat-transfer characteristics of the die and the workpiece.

The finite-element method can be applied to complex workpiece shapes in bulk-deformation and in sheet metal forming. Its accuracy depends on the number, shape, and mathematical complexity of the finite elements, the deformation increment, and the methods of calculation. It gives a detailed picture of the actual stresses and strain distributions throughout the workpiece.

The results of applying the finite-element method of analysis on a solid cylindrical workpiece in impression-die forging are shown in Fig. 6.11. The grid pattern is known as a *mesh*; the distortion of the grid pattern can be numerically predicted from theory.

The finite-element method is also capable of determining the temperature distribution throughout the workpiece, and predicting microstructural changes in the material during hot-working operations and the onset of defects, without the need for experimental data. Modern finite element codes are extremely powerful, with automatic meshing and time step control to ensure stable and accurate solutions. Elaborate constitutive models can be used, including temperature and strain rate effects; fracture predictions are also included. This information is then used to verify and improve die design, for better part quality and robustness.

Deformation-zone geometry. The observations made concerning Fig. 2.23 are helpful in calculating forces in forging, as well as other bulk-deformation processes. As described for hardness testing, the compressive

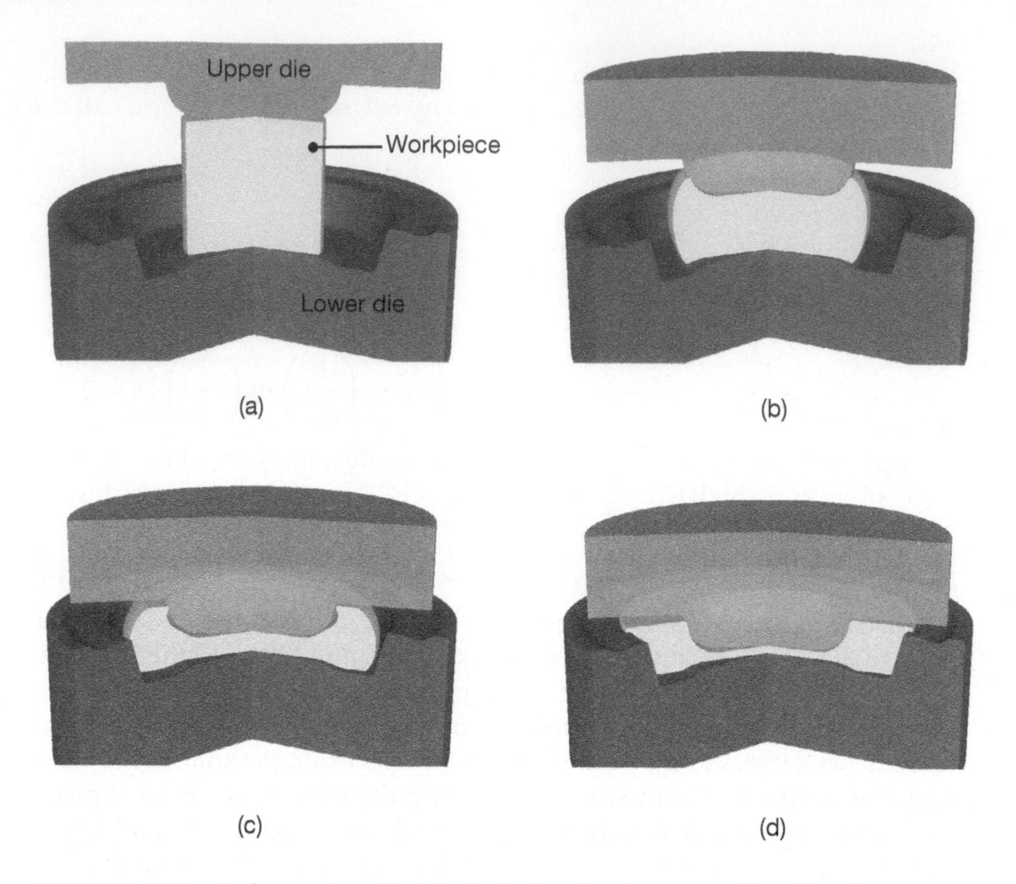

FIGURE 6.11 Deformation of a blank during forging as predicted by the software program DEFORM based on the finite-element method of analysis. *Source:* Courtesy Scientific Forming Technologies Corporation.

stress required for indentation is, theoretically, about three times the yield strength, S_y, required for uniaxial compression (see Section 2.6.8). Recall also that (a) the deformation under the indenter is localized, making the overall deformation highly nonuniform (See Fig. 2.23); and (b) the deformation zone is relatively small compared to the much larger size of the specimen. In contrast, in a simple frictionless compression test, with flat overhanging dies, the top and bottom surfaces of the specimen are always in contact with the dies while the whole specimen undergoes deformation.

6.2.3 Types of Forging

Impression-die forging. In *impression-die forging*, the workpiece acquires the shape of the die cavity (hence the term *impression*) while it is being deformed between the two closing dies. A typical example of such an operation is shown in Fig. 6.12; it can be noted that some of the material flows radially outward, forming a **flash**. Because of its high length-to-thickness ratio (equivalent to a high a/h ratio), the flash is subjected to high pressure, indicating the presence of high frictional resistance encountered as the material flows in the radially outward direction in the flash gap. The flash gap is an important parameter in forging, because high

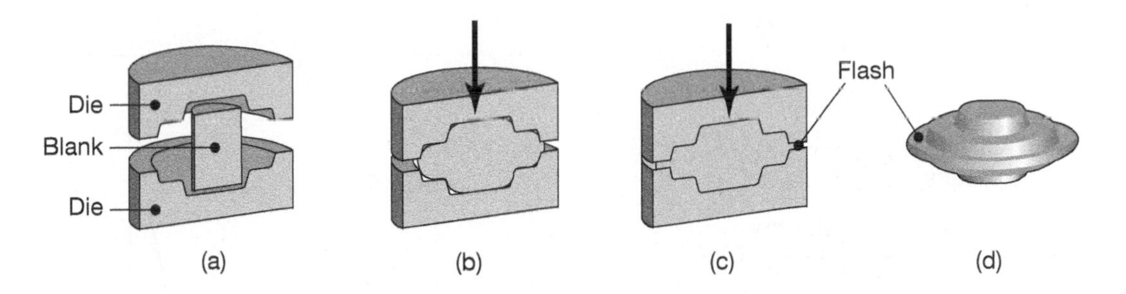

FIGURE 6.12 Schematic illustrations of stages in impression-die forging. Note the formation of a flash, or excess material that subsequently has to be trimmed off.

pressure developed there encourages the filling of the die cavities. Furthermore, if the forging operation is carried out at elevated temperatures (*hot forging*), the flash cools faster than does the bulk of the workpiece (because of the high surface-area-to-thickness ratio of the flash gap). As a result, the cooler flash resists deformation to a greater extent than the bulk, and thus helps filling of the die cavities.

QR Code 6.4 Impression-die forging.
Source: Courtesy of the Forging Industry Association, www.forging.org.

The quality, dimensional tolerances, and surface finish of a forging depend on how well these operations are controlled and performed. Factors that contribute to dimensional inaccuracies are draft angles, radii, fillets, die wear, die closure, and mismatching of the dies. Tolerances generally range between ±0.5% and ±1% of the part dimensions; in hot forging of steel they are typically less than ±6 mm, and in *precision forging* (see below) they can be as low as ±0.25 mm. Surface finish of the forging depends on blank preparation, die surface finish, die wear, and the effectiveness of the lubricant.

Forces in impression-die forging can be difficult to predict, because of the generally complex shapes involved and the fact that each location within the workpiece is typically subjected to different strains, strain rates, and temperatures, as well as variations in coefficient of friction along the die-workpiece contact area. Certain empirical pressure-multiplying factors, K_p, have been developed for use with the expression

$$F = K_p \sigma_f A, \tag{6.22}$$

where F is the forging force, A is the projected area of the forging (including the flash), σ_f is the flow stress of the material at the strain, the strain rate, and temperature to which the material is subjected. Typical values of K_p are given in Table 6.2.

TABLE 6.2 Range of K_p values in Eq. (6.22) for impression-die forging.

Description	K_p
Simple shapes, without flash	3–5
Simple shapes, with flash	5–8
Complex shapes, with flash	8–12

A typical impression-die forging force as a function of the die stroke is shown in Fig. 6.13. Note that, for this axisymmetric part, the force first increases gradually as the cavity is being filled (Fig. 6.13b), and then it increases rapidly as the flash begins to form. As the dies must close further to properly shape the part, an even steeper rise in the force takes place. Also note that the flash has a finite contact length, called **land** (See Fig. 6.24). The land is to ensure that the flash generates sufficient resistance for the

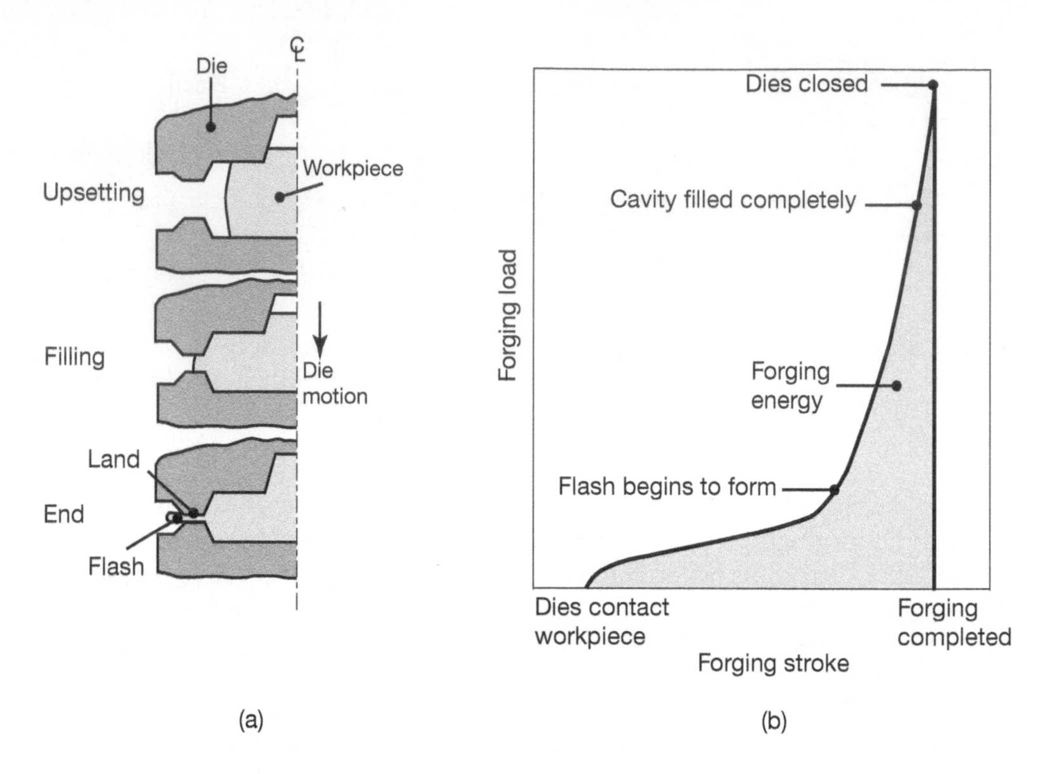

FIGURE 6.13 Typical load-stroke curve for impression-die forging. Note the sharp increase in load when the flash begins to form. *Source:* After T. Altan.

outward flow of the material to aid in die filling, without contributing excessively to the forging load.

Precision forging. In precision forging, special dies are made to a higher accuracy than in typical impression-die forging. This operation requires higher capacity forging equipment than other forging processes, because of the higher stresses required to precisely form the part. Precision forging, as well as similar operations where the part formed is close to the final desired dimensions, are known as **near-net-shape** processes (see Section 1.6). Additional examples are closed-die forging and coining (see below) and powder metallurgy operations of pressing and sintering (Sections 11.3 and 11.4) and metal injection molding (Section 11.3.4).

Aluminum and magnesium alloys are particularly suitable for precision forging, because the forging loads and temperatures required are relatively low, die wear is low, and the forgings have good surface finish. Steels and other alloys are more difficult to precision forge. The choice between conventional forging and precision forging requires an economic analysis, balancing the cost of required special dies against the cost savings associated with reduced machining because the part is closer to or at the specified final shape.

Closed-die forging. In closed-die forging, no flash is formed and the workpiece is completely surrounded by the dies. Proper control of the material volume is essential in order to produce a forging of desired shape

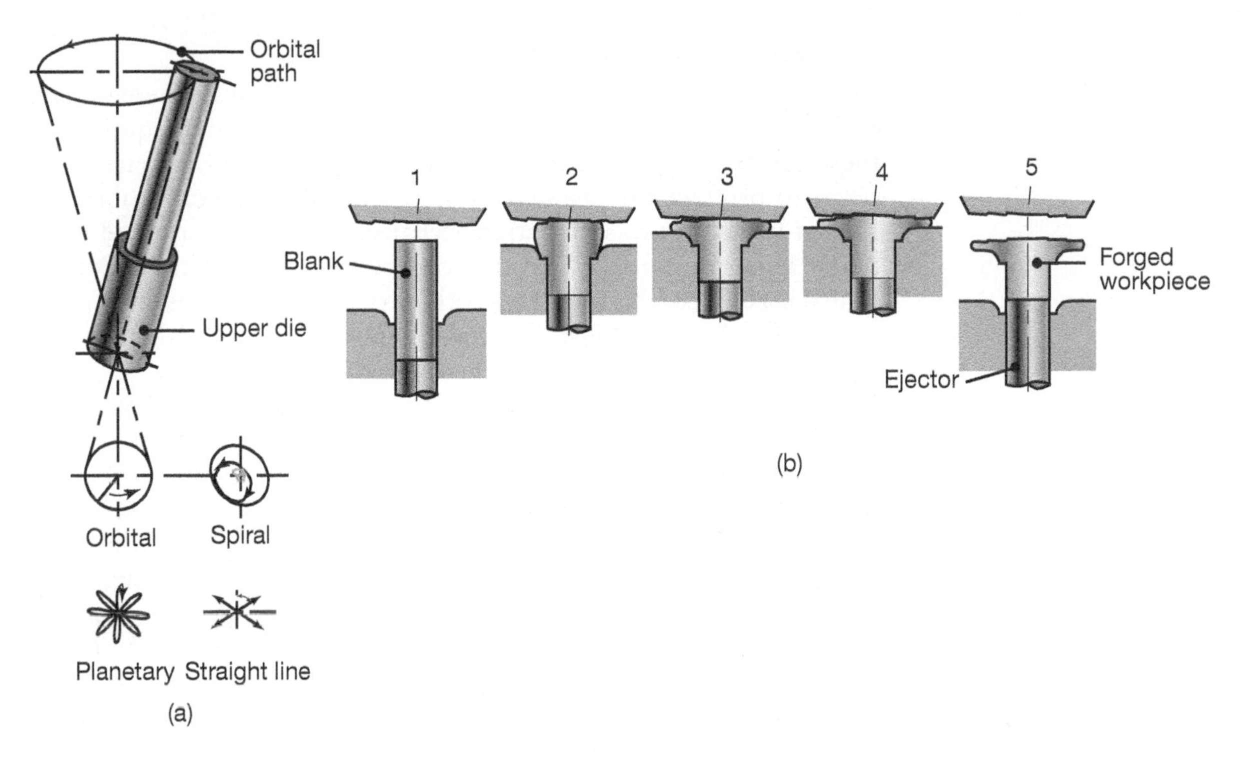

FIGURE 6.14 Schematic illustration of the orbital-forging process. Note that the die is in contact with only a portion of the workpiece surface at a time. Also called *rotary forging, swing forging,* and *rocking-die forging*, this process can be used for forming individual parts such as bevel gears, wheels, and bearing rings.

and dimensions. Undersized blanks will prevent the complete filling of the die cavity, and oversized blanks may cause premature die failure or jamming.

Isothermal forging. Also known as **hot-die forging**, the dies in this operation are heated to the same temperature as that of the hot blank. In this way, cooling of the workpiece is eliminated, the low flow stress of the material is maintained during forging, and material flows easier within the die cavities. The dies are generally made of nickel alloys, and complex parts with good dimensional accuracy can be forged in one stroke, typically in hydraulic presses. Although expensive, isothermal forging can be economical for intricate forgings of costly materials, provided that the quantity required is sufficiently high to justify die costs.

Incremental forging. In this process, the blank is forged into a final shape in several small steps (hence the term incremental), similar to the *cogging* operation (Fig. 6.17). Consequently, incremental forging requires much lower forces compared to conventional forging. **Orbital forging** is an example of incremental forging; the die moves along an orbital path (Fig. 6.14) and forms the part in individual steps. The forces are lower, the operation is quieter, and a family of similar parts can be forged using the same dies.

6.2.4 Miscellaneous Forging Operations

1. **Coining.** An example of coining is the *minting of coins*, where the slug is shaped in a completely closed die cavity. The pressure required can therefore be as high as 5 to 6 times the flow stress of the material, in order to produce the fine details of a coin or medallion. Lubricants cannot be used in this operation, because they can be trapped in die cavities and prevent reproduction of fine die-surface details. Coining is also used in combination with other forging operations to improve surface finish and achieve the desired dimensional accuracy (**sizing**) of the products.
2. **Heading.** This process is basically an upsetting operation, performed at the end of a rod to produce a shape with a larger cross section; common examples include the heads of bolts, screws, and nails (Fig. 6.15). Heading is done on machines called *headers*, which are typically highly-automated horizontal machines with high production rates. An important consideration in heading is the tendency for the bar to *buckle* if its unsupported length-to-diameter ratio is too high. This ratio usually is limited to less than 3:1, but with appropriate dies, it can be higher. For example, higher ratios can be accommodated if the diameter of the die cavity is not more than 1.5 times the bar diameter.
3. **Piercing.** In this process, a punch with a specific shape indents the workpiece and produces a shaped cavity or an impression (Fig. 6.16). The piercing force depends on the cross-sectional area of the punch and its tip geometry, the flow stress of the material, and the friction at the interfaces. Punch pressures may be 3 to 5 times the flow stress of the material. The term *piercing* is also used to describe the process of punching of holes with a punch and die, as shown in Fig. 7.26c.
4. **Hubbing.** In *hubbing*, a hardened punch, with a particular tip geometry, is pressed into a metal workpiece to produce a cavity that is shallower than that produced by piercing. The pressure required to produce a cavity is approximately three times the ultimate tensile strength, S_{ut}, of the material, thus the hubbing force required is

$$\text{Hubbing force} = 3S_{ut}A, \qquad (6.23)$$

QR Code 6.5 Ring preforming presses. *Source:* Courtesy of Erie Press.

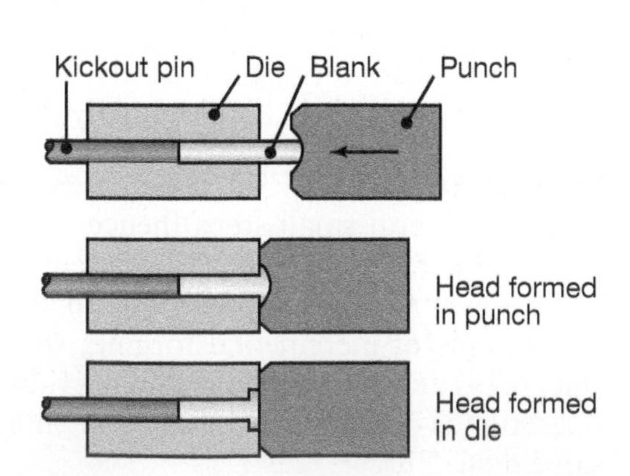

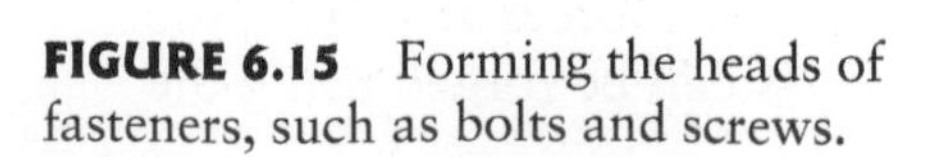

FIGURE 6.15 Forming the heads of fasteners, such as bolts and screws.

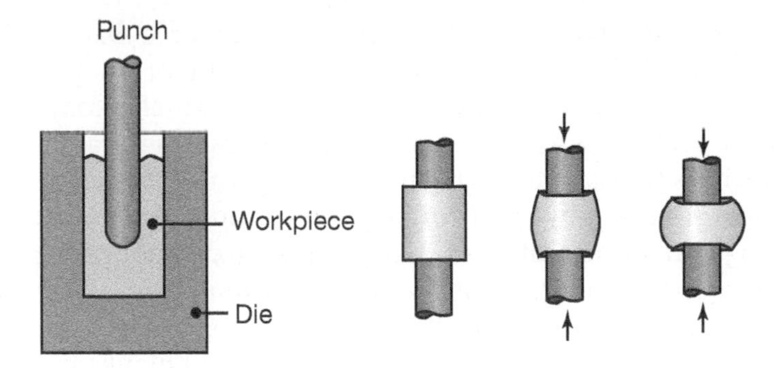

FIGURE 6.16 Examples of piercing operations.

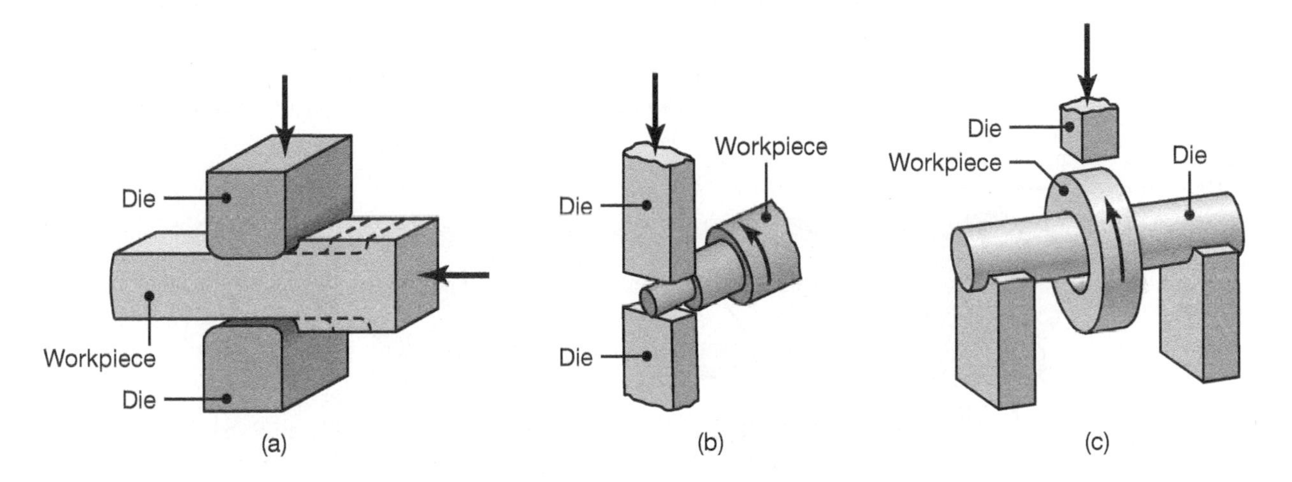

FIGURE 6.17 (a) Schematic illustration of a cogging operation on a rectangular bar. Blacksmiths use a similar procedure to reduce the thickness of parts in small increments by heating the workpiece and hammering it numerous times along the length of the part. (b) Reducing the diameter of a bar by open-die forging; note the movements of the die and the workpiece. (c) The thickness of a ring being reduced by open-die forging.

where A is the projected area of the impression. Note that the factor 3 in this formula is in agreement with the observations made with regard to hardness of materials, as described in Section 2.6.8.

5. **Cogging.** In this operation, also called *drawing out*, the thickness of a bar is reduced by successive steps and at certain intervals (Fig. 6.17). The process is an example of *incremental forming*, in which a long section of a bar, for example, can be reduced in thickness without requiring large dies and forces. The thickness of rings and other parts also can be reduced in a similar operation, as illustrated in Figs. 6.17b and c.
6. **Fullering and edging.** These operations are performed, usually on bar stock, to distribute the material in specific regions prior to forging. In *edging* (See Fig. 6.23) material is gathered into a localized area, whereas in *fullering*, it is distributed away from an area. These operations are thus a method of *preforming* to make material flow easier in forging in die cavities.

7. **Roll forging.** In *roll forging*, the cross-sectional area of a bar is reduced or altered in shape by passing it through a pair or sets of grooved rolls of various shapes (Fig. 6.18). This operation also may be used to produce such parts as tapered shafts and leaf springs, table knives, and numerous tools. Roll forging is also used as a preliminary forming operation, followed by other forging and forming processes, such as in making crankshafts and various automotive components.
8. **Skew rolling.** A process similar to roll forging is *skew rolling*, typically used for making ball bearings. As illustrated in Fig. 6.19a, round wire or rod stock is fed into the roll gap, forming spherical blanks by the continuously rotating rolls. The balls are subsequently ground and polished using special machinery (see Section 9.6). Ball bearings may also be made by first cutting short pieces from a round bar, then upsetting them between a pair of dies, as shown in Fig. 6.19b.

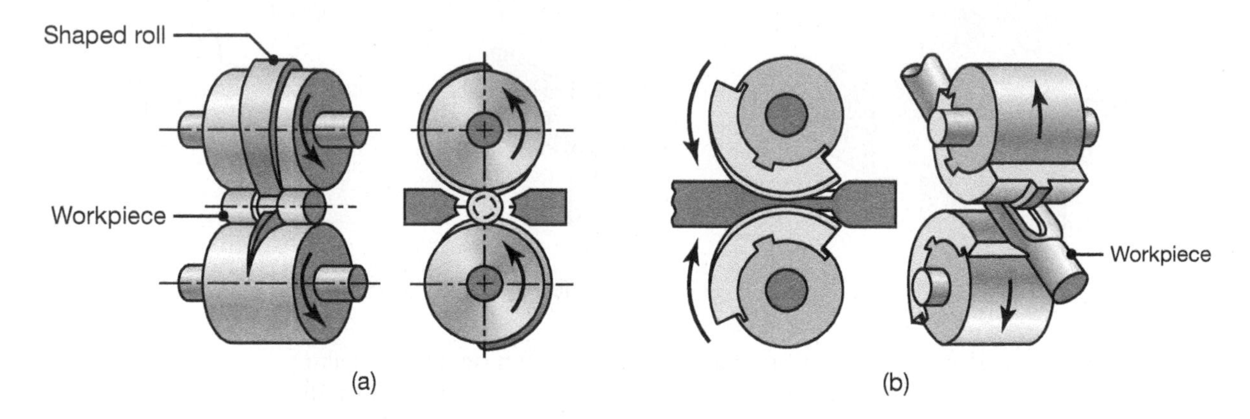

FIGURE 6.18 Two illustrations of roll forging (*cross-rolling*) operations. Tapered leaf springs and knives can be made by this process using specially designed rolls.

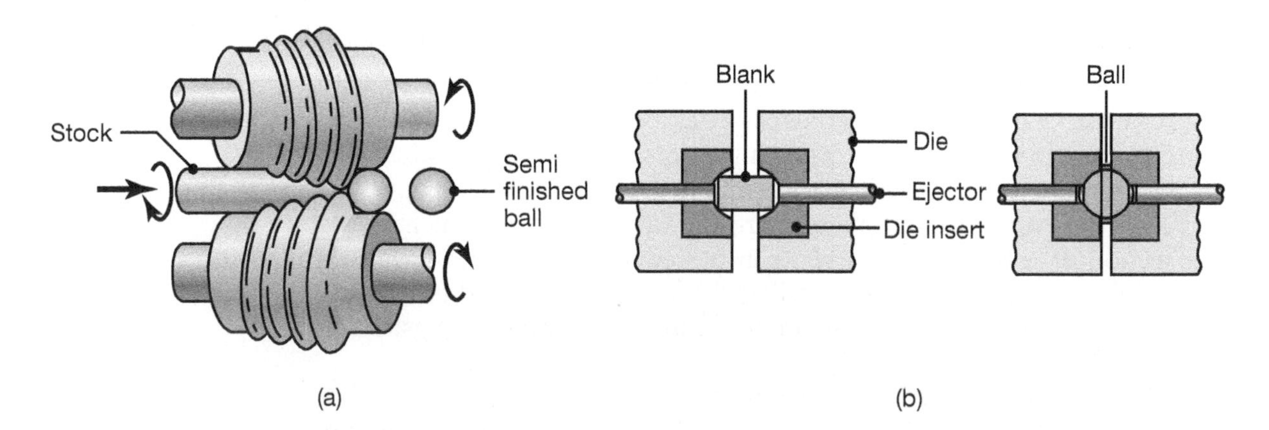

FIGURE 6.19 (a) Production of steel balls for bearings by skew rolling. (b) Production of steel balls by upsetting of a short cylindrical blank; note the formation of flash. The balls are subsequently ground and polished to be used as ball bearings and similar components.

6.2.5 Forgeability

The forgeability of a metal can be defined as its capability to be shaped without cracking or requiring high forces. Recall that impurities in the metal or small changes in the composition of the metal can have a significant effect on the ductility and thus the forgeability of a metal. Various tests have been developed over the years to measure forgeability; two commonly used methods are:

1. **Upsetting test.** A solid cylindrical specimen is upset between two flat dies, and observations are made regarding any cracking on the barreled surfaces (See Fig. 3.21d). The higher the reduction in height prior to cracking, the greater the forgeability of the metal. The cracks on the barreled surface are caused by *secondary tensile stresses*, so called because no external tensile stress is directly applied to the material. Friction at the die-workpiece interfaces has a marked effect on cracking; as friction increases, the specimen cracks at a lower reduction in height.

   It has been observed that the crack may be longitudinal or at a 45° angle, depending on the sign of the axial stress σ_z on the barreled surface (See Fig. 6.8b). If the stress is positive (tensile), the crack is longitudinal; if it is negative (compressive), the crack is at 45°. Surface defects also will adversely affect the results by causing premature cracking. A typical surface defect is a **seam**, which can be a longitudinal scratch, a string of inclusions, or a fold from prior working of the material.

   Upsetting tests can be performed at various temperatures and strain rates. Such tests only serve as guidelines, as in actual forging operations the metal is subjected to a very different state of stress than in simple upsetting.
2. **Hot-twist test.** This is a torsion test (Section 2.4), in which a long, solid round specimen is twisted until it fails. The test is performed at various temperatures and the number of turns that each specimen undergoes before failure is recorded. The hot-twist test is particularly useful in determining the forgeability of steels.

Effect of hydrostatic pressure on forgeability. As described in Section 2.2.8, *hydrostatic pressure* has a significant beneficial effect on the ductility of metals and nonmetallic materials. Experiments have indicated that cracking takes place at higher strain levels if the tests are carried out in an environment of high hydrostatic pressure. In order to take advantage of this phenomenon, special techniques have been developed to forge metals in a high compressive environment; the pressure-transmitting medium is usually a low-strength ductile metal (see also *hydrostatic extrusion*, Section 6.4.3).

Electrically assisted forging. A recent development has been the application of high current during a forging operation in order to increase forgeability, as described in Section 7.5.8. A number of mechanisms through which electrical current increases formability have been suggested, and this may be a thermal effect, but it has been applied to forgings and related bar shearing operations when ductility is limited.

Forgeability of various metals. Based on the results of various tests and observations made during actual forging operations, the forgeability of various metals and alloys has been determined. In general, (a) aluminum, magnesium, copper and their alloys, carbon and low-alloy steels have good forgeability; and (b) high-temperature materials such as superalloys, tantalum, molybdenum, and tungsten and their alloys have poor forgeability.

6.2.6 Forging Defects

In addition to surface cracking (See Fig. 3.21d), several other defects in forging originate in the die cavity.

1. As shown in Fig. 6.20, excess material in the web of a forging can buckle during forging and develop laps.
2. If the web is thick, the excess material flows past the already forged portions and develops internal cracks (Fig. 6.21). These two examples indicate the importance of properly controlling the volume of the blank and its flow in the die cavity.
3. The die radii also can significantly affect the formation of defects. Note in Fig. 6.22, for example, that the material flows more freely

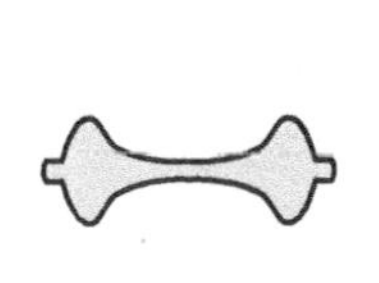

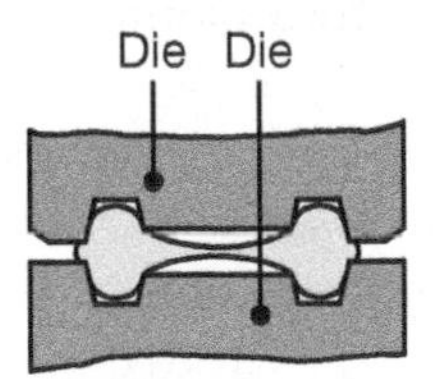

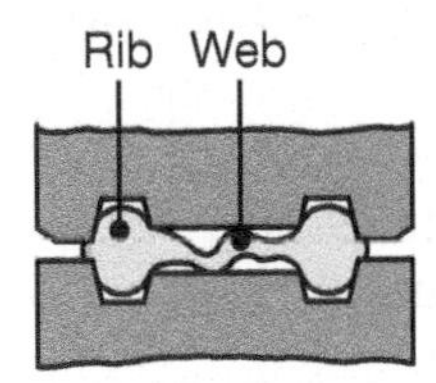

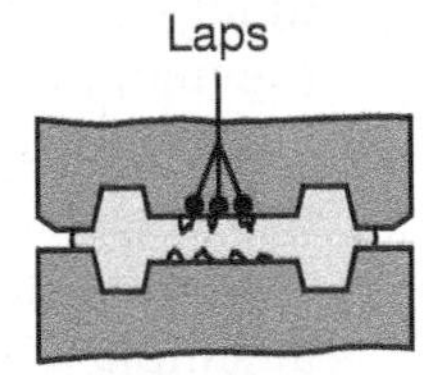

FIGURE 6.20 Stages in lap formation in a part during forging, due to buckling of the web. Web thickness should be increased to avoid this problem.

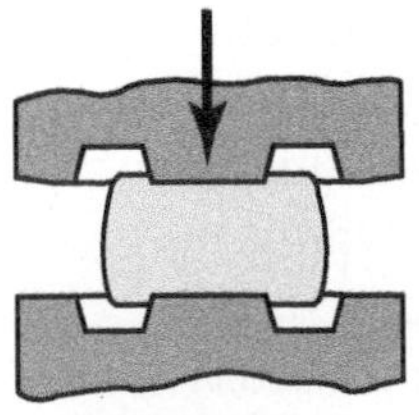

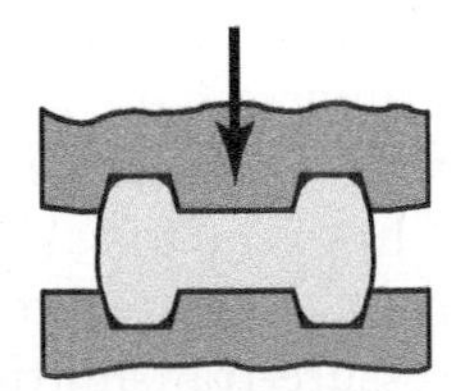

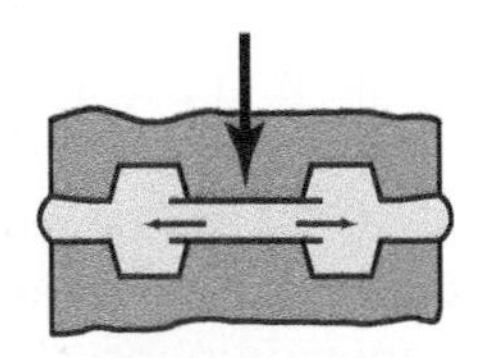

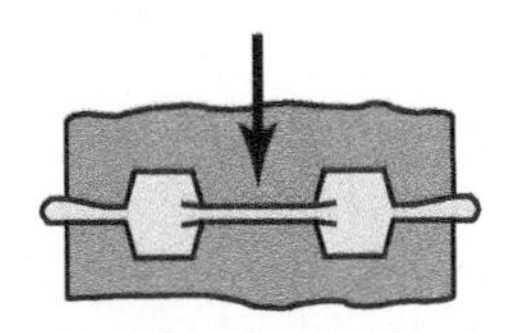

FIGURE 6.21 Stages in internal defect formation in a forging because of an oversized billet. The die cavities are filled prematurely, and the material at the center of the part flows radially outward and past the filled regions as deformation continues.

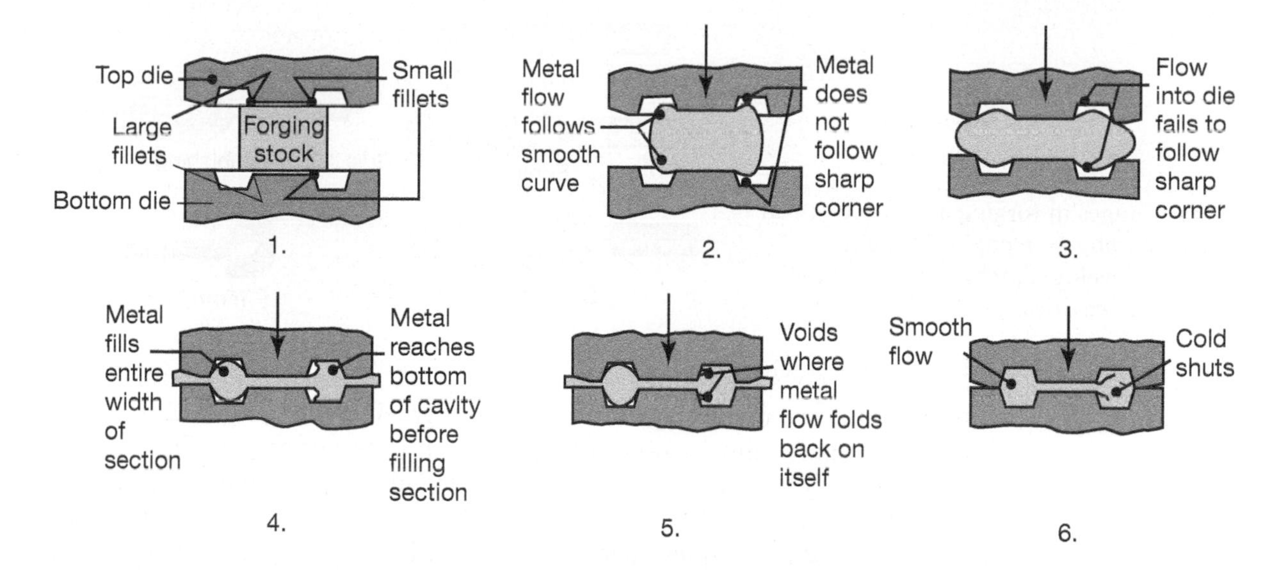

FIGURE 6.22 Effect of fillet radius on defect formation in forging. Note that small fillets (right side of the drawings) lead to defects. *Source:* Aluminum Company of America.

around a large corner radius than it does around a small one. With smaller radii, the material can fold over itself, producing a lap (**cold shut**). Such a defect can then lead to fatigue failure and other difficulties during the service life of a forged component. Inspection of forgings prior to their being placed into service is therefore essential (see Section 4.8).

4. Another important aspect of the quality of a forging is the **grain-flow pattern**. If the grain-flow lines reach a surface perpendicularly, they would expose the grain boundaries directly to the environment (known as **end grains**). End grains are preferentially attacked by the environment (such as salt water, acid rain, or other chemically-active environments), developing rough surfaces, which act as stress raisers. End grains can be avoided by proper selection of grain orientation of the blank with respect to its position in the die cavity and by control of material flow during forging.
5. Because the metal flows in various directions in a forging, and there are also temperature variations within the forging, the properties of a forging are generally *anisotropic* (see Section 3.5). As a result, strength and ductility can vary significantly in different locations and orientations in a forged part.

6.2.7 Die Design

The design of forging dies and selection of die material require knowledge of (a) the strength and ductility of the workpiece material; (b) its sensitivity to strain rate and temperature; (c) its frictional characteristics; and (d) forging temperature. *Die distortion* under high forging loads also is an

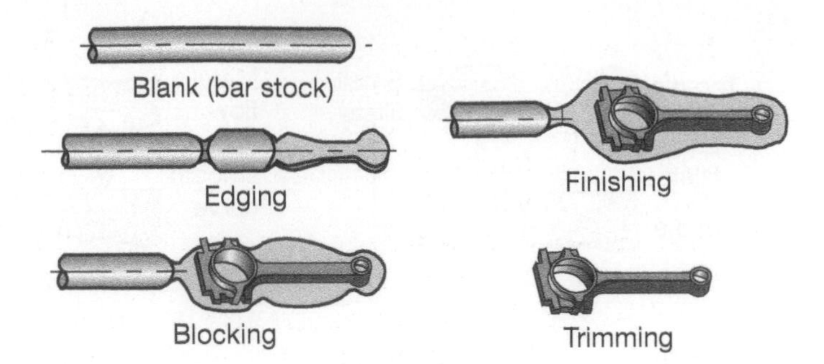

FIGURE 6.23 Stages in forging a connecting rod for an internal combustion engine. Note the amount of flash developed, which is important in properly filling die cavities.

important consideration, particularly if close dimensional tolerances are required. The terminology used in die design is given in Fig. 6.24.

Complex forgings require forming in a number of stages (Fig. 6.23) to ensure proper distribution of the material in the die cavities. Typically, starting with round bar stock, the bar is (a) first **preformed** (*intermediate shape*) using techniques such as fullering and edging (see Section 6.2.4); (b) the part is forged into the final shape in two additional operations; and (c) trimmed. The reason for preforming can best be understood from Eq. (4.6), recognizing that for long die life, wear is to be minimized. In any step of a forging sequence, the die cavity is exposed to either high sliding speeds or high pressures, but not both. Thus, edging, for example, produces large deformations on a relatively thick workpiece, whereas the finishing operation produces fine details, and involves small strains under high pressures. It is also important to accurately calculate the required volume of stock (blank) to ensure that it will properly fill the die cavity. Computer techniques are now widely available to expedite these calculations.

General considerations for die design include the following:

1. The **parting line** or *plane* is where the two dies meet. For simple symmetrical shapes, the parting line is straight and placed at the center of the forging; for more complex shapes, the line may be offset and may not be in a single plane.
2. The significance of **flash** is described in Section 6.2.3. After lateral flow has been sufficiently constrained (by the length of the land), the

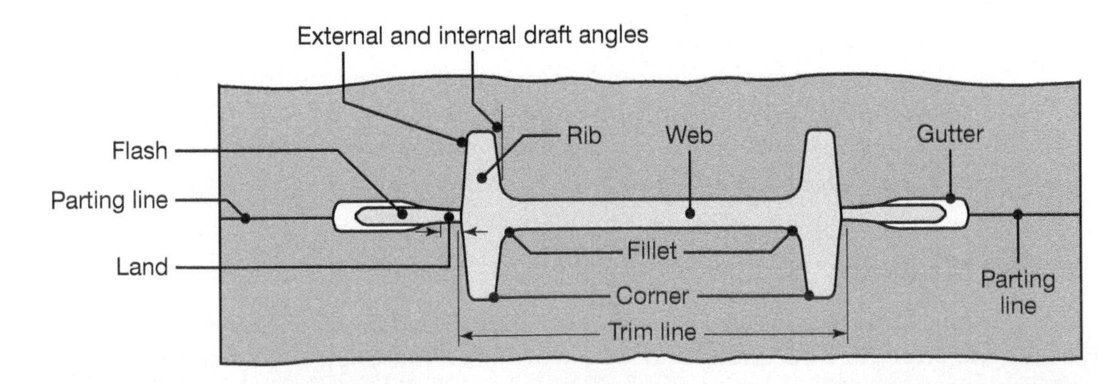

FIGURE 6.24 Standard terminology for various features of a typical forging die.

flash is allowed to flow into a **gutter**; thus, the extra flash does not unnecessarily increase the forging load. A general guideline for flash clearance between the dies is 3% of the maximum thickness of the forging. The length of the land is usually around five times that of the flash clearance.

3. **Draft angles** in dies are necessary in almost all forgings to facilitate removal of the part from the die; they usually are between 3° and 10°. Forgings shrink in all directions as they cool, but because of radial shrinkage, internal draft angles are made larger than external ones. Typically, internal angles are about 7° to 10° and external angles about 3° to 5°.
4. Proper selection of **die radii** for corners and fillets ensures smooth flow of the metal throughout the die cavity and improves die life. Small radii generally are not desirable because of their adverse effect on metal flow and their tendency to wear rapidly from stress concentration and thermal cycling. Small fillet radii can cause fatigue cracking in dies.

Die materials. Because most forgings, particularly large ones, are produced at elevated temperatures, die materials generally must have strength and toughness at elevated temperatures, hardenability, resistance to mechanical and thermal shock, and resistance to wear (especially abrasive wear, because of the presence of scale on heated forgings). Common die materials are tool and die steels containing chromium, nickel, molybdenum, and vanadium (see Section 3.10.4). Factors involved in the selection of die materials are die size, properties of the workpiece, shape complexity, forging temperature, type of forging operation (for example, hammers cause more dynamic loadings than hydraulic presses - see Section 6.2.8), cost of die material, number of forgings to be made, and heat-transfer and distortion characteristics of the die material.

Forging temperature and lubrication. The range of *temperatures* for hot forging are given in Table 6.3. *Lubrication* plays an important role in forging as it affects the flow of metal into the die cavities. Lubricants also serve as a *thermal barrier* between the hot forging and the dies which typically are at room temperature, thus slowing cooling of the workpiece. An additional role of a lubricant is to act as a *parting agent*, that is, to prevent the forging from sticking to the dies and enable the removal of the forging.

TABLE 6.3 Forging temperature ranges for various metals.

Metal	°C	°F
Aluminum alloys	400–450	750–850
Copper alloys	625–950	1150–1750
Nickel alloys	870–1230	1600–2250
Alloy steels	925–1260	1700–2300
Titanium alloys	750–795	1400–1800
Refractory alloys	975–1650	1800–3000

For hot forging, graphite, molybdenum disulfide, and (sometimes) glass are commonly used as lubricants; for cold forging, mineral oils and soaps are common lubricants.

6.2.8 Equipment

Forging equipment of various designs, capacities, and speed-stroke characteristics is available (Fig. 6.25).

QR Code 6.6 Mechanical forging presses.
Source: Courtesy of Erie Press.

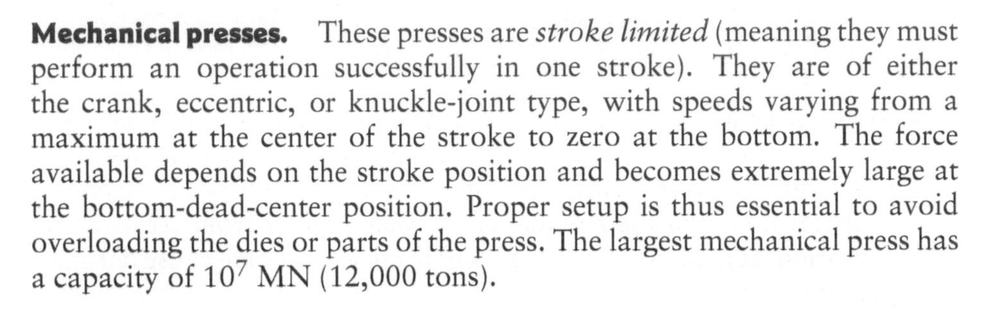

Mechanical presses. These presses are *stroke limited* (meaning they must perform an operation successfully in one stroke). They are of either the crank, eccentric, or knuckle-joint type, with speeds varying from a maximum at the center of the stroke to zero at the bottom. The force available depends on the stroke position and becomes extremely large at the bottom-dead-center position. Proper setup is thus essential to avoid overloading the dies or parts of the press. The largest mechanical press has a capacity of 10^7 MN (12,000 tons).

Screw presses. Deriving their energy from a flywheel, *screw presses* transmit the forging load through a vertical screw; they are *energy limited*, meaning they will deform material until the flywheel energy is expended. Unlike mechanical presses, however, screw presses can apply more than one stroke to a workpiece. These presses can be used for a variety of forging operations, and they are particularly suitable for producing small quantities and for parts requiring precision (such as turbine blades). The largest screw press has a capacity of 280 MN (31,500 tons).

Hydraulic presses. These presses are *load limited* (they can apply only the load developed in their hydraulic cylinder) and have a constant low speed of operation. Large amounts of energy can be transmitted to the workpiece by a constant load that is available throughout the stroke; ram speed can be varied during the stroke. Hydraulic presses are used for both

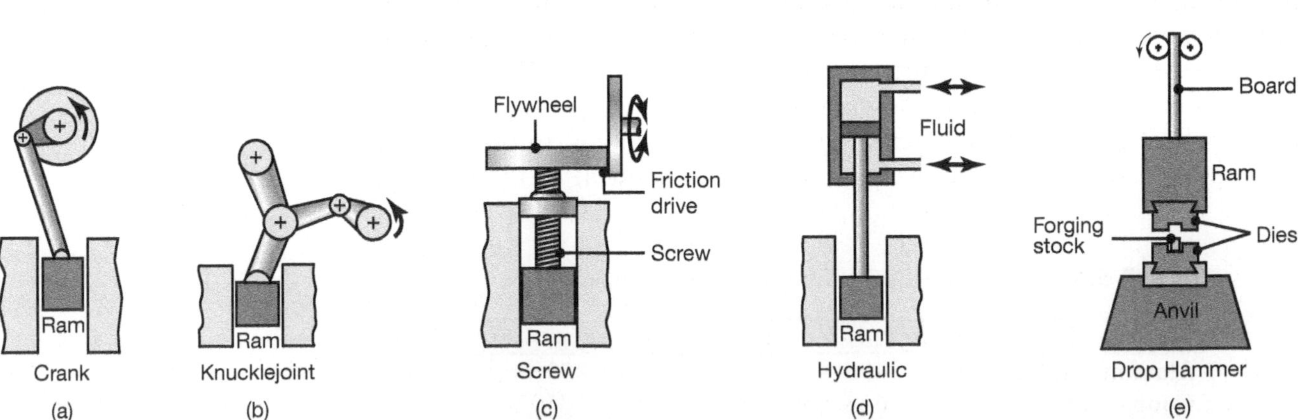

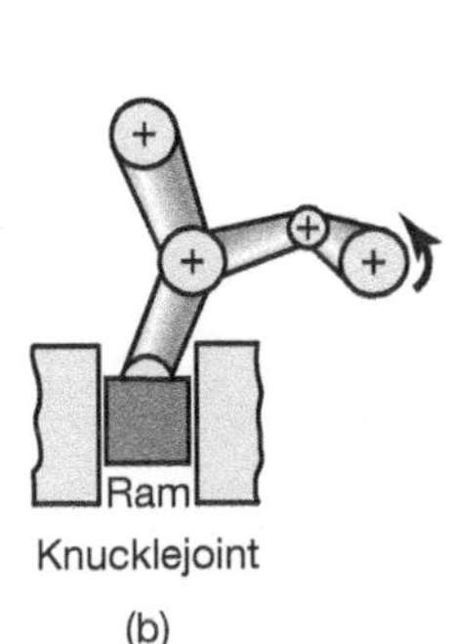

FIGURE 6.25 Schematic illustration of various types of presses used in metalworking. The choice of a press is a major consideration in the overall operation and productivity.

open-die and closed-die forging operations. The largest hydraulic press in existence has a capacity of 730 MN (82,000 tons).

Hammers. *Hammers* derive their energy from the *potential energy* of the ram, which is then converted to kinetic energy; hammers are thus *energy limited.* In *power hammers*, the ram is accelerated in the downstroke by steam or air. Hammer speeds are high, thus minimizing the cooling of the hot forgings, and allowing the forging of complex shapes, particularly with thin and deep recesses (which otherwise would cool rapidly). Several blows may have to be made on the part to finalize its shape. The highest energy available in power hammers is 1150 kJ.

Counterblow hammers. These hammers have two rams; as their name suggests, the hammers simultaneously approach each other to forge the part. They are generally of the mechanical-pneumatic or mechanical-hydraulic type. These machines transmit less vibration to the foundation than other hammers. The largest counterblow hammer has a capacity of 1200 kJ.

High-energy-rate forging (HERF) machines. In this type of machine, the ram is accelerated rapidly by inert gas at high pressure or by expansion associated with a confined explosion (similar to combusting fuel in a car cylinder to move a piston), and the part is forged in one blow at a very high speed. HERF is done on sheet metals through explosive and magnetic pulse forming (see Section 7.5.5), but also bulk shapes. Examples are forming of thick plates in ship building, large dishes, and cladding of plates (see explosive welding, Section 12.12), as well as dynamic compaction of metal powders (see Section 11.3.4).

Servo presses. A recent development is the use of servo presses for forging and stamping applications. These presses utilize servo drives along with linkage mechanisms, as in mechanical, knuckle joint, or screw presses. There are no clutches or brakes; instead, the desired velocity profile is achieved through a servo motor controller. The servo drive thus allows considerable flexibility regarding speeds and stroke heights, which simplifies set up and allows an optimized velocity profile for forging difficult materials or products. In addition, servo presses can produce parts with as little as 10% of the energy consumption of other presses, attributable mainly to their low energy costs when not producing parts (see Section 16.5). Servo presses can develop forces up to 25,000 kN (2800 tons); larger forces can be developed by hybrid machines that combine servo drives with energy storage in a flywheel.

Equipment selection. Selection of forging equipment depends on the size and complexity of the forging, the strength of the material and its sensitivity to strain rate, the extent of deformation involved, production rate, and cost. For hydraulic presses, the number of strokes is typically a few per minute; for power hammers, as many as 300 per minute. Generally, presses are preferred for aluminum, magnesium, beryllium, bronze, and brass, and hammers are preferred for copper, steels, titanium, and refractory alloys.

6.3 Rolling

Rolling is the process of reducing the thickness or changing the cross section of a long workpiece by compressive forces applied through a set of rolls (Fig. 6.26). Rolling, which accounts for about 90% of all metals produced by metalworking operations, was first developed in the late 1500s. The basic rolling operation is called *flat rolling*, or simply *rolling*, where the rolled products are flat plate and sheet.

Plates are generally regarded as having a thickness of more than 6 mm (0.25 in.), and are used for structural applications, such as bridges, girders, ship hulls, machine structures, and nuclear vessels. Plates can be as much as 0.3 m (12 in.) thick for the supports for large boilers, 150 mm thick for reactor vessels, and 100–125 mm thick for large ships and tanks.

Sheets are generally less than 6 mm thick, and are available as flat pieces or as strip in coils shipped to manufacturing facilities for further processing into wide range of products. They are used for automobile bodies, aircraft fuselages, office furniture, appliances, and food and beverage containers. Commercial-aircraft fuselages are typically made of about 1-mm-thick aluminum-alloy sheet, often of the 2000 series, and beverage cans are made of 0.15-mm-thick aluminum-alloy sheet, such as 3104 alloy (Section 3.11.1). Aluminum foil, such as that used to wrap candy and cigarettes, has a thickness of 0.008 mm.

Although the traditional starting form of material for rolling has been ingot, this practice is now rapidly replaced by *continuous casting and rolling* (see Section 5.7.2), with their much higher efficiency and lower cost. Rolling is first carried out at elevated temperatures (hot rolling), wherein the coarse-grained, brittle, and porous *cast* structure of the ingot or continuously cast metal is broken down into a *wrought* structure, with finer grain size and improved properties (Fig. 6.27)

6.3.1 Mechanics of Flat Rolling

The basic flat-rolling process is shown schematically in Fig. 6.28. A strip of thickness h_o enters the roll gap and is reduced to a thickness of h_f by the powered rotating rolls at a surface speed V_r of the roll. Because the volume rate of metal flow is constant, the velocity of the strip must increase as it moves through the roll gap. At the exit of the roll gap, the velocity of the strip is V_f (Fig. 6.29).

Since V_r is constant along the roll gap but the strip velocity increases as it passes through the roll gap, sliding occurs between the roll and the strip. At one location along the arc of contact, however, the two velocities are the same; for this reason, this point is called the **neutral point, neutral plane,** or **no-slip point.** To the left of this point, the roll moves faster than the workpiece, and to the right, the workpiece moves faster than the roll.

Because of the relative motion at the interfaces, the frictional forces (which oppose motion) act on the strip surfaces in the directions shown in Fig. 6.29. In rolling, the frictional force on the left of the neutral point

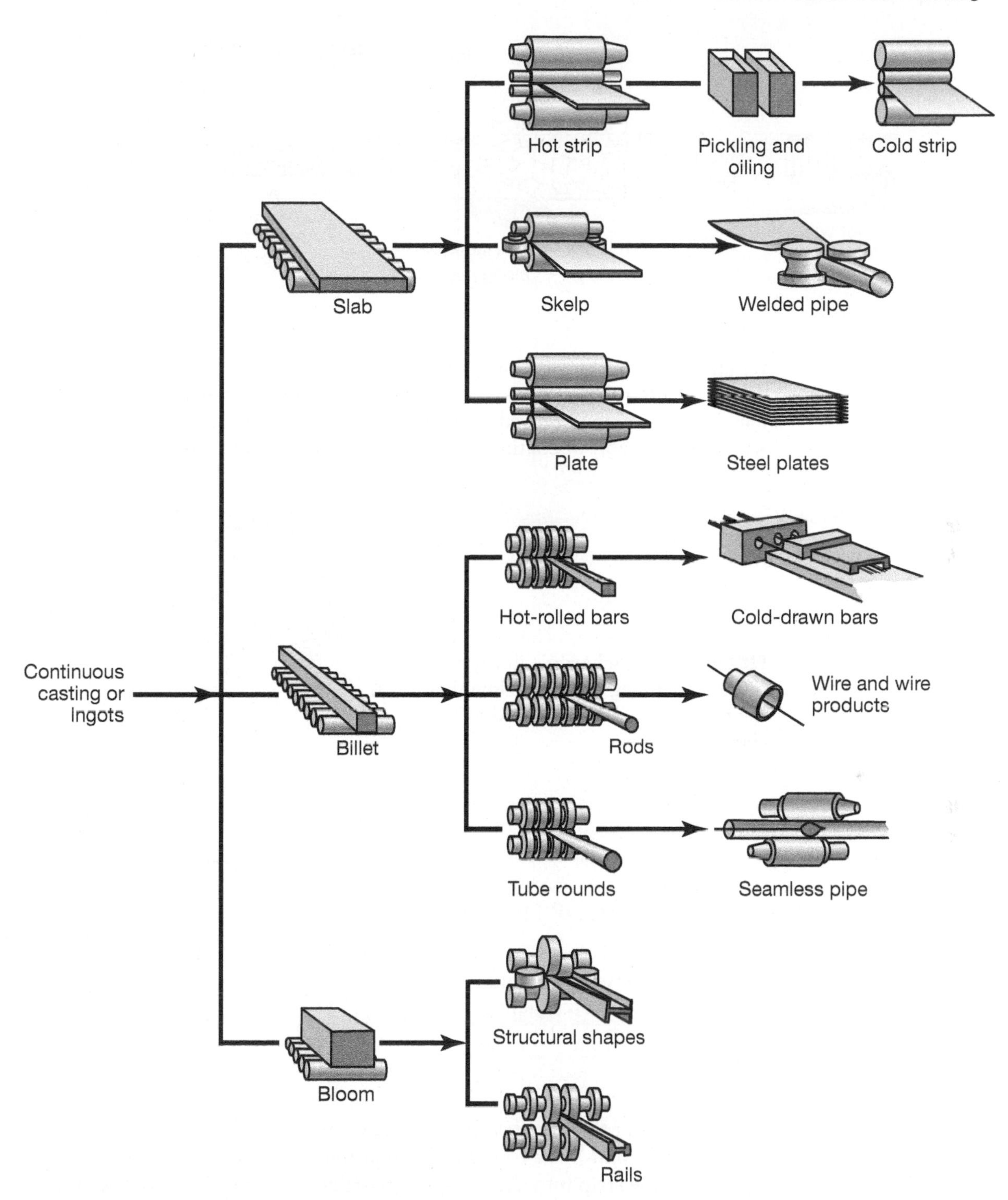

FIGURE 6.26 Schematic outline of various flat-rolling and shape-rolling operations. *Source:* American Iron and Steel Institute.

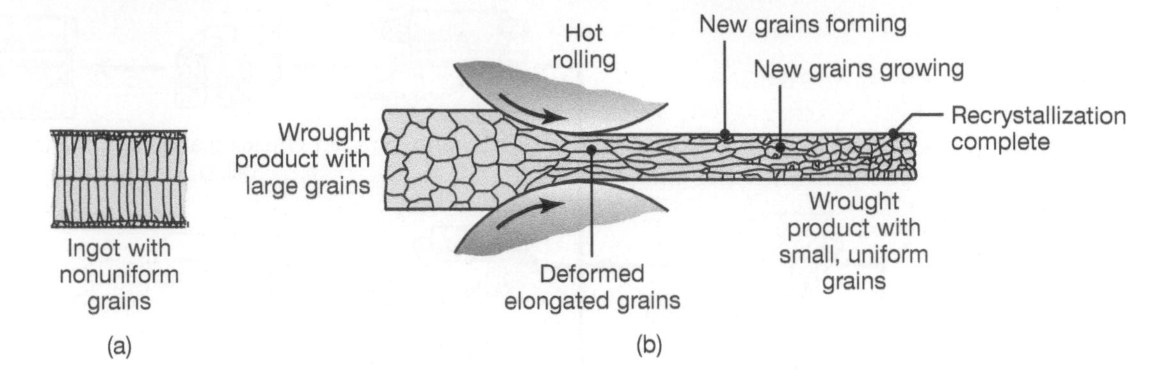

FIGURE 6.27 Changes in the grain structure of metals during hot rolling. This is an effective method to reduce grain size and refine the microstructure in metals, resulting in improved strength and good ductility. In this process cast structures of ingots or continuous castings are converted to a wrought structure.

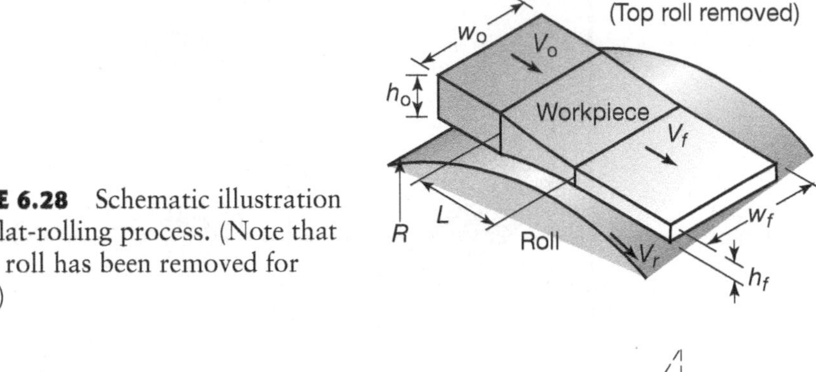

FIGURE 6.28 Schematic illustration of the flat-rolling process. (Note that the top roll has been removed for clarity.)

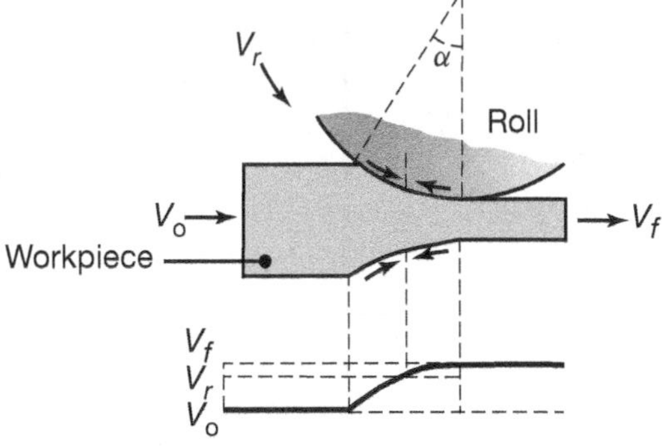

FIGURE 6.29 Relative velocity distribution between roll and strip surfaces. The arrows represent the frictional forces acting along the strip-roll interfaces. Note the difference in their direction in the left and right regions.

must be greater than the frictional force on the right. This difference results in a *net frictional force* to the right, making the rolling operation possible by pulling the strip into the roll gap. In order to supply work to the system, the net frictional force and the surface velocity of the roll must be in the same direction; consequently, the location of the neutral point should be toward the exit in order to satisfy these requirements.

Forward slip in rolling is defined as

$$\text{Forward slip} = \frac{V_f - V_r}{V_r} \tag{6.24}$$

and is a measure of the relative velocities at the exit of the work rolls.

1. **Roll pressure distribution.** It can be noted that the deformation zone in the roll gap is subjected to a state of stress similar to that in upsetting. However, the calculation of forces and stress distribution in rolling is more involved because the contact surfaces are curved. Also, in cold rolling, the material at the exit is strain hardened, and thus the flow stress at the exit is higher than that at the entry.

   The stresses acting on an element in the entry and exit zones, respectively, are shown in Fig. 6.30. Note that the only difference between the two elements is the *direction* of the friction force. Using the *slab method* of analysis for plane strain (described in Section 6.2.2), the stresses in rolling may now be analyzed as follows.

   From the equilibrium of the horizontal forces on the element shown in Fig. 6.30,

$$(\sigma_x + d\sigma_x)(h + dh) - 2pR\,d\phi \sin\phi - \sigma_x h \pm 2\mu pR\,d\phi \cos\phi = 0.$$

   Simplifying and ignoring the second-order terms, this expression reduces to

$$\frac{d(\sigma_x h)}{d\phi} = 2pR(\sin\phi \mp \mu\cos\phi).$$

   In rolling practice, the bite angle α (See Fig. 6.29) is typically only a few degrees; hence, it can be assumed that $\sin\phi = \phi$ and $\cos\phi = 1$. Thus,

$$\frac{d(\sigma_x h)}{d\phi} = 2pR(\phi \mp \mu). \qquad (6.25)$$

   Because the angles involved are very small, p can be assumed to be a *principal stress.* The other principal stress is σ_x. The relationship between these two principal stresses and the flow stress, σ_f, of the material is given by Eq. (2.46) for plane strain, or

$$p - \sigma_x = \frac{2}{\sqrt{3}}\sigma_f = \sigma_f'. \qquad (6.26)$$

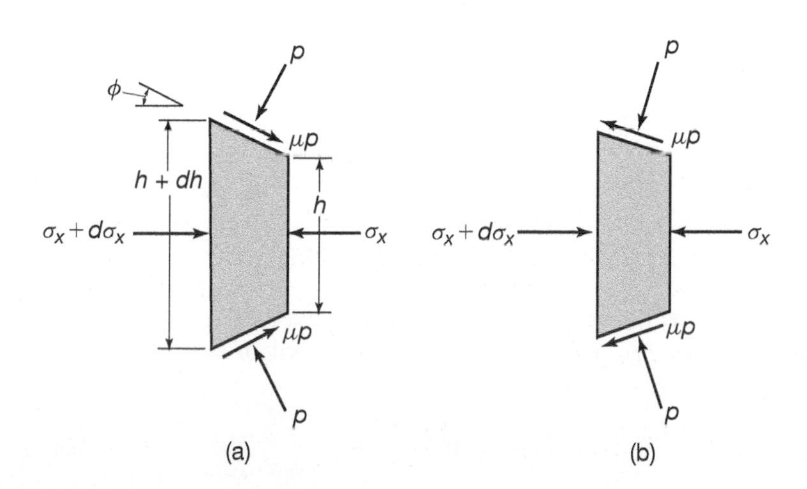

FIGURE 6.30 Stresses acting on an element in rolling: (a) entry zone and (b) exit zone.

Recall that, for a strain-hardening material, the flow stress, σ_f, in these expressions must correspond to the strain that the material has undergone at that particular location in the gap. Rewriting Eq. (6.25),

$$\frac{d\left[\left(p - \sigma_f'\right) h\right]}{d\phi} = 2pR\left(\phi \mp \mu\right),$$

or

$$\frac{d}{d\phi}\left[\sigma_f'\left(\frac{p}{\sigma_f'} - 1\right) h\right] = 2pR\left(\phi \mp \mu\right),$$

which, upon differentiation, becomes

$$\sigma_f' h \frac{d}{d\phi}\left(\frac{p}{\sigma_f'}\right) + \left(\frac{p}{\sigma_f'} - 1\right)\frac{d}{d\phi}\left(\sigma_f' h\right) = 2pR\left(\phi \mp \mu\right).$$

The second term in this expression is very small, because as h decreases, σ_f' increases (due to cold working), thus making the product of σ_f' and h nearly a constant, and thus its derivative is negligibly small. Therefore,

$$\frac{\dfrac{d}{d\phi}\left(\dfrac{p}{\sigma_f'}\right)}{\dfrac{p}{\sigma_f'}} = \frac{2R}{h}\left(\phi \mp \mu\right). \tag{6.27}$$

Letting h_f be the final thickness of the strip being rolled,

$$h = h_f + 2R\left(1 - \cos\phi\right),$$

or, approximately,

$$h = h_f + R\phi^2. \tag{6.28}$$

Substituting this expression for h in Eq. (6.27) and integrating,

$$\ln\frac{p}{\sigma_f'} = \ln\frac{h}{R} \mp 2\mu\sqrt{\frac{R}{h_f}}\tan^{-1}\sqrt{\frac{R}{h_f}}\phi + \ln C$$

or

$$p = C\sigma_f'\frac{h}{R}e^{\mp\mu H},$$

where

$$H = 2\sqrt{\frac{R}{h_f}}\tan^{-1}\left(\sqrt{\frac{R}{h_f}}\phi\right). \tag{6.29}$$

At the entry, $\phi = \alpha$; hence, $H = H_o$ with ϕ replaced by α. At the exit, $\phi = 0$, so that $H = H_f = 0$. Also, at entry and exit, $p = \sigma_f'$. Thus, in the **entry zone**,

$$C = \frac{R}{h_f}e^{\mu H_i}$$

and

$$p = \sigma_f' \frac{h}{h_o} e^{\mu(H_o - H)}. \tag{6.30}$$

In the **exit zone,**

$$C = \frac{R}{h_f},$$

and, hence,

$$p = \sigma_f' \frac{h}{h_f} e^{\mu H}. \tag{6.31}$$

Note that the pressure p at any location in the roll gap is a function of h and its angular position ϕ along the arc of contact. These expressions also indicate that the pressure increases with increasing strength of the material, increasing coefficient of friction, and increasing R/h_f ratio. The R/h_f ratio in rolling is equivalent to the a/b ratio in upsetting, as described in Section 6.2.2.

The dimensionless theoretical pressure distribution in the roll gap is shown in Fig. 6.31. Note the similarity of this curve to that shown in Fig. 6.5 in that both display a friction hill; further note that the neutral point shifts toward the exit as friction decreases. The reason is that with lower friction, a greater extent of the entry zone is needed to draw material through the work rolls.

The effect of reduction in thickness of the strip on the pressure distribution is shown in Fig. 6.32. As reduction increases, the length of contact in the roll gap increases, which, in turn, increases the peak pressure. The curves shown in Fig. 6.32 are theoretically derived; actual pressure distributions, as determined experimentally, are smoother, with rounded peaks.

2. **Determining the location of the neutral point.** The neutral point can be determined by equating Eqs. (6.30) and (6.31); thus, at the neutral point,

$$\frac{h_o}{h_f} = \frac{e^{\mu H_o}}{e^{2\mu H_n}} = e^{\mu(H_o - 2H_n)},$$

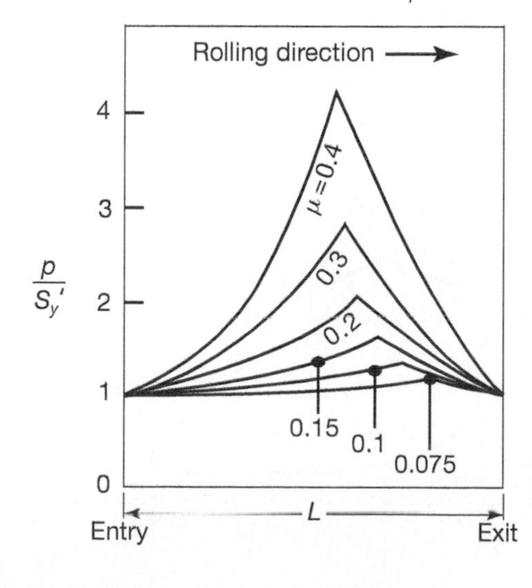

FIGURE 6.31 Pressure distribution in the roll gap as a function of the coefficient of friction. Note that as friction increases, the neutral point shifts toward the entry. Without friction, the rolls will slip, and the neutral point shifts completely to the exit.

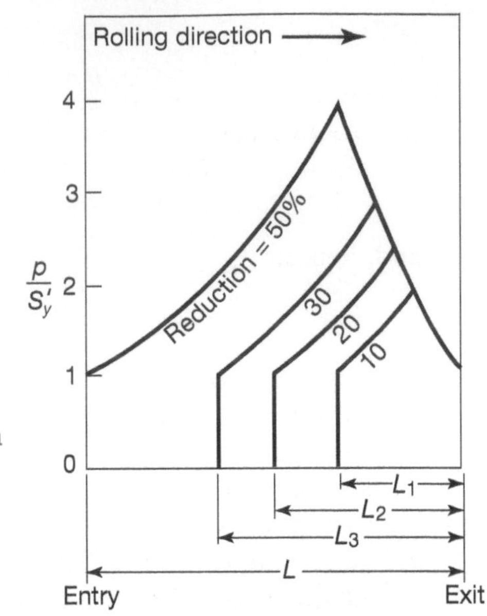

FIGURE 6.32 Pressure distribution in the roll gap as a function of reduction in thickness. Note the increase in the area under the curves with increasing reduction, thus increasing the roll force.

or

$$H_n = \frac{1}{2}\left(H_o - \frac{1}{\mu}\ln\frac{h_o}{h_f}\right). \tag{6.32}$$

Substituting Eq. (6.32) into Eq. (6.29),

$$\phi_n = \sqrt{\frac{h_f}{R}}\tan\left(\frac{H_n}{2}\sqrt{\frac{h_f}{R}}\right). \tag{6.33}$$

3. **Front and back tension.** The roll force, F, can be reduced mainly by (a) lowering friction; (b) using rolls with smaller radii; (c) taking smaller reductions per pass; and (d) raising the workpiece temperature. Another particularly effective method is to reduce the apparent compressive yield strength of the material by applying longitudinal *tension*. Recall from the discussion on yield criteria in Section 2.11 that when a tensile stress is applied to a strip (Fig. 6.33), the yield

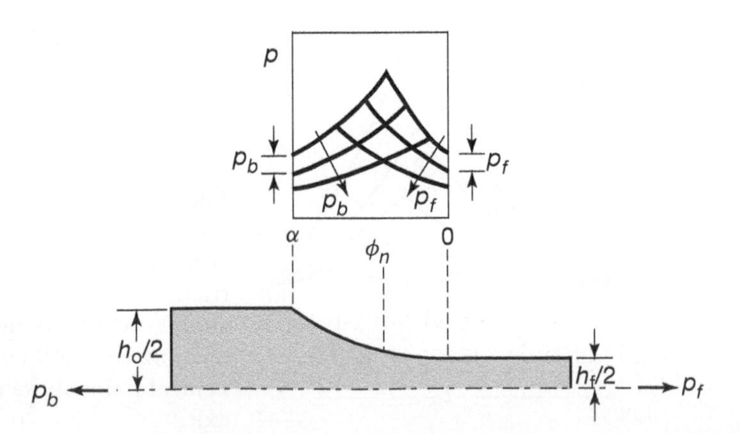

FIGURE 6.33 Pressure distribution as a function of front and back tension in rolling. Note the shifting of the neutral point and the reduction in the area under the curves (hence reduction in the roll force) as tensions increase.

strength normal to the strip surface decreases, and thus the roll pressure and force decrease.

In practice, tensile forces in rolling can be applied either at the entry (**back tension,** p_b) or at the exit (**front tension,** p_f) of the strip, or both. Equations (6.30) and (6.31) can now be modified to include the effect of tension for the entry and exit zones, respectively, as follows:

$$\text{Entry zone: } p = \left(\sigma_f' - p_b\right)\frac{h}{h_o}e^{\mu(H_o-H)} \tag{6.34}$$

and

$$\text{Exit zone: } p = \left(\sigma_f' - p_f\right)\frac{h}{h_f}e^{\mu H}. \tag{6.35}$$

Depending on the relative magnitudes of the tensile stresses applied, the neutral point may *shift*, as shown in Fig. 6.33. As expected, this shift then affects the pressure distribution, torque, and power requirements.

Tensions are particularly important in rolling thin, high-strength materials, because such materials require high roll forces. Front tension in practice is typically controlled by the torque on the coiler (**delivery reel**), around which the rolled sheet is coiled. Back tension is controlled by a braking system in the uncoiler (**payoff reel**). Specialized instrumentation is available for such controls.

EXAMPLE 6.3 Back Tension Required to Cause Slip in Rolling

Given: If the back tension, p_b, is too high the rolls will begin to slip.

Find: Derive an expression for the magnitude of back tension required to make the rolls begin to slip.

Solution: Slipping of the rolls means that the neutral point has moved all the way to the exit in the roll gap. Thus, the whole contact area now becomes the entry zone, and therefore Eq. (6.34) is applicable. Also, when $\phi = 0$, $H = 0$; hence, the pressure at the exit is given by

$$p_{\phi=0} = \left(\sigma_f' - p_b\right)\left(\frac{h_f}{h_o}\right)e^{\mu H_o}.$$

However, the pressure at the exit is equal to σ_f'. Rearranging this equation,

$$p_b = \sigma_f'\left[1 - \left(\frac{h_o}{h_f}\right)\left(e^{-\mu H_o}\right)\right],$$

where H_o is obtained from Eq. (6.29) for the condition of $\phi = \alpha$. Since all the quantities are known, the magnitude of the back tension can be calculated.

4. **Calculating roll forces.** The *roll force*, F, (also called the *roll-separating force*) on the strip is the product of the area under the pressure vs. contact-length curve and the strip width, w. The roll force can then be calculated from the expression,

$$F = \int_0^{\phi_n} wpR\, d\phi + \int_{\phi_n}^{\alpha} wpR\, d\phi. \tag{6.36}$$

A simple method of calculating the roll force is to multiply the contact area by an average contact stress, p_{av},

$$F = Lwp_{\text{av}}, \tag{6.37}$$

where L is the length of contact and can be approximated as the projected length; thus

$$L = \sqrt{R\Delta h}, \tag{6.38}$$

where R is the roll radius and Δh is the difference between the original and final thicknesses of the strip (called **draft**).

The magnitude of p_{av} depends on the h/L ratio, where h is now the *average thickness* of the strip in the roll gap. For large h/L ratios (such as small reductions in strip thickness and/or using large roll diameters), the rolls will act in a manner similar to indenters in a hardness test. Friction is not significant, and p_{av} is obtained from Fig. 2.22. With small h/L ratios (such as large reductions and/or large roll diameters), friction is predominant, and p_{av} is obtained from Eq. (6.15). Note that for strain-hardening materials, the appropriate flow stresses must be calculated and used.

As an approximation and for low frictional conditions, Eq. (6.37) can be simplified to

$$F = Lw\bar{\sigma}'_f, \tag{6.39}$$

where $\bar{\sigma}'_f$ is the average flow stress in plane strain of the material in the roll gap [see Eq. (6.10)]. For higher frictional conditions, an expression similar to Eq. (6.15) can be written as

$$F = Lw\bar{\sigma}'_f\left(1 + \frac{\mu L}{2h_{\text{av}}}\right). \tag{6.40}$$

5. **Roll torque and power.** The *roll torque*, T, for each roll can be calculated from the expression,

$$T = \underset{\text{(entry zone)}}{\int_{\phi_n}^{\alpha} w\mu pR^2\, d\phi} - \underset{\text{(exit zone)}}{\int_0^{\phi_n} w\mu pR^2\, d\phi}. \tag{6.41}$$

Note that the negative sign indicates the change in direction of the friction force at the neutral point. Thus, for example, if the frictional forces are equal to each other, the torque is zero.

The torque in rolling can also be estimated by assuming that the roll force, F, acts in the *middle* of the arc of contact (that is, a length of action of $0.5L$), and that this force is perpendicular to the plane of the strip. (It has been found that whereas $0.5L$ is a good estimate for hot rolling, $0.4L$ is a better estimate for cold rolling.)

Using the hot rolling estimate of length of action for the torque, the **torque per roll** is then

$$T = \frac{FL}{2}. \tag{6.42}$$

The **power required per roll** is

$$\text{Power} = T\omega = \frac{FL\omega}{2} = \frac{\pi FLN}{60}, \tag{6.43}$$

where N is the roll speed in revolutions per minute.

EXAMPLE 6.4 Power Required in Rolling

Given: A 225 mm-wide 6061-O aluminum strip is rolled from a thickness of 25 mm to 20 mm. The roll radius is 300 mm and the roll speed is $N = 100$ rpm.

Find: Estimate the total power required for this operation.

Solution: The length of the arc of contact, L, is obtained from Eq. (6.38) as

$$L = \sqrt{R\Delta h} = \sqrt{(300)(25 - 20)} = 38.73 \text{ mm}$$

For 6061-O aluminum, $K = 205$ MPa and $n = 0.2$ (from Table 2.2). The true strain in this operation is

$$\epsilon_1 = \ln\left(\frac{25}{20}\right) = 0.223.$$

Thus, from Eq. (6.10),

$$\bar{\sigma}_f = \frac{K\epsilon^n}{n+1} = \frac{(205)(0.223)^{0.2}}{1.2} = 126 \text{ MPa}$$

and, from Eq. (2.46),

$$\bar{\sigma}'_f = (1.15)(126) = 145 \text{ MPa}.$$

Therefore, noting that the width is $w = 225$ mm, Eq. (6.39) yields

$$F = Lw\bar{\sigma}'_f = (0.03873)(0.225)(145 \times 10^6) = 1.26 \text{ MN},$$

so that the power per roll is

$$\text{Power} = \frac{(\pi)(1.26 \times 10^6)(0.03873)(100)}{60} = 255 \text{ kW}.$$

Therefore, the power needed for both rolls is 510 kW.

6. **Roll forces in hot rolling.** The calculation of forces and torque in hot rolling presents two difficulties: determining (a) the strain-rate sensitivity of materials at elevated temperatures (see Section 2.2.7); and (b) the coefficient of friction, μ, at elevated temperatures.

   The *average strain rate*, $\dot{\bar{\epsilon}}$, in flat rolling can be obtained by dividing the strain by the time required for an element to undergo this strain in the roll gap. The time can be approximated as L/V_r, hence

$$\dot{\bar{\epsilon}} = \frac{V_r}{L} \ln\left(\frac{h_o}{h_f}\right). \tag{6.44}$$

   The flow stress, σ_f, of the material corresponding to this strain rate must first be obtained [see Eq. (2.18)], and then substituted into the proper equations [Eq. (6.40), for example].

   It has been observed that in cold rolling, the coefficient of friction typically is between 0.02 and 0.3 (see Section 4.4.1), depending on the materials involved and the lubricants used. Also, μ will decrease with the use of effective lubricants and in regimes approaching *hydrodynamic lubrication* (which can, for example, occur in cold rolling of aluminum at high speeds). In hot rolling, μ may range from about 0.2 (with effective lubrication) to as high as 0.7 for sticking, which typically occurs with steels, stainless steels, and high-temperature alloys.

   The maximum possible *draft*, that is, $h_o - h_f$, in flat rolling can be shown to be a function of friction and radius as

$$\Delta h_{\text{max}} = \mu^2 R. \tag{6.45}$$

   Thus, the higher the friction coefficient and the larger the roll radius, the greater the maximum draft. The maximum value of the angle α in Fig. 6.29 (called **angle of acceptance**) can be shown to be related to Eq. (6.45). Based on the simple model of a block sliding down an inclined plane,

$$\alpha_{\text{max}} = \tan^{-1} \mu. \tag{6.46}$$

   If α_{max} is larger than this value, the rolls begin to slip, because the friction is not high enough to pull the material through the roll gap.

7. **Roll deflections and roll flattening.** Roll forces tend to bend the rolls (Fig. 6.34a), resulting in a strip that is thicker at its center than at its edges (**crown**). One method for avoiding this problem is to grind the roll in such a way that its diameter at the center is slightly larger than at the edges. This curvature is known as **camber.** In sheet metal rolling practice, the camber is typically less than 0.50 mm on the roll diameter. In addition, in hot rolling or because of heat generation, rolls can become slightly barrel-shaped. Known as **thermal camber,** this effect can be controlled by varying the location of the coolant on the rolls along their axial direction. When designed properly, such

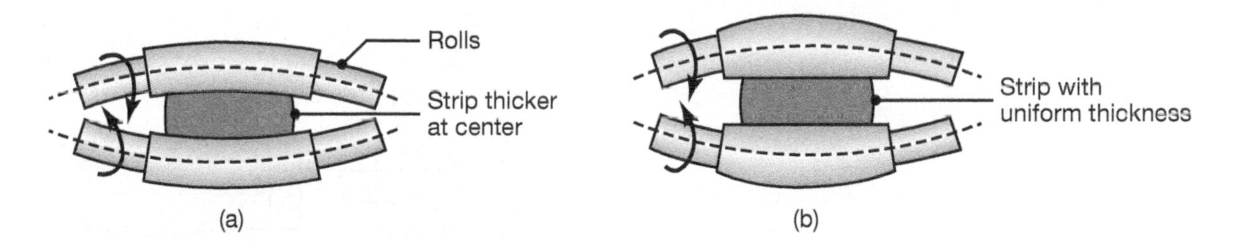

FIGURE 6.34 (a) Bending of straight cylindrical rolls (exaggerated) because of the roll force. (b) Bending of rolls, ground with camber, that produce a sheet of uniform thickness during rolling.

rolls produce flat strips, as shown in Fig. 6.34b; however, a particular camber is correct only for a certain load and width of strip.

Another method of developing camber in a work roll is to apply external forces to the roll to induce bending; the deformation of the rolls due to bending from the workpiece is compensated by the external bending. This approach has the benefit of allowing closed-loop control of crowning, as the applied moment to the roll can be changed based on measurements of thickness distribution.

Roll forces also tend to *flatten* the rolls elastically, much like the flattening of vehicle tires on roads. Flattening increases the roll's radius, resulting in a larger contact area for the same reduction in thickness, and increasing the roll force, F. Because of the combined effects of flattening and the friction hill, reductions in thickness of thin sheets is very difficult to obtain with large work rolls. The usual solution is to roll thin workpieces using a backing roll arrangement (see Section 6.3.4).

It can be shown that the new roll radius, R', is given by the expression

$$R' = R\left(1 + \frac{CF'}{h_o - h_f}\right), \tag{6.47}$$

where the value of C is 2.3×10^{-2} mm^2/kN (1.6×10^{-4} in.^{2}/kip) for steel rolls and 4.57×10^{-2} (3.15×10^{-4}) for cast-iron rolls, and F' is the **roll force per unit width** of strip, expressed in kN/mm. The higher the elastic modulus of the roll material, the less the roll distorts. The magnitude of R' cannot be calculated directly from Eq. (6.47) because the roll force is itself a function of the roll radius; the solution is then obtained iteratively. Note that R in all previous equations should be replaced by R' when significant roll flattening occurs.

8. **Spreading.** Rolling plates and sheets with high strip width-to-thickness ratios is essentially a process of plane strain. At smaller ratios, such as with a square cross section, the width increases considerably during rolling, known as *spreading* (Fig. 6.35). It can be shown that spreading decreases with (a) increasing width-to-thickness ratios

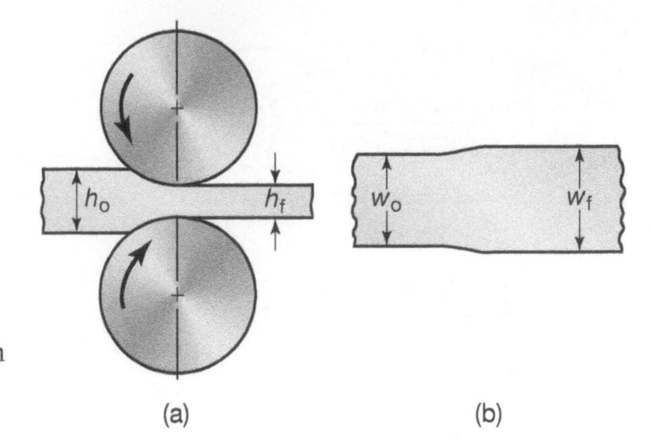

FIGURE 6.35 Increase in the width of a strip (spreading) during flat rolling. (a) Side view; (b) top view. Spreading can be similarly observed when dough is rolled on a flat surface with a rolling pin.

of the entering material; (b) decreasing friction; and (c) increasing ratios of roll-radius to strip-thickness. One technique of prevent spreading is by using a pair of vertical rolls that constrain the edges of the product being rolled, known as an **edger mill.**

6.3.2 Defects in Rolled Products

There may be surface defects on rolled plates and sheets, or structural defects within the material. **Surface defects** may result from inclusions and impurities in the material, scale, rust, dirt, roll marks, and other causes related to the prior treatment and working of the material. In hot rolling of blooms, billets, and slabs, the surface is usually preconditioned by various means, such as by **scarfing** (using a torch).

Structural defects can affect the integrity of the rolled product. Some typical defects are shown in Fig. 6.36. **Wavy edges** are caused by bending of the rolls, whereby the edges of the strip become thinner than at the center. Because the edges then elongate more than the center and are restrained by the bulk of the material from expanding freely, they buckle. The cracks shown in Fig. 6.36b and c are usually caused by low material ductility and barreling of the edges. **Alligatoring** is a complex phenomenon that results from inhomogeneous deformation of the material during rolling or from defects in the original cast ingot, such as pipe (see Section 6.4.4).

Residual stresses. Residual stresses can develop in rolled plates and sheets because of inhomogeneous plastic deformation in the roll gap. Small-diameter rolls and/or small reductions in thickness tend to deform

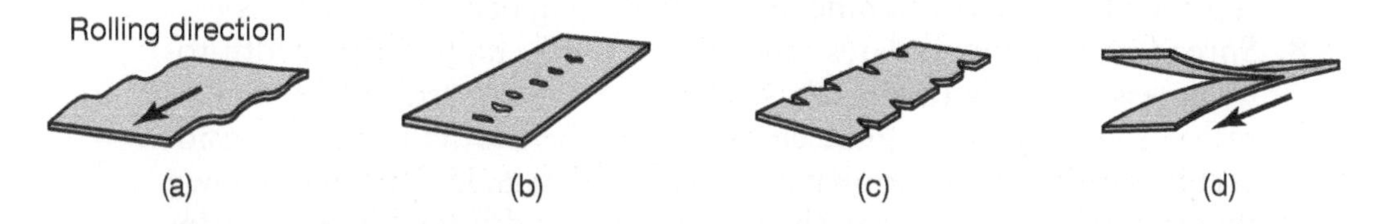

FIGURE 6.36 Schematic illustration of some defects in flat rolling: (a) wavy edges; (b) zipper cracks in the center of strip; (c) edge cracks; (d) alligatoring.

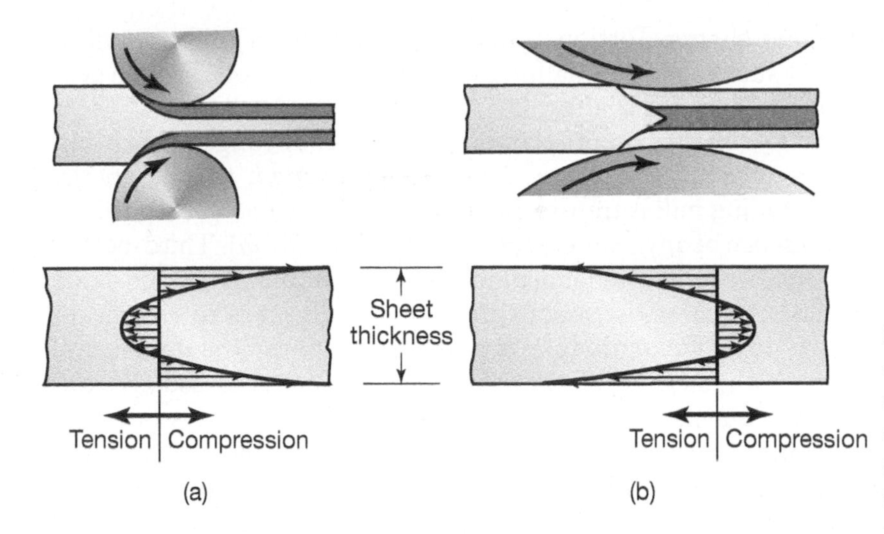

FIGURE 6.37 The effect of roll radius on the type of residual stresses developed in flat rolling: (a) small rolls and/or small reduction in thickness; and (b) large rolls and/or large reduction in thickness.

the metal plastically at its surfaces (similarly to shot peening or roller burnishing as described in Section 4.5.1). This type of deformation generates compressive residual stresses on the surfaces and tensile stresses in the bulk (Fig. 6.37a). Large-diameter rolls and/or high reductions, on the other hand, tend to deform the bulk to a greater extent than the surfaces, because of friction at the surfaces along the arc of contact. This situation generates tensile residual stresses on the surface, as shown in Fig. 6.37b. For many applications, these residual stresses can compromise performance.

6.3.3 Vibration and Chatter in Rolling

Vibration and *chatter* in rolling, as well as other metalworking operations, can have significant effects on product quality and productivity. For example, it has been estimated that modern rolling mills could operate at speeds up to 50% higher were it not for chatter. Considering the very high cost of rolling mills, this issue is a major economic concern. Chatter is generally defined as *self-excited vibration*, and can occur in such metal forming and machining processes as rolling, extrusion, drawing, machining, and grinding, as described in Sections 8.12 and 9.6.8. In rolling, it leads to small random variations in the thickness of the rolled sheet and, significantly, in its surface finish, and could lead to excessive material scrap; it has been found to occur predominantly in *tandem mills* (See Fig. 6.39d).

Vibration and chatter result from interactions between the structure of the mill stand and the dynamics of the rolling operation. Several variables are involved, but rolling speed and lubrication have been found to be the two most important parameters. The vibration modes commonly encountered in rolling are classified as torsional, third-octave, and fifth-octave chatter (octave meaning a measure of musical pitch - it is the interval between one musical pitch and another with half its frequency).

1. **Torsional chatter** is characterized by a low resonant frequency (approximately 5 to 15 Hz); it is usually a result of forced vibration (see Section 8.12), although it can also occur simultaneously with

third-octave chatter. Torsional chatter is usually not significant unless it is due to such factors as malfunctioning speed controls, broken gear teeth, and misaligned shafts.

2. **Third-octave chatter** occurs around the frequency range of the third musical octave (128 to 256 Hz) and is *self-excited*; that is, energy from the rolling mill is transformed into vibratory energy, regardless of the presence of any external forces (*forced vibration*). Third-octave chatter is most serious in tandem mills (see Section 6.3.4). This mode of vibration is the most serious in rolling, as it leads to significant gage variations, fluctuations in strip tension between the stands, and often results in strip breakage. Third-octave chatter is usually controlled by reducing the rolling speed. Although not always practical to implement, this type of chatter can be reduced by (a) increasing the distance between the stands of the tandem mill; (b) increasing the strip width, w; (c) incorporating *dampers* (see Section 8.12) in the roll supports; (d) decreasing the reduction per pass (draft); (e) increasing the roll radius R; and (f) increasing the strip-roll friction coefficient.
3. **Fifth-octave chatter** occurs in the frequency range of 550 to 650 Hz. It has been attributed to chatter marks that develop on work rolls, caused by (a) extended operation at certain critical speeds; (b) surface defects on a backup roll or in the incoming strip; and/or (c) improperly ground rolls. The chatter marks or striations are then imprinted onto the rolled sheet, thus adversely affecting its surface finish and appearance. Moreover, fifth-octave chatter is undesirable because of the objectionable noise it generates. This type of chatter can be controlled by (a) adjusting the mill speed; (b) using backup rolls of progressively larger diameters in the stands of a tandem mill; (c) avoiding chatter in grinding the rolls; and (d) eliminating the external sources of any other vibrations in the mill stand.

6.3.4 Flat-Rolling Practice

The initial breaking down of a cast ingot by hot rolling (at temperature similar to forging, Table 6.3) converts the coarse-grained, brittle, and porous structure into a *wrought* structure (Fig. 6.27) with finer grains and enhanced ductility.

The product of the first hot-rolling operation is called a **bloom** or **slab** (See Fig. 6.26). A bloom usually has a square cross section, at least 150 mm on the side; a slab is rectangular in cross section. Blooms are further processed by *shape rolling* them into structural shapes, such as I-beams and railroad rails; slabs are rolled into plates and sheet. **Billets** are usually square in cross section and their area is smaller than blooms. Billets are shape rolled into various shapes, such as round rods and bars. Hot-rolled round rods are used as the starting material for rod and wire drawing and are called **wire rods**.

In hot rolling of blooms, billets, and slabs, the surface of the material is usually *conditioned* (prepared for a particular operation) prior to rolling. Conditioning is done by such means as scarfing (flame cutting) to remove heavy scale or by rough grinding. Prior to cold rolling, the scale developed during hot rolling (or any other surface defects) may be removed by *pickling* with acids, or by mechanical means such as grinding or blasting with water).

Pack rolling is a flat-rolling operation in which two or more layers of metal are rolled together, thus improving productivity. Aluminum foil, for example, is pack rolled in two layers; it can easily be observed that one side of foil is matte and the other side is shiny. The foil-to-roll side is shiny and bright (because it has been in contact with the polished roll surfaces); the foil-to-foil side is lubricated to prevent cold welding and as a result has a matte and satiny finish due to orange peel.

Mild steel, when stretched during sheet forming operations, undergoes **yield-point elongation**, causing surface irregularities called **stretcher strains** or **Lüder's bands** (see also Section 7.2.1). To avoid this phenomenon, the sheet metal is typically subjected to a light pass of 0.5–1.5% reduction, known as **temper rolling** (*skin pass*).

A rolled sheet may not be sufficiently flat as it leaves the roll gap, because of variations in the material or in the processing parameters during rolling. To improve flatness, the strip is passed through a series of **leveling rolls**. Each roll is usually driven separately with individual electric motors; the strip is flexed in opposite directions as it passes through the sets of rollers. Several different roller arrangements are used (Fig. 6.38).

Sheet thickness is identified by a **gage number**; the smaller the number, the thicker is the sheet. There are various numbering systems for different sheet metals. Rolled sheets of copper and brass are also identified by thickness changes during rolling, such as $\frac{1}{4}$-hard, $\frac{1}{2}$-hard, and so on.

Lubrication. Ferrous alloys are usually hot rolled without a lubricant, although graphite may be used. Aqueous solutions are used to cool the rolls and break up the scale. Nonferrous alloys are hot rolled using a variety of compounded oils, emulsions, and fatty acids. Cold rolling uses low-viscosity lubricants, including mineral oils, emulsions, paraffin, and fatty oils.

Equipment. A variety of rolling equipment is available with several roll arrangements; the major types are shown in Fig. 6.39. Small-diameter rolls are preferable because roll force is lower and roll flattening is less likely. On the other hand, small rolls deflect under the roll forces; they are

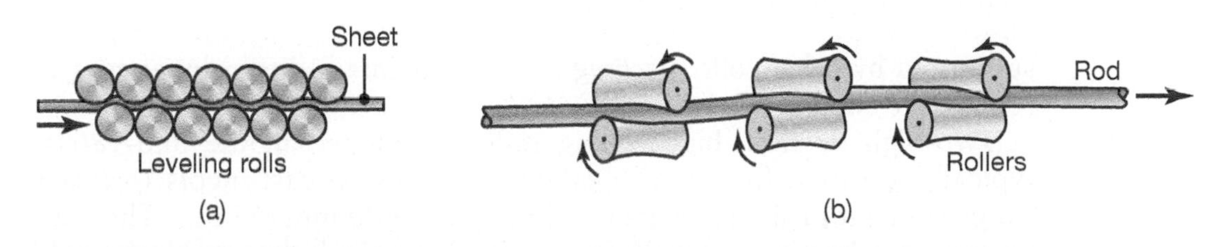

FIGURE 6.38 Schematic illustrations of roller leveling to (a) flatten rolled sheets and (b) straighten round rods.

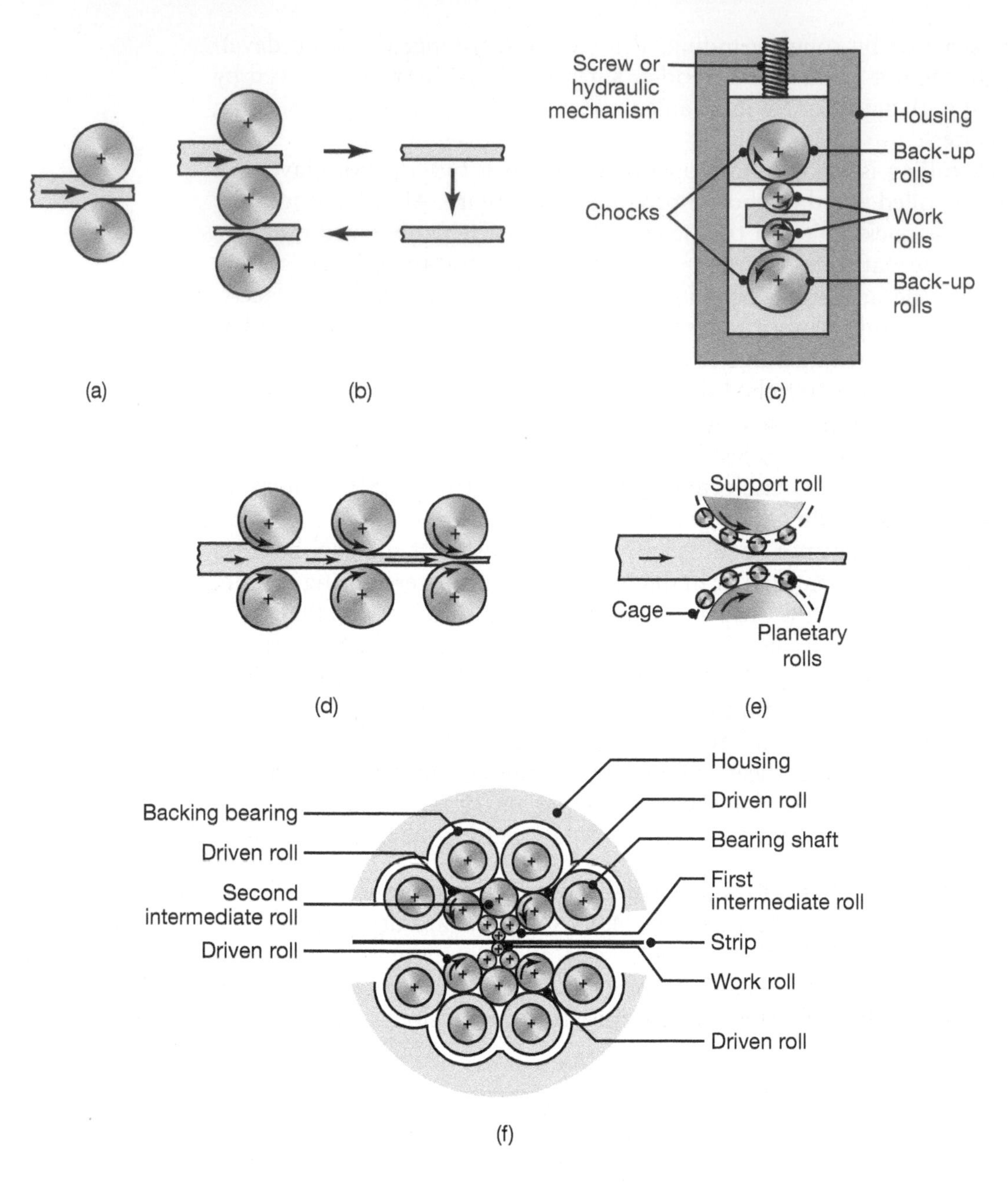

FIGURE 6.39 Schematic illustration of various roll arrangements: (a) two-high mill; (b) three-high mill; (c) four-high mill; (d) tandem rolling, with three stands; (e) planetary mill; and (f) cluster (*Sendzimir*) mill.

supported by other rolls (*backing rolls*) to maintain dimensional control (Fig. 6.39c and e).

Two-high or **three-high** rolling mills (developed in the mid-1800s) typically are used for initial breakdown passes on cast ingots (*primary roughing*), with roll diameters ranging up to 1400 mm (55 in.). The **cluster mill** (**Sendzimir**, or **Z mill**; Fig. 6.39e) is particularly suitable for cold rolling thin sheets or foils. The rolled product obtained in a cluster mill can be as wide as 5000 mm (200 in.) and as thin as 0.0025 mm (0.0001 in.).

The diameter of the work roll (smallest roll) can be as small as 6 mm (0.25 in.), and is usually made of tungsten carbide for rigidity, strength, and wear resistance.

In **tandem rolling** (Fig. 6.39d), the strip is rolled continuously as its passes through a number of **stands**; a group of stands is called a **train**. Control of the gage and of the speed at which the strip travels through each roll gap is critical for proper operation. Flat rolling also can be carried out with front tension only, using idling rolls (**Steckel rolling**).

Stiffness in rolling mills is important for controlling dimensions. The basic requirements for roll materials are mainly strength, stiffness, and resistance to wear. Three common roll materials are cast iron, cast steel, and forged steel. For hot rolling, roll surfaces are generally roughened, and may even have notches or grooves in order to pull the metal through the roll gap at high reductions. Rolls for cold rolling are ground to a fine finish and are polished for special applications, such as aluminum foil.

Minimills. In *minimills*, scrap metal is melted in electric-arc furnaces, cast continuously, and rolled directly into specific lines of products. Each minimill produces essentially one kind of rolled product (such as rod, bar, and structural shapes) from one type of metal, and is often dedicated to markets within the mill's particular geographic area.

Integrated mills. *Integrated mills* are large facilities that involve complete operations, from the production of hot metal in a blast furnace to casting and rolling of finished products that are ready to be shipped to the customer. Nearly all integrated mills specialize in flat-rolled metal or plate. Technically, the only difference between integrated mills and minimills is the type of furnace used; however, the use of blast furnaces implies different economies of scale than electric-arc furnaces that require much larger and more elaborate facilities.

6.3.5 Miscellaneous Rolling Operations

1. **Shape rolling.** Straight structural shapes of various cross sections (such as channels I-beams and railroad rails) are rolled by passing a bloom (See Fig. 6.26) through a series of specially designed rollers (Fig. 6.40). The design of the rolls (**roll-pass design**) requires special care in order to avoid defect formation and to hold dimensional tolerances. For example, in shape rolling a channel, the reduction is different in various locations within the section; thus, elongation is not uniform throughout the cross section, which can cause warping or cracking.
2. **Ring rolling.** In this process, a small-diameter, thick ring is expanded into a larger-diameter, thinner ring, by placing the ring between two rolls, one of which is driven (Fig. 6.41). The ring thickness is reduced by moving the rolls closer as they continue rotating. Because of volume constancy, the reduction in thickness causes an increase in the ring diameter. A variety of cross sections can be ring rolled with variously shaped rolls.

QR Code 6.7 Ring-rolling operations.
Source: Courtesy of the Forging Industry Association, www.forging.org.

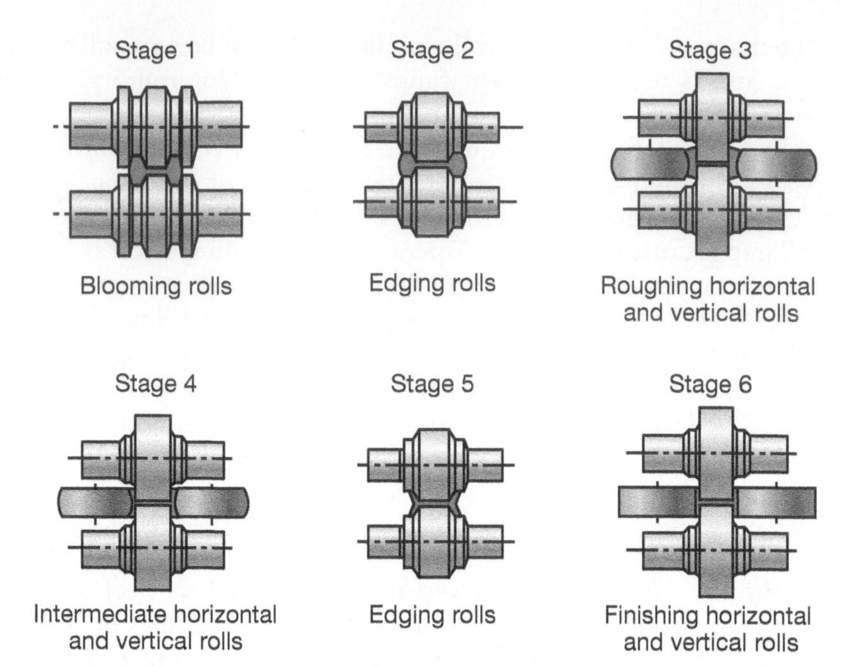

FIGURE 6.40 Stages in shape rolling of an H-section. Several other structural sections, such as channels and rails, also are rolled by this process.

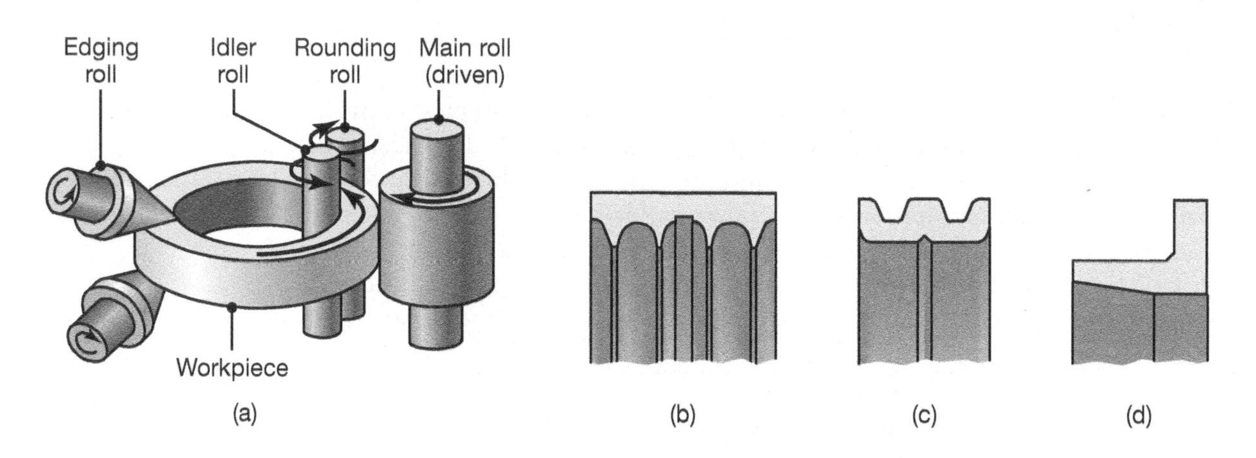

FIGURE 6.41 (a) Schematic illustration of a ring-rolling operation. Reducing the ring thickness results in an increase in its diameter. (b)–(d) Three examples of cross sections that can be produced by ring rolling.

This process can be carried out at room or elevated temperatures, depending on the size and strength of the product. The advantages of ring rolling over competing processes are shorter economical production runs, material savings, close-dimensional tolerances, and favorable grain-flow in the product. Typical parts made include large rings for rockets and turbines, gearwheel rims, ball- and roller-bearing races, flanges and reinforcing rings for pipes, and pressure vessels.

3. **Thread and gear rolling.** This is a cold-forming process in which threads are formed on round parts by passing them between reciprocating or rotating dies (Fig. 6.42a). Typical products made include

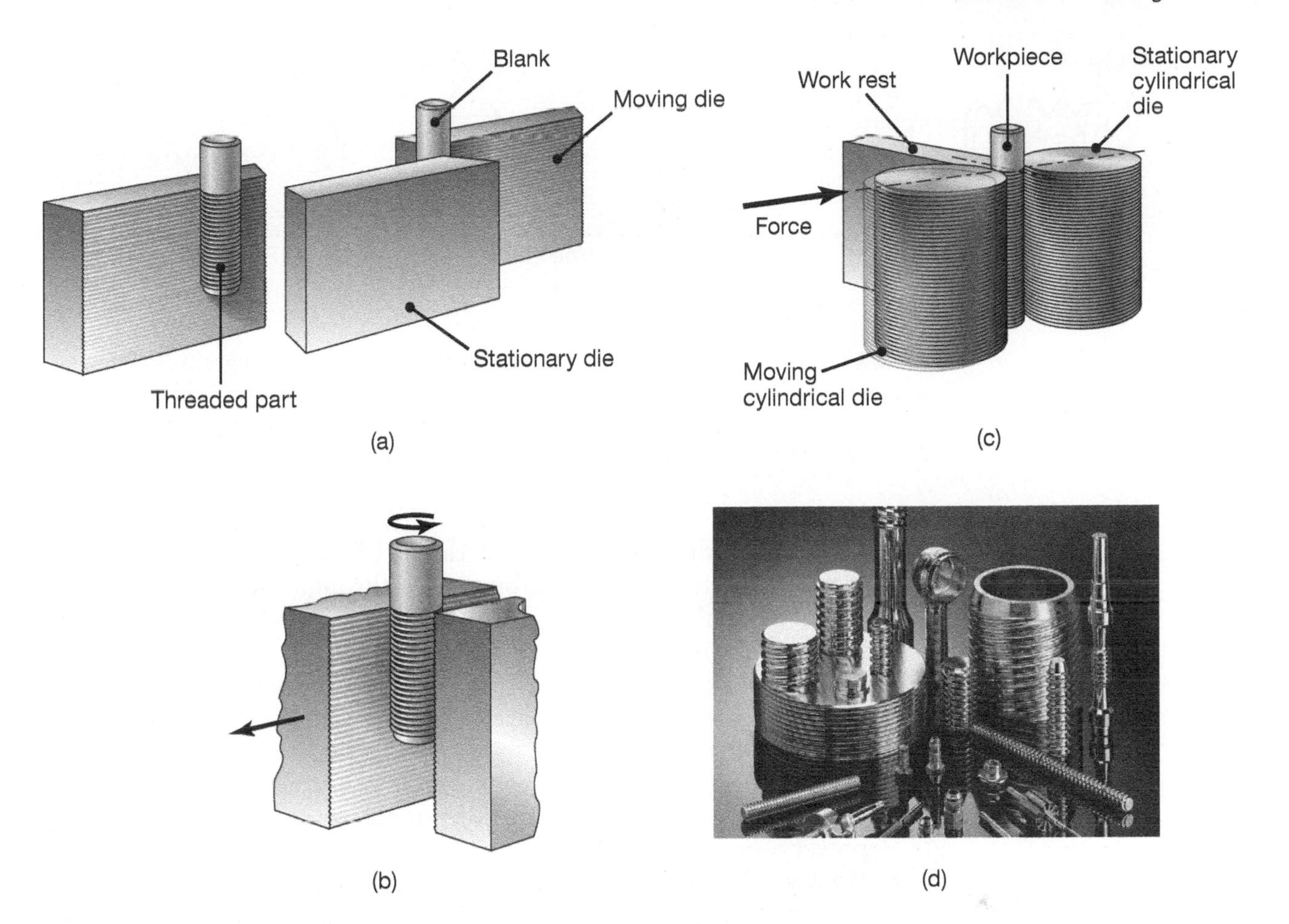

FIGURE 6.42 Thread-rolling processes: (a) and (b) reciprocating flat dies; (c) two-roller dies; and (d) thread-rolled parts, made economically and at high production rates. *Source:* (d) Courtesy of Tesker Manufacturing Corp.

screws, bolts, and similar threaded parts. It is essential that the material have sufficient ductility and that the rod or wire be of proper size. Lubrication is important for good surface finish and to minimize defects. Almost all externally-threaded fasteners are now made by this process. With flat dies, the threads are formed on the rod or wire with each stroke of the reciprocating die. Two- or three-roller thread-rolling machines are available. Threads also can be formed using a rotary die (Fig. 6.42c). Production rates in thread rolling are as high as 30 pieces per second. Thread rolling can also be carried out *internally* with a fluteless forming tap, and produces accurate threads with good strength.

The thread-rolling process produces higher-strength threads without any loss of metal because of the cold working involved. The surface produced is very smooth, and the process induces compressive residual stresses on part surfaces, thus improving fatigue life. Because of volume constancy in plastic deformation, a rolled thread requires a round stock of smaller diameter to produce the same major

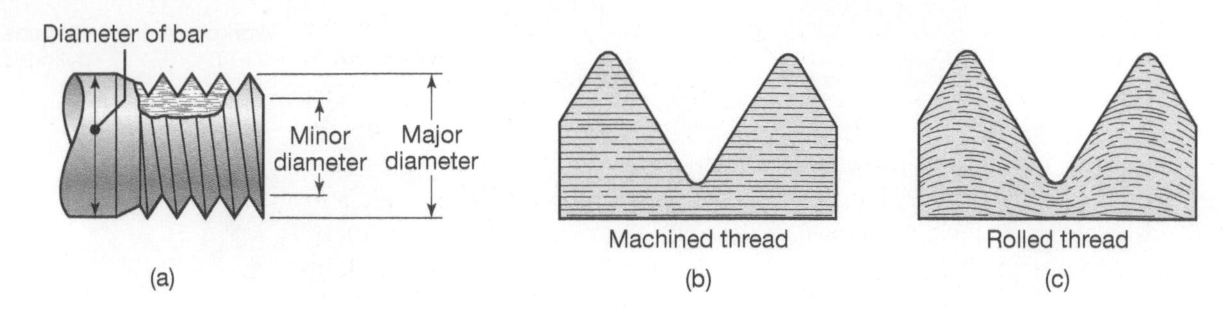

FIGURE 6.43 (a) Schematic illustration of thread features; (b) grain-flow lines in machined and (c) rolled threads. Note that unlike machined threads, which are cut through the grains of the metal, rolled threads follow the grains and because of the cold working involved, they are stronger.

diameter as that of a machined thread (Fig. 6.43). Also, whereas machining removes material by cutting through the grain-flow lines of the material, rolled threads have a grain-flow pattern that improves the strength of the thread.

Spur and helical gears can be produced by processes similar to thread rolling, and may be carried out on solid cylindrical blanks or on precut blanks. Helical gears also can be made by a direct extrusion process (See Fig. 6.46), using specially shaped dies. Cold rolling of gears has numerous applications in automatic transmissions and power tools.

4. **Rotary tube piercing.** Rotary tube piercing is based on the principle that when a round bar is subjected to radial compression in the manner shown in Fig. 6.44a, tensile stresses develop at the center of the roll. When the rod is subjected to cycling compressive stresses, as shown in Fig. 6.44b, a cavity begins to form at the center of the rod.

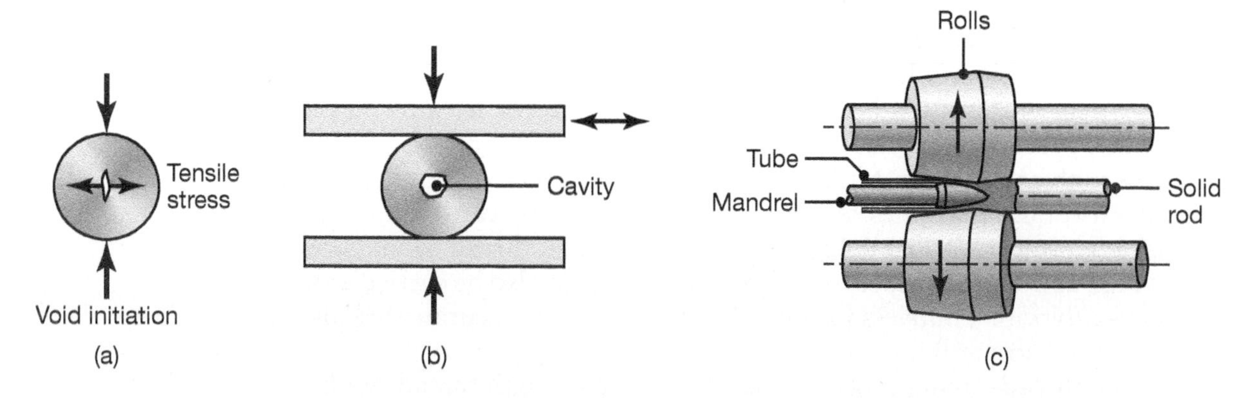

FIGURE 6.44 (a) Cracks developed in a solid round bar due to secondary tensile stresses; (b) simulation of the rotary-tube-piercing process; and (c) the Mannesmann process (mill) for seamless tube making. The mandrel is held in place by a long rod, although techniques also have been developed whereby the mandrel remains in place without using a rod.

Using this principle, the rotary-tube-piercing process, also known as the **Mannesmann process**, is carried out by an arrangement of rotating rolls, as shown in Fig. 6.44c. The axes of the rolls are skewed in order to be able to pull the round bar through the rolls by the longitudinal-force component of their rotary action. A mandrel assists the operation by expanding the hole and sizing the inside diameter of the tube. Because of the severe deformation that the metal undergoes in this process, it is important that the stock be of high quality and defect free.

5. **Tube rolling.** The diameter and thickness of tubes and pipe can be reduced by *tube rolling*, using shaped rolls, either with or without mandrels. In the **pilger mill**, the tube and an internal mandrel undergo a reciprocating motion as the tube is advanced and rotated periodically. Steel tubing of 265 mm (10.6 in.) in diameter have been produced by this process.

6.4 Extrusion

In the extrusion process (Fig. 6.45), developed in the late 1700s for producing lead pipe, a billet is placed in a chamber and forced through a die opening by a ram. The die may be round or of various shapes. Typical parts made are railings for sliding doors, window frames, aluminum

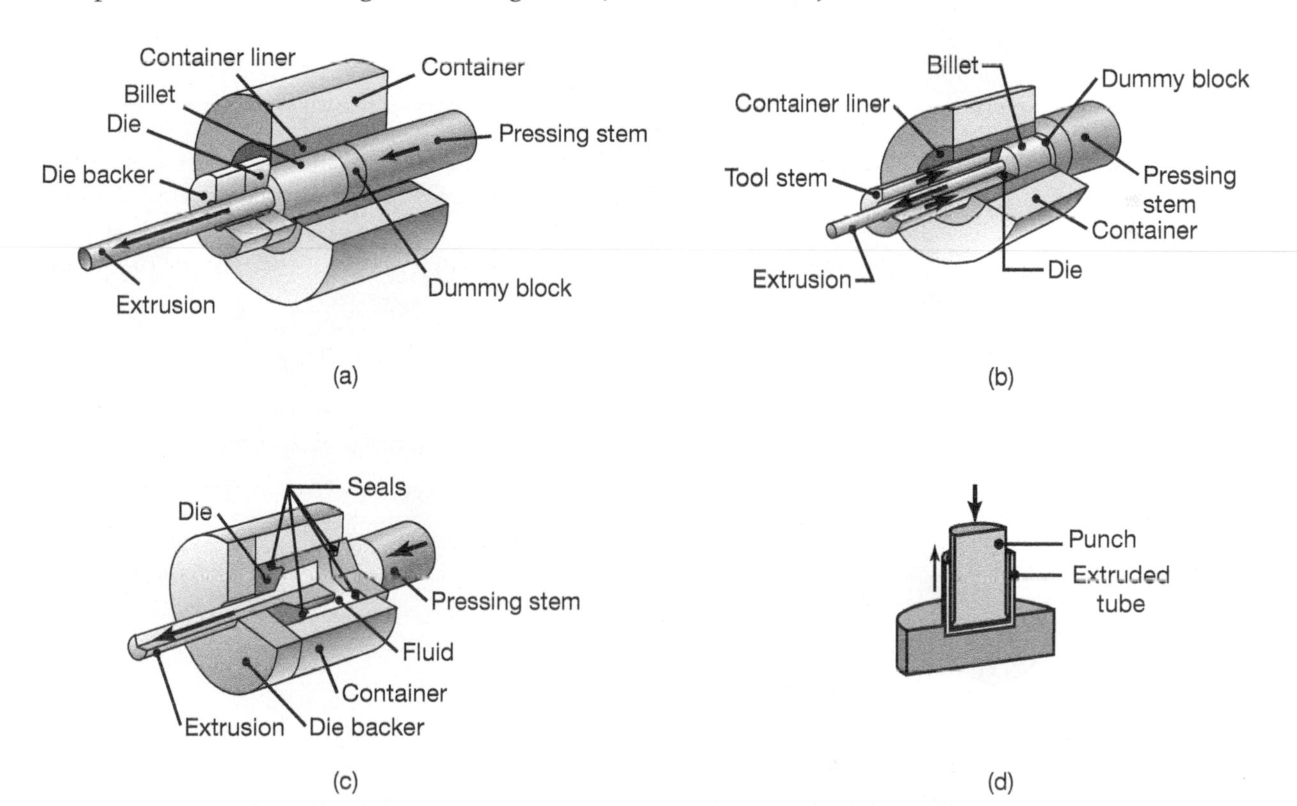

FIGURE 6.45 Types of extrusion: (a) direct; (b) indirect; (c) hydrostatic; (d) impact.

ladders, tubing, and structural and architectural shapes. Extrusion may be carried out cold or at elevated temperatures. Because a chamber with a specific volume is involved, each billet is extruded individually, and thus extrusion is basically a batch operation.

There are four basic types of extrusion, as described below and illustrated in Figs. 6.45 and 6.46.

1. **Direct extrusion** (*forward extrusion*) is similar to forcing toothpaste through the opening of a tube. Note that the billet in this process moves relative to the container wall in the same direction as the extruded product.
2. In **indirect extrusion** (*reverse, inverted,* or *backward extrusion*), the die moves toward the billet and there is no relative motion at the billet-container interface. Indirect extrusion is often used when the workpiece has a high level of friction with the billet, such as hot extrusion of steel.
3. In **hydrostatic extrusion,** the container is filled with a fluid that transmits pressure to the billet, which is then extruded through the die;

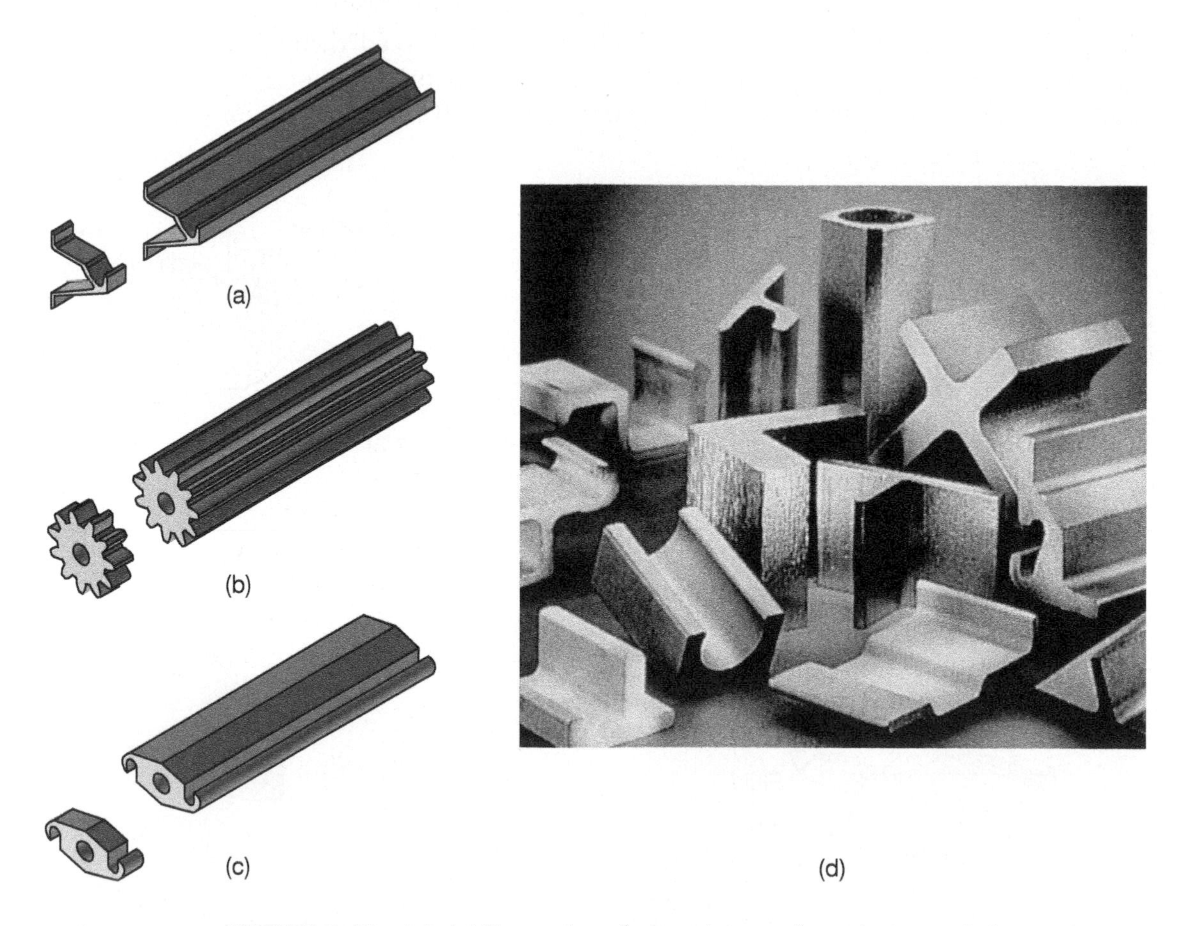

FIGURE 6.46 (a)–(c) Examples of extrusions and products made by sectioning them. *Source:* Kaiser Aluminum. (d) Examples of extruded cross sections. *Source:* (d) Courtesy of Plymouth Extruded Shapes.

there is essentially no friction along the container walls. Hydrostatic stresses in this process greatly increase material ductility (see Section 2.2.8).

4. **Impact extrusion** is a type of indirect extrusion particularly suitable for hollow shapes.

There are several parameters describing the extrusion process. The **extrusion ratio,** R_e, is defined as

$$R_e = \frac{A_o}{A_f}, \tag{6.48}$$

where A_o is the cross-sectional area of the billet and A_f is the area of the extruded product (Fig. 6.45). A geometric feature that describes the shape of the extruded product is the **circumscribing-circle diameter** (CCD), which is the diameter of the smallest circle into which the extruded cross section will fit. For example, the CCD for a square extruded cross section is the diagonal dimension of the cross section, or $1/\cos 45° = 1.44$ times its side dimension. The complexity of an extrusion is described by the term **shape factor,** which is the ratio of the perimeter of the part to its cross-sectional area. Thus, for example, a solid, round extrusion is the simplest shape, having the lowest shape factor.

6.4.1 Metal Flow in Extrusion

The *metal flow pattern* in extrusion is important in the overall process, as it is in all metalworking processes. Figure 6.47 shows three patterns typical of direct extrusion with *square dies*, where the die angle is 90°. From this figure, common flow patterns can be identified:

1. The most homogeneous (uniform) flow pattern (Fig. 6.47a) occurs when there is no friction at the interfaces. This type of flow also occurs in indirect extrusion, where there is no friction at the billet-container interfaces because there is no movement.
2. When friction along all interfaces is high, a **dead-metal zone** develops (Fig. 6.47b). Note the high-shear area as the material flows into the die exit (similar to liquid flowing into a funnel). This configuration may cause the billet surfaces (with their oxide layer and entrained lubricant) to enter this high-shear zone and be extruded, resulting in product defects (see Section 6.4.4).
3. In the third configuration, note that the high-shear zone extends farther back into the billet (Fig. 6.47c). This situation can be due to

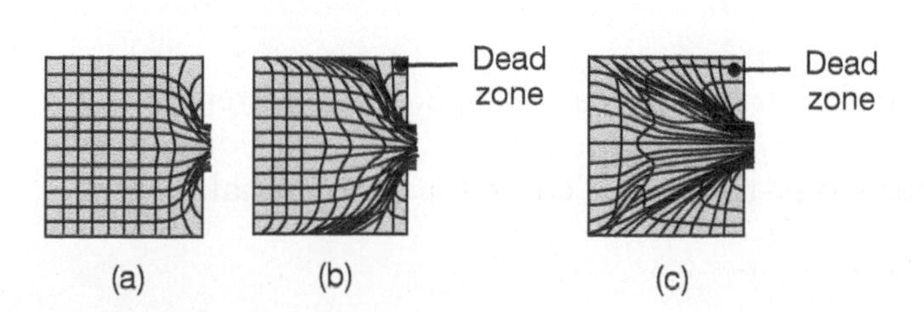

FIGURE 6.47 Schematic illustration of three different types of metal flow in direct extrusion. The die angle in these illustrations is 90°.

high container-wall friction (which retards the flow of the billet) or in extruding materials with a flow stress that drops rapidly with increasing temperature (such as titanium). In hot extrusion, the material near the container walls cools rapidly, thus becoming stronger; as a result, the material in the central regions of the billet flows more easily than that at the outer regions. A large dead-metal zone then forms, and the flow becomes inhomogeneous, leading to a defect known as a *pipe* or *extrusion defect* (Section 6.4.4).

6.4.2 Mechanics of Extrusion

In this section, the extrusion process is analyzed in order to estimate the extrusion force under different conditions of temperature and friction, and how this force can be minimized. The ram or stem force in direct extrusion (Fig. 6.45a) for different conditions can be calculated as follows.

1. **Ideal force, no friction.** Based on the extrusion ratio, the absolute value of the true strain that the material undergoes is

$$\epsilon_1 = \ln\left(\frac{A_o}{A_f}\right) = \ln\left(\frac{L_f}{L_o}\right) = \ln R_e, \tag{6.49}$$

where A_o and A_f and L_o and L_f are the areas and the lengths of the billet and the extruded product, respectively. For a perfectly plastic material with a yield strength of S_y, the energy dissipated in plastic deformation per unit volume, u, is

$$u = S_y \epsilon_1. \tag{6.50}$$

Hence, the work done on the billet is

$$\text{Work} = A_o L_o u. \tag{6.51}$$

The work is supplied by the ram force, F, which travels a distance L_o. Therefore,

$$\text{Work} = F L_o = p A_o L_o, \tag{6.52}$$

where p is the extrusion pressure. Substituting Eq. (6.51) into Eq. (6.52) and further substituting Eqs. (6.49) and (6.50),

$$p = u = S_y \ln\left(\frac{A_o}{A_f}\right) = S_y \ln R_e, \tag{6.53}$$

and the ideal force is $F = pA_o$.

Note that for strain-hardening materials, S_y must be replaced by the *average flow stress*, $\overline{\sigma}_f$. Also note that Eq. (6.53) is equal to the area under the true stress–true strain curve for the material.

2. **Ideal force, with friction.** Based on the slab method of analysis, described in Section 6.2.2, and for small die angles, it can be shown that with friction at the die-billet interface and ignoring the container-wall friction, the extrusion pressure, p, is given by the expression

$$p = S_y\left(1 + \frac{\tan\alpha}{\mu}\right)\left[R_e^{\mu\cot\alpha} - 1\right]. \tag{6.54}$$

If it is assumed that the frictional stress is equal to the shear yield strength k, and that because of the dead zone formed (See Fig. 6.47b and c), the material flows along a 45° "die angle," the pressure can be estimated as

$$p = S_y\left(1.7\ln R_e + \frac{2L}{D_o}\right). \tag{6.55}$$

Note that as the ram travels further toward the die, L decreases, and thus the pressure and force decrease (Fig. 6.48).

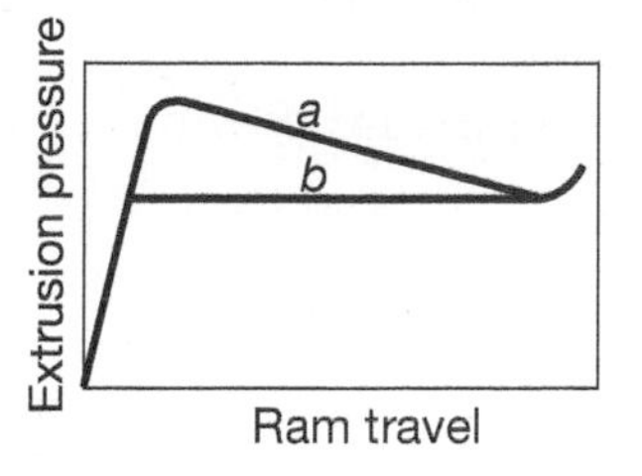

FIGURE 6.48 Schematic illustration of typical extrusion pressure as a function of ram travel: (a) direct extrusion and (b) indirect extrusion. The pressure in direct extrusion is higher because of the frictional resistance at the container-billet interfaces, which decreases as the billet length decreases in the container.

3. **Actual forces.** In extrusion practice, as well as in all metalworking processes, there are difficulties in estimating (a) the coefficient of friction and its variation throughout a process; (b) the flow stress of the material under the actual conditions of temperature and strain rate; and (c) the work involved in inhomogeneous deformation. A simple empirical formula has been developed in the form of

$$p = S_y(a + b\ln R_e), \tag{6.56}$$

where a and b are experimentally determined constants. It has been determined that an approximate value for a is 0.8, and that b ranges from 1.2 to 1.5.

4. **Optimum die angle.** The die angle has an important effect on forces in extrusion, which can be summarized as follows:

   a. The *ideal force* is a function of the strain that the material undergoes, and thus is a function of the extrusion ratio R. Consequently, it is independent of the die angle, as shown by Fig. 6.49.

   b. The force due to *friction* increases with decreasing die angle. This is because, as can be seen in Fig. 6.45, the length of contact along the billet-die interface increases as the die angle decreases. For extrusion, friction is dependent on the area of contact (see Section 4.4.1); thus the work required increases, as shown by curve (d).

   c. An additional force is required for *redundant work* due to inhomogeneous deformation of the material during extrusion (see Section 2.12). This work increases with the die angle.

The *total* extrusion force is the sum of these three components. Note that as shown by curve (a) in Fig. 6.49, there is a specific angle at which this force is a *minimum*, referred to as the *optimum angle*.

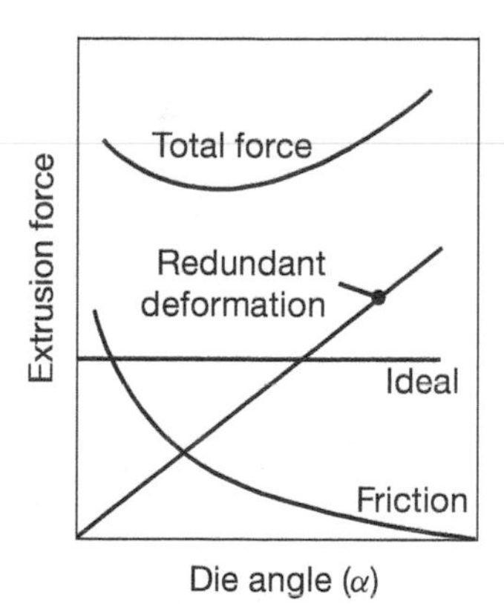

FIGURE 6.49 Schematic illustration of extrusion force as a function of die angle. Note that there is a die angle where the total extrusion force is a minimum (optimum die angle).

EXAMPLE 6.5 Strain Rate in Extrusion

Given: Consider extrusion of a round rod of radius r_o through a conical die.

Find: Determine the true strain rate as a function of distance x from the die entry.

Solution: From the geometry in the die gap,

$$\tan\alpha = \frac{r_o - r}{x}, \tag{6.57}$$

or

$$r = r_o - x\tan\alpha.$$

The incremental true strain can be defined as

$$d\epsilon = \frac{dA}{A},$$

where $A = \pi r^2$. Therefore, $dA = 2\pi r\,dr$, and hence

$$d\epsilon = \frac{2\,dr}{r},$$

where $dr = -\tan\alpha\,dx$. Also,

$$\dot{\epsilon} = \frac{d\epsilon}{dt} = -\left(\frac{2\tan\alpha}{r}\right)\left(\frac{dx}{dt}\right).$$

However, $dx/dt = V$, which is the velocity of the material at any location x in the die. Hence,

$$\dot{\epsilon} = -\frac{2V\tan\alpha}{r}.$$

Because the flow rate is constant (volume constancy), we can write

$$V = \frac{V_o r_o^2}{r^2},$$

and therefore,

$$\dot{\epsilon} = -\frac{2V_o r_o^2\tan\alpha}{r^3} = -\frac{2V_o r_o^2\tan\alpha}{(r_o - x\tan\alpha)^3}. \tag{6.58}$$

The negative sign in the equation is due to the fact that true strain is defined in terms of the cross-sectional area, which decreases as x increases.

5. **Force in hot extrusion.** Because of the strain-rate sensitivity of metals at elevated temperatures (Section 2.2.7), the force in hot extrusion can be difficult to calculate accurately. It can be shown that the average true strain rate, $\dot{\bar{\epsilon}}$, that the material undergoes is given by the expression

$$\dot{\bar{\epsilon}} = \frac{6V_o D_o^2 \tan\alpha}{D_o^3 - D_f^3} \ln R_e, \tag{6.59}$$

where V_o is the ram speed. It can be shown that (a) for high extrusion ratios, that is $D_o \gg D_f$; and (b) a die angle of $\alpha = 45°$, as is the case with a square die and under poor lubrication (thus developing a dead zone), the true strain rate reduces to

$$\dot{\bar{\epsilon}} = \frac{6V_o}{D_o} \ln R_e. \tag{6.60}$$

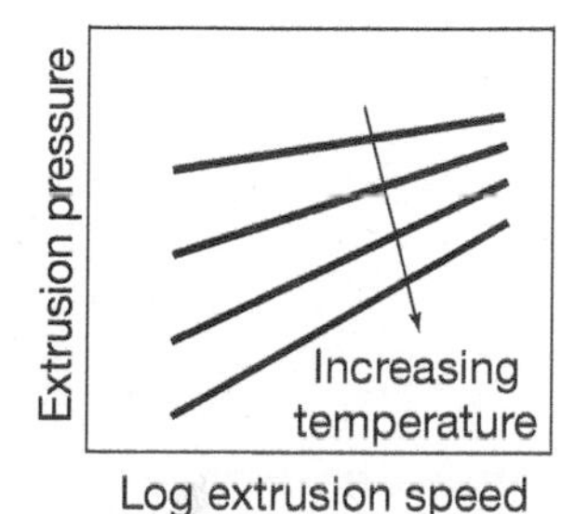

FIGURE 6.50 Schematic illustration of the effect of temperature and ram speed on extrusion pressure. Note the similarity of this figure with Fig. 2.10.

The typical effects of ram speed and temperature on extrusion pressure are shown in Fig. 6.50. As expected, the pressure increases rapidly with ram speed, especially at elevated temperatures due to increased strain-rate sensitivity (see Section 2.2.7). As speed increases, the rate of work done per unit time also increases. Also, the heat generated at higher speeds will not be removed fast enough, thus raising the temperature. The higher temperature can then lead to *incipient melting* of the workpiece material and possibly cause defects. Circumferential surface cracks caused by **hot shortness** also may develop (see Section 3.4.2), a phenomenon known as **speed cracking** (because of the high ram speed involved).

A parameter used to estimate the force in hot extrusion is an experimentally determined *extrusion constant*, K_e, that includes various factors involved in the operation. Thus,

$$p = K_e \ln R_e. \tag{6.61}$$

Some values of K_e for a variety of materials are given in Fig. 6.51.

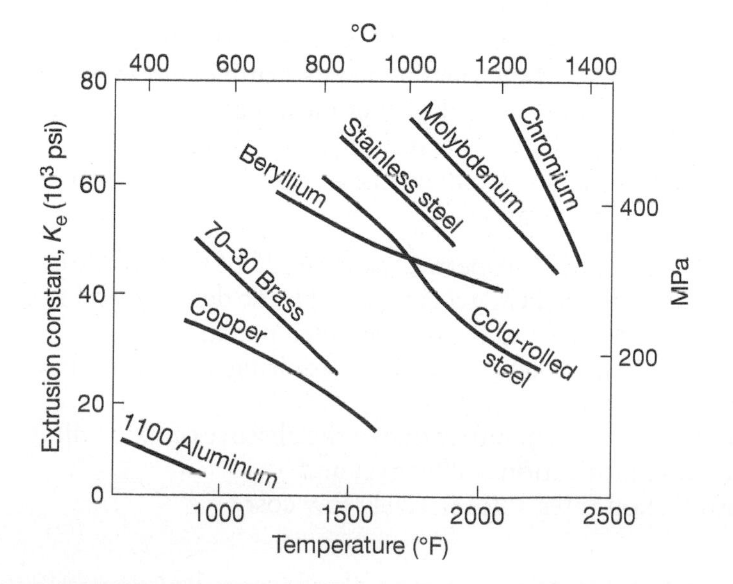

FIGURE 6.51 Extrusion constant, K_e, for various materials as a function of temperature. *Source:* After P. Loewenstein.

EXAMPLE 6.6 Force in Hot Extrusion

Given: A copper billet 125 mm in diameter and 250 mm long is extruded at 800°C at a speed of 250 mm/s.

Find: Using square dies and assuming poor lubrication, estimate the force required in this operation if the extruded diameter is 50 mm.

Solution: The extrusion ratio in this operation is

$$R_e = \frac{125^2}{50^2} = 6.25,$$

and the average true strain rate, from Eq. (6.60), is

$$\dot{\bar{\epsilon}} = \frac{6V_o}{D_o} \ln R_e = \frac{(6)(250)}{(125)} \ln 6.25 = 22 \text{ s}^{-1}.$$

From Table 2.4, use an average value for C of 130 MPa and $m = 0.12$. Hence, from Eq. (2.18),

$$\sigma = C\dot{\epsilon}^m = (130)(22)^{0.12} = 188 \text{ MPa}.$$

Substituting this value for the yield strength, Eq. (6.55) yields

$$p = \sigma\left(1.7 \ln R_e + \frac{2L}{D_o}\right) = (188)\left[(1.7)(\ln 6.25) + \frac{(2)(250)}{(125)}\right] = 1340 \text{ MPa}.$$

Therefore,

$$F = (p)(A_o) = (1340)\frac{(\pi)(0.125)^2}{4} = 16.4 \text{ MN}.$$

6.4.3 Miscellaneous Extrusion Processes

1. **Cold extrusion.** *Cold extrusion* is a general term that often is used to describe a combination of processes, particularly extrusion combined with forging (Fig. 6.52). Many ductile metals can be cold extruded into various configurations, with the billet mostly at room temperature or at a few hundred degrees. Typical parts made are automotive components and gear blanks. Cold extrusion is an important process because of such advantages as:

    a. Improved mechanical properties as a result of strain hardening, provided that the heat generated by plastic deformation and friction does not recrystallize the extruded metal;
    b. Good dimensional tolerances, thus requiring a minimum of finishing operations;
    c. Improved surface finish, partly due to the absence of oxide films, provided that lubrication is effective; and
    d. High production rates and relatively low cost.

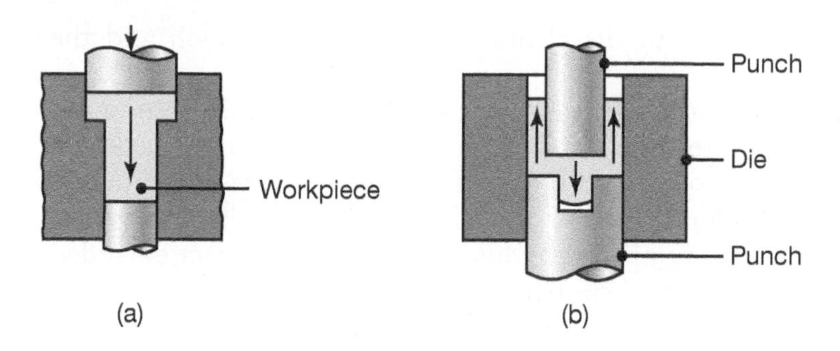

FIGURE 6.52 Two examples of cold extrusion. Arrows indicate the direction of material flow. These parts may also be considered as forgings.

On the other hand, the stresses on tooling and dies in cold extrusion are very high (especially with steel workpieces) and are on the order of the hardness of the material, that is, at least three times its flow stress (see Section 2.6.8). The design of tooling and selection of appropriate tool materials are crucial to success in cold extrusion. Punches are a critical component; they must have sufficient strength, toughness, and resistance to wear and fatigue.

Lubrication in cold extrusion is critical because new surfaces are generated during deformation, possibly causing *seizure* between the workpiece and the tooling. The most effective lubrication is provided by phosphate conversion coatings on the workpiece and with soap or wax as the lubricant. Temperature rise in cold extrusion is an important factor, especially at high extrusion ratios.

2. **Impact extrusion.** A process that often is included in the category of cold extrusion is *impact extrusion* (Fig. 6.53). In this operation, the punch descends at a high speed and strikes the *blank* (*slug*), and extrudes it in the opposite direction to the punch travel. A typical example of impact extrusion is the production of collapsible tubes. Note from the figure that the thickness of the extruded tubular cross

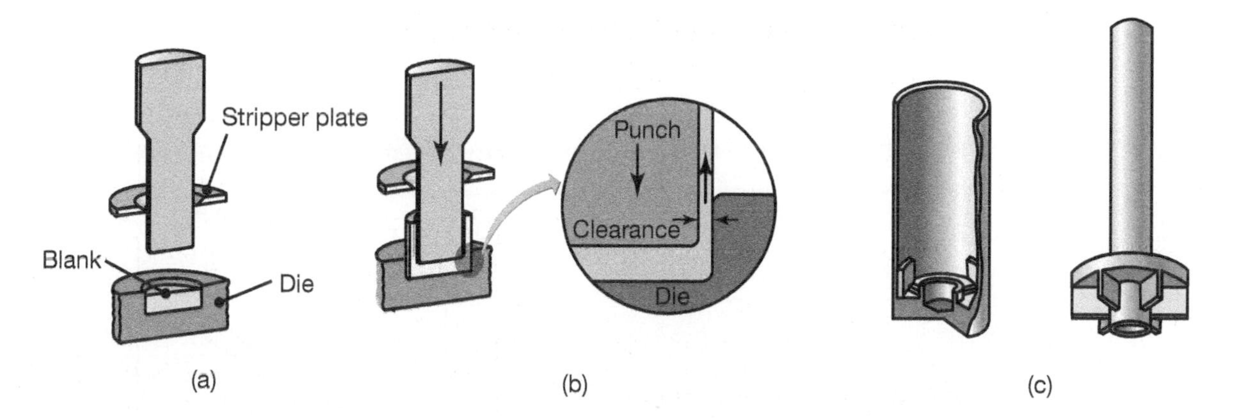

FIGURE 6.53 (a)–(b) Schematic illustration of the impact-extrusion process. The extruded parts are stripped using a stripper plate, as otherwise they may stick to the punch. (c) Two examples of products made by impact extrusion. Collapsible tubes can be produced by impact extrusion, referred to as the Hooker process.

section is a function of the clearance between the punch and the die cavity.

The impact-extrusion process typically produces tubular shapes, with wall thicknesses that are small in relation to their diameters, a ratio that can be as small as 0.005. The concentricity of the punch and the blank is important, as otherwise the wall thickness will not be uniform. A variety of nonferrous metals can be impact extruded into shapes such as those shown in Fig. 6.53b. The equipment used is typically vertical presses, at production rates as high as two parts per second.

3. **Hydrostatic extrusion.** In this process, the pressure required for extrusion is supplied through a fluid medium that surrounds the billet, as shown in Fig. 6.45c; there is no container-wall contact and hence no friction. The high pressure in the chamber also transmits some of the fluid to the die surfaces, significantly reducing friction and forces (Fig. 6.54). Pressures in this process are typically on the order of 1400 MPa (200 ksi). Hydrostatic extrusion can also be carried out by extruding the part into a second pressurized chamber, which is under lower pressure (**fluid-to-fluid extrusion**). Because of the highly pressurized environment, this operation reduces the defects that may otherwise develop in the extruded product. A variety of metals and polymers, solid shapes, tubes and other hollow shapes, and honeycomb and clad profiles can be extruded.

   Hydrostatic extrusion is usually carried out at room temperature, typically using vegetable oils as the fluid, particularly castor oil because its viscosity is not influenced significantly by pressure. For elevated-temperature hydrostatic extrusion, waxes, polymers, and glass are used as the fluid. These materials also serve as thermal insulators and help maintain the billet temperature during extrusion.

4. **Coaxial extrusion.** In this process, coaxial billets are extruded and joined together. *Cladding* is another application of this process; an example is copper clad with silver.

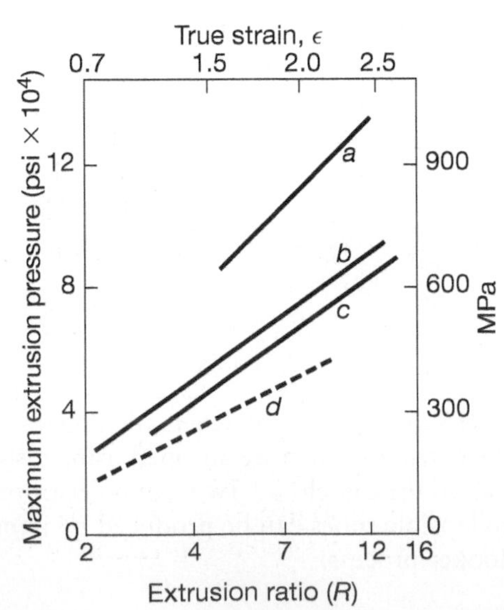

FIGURE 6.54 Extrusion pressure as a function of the extrusion ratio for an aluminum alloy. (a) Direct extrusion, $\alpha = 90°$. (b) Hydrostatic extrusion, $\alpha = 45°$. (c) Hydrostatic extrusion, $\alpha = 22.5°$. (d) Ideal homogeneous deformation, calculated. *Source:* After H. Li, D. Pugh, and K. Ashcroft.

6.4.4 Defects in Extrusion

There are three principal types of defects in extrusion.

1. **Surface cracking.** If the extrusion temperature, speed, and friction are too high, surface temperatures can rise significantly, leading to surface cracking and tearing (*fir-tree cracking* or *speed cracking*). These cracks are intergranular, and are usually the result of hot shortness (Section 3.4.2). They occur especially with aluminum, magnesium, and zinc alloys, but are also observed with other metals, such as molybdenum alloys. Typically, this behavior can be avoided by using lower temperatures and speeds.

   Surface cracking can also occur at low temperatures, and has been attributed to periodic sticking of the extruded product along the die land (**stick-slip**) during extrusion. The causes of stick-slip are as follows: when the product being extruded sticks to the die land, the extrusion pressure rapidly increases. With a high enough pressure, the product moves forward again, and the pressure is relieved; the cycle then repeats. Because of its surface appearance, this defect is known as a **bamboo defect**, and is especially encountered in hydrostatic extrusion where the pressure is sufficiently high to significantly increase the viscosity of the fluid, which then leads to the formation of a thick lubricant film. The billet then surges forward, which, in turn, relieves the pressure of the fluid and increases the friction. Thus, the change in the physical properties of the metalworking fluid is responsible for a stick-slip phenomenon. It has been observed that proper selection of a fluid is critical, and that increasing the extrusion speed also can eliminate stick-slip.
2. **Extrusion defect.** Note that the type of metal flow shown in Fig. 6.47c will tend to draw the surface oxides and impurities toward the center of the billet. Known as *extrusion defect*, **pipe, tailpipe,** and **fishtailing,** this type of defect renders as much as one-third of the length of the extruded material useless. This defect can be reduced by (a) modifying the flow pattern to a more homogeneous one, such as by controlling friction and minimizing the temperature gradient within the metal; (b) machining the billet surface prior to extruding to eliminate scale and impurities; and (c) using a dummy block (Fig. 6.45a) that is slightly smaller in diameter than the container, thus leaving a thin shell (*skull*) along the container wall as extrusion progresses.
3. **Internal cracking.** Cracks, variously known as **centerburst, center cracking, arrowhead fracture,** and **chevron cracking,** can develop at the center of an extruded product, as shown in Fig. 6.55. They are attributed to a state of *hydrostatic tensile stress* (also called *secondary tensile stress*) at the centerline of the deformation zone in the die. Such cracks also have been observed in tube extrusion and in spinning of tubes (Section 7.5.4), but the cracks appear on the inside surfaces for these processes.

   The major variables affecting hydrostatic tension are the (a) die angle; (b) extrusion ratio (or reduction in cross-sectional area); and

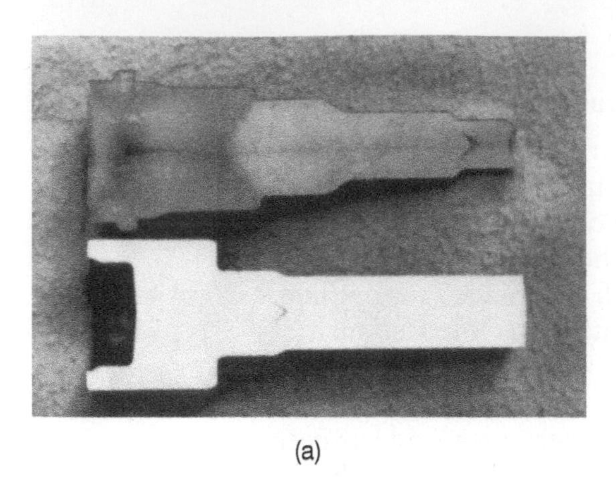

(a)

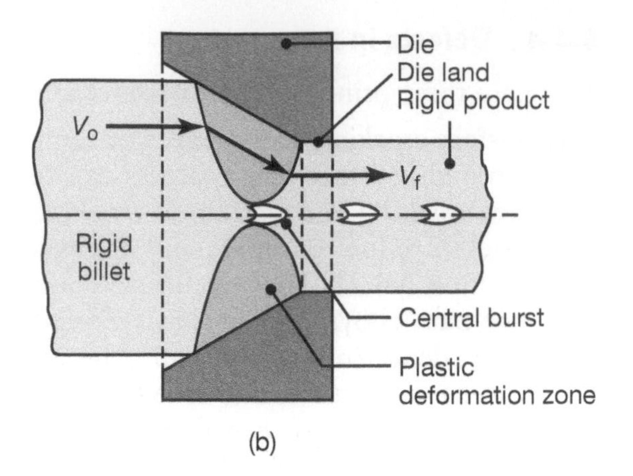

(b)

FIGURE 6.55 (a) Chevron cracking in round steel bars during extrusion. Unless the part is inspected, such internal detects may remain undetected and possibly cause failure of the part in service. (b) Deformation zone in extrusion, showing rigid and plastic zones. Note that the plastic zones do not meet, which then lead to chevron cracking. The same observations are also made in drawing round bars through conical dies and drawing flat sheet or plate through wedge-shaped dies.

(c) friction. The role of these factors can best be understood by observing the extent of inhomogeneous deformation during extrusion. Experimental studies have indicated that, for the same extrusion ratio, the deformation across the part becomes more inhomogeneous as the die angle becomes larger. Another factor is the die contact length, where it can be noted that the smaller the die angle, the longer the contact length. The size and depth of the deformation zone increases with increasing contact length, as illustrated in Fig. 6.55b. The h/L ratio also is an important parameter; the higher this ratio, the more inhomogeneous is the deformation.

Inhomogeneous deformation indicates that the center of the billet is not in a fully plastic state, because the plastic deformation zones under the die-contact lengths do not reach each other (Fig. 6.55a). Likewise, small extrusion ratios and high die angles retard the flow of the material at the surfaces, while the central portions are more free to flow through the opening in the die. High h/L ratios generate hydrostatic tensile stresses in the center of the billet, causing the type of defects shown in Fig. 6.55b. These defects form more readily in materials with impurities, inclusions, and voids, because these imperfections act as nucleation sites for defect formation. High friction in extrusion delays formation of these cracks. These observations are also valid for the drawing of rod and wire, as described in Section 6.5.

6.4.5 Extrusion Practice

A variety of nonferrous and ferrous metals can be extruded to various cross-sectional shapes and dimensions. Extrusion ratios typically range from about 10 to over 100; ram speeds may be up to 0.5 m/s. Generally, slower speeds are preferred for aluminum, magnesium, and copper, and higher speeds for steels, titanium, and refractory alloys. Presses for hot extrusion are typically hydraulic and horizontal, and for cold extrusion, they are usually vertical.

Hot extrusion. In addition to the strain-rate sensitivity of metals at elevated temperatures (Section 2.2.7), hot extrusion requires special considerations. Temperature ranges for hot extrusion are similar to those for hot forging operations, as given in Table 6.3. Cooling of the billet in the container (which normally is not heated) can result in highly inhomogeneous deformation during extrusion. Furthermore, since the billet is heated prior to extrusion, it typically develops an oxide layer, unless heated in an inert atmosphere. The oxide layer affects the frictional properties, can affect the flow of the material, and can produce an extruded part that is covered with an oxide layer.

Lubrication. For steels, stainless steels, and high-temperature materials, glass is an excellent lubricant. It maintains its viscosity at elevated temperatures, has good wetting characteristics, and acts as a thermal barrier between the billet, the container, and the die, thus minimizing billet cooling. In the **Séjournet process**, a circular glass pad is placed in the container at the die entrance. As extrusion progresses, this pad begins to soften and slowly melts; while doing so, it forms an optimal die geometry in order to minimize the energy required. The viscosity-temperature index of the glass is an important factor in its application, as high viscosity will lead to rapid glass depletion at the die entrance. Solid lubricants, such as graphite and molybdenum disulfide, also may be used in hot extrusion. Nonferrous metals are often extruded without a lubricant, although graphite may be used.

For materials that have a tendency to stick to the container and the die, the billet can be enclosed in a jacket (a thin-walled container) of a softer metal, such as copper or mild steel (**jacketing or canning**). Besides providing low friction stresses at the interface, canning prevents contamination of the billet by the environment; similarly it protects the environment from the billet (if it is toxic or radioactive). The canning technique is also used for processing metal powders (Chapter 11).

Die design. As in all metalworking operations, die design and die material selection are major considerations. Dies with angles of 90° (**square dies**, or **shear dies**) can be successfully used for nonferrous metals, particularly aluminum. Dies for tubing typically involve the use of a ram fitted with a mandrel, as shown in Fig. 6.56a. For billets with a pierced hole, the mandrel may simply be attached to the ram; however, if the billet is solid, it must first be pierced. Die materials for hot extrusion are typically hot-work die steels (Section 3.10.4). Various coatings can be applied to dies to extend their life (see Section 4.5).

Tubing and hollow extruded shapes (Fig. 6.57a) can be produced by **welding-chamber methods**, using special dies known as *spider, porthole,* and *bridge dies* (Fig. 6.57b-d). The metal flows around the arms of the

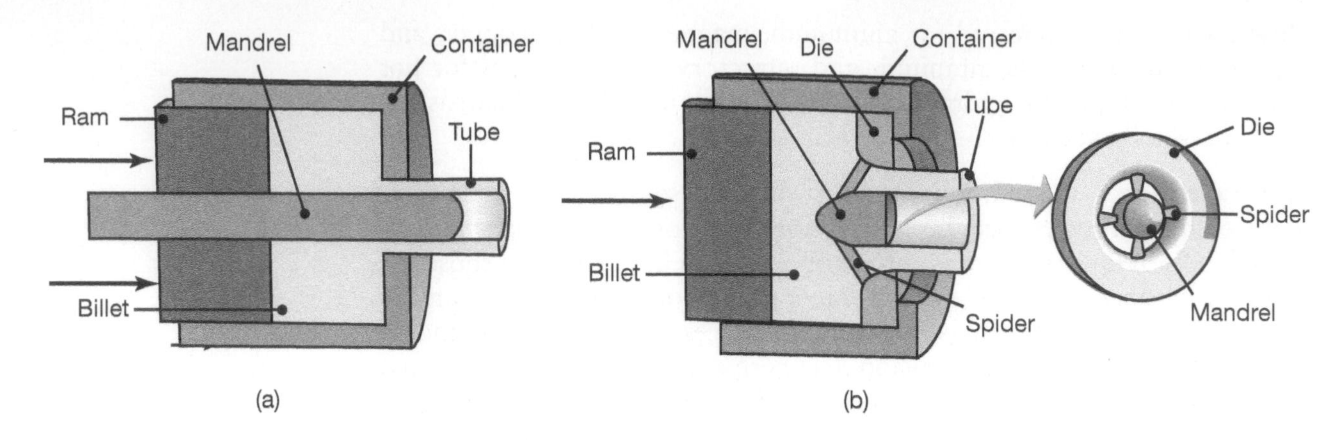

FIGURE 6.56 Extrusion of a seamless tube. (a) Using an internal mandrel that moves independently of the ram. An alternative arrangement has the mandrel integral with the ram. (b) Using a spider die (See Fig. 6.57c) to produce seamless tubing.

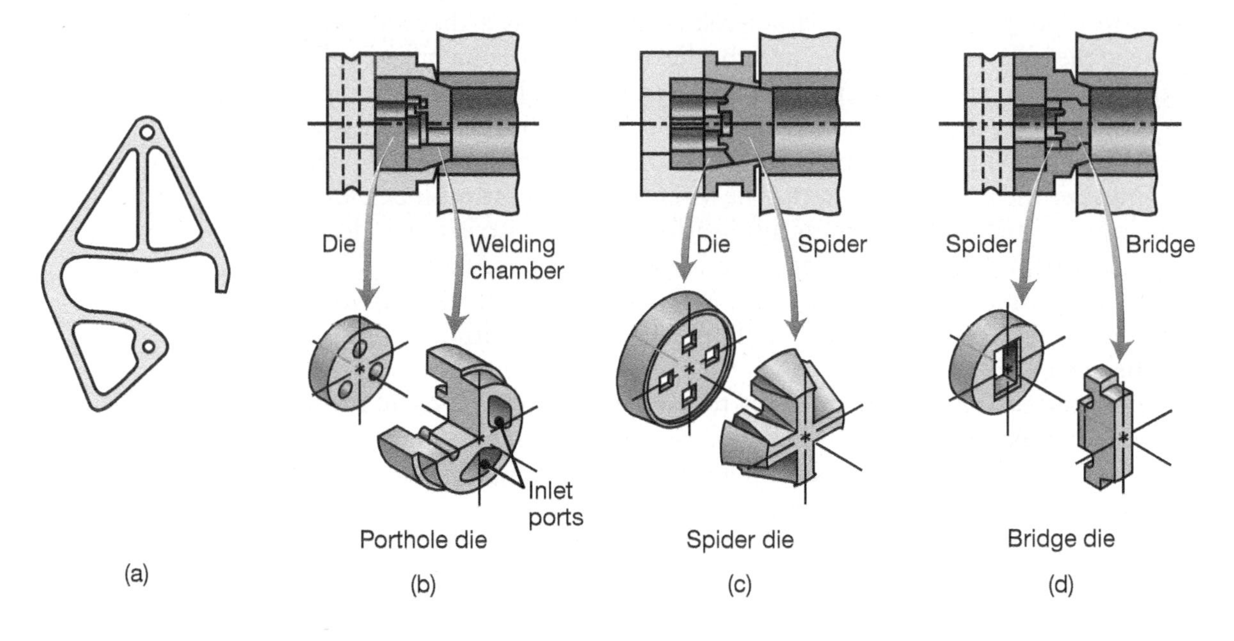

FIGURE 6.57 (a) An extruded 6063-T6 aluminum ladder lock for aluminum extension ladders. This part is thick and is sawed from the extrusion, as also shown in Fig. 6.46a. (b)–(d) Components of various types of dies for extruding intricate hollow shapes.

die into strands, which then reweld under the high pressures present at the die exit. Lubricants cannot be used in these operations because they penetrate interfaces and thus prevent rewelding. The welding-chamber methods are suitable only for aluminum and some of its alloys, because of their capacity for developing strong bonds under high pressure (*pressure welding*); also, die design is especially difficult if the material being extruded has a high strength.

Equipment. The basic equipment for extrusion is a hydraulic press, usually configured horizontally, and can be used for a variety of extrusion operations as well as with different strokes and speeds. The largest hydraulic press for extrusion has a ram force of 160 MN.

6.5 Rod, Wire, and Tube Drawing

Drawing is an operation in which the cross-sectional area of a bar or tube is reduced or changed in shape by pulling it through a converging die (Fig. 6.58). An established art by the 11th century, the drawing process is somewhat similar to extrusion, except that in drawing, the bar is under tension, whereas in extrusion it is under compression.

Rod and wire drawing produce good surface finish and dimensional tolerances and do not generally require finishing processes. The product is either used as produced or is further processed into other shapes by various additional operations. *Rods* are used for such applications as shafts, spindles, and structural members, and as the raw material for making fasteners, such as bolts and screws. *Wire* and wire products have a wide range of applications, such as electrical wiring, cables, springs, musical instruments, paper clips, fencing, welding electrodes, and shopping carts. Wire diameters may be as small as 0.025 mm.

6.5.1 Mechanics of Rod and Wire Drawing

The major variables in determining the drawing stress are reduction in cross-sectional area, die angle, and friction (Fig. 6.58).

1. **Ideal deformation.** The drawing stress, σ_d, for the simplest case of ideal deformation, can be obtained by the same approach as that used for extrusion. Thus,

$$\sigma_d = S_y \ln\left(\frac{A_o}{A_f}\right). \tag{6.62}$$

Note that this expression is the same as Eq. (6.53) and that it also represents the energy per unit volume, u. As emphasized throughout

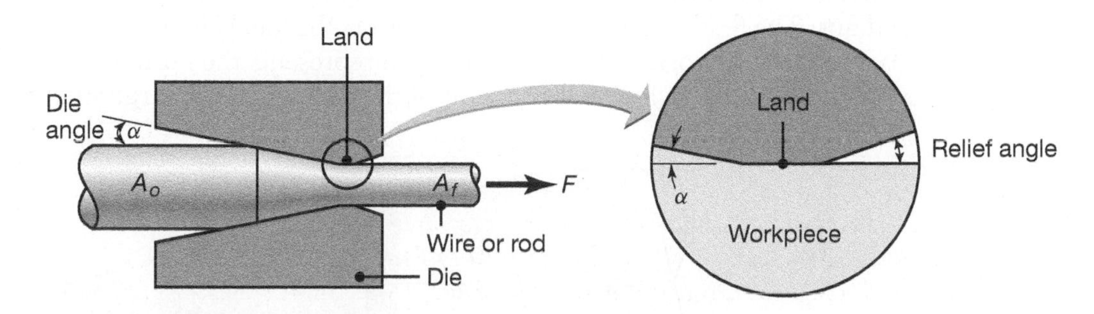

FIGURE 6.58 Variables in drawing round rod or wire.

this chapter, for strain-hardening materials, S_y is replaced by an average flow stress, $\overline{\sigma}_f$, in the deformation zone. Thus, for a material that exhibits the true stress–true strain behavior of

$$\sigma = K\epsilon^n,$$

the quantity $\overline{\sigma}_f$ is obtained from Eq. (6.10). The drawing force, F, is then

$$F = \overline{\sigma}_f A_f \ln\left(\frac{A_o}{A_f}\right). \tag{6.63}$$

Note that the higher the reduction in cross-sectional area and the stronger the material, the higher the drawing force.

2. **Ideal deformation and friction.** Friction at the die-workpiece interface increases the drawing force, because work has to be done to overcome friction. Using the slab method of analysis and on the basis of Fig. 6.59, the following expression for the drawing stress can be obtained:

$$\sigma_d = S_y\left(1 + \frac{\tan\alpha}{\mu}\right)\left[1 - \left(\frac{A_f}{A_o}\right)^{\mu\cot\alpha}\right]. \tag{6.64}$$

With good lubrication, the coefficient of friction, μ, in wire drawing typically ranges from about 0.03 to 0.1 (see Section 4.4.1). Even though Eq. (6.64) does not include the redundant work involved in the process, it is in reasonably good agreement with experimental data for small die angles and for a wide range of reductions in cross section.

3. **Redundant work.** Depending on the die angle and reduction, the material in drawing undergoes *inhomogeneous deformation*, much as it does in extrusion (see Section 6.4). It can be shown that when the redundant work of deformation is included, the expression for the drawing stress becomes

$$\sigma_d = \overline{\sigma}_f\left\{\left(1 + \frac{\tan\alpha}{\mu}\right)\left[1 - \left(\frac{A_f}{A_o}\right)^{\mu\cot\alpha}\right] + \frac{4}{3\sqrt{3}}\alpha^2\left(\frac{1-r}{r}\right)\right\}, \tag{6.65}$$

where r is the fractional reduction of area and α is the die angle, in radians. The first term in Eq. (6.65) represents the ideal and frictional components of work, and the second term represents the redundant-work component, which is a function of the die angle. The larger the

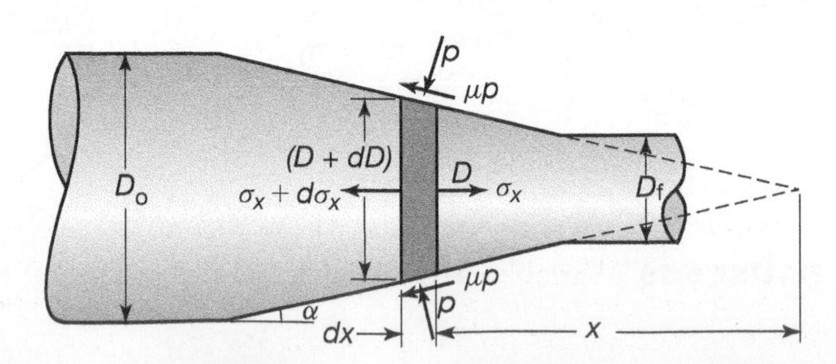

FIGURE 6.59 Stresses acting on an element in drawing of a solid cylindrical rod or wire through a converging conical die.

die angle, the greater the inhomogeneous deformation and, hence, the greater the redundant work.

For small die angles, another expression for the drawing stress that includes all three components of work is

$$\sigma_d = \bar{\sigma}_f \left[\left(1 + \frac{\mu}{\alpha}\right) \ln\left(\frac{A_o}{A_f}\right) + \frac{2}{3}\alpha \right]. \tag{6.66}$$

The last term in this expression is the redundant-work component, assuming that it increases linearly with the die angle, as shown in Fig. 6.49. Because redundant deformation is a function of the aspect ratio, an **inhomogeneity factor**, Φ, for solid round cross sections has been developed and expressed as

$$\Phi = 1 + 0.12\left(\frac{D_f}{L}\right), \tag{6.67}$$

where L is the length of contact between the workpiece and die. A useful expression for the drawing stress is then

$$\sigma_d = \Phi\bar{\sigma}_f \left(1 + \frac{\mu}{\alpha}\right) \ln\left(\frac{A_o}{A_f}\right). \tag{6.68}$$

Equations (6.64) through (6.68) provide reasonable predictions for the stresses required for wire drawing.

4. **Die pressure.** Based on yield criteria and noting that the compressive stresses in the two principal directions are equal to p, the *die pressure* along the die contact length can be obtained from

$$p = \sigma_f - \sigma, \tag{6.69}$$

where σ is the tensile stress in the deformation zone at a particular diameter, and σ_f is the flow stress of the material at that diameter. Note that σ is equal to σ_d at the die exit, and is zero at the die entry. Equation (6.69) indicates that as the tensile stress increases toward the exit, the die pressure drops toward the exit, as also shown qualitatively in Fig. 6.60.

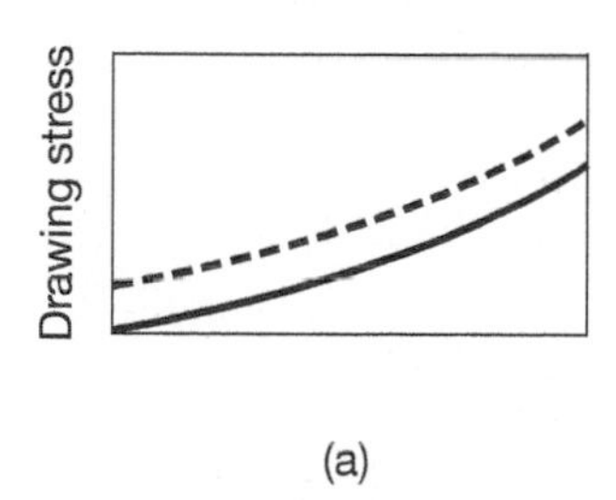

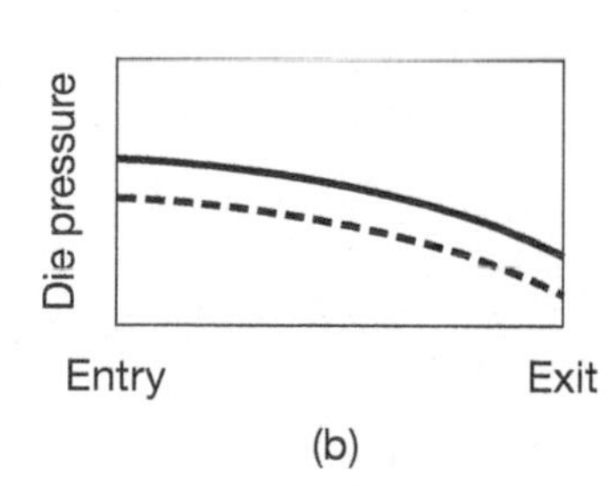

FIGURE 6.60 Variation in the (a) drawing stress and (b) die contact pressure along the deformation zone. Note that as the drawing stress increases, the die pressure decreases (see also *yield criteria*, described in Section 2.11). Note the effect of back tension on the stress and pressure.

EXAMPLE 6.7 Power Required and Die Pressure in Rod Drawing

Given: A round rod of annealed 302 stainless steel is being drawn from a diameter of 10 mm to 8 mm at a speed of 0.5 m/s. Assume that the frictional and redundant work together constitute 40% of the ideal work of deformation.

Find: Calculate the power required in this operation, and the die pressure at the die exit.

Solution: The true strain that the material undergoes in this operation is

$$\epsilon_1 = \ln\left(\frac{10^2}{8^2}\right) = 0.446.$$

From Table 2.2, for this material and condition, $K = 1300$ MPa and $n = 0.30$. Hence,

$$\bar{\sigma}_f = \frac{K\epsilon_1^n}{n+1} = \frac{(1300)(0.446)^{0.30}}{1.30} = 785 \text{ MPa}.$$

From Eq. (6.63), the drawing force is

$$F = \bar{\sigma}_f A_f \ln\left(\frac{A_o}{A_f}\right),$$

where

$$A_f = \frac{(\pi)(0.008)^2}{4} = 5 \times 10^{-5} \text{ m}^2.$$

Therefore,

$$F = (785)\left(5 \times 10^{-5}\right)(0.446) = 0.0175 \text{ MN}$$

and

$$\text{Power} = (F)(V_f) = 8.75 \text{ kW}.$$

As given in the problem statement, the actual power will be 40% higher, or

$$\text{Actual power} = (1.4)(8.75) = 12.25 \text{ kW}.$$

The die pressure can be obtained by noting from Eq. (6.69) that

$$p = \sigma_f - \sigma,$$

where, at the exit of the die, σ_f represents the flow stress of the material and σ refers to the drawing stress. Thus,

$$\sigma_f = K\epsilon_1^n = (1300)(0.446)^{0.30} = 1020 \text{ MPa}.$$

Hence, using the force obtained above,

$$\sigma_d = \frac{F}{A_f} = \frac{(1.4)(0.0175)}{0.00005} = 490 \text{ MPa}.$$

Therefore, the die pressure at the exit is

$$p = 1020 - 490 = 530 \text{ MPa}.$$

5. **Drawing at elevated temperatures.** Recall that the flow stress of metals is a function of the strain rate. In drawing, the *average true strain rate*, $\dot{\bar{\epsilon}}$, in the deformation zone is given by

$$\dot{\bar{\epsilon}} = \frac{6V_o}{D_o} \ln\left(\frac{A_o}{A_f}\right), \tag{6.70}$$

which is the same as Eq. (6.60). After first calculating the average strain rate, the flow stress and the average flow stress, $\overline{\sigma}_f$, of the material can be calculated and substituted in appropriate equations.

6. **Optimum die angle.** Because of the various effects of the die angle on the three components of work (ideal, friction, and redundant), there is, as in extrusion, an *optimum die angle* at which the drawing force is a minimum. Figure 6.61 shows a typical example, where it can be noted that the optimum angle for the minimum drawing force increases with reduction; note also that optimum angles are relatively small.

7. **Maximum reduction per pass.** As reduction increases, the drawing stress increases. There is, however, a limit to the magnitude of the drawing stress. If it reaches the yield strength of the material at the exit, it will yield and fail. An expression for the limiting situation can be obtained based on the fact that, in the ideal case of a perfectly plastic material with a yield strength S_y, the limiting condition is

$$\sigma_d = S_y \ln\left(\frac{A_o}{A_f}\right) = S_y, \tag{6.71}$$

or

$$\ln\left(\frac{A_o}{A_f}\right) = 1,$$

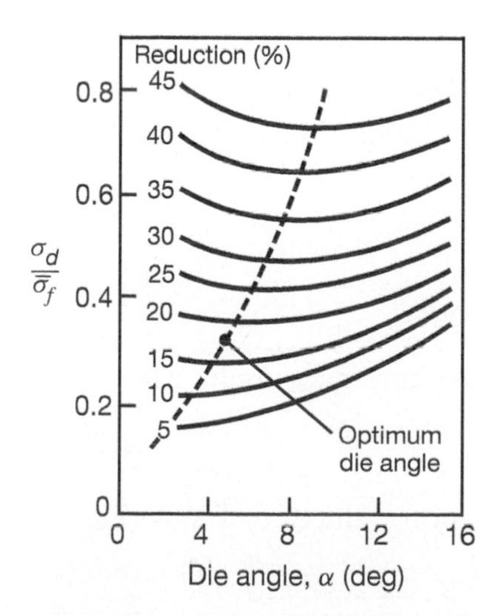

FIGURE 6.61 The effect of reduction in cross-sectional area on the optimum die angle in drawing. *Source:* After J.G. Wistreich.

and therefore,

$$\frac{A_o}{A_f} = e.$$

Thus, the maximum reduction per pass is given by

$$\frac{A_o - A_f}{A_o} = 1 - \frac{1}{e} = 0.63 = 63\%. \tag{6.72}$$

It will be noted that because of strain hardening, the exiting material will be stronger than the rest of the material in the die gap; consequently, the maximum reduction per pass will increase. The effects of friction and die angle on maximum reduction per pass are similar to those shown in Fig. 6.49. Because both friction and redundant deformation contribute to an increase in the drawing stress, the maximum reduction per pass will be lower than in the ideal case. That is, friction and redundant work have a larger effect than strain hardening.

EXAMPLE 6.8 Maximum Reduction Per Pass for a Strain-Hardening Material

Given: A material with a true stress–true strain curve of $\sigma = K\epsilon^n$ is being drawn through conical dies.

Find: Obtain an expression for the maximum reduction per pass, ignoring friction and redundant work.

Solution: From Eq. (6.62),

$$\sigma_d = \overline{\sigma}_f \ln\left(\frac{A_o}{A_f}\right) = \overline{\sigma}_f \epsilon_1,$$

where, from Eq. (6.10),

$$\overline{\sigma}_f = \frac{K\epsilon_1^n}{n+1}$$

and σ_d, for this problem, can have a maximum value equal to the flow stress at ϵ_1, or

$$\sigma_d = K\epsilon_1^n.$$

Equation (6.71) can now be rewritten as (but using flow stress, not yield strength, as the material is strain hardening) as:

$$K\epsilon_1^n = \frac{K\epsilon_1^n}{n+1}\epsilon_1,$$

or

$$\epsilon_1 = n + 1.$$

With $\epsilon_1 = \ln(A_o/A_f)$ and a maximum reduction of $(A_o - A_f)/A_o$, these expressions reduce to

$$\text{Maximum reduction per pass} = 1 - e^{-(n+1)}. \tag{6.73}$$

Note that when $n = 0$ (that is, a perfectly plastic material), this expression reduces to Eq. (6.72), and that as n increases, the maximum reduction per pass increases.

8. **Drawing of flat strip.** The dies in flat strip drawing are *wedge shaped*, and there is little or no change in the width of the strip during drawing. The drawing process is then somewhat similar to that of rolling wide strips, and hence the process can be considered as a **plane-strain** problem, especially at large width-to-thickness ratios. Although not of particular industrial significance, flat drawing is the fundamental deformation mechanism in **ironing**, as described in Section 7.6.

   The approach to determining drawing force and maximum reduction is similar to that for round sections; thus, the drawing stress for the ideal condition is

$$\sigma_d = S'_y \ln\left(\frac{h_o}{h_f}\right), \tag{6.74}$$

   where S'_y is the yield strength of the material in plane strain and h_o and h_f are the original and final thicknesses of the strip, respectively. The effects of friction and redundant deformation in strip drawing also are similar to those for round sections.

   The maximum reduction per pass can be obtained by equating the drawing stress [Eq. (6.74)] to the *uniaxial yield strength* of the material (because the drawn strip is subjected only to simple tension). Thus,

$$\sigma_d = S'_y \ln\left(\frac{h_o}{h_f}\right) = S_y, \quad \ln\left(\frac{h_o}{h_f}\right) = \frac{S_y}{S'_y} = \frac{\sqrt{3}}{2}, \quad \text{and} \quad \frac{h_o}{h_f} = e^{\sqrt{3}/2},$$

   which reduces to

$$\text{Maximum reduction per pass} = 1 - \frac{1}{e^{\sqrt{3}/2}} = 0.58 = 58\%. \tag{6.75}$$

9. **Drawing of tubes.** Tubes produced by extrusion or other processes, such as shape rolling or the Mannesmann process, can be reduced in thickness or diameter (**tube sinking**) by the tube-drawing processes illustrated in Fig. 6.62. Shape changes can also be imparted by using dies and mandrels with various profiles. Drawing forces, die pressure, and the maximum reduction per pass in tube drawing can be calculated by methods similar to those described above.

6.5.2 Defects in Drawing

Defects in drawing are similar to those observed in extrusion (Section 6.4.4), especially center cracking. The factors influencing center cracking are the same, namely, that the tendency for cracking increases with increasing die angle, decreasing reduction per pass, friction, and the presence of inclusions in the material. A type of surface defect in drawing is the formation of **seams**. These are longitudinal scratches or folds in the material that can open up (See, for example, Fig. 3.21d) during subsequent forming operations, such as upsetting, heading, thread rolling, or bending of the rod or wire.

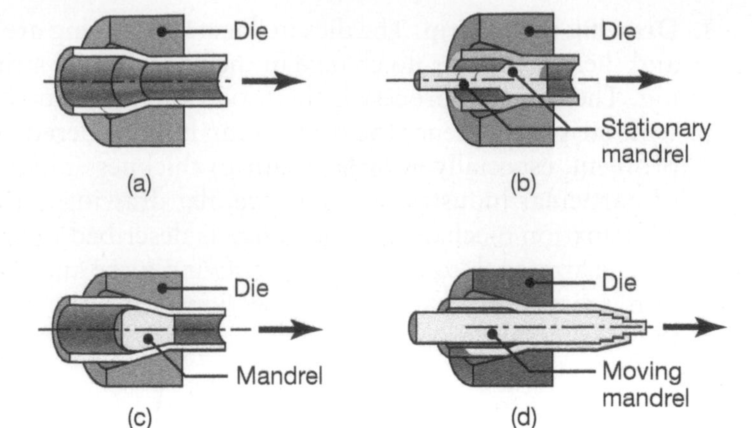

FIGURE 6.62 Examples of tube-drawing operations, with and without an internal mandrel. Note that a variety of diameters and wall thicknesses can be produced from the same tube stock (that has been produced by other processes, such as extrusion).

Because of inhomogeneous deformation, residual stresses usually develop in cold-drawn rod, wire, and tube. Typically, a wide range of residual stresses can be present, and in three principal directions, as shown in Fig. 6.63. For very light reductions, the surface residual stresses are compressive; light reductions are equivalent to shot peening or surface rolling (see Section 4.5.1). Recall that residual stresses can be significant in stress-corrosion cracking metals of over a period of time, as well as in distortion and warping of a part (See Fig. 2.28) when a layer is subsequently removed, such as by machining or grinding.

6.5.3 Drawing Practice

As in all metalworking processes, successful drawing requires proper selection of process parameters. A typical die for drawing and its characteristic

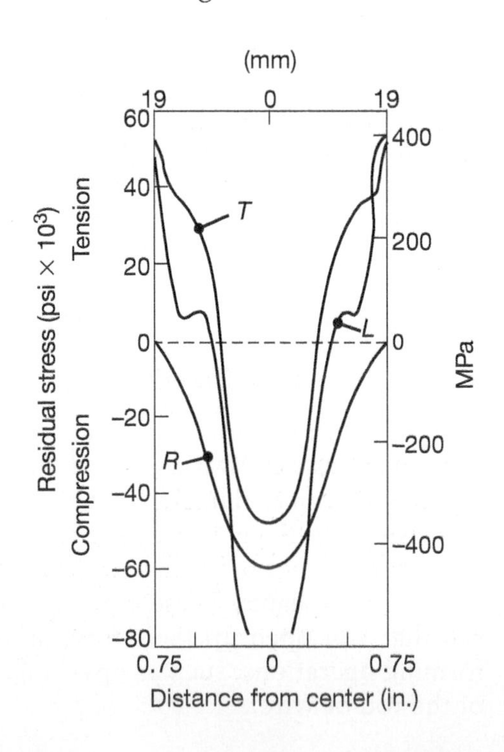

FIGURE 6.63 Residual stresses in cold-drawn 1045 carbon steel round rod: T = transverse direction, L = longitudinal direction and R = radial direction.

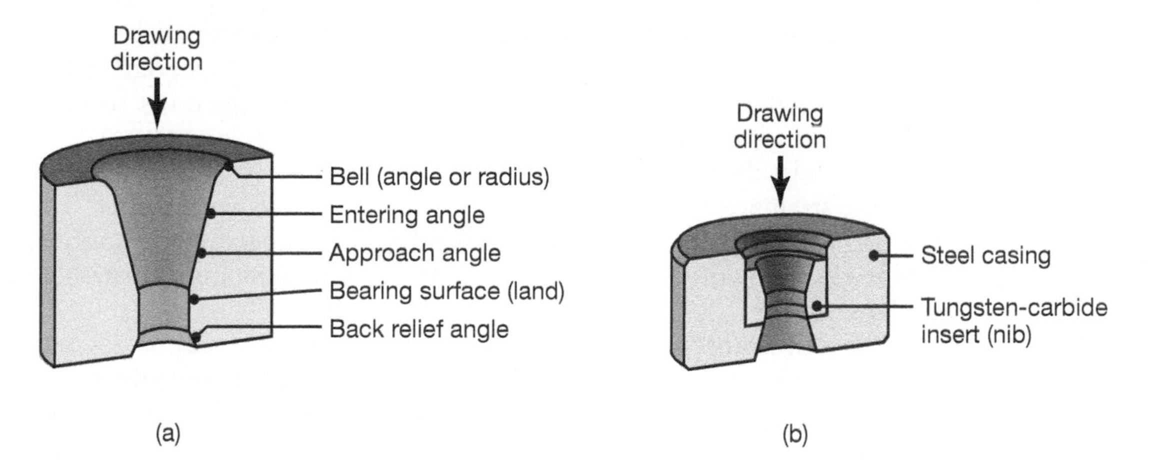

FIGURE 6.64 (a) Terminology for a typical die for drawing round rod or wire. (b) Tungsten-carbide die insert in a steel casing. Diamond dies, used in drawing thin wire, also are encased in a similar manner.

features are shown in Fig. 6.64. The purpose of the land is to size the product, that is, to set its final diameter. Also, because the die is typically reground to extend its life, the land maintains the exit dimension of the die opening.

Die angles typically range from 6° to 15°. Reductions in cross-sectional area per pass range from about 10 to 45%; generally, the smaller the cross section, the lower is the reduction per pass. Light reductions may also be taken on rods (**sizing pass**) to improve dimensional accuracy and surface finish. Reductions per pass greater than 45% may result in breakdown of lubrication and deterioration of the product's surface finish.

In typical drawing practice, a rod or wire is fed into the die by first **pointing** it, such as by *swaging* (forming the tip of the rod into a conical shape, as described in Section 6.6). After it is fed into the die, the tip of the wire is clamped into the jaws of the wire-drawing machine, and the rod is drawn continuously through the die. In most operations, the wire passes through a series of dies (**tandem drawing**). In order to avoid excessive tension in the exiting wire, it is wound one or two turns around a *capstan* (a drum) between each pair of dies (Fig. 6.65). The speed of the capstan

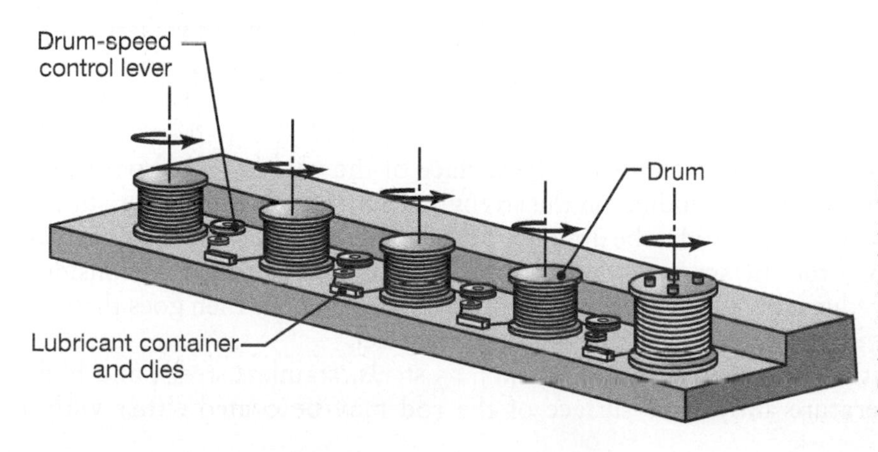

FIGURE 6.65 An illustration of a multistage wire drawing operation typically used to produce copper wire. Shown is a five bull block configuration; wire drawing machines can incorporate 15 or more of these drums, depending on the material and wire size.

is adjusted so that it supplies not only tension, but also a small *back tension* to the wire entering the next die. Depending on the material and its cross-sectional area, drawing speeds may be low for large cross sections and high for very fine wire.

Rods and tubes that are not sufficiently straight, or are supplied as coils, are straightened by passing them through multiple pairs of rolls. The rolls subject the material to a series of bending and unbending operations, similar to the method shown in Fig. 6.38 for straightening rolled sheet or plate.

Because of strain hardening, intermediate annealing between passes may be necessary in cold drawing in order to maintain sufficient ductility of the material. Steel wires for springs and musical instruments are made by a heat-treatment process, called **patenting**, which can be performed either prior to or after the drawing operation. These wires have ultimate tensile strengths as high as 4800 MPa (700,000 psi), and tensile reduction of area of about 20%.

Bundle drawing. In this process, a number of wires (as many as several thousand) are drawn simultaneously as a bundle. To prevent sticking, the wires are separated from each other by a suitable material, usually a viscous lubricant. The cross section of the wires is somewhat polygonal, because of the manner in which the wires are pressed together. The wires produced may be as small as 4 μm (160 μin.) in diameter, and can be made of stainless steels, titanium, and high-temperature alloys.

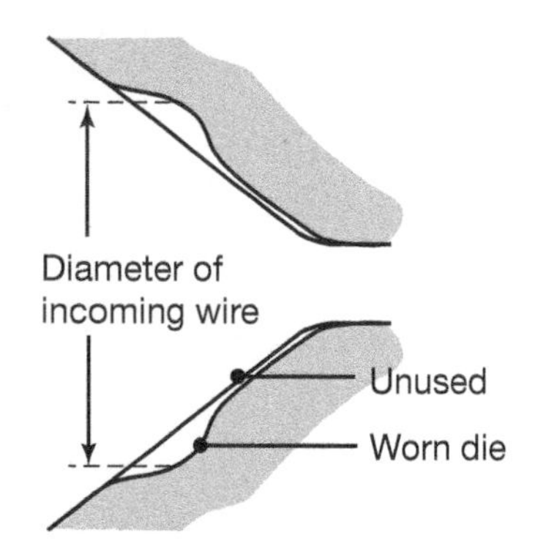

FIGURE 6.66 Schematic illustration of a typical wear pattern in a wire-drawing die.

Dies and die materials. Die materials for drawing are generally tool steels, carbides, or diamond for drawing fine wires, either a single crystal or a polycrystalline diamond in a metal matrix. Carbide and diamond dies are made as inserts or nibs, which are then supported in a steel casing (Fig. 6.64b). A typical wear pattern on a drawing die is shown in Fig. 6.66; note that die wear is highest at the entry. Although the die pressure is highest in this region and is partially responsible for wear, other factors include (a) variations in the diameter of the entering wire; (b) vibration, which subjects the die-entry contact zone to fluctuating stresses; and (c) the presence of abrasive scale on the surface of the entering wire.

A set of idling rolls can also be used in drawing of rods or bars of various shapes. This arrangement, known as a **Turk's head**, is more versatile than ordinary dies, because the rolls can be adjusted to various positions to produce a variety of cross sections.

Lubrication. In **dry drawing**, the surface of the wire is coated with various lubricants, depending on the strength and frictional characteristics of the material. The rod to be drawn is first surface treated by pickling, which removes the surface scale that could lead to surface defects and considerably reduce die life, because of its abrasiveness. The bar then goes through a *stuffing box* filled typically with soap powder.

With high-strength materials, such as steels, stainless steels, and high-temperature alloys, the surface of the rod may be coated either with a softer metal or with a **conversion coating** (see Section 4.5.1). Conversion coatings may consist of sulfate or oxalate which, typically, are then coated with soap as the lubricant. Copper or tin can chemically be deposited (see Section 4.5.1) as a thin layer on the surface of the metal, whereby it acts as a solid lubricant. Polymers also may be used as solid lubricants, such as in drawing titanium.

In **wet drawing**, the dies and rod are completely immersed in a lubricant, such as oils and emulsions containing fatty or chlorinated additives, and various chemical compounds. The technique of **ultrasonic vibration** of the dies and mandrels is also used in drawing. This technique improves surface finish and die life, and reduces drawing forces, thus allowing higher reductions per pass.

Equipment. Two types of equipment are generally used in drawing operations. A **draw bench** is similar to a long horizontal tensile-testing machine but with a hydraulic or chain-drive mechanism; it is used for single draws of straight rods with large cross sections and for tubes with lengths up to 30 m. Smaller cross sections are usually drawn by a **bull block** (Fig. 6.65), which is basically a rotating drum around which the wire is wrapped; the tension in the setup provides the force required to draw the wire.

6.6 Swaging

In *swaging*, also known as **rotary swaging** or **radial forging**, a rod or tube is reduced in diameter by the reciprocating radial movement of two or four dies, as shown in Fig. 6.67. The die movements are obtained from a set of rollers in a cage, with a cam-ended hammer that periodically closes the die when the cam interferes with a roller. The internal diameter and the thickness of the tube can be controlled with or without the use of mandrels (Fig. 6.68). Mandrels can also be made with longitudinal grooves (similar in appearance to a splined shaft), whereby internally-shaped tubes can be swaged, such as those shown in Fig. 6.69a. The *rifling* in gun barrels can be made by swaging a tube over a mandrel with spiral

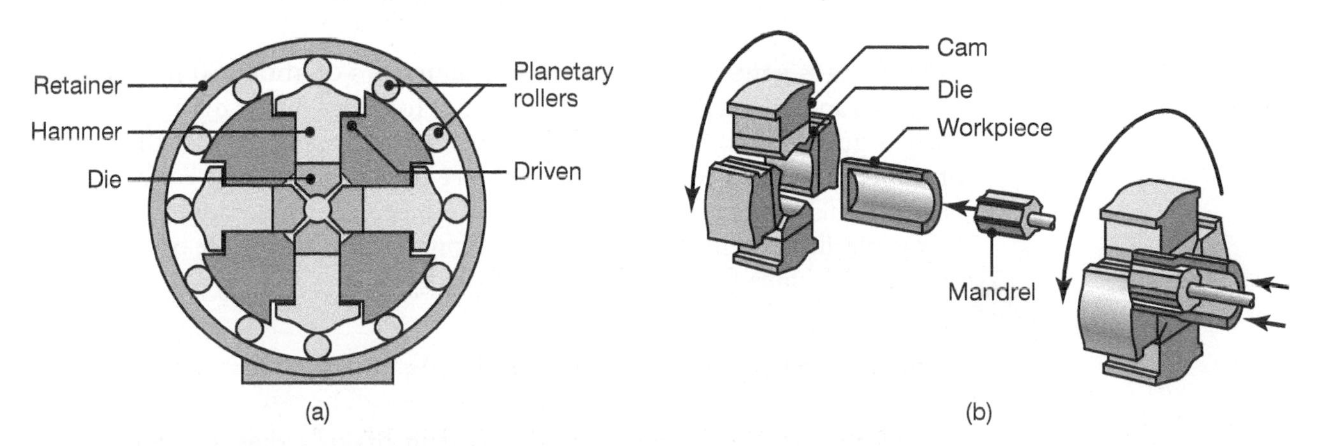

FIGURE 6.67 (a) Schematic illustration of the rotary-swaging process. (b) Forming internal profiles in a tubular workpiece by swaging.

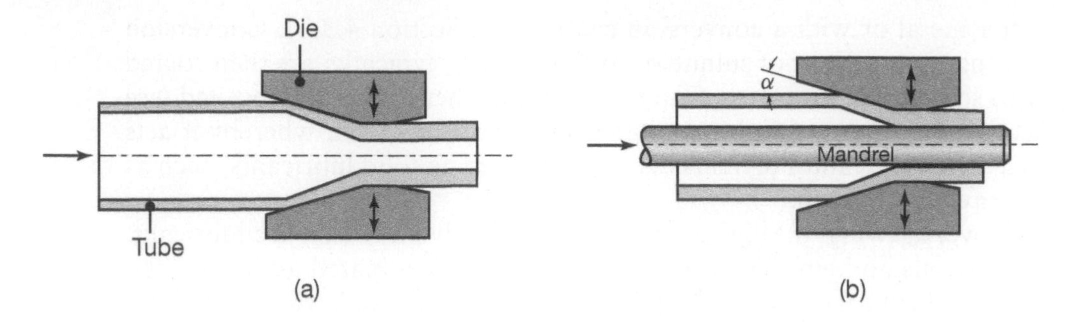

FIGURE 6.68 Reduction of outer and inner diameters of tubes by swaging. (a) Free sinking without a mandrel. The ends of solid bars and wire are tapered (pointing) by this process in order to feed the material into the conical die. (b) Sinking on a mandrel. Coaxial tubes of different materials can also be swaged in one operation.

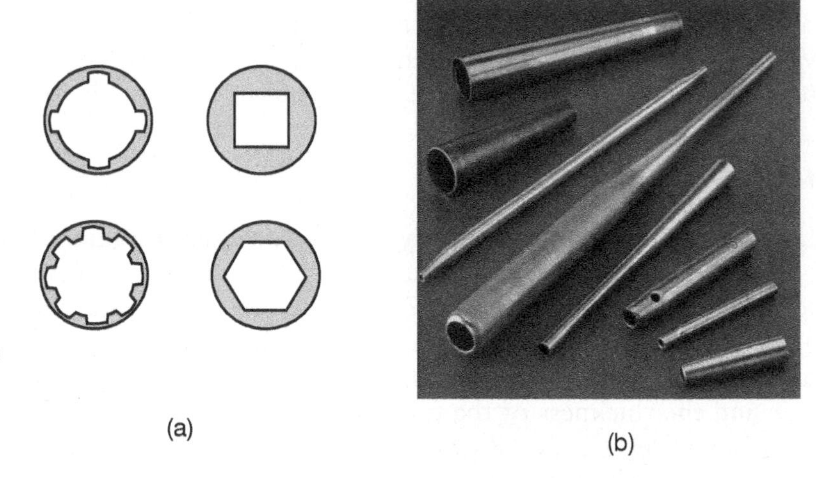

FIGURE 6.69 (a) Typical cross sections produced by swaging tubular blanks with a constant wall thickness on shaped mandrels. Rifling of small gun barrels also can be made by swaging. (b) Typical parts made by swaging. *Source:* Courtesy of J. Richard Industries.

grooves. The swaging process is generally carried out at room temperature. Parts produced by swaging have improved mechanical properties and good dimensional accuracy.

Swaging is generally limited to part diameters of about 50 mm (2 in.), although special machinery have been built to swage larger diameter parts, such as gun barrels. The length of the product is limited only by the length of the mandrel (if needed). Die angles are usually a few degrees and may be compound, that is, the die may have more than one angle for more favorable flow of material during swaging. Lubricants can be used for improved surface finish and to extend die life.

6.7 Die-Manufacturing Methods

Several manufacturing methods are used in making dies and molds for metalworking and various other processes. These methods, which are used either singly or in combinations, include casting, forging, machining, grinding, selective laser sintering, and electrical and electrochemical methods of die sinking. For improved surface finish and dimensional accuracy, dies may be subjected to various finishing operations, such as honing, polishing, and coating. The selection of a die manufacturing method primarily depends on the following parameters:

1. The particular process for which the die will be used;
2. Die shape and size;
3. The surface quality required;
4. Lead time required to produce the die (e.g., large dies may require months to produce);
5. Production run (i.e., die life expected); and
6. Cost.

Economic considerations often dictate the process selected because tool and die costs can be very significant in manufacturing operations. For example, the cost of a set of dies for pressworking sheet metal automotive body panels in Figs. 1.6 and 7.70 may run up to $2 million; even small and simple dies can cost hundreds of dollars. On the other hand, because a large number of parts are typically produced using the same die, the die cost per piece made is only a small fraction of a part's manufacturing cost (Section 16.9).

The processes used for producing dies typically include *casting* and *forging*, although small dies and tools also can be made by powder metallurgy techniques, described in Chapter 11, as well as by *rapid tooling techniques*, described in Section 10.12. Sand casting is used for large dies weighing several tons and shell molding for small dies. Cast steels are generally preferred as die materials, because of their strength and toughness and the ease with which their composition, grain size, and properties can be controlled and modified. Depending on how they solidify in the mold, cast dies, unlike those made of wrought metals, may not have directional properties, thus exhibiting isotropic properties on all working surfaces. However, because of shrinkage, the control of dimensional accuracy can be difficult compared to machined and finished dies, and may require additional finishing operations.

Casting of dies is usually followed by various primary operations, such as forging, rolling, and extrusion, followed by secondary processing, such as machining, grinding, polishing, and coating. Most commonly, dies are machined from cast and forged *die blocks* (called *die sinking*) by milling, turning, grinding, electrical discharge and electrochemical machining, and polishing (Chapters 8 and 9). Conventional machining can be difficult and time consuming for high-strength, hard, tough, and wear-resistant die materials (see also *hard machining* and *hard turning* Section 8.9.2). Dies are commonly machined on *computer-controlled machine tools* and *machining centers*, using various software packages (Fig. 1.10).

Advanced machining processes, especially *electrical-discharge machining*, also are used extensively (Section 9.13), particularly for small and medium-sized dies and for extrusion dies. These processes are generally faster and more economical than traditional machining, and the dies may not require additional finishing operations. It is important to consider any possible adverse effects that these processes may have on the properties of the die (including fatigue life), because of possible surface damage and the development of cracks. Small dies with shallow cavities may be produced by *hubbing* (Section 6.2.4).

One advantage of producing dies or molds through rapid tooling techniques (see Section 10.12) is the ability to use **contour cooling** to extract heat from the molds. Most dies will incorporate some form of cooling channels to extract heat from the die; these are usually deep channels produced by drilling and capping the ends of drilled holes, or through the use of cores when sand casting a mold. These have somewhat crude geometries and are much thicker than required, and usually farther removed from the mold wall than is necessary. With rapid tooling approaches, cooling channels can be designed to follow the profile of the part, making heat extraction far more effective. Cycle times in injection molding and die casting have been reduced by 50% when using contour cooling approaches.

To improve their hardness, wear resistance, and strength, die steels (Section 3.10.4) are usually *heat treated* (Section 5.11). After heat treatment, tools and dies are generally subjected to finishing operations, such as grinding and polishing, to obtain the desired surface roughness and dimensional accuracy. If not controlled properly, the grinding process can cause surface damage by generating excessive heat and inducing detrimental tensile residual stresses on die surfaces, thus reducing fatigue life. Dies may also be subjected to various surface treatments, including *coatings* (Sections 4.5.1 and 8.6.5) for improved frictional and wear characteristics, as described in Chapter 4.

6.8 Die Failures

Failure of dies in metalworking operations generally is due to one or more of the following causes:

1. Improper die design;
2. Defective die materials;
3. Improper heat treatment and finishing operations;
4. Overheating and heat checking;
5. Excessive wear;
6. Improper installation, assembly, and alignment of die components; and
7. Overloading, misuse, and improper handling.

With rapid advances in *computer modeling and simulation* of processes and systems (Section 15.7), die design and optimization has become an advanced technology. Basic die-design guidelines include the following considerations:

1. Dies must have appropriate cross sections and clearances in order to withstand the forces involved.
2. Sharp corners, radii, and fillets, and abrupt changes in cross section must be avoided as they can act as stress raisers.

3. For improved strength, dies may be made in *segments* and may be prestressed during their assembly.
4. Dies can be designed and constructed with *inserts* that can be replaced when worn or cracked.
5. Die surface preparation and finishing are important, especially for die materials such as heat-treated steels, carbides, and diamond because they are susceptible to cracking and chipping from *impact forces* (as in mechanical presses and forging hammers) or from *thermal stresses* caused by temperature gradients within the die (as in hot-working operations).
6. Metalworking fluids can adversely affect tool and die materials; for example, sulfur and chlorine additives in some lubricants and coolants can attack the cobalt binder in tungsten carbide and lower its strength and toughness (called *leaching*, Section 3.9.7); and
7. Overloading can cause premature failure of dies.

6.9 Economics of Bulk Forming

QR Code 6.8 The forging advantage.
Source: Courtesy of the Forging Industry Association, www.forging.org.

Several factors are involved in the cost in bulk forming operations. Depending on the complexity of the part, tool and die costs range from moderate to high. As in other manufacturing operations, this cost is spread over the number of parts produced with that particular tooling. Thus, even though the workpiece material cost per piece made is constant, setup and tooling costs per piece decrease as the manufacturing volume increases (Fig. 6.70).

The size of parts also has an effect on cost. Sizes range from small parts such as utensils and small automotive components, to large ones such as gears and crankshafts and connecting rods for large engines. As part size increases, the share of material cost in the total cost also increases, although at a lower rate. This occurs because (a) the incremental increase in die cost for larger dies is relatively small; (b) the machinery and operations involved are essentially the same regardless of part size; and (c) the labor involved per piece made is not that much higher.

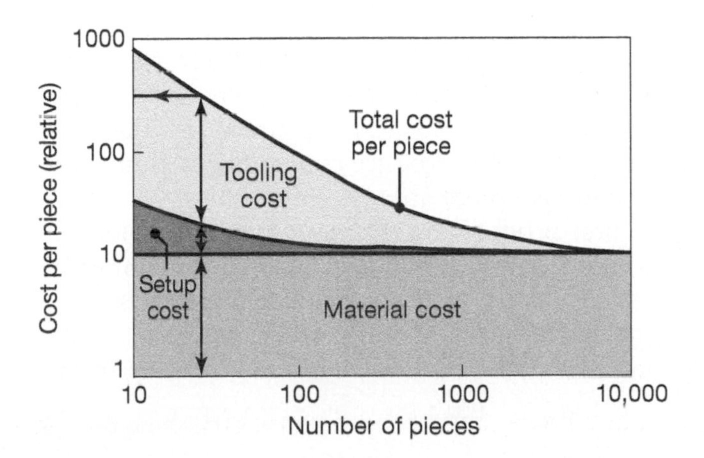

FIGURE 6.70 Typical unit cost (cost per piece) in forging. Note how the setup and the tooling costs per piece decrease as the number of pieces forged increases, if all pieces use the same die.

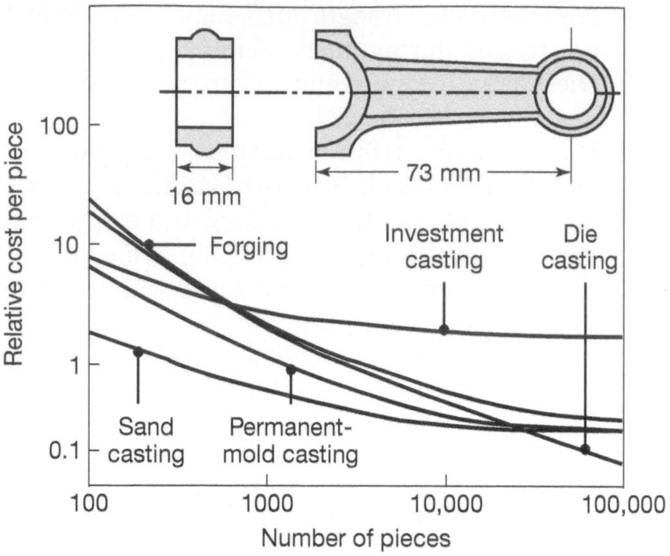

FIGURE 6.71 Relative unit costs of a small connecting rod made by various forging and casting processes. Note that, for large quantities, forging is more economical. Sand casting is the more economical process for fewer than about 20,000 pieces.

Labor costs in bulk forming are generally moderate, because they have been reduced significantly by automation and computer-controlled operations. Computer-aided design and manufacturing techniques (Chapter 15) in die design and manufacturing have resulted in major savings in time and effort in both tooling and part design.

The cost of bulk forming a part compared to that with other operations such as casting, powder metallurgy, machining, etc. is a major consideration (see Chapter 16). For example, all other factors being the same, and depending on the number of pieces required, manufacturing a certain part by, say, expendable-mold casting may well be more economical than producing it by forging (Fig. 6.71). The casting method does not require expensive molds and tooling, whereas forging requires expensive dies. On the other hand, bulk forming operations are often justified because of the improved mechanical properties that often result.

CASE STUDY 6.1 Production of Aluminum Foil

Aluminum foil is commonly used as packaging material, protection of electrical components against static electricity, and for food processing. It combines low cost and safety in a convenient-to-use product. Aluminum foil can be rolled to a thickness as small as 6.3 μm, ten times less than the thickness of human hair. The manufacturing steps that are taken to produce conventional aluminum foil as used in kitchens worldwide are very developed and take place at large scales.

Aluminum is produced from bauxite ore, and is first produced as alumina. The Hall-Héroult process (Fig. 6.72) produces pure aluminum from this alumina in a carbon or graphite lined container, referred to as a reduction pot. In the reduction pot, a small voltage, perhaps as low as 6 volts, is used, but the current can be extremely high (over 150,000 amperes). The Hall-Heroult process uses a carbon anode that combines with the oxygen in the alumina, resulting in molten aluminum forming in the bottom of the pot where it is siphoned periodically.

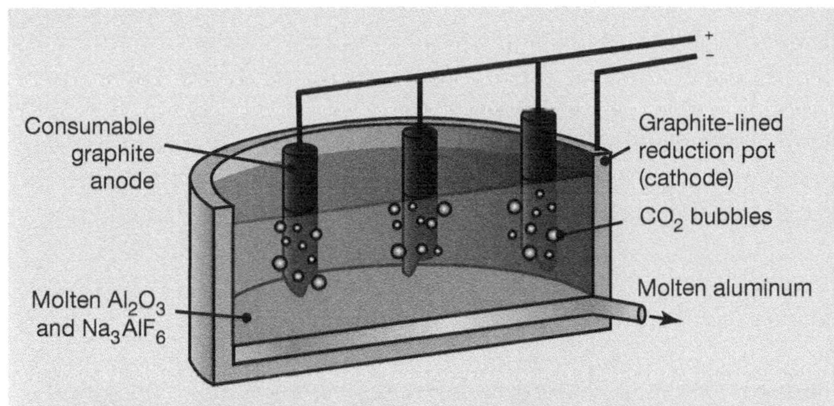

FIGURE 6.72 Schematic illustration of a Hall-Héroult reduction pot, used to produce molten aluminum from aluminum oxide.

Aluminum foil is generally high purity aluminum, even up to 99.9%, to ensure that the workpiece has sufficient ductility to be rolled to the very thin final thickness. Even so, the aluminum coils often need to be annealed to ensure successful rolling. Aluminum foil is usually produced in high volumes, so it is first processed with strip casting (Section 5.7.3) and then coiled and cold rolled. Alternative methods of production exist for different alloys, including the direct casting of strips followed by hot rolling in a tandem mill (Fig. 6.39d), or casting slabs which are then hot rolled in reversing mills followed by tandem mills.

The aluminum is then cold rolled in multiple passes. The temperature at the exit of a cold mill is around 100°C, but this is still less than one-third the homologous temperature of aluminum. Significant coolant/lubricant is needed in this process to maintain thermal equilibrium and ensure good surface finish (Section 4.4.4).

Cold rolling takes place in multi-stand tandem mills (Fig. 6.73), where the strip is reduced in thickness in each stand. At this stage, significant work hardening can occur, and one or two annealing stages may be necessary to achieve the desired mechanical properties and grain structure.

Finally, the cold-rolled strip is flat-rolled on a non-reversing single stand rolling mill. Because of the friction hill, it becomes increasingly difficult to roll the foil; roll flattening results from the high stresses instead. To overcome the friction hill, two strategies are followed: First, the mill that produces

FIGURE 6.73 Aluminum sheet exits a minimill as part of the cold rolling process.
Source: Courtesy of ALCOA Corp.

foil has very small work rolls with larger backing rolls, usually in a four-high configuration (Sendzimer mills are used for harder materials like stainless steel; see Fig. 6.39f.) Second, the foil is rolled two sheets at a time, thereby effectively doubling the foil thickness.

However, care must be taken to avoid cold welding the aluminum sheets, so a viscous lubricant is applied between the foil sheets before cold rolling. As a result, the surface in contact with a roll is impressed with its (polished) surface finish, while the other side experiences orange peel or roughening (Section 3.6).

After foil rolling, the sheet can be up to 1.7 m wide (Fig. 6.74), so it needs to be cut to a desired width. This is performed in coil slitting lines (see Section 7.3.1). The foil is then recoiled and packaged.

FIGURE 6.74 Cold rolled coils ready for slitting or shipment.
Source: Courtesy of ALCOA Corp.

Source: Courtesy of H. Malkani and H. Markman, ALCOA Corp.

SUMMARY

- Bulk deformation processes typically consist of forging, rolling, extrusion, and drawing, and involve major changes in the dimensions of the workpiece. Bulk properties as well as surface characteristics of the material are important considerations. (Section 6.1)
- Forging denotes a family of processes by which deformation of the workpiece is carried out by compressive forces applied through dies. Forging is capable of producing a wide variety of parts, with favorable characteristics of strength, roughness, dimensional accuracy, and reliability in service. Several types of defects can develop, depending on die geometry, quality of the billet materials, and preform shape. Forging machines are available, with various characteristics, sizes, and capabilities. (Section 6.2)

- The commonly used methods of analysis of the stresses, strains, strain rates, temperature distribution, and loads in forging include the slab method and the finite-element method. (Section 6.2)
- Rolling is the process of continuously reducing the thickness or changing the cross section of long stock by compressive forces applied through a set of rolls. Rolled products include plate, sheet, foil, rod, pipe, and tube, as well as shape-rolled products such as I-beams, structural shapes, rails, and bars of various cross sections. The operation involves several material and process variables, including the size of the roll relative to the thickness of the material, the amount of reduction per pass, speed, lubrication, and temperature. (Section 6.3)
- Extrusion is the process of forcing a billet through the opening of a die, and it is capable of producing finite lengths of solid or hollow cross sections. Important factors include die design, extrusion ratio, lubrication, billet temperature, and extrusion speed. Cold extrusion is a combination of extrusion and forging operations, and is capable of economically producing a wide variety of parts with good mechanical properties. (Section 6.4)
- Rod, wire, and tube drawing involve pulling the material through one or more dies. Proper die design and appropriate selection of materials and lubricants are essential to obtaining products of high quality and with good surface finish. The major process variables in drawing are die angle, friction, and amount of reduction per pass. (Section 6.5)
- In swaging, a solid rod or a tube is reduced in diameter by the reciprocating radial movement of two or four dies. This process is suitable for producing short or long lengths of bars or tubing with various internal or external profiles. (Section 6.6)
- Because die failure has a major economic impact, die design, material selection, and the manufacturing methods employed are of major importance. A wide variety of die materials and manufacturing methods is available, including advanced machining methods and subsequent treatment, finishing, and coating operations. (Sections 6.7 and 6.8)

SUMMARY OF EQUATIONS

Forging

Pressure in plane-strain compression, $p = S'_y e^{2\mu(a-x)/h}$

Average pressure in plane-strain compression, $p_{av} \simeq S'_y \left(1 + \frac{\mu a}{h}\right)$

Pressure in axisymmetric compression, $p = S_y e^{2\mu(r-x)/h}$

Average pressure in axisymmetric compression, $p_{av} \simeq Y \left(1 + \frac{2\mu r}{3h}\right)$

Average pressure in plane-strain compression, sticking, $p_{av} = S'_y \left(1 + \frac{a-x}{h}\right)$

Pressure in axisymmetric compression, sticking, $p = S_y \left(1 + \frac{r-x}{h}\right)$

Rolling

Roll pressure in entry zone, $p = \sigma_f' \dfrac{h}{h_o} e^{\mu(H_o - H)}$

with back tension, $p = \left(\sigma_f' - \sigma_b\right) \dfrac{h}{h_o} e^{\mu(H_o - H)}$

Roll pressure in exit zone, $p = \sigma_f' \dfrac{h}{h_f} e^{\mu H}$

with front tension, $p = \left(\sigma_f' - \sigma_f\right) \dfrac{h}{h_f} e^{\mu H}$

Parameter, $H = 2\sqrt{\dfrac{R}{h_f}} \tan^{-1}\left(\sqrt{\dfrac{R}{h_f}}\phi\right)$

Roll force, $F = \int_o^{\phi_n} wpR\, d\phi + \int_{\phi_n}^{\alpha} wpR\, d\phi$

Roll force, approximate, $F = Lw\overline{\sigma}_f' \left(1 + \dfrac{\mu L}{2h_{av}}\right)$

Roll torque, $T = \int_{\phi_n}^{\alpha} w\mu pR^2\, d\phi - \int_o^{\phi_n} w\mu pR^2\, d\phi$

Roll contact length, approximate, $L = \sqrt{R\Delta h}$

Power per roll $= \dfrac{FL\omega}{2} = \dfrac{\pi FLN}{60}$

Maximum draft, $\Delta h_{max} = \mu^2 R$

Maximum angle of acceptance, $\alpha_{max} = \tan^{-1}\mu$

Extrusion

Extrusion ratio, $R = A_o/A_f$

Extrusion pressure, ideal, $p = S_y \ln R$

Extrusion pressure, with friction, $p = S_y\left(1 + \dfrac{\tan\alpha}{\mu}\right)\left[R^{\mu\cot\alpha} - 1\right]$

Rod and wire drawing

Drawing stress, ideal, $\sigma_d = S_y \ln\left(\dfrac{A_o}{A_f}\right)$

Drawing stress, with friction, $\sigma_d = S_y\left(1 + \dfrac{\tan\alpha}{\mu}\right)\left[1 - \left(\dfrac{A_f}{A_o}\right)^{\mu\cot\alpha}\right]$

Die pressure, $p = \sigma_f - \sigma$

BIBLIOGRAPHY

Altan, T., Ngaile, G., and Shen, G., (eds.), *Cold and Hot Forging: Fundamentals and Applications*, ASM International, 2004.

ASM Handbook, Vol. 14A: Metalworking: Bulk Forming, ASM International, 2005.

Boljanovic, V., *Die Design Fundamentals*, 3rd ed., Industrial Press, 2005.

Dieter, G.E., Semiatin, S.L., and Kuhn, H.A., Handbook of Workability and Process Design, ASM International, 2003.

Dixit, U.S., and Narayayanan, R.G., *Metal Forming: Technology and Process Modeling*, McGraw-Hill, 2013.

Ginzburg, V.B., *Flat-Rolled Steel Processes: Advanced Processes*, CRC Press, 2009.

Hosford, W.F., and Caddell, R.M., *Metal Forming, Mechanics and Metallurgy*, 4th ed., Prentice Hall, 2014.

Lenard, J.G., *Primer on Flat Rolling*, Elsevier Science, 2007.

McQueen, H.J., Spigarelli, S., Kassner, M.E., and Evangelista, E., *Hot Deformation and Processing of Aluminum Alloys*, CRC Press, 2011.

Panjkovic, V., *Friction and Hot Rolling of Steel*, CRC Press, 2014.

Pittner, J., and Simaan, M.A., *Tandem Cold Metal Rolling Mill Control: Using Practical Advanced Methods*, Springer, 2010.

Qamar, S.Z., Sheikh, A.K., and Arif, A.F.M., *Modeling and Analysis of Aluminum Extrusion: Process, Tooling, and Defects*, Lambert Academic Publishing, 2011.

Saha, P.K., *Aluminum Extrusion Technology*, ASM International, 2000.

Suchy, I., *Handbook of Die Design*, 2nd ed., McGraw-Hill, 2005.

Tschaetsch, H., *Metal Forming Practice*, Springer, 2006.

Valberg, H.S., *Applied Metal Forming*, Cambridge University Press, 2010.

Verlinden, B., Driver, J., Samajdar, I., and Doherty, R.D., *Thermo-Mechanical Processing of Metallic Materials*, Pergamon Press, 2007.

Wagoner, R.H., and Chenot, J.L., *Metal Forming Analysis*, Cambridge University Press, 2005.

QUESTIONS

Forging

6.1 How can you tell whether a certain part is forged or cast? Describe the features that you would investigate to arrive at a conclusion.

6.2 Why is the control of volume of the blank important in closed-die forging?

6.3 What are the advantages and limitations of a cogging operation? Of die inserts in forging?

6.4 Explain the conditions under which an upsetting test would result in double barreling.

6.5 Explain why there is a variety of forging machines.

6.6 Devise an experimental method whereby you can measure the force required for forging only the flash in impression-die forging. (See Fig. 6.13a.)

6.7 A manufacturer is successfully hot forging a certain part, using material supplied by Company *A*. A new supply of material is obtained from Company *B*, with the same nominal composition of the major alloying elements as that of the material from Company *A*. However, it is found that the new forgings are cracking even though the same procedure is followed as before. What is the probable reason?

6.8 Explain why there might be a change in the density of a forged product as compared to that of the cast blank.

6.9 Since glass is a good lubricant for hot extrusion, would you use glass for impression-die forging as well? Explain.

6.10 Describe and explain the factors that influence spread in cogging operations on square billets.

6.11 Why are end grains generally undesirable in forged products? Give examples of such products.

6.12 Explain why one cannot obtain a finished forging in one press stroke, starting with a blank.

6.13 List the advantages and disadvantages of using a lubricant in forging.

6.14 Explain the reasons why the flash assists in die filling for hot forging.

6.15 By inspecting some forged products (such as a pipe wrench or coins), you can see that the lettering on them is raised rather than sunk. Offer an explanation as to why they are made that way.

6.16 List the implications of conducting a forging operation on a hydraulic press as opposed to a drop forge.

6.17 Figure 6.17(a) shows a cogging operation, used to reduce the thickness of a workpiece. Can this operation be used to produce thin sheets? Explain.

6.18 What is electrically assisted forging?

6.19 What is a gutter? Why is it useful in impression die forging?

6.20 What are the advantages of servo presses?

6.21 Identify casting design rules, described in Section 5.12, that also can be applied to forging.

Rolling

6.22 As stated in Section 6.3.1, three factors that influence spreading in rolling are the width-to-thickness ratio of the strip, friction, and the ratio of the radius of the roll to the thickness of the strip. Explain how these factors affect spreading.

6.23 Explain how you would go about applying front and back tensions to sheet metals during rolling.

6.24 It was noted that rolls tend to flatten under roll forces. Which property(ies) of the roll material can be increased to reduce flattening?

6.25 Describe the methods by which roll flattening can be reduced.

6.26 Is it possible to have a negative forward slip? Explain.

6.27 Explain the technical and economic reasons for taking larger rather than smaller reductions per pass in flat rolling.

6.28 List and explain the methods that can be used to reduce the roll force.

6.29 Explain the advantages and limitations of using small-diameter rolls in flat rolling.

6.30 A ring-rolling operation is successful for the production of bearing races. However, when the bearing race diameter is changed, the operation results in very poor surface finish. List the possible causes and the investigation you would perform to identify the parameters involved and correct the problem.

6.31 Describe the importance of controlling roll speed, roll gap, temperature, and other process variables in a tandem-rolling operation.

6.32 Rolling reduces the thickness of plates and sheets. However, it is also possible to reduce the thickness by simply stretching the material. Would this process be feasible? Explain.

6.33 In Fig. 6.31, explain why the neutral point moves toward the roll-gap entry as friction increases.

6.34 What typically is done to make sure the product in flat rolling is not crowned?

6.35 List the possible consequences of rolling at (a) too high of a speed; and (b) too low of a speed.

6.36 Rolling has sometimes been described as a continuous forging operation. Is this description accurate? Explain.

6.37 Explain, with reference to appropriate equations, why titanium carbide is used as the work roll in Sendzimir mills, but not other rolling mill configurations.

6.38 Make two careful drawings of rolling between two rollers, one with a large reduction in thickness, one with a small reduction in thickness. Sketch the strain distributions in each case.

6.39 It is known that in thread rolling as illustrated in Fig. 6.42c, a workpiece must make roughly six revolutions to form the thread. Under what conditions (process parameters, thread geometry or workpiece properties) can deviation from this rule take place? Explain your answer.

6.40 If a rolling mill encounters chatter, what process parameters would you change, and in what order? Explain your answer.

Extrusion

6.41 The extrusion ratio, die geometry, extrusion speed, and billet temperature all affect the extrusion pressure. Explain why.

6.42 How would you go about avoiding centerburst defects in extrusion? Explain why your methods would be effective.

6.43 How would you go about making a stepped extrusion that has increasingly larger cross sections along its length? Is it possible? Would your process be economical and suitable for high production runs? Explain.

6.44 Note from Eq. (6.53) that, for low values of the extrusion ratio, such as $R_e = 2$, the ideal extrusion pressure p can be lower than the yield strength, S_y, of the material. Explain whether or not this phenomenon is logical.

6.45 In hydrostatic extrusion, complex seals are used between the ram and the container, but not between the extrusion and the die. Explain why.

6.46 What kind of defects may occur in (a) extrusion; and (b) drawing.

6.47 What is a land? What is its function in a die? What are the advantages and disadvantages to having no land?

6.48 Under what circumstances is backwards extrusion preferable to direct extrusion? When is hydrostatic extrusion preferable to direct extrusion?

6.49 What is the purpose of a container liner in direct extrusion (See Fig. 6.45a)? Why is there no container liner in hydrostatic extrusion?

6.50 In extrusion, why does the force increases even as the billet length is decreasing?

Drawing

6.51 In rod and wire drawing, the maximum die pressure is at the die entry. Why?

6.52 Describe the conditions under which wet drawing and dry drawing, respectively, are desirable.

6.53 Name the important variables in drawing, and explain how they affect the drawing process.

6.54 Assume that a rod-drawing operation can be carried out either in one pass or in two passes in tandem. If all die angles are the same and the total reduction is the same, will the drawing forces be different? Explain.

6.55 In Fig. 6.58, assume that the reduction in the cross section is taking place by pushing the rod through the die instead of pulling it. Assuming that the material is perfectly plastic, sketch the die-pressure distribution, for the following situations: (a) frictionless; (b) with friction; (c) frictionless, but with front tension. Explain your answers.

6.56 In deriving Eq. (6.72), no mention is made of the ductility of the original material being drawn. Explain why.

6.57 Explain why the die pressure in drawing decreases toward the exit of the die.

6.58 What is the value of the die pressure at the exit when a drawing operation is being carried out at the maximum reduction per pass?

6.59 Explain why the maximum reduction per pass in drawing should increase as the strain-hardening exponent, n, increases.

6.60 If, in deriving Eq. (6.72), friction is included, will the maximum reduction per pass be the same (63%), higher, or lower? Explain.

6.61 Explain what effects back tension has on the die pressure in wire or rod drawing, and discuss why these effects occur.

6.62 Explain why the inhomogeneity factor ϕ in rod and wire drawing depends on the ratio h/L.

6.63 Describe the reasons for the development of the swaging process.

6.64 Occasionally, steel-wire drawing will take place within a sheath of a soft metal such as copper or lead. Why would this procedure be useful?

6.65 Recognizing that it is very difficult to manufacture a die with a submillimeter diameter, how would you produce a 10 μ-m-diameter wire?

General

6.66 With respect to the topics covered in this chapter, list and explain specifically two examples each where (a) friction is desirable and (b) friction is not desirable.

6.67 Take any three topics from Chapter 2 and, with a specific example for each, show their relevance to the topics covered in this chapter.

6.68 Take any three topics from Chapter 3 and, with a specific example for each, show their relevance to the topics covered in this chapter.

6.69 List and explain the reasons that there are so many different types of die materials used for the processes described in this chapter.

6.70 Explain the reasons why one should be interested in residual stresses developed in workpieces made by the forming processes described in this chapter.

6.71 Make a summary of the types of defects found in the processes described in this chapter. Indicate, for each type, methods of reducing or eliminating the defects.

6.72 Based on the processes in this chapter, list methods of producing ball bearings.

PROBLEMS

Forging

6.73 In the free-body diagram in Fig. 6.4b, the incremental stress $d\sigma_x$ is shown pointing to the left. Yet it would appear that, because of the direction of frictional stresses μp the incremental stress should point to the right in order to balance the horizontal forces. Show that the same answer for the forging pressure is obtained regardless of the direction of this incremental stress.

6.74 Plot the force vs. reduction in height curve in open-die forging of a cylindrical, annealed copper specimen 50 mm high and 25 mm in diameter, up to a reduction of 70%, for the cases of (a) no friction between the flat dies and the specimen, (b) $\mu = 0.25$, and (c) $\mu = 0.5$. Ignore barreling. Use average-pressure formulas.

6.75 Plot the force vs. reduction in height curve in open-die forging of a cylindrical, annealed copper specimen 10 mm high and 25 mm in diameter, up to a reduction of 50%, for the cases of (a) no friction between the flat dies and the specimen, (b) $\mu = 0.10$, and (c) $\mu = 0.30$. Ignore barreling. Use average-pressure formulas. Compare your results to Problem 6.74.

6.76 Use Fig. 6.9b to provide the answers to Problem 6.75.

6.77 Calculate the work done for each case in Problem 6.75.

6.78 Determine the temperature rise in the specimen for each case in Problem 6.75, assuming that the process is adiabatic and the temperature is uniform throughout the specimen.

6.79 Plot the force vs. reduction in height curve in open-die forging of a cylindrical, annealed Ti-6Al-4V specimen that is 10 mm high and 25 mm in diameter, up to a reduction of 50%, for the cases of (a) 20°C with $\mu = 0.2$, and (b) a workpiece preheated to a temperature of 600°C with $\mu = 0.4$.

6.80 Plot the force vs. reduction in height curve in open-die forging of a cylindrical, annealed Ti-6Al-4V specimen that is 10 mm high and 25 mm in diameter, up to a reduction of 50%, with $\mu = 0.2$, for the cases of (a) a hydraulic press with a speed of 0.1 m/s and (b) a mechanical press with a speed of 1 m/s. Assume the temperature is 800°C for both cases.

6.81 To determine forgeability, a hot-twist test is performed on a round bar 25 mm in diameter and 200 mm long. It is found that bar underwent 8 turns before it fractured. Calculate the shear strain at the outer surface of the bar at fracture.

6.82 Derive an expression for the average pressure in plane-strain compression with (a) Tresca friction and (b) sticking friction with $m = 1$.

6.83 What is the value of μ when, for plane-strain compression, the forging load with sliding friction is equal to the load with sticking friction? Use average-pressure formulas.

6.84 Note that in cylindrical upsetting, the frictional stress cannot be greater than the shear yield strength k of the material. Thus, there may be a distance x in Fig. 6.8 where a transition occurs from sliding to sticking friction. Derive an expression for x in terms of r, h, and μ only.

6.85 For the sticking example in Fig. 6.10, derive an expression for the lateral force F required to slide the workpiece to the right while the workpiece is being compressed between flat dies.

6.86 Two solid cylindrical specimens A and B, made of a perfectly plastic material, are being forged with friction and isothermally at room temperature to a reduction in height of 50%. Specimen A has a height of 50 mm and a cross-sectional area of $6.25 \times 10^{-4} m^2$ and specimen B has a height of 25 mm and a cross-sectional area of 0.0025 m^2. Will the work done be the same for the two specimens? Explain.

6.87 Assume that a workpiece is being pushed to the right by a lateral force F while being compressed between flat dies. (See the accompanying figure.) (a) Make a sketch of the die-pressure distribution for the condition for which F is not large enough to slide the workpiece to the right. (b) Make a similar sketch, except with F now being large enough so that the workpiece slides to the right while being compressed.

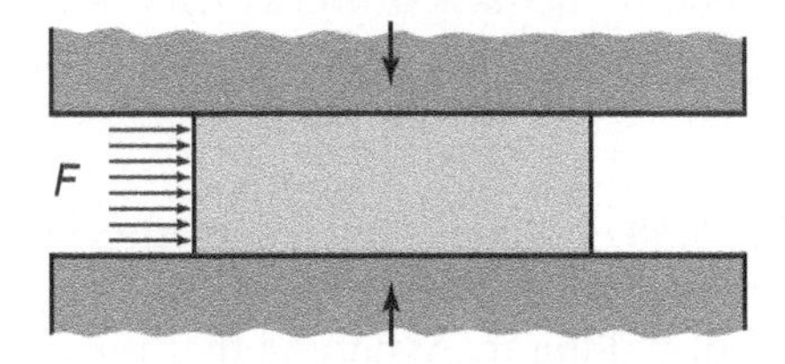

6.88 In Fig. 6.6, does the pressure distribution along the four edges of the workpiece depend on the particular yield criterion used? Explain.

6.89 Under what conditions would you have a normal pressure distribution in forging a workpiece as shown in the accompanying figure? Explain.

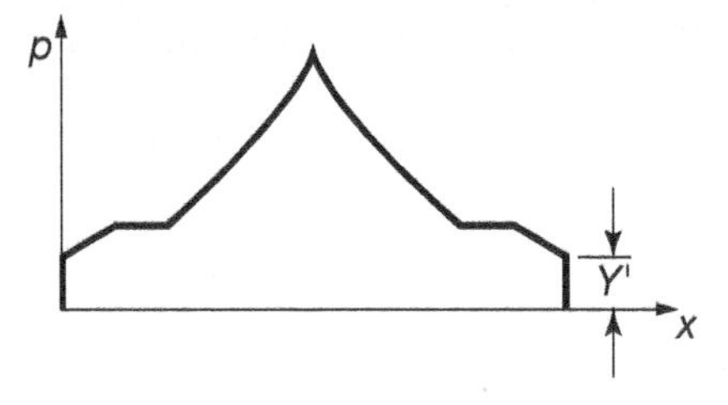

6.90 Derive the average die-pressure formula given by Eq. (6.15). (*Hint:* Obtain the volume under the friction hill over the surface by integration, and divide that quantity by the cross-sectional area of the workpiece.)

6.91 Take two solid cylindrical specimens of equal diameter, but different heights, and compress them (frictionless) to the same percent reduction in height. Show that the final diameters will be the same.

6.92 A rectangular workpiece has the following original dimensions: $2a = 100$ mm, $h = 30$ mm and width = 20 mm. The metal has a strength coefficient of 180 MPa and a strain-hardening exponent of 0.2. It is being forged in plane strain with $\mu = 0.2$. Calculate the force required for the height to be reduced by 20%. Do not use average-pressure formulas.

6.93 Assume that in upsetting a solid cylindrical specimen between two flat dies with friction, the dies are rotated at opposite directions to each other. How, if at all, will the forging force change from that for nonrotating dies? (*Hint:* Note that the dies will now require torque, because of the change in the direction of frictional forces at the die-workpiece interfaces.)

6.94 A solid cylindrical specimen, made of a perfectly plastic material, is being upset between flat dies with no friction. The process is being carried out by a falling weight, as in a drop hammer. The downward velocity of the hammer is at a maximum when it first contacts the workpiece and becomes zero when the hammer stops at a certain height of the specimen. Establish quantitative relationships between workpiece height and velocity, and make a qualitative sketch of the velocity profile of the hammer. (*Hint:* The loss in the kinetic energy of the hammer is the plastic work of deformation; thus, there is a direct relationship between workpiece height and velocity.)

6.95 How would you go about estimating the swaging force acting on each die?

6.96 A mechanical press is powered by a 25 kW motor and operates at 40 strokes per minute. It uses a flywheel, so that the rotational speed of the crankshaft does not vary appreciably during the stroke. If the stroke length is 150 mm, what is the maximum contact force that can be exerted over the entire stroke length? To what height can a 5052-O aluminum cylinder with a diameter of 10 mm and a height of 30 mm be forged before the press stalls?

6.97 Estimate the force required to upset a 5 mm diameter C74500 brass rivet in order to form a 10 mm diameter head. Assume that the coefficient of friction between the brass and the tool-steel die is 0.25 and that the head is 5 mm in thickness. Use $S_y = 175$ MPa.

6.98 Using the slab method of analysis, derive Eq. (6.17).

6.99 A compressor blade is to be forged of Ti-6Al-4V at 900°C, where $K = 140$ MPa and $n = 0.40$. The volume of the compressor blade is 30,000 mm^3, but the blank is oversized so that 20% of the blank volume will go into flash. In the finishing die, the projected area is 3600 mm^2. Use a flash thickness of 5 mm, and recognize that the compressor blade is a simple shape, but has some complexity because of the thin sections and the detail at the mounting end. Estimate the required forging force if the largest strain in the finish forging is $\epsilon = 0.25$.

6.100 A 0.25-m-wide billet of 5052-O aluminum ($K = 210$ MPa, $n = 0.13$) is forged from a thickness of 30 mm to a thickness of 20 mm with a long die with a width of 75 mm. The coefficient of friction for the die/workpiece interface is 0.25. Calculate the maximum die pressure and required forging force.

Rolling

6.101 In Example 6.4, what is the velocity of the strip leaving the rolls if the coefficient of friction is $\mu = 0.3$?

6.102 With appropriate sketches, explain the changes that occur in the roll-pressure distribution if one of the rolls is idling, i.e., power is shut off to that roll.

6.103 It can be shown that it is possible to determine μ in flat rolling without measuring torque or forces. By inspecting equations for rolling, describe an experimental procedure to do so. Note that you are allowed to measure any quantity other than torque or forces.

6.104 Derive a relationship between back tension p_b and front tension p_f in rolling such that when both tensions are increased, the neutral point remains in the same position.

6.105 Take an element at the center of the deformation zone in flat rolling. Assuming that all the stresses acting on this element are principal stresses, indicate the stresses qualitatively, and state whether they are tension or compression stresses. Explain the reasoning behind your devices. Is it possible for all three principal stresses to be equal to each other in magnitude? Explain.

6.106 It has been stated that in rolling a flat strip, the roll force is reduced about twice as effectively by back tension as it is by front tension. Explain this difference, using appropriate sketches. (*Hint:* Note the position of the neutral point.)

6.107 It can be seen that in rolling a strip, the rolls will begin to slip if the back tension, σ_b, is too high. Derive an analytical expression for the magnitude of the back tension in order to make the powered rolls begin to slip. Use the same terminology as applied in the text.

6.108 Prove Eq. (6.45).

6.109 In Steckel rolling, the rolls are idling, and thus there is no net torque, assuming frictionless bearings. Where, then, is the energy coming from to supply the work of deformation in rolling? Explain with appropriate sketches, and state the conditions that have to be satisfied.

6.110 Derive an expression for the tension required in Steckel rolling of a flat sheet, without friction, for a workpiece whose true stress–true strain curve is given by $\sigma = a + b\epsilon$.

6.111 (a) Make a neat sketch of the roll-pressure distribution in ordinary rolling with powered rolls. (b) Assume now that the power to both rolls is shut off and that rolling is taking place by front tension only, i.e., Steckel rolling. Superimpose on your diagram the new roll-pressure distribution, explaining your reasoning clearly. (c) After completing part (b), further assume that the roll bearings are getting rusty and deprived of lubrication while rolling is still taking place by front tension only. Superimpose a third roll-pressure distribution diagram for this condition, explaining your reasoning.

6.112 Derive Eq. (6.28), based on the equation preceding it. Comment on how different the h values are as the angle ϕ increases.

6.113 In Fig. 6.32, assume that $L = 2L_2$. Is the roll force, F, for L twice or more than twice the roll force for L_2? Explain.

6.114 A flat-rolling operation is being carried out where $h_o = 5$ mm, $h_f = 3.75$ mm, $w_o = 250$ mm, $R = 200$ mm, $\mu = 0.25$, and the average flow stress of the material in plane strain is $\overline{\sigma}'_f = 275$ MPa. Estimate the roll force and the torque. Include the effects of roll flattening.

6.115 Estimate the roll force and power for annealed low carbon steel strip ($K = 530$ MPa and $n = 0.26$), 200 mm wide and 10 mm thick, rolled to a thickness of 6 mm. The roll radius is 150 mm, and the roll rotates at 200 rpm. Use $\mu = 0.1$.

6.116 Calculate the individual drafts in each of the stands in the tandem-rolling operation shown.

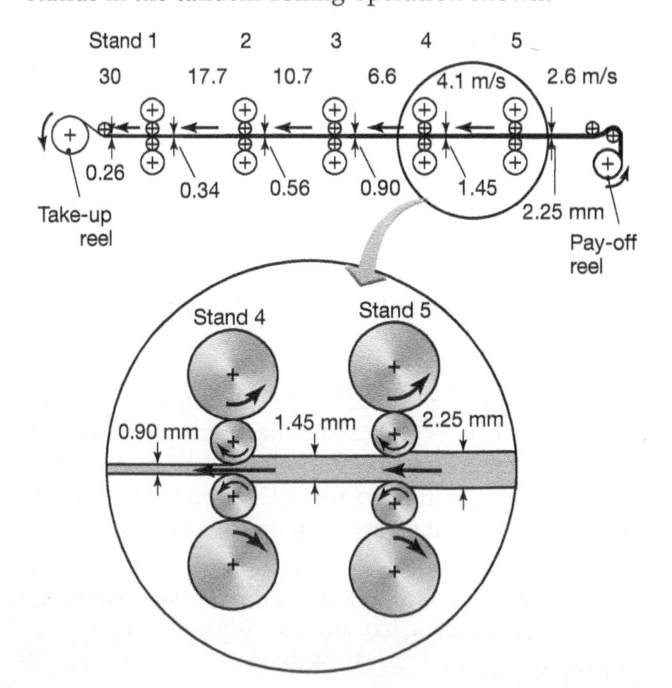

6.117 Calculate the required roll velocities for each roll in Problem 6.116 in order to maintain a forward slip of zero.

6.118 A 0.25-m-wide slab of 5052-O aluminum ($K = 210$ MPa, $n = 0.13$) is rolled from a thickness of 30 mm to a thickness of 20 mm on a cold mill with a roll diameter of 600 mm. The coefficient of friction is 0.25. Calculate the maximum roll pressure and required roll force. Assume the slab width is constant during the process.

6.119 A rolling operation takes place under the conditions shown in the accompanying figure. What is the position x_n of the neutral point? Note that there are a front and back tension that have not been specified.

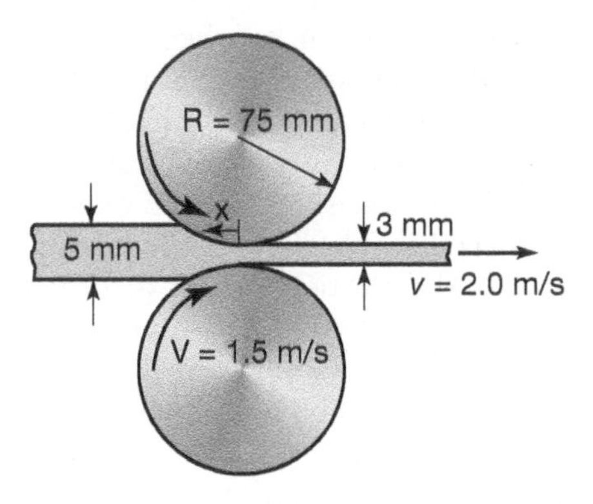

Additional data are as follows: Material is 5052-O aluminum; hardened steel rolls; surface roughness of the rolls = 0.02 μm; rolling temperature = 210°C.

6.120 A U-channel of 85-15 brass will be shape formed, but first must be flat rolled to a thickness of 0.9 mm. The strip has a width of 25 mm and an initial thickness of $h_o = 1.5$ mm. A preliminary process design suggests a 40% reduction in a single pass on a rolling mill with 150 mm-diameter rolls. If the roll surface speed is 1 m/s, and the coefficient of friction is $\mu = 0.1$, calculate the rolling force and power requirements. Repeat the problem if two passes were taken to achieve the desired reduction.

Extrusion

6.121 Estimate the force required in extruding 70–30 brass at 700°C, if the billet diameter is 125 mm and the extrusion ratio is 20.

6.122 Calculate the force required in direct extrusion of 1100-O aluminum from a diameter of 150 mm to a diameter of 100 mm. Assume that the redundant work is 30% of the ideal work of deformation, and the friction work is 25% of the total work of deformation.

6.123 Prove Eq. (6.57).

6.124 Calculate the theoretical temperature rise in the extruded material in Example 6.6, assuming that there is no heat loss. (See Section 3.9 for information on the physical properties of the material.)

6.125 Using the same approach as that shown in Section 6.5 for wire drawing, show that the extrusion pressure is given by the expression

$$p = S_y \left(1 + \frac{\tan\alpha}{\mu}\right)\left[1 - \left(\frac{A_o}{A_f}\right)^{\mu\cot\alpha}\right],$$

where A_o and A_f are the original and final workpiece areas, respectively.

6.126 Derive Eq. (6.55).

6.127 A planned extrusion operation involves steel at 800°C, with an initial diameter of 100 mm and a final diameter of 20 mm. Two presses, one with a capacity of 20 MN and the other of 10 MN, are available for this operation. Obviously, the larger press requires greater care and more expensive tooling. Is the smaller press sufficient for the operation? If not, what recommendations would you make to allow use of the smaller press?

Drawing

6.128 Calculate the power required in Example 6.7 if the workpiece material is annealed 70-30 brass.

6.129 Using Eq. (6.62), make a plot similar to Fig. 6.61 for the following conditions: $K = 100$ MPa; $n = 0.3$, $\mu = 0.04$.

6.130 Using the same approach as that described in Section 6.5 for wire drawing, show that the drawing stress, σ_d in plane-strain drawing of a flat sheet or plate is given by the expression

$$\sigma_d = S_y' \left(1 + \frac{\tan\alpha}{\mu}\right)\left[1 - \left(\frac{h-f}{h_o}\right)^{\mu\cot\alpha}\right],$$

where h_o and h_f are the original and final thickness, respectively, of the workpiece.

6.131 Derive an analytical expression for the die pressure in wire drawing, without friction or redundant work, as a function of the instantaneous diameter in the deformation zone.

6.132 A linearly strain-hardening material with a true stress–true strain curve $\sigma = 35 + 175\epsilon$ MPa is being drawn into a wire. If the original diameter of the wire is 5 mm, what is the minimum possible diameter at the exit of the die? Assume that there is no redundant work and that the frictional work is 15% of the ideal work of deformation. (*Hint:* The yield strength of the exiting wire is the point on the true stress–true strain curve that corresponds to the total strain that the material has undergone.)

6.133 In Fig. 6.63, assume that the longitudinal residual stress at the center is −500 MPa. Using the distortion-energy criterion, calculate the minimum yield strength that this particular steel must have in order to sustain these levels of residual stresses.

6.134 Derive an expression for the die-separating force in frictionless wire drawing of a perfectly plastic material. Use the same terminology as in the text.

6.135 A material with a true stress–true strain curve $\sigma = 10{,}000\epsilon^{0.25}$ is used in wire drawing. Assuming that the friction and redundant work compose a total of 50% of the ideal work of deformation, calculate the maximum reduction in cross-sectional area per pass that is possible.

6.136 Derive an expression for the maximum reduction per pass for a material of assuming that the friction and redundant work contribute a total of 30% to the ideal work of deformation.

6.137 Prove that the true strain rate $\dot{\epsilon}$ in drawing or extrusion in plane strain with wedge-shaped dies is given by the expression

$$\dot{\epsilon} = -\frac{2\tan\alpha V_o h_o}{(h_o - 2x\tan\alpha)^2},$$

where α is the die angle, h_o is the original thickness, and x is the distance from die entry (*Hint:* Note that $d\epsilon = dA/A$).

6.138 In drawing a strain-hardening material with $n = 0.25$ what should be the percentage of friction plus redundant work, in terms of ideal work, so that the maximum reduction per pass is 63%?

6.139 A round wire made of a perfectly plastic material with a yield strength of 200 MPa is being drawn from a diameter of 2.5 to 1.75 mm in a draw die of 15°. Let the coefficient of friction be 0.1. Using both Eqs. (6.62) and (6.66), estimate the drawing force required. Comment on the differences in your answer.

6.140 A 30-mm-diameter rod of 5052-O aluminum ($K = 210$ MPa, $n = 0.13$, $S_y = 90$ MPa) is drawn to a diameter of 20 mm using a die angle of $\alpha = 15°$. The coefficient of friction is 0.25. Calculate the maximum die pressure and required drawing force. Include friction and redundant work in your estimate.

6.141 Assume that you are asked to give a quiz to students on the contents of this chapter. Prepare three quantitative problems and three qualitative questions, and supply the answers.

DESIGN

6.142 Forging is one method of producing turbine blades for jet engines. Study the design of such blades and the relevant technical literature, and then prepare a step-by-step procedure for making the blades. Comment on the potential difficulties that may be encountered.

6.143 In comparing forged parts with cast parts, it has been noted that the same part may be made by either process. Comment on the pros and cons of each process, considering factors such as part size, shape complexity, and design flexibility if a particular design has to be modified.

6.144 Referring to Fig. 6.23, sketch the intermediate steps you would recommend in the forging of a wrench.

6.145 Review the technical literature, and make a detailed list of the manufacturing steps involved in the manufacture of hypodermic needles.

6.146 Figure 6.46a shows examples of products that can be obtained by slicing long extruded sections into small discrete pieces. Name several other products that can be made in a similar manner.

6.147 Make an extensive list of products that either are made of or have one or more components of (a) wire; (b) very thin wire; and (c) rods of various cross sections.

6.148 Survey the technical literature, and describe the design features of the various roll arrangements shown in Fig. 6.39.

6.149 Although extruded products typically are straight, it is possible to design dies whereby the extrusion has a constant radius of curvature. (a) What applications could you think of for such products? (b) Describe your ideas as to the shape that such a die should have.

6.150 The beneficial effects of using ultrasonic vibration to reduce friction in some of the processes was described in this chapter. Survey the technical literature, and offer design concepts to apply such vibrations.

Sheet Metal Processes

This chapter describes the characteristics of sheet metals and the principles of forming them into products. Specifically, the following topics are described:

- Specific material properties that affect formability.
- Principles of various shearing operations.
- Fundamentals of sheet forming operations and the important parameters involved.
- Basic design principles in sheet forming.
- Economic considerations.

Symbols

c	clearance, m
D	diameter, m
e	engineering strain
E	elastic modulus, N/m^2
f	feed, m/rev
F	force, N
K	strength coefficient, N/m^2
K_s	springback factor
L	length, m
L_b	bend allowance, m
LDR	limited drawing ratio
m	strain rate sensitivity
n	strain-hardening exponent
p	pressure, N/m^2
r	reduction in area
R	bend radius, m also, standoff, m also, normal anisotropy ratio
ΔR	planar anisotropy
$\overline{R}$	average normal anisotropy ratio
S	panel stiffness
S_y	yield strength, N/m^2
S_{ut}	ultimate tensile strength, N/m^2
t	thickness, m
u	energy dissipated per volume, N/m^2
V	volume, m^3
W	die dimension, m
α	bend angle, deg
ϵ	true strain
$\dot{\epsilon}$	true strain rate, s^{-1}
ϕ	angle in Mohr's circle plane, deg
γ	shear strain, m/m
σ	normal stress, N/m^2
σ_d	draw stress, N/m^2
$\overline{\sigma}_f$	average flow stress, N/m^2

Subscripts

1,2,3 principal directions
i inner
max maximum
o original, outer
p punch
t thickness direction
w width direction

7.1 Introduction

Sheet metal forming operations produce a wide variety of consumer and industrial products, such as beverage cans, kitchen utensils, metal furniture, appliances, car bodies, and aircraft fuselages. Sheet metal forming, also called **pressworking**, **press forming**, or **stamping**, is among the most important of metalworking processes, dating back to as early as 5000 B.C., when household utensils, jewelry, and other objects were made by hammering and stamping metals such as gold, silver, and copper. Compared to those made by casting or forging, sheet metal parts offer such advantages as light weight, low cost, and product shape versatility.

Sheet forming, unlike bulk-deformation processes, involves workpieces with a high ratio of surface area to thickness, as can be seen by inspecting simple products such as cookie sheets and beverage cans. Sheetmetal is produced by the rolling process, as described in Section 6.3. Sheet thicker than 6 mm is generally called **plate**; If the sheet is thin, it is generally coiled after rolling and is often decoiled and flattened prior to further processing. In a typical forming operation, a blank of suitable thickness and dimensions is first cut from a large sheet. This is usually done by a shearing process, although there are several other methods, as described in Chapters 8 and 9. The blank is then processed by any of the operations outlined in Table 7.1. Each forming process has its own specific characteristics, using a variety of hard tools and dies, as well as soft tooling, such as rubber or polyurethane. The sources of energy typically involve mechanical but may also include hydraulic, magnetic, and explosive means.

7.2 Sheet Metal Characteristics

The mechanics of all sheet forming processes basically consists of *stretching* and *bending*. Recall that in some bulk-deformation processes described in Chapter 6, the thickness or the lateral dimension of the workpiece is intentionally changed to produce the part. In sheet forming, however, any change in thickness is typically due to stretching of the sheet under tensile stresses (*Poisson effect*, see Section 2.2).

7.2.1 Elongation

Recall from Fig. 2.2 that a specimen subjected to tension first undergoes **uniform elongation** up to S_{ut}, after which it begins to neck. The elongation

TABLE 7.1 General characteristics of sheet metal forming processes.

Process	Characteristics
Bending	Wide range of product sizes and shapes; versatile process with high production rates if automated; allows use of generally purpose tooling; low to high production rates
Roll forming	Long parts with constant complex cross sections; good surface finish; high production rates; high tooling costs
Stretch forming	Large parts with shallow contours; suitable for low-quantity production; high labor costs; tooling and equipment costs depend on part size
Deep drawing	Shallow or deep parts with relatively simple shapes; high production rates; high tooling and equipment costs
Ironing	Reduction of wall thickness in deep drawn shapes; high tooling and equipment costs
Stamping	Includes a variety of operations, such as punching, blanking, embossing, bending, flanging, and coining; simple or complex shapes formed at high production rates; tooling and equipment costs can be high, but labor costs are low
Rubber-pad forming	Drawing and embossing of simple or complex shapes; sheet surface protected by rubber membranes; flexibility of operation; low tooling costs
Hydroforming	Drawing and embossing of simple or complex shapes; expansion of tubes; variety of cross sections possible
Spinning	Small or large axisymmetric parts; good surface finish; low tooling costs, but labor costs can be high unless operations are automated
Superplastic forming	Complex shapes, fine detail, and close tolerances; forming times are long, and hence production rates are low; parts not suitable for high-temperature use
Peen forming	Shallow contours on large sheets; flexibility of operation; equipment costs can be high; process is also used for straightening parts
Explosive forming	Very large sheets with relatively complex shapes, although usually axisymmetric; low tooling costs, but high labor costs; suitable for low-quantity production; long cycle times
Magnetic-pulse forming	Shallow forming, bulging, and embossing operations on relatively low-strength sheets; most suitable for tubular shapes; high production rates; requires special tooling
Single point incremental forming	Simple shapes; low tooling costs; no lead time for tooling; suitable for limited production

is then followed by further *nonuniform* elongation (**post-uniform elongation**), until the specimen fractures. Because the sheet is being stretched during forming, high uniform elongation is thus essential for good formability.

It was shown in Section 2.2.3 that, for a material that has a true stress–true strain curve represented by the equation

$$\sigma = K\epsilon^n, \tag{7.1}$$

the strain at which necking begins (**instability**) is given by

$$\epsilon = n. \tag{7.2}$$

Thus, the maximum true uniform strain in a simple stretching operation (that is, uniaxial tension) is numerically equal to the strain-hardening exponent, n, of the material. A high value of n is thus desirable for sheet forming.

Necking of a sheet metal specimen generally takes place at an angle ϕ to the direction of tension, as shown in Fig. 7.1a. For an *isotropic* sheet specimen under simple tension, the Mohr's circle is constructed as shown in Fig. 7.1b. The strain ϵ_1 is the longitudinal strain, and ϵ_2 and ϵ_3 are the two lateral strains. Since the equivalent Poisson's ratio for metals in the plastic range is 0.5, the lateral strains have the value $-\epsilon_1/2$. The narrow band (called **localized necking**) shown in Fig. 7.1a is in **plane strain,** because it is constrained by the material above and below the neck band.

The angle ϕ can be determined from the Mohr's circle by a rotation (either clockwise or counterclockwise) of 2ϕ from the ϵ_1 position (Fig. 7.1b). For isotropic materials, this angle is about 110°, and thus the angle ϕ is about 55°. The angle ϕ will be different for materials that are anisotropic in the plane of the sheet (*planar anisotropy*). Note that although the length of the neck band essentially remains constant during the test, its thickness decreases (because of volume constancy), and the specimen eventually fractures.

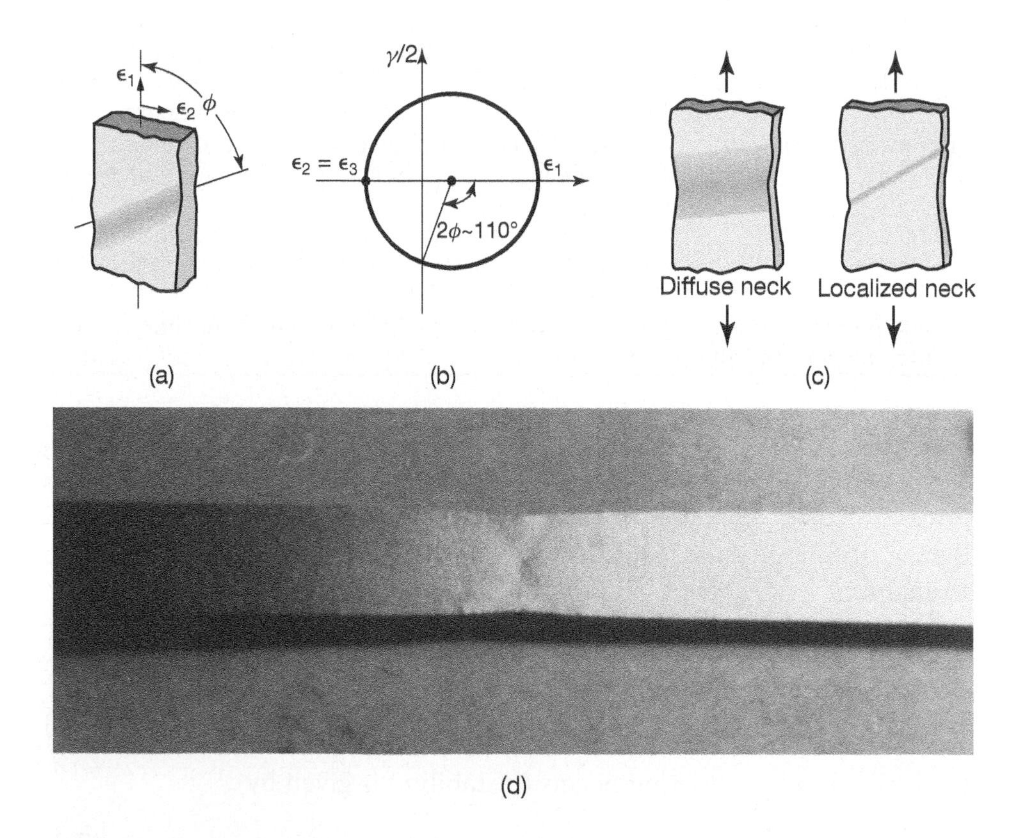

FIGURE 7.1 (a) Localized necking in a sheet metal specimen under tension. (b) Determination of the angle of neck from the Mohr's circle for strain. (c) Schematic illustrations for diffuse and localized necking, respectively. (d) Localized necking in an aluminum strip in tension; note the double neck. *Source:* S. Kalpakjian.

Whether necking is *localized* or is **diffuse** (Fig. 7.1c) depends on the strain-rate sensitivity, m, of the material, as given by the equation

$$\sigma = C\dot{\epsilon}^{m}. \tag{7.3}$$

An example of localized necking on a strip of aluminum in tension is shown in Fig. 7.1d. Note the double localized neck in the specimen; in other words, ϕ can be in either the clockwise or counterclockwise position in Fig. 7.1a, or both (as shown in Fig. 7.1d).

The total elongation of a tension-test specimen, at a gage length of 50 mm, also is a significant factor in formability of sheet metals, where total elongation is the sum of uniform and post-uniform elongation. As stated in Section 2.2.4, uniform elongation is governed by the strain-hardening exponent, n, whereas post-uniform elongation is governed by the strain-rate sensitivity index, m. The higher the m value, the more diffuse the neck is, and hence the greater the post-uniform elongation prior to fracture. Consequently, the total elongation of the material increases with increasing values of both n and m.

The important parameters that affect the sheet metal forming process include:

1. **Yield-point elongation.** Low-carbon steels exhibit a behavior called *yield-point elongation*, and involving *upper* and *lower* yield points, as shown in Fig. 7.2a. Recall also from Section 2.2 that yield-point elongation is usually on the order of a few percent. In yield point elongation, a material yields at a given location, and subsequent yielding occurs in adjacent areas, where the lower yield point is unchanged. When the overall elongation reaches the yield-point elongation, the entire specimen has been deformed uniformly. The magnitude of the yield-point elongation depends on the strain rate (with higher rates,

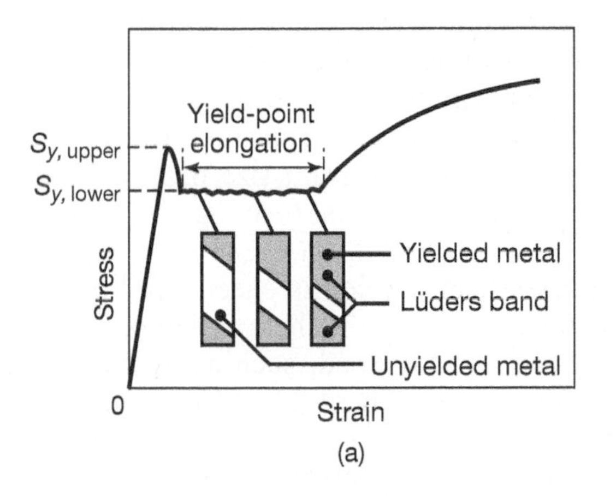

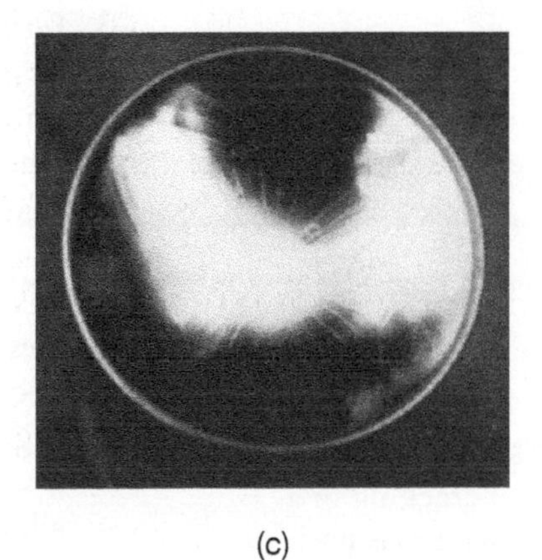

FIGURE 7.2 (a) Yield-point elongation and Lüders bands in tensile testing. (b) Lüder's bands in annealed low-carbon steel sheet. (c) Stretcher strains at the bottom of a steel can for common household products. *Source:* (b) Courtesy of Caterpillar Inc.

the elongation generally increases) and the grain size of the sheet metal (as grain size decreases, yield-point elongation increases).

The behavior of low-carbon steels produces **Lüder's bands** (also called **stretcher strain marks** or *worms*) on the sheet, as shown in Fig. 7.2b. These bands consist of elongated depressions on the surface of the sheet, and can be objectionable in the final product, because of its uneven surface appearance. Also, these bands can subsequently cause difficulties in coating and painting operations of the sheet. Stretcher strain marks can be observed on the curved bottom (dome) of steel cans for common household products, as shown in Fig. 7.2c. Aluminum cans do not exhibit this behavior, because aluminum alloys do not demonstrate yield point elongation.

The usual method of avoiding the formation of these marks is to reduce the thickness of the sheet, by 0.5 to 1.5% by cold rolling, known as **temper rolling** or **skin rolling**. However, because of strain aging (see Section 3.8.1), the yield-point elongation reappears after even a few days at room temperature or after a few hours at higher temperatures. The sheet metal should therefore be formed soon after temper rolling, although the allowable time can be a few weeks for rimmed steel (see Section 5.7.1).

2. **Anisotropy.** *Anisotropy*, or **directionality,** of the sheet metal is acquired during the thermomechanical processing history (mainly rolling) of the sheet. Recall from Section 3.5 that there are two types of anisotropy: **crystallographic anisotropy** (due to preferred grain orientation) and **mechanical fibering** (due to the alignment of impurities, inclusions and voids throughout the thickness of the sheet during processing). Anisotropy may be present not only in the plane of the sheet (**planar anisotropy**), but also in its thickness direction (**normal** or **plastic anisotropy**). These behaviors are particularly important in deep drawing of sheet metals, as described in Section 7.6.
3. **Grain size.** *Grain size* of the sheet metal is important for two reasons: first, because of its effect on the mechanical properties of the material, and secondly, because of the surface appearance of the formed part. The coarser the grain, the rougher the surface appears after deformation (known as **orange peel**); an ASTM grain size of No. 7 or finer is typically preferred for general sheet metal forming operations (see Section 3.4.1).
4. **Residual stresses.** *Residual stresses* can develop in sheet metal parts because of the nonuniform deformation that the sheet undergoes during forming. When these stresses are disturbed, such as by cutting away a portion, the part may distort (See Fig. 2.28). Moreover, tensile residual stresses on surfaces can lead to **stress-corrosion cracking** of the part (Fig. 7.3), unless it is properly stress relieved (see also Sections 3.8.2 and 5.11.4).
5. **Springback.** Because they are thin and are subjected to relatively small strains during forming, sheet metal parts typically display considerable *springback* (see Section 7.4.2). This behavior is particularly significant in simple bending and other forming operations where the bend radius-to-sheet thickness ratio is high.

FIGURE 7.3 Stress-corrosion cracking in a deep-drawn brass part for a light fixture. The cracks have developed over a period of time. Brass and 300-series austenitic stainless steels are particularly susceptible to stress-corrosion cracking.

6. **Wrinkling.** Although in forming operations the sheet metal is typically subjected to biaxial tensile stresses, compressive stresses also can develop in the plane of the sheet. An example is the *wrinkling* of the flange in deep drawing (Section 7.6), because of the circumferential *compressive stresses* that develop in the flange; similar phenomena are called **folding** and **collapsing**. The tendency for wrinkling in sheet metals increases with (a) decreasing thickness; (b) nonuniformity of the thickness of the sheet; and (c) increasing unsupported length or surface area of the sheet. Unevenly distributed lubricants can contribute to the initiation of wrinkling.
7. **Coated sheet.** Sheet metals, especially steels, are often **precoated** with a variety of organic coatings, films, and laminates. Such coatings are primarily for appearance and corrosion resistance. Coatings are applied to the coil stock on continuous lines; thicknesses generally range from 0.0025 to 0.2 mm. Coatings are available with a wide range of characteristics, such as flexibility, durability, hardness, resistance to abrasion and chemicals, color, texture, and gloss. Coated sheet metals are subsequently formed into products such as appliance housing, paneling, shelving, siding, and metal furniture. Zinc is used extensively as a coating for **galvanized** steel (Sections 3.9.7 and 4.5), to protect it from corrosion. Galvanizing of steel may be done by hot dipping, electrogalvanizing, or galvannealing processes (as described in Section 4.5.1).

7.3 Shearing

The *shearing* process involves cutting sheet metal, as well as plates, bars, and tubing of various cross sections, into individual pieces, called *blanks*. This is typically done with a **punch** and a **die** that subject the sheet to shear stresses in the thickness direction, an operation similar to the action of a paper punch (Fig. 7.4). The punch and the die may be of any shape or straight blades (as in a pair of scissors). Important variables in the shearing process are the punch speed, the corner radii of the punch and the die, the punch-die clearance, the edge condition of the sheet (See Figs. 7.5 and 7.6), the punch and die materials, and lubrication condition.

The features of typical sheared surfaces are illustrated in Fig. 7.5; note that the edges are neither smooth nor perpendicular to the plane of the sheet. The **clearance**, c in Fig. 7.4, is the major parameter that determines the shape and quality of the sheared edge. As shown in Fig. 7.6a, as clearance increases, the edges become rougher and the deformation zone in the sheared region becomes larger. Also, with a high clearance, the sheet is pulled more into the cavity, and the sheared edges become more and more rounded. In fact, if the clearance is too large, the sheet metal is bent and thus is subjected to tensile stresses.

In practice, clearances typically range between 2 and 8% of the sheet thickness, but may be as small as 1% in *fine blanking* (see below). In general, clearances are smaller for softer materials; they are higher as the sheet thickness increases. As can be seen in Fig. 7.6b, sheared edges and burrs

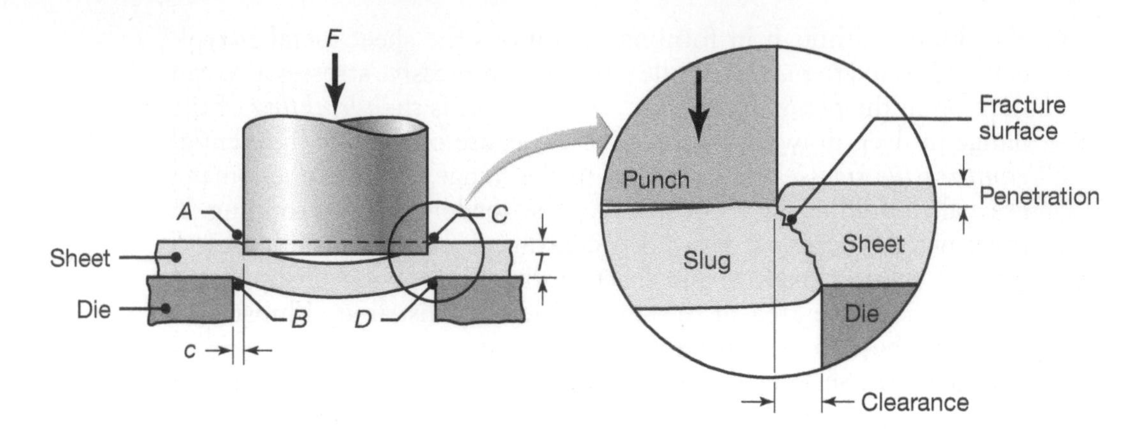

FIGURE 7.4 Schematic illustration of the shearing process with a punch and die, indicating important process variables.

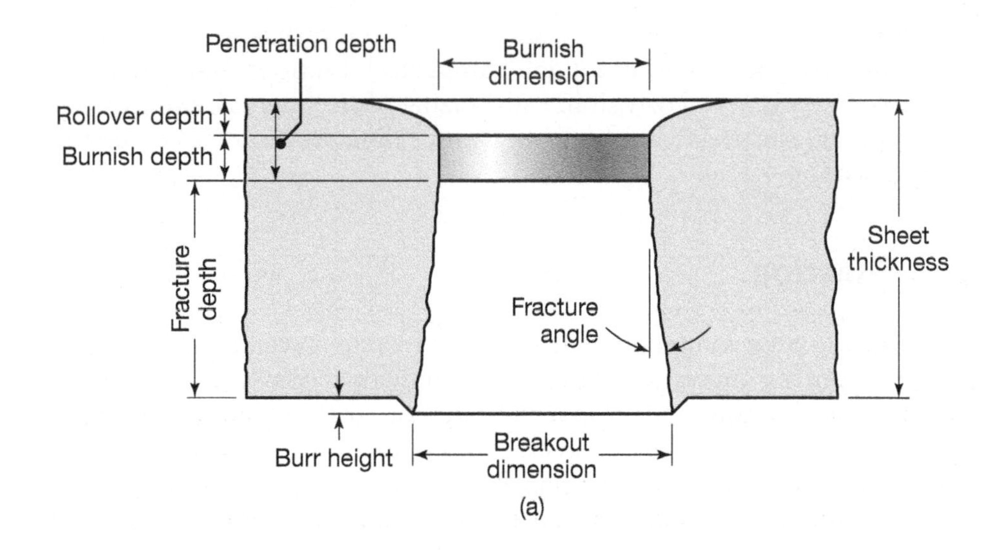

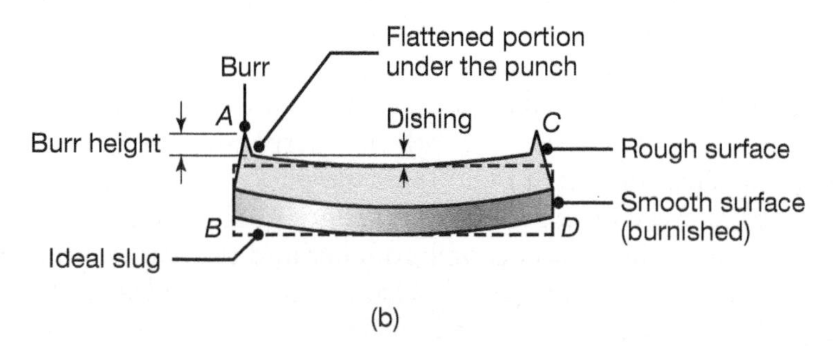

FIGURE 7.5 Characteristic features of (a) a punched hole and (b) a punched slug. Note that the slug has been enlarged with features exaggerated for emphasis.

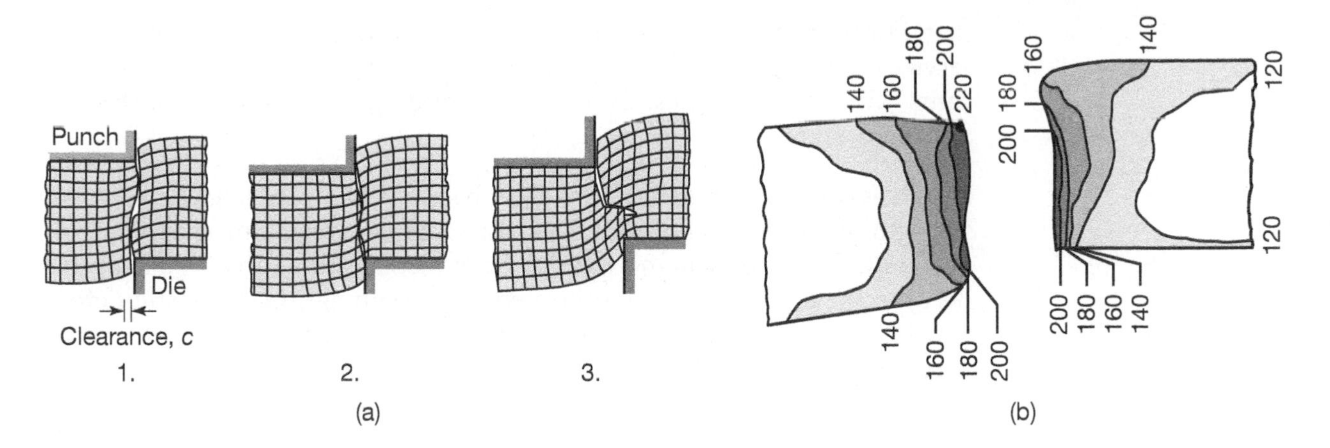

FIGURE 7.6 (a) Effect of clearance, *c*, on the deformation zone in shearing. Note that, as clearance increases, the sheet tends to be pulled into the die, rather than being sheared. (b) Microhardness (HV) contours for a 6.4-mm thick AISI 1020 hot-rolled steel in the sheared region. *Source:* After H.P. Weaver and K.J. Weinmann.

undergo severe cold working, because of the high strains involved. This, in turn, can adversely affect the formability of the sheet during subsequent operations.

It has been observed that shearing typically starts with the formation of cracks on both the top and bottom edges of the sheet (at *A* and *B* in Fig. 7.4). These cracks eventually meet, resulting in complete separation and a **rough fracture surface**. The smooth, shiny, **burnished surfaces** are from the contact and rubbing of the sheared edges against the punch and the die. In the slug shown in Fig. 7.5b, the burnished surface is in the lower region, because this region rubs against the die wall. The burnished surface is on the upper region in the sheared hole and results from rubbing against the punch.

The ratio of the burnished-to-rough areas on the sheared surface (a) increases with increasing ductility of the sheet metal and (b) decreases with increasing sheet thickness and clearance. The distance the punch travels to complete the shearing operation depends on the maximum shear strain that the material can undergo prior to fracture and separation. Thus, a brittle or highly cold-worked material requires little travel of the punch to complete shearing.

Note from Fig. 7.6 that the deformation zone is subjected to high shear strains. The width of this zone depends on the rate of shearing (the punch speed). Also, with increasing punch speed, the heat generated by plastic deformation is confined to a narrower zone; consequently, the sheared surface is smoother.

Note also the formation of a **burr** in Fig. 7.5. *Burr height* increases with increasing clearance and increasing ductility of the metal. Punches and dies with dull edges also contribute to burr formation (see also *slitting*, Section 7.3.1). The height, shape, and size of the burr can significantly affect subsequent forming operations. Moreover, burrs are sharp edges that can be a safety concern; they also may become dislodged during the service life of a part and interfere with the operation or contaminate lubricants with solid particles. Several *deburring* operations are described in Section 9.8.

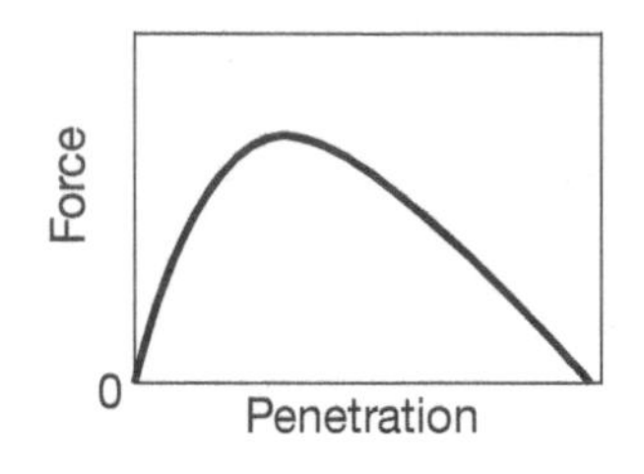

FIGURE 7.7 Typical punch force vs. penetration in shearing. The area under the curve is the work done in shearing. The shape of the curve depends on processing parameters and material properties.

Punch force. The *punch force*, *F*, is basically the product of the shear strength of the sheet metal and the cross-sectional area being sheared; *friction* between the punch and the sheet can increase this force significantly. Because the sheared zone is subjected to a combination of plastic deformation, friction, and development of cracks, the punch-force vs. stroke curves can have a variety of shapes. Figure 7.7 shows one typical curve, for a ductile material. Note also that the area under the curve is the *total work* done in shearing.

An approximate empirical formula for estimating the **maximum punch force,** F_{max}, is given by

$$F_{max} = 0.7 S_{ut} t L, \tag{7.4}$$

where S_{ut} is the ultimate tensile strength of the sheet metal, t is its thickness, and L is the total length of the sheared edge; thus, for example, for a round hole of diameter D, $L = \pi D$. In addition to the punch force, a force is also required to strip the sheet from the punch during its return stroke.

EXAMPLE 7.1 Calculation of Maximum Punch Force

Given: A 25 mm diameter hole will be punched in a 1.6 mm thick, 5052-O aluminum sheet at room temperature.

Find: Estimate the force required.

Solution: The force is estimated from Eq. (7.4), where S_{ut} for this alloy is found in Table 3.5 as 190 MPa. Therefore,

$$F = 0.7\,(0.0016)\,(\pi)(0.025)\left(190 \times 10^6\right) = 16.7 \text{ kN}.$$

7.3.1 Shearing Operations

This section describes various operations that are based on the shearing process. In **punching**, the sheared slug is generally discarded (Fig. 7.8a); in **blanking**, the slug is the part itself and the rest is generally scrap, later to be recycled. The following processes are common shearing operations:

1. **Die cutting.** *Die cutting* refers to the operations shown in Fig. 7.8b, where the parts produced have a wide variety of uses: (a) **perforating** is punching a number of holes in a sheet; (b) **parting** is shearing the sheet into two or more pieces; (c) **notching** is removing pieces of various shapes from the edges of a sheet; (d) **slitting**; and (e) **lancing** involves leaving a tab on the sheet without removing any material.
2. **Fine blanking.** Very smooth and square edges can be produced by *fine blanking*. A basic die design is shown in Fig. 7.9, in which a

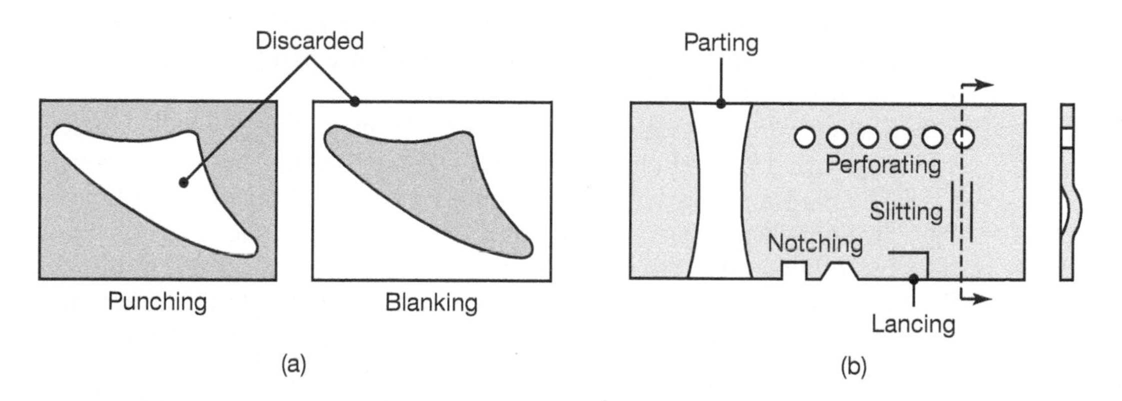

FIGURE 7.8 (a) Punching and blanking. (b) Examples of shearing operations on sheet metal.

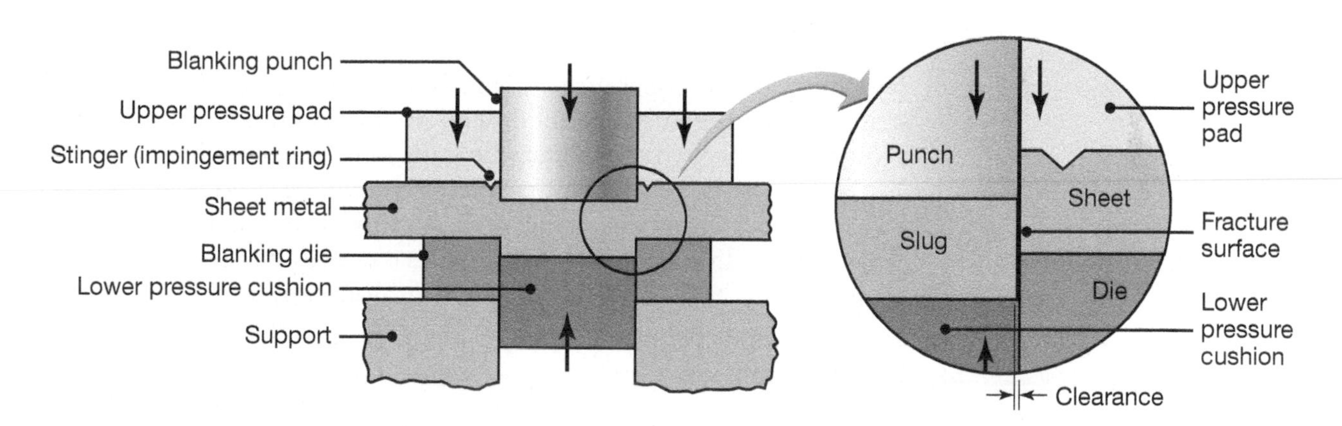

FIGURE 7.9 Schematic illustration of a setup for fine blanking. *Source:* Feintool International Holding.

V-shaped *stinger* (*impingement*) locks the sheet tightly in place and prevents the type of distortion of the material such as that shown in Fig. 7.6. Fine blanking involves clearances on the order of 1% of the sheet thickness, as compared with as much as 8% in ordinary shearing operations. The thickness of the sheet typically ranges from 0.5 to 13 mm, with a dimensional tolerance of ±0.05 mm. Fine blanking usually is carried out on triple-action hydraulic presses, triple meaning that the movements of the punch, pressure pad (See Fig. 7.9), and die are controlled separately.

3. **Slitting**. *Slitting* (Fig. 7.10) is a shearing operation, typically carried out with a pair of circular blades, similar to those on a can opener. The blades follow either a straight line or a circular or curved path. Straight slitting is commonly used in cutting wide sheet, as delivered by rolling mills, into narrower strips for further processing into individual parts. Slitting operations may cause various planar distortions of the slit part or strip. Also, a slit edge typically has a burr (See Fig. 7.5), which can be rolled over the sheet's edge using a set of rolls (see also Section 9.8). There are two basic types of slitting equipment: (a) the *driven* type, in which the blades are powered; and (b) the *pull-through* type, in which the strip is pulled through idling blades.
4. **Steel rules.** Sheets of soft metals, thermoplastics, paper, leather, and rubber can be blanked into various shapes using *steel-rule dies*. Such a die consists of a hardened steel blade that is first bent to the shape to be sheared (similar to a cookie cutter), and then supported on a flat base. The die is pressed against the sheet and cuts it to the shape of the steel rule.
5. **Nibbling.** In this operation, a machine called a *nibbler* moves a straight punch rapidly up and down into a matching die cavity. The sheet is fed through the punch-die gap, making a number of overlapping holes; this operation is similar to making a large elongated hole by successively punching several closely spaced holes with a paper punch. Intricate slots and notches can thus be produced using standard punches; the sheet can be cut along any path by manual or automatic control. The nibbling process is economical for small production runs.

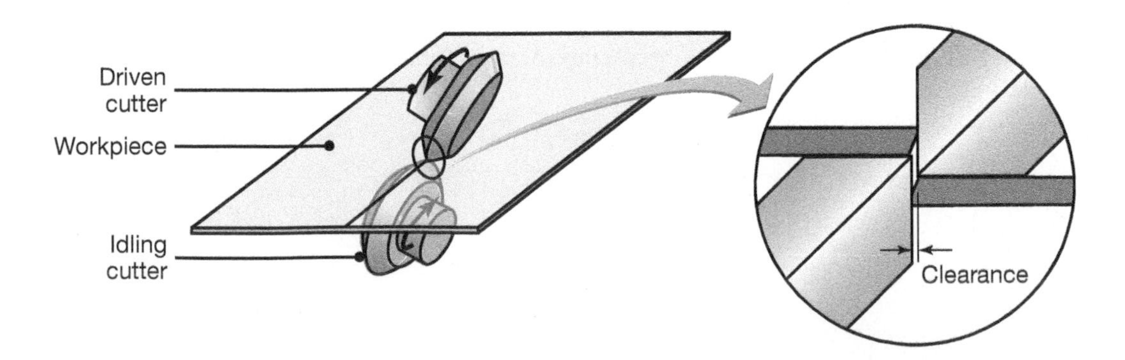

FIGURE 7.10 Slitting with rotary blades, a process similar to opening cans.

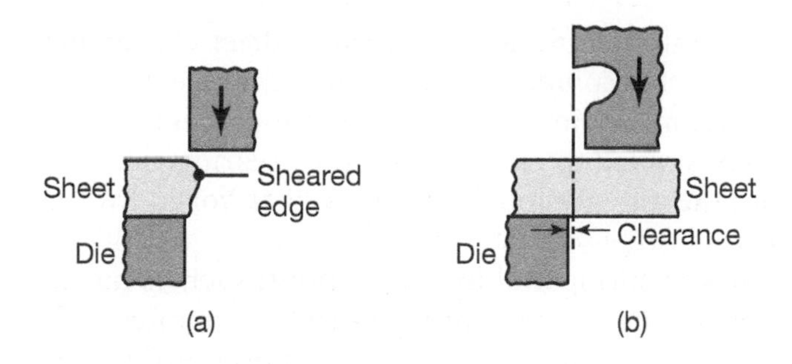

FIGURE 7.11 Schematic illustration of shaving on a sheared edge. (a) Shaving a sheared edge. (b) Shearing and shaving combined in one punch stroke.

Scrap in shearing operations. The amount of *scrap* produced (**trim loss**) in shearing operations can be significant, being as high as 30% of the original sheet for large pieces (See also Table 16.7). Scrap is an important factor in manufacturing costs; it can be reduced significantly by proper arrangement of the shapes on the sheet to be cut, called **layout and nesting** (See Fig. 7.65). Computer-aided design techniques are now available for minimizing scrap, particularly in large-scale operations.

7.3.2 Shearing Dies

The formability of a sheared part can be directly influenced by the quality of its sheared edges. Clearance control is an important factor; generally, the thicker the sheet, the larger the clearance. Recall, however, that the smaller the clearance, the better the quality of the sheared edge. Any extra material from a rough sheared edge can trimmed by **shaving** (Fig. 7.11), which also improves surface quality. Some commonly used shearing dies are described below.

1. **Punch and die shapes.** Note in Fig. 7.4 that the surfaces of the punch tip and the die are flat; the punch force builds up rapidly during shearing, because the entire thickness of the sheet shears at the same time. The area sheared at any instant can be controlled by *beveling* the punch and die surfaces, as illustrated in Fig. 7.12. A beveled shape is particularly suitable for shearing thick sheets, because it reduces the maximum shearing force (since it is an incremental process); it also reduces the noise level during punching. Note also that the punch and the press must have sufficient rigidity in the lateral direction to maintain dimensional tolerances and avoid tool breakage, since there is a net lateral force in a beveled punch (Fig. 7.12b).

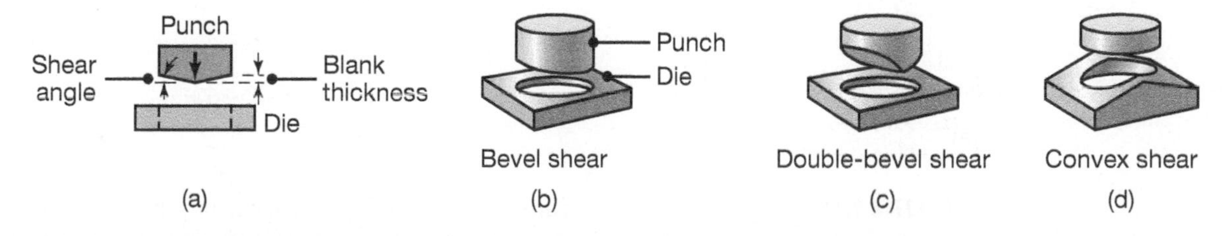

FIGURE 7.12 Examples of the use of shear angles on punches and dies. Compare these designs with that for a common paper punch.

2. **Compound dies.** Several operations on the same sheet can be performed in one stroke with a *compound die*. Although these dies have higher productivity than the simple punch-and-die processes, these operations are usually limited to relatively simple shearing. They are somewhat slow and the dies are more expensive than for individual shearing operations
3. **Progressive dies.** Parts requiring multiple operations, such as punching, bending, and blanking, are made at high production rates using *progressive dies*. A coil strip is fed into the dies and progressively fed with each stroke of the press. A sequence of different operations is performed with each press stroke (Fig. 7.13a). An example of a part made in progressive dies is shown in Fig. 7.13b.
4. **Transfer dies.** In a *transfer die* the sheet undergoes different operations at different stations, which are arranged along a straight line or a circular path. After an operation is completed, the part is transferred (hence the name transfer die) to the next station for another operation, and so on.
5. **Tool and die materials.** Tool and die materials for shearing operations are generally tool steels and, for high production rates, carbides. Lubrication is important for reducing tool and die wear and for improving edge quality.

QR Code 7.1 Production of an automotive clip in progressive dies. *Source:* Courtesy of Scandic Springs.

7.3.3 Miscellaneous Methods of Cutting Sheet Metal

There are several other methods for cutting sheets and plates:

1. The sheet or plate may be cut with a **band saw**, as described in Section 8.10.5.

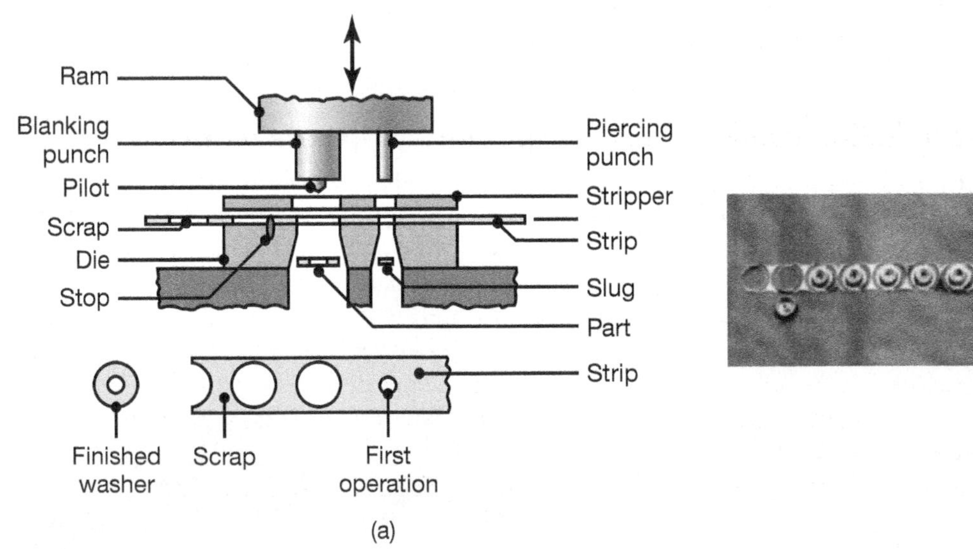

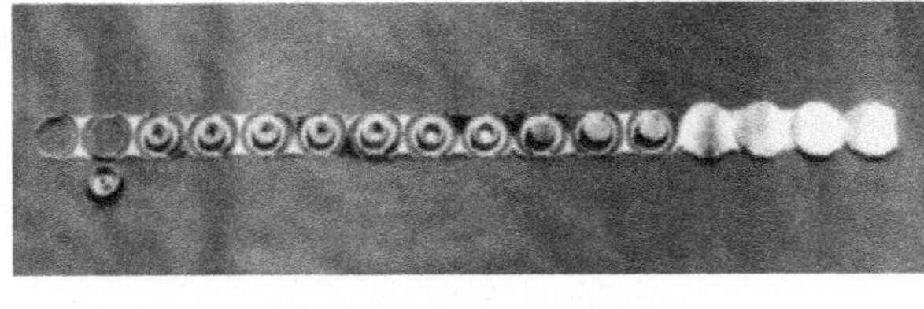

FIGURE 7.13 (a) Schematic illustration of producing a washer in a progressive die. (b) Forming of the top piece of a common aerosol spray can in a progressive die. Note that the part is attached to the strip until the last operation is completed. *Source:* (b) After S. Kalpakjian.

2. **Oxyfuel-gas (flame) cutting** may be employed, particularly for thick plates, as widely used in shipbuilding and heavy-construction industries (see Section 12.6).
3. **Friction sawing** involves the use of a disk or blade, that rubs against the sheet or plate at high surface speed (see Section 8.10.5).
4. **Water-jet cutting** and **abrasive water-jet cutting** are effective operations on sheet metals as well as on nonmetallic materials (see Section 9.15).
5. **Laser-beam cutting** is widely used, consistently cutting a variety of shapes with computer-controlled equipment (see Section 9.14.1). This process also can be combined with shearing processes.

7.3.4 Tailor-Welded Blanks

Note that in sheet metal forming operations the blank has uniform thickness, and is typically supplied in one piece, usually cut from a larger sheet. An important technology, particularly in the automotive industry, involves laser butt welding (see Section 12.5.2) of two or more pieces of sheet with different thicknesses and shapes (called *tailor-welded blanks* or *TWB*). The welded sheet is subsequently formed into a final shape, by using any of the processes described in this chapter.

Each welded piece can have a different thickness (as guided by design considerations), grade of sheet metal, coating, or other characteristics; consequently, these blanks possess the required characteristics in optimal locations of the formed part. As a result, (a) productivity is increased; (b) the need for subsequent spot welding of the product is reduced or eliminated; (c) scrap is reduced; and (d) dimensional control is improved. It should be noted that because the sheet thicknesses involved are small, proper alignment of the sheets prior to butt welding is essential.

CASE STUDY 7.1 Use of Tailor-Welded Blanks in the Automotive Industry

An example of the use of tailor-welded blanks is the production of an automobile outer side panel, shown in Fig. 7.14a. Note that five different pieces are first blanked, then laser butt welded and stamped into the final shape. Thus, the blanks can be tailored for a particular application, including not only sheets with different thicknesses and shapes, but also of different quality and with or without coatings on one or both surfaces of the sheets. This technique of welding and forming of sheet metal pieces allows significant flexibility in product design, structural stiffness and *crashworthiness*, formability, and using different materials in one component, as well as weight savings and cost reduction in materials, scrap, equipment, assembly, and labor.

Tailor-welded blanks are increasingly popular in automotive applications, as shown in Fig. 7.14b, which also exploit the advantages outlined above. For example, the sheet thickness in these components varies, in accordance with loading or required stiffness; this leads to significant weight and cost savings.

Source: After M. Geiger and T. Nakagawa.

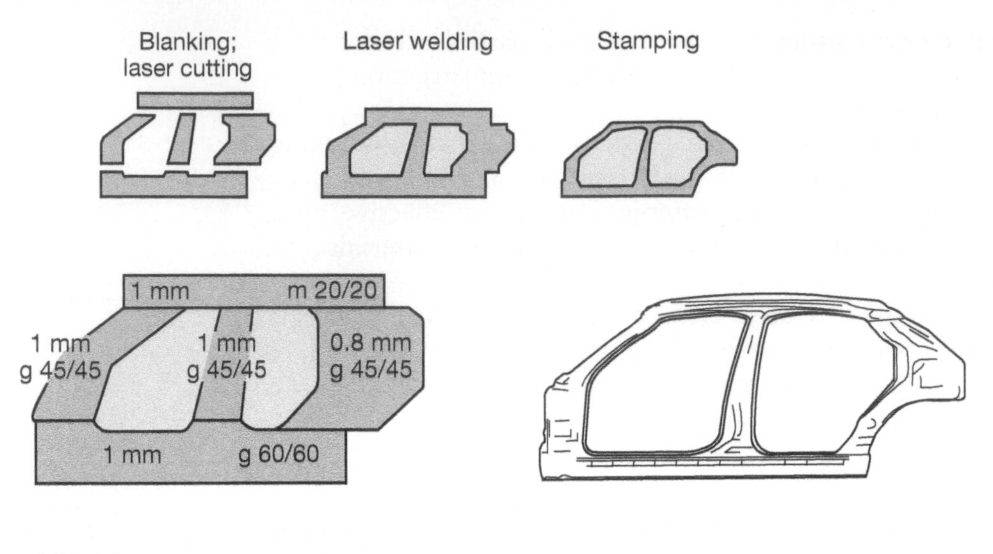

Legend

g 60/60 (45/45) Hot-galvanized alloy steel sheet. Zinc amount: 60/60 (45/45) g/m^2.

m 20/20 Double-layered iron-zinc alloy electroplated steel sheet. Zinc amount 20/20 g/m^2.

(a)

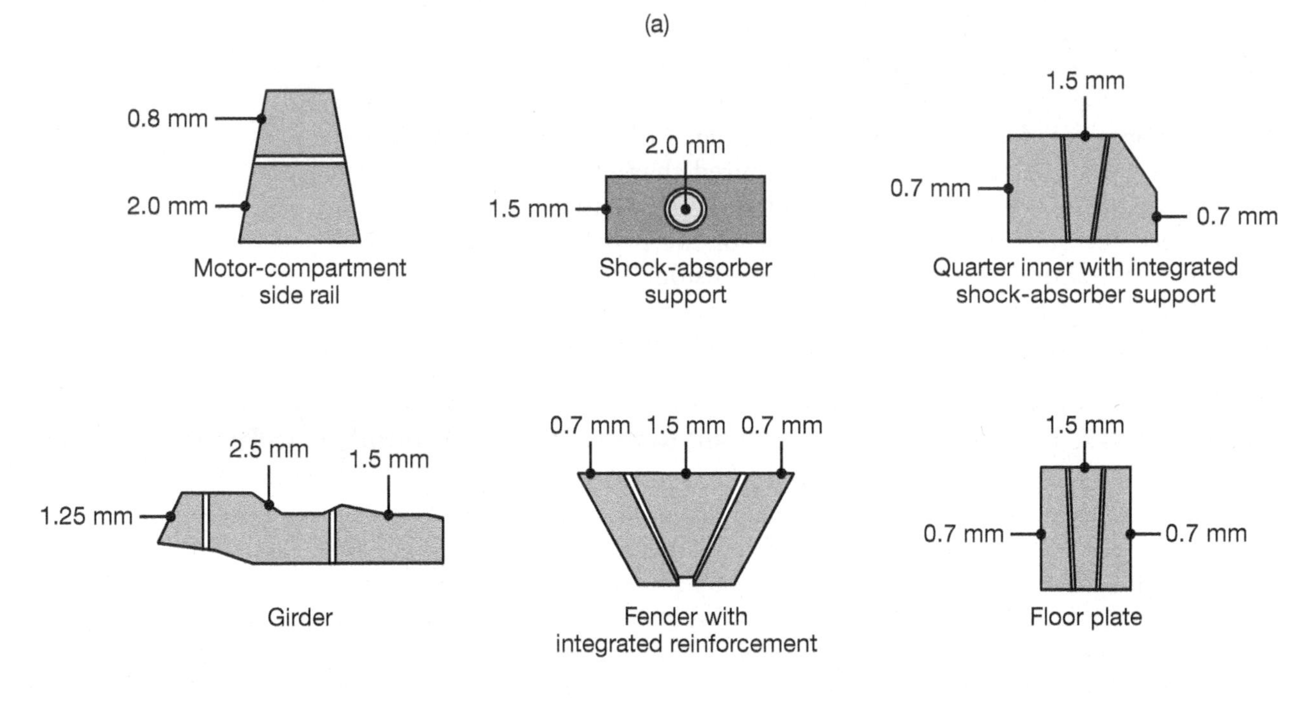

(b)

FIGURE 7.14 Examples of laser-welded and stamped automotive body components. *Source:* After M. Geiger and T. Nakagawa.

7.4 Bending of Sheet and Plate

One of the most common metalworking operations is *bending*, a process that is used not only to form such parts as flanges, curls, seams, and corrugations, but also to impart stiffness to the sheet metal part by increasing its moment of inertia. Figure 7.15a gives the terminology for bending. **Bend allowance** is the length of the *neutral axis* in the bend area, and is used to determine the blank length for a part to be bent. However, as described in texts on the mechanics of solids, the radial position of the neutral axis in bending depends on the bend radius and bend angle. An approximate formula for the bend allowance, L_b, is given by

$$L_b = \alpha\left(R + kt\right),$$

where α is the bend angle (in radians), R is the bend radius, k is a constant, and t is the sheet thickness. For the ideal case, the neutral axis remains at the center, and hence $k = 0.5$. In practice, k usually ranges from 0.33 for $R < 2t$ to 0.5 for $R > 2t$.

QR Code 7.2 Air bending, highlighting compensation for springback. *Source:* Courtesy of the LVD Company.

7.4.1 Minimum Bend Radius

Recall that the outer fibers of a strip being bent are subjected to tension, and the inner fibers to compression. Theoretically, the strains at the outer and inner fibers are equal in magnitude and are given by the equation

$$e_o = e_i = \frac{1}{(2R/t) + 1}. \tag{7.5}$$

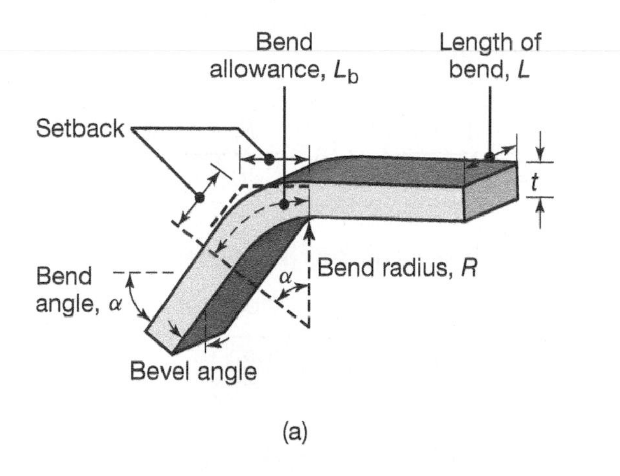

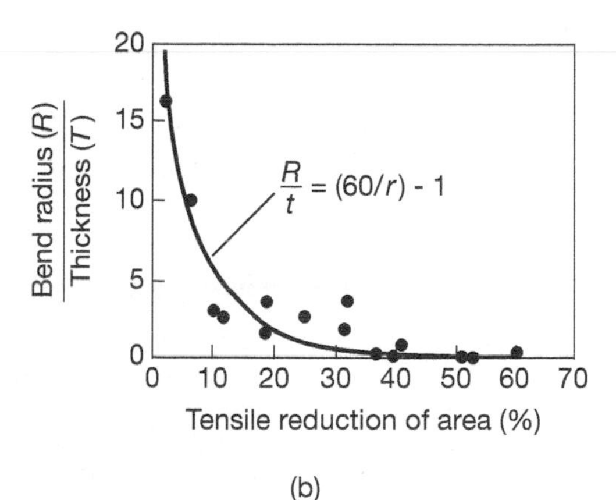

FIGURE 7.15 (a) Bending terminology. Note in the figure that the bend radius is measured to the inner surface of the bend, and that the length of the bend is the width of the sheet being bent. (b) Relationship between the ratio of bend-radius to sheet-thickness and tensile reduction of area for a variety of materials. Note also that sheet metal with a reduction of area of about 50% can be bent and flattened over itself without cracking, similar to folding a piece of paper. *Source:* After J. Datsko and C.T. Yang.

However, due to the *shifting of the neutral axis* toward the inner surface, the *length of bend* (the dimension L in Fig. 7.15a) is smaller in the outer region than in the inner region. This phenomenon can easily be observed by bending a rectangular eraser and noting how the cross section distorts. Consequently, the outer and inner strains are different, with the difference increasing with decreasing R/t ratio.

It can be seen from Eq. (7.5) that as the R/t ratio decreases, the tensile strain at the outer fiber increases; the sheet may thus begin to crack after a certain strain is reached. The radius, R, at which a crack first appears on the outer surface of the bend is called the *minimum bend radius*. This minimum bend radius is generally expressed in terms of its thickness, such as $2t$, $3t$, $4t$, and so on. Thus, for example, a bend radius of $3t$ indicates that the smallest radius to which the sheet can be bent, without cracking, is three times its thickness. The minimum bend radii for various materials have been determined experimentally; some typical results are given in Table 7.2.

Relationships have been established between the minimum R/t ratio and mechanical properties of the sheet material. One such analysis is based on the assumptions that (a) the true strain at cracking on the outer fiber in bending is equal to the true strain at fracture, ϵ_f, of the material in a simple tension test; (b) the material is homogeneous and isotropic; and (c) the sheet is bent in a state of plane stress, that is, its L/t ratio is small.

The true strain at fracture in tension is

$$\epsilon_f = \ln\left(\frac{A_0}{A_f}\right) = \ln\left(\frac{100}{100 - r}\right),$$

where r is the percent reduction of area of the sheet in a tension test. From Section 2.2.2 and Eq. (7.5), the true strain is,

$$\epsilon_o = \ln(1 + e_o) = \ln\left(1 + \frac{1}{(2R/t) + 1}\right) = \ln\left(\frac{R + t}{R + (t/2)}\right).$$

TABLE 7.2 Minimum bend radii for various materials at room temperature.

Material	Material condition: Soft	Hard
Aluminum alloys	0	$6t$
Beryllium copper	0	$4t$
Brass, low leaded	0	$2t$
Magnesium	$5t$	$13t$
Steels		
Austenitic stainless	$0.5t$	$6t$
Low carbon, low alloy, and HSLA	$0.5t$	$4t$
Titanium	$0.7t$	$3t$
Titanium alloys	$2.6t$	$4t$

Equating the two expressions above and simplifying,

$$\text{Minimum}\frac{R}{t} = \frac{50}{r} - 1. \tag{7.6}$$

The experimental data are shown in Fig. 7.15b, where it can be noted that a better curve fit is

$$\text{Minimum}\frac{R}{t} = \frac{60}{r} - 1. \tag{7.7}$$

Note that the R/t ratio approaches zero (*complete bendability*) at a tensile reduction of area of 50%. This percentage is the same value obtained for spinnability of metals (described in Section 7.5.4); that is, a material with 50% reduction of area is found to be completely spinnable.

Factors affecting bendability. The bendability of a metal may be increased by increasing its tensile reduction of area, by such means as *heating* or by the application of *hydrostatic pressure* (see Section 2.2.8). Another technique involves applying a compressive force in the plane of the sheet during bending, to minimize tensile stresses in the outer fibers of the bend area (See Fig. 7.29).

As bend length, L, increases, the state of stress at the outer fibers changes from uniaxial to a *biaxial* stress. The reason for this change is that the bend length tends to become smaller due to stretching of the outer fibers (as in bending a rectangular eraser), but it is constrained by the material around the bend area. Biaxial stretching tends to reduce ductility (strain to fracture); thus, as L increases, the minimum bend radius increases (Fig. 7.16). However, at a bend length of about $10t$, the minimum bend radius increase is negligible, and a *plane-strain condition* is developed. As the R/t ratio decreases, narrow sheets (smaller length of bend) begin to crack at the edges, whereas wider sheets crack at the center (where the biaxial stress is the highest).

The **edge condition** of the sheet being bent significantly affects bendability. Because rough edges have locations of stress concentration, bendability

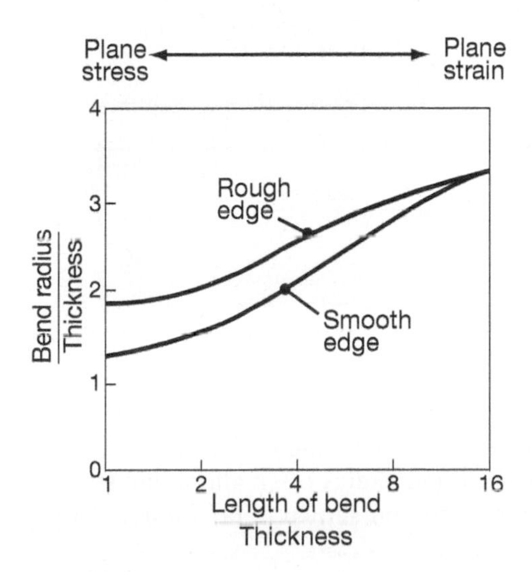

FIGURE 7.16 The effect of length of bend and edge condition on the ratio of bend radius to thickness for 7075-T aluminum sheet. *Source:* After G. Sachs and G. Espey.

decreases as edge roughness increases. Another important factor is the degree of **cold working** that the edges undergo during shearing, as can be observed from microhardness tests in the sheared region, shown in Fig. 7.6b. Removal of the cold-worked regions, by such means as shaving or machining, or by increasing ductility by annealing, greatly improves the resistance to edge cracking during bending.

Another significant factor in edge cracking is the shape and the amount of inclusions in the sheet metal (see also Section 3.3.3). Inclusions in the form of **stringers** are more detrimental than globular-shaped inclusions. Anisotropy of the sheet also is important in bendability; as depicted in Fig. 7.17, cold rolling of sheets results in **anisotropy**, because of the alignment of impurities, inclusions, and voids (*mechanical fibering*). The transverse ductility is thus reduced, as shown in Fig. 7.17c (See also Fig. 3.16). Caution should therefore be exercised in cutting or slitting blanks in the proper direction of the rolled sheet, although this may not always be possible or economical in practice.

7.4.2 Springback

Because all materials have a finite modulus of elasticity, plastic deformation is always followed by **elastic recovery** upon removal of the load; in bending, this recovery is known as *springback*. As shown in Fig. 7.18, the final bend angle after springback is smaller than the angle to which it is bent and the final bend radius is larger than the radius to which it is bent. This phenomenon can easily be observed and verified by bending a piece of wire or a strip of metal.

The amount of springback can be expressed by the **springback factor,** K_s, which is determined as follows. Because the bend allowance

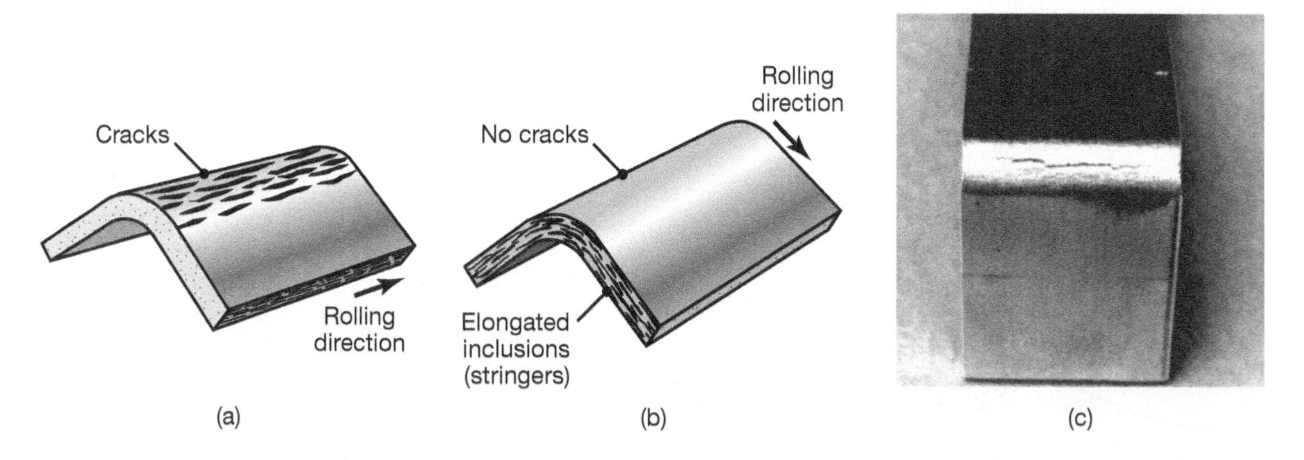

FIGURE 7.17 (a) and (b) The effect of elongated inclusions (stringers) on cracking in sheets as a function of the direction of bending with respect to the original rolling direction. (c) Cracks on the outer radius of an aluminum strip bent to an angle of 90°; compare this part with that shown in (a). *Source:* After S. Kalpakjian.

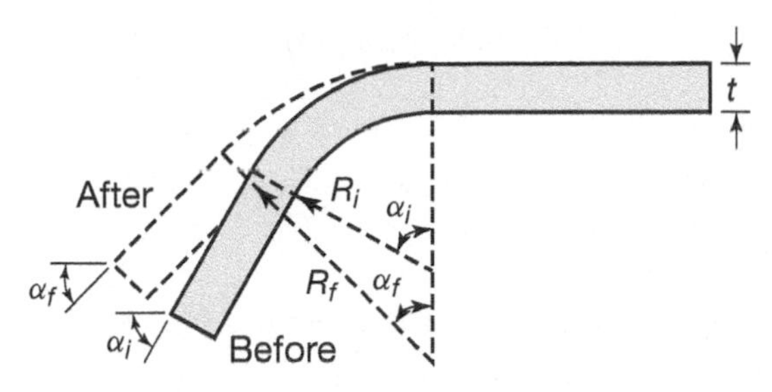

FIGURE 7.18 Terminology for springback in bending. Note that the bend angle has become smaller after the sheet is bent. There are situations whereby the angle becomes larger, called *negative springback* (See Fig. 7.20).

(See Fig. 7.15a) is the same before and after bending, the relationship obtained for pure bending is

$$\text{Bend allowance} = \left(R_i + \frac{t}{2}\right)\alpha_i = \left(R_f + \frac{t}{2}\right)\alpha_f. \qquad (7.8)$$

From this relationship, K_s is defined as

$$K_s = \frac{\alpha_f}{\alpha_i} = \frac{(2R_i/t) + 1}{(2R_f/t) + 1}, \qquad (7.9)$$

where R_i and R_f are the initial and final bend radii, respectively. Note from this expression that K_s depends only on the R/t ratio. The condition of $K_s = 1$ indicates that there is no springback, and $K_s = 0$ indicates that there is complete elastic recovery (Fig. 7.19).

Recall from Fig. 2.4 that elastic recovery increases with the stress level and with decreasing elastic modulus. Therefore, an approximate formula has been developed to estimate springback as

$$\frac{R_i}{R_f} = 4\left(\frac{R_i S_y}{Et}\right)^3 - 3\left(\frac{R_i S_y}{Et}\right) + 1, \qquad (7.10)$$

where S_y is the uniaxial yield stress of the material at 0.2% offset (See Fig. 2.3).

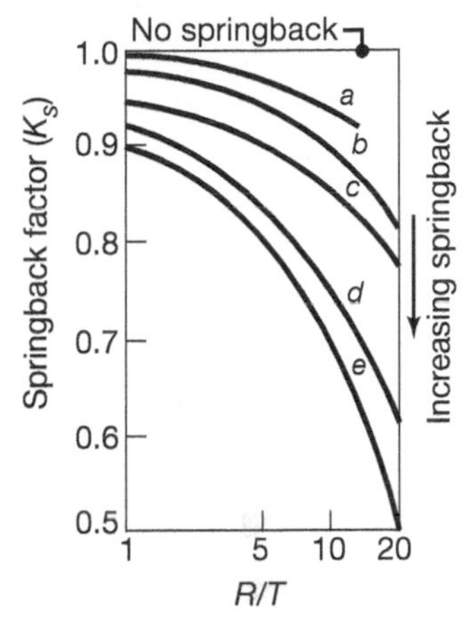

FIGURE 7.19 Springback factor, K_s, for various materials: (a) 2024-0 and 7075-0 aluminum; (b) austenitic stainless steels; (c) 2024-T aluminum; (d) $\frac{1}{4}$-hard austenitic stainless steels; and (e) $\frac{1}{2}$-hard to full-hard austenitic stainless steels. A factor of $K_s = 1$ indicates that there is no springback. *Source:* After G. Sachs.

EXAMPLE 7.2 Estimating Springback

Given: A 1 mm-thick steel sheet with a stiffness of 200 GPa and a yield strength of 275 MPa is bent to a radius of 12 mm.

Find: (a) Calculate the radius of the part after it is bent, and (b) the required bend angle to achieve a 90° bend after springback has occurred.

Solution:

a. The appropriate formula is Eq. (7.10), where

$$R_i = 12 \text{ mm}, \qquad S_y = 275 \text{ MPa}, \qquad E = 200 \times 10^9 \text{ N/m}^2,$$

and $t = 0.001$ mm. Thus,

$$\frac{R_i S_y}{Et} = \frac{(0.012)\,(275 \times 10^6)}{(200 \times 10^9)\,(0.001)} = 0.0165,$$

and

$$\frac{R_i}{R_f} = 4\,(0.0165)^3 - 3\,(0.0165) + 1 = 0.951.$$

Hence,

$$R_f = \frac{12}{0.951} = 12.62 \text{ mm}.$$

b. To calculate the required bend angle,using Eq. (7.9) yields

$$\frac{\alpha_f}{\alpha_i} = \frac{(2R_i/t) + 1}{(2R_f/t) + 1}$$

or

$$\alpha_i = \alpha_f \frac{(2R_f/t) + 1}{(2R_i/t) + 1} = (90^\circ)\,\frac{(2)(0.01262)/(0.001) + 1}{(2)(0.012)/(0.001) + 1} = 94.5^\circ.$$

Negative springback. The springback commonly observed and shown in Fig. 7.18 is called *positive springback*. Under certain conditions, however, *negative springback* also occurs, whereby the bend angle becomes larger after the part is bent. This phenomenon, which generally is associated with V-die bending, is explained by observing the sequence of deformation shown in Fig. 7.20. If the bent part is removed at stage (b), it will undergo positive springback. At stage (c), the ends of the piece are touching the punch; note that between stages (c) and (d), the part is actually being bent in the direction opposite to that between stages (a) and (b).

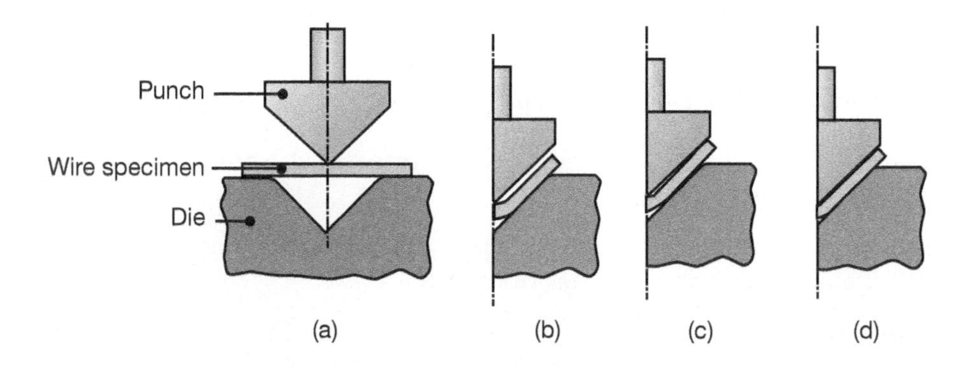

FIGURE 7.20 Schematic illustration of the stages in bending round wire in a V-die. This type of bending can lead to negative springback, which does not occur in air bending (shown in Fig. 7.24a). *Source:* After K.S. Turke and S. Kalpakjian.

Note also that in both stages (b) and (c), there is lack of shape conformity between the punch radius and the inner radius of the part; in stage (d), however, the two radii become the same. Upon unloading, the part in stage (d) will spring back inward, because it is being *unbent* from stage (c), both at the tip of the punch and in the arms of the part. Because of the large strains that the material has undergone in the small bend area in stage (b), the amount of this inward (negative) springback can be greater than the amount of positive springback; the net result is negative springback.

Compensation for springback. In practice, springback is usually compensated using various techniques:

1. **Overbending** the part in the die can compensate for springback. Overbending also can be achieved by the **rotary bending** technique shown in Fig. 7.21e. Note that the upper die consists of a cylindrical rocker (with an angle of $<90°$), and that it is free to rotate as the punch moves downward. The sheet is clamped and is bent by the rocker over the lower die (anvil). A relief angle in the lower die allows overbending of the sheet at the end of the stroke, thus it compensates for springback.
2. **Coining** the bend region by subjecting it to high localized compressive stresses between the tip of the punch and the die surface (Figs. 7.21c and d). This operation is also known as **bottoming** the punch against the die.
3. **Stretch bending**, in which the part is subjected to tension while being bent. In this way, springback, which is due to nonuniform stresses in bending, will decrease. This technique is also used to reduce springback in stretch forming of shallow automotive bodies (see Section 7.5.1).
4. Because springback decreases as yield stress decreases [see Eq. (7.10)], bending may also be carried out at *elevated temperatures* to reduce

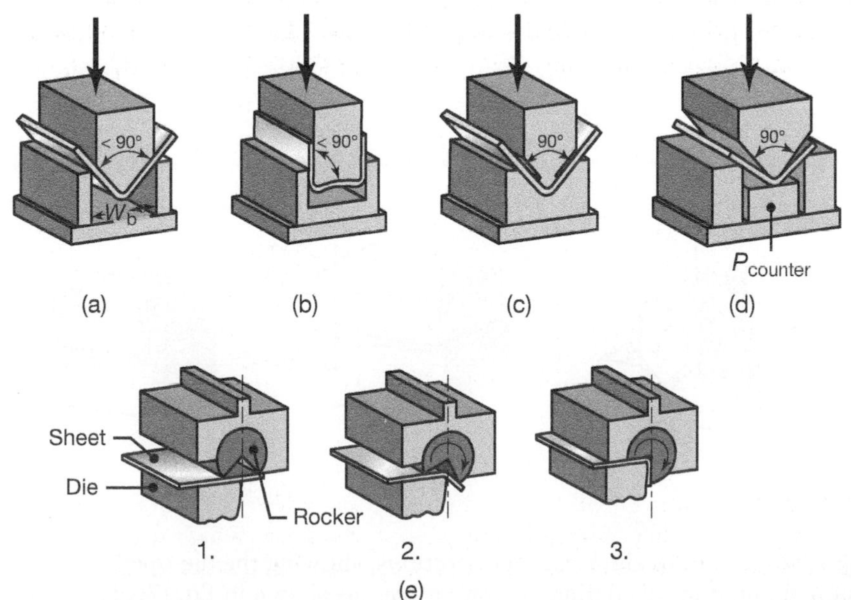

FIGURE 7.21 Methods of reducing or eliminating springback in bending operations.

springback. In practice, this is rarely done because of the complications involved.

7.4.3 Bending Force

The bending force can be estimated by assuming that the process is that of simple bending of a rectangular beam (as described in texts on strength of materials). Thus, the bending force is a function of the material's strength, the length and thickness of the part (L and t, respectively), and the width, W, of the die opening (Fig. 7.22). The general expression for the **maximum bending force,** F_{max}, is

$$F_{max} = k\frac{S_{ut}Lt^2}{W}, \qquad (7.11)$$

QR Code 7.3 Trumpf press brake. *Source:* Courtesy of OGS Industries.

where k includes various factors, and ranges from about 1.2 to 1.33 for a V die, 0.3–0.34 for a wiping die and 2.4–2.6 for V dies. Equation (7.11) works well for situations where the punch radius and sheet thickness are small compared to the size of the die opening, W.

The bending force in die bending (See Fig. 7.22) is also a function of punch travel. It first increases to a maximum value; may then decrease as the bend is completed; and then increases sharply as the punch bottoms. In *air bending* (also called *free bending*, see Fig. 7.24a) the force does not increase again after it begins to decrease, because there is no constraint to the downward movement of the sheet being bent.

7.4.4 Common Bending Operations

QR Code 7.4 Press brake forming. *Source:* Courtesy of OGS Industries.

This section describes common bending operations, some of which are performed on discrete sheet metal parts while others are done continuously, as in roll forming of coiled sheet stock.

1. **Press-brake forming.** Sheet metal or plate can be bent with simple fixtures, using a press (see Section 6.2.8). Parts that are long and relatively narrow are usually bent in a *press brake*. Press brakes are equipped with long dies, are usually mechanically or hydraulically powered, and are suitable for small production runs. The tooling is simple and adaptable to a wide variety of shapes (Fig. 7.23). Die

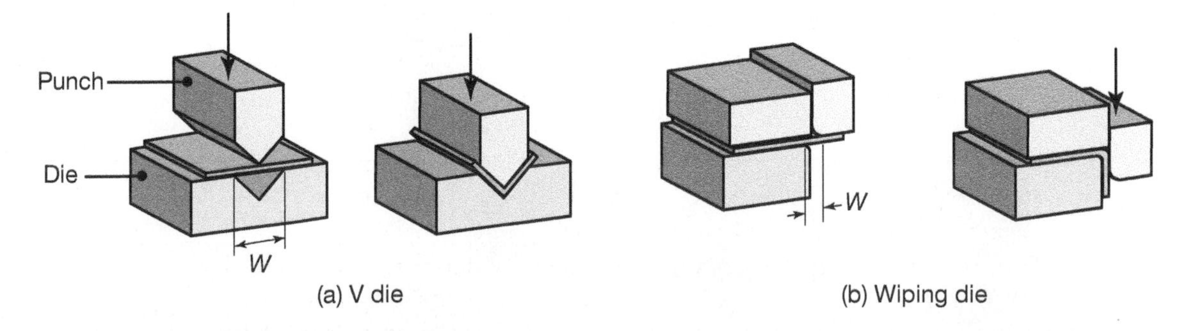

FIGURE 7.22 Common die-bending operations, showing the die-opening dimension W, used in calculating bending forces, as shown in Eq. (7.11).

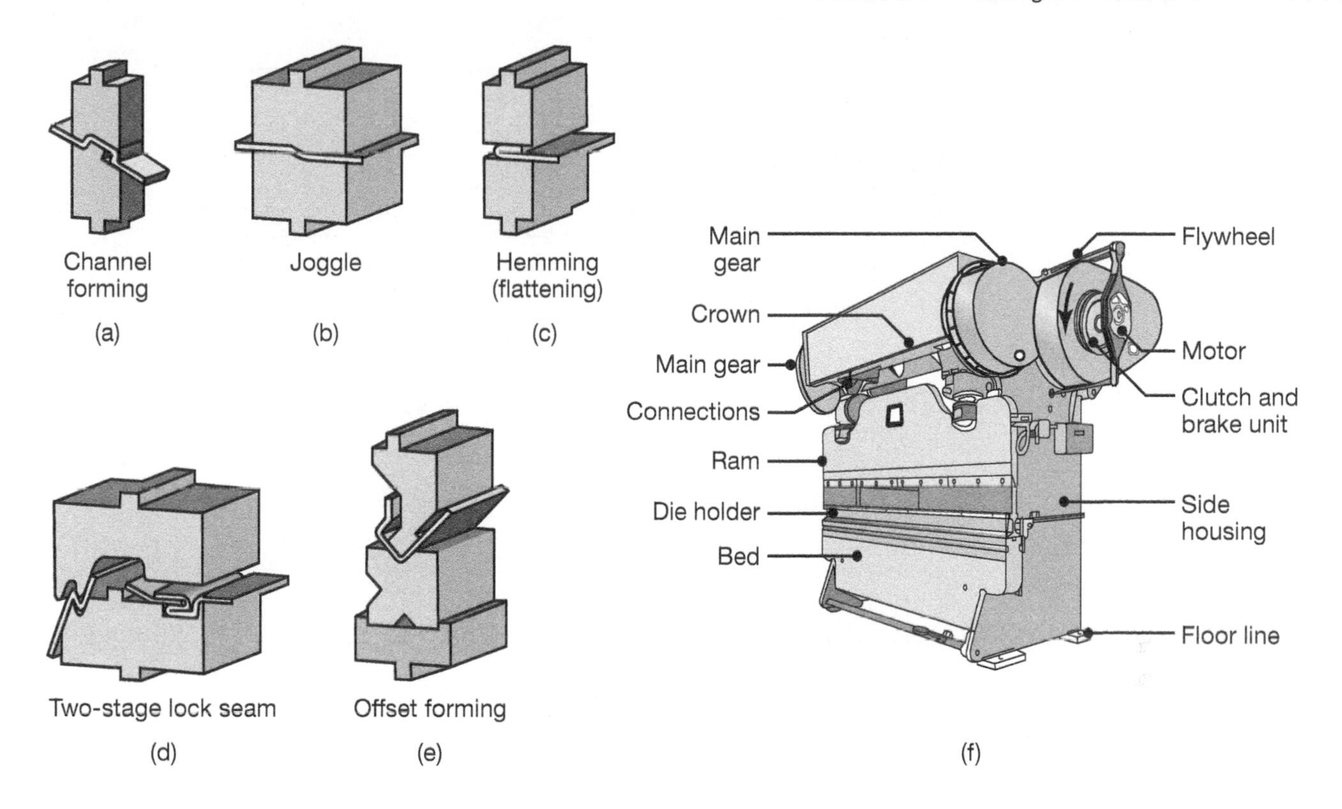

FIGURE 7.23 (a)–(e) Schematic illustrations of various bending operations in a press brake. (f) Schematic illustration of a press brake.

materials for most applications are carbon steel or gray iron, but may include hardwood (for low-strength materials and small production runs) to carbides (for stronger workpieces).

2. **Other bending operations.** Sheet metal may be bent by a variety of processes, as shown in Fig. 7.24. Bending can be carried out with two rolls (**air bending** or *free bending*), the larger one of which is flexible and is typically made of polyurethane. The upper roll pushes the sheet into the flexible lower roll, imparting a curvature to the sheet, the shape of which depends on the degree of indentation into the flexible roll. By controlling the depth of penetration, it is thus possible to form a sheet having different curvatures in the same piece. Bending of relatively short pieces, such as a bushing, can also be done on highly automated **four-slide machines**, a process similar to that shown in Fig. 7.24b.

   Plates are bent, using three rolls, by the **roll bending** process, shown in Fig. 7.24d. Note that by adjusting the distance between the three rolls, various curvatures can be developed.

3. **Beading.** In this operation, the edge of the sheet is bent into the cavity of a die (Fig. 7.25). The bead imparts stiffness to the part by virtue of the higher moment of inertia of the edges. Beading also improves the appearance of the part and eliminates the safety hazard presented by exposed sharp edges.

QR Code 7.5 Press brake with automatic tool changing system. *Source:* Courtesy of the LVD Company.

QR Code 7.6 Tool changer punch press able to punch, form, bend, and tap. *Source:* Courtesy of the LVD Company.

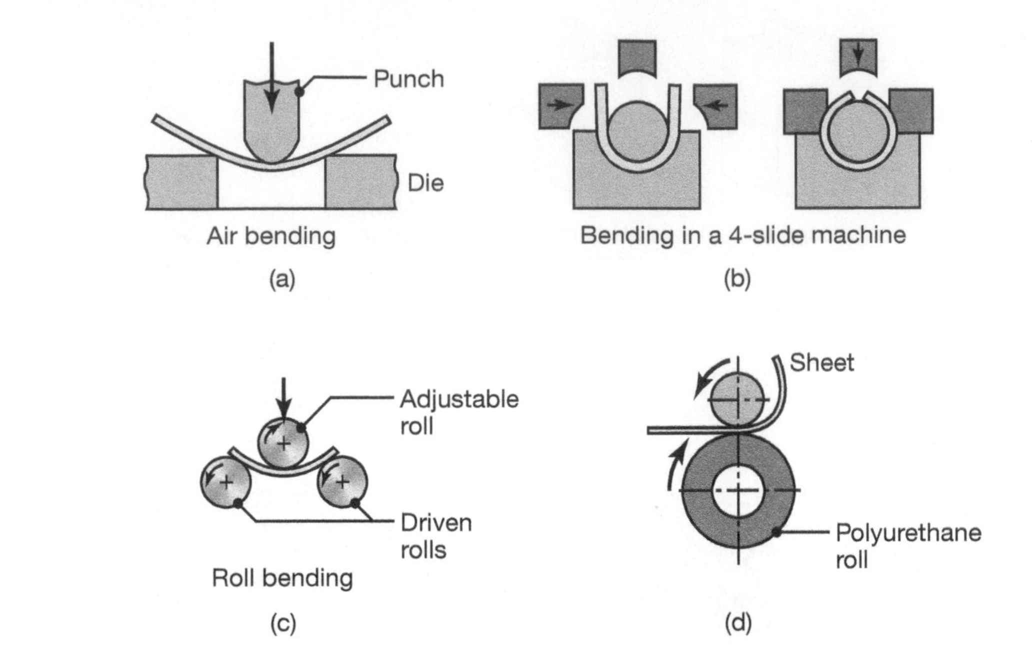

FIGURE 7.24 Examples of various bending operations.

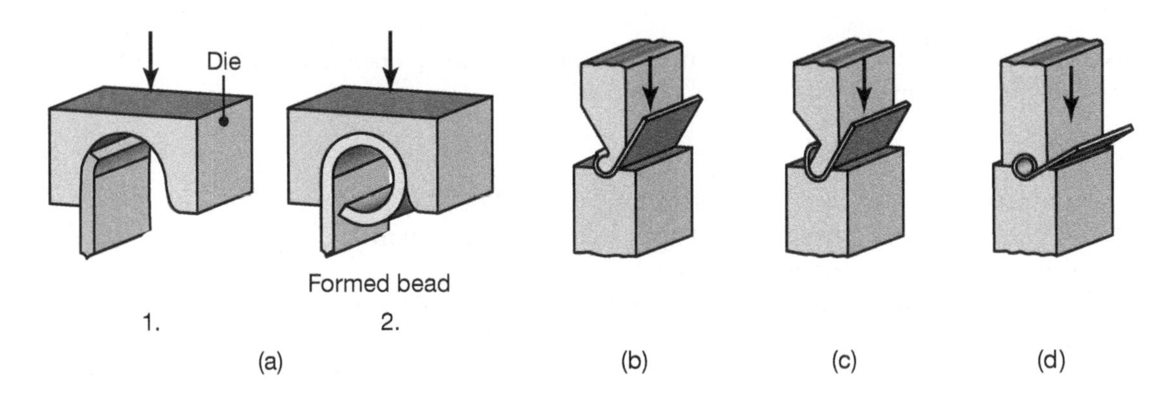

FIGURE 7.25 (a) Bead forming with a single die. (b)-(d) Bead forming with two dies in a press brake.

4. **Flanging.** This is a process of bending the edges of sheet metals, typically to 90°, for the purpose of imparting stiffness, appearance, or for ease of assembly with other components. In **shrink flanging** (Fig. 7.26a), the flange periphery is subjected to compressive hoop stresses that can cause it to wrinkle. In **stretch flanging**, the flange periphery is subjected to tensile stresses, that, if excessive, can lead to cracking and tearing at the edges, similar to that shown in Fig. 7.26c.
5. **Dimpling.** In this operation (Fig. 7.26b), a hole is first punched and then expanded to the shape of a flange. Flanges also may be produced by **piercing,** using a bullet-shaped punch (Fig. 7.26c); the ends of tubes are flanged by a similar process (Fig. 7.26d). When the bend angle is less than 90°, as in fittings with conical ends, the process

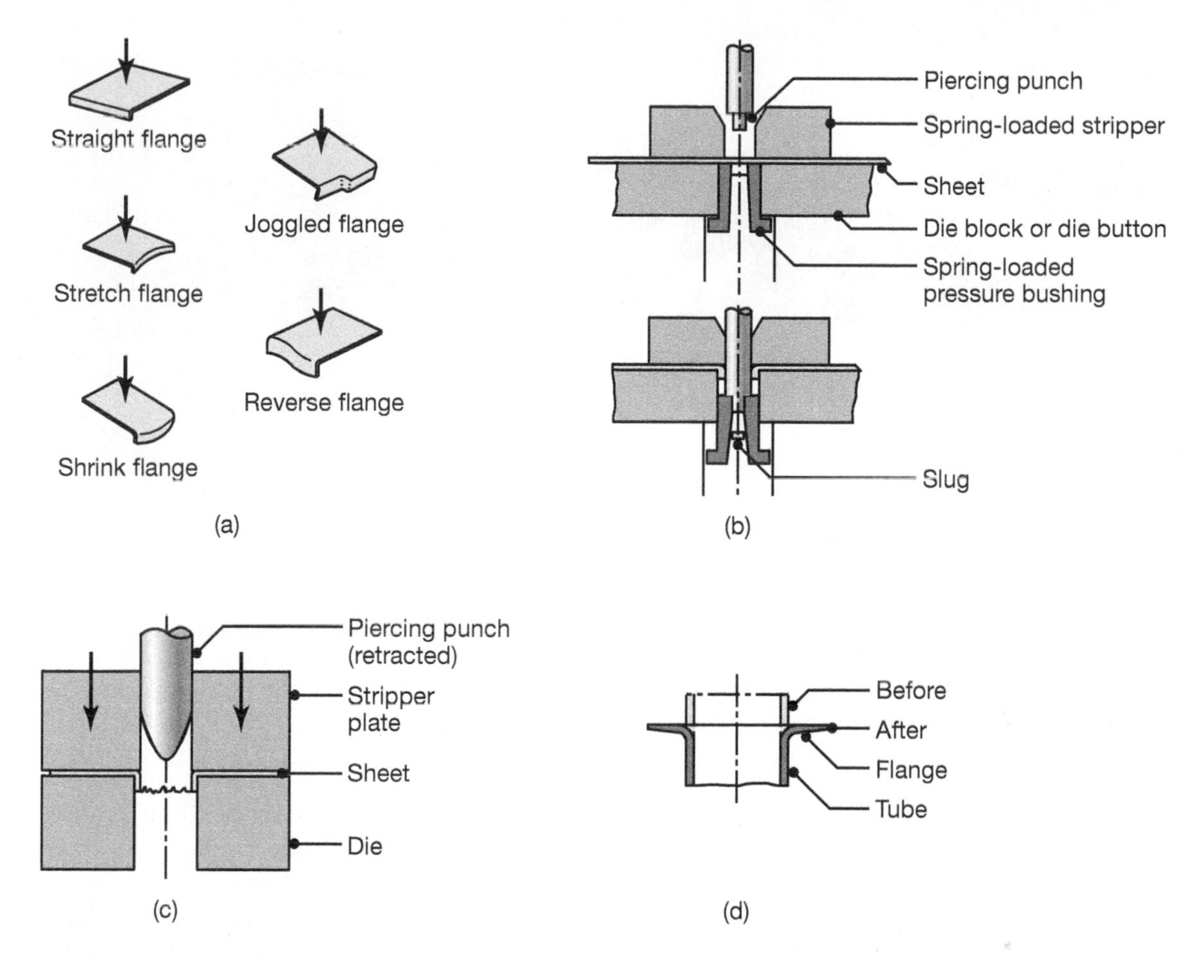

FIGURE 7.26 Illustrations of various flanging operations. (a) Flanges formed on flat sheet. (b) Dimpling. (c) Piercing sheet metal with a punch to form a circular flange. In this operation, a hole does not have to be prepunched; note, however, the rough edges that develop along the circumference of the flange. (d) Flanging of a tube; note the thinning of the periphery of the flange, due to its diametral expansion.

is called **flaring**. The condition of the edges is important in flanging operations. As the ratio of flange diameter-to-hole diameter increases, the tensile strains increase proportionately; thus, the rougher the edge, the greater the tendency for cracking. Sheared or punched edges may be shaved with a sharp tool (See Fig. 7.11), to improve their surface finish and reduce the tendency for cracking.

6. **Hemming.** In thus process, also called **flattening**, the edge of the sheet is folded over itself (See Fig. 7.23c). Hemming increases the stiffness of the part, improves its appearance, and eliminates sharp edges. *Seaming* (Fig. 7.23d) involves joining two edges of sheet metal pieces by hemming (*double seams*). Watertight and airtight joints, such as in some food and beverage containers, are made by a similar operation, using specially shaped rollers.

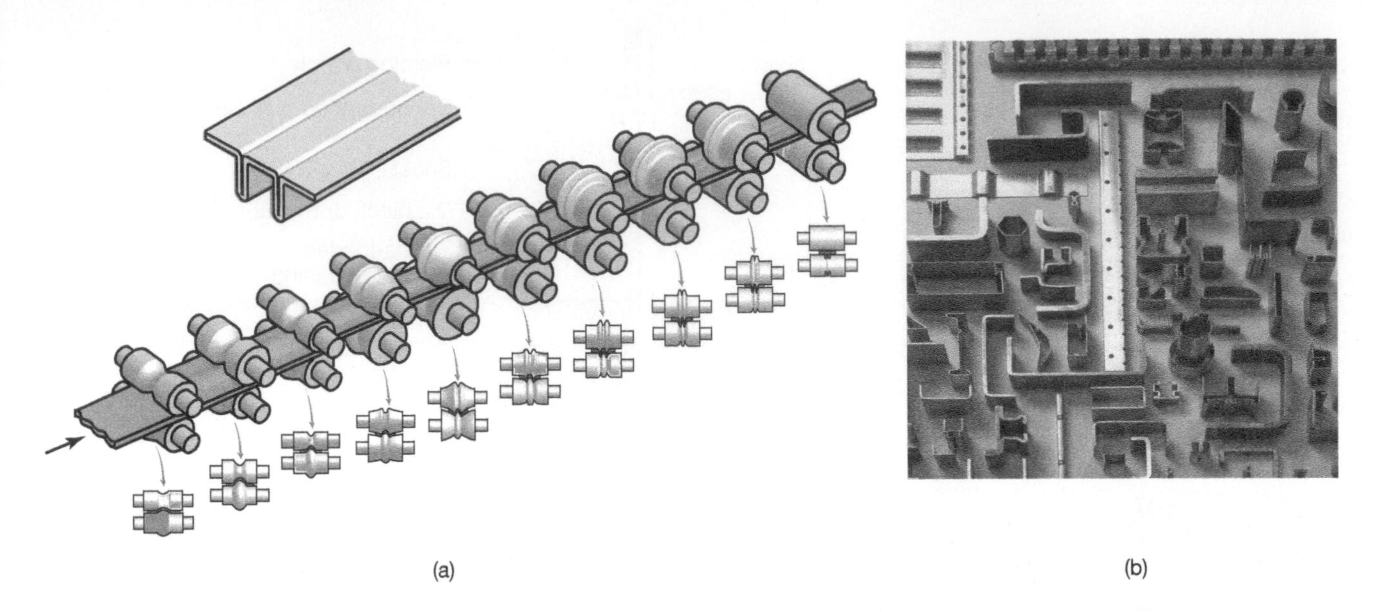

FIGURE 7.27 (a) The roll-forming operation, showing the stages in roll forming of a structural shape. (b) Examples of roll-formed cross sections.
Source: Courtesy of Sharon Custom Metal Forming, Inc.

7. **Roll forming.** This process is used for continuously bending lengths of sheet metal. Also called **contour roll forming** or *cold roll forming*, the strip in this operation is bent in several stages as it passes through a series of rolls (Fig. 7.27a). Typical products made include channels, gutters, siding, panels, frames and pipes and tubing with lock seams (Fig. 7.27b). The length of the part is limited only by the amount of material supplied from the coiled stock; the parts are subsequently sheared and stacked continuously. Forming speeds are generally below 1.5 m/s (300 ft/min), although they can be much higher for special applications.

Dimensional tolerances, springback, tearing and buckling of the strip, and the design and sequencing of the rolls are important considerations. The rolls, which usually are mechanically driven, are generally made of carbon steel or gray iron, and may be chromium plated for improved surface finish of the product and for increased wear resistance of the rolls. Lubricants may be used to improve roll life and product surface finish, as well as to cool the rolls and the workpiece during the forming operation.

7.4.5 Tube Bending

The oldest and simplest method of bending a tube or pipe is to pack it with loose particles (typically sand) and bend it in a suitable fixture. The packing, which prevents the tube from buckling inward while being bent, is shaken out after the tube is bent. Tubes can also be *plugged* with various flexible internal mandrels, such as those shown in Fig. 7.28, which also

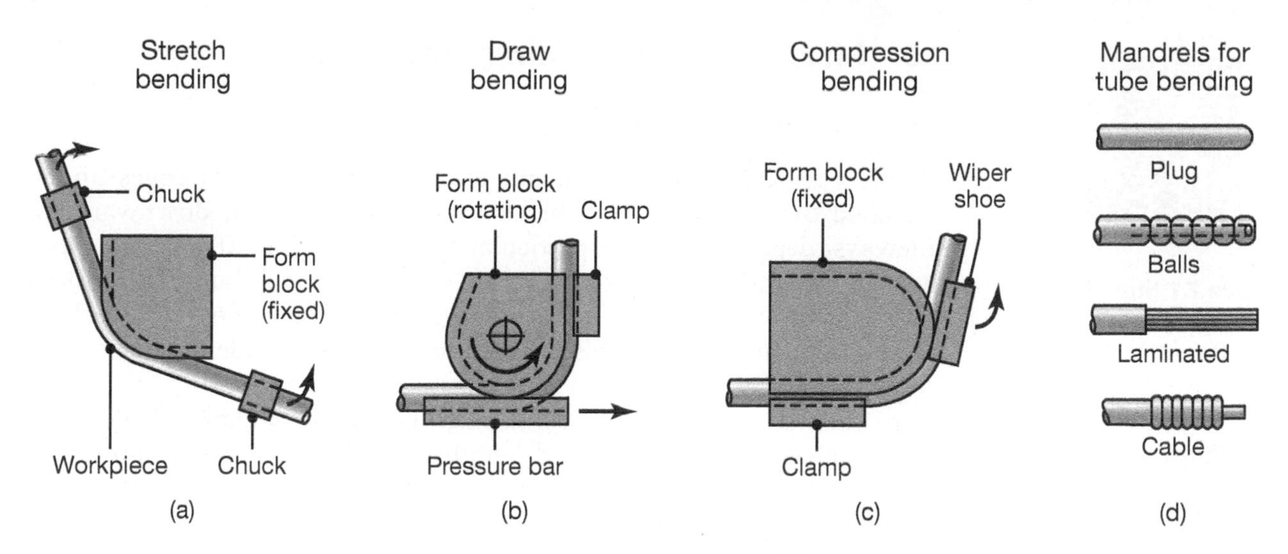

FIGURE 7.28 Methods of bending tubes. Using internal mandrels, or filling tubes with particulate materials, such as sand, prevents the tubes from collapsing during bending. Solid rods and structural shapes are also bent by some of these techniques.

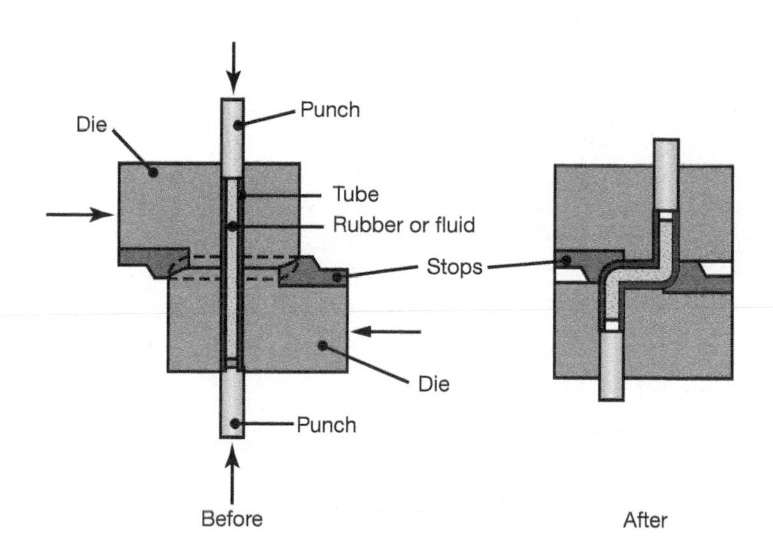

FIGURE 7.29 A method of forming a tube with sharp angles, using an axial compressive force. The tube is supported internally with rubber or a fluid, to avoid collapsing during forming. *Source:* After J.L. Remmerswaal and A. Verkaik.

illustrates various bending methods and fixtures for tubes and sections. A relatively thick tube with a large bend radius can be bent without packing it with particulates or using plugs, since it will have lower a tendency to buckle inward.

The beneficial effect of forming metals under high compressive stresses (see Section 2.2.8) is demonstrated in Fig. 7.29 regarding bending of a tube with relatively sharp corners. Note that the tube is subjected to longitudinal compressive stresses, by mechanical means; these stresses reduce the tensile stresses developed in the outer fibers of the tube in the bend area, thus improving bendability.

7.5 Miscellaneous Forming Processes

7.5.1 Stretch Forming

QR Code 7.7 Stretch forming. *Source:* Courtesy of Erie Press.

QR Code 7.8 Stretch forming machine. *Source:* Courtesy of Triform/Beckwood Press.

In this operation, the sheet metal is clamped around its edges and is stretched over a die or form block that moves upward, downward, or sideways, depending on the particular machine (Fig. 7.30). Stretch forming is used primarily in making aircraft-wing skin panels, automobile door panels, and window frames. Aluminum skins for the Boeing 767 fuselage, for example, are made by this process, using a blank under a tensile force as high as 9 MN (1000 tons).

In most stretch-forming operations, the blank is clamped along its narrower edges and is stretched lengthwise. Controlling the amount of stretching is important to avoid tearing. Stretch forming cannot produce parts with sharp contours or re-entrant corners (depressions of the surface of the die). Dies for stretch forming are generally made of zinc alloys, steel, hard plastics, or wood. Most applications require little or no lubrication. A variety of accessory equipment can be used in conjunction with stretch forming, including additional forming of the sheet metal with both male and female dies while the part is under tension. Although generally used for low-volume production, this is a versatile and economical process.

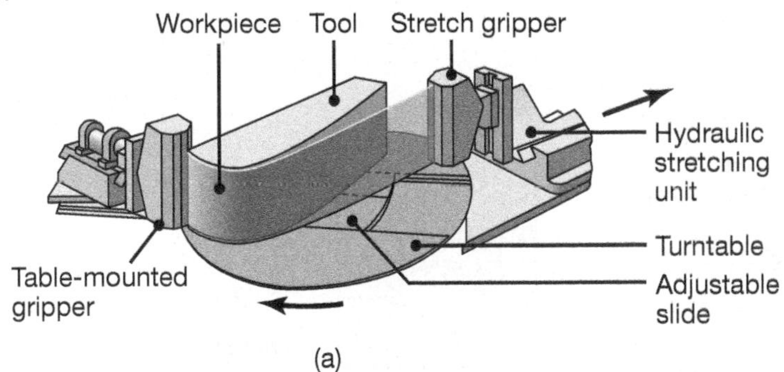

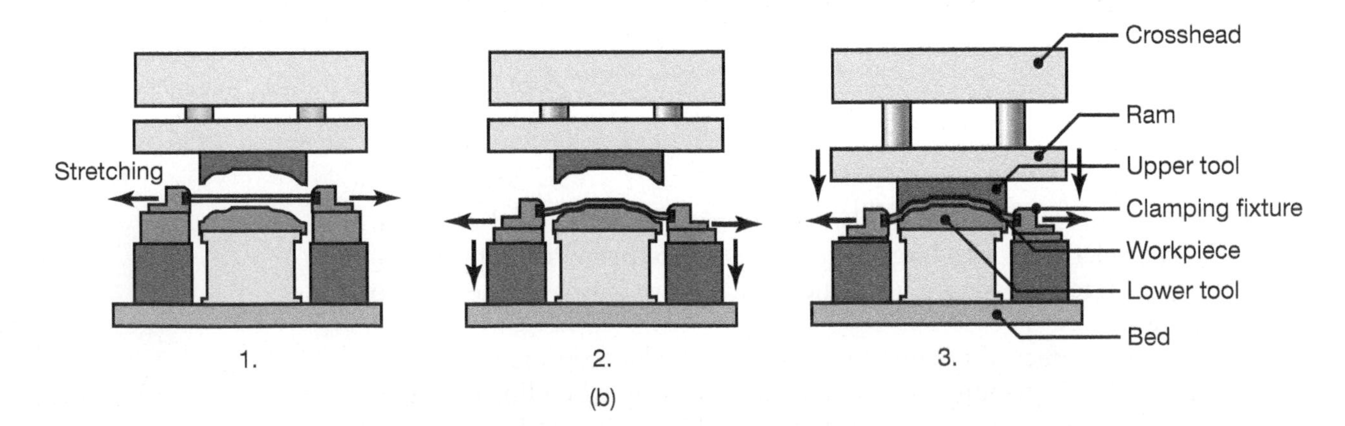

FIGURE 7.30 (a) Schematic illustration of a stretch-forming operation. *Source:* Cyril Bath Co. (b) Stretch forming in a hydraulic press.

EXAMPLE 7.3 Work Done in Stretch Forming

Given: A 375-mm-long sheet with a cross-sectional area of 4×10^{-4} m^2 (See Fig 7.31) is stretched with a force, F, until $\alpha = 20°$. The material has a true stress–true strain curve $\sigma = (650 \text{ MPa})\epsilon^{0.3}$.

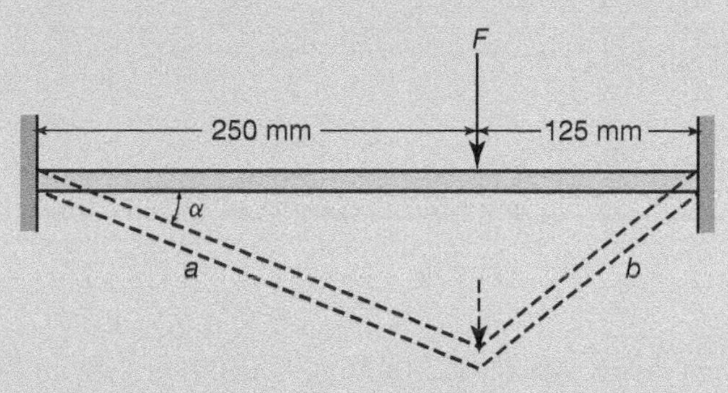

FIGURE 7.31 Workpiece considered in Example 7.3.

Find: (a) Determine the total work done, ignoring end effects and bending. (b) What is α_{max} before necking begins?

Solution:

a. Note that because the cross section is very small compared to the length of the part, this operation is equivalent to stretching a piece of metal from 375 mm to a length of $a+b$. For $\alpha = 20°$, the final length is calculated to be $L_f = 421$ mm and the true strain is

$$\epsilon = \ln\left(\frac{L_f}{L_o}\right) = \ln\left(\frac{421}{375}\right) = 0.116.$$

The work done per unit volume is given by Eq. (2.56) as

$$u = \int_0^{0.116} \sigma \, d\epsilon = \left(650 \times 10^6\right) \int_0^{0.116} \epsilon^{0.3} \, d\epsilon = \left(650 \times 10^6\right) \left[\frac{\epsilon^{1.3}}{1.3}\right]_0^{0.116}$$

or $u = 30.4$ MNm/m^3. Since the volume of the workpiece is

$$V = (0.375)\left(4 \times 10^{-4}\right) = 1.5 \times 10^{-4} \text{ m}^3,$$

the work done is

$$\text{Work} = (u)(V) = 4560 \text{ Nm}.$$

b. The necking limit for uniaxial tension is given by Eq. (7.2). Thus, $\epsilon = n$ and the length at this strain is

$$L_{max} - L_o e^n - 0.375e^{0.3} = 0.506 \text{ m}.$$

Therefore, $a + b = 506$ mm, and from similar triangles,

$$a^2 - 250^2 = b^2 - 125^2,$$

or

$$a^2 = b^2 + 46,875.$$

Hence, $a = 300$ mm and $b = 207$ mm. Thus,

$$\cos\alpha = \frac{250}{300} = 0.833, \qquad \text{or} \qquad \alpha_{max} = 33.6^\circ.$$

7.5.2 Bulging

The *bulging* process involves placing a tubular, conical, or curvilinear hollow part in a split die and expanding it with a rubber or polyurethane plug (Fig. 7.32a). The punch is then retracted, the flexible plug returns to its original shape, and the part is removed by opening the die. Typical products made include bellows, water and coffee pitchers, barrels, and beading on tubular parts. For parts with complex shapes, the plug may be made in specific shapes, in order to be able to apply higher pressure at critical points during forming. Polyurethane plugs are very resistant to

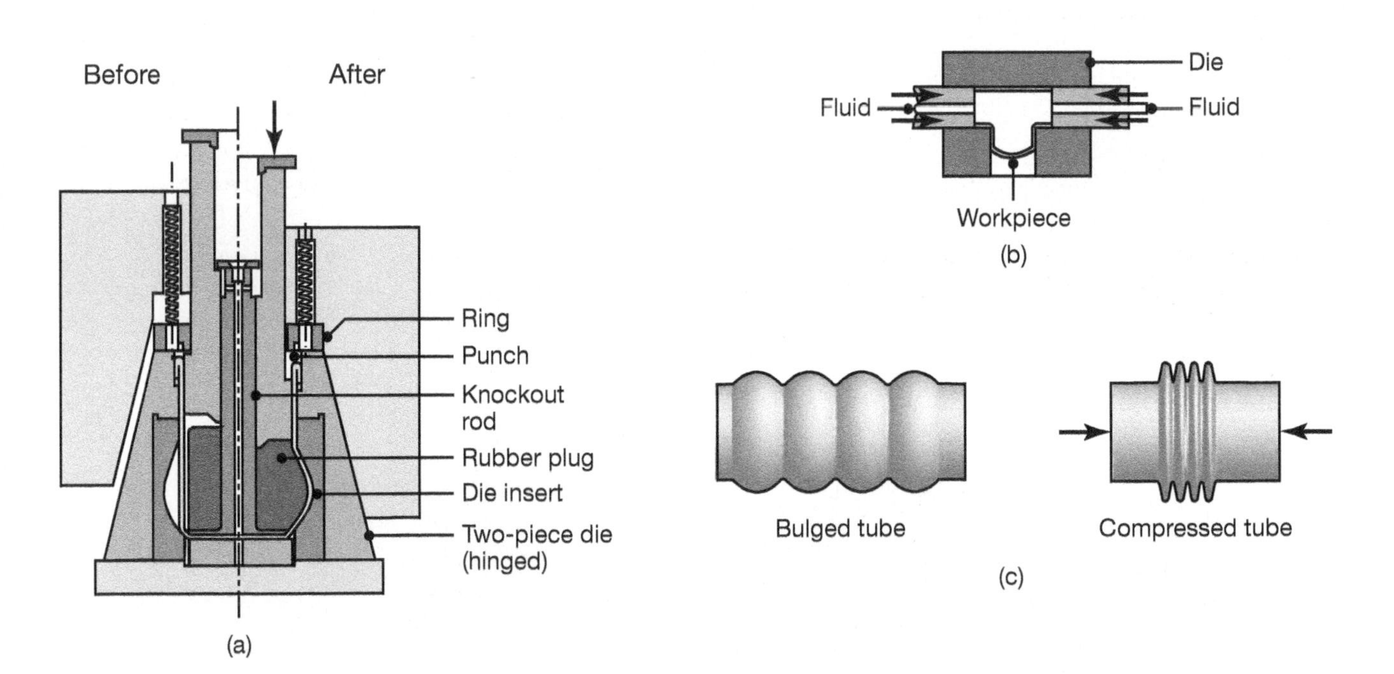

FIGURE 7.32 (a) Bulging of a tubular part with a flexible plug. Water pitchers can be made by this method. (b) Production of fittings for plumbing by expanding tubular blanks with internal pressure; the bottom of the piece is then punched out to produce a "T" section. (c) Sequence involved in manufacturing of a metal bellows.

abrasion, sharp edges, wear, and lubricants, and they do not damage the surface finish of the part being formed (see also Section 7.5.3).

Formability of the sheet metal in bulging operations can be enhanced by applying longitudinal compressive stresses to the parts (see Section 7.4.5). *Hydraulic pressure* may be used, although this technique requires sealing and hydraulic controls (Fig. 7.32b). **Segmented dies**, that are expanded and retracted mechanically, also may be used for bulging operations; the dies are relatively inexpensive and can be used for large production runs. *Bellows* are manufactured by a bulging process, as shown in Fig. 7.32c. After the tube is bulged at several equidistant locations, it is compressed axially to collapse the bulged regions, thus forming bellows.

Embossing. This process consists of forming a number of shallow shapes, such as numbers, letters, or designs, on sheet metal, for decorative as well as specific purposes. Parts may be embossed with male and female dies, or by various means, described throughout this chapter.

7.5.3 Rubber-Pad Forming and Hydroforming

In the processes described in the preceding sections, the dies are typically made of rigid materials; in *rubber-pad forming*, one of the dies in the die set is made of a flexible material, such as a polyurethane or rubber membrane. Polyurethanes are widely used because of their resistance to abrasion, long fatigue life, and resistance to damage by burrs or sharp edges.

In bending and embossing by the rubber-pad forming technique, shown in Fig. 7.33, the female die is replaced with a rubber pad. Note that the outer surface of the sheet is now protected from damage or scratches because it is not in contact with a hard metal surface during forming. Parts can also be formed with laminated sheets consisting of various nonmetallic materials and coatings. The pressures applied are typically on the order of 10 MPa (1450 psi).

In the *hydroforming*, or **fluid-forming process** (Fig. 7.34), the pressure applied over the flexible membrane is controlled throughout the forming

QR Code 7.9 Sheet hydroforming animation. *Source:* Courtesy of Triform/Beckwood Press.

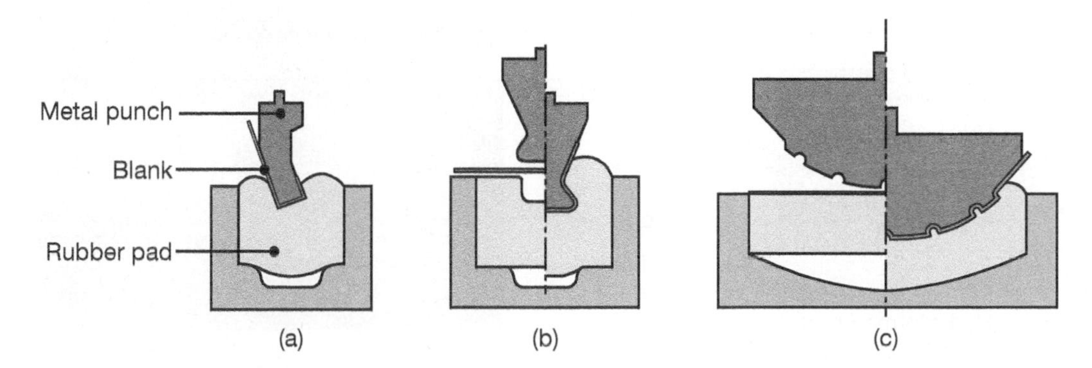

FIGURE 7.33 Examples of bending and embossing sheet metal with a metal punch and a flexible pad serving as the female die. *Source:* Polyurethane Products Corporation.

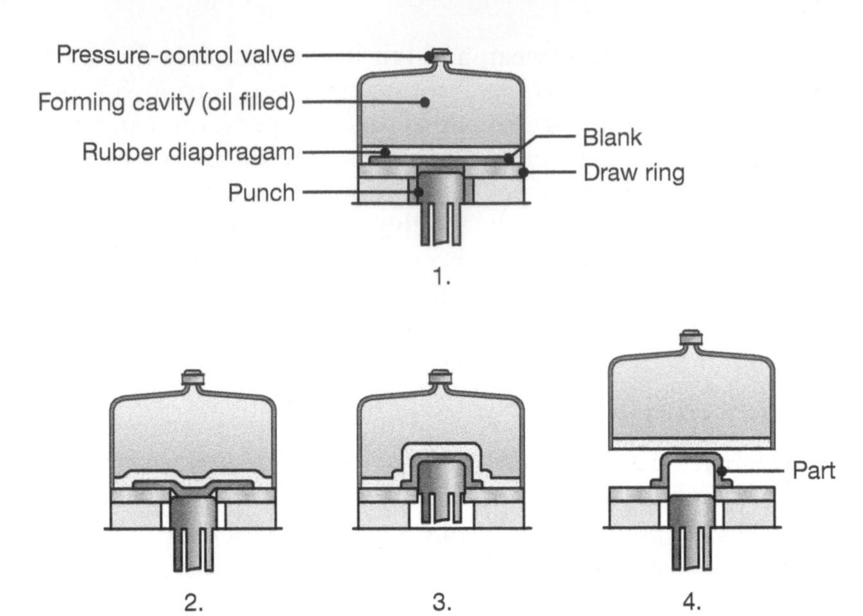

FIGURE 7.34 The principle of the hydroform process, also called fluid forming.

cycle, with maximum pressures reaching 100 MPa. This control procedure allows proper flow of the sheet during the forming cycle and prevents it from wrinkling or tearing. Deeper parts can be made than in conventional deep drawing (described in Section 7.6), because the pressure around the rubber membrane forces the part being formed against the punch. The friction at the punch-cup interface then reduces the longitudinal tensile stresses in the part, thus delaying fracture. The control of frictional conditions in rubber-pad forming, as well as in other sheet forming operations, can be a critical factor in making parts successfully. Selection of proper lubricants and application methods also are important.

In **tube hydroforming** (Fig. 7.35), metal tubing is first bent to form a blank; it is then internally pressurized by a fluid. Simple tubes (Fig. 7.35a)

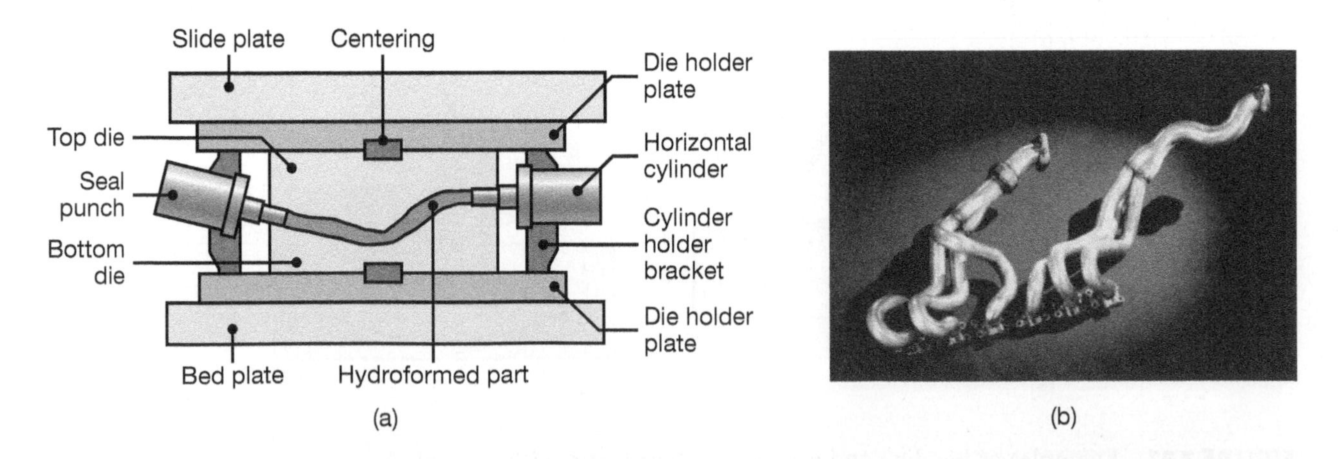

FIGURE 7.35 (a) Schematic illustration of the tube hydroforming process. (b) Examples of tube hydroformed parts. *Source:* Schuler GmbH.

and intricate hollow tubes with varying cross sections (Fig. 7.35b) can be formed by this process. Applications include automotive exhaust pipes, structural components, bicycle frames, and hydraulic and pneumatic fittings.

Rubber-pad forming and hydroforming processes have the advantages of capability to form complex shapes, flexibility and ease of operation, no damage to the surface of the sheet metal, and low tooling cost.

QR Code 7.10 Fluid cell animation.
Source: Courtesy of Triform/Beckwood Press.

7.5.4 Spinning

Spinning involves the forming of axisymmetric parts over a rotating mandrel, using rigid tools or rollers. The equipment is similar to a lathe (see Section 8.9.2), with various special features and computer controls.

Conventional spinning. In *conventional spinning*, a circular blank of flat or preformed sheet metal is held against a rotating mandrel while a rigid tool shapes it over the mandrel (Fig. 7.36a). The operation involves a sequence of passes and requires considerable skill. Typical shapes made by conventional spinning are shown in Fig. 7.37; note that this process is particularly suitable for conical and curvilinear shapes, which would otherwise be difficult or uneconomical to form by other methods. Part diameters may range up to 6 m (20 ft). Although most spinning is performed at room temperature, thick parts or metals with low ductility or high strength require spinning at elevated temperatures. The tools may be actuated either manually or by a hydraulic mechanism. Tooling costs in spinning are relatively low; however, because the operation requires multiple passes to form the final part, it is economical for relatively small production runs only (see Section 7.10).

QR Code 7.11 Sheet hydroforming press.
Source: Courtesy of Triform/Beckwood Press.

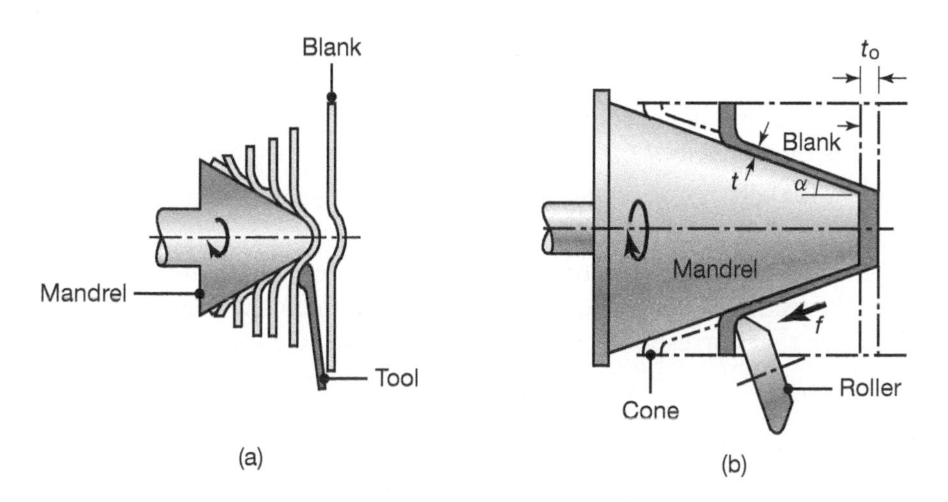

FIGURE 7.36 Schematic illustration of spinning processes: (a) conventional spinning, and (b) shear spinning. Note that in shear spinning, the diameter of the spun part, unlike in conventional spinning, is the same as that of the blank. The quantity f is the feed (in mm/rev).

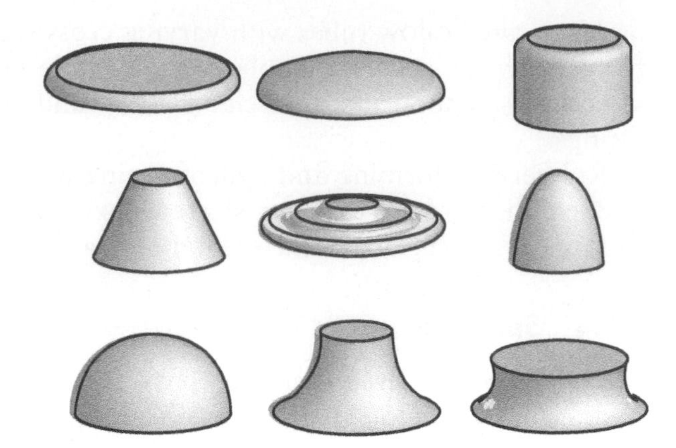

FIGURE 7.37 Typical shapes produced by the conventional spinning process. Circular marks on the external surfaces of the parts made usually indicate that they have been made by spinning, such as aluminum kitchen utensils and light reflectors.

QR Code 7.12 Manual metal spinning. *Source:* Courtesy of Metspin, Ltd.

Shear spinning. In *shear spinning*, also called **power spinning, flow turning, hydrospinning,** and **spin forging**, an axisymmetric conical or curvilinear shape is generated in such a manner that the diameter of the part remains constant during forming (Fig. 7.36b). Typical parts made include rocket-motor casings and missile nose cones. Although a single roller can be used, two rollers are preferred, in order to balance the radial forces acting on the mandrel, to avoid distortions and maintain dimensional accuracy. The spinning operation is completed in a relatively short time.

Parts up to about 3 m (10 ft) in diameter can be spun and to close dimensional tolerances. A wide variety of shapes can be spun, using relatively simple tooling generally made of tool steel. Because of the large plastic deformation involved, the process generates considerable heat, which usually is carried away by a coolant-type fluid (see Section 4.4.4) that can be applied during spinning.

Referring to Fig. 7.36b note that in shear spinning over a conical mandrel, the thickness, t, of the spun part is

$$t = t_o \sin\alpha, \tag{7.12}$$

where t_o is the original sheet thickness. The force that supplies the energy required is the *tangential force*, F_t. For an *ideal* case in shear spinning of a cone, this force is

$$F_t = ut_o f \sin\alpha \tag{7.13}$$

where u is the specific energy of deformation, as obtained from Eq. (2.56), and f is the feed. The specific energy u is the area under the true stress–true strain curve that corresponds to a strain given by the expression

$$\epsilon = \frac{\gamma}{\sqrt{3}} = \frac{\cot\alpha}{\sqrt{3}}. \tag{7.14}$$

The actual force, however, can be as much as 50% higher than that given by Eq. (7.13), because of factors such as redundant work and friction which are difficult to calculate accurately.

An important consideration in shear spinning is the **spinnability** of the metal, defined as the maximum reduction in thickness to which a part

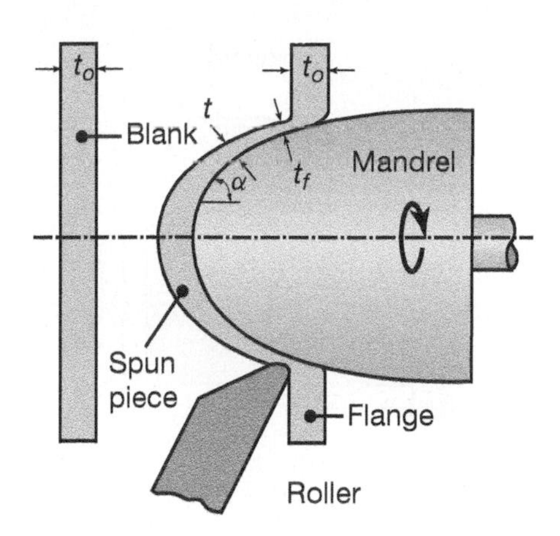

FIGURE 7.38 Schematic illustration of a shear spinnability test. The reduction in thickness at fracture is called the *maximum spinning reduction per pass.*

can be spun without fracture. A simple test method has been developed to determine spinnability (Fig. 7.38), in which a circular blank is spun over an ellipsoid mandrel. As its thickness decreases, the material eventually fractures at some critical thickness. The maximum spinning reduction in thickness is

$$\text{Maximum reduction} = \frac{t_o - t_f}{t_o} \times 100\%. \tag{7.15}$$

It has been observed that ductile metals fail in tension (as is the case in wire and rod drawing; Section 6.5.1), whereas less ductile metals fail in the deformation zone under the roller.

The maximum reduction per pass is then plotted vs. the tensile reduction of area of the material spun (Fig. 7.39). Note that for a material with about 50% reduction of area in a tension test, any further increase in the ductility of the original material does not improve its spinnability. Recall that a similar observation has been made in bending of sheet metals (See Fig. 7.15), in that maximum bendability corresponds to a tensile reduction of area of about 50%.

Tube spinning. In *tube spinning*, tubes or pipes are reduced in thickness by spinning them on a cylindrical mandrel. The operation may be carried out externally or internally (Fig. 7.40), and the part may be spun *forward* or *backward*, similar to a drawing or a backward extrusion process

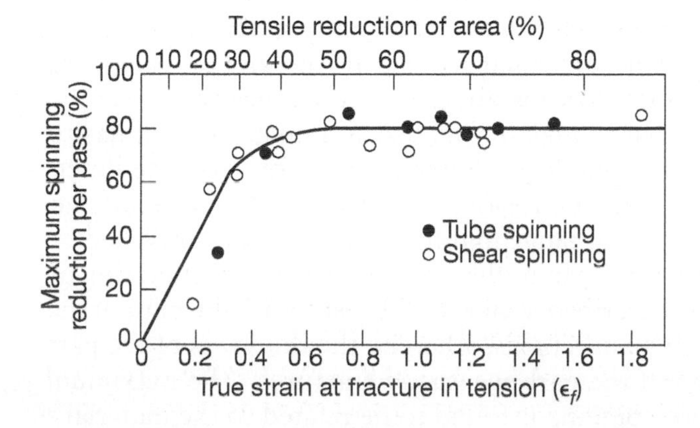

FIGURE 7.39 Experimental data showing the relationship between maximum spinning reduction per pass and the tensile reduction of area of the original material. (See also Fig. 7.15.) *Source:* S. Kalpakjian.

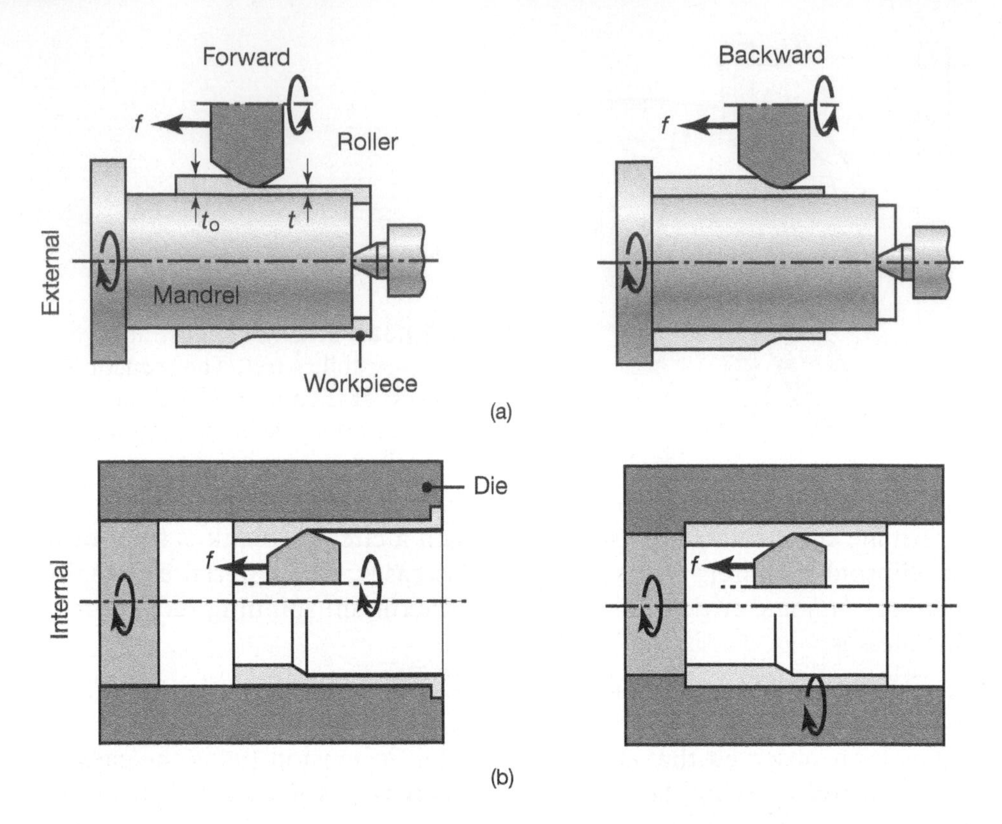

FIGURE 7.40 Examples of (a) external and (b) internal tube spinning, and the process variables involved.

(see Section 6.4). A variety of external and internal profiles can be produced by controlling the path of the roller during its travel along the cylindrical mandrel. This process, which can be also combined with shear spinning, is typically used for making pressure vessels, and rocket, missile, and automotive components.

Based on an approach similar to that for shear spinning, the ideal tangential force, F_t, in forward tube spinning can be expressed as

$$F_t = \overline{\sigma}_f (t_o - t) f, \tag{7.16}$$

where $\overline{\sigma}_f$ is the average flow stress of the material (See Fig. 2.5a). Friction and redundant work of deformation cause the actual force to be about twice as high as that calculated by this formula. For backward spinning, the ideal tangential force is approximately twice that given by Eq. (7.16); the reason is that, unlike in forward spinning, the tool has to travel a shorter distance to complete the operation. Note also the difference between direct and indirect extrusion (see Section 6.4), and recall that the work done is the product of tangential force and distance travelled.

Spinnability in tube spinning can be determined by a test method similar to that for shear spinning described above. In the test setup, the path of the roller is at an angle to the mandrel axis, whereby the thickness of the part is continually reduced, and the part eventually fractures. The maximum reduction per pass in tube spinning is found to be related to the material's tensile reduction of area (See Fig. 7.39), as it is in shear spinning.

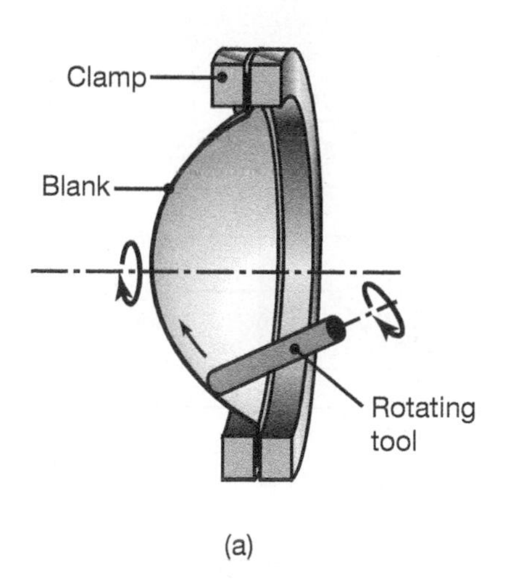

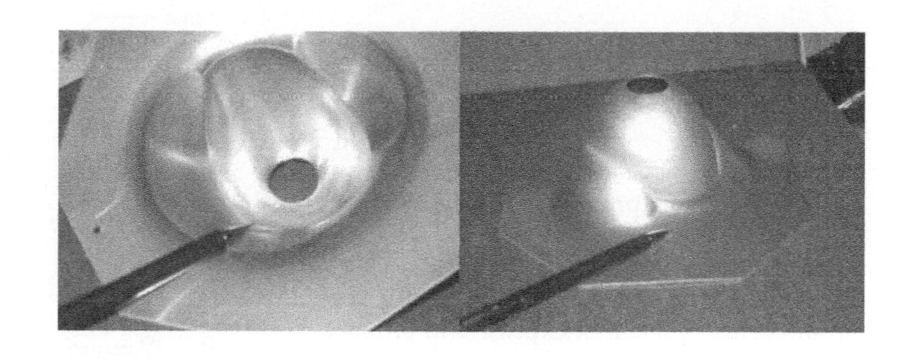

FIGURE 7.41 (a) Illustration of an incremental forming operation. Note that no mandrel is required, and that the final part shape depends on the path of the rotating tool. (b) An automotive headlight reflector produced through CNC incremental forming. *Source:* After J. Jeswiet.

Incremental forming. *Incremental forming* is a term applied to a class of processes that form a material in small increments (see also *incremental forging*, Section 6.2.3). A simple example is *incremental stretch expanding*, shown in Fig. 7.41, wherein a rotating blank is deformed using a steel rod with a smooth hemispherical tip, to produce axisymmetric parts. No special tooling or mandrel is needed; the motion of the rod determines the final part shape, in one or more passes.

CNC incremental forming, also known as *single point incremental forming*, uses a CNC machine tool that is programmed to follow contours, at different depths, across the workpiece surface. The sheet metal blank is clamped and is held stationary, and the tool rotates. Tool paths are calculated in a similar manner as that for metal cutting, using a CAD model of the desired shape as the starting point (See Fig. 10.49). Figure 7.41b depicts an example of a part that has been produced from CNC incremental forming; note that the part does not have to be axisymmetric.

The main advantages of incremental forming are high flexibility in the shapes that can be produced and low tooling costs. CNC incremental forming also has been used for rapid prototyping of sheet metal parts (see Section 10.12), because there are no lead times associated with making hard tooling. The main drawbacks include low production rates and limitations on allowable materials. Most incremental forming is performed on aluminum alloys with very high formability.

QR Code 7.13 Single point incremental forming. *Source:* Courtesy of J. Jeswiet, Queen's University, Ontario.

7.5.5 High-Energy-Rate Forming

This section describes sheet metal forming processes that require chemical, electrical, or magnetic sources of energy. They are called high-energy-rate processes because the energy is released in a very short period of time.

Explosive forming. The most common *explosive forming* operation is shown in Fig. 7.42. The sheet is first clamped over a die, the air in the die cavity is evacuated, and the whole assembly is lowered into a tank filled

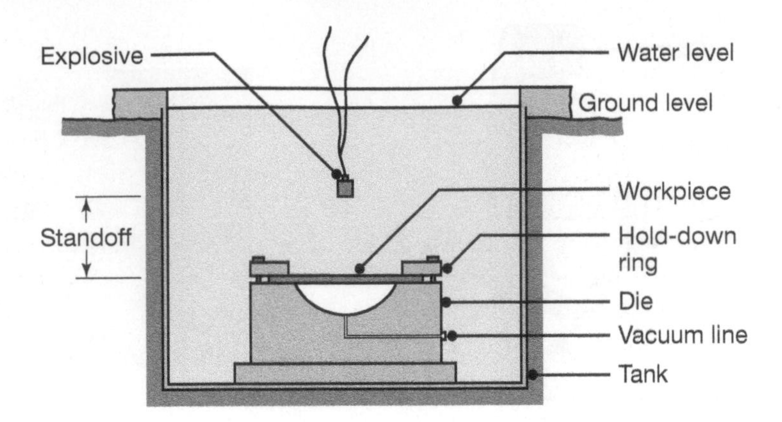

FIGURE 7.42 Schematic illustration of the explosive forming process. Although explosives are typically used for destructive purposes, their energy can be controlled and utilized in forming large sheet metal parts that otherwise would be too difficult or expensive to produce by other methods.

with water. An explosive charge is then placed at a certain distance from the sheet surface, and detonated. The rapid conversion of the explosive into gas generates a shock wave; the pressure of this wave is sufficiently high to force the metal into the die cavity.

The peak pressure generated in water is given by the expression

$$p = K\left(\frac{\sqrt[3]{W}}{R}\right)^{a}, \tag{7.17}$$

where p is the peak pressure in MPa, K is a constant that depends on the type of explosive, e.g., 51.3 for TNT (trinitrotoluene); W is the mass of the explosive, in kg; R is the distance of the explosive from the workpiece (*standoff*), in meters; and a is a constant, generally taken as 1.15.

An important factor in peak pressure is the **compressibility** of the energy-transmitting medium (such as water) and its **acoustic impedance** (defined as the product of mass density and sound velocity in the medium). Thus, the lower the compressibility of the medium and the higher its density, the higher is the peak pressure (Fig. 7.43). Detonation speeds typically are 6700 m/s, and the speed at which the sheet metal is formed is on the order of 30 to 200 m/s.

A variety of shapes can be formed explosively, provided that the material is sufficiently ductile at very high strain rates. Depending on the number of parts to be made, the dies used in this process may be made

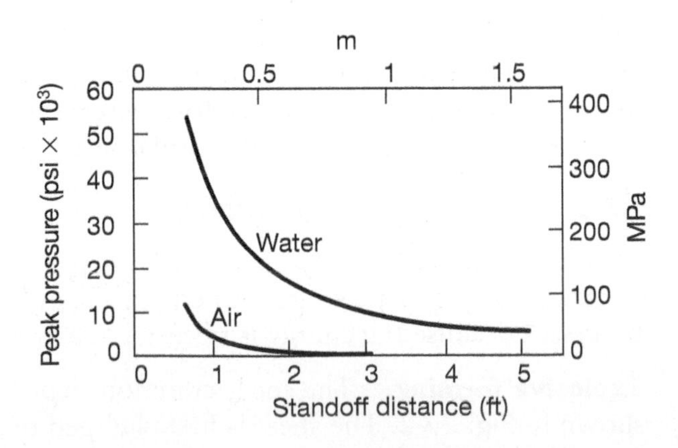

FIGURE 7.43 Effect of the standoff distance and type of energy-transmitting medium on the peak pressure obtained using 1.8 kg of TNT. The pressure-transmitting medium should have a high density and low compressibility. In practice, water is a commonly used medium.

of aluminum alloys, steel, ductile iron, zinc alloys, reinforced concrete, wood, plastics, or composite materials. The final properties of the parts made are basically the same as those made by conventional methods.

Requiring no machinery, only one die, and being versatile, explosive forming is particularly suitable for low-quantity production runs of large parts. Steel plates 25 mm (1 in.) thick and 3.6 m (12 ft) in diameter have been formed by this method. Tubes with a wall thickness of 25 mm (1 in.) have been bulged by explosive-forming techniques. For small parts, a cartridge is used, confined in closed die. Parts made typically involve tubes that are bulged and then expanded.

EXAMPLE 7.4 Peak Pressure in Explosive Forming

Given: 100 grams of TNT is located 250 mm from a workpiece in a water-filled tank.

Find: Calculate the peak pressure. Is this pressure sufficiently high for forming sheet metals?

Solution: Using Eq. (7.17),

$$p = (51.3)\left(\frac{\sqrt[3]{0.1}}{0.25}\right)^{1.15} = 104.5 \text{ MPa}.$$

The pressure is sufficiently high to form sheet metals. Note that in Example 2.5, the pressure required to expand a thin-walled spherical shell of a material similar to soft aluminum alloys is only 3 MPa (435 psi). Also, recall that hydroforming (Section 7.5.3) has a maximum hydraulic pressure of 100 MPa (15,000 psi). Other rubber-pad forming processes utilize pressures ranging from about 10 to 50 MPa (1500 to 7500 psi). Thus, the pressure obtained in this problem is sufficient for most sheet forming processes.

Electrohydraulic forming. In *electrohydraulic forming*, also called **underwater-spark** or **electric-discharge forming**, the source of energy is a spark from two electrodes connected to a thin wire (Fig. 7.44). The energy is stored in a bank of charged condensers; the rapid discharge of this energy, through the electrodes, generates a shock wave, which is strong enough to form the part. Note that this process is basically similar to explosive forming, except that it utilizes a lower level of energy and is used with smaller workpieces; it is also a much safer operation.

Magnetic-pulse forming. In this process, the energy stored in a capacitor bank is discharged rapidly through a magnetic coil. In a typical application, a ring-shaped coil is placed over a tubular workpiece, to be formed

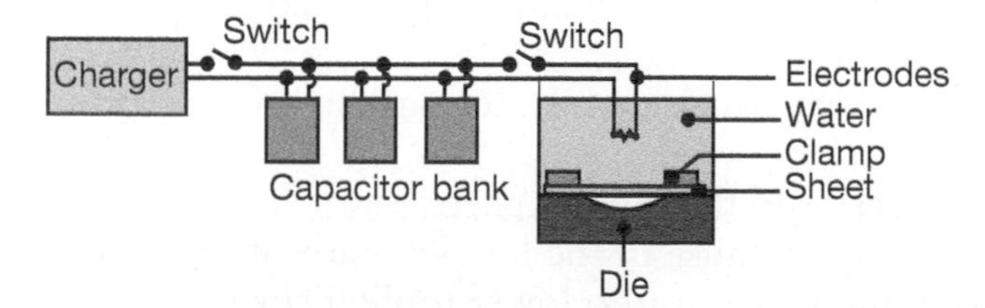

FIGURE 7.44 Schematic illustration of the electrohydraulic forming process.

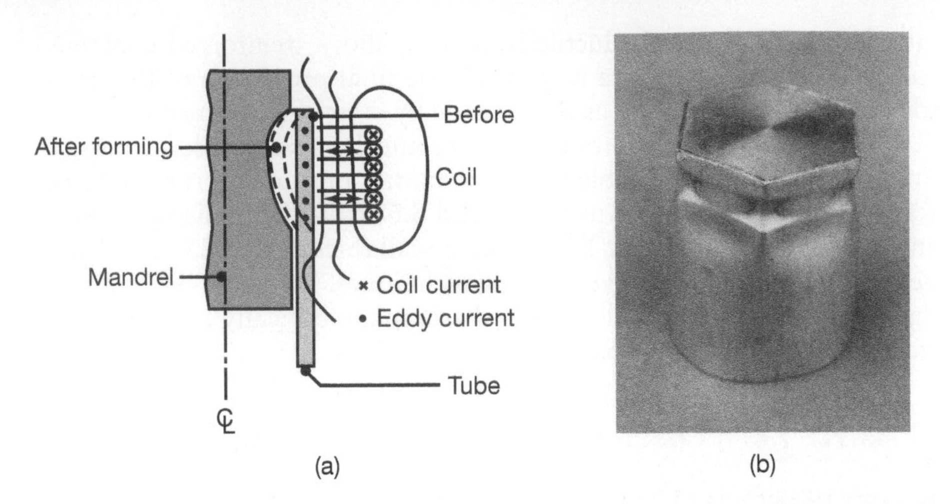

FIGURE 7.45 (a) Schematic illustration of the magnetic-pulse forming process. The part is formed without physical contact with any object, and (b) aluminum tube collapsed over a hexagonal plug by the magnetic-pulse forming process.

over another solid piece (such as a metal plug) and make make an integral part (Fig. 7.45). The transient magnetic field produced by the coil crosses the metal tube, generating *eddy currents* in the tube; this current, in turn, produces its own magnetic field. The forces produced by the two magnetic fields oppose each other. The repelling force between the coil and the tube collapses the tube over the plug. Magnetic-pulse forming can be used for a variety of operations, such as swaging of thin-walled tubes over rods, cables, and plugs, and for bulging and flaring.

7.5.6 Superplastic Forming

In Section 2.2.7, the *superplastic behavior* of some very fine-grained alloys (typically less than 10 to 15 μm) was described, where elongations up to 2000% are obtained at certain temperatures and low strain rates. These alloys, such as zinc-aluminum and titanium, can be formed into complex shapes by employing traditional metalworking or polymer-processing techniques, such as thermoforming, vacuum forming, and blow molding, described in Chapter 10. Die materials typically include low-alloy steels, cast tool steels, ceramics, graphite, and plaster of paris.

The high ductility and relatively low strength of superplastic alloys have the following advantages in superplastic forming operations:

1. Capability of forming one-piece complex shapes, with fine detail, close dimensional tolerances, and the elimination of secondary operations;
2. Weight and material savings, because of the formability of superplastic materials;
3. Little or no residual stresses in the formed parts; and
4. Lower strength of tooling, because of the low strength of the workpiece material at forming temperatures; lower tooling costs.

The limitations are:

1. The material must not be superplastic at service temperatures;
2. Because of the extreme strain-rate sensitivity of the material, it must be formed at sufficiently low rates (typically, at strain rates of 10^{-4}/s to 10^{-2}/s); and
3. Forming times range anywhere from a few seconds to several hours; this can be much longer than in conventional forming processes; superplastic forming is therefore a batch-forming operation.

An important aspect of this process is the ability to fabricate sheet metal structures by combining **diffusion bonding** (see Section 12.13) with superplastic forming (SPF/DB). Typical structures in which flat sheets are diffusion bonded and then superplastically formed are shown in Fig. 7.46. After bonding at selected locations on the sheets, the unbonded regions (*stop off*) are expanded in a mold by air pressure. These structures are relatively thin and have high stiffness-to-weight ratios; consequently, they are particularly important in aerospace applications.

7.5.7 Miscellaneous Forming Operations

Peen forming. This process is used in producing curvatures on thin sheet metals by the technique of **shot peening** only one surface of the sheet (see Section 4.5.1). In peen forming, the surface of the sheet develops compressive stresses, which tend to expand the surface layer. Since the material

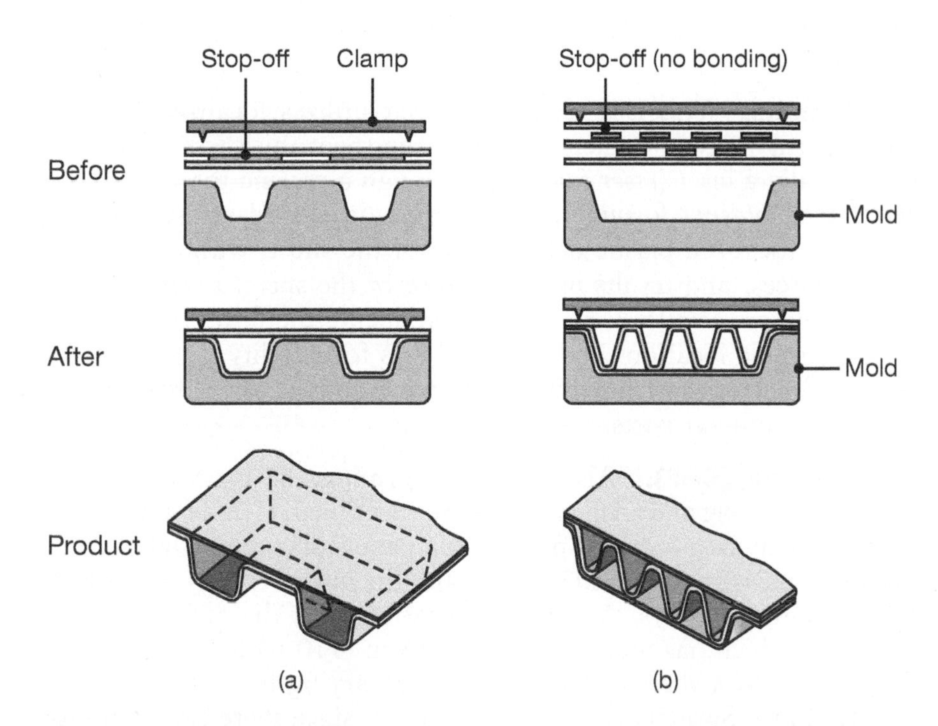

FIGURE 7.46 Two types of structures fabricated by combining diffusion bonding and superplastic forming of sheet metal. Such structures have a high stiffness-to-weight ratio. *Source:* Rockwell Automation, Inc.

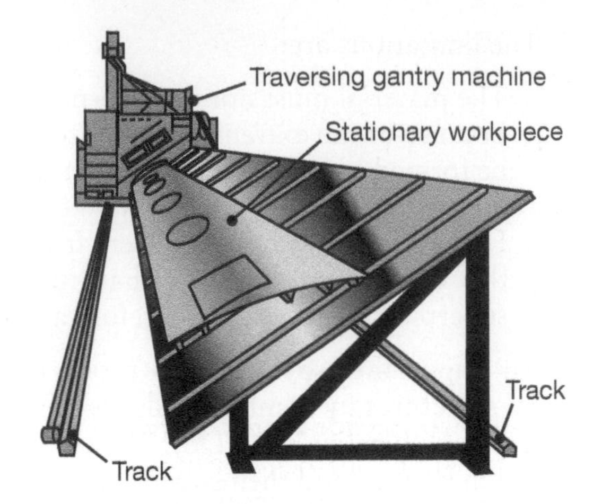

FIGURE 7.47 Schematic illustration of a peen forming machine to shape a large sheet metal part, such as an aircraft-skin panel. Note that the sheet is stationary and the peening head travels along its length. *Source:* Metal Improvement Company.

below the peened surface remains rigid, the surface expansion causes the sheet to develop a curvature. Peening is done with cast-iron or steel shot, discharged either from a rotating wheel or with an air blast from a nozzle. Recall also that shot peening induces compressive surface residual stresses, improving the fatigue strength of the sheet.

The peen forming process is used in the aircraft industry to generate curvatures on aluminum aircraft-wing skins (Fig. 7.47). Cast-steel shot, about 2.5 mm in diameter and at at speeds of 60 m/s, has been used to form wing panels 25 m long. For heavy sections, shot diameters as large as 6 mm may be used. The peen-forming process is also used for straightening twisted or bent parts. Out-of-round rings, for example, can be straightened by this method.

Thermal forming. This is a technology that utilizes localized heating to induce thermal-stress gradients through the sheet thickness. The heat source is usually a laser (*laser forming*, although a plasma torch can also be used, called *plasma forming*). The stresses developed are sufficiently high to cause localized plastic deformation of the sheet, without the use of external forces, and results in a curvature of the sheet. *Laser-assisted forming* uses lasers as a local heat source, reducing the flow stress of the material at specific locations and improving its formability. Applications of this process include straightening, bending, embossing, and forming of complex flat or tubular parts.

Creep age forming (CAF). Also called **age forming**, this process involves simultaneous forming and artificial aging (see Section 5.11.2) of aluminum sheets. It has been used for the top wing-skin panels of commercial aircraft and business jets, made from 2024-T351 and the more recently developed 2022-T8 aluminum alloys. Another application is in the 555-passenger Airbus A380, which has wing skins up to 33 m (110 ft) long and 2.8 m (9.3 ft) wide, with a sheet thickness abruptly varying from 3 to 28 mm (0.12 to 1.12 in.). Springback can be a concern, since there can be up to 80% elastic recovery after forming. With computer modeling and simulation techniques (see Section 7.7.3), the process is capable of economically producing complex multiple-curvatures close to net shapes.

Microforming. *Microforming*, described in greater detail in Chapter 13, involves the use of a variety of metalworking processes to produce very small metallic parts and components (*miniaturized products*). Part sizes typically are in the submillimeter range, and weigh milligrams.

Straightening. Because of the difficulties encountered in controlling all relevant parameters during their production, sheet, plate, or tubular parts may not have sufficient straightness. Several techniques are available to straighten these parts, including those similar to the operations and the tooling illustrated in Figs. 6.38 and 7.28 as well as peen forming.

Manufacturing honeycomb structures. Because of their light weight and high resistance to bending moments, these structures are used for aircraft and aerospace components, as well as buildings and transportation equipment. There are two principal methods of manufacturing honeycomb structures. In the **expansion** process (Fig. 7.48a), the most common method, sheets are cut from a coil, and an adhesive is applied at constant intervals (node lines). The sheets are stacked and cured in an oven, whereby strong bonds develop at the adhesive joints. The block is then cut into slices and stretched to produce a honeycomb structure.

In the **corrugation** process (Fig. 7.48b), the sheets are passed through a pair of specially designed rolls, producing corrugated sheets, that are

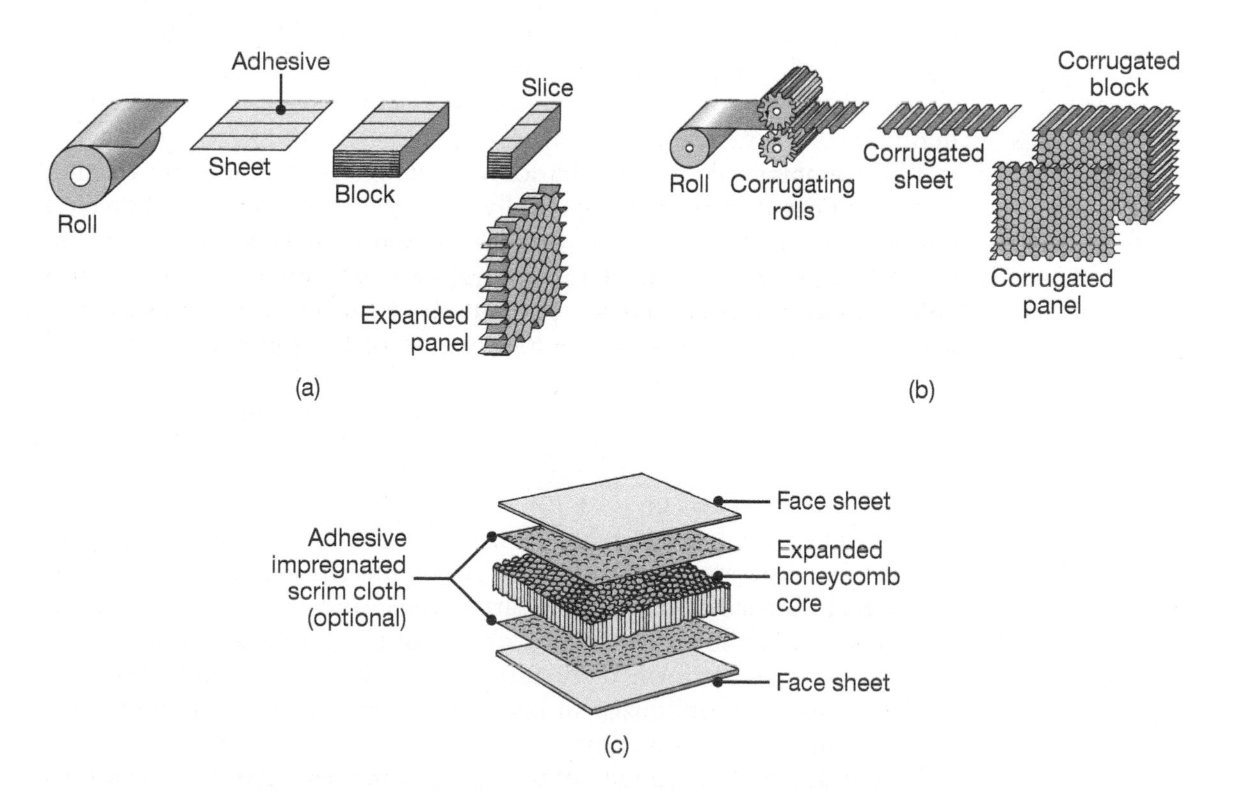

FIGURE 7.48 Methods of making honeycomb structures: (a) expansion process, and (b) corrugation process; and (c) assembling a honeycomb structure into a laminate.

then cut into desired lengths. Adhesive is applied to the node lines, and the block is cured to develop strength; no expansion of the sheets is involved in this process. The honeycomb material, made by either process, is made into a sandwich structure, as shown in Fig. 7.48c. The face sheets are then assembled with adhesives to their top and bottom surfaces. Honeycomb structures are most commonly made from 3000-series aluminum, and can also made of titanium, stainless steels, and nickel alloys.

7.5.8 Hot Stamping

Increasing fuel economy in automobiles has received considerable attention in recent years for both environmental and economic reasons. To achieve increased fuel economy, without compromising performance or safety, manufacturers have increasingly applied advanced materials in automobiles. Die-cast magnesium or extruded aluminum components are examples, but these materials are not sufficiently stiff or as well suited as steel for occupant safety. Thus, there has been a recent trend to consider hot stamping of advanced high-strength steels.

As discussed in Section 3.10.2, high-strength TRIP and TWIP steels have been developed, with yield strengths and ultimate strengths that can exceed 1300 MPa (190 ksi) and 2000 MPa (290 ksi), respectively (See Table 3.3). Conventional sheet metal forming of these materials would be difficult or impossible, because of the high forces required and the excessive springback after forming. For these reasons, the sheet metal is preheated to above 900°C (usually 1000°C to 1200°C) and hot stamped. To extend die life and to quench the material within the die (as discussed below), the tooling is maintained at a much lower temperature, typically 400°C to 500°C.

Hot stamping allows exploitation of steel phases to facilitate forming and maximize part strength. Basically, the steel is maintained at elevated temperatures to form an austenite (see Section 3.10.3), which has a ductile fcc structure at elevated temperatures. When formed and brought into contact with the much cooler tooling, the steel is rapidly quenched to form martensite, which is a very hard and strong but brittle form of steel (Section 3.10.3).

A typical hot-stamping sequence involves the following steps:

1. The material is heated up to the austenization temperature, and allowed to *dwell* or *soak* for a sufficiently long time to ensure that quenching will occur quickly when it contacts the die, but not before. Three basic means are used to heat blanks prior to stamping: roller hearth furnaces, induction heating coils, and resistive heating. The last two methods have the advantage of shorter soak times, but may not lead to uniform temperatures throughout the part. The soak time must be optimized in order to ensure proper quenching while minimizing the cycle time.
2. In order to avoid cooling of the part before forming it, the blank must be transferred to the forming dies as quickly as possible. Forming must be performed quickly, before the beginning of transformation of austenite into martensite.

3. Once the part is formed, the dies remain closed while the part is quenched, which takes from 2 to 10 seconds, depending on sheet thickness, temperature of sheet and die, and workpiece material. The cooling rate must be higher than 27°C/s to obtain martensite. Thus, forming is performed in steel tools that have cooling channels incorporated in them, in order to maintain proper tooling temperature. A complete transformation into martensite results in the high strengths given in Table 3.3. It should be noted that quenching from austenite to martensite results in an increase in volume, which influences the residual stress distribution and workpiece distortion in forming.

A recent development is to use pressurized hot gas (air or nitrogen) as a working media to form the material, similar to hydroforming. This method improves formability, and with proper process control, allows for more uniform blank and tooling temperatures, and thus lower residual stresses and warping.

Because the workpiece is hot and quenching must be done very rapidly, hot stamping is usually performed without a lubricant, and often shot blasting (similar to shot peening described in Section 4.5.1, but with abrasive media) is required after forming to remove scale from part surfaces. The steel may also be coated with an aluminum-silicon layer to prevent oxidation and eliminate the grit blasting step. In such a case, the coating requires a slightly longer soak time, in order to properly bond to the steel substrate.

Hot stamping is not restricted to steels. Magnesium alloys ZEK100, AZ31, and ZE10 are also of great interest, because of their light weight; however, these materials have limited formability at room temperature, and are therefore stamped at up to 300°C. Also, some advanced aluminum-alloy sheets are formed at elevated temperatures in order to attain improved ductility, and even develop superplastic behavior.

Electrically Assisted Forming. A recent technology has involved the simultaneous application of high current through a metallic workpiece during the forming operation, known as *electrically assisted forming* (EAF). This has been applied to bulk forming operations as well, and has been noted to increase the formability of materials. The mechanisms involved may be associated with increased temperature, but it has been suggested that the heating is higher in the vicinity of dislocations and that the malleability associated with electron mobility is increased. EAF has been applied to magnesium, aluminum and titanium alloys where formability at room temperature is limited.

7.6 Deep Drawing

Deep drawing, first developed in the 1700s, is an important forming process; typical parts made include beverage cans, pots and pans, kitchen sinks, and containers of all shapes and sizes. In this process, a flat sheet

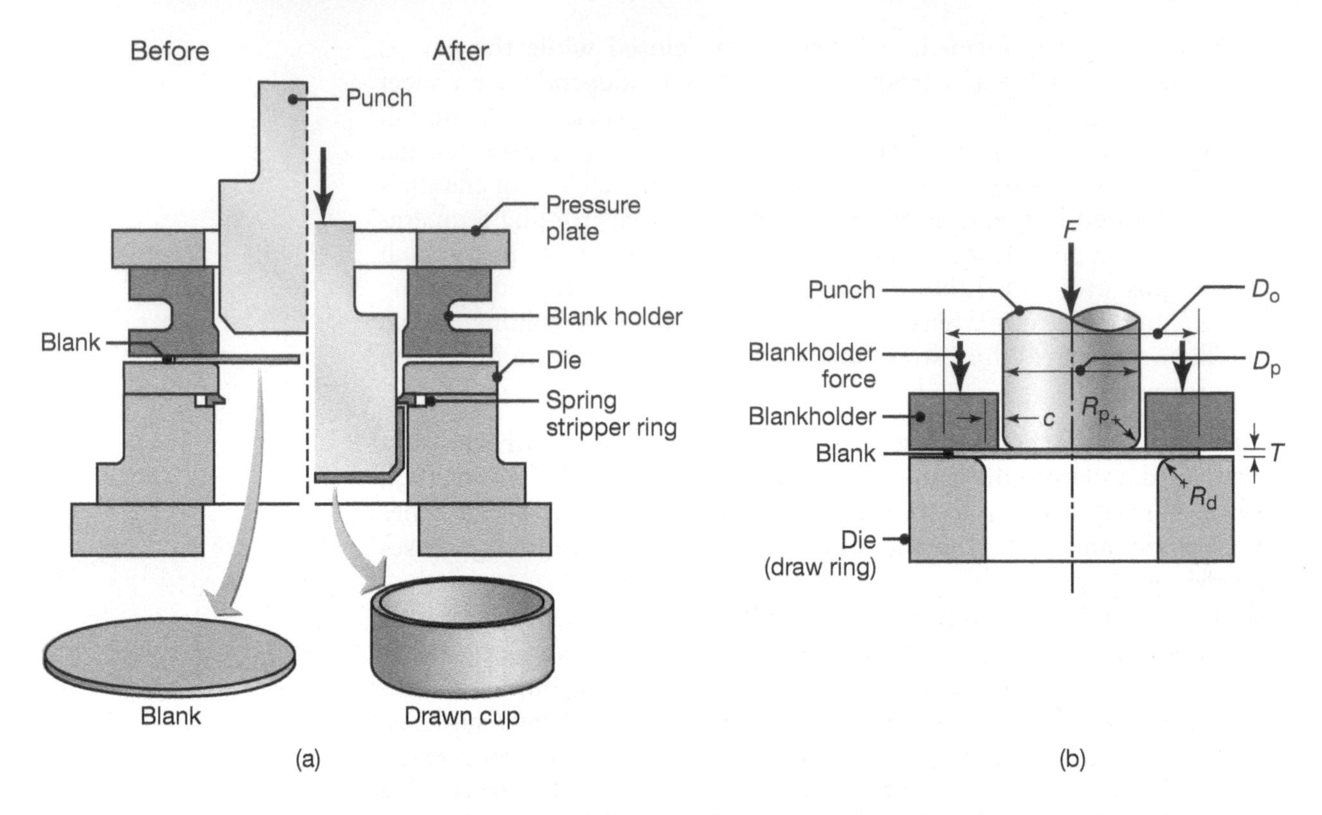

FIGURE 7.49 (a) Schematic illustration of the deep drawing process on a circular sheet metal blank. The stripper ring facilitates the removal of the formed cup from the punch. (b) Variables in deep drawing of a cylindrical cup.

metal blank is formed into a cylindrical or box-shaped part by means of a punch that pushes the blank into the die cavity (Fig. 7.49a). The process is generally called deep drawing (meaning *forming deep parts*), but the basic operation also can produces shallow parts.

The parameters involved in deep drawing a cylindrical cup are shown in Fig. 7.49b. Basically, a circular blank with a diameter D_o and thickness t_o is placed over a die opening, with a corner radius R_d. The blank is held in place with a **blankholder,** or **hold-down ring.** A punch, with a diameter D_p and a corner radius R_p, moves downward and forces the blank into the die cavity, forming a cup. The important variables in deep drawing are

1. Mechanical properties of the sheet metal;
2. Ratio of the blank diameter to the punch diameter;
3. Sheet thickness;
4. Clearance between the punch and the die;
5. Corner radii of the punch and die;
6. Blankholder force;
7. Punch speed; and
8. Friction at the punch, die, and workpiece interfaces.

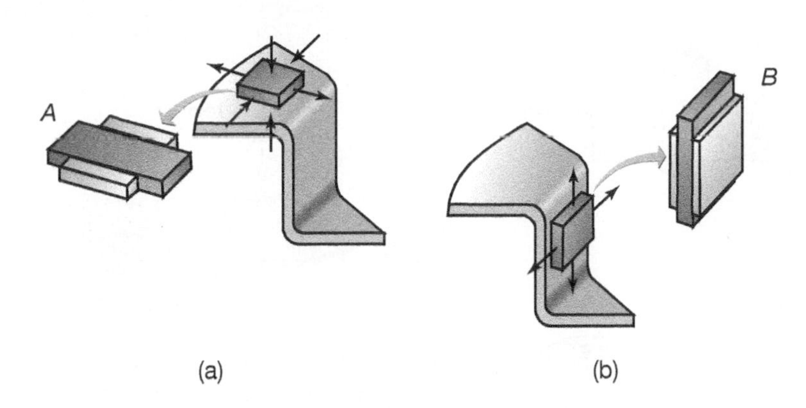

FIGURE 7.50 Deformation of elements in (a) the flange and (b) the cup wall in deep drawing of a cylindrical cup.

At an intermediate stage during the operation, the workpiece is subjected to the states of stress shown in Fig. 7.50. On element A in the blank, the radial tensile stress is due to the blank being pulled into the die cavity; the compressive stress normal to the element is due to the pressure applied on the sheet by the blankholder. With a free-body diagram of the blank along its diameter, it can be shown that the radial tensile stresses cause the compressive hoop stresses on element A. Under this state of stress, the element contracts in the hoop direction and elongates in the radial direction. Note that it is the hoop stress in the flange that tends to cause wrinkling during drawing, thus the necessity for a blankholder under a certain force.

The punch transmits the drawing force, F (See Fig. 7.49b), through the walls of the cup and to the flange (that is being drawn into the die cavity). The cup wall, which is already formed, is subjected principally to a longitudinal tensile stress, as shown in element B in Fig. 7.50. The tensile hoop stress on element B is caused by the cup being held tightly on the punch, because of its contraction under the longitudinal tensile stresses in the cup wall. The diameter of a thin-walled tube becomes smaller when subjected to longitudinal tension, as can be observed from the generalized *flow rule equations*, given by Eq. (2.44). Thus, because it is constrained by the rigid punch, element B does not undergo any width change but it elongates in the longitudinal direction.

An important aspect in this operation is determining how much **pure drawing** and how much **stretching** is taking place (Fig. 7.51). Note that if the blankholder force is low, it will allow the blank to flow freely into the die cavity (pure drawing). Element A in Fig. 7.50a will now tend to increase in thickness as it moves toward the die cavity, because it is being reduced in its diameter. The deformation of the sheet is mainly in the flange, and the cup wall is subjected only to elastic stresses. However, these stresses increase with increasing D_o/D_p ratio, and can eventually lead to failure, because the cup wall cannot support the load required to draw the flange into the die cavity (Fig. 7.51a).

With a large enough blankholder force, or by using **draw beads** (shown in Figs. 7.51b and 7.52), the blank can be prevented from flowing freely into the die cavity. The deformation of the sheet around the bead causes increasing resistance to further deep drawing; the cup eventually begins

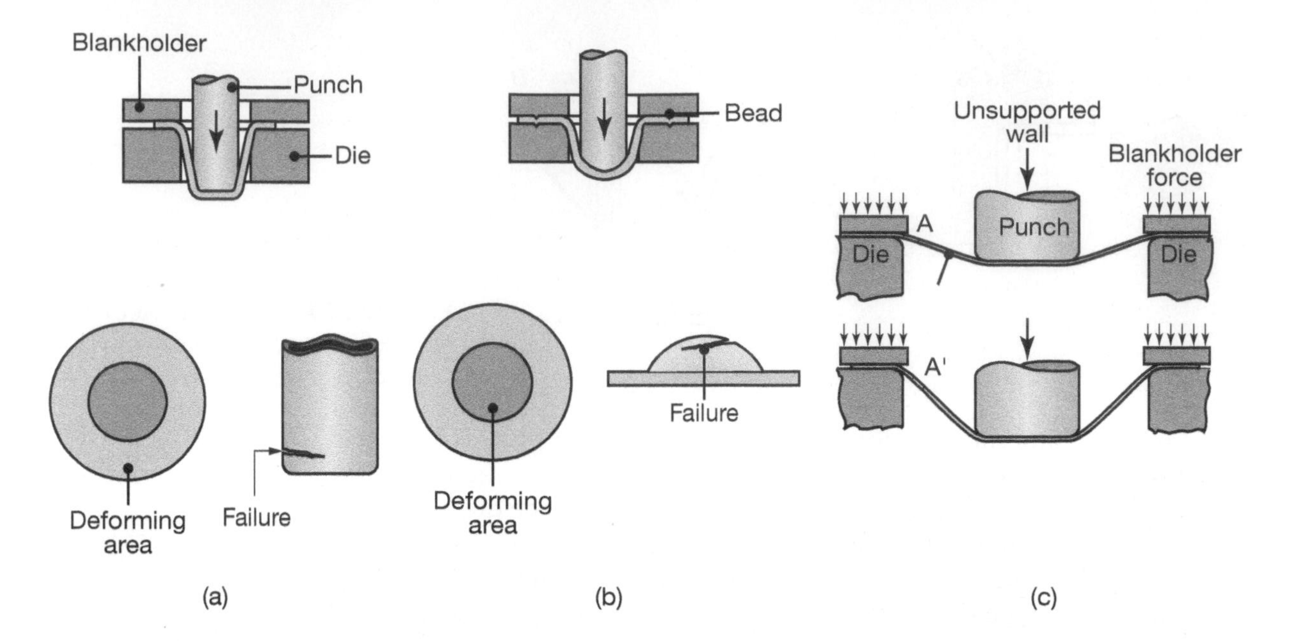

FIGURE 7.51 Examples of (a) pure drawing and (b) pure stretching; the bead prevents the sheet metal from flowing freely into the die cavity. (c) Unsupported wall and possibility of wrinkling of a sheet in drawing. *Source:* After W.F. Hosford and R.M. Caddell.

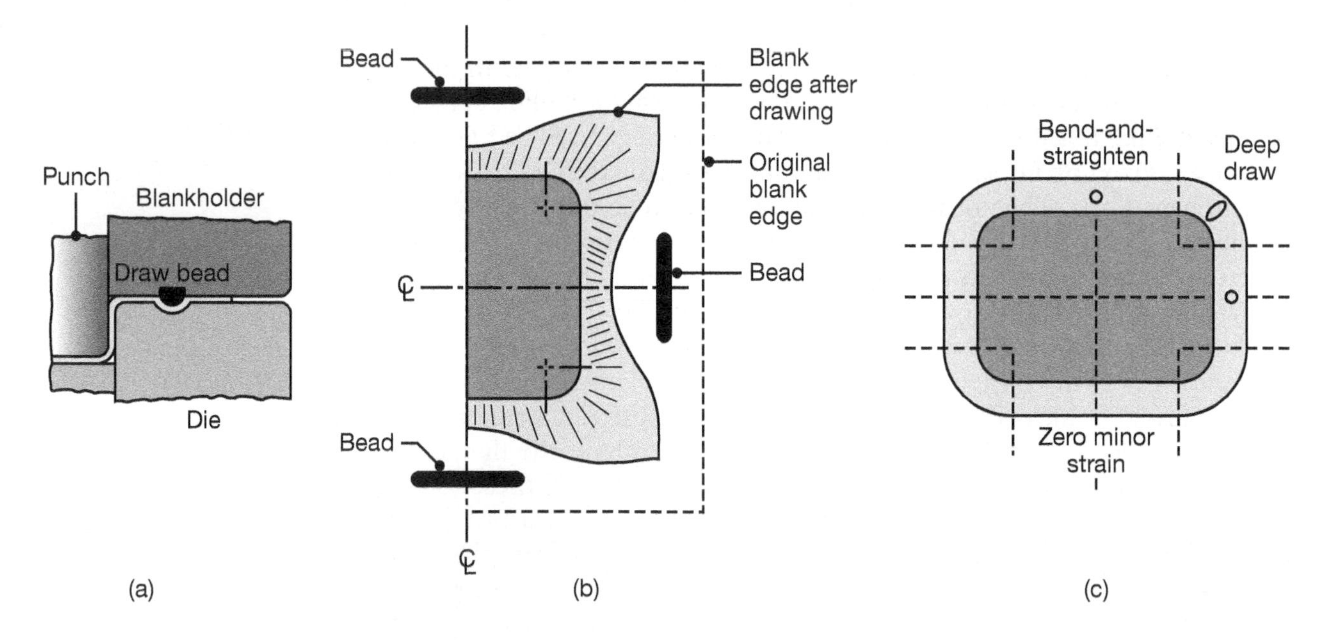

FIGURE 7.52 (a) Schematic illustration of a draw bead. (b) Metal flow during drawing of a box-shaped part, using beads to control the movement of the material. (c) Deformation of circular grids in drawing (see Section 7.7).

to stretch, finally resulting in necking and tearing. Whether necking will be *localized* or *diffuse* depends on (a) the strain-rate sensitivity exponent, m, of the sheet metal (the higher the m value, the more diffuse the neck); (b) punch shape; and (c) lubrication.

The length of the unsupported portion of the sheet (that is, the difference between the die and punch radii) is significant because, if too long, it can lead to **wrinkling** of the sheet. As shown in Fig. 7.51c, element A is pulled into the die cavity as the punch descends; however, as the element moves to position A', the blank becomes smaller in diameter, and the circumference at the element becomes smaller. Thus, at this position, the element is now subjected to circumferential compressive strains and it is unsupported, unlike an element between the blankholder and the die surface. Because the sheet is typically thin and cannot support circumferential compressive stresses to any significant extent, it tends to wrinkle in the unsupported region. This situation is particularly common in pure drawing, but it is less so as the operation approaches pure stretching.

Ironing. If a sheet is thicker than the clearance between the punch and the die, its thickness is reduced, and the process is called *ironing*. By controlling the clearance, a cup with constant wall thickness is produced (Fig. 7.53); thus, ironing can correct earing that occurs in deep drawing (as shown in Fig. 7.56). Note also that because of volume constancy, ironing leads to longer (deeper) cups than deep drawing alone, as described below.

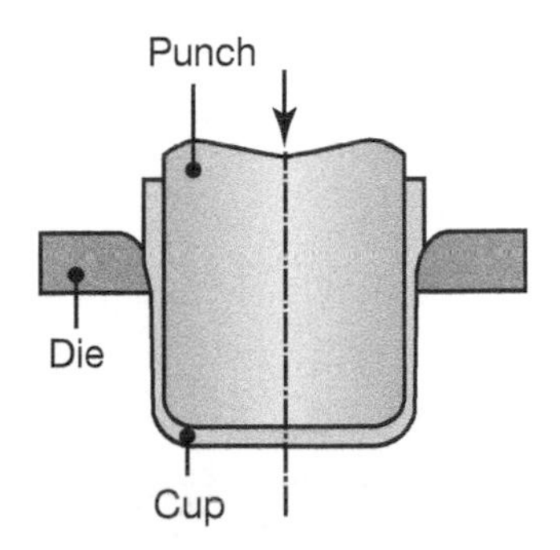

FIGURE 7.53 Schematic illustration of the ironing process. Note that the cup wall is thinner than its bottom. All beverage cans without seams (known as two-piece cans) are ironed, generally in three steps, after being deep drawn into a cup. Cans with separate tops and bottoms are known as three-piece cans.

7.6.1 Deep Drawability (Limiting Drawing Ratio)

An important parameter in drawing is the *limiting drawing ratio* (LDR), defined as the maximum ratio of blank diameter to punch diameter, D_o/D_p, that can be drawn without failure. Generally, failure occurs by *thinning* of the cup wall under high longitudinal tensile stresses. By observing the movement of the round sheet into the die cavity (See Fig. 7.50), it can be noted that the material must be capable of undergoing a reduction in width (being reduced in diameter) but it should resist thinning (under the longitudinal tensile stresses in the cup wall).

The ratio of width strain to thickness strain (Fig. 7.54) is defined as

$$R = \frac{\epsilon_w}{\epsilon_t} = \frac{\ln\left(\frac{w_o}{w_f}\right)}{\ln\left(\frac{t_o}{t_f}\right)}, \tag{7.18}$$

where R is the **normal anisotropy** of the sheet metal, also called **plastic anisotropy** or the **strain ratio.** Subscripts o and f refer to the original and the final dimensions, respectively. An R value of unity indicates that the width and thickness strains are equal to each other; that is, the material is isotropic.

Because errors can be made in the measurement of small thicknesses, Eq. (7.18) is generally modified, based on volume constancy, to

$$R = \frac{\ln\left(\dfrac{w_o}{w_f}\right)}{\ln\left(\dfrac{w_f l_f}{w_o l_o}\right)}, \tag{7.19}$$

where l refers to the gage length of the sheet specimen. The final length and width in a test specimen are usually measured at an elongation of 15 to 20%; for materials with lower ductility, it is measured at less than the elongation at which necking begins.

Rolled sheets generally have **planar anisotropy**, and thus the R value of a specimen, cut from a rolled sheet (Fig. 7.54), will depend on its orientation with respect to the rolling direction of the sheet. An *average* R value, $\overline{R}$, is then calculated as

$$\overline{R} = \frac{R_0 + 2R_{45} + R_{90}}{4}, \tag{7.20}$$

where the subscripts 0, 45, and 90 refer, respectively, to angular orientation (in degrees) of the test specimen with respect to the rolling direction of the sheet. Note that an isotropic material has an $\overline{R}$ value of unity. Typical values are given in Table 7.3. Although hexagonal close-packed metals usually have high $\overline{R}$ values, the low value for zinc, given in the table, is the result of its high c/a ratio in the crystal lattice (See Fig. 3.4).

TABLE 7.3 Typical range of the average normal anisotropy ratio, $\overline{R}$, for various sheet metals.

Material	$\overline{R}$
Zinc alloys	0.4–0.6
Hot-rolled steel	0.8–1.0
Cold-rolled rimmed steel	1.0–1.4
Cold-rolled aluminum-killed steel	1.4–1.8
Aluminum alloys	0.6–0.8
Copper and brass	0.6–0.9
Titanium alloys (α)	3.0–5.0
Stainless steels	0.9–1.2
High-strength low-alloy steels	0.9–1.2

The $\overline{R}$ value also depends on the grain size and texture of the sheet metal. For cold-rolled steels, for example, $\overline{R}$ decreases as grain size decreases. Recall that grain size and grain number are inversely related (See Table 3.1). For hot-rolled sheet steels, $\overline{R}$ is approximately unity, because the texture developed has a random orientation.

It can be seen that deep drawability is enhanced with a high $\overline{R}$ and a low ΔR. Generally, however, sheet metals with a high $\overline{R}$ also have a high ΔR. The controlling parameters have been found to be alloying elements of the metal, processing temperatures, annealing cycles, thickness reduction in rolling, and cross (biaxial) rolling of plates to reduce them to sheets.

Figure 7.55 shows the direct relationship between $\overline{R}$ and LDR, as determined experimentally. In spite of its scatter, no other mechanical

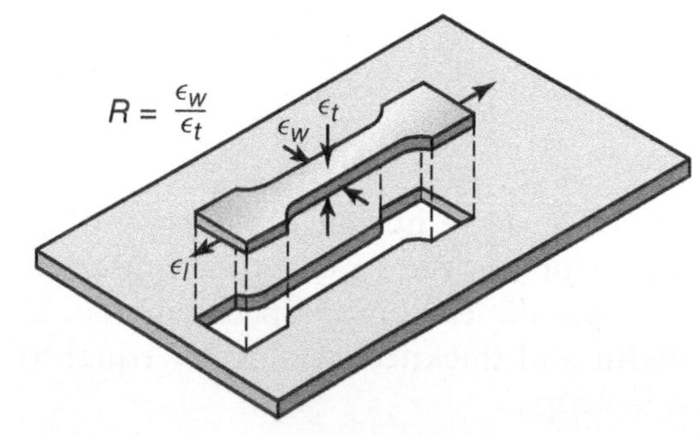

FIGURE 7.54 Definition of the normal anisotropy, R, in terms of width and thickness strains in a tensile-test specimen cut from a rolled sheet. Note that the specimen can be cut in different directions with respect to the length, or rolling direction, of the sheet.

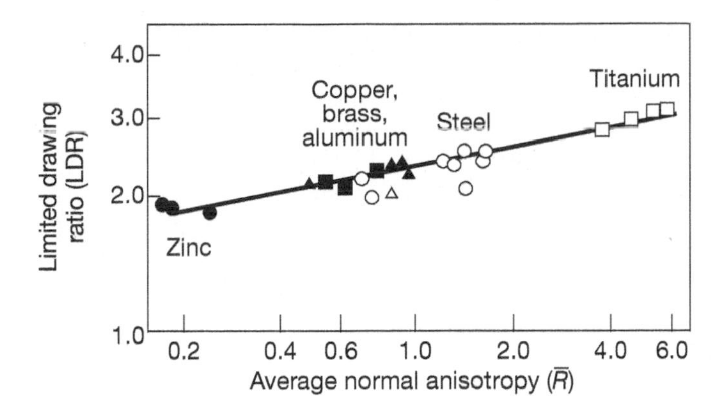

FIGURE 7.55 Effect of average normal anisotropy, $\overline{R}$ on limiting drawing ratio (LDR) for a variety of sheet metals. *Source:* After M. Atkinson.

property of sheet metal has indicated as consistent a relationship with LDR. Note that for an isotropic material, and based on ideal deformation, the maximum LDR is equal to $e = 2.718$.

The **planar anisotropy** of a sheet, ΔR, can also be defined in terms of directional R values as

$$\Delta R = \frac{R_0 - 2R_{45} + R_{90}}{2}, \tag{7.21}$$

which is the difference between the average of the R values in the 0° and 90° directions to rolling and the R value at 45°.

EXAMPLE 7.5 Estimating the Limiting Drawing Ratio

Given: A sheet metal specimen is stretched by 23% in length, resulting in a measured decrease in thickness of 10%.

Find: Estimate the expected limiting drawing ratio (LDR).

Solution: From volume constancy of the test specimen,

$$w_o r_o l_o = w_f t_f l_f \quad \text{or} \quad \frac{w_f t_f l_f}{w_o r_o l_o} = 1.$$

From the information given,

$$\frac{l_f - l_o}{l_o} = 0.23 \quad \text{or} \quad \frac{l_f}{l_o} = 1.23$$

and

$$\frac{t_f - t_o}{t_o} = -0.10 \quad \text{or} \quad \frac{t_f}{t_o} = 0.90.$$

Hence,

$$\frac{w_f}{w_o} = 0.903.$$

From Eq. (7.18),

$$R = \frac{\ln\left(\dfrac{w_o}{w_f}\right)}{\ln\left(\dfrac{t_o}{t_f}\right)} = \frac{\ln 1.107}{\ln 1.111} = 0.965.$$

If the sheet has planar isotropy, then $R = \overline{R}$ and from Fig. 7.55 it can be estimated that

$$\text{LDR} = 2.4.$$

EXAMPLE 7.6 Theoretical Limiting Drawing Ratio

Given: A metal sheet is drawn, and encounters no change in its thickness.

Find: Show that the theoretical limiting drawing ratio is 2.718.

Solution: Reviewing Fig. 7.49b, it can be noted that the diametral change from a blank to a cup involves a true strain of

$$\epsilon_{\max} = \ln\left(\frac{\pi D_o}{\pi D_p}\right) = \ln\left(\frac{D_o}{D_p}\right).$$

The work required for plastic deformation is supplied by the radial stress on element *A* in Fig. 7.50a. From Eq. (6.62), it can then be noted that the drawing stress (the radial stress in Fig. 7.50a), denoted as σ_d, is given by

$$\sigma_d = S_y \epsilon_{\max},$$

where S_y is the yield stress of a perfectly plastic (ideal) material. Since, in the limit, the drawing stress is equal to the yield stress [see also the derivation of Eq. (6.72)],

$$S_y = S_y \epsilon_{\max}$$

or $\epsilon_{\max} = 1$. Consequently, $\ln(D_o/D_p) = 1$ and hence

$$\frac{D_o}{D_p} = 2.718.$$

Earing. Planar anisotropy causes *ears* to form in drawn cups, producing a wavy edge as shown in Fig. 7.56. The number of ears produced can be four, six, or eight, depending on the cystallographic structure of the sheet metal. The height of the ears increases with increasing ΔR; when $\Delta R = 0$, no ears form. Ears are objectionable, because they have to be trimmed off, wasting material which, for a typical drawn aluminum beverage can, is about 1–2% of the length of the can.

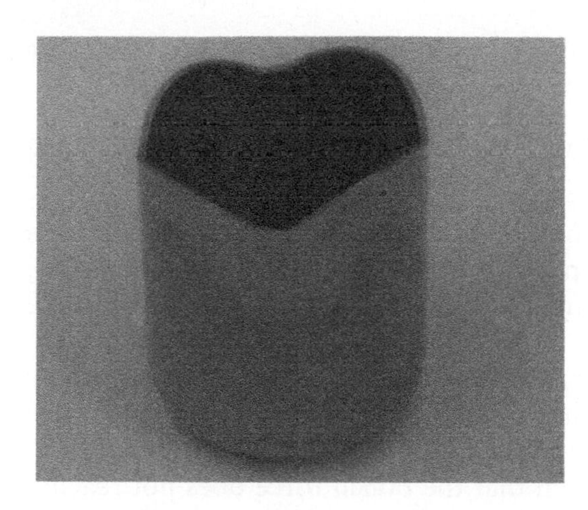

FIGURE 7.56 Typical earing in a drawn steel cup, caused by the planar anisotropy of the sheet metal.

EXAMPLE 7.7 Estimating Cup Diameter and Earing

Given: A steel sheet has R values of 0.9, 1.3, and 1.9 for the 0°, 45° and 90° directions to rolling, respectively.

Find: For a round blank 100 mm in diameter, estimate the smallest cup diameter to which it can be drawn. Will ears form during this operation?

Solution: Substituting the given values into Eq. (7.20),

$$\overline{R} = \frac{0.9 + (2)(1.3) + 1.9}{4} = 1.35.$$

The limiting drawing ratio (LDR) is defined as the maximum ratio of the diameter of the blank to the diameter of the punch that can be drawn without failure, i.e., D_o/D_p. From Fig. 7.55, the LDR for this steel can be estimated to be approximately 2.5. To determine whether or not earing will occur in this operation, the R values are substituted into Eq. (7.21). Thus,

$$\Delta R = \frac{0.9 - (2)(1.3) + 1.9}{2} = 0.1.$$

Ears will not form if $\Delta R = 0$. Since this is not the case here, ears will form in deep drawing this material.

Maximum punch force. The *maximum punch force* in deep drawing, F_{max}, supplies the work required. As in other deformation processes (see Section 2.12), the work consists of the ideal work of deformation, redundant work, frictional work, and, when present, the work required for ironing (Fig. 7.57). Because several variables are involved in deep drawing, and because it is not a steady-state process, accurately calculating the punch force can be difficult. Several expressions have been developed over the years; one simple and approximate formula for the punch force is

$$F_{\max} = \pi D_p t_o S_{\text{ut}} \left(\frac{D_o}{D_p} - 0.7 \right). \tag{7.22}$$

Although this equation does not specifically include such parameters as friction, the corner radii of the punch and die, and the blankholder force, this empirical formula is reasonably accurate for typical ranges of these variables.

If the punch force is excessive, **tearing** will occur, as shown in the lower part of Fig. 7.51a. Note that the cup has been drawn to a considerable depth before failure has occurred. This result can be expected by observing Fig. 7.57, where it can be seen that the punch force does not reach a maximum until *after* the punch has traveled a certain distance. The punch corner radius and die radius do not significantly affect the maximum punch force if they are greater than 10 times the sheet thickness.

7.6.2 Deep Drawing Practice

The parameters that significantly affect the deep-drawing operation are described below.

1. **Clearances and radii.** Generally, *clearances* are 7–14% greater than the original thickness of the sheet. The *corner radii* of the punch and the die also are important; if they are too small, they can cause fracture at the corners (Fig. 7.58), and if they are too large, the unsupported area begins to wrinkle. Wrinkling in this region, as well as in

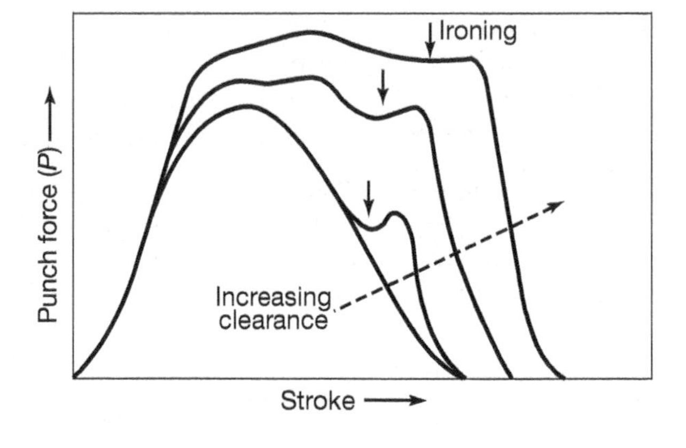

FIGURE 7.57 Schematic illustration of the variation of punch force with stroke in deep drawing. Arrows indicate the initiation of ironing. Note that ironing does not begin until after the punch has traveled a certain distance and the cup is partially formed.

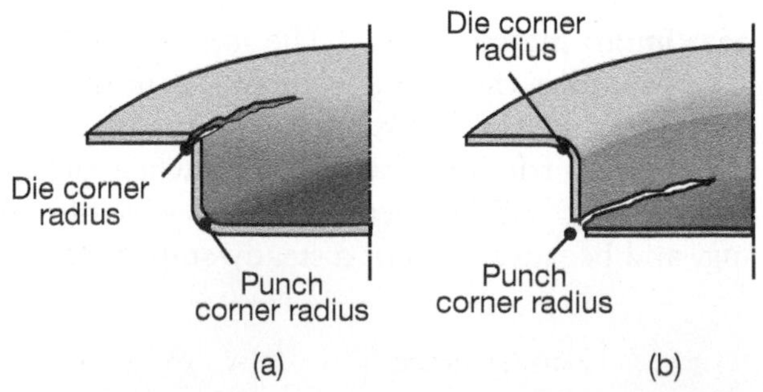

FIGURE 7.58 Effect of die and punch corner radii on fracture in deep drawing of a cylindrical cup. (a) Die corner radius too small; typically, it should be 5 to 10 times the sheet thickness. (b) Punch corner radius too small. Because friction between the cup and the punch aids in the drawing operation, excessive lubrication of the punch is detrimental to drawability.

the flange, causes *puckering* (wrinkling of the portion between the die and the punch).

2. **Draw beads.** Draw beads (See Figs. 7.51b and 7.52) are used to control the flow of the blank into the die cavity. They are particularly important in drawing box-shaped or nonsymmetric parts, because of the nonuniform flow of the sheet into the die cavity. Draw beads also help reduce the blankholder forces required. Draw bead diameters typically range from about 13 to 20 mm.
3. **Blankholder pressure.** Blankholder pressure is generally 0.7 to 1.0% of the sum of the yield and the ultimate tensile strengths of the sheet metal. Too high a blankholder force increases the punch load, because of the higher radial friction forces developed, and also it may cause tearing of the cup wall. On the other hand, if the blankholder force is too low, the flange may wrinkle.

   Since the blankholder force controls the flow of the sheet metal within the die, presses have been designed that are capable of applying a **variable blankholder force.** The blankholder force can thus be programmed and varied throughout the punch stroke. Additional features of such equipment include the control of metal flow by using separate die cushions that allow the *local* blankholder force to be applied at the proper location on the sheet, thus improving drawability.
4. **Redrawing.** Parts that are too difficult to draw in one stroke of the punch are generally *redrawn*, as shown in Fig. 7.59a. In **reverse redrawing** (Fig. 7.59b), the sheet is drawn in the direction opposite to its original configuration. This reversal in deep drawing direction results in **strain softening**, and is another example of the Bauschinger effect, described in Section 2.3.2. The redrawing operation requires

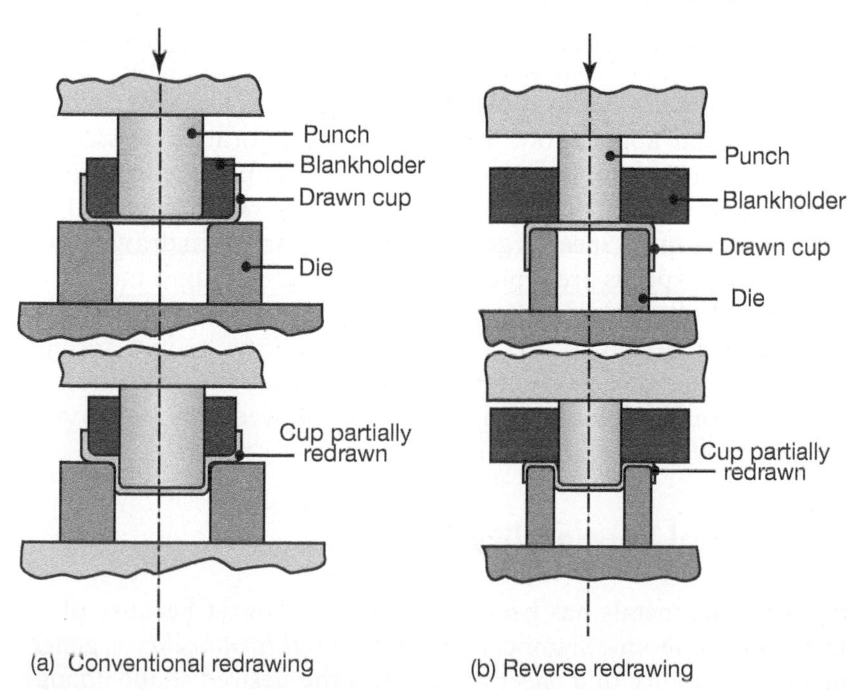

FIGURE 7.59 Reducing the diameter of drawn cups by redrawing operations: (a) conventional redrawing, and (b) reverse redrawing. Small-diameter deep containers may undergo several redrawing operations.

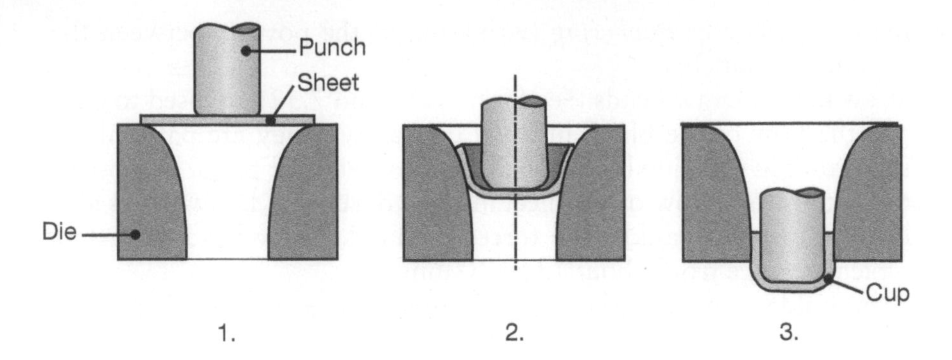

FIGURE 7.60 Stages in deep drawing without a blankholder, using a *tractrix* die profile. The tractrix is a special curve, the construction for which can be found in texts on analytical geometry or in handbooks.

forces lower than those in direct redrawing, and the material behaves in a more ductile manner.

5. **Drawing without a blankholder.** Deep drawing may be carried out without a blankholder, provided that the sheet metal is sufficiently thick to prevent wrinkling. The dies are specially contoured; one example is given in Fig. 7.60. An approximate limit for drawing without a blankholder is given by

$$D_o - D_p < 5t_o. \tag{7.23}$$

It is apparent from Eq. (7.23) that this process can be used with thin sheets for shallow draws. Although the punch stroke is longer than in ordinary drawing, a major advantage of this process is the reduced cost of tooling and equipment.

6. **Tooling and equipment.** The most commonly used tool materials for deep drawing are tool steels and alloyed cast iron; other materials, including carbides and plastics, also may be used, depending on the particular application. Mechanical and hydraulic presses are generally used, although servo presses are becoming more popular (see Section 6.2.8). Punch speeds generally range from 0.1 to 0.3 m/s (4 to 12 in/s). Speed is generally not important in drawability, although lower speeds are typically used for high-strength metals.
7. **Lubrication.** Lubrication in deep drawing operations is important in order to lower punch forces, increase drawability, reduce tooling wear, and reduce defects in drawn parts. In general, lubrication of the punch should be minimized, as friction between the punch and the cup improves drawability.

7.7 Sheet Metal Formability

Formability of sheet metals has been of continued interest because of its technological and economic significance. *Sheet metal formability* is generally defined as the ability of a sheet to undergo the desired shape change without failure, such as by necking, tearing, or splitting. Three factors have a major influence on formability: (a) properties of the sheet metal, as described in Section 7.2; (b) friction and lubrication at various interfaces in the operation; and (c) characteristics of the tools, dies, and the equipment used. Several techniques have been developed over the years to test the formability of sheet metals, including the ability to predict formability by *modeling* the particular forming operation.

7.7.1 Testing Methods for Formability

Several tests are used to determine the formability of sheet metal:

1. **Tension tests.** The *uniaxial tension test* (see Section 2.2) is the most basic and common test used. It evaluates important mechanical properties of the sheet metal, particularly, the total elongation of the sheet specimen at fracture, the strain-hardening exponent, n, the planar anisotropy, ΔR, and the normal anisotropy, R.
2. **Cupping tests.** Because most sheet forming operations basically involve biaxial stretching of the sheet metal, the earliest tests developed to predict formability were *cupping tests*, namely, the **Erichsen** and **Olsen** tests (*stretching*), and the **Swift** and **Fukui** tests (*drawing*). In the Erichsen test, a sheet metal specimen is clamped over a flat die with a circular opening. A 20-mm-diameter (0.8 in.) steel ball is then hydraulically pressed into the sheet, at a load of 1000 kg (2200 lb), until a crack appears on the stretched specimen, or until the punch force reaches a maximum and then begins to decline. The penetration distance, in mm, is the Erichsen number; the greater the value, the better is the formability of the sheet.

   Cupping tests measure the capability of the material to be *stretched* before fracturing, and are relatively easy to perform. Because the stretching under the ball is axisymmetric, however, they do not simulate the exact conditions of actual forming operations.
3. **Bulge test.** The *bulge test* has been used extensively to simulate sheet forming operations. A circular blank is clamped at its periphery and is bulged by *hydraulic* pressure. The deformation of the sheet metal is one of *pure biaxial stretching*, and no friction is involved, as would be the case when using a punch. The bulge limit (depth penetrated prior to failure) is a measure of formability. Bulge tests also can be used to develop effective-stress–effective-strain curves for biaxial loading.
4. **Forming-limit diagrams.** An important development in sheet metal formability is the construction of *forming-limit diagrams* (FLDs). The sheet blank is first marked with a grid pattern of circles, typically 2.5 to 5 mm in diameter, using either chemical-etching or printing techniques. The blank is then stretched over a punch, as shown in Fig. 7.61, and the deformation of the circles is observed and measured in regions where necking and tearing occurs. Note that the sheet is fixed by the draw bead, which prevents it from being drawn into the die cavity. For improved accuracy of measurement, the circles on the sheet are made as small as possible and the grid lines, if used,

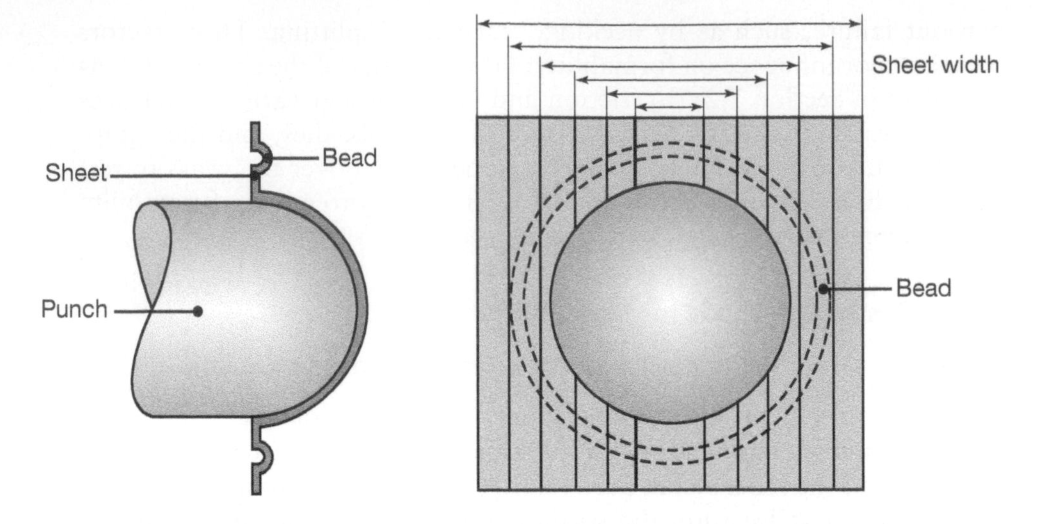

FIGURE 7.61 Schematic illustration of the punch-stretch test on sheet metal specimens with different widths, clamped along the narrower edges. Note that the narrower the specimen, the more uniaxial is the stretching. (See also Fig. 7.64.)

as thin as possible. Because they can significantly influence the test results, lubrication conditions, including dry, are stated in reporting the test results.

Recently, computer-assisted techniques have been developed that print a grid of dots on a sheet metal surface. Optical cameras capture the grid pattern during deformation; specialized software tracks the displacement of the dots during deformation. From the deformations, the planar strains can be calculated.

In the forming-limit diagram, the **major strain** and the **minor strain** are obtained as follows:

a. Note in Figs. 7.62b and 7.63 that after stretching, an original circle has deformed into an ellipse.

b. The strain in the plane of the sheet can be calculated from Eq. (2.10). The major axis of the ellipse represents the major direction of stretching. The minor axis of the ellipse represents the magnitude of *stretching* (positive minor strain) or shrinking (negative minor strain) in the transverse direction.

As an example, if a circle is drawn on the surface of a sheet metal tensile-test specimen and then stretched plastically, the circle becomes narrower; thus, the minor strain is negative. Because of volume constancy in the plastic range [see Eq. (2.49)], and assuming that the normal anisotropy of the sheet is $R = 1$ (that is, the width and thickness strains are equal), then $\epsilon_w = -0.5\epsilon_l$. On the other hand, if a circle is placed on a specimen encountering biaxial tension (an example is a spherical rubber balloon that is inflated), it increases in diameter, indicating that the minor strain is positive and equal to the major strain.

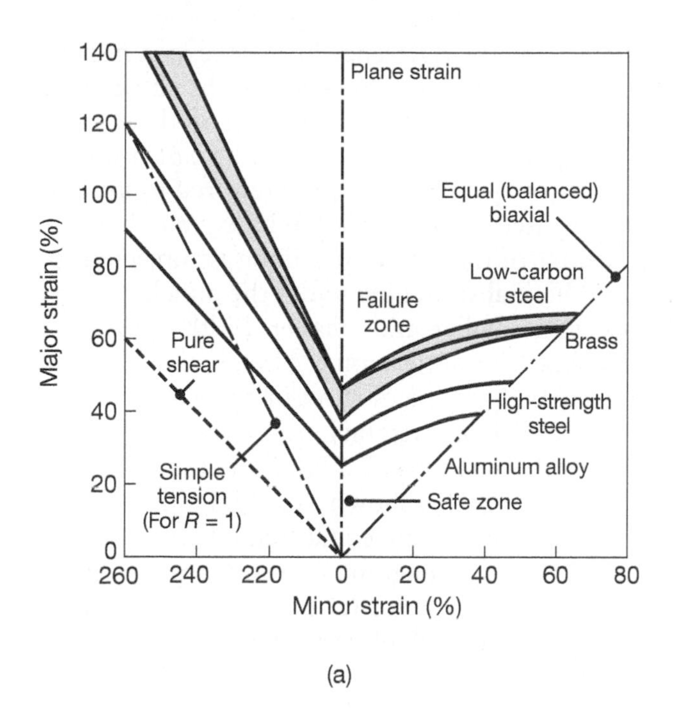

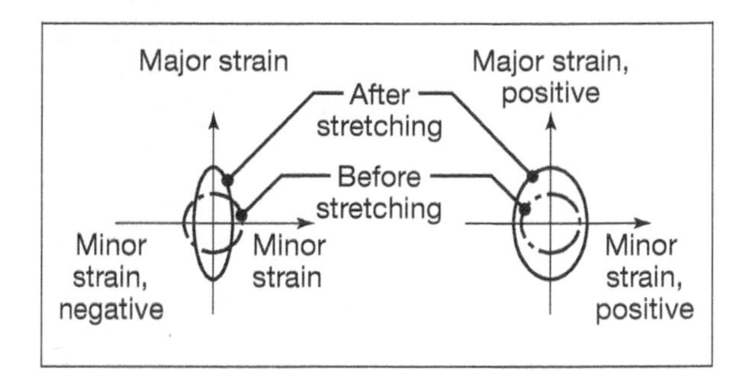

FIGURE 7.62 (a) Forming-limit diagram (FLD) for various sheet metals. Note that the major strain is always positive. The region above the curves is the failure zone; hence, the state of strain in forming must be such that it falls below the curve for a particular material; R is the normal anisotropy. (b) Illustrations of the definition of positive and negative minor strains. If the area of the deformed circle is larger than the area of the original circle, the sheet has become thinner than the original thickness. *Source:* After S.S. Hecker and A.K. Ghosh.

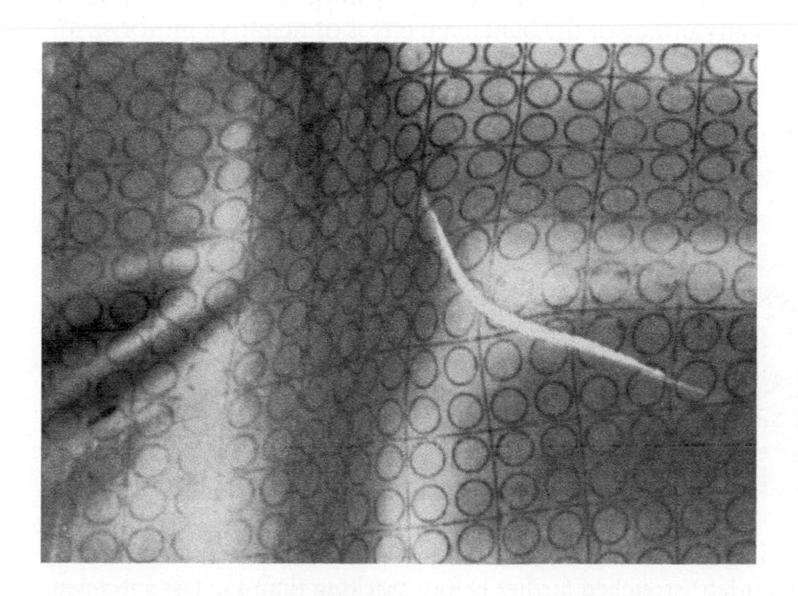

FIGURE 7.63 An example of the use of grid marks (circular and square) to determine the magnitude and direction of surface strains in sheet metal forming. Note that the crack (tear) is generally perpendicular to the major (positive) strain. *Source:* After S.P. Keeler.

For any general stress state, the difference in surface area between the original circle and the ellipse indicates whether the *thickness* of the sheet has changed during deformation. If the area of the ellipse is larger than that of the original circle, the sheet has become thinner.

In order to vary the major and minor strains, referred to as **strain path**, specimens are prepared with varying widths, as shown in Fig. 7.61b. Thus, the center of the square specimen is subjected to a state of **equal (balanced) biaxial stretching** under the punch, whereas rectangular specimens with smaller and smaller width approach a state of **uniaxial stretching** (simple tension). After a series of such tests is performed on a particular type of sheet metal (Fig. 7.64), the boundaries between *safe* and *failed* regions are plotted to obtain a forming-limit diagram, as shown in Fig. 7.62a). Note that this figure also includes the various strain paths, shown as straight lines, including: (a) The line on the right represents **equal biaxial strain**; (b) the vertical line at the center of the diagram represents **plane strain**, because the minor planar strain is zero; (c) the **simple tension** line on the left has a 2:1 slope, because the Poisson's ratio in the plastic range is 0.5, meaning, the minor strain is one-half of the major strain; and (d) the **pure-shear** line has a negative 45° slope.

From Fig. 7.62a, it can be observed that, as expected, different materials have different forming-limit diagrams. Note that the higher the curve, the better the formability of the material. It can also be noted that for the same minor strain, say 20%, a negative (compressive) minor strain is associated with a major strain (before failure) that is higher than that for a positive (tensile) minor strain. In other words, it is desirable for the minor strain to be *negative*, because it allows shrinkage in the minor direction during forming.

Special tooling can be designed for forming sheet metals and tubing to take advantage of the beneficial effect of negative minor strains on extending formability (See, for example, Fig. 7.29). The possible

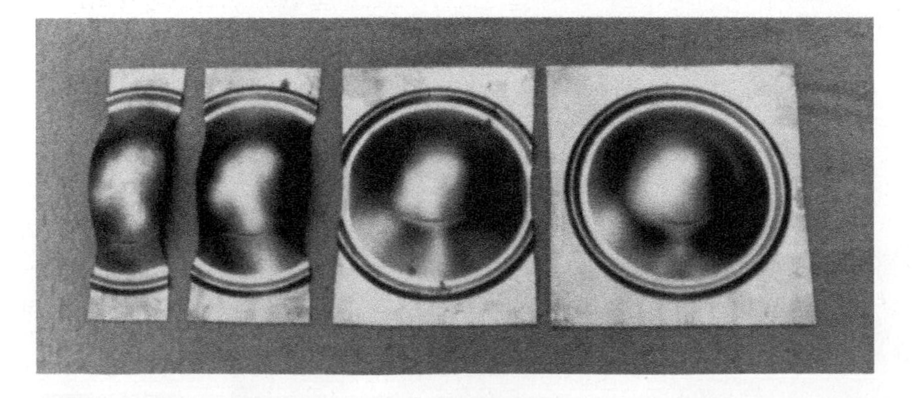

FIGURE 7.64 Bulge test results on steel sheets of various widths. The first specimen (farthest left) stretched farther before cracking than the last specimen. From left to right, the state of stress changes from almost uniaxial to biaxial stretching. *Source:* Courtesy of ArcelorMittal.

effect of the *rate of deformation* on forming-limit diagrams also should be assessed for each material (see also Section 2.2.7).

The effect of sheet metal *thickness* on forming-limit diagrams is to raise the curves in Fig. 7.62a; thus, the thicker the sheet, the higher the curve and the more formable is the sheet. Note, however, that a thick blank may not bend easily around small radii and may develop cracks (see Section 7.4.1).

Friction at the punch, die, and sheet metal interfaces can be significant in test results. With well-lubricated interfaces, the strains are, as expected, more uniformly distributed over the punch. Moreover, depending on the notch sensitivity of the sheet metal, surface scratches, deep gouges, and blemishes can reduce formability, causing premature tearing and failure.

5. **Limiting dome-height test.** The *limiting dome-height* (LDH) test is performed with similarly prepared specimens as in FLD; the *height* of the dome at failure of the sheet, or when the punch force reaches a maximum, is measured. Because the specimens are clamped, the LDH test indicates the capability of the material to *stretch* without failure. It has been shown that high LDH values are related to high n and m values, as well as high total elongation of the sheet metal.

EXAMPLE 7.8 Estimating Diameter of Expansion

Given: A thin-walled spherical shell made of an aluminum alloy, whose forming-limit diagram is shown in Fig. 7.62a, is being expanded by internal pressure.

Find: If the original shell diameter is 200 mm, what is the maximum diameter to which it can safely be expanded?

Solution: Because the material is being stretched in a state of equal (balanced) biaxial tension, Fig. 7.62a indicates that the maximum allowable engineering strain is about 40%. Thus,

$$e = \frac{\pi D_f - \pi D_o}{\pi D_o} = \frac{D_f - 200}{200} = 0.40.$$

Hence,

$$D_f = 280 \text{ mm}.$$

7.7.2 Dent Resistance of Sheet Metal Parts

For automotive body panels, appliances, and office furniture, an important consideration is the *dent resistance* of the sheet metal. A dent is a small but permanent biaxial deformation of a relatively thin metal. The factors that affect dent resistance have been shown to be the yield strength, S_y, the thickness, t, and the shape of the panel. *Dent resistance* is then expressed by a combination of material and geometrical parameters as

$$\text{Dent resistance} \propto \frac{S_y^2 t^4}{S}, \tag{7.24}$$

where S is the panel stiffness, which, in turn, is defined as

$$S = (E)\left(t^a\right)(\text{shape}), \tag{7.25}$$

where the value of a ranges from 1 to 2 for most panels. As for the effect of shape, the flatter the panel, the greater is the dent resistance, because of the sheet's flexibility. Thus, dent resistance (a) increases with increasing strength and thickness of the sheet; (b) decreases with increasing elastic modulus and stiffness; and (c) decreases with decreasing curvature.

Dents are typically caused by *dynamic* forces, such as forces developed by falling objects or other objects that hit the surface of the panel from different directions and angles. In typical automotive panels, impact velocities range up to 45 m/s, thus the *dynamic yield strength* (that is, the yield strength at high strain rates), rather than the static yield strength, is the more significant parameter.

For materials whose yield strength increases with increasing strain rate, denting requires higher energy levels than under static conditions. Also, dynamic forces tend to cause more *localized* dents than do static forces. Because a portion of the energy goes into elastic deformation, the modulus of resilience of the sheet metal [see Eq. (2.5)] is an additional factor to be considered.

7.7.3 Modeling of Sheet Metal Forming Processes

Section 6.2.2 outlined the techniques for studying bulk-deformation processes and described *modeling* of impression-die forging, using the finite-element method (See also Fig. 6.11). Such mathematical modeling can also be applied to sheet forming processes. The ultimate goal is to rapidly analyze stresses, strains, metal flow patterns, wrinkling, and springback as functions of such parameters as material characteristics, sheet anisotropy, speed and temperature of deformation, and friction.

Such interactive analyses can, for example, determine the optimum tool and die geometry in making a specific part, and reducing or eliminating costly die tryouts. Moreover, simulation techniques can be used to determine the size and shape of blanks, intermediate part shapes, required press characteristics, and process parameters that optimize the forming operation. Application of computer modeling are now a cost-effective technique in sheet metal forming operations.

7.8 Equipment for Sheet Metal Forming

The equipment for most pressworking operations generally consists of mechanical, hydraulic, servo mechanical, pneumatic, or pneumatic-hydraulic presses (see also Section 6.2.8 and Fig. 6.25). The traditional *C-frame* press design, with an open front, has been widely used for ease of workpiece accessibility in the press; on the other hand, the *box-type* (O-type) pillar and double-column frame structures are stiffer.

Press selection includes consideration of several factors, including: (a) type of forming operation; (b) size and shape of the parts; (c) length of stroke of the slide(s); (d) number of strokes per minute; (e) press speed; (f) *shut height*, that is, the distance from the top of the press bed to the bottom of the slide (with the stroke down); (g) types and number of slides, such as in double-action and triple-action presses; (h) press capacity and tonnage; (i) type of controls; (j) auxiliary equipment; and (k) safety features.

Because changing dies in presses can involve a significant amount of effort and time, rapid die-changing systems are available. Called **single-minute exchange of dies** (SMED), such systems use automated hydraulic or pneumatic equipment to reduce die-changing times from hours to as little as 10 minutes.

7.9 Design Considerations

As with all other metalworking operations described throughout this text, certain design guidelines and practices for sheet metal forming have evolved with time. The following guidelines identify the most significant design issues of sheet metal forming operations:

1. **Blank design.** Material scrap is a primary concern in blanking operations. Poorly designed parts will not nest well and there will be considerable scrap produced between successive blanking operations (Fig. 7.65).
2. **Bending.** In bending operations, the main concerns are cracking of the material and wrinkling. As shown in Fig. 7.66, a part with a flange to be bent will force the flange to undergo compression, possibly causing buckling. This problem can be controlled by using a relief notch to limit the compressive stresses, as shown in Fig. 7.66. Right-angle bends have similar difficulties; relief notches can be used to avoid tearing (Fig. 7.67).

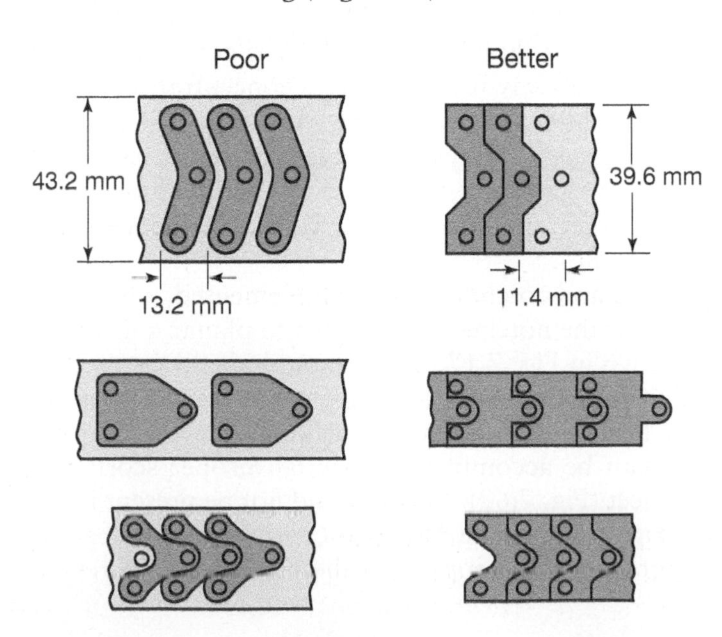

FIGURE 7.65 Efficient nesting of parts for optimum material utilization in blanking. *Source:* Society of Manufacturing Engineers.

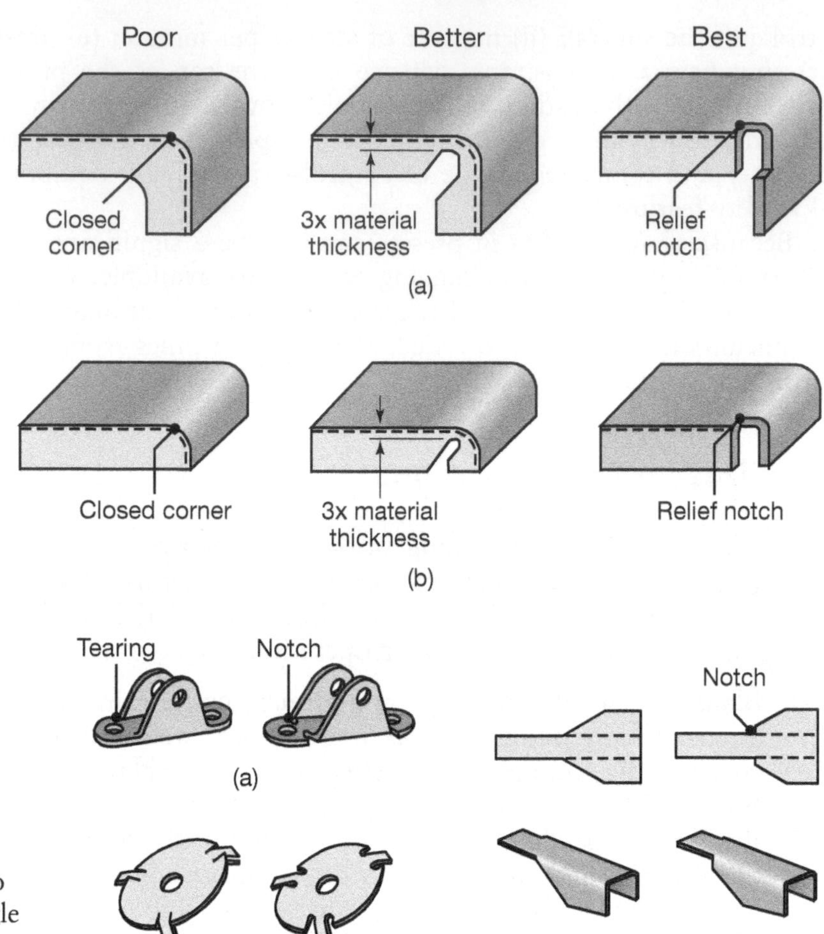

FIGURE 7.66 Control of tearing and buckling of a flange in a right-angle bend. *Source:* Society of Manufacturing Engineers.

FIGURE 7.67 Application of notches to avoid tearing and wrinkling in right-angle bending operations. *Source:* Society of Manufacturing Engineers.

The bend radius is a highly stressed area, and all stress concentrations should be removed as much as possible from that location. One example is the placement of holes near bends; it is generally advantageous to move the hole away from the stress concentration. If this is not possible, a crescent slot or ear can be used (Fig. 7.68a). Similarly, when bending flanges, features such as tabs and notches should be avoided, as they may greatly reduce the part's formability. When tabs are necessary, large radii should be specified to reduce stress concentration (Fig. 7.68b).

When both bending and notches are to be implemented, it is important to properly orient the notches with respect to planar anisotropy of the sheet. As shown in Fig. 7.17, bends should ideally be perpendicular to the rolling direction (or oblique, if perpendicularity is not possible) in order to avoid or minimize cracking. Successful bending at sharp radii can be accomplished through proper scoring or embossing of the sheet (Fig. 7.69). Burrs should not be present in the bend allowance (Fig. 7.15), since they can cause cracking, leading to a stress concentration that propagates the crack into the rest of the sheet.

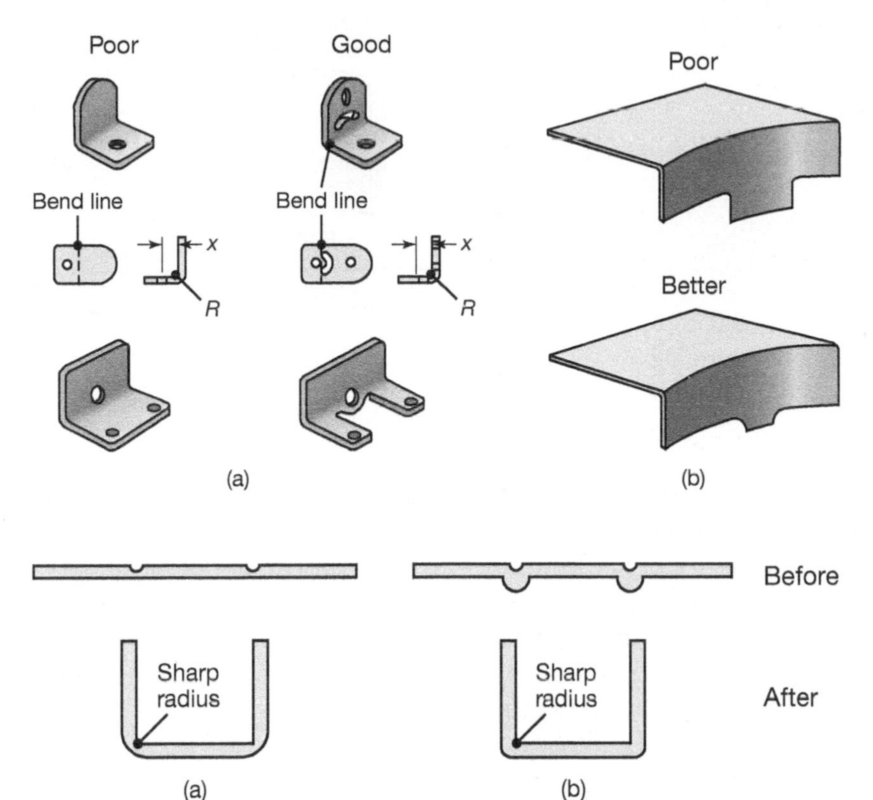

FIGURE 7.68 Stress concentrations near bends. (a) Use of a crescent or ear for a hole near a bend. (b) Reduction of the severity of a tab in a flange. *Source:* Society of Manufacturing Engineers.

FIGURE 7.69 Application of (a) scoring, or (b) embossing to obtain a sharp inner radius in bending. However, unless properly designed, these features can lead to fracture. *Source:* Society of Manufacturing Engineers.

3. **Stamping and progressive-die operations.** In progressive dies (Fig. 7.13), the number of stations and the cost of the tooling are determined by the number and spacing of features on a part. It is advantageous to minimize the number of features on a part. Closely-spaced features may not provide sufficient clearance for punches, requiring additional stations. Narrow cuts and protrusions may be challenging to form with a single punch and die setup.
4. **Deep drawing.** After deep drawing, a cup will tend to spring back to some extent; parts that have a vertical wall can be difficult to draw. Relief angles of at least 3° on each wall improve ease of forming. Cups with sharp internal radii are difficult to draw, and deep cups often require ironing operations.

7.10 Economics of Sheet Metal Forming

Sheet metal forming operations involve economic considerations that are similar to those for other metalworking operations, described in Chapter 6. A complication with sheet forming operations is that the processes are versatile, and several different strategies can be used to produce the same part. For example, a cup-shaped part can be formed by deep drawing, spinning, rubber-pad forming, or explosive forming. Similarly, a cup can be formed by impact extrusion, by casting, or by fabrication involving assembly of several different pieces.

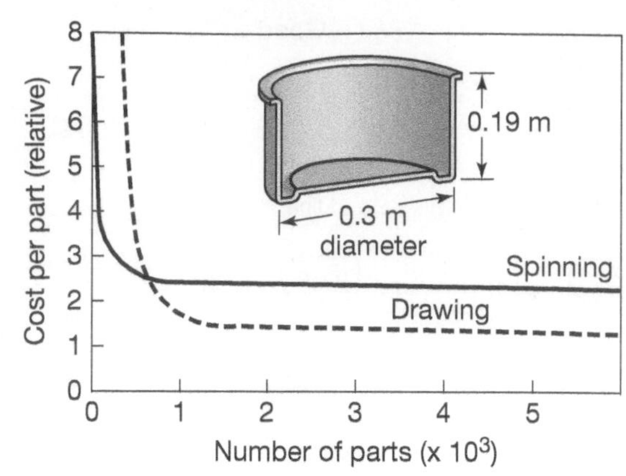

FIGURE 7.70 Cost comparison for manufacturing a cylindrical sheet metal container by conventional spinning and deep drawing. Note that for small quantities, spinning is more economical.

As described in Section 16.6.5, almost all manufacturing processes produce some scrap, ranging up to 60% of the original material for machining operations, and up to 25% for hot forging; sheet forming operations produce scrap on the order of 10 to 25%. While scrap is typically recycled, there is an associated latent energy and handling cost.

As an example, the part shown in Fig. 7.70 can be made either by deep drawing or by conventional spinning; however, the die costs for the two processes are significantly different. Recall that deep-drawing dies have several components and they cost much more than the relatively simple mandrels and tools typically used in spinning. Consequently, the die cost per part in drawing will be high if only few parts are required. On the other hand, this part can be made by deep drawing in a much shorter time (seconds) than by spinning (minutes), even if the latter operation is highly automated. Moreover, deep drawing requires much less skilled labor. Considering all relevant factors, it was found that the break-even point for making this particular part is at about 700 parts; thus, deep drawing is more economical for quantities greater than this number.

CASE STUDY 7.2 Tube Hydroforming of an Automotive Radiator Closure

The conventional assembly used to support an automotive radiator, or radiator closure, is constructed through stamping of the components, which are subsequently welded together. To simplify the design and to achieve weight savings, a hydroformed assembly was designed, as shown in Fig. 7.71 . Note that this design uses varying cross sections, an important feature to reduce weight and provide surfaces to facilitate assembly and mounting of the radiator.

A typical tube hydroforming processing sequence consists of the following steps:

1. Bending of tube to the desired configuration;
2. Tube hydroforming to achieve the desired shape;
3. Finishing operations, such as end shearing and inspection; and
4. Assembly, including welding of components.

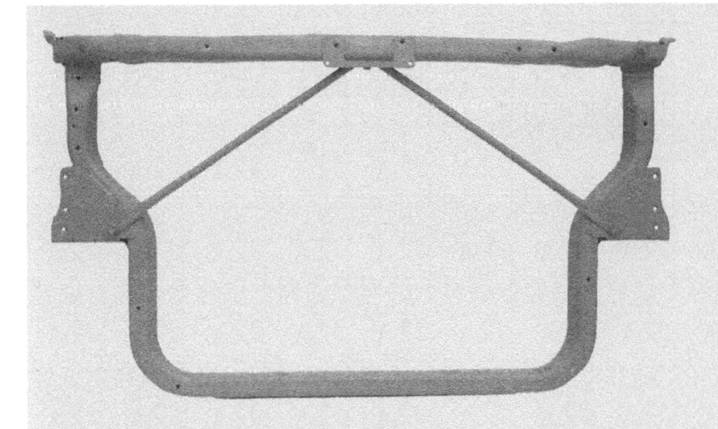

FIGURE 7.71 Hydroformed automotive radiator closure, which serves as a mounting frame for the radiator.

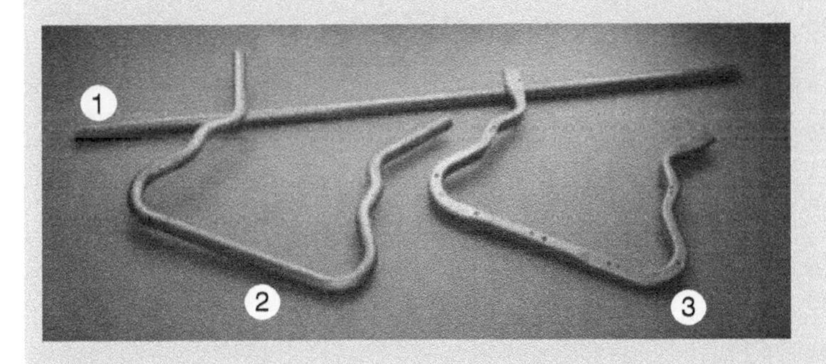

FIGURE 7.72 Sequence of operations in producing a tube-hydroformed component: (1) tube as cut to length; (2) after bending; and (3) after hydroforming.

The operations performed on one of the tube components of the closure is shown in Fig. 7.72. The tube, constructed of steel with a 300 MPa (43.5 ksi) yield strength, is bent to shape (See Fig. 7.28). The bent tube is then placed in a hydroforming press and the end caps are attached.

Conventional hydroforming involves closing the die onto the tube, followed by internal pressurization to force the tube to the desired shape. Figure 7.73a shows a typical cross section. Note that as the tube is expanded, there is significant wall thinning, especially at the corners, because of friction at the tube–die interface. A sequence of pressures that optimize corner formation is therefore used, as shown in Fig. 7.73b.

In this approach, a first pressure stage (prepressure stage) is applied as the die is closing, causing the tube to partially fill the die cavity and form the cross-sectional corners. After the die is completely closed, the internal pressure is increased to lock-in the form and provide support needed for hole piercing. This sequence has the benefit of forming the sharp corners in the cross section by bending, as opposed to pure stretching as in conventional hydroforming. The resulting wall thickness is much more uniform, producing a more structurally sound component.

The assembly shown in Fig. 7.71 has 76 holes that are pierced inside the hydroforming die; the ends are then sheared to length. The 10 components in the hydroformed closure are then assembled through robotic gas-metal arc welding (see Section 12.3.4), using threaded fasteners to aid in serviceability.

Compared to the original stamped design, the hydroformed design has four fewer components, uses only 20 welds as opposed to 174 for the stamped design, and weighs 10.5 kg (23 lb) versus 14.1 kg (31.1 lb). Furthermore, the stiffness of the enclosure and the water cooling areas are both significantly increased.

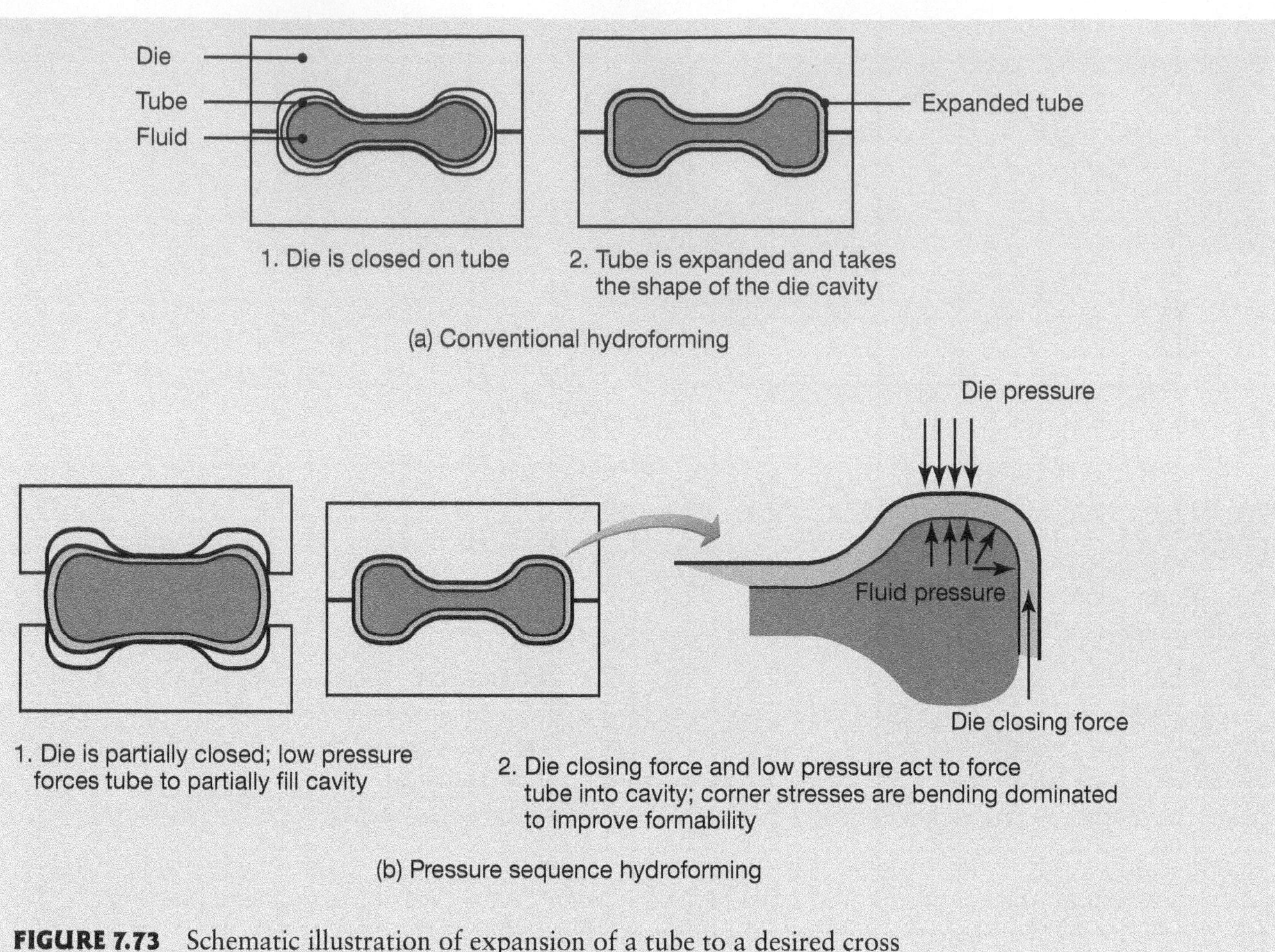

FIGURE 7.73 Schematic illustration of expansion of a tube to a desired cross section through (a) conventional hydroforming and (b) pressure sequence hydroforming.

Source: Courtesy of B. Longhouse, Vari-Form, Inc.

SUMMARY

- Sheet metal forming processes involve workpieces that have a high ratio of surface area to thickness; typically, tensile stresses are applied in the plane of the sheet through various punches and dies. The sheet is generally prevented from being reduced in thickness in order to avoid necking and subsequent tearing and fracture. Material properties and behavior important to sheet forming operations include elongation, yield-point elongation, anisotropy, and grain size. (Sections 7.1, 7.2)
- Blanks for forming operations are generally sheared from large rolled sheets, using a variety of processes. Important parameters include the clearance between the punch and dies, punch and die corner sharpness, and lubrication; the quality of the sheared edge is important in subsequent forming. (Section 7.3)

- Bending of sheet, plate, tubing, and various other cross sections is a common forming operation; important parameters include the minimum bend radius (to avoid cracking in the bend area) and springback (for dimensional accuracy and assembly). Several techniques are available for bending sheet metal and for minimizing springback. (Section 7.4)
- Discrete sheet metal parts can be produced by a variety of methods, including stretch forming, bulging, rubber-pad forming, hydroforming, spinning, and superplastic-forming methods. (Section 7.5).
- Deep drawing operations can produce container-shaped parts, with high depth-to-diameter ratios, economically, and at high rates. Deep drawability has been found to depend on the normal anisotropy of the sheet metal, while earing depends on its planar anisotropy. (Section 7.6)
- Tests have been developed to determine the formability of sheet metals. The most comprehensive test is the punch-stretch (bulge) test, enabling the construction of forming-limit diagrams, indicating safe and failure regions as a function of the planar strains to which the sheet is subjected. Modeling and simulation of sheet metal forming are important tools for predicting the behavior of sheet metals in actual forming operations. (Section 7.7)
- Various types of equipment are available for sheet metal forming operations; their selection depends on several factors concerning workpiece size and shape, equipment characteristics, and control features. Rapid diechanging techniques are important in improving productivity and the overall economics of forming operations. (Section 7.8)
- Design considerations in sheet metal forming are based on such factors as blank shape, the type of operation to be performed, and the nature of the deformation of the sheet metal. Several design rules and practices have been established to eliminate or minimize production problems. (Section 7.9)
- Economic considerations in sheet forming operations often determine the specific processes and practices to be employed. (Section 7.10)

SUMMARY OF EQUATIONS

Maximum punch force in shearing, $F_{\text{max}} = 0.7S_{\text{ut}}tL$

Strain in bending, $e_o = e_i = \dfrac{1}{(2R/t)+1}$

Minimum, $\dfrac{R}{t} = \dfrac{50}{r} - 1$

Springback, $\dfrac{R_i}{R_f} = 4\left(\dfrac{R_iS_y}{Et}\right)^3 - 3\left(\dfrac{R_iS_y}{Et}\right) + 1$

Maximum bending force, $F_{\text{max}} = k\dfrac{S_{\text{ut}}Lt^2}{W}$

Normal anisotropy, $R = \dfrac{\epsilon_w}{\epsilon_t}$

Average, $\overline{R} = \dfrac{R_0 + 2R_{45} + R_{90}}{4}$

Planar anisotropy, $\Delta R = \dfrac{R_0 - 2R_{45} + R_{90}}{2}$

Maximum punch force in drawing, $F_{\max} = \pi D_p t_o S_{\text{ut}} \left(\dfrac{D_o}{D_p} - 0.7 \right)$

Pressure in explosive forming, $p = K \left(\dfrac{\sqrt[3]{W}}{R} \right)^a$

BIBLIOGRAPHY

Altan, T., and Tekkaya, E., (eds.), *Sheet Metal Forming: Fundamentals*, ASM International, 2012.

—, *Sheet Metal Forming: Processes and Applications.*, ASM International, 2012.

ASM Handbook, Vol. 14B: Metalworking: Sheet Forming, ASM International, 2006.

Banabic, D., *Sheet Metal Forming Processes: Constitutive Modeling and Numerical Simulation*, Springer, 2010.

Blazynski, T.Z., *Explosive Welding, Forming and Compaction*, Springer, 2012.

Boljanovic, V., *Sheet Metal Stamping Dies: Die Design and Die-Making Practice*, Industrial Press, 2012.

—, *Sheet Metal Forming Processes and Die Design*, 2nd ed., Industrial Press, 2014.

Davies, G., *Materials for Automobile Bodies*, Elsevier, 2006.

Emmens, W.C., *Formability: A Review of Parameters and Processes that Control, Limit or Enhance the Formability of Sheet Metal*, Springer, 2011.

Fundamentals of Tool Design, 5th ed., Society of Manufacturing Engineers, 2005.

Gillanders, J., *Pipe and Tube Bending Manual*, FMA International, 1994.

Giuliano, G., *Superplastic Forming of Advanced Metallic Materials*, Woodhead, 2011.

Hingole, R.S., *Advances in Metal Forming*, Springer, 2014.

Hosford, W.F., and Caddell, R.M., *Metal Forming, Mechanics and Metallurgy*, 4th ed., Prentice Hall, 2014.

Hu, J., Marciniak, Z., and Duncan, J. *Mechanics of Sheet Metal Forming*, Butterworth-Heinemann, 2002.

Koç, M. (ed.), *Hydroforming for Advanced Manufacturing*, Woodhead, 2008.

Pearce, R., *Sheet Metal Forming*, Springer, 2006.

Rapien, B.L., *Fundamentals of Press Brake Tooling*, Hanser Gardner, 2005.

Suchy, I., *Handbook of Die Design*, 2nd ed., McGraw-Hill, 2005.

Szumera, J., *Metal Stamping Process*, Industrial Press, 2003.

Tschaetch, H., and Koth, A., *Metal Forming Practice: Processes, Machines, Tools*, Springer, 2007.

QUESTIONS

7.1 Take any three topics from Chapter 2, and, with specific examples for each, show their relevance to the topics covered in this chapter.

7.2 Do the same as for Question 7.1, but for Chapter 3.

7.3 Describe (a) the similarities and (b) the differences between the bulk-deformation processes described in Chapter 6 and the sheet metal forming processes described in this chapter.

7.4 Discuss the material and process variables that influence the shape of the curve of punch force vs. stroke for shearing, such as that shown in Fig. 7.7, including its height and width.

7.5 Describe your observations concerning Figs. 7.5 and 7.6.

7.6 Inspect a common paper punch, and comment on the shape of the tip of the punch as compared with those shown in Fig. 7.12.

7.7 Explain how you would go about estimating the temperature rise in the shear zone in a shearing operation.

7.8 As a practicing engineer in manufacturing, why would you be interested in the shape of the curve in Fig. 7.7? Explain.

7.9 Do you think the presence of burrs can be beneficial in certain applications? Give specific examples.

7.10 Explain the reasons that there are so many different types of die materials used for the processes described in this chapter.

7.11 Describe the differences between compound, progressive, and transfer dies.

7.12 It has been stated that the quality of the sheared edges can influence the formability of sheet metals. Explain why.

7.13 Explain why and how various factors influence springback in bending of sheet metals.

7.14 Does the hardness of a sheet metal have an effect on the metal's springback in bending? Explain.

7.15 We note in Fig. 7.16 that the state of stress shifts from plane stress to plane strain as the ratio of length of bend to sheet thickness increases. Explain why.

7.16 Describe the material properties that have an effect on the relative position of the curves shown in Fig. 7.19.

7.17 In Table 7.2, it can be noted that hard materials have higher R/t ratios than soft ones. Explain why.

7.18 Why do tubes buckle when bent? Experiment with a soda straw, and describe your observations.

7.19 Based on Fig. 7.22, sketch the shape of a *U*-die used to produce channel-shaped bends.

7.20 Explain why negative springback does not occur in air bending of sheet metals.

7.21 Give examples of products in which the presence of beads is beneficial or even necessary.

7.22 Assume that you are carrying out a sheet forming operation and you find that the material is not sufficiently ductile. Make suggestions to improve its ductility.

7.23 In deep drawing of a cylindrical cup, is it always necessary for there to be tensile circumferential stresses on the element in the cup wall? (See Fig. 7.50b.) Explain.

7.24 When comparing the hydroforming process with the deep-drawing process, it has been stated that deeper draws are possible in the former method. With appropriate sketches, explain why.

7.25 Note in Fig. 7.50a that element A in the flange is subjected to compressive circumferential (hoop) stresses. Using a simple free-body diagram, explain why.

7.26 From the topics covered in this chapter, list and explain specifically several examples where (a) friction is desirable and (b) friction is not desirable.

7.27 Explain why increasing the normal anisotropy, R, improves the deep drawability of sheet metals.

7.28 What is the reason for the negative sign in the numerator of Eq. (7.21)?

7.29 If you had a choice whereby you could control the state of strain in a sheet forming operation, would you rather work on the left or the right side of the forming-limit diagram? Explain.

7.30 What are the advantages of rubber forming? Which processes does it compete with?

7.31 Comment on the effect of lubrication of the punch on the limiting drawing ratio in deep drawing.

7.32 Why was hot stamping developed? What are the main materials used in hot stamping?

7.33 Comment on the size of the circles placed on the surfaces of sheet metals in determining the metals' formability. Are square grid patterns useful? Explain.

7.34 Make a list of the independent variables that influence the punch force in deep drawing of a cylindrical cup, and explain why and how the variables influence this force.

7.35 Explain why the simple tension line in the forming-limit diagram in Fig. 7.62a states that it is for $R = 1$, where R is the normal anisotropy of the sheet.

7.36 What are the reasons for developing forming-limit diagrams? Do you have any specific criticisms of such diagrams? Explain.

7.37 Explain the reasoning behind Eq. (7.20), for normal anisotropy, and Eq. (7.21), for planar anisotropy, respectively.

7.38 Describe why earing occurs. How would you avoid it?

7.39 In Section 7.7.1 it is stated that the thicker the sheet metal, the higher is the curve in the forming-limit diagram. Explain why.

7.40 Inspect the earing shown in Fig. 7.56, and estimate the direction in which the blank was cut.

7.41 Describe the factors that influence the size and length of beads in sheet metal forming operations.

7.42 It is known that the strength of metals depends on the metals' grain size. Would you then expect strength to influence the R value of sheet metals? Explain.

7.43 Equation (7.23) gives a general rule for dimensional relationships for successful drawing without a blankholder. Explain what happens if this limit is exceeded.

7.44 Explain why the three broken lines (simple tension, plane strain, and equal biaxial stretching) in Fig. 7.62a have those particular slopes.

7.45 Identify specific parts on a typical automobile, and explain which of the processes described in Chapters 6 and 7 can be used to make that part. Then choose only one method for each part, and explain your reasoning.

7.46 It has been seen that bendability and spinnability have a common aspect as far as properties of the workpiece material are concerned. Describe this common aspect.

7.47 Explain the reasons that such a wide variety of sheet forming processes has been developed and used over the years.

7.48 Make a summary of the types of defects found in sheet metal forming processes, and include brief comments on the reason(s) for each defects.

7.49 Which of the processes described in this chapter use only one die? What are the advantages of using only one die?

7.50 It has been suggested that deep drawability can be increased by (a) heating the flange and/or (b) chilling the punch by some suitable means. Comment on how these methods could improve drawability.

7.51 Offer designs whereby the suggestions given in Question 7.50 can be implemented. Would the required production rate affect your designs? Explain.

7.52 In the manufacture of automotive body panels from carbon-steel sheet, stretcher strains (Lüder's bands) are observed, which detrimentally affect surface finish. How can the stretcher strains be eliminated?

7.53 In order to improve its ductility, a coil of sheet metal is placed in a furnace and annealed. However, it is observed that the sheet has a lower limiting drawing ratio than it had before being annealed. Explain the reasons for this behavior.

7.54 What effects does friction have on a forming-limit diagram?

7.55 Why are lubricants generally used in sheet metal forming? Explain, using examples.

7.56 Through changes in clamping, a sheet metal forming operation can allow the material to undergo a negative minor strain. Explain how this effect can be advantageous.

7.57 How would you produce the part shown in Fig. 7.35 other than by tube hydroforming?

7.58 Give three examples of sheet metal parts that (a) can and (b) cannot be produced by incremental forming.

7.59 Due to preferred orientation (see Section 3.5), materials such as iron can have higher magnetism after cold rolling. Recognizing this feature, plot your estimate of LDR vs. degree of magnetism.

7.60 Explain why a material with a fine-grain microstructure is better suited for fine blanking than a coarse grained material.

7.61 If a cupping test were to be performed using a pressurized fluid instead of a spherical die, would you expect the forming limit diagram to change? Why or why not?

7.62 What are the similarities and differences between roll forming discussed in this chapter and shape rolling discussed in Chapter 6?

7.63 Explain how stringers can adversely affect bendability. Do they have similar effect on formability?

7.64 In Fig. 7.55, note that zinc has a high c/a ratio, whereas titanium has a low ratio. Does this have relevance to limited drawing ratio? Explain.

7.65 What is nesting? What is its significance?

7.66 Review Eqs. (7.12) through (7.14) and explain which of these expressions apply to incremental forming.

7.67 What metals do you expect would be best suited for electrically assisted sheet metal forming?

PROBLEMS

7.68 Referring to Eq. (7.5), it is stated that actual values of e_o are considerably higher than values of e_i, due to the shifting of the neutral axis during bending. With an appropriate sketch, explain this phenomenon.

7.69 Note in Eq. (7.11) that the bending force is a function of t^2. Why? (*Hint:* Consider bending-moment equations in mechanics of solids.)

7.70 Estimate the maximum bending force required for a 2-mm-thick and 250-mm-long Ti-6Al-4V titanium alloy, annealed and quenched at 25°C, in a V-die with a width of 150 mm.

7.71 Calculate the minimum tensile true fracture strain that a sheet metal should have in order to be bent to the following R/t ratios: (a) 0.25, (b) 1, and (c) 2. (See Table 7.2.)

7.72 In Example 7.3, calculate the work done by the force-distance method, i.e., work is the integral product of the vertical force, F, and the distance it moves.

7.73 What would be the answer to Example 7.3 if the tip of the force, F, were fixed to the strip by some means, thus maintaining the lateral position of the force? (*Hint:* Note that the left portion of the strip will now be strained more than the right portion.)

7.74 Calculate the magnitude of the force F in Example 7.3 for $\alpha = 30°$.

7.75 How would the force in Example 7.3 vary if the workpiece were made of a perfectly plastic material?

7.76 In Example 7.3, assume that the stretching is done by two equal forces F, each at 150 mm from the ends of the workpiece. (a) Calculate the magnitude of this force for $\alpha = 10°$. (b) If we want the stretching to be done up to $\alpha_{max} = 50°$ without necking, what should be the minimum value of n of the material?

7.77 Calculate the press force needed in punching 0.25-mm-thick 5052-H34 aluminum sheet in the shape of a square hole 30 mm on each side.

7.78 A straight bead is being formed on a 1-mm-thick aluminum sheet in a 20-mm-diameter die, as shown in the accompanying figure. (See also Fig. 7.25a.) Let $S_y = 90$ MPa. Considering springback, calculate the outside diameter of the bead after it is formed and unloaded from the die.

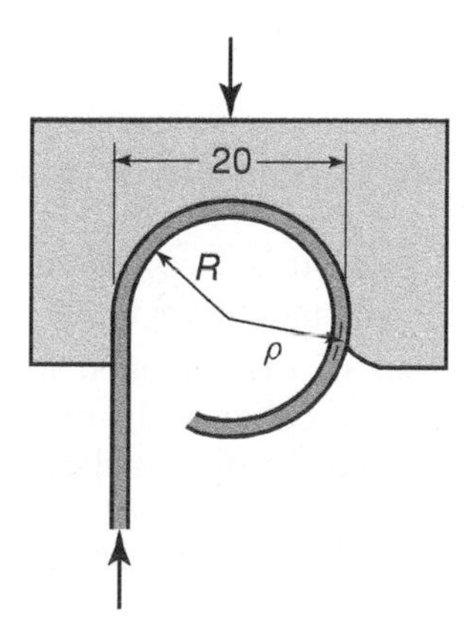

7.79 Calculate and plot the springback in bending 1-mm thick sheet metal around radii from 0.25 to 250 mm for (a) 303 stainless steel; (b) 1100-O aluminum; (c) HK31A magnesium; (d) Ti-6Al-4V.

7.80 Inspect Eq. (7.10), and, substituting in some numerical values, show whether the first term can be neglected without significant error in calculating springback.

7.81 In Example 7.4, calculate the amount of TNT required to develop a pressure of 50 MPa on the surface of the workpiece. Use a standoff of 0.25 m.

7.82 Estimate the limiting drawing ratio (LDR) for the materials listed in Table 7.3.

7.83 For the same material and thickness as in Problem 7.77, estimate the force required for deep drawing with a blank of diameter 250 mm and a punch of diameter 225 mm.

7.84 A cup is being drawn from a sheet metal that has a normal anisotropy of 2.5. Estimate the maximum ratio of cup height to cup diameter that can be drawn successfully in a single draw. Assume that the thickness of the sheet throughout the cup remains the same as the original blank thickness.

7.85 Obtain an expression for the curve shown in Fig. 7.55 in terms of the LDR and the average normal anisotropy, $\overline{R}$ (*Hint:* See Fig. 2.5b).

7.86 A steel sheet has R values of 1.0, 1.5, and 2.0 for the 0°, 45° and 90° directions to rolling, respectively. If a round blank is 100 mm in diameter, estimate the smallest cup diameter to which it can be drawn.

7.87 In Problem 7.86, explain whether ears will form and, if so, why.

7.88 A 1-mm-thick isotropic sheet metal is inscribed with a circle 5 mm in diameter. The sheet is then stretched uniaxially by 25%. Calculate (a) the final dimensions of the circle and (b) the thickness of the sheet at this location.

7.89 Obtain an aluminum beverage can, and cut it in half lengthwise with a pair of tin snips. Using a micrometer, measure the thickness of the bottom of the can and of the wall of the can. Estimate the thickness reductions in ironing and the diameter of the original blank.

7.90 Estimate the percent scrap in producing round blanks if the clearance between blanks is one tenth of

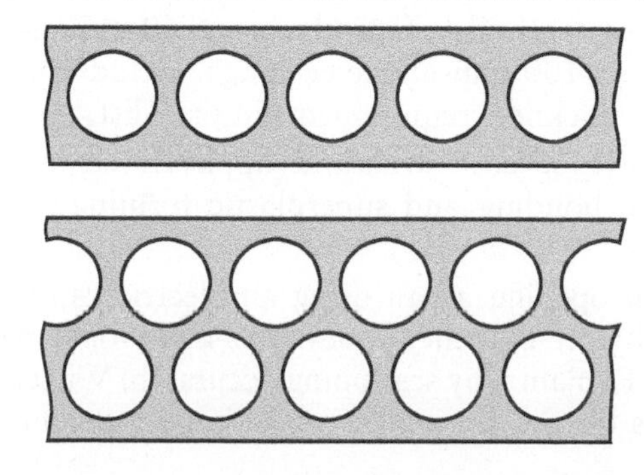

the radius of the blank. Consider single and multiple-row blanking, as sketched in the accompanying figure.

7.91 What is the force required to punch a square hole, 150 mm on each side, from a 0.5-mm-thick 5052-O aluminum sheet, using flat dies? What would be your answer if beveled dies were used instead?

7.92 Plot the final bend radius as a function of initial bend radius in bending for (a) 5052-O aluminum; (b) 5052-H34 Aluminum; (c) C24000 brass and (d) AISI 304 stainless steel.

7.93 Conduct a literature search and obtain the equation for a tractrix curve, as used in Fig. 7.60.

7.94 Derive Eq. (7.5).

7.95 The figure shows a parabolic profile that will define the mandrel shape in a spinning operation. Determine the equation of the parabolic surface. If a spun part will be produced from a 10 mm thick blank, determine the minimum required blank diameter. Assume that the diameter of the profile is 150 mm at a distance of 75 mm from the open end.

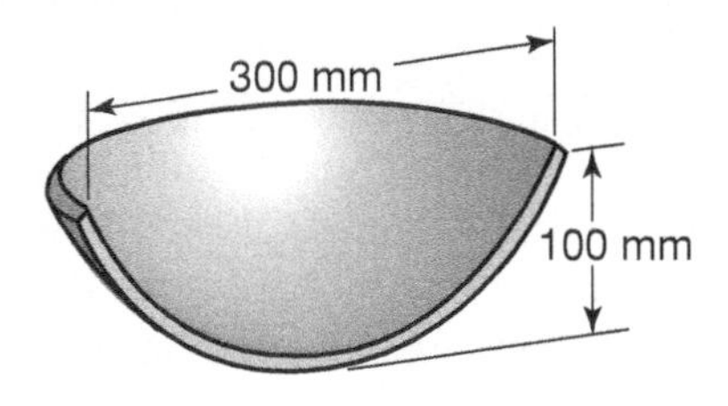

7.96 For the mandrel needed in Problem 7.95, plot the sheet metal thickness as a function of radius if the part is produced by shear spinning. Is this process feasible? Explain.

7.97 Estimate the maximum power in shear spinning a 10-mm thick annealed 304 stainless-steel plate that has a diameter of 300 mm on a conical mandrel of $\alpha = 30°$. The mandrel rotates at 100 rpm and the feed is $f = 1$ mm/rev.

7.98 Circular blanks of 5052-O aluminum, with a diameter of 25 mm and a thickness of 3 mm are to be mass-produced as the starting material for a tube for a paintball gun. The available press has an 800 kN capacity and can take a maximum of 300 mm wide strip. Material utilization improves if more rows are cut from a strip. (a) Determine the force required to blank a single slug. (b) Determine the maximum number of slugs that can be blanked simultaneously by the press. (c) Determine the material utilization if the space around a blanked part needs to be the same as the thickness, or 3 mm.

7.99 A deep-drawing operation will take place on aluminum-killed steel with an LDR of 2.4. A cylindrical cup will be produced with a diameter of 100 mm and a sheet thickness of 2 mm. Find (a) the largest permissible blank diameter; (b) the deepest cup that can be deep drawn; (c) the deepest cup that can be drawn if LDR=2.0. Comment on your results.

7.100 Assume that you are asked to give a quiz to students on the contents of this chapter. Prepare five quantitative problems and five qualitative questions, and supply the answers.

DESIGN

7.101 Consider several shapes to be blanked from a large sheet (such as oval, triangle, *L*-shape, and so forth) by laser-beam cutting, and sketch a nesting layout to minimize scrap.

7.102 List methods of producing a 40-mm-diameter cup with a 100-mm-diameter flange. Include materials and cup thickness requirements in your list.

7.103 Give several structural applications in which diffusion bonding and superplastic forming are used jointly.

7.104 In opening a can using an electric can opener, you will note that the lid develops a scalloped periphery. (a) Explain why scalloping occurs. (b) What design changes for the can opener would you recommend in order to minimize or eliminate, if possible, this scalloping effect? (c) Since lids typically are discarded or recycled, do you think it is necessary or worthwhile to make such design changes? Explain.

7.105 On the basis of experiments, it has been suggested that concrete, either plain or reinforced, can be a suitable material for dies in sheet metal forming operations. Describe your thoughts regarding this suggestion, considering die geometry and any other factors that may be relevant.

7.106 Lay out a roll forming line to produce any three cross sections from Fig. 7.27b.

7.107 Investigate methods for determining optimum shapes of blanks for deep-drawing operations. Sketch the optimally shaped blanks for drawing rectangular cups, and optimize their layout on a large sheet of metal.

7.108 The design shown in the accompanying figure is proposed for a metal tray, the main body of which is made from cold-rolled sheet steel. Noting its features and that the sheet is bent in two different directions, comment on various manufacturing considerations. Include factors such as anisotropy of the rolled sheet, the sheet's surface texture, the bend directions, the nature of the sheared edges, and the way the handle is snapped in for assembly.

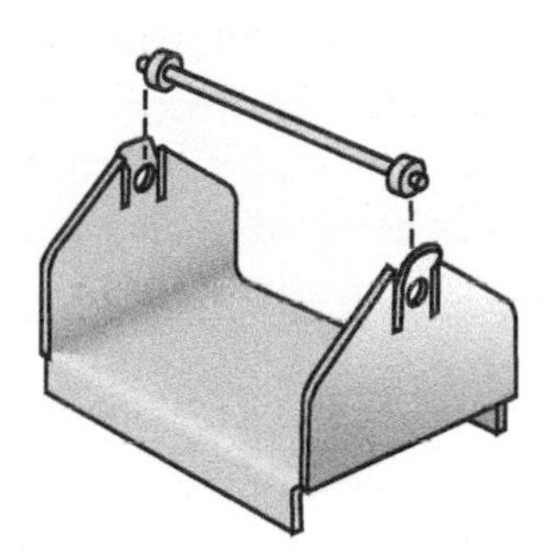

7.109 Metal cans are of either the two-piece variety (in which the bottom and sides are integral) or the three-piece variety (in which the sides, the bottom, and the top are each separate pieces). For a three-piece can, should the seam be (a) in the rolling direction; (b) normal to the rolling direction; or (c) oblique to the rolling direction? Explain your answer, using equations from solid mechanics.

7.110 Design a box that will contain a 100-mm × 150-mm × 75-mm volume. The box should be produced from two pieces of sheet metal and require no tools or fasteners for assembly.

7.111 Repeat Problem 7.110, but design the box from a single piece of sheet metal.

7.112 Sheet metals can now be provided a specially textured surface finish that presents pockets to aid lubricant entrainment. Perform a literature search on this technology, and prepare a short paper on the topic.

7.113 Obtain a few pieces of cardboard and carefully cut the profiles to produce bends as shown in Fig. 7.66. Demonstrate that the designs labeled as "best" are actually the best designs. Comment on the difference in strain states between the designs.

CHAPTER

8 Machining Processes

Machining is a general term describing a group of processes that remove material from a workpiece after it has been produced by the methods described in the preceding chapters. Machining processes are very versatile, and capable of producing almost any shape with good dimensional tolerances and surface finish. The topics covered in this chapter include:

- How chips are produced in machining.
- Force and power requirements.
- Mechanisms of tool wear and failure.
- Types and properties of tool materials.
- Characteristics of machine tools.
- Vibration and chatter in machining operations.
- Design for machining operations.
- Economics of machining.

Symbols

a shear zone thickness, m
A_s shear plane area, m^2
B_m overhead or burden cost
c specific heat, Nm/kg°C
C a constant
C_l loading and unloading cost
C_m machining cost
C_p cost per piece
C_s set up cost
C_t tooling cost
d depth of cut, m
D cutter or drill diameter, m also, diameter of cut, m
D_i depreciation of insert
f feed, m/rev also, feet per tooth, m
F friction force, N
F_c cutting force, N
F_n normal force on shear plane, N
F_r radial force, N
F_s shear force on shear plane, N
F_t thrust force, N
i inclination angle, deg
K strength coefficient, N/m^2
l workpiece length, m
l_c extent of first contact with cutter, m
L_m labor cost
n strain-hardening exponent also, number of inserts or teeth also, exponent in tool life equation

N Normal force, N
also, angular speed, rpm
N_f parts produced per face
N_i parts produced per insert
r cutting ratio
R resultant force, N
also, tool radius, m
R_t peak-to-valley roughness, m
t time or tool life, s
t_c time to change an insert, s
also, chip thickness, m
t_i time to index an insert, s
t_l part loading and unloading time, s
t_m machining time, s
t_o optimum tool life, s
also, depth of cut, m
t_p time to produce one part, s
also, tool life for maximum production, s
T temperature, °C
u_f specific frictional energy, W-s/m^3
u_t energy dissipated per volume, W-s/m^3
v workpiece velocity, m/s
V cutting speed, m/s
V_c chip velocity, m/s
V_o optimum cutting speed, m/s
V_s velocity of shear in shear plane, m/s
w width, m
α rake angle, deg
α_c chip flow angle, deg
α_e effective rake angle, deg
also, edge cutting angle, deg
α_n normal rake angle, deg
α_s side cutting angle, deg
β friction angle, deg
ϵ true strain
ϕ shear angle, deg
γ shear strain, m/m
$\dot{\gamma}$ shear strain rate, s^{-1}
κ thermal diffusivity, m^2/s
μ coefficient of friction
ρ density, kg/m^3
σ normal stress, N/m^2
σ_f flow stress, N/m^2
τ shear stress, N/m^2

Subscripts

avg average
o original, optimum

8.1 Introduction

Machining is the general term used to describe material removal processes and are usually divided into the following broad categories:

- **Cutting**, which generally involves single-point or multipoint cutting tools and processes, such as turning, boring, drilling, tapping, milling, broaching, and sawing. Cutting is the focus of this chapter.
- **Abrasive processes**, such as grinding, honing, lapping, and ultrasonic machining (Sections 9.6 through 9.9).
- **Advanced machining processes**, sometimes referred to as *non-traditional machining processes*, that use electrical, chemical, thermal, hydrodynamic, and optical sources of energy, as well as combinations of these sources, to remove material from the workpiece (Sections 9.10 through 9.15).

As described throughout this chapter, machining processes are preferable or even necessary for the following reasons:

1. Closer **dimensional accuracy** may be required than can be achieved by metalworking or casting processes alone. For example, in a crankshaft, the bearing surfaces cannot be produced with good dimensional accuracy and smooth enough surface finish through forging or sand casting alone.
2. Parts may require external and/or internal **geometric features**, such as sharp corners and internal threads, that cannot be produced by other processes.
3. Some parts are heat treated for improved hardness and wear resistance, but because heat-treated parts may undergo distortion and surface discoloration, they may require additional **finishing operations**.
4. Special **surface characteristics** or **textures** may be required on surfaces of the part that cannot be produced by other means. As an example, copper mirrors with very high reflectivity are typically made by machining with a diamond cutting tool.
5. It may be more **economical** to machine the part than to make it by other processes, particularly if the number of parts required is relatively small. Recall that metalworking processes typically require expensive dies and tooling; the cost of these can only be justified if the number of parts made is sufficiently high.

Against these major advantages, machining has certain limitations:

1. Machining processes inevitably **waste material** and generally require more energy and labor than other metalworking operations.
2. Removing a volume of material from a workpiece generally takes **more time** than other processes.
3. Material-removal processes can have **adverse effects** on the surface integrity of the product, including its fatigue life.

In spite of these limitations, machining operations continue to be indispensable in manufacturing.

As in all manufacturing operations, machining should be viewed as a **system**, consisting of the *workpiece, cutting tool, tool holder, workholding devices, machine tool, and operating personnel*. Machining operations cannot be carried out efficiently and economically without a fundamental knowledge of the often complex interactions among these critical factors, as will be evident throughout this chapter.

8.2 Mechanics of Chip Formation

Machining processes remove material from the surface of a workpiece by producing chips, as shown in Fig. 8.1. The basic mechanics of chip formation are represented by the model shown in Fig. 8.2. A cutting tool moves along the workpiece at a certain velocity (**cutting speed**), V, and a **depth of**

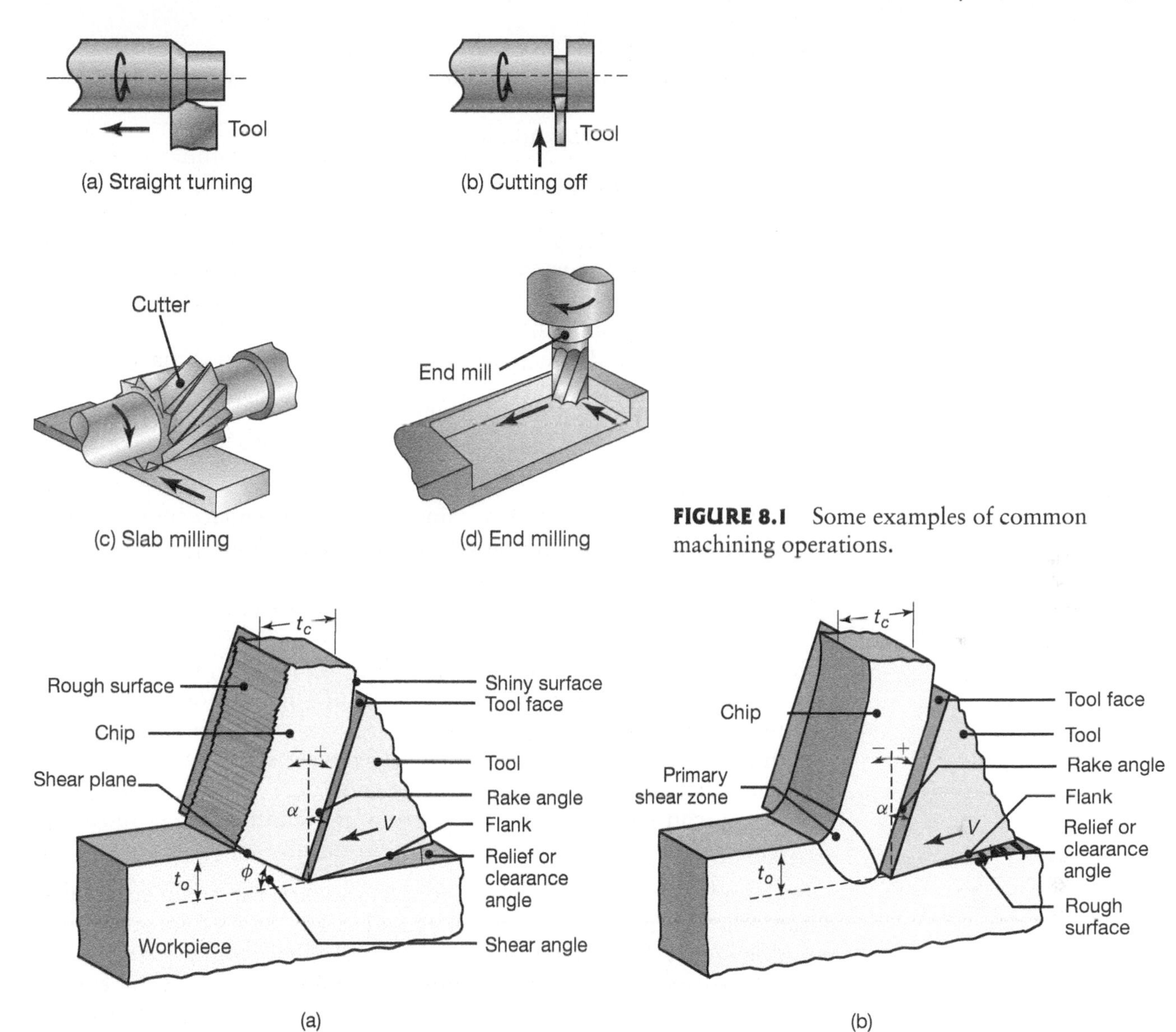

FIGURE 8.1 Some examples of common machining operations.

FIGURE 8.2 Schematic illustration of a two-dimensional cutting process, or orthogonal cutting. (a) Orthogonal cutting with a well-defined shear plane, also known as the Merchant model; and (b) Orthogonal cutting without a well-defined shear plane.

cut, t_o. A **chip** is produced just ahead of the tool by *shearing* the material continuously along the **shear plane.**

In this process, the major *independent variables* are:

- Type of cutting tool and its properties;
- The shape of the tool, its surface finish and sharpness;
- Workpiece material, its properties, and the temperature at which it is machined;
- Cutting conditions, such as speed, feed, and depth of cut;

- Type of cutting fluid, if used;
- Characteristics of the machine tool, particularly its stiffness and damping; and
- Tool holder and workholding devices.

The *dependent variables* are:

- Type of chip produced;
- Force required and energy dissipated in the cutting process;
- Temperature rise in the workpiece, the chip, and the cutting tool;
- Wear, chipping, and failure of the tool; and
- Surface finish and integrity of the workpiece after it is machined.

In order to appreciate the importance of the complex interrelationships among these variables, consider the following commonly encountered situations:

1. If the surface finish of the machined workpiece is unacceptable, which of the independent variables should be modified first?
2. If the workpiece becomes too hot, thus possibly affecting its properties and dimensional accuracy, what modifications should be made to the process parameters?
3. If the cutting tool wears rapidly and becomes dull, what should be changed: the cutting speed, the depth of cut, the tool material, or some other variable?
4. If the dimensional tolerance of the machined part is over the specified limits, what modification should be made?
5. If the cutting tool begins to vibrate and chatter, what should be changed to eliminate or reduce this problem?

Although almost all machining operations are three dimensional in nature, the two-dimensional model shown in Fig. 8.2 is appropriate and useful in studying the basic mechanics of the metal cutting process. This model is known as **orthogonal cutting**, meaning that the cutting edge of the tool is perpendicular (orthogonal) to the cutting direction. The tool has a **rake angle,** α, (positive as shown in the figure) and a **relief,** or **clearance, angle.** Note that the sum of the rake, the relief, and the included angles of the tool is 90°.

Microscopic examinations reveal that metal chips are produced by a **shearing** mechanism, shown in Fig. 8.3a. Shearing takes place along the **shear plane,** which makes an angle ϕ with the workpiece surface, called the *shear angle*. Below the shear plane, the workpiece is deformed elastically, and above the shear plane, the chip is already formed and is moving up the face of the tool as cutting progresses. Because of the relative movement, there is friction involved between the chip and the rake face of the tool.

Note that the thickness of the chip, t_c, can be determined if t_o, α, and ϕ are known. The ratio of t_o to t_c is known as the **cutting ratio,** r, which can be expressed as

$$r = \frac{t_o}{t_c} = \frac{\sin\phi}{\cos(\phi - \alpha)}. \tag{8.1}$$

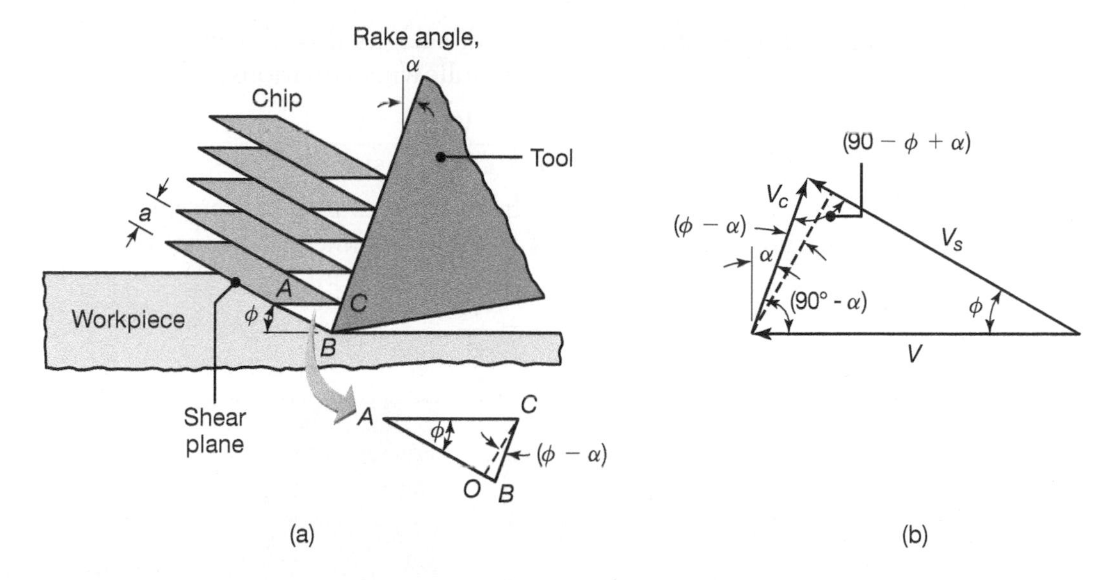

FIGURE 8.3 (a) Schematic illustration of the basic mechanism of chip formation in cutting. (b) Velocity diagram in the cutting zone.

Note that the chip thickness is always greater than the depth of cut (also known as the **undeformed chip thickness**), so that r is always less than unity. The reciprocal of r is known as the **chip compression ratio**, and is a measure of how thick the chip has become compared with the depth of cut; thus the chip compression ratio is always greater than unity.

On the basis of Fig. 8.3a, the **shear strain**, γ, that the material undergoes during cutting can be expressed as

$$\gamma = \frac{AB}{OC} = \frac{AO}{OC} + \frac{OB}{OC}, \tag{8.2}$$

or

$$\gamma = \cot\phi + \tan(\phi - \alpha). \tag{8.3}$$

Note from this equation that high shear strains are associated with low shear angles and low or negative rake angles. Shear strains of 5 or higher have been observed in actual cutting operations. Thus, the chip undergoes greater deformation during cutting than it does in other operations such as forging and shaping operations (Chapter 6), as can also be seen in Table 2.3.

From Fig. 8.2, it can be noted that the undeformed chip thickness and the depth of cut are the same parameter, t_o, in orthogonal cutting. Because the chip thickness, t_c, is greater than the undeformed chip thickness, t_o, the *velocity of the chip*, V_c, must be lower than the cutting speed, V. Since mass continuity has to be maintained,

$$Vt_o = V_c t_c \quad \text{or} \quad V_c = Vr, \tag{8.4}$$

and therefore,

$$V_c = V\frac{\sin\phi}{\cos(\phi - \alpha)}. \tag{8.5}$$

A velocity diagram can be constructed, such as that shown in Fig. 8.3b. From trigonometric relationships, the following equations can be written:

$$\frac{V}{\cos(\phi - \alpha)} = \frac{V_s}{\cos\alpha} = \frac{V_c}{\sin\alpha}, \tag{8.6}$$

where V_s is the velocity at which shearing takes place in the shear plane. The **shear-strain rate** is the ratio of V_s to the thickness, a, of the *shear zone* (Fig. 8.3a), or

$$\dot{\gamma} = \frac{V_s}{a}. \tag{8.7}$$

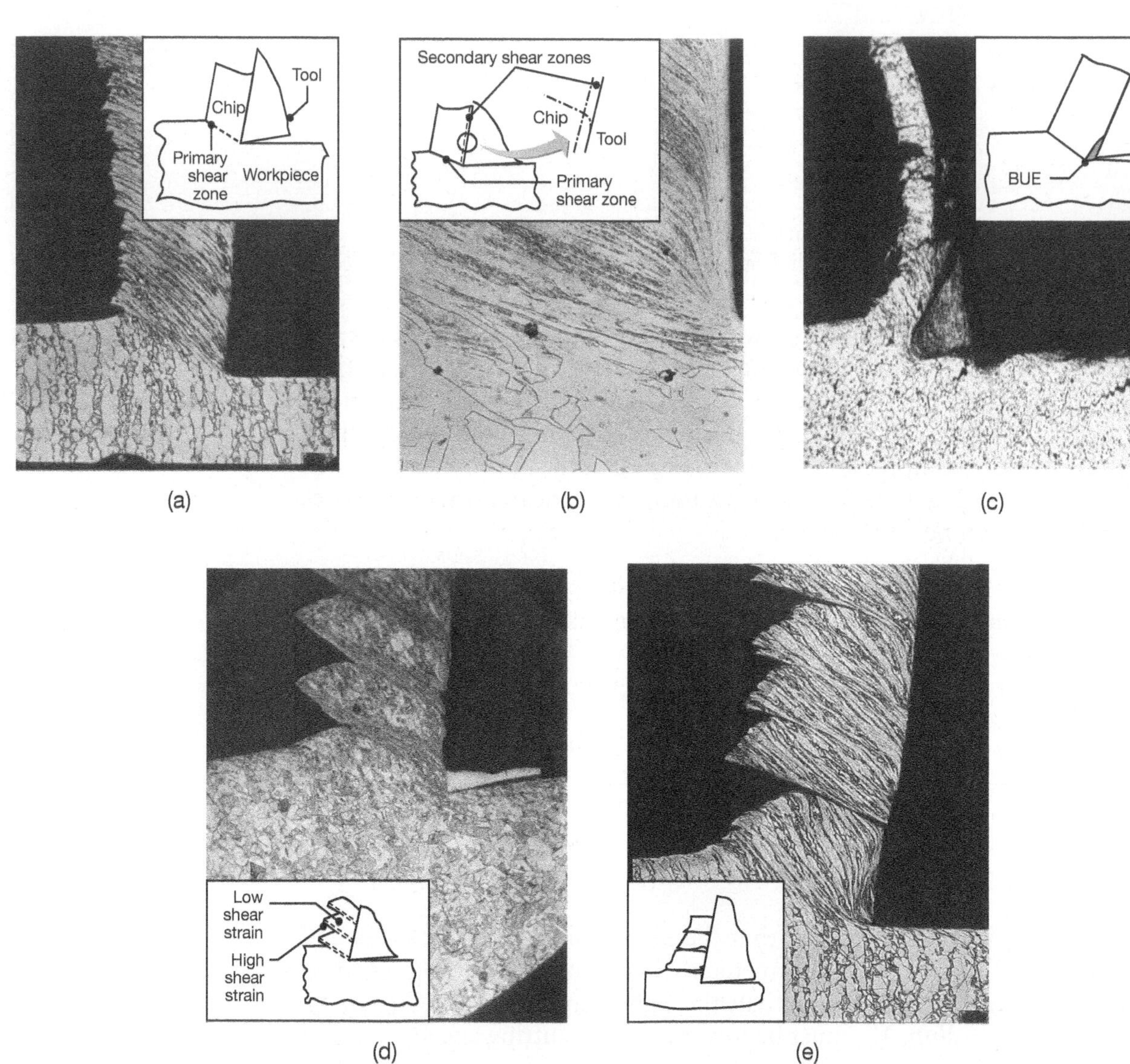

FIGURE 8.4 Basic types of chips produced in metal cutting and their micrographs: (a) continuous chip with narrow, straight primary shear zone; (b) secondary shear zone at the tool-chip interface; (c) continuous chip with built-up edge; (d) segmented or nonhomogeneous chip; and (e) discontinuous chip. *Source:* After M.C. Shaw, P.K. Wright, and S. Kalpakjian.

Experimental evidence indicates that a is typically on the order of 10^{-2} mm to 10^{-3} mm. This range indicates that, even at low cutting speeds, the shear-strain rate is very high, on the order of 10^3/s to 10^6/s. The shear-strain rate is important because of its effects on the strength and ductility of the material (see Section 2.2.7 and Table 2.3), as well as the type of the chip produced.

8.2.1 Chip Morphology

The type of chips produced (*chip morphology*) significantly influences surface finish and integrity and the overall machining operation. When metal chips, produced under different cutting conditions, are observed under a microscope, it will be found that there are significant deviations from the ideal model shown in Figs. 8.2 and 8.3a. Major types of metal chips commonly observed in practice are shown in Fig. 8.4.

Note that a chip has two surfaces: One surface has been in contact with the rake face of the tool, and the other surface is the newly-generated surface of the workpiece. The tool side of the chip surface is shiny or *burnished* (Fig. 8.5), which is caused by rubbing of the chip as it climbs up the tool face of the tool. The other surface of the chip has a jagged, step-like appearance (as can be seen in Fig. 8.4a), which is due to the shearing mechanism of chip formation (See also Figs. 3.5 and 3.7). Note that this surface has not come in contact with any solid body.

FIGURE 8.5 Shiny (burnished) surface on the tool side of a continuous chip produced in turning.

The basic types of metal chips produced in machining are described as follows:

1. **Continuous chips.** *Continuous chips* are typically formed at high cutting speeds and/or high rake angles (Fig. 8.4a). The deformation of the metal takes place along a very narrow shear zone, called the **primary shear zone**. These type of chips also may develop a **secondary shear zone** at the tool-chip interface (Fig. 8.4b), caused by friction; as expected, the secondary zone becomes thicker as the tool-chip friction increases.

   Deformation of continuous chips may also take place along a wide primary-shear zone, with *curved boundaries*, as shown in Fig. 8.2b. Note that the lower boundary of this zone is *below* the machined surface, and thus it has subjected the machined surface to distortion, possible surface damage and induced surface residual stresses. This situation occurs particularly in machining soft metals at low cutting speeds and low rake angles.

   Although they generally produce good surface finish, continuous chips are not always desirable, particularly in computer-controlled machine tools (see Section 8.11), because the chips tend to become tangled around the tool. This situation can be avoided with *chip breaker* features on cutting tools (see below).

   As a result of strain hardening (caused by the shear strain to which it is subjected), a chip generally becomes harder, stronger, and less ductile than the original workpiece material. As the rake angle decreases, the shear strain increases, as can be seen from Eq. (8.3).

2. **Built-up-edge chips.** A *built-up edge* (BUE) may form at the tip of the tool during cutting (Fig. 8.4c); it consists of thin layers of metal from the workpiece that are gradually deposited on the tool (hence the term *built-up*). As it grows larger, the BUE becomes unstable and eventually breaks up; the upper portion of the BUE is carried away on the tool side of the chip and the lower portion is deposited randomly on the machined surface. The process of BUE formation and breakup is repeated continuously during the cutting operation.

   The built-up edge is commonly observed in practice and is one of the significant factors that adversely affects surface finish and integrity in machining, as can be seen in Figs. 8.4 and 8.6. A built-up edge, in effect, changes the geometry of cutting. Note, for example, the large tip radius of the BUE and the rough surface finish it has produced. Because of work hardening and deposition of successive layers of material, BUE hardness increases significantly (Fig. 8.6a). Although BUE is generally undesirable, a thin but stable BUE is generally regarded as desirable, because it protects the tool surface.

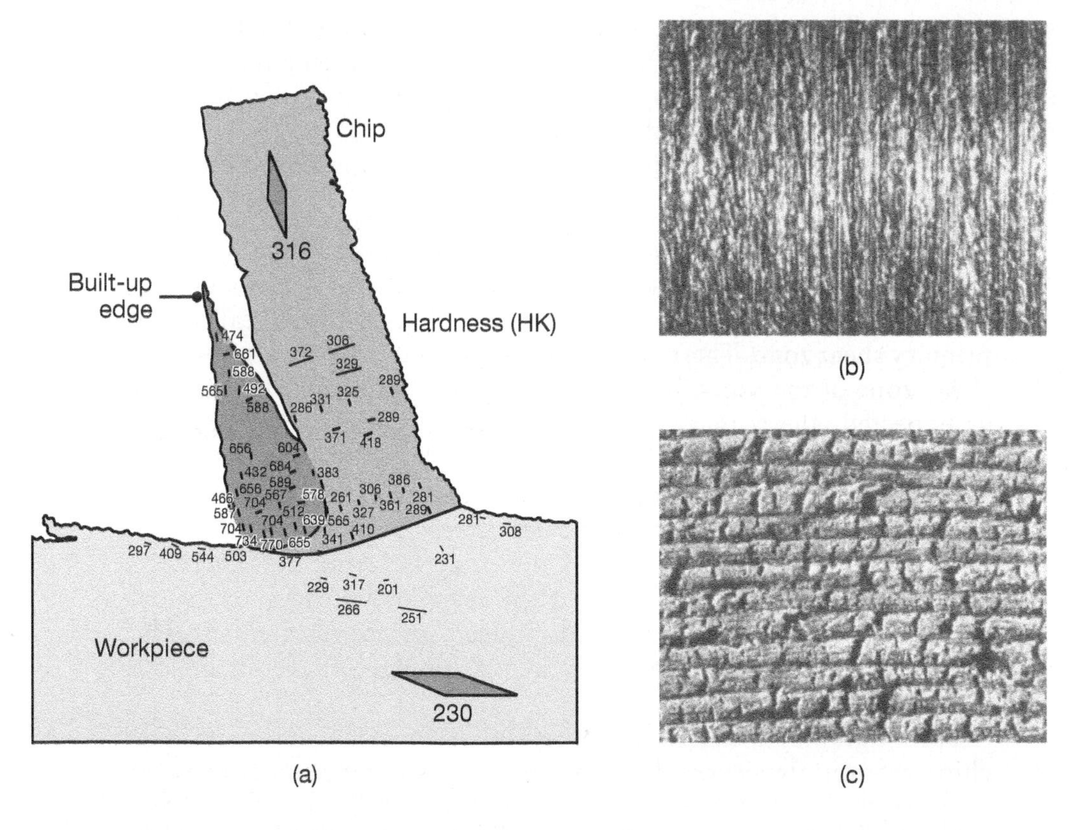

FIGURE 8.6 (a) Hardness distribution in the cutting zone for 3115 steel. Note that some regions in the built-up edge are as much as three times harder than the bulk workpiece. (b) Surface finish in turning 5130 steel with a built-up edge. (c) Surface finish on 1018 steel in face milling. Magnification: 15×. *Source:* Courtesy of Metcut Research Associates, Inc.

BUE formation is affected by (a) adhesion of the workpiece material to the rake face of the tool and the strength of the interfacial bond (see also Section 4.4). Ceramic cutting tools (see Section 8.6), for example, have much lower affinity to form BUE than do tool steels; (b) growth of the successive layers of adhered metal on the tool; (c) tendency of the workpiece material for strain hardening; the higher the strain-hardening exponent, n, the higher the probability for BUE formation.

Generally, the built-up edge decreases or is eliminated (a) as the cutting speed, V, increases; (b) as depth of cut, t_o, decreases; (c) as rake angle, α, increases; (d) as tip radius of the tool decreases; and (e) by using an effective cutting fluid (Section 8.7).

3. **Serrated chips.** *Serrated chips*, also called **segmented** or **nonhomogeneous** chips, are semi-continuous, with zones of low shear and high shear strain (Fig. 8.4d); the chips have the appearance of saw teeth, hence the term serrated. Metals with low thermal conductivity and strength that decreases sharply with temperature, such as titanium, exhibit this behavior.
4. **Discontinuous chips.** *Discontinuous chips* consist of segments that may be either firmly or loosely attached to each other (Fig. 8.4e). These chips usually develop under the following conditions:

   a. The workpiece material is brittle, and cannot undergo the high shear strains involved in cutting.
   b. The workpiece material contains hard inclusions and impurities (see Figs. 3.24 and 3.25) or has a structure such as that of graphite flakes in gray cast iron (Fig. 5.13a). Impurities and hard particles act as sites for internal cracks, thereby producing discontinuous chips. As the depth of cut increases, the probability of such defects being present in the cutting zone increases.
   c. The cutting speed is very low or very high.
   d. The depth of cut (undeformed chip thickness) is large or the rake angle is low.
   e. The machine tool has low stiffness and poor damping.
   f. Lack of an effective cutting fluid.

An additional factor in the formation of discontinuous chips is the magnitude of the compressive stresses on the shear plane [see also Eq. (8.17)]. Recall from Section 2.2.8 that the maximum shear strain at fracture increases with increasing compressive stress; thus, if the normal stress is not sufficiently high, the material will be unable to undergo the large shear strain necessary to form a continuous chip.

Because of the discontinuous nature of chip formation, cutting forces continually vary during machining. Consequently, the stiffness of the cutting-tool holder, the workpiece holding devices, and the machine tool are important factors in cutting with discontinuous or serrated chips. If not sufficiently stiff, the machine tool may begin to vibrate and chatter (see Section 8.12). Such vibration adversely affects the surface finish and dimensional accuracy of the machined component, and may even cause damage or excessive wear of the cutting tool and the machine tool itself.

Chip formation in machining nonmetallic materials. Much of the discussion thus far for metals also is generally applicable to nonmetallic materials. A variety of chips can develop in cutting thermoplastics, depending on the type of polymer (see Chapter 10) and the process parameters, such as depth of cut, cutting speed, and tool geometry. Thermosetting plastics and ceramics tend to produce discontinuous chips, because they generally are brittle (see also *ductile-regime cutting* in Section 8.9.2).

Chip curl. *Chip curl* (Figs. 8.5 and 8.7a) is commonly observed in all machining operations with metals and nonmetallic materials, such as thermoplastics and wood. Factors contributing to chip curl are: (a) the distribution of stresses in the primary and secondary shear zones; (b) thermal gradients in the cutting zone; (c) work-hardening characteristics of the workpiece material; and (d) the shape of the rake face of the tool. Process variables also affect chip curl; generally, the radius of curvature decreases (the chip becomes curlier) with decreasing depth of cut, increasing rake angle, and decreasing friction at the tool-chip interface. The use of cutting fluids and various additives in the workpiece material also influence chip curl.

Chip breakers. Long continuous chips are generally undesirable, because they tend to become entangled and interfere with the machining operation and can become a safety hazard to the operator; this situation is especially troublesome in high-speed automated machinery. The common procedure

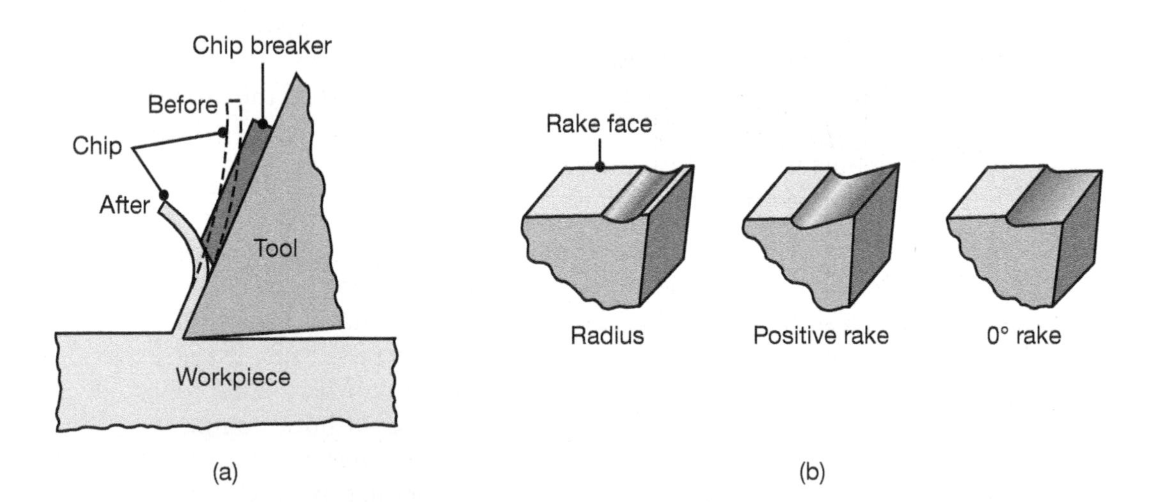

FIGURE 8.7 (a) Schematic illustration of the action of a chip breaker. Note that the chip breaker decreases the radius of curvature of the chip. (b) Grooves on the rake face of cutting tools, acting as chip breakers. Cutting tools inserts generally incorporate built-in chip-breaker features.

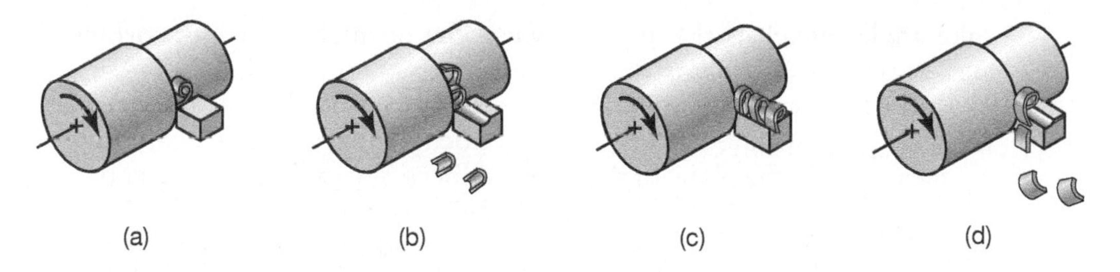

FIGURE 8.8 Various chips produced in turning: (a) tightly curled chip; (b) chip hits workpiece and breaks; (c) continuous chip moving radially outward from workpiece; and (d) chip hits tool shank and breaks off.

to avoid the formation of continuous chips is to break the chip intermittently with a *chip breaker*. Chip breakers are now an integral part of the cutting tool itself (Fig. 8.7). Chips can also be broken by modifying the tool geometry, thus controlling chip flow, as in the turning operations illustrated in Fig. 8.8.

8.2.2 Mechanics of Oblique Cutting

Unlike the two-dimensional cutting described thus far, the majority of machining operations involve tool shapes that are three-dimensional (**oblique**). The basic difference between two-dimensional and oblique cutting are illustrated in Fig. 8.9a. Recall that in orthogonal cutting, the tool cutting edge is perpendicular to the movement of the tool, and that the chip slides straight up the rake face of the tool. In oblique cutting, the cutting edge is at an angle i, called the **inclination angle** (Fig. 8.9b). The chip in Fig. 8.9a flows up the rake face of the tool at an angle α_c (the **chip flow angle**), measured in the plane of the tool face. The angle α_n is the **normal**

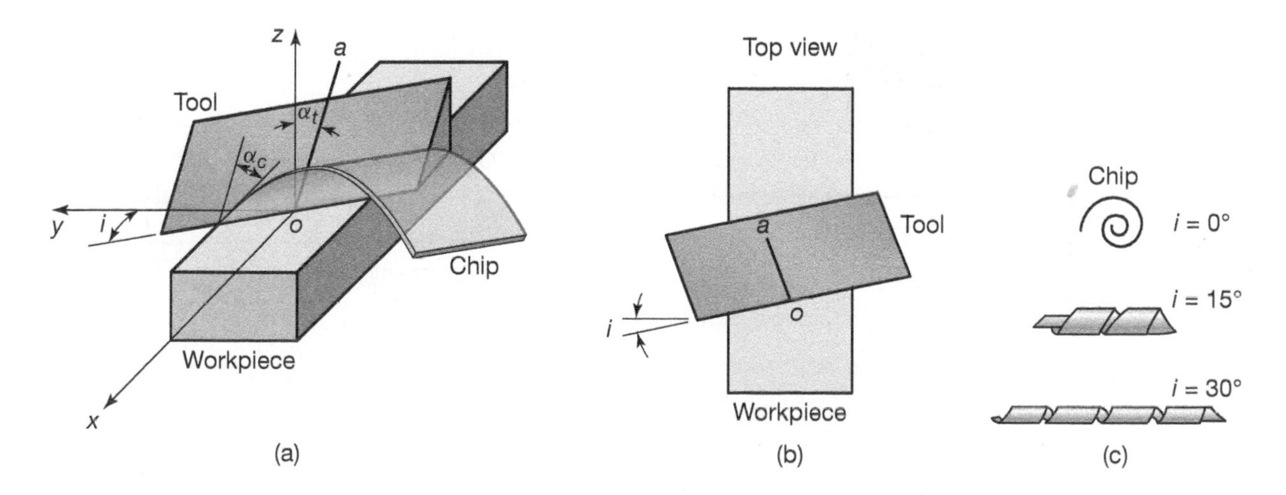

FIGURE 8.9 (a) Schematic illustration of cutting with an oblique tool. (b) Top view, showing the inclination angle, *i*. (c) Types of chips produced with different inclination angles.

rake angle, which is the angle between the normal oz to the workpiece surface and the line oa on the tool face.

As the workpiece material approaches the tool at a velocity V, it leaves the workpiece surface as a chip with a velocity V_c. The effective rake angle, α_e, measured in the plane of these two velocities, can then be calculated as follows: Assuming that the chip flow angle, α_c, is equal to the inclination angle, i (which has been experimentally verified), the effective rake angle, α_e, is

$$\alpha_e = \sin^{-1}\left(\sin^2 i + \cos^2 i \sin\alpha_n\right). \quad (8.8)$$

Since both i and α_n can be measured directly, this equation can be used to calculate the magnitude of the effective rake angle. Note that as i increases, the effective rake angle increases, and thus the chip becomes thinner and longer. The effect of the inclination angle on chip shape is illustrated in Fig. 8.9c.

Figure 8.10 shows a typical single-point turning tool; note the various angles, each of which must be selected properly for efficient machining. Various three-dimensional cutting tools are described in greater detail in Sections 8.8 and 8.9, including those for such operations as drilling, tapping, milling, planing, shaping, broaching, sawing, and filing.

Shaving and skiving. Thin layers of material can be removed from straight or curved surfaces by a process similar to using a *plane* to shave wood. *Shaving* is particularly useful in improving the surface finish and dimensional accuracy of punched slugs or holes (See Fig. 7.11). Parts that are long or have a combination of angles and shapes are shaved by *skiving*, using a specially shaped cutting tool.

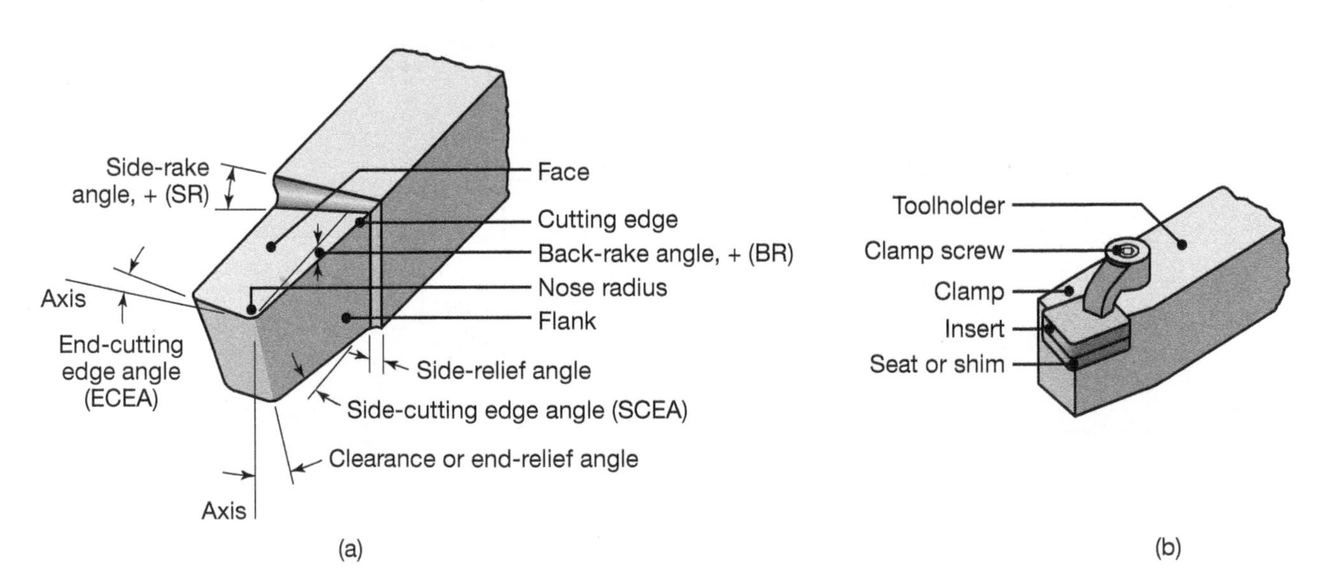

FIGURE 8.10 (a) Schematic illustration of a right-hand cutting tool for turning. Although these tools have traditionally been produced from solid tool-steel bars, they are now replaced by inserts of carbide or other tool materials of various shapes and sizes, as shown in (b).

8.2.3 Forces in Orthogonal Cutting

Determining cutting forces and power requirements in machining operations is essential for the following reasons:

1. **Power requirements** must be known so that a machine tool of suitable capacity can be selected for a particular application;
2. Data on cutting forces are necessary for the proper **design of machine tools** so that they have certain specific characteristics, including stiffness, in order to maintain the desired dimensional accuracy; and
3. The workpiece must be able to withstand the cutting forces without excessive **distortion.**

The factors that significantly influence the forces and power in orthogonal cutting are:

1. **Cutting forces.** The forces acting on the tool in orthogonal cutting are shown in Fig. 8.11. The **cutting force,** F_c, acts in the direction of the cutting speed, V, and supplies the energy required for the machining operation. The **thrust force,** F_t, acts in the direction normal to the cutting velocity, that is, perpendicular to the workpiece. These two forces produce the **resultant force,** R, which can then be resolved into two components on the tool face: a **friction force,** F, along the tool-chip interface and a **normal force,** N, perpendicular to the interface. From Fig. 8.11, it can be shown that the friction force is

$$F = R \sin \beta, \quad (8.9)$$

and the normal force is

$$N = R \cos \beta. \quad (8.10)$$

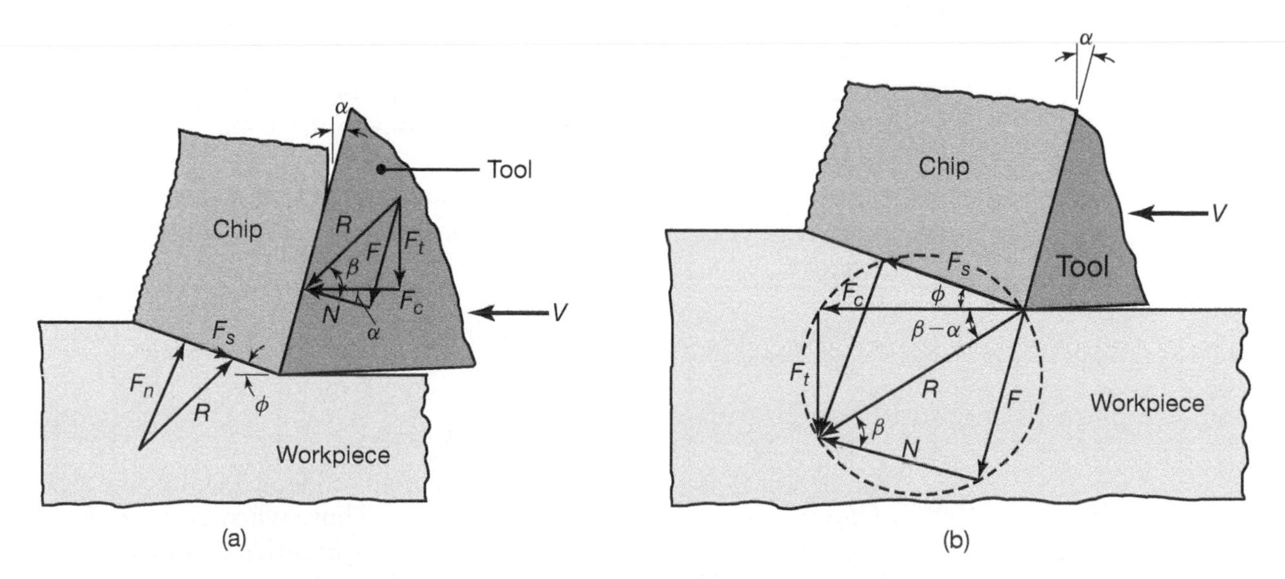

FIGURE 8.11 (a) Forces acting on a cutting tool in two-dimensional cutting. Note that the resultant forces, R, must be collinear to balance the forces. (b) Force circle to determine various forces acting in the cutting zone.

Note also that the resultant force is balanced by an equal and opposite force on the shear plane, and is resolved into a **shear force**, F_s, and a **normal force**, F_n. From Fig. 8.11, the cutting force can be shown to be

$$F_c = R\cos(\beta - \alpha) = \frac{wt_o\tau\cos(\beta - \alpha)}{\sin\phi\cos(\phi + \beta - \alpha)}, \tag{8.11}$$

where τ is the *average shear stress* along the shear plane.

The ratio of F to N is the **coefficient of friction,** μ, at the tool-chip interface (see also Section 4.4.1), and the angle β is known as the **friction angle.** The coefficient of friction can be expressed as

$$\mu = \tan\beta = \frac{F_t + F_c\tan\alpha}{F_c - F_t\tan\alpha}. \tag{8.12}$$

In cutting metals, μ generally ranges from about 0.5 to 2, indicating that the chip encounters considerable frictional resistance in climbing up the rake face of the tool.

The forces in machining operations are generally found to be on the order of a few hundred or thousand newtons. However, the local stresses in the cutting zone and the normal stresses on the rake face of the tool are very high, because the contact areas are very small. The tool-chip contact length (Fig. 8.2), for example, is typically on the order of 1 mm, so that the tool is subjected to very high local stresses.

2. **Thrust force and its direction.** Although the thrust force does not contribute to the energy required in cutting, its magnitude is important because the tool holder, the workholding devices, and the machine tool must be sufficiently stiff to minimize deflections caused by this force. For example, if the thrust force (See Fig. 8.11) is too high and the machine tool is not sufficiently stiff, the tool will be pushed away from the workpiece surface. This deflection will, in turn, reduce the actual depth of cut, leading to loss of dimensional accuracy of the machined part and possibly to vibration and chatter (see Section 8.12).

   Note from Fig. 8.11 that the direction of the thrust force is *downward*. It can be shown, however, that this force can also be upward (negative), by first observing that

$$F_t = R\sin(\beta - \alpha) \tag{8.13}$$

or

$$F_t = F_c\tan(\beta - \alpha). \tag{8.14}$$

The sign of F_c is always positive (as shown in Fig. 8.11), but the sign of F_t can be either positive or negative. Thus, when $\beta > \alpha$, F_t is positive (downward), and for $\beta < \alpha$, it is negative (upward). It is therefore possible to have an upward thrust force when friction at the tool-chip interface is low and/or when the rake angle is high.

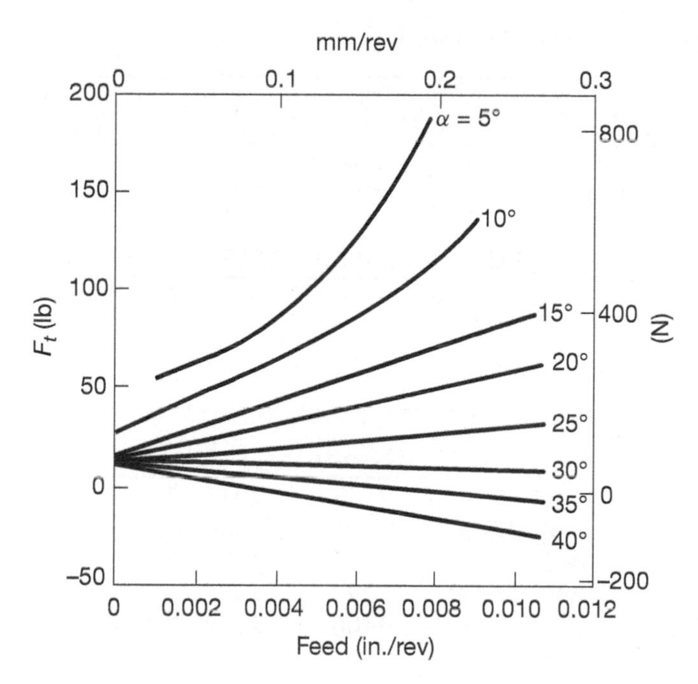

FIGURE 8.12 Thrust force as a function of rake angle and feed in orthogonal cutting of AISI 1112 cold-rolled steel. Note that at high rake angles, the thrust force is negative. A negative thrust force has important implications in the design of machine tools and in controlling the stability of the cutting process.

This situation can be seen in Fig. 8.11 and noting that when $\mu = 0$, $\beta = 0$, so that the resultant force, R, coincides with the normal force, N; thus, F_r will have a thrust-force component that is upward. Also note that for the condition of $\alpha = 0$ and $\beta = 0$, the thrust force will be zero. These observations have been verified experimentally (See Fig. 8.12).

The depth of cut has an important influence; as t_o increases, R must also increase and thus, F_c will increase as well. Note also that additional energy is required in order to remove the extra material associated with the increased depth of cut. The change in the direction and magnitude of the thrust force can play a significant role. It can, for example, lead to *instability* in machining operations, particularly if the machine tool is not sufficiently stiff.

3. **Observations on cutting forces.** In addition to being a function of the strength of the workpiece material, cutting forces are influenced by other variables. Data such as those given in Tables 8.1 and 8.2 indicate that the cutting force increases with increasing depth of cut, decreasing rake angle, and decreasing cutting speed. By reviewing the data given in Table 8.2, the effect of cutting speed can be attributed to the fact that as speed decreases, the shear angle decreases, and the coefficient of friction increases.

   The tip radius of the tool also is an important factor: the larger the radius (hence the duller the tool), the higher the cutting force. Experimental evidence has indicated that, for depths of cut on the order of five times the tip radius or higher, the effect of tool dullness on the cutting forces becomes negligible (see also Section 8.4).

TABLE 8.1 Data on orthogonal cutting of 4130 steel.

α	ϕ	γ	μ	β	F_c (N)	F_t (N)	u_t (MN-m/m^3)	u_s	u_f	u_f/u_t (%)
25°	20.9°	2.55	1.46	56°	1690	996	2206	1441	765	35
35	31.6	1.56	1.53	57°	1130	454	1475	772	703	48
40	35.7	1.32	1.54	57°	1030	316	1344	650	696	52
45	41.9	1.06	1.83	62°	1030	302	1344	517	827	62

$t_o = 0.0625$ mm; $w = 12$ mm; $V = 0.457$ m/s; tool: high-speed steel. *Source:* After E.G. Thomsen.

TABLE 8.2 Data on orthogonal cutting of 9445 steel.

α	V (m/s)	ϕ	γ	μ	β	F_c (N)	F_t (N)	u_t (MN-m/m^3)	u_s	u_f	u_f/u_t (%)
+10°	1.0	17°	3.4	1.05	46°	1646	1214	2758	2013	745	27
	2.0	19°	3.1	1.11	48°	1600	1260	2690	1834	856	32
	3.25	21.5°	2.7	0.95	44°	1463	965	2455	1717	738	30
	6.0	25°	2.4	0.81	39°	1348	747	2260	1551	709	31
−10°	2.0	16.5°	3.9	0.64	33°	1850	1712	3100	2358	742	24
	3.25	19°	3.5	0.58	30°	1708	1450	2861	2150	711	25
	5.9	22°	3.1	0.51	27°	1583	1170	2655	1993	662	25

$t_o = 0.94$ mm; $w = 6.35$ mm; tool: cemented carbide. *Source:* After M.E. Merchant.

4. **Shear and normal stresses in the cutting zone.** The stresses along the shear plane and at the tool-chip interface can be analyzed by first assuming that they are uniformly distributed. The forces in the shear plane can then be resolved into shear and normal forces and stresses. Note that the area, A_s, of the shear plane is

$$A_s = \frac{wt_o}{\sin\phi}, \tag{8.15}$$

and therefore, the *average shear stress* in the shear plane is

$$\tau = \frac{F_s}{A_s} = \frac{F_s \sin\phi}{wt_o}, \tag{8.16}$$

and the *average normal stress* is

$$\sigma = \frac{F_n}{A_s} = \frac{F_n \sin\phi}{wt_o}. \tag{8.17}$$

Some data on average stresses are given in Fig. 8.13, where the following conclusions can be drawn:

a. The shear stress along the shear plane is independent of the rake angle.
b. The normal stress on the shear plane decreases with increasing rake angle.

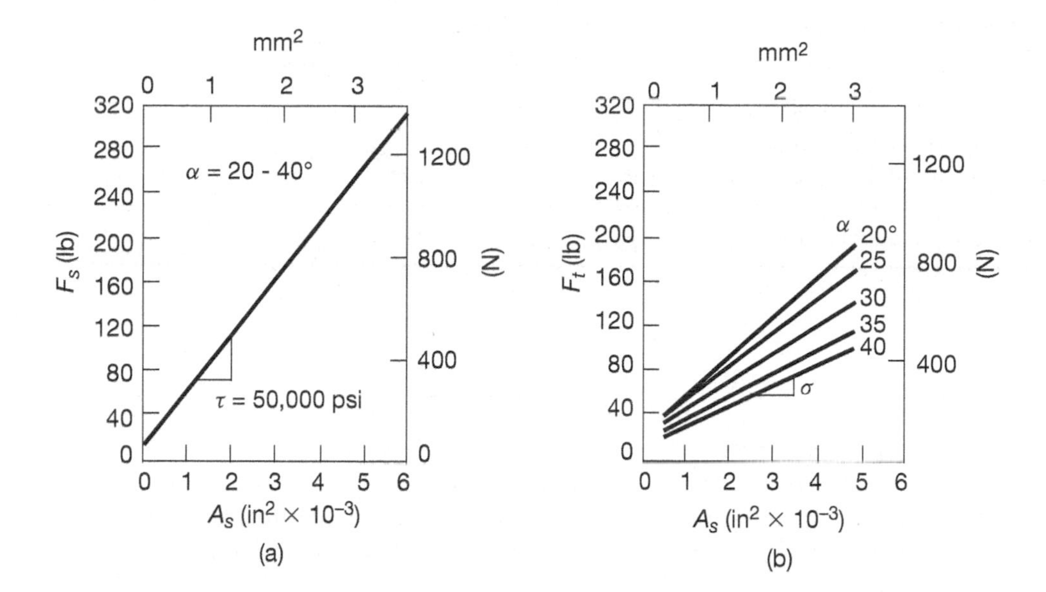

FIGURE 8.13 (a) Shear force and (b) normal force as a function of the area of the shear plane and the rake angle for 85–15 brass. Note that the shear stress in the shear plane is constant, regardless of the magnitude of the normal stress, indicating that the normal stress has no effect on the shear flow stress of the material. *Source:* After S. Kobayashi and E.G. Thomsen.

c. The normal stress in the shear plane has no effect on the magnitude of the shear stress. However, normal stress strongly influences the allowable shear strain in the shear zone prior to fracture. Recall from Section 2.2.8 that the maximum shear strain to fracture increases with the normal compressive stress. For this reason, small or negative rake angles will often be used in machining less ductile materials, in order to promote shearing without fracture.

Determining the stresses on the rake face of the tool presents several difficulties: (a) Accurately determining the length of contact at the tool-chip interface is challenging (Fig. 8.2); the length has been found to increase with decreasing shear angle, indicating that the contact length is a function of rake angle, cutting speed, and friction at the tool-chip interface. (b) The stresses are not uniformly distributed along the rake face; *photoelastic studies* have shown that the actual stress distribution is as shown in Fig. 8.14. Note that the normal stress on the rake face is a maximum at the tool tip, and it decreases rapidly toward the end of the contact length. The shear stress indicates a similar trend, except that it levels off about half-way along the tool-chip contact length. This behavior indicates that *sticking* is taking place (see Section 4.4.1), whereby the shear stress has reached the shear-yield strength of the workpiece material.

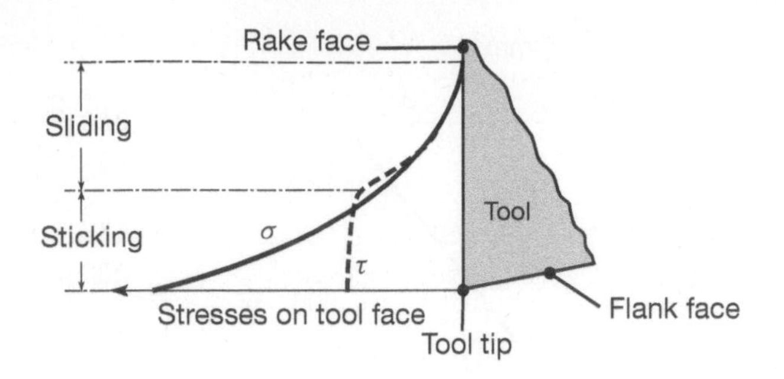

FIGURE 8.14 Schematic illustration of the distribution of normal and shear stresses at the tool-chip interface (rake face). Note that, whereas the normal stress increases continuously toward the tip of the tool, the shear stress reaches a maximum and remains at that value (a phenomenon known as *sticking*; see Section 4.4.1).

5. **Measuring cutting forces.** Cutting forces are typically measured by (a) using **force transducers**, such as piezoelectric crystals; (b) using **force dynamometers**; (c) with resistance-wire strain gages mounted on the tool holder or the workholding device on the machine tool; and (d) calculating it from the power measured during machining, provided that the mechanical efficiency of the machine tool can be determined.

8.2.4 Shear-Angle Relationships

The shear angle and the shape of the shear zone influence the mechanics of cutting; consequently, much effort has been expended in correlating the shear angle to material properties and process variables. One approach is based on the assumption that (a) the shear angle adjusts itself so that the cutting force is a minimum or (b) that the maximum shear stress occurs in the shear plane. From the force diagram given in Fig. 8.11, the shear stress in the shear plane can be expressed as

$$\tau = \frac{F_s}{A_s} = \frac{F_c \sec(\beta - \alpha)\cos(\phi + \beta - \alpha)\sin\phi}{wt_o}. \qquad (8.18)$$

Assuming that β is independent of ϕ, the shear angle corresponding to the maximum shear stress can be determined by differentiating Eq. (8.18) with respect to ϕ and equating it to zero. Thus,

$$\frac{d\tau}{d\phi} = \cos(\phi + \beta - \alpha)\cos\phi - \sin(\phi + \beta - \alpha)\sin\phi = 0, \qquad (8.19)$$

and

$$\tan(\phi + \beta - \alpha) = \cot\phi = \tan\left(90^\circ - \phi\right),$$

or

$$\phi = 45^\circ + \frac{\alpha}{2} - \frac{\beta}{2}. \qquad (8.20)$$

Note from Eq. (8.20) that as the rake angle decreases and/or as the friction at the tool-chip interface increases, the shear angle decreases, and the chip becomes thicker. This is to be expected, because decreasing α and/or increasing β tend to cause greater resistance to the chip as it moves up the rake face of the tool. This makes the chip thicker, indicating that the shear angle is decreasing.

Another method of determining ϕ is based on a slip-line analysis, which gives an expression for the shear angle as

$$\phi = 45^\circ + \alpha - \beta. \tag{8.21}$$

Note that this expression is similar to Eq. (8.20) and indicates the same trends. In another study the following simple relationship has been given:

$$\phi = \alpha \quad \text{for} \quad \alpha > 15^\circ, \tag{8.22}$$

$$\phi = 15^\circ \quad \text{for} \quad \alpha < 15^\circ. \tag{8.23}$$

Several other expressions, based on various models and with different assumptions, have been developed over the years. Equations (8.21) through (8.23) do not agree well with experimental data over a wide range of conditions (Fig. 8.15a). However, the shear angle always decreases with increasing $\beta - \alpha$, as shown in Fig. 8.15a.

8.2.5 Specific Energy

From Fig. 8.11 it can be seen that the **total power** in cutting is

$$\text{Power} = F_c V.$$

The **specific energy**, or *total energy per unit volume of material removed*, u_t, is then

$$u_t = \frac{F_c V}{w t_o V} = \frac{F_c}{w t_o}, \tag{8.24}$$

where w is the width of the cut. Note that u_t is the ratio of the cutting force to the projected area of the cut. From Figs. 8.3 and 8.11 it can be seen that

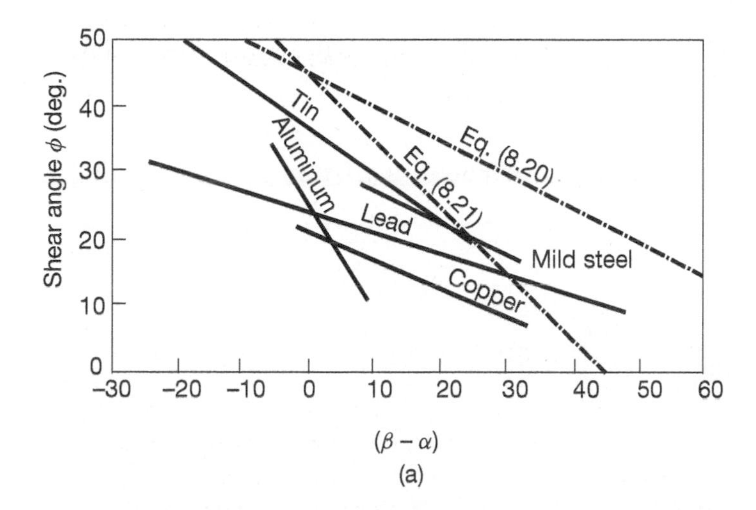

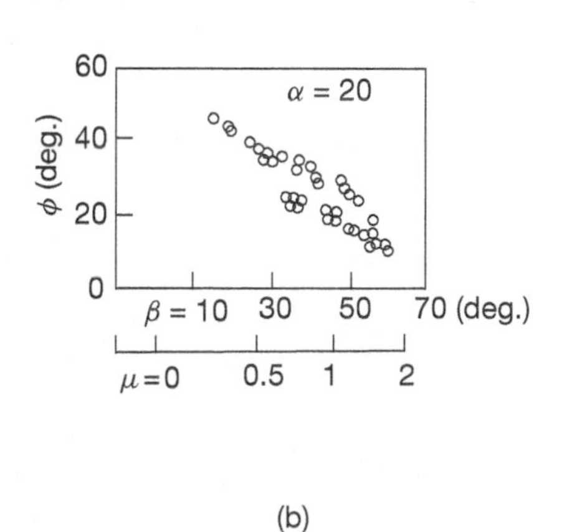

FIGURE 8.15 (a) Comparison of experimental and theoretical shear angle relationships. (b) Relation between the shear angle and the friction angle for various alloys and cutting speeds. *Source:* After S. Kobayashi.

the power required to overcome friction at the tool-chip interface is the product of F and V_c, or, in terms of **specific energy for friction,** u_f, as

$$u_f = \frac{FV_c}{wt_oV} = \frac{Fr}{wt_o} = \frac{(F_c \sin\alpha + F_t \cos\alpha)\, r}{wt_o}. \tag{8.25}$$

Likewise, the power dissipated along the shear plane is the product of F_s and V_s. Thus, the **specific energy for shearing,** u_s, is

$$u_s = \frac{F_s V_s}{wt_o V}. \tag{8.26}$$

The **total specific energy,** u_t, is the sum of the two energies

$$u_t = u_f + u_s. \tag{8.27}$$

Tables 8.1 and 8.2 give some experimental data on specific energies, where it can be noted that as the rake angle increases, the frictional specific energy remains relatively constant, whereas the shear specific energy rapidly decreases. Thus, the ratio u_f/u_t increases significantly as α increases. This trend also can be predicted by developing an expression for the energy ratio, as follows:

$$\frac{u_f}{u_t} = \frac{FV_c}{F_cV} = \frac{R\sin\beta}{R\cos(\beta-\alpha)} \cdot \frac{Vr}{V} = \frac{\sin\beta}{\cos(\beta-\alpha)} \cdot \frac{\sin\phi}{\cos(\phi-\alpha)}. \tag{8.28}$$

Experimental observations have indicated that as α increases, both β and ϕ increase, and inspection of Eq. (8.28) indicates that the ratio u_f/u_t should increase with α.

EXAMPLE 8.1 Relative Energies in Cutting

Given: An orthogonal cutting operation is being carried out in which $t_o = 0.1$ mm, $V = 2$ m/s, $\alpha = 10°$, and the width of cut = 5 mm. It is observed that $t_c = 0.25$ mm, $F_c = 500$ N, and $F_t = 200$ N.

Find: Calculate the percentage of the total energy that is dissipated in friction at the tool-chip interface.

Solution: The percentage of energy can be expressed as

$$\frac{\text{Friction energy}}{\text{Total energy}} = \frac{FV_c}{F_cV} = \frac{Fr}{F_c},$$

where

$$r = \frac{t_o}{t_c} = \frac{0.1}{0.25} = 0.40,$$

$$F = R\sin\beta$$

$$F_c = R\cos(\beta-\alpha)$$

and

$$R = \sqrt{F_t^2 + F_c^2} = \sqrt{200^2 + 500^2} = 538 \text{ N}.$$

Thus,

$$500 = 538 \cos(\beta - 10^\circ),$$

from which we find that

$$\beta = 31.7^\circ \qquad \text{and} \qquad F = 538 \sin 31.7^\circ = 283 \text{ N}.$$

Therefore, the percentage of friction energy is calculated as

$$\text{Percentage} = \frac{(283)(0.40)}{500} = 0.22 = 22\%$$

and similarly, the percentage of shear energy is calculated as 78%.

EXAMPLE 8.2 Comparison of Forming and Machining Energies

Given: Two cylinders of annealed 304 stainless-steel, each with a diameter of 10 mm and a length of 150 mm, are to have their diameters reduced to 9 mm (a) for one piece by *pulling* it in tension and (b) for the other by *machining* it on a lathe (See Fig. 8.8) in one pass.

Find: Calculate the respective amounts of work involved, and explain the reasons for the difference in the energies dissipated.

Solution:

a. The work done in pulling the rod is (see Section 2.12):

$$W_{\text{tension}} = (u)(\text{Volume}),$$

where, from Eq. (2.56),

$$u = \int_0^{\epsilon_1} \sigma \, d\epsilon.$$

The true strain is found from Eq. (2.10) as

$$\epsilon_1 = \ln\left(\frac{D_o}{D}\right)^2 = \ln\left(\frac{10}{9}\right)^2 = 0.105.$$

From Table 2.2, the following values for K and n are obtained for this material:

$$K = 1275 \text{ MPa} \qquad \text{and} \qquad n = 0.45.$$

Thus,

$$u = \frac{K\epsilon_1^{n+1}}{n+1} = \frac{(1275)(0.105)^{1.45}}{1.45} = 33.5 \times 10^6 \text{ Nm/m}^3.$$

and

$$W_{tension} = \left(33.5 \times 10^6\right)(\pi)(0.010)^2(0.15) = 1580 \text{ Nm}.$$

b. From Table 8.3, an average value for the specific energy in machining stainless steels is taken as 4.1 W-s/mm^3. The volume of material machined is

$$\text{Volume} = \frac{\pi}{4}\left[(10)^2 - (9)^2\right](150) = 2240 \text{ mm}^3.$$

Hence, the work done in machining is

$$W_{mach} = (4.1)(2240) = 9180 \text{ Nm}.$$

It will be noted that the work done in machining is over 5 times higher than that for tension. The reasons for the large difference between the two energies are that (a) tension involves very little strain (hence very little work of deformation) and (b) there is no friction. On the other hand, machining involves significant friction, and the material removed (even though relatively small in volume) has undergone much higher strains than it does in the bulk material undergoing tension. Assuming, from Tables 8.1 and 8.2, an average shear strain of 3 [equivalent to an effective strain of 1.7; see Eq. (2.55)], the material removed in machining is subjected to a strain of $1.7/0.105 = 16$ times higher than that in tension for this case.

These differences explain why machining consumes much more energy than bulk deformation. However, it can be shown that as the diameter of the rod decreases and assuming that the same depth of material is involved, the difference between the two energies becomes *smaller*. This result can be explained by noting the changes in the relative volumes involved in machining vs. tension as the diameter of the rod decreases.

TABLE 8.3 Approximate specific energy requirements in machining operations.

	Specific energy*	
Material	W-s/mm^3	hp-min/in^3
Aluminum alloys	0.4–1.1	0.15–0.4
Cast irons	1.6–5.5	0.6–2.0
Copper alloys	1.4–3.3	0.5–1.2
High-temperature alloys	3.3–8.5	1.2–3.1
Magnesium alloys	0.4–0.6	0.15–0.2
Nickel alloys	4.9–6.8	1.8–2.5
Refractory alloys	3.8–9.6	1.1–3.5
Stainless steels	3.0–5.2	1.1–1.9
Steels	2.7–9.3	1.0–3.4
Titanium alloys	3.0–4.1	1.1–1.5

*At drive motor, corrected for 80% efficiency; multiply the energy by 1.25 for dull tools.

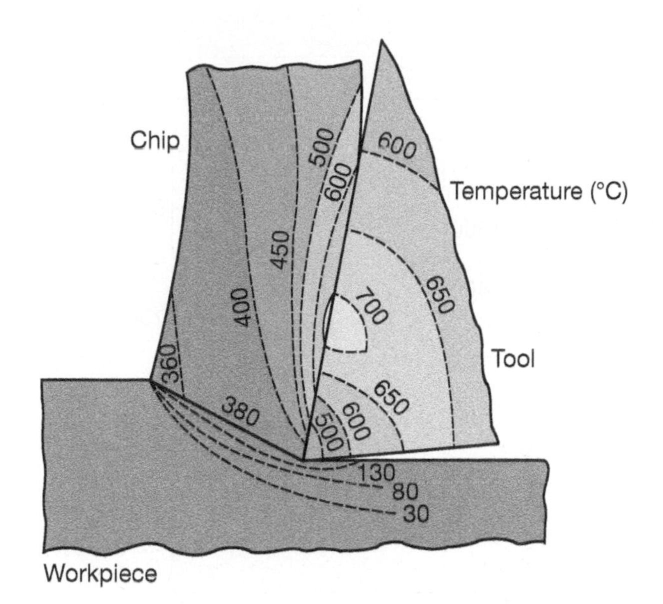

FIGURE 8.16 Typical temperature distribution in the cutting zone. Note the severe temperature gradients within the tool and the chip, and that the workpiece is relatively cool. *Source:* After G. Vieregge.

8.2.6 Temperature

As in all metalworking operations, the energy dissipated in machining operations is converted into heat, which causes a temperature rise in the cutting zone (Fig. 8.16). The rise in temperature is important because:

- High temperatures can adversely affect the strength, hardness, and wear resistance of the cutting tool;
- Control of tolerances can be made difficult by dimensional changes in the workpiece; and
- The machined surface can encounter thermal damage, adversely affecting its properties and service life.

Because of the work done in shearing and in overcoming friction on the rake face of the tool, the principal sources of heat generation are the (a) primary shear zone and (b) friction at the tool-chip interface. Moreover, if the tool is worn, heat is also generated by the dull tool tip rubbing against the machined surface.

Variables affecting temperature. An approximate but simple expression for the *mean temperature* for orthogonal cutting is

$$T = \frac{0.000665\sigma_f}{\rho c}\sqrt[3]{\frac{Vt_o}{\kappa}}, \tag{8.29}$$

where T is the *mean temperature* of the tool-chip interface in K; σ_f is the *flow stress* of the workpiece material (in MPa); V is the *cutting speed* (m/s); t_o is the *depth of cut* (m); ρc is the *volumetric specific heat* of the workpiece (in kJ/m^{3}K); and κ is the *thermal diffusivity* (ratio of thermal conductivity to volumetric specific heat) of the workpiece material (m^2/s). Note that because the material parameters in Eq. (8.29) themselves depend on

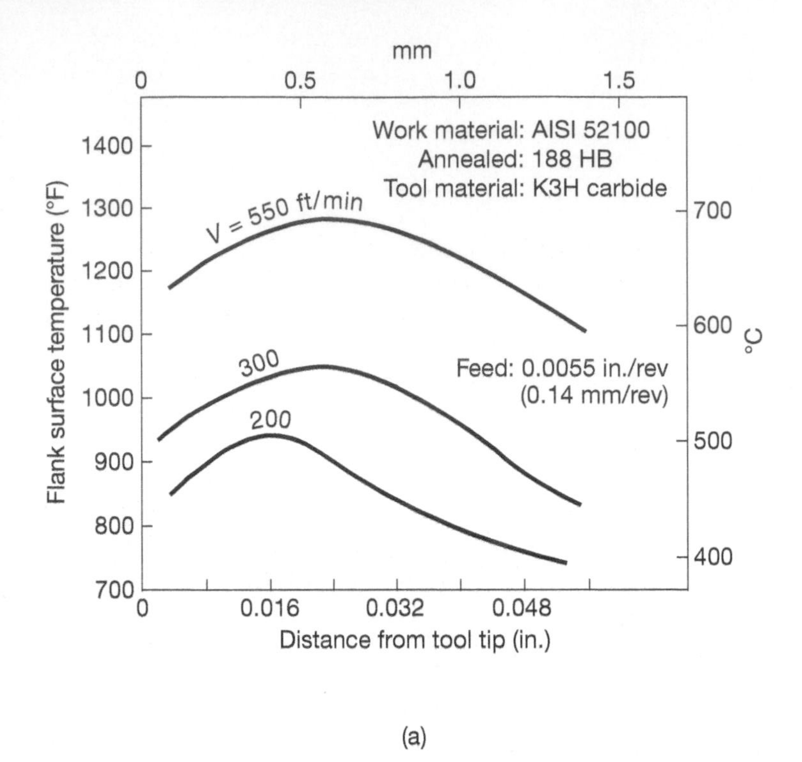

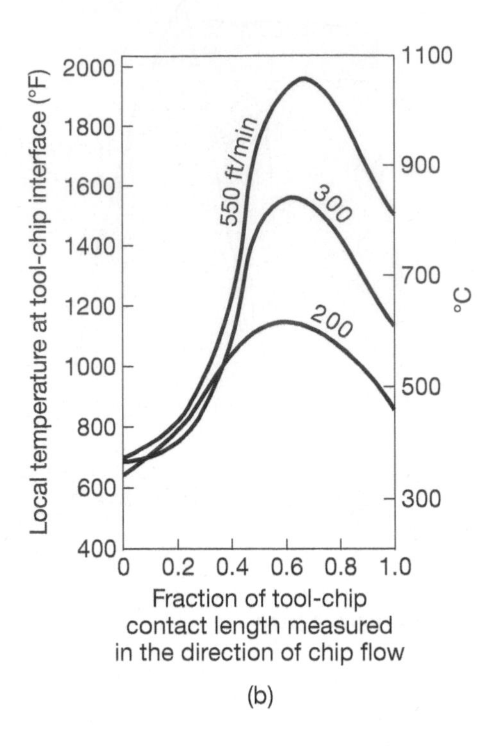

FIGURE 8.17 Temperature distribution in turning as a function of cutting speed: (a) flank temperature; (b) temperature along the tool-chip interface. Note that the rake-face temperature is higher than that at the flank surface. *Source:* After B.T. Chao and K.J. Trigger.

temperature, it is important to substitute appropriate values that are compatible with the *predicted* temperature range. The properties in Eq. (8.29) all pertain to the workpiece. It has been shown that thermal properties of the tool material (see Section 8.6) are relatively unimportant as compared with those of the workpiece material.

The temperature generated in the shear plane is a function of the specific energy for shear, u_s, and the specific heat of the workpiece material. As Eq. (8.29) indicates, temperature rise is highest in machining materials with high strength and low specific heat. The temperature rise at the tool-chip interface is also a function of the coefficient of friction. Moreover, flank wear (see Section 8.3.1 and Fig. 8.20a) is an additional source of heat, caused by rubbing of the tool on the machined surface.

Figure 8.17 shows results from experimental measurements of temperature in a turning operation. Note that (a) the maximum temperature is at a location away from the tool tip and (b) the temperature increases with cutting speed. As the speed increases, there is less time for the heat to be conducted through the workpiece and tool, and hence temperature rises. Also, as speed increases, a larger proportion of the heat generated is carried away by the chip (convective heat transfer), as can be seen in Fig. 8.18. The chip is thus an effective heat sink, in that it carries away most of the heat generated at higher cutting speeds.

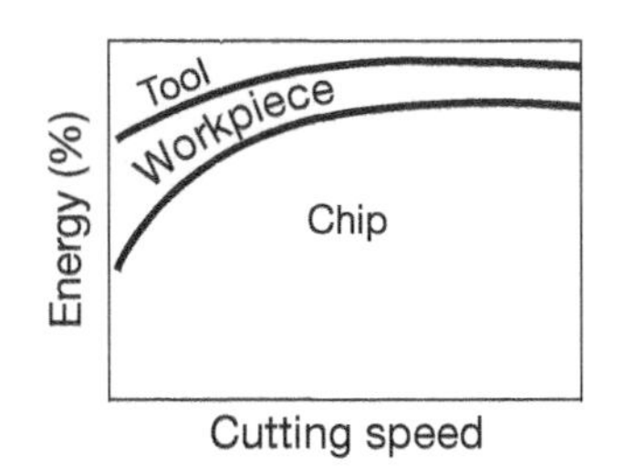

FIGURE 8.18 Proportion of the heat generated in cutting transferred to the tool, workpiece, and chip as a function of the cutting speed. Note that most of the cutting energy is carried away by the chip (in the form of heat), particularly as speed increases.

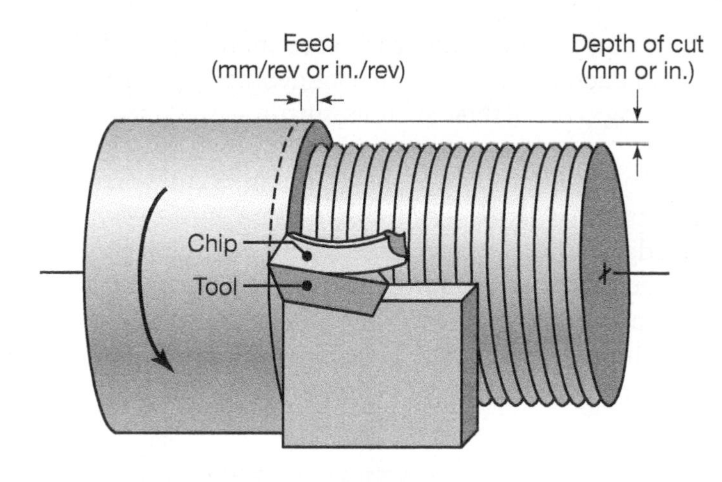

FIGURE 8.19 Terminology used in a turning operation on a lathe, where f is the feed (in mm/rev or in./rev) and d is the depth of cut. Note that feed in turning is equivalent to the depth of cut in orthogonal cutting (See Fig. 8.2), and the depth of cut in turning is equivalent to the width of cut in orthogonal cutting (See also Fig. 8.41).

Equation (8.29) also indicates that the mean temperature increases with (a) the strength of the workpiece material, because of the higher energy required; and (b) the depth of cut, because of the decreasing ratio of surface area to chip thickness. Note also that a thin chip cools faster than a thick chip; however, the depth of cut has been found to have negligible influence on the mean temperature for depths exceeding twice the tool tip radius (See also Fig. 8.27).

Based on Eq. (8.29), another expression has been developed for the mean temperature in turning (See Fig. 8.41), as given by

$$T \propto V^a f^b, \tag{8.30}$$

where a and b are constants, V is the cutting speed, and f is the feed (as shown in Fig. 8.19). For a carbide tool, a is approximately 0.2 and b is 0.125; for a high-speed steel tool, a is around 0.5 and b is 0.375.

Techniques for measuring temperature. Temperature and its distribution in the cutting zone may be determined by several techniques: (a) Using **thermocouples**, embedded in small holes in the tool or in the workpiece; this technique involves considerable effort. (b) Measuring **thermal emf** (electromotive force) at the tool-chip interface, which acts as a hot junction between two different materials (the tool and the chip). (c) Using a *radiation pyrometer*, monitoring the **infrared radiation** from the cutting zone; however, this technique indicates only surface temperatures and the accuracy of the results depends on the *emissivity* of the surfaces, which can be difficult to determine accurately.

8.3 Tool Wear and Failure

Recall that cutting tools are subjected to high stresses, elevated temperatures, and sliding over the machined surface; these conditions all induce wear (see Section 4.4.2). Because of its effects on the quality of the machined surface and the economics of machining, *tool wear* is one of the most important aspects of machining operations. A variety of factors affect tool wear, such as cutting tool and workpiece materials and their physical, mechanical, and chemical properties; tool geometry; cutting fluids (if used); and processing parameters, such as cutting speed, feed, and depth of cut.

Typical wear patterns in cutting tools are shown in Fig. 8.20, where the regions of wear are identified as *flank wear, crater wear, nose wear,* and *chipping* of the cutting edge. Whereas wear is generally a gradual process, chipping of the tool, especially *gross chipping*, is regarded as *catastrophic failure*. In addition to wear, *plastic deformation* of the tool itself also may take place, especially with tool materials that begin to lose their strength and hardness at elevated temperatures (see Section 8.6).

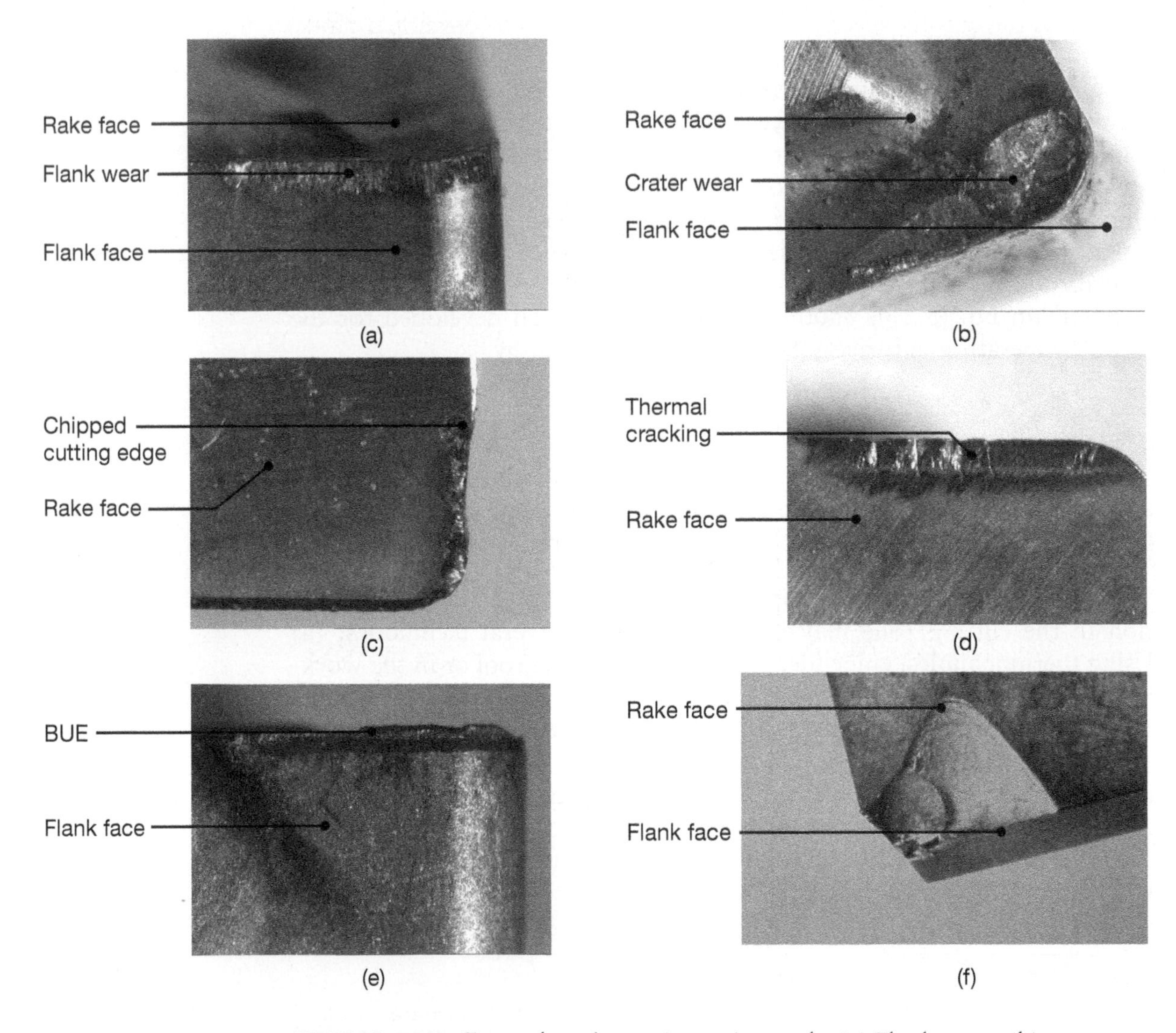

FIGURE 8.20 Examples of wear in cutting tools. (a) Flank wear; (b) crater wear; (c) chipped cutting edge; (d) thermal cracking on rake face; (e) flank wear and built-up edge; and (f) catastrophic failure (fracture). *Source:* Courtesy of Kennametal, Inc.

8.3.1 Flank Wear

Flank wear is generally attributed to

1. Sliding of the tool along the machined surface, causing adhesive and/or abrasive wear of the tool; and
2. Temperature rise, because of its adverse effects on the mechanical properties of the tool material.

Following an extensive study, a tool-wear relationship was established for machining a variety of steels, as

$$Vt^n = C, \tag{8.31}$$

where V is the cutting speed, t is the time that it takes to develop a flank wear land or objectionable surface finish, n is an exponent that depends on workpiece and tool material as well as cutting conditions, and C is a constant. Equation (8.31) is known as the *Taylor tool life equation*, after its developer, F.W. Taylor.

Tool-life curves. *Tool-life curves* are plots of experimental data obtained in machining tests (Fig. 8.21), typically for turning operations. Note that (a) tool life decreases rapidly as cutting speed increases; (b) the condition of the workpiece material has a strong influence on tool life; and (c)

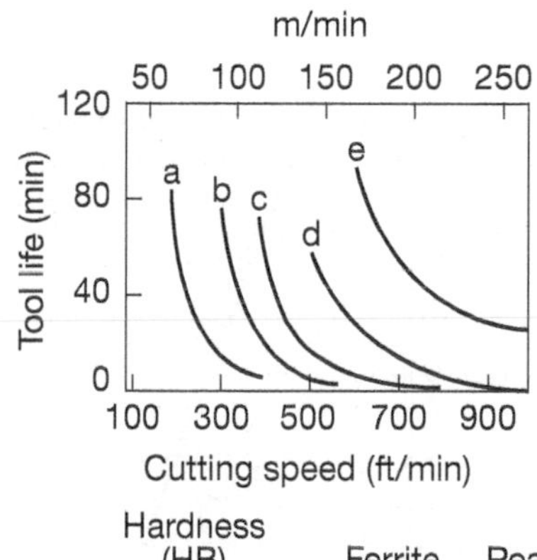

	Hardness (HB)	Ferrite	Pearlite
a. As cast	265	20%	80%
b. As cast	215	40	60
c. As cast	207	60	40
d. Annealed	183	97	3
e. Annealed	170	100	–

(a)

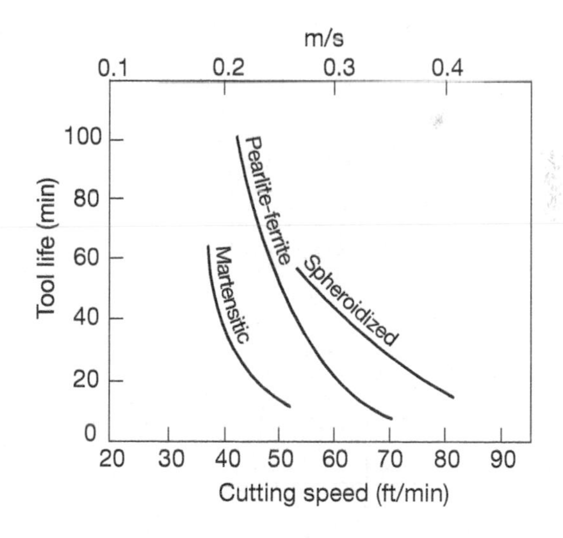

(b)

FIGURE 8.21 Effect of workpiece microstructure on tool life in turning. Tool-Life is hyphenated, is given in terms of the time (in minutes) required to reach a flank wear land of a specified dimension. (a) Ductile cast iron; and (b) steels, with identical hardness. Note in both figures the rapid decrease in tool life as the cutting speed increases.

there is a large difference in tool life for different microstructures of the workpiece. Heat treatment of the workpiece material is important largely because of the increase in hardness. For example, ferrite has a hardness of about 100 HB, pearlite 200 HB, and martensite 300 HB to 500 HB (see Section 5.11). Impurities and hard constituents in the workpiece material also are important, because they reduce tool life due to their abrasive action on the tool (see also Section 4.4.2).

TABLE 8.4 Range of n values for various cutting tools.

Tool material	n
High-speed steels	0.08–0.2
Cast alloys	0.1–0.15
Carbides	0.2–0.5
Ceramics	0.5–0.7

The exponent n is determined from the tool life curves, with a range of typical values given in Table 8.4. It has been observed that although tool-life curves are usually linear over a specific range of cutting speeds, they are rarely so over a wide range. Moreover, the exponent n can become negative at low cutting speeds, indicating that tool-life curves may actually reach a maximum and then curve downward as cutting speed increases. Consequently, caution should be exercised when using tool-life equations beyond the range of cutting speeds at which they were developed.

Because of the major influence of temperature on the physical and mechanical properties of materials (see Section 2.2.6), it would be expected that wear is strongly influenced by temperature. Experimental investigations have shown that there is indeed a direct relationship between flank wear and the temperature generated during machining (Fig. 8.22b). Although cutting speed has been found to be the most significant process

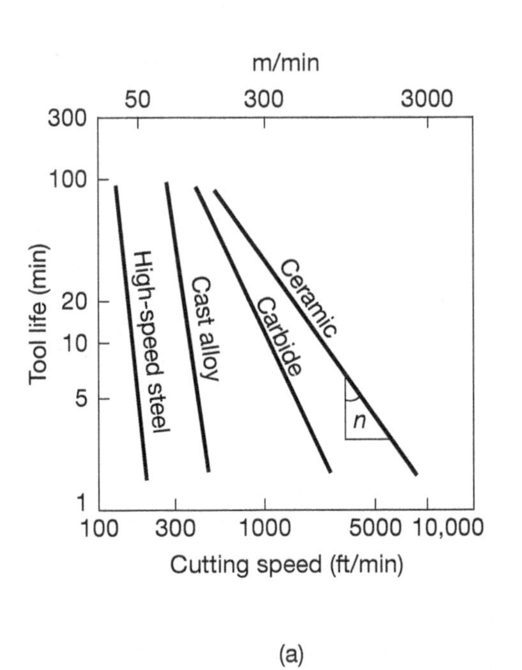

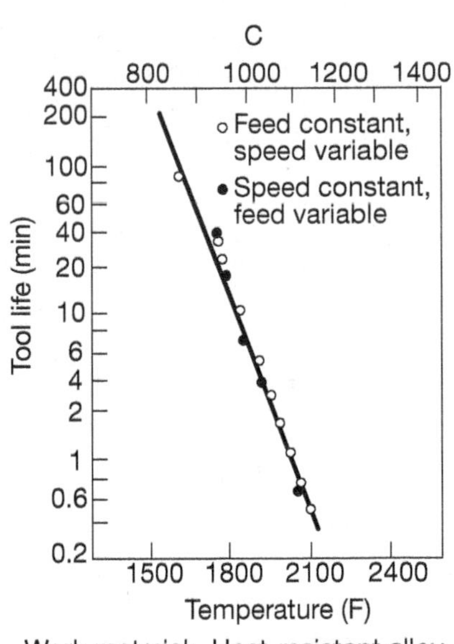

FIGURE 8.22 (a) Tool-life curves for a variety of cutting-tool materials. The negative inverse of the slope of these curves is the exponent n in tool-life equations. (b) Relationship between measured temperature during cutting and tool life (flank wear). Note that high cutting temperatures severely reduce tool life. See also Eq. (8.30). *Source:* After H. Takeyama and Y. Murata.

TABLE 8.5 Allowable average wear land for cutting tools for various operations.

	Allowable wear land (mm)	
Operation	High-speed steels	Carbides
Turning	1.5	0.4
Face milling	1.5	0.4
End milling	0.3	0.3
Drilling	0.4	0.4
Reaming	0.15	0.15

variable in tool life, depth of cut and feed rate also are important; thus, Eq. (8.31) can be modified as

$$Vt^n d^x f^y = C, \tag{8.32}$$

where d is the depth of cut and f is the feed rate (in mm/rev or in./rev) in turning. The exponents x and y must be determined experimentally for each cutting condition. Taking $n = 0.15$, $x = 0.15$, and $y = 0.6$ as typical values encountered in practice, it can be seen that cutting speed, feed rate, and depth of cut are of decreasing order of importance.

Equation (8.32) can be rewritten as

$$t = C^{1/n} V^{-1/n} d^{-x/n} f^{-y/n}, \tag{8.33}$$

or

$$t \simeq C^7 V^{-7} d^{-1} f^{-4}. \tag{8.34}$$

Thus, for a constant tool life, the following observations can be made from Eq. (8.34):

1. If the feed or the depth of cut is increased, the cutting speed must be decreased, and vice versa; and
2. A reduction in the cutting speed will allow an increase in feed and/or depth of cut. Depending on the magnitude of the exponents, this can then result in an increase in the volume of the material removed.

Allowable wear land. Although somewhat arbitrary, typical values of *allowable wear land* for various processing conditions are given in Table 8.5. For improved dimensional accuracy and surface finish, the allowable wear land may be specified at a smaller value than those given in the table. In practice, the recommended cutting speed for a high-speed steel tool is generally the one that gives a tool life of 60–120 min.; for carbide tools, it is 30–60 min (See Table 8.9).

Optimum cutting speed. Recall that as the cutting speed decreases, tool life increases. However, a lower cutting speed means that the rate at which material is being removed will be low; thus, there is an *optimum* cutting speed.

The effect of cutting speed on the volume of metal removed between tool indexing (rotating in its holder) or replacement can be appreciated by analyzing Fig. 8.21a. Assume that the material being machined is in the condition represented by curve "a". If the cutting speed is 1 m/s (200 ft/min), the tool life will be about 40 min. Thus, the tool travels a distance of (1 m/s)(60 s/min)(40 min) = 2400 m before it reaches its life. If the cutting speed is increased to 2 m/s, the tool life will be about 5 min, and the tool travels a distance of (2)(60)(5) = 600 m. Because the volume of material removed is directly proportional to the distance the tool has traveled, it can be seen that by *decreasing* the cutting speed, *more* material will be removed between tool indexing or changes. Note, however, that the lower the cutting speed, the longer it will take to machine the part, thus reducing productivity (see Section 8.15).

EXAMPLE 8.3 Increasing Tool Life by Reducing the Cutting Speed

Given: A tool and material combination has $n = 0.5$ and $C = 400$.

Find: Calculate the percentage increase in tool life when the cutting speed is reduced by 50% using the Taylor equation [Eq. (8.31)].

Solution: Since $n = 0.5$, the Taylor equation can be rewritten as $V\sqrt{t} = 400$. Letting V_1 be the initial speed and V_2 the reduced speed, it can be noted that, for this problem, $V_2 = 0.5V_1$. Because C is a constant,

$$0.5V_1\sqrt{t_2} = V_1\sqrt{t_1}.$$

Simplifying this expression,

$$\frac{t_2}{t_1} = \frac{1}{0.25} = 4.0.$$

This relation indicates that the tool-life change is

$$\frac{t_2 - t_1}{t_1} = \left(\frac{t_2}{t_1}\right) - 1 = 4 - 1 = 3,$$

or that it is increased by 300%. Note that the reduction in cutting speed has resulted in a major increase in tool life.

8.3.2 Crater Wear

Although the factors affecting flank wear also influence *crater wear*, the most significant factors in crater wear are temperature and the level of chemical affinity between the tool and the workpiece materials. Recall that the rake face of the tool is subjected to high localized stress and temperature, as well as sliding of the chip up the rake face at relatively high speeds (See Fig. 8.3b). As shown in Fig. 8.17b, peak temperatures in the cutting zone can be on the order of 1100°C (2000°F). Note that the location of *maximum depth* of crater wear generally coincides with the location of maximum temperature at the tool-chip interface.

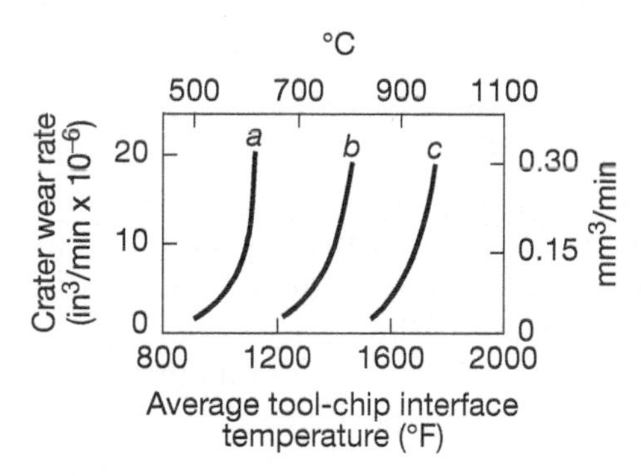

FIGURE 8.23 Relationship between crater-wear rate and average tool-chip interface temperature in turning: (a) high-speed steel tool; (b) C1 carbide; and (c) C5 carbide. Note that crater wear increases rapidly within a narrow range of temperature. *Source:* After K.J. Trigger and B.T. Chao.

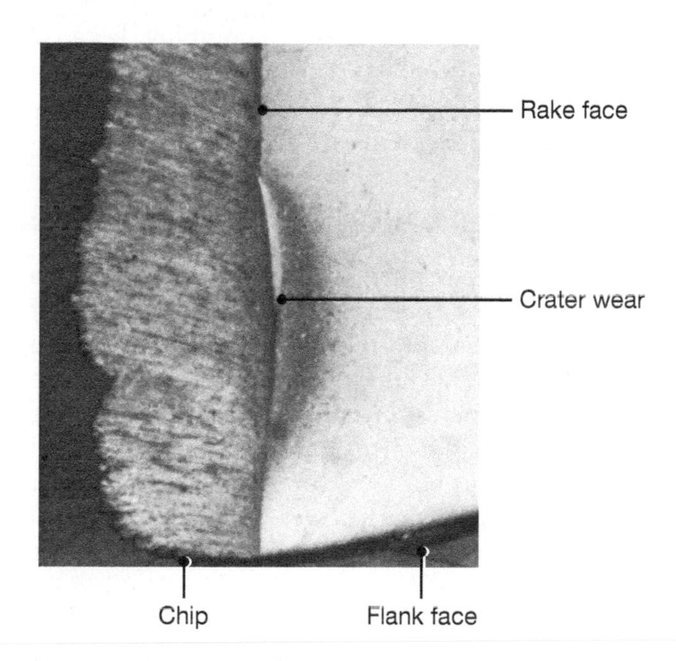

FIGURE 8.24 Interface of chip (left) and rake face of cutting tool (right) and crater wear in cutting AISI 1004 steel at 3 m/s (585 ft/min). Discoloration of the tool indicates the presence of high temperature (loss of temper). Note how the crater-wear pattern coincides with the discoloration pattern. Compare this pattern with the temperature distribution shown in Fig. 8.16. *Source:* Courtesy of P.K. Wright.

Experimental evidence has indicated that a direct relationship exists between crater-wear rate and tool-chip interface temperature (Fig. 8.23). Note, for example, the sharp increase in crater wear after a certain temperature range has been reached. Figure 8.24 shows the cross section of the tool-chip interface in cutting steel at high speeds. Note (a) the location of the crater-wear pattern and the *discoloration* of the tool (called *loss of temper*) as a result of high temperatures to which the tool has been subjected; and (b) the similarity of the discoloration profile to the temperature profile shown in Fig. 8.16.

The effect of temperature on crater wear has been described in terms of a **diffusion** mechanism (the movement of atoms across the tool-chip interface). Diffusion depends on the tool-workpiece material combination and on temperature, pressure, and time; as these quantities increase, the diffusion rate increases. An example of diffusion-induced crater wear can be observed when a diamond cutting tool is used to machine steel. The high solubility of carbon in steel leads to rapid crater wear, and eventually to tool failure.

8.3.3 Chipping

The term *chipping* in machining describes the sudden breaking away of a piece from the cutting edge of the tool. The pieces may be very small (*microchipping* or *macrochipping*), or they may consist of relatively large fragments (*gross chipping* or *fracture*). Two main causes of chipping are **mechanical shock** and **thermal fatigue**, such as seen in interrupted cutting operations as in milling, described in Section 8.10.

Chipping by mechanical shock may occur in a region of a cutting tool where a small crack or defect already exists. High positive rake angles also can contribute to chipping, because of the small included angle of the tool tip (See Fig. 8.32); this is a phenomenon similar to chipping of the tip of a very sharp pencil. Crater wear also may contribute to chipping, because wear progresses toward the tool tip and weakens it. Thermal cracks, which are generally perpendicular to the cutting edge (See Fig. 8.20d), typically are caused by the thermal cycling of the tool in interrupted cutting.

8.3.4 General Observations on Tool Wear

In addition to wear and chipping, other phenomena also occur in tools (Fig. 8.20). The wear **groove** or **notch** has been attributed to the fact that this narrow region is the boundary (the **depth-of-cut line**, DOC) where the chip is no longer in contact with the tool. This boundary oscillates because of inherent variations in the cutting operation, thus accelerating the wear process. Moreover, note that this region on the tool is in contact with the machined surface from the previous cut (See, for example, Fig. 8.19), that may have developed a thin work-hardened layer (depending on tool sharpness and its shape); and this contact would then contribute to the formation of the wear groove.

Because they are hard and abrasive, scale and oxide layers on the surface of a workpiece also increase wear, in which case the depth of cut (See Fig. 8.19) should be greater than the thickness of the oxide film or the work-hardened layer. In other words, light cuts should not be taken on rusted or corroded workpieces, as otherwise the tool will travel through this thin, hard, and abrasive layer.

8.3.5 Tool-Condition Monitoring

With the extensive use of computer-controlled machine tools and implementation of highly automated manufacturing systems, the reliable and repeatable performance of cutting tools is a major consideration. Once programmed properly, machine tools now operate with little direct supervision by an operator; consequently, the failure of a cutting tool will have serious detrimental effects. It is therefore essential to continuously monitor the condition of the cutting tool, such as for wear, chipping, or gross failure.

Techniques for tool-condition monitoring typically fall into two general categories: direct and indirect. The **direct method** involves *optical measurement* of wear, by periodically observing changes in the profile of the tool.

Indirect methods of measuring wear involve correlating the tool condition with variables such as force, power, temperature rise, surface finish, and vibration and chatter. The **acoustic emission** technique utilizes a piezoelectric transducer, attached to a tool holder, which picks up acoustic-emission signals (typically above 100 kHz) that result from the stress waves generated during machining. By analyzing the signals, tool wear and chipping can be monitored. This technique is particularly effective in precision machining operations where, because of the very small amounts of material removed, cutting forces are low.

A similar indirect tool-condition monitoring system consists of **transducers** (installed in an original machine tool or retrofitted on an existing machine) that continually monitor such parameters as spindle *torque* and cutting-tool *forces*. A microprocessor analyzes the signals (which are preamplified) and interprets them. This system is capable of differentiating the signals from tool breakage, tool wear, a missing tool, overloading of the machine, or collision of machine components. It can also automatically compensate for tool wear, so that dimensional accuracy of the part is maintained.

The design of the transducers must be such that they are (a) nonintrusive to the particular machining operation; (b) accurate and repeatable in signal detection; (c) rugged enough for a typical shop-floor environment; and (d) cost effective. *Sensors* (see Section 14.8), including *infrared* and *fiber optic* techniques for temperature measurement, are now being used extensively in a variety of machining operations.

8.4 Surface Finish and Surface Integrity

Surface finish describes the *geometric* features of surfaces, whereas **surface integrity** pertains to *properties* that are strongly influenced by the type of surface produced (see also Section 4.3). The ranges of surface roughness in machining and other processes are given in Fig. 8.25. As can be seen, the processes are generally organized in order of increasing surface quality, which also correspond to increasing cost and machining time (See also Fig. 16.5).

Built-up edge and depth of cut can adversely affect surface finish and integrity. Figure 8.26 shows surfaces obtained in two different machining operations; note the damage to the surfaces from BUE. A shallow depth of cut (or dull tool) can also compromise surface finish. A dull cutting tool has a larger radius along its edges (See Fig. 8.20c), just as a dull pencil or knife does. Figure 8.27 illustrates the relationship between the radius of the cutting edge and depth of cut in orthogonal cutting. Note that at small depths of cut, the rake angle of an otherwise positive-rake tool can effectively become *negative*; the tool may simply ride over the workpiece surface and not remove any material. If the radius is large in relation to the depth of cut, the tool will rub over the machined surface, generating frictional heat, inducing surface residual stresses, and causing surface damage, such as tearing and cracking. In practice, the depth of cut should generally be greater than the radius on the cutting edge.

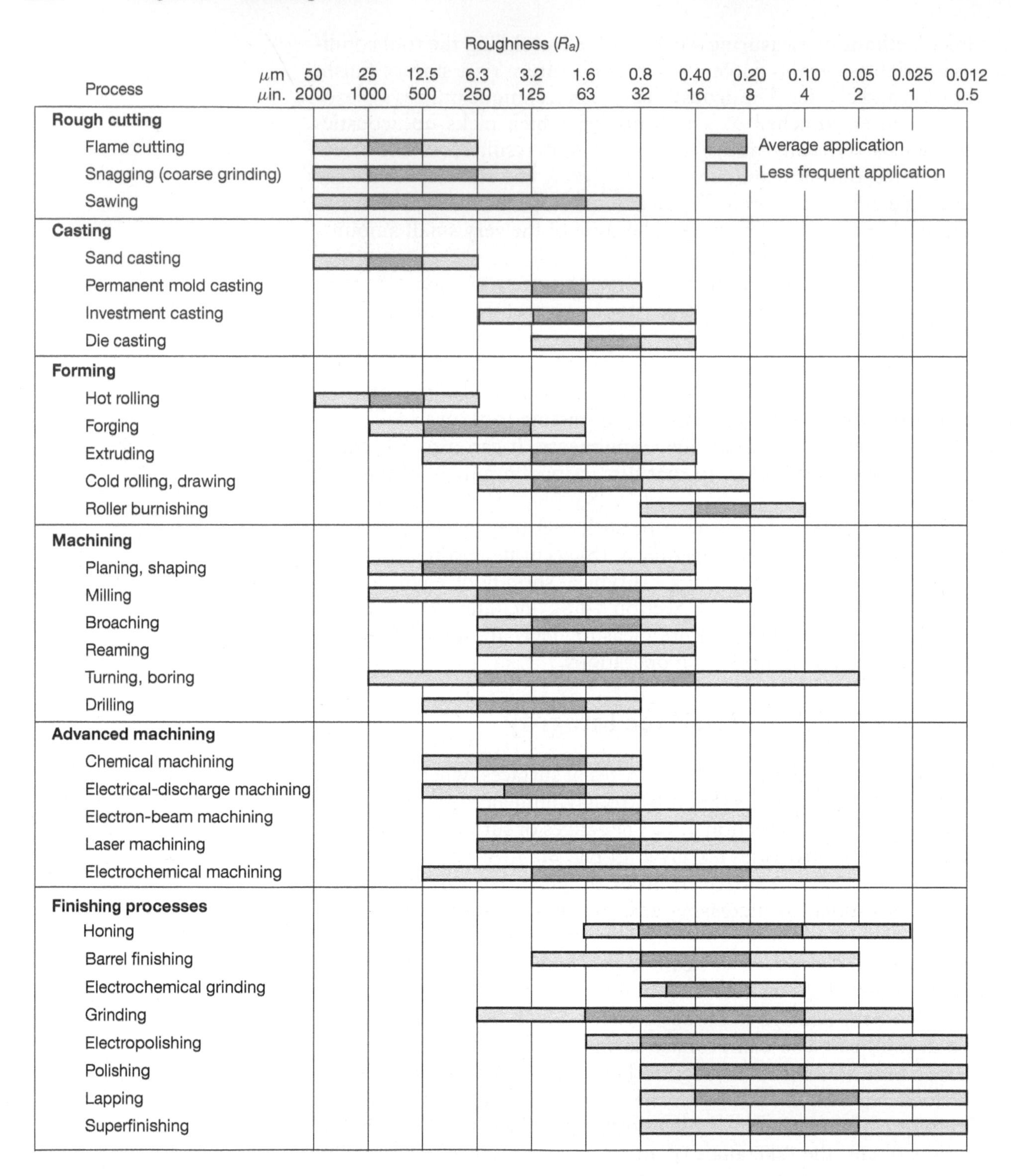

FIGURE 8.25 Range of surface roughness obtained in various machining processes. Note the wide range within each group, especially in turning and boring (See also Fig. 9.27).

(a)

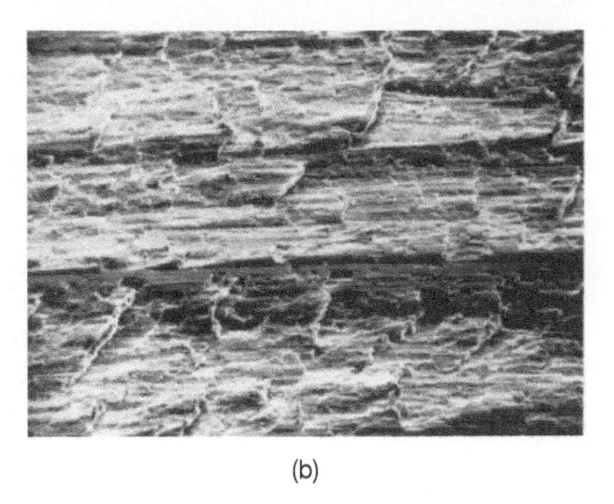

(b)

FIGURE 8.26 Surfaces produced on steel in machining, as observed with a scanning electron microscope: (a) turned surface, and (b) surface produced by shaping. *Source:* JT. Black and S. Ramalingam.

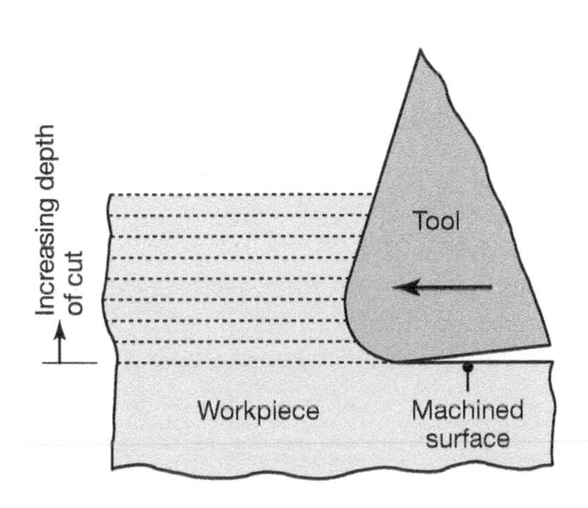

FIGURE 8.27 Schematic illustration of a dull tool in orthogonal cutting (exaggerated). Note that at small depths of cut, the rake angle can effectively become negative. In such cases, the tool may simply ride over the workpiece surface, burnishing it, instead of cutting.

Feed marks. In turning, as in some other machining operations, the cutting tool leaves a spiral profile (*feed marks*) on the machined surface as it moves across the workpiece (See Fig. 8.19). As expected, the higher the feed, f, and the smaller the radius, R, the more prominent are these marks. Although not significant in rough machining operations, feed marks are important in finish machining (see Section 8.9).

The peak-to-valley roughness, R_t, in turning can be expressed as

$$R_t = \frac{f^2}{8R}, \tag{8.35}$$

where f is the feed and R is the nose radius of the tool. For the condition where R is much smaller than f, the roughness is given by the expression

$$R_t = \frac{f}{\tan\alpha_s + \cot\alpha_e}, \tag{8.36}$$

where α_s and α_e are the side and edge cutting angles, respectively (See Fig. 8.40). For face milling (See Fig. 8.56), the roughness is given by

$$R_t = \frac{f^2}{16\left[D \pm (2fn/\pi)\right]}, \tag{8.37}$$

where D is the cutter diameter, f is the feed per tooth, and n is the number of inserts on the cutter. Equations (8.35) to (8.37) are derived considering only geometry; they do not account for such phenomena as cracking of the workpiece surface (See Fig. 8.6c), thermal distortion, and chatter marks.

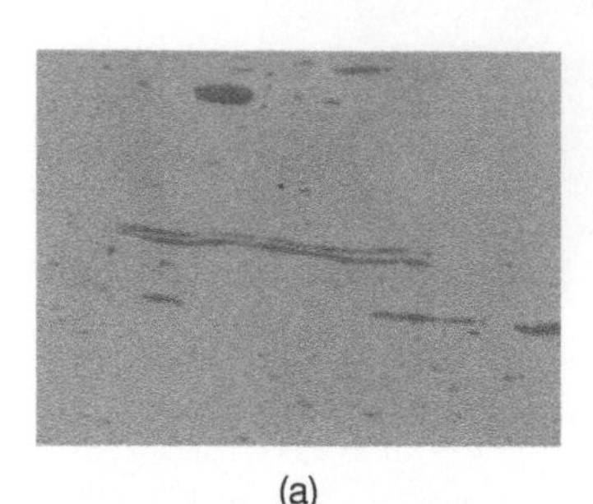

(a)

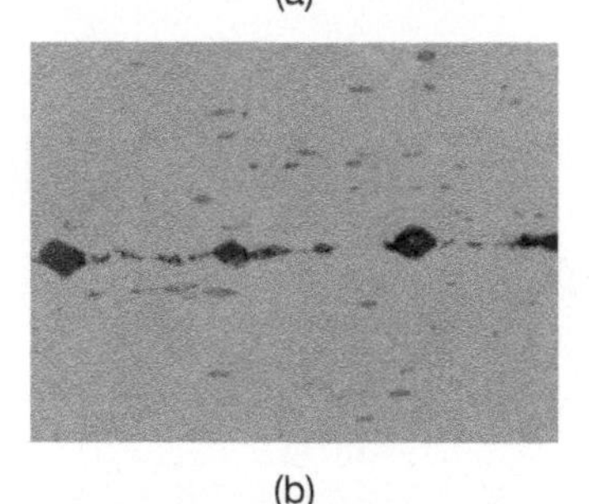

(b)

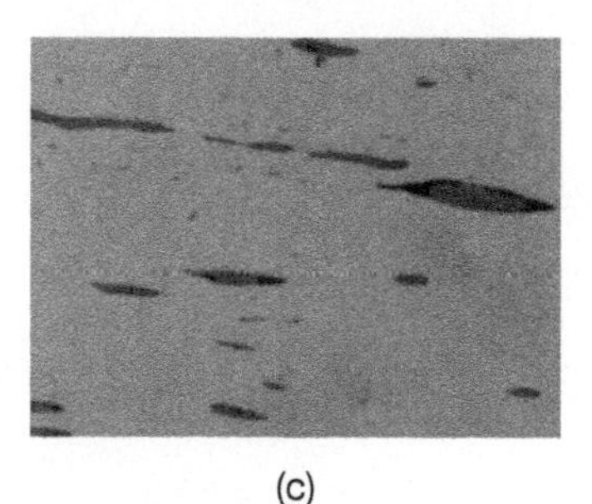

(c)

FIGURE 8.28 Photomicrographs showing various types of inclusions in low-carbon, resulfurized free-machining steels. (a) Manganese-sulfide inclusions in AISI 1215 steel. (b) Manganese-sulfide inclusions and glassy manganese-silicate-type oxide (dark) in AISI 1215 steel. (c) Manganese sulfide with lead particles as tails in AISI 12L14 steel. *Source:* Courtesy of ArcelorMittal.

8.5 Machinability

The *machinability of a material* is generally defined in terms of the following four factors: (a) surface finish and integrity of the machined part; (b) tool life; (c) force and power requirements; and (d) chip control. Thus, good machinability indicates *good surface finish and integrity, long tool life, low force and power requirements,* and type of *chip* produced are easily collected and do not interfere with the machining operation (see Section 8.2.1).

8.5.1 Machinability of Steels

Steels are among the most important and commonly used engineering materials. Their machinability has been studied extensively and has been greatly improved by adding *lead* and *sulfur*, called **free-machining steels.**

Leaded steels. Lead is added to molten steel and takes the form of dispersed fine lead particles (See Fig. 5.2a). During machining, the lead particles are sheared and smeared over the tool-chip interface; because of their low shear strength, the lead particles act as a solid lubricant (see Section 4.4.4). This behavior can be verified from the presence of high concentrations of lead on the tool-side face of the chips when machining leaded steels. In addition, lead lowers the shear stress in the primary shear zone, thus reducing cutting forces and power consumption. Leaded steels are identified by the letter L between the second and third numerals, such as 10L45. Because of its toxicity and environmental concerns, the trend has been toward eliminating the use of lead in favor of such elements as bismuth and tin (*lead-free steels*).

Resulfurized and rephosphorized steels. *Sulfur* in steels forms *manganese-sulfide inclusions* (second-phase particles; see Fig. 8.28). These particles act as stress raisers in the primary shear zone; as a result, the chips produced are small and they break up easily, thus improving machinability. *Phosphorus* in steels improves machinability by virtue of strengthening the ferrite, thereby increasing the hardness of steels and producing less continuous chips.

Calcium-deoxidized steels. In these steels, oxide flakes of *calcium aluminosilicate* (CaO, SiO_2, and Al_2O_3) are formed. These flakes, in turn, lower the strength of the secondary shear zone, thus reducing tool-chip interface friction and wear, and hence lowering the temperature. As a result, these steels develop less crater wear of the tool, especially at high cutting speeds when temperatures are higher.

Effects of other elements on the machinability of steels. (a) The presence of *aluminum* and *silicon* in steels is always harmful, because these elements combine with oxygen to form aluminum oxide and silicates, which are hard and abrasive. (b) *Carbon* and *manganese* have various effects on the machinability of steels, depending on their composition. As the carbon content increases, machinability decreases, although plain low-carbon steels (less than 0.15% C) can produce poor surface finish due to forming a built-up edge. Alloying elements, such as nickel, chromium, molybdenum, and vanadium (which improve the properties of steels) generally reduce machinability. (c) Cast steels have a machinability similar to that of wrought steels. (d) Tool and die steels are very difficult to machine, usually requiring annealing prior to machining them.

Machinability of most steels is generally improved by cold working, which reduces the tendency for built-up edge formation. Austenitic (300 series) stainless steels are generally difficult to machine; chatter can be a problem, requiring the use of machine tools with high stiffness and damping capacity. Ferritic stainless steels (also 300 series) have good machinability. Martensitic (400 series) stainless steels are abrasive, tend to form built-up edge, and require tool materials with high hot hardness and resistance to crater wear. Precipitation-hardening stainless steels are strong and abrasive, and thus require hard and abrasion-resistant tool materials.

8.5.2 Machinability of Various Metals

Aluminum is generally easy to machine, although the softer grades tend to form built-up edge, and thus, poor surface finish. High cutting speeds, rake angles, and relief angles are recommended. Wrought aluminum alloys with high silicon content and cast alloys may be abrasive, thus requiring harder tool materials. Dimensional control may be a challenge in machining aluminum, because of its low elastic modulus and relatively high thermal coefficient of expansion.

Gray-cast irons are generally machinable, although they are abrasive. Free carbides in castings reduce machinability and cause tool chipping and fracture, thus requiring tools with high toughness. Nodular and malleable irons are machinable, using hard tool materials.

Cobalt-base alloys are abrasive and highly work hardening; they require sharp and abrasion-resistant tool materials and low feeds and speeds.

Wrought copper can be difficult to machine, because of built-up edge formation, although cast copper alloys are easy to machine. *Brasses* are easy to machine, especially those containing lead (*leaded free-machining brass*). *Bronzes* are more difficult to machine than brass.

Magnesium is very easy to machine, with good surface finish and long tool life. Care should be exercised when machining magnesium, because of its high rate of oxidation (*pyrophoric*) and the danger of fire.

Molybdenum is ductile and work hardening, indicating that it can produce poor surface finish, although its machinability can be improved using sharp tools.

Nickel-base alloys are work hardening, abrasive, and strong at high temperatures; their machinability is similar to that of stainless steels.

Tantalum is very work hardening, ductile, and soft; it produces a poor surface finish and tool wear is high.

Titanium and its alloys have poor thermal conductivity (the lowest of all metals; see the inside front cover), causing significant localized temperature rise and the formation of built-up edge.

Tungsten is brittle, strong, and very abrasive; its machinability is low, although it improves significantly at elevated temperatures.

8.5.3 Machinability of Nonmetallic Materials

Thermoplastics generally have low thermal conductivity, elastic modulus, and softening temperature. Machining them requires sharp tools with positive rake angle (to reduce cutting forces), large relief angles, small depths of cut and feed, relatively high speeds, and proper support of the workpiece. External cooling of the cutting zone through coolants may be necessary to keep the chips from becoming "gummy" and adhering to the tools.

Thermosetting plastics are brittle and sensitive to thermal gradients during machining; their machinability is generally similar to that of thermoplastics.

Reinforced plastics are generally very abrasive, depending on the type of fiber, and are difficult to machine; fiber tearing and pullout is a significant problem. There are also environmental considerations, as machining these materials requires proper removal of machining debris to avoid human contact with and inhalation of loose fibers.

Metal-matrix and *ceramic-matrix composites* can be difficult to machine, depending on the properties of the individual components in the composite. The reinforcing fibers can be abrasive, and the matrix material may not have sufficient ductility for good machinability.

Ceramics and, especially, *nanoceramics* can have improved machinability with the selection of appropriate machining parameters, such as by *ductile-regime cutting* (see Section 8.9.2).

Graphite is abrasive; it requires hard, abrasion-resistant, and sharp tools.

8.5.4 Thermally Assisted Machining

Materials that are difficult to machine at room temperature may be machined more easily at elevated temperatures. In *thermally assisted machining* (**hot machining**), the source of heat is a torch, or high-energy beam (such as a laser or electron beam) or plasma arc, focused on an area just ahead of the cutting tool. Most applications for hot machining are in turning and milling operations. The process of heating to and maintaining a uniform temperature distribution within the workpiece may be difficult to control, whereby the original microstructure of the workpiece may be altered, adversely affecting the properties of the machined part. Some success has been obtained in laser-assisted machining of silicon-nitride ceramics and *Stellite* (see Section 8.6.3).

8.6 Cutting-Tool Materials

The selection of appropriate cutting-tool materials for a specific application is among the more important considerations in machining operations, as is the selection of mold and die materials for forming and shaping processes. Recall that in machining, the tool is subjected to high localized temperatures, high contact stresses, rubbing on the workpiece surface, and the chip climbing up the rake face of the tool (see Section 8.2). Consequently, a cutting tool must possess the following characteristics:

- **Hardness**, particularly at elevated temperatures (**hot hardness**), so that the strength of the tool is maintained at the temperatures encountered in machining operations (Fig. 8.29);
- **Toughness**, so that impact forces on the tool in interrupted cutting operations, such as milling or turning a splined shaft, do not chip or fracture the tool;
- **Wear resistance**, so that the tool has an acceptable tool life before it is **indexed** or replaced; and

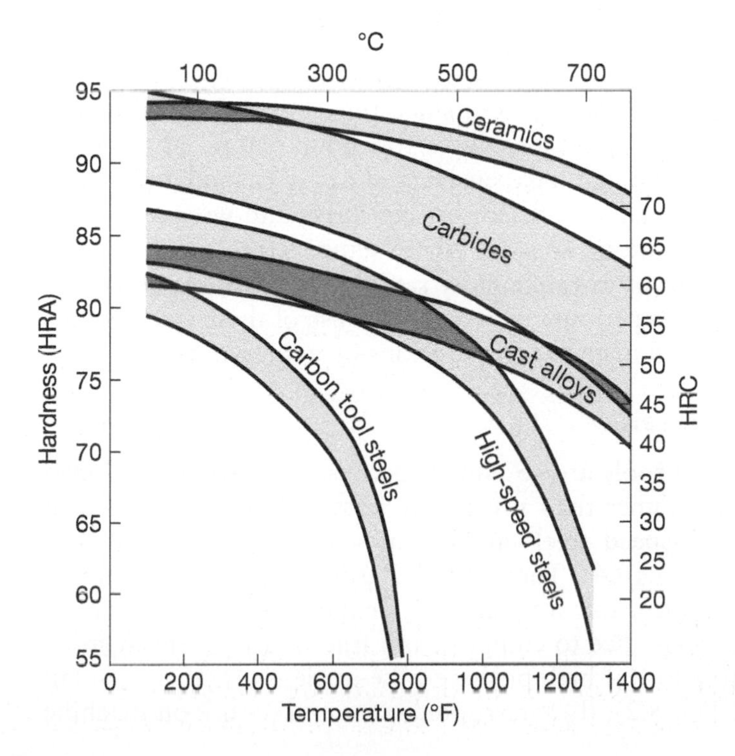

FIGURE 8.29 Hardness of various cutting-tool materials as a function of temperature (hot hardness). The wide range in each group of tool materials results from the variety of compositions and treatments available for that group.

- **Chemical stability** or **inertness** with respect to the workpiece material, so that any adverse reactions that may contribute to tool wear are minimized or avoided.

Several cutting-tool materials with a wide range of these characteristics are available (See Table 8.6). Tool materials are generally divided into the following categories, listed in the approximate chronological order in which they were developed and implemented. Note that many of these materials are also used for dies and molds, as described throughout Chapters 5, 6, 7, 10, and 11.

1. **Carbon and medium-alloy steels;**
2. **High-speed steels;**
3. **Cast-cobalt alloys;**
4. **Carbides;**
5. **Coated tools;**
6. **Alumina-base ceramics;**
7. **Cubic boron nitride;**
8. **Silicon-nitride-base ceramics;**
9. **Diamond;** and
10. **Whisker-reinforced and nanomaterials.**

This section describes the characteristics, applications, and limitations of tool materials. Their characteristics include hot hardness, toughness, impact strength, wear resistance, resistance to thermal shock, costs, and the range of cutting speeds and depth of cut for optimum performance in machining.

8.6.1 Carbon and Medium-Alloy Steels

Carbon steels are the oldest of tool materials, and have been used widely for drills, taps, broaches, and reamers since the 1880s. Low-alloy and medium-alloy steels were developed later for similar applications, but with longer tool life. Although inexpensive and easily shaped and sharpened, these steels do not have sufficient hot hardness and wear resistance for machining at high cutting speeds, where temperature rise is significant. Note in Fig. 8.29, for example, how rapidly the hardness of carbon steels decreases as the temperature increases. The use of these steels is thus limited to very low-speed machining operations or woodworking.

8.6.2 High-Speed Steels

High-speed steel (HSS) tools are so named because they were developed to machine at speeds higher than previously possible. First produced in the early 1900s, high-speed steels are the most highly alloyed of tool steels (see also Section 3.10.4). They can be hardened to various depths, have good wear resistance, and are relatively inexpensive. Because of their high toughness and resistance to chipping and fracture, high-speed steels are especially suitable for (a) high positive-rake-angle tools (that is, small included angle; see Fig. 8.2); (b) interrupted cuts; and (c) use on machine

TABLE 8.6 Typical range of properties of various tool materials.

Property	High-speed steel	Cast alloys	Carbides		Ceramics	Cubic boron nitride	Single crystal diamond*
			WC	TiC			
Hardness	83–86 HRA	82–84 HRA	90–95 HRA	91–93 HRA	91–95 HRA	4000–5000 HK	7000–8000 HK
Compressive strength							
MPa	4100–4500	1500–2300	4100–5850	3100–3850	2750–4500	6900	6900
psi $\times 10^3$	600–650	220–335	600–850	450–560	400–650	1000	1000
Transverse rupture strength							
MPa	2400–4800	1380–2050	1050–2600	1380–1900	345–950	700	1350
psi $\times 10^3$	350–700	200–300	150–375	200–275	50–135	105–200	
Impact strength							
J	1.35–8	0.34–1.25	0.34–1.35	0.79–1.24	<0.1	<0.5	<0.2
in.-lb	12–70	3–11	3–12	7–11	<1	<5	<2
Modulus of elasticity							
GPa	200	—	520–690	310–450	310–410	850	820–1050
psi $\times 10^6$	30	—	75–100	45–65	45–60	125	120–150
Density							
kg/m^3	8600	8000–8700	10,000–15,000	5500–5800	4000–4500	3500	3500
lb/in^3	0.31	0.29–0.31	0.36–0.54	0.2–0.22	0.14–0.16	0.13	0.13
Volume of hard phase (%)	7–15	10–20	70–90	—	100	95	95
Melting or decom-position temperature							
°C	1300	—	1400	1400	2000	1300	700
°F	2370	—	2550	2550	3600	2400	1300
Thermal conductivity, W/mK	30–50	—	42–125	17	29	13	500–2000
Coefficient of thermal expansion, $\times 10^{-6}$/°C	12	—	4–6.5	7.5–9	6–8.5	4.8	1.5–4.8

*The values for polycrystalline diamond are generally lower, except impact strength, which is higher.

tools that, because of their low stiffness, are subject to vibration and chatter. High-speed steels are the most commonly used tool materials, followed closely by various die steels and carbides. They are especially used in machining operations that require complex tool shapes, such as drills, reamers, taps, and gear cutters.

There are two basic types of high-speed steels: **molybdenum** (*M* series) and **tungsten** (*T* series). The *M* series contains up to about 10% molybdenum, with chromium, vanadium, tungsten, and cobalt as alloying elements. The *T* series contains 12 to 18% tungsten, with chromium, vanadium, and cobalt as alloying elements. The *M* series generally has higher abrasion resistance than the *T* series, undergoes less distortion during heat treating, and is less expensive.

High-speed steel tools are available in wrought, cast, and sintered (see *powder-metallurgy*, Chapter 11) conditions. They can be **coated** for improved performance (see Section 8.6.5), and may also be subjected to surface treatments, such as case hardening (Section 4.5.1), for improved hardness and wear resistance.

8.6.3 Cast-Cobalt Alloys

Cast-cobalt alloys have high hardness, typically 58 to 64 HRC, good wear resistance, and maintain their hardness at elevated temperatures. Their composition ranges from 38 to 53% cobalt, 30 to 33% chromium, and 10 to 20% tungsten. Commonly known as *Stellite* tools, these alloys are cast and ground into relatively simple tool shapes. Because they are not as tough as high-speed steels, and are sensitive to impact forces, they less suitable than high-speed steels for interrupted cutting operations.

8.6.4 Carbides

The tool materials described thus far possess sufficient toughness, impact strength, and thermal shock resistance for numerous applications; however, they have significant limitations regarding such important characteristics as strength and hardness, particularly hot hardness. Consequently, they cannot be used as effectively where high cutting speeds, and hence high temperatures, are involved, and their tool life can be relatively short. *Carbides*, also known as **cemented** or **sintered carbides**, were introduced in the 1930s to meet the challenge of higher machining speeds for higher productivity.

Because of their high hardness over a wide range of temperatures (as can be seen in Fig. 8.29), high elastic modulus, high thermal conductivity, and low thermal expansion, carbides are among the most important, versatile, and cost-effective tool and die materials for a wide range of applications. The two basic categories of carbides are *tungsten carbide* and *titanium carbide*. In order to differentiate them from coated tools (see Section 8.6.5), plain carbide tools are usually referred to as **uncoated carbides**.

1. **Tungsten carbide.** *Tungsten carbide* (WC) is a *composite material*, consisting of tungsten-carbide particles bonded together in a *cobalt*

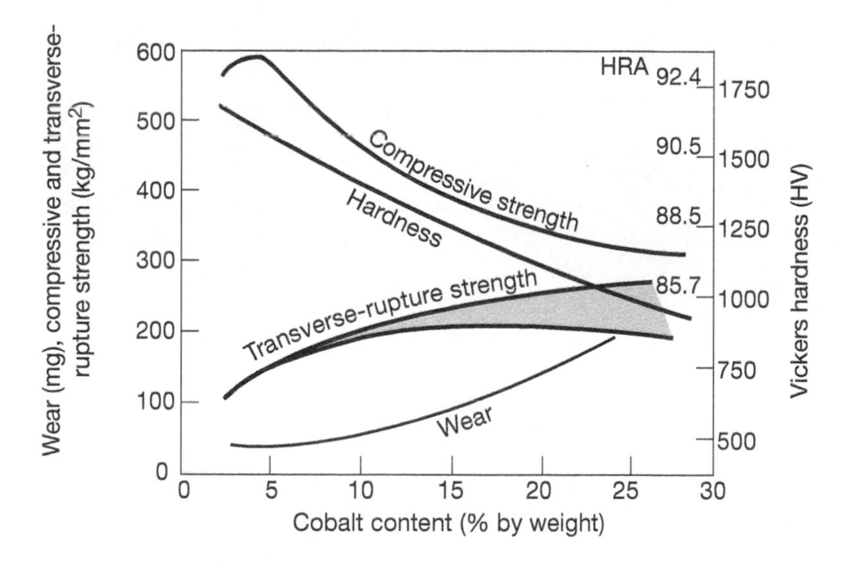

FIGURE 8.30 Effect of cobalt content in tungsten-carbide tools on mechanical properties.

matrix. Tungsten carbide is often compounded with carbides of titanium and niobium to impart special properties to carbide tools and dies. The amount of cobalt significantly affects the properties of carbide tools; as the cobalt content increases, strength, hardness, and wear resistance decrease, while toughness increases (Fig. 8.30). Tungsten-carbide tools are generally used for machining steels, cast irons, and abrasive nonferrous materials. The tools are manufactured by powder-metallurgy techniques.

2. **Titanium carbide.** *Titanium carbide* (TiC) has higher wear resistance than tungsten carbide, but it is not as tough. With a *nickel-molybdenum* alloy as the matrix, TiC is suitable for machining hard materials, mainly steels and cast irons, and for machining at speeds higher than those for tungsten carbide.

Inserts. High-speed steel and carbon-steel cutting tools can be produced in various geometries (See Fig. 8.10), including drills and milling cutters. However, after the cutting edges wear and become dull, the tool has to be removed from its holder and reground, a time-consuming process. The need for a more efficient method led to the development of *inserts*, that are individual cutting tools with a number of cutting edges in various shapes (Fig. 8.31). A square insert, for example, has eight cutting edges, and a triangular insert has six. Inserts also are available with a wide variety of **chip-breaker** features (see Section 8.2.1) for controlling chip flow, reducing vibration, and reducing the heat generated.

Inserts are usually clamped on the tool *shank*, using various locking mechanisms (Figs. 8.31a and b); less frequently, inserts may also be *brazed* (see Section 12.14.1) to the tool shank (See Fig. 8.38). However, because of the difference in thermal expansion between the insert and the tool-shank materials, brazing must be done properly in order to avoid cracking or warping. Clamping is the preferred method because after a cutting edge is worn, it is *indexed* so that a new edge can be used. In addition to those shown in Fig. 8.31, a wide variety of other toolholders also is available

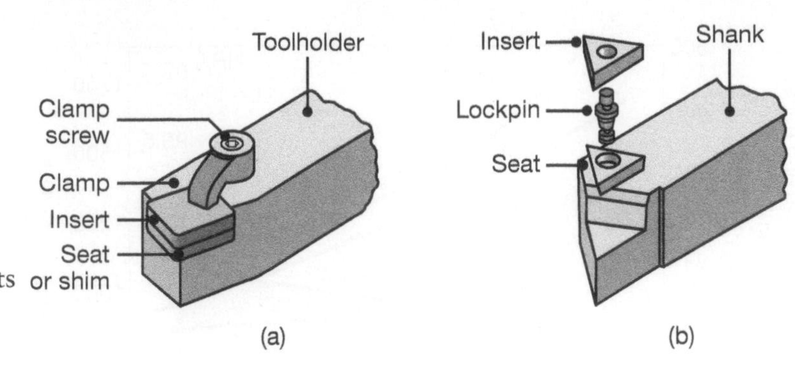

FIGURE 8.31 Methods of mounting inserts on toolholders: (a) clamping, and (b) wing lockpins. *Source:* Courtesy of Valenite.

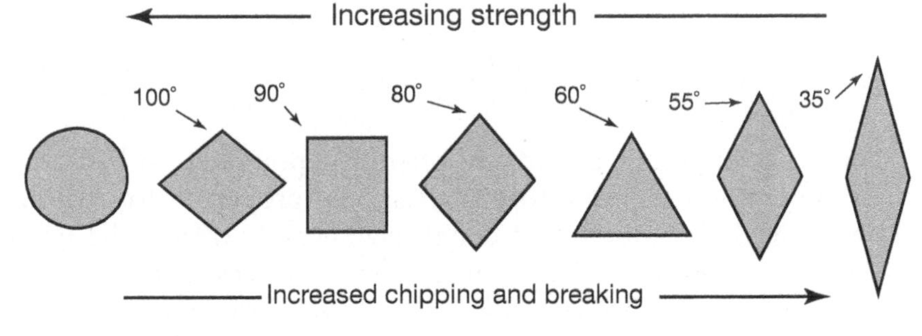

FIGURE 8.32 Relative edge strength and tendency for chipping and breaking of inserts with various shapes. Strength refers to that of the cutting edge shown by the included angles. *Source:* Courtesy of Kennametal, Inc.

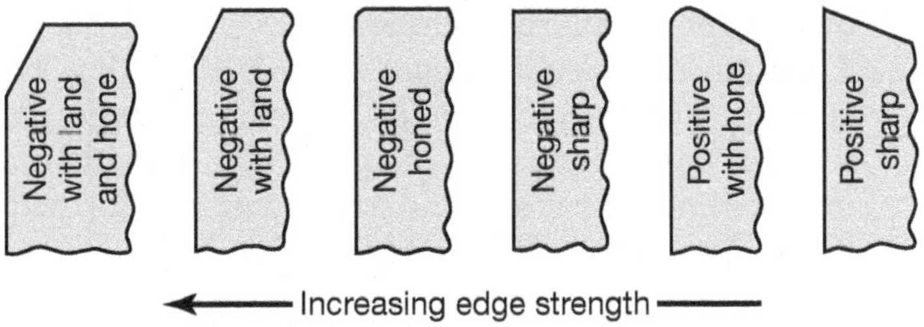

FIGURE 8.33 Edge preparations for inserts to improve edge strength. *Source:* Courtesy of Kennametal, Inc.

for specific applications, including toolholders with quick insertion and removal features, for more efficient operations.

The strength of the cutting edge of an insert depends on its shape; the smaller the included angle of the edge (Fig. 8.32), the lower its strength. In order to further improve edge strength and prevent chipping, insert edges are usually honed, chamfered, or they are made with a negative land (Fig. 8.33). Most inserts are honed to a radius of about 0.025 mm.

8.6.5 Coated Tools

A variety of materials are available as coatings, typically over high-speed steel and carbide tools. Because of their unique properties, **coated tools** can be used at high cutting speeds, thus reducing the machining time, and hence costs. It has been shown that coated tools can improve tool life by an order of magnitude over uncoated tools; note in Fig. 8.34, for example,

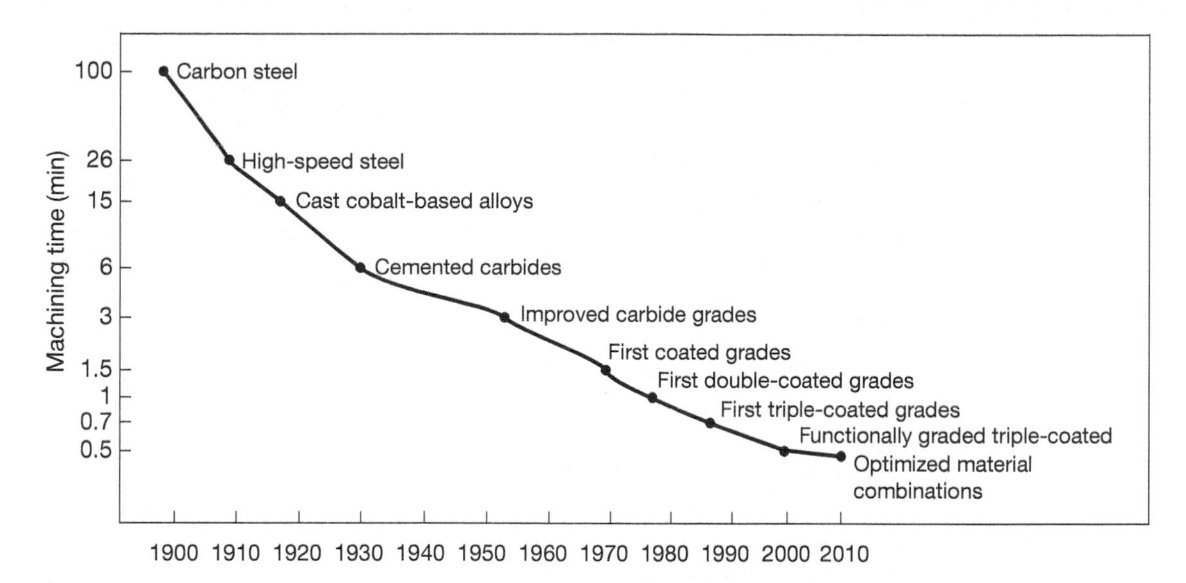

FIGURE 8.34 Relative time required to machine with various cutting-tool materials, with indication of the year the tool materials were introduced. Note that, within one century, machining time has been reduced by two orders of magnitude. *Source:* After Sandvik Coromant.

that the machining time has been reduced by a factor of more than 100 since 1900.

Commonly used *coating materials* include titanium nitride, titanium carbide, titanium carbonitride, and aluminum oxide (Al_2O_3), as described below. Generally in the thickness range of 2–10 μm, coatings are applied by **chemical vapor deposition** (CVD) and **physical-vapor deposition** (PVD) techniques, described in Section 4.5. The CVD process is the most common method for carbide tools with multiphase and ceramic coatings. The PVD-coated carbides with TiN coatings, on the other hand, have higher cutting-edge strength, lower friction, lower tendency to form a built-up edge, and are smoother and more uniform in thickness, which is generally in the range of 2–4 μm. **Medium-temperature chemical-vapor deposition** (MTCVD) provides higher resistance to crack propagation than do CVD coatings.

Coatings must have the following general characteristics:

- High hardness at elevated temperatures;
- Chemical stability and inertness to the workpiece material;
- Low thermal conductivity;
- Good bonding to the substrate; and
- Little or no porosity.

The effectiveness of coatings is enhanced by hardness, toughness, and high thermal conductivity of the substrate, which may be carbide or high-speed steel. Honing (see Section 9.7) of the cutting edges is an important procedure to maintain the strength of the coating and to prevent chipping at sharp edges and corners.

Various coatings are described below.

1. **Titanium nitride.** *Titanium nitride* (TiN) coatings have low coefficient of friction, high hardness, good high temperature properties, and good adhesion to the substrate. These properties greatly improve the life of high-speed steel tools and of carbide tools, drills, and cutters. Titanium-nitride coated tools (gold in color) perform well at higher cutting speeds and feeds; they do not perform as well as uncoated tools at low speeds, because the coating is susceptible to chip adhesion. Using appropriate cutting fluids to discourage chip-tool adhesion is therefore important. Flank wear is significantly lower than for uncoated tools (Fig. 8.35), and flank surfaces can be reground after use without removing the coating on the rake face of the tool.
2. **Titanium carbide.** *Titanium carbide* (TiC) coatings (silver-gray in color) over tungsten-carbide inserts have high resistance to flank wear, especially in machining abrasive materials.
3. **Titanium carbonitride.** *Titanium carbonitride* (TiCN), violet to mauve red in color (depending on carbon content), is deposited by physical-vapor deposition techniques, and is harder and tougher than TiN. It can be used over carbide and high-speed steel tools, and is particularly effective in cutting stainless steels.
4. **Ceramic coatings.** Because of their high-temperature performance, chemical inertness, low thermal conductivity, and resistance to flank and crater wear, ceramics are attractive coating materials. The most commonly used ceramic coating is *aluminum oxide* (Al_2O_3). However, because ceramic coatings are not chemically reactive, oxide coatings generally bond weakly to the substrate and thus they may have a tendency to peel off the tool.
5. **Multiphase coatings.** The desirable properties of various coatings can be combined and optimized by using *multiphase coatings* (Fig. 8.36). Coated carbide tools are available with two or three layers of such coatings, and are particularly effective in machining cast irons and steels.

   In the example shown in Fig. 8.36, the first layer over the tungsten-carbide substrate is TiC, followed by Al_2O_3, and then TiN. It is important for (a) the first layer to bond well to the substrate; (b) the outer layer to resist wear and have low thermal conductivity; and

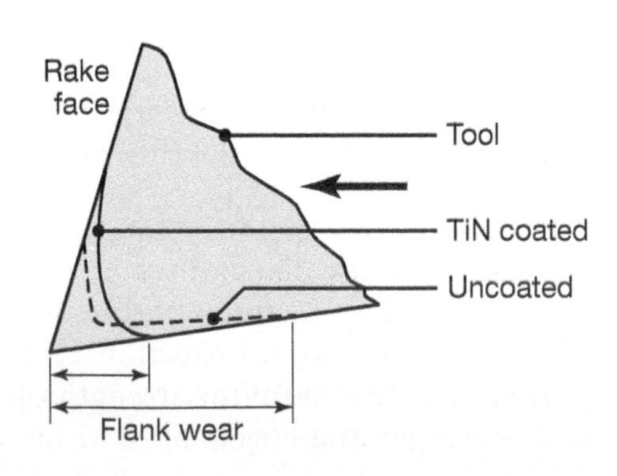

FIGURE 8.35 Wear patterns on high-speed steel uncoated and titanium-nitride-coated cutting tools. Note that flank wear is lower for the coated tool.

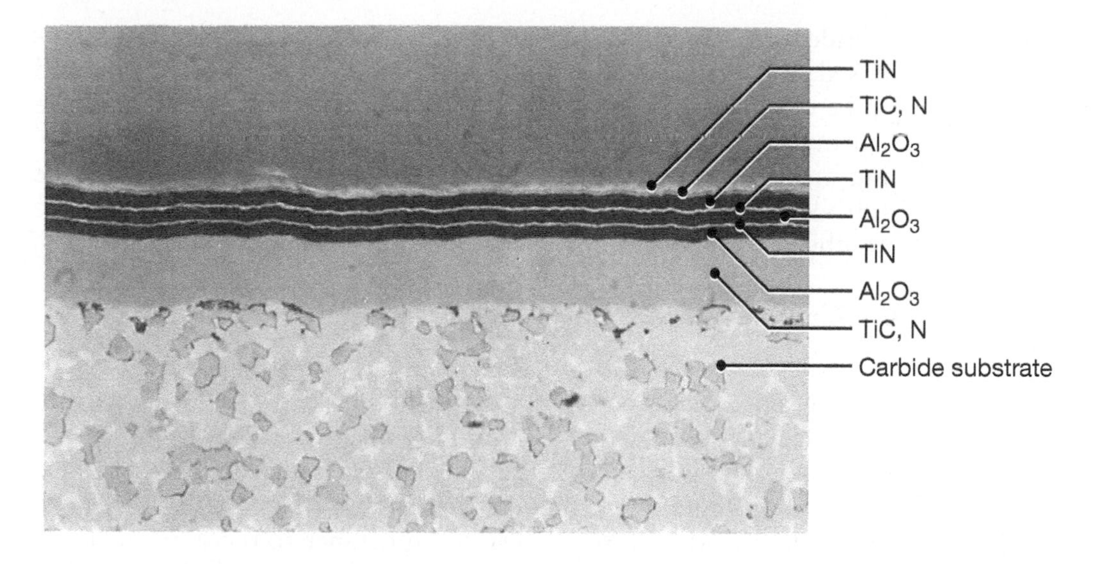

FIGURE 8.36 Multiphase coatings on a tungsten-carbide substrate. Three alternating layers of aluminum oxide are separated by very thin layers of titanium nitride. Inserts with as many as 13 layers of coatings have been made. Coating thicknesses are typically in the range of 2 to 10 μm. *Source:* Courtesy of Kennametal, Inc.

(c) the intermediate layer to bond well and be compatible with both layers.

Typical applications of multiple-coated tools are:

a. High-speed, continuous cutting: TiC/Al_2O_3;
b. Heavy-duty, continuous cutting: $TiC/Al_2O_3/TiN$; and
c. Light, interrupted cutting: TiC/TiC + TiN/TiN.

Coatings consisting of **alternating multiphase layers**, with layers that are thinner than in typical multiphase coatings as shown in Fig. 8.36. The thickness of these layers is in the range of 2–10 μm. The reason for using thinner coatings is that coating hardness increases with decreasing grain size, a phenomenon that is similar to the Hall-Petch effect (see Section 3.4.1).

6. **Diamond coatings.** Polycrystalline diamond is used as a thin coating, particularly over tungsten-carbide and silicon-nitride inserts. The films are deposited on substrates by PVD and CVD techniques, whereas thick films are produced by growing a large sheet of pure diamond, which is then laser cut to shape and brazed to a carbide shank. Diamond-coated tools are particularly effective in machining abrasive materials, such as aluminum-silicon alloys, graphite, and fiber-reinforced and metal-matrix composite materials (see Section 11.14). Improvements in tool life of as much as tenfold have been obtained over other coated tools.
7. **Other coating materials.** Advances are continually being made in developing and testing new coating materials. **Titanium aluminum**

nitride (TiAlN) is effective in machining aerospace alloys. Chromium-based coatings, such as **chromium carbide** (CrC), have been found to be effective in machining softer metals that tend to adhere to the cutting tool, such as aluminum, copper, and titanium. Other coating materials include **zirconium nitride** (ZrN) and **hafnium nitride** (HfN), **nanocoatings** with carbide, boride, nitride, oxide, or some combination, and **composite coatings**, using a variety of materials.

8.6.6 Alumina-Base Ceramics

Ceramic tool materials, introduced in the early 1950s, consist primarily of fine-grained, high-purity **aluminum oxide**. They are pressed into inserts, under high pressure and at room temperature, then sintered (see Section 11.4); they are called **white**, or **cold-pressed, ceramics** (see also Section 11.9.3). Titanium carbide and zirconium oxide can be added to improve such properties as toughness and resistance to thermal shock.

Alumina-base ceramic tools have very high abrasion resistance and hot hardness (Fig. 8.37). Chemically, they are more stable than high-speed steels and carbides, thus they have lower tendency to adhere to metals during machining and hence lower tendency to form a built-up edge. Consequently, good surface finish is obtained, particularly in machining cast irons and steels. However, ceramics lack toughness, which can result in premature tool failure by chipping or fracture (See Fig. 8.20). The shape and setup of ceramic tools also are important; negative rake angles, hence large included angles, are generally preferred in order to avoid chipping. Tool failure can be reduced by increasing the stiffness and damping capacity of machine tools and workholding devices, thus reducing vibration and chatter (see Section 8.12).

Cermets. *Cermets* (from the words *cer*amic and *met*al), also called **black** or **hot-pressed ceramics** (carboxides), typically contain 70% aluminum oxide and 30% titanium carbide. Other cermets may contain molybdenum carbide, niobium carbide, or tantalum carbide. The performance of cermets is between that of ceramics and carbides (See Fig. 8.37).

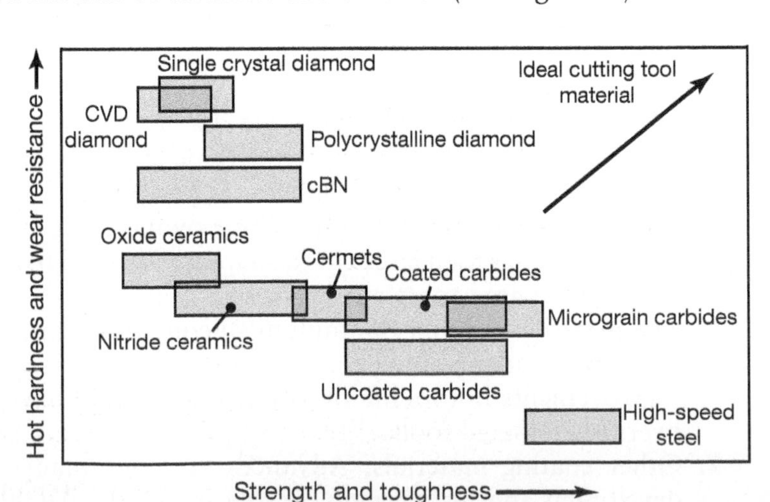

FIGURE 8.37 Ranges of properties for various groups of cutting-tool materials. (See also Tables 8.1 through 8.5.)

8.6.7 Cubic Boron Nitride

Next to diamond, *cubic boron nitride* (cBN) is the hardest material presently available. The cutting tools are made by bonding a 0.5 to 1 mm layer of *polycrystalline cubic boron nitride* to a carbide substrate, by sintering under pressure (Fig. 8.38). While the carbide provides toughness, the cBN layer provides very high wear resistance and cutting-edge strength. Cubic-boron-nitride tools are also made in small sizes without a substrate. At elevated temperatures, cBN is chemically inert to iron and nickel and its resistance to oxidation is high; it is therefore particularly suitable for machining hardened ferrous and high-temperature alloys (see also *hard turning*, in Section 8.9.2). Because cBN tools are brittle, stiffness and damping capacity of the machine tool and fixturing devices is important in order to avoid vibration and chatter. Cubic boron nitride is also used as an abrasive, as described in Section 9.2.

8.6.8 Silicon-Nitride-Base Ceramics

Silicon-nitride-base ceramic (SiN) tool materials consist of silicon nitride with additions of aluminum oxide, yttrium oxide, and titanium carbide. These tools have high toughness, hot hardness, and good thermal-shock resistance. A common example is **sialon**, after the elements *si*licon, *al*uminum, *o*xygen, and *n*itrogen in its composition. It has higher resistance to thermal shock than silicon nitride and is recommended for machining cast irons and nickel-base superalloys, at intermediate cutting speeds. However, because of their chemical affinity to steels, SiN-base tools are not suitable for machining steels.

8.6.9 Diamond

The hardest substance of all known materials is *diamond*, a crystalline form of carbon (see also Section 11.13). As a cutting tool, it has low tool-chip friction, high wear resistance, and thus the ability to maintain a sharp cutting edge. Diamond is used where very fine surface finish and dimensional accuracy are required, particularly in machining abrasive nonmetallic materials and soft nonferrous alloys. *Single-crystal diamond* is used for special applications, such as machining copper-front high-precision optical mirrors. Diamond is brittle, and tool shape and sharpness are thus important; low rake angles and large included angles are normally used for a strong-cutting edge. Wear of diamond tools may occur by microchipping (caused by thermal stresses and oxidation) and transformation to carbon (caused by the high temperatures generated during machining).

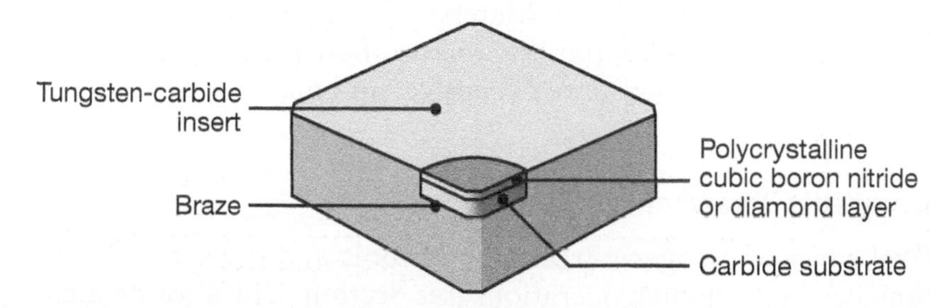

FIGURE 8.38 Construction of polycrystalline cubic-boron-nitride or diamond layer on a tungsten-carbide insert.

Single-crystal single-point diamond tools have been largely replaced by **polycrystalline diamond** tools (called *compacts*), that are also used as wire-drawing dies for fine wire (see Section 6.5). Compacts consist of very small synthetic crystals, fused to a thickness of about 0.5–1 mm, by a high-pressure, high-temperature process, and bonded to a carbide substrate, similar to cBN tools (See Fig. 8.38). The random orientation of the diamond crystals prevents the propagation of cracks through the compact, significantly improving its toughness.

Diamond tools can be used satisfactorily at almost any speed but are suitable mostly for light and uninterrupted finishing cuts. In order to minimize tool fracture, a single-crystal diamond tool must be resharpened as soon as it becomes dull. Because of its strong chemical affinity, diamond is not recommended for machining plain-carbon steels (as is the case with sialon tools) and titanium, nickel, and cobalt-base alloys. Diamond is also used as an abrasive in grinding and polishing operations (Chapter 9), and as a wear-resistant coating (see Section 4.5).

8.6.10 Whisker-Reinforced and Nanocrystalline Tool Materials

In order to further improve the performance and wear resistance of cutting tools, particularly in machining abrasive and hard workpieces, new tool materials are continually being developed with enhanced properties, such as (a) high fracture toughness, (b) resistance to thermal shock, (c) cutting-edge strength, and (d) hot hardness.

Whiskers. *Whiskers* are used as reinforcing fibers in composite tool materials. Examples of **whisker-reinforced** materials include silicon-nitride-base tools reinforced with silicon-carbide (SiC) whiskers, and aluminum-oxide-base tools reinforced with silicon-carbide whiskers, sometimes with the addition of *zirconium oxide* (ZrO_2).

Micrograin carbides. Progress in nanomaterials (Section 3.11.9) has led to the development of cutting tools that are made of very fine-grained (*micrograin*, with grain size in the range of 0.2 to 0.8 μm) carbides of tungsten, titanium, and tantalum. These materials are stronger, harder, and more wear resistant than traditional carbides. In one application, for example, drills with diameters on the order of 100 μm are made from these materials and used in the fabrication of microelectronic circuit boards (see Chapter 13).

Functionally-graded carbides. In these tools, the composition of the carbide in the insert varies through its near-surface depth, instead of being uniform as in carbide inserts. Materials with graded mechanical properties eliminate stress concentrations and promote better tool life and performance; however, they are more expensive, and may not be justified for all applications.

8.6.11 Cryogenic Treatment of Cutting Tools

The beneficial effect of cryogenic treatment of tools and other metals on their performance in machining operations (see Section 5.11.6 for details) continues to be of some interest. In this application, the tool is cooled very slowly to temperatures of about −180°C (−300°F) and slowly returned to room temperature; it is then tempered. Depending on the combinations of tool and workpiece materials involved, tool-life increases of up to 300% can be achieved with this procedure.

8.7 Cutting Fluids

Cutting fluids are used extensively in machining operations to:

- Cool the cutting zone, thus reducing workpiece temperature and distortion, and improving tool life;
- Reduce friction and wear, thus improving tool life and surface finish;
- Reduce forces and energy consumption;
- Wash away chips; and
- Protect the newly machined surfaces from environmental attack.

A cutting fluid predominantly serves as a **coolant** and/or as a **lubricant** (see Section 4.4.4). Its effectiveness in machining operations depends on several factors, such as the method of application, temperature, cutting speed, and type of machining operation. There are situations, however, in which the use of cutting fluids can be detrimental. For example, in interrupted cutting operations, such as milling (Section 8.10), the cooling action of the cutting fluid increases the extent of alternate heating and cooling (*thermal cycling*) to which the cutter teeth are subjected, a condition that can lead to the development of thermal cracks (*thermal fatigue* or *thermal shock*). Moreover, cutting fluids may also cause the chip to become more curled (i.e., smaller radius of curvature), concentrating the stresses on the tool closer to its tip, thus concentrating the heat closer to the tool tip and reduce tool life.

Cutting fluids can present **biological** and **environmental hazards** (see also Section 4.4.4), requiring proper recycling and disposal, and adding to cost. For these reasons, **dry cutting**, or **dry machining**, has become an increasingly important approach, in which no coolant or lubricant is used in the machining operation (see Section 8.7.2). Even though this approach would suggest that higher temperatures and thus more rapid tool wear would occur, some tool materials and coatings maintain an acceptable tool life. Dry cutting has been associated with high-speed machining, because higher cutting speeds transfer a greater amount of heat to the chip (See Fig. 8.18), which is an incentive for reducing the need for a coolant. See also Section 3.9.7 on possible detrimental effects of cutting fluids on some cutting tools, called *selective leaching*, such as in carbide tools with cobalt binders.

8.7.1 Types of Cutting Fluids and Methods of Application

There are four basic types of cutting fluids commonly used in machining operations: **oils, emulsions, semisynthetics,** and **synthetics** (see Section 4.4.4). Cutting-fluid recommendations for specific machining operations are given throughout the rest of this chapter. In selecting an appropriate cutting fluid, considerations should be given to its possible detrimental effects on the workpiece material (such as corrosion, stress-corrosion cracking, staining), machine tool components, biological and environmental effects, and recycling and disposal of chips.

The most common method of applying cutting fluids is **flood cooling**. Flow rates typically range from 10 L/min (2.6 gal/min) for single-point tools, to 225 L/min (60 gal/min) per cutter for multiple-tooth cutters, as in milling. In such operations as gun drilling and end milling (see Section 8.10.1), fluid pressures in the range of 700–14,000 kPa (100–2000 psi) are used to flush away the chips.

Mist cooling involves delivery of fluid as small droplets suspended in air, and is generally used with water-base fluids. Although it requires venting (to prevent inhalation of fluid particles by the machine operator and others nearby) and has limited cooling capacity, mist cooling supplies fluid to otherwise inaccessible areas and provides better visibility of the workpiece being machined. It is particularly effective in grinding operations (see Chapter 9), using air pressures in the range of 70–600 kPa (10–90 psi).

High-pressure refrigerated coolant systems can be used to improve the rate of heat removal from the cutting zone. Pressures on the order of 35 MPa (5 ksi) are used to deliver the fluid, via specially-designed nozzles that aim a powerful jet of fluid to the zone. This action breaks up the chips (thus the fluid also acts as a chip breaker) in situations where the chips produced would otherwise be too long and continuous, and thus interfere with the machining operation.

Through the cutting-tool system. One method to overcome the difficulty of supplying cutting fluids into the cutting zone and flushing away the chips is to provide them through the cutting tool. Narrow passages are produced in the cutting tool and the toolholder, through which the cutting fluid is applied under high pressure.

8.7.2 Near-Dry and Dry Machining

For economic and environmental reasons, there is a continuing trend to minimize or eliminate the use of metalworking fluids. This trend has lead to the practice of *near-dry machining* (NDM), where coolant use is eliminated or reduced significantly. The significance of this approach is apparent when noting that, in the United States alone, millions of liters of metalworking fluids are consumed each year. Moreover, it has been estimated that metalworking fluids constitute about 7 to 17% of the total machining costs. The major benefits of NDM include:

1. Reducing the environmental impact of using cutting fluids, improving air quality in manufacturing plants, and reducing health hazards.
2. Reducing the cost of machining operations, including the cost of maintaining, recycling, and disposing of cutting fluids.

The principle behind near-dry machining is the application of a fine mist of air and fluid mixture, containing a very small amount of cutting fluid.

The mixture is delivered to the cutting zone, through the spindle of the machine tool, typically through a 1-mm diameter nozzle and under a pressure of 600 kPa (90 psi). The fluid is applied at rates of 1 to 100 cc/hr, which is estimated to be, at most, one ten-thousandth of that used in flood cooling; consequently, the process is also known as *minimum quantity lubrication* (MQL).

With continued advances in cutting-tool materials, *dry machining* has been shown to be effective, especially in turning, milling, and gear cutting of steels, alloys steels, and cast irons, although generally not for aluminum alloys.

Recall that one of the functions of a cutting fluid is to flush chips away from the cutting zone. Although it first appears that this may present difficulties in dry machining, tools have been designed to allow the application of *pressurized air*, often through holes in the tool shank (See Fig. 8.48). The compressed air provides limited cooling of the cutting zone, but it is very effective at clearing chips away from the cutting zone.

8.7.3 Cryogenic Machining

In the interest of reducing or eliminating the adverse environmental impact of using metalworking fluids, *liquid nitrogen* can be used as a coolant in machining, as well as in grinding (see Section 9.6.9). With appropriate small-diameter nozzles, liquid nitrogen is injected at a temperature of about −200°C into the tool-workpiece interface, reducing its temperature. As a result, tool hardness is maintained, and tool life is enhanced, thus allowing for higher cutting speeds. Moreover, the chips become more brittle and easier to flush from the cutting zone. Because no fluids are involved and the liquid nitrogen simply evaporates, the chips can be recycled more easily.

8.8 High-Speed Machining

With continuing demands for higher productivity and lower manufacturing costs, much effort has been expended in increasing the cutting speed. Although *high-speed machining* (HSM) is a relative term, an approximate range of cutting speeds are:

1. **High speed:** 600–1800 m/min;
2. **Very high speed:** 1800–18,000 m/min; and
3. **Ultrahigh speed:** >18,000 m/min.

Spindle rotational speeds in machine tools range up to 60,000 rpm. The *spindle power* required is generally on the order of 0.004 W/rpm, which is much lower than in traditional machining which is typically in the range of 0.2 to 0.4 W/rpm. The maximum workpiece speed in high-speed machining is on the order of 1 m/s, and the acceleration of machine-tool components is very high.

Spindles for high rotational speeds require *high stiffness and accuracy*, and generally involve an integral electric motor. The armature is built onto the shaft and the stator is placed in the wall of the spindle housing. The bearings may be rolling element or hydrostatic; the latter is more desirable because it requires less space than the former. Because of *inertial effects* during acceleration and deceleration of machine-tool components, the use of lightweight materials, including ceramics and composite materials, is important. Selection of appropriate cutting-tool materials also is a major consideration. Depending on the workpiece material, multiphase coated carbides, ceramics, cubic boron nitride, and diamond are typical tool materials for high-speed machining.

High-speed machining should be considered primarily for operations where **cutting time** is a significant portion of the time in the overall machining operation. As described in Section 8.15, **non-cutting time** and various other factors (e.g. tool material costs, capital equipment costs, labor costs, etc.) are important considerations in the overall assessment of the benefits of high-speed machining for a particular application. It has, for example, been implemented in machining (a) aluminum structural components for aircraft; (b) submarine propellers of 6 m diameter, made of nickel-aluminum-bronze alloy and weighing 55,000 kg; and (c) automotive engines, with five to ten times the productivity of traditional machining.

Another major factor in the adoption of high-speed machining has been the requirement to further improve dimensional tolerances. As can be seen in Fig. 8.18, as the cutting speed increases, more and more of the heat generated is removed by the chip; thus the tool and, more importantly, the workpiece remain close to ambient temperature. The machine-tool characteristics and special requirements that are important in high-speed machining may be summarized as follows:

1. Spindle design for high stiffness, accuracy, and balance at very high rotational speeds, and workholding devices that can withstand high centrifugal forces;
2. Fast feed drives, bearing characteristics, and effects of inertia of the machine-tool components;
3. Selection of appropriate cutting tools, processing parameters, and their computer control; and
4. effective chip removal systems at very high rates.

8.9 Cutting Processes and Machine Tools for Producing Round Shapes

This section describes the processes that produce parts that are *round in shape*, as outlined in Table 8.7. Typical products machined include parts as small as miniature screws for eyeglass hinges and as large as cylinders, gun barrels, and turbine shafts for hydroelectric power plants. These processes are generally performed by turning the workpiece on a lathe. *Turning* means that the part is rotating while it is being machined using a cutting tool. The starting material is typically a workpiece that has been produced by other processes, such as casting, forging, extrusion, and

TABLE 8.7 General characteristics of machining processes.

Process	Characteristics	Commercial tolerances (±mm)
Turning	Turning and facing operations are performed on all types of materials; requires skilled labor; low production rate, but medium to high rates can be achieved with turret lathes and automatic machines, requiring less skilled labor	Fine: 0.025–0.13 Rough: 0.13 Skiving: 0.025–0.05
Boring	Internal surfaces or profiles, with characteristics similar to those produced by turning; stiffness of boring bar is important to avoid chatter	0.025
Drilling	Round holes of various sizes and depths; requires boring and reaming for improved accuracy; high production rate, labor skill required depends on hole location and accuracy specified	0.075
Milling	Variety of shapes involving contours, flat surfaces, and slots; wide variety of tooling; versatile; low to medium production rate; requires skilled labor	0.13–0.25
Planing	Flat surfaces and straight contour profiles on large surfaces; suitable for low-quantity production; labor skill required depends on part shape	0.08–0.13
Shaping	Flat surfaces and straight contour profiles on relatively small workpieces; suitable for low-quantity production; labor skill required depends on part shape	0.05–0.13
Broaching	External and internal flat surfaces, slots, and contours with good surface finish; costly tooling; high production rate; labor skill required depends on part shape	0.025–0.15
Sawing	Straight and contour cuts on flats or structural shapes; not suitable for hard materials unless the saw has carbide teeth or is coated with diamond; low production rate; requires only low skilled labor	0.8

drawing. Turning processes are very versatile and capable of producing a wide variety of shapes, as outlined in Fig. 8.39.

- **Turning** straight, conical, curved, or grooved workpieces, such as shafts, spindles, pins, handles, and various machine components;
- **Facing**, to produce a flat surface at the end of a part, such as for those that are attached to other components, or to produce grooves for O-ring seats;
- Producing various shapes by **form tools**, for functional purposes and for appearance;
- **Boring**, to enlarge a hole made by a previous process or in a tubular workpiece, or to produce internal grooves;
- **Drilling**, to produce a hole, which may be followed by tapping or by boring to improve the accuracy of the hole and its surface finish;
- **Parting**, also called **cutting off**, to cut a piece from the end of a part, as in making slugs or blanks for subsequent processing into discrete parts;
- **Threading**, to produce external or internal threads in workpieces; and
- **Knurling**, to produce surface characteristics on cylindrical surfaces, as in making knurled knobs.

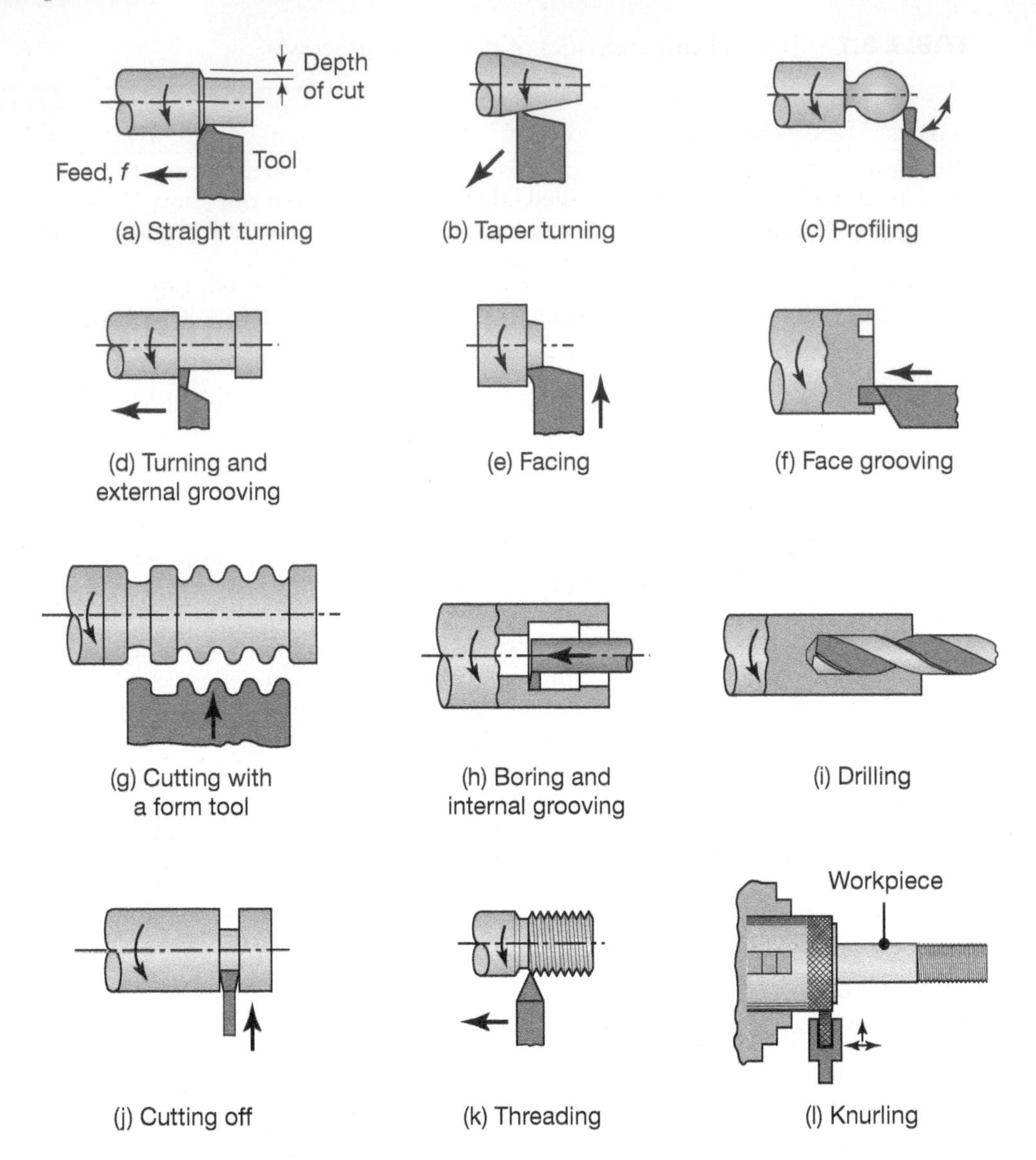

FIGURE 8.39 Examples of the wide variety of machining operations that can be performed on a lathe and similar machine tools.

These machining operations may be performed at various rotational speeds of the workpiece, depths of cut, d, and feed, f (See Fig. 8.19), depending on the workpiece and tool materials, the surface finish and dimensional accuracy required, and the capacity of the machine tool.

Roughing cuts are performed for large-scale material removal, and typically involve depths of cut greater than 0.5 mm (0.02 in.) and feeds on the order of 0.2–2 mm/rev. **Finishing cuts** usually involve smaller depths of cut and feed. Most machining operations consist of roughing cuts to define the part shape, followed by a finishing cut to meet specific dimensional tolerances and surface finish requirements.

8.9.1 Turning Parameters

The majority of turning operations involve the use of *single-point* cutting tools. Figure 8.40 shows the geometry of a typical right-hand cutting tool for turning; such tools are identified by a standardized nomenclature. Each group of tool and workpiece materials has an optimum set of tool angles, that have been developed through many years of industrial experience. Some data on tool geometry can be found in Table 8.8.

1. **Tool geometry**. The various angles on a cutting tool have important functions in machining operations. (a) **Rake angles** are important in controlling the direction of chip flow and in the strength of the tool

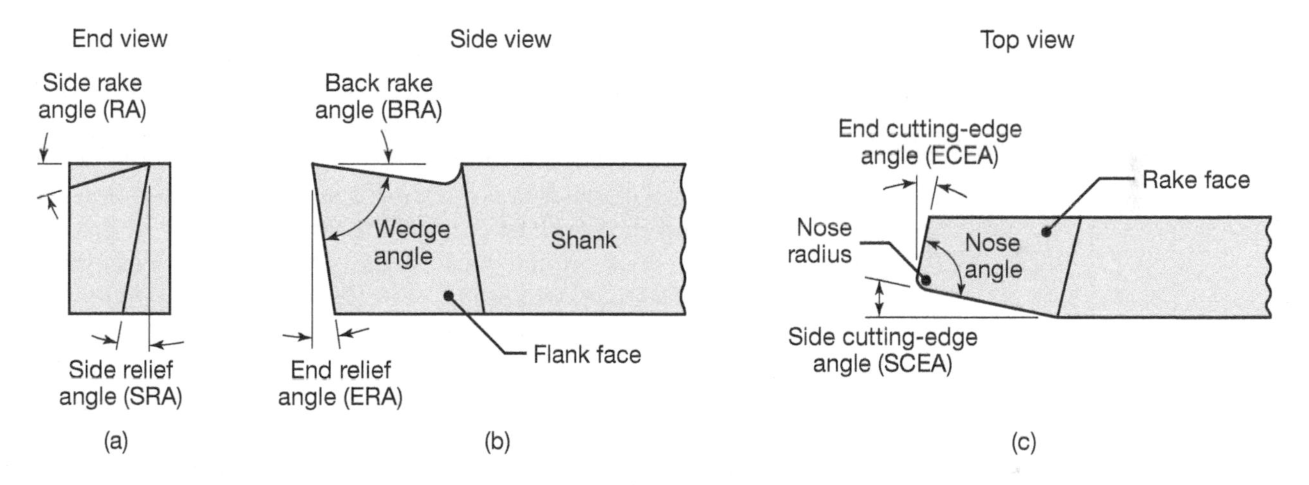

FIGURE 8.40 Designations and symbols for a right-hand cutting tool. The designation "right hand" means that the tool travels from right to left, as shown in Fig. 8.19.

TABLE 8.8 General recommendations for tool angles (in degrees) in turning.

	High-speed steel					Carbide inserts				
Material	Back rake	Side rake	End relief	Side relief	Side and end cutting edge	Back rake	Side rake	End relief	Side relief	Side and end cutting edge
Aluminum and magnesium alloys	20	15	12	10	5	0	5	5	5	15
Copper alloys	5	10	8	8	5	0	5	5	5	15
Steels	10	12	5	5	15	−5	−5	5	5	15
Stainless steels	5	8–10	5	5	15	−5–0	−5–5	5	5	15
High-temperature alloys	0	10	5	5	15	5	0	5	5	45
Refractory alloys	0	20	5	5	5	0	0	5	5	15
Titanium alloys	0	5	5	5	15	−5	−5	5	5	5
Cast irons	5	10	5	5	15	−5	−5	5	5	15
Thermoplastics	0	0	20–30	15–20	10	0	0	20–30	15–20	10
Thermosets	0	0	20–30	15–20	10	0	15	5	5	15

tip. Positive-rake angles improve the cutting operation by reducing forces and temperatures; however, positive angles also have a small included angle of the tool tip (See Fig. 8.2). Depending on the toughness of the tool material, a small included angle may cause premature tool chipping and failure. (b) **Side-rake angle** is more important than **back-rake angle,** although the latter usually controls the direction of chip flow. (c) **Relief angles** control interference and rubbing at the tool-workpiece interface. If the relief angle is too large, the tool tip may chip off, and if it is too small, flank wear may be excessive. (d) **Cutting-edge angles** affect chip formation, tool strength, and cutting forces. (e) **Nose radius** affects surface finish and tool-tip strength. The smaller the radius, the rougher the surface finish of the workpiece, and the lower the strength of the tool; on the other hand, large nose radii can lead to tool chatter (see Section 8.12).

2. **Material-removal rate.** The *material-removal rate* (MRR) is the volume of material removed per unit time, such as mm^3/min. Referring to Fig. 8.41a, note that for each revolution of the workpiece, a ring-shaped layer of material is removed; its cross-sectional area is the product of the axial distance the tool travels in one revolution (known as the feed, f) and the depth of cut, d. The volume of this ring is the product of the cross-sectional area (fd) and the average circumference of the ring (πD_{avg}), where $D_{avg} = (D_o + D_f)/2$. For light cuts on large-diameter workpieces, the average diameter can be replaced by D_o.

   The material-removal rate per revolution is $\pi D_{avg} df$. Since the rotational speed of the workpiece is N revolutions per minute, the removal rate is

$$\text{MRR} = \pi D_{avg} d f N. \qquad (8.38)$$

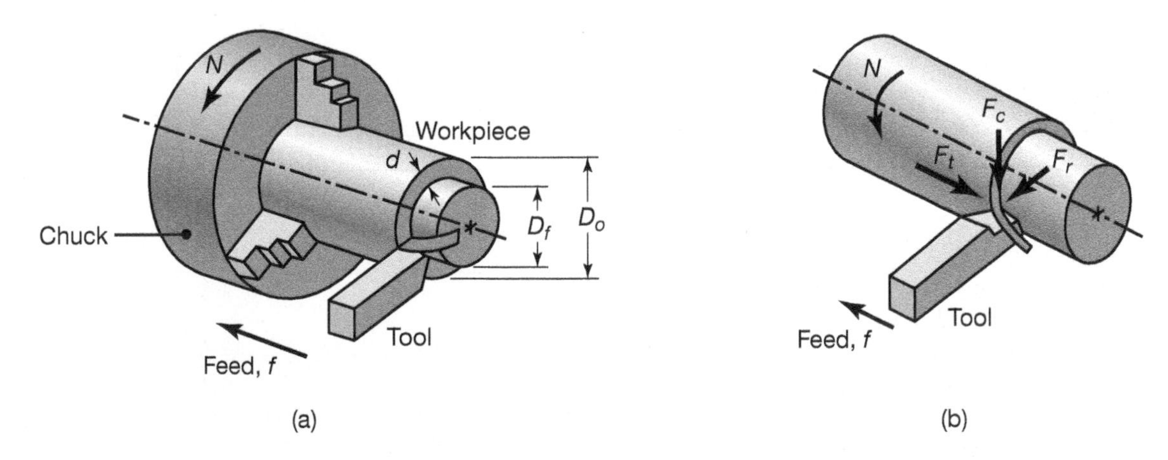

FIGURE 8.41 (a) Schematic illustration of a turning operation, showing depth of cut, d, and feed, f. Cutting speed is the surface speed of the workpiece at the tool tip. (b) Forces acting on a cutting tool in turning. F_c is the cutting force; F_t is the thrust or feed force (in the direction of feed); and F_r is the radial force that tends to push the tool away from the workpiece being machined. Compare this figure with Fig. 8.11 for a two-dimensional cutting operation.

Similarly, the cutting time, t, for a workpiece of length l can be calculated by noting that the tool travels at a *feed rate* of fN = (mm/rev)(rev/min) = mm/min. Since the distance traveled is l mm, the cutting time is

$$t = \frac{l}{fN}. \qquad (8.39)$$

The cutting time does not include (a) the time required for *tool approach*; and (b) *retraction* during the overall machining operation. Using computer controls, modern machine tools are designed and constructed to minimize the *nonproductive time*. A typical method employed in practice is to first quickly move the tool and then to slow it down as the tool engages with the workpiece.

3. **Forces in turning.** The three principal forces acting on a cutting tool in turning are illustrated in Fig. 8.41b. These forces are important in the design of machine tools and in determining the tool deflection, which is especially important in precision machining operations. The *cutting force*, F_c, acts downward on the tool and thus tends to deflect the tool downward; note that the cutting force is the force that supplies the energy required for the cutting operation. As can be seen in Example 8.4, the cutting force can be calculated (a) from the energy per unit volume of material machined, described in Section 8.2.5; and (b) by using the data given in Table 8.3.

   The *thrust force*, F_t, acts in the longitudinal direction; this force is also called the *feed force* because it is in the feed direction. The *radial force*, F_r, is in the radial direction; it tends to push the tool away from the workpiece.

4. **Tool materials, feeds, and cutting speeds.** The general characteristics of cutting-tool materials are described in Section 8.6. Figure 8.42

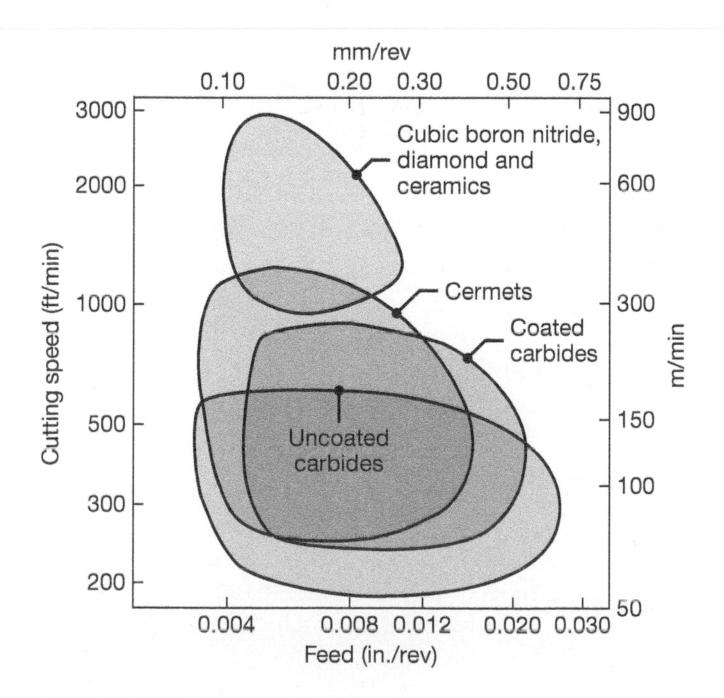

FIGURE 8.42 The range of applicable cutting speeds and feeds for a variety of cutting-tool materials.

TABLE 8.9 Approximate ranges of recommended cutting speeds for turning operations.

	Cutting speed	
Workpiece material	m/min	ft/min
Aluminum alloys	200–1000	650–3300
Cast iron, gray	60–900	200–3000
Copper alloys	50–700	160–2300
High-temperature alloys	20–400	65–1300
Steels	50–500	160–1600
Stainless steels	50–300	160–1000
Thermoplastics and thermosets	90–240	300–800
Titanium alloys	10–100	30–330
Tungsten alloys	60–150	200–500

Note: (a) The speeds given in this table are for carbides and ceramic cutting tools. Speeds for high-speed steel tools are lower than indicated. The higher ranges are for coated carbides and cermets. Speeds for diamond tools are significantly higher than any of the values indicated in the table.
(b) Depths of cut, d, are generally in the range of 0.5–12 mm.
(c) Feeds, f, are generally in the range of 0.15–1 mm/rev.

gives a broad range of cutting speeds and feeds applicable for these tool materials. Specific recommendations for cutting speeds for turning various workpiece materials and for cutting tools are given in Table 8.9.

EXAMPLE 8.4 Material-Removal Rate and Cutting Force in Turning

Given: A 150-mm-long, 10-mm-diameter, 304 stainless-steel rod is being reduced in diameter to 8 mm by turning on a lathe. The spindle rotates at $N = 400$ rpm, and the tool is traveling at an axial speed of 200 mm/min.

Find: Calculate the cutting speed, material-removal rate, cutting time, power dissipated, and cutting force.

Solution: The cutting speed is the tangential speed of the workpiece. The maximum cutting speed is at the outer diameter, D_o, and is obtained from the expression

$$V = \pi D_o N.$$

Thus,

$$V = (\pi)(0.010)(400) = 12.57 \text{ m/min}.$$

The cutting speed at the machined diameter is

$$V = (\pi)(0.008)(400) = 10.05 \text{ m/min}.$$

From the information given, note that the depth of cut is

$$d = \frac{10 - 8}{2} = 1 \text{ mm} = 0.001 \text{ m},$$

and the feed is

$$f = \frac{200}{400} = 0.5 \text{ mm/rev} = 0.0005 \text{ m/rev}.$$

Thus, according to Eq. (8.38), the material-removal rate is

$$\text{MRR} = (\pi)(9)(1)(0.5)(400) = 5655 \text{ mm}^3/\text{min}.$$

The actual time to cut, according to Eq. (8.39), is

$$t = \frac{150}{(0.5)(400)} = 0.75 \text{ min}.$$

The power required can be calculated by referring to Table 8.3 and taking an average value for stainless steel as 4.1 W-s/mm^3. Therefore, the power dissipated is

$$\text{Power} = \frac{(4.1)(5655)}{60} = 386 \text{ W}.$$

The cutting force, F_c, is the tangential force exerted by the tool. Since power is the product of torque, T, and rotational speed in radians per unit time, we have

$$T = \frac{(386)}{(400)(2\pi/60)} = 9.2 \text{ Nm}.$$

Since $T = (F_c)(D_{avg}/2)$,

$$F_c = \frac{(9.2)(2)}{(0.009)} = 2.0 \text{ kN}.$$

8.9.2 Lathes and Lathe Operations

Lathes are generally considered to be the oldest machine tools. Although woodworking lathes were first developed during the period 1000–1 B.C., metalworking lathes, with lead screws, were not built until the late 1700s. The most common type of lathe, shown in Fig. 8.43, was originally called an **engine lathe**, because it was powered with overhead pulleys and belts from nearby engines.

1. **Lathe components.** Lathes are typically equipped with a variety of components and accessories, as shown in Fig. 8.43. The *bed* supports all the other major components of the lathe. The *carriage*, or *carriage assembly*, which slides along the *ways*, consists of an assembly of the *cross slide, tool post*, and *apron*. The cutting tool is mounted on the *tool post*, usually with a *compound rest* that swivels to allow for tool positioning and adjustments. The *headstock*, which is fixed to the bed, is equipped with motors, pulleys, and V-belts that supply power to the *spindle* at various rotational speeds. Headstocks have a hollow spindle to which workholding devices, such as *chucks* and *collets*,

QR Code 8.1 Turning and profiling on a turning center. *Source:* Courtesy of Haas Automation, Inc.

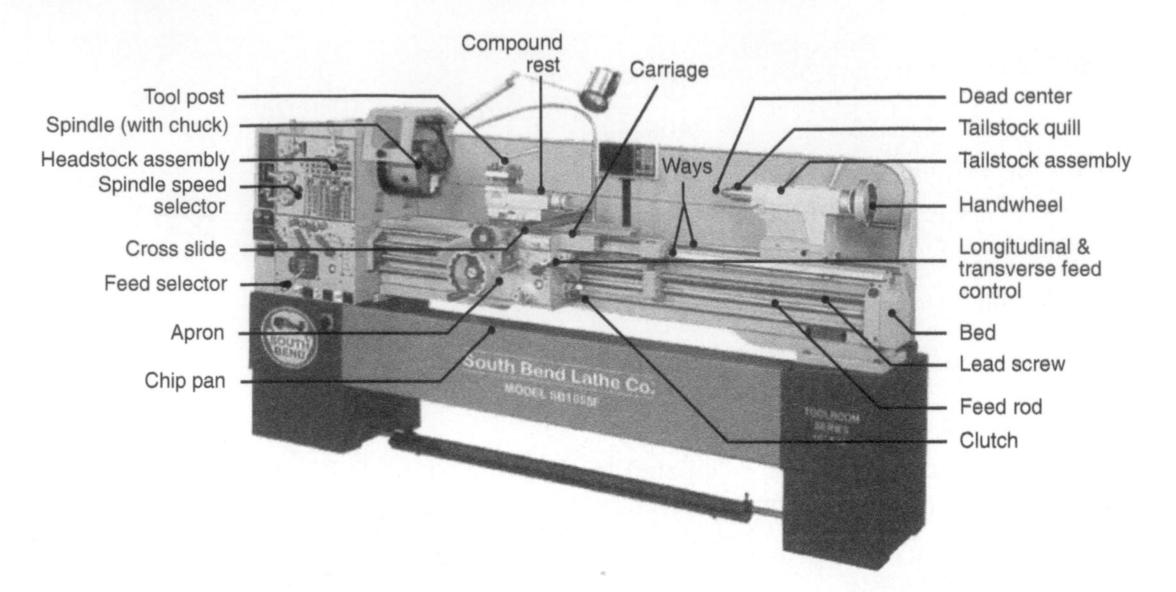

FIGURE 8.43 General view of a typical lathe, showing various components. *Source:* Courtesy of South Bend Lathe Co.

are attached. The *tailstock*, which can slide along the ways and be clamped at any position, supports the opposite end of the workpiece. The *feed rod*, which is powered by a set of gears from the headstock, rotates during operation of the lathe and provides movement to the carriage and the cross slide. The *lead screw*, which is used for accurately cutting threads, is engaged with the carriage by closing a split nut around the lead screw.

A lathe is generally specified by (a) its *swing* (the maximum diameter of the workpiece that can be machined); (b) the maximum distance between the headstock and tailstock centers; and (c) the length of the bed. The wide variety of lathes include *bench lathes, toolroom lathes, engine lathes, gap lathes,* and *special-purpose lathes*.

Tracer lathes, also called *duplicating lathes* or *contouring lathes*, are capable of turning parts to various contours, where the cutting tool follows a path that duplicates the contour of a template. *Automatic lathes*, also called *chucking machines* or *chuckers*, are used for machining individual pieces of regular or irregular shapes. *Turret lathes* are capable of performing multiple machining operations on the same workpiece, such as turning, boring, drilling, thread cutting, and facing. Several cutting tools can be mounted on the hexagonal *main turret*. The lathe usually also has a *square turret* on the cross slide, with as many as four cutting tools mounted on it.

2. **Computer-controlled lathes.** In modern lathes, the movement and control of the machine tool and its components are accomplished by **computer numerical controls** (CNCs); the features of such a lathe are shown in Fig. 8.44. These lathes are typically equipped with one or more turrets; each turret is equipped with a variety of cutting tools and performs several operations on different surfaces of the

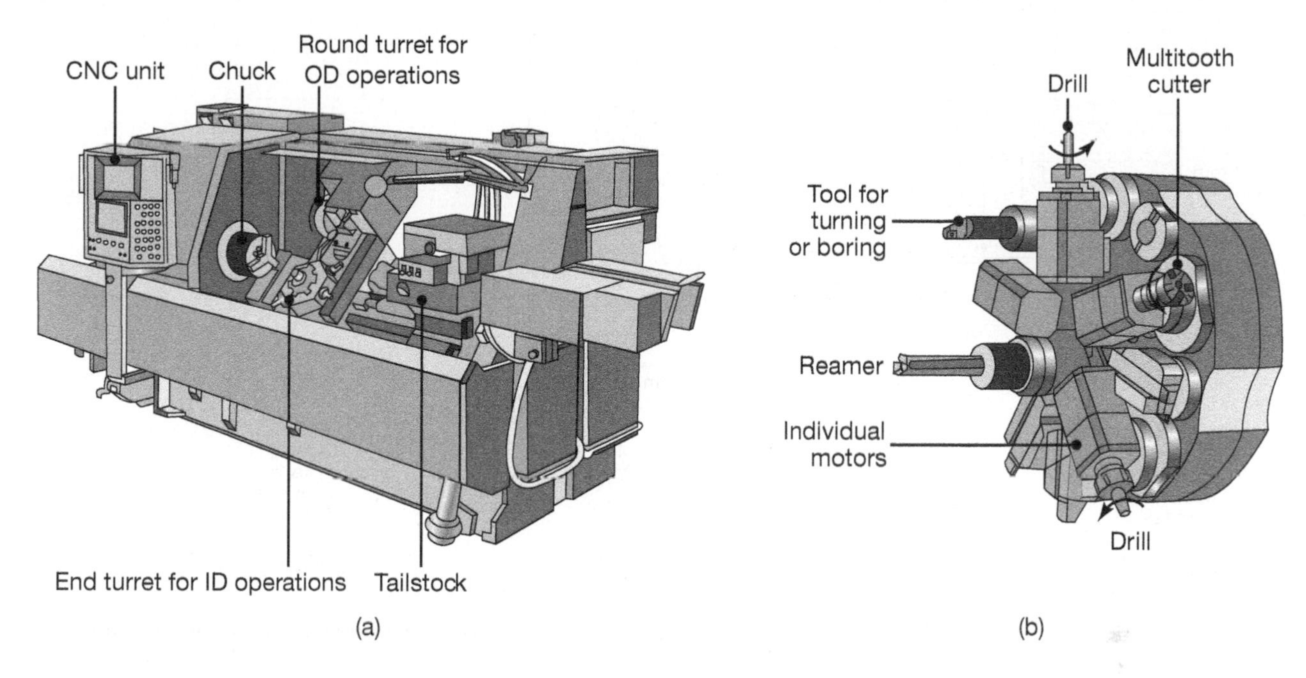

FIGURE 8.44 (a) A computer-numerical-control lathe, with two turrets; these machines have higher power and spindle speed than other lathes in order to take advantage of advanced cutting tools with enhanced properties; (b) a typical turret equipped with 10 cutting tools, some of which are powered.

workpiece. These machine tools are highly automated, the operations are repetitive, they maintain the desired dimensional accuracy, and, once the machine is set up, less skilled labor is required than with common lathes. These lathes are suitable for low to medium volumes of production. Details of the controls are given in Chapters 14 and 15.

EXAMPLE 8.5 Typical Parts Made on CNC Turning Machine Tools

Figure 8.45 illustrates the capabilities of CNC turning machine tools. Workpiece materials, the number of cutting tools used, and machining times are indicated for each part. These parts also can be made on manual or turret lathes, but with more difficulty, higher cost, and less consistency.

Source: Monarch Machine Tool Company.

3. **Turning process capabilities.** Table 8.10 lists relative *production rates* for turning, as well as for other machining operations described in the rest of this chapter. These rates have an important bearing on productivity in machining operations; note that there is a wide range in the production rates of these processes. The differences are due not only to the inherent characteristics of the processes and machine tools, but also to various other factors, such as setup times and the types and sizes of the workpieces involved. The proper selection of a process and the machine tool for a particular product is important

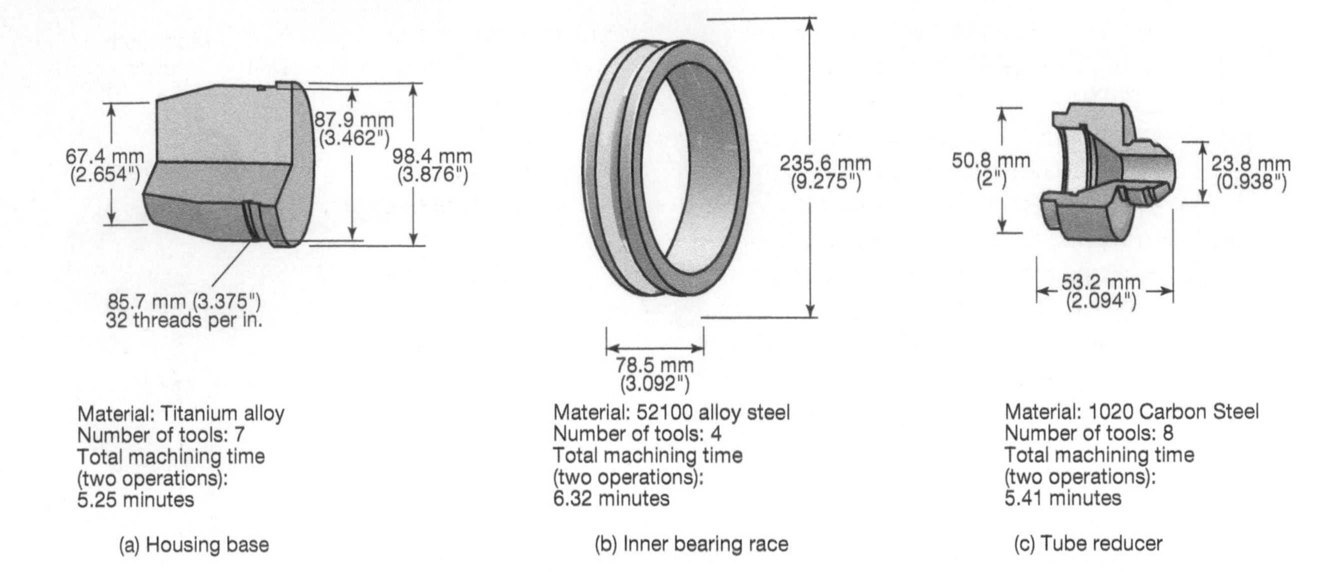

FIGURE 8.45 Typical parts made on computer-numerical-control machine tools.

TABLE 8.10 Typical production rates for various cutting operations.

Operation	Rate
Turning	
Engine lathe	Very low to low
Tracer lathe	Low to medium
Turret lathe	Low to medium
Computer-control lathe	Low to medium
Single-spindle chucker	Medium to high
Multiple-spindle chucker	High to very high
Boring	Very low
Drilling	Low to medium
Milling	Low to medium
Planing	Very low
Gear cutting	Low to medium
Broaching	Medium to high
Sawing	Very low to low

Note: Production rates indicated are relative: *Very low* is about one or more parts per hour; *medium* is approximately 100 parts per hour; *very high* is 1000 or more parts per hour.

for minimizing production costs, as outlined in Section 8.15 and Chapter 16.

The ratings given in Table 8.10 are relative, and there can be significant variations in specific applications. For example, high-carbon cast-steel rolls (for rolling mills; see Section 6.3) can be machined on special lathes at material-removal rates as high as 6000 cm^3/min, using multiple cermet tools. The important factor in this operation (also called **high-removal-rate machining**) is the very high stiffness of the machine tool (to avoid tool breakage due to chatter; see Section 8.12) and its high power, which can be up to 450 kW.

The surface finish and dimensional accuracy obtained in turning and related operations depend on such factors as the characteristics and condition of the machine tool, stiffness, vibration and chatter, processing parameters, tool geometry, tool wear, cutting fluids, machinability of the workpiece material, and operator skill. Consequently, a wide range of surface finishes can be obtained, as shown in Fig. 8.25 (See also Fig. 9.28).

4. **Ultraprecision machining.** There are continued demands for precision manufactured components for computer, electronics, nuclear energy, and defense applications. Examples include optical mirrors and components for optical-systems, with surface finish requirements in the range of tens of nanometers (10^{-9} m or 0.001 μm) and shape accuracies in the μm and sub-μm range. The cutting tool for these *ultraprecision machining* applications is exclusively a single-crystal diamond (hence, the process is also called **diamond turning**), with a polished cutting edge that has a radius as small as a few tens of nanometers; thus, wear of the diamond can be a significant problem.

   The workpiece materials for ultraprecision machining include copper alloys, aluminum alloys, silver, gold, electroless nickel, infrared materials, and plastics (acrylics). The depths of cut involved are in the nanometer range. In this range, hard and brittle materials produce continuous chips (known as **ductile-regime cutting**; see also the discussion of *ductile-regime grinding* in Section 9.5.3); deeper cuts tend to produce discontinuous chips.

   The machine tools for ultraprecision machining are built with very high precision and high machine, spindle, and workholding-device stiffness. These machines, some parts of which are made of structural materials with low thermal expansion and good dimensional stability, are located in a dust-free environment (*clean rooms*) where the temperature is controlled within a fraction of one degree. In **cryogenic diamond turning**, the tooling system is cooled by liquid nitrogen to a temperature of about −120°C (−184°F) (see also Section 8.7.3). Vibrations from external and internal sources must be avoided as much as possible. Feed and position controls are made through laser metrology, and the machines are equipped with highly-advanced computer control systems, which also include thermal and geometric error-compensating features.

5. **Hard turning.** As described in Chapter 9, there are several other processes, particularly grinding, and nonmechanical methods of removing material economically from hard or hardened metals. However, it is still possible to apply traditional machining processes to hard metals and alloys by selecting an appropriate tool material and a machine tool with high stiffness. One common example is finish machining of heat-treated steel (45 to 65 HRC) machine and automotive

components, using polycrystalline cubic-boron-nitride (PcBN) tools. Called *hard turning*, this process produces machined parts with good dimensional accuracy, surface finish, and surface integrity. It has been shown that it can compete successfully with grinding the same components, from both technical and economic aspects. A comparative example of hard turning vs. grinding is given in Example 9.4.

6. **Cutting screw threads.** External threads are produced primarily by (a) *thread rolling* (See Fig. 6.42), but also by (b) cutting, as shown in Fig. 8.39k. When threads are produced externally or internally by cutting, the process is called *thread cutting* or *threading*. When the threads are cut internally with a special threaded tool (*tap*), the process is called *tapping*. External threads may also be cut with a die or by milling. Although it adds considerably to the cost of the operation, threads also may be ground for high accuracy and surface finish.

   Automatic screw machines are designed for machining of screws and similar threaded parts at high production-rates. Because these machines are also capable of producing other components, they are generally called *automatic bar machines*. All operations are performed automatically, with tools clamped to a special turret. The bar stock is automatically fed forward through an opening in the headstock, after each screw or part is machined to finished dimensions and then cut off. The machines may be equipped with single or multiple spindles, and capacities range from 3–150-mm (0.12–6 in.) diameter bar stock.

8.9.3 Boring and Boring Machines

The basic *boring* operation, shown in Fig. 8.39h, consists of producing circular internal profiles in hollow workpieces and enlarging or finishing a hole made by another process, such as drilling. It is carried out using cutting tools that are similar to those used in turning. Note that the boring bar has to reach the full length of the bore, thus tool deflection and maintaining dimensional accuracy can be a concern. The boring bar must also be sufficiently stiff, and is made of a material with high elastic modulus (such as carbides) to minimize deflection and avoid vibration and chatter. Boring bars have been designed with capabilities for damping vibrations (see Section 8.12).

Although boring operations on relatively small workpieces can be carried out on a lathe, **boring mills** are used for large workpieces. These machines are either vertical or horizontal in design and are also capable of performing such operations as turning, facing, grooving, and chamfering. A *vertical boring machine* (Fig. 8.46) is similar to a lathe, but with a vertical axis of workpiece rotation. In *horizontal boring machines*, the workpiece is mounted on a table that can move horizontally, in both axial and radial directions. The cutting tool is mounted on a spindle that rotates in the headstock, and is capable of both vertical and longitudinal movements. Drills, reamers, taps, and milling cutters can also be mounted on the spindle.

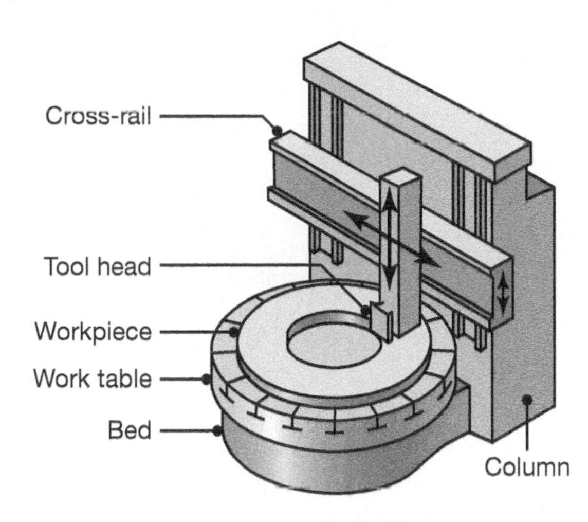

FIGURE 8.46 Schematic illustration of the components of a vertical boring mill.

8.9.4 Drilling, Reaming, and Tapping

One of the most common machining processes is *drilling*. **Drills** typically have a high length-to-diameter ratio (Fig. 8.47) and are capable of producing deep holes. They are, however, somewhat flexible, depending on their length and diameter. Moreover, note that chips produced move through the flutes in a direction opposite to the axial movement of the drill; consequently, chip disposal in drilling and selecting an effective cutting fluid are important.

The most common drill is the standard-point **twist drill** (Fig. 8.47). The main features of the drill point are the *point angle*, the *lip-relief angle*, the *chisel-edge angle*, and the *helix angle*. The geometry of the drill tip is such that the normal rake angle and the velocity of the cutting edge vary with the distance from the center of the drill. Other types of drills include the *step drill, core drill, counterboring* and *countersinking drills, center drill*, and *spade drill* (See Fig. 8.48).

Crankshaft drills have good centering ability, and because the chips they produce tend to break up easily, they are suitable for drilling deep holes. In *gun drilling*, a special drill is used for drilling deep holes; depth-to-diameter ratios produced can be 300 or higher. In the *trepanning* technique, a cutting tool produces a hole by removing a disk-shaped piece of material (*core*), usually from flat plates; the hole is produced without reducing all the material to be removed to chips. The trepanning process can be used to make disks up to 150 mm in diameter from flat sheet or plate.

The *material-removal rate* in drilling can be expressed as

$$\text{MRR} = \frac{\pi D^2}{4} fN, \qquad (8.40)$$

where D is the drill diameter, f is the feed (in mm/revolution), and N is the rpm of the drill. Recommendations for speed and feed in drilling are given in Table 8.11.

The *thrust force* in drilling is the force that acts in the drilling direction; if excessive, this force can cause the drill to bend and break. The thrust

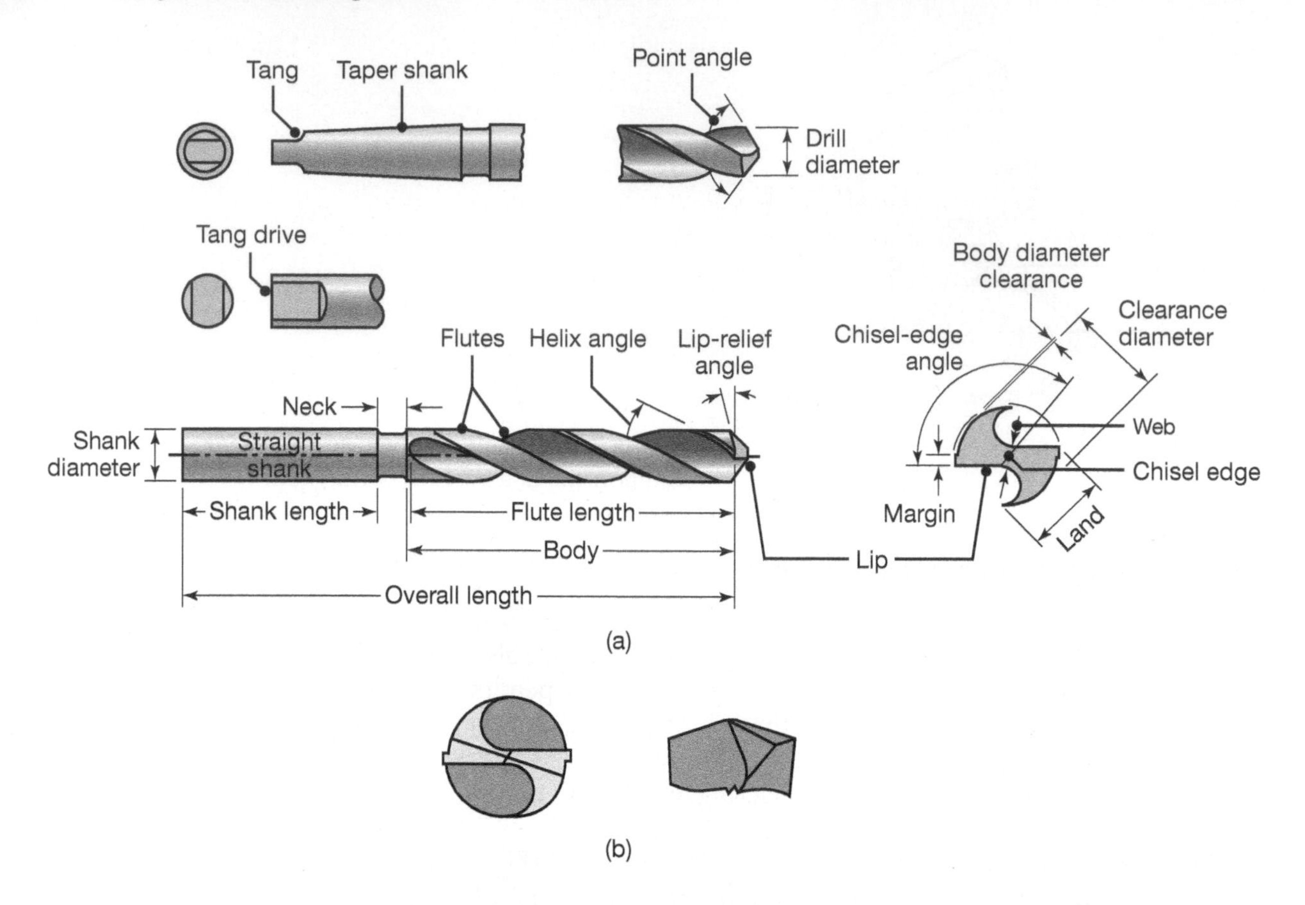

FIGURE 8.47 Two common types of drills: (a) Chisel-point drill. The function of the pair of margins is to provide a bearing surface for the drill against walls of the hole as it penetrates into the workpiece. Drills with four margins (double-margin) are available for improved drill guidance and accuracy. Drills with chip-breaker features are also available. (b) Crankshaft drill. These drills have good centering ability, and because chips tend to break up easily, they are suitable for producing deep holes.

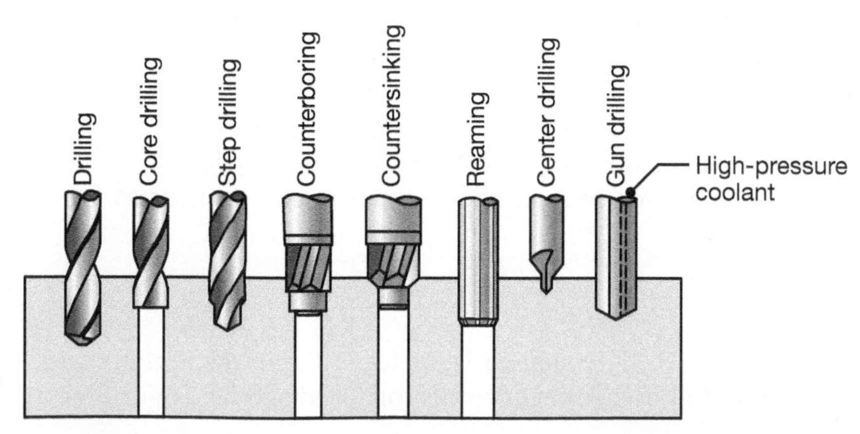

FIGURE 8.48 Various types of drills and drilling operations.

TABLE 8.11 General recommendations for speeds and feeds in drilling.

Workpiece material	Surface speed m/min	Surface speed ft/min	Feed, mm/rev (in./rev) drill diameter 1.5 mm (0.060 in.)	Feed, mm/rev (in./rev) drill diameter 12.5 mm (0.5 in.)	Spindle speed (rpm) drill diameter 1.5 mm (0.060 in.)	Spindle speed (rpm) drill diameter 12.5 mm (0.5 in.)
Aluminum alloys	30–120	100–400	0.025 (0.001)	0.30 (0.012)	6400–25,000	800–3000
Magnesium alloys	45–120	150–400	0.025 (0.001)	0.30 (0.012)	9600–25,000	1100–3000
Copper alloys	15–60	50–200	0.025 (0.001)	0.25 (0.010)	3200–12,000	400–1500
Steels	20–30	60–100	0.025 (0.001)	0.30 (0.012)	4300–6400	500–800
Stainless steels	10–20	40–60	0.025 (0.001)	0.18 (0.007)	2100–4300	250–500
Titanium alloys	6–20	20–60	0.010 (0.0004)	0.15 (0.006)	1300–4300	150–500
Cast irons	20–60	60–200	0.025 (0.001)	0.30 (0.012)	4300–12,000	500–1500
Thermoplastics	30–60	100–200	0.025 (0.001)	0.13 (0.005)	6400–12,000	800–1500
Thermosets	20–60	60–200	0.025 (0.001)	0.10 (0.004)	4300–12,000	500–1500

Note: As hole depth increases, speeds and feeds should be reduced. Selection of speeds and feeds also depends on the specific surface finish required.

force depends on such factors as the strength of the workpiece material, feed, rotational speed, cutting fluids, drill diameter, and drill geometry; thus accurate prediction of the thrust force in drilling has proven to be difficult. Experimental data are available in the technical literature as an aid in the design and use of drills and drilling equipment. Thrust forces in drilling typically range from a few newtons for small drills, to as high as 100 kN in drilling high-strength materials using large drills.

The *torque* during drilling is also difficult to predict accurately, although it can be estimated by using the data in Table 8.3. The power dissipated during drilling is the product of torque and angular velocity. The torque in drilling can range up to 4000 N-m. **Drill life**, as well as tap life, is usually measured by the number of holes drilled before the drill becomes dull and the drilling forces become excessive.

Drilling machines, for drilling holes, tapping, reaming, and other general-purpose, small-diameter boring operations, are generally vertical, the most common type being a **drill press**. The workpiece is placed on an adjustable table, and is clamped directly into the slots and holes on the table, or by holding it in a vise that can be clamped to the table. The drill is then lowered, either manually (by using the hand wheel or by power feed) at preset rates. Drill presses are usually designated by the largest workpiece diameter that can be accommodated on the table; sizes typically range from 150 to 1250 mm.

Reaming and reamers. *Reaming* is an operation that makes an existing hole dimensionally more accurate than can be achieved by drilling alone; it also improves the surface finish of the hole. The most accurate holes are produced by the following sequence of operations: centering, drilling,

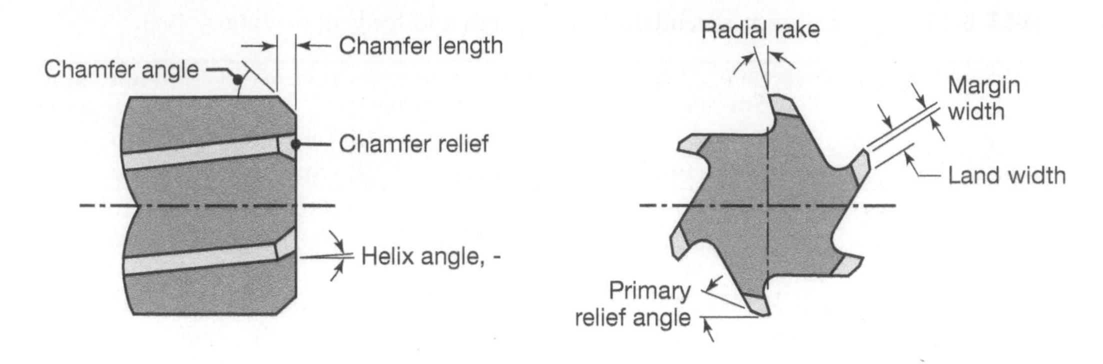

FIGURE 8.49 Terminology for a helical reamer.

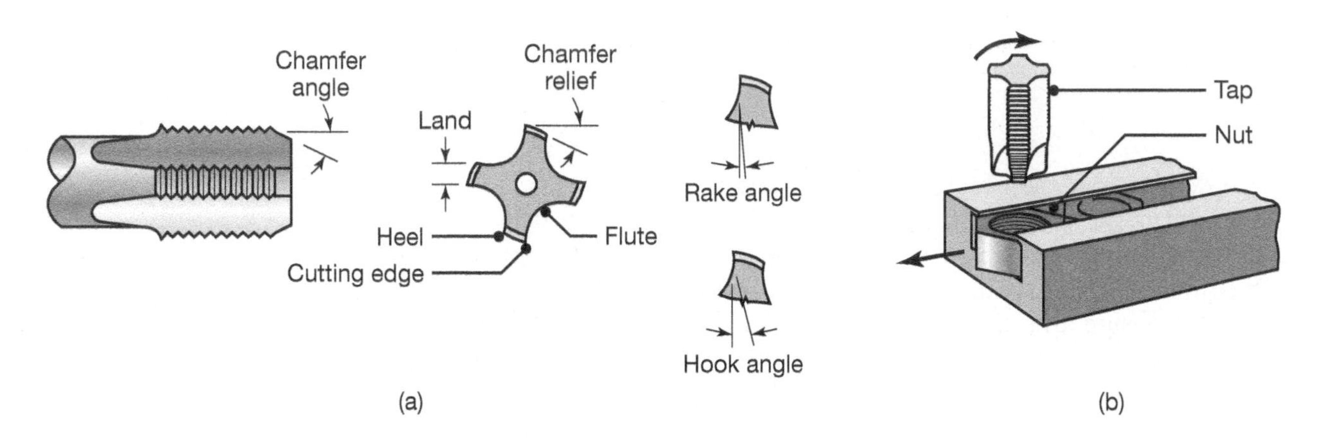

FIGURE 8.50 (a) Terminology for a tap; (b) illustration of tapping of steel nuts in high production.

boring, and reaming. For even better dimensional accuracy and surface finish, holes may be internally *ground* and *honed* (see Section 9.7). A *reamer* (Fig. 8.49) is a multiple-cutting-edge tool with straight or helically fluted edges that removes very little material. The shanks may be straight or tapered, as they are in drills. The basic types of reamers are *hand* and *machine (chucking)* reamers; other types include *rose* reamers, with cutting edges that have wide margins and no relief angle, and *fluted, shell, expansion,* and *adjustable* reamers.

Tapping and taps. Internal threads can be produced by *tapping*; a **tap** is basically a threading tool with multiple cutting teeth (Fig. 8.50). Taps are typically available with three or four flutes; three-fluted taps are stronger because their flute is wider. *Tapered taps* are designed to reduce the torque required for tapping through-holes, and *bottoming taps* are designed for tapping blind holes to their full depth. *Collapsible taps* are used for large-diameter holes; after tapping is completed, the tap is mechanically collapsed and removed from the hole, without having to be rotated. Tap sizes range up to 100 mm.

8.10 Cutting Processes and Machine Tools for Producing Various Shapes

Several machining processes and machine tools are used for producing complex shapes, typically with the use of multitooth cutting tools (Fig. 8.51 and Table 8.7).

8.10.1 Milling Operations

Milling includes a number of versatile machining operations that use a **milling cutter**, a multitooth tool that produces a number of chips per revolution, to machine a wide variety of part geometries. Parts such as the ones shown in Fig. 8.51 can be machined efficiently and repetitively using various milling cutters.

The basic types of milling operations are described as follows.

1. **Slab milling.** In *slab milling*, also called **peripheral milling**, the axis of cutter rotation is parallel to the surface of the workpiece to be machined, as shown in Fig. 8.52a. The cutter, called a *plain mill*, has a number of teeth along its periphery, each tooth acting as a single-point cutting tool. Cutters used in slab milling may have *straight* or *helical* teeth, producing an orthogonal or an oblique cutting action, respectively. Figure 8.1c shows the helical teeth on a milling cutter.

   In **conventional milling**, also called *up milling*, the maximum chip thickness is at the end of the cut (Fig. 8.52b). The advantages of conventional milling are that tooth engagement is not a function of workpiece geometry, and contamination or scale on the surface does not affect tool life. The machining process is smooth, provided that

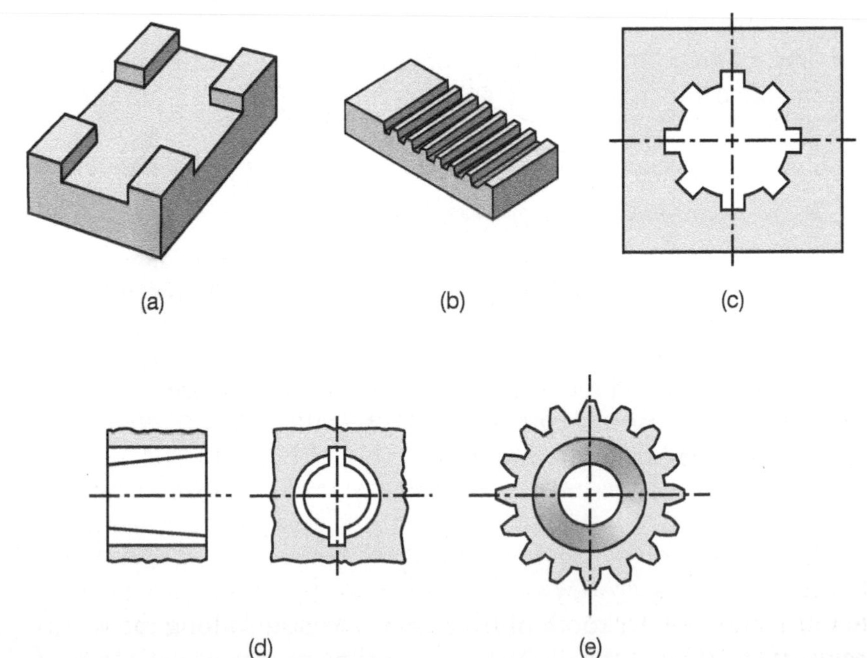

FIGURE 8.51 Typical parts and shapes produced by the machining processes described in Section 8.10.

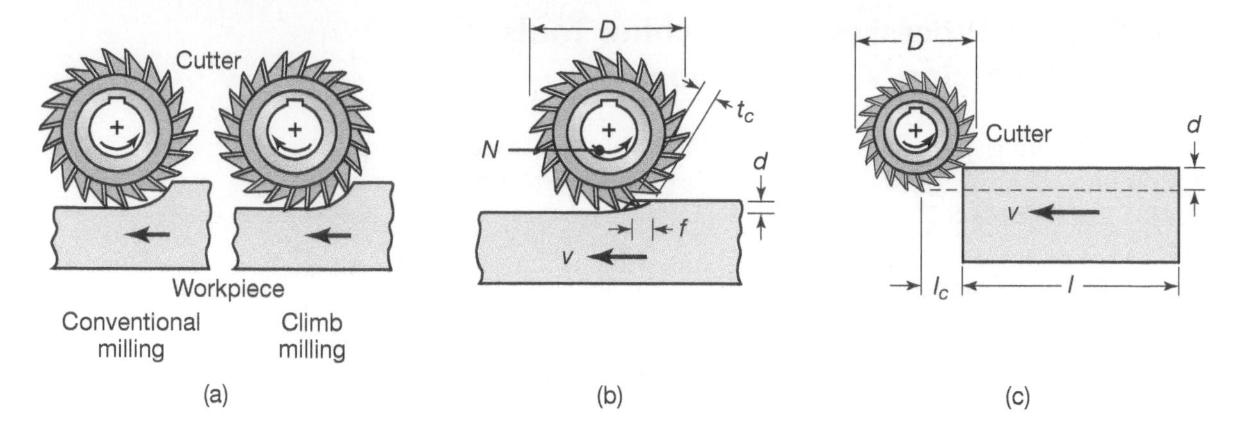

FIGURE 8.52 (a) Illustration showing the difference between conventional milling and climb milling. (b) Slab-milling operation, showing depth of cut, d; feed per tooth, f; chip depth of cut, t_c and workpiece speed, v. (c) Schematic illustration of cutter travel distance, l_c, to reach full depth of cut.

the cutter teeth are sharp. There is, however, a tendency for the tool to chatter, and for the workpiece to be pulled away from the surface, and thus proper clamping is important.

In **climb milling**, also called *down milling*, cutting starts at the thickest location of the chip. The advantage of climb milling is that the downward component of cutting forces holds the workpiece in place, particularly for slender parts. However, because of the resulting high-impact forces when the teeth first engage the workpiece, this operation requires a rigid setup, and backlash in the table feed mechanism must be eliminated. Climb milling is not suitable for machining workpieces that have surface scale, such as hot-worked metals, forgings, and castings; the scale is hard and abrasive, and causes excessive wear and damage to the cutter teeth. In general, climb milling is recommended for applications such as finishing cuts on aluminum workpieces.

The cutting speed in milling, V, is the peripheral speed of the cutter, that is,

$$V = \pi DN, \tag{8.41}$$

where D is the cutter diameter and N is the rotational speed of the cutter (Fig. 8.52b). Note that the thickness of the chip in slab milling varies along its length, because of the relative longitudinal movement between the cutter and the workpiece. For a straight-tooth cutter, the approximate *undeformed chip thickness*, t_c (called *chip depth of cut*), can be determined from the equation

$$t_c = 2f\sqrt{\frac{d}{D}}, \tag{8.42}$$

where f is the feed per tooth of the cutter (measured along the workpiece surface, that is, the distance the workpiece travels per tooth of the cutter, in mm/tooth), and d is the depth of cut. With increasing t_c, the force on the cutter tooth also increases.

Feed per tooth is determined from the equation

$$f = \frac{v}{Nn}, \tag{8.43}$$

where v is the linear speed (feed rate) of the workpiece and n is the number of teeth on the cutter periphery. The dimensional accuracy of this equation can be checked by substituting appropriate units for the individual terms; thus, (mm/tooth) = (m/min)(10^3 mm/m)/(rev/min)(number of teeth/rev), which is correct. The cutting time, t, is given by the expression

$$t = \frac{l + 2l_c}{v}, \tag{8.44}$$

where l is the length of the workpiece (Fig. 8.52c) and l_c is the extent of the cutter's first contact with the workpiece. Based on the assumption that $l_c \ll l$ (although this may not be always reasonable), the *material-removal rate* is

$$\text{MRR} = \frac{lwd}{t} = wdv, \tag{8.45}$$

where w is the width of the cut, which is the same as the workpiece width if it is narrower than the cutter. The distance that the cutter travels in the noncutting-cycle of the operation is an important economic consideration and should be minimized.

2. **Face milling.** In this operation, the cutter is mounted on a spindle with an axis of rotation perpendicular to the workpiece surface, and removes material in the manner shown in Fig. 8.53a. The cutter rotates at a speed of N, and the workpiece moves along a straight path and at a linear speed of v. When the cutter rotates in the direction shown in Fig. 8.53b, the operation is called *climb milling*; when it rotates in the opposite direction (Fig. 8.53c), it is *conventional milling*.

   Because of the relative movement between the cutting teeth and the workpiece, a face-milling cutter leaves *feed marks* on the machined surface, much as in turning operations (See Fig. 8.19). Surface roughness depends on insert corner geometry and feed per tooth [see also Eqs. (8.35) to (8.37)].

   The terminology for a face-milling cutter and its various angles is given in Fig. 8.54; the side view of the cutter is shown in Fig. 8.55. Note that, (a) as in turning operations, the *lead angle* of the insert has a direct influence on the *undeformed chip thickness*; (b) as the lead angle (positive, as shown in the figure) increases, the undeformed chip thickness (thus also the thickness of the actual chip) decreases; (c) the length of contact increases; and (d) the cross-sectional area of the undeformed chip remains constant. The range of lead angles

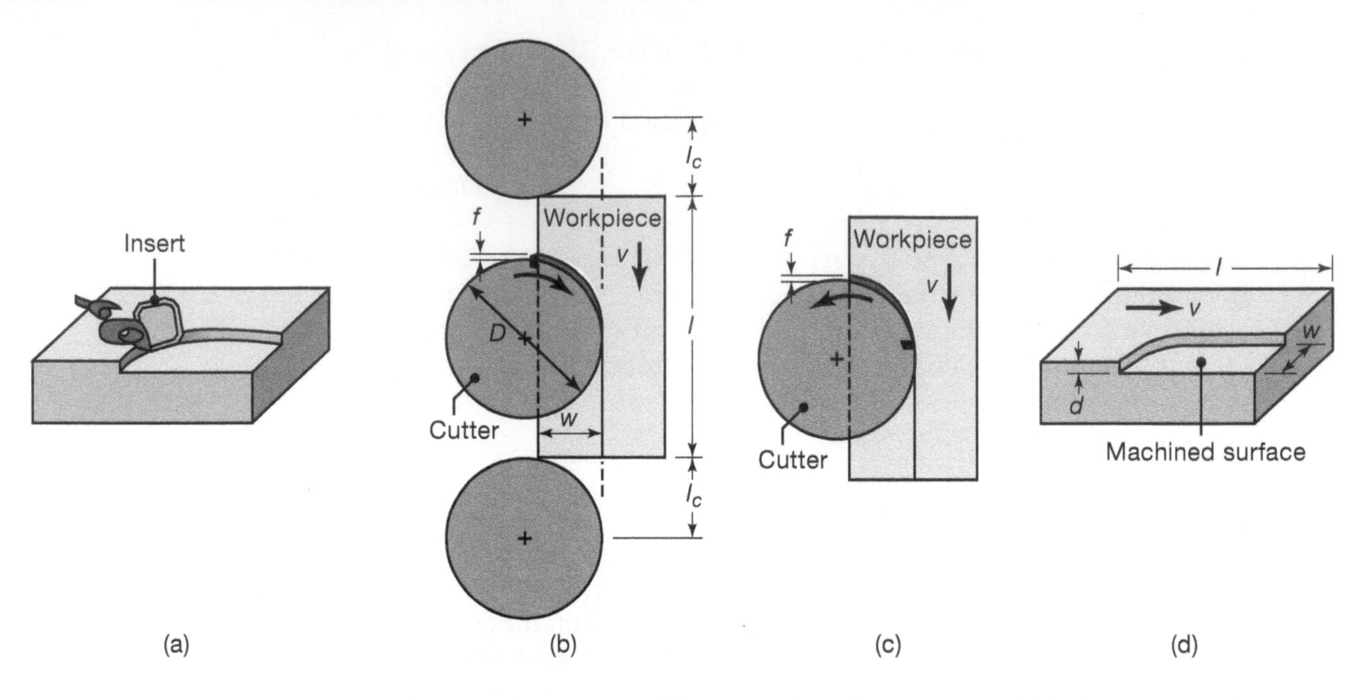

FIGURE 8.53 Face-milling operation showing (a) action of an insert in face milling with the cutter removed for clarity; (b) climb milling; (c) conventional milling; and (d) dimensions in face milling.

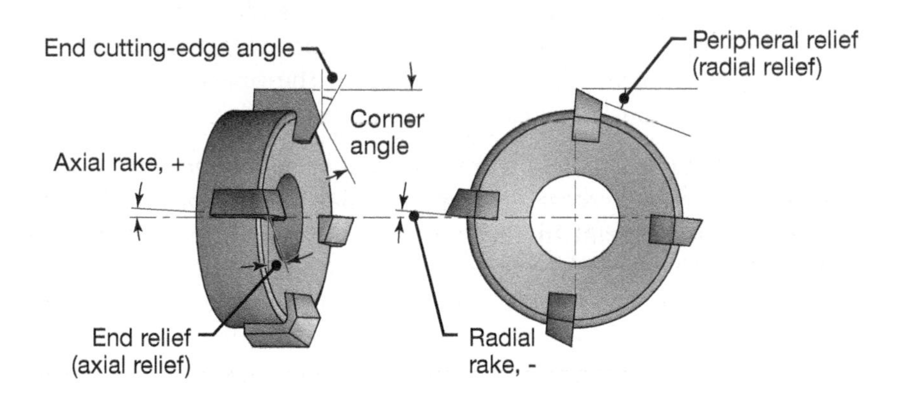

FIGURE 8.54 Terminology for a face-milling cutter.

for most face-milling cutters is typically from 0° to 45°. The lead angle also influences forces in milling; as the lead angle decreases, the vertical force (axial force on the cutter spindle) decreases.

A wide variety of milling cutters is available. The cutter diameter, D, should be chosen so that it will not interfere with fixtures, workholding devices, and other components in the setup of the machine tool. In a typical face-milling operation, the ratio of the cutter diameter to the width of cut should be no less than 3:2. The cutting tools are usually carbide or high-speed steel inserts and are mounted on the cutter body (See Fig. 8.54).

The relationship of cutter diameter and insert angles, and their position relative to the surface to be milled, is important, because

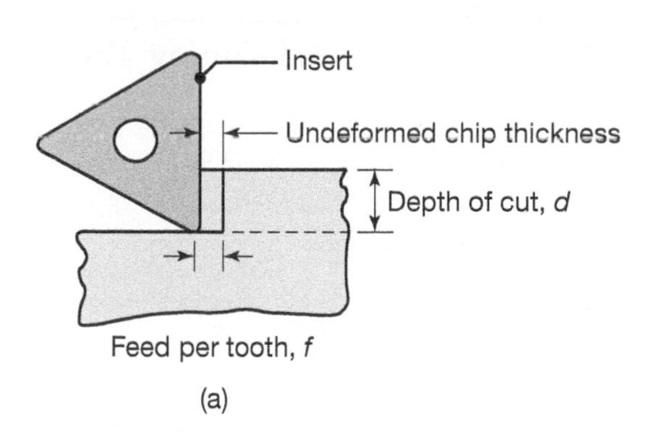

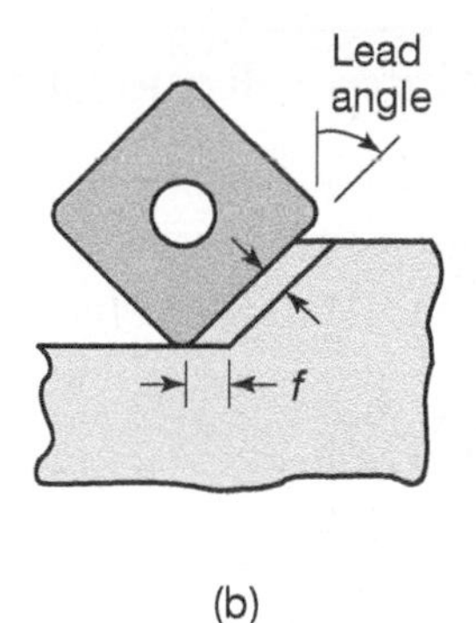

FIGURE 8.55 The effect of lead angle on the undeformed chip thickness in face milling. Note that as the lead angle increases, the undeformed chip thickness (and hence chip thickness) decreases, and the length of contact (and hence the width of the chip) increases. Note that the insert must be sufficiently large to accommodate the increase in contact length.

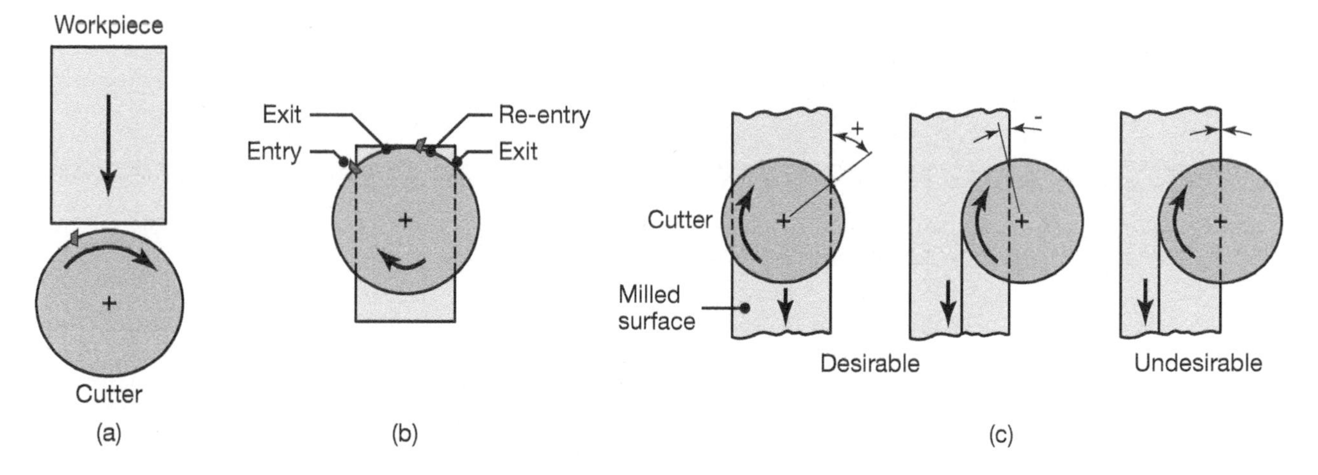

FIGURE 8.56 (a) Relative position of the cutter and the insert as it first engages the workpiece in face milling; (b) insert positions at entry and exit near the end of cut; and (c) examples of exit angles of the insert, showing desirable (positive or negative angle) and undesirable (zero angle) positions. In all figures, the cutter spindle is perpendicular to the page.

they determine the angle at which an insert enters and exits the workpiece. Note in Fig. 8.53b for climb milling that if the insert has zero axial and radial rake angles (See Fig. 8.54), the rake face of the insert engages the workpiece directly, thus subjecting it to a high impact force. However, as can be seen in Fig. 8.56a and b, the same insert will engage the workpiece at different angles, depending on the relative positions of the cutter and the workpiece. In Fig. 8.56a, the edge of the insert makes the first contact, and hence there is potential for the cutting edge to chip off. In Fig. 8.56b, on the other hand, the contacts (at entry, reentry, and the two exits) are at a certain angle and away from the edge of the insert. As a result, there is less of a tendency for the insert to chip off, because the force on the insert increases and decreases gradually. Note from Fig. 8.54 that the radial and axial rake angles also will affect the tendency for the insert material to chip off.

EXAMPLE 8.6 Calculation of Material-Removal Rate, Power Required, and Cutting Time in Face Milling

Given: Referring to Fig. 8.53, assume that $D = 150$ mm, $w = 60$ mm, $l = 500$ mm, $d = 3$ mm, $v = 0.6$ m/min, and $N = 100$ rpm. The cutter has 10 inserts and the workpiece material is a high-strength aluminum alloy.

Find: Calculate the material-removal rate, cutting time, and feed per tooth, and estimate the power required.

Solution: The cross section of the cut is $wd = (60)(3) = 180\ \text{mm}^2$. Since the workpiece speed v is 0.6 m/min = 600 mm/min, the material-removal rate is

$$\text{MRR} = (180)(600) = 108,000\ \text{mm}^3/\text{min}.$$

The cutting time is given by Eq. (8.44) as

$$t = \frac{l + 2l_c}{v}.$$

Note from Fig. 8.53 that for this problem,

$$l_c^2 + \left(\frac{D}{2} - w\right)^2 = \left(\frac{D}{2}\right)^2,$$

so that $l_c = \sqrt{Dw - w^2}$, or 73.5 mm. Thus, the cutting time is

$$t = \frac{[500 + 2(73.5)](60)}{600} = 64.7\ \text{s} = 1.08\ \text{min}.$$

The feed per tooth is obtained from Eq. (8.43). Noting that $N = 100$ rpm $= 1.67$ rev/s,

$$f = \frac{10}{(1.67)(10)} = 0.6\ \text{mm/tooth}.$$

For this material, the unit power can be taken from Table 8.3 to be 1.1 W-s/mm^3; hence, the power can be estimated as

$$\text{Power} = (1.1)(1800) = 1980\ \text{W} = 1.98\ \text{kW}.$$

The exit angles for various cutter positions in face milling are shown in Fig. 8.56c. Note that in the first two examples, the insert exits the workpiece at an angle, whereby the force on the insert diminishes to zero at a slower rate (desirable) than in the third example, where the insert exits suddenly (undesirable).

3. **End milling.** The cutter in *end milling*, shown in Fig. 8.1d, has either a straight or a tapered shank for smaller and larger cutter sizes, respectively. The cutter usually rotates on an axis perpendicular to the workpiece, but it can be tilted to produce inclined surfaces. Some

end mills have cutting teeth on their end faces, allowing the end mill to be used as a drill to start a cavity. End mills are also available with hemispherical ends (called *ball-nose end mills*), for producing curved surfaces, as in making molds and dies. *Hollow end mills* have *internal* cutting teeth and are used for machining the cylindrical surface of solid round workpieces, as in preparing stock with accurate diameters for automatic bar machines.

Section 8.8 described high-speed machining and its typical applications. One of the more common applications is **high-speed milling**, using an end mill and with the same general requirements regarding the stiffness of machines and workholding devices, described in Section 8.12. A typical application is milling aluminum-alloy aerospace components and honeycomb structures (See Fig. 7.48), with spindle speeds on the order of 20,000 rpm. Another application is in *die sinking*, i.e., producing cavities in die blocks. Chip collection and disposal can be a significant problem in these operations, because of the high rate of material removal (see Section 8.8).

4. **Various milling operations and milling cutters.** Several other types of milling operations and cutters are used to machine various surfaces. In *straddle milling*, two or more cutters are mounted on an arbor, and are used to machine two *parallel* surfaces on the workpiece (Fig. 8.57a). *Form milling* is used to produce curved profiles, using cutters with specially shaped teeth (Fig. 8.57b); such cutters are also used for cutting gear teeth (see Section 8.10.7).

   Circular cutters are used for slotting and slitting operations. The teeth may be staggered, as in a saw blade (see Section 8.10.5), to provide clearance for the cutter in making deep slots. *Slitting saws* are relatively thin, usually less than 5 mm. *T-slot cutters* are used to mill T-slots, such as those in machine-tool worktables (See Fig. 8.58). The slot is first milled with an end mill; a T-slot cutter then cuts the complete profile of the slot in one pass. *Key-seat cutters* are used

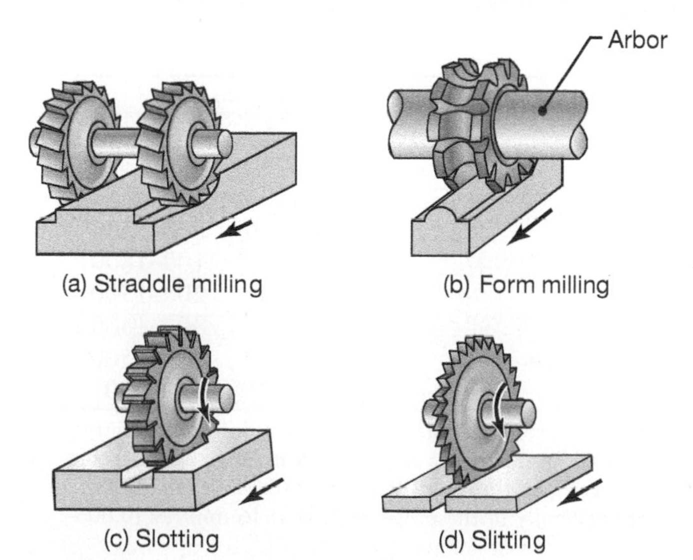

FIGURE 8.57 Cutters for milling operations.

to make the semi cylindrical (Woodruff) key sets for shafts. *Angle-milling cutters*, either with a single angle or double angles, are used to produce tapered surfaces with various angles.

Shell mills are hollow inside and are mounted on a shank, thus allowing the same size of shank to be used for different-sized cutters. Shell mills are used in a manner similar to end mills. Milling with a single cutting tooth, mounted on a high-speed spindle, is known as *fly cutting*, and can used in simple face-milling and boring operations. The cutting tool can be shaped as a single-point tool and can be clamped in various radial positions on the spindle.

5. **Toolholders.** Milling cutters are classified into two basic types. *Arbor cutters* are mounted on an arbor, such as for slab, face, straddle, and form milling. In *shank-type cutters*, the cutter and the shank are in one piece; the most common examples of shank cutters are end mills. Whereas small end mills have straight shanks, larger end mills have tapered shanks for better clamping to resist the higher forces and torque involved. Cutters with straight shanks are mounted in *collet chucks* or special end-mill holders; cutters with tapered shanks are mounted in tapered toolholders. Conventional tapered toolholders have a tendency to wear under the radial forces in milling. In addition to mechanical means, hydraulic tool holders and arbors also are available. The stiffness of cutters and toolholders is important for better surface quality and in reducing vibration and chatter in milling operations (see Section 8.12).

6. **Milling machines and process capabilities.** In addition to the various characteristics of milling processes described thus far, milling capabilities involve parameters such as production rate, surface finish, dimensional tolerances, and cost considerations. Table 8.12 gives the typical conventional ranges of feeds and cutting speeds for milling.

TABLE 8.12 Approximate range of recommended cutting speeds for milling operations.

	Cutting speed	
Workpiece material	m/min	ft/min
Aluminum alloys	300–3000	1000–10,000
Cast iron, gray	90–1300	300–4200
Copper alloys	90–1000	300–3300
High-temperature alloys	30–550	100–1800
Steels	60–450	200–1500
Stainless steels	90–500	300–1600
Thermoplastics and thermosets	90–1400	300–4500
Titanium alloys	40–150	130–500

Note: (a) These speeds are for carbides, ceramic, cermets, and diamond cutting tools. Speeds for high-speed steel tools are lower than those indicated in this table.
(b) Depths of cut, d, are generally in the range of 1–8 mm (0.04–0.3 in.).
(c) Feeds per tooth, f, are generally in the range of 0.08–0.46 mm/rev (0.003–0.018 in./rev).

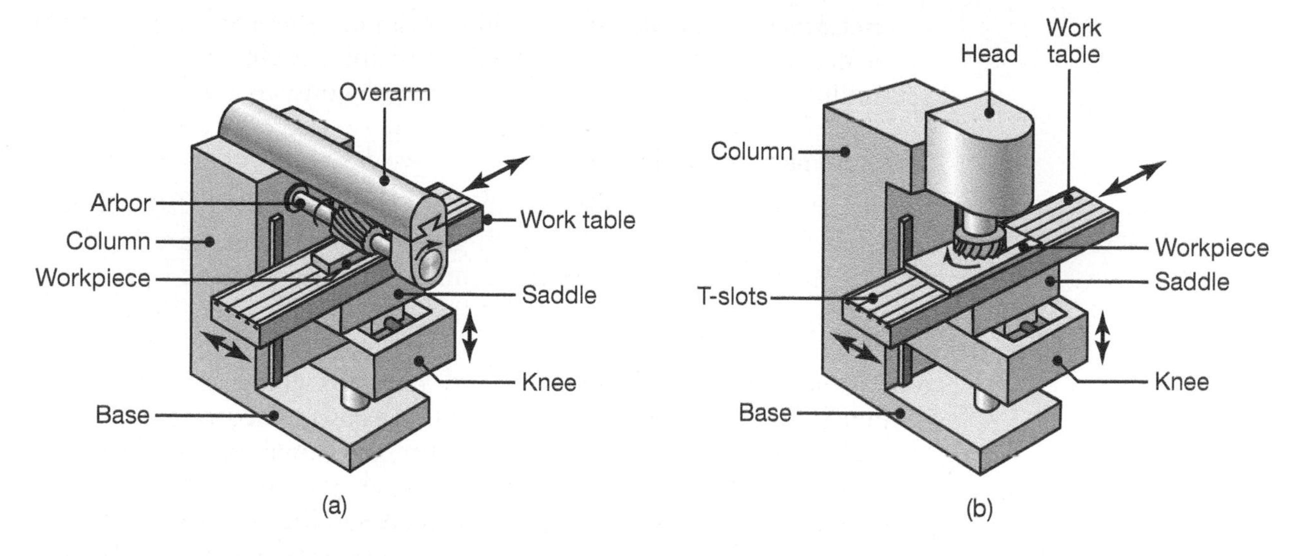

FIGURE 8.58 (a) Schematic illustration of a horizontal-spindle column-and-knee-type milling machine. (b) Schematic illustration of a vertical-spindle column-and-knee-type milling machine.

Cutting speeds vary over a wide range, from 30 to 3000 m/min, depending on workpiece material, cutting-tool material, and process parameters.

Because of their capability to perform a variety of machining operations, milling machines are among the most versatile machine tools, and a wide selection of machines with numerous features is available. Used for general-purpose applications, **column-and-knee-type** machines are the most common milling machines. The spindle, on which the milling cutter is mounted, may be *horizontal* (Fig. 8.58a), for slab milling, or *vertical*, for face milling, end milling, boring, and drilling operations (Fig. 8.58b). The components are moved manually or by power, and are equipped with a variety of CNC controls (see Section 14.3.1).

In **bed-type** machines, the worktable is mounted directly on the bed (which replaces the knee) and can move only longitudinally. These machines are not as versatile as others, but have high stiffness and are used for high-production-rate work. The spindles may be horizontal or vertical and of *duplex* or *triplex* types, that is, with two or three spindles for simultaneously milling two or three workpiece surfaces, respectively. **Planer-type** machines, which are similar to bed-type machines, are equipped with several heads and cutters to machine various surfaces of a workpiece. They are used for heavy workpieces and are more efficient than planers (Section 8.10.2) when used for similar operations.

Rotary-table machines are similar to vertical milling machines and are equipped with one or more heads for face milling. *Profile-milling machines* have five-axis movements. Using tracer fingers, *duplicating machines* (also called copy-milling machines) reproduce parts from a master model. These machine tools are versatile and capable of milling, drilling, boring, and tapping with consistent accuracy.

8.10.2 Planing and Planers

Planing is a relatively simple machining process by which flat surfaces, as well as various cross sections with grooves and notches, are produced along the length of the workpiece. Planing is usually done on large workpieces, as large as 25 m × 15 m (80 ft × 50 ft). In a typical *planer*, the workpiece is mounted on a table that travels along a straight path. A horizontal cross rail, which can be moved vertically along the ways in the column, is equipped with one or more tool heads. The cutting tools are mounted on the heads, and they travel along a straight path. Because of the reciprocating motion of the workpiece, the elapsed noncutting time during the return stroke of the tool heads is significant, both in planing and in shaping (see Section 8.9.3). Planers are generally not efficient or economical, except for low-quantity production.

8.10.3 Shaping and Shapers

Machining by *shaping* is basically the same as in planing. In a *horizontal shaper*, the tool travels along a straight path, and the workpiece is stationary. The cutting tool is mounted on the tool head, which itself is mounted on the ram. The ram has a reciprocating motion, and in most machines, cutting is done during the forward movement of the ram (*push cut*). In others, it is done during the return stroke of the ram (*draw cut*). *Vertical shapers* (called *slotters*) are used for machining notches and keyways. Shapers are capable of producing complex shapes, such as machining helical impellers, where the workpiece is rotated during the cut using a master cam.

8.10.4 Broaching and Broaching Machines

The *broaching* operation is similar to shaping and is used to machine internal and external surfaces (Fig. 8.59), such as holes (circular, square, or irregular in cross section), keyways, teeth of internal gears, multiple spline holes, and flat surfaces. A *broach* (Fig. 8.60) is basically a long multitooth cutting tool that makes successively deeper cuts. Note that the total depth of material removed in one stroke of the broach is the sum of the depths of cut of each tooth. A broach can remove material as deep as 6 mm in one stroke. Broaching can produce parts with good surface finish and dimensional accuracy, thus it competes favorably with other machining processes to produce similar shapes. Although broaches can be expensive, the cost is justified because of their use for high-quantity production runs.

QR Code 8.2 Keyway broaching.
Source: Courtesy of Miles Broaching.

The terminology for a broach is given in Fig. 8.60b. The *rake* (hook) angle depends on the material being machined, and usually ranges between 0° and 20°. The *clearance* angle is typically 1° to 4°; finishing teeth have smaller angles. Too small a clearance angle causes rubbing of the cutter teeth against the broached surface. The *pitch* of the teeth depends on

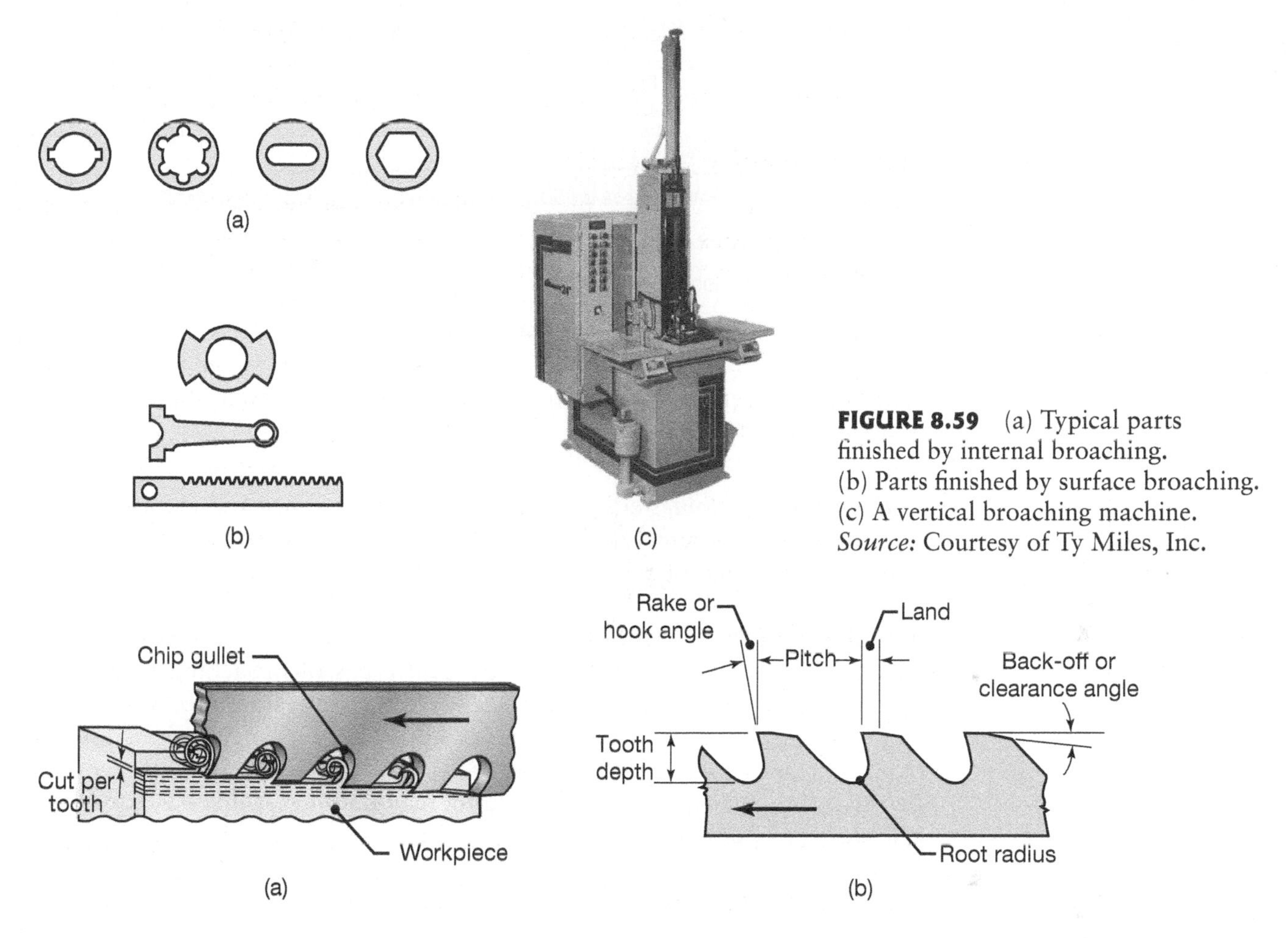

FIGURE 8.59 (a) Typical parts finished by internal broaching. (b) Parts finished by surface broaching. (c) A vertical broaching machine. *Source:* Courtesy of Ty Miles, Inc.

FIGURE 8.60 (a) Cutting action of a broach, showing various features. (b) Terminology for a broach.

factors such as length of the workpiece (length of cut), tooth strength, and size and shape of chips. The tooth depth and pitch must be sufficiently large to accommodate the chips produced during broaching, particularly for long workpieces. At least two teeth should be in contact with the workpiece at all times, similar to sawing (see Section 8.10.5).

Broaches are available with a variety of tooth profiles, including some that incorporate chip breakers (Fig. 8.61). Note that the cutting teeth on broaches have three regions: roughing, semifinishing, and finishing. Round broaches are also available, with circular cutting teeth, and are used to enlarge holes (Fig. 8.61). Irregular internal shapes are usually broached by first starting with a round hole in the workpiece, produced by processes such as drilling or boring, followed by broaching.

In **turn broaching** of crankshafts, the workpiece rotates between centers, the broach is equipped with multiple inserts, and it passes tangentially across the part; the operation is thus a combination of broaching and skiving (see Section 8.2.2). There are also machines that broach a number of locations simultaneously, such as bearing seats for engine crankshafts.

QR Code 8.3 Spline broaching.
Source: Courtesy of Miles Broaching.

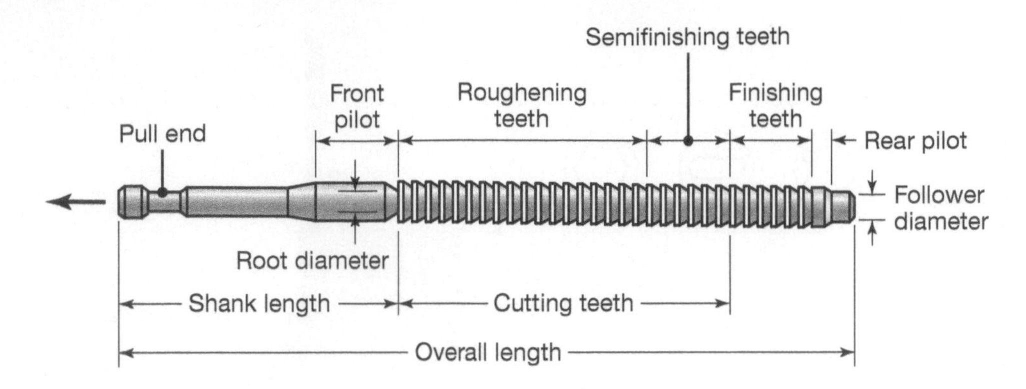

FIGURE 8.61 Terminology for a pull-type internal broach, typically used for enlarging long holes.

Broaching machines either pull or push the broaches and are available in horizontal or vertical designs. *Push broaches* are usually shorter, generally in the range of 150–350 mm (6–14 in.) long. *Pull broaches* tend to straighten a hole, whereas pushing it permits the broach to follow any irregularity of the leader hole. Horizontal machines are capable of longer strokes than are vertical machines. Several types of broaching machines are available, some with multiple heads, allowing a variety of shapes to be produced, including helical splines and rifled gun barrels. Sizes range from machines for making needlelike parts to those used for broaching gun barrels. The capacities of broaching machines are as high as 0.9 MN (100 tons) of pulling force.

8.10.5 Sawing and Saws

Sawing is a cutting operation in which the tool consists of a series of small teeth, each tooth removing a small amount of material. Saws are used for all metallic and nonmetallic machinable materials, and they are also capable of producing various shapes. The width of cut (called **kerf**) in sawing is usually narrow, thus wasting little material. Typical saw-tooth and saw-blade configurations are illustrated in Fig. 8.62. Tooth spacing is usually in the range of 0.08 to 1.25 teeth per mm.

Saw blades are generally made from carbon and high-speed steels; steel blades with brazed tips of carbide or high-speed steel are used for sawing harder materials. In order to prevent the saw from rubbing and binding during sawing, the teeth are alternately set in opposite directions so that the kerf is wider than the blade (Fig. 8.62b). At least two or three teeth should always be engaged with the workpiece in order to prevent *snagging* (catching of the saw tooth on the workpiece). This requirement is the reason that sawing thin materials is difficult or impossible. Cutting speeds in sawing usually range up to 1.5 m/s (300 ft/min), with lower speeds used for high-strength metals. Cutting fluids improve the quality of the cut and extend the life of the saw.

Hacksaws have straight blades and use a reciprocating motion; they may be manually or power operated. *Circular saws (cold saws)* are generally used for high-production-rate sawing of large cross sections.

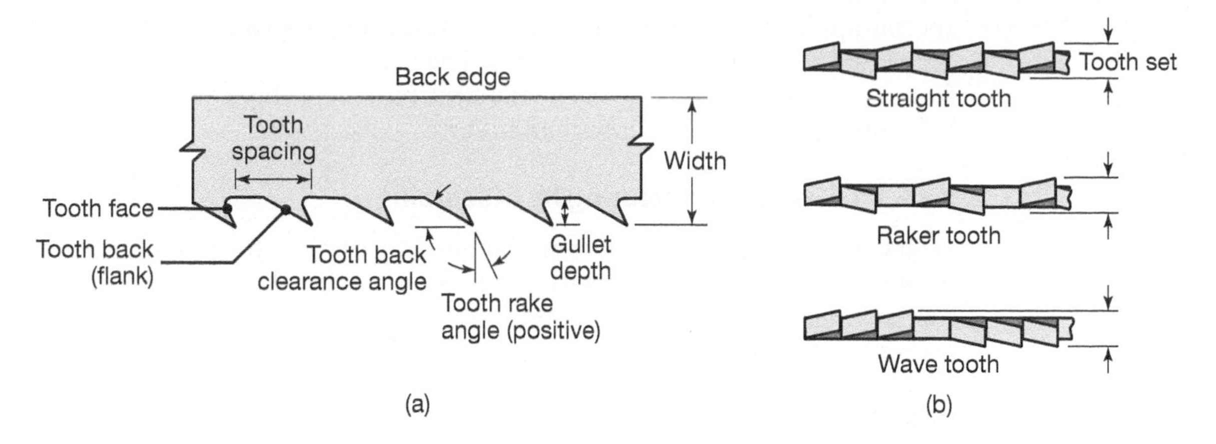

FIGURE 8.62 (a) Terminology for saw teeth. (b) Types of saw teeth, staggered to provide clearance for the saw blade to prevent binding during sawing.

Band saws have long, flexible, and continuous blades, and provide continuous cutting action. Blades and high-strength wire can be coated with diamond powder (*diamond-edged blades* and *diamond saws*); they are suitable for sawing hard metallic, nonmetallic, and composite materials.

Friction sawing. *Friction sawing* is a process in which a mild-steel blade or disk rubs against the workpiece at speeds of up to 125 m/s (25,000 ft/min). The frictional energy is converted into heat, which rapidly softens a narrow zone in the workpiece. The action of the blade or the disk (which can have teeth or notches) pulls and ejects the softened metal from the cutting zone. A *heat-affected zone* (as in welding processes; see Section 12.6) is developed on the cut surfaces, which can adversely affect properties. Because only a small portion of the blade is engaged with the workpiece at any time, the blade cools rapidly. Friction sawing is suitable for hard ferrous metals and fiber-reinforced plastics, but not for nonferrous metals because they have a tendency to stick to the blade. Friction sawing disks as large as 1.8 m in diameter are used to cut off steel structural sections from rolled plate. Friction sawing also can be used to remove flash from castings (see Section 5.12.2).

8.10.6 Filing

Filing is small-scale removal of material from a surface, corner, edge, or hole. First developed about 1000 B.C., files are usually made of hardened steels and are available in a variety of shapes, including flat, round, half round, square, and triangular. Files are available with various tooth forms and grades of coarseness, such as smooth cut, second cut, and intermediate (*bastard*) cut. Although filing is usually done by hand, various machines with automatic features are available for high production rates, with files reciprocating at up to 500 strokes/min.

Band files consist of file segments, each about 75 mm (3 in.) long, that are riveted to flexible steel bands, and used in a manner similar to band saws. *Disk-type files* also are available. *Rotary files* and **burs** are used for special applications; they are usually conical, cylindrical, or spherical in shape and have various tooth profiles. The rotational speed of burs ranges from 1500 rpm, for cutting steel with large burs, to as high as 45,000 rpm, for cutting magnesium using small burs.

8.10.7 Gear Manufacturing by Machining

Gears and gear blanks can be manufactured by casting, forging, extrusion, drawing, rolling, powder metallurgy processes, and sheet metal blanking for making thin gears (see Section 7.3.1). Most gears are machined and ground for better dimensional accuracy and surface finish. Nonmetallic gears are usually made by injection molding and casting (Chapter 10).

In **form cutting** of gears, the cutting tool is similar to a form-milling cutter (See Fig. 8.57b) in the shape of the space between the gear teeth. The cutter travels axially along the length of the gear tooth and at a specific depth, to produce the gear tooth. After each tooth is machined, the cutter is withdrawn, the gear blank is rotated (*indexed*), and the cutter proceeds to cut another tooth; the process continues until all teeth are cut. Broaching also is used to produce gear teeth, and is particularly useful for internal teeth. The broaching process is rapid and produces fine surface finish with high-dimensional accuracy. However, because broaches are expensive and a separate broach is required for each size of gear, this method is suitable mainly for high-quantity production.

In **gear generating**, the tool may consist of the following:

1. *Pinion-shaped cutter.* This type of cutter can be considered as one of the two gears in a conjugate pair of gears and the other is the gear blank (Fig. 8.63a). The cutter is used on **gear shapers** (Fig. 8.63b); it has an axis parallel to that of the gear blank, and rotates slowly with the blank at the same pitch-circle velocity with an axial reciprocating motion. A train of gears provides the required relative motion between the cutter shaft and the gear-blank shaft. Cutting may take place at either the downstroke or the upstroke of the machine. Because the clearance required for cutter travel is small, gear shaping is suitable for gears that are located close to obstructing surfaces, such as flanges in the gear blank shown in Fig. 8.63b. The process can be used for low as well as high-quantity production.
2. *Rack-shaped straight cutter.* Mounted on a **rack shaper**, the generating tool is a segment of a rack (Fig. 8.63c), which reciprocates parallel to the axis of the gear blank. Because it is not practical to have more than 6 to 12 teeth on a rack cutter, the cutter must be disengaged at appropriate intervals and returned to the starting point; the gear blank, meanwhile, remains fixed.
3. *Hob.* A **hob** basically has the shape of a worm or screw, that has been made into a gear-generating tool, by machining a series of longitudinal slots, or gashes, into it to produce cutting teeth (Fig. 8.63d). When hobbing a spur gear, the angle between the hob and gear-blank axes is 90° minus the lead angle at the hob threads. All motions in hobbing are rotary, and the hob and the gear blank both rotate continuously, as two gears in mesh, until all teeth are cut.

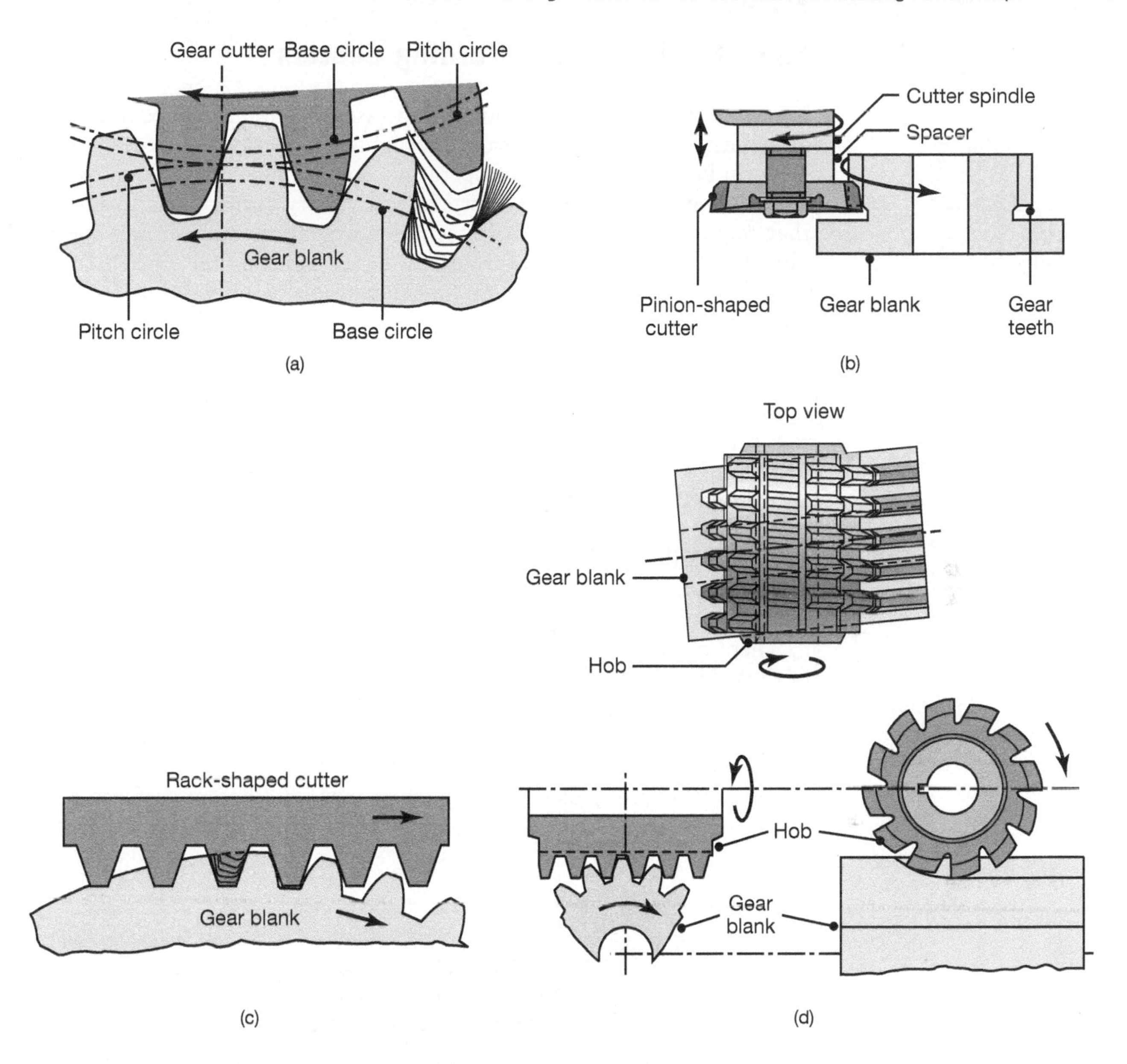

FIGURE 8.63 (a) Schematic illustration of gear generating with a pinion-shaped gear cutter. (b) Schematic illustration of gear generating in a gear shaper, using a pinion-shaped cutter; note that the cutter reciprocates vertically. (c) Gear generating with a rack-shaped cutter. (d) Three views of gear cutting with a hob.

Gears machined by any of the three processes described may, for some applications, not have sufficiently high-dimensional accuracy and surface finish, thus requiring several *finishing* operations, including shaving, burnishing, grinding, honing, and lapping (see Chapter 9). Modern gear-manufacturing machines are all computer controlled, and multiaxis computer-controlled machines have capabilities of machining different types of gears, using indexable cutters.

8.11 Machining and Turning Centers

QR Code 8.4 Turning center demonstration. *Source:* Courtesy of Haas Automation, Inc.

QR Code 8.5 Machining center demonstration. *Source:* Courtesy of Haas Automation, Inc.

In describing individual machining processes thus far, it will be noted that each machine tool, regardless of how highly it is automated, is designed to perform basically one type of operation, such as turning, milling, boring, and drilling. However, many parts have a variety of features and surfaces that require different types of machining operations and certain specific dimensional tolerances and surface finish (See, for example, Figs. 8.25 and 9.28). None of the processes and machine tools described thus far could individually produce these parts.

Traditionally, the required operations have been performed by moving the part from one machine tool to another, and then another, etc., until all operations are completed. This method is a viable manufacturing method that can be highly automated, and it is the principle behind **transfer lines**, consisting of numerous machine tools arranged in a certain sequence (see Section 14.2.4). An engine block, for example, moves from one station to another, with a specific machining operation performed at each station. The block is then transferred automatically to the next machine for another operation, and so on. Transfer lines are commonly used in high-volume or mass production.

There are numerous products, however, for which transfer lines are not feasible or economical, particularly when the types of products to be machined change rapidly due, for example, to changes in market demands. An important concept, developed in the late 1950s, is machining centers. A **machining center** is a computer-controlled machine tool capable of performing a variety of machining operations on different surfaces, locations, and orientations on a workpiece (Fig. 8.64). The development of

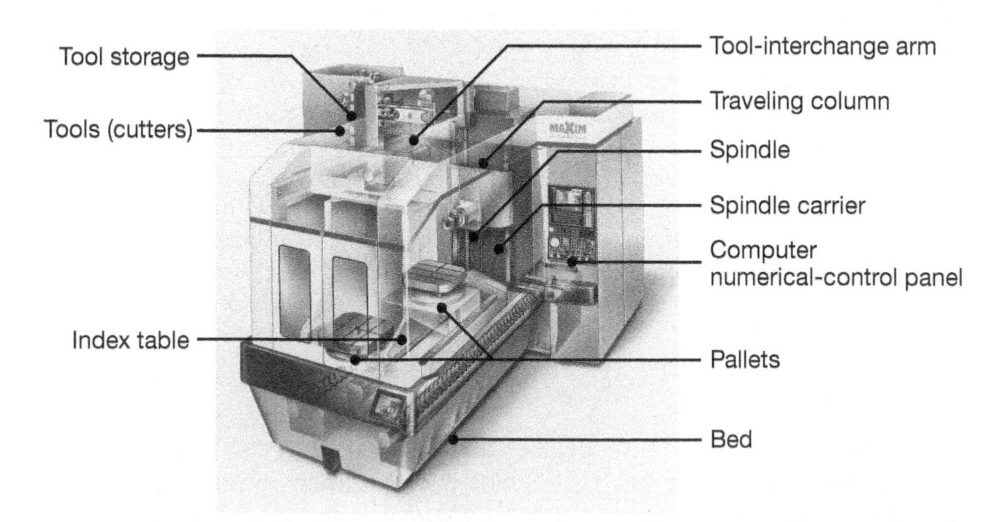

FIGURE 8.64 A horizontal-spindle machining center, equipped with an automatic tool changer. Tool magazines in such machines can store as many as 200 cutting tools, each with its own holder.

machining centers is related to advances in **computer control of machine tools**, details of which are described in Chapter 14.

The workpiece in a machining center is mounted on a **pallet** (Fig. 8.64) that can be oriented in three principal directions, as well as rotated around one or more axes. After a particular machining operation is complete, the workpiece remains on the pallet for another machining operation, such as drilling, reaming, and tapping. Thus, unlike traditional methods, in a machining center the tools and the machine tool are brought to the workpiece. After all machining operations are completed, the pallet automatically moves away, carrying the finished workpiece; another pallet, containing a new workpiece, is then brought into position using an **automatic pallet changer**. All movements in the machine are guided by computer control, with pallet-changing cycle times on the order of 10–30 s. A pallet station can accommodate multiple pallets that serve a machining center, that may also be equipped with various automatic features such as automatic loading and unloading devices for the products being made.

QR Code 8.6 A pallet pool. *Source:* Courtesy of Haas Automation, Inc.

Machining centers also have **programmable automatic tool changers**. Depending on the design, as many as 200 (but usually fewer than 50) cutting tools can be stored in a magazine, drum, or chain (called *tool storage*). For complex machining operations, auxiliary tool storage is available on special machining centers. Cutting tools are automatically selected for each cycle of machining operations, and with random access for the shortest route from storage to the machine spindle. A common design feature in machining centers is a **tool-exchange arm** that swings around to pick up a particular tool from its own toolholder and places it in the spindle. The tools are identified by radio frequency identification (RFID) tags or bar codes attached to the toolholders. Tool changing times are typically on the order of a few seconds.

Machining centers also can be equipped with a **tool-checking** and/or **part-checking station** that feeds information to the computer to compensate for any variations in tool settings or tool wear. **Touch probes** are available that can determine reference surfaces of the workpiece, for selection of tool setting, and for on-line inspection of the parts being machined. Several surfaces can be contacted, and their relative positions are determined and stored in the database of the computer software. The data are then used for programming tool paths and compensating, for example, for tool length and diameter, and tool wear.

8.11.1 Types of Machining and Turning Centers

There are several designs for machining centers, but the basic types are vertical spindle and horizontal spindle; universal machines have the capability of using both axes. The maximum dimensions that the cutting tools can operate in a machining center are known as the *work envelope*, a term also used with industrial robots (see Section 14.7).

1. **Vertical-spindle machining centers**, or *vertical machining centers*, are suitable for performing various machining operations on flat surfaces with deep cavities, such as in mold and die making. Because

the thrust forces in vertical machining are directed downward, these machines have high stiffness and produce parts with good dimensional accuracy; moreover, they are generally less expensive than horizontal-spindle machines.

2. **Horizontal-spindle machining centers**, or *horizontal machining centers*, (See Fig. 8.64) are suitable for large or tall workpieces and requiring machining on a number of their surfaces. The pallet can be rotated on different axes and to various angular positions. Another category of horizontal-spindle machines is **turning centers**, that are computer-controlled *lathes* with several features. Figure 8.65 shows a three-turret computer-numerical-control turning center; note that this machine has two horizontal spindles and three turrets, and are equipped with a variety of cutting tools to perform several operations on a workpiece that rotates.
3. **Universal machining centers** are equipped with both vertical and horizontal spindles. They have a variety of features and are capable of machining all surfaces of a workpiece (vertical, horizontal, and diagonal), hence the term *universal*.

8.11.2 Characteristics and Capabilities of Machining Centers

The major characteristics of machining centers are:

- The machines are capable of handling a wide range of part sizes and shapes, efficiently, economically, and with consistently high dimensional tolerances, on the order of ± 0.0025 mm with proper work holding devices.
- They are versatile, having as many as six axes of linear and angular movements, as well as the capability of quick changeover from one type of product to another. The need for a variety of machine tools and a large amount of floor space is thus significantly reduced.
- The time required for loading and unloading workpieces, changing tools, gaging of workpieces being machined, and troubleshooting is reduced, thus improving productivity, reducing labor requirements (particularly the need for skilled labor), and minimizing overall costs.

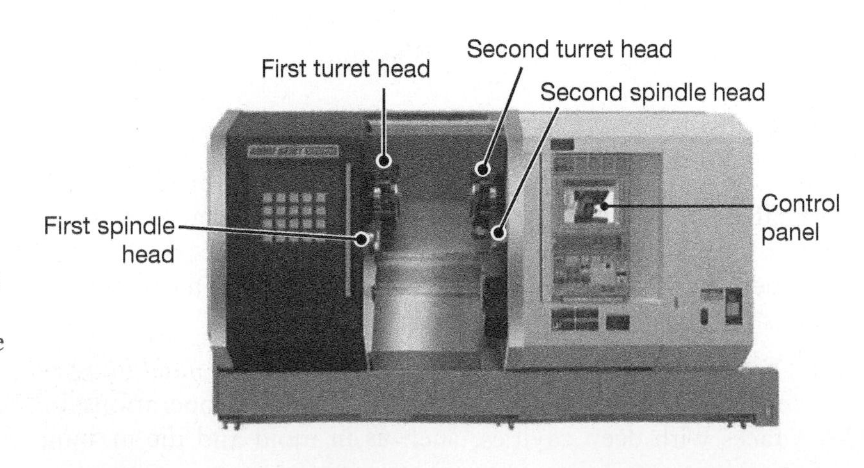

FIGURE 8.65 A computer numerical-control turning center. The two spindle heads and two turret heads make the machine very flexible in its, capabilities; up to three turret heads are commercially available.
Source: Courtesy of Mori Seiki Co., Ltd.

- Machining centers are highly automated and relatively compact, so that one operator often can attend to two or more machines at the same time.
- The machines are equipped with tool-condition monitoring devices (see Section 8.3.5), for detecting tool breakage and wear, as well as probes for compensation for tool wear and for tool positioning.
- In-process and post-process gaging and inspection of machined workpieces are standard features of machining centers.

Machining centers are available in a wide variety of sizes and with a large selection of features; their cost ranges from about $50,000 to $1 million and higher. Typical capacities range up to 75 kW (100 hp), and maximum spindle speeds are typically in the range of 4000–12,000 rpm, but some are as high as 75,000 rpm for special applications and with small-diameter cutters. Some pallets are capable of supporting workpieces that weigh as much as 7,000 kg (15,000 lb); higher capacities are available for special applications.

8.11.3 Reconfigurable Machines and Systems

The need for flexibility of manufacturing operations has led to the concept of *reconfigurable machines*, consisting of various modules to meet different functional requirements. The term *reconfigurable* stems from the fact that, using advanced computer hardware and reconfigurable controllers, and utilizing advances in information management technologies, the machine components can quickly be arranged and rearranged in a number of configurations to meet specific production demands. Based on a typical machine-tool structure of a three-axis machining center, Fig. 8.66 shows an example of how the machine can be reconfigured to become a *modular* machining center. With the resulting flexibility, the machine can perform different machining operations while accommodating a range of

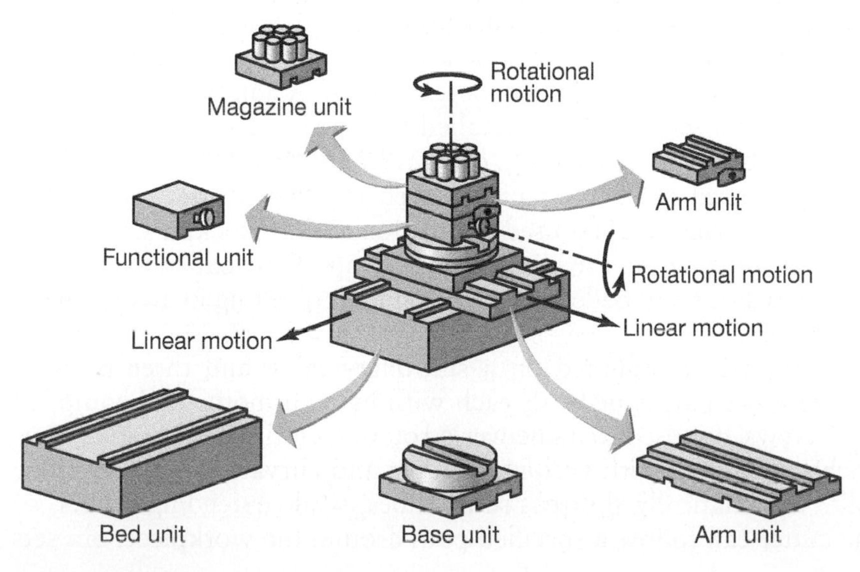

FIGURE 8.66 Schematic illustration of a reconfigurable modular machining center, capable of accommodating workpieces of different shapes and sizes, and requiring different machining operations on their various surfaces. *Source:* After Y. Koren.

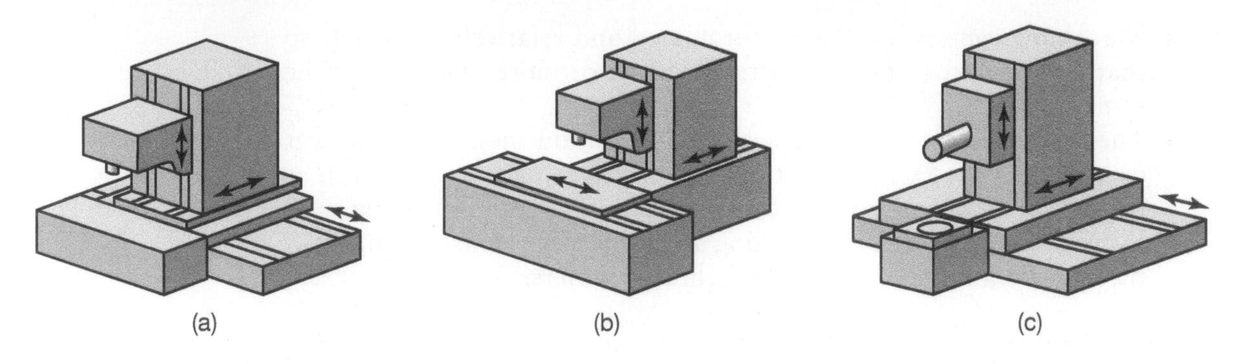

FIGURE 8.67 Schematic illustration of assembly of different components of a reconfigurable machining center. *Source:* After Y. Koren.

workpiece sizes and part geometries. Another example of this flexibility is given in Fig. 8.67, where a five-axis (three linear and two rotational movements) machine can be reconfigured by assembling different modules.

EXAMPLE 8.7 Machining Outer Bearing Races on a Turning Center

Outer bearing races (Fig. 8.68) are machined on a turning center. The starting material is hot-rolled 52100 steel tube, with an outer diameter of 91 mm and an inner diameter of 75.5 mm; the cutting speed is 95 m/min for all operations. The cutting tools are carbide, including the cutoff tool (the last operation performed), which is 3.18 mm wide, instead of 4.76 mm for the high-speed steel cutoff tool that was formerly used. The material that was saved by this change is significant, because the width of the race is small. The turning center was capable of machining these races at high speeds and with repeatable dimensional tolerances of ±0.025 mm.

8.11.4 Hexapod Machines

Developments in the design and materials used for machine-tool structures and components are continually taking place, with important goals of (a) imparting flexibility to machine tools; (b) increasing the machining envelope; and (c) making the machines lighter. A innovative machine-tool structure is a self-contained octahedral (eight-sided) machine frame. Referred to as **hexapods** (Fig. 8.69), or *parallel kinematic linked machines*, the design is based on a mechanism called the *Stewart platform* (after D. Stewart), an invention first used to position aircraft cockpit simulators. The main advantage is that the links in the hexapod are loaded axially and the bending stresses and deflections are minimal, resulting in an extremely stiff structure.

The workpiece is mounted on a stationary table and three pairs of *telescoping tubes* (struts or legs), each with its own motor and equipped with ballscrews, are used to maneuver a rotating cutting-tool holder. During machining a part with various features and curvatures, the machine controller automatically shortens some tubes, while extending others, so that the cutter can follow a specified path around the workpiece. Six sets

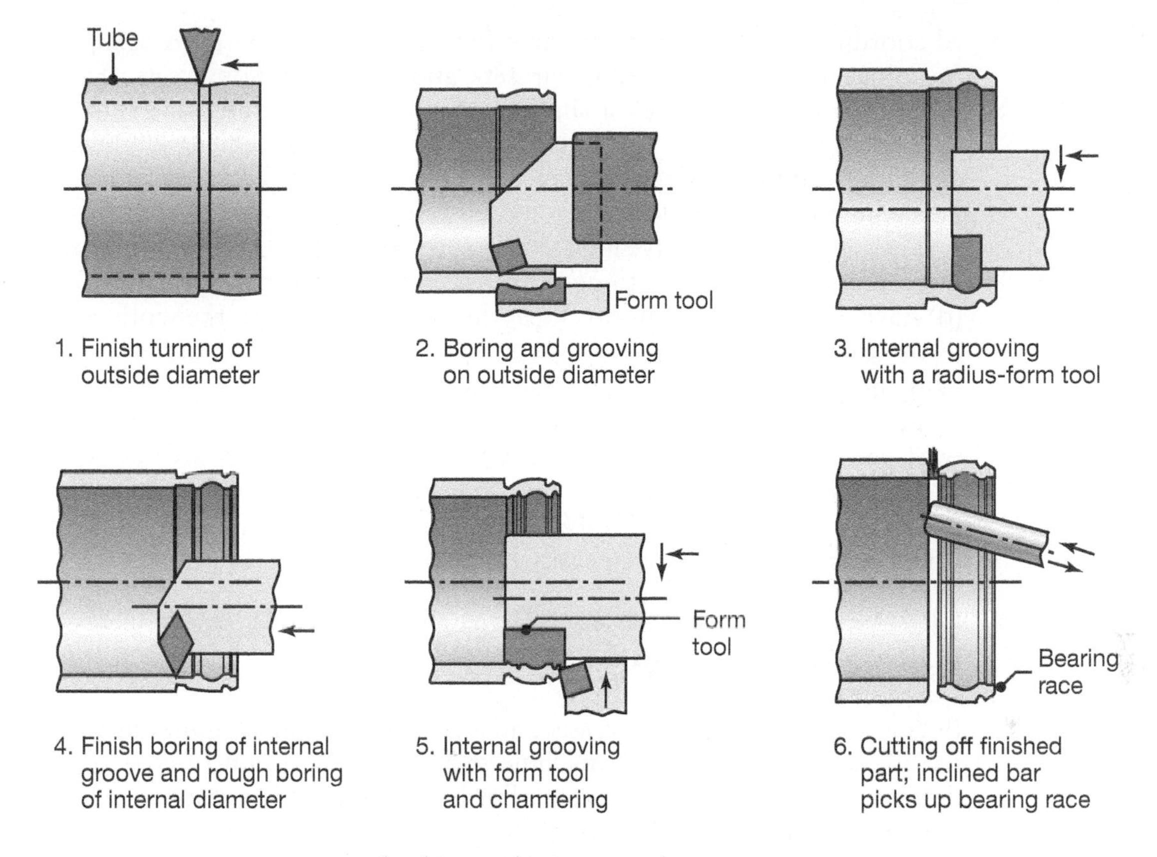

FIGURE 8.68 Sequences involved in machining outer bearing races on a turning center.

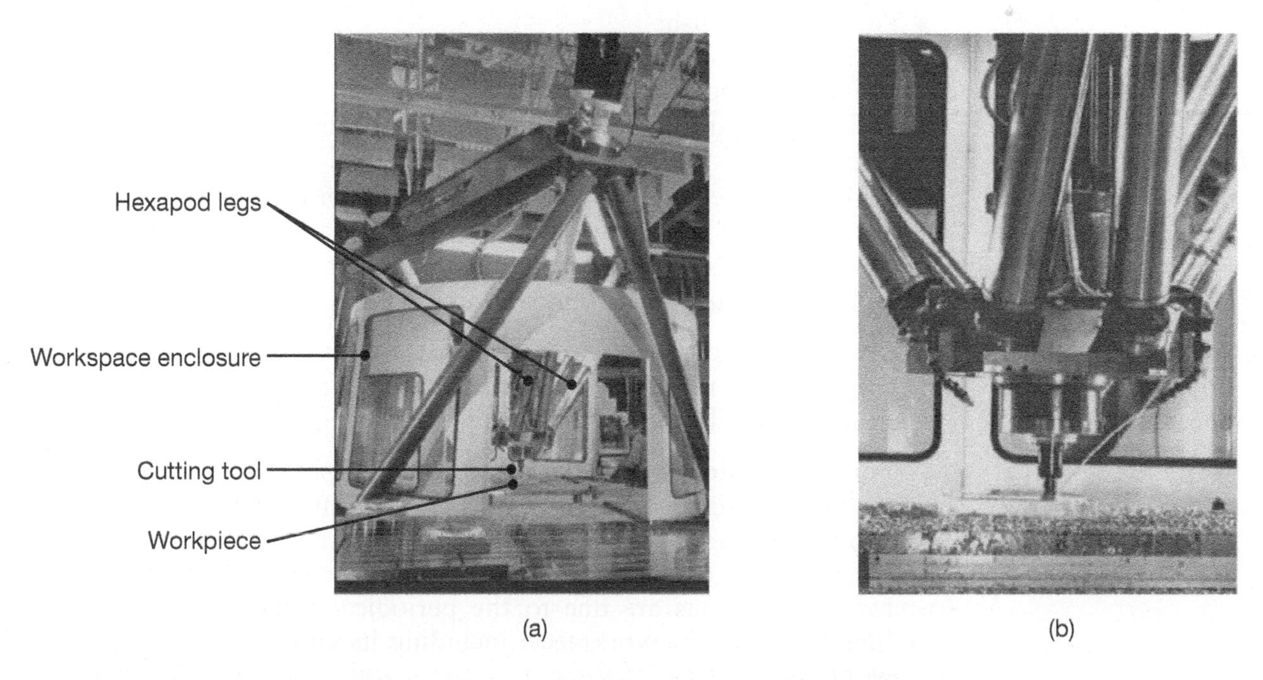

FIGURE 8.69 (a) A hexapod machine tool, showing its major components. (b) Close-up view of the cutting tool and its head in a hexapod machining center. *Source:* National Institute of Standards and Technology.

of coordinates are involved in these machines (hence the term *hexapod*, meaning six legged): three linear sets and three rotational sets. Every motion of the cutter, even a simple linear motion, is translated into six coordinated leg lengths, all moving in real time. The movements of the legs are rapid, and consequently, have high accelerations and decelerations, hence high inertia forces, are involved.

These machines (a) have high stiffness; (b) are not as massive as machining centers; (c) have about one-third fewer parts than machining centers; (d) have a large machining envelope, thus greater access to the work zone; (e) are capable of maintaining the cutting tool perpendicular to the surface being machined, thus improving the machining operation; and (f) with six degrees of freedom, they have high flexibility in the production of parts with various geometries and sizes without having to refixture the work in progress. Unlike most other machine tools, these machines are often portable. In fact, *hexapod attachments* are now available whereby conventional machining centers can easily be converted to a hexapod machine.

8.12 Vibration and Chatter

In describing machining processes and machine tools throughout this chapter, it was pointed out that machine stiffness is important in controlling dimensional accuracy and surface finish of parts. This section describes the adverse effects of low stiffness on product quality and machining operations, and the level of vibration and chatter in cutting tools and machine tools. If uncontrolled, vibration and chatter can result in:

- Poor surface finish;
- Loss of dimensional accuracy of the workpiece;
- Chipping, excessive wear, and failure of the cutting tool, especially critical for brittle tool materials, such as ceramics, some carbides, and diamond;
- Damage to the machine-tool components from excessive vibrations; and
- Noise generated, particularly objectionable if it is of high frequency, such as the squeal in turning brass.

QR Code 8.7 Elimination of chatter through spindle speed variation. *Source:* Courtesy of Haas Automation, Inc.

Because they involve several factors, vibration and chatter in machining are complex phenomena. Machining operations cause two basic types of vibration: forced vibration and self-excited vibration.

1. **Forced vibration** is generally caused by some periodic force present in the machine tool, such as from gear drives, imbalance of the machine components, misalignment, or from motors and pumps. As an example, in milling or in turning a splined shaft, or a shaft with a keyway, forced vibrations are due to the periodic engagement of the cutting tool with the workpiece, including its entry and exit from the workpiece surface.

   A common solution to forced vibrations is to isolate or remove the forcing element. If the forcing frequency is at or near the natural

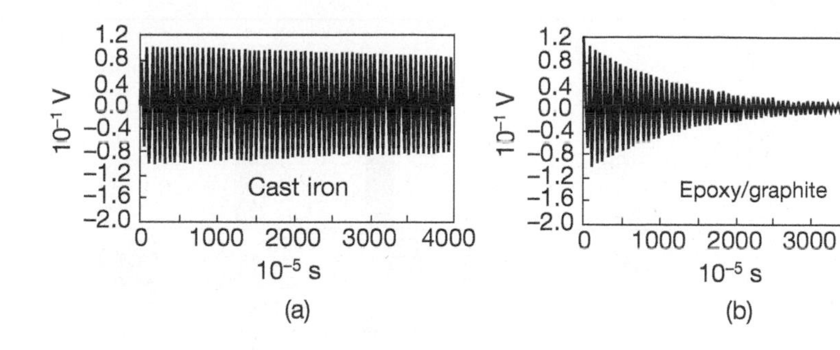

FIGURE 8.70 Relative damping capacity of (a) gray cast iron and (b) epoxy-granite composite material. The vertical scale is the amplitude of vibration and the horizontal scale is time.

frequency of a component of the machine-tool system, one of the frequencies may be raised or lowered. The amplitude of vibration can be reduced by increasing the stiffness or the damping of the system. Although modifying the process parameters generally does not appear to greatly influence forced vibrations, changing the cutting speed and the tool geometry can be helpful.

2. Generally called **chatter**, *self-excited vibration* is caused by the interaction of the machining process with the structure of the machine tool; the vibrations usually have very high amplitude. Chatter typically begins with a disturbance in the cutting zone. The disturbances include lack of homogeneity in the workpiece material or its surface condition, changes in chip morphology during machining (see Section 8.2.1), or a change in frictional conditions at the tool-chip interface, as for example, influenced by cutting fluids and their effectiveness. Self-excited vibrations can generally be controlled by (a) increasing the dynamic stiffness of the system; and (b) by damping. **Dynamic stiffness** is the ratio of the amplitude of the force applied to the amplitude of vibration. Because a machine tool typically has a different stiffness at different frequencies, any changes in cutting parameters, such as cutting speed, also can influence chatter.

   The most important type of self-excited vibration in machining is **regenerative chatter**, which is caused when a tool cuts along a surface that has a roughness or disturbances left from the previous cut. Because the depth of cut varies due to the surface, the resulting cutting force variations cause tool vibration. The process continues repeatedly during machining, hence the term *regenerative*.

Damping. The term *damping* is defined as the rate at which vibrations decay; it is an important factor in controlling vibration and chatter in machine tools.

Internal damping of structural materials. Damping results from the energy loss within materials during vibration. Steels have less damping than gray cast iron, and composite materials have more damping than gray iron (Fig. 8.70). The difference in the damping capacity of materials can be observed by striking them with a gavel and listening to the sound; try, for example, striking pieces of steel, concrete, plastics, fiber-reinforced plastics, and wood.

Joints in the machine-tool structure. Although less significant than internal damping, bolted joints in the structure of a machine tool also are

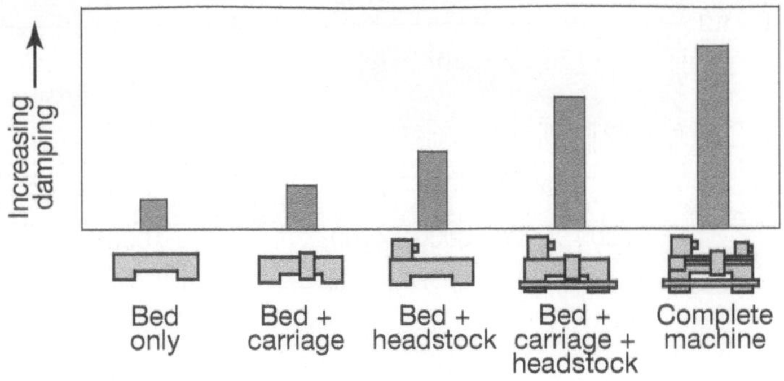

FIGURE 8.71 Damping of vibrations as a function of the number of components on a lathe. Joints dissipate energy; thus, the greater the number of joints, the higher the damping. *Source:* After J. Peters.

a source of damping. Because friction dissipates energy, small relative movements along dry (unlubricated) joints dissipate energy and improve damping. In joints where oil or grease is present, the internal friction of the lubricant layers also dissipates energy, thereby contributing to damping. All machines consist of a number of large and small components, assembled into a structure. Consequently, this type of damping is *cumulative*, owing to the presence of a number of joints. Note in Fig. 8.71, for example, how damping in a lathe increases as the number of its components and their contact areas increase.

External damping. External damping is typically accomplished by using external dampers; these are similar to shock absorbers on automobiles. Special vibration absorbers have been developed to be installed on machine tools for this purpose.

Factors influencing chatter. The tendency for a particular workpiece to chatter during machining has been found to be proportional to the cutting forces and the depth and width of cut. Consequently, because cutting forces increase with strength and hardness of the workpiece material, the tendency to chatter generally increases as hardness increases. For example, aluminum and magnesium alloys have lower tendency to chatter than do martensitic and precipitation-hardening stainless steels, nickel alloys, and high-temperature and refractory alloys. An important factor in chatter is the type of chip produced during machining operations. As noted earlier, continuous chips involve steady cutting forces, and they generally do not cause chatter. Discontinuous chips and serrated chips, on the other hand, are associated with chatter because such chips are produced periodically, and the resulting force variations can cause chatter.

8.13 Machine-Tool Structures

This section describes the *material* and *design* aspects of machine tools, as structures with certain specific characteristics. The proper design of machine-tool structures requires knowledge of the materials available for construction, their forms and various properties, the dynamics of the particular machining process, and the cutting forces involved. The stiffness of a machine tool depends on both the dimensions of its structural components and their elastic modulus and the nature of the joints in the structure. Damping depends on the type of materials used and the number and nature of the joint interfaces.

Materials and design. Traditionally, the base and some of the major components of machine tools have been made of gray or nodular cast iron. These materials have the advantages of low cost and good damping capacity, but they are heavy. *Lightweight* designs are desirable because of their higher natural frequencies, lower inertial forces of the moving components, and ease of transportation. Lightweight designs and design flexibility require such fabrication and assembly processes as (a) mechanical fastening (bolts and nuts) of individual components; and (b) welding; however, because of the preparations involved, this approach increases labor and material costs.

Wrought steels are one choice for such structures because of their low cost, availability in various sizes and shapes (such as channels, angles, and tubes), appropriate mechanical properties, and such favorable characteristics as formability, machinability, and weldability. On the other hand, steels do not have the higher damping capacity of castings and composites.

Another factor that contributes to lack of precision of a machine tool is the *thermal expansion* of its components, that would lead to distortion. The sources of heat may be (a) internal, such as bearings, machine ways, motors, and heat generated from the cutting zone (Section 8.2.5); or (b) external, such as nearby furnaces, space heaters, sunlight, and fluctuations in the ambient and cutting-fluid temperatures. Also important in machine-tool precision are *foundations*, particularly their mass and how the machines are installed in a plant.

There are several options concerning materials for the bases and the components of machine tools. For example, **acrylic concrete**, a mixture of concrete and a polymer (polymethylmethacrylate), can easily be cast into desired shapes for machine bases; various compositions of this material are available. It can also be used for *sandwich construction* with cast iron, thus combining the respective advantages of each type of material. **Granite-epoxy composite** has a typical composition of about 93% crushed granite and 7% epoxy binder. First used in precision grinders in the early 1980s, this material has several favorable properties: (a) good castability, thus allowing for design versatility in machine tools; (b) high stiffness-to-weight ratio; (c) thermal stability; (d) resistance to environmental degradation of the materials used; and (e) good damping capacity (See Fig. 8.70).

8.14 Design Considerations

1. **General requirements.**
   a. Materials should be selected not only to meet design requirements but also for their machinability.
   b. Tolerances should be as wide as possible and surface roughness as high as acceptable. Often, as-cast or as-formed tolerances may be sufficient and machining may not be necessary.

Excessively tight design requirements will require further costly finishing operations, such as grinding and lapping.

c. Design features should be such that standard cutting tools, inserts, and toolholders can be used.
d. Parts should be designed so that they can be securely fixtured on the machines and minimize refixturing; this often requires incorporation of clamping features in the design of the parts. Appropriate spaces should be provided to accommodate fixtures, clamping devices, and clearance for cutters.
e. Whenever possible, all machining operations should be in the same plane or the same diameter, so that they can be performed with a minimum number of part setups.
f. If burrs are unavoidable, space should be provided for burr removal.

2. **Turning.**
   a. Blanks to be machined should be as close to the final dimensions (such as by net or near-net-shape forming) as possible so as to reduce machining time.
   b. Thin and slender workpieces may be difficult to support properly for machining, and may deflect excessively; parts should be as short and stocky as possible.
   c. Sharp corners, tapers, and major dimensional variations along the part should be avoided.
   d. Radii should be as large as permitted and should conform to standard tool nose-radius specifications.

3. **Screw thread cutting.**
   a. Rolled threads are generally preferable to cut threads; whenever practical, cutting threads should be avoided.
   b. Through-holes are preferable to blind holes when machining threads; internal threads in blind holes should have an unthreaded length at the bottom.
   c. Designs should allow the termination of threads before the threads reach a shoulder.
   d. Threaded sections should not be interrupted with holes, slots, or other discontinuities.

4. **Drilling, reaming, boring, and tapping.**
   a. Holes should be specified preferably on flat surfaces and perpendicular to the direction of drill movement.
   b. Excessively deep holes should be avoided; length-to-diameter ratios of three or less should be specified whenever possible, although a ratio of 8:1 is feasible.
   c. When multiple holes are required, they should use the same diameter whenever practical to avoid unnecessary tool changes.
   d. Through holes are usually preferred to blind holes, unless they lead to significantly larger material removal, and blind holes should be drilled deeper than subsequent reaming or tapping

operations would require, typically by an amount at least one-fourth of the hole diameter.

e. The bottoms of blind holes should match standard drill-point angles.

5. **Milling.**

a. Internal corners should have a radius that matches the internal radius of the cutter, whenever possible.

b. For external corners, bevels are preferred over radii, because cutter and setup costs are larger for machining of radii. If the radius and shaft have the same radius, then the transition between them needs to be very accurately machined and is very difficult to produce.

c. Internal cavities and pockets with sharp inner corners should be avoided. When slots or key seats are required, the cutter should be allowed to define both the slot width and the end radii.

d. Small milling cutters can machine any surface, but small cutters are slower and more susceptible to chatter than large cutters.

6. **Broaching.**

a. Features such as keyways and gear teeth have standard sizes, allowing the use of common broaches.

b. Balanced cross sections are preferable to keep the broach from drifting and maintaining tight tolerances.

c. Inverted or dovetail splines should be avoided.

d. Broaching blind holes should be avoided, but when necessary, there must be a relief at the end of the broached area.

7. **Gear machining.**

a. Blank shape is important for proper fixturing and to ease cutting operations. Machining allowances must be incorporated in blanks; if machining is to be followed by other finishing operations, the part must be oversized after machining.

b. Wide gears are more difficult to machine than narrow gears.

c. Spur gears are easier to machine than helical gears, and are easier to machine than bevel and worm gears.

d. Dimensional tolerances and gear shaping tools are specified by standards.

e. A gear quality number should be selected such that the gear has as wide a tolerance range as possible while still meeting performance requirements.

8.15 Economics of Machining

The advantages and limitations of machining should be viewed with respect to the competitive nature of various other manufacturing operations. For example, recall that as compared with forming and shaping processes, machining requires more time and wastes more material. On the other hand, machining operations are much more versatile, and they can produce parts with dimensional accuracy and surface finish that are better than most of those produced by other processes.

This section briefly presents a traditional and basic approach to the *economic* aspects of machining processes, including the important factors involved in determining the *minimum cost per piece* and the *maximum production rate*. The **total cost per piece** basically consists of four components:

$$C_p = C_m + C_s + C_l + C_t, \tag{8.46}$$

where C_p is the cost per piece; C_m is the cost of machining; C_s is the cost of setting up for machining, such as mounting the tool, the cutter, and the fixtures and preparing for the particular operation; C_l is the cost of loading, unloading, and machine handling; and C_t is the tooling cost, which can take different forms depending on the application, but includes factors such as tool or insert changing, indexing, and depreciation of the cutter or insert. The **machining cost** is given by

$$C_m = t_m \left(L_m + B_m\right), \tag{8.47}$$

where t_m is the machining time per piece, L_m is the labor cost of the machine operator per hour, and B_m is the burden rate, or overhead charge, of the machine, including depreciation, maintenance, and indirect labor. The **setup cost** is a fixed figure. The **loading, unloading,** and **machine-handling cost** is given by

$$C_l = t_l \left(L_m + B_m\right), \tag{8.48}$$

where t_l is the time involved in loading and unloading the part, changing speeds, changing feed rates, and so on. The **tooling cost** is expressed as

$$C_t = \frac{1}{N_i}\left[t_c \left(L_m + B_m\right) + D_i\right] + \frac{1}{N_f}\left[t_i \left(L_m + B_m\right)\right], \tag{8.49}$$

where N_i is the number of parts machined per insert, N_f is the number of parts that can be produced per insert face, t_c is the time required to change the insert, t_i is the time required to index the insert, and D_i is the depreciation of the insert in dollars. The **time** required to produce one part is

$$t_p = t_l + t_m + \frac{t_c}{N_i} + \frac{t_i}{N_f}, \tag{8.50}$$

where t_m has to be calculated for each particular operation. For example, in a turning operation, the machining time (see Section 8.9.1) is

$$t_m = \frac{l}{fN} = \frac{\pi l D}{fV}, \tag{8.51}$$

where l is the length of cut, f is the feed, N is the rpm of the workpiece, D is the workpiece diameter, and V is the cutting speed. Note that appropriate units must be used in all these equations. From the tool-life equation [see Eq. (8.31)],

$$Vt^n = C.$$

Hence,

$$t = \left(\frac{C}{V}\right)^{1/n}, \tag{8.52}$$

where t is the time required to reach a flank wear of certain dimension, after which the tool has to be indexed or changed. The number of pieces per insert face is

$$N_f = \frac{t}{t_m}, \tag{8.53}$$

and the number of pieces per insert is

$$N_i = mN_f = \frac{mt}{t_m}. \tag{8.54}$$

Sometimes not all of the faces are used before the insert is discarded or recycled, so it should be recognized that m corresponds to the number of faces that are actually used, not the number of faces in the insert itself. The combination of Eqs. (8.51)–(8.54) gives

$$N_p = \frac{fC^{1/n}}{\pi LDV^{(1/n)-1}}. \tag{8.55}$$

The **cost per piece,** C_p, in Eq. (8.46) can now be defined in terms of several variables. To find the optimum cutting speed and the optimum tool life for **minimum cost,** C_p is differentiated with respect to V and is set to zero. Thus,

$$\frac{\partial C_p}{\partial V} = 0. \tag{8.56}$$

The optimum cutting speed, V_o, is then

$$V_o = \frac{C\,(L_m + B_m)^n}{\left(\frac{1}{n} - 1\right)^n \left\{\frac{1}{m}\left[t_c\,(L_m + B_m) + D_i\right] + t_i\,(L_m + B_m)\right\}^n}, \tag{8.57}$$

and the optimum tool life, t_o, is

$$t_o = \left[\left(\frac{1}{n}\right) - 1\right] \frac{\frac{1}{m}\left[t_c\,(L_m + B_m) + D_i\right] + t_i\,(L_m + B_m)}{L_m + B_m}. \tag{8.58}$$

To determine the optimum cutting speed and the optimum tool life for **maximum production,** t_p is differentiated with respect to V and is set to zero. Thus,

$$\frac{\partial t_p}{\partial V} = 0. \tag{8.59}$$

The **optimum cutting speed** is then found as

$$V_o = \frac{C}{\left[\left(\frac{1}{n} - 1\right)\left(\frac{t_c}{m} + t_i\right)\right]^n}, \tag{8.60}$$

and the **optimum tool life** is

$$t_o = \left(\frac{1}{n} - 1\right)\left(\frac{t_c}{m} + t_i\right). \tag{8.61}$$

Qualitative plots of the minimum cost per piece and the minimum time per piece (meaning maximum production rate) are given in Fig. 8.72. As described in Section 9.18, the cost of a surface machined also depends on the degree of finish required and the cost increases rapidly with finer surface finish.

This brief analysis indicates the importance of (a) identifying all relevant parameters in a machining operation, (b) determining various cost factors, (c) obtaining relevant tool-life curves for the particular operation, and (d) measuring the various time intervals involved in the overall operation. The importance of obtaining accurate data is indicated in Fig. 8.72, as small changes in cutting speed can have a significant effect on the minimum cost and minimum time per piece.

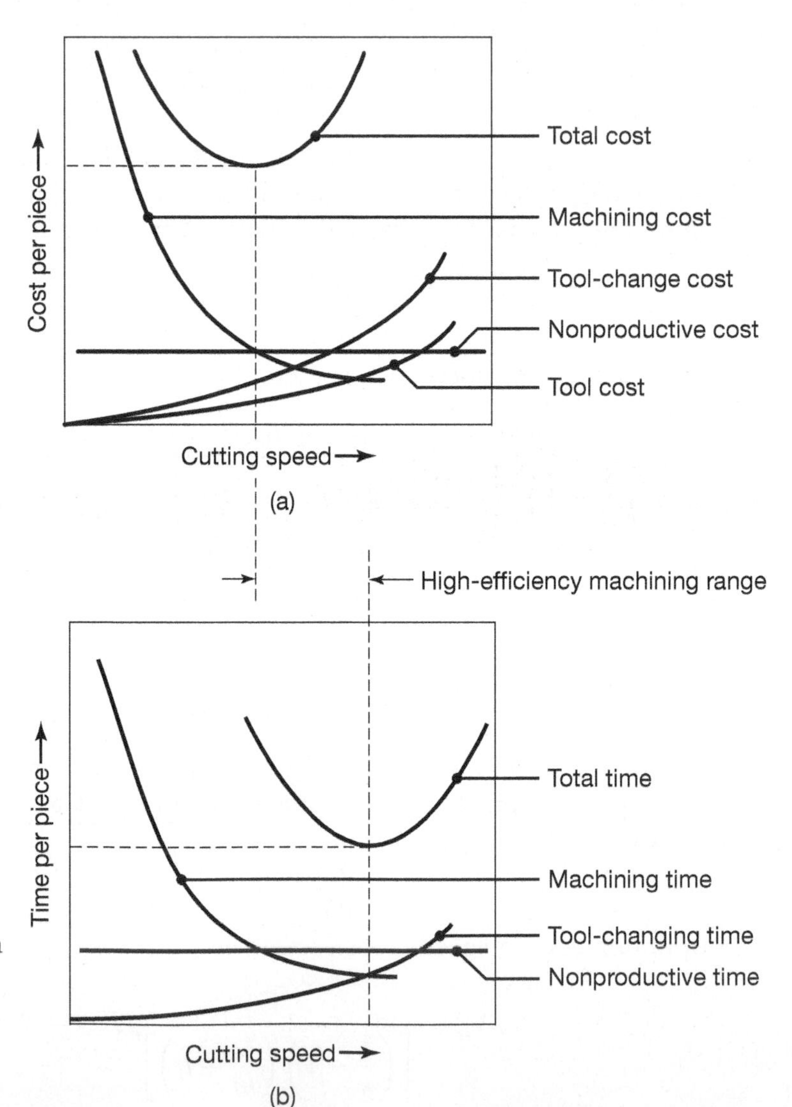

FIGURE 8.72 Qualitative plots showing (a) cost per piece, and (b) time per piece in machining. Note that there is an optimum cutting speed for both cost and time, respectively. The range between the two optimum speeds is known as the *high-efficiency machining range.*

CASE STUDY 8.1 Ping Golf Putters

In their efforts to develop high-end, top performing putters, engineers at Ping Golf, Inc. in Phoenix, Arizona utilized advanced machining practices in their design and production processes for a new style of putter, the Anser® series (Fig. 8.73). Governed by a unique set of design constraints, they had the task and goal of creating putters that were both practical for production quantities and met specific functional and aesthetic requirements.

One of the initial decisions concerned selection of a proper material for the putter to meet its functional requirements. Four types of stainless steel (303, 304, 416, and 17–4 precipitation hardening; see Section 3.10.3) were considered for various property requirements, including machinability, durability, and the sound or feel of the particular putter material (another requirement unique to golf equipment). Among the materials evaluated, 303 stainless steel was chosen because it is free-machining (Section 8.5.1), indicating smaller chips, lower power consumption, better surface finish, and improved tool life, thus allowing for increased machining speeds and higher productivity.

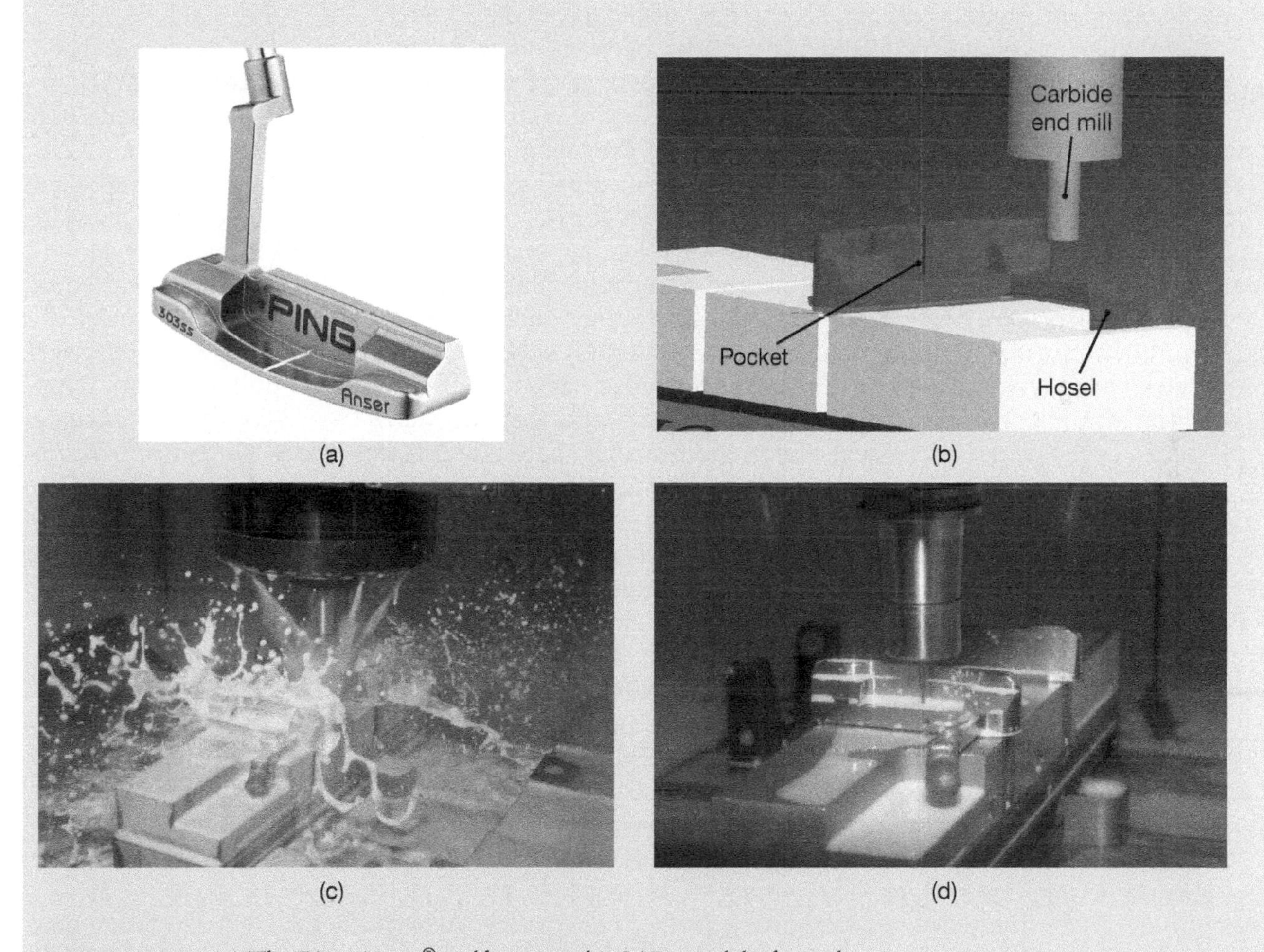

FIGURE 8.73 (a) The Ping Anser® golf putter; (b) CAD model of rough machining of the putter outer surface; (c) rough machining on a vertical machining center; and (d) machining of the lettering in a vertical machining center. The operation was paused to take the photo, as normally the cutting zone is flooded with a coolant. *Source:* Courtesy of Ping Golf, Inc.

The next step of the project involved determining the optimum blank type and the sequence of operations to be performed in its production. In this case, engineers chose to develop a slightly oversized forged blank (Section 6.2). A forging was chosen because it provided a favorable internal grain structure as opposed to a casting, as castings could result in porosity and inconsistent surface finish after machining (Section 5.12.1). The blank incorporated a machining allowance, so dimensions were specified approximately 0.050- 0.075 in. (1.25–1.9 mm) larger in all directions than that of the final part.

The most challenging and longest task of the project was developing the necessary programming and fixturing for each part. Beyond the common requirements of typical machined parts (including tight tolerances and repeatability), putters require an additional set of aesthetic specifications. In this case, both precise machining and the overall appearance of the finished part were imperative. A machining technique known as surfacing or contouring (commonly used in injection mold making; Section 10.10.2) was used to machine most of the finished geometry. Although this operation required additional machining time, it provided a superior finish on all surfaces and allowed machining of more complex geometries, thus adding value to the finished product.

As for all high-volume machined parts, repeatability was essential. Each forged blank was designed with a protrusion across the face of the putter, allowing for the initial locating surfaces (for fixturing). A short machining operation removed a small amount of material around the bar and produced three flat and square surfaces as a reference location for the first primary machining operation.

Each putter required six different operations in order to machine all of its surfaces, and each operation was designed to provide locating surfaces for the next step in the manufacturing process. Several operations were setup using a tombstone loading system (see Section 14.9) on a horizontal-spindle CNC milling machine. This method allowed machine operators to load and unload parts while other parts were being machined, thus significantly increasing the efficiency of the process.

Modular fixturing and tungsten carbide cutting tools coated with TiAlN (Section 8.6.5) allowed the quick change-over between right- and left-handed parts as well as different putter models. After the initial locating operation was complete, the parts were then transferred to a 3-axis VMC to create the putter cavity. Since the forged blanks are near net shape, the maximum radial depth of cut on most surfaces was 0.075 in., but the axial depth of cut of 1.5 in. inside the "cavity" of the putter was the most demanding milling operation (See Figs. 8.73b and c). The putter has small inside radii with a comparatively long depth (7× diameter or greater).

A 4-axis horizontal machining center was used to reduce the number of setups in the operation. The rotary axis was used for creating the relatively complex shape of the hosel (the socket for the shaft of the golf club). Since the hosel is relatively unsupported, chatter was the most complex challenge to overcome. Several iterations of spindle speeds were attempted in conjunction with upfront guidance from a simulation model. Modal analyses were conducted on the fixtured parts in an attempt to identify and avoid the natural frequencies of the part/fixture (see Section 8.12). The machines had spindle speeds ranging from 12,000 to 20,000 rpm, each having 30 horsepower. With the near-net-shape forging, the milling operations were designed to have low depths of cut but high speed.

After each machining operation was completed, a small amount of hand finishing was necessary to produce a superior surface. The putters were then lightly shot blasted (with glass bead media; Section 4.5.1) for the purpose of achieving surface consistency. A black nickel-chrome plating was then applied to all parts to enhance the aesthetic appeal and protect the stainless steel from small dings and dents and corrosion from specific chemicals that may be encountered on a golf course.

Source: Courtesy of D. Jones and D. Petersen, Ping Golf, Inc.

SUMMARY

- Machining processes are often necessary in order to impart the required geometric features, surface finish, and dimensional accuracy to components, particularly those with complex shapes that cannot be produced economically or properly by other shaping techniques. On the other hand, machining processes inevitably waste material in the form of chips, involve longer processing times, and may have adverse effects on the surfaces produced. (Section 8.1)
- Significant process variables in machining include cutting-tool shape and material; cutting conditions, such as speed, feed, and depth of cut; cutting fluids; and the characteristics of the workpiece material and the machine tool. Parameters influenced by these variables include forces and power consumption, tool wear, surface finish and integrity, temperature, and dimensional accuracy of the workpiece. Commonly observed chip types are continuous, built-up edge, discontinuous, and serrated. (Section 8.2)
- Temperature rise is an important consideration, as it can have adverse effects on tool life as well as on dimensional accuracy and surface integrity of the machined part. (Section 8.2)
- Tool wear depends primarily on workpiece and tool-material characteristics, process parameters such as cutting speed and feed, and cutting fluids; machine-tool characteristics also can have an effect. Two major types of wear are flank wear and crater wear. (Section 8.3)
- Surface finish of machined components is an important consideration, as it can adversely affect product integrity. Important variables that affect surface finish include the geometry and condition of the cutting tool, chip morphology, and process variables. (Section 8.4)
- Machinability is generally defined in terms of surface finish, tool life, force and power requirements, and chip type produced. Machinability of materials depends not only on their intrinsic properties and microstructure, but also on proper selection and control of process variables. (Section 8.5)
- A variety of cutting-tool materials are available, the most common ones being high-speed steels, carbides, ceramics, and cubic boron nitride. These materials, including their coatings, have a broad range of mechanical and physical properties, particularly hot hardness, toughness, chemical stability and inertness, and resistance to chipping and wear. (Section 8.6)
- Cutting fluids are important in machining operations, as they reduce friction, forces, and power requirements, and they improve tool life. Generally, slower operations with high tool stresses require a fluid with good lubricating characteristics, whereas in high-speed operations, where temperature rise can be significant, fluids with high cooling capacity are preferred. (Section 8.7)

- The machining processes that produce external and internal circular profiles are turning, boring, drilling, tapping, and thread cutting. Because of the three-dimensional nature of these operations, chip movement and control are important considerations, since chips can interfere with the operation. Optimization of each process requires an understanding of the interrelationships among design and processing parameters. (Section 8.9)
- High-speed machining, ultraprecision machining, and hard turning are among important developments, as they can produce parts with exceptional surface finish and dimensional accuracy, and help reduce machining costs. (Sections 8.8 and 8.9)
- Complex shapes can be machined by slab, face, and end milling, and also by broaching and sawing. These processes use multitooth tools and cutters at various axes with respect to the workpiece. The machine tools are now mostly computer controlled, have a variety of features and attachments, and possess flexibility in operation. (Section 8.10)
- Because of their versatility and capability of performing a variety of cutting operations, machining and turning centers are among the most important advances in machine tools. Their selection depends on such factors as part complexity, the number and type of cutting operations to be performed, the number of cutting tools needed, the dimensional accuracy required, and the production rate. (Section 8.11)
- Vibration and chatter in machining operations are important considerations for dimensional accuracy, surface finish, and tool life. Stiffness and damping capacity of machine tools are important factors in controlling vibration and chatter. (Sections 8.12 and 8.13)
- Design guidelines have been developed over the years for all machining operations. (Section 8.14)
- Analytical methods have been developed for determining the optimum cutting speeds for minimum machining time per piece and for minimum cost per piece, respectively. (Section 8.15)

SUMMARY OF EQUATIONS

Cutting ratio, $r = \frac{t_o}{t_c} = \frac{\sin \phi}{\cos(\phi - \alpha)}$

Shear strain, $\gamma = \cot \phi + \tan(\phi - \alpha)$

Velocity relationships, $\frac{V}{\cos(\phi - \alpha)} = \frac{V_s}{\cos \alpha} = \frac{V_c}{\sin \phi}$

Friction force, $F = R \sin \beta$

Normal force, $N = R \cos \beta$

Coefficient of friction, $\mu = \tan \beta = \frac{F_t + F_c \tan \alpha}{F_c - F_t \tan \alpha}$

Thrust force, $F_t = R\sin(\beta - \alpha) = F_c\tan(\beta - \alpha)$

Shear-angle relationships, $\phi = 45^\circ + \dfrac{\alpha}{2} - \dfrac{\beta}{2}$

$$\phi = 45^\circ + \alpha - \beta$$

Total cutting power, $= F_c V$

Specific energy, $u_t = \dfrac{F_c}{wt_o}$

Frictional specific energy, $u_f = \dfrac{Fr}{wt_o}$

Shear specific energy, $u_s = \dfrac{F_s V_s}{wt_o V}$

Mean temperature, $T = \dfrac{1.2\sigma_f}{\rho c}\sqrt[3]{\dfrac{Vt_o}{K}}$

$$T \propto V^a f^b$$

Tool life, $Vt^n = C$

Material-removal rate (MRR), in turning $= \pi D_{\text{avg}} df N$

in drilling $= \pi\left(\dfrac{D^2}{4}\right)fN$

in milling $= wdv$

BIBLIOGRAPHY

Altintas, Y., *Machining Automation: Metal Cutting Mechanics, Machine Tool Vibrations, and CNC Design*, Cambridge, 2012.

Astakhov, V.P., *Geometry of Single-point Turning Tools and Drills: Fundamentals and Practical Applications*, Springer, 2010.

Astakhov, V.P., and Joksh, S., *Metalworking Fluids for Cutting and Grinding; Fundamentals and Recent Advances*, Widhead, 2012.

Byers, J.P. (ed.), *Metalworking Fluids*, 2nd ed., CRC Press, 2006.

Chang, K., *Machining Dynamics: Fundamentals, Applications and Practices*, Springer, 2010.

Childs, T.H.C., Maekawa, K., Obikawa, T., and Yamane, Y., *Metal Machining: Theory and Applications*, Butterworth-Heinemann, 2000.

Cormier, D., *McGraw-Hill Machining and Metalworking Handbook*, McGraw-Hill, 2005.

Davim, J.P. (ed.), *Machinability of Advanced Materials*, Wiley, 2014.

—, *Machining: Fundamentals and Recent Advances*, Springer, 2010.

—, *Surface Integrity in Machining*, Springer, 2010.

—, *Traditional Machining Processes: Research Advances*, Springer, 2014.

Dudzinski, D., Molinari, A., and Schulz, H., (eds.) *Metal Cutting and High-Speed Machining*, Springer, 2002.

Erdel, B., *High-Speed Machining*, Society of Manufacturing Engineers, 2003.

Gegg, B.C., Suh, S.C., and Luo, C.G., *Machine Tool Vibrations and Cutting Dynamics*, Springer, 2011.

Ito, Y., *Modular Design for Machine Tools*, Springer, 2011.

Jackson, M.J., and Morell, J., *Machining with Nanomaterials*, Springer, 2009.

Joshi, P.H., *Machine Tools Handbook*, McGraw-Hill, 2008.

Knight, M.A., and Boothroyd, G., *Fundamentals of Metal Machining and Machine Tools*, 3rd ed., Dekker, 2006.

Koren, Y., *The Global Manufacturing Revolution: Product-Process and Business Integration and Reconfigurable Systems*, Wiley, 2010.

Krar, S.F., and Check, A.F., *Technology of Machine Tools*, 7th ed., McGraw-Hill, 2010.

Lopez, L.N., and Lamikiz, A., (ed.), *Machine Tools for High Performance Machining*, Springer, 2009.

Mickelson, D., *Hard Milling & High Speed Machining: Tools of Change*, Hanser Gardner, 2005.

Schmitz, T.L., and Smith, K.S., *Machining Dynamics: Frequency Response to Improved Productivity*, Springer, 2008.

Shaw, M.C., *Metal Cutting Principles*, 2nd ed., Oxford, 2005.

Stephenson, D., and Agapiou, J.S., *Metal Cutting: Theory and Practice*, 2nd ed., CRC Press, 2005.

Stout, K.J., Davis, J., and Sullivan, P.J., *Atlas of Machined Surfaces*, Chapman and Hall, 1990.

Trent, E.M., and Wright, P.K., *Metal Cutting*, 4th ed., Butterworth Heinemann, 2000.

Walsh, R.A., *McGraw-Hill Machining and Metalworking Handbook*, McGraw-Hill, 3rd ed., 2006.

Zhang, D., *Parallel Robotic Machine Tools*, Springer, 2009.

QUESTIONS

8.1 Explain why the cutting force, F_c, increases with increasing depth of cut and decreasing rake angle.

8.2 What are the effects of performing a cutting operation with a dull tool tip? A very sharp tip?

8.3 Describe the trends that you have observed in Tables 8.1 and 8.2.

8.4 To what factors would you attribute the large difference in the specific energies within each group of materials shown in Table 8.3?

8.5 Describe the effects of cutting fluids on chip formation. Explain why and how they influence the cutting operation.

8.6 Under what conditions would you discourage the use of cutting fluids?

8.7 Give reasons that pure aluminum and copper are generally classified as easy to machine.

8.8 Can you offer an explanation as to why the maximum temperature in cutting is located at about the middle of the tool-chip interface? (*Hint:* Note that there are two principal sources of heat: the shear plane and the tool-chip interface.)

8.9 State whether or not the following statements are true, explaining your reasons: (a) For the same shear angle, there are two rake angles that give the same cutting ratio. (b) For the same depth of cut and rake angle, the type of cutting fluid used has no influence on chip thickness. (c) If the cutting speed, shear angle, and rake angle are known, the chip velocity can be calculated. (d) The chip becomes thinner as the rake angle increases. (e) The function of a chip breaker is to decrease the curvature of the chip.

8.10 It is generally undesirable to allow temperatures to rise excessively in cutting operations. Explain why.

8.11 Explain the reasons that the same tool life may be obtained at two different cutting speeds.

8.12 Inspect Table 8.6, and identify tool materials that would not be particularly suitable for interrupted cutting operations. Explain your choices.

8.13 Explain the possible disadvantages of a cutting operation in which the type of chip produced is discontinuous.

8.14 It has been noted that tool life can be almost infinite at low cutting speeds. Would you then recommend that all machining be done at low speeds? Explain.

8.15 How would you explain the effect of cobalt content on the properties of carbides?

8.16 Explain why studying the types of chips produced is important in understanding machining operations.

8.17 How would you expect the cutting force to vary for the case of serrated-chip formation? Explain.

8.18 Wood is a highly anisotropic material; that is, it is orthotropic. Explain the effects of orthogonal cutting of wood at different angles to the grain direction on chip formation.

8.19 Describe the advantages of oblique cutting. Which machining processes involve oblique cutting? Explain.

8.20 Explain why it is possible to remove more material between tool resharpenings by lowering the cutting speed.

8.21 Explain the significance of Eq. (8.8).

8.22 How would you go about measuring the hot hardness of cutting tools?

8.23 Describe the reasons for making cutting tools with multiphase coatings of different materials. Describe the properties that the substrate for multiphase cutting tools should have.

8.24 Explain the advantages and limitations of inserts. Why were they developed?

8.25 Make a list of alloying elements used in high-speed steel cutting tools. Explain why they are used.

8.26 What are the purposes of chamfers on cutting tools?

8.27 Why does temperature have such an important effect on cutting-tool performance?

8.28 Ceramic and cermet cutting tools have certain advantages over carbide tools. Why, then, are carbide tools not replaced to a greater extent?

8.29 Why are chemical stability and inertness important in cutting tools?

8.30 What precautions would you take in machining with brittle tool materials?

8.31 Why do cutting fluids have different effects at different cutting speeds? Is the control of cutting-fluid temperature important? Explain.

8.32 Which of the two materials, diamond or cubic boron nitride, is more suitable for cutting steels? Why?

8.33 List and explain the considerations involved in determining whether a cutting tool should be reconditioned, recycled, or discarded after use.

8.34 List the parameters that influence the temperature in metal cutting, and explain why and how they do so.

8.35 List and explain factors that contribute to poor surface finish in machining operations.

8.36 Explain the functions of different angles on a single-point lathe cutting tool. How does the chip thickness vary as the side cutting-edge angle is increased?

8.37 The helix angle for drills is different for different groups of workpiece materials. Why?

8.38 A turning operation is being carried out on a long, round bar at a constant depth of cut. Explain what changes, if any, may occur in the machined diameter from one end of the bar to the other. Give reasons for any changes that may occur.

8.39 Describe the relative characteristics of climb milling and up milling.

8.40 In sawing of difficult materials, high-speed steel cutting teeth are welded to a steel blade. Would you recommend that the whole blade be made of high-speed steel? Explain your reasons.

8.41 Describe the adverse effects of vibrations and chatter in machining.

8.42 Make a list of components of machine tools that could be made of ceramics, and explain why ceramics would be a suitable material for these components.

8.43 In Fig. 8.12, why do the thrust forces start at a finite value when the feed is zero?

8.44 Do you think temperature rise in cutting is related to the hardness of the workpiece material? Explain.

8.45 Describe the effects of tool wear on the workpiece and on the machining operation in general.

8.46 Explain whether or not it is desirable to have a high or low (a) n value and (b) C value in the Taylor tool-life equation.

8.47 Are there cutting operations that cannot be performed on machining centers and/or turning centers? Explain.

8.48 What is the importance of the cutting ratio?

8.49 Emulsion cutting fluids typically consist of 95% water and 5% soluble oil and chemical additives. Why is the ratio so unbalanced? Is the oil needed at all?

8.50 It is possible for the n value in the Taylor tool-life equation to be negative. Explain how.

8.51 Assume that you are asked to estimate the cutting force in slab milling with a straight-tooth cutter, but without running a test. Describe the procedure that you would follow.

8.52 Explain the possible reasons that a knife cuts better when it is moved back and forth. Consider factors such as the material being cut, interfacial friction, and the dimensions of the knife.

8.53 What are the effects of lowering the friction at the tool-chip interface, say with an effective cutting fluid, on the mechanics of cutting operations? Explain, giving several examples.

8.54 Why is it not always advisable to increase cutting speed in order to increase production rate? Explain.

8.55 It has been observed that the shear-strain rate in metal cutting is high even though the cutting speed may be relatively low. Why?

8.56 Note from the exponents in Eq. (8.30) that the cutting speed has a greater influence on temperature than does the feed. Why?

8.57 Describe the consequences of exceeding the allowable wear land (See Table 8.5) for cutting tools.

8.58 Comment on and explain your observations regarding Figs. 8.33, 8.37, and 8.42.

8.59 The tool-life curve for ceramic tools in Fig. 8.22a is to the right of those for other tools. Why?

8.60 In Fig. 8.18, note that the percentage of the energy carried away by the chip increases with cutting speed. Why?

8.61 How would you go about measuring the effectiveness of cutting fluids? Explain.

8.62 Describe the conditions that are critical in using the capabilities of diamond and cubic-boron-nitride cutting tools.

8.63 In Table 8.6, the last two properties listed can be important to the life of the cutting tool. Why? Which of the properties listed are, in your opinion, the least important in machining? Explain.

8.64 In Fig. 8.29, it can be seen that tool materials, especially carbides, have a wide range of hardness at a particular temperature. Why?

8.65 How would you go about recycling used cutting tools? Explain.

8.66 There is a wide range of tool materials available and used successfully today, yet much research and development continues to be carried out on these materials. Why?

8.67 Drilling, boring, and reaming of large holes is generally more accurate than just drilling and reaming. Why?

8.68 A badly oxidized and uneven round bar is being turned on a lathe. Would you recommend a relatively small or large depth of cut? Explain your reasons.

8.69 Does the force or torque in drilling change as the hole depth increases? Explain.

8.70 Explain the advantages and limitations of producing threads by forming and cutting, respectively.

8.71 Describe your observations regarding the contents of Tables 8.8, 8.10, and 8.11.

8.72 The footnote to Table 8.11 states that as the depth of the hole increases, speeds and feeds should be reduced. Why?

8.73 List and explain the factors that contribute to poor surface finish in machining operations.

8.74 Make a list of the machining operations described in this chapter, according to the difficulty of the operation and the desired effectiveness of cutting fluids. (*Example:* Tapping of holes is a more difficult operation than turning straight shafts.)

8.75 Are the feed marks left on the workpiece by a face-milling cutter true segments of a true circle? Explain with appropriate sketches.

8.76 What determines the selection of the number of teeth on a milling cutter? (See, for example, Figs. 8.52 and 8.54.)

8.77 Explain the technical requirements that led to the development of machining and turning centers. Why do their spindle speeds vary over a wide range?

8.78 In addition to the number of components, as shown in Fig. 8.71, what other factors influence the rate at which damping increases in a machine tool? Explain.

8.79 Why is thermal expansion of machine-tool components important? Explain, with examples.

8.80 Would using the machining processes described in this chapter be difficult on nonmetallic or rubberlike materials? Explain your thoughts, commenting on the influence of various physical and mechanical properties of workpiece materials, the cutting forces involved, the parts geometries, and the fixturing required.

8.81 Why is the machinability of alloys difficult to assess?

8.82 What are the advantages and disadvantages of dry machining?

8.83 Can high-speed machining be performed without the use of cutting fluids? Explain.

8.84 If the rake angle is 0°, then the frictional force is perpendicular to the cutting direction and, therefore, does not consume machine power. Why, then, is there an increase in the power dissipated when machining with a rake angle of, say, 20°?

8.85 The accompanying illustration shows a part that is to be machined from a rectangular blank. Suggest the type of operations required and their sequence, and specify the machine tools that are needed.

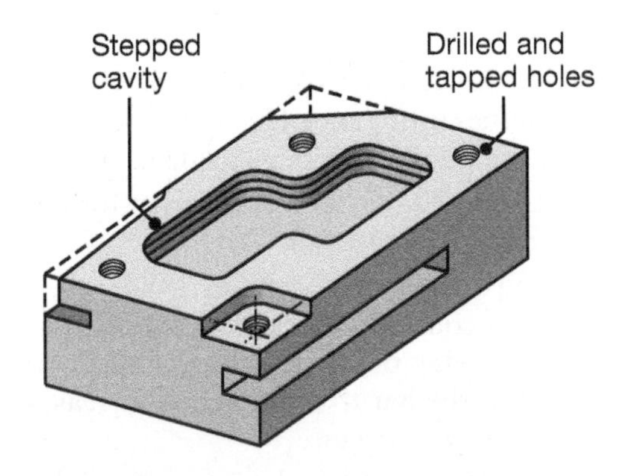

8.86 Select a specific cutting-tool material, and estimate the machining time for the parts shown in the accompanying figure: (a) pump shaft, stainless steel; (b) ductile (nodular) iron crankshaft; and (c) 304 stainless-steel tube with internal rope thread.

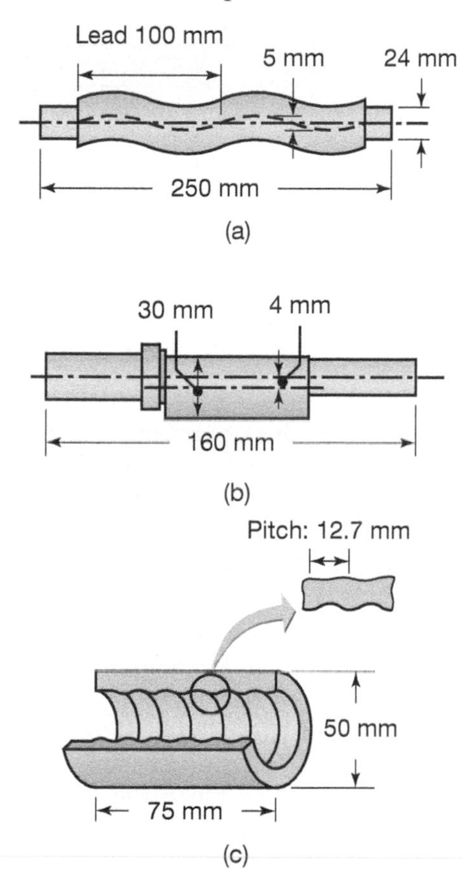

8.87 Would you recommend broaching a keyway on a gear blank before or after the teeth are machined? Explain.

8.88 Given your understanding of the basic metal-cutting process, what are the important physical and chemical properties of a cutting tool?

8.89 Negative rake angles are generally preferred for ceramic, diamond, and cubic boron nitride tools. Why?

8.90 If a drill bit is intended for woodworking applications, what material is it most likely to be made from? (*Hint:* Temperatures rarely rise to 400°C in woodworking.) Are there any reasons why such a drill bit cannot be used to drill a few holes in a metal? Explain.

8.91 What are the consequences of a coating having a different coefficient of thermal expansion than the substrate?

8.92 Discuss the relative advantages and limitations of near-dry machining. Consider all relevant technical and economic aspects.

8.93 In modern manufacturing with computer-controlled machine tools, which types of metal chips are undesirable, and why?

8.94 Explain why hacksaws are not as productive as band saws.

8.95 Describe the parts and conditions under which broaching would be the preferred method of machining.

8.96 With appropriate sketches, explain the differences between and similarities among shaving, broaching, and turn broaching operations.

8.97 Why is it difficult to use friction sawing on non-ferrous metals?

8.98 Review Fig. 8.66 on modular machining centers, and explain workpieces and operations that would be suitable on such machines.

8.99 Describe workpieces that would not be suitable for machining on a machining center. Give specific examples.

8.100 Give examples of forced vibration or self-excited vibration in general engineering practice.

8.101 Tool temperatures are low at low cutting speeds, and high at high cutting speeds, but low at even higher cutting speeds. Explain why.

8.102 Explain the technical innovations that have made high-speed machining advances possible, and the economic motivations for high-speed machining.

8.103 For a constant depth of cut, make a sketch of orthogonal metal cutting with (a) a positive rake angle and (b) a negative rake angle. Identify the shear angle and chip thickness in each.

8.104 Add forces acting on the chip and shear plane in your sketch of Problem 8.103, taking care to use arrows that reflect the size of the forces.

8.105 What are the functions of a cutting fluid?

8.106 Describe the conditions that cause built-up edge. Why is built-up edge undesirable? What can be done to avoid a built-up edge condition?

8.107 Describe the material properties that could lead to (a) segmented and (b) discontinuous chips.

8.108 A turning operation results in a workpiece with an surface finish that is too rough. List and prioritize the steps you would follow to improve the surface finish.

PROBLEMS

8.109 Assume that in orthogonal cutting the rake angle is 15° and the coefficient of friction is 0.15. Using Eq. (8.20), determine the percentage increase in chip thickness when the coefficient of friction is doubled.

8.110 Prove Eq. (8.1).

8.111 Write a simple analytical expression to prove the validity of the last paragraph in Example 8.2.

8.112 Using Eq. (8.3), make a plot of the shear strain, γ, vs. the shear angle, ϕ, with the rake angle, α, as a parameter. Describe your observations.

8.113 Assume that in orthogonal cutting, the rake angle is 15° and the coefficient of friction is 0.25. Determine the percentage change in chip thickness when the friction is doubled.

8.114 Derive Eq. (8.12).

8.115 Determine the shear angle in Example 8.1. Is this calculation exact or an estimate? Explain.

8.116 303 stainless steel, with a Brinnel hardness of 270, is turned with a carbide tool with $\alpha = 8°$. The undeformed chip thickness is 0.3 mm, and the cutting speed is 0.5 m/s. The chip is continuous; a 1-m-long section weighs 5.7 g, and the chip width is the same as the workpiece width and equals 1.5 mm. Calculate the cutting ratio.

8.117 By changing the lubricant, the chip in the previous problem changes from helical to straight. The cutting ratio is now directly measured to be $r = 0.5$. Find the new shear angle and chip velocity.

8.118 The following data are available from orthogonal cutting experiments. In both cases depth of cut (feed) $t_o = 0.13$ mm, width of cut $b = 2.5$ mm, rake angle $\alpha = -5°$, and cutting speed $V = 2$ m/s.

	Workpiece material	
	Aluminum	Steel
Chip thickness, t_c, mm	0.23	0.58
Cutting force, F_c, N	430	890
Thrust force, F_t, N	280	800

Determine the shear angle ϕ [do not use Eq. (8.20)], friction coefficient μ, shear stress τ and shear strain γ on the shear plane, chip velocity V_c and shear velocity V_s, as well as energies u_f, u_s and u_t.

8.119 Estimate the cutting temperatures for the conditions of Problem 8.118 if the following properties apply:

	Workpiece material	
	Aluminum	Steel
Flow stress σ_f, N-mm/mm^3	120	325
Thermal diffusivity, κ, mm^2/s	97	14
Volumetric specific heat, ρc, N/mm^2°C	2.6	3.3

8.120 In a dry cutting operation using a 5° rake angle, the measured forces were $F_c = 1330$ N and $F_t = 740$ N. When a cutting fluid was used, these forces were $F_c = 1200$ N and $F_t = 710$ N. What is the change in the friction angle resulting from the use of a cutting fluid?

8.121 In the dry machining of aluminum with a 10° rake angle, a shear angle of 35° is observed. Determine the new shear angle, if a cutting fluid is applied which decreases the friction coefficient by 20%.

8.122 Taking carbide as an example and using Eq. (8.30), determine how much the feed should be changed in order to keep the mean temperature constant when the cutting speed is doubled.

8.123 With appropriate diagrams, show how the use of a cutting fluid can affect the magnitude of the thrust force, F_t, in orthogonal cutting.

8.124 A 200-mm-diameter stainless-steel bar is being turned on a lathe at 600 rpm and at a depth of cut, $d = 2.5$ mm. If the power of the motor is 5 kW and has a mechanical efficiency of 80%, what is the maximum feed that you can have at a spindle speed of 500 rpm before the motor stalls?

8.125 Using the Taylor equation for tool wear and letting $n = 0.2$ calculate the percentage increase in tool life if the cutting speed is reduced by (a) 30% and (b) 60%.

8.126 Determine the n and C values for the four tool materials shown in Fig. 8.22a.

8.127 Using Eq. (8.30) and referring to Fig. 8.18a, estimate the magnitude of the coefficient a.

8.128 Estimate the machining time required in rough turning a 2.0-m-long, annealed aluminum-alloy round bar that is 75 mm in diameter, using (a) a high-speed steel tool; and (b) a carbide tool. Use a feed of 2 mm/rev.

8.129 A 200-mm-long, 75-mm-diameter titanium-alloy rod is being reduced in diameter to 65 mm by turning on a lathe. The spindle rotates at 400 rpm, and the tool is traveling at an axial velocity of 200 mm/min. Calculate the cutting speed, material removal rate, time of cut, power required, and cutting force.

8.130 Calculate the same quantities as in Example 8.4 for high-strength cast iron at $N = 500$ rpm.

8.131 A 20-mm-diameter drill is being used on a drill press operating at 300 rpm. If the feed is 0.125 mm/rev, what is the material removal rate? What is the MRR if the drill diameter is tripled?

8.132 A hole is being drilled in a block of magnesium alloy with a 15-mm drill at a feed of 0.2 mm/rev. The spindle is running at 500 rpm. Calculate the material removal rate, and estimate the torque on the drill.

8.133 Show that the distance l_c in slab milling is approximately equal to $\sqrt{Dd}$ for situations where $D \gg d$.

8.134 Calculate the chip depth of cut in Example 8.6.

8.135 In Example 8.6, which of the quantities will be affected when the spindle speed is increased to 200 rpm?

8.136 The following flank wear data were collected in a series of machining tests using C6 carbide tools on 1045 steel (HB=192). The feed rate was 0.375 mm/rev, and the width of cut was 0.75 mm. (a) Plot flank wear as a function of cutting time. Using a 0.375 mm wear land as the criterion of tool failure, determine the lives for the two cutting speeds. (b) Plot the life for a 0.375 mm wear land as a function of speed on log-log plot and determine the values of n and C in the Taylor tool life equation. (Assume a straight line relationship.) (c) Using these results, calculate the tool life for a cutting speed of 1.50 m/s.

Cutting speed V, m/s	Cutting time min	Flank wear mm
2.0	0.5	0.0350
	2.0	0.0575
	4.0	0.0750
	8.0	0.1375
	16.0	0.205
	24.0	0.280
	54.0	0.375

Cutting speed V, m/s	Cutting time min	Flank wear mm
3.0	0.5	0.0450
	2.0	0.0875
	4.0	0.150
	8.0	0.250
	13.0	0.3625
	14.0	0.400
4.0	0.5	0.125
	2.0	0.250
	4.0	0.350
	5.0	0.400
5.0	0.5	0.250
	1.0	0.325
	1.8	0.375
	2.0	0.400

8.137 A slab-milling operation is being carried out on a 0.5-m-long, 50-mm-wide high-strength-steel block at a feed of 0.25 mm/tooth and a depth of cut of 3 mm. The cutter has a diameter of 75 mm, has six straight cutting teeth, and rotates at 150 rpm. Calculate the material removal rate and the cutting time, and estimate the power required.

8.138 Referring to Fig. 8.53, assume that $D = 200$ mm, $w = 30$ mm, $l = 600$ mm, $d = 2$ mm, $v = 1$ mm/s, and $N = 200$ rpm. The cutter has 12 inserts, and the workpiece material is 304 stainless steel. Calculate the material removal rate, cutting time, and feed per tooth, and estimate the power required.

8.139 Estimate the time required for face milling a 200-mm-long, 75-mm-wide brass block with an 200-mm-diameter cutter that has 12 high-speed steel teeth.

8.140 A 300-mm-long, 50-mm-thick plate is being cut on a band saw at 50 m/min. The saw has 1 tooth every two millimeters. If the feed per tooth is 0.075 mm, how long will it take to saw the plate along its length?

8.141 A single-thread hob is used to cut 40 teeth on a spur gear. The cutting speed is 70 m/min and the hob has a diameter of 100 mm. Calculate the rotational speed of the spur gear.

8.142 In deriving Eq. (8.20), it was assumed that the friction angle, β, was independent of the shear angle, ϕ. Is this assumption valid? Explain.

8.143 An orthogonal cutting operation is being carried out under the following conditions: depth of cut = 0.10 mm, width of cut = 5 mm, chip thickness = 0.2 mm, cutting speed = 2 m/s, rake angle = 10°, cutting force = 500 N, and thrust force = 200 N. Calculate the percentage of the total energy that is dissipated in the shear plane during cutting.

8.144 An orthogonal cutting operation is being carried out under the following conditions: depth of cut = 0.5 mm, width of cut = 2.5 mm, cutting ratio = 0.3, cutting speed = 1.0 m/s, rake angle = 0°, cutting force = 800 N, thrust force = 600 N, workpiece density = 2700 kg/m^3, and workpiece specific heat = 900 J/kg-K. Assume that (a) the sources of heat are the shear plane and the tool-chip interface; (b) the thermal conductivity of the tool is zero, and there is no heat loss to the environment; and (c) the temperature of the chip is uniform throughout. If the temperature rise in the chip is 70°C, calculate the percentage of the energy dissipated in the shear plane that goes into the workpiece.

8.145 It can be shown that the angle ψ between the shear plane and the direction of maximum grain elongation (See Fig. 8.4a) is given by the expression

$$\psi = 0.5 \cot^{-1}\left(\frac{\gamma}{2}\right),$$

where γ is the shear strain, as given by Eq. (8.3). Assume that you are given a piece of the chip obtained from orthogonal cutting of an annealed metal. The rake angle and cutting speed are also given, but you have not seen the setup on which the chip was produced. Outline the procedure that you would follow to estimate the power required in producing this chip. Assume that you have access to a fully equipped laboratory and a technical library.

8.146 A lathe is set up to machine a taper on a bar stock 120 mm in diameter; the taper is 1 mm per 10 mm. A cut is made with an initial depth of cut of 4 mm at a feed rate of 0.250 mm/rev and at a spindle speed of 200 rpm. Calculate the average metal removal rate.

8.147 Obtain an expression for optimum feed rate that minimizes the cost per piece if the tool life is as described by Eq. (8.34).

8.148 Assuming that the coefficient of friction in cutting is 0.25, calculate the maximum depth of cut for turning a hard aluminum alloy on a 15 kW lathe (mechanical efficiency of 80%) at a width of cut of 5 mm, a rake angle of 0°, and a cutting speed of 1.5 m/s.

8.149 Assume that, using a carbide cutting tool, you measure the temperature in a cutting operation at a speed of 2 m/s and feed of 0.1 mm/rev as 650°C. What is the approximate temperature if the speed is increased by 50%? What speed is required to lower the maximum temperature to 400°C?

8.150 A 75-mm-diameter gray cast-iron cylindrical part is to be turned on a lathe at 500 rpm, with a depth of cut of 5 mm and a feed of 0.5 mm/rev. What should be the minimum power of the lathe? Assume a mechanical efficiency of 80% in the lathe.

8.151 A 150-mm-diameter aluminum cylinder with a length of 300 mm is to have its diameter reduced to 125 mm. (a) Estimate the machining time if an uncoated carbide tool is used. (b) What if a TiN-coated tool is used?

8.152 Calculate the power required for the cases in Problem 8.151.

8.153 Using trigonometric relationships, derive an expression for the ratio of shear energy to frictional energy in orthogonal cutting, in terms of angles α, β, and ϕ only.

8.154 For a turning operation using a ceramic cutting tool, if the speed is increased by 50%, by what factor must the feed rate be modified to obtain a constant tool life? Use $n = 0.5$ and $y = 0.7$.

8.155 Using Eq. (8.35), select an appropriate feed for $R = 1$ mm and a desired roughness of 1 μm. How would you adjust this feed to allow for nose wear of the tool during extended cuts? Explain your reasoning.

8.156 In a drilling operation, a 15-mm drill bit is being used in a low-carbon steel workpiece. The hole is a blind hole which will be tapped to a depth of 25 mm. The drilling operation takes place with a feed of 0.25 mm/rev and a spindle speed of 700 rpm. Estimate the time required to drill the hole prior to tapping.

8.157 Assume that in the face-milling operation shown in Fig. 8.53, the workpiece dimensions are 125 mm by 250 mm. The cutter is 150 mm in diameter, has 8 teeth, and rotates at 300 rpm. The depth of cut is 3 mm and the feed is 0.125 mm/tooth. Assume that the specific energy requirement for this material is 5.5 W-s/mm^3, and that only 75% of the cutter diameter is engaged during cutting. Calculate (a) the power required and (b) the material removal rate.

8.158 A hole in a steel casting (S_{ut} = 475 MPa) is to be bored using a carbide insert. The hole is 130 mm in the as-cast condition; the finished diameter is 138 mm. Suggest a cutting speed and feed, and calculate the required power and cutting force.

8.159 Calculate the ranges of typical machining times for face milling a 250-mm-long workpiece with a 50-mm-wide cutter and a depth of cut of 2.5 mm for the following workpiece materials: (a) low-carbon steel,

(b) titanium alloys, (c) aluminum alloys, and (d) thermoplastics. Assume the cutter has 10 teeth.

8.160 A machining-center spindle and tool extend 300 mm from its machine-tool frame. What temperature change can be tolerated to maintain a tolerance of 0.0025 mm in machining? A tolerance of 0.025 mm? Assume that the spindle is made of steel.

8.161 In the production of a machined valve, the labor rate is \$19.00 per hour, and the general overhead rate is \$15.00 per hour. The tool is a ceramic insert with four faces and costs \$25.00, takes five minutes to change and one minute to index. Estimate the optimum cutting speed from a cost perspective. Use $C = 100$ for V_o in m/min.

8.162 Estimate the optimum cutting speed in Problem 8.161 for maximum production.

8.163 Develop an equation for optimum cutting speed in face milling using a cutter with inserts.

8.164 Develop an equation for optimum cutting speed in turning where the tool is a high speed steel tool that can be reground periodically.

8.165 Aluminum is being machined on a lathe with a high speed steel cutting tool. It is proposed to increase production rates by 25%. Would you accomplish this by increasing speed, feed, or depth of cut? Justify your answer with tool life estimates.

8.166 A 50 mm wide, 500 mm long, 100 mm thick slab of cast iron is milled to a depth of 2 mm using a high-speed steel cutter with seven cutting edges and a 150 mm diameter. (a) Make a sketch of this operation. (b) Select the cutting speed, feed, and feed rate for this material. (c) Calculate the power and time required for this operation.

8.167 Repeat problem 8.166, but if the workpiece is a ductile aluminum alloy using a carbide cutting tool.

8.168 12 holes, each 10 mm in diameter, are to be machined in a 20-mm thick flange made of 304 stainless steel. Estimate the recommended cutting speed, drill angular velocity, feed and power required. How long will it take to produce these holes?

8.169 Repeat problem 8.168 if the material is (a) a soft aluminum alloy or (b) Ti-6Al-4V.

8.170 Assume that you are an instructor covering the topics in this chapter, and you are giving a quiz on the quantitative aspects to test the understanding of the students. Prepare several numerical problems, and supply the answers to them.

DESIGN

8.171 Tool life could be greatly increased if an effective means of cooling and lubrication were developed. Design methods of delivering a cutting fluid to the cutting zone, and discuss the advantages and shortcomings of your design.

8.172 Devise an experimental setup whereby you can perform an orthogonal cutting operation using a round tubular workpiece on a lathe.

8.173 Cutting tools are sometimes designed so that the chip-tool contact length is controlled by recessing it at the rake face some distance away from the tool tip. Explain the possible advantages of such a tool.

8.174 Make a comprehensive table of the process capabilities of the machining processes described in this chapter. Use several columns to describe the machines involved, the type of tools and tool materials used, the shapes of blanks and parts produced, the typical maximum and minimum sizes produced, the surface finish produced, the dimensional tolerances produced, and the production rates achieved.

8.175 The accompanying illustration shows drawings for a cast-steel valve body before (left) and after (right) machining. Identify the surfaces that are to be machined. What type of machine tool would be suitable to machine this part? What type of machining operations are involved, and what should be the sequence of these operations? (Note that not all surfaces are to be machined.)

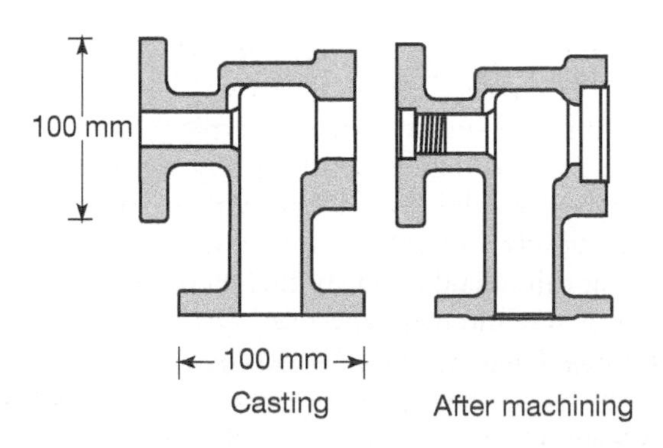

8.176 A large bolt is to be produced from hexagonal bar stock by placing the hex stock into a chuck and machining the cylindrical shank of the bolt by turning on a lathe. List and explain the difficulties that may be presented by this operation.

8.177 Design appropriate fixtures and describe the machining operations required to produce an automotive piston.

8.178 In Figs. 8.16 and 8.17b, it can be seen that the maximum temperature is about halfway up the face of the tool. The text has also described the adverse effects of temperature on various tool materials. Considering the mechanics of cutting operations, describe your thoughts on the technical and economic merits of embedding a small insert, made of materials such as ceramic or carbide, halfway up the rake face of a tool made of a material with lower resistance to temperature than ceramic or carbide.

8.179 Describe your thought on whether chips produced during machining can be used to make useful products. Give some examples of possible products, and comment on their characteristics and differences if the same products were made by other manufacturing processes. Which types of chips would be desirable for this purpose?

8.180 If expanded honeycomb panels (see Section 7.5.7) were to be machined in a form milling operation, what precautions would you take to keep the sheet metal from buckling due to tool forces? Think up as many solutions as you can.

8.181 One of the principal concerns with coolants is degradation due to biological attack by bacteria. To prolong life, chemical biocides are often added, but these biocides greatly complicate the disposal of coolants. Conduct a literature search regarding the latest developments in the use of environmentally benign biocides in cutting fluids.

8.182 Experiments have shown that it is possible to produce thin wide chips, such as 0.08 mm (0.003 in.) thick and 10 mm (4 in.) wide, which would be similar to rolled sheet. Materials have been aluminum, magnesium, and stainless steel. A typical setup would be similar to orthogonal cutting, by machining the periphery of a solid round bar with a straight tool moving radially inward. Describe your thoughts on producing thin metal sheet by this method, its surface characteristics, and its properties.

8.183 Review Fig. 5.37 and reclassify the examples as 'good' or 'poor' based on processes in this chapter. Justify your answers.

8.184 It has been noted that the chips from certain carbon steels are noticeably magnetic, even if the original workpiece is not. Research the reasons for this effect and write a one-page paper explaining the important mechanisms.

8.185 As we have seen, chips carry away the majority of the heat generated during machining. If chips did not have this capacity, what suggestions would you make in order to be able to carry out machining processes without excessive heat? Explain.

8.186 Contact several different suppliers of cutting tools, or search their websites. Make a list of the costs of typical cutting tools as a function of various sizes, shapes, and features.

8.187 List the concerns you would have if you needed to economically machine carbon-fiber reinforced polymers or metal matrix composites with graphite fibers in an aluminum matrix.

8.188 If a bolt breaks in a hole, it typically is removed by first drilling a hole in the bolt shank and then using a special tool to remove the bolt. Inspect such a tool and explain how it functions.

8.189 An important trend in machining operations is the increased use of flexible fixtures. Conduct a search on the Internet regarding these fixtures, and comment on their design and operation.

8.190 Review the following figure, and explain if it would be possible to machine eccentric shafts on the setup illustrated. What if the part is long compared with its cross section? Explain.

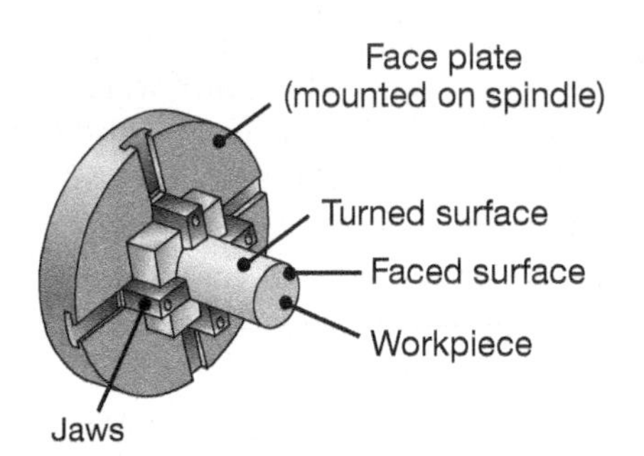

8.191 Boring bars can be designed with internal damping capabilities to reduce or eliminate vibration and chatter during machining. Referring to the technical literature, describe details of designs for such boring bars.

8.192 A large bolt is to be produced from extruded hexagonal bar stock by placing the hex stock into a chuck and machining the shank of the bolt by turning it on a lathe. List and explain the difficulties that may be involved in this operation.

Abrasive and Other Material Removal Processes

CHAPTER

EXAMPLES:

This chapter describes the important features of finishing operations, commonly performed to improve dimensional tolerances and surface finish of products. Among the topics covered are:

- The characteristics of grinding wheels and mechanics of grinding operations.
- Types of grinding machines and advanced abrasive machining processes.
- Abrasive machining operations, including lapping, honing, polishing, chemical mechanical polishing, and the use of coated abrasives.
- Nonmechanical means of material removal, including chemical and electrochemical machining, electrical-discharge machining, laser and electron beam machining, and abrasive jet machining.
- Deburring operations.
- Design and economic considerations for the processes described in this chapter.

Symbols

A area, m^2
b kerf, m
c_o elastic wave speed, m/s
C cutting points per area, m^{-2}
also, electrochemical machining constant, mm^3/A-min
d depth of cut, m
also, laser spot diameter, m
d_w wire diameter, m
D diameter, m
D_w workpiece diameter, m
E cell voltage, V
f feed rate, m/min
F force, N
also, Faraday's constant
G grinding ratio
also, mass, g
h thickness, m
I amperage, A
l undeformed chip length, m
L length ground, m
K electrical conductivity, $\Omega^{-1}mm^{-1}$
also, workpiece material factor, mm^3/A-min
K_p coefficient of loss
m mass, kg
P power, Nm/s
r ratio of chip width to average chip thickness
also, radius, m

R ratio of workpiece to electrode wear
s gap, m
t chip thickness, m; also, cutting depth, m
t_o contact time, s
T_r ratio of workpiece to electrode melting temperature
T_t melting point of electrode, °C
T_w melting point of workpiece, °C
ΔT temperature change, °C
u specific energy, W-s/m^3
v velocity, m/s
V grinding speed, m/s; also, voltage, V
V_f wire feed rate, m/s
V_s grinding wheel penetration, m/s
w width of chip, m
W_t electrode wear rate, mm^3/min
α_n normal rake angle, deg; also, inclination angle, deg
ϕ shear angle, deg
η current efficiency
ρ density, kg/m^3

Subscripts

avg average
c cutting
n normal
o original, optimum

9.1 Introduction

In all the machining operations described in detail in Chapter 8, the cutting tool is made of a certain material and has a clearly defined geometry. Moreover, the machining process involves chip removal, the mechanics of which can be fairly complicated. There are, however, many situations in manufacturing where the workpiece material is either *too hard* or *too brittle*, or its *shape* is difficult to produce with sufficient dimensional accuracy and surface finish by any of the machining methods described previously. One of the best methods for producing such parts is by using **abrasives.** An abrasive is a small, hard particle that has sharp edges and an irregular shape. Abrasives are capable of removing small amounts of material from a surface by a cutting process that produces tiny chips.

Abrasive machining processes are generally among the last operations performed on manufactured products, although they are not necessarily confined to fine or small-scale material removal from workpieces, and they can indeed compete economically with some of the machining processes described in Chapter 8. Because they are hard, abrasives are also suitable for (a) finishing very hard or heat-treated parts; (b) shaping hard nonmetallic materials, such as ceramics and glasses; (c) cutting off lengths of bars, structural shapes, masonry, and concrete; (d) cleaning surfaces, using jets of air or water containing abrasive particles; and (e) removing unwanted weld beads.

In addition to abrasive machining, several **advanced machining processes** have been developed, starting in the 1940s. Also called *nontraditional* or *unconventional* machining, these processes are based on electrical, chemical, fluid, and thermal principles, and are advantageous when:

1. The hardness and strength of the workpiece material is very high, typically above 400 HB.
2. The part is too flexible or slender to support the machining or grinding forces, or it is difficult to clamp in workholding devices.
3. The shape of the part is complex, such as internal and external features or deep small-diameter holes.
4. Surface finish and dimensional accuracy requirements are better than those obtainable by other processes.
5. Temperature rise or residual stresses developed in the workpiece is undesirable or unacceptable.

When selected and applied properly, advanced machining processes offer significant technological and economic advantages over the traditional machining methods.

9.2 Abrasives

The abrasives commonly used in manufacturing are:

1. **Conventional abrasives:**
    - *Aluminum oxide* (Al_2O_3);
    - *Silicon carbide* (SiC).
2. **Superabrasives:**
    - *Cubic boron nitride* (cBN);
    - *Diamond.*

The last two of the four abrasives listed above are the two hardest materials known, hence the term *superabrasives*. Abrasives are significantly harder than conventional cutting-tool materials, as can be seen by comparing Table 8.6 with Table 9.1. In addition to hardness, an important characteristic of an abrasive is **friability**, that is, the ability of an abrasive grain to fracture (break down) into smaller fragments. Friability gives abrasives *self-sharpening characteristics*, important in maintaining the cutting efficiency of the abrasives during use. High friability indicates low strength or low fracture resistance of the abrasive; thus, under the grinding forces, a highly friable abrasive grain fragments more rapidly than an abrasive grain with low friability. Aluminum oxide, for example, has lower friability than silicon carbide.

The **shape** and **size** of an abrasive grain also affect its friability. For example, blocky grains, which are analogous to negative-rake-angle cutting tools (Fig. 8.27), are less friable than plate-like grains. As for grain

TABLE 9.1 Knoop hardness range for various materials and abrasives.

Material	Knoop hardness	Material	Knoop hardness
Common glass	350–500	Titanium nitride	2000
Flint, quartz	800–1100	Titanium carbide	1800–3200
Zirconium oxide	1000	Silicon carbide	2100–3000
Hardened steels	700–1300	Boron carbide	2800
Tungsten carbide	1800–2400	Cubic boron nitride	4000–5000
Aluminum oxide	2000–3000	Diamond	7000–8000

size, because the probability of defects existing in smaller grains is lower (due to the *size effect*; Section 3.8.3), they are stronger and less friable than larger grains. The importance of friability in abrasive machining is described further in Section 9.5.

Types of abrasives. Abrasives found in nature include *emery, corundum (alumina), quartz, garnet,* and *diamond.* However, because natural abrasives contain unknown amounts of impurities and typically have nonuniform properties, their performance is inconsistent and unreliable. Consequently, aluminum oxides and silicon carbides abrasives are made synthetically, in order to produce high-performance abrasives with consistent behavior.

1. Synthetic **aluminum oxide** (Al_2O_3), first made in 1893, is made by fusing bauxite, iron filings, and coke. Aluminum oxide is divided into two groups: fused and unfused. **Fused aluminum oxide** is categorized as *white* (very friable), *dark* (less friable), and *monocrystalline* (single crystal). **Unfused alumina,** also known as *ceramic aluminum oxide,* can be harder than fused alumina. The purest form of fused alumina is **seeded gel.** First introduced in 1987, seeded gel has a particle size on the order of 0.2 μm, which is much smaller than abrasive grains commonly used in industry. Seeded gels are sintered (Section 11.4) to form larger sizes. Because of their hardness and relatively high friability, seeded gels maintain their sharpness and thus are used for difficult-to-grind materials.
2. **Silicon carbide** (SiC), first discovered in 1891, is made with silica sand, petroleum coke, and small amounts of sodium chloride. Silicon carbides are available in *green* (more friable) and in *black* (less friable) types. Silicon carbide generally has higher friability than aluminum oxide, hence a higher tendency to fracture and thus remain sharp longer.
3. **Cubic boron nitride** (cBN) was first produced in the 1970s. Its properties and characteristics are described in Section 11.8.1.
4. **Diamond** was first used as an abrasive in 1955. When produced synthetically, it is called *synthetic* or *industrial diamond.* Its properties and characteristics are described in Sections 8.6.9 and 11.13.2.

Grain size. As used in manufacturing operations, abrasives are generally very small compared with the size of typical cutting tools and inserts described in Section 8.6. Also, abrasives have sharp edges, thus allowing the removal of very small amounts of material from workpiece surfaces, resulting in very fine surface finish and dimensional accuracy (See Figs. 8.25 and 9.28). The size of an abrasive grain is identified by a **grit number**, which is a function of sieve size. The smaller the sieve size, the larger the grit number; for example, grit number 10 is rated as very coarse, 100 as fine, and 500 as very fine. As commonly observed, sandpaper and emery cloth also are identified in this manner, with the grit number printed on the back of the abrasive paper or cloth.

9.3 Bonded Abrasives

Bonded abrasives are typically in the form of a **grinding wheel** (Fig. 9.1). The abrasive grains are held together by a **bonding material**, various types of which are described in Section 9.3.1; the bonding material acts as supporting posts or braces between the grains. Some porosity is essential in bonded abrasives to provide cooling and clearance for the chips being produced, as otherwise the chips would interfere with the grinding operation. Porosity can easily be observed by looking at the surface of any grinding wheel with a magnifying glass or microscope. Other features of the grinding wheel shown in Fig. 9.1 are described in Sections 9.4 and 9.5.

Some of the more commonly used types of grinding wheels are shown in Fig. 9.2 for conventional abrasives, and in Fig. 9.3 for superabrasives. Note that, because of their associated high cost, superabrasives make up only a small portion of the periphery of the wheels. Bonded abrasives are marked with a standardized system of letters and numbers, indicating the type of abrasive, grain size, grade, structure, and bond type. Figure 9.4 shows the marking system for aluminum-oxide and silicon-carbide bonded abrasives, and Fig. 9.5 for diamond and cubic-boron-nitride bonded abrasives.

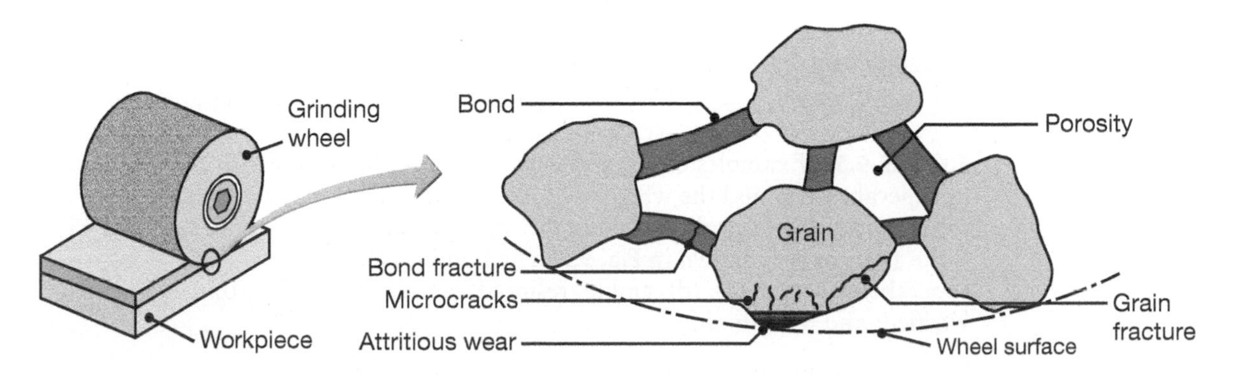

FIGURE 9.1 Schematic illustration of a physical model of a grinding wheel, showing its structure and grain wear and fracture patterns.

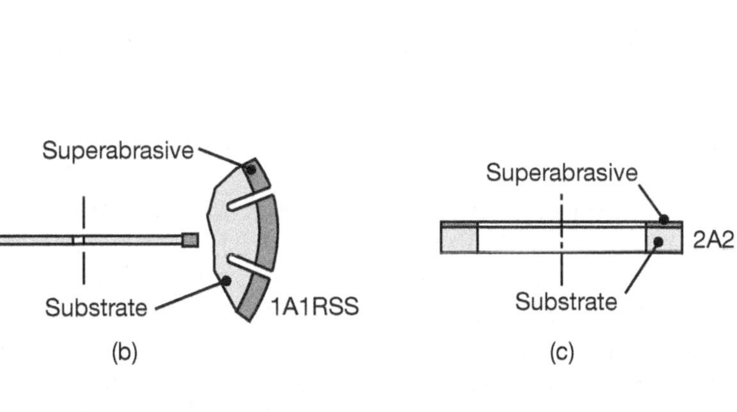

FIGURE 9.2 Some common types of grinding wheels made with conventional abrasives (aluminum oxide and silicon carbide). Note that each wheel has a specific grinding face; grinding on other surfaces is improper and unsafe.

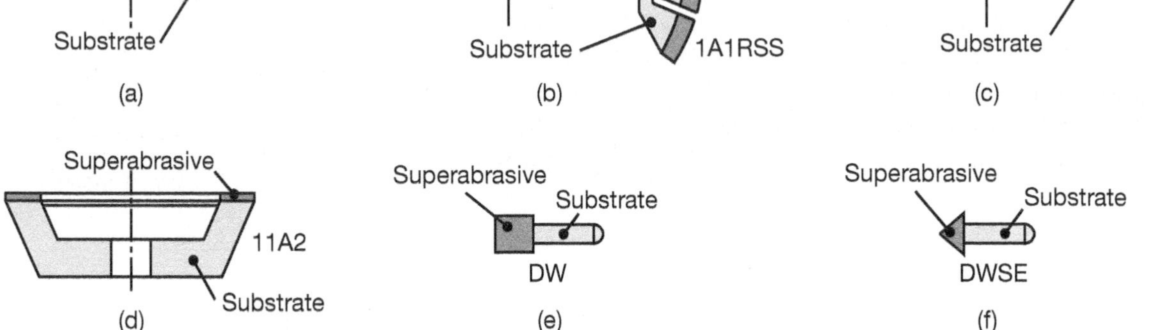

FIGURE 9.3 Examples of superabrasive wheel configurations. The rim consists of superabrasives and the wheel itself (core) is generally made of metal or composites. Note that the basic numbering of wheel types (such as 1, 2, and 11) is the same as that shown in Fig. 9.2. The bonding materials for the superabrasives are: (a), (d), and (e) resinoid, metal, or vitrified; (b) metal; (c) vitrified; and (f) resinoid.

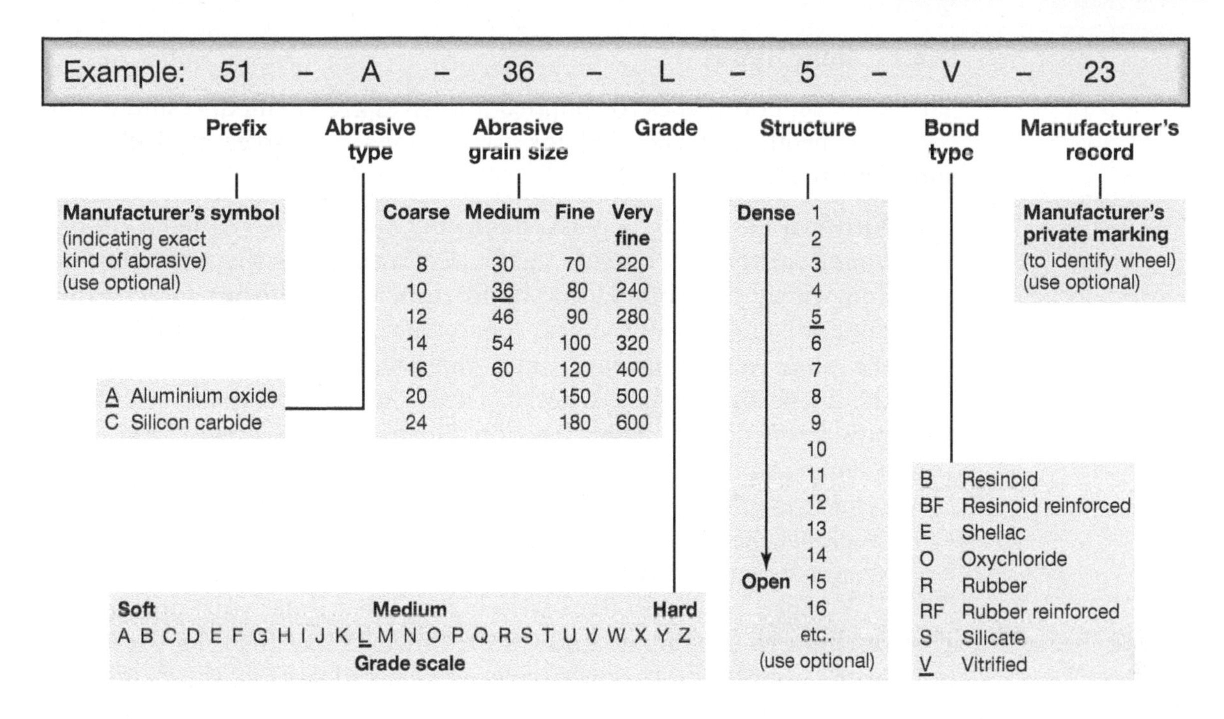

FIGURE 9.4 Standard marking system for aluminum-oxide and silicon-carbide bonded abrasives.

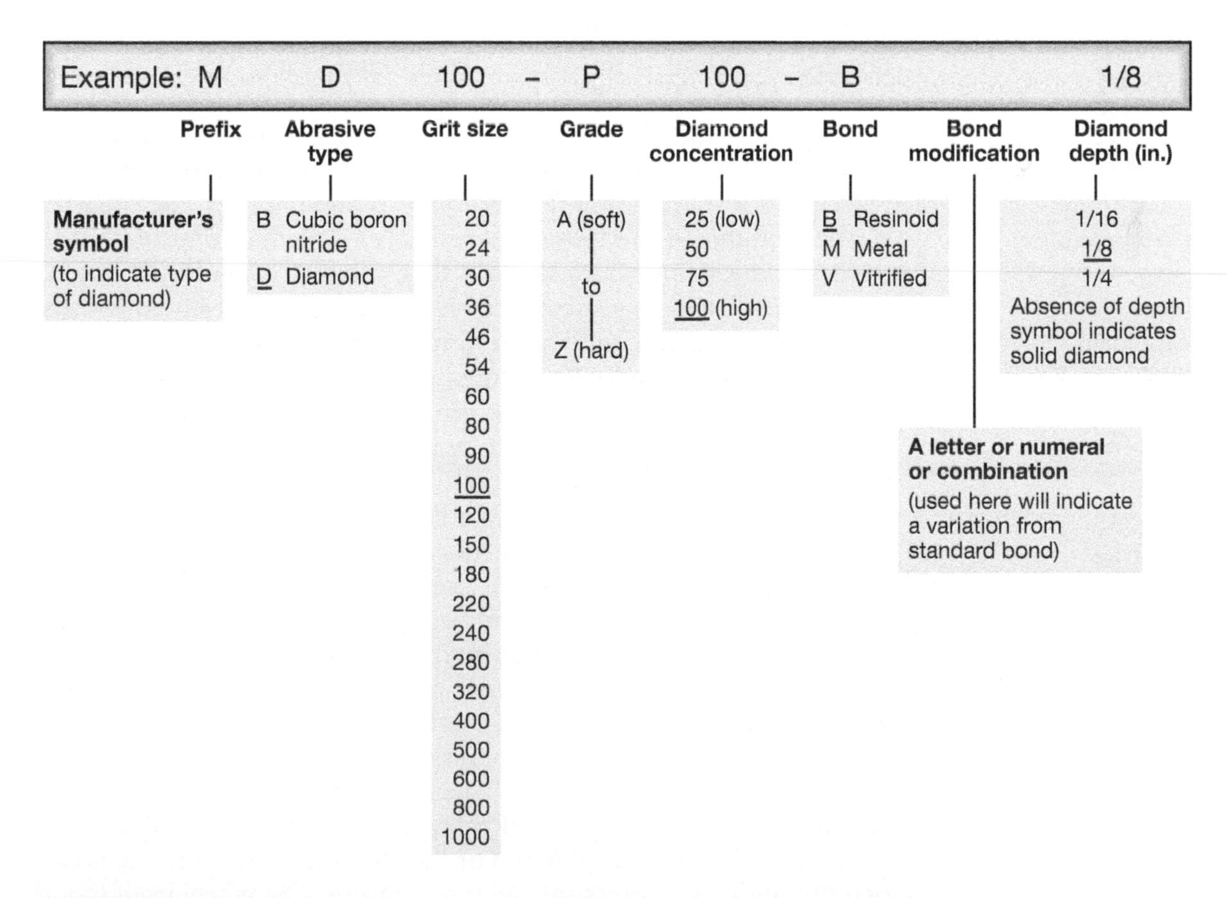

FIGURE 9.5 Standard marking system for diamond and cubic-boron-nitride bonded abrasives.

9.3.1 Bond Types

The common bond types for bonded abrasives are vitrified, resinoid, rubber, and metal, and are used for conventional abrasives as well as for superabrasives.

1. **Vitrified.** Essentially a glass, a *vitrified bond* is also called a *ceramic bond*, particularly outside the United States; it is the most common and widely used bond. The bond consists of feldspar (a crystalline mineral) and various clays. These materials are first mixed with the abrasives, moistened, and then molded under pressure into the shape of grinding wheels. These "green" products, similar to powder-metallurgy parts (Section 11.3), are then slowly fired, up to a temperature of about 1250°C, to fuse the glass and develop structural strength. The wheels are then cooled slowly, to prevent thermal cracking, then finished to size, inspected for quality and dimensional accuracy, and tested for defects.

   Vitrified bonds produce wheels that are strong, stiff, porous, and resistant to oils, acids, and water; however, because the wheels are brittle, they lack resistance to mechanical and thermal shock. Vitrified wheels are also available with steel backing plates or cups for better structural support during their use.

2. **Resinoid.** *Resinoid* bonding materials are *thermosets* (Section 10.4), and are available in a wide range of compositions and properties. Because the bond is an organic compound, these wheels are also called **organic wheels**. The basic manufacturing procedure consists of (a) mixing the abrasive with liquid or powdered phenolic resins and additives; (b) pressing the mixture into the shape of a grinding wheel; and (c) curing it at temperatures of about 175°C. Because the elastic modulus of thermosetting resins is lower than that of glasses, resinoid wheels are more flexible than vitrified wheels. *Polyimide* (Section 10.6) can be a substitute for the phenolic in resinoid wheels; the polymer is tough and has good resistance to high temperatures.

QR Code 9.1 Resin bond grinding wheels. *Source:* Courtesy of Abrasive Technology.

   Reinforced resinoid wheels consist of one or more layers of fiberglass mats of various mesh sizes, providing reinforcement. The main purpose of the reinforcement is to provide strength and prevent catastrophic failure of the wheel. Large-diameter wheels can additionally be supported with one or more internal rings (made of round steel bar) that are inserted during wheel production.

QR Code 9.2 Resin bond diamond grinding wheels. *Source:* Courtesy of Abrasive Technology.

3. **Rubber.** The most flexible bond used in abrasive wheels is *rubber*. The wheels are made by (a) mixing crude rubber, sulfur, and abrasive grains together; (b) rolling the mixture into sheets; (c) cutting out circles; and (d) heating them under pressure to vulcanize the rubber. Thin wheels (called *cut-off blades*) are made in this manner, and are used like saws for cutting-off operations.
4. **Metal bonds.** Abrasive grains, usually diamond or cubic boron nitride, are bonded in a metal matrix to the periphery of a metal disc, typically to depths of 6 mm or less (See Fig. 9.3). Bonding is carried out under high pressure and temperature. The wheel itself (core)

may be made of aluminum, bronze, steel, ceramic, or composite material, depending on special requirements for the wheel, such as strength, stiffness, and dimensional stability. Superabrasive wheels may be layered so that a single abrasive layer is plated or brazed to a metal disc.

5. **Other types of bonds.** In addition to those described above, other types of bonds include *silicate, shellac*, and *oxychloride* bonds. These bonds are far less common, but use the same abrasives as other wheels.

9.3.2 Wheel Grade and Structure

The *grade* of a bonded abrasive is a measure of the strength of the bond, and it includes both the *type* and the *amount* of bond in the wheel. Because strength and hardness are directly related, the grade is also referred to as the *hardness* of a bonded abrasive; thus, a hard wheel has a stronger bond and/or a larger amount of bonding material than a soft wheel. The *structure* is a measure of the *porosity*, the spacing between the grains (Fig. 9.1) of the bonded abrasive. Some porosity is essential to provide space for the grinding chips as otherwise they would interfere with the grinding operation. The structure of bonded abrasives ranges from dense to open (Fig. 9.6).

9.4 Mechanics of Grinding

Grinding is a *chip removal process* in which the cutting tool is an individual abrasive grain. The following are major factors that differentiate the action of a single grain from that of a single-point cutting tool (See Fig. 8.2):

1. Conventional abrasive grains have an irregular geometry and are spaced randomly along the periphery of the wheel (Fig. 9.6). Newly developed shaped abrasives are available that provide less random and more aggressive machining surfaces.
2. The average rake angle of the grains is highly negative, typically -60° and lower, thus the shear angles are very low (Section 8.2.4).

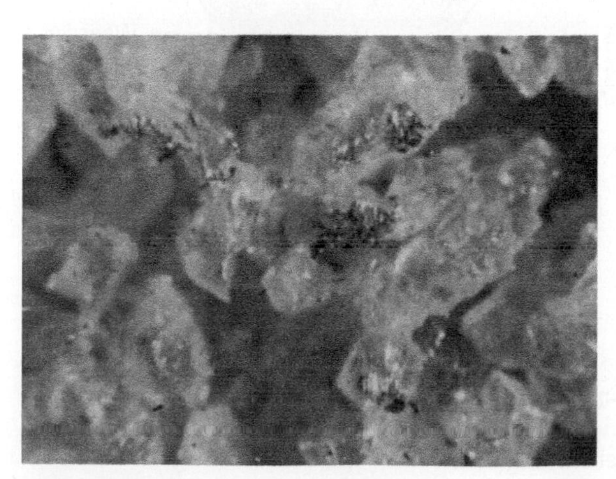

FIGURE 9.6 The grinding surface of an abrasive wheel (A46-J8V), showing grains, porosity, wear flats on grains (See also Fig. 9.7b), and metal chips from the workpiece adhering to the grains. Note the random distribution and shape of the abrasive grains. Magnification: 50×. *Source:* After S. Kalpakjian.

3. Depths of cut are very shallow, and chips are small and always discontinuous.
4. Temperatures are much higher in grinding than in metal cutting; the temperatures can be high enough for some chips to react with oxygen in air, often leading to 'sparks' (see Section 9.4.2).
5. The grains in the periphery of a grinding wheel have different radial positions from the center of the wheel.
6. The cutting speeds of grinding wheels are very high (Table 9.2), typically on the order of 30 m/s (6000 ft/min).

An example of chip formation by an abrasive grain is shown in Fig. 9.7. Note the negative rake angle, the low shear angle, and the very small size of the chip (see also Example 9.1). Grinding chips can easily be collected on an adhesive tape, which is held against the sparks of a grinding wheel; from direct observation of the tape, it will be noted that a variety of metal chips are produced in grinding.

The mechanics of grinding chip formation and the variables involved can best be studied by analyzing the *surface-grinding* operation shown in Fig. 9.9. In this figure, a grinding wheel, with a diameter of D, is removing a layer of material at a depth d, known as the **wheel depth of cut.**

TABLE 9.2 Typical ranges of speeds and feeds for abrasive processes.

Process variable	Conventional grinding	Creep-feed grinding	Buffing	Polishing
Wheel speed (m/min)	1500–3000	1500–3000	1800–3600	1500–2400
Work speed (m/min)	10–60	0.1–1	–	–
Feed (mm/pass)	0.01–0.05	1–6	–	–

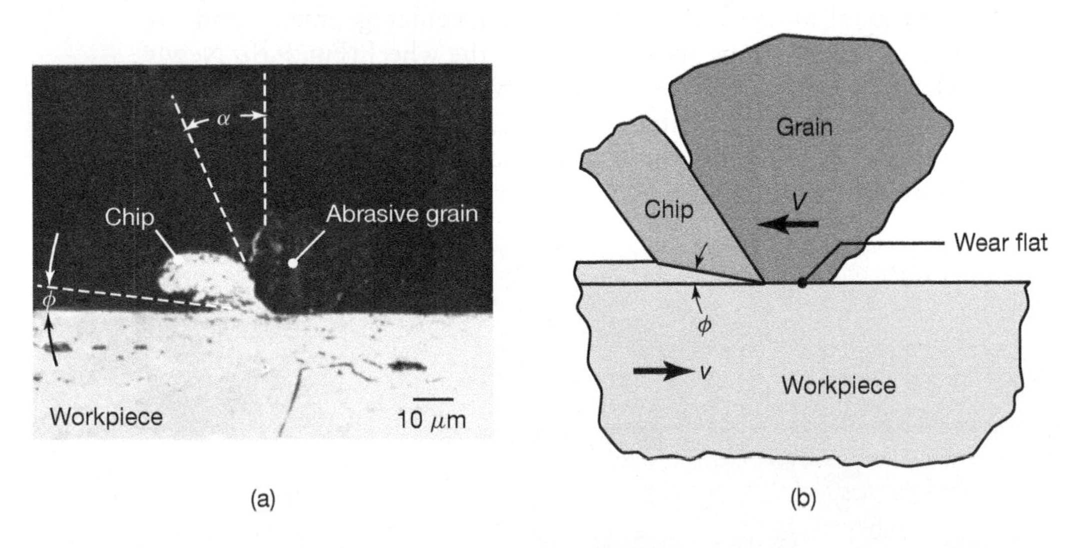

FIGURE 9.7 (a) Grinding chip being produced by a single abrasive grain. Note the large negative rake angle of the grain. *Source:* After M.E. Merchant. (b) Schematic illustration of chip formation by an abrasive grain. Note the negative rake angle, the small shear angle, and the wear flat on the grain.

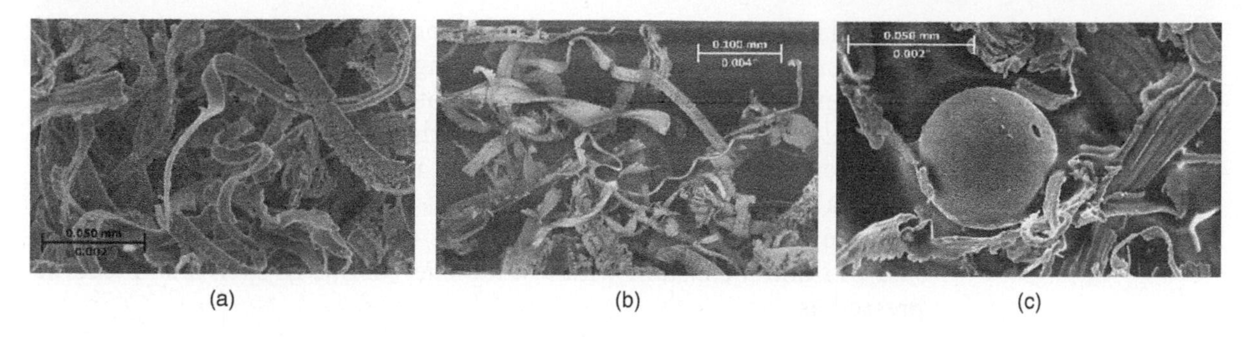

FIGURE 9.8 Typical chips, or swarf, from grinding operations. (a) Swarf from grinding a conventional HSS drill bit; (b) swarf from a tungsten carbide workpiece using a diamond wheel; and (c) swarf of cast iron, showing a melted globule among the chips. *Source:* Courtesy of J. Badger.

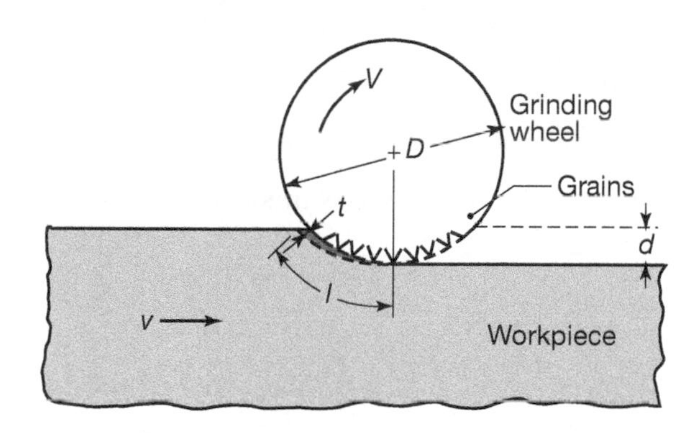

FIGURE 9.9 Basic variables in surface grinding. In actual grinding operations, the wheel depth of cut, d, and contact length, l, are much smaller than the wheel diameter, D. The dimension t is called the *grain depth of cut*.

An individual grain on the periphery of the wheel is rotating at a tangential velocity V; the workpiece is moving at a velocity v. As shown, it is called *up* or *conventional grinding* (see also *milling*, Section 8.10.1). The abrasive grain is removing a chip with an *undeformed thickness* (**grain depth of cut**), t, and an undeformed length, l.

Typical chips from grinding operations are shown in Fig. 9.8. Note that the chips, just as in machining, are thin and long. From geometric relationships, it can be shown that for the condition of $v \ll V$, the *undeformed-chip length*, l, is approximately

$$l \simeq \sqrt{Dd}. \tag{9.1}$$

For *external (cylindrical) grinding* (see Section 9.6),

$$l = \sqrt{\frac{Dd}{1 + (D/D_w)}}, \tag{9.2}$$

and for *internal grinding*,

$$l = \sqrt{\frac{Dd}{1 - (D/D_w)}}, \tag{9.3}$$

where D_w is the diameter of the workpiece.

The relationship between t and other process variables can be derived as follows: Let C be the number of cutting points per unit area of wheel surface, and v and V the surface speeds of the workpiece and the wheel, respectively (Fig. 9.9). Assuming the width of the workpiece to be unity, the number of grinding chips produced per unit time is VC, and the volume of material removed per unit time is vd.

Assume the chip shown in Fig. 9.9 has a triangular cross section with a base of t and a constant width of w. The volume of such a chip can be expressed as

$$\text{Vol}_{\text{chip}} = \frac{wtl}{2} = \frac{rt^2l}{4}, \quad (9.4)$$

where r is the ratio of the chip width, w, to the average chip thickness. The volume of material removed per unit time is the product of the volume of each chip and the number of chips produced per unit time; thus,

$$VC\frac{rt^2l}{4} = vd,$$

and because $l = \sqrt{Dd}$, the undeformed chip thickness in surface grinding will be

$$t = \sqrt{\frac{4v}{VCr}\sqrt{\frac{d}{D}}}. \quad (9.5)$$

Experimental observations have indicated that C is on the order of 0.1 to 10 per mm^2. Note that the finer the grain size of the wheel, the larger is this quantity. The magnitude of r is between 10 and 20 for most grinding operations. Substituting typical values into Eqs. (9.1) through (9.5), it will be found that l and t are very small quantities; typical values of t are in the range of 0.3–0.4 μm.

EXAMPLE 9.1 Chip Dimensions in Grinding

Given: A typical surface grinding operation is being performed with $D = 200$ mm, $d = 0.05$ mm, $C = 2$ per mm^2, and $r = 15$.

Find: Estimate the undeformed-chip length and the undeformed chip thickness.

Solution: The formulas for undeformed length and thickness, respectively, are given by Eqs. (9.1) and (9.5) as

$$l = \sqrt{Dd} \quad \text{and} \quad t = \sqrt{\frac{4v}{VCr}\sqrt{\frac{d}{D}}}.$$

From Table 9.2 the following values are selected:

$$v = 30 \text{ m/min} = 0.5 \text{ m/s} \quad \text{and} \quad V = 1800 \text{ m/min} = 30 \text{ m/s}.$$

Therefore,

$$l = \sqrt{(200)(0.05)} = 3.2 \text{ mm},$$

and

$$t = \sqrt{\frac{(4)(0.5)}{(30)(2)(15)}\sqrt{\frac{0.05}{200}}} = 0.006 \text{ mm}.$$

Note that because of plastic deformation, the *actual* length of the chip is shorter and the thickness greater than these values (See Fig. 9.7).

9.4.1 Grinding Forces

As in machining operations, knowledge of forces is essential not only in the design of grinding machines and workholding devices, but also in determining the deflections that the workpiece and the machine will undergo. Deflections, in turn, adversely affect dimensional accuracy of the workpiece, which is especially critical in precision grinding.

Based on the discussion of cutting force, F_c, in Section 8.2.3 and assuming that the *force* on the grain is proportional to the cross-sectional area of the undeformed grinding chip, it can be shown that the **grain force** is

$$\text{Grain force} \propto \frac{v}{VC}\sqrt{\frac{d}{D}}. \qquad (9.6)$$

The grain force is then the product of the expression in Eq. (9.6) and the strength of the metal being ground.

The **specific energy** consumed in producing a grinding chip consists of three components:

$$u = u_{\text{chip}} + u_{\text{plowing}} + u_{\text{sliding}}, \qquad (9.7)$$

where the quantity u_{chip} is the specific energy required for chip formation by plastic deformation; u_{plowing} is the specific energy required for plowing, which is plastic deformation without chip removal (Fig. 9.10); and the last term, u_{sliding}, can best be understood by observing the grain in Fig. 9.7b. Note that the grain develops a **wear flat**, similar to flank wear in cutting tools (see Section 8.3).

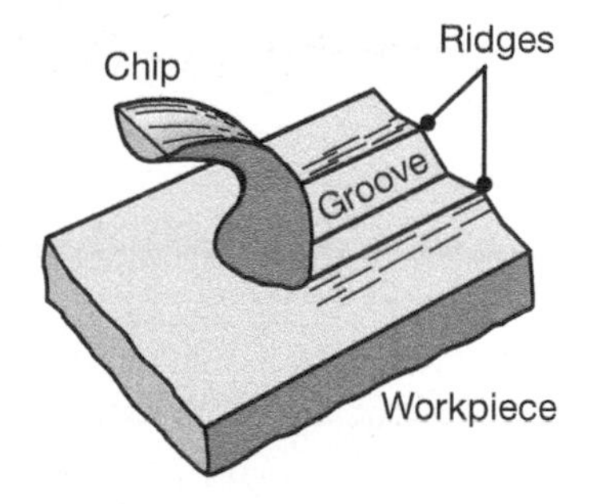

FIGURE 9.10 Chip formation and plowing (plastic deformation without chip removal) of the workpiece surface by an abrasive grain.

Typical specific-energy requirements in grinding are given in Table 9.3. Note that these energy levels are much higher than those in machining operations with single-point tools, as given in Table 8.3. This difference has been attributed to the following three factors:

1. **Size effect.** Recall that the size of grinding chips is very small, as compared with chips produced in machining operations, by about two orders of magnitude. As described in Section 3.8.3, the smaller the size of a metal specimen, the higher is its strength; for this reason, grinding involves higher specific energy than machining operations.

TABLE 9.3 Approximate specific-energy requirements for surface grinding.

Workpiece material	Hardness	Specific energy W-s/mm^3	Specific energy hp-min/in^3
Aluminum	150 HB	7–27	2.5–10
Cast iron (class 40)	215 HB	12–60	4.5–22
Low-carbon steel (1020)	110 HB	14–68	5–25
Titanium alloy	300 HB	16–55	6–20
Tool steel (T15)	67 HRC	18–82	6.5–30

2. **Wear flat.** The wear flat (Fig. 9.7b) dissipates frictional energy, which contributes significantly to the total energy consumed. The size of the wear flat in grinding is much larger than the grinding chip, unlike in metal cutting by a single-point tool, where flank wear land is small compared with the size of the chip (Section 8.3).
3. **Chip morphology.** Recall that the average rake angle of a grain is highly negative (Fig. 9.7), thus the shear strains in grinding are very large. This indicates that the energy required for plastic deformation to produce a grinding chip is higher than in machining processes. Moreover, plowing consumes energy without contributing to grinding chip formation (Fig. 9.10).

EXAMPLE 9.2 Forces in Surface Grinding

Given: Assume that you are performing a surface-grinding operation on a low-carbon steel workpiece using a wheel of diameter $D = 250$ mm that rotates at $N = 4000$ rpm. The width of cut is $w = 25$ mm, depth of cut is $d = 0.05$ mm, and the feed rate of the workpiece is $v = 1.5$ m/s.

Find: Calculate the cutting force, F_c (the force tangential to the wheel), and the thrust force, F_n (the force normal to the workpiece), noting that in general F_n is around 30% higher than F_c.

Solution: We first determine the material removal rate as follows:

$$\text{MRR} = dwv = (0.05)(25)(1500) = 1875 \text{ mm}^3/\text{s}.$$

The power consumed is given by

$$\text{Power} = (u)(\text{MRR}),$$

where u is the specific energy, as obtained from Table 9.3. For low-carbon steel, use an average value of 41 W-s/mm^3. Hence,

$$\text{Power} = (41)(1875) = 76.875 \text{ kW}.$$

Also note that the angular velocity is

$$\omega = (4000)\left(\frac{2\pi}{60}\right) = 418.9 \text{ deg/s}.$$

Since power is defined as

$$\text{Power} = T\omega,$$

where T is the torque and is equal to $(F_c)(D/2)$,

$$76{,}875 = (F_c)\left(\frac{0.25}{2}\right)(418.9),$$

and therefore, $F_c = 1468$ N. The thrust force, F_n, can then be calculated as

$$F_n = (1.3)(1468) = 1908 \text{ N}.$$

9.4.2 Temperature

Temperature rise in grinding is an important consideration, because it can (a) adversely affect workpiece surface properties; (b) cause residual stress; (c) cause distortion and difficulties in controlling dimensional accuracy; and (d) when high, it can cause burning and structural changes. The work expended in grinding is mainly converted into heat. The *surface temperature rise*, ΔT, has been found to be a function of the ratio of the total energy input to the surface area ground. In *surface grinding*, if w is the width and L is the length of the surface area that is ground, then

$$\Delta T \propto \frac{uwLd}{wL} \propto ud. \qquad (9.8)$$

Including the size effect and assuming that u varies inversely with the undeformed-chip thickness, t, the temperature rise is

$$\text{Temperature rise} \propto D^{1/4}d^{3/4}\left(\frac{V}{v}\right)^{1/2}. \qquad (9.9)$$

The *peak temperatures* in chip generation during grinding can be as high as 1650°C (3000°F); however, because the time involved in producing a chip is extremely short (on the order of microseconds), melting of the chip may or may not occur. Because, as in machining, the chips carry away much of the heat generated (See also Fig. 8.18), only a small fraction of the heat generated is conducted into the workpiece (see Section 8.2).

Sparks. The sparks observed in grinding metals are actually glowing chips. The glowing occurs because of the *exothermic reaction* of the hot chips with oxygen in the atmosphere; sparks have not been observed with any metal ground in an oxygen-free environment. The color, intensity, and shape of the sparks depend on the composition of the metal being ground. If the heat generated by exothermic reaction is sufficiently high, the chip may melt and, because of surface tension, solidify as a shiny spherical particle. Scanning electron microscopy has shown that these particles are hollow and have a fine dendritic structure (Fig. 5.8), indicating that they were once molten and have resolidified rapidly. Moreover, some of the spherical particles may also have been formed by plastic deformation and rolling of chips at the grain-workpiece interface during grinding.

9.4.3 Effects of Temperature

The major effects of temperature in grinding are:

1. **Tempering.** Excessive temperature rise caused by grinding can temper (Section 5.11.5) and soften the surfaces of steel components; they are often ground in the heat-treated and hardened state. Tempering can be eliminated by avoiding excessive temperature rise; grinding fluids (Section 9.6.9) also can effectively control temperatures.
2. **Burning.** If the temperature rise is excessive, the workpiece surface may burn, such as a bluish color on steels, which indicates oxidation at high temperatures. A burn may not be objectionable in itself; however, the surface layers may undergo metallurgical transformations, with martensite formation in high-carbon steels from reaustenization, followed by rapid cooling (Section 5.11). This phenomenon is known as *metallurgical burn*, which is especially serious with nickel-base alloys.
3. **Heat checking.** High temperatures in grinding lead to thermal stresses which may cause thermal cracking of the workpiece surface, known as *heat checking* (see also Section 5.10.3). The cracks are usually perpendicular to the grinding direction; under severe grinding conditions, however, parallel cracks may also develop.
4. **Residual stresses.** Temperature gradients within the workpiece are mainly responsible for residual stresses in grinding. Other contributing factors are the physical interactions of the abrasive grain in chip formation and the sliding of the wear flat along the workpiece surface. Two examples of residual stresses in grinding are given in Fig. 9.11, demonstrating the effects of wheel speed and the type of grinding fluid used. The method and direction of the application of grinding fluid also can have a significant effect on the residual stresses developed. Because of the deleterious effect of tensile residual stresses on fatigue strength (see Section 3.8.2), process parameters should be chosen properly. Residual stresses can usually be lowered by (a) using softer grade wheels (called *free-cutting wheels*); (b) lower wheel speeds; and (c) higher work speeds, a procedure known as **low-stress** or **gentle grinding.**

9.5 Grinding-Wheel Wear

Wheel wear is an important consideration because it adversely affects the shape and dimensional accuracy of ground surfaces, a situation similar to the wear of cutting tools (Section 8.3). Grinding wheels wear by three different mechanisms:

1. **Attritious wear.** The cutting edges of a sharp grain become dull by attrition (known as *attritious wear*), developing a *wear flat* (Fig. 9.7b) that is similar to flank wear in cutting tools. This type of wear is caused by the interaction of the grain with the workpiece material,

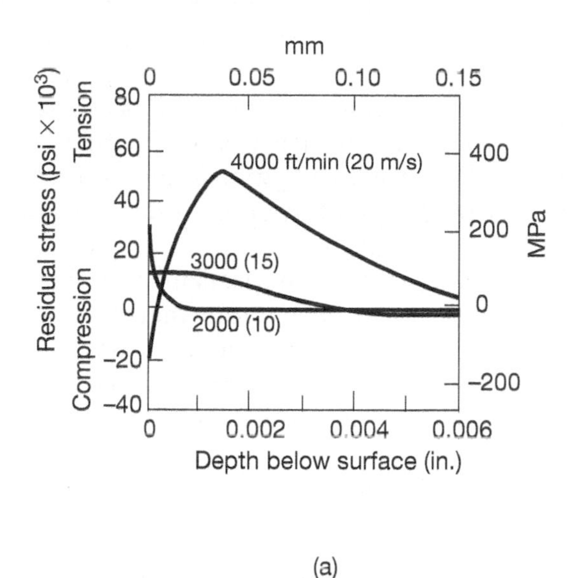

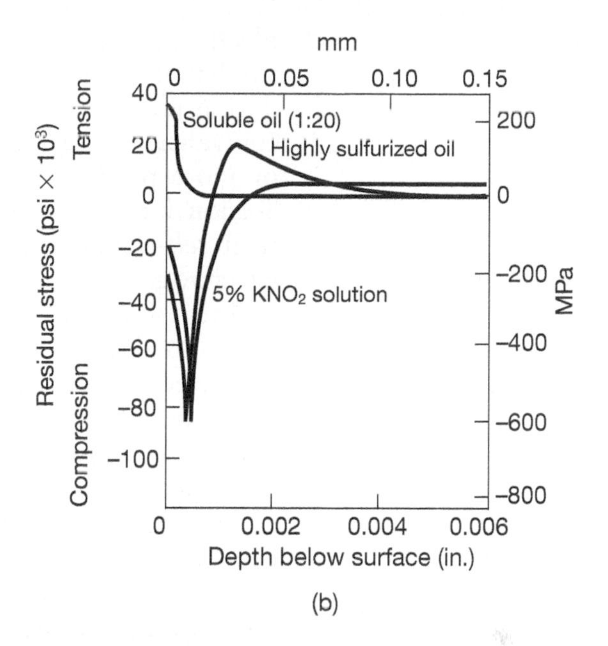

FIGURE 9.11 Residual stresses developed on the workpiece surface in grinding tungsten: (a) effect of wheel speed and (b) effect of type of grinding fluid. Tensile residual stresses on a surface are detrimental to the fatigue life of ground components. The variables in grinding can be controlled to minimize residual stresses, a process known as *low-stress grinding*. *Source:* After N. Zlatin.

resulting in complex physical and chemical reactions between the two. These reactions involve (a) diffusion; (b) chemical degradation or decomposition of the grains; (c) fracture at a microscopic scale; (d) plastic deformation; and (e) melting.

Attritious wear is low when the two materials are chemically inert with respect to each other, much like in cutting tools; the more inert the two materials, the lower the tendency for adhesion to occur between the grain and the workpiece being ground. For example, because aluminum oxide is relatively inert to iron, its rate of attritious wear in grinding steels is much lower than that for silicon carbide and diamond grains. On the other hand, carbon can dissolve in iron, and thus diamond is not suitable for grinding steels. Cubic boron nitride has a higher inertness to steels, and hence it is suitable for use as an abrasive. The selection of the type of abrasive for low attritious wear should thus be based on the reactivity of the grain and the workpiece materials and their relative mechanical properties, especially hardness and toughness. The environment and the type of grinding fluid used also have an influence on grain-workpiece interactions.

2. **Grain fracture.** Because abrasive grains are brittle, their fracture characteristics in grinding are important. If the wear flat caused by attritious wear is excessive, the grain becomes dull and the grinding operation becomes inefficient and produces high temperatures.

Ideally, the grain should fracture or fragment at a moderate rate, so that new and sharp cutting edges are produced continuously during the grinding operation. Note that the fracturing process is equivalent to breaking a piece of rounded and dull chalk into two or more pieces in order to expose new sharp edges. Recall that Section 9.2 has described *friability* of abrasive grains, giving them their *self-sharpening* characteristics, an important consideration in effective grinding.

The selection of grain type and size for a particular application also depends on the attritious-wear rate. Note that a grain-workpiece material combination with high attritious wear and low friability causes dulling of grains and the development of a large wear flat. Grinding then becomes inefficient and surface damage is likely to occur.

The following workpiece material and abrasive combinations are generally recommended:

a. *Aluminum oxide:* steels, ferrous alloys, and alloy steels.
b. *Silicon carbide:* cast iron, nonferrous metals, and hard and brittle materials (such as carbides, ceramics, marble, and glass).
c. *Diamond:* composite materials, ceramics, cemented-carbide ceramics, and some hardened steels.
d. *Cubic boron nitride:* composite materials, steels and cast irons at 50 HRC (such as hardened tool steels) or above, and for high-temperature superalloys.

3. **Bond fracture.** The strength of the bond (*grade*) is a significant parameter in grinding. If the bond is too strong, dull grains cannot be easily dislodged so that other sharp grains, along the circumference of the grinding wheel, can begin to contact the workpiece and remove chips; thus the grinding process then becomes inefficient. On the other hand, if the bond is too weak, the grains can easily be dislodged and wheel wear increases, thus controlling the dimensional accuracy of the workpiece becomes difficult. In general, softer bonds are recommended for harder materials, and for reducing residual stresses and thermal damage to the workpiece. Hard-grade wheels are recommended for softer materials, and for removing large amounts of material at high rates (see also Section 9.5.3).

9.5.1 Dressing, Truing, and Shaping of Grinding Wheels

Dressing is the process of conditioning worn grains on the surface of a grinding wheel in order to produce sharp new grains, but also will true an out-of-round wheel (see below). Dressing is necessary when excessive attritious wear dulls the wheel, called **glazing** because of the shiny appearance of the wheel surface, or when the wheel becomes loaded. **Loading** occurs when the porosities on the grinding surfaces of the wheel (Fig. 9.6) become filled or clogged with grinding chips. Loading can occur (a) when grinding soft workpiece materials; (b) by improper selection of the grinding wheel; and (c) by improper selection of grinding parameters. A loaded wheel grinds very inefficiently, generating much frictional heat and causing surface damage and loss of dimensional accuracy.

Dressing is done by the following techniques:

1. A specially shaped diamond-point tool or a diamond *cluster* is moved across the width of the grinding face of a rotating wheel, removing a very small layer from the wheel surface with each pass across the wheel. This method can be used either dry or wet (using grinding fluids), depending on whether the wheel is to be used dry or wet, respectively, during the grinding operation.
2. A set of star-shaped steel disks is pressed against the rotating grinding wheel, and material is removed from the wheel surface by crushing the grains. This method produces a coarse grinding surface on the wheel and is used only for rough grinding operations on bench or pedestal grinders (Section 9.6.5).
3. *Abrasive sticks* may be held against the grinding surface of the wheel. This is a common method for softer wheels but is not appropriate for precision grinding operations.
4. For metal-bonded diamond wheels, electrical-discharge and electrochemical machining techniques (see Sections 9.11 and 9.13) can be used to erode very small layers of the metal bond, thus exposing new diamond cutting edges.
5. Dressing for *form grinding* involves **crush dressing**, or *crush forming*; the method consists of pressing a metal roll on the surface of the (usually vitrified) grinding wheel. The roll is made of high-speed steel, tungsten carbide, or boron carbide, and has a machined or ground profile, thus reproducing this profile on the surface of the grinding wheel being dressed (Fig. 9.12). Dressing is also done to generate a specific shape on the grinding surface of a wheel for the purpose of grinding profiles on workpieces (see Section 9.6.2).

Dressing techniques and the frequency at which the wheel surface is dressed are significant factors, affecting grinding forces and workpiece surface finish. Modern computer-controlled grinders (Section 9.6) are equipped with automatic dressing features that dress the wheel during the grinding operation. For a typical aluminum-oxide wheel, the depth removed during dressing is on the order of 5 to 15 μm, but for a cBN wheel, it is 2 to 10 μm. Modern dressing systems have a resolution as low as 0.25 to 1 μm.

Truing is an operation by which a wheel is restored to its original shape. A round wheel is *shaped* to make its circumference a true circle, hence the word *truing*. Truing can also produce a desired workpiece shape. The grinding face on the Type 1 straight wheel shown in Fig. 9.2a is cylindrical and thus produces a flat surface; however, modern grinders are equipped with computer-controlled shaping features, whereby the diamond dressing tool automatically traverses the wheel face along a certain prescribed path (Fig. 9.12). Note in this figure that the axis of the diamond dressing tool remains normal to the wheel face at the point of contact. The result of the operation is a profile that can be produced in a workpiece.

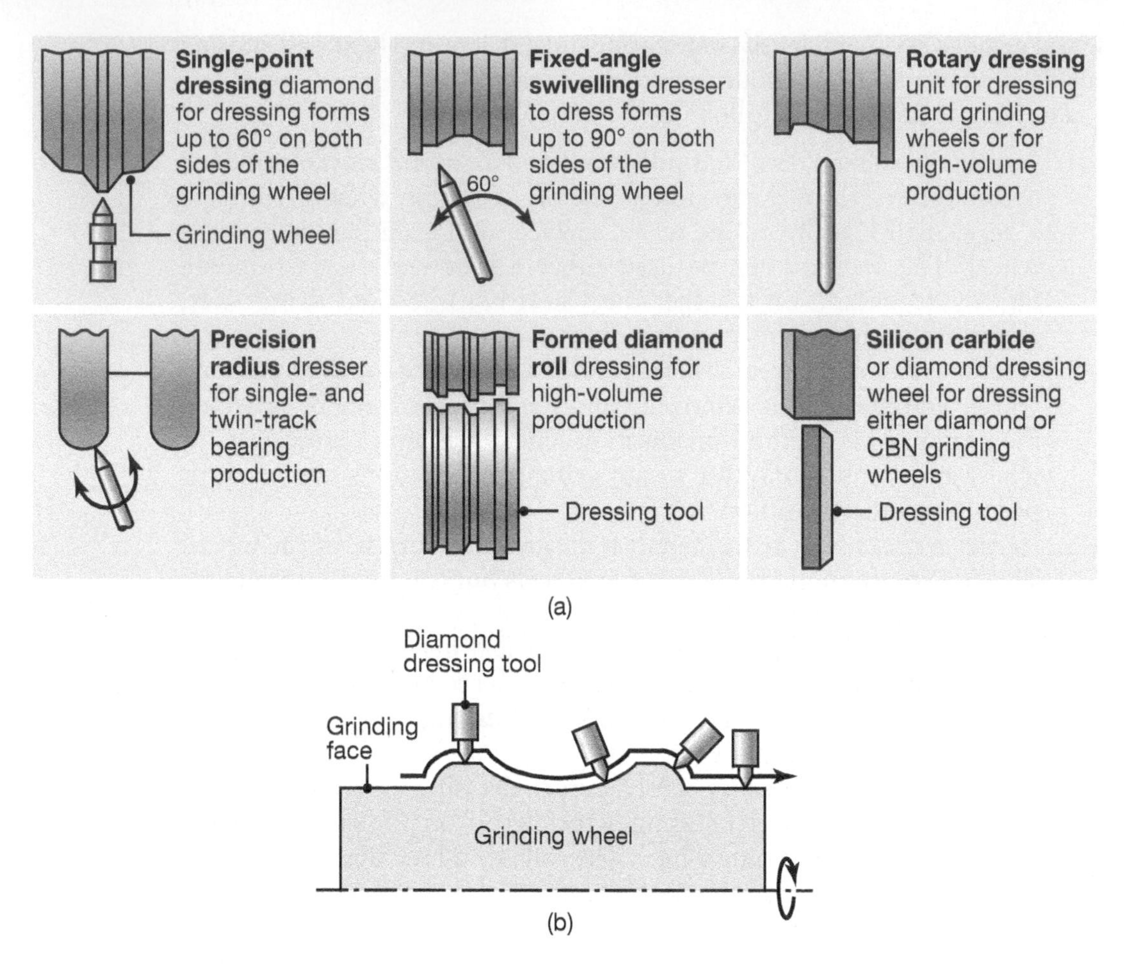

FIGURE 9.12 (a) Methods of grinding wheel dressing. (b) Shaping the grinding face of a wheel by dressing it with computer-controlled shaping features. Note that the diamond dressing tool is normal to the wheel surface at point of contact. *Source:* OKUMA America Corporation.

9.5.2 Grinding Ratio

Grinding-wheel wear is generally correlated to the amount of material ground; it is expressed by the *grinding ratio*, G, which is defined as

$$G = \frac{\text{Volume of material removed}}{\text{Volume of wheel wear}}. \tag{9.10}$$

In practice, grinding ratios vary widely, ranging from 2 to 200 and higher, depending on (a) the type of wheel; (b) workpiece material; (c) grinding fluid; and (d) process parameters, such as depth of cut and speeds of the wheel and the workpiece. Attempting to obtain a high grinding ratio in practice is not necessarily desirable because high ratios may indicate grain dulling, leading to possible surface damage. A lower ratio may well be acceptable if an overall economic analysis justifies it.

Soft-acting or hard-acting wheels. During a grinding operation, a particular wheel may *act soft* (meaning its wear rate is high) or *hard* (wear rate is low), regardless of its grade. This behavior is a function of the force on the grain. The higher the force, the greater the tendency for the grains to undergo fracture or to be dislodged from the wheel surface, hence the higher the wheel wear and the lower the grinding ratio. Equation (9.6) indicates that the grain force increases with the (a) strength of the workpiece material; (b) work speed; and (c) depth of cut, and decreases with increasing (a) wheel speed and (b) wheel diameter.

EXAMPLE 9.3 Action of a Grinding Wheel

Given: A surface-grinding operation is being carried out with the wheel rotating at a constant spindle speed.

Find: Will the wheel act soft or act hard as it wears down over a period of time?

Solution: Referring to Eq. (9.6), it will be noted that the parameters that change with time in this operation are the wheel surface speed, V, and the wheel diameter, D. As both become smaller with time, the relative grain force increases, and therefore the wheel acts softer. Some grinding machines are equipped with variable-speed spindle motors to accommodate these changes and to make provisions for wheels of different diameter.

9.5.3 Wheel Selection and Grindability of Materials

Proper selection of a grinding wheel for a given application greatly influences the quality of surfaces produced and the economics of the operation. Selection involves not only the shape of the wheel with respect to the shape of the part, but also the characteristics of the workpiece material. The *grindability* of materials, like machinability (Section 8.5) or forgeability (Section 6.2.5), is difficult to define precisely. It is a general indication of how easy it is to grind a particular material; it includes such considerations as surface finish, surface integrity, wheel wear, cycle time, and overall economics of the operation.

Specific recommendations for selecting wheels and process parameters can be found in various handbooks. Examples of recommendations are: C60-L6V for cast irons, A60-M6V for steels, C60-I9V or D150-R75B for carbides, A60-K8V for titanium, and D150-N50M (diamond) for ceramics.

Ductile regime grinding. It is possible to obtain continuous chips in grinding ceramics using light passes and rigid machine tools with good damping capacity (Figs. 9.7b and 9.10). Known as *ductile regime grinding*, this technique produces good workpiece surface integrity; however, because ceramic chips are typically 1–10 μm in size, they are more difficult to remove from grinding fluids than are machining chips, requiring fine filtration.

9.6 Grinding Operations and Machines

Grinding operations are typically carried out using a wide variety of wheel-workpiece configurations. The selection of a particular grinding process for a specific application depends on (a) part shape; (b) part size; (c) required tolerances; (d) ease of fixturing, and (e) production rate required. The basic types of grinding operations are surface, cylindrical, internal, and centerless grinding (see below). The movement of the wheel in these operations may be along the surface of the workpiece (*traverse grinding, through feed grinding,* or *cross-feeding*), or it may be radially *into* the workpiece (*plunge grinding*). Surface grinders are the most common machine type, followed by bench grinders (usually with two grinding wheels), cylindrical grinders, and tool grinders. Because grinding wheels are brittle and are operated at high speeds, certain safety procedures must be carefully followed in their handling, storage, and use.

Modern grinding machines are computer controlled, with various features, such as automatic part loading and unloading, clamping, cycling, gaging, dressing, and wheel shaping. Grinders can also be equipped with probes and gages for determining the relative position of the wheel and workpiece surfaces, as well as with tactile sensing features, whereby breakage of the diamond dressing tool, if any, can be detected during the dressing cycle.

9.6.1 Surface Grinding

Surface grinding is one of the most common grinding operations (Fig. 9.13) and basically involves grinding flat surfaces. Typically, the workpiece is secured on a magnetic chuck mounted on the worktable of a **surface**

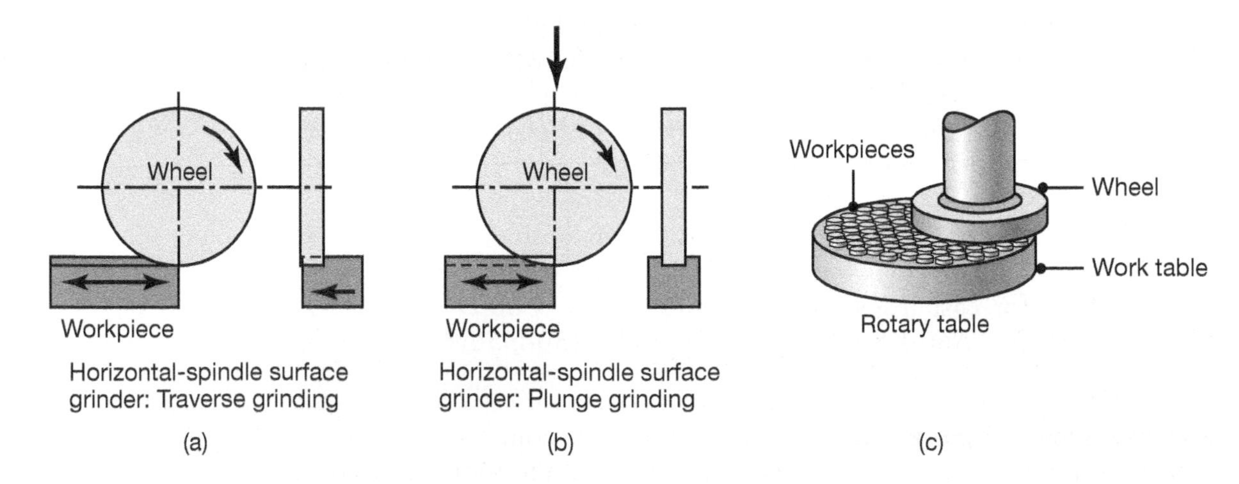

FIGURE 9.13 Schematic illustrations of surface-grinding operations. (a) Traverse grinding with a horizontal-spindle surface grinder. (b) Plunge grinding with a horizontal-spindle surface grinder, producing a groove in the workpiece. (c) Vertical-spindle rotary-table grinder (also known as the *Blanchard-type* grinder).

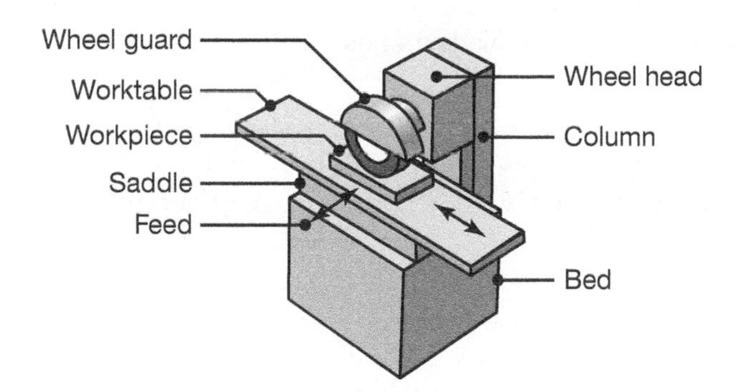

FIGURE 9.14 Schematic illustration of a horizontal-spindle surface grinder.

grinder (Fig. 9.14); nonmagnetic materials generally are held by vises, vacuum chucks, double-sided adhesive tapes, or special fixtures. In this operation, a straight wheel is mounted on the *horizontal spindle* of the grinder; traverse grinding is done as the table reciprocates longitudinally and feeds laterally after each stroke. In *plunge grinding*, the wheel is moved radially into the workpiece, as in grinding a groove illustrated in Fig. 9.13b. The size of a surface grinder is specified by the surface dimensions of length and width that can be ground on the machine. Other types of surface grinders include *vertical spindles* and *rotary tables* (Fig. 9.13c), also called *Blanchard-type* grinders. These configurations allow a number of parts to be ground in one setup, thus improving productivity.

9.6.2 Cylindrical Grinding

In *cylindrical grinding*, also called *center-type grinding*, the external cylindrical surfaces and shoulders of the workpiece are ground, such as crankshaft bearings, spindles, pins, bearing rings, and rolls for rolling mills. The rotating cylindrical workpiece reciprocates laterally along its axis, although in grinders used for large and long workpieces, the grinding wheel reciprocates. The latter design is called a *roll grinder* and is capable of grinding rolls as large as 2 m (80 in.) in diameter, used for metal rolling operations (See Fig. 6.29).

The workpiece in cylindrical grinding is held between *centers*, held in a chuck, or mounted on a faceplate in the headstock of the grinder. For straight cylindrical surfaces, the axes of rotation of the wheel and workpiece are parallel; separate motors drive the wheel and workpiece at different speeds. Long workpieces with two or more diameters are ground on cylindrical grinders. Cylindrical grinding also can produce shapes in which the wheel is dressed to the form to be ground on the workpiece, called *form grinding* or *plunge grinding*. Cylindrical grinders are specified by the maximum diameter and length of the workpiece that can be ground, similar to lathes (Section 8.9.2).

In *universal grinders*, both the workpiece and the wheel axes can be moved and swiveled around a horizontal plane, thus permitting the grinding of tapers and other shapes. Cylindrical grinders can also be equipped with computer control, so that *noncylindrical* parts (such as cams) can be ground on rotating workpieces. The workpiece spindle speed is synchronized with the grinding wheel position such that the distance between the

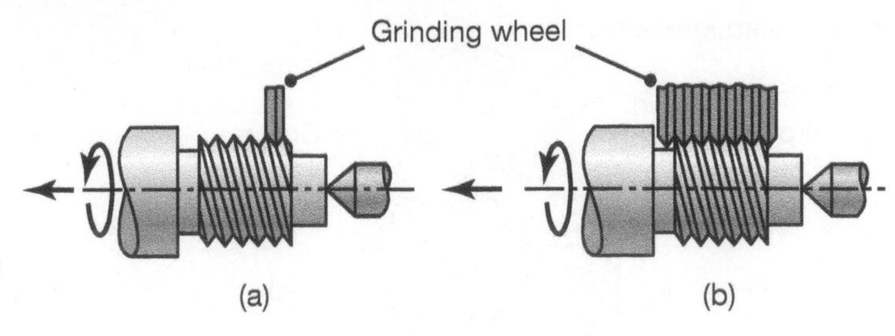

FIGURE 9.15 Threads produced by (a) traverse and (b) plunge grinding.

workpiece and wheel axes is varied continuously to produce a particular shape.

Thread grinding is done on cylindrical grinders, as well as on centerless grinders (Section 9.6.4), with specially dressed wheels that match the shape of the threads (Fig. 9.15). The workpiece and wheel movements are synchronized to produce the proper pitch of the thread, usually in about six passes. Although this operation is costly, it produces more accurate threads than any other manufacturing process, and the threads have a very fine surface finish.

9.6.3 Internal Grinding

In *internal grinding* (Fig. 9.16), a small wheel is used to grind the inside diameter of axisymmetric parts, such as bushings and bearing races. The workpiece is held in a rotating chuck, and the wheel rotates at 30,000 rpm or higher. Internal profiles also can be ground with profile-dressed wheels that move radially into the workpiece. The headstock of internal grinders can be swiveled on a horizontal plane to grind tapered holes.

9.6.4 Centerless Grinding

Centerless grinding is a high-production process for continuously grinding cylindrical surfaces, where the workpiece is supported not by centers (hence the term *centerless*) but by a blade, as shown in Fig. 9.17. Typical parts that are centerless ground include cylindrical roller bearings, piston pins, engine valves, camshafts, and similar components. Parts with diameters as small as 0.1 mm can be ground using this process. Centerless

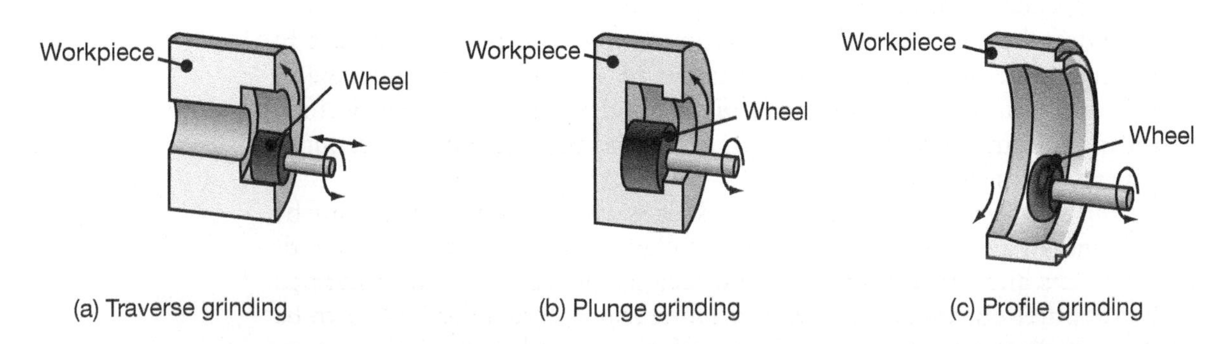

FIGURE 9.16 Schematic illustrations of internal-grinding operations.

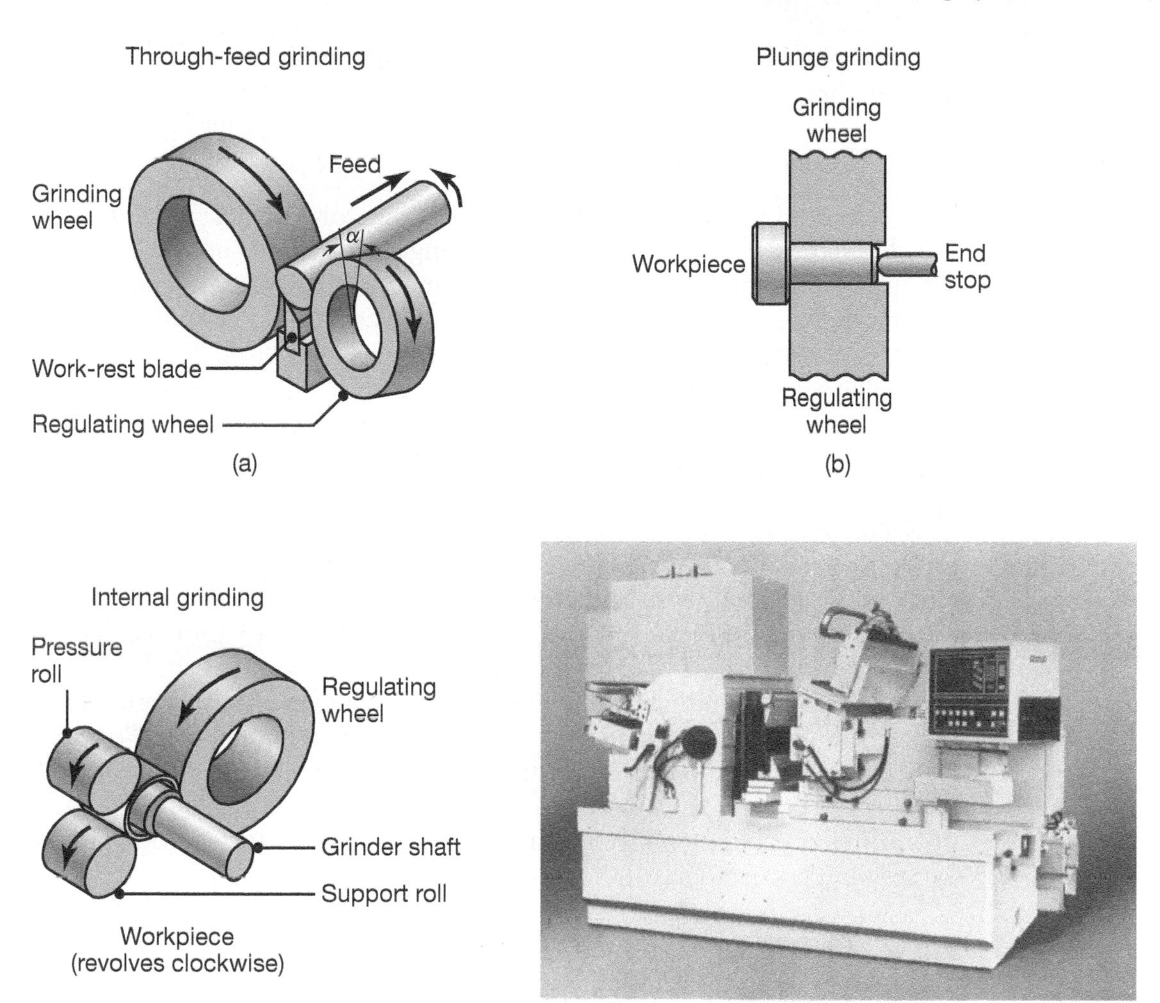

FIGURE 9.17 (a-c) Schematic illustrations of centerless-grinding operations. (d) A computer-numerical-control centerless grinding machine.

grinders are capable of wheel surface speeds on the order of 10,000 m/min, using cubic-boron-nitride abrasive wheels.

In *through-feed grinding* (Fig. 9.17a), the workpiece is supported on a work-rest blade and is ground between two wheels. Grinding is done by the larger wheel, while the smaller wheel regulates the axial movement of the workpiece. The *regulating wheel*, which is rubber bonded, is tilted and runs at a speed of only about $\frac{1}{20}$ of that of the grinding wheel.

Parts with variable diameters, such as bolts, valve tappets, and shafts, can be ground by centerless grinding. Called *infeed grinding* or *plunge grinding* (Fig. 9.17b), the process is similar to plunge or form grinding using cylindrical grinders. Tapered parts are centerless ground by *end-feed grinding*. High-production-rate thread grinding can be done with centerless grinders, using specially dressed wheels. In *internal centerless grinding*, the workpiece is supported between three rolls and is internally ground. Typical applications include sleeve-shaped parts and rings.

9.6.5 Other Types of Grinders

Several special-purpose grinders are available for various applications. *Bench grinders* are used for routine grinding of tools and small parts; they are usually equipped with two wheels; one wheel is usually coarse, for rough grinding, and the other is fine, for finish grinding. *Pedestal*, or *stand grinders* are mounted on stands instead of work benches, and are used similarly to bench grinders.

Universal tool and cutter grinders are used for grinding single-point or multipoint cutting tools and cutters. They are equipped with special workholding devices for accurate positioning of the tools to be ground. *Tool-post grinders* are self-contained units and are usually mounted on the tool post of a lathe (See Fig. 8.44). The workpiece is mounted on the headstock and is ground by moving the tool post. These grinders are versatile, but the ways of the lathe surfaces must be protected from abrasive debris.

Swing-frame grinders are typically used in foundries for grinding large castings. Rough grinding of castings is called *snagging* and is usually done on floorstand grinders, using wheels as large as 0.9 m in diameter. *Portable grinders*, either air or electrically driven, or with a flexible shaft connected to the shaft of an electric motor or gasoline engine, are available for such operations as grinding off weld beads (Fig. 12.5) and *cutting-off* operations using thin abrasive disks.

9.6.6 Creep-Feed Grinding

Although grinding has traditionally been associated with small rates of material removal and fine finishing operations, it can also be used for large-scale removal operations, similar to milling, broaching, and planning (Section 8.10). In *creep-feed grinding*, the wheel depth of cut, d, can be as much as 6 mm, and the workpiece speed is low, typically less than 60 mm/min (Fig. 9.18). The wheels are mostly softer-grade resin bonded, with open structure to keep temperatures low and improve surface finish. Grinders with capabilities for continuously dressing the grinding wheel with a diamond roll also are available. The machines for creep-feed grinding have special features, such as (a) high power of up to 225 kW; (b) high stiffness, because of the high forces due to the larger depth of material removed; (c) high damping capacity; (d) variable and well-controlled spindle and worktable speeds; and (e) ample capacity for grinding fluids.

In view of its overall economics and competitive position with respect to other machining operations, creep-feed grinding has been shown to be economical for specific applications, such as in grinding shaped punches, key seats, twist-drill flutes, the roots of turbine blades and various complex superalloy parts. The wheel is dressed to the shape of the workpiece to be produced; consequently, the workpiece does not have to be previously milled, shaped, or broached. *Near-net-shape* castings and forgings are particularly suitable for creep-feed grinding. Although a single

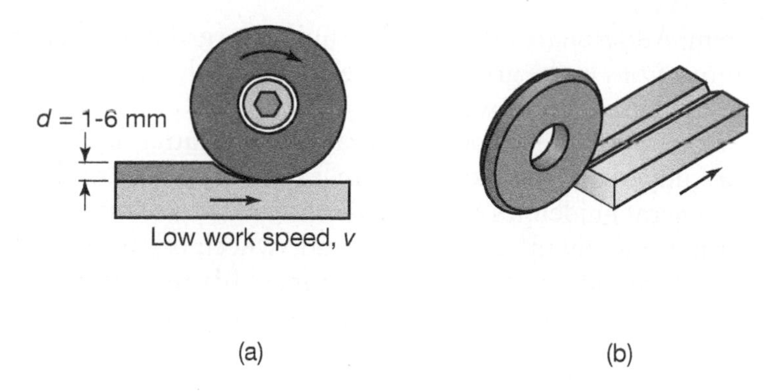

FIGURE 9.18 (a) Schematic illustration of the creep-feed grinding process. Note the large wheel depth of cut. (b) A groove produced on a flat surface in one pass by creep-feed grinding, using a shaped wheel. Groove depth can be on the order of a few mm. *Source:* Courtesy of Blohm, Inc. and Society of Manufacturing Engineers.

pass generally is sufficient, a second pass may be necessary for improved surface finish.

A more recent development is **high-efficiency deep grinding** (HEDG), using workpiece speeds as high as 500 mm/min, depths of cut approaching 10 mm, and superabrasive wheels. Material removal rates in HEDG can be as high as 2000 mm^3/s, compared to under 10 mm^3/s in creep-feed grinding.

9.6.7 Heavy Stock Removal by Grinding

Grinding processes can be used for heavy stock removal, thus competing with machining operations, particularly milling, turning, and broaching, and can be economical for specific applications. Because this process is a rough grinding operation, it can have detrimental effects on the workpiece surface and its integrity. In this operation, surface finish is of secondary importance, and the grinding wheel (or an abrasive belt; Section 9.7) is used to its maximum capacity for minimum cost per piece. The dimensional tolerances are on the same order as those obtained by other machining processes (See Fig. 9.28).

9.6.8 Grinding Chatter

Chatter is particularly significant in grinding because it adversely affects surface finish and wheel performance. Vibrations during grinding may be caused by bearings, spindles, and unbalanced grinding wheels, as well as external sources, such as from nearby machinery. The grinding process also can itself cause regenerative chatter. The analysis of chatter in grinding is similar to that for machining operations (Section 8.12) and involves *self-excited vibration* and *regenerative chatter*. Thus, the important variables are the stiffness of the grinder and of the workholding devices and damping of the system. Additional factors that are unique to grinding chatter include (a) nonuniformities in the grinding wheel itself; (b) the dressing techniques used; and (c) uneven wheel wear.

Because these variables produce characteristic **chatter marks** on ground surfaces, a study of these marks often can lead to the source of the vibration problem. General guidelines have been established to reduce the tendency for chatter in grinding: (a) using a soft-grade wheel; (b) dressing the wheel frequently; (c) modifying dressing techniques; (d) reducing the material removal rate; and (e) supporting the workpiece rigidly.

9.6.9 Grinding Fluids

The functions of grinding fluids are similar to those for cutting fluids, as described in Section 8.7. Although grinding and other abrasive machining processes can be performed dry, the use of a fluid is usually preferred to (a) prevent excessive temperature rise in the workpiece; (b) improve its surface finish and dimensional accuracy; and (c) improve the efficiency of the operation by reducing wheel wear, wheel loading, and power consumption.

Grinding fluids typically are *water-base emulsions* and chemicals and synthetics; oils may be used for thread grinding. The fluids may be applied as a stream (flood) or as mist (a mixture of fluid and air). Because of the high wheel surface speeds in grinding, an *air stream* or *air blanket* develops around the periphery of the wheel, preventing the fluid from reaching the wheel–workpiece interface. Special *nozzles* that conform to the shape of the wheel's grinding surface can be designed for effective application of the grinding fluid under high pressure.

The temperature of water-base grinding fluids can rise significantly during their use, because they remove significant heat from the grinding zone; otherwise the workpiece would expand, making it difficult to control dimensional tolerances. The common method employed to maintain even temperature is to circulate the fluid through a refrigerating system (chiller). As also described for cutting fluids, the biological and ecological aspects, treatment, recycling, and disposal of grinding fluids are among important considerations in their selection and use. Moreover, the practices employed must comply with federal, state, and local laws and regulations.

In **cryogenic grinding**, liquid nitrogen is used as a coolant, largely in the interest of reducing or eliminating the adverse environmental impact of the use of metalworking fluids (Section 8.7.2). With small-diameter nozzles, liquid nitrogen, at around −200°C, is injected into the wheel-workpiece interface thus reducing its temperature. Experimental studies have indicated that, as compared with the use of traditional grinding fluids, cryogenic grinding is associated with a reduced level of metallurgical burn and oxidation, improved surface finish, lower tensile residual surface stress levels, and less loading of the grinding wheel (hence less need for dressing). Cryogenic grinding appears to be particularly suitable for materials such as titanium, which has very low thermal conductivity, low specific heat, and high reactivity.

EXAMPLE 9.4 Grinding vs. Hard Turning

It is evident that grinding and hard turning (Section 8.9.2) can be competitive in specific applications. Hard turning continues to be increasingly competitive with grinding, and the dimensional tolerances and surface finish produced are approaching those for grinding. Furthermore, (a) turning requires much less energy than grinding (See Tables 8.3 and 9.3); (b) thermal and other damage to the workpiece surface is less likely to occur; (c) cutting fluids may not be necessary; and (d) the machine tools are less expensive.

In addition, finishing the part while it is still chucked in the lathe or turning center eliminates the need for material handling and setting the part in the grinder. On the other hand, workholding devices for large and slender workpieces for hard turning can present difficulties because the machining forces are higher than those in grinding and the workpiece will tend to distort. Tool wear and its control in hard turning also can be a significant problem as compared with automatic dressing of grinding wheels.

9.7 Finishing Operations

In addition to the abrasive machining processes described thus far, several processes can be used on workpieces as the final finishing operation, using mainly abrasive grains. However, finishing operations can contribute significantly to production time and product cost, hence they should be specified with due consideration to their economic drawbacks and performance benefits.

Commonly used finishing operations are described below.

1. **Coated abrasives.** Typical examples of *coated abrasives* are sandpaper and emery cloth. The grains in coated abrasives are more pointed than those used for grinding wheels; they are electrostatically deposited on flexible backing materials, such as paper or cloth (Fig. 9.19), with their long axes perpendicular to the plane of the backing. Coated abrasives are available as sheets, belts, and disks, and usually have a much more open structure than the abrasives used in bonded wheels. They are used extensively in finishing flat or curved surfaces of metallic and nonmetallic parts, in finishing metallographic specimens, and in woodworking. The surface finish obtained depends primarily on grain size.

   Coated abrasives are also used as *belts* for high-rate material removal. **Belt grinding** is an important production process and in some cases has replaced conventional grinding operations, such as in

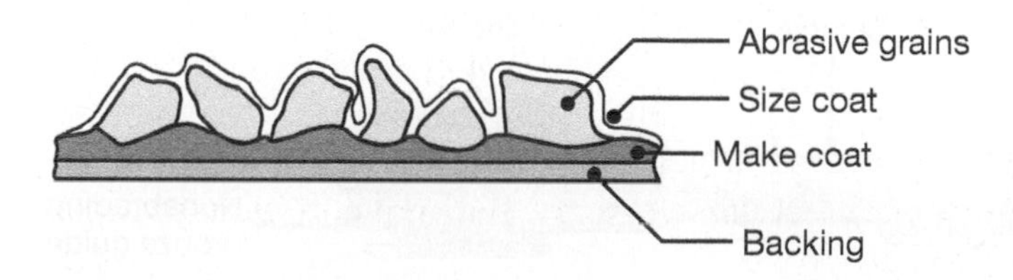

FIGURE 9.19 Schematic illustration of the structure of a coated abrasive. Sandpaper, developed in the 16th century, and emery cloth are common examples of coated abrasives.

grinding camshafts, with 8 to 16 lobes (of approximately elliptic shape) per shaft. Belt speeds are usually in the range of 700–1800 m/min (2300–6000 ft/min). Machines for abrasive-belt operations require proper belt support and rigid construction to minimize vibrations.

In **microreplication**, aluminum-oxide abrasives in the shape of tiny pyramids are placed in a predetermined orderly arrangement on the belt surface. When used on stainless steels and superalloys, their performance is more consistent, and the temperature rise is lower than when using other coated abrasives. Typical applications include surgical implants, turbine blades, and medical and dental instruments.

2. **Wire brushing.** In this process, the workpiece is held against a circular wire brush rotating at high speed, producing longitudinal scratches on the workpiece surface. Wire brushing is used to produce a fine surface texture, but can also serve as a light material-removal process.
3. **Honing.** This is an operation used primarily to produce fine surface finish in holes. The honing tool (Fig. 9.20) consists of a set of aluminum-oxide or silicon-carbide bonded abrasives, called *stones*. The stones are mounted on a mandrel that rotates in the hole, applying a radial force, with a reciprocating axial motion, thus producing a crosshatched pattern. The stones can be adjusted radially for different hole sizes. The surface finish can be controlled by the type and size of abrasive used, the speed of rotation, and the pressure applied. A fluid is used to remove chips and to keep temperatures low. If not implemented properly, honing can produce holes that may not be straight and cylindrical, but rather bellmouthed, wavy, barrel shaped, or tapered. Honing is also used on external cylindrical or flat surfaces, and to remove sharp edges on cutting tools and inserts (See Fig. 8.32).

   In **superfinishing**, the pressure applied is very low, and the motion of the honing stone has a short stroke. The operation is controlled such that the grains do not travel along the same path across the surface of the workpiece. Figure 9.21 shows examples of external superfinishing of a round part.
4. **Electrochemical honing.** This process combines the fine abrasive action of honing with electrochemical action. Although the equipment is costly, the process is as much as 5 times faster than conventional honing, and the tool lasts up to 10 times longer. Electrochemical honing is used primarily for finishing internal cylindrical surfaces.
5. **Lapping.** *Lapping* is a finishing operation on flat or cylindrical surfaces. The *lap* (Fig. 9.22a) is usually made of cast iron, copper, leather,

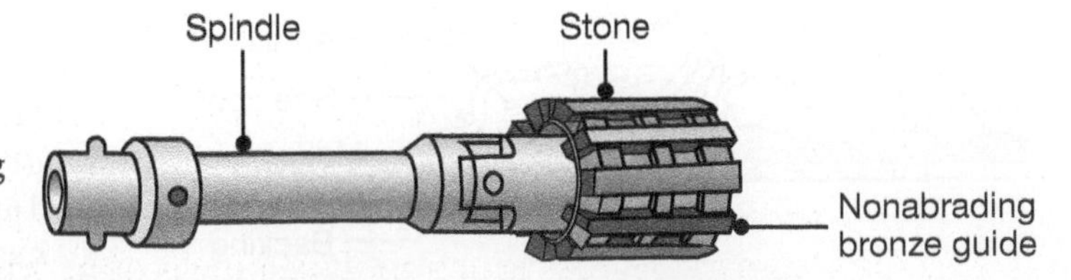

FIGURE 9.20 Schematic illustration of a honing tool to improve the surface finish of bored or ground holes.

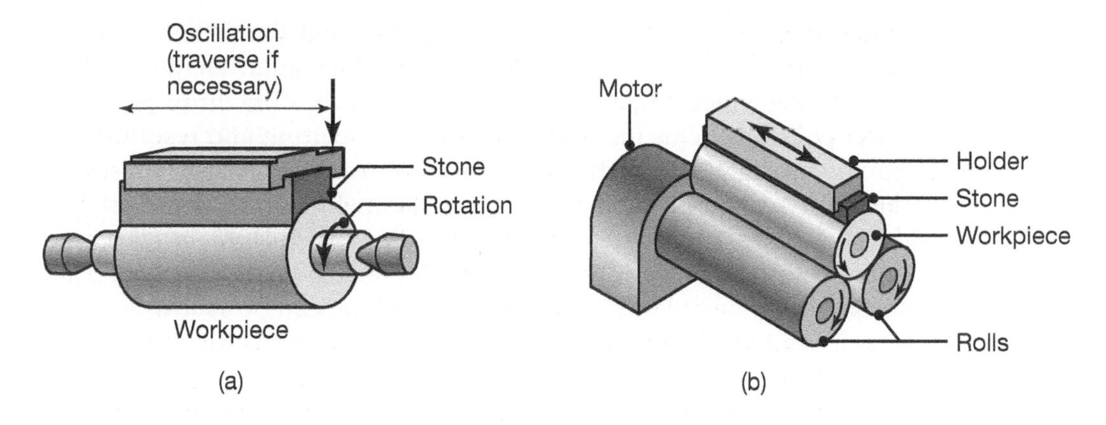

FIGURE 9.21 Schematic illustration of the superfinishing process for a cylindrical part: (a) cylindrical microhoning; and (b) centerless microhoning.

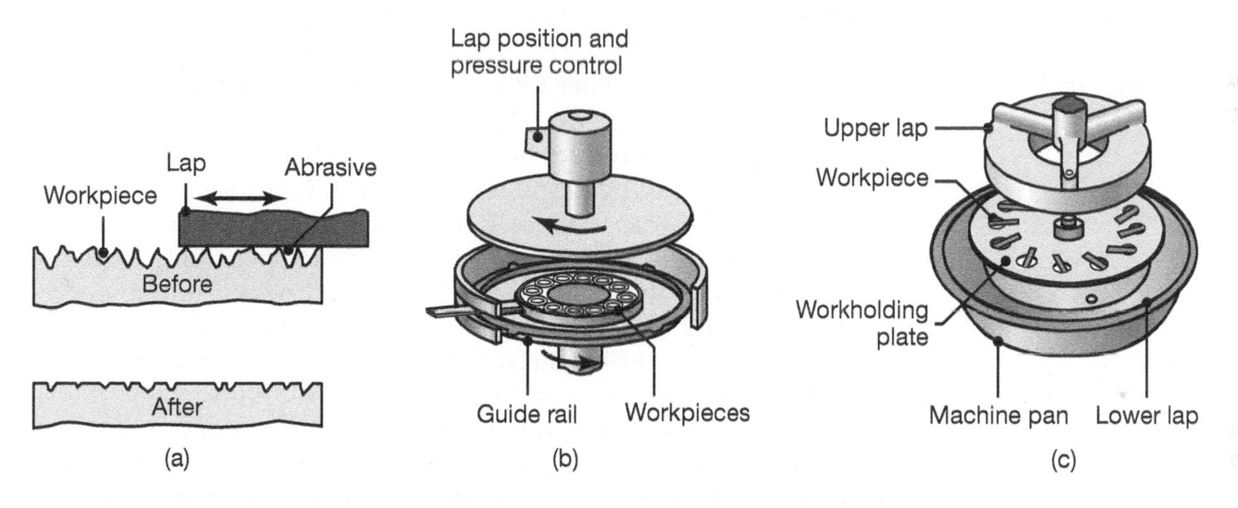

FIGURE 9.22 (a) Schematic illustration of the lapping process. (b) Production lapping on flat surfaces. (c) Production lapping on cylindrical surfaces.

or cloth. The abrasive particles are embedded in the lap or they may be carried through a slurry. Depending on workpiece hardness, lapping pressures range from 7–140 kPa (1–20 psi). Dimensional tolerances on the order of ± 0.4 μm can be obtained with the use of fine abrasives, up to size 900. Surface finish can be as smooth as 0.025–0.1 μm. Production lapping on flat or cylindrical workpieces is done on machines such as those shown in Figs. 9.22b and c. Lapping is also performed on curved surfaces, such as spherical objects and glass lenses, using specially shaped laps. *Running-in* of mating gears also can be done by lapping.

6. **Polishing.** This is a process that produces a smooth, lustrous surface finish. Two basic mechanisms are involved in the polishing process: fine-scale abrasive removal and softening and smearing of surface layers by frictional heating. The shiny appearance of polished surfaces results from the smearing action. Polishing is done with disks or belts,

made of fabric, leather, or felt, that are coated with fine powders of aluminum oxide or diamond. Parts with irregular shapes, sharp corners, deep recesses, and sharp projections are difficult to polish.

7. **Laser polishing.** This method involves rapid melting and resolidification of a surface, at depths of submicrons, using short laser pulses in the range of micro- or nanoseconds. The melting action smoothens the surface by reducing asperity heights (See Fig. 4.4), on ferrous and nonferrous metallic workpieces as well as on glass and diamond. The surface roughness developed by laser polishing can be as low as 0.05 μm (2 μin.) and is suitable for optical use. It also has lower friction, thus making it attractive for such applications as automotive cylinder liners. Unlike traditional polishing operations, this process is much faster and also is applicable to workpieces with uneven surfaces, using programmable controls.
8. **Buffing.** This process is similar to polishing, with the exception that very fine abrasives are used on soft disks typically made of cloth. The abrasive is supplied externally from a stick of abrasive compound. Buffing may be done on polished parts to obtain an even finer surface finish.
9. **Electropolishing.** Mirrorlike finishes can be obtained on metal surfaces by *electropolishing*, a process that is the reverse of electroplating (Section 4.5.1). Because there is no mechanical contact with the workpiece, this process is particularly suitable for polishing irregular shapes. The electrolyte attacks projections and peaks on the workpiece surface and at a higher rate than for the rest of the surface, thus producing a smooth surface. Electropolishing is also used for deburring operations (Section 9.8).
10. **Chemical-mechanical polishing.** Chemical-mechanical polishing (CMP) is very important in the semiconductor industry. The process, shown in Fig. 9.23, uses a suspension of abrasive particles in a water-base solution, with a chemistry selected to cause controlled corrosion. This process removes material from the workpiece surfaces through combined actions of abrasion and corrosion. The result is exceptionally fine surface finish and a workpiece that is essentially flat (plane); for this reason, the process is often referred to as **chemical-mechanical planarization.**

    A major application of this process is the polishing of silicon wafers (Section 13.4), in which the primary function of CMP is to polish a wafer at the micro-level. To evenly remove material across the whole wafer, it is held on a rotating carrier face down and is pressed against a polishing pad attached to a rotating disk (Fig. 9.23). Both the carrier and the pad rotate in order to avoid development of a linear lay (see Section 4.3). The pad contains grooves in order to supply slurry to the wafers.

    Specific abrasive and solution chemistry combinations have been developed for the polishing of copper, silicon, silicon dioxide, aluminum, tungsten, and other metals. For polishing silicon dioxide or silicon, for example, an alkaline slurry of colloidal silica (SiO_2)

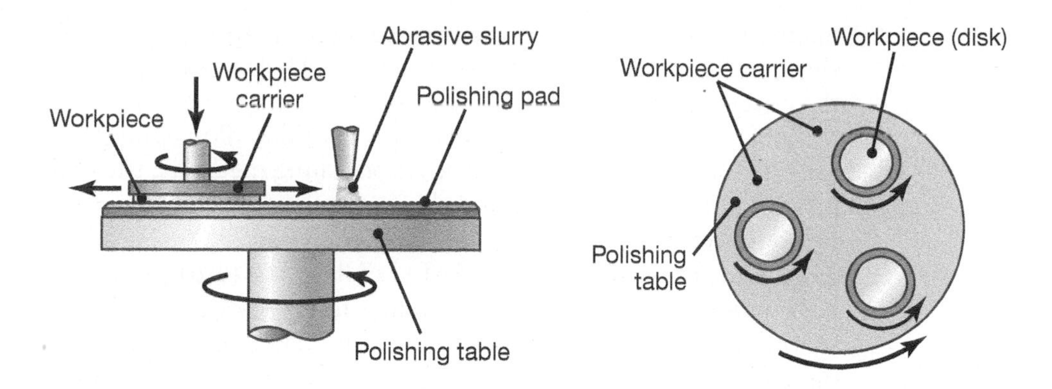

FIGURE 9.23 Schematic illustration of the chemical-mechanical polishing process. This process is widely used in the manufacture of silicon wafers and integrated circuits, where it is known as *chemical-mechanical planarization*. Additional carriers and more disks per carrier also are possible.

particles in a KOH solution or NH_4OH is continuously fed to the pad-wafer interface.

11. **Polishing processes using magnetic fields.** In this technique, an abrasive slurry is supported with and made more effective by magnetic fields. There are two basic methods:

    a. In **magnetic float polishing**, also called *magnetic fluid finishing* (Fig. 9.24a), a magnetic fluid fills the chamber within a guide ring. The magnetic fluid consists of abrasive grains and extremely fine ferromagnetic particles suspended in a carrier, such as water or kerosene. The ceramic balls are located between a drive shaft and a float. The abrasive grains, ceramic balls, and the float (made of a nonmagnetic material) are all suspended by

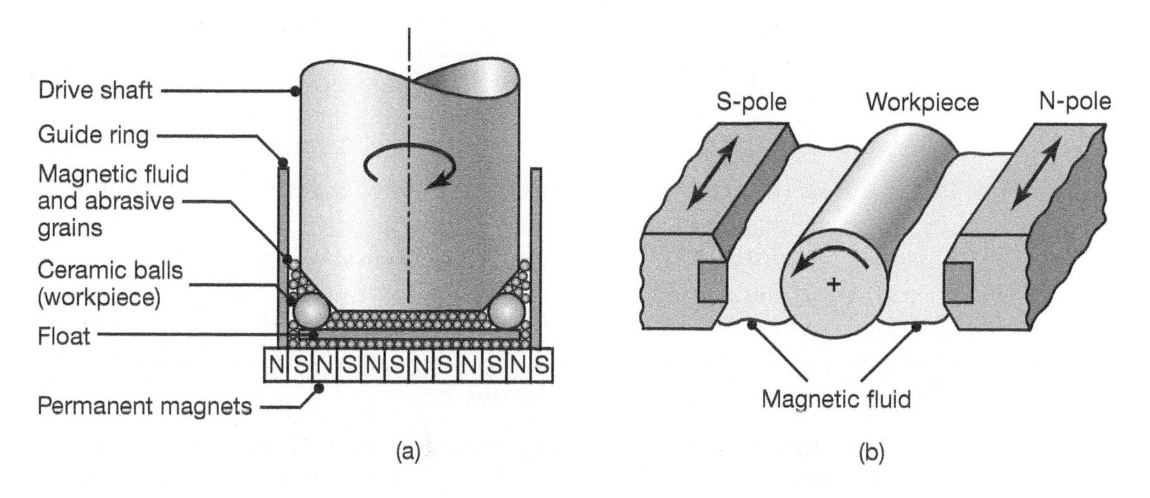

FIGURE 9.24 Schematic illustration of the use of magnetic fields to polish balls and rollers: (a) magnetic float polishing of ceramic balls and (b) magnetic-field-assisted polishing of rollers. *Source:* After R. Komanduri, M. Doc, and M. Fox.

magnetic forces. The balls are pressed against the rotating drive shaft and are polished by abrasive action. The forces applied by the abrasive particles on the balls are extremely small and controllable, hence the polishing action is very fine. Polishing times are much lower than those for other polishing methods, thus the process is very economical and the surfaces produced have little or no defects.

b. **Magnetic abrasive finishing** (MAF) involves fine (1 μm – 2 mm) ferromagnetic particles suspended in a fluid. A workpiece, such as the ceramic roller in Fig. 9.24b, is clamped and rotated on a spindle. The magnetic poles are oscillated, introducing a vibratory motion to the magnetic-abrasive conglomerate. This produces a so-called *magnetic brush* because of the similarities to the response from a wire brush. However, the magnetic brush is more compliant than wire brushes, and geometrically complex surfaces can be accommodated. Bearing steels of 63 HRC have been mirror finished in 30 s by this process.

c. **Magnetorheological finishing** (MRF) is similar to MAF, but involves the use of a fluid whose viscosity increases in the presence of a magnetic field. The fluid, referred to as a *ribbon*, is very viscous next to the workpiece, so that the abrasive particles are pushed against the surface with higher force.

9.8 Deburring

Burrs are thin ridges, usually triangular in shape, that develop along the edges of a workpiece from processes such as shearing of sheet materials (Fig. 7.5), trimming of forgings and castings, and machining operations. Burrs may interfere with the assembly of parts and can cause jamming, misalignment, and short circuiting of electrical components, and may reduce the fatigue life of a part (Section 7.3). Because they are usually sharp, burrs can also be a safety hazard to personnel. The need for deburring may be reduced by adding chamfers to sharp edges on parts.

Several *deburring processes* are available: (a) manually, using files (Section 8.10.6); (b) mechanically, by cutting; (c) wire brushing (Section 9.7); (d) abrasive finishing, using rotary nylon brushes, with the nylon filaments embedded with abrasive grits; (e) abrasive belts (Section 9.7); (f) ultrasonic machining (Section 9.9); (g) electropolishing (Section 9.7); (h) electrochemical machining (Section 9.11); (i) vibratory finishing; (j) shot blasting; (k) abrasive-flow machining; and (l) thermal energy. The last four processes are described below.

1. **Vibratory** and **barrel-finishing** processes are used to improve the surface finish and remove burrs from large numbers of relatively small parts. This is a batch-type operation in which specially-shaped *abrasive pellets* or *media* are placed in a container along with the parts to be deburred. The container is either vibrated or tumbled. The impact of individual abrasives removes sharp edges and burrs from the parts.

Depending on the application, this process is performed dry or wet, and liquid compounds may be added for other requirements, such as degreasing and providing resistance to corrosion.

2. In **shot blasting**, also called *grit blasting*, abrasive particles (usually sand) are propelled by a high-velocity jet of air, or by a rotating wheel, onto the surface of the workpiece. Shot blasting is particularly useful in deburring both metallic and nonmetallic materials and in stripping, cleaning, and removing surface oxides from workpieces. The surface produced by shot blasting has a matte finish. Small-scale polishing and etching also can be done by this process on bench-type units (*microabrasive blasting*).
3. In **abrasive-flow machining**, abrasive grains, such as silicon carbide or diamond, are mixed in a puttylike matrix, which is then forced back and forth through the openings and passageways in the workpiece. The movement of the abrasive matrix under pressure removes burrs and sharp corners and polishes the part. The process is particularly suitable for workpieces with internal cavities that are inaccessible by other means. Pressures applied range from 0.7 to 22 MPa (0.1–3.2 ksi). External surfaces also can be deburred using this process by containing the workpiece within a fixture that directs the abrasive media to the edges and areas to be deburred.
4. The **thermal-energy method** of deburring consists of placing the part in a chamber that is then injected with a mixture of natural gas and oxygen. This mixture is ignited, producing a heat wave with a temperature of 3300°C (6000°F). The burrs instantly heat up and are melted away, while the temperature of the part itself rises only to about 150°C (300°F). The thermal energy method is effective on a wide range of materials including zinc, aluminum, brass, steel, stainless steel, cast iron, and thermoplastics. However, larger burrs or flash from forgings and castings tend to form beads after melting, and the process can distort thin and slender parts. This method does not polish or buff the workpiece surfaces as do many of the other deburring processes.
5. **Robotic deburring.** Deburring and flash removal are now being performed increasingly by programmable robots (Section 14.7), using a force-feedback system for controlling the process. The main advantage of robotic deburring is flexibility, both in terms of the geometries that can be deburred as well as the media that can be applied. This method eliminates tedious and costly manual labor and results in more consistent deburring. The main drawback is the capital equipment cost involved – robotic deburring systems with multiple spindles can cost over $250,000.

9.9 Ultrasonic Machining

In *ultrasonic machining* (UM), material is removed from a workpiece surface by the mechanism of microchipping or erosion with abrasive particles. The tip of the tool (Fig. 9.25a), called a *sonotrode*, vibrates at amplitudes

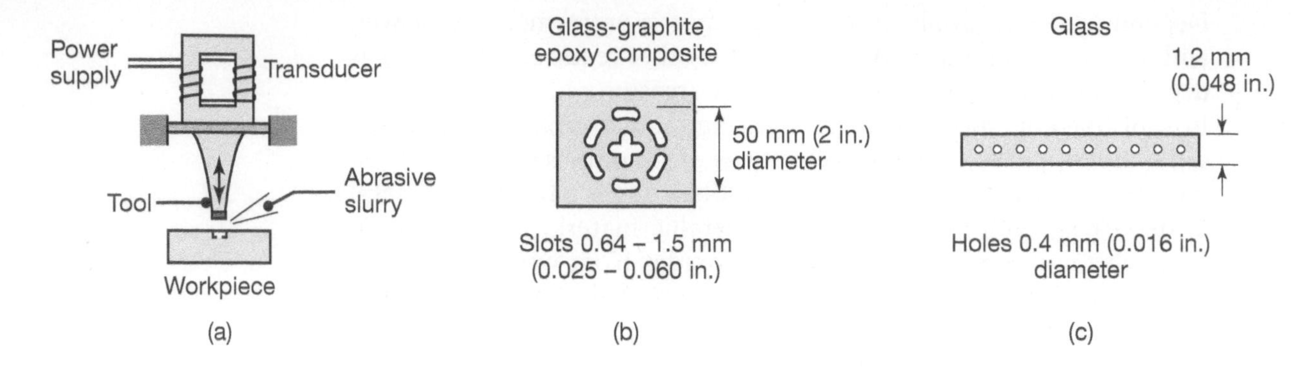

FIGURE 9.25 (a) Schematic illustration of the ultrasonic-machining process; material is removed through microchipping and erosion. (b) and (c) Typical examples of cavities produced by ultrasonic machining. Note the dimensions of cut and the types of workpiece materials.

of 0.05 to 0.125 mm (0.002–0.005 in.) and at a frequency of 20 kHz. The vibration, in turn, transmits a high velocity to fine abrasive grains between the tool and the surface of the workpiece. The grains are in a water slurry with concentrations ranging from 20 to 60% by volume; the slurry also carries away the debris from the cutting area. Although the grains are usually boron carbide, aluminum oxide and silicon carbide are also used, with grain sizes ranging from 100 (for roughing) to 1000 (for finishing).

Ultrasonic machining is best suited for hard and brittle materials, such as ceramics, carbides, glass, precious stones, and hardened steels. The tip of the tool is usually made of low-carbon steel and is attached to a transducer through the toolholder. With fine abrasives, dimensional tolerances of 0.0125 mm (0.0005 in.) or better can be held in this process. Figures 9.25b and c show two applications of ultrasonic machining.

Microchipping in ultrasonic machining is possible because of the high stresses produced by particles striking a solid surface. The contact time between the particle and the surface is very short (10 to 100 μs) and the area of contact is very small. The contact time, t_o, can be expressed as

$$t_o \simeq \frac{5r}{c_o}\left(\frac{c_o}{v}\right)^{1/5}, \tag{9.11}$$

where r is the radius of a spherical particle, c_o is the elastic wave velocity in the workpiece $\left(c_o = \sqrt{E/\rho}\right)$, and v is the velocity with which the particle strikes the surface. The force, F, of the particle on the surface is obtained from the rate of change of momentum; that is,

$$F = \frac{d(mv)}{dt}, \tag{9.12}$$

where m is the mass of the particle. The *average force*, F_{ave}, of a particle striking the surface and rebounding is

$$F_{\text{ave}} = \frac{2mv}{t_o}. \tag{9.13}$$

Substitution of numerical values into Eq. (9.13) indicates that even small particles can exert significant forces and, because of the very small contact area, the contact stresses are very high. In brittle materials, these stresses are sufficiently high to cause microchipping and surface erosion (see also the discussion of *abrasive-jet machining* in Section 9.15).

Rotary ultrasonic machining (RUM). In this process, the abrasive slurry is replaced by a tool with metal-bonded diamond abrasives that have been either impregnated or electroplated on the tool surface. The tool is rotated and ultrasonically vibrated, and the workpiece is pressed against it at a constant pressure; the operation is thus similar to a face milling operation (Section 8.10.1). The rotary ultrasonic machining process is particularly effective in producing deep holes in ceramics at high material removal rates.

9.10 Chemical Machining

The *chemical machining* (CM) process was developed based on the fact that certain chemicals attack metals and etch them, thereby removing small amounts of material from the surface (Table 9.4). The material is removed from a surface by chemical dissolution, using **reagents**, or **etchants**, such as acids and alkaline solutions. Chemical machining is the oldest of the nontraditional machining processes, and has been used for many years for engraving metals and hard stones and in the production of printed-circuit boards and microprocessor chips. Parts can also be deburred by chemical means.

9.10.1 Chemical Milling

In *chemical milling*, shallow cavities are produced on sheets, plates, forgings, and extrusions, either for design requirements or for weight reduction in parts (Fig. 9.26). Chemical milling has been used on a wide variety of metals, with depths of material removal to as much as 12 mm (0.5 in.). Selective attack by the chemical reagent on different areas of the workpiece surfaces is controlled by removable layers of a **maskant** (Fig. 9.27) or by partial immersion in the reagent.

Chemical milling is used in the aerospace industry, particularly for removing shallow layers of material from large aircraft, missile skin panels, and extruded parts for airframes. The process is also used to fabricate microelectronic devices (Sections 13.8 and 13.14), often referred to as **wet etching**. Tank capacities for reagents are as large as 3.7 m × 15 m (12 × 50 ft). Figure 9.28 shows the range of surface finish and dimensional tolerances obtained by chemical machining and other machining processes. Some surface damage may result from chemical milling, because of preferential etching and intergranular attack by the reagents (see Section 3.4), which adversely affect surface properties. Chemical milling of welded and brazed structures may produce uneven material removal because the filler metal or the structure material may machine preferentially.

TABLE 9.4 General characteristics of advanced machining processes.

Process	Characteristics	Process parameters and typical material removal rate or cutting speed
Chemical machining (CM)	Shallow removal (up to 12 mm) on large flat or curved surfaces; blanking of thin sheets; low tooling and equipment cost; suitable for low production runs	0.0025–0.1 mm/min (0.0001–0.004 in./min)
Electrochemical machining (ECM)	Complex shapes with deep cavities; highest rate of material removal; expensive tooling and equipment; high power consumption; medium to high production quantity	V: 5–25 dc; I: 1.5–8 A/mm^2; 2.5–12 mm/min (0.1–0.5 in./min), depending on current density
Electrochemical grinding (ECG)	Cutting off and sharpening hard materials, such as tungsten-carbide tools; also used as a honing process; higher material removal rate than grinding	I: 1–3 A/mm^2; typically 25 mm^3/s (0.0016 in^3/s) per 1000 A
Electrical-discharge machining (EDM)	Shaping and cutting complex parts made of hard materials; some surface damage may result; also used for grinding and cutting; versatile; expensive tooling and equipment	V: 50–380; I: 0.1–500 A; typically 300 mm^3/min (0.02 in^3/min)
Wire EDM	Contour cutting of flat or curved surfaces; expensive equipment	Varies with workpiece material and its thickness
Laser-beam machining (LBM)	Cutting and hole making on thin materials; heat-affected zone; does not require a vacuum; expensive equipment; consumes much energy; extreme caution required in use	0.50–7.5 m/min (1.67–25 ft/min)
Electron-beam machining (EBM)	Cutting and hole making on thin materials; very small holes and slots; heat-affected zone; requires a vacuum; expensive equipment	1–2 mm^3/min (0.004–0.008 in^3/h)
Water-jet machining (WJM)	Cutting all types of nonmetallic materials to 25 mm (1 in.) and greater in thickness; suitable for contour cutting of flexible materials; no thermal damage; environmentally safe process	Varies considerably with workpiece material
Abrasive water-jet machining (AWJM)	Single or multilayer cutting of metallic and nonmetallic materials	Up to 7.5 m/min (25 ft/min)
Abrasive-jet machining (AJM)	Cutting, slotting, deburring, flash removal, etching, and cleaning of metallic and nonmetallic materials; tends to round off sharp edges; some hazard because of airborne particulates	Varies considerably with workpiece material
Laser microjet	Water-jet guided laser uses a 25–100 μm diameter stream to mill or cut; large depth of field; little thermal damage from laser machining	Varies with material; up to 20 mm in silicon, 2 mm in stainless steel; up to 300 mm/s in 50-μm thick silicon

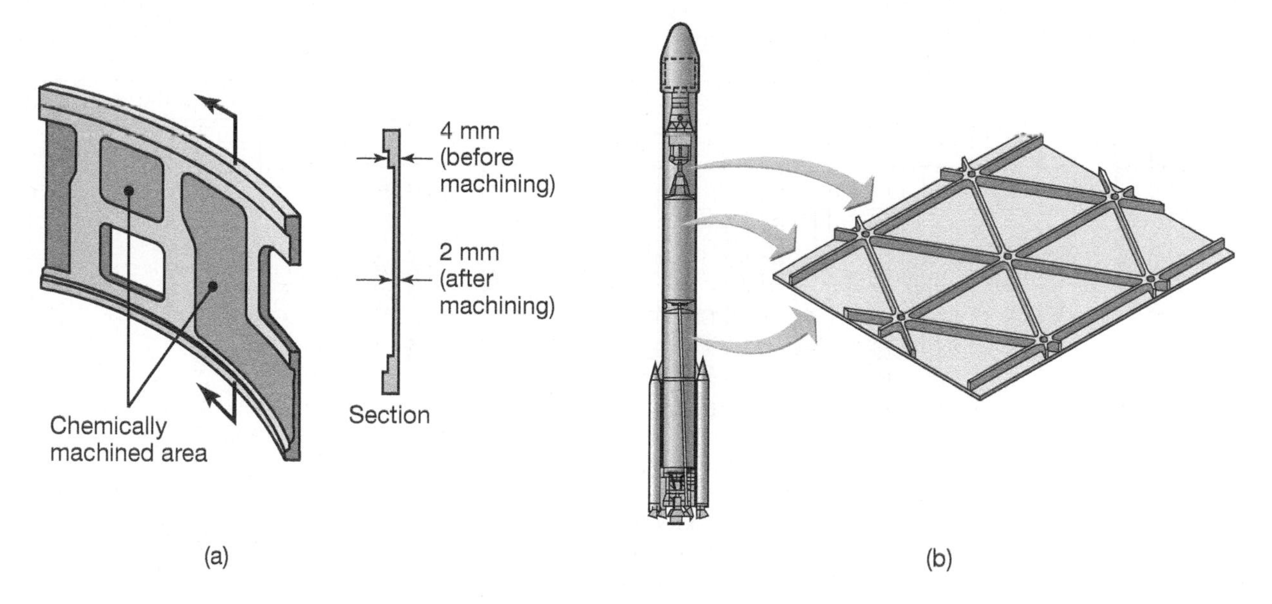

FIGURE 9.26 (a) Missile skin-panel section contoured by chemical milling to improve the stiffness-to-weight ratio of the part. (b) Weight reduction of space launch vehicles by chemical milling of aluminum-alloy plates. These panels are chemically milled after the plates have first been formed into shape, such as by roll forming or stretch forming. *Source:* After ASM International.

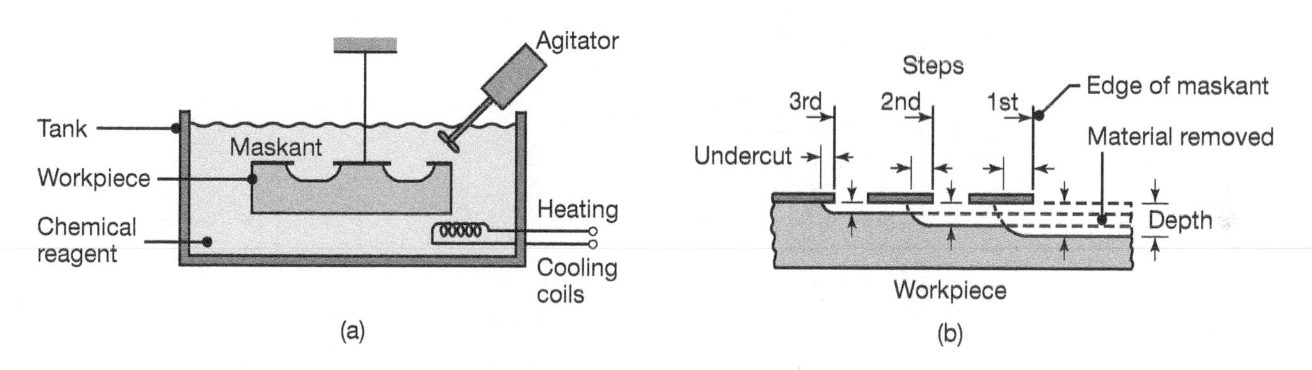

FIGURE 9.27 (a) Schematic illustration of the chemical machining process. Note that no forces are involved in this process. (b) Stages in producing a profiled cavity by chemical machining.

Also, chemical milling of castings may result in uneven surfaces caused by porosity and nonuniformity of the structure.

9.10.2 Chemical Blanking

Chemical blanking is similar to blanking of sheet metal (See Fig. 7.8), in that it is used to produce features that penetrate through the thickness of the material; the material is removed by chemical dissolution. Typical applications include burr-free etching of printed-circuit boards (Fig. 13.33) and the production of decorative panels, thin sheet metal stampings, and small or complex shapes.

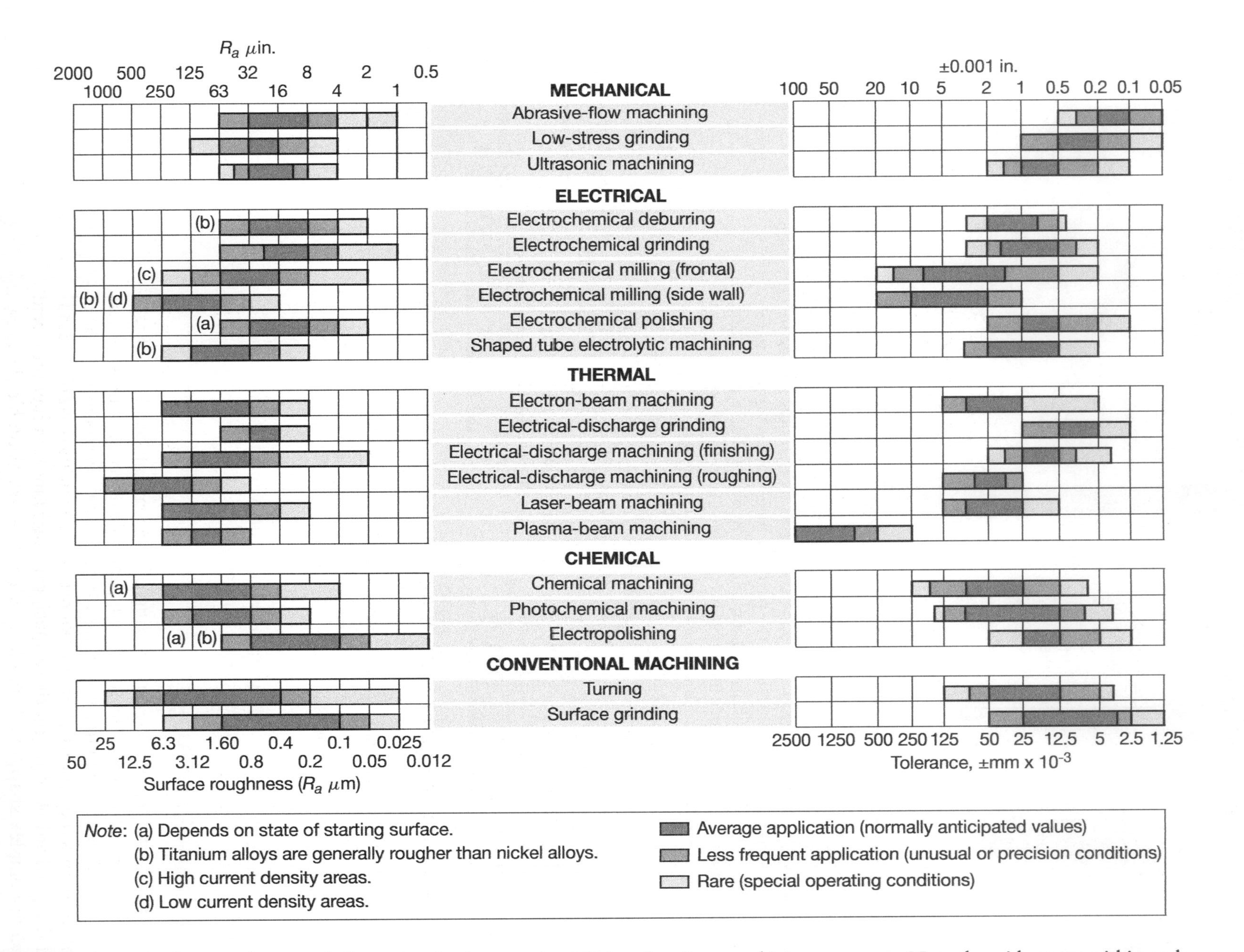

FIGURE 9.28 Surface roughness and dimensional tolerance capabilities of various machining processes. Note the wide range within each process. (See also Fig. 8.25.) *Source: After Machining Data Handbook*, 3rd ed., ©1980. Used by permission of Metcut Research Associates, Inc.

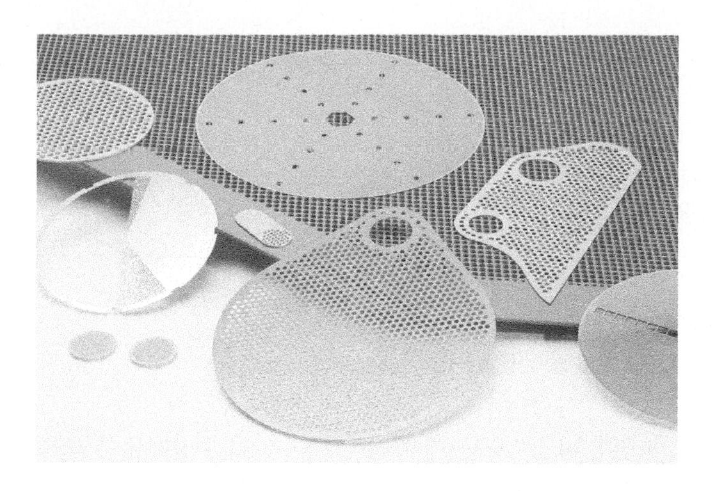

FIGURE 9.29 Typical parts made by chemical blanking; note the fine detail. *Source:* Courtesy of VACCO Industries.

9.10.3 Photochemical Blanking

Also called *photoetching*, *photochemical blanking* is a modification of chemical milling; material is removed by photographic techniques, usually from flat thin sheet. Complex burr-free shapes can be blanked (Fig. 9.29) on metals as thin as 0.0025 mm; the process is also used for etching (see Section 13.8). Typical applications for photochemical blanking include fine screens, printed-circuit boards, electric-motor laminations, and flat springs.

Photochemical blanking is capable of forming very small parts for which traditional blanking dies (Section 7.3) are difficult to make, and the process is effective for blanking fragile workpieces and materials. Handling of chemical reagents requires special safety precautions to protect personnel. Disposal of chemical by-products from this process also is a major consideration, although some by-products can be recycled. Although skilled labor is required, tooling costs are low, the process can be automated, and it is economical for medium to high production volume.

9.11 Electrochemical Machining

Electrochemical machining (ECM) is basically the reverse of electroplating. An **electrolyte** (Fig. 9.30) acts as a current carrier, and the high rate of electrolyte movement in the tool-workpiece gap washes metal ions away from the workpiece (*anode*) before they have a chance to plate onto the tool (*cathode*). Note that the cavity produced is the female mating image of the tool. Modifications of this process are used for operations such as turning, facing, slotting, trepanning, and profiling in which the electrode becomes the cutting tool.

The shaped tool is generally made of brass, copper, bronze, or stainless steel. The electrolyte is a highly conductive inorganic salt solution, such as sodium chloride mixed in water or sodium nitrate, and is pumped at a high rate through the passages in the tool. A dc power supply in the range of 5–25 V maintains current densities, which for most applications are 1.5–8 A/mm^2 of active machined surface. Machines that have current capacities as high as 40,000 A and as small as 5 A are available. Because

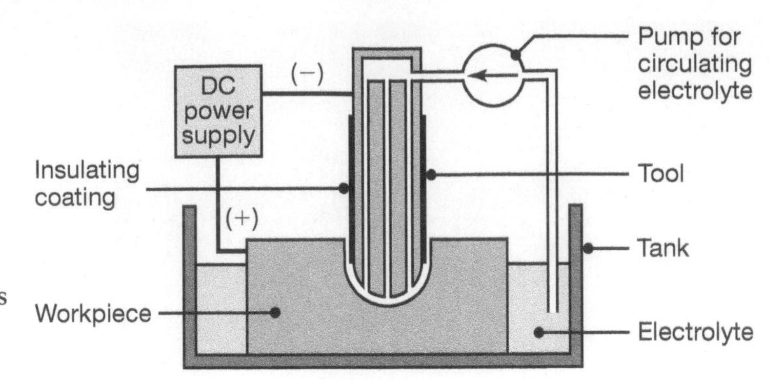

FIGURE 9.30 Schematic illustration of the electrochemical-machining process. This process is the reverse of electroplating, described in Section 4.5.1.

the metal removal rate is only a function of ion exchange rate, it is not affected by the strength, hardness, or toughness of the workpiece (which must be electrically conductive).

The material removal rate by electrochemical machining can be calculated from the equation

$$\text{MRR} = CI\eta, \tag{9.14}$$

where MRR = mm^3/min; I = current in amperes; and η = current efficiency, which typically ranges from 90 to 100%. C is a material constant in mm^3/A-min and, for pure metals, depends on valence; the higher the valence, the lower its value. For most metals, the value of C typically ranges between about 1 and 2. If a cavity of uniform cross-sectional area A_o (in mm^2) is being electrochemically machined, the feed rate, f, in mm/min would be

$$f = \frac{\text{MRR}}{A_o}. \tag{9.15}$$

Note that the feed rate is the speed at which the electrode penetrates the workpiece.

Electrochemical machining is generally used for machining complex cavities in high-strength materials, particularly in the aerospace industry for mass production of turbine blades, jet-engine parts, and nozzles (Fig. 9.31). The process is also used for machining forging-die cavities (*die sinking*) and producing small holes. The ECM process leaves a burr-free surface; in fact, it can also be used as a deburring process. It does not cause any thermal damage to the part, and the lack of tool forces prevents distortion of the part, as would be the case with typical machining operations. Furthermore, there is no tool wear, and the process is capable of producing complex shapes as well as machining hard materials. However, the mechanical properties of components made by ECM should be compared with those produced by other machining methods.

A modification of ECM is the **shaped-tube electrolytic machining** (STEM) process, typically used for drilling small-diameter deep holes, as in turbine blades. The tool is a titanium tube, coated with an electrically insulating resin to prevent removal of material from other regions. Holes as small as 0.5 mm can be drilled, at depth-to-diameter ratios of as high as 300:1.

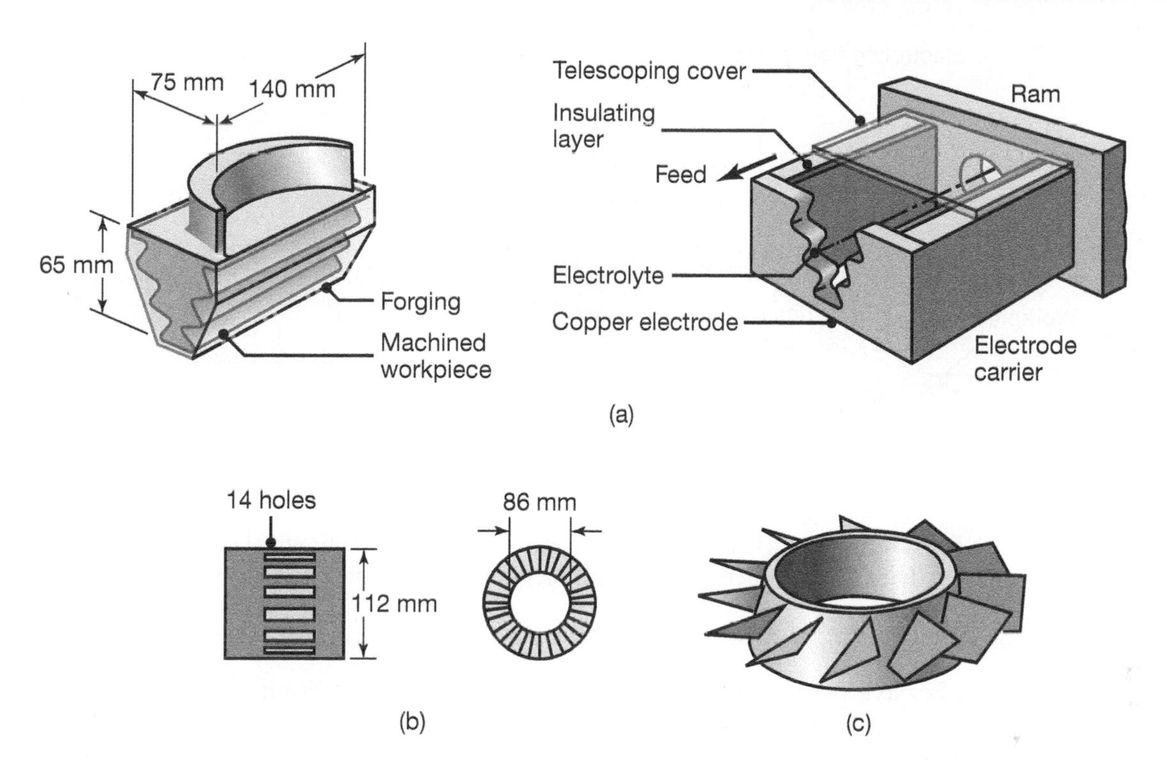

FIGURE 9.31 Typical parts made by electrochemical machining. (a) Turbine blade made of a nickel alloy, 360 HB; the part on the right is the shaped electrode. *Source:* ASM International. (b) Thin slots on a 4340-steel roller-bearing cage. (c) Integral airfoils on a compressor disk.

Electrochemical machining systems are available as *numerically-controlled machining centers*, with the capability of high production rates, high flexibility, and the maintenance of close dimensional tolerances. The ECM process can also be combined with electrical-discharge machining (EDM; Section 9.13) on the same machine. In *hybrid machining systems* ECM and other advanced machining processes are combined to take advantage of the capabilities of each process.

Pulsed electrochemical machining (PECM) is a refinement of ECM. It uses very high current densities (on the order of 100 A/cm^2), but the current is pulsed rather than direct. The purpose of pulsing is to eliminate the need for high electrolyte flow rates, which limits the usefulness of ECM in die and mold making. PECM improves fatigue life over ECM, and the process can be used to eliminate the recast layer left on die and mold surfaces by electrical-discharge machining (Section 9.13).

9.12 Electrochemical Grinding

Electrochemical grinding (ECG) combines electrochemical machining and conventional grinding (Table 9.4), on an equipment that is similar to a conventional grinder, except that the wheel is a rotating cathode, with

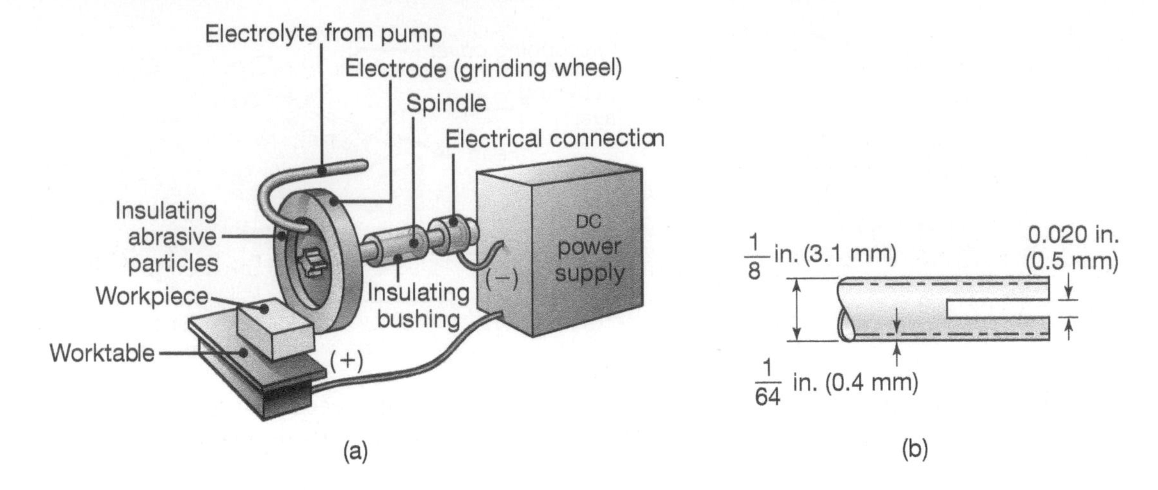

FIGURE 9.32 (a) Schematic illustration of the electrochemical grinding process. (b) Thin slot produced on a round nickel-alloy tube by this process.

abrasive particles (Fig. 9.32a). The wheel is metal bonded with diamond or aluminum-oxide abrasives, and rotates at a surface speed of 1200–2000 m/min (4000–6600 ft/min). Current densities range from 1–3 A/mm^2. The abrasives serve as insulators between the grinding wheel and the workpiece, and mechanically remove electrolytic products from the working area. A flow of electrolyte, usually sodium nitrate, is provided for the electrochemical-machining phase of the operation. The majority of metal removal in ECG is by electrolytic action, and typically less than 5% of metal is removed by the abrasive action of the wheel; consequently, wheel wear is very low. Finishing cuts are usually made by the grinding action, but only to produce good surface finish and dimensional accuracy.

The material removal rate in electrochemical grinding can be calculated from the equation

$$\text{MRR} = \frac{GI}{\rho F}, \tag{9.16}$$

where MRR is in mm^3/min, G = mass in grams, I = current in amperes, ρ = density in g/mm^3, and F = Faraday's constant in Coulombs. The speed of penetration, V_s, of the grinding wheel into the workpiece is given by the equation

$$V_s = \left(\frac{G}{\rho F}\right)\left(\frac{E}{gK_p}\right)K, \tag{9.17}$$

where V_s is in mm^3/min; E = cell voltage in volts; g = wheel-workpiece gap in mm; K_p = coefficient of loss, which is in the range of 1.5 to 3; and K = electrolyte conductivity in Ω^{-1} mm^{-1}.

Electrochemical grinding is suitable for applications similar to those for milling, grinding, and sawing (Fig. 9.32b), and has been successfully applied to carbides and high-strength alloys. It is not adaptable to cavity-sinking operations, such as die making, because deep cavities are not accessible for grinding. The process offers an advantage over traditional diamond-wheel grinding when processing very hard materials, for which wheel wear can be high. ECG machines are available with numerical controls, further improving dimensional accuracy and providing repeatability and increased productivity.

EXAMPLE 9.5 Machining Time in Electrochemical Machining vs. Drilling

Given: A round hole 12.5 mm in diameter is being produced in a titanium-alloy block by electrochemical machining.

Find: Using a current density of 6 A/mm^2, estimate the time required for machining a 20-mm deep hole. Assume that the efficiency is 90%. Compare this time with that required for ordinary drilling.

Solution: Note from Eqs. (9.14) and (9.15) that the feed rate can be expressed by the equation

$$f = \frac{CI\eta}{A_o}.$$

Letting $C = 1.6$ mm^3/A-min and $I/A_o = 6$ A/mm^2, the feed rate is $f = (1.6)(6)(0.9) = 8.64$ mm/min. Since the hole is 20 mm deep,

$$\text{Machining time} = \frac{20}{8.64} = 2.3 \text{ min}.$$

To determine the drilling time, refer to Table 8.11 and note the data for titanium alloys. Selecting the following values for a 12.5-mm drill: rpm = 300 and feed = 0.15 mm/rev. It can be seen that the feed rate is (300 rev/min)(0.15 mm/rev) = 45 mm/min. Since the hole is 20 mm deep,

$$\text{Drilling time} = \frac{20}{45} = 0.45 \text{ min},$$

which is about 20% of the time required for ECM.

9.13 Electrical-Discharge Machining

The principle of *electrical-discharge machining* (EDM), also called *electrodischarge* or *spark-erosion machining*, is based on erosion of metals by spark discharges (Table 9.4). When two current-conducting wires are allowed to touch each other, an arc is produced, eroding away a small portion of the metal and leaving a small crater. The EDM system (Fig. 9.33) consists of a shaped tool (**electrode**) and the workpiece, which are connected to a DC power supply and placed in a **dielectric fluid**.

When a voltage is applied to the tool, a magnetic field causes the suspended particles in the dielectric fluid to concentrate between the electrode and workpiece, eventually forming a bridge for the current to flow to the workpiece. An intense electrical arc is then generated, causing sufficient

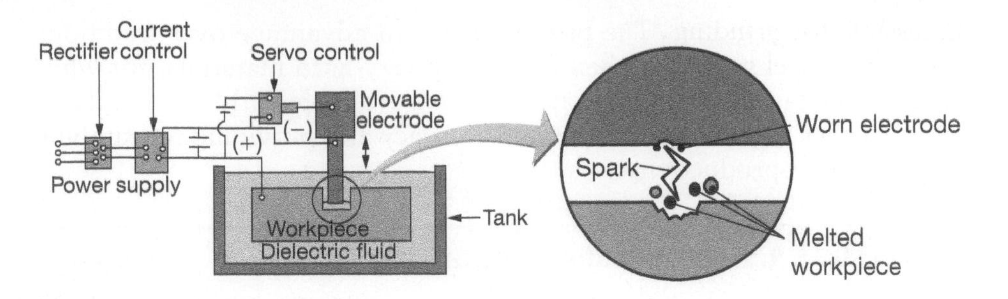

FIGURE 9.33 Schematic illustration of the electrical-discharge-machining process.

heating to melt a portion of the workpiece and, usually, some of the tooling material. The dielectric fluid is heated rapidly, causing evaporation of the fluid in the arc gap. This evaporation, in turn, increases the electrical resistance of the interface, until the arc can no longer be maintained. Once the arc is interrupted, heat is removed from the gas bubble by the surrounding dielectric fluid, and the bubble collapses (*cavitates*). The associated shock wave and flow of dielectric fluid flush debris from the workpiece surface and entrain any molten workpiece material into the dielectric fluid. The capacitor discharge is repeated at rates of between 50 and 500 kHz, with voltages usually ranging between 50 and 380 V and currents from 0.1 to 500 A.

QR Code 9.3 Ram EDM. *Source:* Courtesy of Makino, Inc.

The dielectric fluid (a) acts as an insulator until the electrical potential is sufficiently high; (b) acts as a flushing medium and carries away the debris in the gap; and (c) provides a cooling medium. The gap between the tool and the workpiece (*overcut*) is critical; thus, the downward feed of the tool is controlled by a servomechanism, which automatically maintains a constant gap. The most common dielectric fluids are mineral oils, although kerosene and distilled and deionized water may be used in specialized applications. The machines are equipped with a pump and filtering system for the dielectric fluid.

QR Code 9.4 EDM drilling. *Source:* Courtesy of Makino, Inc.

The EDM process can be used on any material that is an electrical conductor. The melting point and latent heat of melting are important physical properties that determine the volume of metal removed per discharge. As these values increase, the rate of material removal slows. The volume of material removed per discharge is typically in the range of 10^{-6} to 10^{-4} mm^3. Because the process does not involve mechanical energy, the hardness, strength, and toughness of the workpiece material do not necessarily influence the removal rate. The frequency of discharge or the energy per discharge is usually varied to control the removal rate, as are the voltage and current. The rate and surface roughness increase with increasing current density and decreasing frequency of sparks.

Electrodes for EDM are typically made of graphite, although brass, copper, or copper-tungsten alloy may be used. The tools are shaped by forming, casting, powder metallurgy, or machining. Electrodes as small as 0.1 mm in diameter are used. *Tool wear* is an important factor because it adversely affects dimensional accuracy and the shape produced. Tool wear

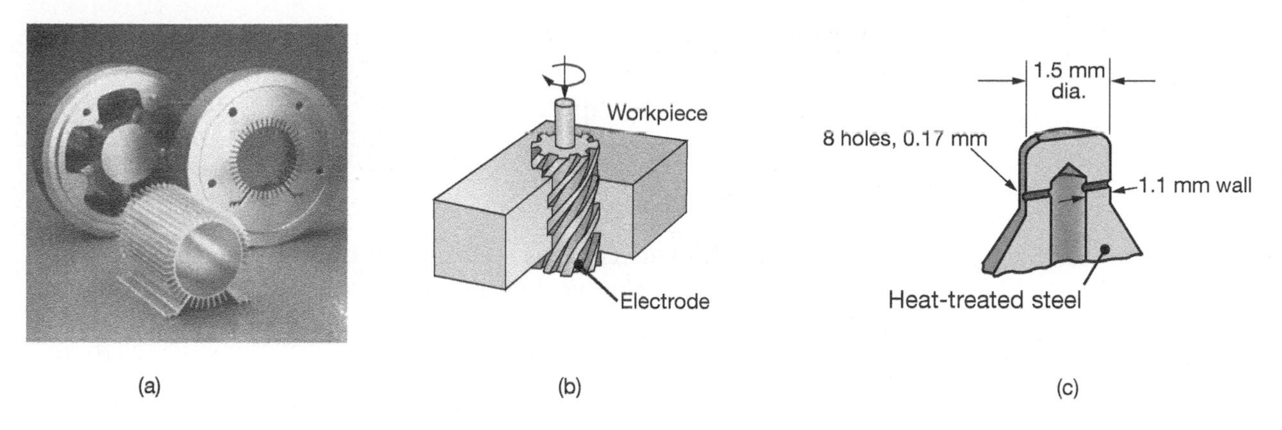

FIGURE 9.34 (a) Examples of shapes produced by the electrical-discharge machining process, using shaped electrodes. The two round parts in the rear are a set of dies for extruding the aluminum piece shown in front; see also Section 6.4. *Source:* Courtesy of GF Machining Solutions. (b) A spiral cavity produced using a shaped rotating electrode. *Source:* American Machinist. (c) Holes in a fuel-injection nozzle produced by electrical-discharge machining.

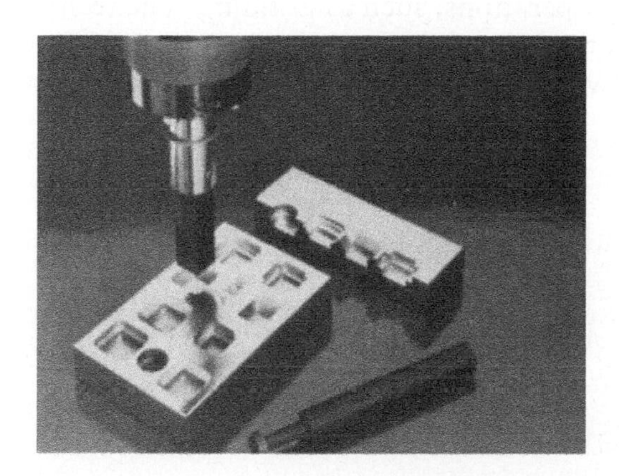

FIGURE 9.35 Stepped cavities produced with a square electrode by EDM. In this operation, the workpiece moves in the two principal horizontal directions, and its motion is synchronized with the downward movement of the electrode to produce these cavities. Also shown is a round electrode capable of producing round or elliptical cavities. *Source:* Courtesy of GF Machining Solutions.

can be minimized by reversing the polarity and using copper tools, a process called **no-wear EDM**.

Electrical-discharge machining has numerous applications, such as producing die cavities for large automotive-body components (*die-sinking machining centers*); narrow slots; turbine blades; various intricate shapes (Figs. 9.34a and b); and small-diameter deep holes (Fig. 9.34c), using tungsten wire as the electrode. Stepped cavities can be produced by controlling the relative movements of the workpiece in relation to the electrode (Fig. 9.35).

The material removal rate in electrical-discharge machining is basically a function of the current and the melting point of the workpiece material, although other process variables, such as temperature and current frequency also have an effect. The metal removal rate in EDM is given by

$$\text{MRR} = 4 \times 10^4 I T_w^{-1.23}, \quad (9.18)$$

where MRR is in mm^3/min, I is the current in amperes, and T_w is the melting point of the workpiece in °C. The wear rate of the electrode, W_t, can be estimated from the empirical equation

$$W_t = \left(1.1 \times 10^{11}\right) I T_t^{-2.38}, \quad (9.19)$$

where W_t is mm^3/min and T_t is the melting point of the electrode material in °C. The wear ratio of the workpiece to electrode, R, can be estimated from the expression

$$R = 2.25 T_r^{-2.38}, \quad (9.20)$$

where T_r is the ratio of workpiece to electrode melting points. In practice, the value of R varies over a wide range, typically between 0.2 and 100. (Note that in abrasive processing, grinding ratios also are in the same range; see Section 9.5.2.)

Metal removal rates in EDM usually range from 2 to 400 mm^3/min. High rates produce a very rough finish, with a molten and resolidified (recast) structure with poor surface integrity and low fatigue properties. Finishing cuts are therefore made at low metal removal rates, or the recast layer is later removed by finishing operations, such as grinding. The technique of using an *oscillating electrode* provides very fine surface finish, requiring significantly less bench work to improve the luster of cavities.

EXAMPLE 9.6 Machining Time in Electrical-Discharge Machining vs. Drilling

Given: Consider the hole in Example 9.5, where the titanium alloy has a melting temperature of 1600°C.

Find: (a) Calculate the machining time for producing the hole by EDM with a current of 100 A, and compare that time to that required for drilling. (b) Calculate the wear rate of the electrode, assuming that the melting point of the electrode is 1100°C.

Solution:

a. Using Eq. (9.18),

$$\text{MRR} = \left(4 \times 10^4\right)(100)\left(1600^{-1.23}\right) = 458 \text{ mm}^3/\text{min}.$$

The volume of the hole is

$$V = \pi \left[\frac{(12.5)^2}{4}\right](20) = 2452 \text{ mm}^3.$$

Hence, the machining time for EDM is 2454/458 = 5.4 min. This time is 2.35 times that for ECM and 11.3 times that for drilling. Note, however, that if the current is increased to 300 A, the machining time for EDM will only be 1.8 minutes, which is less than the time for ECM.

b. The wear rate of the electrode is calculated using Eq. (9.19). Thus,

$$W_t = \left(11 \times 10^3\right)(100)\left(1100^{-2.38}\right) = 0.064 \text{ mm}^3/\text{min}.$$

9.13.1 Electrical-Discharge Grinding

The grinding wheel in *electrical-discharge grinding* (EDG) is made of graphite or brass and contains no abrasives. Material is removed from the workpiece surface by repetitive spark discharges between the rotating wheel and the workpiece. The material removal rate can be estimated from the equation

$$\text{MRR} = KI, \tag{9.21}$$

where MRR is in mm^3/min, I = current in amperes, and K = workpiece material factor in mm^3/A-min; for example, $K = 16$ for steel and 4 for tungsten carbide.

The EDG process can be combined with electrochemical grinding in the **electrochemical-discharge grinding** (ECDG) process. Material is removed by chemical action, with the electrical discharges from the graphite wheel breaking up the oxide film, which is washed away by the electrolyte flow. The process is used primarily for grinding carbide tools and dies, but can also be used for grinding fragile parts, such as surgical needles, thin-walled tubes, and honeycomb structures. The ECDG process is faster than EDG, but power consumption for ECDG is higher.

In **sawing** with EDM, a setup similar to a band or circular saw (without any teeth) is used with the same electrical circuit as in EDM. Narrow slits can be made at high rates of metal removal. Because the cutting forces are negligible, the process can be used on slender components.

9.13.2 Blue Arc Machining

One variation of electrical-discharge machining is the *Blue Arc* process, developed for roughing cuts of difficult-to-machine materials, especially nickel-based superalloys. The geometry of bladed disks (or *blisks*) used in aircraft engines is quite challenging to machine; the Blue Arc process removes most of the material to achieve a rough shape, which is then finish machined through conventional CNC milling. Blue arc uses an electrode and electrical-discharge machining to remove material, but adds high pressure fluid flushing to remove chips from the cutting zone. This approach has been shown to reduce energy consumption by over 30% compared to milling, and has reduced the cycle time to produce blisks from days to hours. Variations of Blue Arc machining exist for turning and grinding.

Because of the molten and resolidified (recast) surface structure developed, high rates of material removal produce a very rough surface finish, with poor surface integrity and low fatigue properties. Finishing cuts are therefore made at low removal rates, or the recast layer is subsequently removed by finishing operations. It has also been shown that surface finish can be improved by *oscillating* the electrode in a planetary motion, at amplitudes of 10 to 100 μm.

9.13.3 Wire EDM

Wire EDM (Fig. 9.36), or *electrical-discharge wire cutting*, is similar to contour cutting with a band saw (Section 8.10.5); a slowly moving wire travels along a prescribed path cutting the workpiece, with the discharge sparks acting like cutting teeth. This process is used to cut plates as thick as 300 mm (12 in.) and for making punches, tools, and dies from hard metals; it can also cut intricate components for the electronics industry (Table 9.4).

The wire is usually made of brass, copper, or tungsten and is typically about 0.25 mm (0.01 in.) in diameter, making narrow cuts possible. Zinc-, brass-, or multi-coated, wires are also used. The wire should have sufficient tensile strength and fracture toughness, and high electrical conductivity. The wire is generally used only once, as it is relatively inexpensive, and is then recycled. It travels at a constant velocity in the range of 0.15 to 9 m/min, and a constant gap (*kerf*; Fig. 9.36) is maintained during the cut.

The cutting speed is generally given in terms of the cross-sectional area cut per unit time. Some typical examples are: 18,000 mm^2/hr (28 in^2/hr) for 50-mm-thick D2 tool steel, and 45,000 mm^2/hr for 150-mm thick aluminum. These material removal rates indicate a linear cutting speed of 18,500/50 = 360 mm/hr = 6 mm/min, and 45,000/150 = 300 mm/hr = 5 mm/min, respectively.

The material removal rate for wire EDM can be obtained from the expression

$$\text{MRR} = V_f h b, \tag{9.22}$$

where MRR is in mm^3/min; V_f is the feed rate of the wire into the workpiece, in mm/min; h is the thickness or height, in mm; and the kerf, b, Fig. 9.36 is denoted as $b = d_w + 2s$, where d_w = wire diameter, in mm and s = gap, in mm, between wire and workpiece during machining.

Multiaxis EDM wire-cutting machining centers are equipped with (a) computer controls to control the cutting path of the wire and its angle with respect to the workpiece plane; (b) automatic self-threading features in case of wire breakage; (c) multiheads for cutting two parts at the same time; (d) automatic controls that prevent wire breakage; and (e) programmed

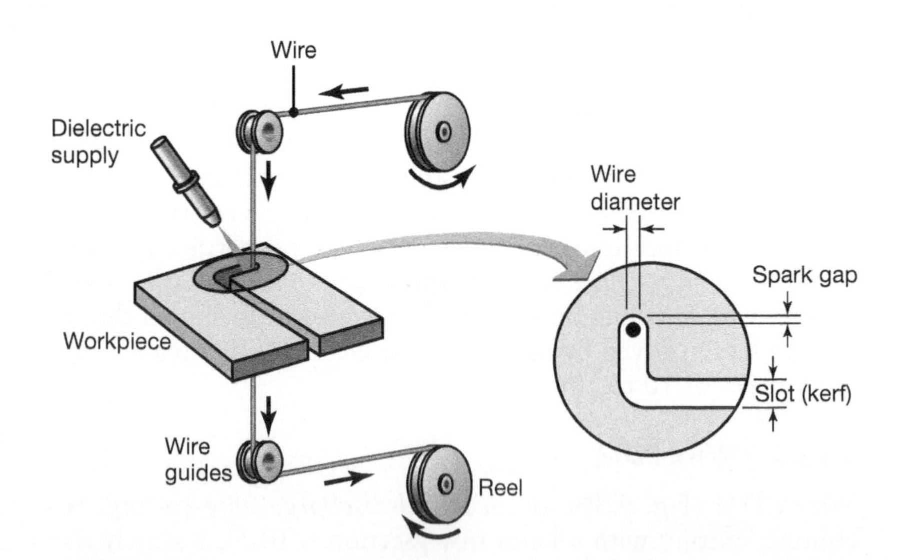

FIGURE 9.36 Schematic illustration of the wire EDM process. As much as 50 hours of machining can be performed with one reel of wire, which is then recycled.

machining strategies. Two-axis computer-controlled machines can produce cylindrical shapes in a manner similar to that of a turning operation or cylindrical grinding.

9.14 High-Energy-Beam Machining

9.14.1 Laser-Beam Machining

In *laser-beam machining* (LBM), the source of energy is a laser (an acronym for *L*ight *A*mplification by *S*timulated *E*mission of *R*adiation), which focuses optical energy on the surface of the workpiece (Fig. 9.37a). The highly focused, high-density energy melts and evaporates small portions of the workpiece in a controlled manner. This process, which does not require a vacuum, is used to machine a variety of metallic and nonmetallic workpieces. There are several types of lasers used in manufacturing operations: CO_2 (pulsed or continuous wave), Nd:YAG (neodymium: yttrium-aluminum-garnet), Nd:glass, ruby, and excimer (from the words *ex*cited and di*mer*, meaning two mers, or two molecules of the same chemical composition). The typical applications of laser-beam machining are outlined in Tables 9.4 and 9.5.

Important physical parameters in LBM are the *reflectivity* and *thermal conductivity* of the workpiece, its specific heat, and latent heats of melting and evaporation; the lower these quantities, the more efficient is the process. The cutting depth, t, may be expressed as

$$t \propto \frac{P}{vd}, \tag{9.23}$$

where P is the power input, v is the cutting speed, and d is the laser-beam spot diameter. The surface produced by LBM is usually rough and has a

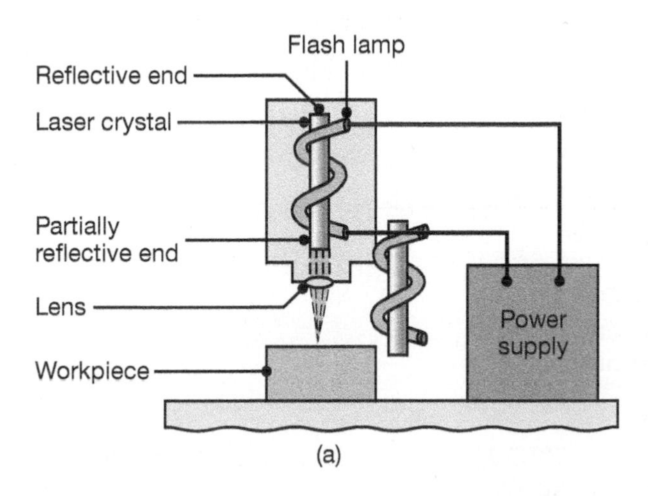

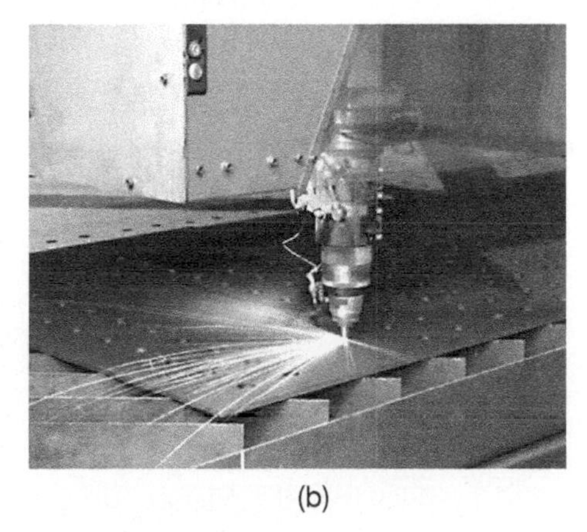

FIGURE 9.37 (a) Schematic illustration of the laser-beam machining process. (b) Cutting sheet metal with a laser beam. *Source:* (b) Courtesy of grigvovan/Fotolia.

heat-affected zone (See Fig. 12.17) that, for critical applications, may have to be removed or heat treated. Kerf width is an important consideration, as it is in other cutting processes, such as sawing, wire EDM, and electron-beam machining.

Laser beams may be used in combination with a gas stream, such as oxygen, nitrogen, or argon (*laser-beam torch*), for cutting thin sheet materials. High-pressure inert-gas-assisted (nitrogen) laser cutting is used for stainless steel and aluminum, as it leaves an oxide-free edge that can improve weldability. Gas streams also have the important function of blowing away molten and vaporized material from the workpiece surface.

Laser-beam machining is used widely in cutting and drilling metals, nonmetals, and composite materials (Table 9.5). The abrasive nature of composite materials and the cleanliness of the operation have made laser-beam machining an attractive alternative to traditional machining methods. Holes as small as 0.005 mm (200 μin.) and with hole depth-to-diameter ratios of 50 to 1 have been produced in various materials, although a more practical minimum is 0.025 mm (0.001 in.). Extreme caution should be exercised with lasers, as even low-power lasers can cause damage to the retina of the eye.

Laser beams are also used for (a) *welding* (Section 12.5); (b) small-scale *heat treating* of metals and ceramics to modify their surface mechanical and tribological properties (Section 4.4); and (c) *marking* of parts with letters, numbers, codes, etc. Marking also can be done by devices such as punches, pins, styli, scroll rolls, or stamps; or by etching. Although the

TABLE 9.5 General applications of lasers in manufacturing.

Application	Laser type
Cutting	
Metals	PCO_2, $CWCO_2$, Nd:YAG, ruby
Plastics	$CWCO_2$
Ceramics	PCO_2
Drilling	
Metals	PCO_2, Nd:YAG, Nd:glass, ruby
Plastics	Excimer
Marking	
Metals	PCO_2, Nd:YAG
Plastics	Excimer
Ceramics	Excimer
Surface treatment	$CWCO_2$
Welding	
Metals	PCO_2, $CWCO_2$, Nd:YAG, Nd:glass, ruby, diode
Plastics	Diode, Nd:YAG
Lithography	Excimer

Note: P = pulsed, CW = continuous wave,
Nd:YAG = neodynmium: yttrium–aluminum–garnet.

equipment is more expensive than for other methods, marking and engraving with lasers has become increasingly common, due to its accuracy, reproducibility, flexibility, ease of automation, and on-line application in a wide variety of manufacturing operations.

The inherent flexibility of the laser-beam cutting process with fiber-optic-beam delivery, simple fixturing, low setup times, and the availability of multi-kW machines and 2D and 3D computer-controlled systems are attractive features. Thus, laser cutting can compete successfully with cutting of sheet metal using the traditional punching processes (Section 7.3), although the two processes can be combined for improved overall efficiency.

Laser Microjet. *Laser microjet,* illustrated in Fig. 9.38, uses a low-pressure laminar water stream to serve as a variable-length fiber optic cable to direct the laser and deliver laser power at the bottom of the kerf. This has an advantage in that the laser focus depth is very large, and large aspect ratio cuts can be produced. The water-jet is produced by a sapphire or diamond nozzle with an opening of 25 to 100 μm (0.004 in.), and exerts a force less than 0.1 N (0.02 lb). In laser microjet machining, material removal is due to the action of the laser, and the water provides cooling (reducing the heat affected zone, see Section 12.7), and prevents weld splatter from attaching to the workpiece. The laser is typically a Nd:Yag laser with micro- or nano-second pulse duration and power between 10 and 200 W.

9.14.2 Electron-Beam Machining and Plasma-Arc Cutting

The source of energy in *electron-beam machining* (EBM) is high-velocity electrons, that strike the surface of the workpiece and generate heat (Fig. 9.39). The applications of this process are similar to those of laser-beam machining, except that EBM requires a vacuum. The machines use voltages in the range of 50–200 kV to accelerate the electrons to speeds of 50 to 80% of the speed of light. Electron-beam machining is used for very accurate cutting of a wide variety of metals. Surface finish is better

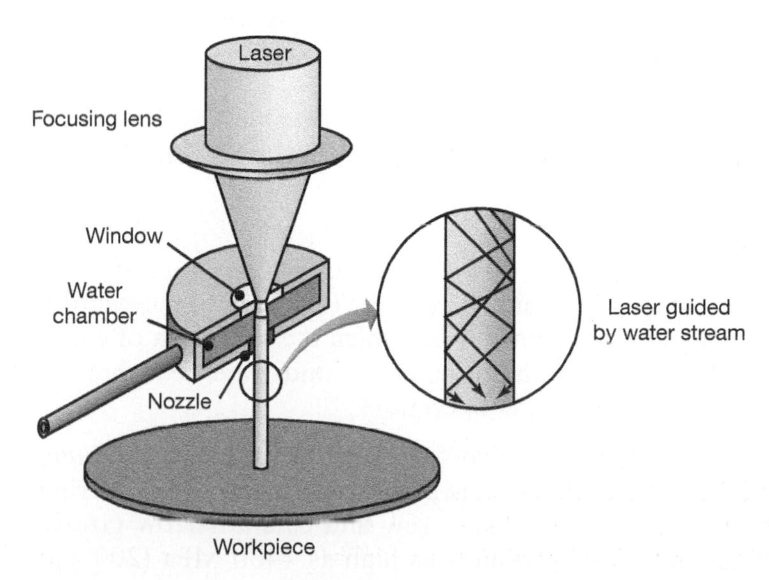

FIGURE 9.38 (a) Schematic illustration of the laser microjet process.

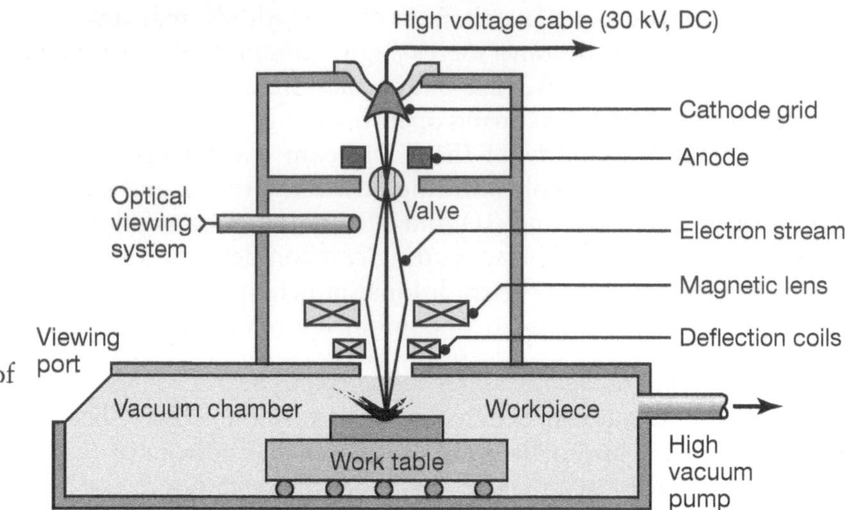

FIGURE 9.39 Schematic illustration of the electron-beam machining process. Unlike LBM, this process requires a vacuum, and hence workpiece size is limited by the chamber size.

and kerf width is narrower than that for other thermal-cutting processes (see also the discussion in Section 12.5 on electron-beam welding). The interaction of the electron beam with the workpiece surface produces hazardous X-rays; consequently, the equipment should be used only by highly trained personnel.

In **plasma-arc cutting** (PAC), *plasma beams* (ionized gas) are used for rapid cutting of nonferrous and stainless-steel plates. The temperatures generated are very high (9400°C in the torch for oxygen as a plasma gas); consequently, the process is fast, kerf width is small, and the surface finish is good. Material removal rates are much higher than in the EDM and LBM processes, and parts can be machined with good reproducibility. Plasma-arc cutting is highly automated through the use of programmable controls.

A more traditional method is **oxyfuel-gas cutting** (OFC), using a torch as in welding (Chapter 12). The operation is particularly useful for cutting steels, cast irons, and cast steels. Cutting occurs mainly by oxidation and burning of the steel, with some melting taking place as well. Kerf widths are usually in the range of 1.5–10 mm (0.06–0.4 in.).

9.15 Water-Jet, Abrasive Water-Jet, and Abrasive-Jet Machining

When a hand is placed across a jet of water or air, there is a considerable concentrated force acting on the hand. This force is a result of momentum change of the stream and is the principle on which the operation of water or gas turbines is based. This is the principle behind water-jet, abrasive water-jet, and abrasive-jet machining operations.

1. **Water-jet machining** (WJM). Also called *hydrodynamic machining* (Fig. 9.40a), the force from the jet is utilized in cutting and deburring operations; the water jet acts like a saw and cuts a narrow groove in the material. Although pressures as high as 1400 MPa (200 ksi)

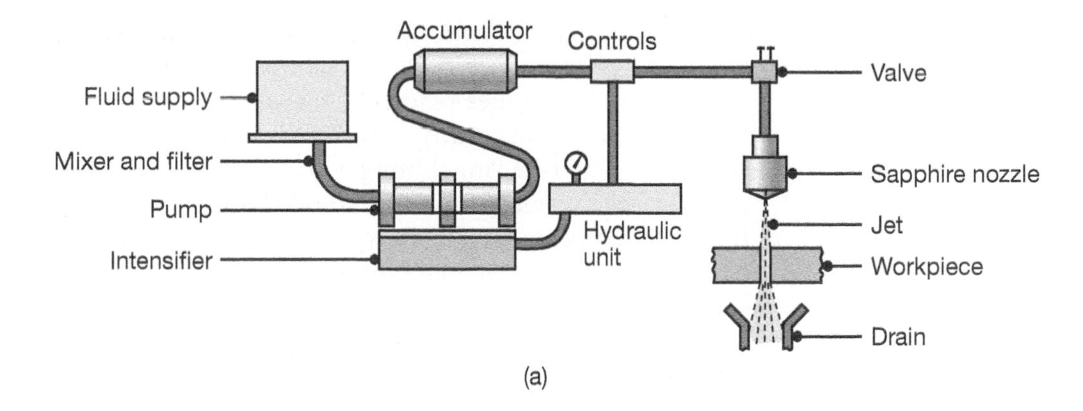

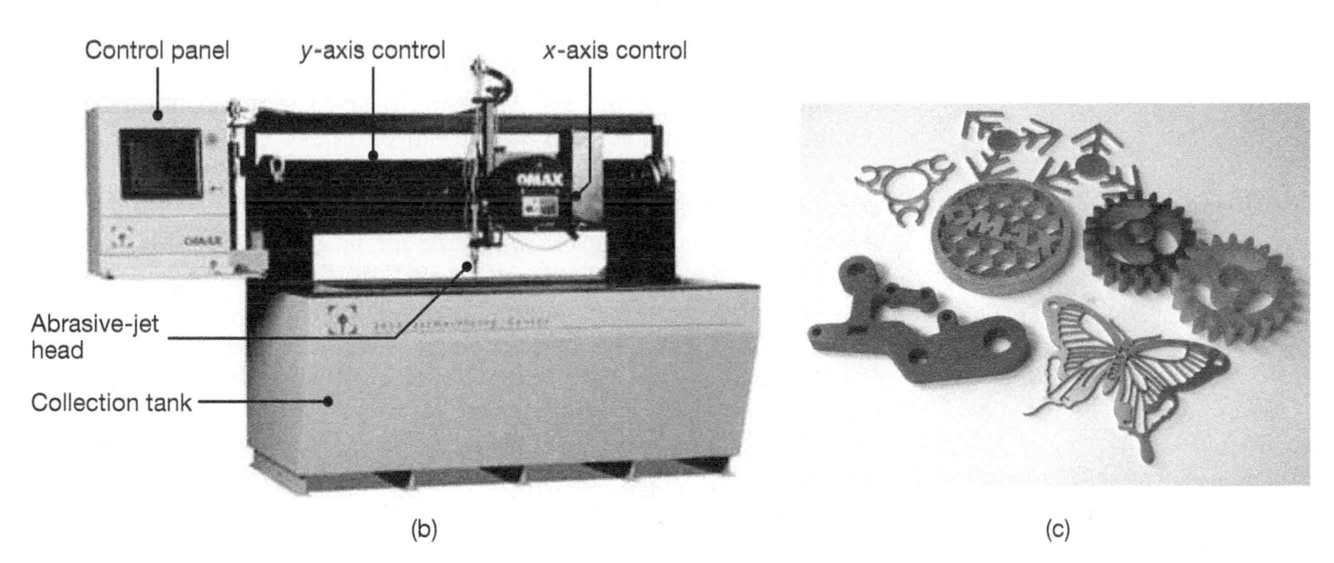

FIGURE 9.40 (a) Schematic illustration of water-jet machining. (b) A computer-controlled water-jet cutting machine. (c) Examples of various nonmetallic parts machined by the water-jet cutting process. *Source:* Courtesy of OMAX Corporation.

can be generated, a pressure level of about 400 MPa (60 ksi) is typically used for efficient operation. Jet-nozzle diameters usually range between 0.05 and 1 mm. A water-jet cutting machine and typical parts produced are shown in Fig. 9.40.

A variety of nonmetallic materials can be cut with this technique, including plastics, fabrics, rubber, wood products, paper, leather, insulating materials, brick, and composite materials (Fig. 9.40c). Thicknesses range to 25 mm or higher. Vinyl and foam coverings for some automobile dashboards are, for example, cut using multiple-axis robot-guided water-jet machining equipment. Because it is an efficient and clean operation as compared with other cutting processes, WJM is also used in the food-processing industry for cutting and slicing food products (Table 9.4).

The advantages of this process are that (a) cuts can be started at any location without the need for predrilled holes; (b) no heat is produced; (c) no deflection of the rest of the workpiece takes place, hence the process is suitable for flexible materials; (d) little wetting of the workpiece takes place; (e) the burr produced is minimal; and (f) it is an environmentally friendly and safe operation.

2. **Abrasive water-jet machining.** In *abrasive water-jet machining* (AWJM), the water-jet contains abrasive particles such as silicon carbide or aluminum oxide, thus increasing the material removal rate over that of water-jet machining. Metallic, nonmetallic, and composite materials of various thicknesses can be cut in single or multiple layers, particularly heat-sensitive materials that cannot be machined by processes in which heat is produced. The minimum hole size that can be produced satisfactorily to date is about 3 mm (0.12 in.), and maximum hole depth is on the order of 25 mm (1 in.). Cutting speeds can be as high as 7.5 m/min (25 ft/min) for reinforced plastics, but are much lower for metals. With multiple-axis and robotic-control machines, complex three-dimensional parts can be machined to finish dimensions. Nozzle life is improved by making it from rubies, sapphires, and carbide-base composite materials.
3. **Abrasive-jet machining.** In *abrasive-jet machining* (AJM), a high-velocity jet of dry air, nitrogen, or carbon dioxide, containing abrasive particles, is aimed at the workpiece surface under controlled conditions (Fig. 9.41). The impact of the particles develops sufficiently high concentrated force (Section 9.9) for operations such as (a) cutting small holes, slots, and intricate patterns in very hard or brittle metallic and nonmetallic materials; (b) deburring or removing small flash from parts; (c) trimming and beveling; (d) removing oxides and other surface films; and (e) general cleaning of components with irregular surfaces.

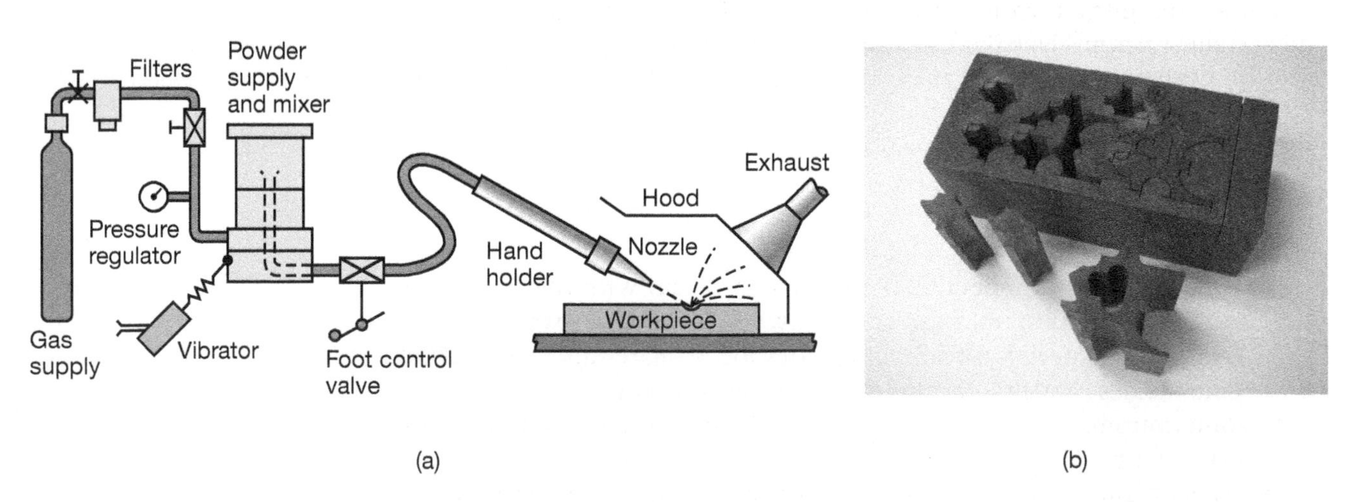

FIGURE 9.41 (a) Schematic illustration of the abrasive-jet machining process. (b) Examples of parts produced by abrasive-jet machining; the parts are 50 mm (2 in.) thick and are made of 304 stainless steel. *Source:* Courtesy of OMAX Corporation.

The gas supply pressure is on the order of 850 kPa (120 psi), and the abrasive-jet velocity can be as high as 300 m/s (1000 ft/s) and is controlled by a valve. The handheld nozzles are usually made of tungsten carbide or sapphire. The abrasive size is in the range of 10–50 μm. Because the flow of the free abrasives tends to round off corners, designs for abrasive-jet machining should avoid sharp corners; holes made in metal parts tend to be tapered. Because of airborne particulates, there is some safety hazard involved in using this process.

9.16 Hybrid Machining Systems

A more recent development in material-removal processes is the concept of *hybrid machining systems*. Two or more individual machining processes are *combined* into one system, thus taking advantage of the capabilities of each process, increasing production speed, and thereby improving the efficiency of the operation. The system is able to handle a variety of materials, including metals, ceramics, polymers, and composites. Examples of hybrid machining systems include combinations and integration of the following processes:

1. Abrasive machining and electrochemical machining;
2. Abrasive machining and electrical-discharge machining;
3. Abrasive machining and electrochemical finishing;
4. Water-jet cutting and wire EDM;
5. High-speed milling, laser ablation, and abrasive blasting, as an example of *three* integrated processes;
6. Machining and blasting;
7. Electrochemical machining and electrical-discharge machining (ECDM), also called electrochemical spark machining (ECSM);
8. Machining and forming processes, such as laser cutting and punching of sheet metal, described in Example 27.1; and
9. Combinations of various other forming, machining, and joining processes.

The implementation of these concepts, and the development of appropriate machinery and control systems, present significant challenges. Important considerations include such factors as:

1. The workpiece material and its manufacturing characteristics;
2. Compatibility of processing parameters among the two or more processes to be integrated, such as speed, size, force, energy, and temperature;
3. Cycle times of each individual operation involved and their synchronization;
4. Safety considerations and possible adverse effects of the presence of various elements, such as abrasives, chemicals, wear particles, chips, and contaminants on the overall operation; and
5. Consequence of a failure in one of the stages in the system, since the operation involves sequential processes.

9.17 Design Considerations

The important design considerations for the processes and operations described in this chapter are outlined below.

9.17.1 Grinding and Abrasive Machining Processes

Typical grinding operations are performed on workpieces that are close to their final shape (near-net shape). They generally are not economical for removing large volumes of material, except in creep-feed grinding. Design considerations for grinding operations are somewhat similar to those for machining, described in Section 8.14. In addition, the following should be given special consideration:

1. Parts to be ground should be designed so that they can be mounted securely in suitable fixtures and workholding devices. Workpieces that are thin, straight, or tubular may distort during grinding, thus requiring special attention.
2. If high dimensional accuracy is required, interrupted surfaces, such as holes and keyways, should be avoided, as they can cause vibrations and chatter.
3. Parts for cylindrical grinding should be balanced, and long and slender designs should be avoided to minimize deflections; fillets and corner radii should be as large as possible.
4. In centerless grinding, short parts may be difficult to grind accurately because of their lack of support on the blade. In through-feed grinding, only the largest diameter on the parts can be ground.
5. Designs requiring precise form grinding should be kept simple to avoid frequent form dressing of the wheel.
6. Deep and small-diameter holes, and blind holes requiring internal grinding, should be avoided or they should include a relief (Fig.9.42).
7. To maintain good dimensional accuracy, designs should preferably allow for all grinding operations to be performed without having to reposition the workpiece. Note that this guideline is also applicable to all manufacturing processes and operations.

9.17.2 Ultrasonic Machining

1. Avoid sharp profiles, corners, and radii, because these features will be eroded by the abrasive slurry.
2. The holes produced will have some taper because of oblique collisions between abrasives and vertical sidewalls.
3. Because of the tendency for chipping of brittle materials at the exit end of holes, the bottom of the parts should have a backup plate (Fig. 9.42b).

9.17.3 Chemical Machining

1. Because the etchant attacks all exposed surfaces continuously, designs involving sharp corners, deep and narrow cavities, large tapers, folded seams, or porous workpiece materials should be avoided.

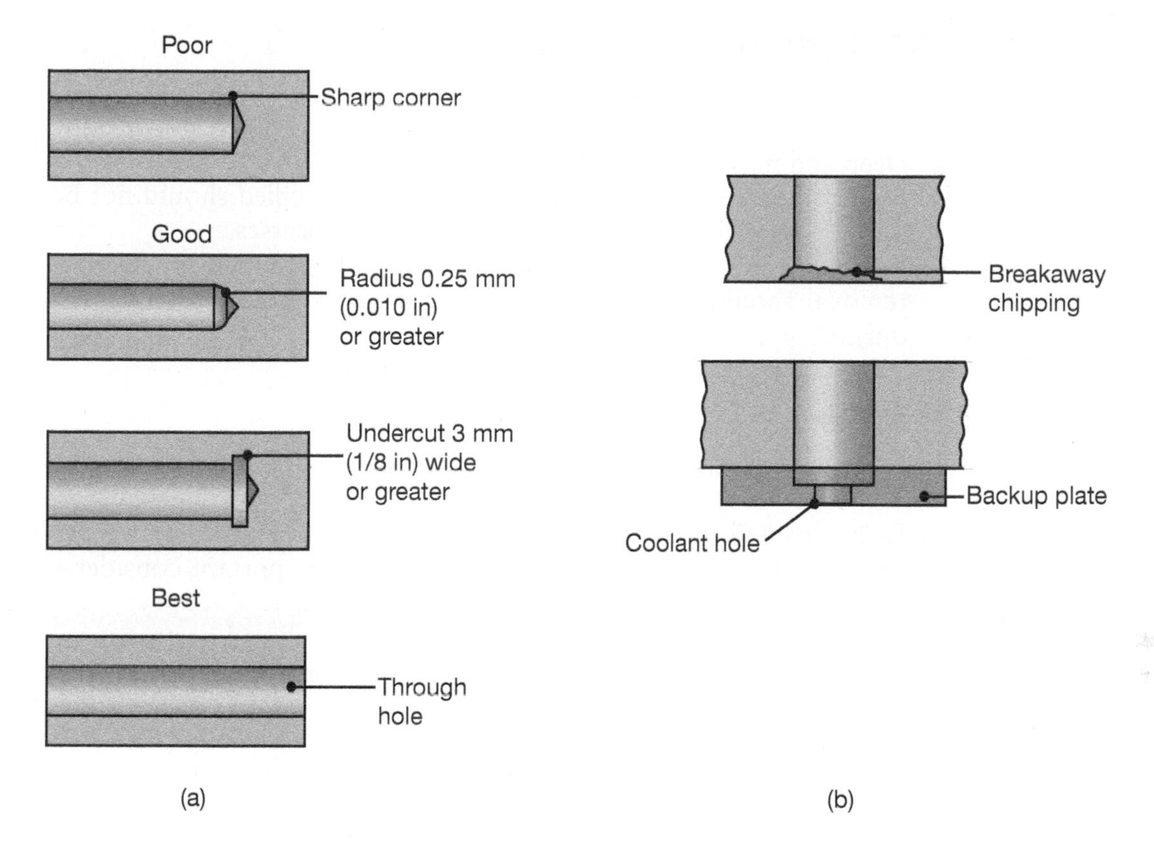

FIGURE 9.42 Design guidelines for internal features, especially as applied to holes. (a) Guidelines for grinding the internal surfaces of holes. These guidelines generally hold for honing as well. (b) The use of a backing plate for producing high-quality through-holes by ultrasonic machining. *Source:* After J. Bralla.

2. Because the etchant attacks the material in both vertical and horizontal directions, *undercuts* may develop (as shown by the areas under the edges of the maskant in Fig. 9.27). Typically, tolerances of ±10% of the material thickness can be maintained in chemical blanking.
3. In order to improve the production rate, the bulk of the workpiece should be preshaped by other processes, such as by machining, prior to chemical machining.

9.17.4 Electrochemical Machining and Grinding

1. Because of the tendency for the electrolyte to erode away sharp profiles, electrochemical machining is not suited for producing sharp square corners or flat bottoms.
2. Controlling the electrolyte flow may be difficult, thus irregular cavities may not be produced to the desired shape with acceptable dimensional accuracy.
3. Designs should make provision for a small taper for holes and cavities to be machined.
4. If flat surfaces are to be produced, the electrochemically ground surface should be narrower than the width of the grinding wheel.

9.17.5 Electrical-Discharge Machining

1. Parts should be designed so that the required electrodes can be shaped properly and economically.
2. Deep and narrow slots should be avoided.
3. For economic production, the surface finish specified should not be too fine, as is the case for all manufacturing processes.
4. In order to achieve a high production rate, the majority of material removal should be done first by conventional processes (roughing out).

9.17.6 Laser and Electron Beam Machining

1. Designs with sharp corners should be avoided since they can be difficult to produce.
2. Deep cuts will produce tapered walls.
3. The reflectivity of the workpiece surface is an important consideration; dull and unpolished surfaces are preferable.
4. Any adverse effects on the properties of the machined parts, caused by high local temperatures and heat-affected zone, should be investigated.
5. Because the vacuum chambers in EBM have limited capacity, part sizes should closely match the size of the chamber for a high-production rate per cycle.
6. If a part requires EBM on only a small section of the workpiece, consideration should be given to manufacturing it as several smaller components and then assembling them.

9.18 Process Economics

This chapter has indicated that the grinding process can be employed both as a finishing operation as well as for large-scale removal operations. The use of grinding as a finishing operation is often necessary because forming and machining processes alone may not produce parts with the desired dimensional accuracy and surface finish (See, for example, Fig. 9.28). Because it is an additional operation, grinding can contribute significantly to product cost. Creep-feed grinding, on the other hand, has proven to be an economical alternative to some machining operations and materials, such as milling, even though wheel wear is high. All finishing operations contribute to product cost. On the basis of the topics described thus far, it can be seen that as the surface finish improves, more operations typically are required, and hence the cost increases. Note in Fig. 9.43, for example, how rapidly cost increases as surface finish is improved by processes such as grinding and honing.

Much progress has been made in automating the equipment involved in finishing operations, including advanced computer controls, the increased use of robotics (Section 14.7), and the availability of powerful and user-friendly software. Production times and labor costs have been reduced,

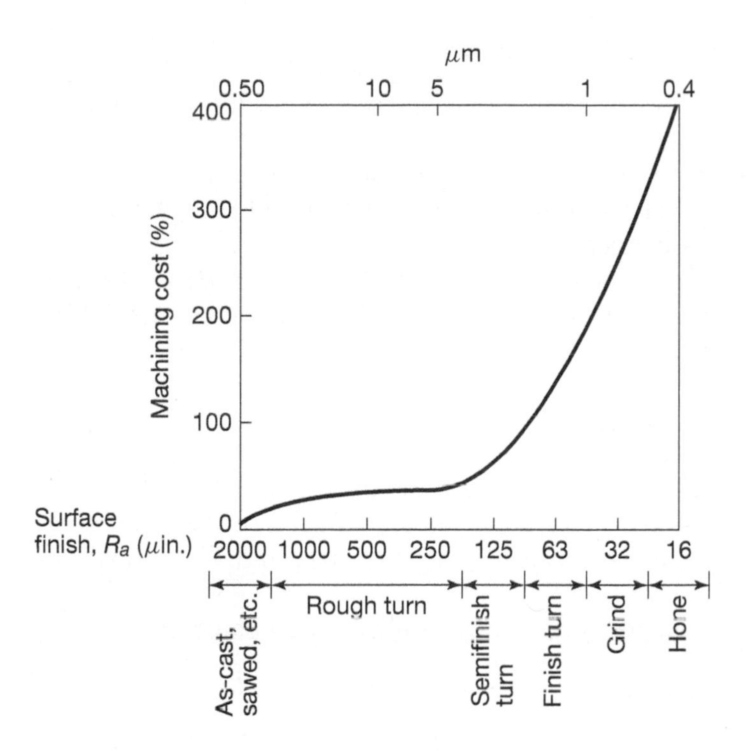

FIGURE 9.43 Increase in the cost of machining and finishing operations as a function of the surface finish required. Note the rapid increase associated with finishing operations.

even though modern machine tools require significant capital investment. If finishing costs are likely to be an important factor in manufacturing a particular product, the conceptual and design stages should include an analysis of the degree of surface finish and dimensional accuracy required. Moreover, all processes that precede finishing operations should be analyzed for their capability to produce a more acceptable surface finish and dimensional accuracy, such as net-shape manufacturing.

The economic production run (Chapter 16) for a particular machining process depends on the cost of tooling and equipment, the material removal rate, operating costs, and the level of operator skill required, as well as the cost of secondary and finishing operations that may be necessary. In chemical machining, for example, the costs of reagents, maskants, and disposal, together with the cost of cleaning the parts, are important factors. In electrical-discharge machining, the cost of electrodes and the need to replace them periodically can be significant.

CASE STUDY 9.1 Gear Grinding with Engineered Abrasives

Grinding in the gear industry has usually focused on finishing steel gear profiles, diameters, bores, etc., and is generally performed on hardened steel. Gear manufacturers need to consider grinding in context, weighing the performance and economic factors of grinding compared to cutting processes such as hard hobbing and power skiving (Section 8.10.7), or to various honing or lapping technologies (Section 9.7).

Grinding offers improved surface quality, dimensional accuracy, and process reliability, giving improvements in, for example, noise reduction and allowable contact stress, but at reduced removal rates, additional cost, and a somewhat higher environmental impact due to the use of coolant.

In general, little attention has been paid to the potential of grinding the rough gear form prior to heat treatment; conventional gear cutting is preferred due the inability of traditional abrasive technologies to provide the necessary metal removal rates and power efficiencies to be competitive.

Consider hobbing, where metal removal rates are of the order of 50–100 mm^3/sec (5–10 in^3/min). Specific power requirements are around 4 J/mm^3 (1.5 hp/in^3). Hobbing is fast and energy-efficient, due to the relatively small numbers of large chips created.

The disadvantages of a process like hobbing lie in the flexibility and reliability of the tooling. The Indiana Tool and Manufacturing Company (ITAMCO) product line involves unique gears, both in terms of size and geometry. For these applications, the cutting tools, although normally long-lived, are costly, are custom to a given gear, and require long lead times to manufacture. Cutting performance and quality also change as the tool wears.

Grinding with conventional abrasives has historically been limited by low grinding ratios (Section 9.5.2) and metal removal rates and high specific grinding energy requirements. Bonded cubic boron nitride (Section 9.2) superabrasive wheels have also been unable to achieve the high stock removal required, even at high wheel speeds; these wheels are also expensive, and, like hobs, are specific to a given gear profile.

Grinding soft steel has also been a problem due to wheel loading unless accompanied by continuous dressing–e.g. continuous dress creep feed grinding (CDCF). CDCF was developed in the 1970s using high porosity wheels to take deep form cuts, where it was found that by continuously dressing with a formed diamond roll dresser to keep the abrasive grains clean and sharp, the specific grinding energy was reduced significantly. In combination with good coolant access through the high wheel porosity, CDCF allowed an order of magnitude increase in stock removal rates. Unfortunately, although this was very effective on tough-to-grind metals such as Inconel, the level of wheel wear from continuous dressing (typically 1 μm /rev at 1000 rpm), resulted in uneconomic wear rates compared to machining. This, combined with the need for a specific form diamond roll, again made CDCF impractical for rough grinding of gears.

Recent advances in wheel technology, especially those related to dressable vitrified bonds utilizing engineered ceramic grains like 3M Cubitron™ II, make grinding feasible for these operations.

Conventional grinding wheels have random shapes, typically with a roughly equiaxed grain, that offers a random cutting edge. The 3M Cubitron™ II uses specially shaped ceramic abrasives that are more aggressive and uniformly oriented on a grinding wheel or disk. Typical grains are shown in Fig. 9.44. Compared to conventional abrasives (Fig. 9.6), the oriented grains cut more efficiently,

FIGURE 9.44 Cubitron™ II abrasives. The shaped particles are much more aggressive than conventional abrasives (Fig. 9.6). *Source:* Courtesy of 3M.

allowing the grinding wheel to operate at lower temperature and higher efficiency, removing up to three times the metal removal rate of conventional grinding wheels.

As one of the largest open gear manufacturers in the world, ITAMCO has developed technologies that can achieve specific grinding energies and metal removal rates approaching those of hobbing and shaping, but with a grinding ratio that allows deep form grinding without continuous dressing. This offers the possibility of grinding in the soft state on standard finish gear grinders. Dressing is performed using a standard CNC contour diamond dress roll instead of a specialty tool.

As an example of the process capabilities, the large gear shown in Fig. 9.45 is a 25 mm (1 in.) diametrical pitch, 600 mm (24 in.) face width, 3 m (120 in.) diameter gear with a 9 degree helix angle weighing over 52,000 pounds. ITAMCO was able to finish the gear to size in 120 hours in one workholding using only two 3M Cubitron™ II grinding wheels, in turn saving time and material.

FIGURE 9.45 Grinding of a large (1.5-m diameter) gear. *Source:* ITAMCO.

The immediate benefits are fast turnaround times since there is no need to wait for specialty hob or shaper manufacture. Also, power requirements are significantly reduced, so that added flexibility is achieved by using grinders that normally would be underpowered for demanding applications.

Source: Courtesy of J. Neidig, ITAMCO.

SUMMARY

- Abrasive machining and advanced machining processes are often necessary and economical when workpiece hardness is high, the materials are brittle or flexible, part shapes are complex, and surface finish and dimensional tolerance requirements are high. (Section 9.1)
- Conventional abrasives basically consist of aluminum oxide and silicon carbide, and superabrasives consist of cubic boron nitride and diamond. Friability of abrasive grains is an important factor, as are their shape and size. (Section 9.2)
- Grinding wheels typically are a combination of abrasive grains and bonding agents. Important characteristics of wheels include the types of abrasive grain and bond, grade, and hardness. Wheels may be reinforced with metal or fibers to maintain wheel integrity in case of fracture. (Section 9.3)
- The mechanics of grinding processes help establish quantitative relationships regarding chip dimensions, grinding forces and energy requirements, temperature rise, residual stresses, and any adverse effects on surface integrity of ground components. (Section 9.4)
- Wheel wear is an important consideration in surface quality. Wear is usually monitored in terms of the grinding ratio, defined as the ratio of material removed to the volume of wheel wear. Dressing and truing of wheels are done by a variety of techniques, with the aid of sensors and computer controls. (Section 9.5)
- A variety of abrasive machining processes and equipment are available for surface, external, and internal grinding, as well as for large-scale material removal processes. The proper selection of abrasives, process variables, and grinding fluids is important in obtaining the desired surface finish and dimensional accuracy and for avoiding workpiece damage, such as burning, heat checking, harmful residual stresses, and chatter. (Section 9.6)
- Several finishing operations are available for improving surface integrity. Because finishing processes can contribute significantly to product cost, proper selection and implementation of finishing operations are important. (Section 9.7)
- Deburring may be necessary for certain finished components. Commonly used deburring methods include vibratory and barrel finishing and shot blasting; thermal-energy processes and other methods also can be used. (Section 9.8)
- Ultrasonic machining removes material by microchipping and is best suited for hard and brittle materials. (Section 9.9)
- Advanced machining processes include chemical and electrical means and high-energy beams, and are particularly suitable for hard materials and complex part shapes. The effects of these processes on surface integrity must be considered as they can damage surfaces and reduce fatigue life. (Sections 9.10–9.14)

- Water-jet, abrasive water-jet, and abrasive-jet machining processes can be used effectively for cutting as well as deburring operations. Because they do not use hard tooling, they have inherent flexibility of operation. (Sections 9.15)
- A recent trend is development of hybrid processes, where two or more individual machining processes are combined into one system, thus taking advantage of the capabilities of each process. (Section 9.16)
- As in all manufacturing processes, certain design guidelines have been established for effective use of advanced machining operations. (Section 9.17)
- The processes described in this chapter have unique capabilities. However, because they involve different types of machinery, controls, process variables, and cycle times, the competitive aspects of each process with respect to a particular product must be studied individually. (Section 9.18)

SUMMARY OF EQUATIONS

Undeformed chip length, l:

Surface grinding, $l = \sqrt{Dd}$

External grinding, $l = \sqrt{\dfrac{Dd}{1 + D/D_w}}$

Internal grinding, $l = \sqrt{\dfrac{Dd}{1 - D/D_w}}$

Undeformed-chip thickness, surface grinding, $t = \sqrt{\dfrac{4v}{VCr}\sqrt{\dfrac{d}{D}}}$

Relative grain force, $\propto \dfrac{v}{VC}\sqrt{\dfrac{d}{D}}$

Temperature rise in surface grinding, $\Delta T \propto D^{1/4} d^{3/4} \left(\dfrac{V}{v}\right)^{1/2}$

Grinding ratio, $G = \dfrac{\text{Volume of material removed}}{\text{Volume of wheel wear}}$

Average force of particle in ultrasonic machining, where $F_{\text{ave}} = \dfrac{2mv}{t_o}$

where $t_o = \dfrac{5r}{c_o}\left(\dfrac{c_o}{v}\right)^{1/5}$

Material removal rate in:

electrochemical machining, $\text{MRR} = CI\eta$

electrochemical grinding, $\text{MRR} = \dfrac{GI}{\rho F}$

electrical-discharge machining, $\text{MRR} = 4 \times 10^4 I T_w^{-1.23}$

electrical-discharge grinding, $\text{MRR} = KI$

wire EDM, $\text{MRR} = V_f h b$, where $b = d_w + 2s$

Penetration rate in electrochemical grinding, $V_s = \left(\frac{G}{dF}\right)\left(\frac{E}{gK_p}\right)K$

Electrode wear rate in EDM, $W_t = \left(1.1 \times 10^{11}\right) I T_t^{-2.38}$

Laser-beam cutting time, $t = \frac{P}{vd}$

BIBLIOGRAPHY

Asibu, E.K., *Principles of Laser Materials Processing*, Wiley, 2009.

Astashev, V.K., and Babitsky, V.I., *Ultrasonic Processes and Machines*, Springer, 2010.

Dahotre, N.B., and Samant, A., *Laser Machining of Advanced Materials*, CRC Press, 2011.

Davim, J.P., *Machining of Advanced Materials*, Springer, 2011.

—, *Machining of Metal Matrix Composites*, Springer, 2014.

El-Hofy, H.A.-G., *Advanced Machining Processes*, McGraw-Hill, 2005.

Gillespie, L.K., *Deburring and Edge Finishing Handbook*, Society of Manufacturing Engineers/American Society of Mechanical Engineers, 2000.

Grzesik, W., *Advanced Machining Processes of Metallic Materials: Theory, Modelling and Applications*, Elsevier, 2008.

Jackson, M.J., and Davim, J.P., *Machining with Abrasives*, Springer, 2010.

Jameson, E.C., *Electrical Discharge Machining*, Society of Manufacturing Engineers, 2001.

Kuchle, A., *Manufacturing Processes 2: Grinding, Honing, Lapping*, Springer, 2009.

Malkin, S., and Guo, C., *Grinding Technology*, 2nd ed., Industrial Press, 2008.

Marinescu, I.D., Uhlmann, E., and Doi, T., *Handbook of Lapping and Polishing*, CRC Press, 2006.

Oliver, M.R., *Chemical Mechanical Planarization of Semiconductor Materials*, Springer, 2004.

Rowe, W.B., *Principles of Modern Grinding Technology*, William Andrew, 2009.

Schaeffer, R., *Fundamentals of Laser Micromachining*, CRC Press, 2012.

Steen, W.M., and Mazumder, J., *Laser Material Processing*, 4th ed., Springer, 2010.

QUESTIONS

9.1 Why are grinding operations necessary for parts that have been machined by other processes?

9.2 Explain why there are so many different types and sizes of grinding wheels.

9.3 Explain the reasons for the large differences between the specific energies involved in grinding (Table 9.3) and in machining (Table 8.3).

9.4 Describe the advantages of superabrasives over conventional abrasives.

9.5 Give examples of applications for the grinding wheels shown in Fig. 9.2.

9.6 Explain why the same grinding wheel may act soft or hard.

9.7 Describe your understanding of the role of friability of abrasive grains on grinding wheel performance.

9.8 Explain the factors involved in selecting the appropriate type of abrasive for a particular grinding operation.

9.9 What are the effects of wear flat on the grinding operation?

9.10 Explain how the grinding ratio, G, depends on the following factors: (a) type of grinding wheel; (b) workpiece hardness; (c) wheel depth of cut; (d) wheel and workpiece speeds; and (e) type of grinding fluid.

9.11 Which terms in Eq. (9.7) would be affected if a grinding operation changed from conventional to precision-shaped abrasives?

9.12 List and explain the precautions you would take when grinding with high precision. Comment on the role of the machine, process parameters, the grinding wheel, and grinding fluids.

9.13 Describe the methods you would use to determine the number of active cutting points per unit surface area on the periphery of a straight (i.e., Type 1; see Fig. 9.2a) grinding wheel. What is the significance of this number?

9.14 Describe and explain the difficulties involved in grinding parts made of (a) thermoplastics; (b) thermosets; and (c) ceramics.

9.15 Explain why ultrasonic machining is not suitable for soft and ductile metals.

9.16 It is generally recommended that a soft-grade grinding wheel be used for hardened steels. Explain why.

9.17 Explain the reasons that the processes described in this chapter may adversely affect the fatigue strength of materials.

9.18 Describe the factors that may cause chatter in grinding operations, and give reasons that these factors cause chatter.

9.19 Outline the methods that are generally available for deburring parts. Discuss the advantages and limitations of each.

9.20 In which of the processes described in this chapter are physical properties of the workpiece material important? Why?

9.21 Give all possible technical and economic reasons that the material-removal processes described in this chapter may be preferred, or even required, over those described in Chapter 8.

9.22 What processes would you recommend for die sinking in a die block? Explain.

9.23 The proper grinding surfaces for each type of wheel are shown in Fig. 9.2. Explain why grinding on other surfaces of the wheel is improper and unsafe.

9.24 Note that wheel (b) in Fig. 9.3 has serrations along its periphery, somewhat similar to circular metal saws. What does this design offer in grinding?

9.25 In Fig. 9.11, we note that wheel speed and grinding fluids can have a major effect on the type and magnitude of residual stresses developed in grinding. Explain the possible reasons for these phenomena.

9.26 Explain the consequences of allowing the workpiece temperature to rise excessively in grinding operations.

9.27 Comment on any observations you have regarding the contents of Table 9.4.

9.28 Why has creep-feed grinding become an important process? Explain.

9.29 There has been a trend in manufacturing industries to increase the spindle speed of grinding wheels. Explain the possible advantages and limitations of such an increase.

9.30 Does built-up edge occur in grinding? Explain your answer.

9.31 Why is preshaping or premachining of parts generally desirable in the advanced machining processes described in this chapter?

9.32 Why are finishing operations sometimes necessary? How could they be minimized? Explain, with examples.

9.33 Why has the wire-EDM process become so widely accepted in industry?

9.34 Make a list of material-removal processes described in this chapter that may be suitable for the following workpiece materials: (a) ceramics; (b) cast iron; (c) thermoplastics; (d) thermosets; (e) diamond; and (f) annealed copper.

9.35 Explain why producing sharp corners and profiles using some of the processes described in this chapter can be difficult.

9.36 How do you think specific energy, u, varies with respect to wheel depth of cut and hardness of the workpiece material? Explain.

9.37 In Example 9.2, it is stated that the thrust force in grinding is about 30% higher than the cutting force. Why is it higher?

9.38 Why should we be interested in the magnitude of the thrust force in grinding? Explain.

9.39 Why is the material removal rate in electrical-discharge machining a function of the melting point of the workpiece material?

9.40 Inspect Table 9.4, and, for each process, list and describe the role of various mechanical, physical, and chemical properties of the workpiece material that may affect performance.

9.41 Which of the processes listed in Table 9.4 would not be applicable to nonmetallic materials? Explain.

9.42 Why does the machining cost increase so rapidly as surface finish requirements become finer?

9.43 Which of the processes described in this chapter are particularly suitable for workpieces made of (a) ceramics, (b) thermoplastics, and (c) thermosets? Why?

9.44 Other than cost, is there a reason that a grinding wheel intended for a hard workpiece cannot be used for a softer workpiece?

9.45 How would you grind the facets on a diamond, as for an engagement ring, since diamond is the hardest material known?

9.46 Define dressing and truing, and describe the difference between them.

9.47 What is heat checking in grinding? What is its significance?

9.48 Explain why parts with irregular shapes, sharp corners, deep recesses, and sharp projections can be difficult to polish.

9.49 Explain the reasons why so many deburring operations have been developed over the years.

9.50 Why does grinding temperature decrease with increasing work speed [Eq. (9.8)]? Does this mean that for a work speed of zero, the temperature is infinite? Explain.

9.51 Describe the similarities and differences in the action of metalworking fluids in machining vs. grinding.

9.52 Are there any similarities among grinding, honing, polishing, and buffing? Explain.

9.53 Is the grinding ratio important in evaluating the economics of a grinding operation? Explain.

9.54 We know that grinding can produce a very fine surface finish on a workpiece. Is this necessarily an indication of the quality of a part? Explain.

9.55 If not done properly, honing can produce holes that are bellmouthed, wavy, barrel-shaped, or tapered. Explain how this is possible.

9.56 Which of the advanced machining processes causes thermal damage? What is the consequence of such damage to workpieces?

9.57 Describe your thoughts regarding laser-beam machining of nonmetallic materials. Give several possible applications, including their advantages as compared to other processes.

9.58 What is kerf? In what processes is kerf important?

9.59 It was stated that graphite is the preferred material for EDM tooling. Would graphite the be useful in wire EDM? Explain.

9.60 What are the functions of the fluid in EDM?

9.61 What is the purpose of the abrasives in electrochemical grinding?

9.62 What are the advantages of precision-shaped abrasives as shown in Fig. 9.44? Are there any disadvantages? Explain.

9.63 List and explain factors that contribute to a poor surface finish in the processes described in this chapter.

9.64 Explain why laser microjet has a large depth of field.

9.65 Explain the principle of hybrid machining.

PROBLEMS

9.66 In a surface-grinding operation, calculate the chip dimensions for the following process variables: $D = 200$ mm, $d = 0.025$ mm, $v = 0.15$ m/s, $V = 25$ m/s, $C = 1$ per mm^2, and $r = 20$.

9.67 If the workpiece strength in grinding is increased by 50%, what should be the percentage decreases in the wheel depth of cut, d, in order to maintain the same grain force, all other variables being the same?

9.68 Taking a thin Type 1 grinding wheel as an example and referring to texts on stresses in rotating bodies, plot the tangential stress, σ_t, and radial stress, σ_r, as a function of radial distance, i.e., from the hole to the periphery of the wheel. Note that because the wheel is thin, this problem can be regarded as a plane-stress problem. How would you determine the maximum combined stress and its location in the wheel?

9.69 Derive a formula for the material removal rate (MRR) in surface grinding in terms of process parameters. Use the same terminology as in the text.

9.70 Assume that a surface-grinding operation is being carried out under the following conditions: $D = 250$ mm, $d = 0.1$ mm, $v = 0.5$ m/s, and $V = 50$ m/s. These conditions are then changed to the following: $D = 150$ mm, $d = 0.1$ mm, $v = 0.3$ m/s, and $V = 25$ m/s. What is the difference in the temperature rise from the initial condition?

9.71 For a surface-grinding operation, derive an expression for the power dissipated in imparting kinetic energy to the chips. Comment on the magnitude of this energy. Use the same terminology as in the text.

9.72 The shaft of a Type 1 grinding wheel is attached to a flywheel only, which is rotating at a certain initial rpm. With this setup, a surface-grinding operation is being carried out on a long workpiece and at a constant workpiece speed, v. Obtain an expression for estimating the linear distance ground on the workpiece before the wheel comes to a stop. Ignore wheel wear.

9.73 Calculate the average impact force on a steel plate by a spherical aluminum-oxide abrasive particle with a 1-mm diameter, dropped from heights of (a) 1 m; (b) 2 m; and (c) 10 m.

9.74 A 50-mm-deep hole, 25 mm in diameter, is being produced by electrochemical machining. A high production rate is more important than the quality of the machined surface. Estimate the maximum current and the time required to perform this operation.

9.75 If the operation in Problem 9.74 were performed on an electrical-discharge machine, what would be the estimated machining time?

9.76 A cutting-off operation is being performed with a laser beam. The workpiece being cut is 5 mm thick and 100 mm long. If the kerf width is 3 mm, estimate the time required to perform this operation.

9.77 Referring to Table 3.3, identify two metals or metal alloys that, when used as workpiece and electrode, respectively, in EDM would give the (a) lowest and (b) highest wear ratios, R. Calculate these quantities.

9.78 In Section 9.5.2, it was stated that, in practice, grinding ratios typically range from 2 to 200. Based on the information given in Section 9.13, estimate the range of wear ratios in electrical-discharge machining, and then compare them with grinding ratios.

9.79 It is known that heat checking occurs when grinding with a spindle speed of 4000 rpm, a wheel diameter of 250 mm, and a depth of cut of 0.0375 mm for a feed rate of 0.25 m/s. For this reason, the spindle speed should be kept at 3500 rpm. If a new, 200-mm-diameter wheel is used, what spindle speed can be employed before heat checking occurs? What spindle speed should be used to keep the same grinding temperatures as those encountered with the existing operating conditions?

9.80 It is desired to grind a hard aerospace aluminum alloy. A depth of 0.075 mm is to be removed from a cylindrical section 200 mm long and with a 75-mm diameter. If each part is to be ground in not more than one minute, what is the approximate power requirement for the grinder? What if the material is changed to a hard titanium alloy?

9.81 A grinding operation is taking place with a 250-mm grinding wheel at a spindle rotational speed of 4000 rpm. The workpiece feed rate is 0.25 m/s, and the depth of cut is 0.050 mm. Contact thermometers record an approximate maximum temperature of 950°C. If the workpiece is steel, what is the temperature if the spindle speed is increased to 5000 rpm? What if it is increased to 10,000 rpm?

9.82 The regulating wheel of a centerless grinder is rotating at a surface speed of 0.125 m/s and is inclined at an angle of 5°. What is the feed rate of material past the grinding wheel?

9.83 Using some typical values, explain what changes, if any, take place in the magnitude of impact force of a particle in ultrasonic machining as the temperature of the workpiece is increased. Assume that the workpiece is made of hardened steel.

9.84 Estimate the percent increase in the cost of the grinding operation if the specification for the surface finish of a part is changed from 1.5 μm to 0.4 μin.

9.85 Assume that the energy cost for grinding an aluminum part, with a specific energy requirement of 8 W-s/mm^3, is \$0.90 per piece. What would be the energy cost of carrying out the same operation if the workpiece material is T15 tool steel?

9.86 Derive an expression for the angular velocity of the wafer as a function of the radius and angular velocity of the pad in chemical mechanical polishing.

9.87 A 25-mm-thick copper plate is being machined through wire EDM. The wire moves at a speed of 0.0125 m/s and the kerf width is 0.5 mm. What is the required power? Note that the latent heat of fusion for copper is 2.05×10^5 J/kg, and the specific heat for copper is 385 J/kgK.

9.88 A 200-mm diameter grinding wheel, 25 mm wide, is used in a surface grinding operation performed on a flat piece of heat-treated 4340 steel. The wheel is rotating with a surface speed $V = 25$ m/s, depth of cut $d = 0.05$ mm/pass, and cross feed $w = 4.0$ mm. The reciprocating speed of the work is $v = 0.1$ m/s, and the operation is performed dry. (a) What is the length of contact between the wheel and the work? (b) What is the volume rate of metal removed? (c) If $C = 0.5$ per mm^2, estimate the number of chips formed per time. (d) What is the average volume per chip? (e) If the tangential cutting force on the workpiece, $F_c = 50$ N, what is the specific energy for the operation?

9.89 A 150 mm diameter tool steel ($u = 60$ W-s/mm^3) work roll for a metal rolling operation is being ground by a 250 mm diameter, 75 mm wide, Type I grinding wheel. Estimate the chip dimensions if $d = 0.04$ mm and $C = 5$ grains per mm^2. If the wheel rotates at $N = 3000$ rpm, estimate the cutting force if the work roll rotates at 1 rpm.

9.90 Estimate the contact time and average force for the following particles striking a steel workpiece at 1 m/s. Comment on your findings. (a) 5 mm diameter steel shot; (b) 10 mm diameter ceramic beads; (c) 3 mm diameter tungsten sphere; (d) 75 mm diameter rubber ball; and (e) 3 mm diameter glass beads.

9.91 Assume that you are an instructor covering the topics in this chapter, and you are giving a quiz on the quantitative aspects to test the understanding of the students. Prepare three quantitative problems, and supply the answers.

DESIGN

9.92 Would you consider designing a machine tool that combines, in one machine, two or more of the processes described in this chapter? Explain. For what types of parts would such a machine be useful? Make preliminary sketches for such machines.

9.93 With appropriate sketches, describe the principles of various fixturing methods and devices that can be used for each of the processes described in this chapter.

9.94 Make a list of machining processes that may be suitable for each of the following materials: (a) ceramics; (b) cast iron; (c) thermoplastics; (d) thermosets; (e) diamond; and (f) annealed copper.

9.95 Describe any workpiece size limitations in advanced machining processes. Give examples.

9.96 Surface finish can be an important consideration in the design of products. Describe as many parameters as you can that could affect the final surface finish in grinding, including the role of process parameters as well as the setup and the equipment used.

9.97 A somewhat controversial subject in grinding is size effect (see Section 9.4.1). Design a setup and a series of experiments whereby size effect can be studied.

9.98 Describe the types of parts that would be suitable for hybrid machining. Consider one such part and make a preliminary sketch for a hybrid machine to produce that part.

9.99 Describe how the design and geometry of the workpiece affects the selection of an appropriate shape and type of a grinding wheel.

9.100 Prepare a comprehensive table of the capabilities of abrasive machining processes, including the shapes of parts ground, types of machines involved, typical maximum and minimum workpiece dimensions, and production rates.

9.101 How would you produce a thin, large-diameter round disk with a thickness that decreases linearly from the center outward?

9.102 Marking surfaces of manufactured parts with letters and numbers can be done not only with labels and stickers, but also by various mechanical and nonmechanical means. Make a list of some of these methods, explaining their advantages and limitations.

9.103 On the basis of the information given in Chapter 8 and this chapter, comment on the feasibility of producing a 10 mm hole 100 mm deep in a copper alloy (a) by conventional drilling, and (b) by internal grinding.

9.104 Make a literature search and explain how observing the color, brightness, and shape of sparks produced in grinding can be a useful guide to identifying the type of material being ground and its condition.

9.105 Visit a large hardware store and inspect the grinding wheels on display. Make a note of the markings on the wheels and, based on the marking system shown in Figs. 9.4 and 9.5, comment on your observations, including the most common types of wheels available in the store.

9.106 Obtain a small grinding wheel or a piece of a large wheel, and (a) observe its surfaces using a magnifier or a microscope, and compare with Fig. 9.6, and (b) rub the abrasive wheel pressing it hard against a variety of flat metallic and nonmetallic materials. Describe your observations regarding the surface produced.

9.107 In reviewing the abrasive machining processes in this chapter, you will note that some use bonded abrasives while others involve loose abrasives. Make two separate lists for these processes, and comment on your observations.

9.108 We have seen that there are several holemaking methods. Based on the topics covered in Chapters 6–9, make a comprehensive table of holemaking processes. Describe the advantages and limitations of each method, comment on the quality and surface integrity of the holes produced, and give examples of specific applications.

9.109 Precision engineering is a term that is used to describe manufacturing high-quality parts with close dimensional tolerances and good surface finish. Based on their process capabilities, make a list of advanced machining processes, with decreasing order of quality of parts produced. Include a brief commentary on each method.

9.110 We have seen that several of the processes described in this chapter can be employed, either singly or in combination, to make or finish dies for metalworking operations. Write a brief technical paper on these methods, describing their advantages and limitations, and typical applications.

9.111 Would the processes described in this chapter be difficult to perform on various nonmetallic or rubberlike materials? Explain your thoughts, commenting on the influence of various physical and mechanical properties of workpiece materials, part geometries, etc.

9.112 Make a list of the processes described in this chapter in which the following properties are relevant: (a) mechanical, (b) chemical, (c) thermal, and (d) electrical. Are there processes in which two or more of these properties are important? Explain.

CHAPTER

10 Polymer Processing and Additive Manufacturing

This chapter covers the characteristics and processing of polymers and reinforced plastics (composite materials), including:

- The structure, properties, and behavior of thermoplastics, thermosets, elastomers, and reinforced plastics, and their manufacturing characteristics.
- The processing methods employed to produce parts, beginning with extrusion and followed by various forming and molding processes, and the processing parameters and the equipment involved.
- Fundamentals of various processes that are unique to producing reinforced plastic parts and the equipment involved.
- The principles of various rapid prototyping operations, where short runs of predominantly plastic parts can be produced quickly.
- Design considerations in forming polymers and reinforced plastics.
- The processes used in additive manufacturing to produce parts quickly from computer descriptions of part geometry, and without the use of tooling.
- The economics of processing these materials into products.

Symbols

A	consistency index
A_s	area, m^2
B	laser spot diameter, m
C	a constant
C_d	cure depth, m
D	screw diameter, m
D_d	die opening diameter,m
D_p	penetration depth, m
E	elastic modulus, N/m^2 also, exposure, N/m also, activation energy, Nm
E_o	exposure at surface, N/m
E_r	viscoelastic modulus, N/m^2
F	force, N
G	shear modulus, N/m^2
H	flight depth, m
k	stiffness, N/m
K	die characteristic coefficient, m^5/Ns
l	length of pumping section, m
l_d	die land, m
L_w	line width, m
n	power law index
N	angular speed, rpm
p	pressure, N/m^2
Q_d	drag flow, m^3/s
Q_p	pressure flow, m^3/s
t	time, s

T temperature, °C
T_g glass transition temperature, °C
T_m melting temperature, °C
v velocity, m/s
w flight width, m
W ribbon width, m
x Cartesian coordinate, m
also, fiber volume fraction
y Cartesian coordinate, m
z Cartesian coordinate, m
ϵ true strain
$\dot{\epsilon}$ true strain rate, s^{-1}
η viscosity, Ns/m^2
κ Boltzmann's constant,
λ rate of decay of stress, s
γ shear strain, m/m
$\dot{\gamma}$ shear strain rate, s^{-1}
ν Poisson's ratio
σ normal stress, N/m^2
τ shear stress, N/m^2
θ flight angle, deg
ω angular velocity, deg/s

Subscripts

c composite
f fiber
m matrix

10.1 Introduction

The word **plastics** was first used around 1909 and is a common synonym for polymers. *Plastics*, from the Greek word *plastikos*, means "it can be molded and shaped." Plastics are one of numerous polymeric materials having extremely large molecules (*macromolecules*). Consumer and industrial products made of polymers include food and beverage containers, packaging, signs, housewares, textiles, medical devices, foams, paints, safety shields, and toys. Compared with metals, polymers are generally characterized by low density, low strength and stiffness (Table 10.1), low electrical and thermal conductivity, good resistance to chemicals, and a high coefficient of thermal expansion.

The useful temperature range for most polymers is generally low, being up to about 350°C, and they are not as dimensionally stable in service as metals. Plastics can be machined, cast, formed, and joined into a wide variety of complex shapes with relative ease; few, if any, additional surface-finishing operations are required, an important advantage over metals. Plastics are commercially available as sheet, plate, film, rods, and tubing of various cross sections.

The word *polymer* was first used in 1866. The earliest polymers were made of **natural organic materials** from animal and vegetable products, *cellulose* being the most common example. With various chemical reactions, cellulose is modified into *cellulose acetate*, used in making sheets for packaging and textile fibers. It is also modified into *cellulose nitrate* for plastics, explosives, rayon (a cellulose textile fiber), and varnishes. The earliest **synthetic polymer** was *phenol formaldehyde*, a thermoset developed in 1906 and called *Bakelite* (a trade name, after L.H. Baekeland, 1863–1944).

The development of modern plastics technology began in the 1920s, when raw materials necessary for making polymers were extracted from coal and petroleum products. *Ethylene* was the first example of such a

TABLE 10.1 Range of mechanical properties for various engineering plastics at room temperature.

Material	Ultimate tensile strength (MPa)	Elastic modulus (GPa)	Elongation (%)	Poisson's ratio, ν
Thermoplastics:				
ABS	28–55	1.4–2.8	75–5	–
ABS, reinforced	100	7.5	–	0.35
Acetal	55–70	1.4–3.5	75–25	–
Acetal, reinforced	135	10	–	0.35–0.40
Acrylic	40–75	1.4–3.5	50–5	–
Cellulosic	10–48	0.4–1.4	100–5	–
Fluorocarbon	7–48	0.7–2	300–100	0.46–0.48
Nylon	55–83	1.4–2.8	200–60	0.32–0.40
Nylon, reinforced	70–210	2–10	10–1	–
Polycarbonate	55–70	2.5–3	125–10	0.38
Polycarbonate, reinforced	110	6	6–4	–
Polyester	55	2	300–5	0.38
Polyester, reinforced	110–160	8.3–12	3–1	–
Polyethylene	7–40	0.1–1.4	1000–15	0.46
Polypropylene	20–35	0.7–1.2	500–10	–
Polypropylene, reinforced	40–100	3.5–6	4–2	–
Polystyrene	14–83	1.4–4	60–1	0.35
Polyvinyl chloride	7–55	0.014–4	450–40	–
Thermosets:				
Epoxy	35–140	3.5–17	10–1	–
Epoxy, reinforced	70–1400	21–52	4–2	–
Phenolic	28–70	2.8–21	2–0	–
Polyester, unsaturated	30	5–9	1–0	–
Elastomers:				
Chloroprene (neoprene)	15–25	1–2	100–500	0.5
Natural rubber	17–25	1.3	75–650	0.5
Silicone	5–8	1–5	100–1100	0.5
Styrene-butadiene	10–25	2–10	250–700	0.5
Urethane	20–30	2–10	300–450	0.5

raw material, as it became the building block for *polyethylene*. It is the product of the reaction between acetylene and hydrogen, and acetylene itself is the product of the reaction between coke and methane. Likewise, commercial polymers, including polypropylene, polyvinyl chloride, polymethylmethacrylate, and polycarbonate, are all made in a similar manner; these materials are known as *synthetic organic polymers*. An outline of the basic structure of various synthetic polymers is given in Fig. 10.1.

An important group of polymeric materials is **polymer-matrix reinforced plastics** (a type of **composite material**), which exhibit a wide range of

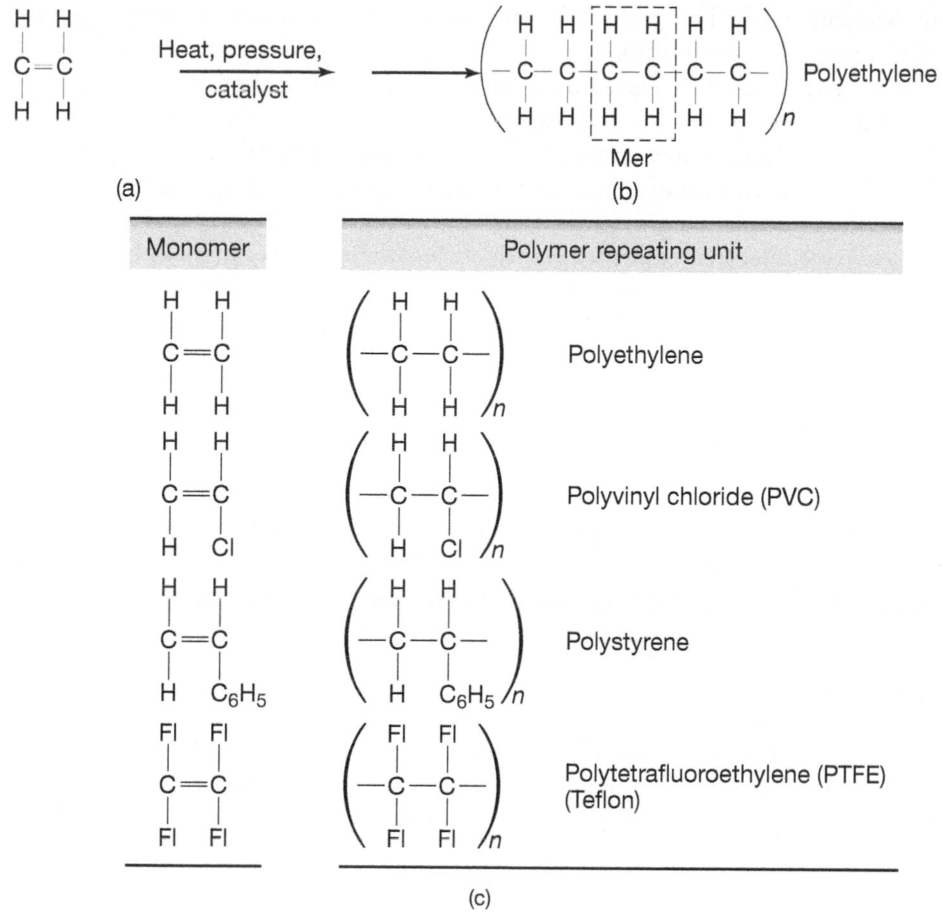

$$H_2C{=}CH_2 \xrightarrow[\text{catalyst}]{\text{Heat, pressure,}} \left(-CH_2-CH_2-\right)_n \quad \text{Polyethylene}$$

Mer

(a) (b)

Monomer	Polymer repeating unit	
$HHC{=}CHH$	$(-CHH-CHH-)_n$	Polyethylene
$HHC{=}CHCl$	$(-CHH-CHCl-)_n$	Polyvinyl chloride (PVC)
$HHC{=}CH(C_6H_5)$	$(-CHH-CH(C_6H_5)-)_n$	Polystyrene
$FlFlC{=}CFlFl$	$(-CFlFl-CFlFl-)_n$	Polytetrafluoroethylene (PTFE) (Teflon)

(c)

FIGURE 10.1 Basic structure of some polymer molecules: (a) ethylene molecule; (b) polyethylene, a linear chain of many ethylene molecules; and (c) molecular structure of various polymers. These molecules are examples of the basic building blocks for plastics.

properties such as stiffness, strength, resistance to creep, and high strength-to-weight and stiffness-to-weight ratios. Their applications include a wide variety of consumer and industrial products, as well as products for the automotive and aerospace industries.

10.2 The Structure of Polymers

A polymer's properties basically depend on (a) the structure of individual polymer molecules; (b) the shape and size of these molecules; and (c) how they are arranged to form a polymer structure.

A **monomer** is the basic building block of polymers. The word **mer**, from the Greek *meros* meaning "part," indicates the smallest repeating unit, similar to *unit cell* in the crystal structures of metals (described in Section 3.2). The term **polymer** means many mers or units, generally repeated in a chainlike structure. Polymers are *long-chain molecules*, also called **macromolecules** or **giant molecules** that are formed by *polymerization*, that is, by linking and cross-linking different monomers. Polymer molecules are characterized by their extraordinary size, a trait that distinguishes them from other organic chemical compositions.

Most monomers are organic materials in which carbon atoms are joined in *covalent* (electron sharing) bonds with other atoms, such as hydrogen, oxygen, nitrogen, fluorine, chlorine, silicon, and sulfur. For example, an ethylene molecule is a simple monomer consisting of carbon and hydrogen atoms (Fig. 10.1a).

10.2.1 Polymerization

Monomers can be linked in repeating units to make longer and larger molecules by a chemical process called *polymerization reaction*. Polymerization processes are complex and are described only briefly here. Although there are several variations, two basic polymerization processes are condensation polymerization and addition polymerization.

1. In **condensation polymerization**, polymers are produced by the formation of bonds between two types of reacting mers. A characteristic of this reaction is that the reaction by-products, such as water, are condensed out, hence the term *condensation*. This process is also known as *step-growth* or *step-reaction polymerization*, because the polymer molecule grows step by step until all of one reactant is consumed.
2. In **addition polymerization**, also known as *chain-growth* or *chain-reaction polymerization*, bonding takes place without reaction by-products. It is called *chain-reaction polymerization* because of the high rate at which long molecules form simultaneously, usually within a few seconds, a rate much higher than that for condensation polymerization. In this reaction, an *initiator* is added to open the double bond between the carbon atoms and begins the linking process by adding more monomers to a growing chain. For example, ethylene monomers (Fig. 10.1a) link to produce polyethylene (Fig. 10.1b); other examples of addition-formed polymers are shown in Fig. 10.1c.

Some basic characteristics of polymers are described below.

1. **Molecular weight.** The sum of the weight of the atoms in the polymer chain is the *molecular weight* of the polymer. The higher the molecular weight in a given polymer, the greater is its chain length. Because polymerization is a random event, the polymer chains produced are not all of equal length, and the chain lengths produced fall into a traditional distribution curve (See Fig. 11.5). The *average* molecular weight of a polymer is determined and expressed statistically as yielding a **molecular weight distribution** (MWD). Average molecular weight and molecular weight distribution have a strong influence on the properties of the polymer. For example, tensile and impact

strength, resistance to cracking, and viscosity in the molten state of a polymer all increase with increasing molecular weight (Fig. 10.2). Most commercial polymers have a molecular weight between 10,000 and 10,000,000.

2. **Degree of polymerization.** It is sometimes more convenient to express the size of a polymer chain in terms of the *degree of polymerization* (DP), defined as the ratio of the molecular weight of the polymer to the molecular weight of the repeating unit. In polymer processing, described in Section 10.10, the higher the DP, the higher is the polymer's viscosity, or resistance to flow (Fig. 10.2), which can affect ease of shaping and the overall cost of processing.
3. **Bonding.** During polymerization, the monomers are linked together in a *covalent bond*, forming a polymer chain. Because of their strength, covalent bonds are also called **primary bonds**. The polymer chains are, in turn, held together by the secondary bonds, such as van der Waals bonds, hydrogen bonds, and ionic bonds. **Secondary bonds** are much weaker than primary bonds, by one to two orders of magnitude. In a given polymer, the increase in strength and viscosity with molecular weight comes, in part, from the fact that the longer the polymer chain, the greater is the energy required to overcome the strength of the secondary bonds. For example, ethylene mers with a DP of 1, 6, 35, 140, and 1350 are, respectively, in the form of gas, liquid, grease, wax, and hard plastic at room temperature.
4. **Linear polymers.** The chainlike polymers shown in Fig. 10.1 are called *linear polymers*, because of their linear structure (Fig. 10.3a). A linear molecule does not, however, mean that it is straight in shape. In addition to those shown in Fig. 10.3, other linear polymers include polyamides (nylon 6,6) and polyvinyl fluoride. Generally, a polymer consists of more than one type of structure; thus, a polymer characterized as linear may still contain some branched and cross-linked chains (see parts 5 and 6 of this list). A polymer's mechanical properties, such as strength and ductility, are strongly dependent on branching and cross-linking.
5. **Branched polymers.** The properties of a polymer depend not only on the type of monomers in the polymer but also on their arrangement in the molecular structure. As shown in Fig. 10.3b, *branched polymers* have side branch chains that are bonded to the main chain during the synthesis of the polymer. Branching interferes with the relative movement of the molecular chains; as a result, resistance to deformation increases and stress cracking resistance is affected. Also, the density of branched polymers is lower than that of linear-chain polymers, because branches interfere with the packing efficiency of polymer chains. The behavior of branched polymers can be compared with that of linear-chain polymers by making an analogy with a pile of tree branches (regarded as branched polymers) and a bundle of straight logs (regarded as linear-chain polymers). It will be noted that it is more difficult to move a branch within the pile of branches than to move a log in a bundle of logs. The three-dimensional entanglements

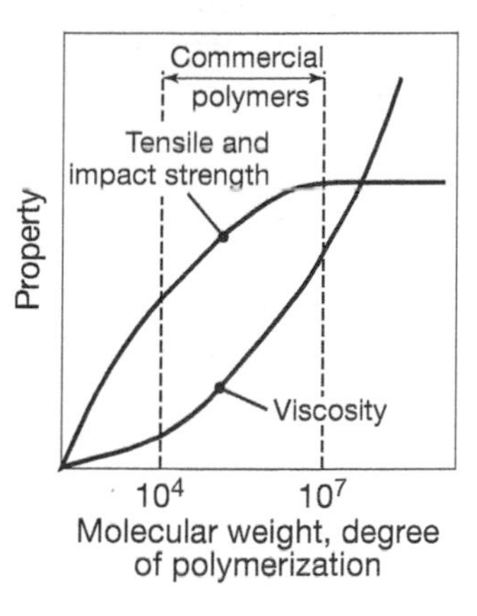

FIGURE 10.2 Effect of molecular weight and degree of polymerization on the strength and viscosity of polymers.

of branches make movements more difficult, a phenomenon akin to increasing strength.

6. **Cross-linked polymers.** Generally three-dimensional in structure, *cross-linked polymers* have adjacent chains linked by covalent bonds (Fig. 10.3c). Polymers with a cross-linked structure are called **thermosets** or **thermosetting plastics**; examples include epoxies, phenolics, and silicones (see Section 10.6). Cross-linking has a major influence on the properties of polymers, generally improving hardness, strength, stiffness, dimensional stability (See Fig. 10.4), but also increasing brittleness. A related process is the **vulcanization** of rubber (Section 10.8).

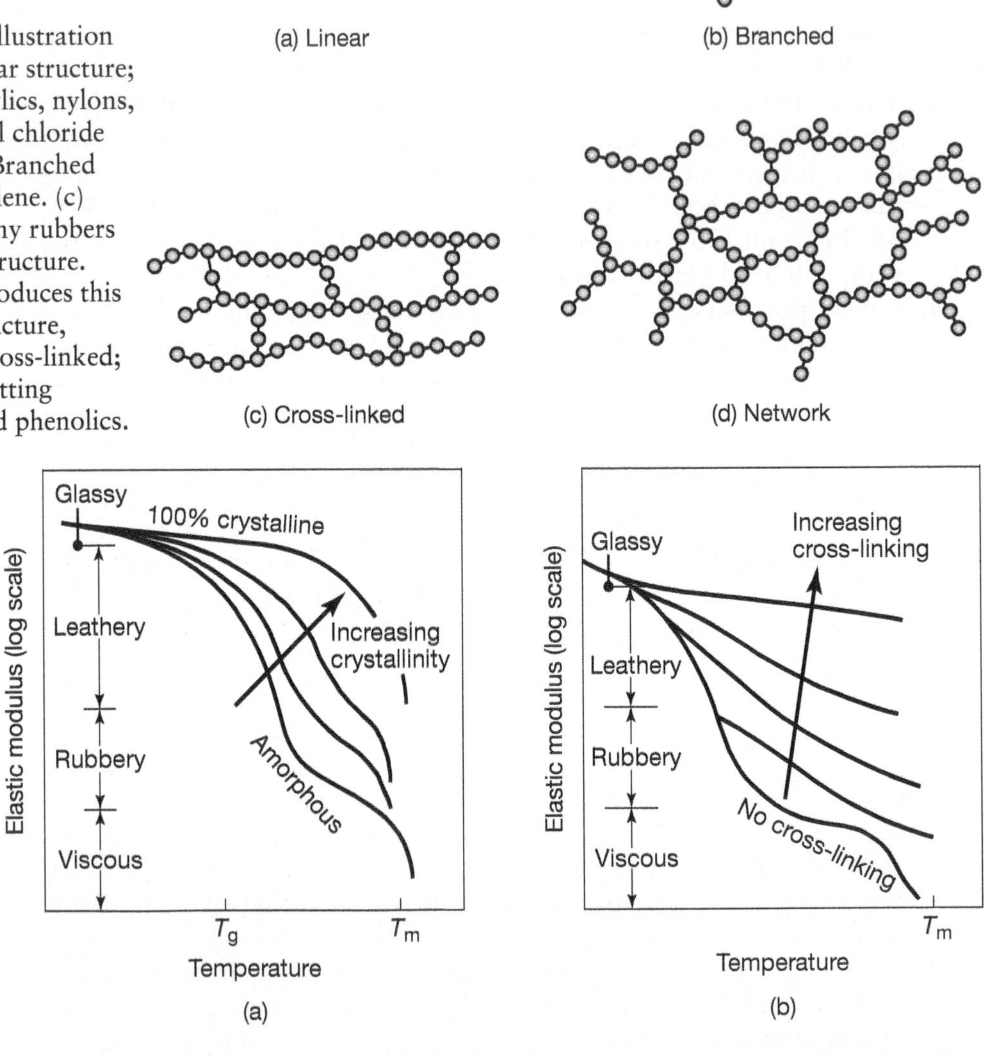

FIGURE 10.3 Schematic illustration of polymer chains. (a) Linear structure; thermoplastics such as acrylics, nylons, polyethylene, and polyvinyl chloride have linear structures. (b) Branched structure, such as polyethylene. (c) Cross-linked structure; many rubbers and elastomers have this structure. Vulcanization of rubber produces this structure. (d) Network structure, which is basically highly cross-linked; examples include thermosetting plastics such as epoxies and phenolics.

FIGURE 10.4 Behavior of polymers as a function of temperature and (a) degree of crystallinity and (b) cross-linking. The combined elastic and viscous behavior of polymers is known as viscoelasticity.

Network polymers consist of three-dimensional networks of active covalent bonds, as shown in Fig. 10.3d; a highly cross-linked polymer is also considered a network polymer.

7. **Copolymers and terpolymers.** If the repeating units in a polymer chain are all of the same type, the molecule is called a **homopolymer.** However, as with solid-solution metal alloys (see Section 5.2.1), two or three different types of monomers can be combined to impart certain special properties to the polymer, such as improvement in strength, toughness, and formability of the polymer. *Copolymers* contain two types of polymers, such as styrene-butadiene, used widely for automobile tires. *Terpolymers* contain three types, such as ABS (acrylonitrile-butadiene-styrene), used for helmets and refrigerator liners.

EXAMPLE 10.1 Degree of Polymerization in Polyvinyl Chloride (PVC)

Given: A polyvinyl chloride (PVC) polymer has an average molecular weight of 50,000.

Find: Determine the molecular weight of a polyvinyl chloride mer. What is the degree of polymerization for the polymer?

Solution: From Fig. 10.1c, note that each PVC mer has three hydrogen atoms, two carbon atoms, and one chlorine atom. Since the atomic numbers of these elements are 1, 12, and 35.5, respectively, the molecular weight of a PVC mer is $(3)(1) + (2)(12) + (1)(35.5) = 62.5$; and the degree of polymerization is $50,000/62.5 = 800$.

10.2.2 Crystallinity

Polymers such as polymethylmethacrylate, polycarbonate, and polystyrene are generally *amorphous*; that is, the polymer chains exist without long-range order (see also the discussion of *amorphous alloys* in Section 3.11.9). The amorphous arrangement of polymer chains is often described as similar to a bowl of spaghetti, or worms in a bucket, all intertwined with each other. In some polymers, however, it is possible to impart some crystallinity and thereby modify their characteristics. This may be done either during the synthesis of the polymer or by deformation (shaping) during its subsequent processing.

The crystalline regions in polymers are called **crystallites** (Fig. 10.5). The crystals are formed when the long molecules arrange themselves in an orderly manner, similar to folding a fire hose in a cabinet or facial tissue in a box. Thus, a partially crystalline (semicrystalline) polymer can be regarded as a two-phase material (see Section 5.2.3), one phase being crystalline and the other amorphous. By controlling the rate of solidification during cooling, it is possible to impart different **degrees of crystallinity** to polymers, although a polymer can never be 100% crystalline.

Crystallinity ranges from an almost complete crystal (up to about 95% by volume in the case of polyethylene) to slightly crystallized but mostly

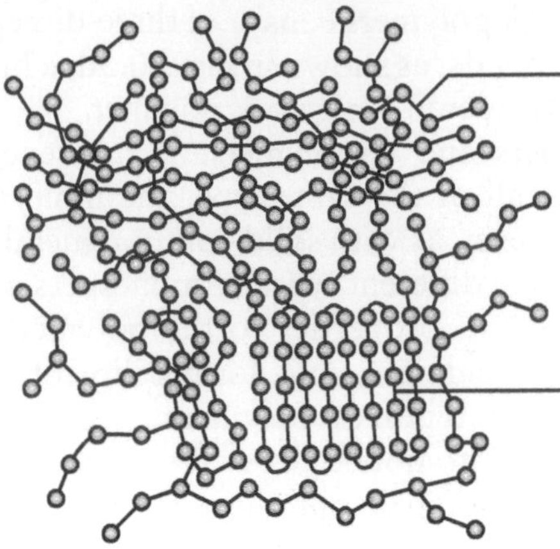

Amorphous region

Crystalline region

FIGURE 10.5 Amorphous and crystalline regions in a polymer. Note that the crystalline region (crystallite) has an orderly arrangement of molecules. The higher the crystallinity, the harder, stiffer, and less ductile is the polymer.

amorphous polymers. The degree of crystallinity is also affected by branching. A linear polymer can become highly crystalline, but a highly branched polymer cannot. Although the latter may develop some low degree of crystallinity, it will never acquire a high crystalline content, because the branches interfere with the alignment of the chains into a regular crystal array.

Effects of crystallinity. The mechanical and physical properties of polymers are greatly influenced by the degree of crystallinity. As crystallinity increases, polymers become stiffer, harder, less ductile, denser, less rubbery, and more resistant to solvents and heat (Fig. 10.4). The increase in density with increasing crystallinity is caused by crystallization shrinkage and a more efficient packing of the molecules in the crystal lattice. For example, the highly crystalline form of polyethylene, known as *high-density polyethylene* (HDPE), has a specific gravity in the range of 0.941 to 0.970 (80% to 95% crystalline) and is stronger, stiffer, tougher, and less ductile than low-density polyethylene (LDPE), which is about 60% to 70% crystalline and has a specific gravity of about 0.910 to 0.925.

Optical properties of polymers also are affected by the degree of crystallinity. The reflection of light from the boundaries between the crystalline and the amorphous regions in the polymer (See Fig. 10.5) causes opaqueness. Furthermore, because the index of refraction is proportional to density, the higher the density difference between the amorphous and the crystalline phases, the greater is the opaqueness of the polymer. Polymers that are completely amorphous can be transparent, such as polycarbonate and acrylics.

10.2.3 Glass-Transition Temperature

Amorphous polymers do not have a specific melting point, but they undergo a distinct change in their mechanical behavior across a narrow range of temperature. The temperature at which this transition occurs is called the *glass-transition temperature*, T_g, or the *glass point* or *glass temperature*. At low temperatures, amorphous polymers are hard, rigid, brittle, and glassy, and at high temperatures, they are rubbery or leathery.

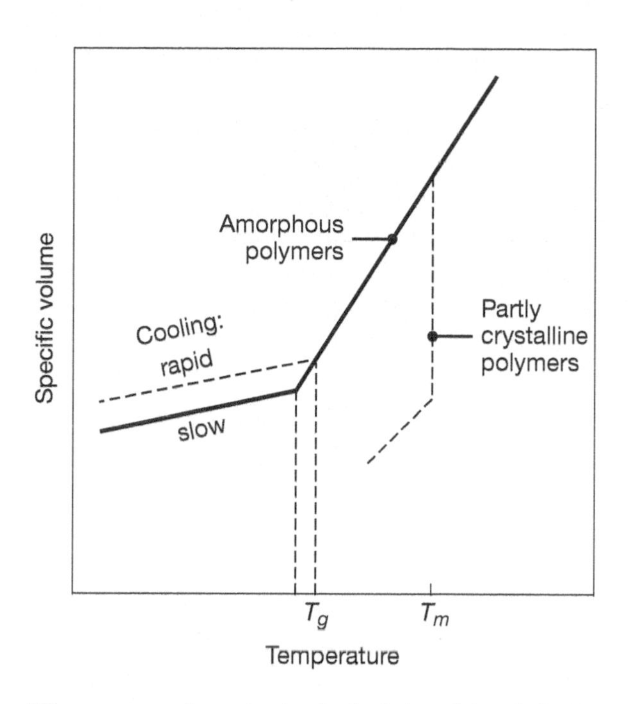

FIGURE 10.6 Specific volume of polymers as a function of temperature. Amorphous polymers, such as acrylic and polycarbonate, have a glass-transition temperature, T_g, but do not have a specific melting point, T_m. Partly crystalline polymers, such as polyethylene and nylons, contract sharply at their melting points during cooling.

The term *glass* is included in this definition, because glasses, which are amorphous solids (see Section 3.11.9), display the same behavior. Although most amorphous polymers exhibit this behavior, there are some exceptions, such as polycarbonate, which is (a) not rigid or brittle below its glass-transition temperature; and (b) tough at ambient temperature and is therefore suitable for safety helmets and shields.

To determine T_g, the specific volume of the polymer is measured and plotted against temperature and the sharp change in the slope of the curve is noted, as can be seen in Fig. 10.6. For highly cross-linked polymers, the slope of the curve changes *gradually* near T_g, thus making it difficult to determine T_g. The glass-transition temperature varies for different polymers (Table 10.2); for example, room temperature is above T_g for some polymers and below it for others. Unlike amorphous polymers, partly crystalline polymers have a distinct melting point, T_m (Fig. 10.6; see also Table 10.2). Because of the structural changes (known as first-order changes) that occur, the specific volume of the polymers drops rapidly as their temperature is reduced.

10.2.4 Polymer Blends

To improve the brittle behavior of amorphous polymers below their glass-transition temperature, they can be blended, usually with small quantities of an elastomer (see Section 10.8), known as **rubber-modified polymers**. These tiny particles are dispersed throughout the amorphous polymer, enhancing its toughness and impact strength by improving its resistance to crack propagation. Blending may involve several components (**polyblends**) to obtain the favorable properties of different polymers. **Miscible blends** involve mixing without separation of two phases (a process similar to alloying of metals) and enable polymer blends to become more ductile. Polymer blends account for about 20% of all polymers produced.

TABLE 10.2 Glass-transition and melting temperatures of selected polymers.

Material	T_g (°C)	T_m (°C)
Nylon 6,6	57	265
Polycarbonate	150	265
Polyester	73	265
Polyethylene		
High density	−90	137
Low density	−110	115
Polymethylmethacrylate	105	–
Polypropylene	−14	176
Polystyrene	100	239
Polytetrafluoroethylene (Teflon)	−90	327
Polyvinyl chloride	87	212
Rubber	−73	–

10.2.5 Additives in Polymers

In order to impart certain specific properties, polymers are usually compounded with *additives*. The additives modify and improve such characteristics as stiffness, strength, color, weatherability, flammability, arc resistance for electrical applications, and ease of subsequent processing of the polymer.

1. **Fillers** are generally wood flour (fine sawdust), silica flour (fine silica powder), clay, powdered mica, and short fibers of cellulose or glass. Depending on their type, fillers improve the strength, hardness, toughness, abrasion resistance, dimensional stability, and stiffness of plastics. As in reinforced plastics (Section 10.9), a filler's effectiveness depends on the nature and strength of the bond between the filler material and the polymer chains. Because of their low cost, fillers are important in reducing the overall cost of polymers.
2. **Plasticizers** are added to some polymers to impart flexibility and softness by lowering their glass-transition temperature. Plasticizers are low-molecular-weight solvents with high boiling points (they are nonvolatile). They reduce the strength of the secondary bonds between the long-chain molecules, thus making the polymer soft and flexible. The most common use of plasticizers is in polyvinyl chloride (PVC), which remains flexible during its many uses. Other applications of plasticizers include thin sheet, film, tubing, shower curtains, and clothing fibers.
3. Most polymers are adversely affected by **ultraviolet radiation** (such as in sunlight) and oxygen, which weaken and break the primary bonds, resulting in the *scission* (splitting) of the long-chain molecules; the polymer then degrades and becomes brittle and stiff.

On the other hand, degradation may be beneficial, as in the disposal of plastic objects by subjecting them to environmental attack, or in the preferred etching of polymethylmethacrylate in LIGA (see Section 13.16 and *biodegradable plastics* in Section 10.7.3.) A typical example of improving resistance to ultraviolet radiation is the compounding of some plastics and rubbers with *carbon black* (soot), which absorbs a high percentage of the ultraviolet radiation. Protection against degradation by oxidation, particularly at elevated temperatures, is achieved by adding *antioxidants* to the polymer. Applying various *coatings* is another method of protecting polymers against environmental attack.

4. The wide variety of colors available in plastics is obtained by adding **colorants** that are either organic (*dyes*) or inorganic (*pigments*). Pigments are dispersed particles and generally have greater resistance than dyes to temperature and light. The selection of a colorant depends on the polymer's service temperature and the length of exposure to light.
5. Most polymers will ignite if the temperature is sufficiently high. The **flammability** of polymers (defined as the ability to support combustion) varies considerably, depending on their composition, such as the chlorine and fluorine content. Flammability can be reduced either by making the polymer from less flammable raw materials or by adding **flame retardants**, such as compounds of chlorine, bromine, and phosphorus. Examples of polymers with different burning characteristics include: (a) Fluorocarbons (such as *Teflon*) which do not burn; (b) carbonate, nylon, and vinyl chloride which burn but are *self-extinguishing*; and (c) acetal, acrylic, ABS, polyester, polypropylene, and styrene which burn and are not self-extinguishing.
6. **Lubricants** may be added to polymers to reduce friction during their subsequent processing into products and to prevent the parts from sticking to molds. Lubrication is also important in preventing thin polymer films from sticking to each other (see also Section 4.4.3).

10.3 Thermoplastics: Behavior and Properties

Recall that the bonds between adjacent long-chain molecules (secondary bonds) are much weaker than the covalent bonds (primary bonds) within each molecule, and that it is the strength of the secondary bonds that determines the overall strength of the polymer. Linear and branched polymers have weak secondary bonds. As the temperature of the polymer is raised above glass-transition temperature or, more commonly, its melting point, certain polymers become easier to form or mold into desired shapes. The increased temperature weakens the secondary bonds, and the adjacent chains can thus move more easily under the shaping forces. If the polymer is then cooled, it returns to its original hardness and strength; in other words, the effects of this process are *reversible*. The repeated heating and cooling of thermoplastics can, however, cause **degradation** (**thermal aging**).

Polymers that exhibit this behavior are known as **thermoplastics**; typical examples include acrylics, cellulosics, nylons, polyethylenes, and polyvinyl chloride. The behavior of thermoplastics depends primarily on temperature and rate of deformation. Below the glass-transition temperature, most polymers are *glassy* (described as rigid, brittle, or hard), and they behave like an elastic solid. If the load exceeds a certain critical value, the polymer fractures, just as a piece of glass does at room temperature. In the glassy region, the relationship between stress and strain is linear, or

$$\sigma = E\epsilon. \tag{10.1}$$

If the polymer is tested in torsion, then

$$\tau = G\gamma. \tag{10.2}$$

The glassy behavior can be represented by a spring with a stiffness equivalent to the modulus of elasticity of the polymer (Fig. 10.7a). Note that the strain is completely recovered when the load is removed at time t_1. When the applied stress is increased, the polymer eventually fractures.

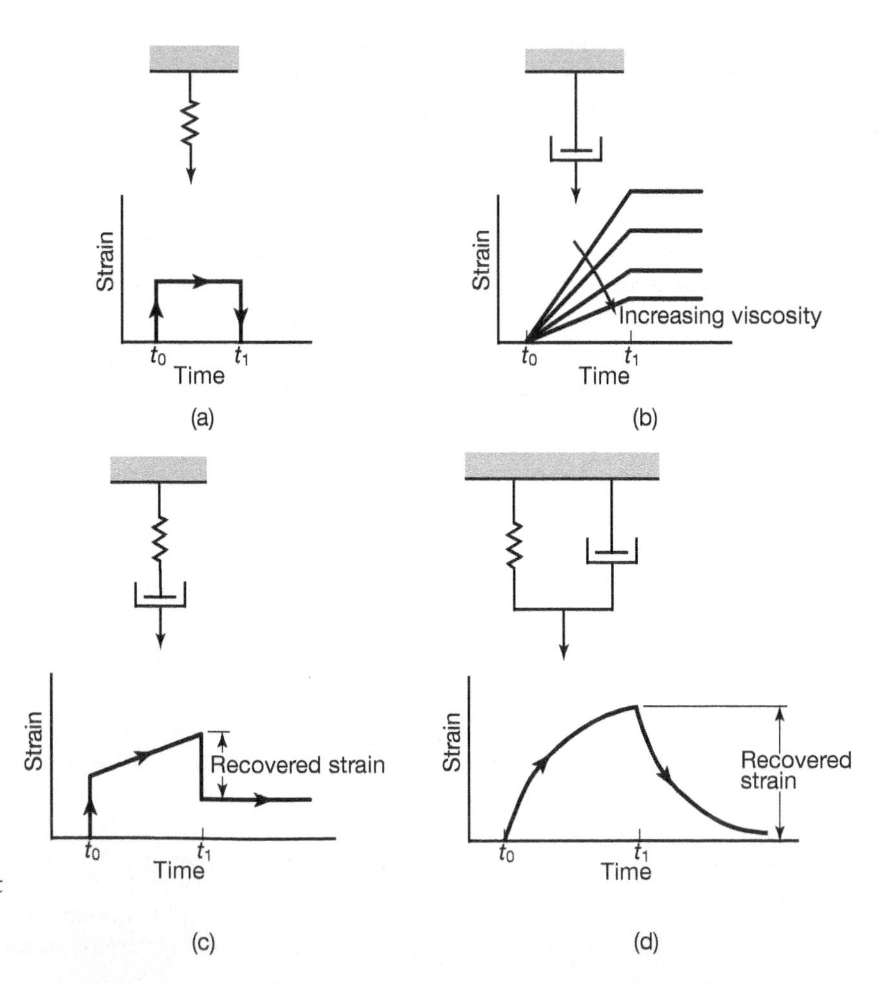

FIGURE 10.7 Various deformation modes for polymers: (a) elastic; (b) viscous; (c) viscoelastic (Maxwell model); and (d) viscoelastic (Voigt or Kelvin model). In all cases, an instantaneously applied load occurs at time t_o, and released at time t_1, resulting in the strain paths shown.

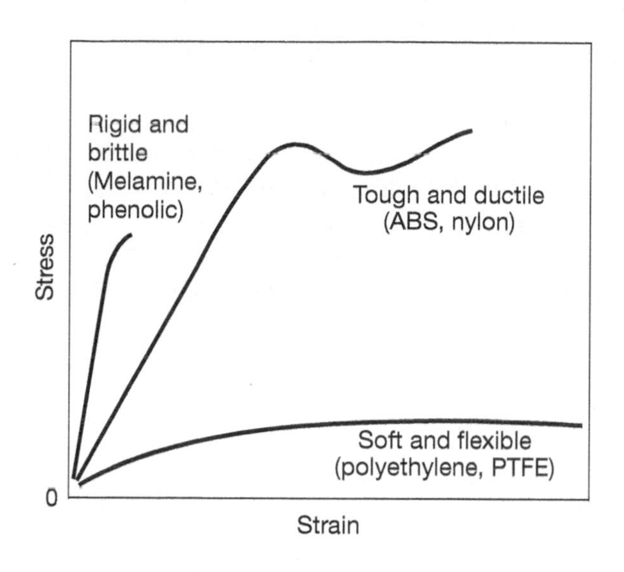

FIGURE 10.8 General terminology describing the behavior of three types of plastics. PTFE is polytetrafluoroethylene (Teflon, a trade name).

The mechanical properties of several polymers listed in Table 10.1 indicate that thermoplastics are about two orders of magnitude less stiff than metals, and their ultimate tensile strength is about one order of magnitude lower than that of metals. Typical stress–strain curves for some thermoplastics and thermosets at room temperature are shown in Fig. 10.8. Note that the plastics shown in the figure exhibit different behaviors, which can be described as rigid, soft, brittle, flexible, and so on. Plastics, like metals, also undergo fatigue and creep phenomena.

Some of the major characteristics of thermoplastics are described below.

1. **Effects of temperature and deformation rate.** The typical effects of temperature on the strength and elastic modulus of thermoplastics are similar to those for metals (see Sections 2.2.6 and 2.2.7). Thus, with increasing temperature, the strength and modulus of elasticity decrease, and the toughness increases (Fig. 10.9). The effect of temperature on impact strength is shown in Fig. 10.10; note the large difference in the impact behavior of various polymers.

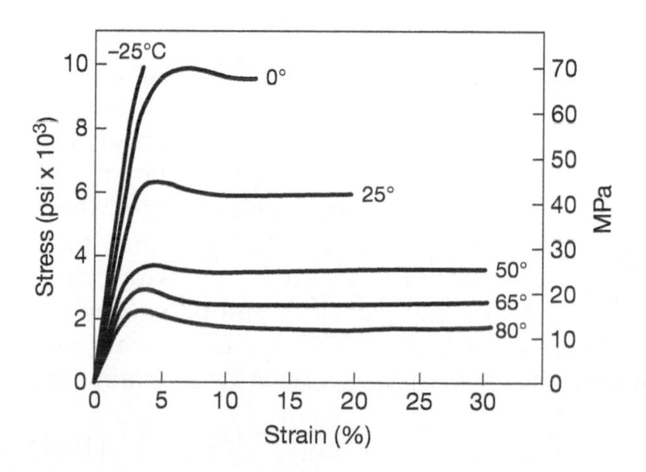

FIGURE 10.9 Effect of temperature on the stress–strain curve for cellulose acetate, a thermoplastic. Note the large drop in strength and increase in ductility with a relatively small increase in temperature.

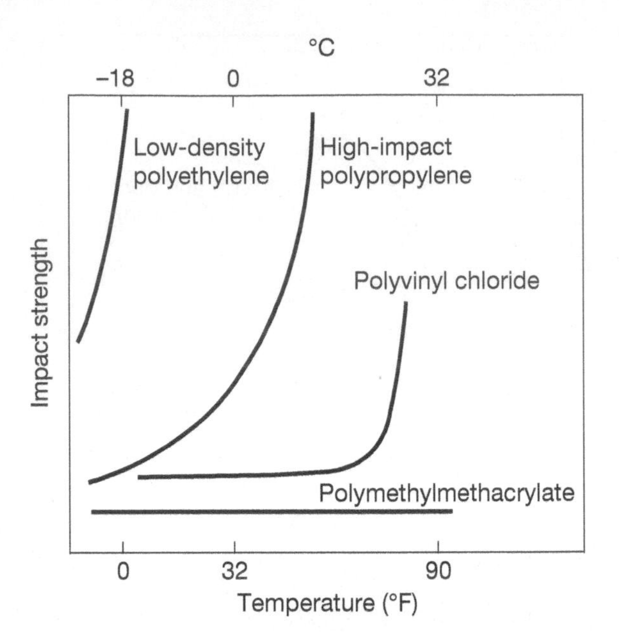

FIGURE 10.10 Effect of temperature on the impact strength of various plastics. Note that small changes in temperature can have a significant effect on impact strength.

If the temperature of a thermoplastic polymer is raised above T_g, it first becomes leathery and then rubbery (Fig. 10.4). Finally, at higher temperatures (above T_m for crystalline thermoplastics) it becomes a viscous fluid, with viscosity decreasing as temperature and strain rate are increased. Thermoplastics display **viscoelastic** behavior, as demonstrated by the spring and dashpot models shown in Figs. 10.7c and d, known as the *Maxwell* and *Kelvin* (or *Voigt*) models, respectively. When a constant load is applied, the polymer first stretches at a high strain-rate and then continues to elongate over a period of time (*creep*, see Section 2.8), because of its viscous behavior. In the models shown in Fig. 10.7 note that the elastic portion of the elongation is reversible (elastic recovery), but the viscous portion is not.

The viscous behavior is expressed by

$$\tau = \eta\left(\frac{dv}{dy}\right) = \eta\dot{\gamma}, \tag{10.3}$$

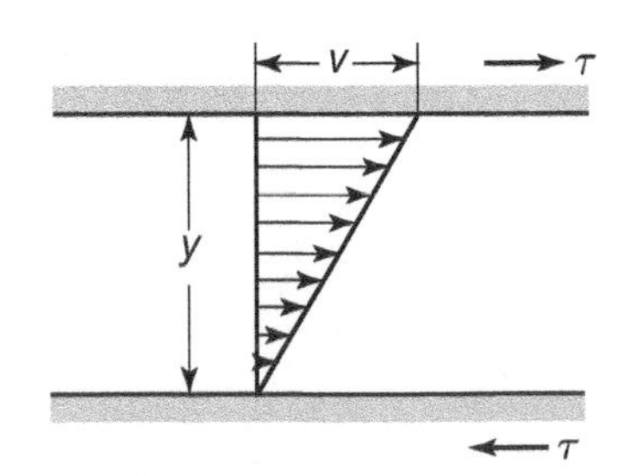

FIGURE 10.11 Parameters used to describe viscosity; see Eq. (10.3).

where η is the *viscosity* and dv/dy is the *shear-strain rate*, $\dot{\gamma}$, as shown in Fig. 10.11. The viscosities of some polymers are given in Fig. 10.12. When the shear stress, τ, is directly proportional to the shear-strain rate, the behavior of the thermoplastic polymer is called *Newtonian*. For many polymers, however, Newtonian behavior is not a good approximation, and Eq. (10.3) will give poor predictions of process performance. For example, polyvinyl chloride, polyethylene (low density or high density), and polypropylene have viscosities that decrease markedly with increasing strain rate (**pseudoplastic behavior**). Their viscosity, as a function of strain rate, can be expressed as

$$\eta = A\dot{\gamma}^{1-n}, \tag{10.4}$$

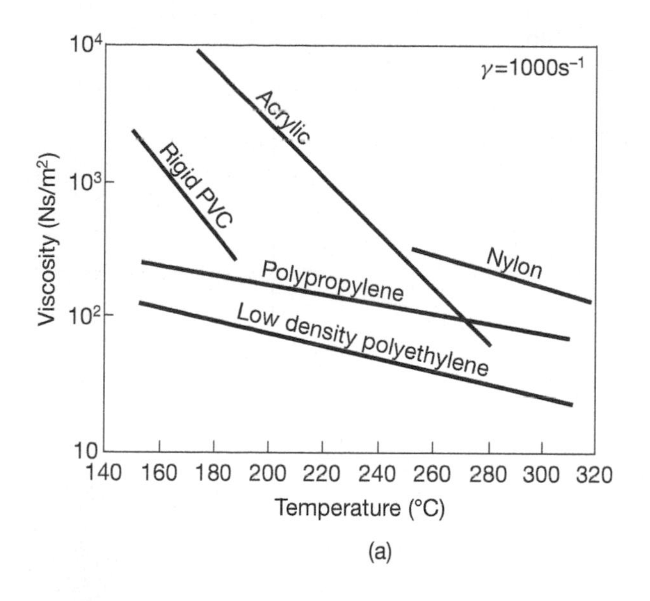

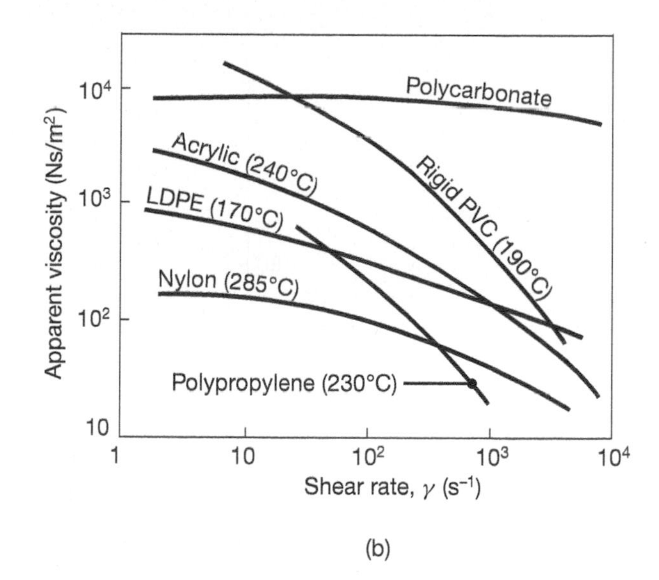

FIGURE 10.12 Viscosity of some thermoplastics as a function of (a) temperature and (b) shear rate.

where A is the *consistency index* and n is the *power-law index* for the polymer.

Note from the foregoing equations that the viscous behavior of thermoplastics is similar to the strain-rate sensitivity of metals (see Section 2.2.7), given by the expression

$$\sigma = C\dot{\epsilon}^m, \tag{10.5}$$

where, for Newtonian behavior, $m = 1$.

Thermoplastics have high m values, indicating that they can undergo large uniform deformations in tension before fracture. Note in Fig. 10.13 how, unlike with ordinary metals, the necked region elongates considerably. This characteristic, which is the same as in superplastic metals (described in Section 2.2.7), enables thermoplastics to be formed into complex shapes, such as bottles for soft drinks, food trays, toys, packaging and frames of all kinds, and lighted signs, as described in Section 10.10. Note also that with increasing rate of loading, the strength of the polymer is increased.

Between T_g and T_m, thermoplastics exhibit leathery and rubbery behavior, depending on their structure and degree of crystallinity, as shown in Fig. 10.4. A term combining the strains caused by elastic behavior (e_e) and by viscous flow (e_v) is called the *viscoelastic modulus*, E_r, and is expressed as

$$E_r = \frac{\sigma}{(e_e + e_v)}. \tag{10.6}$$

This modulus essentially represents a time-dependent elastic modulus.

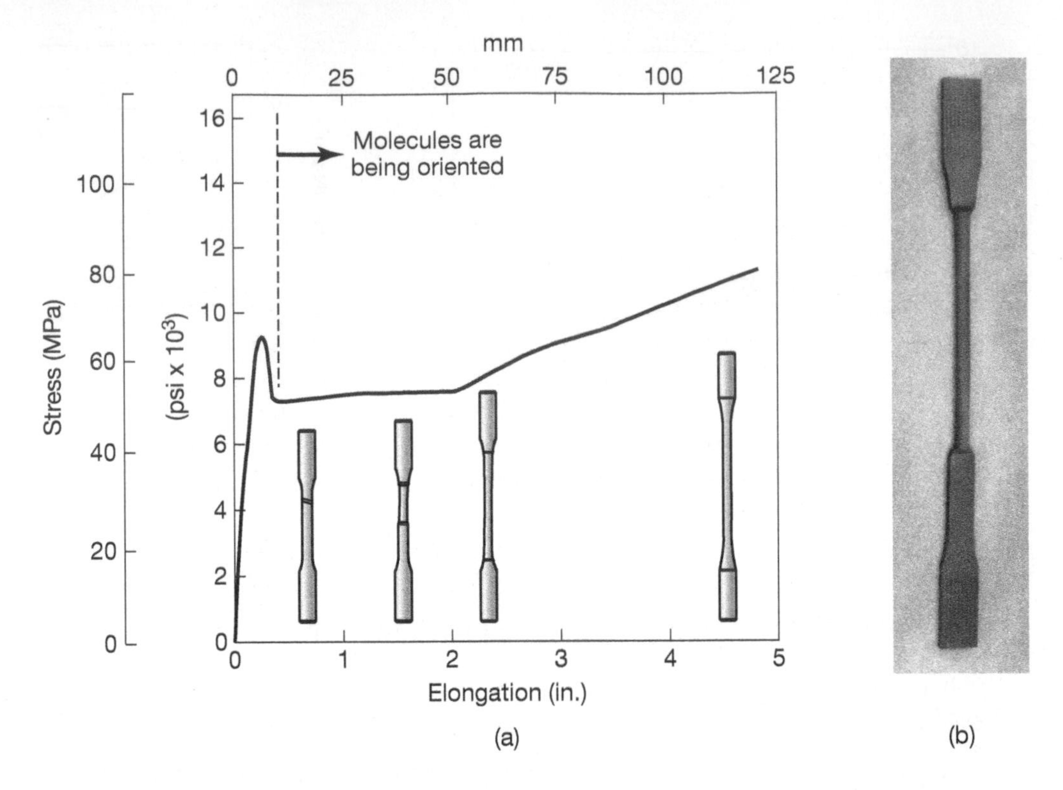

FIGURE 10.13 (a) Load-elongation curve for polycarbonate, a thermoplastic; (b) high-density polyethylene tension-test specimen, showing uniform elongation (the long, narrow region in the specimen). *Source:* After S. Kalpakjian.

The viscosity η of polymers (the ratio of shear stress to shear strain rate) depends on temperature, pressure, and on the polymer's structure and its molecular weight. The effect of temperature can be represented by

$$\eta = \eta_o e^{E/\kappa T}, \tag{10.7}$$

where η_o is a material constant, E is the activation energy (the energy required to initiate a reaction), κ is Boltzmann's constant (the thermal-energy constant, or 13.8×10^{-24} J/K), and T is the temperature (in K).

Thus, as the temperature of the polymer is increased, η decreases. Increasing the molecular weight (and thus increasing the length of the chain) increases η, because of the greater number of secondary bonds present. As the molecular-weight distribution widens, there are more shorter chains and η decreases. With increasing pressure, the viscosity increases, because of reduced *free volume*, or *free space*, defined as the volume in excess of the true volume of the crystal in the crystalline regions of the polymer.

Based on experimental observations that at the glass-transition temperature, T_g, polymers have a viscosity η of about 10^{12} Pa-s, an empirical relationship between viscosity and temperature has been developed for linear thermoplastics:

$$\log \eta = 12 - \frac{17.5\Delta T}{52 + \Delta T}, \qquad (10.8)$$

where $\Delta T = T - T_g$, in K or °C; thus, the viscosity of the polymer at any temperature can be estimated.

2. **Creep and stress relaxation.** Because of their viscoelastic behavior, thermoplastics are particularly susceptible to *creep* and to *stress relaxation* phenomena, described in Section 2.8. Referring back to the dashpot models in Figs. 10.7b - d, note that under a constant load, the polymer undergoes further strain, and thus it creeps. The recovered strain depends on the stiffness of the spring and, hence, the modulus of elasticity.

   According to the Maxwell model of viscoelastic behavior (Fig. 10.7c), the strain caused by an applied load is called the *creep function* and is given by

$$\epsilon(t) = \left(\frac{1}{k} + \frac{1}{\eta}t\right)F, \qquad (10.9)$$

where k is the stiffness (AE/l for a linear spring), η is the coefficient of viscosity for the dashpot, and F is the applied force. For the Voigt model, the creep function is

$$\epsilon(t) = \frac{1}{k}\left[1 - e^{-(k/\eta)t}\right]F. \qquad (10.10)$$

Stress relaxation in polymers occurs over a period of time; recall that stress relaxation refers to a gradual reduction in stress under an applied strain. For the Maxwell model, the *relaxation function* is

$$\sigma(t) = \frac{k\Delta l}{A}e^{-(k/\eta)t} = \frac{k\Delta l}{A}e^{-t/\lambda}, \qquad (10.11)$$

where Δl is the elongation, and A is the instantaneous cross-sectional area. The factor $\lambda = \eta/k$ has the units of time and characterizes the rate of decay of the stress, and hence it is referred to as the *relaxation time*. Because both the stiffness and the viscosity of polymers depend on temperature, to varying degrees, the relaxation time also depends on temperature.

3. **Orientation.** When thermoplastics are permanently shaped, say by stretching, the long-chain molecules align in the general direction of elongation. This process is called *orientation*, and as with metals, the polymer becomes *anisotropic* (see Section 3.5). Moreover, the specimen becomes stronger and stiffer in the elongated (stretched) direction

than in the transverse direction. This behavior is an important technique for enhancing the strength and toughness of polymers; however, orientation weakens the polymer in the transverse direction.

4. **Crazing.** When subjected to tensile stresses or bending, some thermoplastics, such as polystyrene and polymethylmethacrylate, develop localized and wedge-shaped narrow regions of highly deformed material, a phenomenon called *crazing*. Crazes are areas of spongy material, typically containing about 50% voids. With increasing tensile load on the specimen, the voids coalesce to form a crack, eventually leading to fracture. The environment and the presence of solvents, lubricants, and water vapor enhance the formation of crazes (**environmental stress cracking** and **solvent crazing**). Residual stresses in the material also contribute to crazing and cracking, as does radiation exposure.

   A related phenomenon is **stress whitening.** When subjected to tensile stresses, such as by folding or bending, the plastic becomes lighter in color, which is generally attributed to the formation of microvoids in the material. As a result, the material becomes less translucent (transmits less light) or more opaque; this behavior can easily be demonstrated by bending thin plastic components commonly found in household products and toys.

5. **Water absorption.** An important limitation of some polymers, such as nylons, is their ability to absorb water (*hygroscopy*). Water acts as a plasticizing agent (it makes the polymer more ductile and hence formable); thus, in a essence, it lubricates the chains in the amorphous region. Typically, with increasing moisture absorption, the glass-transition temperature, yield stress, and elastic modulus of the polymer are all severely lowered. Dimensional changes also occur because of water absorption, such as in a humid environment.

6. **Thermal and electrical properties.** Compared to metals, plastics generally are characterized by low thermal and electrical conductivity, low specific gravity (ranging from 0.90 to 2.2), and a relatively high coefficient of thermal expansion (about one order of magnitude higher). Because polymers generally have low electrical conductivity, they are commonly used for electrical and electronic components; they can, however, be made electrically conductive, as described in Section 10.7.2.

7. **Shape-memory polymers.** Polymers also can behave in a manner similar to shape-memory alloys, described in Section 3.11.9. The polymers can be stretched or compressed to very large strains, and then, when subjected to heat, light, or a chemical environment, they recover their original shape. The potential applications for these polymers are similar to those for shape-memory metals, such as in opening blocked arteries, probing neurons in the brain, and improving the toughness of spines.

EXAMPLE 10.2 Lowering the Viscosity of a Polymer

Given: In processing a batch of polycarbonate at 170°C to make a certain part, it is found that the polycarbonate's viscosity is twice that desired.

Find: Determine the temperature at which this polymer should be processed.

Solution: From Table 10.2, T_g for polycarbonate is 150°C. Its viscosity at 170°C is determined by Eq. (10.8):

$$\log \eta = 12 - \frac{17.5(20)}{52 + 20} = 7.14.$$

Hence, $\eta = 13.8$ MPa-s, and because this magnitude is twice what is desired, the new viscosity should be 6.9 MPa-s. Substitute this new value to find the new temperature:

$$\log\left(6.9 \times 10^6\right) = 12 - \frac{17.5\,(\Delta T)}{52 + \Delta T}.$$

Therefore, $\Delta T = 21.7$, or the new temperature is $150 + 21.7 = 172$°C. Note that the temperatures are estimates, and that viscosity is very sensitive to temperature.

EXAMPLE 10.3 Stress Relaxation in a Thermoplastic

Given: A long piece of thermoplastic is stretched between two rigid supports at a stress level of 5 MPa. After 30 minutes, the stress level is found to have decayed to one-half of the original level.

Find: How long will it take for the stress to reach one tenth of the original value?

Solution: Substituting these data into Eq. (10.11) as modified for normal stress, we have, noting that at $t = 0$, $\sigma = 5$ MPa,

$$5 = \frac{k\Delta l}{A} e^0 = \frac{k\Delta l}{A}.$$

Therefore, Eq. (10.11) becomes

$$\sigma(t) = 5e^{-t\lambda}.$$

Also, at $t = 30$ minutes, $\sigma = 2.5$ MPa, so that

$$2.5 = 5e^{-(30)/\lambda}; \qquad -\frac{30}{\lambda} = \ln\left(\frac{2.5}{5}\right); \qquad \lambda = 43.3 \text{ min}.$$

Thus, the time needed for $\sigma = 0.5$ MPa is

$$\ln\left(\frac{0.5}{5}\right) = -\frac{t}{43.3}, \text{ or } t = 99.7 \text{ min}.$$

10.4 Thermosets: Behavior and Properties

When the long-chain molecules in a polymer are cross-linked in a three-dimensional (spatial) arrangement, the structure becomes one giant molecule, with strong covalent bonds. Such polymers are called **thermosetting polymers** or *thermosets*, because during polymerization, the network is completed and the shape of the part being formed is permanently set. An important behavior is that the **curing** (*cross-linking*) reaction, unlike that of thermoplastics, is irreversible. The response of a thermosetting plastic to temperature thus can be likened to baking a cake or boiling an egg; once the cake is baked and cooled, or the egg is boiled and cooled, reheating it will not alter its shape. Some thermosets, such as epoxy, polyester, and urethane, cure at room (ambient) temperature, whereby the heat of the exothermic reaction cures the plastic. Thermosetting polymers do not have a sharply defined glass-transition temperature.

The polymerization process for thermosets generally takes place in two stages: (1) The first stage occurs at the chemical plant, where the molecules are partially polymerized into linear chains, and (2) the second stage occurs during subsequent processing, where cross-linking is completed under heat and pressure during the molding and shaping of the part (Section 10.10).

Because of the nature of their bonds, the strength and hardness of thermosets, unlike those of thermoplastics, are not as sensitive to temperature or rate of deformation. Thermosetting plastics (Section 10.6) generally possess better mechanical, thermal, and chemical properties, electrical resistance, and dimensional stability than do thermoplastics. However, if the temperature is increased sufficiently, the thermosetting polymer begins to degrade and char.

10.5 Thermoplastics: General Characteristics and Applications

This section outlines the general characteristics and typical applications of major thermoplastics, particularly as they relate to manufacturing of products and their service life. General recommendations for various plastics applications are given in Table 10.3.

1. **Acetals** (from *ace*tic and *al*cohol) have good strength; stiffness; and resistance to creep, abrasion, moisture, heat, and chemicals. Typical applications include mechanical parts and components where high performance is required over a long period: bearings, cams, gears, bushings, rollers, impellers, wear surfaces, pipes, valves, showerheads, and housings. Common trade names are *Delrin, Duracon, Lupital,* and *Ultraform.*
2. **Acrylics** (such as **polymethylmethacrylate**, or PMMA) have moderate strength, good optical properties, and weather resistance. They are transparent, but can be made opaque, and generally are resistant

TABLE 10.3 General recommendations for plastic products.

Design requirement	Typical applications	Plastics
Mechanical strength	Gears, cams, rollers, valves, fan blades, impellers, pistons	Acetals, nylon, phenolics, polycarbonates, polyesters, polypropylenes, epoxies, polyimides
Wear resistance	Gears, wear strips and liners, bearings, bushings, roller-skate wheels	Acetals, nylon, phenolics, polyimides, polyurethane, ultrahigh-molecular-weight polyethylene
Friction		
High	Tires, nonskid surfaces, footwear, flooring	Elastomers, rubbers
Low	Sliding surfaces, artificial joints	Fluorocarbons, polyesters, polyethylene, polyimides
Electrical resistance	All types of electrical components and equipment, appliances, electrical fixtures	Polymethylmethacrylate, ABS, fluorocarbons, nylon, polycarbonate, polyester, polypropylenes, ureas, phenolics, silicones, rubbers
Chemical resistance	Containers for chemicals, laboratory equipment, components for chemical industry, food and beverage containers	Acetals, ABS, epoxies, polymethylmethacrylate, fluorocarbons, nylon, polycarbonate, polyester, polypropylene, ureas, silicones
Heat resistance	Appliances, cookware, electrical components	Fluorocarbons, polyimides, silicones, acetals, polysulfones, phenolics, epoxies
Functional and decorative features	Handles, knobs, camera and battery cases, trim moldings, pipe fittings	ABS, acrylics, cellulosics, phenolics, polyethylenes, polypropylenes, polystyrenes, polyvinyl chloride
Functional and transparent features	Lenses, goggles, safety glazing, signs, food-processing equipment	Acrylics, polycarbonates, polystyrenes, polysulfones, laboratory hardware
Housings and hollow shapes	Power tools, housings, sport helmets, telephone cases	ABS, cellulosics, phenolics, polycarbonates, polyethylenes, polypropylene, polystyrenes

to chemicals and have good electrical resistance. Typical applications include lenses, lighted signs, displays, window glazing, skylights, automotive windshields, lighting fixtures, and furniture. Common trade names: *Orlon, Plexiglas, Diakon, Zylar,* and *Lucite.*

3. **Acrylonitrile-butadiene-styrene** (ABS) is dimensionally stable and rigid and has good resistance to impact, abrasion, chemicals, and electricity, strength and toughness, and low-temperature properties. Typical applications include pipes, fittings, chrome-plated plumbing supplies, helmets, tool handles, automotive components, boat hulls, luggage, housing, appliances, refrigerator liners, and decorative panels. Common trade names: *Cycolac, Delta, Denka,* and *Magnum.*
4. **Cellulosics** have a wide range of mechanical properties, depending on their composition. They can be made rigid, strong, and tough. However, they weather poorly and are affected by heat and chemicals. Typical applications include tool handles, pens, knobs, frames for eyeglasses, safety goggles, machine guards, helmets, tubing and pipes, lighting fixtures, rigid containers, steering wheels, packaging film, signs, billiard balls, toys, and decorative parts.

5. **Fluorocarbons** have good resistance to temperature, chemicals, weather, and electricity; they also have unique non-adhesive properties and low friction. Typical applications include linings for chemical-processing equipment, nonstick coatings for cookware, electrical insulation for high-temperature wire and cable, gaskets, low-friction surfaces, bearings, and seals. Common trade name: *Teflon*.
6. **Polyamides** (from the words *poly*, *am*ine, and carboxyl ac*id*) are available in two main types:

   a. **Nylons** have good mechanical properties and abrasion resistance. They are self-lubricating and resistant to most chemicals. All nylons are hygroscopic (they absorb water). Moisture absorption reduces mechanical properties and increases part dimensions. Typical applications include gears, bearings, bushings, rollers, fasteners, zippers, electrical parts, combs, tubing, wear-resistant surfaces, guides, and surgical equipment.
   b. **Aramids** (*ar*omatic poly*amides*) have very high-tensile strength and stiffness. Typical applications include fibers for reinforced plastics (composite materials), bulletproof vests, cables, and tires. Common trade name: *Kevlar*.

7. **Polycarbonates** are versatile and have good mechanical and electrical properties; they also have high impact resistance and can be made resistant to chemicals. Typical applications include safety helmets, optical lenses, bullet-resistant window glazing, signs, bottles, food-processing equipment, windshields, load-bearing electrical components, electrical insulators, medical instruments, business-machine components, guards for machinery, and parts requiring dimensional stability. Common trade name: *Lexan, Makrolon,* and *Merlon*.
8. **Polyesters** (thermoplastics; see also Section 10.6) have good mechanical, electrical, and chemical properties, good abrasion resistance, and low friction. Typical applications include gears, cams, rollers, load-bearing members, pumps, and electromechanical components. Common trade names: *Dacron, Mylar*, and *Kodel*.
9. **Polyethylenes** have good electrical and chemical properties. Their mechanical properties depend on their composition and structure. The three major classes of polyethylenes are low density (LDPE), high density (HDPE), and ultrahigh molecular weight (UHMWPE). Typical applications for LDPE and HDPE include housewares, bottles, garbage cans, ducts, bumpers, luggage, toys, tubing, bottles, and packaging material. UHMWPE is used in parts requiring high-impact toughness and abrasive wear resistance; examples include artificial knee and hip joints.
10. **Polyimides** have the structure of a thermoplastic but the non-melting characteristic of a thermoset (see Section 10.6). Typical applications include coatings on flexible electrical cables, medical tubing, and as a high-temperature adhesive. Common trade names: *Avimid, Torlon,* and *Kapton*.
11. **Polypropylenes** have good mechanical, electrical, and chemical properties and good resistance to tearing. Typical applications include

automotive trim and components, medical devices, appliance parts, wire insulation, TV cabinets, pipes, fittings, drinking cups, dairy-product and juice containers, luggage, ropes, and weather stripping. Common trade names: *Fortilene, Oleplate,* and *Olevac.*

12. **Polystyrenes** are inexpensive, have generally average properties, and are somewhat brittle. Typical applications include disposable containers, packaging, foam insulation, appliances, automotive, radio, and TV components, housewares, and toys and furniture parts (as a wood substitute). Common trade names: *Lustrex, Polystrol, Styron,* and *Styrofoam.*
13. **Polysulfones** have excellent resistance to heat, water, and steam and are highly resistant to some chemicals, but are attacked by organic solvents. Typical applications include steam irons, coffeemakers, hot-water containers, medical equipment that requires sterilization, power-tool and appliance housings, aircraft cabin interiors, and electrical insulators. Common trade names: *Mindel* and *Udel.*
14. **Polyvinyl chloride** (PVC) has a wide range of properties, is inexpensive and water resistant, and can be made rigid or flexible. It is not suitable for applications that require strength and heat resistance. *Rigid PVC* is tough and hard and is used for signs and in the construction industry, such as for pipes and conduits. *Flexible PVC* is used in wire and cable coatings, low-pressure flexible tubing and hose, footwear, imitation leather, upholstery, records, gaskets, seals, trim, film, sheet, and coatings. Common trade names: *Saran, Sintra,* and *Tygon.*

10.6 Thermosets: General Characteristics and Applications

This section outlines the general characteristics and typical applications of major thermosetting plastics.

1. **Alkyds** (from *alkyl*, meaning alcohol, and *acid*) have good electrical insulating properties, impact resistance, and dimensional stability and have low water absorption. Typical applications include electrical and electronic components.
2. **Aminos** (**urea** and **melamine**) have properties that depend on composition. Generally, aminos are hard and rigid and are resistant to abrasion, creep, and electrical arcing. Typical applications include small appliance housings, countertops, toilet seats, and handles. Urea is used for electrical and electronic components, and melamine is used for dinnerware.
3. **Epoxies** have excellent mechanical and electrical properties, dimensional stability, strong adhesive properties, and good resistance to heat and chemicals. Typical applications include electrical components that require mechanical strength and high insulation, tools and dies, and adhesives. *Fiber-reinforced epoxies* (Section 10.9) have excellent mechanical properties and are used in pressure vessels, rocket motor casings, tanks, and similar structural components.

4. **Phenolics** are rigid but brittle; they are dimensionally stable and have high resistance to heat, water, electricity, and chemicals. Typical applications include knobs, handles, laminated panels, bond material to hold abrasive grains together in grinding wheels, and electrical components, such as wiring devices, connectors, and insulators.
5. **Polyesters** have good mechanical, chemical, and electrical properties. Polyesters are generally reinforced with glass or other fibers. Typical applications include boats, luggage, chairs, automotive bodies, swimming pools, and material for impregnating cloth and paper. Polyesters are also available as casting resins.
6. **Polyimides** have good mechanical, physical, and electrical properties at elevated temperatures; they also have creep resistance and low friction and wear characteristics. Polyimides have the non-melting characteristics of a thermoset, but the structure of a thermoplastic. Typical applications include pump components (bearings, seals, valve seats, retainer rings, and piston rings); electrical connectors for high-temperature use; aerospace parts; high-strength, impact-resistant structures; sports equipment; and safety vests.
7. **Silicones** have properties that depend on composition. Generally, they weather well, have excellent electrical properties over a wide range of humidity and temperatures, and resist chemicals and heat (see also Section 10.8). Typical applications include electrical components that require strength at elevated temperatures, oven gaskets, heat seals, and waterproof materials.

10.7 High-Temperature Polymers, Electrically Conducting Polymers, and Biodegradable Plastics

10.7.1 High-Temperature Polymers

Polymers and polymer blends for high-temperature applications are available, particularly for the aerospace industry. High-temperature resistance may be for short term at relatively high temperatures, or for long term at lower temperatures. (See, for example, Section 3.11.6 for a similar consideration regarding titanium alloys.) Short-term exposure to high temperatures generally requires *ablative* materials, or materials that wear, melt or vaporize to dissipate heat, such as phenolic-silicone copolymers, used for rocket and missile components, at temperatures of thousands of degrees. Long-term exposure for polymers is presently confined to temperatures around 260°C (500°F). High-temperature thermoplastic polymers include fluorine-containing thermoplastics, polyketones, and polyimides. High-temperature thermosetting polymers include phenolic resins, epoxy resins, silicone-based thermosetting polymers, and phenolic-fiberglass systems.

10.7.2 Electrically Conducting Polymers

The electrical conductivity of some polymers can be increased by **doping**, that is, introducing certain impurities in the polymer, such as metal powder, salts, and iodides. The conductivity of polymers increases with moisture absorption, and their electronic properties can be altered by irradiation. Some electrically conductive polymers have been developed, such as poly(3,4-ethylene dioxitiophene) or PEDOT. When PEDOT is doped with styrene sulfonate (PEDOT:PSS), it becomes a semiconductor. These conducting and semiconducting polymers have applications in roll-to-roll printing of electronic devices (see Section 13.14). Other applications of electrically conducting polymers include microelectronic devices, rechargeable batteries, capacitors, fuel cells, catalysts, fuel-level sensors, de-icer panels, antistatic coatings, and as conducting adhesives for surface-mount technologies (see also Section 12.14.3).

10.7.3 Biodegradable Plastics

About one-third of plastics produced today are in the disposable-products sector, such as beverage bottles, packaging, and garbage disposal bags. These plastics contribute about 10% of municipal solid waste, and on a volume basis, they contribute between two and three times their weight. With the growing use of plastics and increasing concern over environmental issues regarding disposal of plastic products and limited landfills, major efforts are continuing toward development of biodegradable plastics.

Most plastic products have traditionally been made from synthetic polymers that are (a) derived from nonrenewable natural resources; (b) are not biodegradable; and (c) are difficult to recycle. **Biodegradability** means that microbial species in the environment (e.g., microorganisms in soil and water) will degrade a portion of (or even the entire) polymeric material, under the proper environmental conditions and without producing toxic by-products (see also *biological cycle*, Section 1.4). The end products of the degradation of the biodegradable portion of the material are carbon dioxide and water.

Three different *biodegradable plastics* have thus far been developed. They have different degradability characteristics, and they degrade over different periods of time, ranging from a few months to a few years.

1. The **starch-based system** is the farthest along of the three types of bioplastics in terms of production capacity. Starch may be extracted from potatoes, wheat, rice, and corn. In this system, starch granules are processed into a powder, which is heated and becomes a sticky liquid. Various binders and additives are blended in the starch to impart special characteristics to the bioplastic materials. The liquid is then cooled, formed into pellets, and processed in conventional plastic-processing equipment (see Section 10.10).
2. In the **lactic-based system**, fermenting corn or other feedstock produce lactic acid, which is then polymerized to form a polyester resin.
3. In the third system, **organic acids** are added to a **sugar** feedstock. Through a specially developed process, the resulting reaction produces a highly crystalline and very stiff polymer that, after further processing, behaves in a manner similar to polymers developed from petroleum.

Fully biodegradable plastics are continually under development, using feedstocks such as various agricultural waste (*agrowastes*), plant carbohydrates, plant proteins, and vegetable oils. Typical applications include the following:

- Disposable tableware made from a cereal substitute, such as rice grains or wheat flour;
- Plastic parts made from coffee beans and rice hulls, dehydrated and molded under high pressure and temperature;
- Water-soluble and compostable polymers, for medical and surgical applications; and
- Food and beverage containers (made from potato starch, limestone, cellulose, and water) that can dissolve in storm sewers and oceans without affecting wildlife.

The long-range performance of biodegradable plastics, both during their useful life-cycle as products and in landfills, has not yet been fully assessed. There is also concern that emphasis on biodegradability may divert attention from the issue of *recyclability* of plastics and the efforts for *conservation* of materials and energy. A major consideration is the fact that the cost of today's biodegradable polymers is substantially higher than that of synthetic polymers. Consequently, a mixture of agricultural waste, such as hulls from corn, wheat, rice, and soy (as the major component) and biodegradable polymers (as the minor component), is an attractive alternative.

Recycling. Recycled plastics are being used increasingly for a variety of products, including automotive-body components, packaging materials, and architectural structural shapes. Plastic products carry the following numerals within a triangular mark with arrows, meaning that the product is recyclable: 1-PETE (polyethylene therephthalate), 2-HDPE (high-density polyethylene), 3-V (vinyl) or PVC (polyvinyl chloride), 4-LDPE (low-density polyethylene), 5-PP (polypropylene), 6-PS (polystyrene), and 7-others.

10.8 Elastomers (Rubbers): General Characteristics and Applications

The terms *elastomer* and *rubber* are often used interchangeably. Generally, an **elastomer** is defined as being capable of recovering substantially in shape and size after a load has been removed; **rubber** is defined as being capable of quickly recovering from large deformations. Elastomers (from the words *elastic* and *mer*) comprise a large family of amorphous polymers that have (a) a low glass-transition temperature; (b) the characteristic ability to undergo large elastic deformations without rupture; (c) a low hardness; and (d) a low elastic modulus. The structure of these polymers is highly *kinked* (tightly twisted or curled); they stretch, but then return to their original shape after the load is removed (Fig. 10.14). They can be cross-linked, the best known example of which is the elevated-temperature **vulcanization** (after Vulcan, the Roman god of fire) of rubber with sulfur, discovered by C. Goodyear in 1839. Note that an automobile tire, which is one giant molecule, cannot be softened and reshaped.

The hardness of elastomers, which is measured with a *durometer* (see Section 2.6.7), increases with increasing cross-linking of the molecular chains. A variety of additives can be blended with elastomers to impart specific properties, as is done with polymers. Elastomers have a wide range of applications, such as high-friction and nonskid surfaces, protection against corrosion and abrasion, electrical insulation, and shock and vibration insulation. Specific examples include tires, hoses, weather stripping, footwear, linings, gaskets, seals, printing rolls, and flooring.

A characteristic of elastomers is their hysteresis in stretching or compression (Fig. 10.14). The clockwise loop in Fig. 10.14 indicates energy loss, whereby mechanical energy is converted into heat; this property is desirable for absorbing vibrational energy (damping) and sound.

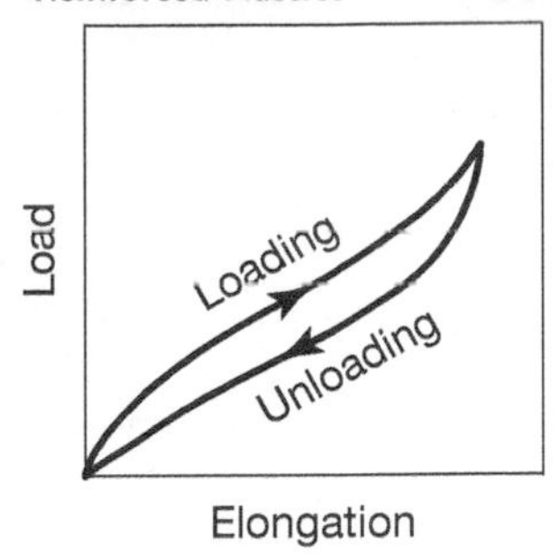

FIGURE 10.14 Typical load-elongation curve for elastomers. The area within the clockwise loop, indicating loading and unloading paths, is the hysteresis loss. Hysteresis gives rubbers the capacity to dissipate energy, damp vibration, and absorb shock loading, as in automobile tires and vibration dampeners for machinery.

The major types of elastomers are described below.

1. **Natural rubber.** The base for natural rubber is **latex**, a milklike sap obtained from the inner bark of a tropical tree. It has good resistance to abrasion and fatigue and high frictional properties, but low resistance to oil, heat, ozone, and sunlight. Typical applications include tires, seals, shoe heels, couplings, and engine mounts.
2. **Synthetic rubbers.** Compared with natural rubbers, synthetic rubbers have improved resistance to heat, gasoline, and chemicals and a higher useful-temperature range. Synthetic rubbers include synthetic natural rubber, butyl, styrene butadiene, polybutadiene, and ethylene propylene. Examples of synthetic rubbers that are resistant to oil are neoprene, nitrile, urethane, and silicone. Typical applications of synthetic rubbers include tires, shock absorbers, seals, and belts.
3. **Silicones.** Silicones (see also Section 10.6) have the highest useful temperature range of all elastomers, up to 315°C, but their other properties such as strength and resistance to wear and oils are generally inferior to those of other elastomers. Typical applications include seals, gaskets, thermal insulation, high-temperature electrical switches, and various electronic components.
4. **Polyurethane.** Polyurethane has very good overall properties of high strength, stiffness, hardness, and exceptional resistance to abrasion, cutting, and tearing. Typical applications include seals, gaskets, cushioning, diaphragms for rubber-pad forming of sheet metals (see Section 7.5.3), and auto-body parts such as bumpers.

10.9 Reinforced Plastics

A major group of important materials are *reinforced plastics* (**composite materials**), as shown in Fig. 10.15. These **engineered materials** are defined as a combination of two or more chemically distinct and insoluble phases whose properties and structural performance are superior to those of the

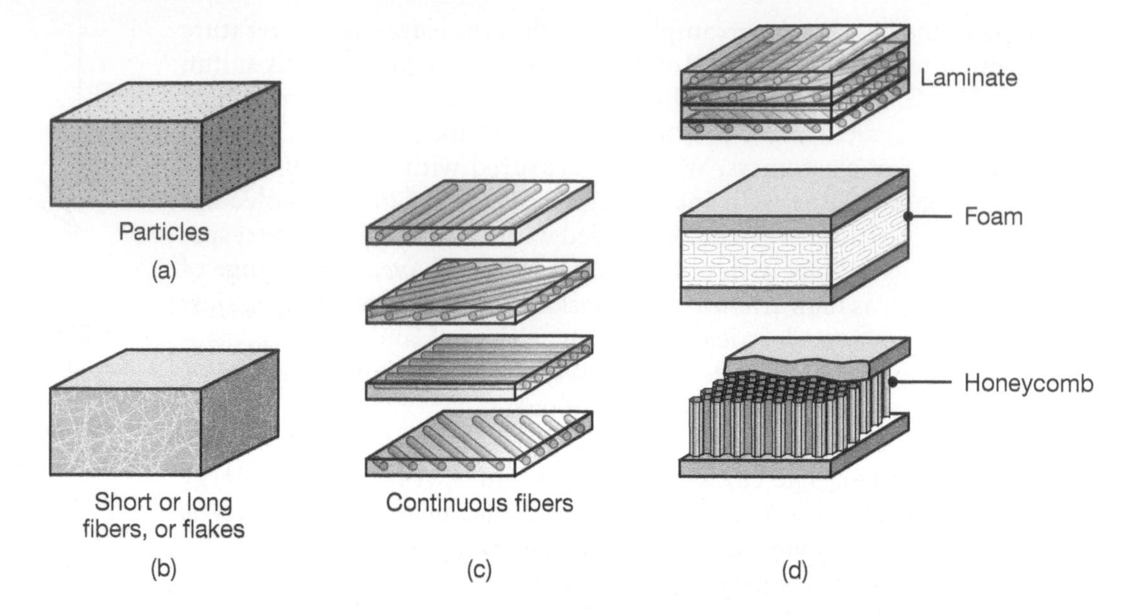

FIGURE 10.15 Schematic illustration of types of reinforcing plastics. (a) Matrix with particles; (b) matrix with short or long fibers or flakes; (c) continuous fibers; and (d) and (e) laminate or sandwich composite structures using a foam or honeycomb core. (See also Fig. 7.48 on making of honeycombs.)

constituents acting independently. Although plastics possess mechanical properties (particularly strength, stiffness, and creep resistance) that are generally inferior to those of metals and alloys, these properties can be greatly improved by embedding reinforcements of various types to produce reinforced plastics. As shown in Table 10.1, reinforcements improve the strength, stiffness, and creep resistance of plastics, and particularly their strength-to-weight and stiffness-to-weight ratios. Reinforced plastics have a wide variety of applications in aircraft (Fig. 10.16), space vehicles, offshore structures, piping, electronics, automobiles, boats, ladders, and sporting goods.

The oldest example of a composite material is the addition of straw to clay for making mud huts and bricks for structural use, dating back to 4000 B.C. The straw acts as the reinforcing fiber, and the clay is the matrix. Another example of a composite material is the reinforcement of masonry and concrete with iron rods, beginning in the 1800s. Note that concrete itself is a composite material, consisting of cement, sand, and gravel. In *reinforced concrete*, steel rods impart the necessary tensile strength to the composite, since concrete is brittle and generally has little or no useful tensile strength.

10.9.1 Structure of Polymer-Matrix-Reinforced Plastics

Reinforced plastics consist of *fibers* (the *discontinuous* or *dispersed* phase) in a polymer **matrix** (the *continuous* phase), as shown in Fig. 10.15. Commonly used fibers include glass, graphite, aramids, and boron (Table 10.4). These fibers are strong and stiff and have a high specific strength (strength-to-weight ratio) and specific modulus (stiffness-to-weight ratio), as shown

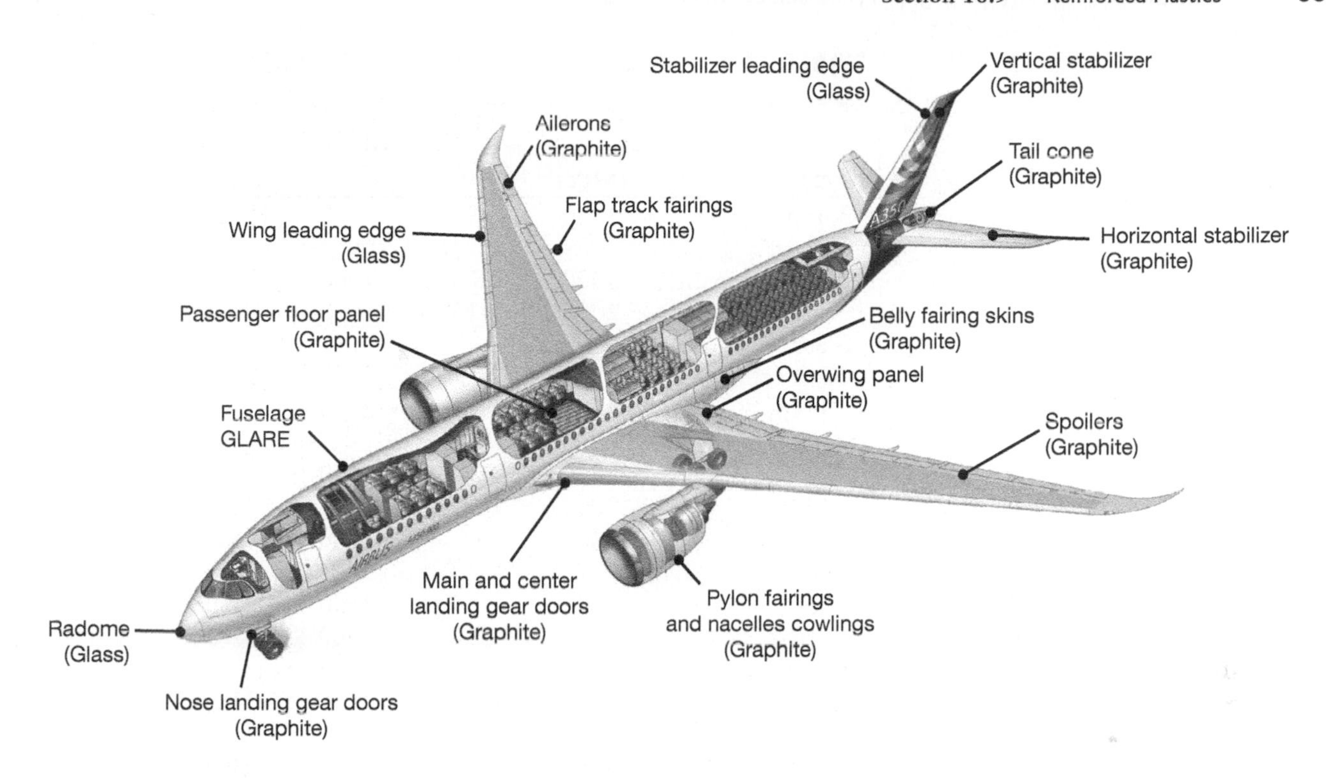

FIGURE 10.16 Application of advanced composite materials in the Airbus 350. The reinforcement type is shown, with the fuselage made of GLARE, a glass-reinforced polymer/aluminum laminate. *Source:* Graphic image Courtesy of Airbus.

in Fig. 10.17. They are, however, generally brittle and abrasive and lack toughness; thus, the fibers, by themselves, have little significance as structural members. The polymer matrix is less strong and less stiff but tougher than the fibers; reinforced plastics thus combine the advantages of each of the two constituents (Table 10.5).

In addition to having high specific strength and specific modulus, reinforced-plastic structures also have higher fatigue resistance, toughness, and creep resistance than unreinforced plastics. The percentage of fibers (by volume) in reinforced plastics usually ranges between 10 and 60%. The highest practical fiber content is 65%; higher percentages generally result in diminished mechanical properties. When a composite material has more than one type of fiber, it is called a **hybrid**, which generally has even better properties than a reinforced plastic using only one type of fiber.

10.9.2 Reinforcing Fibers: Characteristics and Manufacture

The major types of reinforcing fibers and their characteristics are described below.

1. **Polymer fibers.** The most commonly used reinforcing fibers are **aramids** (such as *Kevlar*). They are among the toughest fibers and have very high specific strength (Fig. 10.17 and Table 10.4); they can undergo some plastic deformation before fracture and,

TABLE 10.4 Typical properties of reinforcing fibers.

Type	Tensile strength (MPa)	Elastic modulus (GPa)	Density (kg/m^3)	Relative cost
Boron	3500	380	2380	Highest
Carbon				
High strength	3000	275	1900	Low
High modulus	2000	415	1900	Low
Glass				
E-type	3500	73	2480	Lowest
S-type	4600	85	2540	Lowest
Kevlar				
29	2920	70.5	1440	High
49	3000	112.4	1440	High
129	3200	85	1440	High
Nextel				
312	1700	150	2700	High
610	2770	328	3960	High
Spectra				
900	2270	64	970	High
1000	2670	90	970	High
2000	3240	115	970	High
Alumina (Al_2O_3)	1900	380	3900	High
Silicon carbide	3500	400	3200	High

Note: These properties vary significantly depending on the material and method of preparation.

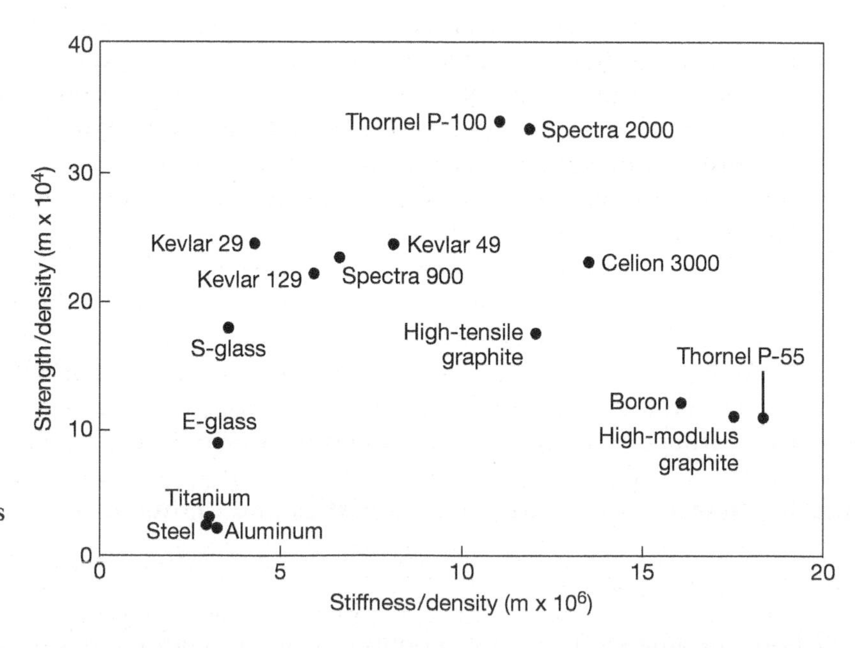

FIGURE 10.17 Specific tensile strength (ratio of tensile strength-to-density) and specific tensile modulus (ratio of modulus of elasticity-to-density) for various fibers used in reinforced plastics. Note the wide range of specific strength and stiffness available.

TABLE 10.5 Types and general characteristics of reinforced plastics and metal-matrix and ceramic-matrix composites.

Material	Characteristics
Fiber	
Glass	High strength, low stiffness, high density; E (calcium aluminoborosilicate) and S (magnesia-aluminosilicate) types are commonly used; lowest cost
Graphite	Available typically as high modulus or high strength; less dense than glass; low cost
Boron	High strength and stiffness; has tungsten filament at its center (coaxial); highest density; highest cost
Aramids (Kevlar)	Highest strength-to-weight ratio of all fibers; high cost
Other	Nylon, silicon carbide, silicon nitride, aluminum oxide, boron carbide, boron nitride, tantalum carbide, steel, tungsten, and molybdenum; see Chapters 3, 8, 9, and 10
Matrix	
Thermosets	Epoxy and polyester, with the former most commonly used; others are phenolics, fluorocarbons, polyethersulfone, silicon, and polyimides
Thermoplastics	Polyetheretherketone; tougher than thermosets, but lower resistance to temperature
Metals	Aluminum, aluminum-lithium alloy, magnesium, and titanium; fibers used are graphite, aluminum oxide, silicon carbide, and boron
Ceramics	Silicon carbide, silicon nitride, aluminum oxide, and mullite; fibers used are various ceramics

consequently, such composites have higher toughness than those containing brittle fibers. However, aramids absorb moisture (as do nylons), which reduces their properties and complicates their application, as *hygrothermal* stresses (see Section 10.3) must be considered. Other polymer reinforcements include rayon, nylon, and acrylics.

A high-performance polyethylene fiber is *Spectra* (a trade name), which has ultrahigh molecular weight and high molecular-chain orientation. Compared to aramid fibers, Spectra has better abrasion resistance and flexural fatigue and at a comparable cost, and also, because of its lower density, it has higher specific strength and specific stiffness. However, it has a low melting point and poor interfacial fiber-matrix adhesion characteristics as compared with other fibers.

Most synthetic fibers used in reinforced plastics are extruded through tiny holes of a device called a *spinneret* (resembling a shower head), producing continuous semi-solid filaments. The extruder forces the polymer through the spinneret, which may have from one to several hundred holes. If the polymers are thermoplastics, they are first melted in the extruder, as described in Section 10.10.1. Thermosetting polymers also can be formed into fibers by first dissolving or chemically treating them, so that they can be extruded. These operations are performed at high production rates and with very high product reliability. As the filaments emerge from the holes in the spinneret, the liquid polymer is converted first to a rubbery state and then solidified.

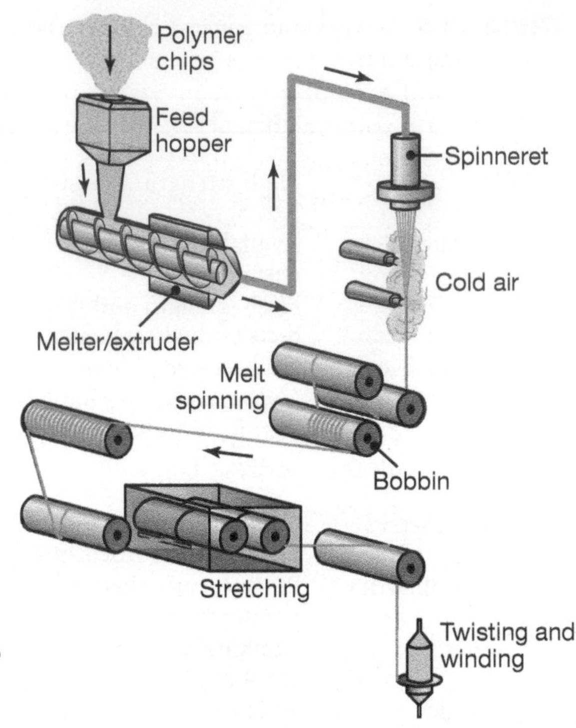

FIGURE 10.18 The melt spinning process for producing polymer fibers. The fibers are used in a variety of applications, including fabrics and as reinforcements for composite materials.

The process of extrusion and solidification of continuous filaments is called **spinning**, a term which is also used for the production of natural textiles (such as cotton or wool), where short pieces of fiber are twisted into yarn. There are four methods of spinning fibers:

a. In **melt spinning**, shown in Fig. 10.18, the polymer is first melted for extrusion through the spinneret and then directly solidified by cooling. A typical spinneret for this operation has about 50 holes of around 0.25 mm (0.01 in.) in diameter, and is about 5 mm (0.2 in.) thick. The fibers that emerge from the spinneret are cooled by forced-air convection and are simultaneously pulled, so that their final diameter becomes much smaller than the spinneret opening. Polymers such as nylon, olefin, polyester, and PVC are produced in this manner. Because of the important applications of nylon and polyester fibers, melt spinning is the most important fiber-manufacturing process.

Melt-spun fibers also can be extruded from the spinneret into various cross sections, such as trilobal (triangle with curved sides), pentagonal, octagonal, as well as hollow shapes. While various cross sections have specific applications, hollow fibers trap air, thus providing additional thermal insulation.

b. **Wet spinning** is the oldest method of fiber production, and is used for polymers that have been dissolved in a solvent. The spinnerets are submerged in a chemical bath, and as the filaments emerge, they precipitate in the chemical bath, producing a fiber that is then wound onto a bobbin. The term wet spinning refers to the use of a precipitating liquid bath, resulting in wet fibers that must be dried before they can be used; acrylic, rayon, and aramid fibers can be produced by this process.

c. **Dry spinning** is used for thermosets carried by a solvent. Instead of precipitating the polymer by dilution, as in wet spinning, solidification is achieved by evaporating the solvent in a stream of air or inert gas. The filaments do not come in contact with a precipitating liquid, thus eliminating the need for drying. Dry spinning may be used for the production of acetate, triacetate, polyether-based elastane, and acrylic fibers.

d. **Gel spinning** is a special process for making fibers with high strength or special properties. The polymer is not completely melted or dissolved in liquid, but the molecules are bonded together at various stages in liquid-crystal form. This operation produces strong inter-chain forces in the filaments produced, thus significantly increasing the tensile strength of the fibers. Moreover, the liquid crystals are aligned along the fiber axis by the strain encountered during extrusion. The filaments emerge from the spinneret with a high degree of orientation relative to each other, further enhancing fiber strength. This process is also called *dry-wet spinning*, because the filaments first pass through air and then are cooled further in a liquid bath. Some high-strength polyethylene and aramid fibers are produced by gel spinning.

A necessary step in the production of most fibers is the application of significant stretching to induce orientation of the polymer molecules in the fiber direction. Orientation is the main reason for the high strength of the fibers as compared to the polymer in bulk form. The stretching can be done while the polymer is still pliable just after it is extruded from the spinneret, or it can be performed as a cold-drawing operation (see Section 6.5). The strain induced can be as high as 800%.

2. **Glass fibers.** These fibers are the most widely used and least expensive of all fibers (see also *glasses*, Section 11.10). The composite material containing glass fibers is called *glass-fiber reinforced plastic* (GFRP), and may contain between 30 and 60% fibers by volume. Glass fibers are made by drawing molten glass through small openings in a platinum die, which are then mechanically elongated, cooled, and wound on a roll. A protective coating (*sizing*) may be applied to facilitate the passage of the glass fibers through machinery. The fibers are treated with **silane** (a silicon hydride) for improved wetting and bonding between the fiber and the matrix.

The principal types of glass fibers are (a) the **E type**, a calcium aluminoborosilicate glass, which is used most often; (b) the **S type**, a magnesia-aluminosilicate glass, which has higher strength and stiffness, but is more expensive than the other two types, and (c) the **E-CR**

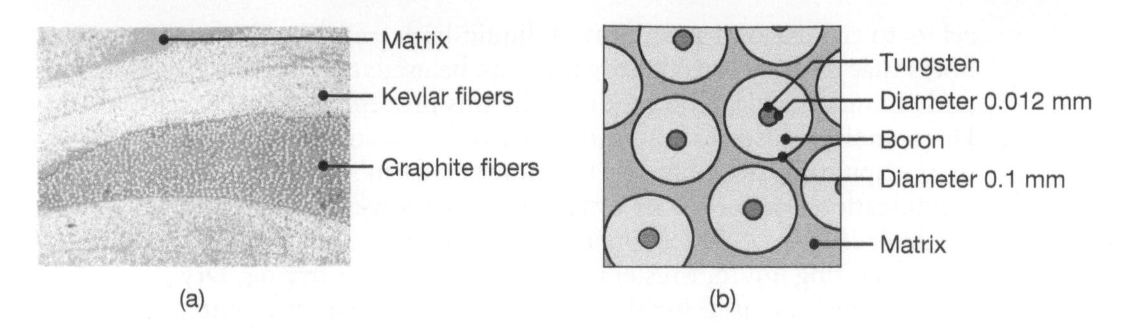

FIGURE 10.19 (a) Cross section of a tennis racket, showing graphite and aramid (Kevlar) reinforcing fibers. *Source:* After J. Dvorak and F. Garrett. (b) Cross section of boron-fiber-reinforced composite material.

type, which offers higher resistance to elevated temperature and acid corrosion than the other two types.

3. **Carbon fibers.** Carbon and graphite fibers (Fig. 10.19a), although more expensive than glass fibers, have the desirable combination of low density, high strength, and high stiffness; the composite is called *carbon-fiber reinforced plastic* (CFRP). Graphite fibers are made by **pyrolysis** of organic **precursors**, commonly polyacrylonitrile (PAN) because of its lower cost. Rayon and pitch (the residue from catalytic crackers in petroleum refining) also can be used as precursors. *Pyrolysis* is the process of inducing chemical changes by heat, such as burning a length of yarn, which becomes carbon and black in color.

   With PAN, the fibers are partially cross-linked, at a moderate temperature, in order to prevent melting during subsequent processing steps, and the fibers are simultaneously elongated. At this stage, the fibers are *carburized* (also known as *oxidizing*); that is, they are exposed to oxygen at elevated temperature to expel the hydrogen (dehydrogenation) and nitrogen (denitrogenation) from the PAN. The temperatures range is up to about 1500°C for carbonizing and up to 3000°C (5400°F) for graphitizing.

   The difference between carbon and graphite, although the terms are often used interchangeably, depends on the temperature of pyrolysis and the purity of the material. Carbon fibers are generally at least 92% carbon, and graphite fibers are usually more than 99% carbon. The fibers are classified by the magnitude of their elastic modulus, which typically ranges from 35 to 800 GPa (5–115 Mpsi); tensile strengths typically range from 250 to 2600 MPa (35–380 ksi).

Conductive graphite fibers. These fibers are produced to enhance the electrical and thermal conductivity of reinforced plastic components. The fibers are coated with a metal (usually nickel), using a continuous electroplating process. The coating is typically 0.5-μm thick on a 7-μm-diameter graphite fiber core. Available in chopped or continuous form, the conductive fibers are incorporated directly into injection-molded plastic parts (see Section 10.10.2). Applications include electromagnetic and radio-frequency shielding and lightning-strike protection.

4. **Boron fibers.** Boron fibers consist of boron deposited by chemical vapor-deposition techniques (see Section 4.5.1), on tungsten fibers (Fig. 10.19b), as well as on carbon fibers. These fibers have favorable properties, such as high strength and stiffness in tension and compression and resistance to high temperatures. However, because of the tungsten, they have high density and are expensive, thus increasing the weight and cost of the reinforced plastic component.
5. **Miscellaneous fibers.** Other fibers that are used for composite materials include silicon carbide, silicon nitride, aluminum oxide, sapphire, steel, tungsten, molybdenum, boron carbide, boron nitride, and tantalum carbide (see also Section 8.6.10).

   Metallic fibers are drawn as described in Section 6.5, although at the smaller diameters, the wires are drawn in bundles (see also Section 6.5.3). **Whiskers** (Section 3.8.3) also are used as reinforcing fibers. Whiskers are tiny needlelike single crystals that grow to 1–10 μm in diameter and have aspect ratios (the ratio of fiber length to diameter) ranging from 100 to 15,000. Because of their small size, the whiskers are either free of imperfections or the imperfections they contain do not significantly affect their strength, which approaches the theoretical strength of the material (see Section 3.3.2).

10.9.3 Fiber Size and Length

The mean diameter of fibers in reinforced plastics is typically less than 0.01 mm. The fibers are very strong and rigid in tension, because the molecules in the fibers are *oriented* in the longitudinal direction, and their cross sections are so small that the probability is low that any significant defects exist in the fiber. Glass fibers, for example, can have tensile strengths as high as 4600 MPa (670 ksi), whereas the strength of glass in bulk form is much lower; in fact, glass fibers are stronger than steel.

Fibers are classified as *short* or *long* fibers, or as *discontinuous fibers* or *continuous fibers*, respectively. Short fibers typically have an *aspect ratio* between 20 and 60, and long fibers between 200 and 500. The short- and long-fiber designations are, in general, based on the following observations: In a given fiber, if the mechanical properties of the composite improve as a result of increasing the fiber length, then the fiber is denoted as a short fiber. When no additional improvement in properties is observed, the fiber is denoted as a long fiber. In addition to the discrete fibers, reinforcements in composites may also be in the form of (a) continuous *roving* (slightly twisted strand of fibers); (b) *woven* fabric (similar to cloth); (c) *yarn* (twisted strand); and (d) *mats* of various combinations. As shown in Fig. 10.15, reinforcing elements also may be in the form of particles and flakes.

10.9.4 Matrix Materials

The matrix in reinforced plastics has three important functions:

1. Support and transfer the stresses to the fibers, which carry most of the load;

2. Protect the fibers from physical damage and the environment; and
3. Reduce propagation of cracks in the composite, by virtue of the ductility and toughness of the plastic matrix.

Matrix materials are typically epoxy, polyester, phenolic, fluorocarbon, polyethersulfone, or silicon. The most common materials are epoxies (80% of all reinforced plastics) and polyesters. Polyimides, which resist exposure to temperatures in excess of 300°C, are used with graphite fibers. A thermoplastic such as polyetheretherketone (PEEK) also is used as a matrix material; it has higher toughness than thermosets, but its resistance to temperature is lower, being limited to the range of 100°C to 200°C (200°F to 400°F).

10.9.5 Properties of Reinforced Plastics

The overall properties of reinforced plastics depend on (a) the type, shape, and orientation of the reinforcing material; (b) the length of the fibers; and (c) the volume fraction (percentage) of the reinforcing material. Short fibers are less effective than long fibers (Fig. 10.20), and their properties are

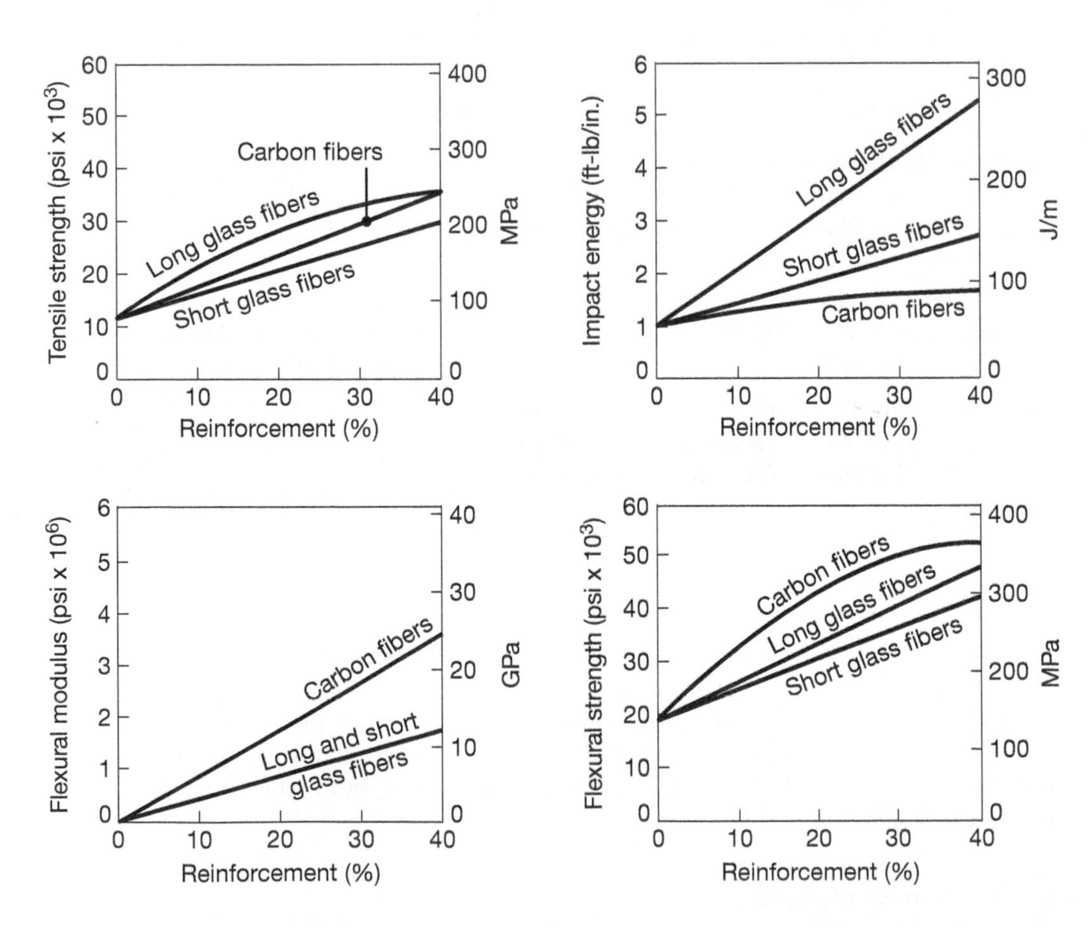

FIGURE 10.20 Effect of the percentage of reinforcing fibers and fiber length on the mechanical properties of reinforced nylon. Note the significant improvement with increasing percentage of fiber reinforcement.

strongly influenced by time and temperature. Long fibers transmit the load through the matrix better and thus are commonly used in critical applications (such as aerospace structural components and pressure vessels), particularly at elevated temperatures.

A critical factor in reinforced plastics is the *strength of the bond* between the fiber and the polymer matrix, since the load is transmitted through the fiber-matrix interface. The importance of proper bonding can be appreciated by inspecting Fig. 10.21, showing the fracture surfaces of a reinforced plastic. Weak bonding in the composite causes **fiber pullout** and **delamination,** particularly under adverse environmental conditions, including temperature and humidity. Adhesion at the interface can be improved by special surface treatments, such as coatings and coupling agents. Glass fibers, for example, are treated with *silane* (see Section 10.9.2) for improved wetting and bonding between the fiber and the matrix.

Generally, the highest stiffness and strength in reinforced plastics is obtained when the fibers are aligned in the direction of the tensile force, which makes the composite highly anisotropic (Fig. 10.22). As a result, other properties of the composite, such as stiffness, creep resistance, thermal expansion, and thermal and electrical conductivity, also are anisotropic. The transverse properties of a unidirectionally reinforced structure are much lower than the longitudinal properties. Note, for

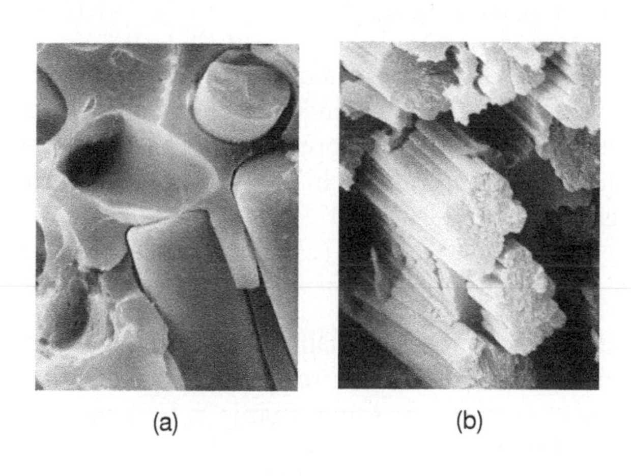

FIGURE 10.21 (a) Fracture surface of glass-fiber-reinforced epoxy composite. The fibers are 10 μm in diameter and have random orientation. (b) Fracture surface of a graphite-fiber-reinforced epoxy composite. The fibers are 9–11 μm in diameter. Note that the fibers are in bundles and are all aligned in the same direction. *Source:* After L.J. Broutman.

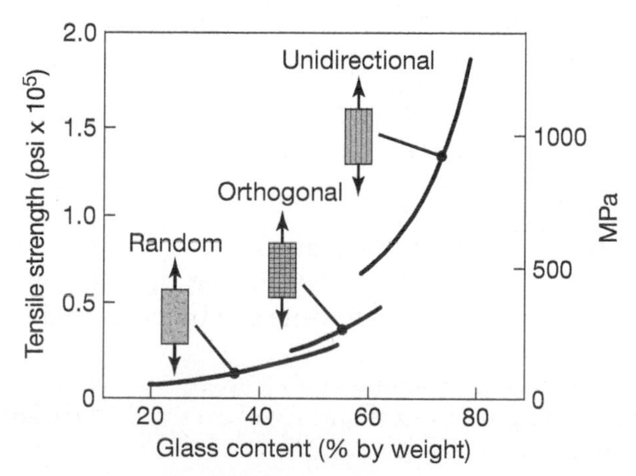

FIGURE 10.22 Tensile strength of glass-reinforced polyester as a function of fiber content and fiber direction in the matrix.

example, how easily fiber-reinforced packaging tape can be split, yet how strong it is when pulled in tension.

For a specific service condition, a reinforced-plastic part can be given an optimal configuration. For example, if the part is to be subjected to forces in *biaxial* directions (such as a thin-walled, pressurized vessel), the fibers are crisscrossed in the matrix (see also the discussion of *filament winding* in Section 10.11.2). Reinforced plastics may also be made with various other materials and shapes of the polymer matrix in order to impart specific properties, such as permeability and dimensional stability, to make processing easier, as well as to reduce costs.

Strength and elastic modulus of reinforced plastics. The strength of a reinforced plastic with longitudinal fibers can be determined in terms of the strength of the fibers and matrix and the volume fraction of fibers in the composite. In the following equations, c refers to the composite, f to the fiber, and m to the matrix. The total load F_c on the composite is shared by the fiber load F_f and the matrix load F_m; thus,

$$F_c = F_f + F_m, \tag{10.12}$$

which can be rewritten as

$$\sigma_c A_c = \sigma_f A_f + \sigma_m A_m, \tag{10.13}$$

where A_c, A_f, and A_m are the cross-sectional areas of the composite, fiber, and matrix, respectively, and $A_c = A_f + A_m$. Define x as the area fraction of the fibers in the composite. (Note that x also represents the volume fraction, because the fibers are uniformly longitudinal in the matrix.) Equation (10.13) can now be rewritten as

$$\sigma_c = x\sigma_f + (1 - x)\sigma_m. \tag{10.14}$$

The fraction of the total load carried by the fibers can now be calculated as follows: First, note that in the composite under a tensile load, the strains sustained by the fibers and the matrix are the same (that is, $e_c = e_f = e_m$), and then recall from Section 2.2 that

$$e = \frac{\sigma}{E} = \frac{F}{AE}.$$

Consequently,

$$\frac{F_f}{F_m} = \frac{A_f E_f}{A_m E_m}. \tag{10.15}$$

Since the relevant quantities for a specific case are known, Eq. (10.12) can be used to determine the fraction F_f/F_c. Then, using the foregoing relationships, the elastic modulus E_c of the composite can be calculated by replacing σ in Eq. (10.14) with E. Thus,

$$E_c = xE_f + (1 - x)E_m. \tag{10.16}$$

EXAMPLE 10.4 Properties of a Reinforced Plastic

Given: Assume that a graphite-epoxy-reinforced plastic with longitudinal fibers contains 20% graphite fibers, which have a strength of 2500 MPa, and elastic modulus of 300 GPa. The strength of the epoxy matrix is 120 MPa, and it has an elastic modulus of 100 GPa.

Find: Calculate (a) the elastic modulus of the composite and (b) the fraction of the load supported by the fibers.

Solution:

a. The data given are $x = 0.2$, $E_f = 300$ GPa, $E_m = 100$ GPa, $\sigma_f = 2500$ MPa, and $\sigma_m =$ 120 MPa. From Eq. (10.16),

$$E_c = 0.2(300)+(1-0.2)100 = 60-80 = 140 \text{ GPa}.$$

b. The load fraction F_f/F_m is obtained from Eq. (10.15):

$$\frac{F_f}{F_m} = \frac{(0.2)(300)}{(0.8)(100)} = 0.75.$$

Using this relation and Eq. (10.12) yields

$$F_c = F_f + \frac{F_f}{0.75} = 2.33F_f,$$

or $F_f = 0.43F_c$. Therefore, the fibers support 43% of the load, even though they occupy only 20% of the cross-sectional area (and hence volume) of the composite.

10.9.6 Applications of Reinforced Plastics

The first application of reinforced plastics (in 1907) was for an acid-resistant storage tank, made of a phenolic resin (as the matrix) with asbestos fibers. *Formica*, commonly used for countertops, was developed in the 1920s. Epoxies were first used as a matrix material in the 1930s. Beginning in the 1940s, boats were made with fiberglass, and reinforced plastics were used for aircraft, electrical equipment, and sporting goods. Major developments in composites began in the 1970s, and these materials are now called **advanced composites**. Glass-fiber or carbon-fiber reinforced hybrid plastics are available for high-temperature applications (up to about 300°C).

Reinforced plastics are typically used in commercial and military aircraft, rocket components, helicopter blades, automobile bodies, leaf springs, drive shafts, pipes, ladders, pressure vessels, sporting goods, helmets, boat hulls, and various other structures and components. About 50% (by weight) of the Boeing 787 Dreamliner is made of composites. By virtue of the resulting weight savings, reinforced plastics have reduced fuel consumption in aircraft by about 2%. The Airbus jumbo jet A380, with a capacity of up to 700 passengers, has horizontal stabilizers, ailerons, wing boxes and leading edges, secondary mounting brackets of the fuselage, and a deck structure made of composites with carbon fibers, thermosetting resins, and thermoplastics. The upper fuselage is made of alternating layers of aluminum and glass-fiber-reinforced epoxy prepregs.

The contoured frame of the Stealth bomber is made of composites, consisting of carbon and glass fibers, epoxy-resin matrices, high-temperature polyimides, and other advanced materials. Boron fiber-reinforced composites are used in military aircraft, golf-club shafts, tennis rackets, fishing rods, and sailboards (Fig. 9.8). Another example is the development of a small, all-composite ship (twin-hull catamaran design) for the US Navy, capable of speeds of 50 knots (58 mph). More recent developments include (a) reinforcing bars (*rebar*) for concrete, replacing steel bars, thus lowering the costs involved due to their corrosion; and (b) rollers for papermaking and similar industries, with lower deflections as compared to traditional steel rollers.

The processing of polymer-matrix-reinforced plastics, described in Section 10.11, presents significant challenges. Several innovative techniques have been developed to manufacture both large and small parts by a combination of processes, such as molding, forming, cutting, and assembly. Careful inspection and testing of reinforced plastics is essential in critical applications, in order to ensure that good bonding between the reinforcing fiber and the matrix has been obtained throughout the structure. It has been shown that, in some cases, the cost of inspection can be as high as one quarter of the total cost of the composite product.

10.10 Processing of Plastics

The processing of plastics involves operations similar to those for forming and shaping metals, described in Chapters 5 through 8. Depending on their type, plastics can be molded, cast, shaped, formed, machined, and joined into many shapes with relative ease and with few or no additional operations required (Table 10.6). Recall that plastics melt (thermoplastics) or cure (thermosets) at relatively low temperatures (Table 10.2) and thus, unlike metals, are easy to handle and require less energy to process. However, because the properties of plastic parts and components are greatly influenced by the method of manufacture and the processing parameters, their proper control is essential for part quality.

Plastics are usually shipped to manufacturing plants as *pellets* or *powders*, and are melted just before the shaping process. Plastics are also available as sheet, plate, rod, and tubing, which can then be formed into a variety of products. *Liquid* plastics, such as polyurethane, are often used to make laminated or cast reinforced-plastic parts.

10.10.1 Extrusion

In *extrusion*, raw thermoplastic materials, in the form of pellets, granules, or powder, are placed into a hopper and fed into the extruder barrel (Fig. 10.23). The barrel is equipped with a *screw* that blends and conveys the pellets down the barrel. The internal friction and shear stresses developed from the mechanical action of the screw, along with heaters around

TABLE 10.6 Characteristics of processing plastics and reinforced plastics.

Process	Characteristics
Extrusion	Long, uniform, solid or hollow, simple or complex cross sections; wide range of dimensional tolerances; high production rates; low tooling cost
Injection molding	Complex shapes of various sizes and with fine detail; good-dimensional accuracy; high production rates; high tooling cost
Structural foam molding	Large parts with high stiffness-to-weight ratio; low production rates; less expensive tooling than in injection molding
Blow molding	Hollow thin-walled parts of various sizes; high production rates and low cost for making beverage and food containers
Rotational molding	Large hollow shapes of relatively simple design; low production rates; low tooling cost
Thermoforming	Shallow or deep cavities; medium production rates; low tooling costs
Compression molding	Parts similar to impression-die forging; medium production rates; relatively inexpensive tooling
Transfer molding	More complex parts than in compression molding, and higher production rates; some scrap loss; medium tooling cost
Casting	Simple or intricate shapes, made with flexible molds; low production rates
Processing of reinforced plastics	Long cycle times; dimensional tolerances and tooling costs depend on the specific process

the extruder's barrel, heats the pellets and liquefies them. The screw action also builds up pressure in the barrel.

Screws have three distinct sections: (1) a *feed section*, which conveys the material from the hopper area into the central region of the barrel; (2) a *melt* or *transition section*, where the heat from the barrel and also generated from shearing of the plastic causes melting to begin; and (3) a *pumping section*, where additional shearing and melting occurs, with pressure buildup at the die. The lengths of these sections can be modified to accommodate the melting characteristics of the type of plastic.

1. **Mechanics of polymer extrusion.** The pumping section of the screw determines the rate of polymer flow through the extruder. Consider

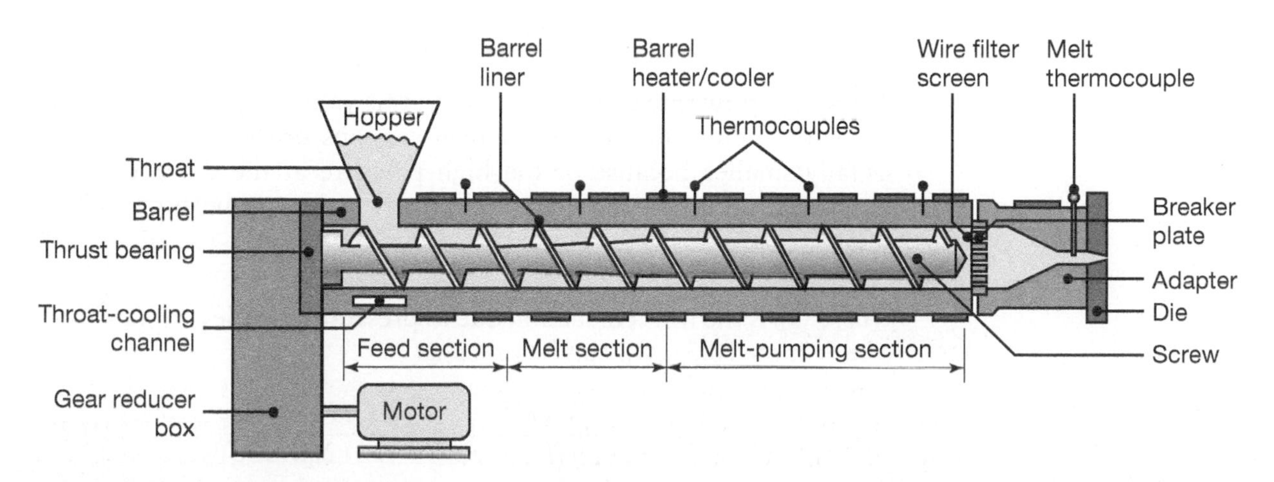

FIGURE 10.23 Schematic illustration of a typical extruder.

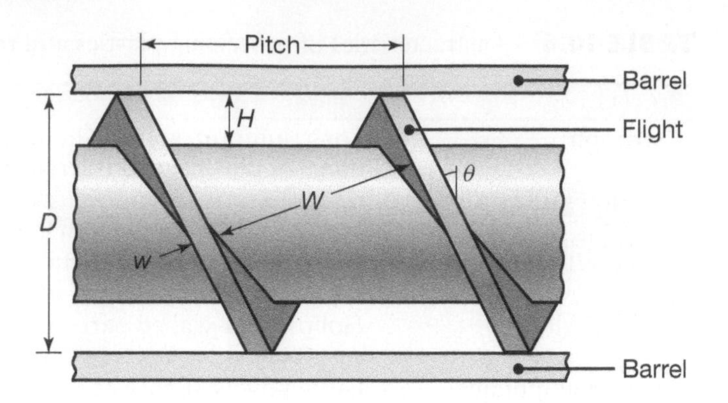

FIGURE 10.24 Geometry of the pumping section of an extruder screw.

a uniform screw geometry with narrow *flights* and small clearance with the barrel (Fig. 10.24). At any point in time, the molten plastic is in the shape of a helical ribbon that is being conveyed toward the extruder outlet by the screw flights. If the pressure is constant along the pumping zone, then the volume flow rate of the plastic out of the extruder (*drag flow*) is given by

$$Q_d = \frac{vHW}{2}, \tag{10.17}$$

where v is the velocity of the flight in the extrusion direction, H is the channel depth, and W is the width of the polymer ribbon. From the geometry defined in Fig. 10.24,

$$v = \omega \cos\theta = \pi DN \cos\theta \tag{10.18}$$

and

$$W = \pi D \sin\theta - w, \tag{10.19}$$

where ω is the angular velocity of the screw, D is the screw diameter, N is the shaft speed (usually in rev/min), θ is the flight angle, and w is the flight width. If the flight width, w, is considered to be negligibly small, then the drag flow can be simplified as

$$Q_d = \frac{\pi^2 HD^2 N \sin\theta \cos\theta}{2}. \tag{10.20}$$

The actual flow rate can be larger than this value if there is significant pressure buildup in the feed or melt sections of the screw, but it is usually smaller because of the high pressure at the die end of the barrel. Thus, the flow rate through the extruder can be expressed as

$$Q = Q_d - Q_p, \tag{10.21}$$

where Q_p is the flow correction due to pressure. For Newtonian fluids (see Section 10.3), Q_p can be taken as

$$Q_p = \frac{WH^3 p}{12\eta\,(l/\sin\theta)} = \frac{p\pi DH^3 \sin^2\theta}{12\eta l}, \tag{10.22}$$

where l is the length of the pumping section. Equation (10.21) then becomes

$$Q = \frac{\pi^2 HD^2 N \sin\theta \cos\theta}{2} - \frac{p\pi DH^3 \sin^2\theta}{12\eta l}. \quad (10.23)$$

Equation (10.23) is known as the **extruder characteristic.** If the power-law index, n from Eq. 10.4, is known, then the extruder characteristic is given by the following approximate relationship:

$$Q = \left(\frac{4+n}{10}\right)\left(\pi^2 HD^2 N \sin\theta \cos\theta\right) - \frac{p\pi DH^3 \sin^2\theta}{(1+2n)4\eta}. \quad (10.24)$$

The die plays a major role in determining the output of the extruder. The **die characteristic** is the expression relating flow to the pressure drop across the die and, in general form, is written as

$$Q_{\text{die}} = Kp, \quad (10.25)$$

where Q_{die} is the flow through the die, p is the pressure at the die inlet, and K is a function of die geometry. The determination of K is usually difficult to develop analytically, although computer-based tools are available for predictions. More commonly, K is determined experimentally. One closed-form solution for extruding solid round cross sections is given by

$$K = \frac{\pi D_d^4}{128\eta l_d}, \quad (10.26)$$

where D_d is the die-opening diameter and l_d is the die land. If the extruder characteristic and the die characteristic are both known, two simultaneous algebraic equations can be written and can be solved for the pressure and flow rate during the operation.

2. **Process characteristics.** Once the extruded product exits the die, it is cooled, either by air or by passing it through a water-filled channel. Controlling the rate and the uniformity of cooling is important for minimizing product shrinkage and distortion. The extruded product can also be drawn (*sized*) by a puller, after it has cooled; the extruded product is then coiled or cut into desired lengths. Complex shapes with constant cross section can be extruded with relatively inexpensive tooling. This process is also used to extrude elastomers (Section 10.10.10).

   Because the material being extruded is still soft as it leaves the die and the pressure is relieved, the cross section of the extruded product is different than the shape of the die opening, an effect known as **die swell** (See Fig. 10.61b). The diameter of a round extruded part, for example, is larger than the die opening, the difference depending on the type of polymer. Modern software can assist in the design of extrusion dies, such as ANSYS Polyflow, that numerically simulates polymer extrusion processes of all kinds, including reverse and coextrusion.

Extruders are generally rated by the diameter and the length-to-diameter (L/D) ratio of the barrel. Typical commercial units are 25 to 200 mm in diameter, with L/D ratios of 5 to 30. Extrusion equipment costs between $30,000 and $150,000, with an additional $30,000 for equipment for downstream cooling and winding of the extruded product. Large production runs are therefore required to justify such an expenditure.

EXAMPLE 10.5 Analysis of a Plastic Extruder

Given: An extruder screw with a thread angle of 20° has a melt-pumping zone that is 1 m long, with a channel depth of 7 mm for a 50-mm diameter barrel. The screw is used to extrude 5 mm-diameter circular nylon bars through a die with a 20-mm land length at 300°C.

Find: If the extruder is operated at 50 rpm = 0.833 rev/s, determine the extruder and die characteristics and obtain the operating flow rate. What is the speed of material leaving the extruder? Ignore die swell.

Solution: Note from Fig. 10.12 that the viscosity of nylon at 300°C is around 300 Ns/m^2, and that the final bar cross-sectional area is $A = \pi D_d^2/4 = 1.96 \times 10^{-5}$ m^2. The extruder characteristic is given by Eq. (10.23) as

$$Q = \frac{\pi^2 HD^2 N \sin\theta \cos\theta}{2} - \frac{p\pi DH^3 \sin^2\theta}{12\eta l}$$

$$= \frac{\pi^2(0.007)(0.050)^2(0.833)\sin 20^\circ \cos 20^\circ}{2} - \frac{\pi(0.050)(0.007)^3 \sin^2 20^\circ}{12(300)(1)}p$$

$$= 2.31 \times 10^{-5} - \left(1.75 \times 10^{-12}\right)p,$$

where Q is in m^3/s and p is in N/m^2. From Eq. (10.26),

$$K = \frac{\pi D_d^4}{128\eta l_d} = \frac{\pi(0.005)^4}{128(300)(0.020)} = 2.56 \times 10^{-12},$$

so that the die characteristic is given by Eq. (10.25) as

$$Q = Kp = \left(2.56 \times 10^{-12}\right)p.$$

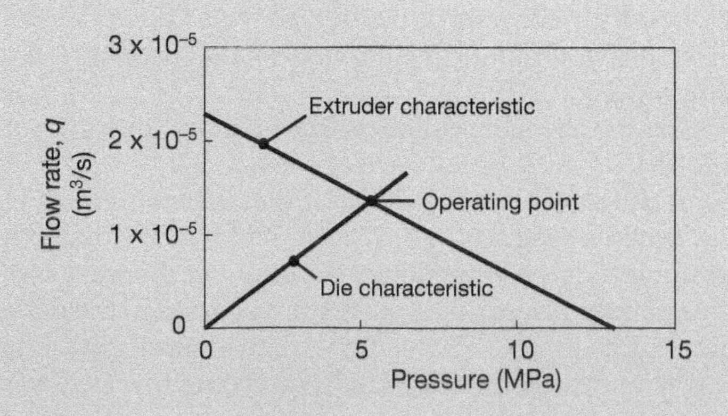

FIGURE 10.25 Extruder and die characteristics for Example 10.5.

The extruder and die characteristics are plotted in Fig. 10.25. Note that for both to be valid, the extruder must operate at the intersection of the two lines. This can be obtained from the graph or directly from the foregoing equations as $p = 5.4$ MPa. At this pressure, the flow rate can be calculated as $Q = 1.37 \times 10^{-5}$ m^3/s. Based on the cross-sectional area of the product, the final velocity is

$$Q = vA; \qquad v = \frac{Q}{A} = \frac{1.37 \times 10^{-5}}{1.96 \times 10^{-5}} = 0.70 \text{ m/s}.$$

3. **Sheet and film extrusion.** Polymer sheet and film can be produced using a flat extrusion die with a thin, rectangular opening (Fig. 10.26). The polymer is extruded by forcing it through a specially designed die, following which the sheet is taken up first on water-cooled rolls and then by a pair of rubber-coated pull-off rolls.

   Thin polymer films are typically made from a tube produced by an extruder (Fig. 10.27). A thin-walled tube is extruded vertically and then expanded (*blown*) into a balloon shape, by blowing warm air through the center of the extrusion die until the desired film thickness

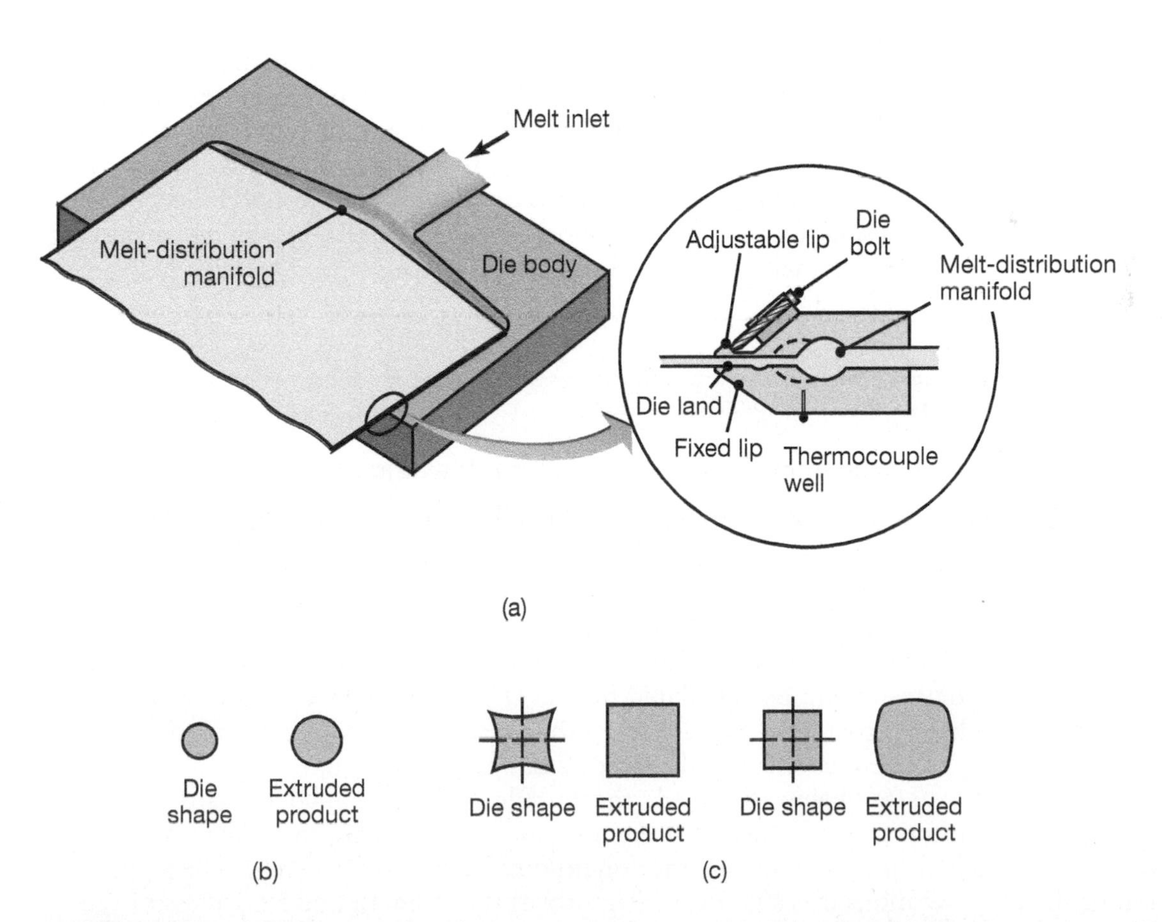

FIGURE 10.26 Common extrusion die geometries: (a) coat-hanger die for extruding sheet; (b) and (c) nonuniform recovery of the part after it exits the die.

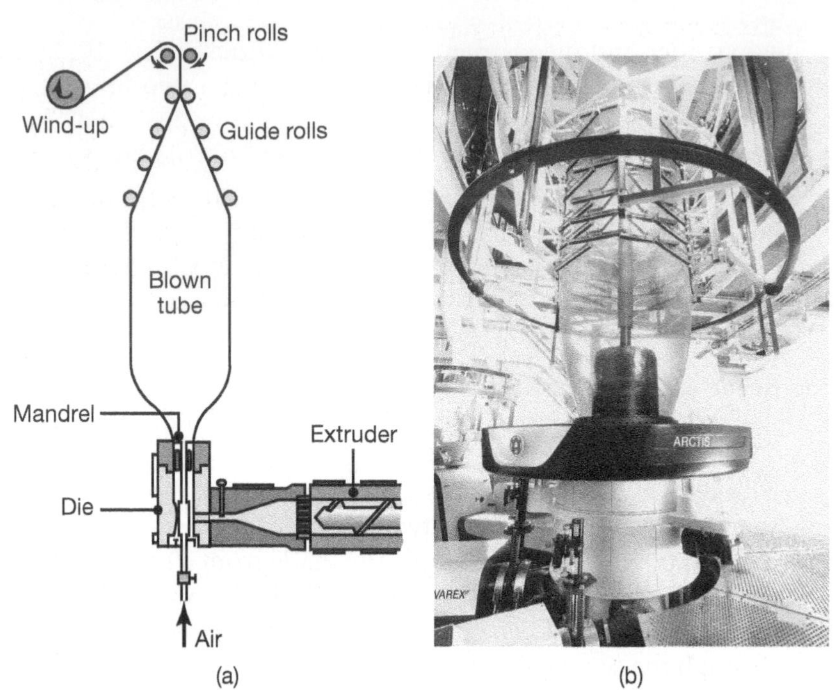

FIGURE 10.27 (a) Schematic illustration of production of thin film and plastic bags from a tube produced by an extruder, and then blown by air. (b) A blown-film operation. *Source:* Courtesy of Windmoeller & Hoelscher Corp.

is reached. The balloon is usually cooled by air from holes in an external cooling ring, that also acts as a barrier to further expansion of the balloon. The **blown film** is sold as wrapping film (after the cooled bubble has been slit) or as bags (where the bubble is pinched and cut off). Film is also produced by *shaving* solid round billets of plastics, especially polytetrafluoroethylene (PTFE), by *skiving*, a shaving process with specially designed knives.

EXAMPLE 10.6 Blown Film

Given: A plastic shopping bag, made from blown film, has a lateral (width) dimension of 400 mm.

Find: (a) What should be the extrusion die diameter? (b) These bags are relatively strong. How is this strength achieved?

Solution:

a. The perimeter of the bag is (2)(400) = 800 mm. Since the original cross section of the film was round, the blown diameter can be calculated from $\pi D = 800$, or $D = 255$ mm. Recall that in this process a tube is expanded 1.5 to 2.5 times the extrusion die diameter. Selecting the maximum value of 2.5, the die diameter is calculated as $255/2.5 = 100$ mm.

b. Note in Fig. 10.27 that, after being extruded, the bubble is being pulled upward by the pinch rolls. Thus, in addition to diametral stretching and the resulting molecular orientation, the film is stretched and oriented in the longitudinal direction as well. The biaxial orientation of the polymer molecules significantly improves the strength and toughness of the blown film.

4. **Miscellaneous extrusion processes**

Pellets, which are used for other plastic-processing methods described throughout the rest of this chapter, are also made by extrusion. The extruded product is a small-diameter rod which is then chopped into short lengths, or *pellets*, as it leaves the die. With some modifications, extruders also can be used to melt plastics for other shaping processes, such as injection molding and blow molding.

Plastic tubes and pipes are produced in an extruder equipped with a *spider die*, as shown in Fig. 10.28a (See also Fig. 6.56 for details). Extrusion of tubes is a first step for related processes, such as extrusion blow molding and blown film (see above). For the production of *reinforced hoses* that can withstand higher pressures, woven fiber or wire reinforcements are coaxially fed through the extruder using specially designed dies.

Coextrusion (Fig. 10.28b) involves simultaneous extrusion of two or more polymers through a single die. The product cross section thus contains different polymers, each with its own characteristics and function. Coextrusion is commonly used for shapes such as sheet, film, and tubes, especially in food packaging where different layers of polymers have different functions. These functions include

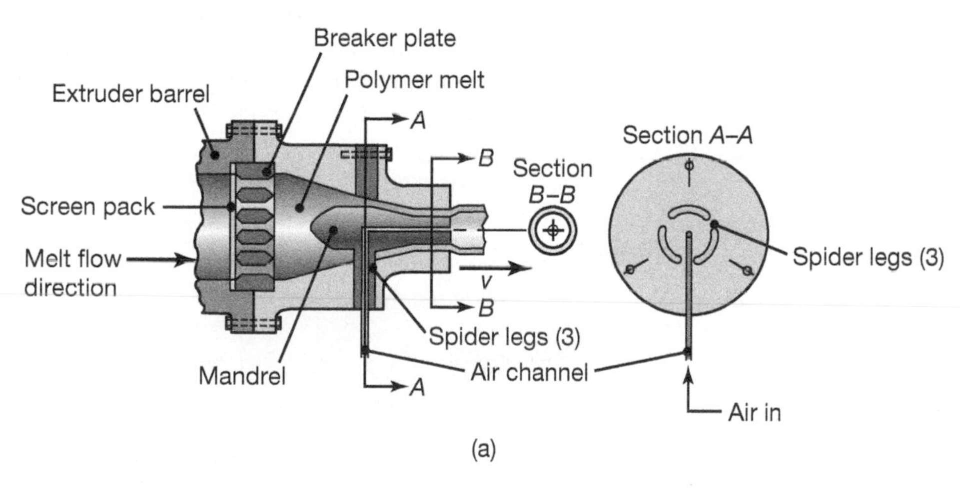

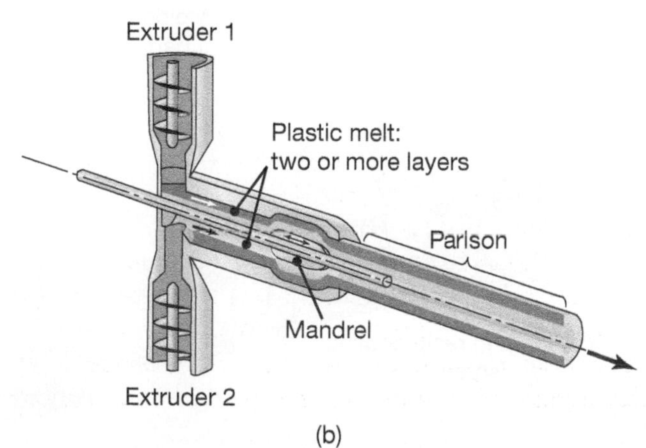

FIGURE 10.28 Extrusion of plastic tubes. (a) Extrusion using a spider die (See also Fig. 6.56) and pressurized air; and (b) coextrusion of tube for producing a bottle.

(a) inertness for food and liquids; (b) serving as barriers to fluids such as water or oil; and (c) labeling of the product.

Plastic-coated electrical wire, cable, and strips also can be extruded and coated by coextrusion. The wire is fed into the die opening at a controlled rate with the extruded plastic in order to produce a uniform coating. To ensure proper electrical insulation, extruded wires are continuously checked for their resistance as they exit the die and are marked automatically with a roller to identify the specific type of wire. *Plastic-coated paper clips* also are made by coextrusion.

10.10.2 Injection Molding

Injection molding is very similar to hot-chamber die casting (Fig. 5.24). The pellets or granules are fed into a heated cylinder, where they are melted, then forced into a split-die chamber (Fig. 10.29a), either by a hydraulic plunger or by the rotating screw of an extruder. Most modern equipment is of the *reciprocating-screw* type (Fig. 10.29b). As the pressure builds up at the mold entrance, the screw starts to move backward and under pressure, to a predetermined distance, thus controlling the volume of material to be injected. The screw then stops rotating and is pushed forward hydraulically, forcing the molten plastic into the mold cavity. Injection-molding pressures usually range from 70 to 200 MPa (10–30 ksi).

Typical injection-molded products include cups, containers, housings, tool handles, knobs, electrical and communication components (such as

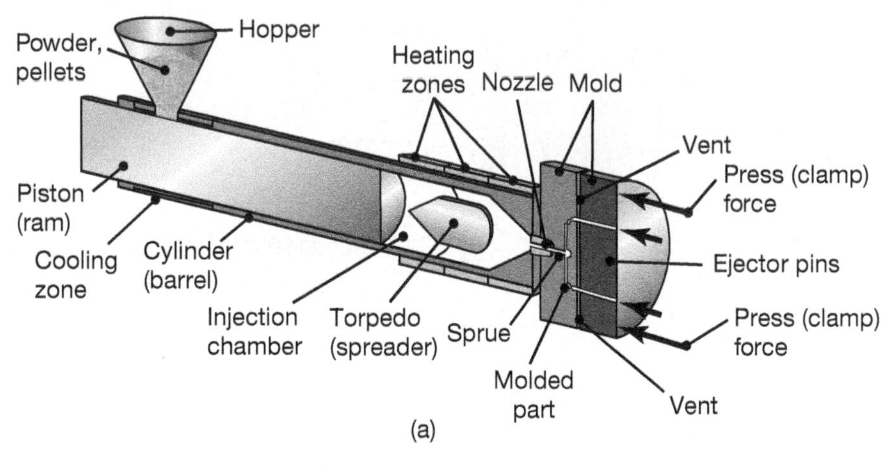

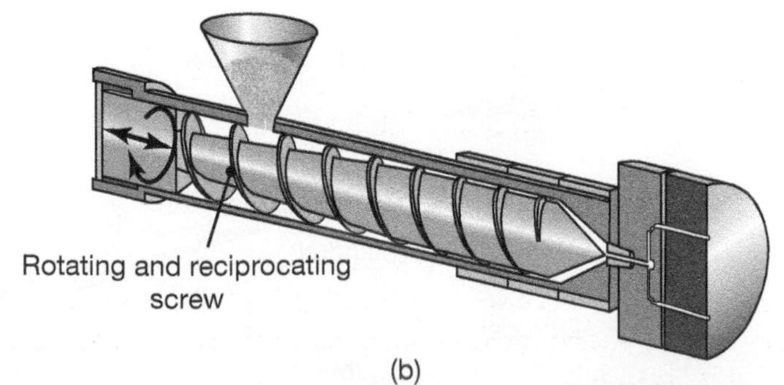

FIGURE 10.29 Injection molding with (a) a plunger and (b) a reciprocating rotating screw. Plumbing fittings, tool handles, and housings are examples of parts made by injection molding.

cell phones housings), toys, and plumbing fittings. Although the molds are relatively cool for thermoplastics, thermosets are molded in heated molds, where *polymerization* and *cross-linking* take place. After the part is sufficiently cooled (for thermoplastics) or set or cured (for thermosets), the molds are opened and the part is ejected. The molds are then closed, and the process is repeated automatically. Elastomers also are injection molded. Molds with moving and unscrewing mandrels are also used; they allow the molding of parts with multiple cavities and internal and external threads.

Because the material is molten when injected into the mold, complex shapes and good dimensional accuracy can be achieved; however, as in metal casting (Chapter 5), the molded part shrinks during cooling. Note also that plastics have a higher thermal expansion coefficient than metals. Linear shrinkage for plastics typically ranges between 0.005 and 0.025 mm/mm, and the volumetric shrinkage is typically in the range of 1.5 to 7% (See also Table 5.1 and Section 11.4). Shrinkage in plastics molding is compensated by oversizing the molds.

Molds. Depending on part design, injection molds have several components, such as runners, cores, cavities, cooling channels, inserts, knockout pins, and ejectors. There are three basic types of molds:

1. *Cold-runner two-plate* mold (Figs. 10.30 and 10.31a), which is the basic and simplest mold design;

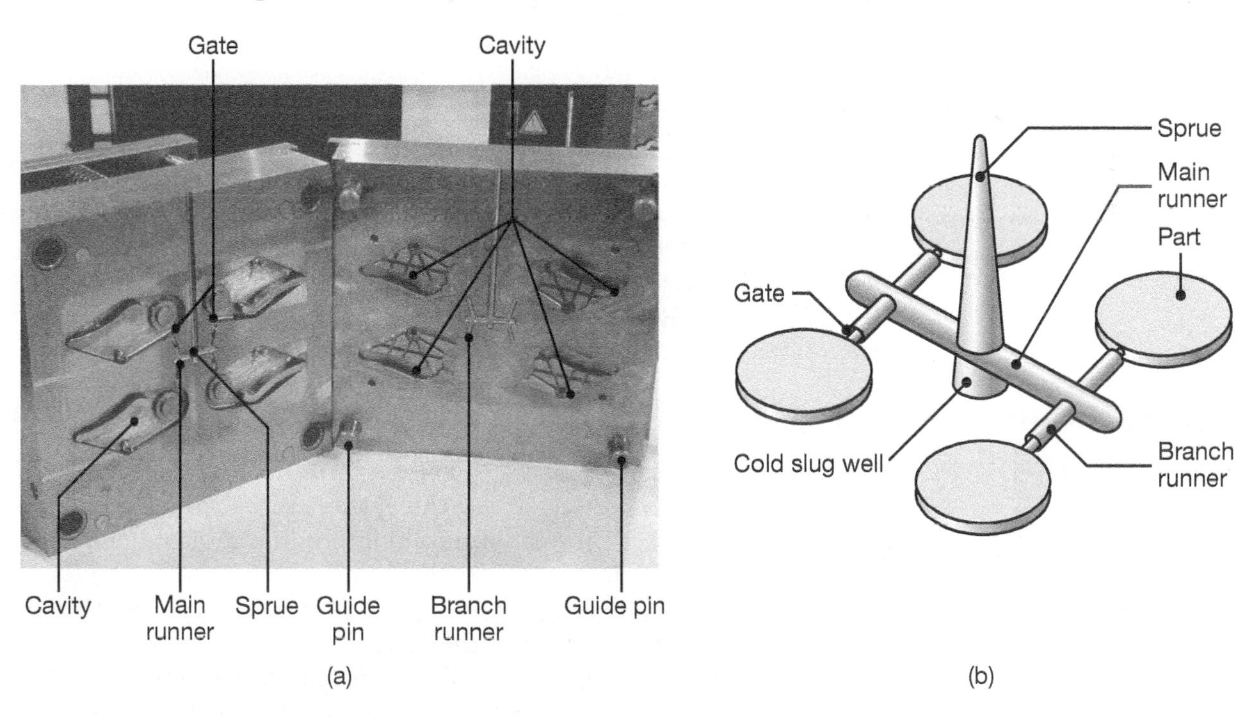

FIGURE 10.30 Illustration of mold features for injection molding. (a) Two-plate mold, with important features identified; and (b) injection molding of four parts, showing details and the volume of material involved. *Source:* Courtesy of Tooling Molds West, Inc.

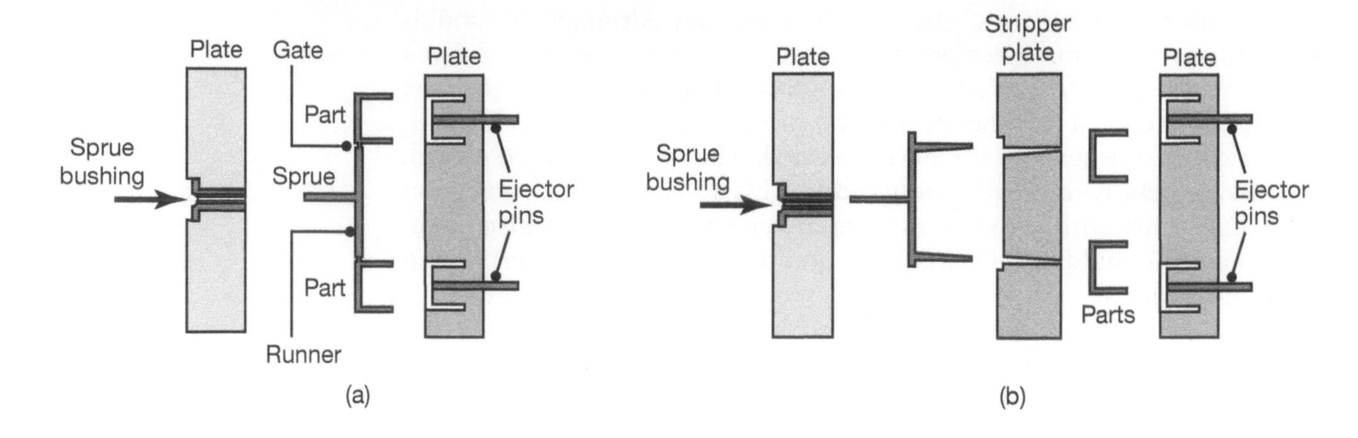

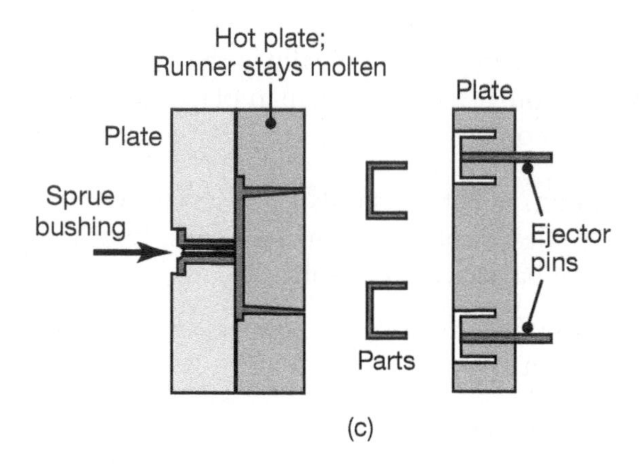

FIGURE 10.31 Types of molds used in injection molding. (a) Two-plate mold, (b) three-plate mold, and (c) hot-runner mold.

2. *Cold-runner three-plate* mold (Fig 10.31b), in which the runner system is separated from the part after the mold is opened; and
3. *Hot-runner* mold, also called *runnerless* mold (Fig. 10.31c), in which the molten plastic is kept hot in a heated runner plate.

FIGURE 10.32 Products made by insert injection molding. Metallic components are embedded in these parts during molding. *Source:* After Rayco Mold & Mfg. LLC.

In cold-runner molds, the solidified plastic in the channels (that connect the mold cavity to the end of the barrel) must be removed, usually by trimming. The scrap is usually chopped and recycled. In hot-runner molds, which are more expensive, there are no gates, runners, or sprues attached to the part. Cycle times are shorter, because only the injection-molded part must be cooled and ejected.

Metallic components, such as screws, pins, and strips, can be placed in the mold cavity before injection of the melted polymer, thereby becoming an integral part of the injection-molded product (**insert molding**; see Fig. 10.32). The most common products made are electrical components and hand tools, such as plastic-handled screwdrivers. *Multicomponent* injection molding, also called *coinjection* or *sandwich* molding, allows the forming of various parts with a combination of colors and shapes. Examples include multicolor molding of rear-light covers for automobiles and ball joints made of different materials. Printed film also can be placed in the mold cavity, thus parts need not be decorated or labeled after molding.

Injection molding is a high-rate production process, with good dimensional control; typical cycle times range from 5 to 60 s, but can be several minutes for thermosetting materials. The molds, generally made of tool steels, aluminum or beryllium-copper alloy, may have *multiple cavities*, so that more than one part can be made in one cycle (as in die casting; Section 5.10.3). Mold design and control of material flow in the die cavities are important factors in the quality of the product. Other factors that also affect part quality include injection pressure, temperature, and condition of the resin. *Computer models* are now available for studying the flow of the material in dies, and thereby improving die design and establishing appropriate process parameters.

Machines. Injection-molding machines are usually horizontal, as shown in Fig. 10.33, and the clamping force on the dies is supplied generally by hydraulic means, although electrical types are also available. Electrically driven models weigh less and are quieter than hydraulic machines. Vertical machines are generally used for making small, close-tolerance parts and for insert molding. Injection-molding machines are rated according to the capacity of the mold and the clamping force. Although in most machines this force generally ranges from 0.9 to 2.2 MN (100–250 tons), the largest machine in operation has a capacity of 75 MN, and can produce parts (large containers, tractor hoods, and similar products) weighing up to 40 kg; however, parts typically weigh 100–600 g. Modern machines used

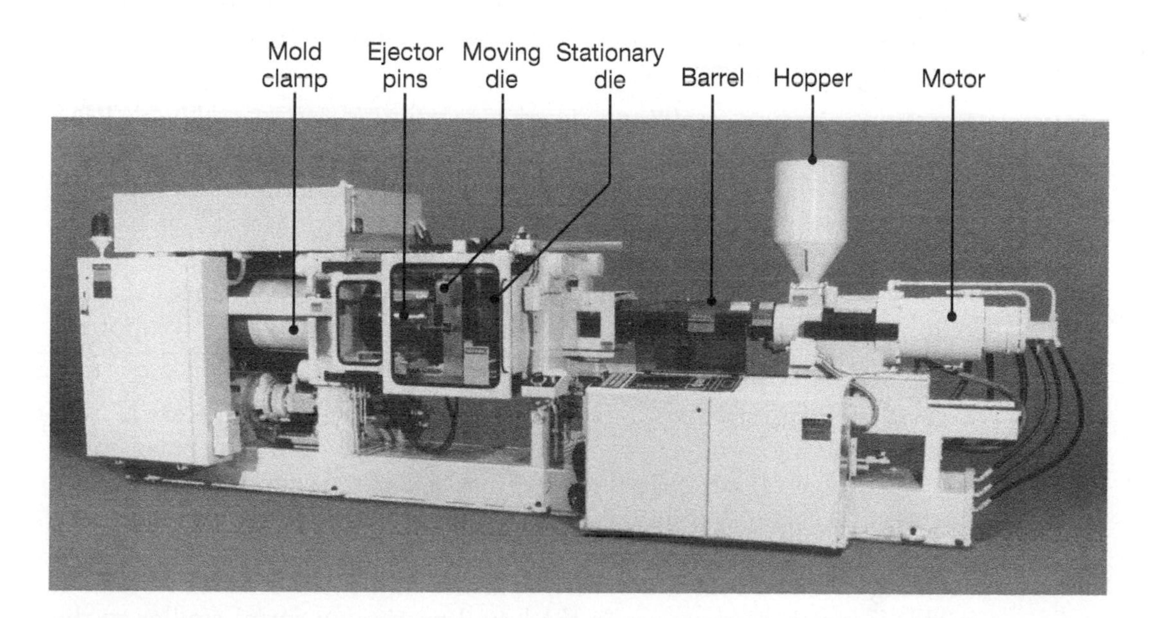

FIGURE 10.33 A 2.2-MN (250-ton) injection-molding machine. The tonnage is the force applied to keep the dies closed during the injection of molten plastic into the mold cavities and hold it there until the parts are cool and stiff enough to be removed from the die.

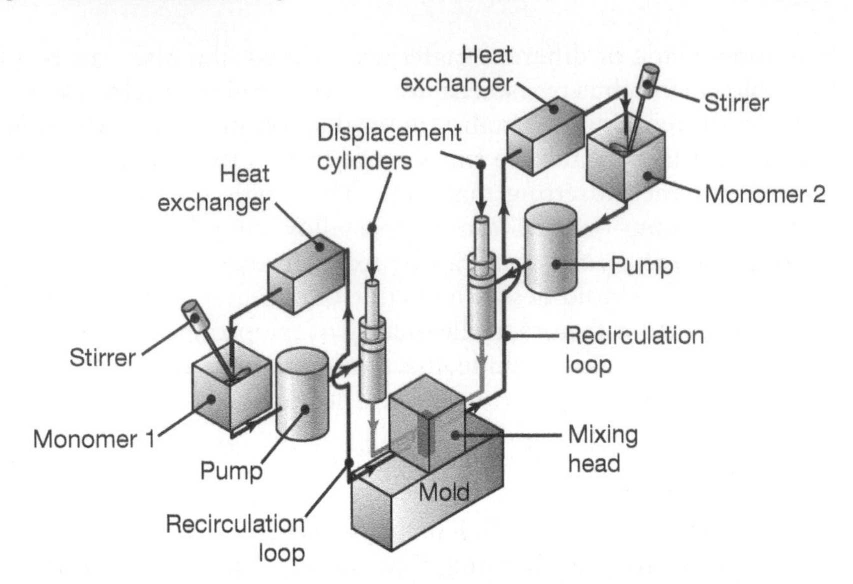

FIGURE 10.34 Schematic illustration of the reaction-injection-molding process.

in injection molding are equipped with microprocessors and microcomputers, and monitor all aspects of the molding operation. Because of the high cost of dies, typically ranging from \$20,000 to \$200,000, high-volume production is required to justify such an expenditure.

Overmolding ice-cold molding. *Overmolding* is a technique for producing, in one operation, hinged joints and ball-and-socket joints. Two different plastics are used to ensure that no bond will form between the two parts, which may interfere with the free movement of the components. *Ice-cold molding* uses the same kind of plastic to form both components, such as in a hinge. The process takes place in one cycle and involves a two-cavity mold, using cooling inserts positioned such that no bond forms between the two pieces.

Reaction-injection molding. In the *reaction-injection-molding* (RIM process), a mixture of two or more reactive fluids is forced into the mold cavity (Fig. 10.34). Chemical reactions then take place rapidly in the mold and the polymer solidifies, producing a thermoset part. Major applications include automotive bumpers and fenders, thermal insulation for refrigerators and freezers, and stiffeners for structural components. Various reinforcing fibers, such as glass or graphite, also may be used to improve the product's strength and stiffness.

Structural foam molding. The *structural-foam-molding* process is used to make plastic products that have a solid skin and a cellular inner structure. Typical products include furniture components, TV cabinets, business-machine housings, and storage-battery cases. Although there are several foam-molding processes, they are basically similar to injection molding or extrusion. Both thermoplastics and thermosets can be foam molded, but thermosets are in the liquid-processing form, similar to the polymers used for reaction-injection molding.

Injection foam molding. In *injection foam molding*, also called *gas-assist molding*, thermoplastics are mixed with a *blowing agent* (usually an inert gas, such as nitrogen, or a chemical agent that produces gas during molding) which expands the polymer. The core of the part is cellular, and the skin is rigid. The thickness of the skin can be as much as 2 mm, and part densities are as low as 40% of the density of the solid plastic. Consequently, parts have a high stiffness-to-weight ratio and can weigh as much as 55 kg.

EXAMPLE 10.7 Injection Molding of Gears

Given: A 2 MN injection-molding machine is used to make 100 mm diameter spur gears with a thickness of 15 mm. The gears have a fine-tooth profile, and require a pressure of 100 MPa to mold properly.

Find: How many gears can be injection molded in one set of molds? Does the thickness of the gears influence the answer?

Solution: The cross-sectional (projected) area of the gear is $\pi(0.1)^2/4 = 0.00785\ \text{m}^2$. Assuming that the parting plane of the two halves of the mold is at the edge of the gear, the force required will be based on this area and is $(0.00785)(100 \times 10^6) = 785$ kN. The capacity of the machine is 2 MN, therefore, the mold can accommodate two cavities, producing two gears per cycle. Because it does not influence the cross-sectional area of the gear, the thickness of the gear does not directly influence the pressures involved, and hence the answer is the same for different gear thicknesses.

10.10.3 Blow Molding

Blow molding is a modified combination of extrusion and injection-molding processes. In **injection blow molding**, a short tubular preform (**parison**) is first injection molded (Fig. 10.35b). The parison can be stored for future molding, or it can be used immediately. If used immediately, the parison molds are opened, and the parison is transferred to a blow-molding die. Hot air is injected into the parison, which expands and fills the mold cavity. Typical products made include plastic beverage bottles and hollow containers.

Multilayer blow molding involves the use of coextruded tubes, or parisons, thus allowing the use of multilayer structures. Typical examples include plastic packaging for food and beverages, with such characteristics as odor and permeation barrier, taste and aroma protection, resistance to scuffing, printing capability, and ability to be filled with hot fluids. Other applications include the products of cosmetics and pharmaceutical industries.

10.10.4 Rotational Molding

Most thermoplastics and some thermosets can be formed into large hollow parts by *rotational molding*. A thin-walled metal mold is made of two pieces (*split female mold*), designed to be rotated about two perpendicular axes (Fig. 10.36). A pre-measured quantity of powdered plastic is placed inside a warm mold. The powder is produced from a polymerization process that precipitates the powder from a liquid. The mold is then heated, usually in a large oven or with heating elements incorporated into the mold, while it rotates about the two axes. This action tumbles the

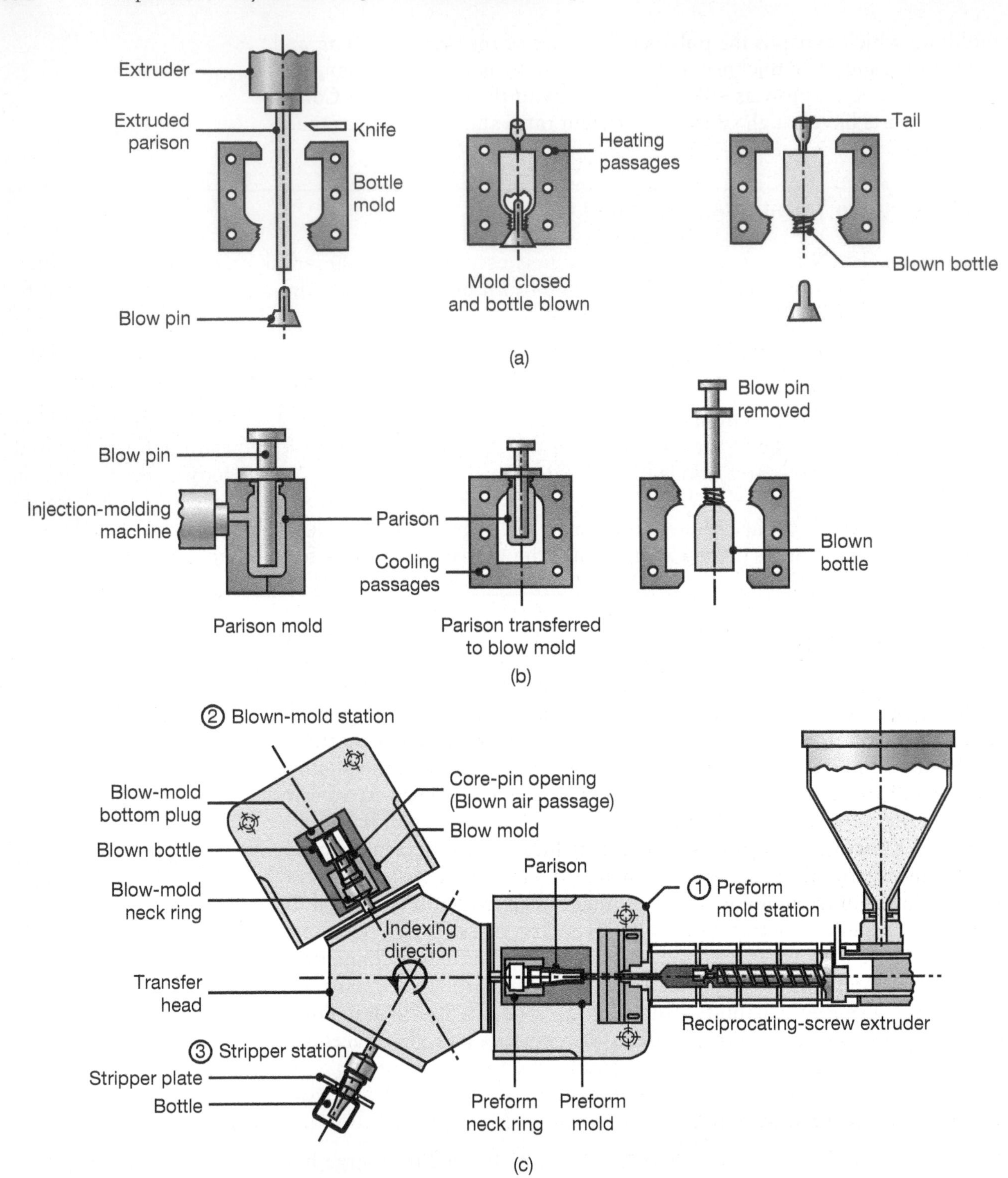

FIGURE 10.35 Schematic illustratons of (a) the blow-molding process for making plastic beverage bottles from tubing; (b) blow molding with a preformed parison; (c) a three-station injection-blow-molding machine.

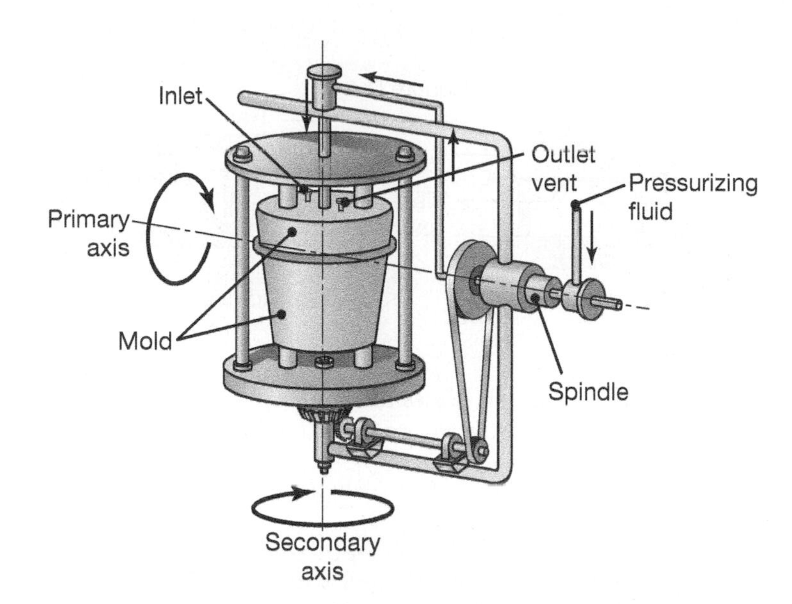

FIGURE 10.36 The rotational molding (rotomolding or rotocasting) process. Trash cans, buckets, carousel horses, and plastic footballs can be made by this process.

powder against the mold, where the heat fuses the powder without melting it. In making some parts, a chemical agent is added to the powder, and cross-linking occurs after the part is formed in the mold by continued heating. Typical parts made include trash cans, boat hulls, buckets, housings, toys, and carrying cases. Various metallic or plastic *inserts* may be molded into the parts made by this process, as in insert molding, described above.

Liquid polymers, called **plastisols** (vinyl plastisols being the most common), can be used in a process called **slush molding**. The mold is simultaneously heated and rotated; the tumbling action forces the plastic particles against the inner walls of the heated mold. Upon contact, the material melts and coats the mold walls. The part is cooled while still rotating, and is then removed by opening the mold.

Rotational molding (**rotomolding**) can produce parts with complex hollow shapes, with wall thicknesses as small as 0.4 mm. Parts as large as 1.8 m × 1.8 m × 3.6 m (6 ft × 6 ft × 12 ft) have been made by this method. Typical examples are storage tanks, containers of all kinds, children's playground slides and structures, rocking horses, and billiard balls. The outer surface finish of the part is a replica of the surface finish of the mold walls. Although cycle times are longer than in other processes, equipment costs are low. Quality-control considerations generally involve the weight of the powder placed in the mold, the rotation of the mold, and the temperature-time relationship during the oven cycle. The most common material in rotomolding is polyethylene, but polycarbonate, nylon, polyvinyl chloride, and many other polymers also are used in this process.

10.10.5 Thermoforming

Thermoforming is a family of processes for forming thermoplastic sheet or film over a mold with the application of heat (Fig. 10.37). A sheet (made by sheet extrusion; see above) is first heated in an oven to the *sag* (softening) *point* but not to the melting point. It is then removed from the oven, placed

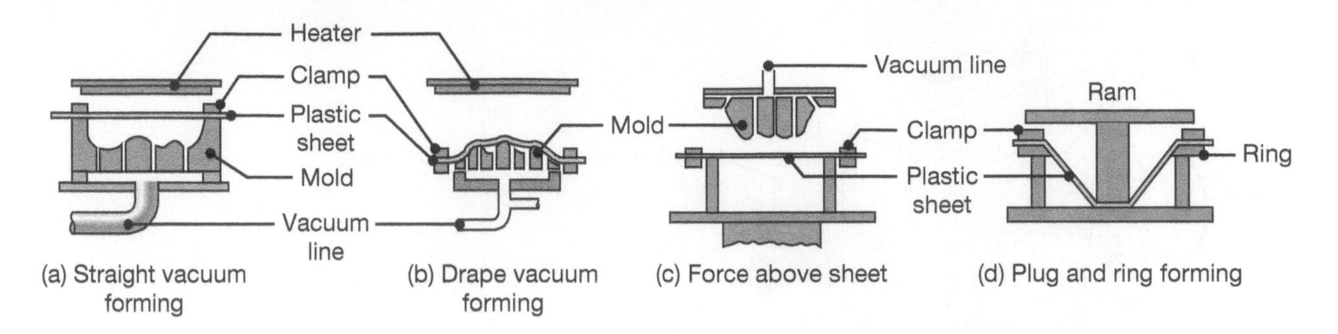

FIGURE 10.37 Various thermoforming processes for thermoplastic sheet. These processes are commonly used in making advertising signs, cookie and candy trays, panels for shower stalls, and packaging.

QR Code 10.1 Thermoforming. *Source:* Courtesy of Shepherd Thermoforming and Packaging, Inc.

over a mold, and forced against the mold by the application of pressure or vacuum. Since the mold is usually at room temperature, the shape of the plastic is set upon contacting the mold.

Typical parts made by this method include packaging, advertising signs, refrigerator liners, appliance housings, and panels for shower stalls. Parts with openings or holes cannot be formed by this method, because the pressure differential required cannot be developed during forming. Because thermoforming is a combination of drawing and stretching operations, much like sheet metal forming (see Chapter 7), the material should exhibit high uniform elongation, as otherwise it will neck and fail. Thermoplastics have a high capacity for uniform elongation by virtue of their high strain-rate sensitivity exponent, m (see Section 2.2.7).

Hollow parts can be made by using twin sheets. In this operation, the two mold halves are brought together with the heated sheets between them. The sheets are drawn to the mold halves through a combination of vacuum from the mold and the compressed air that is introduced between the sheets. The molds also join the sheet around the mold cavity (see *hot-plate welding* in Section 12.17.1).

Molds for thermoforming are usually made of aluminum, since high strength is not a requirement. The holes in the molds are generally less than 0.5 mm in diameter to avoid marks on the formed sheets. Quality considerations include nonuniform wall thickness, tearing, improperly filled molds, and poor surface details (known as part definition).

10.10.6 Compression Molding

In *compression molding*, a preshaped charge of material, a premeasured volume of powder, or a viscous mixture of liquid resin and filler material is first placed directly in a heated mold cavity. Molding is done under pressure with a plug or the upper half of the die, as shown in Fig. 10.38. The flash is removed by trimming or some other means. Typical parts made include dishes, handles, container caps, fittings, electrical and electronic components, washing-machine agitators, and housings. Elastomers and fiber-reinforced parts with long chopped fibers also are formed by this process.

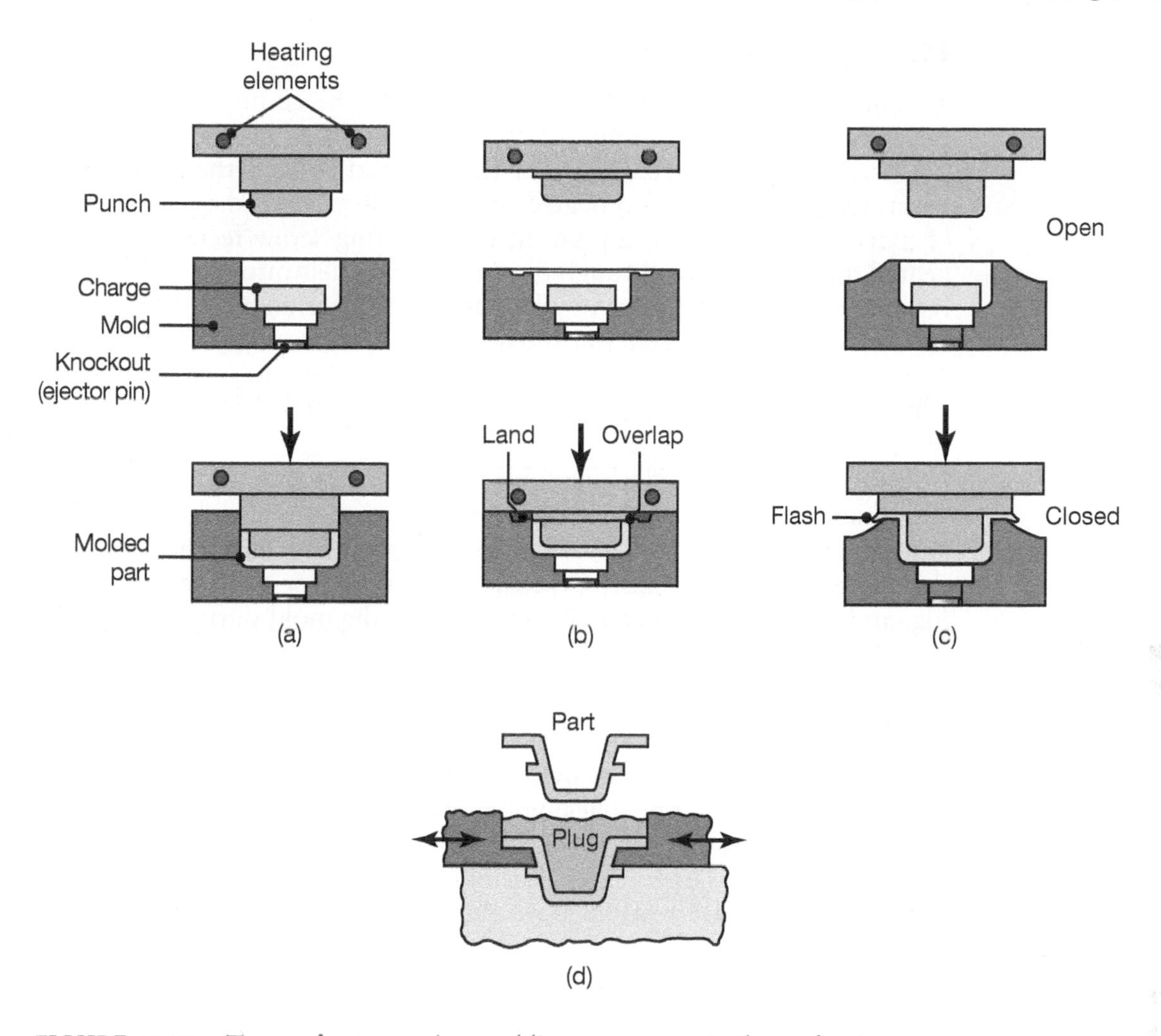

FIGURE 10.38 Types of compression molding, a process similar to forging: (a) positive, (b) semipositive, and (c) flash. The flash in part (c) is trimmed off. (d) Die design for making a compression-molded part with undercuts. Such designs also are used in other molding and shaping operations.

Compression molding is used mainly with thermosetting plastics, with the original material in a partially polymerized state. Cross-linking is completed in the heated die, with curing times typically ranging from 0.5 to 5 min, depending on the polymer, and part geometry and its thickness. The thicker the part, the longer it will take to cure. Because of their relative simplicity, dies for this process generally cost less than those for injection molding. Three types of compression molds are available: (1) **flash type** for shallow or flat parts, (2) **positive** for high-density parts, and (3) **semipositive** for high-quality production. Undercuts in parts are not recommended, but the dies can be designed so as to be opened sideways (Fig. 10.38d) to allow easy removal of the shaped part. In general, the part complexity in this process is less than with injection molding, and the dimensional control is better. Compression molding generally produces polymers that are more crystalline than results from other molding processes; this results in polymers with better strength but lower ductility (see Section 10.2.2).

10.10.7 Transfer Molding

Transfer molding represents a development of the compression-molding process. The uncured thermosetting material is placed in a heated transfer pot or chamber (Fig. 10.39). After the material reaches the proper temperature, it is injected into heated, closed molds. Depending on the type of machine used, a ram, a plunger, or a rotating screw feeder forces the heated material to flow through the narrow channels into the mold cavity. Because of internal friction, the flow generates internal heat, which raises the temperature of the material and homogenizes it; curing then takes place by cross-linking. Because the polymer is molten as it enters the molds, the complexity of the part made and its dimensional control are close to those in injection molding.

The process is particularly suitable for intricate shapes that have varying wall thicknesses. Typical parts made by transfer molding include electrical and electronic components and rubber and silicone parts. The molds for this process tend to be more expensive than those for compression molding, and material is wasted in the channels of the mold during filling. (See also the discussion of *resin transfer molding* in Section 10.11.1.)

10.10.8 Casting

Some thermoplastics, such as nylons and acrylics, and thermosetting plastics, such as epoxies, phenolics, polyurethanes, and polyester, can be *cast* in rigid or flexible molds into a variety of shapes (Fig. 10.40a). Typical parts cast include large gears, bearings, wheels, thick sheets, and components that require resistance to abrasive wear. For thermoplastics, a

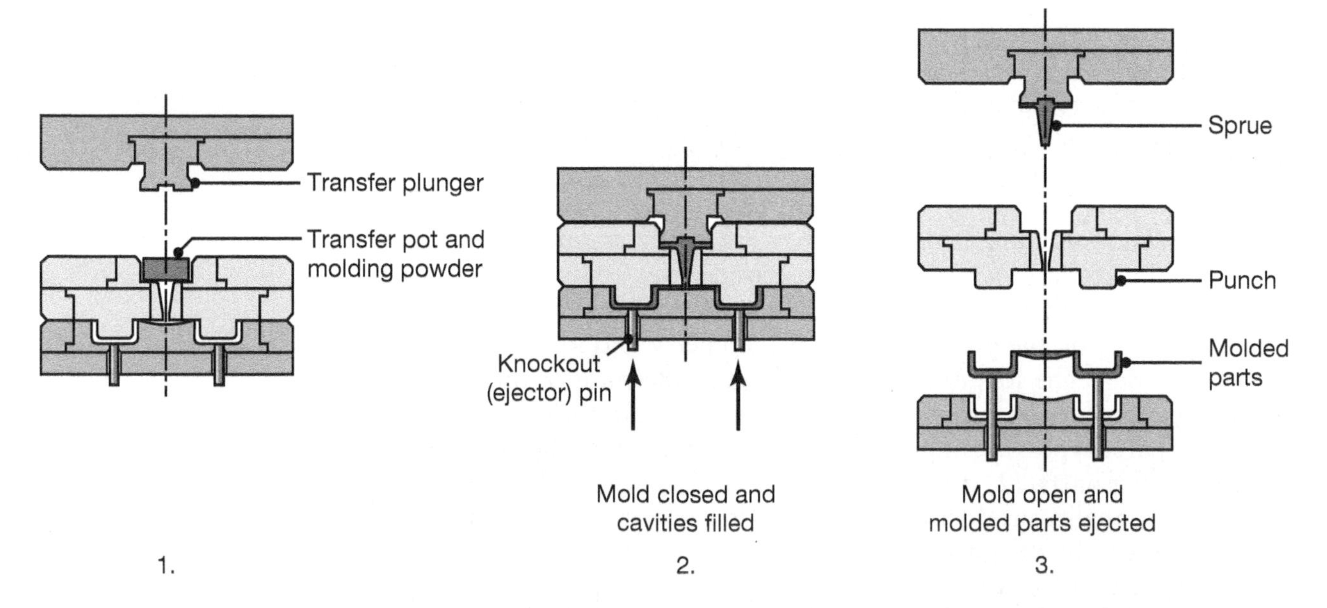

FIGURE 10.39 Sequence of operations in transfer molding of thermosetting plastics. This process is particularly suitable for making intricate parts with varying wall thicknesses.

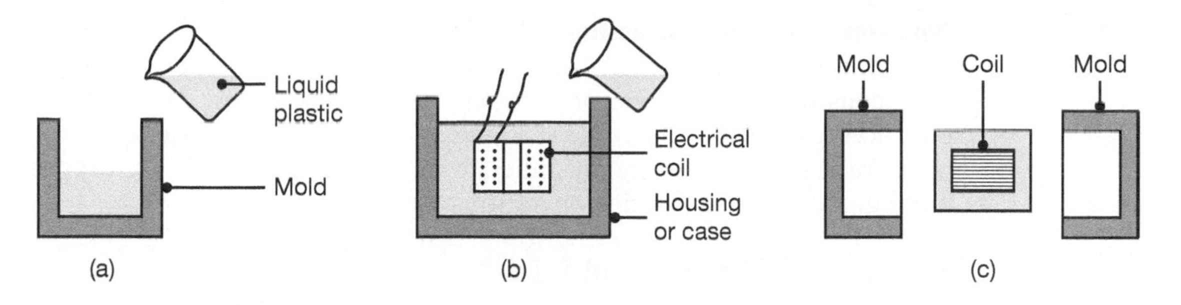

FIGURE 10.40 Schematic illustration of (a) casting, (b) potting, and (c) encapsulation of plastics.

mixture of monomer, catalyst, and various additives is heated and poured into the mold; the part is formed after polymerization takes place at ambient pressure. Intricate shapes can be produced using *flexible molds* (made of polyurethane), which are then peeled off from the solidified cast part. *Centrifugal casting* (see Section 5.10.4) is also used with thermoplastics, including reinforced plastics containing short fibers. Thermosets are cast in a similar manner, and parts are similar to thermoplastic castings.

Potting and encapsulation. A variation of casting, especially important to the electrical and electronics industry, is potting and encapsulation. This involves casting plastic around an electrical component, totally embedding it. *Potting* (Fig. 10.40b) is performed in a housing or case, which becomes an integral part of the product. In *encapsulation* (Fig. 10.40c), the component is covered with a layer of the solidified plastic. In both applications, the plastic serves as a dielectric (nonconductor). Structural members, such as hooks and studs, also may be made by partial encapsulation.

Foam molding. Products such as *styrofoam* cups and food containers, insulating blocks, and shaped packaging materials (such as for computers, appliances, and electronics) are produced by *foam molding*. The material is made of *expandable polystyrene*, made by placing **polystyrene beads** (obtained by polymerization of styrene monomer) containing a blowing agent in a mold, and exposing them to heat, usually by steam. The beads expand to as much as 50 times their original size, and acquire the shape of the mold. The amount of expansion can be controlled through temperature and time. A common method of molding involves the use of *pre-expanded* beads, in which the beads are expanded by steam or hot air, hot water, or an oven in an open-top chamber. The beads are then placed in a storage bin and allowed to stabilize for a period of 3 to 12 hours. They are then molded into shapes.

Polystyrene beads are available in three sizes: (a) small, for products such as cups; (b) medium, for molded shapes such as containers; and (c) large, for molding of insulating blocks; which can then be cut to size. The bead size chosen depends on the minimum wall thickness of the product. Beads can also be colored prior to expansion, or integrally-colored beads can be used.

Polyurethane foam processing. Used for making such products as cushions and insulating blocks, this process involves several steps that basically consist of mixing two or more chemical components. The reaction produces a *cellular structure* that solidifies in the mold. A variety of low-pressure and high-pressure machines, with computer controls for proper mixing, are available for this operation.

10.10.9 Cold Forming and Solid-Phase Forming

Cold-working processes, such as rolling, deep drawing, extrusion, closed-die forging, coining, and rubber forming (described in Chapters 6 and 7), also can be used to form thermoplastics at room temperature (*cold forming*). Typical cold formed polymers include polypropylene, polycarbonate, ABS, and rigid PVC. The major considerations are that (a) the material must be sufficiently ductile at room temperature, hence polystyrenes, acrylics, and thermosets cannot be cold formed; and (b) the material's deformation must be nonrecoverable, in order to minimize springback and creep.

The advantages of cold forming of plastics over other methods of shaping are:

1. Strength, toughness, and uniform elongation of the material are increased;
2. Polymers with high molecular weight can be used to make parts, with superior properties;
3. Forming speeds are not affected by the thickness of the part, because there is no heating or cooling involved; and
4. Typical cycle times are shorter than those for molding processes.

Solid-phase forming is carried out at a temperature about 10°C to 20°C below the melting temperature of the plastic (See Table 10.2), if it is a crystalline polymer and it is formed while still in a solid state. The advantages of solid-phase forming over cold forming are that forming forces and springback are lower for the former. These processes are not as widely used as hot-processing methods and are generally restricted to special applications.

10.10.10 Processing Elastomers

In terms of its processing characteristics, a thermoplastic elastomer is a polymer; in terms of its function and performance, it is a rubber (see Section 10.8). Elastomers can be shaped by many of the same processes used for shaping thermoplastics. Thermoplastic elastomers are commonly shaped by extrusion and injection molding, extrusion being the most economical and the fastest process. These elastomers also can be formed by blow molding and thermoforming. Thermoplastic polyurethane can be shaped by all conventional methods; it can also be blended with thermoplastic rubbers, polyvinyl chloride compounds, ABS, and nylon. For extrusion, the temperatures are in the range of 170° to 230°C (340°F to 450°F), and for molding, they range up to 60°C (140°F). Typical extruded elastomer products include hoses, moldings, and inner tubes.

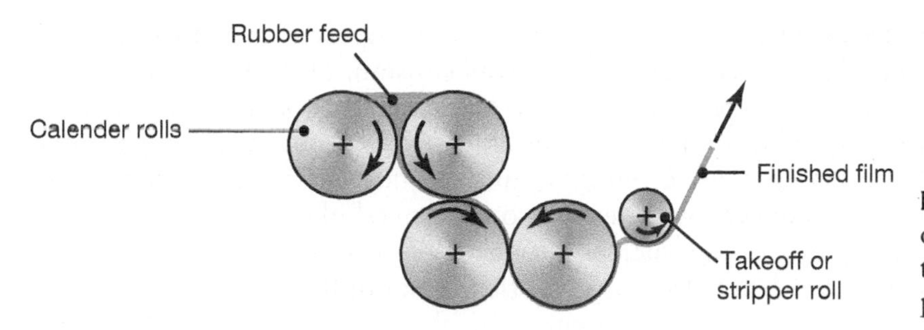

FIGURE 10.41 Schematic illustration of calendaring. Sheets produced by this process are subsequently used in processes such as thermoforming.

Injection-molded products cover a broad range, such as children's toys, wiring harnesses and other components for automobiles and housings and controls for appliances.

Rubber and some thermoplastic sheets are formed by the **calendaring** process (Fig. 10.41), wherein a warm mass of the elastomer feedstock is fed through a series of rolls (called *masticating rolls*) and is then stripped off in the form of a sheet. The rubber may also be shaped between two surfaces of a fabric liner.

Discrete rubber products, such as gloves, balloons, and swim caps, are made by *dipping* or **dip molding**, A solid metal form, such as in the shape of a hand for making gloves, is repeatedly dipped into a liquid compound that adheres to the form. The compound is then vulcanized (cross-linked), usually in steam, and then stripped from the form, thus becoming a discrete product. A typical compound is *latex*, a milk-like sap obtained from the inner bark of a tropical tree.

10.11 Processing of Polymer-Matrix-Reinforced Plastics

As described in Section 10.9, reinforced plastics are among the most important materials that can be engineered to meet specific design requirements, such as high strength-to-weight and stiffness-to-weight ratios and creep resistance. Because of their unique structure and the characteristics of their individual components, reinforced plastics require special methods to shape them into products (Fig. 10.42).

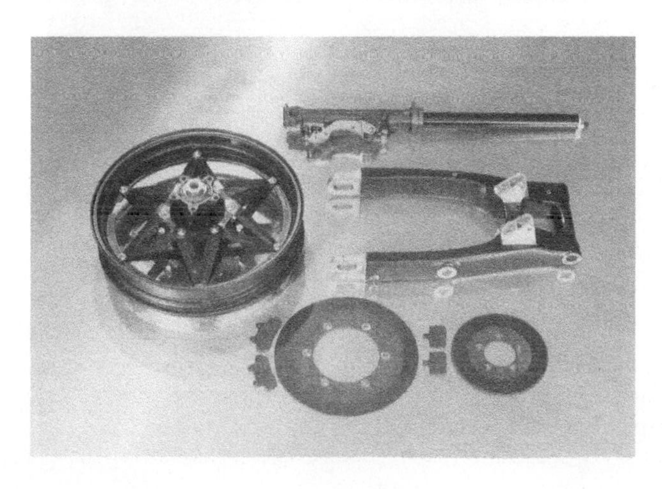

FIGURE 10.42 Reinforced-plastic components for a Honda motorcycle. The parts shown are front and rear forks, a rear swing arm, a wheel, and brake disks.

The care required and the numerous steps involved in manufacturing reinforced plastics make processing costs substantial. Careful assessment and integration of the design and manufacturing processes (see *concurrent engineering*, Section 1.2) is thus essential in order to minimize costs while maintaining product integrity and production rate. An important environmental concern with respect to reinforced plastics is the dust generated during processing, such as airborne carbon fibers that are known to remain in the work area long after fabrication of parts has been completed.

Reinforced plastics are fabricated by the methods described in this chapter, with consideration for the presence of more than one type of material in the composite. As described in Section 10.9, the reinforcement may consist of chopped fibers, woven fabric or mat, roving or yarn (slightly twisted fiber), or continuous lengths of fiber. Short fibers are commonly added to thermoplastics in injection molding, milled fibers can be used in reaction-injection molding, and longer chopped fibers are used primarily in compression molding. In order to obtain good bonding between the fibers and the polymer matrix, as well as to protect the fibers during subsequent processing steps, the fibers are first surface treated by *impregnation* (*sizing*). When the impregnation is done as a separate step, the resulting partially-cured sheets are referred to by various terms:

1. **Prepregs.** The continuous fibers are aligned (Fig. 10.43a) and subjected to surface treatment to enhance their adhesion to the polymer matrix. They are then coated by being dipped in a resin bath and made into a *sheet* or *tape* (Fig. 10.43b). Finally, individual pieces of the sheet are assembled into laminated structures; special computer-controlled tape-laying machines have been developed for this purpose. Typical products made include flat or corrugated architectural paneling, panels for construction and electric insulation, and structural

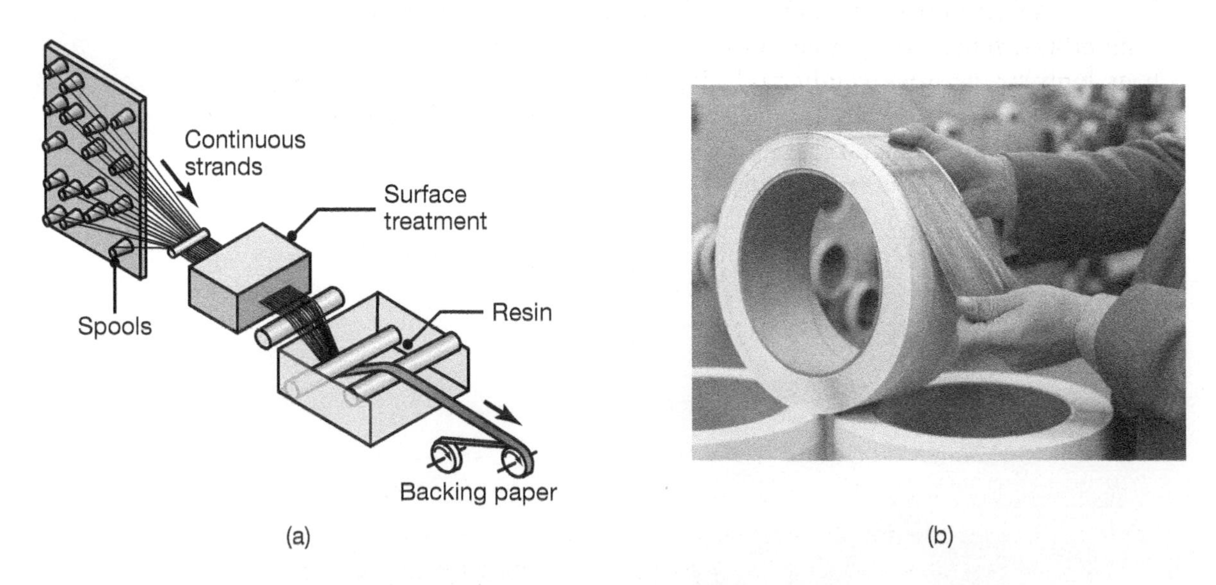

FIGURE 10.43 (a) Manufacturing process for polymer-matrix composite. (b) Boron-epoxy prepreg tape. *Source:* Specialty Materials, Inc.

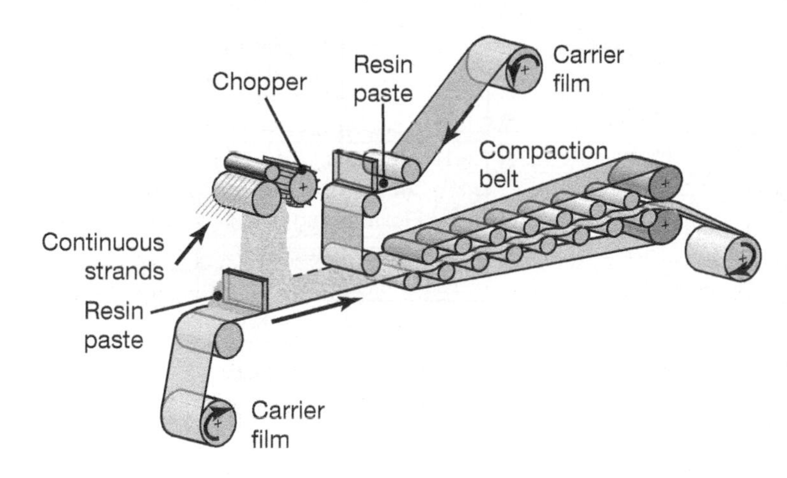

FIGURE 10.44 Manufacturing process for producing reinforced-plastic sheets. The sheet is still viscous at this stage and can later be shaped into various products.

components of aircraft that require good retention of properties and fatigue strength under various severe environmental conditions.

2. **Sheet-molding compound** (SMC). Continuous strands of reinforcing fiber are chopped into short fibers (Fig. 10.44) and deposited over a layer of resin paste, usually a polyester mixture that is carried on a polymer film such as polyethylene. A second layer of resin paste is deposited on top, and the sheet is pressed by passing it through rollers. The product is then gathered into rolls, or placed into containers in layers, and stored until it undergoes a *maturation period*, reaching the desired molding viscosity. The maturing process involves controlled temperature and humidity and usually takes one day. The matured SMC, which has a leatherlike feel, has a shelf life of about 30 days and must be processed within this period. Alternatively, the resin and the fibers may be mixed together only at the time they are placed in the mold.
3. **Bulk-molding compound** (BMC). These compounds are in the shape of billets, generally up to 50 mm in diameter; they are made by extrusion and in the same manner as SMCs. When processed into products, BMCs have flow characteristics similar to that of dough, hence they are called *dough-molding compounds* (DMCs).
4. **Thick-molding compound** (TMC). This compound combines the characteristics of BMCs (lower cost) and SMCs (higher strength); it is usually injection molded, using chopped fibers of various lengths. One of its applications is for electrical components, because of the high dielectric strength of TMCs.

10.11.1 Molding

There are five basic methods for molding reinforced plastics.

1. **Compression molding.** This operation was described in Section 10.10.6. With reinforced plastics, the feedstock may be in bulk form (BMC or DMC), consisting of a viscous, sticky mixture of polymers, fibers, and additives. It is generally shaped into a log, which is then cut into the desired size before being placed in the mold. Fiber lengths

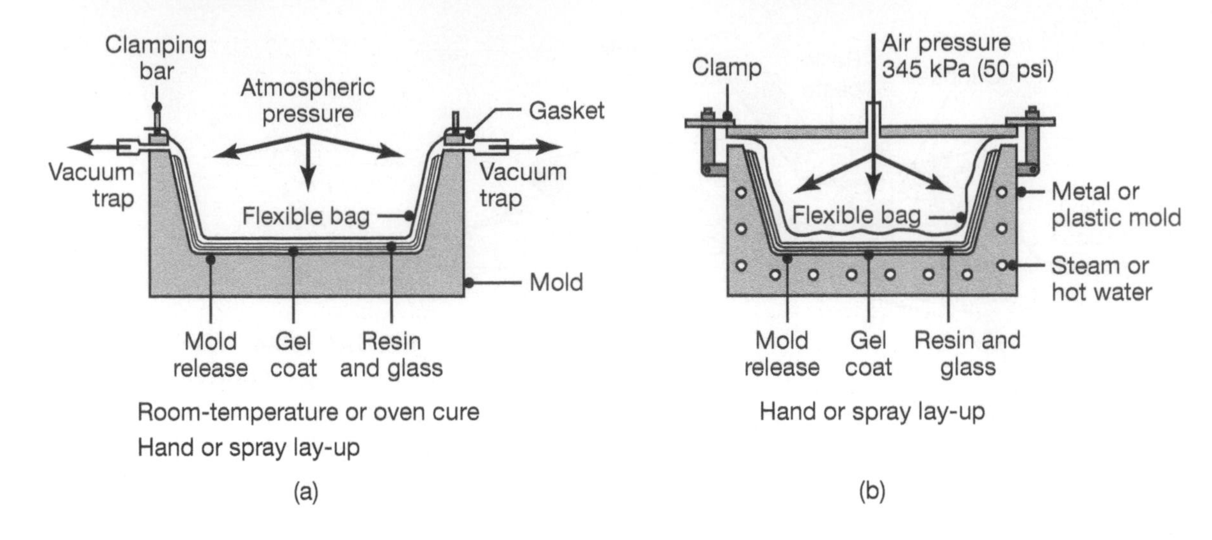

FIGURE 10.45 (a) Vacuum-bag forming. (b) Pressure-bag forming.

typically range from 3 to 50 mm, although fibers 75 mm long also may be used. Sheet-molding compounds (SMC) also can be used in compression molding.

2. **Vacuum-bag molding.** As shown in Fig. 10.45, prepregs are laid in a mold, and the pressure required to form the shape and to develop good bonding is obtained by covering the layup with a plastic bag and developing a vacuum. If additional heat and pressure are necessary, the entire assembly is placed in an *autoclave*. Care should be exercised to maintain fiber orientation, if desired; in materials with chopped fibers, no specific orientation is usually intended. In order to prevent the resin from sticking to the vacuum bag and to facilitate removal of any excess material, several sheets (*release cloth* or *bleeder cloth*) are placed on top of the prepreg. Although the molds can be made of metal, usually aluminum, more often they are made from the same resin (with reinforcement) as the material to be cured; this eliminates any difficulties resulting from differences in thermal expansion between the mold and the part.
3. **Contact-molding.** This process is used in making products with high surface-area-to-thickness ratios, such as swimming pools, boats, tub and shower units, and housings. It uses a single male or female mold (Fig. 10.46), made of materials such as reinforced plastics, wood, or plaster. The contact molding process is a wet method, in that the reinforcement is impregnated with the resin at the time of molding. The simplest method is called **hand lay-up** (Fig. 10.46a). The materials are placed and formed in the mold by hand where the squeezing action expels any trapped air, and compacts the part.

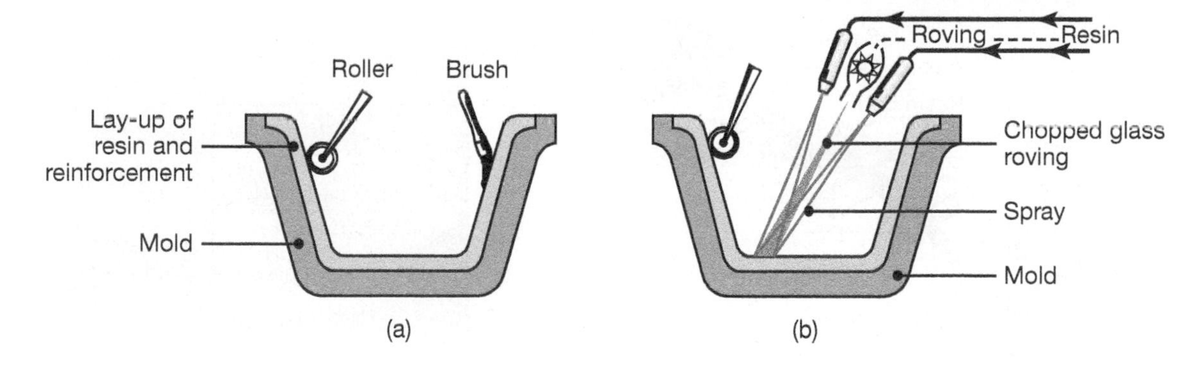

FIGURE 10.46 Manual methods of processing reinforced plastics: (a) hand lay-up and (b) spray-up. These methods are also called *open-mold processing*.

Molding may also be done by spraying (*spray-up*; see Fig. 10.46b). Although the spraying operation can be automated, it is simple, and tooling is inexpensive, it is relatively slow, so labor costs are high, and only the mold-side surface of the part is smooth. Several types of boats can be made by this process.

4. **Resin transfer molding.** This process is based on transfer molding (Section 10.10.7), whereby a resin, mixed with a catalyst, is forced by a piston-type, positive-displacement pump into a mold cavity filled with fiber reinforcement. The process is a viable alternative to hand lay-up, spray-up, and compression molding, particularly for low- or intermediate-volume production.
5. **Transfer/injection molding.** This is an automated operation that combines the processes of compression molding, injection molding, and transfer molding, and thus it takes advantage of each process and produces parts with enhanced properties.

10.11.2 Filament Winding, Pultrusion, and Pulforming

Filament winding. This is a process whereby the resin and the fibers are combined at the time of curing. The reinforcements are impregnated by passing them through a polymer bath (Fig. 10.47a). The reinforcing filament, tape, or roving is wrapped continuously around a rotating mandrel or form. Alternatively, a prepreg (see below) can be wrapped around the mandrel. Axisymmetric parts, such as pipes, gas cylinder pressure vessels, and storage tanks, are produced by this method.

Spools
Separator Combs
Resin bath
Continuous strands
Guide
Nip rollers
Rotating mandrel

(a)

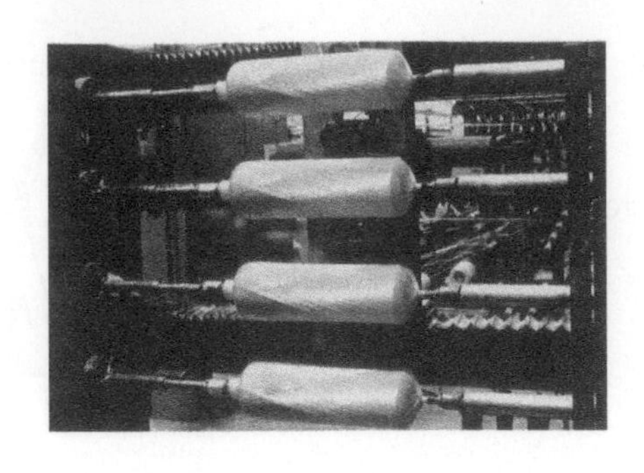

(b)

FIGURE 10.47 (a) Schematic illustration of the filament-winding process. (b) Fiberglass being wound over aluminum liners for slide-raft inflation vessels for the Boeing 767 aircraft. *Source:* Advanced Technical Products Group, Inc., Lincoln Composites.

The products made by filament winding are very strong because of their highly reinforced structure. The process also can be used for strengthening cylindrical or spherical pressure vessels (Fig. 10.47b) made of materials such as aluminum or titanium; the presence of a metal inner lining makes the part impermeable. Filament winding can also be used directly over solid-rocket propellant forms. Seven-axis computer-controlled machines (Section 14.3) have been developed for making asymmetric parts that automatically dispense several unidirectional prepregs. Typical asymmetric parts made include aircraft engine ducts, fuselages, propellers, blades, and struts.

QR Code 10.2 Pultrusion process. *Source:* Courtesy of Strongwell Corporation.

Pultrusion. Parts with high length-to-cross-sectional area ratios and various constant profiles, such as rods, structural profiles, and tubing, are made by the *pultrusion* process. In this process, developed in the early 1950s, the continuous reinforcement (roving or fabric) is pulled through a thermosetting-polymer bath, and then through a long heated steel die (Fig. 10.48). The product is cured during its travel through the die and then cut into desired lengths. The most common material used in pultrusion is polyester with glass reinforcements. Typical products made by this process include golf club shafts, drive shafts, and structural members such as ladders, walkways, and handrails.

Pulforming. Continuously reinforced products with a constant cross section are made by *pulforming*. After being pulled through the polymer bath as in pultrusion, the composite is placed in a die, and cured into a finished product. The part can then be formed in a manner similar to swaging (see Section 6.6) or through pressurization of hollow cross sections. Common examples of products made include glass-fiber-reinforced hammer handles and curved automotive leaf springs.

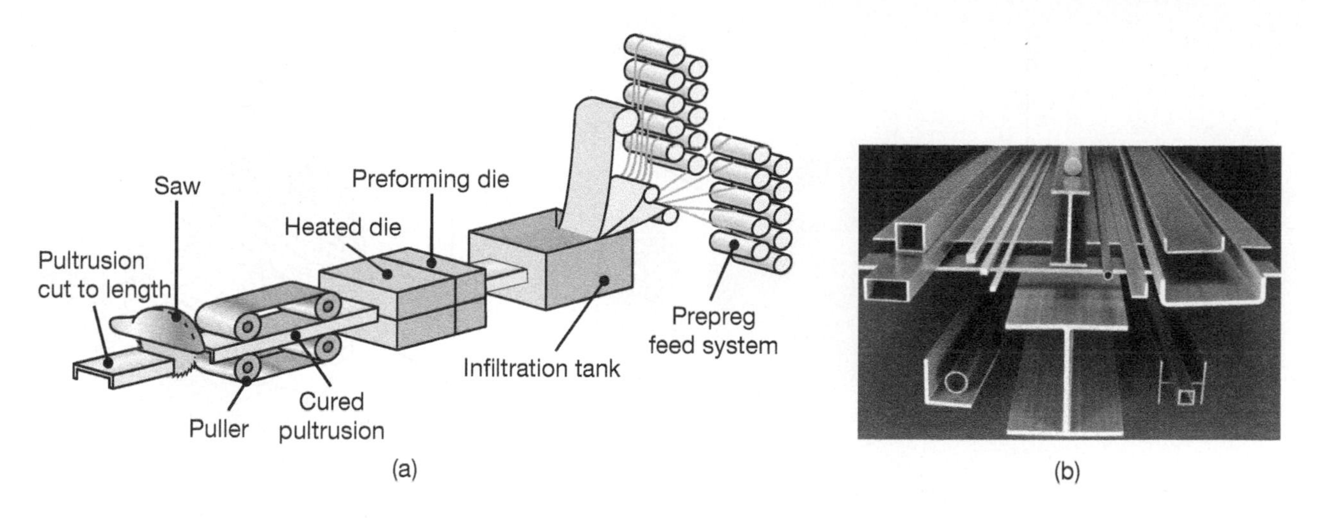

FIGURE 10.48 (a) Schematic illustration of the pultrusion process. (b) Examples of parts made by pultrusion. *Source:* Courtesy of Strongwell Corporation.

10.11.3 Product Quality

The major quality considerations for the processes described thus far principally involve internal voids and gaps between successive layers of material. Volatile gases that develop during processing must be allowed to escape through the vacuum bag to avoid porosity due to trapped gases within the lay-up. Microcracks may also develop due to improper curing or during transportation and handling. These defects can be detected using ultrasonic scanning and other techniques (see Section 4.8.1).

10.12 Additive Manufacturing

Making a *prototype*, the first full-scale model of a product, has traditionally involved flexible manufacturing processes, such as machining operations, using a variety of cutting tools and machines, usually taking several weeks or months. An important advance is *rapid prototyping*, also called **three-dimensional printing** or **additive manufacturing,** a process by which a solid physical model of a part is made directly from a three-dimensional CAD drawing without the use of tools. First developed in the 1980s, additive manufacturing involves several different processes and mechanisms, including resin curing, robot controlled extrusion deposition, melting and solidification, and sintering.

This section describes **additive manufacturing,** whereby parts are *built in layers*. Prototypes also can be produced through **subtractive processes** (basically involving computer-controlled machining operations) or by **virtual prototyping** (involving advanced graphics and software). Additive manufacturing inherently involves integrated computer-driven hardware and software. In order to visualize the methodology used, it is helpful to visualize constructing a loaf of bread by stacking and bonding individual

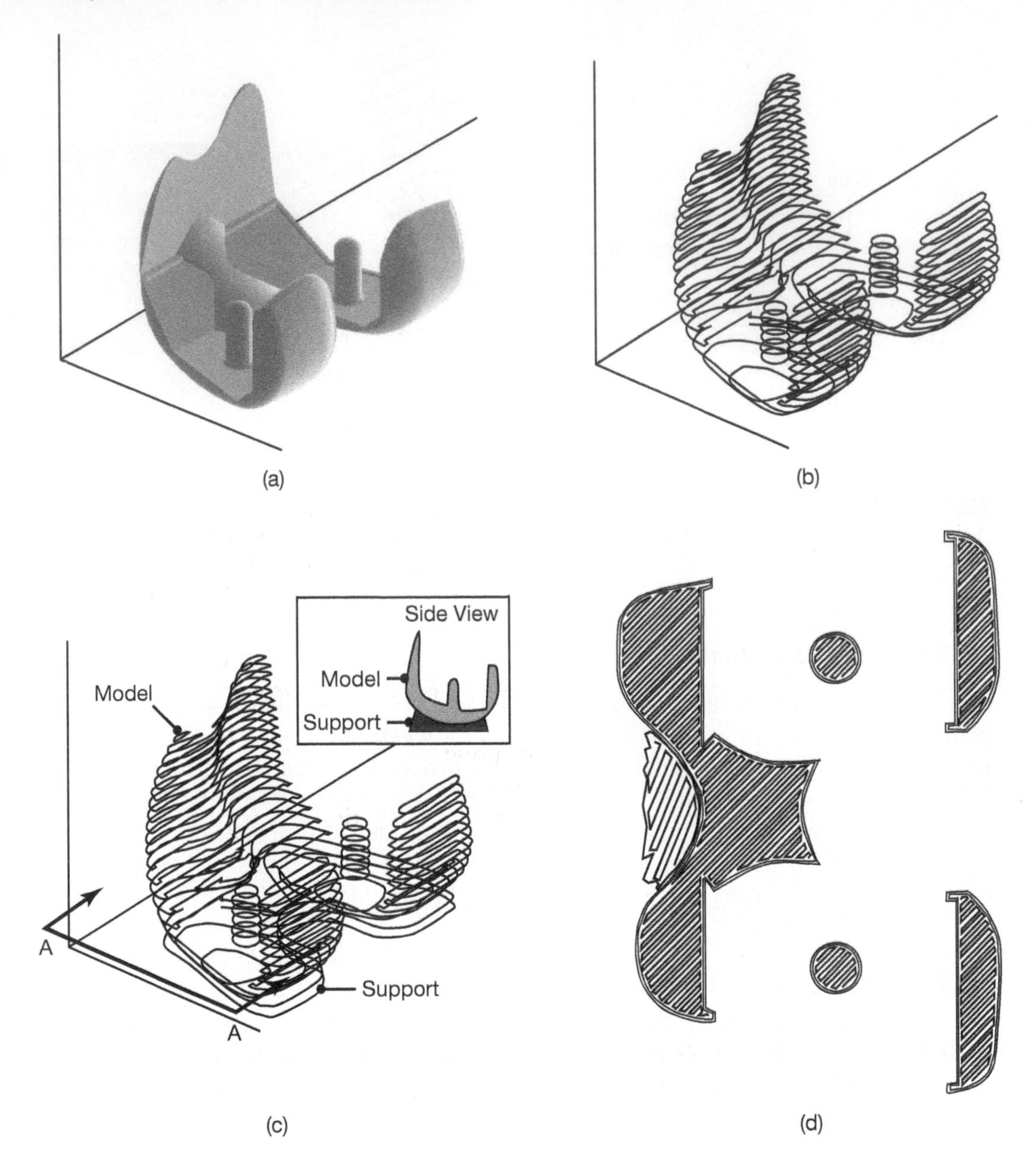

FIGURE 10.49 The computational steps involved in producing a stereolithography file. (a) Three-dimensional description of the part. (b) The part is divided into slices (only 1 in 10 is shown). (c) Support material is planned. (d) A set of tool directions is determined for manufacturing each slice. Shown is the extruder path at section A-A from (c), for a fused-deposition modeling operation.

slices on top of each other. All of the processes described in this section *build parts slice by slice*; Fig. 10.49 shows the computational steps involved in producing a part. The main difference between the various additive processes lies in the method of producing the individual slices, which are typically 0.05 to 0.5 mm thick, but can be higher for some rapid prototyping systems.

Note that Fig. 10.50 shows a **support structure**. Complicated parts may need a support structure; some common support designs are shown in

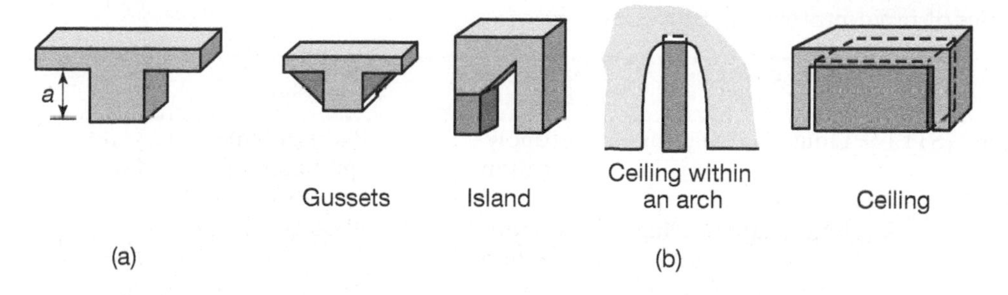

FIGURE 10.50 (a) A part with a protruding section that requires support material. (b) Common support structures used in rapid-prototyping machines.

Fig. 10.50a. Depending on the process, supports may need to be removed from the final part; strategies often employed to facilitate support removal include using a soluble material for the support, producing a weaker support material (such as by using a less dense spacing of material on a layer), or by using unfused powder as a support in powder bed processes (Section 10.12.4).

Part production in rapid prototyping typically takes hours for most parts, although small parts can be produced quicker. Consequently, these processes are not viable for large production runs, especially since the workpiece material is often an expensive polymer, powder, or a laminate. The characteristics of the rapid-prototyping processes are summarized in Table 10.7, and Table 10.8 lists typical material properties that are attained by these methods.

10.12.1 Stereolithography

The *stereolithography* (STL) or *vat photopolymerization* process is based on the principle of curing (hardening) a *liquid photopolymer* into a specific shape. Consider what happens when a laser beam is focused on and translated across the surface of a liquid *photopolymer*. The laser serves to cure the photopolymer by providing the energy necessary for polymerization. Laser energy is absorbed by the polymer; the exposure decreases exponentially with depth according to the rule

$$E(z) = E_o^{-z/D_p}, \tag{10.27}$$

where E is the exposure in energy per area, E_o is the exposure at the resin surface ($z = 0$), and D_p is the penetration depth (in the z-direction) at the laser wavelength and is a property of the resin. At the cure depth, the polymer is exposed to

$$E_c = E_o^{-C_d/D_p}, \tag{10.28}$$

where E_c is the exposure necessary to transform the working liquid to a gel, and C_d is the cure depth. Thus, solving for the cure depth,

$$C_d = D_p \ln\left(\frac{E_o}{E_c}\right), \tag{10.29}$$

TABLE 10.7 Characteristics of rapid-prototyping processes.

Supply phase	Process	Layer creation technique	Phase change type	Materials
Liquid	Stereolithography (STL)	Liquid-layer curing	Photopoly-merization	Photopolymers (acrylates, epoxies, colorable resins, and filled resins)
	Polyjet	Liquid-layer curing	Photopoly-merization	Photopolymers
	Fused-deposition modeling (FDM)	Extrusion of melted plastic	Solidification by cooling	Thermoplastics (ABS, polycarbonate, and polysulfone)
Powder	Three-dimensional printing (3DP)	Binder-droplet deposition onto powder layer	No phase change	Polymer, ceramic and metal powder with binder
	Selective laser sintering (SLS)	Laser-driven	Sintering or melting	Polymers, metals with binder, metals, ceramics, and sand with binder
	Electron beam melting	Electron beam	Melting	Metals
	Laser-engineered net shaping (LENS)	Injection of powder stream into the path of a laser	No phase change	Titanium, stainless steel, and aluminum
	Friction stir processing (FSP)	Solid state fusion of powder	No phase change	Magnesium, aluminum, and copper
Solid	Laminated object manufacturing (LOM)	Laser cutting	No phase change	Paper, polymer films, and metal foil

which represents the thickness at which the resin has polymerized into a gel, although this state does not have particularly high strength. Recognizing this condition, the controlling software slightly overlaps the cured volumes, but additional curing under fluorescent lamps or in an oven is usually required as a finishing operation.

The polymer at the periphery of the laser spot does not receive sufficient exposure to polymerize. It can be shown that the cured line width, L_w, at the surface is given by

$$L_w = B\sqrt{\frac{C_d}{2D_p}}, \tag{10.30}$$

where B is the diameter of the laser-beam spot.

Stereolithography (Fig. 10.51) equipment involves a vat filled with a photocurable liquid *photopolymer*, and contains a mechanism whereby a platform can be lowered and raised vertically. A typical liquid is a mixture of acrylic monomers, oligomers (polymer intermediates), and a *photoinitiator* that starts polymerization when exposed to laser light. When the platform is at its highest position, the layer of liquid above it is shallow. A *laser*, generating an ultraviolet beam, is then focused along a selected surface area of the photopolymer at surface a and moved in the x-y direction.

TABLE 10.8 Mechanical properties of selected materials for rapid prototyping.

Process	Material	Tensile strength (MPa)	Elastic modulus (GPa)	Elongation in 50 mm (%)	Characteristics
Stereolithography	Accura 60	68	3.10	5	Transparent; good general–purpose material for rapid prototyping
	Somos 9920	9	1.35–1.81	15–26	Transparent amber; good chemical resistance; good fatigue properties; used for producing patterns in rubber molding
	WaterClear Ultra	56	2.9	6–9	Optically clear resin with ABS-like properties
	DMX-SL 100	32	2.2–2.6	12–28	Opaque beige; good general-purpose material for rapid prototyping
Polyjet	FC720	60.3	2.87	20	Transparent amber; good impact strength, good paint adsorption, and machinability
	FC830	49.8	2.49	20	White, blue, or black; good humidity resistance; suitable for general-purpose applications
	FC 930	1.4	0.185	218	Semiopaque, gray, or black; highly flexible material used for prototyping of soft polymers or rubber
Fused-deposition modeling	Polycarbonate	52	2.0	3	White; high-strength polymer suitable for rapid prototyping and general use
	ABS-M30i	36	2.4	4	Available in multiple colors, most commonly white; a strong and durable material suitable for general use; biocompatible
	PC	68	2.28	4.8	White; good combination of mechanical properties and heat resistance
Selective laser sintering	WindForm XT	77.85	7.32	2.6	Opaque black polymide and carbon; produces durable heat- and chemical-resistant parts; high wear resistance
	Polyamide PA 3200GF	45	3.3	6	White; glass-filled polyamide has increased stiffness and is suitable for higher temperature applications
	SOMOS 201	–	0.015	110	Multiple colors available; mimics mechanical properties of rubber
	ST-100c	305	137	10	Bronze-infiltrated steel powder
Electron-beam melting	Ti-6Al-4V	970–1030	120	12–16	Can be heat treated by HIP to obtain up to 600-MPa fatigue strength

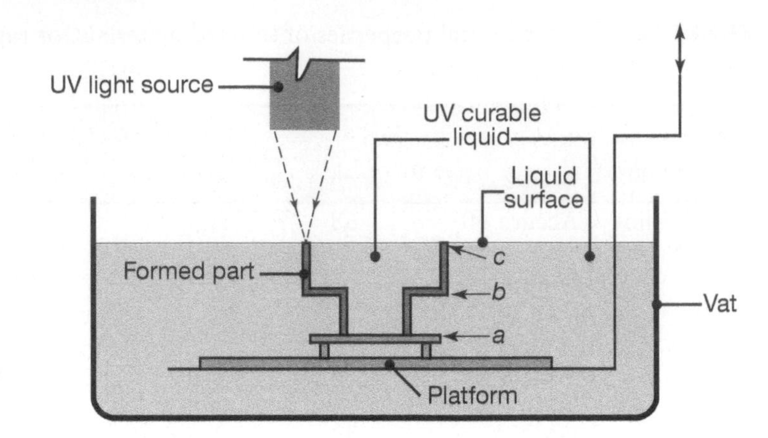

FIGURE 10.51 Schematic illustration of the stereolithography process. *Source:* Courtesy of 3D Systems.

The beam cures that portion of the photopolymer (say, a ring-shaped portion), producing a thin solid body. The platform is then lowered sufficiently to cover the cured polymer with another layer of liquid polymer, and the sequence is repeated. In Fig. 10.51, the process is repeated until level *b* is reached. Note that at level *b*, the platform has been lowered by a vertical distance *ab*, and that there now is a cylindrical part with a constant wall thickness.

At level *b*, the *x*-*y* movements of the beam are wider, so that a flange-shaped piece is built over the previously formed part. For this geometry, a support structure as shown in Fig. 10.50 would be needed. After the desired thickness of the liquid has been cured, the process is repeated, producing another cylindrical section between levels *b* and *c*. Note that the surrounding liquid polymer is still fluid, because it has not been exposed to the ultraviolet beam, and that the part has been produced from the *bottom up* in individual "slices." The unused portion of the liquid polymer can be used again to make another part or a prototype.

The word *stereolithography*, as used in describing this process, comes from the fact that the movements are three-dimensional and that the process is similar to lithography (in which the image to be printed on a flat surface is ink receptive and the blank areas are ink repellent, see Section 13.7). After the operation is completed, the part is removed from the platform, blotted, and cleaned ultrasonically in an alcohol bath. Finally, the part is exposed to UV radiation, for up to a few hours, to fully cure and harden the polymer part.

As shown in Fig. 10.51, the UV light source is directed in raster fashion across the liquid polymer surface. An alternative is to use a *digital light processing* (DLP) device, made up of millions of microscopic mirrors, that directs UV light to expose the entire layer at once. Sometimes called *mask projection stereolithography*, this process has the advantages of much higher rate of part production.

QR Code 10.3 Polyjet. *Source:* Courtsey of Stratasys, Ltd.

10.12.2 Polyjet

The *polyjet process* is similar to inkjet printing, where multiple print heads deposit the photopolymer on the build tray. Ultraviolet bulbs, alongside the jets of photopolymer, immediately cure and harden each layer, thus eliminating the need for any post-modeling curing as is required in stereolithography. The result is a smooth surface of layers as small as 16 μm, that can be handled immediately after the process is completed. Two different materials are used in polyjet: one material is for the actual model, while the second is a gel-like resin used for support, such as those shown in Fig. 10.49. Each material is simultaneously jetted and cured, layer by layer. When the model is completed, the support material is removed using an aqueous solution. Build sizes are fairly large, with an envelope of up to 500 mm × 400 mm × 200 mm (20 in. × 16 in. × 8 in.).

The polyjet process has capabilities similar to those for stereolithography and uses similar resins (Table 10.8). The main advantages are the ability to avoid the necessity for part cleanup, lengthy post-process curing operations, and the much smaller layer thickness, thus allowing for better resolution.

QR Code 10.4 The fused-deposition modeling process. *Source:* Courtesy of Stratasys, Ltd.

10.12.3 Fused-Deposition Modeling

In the *fused-deposition modeling* (FDM) process (Fig. 10.52), also called *roboextrusion*, an extruder head moves in two principal directions over a table. The table can be raised and lowered as needed. A thermoplastic filament is extruded through the small orifice of a heated die. Layers are produced by extruding the filament at a constant rate while

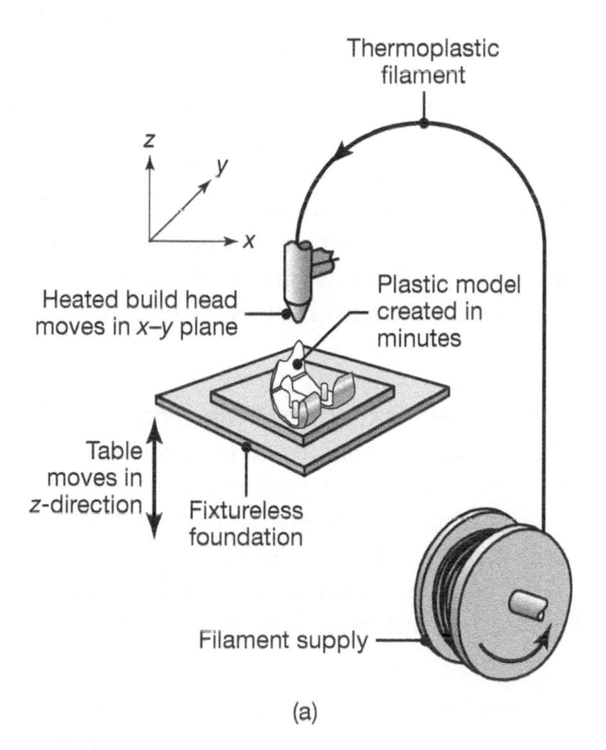

FIGURE 10.52 (a) Schematic illustration of the fused-deposition modeling process. (b) The 450mc Fortus rapid prototyping machine. *Source:* Courtesy of Stratasys, Ltd.

the extruder head follows a predetermined path (Fig. 10.49d). When a layer is completed, the table is lowered so that a subsequent layer can be superposed.

In the FDM process, the extruded layer's thickness is typically 125–325 μm (0.005–0.013 in.). This thickness limits the best achievable dimensional tolerance in the vertical direction. In the *x-y* plane, however, dimensional accuracy can be as fine as 0.025 mm, as long as a filament can be extruded into the feature. Close examination of an FDM-produced part will indicate that a *stepped* surface exists on oblique exterior planes. If the roughness of this surface is unacceptable, chemical-vapor polishing or a heated tool can be used to smoothen the surface; also, a coating can be applied, often in the form of a polishing wax. The overall dimensional tolerances may be compromised, unless care is taken in applying these finishing operations.

An extreme application of FDM is **big area additive manufacturing** (BAAM), which can produce parts as large as 6 m × 2.3 m × 1.8 m (20 ft × 7.5 ft × 6 ft), with a positioning accuracy of 25 μm (0.001 in.). The feedstock in BAAM is injection molding compound (pellets, sometimes with carbon fiber reinforcement) instead of a filament, so that material costs are significantly lower than other additive manufacturing processes.

Fused deposition modeling has come into wide use, and is the main process used in the Maker Movement (see Section 10.12.8); as such, a wide variety of machines with different capacities are now available. Machines cost as little as $1000–2000, and up to $100,000 for large capacity commercial machines, and over $500,000 for a BAAM machine.

10.12.4 Powder Bed Processes

Powder bed processes involve a number of processes that utilize powder as the workpiece material, and where the powder is deposited layer-by-layer in a *bed* or *build chamber* or *cylinder*. A number of powder application systems are used, but usually involve a roller or wiping mechanism. The deposited powder has a limited green strength, and can serve as a support for complicated parts.

Recognizing that powder bed processes generally produce porous parts (electron beam melting and selective laser melting are the exceptions), hot isostatic pressing (HIP; Section 11.3.3) can be performed to remove porosity and improve mechanical properties. The laser power can be reduced by preheating the powder supply and build chamber, thus requiring less incremental energy to produce the phase change. Some less expensive polymer machines therefore can use a DLP micromirror device instead of the galvanometers shown in Fig. 10.53, and use a less powerful and less expensive laser.

Selective laser sintering. (SLS) is a process based on selectively sintering of polymeric or metallic powders into an individual object (see also Section 11.4); the process is sometimes called **direct metal laser sintering** (DMLS). Sintering is usually a misnomer, since some melting is involved for polymers and metals, and therefore the term **selective laser melting** (SLM)

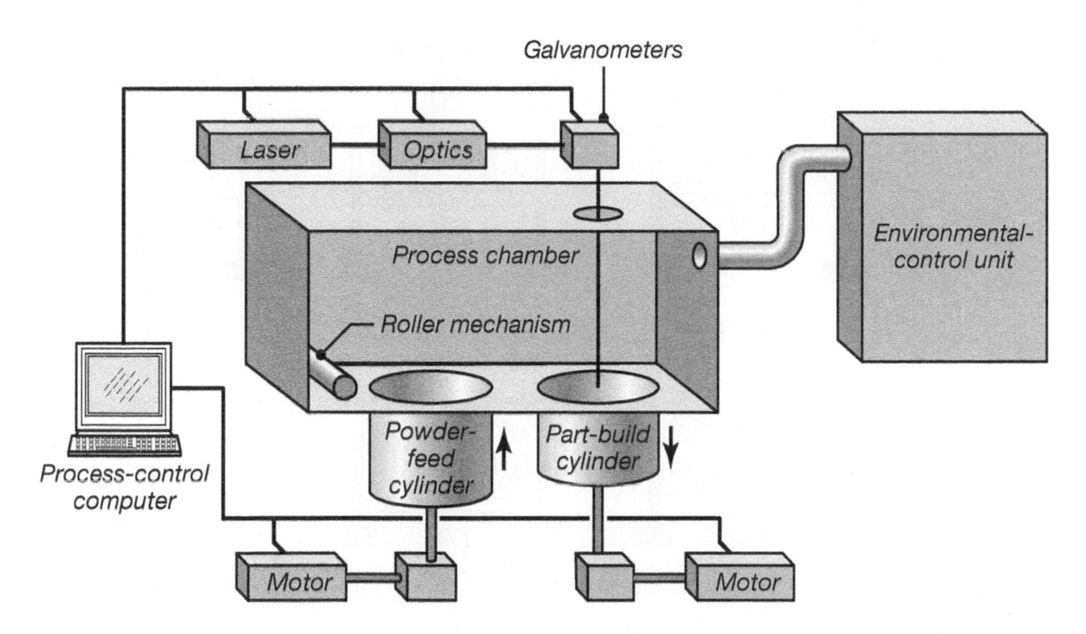

FIGURE 10.53 Schematic illustration of the selective-laser-sintering process. *Source:* After C. Deckard and P.F. McClure.

is sometimes used. The basic components in this process are illustrated in Fig. 10.53. The bottom of the processing chamber is equipped with two cylinders: a *part-build cylinder*, which is lowered incrementally to where the sintered part is being formed, and a *powder-feed cylinder*, which is raised incrementally to supply powder to the part-build cylinder through a roller mechanism or wiper. Some machines use two feed chambers to simplify powder deposition.

In this process, a thin layer of powder is first deposited in the part-build cylinder. A laser beam, guided by a computer (using instructions generated by the 3D CAD program of the desired part), is then focused on that layer, tracing and melting a particular cross section, which then rapidly resolidifies into a solid mass (after the laser beam is moved to another section). The powder in other areas remains loose, but supports the solid portion. Another layer of powder is then deposited, and this cycle is repeated until the entire three-dimensional part has been produced. The completed part is then extracted from the build chamber.

A variety of materials can be used in this process, including polymers (ABS, PVC, nylon, polyester, polystyrene, and epoxy), wax, metals, and ceramics (with appropriate binders). Polymers are the most common because of the smaller, less expensive, and less complicated lasers required for sintering.

Electron beam melting. A process similar to selective laser sintering and electron-beam welding (Section 12.5.1), electron-beam melting (EBM) uses the energy source associated with an electron beam to melt metal powder (such as titanium, stainless steel, or cobalt-chrome) to make metal prototypes. The workpiece is produced in a vacuum, thus part build size is limited to around 200 × 200 × 180 mm (8 in. × 8 in. × 6.3 in.). EBM is up

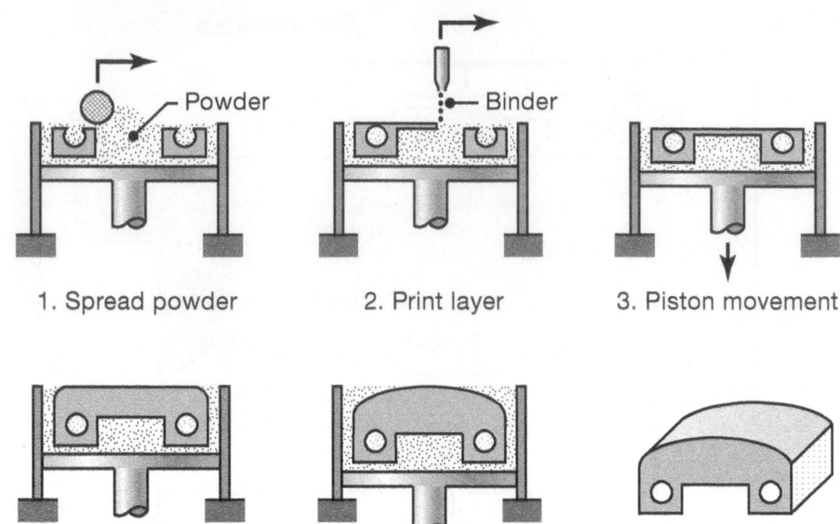

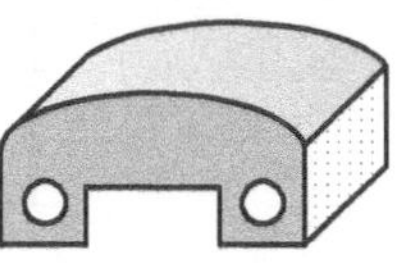

FIGURE 10.54 Schematic illustration of the binder jet printing process. *Source:* After E. Sachs and M. Cima.

to 95% efficient from an energy standpoint, as compared with 10%–20% efficiency for selective laser sintering, so that the metal powder is actually melted, and fully dense parts can be produced. A volume build rate of up to 60 cm^3/hr (3.7 in^3/hr) can be obtained, with individual layer thicknesses of 0.050–0.200 mm (0.002–0.008 in.). Hot isostatic pressing (Section 11.3.3) also can be performed on parts to improve their fatigue strength.

Binder jet printing. In *binder jet printing*, or *binder jetting*, a print head deposits an inorganic binder material onto a layer of nonmetallic or metallic powder, as shown in Fig. 10.54. A piston, supporting the powder bed, is lowered incrementally and, with each step, a layer is deposited and then fused by the binder.

Binder jet printing allows considerable flexibility in the materials used. Commonly materials are blends of polymers and fibers, foundry sand, and metals. Moreover, since multiple binder print heads can be incorporated into one machine, it is possible to produce full-color prototypes by using different color binders (Fig. 10.55). The effect is a three-dimensional analog to printing photographs using three ink colors on an inkjet printer.

The parts produced through the binder jet printing process are somewhat porous and thus may lack strength. Binder jet printing, as well as any other process that produces a porous material, may be combined with HIP (Section 11.3.3) to produce fully-dense parts. With metal powders, sintering and metal infiltration (see Section 11.4) can be performed to produce fully-dense parts, using the sequence shown in Fig. 10.56. The part is produced stated previously, by directing binder onto powders; the build sequence is then followed by sintering, to burn off the binder and partially fuse the metal powders (as in metal injection molding, described in Section 11.3.4). Common metals used in binder jetting include stainless steels, aluminum, and titanium. Infiltration materials typically are copper and bronze, thus providing heat transfer capabilities as well as

FIGURE 10.55 Example of a part produced through inkjet binding. Full color parts also are possible, and the colors can be blended throughout the volume. *Source:* Courtesy of 3D Systems.

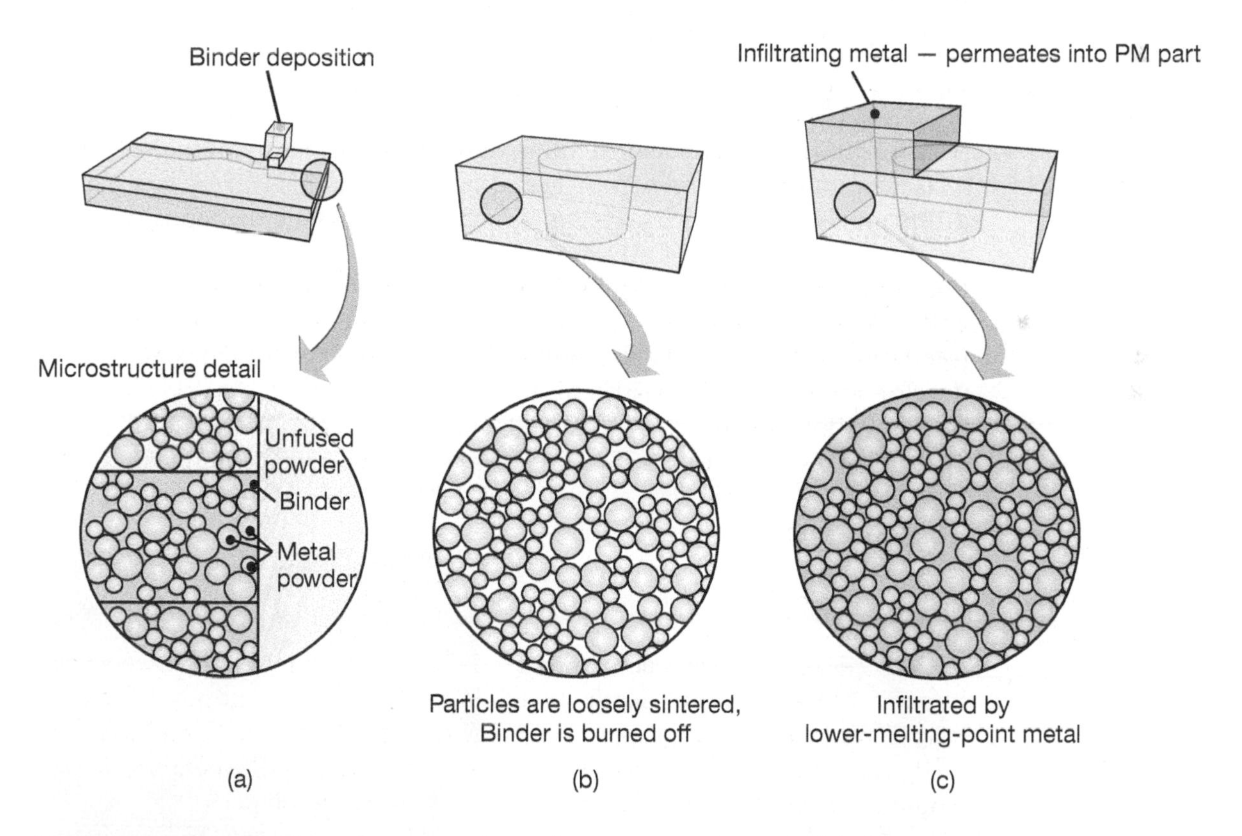

FIGURE 10.56 The three-dimensional printing process: (a) part build; (b) sintering, and (c) infiltration steps to produce metal parts. *Source:* Courtesy of Ex One Corporation.

wear resistance. This approach represents a strategy for *rapid tooling* (see below).

10.12.5 Laminated-Object Manufacturing

Lamination involves laying down layers that are bonded adhesively to one another. Several variations of *laminated-object manufacturing* (LOM) are available. The simplest and least expensive versions of LOM involve using control software and vinyl cutters to produce the prototype; the cutters are simple CNC machines that cut shapes from vinyl or paper sheets. Each sheet has a number of registration holes, which allow proper alignment and placement onto a build fixture. LOM systems are highly economical and are popular in schools and universities, because of the hands-on demonstration of additive manufacturing and production of parts by layers.

LOM systems can also be elaborate, where the more advanced systems use layers of paper or plastic, with a heat-activated glue on one side to bond the layers. The shapes are burned into the sheet with a laser, and the parts are built layer by layer (Fig. 10.57).

10.12.6 Laser-Engineered Net Shaping

A more recent development in additive manufacturing involves the use of a laser beam to melt and deposit metal powders, again layer by layer, over a previously deposited layer (Fig.10.58). The patterns of deposited layers are controlled by a CAD file. This near-net-shaping process is called *laser-engineered net shaping* (LENS) or direct metal deposition (DMD), and is based on the technologies of laser-beam welding and cladding (see Section 12.5). The heat input and cooling are controlled precisely to develop a favorable microstructure.

The deposition process is carried out inside a closed area and in an argon environment, to avoid the adverse effects of oxidation, particularly

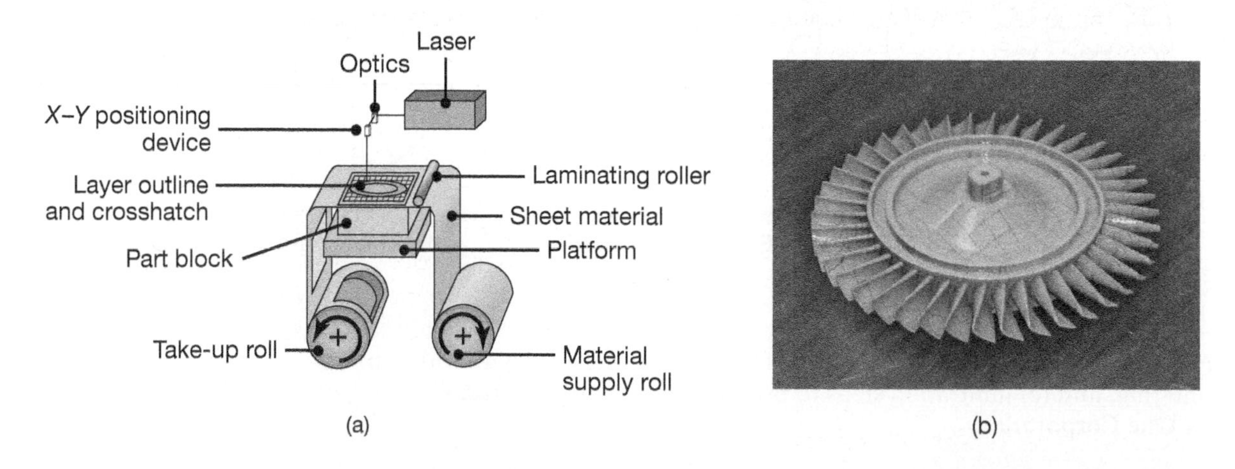

FIGURE 10.57 (a) Schematic illustration of the laminated-object-manufacturing process. (b) Turbine prototype made by LOM. *Source:* Courtesy of M. Feygin, Cubic Technologies, Inc.

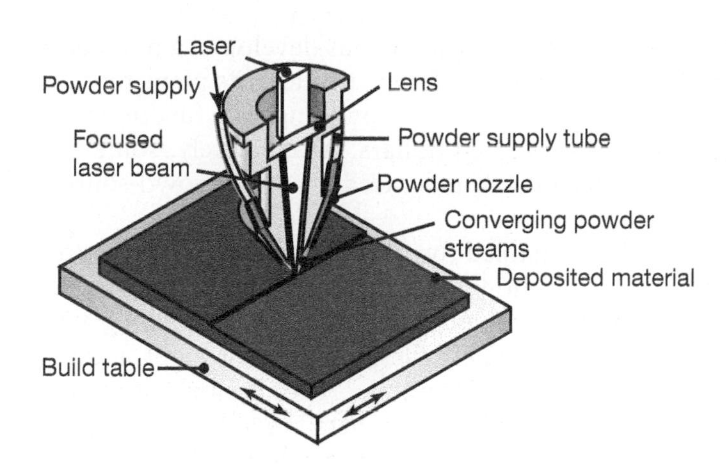

FIGURE 10.58 Schematic illustration of the laser-engineered net shaping (LENS) process.

on aluminum. It is suitable for a wide variety of metals and specialty alloys, for the direct manufacturing of parts, including fully dense tools and molds. It can also be used for repairing thin and delicate components. There are other, similar processing methods using lasers, including *controlled-metal buildup* (CMB) and *precision-metal deposition* (PMD).

LENS has been found suitable for incorporation into *hybrid machines* that conduct both additive and subtractive manufacturing. The advantages are that complex shapes can be quickly produced without refixturing, with high dimensional tolerance and surface finish, and with little material left as scrap. Usually, this involves the incorporation of a LENS deposition head in combination with a machining or turning center (see Section 8.11).

10.12.7 Friction Stir Modeling

The **friction stir modeling** (FSM) process shares many similarities with friction stir welding (Section 12.10). In this process, powder is delivered to a build location by pushing it into a rotating tube; the friction between powders and the substrate are sufficient to densify the powder and develop a solid material. However, because the process is solid state, there is no appreciable heat affected zone.

Friction stir modeling has been successfully applied to magnesium, aluminum, and titanium, and has the advantage of changing the deposited material during a build. For example, a lightweight aluminum part can have an integral hardened surface for wear resistance.

The equipment used for FSM involves conventional CNC milling machines (Section 8.11), modified to deliver the desired powder. Typical layer thickness is around 100 μm, and surface finish is generally poor, requiring machining to achieve smooth surfaces; a machining allowance is therefore necessary.

10.12.8 The Maker Movement

Upon the expiration of the initial patents for fused deposition modeling, a large number of machines based on FDM have been developed. A crowd-sourcing community, known as The Maker Movement, as organized, is linked with Internet communication tools, and has developed a number of so-called *Makerbots*. Some machines are freely available as plans that can be downloaded from the Internet, and used for building fully functional 3D printers at a cost of only a few hundred dollars. Alternatively, some very inexpensive machines have been marketed based on these crowd-sourced designs, such as the system.

The low-cost machines have enabled the development of *maker spaces*, where individual designers (commonly high school students) are given access to FDM equipment, sometimes for a nominal fee. Along with Internet-based services that accept CAD files, this has brought additive manufacturing capabilities to the general public. Also, because of the low cost and availability of these machines, researchers are able to apply new and innovative materials to rapid prototyping machines. Recent novel approaches include printing of food or biological materials for producing medical implants, printing of artificial organs (*bioprinting*), clothing, and shoes.

10.12.9 Direct (Rapid) Manufacturing and Rapid Tooling

Parts produced by various rapid-prototyping operations are useful not only for design evaluation and troubleshooting, but also occasionally for direct application into products or to aid in the direct manufacture of marketable products. Also, it is often desirable, for functional reasons, to use metallic parts, while the best developed and most available rapid-prototyping operations involve polymeric workpieces. Although prototypes can be produced from metal stock, using the machining operations described in Chapter 8, there may be a cost advantage to using other manufacturing approaches. However, additive manufacturing techniques are often incorporated into conventional processes to streamline them and make them more economically competitive.

A valuable method to apply rapid prototyping operations to other manufacturing processes is in the direct production of patterns or molds (see Section 5.8). As an example, Fig. 10.59 shows an approach for investment casting. The individual patterns are first made in a rapid-prototyping operation (in this case, stereolithography), and then are used as the patterns in assembling a *tree* for investment casting. This approach requires a polymer that will melt and burn from the ceramic mold completely; such polymers are available for all forms of polymer rapid-prototyping operations. Moreover, the parts as drawn in CAD programs are usually software modified to account for shrinkage, and it is the modified part that is produced in the rapid-prototyping machinery.

As another example, 3DP can easily produce a ceramic-casting shell (Section 5.8.4), in which an aluminum-oxide or aluminum-silica powder is fused with a silica binder. The molds have to be post-processed in two steps: curing at around 150°C, and then firing them at 1000°C to 1500°C. Such parts are suitable for shell-casting operations (Section 5.8.4); similar methods allow for the direct production of sand molds and injection molding dies.

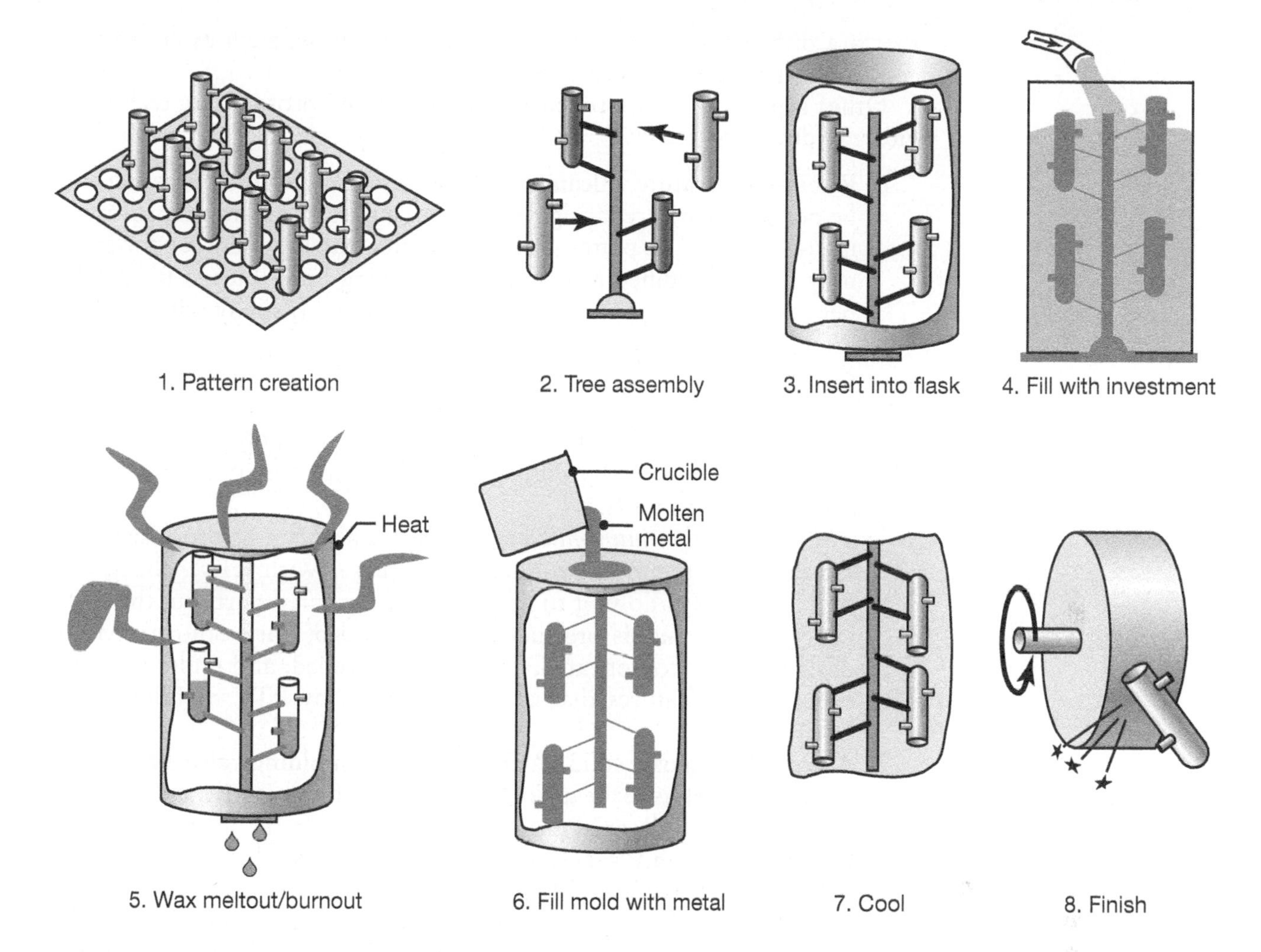

FIGURE 10.59 Manufacturing steps for investment casting that uses rapid-prototyped wax parts as patterns. This approach uses a flask for the investment, but a shell method can also be used. *Source:* 3D Systems, Inc.

Another common application of rapid tooling is injection molding (Section 10.10.2), where the mold or, more typically, a *mold insert* is manufactured by additive manufacturing. This operation may involve metals or polymers, depending on the application and intended tool life and production rates. Molds for slip casting of ceramics (Section 11.9.1) also can be produced in this manner. To produce individual slip casting molds, rapid-prototyping processes are used directly, and the molds are designed to achieve the desired permeability.

The advantage of rapid tooling is its capability to produce a mold, or a mold insert, that can be used to make components without the time lag traditionally required for the procurement of tooling (typically several months). Furthermore, the design is simplified, because the designer needs to only analyze a CAD file of the desired part; software then produces the tool geometry and automatically compensates for shrinkage. Also, features that are difficult or impossible to produce otherwise, such as conformal cooling channels, can be incorporated into a mold.

Other rapid tooling approaches, based on rapid-prototyping technologies, are described below.

1. **Room-temperature vulcanizing (RTV) molding/urethane casting** can be performed by preparing a pattern of a part by any rapid prototyping operation. The pattern is coated with a parting agent, and may or may not be modified to define mold parting lines. Liquid RTV rubber is then poured over the pattern, and cured (usually within a few hours) to produce mold halves. The mold is then used in injection molding operations, using liquid urethanes in injection molding or reaction-injection molding operations (Section 10.10.2). One limitation to this approach is mold life, as curing of the polyurethane in the mold causes progressive damage and the mold may only be suitable for as few as 25 parts.

   Epoxy or aluminum-filled epoxy molds also can be produced, but the design of the mold requires special care. With RTV rubber, the mold flexibility allows it to be peeled off the cured part, but the stiffness of epoxy molds precludes this method of part removal. Mold design is more complicated, as drafts are needed, and undercuts and other design features that can be produced by RTV molding must be avoided.
2. **Acetal clear epoxy solid (ACES) injection molding**, also known as *direct AIM*, refers to the use of additive manufacturing (usually stereolithography) to directly produce molds suitable for injection molding. The molds are shells, with an open end to allow filling with a material, such as epoxy, aluminum-filled epoxy, or low-melting point metals. Depending on the polymer used in injection molding, mold life may be as few as ten parts, although a few hundred parts per mold are possible.
3. **Sprayed metal tooling.** In this process, shown in Fig. 10.60, a pattern is made through rapid prototyping; a metal spray operation (Section 4.5.1) then coats the pattern surface with a zinc-aluminum alloy. The metal coating is placed in a flask and potted with an epoxy or aluminum-filled epoxy material. In some applications, cooling lines can be incorporated into the mold before the epoxy is applied. The pattern is removed; two such mold halves are then suitable for injection molding operations. Mold life is dependent on the material and the temperature, and can vary from a few to thousands of parts.
4. **Keltool process.** In this process, a RTV mold is produced based on a rapid-prototyped pattern as above; the mold is then filled with a mixture of powder A6 tool steel (Section 3.10.4), tungsten carbide, and polymer binder, and allowed to cure. The so-called *green* tool (see Section 11.3) is then fired to burn off the polymer and fuse the steel and tungsten-carbide powders. The tool is then infiltrated with copper in a furnace to produce the final mold. The mold can subsequently be machined or polished to impart superior surface finish

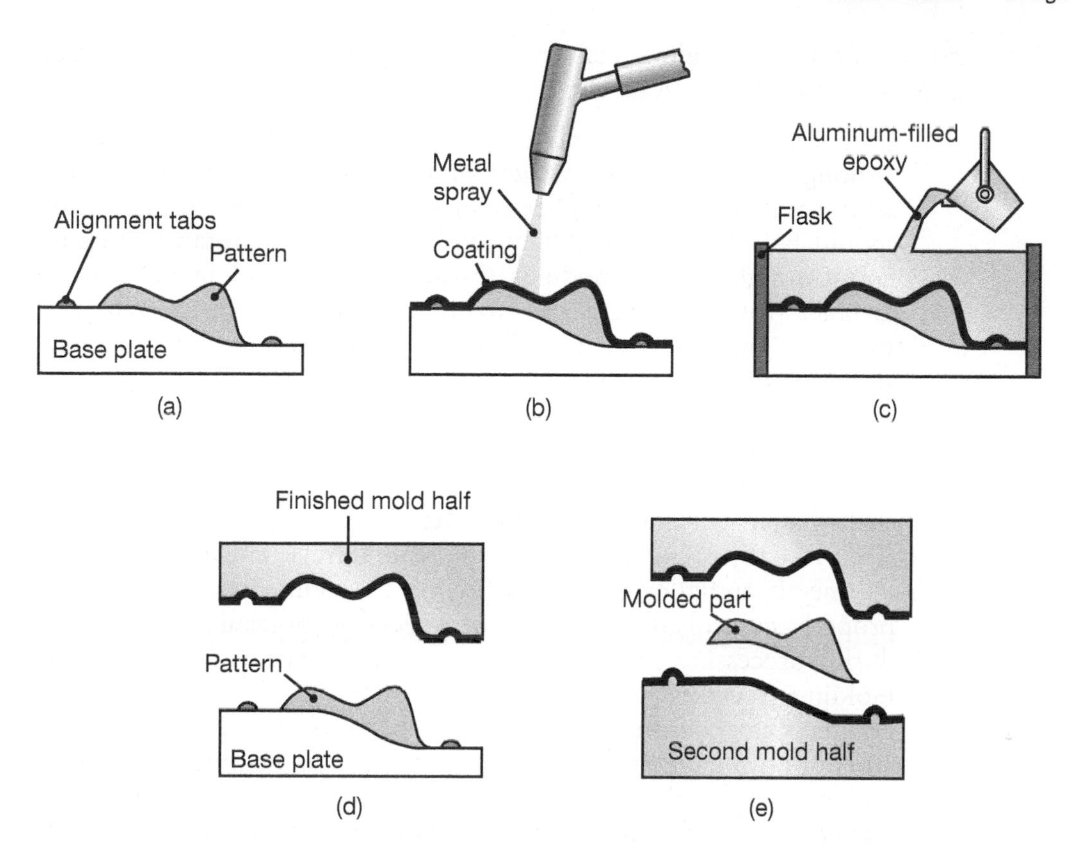

FIGURE 10.60 Production of tooling for injection molding by the sprayed-metal tooling process. (a) A pattern and base plate are prepared through a rapid-prototyping operation; (b) a zinc-aluminum alloy is sprayed onto the pattern (see Section 4.5.1); (c) the coated base plate and pattern assembly is placed in a flask and back-filled with aluminum-impregnated epoxy; (d) after curing, the base plate is removed from the finished mold; and (e) a second mold half suitable for injection molding is prepared.

and dimensional tolerances. Keltool molds are limited to around 150 mm × 150 mm × 150 mm (6 in. × 6 in. × 6 in.), thus typically a mold insert is produced. The process is suitable for high-volume molding operations, and depending on the material and processing conditions, mold life can range from 100,000 to 10 million parts.

10.13 Design Considerations

Design considerations for forming and shaping plastics are similar to those for processing metals. Selection of an appropriate material and available processes, from an extensive list, requires consideration of (a) mechanical and physical properties; (b) service requirements; (c) possible long-range effects of processing on properties and behavior, such as dimensional stability and degradation; (d) manufacturability; (e) economics; and (f) disposal and recycling at the end of the product's life cycle. The following list summarizes the major considerations involved in the design for plastics and for reinforced plastics.

1. Compared with metals, plastics have lower strength and stiffness, although the strength-to-weight and stiffness-to-weight ratios for reinforced plastics are higher than for many metals. Section sizes should be selected accordingly, with a view to maintaining a sufficiently high section modulus (ratio of moment of inertia to the distance from the neutral axis to the part surface) for the required stiffness. Improper design or assembly can lead to warping and shrinking of the part (Fig. 10.61a).
2. The overall part shape often determines the choice of the particular forming or molding process. Even after a particular process is selected, the design of the part, the die or the mold should be such that it will not cause difficulties concerning shape generation (Fig. 10.61b), dimensional control, and surface finish. As in casting metals and alloys, material flow in the mold cavities should be properly controlled, including the effects of molecular orientation during processing, especially in extrusion, thermoforming, and blow molding processes.
3. Polymers can be formed into complex shapes by molding operations; a major advantage of thermoplastics is that a single molded part can replace what would otherwise be an assembly of several parts. Polymers are available in a wide variety of material properties and colors, but usually require fairly large production quantities to justify tooling expenditures (see Section 10.14).
4. Large variations in cross section (Fig. 10.61c) and abrupt changes in geometry should be avoided to achieve better product quality and

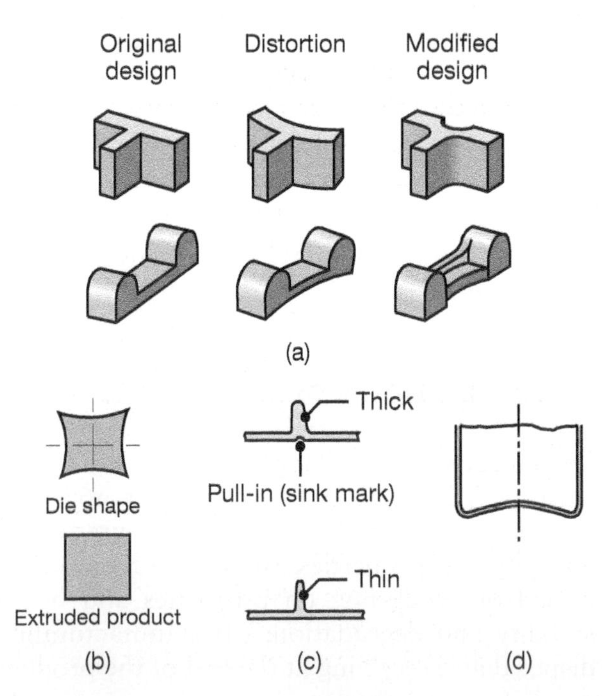

FIGURE 10.61 Examples of design modifications to eliminate or minimize distortion of plastic parts. (a) Suggested design changes to minimize distortion. *Source:* After F. Strasser. (b) Die design (exaggerated) for extrusion of square sections. Without this design modification, product cross sections would not have the desired shape because of the recovery of the material, known as die swell. (c) Design change in a rib to minimize pull-in caused by shrinkage during cooling. (d) Stiffening of the bottom of thin plastic containers by doming, similar to the process used to make the bottoms of aluminum beverage cans and similar containers.

increased mold life. Uniform wall thicknesses should be maintained whenever possible; if changes are required, they should be made gradually. Reinforcing ribs are often necessary to increase stiffness; the ribs should be thinner than the section they reinforce, and they should not be higher than about 3 times the wall thickness. Ribs should also incorporate a draft of 0.5°–1.5°. Sink marks (Fig. 10.61c) can be disguised by surface textures or grooves.

5. Contraction in large cross sections can cause porosity in plastic parts; conversely, because of a lack of stiffness, removing thin sections from molds may be difficult. The low elastic modulus of plastics requires that shapes be properly selected for improved stiffness of the component (Fig. 10.61d), particularly when material saving is important. These considerations are similar to those in designing metal castings and forgings.
6. Holes in thermoplastic parts can be produced in one step, but they present a complication in mold design. Flashing will often develop at the edge of a hole. Through holes are preferable to blind holes when using core pins, because the pin can be supported at both ends.
7. Parts should be designed to facilitate ejection from molds so that side features, such as flanges or holes that would interfere with ejection, should be avoided. Such features can be produced with retractable side cores, although they complicate mold design.
8. Screw threads can be molded, usually using one of three basic approaches: (a) a core containing the thread geometry is used and is rotated to remove the molded part; (b) the part is placed with its axis on the parting line, an approach that leads to a small amount of flash on the threads but eliminates the need for a core; and (c) for flexible polymers, a rounded thread can be molded and then stripped from the mold, but the thread depth must be small and the thread angle must be large for this approach to be successful.
9. Lettering, numbering, and other surface features can be incorporated. Whether these features are raised or depressed depends on the mold making approach; if the mold is machined, it is easier to produce letters into a mold than to remove the surrounding material. Thus, letters should be raised in the part itself and recessed in the mold. On the other hand, if a mold is produced through room-temperature vulcanization (RTV) from a machined blank, it is easier to have raised letters on the mold and depressed letters on the part.

Design for Additive Manufacturing. Several design rules have been developed for additive manufacturing. Since machines are available in a wide variety of capacities and capabilities, detailed design recommendations are manufacturer-specific. The following considerations are generic and are considered good design practice.

1. Additive manufacturing processes tend to warp the part, because of thermal stresses and shrinkage encountered during manufacturing. In general, the design guidelines for plastic parts, given above, are applicable to parts produced through additive manufacturing.

2. The dimensional tolerance standard used (see Section 4.7) should involve symmetric tolerances in order to be easily applied to additive manufacturing.
3. The tolerances within a plane can be much higher than those outside of a plane. Therefore, the part should be oriented to place the critical dimension in the plane of a build, not in its thickness direction.
4. Dimensional tolerances and surface finishes depend on the particular machine, the material, and part size and orientation. In stereolithography, tolerances of ±0.05 to 0.1 mm are achievable, plus 0.001 mm/mm for well-designed parts that do not warp excessively. Typical selective laser sintering of polymers yields tolerances of ±0.4 mm, or 0.1 mm/mm, whichever is greater. For metal selective laser systems, tolerances of 0.05–0.125 mm are commonly quoted, with roughnesses between 5 and 40 μm. To achieve better tolerances, a *machining allowance* of 0.5–1 mm should be provided for post-processing.
5. Steps are noticeable in an inclined plane; generally, the use of flat planes or planes inclined at not less than 20° are producible without noticeable steps.
6. In selective laser sintering of polymers, it is recommended to use a clearance of 0.3–0.5 mm within the plane for surfaces that are not joined together; up to 0.6 mm is required in the build direction.
7. The thinnest wall that can be produced depends on the material and aspect ratio; common ranges are 0.5–1.5 mm for polymers in selective laser sintering. In fused deposition modeling, it is generally recommended that a wall be at least four times wider than the layer thickness.
8. Recognizing that the powder in the build chamber may not be reusable, it is beneficial to fill a build space with as many parts as possible, and nestable parts be used when possible.
9. To reduce costs, the height in the build direction should be low, or stackable parts should be used to increase the amount of powder that is fused in a build chamber.
10. Consideration must be given to the removal of uncured photopolymer or powder when the parts made are hollow.
11. Build time depends on the volume of material that is to be fused in a process. It is therefore beneficial to model an object with solid surfaces, but supported by porous structures or struts instead of a solid bulk. This approach produces designs that can be optimized to minimize weight by carefully designing the supporting structure.

10.14 Economics of Processing Plastics

As in all processes, design and manufacturing decisions are ultimately based on performance and cost, including the costs of equipment, tooling, and production. The final selection of a process or processes also depends greatly on production volume. High equipment and tooling costs in plastics processing can be acceptable only if the production run is large, as is also the case in casting and forging. However, using additive manufacturing operations makes some processes economical for limited production runs, but the tools and molds have limited life. Additive manufacturing operations are suitable for prototypes and even limited production runs, but they require expensive consumables, and are therefore unsuitable for moderate to high production runs. This situation is complicated by the fact that some processes (such as selective laser sintering and electron beam melting) may require the unfused powder in the build chamber to be discarded; thus, if only 10% of the build volume is used, the material cost in the part is ten times the nominal material cost. This is a significant concern; titanium (Ti-6Al-4V), for example, costs over $400/kg for the raw powder. A general guide to economical processing of plastics and composite materials is given in Table 10.9.

Several types of equipment are used in plastics forming and shaping; note that the most expensive is injection-molding machines, with costs being directly proportional to the clamping force. A machine with a 2000-kN (225-ton) clamping force costs about $100,000, and one with a 20,000-kN (2250-ton) clamping force costs about $450,000 (See also Table 16.8). For composite materials, equipment and tooling costs for most molding operations are generally high, and production rates and economic production quantities vary widely.

The optimum number of cavities in the die for making the product in one cycle is an important consideration, as in die casting (Section 5.10.3).

TABLE 10.9 Comparative costs and production volumes for processing of plastics.

Process	Equipment capital cost	Production rate	Tooling cost	Typical production volume, number of parts						
				10	10^2	10^3	10^4	10^5	10^6	10^7
Machining	Med	Med	Low	←	→					
Compression molding	High	Med	High			←	—	—	→	
Transfer molding	High	Med	High			←	—	—	→	
Injection molding	High	High	High				←	—	—	→
Extrusion	Med	High	Low	*						
Rotational molding	Low	Low	Low		←	→				
Blow molding	Med	Med	Med				←	—	—	→
Thermoforming	Low	Low	Low		←	→				
Casting	Low	Very low	Low	←	→					
Foam molding	High	Med	Med				←	—	—	→
Stereolithography	Med	Very low	None	←	→					
Fused-deposition modeling	Low	Very low	None	←	→					
Three-dimensional printing	Med	Very low	None	←	→					

*Continuous process

For small parts, several cavities can be made in a mold, with runners to each cavity; if the part is large, only one cavity may be accommodated. As the number of cavities increases, so does the cost of the die. Larger dies may be considered for larger numbers of cavities, thus increasing die cost even further. On the other hand, more parts will be produced per machine cycle on a larger die, thereby increasing the production rate. A detailed analysis is thus necessary to determine the optimum number of cavities, die size, and machine capacity.

CASE STUDY 10.1 Manufacture of Head Protector® Tennis Racquets

Competitive tennis is a demanding sport, and as a result, there is a strong demand to produce exceptionally lightweight and stiff racquets to improve performance. A tennis racquet consists of a number of regions, as shown in Fig. 10.62. Of particular interest is the sweet spot; when the tennis ball is struck at the sweet spot, the player has optimum control and power, and vibration is minimized. Several innovative racquet-head designs have been developed over the years to maximize the size of the sweet spot. A stiff composite material, typically with high-modulus graphite fibers in an epoxy matrix (see Section 10.9) is used in the manufacture of the racquet head. The orientation of the fibers varies in different locations of the racquet. For example, the main tube for the racquet consists of carbon-epoxy prepreg, and is oriented at $\pm 30°$ from layer to layer.

The advantages to such materials are obvious, in that stiff racquets allow higher forces to be applied to the ball. However, the use of these advanced materials has led to an increased frequency

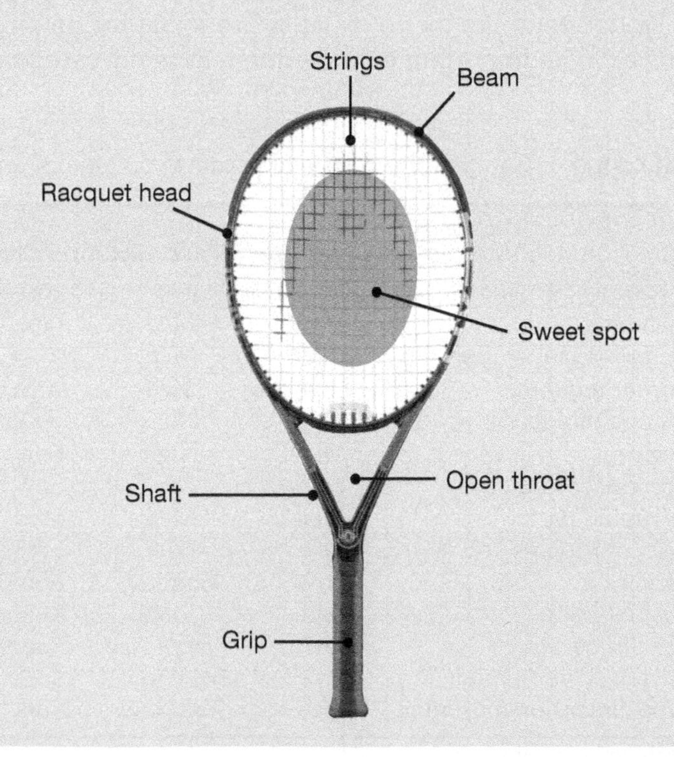

FIGURE 10.62 A Head Protector® tennis racquet. *Source:* Courtesy of Head Sport AG.

of tennis elbow, a painful condition associated with the tendons, that anchor muscles to the bones at the elbow. The condition is due not only to the higher forces involved, but also to the associated greater vibration of the racquet encountered with every stroke, especially when balls are struck away from the sweet spot.

An innovative design for a racquet, the Protector® (made by Head Sport AG) uses lead zirconate titanate (PZT) fibers as an integral layer of the composite racquet frames. PZT is a piezoelectric material (see Section 3.9.6) that produces an electric response when deformed. Modules of the fibers, called Intellifibers®, are integrated into the throat on all sides of the racket, that is, left, right, front, and back). The module consists of about 50 PZT fibers, each approximately 0.3 mm in diameter, sandwiched between two polyamide layers, with printed electrodes for generating the potential difference when the fibers are bent.

During impact, the vibrations constantly excite the Intellifibers®, generating a very high voltage potential but at low current. The energy is stored in coils on the printed circuit board (Chipsystem®) incorporated in the racquet handle in real time, and released back to the Intellifibers®, in the optimal phase and waveform for the most efficient damping. The stored energy is sent back to the Intellifibers® in a phase that causes a mechanical force opposite to the vibration, thereby reducing it. The Chipsystem® is tuned to the first natural frequency of the racket, and can dampen vibrations only within a range of its design frequency.

The manufacture of a Protector® tennis racquet involves a number of steps. First, a carbon-epoxy prepreg is produced, as described in Section 10.11. The prepreg is cut to the proper size and placed on a flat, heated bench to make the matrix material tackier, resulting in better adhesion to adjacent layers. A polyamide sleeve (or bladder) is then placed over a rod, and the prepreg is rolled over the sleeve. When the bar is removed, the result is a tube of carbon-epoxy prepreg with a polyamide sleeve, that can be placed in a mold and internally pressurized to develop the desired cross section.

The throat piece is molded separately by wrapping the prepreg around sand-filled polyamide preforms or expandable foam. Since there is no easy way to provide air pressure to the throat, the preform develops its own internal pressurization, because of the expansion of air during exposure to elevated molding temperatures. If sand is used, it is later removed by drilling holes into the preform during the finishing operation.

Prior to molding, all components are assembled onto a template, and final prepreg pieces are added to strategic areas. The main tube is bent around the template, and the ends are pressed together and wrapped with a prepreg layer to form the handle. The PZT fibers are incorporated as the outer layer in the racquet in the throat area, and the printed electrodes are connected to the Chipsystem®. The racquet is then placed into the mold, internally pressurized, and is allowed to cure. Note that this operation is essentially an internally-pressurized, pressure-bag molding process (See Fig. 10.63b). A racquet as it appears directly after molding is shown in Fig. 10.63a.

The racquet then undergoes a number of finishing operations, including flash removal, drilling of holes to accommodate strings, and finishing of the handle, including wrapping it with a special grip material. A completed Head Protector® racquet is shown in Fig. 10.63b. This design has been found to reduce racquet vibrations by up to 50%, resulting in clinically proven reductions in tennis elbow, without any compromise in performance.

Source: Courtesy of J. Kotze and R. Schwenger, Head Sport AG.

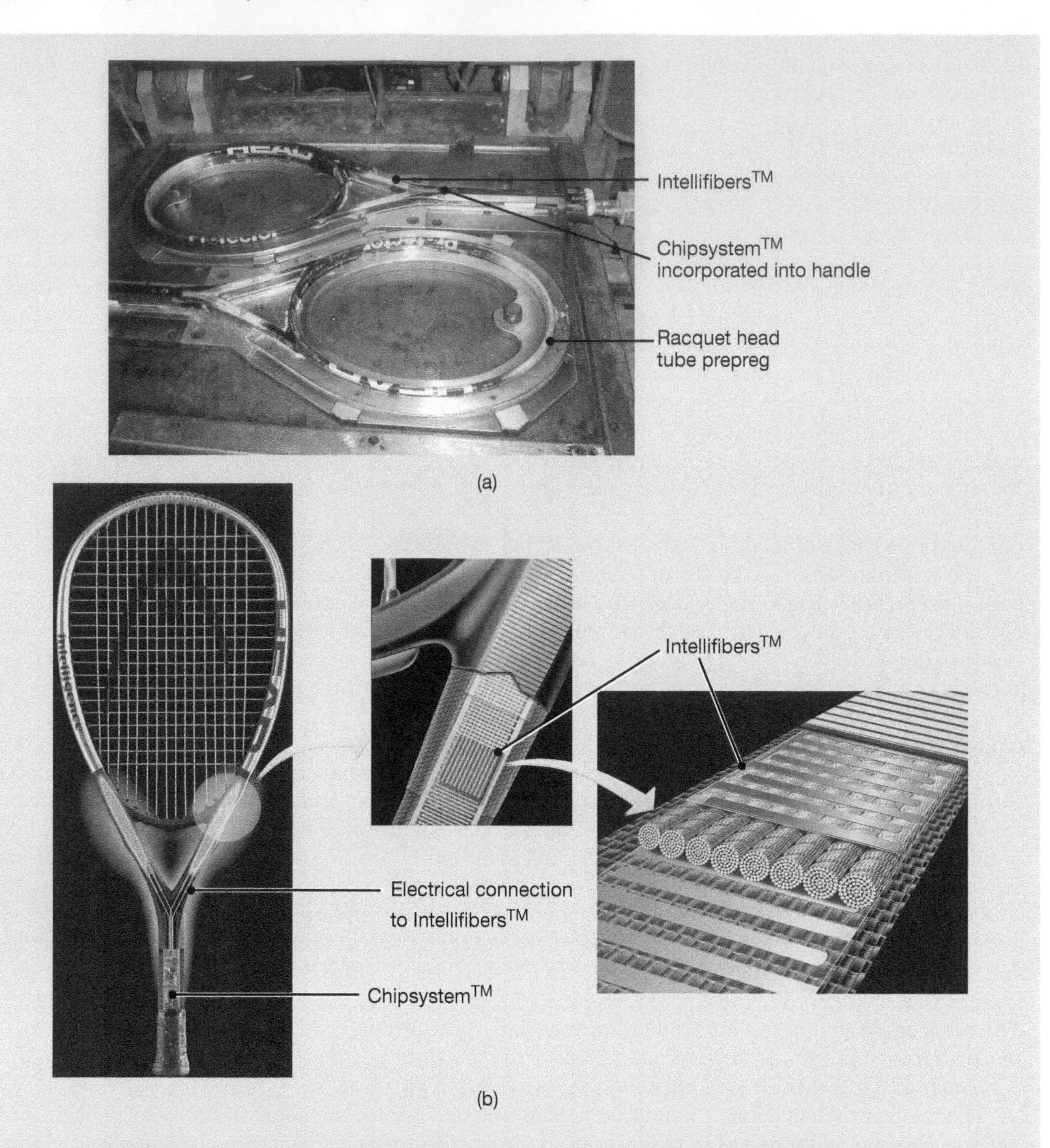

FIGURE 10.63 (a) The composite Head Protector® racquet immediately after molding; (b) a completed Head Protector® racquet, highlighting the incorporation of piezoelectric Intellifibers®. *Source:* Courtesy of Head Sport AG.

SUMMARY

- Polymers are an important class of materials because they possess a very wide range of mechanical, physical, chemical, and optical properties. Compared with metals, plastics generally are characterized by lower density, strength, elastic modulus, and thermal and electrical conductivity, and higher coefficient of thermal expansion. (Section 10.1)
- Plastics are composed of polymer molecules and various additives. The smallest repetitive unit in a polymer chain is called a mer. Monomers are linked by polymerization processes to form larger molecules. The properties of the polymer depend on the molecular structure, the degree of crystallinity, and additives. The glass-transition temperature indicates the brittle and ductile regions of polymers. (Section 10.2)
- Thermoplastics become soft and easy to form at elevated temperatures. Their mechanical behavior can be characterized by various spring and dashpot models, and includes such phenomena as creep and stress relaxation, crazing, and water absorption. (Section 10.3)
- Thermosets are produced by cross-linking polymer chains; they do not become soft to any significant extent with increasing temperature. (Section 10.4)
- Thermoplastics and thermosets have a very wide range of consumer and industrial applications and with a variety of characteristics. (Sections 10.5 and 10.6)
- Among important aspects of polymers is biodegradable plastics; several formulations have been developed. (Section 10.7)
- Elastomers have the characteristic ability to undergo large elastic deformations and return to their original shape when unloaded. Consequently, they have important applications as tires, seals, footwear, hoses, belts, and shock absorbers. (Section 10.8)
- Reinforced plastics, also called composite materials, are an important class of engineered materials that have superior mechanical properties and are lightweight. The reinforcing fibers are typically glass, graphite, aramids, and boron, and epoxies commonly serve as a matrix material. The properties of reinforced plastics primarily depend on their composition and the orientation of the fibers. (Section 10.9)
- Plastics can be formed and shaped by a variety of processes, such as extrusion, molding, casting, and thermoforming. Thermosets are generally molded or cast. (Section 10.10)
- Reinforced plastics are processed into structural components using liquid plastics, prepregs, and bulk- and sheet-molding compounds. Fabricating techniques include various molding methods, filament winding, and pultrusion. (Section 10.11)

- Additive manufacturing has become an increasingly important technology because of its inherent flexibility, low cost, and the much shorter times required for making prototypes. Typical techniques include stereolithography, polyjet, fused-deposition modeling, selective laser sintering, and three-dimensional printing. Rapid production of tooling is a further development of these processes. (Section 10.12)
- Design of plastic parts includes considerations of their relatively low strength and stiffness, high thermal expansion, and generally low resistance to temperature. (Section 10.13)

SUMMARY OF EQUATIONS

Linearly elastic behavior

in tension, $\sigma = E\epsilon$

in shear, $\tau = G\gamma$

Viscous behavior, $\tau = \eta\left(\frac{dv}{dy}\right) = \eta\dot{\gamma}$

Strain-rate sensitivity, $\sigma = C\dot{\epsilon}^m$

Viscoelastic modulus, $E_r = \frac{\sigma}{(e_e + e_v)}$

Viscosity,

$$\eta = \eta_o e^{E/\kappa T}$$

$$\log \eta = 12 - \frac{17.5\Delta T}{52 + \Delta T}$$

Stress relaxation, $\tau = \tau_o e^{-t/\lambda}$

Relaxation time, $\lambda = \frac{\eta}{G}$

Strength of composite, $\sigma_c = x\sigma_f + (1 - x)\sigma_m$

Elastic modulus of composite, $E_c = xE_f + (1 - x)E_m$

Extruder characteristic, $Q = \frac{\pi^2 HD^2 N \sin\theta \cos\theta}{2} - \frac{p\pi DH^3 \sin^2\theta}{12\eta l}$

Die characteristic, $Q_{\text{die}} = Kp$

For round die, $K = \frac{\pi D_d^4}{128\eta l_d}$

Cure depth in stereolithography, $C_d = D_p \ln\left(\frac{E_o}{E_c}\right)$

Line width in stereolithography, $L_w = B\sqrt{\frac{C_d}{2D_p}}$

BIBLIOGRAPHY

Agarwal, B.D., Broutman, L.J., and Chandrashekhara, K., *Analysis and Performance of Fiber Composites*, 3rd ed., Wiley, 2006.

ASM Handbook, Vol. 21: *Composites*, ASM International, 2001.

Barbero, E.J., *Introduction to Composite Materials Design),* 2nd ed., CRC Press, 2010.

Bhowmick, A.K., and Stephens, H.L., *Handbook of Elastomers*, 2nd ed., CRC, 2000.

Campbell, F., (ed.), *Manufacturing Processes for Advanced Composites*, Elsevier, 2003.

Chanda, M., and Roy, S.K., *Plastics Technology Handbook*, 4th ed., Dekker, 2006.

Chawla, K.K., *Composite Materials: Science and Engineering*, 3rd ed., Springer, 2013.

Chua, C.K., and Leong, L.K.F., *Rapid Prototyping: Principles and Applications in Manufacturing*, 3rd. ed., World Scientific Co., 2010.

Daniel, I.M., and Ishai, O., *Engineering Mechanics of Composite Materials*, 2nd ed., Oxford, 2005.

Engelmann, S., *Advanced Thermoforming: Methods, Machines, and Materials, Applications and Automation*, Wiley, 2012.

Erhard, G., *Designing with Plastics*. Hanser Gardner, 2006.

Freakley, P.K., *Rubber Processing and Production Organization*, Springer, 2013.

Gastrow, H., *Injection Molds: 130 Proven Designs*. Hanser Gardner, 2002.

Gebhardt, A., *Understanding Additive Manufacturing: Rapid Prototyping, Rapid Tooling, Rapid Manufacturing*, Hanser, 2012.

Gibson, I., Rosen, D.W., and Stucker, B., *Additive Manufacturing Technologies: Rapid Prototyping to Direct Digital Manufacturing*, Springer, 2009.

Giles, H.F., Wagner, J.R., and Mount, E.M., *Extrusion: the Definitive Processing Guide and Handbook,* Elsevier, 2013.

Grimm, T., *Engineering Design and Rapid Prototyping*, Springer, 2010.

Harper, C.A., *Handbook of Plastic Processes*, Wiley-Interscience, 2006.

Kazmer, D., *Injection Mold Design Engineering*, Hanser, 2007.

Kutz, M., *Applied Plastics Engineering Handbook: Processing and Materials*, William Andrew, 2011.

Larson, E.R.R., *Thermoplastic Material Selection: A Practical Guide*, Elsevier, 2015.

Mallick, P.K. (ed.), *Composites Engineering Handbook*, 3rd ed., Dekker, 2008.

Malloy, R.A., *Plastic Part Design for Injection Molding: An Introduction*, 2nd ed., Hanser-Gardner, 2010.

Mittal, V., (ed.), *High Performance Polymers and Engineering Plastics*, Wiley, 2011.

Osswald, T., and Turng, L-S, *Injection Molding Handbook,* Hanser, 2008.

Pham, D., and Dimov, S.S., *Rapid Manufacturing: The Technologies and Applications of Rapid Prototyping and Rapid Tooling*, Springer, 2011.

Rauwendaal, C., *Polymer Extrusion*, 5th ed., Hanser Gardner, 2014.

Rosato, D.V., and Rosato, D.V., *Injection Molding Handbook*, 3rd ed., Kluwer Academic Publishers, 2012.

Rudin, A., and Choi, P.C., *Elements of Polymer Science and Engineering*, 3rd ed., Academic Press, 2012.

Skotheim, T.A., and Reynolds, J., (eds.), *Handbook of Conducting Polymers*, 2 vols., 3rd ed., Dekker, 2007.

——, *Plastics: Materials and Processing*, 3rd ed., Prentice Hall, 2005.

Tadmor, Z., and Goqos, C., *Principles of Polymer Processing*, 2nd ed., Wiley, 2006.

Venuvinod, P.K., and Ma, W., *Rapid Prototyping: Laser-Based and Other Technologies*, Springer, 2010.

Wang, W., Stoll, H., and Conley, J.G., *Rapid Tooling Guidelines for Sand Casting*, Springer, 2010.

Young, R.J., and Lovell, P.A., *Introduction to Polymers*, 3rd ed., CRC Press, 2011.

QUESTIONS

10.1 Summarize the most important mechanical and physical properties of plastics.

10.2 What are the major differences between the properties of plastics and metals?

10.3 What properties are influenced by the degree of polymerization?

10.4 Give applications for which flammability of plastics would be a major concern.

10.5 An important property for processing of melted polymers is viscosity. Define viscosity, using necessary sketches and equations.

10.6 What properties do elastomers have that thermoplastics, in general, do not have?

10.7 Is it possible for a material to have a hysteresis behavior that is the opposite of that shown in Fig. 10.14, whereby the arrows are counterclockwise? Explain.

10.8 Observe the behavior of the tension-test specimen shown in Fig. 10.13, and state whether the material has a high or low m value (see Section 2.2.7). Explain why.

10.9 Why would we want to synthesize a polymer with a high degree of crystallinity?

10.10 Add more to the applications column in Table 10.3.

10.11 Discuss the significance of the glass-transition temperature, T_g, in engineering applications.

10.12 Why does cross-linking improve the strength of polymers?

10.13 Describe the methods by which optical properties of polymers can be altered.

10.14 Explain the reasons that elastomers were developed. Are there any substitutes for elastomers? Explain.

10.15 Give several examples of plastic products or components for which creep and stress relaxation are important considerations.

10.16 Describe your opinions regarding recycling of plastics versus developing plastics that are biodegradable.

10.17 Explain how you would go about determining the hardness of the plastics described in this chapter.

10.18 Distinguish between composites and alloys.

10.19 Describe the functions of the matrix and the reinforcing fibers in reinforced plastics. What fundamental differences are there in the characteristics of the two materials?

10.20 What products have you personally seen that are made of reinforced plastics? How can you tell that they are reinforced?

10.21 Identify metals and alloys that have strengths comparable with those of reinforced plastics.

10.22 Compare the advantages and disadvantages of metal-matrix composites, reinforced plastics, and ceramic-matrix composites.

10.23 You have studied the many advantages of composite materials in this chapter. What limitations or disadvantages do these materials have? What suggestions would you make to overcome these limitations?

10.24 A hybrid composite is defined as a material containing two or more different types of reinforcing fibers. What advantages would such a composite have over other composites?

10.25 Why are fibers capable of supporting a major portion of the load in composite materials?

10.26 Assume that you are manufacturing a product in which all the gears are made of metal. A salesperson visits you and asks you to consider replacing some of the metal gears with plastic ones. Make a list of the questions that you would raise before making such a decision.

10.27 Review the three curves in Fig. 10.8, and name applications for each type of behavior. Explain your choices.

10.28 Repeat Question 10.27 for the curves in Fig. 10.10.

10.29 Do you think that honeycomb structures could be used in passenger cars? If so, where? Explain.

10.30 Other than those described in this chapter, what materials can you think of that can be regarded as composite materials?

10.31 What applications for composite materials can you think of in which high thermal conductivity would be desirable?

10.32 Make a survey of a variety of sports equipment, and identify the components that are made of composite materials. Explain the reasons for and advantages of using composites for these specific applications.

10.33 We have described several material combinations and structures in this chapter. In relative terms, identify those that would be suitable for applications involving one of the following: (a) very low temperatures; (b) very high temperatures; (c) vibrations; (d) high humidity.

10.34 Explain how you would go about determining the hardness of the reinforced plastics and composite materials described in this chapter. Are hardness measurements for these types of materials meaningful? Does the size of the indentation make a difference in your answer? Explain.

10.35 Describe the advantages of applying traditional metalworking techniques to the formation of plastics.

10.36 Describe the advantages of cold forming of plastics over other processing methods.

10.37 Explain the reasons that some forming processes are more suitable for certain plastics than for others.

10.38 Would you use thermosetting plastics for injection molding? Explain.

10.39 By inspecting plastic containers, such as for baby powder, you can see that the lettering on them is raised rather than sunk. Can you offer an explanation as to why they are molded in that way?

10.40 Give examples of several parts that are suitable for insert molding. How would you manufacture these parts if insert molding were not available?

10.41 What manufacturing considerations are involved in making a metal beverage container versus a plastic one?

10.42 Inspect several electrical components, such as light switches, outlets, and circuit breakers, and describe the process or processes used in making them.

10.43 Inspect several similar products that are made of metals and plastics, such as a metal bucket and a plastic bucket of similar shape and size. Comment on their respective thicknesses, and explain the reasons for their differences, if any.

10.44 Make a list of processing methods used for reinforced plastics. Identify which of the following fiber orientation and arrangement capabilities each has: (a) uniaxial, (b) cross-ply, (c) in-plane random, and (d) three-dimensional random.

10.45 Some plastic products have lids with integral hinges; that is, no other material or part is used at the junction of the two parts. Identify such products, and describe a method for making them.

10.46 Explain why operations such as blow molding and film-bag making are done vertically and why buildings that house equipment for these operations have ceilings 10 m to 15 m (35 ft to 50 ft) high.

10.47 Consider the case of a coffee mug being produced by rapid prototyping. How can the top of the handle be manufactured, since there is no material directly beneath the arch?

10.48 Make a list of the advantages and disadvantages of each of the rapid-prototyping operations.

10.49 Explain why finishing operations are needed for rapid-prototyping operations. If you are making a prototype of a toy automobile, list the post-rapid-prototyping finishing operations you would perform.

10.50 A current topic of research involves producing parts from rapid-prototyping operations and then using them in experimental stress analysis, in order to infer the strength of final parts produced by means of conventional manufacturing operations. List your concerns with this approach, and outline means of addressing these concerns.

10.51 Because of relief of residual stresses during curing, long unsupported overhangs in parts from stereolithography will tend to curl. Suggest methods of controlling or eliminating this curl.

10.52 One of the major advantages of stereolithography and polyjet is that semi- and fully transparent polymers can be used, so that internal details of parts can be readily discerned. List parts for which this feature is valuable.

10.53 Based on the processes used to make fibers, explain how you would produce carbon foam. How would you make a metal foam?

10.54 Die swell in extrusion is radially uniform for circular cross sections, but is not uniform for other cross sections. Recognizing this fact, make a qualitative sketch of a die profile that will produce (a) square and (b) triangular cross sections of extruded polymer.

10.55 What are the advantages of using whiskers as a reinforcing material?

10.56 By incorporating small amounts of blowing agent, it is possible to manufacture polymer fibers with gas cores. List some applications for such fibers.

10.57 With injection-molding operations, it is common practice to remove the part from its runner and then to place the runner into a shredder and recycle the resultant pellets. List the concerns you would have in using such recycled pellets as opposed to so-called virgin pellets.

10.58 What characteristics make polymers attractive for applications such as gears? What characteristics are drawbacks for such applications?

10.59 Can polymers be used to conduct electricity? Explain.

10.60 Why is there so much variation in the stiffness of polymers?

10.61 Explain why thermoplastics are easier to recycle than thermosets.

10.62 Describe how shrink-wrap works.

10.63 List the characteristics required of a polymer for: (a) a total hip replacement insert; (b) a golf ball; (c) an automotive dashboard; (d) clothing; and (e) a child's doll.

10.64 How can you tell whether a part is made of a thermoplastic or a thermoset?

10.65 Describe the features of an extruder screw and their functions.

10.66 An injection-molded nylon gear is found to contain small pores. It is recommended that the material be dried before molding it. Explain why drying will solve this problem.

10.67 What determines the cycle time for (a) injection molding; (b) thermoforming; and (c) compression molding?

10.68 Does the pull-in defect (sink marks) shown in Fig. 10.61 also occur in metal forming and casting processes? Explain.

10.69 List the differences between the barrel section of an extruder and that on an injection molding machine.

10.70 Identify processes that are suitable for making small production runs of plastic parts, such as quantities of 100.

10.71 Identify processes that are capable of producing parts with the following fiber orientations: (a) uniaxial; (b) cross-ply; (c) in-plane random; and (d) three-dimensional random.

10.72 List approaches for quickly manufacturing tooling for injection molding.

10.73 Careful analysis of a rapid-prototyped part indicates that it is made up of layers with a white filament outline visible on each layer. Is the material a thermoset or a thermoplastic? Explain.

10.74 List the advantages of using a room-temperature vulcanized (RTV) rubber mold in injection molding.

10.75 What are the similarities and differences between stereolithography and polyjet?

10.76 Explain how color can be incorporated into rapid prototyped components.

10.77 Many oils, such as olive and palm oil, are paraffinic like polyethylene, but are liquid at room temperature. Explain why.

10.78 For thermoplastics, explain the effects of increasing the following properties on viscosity: (a) temperature; (b) pressure; (c) strain rate; (d) molecular weight; and (e) the presence of side branches.

10.79 Hydrostatic pressure increases the strength and ductility of metals. List the effects of hydrostatic pressure on polymer melts.

10.80 List the additive manufacturing approaches that are suitable for metals.

10.81 Explain why part orientation in rapid prototyping is important.

10.82 Do you expect that materials produced from additive manufacturing will be isotropic? Explain.

10.83 What is the Maker Movement? What is BAAM? Are they related? Explain.

10.84 How are latex gloves produced?

10.85 Explain why part thickness is limited in additive manufacturing.

10.86 Why are build platforms heated in selective laser sintering?

PROBLEMS

10.87 Calculate the areas under the stress–strain curve (toughness) for the material in Fig. 10.9, plot them as a function of temperature, and describe your observations.

10.88 Note in Fig. 10.9 that, as expected, the elastic modulus of the polymer decreases as temperature increases. Using the stress–strain curves in the figure, make a plot of the modulus of elasticity versus temperature.

10.89 Calculate the percentage increase in mechanical properties of reinforced nylon from the data shown in Fig. 10.20.

10.90 A rectangular cantilever beam 100 mm high, 50 mm wide, and 1 m long is subjected to a concentrated force of 100 N at its end. Select two different unreinforced and reinforced materials from Table 10.1, and calculate the maximum deflection of the beam. Then select aluminum and steel, and for the same beam dimensions, calculate the maximum deflection. Compare the results.

10.91 In Sections 10.5 and 10.6, several plastics and their applications are listed. Rearrange this information, respectively, by making a table of products and the type of plastics that can be used to make the products.

10.92 Determine the dimensions of a tubular steel drive shaft required to transmit a given torque, T, for a typical automobile. If you now replace this shaft with shafts made of unreinforced and reinforced plastic, respectively, what should be the shaft's new dimensions to transmit the same torque for each case? Choose the materials from Table 10.1, and assume a Poisson's ratio of 0.4.

10.93 Calculate the average increase in the properties of the plastics listed in Table 10.1 as a result of their reinforcement, and describe your observations.

10.94 In Example 10.4, what would be the percentage of the load supported by the fibers if their strength is 1250 MPa and the matrix strength is 240 MPa? What if the strength is unaffected, but the elastic modulus of the fiber is 200 GPa while the matrix is 50 GPa?

10.95 Estimate the die clamping force required for injection molding ten identical 50-mm-diameter disks in one die. Include the runners of appropriate length and diameter.

10.96 A two-liter plastic beverage bottle is made from a parison with the same diameter as the threaded neck of the bottle and has a length of 125 mm. Assuming uniform deformation during blow molding, estimate the wall thickness of the tubular section.

10.97 Estimate the consistency index and power-law index for the polymers in Fig. 10.12.

10.98 An extruder has a barrel diameter of 125 mm. The screw rotates at 100 rpm, has a channel depth of 6 mm, and a flight angle of 17.5°. What is the largest flow rate of polypropylene that can be achieved?

10.99 The extruder in Problem 10.98 has a pumping section that is 3 m long and is used to extrude round polyethylene rod. The die has a land of 1 mm and a diameter of 5 mm. If the polyethylene is at a mean temperature of 250°C, what is the flow rate through the die? What if the die diameter is 10 mm?

10.100 An extruder has a barrel diameter of 100 mm, a channel depth of 5 mm, a flight angle of 18°, and a pumping zone that is 1.5 m long. It is used to pump a plastic with a viscosity of 80 Ns/m^2. If the die characteristic is experimentally determined as $Q_x = (2.0 \times 10^{-12}$ m^5/Ns$)p$, what screw speed is needed to achieve a flow rate of 10^{-4} m^3/s from the extruder?

10.101 What flight angle should be used on a screw so that a flight translates a distance equal to the barrel diameter with every revolution?

10.102 For a laser providing 12.5 kJ of energy to a spot with diameter of 0.20 mm, determine the cure depth and the cured line width in stereolithography. Use $E_c = 6.36 \times 10^{10}$ J/m^2 and $D_p = 25$ μm.

10.103 Assume the line width in a stereolithography operation is 0.20 mm. Estimate the time required to cure a layer defined by a 40-mm circle if adjacent lines overlap each other by 10% and the laser traverse rate is 0.25 m/s.

10.104 The extruder head in a fused-deposition-modeling setup has a diameter of 1.25 mm and produces layers that are 0.25 mm thick. If the velocities of the extruder head and polymer extrusion are both 50 mm/s, estimate the production time for the generation of a 30-mm solid cube. Assume that there is a 15-s delay between layers as the extruder head is moved over a wire brush for cleaning.

10.105 Using the data for Problem 10.104 and assuming that the porosity of the support material is 50%, calculate the production rate for making a 100-mm high cup with an outside diameter of 88 mm and wall thickness of 5 mm. Consider both the case with the closed end (a) up and (b) down.

10.106 What would the answer to Example 10.5 be if the nylon was recognized to have a power law viscosity with $n = 0.4$? What if $n = 0.2$?

10.107 Referring to Fig. 10.7, plot the relaxation curves, that is, plot the stress as a function of time if a unit strain is applied at time $t = t_o$.

10.108 Derive a general expression for the coefficient of thermal expansion for a continuous fiber-reinforced composite in the fiber direction.

10.109 Estimate the number of molecules in a typical 20 kg automobile tire. Estimate the number of atoms.

10.110 Calculate the elastic modulus and load supported by fibers in a composite with an epoxy matrix (E = 100 GPa), made up of 25% fibers made of (a) high-modulus carbon fiber and (b) Kevlar 29 fibers.

10.111 Calculate the stress in the fibers and in the matrix for Problem 10.110. Assume that the cross-sectional area is 50 mm^2 and $P_c = 2000$ N.

10.112 For a composite material consisting of high modulus carbon fibers (E = 415 GPa) and an epoxy matrix (E = 100 GPa), determine the volume fraction of fibers needed to produce a composite material with a stiffness equal to that of steel.

10.113 Consider a composite consisting of reinforcing fibers (E_f = 300 GPa) in an epoxy matrix (E = 100 GPa). If the allowable fiber stress is 200 MPa and the matrix strength is 75 MPa, what should be the fiber content so that the fibers and matrix fail simultaneously?

10.114 An extrusion operation is being carried out with D = 100 mm, H = 6 mm, θ = 20°, L = 1 m, and N = 200 rpm. A 20 MPa pressure is produced at the die. Acrylic polymer is being extruded at 240° and $\dot{\gamma} = 1000$ s^{-1}. What is the flow rate?

10.115 For the same extruder as in Problem 10.114, the speed is reduced to 150 rpm, so that the strain rate is now 100 s^{-1}. What is the flow rate?

10.116 For the same extruder as in Problem 10.114, with N = 200 rpm and $\dot{\gamma}$ = 1000 s^{-1}, what is the flow rate for low density polyethylene at (a) 280°C and (b) 180°C?

10.117 In an extruder, is it better to increase the diameter or the flight depth if one wishes to increase drag flow? Explain your answer with proper equations.

10.118 A fused deposition modeling machine has a diameter of 0.125 mm, and a land length of 0.1 mm. If nylon is to be extruded at a shear strain rate of 1000 s^{-1}, and a filament feed mechanism is designed to allow the extruder head to move at 50 mm/s, what flow rate is produced? What pressure must be accommodated by the feed mechanism?

10.119 Inspect Table 10.8 and compare the numerical values given with those for metals and other materials, as can be found in Part I of this text. Comment on your observations.

10.120 Assume that you are asked to give a quiz to students on the contents of this chapter. Prepare five quantitative problems and five qualitative questions, and supply the answers.

DESIGN

10.121 Make a survey of the recent technical literature and present data indicating the effects of fiber length on such mechanical properties as the strength, the elastic modulus, and the impact energy of reinforced plastics.

10.122 Discuss the design considerations involved in replacing a metal beverage container with a container made of plastic.

10.123 Using specific examples, discuss the design issues involved in products made of plastics versus reinforced plastics.

10.124 Make a list of products, parts, or components that are not currently made of plastics, and offer some reasons that they are not.

10.125 In order to use a steel or aluminum container to hold an acidic material, such as tomato juice or sauce, the inside of the container is coated with a polymeric barrier. Describe the methods of producing such a container (see also Chapter 7).

10.126 Using the information given in this chapter, develop special designs and shapes for possible new applications of composite materials.

10.127 Would a composite material with a strong and stiff matrix and soft and flexible reinforcement have any practical uses? Explain.

10.128 Make a list of products for which the use of composite materials could be advantageous because of their anisotropic properties.

10.129 Name several product designs in which both specific strength and specific stiffness are important.

10.130 Describe designs and applications in which strength in the thickness direction of a composite is important.

10.131 Design and describe a test method to determine the mechanical properties of reinforced plastics in their thickness direction.

10.132 As described in this chapter, reinforced plastics can be adversely affected by environmental factors, such as moisture, chemicals, and temperature variations. Design and describe test methods to determine the mechanical properties of composite materials under these conditions.

10.133 As with other materials, the mechanical properties of composites are obtained by preparing appropriate specimens and testing them. Explain what problems you might encounter in preparing specimens for testing and in the actual testing process itself.

10.134 Add a column to Table 10.1, which describes the appearance of these plastics, including available colors and opaqueness.

10.135 It is possible to weave fibers in three dimensions, and to impregnate the weave with a curable resin. Describe the property differences that such materials would have compared to laminated composite materials.

10.136 Make a survey of various sports equipment and identify the components made of composite materials. Explain the reason for and the advantages of using composites in these applications.

10.137 Instead of a constant cross section, it is possible to make fibers or whiskers with a varying cross section or a "wavy" fiber. What advantages would such fibers have?

10.138 It has been suggested that polymers (either plain or reinforced) can be a suitable material for dies in sheet metal forming operations described in Chapter 7. Describe your thoughts concerning this suggestion, considering die geometry and any other factors that may be relevant.

10.139 For ease of sorting for recycling, a rapidly increasing number of plastic products are now identified with a triangular symbol with a single-digit number at its center and two or more letters under it. Explain what these numbers indicate and why they are used.

10.140 Obtain different kinds of toothpaste tubes, carefully cut them across, and comment on your observations regarding the type of materials used and how the tube could be produced.

10.141 Design a machine that uses rapid-prototyping technologies to produce ice sculptures. Describe its basic features, commenting on the effect of size and shape complexity on your design.

10.142 A manufacturing technique is being proposed that uses a variation of fused-deposition modeling, where there are two polymer filaments that are melted and mixed before being extruded in order to produce the workpiece. What advantages does this method have?

10.143 It is possible to injection mold or compression mold gears. What design considerations and modifications would you propose for polymer gears compared to steel gears?

10.144 In production of pressure vessels, it is common to begin with a metal container and wrap fiber reinforcement around this container through filament winding. Discuss the purpose of both the metal and the fiber composite, and explain how you would determine the proper thickness of each.

10.145 In selective laser sintering and melting, the unfused powder in the build chamber is usually discarded. Design a research program to evaluate the effects of powder reuse on mechanical properties of SLS materials.

10.146 There is a great desire to increase the speed of additive manufacturing approaches. List three strategies for increasing the speed of a process, along with the advantages and disadvantages of each method. Write a one-page paper on the approach you think is best.

10.147 Two cubes constructed from the same polymer but with different processing history are soaked in water. After a few hours, one of them has increased in weight by 5% from absorbing water, and is warped and larger as a result. The other has hardly absorbed any water. Give an explanation for this observation.

10.148 Bioprinting is a manufacturing process related to fused deposition modeling. Write a two-page paper on the latest trends in bioprinting.

10.149 Perform an Internet search, and produce a presentation that summarizes applications of additive manufacturing. Whenever possible, indicate the material and process used.

CHAPTER

11 Powder Metallurgy and Processing of Ceramics and Glasses

EXAMPLES:

This chapter explores the manufacturing processes and technologies for producing net-shape parts from metal and ceramic powders, as well as glass, diamond, and graphite. Specifically, the following are described:

- Methods of producing metal and ceramic powders.
- Processes used to shape compacts of powders, including the mechanics of powder compaction.
- Finishing operations to improve dimensional tolerances and surface properties, as well as part aesthetics.
- Methods of producing superconductor materials and commercially important forms of carbon, such as graphite and diamond.
- Design and competitive aspects of these processes.

Symbols

D diameter, m

E elastic modulus, N/m^2

H thickness, m

k friction measure in compaction

L length, m

k thermal conductivity, W/mK

L length, m

n exponent

p pressure, N/m^2

P volume fraction of pores

P_d dry porosity

P_f fired porosity

r neck radius, m

R particle radius, m

S_{ut} ultimate tensile strength, N/m^2

V volume, m^3

x Cartesian coordinate, m

μ coefficient of friction

ν Poisson's ratio

ρ density, g/cm^3 also, neck profile radius, m

σ normal stress, N/m^2

11.1 Introduction

In the manufacturing processes described in the preceding chapters, the raw materials used were either in a molten state or in solid form. This chapter describes groups of operations whereby metal powders, ceramics, and glasses are processed into products, as well as those involved in processing composite materials and superconductors. Design considerations in each process are also described.

The **powder metallurgy** (PM) process is capable of making complex parts by compacting metal powders in dies and sintering them (heating without melting) to net- or near-net-shape products. The availability of a wide range of powder compositions, the capability to produce parts to net dimensions, and the economics of the overall operation make this process attractive for many applications. The application of powder metals to additive manufacturing processes, described in Section 10.12, has also led to increased interest in PM.

QR Code 11.1 Powder metallurgy touches your life: Part 1 - Overview. *Source:* Courtesy of the Metal Powder Industries Federation.

This chapter next describes the structure, properties, and processing of **ceramics, glasses, graphite,** and **diamond.** These materials have significantly different mechanical and physical properties compared to metals, including hardness, high-temperature resistance, and electrical and optical properties. Consequently, they have important and unique applications.

The chapter then describes the structure, properties, and processing of **metal-matrix and ceramic-matrix composites.** Recall that polymer-matrix composites are described in Chapter 10. Because of the large number of possible combinations of elements, a great variety of composite materials is now available for a wide range of consumer and industrial applications, particularly in the automotive and aerospace industries.

The chapter ends with a description of how superconductors are processed into products, such as coils for magnetic resonance imaging and other devices that monitor magnetic fields.

QR Code 11.2 Powder metallurgy touches your life Part 2 - Processes. *Source:* Courtesy of the Metal Powder Industries Federation.

11.2 Powder Metallurgy

One of the first uses of powder metallurgy was the tungsten filaments for incandescent light bulbs in the early 1900s. Applications now include gears, cams, bushings, cutting tools, porous products such as filters and oil-impregnated bearings, and automotive components such as piston rings, valve guides, connecting rods, and hydraulic pistons (Fig. 11.1). Recent advances in PM technology also permit structural parts of aircraft, such as landing gear components, engine-mount supports, engine disks, impellers, and engine nacelle frames, to be made from metal powders.

Powder metallurgy has become competitive with processes such as casting, forging, and machining, particularly for relatively complex parts made of high-strength and hard alloys. Nearly 70% of PM part production is for automotive applications. Parts made by this process have good dimensional accuracy, and their sizes range from tiny balls for ball-point pens to parts weighing about 50 kg, although most parts made by PM weigh less

(a)

(b)

FIGURE 11.1 (a) Examples of typical parts made by powder-metallurgy processes. (b) Upper trip lever for a commercial irrigation sprinkler, made by PM. Made of unleaded brass alloy, it replaces a die-cast part, at a 60% cost savings. *Source:* Courtesy of the Metal Powder Industries Federation.

than 2.5 kg. A typical family car now contains, on average, 18 kg of precision metal parts made by powder metallurgy, an amount that has been increasing by about 10% annually.

The powder-metallurgy operation consists of the following steps, in sequence:

1. **Powder production;**
2. **Blending;**
3. **Compaction, forming, or molding;**
4. **Sintering or pressing;** and
5. **Finishing operations.**

The finishing operations may include processes such as coining, sizing, machining, and infiltration for improved quality, dimensional accuracy, and part strength.

(a)

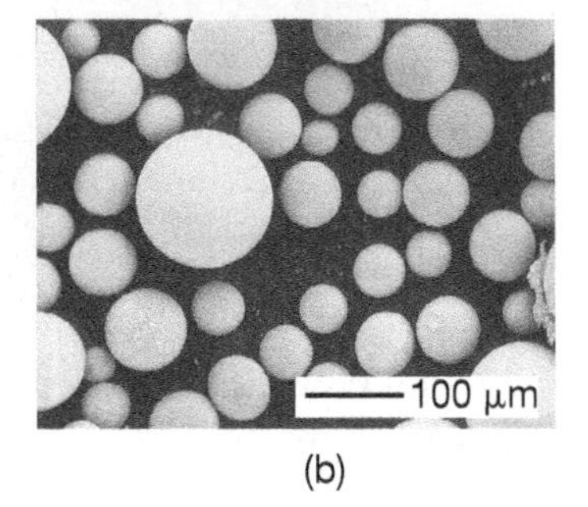

(b)

FIGURE 11.2 (a) Scanning-electron microscope image of iron powder particles made by atomization. (b) Nickel-based superalloy (Udimet 700) powder particles made by the rotating electrode process; see Fig. 11.4d. *Source:* Courtesy of P.G. Nash.

11.2.1 Production of Metal Powders

Metal powders can be produced by several methods, the choice of which depends on the particular requirements of the end product. Sources for metals generally are their bulk form, ores, salts, and other compounds. These forms are then produced, by various methods, into powders. The shape, size distribution, porosity, chemical purity, and bulk and surface characteristics of the powder particles depend on the particular process used (Figs. 11.2 and 11.3). These characteristics are important, because they significantly affect flow during powder compaction and reactivity in subsequent sintering operations. Metal powder production is generally done at a large scale, and then shipped to customers in the forms of powders with particle sizes ranging from 0.1 to 1000 μm.

The methods of powder production are:

1. **Atomization.** *Atomization* produces a liquid-metal stream by injecting molten metal through a small orifice (Fig. 11.4a). The stream is broken up by jets of inert gas, air, or water, known as *gas* or *water atomization*, respectively. The size of the particles formed depends

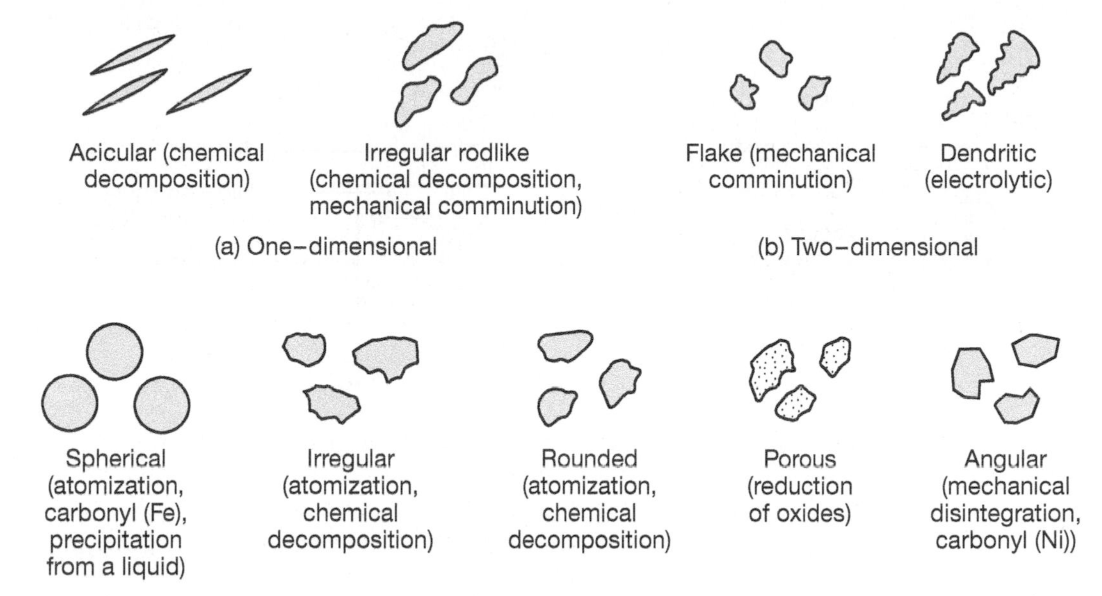

FIGURE 11.3 Particle shapes and characteristics of metal powders and the processes by which they are produced.

on the temperature of the metal, the rate of flow, nozzle size, and jet characteristics. The use of water results in a slurry of metal powder and liquid at the bottom of the atomization chamber. Although the powders must be dried before they can be used, the water allows for more rapid cooling of the particles and thus higher production rates. Gas atomization usually results in more spherical particles, with the rotating-electrode process producing the best powder (See Fig. 11.2b).

Several related melt-atomization methods are used for production of powders for PM. In *centrifugal atomization*, the molten-metal stream drops onto a rapidly rotating disk or cup; the centrifugal force breaks up the stream and generates particles (Fig. 11.4c). In a variation of this method, a consumable electrode is rotated at about 15,000 rev/min in a helium-filled chamber (Fig. 11.4d). The centrifugal force breaks up the molten tip of the electrode into metal particles.

2. **Reduction.** *Reduction* of metal oxides (removal of oxygen) involves gases, such as hydrogen and carbon monoxide, as reducing agents, to produce fine metallic powders from metallic oxide powders. The powders produced are spongy and porous, and have uniformly-sized spherical or angular shapes.
3. **Electrolytic deposition.** This operation uses either aqueous solutions or fused salts; the powders produced are among the purest of all metal powders.
4. **Carbonyls.** *Metal carbonyls*, such as iron carbonyl [$Fe(CO)_5$] and nickel carbonyl [$Ni(CO)_4$], are formed by letting iron or nickel react with carbon monoxide. The reaction products are then decomposed to iron and nickel, producing small, dense, and uniform spherical particles of high purity.

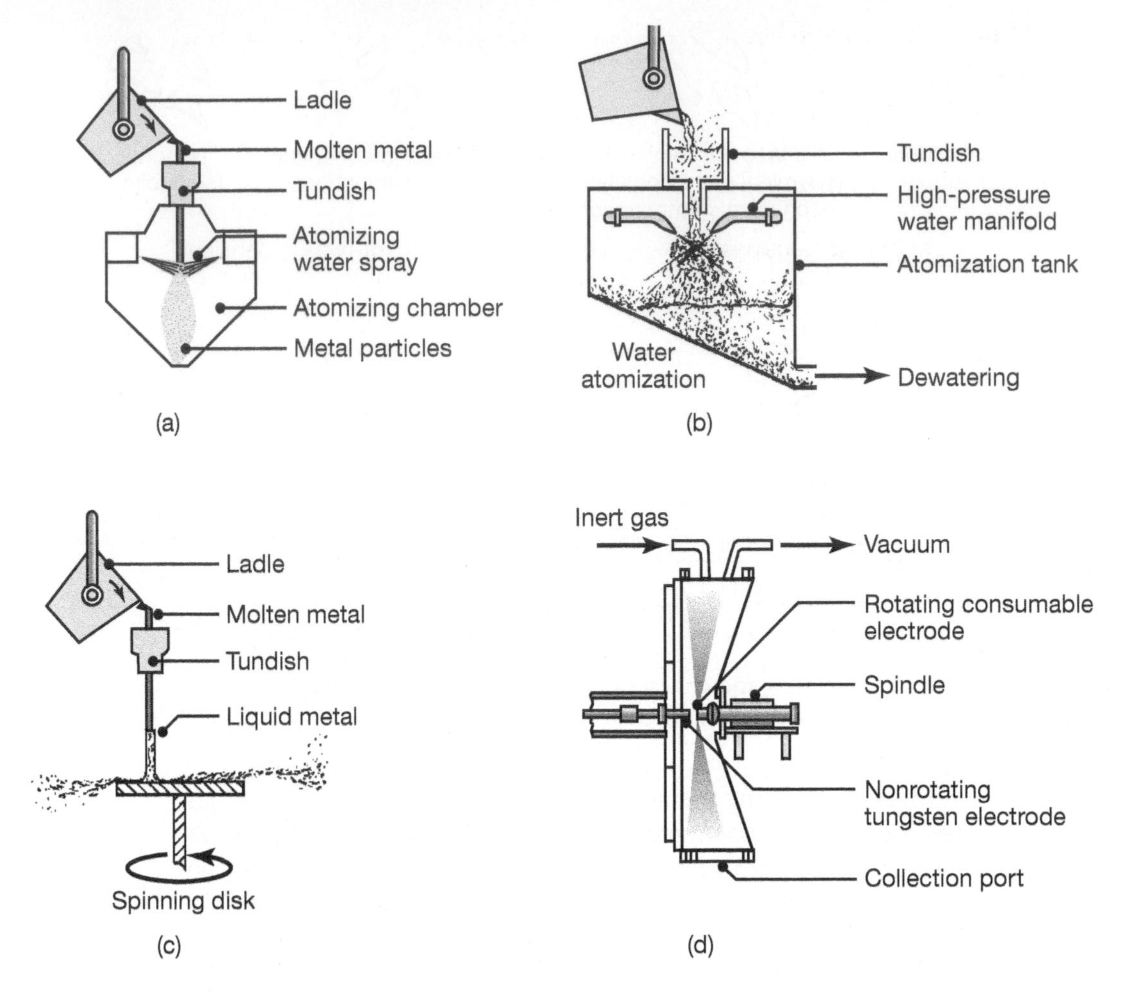

FIGURE 11.4 Methods of metal-powder production by atomization: (a) gas atomization; (b) water atomization; (c) centrifugal atomization with a spinning disk or cup; and (d) atomization with a rotating consumable electrode.

5. **Comminution.** Mechanical *comminution* or *pulverization* involves crushing, milling in a *ball mill*, or grinding brittle or less ductile metals into small particles. A ball mill (See Fig. 11.29) consists of a stirred or rotating hollow chamber that is partly filled with steel or white cast-iron balls. Their successive impact with the smaller powders causes fracture and the production of finer powders. When made from brittle metals, the powder particles have angular shapes, and when made from ductile metals, they are flaky and are not particularly suitable for PM applications.
6. **Mechanical alloying.** In this process, powders of two or more pure metals are mixed in a ball mill; under the impact of the hard balls, the powders repeatedly fracture and bond together by diffusion, producing an alloy. Note that the alloy produced can have concentrations that exceed solubility limits of castings. The dispersed phase can result in strengthening of the particles, or it can impart special electrical or magnetic properties to the powder.

7. **Other methods.** Less commonly used methods of powder production include:

    a. **precipitation** from a chemical solution;
    b. production of fine metal chips by **machining**;
    c. **vapor condensation**;
    d. high-temperature **extractive metallurgy**;
    e. reaction of volatile halides (a compound of halogen and an electropositive element) with liquid metals; and
    f. controlled reduction and reduction/carburization of solid oxides.

Nanopowders of various metals, including copper, aluminum, iron, and titanium are available (see also *nanomaterials*, Section 3.11.9). Such powders have limited applications outside of research, because of the high cost, limited availability, and poor flowability.

Microencapsulated powders. These powders are completely coated with a binder and are compacted by warm pressing (see also *metal injection molding*, Section 11.3.4). In electrical applications, the binder acts like an insulator, preventing electricity from flowing and reducing eddy-current losses.

11.2.2 Particle Size, Distribution, and Shape

Particle size is usually controlled by screening, that is, passing the powder through screens (*sieves*) of various mesh sizes. The screens are stacked vertically, with the mesh size becoming finer as the powder flows downward through the screens. The larger the mesh size, the smaller the opening in the screen. For example, a mesh size of 30 has openings of 600 μm, size 100 has 150 μm, and size 400 has 38 μm. This method is similar to the technique used for numbering abrasive grains; the higher the number, the smaller the size of the abrasive particle (see Section 9.2).

Several other methods are also used for particle-size control and measurement, particularly for powders finer than 45 μm in diameter:

1. **Sedimentation,** which involves measuring the rate at which particles settle in a fluid;
2. **Microscopic analysis,** including the use of transmission and scanning electron microscopy;
3. **Light scattering,** from a *laser* that illuminates a sample consisting of particles suspended in a liquid medium; the particles cause the light to be scattered, which is then focused on a detector that digitizes the signals and computes the particle-size distribution;
4. **Optical means,** such as having particles blocking a beam of light, which is then sensed by a *photocell*; and
5. **Suspension of particles** in a liquid, and subsequent detection of particle size and distribution by electrical *sensors*.

The **size distribution** of particles is important because it affects the processing characteristics of the powder; it is given in terms of a *frequency-distribution* plot, as shown in Fig. 11.5a. (For a detailed description of this

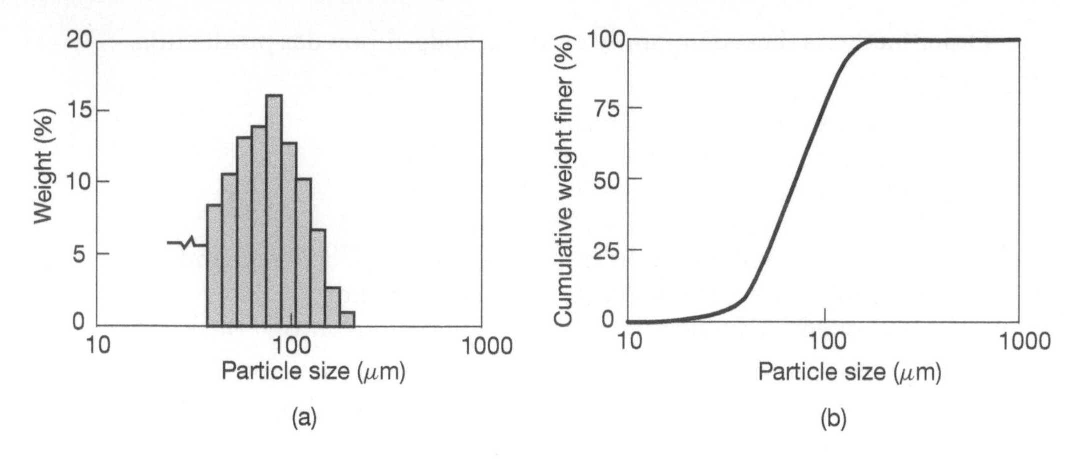

FIGURE 11.5 (a) Distribution of particle size, given as weight percentage; note that the highest percentage of particles have a size between 75 and 90 μm. (b) Cumulative particle-size distribution as a function of weight.

type of plot, see Section 4.9.1 and Fig. 4.20a.) Note in this plot that the highest percentage of the particles (by weight) has a size in the range of 75 to 90 μm; this size is called the *mode size*. The same particle-size data are also plotted in the form of a *cumulative distribution* (Fig. 11.5b). Thus, for example, about 75% of the particles (by weight) have a size finer than 100 μm. Note that the cumulative weight reaches 100% at a particle size of about 200 μm.

The **particle shape** has a major influence on its processing characteristics. The shape is generally described in terms of *aspect ratio* or *shape index*. The aspect ratio is the ratio of the largest dimension to the smallest dimension of the particle; it ranges from 1 for a spherical particle, to about 10 for flakelike or needlelike particles. Shape index, or shape factor (SF), is a measure of the surface area to the volume of the particle with reference to a spherical particle of equivalent diameter. Thus, for example, the shape factor for a flake is higher than that for a sphere.

11.2.3 Blending and Mixing of Metal Powders

Blending or mixing of powders is the second step in PM processing and is carried out for the following purposes:

1. Powders of different metals can be mixed in order to impart specific physical and mechanical properties and certain characteristics to the part. Powders can be made from metal alloys, or mixes of different metal powders can be produced. Proper mixing is essential to ensure uniformity of properties throughout the part.
2. Even when one type of metal is used, the powders may significantly vary in size and shape, hence they must be blended to ensure uniformity from part to part. The ideal mix is one in which all the particles of each material, and of each size and morphology, are distributed uniformly.
3. *Lubricants* can be mixed with the powders to improve their flow characteristics during processing. Lubricants reduce friction between the

metal particles, improve their flow into the dies, and improve die life. Typical lubricants are stearic acid or zinc stearate, in a proportion of from 0.25 to 5% by weight.

4. Various additives, including *binders* (as in sand molds; see Section 5.8.1) are used to develop sufficient *green strength* (see below), and facilitate sintering.

Powder mixing must be carried out under controlled conditions to avoid contamination and deterioration. *Deterioration* is caused by excessive mixing, which may alter the shape of the particles and work-harden them, thus making the subsequent compaction process more difficult. Powders can be mixed in air, in inert atmospheres (to avoid oxidation), or in liquids, which act as lubricants and make the mix more uniform. Several types of blending equipment are available. These operations are controlled by microprocessors to improve and maintain quality.

Because of their high surface-area-to-volume ratio, metal powders are *explosive*, particularly aluminum, magnesium, titanium, zirconium, and thorium powders. Great care must be exercised, both during blending and during storage and handling. Among precautions to be taken are (a) grounding equipment; (b) preventing the generation of sparks, by using nonsparking tools and avoiding the use of friction as a source of heat; and (c) avoiding dust clouds, open flames, and possible chemical reactions.

11.3 Compaction of Metal Powders

Compaction is the step in which the blended powders are pressed into specific shapes (Figs. 11.6a and b), using dies and presses that are either hydraulically or mechanically actuated (See also Fig. 6.25). Pressing is generally carried out at room temperature, although it can be done at elevated temperatures as well. The purposes of compaction are to obtain the required shape, density, and particle-to-particle contact, and to make the part sufficiently strong to be handled and processed further.

The stages of powder compaction are shown in Fig. 11.7. Initially, the powder is loosely packed, and thus there is significant porosity. With low applied pressure, the powder rearranges, filling the voids and producing a denser powder. Continued compaction causes increased contact stress and plastic deformation of the powders, resulting in increased powder-to-powder adhesion. The as-pressed powder is known as a **green compact**. The green parts are very fragile, and can easily crumble or become damaged, a situation that is exacerbated by poor pressing practices or rough handling of the compacted parts. For higher green strength, the powder must be fed properly into the die cavity, and to allow sufficient pressure to be developed throughout the part.

Density is relevant during three different stages in PM processing: (a) as loose powder; (b) as a green compact; and (c) after sintering. The particle shape, average size, and size distribution all affect the packing density of loose powder. An important factor in density is the size *distribution* of the particles. If all of the particles are of the same size, there will always be some porosity when packed together. Theoretically, the porosity is at least

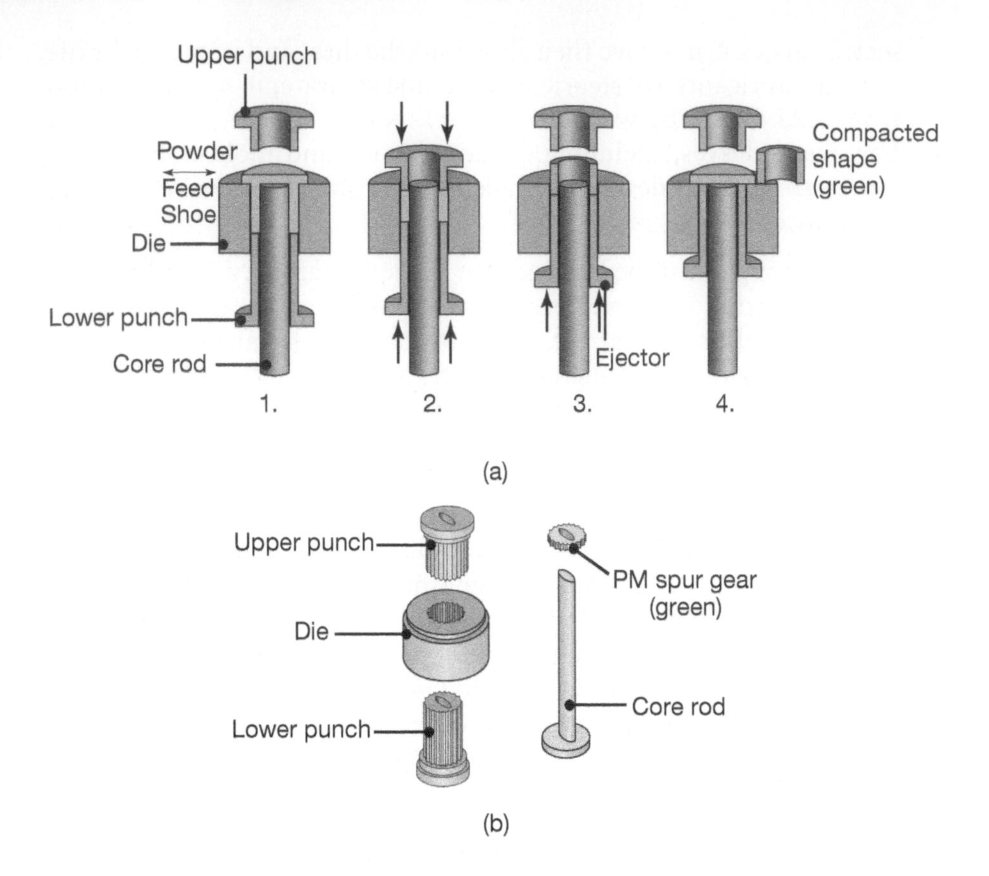

FIGURE 11.6 (a) Compaction of metal powder to produce a bushing. (b) A typical tool and die set for compacting a spur gear. *Source:* After the Metal Powder Industries Federation.

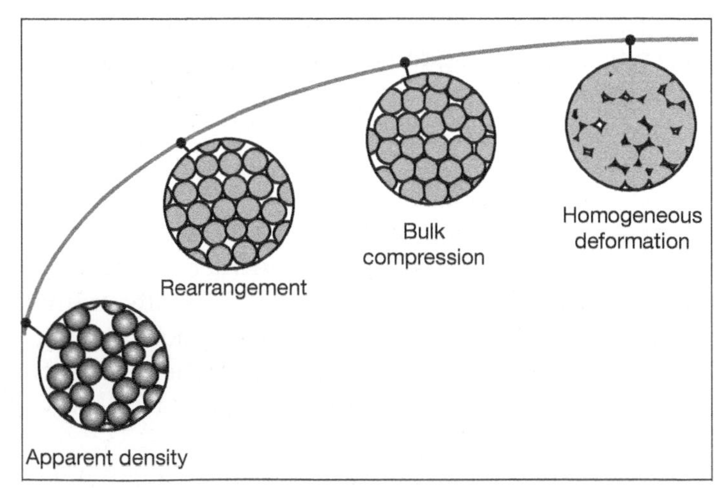

FIGURE 11.7 Compaction of metal powders: at low pressures, the powder rearranges without deforming, leading to a high rate of density increase. Once the powders are more closely packed, plastic deformation occurs at their interfaces, leading to further density increases but at lower rates. At very high densities, the powder behaves like a bulk solid.

24% by volume. (Observe, for example, a box filled with small balls; there are always open spaces between the individual balls.) Introducing smaller particles into the powder mix will fill the spaces between the larger powder particles, and thus result in a higher density of the compact.

The density after compaction (**green density**) depends primarily on the (a) compaction pressure; (b) powder composition; and (c) hardness of the powder (Fig. 11.8a). The higher the compacting pressure and the softer

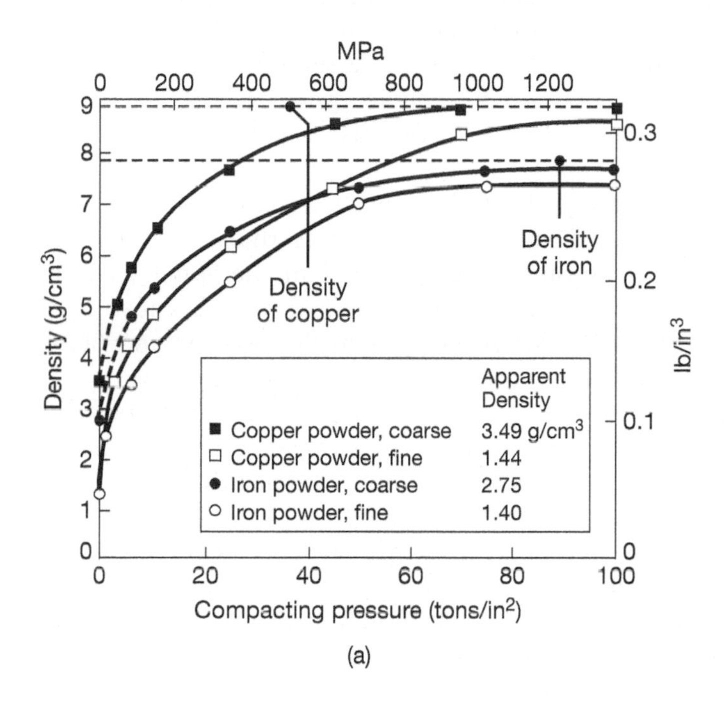

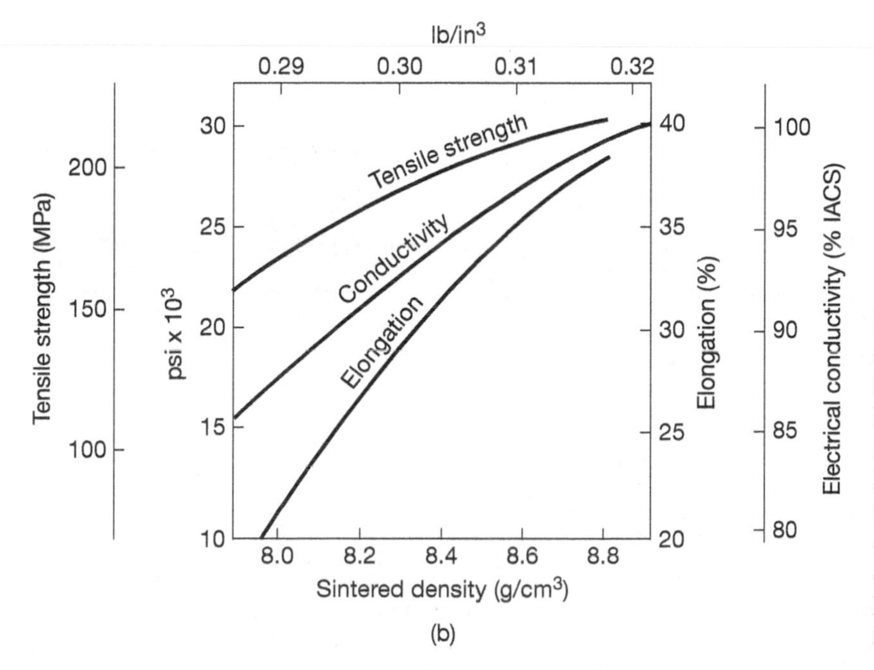

FIGURE 11.8 (a) Density of copper- and iron-powder compacts as a function of compacting pressure. Density greatly influences the mechanical and physical properties of PM parts. *Source:* After F.V. Lenel. (b) Effect of density on tensile strength, elongation, and electrical conductivity of copper powder. (IACS is International Annealed Copper Standard for electrical conductivity.)

the powder, the higher the green density. The density and its uniformity within a compact can be improved with the addition of a small quantity of admixed (blended-in) lubricant.

The effect of particle shape on green density can best be understood by considering two powder grades with the same chemical composition and hardness: one with a spherical particle shape and the other with an irregular shape. The spherical grade will have a higher apparent density (*fill density*), but after compaction under higher pressure, compacts from both grades will have similar green densities. When comparing two similar powders that were pressed under some standard conditions, the powder that gives a higher green density is said to have a higher *compressibility* (see also the discussion of *sintered density* in Section 11.4).

The higher the density, the higher the strength and the elastic modulus of the PM part (Fig. 11.8b). The reason is that the higher the density, the higher the amount of solid metal in the same volume, hence the greater the part's resistance to external forces. Because of the friction present between the metal particles in the powder and between the punches and the die walls, the density can significantly vary *within* the part. This variation can be minimized by proper punch and die design and by controlling friction. Thus, for example, it may be necessary to use *multiple punches*, with separate movements, in order to ensure that the density is more uniform throughout the part (Fig. 11.9). On the other hand, in some compacted parts, such as gears and cams, density variations may be desirable. For example, densities can be increased in critical locations of a part where high strength and wear resistance are important.

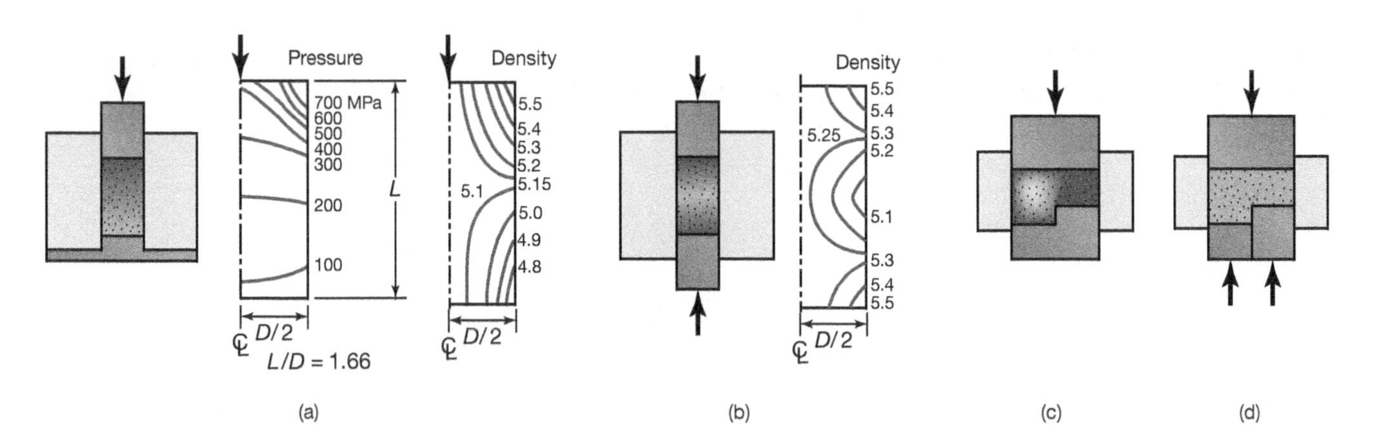

FIGURE 11.9 Density variation in compacting metal powders in various dies. (a) and (c) Single action press; (b) and (d) double action press, where the punches have separate movements. Note the greater uniformity of density in (d) as compared to (c). All pressure contours shown are for compacted copper powder.

EXAMPLE 11.1 Density of Metal-Powder-Lubricant Mix

Given: Zinc stearate is a lubricant that commonly is mixed with metal powders prior to compaction, in proportions up to 2% by weight. Measurements have shown that the density of the lubricant is 1.10 g/cm^3, and the apparent density of the iron powder is 2.75 g/cm^3. The theoretical density of the iron powder is 7.86 g/cm^3.

Find: Calculate the theoretical and apparent densities of an iron powder-zinc stearate mix if 1000 g of iron powder is mixed with 20 g of lubricant.

Solution: The volume of the mixture is

$$V = \left(\frac{1000}{7.86}\right) + \left(\frac{20}{1.10}\right) = 145.4 \text{ cm}^3.$$

The combined weight of the mixture is $1000 + 20 = 1020$ g, and thus its theoretical density is $1020/145.41 = 7.01$ g/cm^3. The apparent density of the iron powder is given as 2.75 g/cm^3, hence its density is $(2.75/7.86)100 = 35\%$ of the theoretical density. Assuming a similar percentage for the mixture, the apparent density of the mix can be estimated as $(0.35)(7.01) = 2.45$ g/cm^3. Note that although g and g/cm^3 are not SI units, they continue to be commonly used in the powder-metallurgy industry.

11.3.1 Pressure Distribution in Powder Compaction

As can be seen in Fig. 11.9, in single-action pressing, the pressure decays rapidly toward the bottom of the compact. The pressure distribution along the length of the compact can be determined by using the slab method of analysis, described in Section 6.2.2. As also done in Fig. 6.4, the operation is first described in terms of its coordinate system, as shown in Fig. 11.10, where D is the diameter of the compact, L is the compact's length, and p_o is the pressure applied by the punch. An element dx thick is then taken, with all the relevant stresses indicated on it, namely, the compacting pressure, p_x, the die-wall pressure, σ_r, and the frictional stress along the die wall, $\mu\sigma_r$. Note that the frictional stresses act upward on the element, because the punch movement is downward.

Balancing the vertical forces acting on this element,

$$\left(\frac{\pi D^2}{4}\right) p_x - \left(\frac{\pi D^2}{4}\right)(p_x + dp_x) - (\pi D)(\mu\sigma_r)\,dx = 0, \qquad (11.1)$$

which can be simplified to

$$D\,dp_x + 4\mu\sigma_r\,dx = 0.$$

This results in one equation, but two unknowns (p_x and σ_r). At this point, a factor k is introduced, which is a measure of the interparticle friction during compaction:

$$\sigma_r = kp_x.$$

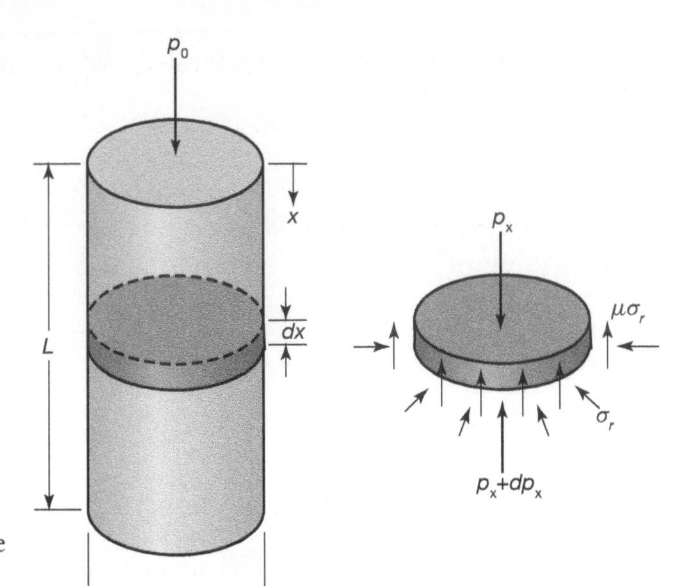

FIGURE 11.10 Coordinate system and stresses acting on an element in compaction of powders. The pressure is assumed to be uniform across the cross section. (See also Fig. 6.4.)

If there is no friction between the particles, $k = 1$, the powder behaves like a fluid, and thus $\sigma_r = p_x$, signifying a state of hydrostatic pressure. Using this expression yields

$$dp_x + \frac{4\mu k p_x \, dx}{D} = 0,$$

or

$$\frac{dp_x}{p_x} = -\frac{4\mu k \, dx}{D}.$$

This expression is similar to that given in Section 6.2.2 for upsetting. In the same manner, this expression can be integrated, noting that the boundary condition in this case is $p_x = p_o$ when $x = 0$:

$$p_x = p_o e^{-4\mu k x/D}. \tag{11.2}$$

It can be seen from this expression that the pressure within the compact decays as the coefficient of friction, the parameter k, and the length-to-diameter ratio increase.

EXAMPLE 11.2 Pressure Decay in Compaction

Given: A powder mix has the values of $k = 0.5$ and $\mu = 0.3$.

Find: At what depth will the pressure in a straight cylindrical compact 10 mm in diameter become (a) zero and (b) one-half the pressure at the punch?

Solution: For case (a), $p_x = 0$. Consequently, from Eq. (11.2)

$$0 = p_o e^{-(4)(0.3)(0.5)x/10}, \qquad \text{or} \qquad e^{-0.06x} = 0.$$

Note that the value of x must approach ∞ for the pressure to decay to 0. For case (b), we have $p_x/p_o = 0.5$. Therefore,

$$e^{-0.06x} = 0.5, \qquad \text{or} \qquad x = 11.55 \text{ mm}.$$

In practice, a pressure drop of 50% is considered severe as the compact density will then be unacceptably low. This example shows that, under the conditions assumed, uniaxial compaction of a cylinder of even a length-to-diameter ratio of about 1.2 will be unsatisfactory.

11.3.2 Equipment

The compacting pressure required for pressing metal powders ranges from 70 MPa (10 ksi) for aluminum to 800 MPa (120 ksi) for high-density iron (Table 11.1). The pressure required depends on the characteristics and shape of the particles, the method of blending, and the lubricant.

Press capacities are on the order of 1.8–2.7 MN (200–300 tons), although presses with much higher capacities are available for special applications; most applications require less than 1 MN (110 tons). For small capacities, crank or eccentric-type mechanical presses are used; for higher capacities, toggle or knucklejoint presses are employed (See Fig. 6.25). Hydraulic presses with capacities as high as 45 MN (5000 tons) can be used for compacting large parts.

The selection of a press depends on part size and its configuration, density requirements, and production rate. Also, the higher the pressing speed, the greater is the tendency for air to be trapped in the die cavity. Proper die design, including provision of vents, are thus important.

TABLE 11.1 Compacting pressures for various metal powders.

	Pressure	
	MPa	ksi
Metal		
Aluminum	70–275	10–40
Brass	400–700	60–100
Bronze	200–275	30–40
Iron	350–800	50–120
Tantalum	70–140	10–20
Tungsten	70–140	10–20
Other materials		
Aluminum oxide	110–140	16–20
Carbon	140–165	20–24
Cemented carbides	140–400	20–60
Ferrites	110–165	16–24

QR Code 11.3 Hot isostatic pressing. *Source:* Courtesy of the Metal Powder Industries Federation.

11.3.3 Isostatic Pressing

For improved compaction, PM parts may be subjected to a number of additional operations, such as *rolling, forging,* and *isostatic pressing.* Because the density of die-compacted powders can vary significantly, powders can be subjected to *hydrostatic pressure* in order to achieve more uniform compaction, similar to pressing cupped hands together when making snowballs.

In **cold isostatic pressing** (CIP), the powder is placed in a flexible rubber mold, made of neoprene rubber, urethane, polyvinyl chloride, or other elastomers (Fig. 11.11). The assembly is then pressurized hydrostatically in a chamber, usually filled with water. The most common pressure is 400 MPa, although pressures of up to 1000 MPa may be used. The applications of CIP and other compacting methods, in terms of size and part complexity, are shown in Fig. 11.12.

In **hot isostatic pressing** (HIP), the container is usually made of a high-melting-point sheet metal, and the pressurizing medium is inert gas or vitreous (glasslike) fluid (Fig. 11.13). A typical operating condition for HIP is 100 MPa at 1100°C (15 ksi at 2000°F), although the trend is toward higher pressures and temperatures. The main advantage of HIP is its ability to produce compacts with essentially 100% density, high bond strength among the particles, and good mechanical properties. Although relatively expensive, typical applications include making superalloy components for the aerospace industry, final densification for tungsten-carbide cutting tools and PM tool steels, and closing internal porosity and improving mechanical properties in superalloy and titanium-alloy castings.

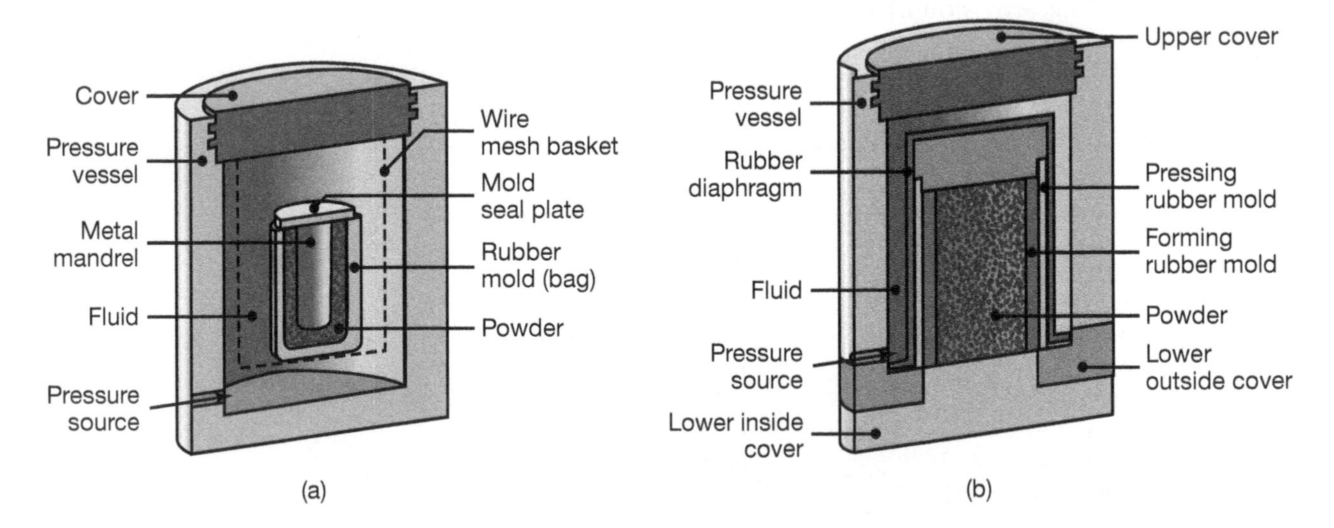

FIGURE 11.11 Schematic illustration of cold isostatic pressing in compaction of a tube. (a) The *wet-bag process*, where the rubber mold is inserted into a fluid that is subsequently pressurized. In the arrangement shown, the powder is enclosed in a flexible container around a solid core rod. (b) The *dry bag process*, where the rubber mold does not contact the fluid, but instead is pressurized through a diaphragm. *Source:* After R.M. German.

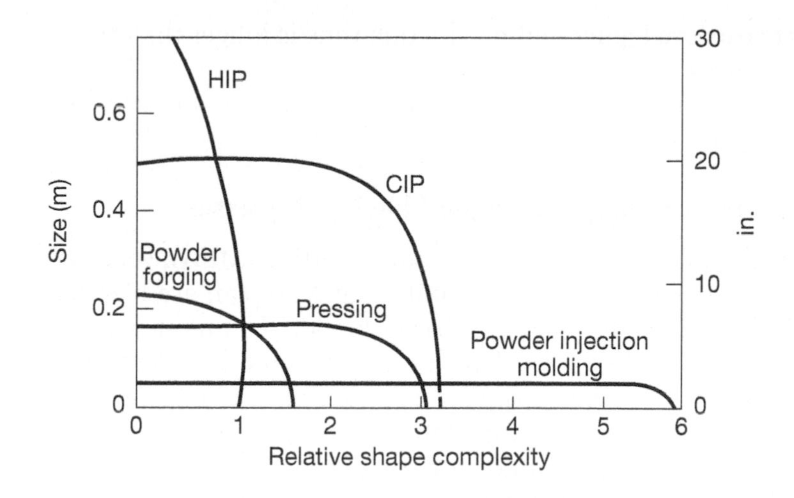

FIGURE 11.12 Process capabilities of part size and shape complexity for various PM operations; PF is powder forging. *Source:* After the Metal Powder Industries Federation.

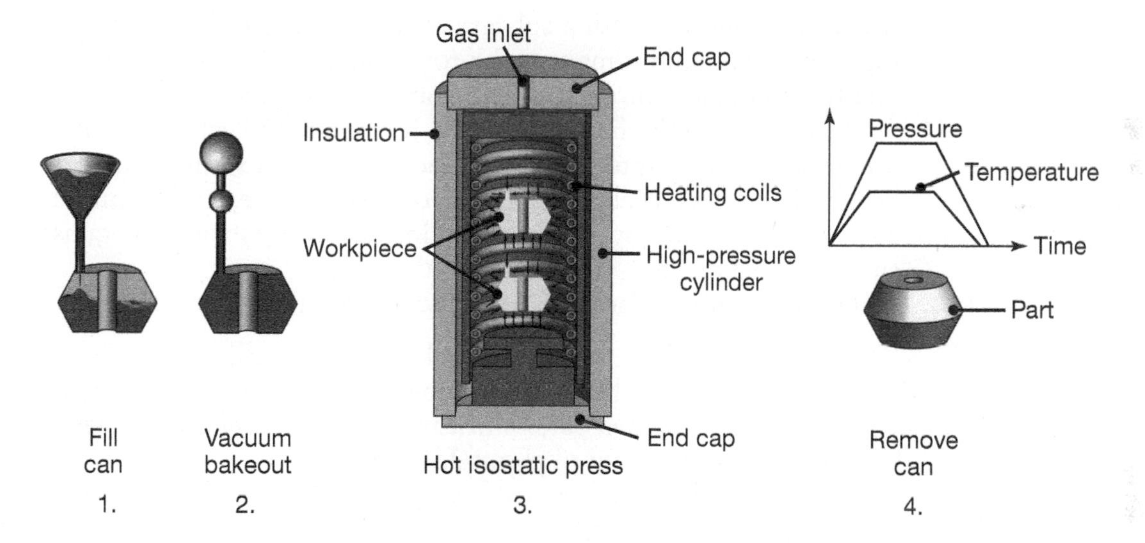

FIGURE 11.13 Schematic illustration of the sequence of steps in hot isostatic pressing. Diagram (4) shows the pressure and temperature variation versus time.

The main advantages of hot isostatic pressing over conventional PM are:

- Because of the uniformity of pressure from all directions and the absence of die-wall friction, it produces fully dense compacts of practically uniform grain structure and density, irrespective of part shape; the properties are thus *isotropic*. Parts with high length-to-diameter ratios can be produced, with very uniform density, strength, toughness, and good surface detail.
- HIP is capable of processing much larger parts than those in other compacting processes.

On the other hand, HIP has some limitations, such as:

- Dimensional tolerances are higher than those in other compacting methods.

- Equipment costs are higher and production time is longer than those in other processes.
- HIP is suitable for relatively small production quantities, typically less than 10,000 parts per year.

11.3.4 Miscellaneous Compacting and Shaping Processes

1. **Metal injection molding.** In *metal injection molding* (MIM), very fine metal powders (generally <45 μm, and often <10 μm) are blended with a 25–45% polymer or a wax-based binder. The term *powder injection molding* (PIM) is used to include ceramic powders. The mixture undergoes a process similar to injection molding of plastics (Section 10.10.2); it is injected into a mold at a temperature of 135°C to 200°C. The molded green parts are then placed in a low-temperature oven to burn off the plastic (*debinding*), or the binder is partially removed with a solvent; some binder is purposely left in the part to maintain some strength for handling. The green parts are finally sintered in a furnace at temperatures as high as 1375°C.

   Generally, metals that are suitable for powder injection molding are those that melt at temperatures above 1000°C; examples are carbon and stainless steels, tool steels, copper, bronze, and titanium. Typical parts made by this method include components for guns, surgical instruments, automobiles, and watches.

   The major advantages of PIM over conventional compaction are:

   a. Complex shapes, with wall thicknesses as small as 5 mm, can be molded, and then easily removed from the dies. Most commercially produced parts fabricated by this method weigh from a fraction of a gram to about 250 g each.
   b. Mechanical properties are nearly equal to those of wrought parts.
   c. Dimensional tolerances are good.
   d. High production rates can be achieved by using multicavity dies (See also Fig. 10.30).
   e. Parts produced compete well against small investment-cast parts (Section 5.9.2), small forgings, and complex machined parts. It does not, however, compete well with zinc and aluminum die casting (Section 5.10.3) and with screw machining (Section 8.9.2).

   The major limitations of PIM are the added complications of burning off binder, and thermal distortion associated with lower green density.

   An example where the advantages of MIM are apparent is in the production of light-duty gears (such as in office equipment, where load and power are low). An inexpensive gear can be produced directly by MIM, instead of producing blanks from sheet metal, casting, or forging followed by machining. Avoiding machining is a significant cost savings.

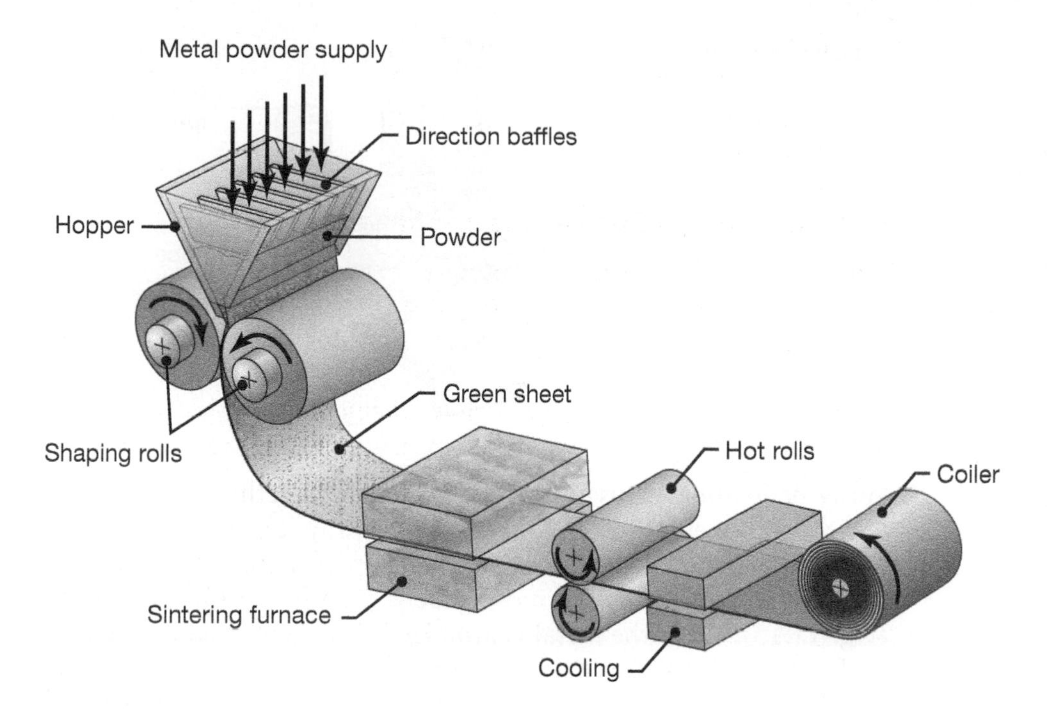

FIGURE 11.14 An example of powder rolling. The purpose of direction baffles in the hopper is to ensure uniform distribution of powder across the width of the strip.

2. **Forging.** In *powder forging* (PF), the part produced from compaction and sintering serves as the *preform* in a hot-forging operation. The forged products are almost fully dense, and have a good surface finish, good dimensional tolerances, and uniform and fine grain size. The superior properties obtained make powder forging particularly suitable for such applications as highly stressed automotive parts, such as connecting rods, and jet-engine components.
3. **Rolling.** In *powder rolling*, also called *roll compaction*, the powder is fed into the roll gap in a two-high rolling mill (See Fig. 6.39a) and is compacted into a continuous strip at speeds of up to 0.5 m/s (100 ft/min), as shown in Fig. 11.14. The process can be carried out at room or at elevated temperatures, depending on the type of powder metal. Sheet metal for electrical and electronic components and for coins also can be made by powder rolling.
4. **Extrusion.** Powders may be compacted by *hot extrusion*, in which the powder is first encased in a metal container and then extruded (see also Section 6.4). Superalloy powders, for example, are hot extruded for improved properties. Preforms made from the extrusions may be reheated and then forged in a closed die to their final shape.
5. **Pressureless compaction.** In this process, the die is filled with metal powder by gravity, and the powder is sintered directly in the die. Because of the resulting low density, pressureless compaction is used principally for porous parts, such as filters.

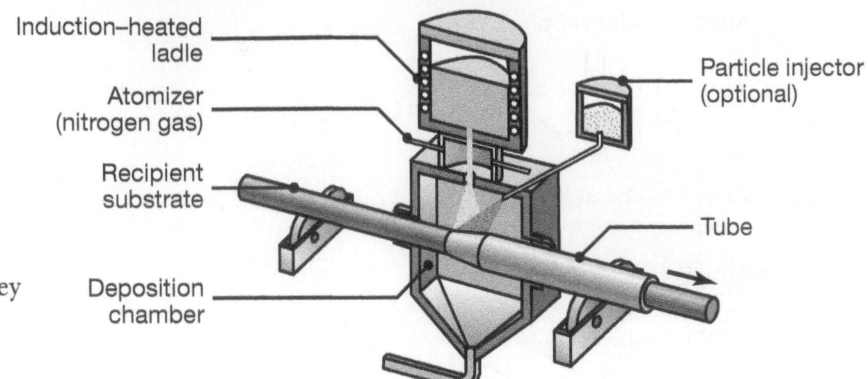

FIGURE 11.15 Spray casting (Osprey process) in which molten metal is sprayed over a rotating mandrel to produce seamless tubing and pipe.

6. **Spray deposition.** This is a shape-generation method in which the basic components are an atomizer, a spray chamber with inert atmosphere, and a mold for producing preforms. Although there are several variations, the most widely used is the *Osprey* process (Fig. 11.15). After the metal is atomized, it is deposited on a cooled preform mold (usually made of copper or ceramic) where it solidifies. The metal particles stick together, developing a density that is typically above 99% of the solid metal density. The mold may have various shapes, such as billets, tubes, disks, and cylinders. The parts made may be subjected to further shaping and consolidation processes, such as forging, rolling, and extrusion. The grain size of the part is fine, and its mechanical properties are comparable to those of wrought products of the same alloy.

Dynamic and Explosive Compaction. Metal powders that are difficult to compact with sufficient green strength can be compacted rapidly to near full density using the setup shown in Fig. 11.16. The explosive drives a mass into a green powder contained in a chamber. This high-speed collision generates a shock wave that develops pressures up to 30 GPa. The shock wave traverses across the part at speeds up to 6 km/s. Preheating of the powder is often practiced to prevent fracture.

Combustion Synthesis. This operation takes advantage of the highly combustible nature of metal powders, by placing a lightly compacted powder into a pressure vessel. An ignition source is then introduced, such as an arc from a tungsten electrode, igniting the powder. The explosion produces a shock wave that travels across the compact, developing heat and pressure sufficient for compaction of the powder metal.

Pseudo-isostatic Pressing. In this process, a preform is preheated, surrounded by hot ceramic or graphite granules, and placed in a container. A mechanical press compacts the granules and the preform. The granules are large and cannot penetrate the pores of the PM part. The compaction is uniaxial, but because of the presence of the ceramic granules, the loading on the preform is basically multi-axial (hence the word pseudo). This technique has cycle times shorter than HIP, but because the pressure

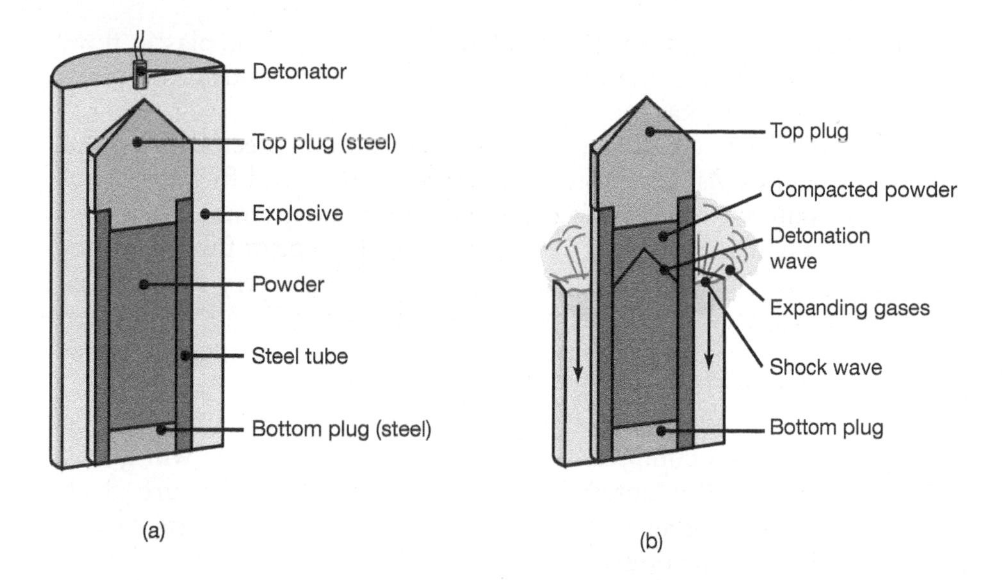

FIGURE 11.16 Schematic illustration of explosive compaction. (a) A tube filled with powder is surrounded by explosive media inside a container, typically cardboard or wood. (b) After detonation, a compression wave follows the detonation wave, resulting in a compacted metal powder part.

is not strictly hydrostatic, dimensional changes during compaction are not uniform.

11.3.5 Punch and Die Materials

The selection of punch and die materials for PM depends on the abrasiveness of the powder metal and the number of parts to be made. The most commonly used die materials are air- or oil-hardening tool steels, such as D2 or D3, with a hardness range of 60–64 HRC (Section 3.10.4). Because of their greater hardness and wear resistance, tungsten-carbide dies are used for more severe applications. Punches are generally made of similar materials as dies. Close control of die and punch dimensions are essential for proper compaction and longer die life. Too large a clearance between the punch and the die will allow the metal powder to enter the gap, interfere with the operation, and produce eccentric parts. Diametral clearances are generally less than 25 μm. Die and punch surfaces must be lapped or polished in the direction of tool movements, for improved die life and overall performance.

11.4 Sintering

Sintering is the process whereby compressed metal powder is heated in a controlled-atmosphere furnace to a temperature below its melting point, but sufficiently high to allow bonding (fusion) of the individual metal particles. Prior to sintering, the compact is brittle and its strength is low. The nature and strength of the bond between the particles, and hence of the sintered compact, depend on the mechanisms of diffusion, plastic flow, evaporation of volatile materials in the compact, recrystallization, grain growth, and pore shrinkage.

The density of a sintered part mainly depends on its green density and on sintering conditions in terms of temperature, time, and furnace atmosphere. The sintered density increases with increasing temperature and time, and usually with the use of a more deoxidizing type of furnace atmosphere. For structural parts, a higher sintered density is very desirable as it leads to better mechanical properties. On the other hand, it is often preferable to minimize the increase in density during sintering, for better dimensional accuracy.

Better properties and dimensional accuracy can be achieved by using a powder with a high compressibility, that is, a powder that will give a high green density and maintain a moderate sintering temperature. Such a powder also has another important benefit in that larger parts can be produced with a specific press capacity.

Sintering temperatures (Table 11.2) are generally within 70 to 90% of the melting point of the metal or alloy. *Sintering times* range from a minimum of about 10 min for iron and copper alloys to as much as 8 hrs for tungsten and tantalum. Continuous-sintering furnaces are used for most production today. These furnaces have three chambers: (1) a burn-off chamber, to volatilize the lubricants in the green compact to improve bond strength and prevent cracking; (2) a high-temperature chamber, for sintering; and (3) a cooling chamber.

Proper control of the furnace atmosphere is essential for successful sintering and obtaining optimum properties. An oxygen-free atmosphere is essential in order to control the carburization and decarburization of iron and iron-base compacts and to prevent oxidation of powders. Oxide inclusions have a detrimental effect on mechanical properties (see Section 3.8). For the same volume of inclusions, the smaller inclusions have a larger effect because there are more of them per unit volume of the part. The gases most commonly used for sintering are hydrogen, dissociated or burned ammonia, partially combusted hydrocarbon gases,

TABLE 11.2 Sintering temperature and time for various metal powders.

Material	Temperature (°C)	Time (min)
Copper, brass, and bronze	760–900	10–45
Iron and iron graphite	1000–1150	8–45
Nickel	1000–1150	30–45
Stainless steels	1100–1290	30–60
Alnico alloys (for permanent magnets)	1200–1300	120–150
Ferrites	1200–1500	10–600
Tungsten carbide	1430–1500	20–30
Molybdenum	2050	120
Tungsten	2350	480
Tantalum	2400	480

and nitrogen. A vacuum is generally used for sintering refractory metal alloys and stainless steels.

Sintering mechanisms. Sintering mechanisms are complex and depend on the composition of the metal particles as well as the processing parameters. As the temperature increases, two adjacent particles begin to form or strengthen a bond by **diffusion** (*solid-state bonding*), as shown in Fig. 11.17a. As a result, the strength, density, ductility, and thermal and electrical conductivities of the compact increase (Fig. 11.18). At the same time, however, the compact shrinks, hence allowances should be made for shrinkage, as is done in casting (see Section 5.12.2).

A second sintering mechanism is **vapor-phase transport**. Because the material is heated very close to its melting temperature, metal atoms can vaporize. At convergent geometries (the interface of two particles), the melting temperature is locally higher, and the vapor resolidifies; thus, the interface grows and strengthens, while each particle shrinks as a whole. If two adjacent particles are of different metals, *alloying* can take place at the interface of the two particles. One of the particles may have a lower melting point than the other, in which case one particle may melt and, because of surface tension, surround the particle that has not melted (**liquid-phase sintering**), as shown in Fig. 11.17b. Stronger and denser parts can be obtained in this way; an example is cobalt in tungsten-carbide tools and dies (Section 8.6.4).

Depending on temperature, time, and processing history, different structures and porosities can develop in a sintered compact. However, porosity

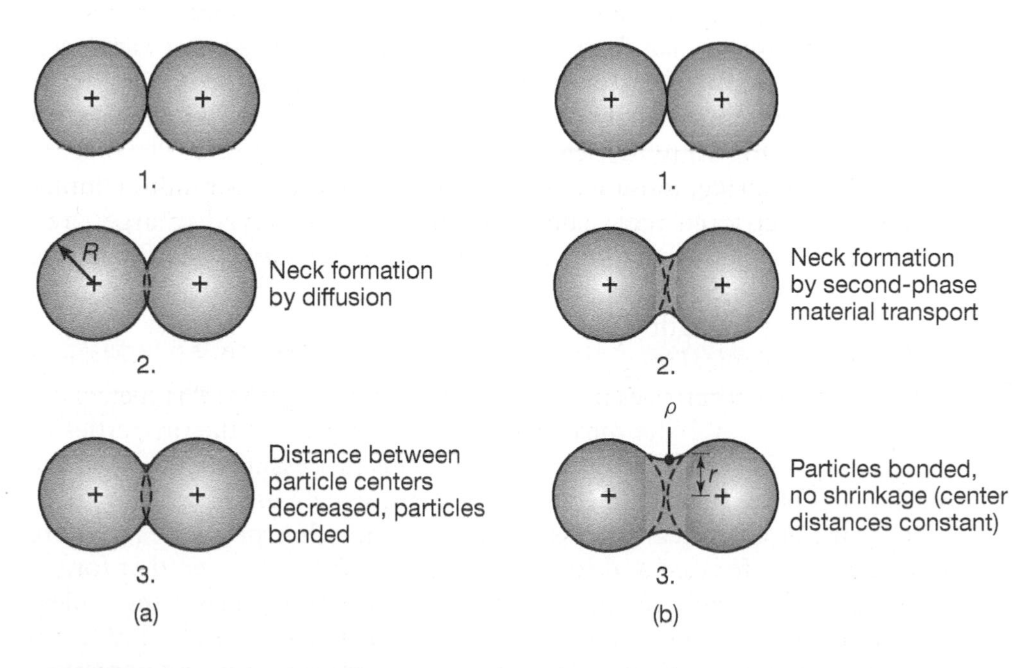

FIGURE 11.17 Schematic illustration of two basic mechanisms in sintering metal powders: (a) solid-state material transport and (b) liquid-phase material transport. R = particle radius, r = neck radius, and ρ = neck profile radius.

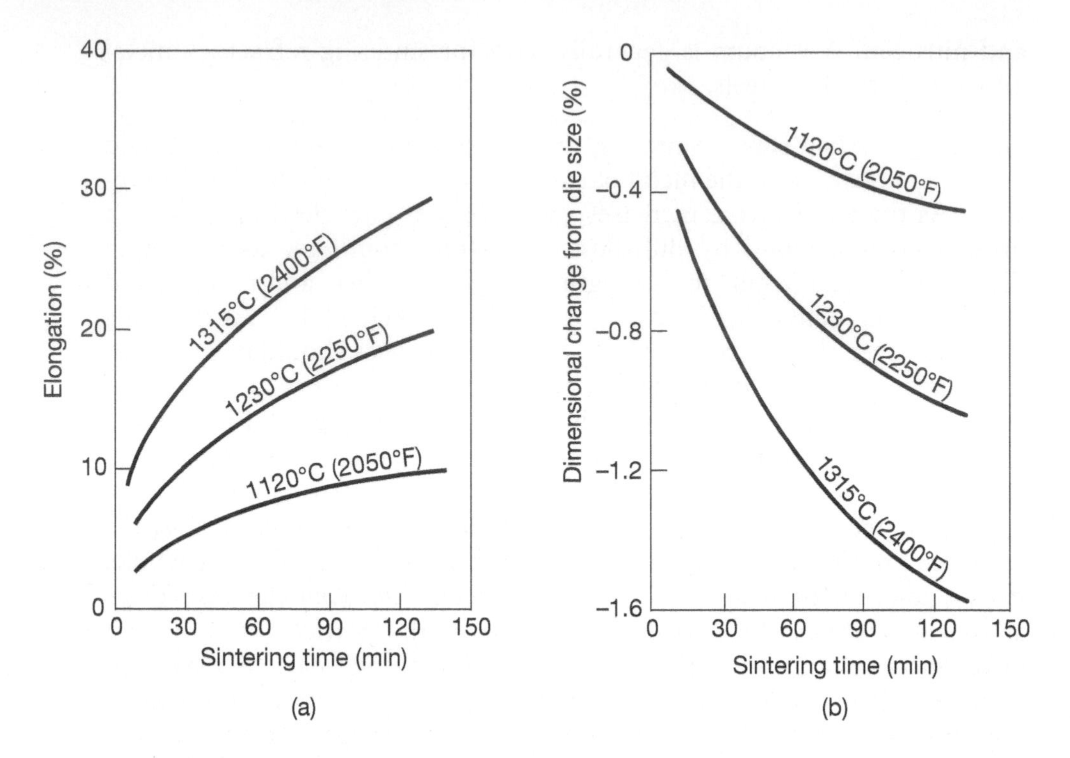

FIGURE 11.18 Effect of sintering temperature and time on (a) elongation and (b) dimensional change during sintering of type 316L stainless steel. *Source:* ASM International.

cannot be completely eliminated, because some voids remain after compaction and gases evolve during sintering and are trapped. Porosities can consist of either a network of interconnected pores or closed holes. Their presence is important in making PM parts such as filters and bearings (with some porosity to entrap lubricants).

In *spark sintering*, loose metal powders are placed in a graphite mold, heated by electric current, subjected to a high-energy discharge, and compacted, all in one step. Another technique is *microwave sintering*, which reduces sintering time, thereby preventing grain growth, which can adversely affect strength.

Typical mechanical properties for several sintered PM alloys are given in Table 11.3. The differences in the properties of wrought vs. PM metals are given in Table 11.4. Note the effect of heat treatment on the properties of metals. Table 11.5 shows the effects of various processing methods on the mechanical properties of Ti-6Al-4V, the most widely used titanium alloy. Note that hot isostatically pressed (HIP) titanium has properties that are similar to those for cast and forged titanium. Recall, however, that forged components are likely to require additional machining processes (unless the components are precision forged to near-net shape), while PM components may not. Consequently, powder metallurgy can be a competitive alternative to casting and forging operations.

TABLE 11.3 Mechanical properties of selected PM materials. See Table 11.5 for titanium.

Material	Yield strength (MPa)	Ultimate tensile strength (MPa)	Elastic modulus (GPa)	Hardness	Elongation in 25 mm (%)	Density (g/cm^3)
Ferrous[1]						
F-0008-20	170	200	85	35 HRB	<1	5.8
F-0008-35	260	390	140	70 HRB	1	7.0
F-0008-55HT		450	115	22 HRC	<1	6.3
F-0008-85HT		660	150	35 HRC	<1	7.1
FC-0008-30[3]	240	240	85	50 HRB	<1	5.8
FC-0008-60[3]	450	520	155	84 HRB	<1	7.2
FC-0008-95[3]		720	150	43 HRC	<1	7.1
FN-0205-20	170	280	115	44 HRB	1	6.6
FN-0205-35	280	480	170	78 HRB	5	7.4
FN-0205-180HT		1280	170	78 HRB	<1	7.4
FX-1005-40	340	530	160	82 HRB	4	7.3
FX-1005-110HT		830	160	38 HRC	<1	7.3
Stainless Steel						
SS-303N1-38	310	470	115	70 HRB	5	6.9
SS-304N1-30	260	300	105	61 HRB	<1	6.4
SS-316N1-25[2]	230	280	105	59 HRB	<1	6.4
SS-316N2-38[2]	310	480	140	65 HRB	<1	6.9
Copper and Copper Alloys						
CZ-1000-9[2]	70	120	80	65 HRH	9	7.6
CZ-1000-11[2]	80	160	100	80 HRH	12	8.1
CZP-3002-14[2]	110	220	90	88 HRH	16	8.0
CT-1000-13[3]	110	150	60	82 HRH	4	7.2
Aluminum Alloys						
Ax 123-T1[2]	200	270	—	47 HRB	3	2.7
Ax 123-T6[2]	390	400	—	72 HRB	<1	2.7
Ax 231-T6[3]	200	220	—	55 HRB	1	2.7
Ax 231-T6[3]	310	320	—	77 HRB	<1	2.7
Ax 431-T6[2]	270	300	—	55 HRB	5	2.8
Ax 431-T6[2]	440	470	—	80 HRB	2	2.8
Superalloys						
Stellite 19	—	1035	—	49 HRC	<1	

Notes: 1, F-0008 is often most cost effective; FX-1005 is copper-infiltrated steel; 2, Suitable for general-purpose structural applications; 3, Wear-resistant alloy.

TABLE 11.4 Comparison of mechanical properties of some wrought and equivalent PM metals (as sintered).

Metal	Condition	Relative density[a] (%)	Ultimate tensile strength[a] (MPa)	Elongation in 50 mm (%)
Aluminum				
2014-T6	Wrought (W)	100	480	20
	PM	94	330	2
6061-T6	W	100	310	15
	PM	94	250	2
Copper, OFHC[b]	W, annealed	100	235	50
	PM	89	160	8
Brass, 260	W, annealed	100	300	65
	PM	89	255	26
Steel, 1025	W, hot rolled	100	590	25
	PM	84	235	2
Stainless steel, 303	W, annealed	100	620	50
	PM	82	360	2

Notes:
[a]The density and strength of PM materials greatly increase with further processing, such as forging, isostatic pressing, and heat treatments.
[b]OFHC = oxygen-free high conductivity.

TABLE 11.5 Mechanical property comparisons for Ti-6Al-4V titanium alloy.

Process	Density (%)	Yield strength (MPa)	Ultimate tensile strength (MPa)	Elongation (%)	Reduction of area (%)
Cast	100	840	930	7	15
Cast and forged	100	875	965	14	40
Powder metallurgy					
Blended elemental (P+S)*	98	786	875	8	14
Blended elemental (HIP)*	>99	—	875	9	17
Realloyed (HIP)	100	880	975	14	26

*P+S = pressed and sintered; HIP = hot isostatically pressed. *Source:* After R.M. German

EXAMPLE 11.3 Shrinkage in Sintering

Given: In solid-state bonding during sintering of a powder-metal green compact, the linear shrinkage is 4%.

Find: If the desired sintered density is 95% of the theoretical density of the metal, what should be the density of the green compact? Ignore the small changes in mass that occur during sintering.

Solution: Linear shrinkage is defined as $\Delta L/L_o$, where L_o is the original length of the part. The volume shrinkage during sintering can then be expressed as

$$V_{\text{sint}} = V_{\text{green}}\left(1 - \frac{\Delta L}{L_o}\right)^3. \qquad (11.3)$$

The volume of the green compact must be larger than that of the sintered part. However, the mass does not change during sintering, this expression can be rewritten in terms of the density, ρ, as

$$\rho_{\text{green}} = \rho_{\text{sint}}\left(1 - \frac{\Delta L}{L_o}\right)^3. \qquad (11.4)$$

Therefore,

$$\rho_{\text{green}} = 0.95(1 - 0.04)^3 = 0.84 \quad \text{or} \quad 84\%.$$

11.5 Secondary and Finishing Operations

In order to further improve the properties of sintered powder-metallurgy products or to impart certain specific characteristics, several additional operations may be carried out following the sintering process.

1. **Coining** and **sizing** are compacting operations, performed under high pressure in presses. The purposes of these operations are to impart dimensional accuracy to the sintered part and to improve its strength and surface finish by further densification.
2. **Forging** involves the use of either unsintered or sintered alloyed-powder preforms that are subsequently hot forged in heated, confined dies to the desired final shapes; the preforms also may be shaped by impact forging (see also *hammers*, Section 6.2.8). The forging process is generally referred to as *powder-metallurgy forging* (PF); when the unsintered preform is later sintered, the process is usually referred to as *sinter forging*. Ferrous and nonferrous powders can be processed in this manner, resulting in *full-density* products, on the order of 99.9% of the theoretical density of the material.

   The deformation of the preform may consist of the following two modes: *upsetting* and *re-pressing*. In upsetting, the material flows laterally outward, as also shown in Fig. 6.1. The preform is thus subjected to compressive and shear stresses during deformation; as a result, any interparticle residual oxide films present are broken up, and the part has improved toughness, ductility, and fatigue strength. In *re-pressing*, the material flow is mainly in the direction of the movement of the punch; as a result, the mechanical properties of the part are not as high as those made by upsetting.

   PF parts have good surface finish and dimensional accuracy, little or no flash, and uniform and fine grain size throughout the part. The superior properties obtained make this technology particularly suitable for such applications as highly stressed components, for use as connecting rods, jet engines, military hardware, and off-road equipment.

3. The porosity of PM components can be taken advantage of by **impregnating** the components with a fluid; a typical application is to impregnate the sintered part with oil, usually done by immersing the part in heated oil. Bearings and bushings that are internally lubricated, with up to 30% oil by volume, are made by this method. Such components thus have a continuous supply of lubricant during their service life. Universal joints are made with grease-impregnated PM techniques, as a result of which these parts no longer requires grease fittings.
4. **Infiltration** is a process whereby a metal slug, with a lower melting point than that of the part, is placed against the sintered part, and the assembly is heated to a temperature sufficiently high to melt the slug. The molten metal infiltrates the pores of the part by *capillary action*, resulting in a relatively pore-free part with good density and strength. The most common application is the infiltration of iron-base compacts with copper.

   The advantages of infiltration are that hardness and tensile strength are improved and the pores of the part are filled, thus preventing moisture penetration, which otherwise could cause corrosion. Infiltration also may be done using a lead slug whereby, because of the low shear strength of lead, the infiltrated part has lower frictional characteristics than the uninfiltrated one (see Section 4.4.4); some bearing materials are formed in this manner.
5. **Densification**, or **roll densification**, is similar to roller burnishing (Section 4.5.1), where a small-diameter hard roll is pressed against a PM part, developing sufficiently high contact pressures to cause plastic deformation of its surface layers. Thus, instead of cold working the part, the effect is to cause an increase in density, hence densification, of the surface layers (Fig. 11.19). PM gears and bearing races are often treated by roll densification; the surface layers are more fatigue resistant and better able to support higher contact stresses than untreated components.

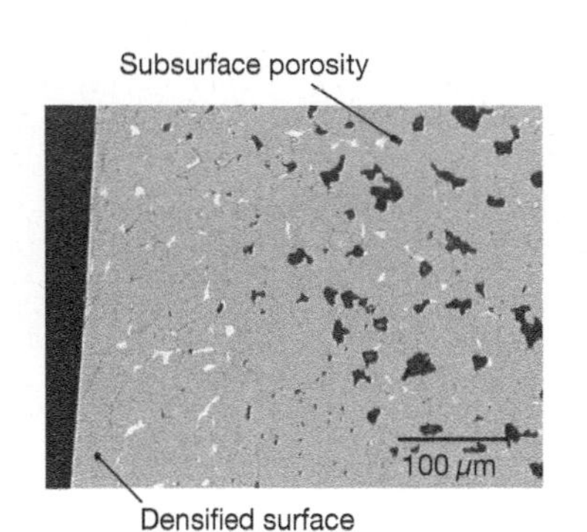

FIGURE 11.19 Micrograph of a PM material surface after roll densification; note the low porosity near the surface, increasing the material's ability to support contact stresses and resist fatigue. *Source:* Courtesy of Capstan Atlantic Corp.

6. PM parts may be subjected to various other operations:

   a. *Heat treating* (quenching and tempering, and steam treating; Section 5.11), for improved strength, hardness, and resistance to wear.
   b. *Machining* (turning, milling, drilling, tapping, and grinding; Chapters 8 and 9), for making undercuts and slots, improving surface finish and dimensional accuracy, and producing threaded holes and other surface features.
   c. *Finishing* (deburring, burnishing, mechanical finishing, plating, and coating), for improved surface characteristics, corrosion and fatigue resistance, and appearance.

CASE STUDY 11.1 Production of Tungsten Carbide for Tools and Dies

Tungsten carbide is an important tool and die material, mainly because of its hardness, strength, and wear resistance over a wide range of temperatures (see Section 8.6.4); it is made by PM techniques. First, powders of tungsten and carbon are blended together in a ball mill or a rotating mixer. The mixture (typically 94% tungsten and 6% carbon, by weight) is heated to approximately 1500°C in a vacuum-induction furnace; as a result, the tungsten is carburized, forming tungsten carbide in a fine powder form. A binding agent (usually cobalt) is then added to the tungsten carbide (with an organic fluid, such as hexane), and the mixture is ball milled to produce a uniform and homogeneous mix. This process can take several hours, or even days.

The mixture is then dried and consolidated, usually by cold compaction, at pressures in the range of 200 MPa (30 ksi). Finally, it is sintered in a hydrogen-atmosphere or a vacuum furnace at a temperature of 1350°C to 1600°C (2500°F to 2900°F), depending on its composition. (Powders may also be hot pressed at the sintering temperature, using graphite dies.) At this temperature, the cobalt is in a liquid phase and acts as a binder for the carbide particles. During sintering, the tungsten carbide undergoes a linear shrinkage of about 16%, corresponding to a volume shrinkage of about 40%; thus, control of size and shape is important for producing tools with accurate dimensions. A combination of other carbides, such as titanium carbide and tantalum carbide, can likewise be produced, using mixtures made by the methods described in this case study.

11.6 Design Considerations for Powder Metallurgy

Because of the unique properties of metal powders and their flow characteristics in a die, and the brittleness of green compacts, certain design principles have been established, as illustrated in Figs. 11.20 through 11.22.

QR Code 11.4 Design guide for powder metallurgy. *Source:* Courtesy of the Metal Powder Industries Federation.

1. The shape of the compact must be kept as simple and as uniform as possible. Features such as sharp changes in contour, thin sections,

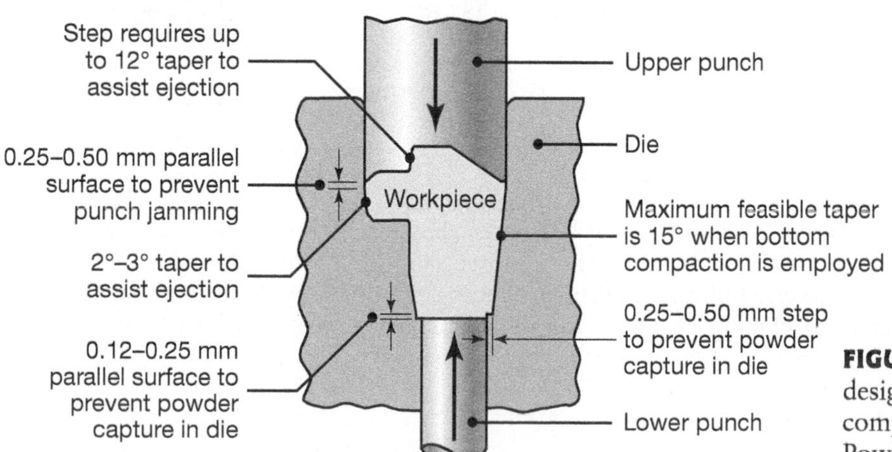

FIGURE 11.20 Die geometry and design features for powder-metal compaction. *Source:* After the Metal Powder Industries Federation.

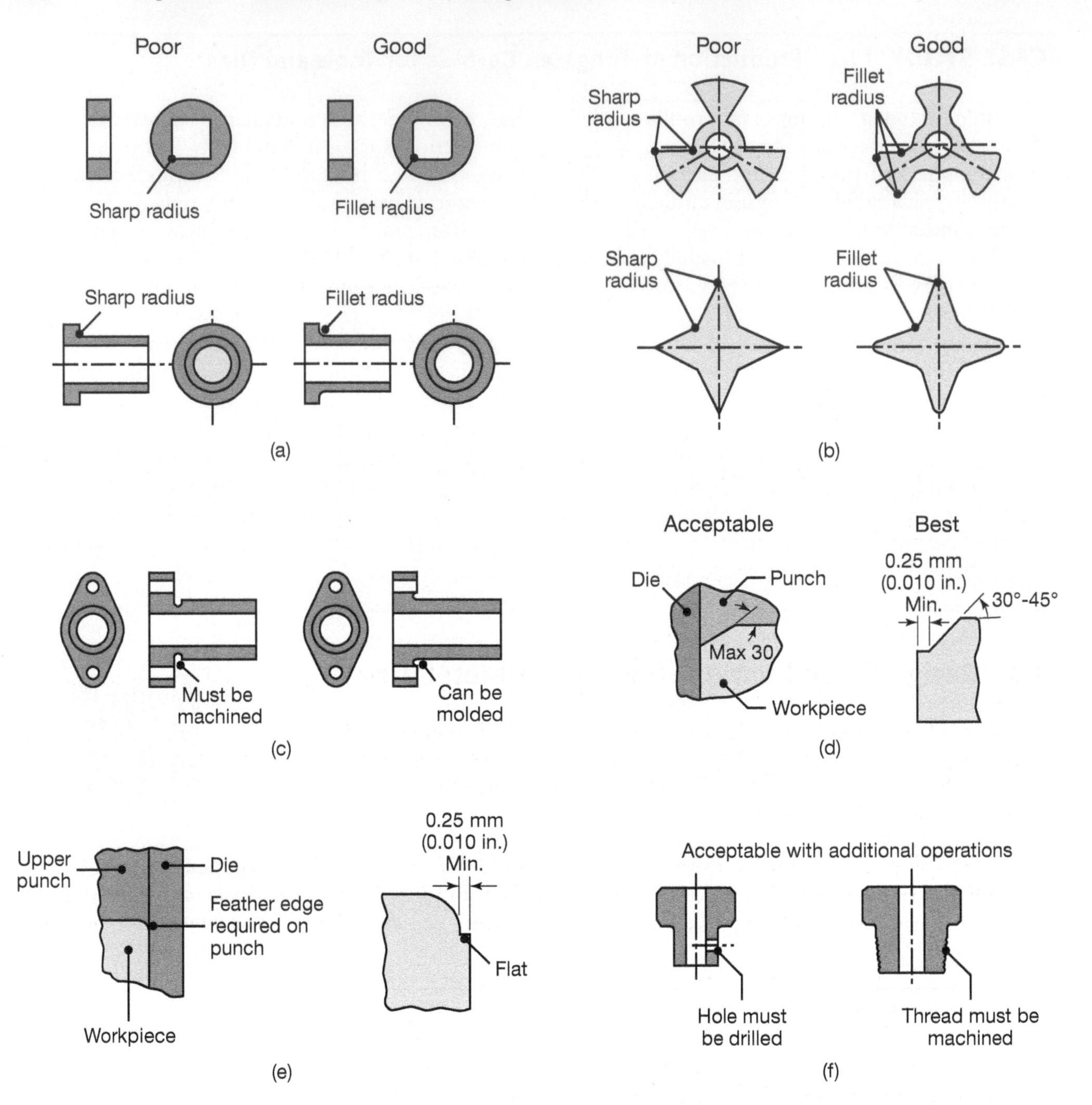

FIGURE 11.21 Examples of PM parts, showing various poor and good designs. Note that sharp radii and reentry corners should be avoided, and that threads and transverse holes have to be produced separately, by additional operations, such as machining or grinding. *Source:* After the Metal Powder Industries Federation.

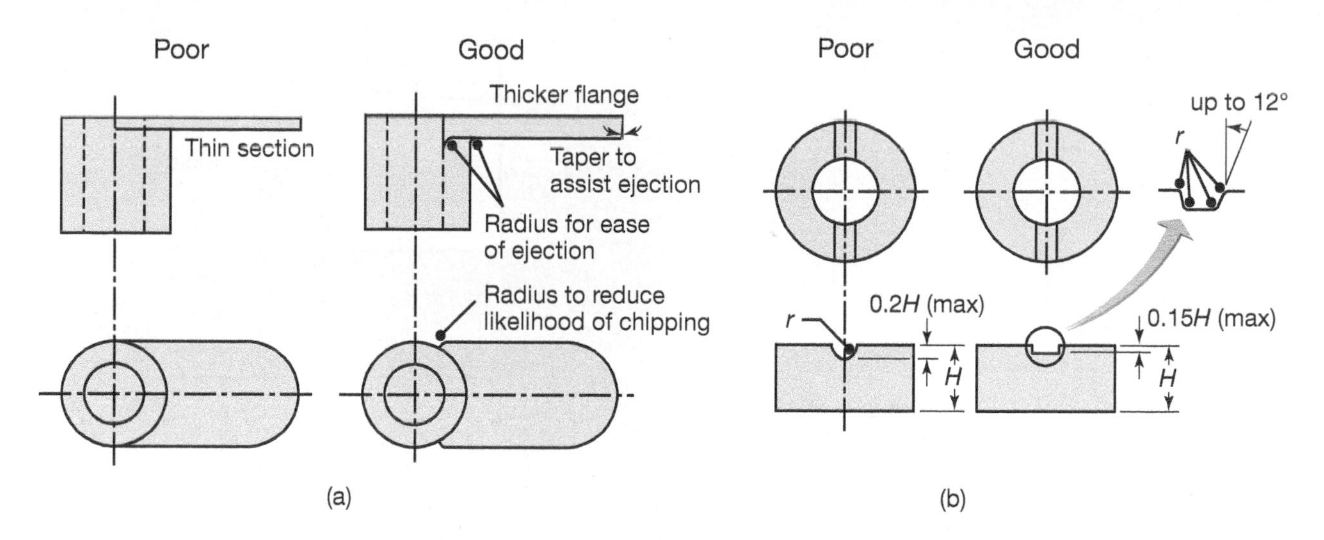

FIGURE 11.22 (a) Design features for use with unsupported flanges. (b) Design features for use with grooves. *Source:* After the Metal Powder Industries Federation.

variations in thickness, and high length-to-diameter ratios should be avoided.

2. Provision must be made for ease of ejection of the green compact from the die without damage; holes or recesses should be parallel to the axis of punch travel. Chamfers should be provided.
3. As is the case for parts made by most other processes, PM parts should be made with the widest dimensional tolerances, consistent with their intended applications, in order to increase tool and die life and to reduce production costs.
4. Generally, part walls should not be less than 1.5 mm thick, although walls as thin as 0.34 mm have been pressed successfully on components 1 mm in length. Walls with length-to-thickness ratios greater than 8:1 can be difficult to press, and density variations are virtually unavoidable.
5. Simple steps in parts can be produced if their size does not exceed 15% of the overall part length. Larger steps can be pressed, but they require more complex, multiple-action tooling (See also Fig. 11.9).
6. Letters and numbers for part identification can be pressed into parts if they are oriented perpendicular to the direction of pressing. Letters can be either raised or recessed; however, raised letters are more susceptible to damage in the green stage and may interfere with proper stacking of parts during sintering.
7. Flanges and overhangs can be produced by a step in the die; however, such protrusions may be damaged during ejection of the green part and may require more elaborate tooling. Also, a long flange should incorporate a draft (see Section 5.12.2) around the flange, a radius at the bottom edge, and a radius at the juncture of the flange and/or component body, to reduce stress concentrations and the likelihood of fracture.

QR Code 11.5 2014 Design Awards. *Source:* Courtesy of the Metal Powder Industries Federation.

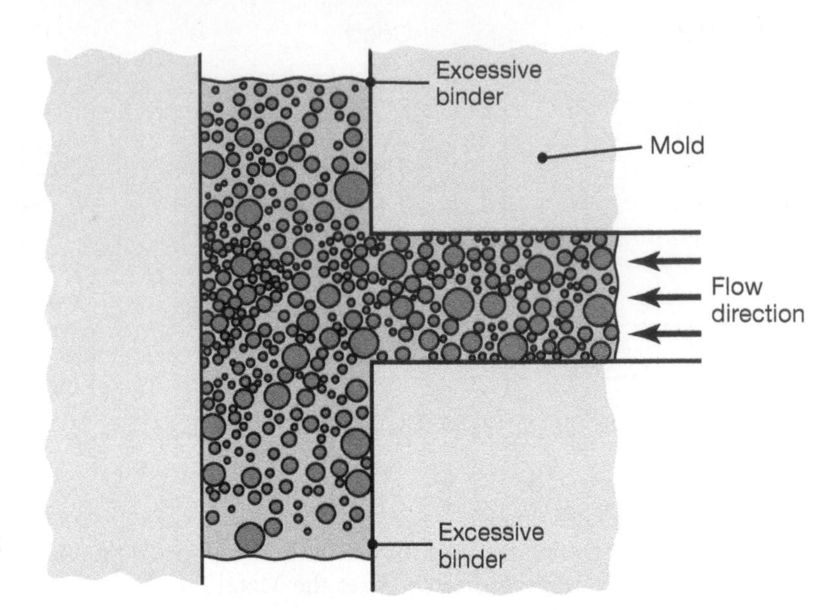

FIGURE 11.23 The use of abrupt transitions in molds for powder injection molding causing non-uniform metal-powder distribution within a part.

8. A true radius cannot be pressed into the edge of a part, because it would require the punch to be feathered (to have a smooth transition) to a zero thickness (Fig. 11.21e). Chamfers or flats are preferred for pressing; a common design approach is to use a 45° angle and 0.25-mm flat (Fig. 11.21d).
9. Features such as keys, keyways, and holes that are used for transmitting torques on gears and pulleys can be shaped during compaction. Bosses (See Fig. 5.35c) can be produced if proper drafts are used and their length is small compared with the overall size of the pressed component.
10. Notches and grooves can be incorporated if they are oriented perpendicular to the pressing direction (Fig. 11.21c). Circular grooves should not exceed a depth of 20% of the overall depth of the component; rectangular grooves should not exceed a depth of 15% of the overall depth of the component (See Fig. 11.22b).
11. Parts made by MIM have design constraints similar to injection molding. Wall thicknesses should be as uniform as possible in order to minimize distortion during sintering. Molds and dies should be designed with smooth transitions, to prevent powder accumulation and to allow uniform distribution of metal powder (Fig. 11.23).
12. Dimensional tolerances of sintered parts are usually on the order of ±0.05–0.1 mm. Tolerances improve significantly with additional operations such as sizing, machining, and grinding (See Fig. 9.28).

11.7 Economics of Powder Metallurgy

Major cost elements in PM parts production are the costs of the powders, compacting dies, and equipment for compacting and sintering. The cost of metal powders per unit weight is much higher (by a factor of about 1.5 to

TABLE 11.6 Competitive features of PM and some other manufacturing processes.

Process	Advantages over PM	Limitations as compared with PM
Casting	Wide range of part shapes and sizes produced; generally low mold and setup cost	Some waste of material in processing; some finishing required; may not be feasible for some high-temperature alloys
Forging (hot)	High production rate of a wide range of part sizes and shapes; high mechanical properties through control of grain flow	Some finishing required; some waste of material in processing; die wear; relatively poor surface finish and dimensional control
Extrusion (hot)	High production rate of long parts; complex cross sections may be produced	Only a constant cross-sectional shape can be produced; die wear; poor dimensional control
Machining	Wide range of part shapes and sizes; short lead time; flexibility; good dimensional control and surface finish; simple tooling	Waste of material in the form of chips; relatively low productivity

7 or even more) than that for molten metal for casting or for wrought bar stock for machining and forming. The cost also depends on the method of powder production, its quality, and the quantity purchased. The least expensive powder metal is iron, followed by, in increasing order, aluminum, zinc, copper, chromium, stainless steel, molybdenum, tungsten, cobalt, niobium, zirconium, and tantalum.

The net- or near-net-shape capability of PM is a highly significant factor because of its cost-effectiveness in numerous applications. However, due to the high cost of punches, dies, and various equipment for PM processing, production volume must be sufficiently high to warrant this major expenditure, typically upwards of 50,000 parts per year. Modern PM equipment is highly automated, and production is ideal for many automotive parts made in millions per year, with the labor cost per part being very low.

Powder metallurgy has become increasingly competitive with cast, forged, extruded, or machined parts (Table 11.6). Many die-pressed and sintered parts are used without the need for any machining, although some may require minor finishing operations. Even in highly complex parts, finish machining of PM parts involves relatively simple operations, such as drilling and tapping holes or grinding some surfaces. Furthermore, an important PM feature is that a single part may replace an assembly of several parts made by various manufacturing processes.

PM forging is used mainly for critical applications in which full density and the accompanying superior fatigue resistance are essential. For specific applications, PM forging competes with processes such as conventional forging and casting, in terms of both product properties and production cost. Automotive connecting rods can be produced by both PM forging and casting. Metal injection molding (MIM) uses finer, and therefore more expensive, powder, and also requires more production steps, such as feedstock preparation, molding, debinding, and sintering. Because of these higher cost elements, MIM is cost effective mainly for small but highly complex parts (generally weighing less than 100 g) required in large quantities.

Relatively large aerospace parts can be produced in small quantities when there are critical property requirements or special metallurgical considerations. For example, some nickel-base superalloys are so highly alloyed that they are prone to segregation during casting (see Section 5.3.3) and can be processed only by PM methods. The powders may first be consolidated by hot extrusion (Section 6.4) and then hot forged (Section 6.2) in dies maintained at a high temperature. All beryllium processing also is based on powder metallurgy, and involves either cold isostatic pressing and sintering or hot isostatic pressing. Many of these materials are difficult to process and may require much finish machining even when the parts are made by PM methods. This is in sharp contrast to the numerous ferrous press-and-sinter PM parts that require practically no finish machining. To a large extent, the special and considerably more expensive PM methods are used for aerospace applications because no competing processes exist.

11.8 Ceramics: Structure, Properties, and Applications

Ceramics are compounds of metallic and nonmetallic elements. The term ceramics refers both to the material and to the ceramic product itself; in Greek, the word *keramos* means potter's clay, and *keramikos* means clay products. Because of the large number of possible combinations of elements, a wide variety of ceramics are now available for numerous consumer and industrial applications.

The earliest use of ceramics was in pottery and bricks, dating back to before 4000 B.C. Ceramics have been used for many years as an electrical insulator and for high-temperature strength; they have become increasingly important materials in heat engines and various other applications (Table 11.7), as well as in tools, dies, and molds. Modern applications of ceramics include cutting tools (Section 8.6), whiteware, tiles for architectural applications, and automotive components, such as exhaust-port liners, coated pistons, and cylinder liners.

Some properties of ceramics are significantly better than those of metals, particularly their hardness and thermal and electrical resistance. Ceramics are available as *single crystal* or in *polycrystalline* form. Grain size has a major influence on the properties of ceramics; the finer the size, the higher is the strength and the toughness (**fine ceramics**). Ceramics are generally divided into the categories of **traditional ceramics** (tiles, brick, pottery, whiteware, and abrasive wheels) and **industrial ceramics**, also called **engineering** or **high-tech ceramics** (heat exchangers, cutting tools, semiconductors, and prosthetics).

TABLE 11.7 Types and general characteristics of ceramics and glasses.

Type	General characteristics
Oxide Ceramics	
Alumina	High hot hardness and abrasion resistance, moderate strength and toughness; most widely used ceramic; used for cutting tools, abrasives, and electrical and thermal insulation
Zirconia	High strength and toughness; resistance to thermal shock, wear, and corrosion; partially-stabilized zirconia and transformation-toughened zirconia have better properties; suitable for heat-engine components
Carbides	
Tungsten carbide	High hardness, strength, toughness, and wear resistance, depending on cobalt binder content; commonly used for dies and cutting tools
Titanium carbide	Not as tough as tungsten carbide, but has a higher wear resistance; has nickel and molybdenum as the binder; used as cutting tools
Silicon carbide	High-temperature strength and wear resistance; used for engines components and as abrasives
Nitrides	
Cubic boron nitride	Second hardest substance known, after diamond; high resistance to oxidation; used as abrasives and cutting tools
Titanium nitride	Used as coatings on tools, because of its low friction characteristics
Silicon nitride	High resistance to creep and thermal shock; high toughness and hot hardness; used in heat engines
Sialon	Consists of silicon nitrides and other oxides and carbides; used as cutting tools
Cermets	Consist of oxides, carbides, and nitrides; high chemical resistance but is somewhat brittle and costly; used in high-temperature applications
Nanophase ceramics	Stronger and easier to fabricate and machine than conventional ceramics; used in automotive and jet-engine applications
Silica	High temperature resistance; quartz exhibits piezoelectric effects; silicates containing various oxides are used in high-temperature, nonstructural applications
Glasses	Contain at least 50% silica; amorphous structure; several types available, with a wide range of mechanical, physical, and optical properties
Glass ceramics	High-crystalline component to their structure; stronger than glass; good thermal-shock resistance; used for cookware, heat exchangers, and electronics
Graphite	Crystalline form of carbon; high electrical and thermal conductivity; good thermal-shock resistance; also available as fibers, foam, and buckyballs for solid lubrication; used for molds and high-temperature components
Diamond	Hardest substance known; available as single-crystal or polycrystalline form; used as cutting tools and abrasives and as die insert for fine wire drawing; also used as coatings

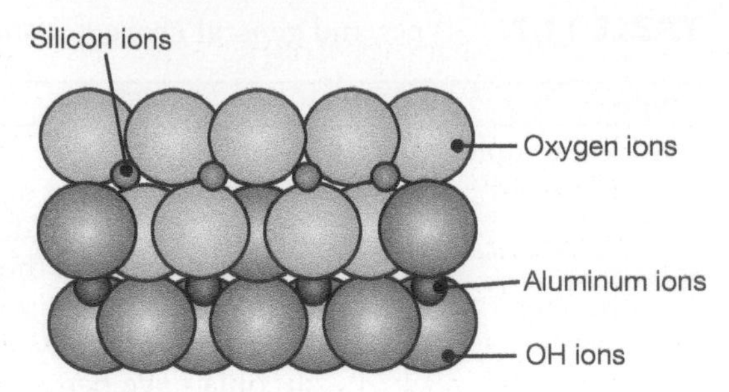

FIGURE 11.24 The crystal structure of kaolinite, commonly known as clay; compare with Figs. 3.2–3.4 for metals.

11.8.1 Structure and Types of Ceramics

The structure of ceramic crystals is among the most complex of all materials, containing various elements of different sizes. The bonding between the atoms is generally *covalent* (electron sharing) and *ionic* (primary bonding between oppositely charged ions). Among the oldest raw materials for ceramics is **clay**, a fine-grained sheetlike structure. The most common example is *kaolinite* (from Kaoling, a hill in China), a white clay, consisting of silicate of aluminum with alternating weakly bonded layers of silicon and aluminum ions (Fig. 11.24). When added to kaolinite, water attaches itself to the layers (*adsorption*), makes them slippery, and gives wet clay its well-known softness and plastic properties (*hydroplasticity*). Other raw materials for ceramics are *flint* (rock of very fine-grained silica, SiO_2) and *feldspar* (a group of crystalline minerals consisting of aluminum silicates, potassium, calcium, or sodium). In their natural state, these raw materials generally contain various impurities which, for reliable performance, have to be removed prior to further processing of the materials into individual products.

The most common types of ceramics are:

1. **Oxide ceramics.**
   Alumina. Also called *corundum* or *emery*, *alumina* (aluminum oxide, Al_2O_3) is the most widely used **oxide ceramic**, either in pure form or as a raw material to be blended with other oxides; it has high hardness and moderate strength. Although alumina exists in nature, it contains varying amounts of impurities and thus possesses nonuniform properties, thus its behavior is unreliable. Aluminum oxide, silicon carbide, and several other ceramics are now almost totally manufactured synthetically, so that its quality can be controlled (see also Section 9.2).

   First made in 1893, *synthetic aluminum oxide* is obtained by the fusion of molten bauxite (an aluminum-oxide ore, the principal source of aluminum), iron filings, and coke in electric furnaces; it is then crushed and graded (by size) by passing the particles through standard screens. Parts made of aluminum oxide are then *cold pressed and sintered* (*white ceramics*). Their properties are further improved by minor additions of other ceramics, such as titanium oxide and titanium carbide. Structures containing various alumina and other oxides are known as *mullite* and *spinel*, and are used as refractory materials for high-temperature applications (see Section 3.11.7). The mechanical and physical properties of alumina are particularly suitable for such applications as electrical and thermal insulation, cutting tools, and abrasives.

Zirconia. *Zirconia* (zirconium oxide, ZrO_2) is white in color and has good toughness and resistance to wear, thermal shock, and corrosion; it also has low thermal conductivity and low friction coefficient. **Partially stabilized zirconia** (PSZ) has high strength and toughness. It is obtained by doping (see Section 13.3) the zirconia with oxides of calcium, yttrium, or magnesium, in a process that forms a ceramic material with fine particles of tetragonal zirconia in a cubic lattice. Typical applications of PSZ include dies for hot extrusion of metals and, as zirconia beads, for grinding and dispersion media for aerospace coatings, automotive primers and topcoats, and fine glossy print on flexible food packaging.

Another important characteristic of PSZ is its coefficient of thermal expansion, which is only about 20% lower than that of cast iron, and its thermal conductivity, which is about one-third that of other ceramics. Consequently, it is very suitable for heat-engine components, cylinder liners, and valve bushings, as they keep the cast-iron engine assembly intact during service (no loosening of components or excessive tightness). Other forms of zirconia include **transformation-toughened zirconia** (TTZ) which, because of dispersed tough phases in the ceramic matrix, has higher toughness than that of PSZ.

2. **Other ceramics.**

Carbides. Typical examples of *carbides* are tungsten and titanium, used as cutting tools and die materials, and silicon carbide, used as abrasives (as in grinding wheels):

a. **Tungsten carbide** (WC) consists of tungsten-carbide particles with cobalt as a binder; the amount of binder has a major influence on the material's properties. Toughness increases with the cobalt content (Fig. 8.31), whereas hardness, strength, and resistance decrease.
b. **Titanium carbide** (TiC) has nickel and molybdenum as the binder, and is not as tough as tungsten carbide.
c. **Silicon carbide** (SiC) has good resistance to wear, thermal shock, and corrosion; it also has a low friction coefficient and retains its strength at elevated temperatures. It is suitable for high-temperature components in heat engines and is also used as an abrasive (Section 9.2). Synthetic silicon carbide is made from silica sand, coke, and small amounts of sodium chloride and sawdust, in a process similar to making synthetic aluminum oxide (see above).

Nitrides. Another important class of ceramics is *nitrides*:

a. **Cubic boron nitride** (cBN) is the second hardest known substance, after diamond, and has applications such as abrasives in grinding wheels and as cutting tools. It does not exist in nature, and was first made synthetically in the 1970s, with techniques similar to those used for making synthetic diamond (Section 11.13.2).

b. **Titanium nitride** (TiN) is used extensively as coatings on cutting tools, improving tool life partly because of its low friction characteristics.

c. **Silicon nitride** (Si_3N_4) has high resistance to creep at elevated temperatures, and a combination of low thermal expansion and high thermal conductivity; hence it resists thermal shock. It is suitable for high-temperature structural applications, such as in automotive engine and gas-turbine components, cam-follower rollers, bearings, sand-blast nozzles, and components for the paper industry.

Sialon. *Sialon* consists of silicon nitride and various proportions of aluminum oxide, yttrium oxide, and titanium carbide (see Section 8.6.8). The word *sialon* is derived from the words *si*licon, *al*uminum, *o*xygen, and *n*itrogen. Sialon has higher strength and thermal-shock resistance than silicon nitride, and is used primarily as a cutting-tool material.

Cermets. *Cermets* are combinations of *cer*amics bonded with a *met*allic phase. Introduced in the 1960s, and also called **black ceramics** or *hot-pressed ceramics*, they combine the high-temperature oxidation resistance of ceramics and the toughness, thermal-shock resistance, and ductility of metals. A common application of cermets is in cutting tools, a typical composition being 70% Al_2O_3 and 30% TiC. Other cermets contain various oxides, carbides, and nitrides, and are developed specifically for high-temperature applications such as nozzles for jet engines and aircraft brakes. Cermets, which can be regarded as composite materials, are used in various combinations of ceramics and metals, bonded by powder-metallurgy techniques (see also Section 11.14).

3. **Silica.** Abundant in nature, *silica* is a polymorphic material (it can have different crystal structures; see Section 3.2). The cubic structure is found in refractory bricks, used for high-temperature furnace applications. Most glasses contain more than 50% silica. The most common form of silica is **quartz**, a hard and abrasive hexagonal crystal, used extensively as oscillating crystals of fixed frequency in communications applications, because it exhibits the *piezoelectric* effect (see Section 3.9.6).

   Silicates are products of the reaction of silica with oxides of aluminum, magnesium, calcium, potassium, sodium, and iron. Examples include clay, asbestos, mica, and silicate glasses. **Lithium aluminum silicate** has very low thermal expansion and thermal conductivity and good thermal shock resistance. However, it has very low strength and fatigue life, and thus it is suitable only for nonstructural applications, such as catalytic converters, regenerators, and heat-exchanger components.

4. **Nanoceramics (nanophase ceramics).** *Nanoceramics* consist of atomic clusters containing a few thousand atoms. They exhibit ductility at significantly lower temperatures than conventional ceramics, and are stronger and easier to fabricate, with fewer flaws. Control of particle size, its distribution, and contamination during processing are important. Applications of nanophase ceramics in the automotive industry include valves, rocker arms, turbocharger rotors, and cylinder liners; they are also used in jet-engine components. Other applications in various stages include coatings, microbatteries, optical filters, very thin capacitors, nanoabrasives for lapping, solar cells, and valves for artificial hearts. Nanocrystalline second-phase particles, on the order of 100 nm or less, and fibers also are used as reinforcement in composites (see Section 11.14). They have enhanced properties such as tensile strength and creep resistance (see also *nanomaterials* in Sections 3.11.9 and 13.19).

11.8.2 General Properties and Applications of Ceramics

Compared with metals, ceramics have the following general characteristics: brittleness, high strength and hardness at elevated temperatures, high elastic modulus, low toughness, low density, low thermal expansion, and low thermal and electrical conductivity. However, because of the wide variety of compositions and grain size, the mechanical and physical properties of ceramics can vary significantly. For example, the electrical conductivity of ceramics can be modified from poor to good, which is the principle behind semiconductors (Section 13.3). Ceramics have a wide range of properties depending also on the presence of defects, surface or internal cracks, impurities, and the particular method of production.

1. **Mechanical properties.** The mechanical properties of several engineering ceramics are given in Table 11.8. Note that, because of their sensitivity to cracks, impurities, and porosity, their strength in tension (transverse rupture strength) is approximately one order of magnitude lower than their compressive strength. Such defects lead to the initiation and propagation of cracks, under tensile stresses, severely reducing tensile strength (see also Section 3.8). Consequently, reproducibility of their properties and reliability (acceptable performance over a specified period of time) of ceramic components are critical aspects in their service life.

   The tensile strength of polycrystalline ceramic parts increases with decreasing grain size and increasing porosity. Common earthenware, for example, has a porosity of 10–15%, whereas the porosity of hard *porcelain* (a white ceramic, composed of kaolin, quartz, and feldspar) is about 3%. An empirical relationship is given by

$$S_{ut} \simeq S_{ut,o} e^{-nP}, \tag{11.5}$$

TABLE 11.8 Approximate range of properties of various ceramics at room temperature.

Material	Symbol	Transverse rupture strength (MPa)	Compressive strength (MPa)	Elastic modulus (GPa)	Hardness (HK)	Density (kg/m^3)
Aluminum oxide	Al_2O_3	140–240	1000–2900	310–410	2000–3000	4000–4500
Cubic boron nitride	cBN	725	7000	850	4000–5000	3480
Diamond	—	1400	7000	830–1000	7000–8000	3500
Silica, fused	SiO_2	—	1300	70	550	—
Silicon carbide	SiC	100–750	700–3500	240–480	2100–3000	3100
Silicon nitride	Si_3N_4	480–600	—	300–310	2000–2500	3300
Titanium carbide	TiC	1400–1900	3100–3850	310–410	1800–3200	5500–5800
Tungsten carbide	WC	1030–2600	4100–5900	520–700	1800–2400	10,000–15,000
Partially stabilized zirconia	PSZ	620	—	200	1100	5800

Note: These properties vary widely, depending on the condition of the material.

where P is the volume fraction of pores in the solid, $S_{ut,o}$ is the tensile strength at zero porosity, and the exponent n ranges between 4 and 7.

The modulus of elasticity is likewise affected by porosity, as given by

$$E \simeq E_o \left(1 - 1.9P + 0.9P^2\right), \tag{11.6}$$

where E_o is the modulus at zero porosity. Equation (11.6) is valid up to 50% porosity.

Unlike most metals and thermoplastics, ceramics generally lack impact toughness and thermal-shock resistance, because of their inherent lack of ductility. In addition to fatigue failure under cyclic loading, ceramics (particularly glasses) exhibit a phenomenon called **static fatigue:** When subjected to a static tensile load over a period of time, these materials may suddenly fail. This phenomenon occurs in environments where water vapor is present. Static fatigue, which does not occur in dry air or a vacuum, has been attributed to a mechanism similar to stress-corrosion cracking of metals (Section 3.8.2).

Ceramic components subjected to tensile stresses in service may be **prestressed,** much like prestressed concrete. Prestressing of shaped ceramic components subjects them to compressive stresses. Methods used for prestressing include (a) heat treatment and chemical tempering (Sections 5.11 and 11.11.2); (b) laser treatment of surfaces; (c) coating with ceramics with different degrees of thermal expansion; and (d) surface-finishing operations, such as grinding, whereby compressive residual stresses are induced on the surfaces. Significant advances have been made in improving the toughness and other properties of ceramics, including **machinable ceramics** (Section 8.5.3).

2. **Physical properties.** Ceramics generally have relatively low specific gravity, ranging from about 3 to 5.8 for oxide ceramics, as compared with 7.86 for iron, and very high melting or decomposition

temperatures. Their thermal conductivity varies by as much as three orders of magnitude, depending on composition, whereas the thermal conductivity of metals varies by one order of magnitude. Also, as with other materials, thermal conductivity decreases with increasing temperature and porosity (because air is a very poor thermal conductor).

The thermal conductivity, k, is related to porosity by

$$k = k_o(1 - P), \tag{11.7}$$

where k_o is the conductivity at zero porosity.

The thermal-expansion characteristics of ceramics are shown in Fig. 11.25. Thermal expansion and thermal conductivity induce thermal stresses that can lead to shock or fatigue (see Section 3.9.5). The tendency for *thermal cracking*, also called *spalling*, (when a piece or a layer from the surface breaks off) is lower, with low thermal expansion and high thermal conductivity. Fused silica, for example, has high thermal-shock resistance, because of its virtually zero thermal expansion.

A common example illustrating the importance of low thermal expansion is heat-resistant ceramics for cookware and flat stove tops; these ceramics can sustain high thermal gradients, from hot to cold and vice versa. The relative thermal expansion of ceramics vs. metals

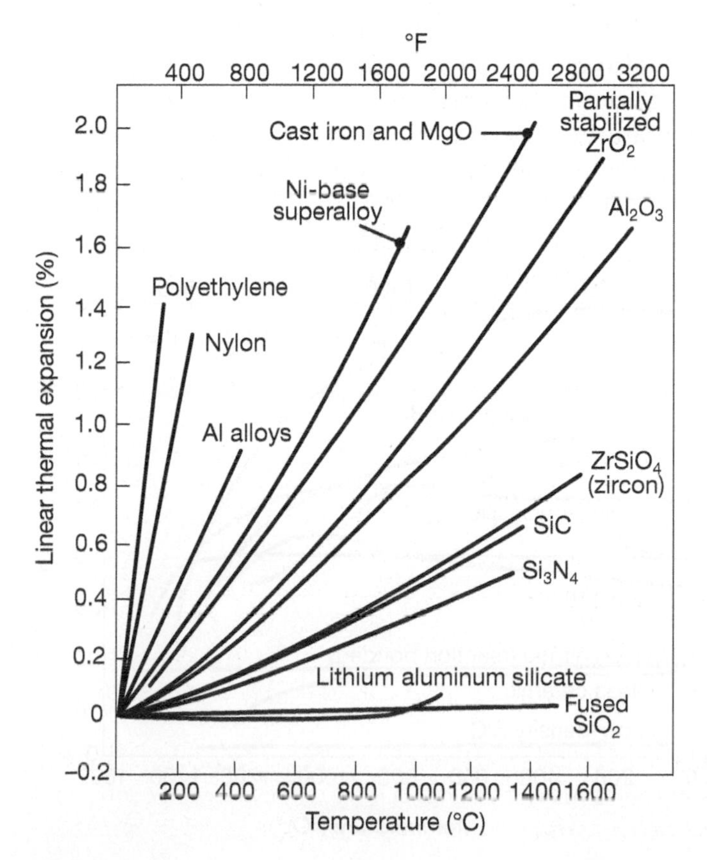

FIGURE 11.25 Effect of temperature on thermal expansion for several ceramics, metals, and plastics. Note that the expansions for cast iron and for partially stabilized zirconia (PSZ) are within about 20%.

is also important in the use of ceramic components in heat engines. The fact that the thermal conductivity of partially stabilized zirconia components is close to that of the cast iron in engine blocks (Fig. 11.25) is a further advantage in the use of PSZ in engines. An additional characteristic, typically exhibited by oxide ceramics, is the **anisotropy of thermal expansion**, whereby thermal expansion varies in different directions of the ceramic. This behavior causes thermal stresses that can lead to cracking of the ceramic component.

The *optical properties* of ceramics depend on their formulations and control of structure, thereby imparting different degrees of transparency and colors. Single-crystal sapphire, for example, is completely transparent, whereas zirconia is white and fine-grained polycrystalline aluminum oxide is a translucent gray. Porosity also influences optical properties, much like trapped air in ice cubes, which makes ice less transparent and gives it a white appearance.

Although ceramics basically are resistors, they can be made *electrically conducting* by alloying them with certain elements (*doping*), thus making them act like a semiconductor or as a superconductor (Section 11.15).

3. **Applications.** As shown in Table 11.7, ceramics have numerous consumer and industrial applications. Several types of ceramics are used in the electrical and electronics industry, because of their high electrical resistivity, dielectric strength, and magnetic properties suitable for applications such as magnets for speakers; an example is *porcelain*. Certain ceramics, such as lead zirconate titanate (PZT) and barium titanate ($BaTiO_3$), also have good *piezoelectric* properties.

   The capability of ceramics to maintain their strength and stiffness at elevated temperatures (Figs. 11.26 and 11.27) makes them very attractive for high-temperature applications. The higher operating

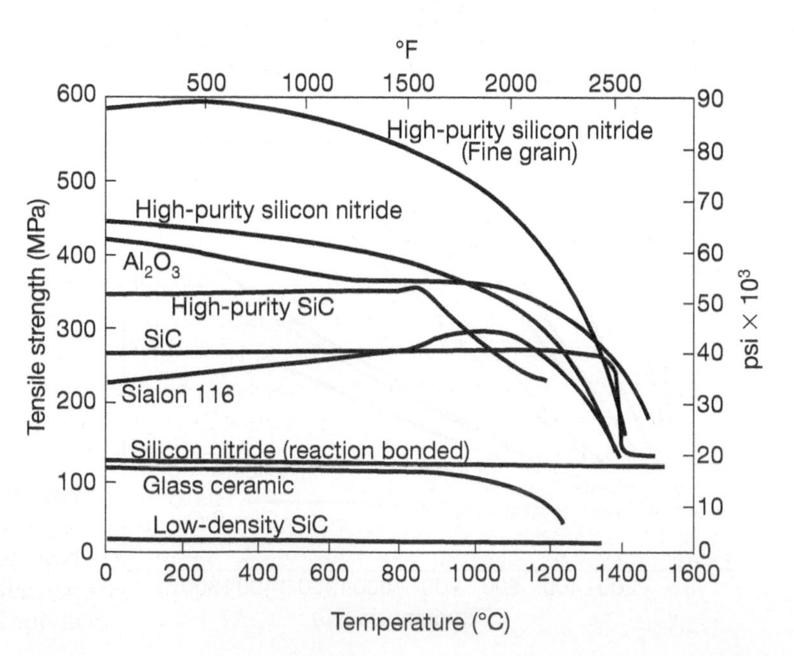

FIGURE 11.26 Effect of temperature on the strength of various engineering ceramics. Note that much of the strength is maintained at high temperatures; compare with Figs. 2.9 and 8.30.

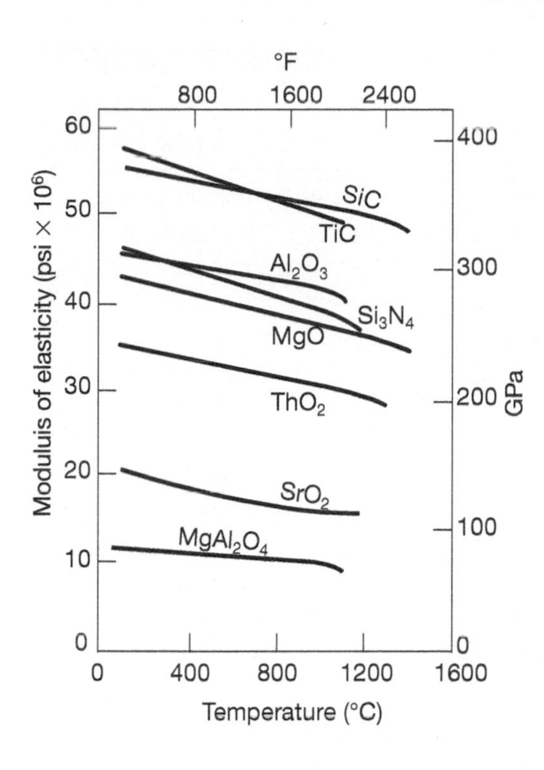

FIGURE 11.27 Effect of temperature on the modulus of elasticity for various ceramics; compare with Fig. 2.9. *Source:* After D.W. Richerson.

temperatures made possible by the use of ceramic components enable more efficient fuel burning and reduced emissions. Internal combustion engines are only about 30% efficient, but with the use of ceramic components, their operating performance can be improved by at least 30%. Ceramics that have been used successfully, especially in gasoline and diesel-engine components and as rotors, are silicon nitride, silicon carbide, and partially stabilized zirconia. Their high resistance to wear also makes them suitable for applications such as cylinder liners, bushings, seals, and bearings. Ceramics are also used as coatings on metal to reduce wear, prevent corrosion, and/or provide a thermal barrier.

The low density and high elastic modulus of ceramics make it possible to reduce engine weight and the inertial forces generated by moving parts. High-speed components for machine tools also are candidates for ceramics, replacing metals; silicon-nitride ceramics are used as ball bearings and rollers in high-speed machines. Their higher elastic modulus also makes them attractive for improving stiffness as well, reducing the tendency for vibration and chatter (Section 8.12), and improving the dimensional accuracy of parts being machined.

Because of their strength and inertness, ceramics are also used as **biomaterials** to replace joints in the human body, as prosthetic devices, and for dental work. Commonly used *bioceramics* include aluminum oxide, silicon nitride, and various compounds of silica. Furthermore, ceramics can be made to be porous, thus allowing bone to grow into its porous surface and develop a strong mechanical bond.

EXAMPLE 11.4 Effect of Porosity on Properties

Given: A fully dense ceramic has the properties of $S_{ut,o} = 100$ MPa, $E_o = 400$ GPa, and $k_o = 0.5$ W/m-K.

Find: What are these properties at 10% porosity? Assume that $n = 5$.

Solution: Using Eqs. (11.5) through (11.7),

$$S_{ut} = 100e^{-(5)(0.1)} = 61 \text{ MPa}$$

$$E = 400\left[1 - (1.9)(0.1) + (0.9)(0.1)^2\right] = 328 \text{ GPa},$$

and

$$k = 0.5(1 - 0.1) = 0.45 \text{ W/m-K}.$$

CASE STUDY 11.2 Ceramic Ball and Roller Bearings

Silicon-nitride ceramic ball and roller bearings are used when high temperature, high speed, or marginally lubricated conditions occur. The bearings can be made entirely from ceramics or just the ball and rollers, in which case they are referred to as hybrid bearings (Fig. 11.28). Examples of machines utilizing ceramic and hybrid bearings include high-performance machine-tool spindles (Section 8.13), metal can seaming heads, and high-speed flow meters.

Ceramic balls have high wear resistance, high fracture toughness, perform well with little or no lubrication, and have low density. The balls have a coefficient of thermal expansion one-fourth that

(a)

(b)

FIGURE 11.28 A selection of ceramic bearings and races. *Source:* Courtesy of Timken, Inc.

of steel, and they can withstand temperatures of up to 1400 °C. The ceramic balls have a diametral tolerance of 0.13 μm and a surface roughness of 0.02 μm. Produced from titanium and carbon nitride by powder-metallurgy techniques, the full-density titanium carbonitride (TiCN) or silicon nitride (Si_3N_4) bearing-grade material can be twice as hard as chromium steel and 40% lighter. Components up to 300 mm in diameter can be produced.

11.9 Shaping Ceramics

Several techniques can be used for shaping ceramics into individual products (Table 11.9). Generally, the procedure involves the following steps: (1) crushing or grinding the raw materials into very fine particles, (2) mixing the particles with additives to impart certain desirable characteristics, and (3) shaping, drying, and firing the material. *Crushing* of the raw materials (also called *comminution* or *milling*) is usually carried out in a ball mill (Fig. 11.29b), either dry or wet. Wet crushing is more effective because it keeps the particles together and prevents the suspension of fine particles in air. The ground particles are then mixed with *additives*, the function of which is one or more of the following:

1. *Binder,* for the ceramic particles;
2. *Lubricant,* for mold release and to reduce internal friction between particles during molding;
3. *Wetting agent,* to improve mixing;
4. *Plasticizer,* to make the mix more plastic and formable;
5. *Deflocculent,* to make the ceramic-water suspension uniform. Typical deflocculents include Na_2CO_3 and Na_2SiO_3 in amounts of less than

TABLE 11.9 General characteristics of ceramics processing methods.

Process	Advantages	Limitations
Slip casting	Large parts; complex shapes; low equipment cost	Low production rate; limited dimensional accuracy
Extrusion	Hollow shapes and small diameters; high production rate	Parts have constant cross section; limited thickness
Dry pressing	Close tolerances; high production rate with automation	Density variation in parts with high length-to-diameter ratios; dies require high abrasive-wear resistance; equipment can be costly
Wet pressing	Complex shapes; high production rate	Limited part size and dimensional accuracy; tooling costs can be high
Hot pressing	Strong, high-density parts	Protective atmospheres required; die life can be short
Isostatic pressing	Uniform density distribution	Equipment can be costly
Jiggering	High production rate with automation; low tooling cost	Limited to axisymmetric parts; limited dimensional accuracy
Injection molding	Complex shapes; high production rate	Tooling costs can be high

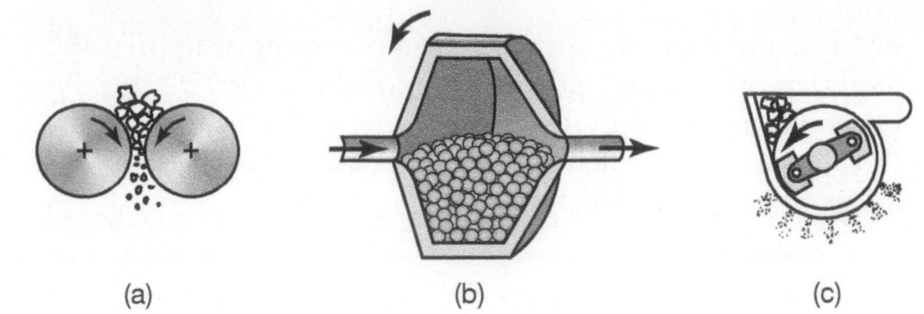

FIGURE 11.29 Methods of crushing ceramics to obtain very fine particles: (a) roll crushing, (b) ball milling, and (c) hammer milling.

1%; they change the electrical charges on the particles of clay so that they repel instead of attract each other. Water is added to make the mixture more pourable and less viscous; and

6. Various agents to control *foaming* and *sintering*.

The three basic shaping processes for ceramics are casting, plastic forming, and pressing.

11.9.1 Casting

The most common casting process is **slip casting**, also called *drain casting* (Fig. 11.30); a *slip* is a suspension of ceramic particles in a liquid, generally water. In this process, the slip is poured into a porous mold made of plaster of paris; the slip must have sufficient fluidity and low viscosity to flow easily into the mold, much like the fluidity of molten metals (Section 5.4.2). After the mold has absorbed some of the water from the outer layers of the suspension, it is inverted, and the remaining suspension is poured out (in making hollow objects, as in slush casting of metals, as illustrated in Fig. 5.11). The top of the part is then trimmed, the mold is opened, and the part is removed.

Large and complex parts, such as plumbing ware, art objects, and dinnerware, can be made by slip casting. Although dimensional control is limited and the production rate is low, mold and equipment costs are also low. Molds for slip casting may consist of several components. Handles for cups and pitchers, for example, are made separately and then joined, using the slip as an adhesive. For solid ceramic parts, the slip is supplied

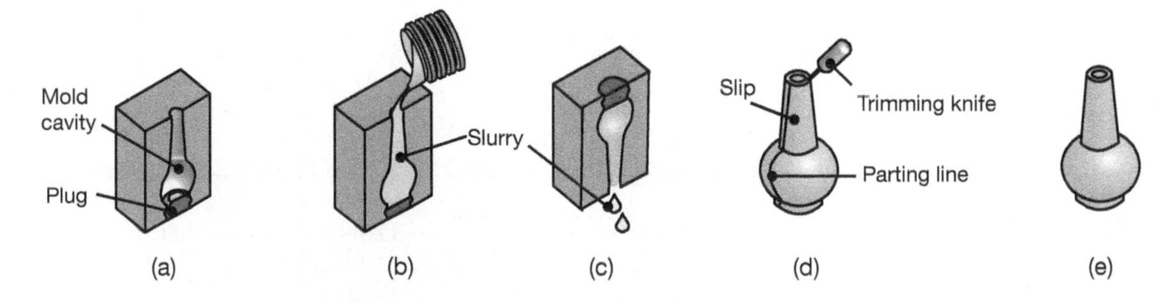

FIGURE 11.30 Sequence of operations in slip casting a ceramic part. After the slip has been poured, the part is dried and fired in an oven to give it strength and hardness. The step in (d) is a trimming operation.

continuously into the mold to replenish the absorbed water, as otherwise the part will shrink. At this stage, the part is described as a soft solid or semirigid. The higher the concentration of solids in the slip, the less water has to be removed. The part, called *green* (as in powder metallurgy), is then fired.

While the ceramic parts are still green, they may be machined to produce certain features or for dimensional accuracy to the parts. Because of the delicate nature of the green compacts, however, machining is usually done manually or with simple tools. The flashing (similar to a flash in forging, shown in Fig. 6.12c) in a slip casting, for example, may be removed gently, with a fine wire brush; features such as holes can be drilled. Operations such as tapping of threads is generally not done on green compacts because warpage, due to firing, makes such machining not viable.

Ceramic sheets with a minimum thickness around 1.5 mm can be made by the **doctor-blade process**. The slip is cast over a moving plastic belt while its thickness is controlled by a blade, as shown in Fig. 11.31. Other flat shaping processes include (a) *rolling* the slip between pairs of rolls, and (b) casting the slip over a paper tape, which is then burned off during firing of the slip.

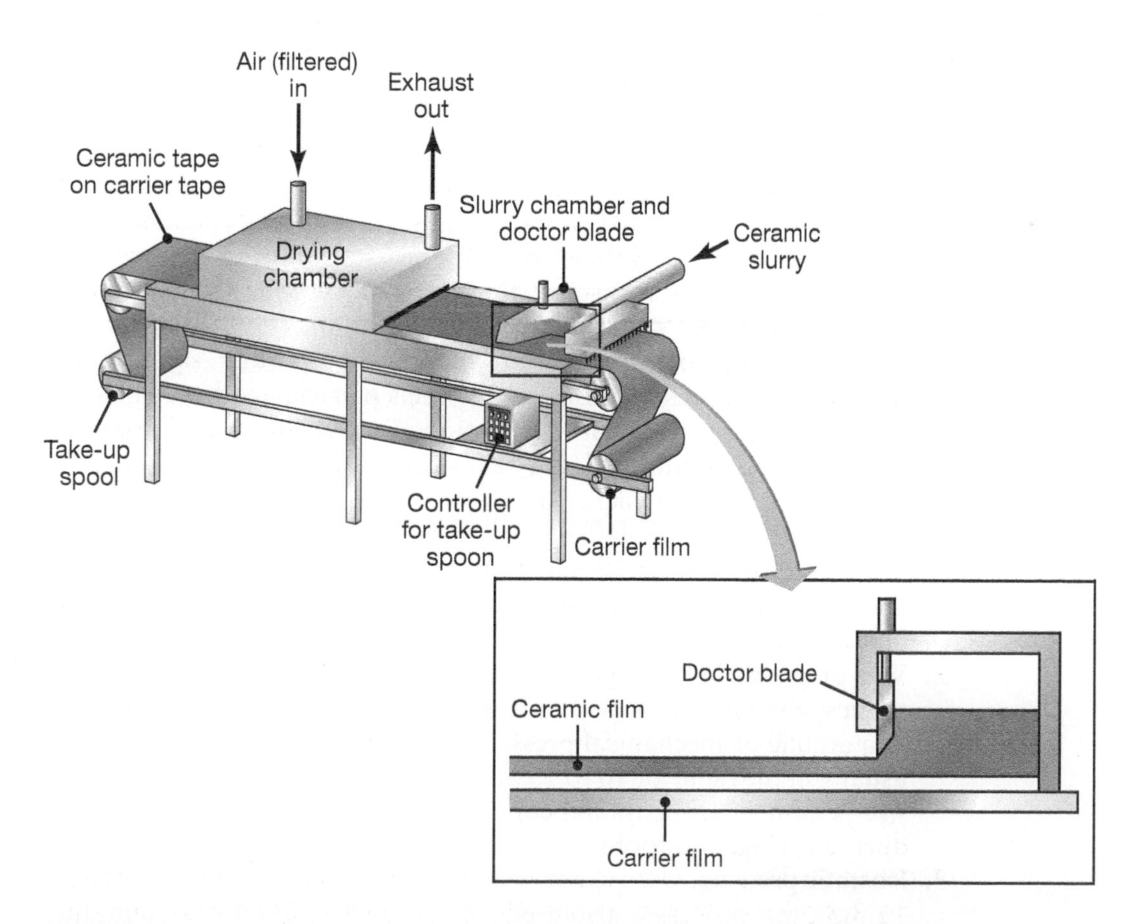

FIGURE 11.31 Production of ceramic sheets through the doctor-blade process.

11.9.2 Plastic Forming

Plastic forming, also called *soft*, *wet*, or *hydroplastic forming*, can be done by various methods, such as extrusion, injection molding, or molding and jiggering (as done on a potter's wheel). Plastic forming tends to orient the layered structure of clays along the direction of material flow (as is the case in metal forming; see Fig. 2.34), which leads to anisotropic behavior of the clay, both in subsequent processing and in the final properties of the ceramic product.

In **extrusion**, the clay mixture, containing 20 to 30% water, is forced through a die opening in screw-type equipment (See, for example, Fig. 10.23). The cross section of the extruded product is constant, and there are limitations on the wall thickness for hollow extrusions. Tooling costs are low and production rates are high.

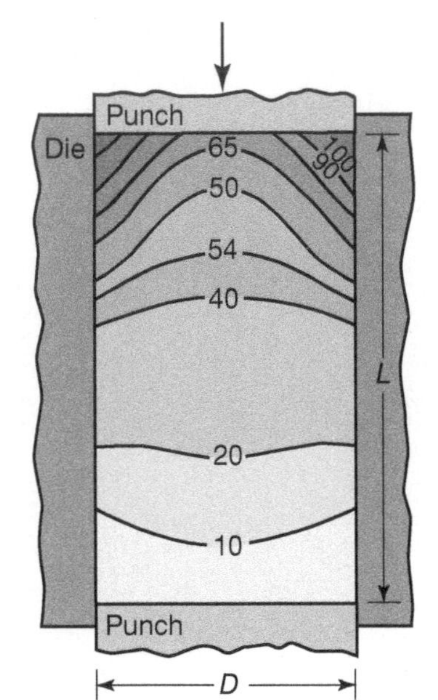

FIGURE 11.32 Density variation in pressed compacts in a single-action press for $L/D = 1.75$. Variation increases with increasing L/D ratio; see also Fig. 11.9e. *Source:* After W.D. Kingery.

11.9.3 Pressing

The various methods of pressing are:

1. **Dry pressing.** Similar to powder-metal compaction, *dry pressing* is used for relatively simple shapes, such as whiteware, refractories, and abrasive products. The process has the same high production rates and close control of dimensional tolerances. The moisture content of the mixture is generally below 4%, but it may be as high as 12%. Organic and inorganic binders, such as stearic acid, wax, starch, and polyvinyl alcohol, are usually added to the mixture; the binders also act as lubricants. The pressing pressure is between 35 and 200 MPa, and presses are highly automated. Dies are usually made of carbides or hardened steel, and must have high wear resistance to withstand the abrasive ceramic particles, and can be expensive.

   The density of dry-pressed ceramics can vary significantly (Fig. 11.32) because of friction between particles and at the mold walls, as in powder-metal compaction (Fig. 11.9). Several methods may be used to minimize density variations. Isostatic pressing reduces density variations, which may cause warping during firing. Warping is particularly severe for parts that have high length-to-diameter ratios; the recommended maximum ratio is 2:1. Vibratory pressing and impact forming may also be used, particularly for nuclear-reactor fuel elements.
2. **Wet pressing.** In this operation, generally used in making intricate shapes, the part is formed in a mold while under high pressure in a hydraulic or mechanical press. The moisture content of the part is usually in the range of 10 to 15%. Production rates are high, but part size is limited. Dimensional control is difficult because of shrinkage during drying, and tooling costs can be high.
3. **Isostatic pressing.** Used extensively in powder metallurgy (see Section 11.3.3), this process is also used for ceramics in order to obtain uniform density throughout the part. Among products made are (a)

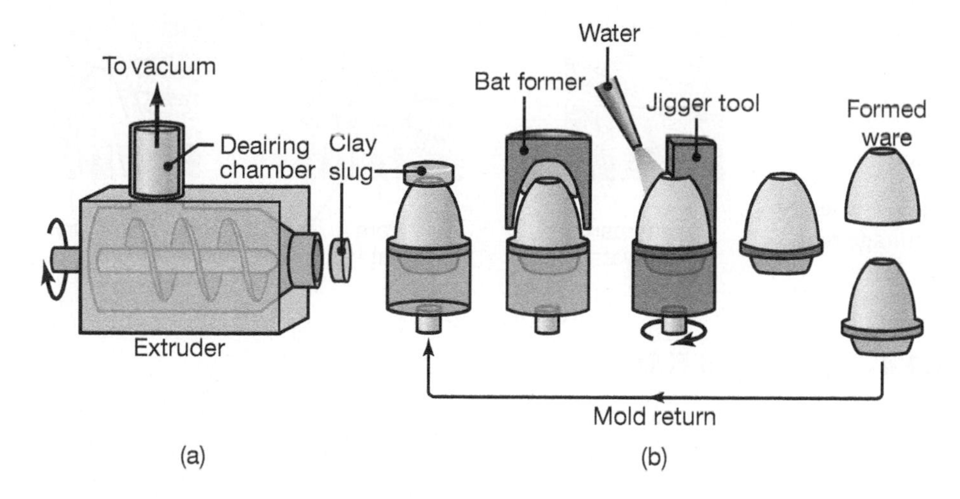

FIGURE 11.33 (a) Extruding and (b) jiggering operations in shaping ceramics. *Source:* After R.F. Stoops.

automotive spark-plug insulators and (b) silicon-nitride vanes for high-temperature use.

4. **Jiggering.** A combination of processes is used to make axisymmetric parts, such as ceramic plates. Clay slugs are first extruded, then formed into a *bat* over a plaster mold, and finally jiggered on a rotating mold (Fig. 11.33). *Jiggering* is an operation in which the clay bat is formed with templates or rollers. The part is then dried and fired. Although production rate is high, the process has limited dimensional accuracy.
5. **Injection molding.** The advantages of *injection molding* of plastics and powder metals were indicated in Sections 10.10.2 and 11.3.4. *Ceramic injection molding* (CIM) is used extensively for *precision forming* of ceramics for high-technology applications, such as in rocket-engine components, piezo-electric scanners, and such medical devices as ultrasonic scalpels. The raw material is mixed with a binder, such as a thermoplastic polymer (polypropylene, low-density polyethylene, ethylene vinyl acetate, or wax). The binder is usually removed by pyrolysis and the part is sintered by firing. Injection molding can produce thin sections, typically less than 10–15 mm, using most engineering ceramics, such as alumina, zirconia, silicon nitride, and silicon carbide. Thicker sections require control of the materials used and processing parameters in order to avoid internal voids and cracks, such as those due to shrinkage.
6. **Hot pressing.** In this operation, also called *pressure sintering*, pressure and temperature are applied simultaneously, thus reducing porosity and making the part denser and hence stronger. *Hot isostatic pressing* (Section 11.3.3) also may be used in this process, particularly to improve the quality of high-technology ceramics. Because of the presence of both pressure and temperature, die life in hot pressing can be shorter than in others processes. Protective atmospheres are usually used, and graphite is a commonly used punch and die material.

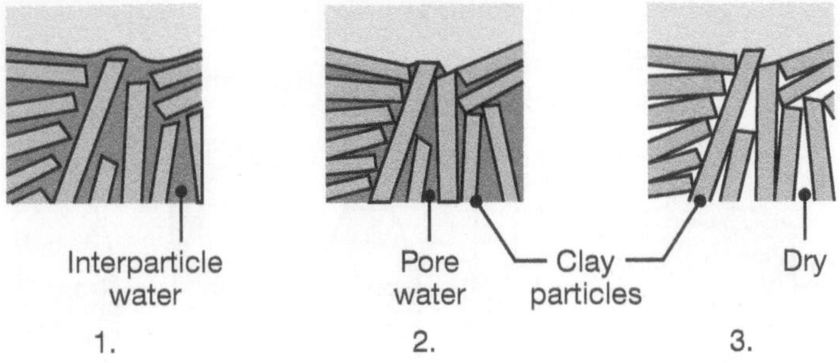

FIGURE 11.34 Shrinkage of wet clay, caused by removal of water during drying; shrinkage may be as much as 20% by volume. *Source:* After F.H. Norton.

11.9.4 Drying and Firing

After a part has been shaped by any of the methods described above, the next step is to dry and fire it to give it strength. *Drying* is a critical stage because of the tendency for the part to warp or crack from variations in moisture content throughout the part, particularly for complex shapes. The control of atmospheric humidity and temperature is thus important.

Loss of moisture results in shrinkage of the part, by as much as 15 to 20% of the original moist size (Fig. 11.34). In a humid environment, the evaporation rate is low, and consequently the moisture gradient across the thickness of the part is lower than that in a dry environment. The low moisture gradient, in turn, prevents a large and uneven shrinkage from the surface to the interior during drying. The dried part, called *green*, can be machined relatively easily at this stage to bring it closer to its final shape, although it must be handled carefully.

Firing, also called *sintering*, involves heating the part to an elevated temperature in a controlled environment, similar to the sintering process in powder metallurgy, and some shrinkage occurs during firing. Firing gives the ceramic part its strength and hardness. The improvement in properties results from (a) development of a strong bond between the complex oxide particles in the ceramic and (b) reduced porosity. Another technology is **microwave sintering** in a furnace operating; the microwaves used have a frequency higher than 2 GHz.

Nanophase ceramics (Sections 11.8.1 and 13.19) can be sintered at lower temperatures than those for conventional ceramics. Nanophase ceramics are easier to fabricate and can be compacted at room temperature to high densities, hot pressed to theoretical density, and formed into net-shape parts without using binders or sintering aids.

EXAMPLE 11.5 Dimensional Changes During Shaping of Ceramic Components

Given: A solid cylindrical ceramic part is to be made, whose final length must be $L = 20$ mm. It has been established that for this material, linear shrinkages during drying and firing are 7% and 6%, respectively, based on the dried dimension L_d.

Find: Calculate (a) the initial length L_o of the part and (b) the dried porosity, P_d, if the porosity of the fired part, P_f, is 3%.

Solution:

a. On the basis of the information given, and remembering that firing is preceded by drying,

$$\frac{L_d - L}{L_d} = 0.06,$$

or

$$L = (1 - 0.06)L_d.$$

Hence,

$$L_d = \frac{20}{0.94} = 21.28 \text{ mm}$$

and

$$L_o = (1+0.07)L_d = (1.07)(21.28) = 22.77 \text{ mm}.$$

b. Since the final porosity is 3%, the actual volume V_a of the ceramic material is

$$V_a = (1 - 0.03)V_f = 0.97V_f,$$

where V_f is the fired volume of the part. Since the linear shrinkage during firing is 6%, the dried volume V_d of the part can be determined as

$$V_d = \frac{V_f}{(1 - 0.06)^3} = 1.2V_f.$$

Hence,

$$\frac{V_a}{V_d} = \frac{0.97}{1.2} = 0.81 = 81\%.$$

Therefore, the porosity P_d of the dried part is 19%.

11.9.5 Finishing Operations

Additional operations may be performed to give a fired part its final shape, remove surface flaws, and improve surface finish and dimensional accuracy. The processes involved are (a) grinding; (b) lapping; (c) ultrasonic machining; (d) electrical-discharge machining; (e) laser-beam machining; (f) abrasive water-jet machining; and (g) tumbling, to remove sharp edges and grinding marks. The selection of a particular process is important in view of the brittle nature of most ceramics and the additional costs involved in these processes. The effect of the finishing operation on the properties of the product also must be considered; because of notch sensitivity of ceramics, the finer the finish, the higher is the part's strength. To improve appearance and strength, and to make them impermeable, ceramic products are often coated with a glaze material (see also Section 4.5.1), which forms a glassy coating after being fired.

11.10 Glasses: Structure, Properties, and Applications

Glass is an *amorphous solid* with the structure of a liquid; it has been *supercooled*, that is, cooled at a rate that is too high for crystals to form. Generally, glass is defined as an inorganic product of fusion that has cooled to a rigid condition without crystallizing. Glass has no distinct melting or freezing point, thus its behavior is similar to that of amorphous polymers (Section 10.2.2). It has been estimated that there are some 750 different types of commercially available glasses today.

The uses of glass range from windows, bottles, and cookware to glasses with special mechanical, electrical, high-temperature, chemical-resistant, corrosion-resistant, and optical characteristics. Special glasses are used in *fiber optics* for communication, with little loss in signal power, and in *glass fibers* with very high strength, for use in reinforced plastics (Section 10.9.2).

All glasses contain at least 50% silica, known as a *glass former*. The composition and properties of glasses, except strength, can be modified by the addition of oxides of aluminum, sodium, calcium, barium, boron, magnesium, titanium, lithium, lead, and potassium. Depending on their function, these oxides are known as *intermediates* or *modifiers*. Glasses are generally resistant to chemical attack, and are also ranked by their resistance to acid, alkali, or water corrosion.

11.10.1 Types of Glasses

Almost all commercial glasses are categorized by type, as shown in Table 11.10:

1. **Soda-lime glass** (the most common type);
2. **Lead-alkali glass;**
3. **Borosilicate glass;**
4. **Aluminosilicate glass;**
5. **96% silica glass;** and
6. **Fused silica.**

Glasses are also classified as colored, opaque (white and translucent), multiform (a variety of shapes), optical, photochromatic (darkens when exposed to light), photosensitive (changes from clear to opaque), fibrous (drawn into long fibers, as in fiberglass), and foam glass or cellular glass (containing air bubbles,). Glasses also are referred to as **hard** or **soft**, usually in the sense of a thermal property rather than a mechanical property (as in hardness). Thus, a soft glass softens at a lower temperature than does a hard glass. Soda-lime and lead-alkali glasses are considered soft, and the rest are hard.

11.10.2 Mechanical Properties

The behavior of glass, as for most ceramics, can be considered as linearly elastic and brittle. The range of elastic modulus for most commercial

TABLE 11.10 General characteristics of various types of glasses.

	Soda-lime glass	Lead glass	Borosilicate glass	Fused glass	96% silica glass
Density	High	Highest	Medium	Low	Lowest
Strength	Low	Low	Moderate	High	Highest
Resistance to thermal shock	Low	Low	Good	Better	Best
Electrical resistivity	Moderate	Best	Good	Good	Good
Hot workability	Good	Best	Fair	Poor	Poorest
Heat treatability	Good	Good	Poor	None	None
Chemicals resistance	Poor	Fair	Good	Better	Best
Impact abrasion resistance	Fair	Poor	Good	Good	Best
Ultraviolet-light transmission	Poor	Poor	Fair	Good	Good
Relative cost	Lowest	Low	Medium	High	Highest

glasses is 55–90 GPa (8–13 Mpsi), and the Poisson's ratio ranges from 0.16 to 0.28. The hardness of glasses, as a measure of resistance to scratching, ranges from 5 to 7 on the Mohs scale (Section 2.6.6), equivalent to a range of approximately 350–500 HK.

In *bulk* form glass has a strength of less than 140 MPa (20 ksi). The relatively low strength is attributed to the presence of small flaws and microcracks on the surface of the glass, some or all of which may be introduced during normal handling of the glass by inadvertent abrading. These defects reduce the strength of glass by two to three orders of magnitude, compared with its ideal (defect-free) strength (Section 3.3.2). Glasses can be strengthened by thermal or chemical treatments to obtain high strength and toughness, as described in Section 11.11.2.

The strength of glass can theoretically reach as high as 35 GPa (5000 ksi). When molten glass is freshly drawn into fibers (*fiberglass*), its tensile strength ranges from 0.2 to 7 GPa (30–1000 ksi), with an average value of about 2 GPa (300 ksi). Glass fibers are thus stronger than steel and are used to reinforce plastics in such applications as boats, automobile bodies, furniture, and sports equipment. The strength of glass is generally measured by bending it (see Section 2.5). The surface of the glass is first thoroughly abraded (roughened) to ensure that the test gives a reliable strength level in actual service under adverse conditions. The phenomenon of *static fatigue* observed in ceramics (Section 11.8.2) is also exhibited by glasses. As a general rule, if a glass item (such as a glass shelf) must withstand a load for 1000 hours or longer, the maximum stress that can be applied to it is approximately one-third the maximum stress that the same item can withstand during the first second of loading.

11.10.3 Physical Properties

Glasses have low thermal conductivity and high dielectric strength. Their thermal expansion coefficient is lower than those for metals and plastics and may even approach zero. For example, titanium-silicate glass, a clear, synthetic high-silica glass and fused silica, a clear, synthetic amorphous silicon dioxide of very high purity both have near-zero coefficient of expansion (Fig. 11.25). Optical properties of glasses, such as reflection, absorption, transmission, and refraction, can be modified by varying their composition and treatment.

11.10.4 Glass Ceramics

Glass ceramics, such as *Pyroceram* (a trade name), contain large proportions of several oxides; hence their properties are a combination of those for glass and ceramics. They have a high-crystalline component to their microstructure and most are stronger than glass. First developed in 1957, glass ceramics are first shaped and then heat treated to cause **devitrification** (recrystallization) of the glass. Unlike most glasses, which are clear, glass ceramics generally are white or gray in color.

The hardness of glass ceramics ranges approximately from 520 to 650 HK. They have a near-zero coefficient of thermal expansion, hence they have good thermal shock resistance. They also have high strength because of the absence of porosity usually found in conventional ceramics, and their properties can be improved by modifying their composition and by heat-treatment techniques. Glass ceramics are suitable for cookware, heat exchangers for gas-turbine engines, radomes (housings for radar antenna), and electrical and electronics applications.

11.11 Forming and Shaping Glass

All forming and shaping processes begin with molten glass, which has the appearance of red-hot viscous syrup, supplied from a melting furnace or tank. Glass products generally can be categorized as:

- **Flat sheet** or **plate**, ranging in thickness from about 0.8 to 10 mm (0.03–0.4 in.), such as window glass, glass doors, and glass tabletops;
- **Rods** and **tubing**, used for chemicals, laboratory glassware, neon lights, and decorative artifacts;
- **Discrete products**, such as bottles, vases, headlights, and television tubes; and
- **Glass fibers**, to reinforce composite materials and for fiber optics.

Flat-sheet glass is made by the *float method*, although it traditionally was made by *drawing* or *rolling* from the molten state; all of these processes are continuous operations. In the **drawing** process, molten glass passes through a pair of rolls. The solidifying glass is squeezed between the rolls, forming a sheet, which is then moved forward over a set of smaller rolls. In the **rolling** method, molten glass is squeezed between rollers, shaping it into a sheet. The surfaces of the glass can be embossed with a pattern by shaping the roller surfaces accordingly. Glass sheet produced by these two processes has a usually rough surface appearance. In making plate glass, both surfaces have to be ground parallel and polished for a smooth appearance.

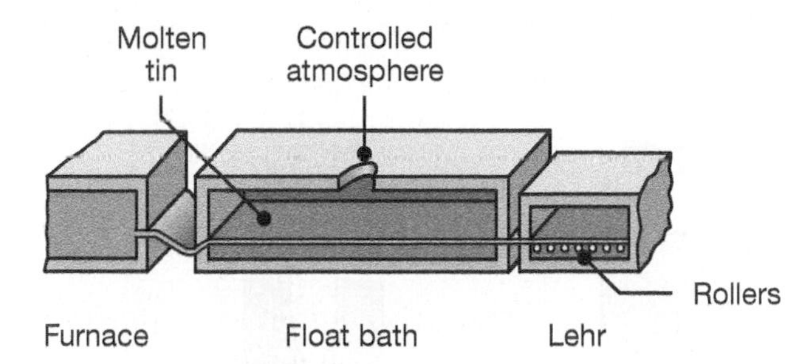

FIGURE 11.35 The float method of forming sheet glass. *Source:* After Corning Glass Works.

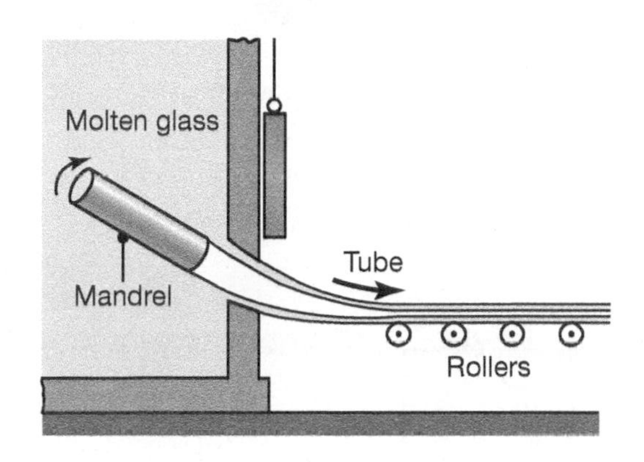

FIGURE 11.36 Continuous manufacturing process for glass tubing. Air is blown through the mandrel to keep the tube from collapsing. *Source:* After Corning Glass Works.

In the **float method** (Fig. 11.35), molten glass from the furnace is fed into a bath of molten *tin*, under controlled atmosphere. The glass floats on the tin bath, then moves over rollers into another chamber (*lehr*), and solidifies. **Float glass** has a smooth (*fire-polished*) surface and needs no further finishing operations.

Glass tubing is manufactured by the process shown in Fig. 11.36. Molten glass is wrapped around a rotating hollow cone-shaped or cylindrical *mandrel*, and is drawn out by a set of rollers; air is blown through the hollow mandrel to keep the glass tube from collapsing. The machines used in this process may be horizontal, vertical, or slanted downward. **Glass rods** are made in a similar manner.

Continuous **fibers** are drawn through multiple (200 to 400) orifices in heated platinum plates, at speeds as high as 500 m/s (1600 ft/s); fibers as small as 2 μm in diameter can be produced by this method. In order to protect their surfaces, the fibers are subsequently coated with a chemical such as silane. Short glass fibers, used as thermal insulating material (**glass wool**) or for acoustic insulation, are made by a *centrifugal spraying process* in which molten glass is ejected (*spun*) from a rotating head. Fiber diameters are typically in the range of 20 to 30 μm.

11.11.1 Manufacture of Discrete Glass Products

Several processes are used for making discrete glass objects, including blowing, pressing, centrifugal casting, and sagging.

Blowing. This process is used to make hollow, thin-walled glass items, such as bottles and flasks, and is similar to blow molding of thermoplastics (Section 10.10.3). The steps involved in the production of an ordinary

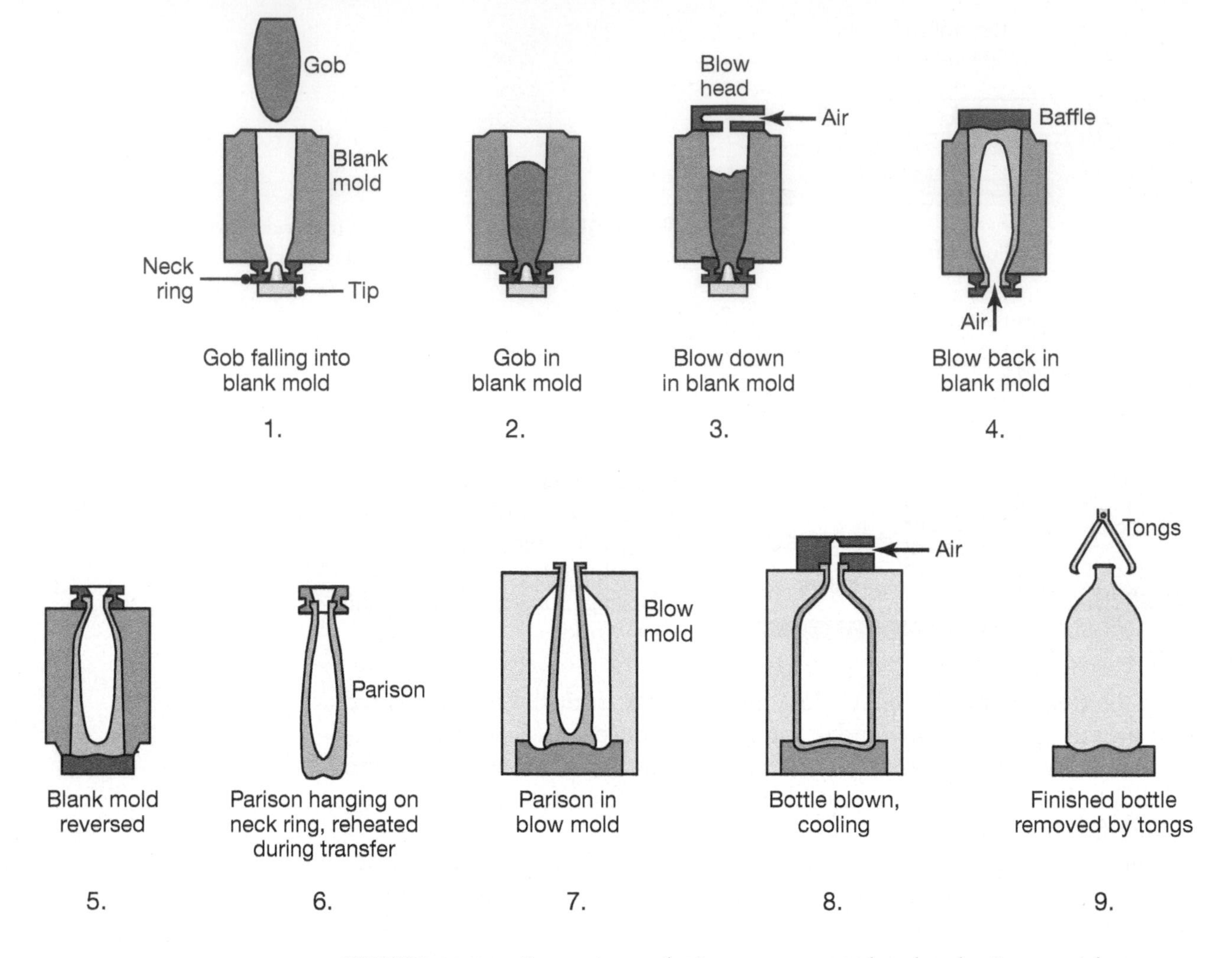

FIGURE 11.37 Stages in producing a common glass bottle. *Source:* After F.H. Norton.

glass bottle by the blowing process are illustrated in Fig. 11.37. Blown air expands a *gob* of heated glass against the inner walls of a mold, which usually are coated with a *parting agent* (such as oil or emulsion) to prevent the glass part from sticking. After shaping, the two halves of the mold are opened and the product is removed. The surface finish of products made by this process is acceptable for most applications. Although it is difficult to control the wall thickness of the product, blowing is used for high rates of production. Bottles made by this process have a typical visible parting line.

Pressing. In this operation, a gob of molten glass is placed in a mold and then pressed into shape using a shaped plunger. The mold may be made in one piece (Fig. 11.38) or it may be a split mold (Fig. 11.39). After pressing, the solidifying glass acquires the shape of the cavity between the mold and the plunger. Because of its confined environment, the product

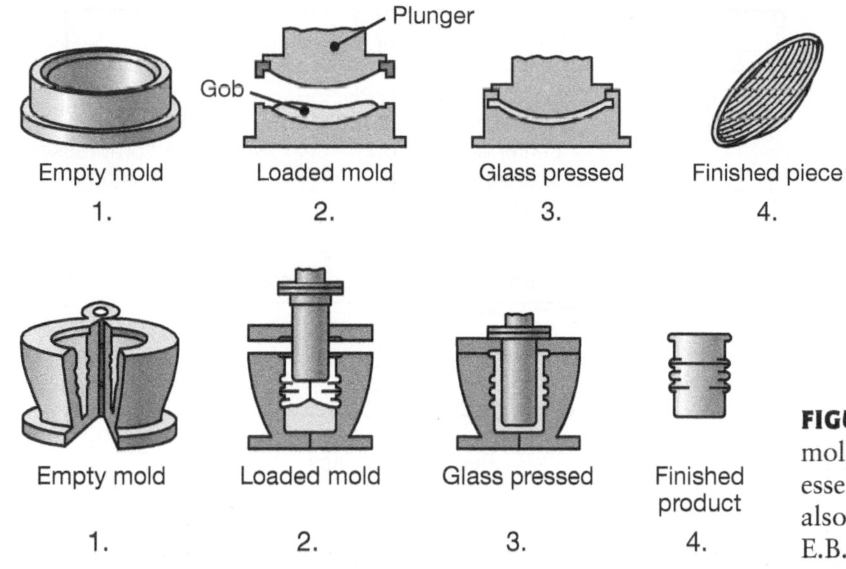

FIGURE 11.38 Steps for producing a glass item by pressing in a mold. *Source:* After Corning Glass Works.

FIGURE 11.39 Pressing glass in a split mold. Note that the use of a split mold is essential to be able to remove the part; see also Figs. 10.35 and 10.38. *Source:* After E.B. Shand.

has higher dimensional accuracy than can be obtained with the blowing method. Pressing cannot be used on thin-walled items or for parts, such as bottles, from which the plunger cannot be retracted.

Centrifugal casting. Also known as *spinning*, this process is similar to that for metals, as described in Section 5.10.4. The centrifugal force pushes the molten glass against the cool mold walls, where it solidifies. Typical products made include curved glass screens and lenses for large telescopes.

Sagging. Shallow, dish shaped or lightly embossed glass parts can be made by this process. A sheet of glass is placed over the mold and is heated, whereby the glass sags by its own weight and takes the shape of the mold cavity. The operation is similar to thermoforming of thermoplastic sheets (Fig. 10.37) but without applying pressure or a vacuum. Typical parts made include dishes, lenses for sunglasses, mirrors for telescopes, and lighting panels.

11.11.2 Techniques for Treating Glass

Glass can be strengthened by thermal tempering, chemical tempering, and laminating; glass products may also be subjected to annealing and other finishing operations.

1. **Thermal tempering.** Also called *physical tempering* or *chill tempering*, the surfaces of the hot glass are cooled rapidly (Fig. 11.40). As a result, the cooler surfaces shrink, and because the bulk is still hot, tensile stresses develop on the surfaces. Then, as the bulk of the glass begins to cool, it contracts, and the solidified surfaces are forced to contract. As a result, residual compressive stresses develop on the outer surfaces while the interior develops tensile stresses. Compressive surface stresses improve the strength of the glass, as they do in other materials. Recall also that the higher the coefficient of thermal

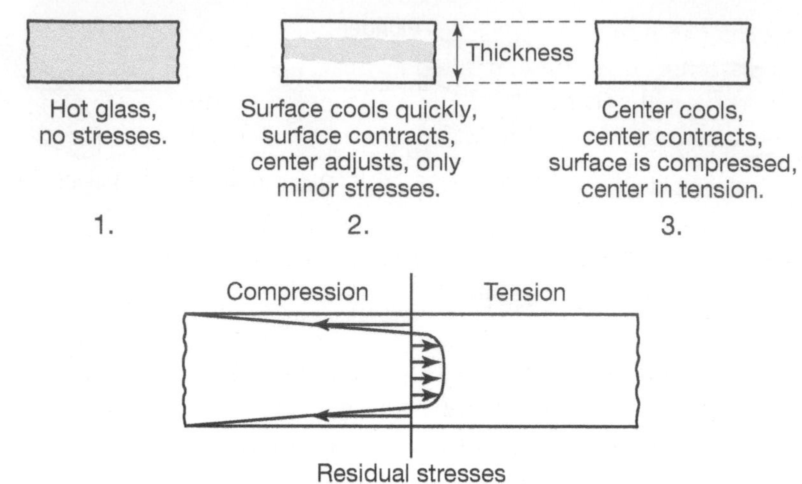

FIGURE 11.40 Stages in the development of residual stresses in tempered glass plate.

expansion of the glass and the lower its thermal conductivity, the higher the level of residual stresses developed and the stronger the glass becomes. Thermal tempering takes a relatively short time (minutes) and can be applied to most glasses. Because of the large amount of energy stored from residual stresses, **tempered glass** shatters into a large number of pieces when broken.

2. **Chemical tempering.** In this operation, the glass is heated in a bath of molten KNO_3, K_2SO_4, or $NaNO_3$, depending on the type of glass, whereby ion exchange takes place, with larger atoms replacing the smaller atoms on the surface of the glass. As a result, residual compressive stresses develop on the surface, a situation similar to forcing a wedge between two bricks in a wall. The time required for chemical tempering is about one hour, significantly longer than for thermal tempering. Chemical tempering may be performed at various temperatures; at low temperatures, distortion of the part is minimal and complex shapes can be treated. At elevated temperatures, there may be some distortion of the part, but the product can then be used at higher temperatures without loss of strength.
3. **Laminated glass.** This is a strengthening method in which two pieces of flat glass are assembled with a thin sheet of tough plastic, such as polyvinyl butyral (PVB), between them, hence the process is also called **laminate strengthening**. When laminated glass is broken, its pieces are held together by the plastic sheet, a phenomenon that can be observed on a shattered automobile windshield.

Finishing operations. Glass products may be subjected to further operations, such as cutting, drilling, grinding, and polishing, to meet specific product requirements. Sharp edges and corners can, as in glass tops for desks and shelves, be smoothened by (a) grinding or (b) holding a torch against the edges. Called **fire polishing**, this operation causes localized softening, resulting in edge rounding (due to surface tension).

As in metal products, residual stresses also can develop in glass products if they are not cooled at a sufficiently low rate. In order to ensure that the product is free from these stresses it is *annealed*, by a process similar to stress-relief annealing of metals (Section 5.11.4). The glass is heated to a certain temperature and then cooled gradually. Depending on the size, thickness, and type of glass, annealing times may range from a few minutes to as long as 10 months, as in the case of a 600-mm mirror for a telescope.

11.12 Design Considerations for Ceramic and Glass Products

Ceramic and glass products require careful selection of composition, processing methods, and finishing operations, as well as methods of assembly into other metallic or nonmetallic components. Material limitations, such as general lack of tensile strength, sensitivity to external and internal defects, and low impact toughness, are important considerations. These limitations have to be balanced against desirable characteristics, such as hardness, resistance to scratching, compressive strength at room and elevated temperatures, and physical properties. Control of processing parameters and the type and level of impurities in the raw materials are also important. Various other factors also should be considered, including the number of parts required and the costs of tooling, equipment, and labor.

The possibilities of dimensional changes, warping, and cracking during processing of these materials are significant factors in selecting methods for shaping them. When a ceramic or a glass component is a component of a larger assembly, its compatibility with other components is another important consideration, particularly thermal expansion (as in seals) and the type of loading.

11.13 Graphite and Diamond

11.13.1 Graphite

Graphite is a crystalline form of carbon, with a *layered structure* of basal planes or sheets of close-packed carbon atoms. Although brittle, graphite has high electrical and thermal conductivity and resistance to thermal shock and high temperature. However, it begins to oxidize at 500°C. Graphite is an important material for applications, such as electrodes, brushes for motors, heating elements, high-temperature fixtures and furnace parts, crucibles for melting metals, molds for casting of metals (See, for example, Fig. 5.21), and seals (because of its low friction and high wear resistance). Unlike other materials, the strength and stiffness of graphite increases with temperature.

The low thermal-neutron-absorption cross section and high scattering cross section make graphite also suitable for nuclear applications. Another characteristic of graphite is its resistance to chemicals, making it also useful as filters for corrosive fluids. An important use of graphite is as **fibers** in composite materials and reinforced plastics, as described in Section 10.9.2.

Because of its layered structure, graphite is weak when sheared along the layers. This characteristic, in turn, gives graphite its low frictional properties as a solid lubricant (Section 4.4.4), although its frictional properties are low only in an environment of air or moisture. Graphite is abrasive and a poor lubricant in a vacuum.

Graphene. A single layer of graphite is known as **graphene**, and has many unique properties. It is the strongest material ever tested, with a reported tensile strength of 130 GPa (19 Mpsi) and elastic modulus of 1000 GPa (145 Mpsi). Thermal conductivity has not been accurately measured, but graphene is thought to be very valuable for temperature control of micro and nano electronics. The commercial applications of graphene have been limited, although it has been suggested that graphene can be used for electronics, filtration, and energy storage applications. When produced as a powder and dispersed in a polymer, graphene is thought to be useful for advanced composites, coatings, capacitors and batteries, energy storage, and other applications. Graphene has been used in small concentrations in a 3D Printing material referred to as *Graphenite*, where a 0.03% graphene added to a polymer material resulted in up to 50% improvements in compressive strength. Graphene's use is still rare in industry, but it remains a material with significant research interest.

Carbon nanotubes. Carbon nanotubes can be thought of as tubular forms of graphene, and are of interest for the development of nanoscale devices (see Section 13.19). Nanotubes are produced by laser ablation of graphite, carbon-arc discharge, and, most often, by chemical vapor deposition (Section 4.5.1). They can be single-walled (SWNTs) or multi-walled (MWNTs), and can be doped with various species.

Carbon nanotubes have exceptional strength, thus making them attractive as reinforcing fibers for composite materials. However, because they have very low adhesion with most materials, delamination with a matrix can limit their reinforcing effectiveness. It is difficult to disperse nanotubes properly because they have a tendency to clump and this limits their effectiveness as reinforcement. A few products have used carbon nanotubes, such as bicycle frames, specialty baseball bats, golf clubs, and tennis racquets. Nanotubes provide only a fraction of the reinforcing material (by volume), with graphite fibers playing the major role.

An additional characteristic of carbon nanotubes is their very high electrical current carrying capability; they can be made as semiconductors or conductors, depending on the orientation of the graphite in the nanotube (See Fig. 11.41). Armchair nanotubes are theoretically capable of carrying a current density higher than 1,000 times that for silver or copper, making them attractive for electrical connections in nanodevices (see Section 13.19). Carbon nanotubes have also been incorporated into polymers to improve their static-electricity discharge capability, especially in fuel lines for automotive and aerospace applications.

Among the numerous potential uses for carbon nanotubes are storage of hydrogen for use in hydrogen-powered vehicles, flat-panel displays,

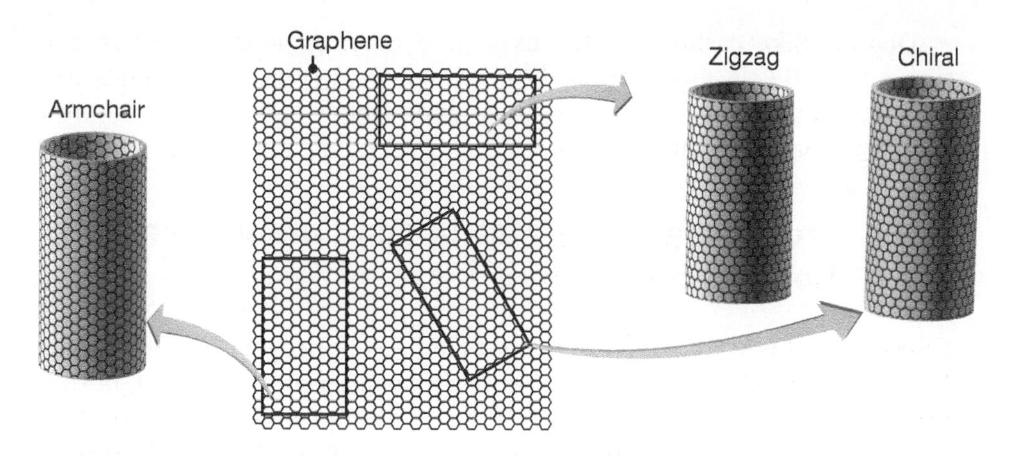

FIGURE 11.41 Forms of carbon nanotubes produced from a section of graphene: armchair, zigzag, and chiral. Armchair nanotubes are noteworthy for their high electrical conductivity, whereas zigzag and chiral nanotubes are semiconductors.

tissue engineering, electrical cables for nano-scale circuitry, catalysts, and X-ray and microwave generators. Highly sensitive sensors using aligned carbon nanotubes are now being developed for detecting deadly gases, such as sarin.

11.13.2 Diamond

A principal form of carbon is *diamond*, which has a covalently bonded structure and is the hardest substance known (7000 to 8000 HK). Diamond is brittle and it begins to decompose in air at about 700°C; in nonoxidizing environments, however, it resists higher temperatures. The high hardness of diamond makes it an important material (a) for cutting tools (Section 8.6.9), either as a single crystal or in polycrystalline form; (b) as an abrasive in grinding wheels (Section 9.2) and for dressing of grinding wheels (see Section 9.5.1); and (c) as a die material for drawing thin wire (Section 6.5.3), with a diameter of less than 0.06 mm (0.0025 in.).

Synthetic or **industrial diamond** was first made in 1955; a principal method is to subject graphite to a hydrostatic pressure of 14 GPa (2000 ksi) and a temperature of 3000°C (5400°F). Synthetic diamond is identical to natural diamond, but has superior properties because of its absence of impurities. It is available in various sizes and shapes, the most common abrasive grain size being 0.01 mm in diameter. Diamond particles can also be *coated* (with nickel, copper, or titanium) for improved performance in grinding operations. *Gem-quality synthetic diamond* has an electrical conductivity 50 times higher than that for natural diamond and 10 times more resistance to laser damage in optical applications. Its applications include heat sinks for computers, in the telecommunications and integrated-circuit industries, and windows for high-power lasers.

Diamondlike carbon (DLC), used as a coating, is described in Section 4.5.1. DLC is widely applied for its high wear resistance; applications include cutting tools, razors for shaving, and high-performance automotive engine components.

11.14 Processing Metal-Matrix and Ceramic-Matrix Composites

Composite materials are continually being developed, with a wide range and form of polymeric, metallic, and ceramic materials being used, both as fibers and as matrix materials, with the aim of improving strength, toughness, stiffness, resistance to high temperatures, and reliability in service, particularly under adverse environmental conditions.

11.14.1 Metal-Matrix Composites

The advantage of a metal matrix over a polymer matrix is its higher resistance to elevated temperatures and higher ductility and toughness; the limitations are higher density and greater difficulty in processing. The matrix materials in these composites are usually aluminum, aluminum-lithium, magnesium, and titanium. The fiber materials typically are graphite, aluminum oxide, silicon carbide, and boron, with beryllium and tungsten as other possibilities. An important consideration is the proper bonding of fibers to the metal matrix.

Because of their high specific stiffness, light weight, and high thermal conductivity, boron fibers in an aluminum-matrix, for example, are used for structural tubular supports in manned satellites. Other applications include bicycle frames, sporting goods, stabilizers for aircraft and helicopters gas turbines, electrical components, and various structural components (See Table 11.11).

TABLE 11.11 Metal-matrix composite materials and typical applications.

Fiber	Matrix	Typical applications
Graphite	Aluminum	Satellite, missile, and helicopter structures
	Magnesium	Space and satellite structures
	Lead	Storage-battery plates
	Copper	Electrical contacts and bearings
Boron	Aluminum	Compressor blades and structural supports
	Magnesium	Antenna structures
	Titanium	Jet-engine fan blades
Alumina	Aluminum	Superconductor restraints in fusion power reactors
	Lead	Storage-battery plates
	Magnesium	Helicopter transmission structures
Silicon carbide	Aluminum, titanium	High-temperature structures
	Superalloy (cobalt base)	High-temperature engine components
Molybdenum, tungsten	Superalloy	High-temperature engine components

There are three principal methods of manufacturing metal-matrix composites into near-net-shape parts.

1. **Liquid-phase processing** basically consists of casting together the liquid matrix and the solid reinforcement, using either conventional casting processes or pressure-infiltration casting techniques. In the latter process, pressurized gas is used to force the liquid matrix metal into a preform (usually as sheet or wire) made of the reinforcing fibers.
2. **Solid-phase processing** involves powder-metallurgy techniques, including cold and hot isostatic pressing. Proper mixing of the constituents for homogeneous distribution of the fibers throughout the matrix is important. An application of this technique, employed in tungsten-carbide tool and die manufacturing with cobalt as the matrix material, is described in Case Study 11.3. In making complex MMC parts, with whisker or fiber reinforcement, die geometry and control of process variables are very important in ensuring proper distribution and orientation of the fibers within the part. MMC parts made by powder-metallurgy processes are subsequently heat treated for optimum properties.
3. **Two-phase processing** consists of *rheocasting* (Section 5.10.6), and *spray atomization and deposition*. In the latter, the reinforcing fibers are mixed with a matrix that contains both liquid and solid phases.

CASE STUDY 11.3 Aluminum-Matrix Composite Brake Calipers

One of the trends in automobile design and manufacture is the increased use of lighter weight designs in order to realize improved performance and/or fuel economy. This can be seen in the development of metal-matrix composite brake calipers. Traditional brake calipers are made from cast iron, and each can weigh about 3 kg in a small car, and up to 14 kg each in a truck. The cast-iron caliper

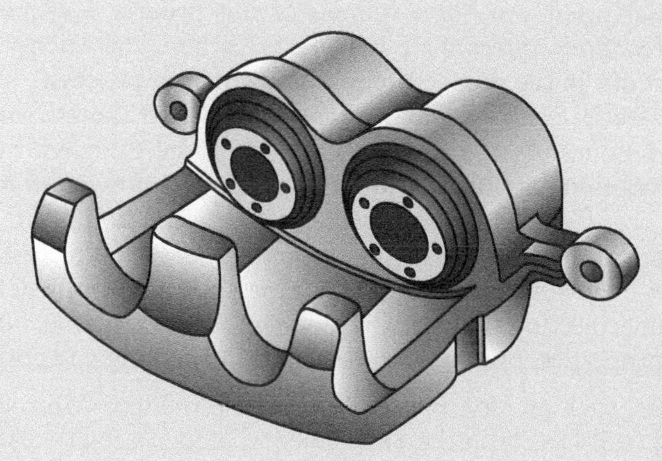

FIGURE 11.42 A rendering of aluminum-matrix composite brake caliper, using nanocrystalline alumina-fiber reinforcement.

could be completely redesigned using aluminum to achieve weight savings, but this would require a larger volume of material, and the space available between the wheel and rotor is highly constrained.

A different brake caliper (Fig. 11.42) uses an aluminum alloy reinforced with precast composite inserts using continuous ceramic fiber. The fiber is nanocrystalline alumina (S_{ut} = 3100 MPa and density = 3.9 g/cm^3), with diameter of 10–12 μm, and fiber volume fraction of 65%. The resulting metal-matrix composite has a tensile strength of 1500 MPa, and a density of 3.48 g/cm^3. Finite-element analysis confirmed that the design exceeded minimum design requirements and matched deflections of cast-iron calipers in a volume-constrained environment. The new brake caliper resulted in a weight savings of 50%, with the additional benefits of corrosion resistance and ease of recyclability.

11.14.2 Ceramic-Matrix Composites

Recall that ceramics are strong and stiff and resist high temperatures, but generally lack toughness. On the other hand, matrix materials, such as silicon carbide, silicon nitride, aluminum oxide, and mullite (a compound of aluminum, silicon, and oxygen), are tough and retain their strength to 1700°C. Carbon-carbon-matrix composites retain much of their strength up to 2500°C, although they lack oxidation resistance at high temperatures.

There are several processes to make ceramic-matrix composites; three common ones are briefly described below.

1. **Slurry infiltration**, the most common process, involves the preparation of a fiber preform that is first hot pressed and then impregnated with a slurry that contains the matrix powder, a carrier liquid, and an organic binder. High strength, toughness, and uniform structure are obtained by this process, but the product has limited high-temperature properties due to the low melting temperature of the matrix materials used. An improvement of this process is *reaction bonding* or *reaction sintering* of the slurry, where the energy of chemical reactions in the powder help in the sintering process.
2. **Chemical synthesis** processes involve the *sol-gel* and the *polymer-precursor* techniques. In the first, a sol (a colloidal fluid with the liquid as the continuous phase), containing fibers, is converted to a gel; it is then heat treated to produce a ceramic-matrix composite.
3. **Chemical vapor infiltration** involves a porous fiber preform, which is infiltrated with the matrix phase, using the chemical vapor deposition technique (Section 4.5.1). The product has very good high-temperature properties, but the process is time consuming and costly.

Other techniques, such as melt infiltration, controlled oxidation, and hot-press sintering (still largely in the experimental stage), are at various stages of development for improving the properties and performance of these composites.

Applications for ceramic-matrix composites are in jet and automotive engines, equipment for deep-sea mining, pressure vessels, and various structural components. Silicon carbide ceramics reinforced by carbon fibers have been used as brake disks for race cars and aircraft, as well as bearings, because of its high wear resistance. CMCs have significant applicability to aerospace engines because of their light weight and strength at high temperatures; envisioned applications include turbine shrouds and turbine blades.

11.14.3 Miscellaneous Composites

Described below are several other composites.

1. Composites may consist of *coatings* on base metals or substrates. Examples include (a) plating of aluminum and other metals over plastics for decorative purposes and (b) enamels, dating to prior to 1000 B.C., or applying similar vitreous (glasslike) coatings on metal surfaces, for various functional or ornamental purposes (see also Section 4.5).
2. Another composite is *glass-reinforced fiber-metal laminate* (GLARE), used extensively in aircraft wings, fuselage sections, tail surfaces, and doors for the Airbus A380 aircraft. GLARE consists of several layers of aluminum interspersed with layers of glass-fiber reinforced epoxy matrix (prepreg), and has such advantages of weight savings and improved impact and fatigue properties.
3. Composites such as *cemented carbides*, usually tungsten carbide and titanium carbide, with cobalt and nickel, respectively, as a binder, are made into tools and dies (Section 8.6).
4. *Grinding wheels* are typically made of aluminum-oxide, silicon-carbide, diamond, or cubic-boron-nitride grains. The abrasive particles are held together with various organic, inorganic, or metallic binders (see Chapter 9).
5. Another composite material is granite particles embedded in an epoxy matrix (Section 8.13). It has high strength and good frictional characteristics and vibration-damping capacity (better than that of gray cast iron). This composite is used as machine-tool beds for some precision grinders.

11.15 Processing Superconductors

Although superconductors (Section 3.9.6) have major energy-saving potential in the generation, storage, and distribution of electrical power, their processing into useful shapes and sizes for practical applications presents significant challenges. Two basic types of superconductors are (a) *metals* (**low-temperature superconductors,** LTSCs), including combinations of niobium, tin, and titanium] and (b) *ceramics* (**high-temperature superconductors,** HTSCs), including various copper oxides. In this application, *high temperature* means closer to *ambient temperature*; consequently, the HTSCs are of more practical use.

Ceramic superconducting materials are available in powder form. The major difficulty in manufacturing them is their inherent brittleness and

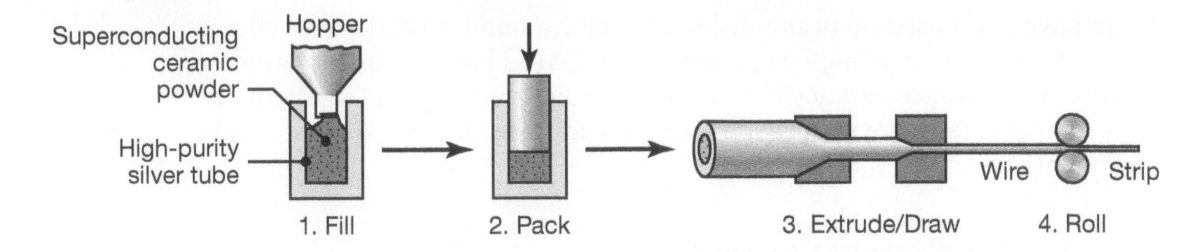

FIGURE 11.43 Schematic illustration of the steps involved in the powder-in-tube process.

anisotropy, which make it difficult to align the grains in the proper direction, for high efficiency. The smaller the grain size, the more difficult it is to align the grains.

The basic manufacturing process for superconductors consists of the following steps:

1. Preparing the *powder*, mixing, and grinding in a ball mill to a grain size of 0.5 to 10 mm;
2. *Shaping*; and
3. *Heat treating*, to improve grain alignment.

The most common shaping process is **oxide powder in tube** (OPIT), shown in Fig. 11.43. In this operation, the powder is first packed into silver tubes (because silver has the highest electrical conductivity of all metals) and sealed at both ends. The tubes are then mechanically worked (such as by swaging, drawing, extrusion, isostatic pressing, and rolling) into final shapes, which may be wire, tape, coil, or bulk.

Other principal superconductor-shaping processes include (a) coating of silver wire with a superconducting material; (b) deposition of superconductor films by laser ablation (ablation refers to laser heating causing layers of material to melt away from a surface, carrying away much of the heat); (c) the doctor-blade process (Section 11.9.1); (d) explosive cladding (Section 12.11); and (e) chemical spraying.

CASE STUDY 11.4 Powder Metal Parts in a Snowblower

Some of the parts in the freewheeling steering system of a commercial snowblower are shown in Fig. 11.44. Among the 16 PM components, the sprocket is the largest piece, at about 140 mm in diameter.

The final assembly incorporates a stamped steel frame, bronze and plastic bearings, and a wrought-steel axle, producing a highly functional and low-cost machine. Unique features, compatible with PM manufacturing, were incorporated into the design of these parts to enhance their functionality.

The PM components in the assembly range from single-level parts, with fixed features on punch faces and core rods, to intricate multilevel parts with complex die geometries, core rods, and transfer punches. These are unique features and they ensure proper and uniform distribution of powder. The clutch pawl, for example, is produced to a net-shape peripheral geometry that is not practical or economical with other manufacturing technologies. The material used is FLC4608-70 steel

FIGURE 11.44 Powder metallurgy parts in a commercial snowblower.

(a prealloyed powder of iron, with 1.9% Ni, 0.56% Mo, and 0.8% C, mixed in with 2% Cu); it has a tensile strength of 500 MPa and a density of 6.8 g/cm^3.

The part numbers are pressed into the face of each component, as a simple means of identifying them. Two of the components are made with especially close tolerances: The pawl latch gear has a 0.15-mm tolerance on the pitch diameter (PD), with 0.11 mm PD to ID run-out and 0.025-mm tolerance on the bore. The 32-tooth sprocket has a thin-walled 57.75 mm ID, with a 0.05-mm tolerance. Both the pawl latch gear and the sprocket achieve a density of 6.7 g/cm^3 and a tensile strength of 690 MPa.

All components shown passed normal life-cycle testing and product-life testing, including shock loading by engaging the drive in reverse, while traveling at the maximum forward speed down an incline. Clutch components, which were also subjected to salt-spray corrosion resistance test, and proper operation in subzero temperatures, suffered no failures. No machining is required on any of these parts, as they are net-shaped components with sufficient dimensional tolerances.

The only additional operations, prior to final assembly, were vibratory deburring and honing of the 32-tooth sprocket, in order to produce a close-tolerance bore and surface finish. The clutch pawls, made of sinter-hardened steel, are quenched in an atmosphere so that the porosity present can be filled with a lubricant, to provide lubricity at the interface of the mating parts (see also Section 4.4.3).

Source: Courtesy of the Metal Powder Industries Federation and Burgess-Norton Manufacturing Co.

SUMMARY

- The powder-metallurgy process is capable of economically producing complex parts in net- or near-net shape to close dimensional tolerances from a wide variety of metal and alloy powders. (Section 11.1)
- The steps in PM processing are powder production, blending, compaction, sintering, and additional processing to improve surface finish, mechanical or physical properties, or appearance. (Sections 11.2–11.5)

- Design considerations in PM include the shape of the compact, the ejection of the green compact from the die without damage, and the acceptable dimensional tolerances of the application. (Section 11.6)
- The PM process is suitable for medium- to high-volume production runs and for relatively small parts, and has competitive advantages over other processing methods. (Sections 11.7)
- Ceramics have such characteristics as high hardness and strength at elevated temperatures, high modulus of elasticity, brittleness, low density, low thermal expansion, and low thermal and electrical conductivity. (Section 11.8)
- The three basic shaping processes for ceramics are casting, plastic forming, and pressing; the part is then dried and fired to give it the proper strength. Finishing operations, such as machining and grinding, may be performed to give the part its final shape, or the part may be subjected to surface treatments to improve specific properties. (Section 11.9)
- Commercial glasses are categorized as one of six types, indicating their composition. Glass in bulk form has relatively low strength, but it can be strengthened by thermal or chemical treatments to impart high strength and toughness. (Section 11.10)
- Continuous methods of glass processing are floating, drawing, and rolling; discrete glass products are made by blowing, pressing, centrifugal casting, and sagging. After initial processing, glasses can be strengthened by thermal or chemical tempering or by laminating. (Section 11.11)
- Design considerations for ceramics and glasses include factors such as general lack of tensile strength and toughness, and sensitivity to external and internal defects. Warping and cracking are important considerations, as are the methods employed for their production and assembly. (Section 11.12)
- Graphite, carbon nanotubes, graphene, and diamond are forms of carbon that display unusual combinations of properties; these materials have several unique applications (Section 11.13).
- Metal-matrix and ceramic-matrix composites possess unique combinations of properties and have increasingly broad applications. Metal-matrix composites are processed through liquid-phase, solid-phase, and two-phase processes; ceramic-matrix composites can be processed by slurry infiltration, chemical synthesis, and chemical-vapor infiltration. (Section 11.14)
- Manufacture of superconductors into useful products is a challenging area because of the anisotropy and inherent brittleness of the materials. Although other processes are being developed, the basic and common practice involves packing the powder into a silver tube and deforming it plastically into desired shapes. (Section 11.15)

SUMMARY OF EQUATIONS

Shape factor of particles, $k = \left(\frac{A}{V}\right) D_{eq}$

Pressure distribution in compaction, $p_x = p_o e^{-4\mu kx/D}$

Volume, $V_{sint} = V_{green}\left(1 - \frac{\Delta L}{L_o}\right)^3$

Density, $\rho_{green} = \rho_{sint}\left(1 - \frac{\Delta L}{L_o}\right)^3$

Tensile strength, $S_{ut} \simeq S_{ut,o}e^{-nP}$

Elastic modulus, $E \simeq E_o\left(1 - 1.9P + 0.9P^2\right)$

Thermal conductivity, $k = k_o(1 - P)$

BIBLIOGRAPHY

Powder Metallurgy

ASM Handbook, Vol. 7: *Powder Metallurgy*, ASM International, 2016.

El-Eskandarany, M.S., *Mechanical Alloying*, 2nd ed., Elsevier, 2015.

German, R.M., *Powder Metallurgy and Particulate Materials Processing*, The Metal Powder Industries Federation, 2005.

——, *Sintering: From Empirical Observations to Scientific Principles*, Butterworth-Heinemann, 2014.

——, *Metal Injection Molding: A Comprehensive MIM Design Guide*, The Metal Powder Industries Federation, 2011.

Heaney, D., *Handbook of Metal Injection Molding*, Woodhead, 2012.

Nasr, G.G., Yule. A.J., and Bendig, L., *Industrial Sprays and Atomization: Design, Analysis and Applications*, Springer, 2010.

Pease III, L.F., and West, W.G., *Fundamentals of Powder Metallurgy*, The Metal Powder Industries Federation, 2002.

Powder Metallurgy Design Manual, 3rd ed., The Metal Powder Industries Federation, 1998.

Upadhyaya, G.S., *Sintering Metallic and Ceramic Materials: Preparation, Properties and Applications*, Wiley, 2000.

Upadhyaya, A., and Udaphyaya, G.S., *Powder Metallurgy: Science, Technology and Materials*, Universities Press, 2011.

Ceramics, Glasses, and Diamond

Advanced Ceramic Technologies & Products, Springer, 2012.

Buchanan, R.C., *Ceramic Materials for Electronics: Processing, Properties, and Applications*, 3rd ed., Dekker, 2004

Carter, C.B. and Norton, M.G., *Ceramic Materials: Science and Engineering*, Springer, 2013.

Holand, W., and Beall, G.H., *Glass Ceramic Technology*, Wiley, 2012.

Rahaman, M.N., *Ceramic Processing Technology and Sintering*, 2nd ed., Dekker, 2003.

Richerson, D.W., *Modern Ceramic Engineering: Properties, Processing, and Use in Design*, 3rd ed., Dekker, 2005.

Shackelford, J.F., and Doremus, R.H. (eds.). *Ceramic and Glass Materials: Structure, Properties and Processing*, Springer, 2008.

Somiya, S., *Modern Ceramic Engineering: Properties, Processing and Use in Design*, 3rd ed., CRC Press, 2005.

——, *Handbook of Advanced Ceramics*, 2nd ed., Academic Press, 2013.

Composites

Agarwal, A., Bakshi, S.R., and Lahiri, D., *Carbon Nanotubes: Reinforced Metal Matrix Composites*, CRC Press, 2010.

ASM Handbook, Vol. 21: *Composites*, ASM International, 2001.

Belitskus, D.L., *Fiber and Whisker Reinforced Ceramics for Structural Applications*, Dekker, 2004.

Chawla, K.K., *Ceramic Matrix Composites*, 2nd ed., Springer, 2003.

____, *Composite Materials*, 3rd ed., Springer, 2013.

____, *Metal Matrix Composites*, 2nd ed., Springer, 2013.

Daniel, I.M., and Ishai, O., *Engineering Mechanics of Composite Materials*, 2nd ed., Oxford, 2005.

Elhajjar, R., and La Saponara, V., *Smart Composites: Mechanics and Design*, CRC Press, 2013.

Gay, D., *Composite Materials Design and Applications*, 3rd ed.,CRC Press, 2014.

Strong, A.B., *Fundamentals of Composites Manufacturing*, Society of Manufacturing Engineers, 2007.

QUESTIONS

Powder Metallurgy

11.1 Explain the advantages of mixing metal powders.

11.2 Make a flow chart describing the production steps involved in powder metallurgy.

11.3 Is green strength important in powder metal processing? Explain.

11.4 Give the reasons that injection molding of metal powders has become an important process.

11.5 Describe the events that occur during sintering.

11.6 What is mechanical alloying, and what are its advantages over conventional alloying of metals?

11.7 It is possible to infiltrate PM parts with various resins, as well as with metals. What possible benefits would result from infiltration?

11.8 What concerns would you have when electroplating PM parts?

11.9 Describe the effects of different shapes and sizes of metal powders in PM processing, commenting on the magnitude of the shape factor (SF) of the particles.

11.10 Comment on the shapes of the curves and their relative positions shown in Fig. 11.9.

11.11 Should green compacts be brought up to the sintering temperature slowly or rapidly? Explain the advantages and limitations of both methods.

11.12 Explain the effects of using fine powders and coarse powders, respectively, in making PM parts.

11.13 Are the requirements for punch and die materials in powder metallurgy different than those for forging and extrusion, described in Chapter 6? Explain.

11.14 Describe the relative advantages and limitations of cold and hot isostatic pressing, respectively.

11.15 Why do mechanical and physical properties depend on the density of PM parts? Explain with appropriate sketches.

11.16 What type of press is required to compact parts by the set of punches shown in Fig. 11.9d? (See Section 6.2.8.)

11.17 Explain the difference between impregnation and infiltration. Give some applications of each.

11.18 Tool steels are now being made by PM techniques. Explain the advantages of this method over traditional methods such as casting and subsequent metal-working techniques.

11.19 Why do the compacting pressure and sintering temperature depend on the type of powder metal used?

11.20 Name the various methods of powder production and describe the morphology of powders produced.

11.21 What hazards are involved in PM processing? Explain their causes.

11.22 What is "screening" of metal powders? Why is it done?

11.23 Why is there density variation in the compaction of metal powders? How is it reduced?

11.24 We have stated that PM can be competitive with processes such as casting and forging. Explain why this is so, giving a range of process parameters and illustrative applications.

11.25 Selective laser sintering was described in Section 10.12.4 as a rapid prototyping operation. What similarities does this process have with the processes described in this chapter?

11.26 Prepare an illustration similar to Fig. 6.28 showing the variety of PM manufacturing options.

11.27 List the similarities and differences between forging and compacting metal powders.

11.28 What will be stronger: a blend of stainless steel and copper powder that is compacted and sintered, or a stainless steel powder that is compacted, sintered, and infiltrated by copper? Explain.

11.29 Describe other methods of manufacturing the parts shown in Fig. 11.1a. Comment on the advantages and limitations of these methods over PM.

11.30 What is the difference between a lubricant and a binder?

Ceramics and Other Materials

11.31 Compare the major differences between ceramics, metals, thermoplastics, and thermosets.

11.32 Explain why ceramics are weaker in tension than in compression.

11.33 Explain why the mechanical and physical properties of ceramics decrease with increasing porosity.

11.34 What engineering applications could benefit from the fact that, unlike metals, ceramics generally maintain their modulus of elasticity at elevated temperatures?

11.35 Explain why the mechanical-property data in Table 11.8 have such a broad range. What is the significance of this range in engineering applications?

11.36 List the factors that you would consider when replacing a metal component with a ceramic component. Give examples of such substitutions.

11.37 How are ceramics made tougher? Explain.

11.38 Describe applications in which static fatigue can be important.

11.39 Explain the difficulties involved in making large ceramic components. What recommendations would you make to improve the process?

11.40 Is there any flash that develops in slip casting? How would you propose to remove such flash?

11.41 Explain why ceramics are effective cutting-tool materials. Would ceramics also be suitable as die materials for metal forming? Explain.

11.42 Describe applications in which the use of a ceramic material with a zero coefficient of thermal expansion would be desirable.

11.43 Give reasons for the development of ceramic-matrix components. Name some present and possible future applications of these components.

11.44 List the factors that are important in drying ceramic components, and explain why they are important.

11.45 It has been stated that the higher the coefficient of thermal expansion of the glass and the lower the glass' thermal conductivity, the higher is the level of residual stresses developed during processing. Explain why.

11.46 What types of finishing operations are performed on ceramics? Why are they done?

11.47 What should be the property requirements for the metal balls used in a ball mill? Explain why these properties are required.

11.48 Which properties of glasses allow them to be expanded and shaped into bottles by blowing?

11.49 What properties should plastic sheet have when used in laminated glass? Why?

11.50 Consider some ceramic products that you are familiar with, and outline a sequence of processes performed to manufacture each of them.

11.51 Explain the difference between physical and chemical tempering of glass.

11.52 What do you think is the purpose of the operation shown in Fig. 11.30d?

11.53 As you have seen, injection molding is a process that is used for plastics and powder metals as well as for ceramics. Why is it used for all these materials?

11.54 Are there any similarities between the strengthening mechanisms for glass and those for other metallic and nonmetallic materials described throughout this text? Explain.

11.55 Describe and explain the differences in the manner in which each of the following flat surfaces would fracture when struck with a large piece of rock: (a) ordinary window glass; (b) tempered glass; and (c) laminated glass.

11.56 Describe the similarities and the differences between the processes described in this chapter and in Chapters 5 through 10.

11.57 What is the doctor-blade process?

11.58 Describe the methods by which glass sheet is manufactured.

11.59 Describe the glass-blowing process. What properties of glass make blowing possible?

11.60 How are glass fibers made?

11.61 Is diamond a ceramic? Why or why not?

11.62 What are the similarities and differences between injection molding, metal injection molding, and ceramic injection molding?

11.63 Aluminum oxide and partially stabilized zirconia are normally white in appearance. Can they be colored? If so, how would you accomplish this?

11.64 It was stated in the text that ceramics have a wider range of strengths in tension than they have in metals. List reasons why this is so, with respect to both the ceramic properties that cause variations and the difficulties in obtaining repeatable results.

11.65 What techniques, other than the powder-in-tube process, could be used to produce superconducting monofilaments?

PROBLEMS

11.66 Estimate the number of particles in a 500-g sample of iron powder, if the particle size is 25 μm.

11.67 Referring to Fig. 11.8a, what should be the volume of loose, fine iron powder in order to make a solid cylindrical compact 20 mm in diameter and 10 mm high?

11.68 Give reasons that HIP-treated PM parts have a more uniform strength compared to cast parts.

11.69 Estimate the compaction force required to compact a brass slug 75 mm. in diameter. Would the height of the slug make any difference in your answer? Explain your reasoning.

11.70 Assume that the surface of a copper particle is covered with a 0.1-μm-thick oxide layer. What is the volume occupied by this layer if the copper particle itself is 50 μm in diameter? What would be the role of this oxide layer in subsequent processing of the powders?

11.71 Determine the shape factor for a flakelike particle with a ratio of surface cross-sectional area to thickness of $15 \times 15 \times 1$, for a cylinder with dimensional ratios 1:1:1, and for an ellipsoid with an axial ratio of $5 \times 2 \times 1$.

11.72 We stated in Section 3.3 that the energy in brittle fracture is dissipated as surface energy. We also noted that the comminution process for powder preparation generally involves brittle fracture. What are the relative energies involved in making spherical powders of diameters 1, 10, and 100 μm, respectively?

11.73 In Fig. 11.9e, we note that the pressure is not uniform across the diameter of the compact. Explain the reasons for this lack of uniformity.

11.74 Plot the family of pressure-ratio p_x/p_o curves as a function of x for the following ranges of process parameters: $\mu = 0$ to 1, $k = 0$ to 1, and $D = 5$ mm to 50 mm.

11.75 Derive an expression similar to Eq. (11.2) for compaction in a square die with dimensions a by a.

11.76 For the ceramic in Example 11.7, calculate (a) the porosity of the dried part if the porosity of the fired part is to be 9%, and (b) the initial length L_o of the part if the linear shrinkages during drying and firing are 8% and 7%, respectively.

11.77 What would be the answers to Problem 11.76 if the quantities given were halved?

11.78 Plot S_{ut}, E, and k values for ceramics as a function of porosity P, and describe and explain the trends that you observe in their behavior.

11.79 Plot the total surface area of a 1-g sample of aluminum as a function of the natural log of particle size.

11.80 How large is the grain size of metal powders that can be produced in atomization chambers? Conduct a literature search to determine the answer.

11.81 A coarse copper powder is compacted in a mechanical press at a pressure of 400 MPa. During sintering, the green part shrinks an additional 9%. What is the final density of the part?

11.82 A gear is to be manufactured from iron powder. It is desired that it have a final density that is 90% of that of cast iron, and it is known that the shrinkage in sintering will be approximately 7%. For a gear 50 mm in diameter and a 20-mm hub, what is the required press force?

11.83 What volume of powder is needed to make the gear in Problem 11.82 if its thickness is 15 mm?

11.84 Coarse iron powder is compacted into a cylinder with a 25-mm diameter and 40-mm height. The green part has a measured mass of 130 g. Calculate (a) the apparent density; (b) the percentage of the theoretical full density; and (c) an estimate of the compacting pressure used.

11.85 The part in Problem 11.84 is to be hot isostatically pressed to full density. If shrinkage is the same in all directions, estimate the final part dimensions.

11.86 Fine iron powder is compressed into a cylinder ($d = 30$ mm, $h = 20$ mm), achieving 70% of theoretical density. It is to be hot forged to a height of 5 mm. What diameter should be planned in order to achieve a 95% final density?

11.87 Determine the shape factor for a (a) spherical particle; (b) cubic particle; and (c) cylindrical particle with a length-to-diameter ratio of 2.

11.88 Estimate the maximum force needed to compress a PM pure copper powder into a 50 mm-diameter cylinder and an 80% relative density. Would the height of the cylinder make any difference in your answer? Explain.

11.89 A green compact of iron is sintered for an hour at 890°C, and is found to have a higher strength than the same material sintered at 1000°, with all other variables being constant. Explain these results.

11.90 The axisymmetric part shown in the accompanying figure is to be produced from fine copper powder and is to have a tensile strength of 175 GPa. Determine the compacting pressure and the initial volume of powder needed.

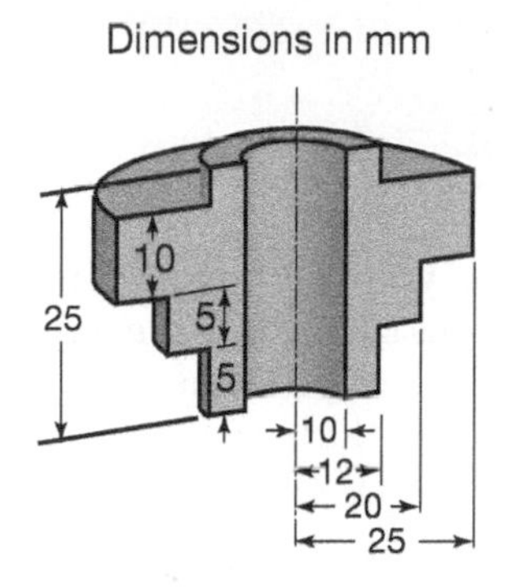

11.91 The part considered in Problem 11.90 is to be compressed in a dual-action press. The density of loose powder is 40%, but it is desired to produce a green compact with 80% of full density. Specify the initial positions of the vertical features in the die.

11.92 A 100-g sample is used to measure the particle size of Ti-6Al-4V powder. If the average size is 100 μm, roughly how many particles are in the sample? Do you think the particle size distribution will be accurate? Explain.

11.93 If a fully dense ceramic has the properties of $S_{ut,o} = 180$ MPa and $E_o = 300$ GPa, what are these properties at 20% porosity for values of $n = 4, 5, 6,$ and 7, respectively?

11.94 Calculate the thermal conductivities for ceramics at porosities of 1%, 5%, 10%, 20%, and 30% for $k_o = 0.7$ W/m-K.

11.95 A ceramic has $k_o = 0.65$ W/m-K. If this ceramic is shaped into a cylinder with a porosity distribution of $P = 0.1(x/L)(1 - x/L)$, where x is the distance from one end of the cylinder and L is the total cylinder length, estimate the average thermal conductivity of the cylinder.

11.96 Plot S_{ut}, E, and k as a function of porosity, up to 10% porosity, for the conditions in Example 11.4.

11.97 A cube of 25-mm dimension is to be produced through metal injection molding. In the metal-powder blend, the binder is 30% by volume, and is completely removed during sintering. Calculate the change in volume and the dimensions of the sintered object.

11.98 A cylinder is being compressed to a final height of 50 mm and a diameter of 25 mm. The pressure at the punch is 200 MPa. What is the pressure at the middle of the cylinder if compaction takes place (a) unlubricated ($k = 0.6, \mu = 0.2$); and (b) lubricated to almost frictionless conditions?

11.99 Assume that you are asked to give a quiz to students on the contents of this chapter. Prepare three quantitative problems and three qualitative questions, and supply the answers.

DESIGN

11.100 Historically, PM was known as "metal ceramics". Does this term make sense? Write a one-page paper expressing your viewpoint.

11.101 List the process parameters and material properties that you expect will influence particle size and shape resulting from gas atomization.

11.102 Can mechanical alloying be used to produce (a) amorphous and (b) polycrystalline powders? Explain.

11.103 Make sketches of several PM products in which density variations would be desirable. Explain why in terms of the function of these parts.

11.104 Compare the design considerations for PM products with those for (a) products made by casting and (b) products made by forging. Describe your observations.

11.105 It is known that in the design of PM gears, the outside diameter of the hub should be as far as possible from the root of the gear. Explain the reasons for this design consideration.

11.106 How are the design considerations for ceramics different, if at all, than those for the other materials described in this chapter?

11.107 Are there any shapes or design features that are not suitable for production by powder metallurgy? By ceramics processing? Explain.

11.108 It is known that a higher green density can be obtained from a blend of two powders, with two different mean particle sizes, than from a single powder. With proper sketches, explain why this would be the case.

11.109 Using the Internet, locate suppliers of metal powders and compare the cost of the powder with the cost of ingots for five different materials.

11.110 It has been noted that PM gears are very common for low-cost office equipment such as the carriage mechanism of inkjet printers. Review the design requirements of these gears and list the advantages of PM manufacturing approaches for these gears.

11.111 What design changes would you recommend for the part shown in Problem 11.87?

11.112 Samples of metal powder from a single batch are sent to two different laboratories for analysis. One reports that the mean particle size is 50 μm, while the other reports 64 μm. List reasons to explain the difference.

11.113 Describe any special design considerations in products that use ceramics with a near-zero coefficient of thermal expansion.

11.114 Prepare an illustration similar to Fig. 13.3, showing the sequence of PM manufacturing steps.

11.115 Construct a table that describes the approach for manufacturing plate from (a) metals; (b) thermoplastics; (c) ceramics; (d) powder metal; and (e) glass. Include descriptions of process capabilities and shortcomings in your descriptions.

11.116 Describe your thoughts regarding designs of internal combustion engines using ceramic pistons.

11.117 List and discuss the factors that you would take into account when replacing a metal component with a ceramic component.

11.118 Assume that in a particular design, a metal beam is to be replaced with a beam made of ceramics. Discuss the differences in the behavior of the two beams, such as with respect to strength, stiffness, deflection, and resistance to temperature and to the environment.

11.119 The axisymmetric parts shown in the figures below are to be produced through PM. Describe the design changes that you would recommend.

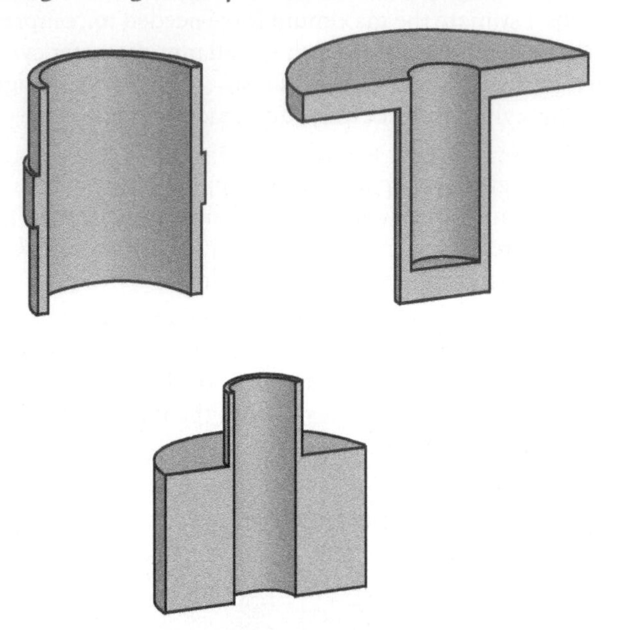

11.120 Assume you are working in technical sales. What applications currently using non-PM parts would you attempt to develop? What would you say to your potential customers during your sales visits? What kind of questions do you think they would ask?

11.121 Describe your thoughts on the processes that can be used to make (a) small ceramic statues, (b) whiteware for bathrooms, (c) common brick, and (d) floor tile.

11.122 Pyrex cookware displays a unique phenomenon: it functions well for a large number of cycles and then shatters into many pieces. Investigate this phenomenon, list the probable causes, and discuss the manufacturing considerations that may alleviate or contribute to such failures.

11.123 It has been noted that the strength of brittle materials such as ceramics and glasses are very sensitive to surface defects such as scratches (notch sensitivity). Obtain some pieces of these materials, make scratches on them, and test them by carefully clamping in a vise and bending them. Comment on your observations.

11.124 Consider some ceramic products with which you are familiar, and outline a sequence of processes that you think were used to manufacture them.

11.125 Make a survey of the technical literature and describe the differences, if any, between the quality of glass fibers made for use in reinforced plastics and that those made for use in fiber-optic communications. Comment on your observations.

11.126 Describe your thoughts on the processes that can be used to make (a) small ceramic statues; (b) whiteware for bathrooms; (c) common brick; and (d) floor tile.

11.127 As we have seen, one method of producing superconducting wire and strip is by compacting powders of these materials, placing them into a tube, and drawing them through dies, or rolling them. Describe your thoughts concerning the steps and possible difficulties involved in each step of this production.

11.128 We have explained briefly the characteristics of bulletproof glass. Describe your own thoughts on possible new designs for this type of glass. Explain your reasoning.

11.129 Review Figure 11.21 and prepare a similar figure for constant thickness parts, as opposed to the axisymmetric parts shown.

CHAPTER

12 Joining and Fastening Processes

This chapter describes the principles, characteristics, and applications of major joining processes and operations. The topics covered include:

- Oxyfuel gas welding and arc welding processes, using nonconsumable and consumable electrodes.
- Solid state welding processes.
- Laser beam and electron beam welding.
- The nature and characteristics of the weld joint and factors involved in weldability of metals.
- Adhesive bonding and applications.
- Mechanical fastening methods and fasteners.
- Economic considerations in joining operations.
- Good joint design practices and process selection.

Symbols

A	area, m^2	l	weld length, m
b	weld dimension, m	R	electrical resistance, ohm
e	efficiency	t	time, s
h	weld dimension, m	T_m	melting temperature, °C
H	heat input, Nm	u	specific energy, W-s/m^3
I	current, A	v	velocity, m/s
K	energy loss factor	V	voltage, V

12.1 Introduction

Joining is a general, all-inclusive term covering numerous processes that are important in manufacturing. The necessity of joining and assembling components can be appreciated by inspecting various products, such as automobiles, bicycles, printed circuit boards, machinery, and appliances. Joining may be preferred or necessary for one or more of the following reasons:

- The product is impossible or uneconomical to manufacture as a single piece;
- The product is easier and less costly to manufacture in individual components, which are then assembled;
- The product may have to be taken apart for repair or maintenance during its service life;
- Different properties may be required for function of the product; for example, surfaces that are subjected to friction and wear, or to corrosion and environmental attack, typically require characteristics different than those of the component's bulk; and
- Transportation of the product in individual components and their subsequent assembly may be easier and more economical than transporting it as a single unit.

There are many categories of joining and fastening processes; this chapter will group them according to their common principle of operation. The joints being produced typically undergo metallurgical and physical changes, which, in turn, have major effects on the properties and performance of welded components.

Fusion welding involves melting and coalescing of materials by means of heat, usually supplied by electrical or other means. The processes consist of *oxyfuel gas welding*, consumable- and nonconsumable-electrode *arc welding*, and *high energy beam welding*. The high energy joining processes, described in Section 9.14, are also used for cutting and machining applications.

Solid-state welding involves joining without fusion; there is no liquid (molten) phase in the joint. The basic categories are *cold, ultrasonic, friction, resistance, explosion welding,* and *diffusion bonding*.

Brazing and **soldering** use filler metals and involve lower temperatures than in welding; the heat required is supplied externally.

Adhesive bonding is an important technology because of its several unique advantages in applications requiring strength, sealing, insulation, vibration damping, and resistance to corrosion between dissimilar or similar metals. Included in this category are *electrically conducting adhesives* for surface-mount technologies.

Mechanical fastening processes typically involve fasteners such as bolts, nuts, screws, and rivets. *Joining nonmetallic materials* can be accomplished by such means as mechanical fastening, adhesive bonding, fusion, diffusion, etc.

The individual groups of joining processes each has several important characteristics, including joint design (Fig. 12.1), size and shape of the parts to be joined, strength and reliability of the joint, cost and maintenance of equipment, and skill level of labor (Tables 12.1 and 12.2).

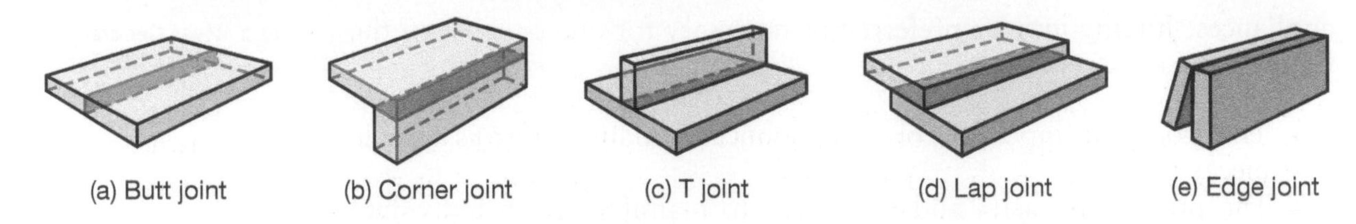

FIGURE 12.1 Examples of welded joints.

TABLE 12.1 Comparison of various joining methods.

Method	Strength	Design variability	Small parts	Large parts	Tolerances	Reliability	Ease of maintenance	Visual inspection	Cost
Arc welding	1	2	3	1	3	1	2	2	2
Resistance welding	1	2	1	1	3	3	3	3	1
Brazing	1	1	1	1	3	1	3	2	3
Bolts and nuts	1	2	3	1	2	1	1	1	3
Riveting	1	2	3	1	1	1	3	1	2
Fasteners	2	3	3	1	2	2	2	1	3
Seaming, crimping	2	2	1	3	3	1	3	1	1
Adhesive bonding	3	1	1	2	3	2	3	3	2

Note: 1, very good; 2, good; 3, poor

TABLE 12.2 General characteristics of joining processes.

Joining process	Operation	Advantage	Skill level required	Welding position	Current type	Distortion*	Cost of equipment
Shielded metal arc	Manual	Portable and flexible	High	All	AC, DC	1 to 2	Low
Submerged arc	Automatic	High deposition	Low to medium	Flat and horizontal	AC, DC	1 to 2	Medium
Gas metal arc	Semiautomatic or automatic	Works with most metals	Low to high	All	DC	2 to 3	Medium to high
Gas tungsten arc	Manual or automatic	Works with most metals	Low to high	All	AC, DC	2 to 3	Medium
Flux-cored arc	Semiautomatic or automatic	High deposition	Low to high	All	DC	1 to 3	Medium
Oxyfuel	Manual	Portable and flexible	High	All	—	2 to 4	Low
Electron beam, laser beam	Semiautomatic or automatic	Works with most metals	Medium to high	All	—	3 to 5	High
Thermit	Manual	Steels	Low	Flat and horizontal	—	2 to 4	Low

*1, highest; 5, lowest

12.2 Oxyfuel Gas Welding

Developed in the early 1900s, *oxyfuel gas welding* (OFW) is a general term describing any welding process that uses a **fuel gas**, combined with *oxygen* to produce a flame as the heat source required to melt the metals at the joint. The most common gas welding process uses *acetylene*; the process is known as oxyacetylene gas welding (OAW) and is typically used for structural fabrication, automotive frames, and various other repair work.

The heat generated in OFW is a result of the combustion of acetylene gas (C_2H_2) in a mixture with oxygen, in accordance with a pair of chemical reactions. The primary combustion, which occurs in the inner core of the flame (Fig. 12.2), involves the reaction

$$C_2H_2 + O_2 \rightarrow 2CO + H_2 + \text{heat}. \tag{12.1}$$

This reaction dissociates the acetylene into carbon monoxide and hydrogen, and produces about one-third of the total heat generated in the flame. The secondary combustion process is

$$2CO + H_2 + 1.5O_2 \rightarrow 2CO_2 + H_2O + \text{heat}. \tag{12.2}$$

This reaction produces about two-thirds of the total heat; note that the reaction also produces water vapor. The temperatures developed in the flame can reach 3300°C (6000°F).

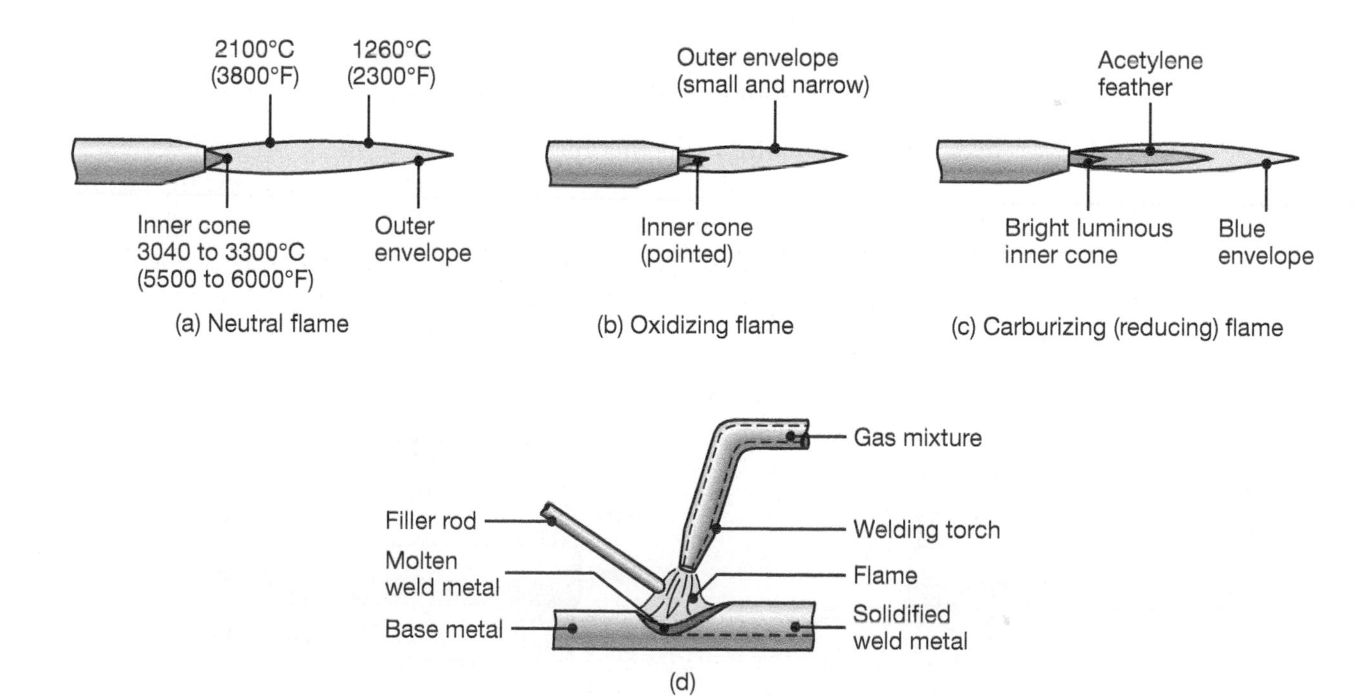

FIGURE 12.2 Three basic types of oxyacetylene flames used in oxyfuel gas welding and cutting operations: (a) neutral flame; (b) oxidizing flame; (c) carburizing, or reducing, flame; and (d) the principle of the oxyfuel gas welding operation.

Flame types. The proportion of acetylene and oxygen in the gas mixture is an important factor in oxyfuel gas welding. At a ratio of 1:1 (when there is no excess oxygen), the flame is *neutral* (Fig. 12.2a). With a greater oxygen supply, the flame is known as **oxidizing** (Fig. 12.2b), and can be harmful, especially for steels because it oxidizes the metal. Only in welding copper and copper-base alloys is an oxidizing flame desirable, because it forms a thin protective layer of *slag* (compounds of various oxides, fluxes, and electrode coating materials) over the molten metal. When the oxygen is insufficient for full combustion, the flame is known as a **reducing** (having excess acetylene) or a **carburizing flame** (Fig. 12.2c); its temperature is lower, thus carburizing flames are suitable for applications requiring low heat, such as in brazing, soldering, and flame hardening operations.

Other fuel gases, such as hydrogen and methylacetylene propadiene, can also be used in oxyfuel gas welding. The temperatures developed by these gases are lower than with acetylene; thus they are used for welding metals with low melting points, such as lead, and for parts that are thin or small. The flame with pure hydrogen gas is colorless, and it is difficult to adjust the flame by eyesight, as is the case with other gases.

Filler metals. These metals are used to supply additional metal to the weld zone during welding; they are available as *filler rods* or *wire* (Fig. 12.2d), and may be bare or coated with flux. The purpose of the **flux** is to retard oxidation of the surfaces being welded, by generating a *gaseous shield* around the weld zone. The flux also helps dissolve and remove oxides and other substances from the weld zone, thus enhancing the development of a stronger joint; the slag protects the molten puddle of metal against oxidation as it cools.

Pressure gas welding. In this method, the interfaces of the two components to be welded are heated by means of a torch, typically using an oxyacetylene gas mixture (Fig. 12.3a). When the interface begins to

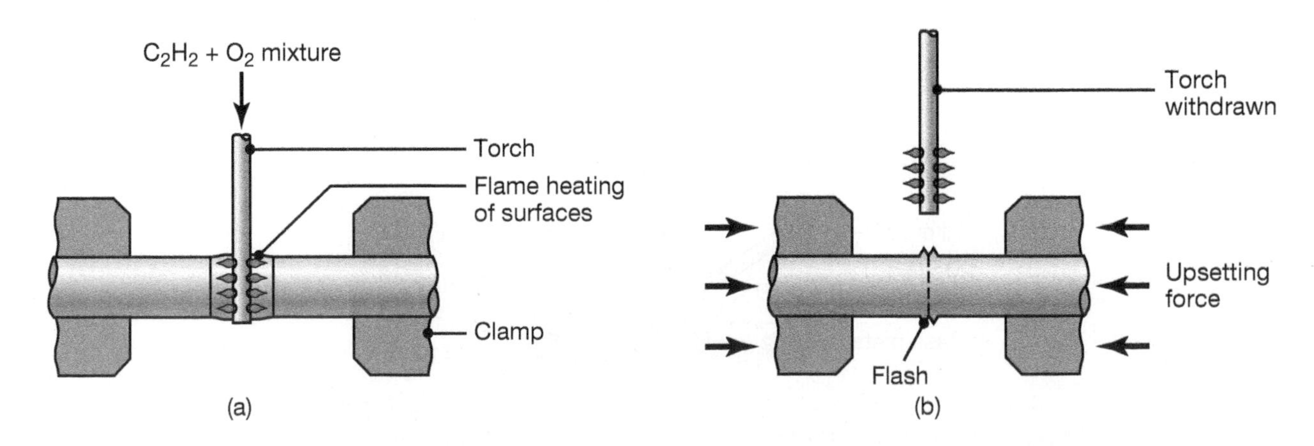

FIGURE 12.3 Schematic illustration of the pressure gas welding process; (a) before, and (b) after. Note the formation of a flash at the joint, which can later be trimmed off.

melt, the torch is withdrawn and an axial force is applied, pressing the two components together (Fig. 12.3b); the force is maintained until the interface solidifies. Note the formation of a flash in Fig. 12.3b, due to the upsetting of the joined ends of the two components.

Thermit welding. Also known as *thermite* or *exothermic welding*, and developed in 1895, this process involves mixing a metal powder with a metal oxide, using a high temperature ignition source to cause an oxidation-reduction reaction (Fig. 12.4). A common arrangement in this process is to use iron oxide (*rust*) powder in combination with aluminum powder; upon ignition with a magnesium fuse, the chemical reaction forms aluminum oxide (Al_2O_3) and iron.

Temperatures can reach 2500°C, melting the iron which subsequently flows into a pouring basin and then into a mold placed around the parts to be welded; the aluminum oxide floats to the slag basin because of its lower density. The features of a thermit welding mold are very similar to a casting mold (See Fig. 5.10). Note from Fig. 12.4 that a heating port is present, a feature that allows insertion of an oxyacetylene torch to preheat the workpieces and prevent the weld from cracking (see Section 12.7).

Several combinations of powder and oxide can be used in thermit welding, but aluminum powder combined with iron oxide is the most common, because of the widespread use of this process in joining railroad rails. Some copper and magnesium oxides are often added to improve flammability. Other applications of this process include welding of large-diameter copper conductors, using copper oxide, and field repair of large equipment, such as locomotive axle frames.

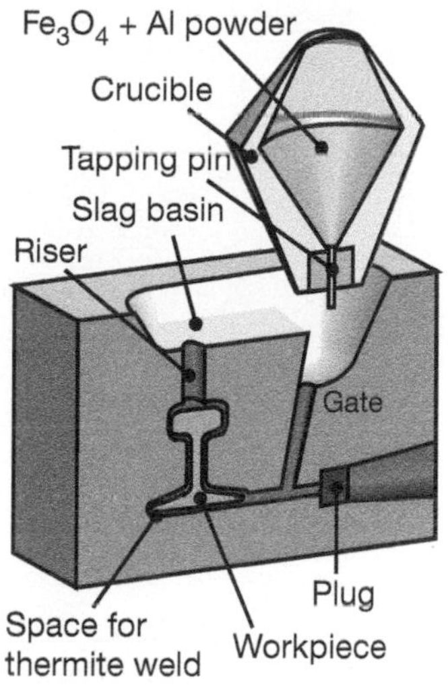

FIGURE 12.4 Schematic illustration of thermite welding.

12.3 Arc Welding Processes: Consumable Electrode

In arc welding, developed in the mid-1800s, the heat required is obtained from electrical energy. Using either a consumable or a nonconsumable rod or wire electrode, an arc is produced between the tip of the electrode and the parts to be welded, using an AC or DC power supply. Arc welding operations include several processes (Table 12.2), as described throughout this section.

12.3.1 Heat Transfer in Arc Welding

The heat input in arc welding can be calculated from the equation

$$\frac{H}{l} = e\frac{VI}{v}, \tag{12.3}$$

where H is the heat input in Joules, l is the weld length, V is the voltage applied, I is the current in amperes, and v is the welding speed. The term e is the efficiency of the process, and varies from around 75% for shielded metal arc welding to 90% for gas metal arc and submerged arc welding. The efficiency is an indication that not all of the available energy is used to melt material, and some of the available heat is conducted through the

TABLE 12.3 Approximate specific energies required to melt a unit volume of commonly-welded metals.

	Specific Energy, u	
Material	J/mm^3	BTU/in^3
Aluminum alloys	2.9	41
Cast irons	7.8	112
Copper	6.1	87
Bronze	4.2	59
Magnesium	2.9	42
Nickel	9.8	142
Steels	9.7	137
Stainless steels	9.4	135
Titanium	14.3	204

workpiece and some is lost by radiation and convection to the surrounding environment.

The heat input, given by Eq. (12.3), melts some material, usually the electrode or the filler metal, and can also be expressed as

$$H = u(\text{Volume}) = uAl, \tag{12.4}$$

where u is the specific energy required for melting and A is the cross section of the weld. Typical values of u are given in Table 12.3. Equations (12.3) and (12.4) allow an expression of the welding speed as

$$v = e\frac{VI}{uA}. \tag{12.5}$$

Although these equations have been developed for arc welding, similar expressions can be obtained for other fusion-welding operations, taking into account differences in weld zone geometry and process efficiency.

EXAMPLE 12.1 Estimation of Welding Speed for Different Materials

Given: A welding operation is being performed with $V = 20$ volts, $I = 200$ A, and the cross-sectional area of the weld bead is 30 mm^2.

Find: Estimate the welding speed if the workpiece and electrode are made of (a) aluminum; (b) carbon steel; and (c) titanium. Use an efficiency of 75%.

Solution: For aluminum, note from Table 12.3 that the specific energy required is $u = 2.9\ J/mm^3$. From Eq. (12.5) then,

$$v = e\frac{VI}{uA} = (0.75)\frac{(20)(200)}{(2.9)(30)} = 34.5\ \text{mm/s}.$$

Similarly, for carbon steel, u is taken from Table 12.3 as 9.7 J/mm^3, leading to $v = 10.3$ mm/s. For titanium, $u = 14.3\ J/mm^3$, so that $v = 7.0$ mm/s.

12.3.2 Shielded Metal Arc Welding

Shielded metal arc welding (SMAW) is one of the oldest, simplest, and most versatile joining processes; currently, about one-half of all industrial and maintenance welding is performed by this process. The electric arc is generated by touching the tip of a coated electrode to the workpiece, then withdrawing it quickly to a distance sufficient to maintain the arc (Fig. 12.5a). The electrodes are in the shape of thin, long sticks (see Section 12.3.8), hence this process is also known as **stick welding**. The heat generated melts a portion of the electrode tip, its coating, and the base metal in the immediate area of the arc. A weld forms after the molten metal (a mixture of the base metal, electrode metal, and electrode coating) solidifies in the weld zone; the electrode coating also deoxidizes and provides a shielding gas to protect the zone from oxygen in the environment.

In this operation, the bare section at one end of the electrode is clamped to one terminal of the power source, while the other terminal is connected to the part being welded (Fig. 12.5b). The current usually ranges between 50 and 300 A, with power generally less than 10 kW. Too low a current causes incomplete fusion, and too high a current can damage the electrode coating and reduce its effectiveness. The current may be AC or DC; for sheet metal welding, DC is preferred, because of the steady arc produced.

The **polarity** (direction of current flow) of the DC current may be important, depending on such factors as the type of electrode, the metals to be welded, and the depth of the heated zone. In **straight polarity**, the workpiece is positive and the electrode negative; this polarity is preferred for sheet metals because of its shallow penetration, and for joints with very wide gaps. In **reverse polarity** the electrode is positive and the workpiece negative, and weld penetration is deeper. When the current is AC, the arc pulsates rapidly, a condition that is suitable for welding thick sections, using larger diameter electrodes and higher electrical current.

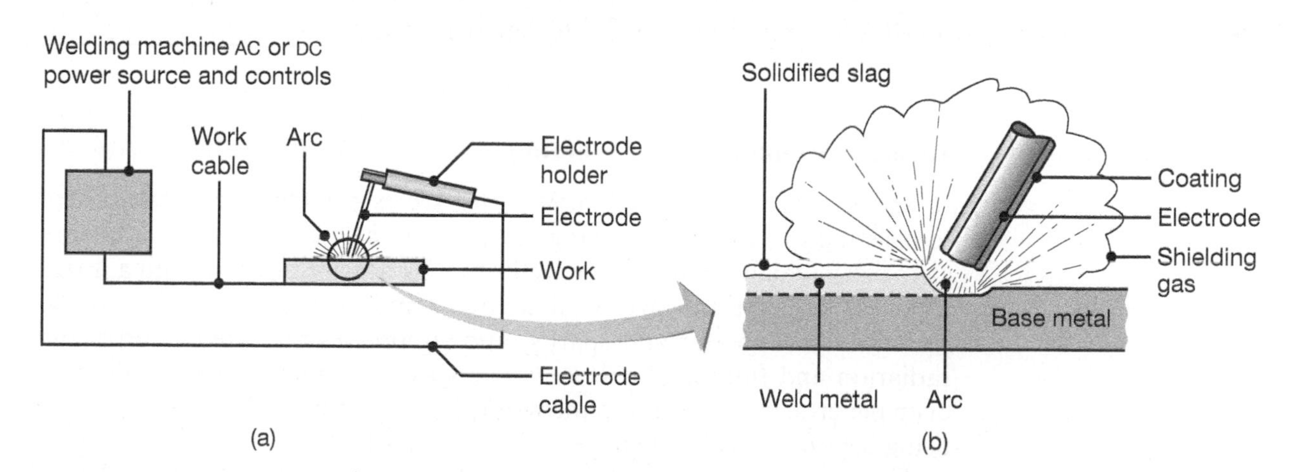

FIGURE 12.5 (a) Schematic illustration of the shielded metal arc welding process; about one-half of all large-scale industrial welding operations use this process. (b) Schematic illustration of the shielded metal arc welding operation.

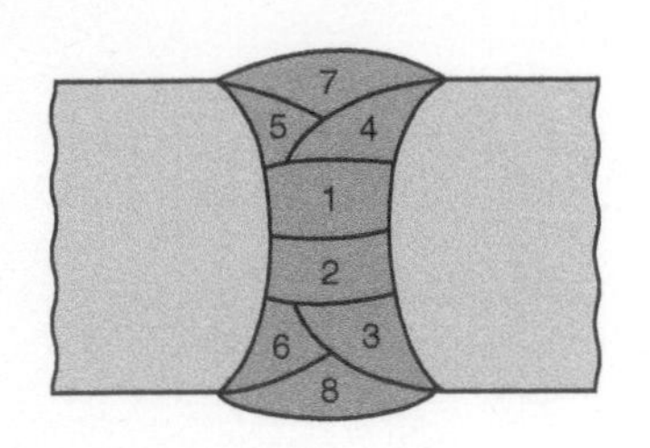

FIGURE 12.6 A weld zone showing the build-up sequence of individual weld beads in deep welds.

The SMAW process requires a relatively small selection of electrodes, and the equipment consists of a power supply, power cables, and an electrode holder. This process is commonly used in general construction, shipbuilding, and for pipelines, as well as for maintenance work because the equipment is portable and can be easily maintained. SMAW is especially useful for work in remote areas, where portable fuel-powered generators can be used as the power supply. The process is best suited for workpiece thicknesses of 3–20 mm (0.12–0.8 in.), although this range can easily be extended using multiple-pass techniques (Fig. 12.6). SMAW requires that slag be cleaned after each weld bead is completed. Unless removed completely, the solidified slag can cause severe corrosion of the weld area and subsequently lead to failure of the weld. Labor costs are high in SMAW, as are material costs.

EXAMPLE 12.2 Heat and Speed of Shielded Metal Arc Welding

Given: A shielded metal arc welding operation takes place on a steel workpiece (with a steel electrode) with a 20 volt power supply.

Find: If a weld with a triangular cross section with a 10 mm leg length is to be produced, estimate the current needed for a welding speed of 10 mm/s. Use an efficiency of 75%.

Solution: The cross-sectional area of the weld is calculated from the given geometry as:

$$A = \frac{1}{2}bh = \frac{1}{2}(10)(10) = 50 \text{ mm}^2.$$

The specific energy needed to melt the steel electrode is taken from Table 12.3 as 10.3 J/mm^2. Therefore, Eq. (12.5) yields:

$$v = e\frac{VI}{uA}; \qquad I = \frac{vuA}{eV} = \frac{(10)(10.3)(50)}{(0.75)(20)} = 343 \text{ A}.$$

12.3.3 Submerged Arc Welding

In *submerged arc welding* (SAW), the arc is shielded by *granular flux* (consisting of lime, silica, manganese oxide, calcium fluoride, and other elements), which is fed into the weld zone by gravity flow through a nozzle (Fig. 12.7). The thick layer of flux covers and protects the molten metal, prevents weld spatter and sparks, and suppresses the intense ultraviolet radiation and fumes. The flux also acts as a thermal insulator, allowing deep penetration of heat into the workpiece. The unfused flux is recovered, using a *recovery tube*, then treated, and reused.

The consumable electrode is a coil of bare wire 1.5–10 mm (0.06–0.4 in.) in diameter and is fed automatically through a tube (**welding gun**). Electric currents usually range between 300 and 2000 A, usually with standard single- or three-phase power supply at a primary rating of up to 440 V.

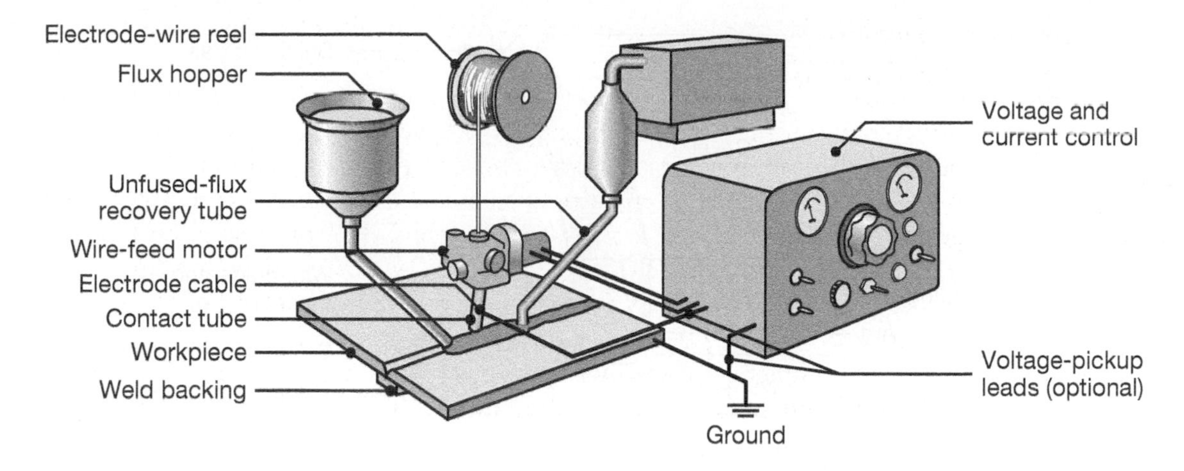

FIGURE 12.7 Schematic illustration of the submerged arc welding process and equipment.

Because the flux is fed by gravity, the SAW process is somewhat limited to welds in a horizontal or flat position, with a backup piece to assure alignment.

The SAW process can be automated and is used to weld a variety of carbon- and alloy-steel and stainless-steel sheet or plate, at speeds as high as 5 m/min (17 ft/min). Typical applications include thick-plate welding for shipbuilding and fabrication of pressure vessels; circular welds on pipes can also be made, provided that the pipes are rotated during the welding operation. The quality of the weld is very high, with good toughness, ductility, and uniformity of properties. The process has very high welding productivity, depositing 4 to 10 times the amount of weld metal per hour compared to the SMAW process.

12.3.4 Gas Metal Arc Welding

In *gas metal arc welding* (GMAW), the weld area is shielded by an external source of gas, such as argon, helium, carbon dioxide, and various gas mixtures (Fig. 12.8a). Deoxidizers are usually present in the electrode metal itself to prevent oxidation of the molten weld puddle. The consumable bare wire is fed automatically through a nozzle into the weld arc (Fig. 12.8b), and multiple weld layers can be deposited at the joint.

Developed in the 1950s, the GMAW process is suitable for a variety of ferrous and nonferrous metals. Used extensively in the metal fabrication industry, the operation is rapid, versatile, and economical, and its welding productivity is double that of the SMAW process. Moreover, it can easily be automated and lends itself readily to robotics and flexible manufacturing systems (Chapters 14 and 15).

Formerly called *MIG welding* (for *metal inert gas*), the weld metal in this process is transferred in one of four ways:

1. **Spray transfer.** Small droplets of molten metal from the electrode are transferred to the weld area, at rates of several hundred droplets per

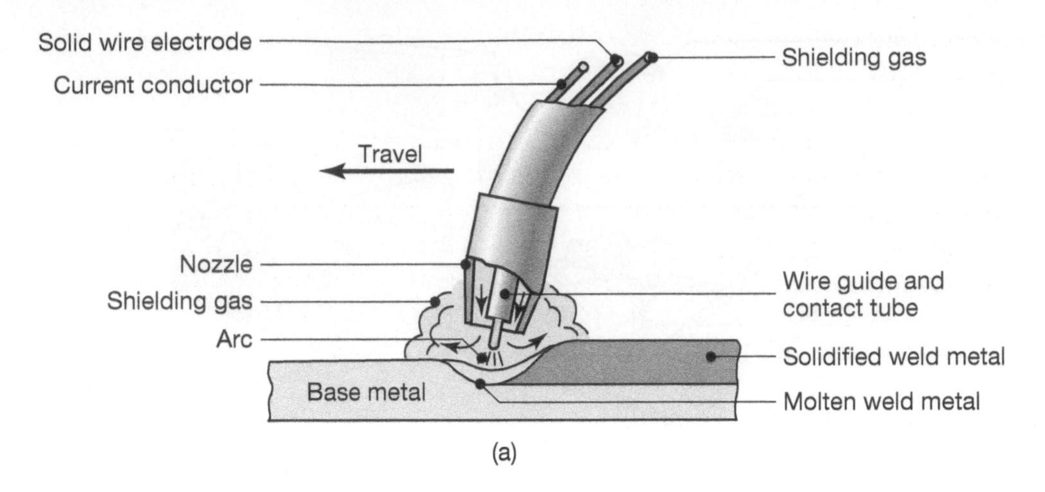

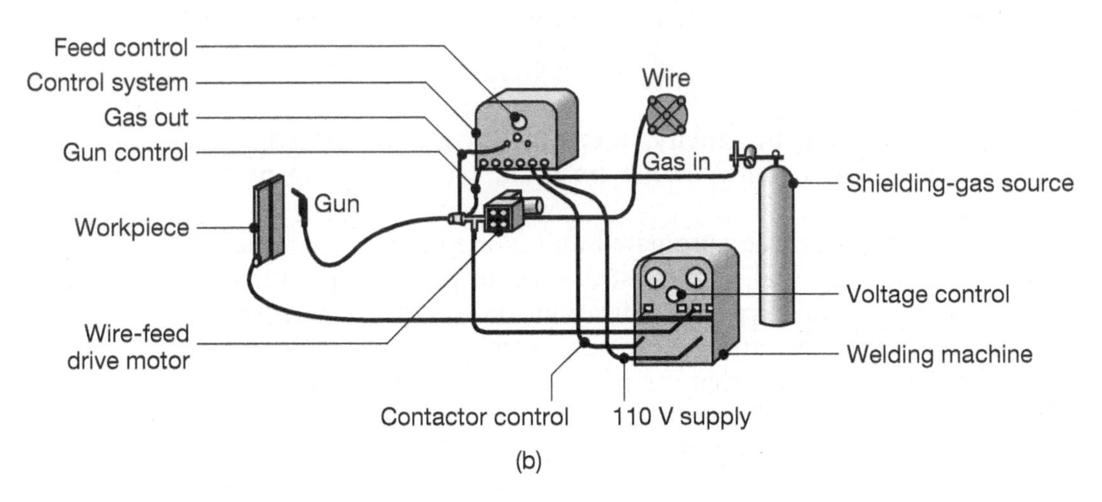

FIGURE 12.8 (a) Gas metal arc welding process, formerly known as MIG welding (for metal inert gas). (b) Basic equipment used in gas metal arc welding operations.

second; the transfer is spatter free and very stable. High DC current, high voltages, and large-diameter electrodes are used, with argon or argon-rich gas mixtures used as the shielding gas. The average current required can be reduced using *pulsed arcs*, which are high-amplitude pulses superimposed over a low, steady current. The process can be used in all welding positions.

2. **Globular transfer.** Using carbon-dioxide rich gases, globules of molten metal are propelled by the force of the electric arc that transfers the metal; the operation produces considerable spatter. Welding currents are high, with greater weld penetration, and welding speed is higher than in spray transfer; heavy steel sections are commonly welded by this method.
3. **Short circuiting.** The metal is transferred in individual droplets, at rates of more than 100/s, as the electrode tip touches the molten weld pool and short circuits. Low currents and voltages are used, with carbon-dioxide rich gases and electrodes made of small-diameter

wire; the power required is about 2 kW. The temperatures developed are relatively low, hence this method is suitable for thin sheets and sections (less than 6 mm thick); with higher thickness, incomplete fusion may occur (See Fig. 12.22). This process is very easy to use and is a common method for welding ferrous metals in thin sections.

4. **Pulsed-spray deposition.** A variation of spray transfer is *pulsed-spray deposition*, where a pulsed current is used to melt the filler wire. Pulsing allows the average current to be lower, leading to a reduction in the size of the heat-affected zone (see Section 12.7). Pulsed-spray deposition can be applied to thin workpieces.

12.3.5 Flux-Cored Arc Welding

The *flux-cored arc welding* (FCAW) process (Fig. 12.9) is similar to gas metal arc welding, with the exception that the electrode is tubular in shape and is filled with flux (hence the term flux-cored). The power required is about 20 kW. Cored electrodes produce a more stable arc, and improve the weld contour and the mechanical properties of the joint. The flux is much more flexible than the brittle coating on SMAW electrodes, thus tubular electrodes can be provided in coiled lengths.

The electrodes are usually 0.5–4 mm (0.02–0.16 in.) in diameter; the smaller electrodes are used for welding thinner materials, as well as making it relatively easy to weld parts in difficult orientations. The flux chemistry enables welding of different base metals. *Self-shielded cored electrodes* are available; they contain emissive fluxes that shield the weld zone against the surrounding atmosphere, and thus do not require external gas shielding.

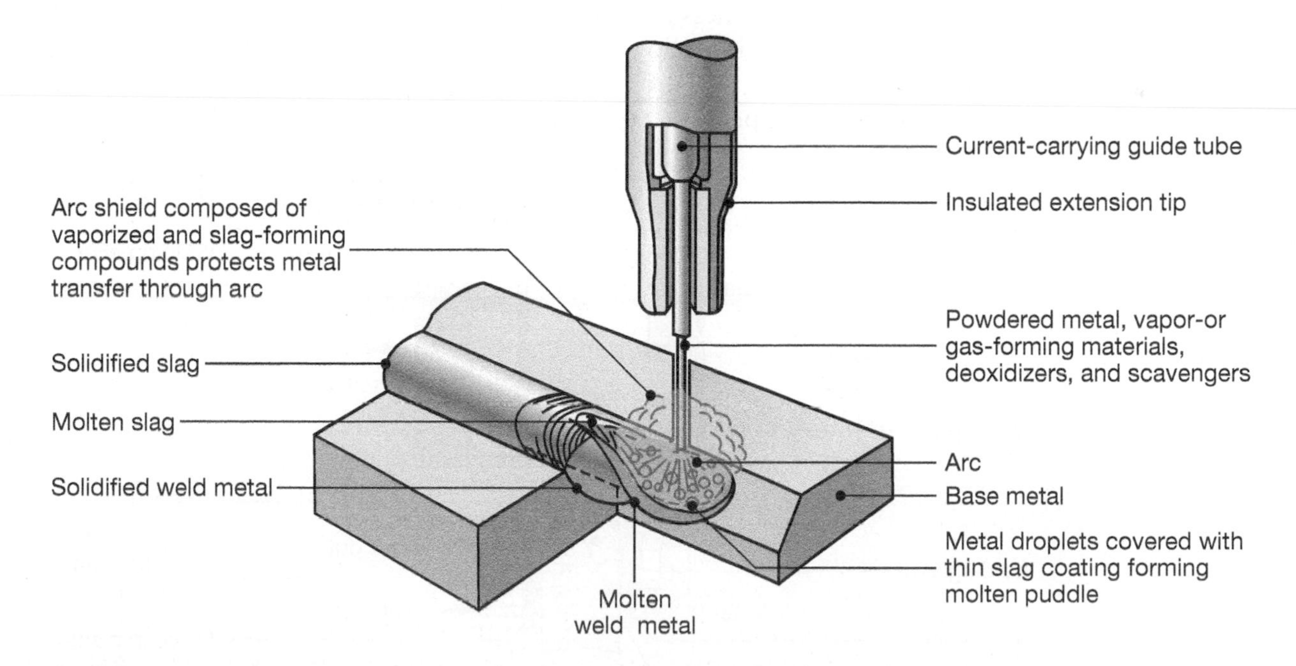

FIGURE 12.9 Schematic illustration of the flux-cored arc welding process; this operation is similar to gas metal arc welding.

The flux-cored arc welding operation combines the versatility of SMAW with the continuous and automatic electrode-feeding feature of GMAW. It is economical and is used for welding a variety of joints with different thicknesses (see also tailor-welded blanks; Section 7.3.4), mainly with steels, stainless steels, and nickel alloys. A major advantage of FCAW is the ease with which specific weld-metal chemistries can be developed; also, by adding alloys to the flux core, virtually any alloy composition can be developed. The process is easy to automate and used with flexible manufacturing systems and robotics.

12.3.6 Electrogas Welding

Electrogas welding (EGW) is primarily used for welding the vertical edges of sections in one pass, with the pieces placed edge to edge (*butt welding*; see Fig. 12.1a). The process is classified as a machine welding process because it requires special equipment (Fig. 12.10). The weld metal is deposited into a weld cavity between the two pieces to be joined; the space is enclosed by two water-cooled copper dams (*shoes*) to prevent the molten slag from running off; mechanical drives move the shoes upward. Circumferential welds, such as on pipes, are also possible, with the workpiece rotating during welding.

Single or multiple electrodes are fed through a conduit, and a continuous arc is maintained, using flux-cored electrodes and currents up to 800 A, or solid electrodes at 400 A; power requirements are about 20 kW. Shielding is provided by an inert gas, such as carbon dioxide, argon, or helium, depending on the type of material being welded. The gas may be provided from an external source or it may be produced from a flux-cored electrode, or both. Weld thickness ranges from 12 to 75 mm (0.5–3 in.) on steels, titanium, and aluminum alloys. Typical applications are in the construction of bridges, ships, pressure vessels, storage tanks, and thick-walled and large-diameter pipes.

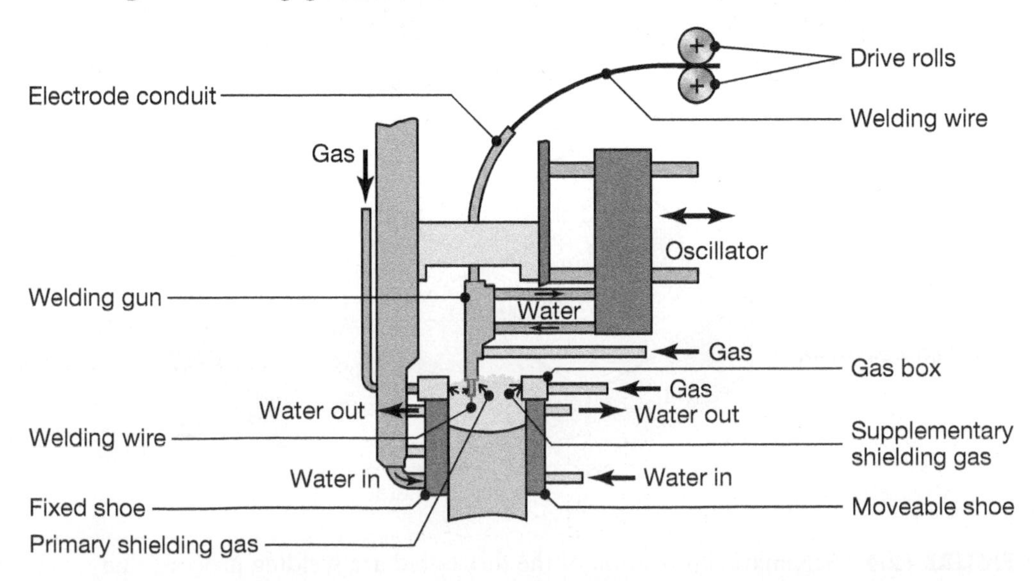

FIGURE 12.10 Schematic illustration of the electrogas welding process.

12.3.7 Electroslag Welding

In *electroslag welding* (ESW), the arc is started between the electrode tip and the bottom of the part to be welded (Fig. 12.11); flux is added and melted by the heat of the arc. After the molten slag reaches the tip of the electrode, the arc is extinguished, while energy is supplied continuously through the electrical resistance of the molten slag. Single or multiple solid or flux-cored electrodes may be used, and the guide tube may be nonconsumable (the conventional method) or consumable.

Electroslag welding has applications that are similar to those for electrogas welding. The process is capable of welding plates with thicknesses ranging from 50 to more than 900 mm (2 in. to more than 3 ft) and welding is usually done in one pass. The current required is about 600 A at 40–50 V, although higher currents are used for thick plates. The travel speed of the weld is 12–36 mm/min (0.5–1.5 in./min). The process is used for heavy structural steel sections, such as heavy machinery and nuclear reactor vessels, and the weld quality is good.

12.3.8 Electrodes for Arc Welding

The *electrodes* for the consumable electrode arc welding processes described thus far are classified according to the strength of the deposited weld metal, the current (AC or DC), and the type of coating. Electrodes are identified by numbers and letters or by color code, particularly if they are too small to imprint with identification. Typical coated electrodes are 150 to 450 mm (6–18 in.) long and 1.5 to 8 mm (0.06–0.32 in.) in diameter.

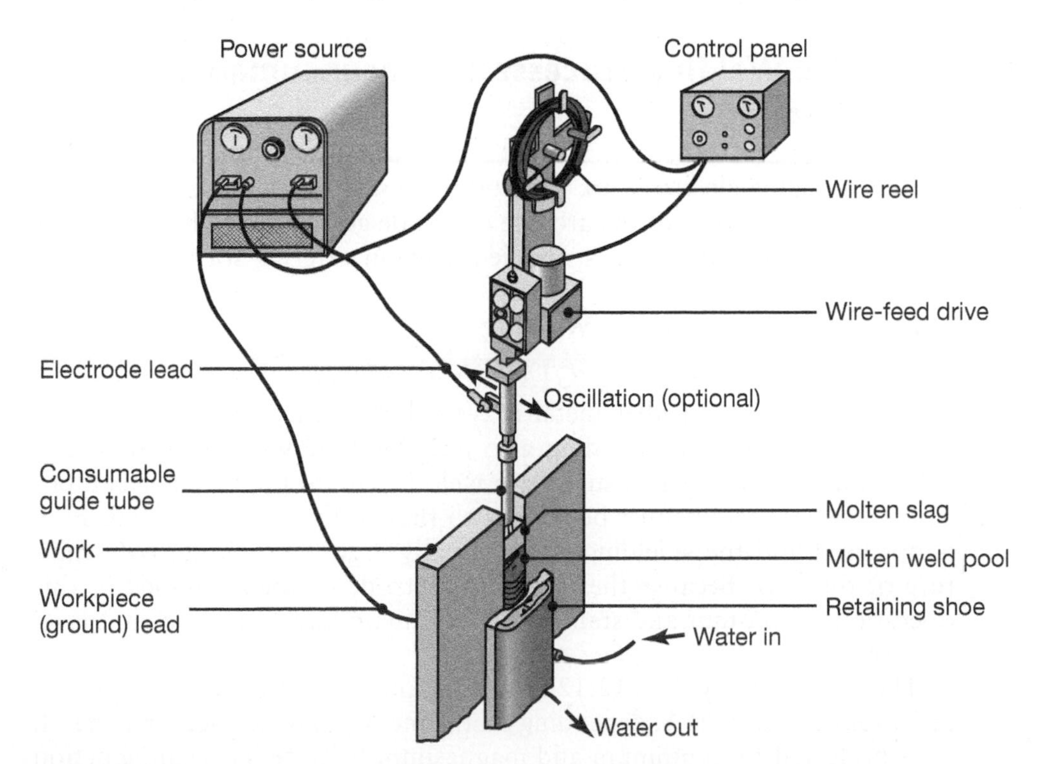

FIGURE 12.11 Equipment used for electroslag welding operations.

The diameter of the electrode decreases with the thickness of the sections to be welded and the current required.

Electrode coatings. Electrodes are *coated* with claylike materials that include silicate binders and powdered materials, such as oxides, carbonates, fluorides, metal alloys, and cellulose (cotton cellulose and wood flour). The coating, which typically is brittle and has complex interactions during welding, has the following basic functions:

1. Stabilizes the arc;
2. Generates gases to act as a shield against the surrounding atmosphere; the gases produced are carbon dioxide and water vapor, and carbon monoxide and hydrogen in small amounts;
3. Controls the rate at which the electrode melts;
4. Acts as a flux to protect the weld against the formation of oxides, nitrides, and other inclusions; the slag that is produced also protects the weld; and
5. Adds alloying elements to the weld zone to enhance the properties of the weld, including deoxidizers to prevent the weld from becoming brittle.

In order to ensure weld quality, the slag must be removed from the weld surfaces after each pass (see also Fig. 12.6); a manual or powered wire brush can be used for this purpose. Bare stainless steel, aluminum alloys electrodes, and wires also are available, and they are used as filler metals in various welding operations.

12.4 Arc Welding Processes: Nonconsumable Electrode

Nonconsumable electrode arc welding processes typically use a **tungsten electrode**. As one pole of the arc, the electrode generates the heat required for welding; a shielding gas is supplied from an external source.

12.4.1 Gas Tungsten Arc Welding

In *gas tungsten arc welding* (GTAW), formerly called *TIG welding* (for *tungsten inert gas*), a filler metal is typically supplied from a **filler wire** (Fig. 12.12a); however, welding also may be done without filler metals (called *autogenous welds*), such as in welding close-fit joints. The composition of filler metals must be similar to that of the metals welded. Flux is not used and the shielding gas is usually argon or helium, or a mixture of the two. Because the tungsten electrode is not consumed in this operation, a constant and stable arc gap is maintained at a constant level of current.

The power supply (Fig. 12.12b) ranges from 8 to 20 kW and is either DC at 200 A or AC at 500 A, depending on the metals to be welded. In general, AC is preferred for aluminum and magnesium, because its cleaning action removes oxides and improves weld quality. The tungsten electrodes may

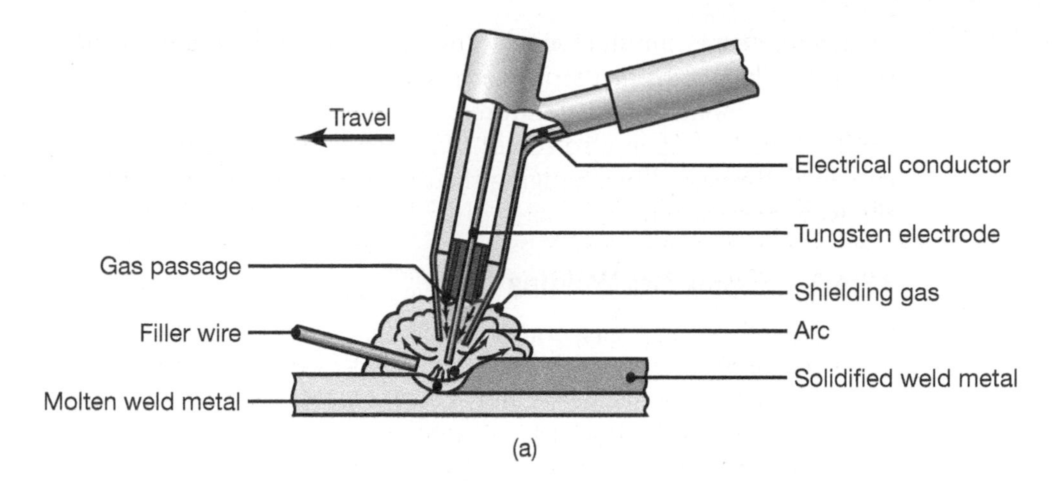

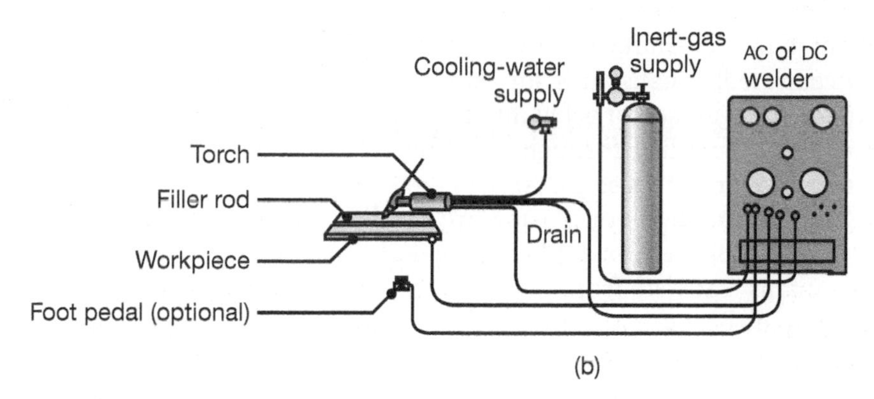

FIGURE 12.12 (a) Gas tungsten arc welding process, formerly known as TIG welding (for tungsten inert gas). (b) Equipment for gas tungsten arc welding operations.

contain thorium or zirconium to improve their characteristics. Contamination of the tungsten electrode by the molten metal can be a significant problem in this process, particularly in critical applications, as contamination can cause discontinuities in the weld, thus contact of the electrode with the molten metal pool must be avoided.

The GTAW process is used for a wide variety of applications and metals, particularly aluminum, magnesium, titanium, and refractory metals; it is especially suitable for thin metals. The cost of the inert gas makes this process more expensive than SMAW, but it provides welds with very high quality and surface finish; however, it is a manual process, requiring skilled labor to produce high quality welds.

12.4.2 Atomic Hydrogen Welding

In *atomic hydrogen welding* (AHW), an arc is generated between two tungsten electrodes in a shielding atmosphere of flowing hydrogen gas. The hydrogen gas is normally diatomic (H_2), but near the arc, where the temperatures are over 6000°C (10,800°F), the hydrogen breaks down into its atomic form, simultaneously absorbing a large amount of heat from the arc. When hydrogen strikes a relatively cold surface, that is, the weld zone, it recombines into its diatomic form and releases the stored heat very rapidly; the hydrogen also serves as a shielding gas. The energy in AHW can easily be varied by changing the distance between the arc stream and the workpiece surface.

12.4.3 Plasma Arc Welding

Plasma is ionized hot gas, consisting of nearly equal numbers of electrons and ions. In *plasma arc welding* (PAW), developed in the 1960s, a concentrated plasma arc is produced and aimed at the weld area. The arc is stable and reaches temperatures as high as 33,000°C; operating currents are usually below 100 A, but they can be higher for special applications. The plasma is initiated between the tungsten electrode and the orifice, using a low-current pilot arc. Unlike the arc in other joining processes, the plasma arc is concentrated, because it is forced through a relatively small orifice. When a filler metal is used, it is fed into the arc, as in GTAW. Arc and weld-zone shielding is supplied through an outer shielding ring, using argon, helium, or a mixture of these gases.

There are two methods of plasma arc welding. In the **transferred arc** method (Fig. 12.13a), the metal being welded is part of the electrical circuit; the arc thus transfers from the electrode to the workpiece, hence the term transferred. Because a transferred arc provides high energy density, this process is commonly used for cutting stainless steels and non-ferrous alloys. In the **nontransferred** method (Fig. 12.13b), the arc is between the electrode and the nozzle, and the heat is carried to the workpiece by the plasma gas. This technique is also used for *thermal spraying*, as described in Section 4.5.1.

Compared with other arc welding processes, plasma arc welding has (a) greater energy concentration, thus deeper and narrower welds can be made; (b) better arc stability; (c) less thermal distortion; and (d) higher

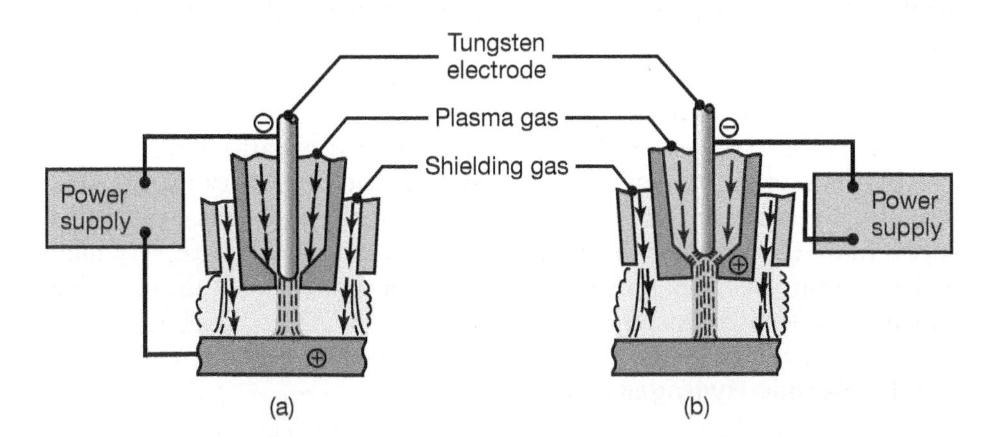

FIGURE 12.13 Two types of plasma arc welding processes: (a) transferred and (b) nontransferred. Deep and narrow welds are made by this process at high welding speeds.

welding speeds, such as 120–1000 mm/min (5–40 in/min). A variety of metals can be welded, with part thicknesses generally less than 6 mm (0.25 in.). The high heat concentration can completely penetrate through the joint (**keyhole technique**), to thicknesses as much as 20 mm (0.8 in.) for some titanium and aluminum alloys. In this technique, the force of the plasma arc displaces the molten metal and produces a hole at the leading edge of the weld pool. Plasma arc welding is often used for butt and lap joints, because of its high energy concentration, better arc stability, and higher welding speeds.

12.5 High Energy Beam Welding

Joining with high energy beams, principally laser-beam and electron-beam welding, has important applications in modern manufacturing, because of its high quality and technical and economic advantages. The general characteristics of high energy beams and their unique applications in machining operations are described in Section 9.14.

12.5.1 Electron-Beam Welding

In *electron-beam welding* (EBW), heat is generated by high-velocity, narrow-beam electrons; the kinetic energy of the electrons is converted into heat as the electrons strike the workpiece. This process requires special equipment to focus the beam on the workpiece, in a vacuum; the capacities of electron-beam guns range up to 100 kW. The higher the vacuum, the deeper the beam penetrates the part and the greater is the depth-to-width ratio that can be obtained. The level of vacuum is specified as HV (high vacuum) or MV (medium vacuum), both of which are commonly used. NV (non vacuum) can be effective on some materials, mainly copper. Almost any similar or dissimilar metals can be butt or lap welded, with thicknesses ranging from foil to 150-mm (6-in.) thick plates. The intense energy is also capable of producing holes (see Section 9.14), and is the basis for electron beam melting (see Section 10.12.4). Generally, no shielding gas, flux, or filler metal are required.

Developed in the 1960s, the EBW process produces high-quality, deep and narrow welds that are almost parallel sided and have small heat-affected zones (see Section 12.7); depth-to-width ratios range between 10 and 30 (Fig. 12.14). Welding parameters can be precisely controlled, at speeds as high as 12 m/min (40 ft/min). Distortion and shrinkage in the weld area are minimal, and the weld has very high purity. Typical applications include welding of aircraft, missile, nuclear, and electronic components, as well as gears and shafts for the automotive industry. Because EBW equipment generates X-rays, proper monitoring and periodic maintenance are essential.

12.5.2 Laser-Beam Welding

Laser-beam welding (LBW) uses a high-power laser as the source of heat (See Fig. 9.37 and Table 9.5) to produce a fusion weld. Because the laser

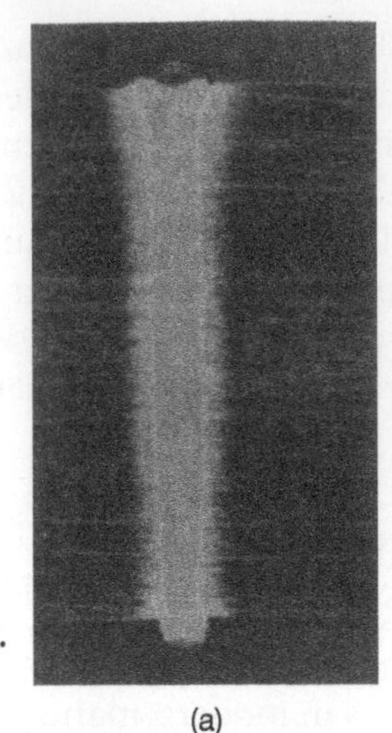
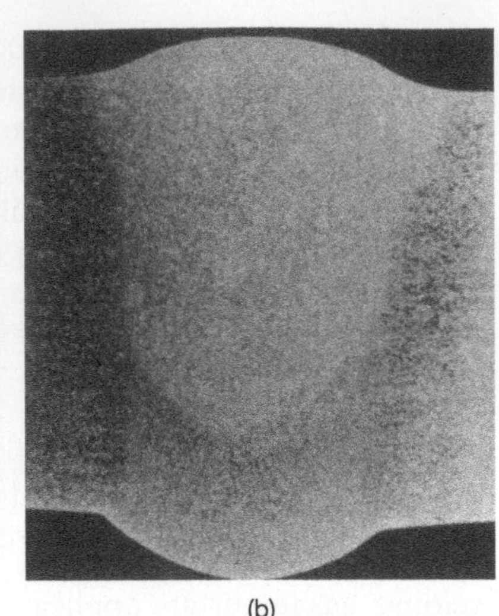
(a) (b)

FIGURE 12.14 Comparison of the size of weld beads in (a) electron-beam or laser-beam welding with that in (b) conventional (tungsten arc) welding. *Source:* American Welding Society, *Welding Handbook*, 8th ed., 1991.

QR Code 12.1 Laser cutting system. *Source:* Courtesy of the LVD Company.

beam can be highly focused to as small a diameter as 10 μm, it has high energy density and deep penetrating capability. This process is thus particularly suitable for making deep and narrow joints (Fig. 12.14a), with depth-to-width ratios typically ranging from 4 to 10. The laser beam may be **pulsed** (milliseconds) for such applications as spot welding of thin materials, with power levels up to 100 kW. **Continuous** multi-kW laser systems are used for deep welds on thick sections. The efficiency of this process decreases with increasing *reflectivity* of the workpiece materials. To further improve performance, oxygen may be used for welding steels and inert gases for nonferrous metals.

The LBW process can be automated and used on a variety of materials, with thicknesses up to 25 mm (1 in.); it is particularly effective on thin workpieces. Typical metals and alloys welded include aluminum, titanium, ferrous metals, copper, superalloys, and refractory metals. Welding speeds range from 2.5 m/min (100 in./min) to as high as 80 m/min (270 ft/min) for thin metals. Because of the nature of the process, welding can be done in otherwise inaccessible locations. Safety is particularly important because of the extreme hazards to the eye and the skin; solid state (YAG) lasers are particularly harmful.

Laser-beam welding produces welds of good quality, with minimum shrinkage and distortion; they have good strength and generally are ductile and free of porosity. In the automotive industry, welding of transmission components is a widespread application of this process; among numerous other applications is the welding of thin parts for electronic components, hermetic welding of pacemakers, and transmission shaft components (where it competes with *friction welding*, described in Section 12.10). As shown in Fig 7.14, another important application of laser welding is the forming of butt-welded sheet metal, particularly for automotive body panels (see *tailor-welded blanks* in Section 7.3.4).

The major advantages of LBW over EBW are:

- The laser beam can be transmitted through air, hence a vacuum is not required as is in electron-beam welding.

- Because laser beams can be shaped, manipulated, and focused optically (using optical fibers) the process can easily be automated.
- Laser beams do not generate X-rays, unlike electron beams.
- The quality of the weld is better than EBW, with less tendency for incomplete fusion, spatter, porosity, and distortion.

QR Code 12.2 Adaptive CO_2 laser cutting system. *Source:* Courtesy of the LVD Company.

Laser GMAW. This is a *hybrid welding* technology that combines the narrow heat-affected zone of laser welding with the high deposition rate of gas-metal arc welding. In this process, shown in Fig. 12.15, the laser is focused on the workpiece ahead of the GMAW arc, heating the workpiece, resulting in deep penetration and allowing high travel speeds. Furthermore, the process is able to bridge larger gaps than traditional laser welding, and the metallurgical quality of the weld is improved because of the presence of the shielding gas.

12.6 Cutting

In addition to being cut by mechanical means, material can be cut into various contours using a heat source that melts and removes a narrow zone in the workpiece. The sources of heat can be torches, electric arcs, or lasers.

Oxyfuel-gas cutting. *Oxyfuel–gas cutting* (OFC) is similar to oxyfuel–gas welding (Section 12.2), but the heat source is used to *remove* a narrow zone from a metal plate or sheet (Fig. 12.16a). This process is suitable particularly for steels, where the basic reactions are

$$\mathrm{Fe} + \mathrm{O} \rightarrow \mathrm{FeO} + \mathrm{Heat}, \tag{12.6}$$

$$3\mathrm{Fe} + 2\mathrm{O_2} \rightarrow \mathrm{Fe_3O_4} + \mathrm{Heat}, \tag{12.7}$$

and

$$4\mathrm{Fe} + 3\mathrm{O_2} \rightarrow 2\mathrm{Fe_2O_3} + \mathrm{Heat}. \tag{12.8}$$

The greatest heat is generated by the second reaction, with temperatures rising to about 870°C (1600°F). However, because this temperature is not sufficiently high, the workpiece is *preheated* with fuel gas, and oxygen is introduced later, as can be seen from the nozzle cross section in Fig. 12.16a.

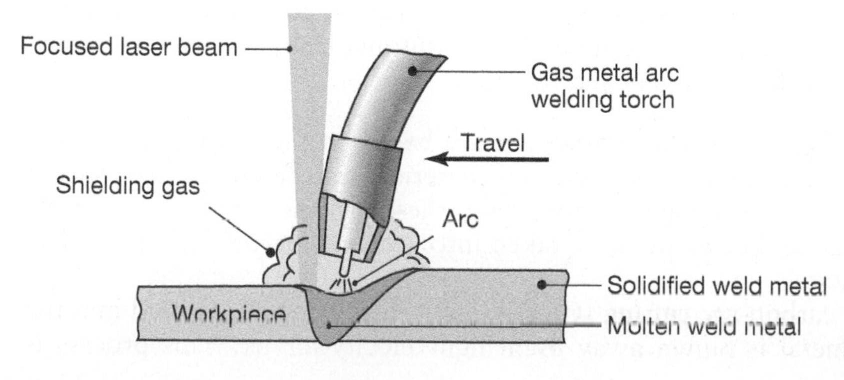

FIGURE 12.15 Schematic illustration of the Laser GMAW hybrid welding process.

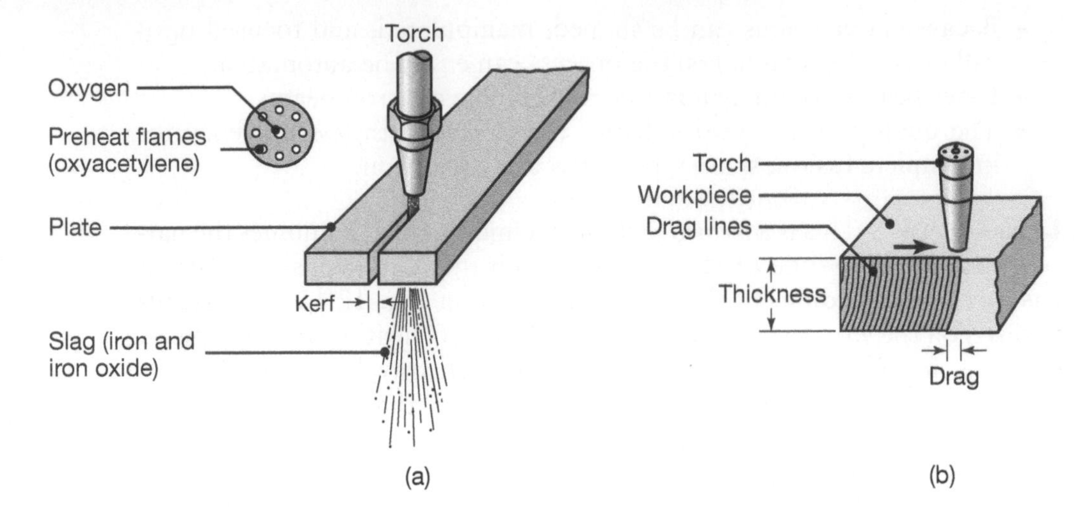

FIGURE 12.16 (a) Flame cutting of a steel plate with an oxyacetylene torch, and a cross section of the torch nozzle. (b) Cross section of a flame-cut plate, showing drag lines.

The higher the carbon content of the steel, the higher the preheating temperature required. Cutting takes place mainly by the oxidation (burning) of the steel; some melting also takes place. Cast irons and steel castings also can be cut by this method. The process generates a **kerf**, similar to that produced in sawing with a saw blade or by wire electrical-discharge machining (See Fig. 9.36). Kerf widths range from about 1.5 to 10 mm (0.06 to 0.4 in.), with good control of dimensional tolerances. However, distortion caused by uneven temperature distribution can be a problem in OFC.

The maximum thickness that can be cut by OFC depends mainly on the gases used. With oxyacetylene gas, for example, the maximum thickness is about 300 mm (12 in.); with oxyhydrogen, it is about 600 mm (24 in.). The flame leaves **drag lines** on the cut surface (Fig. 12.16b), resulting in a rougher surface than that produced by processes such as sawing, blanking, or other similar operations that use mechanical cutting tools. *Underwater cutting* is done with specially designed torches that produce a blanket of compressed air between the flame and the surrounding water. Although long used for salvage and repair work, OFC can be used in manufacturing as well. Torches may be guided along specified paths either manually, mechanically, or automatically by machines using programmable controllers and robots.

Arc cutting. *Arc-cutting* processes are based on the same principles as arc-welding processes. A variety of materials can be cut at high speeds by arc cutting, although as in welding, these processes also leave a heat-affected zone that needs to be taken into account, particularly in critical applications.

In **air carbon-arc cutting** (CAC-A), a carbon electrode is used and the molten metal is blown away by a high-velocity air jet. This process is used especially for gouging and scarfing (removal of metal from a surface). However, the process is noisy, and the molten metal can be blown substantial distances and cause safety hazards.

Plasma-arc cutting (PAC) produces the highest temperatures, and is used for the rapid cutting of nonferrous and stainless-steel plates. The cutting productivity of this process is higher than that of oxyfuel–gas methods. PAC produces a good surface finish and narrow kerfs, and is the most common cutting process utilizing programmable controllers employed in manufacturing today. **Electron beams** and **lasers** also are used for very accurately cutting a wide variety of metals, as was described in Sections 12.5.1 and 12.5.2. The surface finish is better than that of other thermal cutting processes, and the kerf is narrower.

12.7 The Fusion Welded Joint

This section describes the following aspects of fusion welding processes:

- The nature, properties, and quality of the *welded joint*;
- *Weldability* of metals; and
- *Testing* welds.

A typical fusion welded joint is shown in Fig. 12.17, where three distinct zones are identified:

1. The **base metal**, the metal to be joined;
2. The **heat-affected zone** (HAZ); and
3. The **fusion zone**, the region that melts during welding.

The metallurgy and properties of the second and third zones depend greatly on (a) the metals being joined (whether they are single phase, two phase, or dissimilar); (b) the particular welding process; (c) the filler metals used

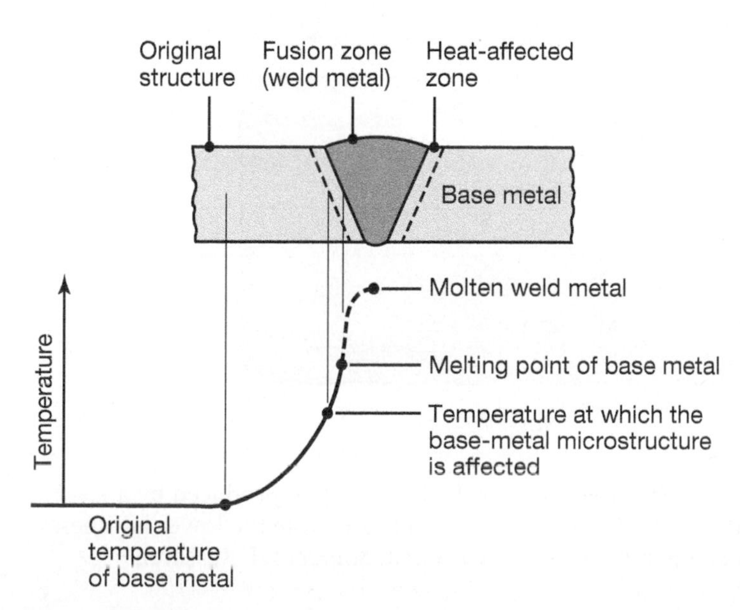

FIGURE 12.17 Characteristics of a typical fusion weld zone in oxyfuel gas welding and arc welding processes.

(if any); and (d) the process variables. A joint basically consists of the *resolidified* base metal; if a filler metal is used, there is a central zone, called the **weld metal**, composed of a mixture of the base and filler metals.

The mechanical properties of a welded joint depend on several factors: (a) the rate of heat application and the thermal properties of metals, because they control the magnitude and distribution of temperature in a joint during welding; (b) the microstructure and grain size of the joint, as they depend on the magnitude of heat applied, the temperature rise, the degree of prior cold work of the metals, and the rate of cooling after the weld is completed; and (c) the weld quality, depending on such factors as the geometry of the weld bead, the presence of any cracks, residual stresses, inclusions, and oxide films.

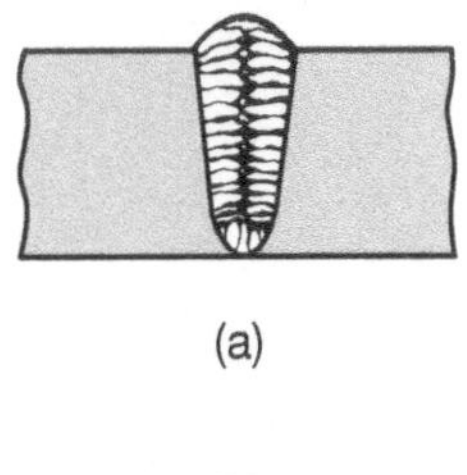

(a)

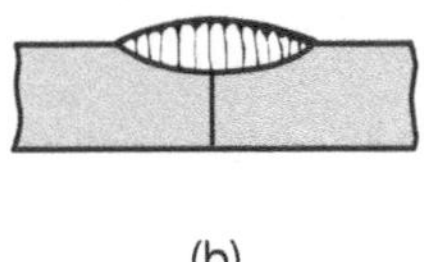

(b)

FIGURE 12.18 Grain structure in (a) a deep weld and (b) a shallow weld. Note that the grains in the solidified weld metal are perpendicular to their interface with the base metal.

Solidification of the weld metal. After heat has been applied and the filler metal (if any) has been introduced into the weld zone, the molten weld begins to solidify and cool. The *solidification process* is similar to that in casting (Section 5.3) and begins with the formation of columnar (dendritic) grains; these grains are relatively long and form parallel to the heat flow (See Fig. 5.5). Because metals are much better heat conductors than the surrounding air, the grains lie parallel to the plane of the two plates or sheets being welded (Fig. 12.18a). The grains developed in a *shallow* weld are shown in Fig. 12.18b; the final structure of the grains and their size depend on the specific alloy, the particular welding process, and the filler metal.

The weld metal has a *cast structure*, and because it has cooled rather slowly, it typically has coarse grains; consequently, the metal has relatively low strength, hardness, toughness, and ductility (Fig. 12.19). These properties can, however, be improved with proper selection of filler metal composition and with subsequent heat treatment of the joint. The result depends on the particular alloy, its composition, and the thermal cycling

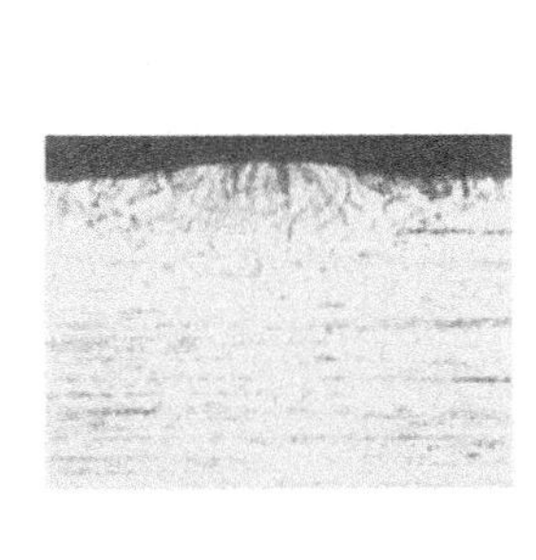

(a)

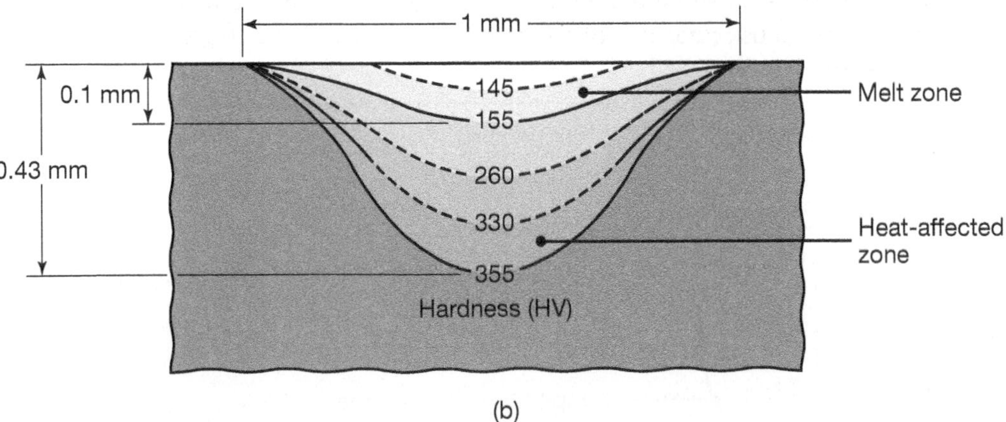

(b)

FIGURE 12.19 (a) Weld bead on a cold-rolled nickel strip produced by a laser beam. (b) Microhardness profile across the weld bead. Note the lower hardness of the weld bead as compared with the base metal. *Source:* IIT Research Institute.

to which the joint has been subjected. For example, cooling rates may be controlled and reduced by *preheating* the weld zone prior to welding. Preheating is particularly important for metals with high thermal conductivity, such as aluminum and copper, as otherwise the heat can rapidly dissipate during welding.

Heat-affected zone. The heat-affected zone is within the base metal itself, as can be seen in Fig. 12.17. It has a microstructure different from that of the base metal prior to welding, because it has been subjected to elevated temperatures for a period of time during the welding process. The regions of the base metal that are far away from the heat source do not undergo any significant structural changes during welding. The properties and microstructure of the HAZ depend on (a) the rates of heat input and of cooling; (b) the temperature to which the zone has been raised during welding; and (c) the original grain size, grain orientation, degree of prior cold work, and the specific heat and thermal conductivity of the metals.

The strength and hardness of the heat-affected zone also depend partly on how the original strength and hardness of the particular alloy had been developed prior to welding. As described in Chapters 3 and 5, they may, for example, have been developed by cold working, solid solution strengthening, precipitation hardening, or various other thermal treatments. Among metals strengthened by these methods, the simplest to analyze is the base metal that has been cold worked, such as by cold rolling or cold forging. The heat applied during welding *recrystallizes* the elongated grains of the cold-worked base metal. The grains that are away from the weld metal will recrystallize into fine equiaxed grains, but those close to the weld metal, having been subjected to elevated temperatures for a longer period of time, will grow (see also Section 3.6). Grain growth will result in a region that is softer and has lower strength; such a joint will be weakest in its heat-affected zone. The grain structure of such a weld, exposed to corrosion by chemical reaction, is shown in Fig. 12.20; the central vertical line is the juncture of the two workpieces. The effects of heat during welding on the HAZ for joints made with *dissimilar* metals and for alloys strengthened by other mechanisms are complex, and beyond the scope of this text.

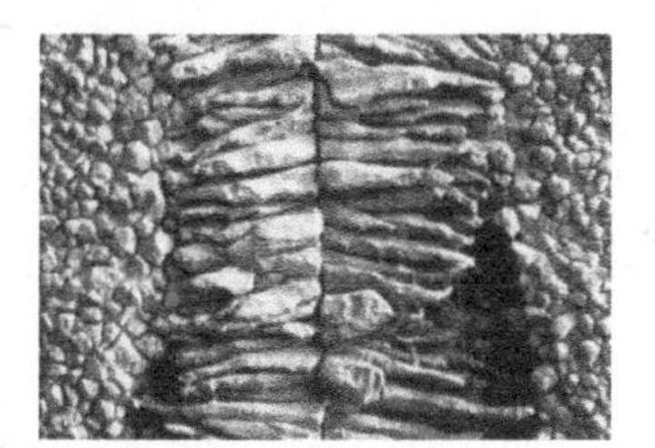

FIGURE 12.20 Intergranular corrosion of a weld joint in ferritic stainless-steel welded tube, after exposure to a caustic solution. The weld line is at the center of the photograph.

12.7.1 Weld Quality

Because of its past history of thermal cycling and the attendant microstructural changes, a welded joint may develop *discontinuities* that can be caused by inadequate or careless application of established welding procedures or poor operator training. The major discontinuities that affect weld quality are described below.

1. **Porosity** in welds may be caused by (a) *trapped gases* that are released during melting of the weld zone but trapped during solidification of the weld zone; (b) *chemical reactions* that occur during welding; or (c) *contaminants* present in the weld zone. Most welded joints contain some porosity, generally spherical in shape or as elongated

cavities (see also Section 5.12.1). The *distribution* of porosity may be random or it may be concentrated in a certain region. Particularly important is the presence of *hydrogen*, which may be due to the use of damp fluxes or to environmental humidity; examples are porosity in aluminum-alloy welds and hydrogen embrittlement in steels (see Section 3.8.2).

Porosity in welds can be reduced by the following methods:

a. Proper selection of electrodes and filler metals;
b. Preheating the weld area or increasing the rate of heat input;
c. Cleaning of the weld zone and prevention of contaminants from entering the weld zone; and
d. Lowering the welding speed, to allow time for gases to escape.

2. **Slag inclusions.** These inclusions may be trapped in the weld zone. If the shielding gases used are not effective, contamination from the environment may contribute to slag inclusions. Maintenance of proper welding conditions also is important; for example, when proper techniques are used, the slag will float to the surface of the molten weld metal and not be entrapped. Slag inclusions or *slag entrapment* may be prevented by the following methods:

   a. Cleaning the weld-bead surface before the next layer is deposited (See Fig. 12.6), typically using a hand or power wire brush;
   b. Providing adequate shielding gas; and
   c. Redesigning the joint (Section 12.18) to permit sufficient space for proper manipulation of the puddle of molten weld metal.

3. **Incomplete fusion and incomplete penetration.** Incomplete fusion (*lack of fusion*) produces poor weld beads, such as those shown in Fig. 12.21. A better weld can be obtained by the following methods:

   a. Raising the temperature of the base metal;
   b. Cleaning the weld area prior to welding;
   c. Modifying the joint design;
   d. Changing the type of electrode; and
   e. Providing adequate shielding gas.

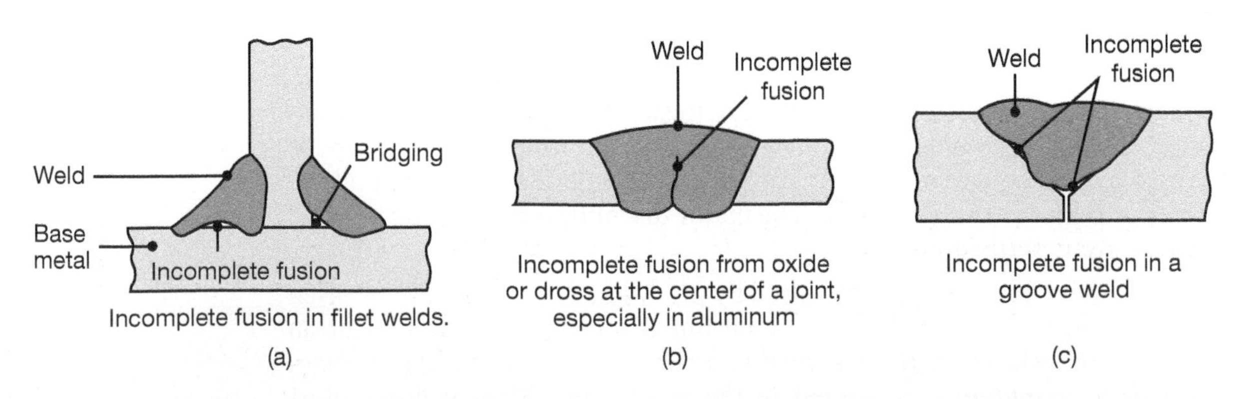

FIGURE 12.21 Examples of various incomplete fusion in welds.

Incomplete penetration occurs when the depth of the welded joint is insufficient. Penetration can be improved by the following methods:

a. Increasing the heat input;
b. Lowering travel speed during welding;
c. Modifying the design of the joint; and
d. Ensuring that the surfaces to be joined fit properly.

4. **Weld profile.** Weld profile is important not only because of its effects on the strength and appearance of the weld but also because it can indicate incomplete fusion or the presence of slag inclusions in multiplayer welds.

   Underfilling, shown in Fig. 12.22a, is a result of the joint not being filled with the proper amount of weld metal.

   Undercutting results from melting away of the base metal and the subsequent development of a groove in the shape of a sharp recess or notch. If deep or sharp, an undercut can act as a stress raiser and reduce the fatigue strength of the joint.

   Overlap (Fig. 12.22b) is a surface discontinuity, generally caused by poor welding practice and improper selection of the materials involved. A proper weld bead, without these defects, is shown in Fig. 12.22c.
5. **Cracks.** *Cracks* may develop at various locations and directions in the weld zone. Typical types are longitudinal, transverse, crater, underbead, and toe cracks (Fig. 12.23). Cracks generally result from a combination of the following factors:

   a. Temperature gradients, causing thermal stresses in the weld zone;
   b. Variations in the composition of the weld zone, causing different contractions;

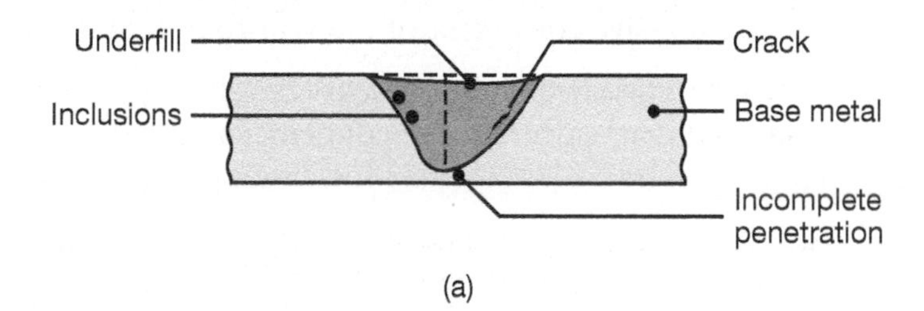

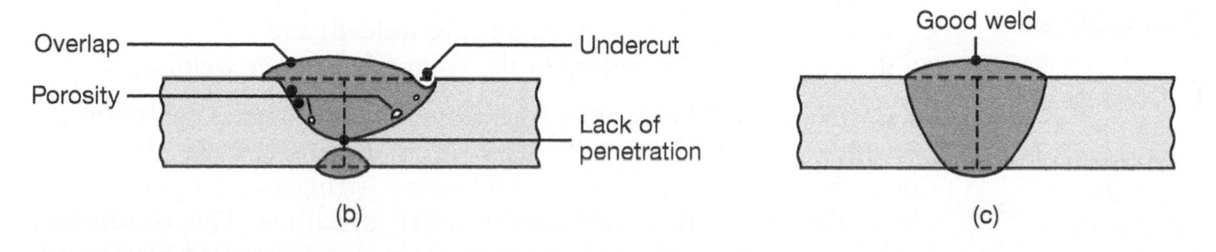

FIGURE 12.22 Examples of various defects in fusion welds.

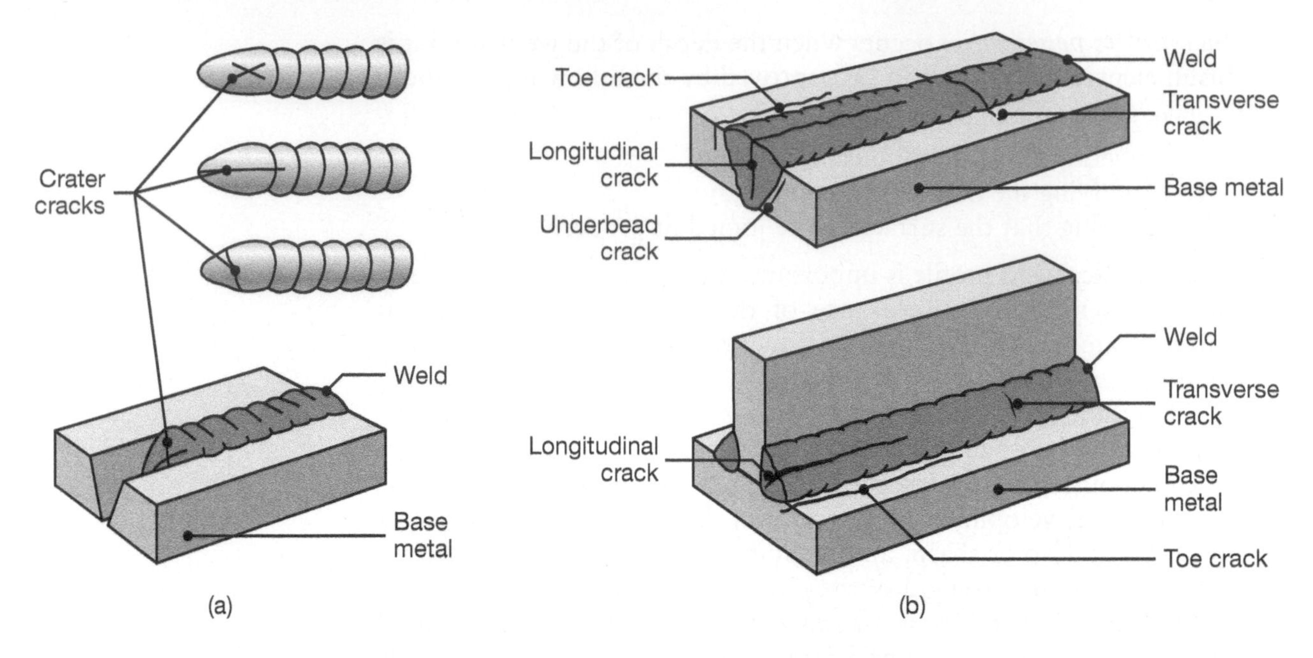

FIGURE 12.23 Types of cracks in welded joints. The cracks are caused by thermal stresses that develop during solidification and contraction of the weld bead and the welded structure: (a) crater cracks and (b) various types of cracks in butt and T joints.

c. Embrittlement of grain boundaries by segregation of elements such as as sulfur to the grain boundaries (see Section 3.4.2), as the solid-liquid boundary moves when the weld metal begins to solidify;
d. Hydrogen embrittlement (Section 3.8.2); and
e. Inability of the weld metal to contract during cooling (Fig. 12.24), a situation similar to the development of *hot tears* in castings, due to excessive constraint of the workpiece (Section 5.12.1).

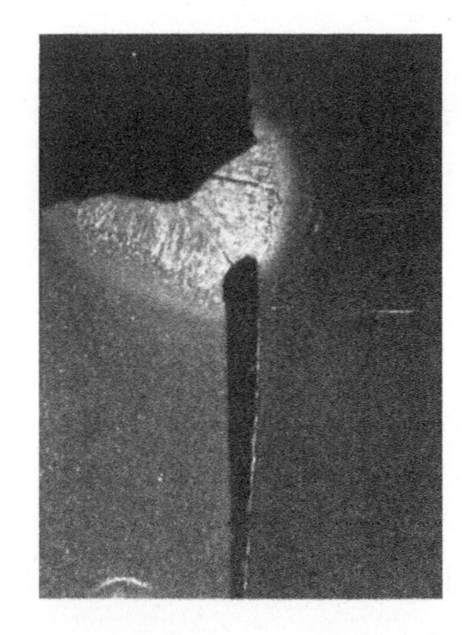

FIGURE 12.24 Crack in a weld bead, due to the two components not being allowed to contract after the weld was completed. *Source:* Courtesy of Packer Engineering.

Cracks are classified as **hot cracks**, developing while the joint is still at elevated temperatures, and **cold cracks**, that develop after the weld metal has solidified. Measures for crack prevention are:

a. Modifying the joint design to minimize thermal stresses due to shrinkage during cooling;
b. Changing welding process parameters, procedures, and sequence of steps taken;
c. Preheating the components being welded; and
d. Avoiding rapid cooling of the components after welding.

6. **Lamellar tears.** In describing the anisotropy of plastically deformed metals (Section 3.5), it was stated that, because of the alignment of nonmetallic impurities and inclusions (stringers), the workpiece is weaker in one direction than in other directions. This condition, called *mechanical fibering*, is particularly found in rolled plates and structural shapes. In welding such components, lamellar tears may

develop, because of shrinkage of the restrained members in the structure during cooling. These tears can be avoided by providing for shrinkage of the members or by modifying the joint design to make the weld bead penetrate the weaker member to a greater extent.

7. **Surface damage.** During welding, some of the metal may spatter and be deposited as small, solid droplets on adjacent surfaces. Also, in arc welding, the electrode may inadvertently be allowed to contact the parts being welded (**arc strikes**), at locations outside of the weld zone. Such surface discontinuities may be objectionable for reasons of appearance or in the subsequent use of the welded part. If severe, they may also adversely affect the properties of the welded structure, particularly for notch-sensitive metals (Section 2.9).
8. **Residual stresses.** Because of localized heating and cooling during welding, expansion and contraction of the weld area develop *residual stresses* (Section 2.10). Residual stresses can cause detrimental effects, such as:

   a. Distortion, warping, and buckling of the welded parts (Fig. 12.25);
   b. Stress-corrosion cracking (Section 3.8.2);
   c. Further distortion if a portion of the welded structure is subsequently removed, say, by machining or sawing (See Fig. 2.28); and
   d. Reduced fatigue life.

   The type and distribution of residual stresses in welds can be best described by reference to Fig. 12.26a. When two plates are being welded, a long, narrow region is subjected to elevated temperatures, whereas the plates as a whole are essentially at ambient temperature. As the weld is completed and time elapses, the heat from the weld area dissipates laterally into the plates; the plates then begin to expand longitudinally while the welded length begins to contract. These two opposing effects cause residual stresses that typically are distributed as illustrated in Fig. 12.26b. Note that, as expected, the

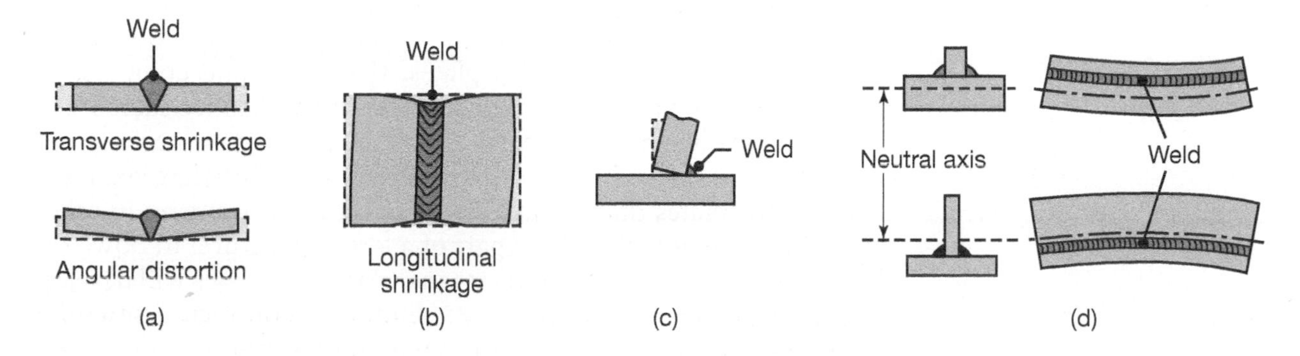

FIGURE 12.25 Distortion and warping of parts after welding, caused by differential thermal expansion and contraction of different regions of the welded assembly. Warping can be reduced or eliminated by proper weld design and fixturing prior to welding.

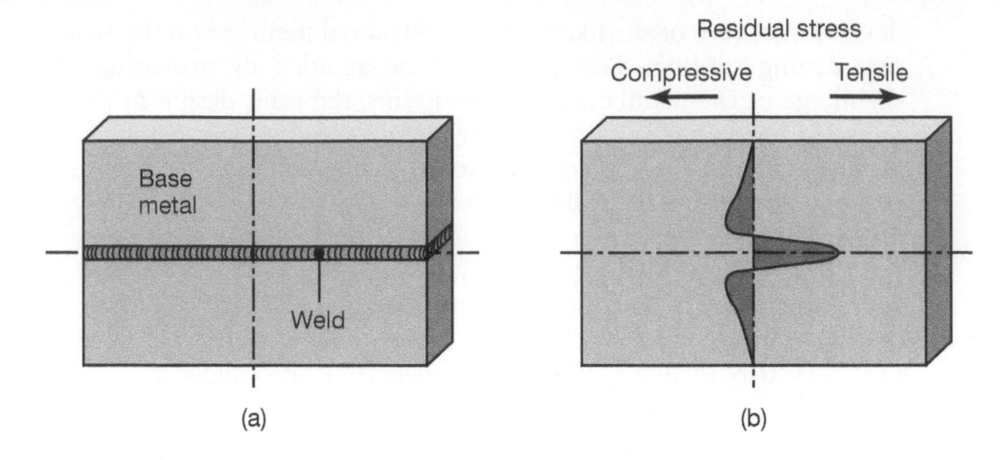

FIGURE 12.26 Residual stresses developed in a straight butt joint.

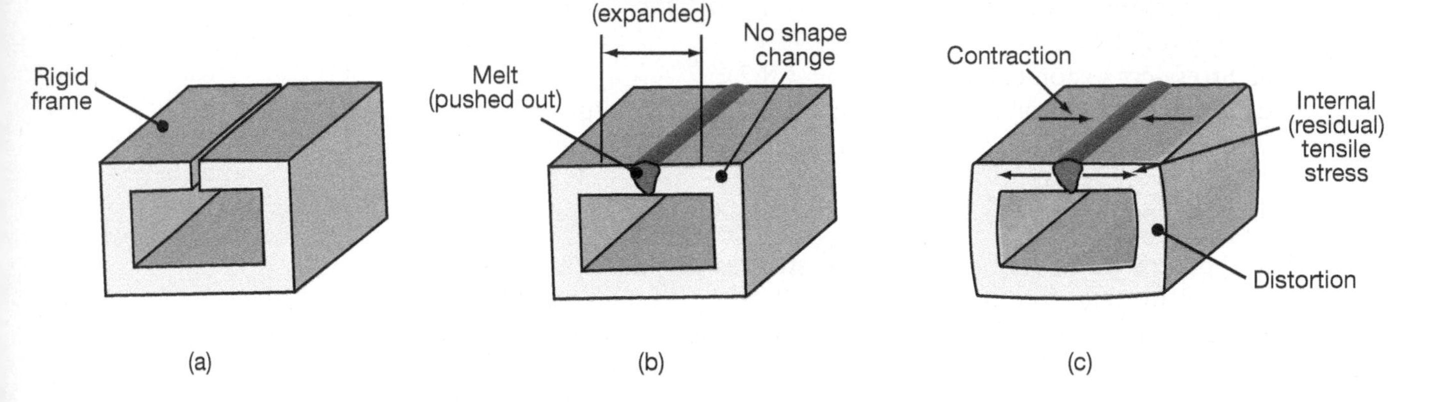

FIGURE 12.27 Distortion of a welded structure. (a) Before welding; (b) during welding, with weld bead placed in joint; and (c) after welding, showing distortion in the structure. *Source:* After J.A. Schey.

magnitude of compressive residual stresses in the plates diminishes to zero at a location away from the weld area. Because no external forces are acting on the welded plates, the tensile and compressive forces represented by these residual stresses must balance each other.

In complex welded structures, residual stress distributions are three dimensional and difficult to analyze. The preceding example involves two plates that are not restrained from movement; in other words, they are not an integral part of a larger structure. If, however, the plates are restrained by some means, reaction stresses will develop because the plates are not free to expand or to contract, a situation that occurs particularly in structures with high stiffness.

Events leading to the *distortion* of a welded structure are shown in Fig. 12.27. Prior to welding, the structure is stress-free, as shown in Fig. 12.27a; it may be rigid and fixtures may be used to support the structure. When the weld bead is being placed, the molten metal fills the gap between the surfaces to be joined, and flows outwards to complete the weld bead; at this stage, the weld is not under any stress. The weld bead then solidifies, and both the weld bead and the surrounding material begin to cool to room temperature; as they cool, they contract but are constrained by the bulk of the surrounding structure. As result, the structure distorts (Fig. 12.27c) and residual stresses develop.

Stress relieving of welds. The effects of residual stresses, such as distortion, buckling, and cracking, can be reduced by *preheating* the base metal or the parts to be welded. Preheating reduces distortion by lowering the cooling rate and the magnitude of thermal stresses (by reducing the elastic modulus of the metals being welded). This method also reduces shrinkage and the tendency for cracking of the joint. Preheating also is effective in welding hardenable steels (Section 5.11) for which rapid cooling of the weld area would otherwise be harmful. The parts may be heated in a furnace or electrically or inductively; thin sections may be heated by radiant lamps or a hot-air blast. For optimum results, preheating temperatures and cooling rates must be controlled for acceptable strength and toughness in the welded structure.

Residual stresses can be reduced by *stress relieving* or *stress-relief annealing* of the welded structure (Section 5.11.4). The temperature and the time required for stress relieving depend on the type of material and the magnitude of the residual stresses. Other methods of stress relieving include peening, hammering, and surface rolling the weld bead area (see Section 4.5.1). For multilayer welds, the first and last layers should not be peened, because peening may cause damage to those layers.

Plastically deforming the structure by a small amount also can reduce or relieve residual stresses (See Fig. 2.27). This technique can be used in welded pressure vessels by pressurizing the vessels internally (called *proof stressing*), in turn producing axial and hoop stresses (Section 2.11). To reduce the possibility of sudden fracture under high internal pressure, the weld must be free from significant notches or discontinuities, which could act as points of stress concentration.

Weld may also be *heat treated* in order to modify and improve mechanical properties, such as strength and toughness. The techniques employed include (a) annealing, normalizing, or quenching and tempering of steels, and (b) solution treatment and aging of precipitation-hardenable alloys (Section 5.11).

12.7.2 Weldability

Weldability is generally defined as (a) a metal's capability to be welded into a specific structure with specific properties and characteristics and (b) the ability of the welded structure to satisfactorily meet service requirements. Weldability involves several variables; material characteristics, such as alloying elements, impurities, inclusions, grain structure, and processing history of the base metal and filler metal, are all important. Other factors are strength, toughness, ductility, notch sensitivity, elastic modulus, specific heat, melting point, thermal expansion, surface-tension characteristics of the molten metal, and corrosion.

Surface preparation is important, as are the nature and properties of surface oxide films and adsorbed gases. Other factors are shielding gases, fluxes, moisture content of the coatings on electrodes, welding speed, welding position, cooling rate, preheating, and postwelding techniques (such as stress relieving and heat treating).

The following list, in alphabetical order, briefly summarizes the weldability of specific groups of metals. Their weldability can vary significantly, with some requiring special welding techniques and proper control of all processing parameters.

1. *Aluminum alloys:* weldable at a high rate of heat input; alloys containing zinc or copper generally are considered unweldable.
2. *Cast irons:* generally weldable.
3. *Copper alloys:* similar to that of aluminum alloys.
4. *Lead:* weldable.
5. *Magnesium alloys:* weldable with the use of protective shielding gas and fluxes.
6. *Molybdenum:* weldable under well-controlled conditions.
7. *Nickel alloys:* weldable.
8. *Niobium (columbium):* weldable under well-controlled conditions.
9. *Stainless steels:* weldable.
10. *Steels, galvanized and prelubricated:* The presence of zinc coating and lubricant layer adversely affects weldability.
11. *Steels, plain-carbon:* (a) excellent weldability for low-carbon steels; (b) fair to good weldability for medium-carbon steels; and (c) poor weldability for high-carbon steels.
12. *Steels, low-alloy:* fair to good weldability.
13. *Steels, high-alloy:* generally good weldability under well-controlled conditions.
14. *Tantalum:* weldable under well-controlled conditions.
15. *Tin:* weldable.
16. *Titanium alloys:* weldable with the use of proper shielding gases.
17. *Tungsten:* weldable under well-controlled conditions.
18. *Zinc:* difficult to weld; soldering preferred.
19. *Zirconium:* weldable with the use of proper shielding gases.

12.7.3 Testing Welded Joints

The *quality* of a welded joint is generally established by testing the joint itself. Several standardized tests and procedures have been established by organizations such as the American Society for Testing and Materials (ASTM), the American Welding Society (AWS), the American Society of Mechanical Engineers (ASME), the American Society of Civil Engineers (ASCE), and various federal agencies. Welded joints may be tested *destructively* or *nondestructively* (Section 4.8). Each technique has its capabilities, limitations, reliability, and need for special equipment and operator skill.

1. **Destructive testing techniques.**

   a. **Tension test.** Longitudinal and transverse tension tests are performed on specimens removed from welded joints and from the weld-metal area; stress–strain curves are then obtained (Section 2.2). These curves indicate the yield strength (S_y), ultimate tensile strength (S_{ut}), and ductility of the welded joint in different locations and directions. Ductility is measured in terms of elongation and reduction of area. Weld hardness may also be used as an indication of weld strength and microstructure in the weld zone.

   b. **Tension-shear test.** The specimens in this test (Fig. 12.28a) are prepared, then subjected to tension to determine the shear strength of the weld metal and the location of fracture.

   c. **Bend test.** Several bend tests can be used to determine the ductility and strength of welded joints. In one test, the welded specimen is bent around a fixture (called *wraparound bend test*; Fig. 12.28b). In another test, the specimens are tested in *three-point transverse bending* (See Figs. 2.19 and 12.28c).

   d. **Fracture toughness test.** This test commonly uses impact-testing techniques (Section 2.9). Charpy V-notch specimens are prepared and tested for toughness; other tests include the drop-weight test, in which the energy is supplied by a falling weight.

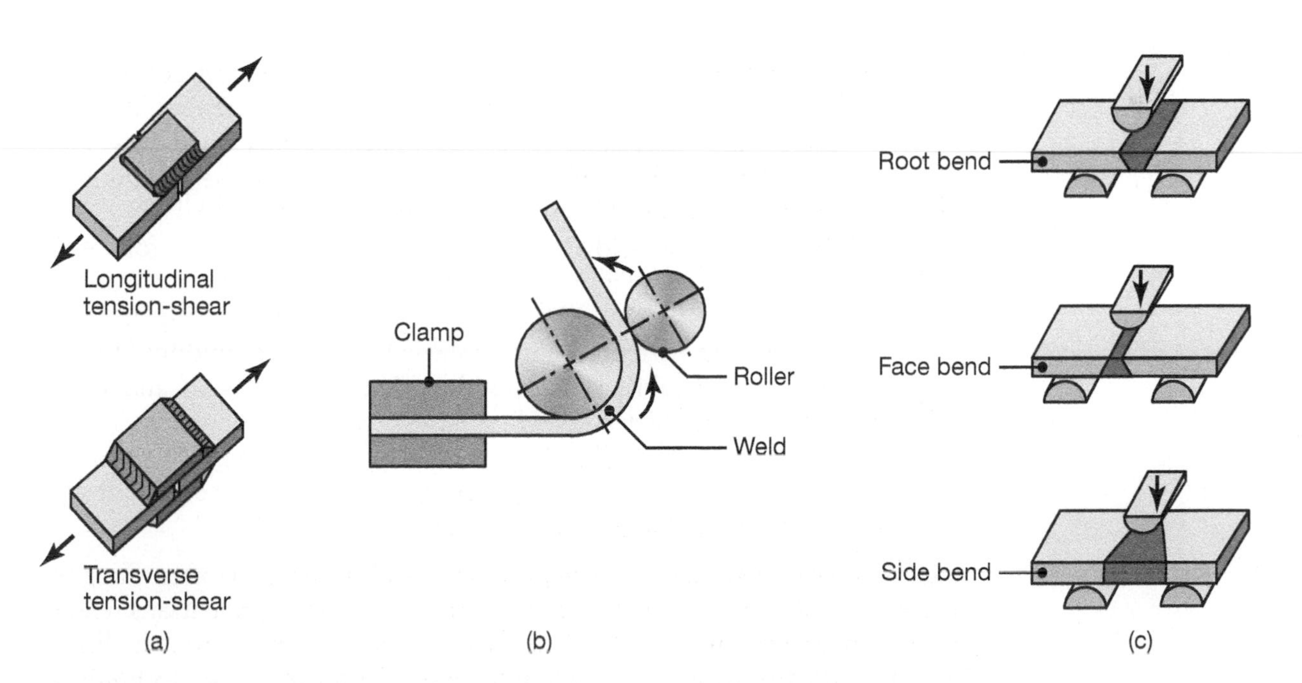

FIGURE 12.28 (a) Types of specimens for tension-shear testing of welds. (b) Wraparound bend test method. (c) Three-point bending of welded specimens. (See also Fig. 2.19.)

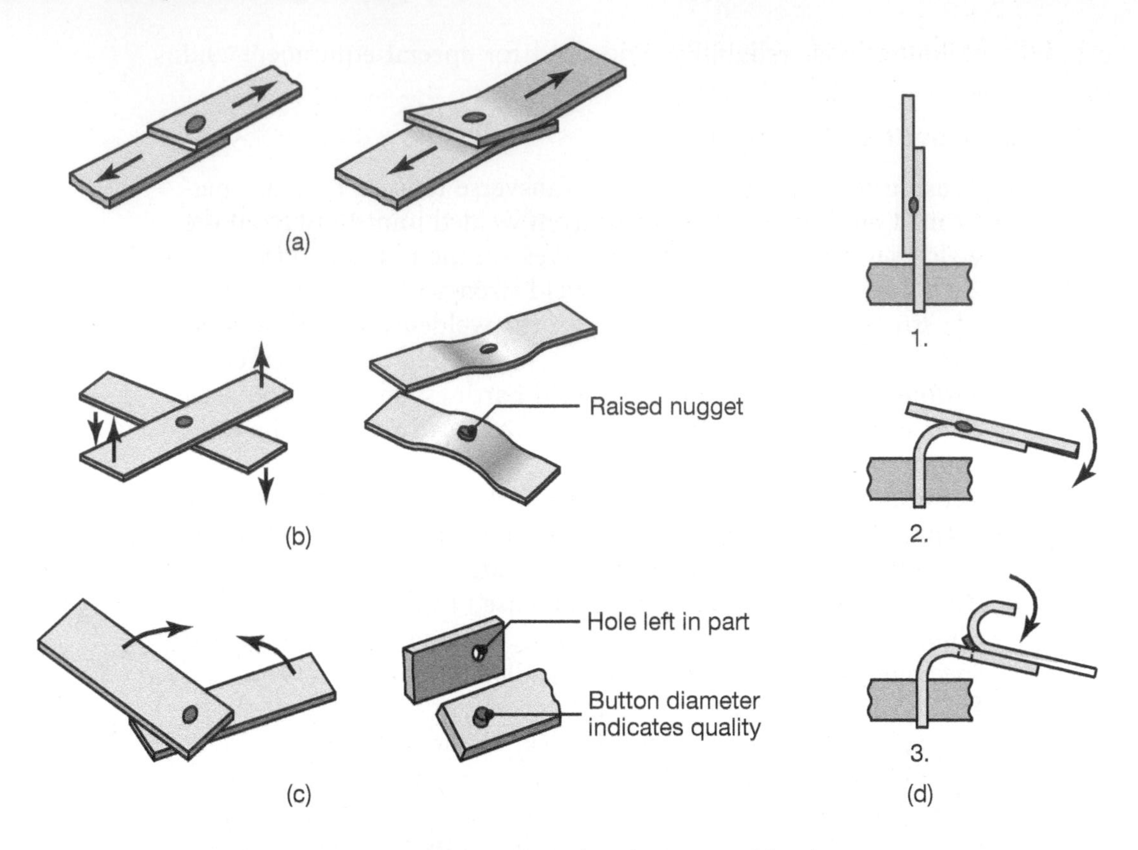

FIGURE 12.29 (a) Tension-shear test for spot welds; (b) cross-tension test; (c) twist test; and (d) peel test.

e. **Corrosion and creep tests.** Because of the differences in the composition and microstructure of the materials in the weld zone, preferential corrosion may take place in this zone (See Fig. 12.20). Creep tests are essential in determining the behavior of welded joints at elevated temperatures.

f. **Testing of spot welds.** Spot-welded joints may be tested for weld-nugget strength (Fig. 12.29) using the (a) *tension-shear*, (b) *cross-tension*, (c) *twist*, and (d) *peel tests*. Because they are inexpensive and easy to perform, tension-shear tests are commonly used in sheet metal fabricating facilities. The cross-tension and twist tests are capable of revealing flaws, cracks, and porosity in the weld zone. The peel test is commonly used for thin sheets; after the joint has been bent and peeled, the shape and size of the torn-out weld nugget are evaluated.

2. **Nondestructive testing techniques.** Welded structures often have to be tested nondestructively, particularly for critical applications where weld failure can be catastrophic, such as in pressure vessels, pipelines, and load-bearing structural members. Nondestructive techniques for welded joints usually consist of visual, radiographic, magnetic-particle, liquid-penetrant, and ultrasonic testing methods, as described in detail in Section 4.8.1.

12.7.4 Welding Process Selection

In addition to the role of material characteristics, selection of a welding process for a particular application involves the following considerations:

1. Configuration of the parts or structure to be welded and their shape, thickness, and size;
2. Methods used to manufacture component parts;
3. Service requirements, such as the type of loading and stresses developed;
4. Location, accessibility, and ease of welding;
5. Effects of distortion and discoloration;
6. Joint appearance; and
7. Costs involved in various operations, such as edge preparation, welding, and postprocessing of the weld, including machining and finishing operations.

12.8 Cold Welding

In *cold welding* (CW), pressure is applied to the mating interfaces of the parts through dies or rolls. Because plastic deformation is involved, it is essential that at least one, but preferably both, of the mating parts be sufficiently ductile. The interface is usually cleaned by wire or power brushing prior to welding. In joining two dissimilar metals that are mutually soluble; however, brittle *intermetallic compounds* may form (Section 5.2.2), resulting in a weak and brittle joint. An example is the bonding of aluminum and steel, where a brittle intermetallic compound is formed at the interface. The best bond strength and ductility are obtained in cold welding of two similar materials. The process can be used to join small workpieces made of soft, ductile metals. Applications include electrical connections, wire stock, and sealing of heat sensitive containers, such as those containing explosives.

Roll bonding. Also called *roll welding* (ROW), the pressure required for cold welding is applied through a pair of rolls (Fig. 12.30). A common application is in manufacturing some bimetallic US coins, using continuous strips of metal (see Case Study 12.1). Another common application is the production of bimetallic strips for thermostats and similar controls, using two layers of materials with different coefficients of thermal expansion (see also Section 12.9). Surface preparation is important for interfacial

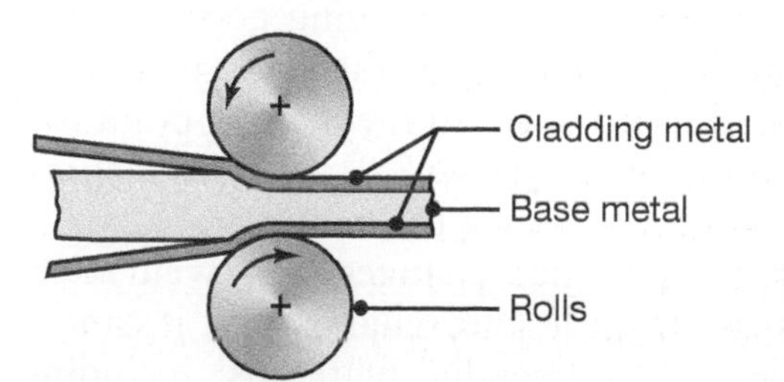

FIGURE 12.30 Schematic illustration of the roll-bonding, or cladding, process.

strength. Bonding in only selected regions in an interface can be achieved by depositing a *parting agent*, such as graphite or ceramic, called *stop-off*. A use of this technique in *superplastic forming* of sheet metal structures is shown in Fig. 7.46.

Roll bonding can also be carried out at elevated temperatures (*hot roll bonding*) for improved interfacial strength. Typical examples include *cladding* (Section 4.5.1) pure aluminum over precipitation-hardened aluminum-alloy sheet and stainless steel over mild steel for corrosion resistance.

CASE STUDY 12.1 Roll Bonding of the US Quarter

The technique used for manufacturing composite US quarters, dimes and other such coins is roll bonding. With quarters, two outer layers are 75% Cu-25% Ni (*cupronickel*), each 1.2 mm thick, with an inner layer of pure copper 5.1 mm thick. For good bond strength, the mating surfaces (*faying surfaces*) are first chemically cleaned and wire brushed. The strips are then rolled to a thickness of 2.29 mm. A second rolling operation reduces the final thickness to 1.36 mm; The strips thus undergo a total reduction in thickness of 82%.

Because volume constancy is maintained in plastic deformation (Section 2.11.5), there is a major increase in the surface area between the layers, generating clean interfacial surfaces and thus better bonding. The extension in surface area under the high pressure of the rolls, combined with the solid solubility of nickel in copper (Section 5.2.1), produce a strong bond between the layers.

12.9 Ultrasonic Welding

In *ultrasonic welding* (USW), the faying surfaces of the two members to be joined are subjected to a static normal force and oscillating shearing (tangential) stresses. The shearing stresses are applied by the tip of a **transducer** (Fig. 12.31a), similar to that used for ultrasonic machining (See Fig. 9.25a). The frequency of oscillation generally ranges from 10 to 75 kHz. The energy required in this operation increases with the thickness and hardness of the materials being joined. Proper coupling between the transducer and the tip (called a *sonotrode*, from the words *son*ic and elec*trode*) is important for effective bonding. The welding tip can be replaced with rotating disks (Fig. 12.31b) for seam welding (similar to those shown in Fig. 12.37 for *resistance seam welding*), one component of which can be a sheet or foil.

The shearing stresses cause small-scale plastic deformation at the interface, breaking up oxide films and contaminants, allowing good contact and producing a strong solid-state bond. The temperature generated in the weld zone is in the range of one-third to one-half the melting point (on the absolute scale) of the metals being joined (Table 3.2), thus no melting and fusion take place. In some situations, however, the temperature can be sufficiently high to cause significant metallurgical changes in the weld zone.

The ultrasonic welding process is versatile and reliable, and it can be used with of a variety of metallic and nonmetallic materials, including

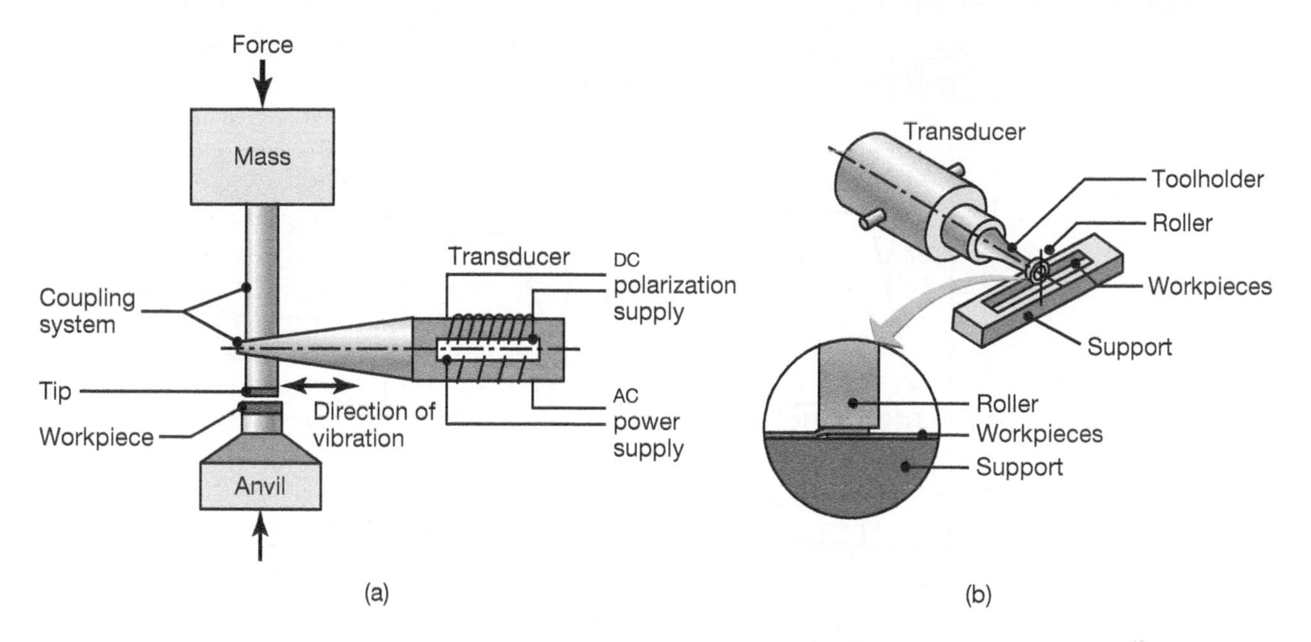

FIGURE 12.31 (a) Components of an ultrasonic welding machine for lap welds. (b) Ultrasonic seam welding using a roller.

dissimilar metals in making bimetallic strips. It is used extensively in joining plastics and in the automotive and consumer electronics industries, for lap welding of sheet, foil, and thin wire, and in packaging with foils. The mechanism of joining thermoplastics (Section 12.17.1) by ultrasonic welding is different from that for metals, and interfacial melting takes place, due to the much lower melting temperature of plastics (See Table 10.2).

12.10 Friction Welding

Note that thus far the energy required for welding is supplied externally. In *friction welding* (FRW), the heat required for welding is generated through *friction* at the interface of the two members being joined. In the friction welding process, one of the members is placed in a chuck, collet, or a similar fixture (See Fig. 8.41), and rotated at a high constant speed; the other member remains stationary. The surface speed of rotation may be as high as 900 m/min. The two members are then brought into contact under an axial force (Fig. 12.32). The members must be clamped securely to resist both torque and axial forces without slipping in their fixtures. After sufficient interfacial contact is established, (a) the rotating member is brought to a sudden stop, so that the weld is not destroyed by subsequent shearing action; and (b) the axial force is increased to upset the members and force intimate contact.

The weld zone is usually confined to a narrow region, depending on (a) the amount of heat generated; (b) the thermal conductivity of the materials; (c) the mechanical properties of the materials at elevated temperature;

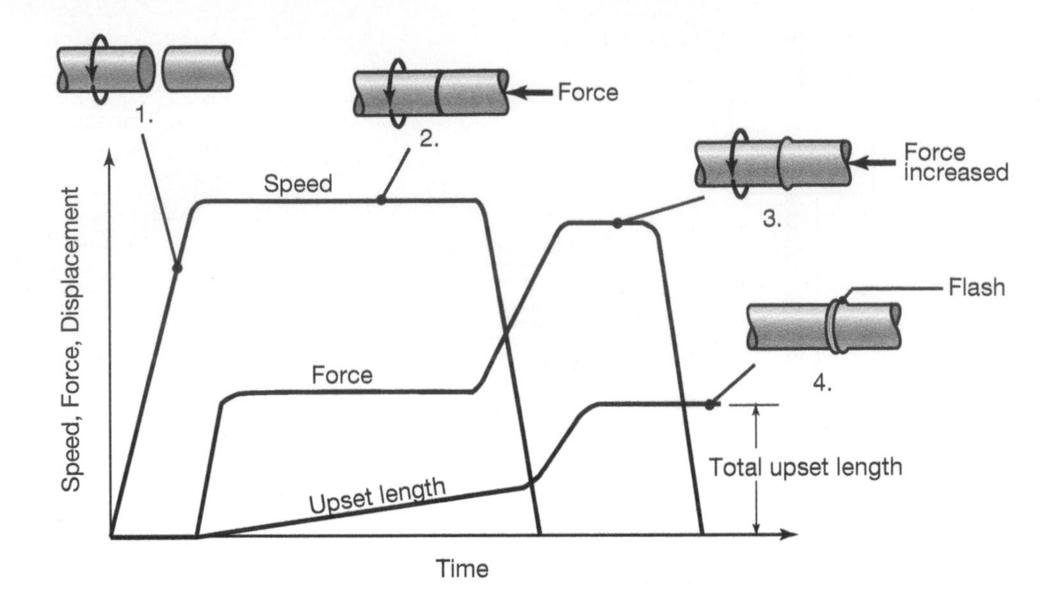

FIGURE 12.32 Sequence of operations in the friction welding process. (1) The part on the left is rotated at high speed. (2) The part on the right is brought into contact under an axial force. (3) The axial force is increased, and the part on the left stops rotating; flash begins to form. (4) After a specified upset length or distance is achieved, the weld is completed. The upset length is the distance the two pieces move inward during welding after their initial contact; thus, the total length after welding is less than the sum of the lengths of the two pieces.

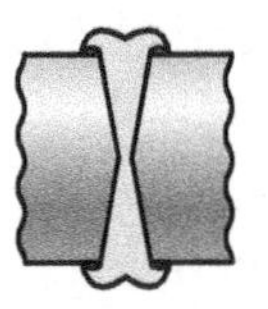

(a) High pressure or low speed

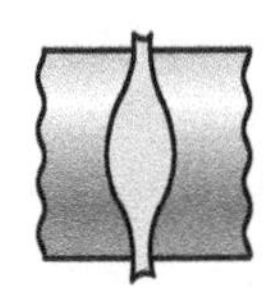

(b) Low pressure or high speed

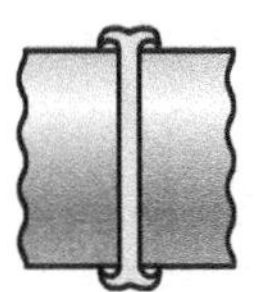

(c) Optimum

FIGURE 12.33 Shapes of the fusion zone in friction welding as a function of the force applied and the rotational speed.

QR Code 12.3 Demonstration of inertia friction welding. *Source:* Courtesy of Manufacturing Technologies, Inc.

and (d) the welding time. The shape of the joint depends on the rotational speed and the axial force applied, as can be seen in Fig. 12.33. These factors must be controlled to obtain a uniformly strong joint. The radially-outward movement of the hot metal at the interface (*flash*) pushes oxides and other contaminants out of the interface, further increasing joint strength.

Developed in the 1940s, friction welding can be used to join a wide variety of materials, provided that one of the components has some rotational symmetry. Solid and tubular parts can be joined, with good joint strength. Solid steel bars up to 100 mm in diameter and pipes up to 500 mm outside diameter have been welded successfully. Because of the combined heat and pressure, the interface in FRW develops a flash by plastic deformation (*upsetting*) of the heated zone. If objectionable, this flash can easily be removed by subsequent machining or grinding (see also the Case Study at the end of this chapter).

There are several variations on the friction welding process.

1. **Inertia friction welding (IFW).** Although this term has been used interchangeably with *friction welding*, the energy required in this process is supplied through the kinetic energy of a flywheel. The operation consists of the following sequence: (a) the flywheel is accelerated to the proper speed; (b) the two members are brought into contact; (c) an axial force is applied; (d) as the friction at the interface begins to slow down the flywheel, the axial force is increased; and (e) the weld is completed when the flywheel comes to a stop. The timing of this sequence is important for weld quality. The rotating mass in inertia friction welding machines, hence the energy, can be adjusted depending on workpiece cross section and the properties of the materials to be joined.

QR Code 12.4 Demonstration of linear friction welding. *Source:* Courtesy of Manufacturing Technologies, Inc.

2. **Linear friction welding (LFW).** This operation involves subjecting the interface of the two parts to be joined to a reciprocating linear motion of at least one of the components to be joined; the parts do not have to be circular or tubular in cross section. In this operation, one part is moved across the face of the other part, using a balanced reciprocating mechanism. The process is capable of joining square or rectangular components, as well as round parts, made of metals or plastics. In one application, a rectangular titanium-alloy part is friction welded at a linear frequency of 25 Hz, an amplitude of ± 2 mm, and a pressure of 100 MPa, acting on a 240 mm^2 interface. Various other metal parts have been welded successfully, with rectangular cross sections as large as 100 mm $\times$ 25 mm.

   Although this process was originally intended for welding aerospace alloys, especially for aluminum extrusions and titanium engine components, current applications include welding of polymers and composite materials. Significant progress is being made in linear friction welding to bond dissimilar materials, such as aluminum to steel, in automotive applications.

QR Code 12.5 Demonstration of friction stir welding. *Source:* Courtesy of Manufacturing Technologies, Inc.

3. **Friction stir welding.** Whereas in conventional friction welding, heating of interfaces is achieved through friction, in *friction stir welding* (FSW), a rotating tool, about 5 to 6 mm in diameter and 5 mm long, is plunged into the joint (Fig. 12.34). The contact pressures and the rotation cause frictional heating, raising the temperature to the range of 230°C to 260°C (450°F to 500°F) at the tip of the rotating tool, inducing heating and stirring of the material in the joint.

   The thickness of the welded parts can be as small as 1 mm and as much as 30 mm. The weld quality is high, with uniform material structure, and with minimal pores. The welds are produced with low heat input and thus have low distortion and little microstructural changes; moreover, there are no fumes or spatter produced. The process is relatively easy to implement and to be automated.

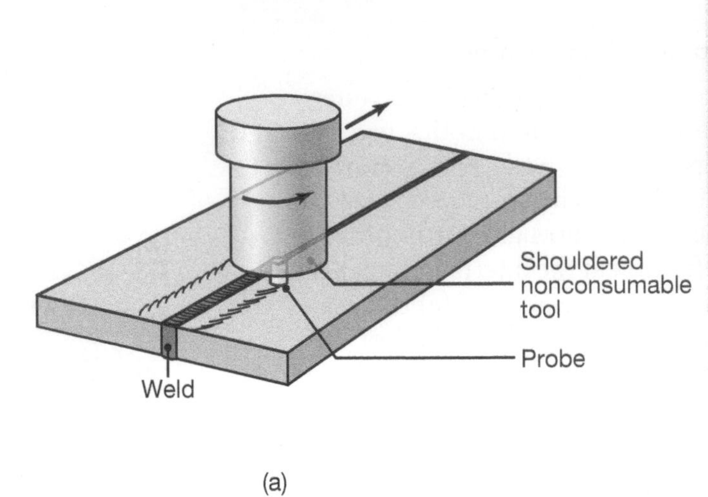

FIGURE 12.34 The friction stir welding process. (a) Schematic illustration of friction stir welding. Aluminum-alloy plates up to 75 mm (3 in.) thick have been welded by this process. (b) Multi-axis friction stir welding machine for large workpieces, such as aircraft wing and fuselage structures, that can develop 67 kN (15,000 lb) axial forces, is powered by a 15 kW (20 hp) spindle motor, and can achieve welding speeds up to 1.8 m/s. *Source:* (b) Courtesy of Manufacturing Technology, Inc.

12.11 Resistance Welding

QR Code 12.6 Friction stir spot welding. *Source:* Courtesy of Manufacturing Technologies, Inc.

Resistance welding (RW) operations include several processes in which the heat required for welding is produced by means of the *electrical resistance* between the two members to be joined. The major advantages include the fact that they do not require any consumable electrodes, shielding gases, or flux. The heat generated in resistance welding is given by the general expression

$$H = I^2Rt, \tag{12.9}$$

where H is the heat generated in joules (watt-seconds); I is the current in amperes; R is the resistance in ohms; and t is the flow time of current in seconds. This equation is generally modified to determine the actual heat energy available, by including a factor K that represents the energy losses through radiation and conduction. Equation (12.9) then becomes $H = I^2RtK$, where the value of K is less than unity.

The total electrical resistance in these processes, such as in *resistance spot welding* shown in Fig. 12.35, is the sum of the following four components:

1. Resistance of the electrodes;
2. Electrode-workpiece contact resistances;
3. Resistances of the individual parts to be welded; and
4. Workpiece-workpiece contact resistances (faying surfaces).

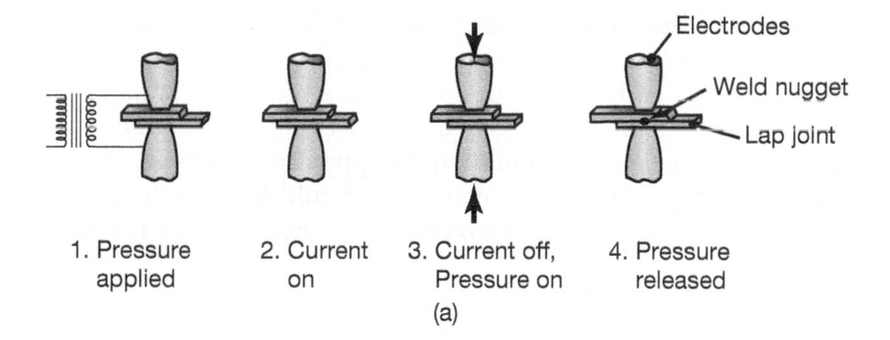

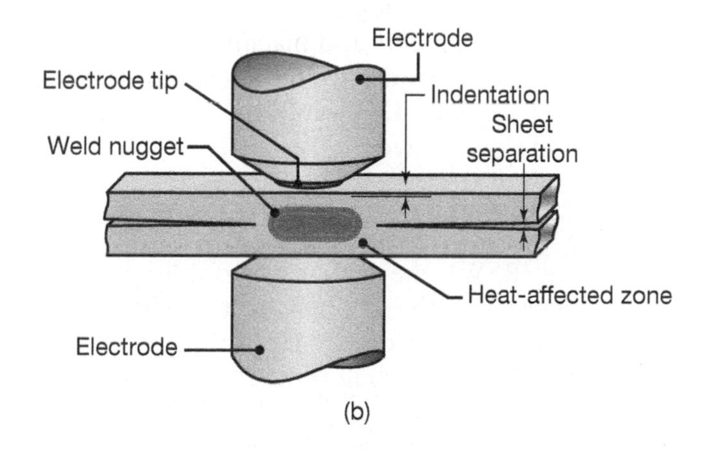

FIGURE 12.35 (a) Sequence in the resistance spot welding operation. (b) Cross section of a spot weld, showing weld nugget and light indentation by the electrode on sheet surfaces.

EXAMPLE 12.3 Heat Generated in Resistance Spot Welding

Given: Two 1-mm-thick steel sheets are being spot welded at a current of 5000 A and current-flow time of $t = 0.1$ s. The electrodes are 5 mm in diameter.

Find: Estimate the amount of heat generated and its distribution in the weld zone. Use an effective resistance of 200 $\mu\Omega$.

Solution: According to Eq. (12.9),

$$\text{Heat} = (5000)^2\,(0.0002)(0.1) = 500 \text{ J}.$$

If it is assumed that the material below the electrode is heated enough to melt and fuse, the weld nugget volume can be calculated as

$$V = \left(\frac{\pi}{4}d^2\right)(t) = \frac{\pi}{4}(5)^2(2) = 39.3 \text{ mm}^3.$$

From Table 12.3, u for steel is 9.7 J/mm^3. Therefore, the heat required to melt the weld nugget is, from Eq. (12.4),

$$H = u(\text{Volume}) = (9.7)(39.3) = 381 \text{ J}.$$

Consequently, the remaining heat (119 J), or 24%, is dissipated into the metal surrounding the nugget.

The magnitude of the current in resistance welding operations may be as high as 100,000 A, and the voltage is typically only 0.5–10 V. The actual temperature rise at the joint depends on the specific heat and thermal conductivity of the metals to be joined. Because they have high thermal conductivity, metals such as aluminum and copper, for example, require high heat concentrations. Electrode materials should have high thermal conductivity and strength at elevated temperatures; they are typically made of copper alloys.

Developed in the early 1900s, resistance welding processes require specialized machinery, and are now operated with programmable computer control. The machinery is generally not portable, and the process is most suitable for use in manufacturing plants and machine shops.

There are five basic types of resistance welding operations: spot, seam, projection, flash, and upset welding. Lap joints are used in the first three processes and butt joints in the last two.

12.11.1 Resistance Spot Welding

In *resistance spot welding* (RSW), the tips of two opposing solid, round electrodes contact the surfaces of the lap joint of two sheet metals, and resistance heating produces a spot weld (Fig. 12.35a). The current usually ranges from 3000 to 40,000 A, depending on the materials being welded and their thickness. In order to obtain a good bond in the **weld nugget** (Fig. 12.35b), pressure is continually applied until the current is turned off. Accurate control and timing of the electric current and pressure are thus essential.

The weld nugget is generally up to 10 mm in diameter and the surface of the weld spot has a slightly discolored indentation. The strength of the joint depends on the surface roughness and cleanliness of the mating surfaces. Oil, paint, and thick oxide layers should be removed before welding.

Spot welding is the simplest and most commonly used resistance welding process. The operation may be performed by means of single or multiple electrodes, and the pressure required is supplied through mechanical or pneumatic means. The *rocker-arm type* machines are generally used for smaller parts, whereas the *press type* machines are used for welding larger workpieces. The shape and surface condition of the electrode tip and accessibility of the weld area are important factors in spot welding. A variety of electrode shapes are used to weld areas that are difficult to reach (Fig. 12.36).

Modern equipment for spot welding is computer controlled for optimum timing of current and pressure, while the welding guns are manipulated by programmable robots (Section 14.7). Spot welding is widely used for fabricating sheet metal products, with applications ranging from attaching metal handles to cookware to rapid spot welding of automobile bodies, using multiple electrodes. Some automobiles have as many as 5,000 spot welds.

12.11.2 Resistance Seam Welding

Resistance seam welding (RSEW) is a modification of resistance spot welding wherein the electrodes are replaced by wheels or rollers (Fig. 12.37).

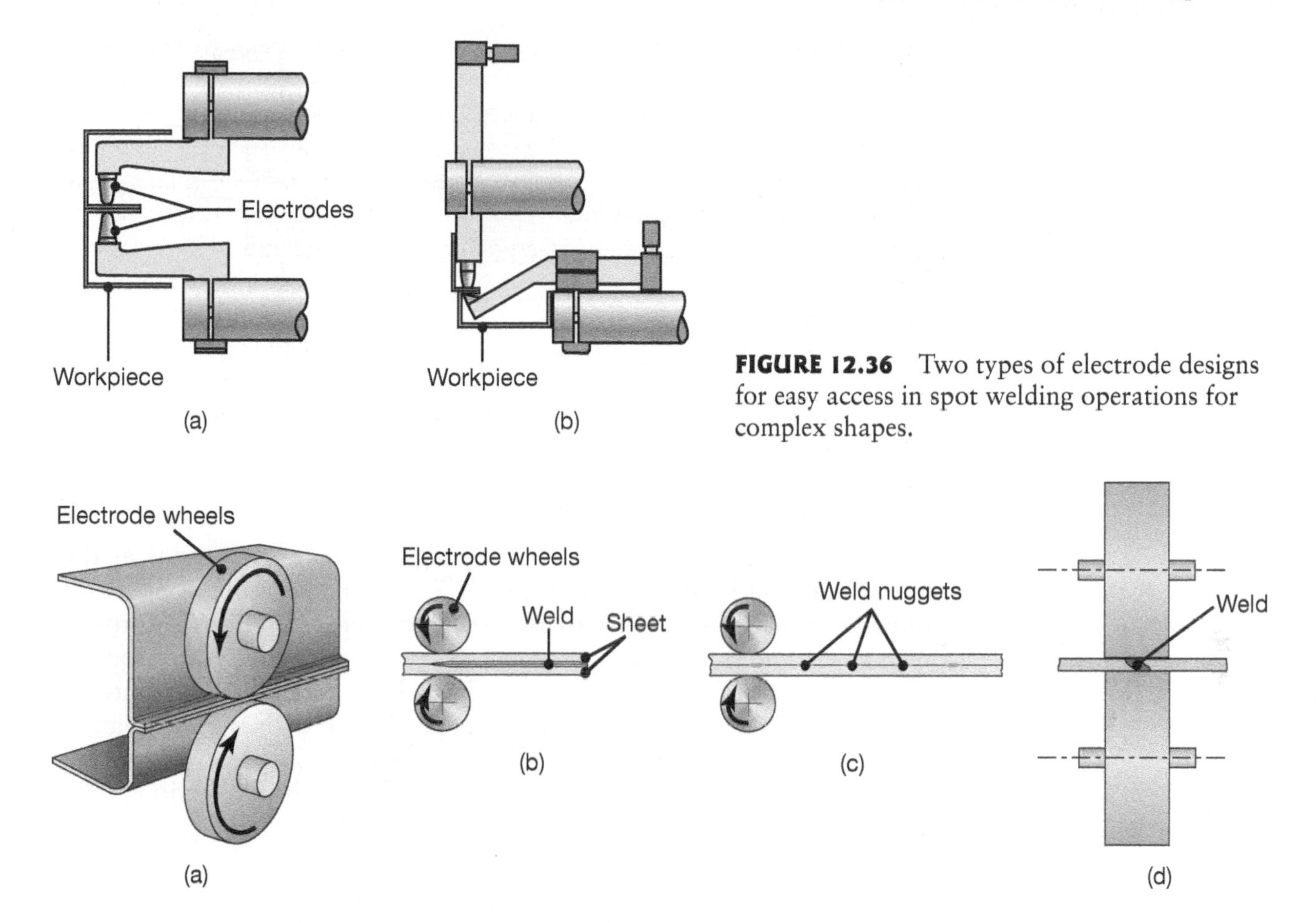

FIGURE 12.36 Two types of electrode designs for easy access in spot welding operations for complex shapes.

FIGURE 12.37 (a) Illustration of the seam welding process, with rolls acting as electrodes. (b) Overlapping spots in a seam weld. (c) Cross section of a roll spot weld. (d) Mash seam welding.

With a continuous AC power supply, the electrically conducting and rotating electrodes produce continuous spot welds whenever the current reaches a sufficiently high level in the AC cycle. The welds produced actually are overlapping spot welds, and can produce a joint that is liquid and gas tight (Fig. 12.37b). The typical welding speed is 1.5 m/min (5 ft/min) for thin sheet. With *intermittent* application of current, a series of spot welds at various intervals can be made along the length of the seam (Fig. 12.37c), a procedure called **roll spot welding.**

The RSEW process is used in making the longitudinal (side) seam of cans for household products, mufflers, gasoline tanks, and other containers. In *mash seam welding*, the overlapping welds are about one to two times the sheet thickness. Tailor-welded blanks (Section 7.3.4) also may be made by this process.

High frequency resistance welding (HFRW) is similar to resistance seam welding, with the exception that a high frequency current of up to 450 kHz is employed. Applications include butt-welded tubing, spiral pipe and tubing, finned tubes for heat exchangers, wheel rims, and structural sections such as I-beams. In **high frequency induction welding** (HFIW), the tube or

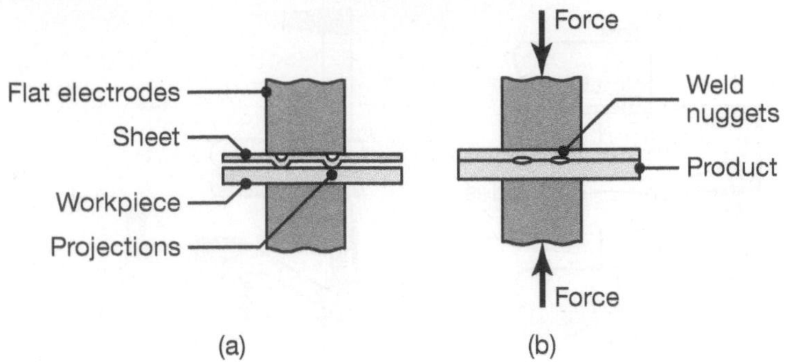

FIGURE 12.38 Schematic illustration of resistance projection welding: (a) before and (b) after. The projections on sheet metal are produced by embossing operations, as described in Section 7.5.2.

pipe to be welded is subjected to high frequency induction heating, using an induction coil ahead of the rollers.

12.11.3 Resistance Projection Welding

In *resistance projection welding* (RPW), high electrical resistance at the joint is developed by embossing (Section 7.5.2) one or more projections (dimples) on one of the surfaces to be welded (Fig. 12.38). The projections may be round or oval for specific design or part strength purposes. Because of the small contact areas involved, high localized temperatures are generated at the projections that are in contact with the flat mating part. The weld nuggets are similar to those produced in spot welding, and are formed as the electrodes exert pressure to flatten the projections. The electrodes generally are made of copper-base alloys; they have flat tips and are water-cooled to keep their temperature low.

Although embossing workpieces is an added operation and cost in RPW, it produces a number of welds in one stroke, extending electrode life, and it is capable of welding metals with different thicknesses. Fasteners such as nuts and bolts also can be welded to a sheet or plate, with projections that can be produced by machining or forging. Joining a network of wires, such as metal baskets, grills, oven racks, and shopping carts, also is considered to be resistance projection welding, because of the small contact area between crossing wires (*grids*).

12.11.4 Flash Welding

In *flash welding* (FW), also called **flash butt welding**, heat is generated from the arc as the ends of the two members begin to make contact, developing an electrical resistance at the joint (Fig. 12.39). Because of the electrical arc, this process is also classified as arc welding. After the temperature rises and the interface begins to soften, an axial force is applied at a controlled rate, and a weld is formed by plastic deformation of the joint. Note that this is a process of *hot upsetting* (See Fig. 6.1), hence also the term **upset welding** (UW). A significant amount of metal may be expelled from the joint as a shower of sparks during the flashing process; the flash may later be removed to improve joint appearance. Because impurities and contaminants also are squeezed out during this operation, the quality of the weld is good. The FW machines are large and usually automated, with power supplies ranging from 10 to 1500 kVA.

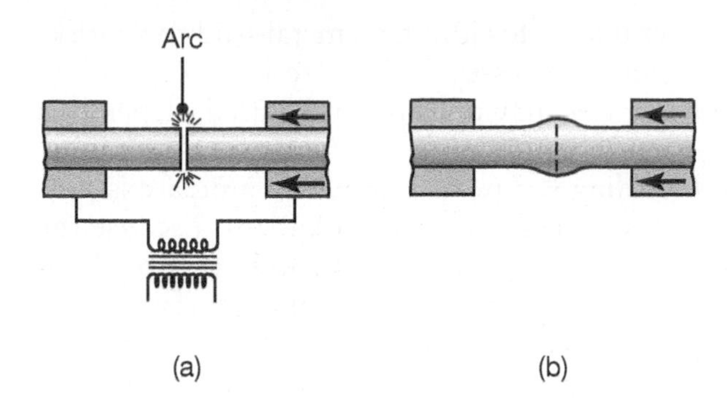

FIGURE 12.39 Flash welding process for end-to-end welding of solid rods or tubular parts. (a) Before and (b) after.

The flash welding process is suitable for end-to-end or edge-to-edge joining of similar or dissimilar metals, 1–75 mm (0.04–3 in.) in diameter, and sheet and bars 0.2–25 mm (0.008–1 in.) thick. Thinner sections have a tendency to buckle under the axial force applied during the process. Rings, made by forming processes such as those shown in Figs. 7.24b and c, also can be flash butt welded. This technique is also used to repair broken band-saw blades (Section 8.10.5), with fixtures that are attached to the saw frame. Typical applications of flash welding include (a) joining pipe and tubular shapes, for metal furniture and windows; (b) welding high-speed steels to steel shanks; (c) joining sections of steel rails; and (d) welding the ends of coils of sheet for continuous operation of rolling mills and wire-drawing equipment (Chapter 6).

12.11.5 Stud Arc Welding

Stud arc welding (SW) is similar to flash welding; the stud, such as a bolt or threaded rod, hook, or hanger, serves as one of the electrodes while being joined to another member, which typically is a flat plate (Fig. 12.40). In order to concentrate the heat generated, prevent oxidation, and retain the molten metal in the weld zone, a disposable ceramic ring (*ferrule*) is placed around the joint. The equipment for stud welding can be automated, with various controls for arcing and applying pressure; portable equipment is also available. The process has numerous applications in the automotive, construction, appliance, electrical, and shipbuilding industries.

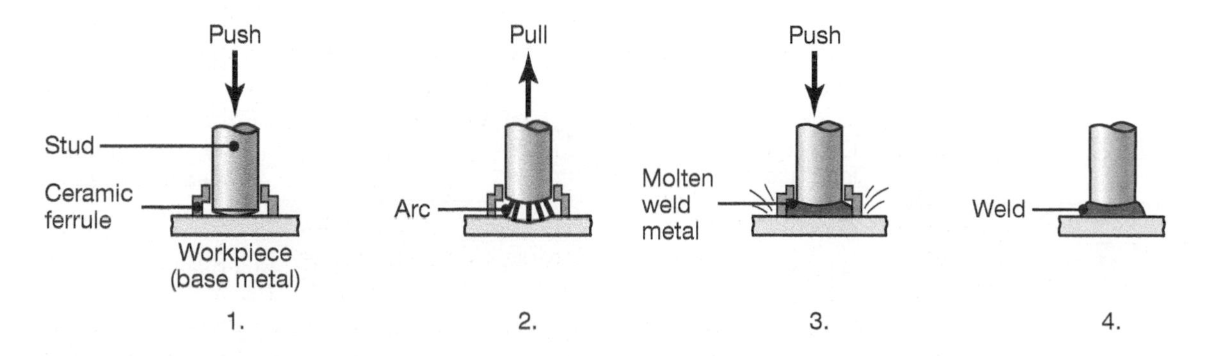

FIGURE 12.40 Sequence of operations in stud arc welding, used for welding bars, threaded rods, and various fasteners on metal plates.

Stud welding is a more general term for joining a metal stud to a workpiece; it can be done by various processes such as resistance welding, friction welding, etc. Shielding gases may or may not be used, depending on the particular application.

Capacitor discharge stud welding is a process in which a DC arc is produced from a capacitor bank. No ferrule or flux is required, because the welding time is very short (1–6 milliseconds). The process is capable of stud welding on thin metal sheets with coated or painted surfaces. The choice between this process and stud arc welding depends on such factors as the type of metals to be joined, part thickness, stud diameter, and shape of the joint.

12.11.6 Percussion Welding

The electrical energy for welding may also be stored in a *capacitor*. *Percussion welding* (PEW) uses this technique, in which the power is discharged within 1–10 milliseconds, developing highly localized heat at the joint. Pressure is applied to the joint immediately after the discharge. This process is useful where heating of the components adjacent to the joint is to be avoided, such as in electronic components.

12.12 Explosion Welding

In *explosion welding* (EXW), extremely high contact pressure is developed by detonating a layer of explosive placed over one of the members being joined (*flyer plate*; Fig. 12.41). The pressure is on the order of 10 GPa (1450 ksi), and the kinetic energy of the flyer plate striking the mating member produces a turbulent, wavy interface. The impact mechanically interlocks the two mating surfaces (Fig. 12.42); in addition, cold pressure welding by plastic deformation also takes place (Section 4.4). The flyer plate is placed at an angle, so that any oxide films present at the interface are broken up and propelled from the interface; as a result, bond strength is very high.

The explosive may be in the form of flexible plastic sheet, cord, granular solid, or a liquid that is cast or pressed onto the flyer plate. Detonation

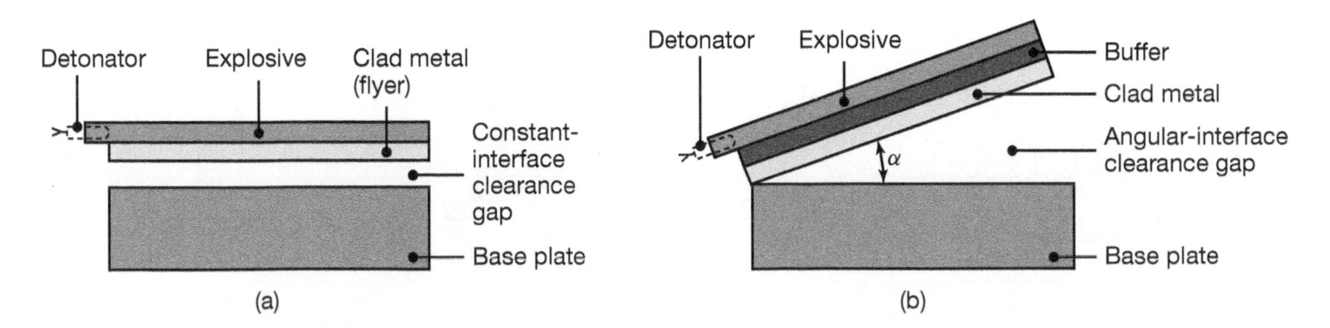

FIGURE 12.41 Schematic illustration of the explosion welding process: (a) constant interface clearance gap and (b) angular interface clearance gap.

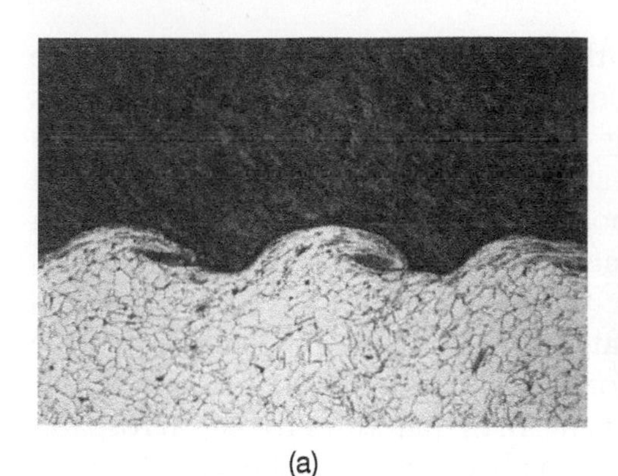

(a)

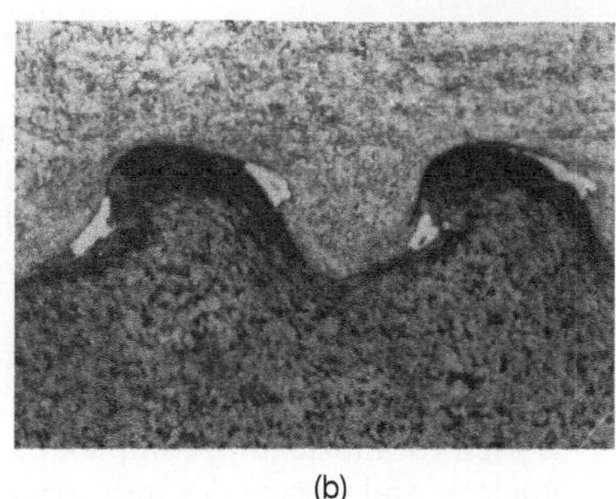

(b)

FIGURE 12.42 Cross sections of explosion welded joints: (a) titanium (top) on low-carbon steel (bottom) and (b) Incoloy 800 (iron-nickel-base alloy) on low-carbon steel. The wavy interfaces shown improve the shear strength of the joint. Some combinations of metals, such as tantalum and vanadium, produce a much less wavy interface. If the two metals have little metallurgical compatibility, an interlayer may be added that has compatibility with both metals.

speeds are typically 2400–3600 m/s (8000–12,000 ft/sec), depending on the type of explosive and the thickness of the explosive layer. Detonation is carried out using a commercial blasting cap.

Explosion welding is suitable for cladding plates and slabs with dissimilar metals, particularly for the chemical industry and military applications such as armor. Plates as large as 6 × 2 m (20 × 6.6 ft) have been explosively clad; the plates may then be rolled into thinner sections. Tube and pipe are often joined to the holes in header plates of boilers and tubular heat exchangers by this method. The explosive is placed inside the tube, and the pressure developed expands the tube and seals it tightly against the holes in the header plate.

12.13 Diffusion Bonding

Diffusion bonding, also called **diffusion welding** (DFW), is a solid-state joining process in which the strength of the joint is primarily developed from diffusion (movement of atoms across the interface) and, to a lesser extent, from the plastic deformation of the faying surfaces. This process requires temperatures of about 0.5 T_m (where T_m is the melting point of the metal, on the absolute scale) in order to have a sufficiently high diffusion rate between the parts to be joined. The bonded interface has essentially the same physical and mechanical properties as those of the base metal. Bond strength depends on pressure, temperature, time of contact, and the cleanliness of the faying surfaces.

In diffusion bonding, the two parts are usually heated in a furnace or by electrical resistance. The pressure required may then be applied by (a) using dead weights; (b) a press; (c) differential gas pressure; or (d) from the relative thermal expansion of the parts to be joined. High-pressure autoclaves may also be used for bonding complex parts. The process is generally most suitable for joining dissimilar metal pairs, although it is also used for reactive metals, such as titanium, beryllium, zirconium, and refractory metal alloys. Diffusion bonding is also important in sintering in powder metallurgy (Section 11.4) and for processing composite materials (Section 11.14).

Because diffusion involves migration of the atoms across the joint, the process is significantly slower than other welding methods. Although DFW is used for fabricating complex parts in small quantities for the aerospace, nuclear, and electronics industries, it has been automated to make it suitable and economical for moderate volume production as well, such as in bonding sections of orthopedic implants and sensors (See Fig. 13.48).

The practice of diffusion bonding dates back centuries, when goldsmiths bonded gold over copper. Called **filled gold**, a thin layer of gold foil is first made by hammering; the foil is placed over copper, and a weight is placed on top of it. The assembly is then placed in a furnace and left there until a good bond is developed; this process is also called *hot pressure welding* (HPW).

Diffusion bonding/superplastic forming. Beginning in the 1970s, complex sheet metal structures began to be fabricated by combining *diffusion bonding* and *superplastic forming* (DB/SPF; see also Section 7.5.6). After selected locations of the sheets are diffusion bonded, the unbonded regions (called *stop-off*) are expanded, by air pressure, into a mold (Fig. 12.43). The structures are thin and have high stiffness-to-weight ratios (Section 3.9.1), which is particularly important in aircraft and aerospace applications.

This technology improves productivity by (a) eliminating mechanical fasteners; (b) reducing the number of parts required; (c) producing

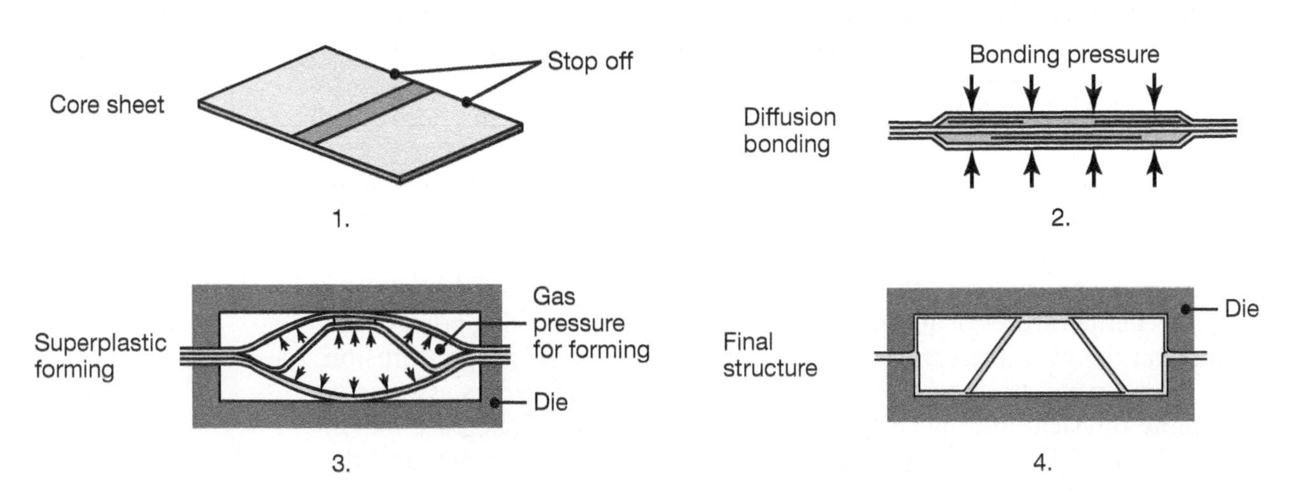

FIGURE 12.43 Sequence of operations in diffusion bonding and superplastic forming of a structure with three flat sheets; see also Fig. 7.46. *Source:* After D. Stephen and S.J. Swadling.

assemblies with good dimensional accuracy and low residual stresses; and (d) reducing labor costs and lead times. The process is used for titanium structures (typically Ti-6Al-4V) and aluminum alloys and various other alloys for aerospace structural applications.

12.14 Brazing and Soldering

Two joining processes that involve lower temperatures than those required for welding are *brazing* and *soldering*; they are arbitrarily distinguished by temperature, although brazing temperatures are higher than those for soldering.

12.14.1 Brazing

In *brazing*, a process first used as far back as 3000–2000 BC, a filler metal is first placed at or between the surfaces to be joined; the temperature is then raised to melt the filler metal, but not the parts (Fig. 12.44a). The molten metal fills the closely fitting space by *capillary action*. Upon cooling and solidification of the filler metal, a strong joint is obtained. In **braze welding**, the filler metal is deposited at the joint, as illustrated in Fig. 12.44b.

Filler metals for brazing generally melt above 450°C (850°F), below the melting point (*solidus temperature*) of the metals to be joined (See Fig. 5.3). Note that this process is unlike liquid-state welding, in which the workpieces must melt in the weld area for fusion to take place; any difficulties associated with heat-affected zones (Section 12.7), warping, and residual stresses are thus reduced in the brazing process. The strength of a brazed joint depends on joint design and the bond strength at the interfaces of the workpiece and filler metal. The surfaces to be brazed should be chemically or mechanically cleaned to ensure full capillary action; the use of a flux is thus important.

The *clearance* between the mating surfaces is an important parameter, as it directly affects the strength of the joint (Fig. 12.45). Note in the figure that the smaller the gap, the higher the *shear strength* of the joint; also, there is an optimum gap to achieve maximum *tensile strength*. The typical joint clearance ranges from 0.025 to 0.2 mm, and because clearances are very small, surface roughness of the mating surfaces is also important (Section 4.3).

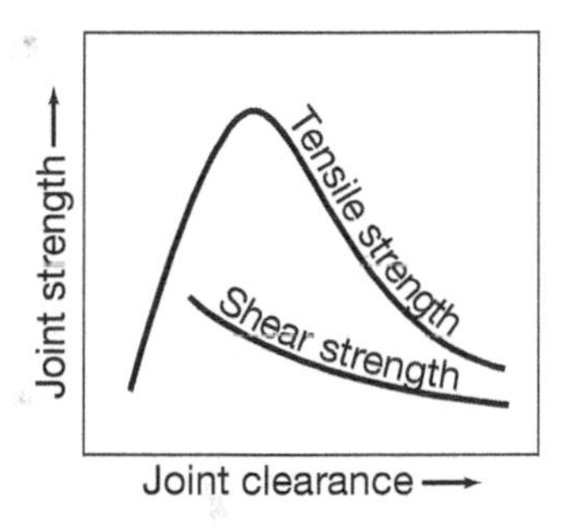

FIGURE 12.45 The effect of joint clearance on tensile and shear strength of brazed joints. Note that unlike tensile strength, shear strength continually decreases as clearance increases.

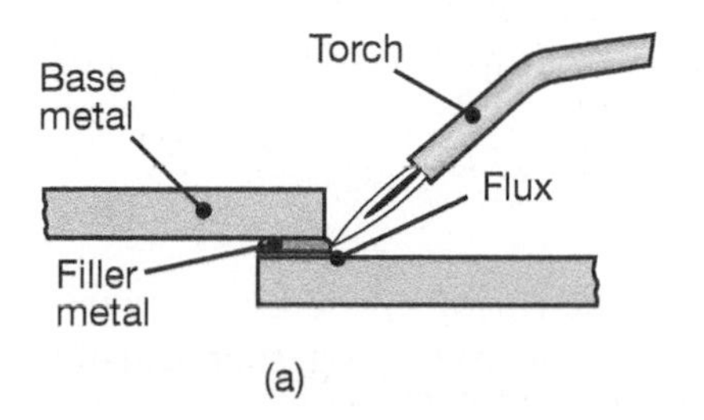

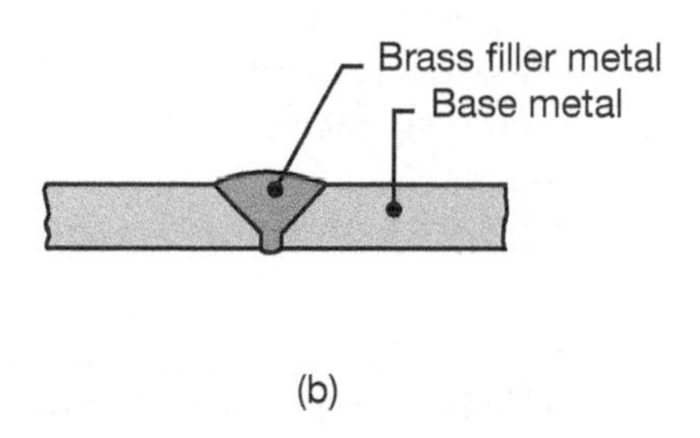

FIGURE 12.44 (a) Brazing and (b) braze welding operations.

TABLE 12.4 Typical filler metals for brazing various metals and alloys.

Base metal	Filler metal	Brazing temperature (°C)
Aluminum and its alloys	Aluminum-silicon	570–620
Magnesium alloys	Magnesium-aluminum	580–625
Copper and its alloys	Copper-phosphorus	700–925
Ferrous and nonferrous alloys (except aluminum and magnesium)	Silver and copper alloys, copper-phosphorus	620–1150
Iron-, nickel-, and cobalt-base alloys	Gold	900–1100
Stainless steels, nickel- and cobalt-base alloys	Nickel-silver	925–1200

Filler metals. Several filler metals (**braze metals**) are available, with a range of brazing temperatures (Table 12.4) and as wire, strip, rings, shims, preforms, filings, and powder. Note that unlike those for other welding operations, filler metals for brazing generally have significantly different compositions than those of the metals to be joined. The choice of a filler metal and its composition is important to avoid (a) *embrittlement* of the joint (by grain-boundary penetration of liquid metal; see Section 3.4.2); (b) formation of brittle intermetallic compounds at the joint; and (c) galvanic corrosion in the joint.

Because of diffusion between the filler metal and the base metal, mechanical and metallurgical properties of joints can undergo changes during the service life of brazed components or in subsequent processing of the brazed parts. For example, when titanium is brazed with pure tin filler metal, it is possible for the tin to completely diffuse into the titanium base metal by aging; when that happens, a well-defined interface no longer exists.

Fluxes. Flux is essential in brazing in order to prevent oxidation and to remove oxide films from the surfaces to be joined. Generally made of borax, boric acid, borates, fluorides, and chlorides, fluxes are available as paste, slurry, or powder. *Wetting agents* may be added to improve both the wetting characteristics of the molten filler metal and its capillary action. Because they are corrosive, fluxes should be removed after brazing, especially in hidden crevices, generally by washing vigorously with hot water.

Surfaces to be brazed must be clean and free from rust, oil, lubricants, and other contaminants. Clean surfaces (Section 4.5.2) are necessary for proper wetting and spreading characteristics of the molten filler metal in the joint and for developing maximum bond strength. Sand blasting and other processes (Section 9.9) also may be used to improve surface finish of faying surfaces.

12.14.2 Brazing Methods

The *heating methods* for brazing also identify the various processes. A variety of special fixtures may be used to hold the parts together during brazing, some with a provision for allowing thermal expansion and contraction.

1. **Torch brazing (TB).** The source of heat in this operation is oxyfuel gas with a carburizing flame (Fig. 12.2c). Brazing is performed by first heating the joint with the torch, and then depositing the brazing rod or wire in the joint. More than one torch may be used in the operation. Part thicknesses are usually in the range of 0.25–6 mm (0.01–0.25 in.). Although it can be automated as a production process, torch brazing is difficult to control and requires skilled labor.
2. **Furnace brazing (FB).** The parts are precleaned and preloaded with brazing metal, in appropriate configurations, then placed in a furnace (Fig. 12.46); the whole assembly is heated uniformly in the furnace. Furnaces (Section 5.5) may be batch type for complex shaped parts or continuous type for high production runs, especially for small parts with simple joint designs. *Vacuum furnaces* or *neutral atmospheres* are used for metals that react with the environment, such as stainless steels where the passivation layer of chromium oxide (Section 3.10.3) can adversely affect joint strength.
3. **Induction brazing (IB).** The source of heat is induction heating by high-frequency AC current (Section 5.5). Parts are first preloaded with filler metal, then placed near the induction coils for rapid heating. Unless there is a protective atmosphere, fluxes are generally required. The process is particularly suitable for continuous brazing of parts that are usually less than 3 mm (0.12 in.) thick.
4. **Resistance brazing (RB).** In this process, the source of heat is the electrical resistance of the components to be brazed; electrodes are used for this purpose, as in resistance welding. Parts are either preloaded with filler metal or the filler metal is supplied externally during brazing. Parts commonly brazed have a thickness of 0.1–12 mm (0.004–0.5 in.). As in induction brazing, the process is rapid, heating zones can be confined to very small areas, and the process can be automated to produce uniform quality.
5. **Dip brazing (DP).** Dip brazing is carried out by dipping the assemblies to be brazed into a molten filler metal bath, at a temperature just above the melting point of the filler metal. All workpiece surfaces are

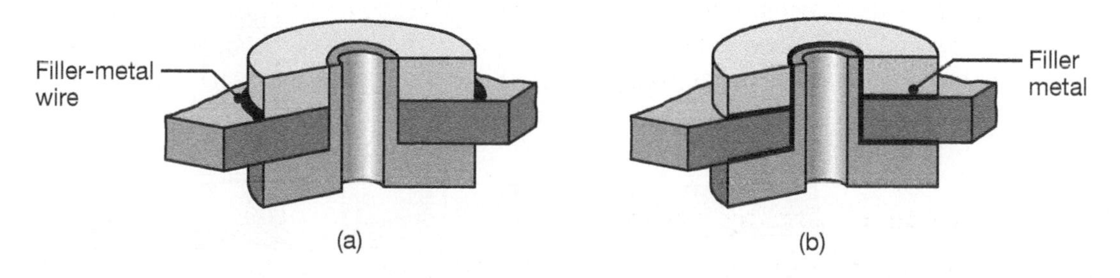

FIGURE 12.46 An application of furnace brazing: (a) before and (b) after. Note that the filler metal is a shaped wire.

thus coated with the filler metal. Alternatively, a clamped assembly incorporating braze can be dipped into a molten salt bath. Dip brazing in metal baths is used only for small parts, such as fittings, usually less than 5 mm (0.2 in.) in thickness or diameter. Molten salt baths, which also act as fluxes, are used for complex assemblies of parts with various thicknesses. As many as 1000 joints can be made at one time by dip brazing.

6. **Infrared brazing (IRB).** The heat source in this process is a high-intensity quartz lamp. This process is particularly suitable for brazing very thin (usually less than 1 mm) components, including honeycomb structures (Fig. 7.48). The radiant energy of the lamp is focused on the joint, and the process can be carried out in a vacuum; a related method is *microwave heating*.
7. **Diffusion brazing (DFB).** This process is carried out in a furnace in which, with proper control of temperature and time, the filler metal diffuses into the faying surfaces of the components to be joined. The brazing time may range from 30 min to 24 hrs. Diffusion brazing is used for strong lap or butt joints and for difficult joining operations. Because the rate of diffusion at the interface does not depend on the thickness of the components, part thicknesses may range from foil to as much as 50 mm (2 in.).
8. **High energy beams.** For specialized high-precision applications and with high-temperature metals and alloys, *electron beam* and *laser beam* heating may also be used, as described in Section 12.5.

Braze welding. The joint is prepared as in fusion welding. Using an oxyacetylene torch with an oxidizing flame, filler metal is then deposited at the joint; considerably more filler metal is thus used as compared to brazing. Temperatures in braze welding are generally lower than in fusion welding and part distortion is minimal. The use of a flux is essential in this process to develop proper joint strength.

Examples of typical brazed joints are given in Fig. 12.47. In general, dissimilar metals can be assembled with good joint strength, including carbide drill bits (*masonry drill*) and carbide inserts on steel shanks (See Fig. 8.38). The shear strength of brazed joints can reach 800 MPa (115 ksi), using

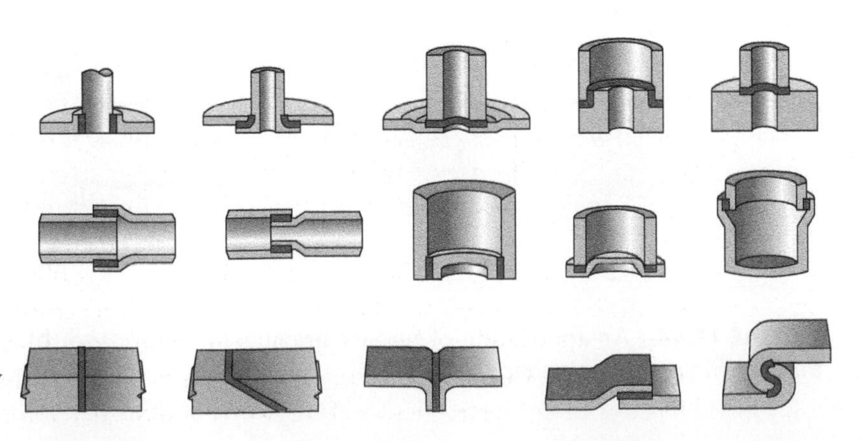

FIGURE 12.47 Joint designs commonly used in brazing operations.

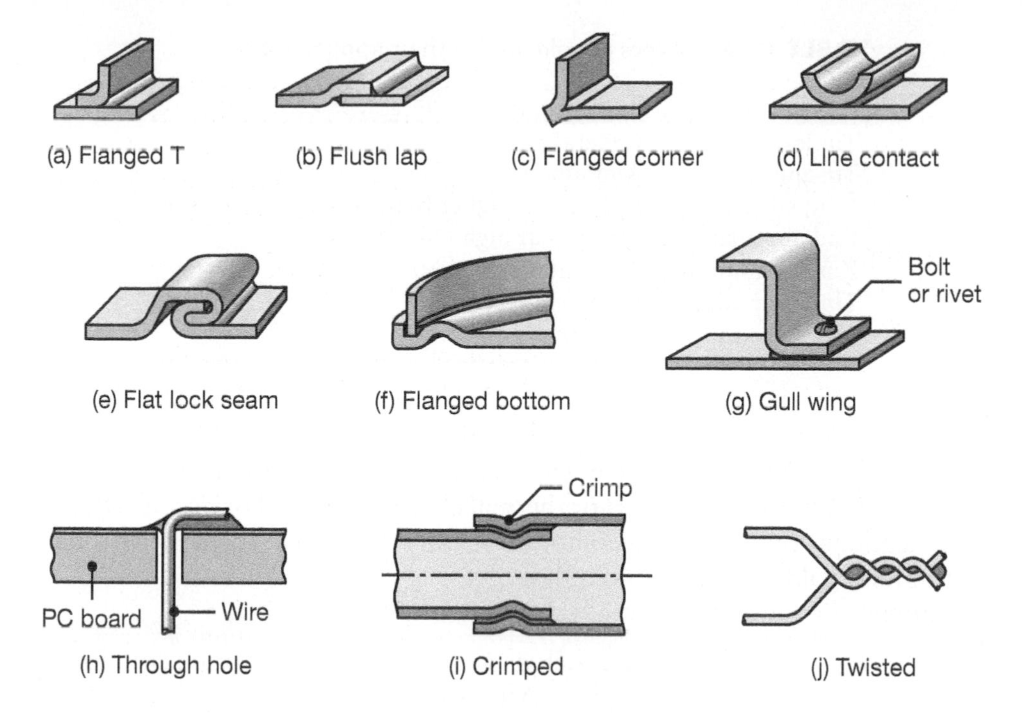

FIGURE 12.48 Joint designs commonly used for soldering.

brazing alloys containing silver (*silver solder*). Intricate and lightweight parts can be joined rapidly and with little distortion.

12.14.3 Soldering

In soldering, the filler metal (**solder**), which typically melts below 450°C (850°F), fills the joint by capillary action between closely-fitting or closely-spaced components (Fig. 12.48). Soldering with copper-gold and tin-lead alloys was first practiced in 4000–3000 BC Heat sources for soldering typically are soldering irons, torches, or ovens. Soldering can be used to join various metals and part thicknesses, and is used extensively in the electronics industry, such as for printed circuit boards. Although manual soldering operations require skill and are time consuming, soldering speeds can be high with automated equipment.

Unlike in brazing, temperatures in soldering are relatively low; consequently, a soldered joint has very limited use at elevated temperatures. Moreover, because solders do not generally have much strength, they are not used for load-bearing structural members. Butt joints are rarely made with solders because of the small faying surfaces involved. Joint strength can be improved by *mechanical interlocking* of the soldered joint (see also Section 12.18).

Solders are generally tin-lead alloys, in various proportions; for higher joint strength and for special applications, other solder compositions include tin-zinc, lead-silver, cadmium-silver, and zinc-aluminum alloys (Table 12.5). Since the European Union prohibited the intentional addition of lead to consumer electronics in 2006, tin-silver-copper solders have

TABLE 12.5 Types of solders and their applications.

Solder	Typical application
Tin-lead	General purpose
Tin-zinc	Aluminum
Lead-silver	Strength at higher than room temperature
Cadmium-silver	Strength at high temperatures
Zinc-aluminum	Aluminum; corrosion resistance
Tin-silver-copper	Electronics
Tin-bismuth	Electronics

come into wide use. A typical composition is 96.5% tin, 3.0% silver, and 0.5% copper; a fourth element, such as zinc or manganese, is often added to provide desired mechanical or thermal characteristics. For non-electrical applications, several other compositions are available and may include cadmium, gold, bismuth, and indium.

Fluxes for soldering generally are of two types:

1. *Inorganic acids* or *salts*, such as zinc ammonium chloride solutions, that clean the surface rapidly. After soldering, the flux residues should be removed by washing thoroughly with water to avoid subsequent corrosion.
2. *Noncorrosive resin-based fluxes*, used typically in electrical applications.

Solderability. Solderability can be defined in a manner similar to weldability (Section 12.7.2), as the capability to be soldered and meet service requirements. Solderability can be improved by first coating the metals; a common example is **tinplate**, which is steel coated with tin, used widely for food containers.

The solderability of some common materials are:

a. *Copper, silver,* and *gold* are easy to solder;
b. *Iron* and *nickel* are more difficult to solder;
c. *Aluminum* and *stainless steels* are difficult to solder, because of their strong, thin oxide film (Section 4.2); these and some other metals can be soldered using special fluxes that modify surfaces; and
d. *Cast irons, magnesium*, and *titanium*, and nonmetallic materials, such as *graphite* and *ceramics*, can be soldered by first plating the parts with metallic elements (see also Section 12.17.3 for a similar technique for joining ceramics).

Soldering methods. There are several soldering methods, that are similar to the brazing methods described in Section 12.14.2:

1. **Torch soldering** (TS);
2. **Furnace soldering** (FS);
3. **Iron soldering** (INS), using a soldering iron;
4. **Induction soldering** (IS);
5. **Resistance soldering** (RS);

6. **Dip soldering** (DS);
7. **Infrared soldering** (IRS);
8. **Ultrasonic soldering,** in which a transducer subjects the molten solder to ultrasonic cavitation that removes the oxide films from the surfaces to be joined; the need for a flux is eliminated;
9. **Wave soldering** (WS), used for automated soldering of printed circuit boards; and
10. **Reflow (paste) soldering** (RS).

The last two techniques listed above are significantly different from the other methods, as described below.

1. **Reflow (paste) soldering.** Semisolid in behavior, solder pastes consist of solder metal particles bound together by flux and wetting agents. They have high viscosity, but maintain a solid shape for relatively long periods of time, similar to the behavior of greases. In this operation, the paste is placed directly onto the joint or on flat objects for finer detail, and is applied using a *screening* or stenciling process (Fig. 12.49). This is common operation for mounting electrical components onto printed circuit boards (Section 13.13). An additional benefit is that the surface tension of the paste helps keep surface-mount packages aligned on their pads, improving the reliability of solder joints.

   Once the paste and the components are placed, the assembly is heated in a furnace, and reflow soldering takes place; the product must be heated in a controlled manner so that the following sequence of events takes place:

   a. Solvents present in the paste are evaporated.
   b. The flux in the paste is activated, and fluxing action occurs.
   c. The components are slowly preheated.
   d. The solder particles melt and wet the joint.
   e. The assembly is cooled, at a low rate, to prevent thermal shock and possible fracture of the joint.

   There are several process variables in each stage, and proper control of temperature and exposure time must be maintained at each

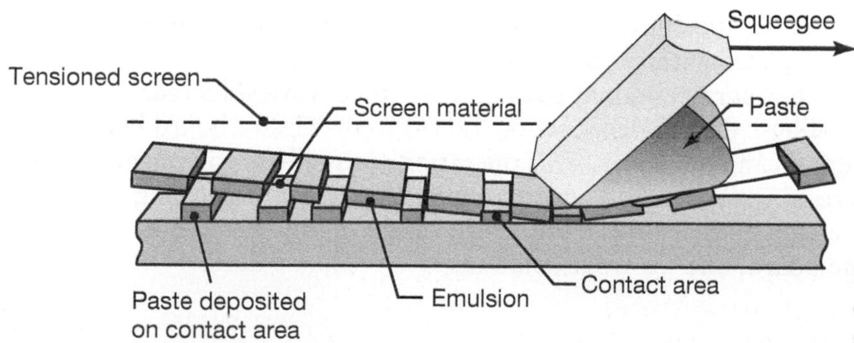

FIGURE 12.49 Screening solder paste onto a printed circuit board in reflow soldering. *Source:* After V. Solberg.

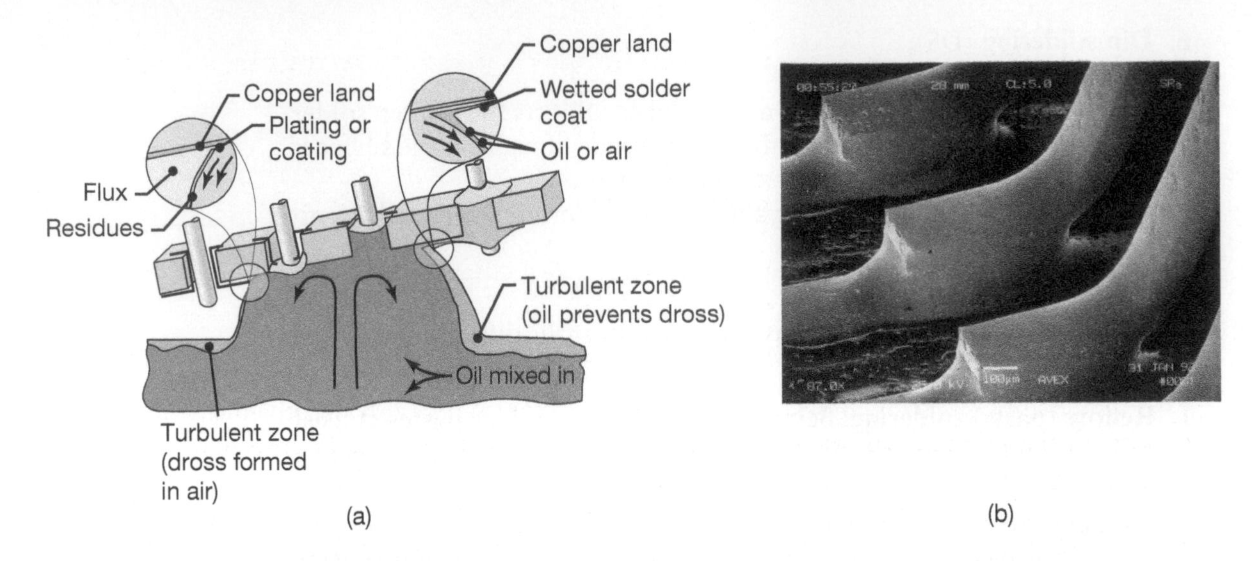

FIGURE 12.50 (a) Schematic illustration of the wave soldering process. (b) SEM image of a wave soldered joint on a surface-mount device (see also Section 13.13).

stage to ensure joint strength. Reflow soldering is the most common soldering method in circuit board manufacturing.

2. **Wave soldering.** This is a common method for soldering circuit components to their boards (Section 13.13). In wave soldering, the molten solder does not wet all surfaces; in fact, solder will not bond to most polymer surfaces and can easily be removed while it is still in the molten state. Also, the solder wets metal surfaces and forms a good bond only when the metal is preheated to a certain temperature. Wave soldering thus requires separate fluxing and preheating operations before it can be applied successfully.

   Wave soldering is schematically illustrated in Fig. 12.50a. A standing laminar wave of molten solder is first generated by a pump; preheated and prefluxed circuit boards are then conveyed over the wave. The solder wets the exposed metal surfaces, but it does not remain attached to the polymer package for integrated circuits and it does not bond to the polymer-coated circuit boards. An *air knife* (basically a high-velocity jet of hot air) blows any excess solder from the joint, thus preventing bridging between adjacent leads. Figure 12.50b shows a scanning-electron microscope photograph of a completed surface-mount joint.

   When surface-mount packages are to be wave soldered, they must be adhesively bonded (Section 12.15) to the circuit board before soldering can commence. This operation is usually accomplished by (1) screening or stenciling epoxy onto the boards; (2) placing the components in their proper locations; (3) curing the epoxy; (4) inverting the board; and (5) wave soldering.

CASE STUDY 12.2 Soldering of Components onto a Printed Circuit Board

Computer and consumer electronics industries place extremely high demands on electronic components, as they are expected to function reliably for extended periods of time during which they can be subjected to significant temperature variations and to vibration. It is thus essential that the solder joints are sufficiently strong and very reliable.

Moreover, the trend continues toward reduction of chip size and increasing compactness of circuit boards. Space savings are achieved by mounting integrated circuits into surface-mount packages, allowing tighter packing on a circuit board and, more importantly, mounting of components on *both sides* of a circuit board.

A challenging problem arises when a printed circuit board (a) has both surface-mount and in-line circuits on the same board. In such a case, all in-line circuits should be inserted from one side of the board for easy assembly.

The basic steps in soldering the connections on a board are:

1. Apply solder paste to one side;
2. Place the surface-mount packages onto the board; insert in-line packages through the primary side of the board;
3. Reflow soldering;
4. Apply adhesive to the secondary side of the board;
5. Attach the surface mount devices on the secondary side;
6. Cure the adhesive; and
7. Wave soldering on the secondary side (to electrically bond the surface mounts and the in-line circuits to the board).

Solder paste is applied with chemically etched stencils or screens, whereby the paste is placed only onto the designated areas of a circuit board. Stencils are more widely used for fine-pitch devices, and produce a more uniform paste thickness. Surface-mount circuit components are then placed on the board, and the board is heated in a furnace to around 200°C (400°F), to reflow the solder and produce strong connections between the surface mount and the circuit board.

At this point the components, with leads, are inserted into the primary side of the board, their leads are crimped, and the board is flipped over. An adhesive pattern is printed onto the board, using a dot of epoxy at the center of a surface-mount component location. The surface-mount packages are then placed onto the adhesive by high speed computer controlled systems. The adhesive is cured, the board is flipped, and wave soldered. This operation simultaneously joins the surface-mount components to the secondary side and solders the leads of the in-line components from the board's primary side. The board is then cleaned and inspected prior to performing electronic quality checks.

12.15 Adhesive Bonding

Adhesive bonding has been a common method of joining and assembling in numerous applications, such as bookbinding, labeling, packaging, home furnishings, and footwear. Plywood, developed in 1905, is an example of adhesive bonding of several layers of wood. Adhesive bonding has been gaining increased acceptance in manufacturing ever since

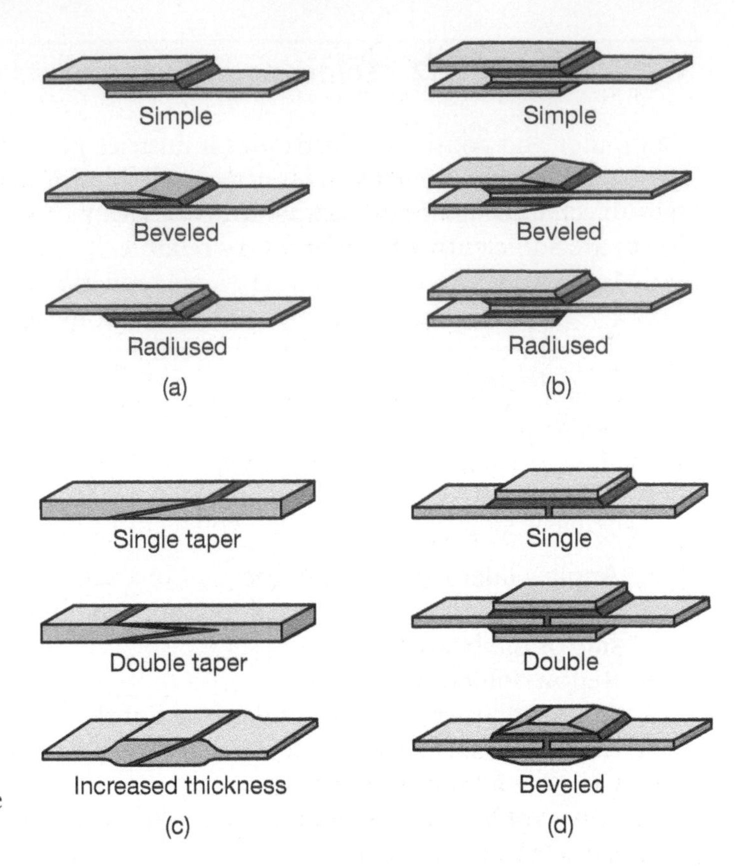

FIGURE 12.51 Various configurations for adhesively bonded joints: (a) single lap, (b) double lap, (c) scarf, and (d) strap.

its use in assembling load-bearing components in military aircraft during World War II (1939–1945). It became a common industrial technique beginning in the 1960s. Various configurations of adhesively bonded joints are shown in Fig. 12.51. Adhesives are available as liquids, pastes, solutions, emulsions, powder, tape, and film; when applied, adhesives typically are on the order of 0.1 mm thick.

Most industries now use adhesive bonded products, including wide acceptance in the aerospace, automotive, appliances, and building products industries. Applications are as varied as attaching rearview mirrors to windshields, automotive brake-lining assemblies, laminated windshield glass (Section 11.11.2), home appliances, helicopter blades, honeycomb structures, and aircraft fuselage and control surfaces.

12.15.1 Types of Adhesives

Several types of adhesives are now available, providing good joint strength, fatigue strength, and resistance to environmental attack. Three basic types of adhesives are:

1. **Natural adhesives:** starch, dextrin (a gummy substance derived from starch), soya flour, and animal products;
2. **Inorganic adhesives:** sodium silicate and magnesium oxychloride; and

3. **Synthetic organic adhesives**: thermoplastics (for nonstructural and some structural bonding) or thermosetting polymers (primarily for structural bonding).

Because of their high cohesive strength, synthetic organic adhesives are the most important in manufacturing, particularly for load-bearing applications (called *structural adhesives*). They are generally classified as follows:

1. **Chemically reactive**: polyurethanes, silicones, epoxies, cyanoacrylates, modified acrylics, phenolics, and polyimides; also included are anaerobics, that cure in the absence of oxygen (Loctite® for threaded fasteners);
2. **Pressure sensitive**: natural rubber, styrene-butadiene rubber, butyl rubber, nitrile rubber, and polyacrylates;
3. **Hot melt**: thermoplastics, such as ethylene-vinyl-acetate copolymers, polyolefins, polyamides, polyester, and thermoplastic elastomers; **reactive hot melt** adhesives have a thermoset portion based on urethane's chemistry, with improved properties;
4. **Evaporative** or **diffusion**: vinyls, acrylics, phenolics, polyurethanes, synthetic rubbers, and natural rubbers;
5. **Film** and **tape**: nylon-epoxies, elastomer-epoxies, nitrile-phenolics, vinyl-phenolics, and polyimides;
6. **Delayed tack**: styrene-butadiene copolymers, polyvinyl acetates, polystyrenes, and polyamides; and
7. **Electrically** and **thermally conductive**: epoxies, polyurethanes, silicones, and polyimides (see Section 10.7.2).

Adhesive systems may be classified based on their specific chemistries:

1. **Epoxy based systems**: These systems have high strength and high-temperature properties, to as high as 200°C (400°F); typical applications include automotive brake linings and as bonding agents for sand molds for casting (Section 5.8.1);
2. **Acrylics**: Suitable for applications on substrates that are not clean or are contaminated;
3. **Anaerobic systems**: Curing is by oxygen deprivation, and the bond is usually hard and brittle; curing times can be reduced by heating externally or by ultraviolet (UV) radiation;
4. **Cyanoacrylate**: The bond lines are thin, and the bond sets within 5 to 40 s;
5. **Urethanes**: High toughness and flexibility at room temperature; widely used as sealants; and
6. **Silicones**: Highly resistant to moisture and solvents, silicones have high impact and peel strength; curing times are typically in the range of one to five days.

Many of these adhesives can be combined to optimize their properties; examples include epoxy-silicon, nitrile-phenolic, and epoxy-phenolic. The least expensive adhesives are epoxies and phenolics, followed by

TABLE 12.6 Typical properties and characteristics of chemically reactive structural adhesives.

	Epoxy	Polyurethane	Modified acrylic	Cyanocrylate	Anaerobic
Impact resistance	Poor	Excellent	Good	Poor	Fair
Tension-shear strength, MPa (ksi)	15–22 (2.2–3.2)	12–20 (1.7–2.9)	20–30 (2.9–4.3)	18.9 (2.7)	17.5 (2.5)
Peel strength[a], N/m (lb/in.)	<523 (3)	14,000 (80)	5250 (30)	<525 (3)	1750 (10)
Substrates bonded	Most	Most smooth, nonporous	Most smooth, nonporous	Most non-porous metals or plastics	Metals, glass, thermosets
Service temperature range, °C (°F)	−55 to 120 (−70 to 250)	−40 to 90 (−250 to 175)	−70 to 120 (−100 to 250)	−55 to 80 (−70 to 175)	−55 to 150 (−70 to 300)
Heat cure or mixing required	Yes	Yes	No	No	No
Solvent resistance	Excellent	Good	Good	Good	Excellent
Moisture resistance	Good-Excellent	Fair	Good	Poor	Good
Gap limitation, mm (in.)	None	None	0.5 (0.02)	0.25 (0.01)	0.60 (0.025)
Odor	Mild	Mild	Strong	Moderate	Mild
Toxicity	Moderate	Moderate	Moderate	Low	Low
Flammability	Low	Low	High	Low	Low

Note: (a) Peel strength varies widely depending on surface preparation and quality.

polyurethanes, acrylics, silicones, and cyanoacrylates; the most expensive are adhesives for high-temperature applications, at up to about 260°C (500°F), such as polyimides and polybenzimidazoles.

Depending on the particular application, an adhesive generally must have one or more of the following properties (Table 12.6):

- Capability to wet the surfaces to be bonded;
- Strength (shear and peel);
- Toughness;
- Resistance to various fluids and chemicals; and
- Resistance to environmental degradation, including heat and moisture.

Adhesive joints are designed to withstand shear, compressive, and tensile forces, but they should not be subjected to peeling forces (Fig. 12.52); note, for example, how easily an adhesive tape can be peeled from a surface. During peeling, the behavior of an adhesive may be described as *brittle* or as *ductile* and *tough* (thus requiring large forces to peel it).

QR Code 12.7 Peel test of an adhesive.
Source: Courtesy of Instron.

12.15.2 Surface Preparation and Application

Surface preparation is very important in adhesive bonding because joint strength depends greatly on the absence of dirt, dust, oil, and various other contaminants; note, for example, that it is virtually impossible to apply an adhesive tape over a dusty or oily surface. Contaminants also affect the *wetting* ability of the adhesive, preventing the spreading of the

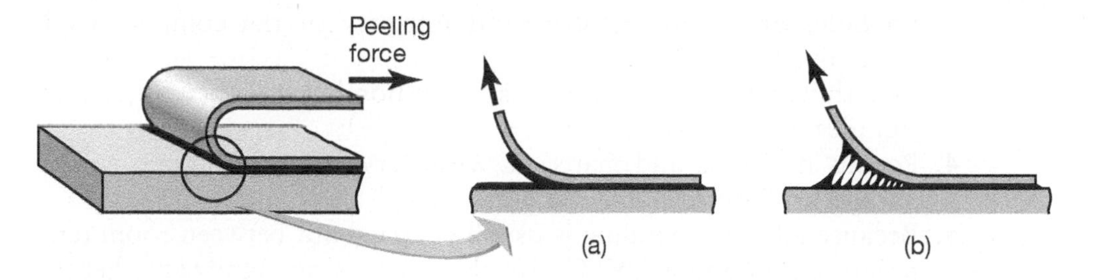

FIGURE 12.52 Characteristic behavior of (a) brittle and (b) tough and ductile adhesives in a peeling test. This test is similar to peeling adhesive tape from a solid surface.

adhesive evenly over the interface; thick, weak, or loose oxide films on workpieces also are detrimental in adhesive bonding. On the other hand, a porous or thin and strong oxide film can be desirable, particularly one with some surface roughness that would improve adhesion at the interfaces (by mechanical locking). The roughness must not be too high, however, as air may be trapped, reducing joint strength. Various compounds and primers are now available to modify surfaces to increase the strength of adhesive bonds.

12.15.3 Process Capabilities

A wide variety of similar and dissimilar metallic and nonmetallic materials and components with different shapes, sizes, and thicknesses can be bonded to each other by adhesives. Adhesive bonds for structural applications are rarely suitable for service above 250°C (500°F), but they can be combined with *mechanical fastening* methods (Section 12.16) to further improve bond strength. Joint design and bonding methods require skill and special equipment, such as fixtures, presses, various tooling, and autoclaves and ovens for curing. An important consideration in the use of adhesives is curing time, which can range from a few seconds at high temperatures to several hours at room temperature, particularly for thermosetting adhesives.

Nondestructive inspection of the quality and strength of adhesively bonded components can be difficult; some of the techniques described in Section 4.8.1, such as acoustic impact (tapping), holography, infrared detection, and ultrasonic testing, can be effective.

The *advantages* of adhesive bonding include the following:

1. Provides a bond at the interface, either for structural strength or for nonstructural applications (sealing, insulating, preventing electrochemical corrosion between dissimilar metals, and reducing vibration and noise through internal damping at the joints).
2. Distributes the load at an interface evenly, eliminating localized stresses that typically result from joining the components with welds or mechanical fasteners, such as bolts and screws. Moreover, because

no holes are required, structural integrity of the components is maintained.

3. Very thin and fragile components can be bonded without contributing significantly to weight.
4. Porous materials and materials with very different properties and sizes can be joined.
5. Because adhesive bonding is usually carried out between room temperature and about 200°C (400°F), there is no significant thermal distortion of the components or change in their original properties, a factor that is particularly important for heat-sensitive materials and components.
6. The appearance of the joined components is unaffected.

The *limitations* of adhesive bonding are:

1. Service temperatures are relatively low.
2. Bonding time can be long.
3. Surface preparation is critical.
4. Joints are difficult to test nondestructively, particularly with large structures.
5. Reliability of structures during their service life and under hostile environmental conditions (such as *degradation* by elevated temperature, oxidation, radiation, stress corrosion, and dissolution) can be a significant concern.

12.15.4 Electrically Conducting Adhesives

Although the majority of adhesive bonding is for mechanical strength and structural integrity, an important type is *electrically conducting adhesives*. Applications include calculators, remote controls, control panels, electronic assemblies, liquid-crystal displays, and electronic games.

These adhesives require curing temperatures that are lower than those for soldering. Electrical conductivity in adhesives is achieved by adding electrically conducting *fillers*, such as silver, copper, aluminum, nickel, gold, and graphite (see also Section 10.7.2). Because of its very high electrical conductivity, silver is the most commonly used metal as a filler, with up to 85% silver content. There is a minimum concentration of fillers to make the adhesive electrically conducting, typically in the range of 40–70%. Fillers generally are in the form of flakes or particles, and also include polymeric particles, such as polystyrene, that are coated with thin films of silver or gold. Matrix materials are generally epoxies, although various thermoplastics are also used.

The size, shape, and distribution of the particles and the nature of their contact between the individual particles and how heat and pressure are applied, can be controlled to impart isotropic or anisotropic electrical conductivity to the adhesive. Note that fillers that improve electrical conductivity also improve thermal conductivity of the adhesively bonded joints (see Sections 3.9.4 and 3.9.6).

12.16 Mechanical Fastening

Countless products are fastened *mechanically*, whereby two or more components are assembled in such a way that they can be taken apart during the product's service life or life cycle (see Section 14.10). Mechanical fastening may be preferred over other methods because of (a) ease of manufacturing, assembly and transportation; (b) part disassembly for replacement, maintenance, and repair; (b) designs requiring movable joints and adjustable components; and (f) lower production cost.

The most common **fasteners** are bolts, nuts, screws, rivets, and pins. They typically require *holes* through which these fasteners are inserted and secured.

12.16.1 Hole Preparation

Hole preparation is an important aspect of mechanical fastening. Depending on the type of material, its properties, and its thickness, a hole in a solid body can be produced by such processes as punching, drilling, chemical and electrical means, and high energy beams (Chapters 7, 8, and 9). Recall also from Chapters 5, 6, and 11 that holes can also be made as an *integral* part of the product, such as during casting, forging, extrusion, and powder metallurgy; additional operations are thus avoided and product costs are reduced. For improved dimensional accuracy and surface finish, many of these holemaking operations may also be followed by such finishing operations as deburring, reaming, shaving, and honing.

Because of their fundamental differences, each type of holemaking operation produces a hole with different surface finish, texture, properties, and dimensional characteristics. For example, a punched hole will have an axial *lay* (see Sections 4.3 and 7.3), whereas a drilled hole will have a circumferential lay. The most significant influence of a hole in a solid body is its stress concentration, because its presence can reduce the component's fatigue life. Fatigue life can best be improved by inducing compressive residual stresses on the cylindrical surface of the hole; a common technique is to push a round rod (*drift pin*) through the hole and expand it by a very small amount. This operation plastically deforms the surface layers of the hole in a manner similar to shot peening or roller burnishing (Section 4.5.1).

12.16.2 Threaded Fasteners

Bolts, screws, and studs are among the most commonly used *threaded fasteners*. References on machine design describe in detail numerous standards and specifications, including thread dimensions, tolerances, pitch, strength, and the quality of materials used to make these fasteners. Bolts and screws may be secured either with nuts or threaded holes, or they may be *self-tapping*, whereby the screw either cuts or forms the thread into the part to be fastened; the latter method is particularly effective and economical for plastic products. If the joint is to be subjected to vibration, such as in aircraft and various machinery and engines, several specially designed

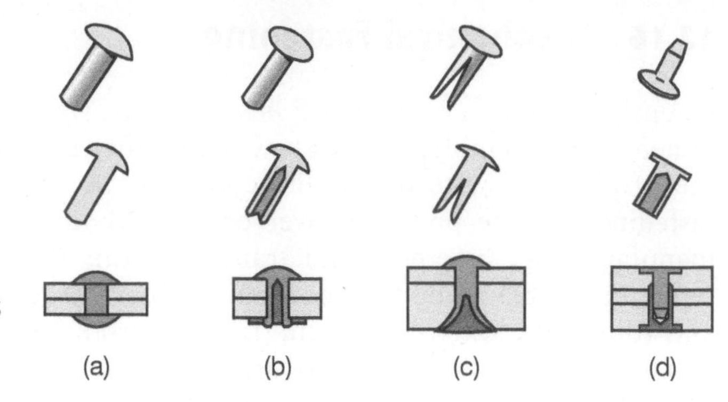

FIGURE 12.53 Examples of rivets: (a) solid; (b) tubular; (c) split; or bifurcated; and (d) compression.

nuts and lock washers are available; they increase the frictional resistance, thus preventing loosening of the fasteners.

12.16.3 Rivets

The most common method of permanent or semipermanent mechanical joining or assembly is by *riveting* (Fig. 12.53). The basic types of rivets are solid, hollow, and blind (inserted from one side only). Installing a rivet basically consists of placing the rivet in the hole and deforming the end of its shank by upsetting it (similar to the *heading* process shown in Fig. 6.15); this operation can be performed by hand hammer or by mechanized means, including the use of robots. Riveting may be done either at room temperature or high temperatures, or using explosives placed in the cavity of the rivet.

12.16.4 Various Methods of Fastening

Metal stitching or **stapling** (Fig. 12.54) is much like that of ordinary stapling of papers, an operation that is fast and particularly suitable for joining thin metallic and nonmetallic materials. A common example is the stapling of cardboard boxes and containers. **Clinching** is illustrated in Fig. 12.54b; note that the fastener is subjected to sharp bends.

Seaming is based on the same principle as folding two thin pieces of material together; common examples are the lids of beverage cans and containers for food and household products (*lock seams*, Fig. 12.55). The materials must be capable of undergoing bending and folding at very small

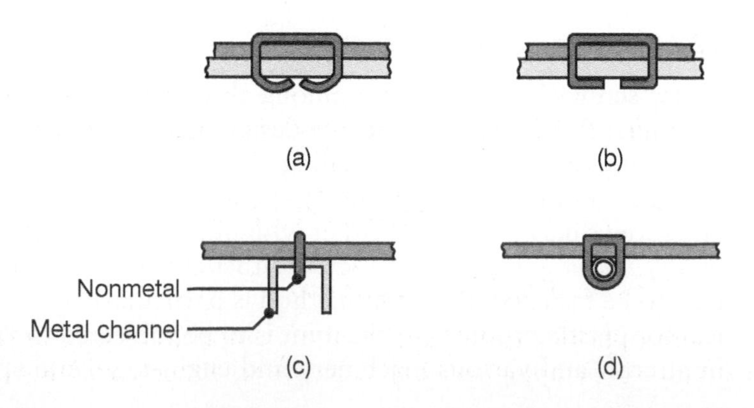

FIGURE 12.54 Examples of various fastening methods. (a) Standard loop staple; (b) flat clinch staple; (c) channel strap; and (d) pin strap.

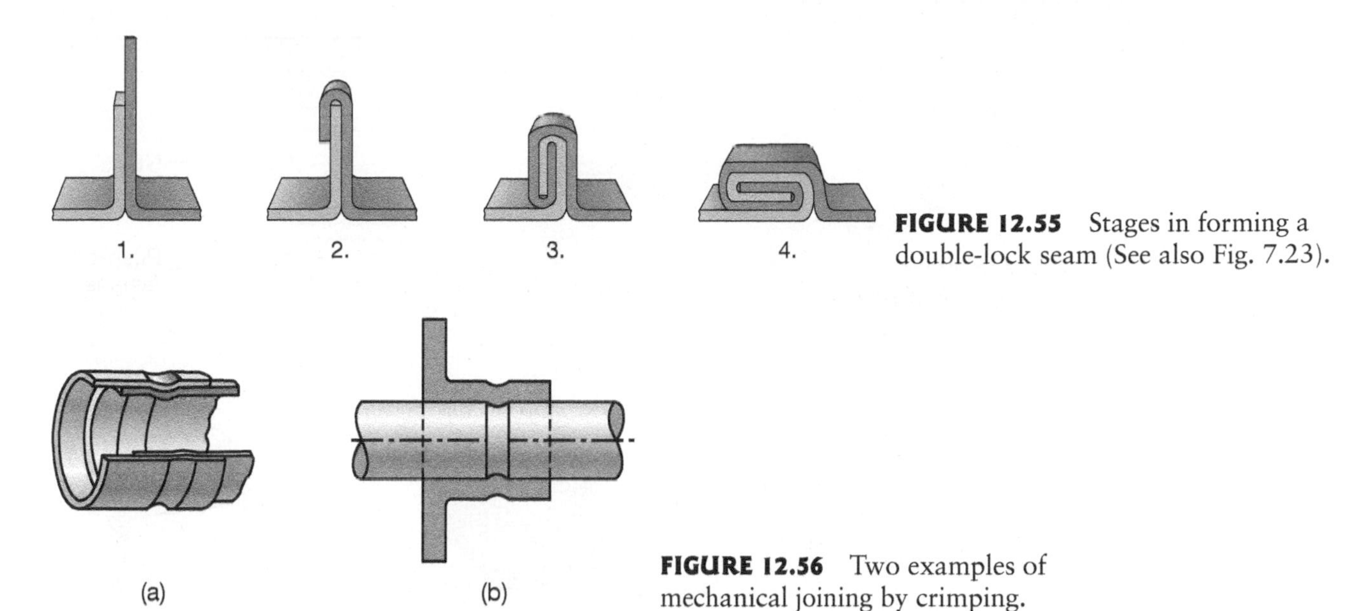

FIGURE 12.55 Stages in forming a double-lock seam (See also Fig. 7.23).

FIGURE 12.56 Two examples of mechanical joining by crimping.

radii (see Section 7.4.1 and Fig. 7.15). The performance and airtightness of lock seams may be improved with adhesives, coatings, and polymeric materials at the interfaces or by soldering, as is done in some steel cans for household products.

Crimping is a method of joining without using fasteners; caps on glass bottles, for example, are assembled by crimping, as are some connectors for electrical wiring. Crimping can be used on both tubular and flat parts or it can be done with beads or dimples (Fig. 12.56), which can be produced by *shrinking* (Section 7.4.4) or swaging operations (Section 6.6).

Shape-memory alloys. The characteristics of these materials were described in Section 3.11.9. Because of their unique capability to recover their shape, shape-memory alloys can be used for fasteners. Several advanced applications include their use as couplings in the assembly of titanium-alloy tubing for aircraft.

Snap-in fasteners are shown in Fig. 12.57. They are widely used in the assembly of automobile bodies and household appliances; they are economical and permit easy and rapid assembly and disassembly of components.

Shrink fits are based on the principle of the differential thermal expansion and contraction of two components. Typical applications include the assembly of die components and mounting of gears and cams on shafts. In **press fitting**, one component is forced over another, resulting in high joint strength; examples include shrink fit toolholders and hubs on shafts.

12.17 Joining Nonmetallic Materials

12.17.1 Joining Thermoplastics

Thermoplastics soften and melt as temperature increases, thus they can be joined by applying heat to the joint, either through external means or

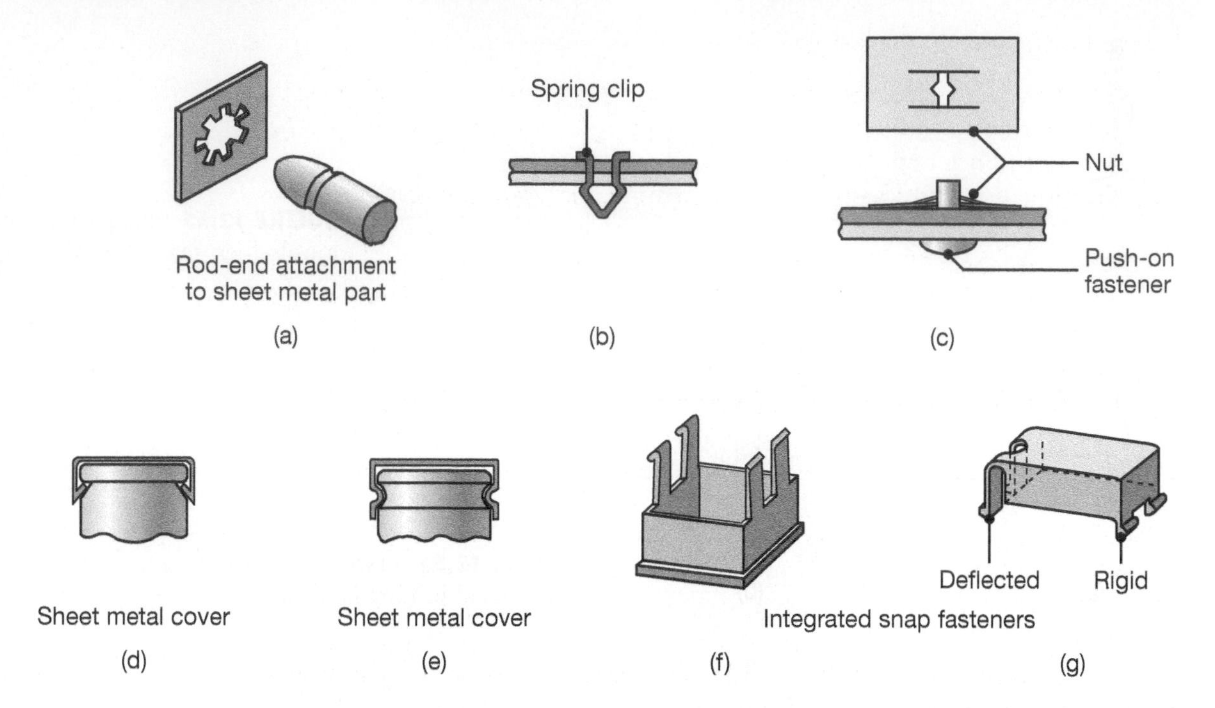

FIGURE 12.57 Examples of spring and snap-in fasteners to facilitate assembly.

internally. The heat softens the thermoplastic at the interface, then pressure is applied, allowing *fusion* to take place, ensuring a good bond. Filler materials of the same type of polymer may be used for stronger bonds.

Oxidation can be a consideration in joining some polymers, such as polyethylene, because of degradation; an inert shielding gas, such as nitrogen, is typically used to prevent oxidation. Moreover, because of the low thermal conductivity of thermoplastics, the heat source may burn or char the joint if heat input is not controlled properly.

External heat sources may consist of the following, depending on the compatibility of the polymers to be joined:

1. Hot air or gases, or infrared radiation with high-intensity focused quartz heat lamps;
2. Heated tools and dies that contact and heat the surfaces to be joined. Known as *hot-tool* or *hot-plate welding*, this is a common practice in butt welding of pipe and tubing;
3. Radio-frequency or dielectric heating, particularly useful for thin films;
4. Lasers, using defocused beams at low power, to prevent degradation of the polymer; and
5. Electrical resistance wire or braids, or carbon-based tapes, sheet, and ropes; they are placed at the interface and allow passing of electrical current (*resistive implant welding*). The elements at the interface, which may also be subjected to a radio-frequency field (*induction welding*), must be compatible with the intended use of the joined product because they remain in the weld zone.

Internal heat sources consist of the following:

1. Ultrasonic welding (Section 12.9), the most commonly used process for thermoplastics, particularly amorphous polymers such as ABS and high-impact polystyrene;
2. Friction welding (Section 12.10), also called *spin welding* for polymers; it includes linear friction welding, also called *vibration welding*, which is particularly useful for joining polymers with a high degree of crystallinity (Section 10.2.2), such as acetal, polyethylene, nylon, and polypropylene; and
3. *Orbital welding*, which is similar to friction welding except that the rotary motion of one part is in an orbital path (See also Fig. 6.14a).

Other joining methods. *Adhesive bonding* of polymers is a versatile process, commonly used for joining sections of PVC or ABS pipe. Liquid adhesive is applied to the connecting sleeve and pipe surfaces, sometimes using a primer to improve adhesion. Adhesive bonding of polyethylene, polypropylene, and polytetrafluoroethylene (Teflon) can be difficult, because these polymers have low surface energy, and thus adhesives do not readily bond to their surfaces. The surfaces generally have to be treated chemically to improve bond strength. The use of adhesive *primers* or *double-sided adhesive tapes* can also be effective. Solvents may also be used for bonding, called *solvent bonding*.

Coextruded multilayer food wrappings consist of different types of films, bonded by the *heat* generated during *extrusion* (Section 10.10.1). Each film has a specific function, such as to keep out moisture and oxygen or to facilitate heat sealing during packaging. Some wrappings have as many as seven layers, all bonded together by *cocuring* during production of the film.

Thermoplastics may also be joined by *mechanical* means, such as fasteners and self-tapping screws. The strength of the joint depends on the particular method used and the inherent toughness and resilience of the polymer to resist tearing at the holes that exist for mechanical fastening. *Integrated snap fasteners* (Figs. 12.57f and g) are gaining wide acceptance for their simplified and cost effective assembly.

Bonding of polymers may also be achieved by *electromagnetic bonding* whereby tiny (on the order of 1 μm) metallic particles are embedded in the polymer. A high frequency field then causes induction heating of the polymer and melts it at the interfaces to be joined.

12.17.2 Joining Thermosets

Because they do not soften or melt with increasing temperature, thermosetting plastics are usually joined using (a) threaded or molded-in inserts (See Fig. 10.32); (b) mechanical fasteners; and (c) solvent bonding. Bonding with solvents typically involves the following steps: (1) roughening the surfaces of the parts with an abrasive cloth or paper; (2) wiping both surfaces with a solvent; and (3) pressing the surfaces together and maintaining the pressure until sufficient bond strength is developed.

12.17.3 Joining Ceramics and Glasses

Ceramics and glasses can be assembled, either with the same type of material or with different materials using (a) adhesive bonding, (b) mechanical means, and (c) shrink or press fitting. If the assembly is to be subjected to elevated temperature during its service life, the *relative thermal expansion* of the two materials should be taken into consideration to prevent damage to or loosening of the joint (see Section 3.9.5).

Ceramics. A technique, effective in joining difficult-to-bond material combinations, consists of first applying a coating of a material that bonds itself easily to one or both parts, thus acting as a bonding agent. The surface of alumina ceramics, for example, can be *metallized* (Section 4.5.1); called the *Mo-Mn process*, the ceramic part is first coated with a slurry that, after being fired, forms a glassy layer. This layer is then plated with nickel; because the part now has a metallic surface, it can be brazed to a metal. Depending on their particular structure, ceramics can also be joined on to metals by diffusion bonding (Section 12.13); it may be necessary to first place a metallic layer at the joint to make it stronger.

As described in Section 8.6.4, tungsten carbide has a matrix (binder) of cobalt, and titanium carbide has matrix of nickel-molybdenum alloy; consequently, with both binders being metal, carbides can easily be *brazed* to other metals. Common applications include brazing carbide tips to masonry drills and cubic boron nitride tips to carbide inserts (See Fig. 8.39).

Glasses. Glasses can easily be bonded to each other, as evidenced by the availability of numerous complex ornamental or other glass objects. This is accomplished by first heating and softening the glass surfaces to be joined, then pressing the two pieces together. Glasses can be bonded to metals also, because of the diffusion of metal ions into the amorphous glass (see also Section 11.11.2).

12.18 Design Considerations in Joining

12.18.1 Design for Welding

As in all manufacturing processes, the optimum design for welding is one that satisfies all design requirements, and specifications at a minimum cost. Examples of various weld guidelines are given in Figs. 12.58 and 12.59, but the following rules are also important.

1. Product design should minimize the number of welds.
2. Weld locations should be selected so as not to interfere with subsequent processing of the part and its appearance.
3. Parts should fit properly prior to welding; the method employed to produce the edges can affect weld quality.
4. Weld bead size should be kept to a minimum to conserve weld metal as well as for appearance.
5. Mating surfaces for some joining processes may require uniform cross section at the joint.

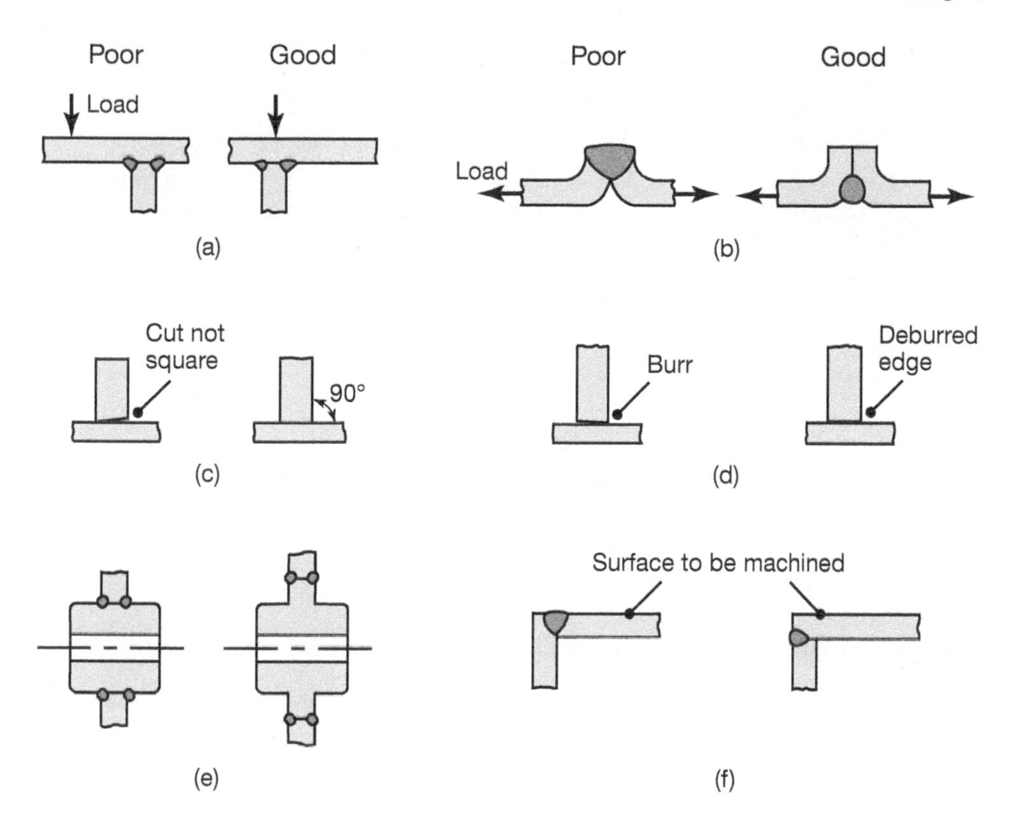

FIGURE 12.58 Design guidelines for welding.

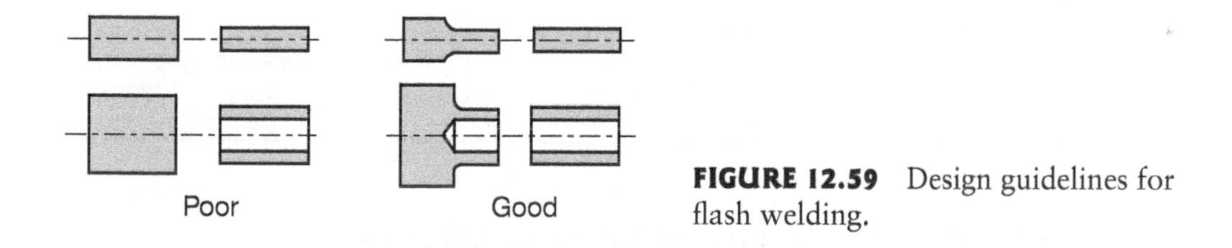

FIGURE 12.59 Design guidelines for flash welding.

EXAMPLE 12.4 Weld Design Selection

Three different types of weld designs are illustrated in Fig. 12.60. As shown in Fig. 12.60a, the two vertical joints can be welded either externally or internally. Note that full-length external welding takes considerable time and requires more weld material than the alternative design, which consists of intermittent internal welds. Moreover, in the alternative method, the appearance is improved and distortion is reduced because of the lower energy content in the structure (See Fig. 12.27).

Although both designs require the same amount of weld material and welding time (Fig. 12.60b), further analysis will show that the design on the right can carry three times the moment M of the design on the left. In Fig. 12.60c, the weld on the left requires about twice the amount of weld material than that required by the design on the right. Note also that because more material must be machined in the single V groove, the design on the left will require more time for joint preparation and more base metal will be wasted.

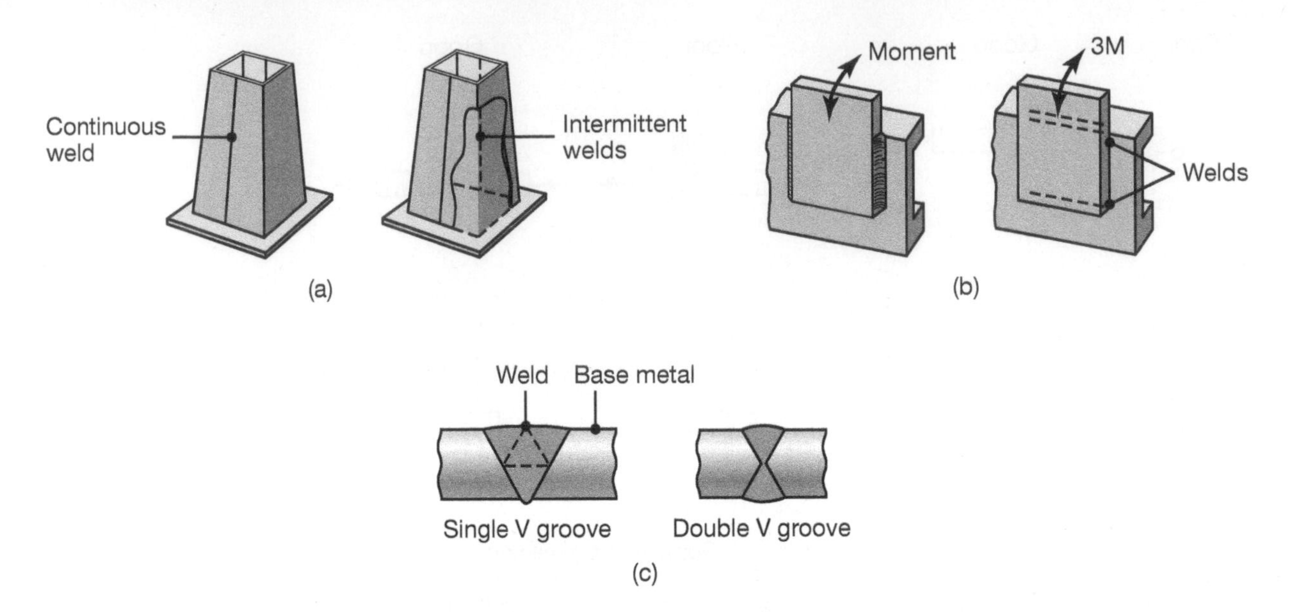

FIGURE 12.60 Weld designs for Example 12.4.

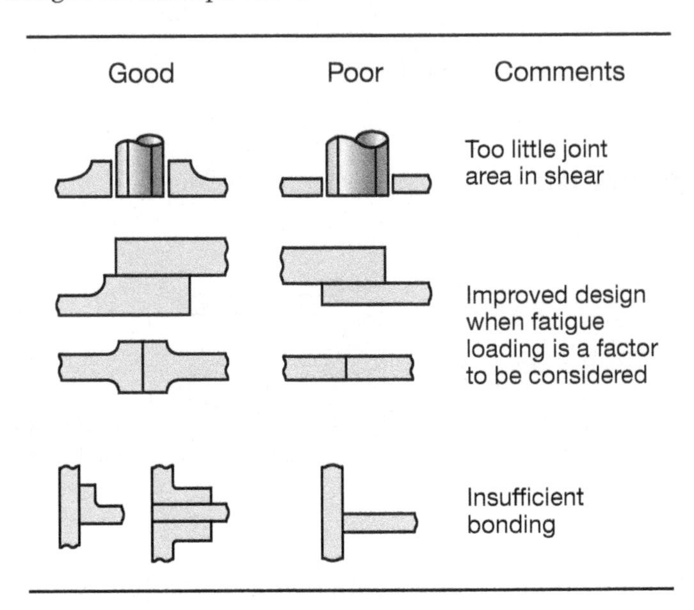

FIGURE 12.61 Examples of good and poor designs for brazing.

12.18.2 Design for Brazing and Joining

General design guidelines for brazing are briefly given in Fig. 12.61. Note that strong joints require a larger contact area for brazing than for welding. Design guidelines for soldering are similar to those for brazing; Figs. 12.47 and 12.48 show some examples of frequently used joint designs. Note again the importance of large contact surfaces to develop sufficient joint strength.

12.18.3 Design for Adhesive Bonding

Several joint designs for adhesive bonding are given in Fig. 12.62; note that they vary considerably in joint strength. Well-designed adhesive bonded joints should be loaded in tension, compression or shear, but not peeling

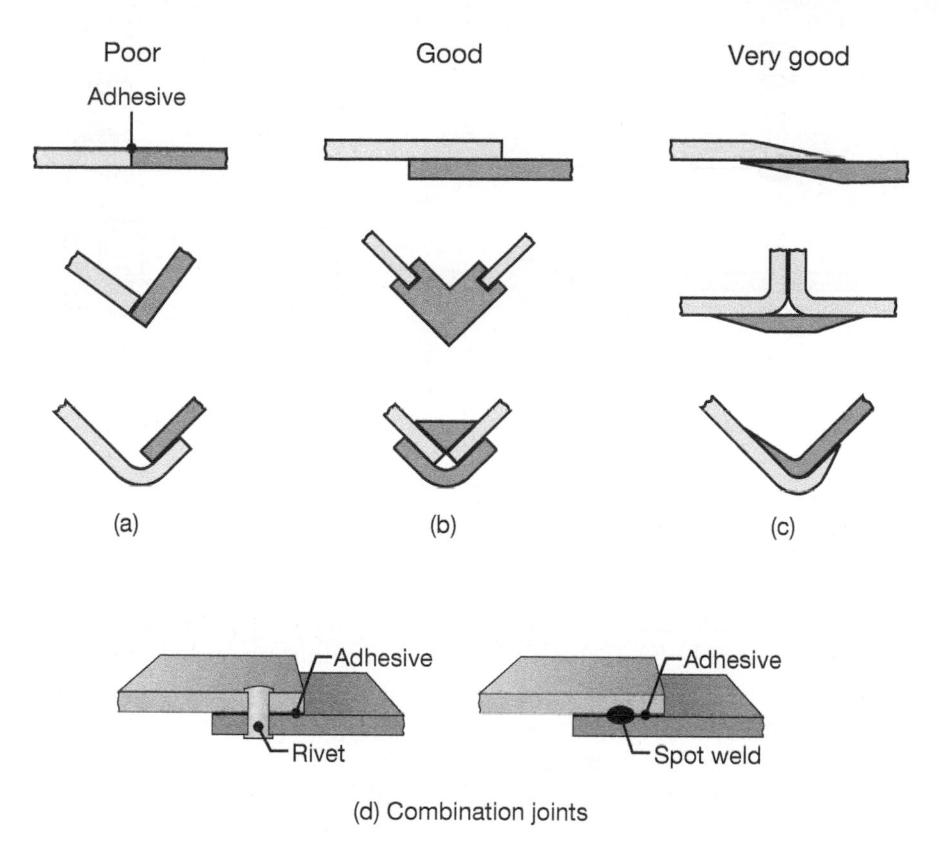

FIGURE 12.62 Various joint designs in adhesive bonding. Note that good designs require large contact areas for better joint strength.

or cleavage (Fig. 12.52). Selection of an appropriate design should include considerations of the type of loading and the environment to which the bonded structure will be subjected, particularly over its service life. Butt joints require large bonding surfaces; note also that because of the force couple developed at the joint, lap joints tend to distort under a tensile force. The coefficients of thermal expansion of the two components to be bonded should preferably be close, to avoid the development of internal stresses during adhesive bonding. Note also that thermal cycling can cause differential movement across the joint, possibly reducing joint strength.

12.18.4 Design for Mechanical Fastening

Some design guidelines for riveting are given in Fig. 12.63. The design of mechanical joints requires basic considerations of (a) the type of loading to which the structure will be subjected; (b) the size and spacing of the holes; and (c) compatibility of the fastener material with the components to be joined. Incompatibility may lead to galvanic corrosion, also known as *crevice corrosion*. For example, if using a steel bolt or rivet to fasten copper sheets, the bolt is anodic and the copper plate is cathodic, thus resulting in rapid corrosion and loss of joint strength. Similar reactions take place when aluminum or zinc fasteners are used on copper products.

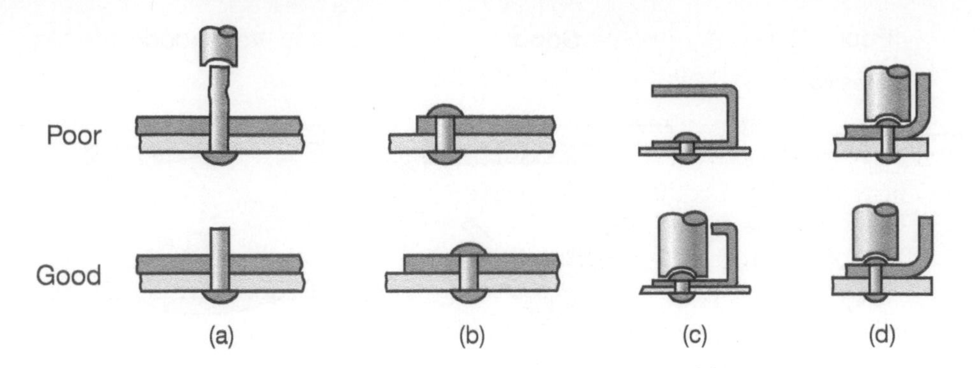

FIGURE 12.63 Design guidelines for riveting. *Source:* After J.G. Bralla.

Other design guidelines for mechanical joining include the following (see also Section 14.10):

1. Fasteners of standard size should be used whenever possible.
2. Assembly should be accomplished with a minimum number of fasteners.
3. It is generally less costly to use fewer but larger fasteners than a large number of smaller fasteners.
4. The fit between parts to be joined should be as loose as permissible to reduce costs and facilitate the assembly operation.
5. Holes should not be too close to edges or to corners, to avoid tearing of the material when subjected to external forces.

12.19 Economic Considerations

The basic economic considerations for particular joining processes have been described individually in this chapter. It will be noted that because of the wide variety of the factors involved, it is difficult to make generalizations regarding costs. The overall relative costs have been included in the last columns in Tables 12.1 and 12.2. Note in Table 12.1, for example, that because of the preparations involved, brazing and mechanical fastening can be the most costly methods. On the other hand, because of their highly automated nature, resistance welding and seaming and crimping processes are the least costly of the methods described. Although the overall cost of adhesive bonding depends on the particular application, the overall economics of the process makes adhesive bonding an attractive alternative, and sometimes, it is the only process that is feasible or practical for a specific application.

Table 12.2 lists the cost of equipment involved for major categories of joining processes. Note that because of the specialized equipment involved, laser-beam welding often is the most expensive process, whereas the traditional oxyfuel gas welding and shielded metal arc welding are the least expensive processes. Typical ranges of equipment costs for some processes are listed below (in alphabetical order). The cost of some equipment can, however, significantly exceed these values, depending on the size and the level of automation and controls implemented:

Electron-beam welding: $90,000 to over $1 million;
Electroslag welding: $15,000 to $25,000;
Flash welding: $5000 to $1 million;
Friction welding: $75,000 to over $1 million;
Furnace brazing: $2000 to $300,000;
Gas metal arc and flux-cored arc welding: $1000 to $3000;
Gas tungsten arc welding: $1000 to $5000;
Laser-beam welding: $30,000 to $1 million;
Plasma arc welding: $1500 to $6000;
Resistance welding: $20,000 to $50,000; and
Shielded metal arc welding: $300 to $2000.

As in all manufacturing operations, an important aspect of the economics of joining is in process automation and optimization (see Chapters 14 and 15). Particularly important is the extensive and effective use of industrial robots, programmed in such a manner that they can accurately track complex weld paths (such as in *tailor-welded blanks*; Section 7.3.4), using *machine vision* systems and closed-loop controls, greatly improving the repeatability and accuracy of joining operations.

CASE STUDY 12.3 Linear Friction Welding of Blanes and Blisks

Ti-6Al-4V titanium alloy bladed vanes (or *blanes*) and bladed disks (*blisks*) are integral components of modern jet engines. Figure 12.64 shows a typical jet engine with casings removed to highlight the blisks; Fig. 12.65 shows details of a typical blisk. Note that there are many blades mounted in close proximity to each other, and that very strict tolerances must be maintained for operating efficiency. Further, the environment in a jet engine is very demanding; temperatures can easily exceed 1000°C, and loadings are unsteady, so that fatigue is an issue.

Blanes and blisks traditionally required skilled machinists to attach blades to a central hub using mechanical fasteners. However, this approach was time-consuming and expensive, and the quality of the product was difficult to control. Beginning in the 1990s, laser welding started to be used to fasten blades to disks, and significant economic and performance improvements resulted. However, blade failures in the heat affected zone still occurred.

Linear friction welded (LFW) blanes and blisks started appearing in aerospace applications in 2001, and have seen steadily increasing use ever since. LFW involves reciprocating sliding motion under controlled pressure. The oscillation frequency is between 30 and 50 Hz and the amplitude is

FIGURE 12.64 Photograph of a jet engine, highlighting a blisk. (Shutterstock)

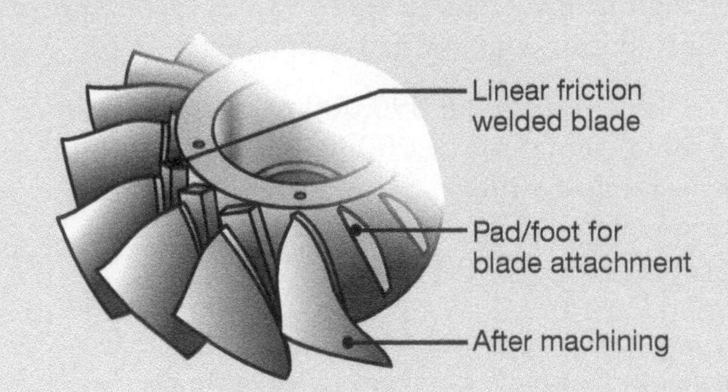

FIGURE 12.65 Detailed view of blades that have been attached to a compressor disk through linear friction welding. During linear friction welding, the parts encounter plastic deformation; the block shown is removed by machining.

2.5–5.0 mm. As the temperature increases, the load is also increased, resulting in a pressure of around 100 MPa, sufficient to cause plastic deformation at the interface between parts to be joined. This deformation removes surface oxides and other defects from the joint. When the desired deformation is achieved, the relative motion between parts is stopped, resulting in a strong diffusion-based joint.

Since the part cools fairly quickly, the joint is cold worked and has an advantageous microstructure for fatigue resistance. Linear friction welding has a number of advantages for this application:

1. The properties of linear friction welded joints are superior to traditional fusion based welded methods since the friction welding process does not actually melt the parent material. Melting causes a drastic change in a material's properties in the weld zone. The heat-affected zone (HAZ) of a friction welded joint is narrow and fine grained with a smooth transition to the unaffected base material.
2. Complex geometries result in forged-quality across the entire butt-weld area.
3. The welding process is very fast – between 2 and 100 times faster than competing processes. Further, it is possible to weld more than one blade at a time, further reducing cycle times.
4. By welding blades onto a disk, significant material cost savings can be achieved compared to designs involving machining from a single billet or block.
5. The process is energy efficient. Power requirements for LFW are as much as 20% lower than those for conventional welding processes. Further, it is environmentally friendly in that it requires no flux, filler metal, or shielding gases, and it doesn't emit any smoke, fumes or gases.
6. LFW is extremely repeatable; since the process is steady state, there is essentially no porosity, segregation or slag inclusions.

As can be seen in Fig. 12.65, the blades are produced with a relatively large block, and the disks have a prepared shoe or pad for the blades. After welding, the block and flash need to be removed by machining, resulting in the high-quality blisks required in modern aircraft. The blades are attached with a more fatigue-resistant weld zone without a heat affected zone, and the blanes and blisks are more reliable as a result. LFW installations are quite extensive; a schematic of a LFW setup for blisks is shown in Fig. 12.66.

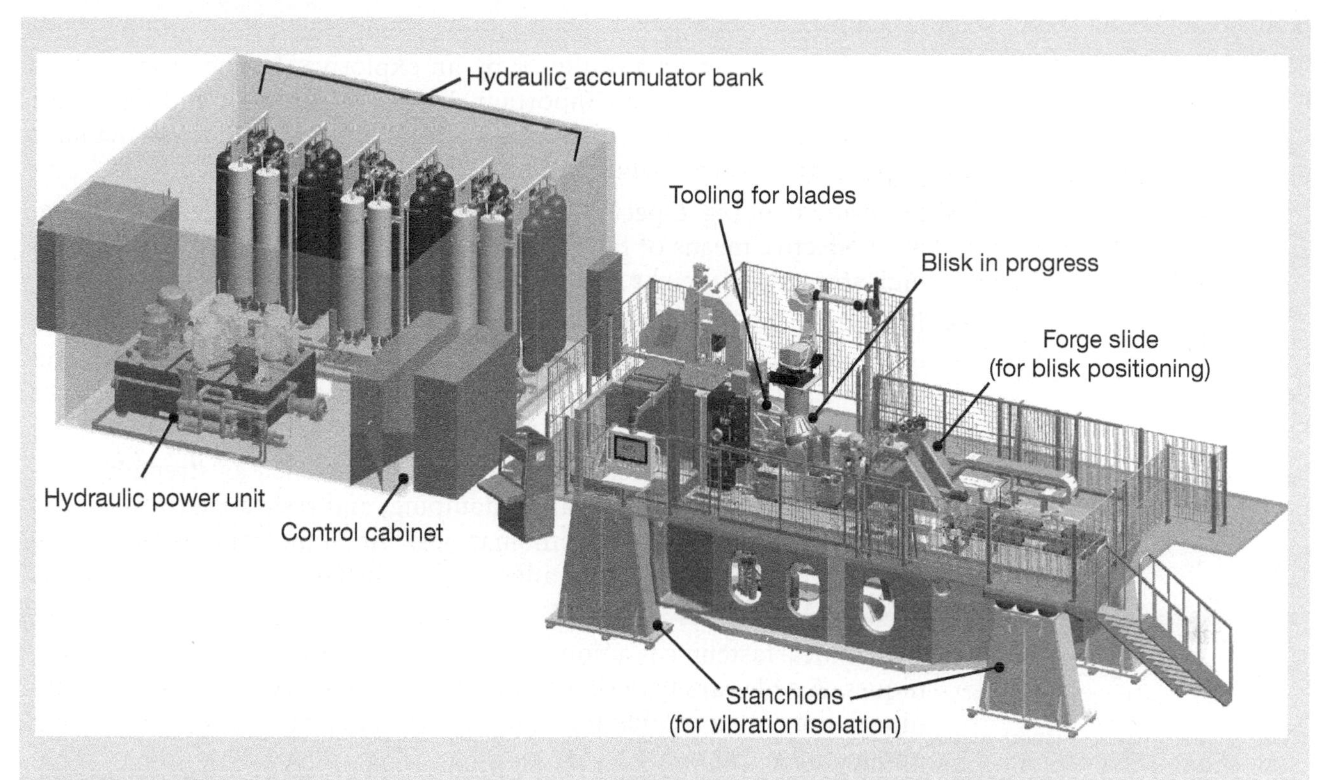

FIGURE 12.66 Layout of a linear friction welding line for the production of blisks.

Source: D. Adams, Manufacturing Technology, Inc.

SUMMARY

- Joining and fastening processes are an important aspect of manufacturing, as well as of servicing and transportation. (Section 12.1)
- A major category of joining processes is fusion welding, in which the two pieces to be joined melt together at their interface and coalesce by means of heat; filler metals may or may not be used. Two common methods are consumable electrode arc welding and nonconsumable electrode arc welding. Selection of a particular method depends on several factors, including workpiece material, shape complexity, size, and thickness, and type of joint. (Sections 12.2–12.4)
- Electron-beam and laser-beam welding are major categories of joining using high energy beams. They produce small and high-quality weld zones, thus have important and unique applications in manufacturing, including cutting. (Sections 12.5–12.6)
- Because it undergoes important metallurgical and physical changes, the nature, properties, and quality of a weld joint are important factors to consider. Weldability of metals, weld design, and welding process selection are factors that must be considered. (Section 12.7)

- In solid state welding, joining takes place without fusion, and pressure is applied either mechanically or by an explosive. Surface preparation and its cleanliness can be important. Ultrasonic welding and resistance welding are two major examples of solid state welding, particularly important for sheet metals and foil. (Sections 12.8–12.12)
- Diffusion bonding, especially when combined with superplastic forming, is an effective means of fabricating complex sheet metal structures with high strength-to-weight and stiffness-to-weight ratios. (Section 12.13)
- Brazing and soldering involve the use of filler metals at the interfaces to be joined. These processes require lower temperatures than those in welding and are capable of joining dissimilar metals with intricate shapes and various thicknesses. (Section 12.14)
- Adhesive bonding has favorable characteristics, such as strength, sealing, thermal insulation, vibration damping, and resistance to corrosion between dissimilar metal combinations. An important aspect is the use of electrically conducting adhesives for surface-mount technologies. (Section 12.15)
- Mechanical fastening is among the oldest and most commonly used techniques. A wide variety of shapes and sizes of fasteners and fastening techniques are now available for numerous permanent and semipermanent applications. (Section 12.16)
- Several joining techniques have been developed for joining thermoplastics and thermosetting plastics, as well as for various types of ceramics and glasses. (Section 12.17)
- General design guidelines have been established for joining, some of which are applicable to a variety of processes whereas others require special considerations, depending on the particular application. (Section 12.18)
- The economics of joining operations involves such factors as equipment cost, labor cost, and skill level required, as well as consideration of processing parameters, such as time required, joint quality, and the necessity to meet specific requirements. The implementation of automation, process optimization, and computer controls has a major impact on costs involved. (Section 12.19)

SUMMARY OF EQUATIONS

Heat input in welding, $\frac{H}{l} = e\frac{VI}{v}$

or $H = uAl$

Welding speed, $v = e\frac{VI}{uA}$

Heat input in resistance welding, $H = I^2Rt$

BIBLIOGRAPHY

Adams, R.D., (ed.), *Adhesive Bonding*, CRC Press, 2005.

ASM Handbook, Vol. 6A: *Welding Fundamentals*, ASM International, 2011.

Bath, J. (ed.), *Lead-Free Soldering*, Springer, 2007.

Benedek, I., *Pressure-Sensitive Adhesives and Applications*, 2nd ed., CRC Press, 2004.

Bickford, J.H., and Nassar, S., (eds.), *Introduction to the Design and Behavior of Bolted Joints*, 4th ed. Dekker, 2012.

Bowditch, K.E., and Bowditch, M.A., *Oxyfuel Gas Welding*, 7th ed., Goodheart-Wilcox, 2011.

Brockman, W., Geiss, P.L., Klingen, J., Schroeder, K.B., and Mikhail, B., *Adhesive Bonding: Adhesives, Applications and Processes*, Wiley, 2009.

Cary, H.B., and Helzer, S., *Modern Welding Technology*, 6th ed., Prentice Hall, 2004.

Habenicht, G., *Applied Adhesive Bonding: A Practical Guide for Flawless Results*, Wiley, 2009.

Humpston, G., and Jacobson, D.M., *Principles of Soldering*, ASM International, 2004.

Jacobson, D.M., and Humpston, G., *Principles of Brazing*, ASM International, 2005.

Jeffus, L.F., *Welding: Principles and Applications*, 7th ed., Delmar, 2011.

Katayama, S. (ed.), *Handbook of Laser Welding Technologies*, Woodhead Publishing, 2013.

Kou, S., *Welding Metallurgy*, 2nd ed., Wiley, 2002.

Lippold, J.C., *Welding Metallurgy and Weldability*, Wiley, 2014.

Mandal, N.R., *Aluminum Welding*. 2nd ed., ASM International, 2006.

Minnick, W.H., *Gas metal arc Welding Handbook*, 5th ed., Goodheart-Wilcox, 2007.

Mishra, R.S., De, P.S., and Kumar, N., *Friction Stir Welding and Processing: Science and Engineering*, Springer, 2014.

Petrie, E.M., *Handbook of Adhesives and Sealants*, 2nd ed., McGraw-Hill, 2006.

Powell, J., *CO_2 Laser Cutting*, 2nd ed., Springer, 2011.

Rotheiser, J., *Joining of Plastics: Handbook for Designers and Engineers*, 3rd ed., Hanser, 2009.

Schwartz, M.M., *Brazing*, 2nd ed., ASM International, 2003.

Speck, J.A., *Mechanical Fastening, Joining, and Assembly*, 2nd ed., CRC Press, 2015.

Steen, W.M., *Laser Material Processing*, 4th ed., Springer, 2010.

Tres, P.A., *Designing Plastic Parts for Assembly*, 7th ed., Hanser-Gardner, 2014.

Weman, K., *Welding Processes Handbook*, 2nd ed., Wileman, 2011.

Yilbas, B.S., and Sahin, A.Z., *Friction Welding: Thermal and Metallurgical Characteristics*, Springer, 2014.

QUESTIONS

12.1 Explain the reasons that so many different welding processes have been developed.

12.2 List the advantages and disadvantages of mechanical fastening as compared with adhesive bonding.

12.3 What are the similarities and differences between consumable and nonconsumable electrodes?

12.4 Explain the features of neutralizing, reducing, and oxidizing flames. Why is a reducing flame so called?

12.5 What determines whether a certain welding process can be used for workpieces in horizontal, vertical, or upside-down positions, or for all types of positions? Explain, giving appropriate examples.

12.6 Comment on your observations regarding Fig. 12.6.

12.7 Discuss the need for and role of fixtures in holding workpieces in the welding operations described in this chapter.

12.8 Describe the factors that influence the size of the two weld beads in Fig. 12.14.

12.9 Why is the quality of welds produced by submerged arc welding very good?

12.10 Explain the factors involved in electrode selection in arc welding processes.

12.11 Explain why the electroslag welding process is suitable for thick plates and heavy structural sections.

12.12 What are the similarities and differences between consumable and nonconsumable electrode arc welding processes?

12.13 In Table 12.2, there is a column on the distortion of welded components, ranging from lowest to highest. Explain why the degree of distortion varies among different welding processes.

12.14 What keeps the weld bead on a steel surface from oxidizing (rusting) during welding?

12.15 Explain why the grains in Fig. 12.18 grow in the particular directions shown.

12.16 Prepare a table listing the processes described in this chapter and providing, for each process, the range of welding speeds as a function of workpiece material and thickness.

12.17 Explain what is meant by *solid state welding*.

12.18 Describe your observations concerning Figs. 12.21, 12.22, and 12.23.

12.19 What materials can be welded by Laser GMAW hybrid welding?

12.20 What advantages does friction welding have over the other joining methods described in this chapter?

12.21 Why is diffusion bonding, when combined with superplastic forming of sheet metals, an attractive fabrication process? Does it have any limitations?

12.22 Can roll bonding be applied to various part configurations? Explain.

12.23 Comment on your observations concerning Fig. 12.43.

12.24 If electrical components are to be attached to both sides of a circuit board, what soldering process(es) would you use? Explain.

12.25 Discuss the factors that influence the strength of (a) a diffusion bonded component and (b) a cold welded component.

12.26 Describe the difficulties you might encounter in applying explosion welding in a factory environment.

12.27 Inspect the edges of a US quarter, and comment on your observations. Is the cross section, that is, the thickness of individual layers, symmetrical? Explain.

12.28 What advantages do resistance welding processes have over others described in this chapter?

12.29 What does the strength of a weld nugget in resistance spot welding depend on?

12.30 Explain the significance of the magnitude of the pressure applied through the electrodes during resistance welding operations.

12.31 Which materials can be friction stir welded, and which cannot? Explain your answer.

12.32 List the joining methods that would be suitable for a joint that will encounter high stresses and will need to be disassembled several times during the product life, and rank the methods.

12.33 Inspect Fig. 12.33, and explain why the particular fusion-zone shapes are developed as a function of pressure and speed. Comment on the influence of the properties of the material.

12.34 Which applications could be suitable for the roll spot welding process shown in Fig. 12.37c? Give specific examples.

12.35 Give several examples concerning the bulleted items listed at the beginning of Section 12.1.

12.36 Could the projection welded parts shown in Fig. 12.38 be made by any of the processes described in other parts of this text? Explain.

12.37 Describe the factors that influence flattening of the interface after resistance projection welding takes place.

12.38 What factors influence the shape of the upset joint in flash welding, as shown in Fig. 12.39b?

12.39 Explain how you would fabricate the structures shown in Fig. 12.43 with methods other than diffusion bonding and superplastic forming.

12.40 Make a survey of metal containers used for household products and foods and beverages. Identify those that have utilized any of the processes described in this chapter. Describe your observations.

12.41 Which process uses a solder paste? What are the advantages to this process?

12.42 Explain why some joints may have to be preheated prior to welding.

12.43 What are the similarities and differences between casting of metals (Chapter 5) and fusion welding?

12.44 Explain the role of the excessive restraint (stiffness) of various components to be welded on weld defects.

12.45 Discuss the weldability of several metals, and explain why some metals are easier to weld than others.

12.46 Must the filler metal be of the same composition as that of the base metal to be welded? Explain.

12.47 Describe the factors that contribute to the difference in properties across a welded joint.

12.48 How does the weldability of steel change as the steel's carbon content increases? Why?

12.49 Are there common factors among the weldability, solderability, castability, formability, and machinability of metals? Explain, with appropriate examples.

12.50 Rate lap, butt and scarf joints in terms of joint strength. Explain your answers.

12.51 Explain why hydrogen welding can be used to weld tungsten without melting the tungsten electrode.

12.52 Assume that you are asked to inspect a weld for a critical application. Describe the procedure you would follow. If you find a flaw during your inspection, how would you go about determining whether or not this flaw is important for the particular application?

12.53 Do you think it is acceptable to differentiate brazing and soldering arbitrarily by temperature of application? Comment.

12.54 Loctite® is an adhesive used to keep metal bolts from vibrating loose; it basically glues the bolt to the nut once the nut is inserted in the bolt. Explain how this adhesive works.

12.55 List the joining methods that would be suitable for a joint that will encounter high stresses and cyclic (fatigue) loading, and rank the methods in order of preference.

12.56 Rank the processes described in this chapter in terms of (a) cost and (b) weld quality.

12.57 Why is surface preparation important in adhesive bonding?

12.58 Why have mechanical joining and fastening methods been developed? Give several specific examples of their applications.

12.59 Explain why hole preparation may be important in mechanical joining.

12.60 What precautions should be taken in mechanical joining of dissimilar metals?

12.61 What difficulties are involved in joining plastics? What about in joining ceramics? Why?

12.62 Comment on your observations concerning the numerous joints shown in the figures in Section 12.18.

12.63 How different is adhesive bonding from other joining methods? What limitations does it have?

12.64 Soldering is generally applied to thinner components. Why?

12.65 Explain why adhesively bonded joints tend to be weak in peeling.

12.66 Inspect various household products, and describe how they are joined and assembled. Explain why those particular processes were used.

12.67 Name several products that have been assembled by (a) seaming; (b) stitching; and (c) soldering.

12.68 Suggest methods of attaching a round bar made of thermosetting plastic perpendicularly to a flat metal plate.

12.69 Describe the tooling and equipment that are necessary to perform the double-lock seaming operation shown in Fig. 12.55, starting with flat sheet (See also Fig. 7.23).

12.70 What joining methods would be suitable to assemble a thermoplastic cover over a metal frame? Assume that the cover has to be removed periodically.

12.71 Repeat Question 12.70, but for a cover made of (a) a thermosetting plastic; (b) metal; and (c) ceramic. Describe the factors involved in your selection of methods.

12.72 Do you think the strength of an adhesively bonded structure is as high as that obtained by diffusion bonding? Explain.

12.73 Comment on workpiece size limitations, if any, for each of the processes described in this chapter.

12.74 Describe part shapes that cannot be joined by the processes described in this chapter. Gives specific examples.

12.75 Give several applications of electrically conducting adhesives.

12.76 Give several applications for fasteners in various household products, and explain why other joining methods have not been used instead.

12.77 Comment on workpiece shape limitations, if any, for each of the processes described in this chapter.

12.78 List and explain the rules that must be followed to avoid cracks in welded joints, such as hot tearing, hydrogen-induced cracking, lamellar tearing, etc.

12.79 If a built-up weld is to be constructed (See Fig. 12.6), should all of it be done at once, or should it be done a little at a time, with sufficient time allowed for cooling between beads?

12.80 Describe the reasons that fatigue failure generally occurs in the heat-affected zone of welds instead of through the weld bead itself.

12.81 If the parts to be welded are preheated, is the likelihood that porosity will form increased or decreased? Explain.

12.82 Which of the processes described in this chapter are not portable? Can they be made so? Explain.

12.83 What is the advantage of electron-beam and laser-beam welding, as compared to arc welding?

12.84 Describe the common types of discontinuities in welds, explain the methods by which they can be avoided.

12.85 What are the sources of weld spatter? How can spatter be controlled?

12.86 Describe the functions and characteristics of electrodes. What functions do coatings have? How are electrodes classified?

12.87 Describe the advantages and limitations of explosion welding.

12.88 Explain the difference between resistance seam welding and resistance spot welding.

12.89 What are the advantages of integrated snap fasteners?

12.90 Could you use any of the processes described in this chapter to make a large bolt by welding the head to the shank (See Fig. 6.15)? Explain the advantages and limitations of this approach

12.91 Describe wave soldering. What are the advantages and disadvantages to this process?

12.92 What are the similarities and differences between a bolt and a rivet?

12.93 It is common practice to tin plate electrical terminals to facilitate soldering. Why is it tin that is used?

12.94 Review Table 12.3 and explain why some materials require more heat than others to melt a given volume.

12.95 Review the electric arc welding and hybrid processes in this chapter. Which of these processes are suitable for the use of more than one electrode at a time? Explain.

PROBLEMS

12.96 Two 1 mm thick, flat copper sheets are being spot welded using a current of 3000 A and a current flow time of $t = 0.18$ s. The electrodes are 5 mm in diameter. Estimate the heat generated in the weld zone.

12.97 Calculate the temperature rise in Problem 12.96, assuming that the heat generated is confined to the volume of material directly between the two electrodes and that the temperature distribution is uniform.

12.98 Calculate the range of allowable currents for Problem 12.96, if the temperature should be between 0.7 and 0.85 times the melting temperature of copper. Repeat this problem for carbon steel.

12.99 A resistance projection welding machine is used to join two 1-mm thick sheets with eight 5-mm diameter spot welds produced simultaneously. If ten seconds are needed for the welding operation, determine (a) the welding current; (b) the required kVA if the applied voltage is 10 V; and (c) the electrical energy consumption for each weld.

12.100 In Fig. 12.26, assume that most of the top portion of the top piece is cut horizontally with a sharp saw. Thus, the residual stresses will be disturbed, and, as described in Section 2.10, the part will undergo shape change. For this case, how will the part distort? Explain.

12.101 The accompanying figure shows a metal sheave that consists of two matching pieces of hot-rolled, low-carbon-steel sheets. These two pieces can be joined either by spot welding or by V-groove welding. Discuss the advantages and limitations of each process for this application.

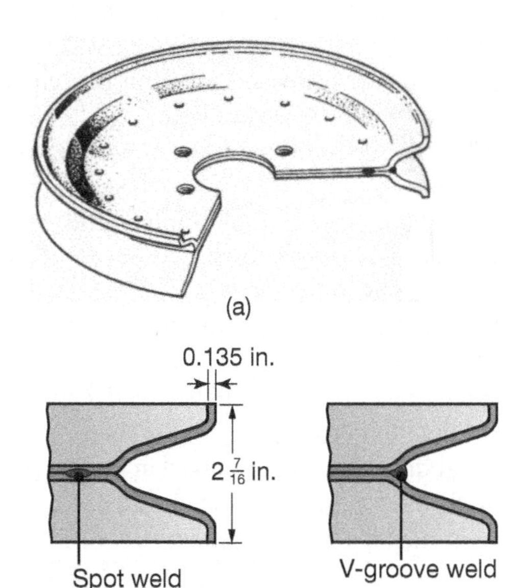

12.102 An arc welding operation is taking place on carbon steel. The desired welding speed is around 20 mm/sec. If the power supply is 10 V, what current is needed if the weld width is to be 5 mm?

12.103 In oxyacetylene, arc, and laser-beam cutting, the processes basically involve melting of the workpiece. If a 50 mm. diameter hole is to be cut from a 250-mm-diameter, 12-mm-thick plate, plot the mean temperature rise in the blank as a function of kerf. Assume that one-half of the energy goes into the blank.

12.104 A welding operation takes place on an aluminum-alloy plate. A pipe 50 mm in diameter, with a 5 mm wall thickness and a 50 mm length, is butt welded onto an angle iron 150 mm by 150 mm by 5 mm thick. The angle iron is of an L cross section and has a length of 1 m. If the weld zone in a gas tungsten arc welding process is approximately 10 mm wide, what would be the temperature increase of the entire structure due to the heat input from welding only? What if the process were an electron-beam welding operation, with a bead width of 2 mm? Assume that the electrode requires 1500 J and the aluminum alloy requires 1200 J to melt one gram.

12.105 The energy applied in friction welding is given by the formula $E = IS^2/C$, where I is the moment of inertia of the flywheel, S is the spindle speed in rpm, and C is a constant of proportionality. ($C = 5873$ when the moment of inertia is given in lb-ft^2.) For a spindle speed of 600 rpm and an operation in which a steel tube with a 3.5 in. outside diameter and a 0.25 in. wall thickness is welded to a flat frame, what is the required moment of inertia of the flywheel if all of the energy is used to heat the weld zone, approximated as the material $\frac{1}{4}$ in. deep and directly below the tube? Assume that 1.4 ft-lbm is needed to melt the electrode.

12.106 Refer to the simple butt and lap joints shown in Fig. 12.1. (a) Assuming the area of the butt joint is 3 mm × 25 mm and referring to the adhesive properties given in Table 12.6, estimate the minimum and maximum tensile force that this joint can withstand. (b) Estimate these forces for the lap joint assuming its area is 25 mm × 25 mm.

12.107 As shown in Fig. 12.63, a rivet can buckle if it is too long. Using information from solid mechanics, determine the length-to-diameter ratio of a rivet that will not buckle during riveting.

12.108 Repeat Example 12.2 if the workpiece is (a) aluminum or (b) titanium.

12.109 A submerged arc welding operations takes place on 10 mm thick stainless steel, producing a butt weld as shown in Fig. 12.22c. The weld geometry can be approximated as a trapezoid with 15 mm and 10 mm as the top and bottom dimensions, respectively. If the voltage provided is 40 V at 300 A, estimate the welding speed if a stainless steel filler wire is used.

12.110 The energy required in ultrasonic welding is found to be related to the product of workpiece thickness and hardness. Explain why this relationship exists.

12.111 6061 aluminum plates with a 2-mm thickness are to be butt-welded by GMAW using a 1-mm diameter electrode. The applied voltage is 20 V, the current is 120A, and the arc travel speed is 15 mm/s. Calculate the power, the deposition rate of electrode material, and the required electrode feed rate.

12.112 Assume that you are asked to give a quiz to students on the contents of this chapter. Prepare three quantitative problems and three qualitative questions, and supply the answers.

DESIGN

12.113 Design a machine that can perform friction welding of two cylindrical pieces, as well as remove the flash from the welded joint (See Fig. 12.32).

12.114 How would you modify your design in Problem 12.113 if one of the pieces to be welded is noncircular?

12.115 Describe product designs that cannot be joined by friction welding processes.

12.116 Make a comprehensive outline of joint designs relating to the processes described in this chapter. Give specific examples of engineering applications for each type of joint.

12.117 Arc Blow is a phenomenon where the magnetic field induced by the welding current passing through the electrode and workpiece in shielded metal arc welding interacts with the arc and causes severe weld splatter. Identify the variables that you feel are important in arc blow. When arc blow is a problem, would you recommend minimizing it using AC or DC power?

12.118 Review the two weld designs in Fig. 12.60a, and, based on the topics covered in courses on the strength of materials, show that the design on the right is capable of supporting a larger moment, as shown.

12.119 In the building of large ships, there is a need to weld large sections of steel together to form a hull. For this application, consider each of the welding operations described in this chapter, and list the benefits and drawbacks of that operation for this product. Which welding process would you select? Why?

12.120 Examine various household products, and describe how they are joined and assembled. Explain why those particular processes are used for these applications.

12.121 A major cause of erratic behavior (hardware bugs) and failures of computer equipment is fatigue failure of the soldered joints, especially in surface-mount devices and devices with bond wires (See Fig. 12.50). Design a test fixture for cyclic loading of a surface-mount joint for fatigue testing.

12.122 Using two strips of steel 25 mm wide and 200 mm long, design and fabricate a joint that gives the highest strength in a tension test in the longitudinal direction.

12.123 Make an outline of the general guidelines for safety in welding operations. For each of the operations described in this chapter, prepare a poster which effectively and concisely gives specific instructions for safe practices in welding (or cutting). Review the various publications of the National Safety Council and other similar organizations.

12.124 A common practice for repairing expensive broken or worn parts, such as may occur when, for example, a fragment is broken from a forging, is to fill the area with layers of weld bead and then to machine the part back to its original dimensions. Make a list of the precautions that you would suggest to someone who uses this approach.

12.125 In the roll bonding process shown in Fig. 12.30, how would you go about ensuring that the interfaces are clean and free of contaminants, so that a good bond is developed? Explain.

12.126 Alclad stock is made from 5182 aluminum alloy, and has both sides coated with a thin layer of pure aluminum. The 5182 provides high strength, while the outside layers of pure aluminum provide good corrosion resistance, because of their stable oxide film. Alclad is commonly used in aerospace structural applications for these reasons. Investigate other common roll bonded materials and their uses, and prepare a summary table.

12.127 Obtain a soldering iron and attempt to solder two wires together. First, try to apply the solder at the same time as you first put the soldering iron tip to the wires. Second, preheat the wires before applying the solder. Repeat the same procedure for a cool surface and a heated surface. Record your results and explain your findings.

12.128 Perform a literature search to determine the properties and types of adhesives used to affix artificial hips onto the human femur.

12.129 Using the Internet, investigate the geometry of the heads of screws that are permanent fasteners, that is, ones that can be screwed in but not out.

12.130 Obtain an expression similar to Eq. (12.9), but for electron beam and laser welding.

12.131 Lattice booms for cranes are constructed from extruded cross sections (See Fig. 7.73) that are welded together. Any warpage that causes such a boom to deviate from straightness will severely reduce its lifting capacity. Conduct a literature search on the approaches used to minimize distortion due to welding and how to correct it, specifically in the construction of lattice booms.

12.132 Two 0.5-mm thick, 304 stainless steel sheets need to be butt welded. (a) List the processes described in this chapter that are suitable for this operation, in order of resultant butt weld strength; (b) List the processes in order of decreasing speed.

12.133 Repeat Problem 12.132 if the sheets were allowed to overlap.

12.134 Welding of aluminum to steel is seen as a critical technology for automotive lightweighting. Conduct an Internet search and determine the suitability of the following processes for this application: (a) cold welding; (b) diffusion bonding; (c) explosion welding; (d) friction welding; (e) spot welding; (f) electron beam welding; (g) flash butt welding. Justify your answer.

12.135 Consider a butt joint that is to be welded. Sketch the weld shape you would expect for (a) SMAW; (b) laser welding; and (c) Laser-SMAW hybrid welding. Indicate the size and shape of the heat-affected zone you would expect. Comment on your observations.

12.136 Sketch the microstructure you would expect if a butt joint were created by (a) linear friction welding; (b) friction stir welding; (c) mash seam welding and (d) flash welding.

12.137 Design a joint to connect two 25 mm wide, 5 mm thick steel members. The overlap may be as much as 25 mm, and any one approach described in this chapter can be used.

12.138 For the same members in Problem 12.137, design a joint using threaded fasteners arranged in one row. Do you advise the use of one large fastener or many small fasteners? Explain.

12.139 For the same members in 12.137, design a joint using a *combination* of joining techniques.

Micro- and Nano-manufacturing

CHAPTER 13

This chapter presents the science and technology of the production of microscopic devices, microelectronic and microelectromechanical systems, and the materials commonly used with these products.

- The unique properties of silicon that make it ideal for producing oxide and dopants and complementary metal-on-oxide semiconductor devices.
- Treatment of a cast ingot and machining operations to produce a wafer.
- Lithography, etching, and doping processes.
- Wet and dry etching for circuit manufacture.
- Electrical connections at all levels, from a transistor to a computer.
- Packages for integrated circuits and manufacturing methods for printed circuit boards.
- Specialized processes for manufacturing MEMS devices.
- Nanomanufacturing.

Symbols

AR	anisotropy ratio	α	spinning constant
C	concentration, kg/m^3	β	spinning constant
E	etch rate, m/s	η	viscosity, Ns/m^2
k	spinning constant	ω	angular velocity, rad/s
t	thickness, m		

13.1 Introduction

Micromanufacturing refers to manufacturing on a microscopic scale, thus not visible to the naked eye (See Fig. 1.9). The terms micromanufacture, microelectronics, and microelectromechanical systems (MEMS) are not strictly limited to such small length scales but suggest a material and manufacturing strategy. Generally, this type of manufacturing relies heavily on lithography, wet and dry etching, and coating techniques. Micromanufacturing of *semiconductors* exploits the unique ability of silicon to form oxides and **complementary metal-on-oxide semiconductors** (CMOS). Examples of products that rely upon micromanufacturing technologies include a wide variety of sensors and probes, inkjet printing heads, microactuators and associated devices, magnetic hard-drive heads, and microelectronic devices, such as computer processors and memory chips.

Although semiconducting materials have been used in electronics for several decades, it was the invention of the *transistor*, in 1947, that set the stage for what would become one of the greatest technological advancements in all of history. Microelectronics has played an ever-increasing role since **integrated circuit** (IC) technology (Fig. 13.1) became the foundation for personal computers, cellular telephones, information systems, automotive control, and telecommunications.

The major advantages of today's ICs are their very small size and low cost. As their fabrication technology continues to become more advanced, the size and cost of such devices as transistors, diodes, resistors, and capacitors continue to decrease, while the global market becomes highly competitive. More and more components can now be put onto a **chip**, a semiconducting material on which the circuit is fabricated.

Typical chips produced today have sizes that are as small as 0.5 mm × 0.5 mm (0.02 in. × 0.2 in.) and, in rare cases, can be more than 20 mm × 20 mm (0.8 in. × 0.8 in.), if not an entire wafer. New technologies now allow densities in the range of 10 million devices per chip (Fig. 13.1), a magnitude that has been termed **very large scale integration** (VLSI). Some of the advanced ICs may contain more than 100 million devices, called **ultralarge-scale integration** (ULSI). The Intel Itanium® processors, for example, has surpassed 2 billion transistors, and the Advanced Micro Devices Tahiti® graphic processing unit has surpassed 4.3 billion transistors.

Among more recent advances is **wafer-scale integration** (WSI), in which an entire silicon wafer is used to build a single device. This approach has been of greatest interest in the design of massively parallel supercomputers, including **three-dimensional integrated circuits** (3DICs) that use multiple layers of active circuits, maintaining both horizontal and vertical connections.

The basic building block of a complex IC is the transistor. A **transistor** (Fig. 13.2) is a three-terminal device that acts as a simple on-off switch: If a positive voltage is applied to the **gate** terminal, a conducting channel is formed between the **source** and **drain** terminals, allowing current to flow between those two terminals (switch closed). If no voltage is applied to the gate, no channel is formed, and the source and drain are isolated from each other (switch opened). Figure 13.3 shows how the basic processing steps, described in this chapter, are combined to form a **metal-oxide-semiconductor field effect transistor** (MOSFET).

In addition to the metal-oxide semiconductor structure, the **bipolar junction transistor** (BJT) also is used, but to a lesser extent. While the actual fabrication steps for these transistors are very similar to those for both the MOSFET and MOS technologies, their circuit applications are different. Memory circuits, such as **random access memory** (RAM), and microprocessors consist primarily of MOS devices, whereas linear circuits,

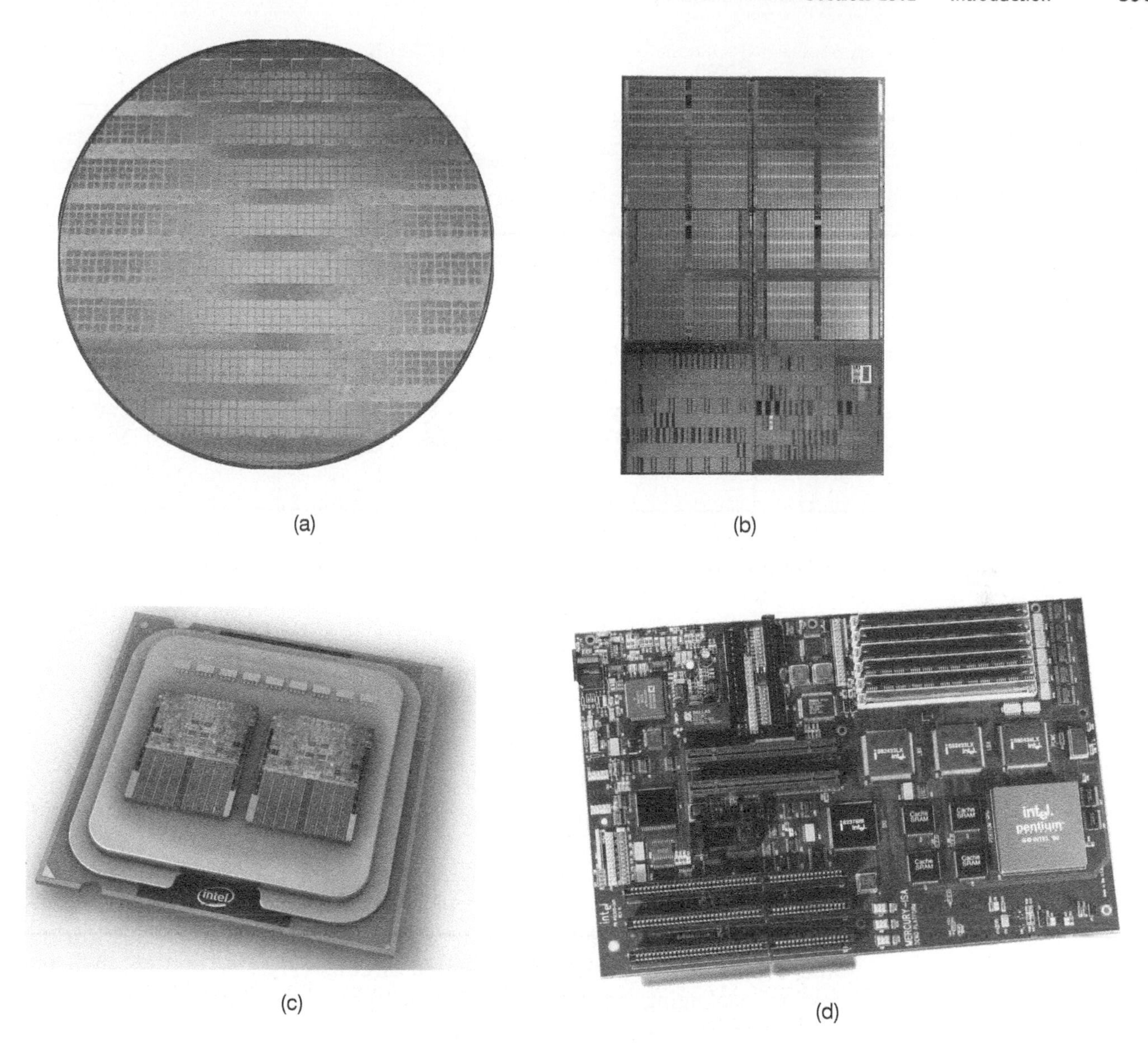

FIGURE 13.1 (a) A 300-mm wafer with a large number of dies fabricated onto its surface; (b) Detail view of an Intel 45-nm chip including a 153 Mbit SRAM (static random access memory) and logic test circuits; (c) Image of the Intel® Itanium® 2 processor; and (d) Pentium® processor motherboard.
Source: Courtesy of Intel Corporation.

such as amplifiers and filters, are more likely to contain bipolar transistors. Other differences between these two types of devices include the higher operating speeds, breakdown voltage, and current needed (with a lower current for the MOSFET).

The major advantage of today's ICs is their high degree of complexity, reduced size, and low cost. As fabrication technologies becomes more advanced, the size of devices decreases, and, consequently, more components can be placed onto a **chip** (a small slice of semiconducting material

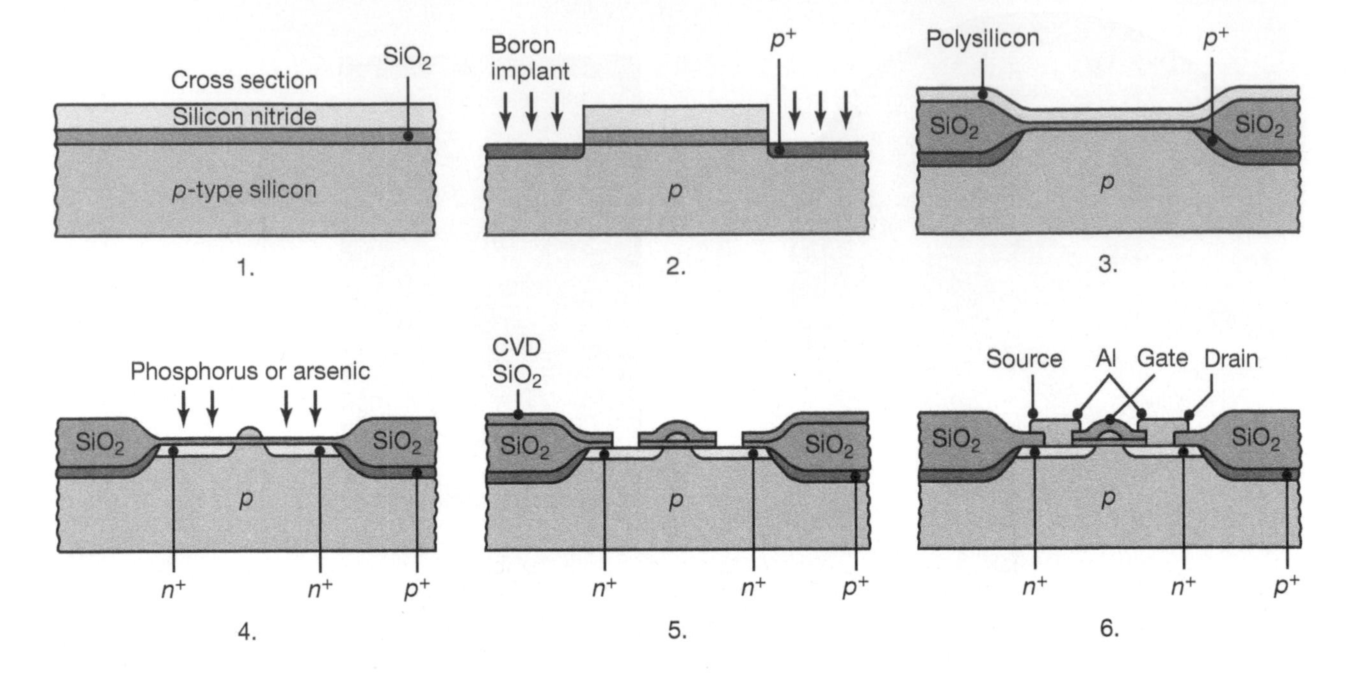

FIGURE 13.2 Cross-sectional views of the fabrication of a metal oxide semiconductor (MOS) transistor. *Source:* After R.C. Jaeger.

on which the circuit is fabricated). In addition, mass processing and automation have greatly helped reduce the cost of each completed circuit. The components fabricated include transistors, diodes, resistors, and capacitors.

This chapter begins with a description of the properties of common semiconductors, such as silicon, gallium arsenide, and polysilicon. It then describes, in detail, the current processes employed in fabricating microelectronic devices and integrated circuits (Fig. 13.3), including IC testing, packaging, and reliability. The chapter also describes a potentially more important development concerning the manufacture of **microelectromechanical systems** (MEMS). These are combinations of electrical and mechanical systems with characteristic lengths of less than 1 mm. These devices utilize many of the batch-processing technologies used for the manufacture of electronic devices, although several other, unique processes have been developed as well.

Although MEMS is a term that came into use around 1987, it has been applied to a wide variety of applications, including precise and rapid sensors, microrobots for nanofabrication, medical delivery systems, and artificial organs. Currently, there are relatively few microelectromechanical systems in use, such as accelerometers and pressure sensors with on-chip electronics. Often, MEMS is a label applied to micromechanical and microelectromechanical devices, such as pressure sensors, valves, and micromirrors. This is not strictly accurate, since MEMS require an integrated microelectronic controlling circuitry; even so, MEMS sales

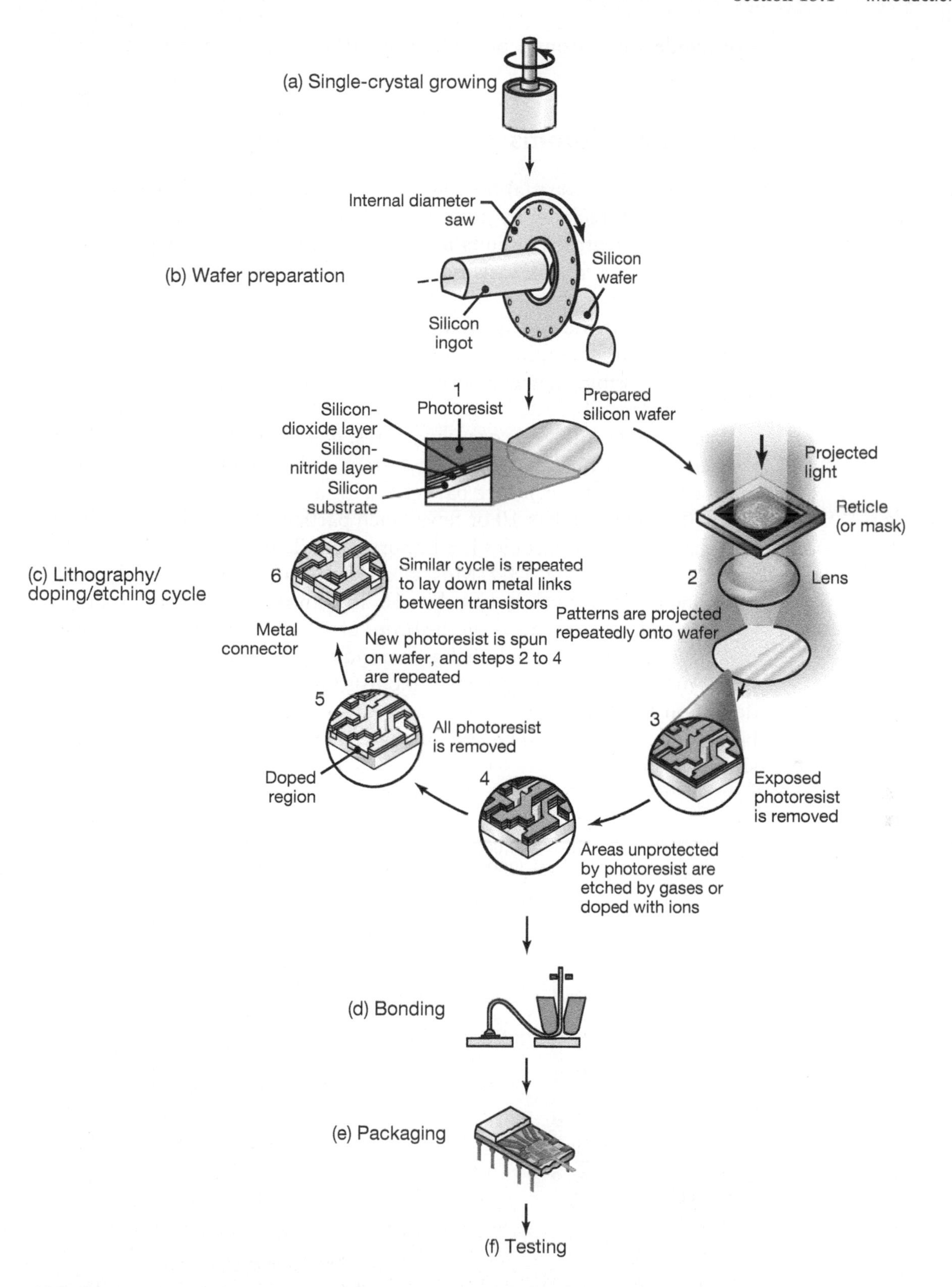
(a) Single-crystal growing
Internal diameter saw
(b) Wafer preparation
Silicon wafer
Silicon ingot
1
Photoresist
Prepared silicon wafer
Silicon-dioxide layer
Silicon-nitride layer
Silicon substrate
Projected light
Reticle (or mask)
(c) Lithography/ doping/etching cycle
6
Similar cycle is repeated to lay down metal links between transistors
2
Lens
Patterns are projected repeatedly onto wafer
Metal connector
New photoresist is spun on wafer, and steps 2 to 4 are repeated
5
All photoresist is removed
3
Doped region
4
Exposed photoresist is removed
Areas unprotected by photoresist are etched by gases or doped with ions
(d) Bonding
(e) Packaging
(f) Testing

FIGURE 13.3 General fabrication sequences for integrated circuits.

worldwide were around \$8 billion in 2014, and are projected to grow by 15% annually.

13.2 Clean Rooms

Clean rooms are essential for the production of integrated circuits, a fact that can be appreciated by noting the scale of manufacturing that is to be performed. Integrated circuits are typically a few mm in length, and the smallest features in a transistor on the circuit may be as small as a few tens of nanometers (nano = 10^{-9}). This size range is smaller than particles that normally are not considered as harmful, such as dust, smoke, perfume, and bacteria; if, however, these contaminants are present on a silicon wafer during processing, they can severely compromise the performance of the whole device.

There are several levels of clean rooms, defined by the **class** of the clean rooms. A traditional classification system, still in wide use, refers to the number of 0.5 μm or larger particles within a cubic foot of air. Thus, a class 10 clean room has 10 or fewer such particles per cubic foot. The size and the number of particles are important in defining the class of a clean room, as shown in Fig. 13.4. Most clean rooms for microelectronics manufacturing range from class 1 to class 10. In comparison, the contamination level in modern hospitals is on the order of 10,000 particles per cubic foot of air, while normal room air has around one million particles per cubic foot. An alternate International Standards Organization (ISO) clean room definition has been standardized and refers to the number of 0.1 μm or larger per cubic meter. The approximate ISO clean room designations are shown in Fig. 13.4.

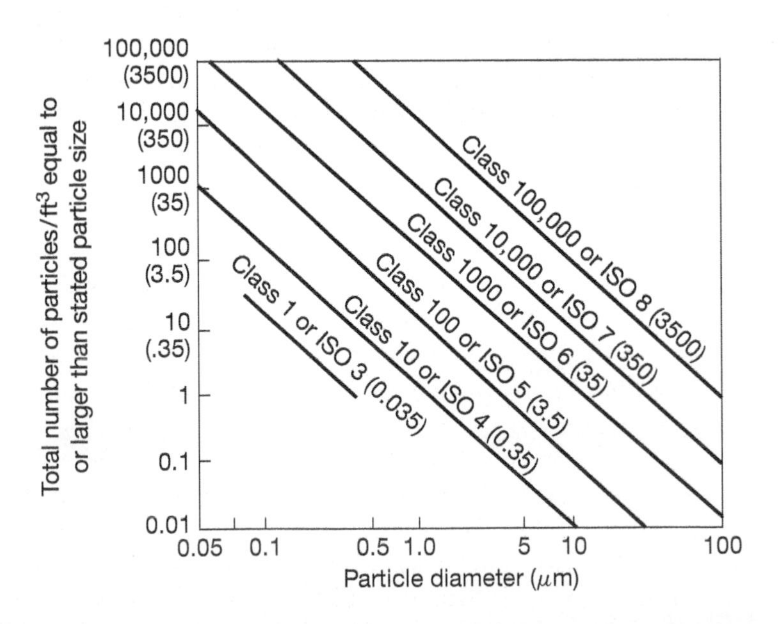

FIGURE 13.4 Allowable particle size concentrations for different clean-room classes. The numbers in parentheses refers to the number of 0.5 μm or larger particles per cubic meter.

To obtain controlled atmospheres that are free from particulate contamination, all ventilating air is passed through a *high-performance particulate air* (HEPA) filter. In addition, the air is usually conditioned so that it is at 21°C and 45% relative humidity. Clean rooms are designed such that the cleanliness at critical processing locations is greater than in the clean room in general. This is accomplished by directing the filtered ventilating air such that it displaces ambient air and directs dust particles away from the operation; it can be facilitated by laminar-flow hooded work areas.

The largest source of contaminants in a clean room is people themselves. Skin particles, hair, perfume, makeup, clothing, bacteria, and viruses are given off naturally by people; they are in sufficiently large numbers to quickly compromise a class 100 clean room. For these reasons, most clean rooms require special coverings, such as white laboratory coats, gloves, and hair nets, as well as avoiding perfumes and makeup. The most stringent clean rooms require full-body coverings, called *bunny suits*. There are other stringent precautions as well; for example, using a ball point pen instead of a pencil (as it can produce objectionable graphite particles), and the paper used is a special clean room paper.

13.3 Semiconductors and Silicon

Semiconductor materials have electrical properties that are between those of conductors and insulators; they exhibit resistivities between 10^{-3} and 10^{8} Ω-cm. Semiconductors are the foundation for electronic devices, because their electrical properties can be altered by adding controlled amounts of selected impurity atoms into their crystal structures. The impurity atoms, known as **dopants**, either have one more valence electron (n-type, or *negative dopant*) or one fewer valence electron (p-type, or *positive dopant*) than the atoms in the semiconductor lattice. For silicon, a Group IV element in the periodic table, typical n-type and p-type dopants include phosphorus and arsenic (Group V) and boron (Group III), respectively. The electrical operation of semiconductor devices is controlled by creating regions of different doping types and concentrations.

Although the earliest electronic devices were fabricated on *germanium*, **silicon** is the industry standard. The abundance of silicon in its alternative forms in the earth's crust is second only to that of oxygen, thus making it economically attractive. Silicon's main advantage over germanium is its larger energy gap (1.1 eV) compared with that of germanium (0.66 eV). This larger energy gap allows silicon-based devices to operate at temperatures about 150°C higher than the operating temperatures of devices fabricated on germanium (about 100°C).

Another important processing advantage of silicon is that its oxide (*silicon dioxide*, SiO_2) is an excellent electrical *insulator* and can be used for both isolation and passivation (Section 3.9.7) purposes. In contrast, germanium oxide is water soluble, thus, it is unsuitable for electronic devices. Moreover, the oxidized form of silicon allows the production of **metal-oxide-semiconductor** (MOS) devices, which are the basis for MOS transistors. These materials are used in memory devices, processors, and various other devices, and are, by far, the largest volume of semiconductor material produced worldwide.

The *crystallographic structure* of silicon is a diamond-type fcc structure, as shown in Fig. 13.5, along with the Miller indices of an fcc material. (Miller indices are a notation for identifying planes and directions within a unit cell.) A crystallographic plane is defined by the reciprocal of its intercepts on the three principal axes. Since anisotropic etchants (Section 13.8.1) preferentially remove material in specific crystallographic planes, the orientation of the silicon crystal in a *wafer* (Section 13.4) is important.

Silicon has some limitations, which has encouraged the development of compound semiconductors, specifically **gallium arsenide** (GaAs). Its major advantage over silicon is its capability for light emission, thus allowing fabrication of devices such as lasers and *light-emitting diodes* (LEDs). Also, its larger energy gap (1.43 eV) allows for higher maximum operating temperature, to about 200°C. Devices fabricated on gallium arsenide also have much higher operating speeds than those fabricated on silicon. Some disadvantages of gallium arsenide include its considerably higher cost, greater processing complications, and the difficulty of growing high-quality oxide layers.

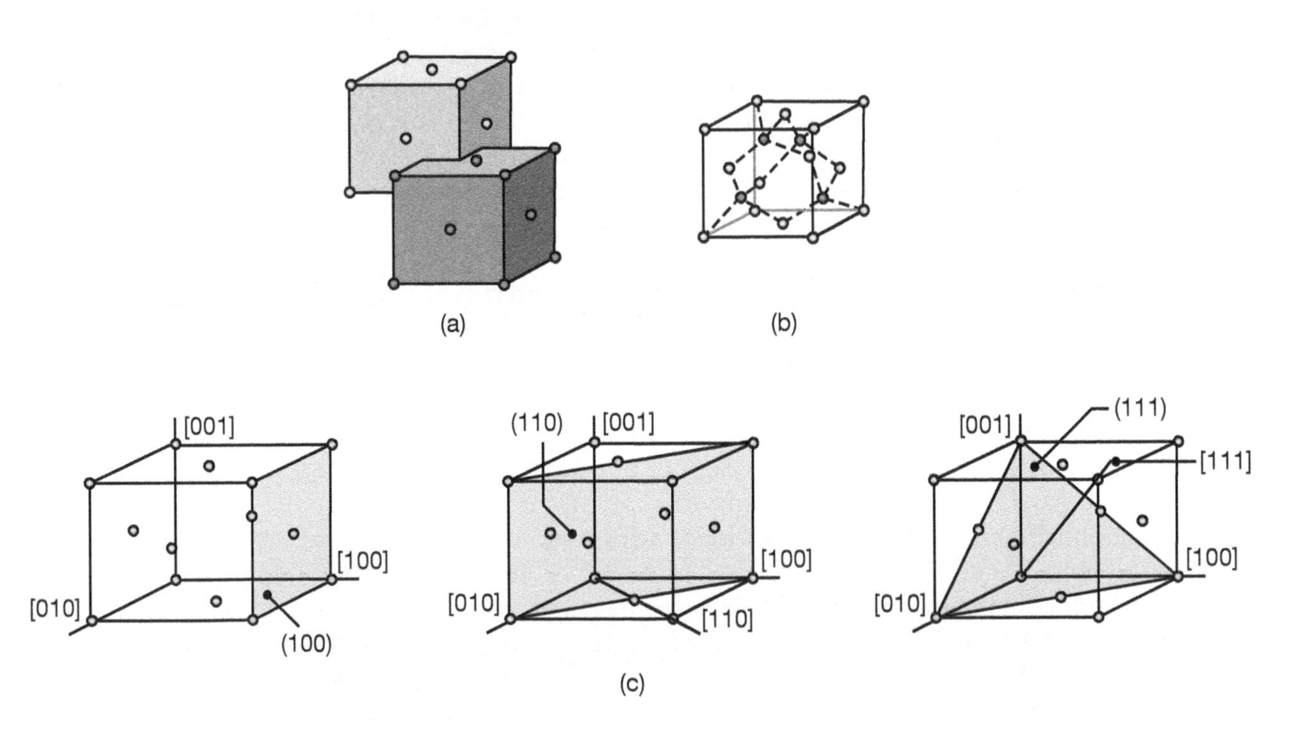

FIGURE 13.5 Crystallographic structure and Miller indices for silicon. (a) Construction of a diamond-type lattice from interpenetrating face-centered cubic cells (one of eight penetrating cells shown). (b) The diamond-type lattice of silicon. The interior atoms have been shaded darker than the surface atoms. (c) Miller indices for a cubic lattice.

13.4 Crystal Growing and Wafer Preparation

Silicon, which occurs naturally in the forms of silicon dioxide and various silicates, must undergo a series of purification steps in order to become a high-quality, defect-free single-crystal material required for semiconductor-device fabrication. The purification process begins by first heating silica and carbon together in an electric furnace, resulting in 95 to 98% pure polycrystalline silicon. This material is then converted to an alternative form, commonly trichlorosilane, which in turn is purified and decomposed in a high-temperature hydrogen atmosphere. The result is an extremely high-quality **electronic-grade silicon** (EGS).

Single-crystal silicon is usually obtained by using the **Czochralski, or CZ, process** (See Fig. 5.28). This method begins with a seed crystal that is dipped into a silicon melt, and then slowly pulled out while being rotated; typical pull rates are on the order of 20 μm/s. At this stage, controlled amounts of impurities can be added to the system for a uniformly doped crystal. The growing technique produces a cylindrical single-crystal ingot, typically over 1 m (40 in.) long and 100–300 mm (4–12 in.) in diameter. Because this technique does not allow for exact control of the ingot diameter, ingots are grown a few millimeters larger than the required size, and are ground to a desired diameter. Silicon **wafers** then are produced

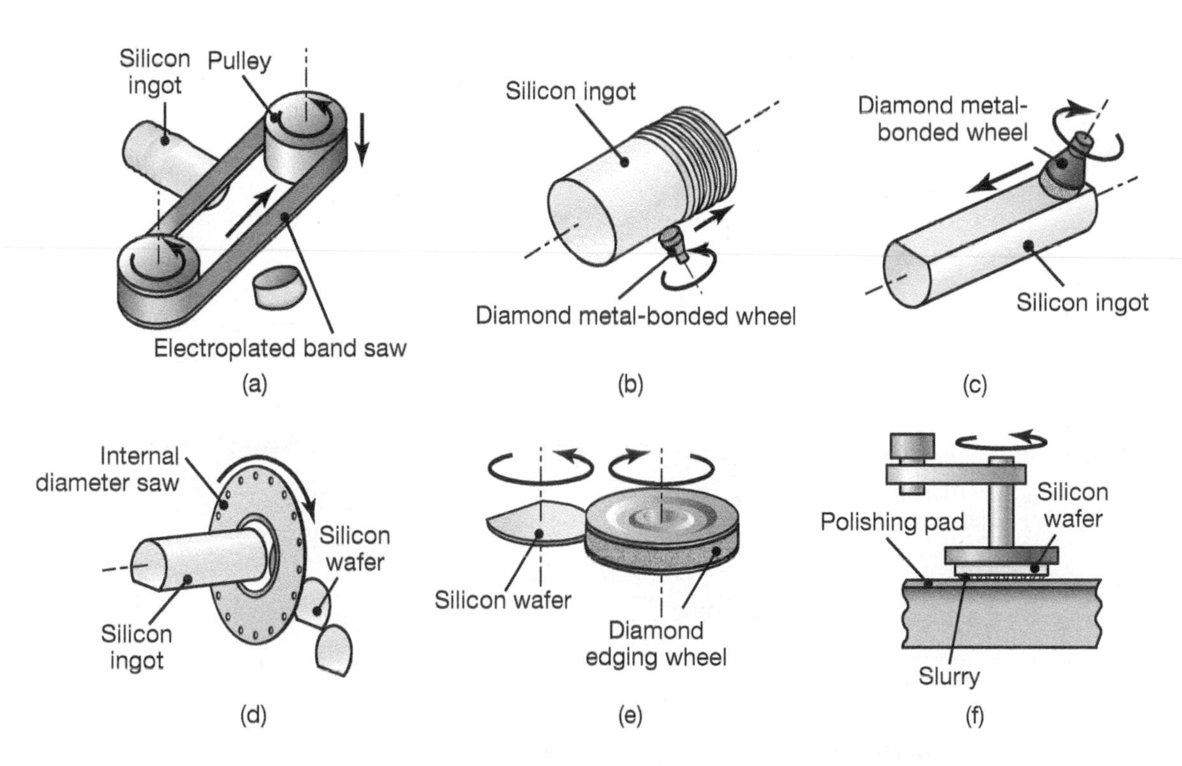

FIGURE 13.6 Operations on a silicon ingot to produce wafers. (a) Electroplated band sawing the ends off the ingot; (b) grinding of the end and cylindrical surfaces of a silicon ingot; (c) machining of a notch or flat; (d) slicing of wafers; (e) end grinding of wafers; and (f) chemical–mechanical polishing of wafers.

from silicon ingots, by a sequence of machining and finishing operations, illustrated in Fig. 13.6.

A notch or a flat is often machined into the silicon cylinder to identify its crystal orientation (Fig. 13.6c). The crystal is then sliced into individual **wafers** using an inner-diameter blade (Fig. 13.6d). While the substrate depth required for most electronic devices is no more than several microns, wafers are typically cut to a thickness of about 0.5 mm to provide the necessary mass to withstand temperature variations and the mechanical support during subsequent fabrication steps.

The wafer is then ground along its edges with a diamond wheel; this operation gives the wafer a rounded profile that is more resistant to chipping. Finally, the wafers are polished and cleaned to remove any surface damage caused by the sawing process, commonly performed by *chemical mechanical polishing*, also referred to as *chemical–mechanical planarization* (CMP, described in Section 9.7).

In order to properly control the manufacturing process, it is important to determine the orientation of the crystal in a wafer; wafers therefore have notches or flats machined into them for identification, as described above and shown in Fig. 13.7. Most commonly, the (100) or (111) plane of the crystal defines the wafer surface, although (110) surfaces also can be used for micromachining applications. Wafers are also identified by a laser scribe marking by the manufacturer. Laser scribing of information may take place on the front or on the back side of the wafer. The front side of some wafers has an exclusion edge area, 3 to 10 mm (0.12 to 0.4 in.) in size,

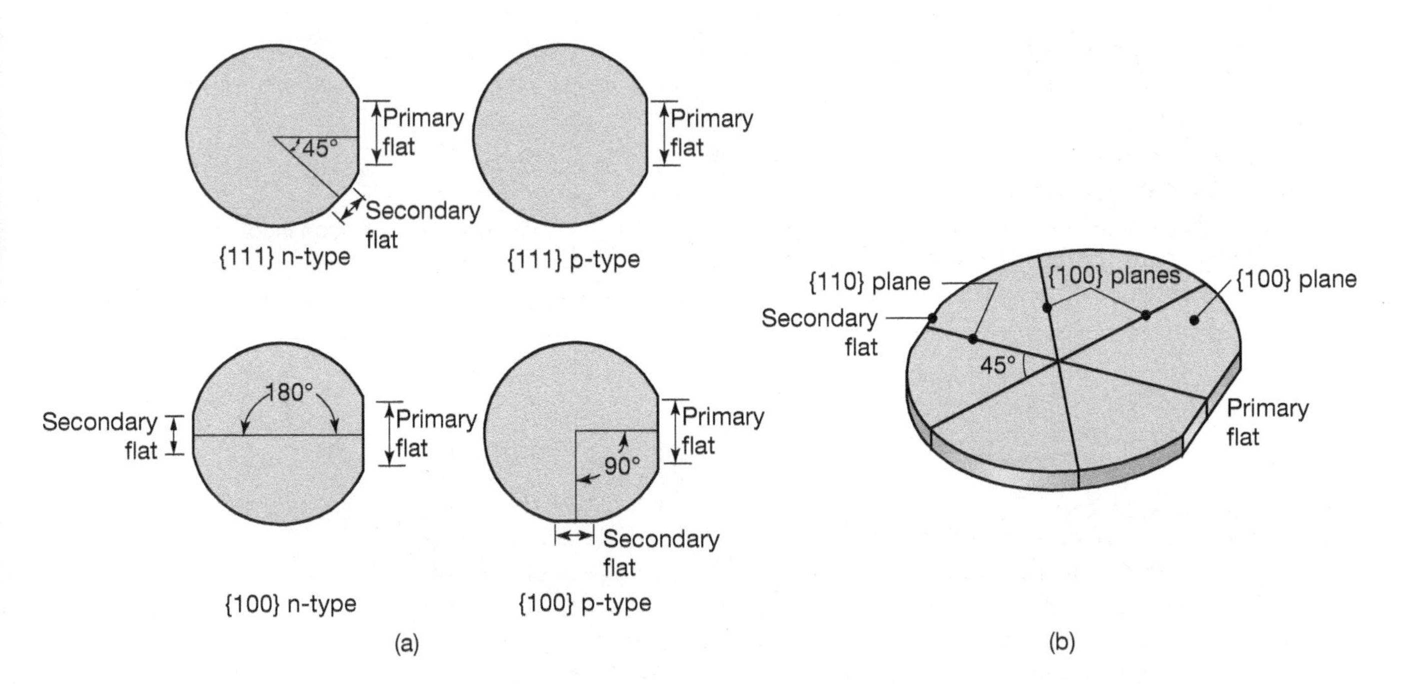

FIGURE 13.7 Identification of single-crystal wafers of silicon. This identification scheme is common for 150-mm diameter wafers, but notches are more common for larger wafers.

reserved for the scribe information, including lot numbers, orientation, and a unique wafer identification code.

Device fabrication takes place over the entire wafer surface. Wafers are typically processed in lots of 25 or 50 wafers with 150–200 mm diameters each, or lots of 12–25 wafers with 300 mm diameters each. In this way, they can be easily handled and automatically transferred during processing. Because the number of chips that can be produced is dependent on the cross-sectional area of the wafer, advanced-circuit manufacturers have moved toward using larger single-crystal solid cylinders, and 300-mm (12 in.) diameter wafers now are common, with 450 mm (18 in.) diameter wafers demonstrated but not in general use. Once processing is completed, the wafer is sliced into individual **chips**, each containing one complete *integrated circuit*.

13.5 Films and Film Deposition

Films of many different types, particularly insulating and conducting, are used extensively in microelectronic-device processing. Common deposition films include polysilicon, silicon nitride, silicon dioxide, tungsten, titanium, and aluminum. In some cases, single-crystal silicon wafers merely serve as a mechanical support on which custom *epitaxial layers* are grown (see Section 13.6). Because these silicon epitaxial films are of the same lattice structure as the substrate, they are also single-crystal materials. The advantages of processing on these deposited films, instead of on the actual wafer surface, include the presence of fewer impurities (notably carbon and oxygen), improved device performance, and the production of tailored material properties not obtainable on the wafers themselves.

Some of the major functions of deposited films include **masking** for diffusion or implantation and protection of the semiconductor's surface. In masking applications, the film must effectively inhibit the passage of dopants while also being able to be etched into patterns of high resolution. Upon completion of device fabrication, films also are applied to protect the underlying circuitry. Films used for masking and protection include silicon dioxide, phosphosilicate glass (PSG), borophosphosilicate glass (BPSG), and silicon nitride. Each of these materials has distinct advantages, and they often are used in combination.

Other films contain dopant impurities and are used as doping sources for the underlying substrate. Conductive films are used primarily for device interconnection; they must have a low resistivity, be capable of carrying large currents, and be suitable for connecting to terminal-packaging leads with wire bonds. Aluminum and copper are generally used for this purpose. To date, aluminum is the more common material, because it can be dry etched more easily. However, copper, with its low resistivity, is used for high-performance processors; the main drawback to copper is associated with difficulties in plasma etching, requiring development of alternate deposition and patterning techniques (see *additive patterning*, Section 13.10). Increasing circuit complexity has required up to six levels of conductive layers, which must all be separated by insulating films.

Films may be deposited by a number of techniques, involving a range of pressures, temperatures, and vacuum systems (see also Section 4.5.1):

1. **Evaporation.** One of the simplest and oldest methods of film deposition is *evaporation*, used primarily for depositing metal films. In this process, the metal is heated in a vacuum until it evaporates; the metal then condenses into a thin layer on the cooler workpiece surface. The heat of evaporation is usually provided by a heating filament or electron beam.
2. **Sputtering.** Another method of metal deposition is *sputtering*, which involves bombarding a target in a vacuum with high-energy ions, usually argon (Ar^+). Sputtering systems usually include a DC power source to produce the energized ions. As the ions impinge on the target, atoms are knocked off and are subsequently deposited on wafers mounted within the system. Although some argon may be trapped within the film, this technique provides very uniform coverage. Advanced sputtering techniques include use of a radio-frequency power source (**RF sputtering**) and introduction of magnetic fields (**magnetron sputtering**).
3. **Chemical vapor deposition.** In *chemical vapor deposition* (CVD), film deposition is achieved by the reaction and/or decomposition of gaseous compounds (see also Section 4.5.1) whereby silicon dioxide is deposited by the oxidation of silane or a chlorosilane. Figure 13.8a shows a continuous CVD reactor that operates at atmospheric pressure.
4. **Low-pressure chemical vapor deposition.** Figure 13.8b shows the apparatus for a method similar to CVD that operates at lower pressures, referred to as *low-pressure chemical vapor deposition* (LPCVD). Capable of coating hundreds of wafers at a time, this

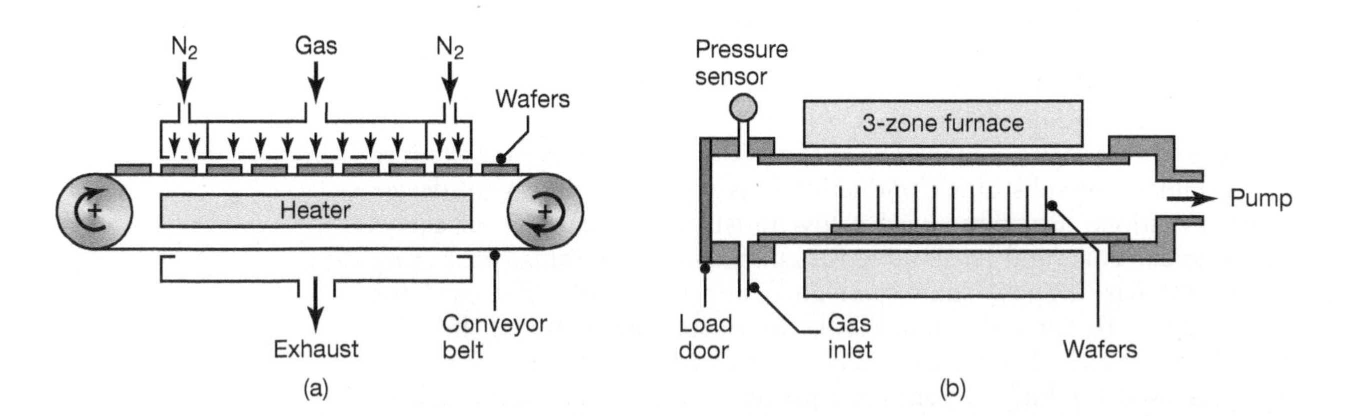

FIGURE 13.8 Schematic diagrams of a (a) continuous, atmospheric-pressure CVD reactor and (b) low-pressure CVD reactor. *Source:* After S.M. Sze.

method has a much higher production rate than that of atmospheric-pressure CVD, and provides superior film uniformity with less consumption of carrier gases. This technique is commonly used for depositing polysilicon, silicon nitride, and silicon dioxide.

5. **Plasma-enhanced chemical vapor deposition.** The *plasma-enhanced chemical vapor deposition* (PECVD) process involves placing wafers in radio-frequency (RF) plasma containing the source gases and has the advantage of maintaining low wafer temperature during deposition. The films deposited generally include hydrogen and are of lower quality than those deposited by other methods.

Silicon **epitaxy** layers, in which the crystalline layer is formed using the substrate as a seed crystal, can be grown by a variety of methods. If the silicon is deposited from the gaseous phase, the process is known as **vapor-phase epitaxy** (VPE). In **liquid-phase epitaxy** (LPE), the heated substrate is brought into contact with a liquid solution containing the material to be deposited.

Another high-vacuum process, called **molecular-beam epitaxy** (MBE), uses evaporation to produce a thermal beam of molecules that deposit on the heated substrate; it offers a very high degree of purity. Moreover, because the films are grown one atomic layer at a time, excellent control of doping profiles is achieved, especially important in gallium-arsenide technology. However, MBE suffers from relatively low growth rates compared with those of other conventional film-deposition techniques.

13.6 Oxidation

The term *oxidation* refers to the growth of an oxide layer by the reaction of oxygen with the substrate material; oxide films also can be formed by the deposition techniques described in Section 13.5. Thermally grown oxides, described in this section, display a higher level of purity than deposited oxides, because the former are grown directly from the high-quality substrate. Deposition methods must be used if the composition of the desired film is different from that of the substrate material.

Silicon dioxide is the most widely used oxide in integrated-circuit technology, and its superior manufacturing characteristics are one of the major reasons for the widespread use of silicon. Aside from its functions of dopant masking and device isolation, silicon dioxide's most critical role is that of the *gate-oxide* material in MOSFETs. Silicon surfaces have an extremely high affinity for oxygen, and a freshly sawed silicon slice will quickly grow a native oxide of 3 to 4 nm in thickness. Modern IC technology requires oxide thicknesses from a few to a few hundred nanometers.

1. **Dry oxidation.** Dry oxidation is a relatively simple process, accomplished by elevating the substrate temperature, typically to 750°C to 1100°C, in an oxygen-rich environment with variable pressure. Silicon dioxide is produced according to the chemical reaction

$$Si + O_2 \rightarrow SiO_2. \quad (13.1)$$

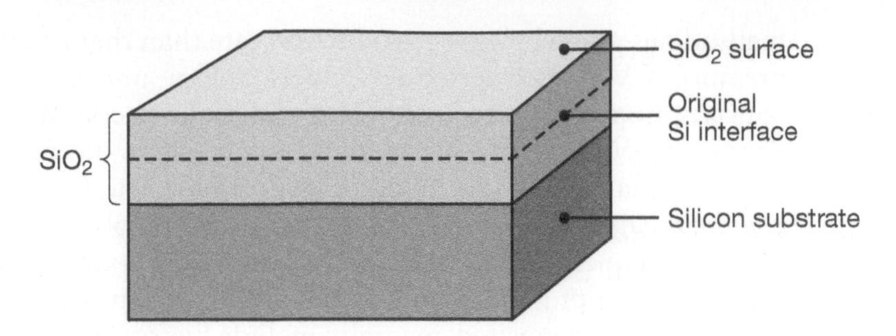

FIGURE 13.9 Growth of silicon dioxide, showing consumption of silicon.

Most oxidation is carried out in a batch process where up to 150 wafers are placed in a furnace. **Rapid thermal processing** (RTP) is a related process combining **rapid thermal oxidation** (RTO) with an anneal step, called **rapid thermal annealing** (RTA), and is used to produce thin oxides on a single wafer.

As a layer of oxide forms, the oxidizing agents must be able to pass through the oxide layer and reach the silicon subsurface, where the actual reaction takes place. Thus, an oxide layer does not continue to grow on top of itself, but rather it grows from the silicon–silicon-dioxide interface outward. Some of the silicon substrate is consumed in the oxidation process (Fig. 13.9). The ratio of oxide thickness to the amount of silicon consumed is found to be 1:0.44. Thus, for example, to obtain an oxide layer 100 nm thick, approximately 44 nm of silicon will be consumed.

One important effect of silicon consumption is the rearrangement of dopants in the substrate near the interface. Some dopants deplete away from the oxide interface while others pile up, hence processing parameters must be adjusted to compensate for this effect.

2. **Wet oxidation.** Another oxidizing technique uses a water-vapor atmosphere as the agent, appropriately called *wet oxidation*. The chemical reaction involved in wet oxidation is

$$Si + 2H_2O \rightarrow SiO_2 + 2H_2. \tag{13.2}$$

This method offers a considerably higher growth rate than that of dry oxidation, but it suffers from a lower oxide density and a lower dielectric strength. The common practice is to combine both dry and wet oxidation methods, growing an oxide in a three-part layer: dry, wet, and dry. This approach combines the advantages of the much higher growth rate in wet oxidation and the high quality obtained in dry oxidation.

3. **Selective oxidation.** The two oxidation methods described above are useful primarily for coating the entire silicon surface with oxide, but it is also necessary to oxidize only certain portions of the surface. The procedure used for this task, called *selective oxidation*, uses silicon nitride (Sections 8.6 and 11.8.1), which inhibits the passage of oxygen and water vapor. Thus, by masking certain areas with silicon nitride, the silicon under these areas remains unaffected while the uncovered areas are oxidized.

13.7 Lithography

Lithography is the process by which the geometric patterns that define devices are transferred from a **reticle** (also called a **photomask** or **mask**) to the substrate surface. A summary of lithographic techniques is given in Table 13.1 and a comparison in Fig. 13.10. There are several forms of lithography, but the most common form used today is *photolithography*. *Electron-beam* and *X-ray lithography* are of great interest because of their ability to transfer patterns of higher resolution, which is an essential feature for increased miniaturization of integrated circuits. However, most integrated circuit applications can be successfully manufactured with photolithography.

A **reticle** is a glass or quartz plate with a pattern of the chip deposited onto it, usually with a chromium film; although, iron oxide and emulsion also are used. The reticle image can have the same size as the desired structure on the chip, but it is often an enlarged image (4× to 20× larger, although 10× magnification is most common). The enlarged images are then focused onto a wafer through a lens, a process known as *reduction lithography*.

In current practice, the lithographic process is applied to microelectronic circuits several times (modern IC devices can require up to 40 lithography steps), each time using a different reticle to define the different areas of the devices and the interconnect. Designed typically at several thousand times their final size, reticle patterns undergo a series of reductions before being applied permanently to a defect-free quartz plate. Computer-aided design (Section 15.4) continues to have a major impact on reticle design and generation. Cleanliness is especially important in lithography, and robots and specialized wafer-handling apparatus are used in order to minimize dust and dirt contamination.

Once the film deposition process is completed and the desired reticle patterns have been generated, the wafer is cleaned and coated with an organic **photoresist** (PR). A photoresist consists of three principal components: (a) a polymer that changes its structure when exposed to radiation; (b) a sensitizer that controls the reactions in the polymer; and (c) a solvent, necessary to deliver the polymer in liquid form.

The wafer is then placed inside a resist spinner, and the photoresist is applied as a viscous liquid onto the wafer (Fig. 13.11). Photoresist layers of

TABLE 13.1 General characteristics of lithography techniques.

Method	Wavelength (nm)	Finest feature size (nm)
Ultraviolet (Photolithography)	365	350
Deep UV	193	190
Extreme UV	10–20	30–100
X-ray	0.01–1	20–100
Electron beam	–	80
Immersion	193	11

Source: After P.K. Wright

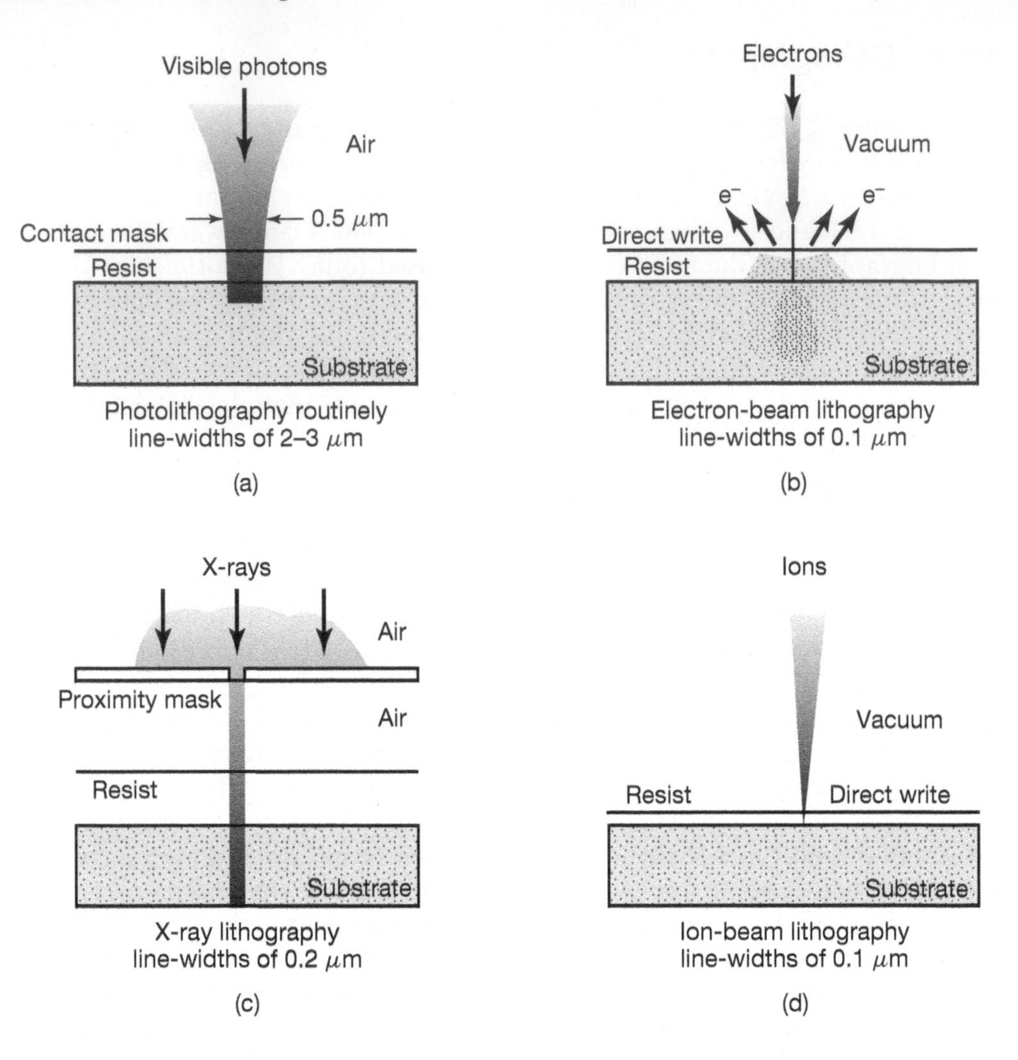

FIGURE 13.10 Comparison of four different lithography techniques. (a) photolithography; (b) electron-beam lithography; (c) X-ray lithography; and (d) ion-beam lithography.

0.5–2.5 μm thick are obtained by spinning them at several thousand rpm for 30 to 60 seconds to give uniform coverage. Proper control of the resist layer is essential to ensure proper performance of subsequent lithography operations. The thickness of the photoresist is given by the expression

$$t = \frac{kC^{\beta}\eta^{\gamma}}{\omega^{\alpha}}, \tag{13.3}$$

where t is the thickness, C is the polymer concentration in mass per volume, η is the viscosity of the polymer, ω is the angular velocity during spinning, and k, α, and β are constants for the particular spinning system. Where masking levels are considered critical, a **barrier antireflective layer** (BARL) or **barrier antireflective coating** (BARC) is applied either beneath or on top of the photoresist to provide **linewidth** control, especially over aluminum.

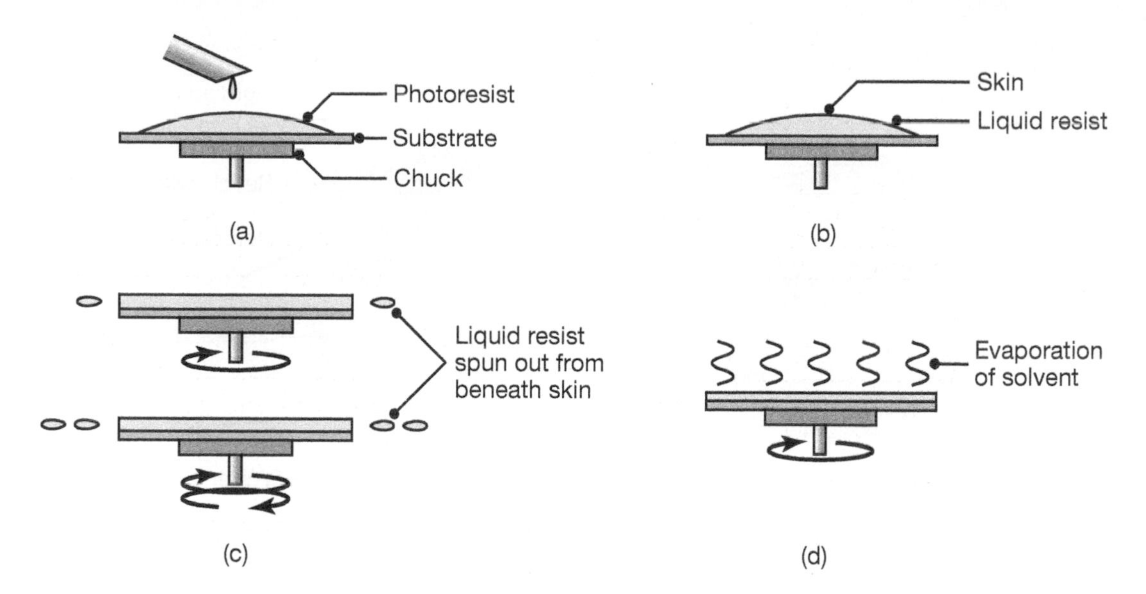

FIGURE 13.11 Spinning of an organic coating on a wafer. (a) Liquid dispensed; (b) liquid is spread over the wafer surface by spinning at low speed; (c) speed is increased, developing a uniform coating thickness and expelling excess liquid; and (d) evaporation of solvent at final spin speed to obtain organic coating.

Linewidth refers to the width of the smallest feature obtainable on the silicon surface and is also called **critical dimension** (CD).

The next step in lithography is **prebaking** the wafer to remove the solvent from the photoresist and harden it. This step is carried out in a convection oven or hot plate, at around 100°C for a period of 10 to 30 min. The pattern is then transferred to the wafer through **stepper** or **step-and-scan** systems. With wafer steppers (Fig. 13.12a), the full image is first exposed in one flash, and then the reticle pattern is refocused onto another adjacent section of the wafer. With *step-and-scan* systems (Fig. 13.12b), the exposing light source is focused into a line, and the reticle and wafer are translated simultaneously in opposite directions to transfer the pattern.

The wafer must be carefully aligned; in this crucial step, called **registration**, the reticle must be aligned correctly with the previous layer on the wafer. Upon development and removal of the exposed photoresist, a duplicate of the reticle pattern will appear in the PR layer. As can be seen in Fig. 13.13, the reticle can be either a negative or a positive image of the desired pattern. A positive reticle uses the UV radiation to break down the chains in the organic film, so that these chains are preferentially removed by the developer. Positive masking is more commonly used than negative masking because, with negative masking, the photoresist can swell and distort, making it unsuitable for small geometries, although newer negative photoresist materials do not have this problem.

Following the exposure and development sequence, **postbaking** the wafer is performed to drive off solvent and toughen and improve the adhesion of the remaining resist. Also, a deep UV treatment, which consists of

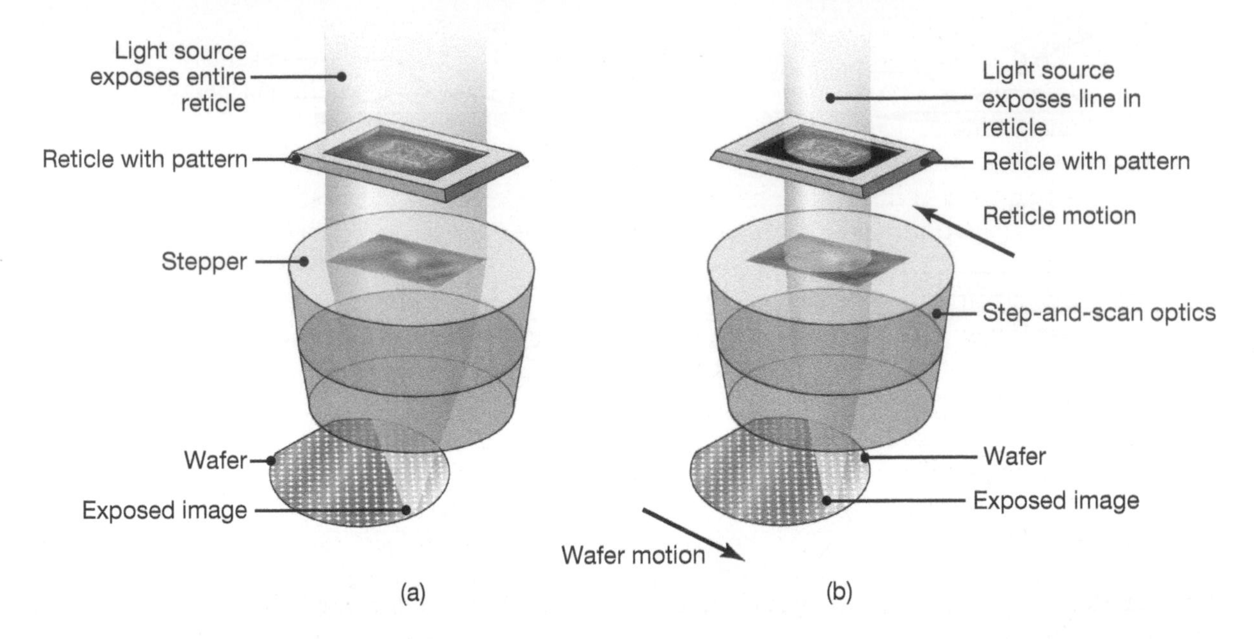

FIGURE 13.12 Schematic illustration of (a) wafer stepper technique for pattern transfer and (b) step-and-scan technique.

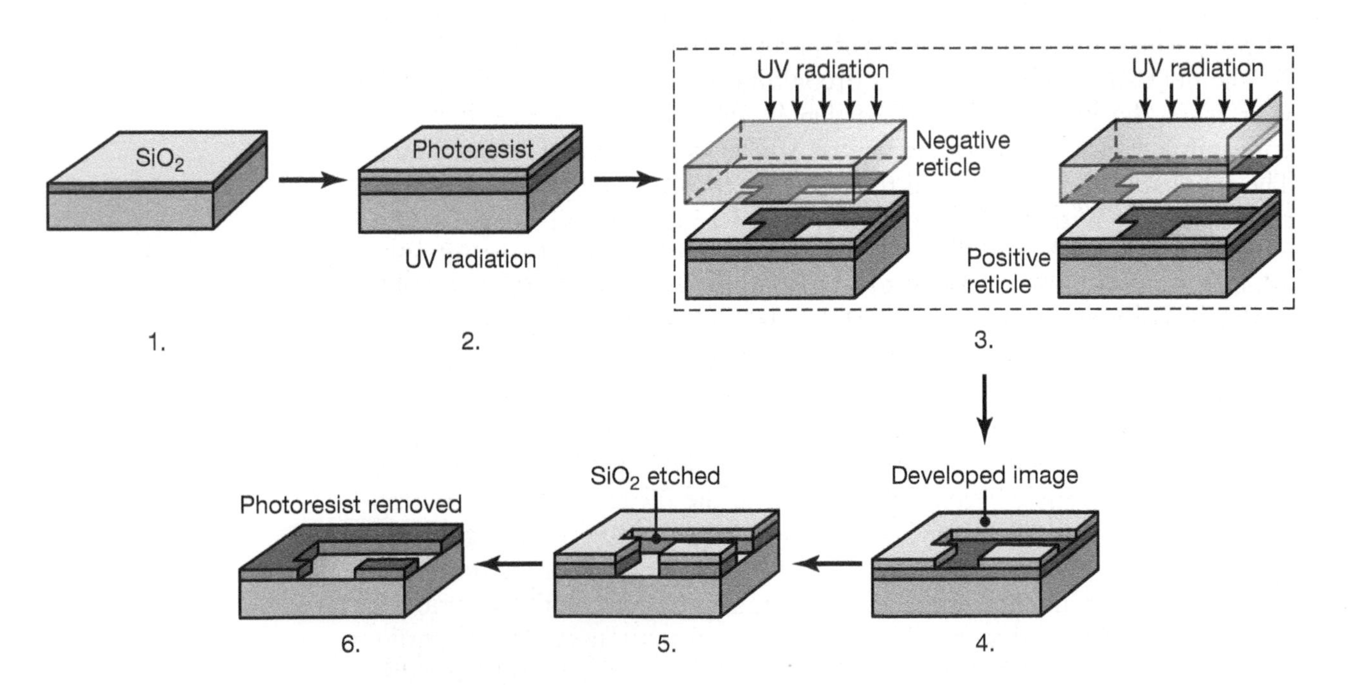

FIGURE 13.13 Pattern transfer by lithography. Note that the mask in Step 3 can be either a positive or a negative image of the pattern. *Source:* After W.C. Till and J.T. Luxon.

baking the wafer to 150°C to 200°C in ultraviolet light, can be used to further strengthen the resistance against high-energy implants and dry etches. The underlying film not covered by the PR is then implanted or etched away (Sections 13.8 and 13.9).

After completion of lithography, the developed photoresist must be removed, in a process called **stripping**. In *wet stripping*, the photoresist is dissolved by such solutions as acetone or strong acids; in this method, the solutions tend to lose potency in use. *Dry stripping* involves exposing the photoresist to an oxygen plasma, referred to as **ashing**. Dry stripping has become more common, because it does not involve the disposal of consumed hazardous chemicals, is easier to control, and can produce exceptional surfaces.

As circuit densities have increased over the years, device sizes and features have become smaller and smaller. Today, minimum commercially feasible linewidths have been reduced to 0.014 μm (14 nm) with considerable research being conducted at smaller linewidths. The Apple A9 and Intel Broadwell processors, for example, both use linewidths of 14 nm.

As pattern resolution, and hence device miniaturization, is limited by the wavelength of the radiation source used, the need has arisen to move to wavelengths shorter than those in the ultraviolet range, such as "deep" UV wavelengths, "extreme" UV wavelengths, electron beams, and X-rays (See Table 13.1). In these technologies, the photoresist is replaced by a similar resist that is sensitive to a specific range of shorter wavelengths.

Pitch splitting. Multiexposure techniques have been developed to obtain higher resolution images than can be attained through conventional single-exposure lithography, and has been applied for sub-32 nm feature developments. *Pitch splitting* is shown in Fig. 13.14, using conventional lithography in multiple stages. Recognizing that the spacing between features is the limiting dimension, pitch splitting involves breaking up the desired pattern into two complementary portions and creating corresponding masks. Using two imaging steps, features can be developed in the substrate with twice the resolution of a single imaging step.

There are two forms of pitch splitting. In **double exposure** (DE), a mask exposes some of the desired trenches or regions in the photoresist, then a second mask is used to expose the remaining features (Fig. 13.14a). The photoresist is then exposed and the substrate is etched. **Double patterning** (DP) involves two sequential lithography and etch steps, so that it is sometimes referred to as the *LELE* (lithography-etch-lithography-etch) *process*.

Immersion lithography. The resolution of lithography systems can be increased by inserting a fluid with a high refractive index between the final lens and the wafer, a technique called *immersion lithography*. Water has been mainly used to date and has been the main approach used to attain feature sizes below 45 nm. Fluids with a refractive index higher than that of water also are being investigated to increase the resolutions. Immersion lithography requires careful process controls, especially thermal controls,

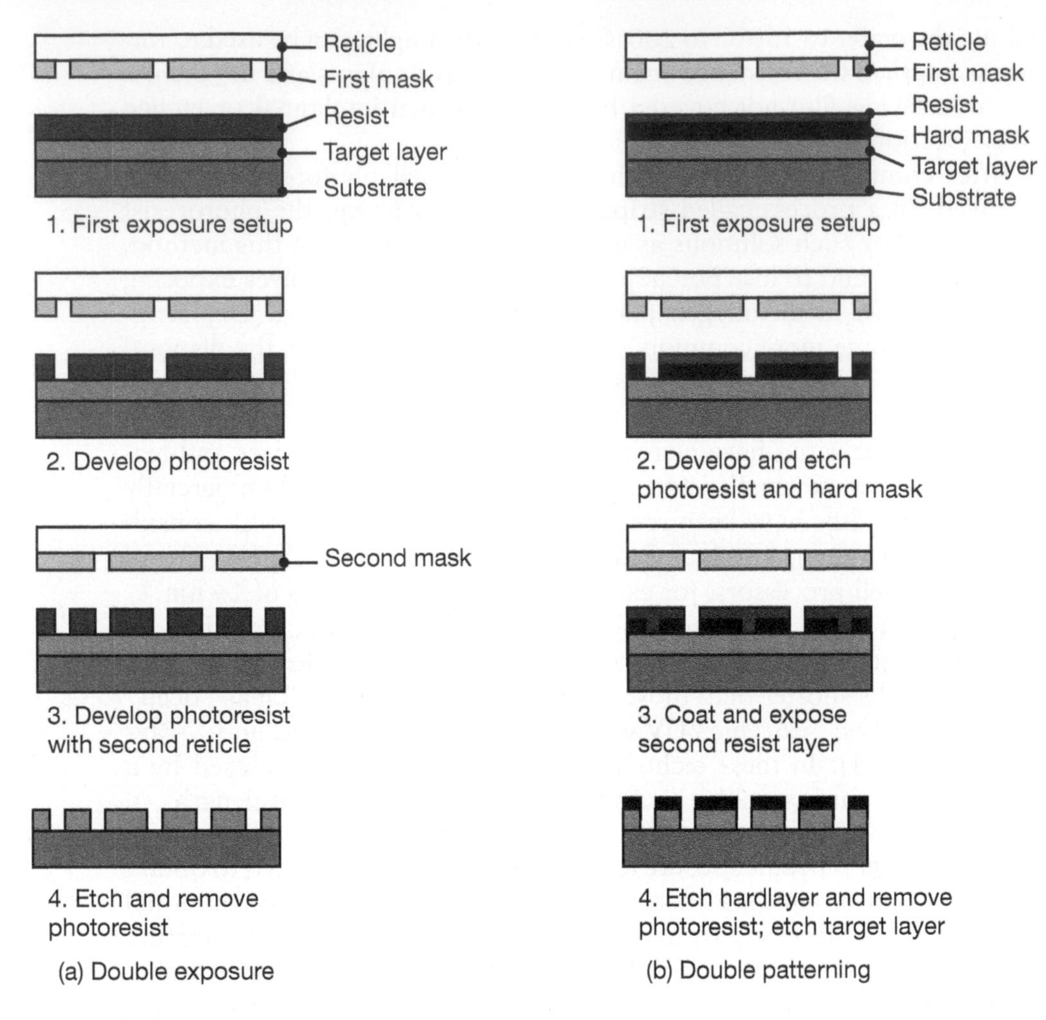

FIGURE 13.14 Pitch splitting lithography. (a) Double exposure (DE) process; and (b) double patterning (DP) process, also known as the LELE (lithography-etch-lithography-etch) process.

because any bubbles that develop in the water will result in defects due to distortion of the light source.

Extreme-ultraviolet lithography. The pattern resolution in photolithography is limited by light diffraction; one method of reducing this effect is to use ever shorter wavelengths. *Extreme-ultraviolet* (EUV) *lithography* uses light at a wavelength of 13 nm in order to obtain features around 30 to 100 nm in size. The waves are focused by highly-reflective molybdenum/silicon mirrors (instead of glass lenses that absorb EUV light) through the mask to the wafer surface.

X-ray lithography. Although photolithography is the most widely used lithography technique, it has fundamental resolution limitations associated with light diffraction. *X-ray lithography* is superior to photolithography, because of the shorter wavelength of the radiation and the very large depth of focus involved. This characteristic allows much finer patterns to be resolved, and X-ray lithography is far less susceptible to dust. Moreover, the aspect ratio (defined as the ratio of depth to lateral dimension) can be more than 100 with X-ray lithography, but is limited to around 10 with photolithography.

In order to achieve this benefit, synchrotron radiation is required, which is expensive and available at only a few research laboratories. Given the very large capital investment required for a manufacturing facility, industry has preferred to refine and improve optical lithography, instead of investing new capital into X-ray based production. X-ray lithography is currently not widespread, although the LIGA process (Section 13.16) fully exploits the benefits of X-ray lithography.

Electron-beam and ion-beam lithography. With respect to the resolutions attainable *electron-beam* (e-beam) and *ion-beam* (i-beam) *lithography* are superior to photolithography. These two methods involve high current density in narrow electron or ion beams (called *pencil sources*) that scan a pattern onto a wafer, one pixel at a time. Masking is done by controlling the point-by-point transfer of the stored pattern and is performed by software.

These techniques have the advantages of accurate control of exposure over small areas of the wafer, large depth of focus, and low defect densities. Resolutions are limited to around 10 nm, because of electron scatter, although 2-nm resolutions have been reported for some materials. It should be noted, however, that the scan time increases significantly as the resolution increases, because more highly-focused beams are required. The main drawback of these techniques is that electron and ion beams have to be maintained in a vacuum, significantly increasing equipment complexity and production cost. Also, the scan time for a wafer for these techniques is much slower than that for other lithographic methods.

SCALPEL. In the SCALPEL process, from Scattering with Angular Limitation Projection Electron-Beam Lithography (Fig. 13.15), a mask is produced from a roughly 0.1-μm thick membrane of silicon nitride, and is patterned with an approximately 50-nm thick coating of tungsten. High-energy electrons pass through both the silicon nitride and the tungsten, but the tungsten scatters the electrons widely, whereas the silicon nitride results in very little scattering. An aperture blocks the scattered electrons, resulting in a high-quality image at the wafer. The limitation is the small sized masks that are currently in use, but the process has high potential. Its most significant advantage is that energy does not have to be absorbed by the reticle, but instead, it is blocked by the aperture, which is not as fragile or expensive as the reticle.

Soft lithography. *Soft lithography* refers to a number of processes for pattern transfer. All processes require that a master mold must first be created by standard lithography techniques, as described above. The master mold is then used to produce an elastomeric pattern or stamp, as shown in Fig. 13.16. An elastomer that has been commonly used for the stamp

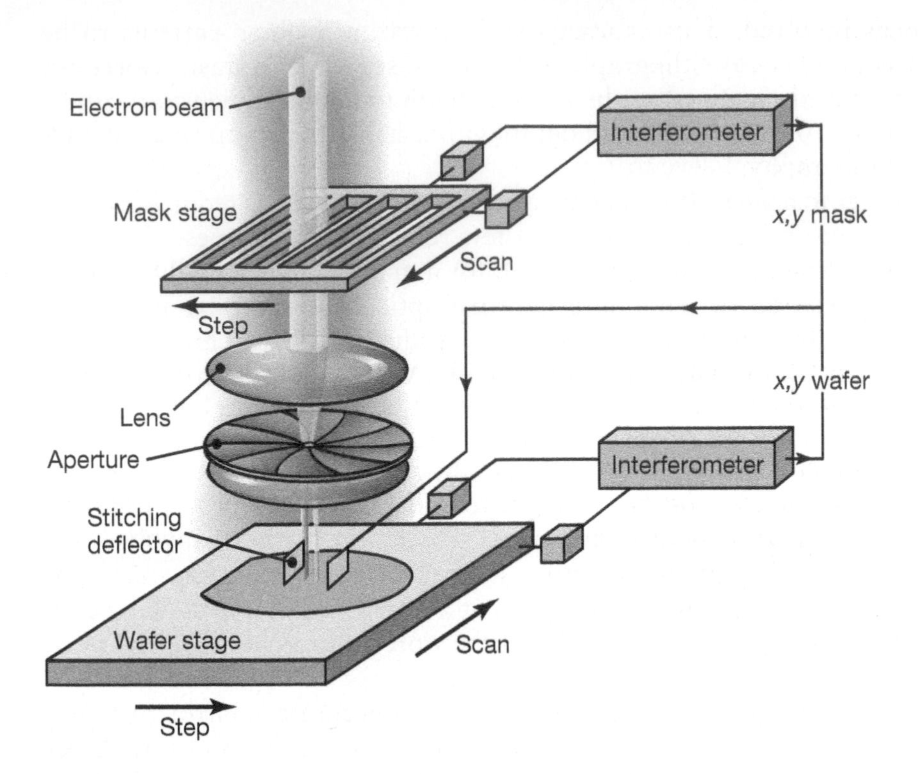

FIGURE 13.15 Schematic illustration of the SCALPEL process.

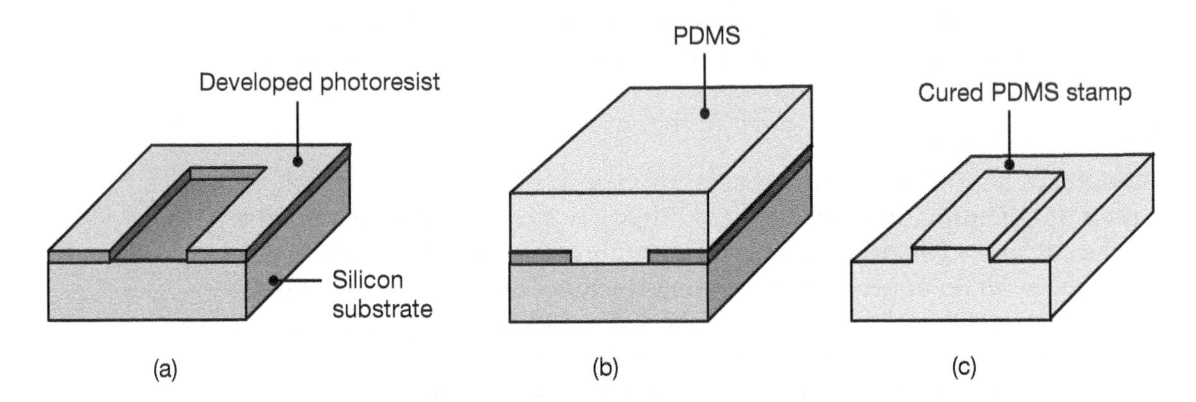

FIGURE 13.16 Production of a polydimethylsiloxane (PDMS) mold for soft lithography. (a) A developed photoresist is produced through standard lithography (See Fig. 13.13); (b) a PDMS stamp is cast over the photoresist; and (c) the PDMS stamp is peeled off the substrate to produce a stamp. The stamp shown in the figure has been rotated to emphasize replication of surface features; the master pattern can be used several times. *Source:* After Y. Xia and G.M. Whitesides.

is silicone rubber, or polydimethylsiloxane (PDMS), because it is chemically inert and not hygroscopic (it does not swell with humidity), has good thermal stability, strength, durability, and surface properties.

Several PDMS stamps can be produced from the same pattern, and each stamp can be used several times. Some of the common soft lithography processes are:

- **Micro-contact printing** (μCP). In this process the PDMS stamp is coated with an "ink" and then pressed against a surface. The peaks of the pattern are in contact with the opposing surface and a thin layer of the ink is transferred, often only one molecule thick (self-assembled monolayer or boundary film, Section 4.4.3). This thin film can serve as a mask for selective wet etching, described below, or it can be used to impart a desired chemistry onto the surface.
- **Micro-transfer molding** (μTM). In this process, shown in Fig. 13.17a, the recesses of the PDMS mold are filled with a liquid polymer precursor, and then pushed against a surface. After the polymer has cured, the mold is peeled off, leaving behind a pattern suitable for further processing.
- **Micromolding in capillaries** (MIMIC). In the MIMIC technique, shown in Fig. 13.17b, the PDMS stamp pattern consists of channels that use capillary action to wick a liquid into the stamp, either from the side of the stamp or from reservoirs within the stamp. The liquid can be a thermosetting polymer, a ceramic sol-gel, or suspensions of solids within liquid solvents. Good pattern replication occurs as long as the channel aspect ratio is moderate and depending on the liquid used. The MIMIC process has been used to produce all-polymer field effect transistors and diodes and has various applications in sensors.

13.8 Etching

Etching is the process by which entire films, or particular sections of films or the substrate, are removed. One of the most important criteria in this important process is **selectivity**, which refers to the ability to etch one material without etching another. Tables 13.2 and 13.3 provide a summary of etching processes. In silicon technology, an etching process must effectively etch the silicon-dioxide layer, with minimal removal of the underlying silicon or the resist material. In addition, polysilicon and metals must be etched into high-resolution lines with vertical wall profiles and with minimal removal of the underlying insulating film. Typical etch rates range from tens to several thousands of nm/min, and *selectivities* (defined as the ratio of the etch rates of the two films) can range from 1:1 to 100:1.

13.8.1 Wet Etching

Wet etching involves immersing the wafers in a liquid solution, usually acidic. The main drawback to most wet-etching operations is that they

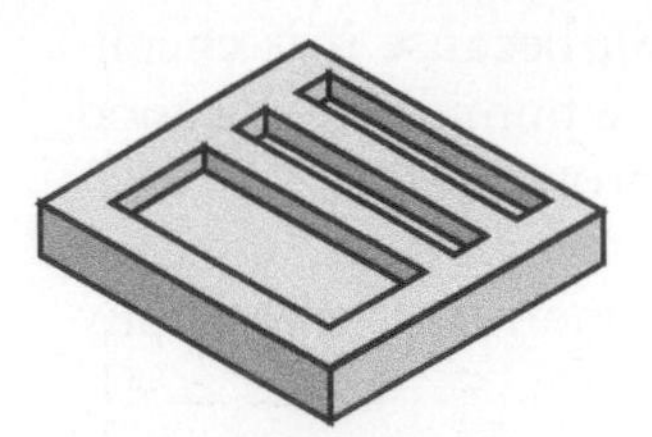

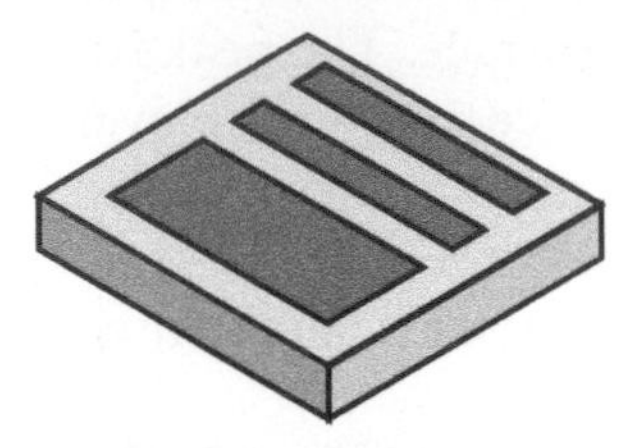

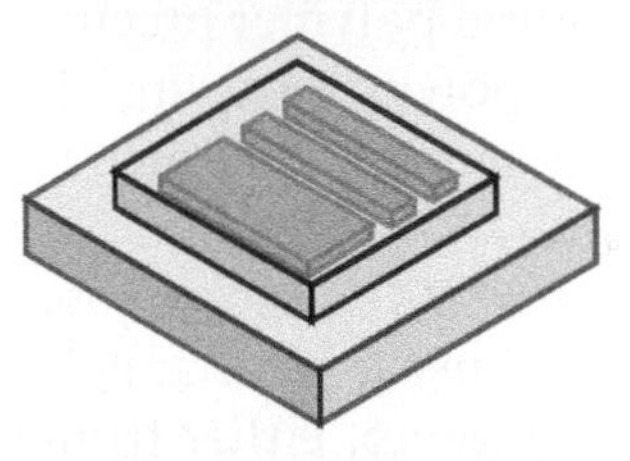

3. Press stamp against surface; allow precursor to cure

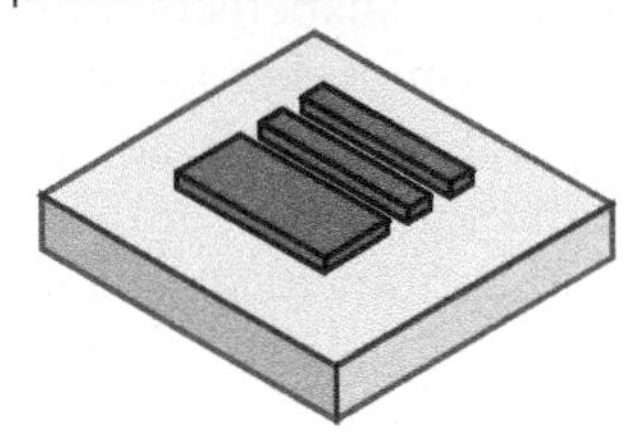

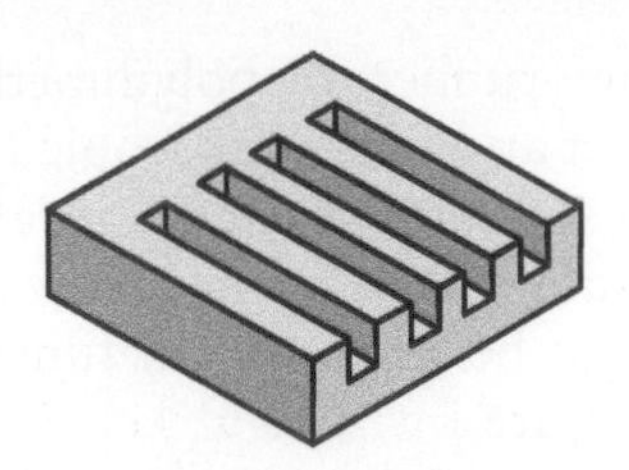

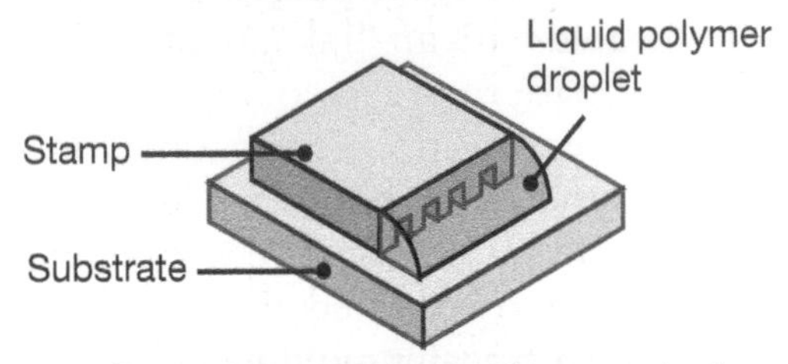

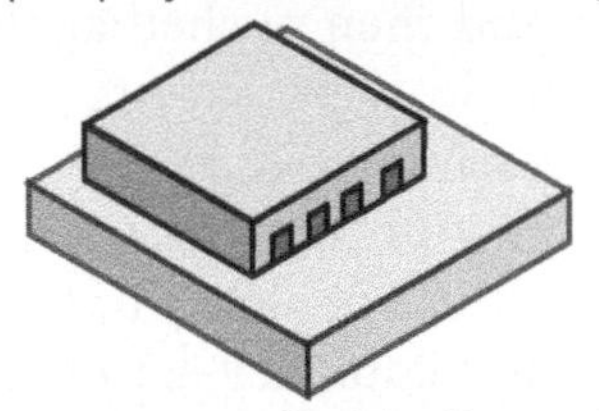

3. Remove excess liquid; allow polymer to cure

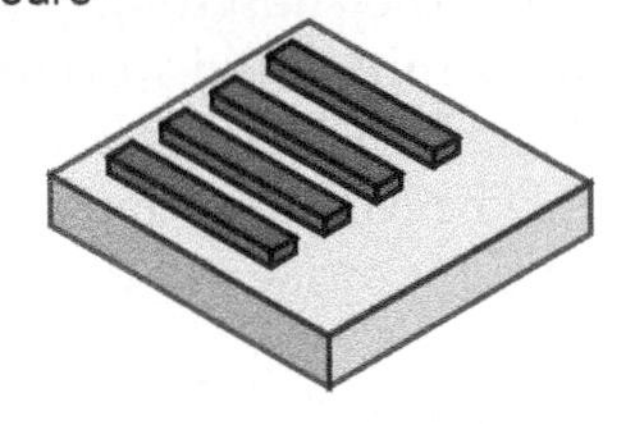

4. Peel off stamp

4. Peel off stamp

FIGURE 13.17 Soft lithography techniques. (a) Micro-transfer molding (μTM); and (b) micromolding in capillaries (MIMIC).

are *isotropic*, that is, they etch in all directions of the workpiece at the same rate. This condition results in *undercuts* beneath the mask material (See, for example, Fig. 13.18a) and thus limits the resolution of geometric features in the substrate.

Effective etching requires the following conditions:

1. Transport of etchant to the surface;
2. A chemical reaction to remove material;
3. Transporting reaction products away from the surface; and
4. Ability to rapidly stop the etching process (*etch stop*) in order to obtain superior pattern transfer, usually using an underlying layer with high selectivity.

TABLE 13.2 Comparison of etch rates.

Etchant	Target material	Etch rate (nm/min)[a]							
		Poly-silicon n^+	Poly-silicon, undoped	SiO_2	SiN	Phospho-silicate glass, annealed	Alum-inum	Tita-nium	Photo-resist (OCG-820PR)
Wet etchants									
Concentrated HF (49%)	Silicon oxides	0	—	2300	14	3600	4.2	>1000	0
25:1 $HF{:}H_2O$	Silicon oxides	0	0	9.7	0.6	150	—	—	0
5:1 BHF[b]	Silicon oxides	9	2	100	0.9	440	140	>1000	0
Silicon etchant (126 HNO_3:60H_2O:5NH_4F)	Silicon	310	100	9	0.2	170	400	300	0
Aluminum etchant (16H_3PO_4:1HNO_3:1HAc:2H_2O)	Aluminum	<1	<1	0	0	<1	660	0	0
Titanium etchant (20 H_2O:1 H_2O_2:1HF)	Titanium	1.2	—	12	0.8	210	>10	880	0
Piranha (50 H_2SO_4:1H_2O_2)	Cleaning off metals and organics	0	0	0	0	0	180	240	>10
Acetone (CH_3COOH)	Photoresist	0	0	0	0	0	0	0	>4000
Dry etchants									
CF_4+CHF_3+He, 450W	Silicon oxides	190	210	470	180	620	—	>1000	220
SF_6+He, 100W	Silicon nitride	73	67	31	82	61	—	>1000	69
SF_6m 125 W	Thin silicon nitrides	170	280	110	280	140	—	>1000	310
O_2, 400W	Ashing photoresist	0	0	0	0	0	0	0	340

Notes: (a) Results are for fresh solutions at room temperature unless otherwise noted. Actual etch rates will vary with temperature and prior use of solution, area of exposure of film, other materials present, and film impurities and microstructure.
(b) Buffered hydrofluoric acid, 33% NH4F and 8.3% HF by weight. *Source:* After K. Williams and R. Muller.

TABLE 13.3 General characteristics of silicon etching operations.

	Temperature (°C)	Etch rate (μm/min)	{111}/{100} selectivity	Nitride etch rate (nm/min)	SiO_2 etch rate (nm/min)	p^{++} etch stop
Wet etching						
$HF:HNO_3:CH_3COOH$	25	1–20	—	Low	10–30	No
KOH	70–90	0.5–2	100:1	<1	10	Yes
Ethylene-diamine pyrochatechol (EDP)	115	0.75	35:1	0.1	0.2	Yes
$N(CH_3)_4OH$ (TMAH)	90	0.5–1.5	50:1	<0.1	<0.1	Yes
Dry (plasma) etching						
SF_6	0–100	0.1–0.5	—	200	10	No
SF_6/C_4F_8 (DRIE)	20–80	1–3	—	200	10	No

Source: After N. Maluf.

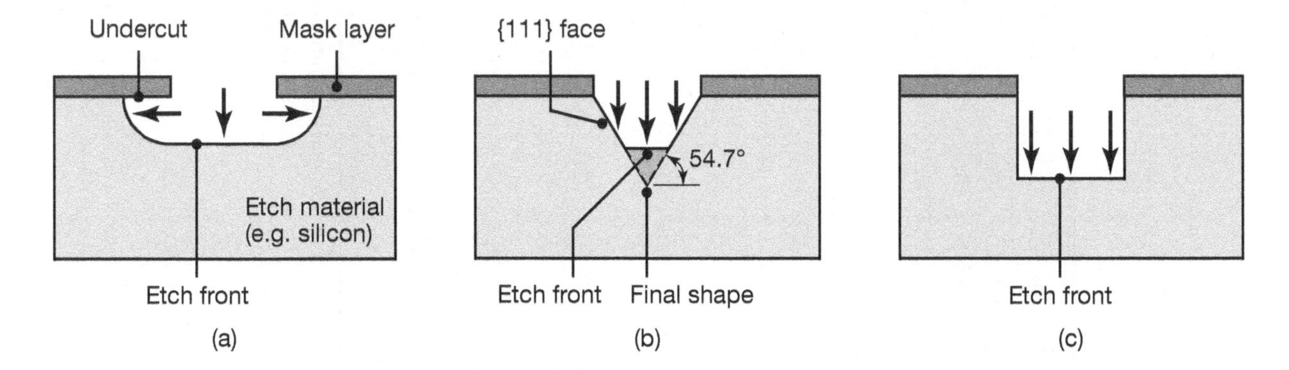

FIGURE 13.18 Etching directionality. (a) Isotropic etching: etch proceeds vertically and horizontally at approximately the same rate; note the significant mask undercut. (b) Orientation-dependent etching (ODE): etch proceeds vertically, terminating on {111} crystal planes, with little mask undercut. (c) Vertical etching: etch proceeds vertically, with little mask undercut. *Source:* After K.R. Williams.

If the first or third steps above limit the speed of the process, agitation or stirring of the solution can increase etching rates (See also Fig. 9.27a). If the second step limits the speed of the process, the etching rate will strongly depend on temperature, etching material, and the composition of the solution. Reliable etching therefore requires both good temperature control and repeatable stirring capability.

Isotropic etchants are widely used for the following procedures:

1. Removal of damaged surfaces;
2. Rounding of sharp, etched corners to avoid stress concentrations;
3. Reduction of roughness following anisotropic etching;
4. Development of structures in single-crystal slices; and
5. Evaluation of defects.

Microelectronic devices and MEMS (Sections 13.15 through 13.17) require accurate machining of structures, a task that is done through masking, which is a challenge with isotropic etchants. The strong acids used will (a) etch aggressively, at a rate of up to 50 μm/min with an etchant of 66% nitric acid (HNO_3) and 34% hydrofluoric acid (HF), although etch rates of 0.1–1 μm/s are more typical; and (b) produce rounded cavities. The etch rate is very sensitive to agitation and, therefore, lateral and vertical features are difficult to control.

The size of the features in an integrated circuit determines its performance, and for this reason, there is a strong need to produce well-defined, extremely small structures. Such small features cannot be attained through isotropic etching, because of the poor definition that results from undercutting of masks.

Anisotropic etching occurs when etching is strongly dependent on compositional or structural variations in the material. There are two basic kinds of anisotropic etching: *orientation-dependent etching* and *vertical etching*. Most vertical etching is done with dry plasmas (Section 13.8.2). Orientation-dependent etching commonly occurs in a single crystal when etching takes place at different rates at different directions, as shown in Fig. 13.18b. When orientation-dependent etching is performed properly, the etchants produce geometric shapes with walls defined by the crystallographic planes that resist the etchants. Figure 13.19 shows the vertical etch rate for silicon as a function of temperature. As can be seen, etching rate is more than one order of magnitude lower in the [111] crystal direction than in other directions, hence well-defined walls can be obtained along the [111] crystal direction.

The **anisotropy ratio** (AR) for etching is defined by

$$AR = \frac{E_1}{E_2}, \tag{13.4}$$

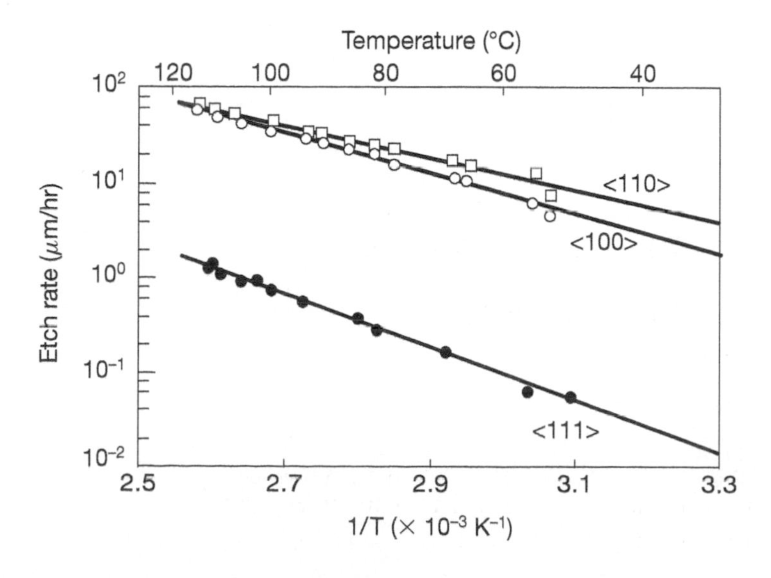

FIGURE 13.19 Etch rates of silicon at different crystallographic orientations, using ethylene-diamine/pyrocatechol-in-water as the solution. *Source:* After H. Seidel, et al., *J. Electrochemical Society*, 1990, pp. 3612–3626.

where E is the etch rate and the subscripts refer to two crystallographic directions of interest. Recall that selectivity refers to the etch rates between the materials of interest. The anisotropy ratio is unity for isotropic etchants and can be as high as 400:200:1 for (110)/(100)/(111) silicon. The {111} planes always etch the slowest, but the etch rates for the {100} and {110} planes can be controlled through etchant chemistry.

Masking is also a concern in anisotropic etching, but silicon oxide is less valuable as a mask material for different reasons than in isotropic etching. Anisotropic etching is slower than isotropic etching (typically 3 μm/m), and thus anisotropic etching through a wafer may take several hours. Silicon oxide may etch too rapidly to be used as a mask, and hence a high-density silicon-nitride mask may be required.

Often, it is important to rapidly halt the etching process (etch stop), a situation that is typically the case when thin membranes are to be manufactured or when features with very precise thicknesses are required. Conceptually, this task can be accomplished by removing the wafer from the etching solution; however, etching depends to a great extent on the ability to circulate fresh etchants to the desired locations. Since the circulation varies across the wafer surface, this strategy for halting the etching process would lead to large variations in etch depth.

The most common approach for producing uniform feature sizes across a wafer is to use a *boron etch stop*, where a boron layer is diffused or implanted into silicon. Another common etch stop is the placement of silicon oxide (SiO_2) beneath silicon nitride (Si_3N_4). Because anisotropic etchants do not attack boron-doped silicon as aggressively as they do undoped silicon, surface features or membranes can be created by **back etching**. Figure 13.20 shows an example of the boron etch-stop approach.

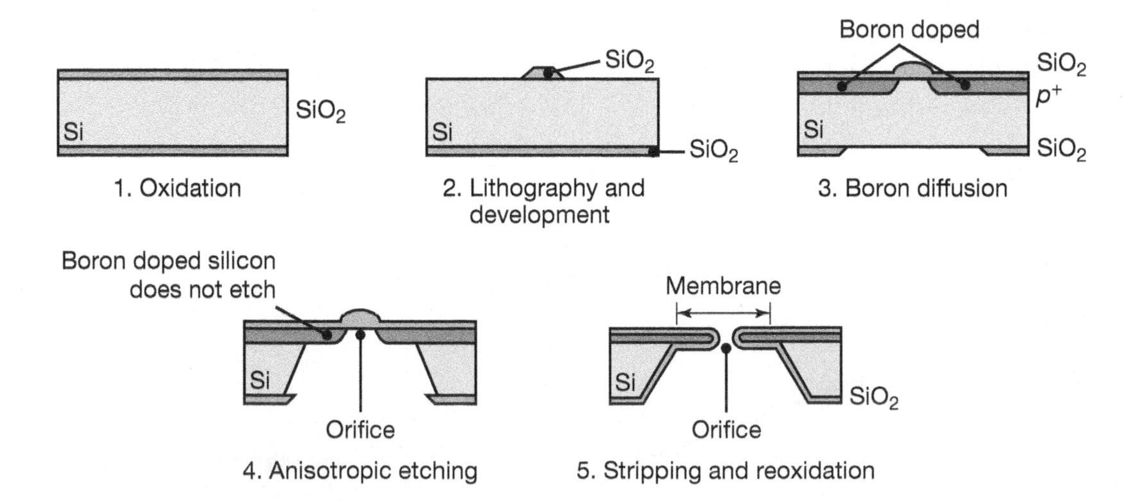

FIGURE 13.20 Application of a boron etch stop and back etching to form a membrane and orifice. *Source:* After I. Brodie and J.J. Murray.

Numerous etchant formulations have been developed over the years; some of the more common wet etchants are described below.

1. Silicon dioxide is commonly etched with hydrofluoric (HF) acid solutions. The driving chemical reaction in pure HF etching is

$$SiO_2 + 6HF \rightarrow H_2SiF_6 + 2H_2O. \tag{13.5}$$

It is, however, rare that silicon dioxide is etched purely through the reaction in Eq. (13.5). Hydrofluoric acid is a weak acid, and it does not completely dissociate into hydrogen and fluorine ions in water. HF_2^- is an additional ion that exists in hydrofluoric acid, and HF_2^- attacks silicon oxide about 4.5 times faster than does HF alone. The reaction involving the HF_2^- ion is

$$SiO_2 + 2HF_2^- + H^+ \rightarrow SiF_6^{2-} + 2H_2O. \tag{13.6}$$

The pH value of the etching solution is critical, because acidic solutions have sufficient hydrogen ions to dissociate the HF_2^- ions into HF ions. As HF and HF_2^- are consumed, the etch rate decreases; for that reason, a buffer of ammonium fluoride (NH_4F) is used to maintain the pH and, thus, keep the concentrations of HF and HF_2^- constant, stabilizing the etch rate. Such an etching solution is referred to as a *buffered hydrofluoric acid* (BHF) or *buffered oxide etch* (BOE) and involves the reaction

$$SiO_2 + 4HF + 2NH_4F \rightarrow (NH_4)_2\,SiF_6 + 2H_2O. \tag{13.7}$$

2. Silicon nitride is etched with phosphoric acid (H_3PO_4), usually at an elevated temperature, typically 160°C (320°F). The etch rate of phosphoric acid decreases with water content, thus a *reflux system* is used to return condensed water vapor to the solution to maintain a constant etch rate.
3. Etching silicon often involves mixtures of nitric acid (HNO_3) and hydrofluoric acid (HF); water can be used to dilute these acids. The preferred buffer is acetic acid because it preserves the oxidizing power of HNO_3; this system is referred to as a *HNA etching system*. A simplified description of this etching process is that the nitric acid oxidizes the silicon and then the hydrofluoric acid removes the silicon oxide. This two-step process is a common approach for chemical machining of metals (see Section 9.10). The overall reaction is

$$18HF + 4HNO_3 + 3Si \rightarrow 3H_2SiF_6 + 4NO + 8H_2O. \tag{13.8}$$

The etch rate is limited by the silicon-oxide removal; a buffer of ammonium fluoride is therefore usually used to maintain etch rates.
4. Anisotropic etching, or orientation-dependent etching, of single-crystal silicon can be done with solutions of potassium hydroxide, although other etchants have can be used. The reaction is

$$Si + 2OH^- + 2H_2O \rightarrow SiO_2\,(OH)_2^{2-} + 2H_2. \tag{13.9}$$

Note that this reaction does not require potassium to be the source of the OH ions; KOH attacks {111}-type planes much more slowly than other planes. Isopropyl alcohol is sometimes added to KOH solutions to reduce etch rates and increase the uniformity of etching. KOH also is extremely valuable in that it stops etching when it contacts a very heavily doped *p*-type material (*boron etch stop*; see Fig. 13.20).

5. Aluminum is etched through a solution typically consisting of 80% phosphoric acid (H_3PO_4), 5% nitric acid (HNO_3), 5% acetic acid (CH_3COOH), and 10% water. The nitric acid first oxidizes the aluminum, and the oxide is then removed by the phosphoric acid and water; this solution can be masked with a photoresist.
6. Wafer cleaning is accomplished through *Piranha solutions*, which have been in use for decades. These solutions consist of hot mixtures of sulfuric acid (H_2SO_4) and peroxide (H_2O_2). Piranha solutions strip photoresist and other organic coatings and remove metals on the surface, but do not adversely affect silicon dioxide or silicon nitride, thus making it an ideal cleaning solution. Bare silicon forms a thin layer of hydrous silicon oxide, which later is removed through a short dip in hydrofluoric acid.
7. Although photoresist can be removed through Piranha solutions, acetone is commonly used for this purpose instead. Acetone dissolves the photoresist, but if the photoresist is excessively heated during a process step, it will significantly be more difficult to remove with acetone. In such a case, the photoresist can be removed through an ashing plasma.

EXAMPLE 13.1 Processing of a *p*-Type Region in *n*-Type Silicon

Given: Assume that it is desired to create a *p*-type region within a sample of *n*-type silicon.

Find: Draw cross sections of the sample at each processing step in order to accomplish this task.

Solution: See Fig. 13.21. This simple device is known as a *pn junction diode*, and the physics of its operation is the foundation for most semiconductor devices.

13.8.2 Dry Etching

Modern integrated circuits are etched exclusively through *dry etching*, which involves the use of chemical reactants in a low-pressure system. In contrast to the wet process, described above, dry etching can have a high degree of directionality, resulting in highly anisotropic etch profiles (Fig. 13.18c). Moreover, the dry etching process requires only small amounts of the reactant gases, whereas the solutions used in wet etching have to be refreshed periodically. Dry etching usually involves a plasma or discharge in areas of high electric and magnetic fields; any gases that are present are dissociated to form ions, electrons, or highly reactive molecules.

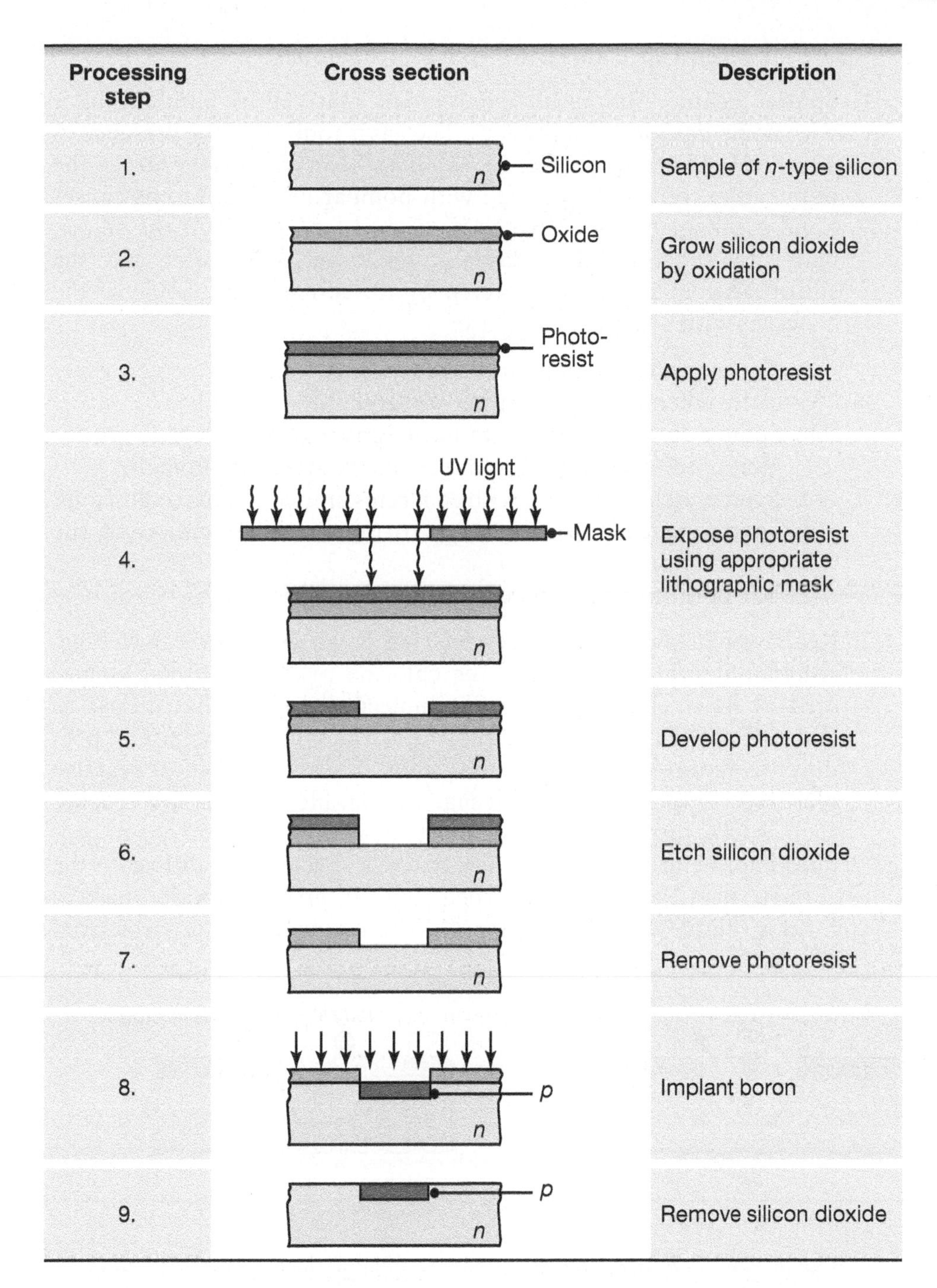

Processing step
Cross section
Description
1.
Silicon
n
Sample of n-type silicon
2.
Oxide
Grow silicon dioxide by oxidation
3.
Photo-resist
Apply photoresist
4.
UV light
Mask
Expose photoresist using appropriate lithographic mask
5.
Develop photoresist
6.
Etch silicon dioxide
7.
Remove photoresist
8.
p
Implant boron
9.
Remove silicon dioxide

FIGURE 13.21 Sequence in processing of a p-type region in n-type silicon.

There are several specialized dry-etching techniques.

1. **Sputter etching.** *Sputter etching* removes material by bombarding it with noble-gas ions, usually Ar^+. The gas is ionized in the presence of a cathode and an anode (Fig. 13.22). If a silicon wafer is the target, the momentum transfer associated with bombardment of atoms causes bond breakage and material to be ejected (sputtered). If the silicon chip is the substrate, the material in the target is deposited onto the silicon after it has been sputtered by the ionized gas. Some of the concerns with sputter etching are:

    a. The ejected material can be redeposited onto the target, especially when the aspect ratios are large;
    b. Sputter etching is not material selective; most materials sputter at about the same rate, and masking is therefore difficult;
    c. Sputter etching is slow, with etch rates limited to tens of nm/min;
    d. Sputtering can cause damage to or excessive erosion of the material; and
    e. The photoresist is difficult to remove.

2. **Reactive plasma etching.** Also referred to as *dry chemical etching*, *reactive plasma etching* involves chlorine or fluorine ions (generated by RF excitation) and other molecular species that diffuse to and chemically react with the substrate. The volatile compound produced is removed by a vacuum system. The mechanism of reactive plasma etching is illustrated in Fig. 13.23a. Here, a reactive species, such as CF_4, dissociates upon impact with energetic electrons to produce fluorine atoms (step 1). The reactive species then diffuse to the surface (step 2), become adsorbed (step 3), and chemically react to form a volatile compound (step 4). The reactant then desorbs from

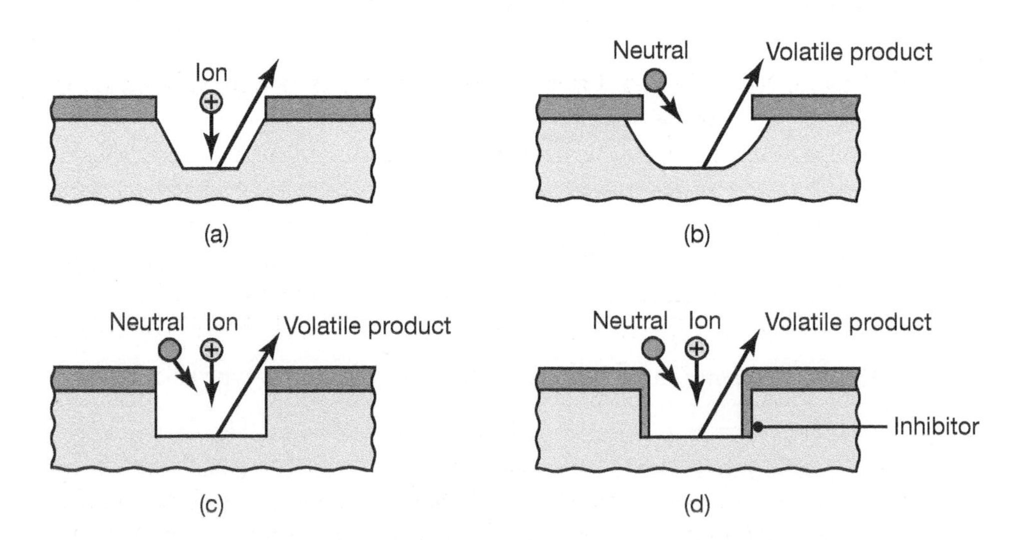

FIGURE 13.22 Machining profiles associated with different dry-etching techniques. (a) sputtering; (b) chemical; (c) ion-enhanced energetic; and (d) ion-enhanced inhibitor. *Source:* After M. Madou.

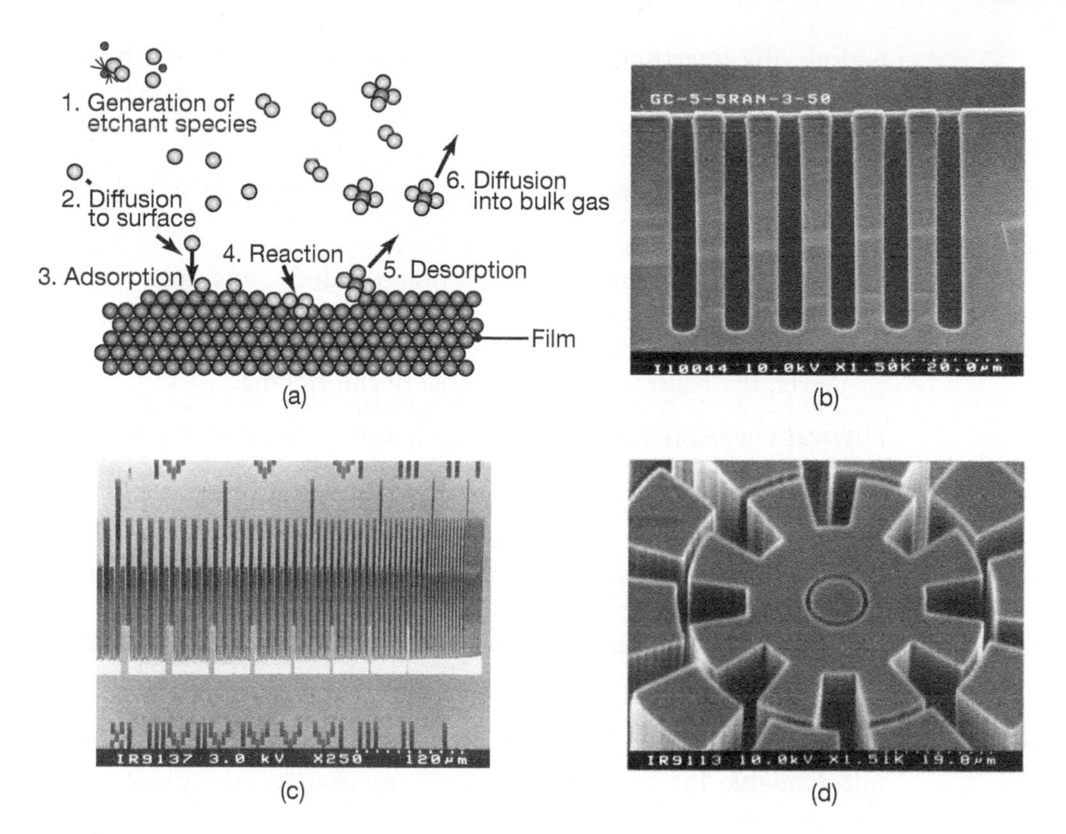

FIGURE 13.23 (a) Schematic illustration of reactive plasma etching. *Source:* After M. Madou. (b) Example of deep reactive ion etched trench; note the periodic undercuts, or scalloping. (c) Near vertical sidewalls produced through DRIE with an anisotropic etching process. (d) An example of cryogenic dry etching, showing a 145 μm deep structure etched into Si using a 2.0 μm thick oxide masking layer. The substrate temperature was −140°C during etching. *Source:* for (b) to (d): R. Kassing and I.W. Rangelow.

the surface (step 5) and diffuses into the bulk gas, where it is removed by the vacuum system.

Some reactants polymerize on the surface, and require additional removal, either with oxygen in the plasma reactor or by an external ashing operation. The electrical charge of the reactive species is not high enough to cause damage through impact on the surface, so that no sputtering occurs. Thus, the etching is isotropic and undercutting of the mask takes place (Fig. 13.18a). Table 13.2 lists some of the more commonly used dry etchants, their target materials, and their typical etch rates.

3. **Physical-chemical etching.** Processes such as *reactive ion-beam etching* (RIBE) and *chemically assisted ion-beam etching* (CAIBE) combine the advantages of physical and chemical etching. Although these processes use chemically-reactive species to remove material, these processes are physically assisted by the impact of ions onto the surface. In RIBE, also known as *deep reactive ion etching* (DRIE), vertical trenches hundreds of micrometers deep can be produced by

periodically interrupting the etch process and depositing a polymer layer. When performed with an isotropic dry etching process, this operation results in scalloped sidewalls, as shown in Fig. 13.23b. Anisotropic DRIE can produce near-vertical sidewalls (Fig. 13.23c).

In CAIBE, ion bombardment can assist dry chemical etching by

a. Making the surface more reactive;
b. Clearing the surface of reaction products and allowing the chemically reactive species access to the cleared areas; and
c. Providing the energy to drive surface chemical reactions; however, the neutral species do most of the etching.

Physical-chemical etching is extremely useful, because the ion bombardment is directional, so that etching is anisotropic. Also, the ion bombardment energy is low and does not contribute much to mask removal. This factor allows generation of near-vertical walls with very large aspect ratios. Since the ion bombardment does not directly remove material, masks can be used.

4. **Cryogenic dry etching.** This procedure is an approach to obtain very deep features with vertical walls. The workpiece is lowered to cryogenic temperatures, and chemically-assisted ion-beam etching takes place. The low temperatures involved ensure that sufficient energy is not available for a surface chemical reaction to take place, unless the ion bombardment direction is normal to the surface. Oblique impacts, such as occur on side walls in deep crevices, cannot drive the chemical reactions; therefore, very smooth vertical walls can be produced (Fig. 13.23d).

   Because dry etching is not selective, etch stops cannot be directly applied; dry-etch reactions must be terminated when the target film is removed. Optical emission spectroscopy is often used to determine the "end point" of a reaction; filters can be used to capture the wavelength of light emitted during a particular reaction.

EXAMPLE 13.2 Comparison of Wet and Dry Etching

Given: Consider a case where a ⟨100⟩ wafer has an oxide mask placed on it in order to produce square or rectangular holes. The sides of the square mask are precisely oriented with the ⟨110⟩ direction (See Fig. 13.7) of the wafer surface, as shown in Fig. 13.24.

Find: Describe the possible hole geometries that can be produced by etching.

Solution: Isotropic etching results in the cavity shown in Fig. 13.24a. Since etching occurs at constant rates in all directions, a rounded cavity is produced that undercuts the mask. An orientation-dependent etchant produces the cavity shown in Fig. 13.24b. Since etching is much faster in the ⟨100⟩ and ⟨110⟩ directions than in the ⟨111⟩ direction, sidewalls defined by the {111} plane are generated. For silicon, for example, these sidewalls are at an angle of 54.74° to the surface.

The effect of a larger mask or shorter etch time is shown in Fig. 13.24c. The resultant pit is defined by ⟨111⟩ sidewalls and a bottom in the ⟨100⟩ direction parallel to the surface. A rectangular

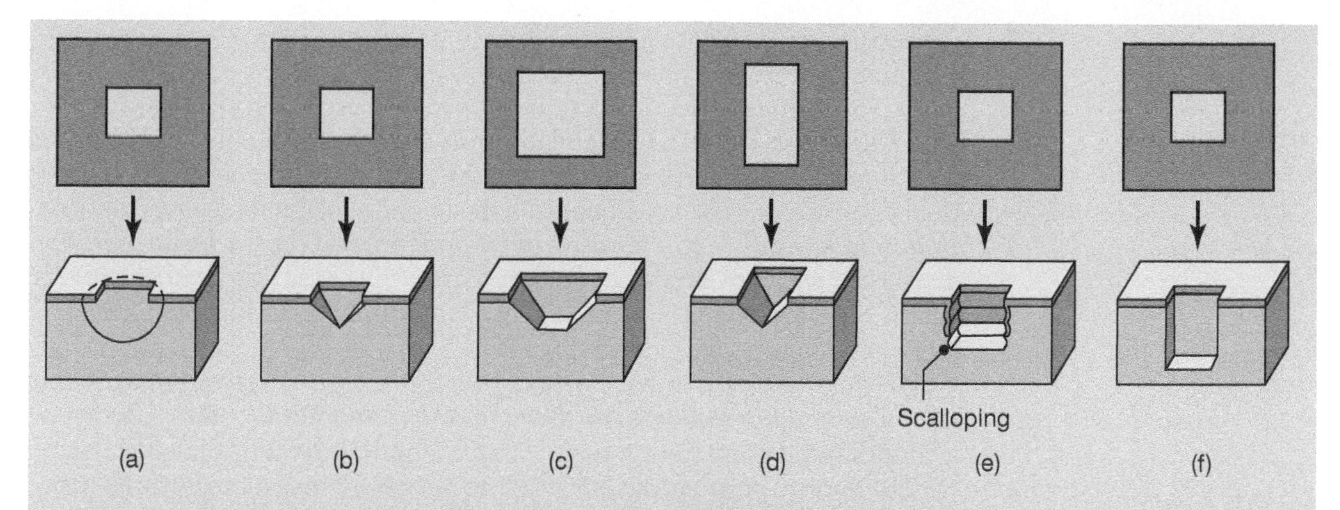

FIGURE 13.24 Various types of holes generated from a square mask in (a) isotropic (wet) etching; (b) orientation-dependent etching (ODE); (c) ODE with a larger hole; (d) ODE of a rectangular hole; (e) deep reactive ion etching; and (f) vertical etching. *Source:* After M. Madou.

mask and the resultant pit are shown in Fig. 13.24d. Deep reactive ion etching is depicted in Fig. 13.24e; note that a polymer layer is periodically deposited onto the sidewalls of the hole to allow for deep pockets, but scalloping (greatly exaggerated in the figure) is unavoidable. A hole produced from chemically reactive ion etching is shown in Fig. 13.24f.

13.9 Diffusion and Ion Implantation

As stated in Section 13.3, the operation of microelectronic devices often depends on regions of different doping types and concentrations. The electrical character of these regions can be altered by introducing dopants into the substrate, accomplished by *diffusion* and *ion implantation* processes. Because many different regions of microelectronic devices must be doped, this step in the fabrication sequence is repeated several times.

In the diffusion process, the movement of atoms results from thermal excitation. Dopants can be introduced to the substrate in the form of a deposited film, or the substrate can be exposed to a vapor containing the dopant source; this operation takes place at elevated temperatures, usually 800°C to 1200°C. Dopant movement within the substrate is a function of temperature, time, and the diffusion coefficient (or diffusivity) of the dopant species, as well as the type and quality of the substrate material. Because of the nature of diffusion, dopant concentration is very high at the surface and drops off sharply away from the surface.

To obtain a more uniform concentration within the substrate, the wafer is further heated to drive in the dopants, a process called **drive-in diffusion.** The fact that diffusion, desired or undesired, will always occur at high temperatures is invariably taken into account during subsequent processing steps. Although the diffusion process is relatively inexpensive, it is highly isotropic.

Ion implantation is a much more extensive process and requires specialized equipment (Fig. 13.8). Implantation is accomplished by accelerating ions through a high-voltage beam (of as much as one million volts) and then choosing the desired dopant by means of a magnetic mass separator. In a manner similar to that used in cathode-ray tubes, the beam is swept across the wafer by sets of deflection plates, thus ensuring uniform coverage of the substrate. The complete implantation system must be operated in a vacuum.

The high-velocity impact of ions on the silicon surface damages the lattice structure, resulting in lower electron mobilities. This condition is undesirable, but the damage can be repaired by an annealing step, which involves heating the substrate to relatively low temperatures, usually 400°C to 800°C, for 15–30 min. This process provides the energy that the silicon lattice requires to rearrange and mend itself. Another important function of annealing is to allow the dopant to move from the interstitial to substitutional sites (See Fig. 3.9), where they are electrically active.

13.10 Metallization and Testing

Producing a complete and functional integrated circuit requires that devices be interconnected; this task must take place at a number of levels, as illustrated in Fig. 13.25. **Interconnections** are made of metals that exhibit low electrical resistance and good adhesion to dielectric surfaces. Aluminum and aluminum-copper alloys remain the most commonly used materials for this purpose in VLSI (*very large scale integration*) technology. However, as device dimensions continue to shrink, electromigration has become more of a concern with aluminum interconnects.

Electromigration is the process by which metal atoms are physically moved by the impact of drifting electrons under high-current conditions; low-melting-point metals such as aluminum are especially prone to electromigration. In extreme cases, this condition can lead to severed and/or shorted metal lines. Solutions to this problem include the addition of sandwiched metal layers, such as tungsten and titanium, and the use of pure copper, which displays lower resistivity and significantly less electromigration than aluminum.

Metals are deposited by standard techniques and interconnection patterns are generated by lithographic and etching processes. The exception is copper, which is used for high performance ICs as an interconnect material. Because copper is difficult to pattern, an approach called **additive patterning** or the *Damascene* process is used, where the underlying semiconductor is patterned to produce open trenches. Copper is applied to the surface, filling the trenches and establishing electrical connections; the excess copper is removed in planarization.

Modern ICs typically have one to thirteen layers of metallization, each layer of metal being insulated by a dielectric, either silicon oxide or borophosphosilicate glass. **Planarization** (producing a planar surface) of

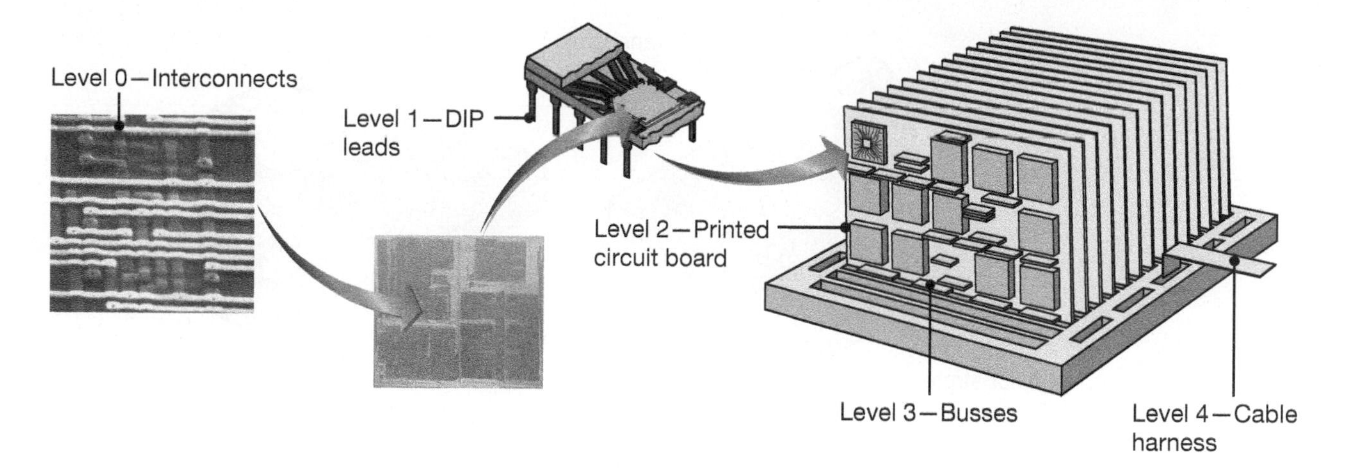

Level	Element example	Interconnection method
Level 0	Transistor within an IC	IC metallization
Level 1	ICs, other discrete components	Package leads or module interconnections
Level 2	IC packages	Printed circuit board
Level 3	Printed circuit boards	Connectors (busses)
Level 4	Chassis or box	Connectors/cable harnesses
Level 5	System, e.g., computer	

FIGURE 13.25 Connections between elements in the hierarchy for integrated circuits.

these interlayer dielectrics is critical to reducing metal shorts and linewidth variation of the interconnect. A common method for achieving a planar surface has been a uniform oxide etch process that produces a smooth surface on the dielectric layer.

Superior planar surfaces for high-density interconnects are produced through **chemical-mechanical polishing** (Section 9.7), also called *chemical-mechanical planarization*. A typical CMP process combines an abrasive medium with a polishing compound or slurry and can polish a wafer to within 0.03 μm of being perfectly flat, with a R_q roughness (Section 4.3) on the order of 0.1 nm for a new, bare silicon wafer.

Different layers of metal are connected together by **vias**, and access to the devices on the substrate is achieved through **contacts** (Fig. 13.26). As devices have become smaller and faster the size and speed of some chips have become dominated by the resistance of the metallization process itself and by the capacitance of the dielectric and the transistor gate. Wafer processing is completed upon application of a *passivation layer*, usually silicon

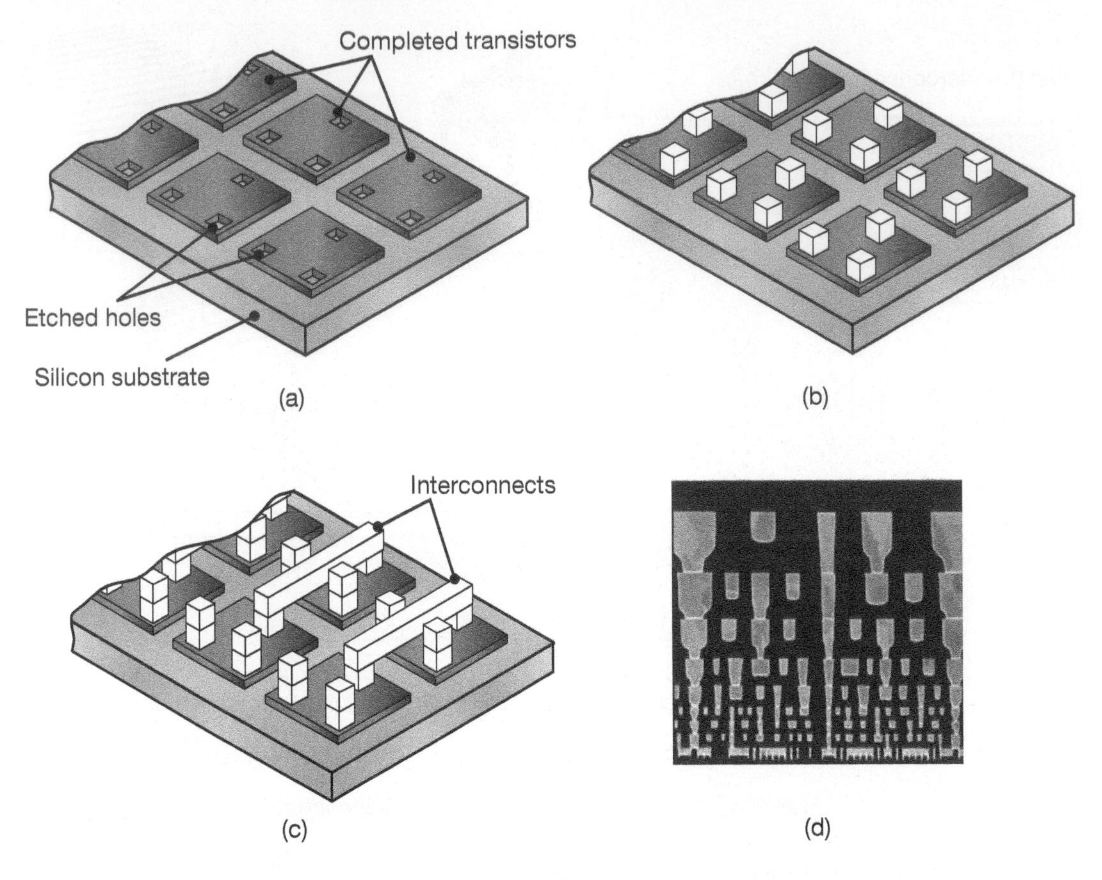

FIGURE 13.26 Production of metal interconnects. (a) Wafer with fabricated transistors (See Fig. 13.21). (b) First metallization layer, produced by electroplating copper. The dielectric has been removed for clarity. Note that the copper extends into the holes in the transistors to produce an electrical connection. (c) After the second layer, showing an interconnect between transistors. Over 20 layers can be required to fully interconnect an IC, although most applications require less than thirteen. (d) Cross section of a nine-layer interconnect structure; each layer is between 0.25 and 1 μm thick.
Source: (d) Courtesy of Intel Corporation.

nitride (Si_3N_4), which acts as an ion barrier for sodium ions and also provides excellent scratch resistance.

The next step in production is to test each of the individual circuits on the wafer (Fig. 13.27). Each chip, also referred to as a **die**, is tested with a computer-controlled platform which contains needlelike probes to access the bonding pads on the die. The probes are of two forms:

1. **Test patterns or structures.** The probe measures test structures, often outside of the active die, placed in the so-called *scribe line* (the vacant space between dies). These structures consist of transistors and interconnect structures that measure various quantities such as resistivity, contact resistance, and electromigration.
2. **Direct probe.** This approach uses 100% testing on the bond pads of each die.

FIGURE 13.27 A probe (top center) checking for defects in a wafer; an ink mark is placed on each defective die. *Source:* Courtesy of Intel Corporation.

The platform indexes across the wafer, and using computer-generated timing wave forms, tests whether each circuit functions properly. If a chip is defective, it is marked with a drop of ink. Up to one third of the cost of a micromachined part can be incurred during this testing.

After the wafer-level testing is completed, *back grinding* may be done to remove a large amount of the original substrate on the opposite side of the wafer from the fabricated integrated circuit. Final die thickness depends on the packaging requirement, but anywhere from 25 to 75% of the wafer thickness may be removed. After back grinding, each die is separated from the wafer. Diamond sawing is a commonly used separation technique and results in very straight edges, with minimal chipping and cracking damage. The chips are then sorted, with the inked dice discarded.

13.11 Wire Bonding and Packaging

To ensure reliability, the working dice must be attached to a rugged foundation. One simple method is to *bond* a die to its packaging material with an *epoxy cement* (Section 12.15). Another method uses a *eutectic bond*, made by heating metal-alloy systems; one widely used mixture is 96.4% gold and 3.6% silicon, which has a eutectic point (See Fig. 5.4) at 370°C.

QR Code 13.1 Wire bonding demonstration. *Source:* Courtesy of Palomar Technologies.

Once the chip has been bonded to its substrate, it must be electrically connected to the package leads. This task is accomplished by **wire bonding** very thin (25 μm diameter) gold wires either from the package leads to bonding pads located around the perimeter or down the center of the

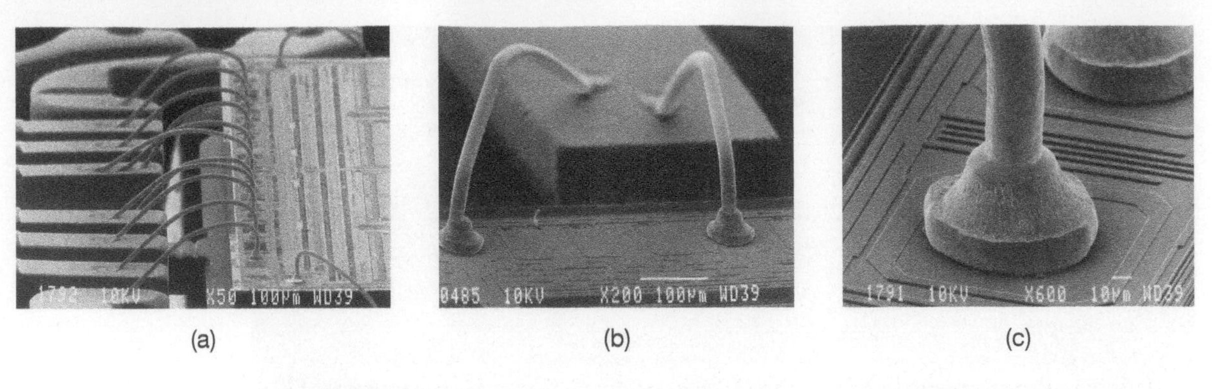

FIGURE 13.28 (a) SEM photograph of wire bonds connecting package leads (left-hand side) to die bonding pads. (b) and (c) Detailed views of (a). *Source:* Courtesy of Micron Technology, Inc.

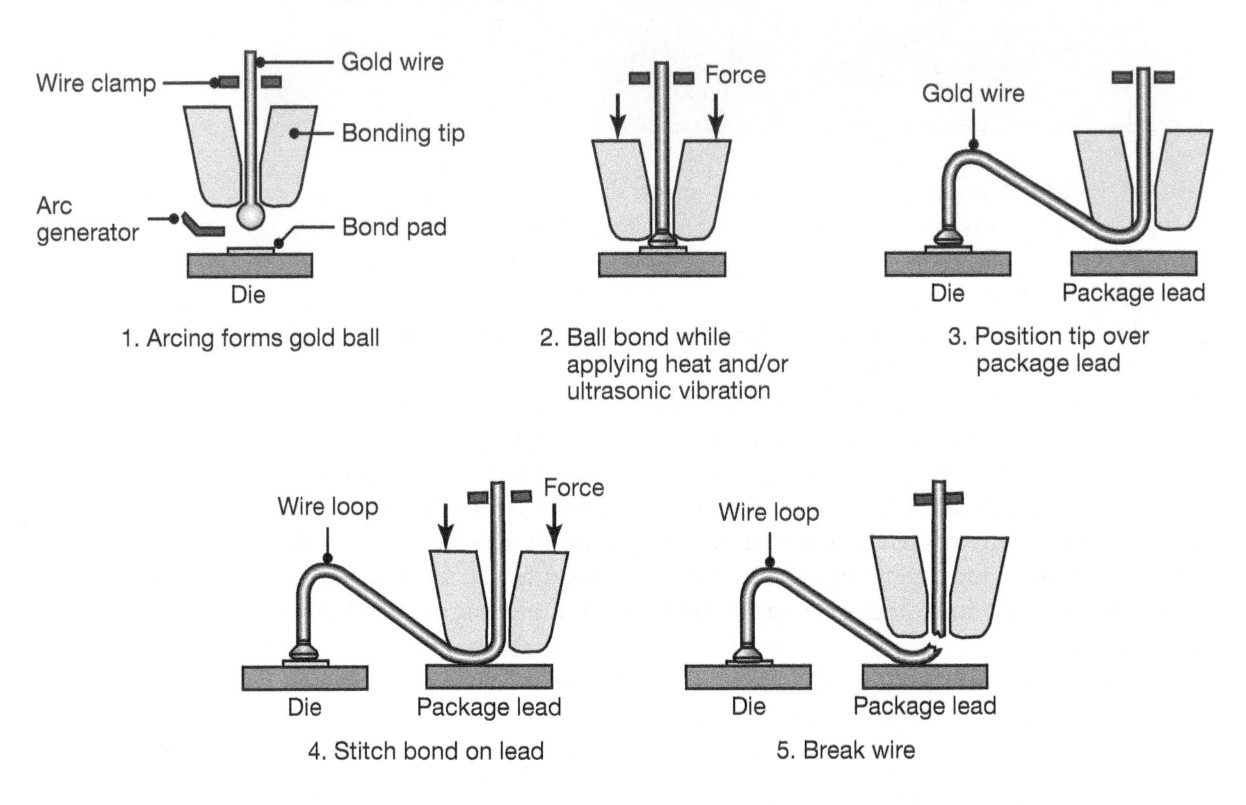

FIGURE 13.29 Schematic illustration of the thermosonic ball and stitch process. *Source:* After N. Maluf.

die (Fig. 13.28a). The bonding pads on the die are typically 50 μm or more per side, and the bond wires are attached using thermocompression, ultrasonic, or thermosonic techniques (Fig. 13.29).

The connected circuit is now ready for final **packaging**. The packaging operation largely determines the overall cost of each completed IC, since the circuits are mass produced on the wafer, but are then packaged individually. Packages are available in a variety of styles (Table 13.4),

TABLE 13.4 Summary of molded-plastic IC packages.

Package	Abbreviation	Pins Min.	Pins Max.	Description
Through-hole mount				
Dual in-line	DIP	8	64	Two in-line rows of leads
Single in-line	SIP	11	40	One in-line row of leads
Zigzag in-line	ZIP	16	40	Two rows with staggered leads
Quad in-line package	QUIP	16	64	Four in-line rows of staggered leads
Surface mount				
Small-outline IC	SOIC	8	28	Small package with leads on two sides
Thin small-outline package	TSOP	26	70	Thin version of SOIC
Small-outline J-lead	SOJ	24	32	Same as SOIC, with leads in a J-shape
Plastic leaded chip carrier	PLCC	18	84	J-shaped leads on four sides
Thin quad flat pack	TQFP	32	256	Wide but thin package with leads on four sides

and selection of the appropriate one must take into account operating requirements. These include consideration of chip size, the number of external leads, operating environment, heat dissipation, and power requirements. ICs that are used for military and industrial applications, for example, require packages with particularly high strength, toughness, hermeticity, and high-temperature resistance.

Packages are produced from polymers, metals, or ceramics. Metal containers are produced from alloys such as Kovar (an iron-cobalt-nickel alloy with a low coefficient of thermal expansion; Section 3.9.5) and provide a hermetic seal and good thermal conductivity, but are limited in the number of leads that can be accommodated. Ceramic packages are usually produced from Al_2O_3, are hermetic, and have good thermal conductivity, with higher lead counts than metal packages; they are, however, more expensive than metal packages. Plastic packages are inexpensive, with high lead counts, but they have high thermal resistance and are not hermetic.

A common style of packaging is the **dual in-line package** (DIP), shown schematically in Fig. 13.30a. Characterized by low cost (if using plastic) and ease of handling, DIP packages are made of thermoplastic, epoxy, or ceramic, and they can have from 2 to 500 external leads. Ceramic packages are designed for use over a broader temperature range, for high-performance, and military applications; they cost considerably more than plastic packages. A flat ceramic package is shown in Fig. 13.30b, in which the package and all the leads are in the same plane. Because this package style does not offer the ease of handling or the modular design of the DIP package, it is usually affixed permanently to a multiple-level circuit board in which the low profile of the flat pack is essential.

Surface-mount packages are the standard for modern integrated circuits. In some common examples, shown in Fig. 13.30c, it can be noted that the main difference among them is in the shape of the connectors. The DIP connection to the surface board is by way of prongs that are inserted

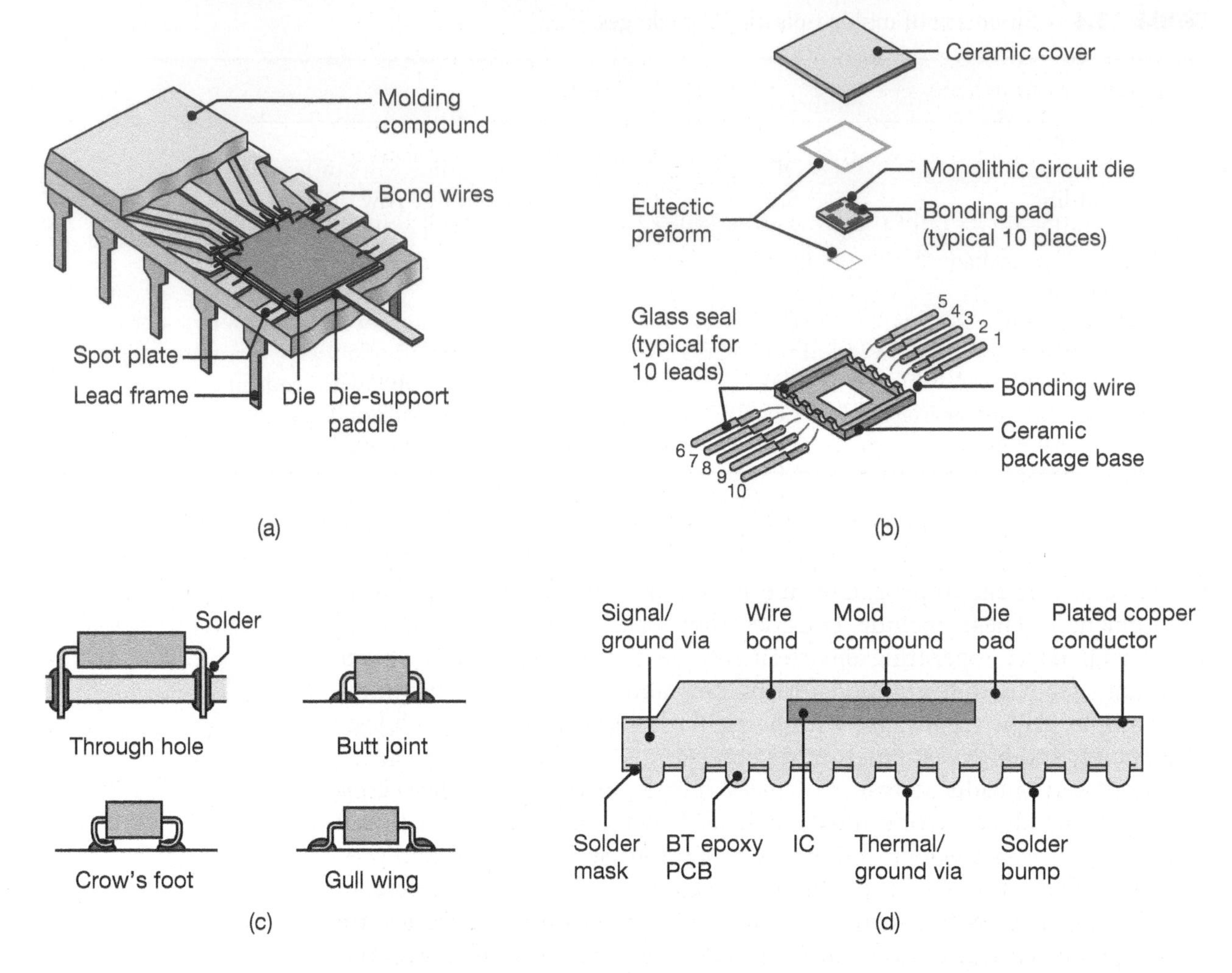

FIGURE 13.30 Schematic illustration of various IC packages. (a) dual in-line (DIP); (b) ceramic flat pack; (c) common surface-mount configurations; and (d) ball-grid array (BGA). *Source:* After R.C. Jaeger, A.B. Glaser, and G.E. Subak-Sharpe.

into corresponding holes, whereas a surface mount is soldered onto specially fabricated pad or land designs. Package size and land layouts are selected from standard patterns and usually require adhesive bonding of the package to the board, followed by wave soldering of the connections (Section 12.14.3).

Faster and more versatile chips require increasingly closely-spaced connections. **Pin-grid arrays** (PGAs) use tightly packed pins that connect by way of through-holes onto printed circuit boards. However, PGAs and other in-line and surface-mount packages are extremely susceptible to plastic deformation of the wires and legs, especially with small-diameter, closely-spaced wires. One way to achieve tight spacing and avoid the difficulties of slender connections is through **ball-grid arrays** (BGAs), shown in Fig. 13.30d. Such arrays have a plated solder coating on a number of closely spaced metal balls on the underside of the package. The spacing between the balls can be as small as 50 μm, but more commonly, the spacing is standardized as 1.0 mm, 1.27 mm, or 1.5 mm.

BGAs can be designed with over 1000 connections, but such high numbers of connections are extremely rare; usually, 200–300 connections are sufficient for demanding applications. By using reflow soldering, the solder serves to center the BGAs by surface tension, resulting in well-defined electrical connections for each ball.

Chip on board. *Chip on board* (COB) designs refer to the direct placement of chips onto an adhesive layer on a circuit board. Electrical connections are then made by wire bonding the chips directly to the pads on the circuit board. After wire bonding, final encapsulation with an epoxy is necessary, not only to attach the IC package more securely to the printed circuit board but also to transfer heat evenly during its operation.

Flip-chip on board. The flip-chip on board (FOB) technology, illustrated in Fig. 13.31, involves the direct placement of a chip with solder bumps onto an array of pads on the circuit board. The main advantage to flip chips and ball-grid array packages, is that the space around the package, normally reserved for bond pads, is saved; Thus, a higher level of miniaturization can be achieved.

System in package. A trend that allows for more compact devices involves incorporating more than one integrated circuit into a package. Figure 13.32 illustrates the major categories of *silicon in package* (SiP) designs. Although these packages can be integrated horizontally, vertical integration, through stacked or embedded structures (Figs. 13.32b and 13.32c), has the advantage of achieving performance increases over conventional packages. These benefits have been described as "more than Moore," referring to the famous "Moore's Law" that professed that the number of chips per area would double annually. SiPs also have other advantages, such as (a) they present reduced size and less noise; (b) cross talk between chips can be better isolated; and (c) individual chips can be upgraded more easily. On the other hand, these packages are more complex, require higher power density and associated heat extraction, and are more expensive than conventional packages.

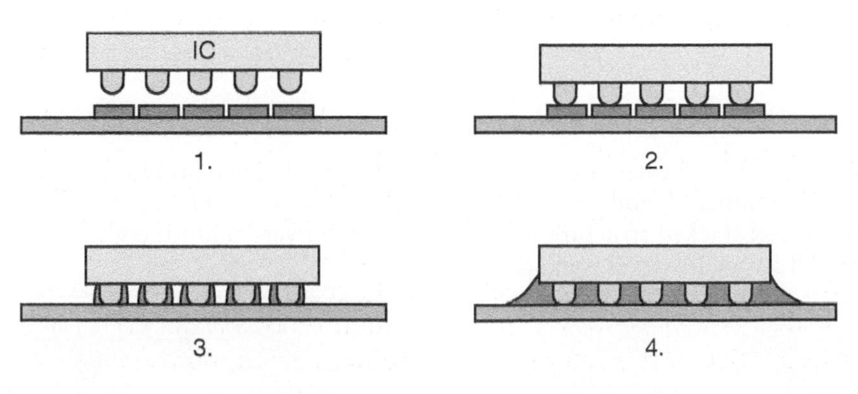

FIGURE 13.31 Illustration of flip-chip technology. Flip-chip package with 1. solder-plated metal balls and pads on the printed circuit board; 2. flux application and placement; 3. reflow soldering; and 4. encapsulation.

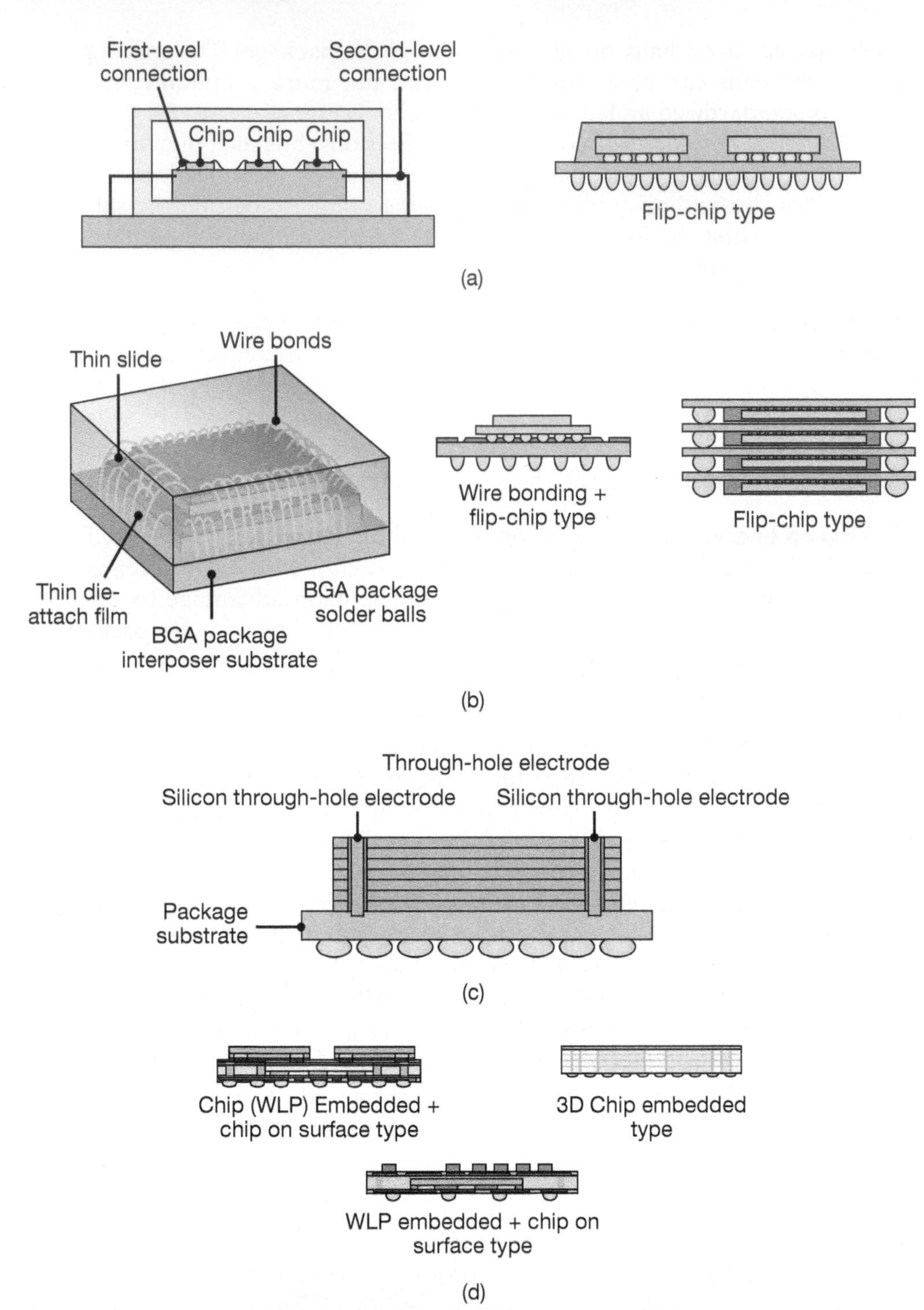

FIGURE 13.32 Major categories of system-in-package designs. (a) Horizontal placement, or multichip modules (MCMs); (b) interposer-type stacked structure; (c) interposerless stacked structure with through-silicon vias; and (d) embedded structure. WLP = wafer level package.

SiP packages can, however, be made very simple by incorporating more than one chip inside a single package, as shown in Fig. 13.32a. To preserve area on a circuit board, chips and/or flip-chips can be stacked and bonded to a circuit board to produce three-dimensional integrated circuits, as illustrated in Fig. 13.32b. Here, an interposing layer, commonly an adhesive, separates the chips and electrically-isolates adjacent layers. An alternative is to employ a so-called *interposerless* structure, using **through-silicon vias** (TSVs) instead of wire bonding, to provide electrical connections to all layers. TSVs are sometimes considered a packaging feature, but it has been noted that this is perhaps a case of 3D integration of a wafer, as shown in Fig. 13.32c.

13.12 Yield and Reliability of Chips

Yield is defined as the ratio of functional chips to the total number of chips produced. The overall yield of the total IC manufacturing process is the product of the wafer yield, bonding yield, packaging yield, and test yield. This value can range from only a few percent, for new processes, to above 90%, for mature manufacturing lines. Most loss of yield occurs during wafer processing due to its complex nature; in this stage, wafers are commonly separated into regions of good and bad chips. Failures at this stage can arise from point defects (such as oxide pinholes), film contamination, or metal particles, as well as from area defects, such as uneven film deposition or etch nonuniformity.

A major concern about completed ICs is their **reliability** and **failure rate.** Since no device has an infinite lifetime, statistical methods are used to characterize the expected lifetimes and failure rates of microelectronic devices (see also *Six Sigma*, Section 4.9.1). The unit for failure rate is the FIT (*failure in time*), defined as the number of failures per one billion device-hours. However, complete systems may have millions of devices, so the overall failure rate in entire systems is correspondingly higher; failure rates higher than 100 FIT are generally unacceptable.

Equally important in failure analysis is the determination of the failure mechanism, that is, the actual incident or component that causes the device to fail. Common failures due to processing involve (a) diffusion regions, leading to nonuniform current flow and junction breakdown; (b) oxide layers (dielectric breakdown and accumulation of surface charge); (c) lithography (uneven definition of features and mask misalignment); and (d) metal layers (poor contact and electromigration, resulting from high current densities). Other failures can originate from improper chip mounting, degradation of wire bonds, and loss of package hermeticity. Wire-bonding and metallization failures account for over one-half of all integrated-circuit failures.

Because device lifetimes are very long (10 years or more), it is impractical to study device failure under normal operating conditions. One method of studying failures efficiently is by **accelerated life testing**, which involves accelerating the conditions whose effects are known to cause device breakdown. Cyclic variations in temperature, humidity, voltage, and current are used to stress the components. Statistical data taken from these tests are then used to predict device failure modes and device life under normal operating conditions. Chip mounting and packaging are subjected to cyclical temperature variations.

13.13 Printed Circuit Boards

Packaged integrated circuits are seldom used alone; they are usually combined with other ICs to serve as building blocks of a yet larger system. A *printed circuit board* (PCB) is the *substrate* for the final interconnections among all completed chips, and serves as the communication link between the busses to other PCBs and the microelectronic circuitry within each packaged IC (See Fig. 13.25). In addition to the ICs, circuit boards also usually contain discrete circuit components, such as resistors and capacitors, that would (a) take up too much "real estate" on the limited silicon surface; (b) have special power dissipation requirements; or (c) cannot be implemented on a chip. Other common discrete components include inductors, which cannot be integrated onto the silicon surface, and high-performance transistors, large capacitors, precision resistors, and crystals for frequency control.

A printed circuit board is basically a resin material, containing several layers of copper foil (Fig. 13.33). *Single-sided* PCBs have copper tracks on only one side of an insulating substrate, while *double-sided* boards have copper tracks on both sides. Multilayer boards also can be constructed from alternating copper and insulator layers. Single-sided boards are the simplest form of circuit board; double-sided boards usually have locations where electrical connectivity is established between the features on both

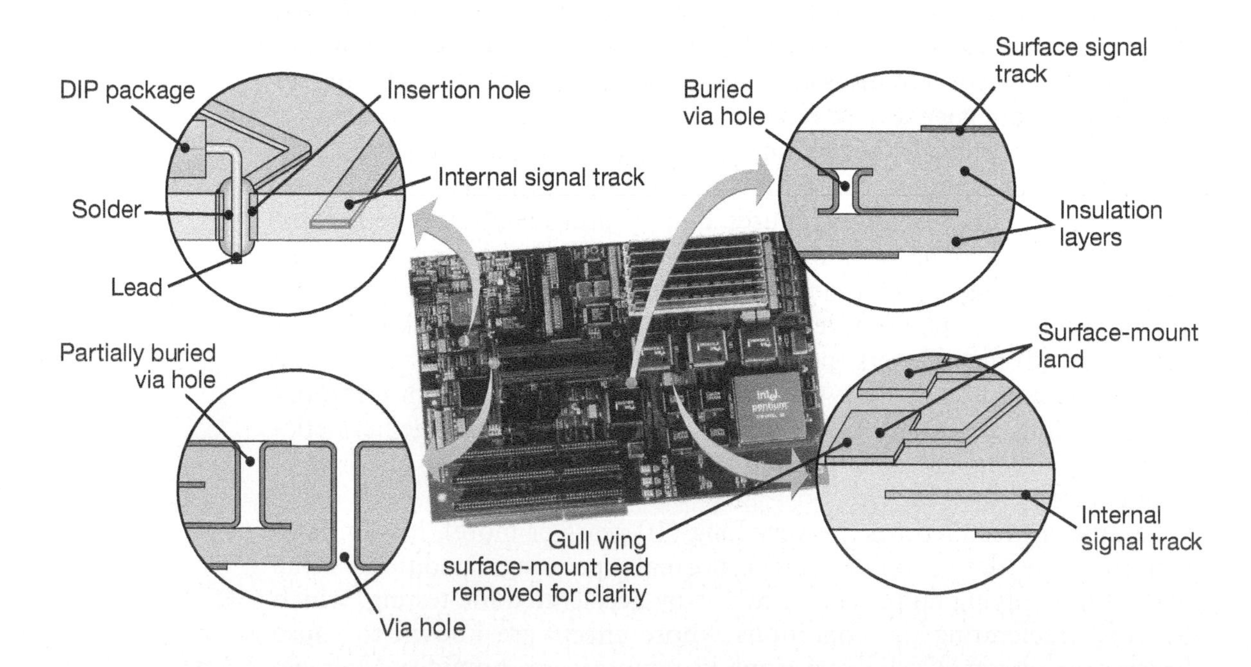

FIGURE 13.33 Printed circuit board structures and design features.

sides of the board. This structure is accomplished with *vias*, as shown in Fig. 13.33. Multilayer boards can have partial, buried, or through-hole vias to allow for flexible PCBs. Double and multilayer boards have advantage in that IC packages can be bonded to both sides of the board, allowing more compact designs.

The insulating material of the board is usually an epoxy resin 0.25 to 3 mm thick, reinforced with an epoxy/glass fiber, referred to as E-glass (see Section 10.9.2). They are produced by impregnating sheets of glass fiber with epoxy and then pressing the layers together between hot plates or rolls. The heat and pressure cure the board, resulting in a stiff and strong basis for the printed circuit boards. Boards are sheared to a desired size, and approximately 3-mm-diameter locating holes are then drilled or punched into the board corners to permit alignment and proper location of the board within chip-insertion machines. Holes for vias and connections are punched or produced through CNC drilling (Section 14.3); stacks of boards can be drilled simultaneously to increase production rates.

The conductive patterns on circuit boards are defined by lithography, although originally they were produced through screen-printing technologies, hence the terms *printed circuit board* and *printed wiring board* (PWB). In the *subtractive method*, a copper foil is first bonded to the circuit board; the desired pattern on the board is then defined by a positive mask developed through photolithography, and the remaining copper is removed through wet etching. In the *additive method*, a negative mask is placed directly onto an insulator substrate to define the desired shape. Electroless plating and electroplating of copper (Section 4.5.1) serve to define the connections, tracks, and lands on the circuit board.

The ICs and other discrete components are then fastened to the board by soldering. This procedure is the final step in making the integrated circuits (and the microelectronic devices they contain) accessible. *Wave soldering* and *reflow paste soldering* (Section 12.14.3) are the preferred methods of soldering ICs onto circuit boards.

Basic design considerations in laying out PCBs are:

1. Wave soldering should be used only on one side of the board, thus all through-hole mounted components should be inserted from the same side of the board. Surface-mount devices placed on the insertion side of the board must be reflow soldered in place, because this side is not exposed to the solder. Surface-mount devices on the leg side can be wave soldered.
2. To allow good solder flow in wave soldering, IC packages should carefully be laid out on the printed circuit board. Inserting the packages in the same direction is advantageous for automated placing, whereas random orientations can cause difficulties in flow of solder across all of the connections.
3. The spacing of ICs is determined mainly by the need to remove heat during operation. Sufficient clearance between packages and adjacent boards is required to allow forced air flow and heat convection.
4. There should be sufficient space around each IC package to allow for rework and repair without disturbing adjacent devices.

13.14 Roll-to-Roll Printing of Flexible Electronics

QR Code 13.2 Roll-to-roll manufacturing.
Source: Courtesy of Oak Ridge National Laboratory.

QR Code 13.3 Flexible Hybrid Electronics Innovation Institute.
Source: Courtesy of the US Department of Defense.

A relatively new development for electronic device manufacturing is the use of **roll-to-roll** printing, also known as *R2R processing*. This approach uses various rotary printing processes to transfer functional ink to a sheet or continuous feed of a flexible substrate (usually metal, paper or plastic), which is then wound into a roll. While the resolution that can be achieved in photolithography is much better than that for R2R, the main advantage of R2R is that large areas can be printed quickly and relatively inexpensively. For example, printing speeds of 10 m^2/s are routine with R2R, while soft lithography generally reaches speeds around 1×10^{-4} m^2/s. R2R has been widely applied in flexible solar panels, *organic light emitting diodes* (OLEDs) used in curved displays, and radio-frequency identification (RFID) circuits, where the large aerials can be printed at the same time as the circuit. There are numerous applications under development, including fuel cells, advanced batteries, multilayer capacitors, X-ray and other radiation detectors, and membranes for chemical processing industries.

Roll-to-roll processing requires the use of special inks, which can be either dissolved solutions of ink in a carrier fluid, an emulsion of one liquid suspended in the other in the form of micro- or nanoscale droplets, and suspensions of particles in a carrier fluid. There are several different inks that have been developed, of which the following are particularly important:

- **Silver nanoparticles** can be suspended in water or other carriers. Deposited silver is useful as an electrode material. Nanoparticle suspensions of gold can also be inkjetted and used as a conductor.
- **Indium tin oxide** (ITO) is a conductor that is transparent in thin layers.
- Organic polymers can be produced with a wide variety of attractive properties, including semiconducting, conducting, photovoltaic, and electroluminescent formulations. For example, conductive polymers are based on poly(3, 4-ethylene dioxitiophene) or PEDOT, while semiconductors are based on this polymer doped with styrene sulfonate or PEDOT:PSS. Special formulations are available for each of these functions.
- Inorganic semiconductors, such as *copper indium gallium selenide* (*CIGS*), can be deposited through roll-to-roll techniques, and is a common material used in solar cells.
- *Electroluminescent* materials can be produced from inorganic materials, such as copper-doped phosphor on a plastic film.

Substrates can be glass or silicon (for inkjet printing), but they can also be polymer films, often poly(ethylene terephthalate) (PET) for roll-to-roll processing, or paper. A flexible substrate allows continuous printing, significantly increasing speed and lowering costs. Furthermore, flexible electronics enable designs such as flexible displays, solar panels that conform to an automobile roof, and wearable electronics.

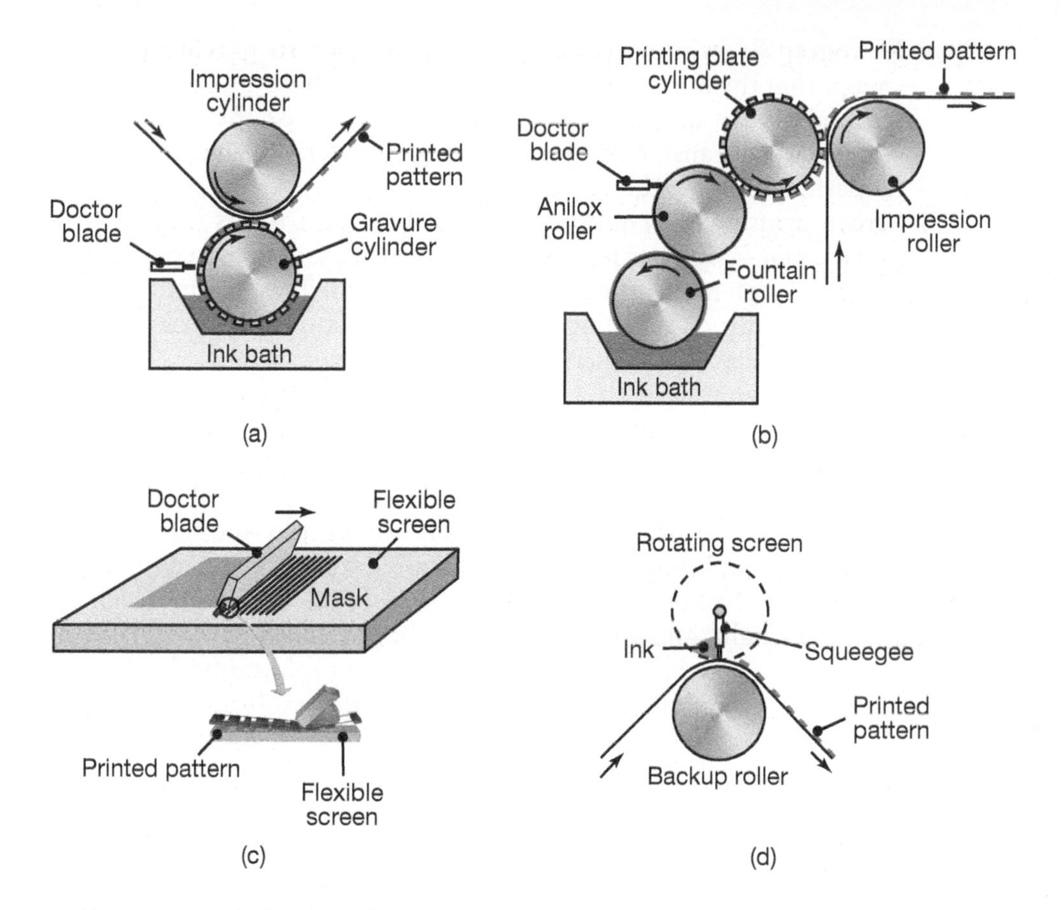

FIGURE 13.34 Schematic illustration of roll-to-roll processing approaches. (a) Gravure or Rotogravure, using an engraved roller or rotograve with a rubber doctor blade to transfer a desired ink pattern to the flexible web. (b) Flexography, where ink is transferred from an *anilox roller* (a hard cylinder, coated with a ceramic that has dimples or micro cavities embedded into its surface) to raised portions of a printing plate that is then transferred to a flexible substrate. (c) Flat bed screen printing (See also Fig. 12.49); and (d) Rotary screen printing.

The commonly used printing techniques in R2R are shown in Fig. 13.34, and include the following:

- **Inkjet printing** allows the deposition of a wide variety of inks as small droplets whose position can be precisely controlled.
- **Gravure**, or *rotogravure*, printing, is commonly used in the printing of magazines and catalogs. The technique requires the transfer of ink from cavities in the engraved or gravure cylinder to the web. Ink is continuously fed into the rotating gravure cylinder, using a doctor blade or squeegee to remove excess ink. The gravure cylinder can have varying depths to control the amount of deposited ink over the printed image.

- **Flexographic printing** (Fig. 13.34b) is similar to gravure printing, except that the ink is transferred to an intermediate anilox roller. This method allows uniform distribution of ink across the anilox roller, which can be transferred to the printing cylinder, which then transfers its pattern to the web.
- **Screen printing** or *screening* is available in two forms (Figs. 13.34c and d). *Flat bed screen printing* uses a moving squeegee to push an ink paste through a screen with a specific pattern. This technique allows the deposition of very thick layers, and is a preferred technique for producing electrodes where high conductivity is needed.
- **Rotary screen printing** is a continuous form of screen printing, where the ink is inside the rotating screen and is pushed outwards by a stationary squeegee.

Another important process in R2R is soft lithography and referred to as **self-aligned imprint lithography** (SAIL). Flexible stamps are used to deposit multiple layers of photopolymers (see also Section 10.12.1), which are cured with UV light, allowing the stamp to be peeled off easily. SAIL consists of deposition of base films, followed by multiple stamp layers, and completed by wet and dry etching to produce high-resolution patterns.

R2R involves printing layers one at a time, as in the approach for integrated circuits shown in Fig. 13.3, but the layers are printed on a continuous web at very high rates. Each layer requires its own printing steps, just as a color magazine requires layers of red, blue, yellow and black ink. Once a multiple-layer device has been printed, it is laminated to protect the circuit and to provide operational stability. Cold lamination involves application of a polymer film with a pressure-sensitive adhesive; hot melt lamination uses a material that becomes adhesive when heated. Other forms of lamination are also in use.

Roll-to-roll printing is not in common use but is being developed rapidly. It is expected that R2R will allow the mass production of inexpensive circuits, enabling such innovations as the Internet of Things (see Section 15.14.2).

13.15 Micromachining of MEMS Devices

The topics described thus far have dealt with the manufacture of integrated circuits and products that operate based purely on electrical or electronic principals. These processes also are suitable for manufacturing devices that incorporate mechanical elements or features as well.

The following types of devices can be made through the approach described in Fig. 13.3:

1. **Microelectronic devices.** These semiconductor-based devices often have the common characteristic of extreme miniaturization and use electrical principles in their design.
2. **Micromechanical devices.** This term refers to a product that is purely mechanical in nature, and has dimensions between atomic

length scales and a few millimeters, such as some very small gears and hinges.

3. **Microelectromechanical devices** are products that combine mechanical and electrical or electronic elements at very small length scales' most sensors are examples of microelectromechanical devices.
4. **Microelectromechanical systems** (MEMS). These are microelectromechanical devices that also incorporate an integrated electrical system in one product. Microelectromechanical systems are rare compared to microelectronic, micromechanical, or microelectromechanical devices, typical examples being airbag sensors and digital micromirror devices.

Microelectronic devices are semiconductor-based, whereas microelectromechanical devices and portions of MEMS do not have this material restriction. This characteristic allows the use of many more materials and the development of processes suitable for these materials. Regardless, silicon is the material often used because several highly advanced and reliable manufacturing processes have been developed for microelectronic applications.

The production of features from micrometers to millimeters in size is called **micromachining**. MEMS devices have been constructed from *polycrystalline silicon (polysilicon)* and *single-crystal silicon*, because the technologies for integrated-circuit manufacture are well developed and exploited for these devices, and other, new processes have been developed that are compatible with the existing processing steps. The use of anisotropic etching techniques allows the fabrication of devices with well-defined walls and high aspect ratios, and for this reason, some single-crystal silicon MEMS devices have been fabricated.

One of the recognized difficulties associated with using silicon for MEMS devices is the high adhesion encountered at small length scales and the associated rapid wear. Most commercial devices are designed to avoid friction by, for example, using flexing springs instead of bushings. However, this approach complicates designs and makes some MEMS devices unfeasible. Consequently, significant research is being conducted to identify materials and lubricants that provide reasonable life and performance.

Silicon carbide, diamond, and metals such as aluminum, tungsten, and nickel have been investigated as potential MEMS materials. In addition, lubrication remains a pressing concern. It is known that surrounding the MEMS device in a silicone oil, for example, practically eliminates adhesive wear (Section 4.4.2), but it also limits the performance of the device. Self-assembling layers of polymers are being investigated, as well as novel and new materials with self-lubricating characteristics. However, the tribology of MEMS devices remains a main technological barrier to expansion of their already widespread use.

MEMS and MEMS devices are rapidly advancing, and new processes or variations on existing processes are continually being developed. It has been suggested that MEMS technology can have widespread industrial applications; however, only a few industries have exploited MEMS thus far, such as the computer, medical, and automotive industries.

13.15.1 Bulk Micromachining

Until the early 1980s, bulk micromachining was the most common form of machining at micrometer scales, using orientation-dependent etches on single-crystal silicon (See Fig. 13.18b). The approach is based on etching down into a surface, stopping on certain crystal faces, the doped regions, and the etchable films to develop the required structure. As an example of this process, consider the fabrication of the silicon cantilever shown in Fig. 13.35. Using masking techniques, described in Section 13.7, a rectangular patch of the *n*-type silicon substrate is changed to *p*-type silicon through boron doping. Recall that etchants for orientation-dependent etching, such as potassium hydroxide, will not be able to etch heavily boron-doped silicon, hence this patch will not be etched.

A mask is then produced, such as with silicon nitride on silicon. When etched with potassium hydroxide, the undoped silicon will be removed rapidly, while the mask and the doped patch will essentially be unaffected. Etching progresses until the (111) planes are exposed in the *n*-type silicon substrate, and they undercut the patch, leaving a suspended cantilever, as shown in the figure.

13.15.2 Surface Micromachining

Although bulk micromachining is useful for producing simple shapes, it is restricted to single-crystal materials, because polycrystalline materials will not machine at different rates in different directions when using wet etchants. Many MEMS applications require the use of other materials, so that alternatives to bulk micromachining are needed. One such method is *surface micromachining*, the basic steps of which are illustrated in Fig. 13.36 for silicon devices. A spacer or sacrificial layer is deposited onto a silicon substrate coated with a thin dielectric layer (called an *isolation* or *buffer layer*).

Phosphosilicate glass, deposited by chemical vapor deposition, is the most common material for a spacer layer, because it etches very rapidly

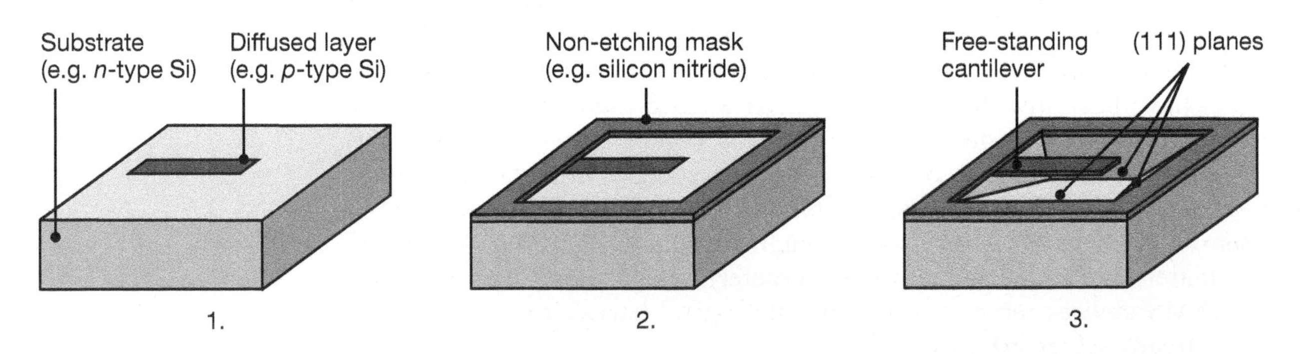

FIGURE 13.35 Schematic illustration of the steps in bulk micromachining. (1) Diffuse dopant in desired pattern, (2) deposit and pattern masking film, and (3) orientation-dependent etch, leaving behind a freestanding structure. *Source:* After K.R. Williams.

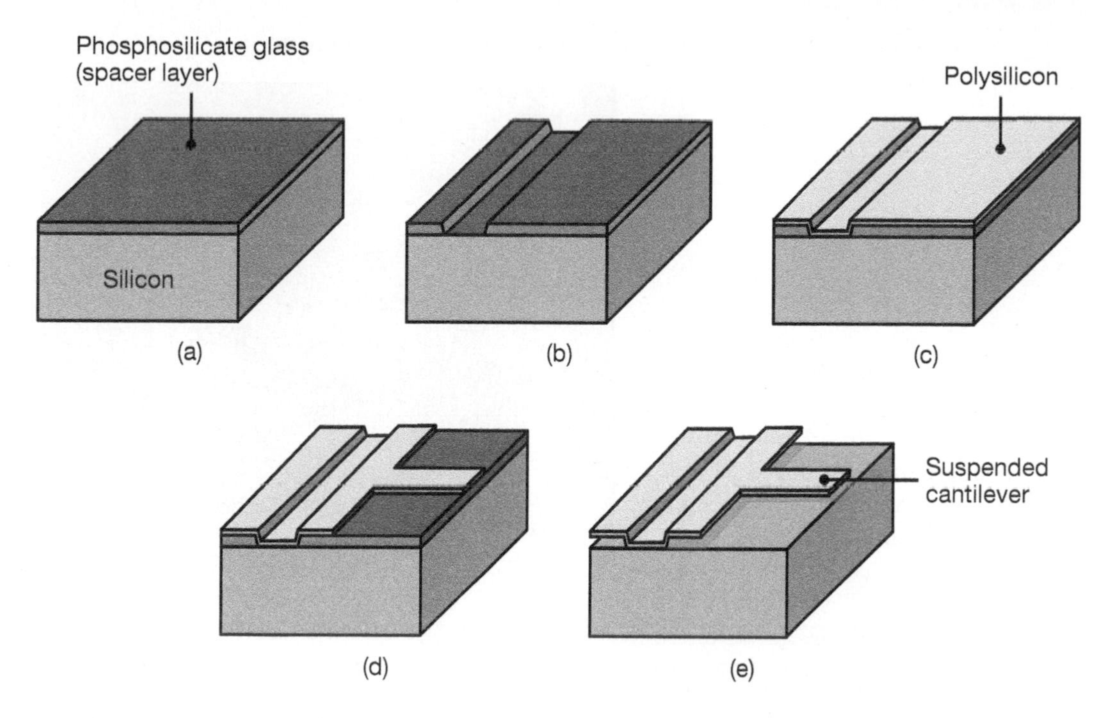

FIGURE 13.36 Schematic illustration of the steps in surface micromachining. (a) Deposition of a phosphosilicate glass (PSG) spacer layer; (b) etching of spacer layer; (c) deposition of polysilicon; (d) etching of polysilicon; and (e) selective wet etching of PSG, leaving the silicon substrate and the deposited polysilicon unaffected.

in hydrofluoric acid. Figure 13.36b shows the spacer layer after the application of masking and etching. At this stage, a structural thin film is deposited onto the spacer layer; the film can be polysilicon, metal, metal alloy, or a dielectric (Fig 13.36c). The structural film is then patterned, usually through dry etching in order to maintain vertical walls and tight dimensional tolerances. Wet etching of the sacrificial layer leaves a free-standing, three-dimensional structure, as shown in Fig. 13.36e. It should be noted that the wafer must be annealed to remove the residual stresses in the deposited metal before it is patterned, as otherwise the structural film will severely warp once the spacer layer is removed.

Figure 13.37 shows a microlamp that emits a white light when current is passed through it; this part was produced through a combination of surface and bulk micromachining. The top patterned layer is 2.2-μm thick plasma-etched tungsten, forming a meandering filament and bond pad. The rectangular overhang is dry-etched silicon nitride; the steeply sloped layer is wet-HF-etched phosphosilicate glass, and the substrate is silicon, which is orientation dependent etched (ODE).

The etchant used to remove the spacer layer must be carefully chosen. It must preferentially dissolve the spacer layer while leaving the dielectric, the silicon, and the structural film as intact as possible. With large features

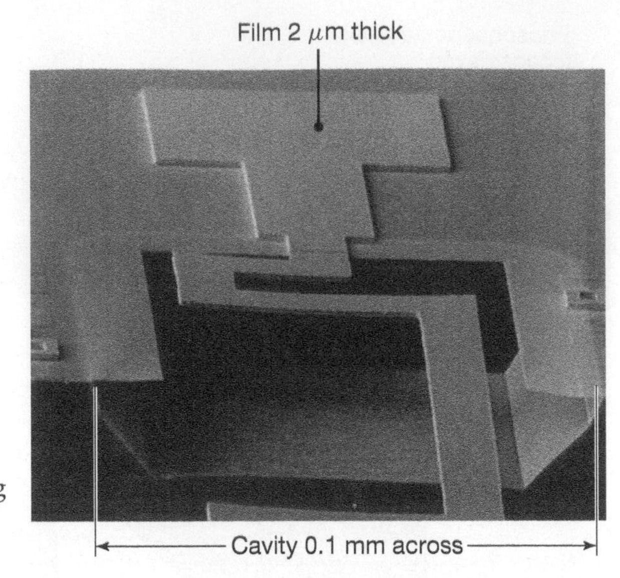

FIGURE 13.37 A microlamp produced by a combination of bulk and surface micromachining processes. *Source:* K.R. Williams, Agilent Technologies.

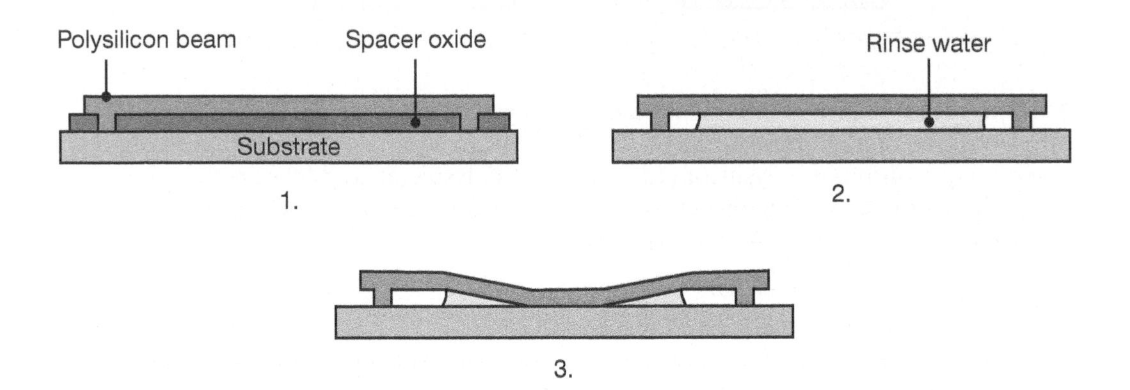

FIGURE 13.38 Stiction after wet etching. (1) Unreleased beam, (2) released beam before drying, and (3) released beam pulled to the surface by capillary forces during drying. Once contact is made, adhesive forces prevent the beam from returning to its original shape. *Source:* After B. Bhushan.

and narrow spacer layers, this task becomes very difficult to do and etching can take several hours. To reduce the etch time, additional etch holes can be designed into the microstructures to increase access of the etchant to the spacer layer.

Another difficulty that must be overcome in this operation is **stiction** after wet etching, described best by considering the situation illustrated in Fig. 13.38. After the spacer layer has been removed, the liquid etchant is dried from the wafer surface. A meniscus formed between the layers then results in capillary forces that can deform the film and causes the substrate to contact as the liquid evaporates. Since adhesion forces are more significant at small length scales, it is then possible for the film to permanently *stick* to the surface, whereby the desired three-dimensional features will not be produced.

EXAMPLE 13.3 Surface Micromachining of a Hinge for a Mirror Actuation System

Surface micromachining is a very common technology for the production of microelectromechanical systems; applications include accelerometers, pressure sensors, micropumps, micromotors, actuators, and microscopic locking mechanisms. These devices often require very large vertical walls, which cannot be directly manufactured because the high vertical structure is difficult to deposit. This difficulty is overcome by machining large flat structures horizontally, and then rotating or folding them into an upright position, as shown in Fig. 13.39.

Figure 13.39a shows a micromirror that has been inclined with respect to the surface on which it was manufactured. Such systems can be used for reflecting light that is oblique to a surface onto detectors or toward other sensors. It is apparent that a device that has such depth, and has the aspect ratio of the deployed mirror, is very difficult to machine directly. Instead, it is easier to surface micromachine the mirror along with a linear actuator, and then fold the mirror into a deployed position. In order to do so, special hinges, as shown in Fig. 13.39b, are integrated into the design.

Figure 13.40 shows the hinge during its manufacture through the following steps:

1. A 2-μm thick layer of phosphosilicate glass (PSG) is deposited onto the substrate material.
2. A 2-μm thick layer of polysilicon (Poly1 in Fig. 13.40a) is deposited onto the PSG, and patterned by photolithography. It is then dry etched to form the desired structural elements, including the hinge pins.
3. A second layer of sacrificial PSG, with a thickness of 0.5 μm, is deposited (Fig. 13.40b).
4. The connection locations are etched through both layers of PSG (Fig. 13.40c).
5. A second layer of polysilicon (Poly2 in Fig. 13.40d) is deposited, patterned, and etched.
6. The sacrificial layers of PSG are then removed through wet etching.

Hinges such as these have very high friction; consequently, if the mirrors, as shown, are manually and carefully manipulated with probe needles, they will remain in position. Often, such mirrors are combined with linear actuators as described here to precisely control their deployment.

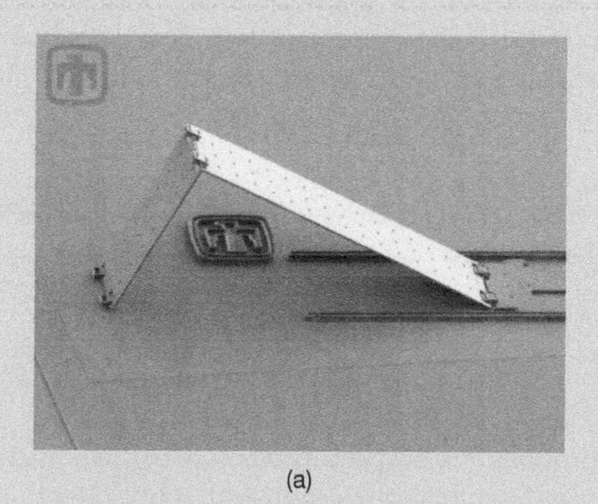

(a)

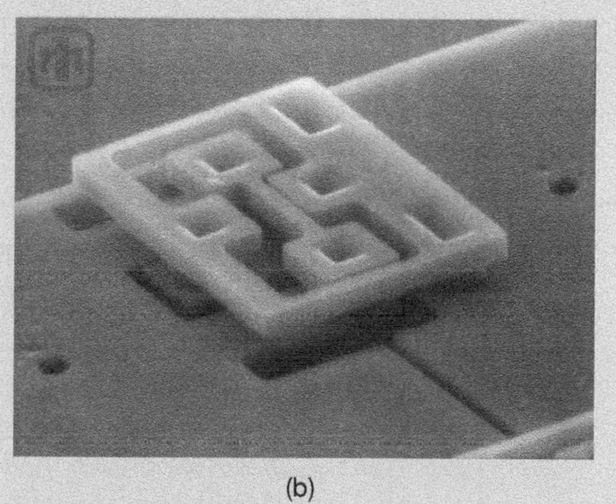

(b)

FIGURE 13.39 (a) SEM image of a deployed micromirror and (b) detail of the micromirror hinge. *Source:* Sandia National Laboratories.

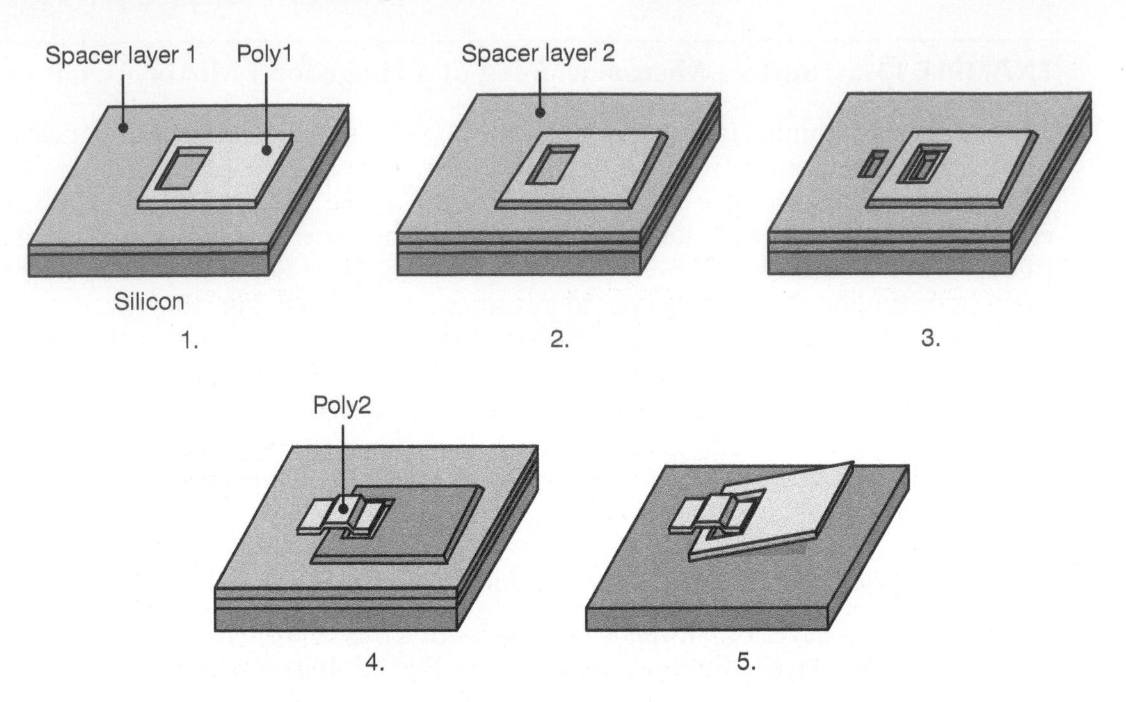

FIGURE 13.40 Schematic illustration of the steps in manufacturing a hinge. (1) Deposition of a phosphosilicate glass (PSG) spacer layer and polysilicon layer (See Fig. 13.36), (2) deposition of a second spacer layer, (3) selective etching of the PSG, (4) deposition of polysilicon to form a staple for the hinge, and (5) after selective wet etching of the PSG, the hinge can rotate.

SCREAM. Another approach for making very deep MEMS structures is the *single-crystal silicon reactive etching and metallization* (SCREAM) process, depicted in Fig. 13.41. In this technique, standard lithography and etching processes produce trenches 10–50 μm deep, which are then protected by a layer of chemically vapor-deposited silicon oxide. An anisotropic etch step removes the oxide only at the bottom of the trench, and the trench is then extended through dry etching. An isotropic etch, using sulfur hexafluoride (SF_6), laterally etches the exposed sidewalls at the bottom of the trench. This undercut, when it overlaps adjacent undercuts, releases the machined structures.

SIMPLE. An alternative to SCREAM is the *silicon micromachining by single-step plasma etching* (SIMPLE) technique, as depicted in Fig. 13.42. This technique uses a chlorine-gas-based plasma etch process that machines *p*-doped or lightly doped silicon anisotropically, but heavily *n*-doped silicon isotropically. A suspended MEMS device can thus be produced in a single plasma etching step, as shown in the figure.

Some of the concerns regarding the SIMPLE process are:

1. The oxide mask is machined, although at a slower rate, by the chlorine-gas plasma; therefore, relatively thick oxide masks are required.

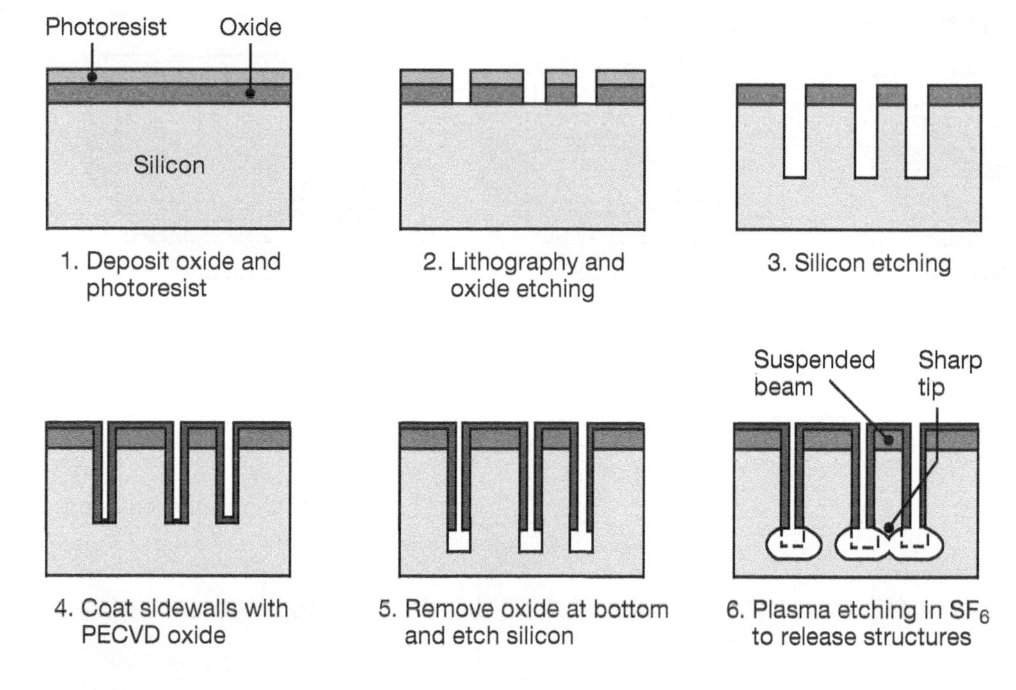

FIGURE 13.41 Steps involved in the SCREAM process. *Source:* After N. Maluf.

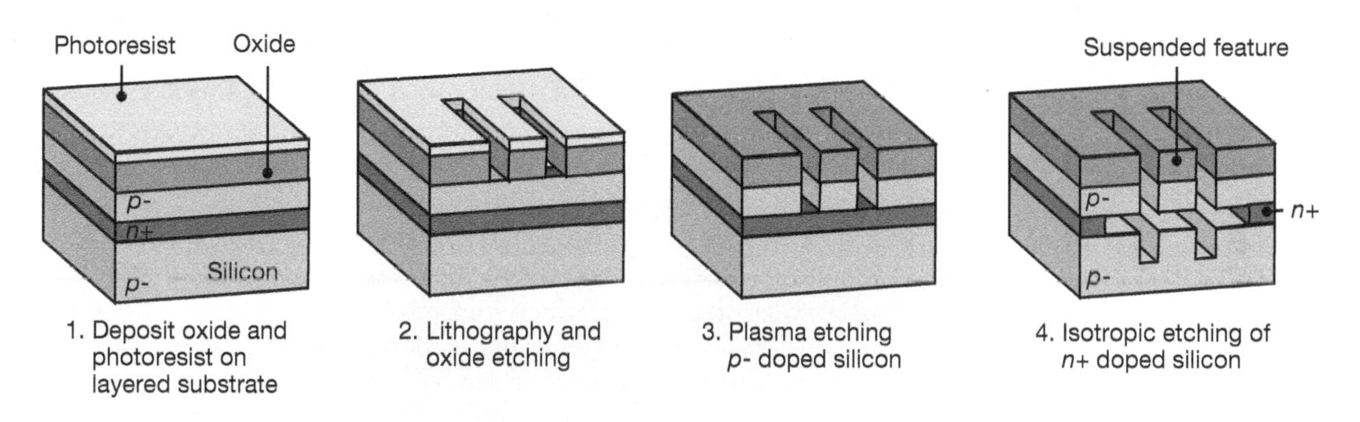

FIGURE 13.42 Schematic illustration of silicon micromachining by the single-step plasma etching (SIMPLE) process.

2. The isotropic etch rate is low, typically 50 nm/min; consequently, this process is very slow.
3. The layer beneath the structures will contain deep trenches, which may affect the motion of free-hanging structures.

Etching combined with diffusion bonding. The terms fusion bonding and diffusion bonding (Section 12.13) are interchangeable, but fusion bonding is the preferred term for MEMS applications. Very tall structures can be produced in single crystal silicon through a combination of *silicon fusion bonding and deep reactive ion etching* (SFB-DRIE), as illustrated in Fig. 13.43. First, a silicon wafer is prepared with an insulating oxide

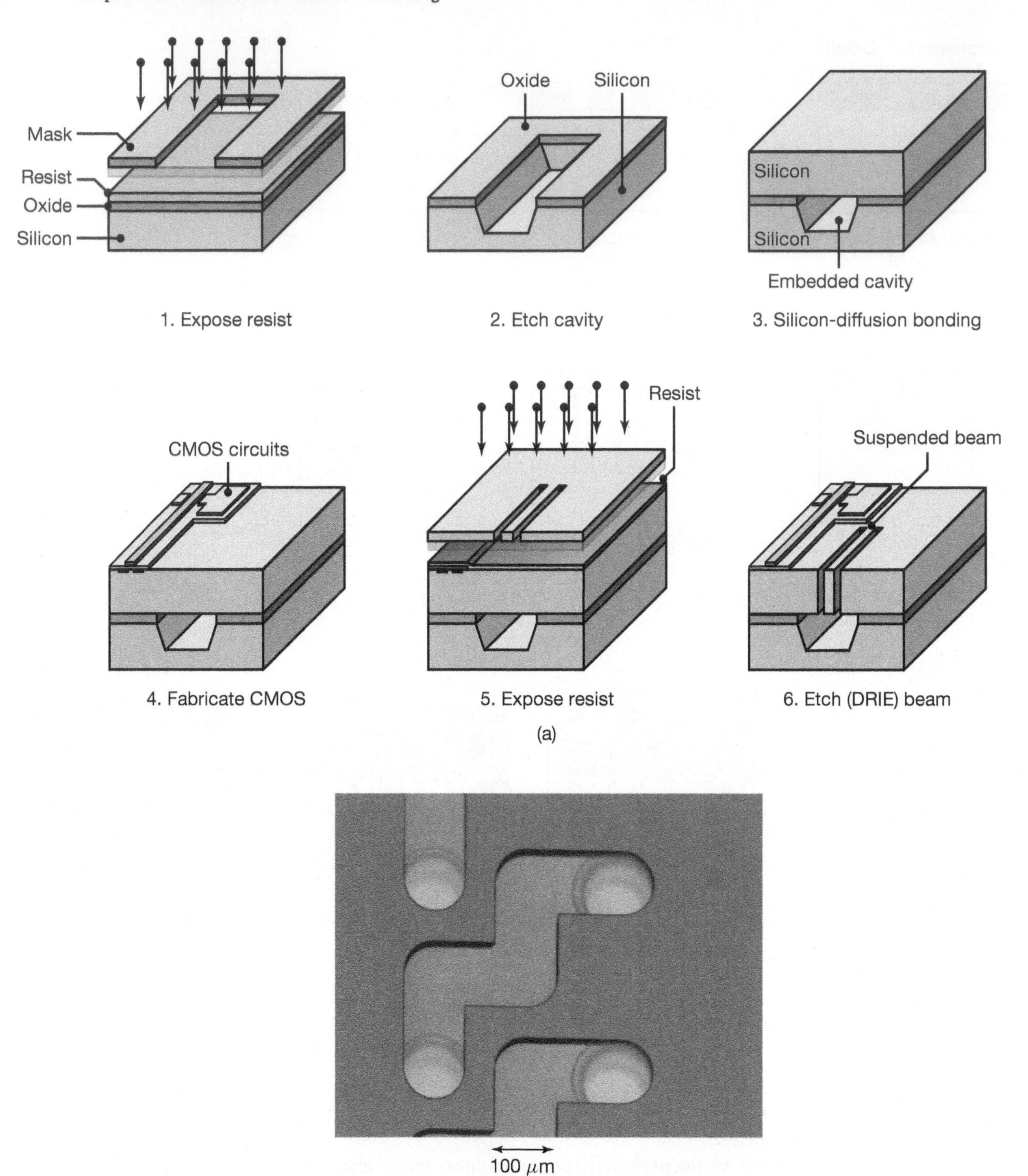

FIGURE 13.43 (a) Schematic illustration of silicon fusion bonding combined with deep reactive ion etching to produce large suspended cantilevers. *Source:* After N. Maluf. (b) A micro-fluid-flow device manufactured by applying the DRIE process to two separate wafers and then aligning and silicon fusion bonding them together. Afterward, a Pyrex layer (not shown) is anodically bonded over the top to provide a window for observing fluid flow. *Source:* After K.R. Williams.

layer, with the deep trench areas defined by a standard lithography procedure; this step is followed by conventional wet or dry etching to form a large cavity. A second layer of silicon is fusion bonded to this layer, which can then be ground and polished to the desired thickness, if necessary. At this stage, integrated circuitry is manufactured through the steps outlined in Fig. 13.3. A protective resist is applied and exposed, and the desired trenches are etched by deep reactive ion etching through to the cavity in the first layer of silicon.

EXAMPLE 13.4 Operation and Fabrication Sequence for a Thermal Ink-Jet Printer

Thermal ink-jet printers are among the most successful applications of MEMS to date, and the mechanisms are also used in applications as varied as binder-jet printing (Fig. 10.54) and production of flexible electronics (Section 13.14). These mechanisms operate by ejecting nano- or picoliters (10^{-12} liters) of ink from a nozzle toward paper. Ink-jet printers use a variety of designs, but silicon machining technology is most applicable to high-resolution printers. It should be noted that a resolution of 50 dots per mm requires a nozzle pitch of approximately 20 μm.

The operation of an ink-jet printer is illustrated in Fig. 13.44. When an ink droplet is to be generated and expelled, a tantalum resistor below a nozzle is heated. The resistor heats a thin film of ink so that a bubble forms within five microseconds. The bubble expands rapidly, with internal pressures reaching 1.4 MPa, and as a result, fluid is forced out of the nozzle;

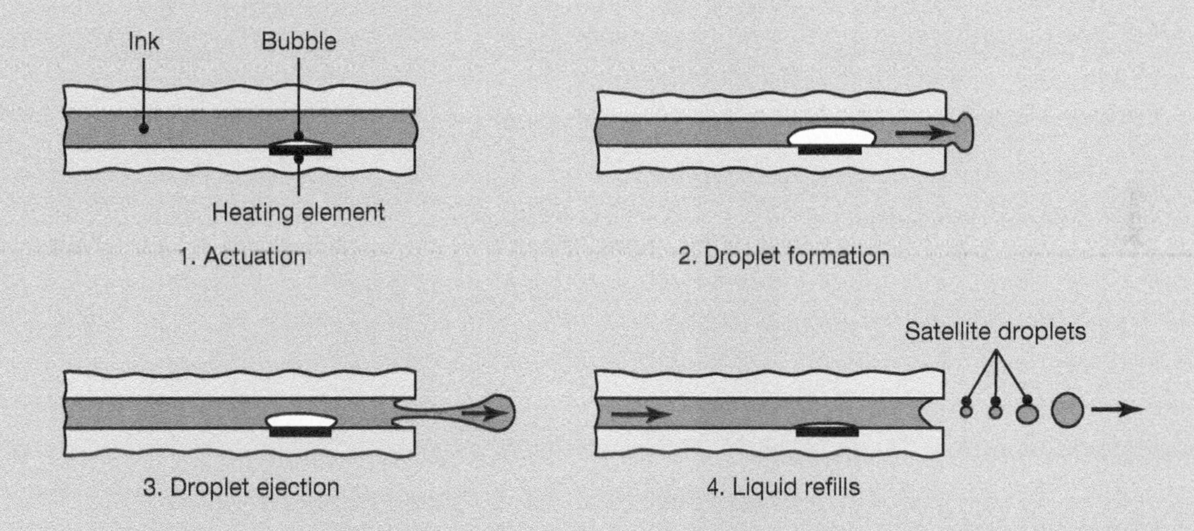

FIGURE 13.44 Sequence of operations of a thermal ink-jet printer. (1) Resistive heating element is turned on, rapidly vaporizing the ink and forming a bubble. (2) Within five microseconds, the bubble has expanded and displaced liquid ink from the nozzle. (3) Surface tension breaks the ink stream into a bubble, which is discharged at high velocity. The heating element is turned off at this time, so that the bubble collapses as heat is transferred to the surrounding ink. (4) Within 24 microseconds, an ink droplet (and any undesirable satellite droplets) are ejected, and surface tension of the ink draws more liquid from the reservoir. *Source:* After F.G. Tseng.

within 24 microseconds, the tail of the ink droplet separates because of surface tension. The heat source is then turned off, and the bubble collapses inside the nozzle. Within 50 microseconds, sufficient ink has been drawn from a reservoir into the nozzle to form the desired meniscus for the next droplet.

Traditional ink-jet printer heads were produced with electroformed nickel nozzles, fabricated separately from the integrated circuitry, and required a bonding operation to attach these two components. With increasing printer resolution, however, it is more difficult to bond the components with a tolerance under a few micrometers; for this reason, single-component (monolithic) fabrication is of interest.

The fabrication sequence for a monolithic ink-jet printer head is shown in Fig. 13.45. A silicon wafer is prepared and coated with a phosphosilicate-glass (PSG) pattern and low-stress silicon-nitride coating. The ink reservoir is obtained by isotropically etching the back side of the wafer, followed by PSG removal and then enlargement of the reservoir. The required CMOS (complementary metal-oxide-semiconductor) controlling circuitry is then produced (this step is not shown in Fig. 13.45), and a tantalum heater pad is deposited. The aluminum interconnection between the tantalum pad and the CMOS circuit is formed, and the nozzle is produced through laser machining. An array of such nozzles can be placed inside an ink-jet printing head, and resolutions of 100 dots per mm or higher can be achieved.

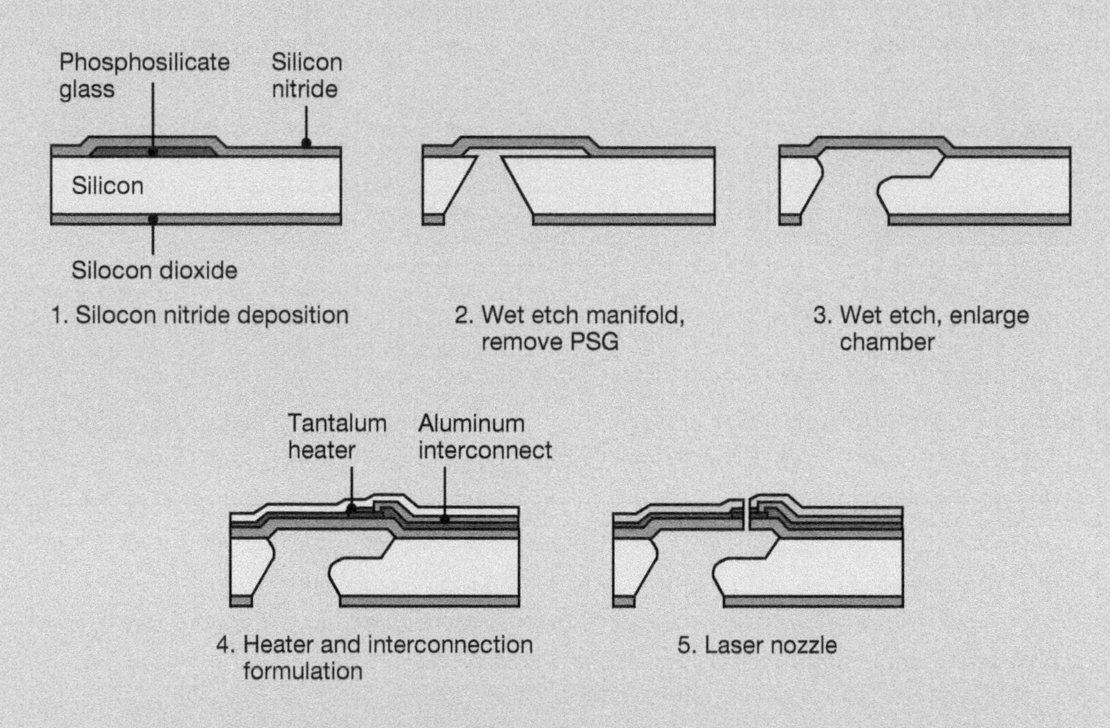

FIGURE 13.45 The manufacturing sequence for producing thermal ink-jet printer heads. *Source:* After F.G. Tseng.

13.16 LIGA and Related Microfabrication Processes

LIGA is a German acronym for the combined processes of X-ray lithography, electrodeposition, and molding (**X-ray Lithographie, Galvanoformung und Abformung**).

The LIGA process, illustrated in Fig. 13.46, involves the following steps:

1. A very thick (up to hundreds of microns) resist layer of polymethylmethacrylate (PMMA) is deposited onto a primary substrate;
2. The PMMA is exposed to collimated X-rays and developed;
3. Metal is electrodeposited onto the primary substrate;
4. The PMMA is removed or stripped, resulting in a freestanding metal structure; and
5. Plastic injection molding is done in the metal structure, which acts as a mold.

Depending on the application, the final product from a LIGA process may be:

1. A freestanding metal structure, resulting from the electrodeposition process;
2. A plastic injection-molded structure;
3. An investment-cast metal part, where the injection molded structure was used as a blank; and
4. A slip-cast ceramic part, produced with the injection-molded parts as the molds.

The substrate used in LIGA is an electrical conductor or a conductor-coated insulator. Examples of primary substrate materials include austenitic steel plate, silicon wafers with a titanium layer, and copper plated with gold, titanium, or nickel; metal-plated ceramic and glass also have been used. The surface may be roughened by grit blasting to encourage good adhesion of the resist material.

Resist materials must have high X-ray sensitivity, dry- and wet-etching resistance when unexposed, and thermal stability. The most common resist material is polymethylmethacrylate, which has a very high molecular weight (more than 10^6 g per mole; see Section 10.2.1). The X-rays break the chemical bonds, leading to the generation of free radicals and significantly reduced molecular weight in the exposed region. Organic solvents then preferentially dissolve the exposed PMMA in a wet-etching process. After development, the remaining three-dimensional structure is rinsed and dried, or it is spun and blasted with dry nitrogen.

Two newer forms of LIGA are **UV-LIGA** and **Silicon-LIGA**. In *UV-LIGA*, special photoresists are used, instead of PMMA, and they are exposed through ultraviolet lithography (Section 13.7). *Silicon-LIGA* uses deep reactive-ion-etched silicon (Section 13.8.2) as a preform for further operations. These processes, like the traditional X-ray-based LIGA, are used to replicate MEMS devices, but, unlike LIGA, they do not require the expensive columnated X-ray source for developing their patterns.

Electrodeposition of metal usually involves electroplating of nickel (Section 4.5.1 and Fig. 4.17). The nickel is deposited onto exposed areas of the substrate; it fills the PMMA structure and can even coat the resist (Fig. 13.46a). Nickel is the material of choice because of the relative

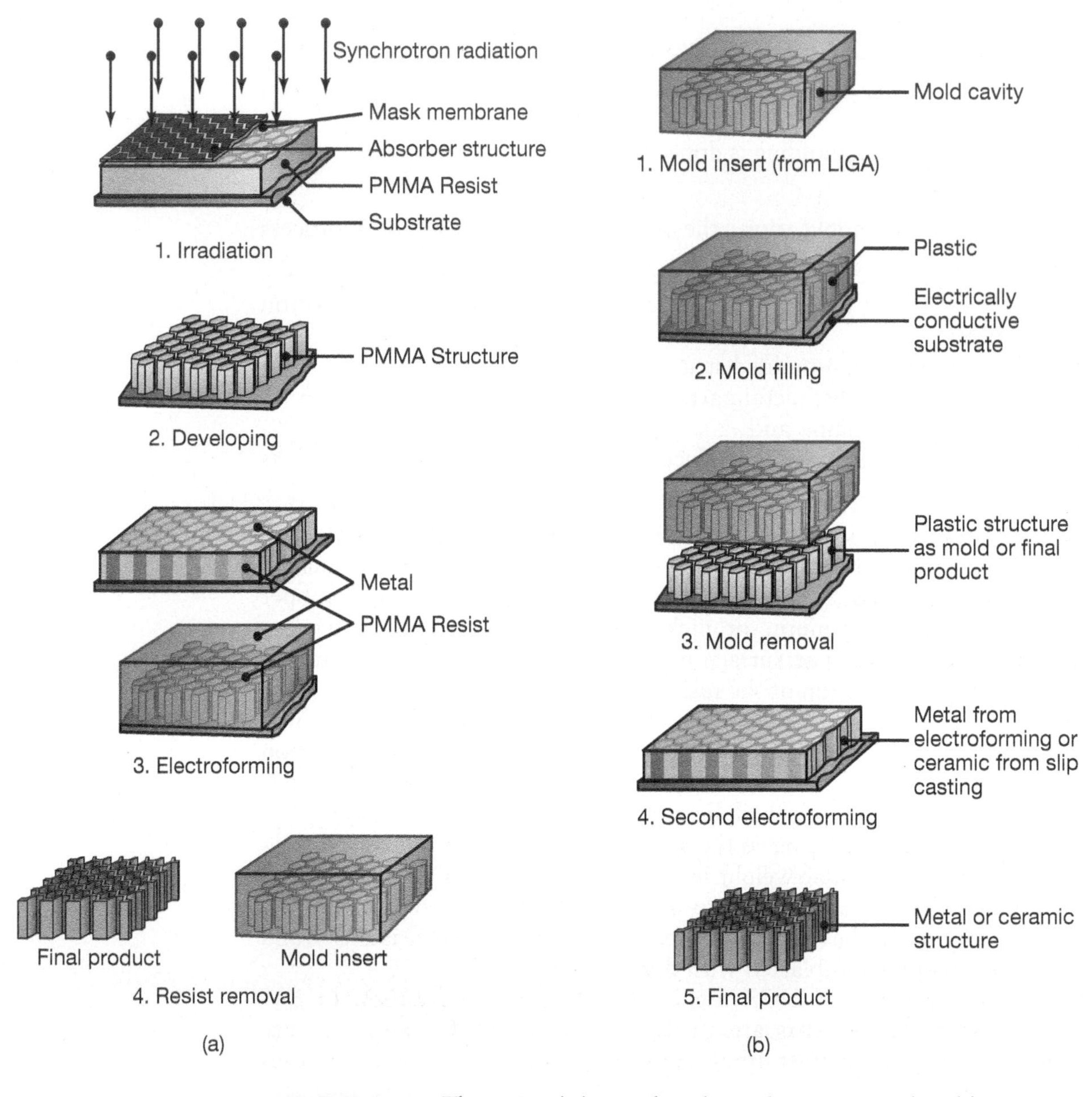

FIGURE 13.46 The LIGA (lithography, electrodeposition, and molding) technique. (a) Primary production of a metal product or mold insert. (b) Use of the primary part for secondary operations, or replication. *Source:* Courtesy of IMM Institut für Mikrotechnik.

ease in electroplating with well-controlled deposition rates and residual stress control. Electroless plating of nickel (Section 4.5.1) is also possible, where the nickel can be deposited directly onto electrically insulating substrates. However, because nickel displays high wear rates in MEMS, significant research continues to be directed toward the use of other materials or coatings.

After the metal structure has been deposited, precision grinding removes either the substrate material or a layer of the deposited nickel, a process called *planarization* (see also Section 13.10). The need for planarization is obvious when it is recognized that three-dimensional MEMS devices require micrometer tolerances on layers several hundreds of micrometers thick. Planarization is difficult to achieve. Conventional lapping (Section 9.7) leads to preferential removal of the soft PMMA and smearing of the metal, so planarization is usually accomplished with a diamond lapping procedure (*nanogrinding*). Here, a diamond slurry loaded, soft-metal plate is used to remove material in order to maintain flatness within 1 μm over a 75-mm diameter substrate.

If cross linked (see Section 10.2.1), the PMMA resist is then exposed to synchrotron X-ray radiation and removed by exposure to an oxygen plasma or through solvent extraction. The result is a metal structure, which may be used for further processing. Examples of freestanding metal structures, produced through electrodeposition of nickel, are shown in Fig. 13.47.

Although the processing steps for making freestanding metal structures are extremely time consuming and expensive, the main advantage of LIGA is that these structures serve as molds for the rapid replication of submicron features through molding operations. Table 13.5 lists and compares the processes that can be used for producing micromolds, where it can be seen that LIGA provides some distinct advantages. Reaction injection molding, injection molding, and compression molding processes (Section 10.10) also have been used to make micromolds.

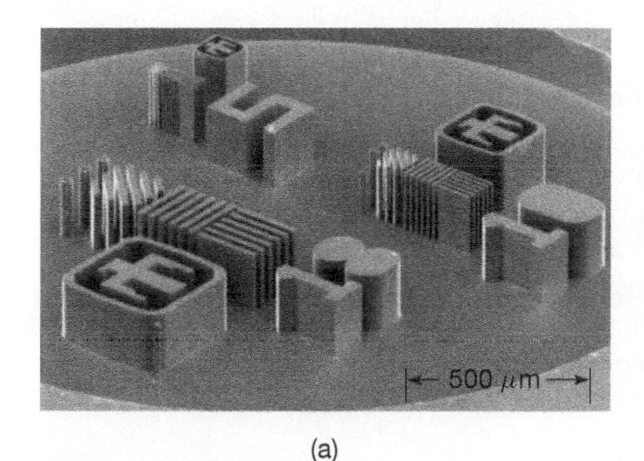

(a)

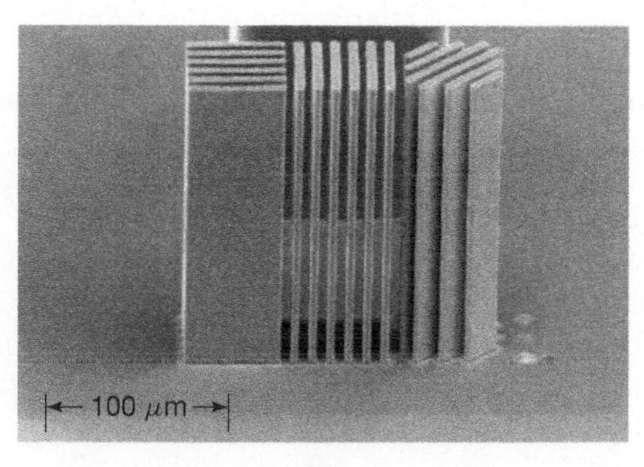

(b)

FIGURE 13.47 (a) Electroformed nickel structures. (b) Detail of nickel lines and spaces. *Source:* After T. Christenson, Sandia National Laboratories.

TABLE 13.5 Comparison of micromold manufacturing techniques.

	Production technique		
	LIGA	Laser machining	EDM
Aspect ratio	10–50	10	up to 100
Surface roughness	<50 nm	100 nm	0.3–1 μm
Accuracy	<1 μm	1–3 μm	1–5 μm
Mask required?	Yes	No	No
Maximum height	1–500 μm	200–500 μm	μm to mm

Source: L. Weber, W. Ehrfeld, H. Freimuth, M. Lacher, M. Lehr, and P. Pech, *SPIE Micromachining and Microfabrication Process Technology II*, Austin, TX, 1996.

TABLE 13.6 Comparison of properties of permanent-magnet materials.

Material	Energy product (Gauss-Oersted $\times 10^{-6}$)
Carbon steel	0.20
36% cobalt steel	0.65
Alnico I	1.4
Vicalloy I	1.0
Platinum-cobalt	6.5
$Nd_2Fe_{14}B$, fully dense	40
$Nd_2Fe_{14}B$, bonded	9

EXAMPLE 13.5 Production of Rare-Earth Magnets

A number of scaling issues in electromagnetic devices indicate that there is an advantage in using rare-earth magnets from the samarium cobalt (SmCo) and neodymium iron boron (NdFeB) families. These materials are available in powder form and are of interest because they can produce magnets that are an order of magnitude more powerful than conventional ones (Table 13.6). Thus, these materials can be used when miniature electromagnetic transducers are to be produced, as are commonly incorporated into in-ear headphones.

The processing steps used to manufacture these magnets are shown in Fig. 13.48. The polymethylmethacrylate mold is produced by exposure to X-ray radiation and solvent extraction. The rare-earth powders are mixed with an epoxy binder and applied to the PMMA mold through a combination of calendering (See Fig. 10.41) and pressing. After curing in a press at a pressure around 70 MPa, the substrate is planarized. The substrate is then subjected to a magnetizing field of at least 35 kilooersteds in the desired orientation. Once the material has been magnetized, the PMMA substrate is dissolved, leaving behind the rare-earth magnets, as shown in Fig. 13.49.

Source: After T. Christenson, Sandia National Laboratories.

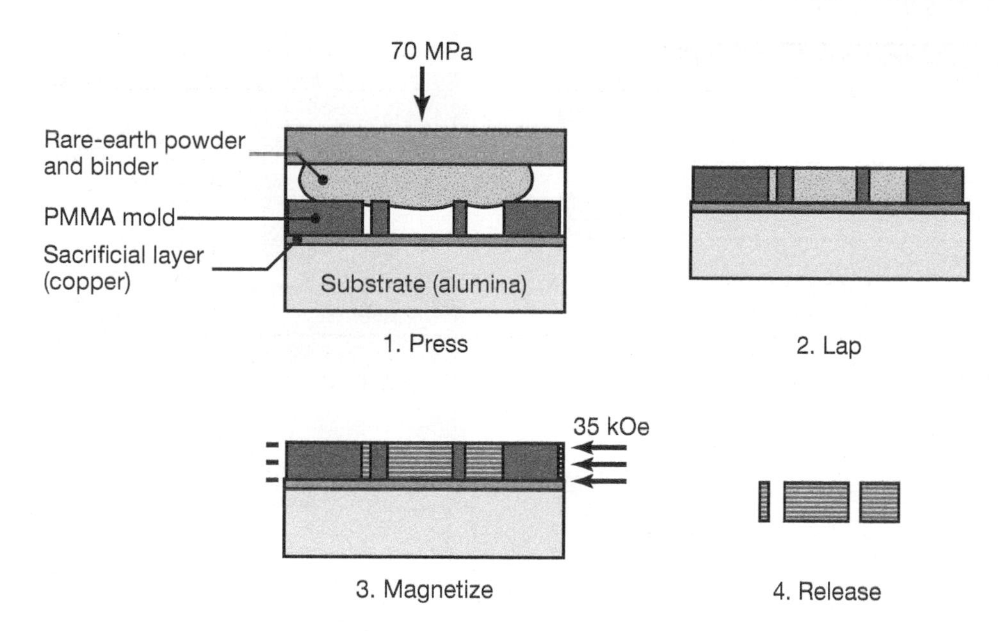

FIGURE 13.48 Fabrication process used in producing rare-earth magnets for microsensors. *Source:* After T. Christenson, Sandia National Laboratories.

Multilayer X-ray lithography. The LIGA technique is very suitable for producing MEMS devices with large aspect ratios and reproducible shapes. Often, however, a multilayer stepped structure that cannot be made directly through LIGA is required. For nonoverhanging geometries, direct plating can be applied. In this technique, a layer of electrodeposited metal with surrounding PMMA is produced as described for LIGA above. A second layer of PMMA resist is then bonded to this structure and X-ray exposed with an aligned X-ray mask.

It is often useful to have overhanging geometries within complex MEMS devices. A batch diffusion bonding and release procedure has been developed for this purpose, schematically illustrated in Fig. 13.50a. This process involves the preparation of two PMMA patterned and electroformed layers, with the PMMA being subsequently removed. The wafers are then aligned, face to face, with guide pins that press-fit into complementary structures on the opposite surface. Next, the substrates are joined in a hot press, and a sacrificial layer on one substrate is etched away, leaving one layer bonded to the other. An example of such a structure is shown in Fig. 13.50b.

(a)

FIGURE 13.49 SEM images of $Nd_2Fe_{14}B$ permanent magnets. Powder particle size ranges from 1 to 5 μm; the binder is a methylene-chloride resistant epoxy. Mild distortion is present in the image due to magnetic perturbation of the imaging electrons. Maximum-energy products of 9 MGOe have been obtained. *Source:* T. Christenson, Sandia National Laboratories.

HEXSIL. The HEXSIL process combines HEXagonal honeycomb structures, SILicon micromachining, and thin-film deposition to produce high aspect-ratio, free-standing structures. HEXSIL, illustrated in Fig. 13.51, can produce tall structures with a shape definition that rivals LIGA's capabilities.

In HEXSIL, a deep trench is first produced in single-crystal silicon, using dry etching, followed by a shallow wet etch to make the trench walls

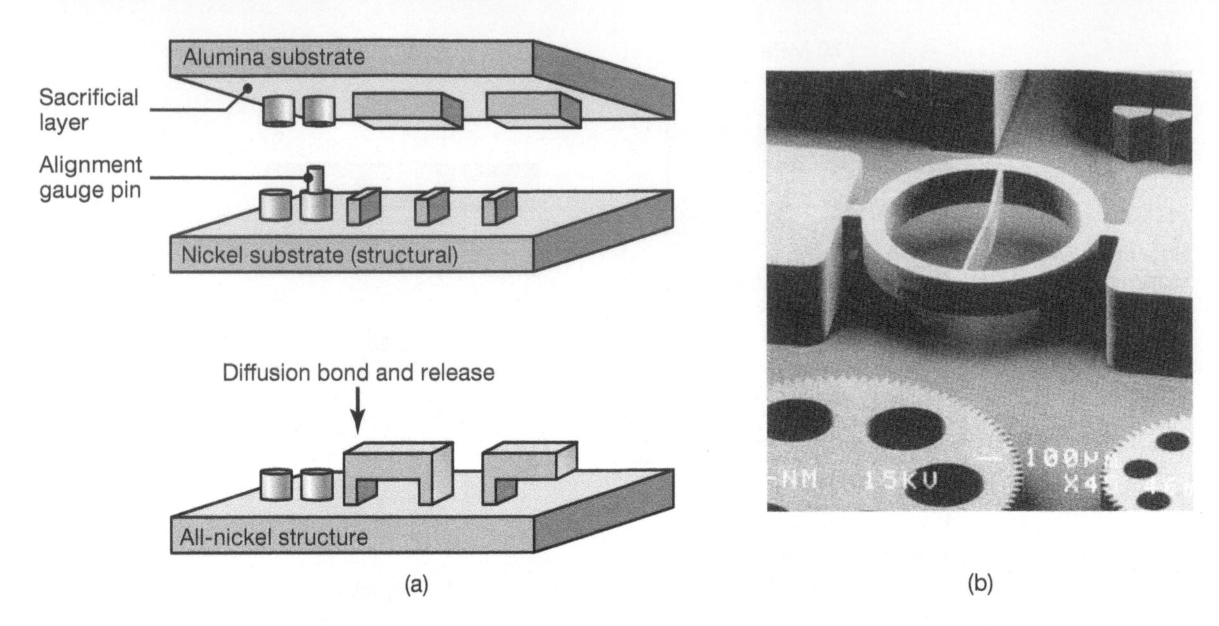

FIGURE 13.50 (a) Multilevel MEMS fabrication through wafer-scale diffusion bonding. (b) A suspended ring structure, for measurement of tensile strain, formed by two-layer wafer-scale diffusion bonding. *Source:* After T. Christenson, Sandia National Laboratories.

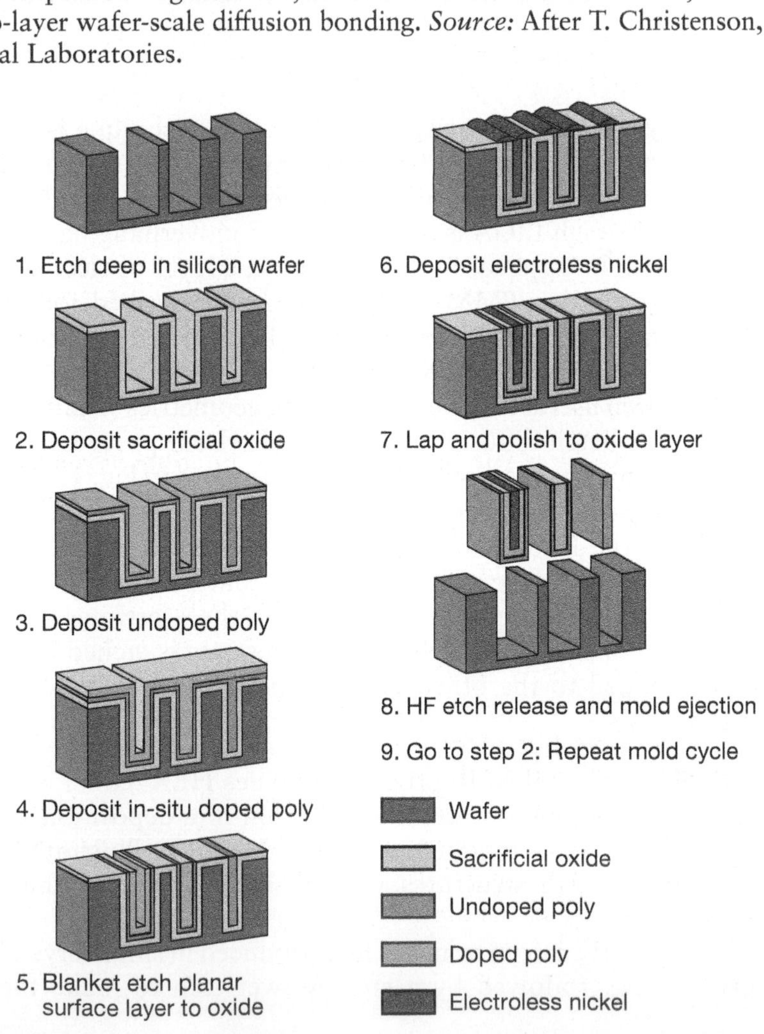

FIGURE 13.51 Illustrations of the HEXagonal honeycomb structure, SILicon micromachining, and thin-film deposition (HEXSIL process).

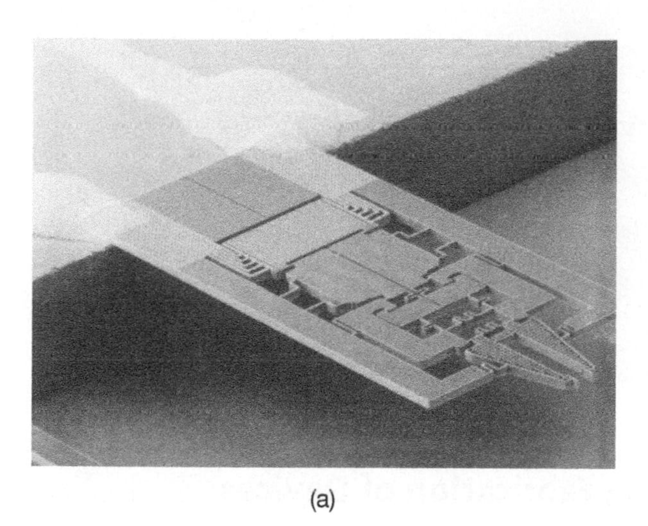

(a)

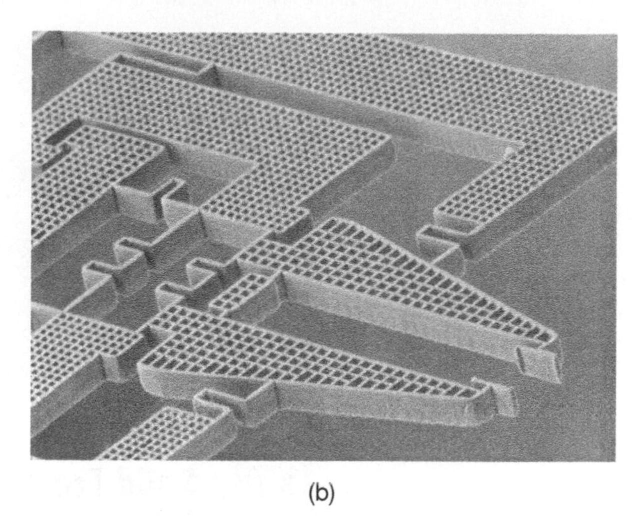

(b)

FIGURE 13.52 (a) SEM image of micro-scale tweezers, used in microassembly and microsurgery. (b) Detailed view of the tweezers. *Source:* Courtesy of MEMS Precision Instruments.

smoother. The depth of the trench matches the desired structure height, and is practically limited to around 100 μm. An oxide layer is then grown or deposited onto the silicon, followed by an undoped polycrystalline silicon layer, which leads to good mold filling and good shape definition. A doped silicon layer is then deposited, to provide for a resistive portion of the microdevice. Electroplated or electroless nickel plate is then deposited. Fig. 13.51 shows various trench widths to demonstrate the different structures that can be produced in HEXSIL. Figure 13.52 shows microtweezers produced through the HEXSIL process. A thermally-activated bar is used to activate the tweezers, which have been used for microassembly and microsurgery applications.

MolTun. The MolTun process (short for *mol*ding of *tun*gsten) was developed in order to utilize the higher mass of tungsten in micromechanical devices and systems. In MolTun, a sacrificial oxide is patterned through lithography and then etched, but instead of electroforming, a layer of tungsten is deposited through chemical vapor deposition. Excess tungsten is then removed by chemical mechanical polishing, which also ensures good control over layer thickness. Multiple layers of tungsten can be deposited to develop intricate geometries (Fig. 13.53).

MolTun has been used for micro mass-analysis systems and a large number of micro-scale latching relays, which take advantage of tungsten's higher strength compared to other typical MEMS materials. The depth of MolTun structures can be significantly larger than those produced through silicon micromachining; the mass analysis array in Fig. 13.53, for example, has a total thickness of about 25 μm.

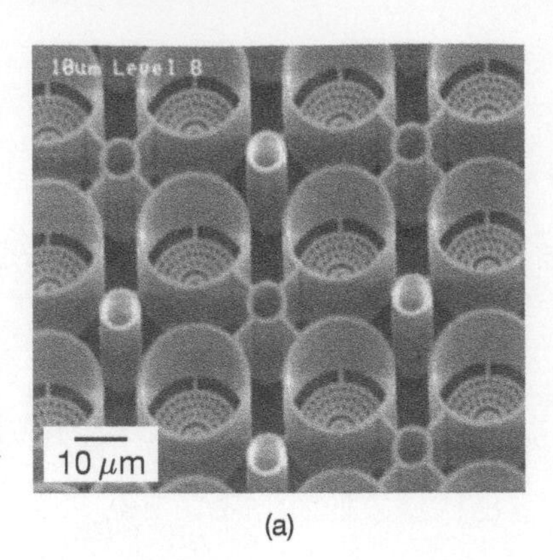

(a)

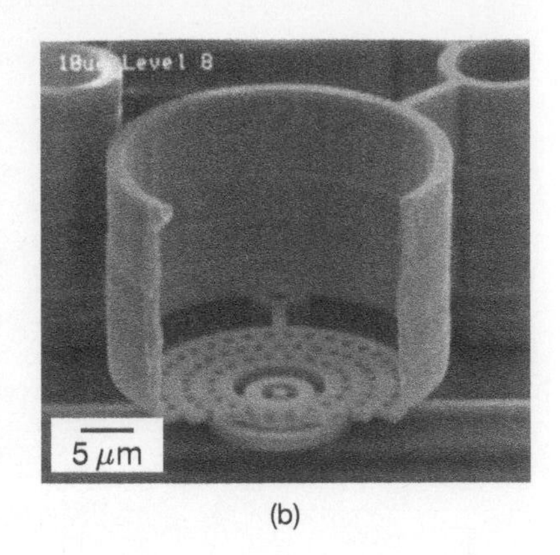

(b)

FIGURE 13.53 (a) An array of micro-mass analysis systems consisting of cylindrical ion traps, constructed of 14 layers of molded tungsten, including 8 layers for the ring electrode. (b) Detail of a since ion trap. *Source:* Courtesy of Sandia National Laboratories.

13.17 Solid Freeform Fabrication of Devices

Solid freeform fabrication is another term for additive manufacturing, described in Section 10.12. Many of the advances in additive manufacturing also are applicable to MEMS manufacture.

Microstereolithography. Recall that *stereolithography* involves curing of a liquid thermosetting polymer, using a photoinitiator and a highly focused light source. Conventional stereolithography uses layers between 75 and 500 μm in thickness, with a laser dot focused to a 0.25-mm diameter.

The **microstereolithography** process uses the same approach, but the laser is more highly focused, to a diameter as small as 1 μm, and layer thicknesses are about 10 μm. This technique has a number of economic advantages, but the MEMS devices are difficult to integrate with the controlling circuitry because stereolithography produces polymer structures that are nonconducting.

Instant masking is another technique for producing MEMS devices (Fig. 13.54). The solid freeform fabrication of MEMS devices by instant masking is also known as **electrochemical fabrication** (EFAB). A mask of elastomeric material is first produced through conventional photolithography techniques (see Section 13.7). The mask is pressed against the substrate in an electrodeposition bath, such that the elastomer conforms to the substrate and excludes plating solution in contact areas. Electrodeposition takes place in areas that are not masked, eventually producing a mirror image of the mask. Using a sacrificial filler, made of a second material, complex three-dimensional shapes can be produced, complete with overhangs, arches, and other features.

13.18 Mesoscale Manufacturing

Conventional manufacturing processes, as those described in Chapters 5 through 12, typically produce parts that can be described as visible to the naked eye. Such parts are generally referred to as *macroscale*, the word

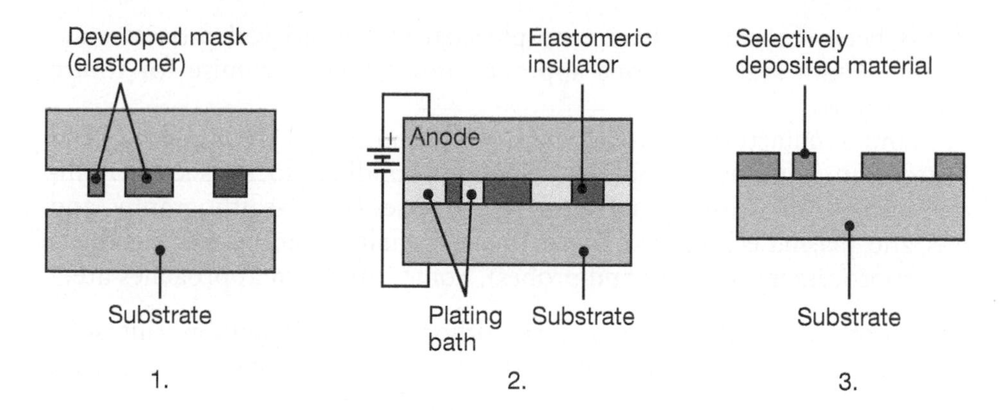

FIGURE 13.54 The instant masking process: (1) bare substrate, (2) during deposition, with the substrate and instant mask in contact, and (3) the resulting pattern deposit. *Source:* After A. Cohen, MEMGen Corporation.

macro being derived from *makros* in Greek, meaning long. The processing of such parts is known as **macromanufacturing**, and is the most developed and best understood size range from a design and manufacturing perspective.

In contrast, examples of **mesomanufacturing** are components for miniature devices, such as hearing aids, medical devices such as stents and valves, mechanical watches, and extremely small motors and bearings. Note that mesomanufacturing overlaps both macro- and micro-manufacturing in Fig. 1.7.

Two general approaches have been used for mesomanufacturing: scaling macromanufacturing processes *down*, and scaling micromanufacturing processes *up*. Examples of the former are a lathe with a 1.5-W motor that measures 32 mm × 25 mm × 30.5 mm and weighs 100 g. Such a lathe can machine brass to a diameter as small as 60 μm and with a surface roughness of 1.5 μm. Similarly miniaturized versions of milling machines, mechanical presses, and various other machine tools have been demonstrated.

Scaling up of micromanufacturing processes is similarly performed; LIGA can produce micro-scale devices, but the largest parts that can be made are meso-scale. For this reason, mesomanufacturing processes are often indistinguishable from their micromanufacturing implementations.

13.19 Nanoscale Manufacturing

In **nanomanufacturing**, parts are produced at nanometer length scales. The term usually refers to manufacturing strategies below the micrometer scale, or between 10^{-6} and 10^{-9} m in length. Many of the features in integrated circuits are at this length scale, but very little else with manufacturing relevance. Molecularly-engineered medicines and other forms of *biomanufacturing* are the only present commercial examples. However, it has been recognized that many physical and biological processes act at this length scale, and this approach holds much promise for future innovations.

Nanomanufacturing has taken two basic approaches: *top-down* and *bottom-up*. Top-down approaches use large building blocks (such as a silicon wafer) and various manufacturing processes (such as lithography, and wet and plasma etching) to construct ever smaller features and products (microprocessors, sensors, and probes). Some top-down approaches are:

- Photolithography, electron-beam lithography, and **nanoimprint (soft) lithography**, are capable of top-down manufacture of structures, with resolution under 100 nm, as discussed in Section 13.7.
- **Nanolithography.** The probes used in atomic-force microscopy vary greatly in size, materials, and capabilities. Diamond-tipped stainless steel cantilevers have a diamond with a tip radius of around 10 nm. By contacting and plowing across a surface, it can produce grooves up to a few μm thick. The spacing between lines depends on the groove depth needed.

At the other extreme, bottom-up approaches use small building blocks (such as atoms, molecules or clusters of atoms and molecules) to build up a structure. In theory, this is similar to additive manufacturing technologies, described in Section 10.12. When placed in the context of nanomanufacturing, however, bottom-up approaches suggest the manipulation and construction of products on an atomistic or molecular scale.

Bottom-up approaches include:

- *Microcontact printing* uses soft-lithography approaches, to deposit material on surfaces from which nanoscale structures can be produced.
- *Scanning tunneling microscopy* can be used to manipulate an atom on an atomically smooth surface (usually cleaved mica or quartz).

It has been pointed out that bottom-up approaches are widely used in nature (building cells is a fundamental bottom-up approach), whereas human manufacturing has, for the most part, consisted of top-down approaches. In fact, there are presently very few bottom-up products that have demonstrated commercial viability. Some prospects that are most applicable to engineering materials are described below.

1. **Carbon nanotube devices.** Carbon nanotubes were described in Section 11.13.1. These are of great research interest because of the exceptional mechanical properties of carbon nanotubes, as well as the high electrical conductivity that can be achieved in armchair nanotubes (see Fig. 11.41).

   Carbon nanotubes are extremely strong (strengths in excess of 150 GPa have been measured) in tension; in compression the nanotubes are not nearly as strong because of buckling. The elastic modulus of carbon nanotubes has been measured at around 900 GPa. Carbon nanotubes are grown from surfaces using chemical vapor deposition approaches, with typical dimensions of 2-100 nm in diameter and 1-15 micrometers in length. However, the longest nanotube reported is over 500 mm long.

Carbon nanotubes are seen as a potential solution to the problems of higher energy density associated with ever smaller microelectronic and MEMS devices. Because of carbon's high melting temperature, it is thought that carbon nanotubes will be the connecting material for future circuits, and this remains a material of significant research interest.

Carbon nanotubes have been produced with single, double, or multiple walls; the added walls provide increased chemical resistance. Also, doping can occur on an outer wall without compromising the strength of an inner wall. A taurus-shaped nanotube has been demonstrated, with outstanding magnetic properties.

QR Code 13.4 Integrated Photonics Institute for Manufacturing Innovation. *Source:* Courtesy of the US Department of Defense.

2. **Nanophase ceramics.** Nanophase ceramics (see also Section 11.8.1) have received interest because by using nanoscale particles in producing the ceramic material, there can be a simultaneous and major increase in both the strength and ductility of the ceramic. Nanophase ceramics are utilized for catalysis because of their high surface-to-area ratios, and nanophase particles can be used as a reinforcement, such as with SiC particles in an aluminum matrix (Section 11.14).

CASE STUDY 13.1 Photonic Integrated Circuits

There is much debate about whether conventional electronic circuitry has reached the limit of miniaturization, also known as "Moore's law." Traditional scaling in nano-electronics assumes that all relevant dimensions of a circuit can be reduced by 0.7× from generation to generation, while keeping the power consumption of the chip identical by area. However, to produce ever smaller features in a nano-electronics, such as transistors or wires connecting the transistors, requires that even more stringent manufacturing tolerances be achieved. Some transistor dimensions have reached their physical limit: scaling the insulator in the gate can only be done down to about 15–20 A, without incurring substantial leakage current due to electron tunneling through the insulator. In addition, the current that each individual transistor can provide scales with the dimension of the transistor, and the resistance that wires exhibit scales with the square root of the cross section dimensions.

The nano-electronics industry has dealt with the transistor-based challenges with materials innovation, such as replacing the SiO_2 gate insulator with improved dielectrics, and architectural innovation. However, the gate-switching energy of transistors has not been reduced at the same rate as their size; the result is increased power consumption in integrated circuits. The metal wire resistance problem has offered fewer options for radical innovation and hence contributes substantially to power losses.

These effects have serious consequences, in that the power consumption per unit area of chip in modern microprocessors can reach the energy density of a clothing iron. Obviously, it is very difficult

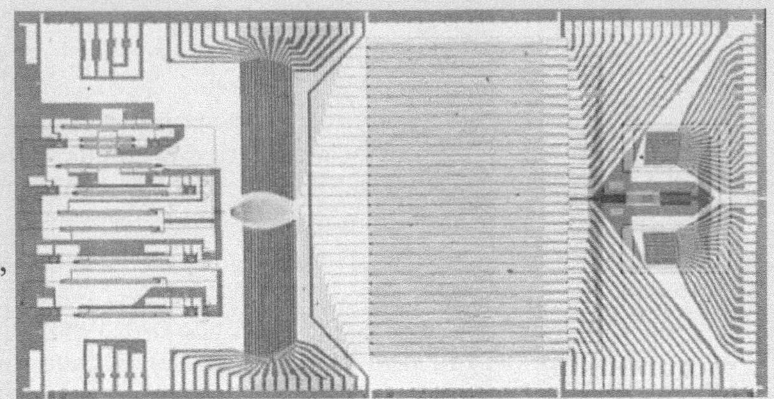

FIGURE 13.55 Two-dimensional scanning chip containing four tunable lasers, 32 phase shifters, 32 optical amplifiers, and 32 photodetectors. It uses heterogeneous integration to combine lasers, amplifiers, and phase shifters in a single chip that emits a beam that can be scanned in two dimensions.
Source: J.C. Hulme, J.K. Doylend, M.J.R. Heck, J.D. Peters, M.L. Davenport, J.T. Bovington, L.A. Coldren, and J.E. Bowers, *Optics Express*, Vol. 23, No. 5, pp. 5861–5874, 2015. Used with permission of J.E. Bowers.

to dissipate the heat produced and the power lost to heat contributes to the power budget. This is a particular concern for large server farms used to power the Internet.

A potential solution is to use photonics-based circuitry, such as the advanced device shown in Fig. 13.55. Light does not have the same scaling issues that exist with current flow. Light can be modulated at very high frequencies and be transmitted instantaneously with little to no power loss. Also, since light can carry multiple signals simultaneously, it has a much higher bandwith.

The manufacture of photonic integrated circuits is very similar to conventional electronics, in that the process involves a cycle as shown in Fig. 13.3. The main departure from semiconductor electronics involves using components and features that modulate light instead of electric current.

The essential elements of a photonic integrated circuit are:

1. **Light sources.** Usually, these are light-emitting diodes (LEDs) or lasers, although a number of other light-emitting devices are under development for various applications. For long-distance optical communications, Indium Phosphate (InP)-based lasers are part of InP-based Photonic Integrated Circuits (PICs), whereas data-center communication uses mainly silicon-based PICs with external indium-gallium-arsenide (InGaAs) lasers.
2. **Transmission media.** Light can be transmitted through any transparent media, including air, which is the medium of choice for sensor applications. Glass fiber can be used to transmit data along a desired path; modern fiber-optic cables allow transmission distances of over 100 km without amplification. Glass fiber can be miniaturized on a wafer where it becomes a wave guide.
3. **Detectors.** Based on photodiodes, detectors recognize the light signals and then transmit an associated electrical current. Detectors in data-center applications can be made of Ge and integrated into a nano-electronics flow.
4. **Modulation.** Information is encoded in a light signal by modulation. A simple form of modulation is to turn a light on or off to transmit a message through Morse code. A similar approach is used to modulate light signals going through a fiber or wave guide, although more advanced approaches are also being used.
5. **Amplifiers.** When needed, optical amplifiers will increase the light signal in an optical fiber. Being active elements, like lasers, amplifiers are more difficult to integrate in a Si-based semiconductor process.

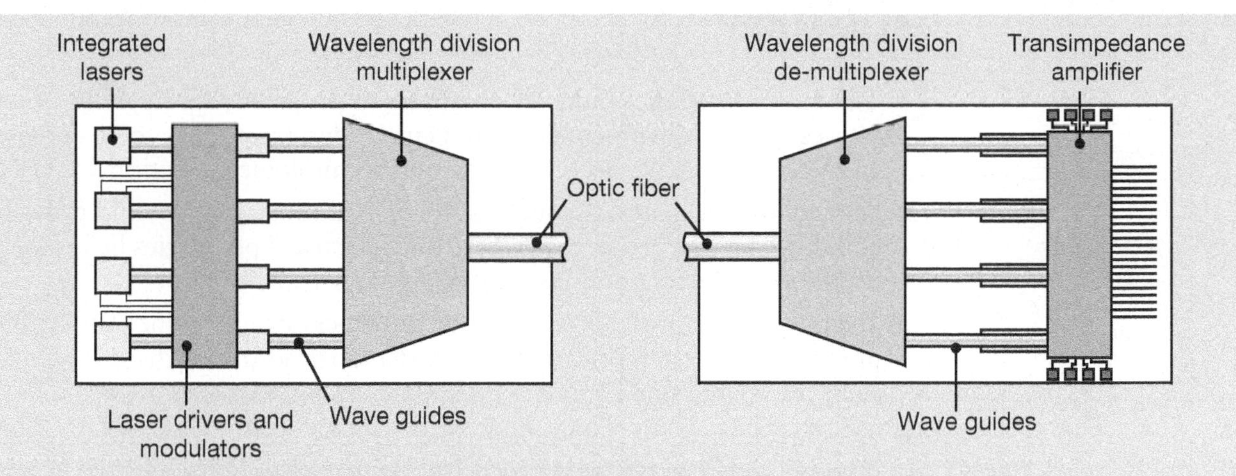

FIGURE 13.56 Schematic illustration of a photonic integrated circuit for data transmission. Numerous lasers, each producing a different wavelength of light, has its output signal modulated and combined into one fiber-optic cable by the wavelength division multiplexer (WDM Mux). The cable can extend thousands of kilometers to the receiving circuit, where the signal is separated into its components by a demultiplexer (WDM Demux), and then detected and amplified.

Improving Internet speed and performance is ultimately a problem of increasing bandwith, or the rate at which information can be transferred, and reducing the power per bit transferred. Bandwidths up to 2 terabytes/s are achievable with conventional electronics. However, the capabilities of photonics circuits are dramatically superior: more than 23 terabytes/s can be achieved in a single optical fiber, while reducing size and power consumption per gigabyte by around 33%. Photonic circuits are intended to drive communications traffic at 400 gigabytes/s, a significant improvement over conventional electronic circuits.

Photonics is able to achieve much higher transmission rates because of Wavelength Division Multiplexing (WDM), allowing multiple channels of light, each with a different frequency, to be transmitted over a single optical fiber. This approach is shown in Fig. 13.56, where four light sources with different frequencies are combined, transmitted over a fiber-optic cable that can extend up to thousands of km (or just the distance between blades or racks in a server farm), and then separated back into its source signals. The light is then detected and the resulting current is amplified as needed.

Photonic circuits are already widely used for long-haul data transmission. Currently, data centers exploit fiber technology and photonic transceivers. The use of PICs and photonic switches have the potential to revolutionize data-center architecture in the next few years, and provide a high-volume application for PICs, further reducing manufacturing costs; it is often said that "photonics will power the Internet." It is expected that photonics will find increased applications in other domains, such as the optical equivalent of radar and the sensors that will power the Internet of Things (see Section 15.14.2).

Source: Courtesy Michael Liehr, Integrated Photonics Institute for Manufacturing Innovation.

SUMMARY

- The microelectronics industry continues to progress rapidly, with possibilities for new device concepts and circuit designs appear to be endless; MEMS devices have become important technologies. (Section 13.1)
- Semiconductor materials, such as silicon, gallium arsenide, and polysilicon, have unique properties, where their electrical properties lie between those of conductors and insulators. (Section 13.3)
- Techniques are well developed for single-crystal growing and wafer preparation; after fabrication, the wafer is sliced into individual chips, each containing one complete integrated circuit. (Section 13.4)
- The fabrication of microelectronic devices and integrated circuits involves different types of processes. After bare wafers have been prepared, they undergo repeated oxidation, film deposition, lithographic, and etching steps to open windows in the oxide layer, in order to provide access to the silicon substrate. (Sections 13.5–13.8)
- Dopants are introduced into various regions of the silicon structure to alter their electrical characteristics, a task involving diffusion and ion implantation. (Section 13.9)
- Microelectronic devices are interconnected by multiple metal layers, and the completed circuit is packaged and made accessible through electrical connections. (Sections 13.10–13.11)
- Yield in devices is significant for economic considerations, and reliability has become increasingly important, because of the very long expected life of these devices. (Section 13.12)
- Requirements for more flexible and faster integrated circuits require packaging that places many closely spaced contacts on a printed circuit board. (Section 13.13)
- New approaches have been developed to allow the production of flexible electronic devices, using advanced printing techniques. These approaches are referred to as roll-to-roll printing. (Section 13.14)
- MEMS devices are manufactured through techniques and with materials that, for the most part, have been pioneered in the microelectronics industry. Bulk and surface micromachining, LIGA, SCREAM, HEXIL, MolTun and fusion bonding of multiple layers are the most prevalent methods. (Sections 13.15–13.17)
- Mesomanufacturing and nanomanufacturig are areas of great promise and rapid change, using top-down and bottom-up techniques for making products at very small length scales. (Sections 13.18–13.19)

BIBLIOGRAPHY

Adams, T.M. and Layton, R.W., *Introductory MEMS: Fabrication and Applications*, Springer, 2009.

Allen, J.J., *Microelectromechanical System Design*, CRC Press, 2006.

Baker, R.J., *CMOS Circuit Design, Layout, and Simulation*, 3rd ed., IEEE Press, 2011.

Cabrini, S., and Kawata, S., (eds.) *Nanofabrication Handbook*, CRC Press, 2012.

Caironi, M., and Noh, Y.-Y., *Large Area and Flexible Electronics*, Wiley, 2016.

Coombs, C., and Holden, H., Printed Circuits Handbook, 7th ed., McGraw-Hill, 2016.

Elwenspoek, M., and Jansen, H., *Silicon Micromachining*, Cambridge University Press, 2004.

Elwenspoek, M., and Wiegerink, R., *Mechanical Microsensors*, Springer, 2001.

Fraden, J., *Handbook of Modern Sensors: Physics, Designs, and Applications*, 4th ed., Springer, 2010.

Franssila, S., *Introduction to Microfabrication*, 2nd ed., Wiley, 2011.

Gardner, J.W., and Udrea, F., *Microsensors: Principles and Applications*, 2nd ed., Wiley, 2016.

Ghodssi, R. and Ghodssi, L.P., (eds.), *MEMS Materials and Processes Handbook*, Springer, 2011.

Harper, C.A. (ed.), *Electronic Packaging and Interconnection Handbook*, 4th ed., McGraw-Hill, 2004.

Hsu, T.-R., *MEMS & Microsystems: Design, Manufacture, and Nanoscale Engineering*, 2nd ed., Wiley, 2008.

Jha, A.R., *MEMS and Nanotechnology-Based Sensors and Devices for Communications, Medical and Aerospace Applications*, CRC Press, 2008.

Kempe, V., *Inertial MEMS: Principles and Practice*, Cambridge, 2011.

Korvink, J., and Oliver, P. (eds.), *MEMS: A Practical Guide to Design, Analysis, and Applications*, William Andrew, 2005.

Lee, T.-K., and Bieler, T.R., *Fundamentals of Lead-Free Solder Interconnect Technology*, Springer, 2015.

Liu, C., *Foundations of MEMS*, 2nd ed., Prentice Hall, 2011.

Madou, M.J., *Fundamentals of Microfabrication and Nanotechnology*, 3rd ed., CRC Press, 2012.

Maluf, N., and Williams, K., *An Introduction to Microelectromechanical Systems Engineering*, 2nd ed., Artech House, 2004.

May, G.S., and Spanos, C.J., *Fundamentals of Semiconductor Manufacturing and Process Control*, Wiley, 2006.

Natelson, D., *Nanostructures and Nanotechnology*, Cambridge University Press, 2016.

Rizvi, S., *Handbook of Photomask Manufacturing Technology*, CRC Press, 2005

Rockett, A., *The Materials Science of Semiconductors*, Springer, 2008.

Saleh, B.E.A., *Fundamentals of Photonics*, 2nd ed., Wiley, 2013.

Schaeffer, R., *Fundamentals of Laser Micromachining*, Taylor & Francis, 2012.

Sze, S.M. (ed.), *Semiconductor Devices: Physics and Technology*, 3rd ed., Wiley, 2012.

Ulrich, R.K., and Brown, W.D., (eds.), *Advanced Electronic Packaging*, Wiley, 2006.

van Zant, P., *Microchip Fabrication: A Practical Guide to Semiconductor Processing*, 6th ed., McGraw-Hill, 2015.

QUESTIONS

13.1 Define the terms *wafer, chip, device, integrated circuit,* and *surface mount*.

13.2 Why is silicon the most commonly used semiconductor in IC technology?

13.3 What do VLSI, IC, CVD, CMP, and DIP stand for?

13.4 How do *n*-type and *p*-type dopants differ?

13.5 How is epitaxy different than other forms of film deposition?

13.6 Compare wet and dry etching.

13.7 How is silicon nitride used in oxidation?

13.8 What are the purposes of prebaking and postbaking in lithography?

13.9 Define selectivity and isotropy and their importance in relation to etching.

13.10 What do the terms *linewidth* and *registration* refer to?

13.11 Compare diffusion and ion implantation.

13.12 What is the difference between evaporation and sputtering?

13.13 What is the definition of yield? How important is yield?

13.14 What is accelerated life testing? Why is it practiced?

13.15 What do BJT and MOSFET stand for?

13.16 Explain the basic processes of (a) surface micromachining and (b) bulk micromachining.

13.17 What is LIGA? What are its advantages?

13.18 What is the difference between isotropic and anisotropic etching?

13.19 What is a mask? Of what materials is it composed?

13.20 What is the difference between chemically reactive ion etching and dry plasma etching?

13.21 Which process(es) in this chapter allow(s) fabrication of products from polymers?

13.22 What is a PCB?

13.23 Explain the process of thermosonic stitching.

13.24 What is the difference between a die, a chip, and a wafer?

13.25 Why are flats or notches machined onto silicon wafers?

13.26 What is a via? What is its function?

13.27 Describe the procedures of image splitting lithography and immersion lithography.

13.28 What is a flip chip? What are its advantages over a surface-mount device?

13.29 Explain how IC packages are attached to a printed circuit board if both sides will contain ICs.

13.30 In a horizontal epitaxial reactor (see the accompanying figure), the wafers are placed on a stage (susceptor) that is tilted by a small amount, usually 1° to 3°. Why is this procedure done?

13.31 The accompanying table describes three changes in the manufacture of a wafer: increase of the wafer diameter, reduction of the chip size, and increase of the process complexity. Complete the table by filling in the words "increase," "decrease," or "no change" to indicate the effect that each change would have on the wafer yield and on the overall number of functional chips.

Effects of manufacturing changes		
Change	Wafer yield	Number of functional chips
Increase wafer diameter		
Reduce chip size		
Increase process complexity		

13.32 What is MolTun? What are its main advantages?

13.33 The speed of a transistor is directly proportional to the width of its polysilicon gate, with a narrower gate resulting in a faster transistor, and a wider gate resulting in a slower transistor. With the knowledge that the manufacturing process has a certain variation for the gate width, say $\pm 0.1\ \mu m$, how might a designer alter the gate size of a critical circuit in order to minimize its speed variation? Are there any penalties for making this change? Explain.

13.34 A common problem in ion implantation is channeling, in which the high-velocity ions travel deep into the material through channels along the crystallographic planes before finally being stopped. What is one simple way to stop this effect?

13.35 The MEMS devices discussed in this chapter apply macroscale machine elements, such as spur gears, hinges, and beams. Which of the following machine elements can and cannot be applied to MEMS, and why?

1. ball bearing;
2. helical springs;
3. bevel gears;
4. rivets;
5. worm gears;
6. bolts; and
7. cams.

13.36 Figure 13.7b shows the Miller indices on a wafer of (100) silicon. Referring to Fig. 13.5, identify the important planes for the other wafer types illustrated in Fig. 13.7a.

13.37 Referring to Fig. 13.24, sketch the holes generated from a circular mask.

13.38 Explain how you would produce a spur gear if its thickness were one tenth its diameter and its diameter were (a) 10 μm, (b) 100 μm, (c) 1 mm, (d) 10 mm, and (e) 100 mm.

13.39 What is cleaner, a Class-10 or a Class-1 clean room?

13.40 Describe the difference between a microelectronic device, a micromechanical device and MEMS.

13.41 Why is silicon often used with MEMS and MEMS devices?

13.42 What is the purpose of a spacer layer in surface micromachining?

13.43 What do SIMPLE and SCREAM stand for?

13.44 Which process(es) in this chapter allow the fabrication of products from polymers?

13.45 What is HEXSIL?

13.46 What are the differences between stereolithography and microstereolithography?

13.47 Lithography produces projected shapes, so that true three-dimensional shapes are more difficult to produce. What processes in this chapter are best able to produce three-dimensional shapes, such as lenses?

13.48 List the advantages and disadvantages of surface micromachining compared to bulk micromachining.

13.49 What are the main limitations to the LIGA process?

13.50 What other process(es) can be used to make the microtweezers shown in Fig. 13.52 other than HEXSIL? Explain.

13.51 The MEMS devices described in this chapter are applicable to macroscale machine elements, such as spur gears, hinges, and beams. Which of the following machine elements can or cannot be applied to MEMS, and why? (a) ball bearings; (b) bevel gears; (c) worm gears; (d) cams; (e) helical springs; (f) rivets; and (g) bolts.

13.52 Explain how you would produce a spur gear if its thickness was one-tenth of its diameter and its diameter was (a) 1 μm; (b) 100 μm; (c) 1 mm; (d) 10 mm; and (e) 100 mm.

13.53 List the advantages and disadvantages of surface micromachining compared with bulk micromachining.

13.54 What are the main limitations to the LIGA process? Explain.

13.55 Other than LIGA, what process can be used to make the micromagnets shown in Fig. 13.49? Explain.

13.56 Is there an advantage to using the MolTun process for other materials? Explain.

PROBLEMS

13.57 A certain wafer manufacturer produces two equal-sized wafers, one containing 500 chips and the other containing 300 chips. After testing, it is observed that 50 chips on each wafer are defective. What are the yields of the two wafers? Can any relationship be drawn between chip size and yield?

13.58 A chlorine-based polysilicon etch process displays a polysilicon:resist selectivity of 4:1 and a polysilicon:oxide selectivity of 50:1 How much resist and exposed oxide will be consumed in etching 350 nm of polysilicon? What would the polysilicon:oxide selectivity have to be in order to lose only 4 nm of exposed oxide?

13.59 During a processing sequence, four silicon-dioxide layers are grown by oxidation: 400 nm, 150 nm, 40 nm, and 15 nm. How much of the silicon substrate is consumed?

13.60 A certain design rule calls for metal lines to be no less than 2 μm wide. If a 1 μm-thick metal layer is to be wet etched, what is the minimum photoresist width allowed (assuming that the wet etch is perfectly isotropic)? What would be the minimum photoresist width if a perfectly anisotropic dry-etch process were used?

13.61 Obtain mathematical expressions for the etch rate as a function of temperature, using Fig. 13.19.

13.62 If a square mask of side length 100 μm is placed on a {100} plane and oriented with a side in the ⟨110⟩ direction, how long will it take to etch a hole 4 μm deep at 80°C using thylene-diamine/pyrocatechol? Sketch the resulting profile.

13.63 Obtain an expression for the width of the trench bottom as a function of time for the mask shown in Fig. 13.18b.

13.64 Estimate the time of contact and average force when a fluorine atom strikes a silicon surface with a velocity of 1 mm/s. *Hint:* See Eqs. (9.11) and (9.13).

13.65 Calculate the undercut in etching a 10-μm-deep trench if the anisotropy ratio is (a) 200, (b) 2, and (c) 0.5. What is the sidewall slope for these cases?

13.66 Calculate the undercut in etching a 10-μm-deep trench for the wet etchants listed in Table 13.3.

What would the undercut be if the mask were silicon oxide?

13.67 Estimate the time required to etch a spur-gear blank from a 75-mm-thick slug of silicon.

13.68 A resist is applied in a resist spinner spun operating at 2000 rpm, using a polymer resist with viscosity of 0.05 N-s/m. The measured resist thickness is 1.5 μm. What is the expected resist thickness at 6000 rpm? Assume that $\alpha = 1.0$ in Eq. (13.3).

13.69 Examine the cavity profiles in the figure below and explain how they might be produced.

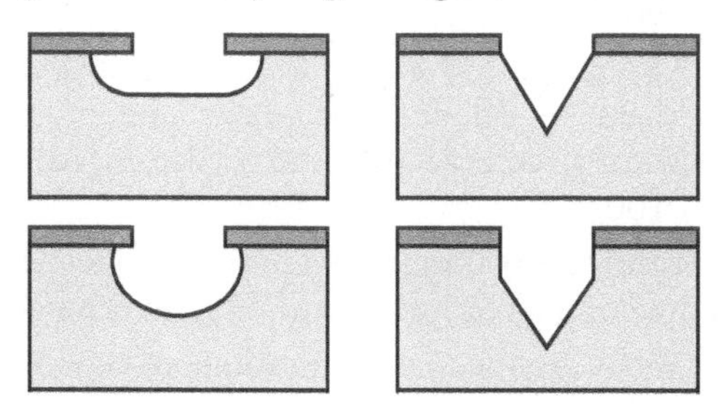

13.70 A polyimide photoresist needs 100 mJ/cm^2 per micrometer of thickness in order to develop properly. How long does a 150 μm film need to develop when exposed by a 1000 W/m^2 light source?

13.71 How many levels are needed to produce the micromotor shown in Fig. 13.23d?

13.72 It is desired to produce a 500 μm by 500 μm diaphragm, 25 μm thick, in a silicon wafer 250 μm thick. Given that you will use a wet etching technique with KOH in water with an etch rate of 1 μm/min, calculate the etching time and the dimensions of the mask opening that you would use on a (100) silicon wafer.

13.73 If the Reynolds number for water flow through a pipe is 2000, calculate the water velocity if the pipe diameter is (a) 10 mm; (b) 100 μm. Do you expect flow in MEMS devices to be turbulent or laminar? Explain.

DESIGN

13.74 The accompanying figure shows the cross-section of a simple *npn* bipolar transistor. Develop a process flow chart to fabricate this device.

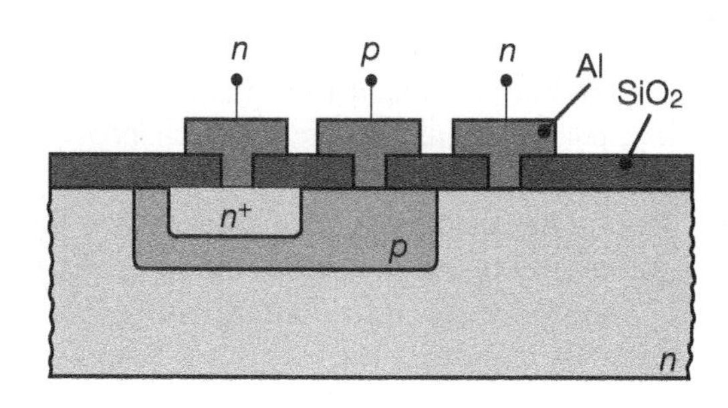

13.75 Referring to Fig. 13.38, design an experiment to find the critical dimensions of an overhanging cantilever that will not stick to the substrate.

13.76 Explain how you would manufacture the device shown in Fig. 13.37.

13.77 Inspect various electronic and computer equipment, take them apart as much as you can, and identify components that may have been manufactured by the techniques described in this chapter.

13.78 Do any aspects of this chapter's contents and the processes described bear any similarity to the processes described throughout previous parts of this book? Explain and describe what they are.

13.79 Describe your understanding of the important features of clean rooms, and how they are maintained.

13.80 Describe products that would not exist without the knowledge and techniques described in this chapter. Explain.

13.81 Review the technical literature and give more details regarding the type and shape of the abrasive wheel used in the wafer-cutting process shown in Fig. 13.6b.

13.82 As you know, microelectronic devices may be subjected to hostile environments (such as high temperature, humidity, and vibration) as well as physical abuse (such as being dropped on a hard surface) Describe your thoughts on how you would go about testing these devices for their endurance under these conditions.

13.83 Conduct a literature search and determine the smallest diameter hole that can be produced by (a) drilling; (b) punching; (c) water-jet cutting; (d) laser machining; (e) chemical etching and (f) EDM.

13.84 Conduct a literature review and summarize the common designs for MEMS-based accelerometers. Outline the processing sequence for such a device using the (a) SCREAM process and (b) HEXSIL processes.

13.85 Conduct a literature search and write a one-page summary of applications in biomems.

13.86 Describe the crystal structure of silicon. How does it differ from FCC? What is the atomic packing factor?

13.87 Referring to the MOS transistor cross section in the figure below and the given table of design rules, what is the smallest obtainable transistor size W? Which design rules, if any, have no impact on W?

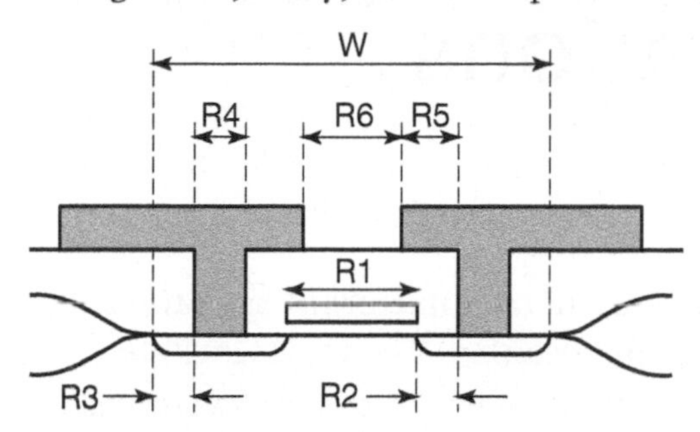

Rule no.	Rule name	Value (μm)
R1	Minimum polysilicon width	0.50
R2	Minimum poly-to-contact spacing	0.15
R3	Minimum enclosure of contact by diffusion	0.10
R4	Minimum contact width	0.60
R5	Minimum enclosure of contact by metal	0.10
R6	Minimum metal-to-metal spacing	0.80

CHAPTER

14 Automation of Manufacturing Processes and Operations

This chapter describes automation of manufacturing operations, including the use of robots, sensors, and fixturing. The topics include:

- Characteristics of hard and soft automation for different production quantities.
- Numerical control of machines for improved productivity and increased flexibility.
- Control strategies, including open-loop, closed-loop, and adaptive control.
- Industrial robots in manufacturing operations.
- Sensors for monitoring and controlling manufacturing processes.
- Design considerations for product assembly and disassembly.
- The impact of automation on product design and process economics.

Symbols

a	acceleration, m/s^2	t	time, s
F	force, N	x	Cartesian coordinate, m
m	mass, kg		

14.1 Introduction

Until the early 1950s, most manufacturing operations were carried out on traditional machinery, which lacked flexibility and required skilled labor. Each time a different product was manufactured, (a) the machinery had to be retooled; (b) the movement of materials had to be rearranged; (c) new products and parts with complex shapes required numerous attempts by the operator to set the proper processing parameters; and (d) because of human involvement, making a large number of parts exactly alike was difficult and time consuming. These circumstances meant that processing methods were generally inefficient and that labor costs were a significant portion of the overall production costs. The necessity for reducing the labor share of product cost became apparent, as did the need to improve the efficiency and flexibility of manufacturing operations.

Also, **productivity** became a major concern. Defined as output per employee, productivity basically measures *operating efficiency* in a plant. **Mechanization** of machinery and operations had, by and large, reached its peak by the 1940s, but with continuous advances in the science and technology of manufacturing, the efficiency of manufacturing operations began to improve, and the percentage of total cost attributed to labor began to decline.

The next step in improving productivity was **automation**, from the Greek *automatos*, meaning self-acting. This term was coined in the mid-1940s by the US automobile industry to indicate *automatic handling and processing* of parts in production machines. During the past decades, major advances in the types and the extent of automation have been taking place, made possible through rapid advances in the capacity and sophistication of **computers and control systems.**

This chapter follows the outline shown in Fig. 14.1. It first reviews the history and principles of automation and how they have helped integrate various operations and activities in manufacturing plants. It then introduces the concept of control of machines and systems through **numerical control** and **adaptive-control** techniques. Next described is an essential aspect of manufacturing, namely **material handling,** which has been developed into various systems, particularly those that include the use of **industrial robots. Sensor technology** is then described; this technology is an essential element in the control and optimization of machinery, processes, and systems. Other developments described include **flexible fixturing** and **assembly operations**, taking full advantage of advanced manufacturing technologies, particularly **flexible manufacturing systems**. Finally, the chapter presents developments in **computer-integrated manufacturing systems** and their impact on all aspects of manufacturing operations.

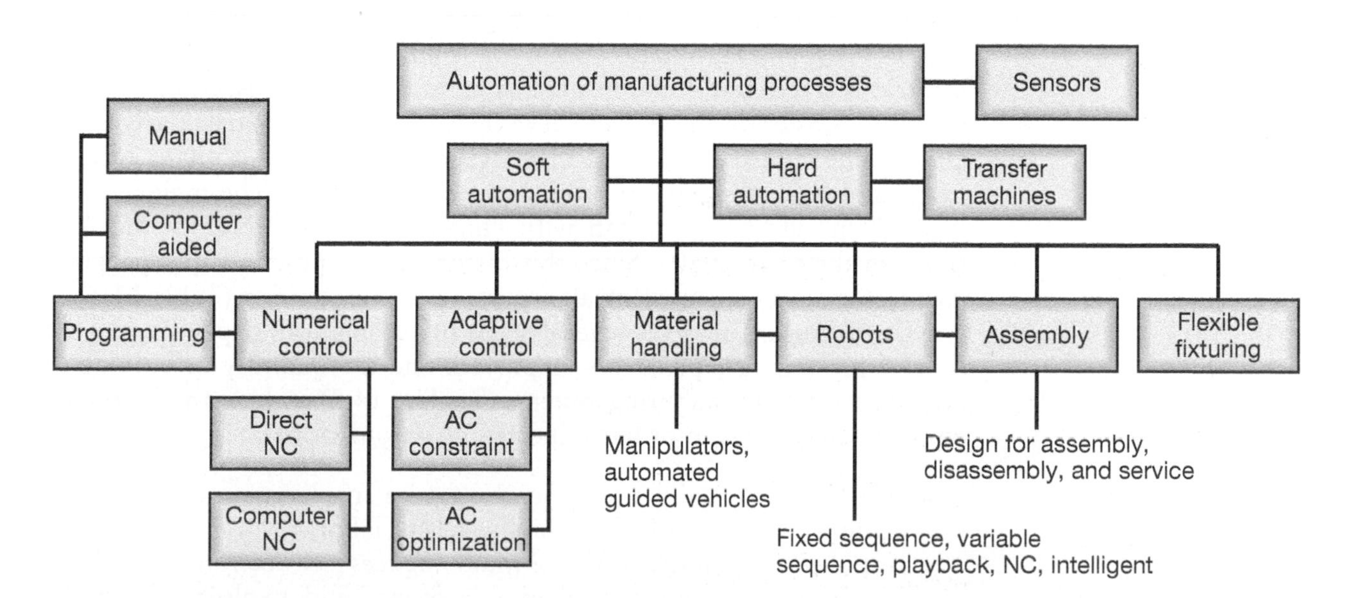

FIGURE 14.1 Outline of topics described in this chapter.

14.2 Automation

Automation is generally defined as the process of having machines follow a predetermined sequence of operations with little or no human involvement, using specialized equipment and devices that perform and control manufacturing processes. As described in Section 14.8 and in Chapter 15, automation is achieved through the use of a variety of devices, sensors, actuators, and specialized equipment that are capable of (a) continuously observing all aspects of the manufacturing operation; (b) rapidly making decisions concerning the changes that should be made; and (c) controlling and optimizing all aspects of the operation.

Automation is an *evolutionary* rather than a revolutionary concept, and it has been implemented successfully in the following basic areas of activity:

- **Manufacturing processes and operations.** Machining, grinding, forging, cold extrusion, casting, and plastics molding are examples of processes that have been extensively automated.
- **Material handling.** Materials and parts in various stages of completion are moved throughout a plant by computer-controlled equipment, including the use of robots, with little or no human guidance.
- **Inspection.** While they are being made, parts are automatically inspected for dimensional accuracy, surface finish, and defects.
- **Assembly.** Individual parts are automatically assembled into subassemblies and into products.
- **Packaging.** Products are packaged automatically.

14.2.1 Evolution of Automation

Although metalworking processes began to be developed as early as 4000 BC (See Table 1.1), it was at the beginning of the Industrial Revolution, in the 1750s, that automation began to be introduced. Machine tools were in various stages of development starting in the late 1890s. Mass production and transfer machines were developed in the 1920s, particularly in the automobile industry; however, these systems had *fixed* automatic mechanisms and were designed to produce specific products. The major breakthrough in automation began with *numerical control* (NC) of machine tools, in the early 1950s. Since this historic development, rapid progress has been made in automating all aspects of manufacturing (Table 14.1).

Manufacturing involves several levels of automation, depending on the type of product, the processes employed, and the quantity or volume to be produced. Manufacturing systems (See Fig. 14.2) include the following basic classifications, in order of increasing automation:

- **Job shops** typically use general-purpose machines and machining centers (Section 8.11), with high levels of labor.
- **Rapid-prototyping facilities** or *Maker Spaces* (see Section 10.12) have less human involvement than job shops, and exploit the unique

TABLE 14.1 Developments in the history of automation and control of manufacturing processes. (See also Table 1.1.)

Date	Development
1500–1600	Water power for metalworking; rolling mills for coinage strips
1600–1700	Hand lathe for wood; mechanical calculator
1700–1800	Boring, turning, and screw cutting lathe, drill press
1800–1900	Copying lathe, turret lathe, universal milling machine; advanced mechanical calculators
1808	Sheet metal cards with punched holes for automatic control of weaving patterns in looms
1863	Automatic piano player (Pianola)
1900–1920	Geared lathe; automatic screw machine; automatic bottle-making machine
1920	First use of the word robot
1920–1940	Transfer machines; mass production
1940	First electronic computing machine
1943	First digital electronic computer
1945	First use of the word automation
1947	Invention of the transistor
1952	First prototype numerical control machine tool
1954	Development of the symbolic language APT (Automatically Programmed Tool); adaptive control
1957	Commercially available NC machine tools
1959	Integrated circuits; first use of the term group technology
1960	Industrial robots
1965	Large-scale integrated circuits
1968	Programmable logic controllers
1970s	First integrated manufacturing system; spot welding of automobile bodies with robots; microprocessors; minicomputer-controlled robot; flexible manufacturing system; group technology
1980s	Artificial intelligence; intelligent robots; smart sensors; untended manufacturing cells
1990–2000s	Integrated manufacturing systems; intelligent and sensor-based machines; telecommunications and global manufacturing networks; fuzzy-logic devices; artificial neural networks; Internet tools; virtual environments; high-speed information systems
2010s	Cloud-based storage and computing; MTConnect for information retrieval; three-dimensional geometry files; STEP-NC and autogenerated G-code

capabilities of computer-driven additive manufacturing machinery, with limited production runs.

- **Stand-alone NC production,** using **numerically controlled machines** (Section 14.3) but still with significant operator/machine interaction.
- **Manufacturing cells** (Section 15.9) consist of clusters of machines with integrated computer control and flexible material handling, often with industrial robots (Section 14.7).
- **Flexible manufacturing cells** (Section 15.10) use computer control of all aspects of manufacturing, the simultaneous incorporation of several manufacturing cells, and automated material handling systems.

- **Flexible manufacturing** lines involve computer-controlled machinery in production lines, instead of cells; part transfer is through hard automation and product flow is more limited than in flexible manufacturing systems. However, the throughput is higher, resulting in higher production quantities.
- **Flowlines** and **transfer lines** consist of organized groupings of machinery with automated material handling between machines; the manufacturing line is designed for limited or no flexibility, because the goal is to produce a single part in large quantities.

14.2.2 Goals of Automation

Automation has the following primary goals:

1. **Integrate** various aspects of manufacturing operations so as to improve product quality and uniformity, minimize cycle times and effort involved, and reduce labor costs.
2. **Improve productivity,** by reducing manufacturing costs through better control of production. Raw materials and parts in various stages of completion are loaded and unloaded on machines, faster and more efficiently; machines are used more effectively; and production is organized more efficiently.
3. **Improve quality,** by improving the repeatability of manufacturing processes.
4. **Reduce human involvement,** boredom, and the possibility of human error.
5. **Reduce workpiece damage** caused by manual handling of parts.
6. **Economize on floor space** in the plant, by arranging machines, material handling and movement, and auxiliary equipment more efficiently.
7. **Raise the level of safety** for personnel, especially under hazardous working conditions.

Automation and production quantity. Production quantity or volume is crucial in determining the type of machinery and the level of automation required to produce the number of parts. **Total production quantity** is defined as the total number of parts to be produced, in various **lot sizes.** Lot size greatly influences the economics of production, as described in Chapter 16. **Production rate** is defined as the number of parts produced per unit time. The approximate and generally accepted ranges of production quantities for some typical applications are shown in Table 14.2. Note that, as expected, **experimental** and **prototype** products represent the lowest volume (see also Section 10.12).

Small quantities can be manufactured in *job shops* (Fig. 14.2). These operations have high part variety, meaning that different parts can be produced in a short time and without extensive changes in tooling and in operations. On the other hand, machinery in job shops generally requires skilled labor, and the production quantity and rate are low; as a result, the cost per part can be high (Fig. 14.3). When products involve a large

TABLE 14.2 Approximate annual quantity of production.

Type of production	Numbers produced	Typical products
Experimental or prototype	1–10	All types
Piece or small batch	<5000	Aircraft, missiles, special machinery, dies, jewelry, and orthopedic implants
Batch or high quantity	5000–100,000	Trucks, agricultural machinery, jet engines, diesel engines, computer components, and sporting goods
Mass production	100,000+	Automobiles, appliances, fasteners, and food and beverage containers

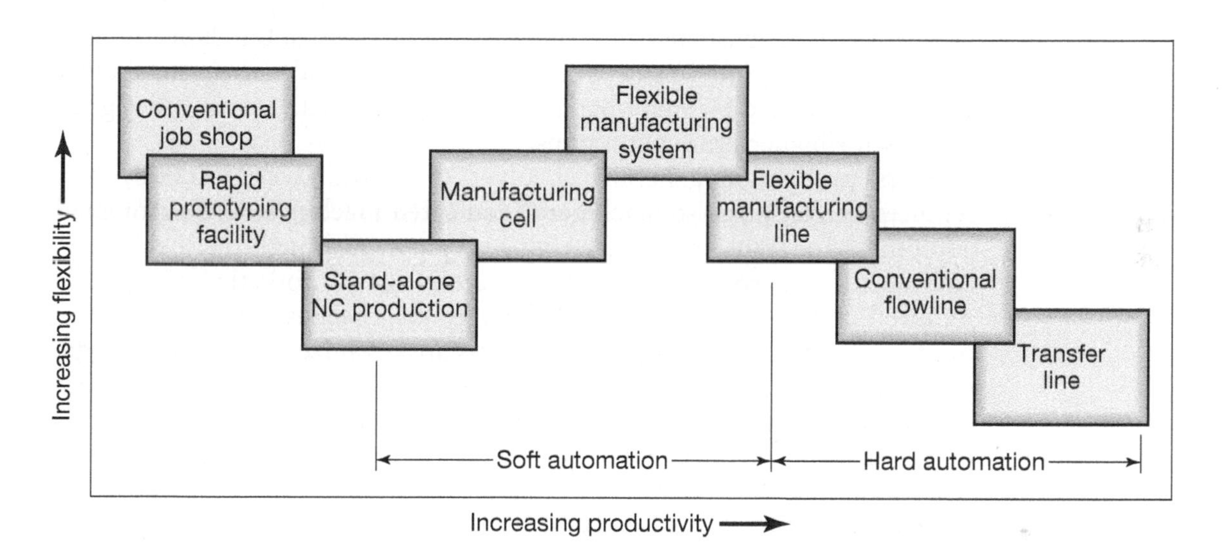

FIGURE 14.2 Flexibility and productivity of various manufacturing systems. Note the overlap between the systems, which is due to the various levels of automation and computer control that are applicable in each group (see also Chapter 15 for more details).

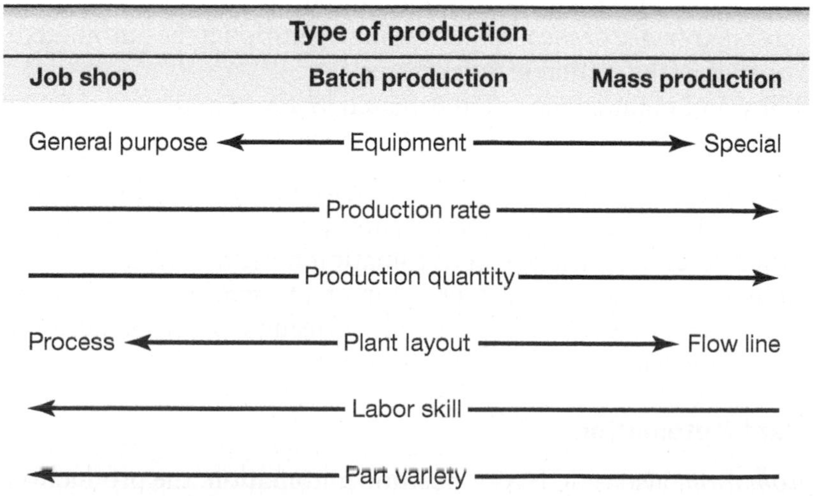

Type of production
Job shop | Batch production | Mass production
General purpose ← Equipment → Special
Production rate →
Production quantity →
Process ← Plant layout → Flow line
← Labor skill
← Part variety

FIGURE 14.3 General characteristics of three types of production methods: job shop, batch production, and mass production.

labor component, the operation is referred to as **labor intensive** (see also Section 14.12).

Additive manufacturing has significantly transformed low-volume production. Along with the development of computer software with three-dimensional geometric modeling ability, piece parts can now be designed and manufactured with less effort and cost than previously possible. Operator/programmer skill is still quite high, but rapid prototyping facilities can achieve almost the same flexibility as job shops, being limited only in the variety of materials that can be effectively processed.

Piece-part production usually involves small quantities and is suitable for job shops. The majority of piece-part production is in lot sizes of 50 or less. Quantities for **small-batch production** typically range from 10 to 100, using general-purpose machines and machining centers. **Batch production** typically involves lot sizes between 100 and 5000, and utilizes machinery similar to that used for small-batch production, but with specially designed fixtures for higher production rates.

Mass production generally involves quantities over 100,000, and requires special-purpose machinery (**dedicated machines**) and automated equipment for transferring materials and parts. Although the machinery, equipment, and specialized tooling are expensive, both the labor skills required and the labor costs are relatively low, because of the high level of automation. These production systems are organized for a specific type of product and thus they lack flexibility. Most manufacturing facilities operate with a variety of machines *in combination* and with various levels of automation and computer controls.

14.2.3 Applications of Automation

Automation can be applied to the manufacture of all types of materials, products, and production methods. The decision to automate a new or existing facility requires that the following additional considerations be taken into account:

- Type of product manufactured;
- Quantity and the rate of production required;
- The particular phase of the manufacturing operation to be automated;
- Level of skill in the available workforce;
- Reliability and maintenance of automated systems; and
- Economics.

Because automation involves a high initial cost and requires knowledge of the principles of operation and maintenance, a decision regarding the implementation of even low levels of automation must involve a study of the needs of an organization. Implementation of automation also changes the nature of a factory's workforce, generally causing greater demand for highly skilled *technicians* in certain specific areas.

14.2.4 Hard Automation

In *hard automation*, also called **fixed-position automation**, the production machines are designed to produce a standardized product, such as an engine block, valve, gear, or spindle. Although product size and processing parameters can be changed, these machines are specialized and lack flexibility; they cannot be modified to any significant extent to accommodate a variety of products with different shapes and dimensions (see also *group technology*, Section 15.8). Because these machines are expensive, their economical use requires the production of parts in very large quantities.

The machines used in hard-automation applications are usually fabricated on the **building-block (modular) principle.** Generally called **transfer machines,** they consist of two major components: power-head production units and transfer mechanisms. **Power-head production units** consist of a frame or bed, electric drive motors, gearboxes, and tool spindles, and are self-contained. Because of their intrinsic modularity, these units can easily be regrouped to produce a different part. Transfer machines consisting of two or more power-head units can be arranged on the shop floor in *linear, circular,* or *U patterns.* **Buffer storage** features are often incorporated in these systems to permit continued operation in the case of tool failure or individual machine breakdown.

Transfer mechanisms and transfer lines are used to move the workpiece from one station to another station in the machine, or from one machine to another, usually controlled by sensors and other devices. Workpieces in various stages of completion are transferred by several methods: (a) *rails* along which the parts, usually placed on pallets, are pushed or pulled by various mechanisms (Fig. 14.4a); (b) *rotary indexing tables* (Fig. 14.4b); and (c) *overhead conveyors.* Tools on transfer machines can easily be changed in toolholders with quick-change features (See, for example, Fig. 8.64). These machines are also equipped with various automatic gaging and inspection systems, to ensure that a part produced in one station is within acceptable dimensional tolerances before it is transferred to the next station. Transfer machines are used extensively in automated assembly (Section 14.10).

Figure 14.5 shows the **transfer lines,** or **flow lines,** in a very large system for producing cylinder heads for engine blocks. Consisting of a number of transfer machines, this system is capable of producing 100 cylinder heads per hour. Note the various machining operations being performed: milling, drilling, reaming, boring, tapping, and honing, as well as cleaning and gaging.

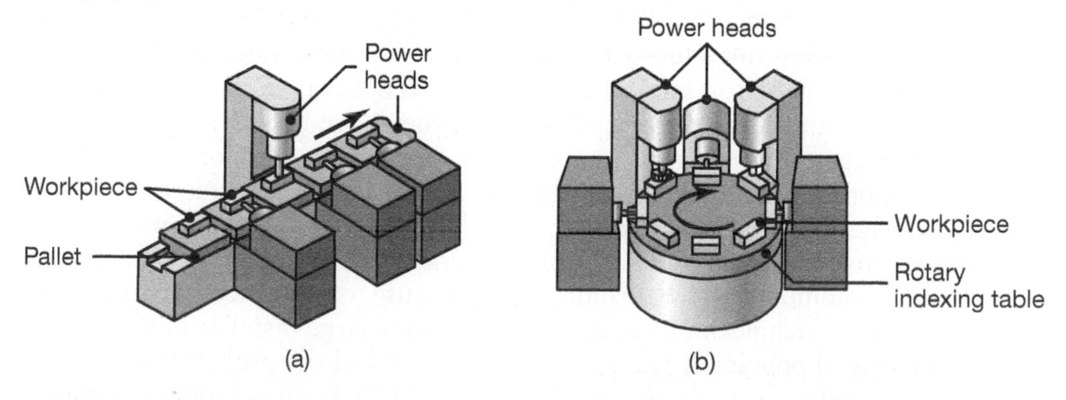

FIGURE 14.4 Two types of transfer mechanisms: (a) straight and (b) circular patterns.

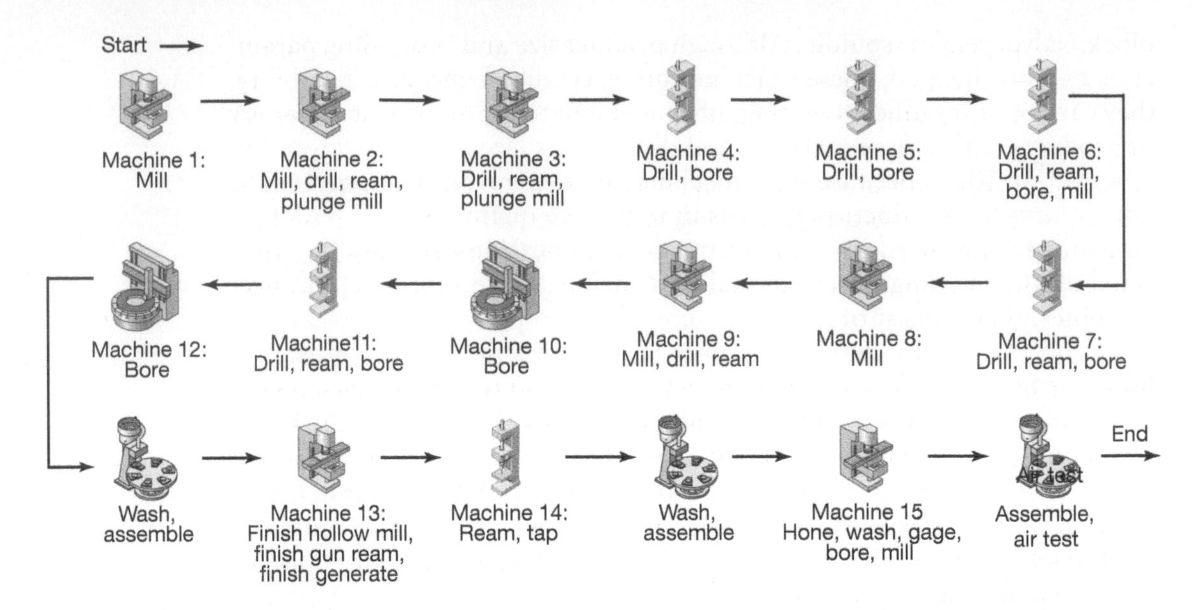

FIGURE 14.5 A traditional transfer line for producing engine blocks and cylinder heads.

14.2.5 Soft Automation

In contrast to hard automation, *soft automation*, also called *flexible automation* and *programmable automation*, has greater flexibility because of the use of computer control of the machinery (Sections 14.3 and 14.4 for details). The machines can easily be reprogrammed to produce a part that has a different shape or dimensions. Flexible automation is the principle behind **flexible manufacturing systems** (Section 15.10), which have high levels of efficiency and productivity.

14.2.6 Programmable Controllers

The control of a manufacturing operation has traditionally been performed by devices such as timers, switches, relays, counters, and similar hard-wired devices that are based on mechanical, electromechanical, and pneumatic principles. Beginning in 1968, **programmable logic controllers** (PLCs) were introduced to replace these devices. Because PLCs eliminate the need for relay control panels and because they can easily be reprogrammed and take up less space, they have been widely adopted in manufacturing systems and operations. Their basic functions are (a) on-off motion; (b) sequential operations; and (c) feedback control.

These controllers perform reliably in industrial environments and improve the overall efficiency of an operation. Although they have become less common in new installations (because of advances in numerical-control machines), PLCs still represent a very large installation base. Their sustained popularity is due to their low cost and the proliferation of powerful software to allow programming of PLCs from personal computers, which upload control programs via Ethernet or wireless communications. Modern PLCs are often programmed in specialized versions of BASIC or C programming languages. *Microcomputers* are now used more often because they are less expensive and are easier to program and to network. PLCs are also used in control of systems, with high-speed digital-processing and communication capabilities.

Programmable automation controllers is a term often used interchangeably with programmable logic controller, but generally refers to a higher performance product with more advanced communications tools, greater memory, and ability for modular architecture to be integrated with other devices and networks.

Microcontrollers. A number of low-cost but capable systems have been developed, notably the Arduino™ and Raspberry Pi™ platforms that are widely used in the Maker Movement (see Section 10.12). While not as robust as PLCs, these systems use a programming environment that allows rapid incorporation of control logic. These devices use proprietary, relatively simple programming languages, and are being increasingly applied in manufacturing.

14.2.7 Total Productive Maintenance

The management and maintenance of a wide variety of machines, equipment, and systems are among the significant aspects that affect productivity in a manufacturing organization. Consequently, the concepts of *total productive maintenance* (TPM) and *total productive equipment management* (TPEM) have become important. These activities include continuous analysis of such factors as (a) equipment breakdown and equipment problems; (b) monitoring and improving equipment productivity; (c) implementation of preventive and predictive maintenance; (d) reduction of setup time, idle time, and cycle time of machine; (e) full utilization of machinery and equipment and improvement of their effectiveness; and (f) reduction of defective products. Teamwork is an essential aspect of this activity and involves the full cooperation of machine operators, maintenance personnel, engineers, and management of the organization.

14.3 Numerical Control

Numerical control (NC) is a method of controlling the movements of machine components by directly inserting coded instructions, in the form of a program, into the system. The system automatically interprets the program and controls various machine components, such as, for example, (a) turning spindles on and off; (b) changing tools; (c) moving the workpiece or the tools along specific paths; and (d) turning cutting fluids on and off.

Numerical control machines are capable of producing parts repeatedly and accurately by executing part programs, as described in Section 14.4. Data concerning all aspects of the machining operation can be stored and retrieved for later analysis and review. Complex operations, such as turning a part that has various contours or die sinking in a milling machine, can easily be carried out.

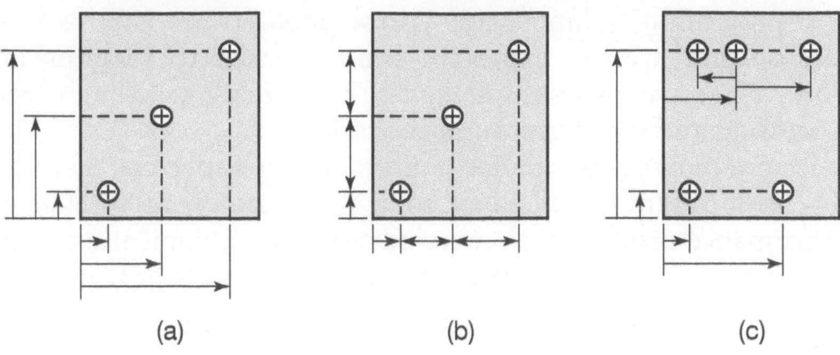

FIGURE 14.6 Positions of drilled holes in a workpiece. Three methods of measurements are shown: (a) absolute dimensioning, referenced from one point at the lower left of the part; (b) incremental dimensioning, made sequentially from one hole to another; and (c) mixed dimensioning, a combination of both methods.

14.3.1 Computer Numerical Control

In *computer numerical control* (CNC), the control hardware (mounted on the NC machine) follows directions received from *local* computer software. There are two types of computerized systems: direct numerical control and computer numerical control.

In **direct numerical control** (DNC), several machines are directly controlled, step by step, by a central computer, usually because the individual CNC controllers lack memory for especially complex parts or surface contours. In this case, the central computer sends blocks of information to the CNC controller, as needed; sequential execution of the code blocks produces the desired part. In this system, the status of all machines in a manufacturing facility can be monitored and assessed from the central computer. However, DNC has a significant disadvantage: If, for some reason, the computer shuts down, all the machines become inoperative. **Distributed numerical control** involves the use of a central computer, serving as the control system over a number of individual CNC machines with dedicated microcomputers. This system provides large memory and computational capabilities, and offers flexibility while overcoming the disadvantage of direct numerical control. Wireless DNC is now widely used instead of hard-wired computers.

Computer numerical control (CNC) is a system in which a control microcomputer is an integral part of a machine or equipment. The part program may be prepared at a remote site often by a software package, incorporating information obtained from CAD and machining simulations. CNC systems are widely used today because of the availability of inexpensive computers with large memory, microprocessors, and program-editing capabilities, as well as increased flexibility, accuracy, and versatility.

14.3.2 Principles of NC Machines

Fig. 14.7 depicts the basic elements and operation of a typical NC machine. The functional elements involved are:

1. **Data input:** The numerical information is read and stored in computer memory.
2. **Data processing:** The programs are read into the machine control unit for processing.

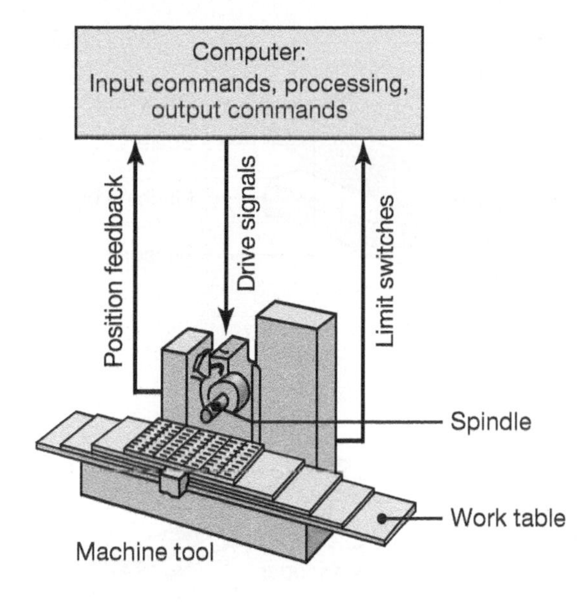

FIGURE 14.7 Schematic illustration of the major components of a numerical control machine tool.

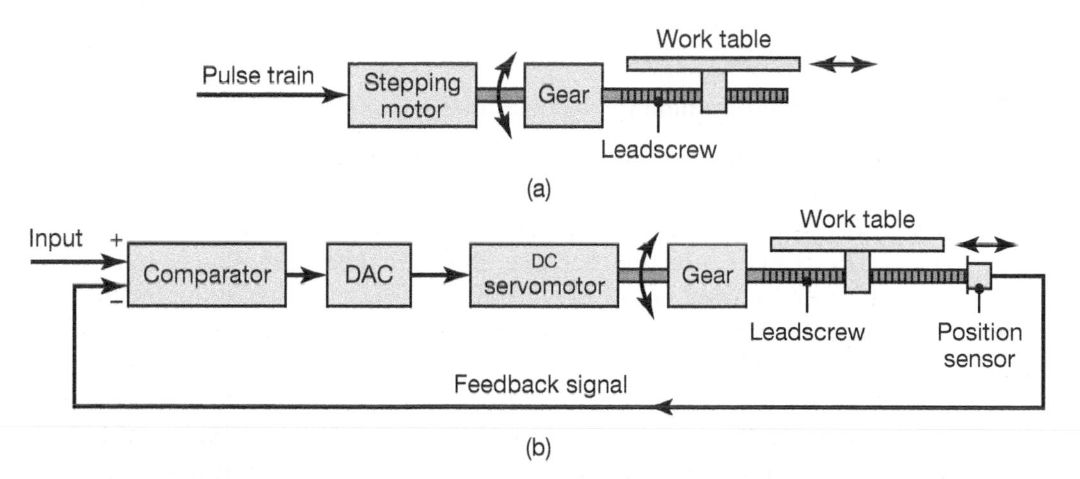

FIGURE 14.8 Schematic illustration of the components of (a) an open-loop, and (b) a closed-loop control system for a numerical control machine. (DAC is digital-to-analog converter.)

3. **Data output:** The information is translated into commands (typically pulsed commands) to the servomotor (Fig. 14.8), which then moves the machine table to specific positions by means of stepping motors, lead screws, and other devices.

Types of control. In the **open-loop system** (Fig. 14.8a), the signals are sent to the servomotor by the controller, but the movements and final positions of the worktable are not checked for accuracy. In contrast, the **closed-loop** system (Fig. 14.8b) is equipped with various transducers, sensors, and counters that accurately measure the position of the worktable. Through **feedback control**, the position of the worktable is compared against the signal, and table movements terminate when the proper coordinates are reached.

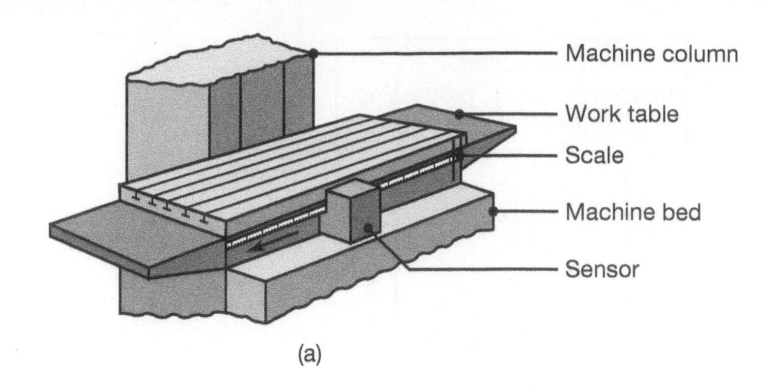

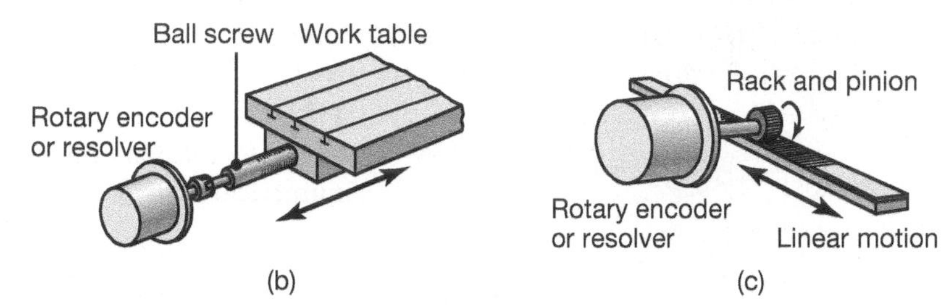

FIGURE 14.9 (a) Direct measurement of the linear displacement of a machine-tool worktable. (b) and (c) Indirect measurement methods.

Position measurement in NC machines is accomplished by two methods:

1. In the *direct measuring system*, a sensing device reads a graduated scale on the machine table or a slide for linear movement (Fig. 14.9a). This system is more accurate than indirect methods, because the scale is built into the machine, and *backlash* (the play between two adjacent mating gear teeth) in the mechanisms is not significant.
2. In the *indirect measuring system*, **rotary encoders** or **resolvers** (Figs. 14.9b and c) convert rotary movement to translation movement, but worktable backlash can significantly affect measurement accuracy. Position feedback mechanisms use various sensors that are based mainly on magnetic and photoelectric principles.

EXAMPLE 14.1 Comparison of Open-Loop and Closed-Loop Controls

Given: Consider the simple, one-dimensional open-loop and closed-loop control systems illustrated in Fig. 14.8. Assume that the worktable mass, m, is known and that control is achieved by changing the force applied to the worktable through a torque to the gear.

Find: If the system is frictionless, develop equations for the force required to obtain a new position for (a) open-loop control; and (b) closed-loop control, where both position and velocity are measured.

Solution: For the open-loop control system, the new position is determined by the equations of motion:

$$x - x_0 = \frac{1}{2}at^2,$$

where it is assumed that the initial velocity is zero (at time $t = 0$) and the initial position is x_0. By applying $F = ma$, but realizing that the mass has to accelerate for the first half of the desired motion and decelerate for the second half, the force as a function of time is given by

$$F = ma \quad \text{for} \quad 0 < t < \frac{x - x_0}{a}$$

and

$$F = -ma \quad \text{for} \quad \frac{x - x_0}{a} < t < \frac{2(x - x_0)}{a}.$$

In theory, the acceleration selected for this system could be defined by the maximum torque that can be developed by the motor to minimize the time required to attain the new position. However, motors cannot attain their maximum torque instantaneously and, moreover, the frictionless situation is not realistic. Note also that when external forces act upon the worktable, it will move; there is no provision in an open-loop control system to ensure that the desired position will be achieved or maintained.

For the closed-loop control system, the force is given by

$$F = -k_v\dot{x} + k_p(x - x_0),$$

where $\dot{x}$ is the instantaneous velocity, x is the desired position, and k_v and k_p are the velocity and position *gains* of the worktable, respectively. If the gains are selected properly, the closed-loop control system can maintain the desired position and be stable even if external forces are applied.

14.3.3 Types of Control Systems

There are two basic types of control systems in numerical control: point-to-point and contouring.

1. In the **point-to-point system,** also called the **positioning system,** each axis of the machine is driven separately by lead screws and, depending on the type of operation, at different speeds. The machine initially moves at maximum velocity (in order to reduce nonproductive time) but decelerates as the tool approaches its numerically defined position. Thus, for example, in an operation such as drilling, the positioning and drilling take place *sequentially* (Fig. 14.10a). After the hole is drilled, the tool retracts and moves rapidly to another position, and the operation is repeated. Punching is another application of this type of system.
2. In the **contouring system,** also known as the **continuous-path system,** the positioning and the machining are both performed along controlled paths, as in milling (Fig. 14.10b). Because the tool is operating as it travels along a prescribed path, accurate control, and synchronization of velocities and movements are important. Other applications of the contouring system include lathes, grinders, welding machinery, and machining centers.

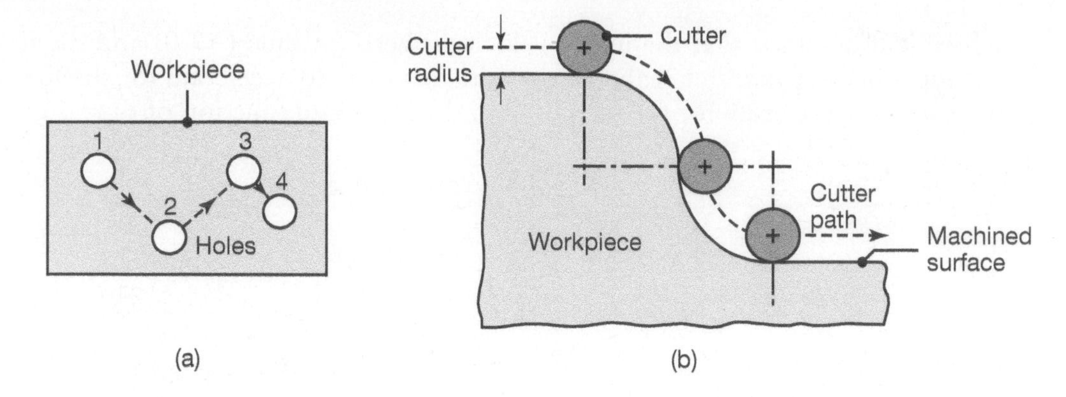

FIGURE 14.10 Movement of tools in numerical control machining. (a) Point-to-point system: The drill bit drills a hole at position 1, is then retracted and moved to position 2, and so on. (b) Continuous path by a milling cutter; note that the cutter path is compensated for by the cutter radius and it can also compensate for cutter wear.

Interpolation. Movement along a path (*interpolation*) occurs incrementally by one of several basic methods (Fig. 14.11). Figure 14.12 illustrates examples of actual paths in drilling, boring, and milling operations. In all interpolations, the tool path controlled is that of the *center of rotation* of the tool. Compensation for different types of tools, different tool diameters, or for tool wear during machining can be made in the NC program.

1. In **linear interpolation**, the tool moves in a *straight* line, from start to end (Fig. 14.11a), on two or three axes. Theoretically, all types of profiles can be produced by this method, by making the increments between the points small (Fig. 14.11b); however, a large amount of data has to be processed in this type of interpolation.
2. In **circular interpolation** (Fig. 14.11c), the inputs required for the tool path are (a) the coordinates of the end points; (b) the coordinates of the center of the circle and its radius; and (c) the direction of the tool along the arc.
3. In **parabolic interpolation** and **cubic interpolation**, the tool path is approximated by *curves*, using higher order mathematical equations. This method is effective in five-axis machines (Section 8.11.2) and is

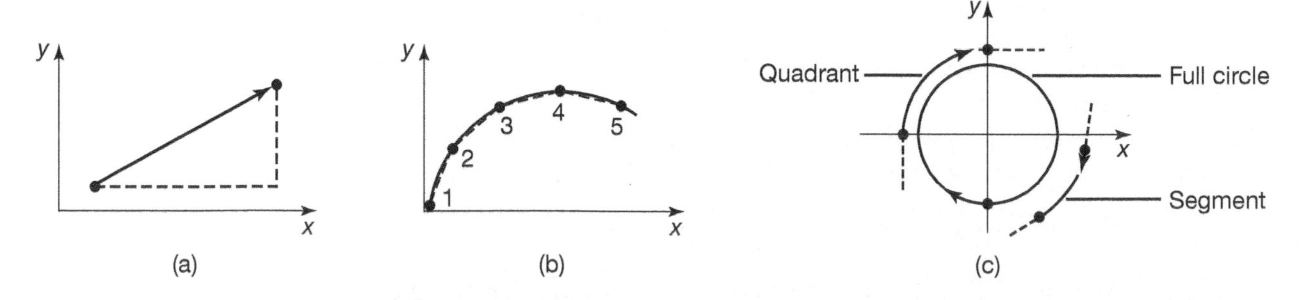

FIGURE 14.11 Types of interpolation in numerical control: (a) linear; (b) continuous path approximated by incremental straight lines; and (c) circular.

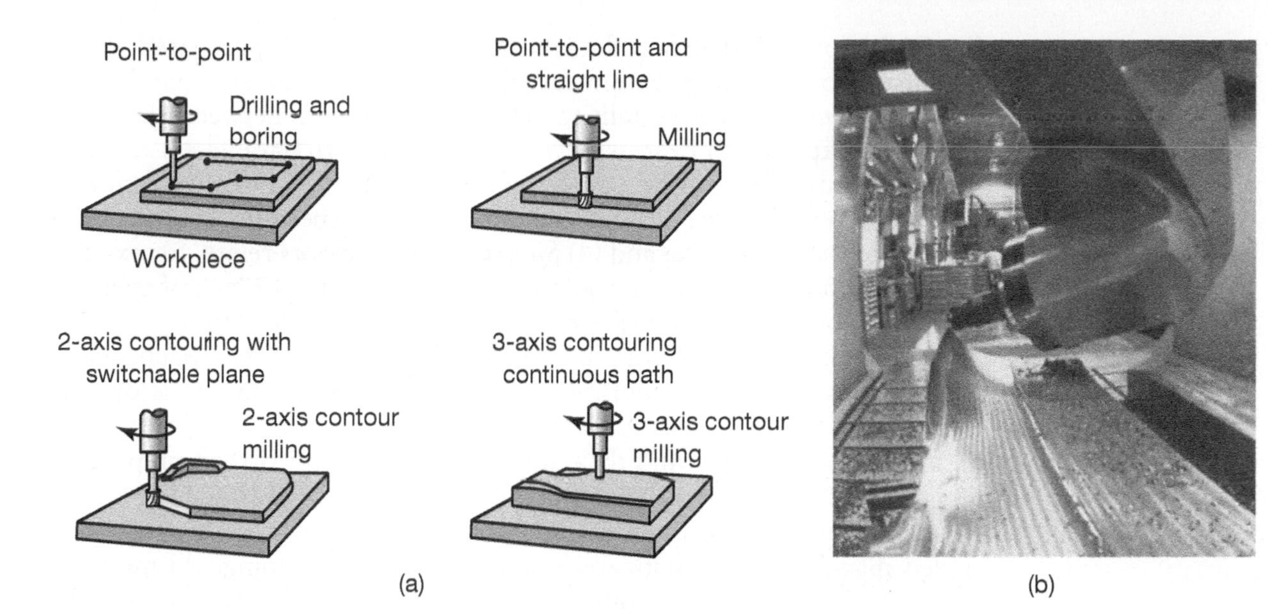

FIGURE 14.12 (a) Schematic illustration of drilling, boring, and milling operations with various cutter paths. (b) Machining a sculptured surface on a five-axis numerical control machine. *Source:* The Ingersoll Milling Machine Co.

particularly useful in die-sinking operations, such as for sheet forming of automotive bodies. These interpolations are also used for the movements of industrial robots (see Section 14.7).

14.3.4 Positioning Accuracy of NC Machines

Positioning accuracy is defined by how accurately the machine can be positioned to a certain coordinate system. An NC machine, for example, generally has a positioning accuracy of at least ± 3 μm. **Repeatability** is defined as the closeness of agreement of repeated position movements under the same operating conditions of the machine; it is usually about ± 8 μm. **Resolution** is defined as the smallest increment of motion of the machine components; it is usually about 2.5 μm.

The **stiffness** of the machine tool (Section 8.13) and elimination of the *backlash* in its gear drives and lead screws are important for dimensional accuracy. Backlash in modern machines is eliminated by means such as using preloaded ball screws. Rapid response of the machine tool to command signals requires that friction and inertia effects be minimized, for example, by reducing the mass of moving components of the machine.

14.4 Programming for Numerical Control

A *program* for numerical control consists of a sequence of directions that instructs the machine to carry out a series of specific operations. *Programming for NC* may be accomplished with the same software that is used for CAD, done on the shop floor, or it can be a service purchased from an outside source. The program contains instructions and commands: (1) *Geometric instructions* pertain to relative movements between the tool and the workpiece; (2) *Processing instructions* concern parameters such as spindle speeds, feeds, cutting tools, and cutting fluids; (3) *Travel instructions* pertain to the type of interpolation and to the speed of movement of the tool or the worktable; and (4) *Switching instructions* relate, for example, to the on-off position for coolant supplies, direction or lack of spindle rotation, tool changes, workpiece feeding, and workpiece clamping.

Manual part programming consists first of calculating the dimensional relationships of the tool, workpiece, and worktable, on the basis of the engineering drawings of the part (including CAD; Section 15.4), the operations to be performed and their sequence. A program is then prepared, detailing the information necessary for carrying out the particular task. Because they are familiar with machine tools and process capabilities, skilled machinists can, with some training in programming, do manual programming, but it has become less popular as CAD software capabilities have improved.

Computer-aided part programming involves special symbolic **programming languages** that determine the coordinate points of corners, edges, and surfaces of the part. A programming language is a means of communicating with the computer, and involves the use of symbolic characters. The programmer describes the component to be processed in this language, and the computer converts that description to commands for the NC machine.

Complex parts are machined using graphics-based, computer-aided machining programs. Standardized programming languages such as STEP-NC and the older but still common G-Code are used for communicating machining instructions to the CNC hardware. The STEP-NC software is a standardized language that has been extended beyond machine tools, and incorporates models for milling, turning, and EDM. Plasma cutting and laser welding and cutting systems also have been developed for use with STEP-NC.

In *shop-floor programming*, CNC programming software is used directly on the machine tool controller. This allows higher-level geometry and processing information to be sent to the CNC controller instead of G-code. G-code is then developed by the dedicated computer under the control of the machine operator. The advantage of this method is that any changes that are made to the machining program is sent back to the programming group and is stored as a shop-proven design iteration; it can be re-used or standardized across a family of parts.

14.5 Adaptive Control

In *adaptive control* (AC), the operating parameters automatically adapt themselves to conform to new circumstances, such as changes in the dynamics of the particular process and any disturbances that may arise; this approach is basically a feedback system. Human reactions to everyday occurrences in life already contain dynamic feedback control. For example, driving a car on a smooth road is relatively easy, and one needs to make few, if any, adjustments. However, on a rough road, one has to steer to avoid potholes by observing the condition of the road. Also, one feels the car's rough movements and vibrations, then reacts by changing the direction and the speed of the car to minimize the effects of the rough road and to increase the comfort of the ride. Similarly, an **adaptive controller** would check these conditions, adapt an appropriate desired braking profile, for example, antilock brake system and traction control, and then use feedback to implement it.

Several adaptive-control systems are commercially available for applications such as ship steering, chemical-reactor control, rolling mills, and medical technology. Adaptive control has been widely used in continuous processing in the chemical industry and in oil refineries. In manufacturing operations, adaptive control helps (a) optimize *production rate*; (b) optimize *product quality*; and (c) minimize *production costs*. Application of AC in manufacturing is particularly important in situations where incoming workpiece dimensions and quality are not uniform, such as a poor casting or an improperly heat-treated part.

Gain scheduling is the simplest form of adaptive control, where a different gain is selected depending on the measured operating conditions and a different gain is assigned to each region of the system's operating space. With advanced adaptive controllers, the gain may vary continuously with changes in operating conditions. Adaptive control is a logical extension of CNC systems. As described in Section 14.4, the part programmer sets the processing parameters, based on the existing knowledge of the workpiece material and various data on the particular manufacturing process. Whereas in CNC machines, these parameters are held constant during a particular processing cycle, in adaptive control the system is capable of automatic adjustments *during* processing through closed-loop feedback control (Fig. 14.13).

Principles and applications of adaptive control. The basic functions common to adaptive-control systems are:

1. Determine the operating conditions of the process, including measures of performance. This information is typically obtained by using sensors that measure process parameters, such as force, torque, vibration, and temperature.
2. Configure the process control in response to the operating conditions. Large changes in the operating conditions may provoke a decision to make a major switch in control strategy; more modest alterations may include the modification of process parameters, such as the speed of the operation or the feed in machining.
3. Continue to monitor the operation, making further changes in the controller as needed.

In processes such as turning on a lathe (Section 8.9.2), for example, the adaptive-control system would sense, in real-time, cutting forces, torque, temperature, tool-wear rate, tool chipping or tool fracture, and surface

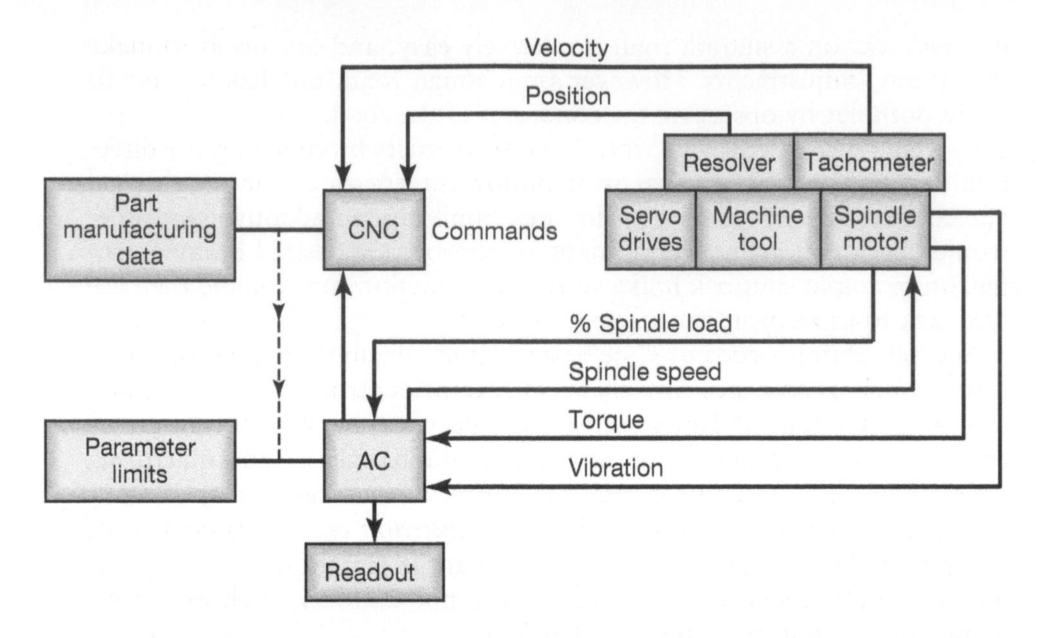

FIGURE 14.13 Schematic illustration of the application of adaptive control (AC) for a turning operation. The system monitors such parameters as cutting force, torque, and vibrations; if they are excessive, AC modifies process variables, such as feed and depth of cut, to bring them back to acceptable levels.

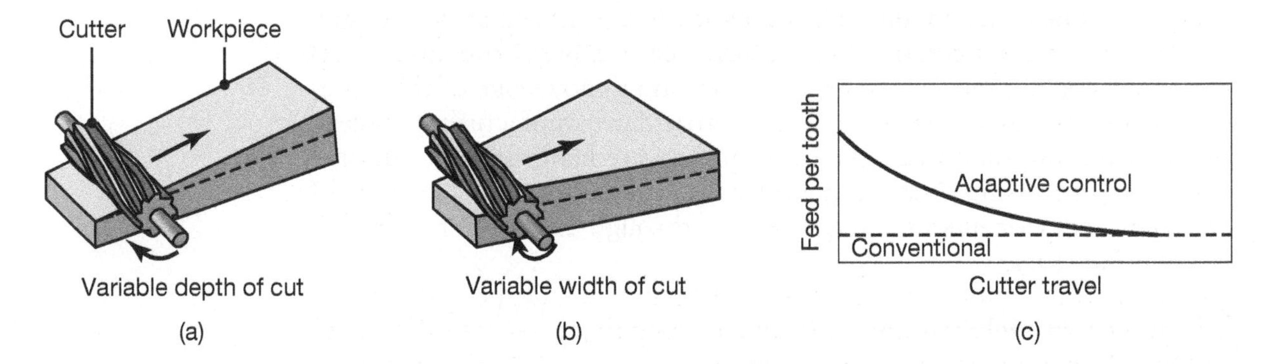

FIGURE 14.14 An example of adaptive control in slab milling. As the depth of cut or the width of cut increases, the cutting forces and the torque increase; the system senses this increase and automatically reduces the feed to avoid excessive forces or tool breakage. *Source:* After Y. Koren.

finish of the workpiece. The system then converts this information into commands that modify the process parameters on the machine tool to hold the parameters constant (or within certain limits) or to optimize the machining operation.

Systems that place a constraint on a process variable are called **adaptive-control constraint** (ACC) systems. Thus, if the thrust force and the cutting force (hence the torque) increase excessively because, for instance, of the presence of a hard region in a casting, the AC system changes the cutting speed or the feed to lower the cutting force to an acceptable level (Fig. 14.14). Note that without AC or the direct intervention of the operator, high cutting forces may cause the tools to chip or break or cause the tool or workpiece to deflect excessively; as a result, the dimensional accuracy and surface finish of the part deteriorate.

Systems that optimize an operation are called **adaptive-control optimization** (ACO) systems. Optimization may, for example, involve maximizing material-removal rate between tool changes or indexing, or improving surface finish. Most systems are based on ACC, because the development and proper implementation of ACO is more complex.

Response time during the operation must be short for AC to be effective, particularly in high-speed machining operations (Section 8.8). Assume, for example, that a turning operation is being performed on a lathe at a spindle speed of 1000 rpm; the tool suddenly chips off, adversely affecting the surface finish and dimensional accuracy of the part being machined. In order for the AC system to be effective, the sensing system must respond within a very short time, as otherwise the damage to the workpiece will be extensive.

For adaptive control to be effective, *quantitative* relationships must be established and programmed. For example, if the tool-wear rate in a machining operation is excessive, the computer must be able to determine how much of a change (increase or decrease) in cutting speed or feed is necessary in order to reduce the wear rate to an acceptable level. The system must also be able to compensate for dimensional changes in the workpiece due to such causes as tool wear or temperature rise (Fig. 14.15). If the operation is grinding, the computer software must contain the desired quantitative relationships among process variables and such parameters as wheel wear, dulling of abrasive grains, grinding forces, temperature, surface finish, and part deflections. Likewise, in bending of a sheet in a V-die, data on the dependence of springback on punch travel and on other material and process variables must be stored in the computer memory.

Because of the numerous factors involved, mathematical equations for such quantitative relationships in manufacturing processes are difficult to establish. Compared to the various other parameters involved, forces and torque in machining have been found to be the easiest to monitor by AC. Several solid-state power controls are commercially available, in which power is displayed or interfaced with data-acquisition systems.

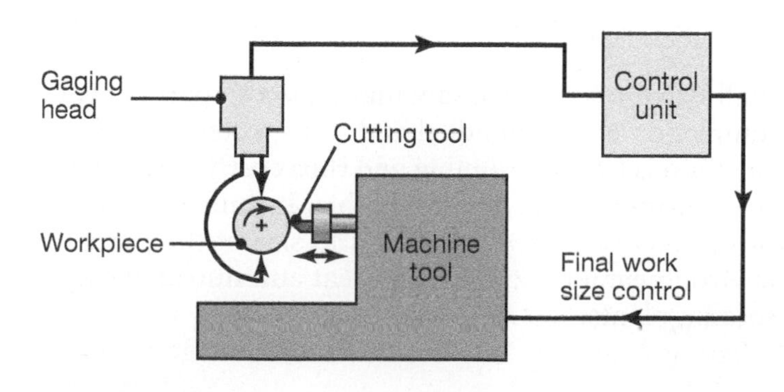

FIGURE 14.15 In-process inspection of workpiece diameter in a turning operation. The system automatically adjusts the radial position of the cutting tool in order to machine the correct diameter.

14.6 Material Handling and Movement

Material handling is defined as the functions and systems associated with the transportation, storage, and control of materials and parts in the total manufacturing cycle of a product. During this cycle, raw materials and parts in various stages of completion (*work-in-progress, or WIP*) are typically moved from storage to machines, from machine to machine, from inspection to assembly and to inventory (see also *just-in-time*, Section 15.12), and finally to shipment. For example, (a) a forging is mounted on a milling-machine bed (Fig. 8.66) for finish machining for better dimensional accuracy; (b) the machined forging may subsequently be ground for even better surface finish and dimensional accuracy; and (c) the part is inspected prior to being assembled into a product. Similarly, cutting tools have to be mounted on toolholders on lathes, dies have to be placed in presses or hammers, grinding wheels have to be mounted on spindles, and parts have to be placed in special fixtures for dimensional measurement and inspection.

Plant layout is crucial for the orderly flow of materials and components throughout the manufacturing cycle. Some plant layout considerations are (a) distances required for moving raw materials and parts in progress should be minimized, (b) storage areas and service centers should be organized accordingly, and for parts requiring multiple operations, equipment should be grouped around the operator or the industrial robot (see also *cellular manufacturing*, Section 15.9).

Important aspects of material handling include:

1. **Methods of material handling.** Several factors must be considered in selecting an appropriate material-handling method for a particular manufacturing operation:

   a. Shape, size, weight, and characteristics of the parts;
   b. Distances involved and the position and orientation of the parts during movement through the machines, and at their final destination;
   c. Conditions of the path and any obstructions along which the parts are to be transported;
   d. Level of automation and control, and their integration with other equipment in the system;
   e. Operator skill required; and
   f. Economic considerations.

   Although for small batch operations, raw materials and parts in various stages of completion can be handled and transported manually, this traditional method is time consuming and thus costly. Moreover, this practice can be unpredictable and unreliable. It can be unsafe to the operator, for reasons such as the weight and size of the parts to be moved and environmental factors such as heat and smoke in older foundries and forging plants.

2. **Equipment.** Several types of equipment can be used to move materials: conveyors, rollers, carts, forklift trucks, automated guided vehicle, self-powered monorails, and various mechanical, electrical, magnetic, pneumatic, and hydraulic devices and manipulators. **Manipulators** are designed to be controlled directly by the operator, or they can be automated for repetitive movements, such as loading and unloading of parts from machine tools, presses, dies, and furnaces. They are capable of gripping and moving heavy parts and orienting them as necessary. Machinery combinations that have the capability of conveying parts without the need for additional equipment are called **integral transfer devices**.

   Warehouse space must be used efficiently by using **automated storage/retrieval systems** (AS/RS); there are several such systems available, with varying degrees of automation and complexity. However, these systems are not always desirable because of the importance of *minimal or zero inventory* and on *just-in-time production* methods (Section 15.12).
3. **Automated guided vehicles.** Flexible material handling and movement, with real-time control, is an integral part of modern manufacturing. First developed in the 1950s, *automated guided vehicles* (AGVs) are used extensively in flexible manufacturing (Fig. 14.16). This transport system has high flexibility and is capable of efficient delivery to different workstations; the movements of AGVs can be planned so that they also interface with automated storage/retrieval systems.

   There are several types of AGVs, with a variety of designs, load-carrying capacities, and features for specific applications. Among these are: (a) unit-load vehicles with pallets that can be handled from both sides of the aisle; (b) light-load vehicles, equipped with bins or trays for light manufacturing; (c) tow vehicles (Fig. 14.16a);

(a)

(b)

FIGURE 14.16 (a) A self-guided vehicle (Tugger type). This vehicle can be arranged in a variety of configurations to pull caster-mounted cars; it has a laser sensor to ensure that the vehicle operates safely around people and various obstructions. (b) A self-guided vehicle configured with forks for use in a warehouse. *Source:* Courtesy of Egemin, Inc.

(d) forklift-style vehicles (Fig 14.16b); and (e) assembly-line vehicles to carry subassemblies to final assembly.

AGVs are guided automatically along pathways with in-floor wiring (for *magnetic guidance*) or tapes or fluorescent-painted strips (for *optical guidance*, called *chemical guide path*). *Autonomous guidance* involves no wiring or tapes, using various optical, ultrasonic, or inertial techniques with onboard controllers. *Routing* of the AGV can be monitored and controlled from a central computer, such that the system optimizes the movement of materials and parts in case of congestion around workstations, machine breakdown, or the failure of one section of the manufacturing system. Traffic management on the factory floor is important; sensors and various controls on the vehicles are designed to avoid collisions with other AGVs or machinery on the plant floor.

4. **Coding systems.** Various coding systems have been developed to locate and identify parts and subassemblies throughout the production system and to correctly transfer them to their appropriate stations:

   a. **Bar coding**, including **QR barcodes** (as also used in the margins this book), is the most widely used and least costly system. The codes are printed on labels attached to the parts themselves, and read by fixed or portable bar code readers or handheld scanners.
   b. **Magnetic strips** are the second most common coding system.
   c. **Radio-frequency (RF) tags** are the third system. Although more expensive than bar codes, the tags do not require the clear line of sight necessary for the first two systems; they have a long range, from hundreds of meters for conventional RF tags to around 10–30 m for Bluetooth systems (Section 15.14), and are rewritable.
   d. **Acoustic waves, optical character recognition,** and **machine vision** are other identification methods (see Section 14.8.1).

14.7 Industrial Robots

QR Code 14.1 Ten most popular applications for robots. *Source:* Courtesy of ABB Robotics.

The word *robot* was coined in 1920 by K. Čapek in his play *R.U.R.* (Rossum's Universal Robots); it is derived from the Czech word *robota*, meaning worker. An *industrial robot* has been defined as a reprogrammable multifunctional manipulator designed to move materials, parts, tools, or other devices by means of variable programmed motions and to perform a variety of other tasks. In a broader context, the term *robot* also includes manipulators that are activated directly by an operator. Today, industrial robots are critical components in manufacturing operations, and have contributed to productivity and product quality improvements, and have significantly reduced labor costs.

14.7.1 Robot Components

The functions of robot components and their capabilities can readily be appreciated by observing the flexibility and capability of the diverse

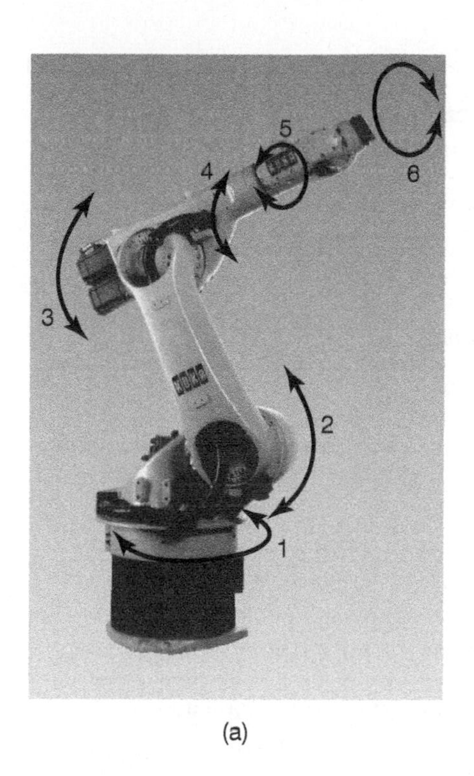

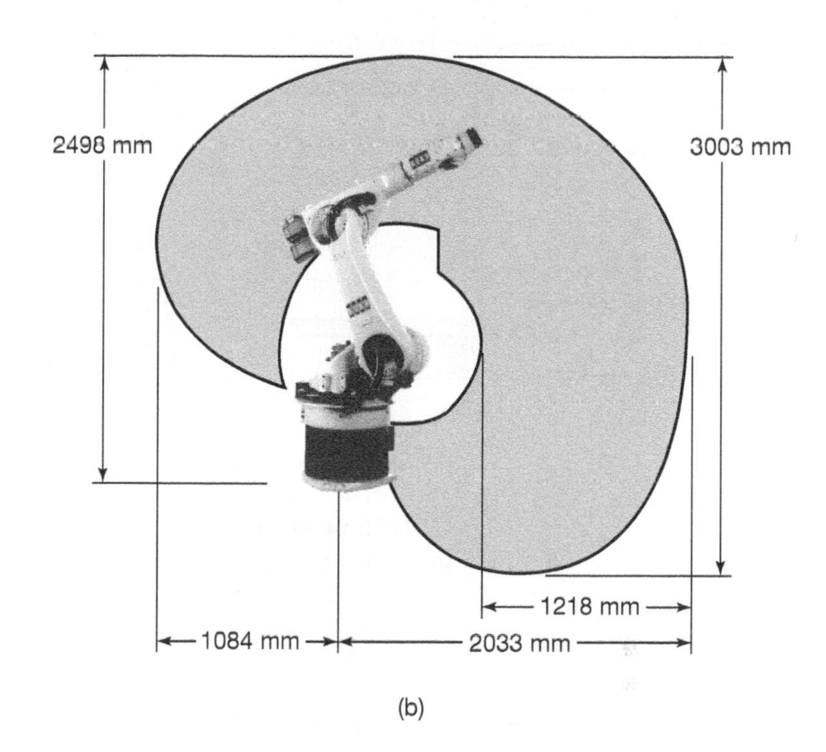

FIGURE 14.17 (a) Schematic of a six-axis KR-30 KUKA robot; the payload at the wrist is 30 kg (65 lbs) and repeatability is ±0.15 mm (±0.006 in.). The robot has mechanical brakes on all of its axes. (b) The work envelope of the KUKA robot, as viewed from the side. *Source:* Courtesy of KUKA Robotics.

movements of the human fingers, hands, wrists, and arms. The basic components of an industrial robot are (Fig. 14.17a):

QR Code 14.2 A new era of robotics. *Source:* Courtesy of ABB Robotics.

1. **Manipulator.** Also called **arm and wrist**, the *manipulator* is a mechanical unit that provides motions (trajectories) similar to those of a human arm and hand, using various devices such as linkages, gears, and joints. The end of the wrist can reach a point in space with a specific set of coordinates and in a specific orientation. Most robots have six rotational joints (See Fig. 14.17a). There are also four-degrees-of-freedom (d.o.f.) and five-d.o.f. robots, but these kinds are not fully dexterous, because full dexterity, by definition, requires six-d.o.f. Seven-d.o.f. (or *redundant*) robots for special applications also are available.
2. **End effector.** The end of the wrist in a robot is equipped with an *end effector*. Also called *end-of-arm tooling*, end effectors can be custom made to meet special handling requirements. *Mechanical grippers* are the most commonly used end effectors and are equipped with two or more fingers. The selection of an appropriate end effector for a specific application depends on such factors as the *payload* (weight of the object to be lifted and moved), environment, reliability, and cost. Depending on the type of operation, end effectors may be equipped with any of the following:

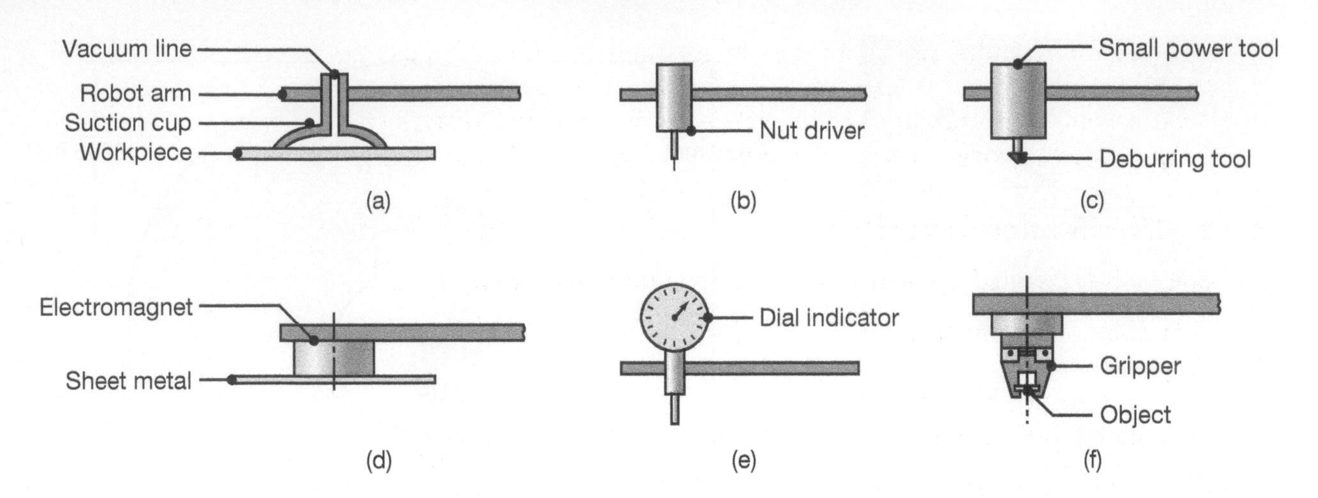

FIGURE 14.18 Devices and tools that can be attached to end effectors to perform a variety of operations.

a. Grippers, hooks, scoops, electromagnets, vacuum cups, and adhesive fingers, for material handling (Fig. 14.18);
b. Various attachments, such as for spot and arc welding and for arc cutting;
c. Power tools, such as drills, nut drivers, burrs, and sanding belts;
d. Measuring instruments, such as dial indicators and laser or contact probes; and
e. Spray guns, for painting.

Compliant end effectors are designed to handle fragile items or to facilitate assembly; they can include elastic mechanisms to limit the force that can be applied to a workpiece or part (such as an egg), and can be designed with a specific required stiffness. For example, end effectors can be designed to be stiff in one direction, but compliant in another direction. This arrangement also prevents damage to parts when slight misalignment occur.

3. The **control system** is the *brain* of a robot. Also known as the **controller**, the control system is a communications and information-processing system that gives commands for the movements of the robot; it stores data to initiate and to terminate movements of the manipulator.

Feedback devices are an important part of a robot's control system. Robots with a *fixed set of motions* have **open-loop control**; in this system, commands are given, and the robot arm goes through its motions. However, accuracy of the movements is not monitored and the system does not have a self-correcting capability (see also Section 14.3.1 and Fig. 14.8). In **closed-loop systems**, positioning feedback has better accuracy.

As in numerical control machines, the types of control in industrial robots are *point to point* or *continuous path* (Section 14.3.3). Depending on the particular task, the *positioning repeatability* required for an industrial robot may be as small as 0.050 mm, as in assembly operations for electronic printed circuitry (Section 13.13). Accuracy and repeatability vary greatly with payload and with position within the *work envelope* (Section 14.7.2).

14.7.2 Classification of Robots

Robots may be classified by their basic types, as illustrated in Fig. 14.19.

1. **Cartesian,** or **rectilinear;**
2. **Cylindrical;**
3. **Spherical,** or **polar;** and
4. **Articulated, revolute, jointed,** or **anthropomorphic.**

Robots may be attached permanently to the plant floor, move along overhead rails (**gantry robots**), or they may be equipped with wheels and move along the factory floor (**mobile robots**). Robots can also be classified according to their operation and programming.

1. **Fixed-sequence and variable-sequence robots.** Also called a **pick-and-place robot**, a fixed-sequence robot is programmed for a specific sequence of operations movements, which are from point to point; the cycle is repeated continuously. These robots are simple and relatively inexpensive. The *variable-sequence robot* can be programmed for multiple specific operation sequences, any of which it can execute when given the proper cue, such as a signal from a controlling or scheduling computer, a bar code, or a signal from an inspection station.
2. **Playback robot.** An operator leads or walks the playback robot and its end effector through the desired path; in other words, the operator teaches the robot by showing it what to do. The robot memorizes and records the path and sequence of motions; it can then repeat them continually without any further action or guidance. The robot has a *teach pendant*, which uses handheld button boxes that are connected

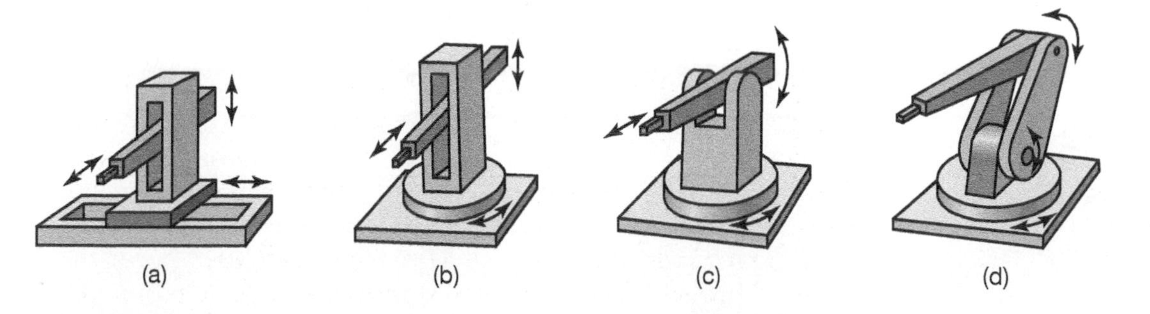

FIGURE 14.19 Four types of industrial robots: (a) Cartesian (rectilinear); (b) cylindrical; (c) spherical (polar); and (d) articulated (revolute, jointed, or anthropomorphic); some modern robots are *anthropomorphic*, meaning that they resemble humans in shape and in movement. The complex mechanisms in robots are made possible by powerful computer processors and fast motors that can maintain a robot's balance and accurate movement control.

to the control panel; these boxes are used to control and guide the robot through the work to be performed. These movements are registered in the memory of the controller, and are automatically reenacted by the robot whenever required.

3. **Numerically controlled robot.** This type of robot is programmed and operated much like a numerically controlled machine; the robot is *servocontrolled* by digital data, and its sequence of movements can be changed with relative ease. There are two basic types of controls: point to point and continuous path, as in NC machines. *Point-to-point* robots are easy to program and have a higher payload and a larger **work envelope** (also called the **working envelope**), the maximum extent or reach of the robot in all directions, as illustrated in Figs. 14.17b and 14.20. *Continuous-path* robots have a higher accuracy than point-to-point robots, but they have a lower payload. More advanced robots have a complex system of path control, enabling high-speed movements with high accuracy.
4. **Intelligent (sensory) robot.** The robot is capable of performing some of the functions and tasks carried out by people. It is equipped with a variety of sensors with visual (*computer vision*, Section 14.8) and *tactile* (touching) capabilities. Much like humans, the robot observes and evaluates the immediate environment and its proximity to other objects in its path, by *perception* and *pattern recognition*. The robot then makes appropriate decisions for the next movement and proceeds accordingly. Because its operation is complex, powerful computers are required to control this type of robot.

   Developments in intelligent robots include:

   - Behaving more like humans and performing tasks, such as moving among a variety of machines and equipment on the shop floor and avoiding collisions.
   - Recognizing, selecting, and gripping the correct raw material or workpiece for further processing.
   - Transporting a part from machine to machine.

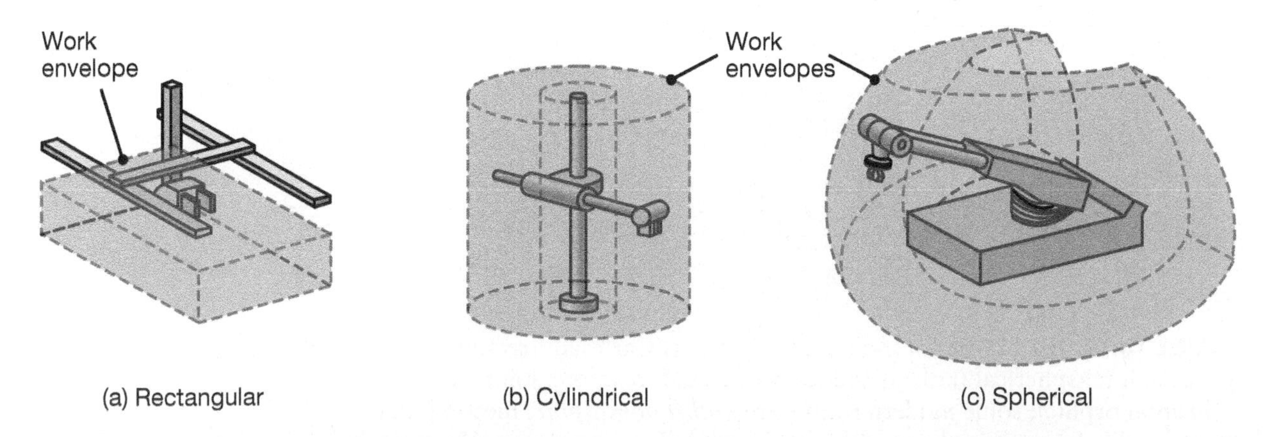

FIGURE 14.20 Work envelopes for three types of robots. The selection depends on the particular application. (See also Fig. 14.17b.)

- Assembling components into subassemblies or into a final product.

5. **Cobot.** Robots are typically programmed to perform a certain task without human guidance and are then isolated behind a fence or guard for the protection of operating personnel. *Cobots*, or *collaborative robots*, are designed to interact with plant personnel and have sensors to aide interaction with humans. Most applications to date have been in assembly or in materials handling; depending on the definition used, modern automobiles may indeed be considered to be cobots.

14.7.3 Applications and Selection of Robots

Major applications of industrial robots include:

1. *Material handling*, such as (a) casting and molding operations, in which molten metal, raw materials, and parts in various stages of completion are handled without operator interference; (b) heat treating, in which parts are loaded and unloaded from furnaces and quench baths; and (c) forming and shaping operations, in which parts are loaded and unloaded from presses.
2. *Spot welding*, especially for automobile and truck bodies, producing reliable welds of high quality (Fig. 14.21), including arc welding, arc cutting, and riveting.
3. *Finishing operations*, such as grinding, deburring, and polishing, using appropriate tools attached to end effectors.
4. *Applying adhesives and sealants*, such as in the automobile frame shown in Fig. 14.22.
5. *Spray painting*, particularly of parts with complex shapes, and cleaning operations.

FIGURE 14.21 Spot welding automobile bodies with industrial robots. *Source:* Marin Tomas/ Getty Images.

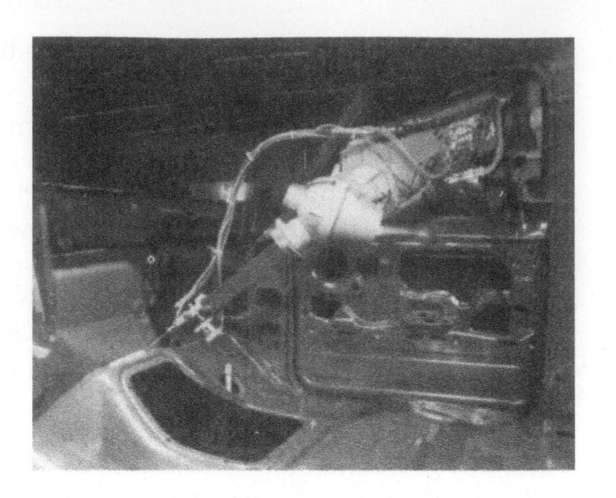

FIGURE 14.22 Sealing joints of an automobile body with an industrial robot.

6. *Automated assembly*, performing repetitive operations (Fig. 14.23) (see also Section 14.10).
7. *Inspection and gaging*, in various stages of manufacture and at speeds much higher than can be done manually.

Robot selection. Factors influencing selection of robots in manufacturing operations are: (a) payload; (b) speed of movement; (c) reliability; (d) repeatability; (e) arm configuration; (f) the degrees of freedom; (g) control system; (h) program memory; (i) work envelope; and (j) cost. Robots are rarely off-the-shelf items; they must be integrated with controllers, end effectors, and their environment.

14.8 Sensor Technology

A *sensor* is a device that produces a signal in response to its detecting or measuring a specific quantity or a property, such as position, force, torque, pressure, temperature, speed, acceleration, or vibration. Traditionally,

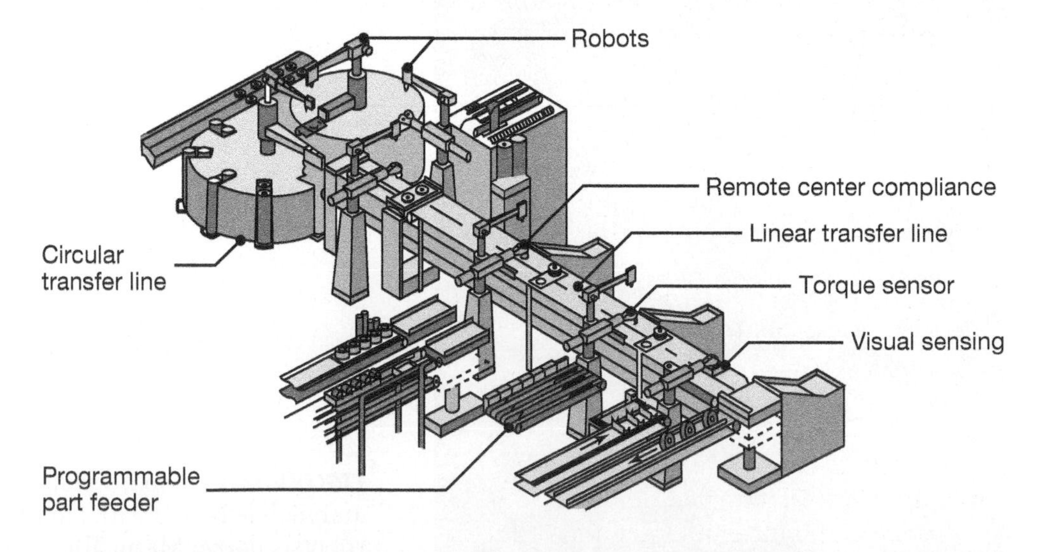

FIGURE 14.23 An example of automated assembly operations using industrial robots and circular and linear transfer lines.

sensors, actuators, and switches have been used to set limits on the performance and movements of machines, such as stops on machine-tool slideways to restrict worktable movements, pressure and temperature gages with automatic shutoff features, and governors on engines to prevent excessive speed of operation. Modern uses of sensors include the monitoring of manufacturing processes to provide feedback or confirmation of operating conditions.

Sensor technology is an important aspect of modern manufacturing, and is essential for data acquisition, monitoring, communication, and computer control. Low-cost sensors, often based on *flexible electronics* (Section 13.14), are being increasingly applied to process monitoring and control. The widespread use of sensors and the large amount of data they generate is the foundation for the application of **Big Data** and the **Internet of Things** approaches in manufacturing (see Section 15.14.2).

Because they convert one quantity to another, sensors are often also referred to as **transducers. Analog sensors** produce a signal that is proportional to the measured quantity. **Digital sensors** have numeric outputs that can be directly transferred to computers. **Analog-to-digital converters** (ADCs) are used for interfacing analog sensors with computers.

14.8.1 Sensor Classification

Sensors that are of interest in manufacturing processes and their capabilities may be classified as:

1. *Mechanical* sensors: position, shape, velocity, force, torque, pressure, vibration, strain, and mass;
2. *Electrical* sensors: voltage, current, charge, and conductivity;
3. *Magnetic* sensors: magnetic field, flux, and permeability;
4. *Thermal* sensors: temperature, conductivity, specific heat, and flux; and
5. *Acoustic, ultrasonic, chemical, optical, radiation, laser,* and *fiber-optic* sensors.

Depending on its application, a sensor may be made of metallic, nonmetallic, organic, or inorganic materials, semiconductors, fluids, gases, or plasmas. Using the special characteristics of these materials, sensors convert the quantity or the property measured to analog or digital outputs.

Common sensor classes are the following:

1. **Tactile sensors** continuously sense varying contact forces, commonly by an *array of sensors*, and are capable of performing within an arbitrary three-dimensional space. Fragile objects, such as glass bottles, eggs, and electronic devices, can be handled with **compliant (smart) end effectors.** They can sense the force applied to the object being handled by such means as piezoelectric devices, strain gages, magnetic induction, ultrasonic, and optical systems such as fiber optics and light-emitting diodes. Tactile sensors capable of measuring and controlling gripping forces and moments in three axes have been developed (Fig. 14.24). The *gripping force* of an end effector

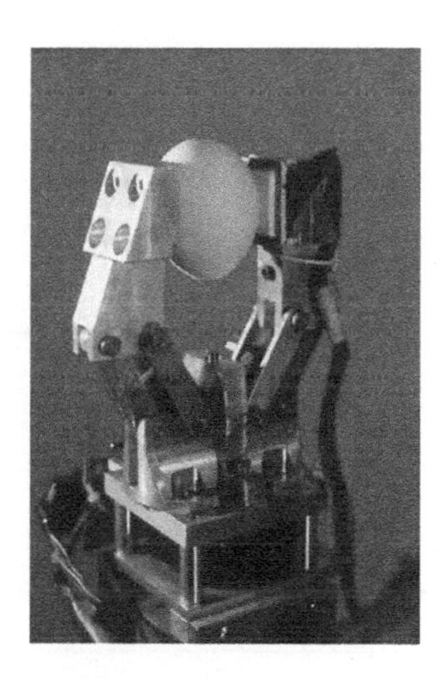

FIGURE 14.24 A robot gripper with tactile sensors. In spite of their capabilities, tactile sensors are now being used less frequently, because of their high cost and low durability (lack of robustness) in industrial applications.
Source: Hank Morgan/Getty Images.

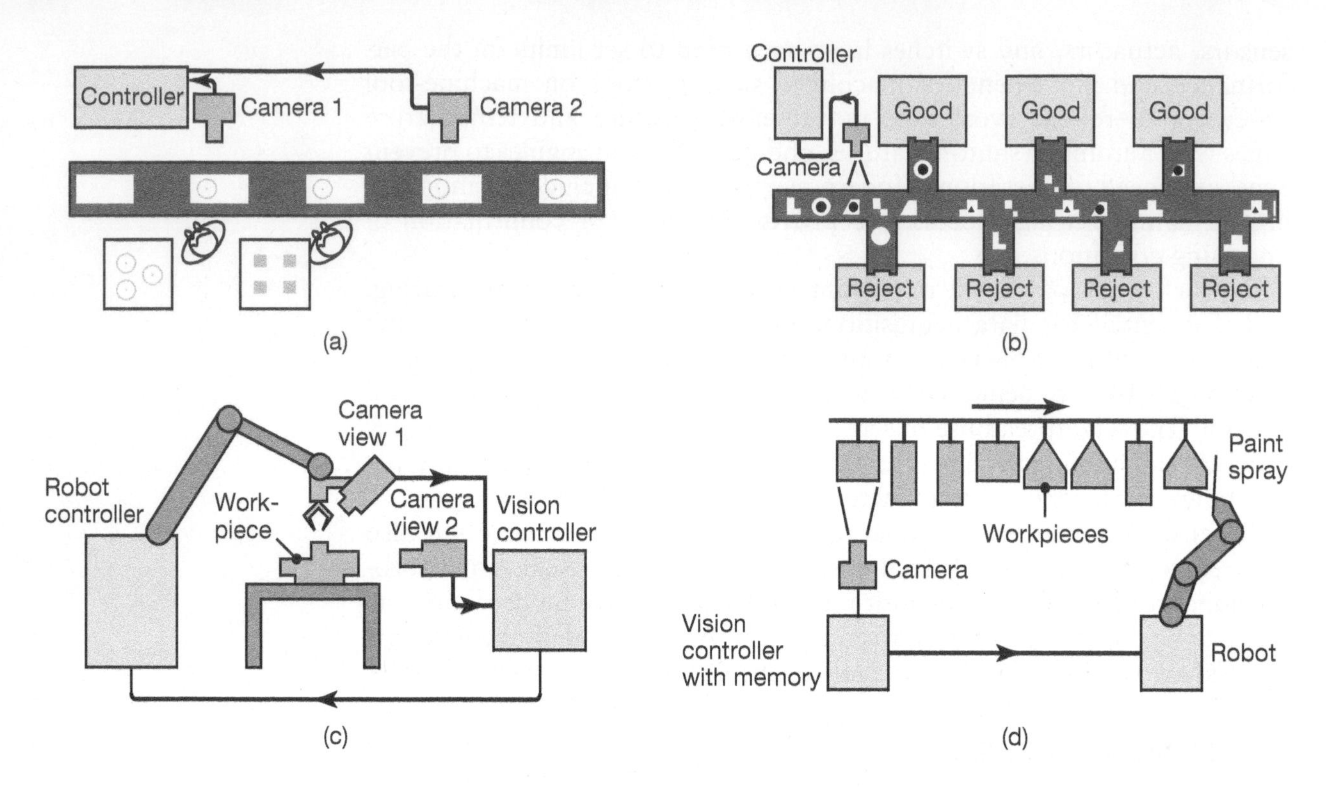

FIGURE 14.25 Examples of machine vision applications. (a) In-line inspection of parts. (b) Identifying parts with various shapes, and inspection and rejection of defective parts. (c) Using cameras to provide positional input to a robot relative to the workpiece. (d) Painting of parts with different shapes by means of input from a camera; the system's memory allows the robot to identify the particular shape to be painted and to proceed with the correct movements of a paint spray nozzle attached to the end effector.

is sensed, monitored, and controlled through closed-loop feedback devices. *Anthropomorphic end effectors* are designed to simulate the human hand and fingers and have the capability of sensing touch, force, movement, and patterns.

In **visual sensing**, cameras scan an image, and the software processes the data; its most important applications in manufacturing are *pattern recognition* (Fig. 14.25), edge detection, and the transfer of information such as from barcodes. Machine vision commonly uses digital cameras that communicate with a computer through wireless, bluetooth, or USB connections. Scanning can take place in (a) one direction (line scan), as with bar codes; (b) 2D scanning, as with QR barcodes; or (c) 3D scanning, as with CT scanning or confocal cameras (Fig. 14.26). Three-dimensional scanning has become more common, with powerful software available to allow simple digital cameras on smart phones to take three-dimensional data.

Machine vision is particularly suitable for parts with inaccessible features, in hostile manufacturing environments, measuring a large number of small features, and in situations where physical contact

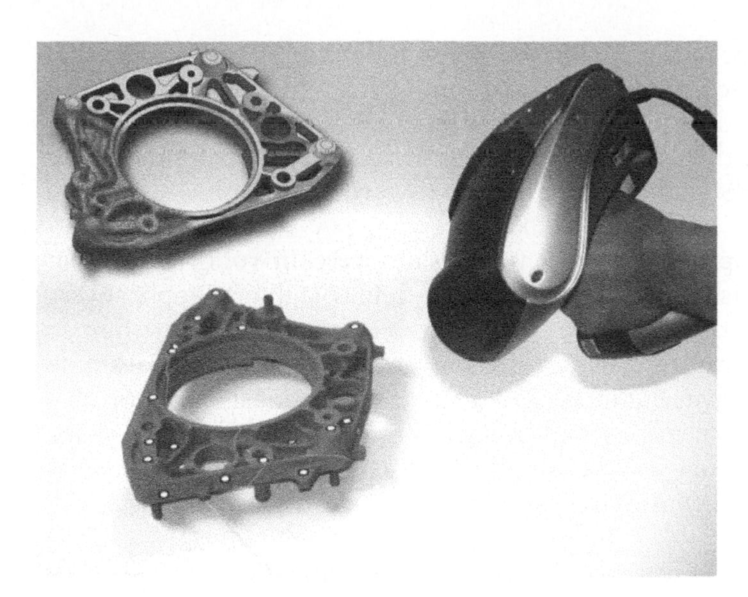

FIGURE 14.26 The use of a 3D scanner to digitize the geometry of a casting (below) to generate a 3D data file describing the geometry (above). The data file can then be used for quality control, or can it be sent to a 3D printer to produce a part with the same geometry. *Source:* EMS-USA.

with the part may cause damage to the part. Sensors for machine tools can now sense tool breakage, verify part placement and fixturing, and monitor surface finish (See also Fig. 14.25). Machine vision is capable of in-line identification and inspection of parts and rejecting defective parts. With visual sensing capabilities, end effectors can pick up parts and grip them in the proper orientation and location.

2. **Smart sensors** can perform a logic function, conduct two-way communications, and make decisions and take appropriate actions. The necessary input and the knowledge required to make decisions can be built into a smart sensor; a computer chip with sensors, for example, can be programmed to turn a machine tool off in the event that a cutting tool fails. Likewise, by sensing distance, heat, and noise, a smart sensor can stop a mobile robot or a robot arm from accidentally colliding with a machine or person.

Sensor selection. The selection of a sensor depends on such factors as (a) the particular quantity to be measured or objects to be sensed; (b) its interaction with other components in the system; (c) its expected service life; (d) the required level of performance; (e) difficulties associated with use of the sensor; and (f) cost. An important consideration is the environment in which the sensor is to be used. **Robust** sensors are designed to withstand extremes of temperature, shock, vibration, humidity, corrosion, dust, fluids, electromagnetic radiation, and any other interference present in industrial environments (see also *robust design*, Section 16.2.3).

14.8.2 Sensor Fusion

Sensor fusion involves the *integration* of several different sensors in a manner such that the individual data from each of the sensors are combined to provide a higher level of information and reliability. A simple and common example of sensor fusion occurs when drinking from a cup of hot fluid.

Although such a practice is taken for granted, it can readily be seen that this process involves simultaneous data input from a person's eyes, lips, tongue, fingers, and hands. Through the five senses (sight, hearing, smell, taste, and touch), there is real-time monitoring; thus, for example, if the coffee is too hot, the movement of the cup toward the mouth is controlled accordingly.

The earliest applications of sensor fusion were in robot movement control and in missile-flight tracking and similar military applications, primarily because these activities involve movements that mimic human behavior. An example of sensor fusion is a machining operation in which several different, but integrated, sensors continuously monitor such quantities as (a) dimensions and surface finish of the workpiece being machined; (b) cutting forces, vibration, and tool wear and fracture; (c) the temperature in various regions of the tool-workpiece interface; and (d) spindle power.

An essential aspect in sensor fusion is **sensor validation**, in which the failure of one particular sensor is detected so that the control system retains high reliability; in validation, the receipt of redundant data from different sensors is essential. Although complex and relatively expensive, sensor fusion and validation are now possible due to the advances made in sensor size, cost, quality, and technology, and continued developments in control systems, artificial intelligence, expert systems, and artificial neural networks, described in Chapter 15.

14.9 Flexible Fixturing

In manufacturing operations, the words *fixture, clamp,* and *jig* are often used interchangeably and sometimes in pairs, such as *jigs and fixtures*. Common workholding devices include chucks, collets, and mandrels. Although many of these devices are operated manually, particularly in job shops, other workholding devices are designed and operated at various levels of mechanization and automation, such as *power chucks*, which are driven by mechanical, hydraulic, or electrical means. **Fixtures** are generally designed for specific purposes; **clamps** are simple multifunctional devices; **jigs** have reference surfaces and points for accurate alignment of parts and tools, and are widely used in mass production. These devices may be used for actual manufacturing operations (in which case the forces exerted on the part must maintain the part's position in the machine tool without slipping or distortion), or they may be used to hold workpieces for purposes of measurement and inspection, where the part is not subjected to any external forces.

Workholding devices have specific ranges of capacity; for example, (a) a particular collet can accommodate rods or bars only within a certain range of diameters; (b) four-jaw chucks can accommodate square or prismatic workpieces of various sizes; and (c) other devices and fixtures that are made for specific workpiece shapes and dimensions and for specific tasks, called **dedicated fixtures**. It is simple to reliably fixture a rectangular bar, such as by clamping it between the parallel jaws of a four-jaw vise.

If the part has curved surfaces, it is possible to shape the contacting surfaces of the jaws themselves by machining them (called *machinable jaws*) to conform to the workpiece surfaces; this method is particularly valuable for machining large quantities of parts.

Flexible manufacturing systems (Section 15.10) require the design and use of workholding devices and fixtures that have **built-in flexibility**. There are several methods of *flexible fixturing*, based on different principles (also called **intelligent fixturing systems**), although the term itself has been defined somewhat loosely. These devices are capable of quickly accommodating a range of part shapes and sizes without the necessity of making extensive changes and adjustments, which would adversely affect productivity.

1. **Modular fixturing.** *Modular fixturing* is often used for small or moderate lot sizes (Fig. 14.27), especially when the cost of dedicated fixtures and the time required to make them are difficult to justify. Complex workpieces can be accommodated within machines through fixtures produced quickly from standard components, and can be disassembled when a production run is completed. Modular fixtures are usually constructed on tooling plates or blocks configured with grid holes or T-slots.

   Several other standard components, such as locating pins, adjustable stops, workpiece supports, V-blocks, clamps, and springs, can be mounted onto the base plate or block to quickly produce a fixture. By computer-aided fixture planning for specific situations, such fixtures can be assembled and modified using robots. As compared with dedicated fixturing, modular fixturing has been shown to be low in cost, have a shorter lead time, provide easier repair of damaged components, and offer a more intrinsic flexibility of application.
2. **Tombstone fixtures.** Also referred to as *pedestal-type fixtures*, *tombstone* fixtures have between two and six vertical faces (hence resembling tombstones) onto which parts can be mounted. These fixtures

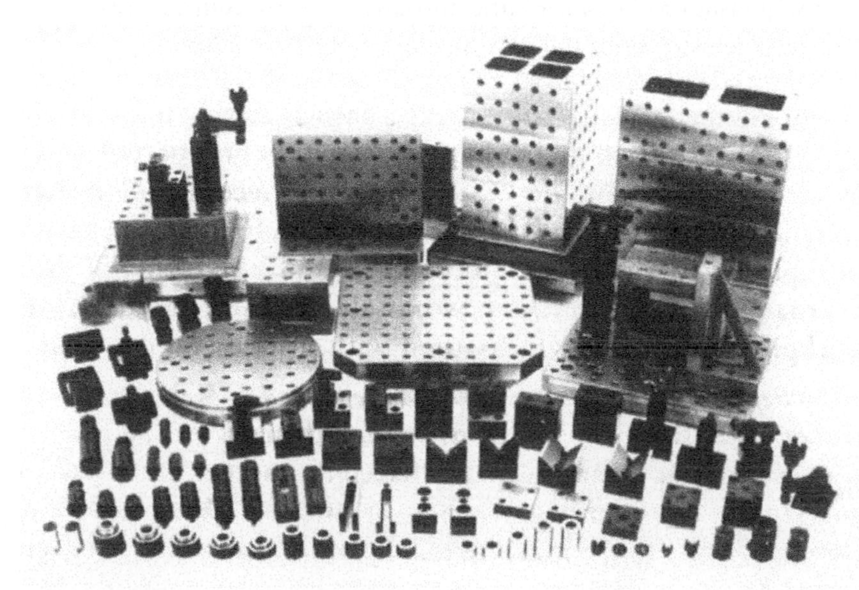

FIGURE 14.27 Components of a modular workholding system. *Source:* Carr Lane Manufacturing Co.

are typically used in automated or robot-assisted manufacturing; the machine tool performs the desired operations on the part or parts on one face, then flips or rotates the tombstone to begin work on other faces. The fixtures allow feeding more than one part into a machine, but are not as flexible as other fixturing systems. They are commonly used for higher volume production, typically in the automotive industry.

3. **Bed-of-nails device.** This fixture consists of a series of air-actuated pins that conform to the shape of the external part surfaces. Each pin moves as necessary to conform to the shape at its point of contact with the part; the pins are then mechanically locked against the part. The fixture is compact, has high stiffness, and is reconfigurable.
4. **Adjustable-force clamping.** Figure 14.28 shows a schematic illustration of another flexible fixturing system. Referred to as an *adjustable-force clamping system*, a strain gage mounted on the clamp senses the magnitude of the clamping force; the system then adjusts this force to ensure the workpiece is securely held during machining.
5. **Phase-change materials.** There are two basic methods to hold irregularly shaped workpieces in a solid medium.

   a. In the first, and older, method, a *low-melting-point metal* is used as the clamping medium. An irregularly shaped workpiece is partially dipped into molten metal (basically a lead-free solder) and allowed to solidify. The process is similar to a wooden stick in an ice-cream bar; see *insert molding* (Section 10.10.2 and Fig. 10.32); after setting, the assembly is clamped in a simple fixture.

      An application of this method is machining of honeycomb structures (Fig. 7.48); note that because the walls of the hollow structure are very thin, the forces exerted by the cutting tool would easily distort and damage them. One method of stiffening this structure is to fill the cavities with water and freeze it. The hexagonal cavities are now supported by ice, whose strength is sufficient to resist the cutting forces. After machining, the ice is allowed to melt away.

   b. In another method, still in experimental stages, the supporting medium is a *magnetorheological* (MR) or *electrorheological* (ER) fluid. In MR, magnetic particles (micrometer size or

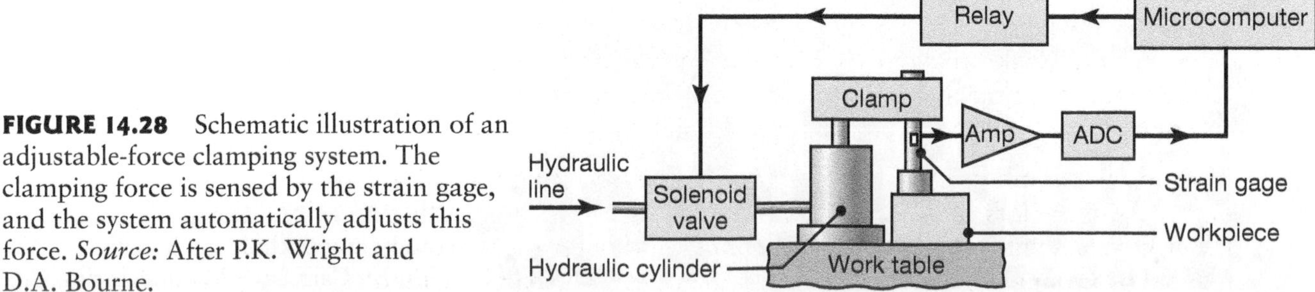

FIGURE 14.28 Schematic illustration of an adjustable-force clamping system. The clamping force is sensed by the strain gage, and the system automatically adjusts this force. *Source:* After P.K. Wright and D.A. Bourne.

nanoparticles (Sections 3.11.9, 11.2.1, and 11.8.1) are suspended in a nonmagnetic fluid. Surfactants are added to maintain dispersal of powders. After the workpiece is immersed in the fluid, an external magnetic field is applied, whereby the particles are polarized, and the behavior of the fluid changes from a liquid to a solid. After the part is processed, the external magnetic field is removed and the part is retrieved. In the ER application, the fluid is a suspension of fine dielectric particles in a liquid with a low dielectric constant. After applying an electrical field, the liquid becomes a solid.

14.10 Assembly, Disassembly, and Service

Some products are simple and have only two or three components to *assemble*; examples include a hammer, cookware with a separate handle, and aluminum beverage can. The vast majority of products consist of numerous parts (see Section 1.1). Traditionally, assembly has typically involved much *manual labor*, contributing significantly to product cost. Depending on the type of product, assembly costs can vary widely; for example, Apple iPhones cost $12.50–30.00 in total labor, with a total cost of $200–600; the assembly cost for automobiles is around 10% of the sales price. Assembly costs are 10–50% of the total cost of manufacturing, with the percentage of workers involved in assembly operations ranging from 20 to 60%. In developed countries, with high productivity and associated automation, the number of workers involved in assembly is on the low end of this range; in countries with inexpensive labor, the percentage is higher. As production costs and quantities of products to be assembled began to increase, the necessity for **automated assembly** became obvious.

Beginning with the hand assembly of US muskets in the late 1700s, and the introduction of **interchangeable parts** in the early 1800s, (see *Eli Whitney*, Section 4.9.1), assembly methods have been vastly improved over the years. The first application of large-scale modern assembly was for the flywheel magnetos for the Model T Ford; this activity eventually led to **mass production** of automobiles.

The choice of an assembly method and system depends on the required production rate, the total quantity to be produced, the product's market life, labor availability, and cost. As indicated throughout this text, parts are manufactured within certain dimensional tolerance ranges (see also Section 4.7). Taking ball bearings as an example, it is well known that, although they all have the same *nominal* dimensions, some balls in a particular lot will be a little smaller or larger than others and some inner or outer races made will be a little smaller or larger than others in the lot.

There are two methods of assembly for such high-volume products:

1. In **random assembly**, parts are put together by selecting them randomly from the lots produced.
2. In **selective assembly**, the balls and the inner and outer races are segregated by groups of sizes, from smallest to largest. The parts are then

selected to mate in a manner whereby the smallest diameter balls are mated with inner races with the largest outside diameters and with outer races with the smallest inside diameters.

14.10.1 Assembly Systems

There are three basic methods of assembly: *manual, high-speed automatic,* and *robotic.* These methods can be used individually or, as is the case for most applications in practice, in combination. The first step in designing an assembly system is to perform an analysis of the product design (Fig. 14.29) to determine the most appropriate and economical method of assembly.

1. **Manual assembly** uses simple tools and is generally economical for relatively small lots. Because of the dexterity of the human hand and fingers, workers can assemble even complex parts without much difficulty. For example, aligning and placing a square peg into a square hole involving small clearances is a simple manual operation, but it can be difficult in automated assembly. Note, however, the potential problems of *cumulative trauma disorders* (*carpal tunnel syndrome*) associated with manual assembly.
2. **High-speed automated assembly** uses *transfer mechanisms*, designed specially for assembly operations. Two examples of individual assembly are shown in Fig. 14.30, in which products are *indexed* for proper positioning during assembly. In **robotic assembly**, one or two general-purpose robots operate at a single workstation, or the robots operate at a multistation assembly system.

   There are three basic types of automated assembly systems: synchronous, nonsynchronous, and continuous.

   a. **Synchronous systems.** Also called **indexing systems**, individual parts and components are supplied and assembled at a constant rate at fixed individual stations. The rate of movement is based on the station that requires the longest time to complete its portion of the assembly. This system is used primarily for high-volume, high-speed assembly of small products.

   Transfer systems move the partial assemblies from workstation to workstation by various mechanical means, two typical

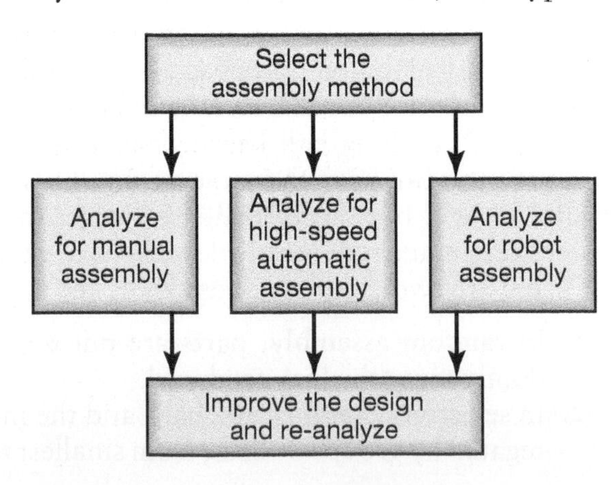

FIGURE 14.29 Stages in the design-for-assembly analysis. *Source:* After G. Boothroyd and P. Dewhurst.

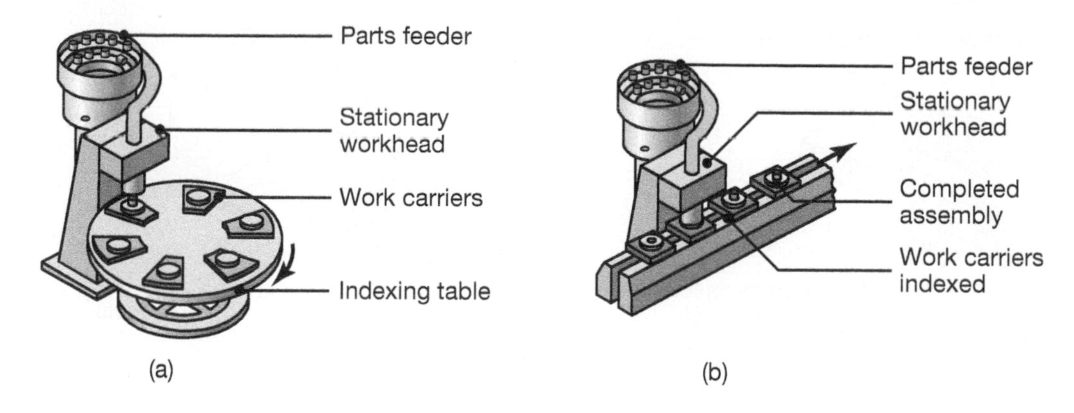

FIGURE 14.30 Transfer systems for automated assembly: (a) rotary indexing machine, and (b) in-line indexing machine. *Source:* After G. Boothroyd.

transfer systems being **rotary indexing** and **in-line indexing** (Fig. 14.30). These systems can operate in either a fully automatic mode or a semiautomatic mode; however, a breakdown of one station will shut down the whole assembly operation. The part feeders supply the individual parts to be assembled and place them on other components that are secured on work carriers or fixtures. The feeders move the individual parts, by vibratory or other means, through delivery chutes and ensure their proper orientation (Fig. 14.31) in order to avoid jamming.

b. **Nonsynchronous systems.** Each station operates independently, and any imbalance in product flow is accommodated in storage (**buffer**) between stations. The station continues operating until the next buffer is full or the previous buffer is empty. If, for some reason, one station becomes inoperative, the assembly line continues to operate until all the parts in the buffer have been used up. Nonsynchronous systems are suitable for products with several parts to be assembled, such as motors, brake components, circuit boards, etc. Note that for products in which the times required for individual assembly operations vary widely, the rate of output will be constrained by the slowest station.

c. **Continuous systems.** The product is assembled while moving at a constant speed, on pallets or similar workpiece carriers. The parts to be assembled are brought to the product by various workheads, and their movements are synchronized with the continuous movement of the product. Typical applications of this system are in bottling and packaging plants, although it is also used on mass-production lines for automobiles and appliances.

Although assembly systems are generally set up for a specific product line, they can be modified for increased flexibility for lines that have a variety of models. Called **flexible assembly systems** (FAS), these systems use computer controls, interchangeable and programmable workheads and feeding devices, coded pallets, and automated guiding devices.

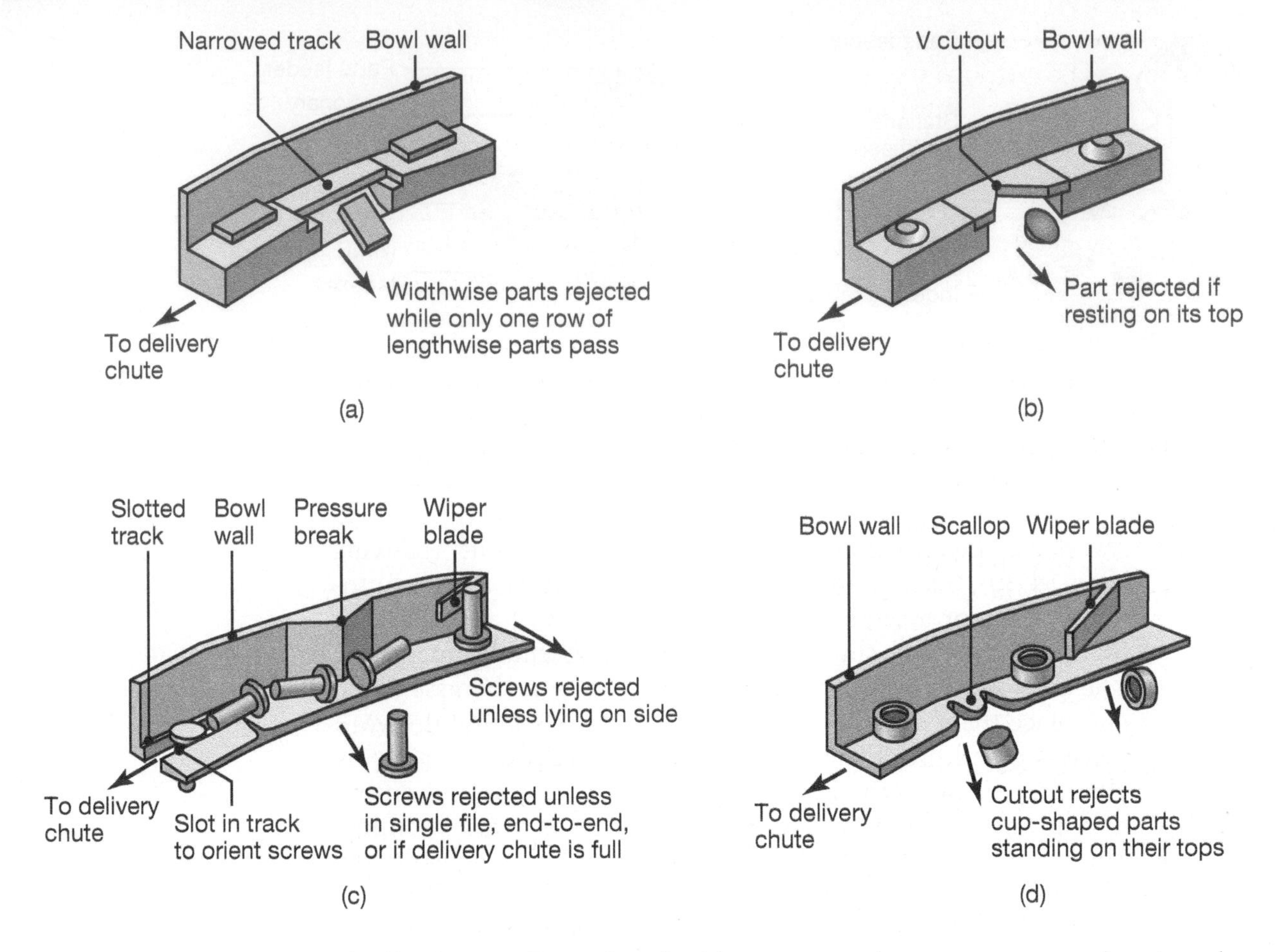

FIGURE 14.31 Examples of guides to ensure that parts are properly oriented for automated assembly. *Source:* After G. Boothroyd.

14.11 Design Considerations

14.11.1 Design for Fixturing

The design, construction, and operation of flexible workholding devices and fixtures are essential to the efficient operation of advanced manufacturing systems. The following is a list of the major design considerations involved:

1. Workholding devices must be able to position the workpiece automatically, accurately, and repeatedly, and must maintain its position with sufficient clamping force to withstand the forces involved in the particular manufacturing operation.
2. Fixtures must have sufficient stiffness to withstand, without excessive distortion, the normal and shear stresses developed at the fixture-workpiece interfaces.
3. The presence of loose chips and other debris between the locating surfaces of the workpiece and the fixture can be a serious problem. Chips

are most likely to be present where cutting fluids are used, because they tend to stick to wet surfaces, by virtue of surface-tension forces.

4. A flexible fixture must accommodate parts to be made by different processes and for situations in which dimensions and surface features vary from part to part. This is even more important when the workpiece (a) is fragile or made of a brittle material, such as a ceramic; (b) is made of a relatively soft or flexible material, such as a thermoplastic or rubber; or (c) has a relatively soft coating on its contacting surfaces.
5. Fixtures and clamps should avoid collision with cutting tools; collision avoidance also is an important factor in programming tool paths in manufacturing operations (see Sections 14.3 and 14.4).
6. Flexible fixturing must meet special requirements to function properly in manufacturing cells and flexible manufacturing systems. The time required to load and unload parts on machinery should be minimal in order to reduce cycle times.
7. Parts should be designed to allow for easy locating and clamping within a fixture. Flanges, flats, or other locating surfaces should be incorporated into the part design, in order to simplify fixture design and to aid in part transfer among different machinery.

14.11.2 Design for Assembly, Disassembly, and Service

Design for assembly. Design for assembly (DFA) continues to attract special attention, particularly design for automated assembly, because of the need to reduce assembly costs. In manual assembly, a major advantage is that humans can easily pick the correct part from a supply of parts or pick one from a supply of identical parts from a nearby bin. Human vision, intelligence, and dexterity allow for proper orientation and assembly of very complex systems. In *high-speed automated assembly*, however, automatic handling generally requires that parts be separated from the bulk, conveyed by hoppers or vibratory feeders (See Fig. 14.30), and assembled in proper locations and orientations.

Based on analyses of assembly operations as well as on experience, several guidelines for DFA have been developed through the years (See also Fig. 1.4):

1. The number and variety of parts in a product should be minimized and, as much as possible, multiple functions should be incorporated into a single part; also, subassemblies that could serve as modules should be considered.
2. Parts should have a high degree of symmetry (such as round or square) or a high degree of asymmetry (such as oval or rectangular), so that they cannot be installed incorrectly and that they do not require location, alignment, or adjustment. Parts should be designed for easy insertion into other components.
3. Designs should allow parts to be assembled without obstructions and with a direct line of sight.

4. Designs should, as much as possible, avoid the need for fasteners, such as bolts, nuts, and screws; other methods such as snap fits should be considered (See Fig. 12.57). If fasteners are required, their variety should be minimized, and they should be located and spaced in such a manner that different tools can be used without obstruction.
5. Part designs should consider such factors as size, shape, weight, flexibility, abrasiveness, and tangling with other parts.
6. As much as possible, assemblies should not be turned over for insertion of parts, so that they can be inserted from only *one* direction, preferably vertically and from above to take advantage of gravity (see also Section 13.13).
7. Products should be designed, or existing products redesigned, so that there are no physical obstructions to the free movement of the parts during assembly (See Fig. 1.5). Thus, for example, sharp external and internal corners should be replaced with chamfers, tapers, or radii.
8. Parts that may appear to be similar but are different should be color coded.

Design guidelines for *robotic assembly* have rules similar to those for manual and high-speed automated assembly. *Compliant end effectors* and *dexterous manipulators* have greatly increased the flexibility of robots. Some additional guidelines are:

1. Parts should be designed so that they can be gripped and manipulated by the same gripper (end effector) of the robot (Fig. 14.18), thus avoiding the need for different grippers. Parts should be made available to the gripper in the proper orientation.
2. Assembly that involves threaded fasteners, such as bolts, nuts, and screws, may be difficult for robots to perform, but they easily can handle self-threading screws (for sheet metal, thermoplastics, and wood), snap fits, rivets, welds, and adhesives.

Evaluating assembly efficiency. Significant effort has been directed toward the development of analytical and computer-based tools for estimating the efficiency of assembly operations. The tools provide a basis for comparisons of different product designs and help in the selection of design attributes that make assembly easier.

To assess assembly efficiency, each component is evaluated with respect to its features that can affect assembly and provides a baseline estimated time required for assembling the part. Note that such an evaluation also can be made for existing products. Assembly efficiency, ν, is expressed as

$$\nu = \frac{Nt}{t_{tot}}, \tag{14.1}$$

where N is the number of parts, and t_{tot} is the total assembly time. t is the ideal assembly time for a small part that presents no difficulties in handling, orientation, or assembly; t is commonly taken as three seconds. Using Eq. (14.1), competing designs can be evaluated with respect to design for assembly. It has been noted that products that are in need of redesign to facilitate assembly usually have assembly efficiencies of around 5–10%, whereas well designed parts have efficiencies of around 25%. It should also be noted that assembly efficiencies near 100% are unlikely to be achieved in practice, because the three second baseline is usually not practical.

Design for disassembly. The manner and ease with which a product can be taken apart for maintenance or replacement of its parts is an important consideration in product design. Although there is no established set of guidelines, the general approach to design for disassembly requires consideration of factors that are similar to those for design for assembly. For example, analysis of computer or physical models of products and their components can indicate potential problems in disassembly, such as obstructions, lack of line of sight, narrow and long passageways, and difficulty of firmly gripping and guiding objects.

After its life cycle, how a product can be be taken apart for *recycling*, especially with respect to its more valuable components, continues to be an important consideration. Recall, for example, that fasteners can be difficult to disassemble, depending on (a) their design and location; (b) the type of tools required for disassembly; and (c) whether the tools are manual or powered. In general, rivets or welds will take longer to remove than screws or snap fits, and a bonded layer of valuable material on a component could be difficult, if not impossible, and thus uneconomical for recycling or reuse. Consequently, the time required for disassembly has to be studied and measured. Although it depends on the manner in which disassembly takes place, some examples of the time required are as follows: cutting wire, 0.25 s; disconnecting wire, 1.5 s; removing snap fits and clips, 1–3 s; and removing screws and bolts, 0.15–0.6 s per revolution.

Design for service. Design for assembly and disassembly must include the ease with which a product can be serviced and, if justified, repaired. *Design for service* essentially is based on the concept that the elements that are most likely to require servicing should be placed at the outer layers of a product.

14.12 Economic Considerations

There are numerous considerations involved in determining the overall economics of manufacturing operations. Important factors influencing the final decision include the type of product, the machinery involved, its cost of operation, the skill level of labor required, lot size, and production rate. Small quantities per year can be produced in job shops, which generally require skilled laborers. Moreover, because production volume and production rate in job shops are typically low, cost per part can be high (See Fig. 14.3).

At the other extreme is the production of very large quantities of parts, using conventional flow lines and transfer lines, and involving special-purpose machinery, equipment, tooling, and computer control systems. Although these components constitute major investments, the level of labor skill required and the labor costs are both relatively low because of the high level of automation implemented. Recall also that these production systems are organized for a specific type of product and hence lack flexibility.

Because most manufacturing operations fall within these two extremes, an appropriate decision must be made regarding the optimum level of automation to be implemented. In many situations, selective automation rather than total automation of a facility has been found to be cost effective. Generally, the higher the level of skill available in the workforce, the lower is the need for automation, provided that the higher labor costs involved are justified and assuming that there is a sufficient number of qualified workers available. Conversely, if a manufacturing facility already has been automated, the skill level required is lower.

In addition, the manufacture of some products may have a large labor component, and thus their production is *labor intensive*; this is especially the case with products that require extensive assembly. Examples of labor-intensive products include aircraft, locomotives, bicycles, pianos, furniture, toys, shoes, and garments. High labor requirement is a major reason that so many household as well as high-tech products are now made or are assembled in countries where labor costs are low (see also Sections 1.10 and 16.10).

SUMMARY

- Automation has been successfully implemented, to varying degrees, in all manufacturing processes, material handling, inspection, assembly, and packaging. Production volume and production rate are major factors in selecting the most economic level of automation for a process or operation. (Sections 14.1, 14.2)
- True automation began with the numerical control of machines, achieving flexibility of operation, lower cost, and ease of making different parts with lower operator skill required. (Sections 14.3, 14.4)
- Manufacturing operations can be optimized by adaptive-control techniques, which continuously monitor the operation and automatically make appropriate adjustments in the processing parameters. (Section 14.5)
- Advances in material handling include the implementation of industrial robots and automated guided vehicles. (Sections 14.6, 14.7)
- Sensors are essential in the implementation of modern manufacturing technologies and computer-integrated manufacturing. A wide variety of sensors, based on various principles, has been developed and successfully installed. (Section 14.8)

- Flexible fixturing and automated assembly techniques greatly reduce the need for worker intervention and thus lower manufacturing costs. Effective and economic implementation of these techniques requires that design for assembly, disassembly, and servicing be recognized as an important factor in the total design as well as in manufacturing operations. (Sections 14.9, 14.10)
- As in all manufacturing processes, there are certain design considerations and guidelines regarding the implementation of the topics described in this chapter. (Section 14.11)
- Economic considerations in automation include decisions regarding the appropriate level of automation to be implemented; such decisions, in turn, involve parameters such as type of product, production volume and rate, and availability of labor. (Section 14.12)

BIBLIOGRAPHY

Batchelor, B.G. (ed.), *Machine Vision Handbook*, Springer, 2012.

Blum, R.S., and Liu, Z., *Multi-Sensor Image Fusion and its Applications*, CRC, 2005.

Boothroyd, G., *Assembly Automation and Product Design*, 2nd ed., Dekker, 2005.

Boothroyd, G., Dewhurst, P., and Knight, W., *Product Design for Manufacture and Assembly*, 3rd ed., Dekker, 2010.

Corke, P., *Robotics, Vision and Control*, Springer, 2013.

Craig, J.J., *Introduction to Robotics: Mechanics and Control*, 3rd ed., Prentice Hall, 2003.

Davies, E.R., *Computer and Machine Vision*, 4th ed., Academic Press, 2012.

Fraden, J., *Handbook of Modern Sensors: Physics, Designs, and Applications*, 5th ed., Springer, 2015.

Hornberg, A., *Handbook of Machine Vision*, Wiley, 2006.

Kandray, D., *Programmable Automation Technologies: An Introduction to CNC, Robotics and PLCs*, Industrial Press, 2010.

Kurfess, T.R. (ed.), *Robotics and Automation Handbook*, CRC Press, 2004.

Mitchell, H.B., *Multi-Sensor Data Fusion: An Introduction*, Springer, 2007.

Pratt, W.K., *Introduction to Digital Image Processing*, CRC Press, 2013.

Quesada, R., *Computer Numerical Control Machining and Turning Centers*, Prentice Hall, 2004.

Rehg, J.A., *Introduction to Robotics in CIM Systems*, 5th ed., Prentice Hall, 2002.

Rekiek, B., and Delchambre, A., *Assembly Line Design*, Springer, 2005.

Ripka, P., and Tipek, A., *Modern Sensors Handbook*, ISTE Publishing Co., 2007.

Snyder, W.E., and Qi, H., *Machine Vision*, Cambridge, 2004.

Soloman, S., *Sensors Handbook*, McGraw-Hill, 2009.

Stenerson, J., and Curran, K.S., *Computer Numerical Control: Operation and Programming*, 3rd ed., Prentice Hall, 2005.

Ulsoy, A.G., Peng, H., and Cakmakci, M., *Adaptive Control Systems*, Cambridge University Press, 2012.

Umbaugh, S.E., *Computer Imaging*, CRC, 2005.

Valentino, J.V., and Goldenberg, J., *Introduction to Computer Numerical Control*, 5th ed., Prentice Hall, 2012.

Van Doren, V., *Techniques for Adaptive Control*, Butterworth-Heinemann, 2002.

Wilson, J., *Sensor Technology Handbook*, Newnes, 2004

Zhang, J., *Practical Adaptive Control: Theory and Applications*, VDM Verlag, 2008.

QUESTIONS

14.1 Describe the differences between mechanization and automation. Give several specific examples for each.

14.2 Why is automation generally regarded as evolutionary rather than revolutionary?

14.3 Are there activities in manufacturing operations that cannot be automated? Explain, and give specific examples.

14.4 Explain the difference between hard and soft automation. Why are they named as such?

14.5 Describe the principle of numerical control of machines. What factors led to the need for and development of numerical control? Name some typical applications of NC.

14.6 Explain the differences between direct numerical control and computer numerical control. What are their relative advantages?

14.7 Describe open-loop and closed-loop control circuits.

14.8 What are the advantages of computer-aided NC programming?

14.9 Describe the principle and purposes of adaptive control. Give some examples of present applications in manufacturing and other areas that you think can be implemented.

14.10 What factors have led to the development of automated guided vehicles? Do automated guided vehicles have any limitations? Explain your answers.

14.11 List and discuss the factors that should be considered in choosing a suitable material-handling system for a particular manufacturing facility.

14.12 Make a list of the features of an industrial robot. Why are these features necessary?

14.13 Discuss the principles of various types of sensors, and give two applications for each type.

14.14 Describe the concept of design for assembly. Why has it become an important factor in manufacturing?

14.15 Is it possible to have partial automation in assembly operations? Explain.

14.16 Describe your thoughts on adaptive control in manufacturing operations.

14.17 What are the two kinds of robot joints? Give applications for each.

14.18 What are the advantages of flexible fixturing over other methods of fixturing? Are there any limitations to flexible fixturing? Explain.

14.19 How are robots programmed to follow a certain path?

14.20 Giving specific examples, discuss your observations concerning Fig. 14.2.

14.21 What are the relative advantages and limitations of the two arrangements for power heads shown in Fig. 14.4?

14.22 Discuss methods of on-line gaging of workpiece diameters in turning operations other than that shown in Fig. 14.15. Explain the relative advantages and limitations of the methods.

14.23 Are drilling and punching the only applications for the point-to-point system shown in Fig. 14.10a? Are there others? Explain.

14.24 Describe possible applications for industrial robots not discussed in this chapter.

14.25 What determines the number of robots in an automated assembly line such as that shown in Fig. 14.23?

14.26 Describe situations in which the shape and size of the work envelope of a robot (See Fig. 14.20) can be critical.

14.27 Explain the functions of each of the components of the robot shown in Fig. 14.17a. Comment on their degrees of freedom.

14.28 Explain the difference between an automated guided vehicle and a self-guided vehicle.

14.29 It has been commonly acknowledged that, at the early stages of development and implementation of industrial robots, the usefulness and cost effectiveness of the robots were overestimated. What reasons can you think of to explain this situation?

14.30 Describe the type of manufacturing operations (See Fig. 14.2) that are likely to make the best use of a machining center (see Section 8.11). Comment on the influence of product quantity and part variety.

14.31 Give a specific example of a situation in which an open-loop control system would be desirable, and give a specific example of a situation in which a closed-loop system would be desirable. Explain.

14.32 Why should the level of automation in a manufacturing facility depend on production quantity and production rate?

14.33 Explain why sensors have become so essential in the development of automated manufacturing systems.

14.34 Why is there a need for flexible fixturing for holding workpieces? Are there any disadvantages to such flexible fixturing? Explain.

14.35 Describe situations in manufacturing for which you would not want to apply numerical control. Explain your reasons.

14.36 Table 14.2 shows a few examples of typical products for each category of production by volume. Add several other examples to this list.

14.37 Describe situations for which each of the three positioning methods shown in Fig. 14.6 would be desirable.

14.38 Describe applications of machine vision for specific parts, similar to the examples shown in Fig. 14.25.

14.39 Add examples of guides other than those shown in Fig. 14.31.

14.40 Sketch the work envelope of each of the robots in Fig. 14.19. Describe its implications in manufacturing operations.

14.41 Give several applications for the types of robots shown in Fig. 14.19.

14.42 Name some applications for which you would not use a vibratory feeder. Explain why vibratory feeding is not appropriate for these applications.

14.43 Give an example of a metal-forming operation (from Chapters 6 and 7) that is suitable for adaptive control similar to that shown in Fig. 14.14.

14.44 Give some applications for the systems shown in Fig. 14.25a and c.

14.45 Comment on your observations regarding the system shown in Fig. 14.18b.

14.46 Give examples for which tactile sensors would not be suitable. Explain why tactile sensors are unsuitable for these applications.

14.47 Give examples for which machine vision cannot be applied properly and reliably. Explain why machine vision is inappropriate for these applications.

14.48 Comment of the effect of cutter wear on the profiles produced, such as those shown in Figs. 14.10b and 14.12.

14.49 Although future trends are always difficult to predict with certainty, describe your thoughts as to what new developments in the topics covered in this chapter could possibly take place as we move through the early 2000s.

14.50 Describe the circumstances under which a dedicated fixture, a modular fixture, and a flexible fixture would be preferable.

14.51 What is an anthropomorphic robot?

PROBLEMS

14.52 A spindle/bracket assembly uses the following parts: a steel spindle, two nylon bushings, a stamped steel bracket, and six screws and six nuts to attach the nylon bushings to the steel bracket and thereby support the spindle. Compare this to the spindle/bracket assembly shown in Problem 7.108 and estimate the assembly efficiency for each design.

14.53 Disassemble a simple ball point pen. Carefully measure the time needed to reassemble the pen and calculate the assembly efficiency. Repeat the exercise for a mechanical pencil.

14.54 Examine Figure 14.11b and obtain an expression for the maximum error in approximating a circle with linear increments as a function of the circle radius and number of increments on the circle circumference.

14.55 Assume that you are asked to give a quiz to students on the contents of this chapter. Prepare five quantitative problems and five qualitative questions, and supply the answers.

14.56 Review Example 14.1, and develop open- and closed-loop control system equations for the force if the coefficient of friction is μ.

14.57 Develop open- and closed-loop control system equations in order to have the worktable achieve a sinusoidal position given by $x = \sin \omega t$, where t is time.

14.58 A 35 kg worktable is controlled by a motor/gear combination that can develop a maximum force of 500 N. It is desired to move the worktable from it's current location ($x = 0$) to a new location ($x = 100$ mm). Plot the resultant force and position as a function of time for (a) an open-loop control system and (b) a closed loop control system for $k_p = 5$ N/mm and $k_v = 4.5$ Ns/mm.

DESIGN

14.59 Design two different systems of mechanical grippers for widely different applications.

14.60 For a system similar to that shown in Fig. 14.27, design a flexible fixturing setup for a lathe chuck (See Figs. 8.41 and 8.44).

14.61 Add other examples to those shown in Fig. 1.4.

14.62 Give examples of products that are suitable for the type of production shown in Fig. 14.3.

14.63 Choose one machine each from Chapters 6 through 12, and design a system for the machine in which sensor fusion can be used effectively. How would you convince a prospective customer of the merits of such a system? Would the system be cost effective?

14.64 Does the type of material (metallic or nonmetallic) used for the parts shown in Fig. 14.31 have any influence on the effectiveness of the guides? Explain.

14.65 Think of a product, and design a transfer line for it similar to that shown in Fig. 14.5. Specify the types of and the number of machines required.

14.66 Section 14.9 has described the basic principles of flexible fixturing. Considering the wide variety of parts made, prepare design guidelines for flexible fixturing. Make simple sketches illustrating each guideline for each type of fixturing, describing its ranges of applications and limitations.

14.67 Describe your thoughts on the usefulness and applications of modular fixturing, consisting of various individual clamps, pins, supports, and attachments mounted on a base plate.

14.68 Inspect several household products, and describe the manner in which they have been assembled. Comment on any design changes you would make so that assembly, disassembly, and servicing are simpler and faster.

14.69 Review the last design shown in Fig. 14.18, and design grippers that would be suitable for gripping the following products: (a) an egg; (b) an object made of soft rubber; (c) a metal ball with a very smooth and shiny surface; (d) a newspaper; and (e) tableware, such as forks, knives, and spoons.

14.70 Obtain an old toaster, and disassemble it. Make recommendations on its redesign, using the guidelines given in Section 14.11.2.

14.71 Comment on the design and materials used for the gripper shown in Fig. 14.24. Why do such grippers have low durability on the shop floor?

14.72 Comment on your observations regarding Fig. 14.28, and offer designs for similar applications in manufacturing. Also comment on the usefulness of the designs in actual production on the shop floor.

14.73 Review the various toolholders used in the machining operations described in Chapter 8, and design sensor systems for them. Comment on the features of the sensor systems and discuss any problems that may be associated with their use on the factory floor.

14.74 Design a guide that operates in the same manner as those shown in Fig. 14.31, but to align U-shaped parts so that they are inserted with the open end down.

14.75 Design a flexible fixture that uses powered workholding devices for a family of parts, as in the case study, but that allows for a range of diameters and thicknesses.

14.76 Design a modular fixturing system for the part shown in the figure accompanying Problem 8.160.

14.77 Review the specifications of various numerical-control machines and make a list of typical numbers for their (a) positioning accuracy, (b) repeat accuracy, and (c) resolution. Comment on your observations.

14.78 Consider the automated guided vehicles in Fig. 14.16. Is it possible for such systems to work in multiple buildings? What additional difficulties would you expect that would need to be overcome?

Computer-Integrated Manufacturing Systems

CHAPTER 15

This chapter describes how computers are integrated into the whole manufacturing environment, using hardware, software, and communications networks and protocols. Topics covered include:

- Computer-aided design, by assisting in graphical descriptions of parts and their analysis.
- The use of computers to simulate manufacturing processes and systems.
- Group technology and database approaches, to allow the rapid recovery of previous design and manufacturing experience and apply it to new situations.
- The principles of attended and unattended (unmanned) manufacturing cells.
- The concept of holonic manufacturing and its applications.
- The principles of just-in-time and lean manufacturing and their benefits.
- The types and features of communication systems.
- Applications of artificial intelligence and expert systems in manufacturing.

15.1 Introduction

This chapter addresses the *integration of all manufacturing activities*, whereby processes, machinery, equipment, operations, and their management are treated as a **manufacturing system**. Such a system allows the *total control* of the manufacturing facility, increasing productivity, product quality and reliability, and reducing manufacturing costs.

In **computer-integrated manufacturing** (CIM), the traditionally separate functions of product design, research and development, planning, production, assembly, inspection, and quality control are all interlinked. Integration requires that quantitative relationships among design, materials, manufacturing processes, process and equipment capabilities, and related activities be well understood and established. In this way, changes in, for example, materials, product types, production methods, or response to market demand can be properly and effectively accommodated.

Recall the statements that (a) quality must be built *into* a product; (b) higher quality does *not* necessarily indicate higher cost; and (c) marketing poor-quality products can indeed be *very costly* to the manufacturer. High quality is far more attainable and less expensive through proper *integration* of design and manufacturing than if the two were separate activities. As illustrated throughout this chapter, integration can effectively and successfully be accomplished through **computer-aided design, engineering, manufacturing, process planning,** and **simulation of processes and systems.**

This chapter also describes and emphasizes how **flexibility** in machines, tooling, equipment, and production operations greatly enhances the ability to respond to market demands and product changes and ensures **on-time delivery** of high-quality products. Important developments during the past fifty years or so (See Table 1.1) have had a major impact on modern manufacturing, especially in an increasingly competitive global marketplace. Among these important advances are **group technology, cellular manufacturing, flexible manufacturing systems,** and **just-in-time production** (also called **zero inventory, stockless production,** or **demand scheduling**). Moreover, because of the extensive use of **computer controls** and hardware and software, the planning and effective implementation of **communication networks** have become an essential component of these activities.

The chapter concludes with a review of **artificial intelligence**, expert systems, natural-language processing, machine vision, artificial neural networks, and fuzzy logic, and how these important developments can impact manufacturing activities.

15.2 Manufacturing Systems

As described throughout this text, manufacturing requires several interdependent components, such as materials, tools, machines, computer controls, and people. Consequently, manufacturing should be regarded as a large and complex *system*, consisting of numerous diverse physical and human elements. Some of these factors are difficult to predict and to control, such as the supply and demand of raw materials, changes in market conditions, global economic conditions, as well as human behavior and performance.

A system should ideally be represented by **mathematical and physical models** that identify the nature and extent of the interdependence of the relevant variables involved. In a manufacturing system, a change or disturbance anywhere in the system requires that the system adjust itself in order to continue functioning effectively. For example, if the supply of a particular raw material decreases (such as due to geopolitical issues, wars, or strikes) and, consequently, its cost increases, alternative materials must be investigated and selected.

Likewise, the demand for a particular product may fluctuate randomly and rapidly due, for example, to its style, size, or capacity. Note, for example, the downsizing of automobiles during the 1980s in response to fuel shortages, the popularity of lower fuel efficiency sport-utility vehicles in the 1990s, and the current trend of vehicle electrification due to corporate averaged fuel economy (CAFE) standards, as well as increased concerns for the environment. The manufacturing system must also be capable of producing the modified product on short **lead time** and, preferably, with relatively small major capital investment in machinery and tooling. Lead time is defined as the length of time between the creation of the product as a concept, or receipt of an order for a product and the time that the product first becomes available in the marketplace.

Computer simulation and modeling of such a complex system can be difficult and time-consuming, because of the lack of comprehensive and reliable data on some of the numerous variables involved. Moreover, it is generally not easy to correctly predict and control some of these variables. Some examples of the problems that may be encountered are:

1. The characteristics of a particular machine tool, its performance, and its response to random external disturbances cannot always be modeled precisely (see also Section 15.7).
2. Raw-material costs and their properties may vary over time and be difficult to predict accurately.
3. Sensors may lack sufficient robustness to allow closed-loop control under all conditions in a manufacturing facility.
4. Market demands and human behavior and performance are difficult to model reliably.

15.3 Computer-Integrated Manufacturing

The various levels of automation in manufacturing operations, as described in Chapter 14, have been extended further through **information technology** (IT) using an extensive network of interactive computers. *Computer-integrated manufacturing* (CIM) is a broad term to describe the computerized integration of product design, planning, production, distribution, and management; CIM is a *methodology* and a *goal*, rather than an assemblage of equipment and computers.

The effectiveness of CIM greatly depends on **integrated communications systems**, involving computers, machines, equipment, and their controls; the difficulties that may arise in such systems are described in Section 15.14. Furthermore, because it ideally should involve the total operation of a company, CIM requires an extensive database containing technical as well as business information.

Implementation of CIM in existing manufacturing plants may begin with the use of modules in *selected phases* of a company's operation. For new plants, comprehensive and long-range strategic planning, covering all phases of the operation, is essential. Such plans must take into account (a) the mission, goals, and culture of the organization; (b) the availability of resources; (c) emerging technologies that are relevant to the products made; and (d) the level of integration. It is apparent that if planned and implemented all at once, CIM can be prohibitively expensive.

Subsystems. Computer-integrated manufacturing systems consist of *subsystems* integrated into a whole (Fig. 15.1); the subsystems consist of the following:

1. Business planning and support;
2. Product design;
3. Manufacturing process planning;
4. Process automation and control; and
5. Factory-floor monitoring systems.

The subsystems are designed, developed, and implemented in such a manner that the output of one subsystem serves as the input to another subsystem, as shown by the various arrows in Fig. 15.1. Organizationally, the subsystems are generally divided into two functions: (a) **business-planning functions,** which include activities such as forecasting,

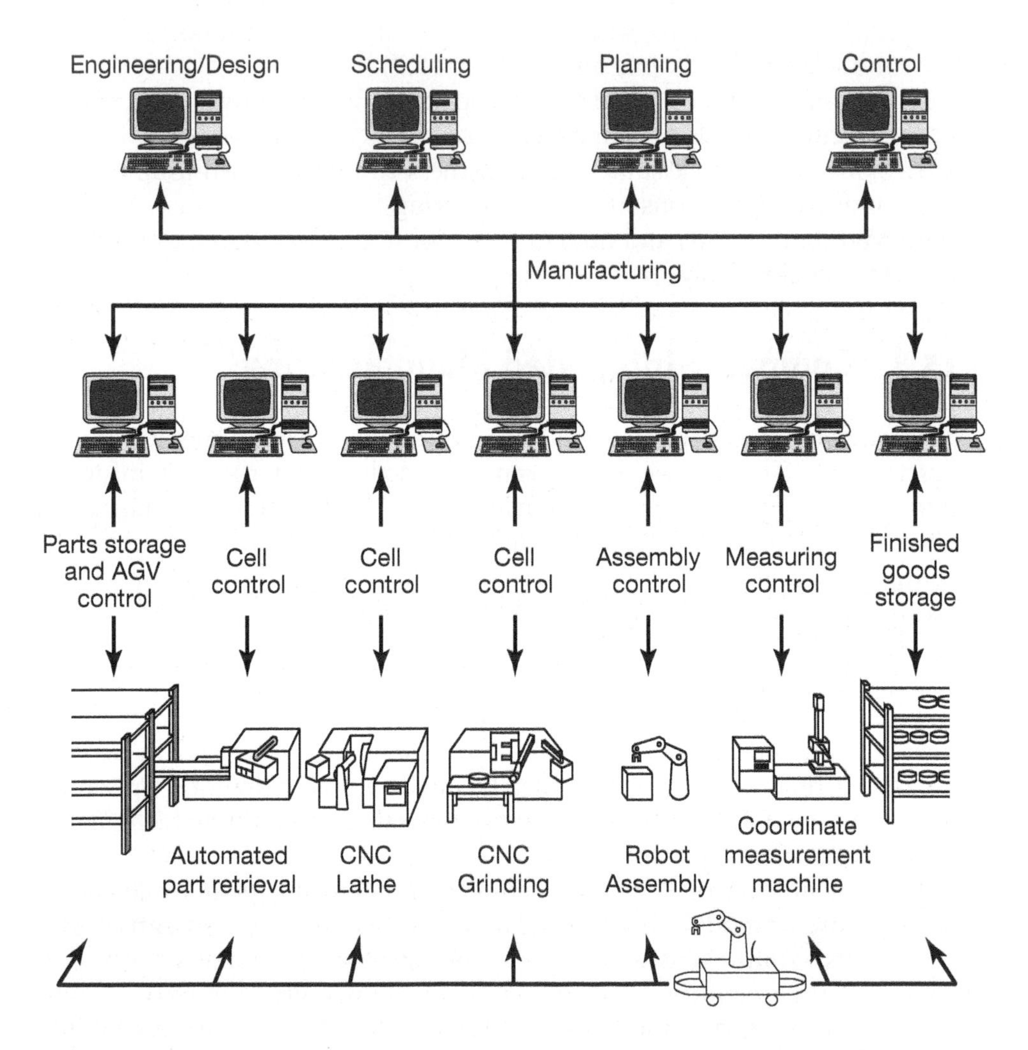

FIGURE 15.1 A schematic illustration of a computer-integrated manufacturing system. *Source:* After U. Rembold.

scheduling, material-requirements planning, invoicing, and accounting, and (b) **business-execution functions**, which include production and process control, material handling, testing, and inspection.

Benefits. The benefits of CIM include the following:

1. Emphasis on *product quality* and *uniformity* through better process control.
2. *Efficient* use of materials, machinery, and personnel, and major reduction in work-in-progress inventory, to improve productivity and lower product cost.
3. *Total control* of production, scheduling, and management of the entire manufacturing operation.
4. *Responsiveness* to shorter product life cycles, changing market demands, and global competition.

15.3.1 Databases

An effective CIM system requires a single, large database that is shared by the entire organization. *Databases* consist of real-time, detailed, and accurate data on product designs, machines, processes, materials, production, finances, purchasing, sales, marketing, and inventory. This vast array of data is stored in computer memory or on a server and recalled or modified, as necessary, either by individuals in the organization or by the CIM system itself. Data storage on remote servers using Internet tools for data retrieval is referred to as *cloud storage*, and is increasingly popular for the large databases typical of large organizations. However, cybersecurity and protection of the data remains a serious concern.

A database typically consists of the following:

1. **Product data** (part shape, dimensions, tolerances, specifications);
2. **Data management attributes** (creator, revision level, part number, etc.);
3. **Production data** (manufacturing processes used in making parts and products);
4. **Operational data** (scheduling, lot sizes, assembly requirements); and
5. **Resources data** (capital, machines, equipment, tooling, and personnel, and specific capabilities of resources).

Databases are built by individuals and through the use of various sensors in the machinery and equipment employed in production. Data from the latter are automatically collected by a **data acquisition system** (DAS) that can report, for example, the number of parts being produced per unit time, weight, dimensional accuracy, and surface finish, at specified rates of sampling. The components of DAS include microprocessors, transducers, and analog-to-digital converters (ADC). Data acquisition systems are also capable of analyzing and transferring the data to other computers, for such purposes as data presentation, statistical analysis, and forecasting of product demand.

Several factors are important in the use and implementation of databases:

1. They should be timely, accurate, user friendly, and easily accessible and shared.
2. Because they are used for a variety of purposes and by many people or groups in an organization, databases must be flexible and responsive to the needs of different users.
3. CIM systems are accessed by designers, manufacturing engineers, process planners, financial officers, and the management of the company through appropriate access codes; companies must protect data against tampering or unauthorized use.
4. If difficulties arise with accuracy or loss of data, the correct data has to be recovered and restored.

15.4 Computer-Aided Design and Engineering

Computer-aided design (CAD) involves the use of software to create geometric representations of products and components (See also Fig. 1.10a) and is associated with **interactive computer graphics**, known as a **CAD system**. *Computer-aided engineering* (CAE) simplifies the creation of the database by allowing several applications to share the information in the database. These applications include, for example, (a) finite-element analysis of stresses, strains, deflections, and temperature distribution in structures and load-bearing members; (b) the generation, storage, and retrieval of NC data; and (c) the design of integrated fixtures and machinery used to produce the design.

In CAD, the user can generate drawings or sections of a drawing on the computer. The design can be printed if desired, but often it is stored as a digital file to be accessed as needed, often on networked computers throughout the organization. When using a CAD system, the designer can conceptualize the object to be designed, can consider alternative designs, or quickly modify a particular design to meet specific requirements.

Through the use of powerful software, such as SolidWorks, ProEngineer, CATIA, AutoCAD, Solid Edge, and VectorWorks, the design can be subjected to detailed *engineering analysis*, which can identify potential problems, such as excessive load or deflection, or interference at mating surfaces during assembly. In addition to geometric and dimensional features, other information, such as a list of materials, specifications, and manufacturing instructions, is stored in the CAD database. Using such information, the designer can also analyze the economics of alternative designs using other software.

Modern CAD software allows the use of parametric design, as described in Section 15.4.2. In a parametric design, instead of specifying dimensions explicitly, relations between dimensions are set, so that a different sized part of the same design can easily be generated.

15.4.1 Exchange Specifications

Because of the availability of a wide variety of CAD systems with different characteristics and supplied by different vendors, proper communication and exchange of data between these systems is essential (see also Section 15.14). **Drawing exchange format** (DFX) was developed for use with Autodesk™ and has become a de facto standard, because of the long-term success of this particular software package. However, a shortcoming of DFX is that it is limited to transferring geometry information only. Similarly, **STL** (*ST*ereo *L*ithography; see Section 10.12.1) formats are used to export 3D geometries, initially to rapid prototyping systems; recently, however, STL has also become a format for data exchange between CAD systems.

The need for a single neutral format for better compatibility and for the transfer of more information than geometry alone is filled by a number of file formats, such as the **Initial Graphics Exchange Specification** (IGES). Vendors only have to provide translators for their own systems, to preprocess outgoing data into the neutral format, and to postprocess incoming data from the neutral format into their system. IGES is used for translation in two directions (into and out of a system), and is also used widely for translation of 3D line and surface data.

Another exchange specification is a solid-model-based standard, called **Product Data Exchange Specification** (PDES), which is based on the *Standard for the Exchange of Product Model Data* (STEP), developed by the International Standards Organization. PDES allows information on shape, design, manufacturing, quality assurance, life cycle, testing, and maintenance to be transferred between CAD systems.

15.4.2 Elements of CAD Systems

The design process in a CAD system consists of four stages:

1. **Geometric modeling** where a physical object or any of its parts is described mathematically or analytically. The designer first constructs a geometric model by giving commands that create or modify lines, surfaces, solids, dimensions, and text that, together, compose an accurate and complete two- or three-dimensional representation of the object. The results of these commands are then displayed, where images can be manipulated on the screen and any section can be magnified to view its details. The models are stored in the database.

   The models can be presented as follows.

   a. In **line representation**, also called **wire-frame representation** (Fig. 15.2), all edges of the model are visible as solid lines. This type of image can, however, be ambiguous because line resolution may be a problem, particularly for complex shapes; hence various colors are generally used for different parts of the object. The three types of wire-frame representations are 2D, $2\frac{1}{2}$-D,

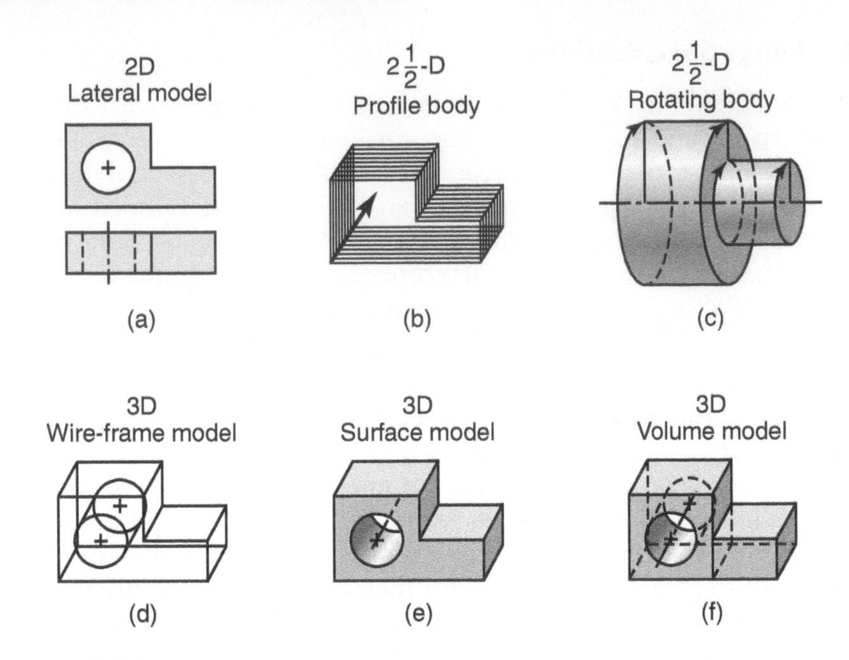

FIGURE 15.2 Types of modeling for CAD.

and 3D. A 2D image shows the profile of the object, and a $2\frac{1}{2}$-D image can be obtained by *translational sweep*, that is, moving the 2D object along the z-axis. For round objects, a $2\frac{1}{2}$-D model can be generated by simply *rotating* a 2D model around its axis.

b. In the **surface model**, all visible surfaces are shown, defining surface features and edges of objects. For surface modeling, CAD programs use *Bezier curves*, *B-splines*, or *nonuniform rational B-splines* (NURBS). Each of these methods uses control points to define a polynomial curve or a surface. A Bezier curve passes through the first and last vertex and uses the other control points to generate a blended curve. The drawback to Bezier curves is that modification of one control point will affect the entire curve. B-splines are a blended, piecewise polynomial curve, where modification of a control point affects the curve only in the area of the modification. Examples of two-dimensional Bezier curves and B-splines are given in Fig. 15.3. A NURBS is a special type of B-spline where each control point has a weight associated with it.

c. In the **solid model**, all surfaces are shown but the data describe the interior volume as well. Solid models can be constructed from *swept volumes* (Figs. 15.2b and c) or by the techniques shown in Fig. 15.4. In **boundary representation** (BREP), surfaces are combined to develop a solid model (Fig. 15.4a). In **constructive solid geometry** (CSG), simple shapes such as spheres, cubes, blocks, cylinders, and cones (called **primitives of solids**) are combined to develop a solid model (Fig. 15.4b). The user selects any combination of primitives and their sizes, and combines them into the desired solid model. Although solid models have advantages such as ease of design analysis and ease of preparation for

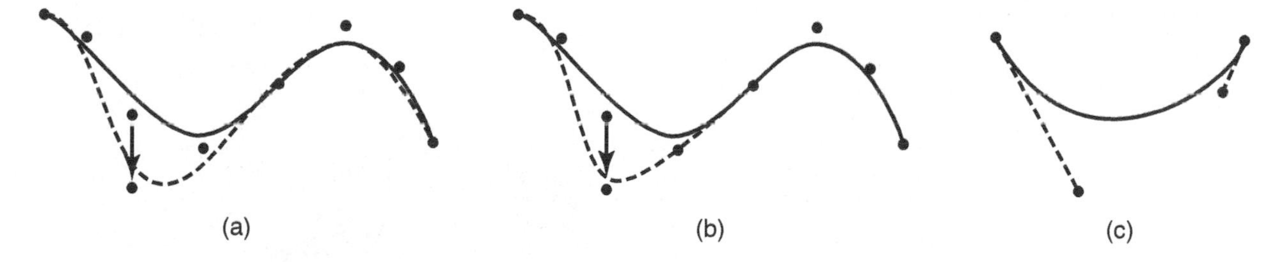

FIGURE 15.3 Types of splines. (a) A Bezier curve passes through the first and last control point and generates a curve from the other points; changing a control point modifies the entire curve. (b) A B-spline is constructed piecewise, so that changing a vertex affects the curve only in the vicinity of the changed control point. (c) A third-order piecewise Bezier curve constructed through two adjacent control points, with two other control points defining the curve slope at the end points; although a third-order piecewise Bezier curve is continuous, its slope may be discontinuous.

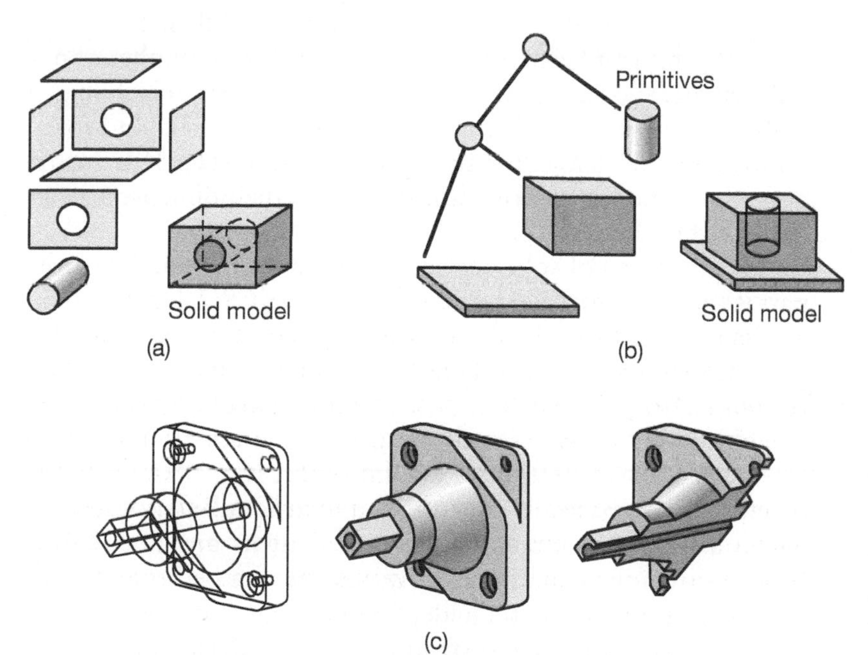

FIGURE 15.4 (a) Boundary representation of solids, showing the enclosing surfaces and the generated solid model. (b) A solid model, represented as compositions of solid primitives. (c) Three different representations of the same part by CAD. *Source:* After P. Ranky.

part manufacturing, they require more computer memory and processing time than the wire-frame and surface models.

The standard for rapid prototyping machinery, the **STL file format** (an abbreviation for *stereolithography*, and also called *Standard Tessellation Language*), allows for three-dimensional part descriptions. Basically, an STL file consists of a number of triangles that define the exterior surface of the part (Fig. 15.5). With a sufficiently large number of triangles, the surface can be defined within a prescribed tolerance, although requiring a larger file size. With additive manufacturing, a part cross section can be obtained at any height; the resulting polygon is then used to plan the part (See Fig. 10.49). The popularity of STL in

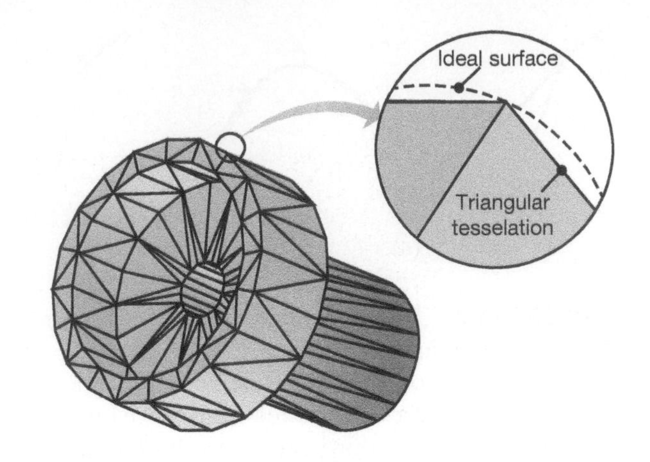

FIGURE 15.5 An example of an STL part description. Note that the surface is defined by a tessellation of triangles, and that there is an inherent error of form that occurs with curved surfaces; however, this can be brought down to any desired tolerance by incorporating more triangles in the surface.

rapid prototyping, along with its easy implementation, has led to using this format in other applications as well, such as computer graphics and general CAD data transfer. A number of new file formats, such as 3MF, are under development, which in general migrate away from triangular tessellations and include other information, such as color and surface texture. The 3MF format also allows the transfer of additional data, including proprietary data, within the same CAD file.

A special type of solid model is a **parametric model**, where a part is not only stored in terms of a BREP or CSG definition, but also is derived from the dimensions and constraints that define the features (Fig. 15.6). Whenever a change is made, the part is re-created from these definitions, a feature that allows simple and straightforward updates and changes to be made to the models.

d. The **octree representation** of a solid object (Fig. 15.7) is a type of model that basically is a three-dimensional analog to pixels on a television screen or monitor. Just as any area can be broken down into quadrants, any volume can be broken down into *octants*, which are then identified as solid, void, or partially filled. Partially filled *voxels* (from *vo*lume pi*xels*) are broken into smaller octants and then reclassified. With increasing resolution, exceptional part detail can be achieved. This process may appear to be somewhat cumbersome, but it allows for accurate description of complex surfaces, and is used particularly in biomedical applications such as modeling bone or organ shapes.

e. A **skeleton** (Fig. 15.8) is commonly used for kinematic analysis of parts or subassemblies. A skeleton is a family of lines, planes, and curves that describe a part without the detail of surface models. Conceptually, a skeleton can be constructed by fitting the largest circles (or spheres for three-dimensional objects) within the geometry. The skeleton is the set of points that connect the centers of the circles (or spheres); the circle radius data at each point also is stored.

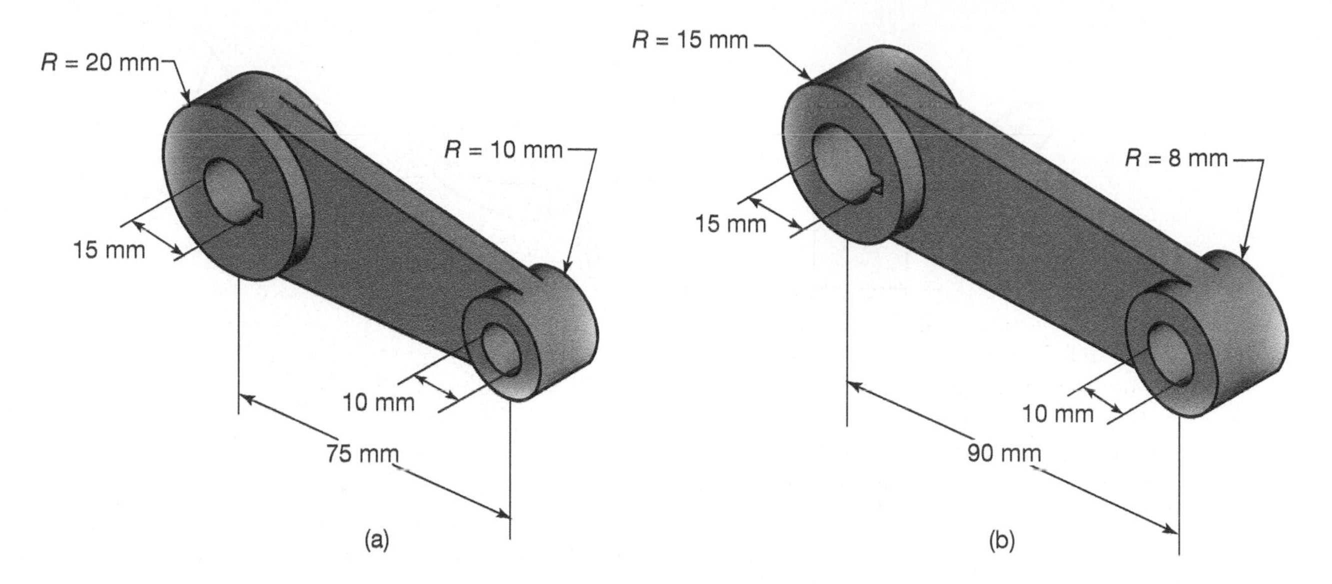

FIGURE 15.6 An example of parametric design. (a) Original design; (b) modified design, produced by modifying parameters in the data file of the part in (a). Dimensions of part features can easily be modified to quickly produce an updated solid model.

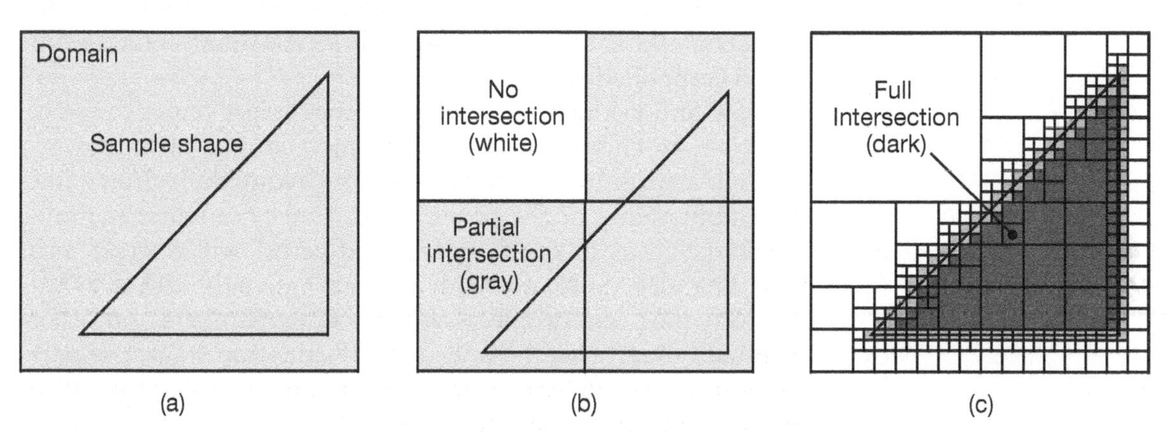

FIGURE 15.7 The octree representation of a solid object. (a) After one iteration; (b) two iterations; (c) six iterations. Any volume can be broken down into octants, which are then identified as solid, void, or partially filled. Shown is two-dimensional, or quadtree, version, for representation of shapes in a plane.

2. **Design analysis and optimization.** After its geometric features have been determined, the design is subjected to an *engineering analysis*, a phase which may consist, for example, of analyzing stresses, strains, deflections, vibrations, heat transfer, temperature distribution, or dimensional tolerances. Several software packages are available with the capabilities to compute accurately and rapidly these quantities, such as the finite element based programs ABAQUS, ANSYS, NASTRAN, LS-DYNA, MARQ, and ALGOR, each having the capability to compute these quantities accurately and rapidly. Because of the relative ease with which such analyses can now be

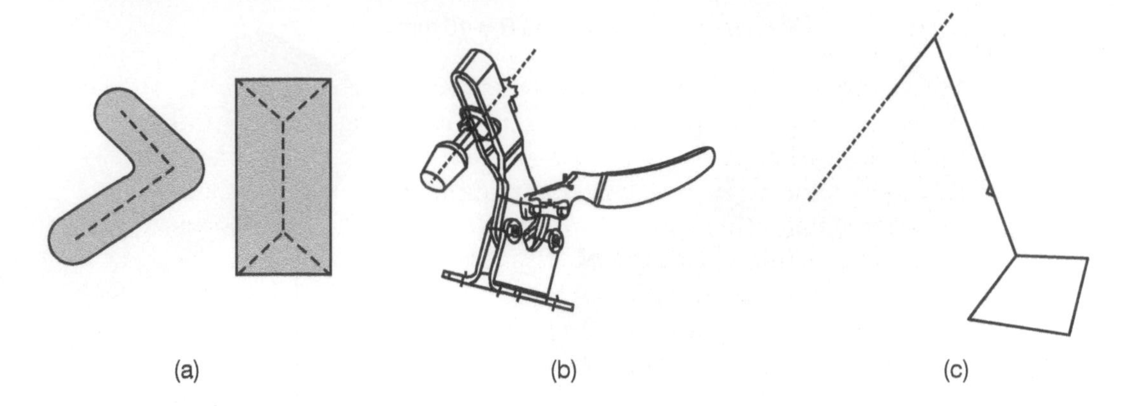

FIGURE 15.8 (a) Illustration of the skeleton data structure for two different solid objects; the skeleton is the dashed lines in the interior of the objects. (b) A general view of a clamp. (c) A skeleton model used for kinematic analysis of the clamp. *Source:* S.D. Lockhart and C.M. Johnson, *Engineering Design Communication*, Prentice Hall, 2012.

performed, designers can study a design more thoroughly before it moves on to production. Experiments and measurements in a working environment may nonetheless still be essential to determine and verify the actual effects of loads, temperature, and other variables on the designed components.

3. **Design review and evaluation.** An important design stage is *review and evaluation*, to check for any interference among various components. This step is necessary in order to avoid difficulties later during assembly or use of the part and to determine whether moving members (such as linkages, gears, and cams) will operate and function as intended. Software with animation capabilities is available to identify potential problems with moving members and other dynamic situations. During the review and evaluation stage, the part is also precisely dimensioned and set within the full range of tolerance required for its production (Section 4.7).
4. **Documentation.** After the preceding stages have been completed, a paper copy of the design may be produced for documentation and reference. At this stage, detail and working drawings can also be developed and printed; however, this practice has become increasingly rare. Companies now often rely only on the databases that have been developed, especially since databases allow part drawing retrieval from any computer on the network.
5. **Database.** The components in products are typically either standard components that are mass produced according to a given design specification, such as bolts or gears, or are identical to those used in previous designs. CAD systems have a built-in database management system that allows designers to identify, view, and access parts from a library of stock parts. These parts can be parametrically modeled to allow cost-effective updating of their geometry. Some databases with extensive parts libraries are now commercially available; many vendors make their part libraries available on the Internet.

15.5 Computer-Aided Manufacturing

Computer-aided manufacturing (CAM) involves the use of computer technology to assist in all phases of manufacturing, including process and production planning, scheduling, manufacture, quality control, and management. Because of their obvious benefits, computer-aided design and computer-aided manufacturing are often combined into **CAD/CAM systems.** This combination allows information transfer from the design stage to the planning stage for the manufacture of a product without the need to manually reenter the data on part geometry. The database developed is stored, and then processed further by CAM into the necessary data and instructions for operating and controlling production machinery and material-handling equipment, as well as for performing automated testing and inspection for maintaining product quality (Section 4.8.3).

The emergence of CAD/CAM has had a major impact on manufacturing operations by standardizing product development and by reducing design effort, evaluation, and prototype work; it also has made possible significant cost reductions and improved productivity. The Boeing 777 airplane, for example, was designed completely by computer (known as **paperless design**), with 2000 workstations linked to eight computers. The plane was constructed directly from the CAD/CAM software developed (an enhanced CATIA system), and no prototypes or mock-ups were built, such as were required for previous models. The development cost for this aircraft was on the order of $6 billion.

An example of an important feature of CAD/CAM in machining is its ability to describe the *cutting-tool path* for such operations as NC turning, milling, and drilling (see Sections 8.10 and 14.4). The instructions (*programs*) are computer generated and can be modified by the programmer to optimize the tool path. The engineer or technician can then display and visually check the tool path for the possibility of tool collisions with clamps or fixtures. The tool path can be modified at any time to accommodate other part shapes or features. CAD/CAM systems are capable of *coding and classifying parts* into groups that have similar shapes, using alphanumeric coding (see the discussion of *group technology* in Section 15.8).

15.6 Computer-Aided Process Planning

For a manufacturing operation to be efficient, all of its diverse activities must be planned and coordinated. *Process planning* involves selecting methods of production, machinery, tooling, fixtures, the sequence of operations, the standard processing time for each operation, and methods of assembly. These choices were traditionally documented on a **routing sheet**, such as that shown in Fig. 15.9. *Computer-aided process planning* (CAPP) accomplishes this complex task by viewing the total operation as an integrated system, so that the individual operations and steps involved in making each part are coordinated with each other.

CAPP is thus an essential adjunct to CAD and CAM. Although it requires extensive software and coordination with CAD/CAM and various

ROUTING SHEET		
CUSTOMER'S NAME: Midwest Valve Co.		PART NAME: Valve body
QUANTITY: 15		PART NO.: 302
Operation No.	Description of operation	Machine
10	Inspect forging, check hardness	Rockwell tester
20	Rough machine flanges	Lathe No. 5
30	Finish machine flanges	Lathe No. 5
40	Bore and counter bore hole	Boring mill No. 1
50	Turn internal grooves	Boring mill No. 1
60	Drill and tap holes	Drill press No. 2
70	Grind flange end faces	Grinder No. 2
80	Grind bore	Internal grinder No. 1
90	Clean	Vapor degreaser
100	Inspect	Ultrasonic tester

FIGURE 15.9 A simple example of a routing sheet. These sheets may include additional information on materials, tooling, estimated time for each operation, processing parameters, and various other details. Routing sheets travel with the part from operation to operation. Current practice is to store all relevant data in computers and to affix to the part a bar code, radio-frequency ID, or equivalent label that serves as a key into the database of parts information.

other aspects of integrated manufacturing systems, CAPP is a powerful tool for efficiently planning and scheduling manufacturing operations. It is particularly effective in small-volume, high-variety parts production involving machining operations.

15.6.1 Elements of CAPP Systems

There are two types of computer-aided process-planning systems: variant and generative.

1. In the **variant system**, also called the **derivative system**, the computer files contain *a standard process plan* for a particular part to be made. The search for a standard plan is made in the database, using a specific code number for the part. The plan is based on the part's shape and its manufacturing characteristics (see Section 15.8). The plan can be retrieved, displayed for review, and printed or stored as that part's

routing sheet. The process plan includes information such as the types of tools and machines to be used, the sequence of operations to be performed, the cutting speeds and feeds, and the time required for each sequence. Minor modifications to an existing process plan also can be made. If the standard plan for a particular part is not available, a plan that is similar to it and one that has a similar code number and an existing routing sheet is retrieved. If a routing sheet does not exist, then one is prepared for the new part and stored.

2. In the **generative system**, a process plan is automatically generated on the basis of the same logical procedures that would be followed by a traditional process planner. The generative system is complex, because it must contain comprehensive and detailed knowledge of the part's shape and dimensions, process capabilities, the selection of manufacturing methods, machinery, tools, and the sequence of operations to be performed. The system is capable of creating a new plan instead of having to use or to modify an existing plan, as the variant system must do.

   The generative system has such advantages as (a) flexibility and consistency in process planning for new parts and (b) higher overall quality of planning, because of the capability of the decision logic in the system to optimize planning using up-to-date manufacturing technology. However, the generative system requires the ready availability of accurate cost data for the processes under consideration.

Process-planning capabilities of computers can be integrated into the planning and control of production systems as a subsystem of CIM (see Section 15.3.) Several functions can be performed using these activities, such as *capacity planning* for plants to meet production schedules, control of *inventory, purchasing,* and *production scheduling.*

15.6.2 Material-Requirements Planning Systems and Manufacturing Resource Planning Systems

Computer-based systems for managing inventories and delivery schedules of raw materials and tools are called *material-requirements planning* (MRP) systems. Sometimes regarded as a method of inventory control, this activity involves keeping complete records of inventories of materials, supplies, parts in various stages of production (*work in progress or WIP*), orders, purchasing, and scheduling. Several types of data are typically involved in a master production schedule. These data pertain to the raw materials needed (**bill of materials**), product structure levels such as individual items in a product (its components, subassemblies, and assemblies), and scheduling.

In *manufacturing resource planning* (MRP-II) systems, all aspects of manufacturing planning are controlled through a feedback mechanism. Although the system is complex, MRP-II is capable of final production scheduling, monitoring actual results in terms of performance and output, and comparing those results against a master production schedule.

15.6.3 Enterprise Resource Planning

Beginning in the 1990s, *enterprise resource planning* (ERP) became important; it basically is an extension of MRP-II. Although there are variations, ERP generally has been defined as a method for effective planning and control of all the resources required in a business enterprise: to take orders for products, produce them, ship them to the customer, develop traceability data, and service the products; accounting and billing are also incorporated in ERP. Thus, ERP coordinates, optimizes, and dynamically integrates all information sources and the widely diverse technical and financial activities in a manufacturing organization, often using **cloud computing** (see Section 15.14.2). Its major goals are to improve productivity, reduce manufacturing cycle times, and optimize processes, thus benefiting not only the organization but the customer as well.

Modern FMS packages plan and manage availability and utilization of employees, machines, tools, fixtures, and many other resources in order to maximize shop-floor efficiency. Moreover, ERP systems are increasingly sophisticated regarding the management of global supply chains. It has therefore been suggested that a fully functional ERP will be able to activate a whole factory as soon as an order comes in to the company's website, thereby enabling *pull* (see Section 15.12). Every ERP is integrated with a **manufacturing execution system** (MES), which gives specific scheduling and routing instructions to the factory, organizes production, and even generates a CNC plan for each part.

An effective implementation of ERP is a challenging task because:

1. Difficulties are encountered in attaining timely, effective, and reliable communication among all parties involved, especially in a global business enterprise, indicating the importance and necessity of teamwork.
2. There is a need for modifying business practices in an age where information systems and e-commerce have become very important to the success of a business organization.
3. Extensive and specific hardware and software requirements must be provided to implement ERP. ERP-II is a more recent development that uses web-based tools to perform the tasks of ERP; these systems are intended to extend the ERP capabilities beyond the host organization to allow interaction and coordination across corporate entities.

15.7 Computer Simulation of Manufacturing Processes and Systems

With increasing power and sophistication of computer hardware and software, *computer simulation* of manufacturing processes and systems has grown rapidly. Process simulation takes two basic forms:

1. A model of a specific manufacturing operation, that is intended to determine the viability of a process or to optimize or improve its performance, and

2. A model of multiple processes and their interactions, to help process planners and plant designers during the layout of machinery and various facilities.

Individual processes have been modeled using various mathematical schemes (Section 6.2.2). Finite-element analyses in software packages (**process simulation**) are commercially available and inexpensive. Typical problems addressed by such models include **process viability** (such as assessing the formability and behavior of a particular type of sheet metal in a pressworking operation) and **process optimization**, such as (a) analyzing the metal-flow pattern in forging a blank (See Fig. 6.11), to identify potential defects, or (b) improving mold design in a casting operation, to reduce or eliminate hot spots and minimize defects by promoting uniform cooling.

Mathematical models can take several forms. Finite element simulation can involve stresses and strains but may ignore accelerations (*quasi-static*), or simulations can involve vibrations and chatter. Similarly, a *design of experiments* approach can be used to experimentally characterize a system. A current trend is to incorporate process simulations into machine control programs to monitor and, if necessary, modify operations in real time.

15.8 Group Technology

Group technology (GT) is a methodology that seeks to take advantage of the **design and processing similarities** among the parts to be produced. As illustrated in Fig. 15.10, these characteristics clearly suggest that major benefits can be obtained by **classifying** and **coding** the parts into *families*. One company found that by disassembling each product into its individual components and then identifying the similar parts, 90% of the 3000 parts made fell into only five major families.

A pump, for example, can be broken down into its basic components, such as the motor, housing, shaft, flanges, and seals. Each of these components is basically the same in terms of its design and manufacturing characteristics; consequently, all shafts, for example, can be placed in one family of shafts or use a family of seals. Group technology becomes especially attractive because of the ever-growing variety of products, which are often produced in batches. Nearly 75% of manufacturing today is batch production.

The group-technology approach becomes even more attractive in view of consumer demand for an ever-larger variety of products, each in smaller quantities, thus involving *batch production* (see Section 14.2.2). Maintaining high efficiency in batch operations can, however, be difficult and overall manufacturing efficiency can therefore be compromised by a reduction in production volume. A traditional *product flow* in a batch manufacturing operation is shown in Fig. 15.11a. Note that machines of the same type are arranged in groups: groups of lathes, milling machines, drill presses, and grinders. In such a layout, called **functional layout**, there is considerable random movement on the production floor, as shown by the arrows in the figure, indicating movement of materials and parts.

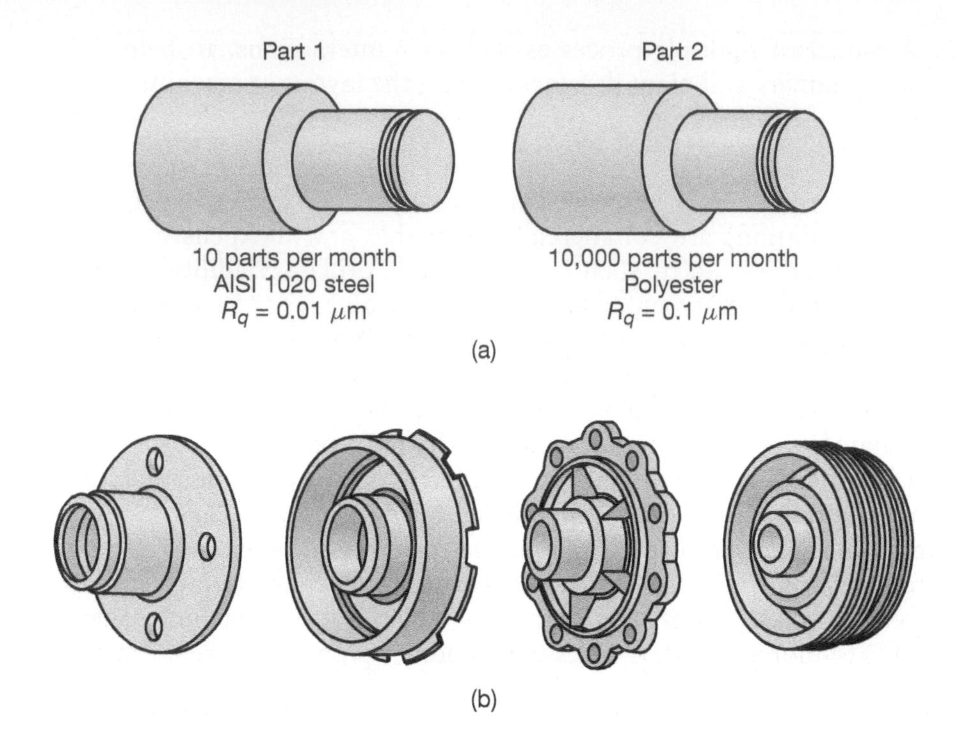

FIGURE 15.10 Grouping parts according to their (a) geometric similarities and (b) manufacturing attributes.

The extreme case of customer-driven variety can result in batches of one product and is referred to as *mass customization*. While this is the typical case for additive manufacturing processes, mass customization requires effective ERP and MES systems, and a fairly high level of automation, generally involving manufacturing cells.

The machines in *cellular manufacturing* (Section 15.9) are arranged in an efficient product flow line, called **group layout** (Fig. 15.11b). The manufacturing cell layout depends on the common features in parts, thus group technology is an essential feature for designing cells and therefore for implementing lean manufacturing (see Section 15.13).

15.8.1 Classification and Coding of Parts

In group technology, parts are identified and grouped into families by **classification and coding (C/C) systems.** This process is a critical and complex first step in GT, and it is done according to the part's design and manufacturing attributes (See Fig. 15.10):

1. Design attributes pertain to *similarities in geometric features* and consist of the following:
    a. External and internal shapes and dimensions;
    b. Aspect ratio (length-to-width ratio or length-to-diameter ratio);
    c. Dimensional tolerance;
    d. Surface finish; and
    e. Part function.

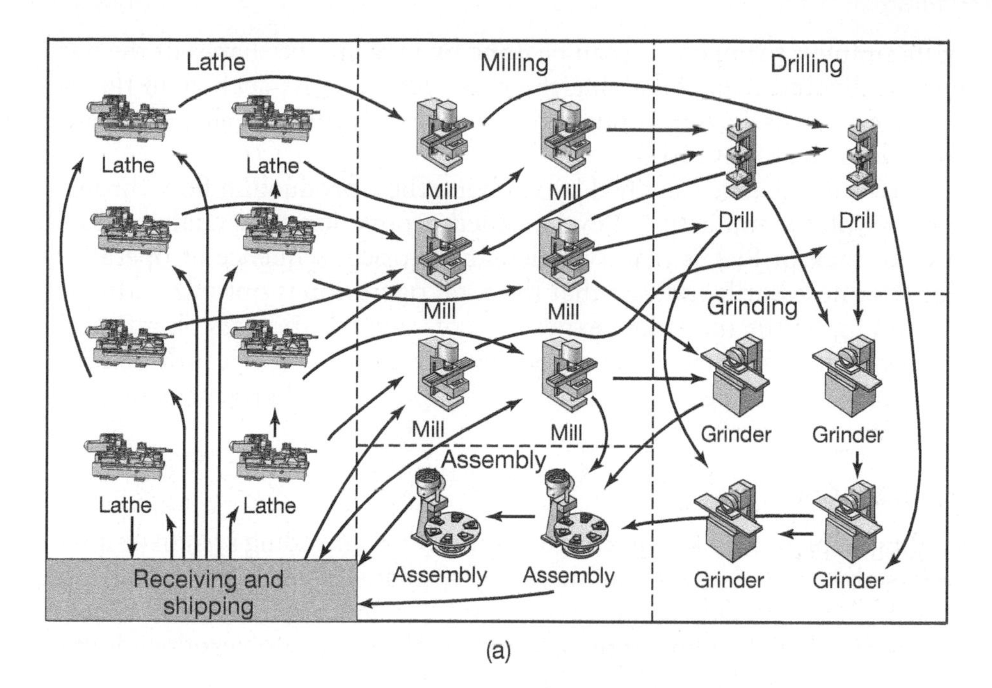

(a)

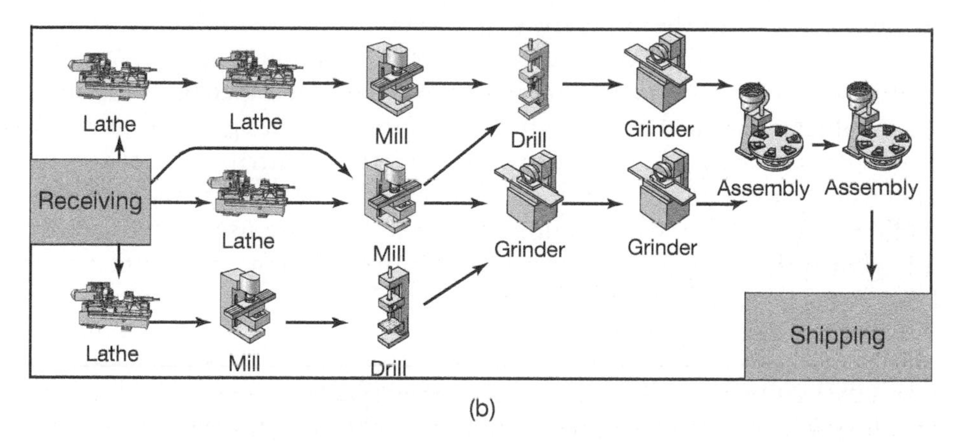

(b)

FIGURE 15.11 (a) Functional layout of machine tools in a traditional plant; arrows indicate the flow of materials and parts in various stages of completion. (b) Group-technology (cellular) layout.

2. **Manufacturing attributes** pertain to *similarities in the methods and the sequence of the operations* performed on the part. The manufacturing attributes of a particular part or component consist of the following:

    a. Primary processes;
    b. Secondary processes and finishing operations;
    c. Process capabilities, such as dimensional tolerances and surface finish;
    d. Sequence of operations performed;
    e. Tools, dies, fixtures, and machinery; and
    f. Production volume and production rate.

In its simplest form, coding can be done by viewing the shapes of the parts generically and then classifying the parts accordingly, such as in the categories of parts having rotational symmetry, rectilinear shape, or large surface-to-thickness ratios.

Parts may also be classified by studying their production flow throughout the total manufacturing cycle, called **production-flow analysis** (PFA). A drawback of PFA is that a particular historical sequence of operations does not necessarily indicate that the total operation is optimized. In fact, depending on the individual experience of, say, two process planners, the strategies for manufacturing the same part can be significantly different from each other. The beneficial role of computer-aided process planning in avoiding or minimizing such problems are thus obvious.

15.8.2 Coding

Coding of parts may be based on a company's own coding system or it may be based on one of several commercially-available classification and coding systems (C/C system). Because product lines and organizational requirements vary widely, none of the C/C systems has been universally adopted. Whether it has been developed in-house or is purchased, the system must be compatible with the company's other systems, such as NC machinery and CAPP systems. The **code structure** for part families typically consists of numbers, letters, or a combination of the two. Each specific component of a product is assigned a code, which may pertain to design attributes only (generally less than 12 digits) or to manufacturing attributes only. Most advanced systems include both attributes and use as many as 30 digits. Coding may be done without input from the software user and displayed only if the information is requested. Commonly, design or manufacturing data retrieval can be based on keyword searches.

The three basic **levels of coding** are:

1. **Hierarchical coding.** In this type of coding, also called **monocode**, the interpretation of each succeeding digit depends on the value of the preceding digit. Each symbol amplifies the information contained in the preceding digit, so that a single digit in the code cannot be interpreted alone. The advantage of hierarchical coding is that a short code can contain a large amount of information; this method is, however, difficult to apply in a computerized system.
2. **Polycodes.** Each digit in this code, also known as **chain-type code**, has its own interpretation and does not depend on the preceding digit. This structure tends to be relatively long, but it allows the identification of specific part **design and manufacturing** attributes and is well suited to computer implementation.
3. **Decision-tree coding.** This system, also called **hybrid code**, combines both design and manufacturing attributes (Fig. 15.12).

15.8.3 Coding Systems

There are numerous coding systems in use, including Opitz, Multiclass, KK-3, Brisch, CODE, CUTPLAN, and DCLASS. Each has its particular

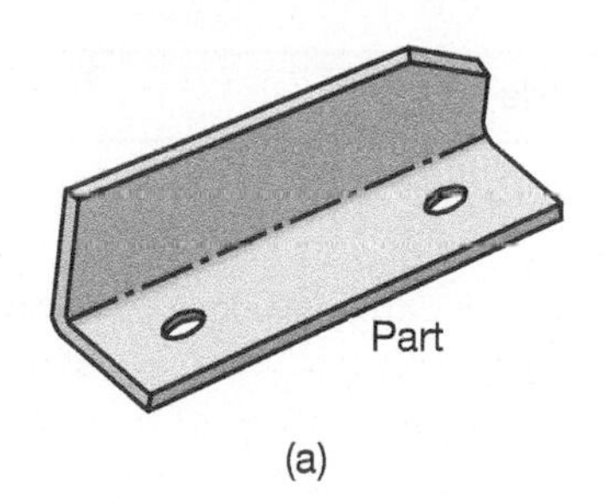

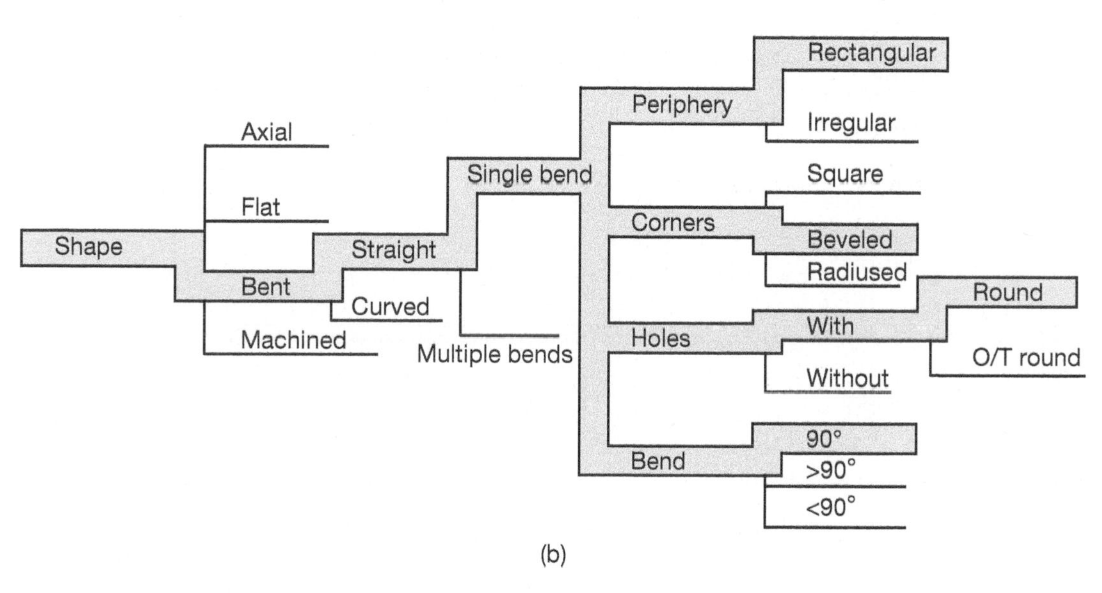

FIGURE 15.12 Decision-tree classification for a sheet metal bracket.
Source: After G.W. Millar.

strengths and have more or less integration with commercial database and CAD software. Three of the most common industrial coding systems are:

1. The **Opitz** system, developed in the 1960s in Germany by H. Opitz, was the first comprehensive coding system presented. The basic code consists of nine digits (in the format 12345 6789) that represent design and manufacturing data (Fig. 15.13); four additional codes (in the format ABCD) also may be used to identify the type and sequence of production operations. This system has two drawbacks: (a) It is possible to have different codes for parts that have similar manufacturing attributes, and (b) several parts with different shapes can have the same code.
2. The **MultiClass system,** illustrated in Fig. 15.14, was originally developed under the name MICLASS (Metal Institute Classification System of the Netherlands Organization for Applied Scientific Research) for the purpose of helping automate and standardize several design, production, and management functions. It involves up to 30 digits, used interactively with a computer that asks the user a number of questions. On the basis of the answers given, the computer automatically assigns a code number to the part.

		Form Code					Supplementary Code			
		1st digit Part class	2nd digit Main shape	3rd digit Rotational surface machining	4th digit Plane surface machining	5th digit Auxiliary holes, gear teeth, forming	Digit 1	Digit 2	Digit 3	Digit 4
0	Rotational parts	$\frac{L}{D} \leq 0.5$	External shape, external shape elements	Internal shape, internal shape elements	Plane surface machining	Auxiliary holes	Dimension	Material	Original shape of raw material	Accuracy
1		$0.5, \frac{L}{D} < 3$								
2		$\frac{L}{D} \geq 3$				Gear teeth				
3		$\frac{L}{D} \leq 2$ with deviation	Main shape	Rotational machining, internal and external shape elements		Auxiliary holes, gear teeth, and forming				
4		$\frac{L}{D} > 2$ with deviation								
5		Special								
6	Nonrotational parts	$\frac{A}{B} \leq 3, \frac{A}{C} \geq 4$ Flat parts	Main shape	Principal bores	Plane surface machining	Auxiliary holes, gear teeth, and forming				
7		$\frac{A}{B} > 3$ Long parts	Main shape							
8		$\frac{A}{B} \leq 3, \frac{A}{C} \geq 4$ Cubic parts	Main shape							
9		Special								

FIGURE 15.13 A classification and coding scheme using the Opitz system, consisting of five digits and a supplementary code of four digits, as shown along the top of the figure. *Source:* After H. Opitz.

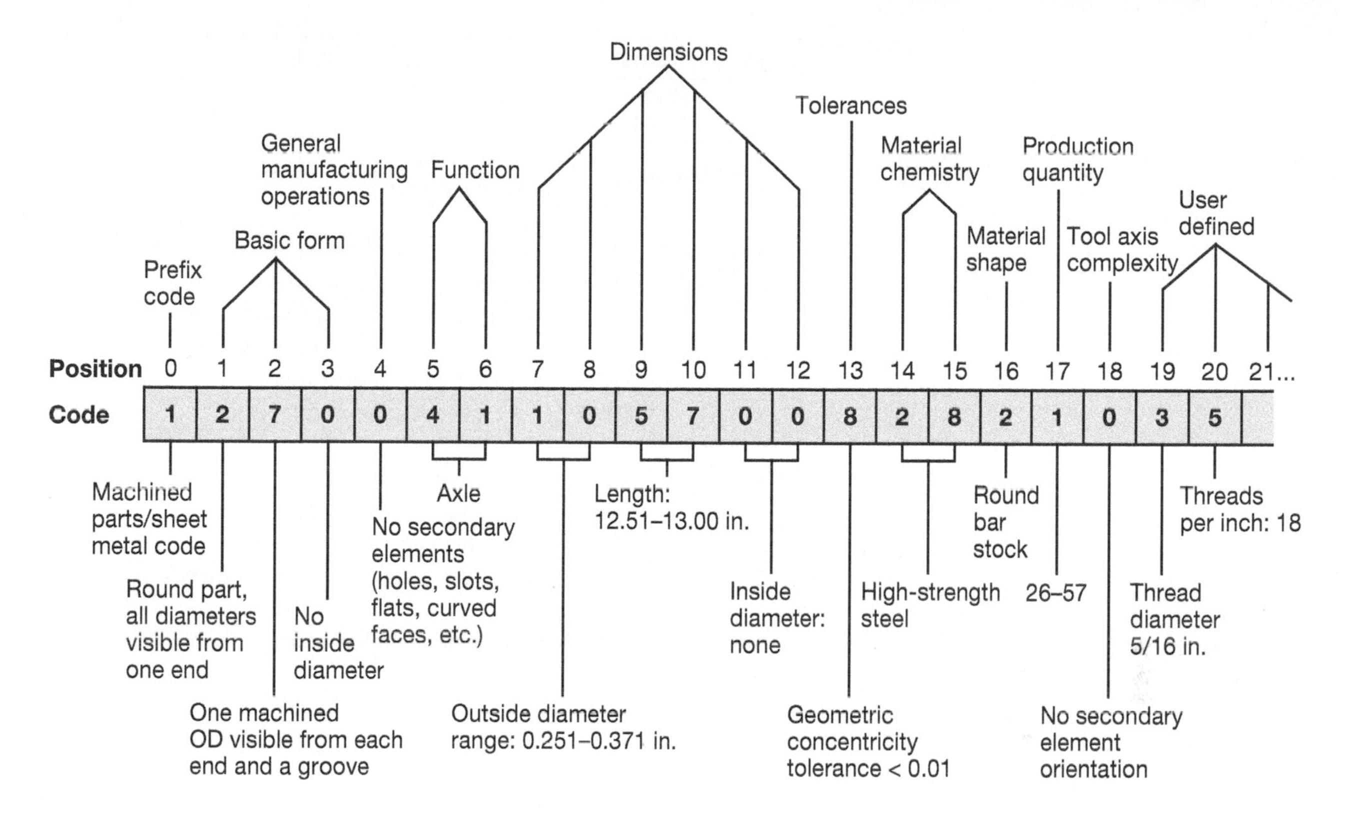

FIGURE 15.14 Typical MultiClass code for a machined part.
Source: Organization for Industrial Research.

3. The **KK-3 system** is a general-purpose system for parts that are to be machined or ground; it uses a 21-digit decimal system. This system is much longer than the two coding systems described above, but it also classifies dimensions and dimensional ratios, such as the length-to-diameter ratio of the part. The structure of a KK-3 system for rotational components is shown in Fig. 15.15.

Production Flow Analysis. One of the main benefits of GT is in the design of manufacturing cells (Section 15.9). Consider the situation in Fig. 15.16, which is intended to show a highly simplified list of parts that are to be produced and the required machinery. As it can be seen, the types of machines and the variety of parts do not lend themselves to groupings of machines. Using GT, parts can be classified and codified, then sorted into logical groupings. Combined with production requirements, this approach allows the design of manufacturing cells to achieve the associated benefits of flexibility and utility.

For example, Fig. 15.16a shows a spreadsheet on a collection of parts to be produced by a pump manufacturer and the machines required to produce them. From this spreadsheet, no intuitive machine groupings are apparent; however, with a reorganization of the data, the parts suggest logical groupings of machines, as shown in Fig, 15.16b. These figures are highly simplified in order to demonstrate the significance of group

Digit	Items		(Rotational component)
1	Parts name		General classification
2			Detail classification
3	Materials		General classification
4			Detail classification
5	Major dimensions		Length
6			Diameter
7	Primary shapes and ratio of major dimensions		
8	Shape details and kinds of processes	External surface	External surface and outer primary shape
9			Concentric screw-threaded parts
10			Functional cut-off parts
11			Extraordinary shaped parts
12			Forming
13			Cylindrical surface
14		Internal surface	Internal primary shape
15			Internal curved surface
16			Internal flat and cylindrical surface
17		End surface	
18		Nonconcentric holes	Regularly located holes
19			Special holes
20		Noncutting process	
21	Accuracy		

FIGURE 15.15 The structure of a KK-3 system for rotational components. *Source:* Japan Society for the Promotion of Machine Industry.

Part	Broach	CNC Lathe	Hor. Machining Center	Hob	Manual Lathe	Hor. Lathe	Screw machine
Cover bearing		X					
Pinion gear, 56 teeth, 8 pitch	X	X		X			
Shaft coupling							X
Impellor	X	X					
6" Gland					X		
Driven gear, 26 teeth, 8 pitch	X	X		X			
Bushing							X
Body casing			X			X	
Relief valve			X				
Bearing spacer							X
Intake manifold			X		X	X	
Yoke bar							X
Spring seat							X
Yoke		X		X			
Pump head					X	X	

(a)

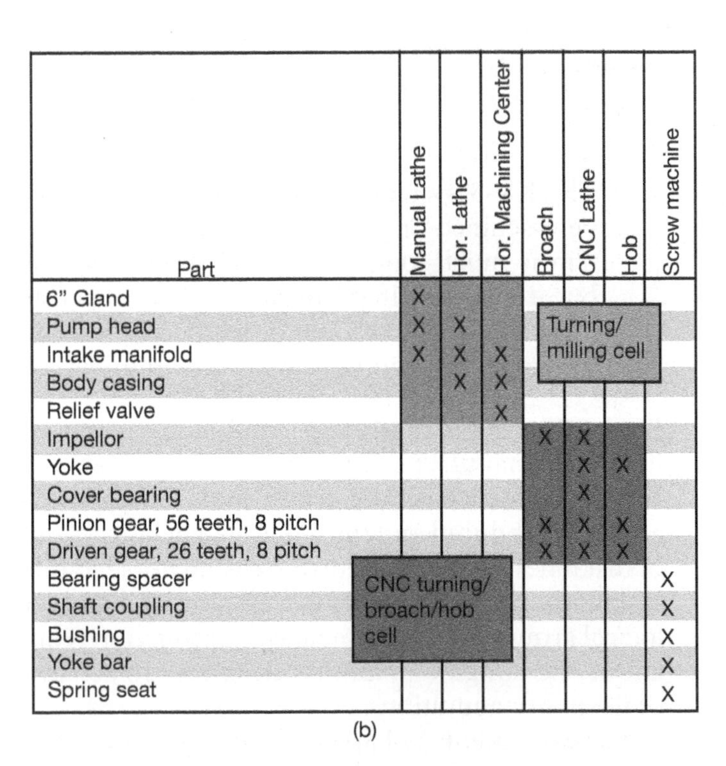

Part	Manual Lathe	Hor. Lathe	Hor. Machining Center	Broach	CNC Lathe	Hob	Screw machine
6" Gland	X						
Pump head	X	X					
Intake manifold	X	X	X				
Body casing		X	X				
Relief valve			X				
Impellor				X	X		
Yoke					X	X	
Cover bearing					X		
Pinion gear, 56 teeth, 8 pitch				X	X	X	
Driven gear, 26 teeth, 8 pitch				X	X	X	
Bearing spacer							X
Shaft coupling							X
Bushing							X
Yoke bar							X
Spring seat							X

(b)

FIGURE 15.16 The use of group technology to organize manufacturing cells. (a) Original spreadsheet of parts; no logical organization is apparent. (b) After grouping parts based on manufacturing attributes, logical machinery groupings are discernible.

technology in organizing manufacturing cells; usually there are many more parts and processes, including metrology, that must be considered.

15.8.4 Advantages of Group Technology

The major advantages of group technology are:

1. Group technology makes possible the standardization of part designs and minimization of design duplication; new part designs can be developed using already existing and similar designs, thus saving a significant amount of time and effort. The product designer can quickly determine whether data on a similar part already exist in computer files.
2. Data that reflect the experience of the designer and the process planner are stored in the database; a new or less experienced engineer can then quickly benefit from that experience.
3. Manufacturing costs can more easily be estimated, and the relevant statistics on materials, processes, number of parts produced, and other factors can easily be obtained.
4. Process plans are standardized and scheduled more efficiently, orders are grouped for more efficient production, machine utilization is improved, setup times are reduced, and parts are produced more efficiently and with better and more consistent product quality; similar tools, fixtures, and machinery are shared in the production of a family of parts.
5. With the implementation of CAD/CAM, cellular manufacturing, and CIM, group technology is capable of greatly improving productivity and reducing costs in small-batch production, approaching those of mass production. Depending on the level of implementation, potential savings in each of the various design and manufacturing phases have been estimated to range from 5 to 75%.

15.9 Cellular Manufacturing

A **manufacturing cell** is a small unit consisting of one or more workstations. A workstation typically contains one machine (**single-machine cell**) or several machines (**group-machine cell**), each of which performs a different operation on the part. The machines can be modified, retooled, and regrouped for different product lines within the same family of parts. Manufacturing cells are particularly effective in producing families of parts for which there is a relatively constant demand.

Cellular manufacturing has thus far been used primarily in machining and grinding (Chapters 8 and 9), sheet metal forming operations (Chapter 7), and in polymer processing involving injection or compression molding. Some form of metrology (Section 4.6) is also generally included. The machine tools commonly used are lathes, milling machines, drills, grinders, electrical-discharge machines, and machining centers. For sheet metal forming, they generally consist of shearing, punching, bending, and other forming machines. Cellular manufacturing has some degree of automatic control for the following types of operations:

1. Loading and unloading blanks and workpieces at workstations;
2. Changing tools and dies at workstations;
3. Transferring workpieces and tools and dies between workstations; and
4. Scheduling and controlling the total operation in the manufacturing cell.

Central to these activities is a **material-handling system** for transferring materials and parts among workstations. In attended (manned) machining cells, materials can be moved and transferred manually by the operator or, more often, by an industrial robot (Section 14.7), located centrally in the cell. Automated inspection and testing equipment are components of such a cell. The major benefits of cellular manufacturing are (a) the economics of reduced work in progress; (b) improved productivity; and (c) the ability to readily detect product quality problems right away.

Manufacturing cell design. Because of their unique features, the design and implementation of manufacturing cells in traditional plants requires a reorganization of the plant and rearrangement of existing product flow lines. The machines may be arranged along a line, a *U* shape, an *L* shape, or in a loop (See also Fig. 14.23). For a group-machine cell, where the materials are handled by the operator, the *U*-shaped arrangement is convenient and efficient, because the operator can easily reach various machines. For mechanized or automated material handling, the linear arrangement or the loop layout are more efficient because operator access is not required. The optimum arrangement of machines and material-handling equipment in a cell also involves consideration of such factors as the production rate and the type of product, its shape, size, and weight.

Flexible manufacturing cells. With rapid changes in market demand and the need for more product variety in smaller quantities, the need for *flexibility* of manufacturing operations is essential. Manufacturing cells can be made flexible using CNC machines and machining centers (see Section 8.11) and industrial robots and other mechanized systems (Sections 14.6 and 14.7). An example of an attended *flexible manufacturing cell* (FMC) involving machining operations is illustrated in Fig. 15.17. Note that an AGV moves parts between machines and inspection stations; equipped with automatic tool changers and tool magazines, machining centers have the ability to perform a wide variety of operations (see Section 8.11).

A computer-controlled inspection station, with a coordinate measurement machine, can similarly inspect dimensions on a wide variety of parts; the proper organization of these machines into a cell can allow the manufacture of very different parts. With computer integration, such a cell can manufacture parts in batch sizes as small as one, with negligible delay between parts and requiring only the time to download new machining instructions.

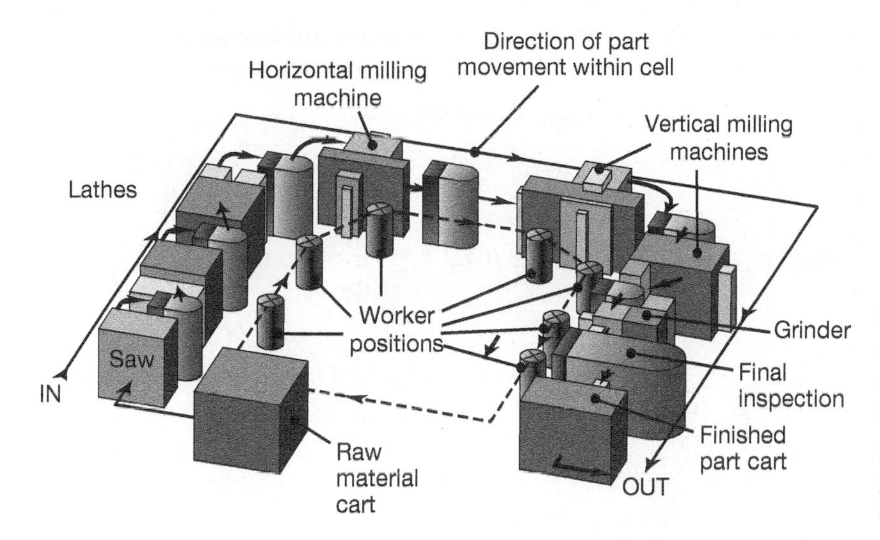

FIGURE 15.17 Schematic view of an attended flexible manufacturing cell, showing various machine tools and an inspection station. Note the worker positions and the flow of parts in progress from machine to machine. *Source:* After J T. Black.

Flexible manufacturing cells are usually unattended; their design and operation are therefore more exacting than those for other cells. The selection of machines and robots, including the types and capacities of end effectors and their control systems, is critical to the proper functioning of FMC. The likelihood of significant changes in demand for part families should be considered during design of the cell to ensure that the equipment involved has the necessary flexibility and capacity. The cost of flexible manufacturing cells is very high; however, this can be outweighed by increased productivity (at least for batch production), flexibility, and controllability.

15.10 Flexible Manufacturing Systems

A **flexible manufacturing system** (FMS) integrates all of the major elements of production into a highly automated system (Fig. 15.18); a general view of an FMS installation in a plant is shown in Fig. 15.19. First implemented in the late 1960s, an FMS basically consists of (a) a number of manufacturing cells, each with an industrial robot serving several CNC machines, and (b) an automated material-handling system. Note in Fig. 15.18 that an automated guided vehicle moves parts between different machines and inspection stations, which can easily involve an industrial robot or an integral transfer device. All of these activities are *interfaced* with a central computer, and different instructions can be downloaded for each successive part passing through a particular workstation. The system can handle a *variety of part configurations* and produce them *in any order*.

This highly automated system is capable of optimizing each step of the total manufacturing operation. These steps may specifically involve one or more processes and operations, such as handling incoming materials, machining, grinding, cutting, forming, powder metallurgy, heat treating, finishing, part inspection, and assembly. The most common applications of FMS to date have been in machining and assembly operations.

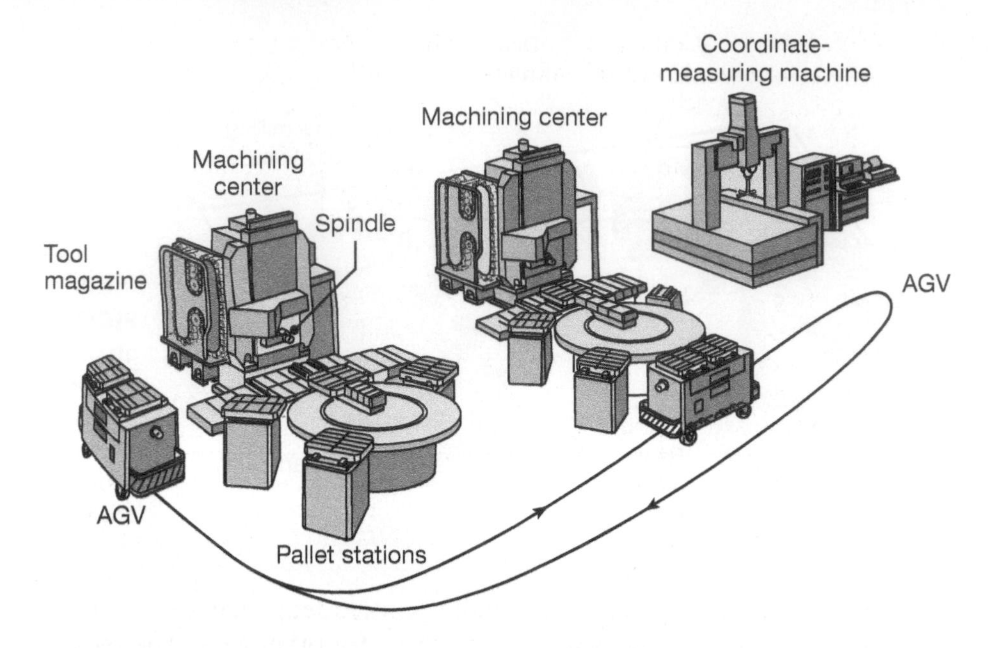

FIGURE 15.18 A schematic illustration of a flexible manufacturing system, showing two machining centers, a coordinate measuring machine, and two automated guided vehicles. *Source:* After J T. Black.

FIGURE 15.19 A general view of a flexible manufacturing system, showing several machining centers and an automated guided vehicle.

Flexible manufacturing systems represent the highest level of efficiency and productivity that has been achieved in manufacturing plants (Figs. 1.11 and 15.19).

FMS can be regarded as a system that combines the benefits of (a) the highly productive, but inflexible, transfer lines (Section 14.2.4); and (b) job-shop production, that can fabricate a large variety of products on stand-alone machines, but is inefficient. Table 15.1 compares some of

TABLE 15.1 Comparison of the characteristics of transfer lines and flexible manufacturing systems.

Characteristic	Transfer line	FMS
Types of parts made	Generally few	Infinite
Lot size	> 100	1–50
Part changing time	$\frac{1}{2}$ to 8 hr	1 min
Tool change	Manual	Automatic
Adaptive control	Difficult	Available
Inventory	High	Low
Production during breakdown	None	Partial
Efficiency	60–70%	85%
Justification for capital expenditure	Simple	Difficult

the characteristics of transfer lines and FMS. Note that in FMS, the time required for changeover to a different part is very short, thus it can be highly responsive to product and market-demand variations.

Elements of FMS. The basic elements of FMS are (a) computers and software for tracking of parts and materials through a plant and for control of supply chains; (b) automated handling and transport of materials and parts; and (c) control systems. The types of machines depend on the type of product. For machining operations, they typically consist of a variety of three- to five-axis machining centers, CNC lathes, milling machines, drill presses, and grinders. Also included are equipment for automated inspection (including coordinate measuring machines), assembly, and cleaning operations. Other manufacturing operations suitable for FMS include sheet metal forming, punching and shearing, and forging systems for these operations incorporate furnaces, forging machines, trimming presses, heat-treating facilities, and cleaning equipment.

Material handling in FMS is controlled by a central computer and performed by automated guided vehicles, conveyors, and various transfer mechanisms. The system is capable of transporting raw materials, blanks, and parts in various stages of completion to any machine, in any order, and at any time. Prismatic parts are usually moved on specially designed **pallets**, and parts with rotational symmetry are moved by various mechanical devices and robots.

Scheduling. Efficient machine utilization is essential in FMS; machines must not stand idle, thus proper scheduling and process planning are crucial. Unlike that in job shops, where a relatively rigid schedule is followed to perform a set of operations, scheduling for FMS is *dynamic*, meaning that it is capable of responding to rapid changes in product type; thus it is responsive to real-time decisions. The scheduling system (a) specifies the types of operations to be performed on each part; and (b) identifies the machines or the manufacturing cells on which these operations are to be performed. Setup time is not required in switching between different manufacturing operations; however, the characteristics, performance, and reliability of each unit in the system must be monitored to ensure that parts are of acceptable quality and dimensional accuracy before they move on to the next workstation.

15.11 Holonic Manufacturing

Holonic organizational systems have been studied since the 1960s, and there are a number of examples in biological systems. These systems are based on three fundamental observations:

1. Complex systems will evolve from simple systems much more rapidly if there are stable, intermediate forms than if there are none; also, stable and complex systems require a hierarchical system for evolution.
2. Holons are simultaneously self-contained wholes (to their subordinated parts) and the dependent parts of other systems. Holons are autonomous and self-reliant units, which have a degree of independence and can handle contingencies without asking higher levels in the hierarchical system for instructions. At the same time, holons are subject to control from multiple sources of higher system levels.
3. A holarchy consists of (a) autonomous wholes in charge of their parts, (b) dependent parts controlled by higher levels of a hierarchy, and (c) coordinate according to their local environment.

In biological systems, holarchies have the characteristics of (a) stability regardless of disturbances; (b) optimum use of available resources; and (c) a high level of flexibility when their environment changes. A **manufacturing holon** is an autonomous and cooperative building block of a manufacturing system for production, storage, and transfer of objects or information. It consists of a control part and an optional physical-processing part; for example, a holon may be a combination of a CNC milling machine and an operator interacting via a suitable interface. A holon also may consist of other holons that provide the necessary processing, information, and human interfaces to the outside world, such as a group of manufacturing cells. Holarchies can be created and dissolved dynamically, depending on the current needs of a particular manufacturing operation.

A holonic-systems view is one of creating a working manufacturing environment from the bottom-up. Maximum flexibility can be achieved by providing intelligence within holons to (a) support all production and control functions required to complete production tasks, and (b) manage the underlying equipment and systems. The manufacturing system will dynamically reconfigure into operational hierarchies to optimally produce the desired products, with holons or elements being added or removed as needed.

Holarchical manufacturing systems rely on rapid and effective communication between holons, as opposed to traditional hierarchical control where individual processing power is essential (such as adaptive control or machine vision programming). Numerous specific arrangements and software algorithms have been proposed for holarchical systems. Although a detailed description of these is beyond the scope of this book, the general sequence of events can be outlined as follows:

1. A factory consists of a number of resource holons, available as separate entities in a resource pool. For example, available holons may consist of a (a) CNC milling machine and an operator; (b) CNC grinder and operator; and (c) CNC lathe and operator.
2. Upon receipt of an order or directive from higher levels in the factory hierarchical structure, an order holon is formed and it begins communicating and negotiating with the available resource holons.
3. The negotiations lead to a self-organized grouping of resource holons, assigned on the basis of product requirements, resource holon availability, and customer requirements. For example, a given product may require the use of a CNC lathe, a CNC grinder, and an automated inspection station to be organized into a production holon.
4. In case of machine breakdown, lack of machine availability, or changing customer requirements, other holons from the resource pool can be added or removed as needed, allowing a reorganization of the production holon. Production bottlenecks can be identified and eliminated through communication and negotiation between the holons in the resource pool. This last step is referred to as *plug and play*, a term borrowed from the computer industry where hardware components seamlessly integrate into a system.

15.12 Just-in-Time Production

The *just-in-time production* (JIT) concept was originated in the United States but was first implemented on a large scale in 1953 at the Toyota Motor Company in Japan to eliminate waste of materials, machines, capital, manpower, and inventory throughout the whole manufacturing system. The JIT concept has the following goals:

1. Receive supplies just in time to be used;
2. Produce parts just in time to be made into subassemblies;
3. Produce subassemblies just in time to be assembled into finished products; and
4. Produce and deliver finished products just in time to be sold.

In traditional manufacturing operations, parts or subassemblies are made in batches, placed in inventory, and used when required. In this approach, known as a **push system**, parts are made according to a schedule and are in inventory to be used if and when they are needed. In contrast, just-in-time manufacturing is a **pull system**, meaning that parts are produced to order and the production is matched with demand. Consequently, there are no stockpiles and the extra expenses involved in stockpiling parts and then retrieving them from storage are eliminated.

Note that the ideal production quantity in JIT is one, hence this system is also called **zero inventory, stockless production,** or **demand scheduling.**

Moreover, parts are inspected in real-time, either automatically or by the worker as they are being manufactured and are used within a short period of time. In this way, control is maintained continuously during production, immediately identifying defective parts or process variations; any necessary adjustments are made rapidly (see also Section 16.3).

The capability of JIT to detect production problems has been likened to the level of water (representing inventory levels) in a lake covering a bed of boulders (representing production problems). When the water level is high (the high inventories associated with *push* production), the boulders are not exposed. When the water level is low (low inventories associated with *pull* production), the boulders are exposed, and the problems then can readily be identified and solved. This analogy indicates that high inventory levels can mask quality and production problems involving parts that already have been stockpiled.

Implementation of the JIT concept requires that all aspects of manufacturing operations be continuously monitored and reviewed, so that all operations and resources that *do not add value* are eliminated. This approach emphasizes (a) worker dedication and pride in producing high-quality products; (b) elimination of idle resources; and (c) teamwork among workers, engineers, and management to quickly solve any problems that arise *during* production or assembly.

An important aspect of JIT is the delivery of supplies and parts from external sources or from other divisions of the company, so as to significantly reduce in-plant inventory. As a result, major reductions in building storage facilities have taken place; in fact, the concept of building large warehouses for parts is now obsolete. Suppliers are expected to deliver, often on a daily basis, pre-inspected parts as they are needed for production. This approach requires reliable suppliers, close cooperation and trust among a company and its vendors, and reliable transportation infrastructure. Also important for smoother operations is the reduction of the number of suppliers.

Kanban. *Kanban* means visible record, typically consisting of two types of cards (kanbans). The **production card** authorizes the *production* of one container or cart of identical, specified parts at a workstation. The **conveyance card**, or **move card**, authorizes the transfer of one container or cart of parts from that particular workstation to the workstation where the parts will next be used. The cards, which are now bar-coded plastic tags or other devices, contain such information as the place of issue, part type, part number, and the number of items in that particular container. The number of containers in circulation at any time is completely controlled and can be scheduled as desired for maximum production efficiency.

Advantages of JIT. The advantages of just-in-time production may be summarized as follows:

- Low inventory-carrying costs;
- Rapid detection of defects in production or in delivery of supplies, hence low scrap;

- Reduced need for inspection and reworking of parts; and most importantly; and
- Production of high-quality parts at low cost.

Although there are significant variations, implementation of just-in-time production has resulted in reductions estimated at (a) 20 to 40% in product cost; (b) 60 to 80% in inventory; (c) up to 90% in rejection rates; (d) 90% in lead times; (e) 50% in scrap, rework, and warranty costs; and (f) 30 to 50% increase in direct labor productivity and of 60% in indirect labor productivity.

15.13 Lean Manufacturing

In a modern manufacturing environment, companies must be responsive to the needs of the customers and their specific requirements and to fluctuating global market demands. At the same time, the manufacturing enterprise must be conducted with a minimum waste of resources. This realization has lead to *lean manufacturing* or *lean production* strategies.

Lean manufacturing generally involves the following steps:

1. **Identify value.** The critical starting point for lean thinking is the recognition of **value**, which can be done only by a customer and considering a customer's product (see also Section 16.3.4). The goal of any manufacturing organization is to produce a product that a customer needs and wants, at a desired price, location, time, and volume. Providing the wrong goods or services produces waste, even if it has been delivered efficiently. It is important to identify all of a manufacturer's activities from the *viewpoint of the customer* and to *optimize* processes and operations in order to maximize added value. This viewpoint is critically important because it helps identify whether or not a particular activity:
   a. Clearly adds value;
   b. Adds no value but cannot be avoided; and
   c. Adds no value and can be avoided.
2. **Identify value streams**, meaning the set of all actions required to produce a product, including
   a. Product design and development tasks, involving all actions from concept, to detailed design, and to production launch;
   b. Information management tasks, involving order taking, detailed scheduling, and delivery; and
   c. Physical production tasks, by means of which raw materials progress to a finished product delivered on time to the customer.
3. **Make the value stream flow**, achieved when parts encounter a minimum of idle time between any two successive operations. It has been observed that flow is easiest to achieve in mass production, and that it is more difficult for small-lot production. However,

production in batches inherently involves part idle time; thus, just-in-time approaches are essential. In such cases, the solution is to implement manufacturing cells, where minimum time and effort are required to switch from one product to another.

In addition to JIT approaches, establishing product flow through factories requires:

- Eliminating waiting time, which may be caused by unbalanced workloads, unplanned maintenance, or quality-control problems; the efficiency of workers must be maximized at all times.
- Leveling or balancing production, as they may vary at different times of the day or a day of the week. Uneven production occurs when machines are underutilized, invariably leading to longer waiting time and waste. When production is leveled, inventory is reduced, productivity increases, and process flow becomes easier to achieve. It should also be noted that balancing can occur at cell, line, and plant scales.
- Eliminating unnecessary or additional operations and steps.
- Minimizing or eliminating movements of parts or products in plants, because it represents an activity that adds no value. Such waste can be eliminated or minimized by, for example, using machining cells or better plant layouts.
- Performing motion and time studies, to avoid unnecessary part or product movements in a plant or to identify inefficient workers.
- Eliminating product defects.
- Avoiding a single-source supplier, especially in unforeseen events of natural disasters and regional conflicts.

4. **Establish pull.** It has been observed that once value streams are flowing, significant savings are gained in terms of inventory reduction, product development, order processing, and production. In some cases, up to 90% savings have been obtained in actual production. It is possible to establish *pull manufacturing*, where products are made upon order by a customer or an upstream machine, and not in batches that ultimately are unwanted and do not create value. The classic examples of unwanted batch-manufactured goods are electronic devices that are superseded by a new model or clothing that has gone out of fashion; if these products could be produced using pull, then they could be sold with a higher margin and therefore create more value.
5. **Achieve perfection.** The Japanese term **kaizen** is used to signify continuous improvement; clearly, there is a challenge for continuous improvement in all organizations. With lean manufacturing approaches, it has been found that continuous improvement can be accelerated, so that production without waste is possible. Moreover, upon adoption of lean manufacturing principles, companies encounter an initial benefit, referred to as *kaikaku* or "radical improvement."

15.14 Communications Networks in Manufacturing

In order to maintain a high level of coordination and efficiency of operation in integrated manufacturing, an extensive, high-speed, and interactive **communications network** is essential. The Internet has profoundly impacted business, as well as the personal lives of people. Often, however, the Internet is not suitable for the manufacturing environment, because of security and performance issues. **Local area networks** (LANs) have been a major advance in communications technology, in which logically related groups of machines and equipment communicate with each other. The network links these groups to each other, bringing different phases of manufacturing into a unified operation.

A local area network can be very large and complex, linking hundreds or even thousands of machines and devices in several buildings. Various network layouts (Fig. 15.20) of fiber-optic or copper cables are typically used, over distances ranging from a few meters to as much as 30 km; for longer distances, **wide area networks** (WANs) are used. Different types of networks can be linked or integrated through "gateways" and "bridges," often using secure **file transfer protocols** (FTP) over Internet connections. A number of advanced network protocols have been developed, such as Internet protocol version 6 (ipV6) and Internet2 (see Section 15.14.2).

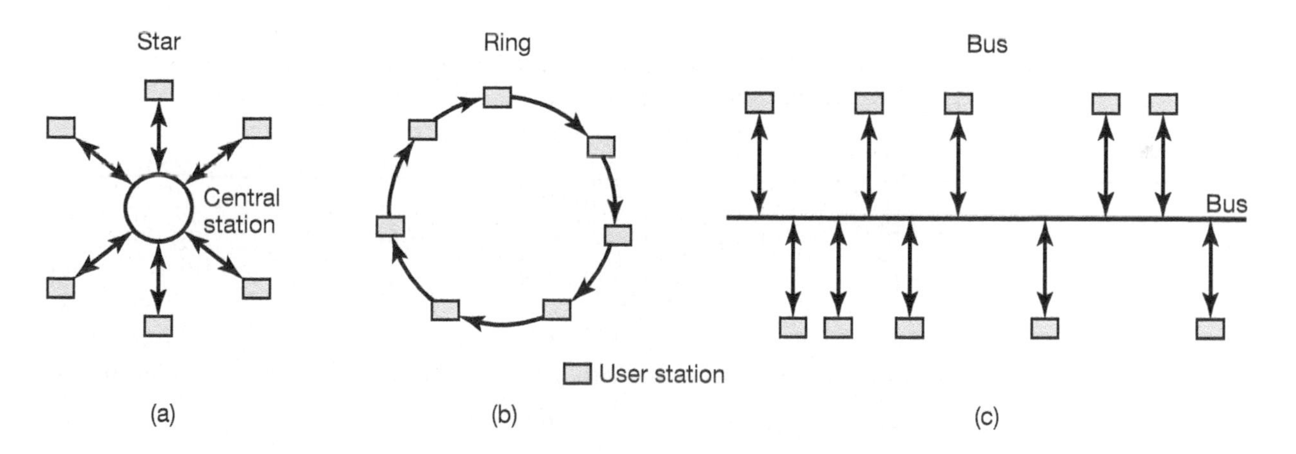

FIGURE 15.20 Three basic types of topology for a local area network (LAN). (a) The *star topology* is suitable for situations that are not subject to frequent configuration changes; all messages pass through a central station. (b) In the *ring topology*, all individual user stations are connected in a continuous ring; the message is forwarded from one station to the next, until it reaches its assigned destination. Although the wiring is relatively simple, the failure of one station shuts down the entire network. (c) In the *bus topology*, all stations have independent access to the bus; this system is reliable and easier to service than the previous two. Because its arrangement is similar to the layout of the machines in a factory, its installation is relatively easy, and it can easily be rearranged when the machines are rearranged.

Conventional LANs require routing of wires, often through masonry walls or other permanent structures, and require computers or machinery to remain stationary. **Wireless local area networks** (WLAN) allow equipment such as mobile test stands or data collection devices (such as bar code readers) to easily maintain a network connection. A communication standard (IEEE 802.11) defines frequencies and specifications of signals, and two radio-frequency and one infrared methods. Wireless networks are slower than those that are hard wired, but their flexibility makes them desirable, especially where slow tasks, such as machine monitoring, are the main application.

Personal area networks (PAN) are based on communications standards, such as Bluetooth, IrDA, and HomeRF, and are designed to allow data and voice communication over short distances. For example, a short-range Bluetooth device will allow communication over a 10 m distance. This technology has been integrated into personal computers and cellular phones, and occasionally for manufacturing applications. PANs are undergoing major changes, and communications standards are continually being refined.

15.14.1 Communications Standards

Typically, one manufacturing cell is built with machines and equipment purchased from one vendor, another cell with machines purchased from another vendor, and a third, purchased from yet another vendor. As a result, a variety of programmable devices are involved, driven by several computers and microprocessors purchased at various times from different vendors and having various capacities and levels of sophistication.

In 2008, **MTConnect** was demonstrated for the first time; it is a *machine tool communications protocol*, developed by the Association for Manufacturing Technology, which has quickly become an industry standard, aided by its royalty-free availability. MTConnect uses *hypertext transfer protocols* (http) for communication of data, suitable for all machine-tool manufacturers. Software is widely available for retrieving data from machine tools, called *agents*, and is also available for tablet computers, smart phones, and specially-designed Bluetooth-based devices, such as iBlue. Collection of manufacturing data, tracking of machine utilization, production rates, and other forms of data are greatly enabled by this protocol, allowing for real-time plant management.

15.14.2 The Internet of Things

The **Internet of Things** is a concept that refers to objects being connected over the Internet; it has already progressed significantly in several applications. A person may have, for example, a number of devices that communicate either with each other or with other devices on the Internet; cell phones, laptop computers, television sets, and gaming systems are well-known to be Internet capable. Automobiles, security cameras or web cams, and building environmental controls are also examples where Internet communication capability is now commonplace, but is not usually noticed; such integration is said to be *transparent*.

The *Industrial Internet of Things* refers to a state where all machines, sensors, fixtures, and even manufactured products are able to communicate with each other or with computers. A number of technological advances have been combined to enable much greater integration of design and manufacturing:

1. Communications protocols, such as MTConnect, have enabled machine tools to communicate with each other and with computers through Internet protocols.
2. Remote servers are available (**cloud-based storage**) to store data (See Fig. 15.21). Cloud-based storage can combine multiple storage devices as needed, providing essentially unlimited storage capability.
3. Sensors of all types have been integrated into machinery to allow monitoring of manufacturing processes. Roll-to-roll printing of integrated circuits and sensors (see Section 13.14) promises to speed the proliferation of sensors with integrated communication capability.
4. The development of Internet protocol version 6 has resulted in 2^{128} discrete web addresses, allowing hundreds of trillions of web addresses for every person and company in the world.
5. *Big Data* approaches have been developed to *mine* information from large data files. Internet search engines, such as Google, exemplify big data mining; it is common to find information using a few search terms, and to have valuable and relevant results based on a search of the entire Internet, produced almost immediately.

Combining these technologies has led to a manufacturing environment where every machine, part, tool, and fixture are connected with communications tools, and data from the entire manufacturing enterprise (including

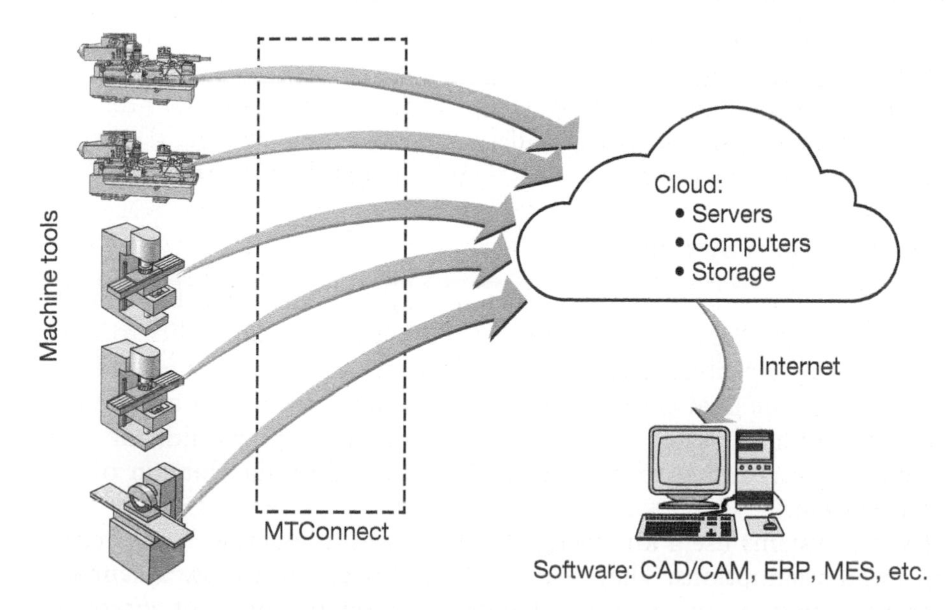

FIGURE 15.21 Schematic illustration of the use of cloud computing to capture manufacturing data and make it available wherever needed. A critical technology for cloud computing remains data security.

historical and real-time) is available to anyone in an organization from cloud servers. This leads to an unprecedented access to information that can help guide management decisions, and has been referred to as the fourth industrial revolution or **Industry 4.0**.

The availability of data also allows new approaches to machine control, quality control, and process validation, as well as extended capabilities for supply-chain management and ERP. Although The Internet of Things in manufacturing is still in its infancy, it continues to be developed at a rapid pace.

15.15 Artificial Intelligence in Manufacturing

Artificial intelligence (AI) is an area of computer science concerned with systems that exhibit some of the characteristics usually associated with intelligence in human behavior, such as learning, reasoning, problem solving, and the understanding of language. The goal of AI is to simulate such human endeavors on the computer; the art of bringing relevant principles and tools of AI to bear on difficult applications is known as **knowledge engineering**.

Artificial intelligence has had a major impact on the design, automation, and overall economics of manufacturing operations, due, in large part, to major advances in computer-memory expansion (VLSI chip design), cloud storage and *cloud computing*, and decreasing costs. AI packages, costing on the order of a few thousand dollars, have been developed, that can run on personal computers, thus making AI readily accessible.

Artificial-intelligence applications in manufacturing generally encompass the following:

1. Expert systems;
2. Natural language;
3. Machine (computer) vision;
4. Artificial neural networks; and
5. Fuzzy logic.

15.15.1 Expert Systems

Also called a **knowledge-based system**, an *expert system* (ES) is generally defined as an intelligent computer program that has the capability to solve difficult real-life problems using **knowledge-base** and **inference** procedures (Fig. 15.22). The goal is to develop the capability to conduct an intellectually demanding task in the way that a human expert would; the field of knowledge required to perform this task is called the **domain** of the expert system.

Expert systems use a knowledge base that contains facts, data, definitions, and assumptions. They also have the capacity to follow a **heuristic** approach, that is, to make good judgments on the basis of *discovery* and *revelation* and to make *high-probability* guesses, just as a human expert would. The knowledge base is expressed in computer codes, usually

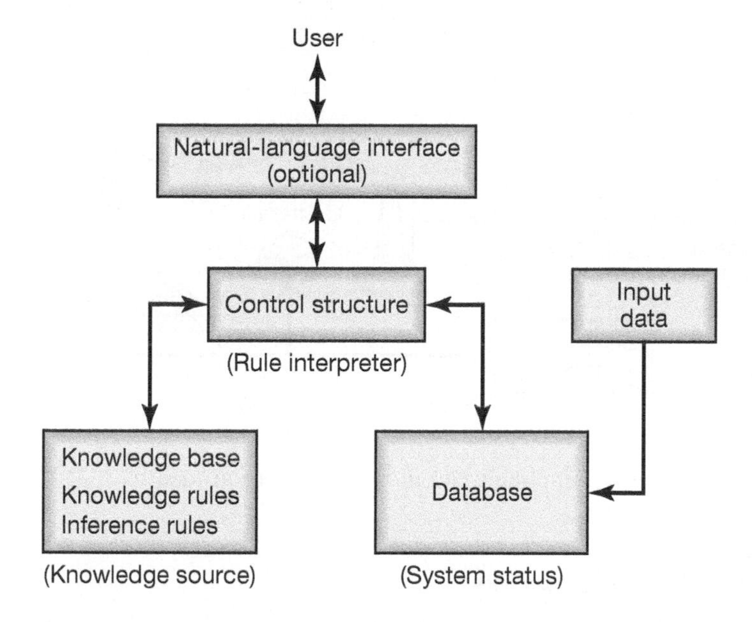

FIGURE 15.22 The basic structure of an expert system. The knowledge base consists of knowledge rules (general information about the problem) and inference rules (the approach used to reach conclusions). The results are communicated to the user through the natural-language interface. *Source:* After K.W. Goff.

in the form of **if-then rules**, and can generate a series of questions. The mechanism for using these rules to solve problems is called an **inference engine**. Expert systems can also communicate with other computer software packages.

To construct expert systems for solving complex design and manufacturing problems, one needs (a) knowledge and (b) a mechanism for manipulating this knowledge to develop solutions. Knowledge-based systems require much time and effort to develop because of (a) the difficulty involved in accurately modeling the years of experience of an expert or a team of experts, and (b) the complex inductive reasoning and decision-making capabilities of humans, including their capacity to learn from past mistakes.

Several expert systems have been developed and used for specialized applications, such as:

1. Problem diagnosis in machines and equipment, and determination of the corrective actions to take;
2. Modeling and simulation of production facilities;
3. Computer-aided design, process planning, and production scheduling; and
4. Management of a company's manufacturing strategy.

15.15.2 Machine Vision

The basic features of *machine vision* are described and illustrated in Section 14.8 and Fig 14.26. In systems that incorporate machine vision (a) computers and software implementing artificial intelligence are combined with cameras and various optical sensors and (b) then perform such operations as inspecting, identifying, and sorting parts, and guiding robots (*intelligent robots*; Fig. 15.23).

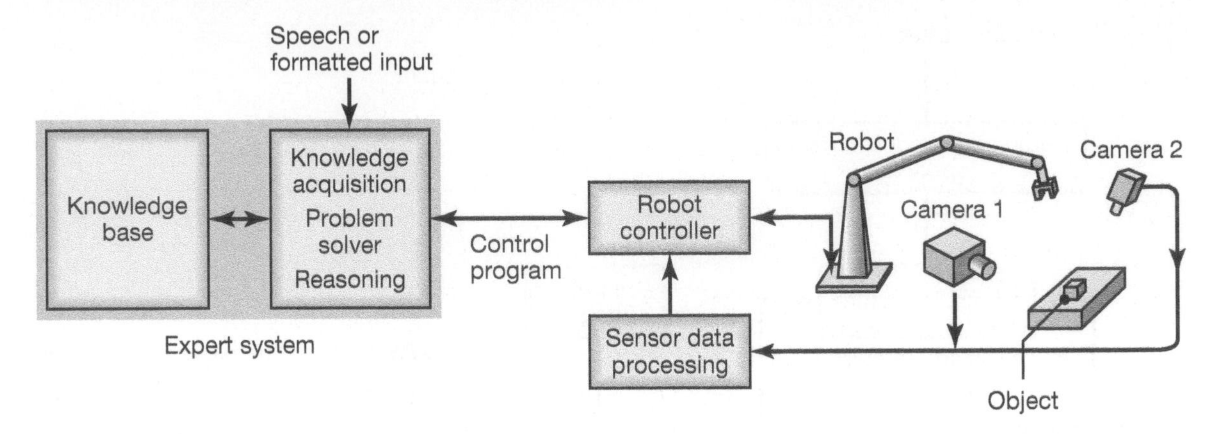

FIGURE 15.23 Expert system as applied to an industrial robot guided by machine vision using two cameras.

15.15.3 Artificial Neural Networks

Although computers are much faster than the human brain at *sequential* tasks, humans are much better at pattern-based tasks that can be attacked with *parallel processing*, such as recognizing features (faces and voices even under noisy conditions), assessing situations quickly, and adjusting to new and dynamic conditions. These advantages are also due partly to the ability of humans to use, in real time, their five **senses** (sight, hearing, smell, taste, and touch) simultaneously, a process called *data fusion*. **Artificial neural networks** (ANN), a branch of artificial intelligence, attempts to gain some of these human capabilities through computer imitation of the way data are processed by the human brain.

Artificial neural networks include such applications as (a) noise reduction (in telephones); (b) speech recognition; and (c) process control in manufacturing operations. They can, for example, also be used to predict the surface finish of a workpiece obtained by end milling (See Fig. 8.1d) based only on input parameters such as cutting force, torque, acoustic emission, and spindle speed.

The human brain has about 100 billion linked **neurons** (cells that are the fundamental functional units of nerve tissue) and more than a 100 trillion *connections*. Each neuron performs only one simple task: It receives input signals from a fixed set of neurons; when the input signals are related in a certain way (specific to that particular neuron), it generates an electrochemical output signal that is sent to a fixed set of neurons. It is now believed that human learning is accomplished by changes in the strengths of these signal connections between neurons.

The most common type of ANN is built from several layers of processing elements (*simulating neurons*). The elements in the first (input) layer are fed with input data, such as force, velocity, or voltage. Each element sums up all its inputs, one per element, in the input layer, and many inputs per element in succeeding layers. Each element in a layer then transfers the data (according to a transfer function) to all the elements in the next layer. Each element in the next layer, however, receives a different signal, because of the different connection weights between the elements.

15.15.4 Fuzzy Logic

An element of artificial intelligence, with important applications in control systems and pattern recognition, is *fuzzy logic*, using methods that deal with reasoning and decision making at a level higher than do neural networks. Based on the observation that people can make good decisions on the basis of imprecise and nonnumeric information, *fuzzy models* were first introduced in 1965 as mathematical means of representing information that is *vague* and *imprecise*, hence the term fuzzy. These models have the capability to recognize, represent, manipulate, interpret, and use data and information that are vague or lack precision. Linguistic examples are such words and terms as *few, almost all, very, more or less, small, medium,* and *extremely*.

Fuzzy-logic technologies and devices have been developed and successfully applied in areas such as robotics and motion control, image processing and machine vision, machine learning, and the design of intelligent systems. Some applications of fuzzy logic include: (a) automatic transmissions of automobiles; (b) washing machines that automatically adjusts the washing cycle for load size, fabric type, and amount of dirt; and (c) helicopters that obey vocal commands to go forward, up, left, and right; to hover; and to land.

SUMMARY

- Computer-integrated manufacturing systems are the most effective means of improving productivity, responding to changes in market conditions and better controlling manufacturing operations and management functions. (Sections 15.1–15.3)
- With rapid advances in software, together with the capability for computer simulation and analysis, computer-aided product design and engineering are now much more efficient, detailed, and comprehensive. (Section 15.4)
- Computer-aided manufacturing is often combined with computer-aided design, to transfer information from product design to the planning stage and to production. (Section 15.5)
- Continuing advances in manufacturing operations, such as computer-aided process planning, computer simulation of manufacturing processes and systems, group technology, cellular manufacturing, flexible manufacturing systems, and just-in-time manufacturing, are all essential elements in improving productivity. (Sections 15.6–15.12)
- Lean manufacturing is intended to identify and eliminate waste, leading to improvements in quality and customer satisfaction and decreasing product cost. A key concept is to establish a pull system of inventory control using JIT principles. (Section 15.12–15.13)
- Communications networks and their global standardization are critical to CIM strategies. Unprecedented levels of sensor integration and communication have led to the concept of the Internet of Things. In manufacturing it refers to machinery and associated sensor data

being available across an enterprise, helping make better decisions. (Section 15.14)

- Artificial intelligence has increasing applications in all aspects of manufacturing science, engineering, and technology. (Section 15.15)

BIBLIOGRAPHY

Amirouche, F.M.L., *Principles of Computer-Aided Design and Manufacturing*, 2nd ed., Prentice Hall, 2004.

Beard, C., and Stallings, W., *Wireless Communications Networks and Systems*, Pearson, 2015.

Black, J T., and Hunter, S.L., *Lean Manufacturing: Systems and Cell Design*, Society of Manufacturing Engineers, 2003.

Chang, T.-C., Wysk, R.A., and Wang, H.P., *Computer-Aided Manufacturing*, 3rd ed., Prentice Hall, 2005.

Curry, G.L., and Feldman, R.M., *Manufacturing Systems Modeling and Analysis*, 2nd ed., Springer, 2011.

Davis, J.W., *Progressive Kaizen: The Key to Gaining a Global Competitive Advantage*, Productivity Press, 2011.

Ertel, W., and Black, R.T., *Introduction to Artificial Intelligence*, Springer, 2011.

Forouzan, B.A., *Data Communications and Networking*, 5th ed., McGraw-Hill, 2012.

Groover, M., *Automation, Production Systems, and Computer-Integrated Manufacturing*, 4th ed., Pearson, 2014.

Haykin, S.O., *Neural Networks and Learning Machines*, 3rd ed., Prentice Hall, 2008.

Imai, M., *Gemba Kaizen: A Commonsense Approach to Continuous Improvement Strategy*, 2nd ed., McGraw-Hill, 2012.

Leon-Carcia, L., and Widjaja, I., *Communications Networks*, 2nd ed., McGraw-Hill, 2003.

Luger, G.F., *Artificial Intelligence*, 6th ed., Pearson, 2008.

Marik, V., *Holonic and Multi-Agent Systems for Manufacturing*, Springer-Verlag, 2011.

Mir, N.F., *Computer and Communications Networks*, 2nd ed., Prentice Hall. 2014.

Monden, Y., *Toyota Production System: An Integrated Approach to Just-in-time*, 4th ed., Productivity Press, 2011.

Parsaei, H., Leep, H., and Jeon, G., *The Principles of Group Technology and Cellular Manufacturing Systems*, Wiley, 2006.

Rehg, J.A., *Introduction to Robotics in CIM Systems*, 5th ed., Prentice Hall, 2002.

Rehg, J.A., and Kraebber, H.W., *Computer-Integrated Manufacturing*, 3rd ed., Prentice Hall, 2005.

Russell, S., and Norvig, P., *Artificial Intelligence: A Modern Approach*, 3rd ed., Pearson, 2009.

Vollmann, T.E., Berry, W.L., Whybark, D.C., and Jacobs, F.R., *Manufacturing Planning and Control for Supply Chain Management*, McGraw-Hill, 2004.

Wilson, L., *How to Implement Lean Manufacturing*, 2nd ed., McGraw-Hill, 2015.

QUESTIONS

15.1 In what ways have computers had an impact on manufacturing?

15.2 What advantages are there in viewing manufacturing as a system? What are the components of a manufacturing system?

15.3 Discuss the benefits of computer-integrated manufacturing operations.

15.4 What is a database? Why is it necessary in manufacturing? Why should the management of a manufacturing company have access to databases?

15.5 Explain how a CAD system operates.

15.6 What are the advantages of CAD systems over traditional methods of design? Do CAD systems have any limitations?

15.7 What is a NURB?

15.8 What are the advantages of a third-order piecewise Bezier curve over a B-spline or conventional Bezier curve?

15.9 Describe your understanding of the octree representation in Fig. 15.7.

15.10 Describe the purposes of process planning. How are computers used in such planning?

15.11 Explain the features of two types of CAPP systems.

15.12 Describe the features of a routing sheet. Why is a routing sheet necessary in manufacturing?

15.13 What is group technology? Why was it developed? Explain its advantages.

15.14 What does classification and coding mean with respect to group technology?

15.15 What is a manufacturing cell? Why was it developed?

15.16 Describe the principle of flexible manufacturing systems. Why do they require major capital investment?

15.17 Why is a flexible manufacturing system capable of producing a wide range of lot sizes?

15.18 What are the benefits of just-in-time production? Why is it called a pull system?

15.19 Explain the function of a local area network.

15.20 What are the advantages of a communications standard?

15.21 What is meant by the term factory of the future?

15.22 What is meant by the term holonic manufacturing?

15.23 What are the differences between ring and star networks? What is the significance of these differences?

15.24 What is kanban? Why was it developed?

15.25 What is an FMC, and what is an FMS? What are the differences between them?

15.26 Describe the elements of artificial intelligence. Why is machine vision a part of it?

15.27 Explain why humans will still be needed in the factory of the future.

15.28 How would you describe the principle of computer-aided manufacturing to an older worker in a manufacturing facility who is not familiar with computers?

15.29 Give examples of primitives of solids other than those shown in Fig. 15.4b.

15.30 Explain the logic behind the arrangements shown in Fig. 15.11b.

15.31 Describe your observations regarding Fig. 15.19.

15.32 What should be the characteristics of an effective guidance system for an automated guided vehicle?

15.33 Give examples in manufacturing for which artificial intelligence could be effective.

15.34 Describe your opinions concerning the voice-recognition capabilities of future machines and controls.

15.35 Would machining centers be suitable for just-in-time production? Explain.

15.36 Give an example of a push system and of a pull system, to clarify the fundamental difference between the two methods.

15.37 Give a specific example in which the variant system of CAPP is desirable and an example in which the generative system is desirable.

15.38 Artificial neural networks are particularly useful where the problems are ill defined and the data are fuzzy. Give examples in manufacturing where this is the case.

15.39 Is there a minimum to the number of machines in a manufacturing cell? Explain.

15.40 List as many three-letter acronyms used in manufacturing (such as CNC) as you can, and give a brief definition of each, for your future reference.

15.41 What are the disadvantages of zero inventory?

15.42 Why are robots a major component of an FMC?

15.43 Is it possible to exercise JIT in global companies?

15.44 A term sometimes used to describe factories of the future is untended factories. Can a factory ever be completely untended? Explain your answer.

15.45 What are the advantages of hierarchical coding?

15.46 Assume that you are asked to rewrite Section 15.15, on artificial intelligence. Briefly outline your thoughts regarding this topic.

15.47 Assume that you own a manufacturing company and that you are aware that you have not taken full advantage of the technological advances in manufacturing. Now, however, you would like to do so, and you have the necessary capital. Describe how you would go about analyzing your company's needs and how you would plan to implement these technologies. Consider technical as well as human aspects.

15.48 With specific examples, describe your thoughts concerning the state of manufacturing in the United States as compared with its state in other industrialized countries.

15.49 It has been suggested by some that artificial intelligence systems will ultimately be able to replace the human brain. Do you agree? Explain your response.

15.50 Make a list of the number of devices you own that are connected to the Internet. How many of these existed ten years ago? Make a list of the devices you think you will own in ten years that will be connected to the Internet.

PROBLEMS

15.51 Sketch skeletons for (a) a circle; (b) a square; and (c) an equilateral triangle. Indicate the radius information at every node and endpoint of the skeleton if the shapes have an area of 1 in^2.

15.52 List the primitives needed to describe the part in Figure 15.12.

15.53 Assume that you are asked to give a quiz to students on the contents of this chapter. Prepare one quantitative problems and three qualitative questions, and supply the answers.

DESIGN

15.54 Review various manufactured parts described in this text, and group them in a manner similar to that shown in Fig. 15.10.

15.55 Evaluate a process from a lean production perspective. For example, closely observe the following and identify, eliminate (when possible), or optimize the steps that produce waste when:

1. Preparing breakfast for a group of eight.
2. Washing clothes or cars.
3. Using Internet browsing software.
4. Studying for an exam, or writing a report or a term paper.

15.56 Consider a routing sheet for a product as discussed in Section 15.8. If a holonic manufacturing system is to automatically generate the manufacturing sequence, (a) what is the likelihood that it will be the same as produced manually? (b) what is the likelihood that a holonic system will produce the same sequence every time? Explain.

15.57 Think of a specific product, and make a decision-tree chart similar to that shown in Fig. 15.12.

15.58 Describe the trends in product designs and features that have had a major impact on manufacturing.

15.59 Think of a specific product, and design a manufacturing cell for making it (See Fig. 15.17), describing the features of the machines and equipment involved. Explain how the cell arrangement would change, if at all, if design changes are made to the product.

15.60 Surveys have indicated that 95% of all the different parts made in the United States are produced in lots of 50 or less. Comment on this observation, and describe your thoughts regarding the implementation of the technologies outlined in Chapters 14 and 15.

15.61 Think of a simple product, and make a routing sheet for its production, similar to that shown in Fig. 15.9. If the same part is given to another person, what is the likelihood that the routing sheet developed will be the same? Explain.

15.62 Review Fig. 15.9, and then suggest a routing sheet for the manufacture of each of the following: (a) an automotive connecting rod; (b) a compressor blade; (c) a glass bottle; (d) an injection-molding die; and (e) a bevel gear.

15.63 What types of production machines would not be suitable for a manufacturing cell? What design or production features make them unsuitable? Explain.

15.64 Conduct a literature and Internet review, and summarize the basic principles of file encryption and cybersecurity.

15.65 In the Internet of Things, how you would you connect people? Explain.

Competitive Aspects of Product Design and Manufacturing

CHAPTER 16

Manufacturing world-class products at low cost requires an understanding of the often complex relationships among several factors. Those that have a major impact on product design and manufacture, covered in this chapter are:

- Quality in design and the design of robust products.
- Product life cycle and design for sustainability.
- Amount of energy required to produce a particular material and process it into a complete part.
- Importance of material and process selection on product design.
- Comparative economics of various manufacturing operations and description of significant factors in product costs.

Only by considering competitive aspects of manufacturing can one make intelligent decisions about material and process selection. An ever-present concern in the modern manufacturing enterprise is continuous improvement, but without increasing costs. Indeed, it is expected that costs will decrease over time, as process and organizational efficiencies are realized.

Symbols

k	constant	USL	upper specification limit
LSL	lower specification limit	Y	mean value
T	target value	σ	standard deviation

16.1 Introduction

In an increasingly competitive global marketplace, manufacturing high-quality products at the lowest cost requires an understanding of the often complex relationships among numerous factors. Throughout this text, it has been pointed out that product design and selection of materials and manufacturing processes are all interrelated, and that designs are periodically modified to:

1. Improve product performance;
2. Take advantage of available new materials or less expensive materials;
3. Make products easier and faster to manufacture and assemble; and
4. Strive for zero-based rejection and waste.

In view of the extensive variety of materials and manufacturing processes now available, the task of producing a high-quality product by selecting the best materials and the best processes, while minimizing costs, continues to be a major challenge as well as an opportunity in the global marketplace. The term **world-class** is widely used to indicate a high level of product quality, signifying the fact that products must meet international standards and be marketable and acceptable to customers worldwide. Note also that the designation of world-class, like product quality, is not a fixed target for a company to reach, but a *moving target*, rising to higher and higher levels, generally referred to as **continued improvement.**

This final chapter begins with considerations in **product design, robust design**, and the methods of making **high-quality, low-cost products.** Design involves numerous factors not only in the basic design itself, but also the ease of manufacturing and assembling that product and the time and energy required to produce it. There are always opportunities to improve designs, reduce the number of their components and the size and weight to reduce costs.

Product quality and **life expectancy** are then described, outlining the relevant parameters, including the concept of *return on quality*. Increasingly important are **life-cycle assessment** and **life-cycle engineering** of products, services, and systems, particularly regarding their possible adverse impact on the environment. The major emphasis in **sustainable manufacturing** is to reduce or eliminate any adverse effects of material or manufacturing operation on the quality of life, while recognizing the fact that a company must be profitable to survive.

Selection of materials for products traditionally has required considerable experience to meet specific performance requirements. Several software packages are now available to facilitate the selection process, especially when numerous opportunities exist for material substitution.

In the production phase of a product, it is imperative that the *capabilities of manufacturing processes* and the latest technological advances be properly assessed as an essential guide to their final selection. As described throughout this text, and to be further demonstrated in this chapter, there usually is more than one method of manufacturing the individual components and subassemblies of a product.

Although the all-important *economics* of various types of manufacturing processes have briefly been described throughout this book, this chapter takes a broader view and summarizes the important overall manufacturing cost factors. The **cost** of a product often, but not always, determines its marketability and customer acceptance and satisfaction.

Meeting this challenge also requires knowledge of economic factors as well as innovative approaches to product design and manufacturing.

16.2 Product Design and Robust Design

Although it is beyond the scope of this book to describe in detail all product design principles, various chapters have highlighted manufacturing aspects that are relevant to design. The references listed in Table 16.1 give design guidelines for specific manufacturing processes and materials.

Major advances continue to be made in *design for manufacture and assembly*, for which software packages are widely available. The software helps designers to quickly develop products that require fewer components, reduced manufacturing and assembly time, and reduced total product cost.

16.2.1 Product Design Considerations

Engineers pursuing new designs or revisions to existing products or components must weigh many simultaneous considerations. Some of the common general product design guidelines, stated in the form of questions that require solutions, are summarized as:

1. Can the product design be simplified and the number of its components reduced without adversely affecting its intended functions and performance?
2. Have environmental considerations been considered and incorporated into material and process selection?
3. Have all potential material and process considerations been considered, and are design opportunities available if alternative materials and processes were selected?
4. Have all alternative designs been investigated?
5. Are there any unnecessary features of the product, or some of its components, that can be eliminated or combined with other features?
6. Have modular design and building-block concepts been considered?
7. Can the design be made simpler and lighter?
8. Are the specified dimensional tolerances and surface finish excessively stringent; can they be relaxed without any significant adverse effects?
9. Will the product be difficult or time consuming to assemble and disassemble, for maintenance, servicing, or recycling?
10. Have the use of fasteners, and their quantity and variety, been minimized?
11. Are some of the components commercially available?
12. Is the product safe for its intended application?
13. Have environmental considerations all been taken into account?
14. Have green design and life-cycle engineering principles been applied, including recycling and cradle-to-cradle principles?
15. Can any of the design or manufacturing activities be outsourced, and can any of these be reshored?

TABLE 16.1 References to various topics in this text.

Processes:	Design considerations
Metal casting	Section 5.12
Bulk deformation	Various sections in Ch. 6
Sheet metal forming	Section 7.9
Machining	Section 8.14
Abrasives	Section 9.17
Polymers	Section 10.13
Powder metallurgy and ceramics	Sections 11.6 and 11.12
Joining	Section 12.18
Material properties	**Manufacturing characteristics of materials**
Tables 2.2 and 2.4; Figs. 2.6, 2.13, and 2.24	Tables 5.2 and 5.3
Tables 3.3, 3.4, 3.5, and 3.7 through 3.12	Table 7.1
Tables 5.4, 5.5, and 5.6; Fig. 5.12	Section 8.5
Table 7.3	Tables 11.6, 11.9, and 11.10
Table 8.6; Figs. 8.29, 8.30	Table 16.8
Table 9.1	
Tables 10.1, 10.4, and 10.8; Fig. 10.17	
Tables 11.3, 11.4, 11.5, 11.8 and 11.10; Figs. 11.25, 11.26, and 11.27	
Table 12.6	
Capabilities of manufacturing processes	**Dimensional tolerances and surface finish**
Fig. 4.19	Fig. 4.19
Tables 5.2, 5.7, and 5.8	Table 5.2
Table 6.1	Table 8.7 and Fig. 8.29
Table 7.1	Fig. 9.28
Table 8.7; Fig. 8.25	Fig. 16.7
Table 9.4; Fig. 9.28	
Tables 10.6, 10.7, and 10.9	
Tables 11.6 and 11.9	
Tables 12.1, 12.2, and 12.5	
Tables 13.1, 13.2 and 13.5	
Fig. 14.3; Table 15.1; Fig. 16.6	
General costs	**Material costs**
Table 5.2; Fig. 5.38	Table 10.4
Fig. 7.70	Table 11.10
Fig. 9.43	Table 16.6
Table 10.9	
Table 11.9	
Tables 12.1 and 12.2	
Table 16.9; Fig. 16.4	

As examples of the foregoing considerations, note how (a) the size of products, such as batteries, electronic equipment, cameras, and laptop computers, has been and continue to be greatly reduced; (b) repair of products is now often eliminated by simply replacing subassemblies and modules; (c) there are now fewer traditional fasteners and more snap-on types of assemblies used in products; (d) additive manufacturing has been used to produce optimum geometries that are difficult to produce otherwise; and (e) automobile frames have been made lighter and stronger through application of advanced steels and aluminum alloys.

16.2.2 Product Design and Quantity of Materials

With high production rates and reduced labor, made possible by automation and computer integration systems, the cost of materials can often become a significantly higher proportion of a product's cost. Although the cost of material cannot be reduced below their market level, reductions can be made in the *quantity* of materials used. The overall shape of the product is usually optimized during the design and prototype stages (see also *additive manufacturing*, Section 10.12) and through such techniques as *minimum-weight design, design optimization*, and *computer-aided design and manufacturing*. These methods have greatly facilitated design analysis, material selection, material usage, and overall optimization of production.

Reductions in the *amount* of material used can be achieved by decreasing the component's volume, an approach that would typically require selection of materials with high strength-to-weight or stiffness-to-weight ratios (see Section 3.9.1). Higher ratios can also be obtained by improving component design and selecting different cross sections, such as those that have a high moment of inertia (as in I-beams) or using tubular or hollow components instead of solids.

Note that design changes and minimizing the amount of materials used can lead to *thinner* cross sections, thus presenting significant difficulties in manufacturing the parts. Consider, for example, the following examples:

1. Casting or molding of thin sections (Section 5.12) can present difficulties in mold filling and in maintaining the required dimensional accuracy and surface finish.
2. Forging of thin sections (Section 6.2.3) requires higher forces, due to friction and chilling of thin sections in impression die forging.
3. Impact extrusion (Section 6.4.3) of thin-walled parts can be difficult, especially when high dimensional accuracy is required.
4. Formability of sheet metals (Section 7.7) typically is reduced as their thickness decreases; also, reducing sheet thickness can lead to wrinkling due to compressive stresses developed in the plane of the sheet during forming (Section 7.6 and Fig. 7.50).
5. Machining and grinding (Sections 8.14 and 9.6) of thin workpieces can present such difficulties as part distortion, poor dimensional accuracy, and chatter; thus, advanced machining processes may have to be considered.

6. Thin sections in additive manufacturing (Section 10.13) may have very low strength and high warping.
7. Welding of thin sheets or slender structures (Section 12.17) can cause distortion due to thermal gradients developed during welding.

Conversely, making parts with *thick* cross sections also can have adverse effects. Some examples are as follows:

1. In die casting (Section 5.10.3) and injection molding (Section 10.10.2), the production rate can be lower for parts with thick cross sections, because of the increased cycle time required to allow the part to cool in the mold cavity prior to its removal.
2. Unless controlled, porosity can develop in the thicker regions of castings (Fig. 5.35).
3. The bendability of sheet metals decreases as their thickness increases (Fig. 7.15b and Table 7.2).
4. In powder metallurgy (Fig. 11.9), there can be significant variations in density and mechanical properties throughout thicker parts.
5. Thick cross sections and large parts will take more time to be produced in additive manufacturing if they can be produced at all because of part size limitations (Section 10.13).
6. Welding of thick sections can present problems that complicate weld strength, such as residual stresses and incomplete penetration (Section 12.7).
7. In die-cast parts (Section 5.10.3), thinner sections will have higher strength per unit thickness than will thicker sections, because of the smaller grain size developed in the thinner sections.
8. Thick cross sections produced by hot rolling (Section 6.3) will have lower strength and dimensional accuracy and higher roughness than cold-rolled thinner sections.
9. Processing of polymer parts (Section 10.13) requires increased cycle times as their thickness or volume increases, because of the longer time required for the parts to cool sufficiently to be removed from the molds.

16.2.3 Robustness and Robust Design

An important consideration in product design is *robustness*. Originally introduced by G. Taguchi (Section 16.3.5), robustness is defined as a design, a process, or a system that continues to function within acceptable parameters *despite variations* in its environment. The variations, called **noise**, are defined as those factors that are impossible or difficult to control. Examples of noise in manufacturing operations include (a) variations in ambient temperature and humidity in a production facility; (b) random and unanticipated vibrations of plant floors; (c) variations in the dimensions and surface and bulk properties of incoming batches of raw materials; and (d) the performance of different operators and machines at different times during the day or on different days. A robust process is unaffected by the presence or changes in noise; in a robust design, the part will function well even if unanticipated events occur.

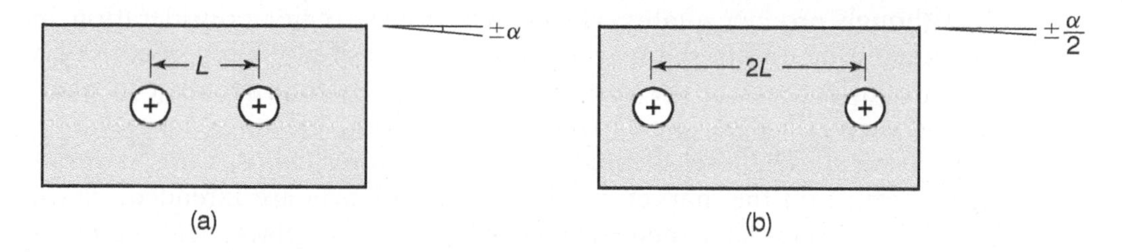

FIGURE 16.1 An example of robust design. (a) Location of two mounting holes on a sheet metal bracket, where the deviation of the top and bottom surfaces of the bracket from being perfectly horizontal is $\pm\alpha$. (b) New location of holes in a robust design, whereby the deviation is reduced to $\pm\alpha/2$.

As an illustration, consider a sheet metal mounting bracket to be attached to a wall with two bolts (Fig. 16.1a). The positioning of the two mounting holes will, as expected, include some error, perhaps caused by the particular manufacturing process and machine involved, such as an error in punching, drilling, or piercing. This error may, in turn, prevent the top edge of the bracket from being horizontal. In a more robust design (Fig. 16.1b), the mounting holes are now twice as far apart from each other. Even though the same production method is used and the product cost will remain basically the same, the robust design has a variability one-half of that of the original design. In an even more robust design, the type of bolts used would be such that they will not loosen in case of vibration or during their use over a period of time (see *mechanical fastening*, Section 12.16).

16.3 Product Quality as a Manufacturing Goal

Product quality and the techniques employed in quality assurance and control are described in Section 4.9. The word *quality* is difficult to define precisely, because it includes not only well-defined technical considerations but also human, and thus subjective, opinions. As defined by Taguchi (Section 16.3.4), manufacturers are charged with the task of providing products that *delight* their customers. To do so, manufacturers must provide products with the following specific characteristics, that also describe a product with *high quality*:

1. High reliability;
2. Perform the required functions well and safely;
3. Good appearance;
4. Inexpensive;
5. Upgradeable;
6. Available in the quantities desired when needed; and
7. Robust over their intended life.

These characteristics clearly are the goals of manufacturers striving to provide high-quality products. Although it is very challenging to actually provide all of these characteristics, *excellence in manufacturing* is undeniably a prerequisite.

Although product quality always has been a major consideration in manufacturing and in the marketplace, in view of global competition, a major consideration is the concept of **continuous improvement in quality**, as exemplified by the Japanese term **kaizen**, meaning *never-ending improvement*. The level of quality a manufacturer chooses for its products depends on the market for which the products are intended. Thus, low-cost products have their own niche in the marketplace, just as there is a market for high-quality products, such as automobiles, watches, audio equipment, and sporting goods.

Because of the significant costs that can be incurred in manufacturing, the concept of **return on quality** (ROQ) is an important consideration, the components of which are:

1. Because of its major influence on customer satisfaction, quality should be viewed as an *investment*.
2. Any improvement in quality must be investigated with respect to the additional costs involved.
3. The specific area for which the expenditure should be made toward quality improvement must be properly assessed.
4. The incremental improvement in quality must be carefully reviewed, especially since quality measures can be subjective.

High-quality products do not necessarily cost more. In most industries, such as automotive, aerospace and medical products, the ROQ is minimized at a value of *zero defects*, while in others, such as integrated circuits, the cost of *eliminating* the final few defects is very high, leading to approaches such as *fault-tolerant design*. Regardless of the direct cost of eliminating the final defects, indirect factors still may encourage pursuit of zero defects. For example, although customer satisfaction is a qualitative factor, thus difficult to include in calculations, satisfaction is increased and customers are more likely to be retained when there are no defects in the products they purchase. Although there can be significant variations, it has been estimated that the *relative costs* involved in identifying and repairing defects in products grow by orders of magnitude, called the *rule of ten* (see also Section 16.10.1), as shown in Table 16.2. Hence, it can be very costly and unsatisfactory if repairs are the burden of the customer.

TABLE 16.2 The cost of repair at different stages in a product lifecycle.

Stage	Relative cost of repair
Fabrication of the part	1
Subassembly	10
Final assembly	100
At product distributor	1000
At the customer	10,000

16.3.1 Total Quality Management

Total quality management (TQM) is a management system that emphasizes the concept that quality must be designed and *built into a product*. It is a systems approach, in that management and employees must *both* make concerted efforts to consistently manufacture high-quality products. Defect *prevention*, rather than defect *detection*, is the major goal in TQM.

Leadership and *teamwork* in the organization are essential. They ensure that the goal of **continuous improvement** in *all* aspects of manufacturing operations is imperative, because they reduce product variability and improve customer satisfaction. The TQM concept also requires the control of *processes* and not of *parts produced*, so that process variability is reduced and no defective parts are allowed to continue through the production line.

Quality circle. This concept, first implemented in 1979, consists of regular meetings by groups of employees (workers, supervisors, managers) who discuss how to improve and maintain product quality at all stages of the manufacturing operation; worker involvement, responsibility, creativity, and team effort are all emphasized. Comprehensive training is provided so that the worker (a) can become aware of the importance of quality; (b) be capable of analyzing statistical data to identify causes of poor quality; and (c) take immediate action to correct the problem. Experience has indicated that quality circles are especially effective in lean manufacturing environments (Section 15.13).

Quality engineering. Experts in quality control have put many of the quality-control concepts and methods into a larger perspective. Notable among these experts are Deming, Juran, and Taguchi, whose philosophies of quality and product cost have had a major impact on modern manufacturing.

16.3.2 Deming Methods

During World War II, W.E. Deming (1900–1993) and several others developed new methods of **statistical process control** (Section 4.9.2) for war-industry manufacturing plants. The methods of statistical control arose from the recognition that there were *variations* in (a) the performance of machines and people; and (b) the quality and dimensions of incoming raw materials. His efforts involved not only statistical methods of analysis, but also a new way of looking at manufacturing operations from the perspective of improving quality while lowering costs.

Deming recognized that manufacturing organizations are *systems* consisting of management, workers, machines, and products; his basic ideas are summarized in the well-known 14 Points (Table 16.3). These points are not to be viewed as a checklist or menu of tasks to be performed; they

TABLE 16.3 Deming's 14 points.

1.	Create constancy of purpose toward improvement of product and service, with the goal of improving competitiveness and providing jobs.
2.	Adopt the new philosophy. Management must be leaders and assume responsibilities of leadership.
3.	Cease dependence on mass inspection to achieve quality. Instead, build quality into parts in the first place.
4.	End the practice of awarding business on the basis of price tag. Minimize total cost, and include considerations of quality.
5.	Improve constantly and forever the system of production and service, to improve quality and productivity, and thus constantly decrease cost.
6.	Institute training for the requirements of a particular task, and document it for future training.
7.	Institute leadership, as opposed to supervision.
8.	Drive out fear so that everyone can work effectively.
9.	Break down barriers between departments. Institute teamwork.
10.	Eliminate slogans, exhortations and targets for zero defects and new levels of productivity. Eliminate quotas and management by numbers, and eliminate numerical goals. Substitute leadership.
11.	Remove barriers that rob the hourly worker of pride of workmanship.
12.	Remove barriers that rob people in management and engineering of their right to pride of workmanship. Abolish annual ratings and management by objective.
13.	Institute a vigorous program of education and self-improvement.
14.	Put everybody in the company to work to accomplish the transformation.

are the *characteristics* that Deming recognized in companies that produce high-quality products. He placed great emphasis on communication, direct worker involvement, and education in statistics and advanced manufacturing technology; his ideas have been accepted worldwide since the end of World War II.

16.3.3 Juran Methods

A contemporary of Deming, J.M. Juran (1904–2008), an electrical engineer and management consultant, emphasized the importance of:

- Recognizing quality at all levels of an organization, including upper management;
- Fostering a responsive corporate culture; and
- Training all personnel in how to plan, control, and improve quality.

According to Juran, the main concern of the top management in an organization is business and management, whereas those in quality control are basically concerned with technology. These different areas have often been at odds and their conflicts have led to quality problems. Planners determine who the customers are and their needs. Customers may be external (end users who purchase products or services) or they may be internal (different parts of an organization that rely on other segments of the organization to supply them with products and services). The planners then develop products and process designs to respond to customers' needs. The plans are turned over to those in charge of operations, who then become responsible for implementing both quality control and continued improvements in quality.

16.3.4 Taguchi Methods

In the G. Taguchi (1924–2012; an engineer and a statistician) methods, high quality and low costs are achieved by combining engineering and statistical methods to optimize product design and manufacturing processes. Taguchi methods refer to the approaches he developed for high-quality products. One fundamental viewpoint he put forward was the quality challenge facing manufacturers: Provide products that delight customers. To do so, manufacturers must offer products with the characteristics described in Section 16.3.

Taguchi also contributed to approaches used to document quality, emphasizing that any deviation from the optimum state of a product represents a *financial loss*, because of such factors as reduced product life, performance, and economy. He defined *loss of quality* as the financial loss to society after the product is shipped, resulting in the following:

1. Poor quality leads to customer dissatisfaction.
2. Costs are incurred in servicing and repairing defective products, some of which are in the field.
3. The manufacturer's credibility in the marketplace is diminished,
4. The manufacturer eventually loses its share of the market.

The Taguchi methods of *quality engineering* emphasize the importance of:

- **Enhancing cross-functional team interaction,** whereby engineers in design and in manufacturing communicate with each other in a common language; they quantify the relationships between design requirements and manufacturing process selection.
- **Implementing experimental design,** in which the factors involved in a process or operation and their interactions are studied simultaneously.

In *experimental design*, the effects of controllable and uncontrollable variables are identified; this approach minimizes variations in product dimensions and characteristics, and also brings the mean to the desired level (*on target*). The methods used for experimental design are complex; they involve the use of *factorial design* and *orthogonal arrays*, that reduce the number of experiments required. These methods are also capable of identifying the effects of variables that cannot be controlled (called *noise*), such as changes in environmental conditions and normal property variations of incoming raw materials.

Implementation of these methods leads to (a) rapid identification of the controlling variables (observing *main effects*), and (b) ability to determine the best method of process control. Thus, for example, variables affecting dimensional tolerances in machining a particular part can readily be identified and, whenever possible, the proper cutting speed, feed, cutting tool, and cutting fluids can be specified. Controlling these variables may require new equipment or major modifications to existing equipment.

16.3.5 Taguchi Loss Function

An important concept, introduced by Taguchi, is that any deviation from a design objective constitutes a loss in quality. Consider, for example, the tolerancing standards given in Fig. 4.18; note that there is a *range* of dimensions over which a part is acceptable. The loss function calls for a *minimization of deviation* from the design objective. Thus, using Fig. 4.18a as an example, a shaft with a diameter of 40.030 mm would normally be considered acceptable and would pass inspections; however, a shaft having this diameter represents a deviation from the design objective. Such deviations generally reduce the performance and robustness of products, especially in complex systems.

The *Taguchi loss function* was introduced because traditional accounting practices had no established methodologies of calculating losses on parts that meet design specifications. In the traditional accounting approach, a part is defective and incurs a loss to the company when it exceeds its design tolerances. The loss function is a tool for comparing quality, based on minimizing variations. It calculates an increasing loss to the company when a particular component of a product is further from the design objective. This function is defined as a parabola, where one point is the cost of replacement (including shipping, scrapping, and handling costs) at an extreme of the tolerances, while a second point corresponds to zero loss at the design objective.

The cost of loss can be described mathematically as

$$\text{Loss cost} = k\left[(Y - T)^2 + \sigma^2\right], \tag{16.1}$$

where Y is the mean value from manufacturing, T is the target value from design, σ is the standard deviation, and k is a constant defined as

$$k = \frac{\text{Replacement cost}}{(\text{LSL} - T)^2}, \tag{16.2}$$

where LSL is the lower specification limit (see also Section 4.9). When the lower and upper specification limits (LSL and USL, respectively) are the same (that is, the tolerances are balanced), either of the specification limits can be used in this equation.

Note that LSL and USL are not necessarily related to lower and upper control limits, described in Section 4.9.2. LSL and USL are *design requirements*, which often do not take manufacturing capabilities into account. Often, designers choose specification limits or tolerances based on experience or as a maximum value that can be allowed and still have a functioning design; the purpose is to allow manufacturing engineers wide tolerances to ease manufacturing. On the other hand, LCL and UCL do not necessarily reflect the designer's intention regarding tolerances. Ideally, control limits and specification limits, considered here, should not be related; instead, it should be the goal of manufacturers to meet design targets, as implied by Eq. (16.1).

EXAMPLE 16.1 Production of Polymer Tubing

Given: High-quality polymer tubes are being produced for medical applications, where the target wall thickness is 2.6 mm, with a USL of 3.2 mm and a LSL of 2.0 mm (2.6 ± 0.6 mm). If the units are defective, they are replaced at a cost of $10.00 including shipping. The current process produces parts with a mean of 2.6 mm and a standard deviation of 0.2 mm. The current volume is 10,000 tubes per month. An improvement is being considered for the extruder heating system, that will cut the variation in half, but it will cost $50,000.

Find: Determine the Taguchi loss function and the payback period for the original production process and incorporating the process improvement.

Solution: The quantities involved can be identified as follows: USL = 3.2 mm, LSL = 2.0 mm, $T = 2.6$ mm, $\sigma = 0.2$ mm, and $Y = 2.6$ mm. The quantity k is given by Eq. (16.2) as

$$k = \frac{(\$10.00)}{(3.2 - 2.6)^2} = \$27.28.$$

Thus, the loss cost is

$$\text{Loss cost} = (27.78)\left[(2.6 - 2.6)^2 + 0.2^2\right] = \$1.11 \text{ per unit.}$$

After the improvement, the standard deviation is cut in half to 0.1 mm; thus,

$$\text{Loss cost} = (27.78)\left[(2.6 - 2.6)^2 + 0.1^2\right] = \$0.28 \text{ per unit.}$$

The savings are ($1.11 − $0.28)(10,000) = $8300 per month and the payback period for the investment is $50,000/($8300/month) = 6.02 months.

CASE STUDY 16.1 Manufacture of Television Sets by Sony Corporation

In the mid-1980s, Sony Corporation executives found a confusing situation: The television sets manufactured in Japanese production facilities sold faster than those produced in a San Diego, CA, facility, even though (a) they were produced from identical designs; (b) there were no identifications to distinguish the sets made in Japan from those made in the United States; and (c) there was no apparent reason for this discrepancy. However, investigations revealed that the sets produced in Japan were superior to the US versions; color sharpness was better and hues were more brilliant. Because they were all on display in the same stores, consumers could easily detect the difference and purchase the model with the best picture quality.

The reasons for the difference were not clear. Another point of confusion was the constant assurances that the San Diego facility had a total quality-control program in place, and that the plant was maintaining quality-control standards, so that no defective parts were being produced. Although the Japanese facility did not have a total quality program, there was an emphasis on reducing variations from part to part.

Further investigations found a typical pattern in an integrated circuit in the sets that was critical in affecting color density. The distribution of parts that met the color-design objective is shown in Fig. 16.2a, and the Taguchi loss function for these parts is shown in Fig. 16.2b.

In the San Diego facility, where the number of defective parts was minimized (to zero in this case), a uniform distribution within the specification limits was achieved. The Japanese facility actually produced parts outside of the design specifications, but the standard deviation about the mean was lower. Using the Taguchi loss-function approach (see Example 16.1) made it clear that the San Diego facility lost about \$1.33 per unit while the Japanese facility lost \$0.44 per unit.

Traditional quality viewpoints would find a uniform distribution without defects to be superior to a distribution in which a few defects are produced, but the majority of parts are closer to the design target values. Consumers, however, can readily detect which product is superior, and the marketplace proves that minimizing deviations is indeed a financially worthwhile quality goal.

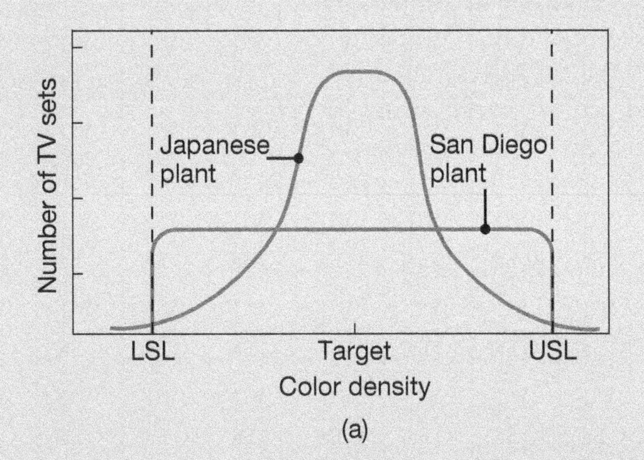

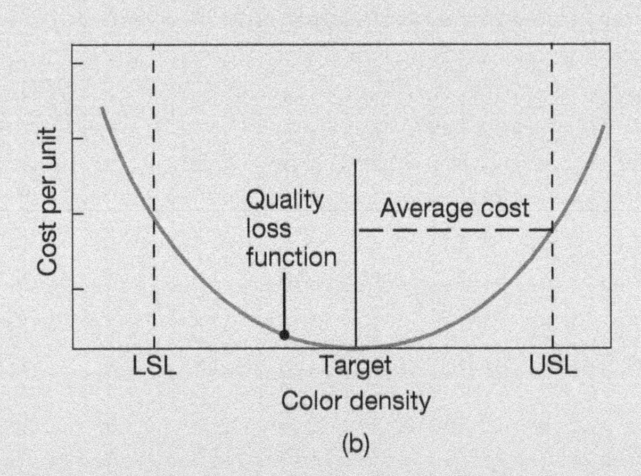

FIGURE 16.2 (a) Objective-function value distribution of color density for television sets. (b) Taguchi loss function, showing the average replacement cost per unit to correct quality problems. *Source:* After G. Taguchi.

Source: After D.M. Byrne and G. Taguchi.

16.3.6 The ISO and QS Standards

With increasing international trade and global manufacturing, economic competitiveness has become very challenging to manufacturers. Customers are demanding *high-quality products and services* and at *low prices*, and they are looking for suppliers that can respond to this demand consistently and reliably. This trend has, in turn, created the need for international conformity and consensus regarding the establishment of methods for quality control, reliability, and safety of products. Equally important are concerns regarding the environment and quality of life, with new international standards being applied.

The ISO 9000 Standard. First issued in 1987, and most recently revised in 2015, the ISO 9000 standard (*Quality Management and Quality Assurance Standards*) is a generic series of quality-system management standards. This standard, which has become the world standard for quality, has permanently influenced the manner in which manufacturing companies conduct business in world trade.

The ISO 9000 series includes the following standards:

- ISO 9001 – *Quality systems: Model for quality assurance in design/development, production, installation, and servicing.*
- ISO 9002 – *Quality systems: Model for quality assurance in production and installation.*
- ISO 9003 – *Quality systems: Model for quality assurance in final inspection and test.*
- ISO 9004 – *Quality management and quality system elements: Guidelines.*

Companies voluntarily register for these standards and are issued certificates. Registration may be sought generally for ISO 9001 or 9002, and some companies have registration up to ISO 9003. The 9004 standard is a guideline and not a model or a basis for registration. For certification, a company's plants are first visited and audited by accredited and independent third-party teams to certify that the standard's 20 key elements are in place and are functioning as required.

Depending on the extent to which a company fails to meet the requirements of the standard, registration may or may not be recommended at that time. The audit team does not advise or consult with the company on how to reconcile discrepancies, but merely describes the nature of the noncompliance. Periodic audits are required to maintain certification. The certification process can take from six months to a year or more, and can cost tens of thousands of dollars, depending on the company's size, number of its plants, and product lines.

The ISO 9000 standard is not a product certification but a **quality process certification.** Companies establish their own criteria and practices for quality, but the documented quality system must be in compliance with the ISO 9000 standard. Thus, a company cannot write into the system any criterion that opposes the intent of the standard.

Registration indicates a company's commitment to consistent practices, as specified by the company's own quality system (such as quality in design, development, production, installation, and servicing), including proper documentation of such practices. In this way customers, including also government agencies, are assured that the supplier of the product or service (which may or may not be within the same country) is following specified practices. In fact, manufacturing companies are themselves assured of such practices regarding their own suppliers who have ISO 9000 registration; therefore, suppliers also must be registered.

The QS 9000 Standard. Jointly developed by Chrysler, Ford, and General Motors, this standard was first published in 1994. Prior to the development of QS 9000, each of the Big Three automotive companies had its own standard for quality system requirements. Tier I suppliers have been required to obtain third-party registration to QS 9000; often, this standard has been described as an "ISO 9000 chassis with a lot of extras."

The ISO 14000 Standard. This is a family of standards, first published in 1996, pertaining to the international *Environmental Management Systems* (EMS), and it concerns how an organization's activities affect the environment throughout the life of its products (see also Section 16.4). These activities (a) may be internal or external to the organization; (b) range from production to ultimate disposal of the product after its useful life; and (c) include any effects on the environment, such as pollution, waste generation and disposal, noise, depletion of natural resources, and energy use.

The ISO 14000 has several sections: Guidelines for Environmental Auditing, Environmental Assessment, Environmental Labels and Declarations, and Environmental Management. ISO 14001: Environmental Management System Requirements consists of sections on General Requirements, Environmental Policy, Planning, Implementation and Operation, Checking and Corrective Action, and Management Review.

16.4 Life-Cycle Assessment and Sustainable Manufacturing

Life-cycle assessment (LCA) is defined, according to the ISO 14000 standard, as "a systematic set of procedures for compiling and examining the inputs and outputs of materials and energy, and the associated environmental impacts or burdens directly attributable to the functioning of a product, process, or service system throughout its entire life cycle." **Life cycle** consists of consecutive and interlinked stages of a product or a service, from the very beginning of design and manufacture to its recycling or disposal; it includes:

1. Extraction of natural resources;
2. Processing of raw materials;
3. Manufacturing of products;

4. Transportation and distribution of the product to the customer;
5. Use, maintenance, and reuse of the product; and
6. Recovery, recycling, and reuse of components of the product or their disposal, which also include metalworking fluids, cleaning solvents, and various liquids used in heat-treating and plating processes.

Life-cycle engineering (LCE) is concerned with environmental factors, especially as they relate to design, optimization, and various technical considerations regarding each component of a product or a process life cycle. A major aim of LCE is to consider the reuse and recycling of products from their earliest stage of the design process, known as *green design*, or *green engineering*. Software has been developed to expedite these analyses, especially for the manufacturing and chemical industries, because of the higher potential for environmental and ecological impact that they have. Examples of such software include FeaturePlan and Teamcenter, which runs in a ProEngineer environment.

Cradle-to-grave. A traditional product life cycle, referred to as the *cradle-to-grave* model, can be defined as consecutive and interlinked stages of a product or a service system that includes:

1. Extraction of natural resources, including raw materials and energy;
2. Processing of raw materials;
3. Manufacture of products;
4. Transportation and distribution of the product to the customer;
5. Use, reuse, and maintenance of the product; and
6. Ultimate disposal of the product.

Life-cycle analysis and engineering is a comprehensive, powerful, and necessary tool, although its total implementation can be expensive and time consuming. This is largely because of uncertainties in input data regarding materials, processes, long-term effects, and costs, as well as the time required to gather reliable data to properly assess the often-complex interrelationships among numerous components of the whole system. A number of software packages are available to expedite such analysis in some specific industries, particularly the chemical and process industries, because of the higher potential for environmental damage in their operations.

Sustainable manufacturing. In recent years, it has become increasingly obvious that natural resources are limited, clearly necessitating the urgent need for conservation of materials and energy. The term *sustainable manufacturing* emphasizes the necessity for conserving these resources, especially through maintenance and reuse. The importance of product life cycle considerations and the differences between cradle-to-grave and cradle-to-cradle approaches and between biological and industrial recycling were described in Section 1.4.

Recycling often requires that individual components in a product be taken apart, but if much effort and time has to be expended in doing so, then recycling can become prohibitively expensive. The general guidelines to facilitate recycling are:

1. Do not design products, instead design life cycles, where all of the material and energy inputs are considered, as well as the product's destination when its design life is over.
2. Use modular design to facilitate disassembly.
3. Whenever possible, use materials that can easily be recycled.
4. Reduce the number of parts and types of materials in products, and use as little material as possible.
5. Avoid mixing materials that are biologically recyclable with those that can be part of an industrial recycling life cycle; for plastic parts, use a single type of polymer as much as possible.
6. Mark plastic parts for ease of identification, as is already done with plastic food containers and beverage bottles.
7. Avoid using coatings, paints, and plating; use molded-in colors in plastic parts whenever possible.
8. Avoid using adhesives, rivets, and other permanent joining methods in assembly; instead, use fasteners, especially snap-in fasteners.

CASE STUDY 16.2 Sustainable Manufacturing in the Production of Nike Athletic Shoes

Nike athletic shoes are assembled using adhesives (Section 12.15). Up until about 1990, the adhesives used contained petroleum-based solvents, which pose health hazards to humans and contribute to petrochemical smog. The company worked with chemical suppliers to successfully develop water-based adhesive technology, which is now used for the majority of the shoe assembly operations. As a result, solvent use in all manufacturing operations in Nike's subcontracted facilities in Asia was reduced by 67% since 1995. Specifically, in 1997, 834,000 gallons of hazardous solvents were replaced with 1290 tons of water-based adhesives.

The rubber outsoles of the athletic shoe were made by a process that resulted in significant amounts of extra rubber from around the periphery of the sole, called *flashing*, similar to the flash shown in Figs. 6.14c and 10.38c. With about 40 factories using thousands of molds and producing over a million outsoles a day, the flashing constituted the largest source of waste in making these shoes. To reduce waste, the company developed a technology that grinds the flashing into rubber powder, which is then added to the rubber mixture used to make outsoles. As a result, waste was reduced by 40%; also it was found that the mixed rubber had better abrasion resistance, durability, and overall performance than the highest premium rubber.

16.5 Energy Consumption in Manufacturing

The global manufacturing sector consumes approximately 25% of annual energy production. This number has fallen after peaking at around 50% in the 1970s, as a result of major efforts to reduce waste and improve the efficiency of machinery and manufacturing operations. The most common source of energy in manufacturing is electrical, produced from oil, natural gas, biofuels, coal, nuclear, wind, wave, and solar. Given such a large percentage and varied sources of energy, societal concerns regarding energy availability and conservation must be considered and addressed in manufacturing. Indeed, it is not likely that viable national energy policies can be developed and implemented without a central consideration of the manufacturing sector.

Process Energy Demand. The energy required to make a particular part or component is determined basically by its design, materials, and the manufacturing operation, as well as the quality, condition, and age of the machinery and equipment. The energy requirement is relatively easy to calculate for a manufacturing process (See Table 16.4), but becomes difficult when ancillary equipment, such as pumps, fans, blowers, furnaces, and lights, are included in the final calculation. Moreover, the efficiency of a manufacturing operation typically varies, depending on a particular plant's practices and procedures or from company to company. For example:

- Some manufacturing operations are more demanding from an energy standpoint than others, as shown in Fig. 16.3.
- Each manufacturing process has a range of performance; processing rates can vary greatly depending on factors such as the specified tolerances and surface finish, with tighter tolerances and smoother surfaces being the most time-consuming and energy intensive (Figs. 16.4 and 16.5).

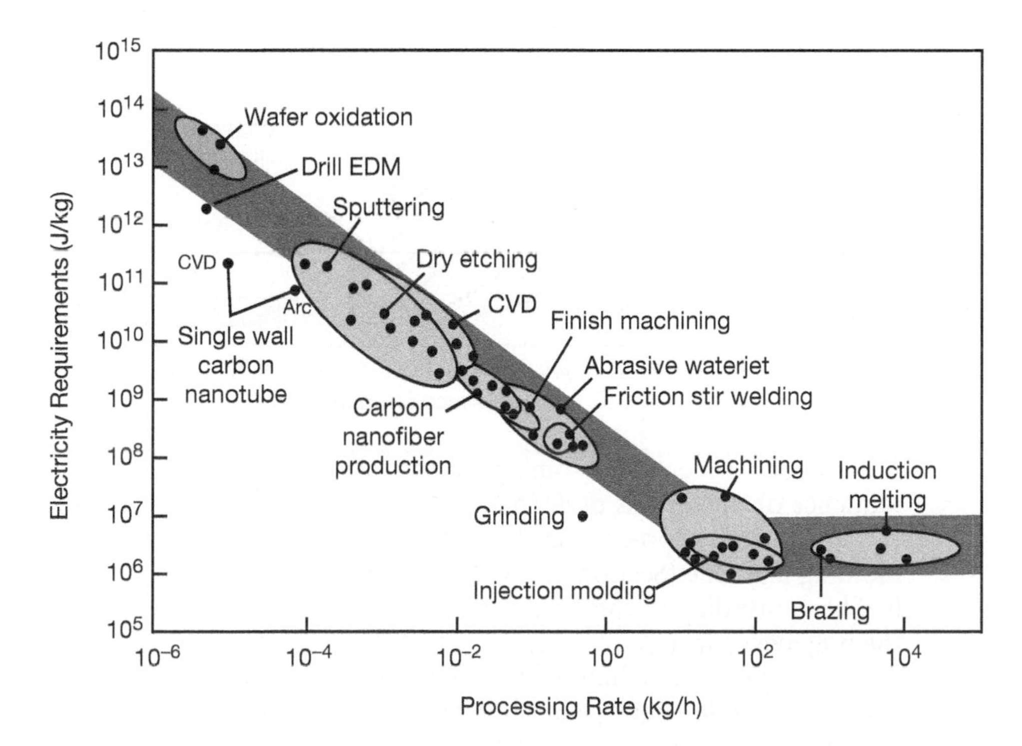

FIGURE 16.3 Specific energy requirements for various manufacturing processes, including ancillary equipment. Note that most manufacturing processes lie within the dark band. Also note the increased energy intensity for slower processes. *Source:* After T. Gutowski.

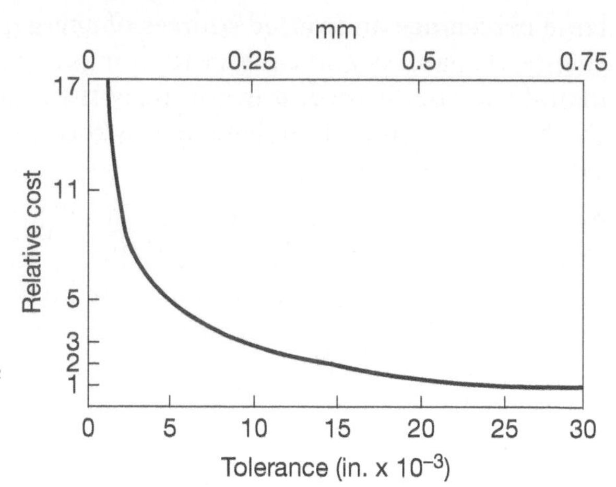

FIGURE 16.4 Relationship between relative manufacturing cost and dimensional tolerance; note how rapidly cost increases as tolerance decreases.

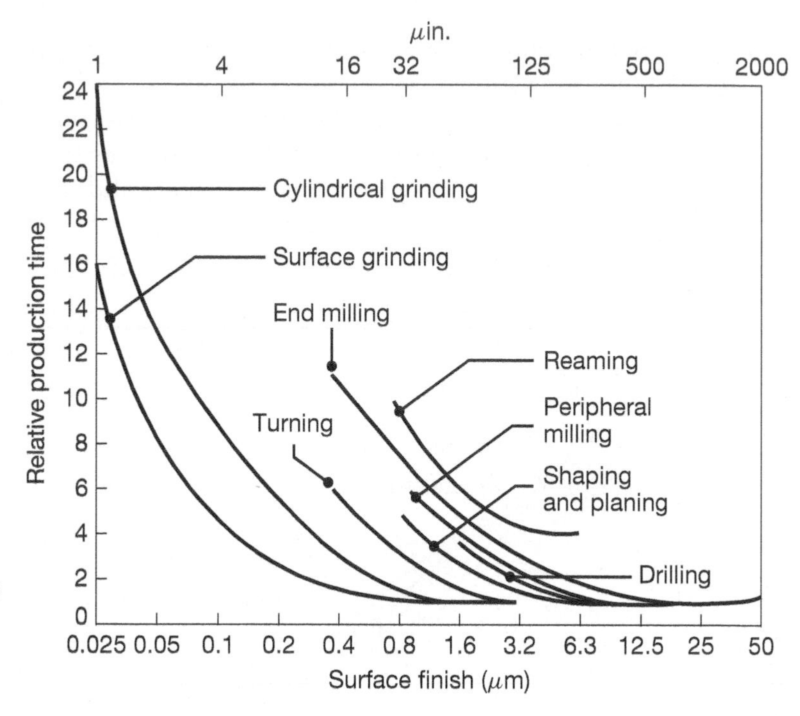

FIGURE 16.5 Relative production time as a function of surface finish produced by various manufacturing processes. (See also Fig. 9.43.)

- Energy requirements for some processes are strongly related to the sequence of operations performed, as when a machine is 'ready' but is not yet actively processing the material. For example, an injection molding machine (Section 10.10.2) is not productive when the mold has been installed in the machine but it has not yet been filled with the polymer.
- Some production machinery have hydraulic pumps that operate continuously, while others will shut the pumps off during periods of inactivity.
- Workpieces in hot-working operations can be cooled by blowing cool air over them; in practice, the hot air is often vented to the atmosphere but it can be used to preheat stock for processing, heat the facility, or provide hot water.

Effect of Materials. Whether a material is mined, refined, cast, synthesized from chemicals, or recycled from discarded products, energy is still required to produce stock materials into forms that can be processed further. In many applications, certain materials have a performance advantage over others. For example, titanium and aluminum alloys are obviously preferred to steel for aircraft, primarily because of their higher strength-to-weight and stiffness-to-weight ratios, allowing for lighter designs and associated fuel savings.

Different materials have very different energy requirements to process them, although recycled materials have significantly less embedded energy. Table 16.4 summarizes the energy required to produce a variety of materials, presenting the data by mass and by volume. It has been noted that if energy or *carbon footprint* is divided by mass, there is a natural benefit to using heavier materials that otherwise may not be justified in weight-constrained applications. On the other hand, if volume is a design constraint, the energy per unit weight may then be a more accurate measure.

Note in Table 16.4 that metals require significant energy to be produced, as they are generally reduced or extracted from their oxides, a

TABLE 16.4 Energy content of selected materials.

Material	Energy content	
	MJ/kg	GJ/m^3
Metals		
Aluminum		
From bauxite	300	810
Recycled	42.5	115
Cast Iron	30–60	230–460
Copper		
From ore	105	942
Recycled	55.4	497
Lead	30	330
Magnesium	410	736
Steel		
From ore	55	429
Recycled	9.8	76.4
Zinc	70	380
Polymers		
Nylon 6,6	175	200
Polyethylene		
High density	105–120	100–115
Low density	80–100	75–95
Polystyrene	95–140	95–150
Polyvinyl chloride	67–90	90–150
Ceramics	1–50	4–100
Glasses	10–25	30–60
Wood	1.8–4.0	1.2–3.6

process that is energy intensive. It has been estimated that 5% of the total energy consumption in the United States is in producing aluminum. It is not uncommon for the energy required to produce a material to be three to four orders of magnitude greater than the energy required to shape it into its final shape.

It is thus understandable why using aluminum in automobiles, to reduce energy intensity, is indeed difficult to justify without the sustained implementation of recycling it. Energy savings from weight reductions of an automobile may not be appreciably higher, and may even be lower, than the energy required to produce the aluminum itself.

16.6 Selection of Materials for Products

Although the general criteria for selecting materials have been described in Section 1.5, this section describes them in greater detail.

16.6.1 General Properties of Materials

As described throughout Chapter 2, *mechanical properties* include (a) strength; (b) toughness; (c) ductility; (d) hardness; and (e) resistance to fatigue, creep, and impact. Moreover, such characteristics as (a) stiffness depend not only on the elastic modulus of the material but also on the geometric features of a part, and (b) friction and wear depend on a combination of several factors, as described in Section 4.4.

Physical properties (Section 3.9) include density, melting point, specific heat, thermal and electrical conductivity, thermal expansion, and magnetic properties. The *chemical properties* (Section 3.9.7) that are of primary concern in manufacturing are oxidation and corrosion. The relevance of these properties to product design and manufacturing has been described in various chapters, and include several tables relating to properties of metallic and nonmetallic materials. Table 16.1 lists the sections relevant to various material properties.

Selection of materials has now become much easier and faster, because of the availability of software (*smart databases*) that provide much greater accessibility to information. Also, expert systems are now capable of rapidly identifying appropriate materials for a particular application.

Regardless of the method employed, the following considerations are essential in materials selection:

1. Do the materials selected have properties that satisfy but do not unnecessarily exceed minimum requirements and specifications?
2. Can some of the materials be replaced by others that are less expensive or more environmentally friendly?
3. Do the materials selected possess the appropriate manufacturing characteristics?
4. Are the raw materials (stock) to be ordered available in standard shapes, dimensions, tolerances, and surface finish?
5. How reliable is the material supply?

6. Are there likely to be significant price increases or market fluctuations for the selected materials?
7. Can the materials be obtained in the required quantities in the desired time frame?

16.6.2 Shapes of Commercially Available Materials

Materials are generally available as a casting, extrusion, forging, bar, plate, rod, wire, sheet, foil, and powders (Table 16.5). Their cost per unit weight or volume can vary widely, depending on the particular process used, as well as the amount purchased (such as bulk discount). Characteristics such as overall material quality, surface texture, dimensional tolerances, straightness, and waviness also must be taken into account, as any improvements on these during processing them will indicate additional costs.

Consider, for example, producing a simple shaft with good dimensional accuracy, surface finish, roundness, and straightness. One could purchase round bars that are already turned and centerless ground (Chapters 8 and 9). Unless a company's facilities are capable of producing round bars with these characteristics and economically, it is generally cheaper to purchase them from outside sources. If the part to be made is a stepped shaft (having different diameters along its length), one could purchase a round bar with a diameter at least equal to the largest diameter of the stepped shaft, then turn it on a lathe. If the stock purchased has broad dimensional tolerances, is warped, or is out-of-round, one could purchase a larger diameter, to ensure proper dimensions of the final shaft after machining.

As indicated throughout various chapters, each manufacturing operation produces parts that have their own specific range of geometric

TABLE 16.5 Commercially available shapes of materials; lowercase letters indicate limited availability. Most of the metals are also available in powder form, including prealloyed powders.

Material	Available as
Aluminum	B, F, I, P, S, T, W
Ceramics	B, p, s, T
Copper and brass	B, f, I, P, s, T, W
Elastomers	b, P, T
Glass	B, P, s, T, W
Graphite	B, P, s, T, W
Magnesium	B, I, P, S, T, w
Plastics	B, f, P, T, w
Precious metals	B, F, I, P, t, W
Steels and stainless steels	B, I, P, S, T, W
Titanium	B, f, I, P, S, w
Zinc	F, I, P, W

Note: B = bar and rod; F = foil; I = ingots; P = plate and sheet; S = structural shapes; T = tubing; W = wire

features, shapes, dimensional tolerances, and surface finish and texture. Consider also the following specific examples:

1. Castings generally have less dimensional accuracy and rougher surface finish than parts made by cold extrusion or powder metallurgy (Chapters 5, 6, and 11).
2. Hot-rolled and hot-drawn products have a wider dimensional tolerances and rougher surface finish than cold-rolled and cold-drawn products (Chapter 6).
3. Extrusions have smaller cross-sectional dimensional tolerances than parts made by roll forming (Chapters 6 and 7).
4. Seamless tubing made by the tube-rolling process have more thickness variation than that of roll-formed and welded tubing (Chapters 6 and 12).
5. Round bars turned on a lathe have a rougher surface finish and wider dimensional tolerances than bars finished on cylindrical or centerless grinders (Chapters 8 and 9).

16.6.3 Manufacturing Characteristics of Materials

Manufacturing characteristics typically include castable, workable, formable, machinable, grindable, weldable, and hardenable by heat treatment. Because raw materials have to be made into individual components having specific shapes, dimensions, and surface finish, these properties are crucial to the proper selection of materials. Recall also that the quality of the incoming stock can greatly influence its manufacturing characteristics.

Consider the following examples:

1. A rod or bar with a longitudinal seam (lap) is likely to develop cracks during simple upsetting or heading operations.
2. Bars with internal defects, such as inclusions, may develop cracks during seamless-tube production.
3. Porous castings will produce poor and irregular surface finish when machined or ground.
4. Solid blanks that are not heat treated uniformly and bars that are not stress relieved will distort during subsequent machining or drilling operations.
5. Incoming stock that has variations in its composition and microstructure cannot be uniformly heat treated or machined.
6. Sheet metal stock that has variations in its thickness, will exhibit uneven springback during bending and forming operations.
7. If prelubricated sheet metal blanks have uneven lubricant distribution over their surfaces, their formability, surface finish, dimensional accuracy, and overall quality may be adversely affected.

16.6.4 Reliability of Material Supply

Several factors influence the *reliability of material supplies*: shortages of materials, geopolitics, and the reluctance of some suppliers to produce materials in a particular shape or quality. Moreover, even though raw materials may generally be available throughout a country as a whole, they may not readily be available at a particular location or a particular plant.

Recycling Considerations. *Recycling* may involve relatively simple parts, such as plastic bottles, aluminum cans, and scrap sheet metal. Most often, however, the individual components of a product must be taken apart and separated into types for individual recycling. It will be appreciated that if much effort, energy, and time has to be expended in doing so, then recycling can become uneconomical. General guidelines to facilitate the recycling process were given in Section 16.4.

16.6.5 Material and Processing Costs

The *unit cost* (cost per unit weight or volume) of a raw material depends on the material itself, its processing history, shape, condition, and the quantity purchased (*bulk discount*). For example, because more operations are involved in producing thin wire than making round rods (Section 6.5), the unit cost of wire is much higher; similarly, powder metals generally are more expensive than bulk metals. Note also that the cost of a particular material is also subject to fluctuations caused by other factors, such as supply and demand and geopolitics. The cost per unit volume for wrought metals and plastics relative to carbon steel are given in Table 16.6.

If a product is no longer cost competitive, alternative and less costly materials can be selected. For example, the copper shortage in the 1940s led the US government to mint pennies from zinc-plated steel. Similarly, when the price of copper increased substantially during the 1960s, the copper wiring traditionally installed in homes was, for a time, made of aluminum. However, this substitution also required the redesign of switches and outlets to avoid excessive heating at the junctions.

TABLE 16.6 Approximate cost per unit volume for wrought metals and plastics relative to the cost of carbon steel. Data is representative of 25-mm diameter bar stock.

Material	Cost	Material	Cost
Gold	70,000	Carbon steel	1
Silver	680	Magnesium alloys	2–4
Molybdenum alloys	200–250	Aluminum alloys	1.5–3
Nickel	40	Gray cast iron	1.2
Titanium alloys	25–40	Nylons, acetals, and silicon	1.1–2
Copper alloys	5–9	Rubber*	0.2–1
Stainless steels	2–9	Other plastics and elastomers*	0.2–2
High-strength low-alloy steels	1.4		

*As molding compounds

Note: Costs vary significantly with the quantity of purchase, supply and demand, size and shape, and various other factors.

TABLE 16.7 Typical scrap produced in various manufacturing processes.

Process	Scrap (%)	Process	Scrap (%)
Machining	10–60	Permanent-mold casting	10
Closed-die forging, hot	20–25	Powder metallurgy	<5
Sheet metal forming	10–25	Rolling and ring rolling	<1
Extrusion, hot	15	Additive manufacturing	50–80

When scrap is produced in manufacturing operations, as in sheet metal fabrication, forging, and machining (Table 16.7), the value of the scrap is deducted from the material's cost to calculate net material cost. In machining, the amount of scrap (in the form of chips of various shapes; Section 8.2.1) can be high, whereas shape rolling, ring rolling, and powder metallurgy, all of which are generally regarded as *net-shape* or *near-net-shape processes*, produce the least scrap.

The value of the scrap depends on the type of material and the demand for the scrap; typically, it is between 10 and 40% of the original cost of the material. Another factor in the value of scrap is whether or not the material is contaminated; thus, for example, metal chips collected from operations where cutting fluids have been used have less value than those produced in dry machining.

16.7 Substitution of Materials in Products

Although new products continually appear in the global marketplace, the majority of the design and manufacturing efforts are generally concerned with improving existing products. Significant improvements can be made by (a) substituting materials; (b) implementing new design concepts, technologies, and processing techniques; (c) better control of processing parameters; and (d) increased automation of plant operations.

Automotive and aircraft manufacturing are examples of major industries in which substitution of materials is an important and ongoing activity; other such industries are sporting goods and medical products. There are several reasons for substituting materials in products:

1. Reduce the costs of materials and processing.
2. Improve assembly and conversions to automated assembly.
3. Improve the performance of products, such as by reducing weight and improving characteristics such as resistance to wear, fatigue, and corrosion.
4. Increase the stiffness-to-weight and strength-to-weight ratios of product components.
5. Reduce the need for maintenance and repair.
6. Reduce vulnerability to the unreliability of domestic and overseas supply of certain materials.
7. Improve the appearance of a product.
8. Reduce performance variations in products, such as by improving *robustness*.

Substitution of materials in the automotive industry. Trends in the automobile industry provide several examples of effective substitution of materials, to achieve one or more of the foregoing objectives:

1. Numerous components in metal bodies replaced with plastic or reinforced-plastic parts.
2. Metal bumpers, fuel tanks, housings, clamps, and various other components replaced with plastic substitutes.
3. Engine components replaced with ceramic and reinforced-plastic parts.
4. Cast-iron engine blocks replaced with cast-aluminum blocks; forged crankshafts with cast crankshafts; and forged connecting rods with cast, powder-metallurgy, or composite-material connecting rods, and some cast-aluminum pistons have replaced forged steel pistons.
5. Wider use of magnesium and aluminum to improve fuel efficiency.
6. Steel structural elements replaced with extruded aluminum sections, and rolled sheet-steel body panels replaced rolled aluminum sheet.
7. Occupant crash zones, where strength and stiffness are essential, are being made of hot-stamped, advanced high-strength steels (see Sections 3.10.2 and 7.5.8).

Substitution of materials in the aircraft industry.

1. Conventional aluminum alloys (2000 and 7000 series) for some components replaced with aluminum-lithium alloys, titanium alloys, and composite materials, particularly because of their higher strength-to-weight ratios.
2. Forged parts replaced with powder-metallurgy parts made with better control of impurities and microstructure; these parts require less machining and finishing operations, producing less scrap of expensive materials.
3. Advanced composite materials and honeycomb structures replacing traditional aluminum airframe components (Fig. 7.48).
4. Metal-matrix composites replacing some of the aluminum and titanium parts previously used in structural components.

16.8 Capabilities of Manufacturing Processes

As indicated throughout this text, each manufacturing process has its specific advantages and limitations. References to the general characteristics and capabilities of manufacturing processes are listed in Table 16.1. Note that casting and injection molding, for example, can generally produce more complex shapes than can forging and powder metallurgy. On the other hand, forgings have toughness and fatigue resistance that generally are superior to that of castings and PM products; moreover, they can be made into more complex shapes by subsequent machining and finishing operations.

Dimensional tolerances and surface finish. The *dimensional tolerances* and *surface finish* produced are important not only for the functioning of

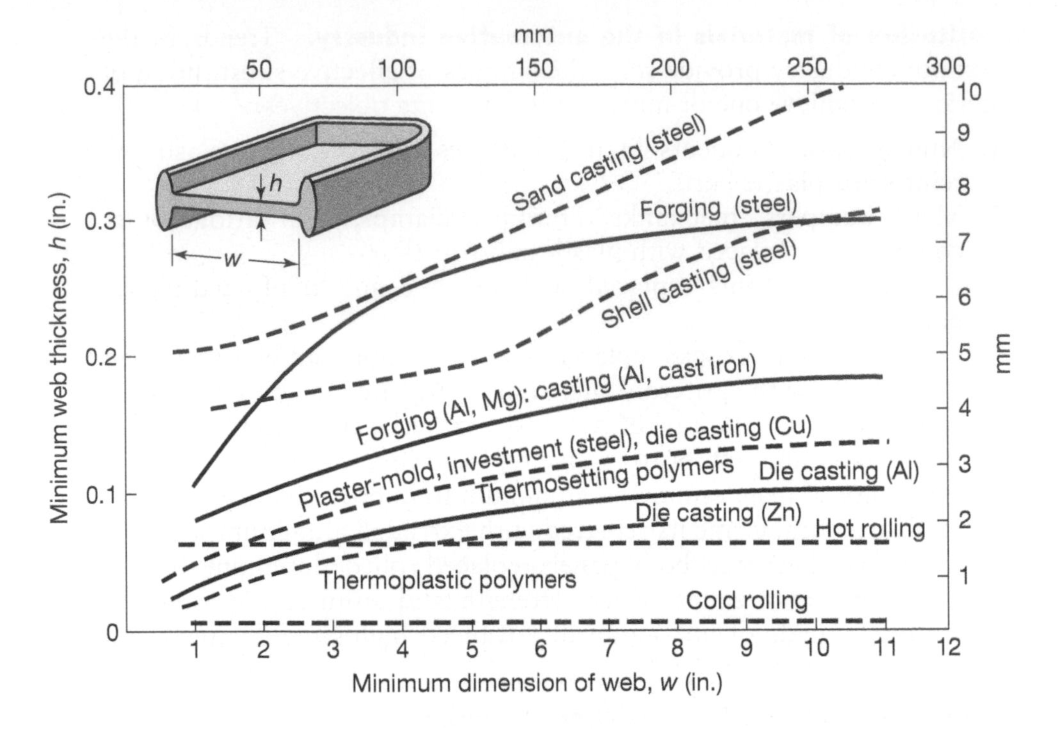

FIGURE 16.6 Minimum part dimensions obtainable by various manufacturing processes. *Source:* After J.A. Schey.

parts made, but also in subsequent sub-assembly and assembly. The capabilities of various manufacturing processes in this regard are qualitatively illustrated in Fig. 16.7 and Table 16.1.

Closer dimensional tolerances and better surface finish can be obtained by additional finishing operations, better control of processing parameters, and the use of higher quality equipment and controls. However, the closer the tolerance and the finer the surface finish specified, the higher is the cost of manufacturing, because of the greater number of additional processes required and the longer the manufacturing time involved (Figs. 9.43, 16.4, and 16.5). In machining aircraft structural members made of titanium alloys, for example, as much as 60% of the cost of machining the part may be expended in the final machining pass in order to maintain specified tolerances and surface finish.

Production quantity or volume. Depending on the type of product, the *production quantity*, or *volume* (*lot size*), can vary widely. Bolts, nuts, washers, bearings, and ballpoint pens are produced in very large quantities, whereas jet engines for commercial aircraft, diesel engines for locomotives, machine tools, and propellers for cruise ships are manufactured in limited quantities. Production quantity also plays a significant role in process and equipment selection; an entire manufacturing discipline, called the *economic order quantity* is devoted to determining the optimum production quantity.

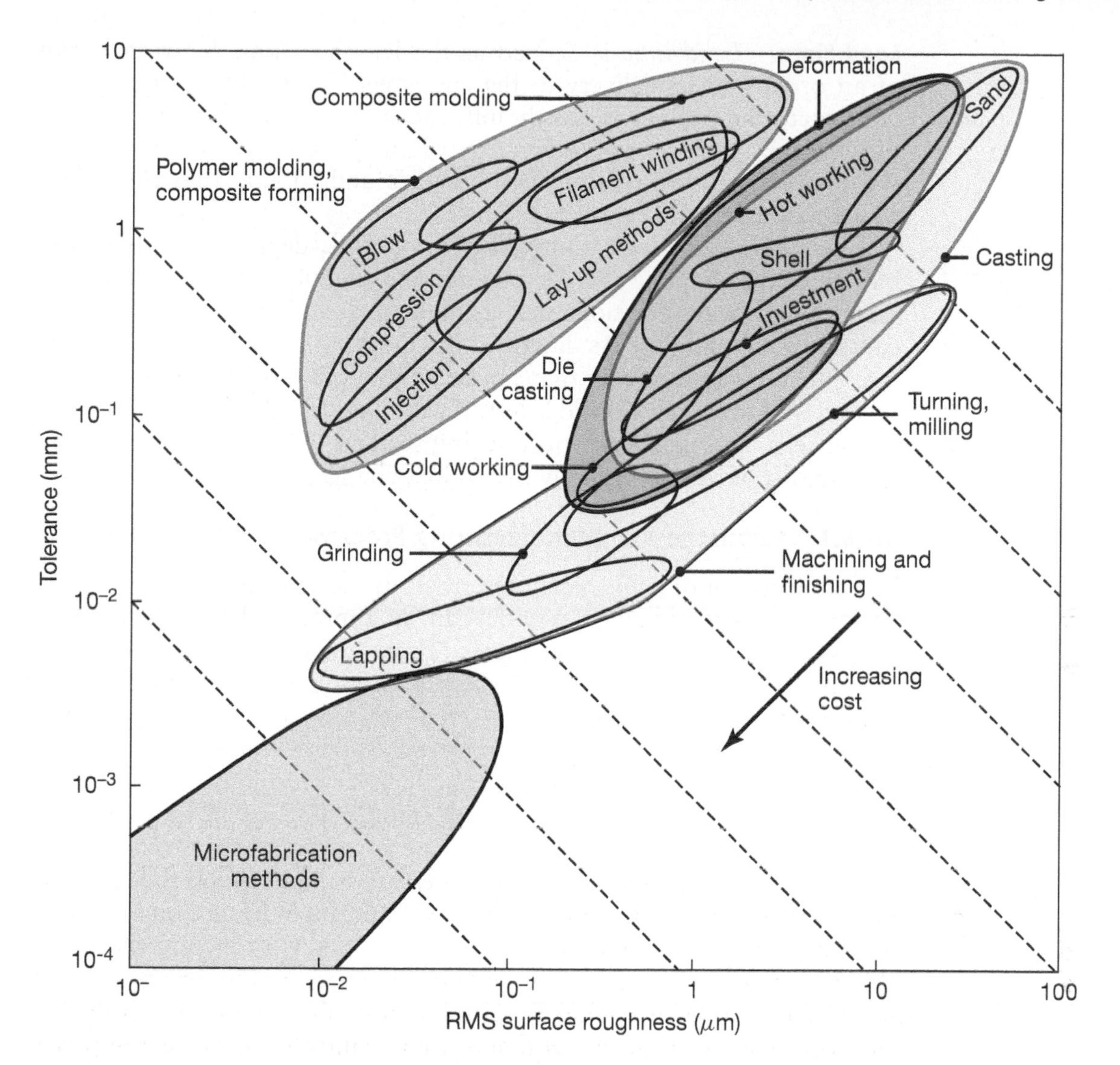

FIGURE 16.7 A plot of achievable dimensional tolerance versus surface roughness for various manufacturing operations; the dashed lines indicate cost factors. An increase in precision corresponding to the separation of two neighboring lines corresponds to a two-fold increase in cost. *Source:* After M.F. Ashby.

Production rate. A major factor in manufacturing process selection is the *production rate*, defined as the number of pieces produced per unit of time. Die casting, deep drawing, roll forming, and powder metallurgy, for example, are high-production-rate operations. In contrast, sand casting, conventional and electrochemical machining, spinning, superplastic forming, adhesive and diffusion bonding, and the processing of composite materials are relatively slow operations. Production rates can of course be increased by such means as automation, computer controls, and using multiple machines; they could be considered after a thorough cost-benefit analysis.

Lead time. *Lead time* is defined as the length of time between receipt of an order and its delivery to the customer. The selection of a specific manufacturing process is greatly influenced by the time required to start production. Processes as forging, extrusion, die casting, roll forming, and sheet metal forming typically require dies and tooling that can take a considerable length of time to produce.

Lead times can range from weeks to months, depending on the complexity of the shape of the die, its size, and the die material. In contrast, material-removal operations such as machining and grinding have significant built-in flexibility. As described in Chapters 8 and 9, these processes use machinery and tooling that can be adapted to most requirements in a relatively short time. Consider also machining centers, flexible manufacturing cells, and flexible manufacturing systems (Chapters 8 and 15) that are capable of respond rapidly to product changes.

16.8.1 Robustness in Manufacturing Processes and Machinery

The importance of *robustness*, described in Section 16.2.3 in terms of a design, or a process, or a system, can be appreciated by considering a simple injection-molded plastic gear. There are several well-understood variables in injection molding of plastics (Section 10.10.2), including the quality of the pellets (raw material), temperature, and time, all of which can be controlled. As noted in Section 16.2.3, there are certain other variables that can influence the operation, such as variations in the ambient temperature and humidity in a plant, dust in the air entering the plant from an open door, and the variability in the performance of the machine itself.

For sustained good quality, it is necessary to understand the effects of each variable on product quality. For example: (a) Why and how does the ambient temperature and humidity affect the quality of the molded gears? (b) Why and how does even a light dust coating on the pellets affect its performance in the machine? Responding to such questions help establish new operating parameters so that such variations do not adversely and significantly affect gear quality, thus resulting in a robust system.

16.9 Selection of Manufacturing Processes

Materials have a wide range of properties, yet no single material has all the desirable manufacturing characteristics, as can be noted in Table 16.8. Thus, a particular material may be cast or forged with relative ease, yet it may subsequently present difficulties in finishing to final dimensions.

As an illustration of the important factors involved in process selection consider the following questions:

1. Have all alternative manufacturing processes been investigated?
2. Are the processing methods under consideration economical for the type of material, the shape to be produced, the specifications to be met, and the required production rate?

TABLE 16.8 General characteristics of manufacturing processes for various metals and alloys.

	Carbon steels	Alloy steels	Stainless steels	Tool and die steels	Aluminum alloys	Magnesium alloys	Copper alloys	Nickel alloys	Titanium alloys	Refractory alloys
Casting										
Sand	1	1	1	2	1	1	1	1	2	1
Plaster	3	3	3	3	1	1	1	3	3	–
Ceramic	1	1	1	1	2	2	1	1	2	1
Investment	1	1	1	3	1	2	1	1	1	1
Permanent	2	2	3	3	1	1	1	3	3	–
Die	3	3	3	3	1	1	1	3	3	–
Forging										
Hot	1	1	1	1	1	1	1	1	1	1
Cold	1	1	1	3	1	2	1	3	3	–
Extrusion										
Hot	1	1	1	2	1	1	1	1	1	1
Cold	1	2	1	3	1	3	1	2	3	3
Impact	3	3	3	3	1	1	1	3	3	3
Rolling	1	1	1	3	1	1	1	1	1	2
Powder metallurgy	1	1	1	1	1	1	1	1	1	1
Sheet metal forming	1	1	1	3	1	1	1	1	1	1
Machining	1	1	1	3	1	1	1	2	1	2
Chemical	1	2	1	2	1	1	1	2	2	2
ECM	3	1	2	1	3	3	2	1	1	1
EDM	3	2	2	1	3	3	2	2	2	1
Grinding	1	1	1	1	1	1	1	1	1	1
Welding	1	1	1	3	1	1	1	1	1	1

Note: 1 = Generally processed by this method; 2 = Can be processed by this method; 3 = Usually not processed by this method

3. Can the requirements for dimensional tolerances, surface finish, and product quality be met consistently?
4. Can the part be produced to final dimensions without requiring additional finishing operations?
5. Is the tooling required available in the plant? Can it be purchased as a standard item?
6. Is scrap produced, and if so, can it be minimized?
7. Have all the automation and computer-control possibilities been explored for each phase of the manufacturing cycle?
8. Are inspection techniques and quality control being properly implemented?

9. Does each part of the product have to be manufactured or are some parts commercially available?
10. What is the ecological impact of the materials and processes to be employed?

16.10 Manufacturing Costs and Cost Reduction

The total cost of a product consists of several categories, including material costs, tooling costs, fixed costs, variable costs, and direct and indirect labor costs. References to some cost data are listed in Table 16.1. Manufacturers use several methods of cost accounting; the procedures can be complex and even controversial. Trends in costing systems (*cost justification*) include considerations of (a) the intangible benefits of quality improvements; (b) life-cycle costs; (c) changes in machine usage; and (d) the financial risks involved in implementing more automation and new technologies becoming available.

The major cost factors in manufacturing are:

1. **Material costs.** These costs are described throughout various sections in this text; see also Table 16.1.
2. **Tooling costs.** These are the costs involved in tools, dies, molds, patterns, and special jigs and fixtures necessary for manufacturing a particular product. Tooling costs greatly depend on the manufacturing process itself; high tooling costs can be justified in high-volume production of the same part (*die cost per piece*). The examples below briefly indicate various considerations involved in tooling costs.
   a. The cost for die casting is much higher than that for sand casting.
   b. Machining and grinding costs are much lower than those for such processes powder metallurgy, forging, or extrusion.
   c. In machining operations, carbide tools are more expensive than are high-speed steel tools, but the life of carbide tools is much longer.
   d. If a part is to be made by spinning, the tooling cost for conventional spinning is much lower than that for shear spinning.
   e. Tooling for rubber-pad forming processes is less expensive than that of the male and female die sets required for deep drawing or stamping of sheet metals.
3. **Fixed costs.** These costs include electric power, fuel, real estate taxes, rent, insurance, and capital (including depreciation and interest). They have to be met regardless of the number of parts made; consequently, fixed costs are not sensitive to production quantity.
4. **Capital costs.** These are major expenditures representing investments in buildings, land, machinery, tooling, and equipment. Note in Table 16.9 the wide range of costs in each category and that some machines can cost millions of dollars. Consequently, high production volumes are necessary to justify such large expenditures. Periodic

TABLE 16.9 Relative costs for machinery and equipment.

Automatic screw machine	M-H	Lathes	L-M
Boring mill, horizontal	M-H	Machining center	L-M
Broaching	M-H	Mechanical press	L-M
Deep drawing	M-H	Milling	L-M
Die casting	M-H	Powder-injection molding	M-H
Drilling	L-M	Powder metallurgy	L-M
EDM	L-M	Powder metallurgy, HIP	M-H
Electron-beam welding	M-H	Resistance spot welding	L-M
Extruder, polymer	L-M	Ring rolling	M-H
Extrusion press	M-H	Robots	L-M
Flexible manufacturing cell and system	H-VH	Roll forming	L-M
Forging	M-H	Rubber forming	L-M
Fused deposition modeling	L	Sand casting	L-M
Gas tungsten-arc welding	L	Spinning	L-M
Gear shaping	L-H	Stereolithography	L-M
Grinding	L-H	Stamping	L-M
Headers	L-M	Stretch forming	M-H
Honing, lapping	L-M	Transfer lines	H-VH
Injection molding	M-H	Ultrasonic welding	L-M
Laser-beam welding	M-H		

Note: L = low; M = medium; H = high; VH = very high. Costs vary greatly, depending on size, capacity, options, and level of automation and computer controls (see also the sections on economics in various chapters).

equipment maintenance is also essential to ensure consistently high productivity (Section 14.2.7), as any breakdown of machinery will lead to significant downtime, typically costing from a few hundred to thousands of dollars per hour.

5. **Labor costs.** These costs generally consist of direct and indirect costs. Direct labor costs are for the labor directly involved in manufacturing, called *productive labor*. They include the cost of all labor, from the time raw materials are first handled to the time the product is finished, generally referred to as *floor-to-floor time*. Direct labor costs are calculated by multiplying the labor rate (hourly wage, including benefits) by the time spent in producing the part. Note that the time required for producing a particular part depends on numerous factors, such as the material and its manufacturing characteristics, part size, shape complexity, dimensional accuracy, and surface finish. For example, as can be seen in Table 8.9, the highest recommended cutting speeds for high-temperature alloys are lower than those for aluminum, cast iron, or copper alloys; thus, the cost of machining aerospace materials is higher than that for machining the more common metals and alloys.

   Labor costs vary greatly from country to country, as described in Section 1.10 and shown in Table 1.4. For labor-intensive industries, such as casting, assembly of electronic devices, and clothing and textiles, many manufacturers have moved production to countries with a lower labor rate, a practice known as **outsourcing**. While this approach can be financially justifiable, the cost savings anticipated may not always be realized, because of the following hidden costs associated with outsourcing:

- *International shipping* is far more involved and time consuming than domestic shipping. For example, it takes about four to six weeks for a container ship to transport a product from China to the United States or to Europe, a time interval that has continued to increase because of homeland security concerns. Moreover, shipping costs can fluctuate significantly and in an unpredictable manner; it is thus not possible to reliably predict or to budget shipping costs.
- *Lengthy shipping schedules* indicate that the benefits of the just-in-time manufacturing approach and its associated cost savings may not be easily realized. Also, because of the long shipping times, schedules are rigid, product design modifications cannot be made easily, and companies cannot readily address changes in the market or in product demand. Thus, companies that outsource can lose agility and may also have difficulties in adopting and following lean-manufacturing approaches.
- *Legal systems* are not as well established in countries with lower labor rates as they are in more industrialized countries. Procedures that are common in the United States and the European Union, such as accounting audits, protection of patented product designs and intellectual property, and conflict resolutions, are more difficult to enforce in other countries.
- *Payments* typically are expected on the basis of units completed and sold, thus the consequences of defective products and their rate of occurrence can be come significant.
- There are various other *hidden costs* in outsourcing, such as increased paperwork and documentation, lower productivity from employees, and difficulties in communication among different nationalities.

For all these reasons, manufacturers that have outsourced manufacturing activities have not always been able to realize the hoped-for benefits; consequently, the trend in the United States is now toward *reshoring* manufacturing.

Manufacturing costs and production quantity. A significant cost factor in manufacturing is *production quantity*. Large production volumes typically require high production rates which, in turn, requires the implementation of mass-production techniques, typically involving *dedicated machinery* and employing proportionally less direct labor (see Section 14.2.2). On the other hand, small production volume usually indicates a larger direct labor involvement.

Small batch production usually requires general-purpose machines, such as lathes, milling machines, and hydraulic presses. The equipment is versatile, and parts with different shapes and sizes can be produced by appropriate changes in the tooling. Direct labor costs are high because the machines are usually operated by skilled labor. For larger quantities (*medium-batch production*), these same general-purpose machines are computer controlled; to reduce labor costs further, machining centers and flexible manufacturing systems are important alternatives. For production quantities of 100,000 or more, the machines are generally designed for specific purposes; they perform a variety of specific operations with very little direct labor.

16.10.1 Cost Reduction

The approach to *cost reduction* requires an assessment of how each of the costs, described above, are incurred and interrelated, with relative costs depending on numerous factors described throughout this chapter. The unit cost of a particular product can vary widely, depending on design and manufacturing characteristics. For example, some parts may be made from expensive materials but may require very little processing, such as minted gold coins, where the cost of the material relative to that of direct labor is high. By contrast, some products made of relatively inexpensive materials, such as carbon steels, may require several complex and expensive production steps to process. An electric motor, for example, is made of relatively inexpensive materials, but several different manufacturing processes are involved in the production of its components, such as housing, rotor, bearings, brushes, and wire windings.

An approximate breakdown of costs in manufacturing is given in Table 16.10. In the 1960s, labor accounted for as much as 40% of the total production cost; today, it can be as low as 5%, depending on the type of product and the level of automation. Such a small share is basically due to using highly-automated equipment and computer-controlled operations; it also indicates that moving production to countries with low wages may not now be an economically viable strategy, based on labor costs alone.

Note in Table 16.10 the very small contribution of the design phase, yet it is this phase that generally has the greatest influence on the overall cost of the product and its success in the marketplace. The engineering changes that often are made in the development of products also can have a significant influence on costs. The stage at which changes are made is very significant, as the cost of engineering changes, from design stage to final production, increases by orders of magnitude (*rule of ten*), as clearly indicated in Table 16.2.

TABLE 16.10 Approximate breakdown of costs in products.

Design	5%
Material	55%
Direct labor	10%
Overhead	30%

Cost reductions can be achieved by a thorough analysis of all the costs incurred in each step in manufacturing a product, by such means as follows:

1. Considering alternative methods of manufacturing;
2. Simplifying product design;
3. Reducing the number of subassemblies;
4. Reducing the amount of materials used for each part;

5. Using less expensive materials; and
6. Specifying broader dimensional tolerances and surface finish.
7. Implementing more automation and computer controls.

The last item above requires a thorough **cost-benefit analysis**, with reliable data input and considerations of all technical and human factors involved. Because the implementation of advanced technology and computer-control machinery can be very expensive, the importance of the concept of *return on investment* (ROI) becomes self-evident (see also **return on quality**, Section 16.3).

SUMMARY

- Competitive aspects of production methods and costs are among the most significant considerations in manufacturing; regardless of how well a product meets design specifications and quality standards, it must also meet economic criteria in order to be competitive in the global marketplace. (Section 16.1)
- Various guidelines for designing products for economical production have been established. (Section 16.2)
- Product quality and life expectancy are important concerns because of their impact on customer satisfaction and the marketability of the product. (Section 16.3)
- The management philosophies of Deming, Juran, and Taguchi provide a framework for improving quality and designing robust products; the Taguchi quality loss function is a valuable tool to evaluate product quality. (Section 16.3)
- Life cycle assessment and life cycle engineering are among increasingly important considerations in manufacturing, particularly reducing any adverse impact on the environment. Sustainable manufacturing aims at reducing waste of natural resources, including materials and energy. (Section 16.4)
- Selecting an appropriate material from among numerous candidates is a challenging aspect of manufacturing. (Section 16.6)
- Substitution of materials, modifications of product design, and relaxing of dimensional tolerance and surface finish requirements are among important considerations for reducing costs. (Section 16.7)
- Because the capabilities of manufacturing processes vary widely, their proper selection for making a particular product meeting specific design and functional requirements is critical. (Sections 16.8 and 16.9)
- The total cost of a product includes several elements; with software programs, the least expensive material can now be identified, without compromising design, service requirements, and quality. Labor costs are becoming an increasingly smaller percentage of production costs through implementation of highly automated and computer-controlled machinery. (Sections 16.10)

SUMMARY OF EQUATIONS

Taguchi loss function, Loss cost $= k\left[(Y - T)^2 + \sigma^2\right]$

where $k = \dfrac{\text{Replacement cost}}{(\text{LSL} - T)^2}$

BIBLIOGRAPHY

Anderson, D.M., *Design for Manufacturability*, CRC Press, 2014.

Andrae, A.S.G., *Global Life Cycle Assessments of Material Shifts*, Springer, 2009.

Ashby, M.F., *Materials Selection in Mechanical Design*, 4th ed., Pergamon, 2010.

Boothroyd, G., Dewhurst, P., and Knight, W. *Product Design for Manufacture and Assembly*, 3rd ed., Dekker, 2010.

Cha, J., Jardim-Gonclaves, R., and Steiger-Garcao, A., *Concurrent Engineering*, Taylor & Francis, 2003.

Cook, H.F., and Wissmann, L.A., *Value Driven Product Planning and Systems Engineering*, Springer, 2010.

Crowson, R., *Product Design and Factory Development*, 2nd ed., CRC Press, 2005.

Fiksel, J., *Design for the Environment*, 2nd ed., McGraw-Hill, 2011.

Giudice, F., La Rosa, G., and Risitano, A., *Product Design for the Environment*, CRC Press, 2006.

Kutz, M., *Environmentally Conscious Manufacturing*, Wiley, 2007.

Magrab, E.B., *Integrated Product and Process Design and Development: The Product Realization Process*, 2nd ed., CRC Press, 2009.

McDonough, W., and Braungart, M., *Cradle to Cradle: Rethinking the Way We Make Things*, North Point Press, 2002.

Mori, T., *Taguchi Methods: Benefits, Impacts, Mathematics, Statistics, Applications*, ASME Press, 2011.

Pyzdek, T., and Keller, P., *The Six Sigma Handbook*, 4th ed., McGraw-Hill, 2014.

Roy, R.K., *A Primer on the Taguchi Method*, 2nd ed., Society of Manufacturing Engineers, 2010.

Seliger, G., (ed.), *Sustainability in Manufacturing*, Springer, 2010.

Stjepandic, J., Wognum, N., and Verhagen, W.J.C., (eds.) *Concurrent Engineering in the 21st Century: Foundations, Developments, and Challenges*, Springer, 2015.

Swift, K.G., and Booker, J.D., *Process Selection: From Design to Manufacture*, 2nd ed., Butterworth-Heinemann, 2003.

Taguchi, G., Chowdhury, S., and Wu, Y., *Taguchi's Quality Engineering Handbook*, Wiley, 2004.

Ulrich, K., and Eppinger, S., *Product Design and Development*, 5th ed., McGraw-Hill, 2011.

Walker, J.M. (ed.), *Handbook of Manufacturing Engineering*, 2nd ed., Dekker, 2006.

Wang, J.X., *Engineering Robust Designs with Six Sigma*, Prentice Hall, 2005.

Wenzel, H., Hauschild, M., and Alting, L., *Environmental Assessment of Products*, Vol. 1, Springer, 2003.

Wu, Y., and Wu, A., *Taguchi Methods for Robust Design*, American Society of Mechanical Engineers, 2000.

QUESTIONS

16.1 List and describe the major considerations involved in selecting materials for products.

16.2 Why is a knowledge of available shapes of materials important? Give five different examples.

16.3 Describe what is meant by the manufacturing characteristics of materials. Give three examples demonstrating the importance of this information.

16.4 Why is material substitution an important aspect of manufacturing engineering? Give five examples from your own experience or observations.

16.5 Why has material substitution been particularly critical in the automotive and aerospace industries?

16.6 What factors are involved in the selection of manufacturing processes? Explain why these factors are important.

16.7 What is meant by process capabilities? Select four different, specific manufacturing processes, and describe their capabilities.

16.8 Is production volume significant in process selection? Explain your answer.

16.9 Discuss the advantages of long lead times, if any, in production.

16.10 What is meant by an economic order quantity?

16.11 Describe the costs involved in manufacturing. Explain how you could reduce each of these costs.

16.12 What is value analysis? What are its benefits?

16.13 What is meant by trade-off? Why is it important in manufacturing?

16.14 Explain the difference between direct labor cost and indirect labor cost.

16.15 Explain why the larger the quantity per package of food products, the lower the cost per unit weight.

16.16 Explain why the value of the scrap produced in a manufacturing process depends on the type of material.

16.17 Comment on the magnitude and range of scrap shown in Table 16.6.

16.18 Describe your observations concerning the information given in Table 16.6.

16.19 Other than the size of the machine, what factors are involved in the range of prices in each machine category shown in Table 16.7?

16.20 Explain how the high cost of some of the machinery listed in Table 16.7 can be justified.

16.21 Explain the reasons for the relative positions of the curves shown in Fig. 16.6.

16.22 What factors are involved in the shape of the curve shown in Fig. 16.4?

16.23 Make suggestions as to how to reduce the dependence of production time on surface finish (shown in Fig. 16.5).

16.24 Is it always desirable to purchase stock that is close to the final dimensions of a part to be manufactured? Explain your answer, and give some examples.

16.25 What course of action would you take if the supply of a raw material selected for a product line became unreliable?

16.26 Describe the potential problems involved in reducing the quantity of materials in products.

16.27 Present your thoughts concerning the replacement of aluminum beverage cans with steel ones.

16.28 There is a period, between the time that an employee is hired and the time that the employee finishes with training, during which the employee is paid and receives benefits, but produces nothing. Where should such costs be placed among the categories given in this chapter?

16.29 Why is there a strong desire in industry to practice near-net-shape manufacturing? Give several examples.

16.30 Estimate the position of the following processes in Fig. 16.7: (a) centerless grinding; (b) electrochemical machining; (c) chemical milling; and (d) extrusion.

16.31 From your own experience and observations, comment on the size, shape, and weight of specific products as they have changed over the years.

16.32 In Section 16.10.1, a breakdown of costs in today's manufacturing environment suggests that design costs contribute only 5% to total costs. Explain why this suggestion is reasonable.

16.33 Make a list of several (a) disposable and (b) reusable products. Discuss your observations, and explain how you would go about making more products that are reusable.

16.34 Describe your own concerns regarding life-cycle assessment of products.

PROBLEMS

16.35 A manufacturer is ring rolling ball-bearing races (See Fig. 6.41). The inner surface has a surface roughness specification of 0.10 ± 0.06 μm. Measurements taken from rolled rings indicate a mean roughness of 0.112 μm with a standard deviation of 0.02 μm. 50,000 rings per month are manufactured and the cost of discarding a defective ring is \$5.00. It is known that by changing lubricants to a special emulsion, the mean roughness could be made essentially equal to the design specification. What additional cost per month can be justified for the lubricant?

16.36 For the data of Problem 16.35, assume that the lubricant change can cause the manufacturing process to achieve a roughness of 0.10 ± 0.01 μm. What additional cost per month for the lubricant can be justified? What if the lubricant did not add any new cost?

16.37 Assume that you are asked to give a quiz to students on the contents of this chapter. Prepare three quantitative problems and three qualitative questions, and supply the answers

DESIGN

16.38 As you can see, Table 16.7 on manufacturing processes includes only metals and their alloys. On the basis of the information given in this book and other sources, prepare a similar table for nonmetallic materials, including ceramics, plastics, reinforced plastics, and metal-matrix and ceramic-matrix composite materials.

16.39 Review Fig. 1.3, and present your thoughts concerning the flowchart. Would you want to make any modifications, and if so, what would the modifications be and why?

16.40 Over the years, numerous consumer products have become obsolete, or nearly so, such as rotary-dial telephones, analog radio tuners, turntables, and vacuum tubes. By contrast, many new products have entered the market. Make a comprehensive list of obsolete products and one of new products. Comment on the possible reasons for the changes you have observed. Discuss how different manufacturing methods and systems have evolved in order to make the new products.

16.41 Select three different household products, and make a survey of the changes in their prices over the past 10 years. Discuss the reasons for the changes.

16.42 Figure 2.2a shows the shape of a typical tension-test specimen having a round cross section. Assuming that the starting material (stock) is a round rod and that only one specimen is needed, discuss the processes and the machinery by which the specimen can be made, including their relative advantages and limitations. Describe how the process you selected can be changed for economical production as the number of specimens required increases.

16.43 Table 16.5 lists several materials and their commercially available shapes. By contacting suppliers of materials, extend the list to include (a) titanium; (b) superalloys; (c) lead; (d) tungsten; and (e) amorphous metals.

16.44 Select three different products commonly found in homes. State your opinions about (a) what materials were used in each product, and why, and (b) how the products were made, and why those particular manufacturing processes were used.

16.45 Inspect the components under the hood of your automobile. Identify several parts that have been produced to net-shape or near-net-shape condition. Comment on the design and production aspects of these parts and on how the manufacturer achieved the near-net-shape condition for the parts.

16.46 Comment on the differences, if any, between the designs, the materials, and the processing and assembly methods used to make such products as hand tools, ladders for professional use, and ladders for consumer use.

16.47 Other than powder metallurgy, which processes could be used (singly or in combination) in the making of the parts shown in Fig. 11.1? Would they be economical?

16.48 Discuss production and assembly methods that can be employed to build the presses used in sheet metal forming operations described in Chapter 7.

16.49 The shape capabilities of some machining processes are shown in Fig. 8.39. Inspect the various shapes produced, and suggest alternative processes for producing them. Comment on the properties of workpiece materials that would influence your suggestions.

16.50 Consider the internal combustion engine in the tractor shown in Fig. 1.1. On the basis of the topics covered in this text, select any three individual components of such an engine, and describe the materials and processes that you would use in making those components. Remember that the parts must be manufactured at very large numbers and at minimum cost, yet maintain their quality, integrity, and reliability during service.

16.51 Discuss the trade-offs involved in selecting between the two materials for each of the applications listed:

- **a.** Sheet metal vs. reinforced plastic chairs;
- **b.** Forged vs. cast crankshafts;
- **c.** Forged vs. powder-metallurgy connecting rods;
- **d.** Plastic vs. sheet metal light-switch plates;
- **e.** Glass vs. metal water pitchers;
- **f.** Sheet metal vs. cast hubcaps;
- **g.** Steel vs. copper nails; and
- **h.** Wood vs. metal handles for hammers.

Also, discuss the typical conditions to which these products are subjected in their normal use.

16.52 Discuss the manufacturing process or processes suitable for making the products listed in Question 16.51. Explain whether the products would require additional operations (such as coating, plating, heat treating, and finishing). If so, make recommendations and give the reasons for them.

16.53 Discuss the factors that influence the choice between the following pairs of processes to make the products indicated:

- **a.** Sand casting vs. die casting of a fractional electric-motor housing;

b. Machining vs. forming of a large-diameter bevel gear;
c. Forging vs. powder-metallurgy production of a cam;
d. Casting vs. stamping a sheet metal frying pan;
e. Making outdoor summer furniture from aluminum tubing vs. cast iron;
f. Welding vs. casting of machine-tool structures; and
g. Thread rolling vs. machining of a bolt for high-strength application;

16.54 The following figure shows a sheet metal part made of steel. Discuss how this part could be made and how your selection of a manufacturing process may change (a) as the number of parts required increases from 10 to thousands and (b) as the length of the part increases from 2 m to 20 m.

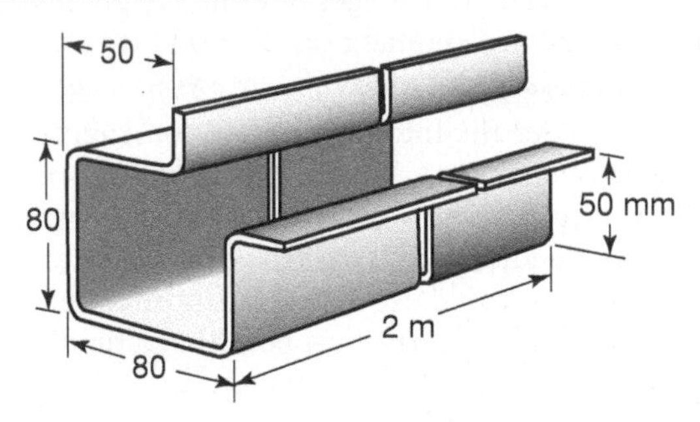

16.55 Many components in products have minimal effect on part robustness and quality. For example, the hinges in the glove compartment of an automobile do not really impact the owner's satisfaction, and the glove compartment is opened so few times that a robust design is easy to achieve. Would you advocate using Taguchi methods like the loss function on this type of component? Explain.

16.56 The part shown in the following figure is a carbon-steel segment (partial) gear:

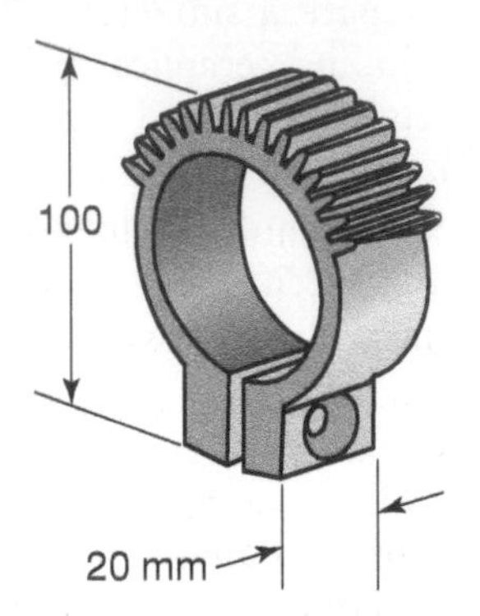

The smaller hole at the bottom is for clamping the part onto a round shaft, using a screw and a nut. Suggest a sequence of manufacturing processes to make this part. Consider such factors as the influence of the number of parts required, dimensional tolerances, and surface finish. Discuss such processes as machining from a bar stock, extrusion, forging, and powder metallurgy.

16.57 Review the automotive brake designs illustrated below and describe your thoughts on (a) the materials that could be used, your own selection, and your reason for it, (b) manufacturing processes, and why you would select them and (c) based on your review, any design changes that you would like to recommend.

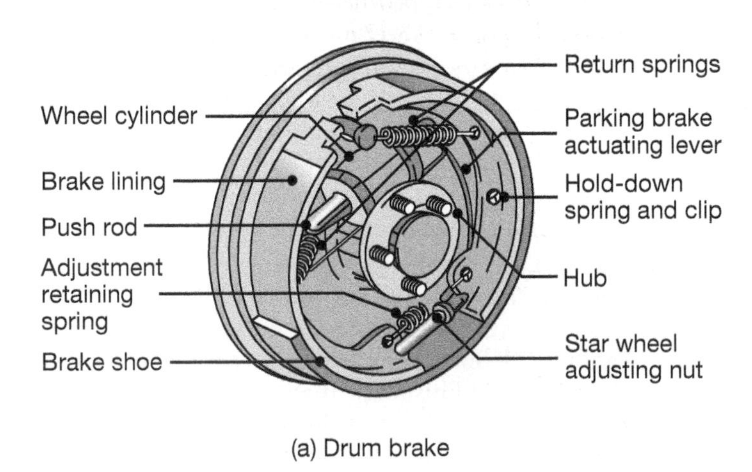

(a) Drum brake

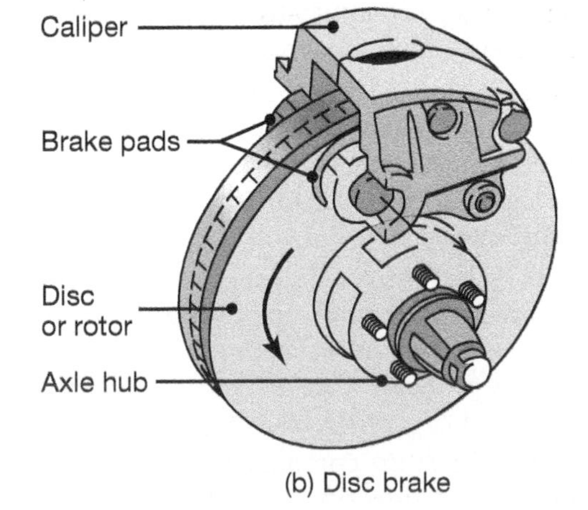

(b) Disc brake

Answers to Selected Problems

Note that the answers given here are for the guidance of students. Often, where a range of values is available in a table or chart, the mean value has been taken. In many problems, students need to develop methodologies and may rely on values taken from sources external to the book. In such cases, it should be straightforward to achieve close to the answers given here, but perhaps not the exact value.

2.50 $e_d = -0.933$; $\epsilon_l = 5.412$.
2.51 $K = 589$ MPa.
2.53 $S_{ut} = 165$ MPa.
2.56 $P = 17.1$ kN.
2.64 $P = 98$ kN.
2.65 $l_f = 500.93$ mm.
2.69 $l_f = 1.0049$ m.
2.83 For MSST, $P = 696$ kN.
2.87 $\epsilon_l = 0.6732$.
2.89 $h_f = 8.02$ mm.
2.93 $S_{ut} = 65.6$ MPa.
2.100 $t = 0.318$ mm.
3.43 For Al, $\tau = 4.6$ GPa, $\sigma = 7.9$ GPa.
3.45 $N = 1.16$ million.
3.48 For 2024-T4, $d = 25.2$ mm.
3.51 $k = 458$ MPa-$\sqrt{\mu\text{m}}$.
3.54 For 303SS, $W = 1.24$ N.
4.58 For sine wave, $R_a/R_q = 0.90$.
4.61 $\text{ID}_{\text{final}} = 8.25$ mm.
4.63 $h = 36.7\ \mu$m.
4.66 $\text{UCL}_{\bar{x}} = 51.685$, $\text{LCL}_{\bar{x}} = 48.315$.
4.67 $\text{UCL}_{\bar{x}} = 41.455$, $\text{LCL}_{\bar{x}} = 39.545$.
4.69 $\text{UCL}_{\bar{x}} = 0.7243$, $\text{LCL}_{\bar{x}} = 0.6223$.
4.70 $\bar{x} = 0.6733$.
5.62 At 728°C, $\%\alpha = 23\%$.
5.64 $T_{\text{round}}/T_{\text{ellipse}} = 2.67$.
5.65 $P = 225$ N.
5.68 $t = 1.88$ min.
5.75 For tripled height, $t = 1.0$ min.
5.77 For Al, $L = 253.2$ mm.
5.80 $d = 6.2$ mm.
5.81 $N = 354$ rpm.
5.85 $d = 0.5$ mm.
5.91 $N = 1640$ rpm.
6.77 For $\mu = 0$, $W = 571$ Nm.
6.78 For $\mu = 0$, $\Delta T = 33.7$°C.
6.81 $\gamma = 3.14$.
6.92 $P = 673$ kN.
6.96 $h_f = 9.54$ mm.
6.97 $P = 16.0$ kN.
6.99 $P = 3.26$ MN.
6.101 $v = 3.25$ m/s.
6.115 $P = 2.295$ MN.
6.118 $P = 3.32$ MN.
6.119 $x_n = 8.64$ mm.
6.121 $P = 8.27$ MN.
6.128 $P = 6.4$ kW.
6.140 $P = 100$ kN.
7.70 $P = 16.67$ kN.
7.71 For $R/t = 0.25$, $\epsilon = 0.654$.
7.77 $P = 5460$ N.
7.78 $d_f = 20.65$ mm.
7.84 $h/D_p = 0.8525$.
7.90 For single row, scrap = 32%.
7.91 $P_{\text{max}} = 39.9$ kN.
7.95 $d_b = 350$ mm.
7.99 $D_{b,\text{max}} = 240$ mm.
8.116 $r = 0.634$.
8.117 $V_c = 0.5$ m/s.
8.120 $\Delta\beta = 1.4°$.
8.121 $\phi = 37.65°$.
8.129 MRR = 3660 mm^3/s.
8.131 For original diameter, MRR = 11,780 mm^3/min.
8.136 $n = 0.2726$, $C = 5.942$.
8.137 MRR = 5625 mm^3/s.
8.139 $t = 22.5$ s.
8.140 $t = 9.6$ s.

8.143 70.5%.
8.146 MRR $= 37,070$ mm^3/min.
8.148 $t_c = 4.2$ mm.
8.156 $t = 10$ s.
8.157 MRR $= 16,875$ mm^3/s.
8.161 $V_o = 212$ m/min.
9.66 $t = 0.0037$ mm.
9.73 For 1 m, $F_{ave} = 9.8$ N.
9.74 $t = 4.17$ min.
9.80 For Al, Power= 1000 W.
9.88 $l = 3.162$ mm, $u = 10$ Ws/mm^3.
9.89 Using $r = 15$, $t = 0.469$ μm; $F_c = 36$ N.
9.90 For steel, $F_{ave} = 56.5$ N.
10.98 $Q_d = 221,000$ mm^3/s.
10.100 $N = 190$ rpm.
10.101 $\theta = 17.6°$.
10.110 For high-modulus carbon, $E_c = 178$ GPa.
10.112 $x = 31.7\%$.
11.66 $N = 7.78$ billion.
11.67 $V = 17.6$ cm^3.
11.71 For the flake, SF $= 17.1$.
11.82 $P = 528$ kN.
11.83 $V = 63.6$ cm^3.
11.86 $d = 5.15$ cm.
11.98 For $\mu = 0.2$, $p = 123.7$ MPa.
12.96 $H = 324$ J.
12.102 $I = 129$ A.
12.104 For GTAW, $\Delta T = 8.25°$C.
12.108 For Al, $I = 54.3$ A.
13.59 266 nm.
13.62 $t = 10.9$ min.
13.67 $t = 3$ days.
13.68 $t = 0.5$ μm.
16.35 Justified additional monthly cost is $10,000.
16.36 Justified additional monthly cost is $31,000.

Index

C

Q

R

T